The Elements

Element	Symbol	Atomic Number	Atomic Mass*	Element	Symbol	Atomic Number	Atomic Mass*
Actinium	Ac	89	(227)	Molybdenum	Mo	42	95.94
Aluminum	Al	13	26.98	Neodymium	Nd	60	144.2
Americium	Am	95	(243)	Neon	Ne	10	20.18
Antimony	Sb	51	121.8	Neptunium	Np	93	(244)
Argon	Ar	18	39.95	Nickel	Ni	28	58.70
Arsenic	As	33	74.92	Niobium	Nb	41	92.91
Astatine	At	85	(210)	Nitrogen	N	7	14.01
Barium	Ba	56	137.3	Nobelium	No	102	(253)
Berkelium	Bk	97	(247)	Osmium	Os	76	190.2
Beryllium	Be	4	9.012	Oxygen	O	8	16.00
Bismuth	Bi	83	209.0	Palladium	Pd	46	106.4
Bohrium	Bh	107	(262)	Phosphorus	P	15	30.97
Boron	B	5	10.81	Platinum	Pt	78	195.1
Bromine	Br	35	79.90	Plutonium	Pu	94	(242)
Cadmium	Cd	48	112.4	Polonium	Po	84	(209)
Calcium	Ca	20	40.08	Potassium	K	19	39.10
Californium	Cf	98	(249)	Praseodymium	Pr	59	140.9
Carbon	C	6	12.01	Promethium	Pm	61	(145)
Cerium	Ce	58	140.1	Protactinium	Pa	91	(231)
Cesium	Cs	55	132.9	Radium	Ra	88	(226)
Chlorine	Cl	17	35.45	Radon	Rn	86	(222)
Chromium	Cr	24	52.00	Rhenium	Re	75	186.2
Cobalt	Co	27	58.93	Rhodium	Rh	45	102.9
Copper	Cu	29	63.55	Rubidium	Rb	37	85.47
Curium	Cm	96	(247)	Ruthenium	Ru	44	101.1
Dubnium	Db	105	(262)	Rutherfordium	Rf	104	(261)
Dysprosium	Dy	66	162.5	Samarium	Sm	62	150.4
Einsteinium	Es	99	(254)	Scandium	Sc	21	44.96
Erbium	Er	68	167.3	Seaborgium	Sg	106	(266)
Europium	Eu	63	152.0	Selenium	Se	34	78.96
Fermium	Fm	100	(253)	Silicon	Si	14	28.09
Fluorine	F	9	19.00	Silver	Ag	47	107.9
Francium	Fr	87	(223)	Sodium	Na	11	22.99
Gadolinium	Gd	64	157.3	Strontium	Sr	38	87.62
Gallium	Ga	31	69.72	Sulfur	S	16	32.07
Germanium	Ge	32	72.61	Tantalum	Ta	73	180.9
Gold	Au	79	197.0	Technetium	Tc	43	(98)
Hafnium	Hf	72	178.5	Tellurium	Te	52	127.6
Hassium	Hs	108	(265)	Terbium	Tb	65	158.9
Helium	He	2	4.003	Thallium	Tl	81	204.4
Holmium	Ho	67	164.9	Thorium	Th	90	232.0
Hydrogen	H	1	1.008	Thulium	Tm	69	168.9
Indium	In	49	114.8	Tin	Sn	50	118.7
Iodine	I	53	126.9	Titanium	Ti	22	47.88
Iridium	Ir	77	192.2	Tungsten	W	74	183.9
Iron	Fe	26	55.85	Uranium	U	92	238.0
Krypton	Kr	36	83.80	Vanadium	V	23	50.94
Lanthanum	La	57	138.9	Xenon	Xe	54	131.3
Lawrencium	Lr	103	(257)	Ytterbium	Yb	70	173.0
Lead	Pb	82	207.2	Yttrium	Y	39	88.91
Lithium	Li	3	6.941	Zinc	Zn	30	65.39
Lutetium	Lu	71	175.0	Zirconium	Zr	40	91.22
Magnesium	Mg	12	24.31			110**	(269)
Manganese	Mn	25	54.94			111	(272)
Meitnerium	Mt	109	(266)			112	(277)
Mendelevium	Md	101	(256)			114	(285)
Mercury	Hg	80	200.6			116	(292)

*All atomic masses are given to four significant figures. Values in parentheses represent the mass number of the most stable isotope.

**The names and symbols for elements 110–112, 114, and 116 have not been chosen.

IMPORTANT:

HERE IS YOUR REGISTRATION CODE TO ACCESS
YOUR PREMIUM McGRAW-HILL ONLINE RESOURCES.

For key premium online resources you need THIS CODE to gain access. Once the code is entered, you will be able to use the Web resources for the length of your course.

If your course is using **WebCT** or **Blackboard**, you'll be able to use this code to access the McGraw-Hill content within your instructor's online course.

Access is provided if you have purchased a new book. If the registration code is missing from this book, the registration screen on our Website, and within your WebCT or Blackboard course, will tell you how to obtain your new code.

Registering for McGraw-Hill Online Resources

TO gain access to your McGraw-Hill web resources simply follow the steps below:

1. USE YOUR WEB BROWSER TO GO TO: **http://www.mhhe.com/silberberg3/**

2. CLICK ON **FIRST TIME USER**.

3. ENTER THE REGISTRATION CODE* PRINTED ON THE TEAR-OFF BOOKMARK ON THE RIGHT.

4. AFTER YOU HAVE ENTERED YOUR REGISTRATION CODE, CLICK **REGISTER**.

5. FOLLOW THE INSTRUCTIONS TO SET-UP YOUR PERSONAL UserID AND PASSWORD.

6. WRITE YOUR UserID AND PASSWORD DOWN FOR FUTURE REFERENCE. KEEP IT IN A SAFE PLACE.

REGISTRATION CODE

organicism-99090881

TO GAIN ACCESS to the McGraw-Hill content in your instructor's **WebCT** or **Blackboard** course simply log in to the course with the UserID and Password provided by your instructor. Enter the registration code exactly as it appears in the box to the right when prompted by the system. You will only need to use the code the first time you click on McGraw-Hill content.

Thank you, and welcome to your McGraw-Hill online Resources!

Mc Graw Hill Higher Education

Mc Graw Hill Higher Education

* YOUR REGISTRATION CODE CAN BE USED ONLY ONCE TO ESTABLISH ACCESS. IT IS NOT TRANSFERABLE.

0-07-239681-4 SILBERBERG: CHEMISTRY, 3E

Martin S. Silberberg

Third Edition

CHEMISTRY

The Molecular Nature of Matter and Change

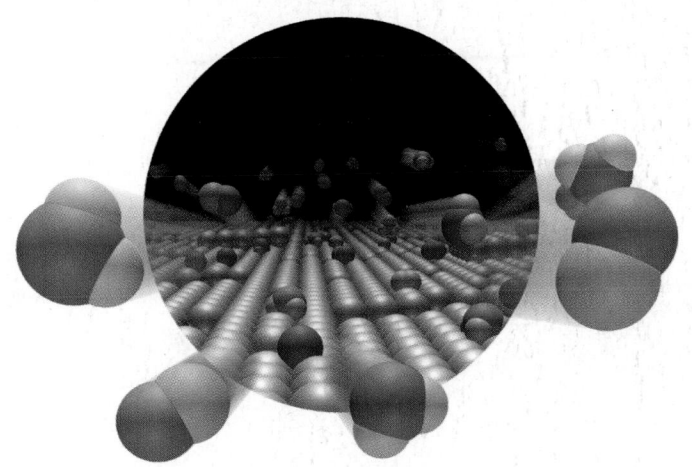

Consultants

Randy Duran
University of Florida—Gainesville

L. Peter Gold (emeritus)
Pennsylvania State University

Charles G. Haas (emeritus)
Pennsylvania State University

Arlan D. Norman
University of Colorado—Boulder

McGraw Hill

Boston   Burr Ridge, IL   Dubuque, IA   Madison, WI   New York   San Francisco   St. Louis
Bangkok   Bogotá   Caracas   Kuala Lumpur   Lisbon   London   Madrid   Mexico City
Milan   Montreal   New Delhi   Santiago   Seoul   Singapore   Sydney   Taipei   Toronto

McGraw-Hill Higher Education ⚛

*A Division of The **McGraw-Hill** Companies*

CHEMISTRY: THE MOLECULAR NATURE OF MATTER AND CHANGE
THIRD EDITION

Published by McGraw-Hill, a business unit of The McGraw-Hill Companies, Inc., 1221 Avenue of the Americas, New York, NY 10020.
Copyright © 2003, 2000, 1996 by The McGraw-Hill Companies, Inc. All rights reserved. No part of this publication may be reproduced
or distributed in any form or by any means, or stored in a database or retrieval system, without the prior written consent of The
McGraw-Hill Companies, Inc., including, but not limited to, in any network or other electronic storage or transmission, or broadcast for
distance learning.

Some ancillaries, including electronic and print components, may not be available to customers outside
the United States.

This book is printed on acid-free paper.

International   2 3 4 5 6 7 8 9 0 VNH/VNH 0 9 8 7 6 5 4 3
Domestic   3 4 5 6 7 8 9 0 VNH/VNH 0 9 8 7 6 5 4 3

ISBN 0–07–239681–4
ISBN 0–07–119894–6 (ISE)

Publisher: *Kent A. Peterson*
Developmental editor: *Joan M. Weber*
Senior development manager: *Kristine Tibbetts*
Marketing manager: *Thomas D. Timp*
Lead project manager: *Peggy J. Selle*
Lead production supervisor: *Sandra Hahn*
Design manager: *Stuart D. Paterson*
Cover/interior designer: *Jamie E. O'Neal*
Cover image: *Federico/Goodman Studios*
Page layout/special features designer: *Ruth Melnick*
Illustrations: *Federico/Goodman Studios*
Senior photo research coordinator: *Lori Hancock*
Photo research: *Feldman and Associates, Inc.*
Supplement producer: *Brenda A. Ernzen*
Senior media project manager: *Stacy A. Patch*
Media technology lead producer: *Steve Metz*
Compositor: *GTS Graphics, Inc.*
Typeface: *10.5/12 Times Roman*
Printer: *Quebecor World Kingsport*

COVER IMAGE: A molecular view at the beginning of the first step in the synthesis of nitric acid shows ammonia and oxygen landing
on the surface of a platinum catalyst and forming water and nitric oxide, which then leave the surface. In a few moments, the platinum
will become covered with reacting molecules. Catalysts speed up reactions and are essential parts of many industrial and virtually all
biological processes.

The credits section for this book begins on page C-1 and is considered an extension of the copyright page.

Library of Congress Cataloging-in-Publication Data

Silberberg, Martin S. (Martin Stuart), 1945–
    Chemistry : the molecular nature of matter and change / Martin S. Silberberg. — 3rd ed.
       p.   cm.
    Includes index.
    ISBN 0–07–239681–4 (acid-free paper) — ISBN 0–07–119894–6 (ISE : acid-free paper)
    1. Chemistry. I. Title.

QD33.2 .S55   2003
540—dc21   2002016642
                                                                             CIP

INTERNATIONAL EDITION ISBN 0–07–119894–6
Copyright © 2003. Exclusive rights by The McGraw-Hill Companies, Inc., for manufacture and export. This book cannot be re-exported
from the country to which it is sold by McGraw-Hill. The International Edition is not available in North America.

www.mhhe.com

With all my love,

For Ruth,
who lives it with me every day,

and

For Daniel,
who makes it all worthwhile

BRIEF CONTENTS

1 Keys to the Study of Chemistry 1

2 The Components of Matter 40

3 Stoichiometry: Mole-Mass-Number Relationships in Chemical Systems 86

4 The Major Classes of Chemical Reactions 131

5 Gases and the Kinetic-Molecular Theory 173

6 Thermochemistry: Energy Flow and Chemical Change 220

7 Quantum Theory and Atomic Structure 254

8 Electron Configuration and Chemical Periodicity 288

9 Models of Chemical Bonding 326

10 The Shapes of Molecules 357

11 Theories of Covalent Bonding 392

12 Intermolecular Forces: Liquids, Solids, and Phase Changes 419

13 The Properties of Mixtures: Solutions and Colloids 484

Interchapter: A Perspective on the Properties of the Elements 531

14 Periodic Patterns in the Main-Group Elements: Bonding, Structure, and Reactivity 542

15 Organic Compounds and the Atomic Properties of Carbon 606

16 Kinetics: Rates and Mechanisms of Chemical Reactions 663

17 Equilibrium: The Extent of Chemical Reactions 713

18 Acid-Base Equilibria 756

19 Ionic Equilibria in Aqueous Systems 805

20 Thermodynamics: Entropy, Free Energy, and the Direction of Chemical Reactions 855

21 Electrochemistry: Chemical Change and Electrical Work 892

22 The Elements in Nature and Industry 950

23 The Transition Elements and Their Coordination Compounds 998

24 Nuclear Reactions and Their Applications 1040

Appendix A Common Mathematical Operations in Chemistry A-1

Appendix B Standard Thermodynamic Values for Selected Substances at 298 K A-5

Appendix C Solubility-Product Constants (K_{sp}) of Slightly Soluble Ionic Compounds at 298 K A-8

Appendix D Standard Electrode (Half-Cell) Potentials at 298 K A-9

Appendix E Answers to Selected Problems A-10

Glossary G-1

Photo Credits C-1

Index I-1

DETAILED CONTENTS

Summary List of Special Features: Chemical Connections;
    Tools of the Laboratory; Galleries; Animations; and
    Margin Notes xiv
About the Author and Consultants xvii
Preface xix
Guided Tour xxii
Acknowledgments xxix
A Note to the Student: How to Do Well in This
    Course xxxii

CHAPTER 1

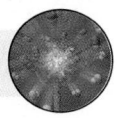

Keys to the Study of Chemistry 1

1.1 Some Fundamental Definitions 3
The Properties of Matter 3
The Three States of Matter 4
The Central Theme in Chemistry 6
The Importance of Energy in the Study of Matter 6

**1.2 Chemical Arts and the Origins
of Modern Chemistry 8**
Prechemical Traditions 8
The Phlogiston Fiasco and the Impact of Lavoisier 10

**1.3 The Scientific Approach: Developing
a Model 11**

1.4 Chemical Problem Solving 13
Units and Conversion Factors in Calculations 13
A Systematic Approach to Solving Chemistry
    Problems 15

1.5 Measurement in Scientific Study 17
General Features of SI Units 17
Some Important SI Units in Chemistry 18

**1.6 Uncertainty in Measurement: Significant
Figures 27**
Determining Which Digits Are Significant 28
Working with Significant Figures in Calculations 28
Precision, Accuracy, and Instrument Calibration 31

Chemical Connections CHEMISTRY PROBLEM
SOLVING IN THE REAL WORLD 32

Chapter Perspective 33
For Review and Reference 34   Problems 35

CHAPTER 2

The Components of Matter 40

**2.1 Elements, Compounds, and Mixtures:
An Atomic Overview 41**

**2.2 The Observations That Led to an Atomic View
of Matter 43**
Mass Conservation 43
Definite Composition 44
Multiple Proportions 45

2.3 Dalton's Atomic Theory 46
Postulates of the Atomic Theory 46
How the Theory Explains the Mass Laws 46
The Relative Masses of Atoms 47

**2.4 The Observations That Led to the Nuclear
Atom Model 48**
Discovery of the Electron and Its Properties 48
Discovery of the Atomic Nucleus 50

2.5 The Atomic Theory Today 51
Structure of the Atom 51
Atomic Number, Mass Number, and Atomic Symbol 52
Isotopes and Atomic Masses of the Elements 52

Tools of the Laboratory MASS SPECTROMETRY 54

A Modern Reassessment of the Atomic Theory 55

2.6 Elements: A First Look at the Periodic Table 56

2.7 Compounds: Introduction to Bonding 59
The Formation of Ionic Compounds 59
The Formation of Covalent Compounds 62
Polyatomic Ions: Covalent Bonds Within Ions 63

2.8 Compounds: Formulas, Names, and Masses 63
Types of Chemical Formulas 64
Some Advice About Learning Names and Formulas 64
Names and Formulas of Ionic Compounds 65
Names and Formulas of Binary Covalent Compounds 70
Molecular Masses from Chemical Formulas 71

Gallery PICTURING MOLECULES 73

2.9 Mixtures: Classification and Separation 74

Tools of the Laboratory BASIC SEPARATION
TECHNIQUES 75

Chapter Perspective 77
For Review and Reference 78   Problems 79

CHAPTER 3

Stoichiometry: Mole-Mass-Number Relationships in Chemical Systems 86

3.1 The Mole 87
Defining the Mole 87
Molar Mass 89
Interconverting Moles, Mass, and Number of Chemical
    Entities 90
Mass Percent from the Chemical Formula 93

3.2 Determining the Formula of an Unknown Compound 95
Empirical Formulas 95
Molecular Formulas 96
Combustion Analysis 98
Chemical Formulas and the Structures of Molecules 99

3.3 Writing and Balancing Chemical Equations 101

3.4 Calculating Amounts of Reactant and Product 105
Stoichiometrically Equivalent Molar Ratios from the Balanced Equation 106
Chemical Reactions That Occur in a Sequence 108
Chemical Reactions That Involve a Limiting Reactant 110
Chemical Reactions in Practice: Theoretical, Actual, and Percent Yields 112

3.5 Fundamentals of Solution Stoichiometry 114
Expressing Concentration in Terms of Molarity 114
Mole-Mass-Number Conversions Involving Solutions 115
Preparing and Diluting Molar Solutions 116
Stoichiometry of Chemical Reactions in Solution 118

Chapter Perspective 120
For Review and Reference 120   Problems 123

CHAPTER 4

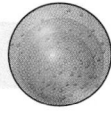

The Major Classes of Chemical Reactions 131

4.1 The Role of Water as a Solvent 132
The Solubility of Ionic Compounds 132
The Polar Nature of Water 134

4.2 Writing Equations for Aqueous Ionic Reactions 137

4.3 Precipitation Reactions 138
The Driving Force for a Precipitation Reaction 138
Predicting Whether a Precipitation Reaction Will Occur 139

4.4 Acid-Base Reactions 140
The Driving Force and Net Change: Formation of H_2O from H^+ and OH^- 141
Acid-Base Titrations 143
Acid-Base Reactions as Proton-Transfer Processes 144

4.5 Oxidation-Reduction (Redox) Reactions 146
The Driving Force for Redox Processes 147
Some Essential Redox Terminology 148
Using Oxidation Numbers to Monitor the Movement of Electron Charge 148
Balancing Redox Equations 150
Redox Titrations 152

4.6 Elemental Substances in Redox Reactions 154

4.7 Reversible Reactions: An Introduction to Chemical Equilibrium 162

Chapter Perspective 164
For Review and Reference 164   Problems 166

CHAPTER 5

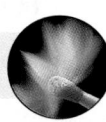

Gases and the Kinetic-Molecular Theory 173

5.1 An Overview of the Physical States of Matter 174

5.2 Gas Pressure and Its Measurement 176
Laboratory Devices for Measuring Gas Pressure 176
Units of Pressure 178

5.3 The Gas Laws and Their Experimental Foundations 180
The Relationship Between Volume and Pressure: Boyle's Law 180
The Relationship Between Volume and Temperature: Charles's Law 181
The Relationship Between Volume and Amount: Avogadro's Law 183
Gas Behavior at Standard Conditions 184
The Ideal Gas Law 185
Solving Gas Law Problems 186

5.4 Further Applications of the Ideal Gas Law 189
The Density of a Gas 189
The Molar Mass of a Gas 191
The Partial Pressure of a Gas in a Mixture of Gases 192

5.5 The Ideal Gas Law and Reaction Stoichiometry 195

5.6 The Kinetic-Molecular Theory: A Model for Gas Behavior 197
How the Kinetic-Molecular Theory Explains the Gas Laws 197
Effusion and Diffusion 201
The Chaotic World of Gases: Mean Free Path and Collision Frequency 203

Chemical Connections Chemistry in Planetary Science: STRUCTURE AND COMPOSITION OF THE EARTH'S ATMOSPHERE 204

5.7 Real Gases: Deviations from Ideal Behavior 207
Effects of Extreme Conditions on Gas Behavior 207
The van der Waals Equation: The Ideal Gas Law Redesigned 209

Chapter Perspective 210
For Review and Reference 210   Problems 212

CHAPTER 6

Thermochemistry: Energy Flow and Chemical Change 220

6.1 Forms of Energy and Their Interconversion 221
The System and Its Surroundings 221
Energy Flow to and from a System 222
Heat and Work: Two Forms of Energy Transfer 223
The Law of Energy Conservation 225

Units of Energy 225
State Functions and the Path Independence of the Energy
  Change 226

**6.2 Enthalpy: Heats of Reaction and Chemical
      Change 228**
The Meaning of Enthalpy 228
Comparing ΔE and ΔH 228
Exothermic and Endothermic Processes 229
Some Important Types of Enthalpy Change 230
Changes in Bond Strengths, or Where Does the Heat of
  Reaction Come From? 230

**6.3 Calorimetry: Laboratory Measurement of Heats
      of Reaction 233**
Specific Heat Capacity 233
The Practice of Calorimetry 234

**6.4 Stoichiometry of Thermochemical
      Equations 236**

6.5 Hess's Law of Heat Summation 238

6.6 Standard Heats of Reaction (ΔH^0_{rxn}) 240
Formation Equations and Their Standard Enthalpy
  Changes 240
Determining ΔH^0_{rxn} from ΔH^0_f Values of Reactants and
  Products 241

Chemical Connections   Chemistry in Environmental
  Science: THE FUTURE OF ENERGY USE 243

Chapter Perspective 245
For Review and Reference 246   Problems 247

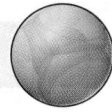

CHAPTER 7

Quantum Theory and Atomic
Structure 254

7.1 The Nature of Light 255
The Wave Nature of Light 256
The Particle Nature of Light 260

7.2 Atomic Spectra 262
The Bohr Model of the Hydrogen Atom 263
Limitations of the Bohr Model 264
The Energy States of the Hydrogen Atom 264

Tools of the Laboratory   SPECTROPHOTOMETRY IN
  CHEMICAL ANALYSIS 267

**7.3 The Wave-Particle Duality of Matter and
      Energy 269**
The Wave Nature of Electrons and the Particle Nature
  of Photons 269
The Heisenberg Uncertainty Principle 272

**7.4 The Quantum-Mechanical Model of the
      Atom 273**
The Atomic Orbital and the Probable Location of the
  Electron 273
Quantum Numbers of an Atomic Orbital 275
Shapes of Atomic Orbitals 278
Energy Levels of the Hydrogen Atom 281

Chapter Perspective 281
For Review and Reference 281   Problems 283

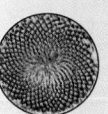

CHAPTER 8

Electron Configuration and
Chemical Periodicity 288

8.1 Development of the Periodic Table 289

8.2 Characteristics of Many-Electron Atoms 290
The Electron-Spin Quantum Number 290
The Exclusion Principle 291
Electrostatic Effects and the Splitting of Energy
  Levels 292

**8.3 The Quantum-Mechanical Model and the
      Periodic Table 295**
Building Up Periods 1 and 2 295
Building Up Period 3 298
Electron Configurations Within Groups 299
The First d-Orbital Transition Series: Building Up
  Period 4 299
General Principles of Electron Configurations 301
Complex Patterns: The Transition and Inner Transition
  Elements 302

**8.4 Trends in Some Key Periodic Atomic
      Properties 304**
Trends in Atomic Size 304
Trends in Ionization Energy 307
Trends in Electron Affinity 310

**8.5 The Connection Between Atomic Structure and
      Chemical Reactivity 311**
Trends in Metallic Behavior 311
Properties of Monatomic Ions 314

Chapter Perspective 320
For Review and Reference 320   Problems 321

CHAPTER 9

Models of Chemical Bonding 326

9.1 Atomic Properties and Chemical Bonds 327
Types of Chemical Bonding 327
Lewis Electron-Dot Symbols: Depicting Atoms in
  Chemical Bonding 329

9.2 The Ionic Bonding Model 330
Energy Considerations in Ionic Bonding: The Importance
  of Lattice Energy 331
Periodic Trends in Lattice Energy 333
How the Model Explains the Properties of Ionic
  Compounds 335

9.3 The Covalent Bonding Model 337
The Formation of a Covalent Bond 337
The Properties of a Covalent Bond: Bond Energy and
  Bond Length 338
How the Model Explains the Properties of Covalent
  Compounds 341

Tools of the Laboratory   INFRARED
  SPECTROSCOPY 343

9.4   Between the Extremes: Electronegativity and Bond Polarity 344
Electronegativity 344
Polar Covalent Bonds and Bond Polarity 346
The Partial Ionic Character of Polar Covalent Bonds 347
The Continuum of Bonding Across a Period 348

9.5   An Introduction to Metallic Bonding 349
The Electron-Sea Model 349
How the Model Explains the Properties of Metals 350

Chapter Perspective 351
For Review and Reference 352   Problems 353

CHAPTER 10

The Shapes of Molecules 357

10.1   Depicting Molecules and Ions with Lewis Structures 358
Using the Octet Rule to Write Lewis Structures 358
Resonance: Delocalized Electron-Pair Bonding 362
Formal Charge: Selecting the Best Resonance Structure 364
Lewis Structures for Exceptions to the Octet Rule 365

10.2   Using Lewis Structures and Bond Energies to Calculate Heats of Reaction 368

10.3   Valence-Shell Electron-Pair Repulsion (VSEPR) Theory and Molecular Shape 370
Electron-Group Arrangements and Molecular Shapes 371
The Molecular Shape with Two Electron Groups (Linear Arrangement) 372
Molecular Shapes with Three Electron Groups (Trigonal Planar Arrangement) 372
Molecular Shapes with Four Electron Groups (Tetrahedral Arrangement) 373
Molecular Shapes with Five Electron Groups (Trigonal Bipyramidal Arrangement) 375
Molecular Shapes with Six Electron Groups (Octahedral Arrangement) 376
Using VSEPR Theory to Determine Molecular Shape 377
Molecular Shapes with More Than One Central Atom 378

Gallery   Molecular Beauty: ODD SHAPES WITH USEFUL FUNCTIONS 380

10.4   Molecular Shape and Molecular Polarity 381
Bond Polarity, Bond Angle, and Dipole Moment 381
The Effect of Molecular Polarity on Behavior 383

Chapter Perspective 383

Chemical Connections   Chemistry in Sensory Physiology: MOLECULAR SHAPE, BIOLOGICAL RECEPTORS, AND THE SENSE OF SMELL 384

For Review and Reference 386   Problems 387

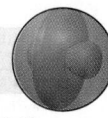

CHAPTER 11

Theories of Covalent Bonding 392

11.1   Valence Bond (VB) Theory and Orbital Hybridization 393
The Central Themes of VB Theory 393
Types of Hybrid Orbitals 394

11.2   The Mode of Orbital Overlap and the Types of Covalent Bonds 400
The VB Treatment of Single and Multiple Bonds 400
Orbital Overlap and Molecular Rotation 403

11.3   Molecular Orbital (MO) Theory and Electron Delocalization 404
The Central Themes of MO Theory 404
Homonuclear Diatomic Molecules of the Period 2 Elements 407
MO Description of Some Heteronuclear Diatomic Molecules 412
MO Descriptions of Ozone and Benzene 413

Chapter Perspective 414
For Review and Reference 414   Problems 416

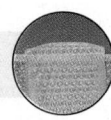

CHAPTER 12

Intermolecular Forces: Liquids, Solids, and Phase Changes 419

12.1   An Overview of Physical States and Phase Changes 420

12.2   Quantitative Aspects of Phase Changes 423
Heat Involved in Phase Changes: A Kinetic-Molecular Approach 423
The Equilibrium Nature of Phase Changes 425
Phase Diagrams: The Effect of Pressure and Temperature on Physical State 430

12.3   Types of Intermolecular Forces 431
Ion-Dipole Forces 432
Dipole-Dipole Forces 432
The Hydrogen Bond 434
Polarizability and Charge-Induced Dipole Forces 436
Dispersion (London) Forces 436

12.4   Properties of the Liquid State 439
Surface Tension 439
Capillarity 439
Viscosity 440

Gallery   PROPERTIES OF LIQUIDS   441

12.5   The Uniqueness of Water 442
Solvent Properties of Water 442
Thermal Properties of Water 442
Surface Properties of Water 443
The Density of Solid and Liquid Water 443

12.6 The Solid State: Structure, Properties, and Bonding 445
Structural Features of Solids 445

Tools of the Laboratory   X-RAY DIFFRACTION ANALYSIS AND SCANNING TUNNELING MICROSCOPY 451
Types and Properties of Crystalline Solids 452
Amorphous Solids 456
Bonding in Solids: Molecular Orbital Band Theory 456

12.7 Advanced Materials 460
Electronic Materials 460
Liquid Crystals 462
Ceramic Materials 465
Polymeric Materials 468
Nanotechnology: Designing Materials Atom by Atom 473
Chapter Perspective 475
For Review and Reference 476   Problems 477

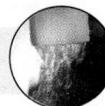

CHAPTER 13

The Properties of Mixtures: Solutions and Colloids 484

13.1 Types of Solutions: Intermolecular Forces and Predicting Solubility 486
Intermolecular Forces in Solution 486
Liquid Solutions and the Role of Molecular Polarity 487

Chemical Connections   Chemistry in Pharmacology: THE MODE OF ACTION OF SOAPS AND ANTIBIOTICS 490
Gas Solutions and Solid Solutions 492

13.2 Energy Changes in the Solution Process 493
Heats of Solution and Solution Cycles 493
Heats of Hydration: Ionic Solids in Water 494
The Solution Process and the Tendency Toward Disorder 496

13.3 Solubility as an Equilibrium Process 497
Effect of Temperature on Solubility 498
Effect of Pressure on Solubility 500

13.4 Quantitative Ways of Expressing Concentration 501
Molarity and Molality 501
Parts of Solute by Parts of Solution 502
Converting Units of Concentration 504

13.5 Colligative Properties of Solutions 506
Colligative Properties of Nonvolatile Nonelectrolyte Solutions 506

Gallery   COLLIGATIVE PROPERTIES IN INDUSTRY AND BIOLOGY 512
Using Colligative Properties to Find Solute Molar Mass 514
Colligative Properties of Volatile Nonelectrolyte Solutions 515
Colligative Properties of Electrolyte Solutions 516

13.6 The Structure and Properties of Colloids 517
Chapter Perspective 519

Chemical Connections   Chemistry in Sanitary Engineering: SOLUTIONS AND COLLOIDS IN WATER PURIFICATION 520
For Review and Reference 522   Problems 524

INTERCHAPTER

A Perspective on the Properties of the Elements 531

Topic 1   **The Key Atomic Properties 532**
Topic 2   **Characteristics of Chemical Bonding 534**
Topic 3   **Metallic Behavior 536**
Topic 4   **Acid-Base Behavior of the Element Oxides 537**
Topic 5   **Redox Behavior of the Elements 538**
Topic 6   **Physical States and Phase Changes 540**

CHAPTER 14

Periodic Patterns in the Main-Group Elements: Bonding, Structure, and Reactivity 542

14.1 Hydrogen, the Simplest Atom 543
Where Does Hydrogen Fit in the Periodic Table? 543
Highlights of Hydrogen Chemistry 544

14.2 Trends Across the Periodic Table: The Period 2 Elements 545

14.3 Group 1A(1): The Alkali Metals 548
Why Are the Alkali Metals Soft, Low Melting, and Lightweight? 548
Why Are the Alkali Metals So Reactive? 548
The Anomalous Behavior of Lithium 549

14.4 Group 2A(2): The Alkaline Earth Metals 552
How Do the Physical Properties of the Alkaline Earth and Alkali Metals Compare? 552
How Do the Chemical Properties of the Alkaline Earth and Alkali Metals Compare? 552
The Anomalous Behavior of Beryllium 553
Diagonal Relationships: Lithium and Magnesium 553
Looking Backward and Forward: Groups 1A(1), 2A(2), and 3A(13) 553

14.5 Group 3A(13): The Boron Family 556
How Do the Transition Elements Influence Group 3A(13) Properties? 556
What New Features Appear in the Chemical Properties of Group 3A(13)? 556
Highlights of Boron Chemistry 560
Diagonal Relationships: Beryllium and Aluminum 562

14.6 Group 4A(14): The Carbon Family 562
How Does the Bonding in an Element Affect Physical
Properties? 562
How Does the Type of Bonding Change in Group 4A(14)
Compounds? 566
Highlights of Carbon Chemistry 566
Highlights of Silicon Chemistry 568
Diagonal Relationships: Boron and Silicon 569
Looking Backward and Forward: Groups 3A(13), 4A(14),
and 5A(15) 569

Gallery   SILICATE MINERALS AND SILICONE
POLYMERS 570

14.7 Group 5A(15): The Nitrogen Family 573
What Accounts for the Wide Range of Physical Behavior
in Group 5A(15)? 573
What Patterns Appear in the Chemical Behavior of
Group 5A(15)? 576
Highlights of Nitrogen Chemistry 577
Highlights of Phosphorus Chemistry: Oxides and
Oxoacids 580

14.8 Group 6A(16): The Oxygen Family 581
How Do the Oxygen and Nitrogen Families Compare
Physically? 581
How Do the Oxygen and Nitrogen Families Compare
Chemically? 584
Highlights of Oxygen Chemistry: Range of Oxide
Properties 586
Highlights of Sulfur Chemistry: Oxides, Oxoacids, and
Sulfides 586
Looking Backward and Forward: Groups 5A(15), 6A(16),
and 7A(17) 588

14.9 Group 7A(17): The Halogens 588
What Accounts for the Regular Changes in the Halogens'
Physical Properties? 588
Why Are the Halogens So Reactive? 588
Highlights of Halogen Chemistry 592

14.10 Group 8A(18): The Noble Gases 595
How Can Noble Gases Form Compounds? 595
Looking Backward and Forward: Groups 7A(17), 8A(18),
and 1A(1) 595

Chapter Perspective 597
For Review and Reference 597   Problems 598

CHAPTER 15

**Organic Compounds and the
Atomic Properties of Carbon 606**

**15.1 The Special Nature of Carbon and the
Characteristics of Organic Molecules 607**
The Structural Complexity of Organic Molecules 608
The Chemical Diversity of Organic Molecules 608

**15.2 The Structures and Classes
of Hydrocarbons 610**
Carbon Skeletons and Hydrogen Skins 610
Alkanes: Hydrocarbons with Only Single Bonds 613
Constitutional Isomerism and the Physical Properties
of Alkanes 615

Chiral Molecules and Optical Isomerism 617
Alkenes: Hydrocarbons with Double Bonds 618

Chemical Connections   Chemistry in Sensory
Physiology: GEOMETRIC ISOMERS AND THE
CHEMISTRY OF VISION 620
Alkynes: Hydrocarbons with Triple Bonds 621
Aromatic Hydrocarbons: Cyclic Molecules with
Delocalized π Electrons 622
Variations on a Theme: Catenated Inorganic
Hydrides 623

Tools of the Laboratory   NUCLEAR MAGNETIC
RESONANCE (NMR) SPECTROSCOPY 624

**15.3 Some Important Classes of Organic
Reactions 624**
Types of Organic Reactions 624
The Redox Process in Organic Reactions 627

**15.4 Properties and Reactivities of Common
Functional Groups 627**
Functional Groups with Single Bonds 628
Functional Groups with Double Bonds 633
Functional Groups with Both Single and
Double Bonds 635
Functional Groups with Triple Bonds 640

**15.5 The Monomer-Polymer Theme I: Synthetic
Macromolecules 641**
Addition Polymers 642
Condensation Polymers 643

**15.6 The Monomer-Polymer Theme II: Biological
Macromolecules 644**
Sugars and Polysaccharides 644
Amino Acids and Proteins 646
Nucleotides and Nucleic Acids 650

Chapter Perspective 654
For Review and Reference 654   Problems 656

CHAPTER 16

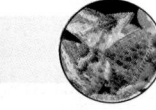

**Kinetics: Rates and Mechanisms
of Chemical Reactions 663**

16.1 Factors That Influence Reaction Rate 665

16.2 Expressing the Reaction Rate 667
Average, Instantaneous, and Initial Reaction Rates 668
Expressing Rate in Terms of Reactant and Product
Concentrations 669

16.3 The Rate Law and Its Components 671

Tools of the Laboratory   MEASURING REACTION
RATES 672
Determining the Initial Rate 672
Reaction Order Terminology 672
Determining Reaction Orders 675
Determining the Rate Constant 677

16.4 Integrated Rate Laws: Concentration Changes over Time 677
Integrated Rate Laws for First-, Second-, and Zero-Order Reactions 677
Determining the Reaction Order from the Integrated Rate Law 679
Reaction Half-Life 680

16.5 The Effect of Temperature on Reaction Rate 682

16.6 Explaining the Effects of Concentration and Temperature 685
Collision Theory: Basis of the Rate Law 685
Transition State Theory: Molecular Nature of the Activated State 688

16.7 Reaction Mechanisms: Steps in the Overall Reaction 691
Elementary Reactions and Molecularity 692
The Rate-Determining Step of a Reaction Mechanism 693
Correlating the Mechanism with the Rate Law 694

16.8 Catalysis: Speeding Up a Chemical Reaction 697
Homogeneous Catalysis 698
Heterogeneous Catalysis 699

Chemical Connections   Chemistry in Enzymology: KINETICS AND FUNCTION OF BIOLOGICAL CATALYSTS 700

Chemical Connections   Chemistry in Atmospheric Science: DEPLETION OF THE EARTH'S OZONE LAYER 702

Chapter Perspective 703
For Review and Reference 703   Problems 705

CHAPTER 17

Equilibrium: The Extent of Chemical Reactions 713

17.1 The Dynamic Nature of the Equilibrium State 714

17.2 The Reaction Quotient and the Equilibrium Constant 717
Writing the Reaction Quotient 718
Variations in the Form of the Reaction Quotient 719

17.3 Expressing Equilibria with Pressure Terms: Relation Between K_c and K_p 724

17.4 Reaction Direction: Comparing Q and K 725

17.5 How to Solve Equilibrium Problems 727
Using Quantities to Determine the Equilibrium Constant 727
Using the Equilibrium Constant to Determine Quantities 730

17.6 Reaction Conditions and the Equilibrium State: Le Châtelier's Principle 736
The Effect of a Change in Concentration 737
The Effect of a Change in Pressure (Volume) 739
The Effect of a Change in Temperature 741
The Lack of Effect of a Catalyst 743

Chemical Connections   Chemistry in Industrial Production: THE HABER PROCESS FOR THE SYNTHESIS OF AMMONIA 744

Chemical Connections   Chemistry in Cellular Metabolism: DESIGN AND CONTROL OF A METABOLIC PATHWAY 745

Chapter Perspective 747
For Review and Reference 747   Problems 749

CHAPTER 18

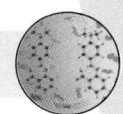

Acid-Base Equilibria 756

18.1 Acids and Bases in Water 758
Proton or Hydroxide Ion Release and the Classical Acid-Base Definition 759
Variation in Acid Strength: The Acid-Dissociation Constant (K_a) 759
Classifying the Relative Strengths of Acids and Bases 762

18.2 Autoionization of Water and the pH Scale 764
The Equilibrium Nature of Autoionization: The Ion-Product Constant for Water (K_w) 764
Expressing the Hydronium Ion Concentration: The pH Scale 765

18.3 Proton Transfer and the Brønsted-Lowry Acid-Base Definition 768
The Conjugate Acid-Base Pair 769
Relative Acid-Base Strength and the Net Direction of Reaction 770

18.4 Solving Problems Involving Weak-Acid Equilibria 772
Finding K_a Given Concentrations 773
Finding Concentrations Given K_a 775
The Effect of Concentration on the Extent of Acid Dissociation 776
The Behavior of Polyprotic Acids 776

18.5 Weak Bases and Their Relation to Weak Acids 779
Molecules as Weak Bases: Ammonia and the Amines 779
Anions of Weak Acids as Weak Bases 782
The Relation Between K_a and K_b of a Conjugate Acid-Base Pair 782

18.6 Molecular Properties and Acid Strength 784
Trends in Acid Strength of Nonmetal Hydrides 784
Trends in Acid Strength of Oxoacids 785
Acidity of Hydrated Metal Ions 786

18.7 Acid-Base Properties of Salt Solutions 787
Salts That Yield Neutral Solutions 787
Salts That Yield Acidic Solutions 787
Salts That Yield Basic Solutions 788
Salts of Weakly Acidic Cations and Weakly Basic
Anions 789

**18.8 Generalizing the Brønsted-Lowry Concept:
The Leveling Effect 790**

**18.9 Electron-Pair Donation and the Lewis Acid-Base
Definition 791**
Molecules as Lewis Acids 792
Metal Cations as Lewis Acids 793
An Overview of Acid-Base Definitions 794

Chapter Perspective 795
For Review and Reference 795   Problems 797

CHAPTER 19

Ionic Equilibria in Aqueous Systems 805

19.1 Equilibria of Acid-Base Buffer Systems 806
How a Buffer Works: The Common-Ion Effect 806
The Henderson-Hasselbalch Equation 811
Buffer Capacity and Buffer Range 812
Preparing a Buffer 813

19.2 Acid-Base Titration Curves 815
Monitoring pH with Acid-Base Indicators 815
Strong Acid–Strong Base Titration Curves 816
Weak Acid–Strong Base Titration Curves 818
Weak Base–Strong Acid Titration Curves 821
Titration Curves for Polyprotic Acids 822
Amino Acids as Biological Polyprotic Acids 823

**19.3 Equilibria of Slightly Soluble Ionic
Compounds 824**
The Ion-Product Expression (Q_{sp}) and the
Solubility-Product Constant (K_{sp}) 824
Calculations Involving the Solubility-Product
Constant 826
The Effect of a Common Ion on Solubility 828
The Effect of pH on Solubility 829
Predicting the Formation of a Precipitate: Q_{sp} vs. K_{sp} 830

Chemical Connections   Chemistry in Geology:
CREATION OF A LIMESTONE CAVE 831

Chemical Connections   Chemistry in Environmental
Science: THE ACID-RAIN PROBLEM 833

19.4 Equilibria Involving Complex Ions 835
Formation of Complex Ions 835
Complex Ions and the Solubility of Precipitates 837
Complex Ions of Amphoteric Hydroxides 838

**19.5 Application of Ionic Equilibria to Chemical
Analysis 840**
Selective Precipitation 840
Qualitative Analysis: Identifying Ions in Complex
Mixtures 841

Chapter Perspective 846
For Review and Reference 846   Problems 848

CHAPTER 20

Thermodynamics: Entropy, Free Energy, and the Direction of Chemical Reactions 855

**20.1 The Second Law of Thermodynamics: Predicting
Spontaneous Change 856**
Limitations of the First Law of Thermodynamics 857
The Sign of ΔH Cannot Predict Spontaneous Change 858
Disorder and Entropy 859
Entropy and the Second Law of Thermodynamics 861
Standard Molar Entropies and the Third Law 861

**20.2 Calculating the Change in Entropy of a
Reaction 866**
Entropy Changes in the System: The Standard Entropy
of Reaction (ΔS°_{rxn}) 866
Entropy Changes in the Surroundings: The Other Part
of the Total 867

Chemical Connections   Chemistry in Biology:
DO LIVING THINGS OBEY THE LAWS
OF THERMODYNAMICS? 869
The Entropy Change and the Equilibrium State 870
Spontaneous Exothermic and Endothermic Reactions: A
Summary 870

20.3 Entropy, Free Energy, and Work 872
Free Energy Change and Reaction Spontaneity 872
Calculating Standard Free Energy Changes 873
ΔG and the Work a System Can Do 874
The Effect of Temperature on Reaction Spontaneity 876
Coupling of Reactions to Drive a Nonspontaneous
Change 878

Chemical Connections   Chemistry in Biological
Energetics: THE UNIVERSAL ROLE OF ATP 879

**20.4 Free Energy, Equilibrium, and Reaction
Direction 880**

Chapter Perspective 884
For Review and Reference 884   Problems 886

CHAPTER 21

Electrochemistry: Chemical Change and Electrical Work 892

21.1 Half-Reactions and Electrochemical Cells 893
A Quick Review of Oxidation-Reduction Concepts 893
Half-Reaction Method for Balancing Redox
Reactions 894
An Overview of Electrochemical Cells 898

**21.2 Voltaic Cells: Using Spontaneous Reactions to
Generate Electrical Energy 900**
Construction and Operation of a Voltaic Cell 900
Notation for a Voltaic Cell 903
Why Does a Voltaic Cell Work? 904

21.3 Cell Potential: Output of a Voltaic Cell 905
Standard Cell Potentials 906
Relative Strengths of Oxidizing and Reducing
Agents 908

21.4 Free Energy and Electrical Work 914
Standard Cell Potential and the Equilibrium Constant 914
The Effect of Concentration on Cell Potential 916
Cell Potential and the Relation Between Q and K 918
Concentration Cells 919

21.5 Electrochemical Processes in Batteries 922

Gallery   BATTERIES AND THEIR APPLICATIONS 923

**21.6 Corrosion: A Case of Environmental
Electrochemistry 926**
The Corrosion of Iron 926
Protecting Against the Corrosion of Iron 927

**21.7 Electrolytic Cells: Using Electrical Energy to
Drive a Nonspontaneous Reaction 929**
Construction and Operation of an Electrolytic Cell 929
Predicting the Products of Electrolysis 931
The Stoichiometry of Electrolysis: The Relation Between
Amounts of Charge and Product 935

Chemical Connections   Chemistry in Biological
Energetics: CELLULAR ELECTROCHEMISTRY
AND THE PRODUCTION OF ATP 937

Chapter Perspective 939
For Review and Reference 939   Problems 942

CHAPTER 22

The Elements in Nature and Industry 950

22.1 How the Elements Occur in Nature 951
Earth's Structure and the Abundance of the Elements 951
Sources of the Elements 955

**22.2 The Cycling of Elements Through the
Environment 956**
The Carbon Cycle 956
The Nitrogen Cycle 958
The Phosphorus Cycle 960

**22.3 Metallurgy: Extracting a Metal from
Its Ore 963**
Pretreating the Ore 963
Converting Mineral to Element 964
Refining and Alloying the Element 967

**22.4 Tapping the Crust: Isolation and Uses of the
Elements 968**
Producing the Alkali Metals: Sodium and Potassium 968
The Indispensable Three: Iron, Copper, and
Aluminum 970
Mining the Sea: Magnesium and Bromine 977
The Many Sources and Uses of Hydrogen 978
A Group at a Glance: Sources, Isolation, and Uses of the
Elements 981

22.5 Chemical Manufacturing: Two Case Studies 987
Sulfuric Acid, the Most Important Chemical 987
The Chlor-Alkali Process 990

Chapter Perspective 991
For Review and Reference 992   Problems 993

CHAPTER 23

The Transition Elements and Their Coordination Compounds 998

23.1 Properties of the Transition Elements 1000
Electron Configurations of the Transition Metals and
Their Ions 1000
Atomic and Physical Properties of the Transition
Elements 1002
Chemical Properties of the Transition Metals 1003

23.2 The Inner Transition Elements 1006
The Lanthanides 1006
The Actinides 1007

23.3 Highlights of Selected Transition Metals 1008
Chromium 1008
Manganese 1009
Silver 1010
Mercury 1012

23.4 Coordination Compounds 1013
Structures of Complex Ions: Coordination Numbers,
Geometries, and Ligands 1014
Formulas and Names of Coordination Compounds 1016
A Historical Perspective: Alfred Werner and Coordination
Theory 1018
Isomerism in Coordination Compounds 1020

**23.5 Theoretical Basis for the Bonding and
Properties of Complexes 1023**
Application of Valence Bond Theory to
Complex Ions 1023
Crystal Field Theory 1025

Chapter Perspective 1031

Chemical Connections   Chemistry in Nutritional
Science: TRANSITION METALS AS ESSENTIAL
DIETARY TRACE ELEMENTS 1032
For Review and Reference 1034   Problems 1035

CHAPTER 24

Nuclear Reactions and Their Applications 1040

24.1 Radioactive Decay and Nuclear Stability 1042
The Components of the Nucleus: Terms and
Notation 1042
The Discovery of Radioactivity and the Types of
Emissions 1042
Types of Radioactive Decay; Balancing Nuclear
Equations 1044
Nuclear Stability and the Mode of Decay 1046

24.2 The Kinetics of Radioactive Decay 1050
The Rate of Radioactive Decay 1050

Tools of the Laboratory COUNTERS FOR THE DETECTION OF RADIOACTIVE EMISSIONS 1051
Radioisotopic Dating 1053

24.3 Nuclear Transmutation: Induced Changes in Nuclei 1055
Early Transmutation Experiments; Discovery of the Neutron 1055
Particle Accelerators and the Transuranium Elements 1056

24.4 The Effects of Nuclear Radiation on Matter 1058
The Effects of Radioactive Emissions: Excitation and Ionization 1058
Effects of Ionizing Radiation on Living Matter 1058

24.5 Applications of Radioisotopes 1062
Radioactive Tracers: Applications of Nonionizing Radiation 1062
Applications of Ionizing Radiation 1065

24.6 The Interconversion of Mass and Energy 1066
The Mass Defect 1066
Nuclear Binding Energy 1067

24.7 Applications of Fission and Fusion 1069
The Process of Nuclear Fission 1069
The Promise of Nuclear Fusion 1073

Chemical Connections Chemistry in Cosmology: ORIGIN OF THE ELEMENTS IN THE STARS 1074

Chapter Perspective 1076
For Review and Reference 1077   Problems 1079

Appendix A Common Mathematical Operations in Chemistry A-1
Manipulating Logarithms A-1
Using Exponential (Scientific) Notation A-2
Solving Quadratic Equations   A-3
Graphing Data in the Form of a Straight Line A-4

Appendix B Standard Thermodynamic Values for Selected Substances at 298 K   A-5

Appendix C Solubility-Product Constants (K_{sp}) of Slightly Soluble Ionic Compounds at 298 K   A-8

Appendix D Standard Electrode (Half-Cell) Potentials at 298 K A-9

Appendix E Answers to Selected Problems A-10

Glossary G-1
Credits C-1
Index I-1

SUMMARY LIST OF SPECIAL FEATURES

Chemical Connections

Chemistry Problem Solving in the Real World 32
Chemistry in Planetary Science: Structure and Composition of the Earth's Atmosphere 204
Chemistry in Environmental Science: The Future of Energy Use 243
Chemistry in Sensory Physiology: Molecular Shape, Biological Receptors, and the Sense of Smell 384
Chemistry in Pharmacology: The Mode of Action of Soaps and Antibiotics 490
Chemistry in Sanitary Engineering: Solutions and Colloids in Water Purification 520
Chemistry in Sensory Physiology: Geometric Isomers and the Chemistry of Vision 620
Chemistry in Enzymology: Kinetics and Function of Biological Catalysts 700
Chemistry in Atmospheric Science: Depletion of the Earth's Ozone Layer 702
Chemistry in Industrial Production: The Haber Process for the Synthesis of Ammonia 744
Chemistry in Cellular Metabolism: Design and Control of a Metabolic Pathway 745
Chemistry in Geology: Creation of a Limestone Cave 831
Chemistry in Environmental Science: The Acid-Rain Problem 833
Chemistry in Biology: Do Living Things Obey the Laws of Thermodynamics? 869
Chemistry in Biological Energetics: The Universal Role of ATP 879
Chemistry in Biological Energetics: Cellular Electrochemistry and the Production of ATP 937
Chemistry in Nutritional Science: Transition Metals as Essential Dietary Trace Elements 1032
Chemistry in Cosmology: Origins of the Elements in the Stars 1074

Tools of the Laboratory

Mass Spectrometry 54
Basic Separation Techniques 75
Spectrophotometry in Chemical Analysis 267
Infrared Spectroscopy 343
X-Ray Diffraction Analysis and Scanning Tunneling Microscopy 451
Nuclear Magnetic Resonance (NMR) Spectroscopy 624
Measuring Reaction Rates 672
Counters for the Detection of Radioactive Emissions 1051

Galleries

Picturing Molecules 73
Molecular Beauty: Odd Shapes with Useful Functions 380
Properties of Liquids 441
Colligative Properties in Industry and Biology 512
Silicate Minerals and Silicone Polymers 570
Batteries and Their Applications 923

Animations and Other Media
This icon in the margin indicates a related media presentation at www.mhhe.com/silberberg3.

Chapter 1
The three states of matter 4

Chapter 2
Rutherford's experiment 51
Formation of an ionic compound 60

Chapter 3
Limiting reactant 111
Making a solution 116

Chapter 4
Dissolution of an ionic compound and a covalent compound 135
Precipitation reactions 139

Chapter 5
Properties of gases 184

Chapter 6
Energy flow 229

Chapter 7
Emission spectra 264
Atomic line specta 281

Chapter 8
Isoelectronic series 319

Chapter 9
Formation of an ionic compound 328
Formation of a covalent bond 341
Ionic vs. covalent bonding 346

Chapter 10
VSEPR theory and the shapes of molecules 376
VSEPR 376
Influence of shape on polarity 382
Polarity of molecules 382

Chapter 11
Molecular shapes and orbital hybridization 398

Chapter 12
Vapor pressure 426
Phase diagrams and the states of matter 431
Cubic unit cells and their origins 454

Chapter 17
Le Châtelier's principle 741

Chapter 18
Dissociation of strong and weak acids 761

Chapter 19
Acid-base titration 818

Chapter 21
Galvanic cell 903
Operation of a voltaic cell 903

Chapter 22
Thermite reaction 966
Iron smelting 971
Aluminum production 975

Chapter 23
Vanadium reduction 1004

Chapter 24
Radioactive decay 1043
Half-life 1053
Nuclear power 1073

Margin Notes
This icon in the text indicates a related application, study aid, or historical note appearing in the margin.

Chapter 1
The Incredible Range of Physical Change 5
Scientific Thinker Extraordinaire [Lavoisier] 10
A Great Chemist Yet Strict Phlogistonist [Priestley] 11
Everyday Scientific Thinking 13
How Many Barleycorns from His Majesty's Nose to His Thumb? [Inexact units] 17
How Long Is a Meter? 19
Don't Drop That Kilogram! 21
Central Importance of Measurement in Science [Lord Kelvin] 27

Chapter 2
Immeasurable Changes in Mass 44
Dalton's Revival of Atomism 46
Atoms? Humbug! [Famous skeptics] 47
Familiar Glow of Colliding Particles [Signs, aurora, and TV] 48
The "Big Three" Subatomic Particles 51
Naming an Element 52
The Heresy of Radioactive "Transmutation" 56

Chapter 3
Imagine a Mole of . . . [Amazing comparisons] 87
A Rose by Any Other Name [Natural product formulas] 95
Limiting Reactants in Everyday Life 111

Chapter 4
Solid Solvents for Ions 135
Displacement Reactions Inside You [Protein metabolism] 142
Space-Age Combustion Without a Flame [Fuel cells] 160

Chapter 5
Atmosphere-Biosphere Redox Interconnections 174
POW! P-s-s-s-t! POP! [Familiar effects of gas behavior] 175
Snowshoes and the Meaning of Pressure 176
The Mystery of the Suction Pump 177
Breathing and the Gas Laws 184
Gas Density and Human Disasters 189
Up, Up, and Away! [Hot-air balloons] 190
Preparing Nuclear Fuel 202
Danger on Molecular Highways [Molecular motion] 203

Chapter 6
Wherever You Look There Is a System 221
Thermodynamics in the Kitchen 223
The Tragic Life of the First Law's Discoverer [von Mayer] 225
Your Personal Financial State Function [Checkbook analogy] 227
Imagine an Earth Without Water [Specific heat capacity] 236

Chapter 7
Hooray for the Human Mind [Major events around 1900] 255
Electromagnetic Emissions Everywhere 257
Rainbows and Diamonds 259
Ping-Pong Photons [Analogy for photoelectric effect] 261
What Are Stars Made Of? 266
"He'll Never Make a Success of Anything" [Einstein] 269
The Electron Microscope 270
Uncertainty Is Unacceptable? [Famous skeptics] 273
A Radial Probability Distribution of Apples 275

Chapter 8
Mendeleev's Great Contribution 289
Moseley and Atomic Number 290
Baseball Quantum Numbers [Analogy with stadium seat] 291

Periodic Memory Aids 302
Packing 'Em In [Nuclear charge and atomic size] 306

Chapter 9
The Remarkable Insights of G. N. Lewis 330
The Amazing Malleability of Gold 351

Chapter 10
A Purple Mule, Not a Blue Horse and a Red Donkey
        [Resonance hybrid] 362
Deadly Free-Radical Activity 365

Chapter 12
Environmental Flow [Solid, liquid, and gas flow] 421
Frozen Gold 421
Cooling Phase Change [Sweating and panting] 422
Cooking Under Low or High Pressure 429
The Remarkable Behavior of a Supercritical Fluid (SCF) 431
A Diamond Film on Every Pot 456
Solar Cells 460
One Strand or Many Pieces? 468

Chapter 13
Waxes for Home and Auto 493
Hot Packs, Cold Packs, and Self-Heating Soup 496
A Saturated Solution Is Like a Pure Liquid and Its Vapor 498
Scuba Diving and Soda Pop 500
Unhealthy Ultralow Concentrations [Pollutants] 503
"Soaps" in Your Small Intestine [Bile salts] 518
From Colloid to Civilization [River deltas] 519

Chapter 14
Fill 'Er Up with Hydrogen? Not Likely 545
Versatile Magnesium 552
Lime: The Most Useful Metal Oxide 552
Gallium Arsenide: The Next Wave of Semiconductors 556
Borates in Your Labware 560
CFCs: The Good, the Bad, and the Strong 567
Hydrazine, Nitrogen's Other Hydride 576
Nitric Oxide: A Biochemical Surprise 578
The Countless Uses of Phosphates 580
Match Heads, Bug Sprays, and O-Rings 581
Selenium and Xerography 584
Hydrogen Peroxide: Hydrazine's Cousin 585
Acid from the Sky 586
HF: Unusual Structure, Familiar Uses 592
Pyrotechnic Perchlorates 594

Chapter 15
"Organic Chemistry Is Enough to Drive One Mad"
        [Wohler] 607
Chiral Medicines 618
Aromatic Carcinogens 623
Pollutants in the Food Chain [PCBs and DDT] 630
A Pungent, Pleasant Banquet [Carboxylic acids and esters] 638
Polysaccharide Skeletons of Lobsters and Roaches 645

Chapter 16
The Significance of R [Dimensional analysis] 683
Sleeping Through the Rate-Determining Step 693
Catalytically Cleaning Your Car's Exhaust 699

Chapter 17
The Universality of Le Châtelier's Principle 737
Temperature-Dependent Systems [Similar math expressions] 743
Catalyzed Perpetual Motion? 743

Chapter 18
Pioneers of Acid-Base Chemistry 757
Logarithmic Scales in Sound and Seismology 765
Ammonia's Picturesque Past 779

Chapter 20
Vital Orderly Information [DNA and gene repair] 859
Poker and Probability 860
A Checkbook Analogy for Heating the Surroundings 867
Greatness and Obscurity of J. Willard Gibbs 872
The Wide Range of Energy Efficiency 875

Chapter 21
The Electrochemical Future Is Here 893
Which Half-Reaction Occurs at Which Electrode? 898
Electron Flow and Water Flow 904
The Pain of a Dental Voltaic Cell 913
Walther Hermann Nernst (1864–1941) 917
Concentration Cells in Your Nerve Cells 920
Minimicroanalysis 922
Father of Electrochemistry and Much More
        [Michael Faraday] 935

Chapter 22
Phosphorus from Outer Space [Meteorite sources] 960
Phosphorus Nerve Poisons 962
Panning and Fleecing for Gold 964
A Plentiful Oceanic Supply of NaCl 969
Was It Slag That Made the Great Ship Go Down? [Titanic] 971
The Dawns of Three New Ages [Copper, bronze, and brass] 972
Energy Received and Returned [Aluminum batteries] 975

Chapter 23
A Remarkable Laboratory Feat [Isolating lanthanides] 1006
Sharing the Ocean's Wealth [Manganese nodules] 1010
Mad as a Hatter [Mercury poisoning] 1013
Grabbing Ions [Chelates] 1016
Anticancer Geometric Isomers 1021

Chapter 24
The Remarkably Tiny, Massive Nucleus 1042
Her Brilliant Career [Marie Curie] 1043
The Little Neutral One [Neutrinos] 1045
The Case of the Shroud of Turin 1054
How Old Is the Solar System? 1055
The Powerful Bevatron 1057
Naming Transuranium Elements 1058
A Tragic Way to Tell Time in the Dark [Painting
        watch dials] 1059
The Risk of Radon 1061
Modeling Radiation Risk 1062
The Force That Binds Us [Strong force] 1067
Lise Meitner (1878–1968) 1069
"Breeding" Nuclear Fuel 1073

ABOUT THE AUTHOR AND CONSULTANTS

Martin S. Silberberg received his B.S. in chemistry from the City University of New York in 1966 and his Ph.D. in chemistry from the University of Oklahoma in 1971. He then accepted a research position at the Albert Einstein College of Medicine, where he studied the chemical nature of neurotransmission and Parkinson's disease. In 1977, Dr. Silberberg joined the faculty of Simon's Rock College of Bard (Massachusetts), a liberal arts college known for its excellence in teaching small classes of highly motivated students. As Head of the Natural Sciences Major and Director of Premedical Studies, he taught courses in general chemistry, organic chemistry, biochemistry, and nonmajors chemistry. The close student contact afforded him insights into how students learn chemistry, where they have difficulties, and what strategies can help them succeed. In 1983, Dr. Silberberg decided to apply these insights in a broader context and established a text writing and editing company. Before writing his own text, he worked on chemistry, biochemistry, and physics texts for several major college publishers. He resides with his wife and child in Massachusetts. For relaxation, he cooks, sings, and walks in the woods.

Randy Duran is a Professor in Chemistry within the Butler Polymer Laboratory and Adjunct Professor in Materials Science at the University of Florida. He obtained a B.S. degree in polymer engineering from Case Western Reserve University and a Ph.D. in polymer physical chemistry from the University Louis Pasteur, Strasbourg, France. In addition to teaching general chemistry and advanced courses in physical and polymer chemistry, Dr. Duran maintains an active research program in the area of polymer surfaces and interfaces. Dr. Duran is Florida's Executive Committee Member, coordinating activities at the Material Research Collaborative Access Team Beamline at the Advanced Photon Source Synchrotron at Argonne National Laboratories. He also directs the National Science Foundation Research Experiences for Undergraduates site in chemistry at the University of Florida. His hobbies include travel, snow skiing, and gardening.

L. Peter Gold, emeritus Professor of Chemistry at The Pennsylvania State University, grew up in Massachusetts and obtained his undergraduate and graduate education at Harvard University. He taught general chemistry to over 10,000 students, as well as physical chemistry and physical chemistry lab. His hobbies included music, both as a performer and a listener, computers, and omnivorous reading. **Editor's Note:** Known as a fine teacher and mentor, Peter was actively involved as a consultant since the first edition. He passed away near the completion of the third edition.

Charles G. Haas, emeritus Professor of Chemistry at The Pennsylvania State University, earned his Ph.D. at the University of Chicago under the supervision of Norman Nachtrieb. During his 38-year career at Penn State, he taught general chemistry, undergraduate and graduate inorganic chemistry, and courses in chemical education for public school teachers at all levels. He was honored with both the College of Science Noll Award and the University Amoco Award for teaching. His research focus is on transition metals chemistry and coordination compounds. Since his retirement, he has spent much of his time reading, traveling worldwide, and enjoying collegiate athletics.

Arlan D. Norman conducted undergraduate studies at the University of North Dakota, graduate work at Indiana University, and postdoctoral research at the University of California—Berkeley; he is a Distinguished Alumnus of the University of North Dakota. Dr. Norman is Professor of Chemistry and Biochemistry at the University of Colorado—Boulder, where he has taught general and inorganic chemistry for the past 33 years, concentrating on the teaching of molecular graphics, modeling, and visualization techniques. He is currently the Associate Dean for Natural Sciences at the University of Colorado—Boulder. He is a main-group element synthetic chemist, with research interests in new materials applications of phosphorus compounds, for which he has been awarded Alfred P. Sloan and University of Colorado Council of Research and Creative Work fellowships. Dr. Norman is a Nordic skier and an avid cyclist: he has cycled extensively in Italy, New Zealand, and Ireland.

For the First and Second Editions
Robert L. Loeschen earned a B.S. in Chemistry from the University of Illinois and a Ph.D. from the University of Chicago. He joined the faculty at California State University—Long Beach in 1969. Trained as an organic chemist, he teaches a wide variety of courses, including general chemistry for science majors, chemistry for nursing majors (he coauthored a text for this subject), organic chemistry for nonmajors, organic chemistry for majors, and occasionally a graduate course in his research specialty, organic photochemistry. At present, he spends half his time in the chemistry department and half his time as Associate Dean for the College of Natural Sciences and Mathematics. When he is not teaching or deaning, he plays golf or putters in his woodworking shop.

PREFACE

Sometimes, when a new edition is in the works, a friend will ask, with a disbelieving tone, "Is there really anything new in chemistry?" What a question! As in any dynamic, modern science, theories in chemistry are refined to reflect new data, established ideas are applied to new systems, and connections are forged with other sciences to uncover new information. But chemistry, as the science of matter and its changes, is central to so many sciences—physical, biological, environmental, medical, and engineering—that it must evolve continuously to allow their progress. Designing safer, "greener" ways to make medicines, fuels, and other commodities; modeling our atmosphere and oceans to predict changes and their effects; and synthesizing new materials with revolutionary properties are among the countless areas in which chemistry is evolving.

In fact, just since the *Second Edition* of this text, hybrid gasoline-electric cars are already on the roads, and cars powered by hydrogen-based fuel cells are being developed by every automobile company. Numerous university and industrial web sites detail research efforts in the amazing field of nanotechnology, exploring the development of molecular-scale computers and biosensors. And, behold, our genes have now been mapped, and the clues they hold to disease, aging, and the miracle of our biology are there to be uncovered.

On the other hand, the basic concepts of chemistry still form the essence of the course. The mass laws and the mole concept still inform the amounts of substances in a chemical reaction; atomic properties, and the periodic trends and types of bonding emerging from them, still determine molecular structure, which in turn still governs the forces between molecules and the resulting physical behavior of substances; and the central concepts of kinetics, equilibrium, and thermodynamics still account for the dynamic aspects of chemical change.

The challenge for a modern chemistry text, then, is to do two jobs at once: to present the fundamental principles clearly and to apply them to the emerging areas of chemistry today. Like chemistry itself, the *Third Edition* of *Chemistry: The Molecular Nature of Matter and Change* has evolved in important ways to meet this challenge. This Preface explains these changes, and the Guided Tour that follows shows actual pages from the book that demonstrate its features.

OVERALL APPROACH TO TEACHING CHEMISTRY

As a species evolves, most of the structures that work well and keep it thriving and successful stay the same. And so it is with an evolving textbook. The three essential themes developed in the first two editions—visualizing chemical models, thinking through a quantitative problem, and demonstrating the amazing relevance of chemistry to society—continue to help students learn chemistry.

Visualizing Chemical Models

Because chemistry deals with observable changes in the world around us that are caused by unobservable atomic-scale events, a size gap of mind-boggling proportions must be spanned. Throughout the text, concepts are explained at the macroscopic level and then from a molecular point of view, with the text's well-known, ground-breaking illustrations placed next to the discussion to bring the point home to today's visually oriented students.

Thinking Logically to Solve Problems

The problem-solving approach, based on a widely accepted, four-step method, is introduced in Chapter 1 and employed consistently throughout the text. It encourages students to first plan a logical approach to a problem, and only then proceed to solve it quantitatively. The Check, a step unique to this text and universally recommended by instructors, fosters the habit of assessing the reasonableness and magnitude of the answer. For practice and reinforcement, each worked problem is followed immediately by a similar one, for which an abbreviated solution is given at the end of the chapter.

Applying Ideas and Skills to the Real World

An understanding of modern chemistry influences a person's attitudes about public policy issues, such as the environment, health care, and energy use, while at the same time explains everyday phenomena, such as the spring in a running shoe, the workings of a ballpoint pen, and the fragrance of a rose. Today's students may enter one of the emerging chemically related, hybrid fields—biomaterials science, nanotechnology, or planetary geochemistry, for example—and their text should keep them abreast of such career directions. But this content is only useful if it advances understanding of a principle being discussed. In addition to countless passages in the main text, four key displayed features seen in the previous two editions—Chemical Connections, Tools of the Laboratory, Galleries, and Margin Notes—provide relevant handles for what may seem abstract ideas.

INNOVATIVE TOPIC TREATMENT

A look at the Detailed Contents shows another aspect of this evolving text that has helped it thrive and, thus, has not changed: a topic order common to most general chemistry courses that incorporates flexibility for instructors to customize their approach. Innovative topic treatments appear in each chapter, but the presentation of the chemistry of the elements, organic chemistry, and biochemistry are especially novel. Rather than leaving these important topics for the end of the course, they are optimally placed for relating principles just learned.

The Interchapter and Chapter 14 apply principles from Chapters 7–13 (atomic structure, periodicity, bonding, molecular shape and polarity, and physical states) to all the main-group elements, thus emphasizing the gradation in element properties, rather than fostering misleading divisions between metals and nonmetals. Chapter 15 is a natural extension of descriptive chemistry, showing how the chemistry of organic and biological compounds arises from the atomic properties of carbon and its few bonding partners. Chapter 22 follows the example of Chapter 14 by applying the principles of kinetics, equilibrium, thermodynamics, and electrochemistry from Chapters 16–21 to the geochemistry, environmental chemistry, and industrial chemistry of the elements. The extensive coverage of biochemistry, more than in any other mainstream text, forms a major portion of Chapter 15 and is integrated into many other chapters in the text, margin notes, and boxed essays. Topics explore molecular shape in physiology, solubility factors in the structures of cell membranes and the action of antibiotics, principles of catalysis that apply to enzymes, principles of equilibrium that relate to metabolic control, electrochemical processes that produce and utilize ATP, and many more.

WHAT'S NEW IN THE *THIRD EDITION*?

This edition evolved from extensive and very positive reviewer feedback, which indicated no need for major structural change. Nevertheless, to improve the overall usability for both student and instructor, several changes were made to improve the pedagogy and enhance the content.

Improving the Pedagogy

My guiding principle throughout the conception, writing, and illustrating of all three editions has been to create a "teaching" text, one with thorough explanations that foresee student confusion before it arises. In addition, the text is replete with learning aids, which are highlighted in the Guided Tour and in the comments to the student that follow this Preface. This edition has these improvements:

- Every paragraph was examined for clarity and directness.
- A cleaner, more open page layout improves readability. Many figures now appear in the text column to help clear the margins.
- The Plan sections of the sample problems are designed to simulate an interchange between student and instructor as they think through the solution. To clarify the process, every Plan now begins explicitly with the known, incorporating data from the problem statement, and points toward the unknown.
- More challenging problems have been added to each end-of-chapter problem set.
- Every figure or table is placed as close to the related text as possible; in only one or two instances must a student turn a page to see a figure being discussed.
- Many new figures that depict the observable and molecular levels simultaneously have been added, and many more molecular models have been included.
- All chapter end matter is now keyed to the text pages on which the items appear.
- Unit canceling is now color-coded for clarity.
- The worked sample problems are now attractively set off to delineate them clearly.
- Nearly every chapter now includes multimedia features—animation, demonstration, or movie—indicated by a margin icon.

Multimedia Icon

An icon indicates a multimedia feature is available. Click on www.mhhe.com/silberberg3. Multimedia features are available for instructors on the Chemistry Animations Library 2003 CD-ROM.

Enhancing the Content

Many detailed changes have been made to achieve the highest standards of accuracy and pedagogy, ranging from clarifying a definition to simplifying a calculation step to correcting a mineral source. But several significant changes were made in order to emphasize a concept, include a topic that was lacking, or make coverage more consistent and up-to-date. Here are the most extensive changes:

- Chapter 4 has been redesigned. Following a presentation of the polar nature of water, the chapter covers ionic equations and then devotes a section to each of the three reaction types—precipitation, acid-base, and redox. Focusing on elements as reactants or products allows a discussion of types of redox reactions and greater emphasis on activity series. The brief introduction to equilibrium now includes the idea of a constant ratio of products to reactants.
- Chapter 12 includes two topics in advanced materials. The first is the physical behavior of polymers, and the discussion highlights their mass, shape, crystallinity, and viscosity—quantitative concepts that allow meaningful homework problems. The second is an overhaul of the earlier coverage of nanotechnology, this time based on the latest material from government, industrial, and academic research labs.
- Chapter 16 now includes a more complete treatment of reaction order. Because of their importance in catalyzed processes, zero-order reactions are now covered alongside first- and second-order reactions.
- Chapter 20 now includes an exceptionally consistent treatment of entropy, and the concept of reversibility is used to clarify the relationship between free energy and work.
- Chapter 21 now employs the most widely approved method for calculating cell potential, based consistently on the half-cell potentials of the cathode and anode compartments.
- Discussions of ozone depletion (Chapter 16), acid rain (Chapter 19), batteries (Chapter 21), and radioactive tracers (Chapter 24) have been thoroughly updated using input from experts.

GUIDED TOUR

This Guided Tour to Chemistry: The Molecular Nature of Matter and Change, Third Edition, *has been designed to walk you through the features of a chapter. As you examine them, note the following:*

- *Each chapter begins by orienting you to the topic flow.*
- *The multipart sample problems help you plan, execute, and check the solution.*
- *Illustrations are placed very close to the related text.*

- *The art depicts a chemical reaction on several levels of reality simultaneously.*
- *The relevance of chemistry is clearly demonstrated.*
- *The end-of-chapter material provides many ways to review and practice the concepts and skills covered.*

Chapter Outline
The outline details the topic flow of the chapter by listing the major topics and the subtopics within them.

Concepts and Skills to Review
This unique feature helps you prepare for the current chapter by listing important material from previous chapters that you should understand *before* you start reading.

CHAPTER 6

THERMOCHEMISTRY: ENERGY FLOW AND CHEMICAL CHANGE

CHAPTER OUTLINE

6.1 Forms of Energy and Their Interconversion
System and Surroundings
Energy Flow to and from a System
Heat and Work
Energy Conservation
Units of Energy
State Functions

6.2 Enthalpy: Heats of Reaction and Chemical Change
The Meaning of Enthalpy
Comparing ΔE and ΔH
Exothermic and Endothermic Processes
Types of Enthalpy Change
Bond Strength and Heat of Reaction

6.3 Calorimetry: Laboratory Measurement of Heats of Reaction
Specific Heat Capacity
The Practice of Calorimetry

6.4 Stoichiometry of Thermochemical Equations

6.5 Hess's Law of Heat Summation

6.6 Standard Heats of Reaction (ΔH°_{rxn})
Formation Equations
Determining ΔH°_{rxn} from ΔH°_f

Figure: The wonder of a burning match. When a match burns, the chemical reaction that occurs releases energy, as do many other processes. Still others absorb energy. Our ability to understand and measure changes in matter and energy is crucial to the ultimate fate of modern society, and in this chapter, we begin an exploration of the factors that govern these changes.

220

All changes in matter, whether chemical or physical, are accompanied by changes in the energy content of the matter. In the inferno of a forest fire, as wood is converted to ash and gases, its energy content changes, and that difference in energy before and after the change is *released* as heat and light. In contrast, some of the energy in a flash of lightning is *absorbed* when atmospheric N_2 and O_2 react to form NO. Energy is *absorbed* when snow melts and is *released* when water vapor condenses to rain.

The production and usage of energy in its many forms have an enormous impact on society. Some of the largest industries manufacture products that release, absorb, or limit the flow of energy. Common fuels—oil, wood, coal, and natural gas—release chemical energy for heating and to power combustion engines and steam turbines. Fertilizers help crops absorb solar energy and convert it to the chemical energy of food, which our bodies convert into other forms. Many plastic, fiberglass, and ceramic materials serve as insulators that limit energy flow.

In this chapter, we investigate the heat, or *thermal energy*, associated with changes in matter. We first examine some basic ideas of **thermodynamics**, the study of heat and its transformations. Then, we discuss **thermochemistry**, the branch of thermodynamics that deals with the heat involved in chemical reactions. We explore changes at the molecular level to find out where the heat change of a reaction comes from. Then we learn how heat is measured in order to focus on the key value: the quantity of heat released or absorbed in a reaction. The chapter ends with an overview of current and future energy sources and the conflicts between energy demand and environmental quality.

Thermodynamics is a fascinating field, a rigorously logical, highly mathematical branch of science that is as relevant in everyday life as it is in the laboratory or the environment. Our discussion of thermodynamics here, and later in Chapter 20, is confined mostly to chemical applications, but also tries to show how widely these ideas can be applied.

6.1 FORMS OF ENERGY AND THEIR INTERCONVERSION

As we discussed in Chapter 1, all energy is either potential or kinetic, and these forms are convertible from one to the other. An object has potential energy by virtue of its position and kinetic energy by virtue of its motion. The potential energy of a weight raised above the ground is converted to kinetic energy as it falls (see Figure 1.3). When the weight hits the ground, it transfers some kinetic energy to the soil and pebbles, causing them to move, thereby doing *work*. In addition, some of the transferred kinetic energy appears as *heat*, as it slightly warms the soil and pebbles. Thus, the potential energy of the weight is converted to kinetic energy, which is transferred to the ground as work and as heat.

Modern atomic theory allows us to consider other forms of energy—solar, electrical, nuclear, and chemical—as examples of potential and kinetic energy on the atomic and molecular scales. No matter what the details of the situation, *when energy is transferred from one object to another, it appears as work and/or as heat*. In this section, we examine this idea in terms of the loss or gain of energy that takes place during a chemical or physical change.

The System and Its Surroundings

To make a meaningful observation and measurement of a change in energy, we must first define the **system**, that part of the universe whose change we are going to observe. The moment we define the system, everything else relevant to the change is defined as the **surroundings**.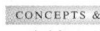

Wherever You Look, There Is a System In the example of the weight hitting the ground, if we define the falling weight as the system, the soil and pebbles that are moved and warmed are the surroundings. An astronomer may define a galaxy as the system and nearby galaxies as the surroundings. An ecologist studying African wildlife can define a zebra herd as the system and other animals, plants, and water supplies as the surroundings. A microbiologist may define a certain cell as the system and the extracellular solution as the surroundings. Thus, in general, it is the experiment and the experimenter that define the system and the surroundings.

to review before you study this chapter
- energy and its interconversion (Section 1.1)
- distinction between heat and temperature (Section 1.5)
- nature of chemical bonding (Section 2.7)
- calculations of reaction stoichiometry (Section 3.4)
- properties of the gaseous state (Section 5.1)
- relation between kinetic energy and temperature (Section 5.6)

Chapter Opener
The opener provides a thought-provoking figure and legend that relate to a main topic of the chapter.

CONSISTENT PROBLEM-SOLVING APPROACH

Sample Problem
A multipart worked-out problem appears when an important new concept or skill is introduced. A step-by-step approach is used for *every* sample problem presented in the text.

Steps

Plan
analyzes steps needed to take what is known and find what is unknown. This approach helps you think through the solution *before* performing calculations.

Solution
shows the calculation steps in the same order as detailed in the Plan and Roadmap.

Check
fosters the habit of quickly going over your work to make sure the answer is chemically and mathematically reasonable—a great way to avoid careless errors on exams.

Comment
provides an additional insight, alternative approach, or common mistake to avoid.

Follow-up Problem
gives you immediate practice with the new skill by presenting a similar problem.

SAMPLE PROBLEM 13.7 Determining the Boiling Point Elevation and Freezing Point Depression of a Solution

Problem You add 1.00 kg of ethylene glycol antifreeze ($C_2H_6O_2$) to your car radiator, which contains 4450 g of water. What are the boiling and freezing points of the solution?
Plan To find the boiling and freezing points of the solution, we first find the molality by converting the given mass of solute (1.00 kg) to amount (mol) and dividing by mass of solvent (4450 g). Then we calculate ΔT_b and ΔT_f from Equations 13.11 and 13.12 (using constants from Table 13.6). We add ΔT_b to the solvent boiling point and subtract ΔT_f from the solvent freezing point. The roadmap shows the steps.
Solution Calculating the molality:

$$\text{Moles of } C_2H_6O_2 = 1.00 \text{ kg } C_2H_6O_2 \times \frac{10^3 \text{ g}}{1 \text{ kg}} \times \frac{1 \text{ mol } C_2H_6O_2}{62.07 \text{ g } C_2H_6O_2} = 16.1 \text{ mol } C_2H_6O_2$$

$$\text{Molality} = \frac{\text{mol solute}}{\text{kg solvent}} = \frac{16.1 \text{ mol } C_2H_6O_2}{4450 \text{ g } H_2O \times \frac{1 \text{ kg}}{10^3 \text{ g}}} = 3.62 \; m \; C_2H_6O_2$$

Finding the boiling point elevation and $T_{b(\text{solution})}$, with $K_b = 0.512°C/m$:

$$\Delta T_b = \frac{0.512°C}{m} \times 3.62 \; m = 1.85°C$$

$$T_{b(\text{solution})} = T_{b(\text{solvent})} + \Delta T_b = 100.00°C + 1.85°C = \boxed{101.85°C}$$

Finding the freezing point depression and $T_{f(\text{solution})}$, with $K_f = 1.86°C/m$:

$$\Delta T_f = \frac{1.86°C}{m} \times 3.62 \; m = 6.73°C$$

$$T_{f(\text{solution})} = T_{f(\text{solvent})} - \Delta T_f = 0.00°C - 6.73°C = \boxed{-6.73°C}$$

Check The changes in boiling and freezing points should be in the same proportion as the constants used. That is, $\Delta T_b/\Delta T_f$ should equal K_b/K_f: 1.85/6.73 = 0.275 = 0.512/1.86.
Comment These answers are only approximate because the concentration far exceeds that of a *dilute* solution, for which Raoult's law is most useful.

FOLLOW-UP PROBLEM 13.7 What is the minimum concentration (molality) of ethylene glycol solution that will protect the car's cooling system from freezing at 0.00°F? (Assume the solution is ideal.)

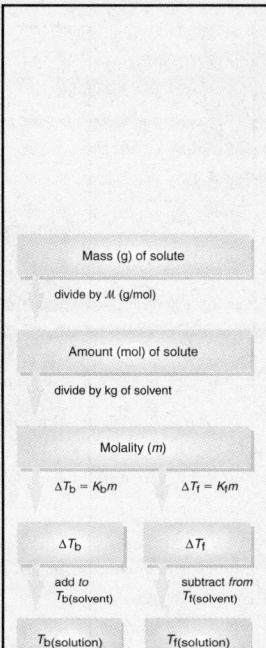

Mass (g) of solute

divide by $\mathcal{M}$ (g/mol)

Amount (mol) of solute

divide by kg of solvent

Molality (m)

$\Delta T_b = K_b m$ $\Delta T_f = K_f m$

ΔT_b ΔT_f

add *to* $T_{b(\text{solvent})}$ subtract *from* $T_{f(\text{solvent})}$

$T_{b(\text{solution})}$ $T_{f(\text{solution})}$

Problem-Solving Roadmaps
are included in many worked exercises. They are block diagrams, *specific* to the problem, that visually lead you through the steps.

Brief Solutions to Follow-up Problems *(continued)*

13.5 Mass % HCl $= \dfrac{\text{mass of HCl}}{\text{mass of soln}} \times 100$

$= \dfrac{\dfrac{11.8 \text{ mol HCl}}{1 \text{ L soln}} \times \dfrac{36.46 \text{ g HCl}}{1 \text{ mol HCl}}}{\dfrac{1.190 \text{ g}}{1 \text{ mL soln}} \times \dfrac{10^3 \text{ mL}}{1 \text{ L}}} \times 100$

$= 36.2$ mass % HCl

Mass (kg) of soln $= 1 \text{ L soln} \times \dfrac{1.190 \times 10^{-3} \text{ kg soln}}{1 \times 10^{-3} \text{ L soln}}$
$= 1.190$ kg soln

Mass (kg) of HCl $= 11.8 \text{ mol HCl} \times \dfrac{36.46 \text{ g HCl}}{1 \text{ mol HCl}} \times \dfrac{1 \text{ kg}}{10^3 \text{ g}}$
$= 0.430$ kg HCl

Molality of HCl $= \dfrac{\text{mol HCl}}{\text{kg water}} = \dfrac{\text{mol HCl}}{\text{kg soln} - \text{kg HCl}}$
$= \dfrac{11.8 \text{ mol HCl}}{0.760 \text{ kg } H_2O} = 15.5 \; m$ HCl

$X_{\text{HCl}} = \dfrac{\text{mol HCl}}{\text{mol HCl} + \text{mol } H_2O}$
$= \dfrac{11.8 \text{ mol}}{11.8 \text{ mol} + \left(760 \text{ g } H_2O \times \dfrac{1 \text{ mol}}{18.02 \text{ g } H_2O}\right)} = 0.219$

13.6 $\Delta P = X_{\text{aspirin}} \times P^0_{\text{methanol}}$
$= \dfrac{\dfrac{2.00 \text{ g}}{180.15 \text{ g/mol}}}{\dfrac{2.00 \text{ g}}{180.15 \text{ g/mol}} + \dfrac{50.0 \text{ g}}{32.04 \text{ g/mol}}} \times 101 \text{ torr}$
$= 0.713$ torr

13.7 $m = \dfrac{(0.00°F - 32°F)\left(\dfrac{5°C}{9°F}\right)}{1.86°C/m} = 9.56 \; m$

13.8 $\Pi = MRT$
$= (0.30 \text{ mol/L})\left(0.0821 \dfrac{\text{atm·L}}{\text{mol·K}}\right)(37°C + 273.15)$
$= 7.6$ atm

Brief Solutions to Follow-up Problems
These provide brief solutions at the end of the chapter, not just a numerical answer at the back of the book. This fuller treatment effectively *doubles* the number of worked-out problems and is an excellent way for you to reinforce skills.

ILLUSTRATIONS AND PAGE LAYOUT
DESIGNED TO MAXIMIZE UNDERSTANDING

Three-Level Illustrations

A Silberberg trademark, these illustrations provide macroscopic and molecular views of a process, so you learn to connect these two levels of reality with each other and with the chemical equation that describes the process in symbols.

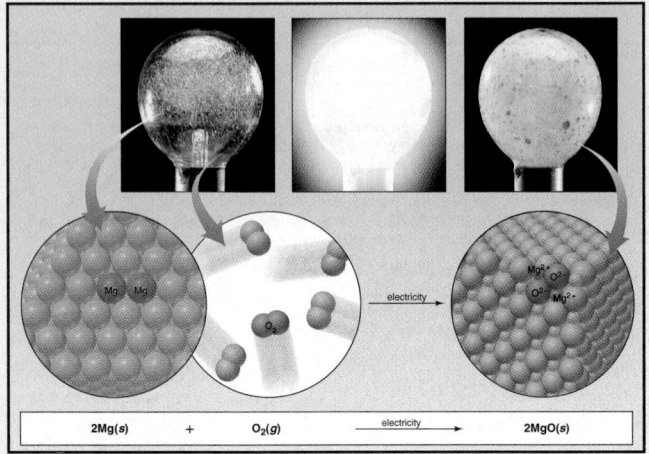

$$2Mg(s) \quad + \quad O_2(g) \xrightarrow{\text{electricity}} 2MgO(s)$$

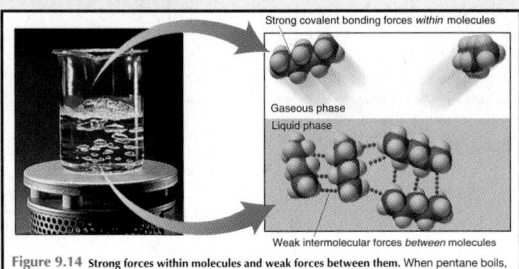

Figure 9.14 Strong forces within molecules and weak forces between them. When pentane boils, weak forces *between* molecules (intermolecular forces) are overcome, but the strong covalent bonds holding the atoms together within each molecule remain unaffected. Thus, the pentane molecules leave the liquid phase as intact units.

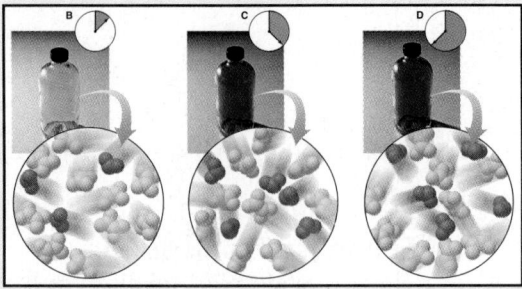

Cutting-Edge Molecular Models

Author and artist worked side by side and employed the most advanced computer-graphic software to provide accurate molecular models and vivid molecular-level scenes. Included in the third edition are over 50 new pieces of art.

Page Layout

Author and pager worked side by side to create a book of two-page spreads that place figures as close as possible to their related text. In some cases, art is even wrapped by text, so you can read concepts and see them depicted simultaneously.

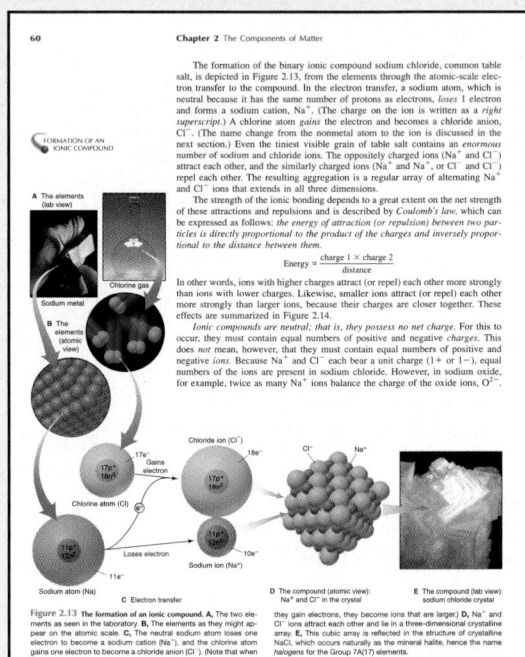

Margin Notes

Over 135 short, lively explanations apply ideas in the related text. You'll explore how your body and the Earth control their temperatures, how crime labs track illegal drugs, how your lungs work, how fat-free chips and decaf coffee are made, the principle of a winning poker hand, the risks of nuclear radiation, handy tips for memorizing relationships, and much more.

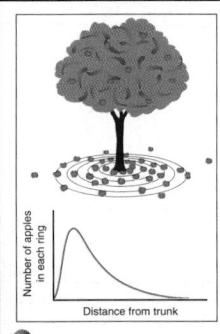

A Radial Probability Distribution of Apples An analogy might clarify why the curve in the radial probability distribution plot peaks and then falls off. Picture the fallen apples around the base of an apple tree: the density of apples is greatest near the trunk and decreases with distance. Divide the ground under the tree into foot-wide concentric rings and collect the apples within each ring. Apple density is greatest in the first ring, but the area of the second ring is larger, so it contains a greater *total* number of apples. Farther out near the edge of the tree, rings have more area but lower apple "density," so the total number of apples decreases. A plot of "number of apples within a ring" vs. "distance of ring from trunk" shows a peak at some close distance from the trunk, as in Figure 7.16D.

APPLYING CHEMISTRY TO THE REAL WORLD

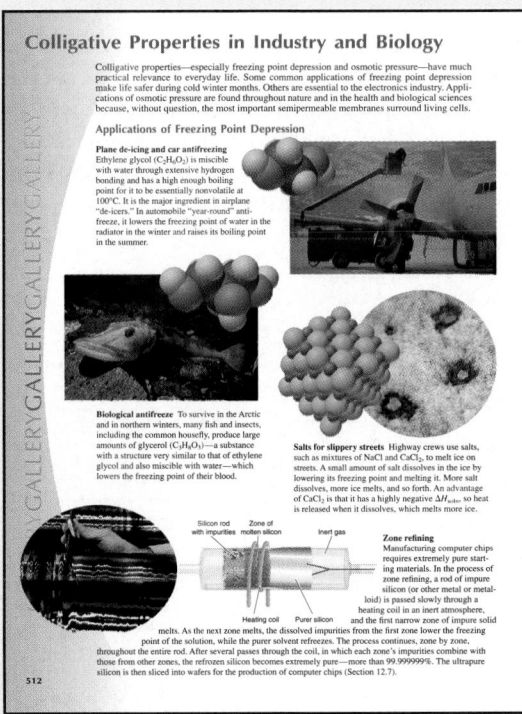

Tools of the Laboratory
These essays describe the key instruments and techniques that chemists use in modern practice to obtain the facts that underlie their theories.

Chemical Connections
These essays show the interdisciplinary nature of chemistry by applying chemical principles to related scientific fields, including physiology, geology, biochemistry, engineering, and environmental science.

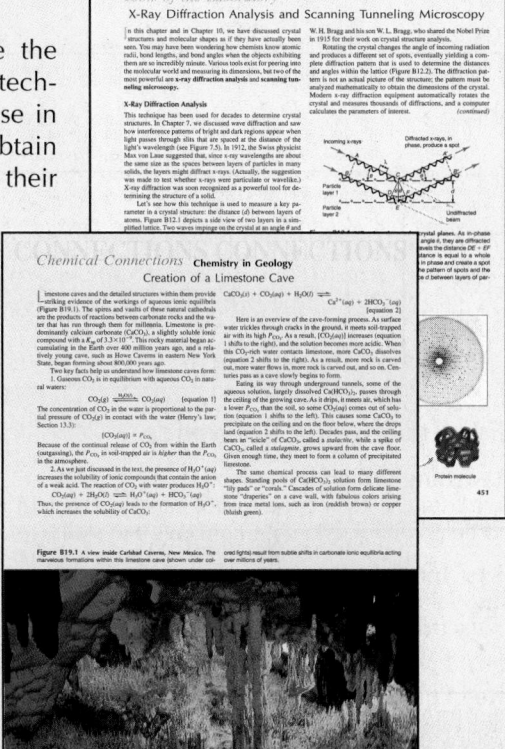

Galleries
These illustrated summaries of applications show how common and unusual substances and products relate to chemical principles. You'll learn how a ballpoint pen works, why bubbles in a drink are round, why contact-lens rinse must have a certain concentration, how disposable and rechargeable batteries differ, and many other intriguing applications.

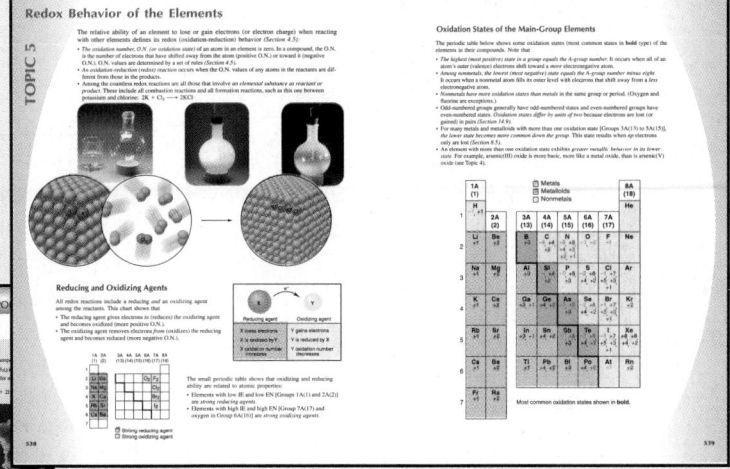

Family Portraits
These two-page illustrated summaries appear in Chapter 14 and detail the atomic, physical, and chemical properties of each main group.

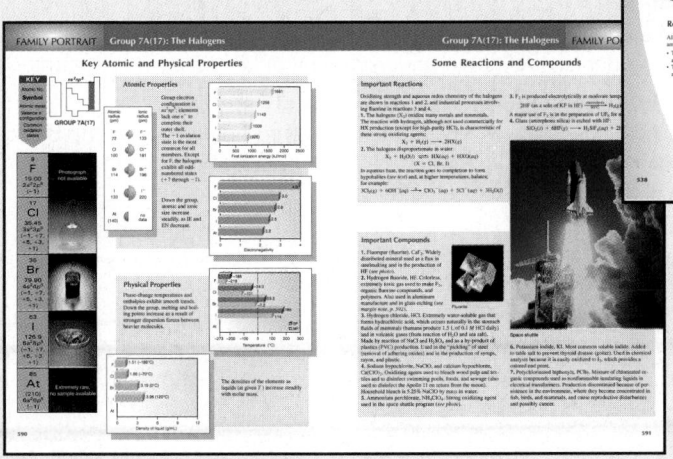

Interchapter
This multipage, illustrated Perspective on the Properties of the Elements reviews the major points in Chapters 7–13 that deal with atomic properties and their resulting effects on element behavior.

TOOLS THAT REINFORCE
THE CONCEPTS

Chapter Perspective

This chapter is the last of three that explore the nature and variety of equilibrium systems. In Chapter 17, we discussed the central ideas of equilibrium in the context of gaseous systems. In Chapter 18, we extended our understanding to acid-base equilibria. In this chapter, we highlighted three types of aqueous ionic systems and examined their role in the laboratory and the environment. The equilibrium constant, in all its forms, is a number that provides a limit to changes in a system, whether a chemical reaction, a physical change, or the dissolution of a substance. You now have the skills to predict whether a change will take place and to calculate its result, but you still do not know why the change occurs in the first place, or why it stops when it does. In Chapter 20, you'll find out.

Section Summaries and Chapter Perspective

Concise paragraphs conclude each section, immediately restating the major ideas just covered. Each chapter ends with a brief overview that places it in the context of previous and upcoming topics.

For Review and Reference

Learning Objectives
are listed, with section and/or sample problem numbers, to help you focus on key concepts and skills.

Key Terms
are boldfaced within the chapter. They are arranged here by section (with page number) and defined again in the Glossary.

Key Equations and Relationships
are screened and numbered within the chapter and listed here with page number.

Highlighted Figures and Tables
are listed by page number, so you can find them easily and review their essential content.

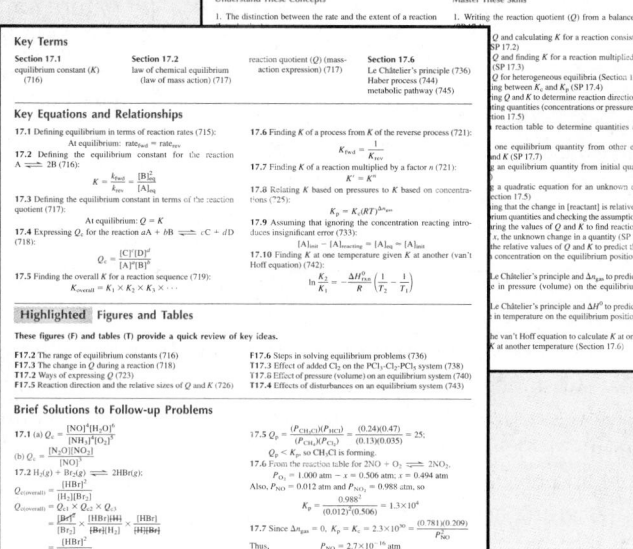

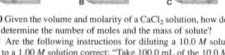

End-of-Chapter Problems

An exceptionally large number of problems follow each chapter. Three types of problems are keyed by chapter section.

Concept Review Questions
These test students' qualitative understanding of key ideas in the chapter.

Skill-Building Exercises
Written in pairs, with one of each pair answered in the back of the book, these exercises begin simply and increase in difficulty, gradually weaning the student from multistep directions.

Problems in Context
These problems apply the learned skills to interesting scenarios, including examples from industry, medicine, and the environment.

Comprehensive Problems
Following the section-based problems is a large group of more challenging problems based on concepts and skills from any section and/or earlier chapter.

SUPPLEMENTS FOR
THE INSTRUCTOR

Multimedia Supplements

Digital Content Manager

This multimedia collection of visual resources allows instructors to utilize artwork from the text in multiple formats to create customized classroom presentations, visually based tests and quizzes, dynamic course website content, or attractive printed support materials. The digital assets on this cross-platform CD-ROM are grouped by chapter within easy-to-use folders.

Chemistry Animations Library 2003 CD-ROM

This CD-ROM contains over 300 animations, several authored by Martin Silberberg. With an easy-to-use application, the CD enables users to quickly view the animations and import them into PowerPoint to create multimedia presentations.

Online Learning Center
(Instructor Center)

The Instructor Center is an online repository for teaching aids. It houses downloadable and printable versions of traditional ancillaries plus a wealth of online content.

All Online Learning Center material is available at WebCT and Blackboard. All end-of-chapter problems from the text are included within these course delivery systems.

Brownstone's DIPLOMA©

With its Test Generator, On-Line Testing Program, Internet Testing, and Grade Management Systems, Brownstone's Diploma is an invaluable instructor resource. This user-friendly software's testing capability is consistently ranked number one over other products.

Printed Supplements

Instructor's Solutions Manual

This printed supplement contains complete, worked-out solutions for all the end-of-chapter problems in the text.

Transparencies

This boxed set of 300 full-color transparency acetates features images from the text that are modified to ensure maximum readability in both small and large classroom settings.

SUPPLEMENTS FOR THE STUDENTS

Multimedia Supplements

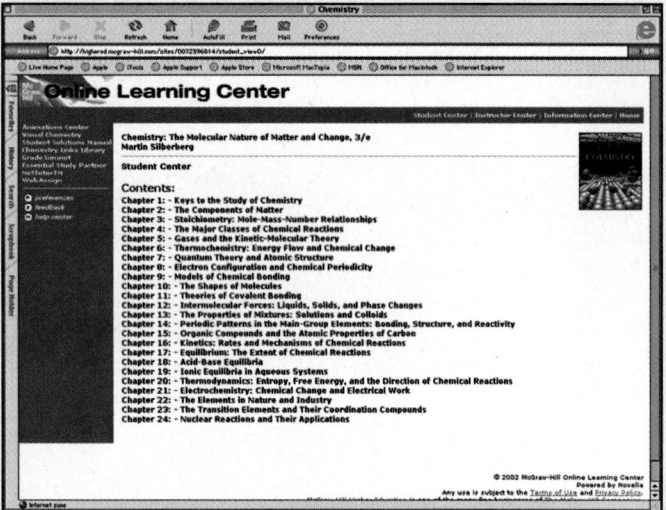

Online Learning Center (Student Center)

The Student Center of the OLC features quizzes, interactive learning games, and study tools tailored to coincide with each chapter of the text.

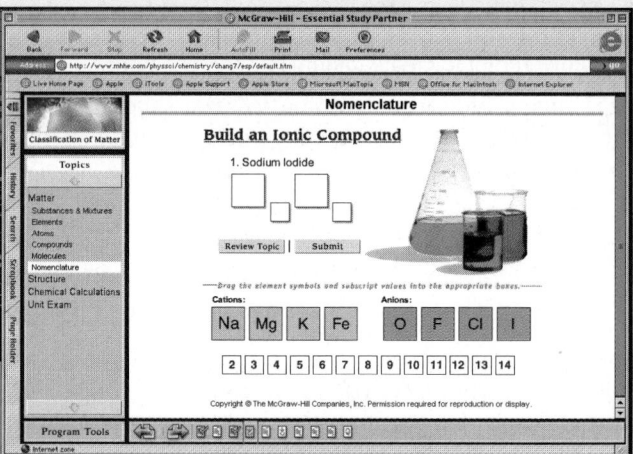

Essential Study Partner

Located within the OLC, this online resource consists of a collection of interactive study modules containing hundreds of animations, learning activities, and quizzes designed to help students grasp complex concepts.

ChemSkill Builder

This incredible online tool contains more than 1500 algorithmically generated questions, each with tutorial feedback. There is a direct correlation between student time investment in this program and increased problem-solving ability. A record of student work is maintained in an online gradebook so that homework can be done at home, in a dorm room, or in a university lab.

NetTutor™

Located within the OLC, NetTutor is a revolutionary online environment that allows users to participate in real-life group learning sessions. Students can communicate with live tutors, submit specific questions, or browse archives of previously asked questions and their answers.

Printed Supplements

Student Study Guide

This excellent resource reinforces the most important points in every chapter of the book.

Student Solutions Manual

This manual contains complete worked-out solutions to all follow-up problems and nearly half the end-of-chapter problems in the text.

ACKNOWLEDGMENTS

A small army of exceptional people from publishing, academia, and industry helped give life to the *Third Edition*. Deepest thanks go, as always, to my team of consultants—Randy Duran, Peter Gold, Chuck Haas, and Arlan Norman—for their continued support and expert advice on countless points of content. I am extremely grateful to Bob Loeschen of California State University—Long Beach for his superb consulting on past editions.

In the final days of this revision, I was deeply saddened to learn that Peter Gold had passed away. From the beginning, Peter was a guiding force in shaping the content and pedagogy of the text; his insight, wit, and friendship will be sorely missed.

Many other professors played essential roles in creating this edition. Most fortunately, I again had the exceptional talents of Dorothy B. Kurland for the project, this time as in-depth reviewer of the entire manuscript. And I was privileged to have Jim Horvath of the University of Florida supply excellent advice and comments for three-quarters of the chapters. Expert contributors helped me keep key topics as up-to-date as possible: Randy Duran for the discussion of polymers (Ch. 12), Robert M. Metzger of the University of Alabama for the nanotechnology discussion (Ch. 12), Jonathan Kurland of Dow Chemical Company for the boxed essays on ozone depletion (Ch. 16) and acid rain (Ch. 19), Perla B. Balbuena of the University of South Carolina and John S. Newman of the University of California—Berkeley for the gallery on batteries (Ch. 21), and Sherry Yennello of Texas A&M University for the discussion of radioactive tracers (Ch. 24). The wide range of homework problems has been universally praised by users of the first two editions, and Joseph Bularzik of Purdue University—Calumet, S. Walter Orchard of Tacoma Community College, and Royale S. Underhill of the University of Florida helped me generate many excellent new ones for this edition. Thanks also to Scott Perry of the University of Houston for his general advice and guidance on media issues.

Modern texts are served by a battery of supplements, and this one is very lucky to have supplement authors so committed to accuracy and clarity for student and instructor. Once again, Elizabeth Bent Weberg wrote a superb *Student Study Guide*. Deborah Wiegand of the University of Washington and Tristine Samberg devoted themselves to creating an excellent *Student Solutions Manual*. Then, Brenda Woodward and Marcia Gillette of Indiana University—Kokomo created the *Instructor's Solutions Manual*, interacting continuously with the student manual authors to ensure consistency.

I am especially grateful for the support of the Board of Advisors, a select group of chemical educators dedicated to helping make this text the optimum teaching tool. The Board was formed during the revision and contributed insightful comments that shaped the edition in many significant ways. I look forward to working together on upcoming editions:

Board of Advisors

Michael R. Abraham *University of Oklahoma*
Margaret Asirvatham *University of Colorado—Boulder*
Christina A. Bailey *California Polytechnic State University—San Luis Obispo*
David S. Ballantine, Jr. *Northern Illinois University*
Richard Bauer *Arizona State University*
Dominick J. Casadonte, Jr. *Texas Tech University*
David K. Erwin *Rose-Hulman Institute of Technology*
Deborah Berkshire Exton *University of Oregon*
David Frank *California State University—Fresno*
Russell Geanangel *University of Houston*
John I. Gelder *Oklahoma State University*
David W. Grainger *Colorado State University*
Jeffrey L. Krause *University of Florida*
Richard H. Langley *Stephen F. Austin State University*
Amy Harris Lindsay *University of New Hampshire*
Michael E. Lipschutz *Purdue University*
Mark E. Noble *University of Louisville*
Frazier W. Nyasulu *University of Washington*
Louis H. Pignolet *University of Minnesota*
James P. Reilly *Indiana University*
Jill K. Robinson *University of Wyoming*
Reva A. Savkar *Northern Virginia Community College*
W. Robert Scheidt *University of Notre Dame*
Diane K. Smith *San Diego State University*
Douglas Tobias *University of California—Irvine*
John B. Vincent *University of Alabama*
Sharon Vincent *Shelton State Community College*
Tom Webb *Auburn University*

My deepest appreciation goes as well to the reviewers who contributed invaluably to the text's final content:

Susmita Acharya *Cardinal Stritch University*
Alexander Altman *St. Josephs College—Suffolk*
Christopher Anukwuem *Prairie View A&M University*
Marsi Archer *Missouri Southern College*
Gabriele Backes *Portland Community College—Sylvania*
Satinder Bains *Arkansas State University—Beebe*
David Baker *Delta College*
Louis Barash *New Jersey Institute of Technology*
Mufeed Basti *North Carolina A&T University*
Partha Basu *Duquesne University*
Danny Bedgood *Arizona State University—Tempe*
Jean Bowdan *Somerset Community College*
Edward Brown *Lee University*
Stuart Burris *Belmont University*
Lynn Carlson *University of Wisconsin—Kenosha*
Donna L. Chevalier *Bossier Parish Community College*
Wheeler Conover *Southeast Community College*
Mary Cook *State Technical Institute at Memphis*
Felicia Corsaro-Barbieri *Gwynedd—Mercy College*
Philip Crawford *Southeast Missouri State University*
Bill Davis *University of Texas at Brownsville*
Son Do *University of Louisiana—Lafayette*
Kelley Donaghy *American University*
Veljko Dragojlovic *Nova Southeastern University*
David Easter *Southwest Texas State University*
Amina K. El-Ashmawy *Collin County Community College—Plano*
Richard Fiore *GateWay Community College*
Paul Flowers *University of North Carolina—Pembroke*
Kathy Flynn *College of the Canyons*
Cheryl Frech *University of Central Oklahoma*
Andrea Gorczyca *Brookhaven College*
Lois Hansen-Polcar *Cuyahoga Community College Western—Parma*

Ryan Harden *Central Lakes College*
Steven Heninger *Allegany College of Maryland*
Narayan Hosmane *Northern Illinois University*
Michael Hurst *Georgia Southern University*
Ramee Indralingam *Stetson University*
Thomas Jeffries *Combellsville University*
Martha Joseph *Westminster College*
David Karpovich *Saginaw Valley State University*
Paul Karr *Wayne State College*
Igor Khudyakov *Catawba College*
Pamela S. Kimbrough *Chaffey College*
Robert Kolodny *Armstrong Atlantic State University*
Lawrence Krebaum *Missouri Valley College*
Brian Lamp *Truman State University*
Andrew Langrehr *Jefferson College*
Pamela Leggett-Robinson *North Greenville College*
Zhaohui Li *University of Wisconsin—Kenosha*
David Lippmann *Southwest Texas State University*
Kirby Lowery *Wharton County Junior College*
Lawrence Madsen *Mesa State College*
Jeffry Madura *Duquesne University*
Ruhulla Massoudi *South Carolina State University*
Mark Masthay *Murray State University*
Lyle McAfee *The Citadel*
Michael McClure *Hopkinsville Community College*
Dan McInnes *East Central University*
Charles W. McLaughlin *University of Nebraska—Lincoln*
Therese Michels *Dana College*
Abdul Mohammed *North Carolina A&T University*
Michael Moran *West Chester University of Pennsylvania*
Richard Nafshun *Oregon State University*
Terry Nagel *Albertson College of Idaho*
William Oliver *Northern Kentucky University*
Mark Ott *Western Wyoming College*
H. Blake Otwell *Jacksonville State University*
David Peitz *Wayne State College*
Suzette Polson *Kentucky State University*
Walda Powell *Meredith College*

Anne Quinn *Delta College*
Carol Quinn *Dominican College*
Don Rufus Ranatunga *Oakwood College*
Jeanette Rice *Georgia Southern University*
Stanley Rice *Southeastern Oklahoma State University*
Rene Rodriguez *Idaho State University*
Michael Russell *Mt. Hood Community College*
Matthew Ryan *Purdue University—Calumet-Hammond*
John E. Schultz *Westminster College*
Amit Sharma *Columbia Basin College*
Joseph Sinski *Bellarmine College*
Joseph Sneddon *McNeese State University*
Gordon Sproul *University of South Carolina—Beaufort*
Max Stappler *Queensborough Community College*
James Stickler *Allegany College of Maryland*
Sam Subramaniam *Miles College*
Richard G. Summers *Lawrence University*
Paris Svoronos *Queensborough Community College*
Erach Talaty *Wichita State University*
Michael Tessmer *Southwestern College*
Mark Thomson *Xavier University*
Tanya To *Peninsula College*
Richard Treptow *Chicago State University*
Phillip Voegel *Midwestern State University*
Elizabeth Mahan Wallace *Western Oklahoma State College*
Randall Wanke *Augustana College*
Susan Weiner *West Valley College*
Neil Weinstein *Santa Fe Community College*
Michael Wells *Campbell University*
Teresa Wenda *Centralia College*
Lane Whitesell *St. Gregory's University*
Daniel Williams *Kennesaw State University*
Anthony Wren *Butte College*
Carolyn Yoder *Millersville University*
Robert Yolles *De Anza College*
Chewki Ziana-Cherif *Florida Community College*
Susan Moyer Zirpoli *Slippery Rock University of Pennsylvvania*
Lisa Zuraw *The Citadel*

Students are, of course, the end users of the text, the final testing ground where the "rubber hits the road." For that reason, I was very fortunate to have extremely useful feedback from students at three institutions:

University of Minnesota
Instructors: Lou Pignolet, Barb Edgar, Ken Leopold, Jeff Roberts. Students: John Arend, Alyssa Bakke, Corrie Bastian, Elizabeth Bentzier, Predrag Bjelogrlic, Matthew Durant, Shelley Gamm, Scott Graupner, Kim Ha, Kelly Hiatt, Melanie Howard, Henry Koeller, Kerry Landry, Britt Lindsay, Benjamin Linstrom, Paul Lobitz, Kristin Martin, Sarah Nelson, Alex Nester, Bayna Oschwald, Andre Reed, Leslie Schultz, Elena Tamura, Paul Wanless III, Emily Whipple, Kaydee Wick, Mohd S. Yahya, Dustin Zastera

University of California—Irvine
Instructor: Pat Rogers. Students: Jason Berg, Anh Dao, Emily Harris, Mat Kladney, Mypha Ninh, Kevin Olson, Dan Trinh

University of Missouri—Columbia
Instructor: Steven Keller and members of his Chemistry 32 class.

And my gratitude extends to those reviewers who helped so much with the previous edition:

Hugh Akers *Lamar University—Beaumont*
Robert Allendoerfer *SUNY—Buffalo*
Michael Batley *Macquarie University*
G. Ronald Brown *University of Northern British Columbia*
Elaine Carter *Los Angeles City College*
Thomas Chasteen *Sam Houston State University*
Nadia Cordero de Figuero *University of Puerto Rico*
Richard Daake *Bartlesville Wesleyan College*
Nordulf Debye *Towson University*
George Fleck *Smith College*
Dennis Flentge *Cedarville College*
Ron Garber *California State University—Long Beach*
John Garland *Washington State University*

Marcia Gillette *Indiana University—Kokomo*
Arthur Grosser *McGill University*
Tandy Grubbs *Stetson University*
Alton Hassell *Baylor University*
Stephen Hawkes *Oregon State University*
Ronald Johnson *Emory University*
Michael W. Jones *New Mexico Military Institute*
Silvia Kölchens *Pima Community College*
Robert Kren *University of Michigan—Flint*
Julie Kuehn *William Rainey Harper College*
Joseph D. Laposa *McMaster University*
Thornton Lipscomb *University of Louisville*
Ian McNaught *University of Sydney*
Richard S. Mitchell *Arkansas State University*

Pamela New *University of Central Oklahoma*
Larry Pederson *College of Misericordia*
Joann Pfeiffer *University of Wyoming*
James Purcell *Manor Junior College*
Melvin Roberts *Columbia University*
Theodore Sakano *Rockland City Community College*
Karl Seff *University of Hawaii—Honolulu*
Mary Jane Shultz *Tufts University*
Scott Sinex *Prince Georges Community College*
Barbara Spohr *Dodge City Community College*
Peter John Stiles *Macquarie University*
Bruce Storhoff *Ball State University*
Howard Wiley, Jr. *Middlesex Community Technical College*
Thao Yang *University of Wisconsin—Eau Claire*

No author could hope for a more supportive group of publishing professionals than the book team at McGraw-Hill Higher Education: Kent Peterson, Publisher—and friend—who led the team and championed the project with consummate skill, tireless effort, and good humor; Joan Weber, Developmental Editor, who managed the supplements and was always available to help in any way I asked; Thomas Timp, Marketing Manager, who mounted a vigorous campaign to present the book to professors and students; and Peggy Selle, Senior Project Manager, who coordinated the in-house production group with patient expertise.

Jane Hoover copyedited the manuscript superbly, making the writing more clear and direct in countless places. My friend Michael Goodman created many more of his amazing pieces of molecular art as well as the astounding cover. Once again, with friendly insistence, Lori Johnson coordinated the innumerable details of production at GTS Graphics. And maneuvering this ship through storms and shallows to arrive at port on time was Karen Pluemer, the freelance developmental editor who, while attending to family, building a new home, and teaching college English, somehow managed to handle the endless needs of book and author with calm, competence, and warmth; I can't thank her enough.

With this edition, my wonderful son Daniel, now 13, could help sketch art layouts, choose new photos, and organize my literature sources. But, most important, he never failed to cheer me on with an admiring smile and a loving hug. As always, Ruth's experienced hand is visible everywhere. In addition to a million other details, she collaborated with the production department on style and design, edited the art manuscript, and laid out the entire book so that the student can read a paragraph and see the accompanying figure without needing to turn the page. Her love and boundless support are my foundation.

A NOTE TO THE STUDENT: HOW TO DO WELL IN THIS COURSE

Why are you taking this course? A requirement, a prerequisite, or maybe just a chance to learn more about chemistry? Whatever your reason, one of the best ways to do well in it is to make good use of this textbook. Think of it as a "how-to-study" manual because it was designed for you! Use the following features (which are illustrated in the Guided Tour of this *Preface*) to help you learn chemistry:

Chapter Outline Before going to class, look at the list of topics and read the introduction. Its last paragraph points out the topics again.

Concepts and Skills to Review These are the topics you must know from earlier chapters in order to understand this one.

Figures, Tables, and Legends As you read, look at the figures and tables; they're placed as close as possible to their description in the text, so you don't have to turn the page back and forth. Read the figure legend, too, because it restates the main idea. Especially important pieces have a number highlighted with a green screen and are listed at the end of the chapter. Review them when you study.

Sample Problems Carefully read through each part of these worked problems. They present a sound problem-solving strategy, helping you devise a Plan before you start calculating the Solution. Each answer is followed by a Check—a great habit for you to get into. Try the Follow-up Problem as soon as you finish studying the worked problem; its Brief Solution at the end of the chapter can help you find where you may have made an error.

Key Terms and Equations Boldfaced terms and screened numbered equations in the text are also listed at the end of the chapter. Focus on them when you study. (The key terms also appear in the Glossary.)

Section Summary Read over the Summary as you finish each section to check whether you understand the most important points. If any are unclear, you can see immediately where you need to review.

Chapter Perspective These brief paragraphs tell you how the various topics in the course fit together. You'll see where you've just been and where you're going in the next chapter.

Learning Objectives Examine this list of concepts and skills to make sure you understand them. Go back to any section or sample problem that you're unsure of. Or use the list as you read the chapter to pick up on the main points to learn.

Problems Keeping up with homework is the key to success. Each text section is represented by a group of problems that includes Concept Review Questions that test your understanding of ideas, paired Skill-Building Exercises that test your ability to apply mathematics in a chemical setting, and Problems in Context that test your ability to apply these skills in realistic scenarios. After you work these, challenge yourself with Comprehensive Problems that apply material from any section or previous chapter (tougher ones have an asterisk). Green-numbered problems are answered in Appendix E.

As the course unfolds, don't forget to take a breath every once in a while and let some of the amazing ideas that you're learning sink in. Seeing the world from a chemical point of view will be an awesome experience. Let me know how the text helps and how to make it better.

Martin S. Silberberg

KEYS TO THE STUDY OF CHEMISTRY

CHAPTER OUTLINE

1.1 Some Fundamental Definitions
Properties of Matter
States of Matter
Central Theme in Chemistry
Importance of Energy

1.2 Chemical Arts and the Origins of Modern Chemistry
Prechemical Traditions
Impact of Lavoisier

1.3 The Scientific Approach: Developing a Model

1.4 Chemical Problem Solving
Units and Conversion Factors
Solving Chemistry Problems

1.5 Measurement in Scientific Study
Features of SI Units
SI Units in Chemistry

1.6 Uncertainty in Measurement: Significant Figures
Which Digits Are Significant
Significant Figures in Calculations
Precision and Accuracy

Figure: A Different View of an Everyday Scene. With molecular sight, a typical scene in the lab, the flame of a Bunsen burner, becomes a source of hurtling particles. Molecules of methane and oxygen react near the burner tip to form carbon dioxide and water. The study of chemistry reveals the world we observe as the result of a hidden, atomic reality—one we can measure and change—and this chapter opens the door to it.

CONCEPTS & SKILLS

to review before you study this chapter
• exponential (scientific) notation
  (Appendix A)

The new millenium has arrived, and, as always, the science of chemistry stands at the forefront of change: creating new batteries and fuel cells to power society and sustain the environment, designing medicines based on our new understanding of the human genetic makeup to prevent the scourge of old and new diseases, and even researching the origin of life as we explore our solar system and nearby systems for signs of it elsewhere. Addressing these and countless other challenges and opportunities depends on an understanding of the concepts you will learn in this course.

You probably already know that chemistry has an impact on your daily life, but the extent of that impact can be mind-boggling. Consider the beginning of a typical day—perhaps this one—from a chemical point of view. Molecules align in the liquid crystal display of your clock, electrons flow through its circuitry to create a rousing sound, and you throw off a thermal insulator of manufactured polymer. You jump in the shower to emulsify fatty substances on your skin and hair with chemically treated water and formulated detergents. You adorn yourself in an array of processed chemicals—pleasant-smelling pigmented materials suspended in cosmetic gels, dyed polymeric fibers, synthetic footwear, and metal-alloyed jewelry. Today, breakfast is a bowl of nutrient-enriched, spoilage-retarded cereal and milk, a piece of fertilizer-grown, pesticide-treated fruit, and a cup of a hot aqueous solution of neurally stimulating alkaloid. After brushing your teeth with artificially flavored, dental-hardening agents dispersed in a colloidal abrasive, you're ready to leave, so you collect some books—processed cellulose and plastic, electronically printed with light- and oxygen-resistant inks—hop in your hydrocarbon-fueled, metal-vinyl-ceramic vehicle, electrically ignite a synchronized series of controlled gaseous explosions, and you're off to class!

The influence of chemistry extends to the natural environment as well. The air, water, and land and the organisms that thrive there form a remarkably complex system of chemical interactions. While modern chemical products have enhanced the quality of our lives, their manufacture and use also pose increasing dangers, such as toxic wastes, acid rain, global warming, and ozone depletion. If our careless application of chemical principles has led to some of these problems, our careful application of the same principles will help to solve them.

Perhaps the significance of chemistry is most profound when you contemplate the chemical nature of biology. Molecular events taking place within you right now allow your eyes to scan this page and your brain cells to translate fluxes of electric charge into thoughts. The most vital biological questions—How did life arise and evolve? How does an organism reproduce, grow, and age? What is the essence of health and disease?—ultimately have chemical answers.

This course comes with a bonus—the development of two mental skills you can apply to any science-related field. The first skill, common to all science courses, trains you to approach problems systematically, especially in planning a solution and then executing it. The second skill is specific to chemistry, for as you comprehend its ideas, your mind's eyes will learn to see a hidden level of the universe, one filled with incredibly minute particles hurtling at fantastic speeds and colliding billions of times a second as they link and relink in different combinations. It is these particles and their interactions that result in everything inside and outside of you.

This first chapter holds the keys to help you enter this new world. It begins with some fundamental definitions and concepts and then examines the historical origins of chemistry and the approach chemists take to understand nature. We discuss modern systems of measurement and calculation and consider how to solve chemistry problems. Finally, we look at one of many examples of chemistry and other sciences working together for society's benefit. If, at this point, you are wondering if chemistry is important to you, soon you'll be wondering how any educated person can function today without knowing chemistry!

1.1 SOME FUNDAMENTAL DEFINITIONS

The science of chemistry deals with the makeup of the entire physical universe. A good place to begin is with the definition of a few central ideas, some of which may be familiar to you. **Chemistry** is *the study of matter and its properties, the changes that matter undergoes, and the energy associated with those changes.*

The Properties of Matter

Matter is the "stuff" of the universe: air, glass, planets, students—*anything that has mass and volume.* (In Section 1.5, we discuss the meanings of mass and volume in terms of how they are measured.) Chemists are particularly interested in the **composition** of matter, *the types and amounts of simpler substances that make it up.* A *substance* is a type of matter that has a defined, fixed composition.

We learn about matter by observing its **properties,** *the characteristics that give each substance its unique identity.* To identify a person, we observe such properties as height, weight, hair and eye color, fingerprints, blood type, and so on, until we arrive at a unique identification. To identify a substance, chemists observe two distinct types of properties, physical and chemical, which are closely related to two types of change that matter undergoes. **Physical properties** are those that a substance shows *by itself, without changing into or interacting with another substance.* Some physical properties are color, melting point, electrical conductivity, and density. It follows that a **physical change** occurs when a substance *alters its physical form, **not** its composition.* A physical change results in different physical properties. For example, when ice melts, several *physical* properties have changed, such as hardness, density, and ability to flow. But the sample has *not* changed its composition: it is still water. The photo in Figure 1.1A shows this change the way you would see it in everyday life. In your imagination, you can see a magnified view of the particles that appears in the "blow-up" circles. Note that the same particles appear in solid and liquid water:

Physical change (same substance before and after):

Water (solid form) $\longrightarrow$ water (liquid form)

On the other hand, **chemical properties** are those that a substance shows *as it changes into or interacts with another substance (or substances).* Examples of chemical properties include flammability, corrosiveness, and reactivity with acids. A **chemical change,** also called a **chemical reaction,** occurs when *a substance (or substances) is converted into a different substance (or substances).* Figure 1.1B shows the chemical change (reaction) that occurs when you pass an electric current through water: the water decomposes (breaks down) into two other substances, hydrogen and oxygen, each with physical and chemical properties different from each other *and* from water. The sample *has* changed its composition: it is no longer water, as you can see from the particles in the magnified view:

Chemical change (different substances before and after):

$$\text{Water} \xrightarrow{\text{electric current}} \text{hydrogen gas} + \text{oxygen gas}$$

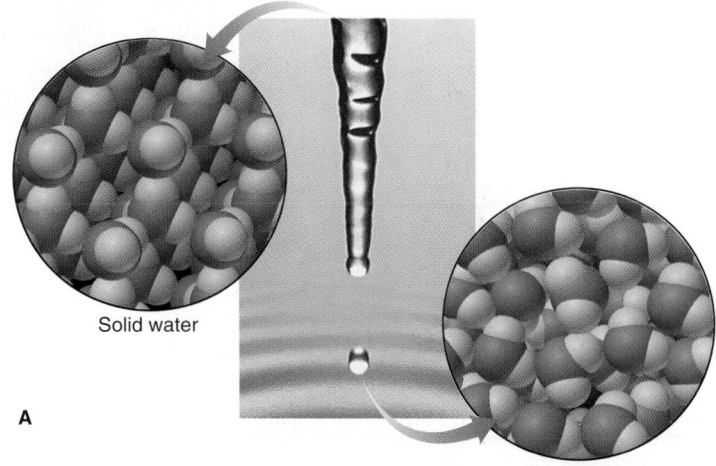

Solid water

Liquid water

A

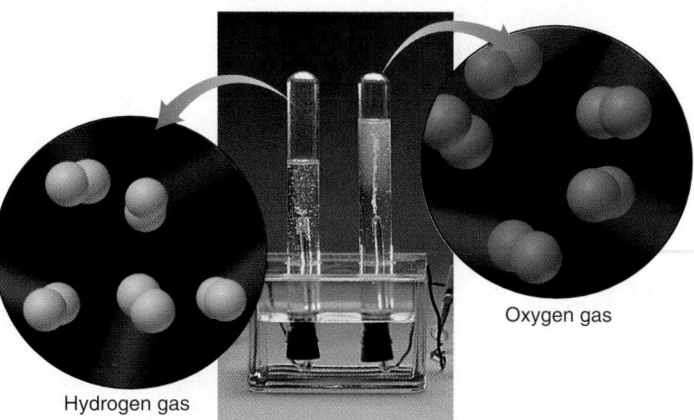

Hydrogen gas

Oxygen gas

B

Figure 1.1 The distinction between physical and chemical change.
A, A physical change occurs when the solid form of water becomes the liquid form. As the magnified views show, the composition of the particles does *not* change. **B,** A chemical change occurs when an electric current passes through water. The magnified views show that there *is* a change in composition as water decomposes to hydrogen and oxygen.

Table 1.1 Some Characteristic Properties of Copper

Physical Properties	Chemical Properties	
Reddish brown, metallic luster		
Easily shaped into sheets (malleable) and wires (ductile) 	Slowly forms a blue-green carbonate in moist air	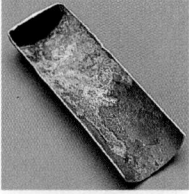
Good conductor of heat and electricity	Reacts with nitric acid (photo) and sulfuric acid	
Can be melted and mixed with zinc to form brass 		
Density = 8.95 g/cm^3 Melting point = 1083°C Boiling point = 2570°C	Slowly forms deep-blue solution in aqueous ammonia	

A substance is identified by its own set of physical and chemical properties. Some properties of copper appear in Table 1.1. When we observe a substance with this set of properties, we identify it as copper.

The Three States of Matter

THE THREE STATES OF MATTER

Matter occurs commonly in three physical forms called **states:** solid, liquid, and gas. As shown in Figure 1.2 for a general substance, each state is defined by the way it fills a container. A **solid** *has a fixed shape that does not conform to the container shape.* Solids are *not* defined by rigidity or hardness: solid iron is rigid, but solid lead is flexible and solid wax is soft. A **liquid** *conforms to the container shape but fills the container only to the extent of the liquid's volume;* thus, a liquid forms a *surface.* A **gas** *conforms to the container shape also, but it fills the entire container;* thus, a gas does *not* form a surface. Now, look at the views within the blow-up circles of the figure. The particles in the solid lie next to each other in a regular, three-dimensional array with a definite shape. Particles in the liquid also lie together but move randomly around one another. Particles in the gas usually have great distances between one another, as they move randomly throughout the container.

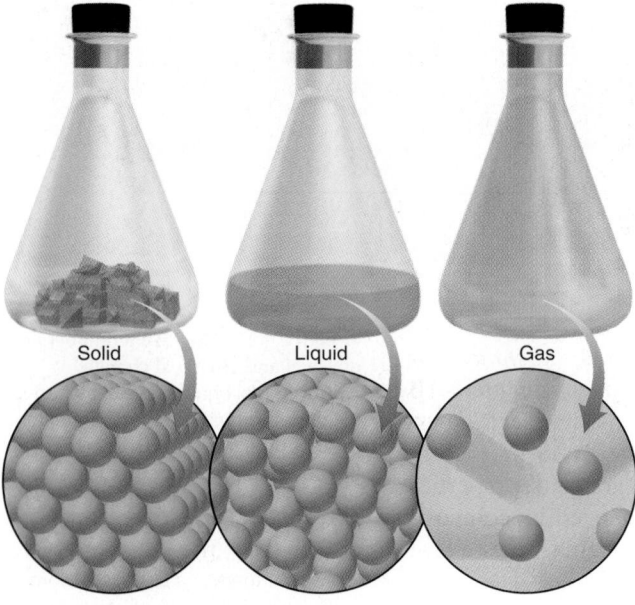

Figure 1.2 **The physical states of matter.** The magnified (blow-up) views of the states show the atomic-scale arrangement of the particles—close together and highly organized in the solid, close together but disorganized in the liquid, and far apart and disorganized in the gas.

Depending on the temperature and pressure of the surroundings, many substances can exist in each of the three physical states and undergo changes in state as well. As the temperature increases, solid water melts to liquid water, which boils to gaseous water (also called *water vapor*). Similarly, with decreasing temperature, water vapor condenses to liquid water, and with further cooling, the liquid freezes to ice. Many other substances behave in the same way: solid iron melts to liquid (molten) iron and then boils to iron gas at a higher temperature. Cooling the iron gas changes it to liquid and then to solid iron.

You can see that changes in state are physical, not chemical, changes. Moreover, a physical change caused by heating can generally be reversed by cooling, and vice versa. This is *not* generally true for a chemical change. For example, heating iron in moist air causes a chemical reaction that yields a brown, crumbly substance known as rust. Cooling does not reverse this change; rather, another chemical change (or series of them) is required.

To summarize the key distinctions:

- A physical change leads to a different form of the same substance (same composition), whereas a chemical change leads to a different substance (different composition).
- A physical change caused by a temperature change can generally be reversed by the opposite temperature change, but this is not generally true of a chemical change.

The following sample problem provides more examples that distinguish between these types of change.

The Incredible Range of Physical Change Scientists often study physical change in remarkable settings, using instruments that allow observation far beyond the confines of the laboratory. Instruments aboard the *Voyager* and *Galileo* spacecrafts and the Hubble Space Telescope have measured temperatures on Jupiter's moon Io (shown here) that are hot enough to maintain lakes of molten sulfur and cold enough to create vast snowfields of sulfur dioxide and polar caps swathed in hydrogen sulfide frost. (On Earth, sulfur dioxide is one of the gases released from volcanoes and coal-fired power plants, and hydrogen sulfide occurs in swamp gas.)

SAMPLE PROBLEM 1.1 Distinguishing Between Physical and Chemical Change

Problem Decide whether each of the following processes is primarily a physical or a chemical change, and explain briefly:
(a) Frost forms as the temperature drops on a humid winter night.
(b) A cornstalk grows from a seed that is watered and fertilized.
(c) Dynamite explodes to form a mixture of gases.
(d) Perspiration evaporates when you relax after jogging.
(e) A silver fork tarnishes in air.
Plan The basic question we ask to decide whether a change is chemical or physical is, "Does the substance change composition or just change form?"
Solution (a) Frost forming is a physical change: the temperature changes water vapor (gaseous water) in humid air to ice crystals (solid water).
(b) A seed growing is a chemical change: the seed, air, fertilizer, soil, and water use energy from sunlight to undergo complex changes in composition.
(c) Dynamite exploding is a chemical change: the dynamite is converted into other substances.
(d) Perspiration evaporating is a physical change: the water in sweat changes its form, from liquid to gas, but not its composition.
(e) Tarnishing is a chemical change: silver changes to silver sulfide by reacting with sulfur-containing substances in the air.

FOLLOW-UP PROBLEM 1.1 Decide whether each of the following processes is primarily a physical or a chemical change, and explain briefly:
(a) Purple iodine vapor appears when solid iodine is warmed.
(b) Gasoline fumes are ignited by a spark in an automobile engine cylinder.
(c) A scab forms over an open cut.

The Central Theme in Chemistry

Understanding the properties of a substance and the changes it undergoes leads to the central theme in chemistry: *macroscopic* properties and behavior, those we can see, are the results of *submicroscopic* properties and behavior that we cannot see. The distinction between chemical and physical change is defined by composition, which we study macroscopically. But it ultimately depends on composition at the atomic scale, as depicted in the magnified views of Figure 1.1. Similarly, the defining properties of the three states of matter are macroscopic, but they arise from the submicroscopic behavior depicted in the magnified views in Figure 1.2. Throughout this text, we return to this basic idea: we study *observable* changes in matter to understand their *unobservable* causes. What is really happening when water boils? Why do iron and copper melt at different temperatures? What events occur in the invisible world of minute particles that cause dynamite to explode, a neon light to glow, or a nail to rust?

The Importance of Energy in the Study of Matter

In general, physical and chemical changes are accompanied by energy changes. **Energy** is often defined as *the ability to do work*. Essentially, all work involves moving something. Work is done when your arm lifts a book, when an engine moves a car's wheels, or when a falling rock moves the ground as it lands. The object doing the work (arm, engine, rock) transfers some of the energy it possesses to the object on which the work is done (book, wheels, ground).

The total energy an object possesses is the sum of its potential energy and its kinetic energy. **Potential energy** is the *energy due to the **position** of the object*. **Kinetic energy** is the *energy due to the **motion** of the object*. Let's examine four systems that illustrate the relationship between these two forms of energy: (1) a weight raised above the ground, (2) two balls attached by a spring, (3) two electrically charged particles, and (4) a fuel and its waste products. A key concept illustrated by all four cases is that *energy is conserved: it may be converted from one form to the other, but it is not destroyed.*

Look at Figure 1.3A and consider a weight you lift above the ground. The energy you use to move the weight against the gravitational attraction of the Earth increases the weight's potential energy (energy due to its position). When the weight is dropped, this change in potential energy is converted to kinetic energy (energy due to motion). Some of this kinetic energy is transferred to the ground as the weight does work, such as driving a stake or simply moving dirt and pebbles. As you can see, the change in potential energy is not destroyed: it is converted to kinetic energy.

In nature, situations of lower energy are typically favored over those of higher energy: because the weight has less potential energy (and thus less total energy) at rest on the ground than held in the air, it will fall when released. We speak of the situation with the weight elevated and higher in potential energy as being *less stable,* and the situation after the weight has fallen and is lower in potential energy as being *more stable.*

To bring the concept somewhat closer to chemistry, consider the two balls attached by a relaxed spring in Figure 1.3B. When you pull the balls apart, the energy you exert to stretch the spring increases its potential energy. This change in potential energy is converted to kinetic energy when you release the balls and they move closer together. In the stretched position, the system of balls and spring is less stable (has more potential energy) than when the spring is relaxed. You can also increase the spring's potential energy by pushing the balls together and compressing the spring. The balls move apart when released, once again converting the change in potential energy into kinetic energy. Similarly, when you wind a clock, the potential energy in the mainspring increases as the spring is

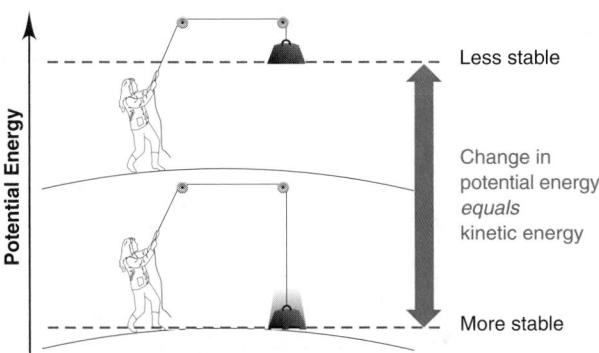

A A gravitational system. The potential energy gained when a weight is lifted is converted to kinetic energy as the weight falls.

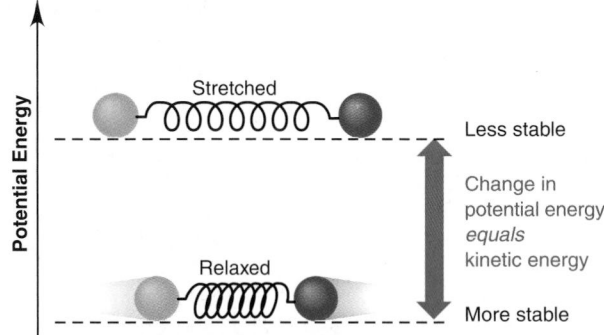

B A system of two balls attached by a spring. The potential energy gained when the spring is stretched is converted to the kinetic energy of the moving balls when it is released.

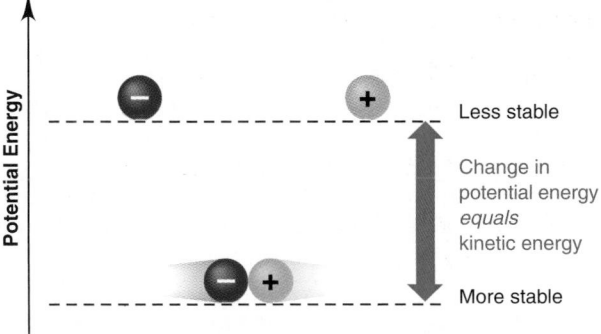

C A system of oppositely charged particles. The potential energy gained when the charges are separated is converted to kinetic energy as the attraction pulls them together.

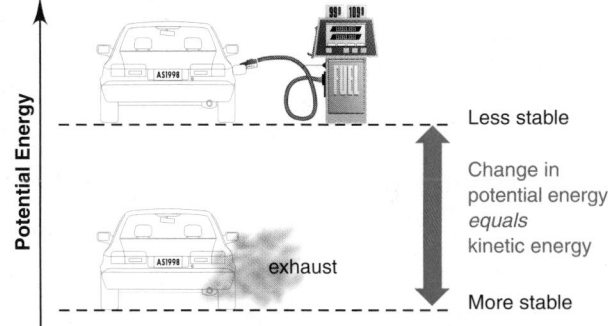

D A system of fuel and exhaust. A fuel is higher in chemical potential energy than the exhaust. As the fuel burns, some of its potential energy is converted to the kinetic energy of the moving car.

Figure 1.3 **Potential energy is converted to kinetic energy.** In all four parts of the figure, the dashed horizontal lines indicate the potential energy of the system in each situation.

compressed. As the spring relaxes, its potential energy is slowly converted into the kinetic energy of the moving gears and hands.

There are no springs in a chemical substance, of course, but the following situation is similar in terms of energy. Much of the matter in the universe is composed of positively and negatively charged particles. A well-known behavior of charged particles (similar to the behavior of the poles of magnets) results from interactions known as *electrostatic forces: opposite charges attract each other and like charges repel each other.* When work is done to separate a positive particle from a negative one (like stretching a spring), the potential energy of the particles increases. As Figure 1.3C shows, that change in potential energy is converted to kinetic energy when the particles move together again. Also, when two positive (or two negative) particles are pushed toward each other (like compressing a spring), their potential energy increases, and when they are allowed to move apart, that increase in potential energy is changed into kinetic energy. Like the weight above the ground and the balls connected by a spring, charged particles move naturally toward a position of lower energy, one that is more stable.

The chemical potential energy of a substance results from the relative positions and the attractions and repulsions among all its particles. Some substances are richer in this chemical potential energy than others. Fuels and foods, for example, contain more potential energy than the waste products they form. Figure 1.3D shows that when gasoline burns in a car engine, substances with higher chemical potential energy (gasoline and air) form substances with lower potential energy (exhaust gases). This change in potential energy is converted into the kinetic energy that moves the car, heats the passenger compartment, makes the lights shine, and so forth. Similarly, the difference in potential energy between the food and air we take in and the waste products we excrete is used to move, grow, keep warm, study chemistry, and so on. Note again the essential point: *energy is neither created nor destroyed—it is always conserved as it is converted from one form to the other*.

SECTION SUMMARY

Chemists study the composition and properties of matter and how they change. Each substance has a unique set of physical properties (attributes of the substance itself) and chemical properties (attributes of the substance as it interacts with or changes to other substances). Changes in matter can be physical (different form of the same substance) or chemical (different substance). Matter exists in three physical states— solid, liquid, and gas. The observable features that distinguish these states reflect the arrangement of their particles. A change in physical state brought about by heating may be reversed by cooling. A chemical change can be reversed only by other chemical changes. Macroscopic changes result from submicroscopic changes.

Changes in matter are accompanied by changes in energy. An object's potential energy is due to its position; an object's kinetic energy is due to its motion. Energy used to lift a weight, stretch a spring, or separate opposite charges increases the system's potential energy. Chemical potential energy arises from the positions and interactions of the particles in a substance. Higher energy substances are less stable than lower energy ones. Through physical or chemical change, a less stable substance changes to a more stable form or a more stable substance as some potential energy is converted into kinetic energy, which can do work.

1.2   CHEMICAL ARTS AND THE ORIGINS OF MODERN CHEMISTRY

Chemistry has a rich, colorful history. Even some concepts and discoveries that led temporarily along a confusing path have contributed to the heritage of chemistry. This brief overview of early breakthroughs and false directions provides some insight into how modern chemistry arose and how science progresses.

Prechemical Traditions

Chemistry has its origin in a prescientific past that incorporated three overlapping traditions: alchemy, medicine, and technology.

The Alchemical Tradition The occult study of nature practiced in the 1st century AD by Greeks living in northern Egypt later became known by the Arabic name **alchemy.** Its practice spread through the Near East and into Europe, where it dominated Western thinking about matter for more than 1500 years! Alchemists were influenced by the Greek idea that matter naturally strives toward perfection, and they searched for ways to change less valued substances into precious ones. What started as a search for spiritual properties in matter evolved over a thousand years into an obsession with potions to bestow eternal youth and elixirs to transmute "baser" metals, such as lead, into "purer" ones, such as gold. Greed prompted some later European alchemists to paint lead objects with a thin coating of gold to fool wealthy patrons.

Alchemy's legacy to chemistry is mixed at best. The confusion arising from alchemists' use of different names for the same substance and from their belief that matter could be altered magically was very difficult to eliminate. Nevertheless, through centuries of laboratory inquiry, alchemists invented the chemical methods of distillation, percolation, and extraction, and they devised apparatus that today's chemists still use routinely (Figure 1.4). Most important, alchemists

Figure 1.4 An alchemist at work. The alchemical tradition influenced ideas about matter and its changes for 1500 years. This is a portion of a painting by the Englishman Joseph Wright entitled "The Alchymist, in search of the Philosopher's Stone, Discovers Phosphorus, and prays for the successful Conclusion of his Operation as was the custom of the Ancient Chymical Astrologers." Some suggest it portrays the German alchemist Hennig Brand in his laboratory, lit by the glow of phosphorus, which he discovered in 1669. A great legacy of the alchemists was their invention of several major laboratory processes. The apparatus shown here is used for distillation, a process still commonly used to separate substances.

encouraged the widespread acceptance of observation and experimentation, which replaced the Greek approach of studying nature solely through reason.

The Medical Tradition Alchemists greatly influenced medical practice in medieval Europe. Since the 13th century, distillates of roots, herbs, and other plant matter have been used as sources of medicines. Paracelsus (1493–1541) was an active alchemist and important physician of the time. His writings are difficult to comprehend, but he seems to have considered the body to be a chemical system whose balance of substances could be restored by medical treatment. His followers introduced mineral drugs into 17th-century pharmacy. Although many of these drugs were useless and some harmful, later practitioners employed other mineral prescriptions with increasing success. Thus began an alliance between medicine and chemistry that thrives today.

The Technological Tradition For thousands of years, people have developed technological skills to carry out changes in matter. Pottery making, dyeing, and especially metallurgy (begun about 7000 years ago) contributed greatly to experience with the properties of materials. During the Middle Ages and the Renaissance, such technology flourished. Books describing how to purify, assay, and coin silver and gold and how to use balances, furnaces, and crucibles were published and regularly updated. Other writings discussed making glass, pottery, dyes, and gunpowder. Some of these even introduced quantitative measurement, which had been lacking in the alchemical literature.

Many creations of these early artisans are still unsurpassed today. Yet, while their working knowledge of substances was expert, their approach to understanding matter was different from ours: techniques were discovered and developed by trial and error, and writings show little interest in exploring *why* a substance changes or *how to predict* the behavior of matter.

The Phlogiston Fiasco and the Impact of Lavoisier

Chemical investigation in the modern sense—inquiry into the *causes* of changes in matter—began in the late 17th century. Alchemical influences persisted, however, and understanding was hampered by an incorrect theory of **combustion,** the process of burning. Even though fire had been used since prehistoric times, no one could explain satisfactorily why some things burn and others do not, or what actually happens when a substance burns.

The English scientist Robert Hooke suggested that burning substances combine with air, but most scientists rejected this idea to embrace the **phlogiston theory,** which held sway for the next hundred years. According to this theory, combustible materials contain *phlogiston*, an undetectable substance that is released when the material burns. A highly combustible material such as charcoal contains large amounts of phlogiston, which is released when the charcoal burns and forms charcoal ash (charcoal without phlogiston):

$$\text{Charcoal} \longrightarrow \text{charcoal ash} + \text{phlogiston (large amount)}$$

Slightly combustible materials such as metals contain small amounts of phlogiston. When a metal burns in air, it forms its *calx* (as metal oxides were then called), which is the metal without phlogiston:

$$\text{Metal} \longrightarrow \text{metal calx} + \text{phlogiston (small amount)}$$

Consider how the theory was used to explain *smelting*, an industrial process that converts a metal calx to the metal by heating the calx with charcoal: when phlogiston-rich charcoal burns, it transfers phlogiston to phlogiston-poor calx, and this transfer converts the charcoal to charcoal ash and the calx to the pure metal:

$$\text{Metal calx} + \text{charcoal} \xrightarrow{\text{phlogiston}} \text{metal} + \text{charcoal ash}$$

These schemes seemed logical, but they ignored certain key observations. Why, the theory's critics asked, is air needed for combustion, and why does charcoal burn for only a short time in a closed vessel? Phlogistonists responded that air is needed to "attract" the phlogiston out of the charcoal, and that burning in a vessel stops when the air is "saturated" with phlogiston. Since a calx weighs more than the metal from which it is formed, critics asked how the *loss* of phlogiston when a metal burns could cause a *gain* in mass. Phlogistonists either dismissed the importance of weighing or proposed that phlogiston had negative mass! These responses seem somewhat ridiculous now, but they point out that the pursuit of science, like any other human endeavor, is subject to human failings; even today, it is easier to dismiss conflicting evidence than to give up an established idea.

Into this chaos of "explanations" entered the young French chemist Antoine Lavoisier (1743–1794), who demonstrated the true nature of combustion. In a series of careful measurements that emphasized the importance of mass, Lavoisier heated mercury calx (mercuric oxide), decomposing it into mercury and a gas. Their combined masses equaled the starting mass of calx. The reverse experiment—heating mercury with the gas—re-formed the mercury calx, and again, *the total mass remained constant*. Lavoisier proposed that when a metal forms its calx, it does not lose phlogiston but rather combines with this gas, which must be a component of air.

In a test of his proposal, Lavoisier heated mercury in a measured volume of air until no further change occurred. Mercury calx formed, and four-fifths of the air volume remained, confirming that the gas was a component of air. A burning candle placed in the remaining volume of air was quickly extinguished, showing that the gas that had combined with the mercury was necessary for combustion. Lavoisier named the gas *oxygen* and called metal calxes *metal oxides*.

The new theory of combustion made sense of the earlier confusion. A combustible substance such as charcoal stops burning in a closed vessel once it has combined with all the available oxygen. A metal oxide weighs more than the metal

Scientific Thinker Extraordinaire Lavoisier's fame would be widespread, even if he had never performed a chemical experiment. A short list of his other contributions: He improved the production of French gunpowder, which became a key factor in the success of the American Revolution. He established on his farm a scientific balance between cattle, pasture, and cultivated acreage to optimize crop yield. He developed public assistance programs for widows and orphans. He quantified the relation of fiscal policy to agricultural production. He proposed a system of free public education and of societies to foster science, politics, and the arts. He sat on the committee that unified weights and measures in the new metric system. His research into combustion clarified the essence of respiration and metabolism. To support these pursuits, he joined a firm that collected taxes for the king, and only this role was remembered during the French Revolution. Despite his devotion to French society, the father of modern chemistry was guillotined at the age of 50.

because it contains the added mass of oxygen. The true nature of smelting is that when charcoal and a metal oxide are heated together, the carbon in charcoal combines with the oxygen in the metal oxide to produce the metal, charcoal ash, and carbon dioxide gas:

$$\text{Metal calx (metal oxide)} + \text{charcoal (carbon)} \xrightarrow{\text{oxygen}} \text{metal} + \text{charcoal ash} + \text{carbon dioxide}$$

Lavoisier's theory of combustion triumphed because it explained observations more clearly and consistently than any other theory. Most important, the new theory relied on *quantitative, reproducible measurements,* not on strange properties of undetectable substances. Because this approach is at the heart of science, many propose that the *science* of chemistry began with Lavoisier.

SECTION SUMMARY

Alchemy, medicine, and technology established processes that have been important to chemists since the 17th century. These prescientific traditions placed little emphasis on objective experimentation, focusing instead on practical experience or mystical explanations. The phlogiston theory dominated thinking about combustion for almost 100 years, but in the mid-1770s, Lavoisier showed that oxygen, a component of air, is required for combustion and combines with a substance as it burns.

1.3 THE SCIENTIFIC APPROACH: DEVELOPING A MODEL

The principles of chemistry have been modified through time and are still evolving. Imagine how differently people learned about the material world tens of thousands of years ago. At the dawn of human experience, our ancestors survived through knowledge acquired by *trial and error:* which types of stone were hard enough to shape others, which types of wood were rigid and which were flexible, which hides could be treated to make clothing, which plants were edible and which were poisonous. Today, the science of chemistry, with its powerful *quantitative theories,* helps us understand the essential nature of materials to make better use of them and create new ones: specialized steels, advanced composites, synthetic polymers, and countless other new materials (Figure 1.5).

A Great Chemist Yet Strict Phlogistonist Despite the phlogiston theory, chemists made key discoveries during the years it held sway. Many were made by the English clergyman Joseph Priestley (1733–1804), who systematically studied the physical and chemical properties of many gases (inventing "soda water," carbon dioxide dissolved in water, along the way). The gas obtained by heating mercury calx was of special interest to him. In 1775, he wrote to his friend Benjamin Franklin: "Hitherto only two mice and myself have had the privilege of breathing it." Priestley also demonstrated that the gas supports combustion, but he drew the wrong conclusion about it. He called the gas "dephlogisticated air," air devoid of phlogiston, and thus ready to attract it from a burning substance. Priestley's contributions make him one of the great chemists of all time. Priestley was a liberal thinker, favoring freedom of conscience and supporting both the French and American Revolutions, positions that caused severe personal problems throughout his later life. But, scientifically, he remained a conservative, believing strictly in phlogiston and refusing to accept the new theory of combustion.

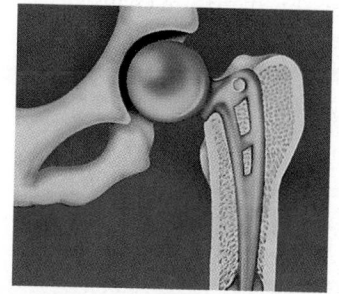

A

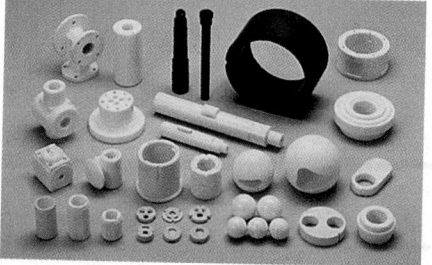

C

B

Figure 1.5 Modern materials in a variety of applications. A, High-tension polymers in synthetic hip joints. **B,** Specialized steels in bicycles; synthetic polymers in clothing and helmets. **C,** Car engine parts made of new ceramics. **D,** Liquid crystal displays in electronic devices.

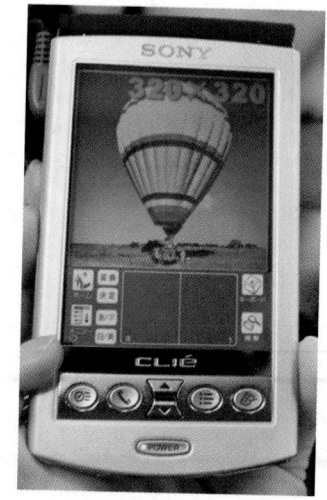

D

Is there something special about the way scientists think? If we could break down a "typical" modern scientist's thought processes, we could organize them into an approach called the **scientific method.** This approach is not a stepwise checklist, but rather a flexible process of creative thinking and testing aimed at objective, verifiable discoveries of how nature works. It is very important to realize that there is no typical scientist and no single method, and that luck can and often has played a role in scientific discovery. In general terms, the scientific approach includes the following parts (see Figure 1.6):

1. *Observations.* These are the facts that our ideas must explain. Observation is basic to scientific thinking. The most useful observations are quantitative because they can be compared and allow trends to be seen. Pieces of quantitative information are **data.** When the same observation is made by many investigators in many situations with no clear exceptions, it is summarized, often in mathematical terms, and called a **natural law.** The observation that mass remains constant during chemical change—made by Lavoisier and numerous experimenters since—is known as the law of mass conservation, discussed in Chapter 2.

2. *Hypothesis.* Whether derived from actual observation or from a "spark of intuition," a hypothesis is a proposal made to explain an observation. A valid hypothesis need not be correct, but it must be testable. Thus, a hypothesis is often the reason for performing an experiment. If the hypothesis is inconsistent with the experimental results, it must be revised or discarded.

3. *Experiment.* An experiment is a clear set of procedural steps that tests a hypothesis. Experimentation is the connection between our hypotheses about nature and nature itself. Often, hypothesis leads to experiment, which leads to revised hypothesis, and so forth. Hypotheses can be altered, but the results of an experiment cannot.

An experiment typically contains at least two **variables,** quantities that can have more than a single value. A well-designed experiment is **controlled** in that it measures the effect of one variable on another while keeping all others constant. For experimental results to be accepted, they must be *reproducible,* not only by the person who designed the experiment, but also by others. Both skill and creativity play a part in great experimental design.

4. *Model.* Formulating conceptual models, or **theories,** *based on experiments* is what distinguishes scientific thinking from speculation. As hypotheses are revised according to experimental results, a model gradually emerges that describes how the observed phenomenon occurs. A model is not an exact representation of nature, but rather a simplified version of nature that can be used to make *predictions* about related phenomena. Further investigation refines a model by testing its predictions and altering it to account for new facts.

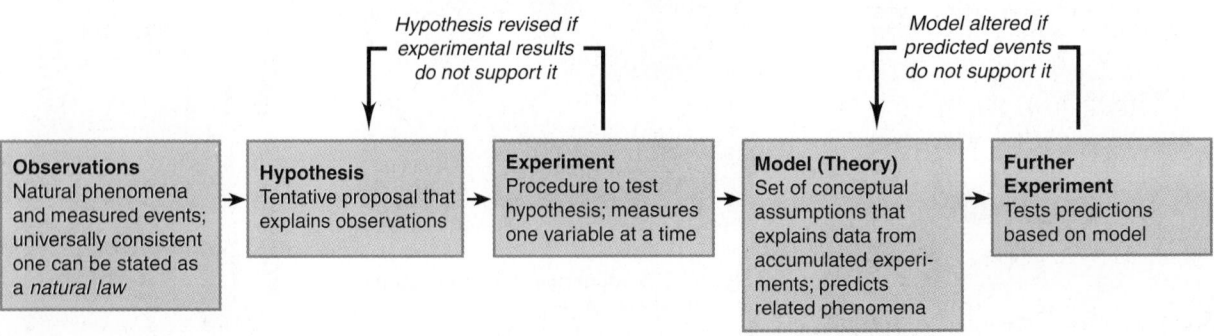

Figure 1.6 **The scientific approach to understanding nature.** Note that hypotheses and models are mental pictures that are changed to match observations and experimental results, *not* the other way around.

Lavoisier's overthrow of the phlogiston theory demonstrates the scientific approach. Observations of burning and smelting led some to hypothesize that combustion involved the loss of phlogiston. Experiments by others showing that air is required for burning and that a metal gains mass during combustion led Lavoisier to propose a new hypothesis, which he tested repeatedly with quantitative experiments. Accumulating evidence supported his developing model (theory) that combustion involves combination with a component of air (oxygen). Innumerable predictions based on this theory have supported its validity. A sound theory remains useful even when minor exceptions appear. An unsound one, such as the phlogiston theory, eventually crumbles under the weight of contrary evidence and absurd refinements.

SECTION SUMMARY

The scientific method is not a rigid sequence of steps, but rather a dynamic process designed to explain and predict real phenomena. Observations (sometimes expressed as natural laws) lead to hypotheses about how or why something occurs. Hypotheses are tested in controlled experiments and adjusted if necessary. If all the data collected support a hypothesis, a model (theory) can be developed to explain the observations. A good model is useful in predicting related phenomena but must be refined if conflicting data appear. ●

1.4 CHEMICAL PROBLEM SOLVING

In many ways, learning chemistry is learning how to solve chemistry problems, not only those on exams or in homework, but also more complex ones in society and professional life. (The Chemical Connections essay at the end of this chapter provides an example.) This text has been designed to help you strengthen your problem-solving skills. Almost every chapter contains sample problems that apply newly introduced ideas and skills and are worked out in detail. In this section, we discuss the approach used in the text for solving them. Because many include calculations, let's first go over some important ideas about measured quantities.

Units and Conversion Factors in Calculations

All measured quantities consist of a number *and* a unit; a person's height is "6 feet," not "6." Ratios of quantities have ratios of units, such as miles/hour. (We discuss the most important units in chemistry in the next section.) To minimize errors, try to make a habit of including units in all calculations.

The arithmetic operations used with measured quantities are the same as those used with pure numbers; in other words, units can be multiplied, divided, and canceled:

- A carpet measuring 3 feet (ft) by 4 ft has an area of
$$\text{Area} = 3 \text{ ft} \times 4 \text{ ft} = (3 \times 4)(\text{ft} \times \text{ft}) = 12 \text{ ft}^2$$

- A car traveling 350 miles (mi) in 7 hours (h) has a speed of
$$\text{Speed} = \frac{350 \text{ mi}}{7 \text{ h}} = \frac{50 \text{ mi}}{1 \text{ h}} \quad (\text{often written } 50 \text{ mi·h}^{-1})$$

- In 3 hours, the car travels a distance of
$$\text{Distance} = 3 \text{ h} \times \frac{50 \text{ mi}}{1 \text{ h}} = 150 \text{ mi}$$

Conversion factors are ratios used to express a measured quantity in different units. Suppose we want to know the distance of that 150-mile car trip in feet. To convert the distance between miles and feet, we use equivalent quantities to

●Everyday Scientific Thinking In an informal way, we often use a scientific approach in daily life. Consider this familiar scenario. While listening to an FM broadcast on your stereo system, you notice the sound is garbled (observation) and assume it is caused by poor reception (hypothesis). To isolate this variable, you play a CD (experiment): the sound is still garbled. If the problem is not poor reception, perhaps the speakers are at fault (new hypothesis). To isolate this variable, you play the CD and listen with headphones (experiment): the sound is clear. You conclude that the speakers need to be repaired (model). The repair shop says the speakers check out fine (new observation), but the power amplifier may be at fault (new hypothesis). Replacing a transistor in the amplifier corrects the garbled sound (new experiment), so the power amplifier was the problem (revised model). Approaching a problem scientifically is a common practice, even if you're not aware of it.

construct the desired conversion factor. The equivalent quantities in this case are 1 mile and the number of feet in 1 mile:

$$1 \text{ mi} = 5280 \text{ ft}$$

We can construct two conversion factors from this equivalency. Dividing both sides by 5280 ft gives one conversion factor (shown in blue):

$$\frac{1 \text{ mi}}{5280 \text{ ft}} = \frac{5280 \text{ ft}}{5280 \text{ ft}} = 1$$

And, dividing both sides by 1 mi gives the other conversion factor (blue):

$$\frac{1 \text{ mi}}{1 \text{ mi}} = \frac{5280 \text{ ft}}{1 \text{ mi}} = 1$$

It's very important to see that, since the numerator and denominator of a conversion factor are equal, multiplying by a conversion factor is the same as multiplying by one. Therefore, *even though the number and unit of the quantity change, the size of the quantity remains the same.*

In our example, we want to convert the distance in miles to the equivalent distance in feet. Therefore, we choose the conversion factor with units of feet in the numerator, because it cancels units of miles and gives units of feet:

$$\text{Distance (ft)} = 150 \text{ mi} \times \frac{5280 \text{ ft}}{1 \text{ mi}} = 792{,}000 \text{ ft}$$
$$\text{mi} \implies \text{ft}$$

Choosing the correct conversion factor is not a matter of chance or memorization. The main goal is that *the chosen conversion factor will cancel all units except those required for the answer.* Set up the calculation so that the unit you are converting *from* (beginning unit) is in the *opposite position in the conversion factor* (numerator or denominator). It will then cancel and leave the unit you are converting *to* (final unit):

$$\text{beginning unit} \times \frac{\text{final unit}}{\text{beginning unit}} = \text{final unit} \qquad \text{as in} \qquad \text{mi} \times \frac{\text{ft}}{\text{mi}} = \text{ft}$$

Or, in cases that involve units raised to a power,

$$(\text{beginning unit} \times \text{beginning unit}) \times \frac{\text{final unit}^2}{\text{beginning unit}^2} = \text{final unit}^2$$
$$\text{as in} \qquad (\text{ft} \times \text{ft}) \times \frac{\text{mi}^2}{\text{ft}^2} = \text{mi}^2$$

Or, in cases that involve a ratio of units,

$$\frac{\text{beginning unit}}{\text{final unit}_1} \times \frac{\text{final unit}_2}{\text{beginning unit}} = \frac{\text{final unit}_2}{\text{final unit}_1} \qquad \text{as in} \qquad \frac{\text{mi}}{\text{h}} \times \frac{\text{ft}}{\text{mi}} = \frac{\text{ft}}{\text{h}}$$

It is also important that you think through the calculation to decide whether the answer expressed in the new units should have a larger or smaller number and whether the conversion factor you have chosen will accomplish this change. In the earlier case, since a foot is *smaller* than a mile, the distance in feet should have a *larger* number (792,000) than the distance in miles (150). Note that the conversion factor we chose has the larger number (5280) in the numerator, so it gave a larger number in the answer.

We use the same procedure to convert between systems of units, for example, between the English (or American) unit system and the International System (a revised metric system discussed fully in the next section). Suppose we know the height of Angel Falls in Venezuela (Figure 1.7) to be 3212 ft and we find its height in miles as

$$\text{Height (mi)} = 3212 \text{ ft} \times \frac{1 \text{ mi}}{5280 \text{ ft}} = 0.6083 \text{ mi}$$
$$\text{ft} \implies \text{mi}$$

Figure 1.7 Angel Falls. The world's tallest waterfall is 3212 ft high.

Now, we want its height in kilometers (km). The equivalent quantities are

$$1.609 \text{ km} = 1 \text{ mi}$$

Since we are converting from miles to kilometers, we use the conversion factor with kilometers in the numerator in order to cancel miles:

$$\text{Height (km)} = 0.6083 \; \cancel{\text{mi}} \times \frac{1.609 \text{ km}}{1 \; \cancel{\text{mi}}} = 0.9788 \text{ km}$$

$$\text{mi} \implies \text{km}$$

Notice that, since kilometers are *smaller* than miles, this conversion factor gave us a *larger* number (0.9788 is larger than 0.6083).

If we want the height of Angel Falls in meters (m), we use the equivalent quantities 1 km = 1000 m to construct the conversion factor:

$$\text{Height (m)} = 0.9788 \; \cancel{\text{km}} \times \frac{1000 \text{ m}}{1 \; \cancel{\text{km}}} = 978.8 \text{ m}$$

$$\text{km} \implies \text{m}$$

In longer calculations, we often string together several conversion steps:

$$\text{Height (m)} = 3212 \; \cancel{\text{ft}} \times \frac{1 \; \cancel{\text{mi}}}{5280 \; \cancel{\text{ft}}} \times \frac{1.609 \; \cancel{\text{km}}}{1 \; \cancel{\text{mi}}} \times \frac{1000 \text{ m}}{1 \; \cancel{\text{km}}} = 978.8 \text{ m}$$

$$\text{ft} \implies \text{mi} \implies \text{km} \implies \text{m}$$

The use of conversion factors in calculations is known by various names, such as the factor-label method or **dimensional analysis** (because units represent physical dimensions). We use this method in quantitative problems throughout the text.

A Systematic Approach to Solving Chemistry Problems

The approach we use in this text provides a systematic way to work through a problem. It emphasizes reasoning, not memorizing, and is based on a very simple idea: plan how to solve the problem *before* you go on to solve it, and then check your answer. Try to develop a similar approach on homework and exams. In general, the sample problems consist of several parts:

1. **Problem.** This part states all the information you need to solve the problem (usually framed in some interesting context).
2. **Plan.** The overall solution is broken up into two parts, *plan* and *solution,* to make a point: *think* about how to solve the problem *before* juggling numbers. If a plan is clear in your mind, the actual calculations will make sense. Since there is often more than one way to solve a problem, the plan shown in a given text problem is just one possibility. The plan will
   - Clarify the known and unknown. (What information do you have, and what are you trying to find?)
   - Suggest the steps from known to unknown. (What ideas, conversions, or equations are needed to solve the problem?)
   - Present a "roadmap" of the solution for many problems in early chapters (and in some later ones). The roadmap is a visual summary of the planned steps. Each step is shown by an arrow labeled with information about the conversion factor or operation needed.
3. **Solution.** In this part, the steps appear in the same order as they were planned.
4. **Check.** In most cases, a quick check is provided to see if the results make sense: Are the units correct? Does the answer seem to be the right size? Did the change occur in the expected direction? Is it reasonable chemically? We often do a rough calculation to see if the answer is "in the same ballpark" as the calculated result, just to make sure we didn't make a large error. *Always* check your answers, especially in a multipart problem, where an error in an early step can affect all later steps.

5. **Comment.** This part is included occasionally to provide additional information, such as an application, an alternative approach, a common mistake to avoid, or an overview.

6. **Follow-up Problem.** This part consists of a problem statement only and provides practice by applying the same ideas as the sample problem. Try to solve it *before* you look at the brief worked-out solution at the end of the chapter.

Of course, you can't learn to solve chemistry problems, any more than you can learn to swim, by reading about an approach. Practice is the key to mastery. Here are a few suggestions that can help:

- Follow along in the sample problem with pencil, paper, and calculator.
- Do the follow-up problem as soon as you finish studying the sample problem. Check your answer against the solution at the end of the chapter.
- Read the sample problem and text explanations again if you have trouble.
- Work on as many of the problems at the end of the chapter as you can. They review and extend the concepts and skills in the text. Answers are given in the back of the book for problems with a colored number, but try to solve the problem yourself first. Let's apply this approach in a unit-conversion problem.

SAMPLE PROBLEM 1.2 Converting Units of Length

Problem What is the price of a piece of copper wire 325 centimeters (cm) long that sells for $0.15/ft?

Plan We know the length of wire in centimeters and the cost in dollars per foot ($/ft). We can find the unknown price of the piece of wire by converting the length from centimeters to inches (in) and from inches to feet. Then the cost (1 ft = $0.15) gives us the equivalent quantities to construct the factor that converts feet of wire to price in dollars. The roadmap starts with the known and moves through the calculation steps to the unknown.

Solution Converting the known length from centimeters to inches: The equivalent quantities alongside the roadmap arrow are the ones needed to construct the conversion factor. We choose 1 in/2.54 cm, rather than the inverse, because it gives an answer in inches:

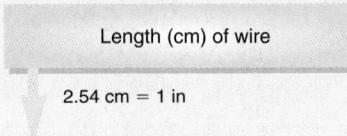

$$\text{Length (in)} = \text{length (cm)} \times \text{conversion factor} = 325 \text{ cm} \times \frac{1 \text{ in}}{2.54 \text{ cm}} = 128 \text{ in}$$

Converting the length from inches to feet:

$$\text{Length (ft)} = \text{length (in)} \times \text{conversion factor} = 128 \text{ in} \times \frac{1 \text{ ft}}{12 \text{ in}} = 10.7 \text{ ft}$$

Converting the length in feet to price in dollars:

$$\text{Price (\$)} = \text{length (ft)} \times \text{conversion factor} = 10.7 \text{ ft} \times \frac{\$0.15}{1 \text{ ft}} = \boxed{\$1.60}$$

Check The units are correct for each step. The conversion factors make sense in terms of the relative unit sizes: the number of inches is *smaller* than the number of centimeters (an inch is *larger* than a centimeter), and the number of feet is *smaller* than the number of inches. The total price seems reasonable: a little more than 10 ft of wire at $0.15/ft should cost a little more than $1.50.

Comment 1. We could also have strung the three steps together:

$$\text{Price (\$)} = 325 \text{ cm} \times \frac{1 \text{ in}}{2.54 \text{ cm}} \times \frac{1 \text{ ft}}{12 \text{ in}} \times \frac{\$0.15}{1 \text{ ft}} = \$1.60$$

2. There are usually alternative sequences in unit-conversion problems. Here, for example, we would get the same answer if we first converted the cost of wire from $/ft to $/cm and kept the wire length in cm. Try it yourself.

FOLLOW-UP PROBLEM 1.2

A furniture factory needs 31.5 ft^2 of fabric to upholster one chair. Its Dutch supplier sends the fabric in bolts of exactly 200 m^2. What is the maximum number of chairs that can be upholstered by 3 bolts of fabric (1 m = 3.281 ft)?

SECTION SUMMARY
A measured quantity consists of a number and a unit. Conversion factors are used to express a quantity in different units and are constructed as a ratio of equivalent quantities. The problem-solving approach used in this text usually has four parts: (1) devise a plan for the solution, (2) put the plan into effect in the calculations, (3) check to see if the answer makes sense, and (4) practice with similar problems.

1.5 MEASUREMENT IN SCIENTIFIC STUDY

Almost everything we own—clothes, house, food, vehicle—is manufactured with measured parts, sold in measured amounts, and paid for with measured currency. Measurement is so commonplace that it's easy to take for granted, but it has a fascinating history characterized by the search for *exact, invariable standards.*

Our current system of measurement began in 1790, when the newly formed National Assembly of France, of which Lavoisier was a member, set up a committee to establish consistent unit standards. This effort led to the development of the *metric system.* In 1960, another international committee met in France to establish the International System of Units, a revised metric system now accepted by scientists throughout the world. The units of this system are called **SI units,** from the French Système International d'Unités.

General Features of SI Units

As Table 1.2 shows, the SI system is based on a set of seven **fundamental units,** or **base units,** each of which is identified with a physical quantity. All other units, which are called **derived units,** are combinations of these seven base units. For example, the derived unit for speed, meters per second (m/s), is the base unit for length (m) divided by the base unit for time (s). (Many derived units can be used as conversion factors because they occur as a ratio of two or more base units.)

How Many Barleycorns from His Majesty's Nose to His Thumb? Systems of measurement began thousands of years ago as trade, building, and land surveying spread throughout the civilized world. For most of that time, however, measurement was based on inexact physical standards. For example, an inch was the length of three barleycorns (seeds) placed end to end; a yard was the distance from the tip of King Edgar's nose to the tip of his thumb with his arm outstretched; and an acre was the area tilled by one man working with a pair of oxen in a day.

Table 1.2 SI Base Units

Physical Quantity (Dimension)	Unit Name	Unit Abbreviation
Mass	kilogram	kg
Length	meter	m
Time	second	s
Temperature	kelvin	K
Electric current	ampere	A
Amount of substance	mole	mol
Luminous intensity	candela	cd

For quantities that are much smaller or much larger than the base unit, we use decimal prefixes and exponential (scientific) notation. Because these prefixes are based on powers of 10, SI units are easier to use in calculations than English units such as pounds and inches. Table 1.3 on the next page shows the most important prefixes.

Table 1.3 Common Decimal Prefixes Used with SI Units

Prefix*	Prefix Symbol	Meaning — Number	Meaning — Word	Multiple†
tera	T	1,000,000,000,000	trillion	10^{12}
giga	G	1,000,000,000	billion	10^{9}
mega	M	1,000,000	million	10^{6}
kilo	k	1,000	thousand	10^{3}
hecto	h	100	hundred	10^{2}
deka	da	10	ten	10^{1}
—	—	1	one	10^{0}
deci	d	0.1	tenth	10^{-1}
centi	c	0.01	hundredth	10^{-2}
milli	m	0.001	thousandth	10^{-3}
micro	μ	0.000001	millionth	10^{-6}
nano	n	0.000000001	billionth	10^{-9}
pico	p	0.000000000001	trillionth	10^{-12}
femto	f	0.000000000000001	quadrillionth	10^{-15}

*The prefixes most frequently used by chemists appear in bold type.
†Many of the calculations you'll perform involve numbers written in exponential notation, which include a multiple written as a power of 10. If you are not familiar with this method of expressing numbers or just need a review, be sure to read Appendix A.

Some Important SI Units in Chemistry

Let's discuss some of the SI units for quantities that we use early in the text: length, volume, mass, density, temperature, and time. Units for other quantities are presented later. Table 1.4 shows some useful SI-English equivalent quantities for length, volume, and mass.

Table 1.4 Common SI-English Equivalent Quantities

Quantity	SI	SI Equivalents	English Equivalents	English to SI Equivalent
Length	1 kilometer (km)	1000 (10^3) meters	0.6214 mile (mi)	1 mile = 1.609 km
	1 meter (m)	100 (10^2) centimeters	1.094 yards (yd)	1 yard = 0.9144 m
		1000 millimeters (mm)	39.37 inches (in)	1 foot (ft) = 0.3048 m
	1 centimeter (cm)	0.01 (10^{-2}) meter	0.3937 inch	1 inch = 2.54 cm (exactly)
Volume	1 cubic meter (m^3)	1,000,000 (10^6) cubic centimeters	35.31 cubic feet (ft^3)	1 cubic foot = 0.02832 m^3
	1 cubic decimeter (dm^3)	1000 cubic centimeters	0.2642 gallon (gal)	1 gallon = 3.785 dm^3
			1.057 quarts (qt)	1 quart = 0.9464 dm^3
				1 quart = 946.4 cm^3
	1 cubic centimeter (cm^3)	0.001 dm^3	0.03381 fluid ounce	1 fluid ounce = 29.57 cm^3
Mass	1 kilogram (kg)	1000 grams	2.205 pounds (lb)	1 pound = 0.4536 kg
	1 gram (g)	1000 milligrams (mg)	0.03527 ounce (oz)	1 ounce = 28.35 g

Length The SI base unit of length is the **meter (m).** The standard meter is now based on two quantities, the speed of light in a vacuum and the second. A meter is a little longer than a yard (1 m = 1.094 yd); a centimeter (10^{-2} m) is about two-fifths of an inch (1 cm = 0.3937 in; 1 in = 2.54 cm). Biological cells are often measured in micrometers (1 μm = 10^{-6} m). On the atomic-size scale, nanometers and picometers are used (1 nm = 10^{-9} m; 1 pm = 10^{-12} m). An older unit still in use is the angstrom (1 Å = 10^{-10} m = 0.1 nm = 100 pm).

Volume Any sample of matter has a certain **volume (V),** the amount of space it occupies. The SI unit of volume is the **cubic meter (m^3).** In chemistry, the most important volume units are non-SI units, the **liter (L)** and the **milliliter (mL)** (note the uppercase L). Physicians and other medical practitioners measure body fluids in cubic decimeters (dm^3), which is equivalent to liters:

$$1 \text{ L} = 1 \text{ dm}^3 = 10^{-3} \text{ m}^3$$

As the prefix *milli-* indicates, 1 mL is $\frac{1}{1000}$ of a liter, and it is equal to exactly 1 cubic centimeter (cm^3):

$$1 \text{ mL} = 1 \text{ cm}^3 = 10^{-3} \text{ dm}^3 = 10^{-3} \text{ L} = 10^{-6} \text{ m}^3$$

A liter is slightly larger than a quart (qt) (1 L = 1.057 qt; 1 qt = 946.4 mL); 1 fluid ounce ($\frac{1}{32}$ of a quart) equals 29.57 mL (29.57 cm^3). Figure 1.8 is a life-size depiction of the 1000-fold decreases in volume from the cubic decimeter to the cubic millimeter. The edge of a cubic meter would be about 2.5 times the width of the open textbook.

How Long Is a Meter? The history of the meter points up the ongoing drive to define units based on unchanging standards. The French scientists who set up the metric system defined the meter as 1/10,000,000, the distance from the equator (through Paris!) to the North Pole. The meter was later redefined as the distance between two fine lines engraved on a corrosion-resistant metal bar kept at the International Bureau of Weights and Measures in France. Fear that the bar would be damaged by war led to the adoption of an exact, unchanging, universally available atomic standard: 1,650,763.73 wavelengths of orange-red light from electrically excited krypton atoms. The current standard is even more reliable: 1 meter is the distance light travels in a vacuum in 1/299,792,458 second.

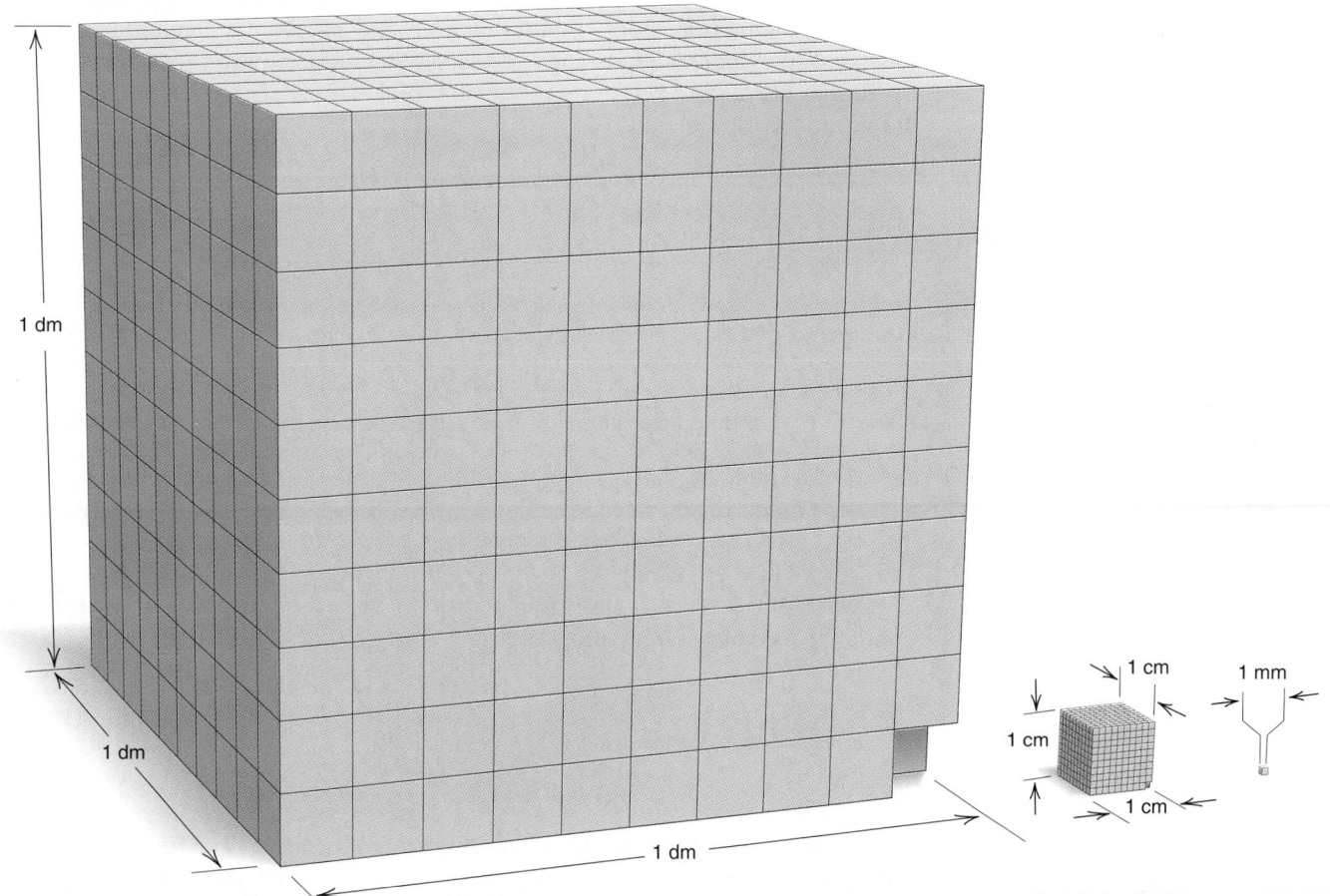

Figure 1.8 **Some volume relationships in SI.** The cube on the left is 1 dm^3; each side is 1 dm long and is divided into ten 1-cm segments. There are 1000 dm^3 in 1 m^3. The middle cube is 1 cm^3; each side is 1 cm long and is divided into ten 1-mm segments; 1000 cm^3 = 1 dm^3 = 1 L = 1000 mL, so 1 cm^3 = 1 mL. The right cube is 1 mm^3; each side is 1 mm long; 1 mm^3 = 1 μL. There are 10^3 μL in 1 cm^3 and 10^6 μL in 1 dm^3 (1 L).

Figure 1.9 Common laboratory volumetric glassware. A, From left to right are two graduated cylinders, a pipet being emptied into a beaker, a buret delivering liquid to an Erlenmeyer flask, and two volumetric flasks. **Inset,** In contact with glass, this liquid forms a concave meniscus (curved surface). **B,** Automatic pipets deliver a given volume of liquid.

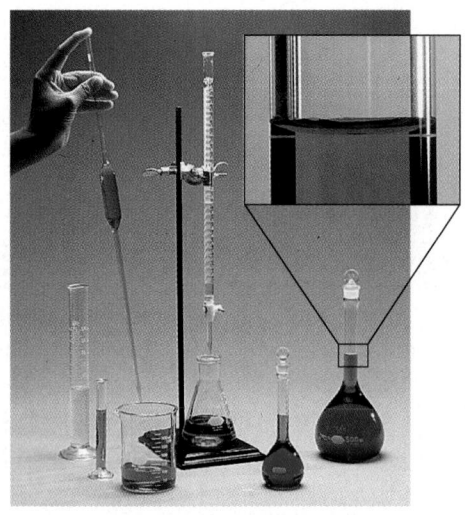

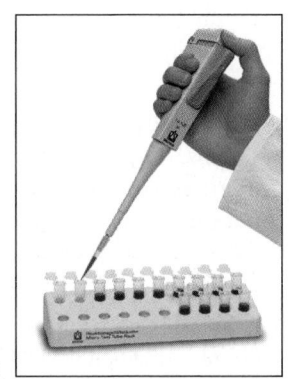

A   B

Figure 1.9 shows some of the types of laboratory glassware designed to contain liquids or measure their volumes. Many come in sizes from a few milliliters to a few liters. Erlenmeyer flasks and beakers are used to contain liquids. Graduated cylinders, pipets, and burets are used to measure and transfer liquids. Volumetric flasks and many pipets have a fixed volume indicated by a mark on the neck. In quantitative work, liquid solutions are prepared in volumetric flasks, measured in cylinders, pipets, and burets, and then transferred to beakers or flasks for further chemical operations. Automatic pipets transfer a given volume of liquid accurately and quickly.

SAMPLE PROBLEM 1.3 Determining the Volume of a Solid by Displacement of Water

---

Problem The volume of an irregularly shaped solid can be determined from the volume of water it displaces. A graduated cylinder contains 19.9 mL of water. When a small piece of galena, an ore of lead, is added, it sinks and the volume increases to 24.5 mL. What is the volume of the piece of galena in cm^3 and in L?

Plan We have to find the volume of the galena from the change in volume of the cylinder contents. The volume of galena in mL is the difference in the known volumes before and after adding it. The mL and cm^3 units represent identical volumes, so the volume of the galena in mL equals the volume in cm^3. We construct a conversion factor to convert the volume from mL to L. The calculation steps are shown in the roadmap.

Solution Finding the volume of galena:

Volume (mL) = volume after − volume before = 24.5 mL − 19.9 mL = 4.6 mL

Converting the volume from mL to cm^3:

$$\text{Volume (cm}^3) = 4.6 \text{ mL} \times \frac{1 \text{ cm}^3}{1 \text{ mL}} = \boxed{4.6 \text{ cm}^3}$$

Converting the volume from mL to L:

$$\text{Volume (L)} = 4.6 \text{ mL} \times \frac{10^{-3} \text{ L}}{1 \text{ mL}} = \boxed{4.6 \times 10^{-3} \text{ L}}$$

Check The units and magnitudes of the answers seem correct. It makes sense that the volume expressed in mL would have a number 1000 times larger than the volume expressed in L, because a milliliter is $\frac{1}{1000}$ of a liter.

FOLLOW-UP PROBLEM 1.3 If you submerge a steel cube 15.0 mm on each side in 31.8 mL of water in a graduated cylinder, what is the total final volume (in L) of the cylinder contents?

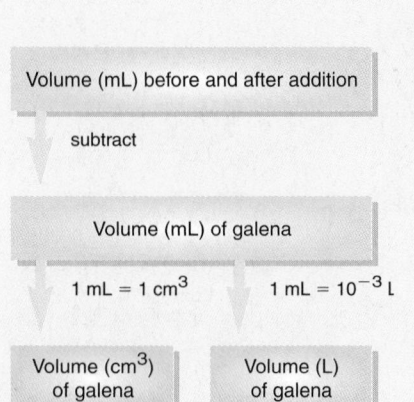

Mass The **mass** of an object refers to the quantity of matter it contains. The SI unit of mass is the **kilogram (kg),** the only base unit whose standard is a physical object—a platinum-iridium cylinder kept in France. It is also the only base unit whose name has a prefix. (In contrast to the practice with other base units, however, we attach prefixes to the word "gram," as in "microgram," rather than to the word "kilogram"; thus, we never say "microkilogram.")

The terms *mass* and *weight* have distinct meanings. Since a given object's quantity of matter cannot change, its *mass is constant.* Its **weight,** on the other hand, depends on its mass *and* the strength of the local gravitational field pulling on it. Because the strength of this field varies with height above the Earth's surface, the object's weight varies also. For instance, you actually weigh slightly less on a high mountaintop than at sea level.

Does this mean that if you weighed an object on a laboratory balance in Miami (sea level) and in Denver (about 1.7 km above sea level), you would obtain different results? Fortunately not. Such balances are designed to measure mass rather than weight, so this chaotic situation does not occur. (We are actually "massing" an object when we weigh it on a balance, but we rarely use that term.) Mechanical balances compare the object's unknown mass with known masses built into the balance, so the local gravitational field pulls equally on them. Electronic (analytical) balances determine mass by generating an electric field that counteracts the local gravitational field. The magnitude of the current needed to restore the pan to its zero position is then displayed as the object's mass. Therefore, an electronic balance must be readjusted with standard masses when it is moved to a different location.

Don't Drop That Kilogram! The U.S. copy of the master kilogram is kept in a vault at the National Institute of Standards and Technology near Washington, DC. Since 1889, it has been taken to France only twice for comparison with the master kilogram, which is removed from its vault only once a year for such purposes. Two people are always present when it is moved: one to carry it with forceps, the other to catch it if the first person stumbles.

SAMPLE PROBLEM 1.4 Converting Units of Mass

Problem Some international computer communications are now carried by optical fibers in cables laid along the ocean floor. If one strand of optical fiber weighs 1.19×10^{-3} lb/m, what is the mass (in kg) of a cable made of six strands of optical fiber, each long enough to link New York and Paris (8.84×10^3 km)?

Plan We have to find the mass of cable (in kg) from the given mass/length of fiber, number of fibers/cable, and the length (distance from New York to Paris). One way to do this (as shown in the roadmap) is to first find the mass of one fiber and then find the mass of cable. We convert the length of one fiber from km to m and then find its mass (in lb) by using the lb/m factor. The cable mass is six times the fiber mass, and finally we convert lb to kg.

Solution Converting the fiber length from km to m:

$$\text{Length (m) of fiber} = 8.84 \times 10^3 \text{ km} \times \frac{10^3 \text{ m}}{1 \text{ km}} = 8.84 \times 10^6 \text{ m}$$

Converting the length of one fiber to mass (lb):

$$\text{Mass (lb) of fiber} = 8.84 \times 10^6 \text{ m} \times \frac{1.19 \times 10^{-3} \text{ lb}}{1 \text{ m}} = 1.05 \times 10^4 \text{ lb}$$

Finding the mass of the cable (lb):

$$\text{Mass (lb) of cable} = \frac{1.05 \times 10^4 \text{ lb}}{1 \text{ fiber}} \times \frac{6 \text{ fibers}}{1 \text{ cable}} = 6.30 \times 10^4 \text{ lb/cable}$$

Converting the mass of cable from lb to kg:

$$\text{Mass (kg) of cable} = \frac{6.30 \times 10^4 \text{ lb}}{1 \text{ cable}} \times \frac{1 \text{ kg}}{2.205 \text{ lb}}$$

$$= \boxed{2.86 \times 10^4 \text{ kg/cable}}$$

Length (km) of fiber

1 km = 10^3 m

Length (m) of fiber

1 m = 1.19×10^{-3} lb

Mass (lb) of fiber

6 fibers = 1 cable

Mass (lb) of cable

2.205 lb = 1 kg

Mass (kg) of cable

Check The units are correct. Let's think through the relative sizes of the answers to see if they make sense: The number of m should be 10^3 larger than the number of km. If 1 m of fiber weighs about 10^{-3} lb, about 10^7 m should weigh about 10^4 lb. The cable mass should be six times as much, or about 6×10^4 lb. Since 1 lb is about $\frac{1}{2}$ kg, the number of kg should be about half the number of lb.

Comment Actually, the pound (lb) is the English unit of *weight,* not *mass.* The English unit of mass, called the *slug,* is rarely used.

FOLLOW-UP PROBLEM 1.4 If a raindrop weighs 65 mg on average and 5.1×10^5 raindrops fall on a lawn every minute, what mass (in kg) of rain falls in 1.5 h?

Figure 1.10 shows the ranges of some common lengths, volumes, and masses.

Density The **density** (*d*) of an object is its mass divided by its volume:

$$\text{Density} = \frac{\text{mass}}{\text{volume}} \qquad (1.1)$$

Whenever needed, you can isolate mathematically each of the component variables by treating density as a conversion factor:

$$\text{Mass} = \text{volume} \times \text{density} = \text{volume} \times \frac{\text{mass}}{\text{volume}}$$

Or,

$$\text{Volume} = \text{mass} \times \frac{1}{\text{density}} = \text{mass} \times \frac{\text{volume}}{\text{mass}}$$

Figure 1.10 **Some interesting quantities of length (A), volume (B), and mass (C).** Note that the scales are exponential.

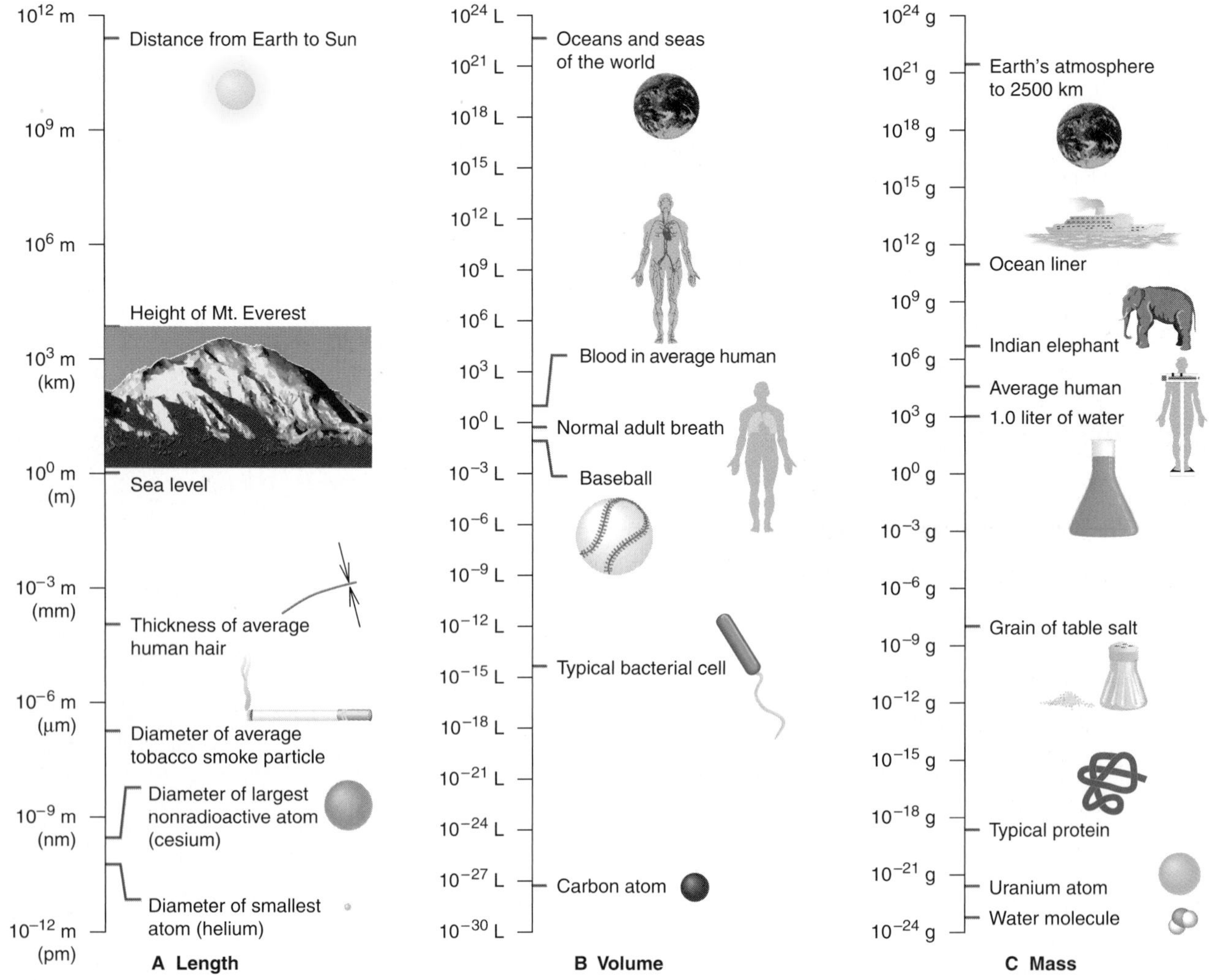

A Length

10^{12} m — Distance from Earth to Sun

10^9 m

10^6 m

Height of Mt. Everest

10^3 m (km)

10^0 m (m) — Sea level

10^{-3} m (mm) — Thickness of average human hair

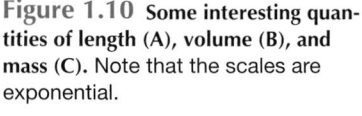

10^{-6} m (μm) — Diameter of average tobacco smoke particle

10^{-9} m (nm) — Diameter of largest nonradioactive atom (cesium)

10^{-12} m (pm) — Diameter of smallest atom (helium)

B Volume

10^{24} L — Oceans and seas of the world

10^{21} L

10^{18} L

10^{15} L

10^{12} L

10^9 L

10^6 L — Blood in average human

10^3 L

10^0 L — Normal adult breath

10^{-3} L — Baseball

10^{-6} L

10^{-9} L

10^{-12} L

10^{-15} L — Typical bacterial cell

10^{-18} L

10^{-21} L

10^{-24} L

10^{-27} L — Carbon atom

10^{-30} L

C Mass

10^{24} g

10^{21} g — Earth's atmosphere to 2500 km

10^{18} g

10^{15} g

10^{12} g — Ocean liner

10^9 g — Indian elephant

10^6 g — Average human

10^3 g — 1.0 liter of water

10^0 g

10^{-3} g

10^{-6} g

10^{-9} g — Grain of table salt

10^{-12} g

10^{-15} g

10^{-18} g — Typical protein

10^{-21} g — Uranium atom

10^{-24} g — Water molecule

Under given conditions of temperature and pressure, *density is a characteristic physical property of a substance* and has a specific value, even though the separate values of mass and volume vary. Mass and volume are examples of **extensive properties,** those dependent on the amount of substance present. Density, on the other hand, is an **intensive property,** one that is independent of the amount of substance present. For example, the mass of a gallon of water is four times the mass of a quart of water, but its volume is also four times greater; therefore, the density of the water, the *ratio* of its mass to its volume, is constant at a particular temperature and pressure, regardless of the sample size.

The SI unit of density is the kilogram per cubic meter (kg/m^3), but in chemistry, density is typically given in units of g/L (g/dm^3) or g/mL (g/cm^3). For example, the density of liquid water at ordinary pressure and room temperature (20°C) is 1.0 g/mL. The densities of some common substances are given in Table 1.5. As you might expect from the magnified views of the physical states (see Figure 1.2), the densities of gases are much lower than those of liquids or solids.

Table 1.5 Densities of Some Common Substances*

Substance	Physical State	Density (g/cm^3)
Hydrogen	gas	0.0000899
Oxygen	gas	0.00133
Grain alcohol	liquid	0.789
Water	liquid	0.998
Table salt	solid	2.16
Aluminum	solid	2.70
Lead	solid	11.3
Gold	solid	19.3

*At room temperature (20°C) and normal atmospheric pressure (1 atm).

SAMPLE PROBLEM 1.5 Calculating Density from Mass and Length

Problem Lithium is a soft, gray solid that has the lowest density of any metal. It is an essential component of some advanced batteries. If a small rectangular slab of lithium weighs 1.49×10^3 mg and has sides that measure 20.9 mm by 11.1 mm by 11.9 mm, what is the density of lithium in g/cm^3?

Plan To find the density in g/cm^3, we need the mass of lithium in g and the volume in cm^3. The mass is given in mg, so we convert mg to g. Volume data are not given, but we can convert the given side lengths from mm to cm, and then multiply them to find the volume in cm^3. Finally, we divide mass by volume to get density. The steps are shown in the roadmap.

Solution Converting the mass from mg to g:

$$\text{Mass (g) of lithium} = 1.49 \times 10^3 \text{ mg} \left(\frac{10^{-3} \text{ g}}{1 \text{ mg}} \right) = 1.49 \text{ g}$$

Converting side lengths from mm to cm:

$$\text{Length (cm) of one side} = 20.9 \text{ mm} \times \frac{1 \text{ cm}}{10 \text{ mm}} = 2.09 \text{ cm}$$

Similarly, the other side lengths are 1.11 cm and 1.19 cm.
Finding the volume:

$$\text{Volume (cm}^3) = 2.09 \text{ cm} \times 1.11 \text{ cm} \times 1.19 \text{ cm} = 2.76 \text{ cm}^3$$

Calculating the density:

$$\text{Density of lithium} = \frac{1.49 \text{ g}}{2.76 \text{ cm}^3} = \boxed{0.540 \text{ g/cm}^3}$$

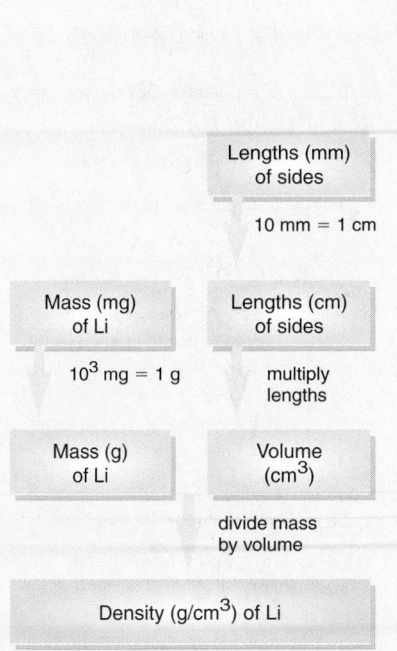

Lengths (mm) of sides

10 mm = 1 cm

Mass (mg) of Li   Lengths (cm) of sides

10^3 mg = 1 g   multiply lengths

Mass (g) of Li   Volume (cm^3)

divide mass by volume

Density (g/cm^3) of Li

Lithium floating in oil floating on water.

Check Since 1 cm = 10 mm, the number of cm in each length should be $\frac{1}{10}$ the number of mm. The units for density are correct, and the size of the answer ($\sim$0.5 g/cm^3) seems correct since the number of g (1.49) is about half the number of cm^3 (2.76). As the photo (see margin) of lithium floating in oil that is floating on water (d = 1.0 g/cm^3) indicates, this answer makes sense.

FOLLOW-UP PROBLEM 1.5 The piece of galena in Sample Problem 1.3 has a volume of 4.6 cm^3. If the density of galena is 7.5 g/cm^3, what is the mass (in kg) of that piece of galena?

Temperature There is a common misunderstanding about heat and temperature. **Temperature (T)** is a measure of how hot or cold a substance is *relative to another substance*. **Heat** is the energy that flows between objects that are at different temperatures. Temperature is related to the *direction* of that energy flow: when two objects at different temperatures touch, energy flows from the one with the higher temperature to the one with the lower temperature until their temperatures are equal. When you hold an ice cube, its "cold" seems to flow *into* your hand; actually, heat flows *from* your hand into the ice. (In Chapter 6, we will see how heat is measured and how it is related to chemical and physical change.) Energy is an *extensive* property (as is volume), but temperature is an *intensive* property (as is density): a vat of boiling water has more energy than a cup of boiling water, but the temperatures of the two water samples are the same.

In the laboratory, the most common means for measuring temperature is the **thermometer,** a device that contains a fluid that expands when it is heated. When the thermometer's fluid-filled bulb is immersed in a substance hotter than itself, heat flows from the substance through the glass and into the fluid, which expands and rises in the thermometer tube. If a substance is colder than the thermometer, heat flows outward from the fluid, which contracts and falls within the tube.

The three temperature scales most important for us to consider are the Celsius (°C), formerly called centigrade, the Kelvin (K), and the Fahrenheit (°F) scales. The SI base unit of temperature is the **kelvin (K);** note that the kelvin has no degree sign (°). The Kelvin scale, also known as the *absolute scale,* is preferred in all scientific work, although the Celsius scale is used frequently. In the United States, the Fahrenheit scale is still used for weather reporting, body temperature, and other everyday purposes. *The three scales differ in the size of the unit and/or the temperature of the zero point.* Figure 1.11 shows some interesting temperatures. Figure 1.12 shows the freezing and boiling points of water in the three scales.

The **Celsius scale,** devised in the 18th century by the Swedish astronomer Anders Celsius, is based on changes in the physical state of water: 0°C is set at water's freezing point, and 100°C is set at its boiling point (at normal atmospheric pressure). The **Kelvin (absolute) scale** was devised by the English physicist William Thomson, known as Lord Kelvin, in 1854 during his experiments on the expansion and contraction of gases. *The Kelvin scale uses the same size degree unit as the Celsius scale—$\frac{1}{100}$* of the difference between the freezing and boiling points of water—*but it differs in zero point.* The zero point in the Kelvin scale, 0 K, is called *absolute zero* and equals −273.15°C. In the Kelvin scale, *all temperatures have positive values.* Water freezes at +273.15 K (0°C) and boils at +373.15 K (100°C). We can convert between the Celsius and Kelvin scales by remembering the difference in zero points: since 0°C = 273.15 K,

$$T \text{ (in K)} = T \text{ (in °C)} + 273.15 \qquad (1.2)$$

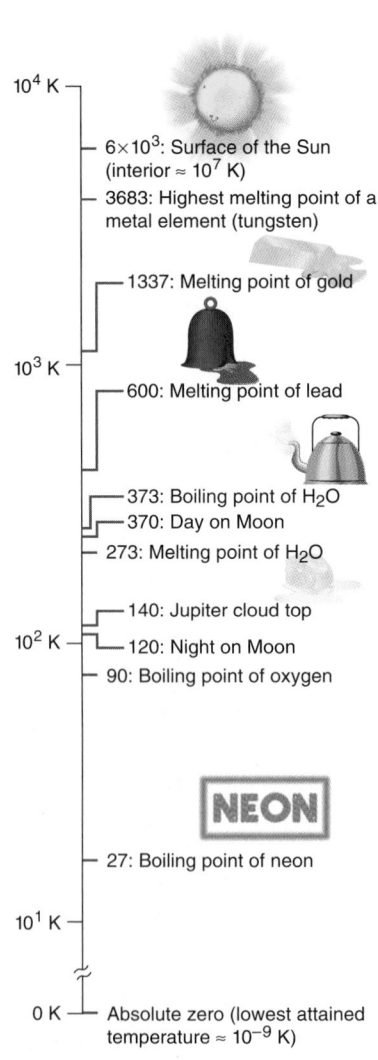

10^4 K —

6×10^3: Surface of the Sun (interior ≈ 10^7 K)

3683: Highest melting point of a metal element (tungsten)

1337: Melting point of gold

10^3 K —

600: Melting point of lead

373: Boiling point of H$_2$O
370: Day on Moon
273: Melting point of H$_2$O

140: Jupiter cloud top

10^2 K —
120: Night on Moon
90: Boiling point of oxygen

NEON

27: Boiling point of neon

10^1 K —

0 K — Absolute zero (lowest attained temperature ≈ 10^{-9} K)

Figure 1.11 **Some interesting temperatures.**

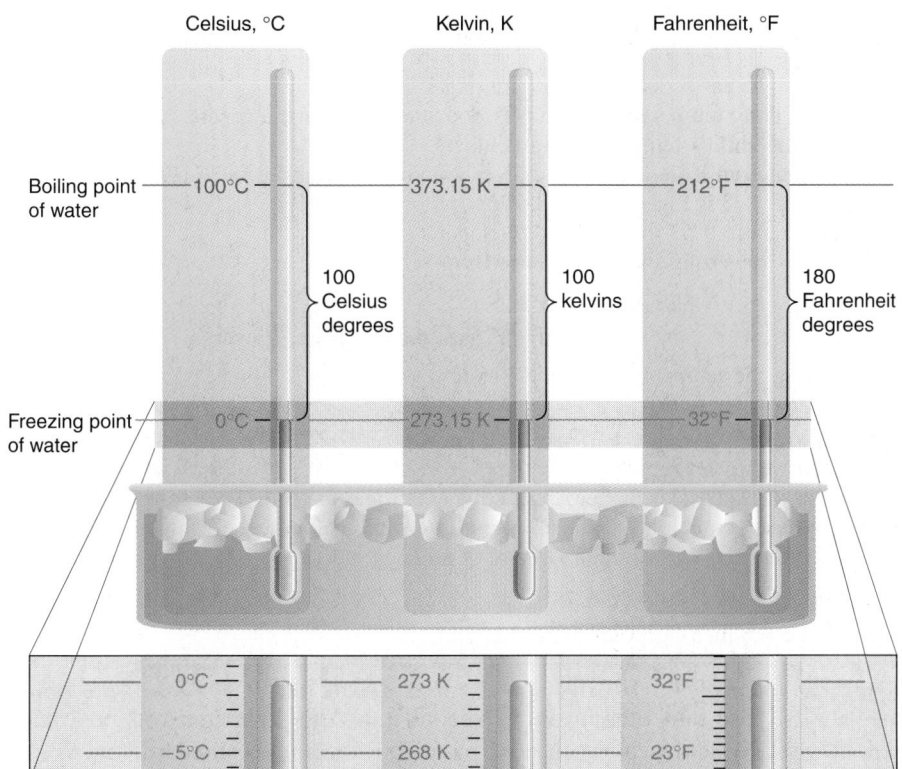

Figure 1.12 **The freezing point and the boiling point of water in the Celsius, Kelvin (absolute), and Fahrenheit temperature scales.** As you can see, this range consists of 100 degrees on the Celsius and Kelvin scales, but 180 degrees on the Fahrenheit scale. At the bottom of the figure, a portion of each of the three thermometer scales is expanded to show the sizes of the units. A Celsius degree (°C; *left*) and a kelvin (K; *center*) are the same size, and each is $\frac{9}{5}$ the size of a Fahrenheit degree (°F; *right*).

Solving Equation 1.2 for T (in °C) gives

$$T \text{ (in °C)} = T \text{ (in K)} - 273.15 \qquad (1.3)$$

The Fahrenheit scale differs from the other scales in its zero point *and* in the size of its unit. Water freezes at 32°F and boils at 212°F. Therefore, 180 Fahrenheit degrees (212°F − 32°F) represents the same temperature change as 100 Celsius degrees (or 100 kelvins). Because 100 Celsius degrees equal 180 Fahrenheit degrees,

$$1 \text{ Celsius degree} = \tfrac{180}{100} \text{ Fahrenheit degrees} = \tfrac{9}{5} \text{ Fahrenheit degrees}$$

To convert a temperature in °C to °F, first change the degree size and then adjust the zero point:

$$T \text{ (in °F)} = \tfrac{9}{5}T \text{ (in °C)} + 32 \qquad (1.4)$$

To convert a temperature in °F to °C, do the two steps in the opposite order; that is, first adjust the zero point and then change the degree size. In other words, solve Equation 1.4 for T (in °C):

$$T \text{ (in °C)} = [T \text{ (in °F)} - 32]\tfrac{5}{9} \qquad (1.5)$$

(The only temperature with the same numerical value in the Celsius and Fahrenheit scales is −40°; that is, −40°F = −40°C.)

SAMPLE PROBLEM 1.6 Converting Units of Temperature

Problem A child has a body temperature of 38.7°C.
(a) If normal body temperature is 98.6°F, does the child have a fever?
(b) What is the child's temperature in kelvins?
Plan (a) To find out if the child has a fever, we convert from °C to °F (Equation 1.4) and see whether 38.7°C is higher than 98.6°F.
(b) We use Equation 1.2 to convert the temperature in °C to K.
Solution (a) Converting the temperature from °C to °F:

$$T \text{ (in °F)} = \tfrac{9}{5}T \text{ (in °C)} + 32 = \tfrac{9}{5}(38.7°C) + 32$$

$$= \boxed{101.7°F; \text{ yes, the child has a fever.}}$$

(b) Converting the temperature from °C to K:

$$T \text{ (in K)} = T \text{ (in °C)} + 273.15 = 38.7°C + 273.15 = \boxed{311.8 \text{ K}}$$

Check (a) From everyday experience, you know that 101.7°F is a reasonable temperature for someone with a fever.
(b) We know that a Celsius degree and a kelvin are the same size. Therefore, we can check the math by approximating the Celsius value as 40°C and adding 273: 40 + 273 = 313, which is close to our calculation, so there is no large errror.

FOLLOW-UP PROBLEM 1.6 Mercury melts at 234 K, lower than any other pure metal. What is its melting point in °C and °F?

Time The SI base unit of time is the **second (s).** Although time was once measured by the day and year, it is now based on an atomic standard: microwave radiation absorbed by cesium atoms (Figure 1.13). In the laboratory, we study the speed of a reaction by measuring the time it takes a fixed amount of substance to undergo a chemical change. The range of reaction speed is enormous: a fast reaction may be over in less than a nanosecond (10^{-9} s), whereas slow ones, such as rusting or aging, take years. Chemists now use lasers to study changes that occur in a few picoseconds (10^{-12} s) or even femtoseconds (10^{-15} s).

SECTION SUMMARY

SI units consist of seven base units and numerous derived units. Exponential notation and prefixes based on powers of 10 are used to express very small and very large numbers. The SI base unit of length is the meter (m). Length units on the atomic

Figure 1.13 The cesium atomic clock. The accuracy of the best pendulum clock is to within 3 seconds per year and that of the best quartz clock is 1000 times greater. The most recent version of the atomic clock, NIST-F1, developed by the Physics Laboratory of the National Institute of Standards and Technology, is over 6000 times more accurate still, to within 1 second in 20 million years! Rather than using the oscillations of a pendulum, the atomic clock measures the oscillations of microwave radiation absorbed by gaseous cesium atoms: 1 second is defined as 9,192,631,770 of these oscillations. This new clock cools the cesium atoms with infrared lasers to around 10^{-6} K above absolute zero, which allows much longer observation times of the atoms, and thus much greater accuracy.

scale are the nanometer (nm) and picometer (pm). Volume units are derived from length units; the most important volume units in chemistry are the cubic meter (m^3) and the liter (L). The mass of an object, a measure of the quantity of matter present in it, is constant. The SI unit of mass is the kilogram (kg). The weight of an object varies with the gravitational field influencing it. Density *(d)* is the ratio of mass to volume of a substance and is one of its characteristic physical properties. Temperature *(T)* is a measure of the relative hotness of an object. Heat is energy that flows from an object at higher temperature to one at lower temperature. Temperature scales differ in the size of the degree unit and/or the zero point. In chemistry, temperature is measured in kelvins (K) or degrees Celsius (°C). Extensive properties, such as mass, volume, and energy, depend on an object's size (extent). Intensive properties, such as density and temperature, are independent of size.

The Central Importance of Measurement in Science It's important to keep in mind *why* scientists measure things: "When you can measure what you are speaking about, and express it in numbers, you know something about it; but when you cannot measure it, . . . your knowledge is of a meager and unsatisfactory kind; it may be the beginning of knowledge, but you have scarcely, in your thoughts, advanced to the stage of science" (William Thomson, Lord Kelvin, 1824–1907).

1.6   UNCERTAINTY IN MEASUREMENT: SIGNIFICANT FIGURES

Because measuring devices are manufactured to limited specifications, and because we use our imperfect senses and skills to read them, we can never measure something exactly or know a quantity with absolute certainty. In other words, every measurement we make includes some **uncertainty.**

The particular measuring device we choose in a given situation depends on how much uncertainty in a measurement we are willing to accept. When you buy potatoes, a supermarket scale that measures in 0.1-kg increments is perfectly acceptable; it tells you that the mass is, for example, 2.0 ± 0.1 kg. The term "± 0.1 kg" expresses the uncertainty in the measurement: the potatoes weigh between 1.9 and 2.1 kg. For a large-scale reaction, a chemist might need a balance that measures in 0.001-kg increments in order to obtain 2.036 ± 0.001 kg of a chemical, that is, between 2.035 and 2.037 kg. The greater number of digits in the mass of the chemical indicates that we know its mass with *more certainty* than we know the mass of the potatoes. We *always estimate the rightmost digit* when reading a measuring device. The uncertainty can be expressed with the ± sign, but generally we drop the sign and *assume an uncertainty of one unit in the rightmost digit.* The digits we record in a measurement, both the certain and the uncertain ones, are called **significant figures.** There are four significant figures in 2.036 kg and two in 2.0 kg. *The greater the number of significant figures in a measurement, the greater is the certainty.* Figure 1.14 shows this point for two thermometers.

Figure 1.14 The number of significant figures in a measurement depends on the measuring device. A, Two thermometers measuring the same temperature are shown with expanded views. The thermometer on the left is graduated in 0.1°C and reads 32.33°C; the one on the right is graduated in 1°C and reads 32.3°C. Therefore, a reading with more significant figures (more certainty) can be made with the thermometer on the left. **B,** This modern electronic thermometer measures the resistance through a fine platinum wire in the probe to determine temperatures to the nearest microkelvin (10^{-6} K).

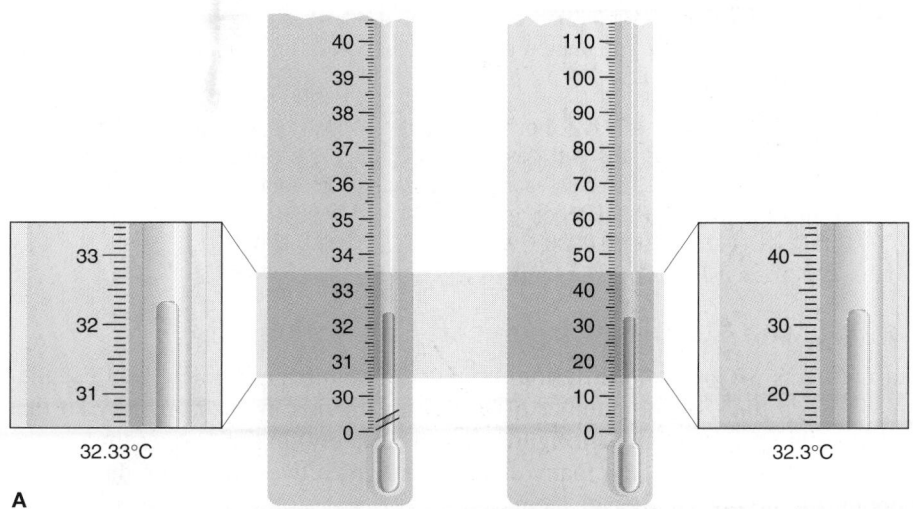

Determining Which Digits Are Significant

When you take measurements or use them in calculations, you must know the number of digits that are significant. In general, *all digits are significant, except zeros that are not measured but are used only to position the decimal point.* Here is a simple procedure that applies this generalization:

1. Make sure that the measured quantity has a decimal point.
2. Start at the left of the number and move right until you reach the first nonzero digit.
3. Count that digit and every digit to its right as significant.

Sometimes, there may be a slight complication with zeros that *end* a number. Zeros that end a number and lie either after or before the decimal point *are* significant; thus, 1.030 mL has four significant figures, and 5300. L has four significant figures also. The problem arises if there is no decimal point, as in 5300 L. In such cases, we would assume that the zeros are *not* significant, so exponential notation is needed to show which of the zeros, if any, were measured and therefore are significant. Thus, 5.300×10^3 L has four significant figures, 5.30×10^3 L has three, and 5.3×10^3 L has only two. In this and other modern texts (and research articles), a terminal decimal point is used when necessary to clarify the situation; thus, 500 mL has one significant figure, but 5.00×10^2 mL, 500. mL, and 0.500 L have three.

SAMPLE PROBLEM 1.7 Determining the Number of Significant Figures

Problem For each of the following quantities, underline the zeros that are significant figures (sf), and determine the number of significant figures in each quantity. For **(d)** to **(f)**, express each in exponential notation first.

(a) 0.0030 L   (b) 0.1044 g   (c) 53,069 mL
(d) 0.00004715 m   (e) 57,600. s   (f) 0.0000007160 cm³

Plan We determine the number of significant figures by counting digits, as just presented, paying particular attention to the position of zeros in relation to the decimal point.

Solution (a) 0.0030 L has 2 sf

(b) 0.1044 g has 4 sf

(c) 53,069 mL has 5 sf

(d) 0.00004715 m, or 4.715×10^{-5} m, has 4 sf

(e) 57,600. s, or 5.7600×10^4 s, has 5 sf

(f) 0.0000007160 cm³, or 7.160×10^{-7} cm³, has 4 sf

Check Be sure that every zero counted as significant comes after nonzero digit(s) in the number.

FOLLOW-UP PROBLEM 1.7 For each of the following quantities, underline the zeros that are significant figures and determine the number of significant figures (sf) in each quantity. For **(d)** to **(f)**, express each in exponential notation first.

(a) 31.070 mg   (b) 0.06060 g   (c) 850.°C
(d) 200.0 mL   (e) 0.0000039 m   (f) 0.000401 L

Working with Significant Figures in Calculations

Measurements may contain differing numbers of significant figures. In a calculation, we keep track of the number of significant figures in each quantity so that we don't claim more significant figures (more certainty) in the answer than in the original data. If we have too many significant figures, we **round off** the answer to obtain the proper number of them.

The general rule for rounding is that *the least certain measurement sets the limit on certainty for the entire calculation and determines the number of significant figures in the final answer.* Suppose you want to find the density of a new ceramic material. You measure the mass of a piece on an analytical balance and obtain 3.8056 g; you measure its volume as 2.5 mL by displacement of water in a graduated cylinder. The mass measurement has five significant figures, but the volume measurement has only two. Should you report the density as 3.8056 g/2.5 mL = 1.5222 g/mL or as 1.5 g/mL? The answer with five significant figures implies more certainty in *all* the measurements than the answer with two. But you didn't measure the volume to five significant figures, so you can't know the density with that much certainty. Therefore, you report the answer as 1.5 g/mL.

Significant Figures and Arithmetic Operations The following two rules tell how many significant figures to show based on the arithmetic operation:

1. *For multiplication and division.* The answer contains the same number of *significant figures* as in the measurement with the fewest significant figures. Suppose you want to find the volume of a sheet of a new graphite composite. The length (9.2 cm) and width (6.8 cm) are obtained with a meterstick and the thickness (0.3744 cm) with a set of fine calipers. The volume calculation is

$$\text{Volume (cm}^3) = 9.2 \text{ cm} \times 6.8 \text{ cm} \times 0.3744 \text{ cm} = 23 \text{ cm}^3$$

The calculator may show 23.4225 cm^3, but the answer should be reported as 23 cm^3, with two significant figures, because the length and width measurements contain only two significant figures.

2. *For addition and subtraction.* The answer has the same number of *decimal places* as there are in the measurement with the fewest decimal places. Suppose you measure 83.5 mL of water in a graduated cylinder and add 23.28 mL of protein solution from a buret. The total volume is

$$\text{Volume (mL)} = 83.5 \text{ mL} + 23.28 \text{ mL} = 106.8 \text{ mL}$$

Here the calculator shows 106.78 mL, but you report the volume as 106.8 mL, with one decimal place, because this is the number in the measurement with the fewest decimal places (83.5 mL).

Rules for Rounding Off In most calculations, you need to round off the answer to obtain the proper number of significant figures or decimal places. Notice that in calculating the volume of the graphite composite above, we removed the extra digits, but in calculating the total solution volume, we removed the extra digit and increased the last digit by one. Here is a set of rules for rounding off:

1. If the digit removed is *more than 5,* the preceding number is increased by 1: 5.379 rounds to 5.38 if three significant figures are retained and to 5.4 if two significant figures are retained.

2. If the digit removed is *less than 5,* the preceding number is unchanged: 0.2413 rounds to 0.241 if three significant figures are retained and to 0.24 if two significant figures are retained.

3. If the digit removed *is 5,* the preceding number is increased by 1 if it is odd and remains unchanged if it is even: 17.75 rounds to 17.8, but 17.65 rounds to 17.6. If the 5 is followed only by zeros, rule 3 is followed; if the 5 is followed by nonzeros, rule 1 is followed: 17.6500 rounds to 17.6, but 17.6513 rounds to 17.7.

4. *Always carry one or two additional significant figures through a multistep calculation and round off the final answer* **only.** Don't be concerned if you string together a calculation to check a sample or follow-up problem and find that your answer differs in the last decimal place from the one in the book. In text calculations, we round off intermediate steps to show you the correct number of significant figures, and this process may sometimes change the last digit.

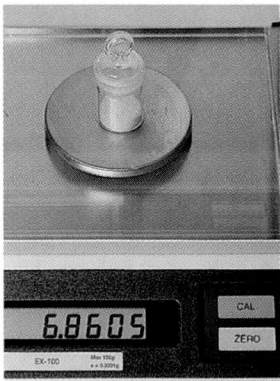

Figure 1.15 Significant figures and measuring devices. The mass (6.8605 g) measured with an analytical balance *(top)* has more significant figures than the volume (68.2 mL) measured with a graduated cylinder *(bottom)*.

Significant Figures and Electronic Calculators A calculator usually gives answers with too many significant figures. For example, if your calculator displays ten digits and you divide 15.6 by 9.1, it will show 1.714285714. Obviously, most of these digits are not significant; the answer should be rounded off to 1.7 so that it has the same number of significant figures as in 9.1. A good way to prove to yourself that the additional digits are not significant is to perform two calculations, indicated by the uncertainty in each of the last digits, to obtain the highest and lowest possible answers. For $(15.6 \pm 0.1)/(9.1 \pm 0.1)$,

The highest answer is $\dfrac{15.7}{9.0} = 1.744444\ldots$ The lowest answer is $\dfrac{15.5}{9.2} = 1.684782\ldots$

No matter how many digits the calculator displays, the values differ in the first decimal place, so the answer has two significant figures and should be reported as 1.7. Many calculators have a FIX button that allows you to set the number of digits displayed.

Significant Figures and Choice of Measuring Device The measuring device you choose determines the number of significant figures you can obtain. Suppose you are doing an experiment that requires mixing a liquid with a solid. You weigh the solid on the analytical balance and obtain a value with five significant figures. It would make sense to measure the liquid with a buret or pipet, which measures volumes to more significant figures than a graduated cylinder. If you chose the cylinder, you would have to round off more digits in the calculations, so the certainty in the mass value would be wasted (Figure 1.15). As you gain experience in the laboratory, you will learn to choose a measuring device based on the number of significant figures you need in the final answer.

Exact Numbers Some numbers are called **exact numbers** because they have no uncertainty associated with them. Some exact numbers are part of a unit definition: there are 60 minutes in 1 hour, 1000 micrograms in 1 milligram, and 2.54 centimeters in 1 inch. Other exact numbers result from actually counting individual items: there are exactly 3 quarters in my hand, 26 letters in the English alphabet, and so forth. Since they have no uncertainty, *exact numbers do not limit the number of significant figures in the answer.* Put another way, exact numbers have as many significant figures as a calculation requires.

SAMPLE PROBLEM 1.8 Significant Figures and Rounding

Problem Perform the following calculations and round the answer to the correct number of significant figures:

(a) $\dfrac{16.3521 \text{ cm}^2 - 1.448 \text{ cm}^2}{7.085 \text{ cm}}$

(b) $\dfrac{(4.80 \times 10^4 \text{ mg})\left(\dfrac{1 \text{ g}}{1000 \text{ mg}}\right)}{11.55 \text{ cm}^3}$

Plan We use the rules just presented in the text. In (a), we subtract before we divide. In (b), we note that the unit conversion involves an exact number.

Solution (a) $\dfrac{16.3521 \text{ cm}^2 - 1.448 \text{ cm}^2}{7.085 \text{ cm}} = \dfrac{14.904 \text{ cm}^2}{7.085 \text{ cm}} = \boxed{2.104 \text{ cm}}$

(b) $\dfrac{(4.80 \times 10^4 \text{ mg})\left(\dfrac{1 \text{ g}}{1000 \text{ mg}}\right)}{11.55 \text{ cm}^3} = \dfrac{48.0 \text{ g}}{11.55 \text{ cm}^3} = \boxed{4.16 \text{ g/cm}^3}$

Check Note that in (a) we lose a decimal place in the numerator, and in (b) we retain 3 sf in the answer because there are 3 sf in 4.80. Rounding to the nearest whole number is always a good way to check: (a) $(16 - 1)/7 \approx 2$; (b) $(5 \times 10^4 / 1 \times 10^3)/12 \approx 4$.

FOLLOW-UP PROBLEM 1.8 Perform the following calculation and round the answer to the correct number of significant figures: $\dfrac{25.65 \text{ mL} + 37.4 \text{ mL}}{73.55 \text{ s}\left(\dfrac{1 \text{ min}}{60 \text{ s}}\right)}$

Precision, Accuracy, and Instrument Calibration

Precision and accuracy are two aspects of certainty. We often use these terms interchangeably in everyday speech, but in scientific measurements they have distinct meanings. **Precision,** or *reproducibility,* refers to how close the measurements in a series are to each other. **Accuracy** refers to how close a measurement is to the actual value.

The concepts of precision and accuracy are linked with two common types of error:

1. **Systematic error** produces values that are *either* all higher or all lower than the actual value. Such error is part of the experimental system, perhaps caused by a faulty measuring device or by a consistent mistake in taking a reading.
2. **Random error,** in the absence of systematic error, produces some values that are higher *and* some that are lower than the actual value. Random error always occurs, but its size depends on the measurer's skill and the instrument's precision.

Precise measurements have low random error, that is, small deviations from the average. *Accurate measurements have low systematic error and, generally, low random error as well.* In some cases, when many measurements are taken that have a high random error, the *average* may still be accurate.

Suppose each of four students measures 25.0 mL of water in a graduated cylinder and then weighs the water in the cylinder on a balance. If the density of water is 1.00 g/mL at the temperature of the experiment, the actual mass of 25.0 mL of water is 25.0 g. Each student performs the operation four times, subtracts the mass of the empty cylinder, and obtains one of the four graphs in Figure 1.16. In graphs A and B, the random error is small; that is, the precision is high (the weighings are reproducible). In A, however, the accuracy is high as well (all the values are close to 25.0 g), whereas in B the accuracy is low (there is a systematic error). In graphs C and D, there is a large random error; that is, the precision is low. Large random error is often called large *scatter.* Note, however, that in D there is also a systematic error (all the values are high), whereas in C the average of the values is close to the actual value.

Systematic error can be avoided, or at least taken into account, through **calibration** of the measuring device, that is, by comparing it with a known standard. The systematic error in graph B, for example, might be caused by a poorly manufactured cylinder that reads "25.0" when it actually contains about 27 mL. If you detect such an error by means of a calibration procedure, you could adjust all volumes measured with that cylinder. Instrument calibration is an essential part of careful measurement.

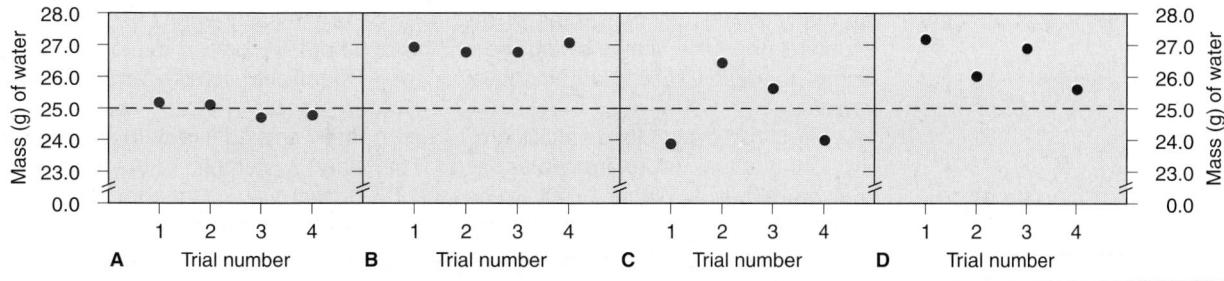

Figure 1.16 Precision and accuracy in a laboratory calibration. Each graph represents four measurements made with a graduated cylinder that is being calibrated (see text for details). **A,** High preci- sion, high accuracy. **B,** High precision, low accuracy (systematic er- ror). **C,** Low precision, average value close to actual. **D,** Low preci- sion, low accuracy.

Chemical Connections

Chemistry Problem Solving in the Real World

Working through the problems in this text will greatly improve your chances of doing well in the course, but that is not really the final goal. Even though you may not choose to become a chemist, learning chemistry is essential to many fields, including medicine, engineering, and environmental science. It is also essential to an understanding of complex science-related issues, such as the recycling of plastics, the reduction of urban smog, and the potential of genetic cloning—to mention just three of many.

A major scientific discipline such as chemistry consists of several subdisciplines that form connections with other sciences to spawn new fields. Traditionally, chemistry has five main branches—organic, inorganic, analytical, physical, and biological chemistry—but these long ago formed interconnections, such as physical organic and bioinorganic chemistry. Solving the problems of today and tomorrow requires further connections, such as ecological chemistry, materials science, planetary chemistry, and molecular biology. The more complex the system that is under study, the greater is the need for interdisciplinary scientific thinking.

Environmental issues are especially complex, and one of the most intractable is the acid rain problem. Let's see how it is being approached by chemists interacting with scientists in related fields. Acid rain results in large part from burning high-sulfur coal, a major fuel used throughout much of North America and Europe. As the coal burns, the gaseous products, including an oxide of its sulfur impurities, are carried away by prevailing winds. In contact with oxygen and rain, this sulfur oxide undergoes chemical changes, yielding acid rain. (We discuss the chemical details in later chapters.) In the northeastern United States and adjacent parts of Canada, acid rain has killed fish, decimated forests, injured crops, and released harmful substances into the soil. Acid rain has severely damaged many forests and lakes in Germany, Sweden, Norway, and several countries in central and eastern Europe. Acidic precipitation has even been observed on both Poles!

Chemists and other scientists are currently working together to solve this problem (Figure B1.1). As geochemists search for low-sulfur coal deposits, their engineering colleagues design better ways of removing sulfur oxides from smokestack gases. Atmospheric chemists and meteorologists track changes through the affected regions, develop computer models to predict the changes, and coordinate their findings with environmental chemists at ground stations. Ecological chemists and aquatic biologists monitor the effects of acid rain on insects, birds, and fish. Agricultural chemists and agronomists study ways to protect crop yields. Biochemists and genetic engineers develop new, more acid-resistant crop species. Soil chemists measure changes in mineral content, sharing their data with forestry scientists to save valuable timber and recreational woodlands. Organic chemists and chemical engineers convert coal to cleaner fuels. Superimposed on this intense experimental activity are scientifically trained economic and policy experts who provide business and government leaders with the information to make their decisions. With all this input, interdisciplinary understanding of the acid rain problem has increased enormously and certainly will continue to do so.

These professions are just a few of those involved in studying a single chemistry-related issue. Chemical principles apply to many other specialties, from medicine and pharmacology to art restoration and criminology, from genetics and space research to archaeology and oceanography. Chemistry problem solving has far-reaching relevance to many aspects of your daily life and your future career as well.

SECTION SUMMARY

Because the final digit of a measurement is estimated, all measurements have a limit to their certainty, which is expressed by the number of significant figures. The certainty of a calculated result depends on the certainty of the data, so the answer has as many significant figures as in the least certain measurement. Excess digits are rounded off in the final answer. The choice of laboratory device depends on the certainty needed. Exact numbers have as many significant figures as the calculation requires.

Precision (how close values are to each other) and accuracy (how close values are to the actual value) are two aspects of certainty. Systematic errors result in values that are either all higher or all lower than the actual value. Random errors result in some values that are higher and some values that are lower than the actual value. Precise measurements have low random error; accurate measurements have low

Figure B1.1 The central role of chemistry in solving real-world problems. Researchers in many chemical specialties join with those in other sciences to investigate complex modern issues such as acid rain. **A,** The sulfur oxide in power plant emissions will be reduced by devices that remove them from flue gases. **B,** Ecologists sample lake, pond, and river water and observe wildlife to learn the effects of acidic precipitation on aquatic environments. **C,** Atmospheric chemists study the location and concentration of air pollutants with balloons that carry monitoring equipment aloft. **D,** Agronomists and other soil scientists use miniaturized equipment to measure acidity in plant and soil samples.

systematic error and often low random error. The size of random errors depends on the skill of the investigator and the precision of the instrument. A systematic error, however, is often caused by faulty equipment and can be compensated for by calibration.

Chapter Perspective

This chapter has provided several keys for you to use repeatedly in your study of chemistry: descriptions of some essential concepts; insight into how scientists think; the units of modern measurement and the mathematical skills to apply them; and a systematic approach to solving problems. You can begin using these keys in the next chapter, where we discuss the components of matter and their classification and see the winding path of scientific discovery that led to our current model of atomic structure.

For Review and Reference (Numbers in parentheses refer to pages, unless noted otherwise.)

Learning Objectives

Relevant section and/or sample problem (SP) numbers appear in parentheses.

Understand These Concepts

1. The distinction between physical and chemical properties and changes (1.1; SP 1.1)
2. The defining features of the states of matter (1.1)
3. The nature of potential and kinetic energy and their interconversion (1.1)
4. The process of approaching a phenomenon scientifically, and the distinctions between observation, hypothesis, experiment, and model (1.3)
5. The common units of length, volume, mass, and temperature and their numerical prefixes (1.5)
6. Distinctions between mass and weight, heat and temperature, intensive and extensive properties (1.5)

7. The meaning of uncertainty in measurements and the use of significant figures and rounding (1.6)
8. The distinction between accuracy and precision and the types of errors (1.6)

Master These Skills

1. Using conversion factors in calculations (1.4; SPs 1.2–1.4)
2. Finding density from mass and volume (SP 1.5)
3. Converting among the Kelvin, Celsius, and Fahrenheit scales (SP 1.6)
4. Determining the number of significant figures (SP 1.7) and rounding to the correct number of digits (SP 1.8)

Key Terms

Section 1.1
chemistry (3)
matter (3)
composition (3)
property (3)
physical property (3)
physical change (3)
chemical property (3)
chemical change (chemical reaction) (3)
state of matter (4)
solid (4)
liquid (4)
gas (4)
energy (6)
potential energy (6)
kinetic energy (6)

Section 1.2
alchemy (8)
combustion (10)
phlogiston theory (10)

Section 1.3
scientific method (12)
observation (12)
data (12)
natural law (12)
hypothesis (12)
experiment (12)
variable (12)
controlled experiment (12)
model (theory) (12)

Section 1.4
conversion factor (13)
dimensional analysis (15)

Section 1.5
SI unit (17)
base (fundamental) unit (17)
derived unit (17)
meter (m) (19)
volume (V) (19)
cubic meter (m^3) (19)
liter (L) (19)
milliliter (mL) (19)
mass (21)
kilogram (kg) (21)
weight (21)
density (d) (22)
extensive property (23)
intensive property (23)
temperature (T) (24)
heat (24)

thermometer (24)
kelvin (K) (24)
Celsius scale (24)
Kelvin (absolute) scale (24)
second (s) (26)

Section 1.6
uncertainty (27)
significant figures (27)
round off (28)
exact number (30)
precision (31)
accuracy (31)
systematic error (31)
random error (31)
calibration (31)

Key Equations and Relationships

1.1 Calculating density from mass and volume (22):

$$\text{Density} = \frac{\text{mass}}{\text{volume}}$$

1.2 Converting temperature from °C to K (24):

$$T \text{ (in K)} = T \text{ (in °C)} + 273.15$$

1.3 Converting temperature from K to °C (25):

$$T \text{ (in °C)} = T \text{ (in K)} - 273.15$$

1.4 Converting temperature from °C to °F (25):

$$T \text{ (in °F)} = \tfrac{9}{5}T \text{ (in °C)} + 32$$

1.5 Converting temperature from °F to °C (25):

$$T \text{ (in °C)} = [T \text{ (in °F)} - 32]\tfrac{5}{9}$$

Highlighted Figures and Tables

These figures (F) and tables (T) provide a quick review of key ideas. Entries in color contain frequently used data.

F1.1 The distinction between physical and chemical change (3)
F1.2 The physical states of matter (4)
F1.3 Potential energy and kinetic energy (7)
F1.6 The scientific approach (12)

T1.2 SI base units (17)
T1.3 Decimal prefixes used with SI units (18)
T1.4 SI-English equivalent quantities (18)
F1.8 Some volume relationships in SI (19)

Brief Solutions to Follow-up Problems

1.1 (a) Physical. Solid iodine changes to gaseous iodine.
(b) Chemical. Gasoline burns in air to form different substances.
(c) Chemical. In contact with air, torn skin and blood react to form different substances.

1.2 No. of chairs

$$= 3 \; \cancel{bolts} \times \frac{200 \; \cancel{m^2}}{1 \; \cancel{bolt}} \times \frac{3.281 \; \cancel{ft}}{1 \; \cancel{m}} \times \frac{3.281 \; \cancel{ft}}{1 \; \cancel{m}} \times \frac{1 \; chair}{31.5 \; \cancel{ft^2}}$$

$$= 205 \; chairs$$

1.3 Volume (mL) $= 15.0^3 \; \cancel{mm^3} \times \dfrac{1^3 \; \cancel{cm^3}}{10^3 \; \cancel{mm^3}} \times \dfrac{1 \; mL}{1 \; \cancel{cm^3}}$

$$= 3.38 \; mL$$

Total volume (L) $= (31.8 \; \cancel{mL} + 3.38 \; \cancel{mL}) \times \dfrac{10^{-3} \; L}{1 \; \cancel{mL}}$

$$= 3.52 \times 10^{-2} \; L$$

1.4 Mass (kg) of rain $= 1.5 \; \cancel{h} \times \dfrac{60 \; \cancel{min}}{1 \; \cancel{h}} \times \dfrac{5.1 \times 10^5 \; \cancel{drops}}{1 \; \cancel{min}}$

$$\times \frac{65 \; \cancel{mg}}{1 \; \cancel{drop}} \times \frac{1 \; \cancel{g}}{10^3 \; \cancel{mg}} \times \frac{1 \; kg}{10^3 \; \cancel{g}}$$

$$= 3.0 \times 10^3 \; kg$$

1.5 Mass (kg) of sample $= 4.6 \; \cancel{cm^3} \times \dfrac{7.5 \; \cancel{g}}{1 \; \cancel{cm^3}} \times \dfrac{1 \; kg}{10^3 \; \cancel{g}}$

$$= 0.034 \; kg$$

1.6 T (in °C) $= 234 \; K - 273.15 = -39°C$

T (in °F) $= \frac{9}{5}(-39°C) + 32 = -38°F$

Answer contains two significant figures (see Section 1.6)

1.7 (a) 31.0̲70 mg, 5 sf　　(b) 0.06060̲ g, 4 sf
(c) 850̲.°C, 3 sf　　(d) 2.0̲00×10² mL, 4 sf
(e) 3.9×10⁻⁶ m, 2 sf　　(f) 4.0̲1×10⁻⁴ L, 3 sf

1.8 $\dfrac{25.65 \; mL + 37.4 \; mL}{73.55 \; s \left(\dfrac{1 \; min}{60 \; s} \right)} = 51.4 \; mL/min$

Problems

Problems with **colored** numbers are answered at the back of the text. Sections match the text and provide the number(s) of relevant sample problems. Most offer Concept Review Questions, Skill-Building Exercises (in similar pairs), and Problems in Context. Then Comprehensive Problems, based on material from any section, follow.

Some Fundamental Definitions

(Sample Problem 1.1)

● **Concept Review Questions**

1.1 Scenes A and B represent changes in matter at the atomic scale:

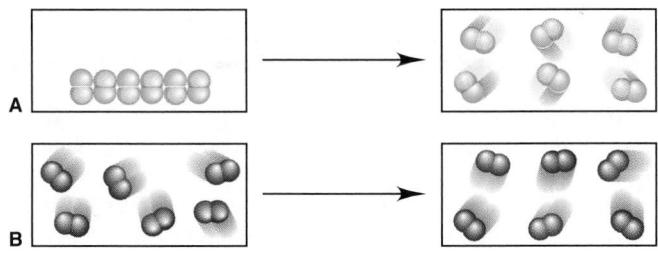

(a) Which show(s) a physical change?
(b) Which show(s) a chemical change?
(c) Which result(s) in different physical properties?
(d) Which result(s) in different chemical properties?
(e) Which result(s) in a change in state?

● **Skill-Building Exercises** *(paired)*

1.2 Describe solids, liquids, and gases in terms of how they fill a container. Use your descriptions to identify the physical state (at room temperature) of the following: (a) helium in a toy balloon; (b) mercury in a thermometer; (c) soup in a bowl.

1.3 Use your descriptions in the previous problem to identify the physical state (at room temperature) of the following: (a) air in your room; (b) vitamin tablets in a bottle; (c) sugar in a packet.

1.4 Define physical property and chemical property. Identify each type of property in the following statements:
(a) Yellow-green chlorine gas attacks silvery sodium metal to form white crystals of sodium chloride (table salt).
(b) A magnet separates a mixture of black iron shavings and white sand.

1.5 Define physical change and chemical change. State which type of change occurs in each of the following statements:
(a) Heating orange mercuric oxide powder produces silvery liquid mercury and colorless gaseous oxygen.
(b) The iron in discarded automobiles slowly forms reddish brown, crumbly rust.

1.6 Which of the following is a chemical change? Explain your reasoning: (a) boiling canned soup; (b) toasting a slice of bread; (c) chopping a log; (d) burning a log.

1.7 Which of the following changes can be reversed by changing the temperature (that is, which are physical changes)?
(a) dew condensing on a leaf; (b) an egg turning hard when it is boiled; (c) ice cream melting; (d) a spoonful of batter cooking on a hot griddle.

1.8 For each pair, which has higher potential energy?
(a) The fuel in your car or the products in its exhaust
(b) Wood in a fireplace or the ashes in the fireplace after the wood burns

1.9 For each pair, which has higher kinetic energy?
(a) A sled resting at the top of a hill or the sled sliding down the hill
(b) Water above a dam or water falling over the dam

Chemical Arts and the Origins of Modern Chemistry

● **Concept Review Questions**

1.10 The alchemical, medical, and technological traditions were precursors to chemistry. State a contribution that each made to the development of the science of chemistry.

1.11 How did the phlogiston theory explain combustion?

1.12 One important observation that supporters of the phlogiston theory had trouble explaining was that the calx of a metal weighs more than the metal itself. Why was that observation important? How did the phlogistonists respond?

1.13 Lavoisier developed a new theory of combustion that overturned the phlogiston theory. What measurements were central to his theory, and what key discovery did he make?

The Scientific Approach: Developing a Model

● **Concept Review Questions**

1.14 How are the key elements of scientific thinking used in the following scenario? While making your breakfast toast, you notice it fails to pop out of the toaster. Thinking the spring mechanism is stuck, you notice that the bread is unchanged. Assuming you forgot to plug in the toaster, you check and find it is plugged in. When you take the toaster into the dining room and plug it into a different outlet, you find the toaster works. Returning to the kitchen, you turn on the switch for the overhead light and nothing happens.

1.15 Why is a quantitative observation more useful than a non-quantitative one? Which of the following are quantitative?
(a) The sun rises in the east.
(b) An astronaut weighs one-sixth as much on the Moon as on Earth.
(c) Ice floats on water.
(d) An old-fashioned hand pump cannot draw water from a well more than 34 ft deep.

1.16 Describe the essential features of a well-designed experiment.

1.17 Describe the essential features of a scientific model.

Chemical Problem Solving
(Sample Problem 1.2)

● **Concept Review Questions**

1.18 When you convert feet to inches, how do you decide which portion of the conversion factor should be in the numerator and which in the denominator?

● **Skill-Building Exercises** *(paired)*

1.19 Write the conversion factor(s) for each of the following:
(a) in^2 to cm^2; (b) km^2 to m^2; (c) mi/h to cm/s; (d) lb/ft^3 to g/cm^3.
1.20 Write the conversion factor(s) for each of the following:
(a) cm/min to in/min; (b) m^3 to in^3; (c) m/s^2 to km/h^2;
(d) gallons/h to L/s.

Measurement in Scientific Study
(Sample Problems 1.3 to 1.6)

● **Concept Review Questions**

1.21 Describe the difference between intensive and extensive properties. Which of the following properties are intensive:
(a) mass; (b) density; (c) volume; (d) melting point?

1.22 Explain the difference between mass and weight. Why is your weight on the Moon one-sixth that on Earth?

1.23 For each of the following cases, state whether the density of the object increases, decreases, or remains the same:
(a) A sample of chlorine gas is compressed.
(b) A lead weight is carried from sea level to the top of a high mountain.
(c) A sample of water is frozen.
(d) An iron bar is cooled.
(e) A diamond is submerged in water.

1.24 Explain the difference between heat and temperature. Does 1 L of water at 65°F have more, less, or the same quantity of energy as 1 L of water at 65°C?

1.25 Why can't a temperature in the Celsius scale be converted into one in the Fahrenheit scale with a simple one-step conversion factor?

● **Skill-Building Exercises** *(paired)*

1.26 The radius of a copper atom is 128 pm. What is its radius in nanometers (nm)?

1.27 The radius of a barium atom is 2.22×10^{-10} m. What is its radius in angstroms (Å)?

1.28 What is the length in inches (in) of a 100.-m soccer field?

1.29 The center on your basketball team is 6 ft 10 in tall. How tall is the player in centimeters (cm)?

1.30 A small hole in the wing of a space shuttle requires a 17.7-cm^2 patch. (a) What is the patch's area in square kilometers (km^2)? (b) If the patching material costs NASA $3.25/$in^2$, what is the cost of the patch?

1.31 The area of a telescope lens is 6322 mm^2. (a) What is the area of the lens in square feet (ft^2)? (b) If it takes a technician 45 s to polish 135 mm^2, how long does it take her to polish the entire lens?

1.32 Express your body weight in kilograms (kg).

1.33 There are 2.60×10^{15} short tons of oxygen in the atmosphere (1 short ton = 2000 lb). How many metric tons of oxygen are present (1 metric ton = 1000 kg)?

1.34 The average density of the Earth is 5.52 g/cm^3. What is its density in (a) kg/m^3? (b) lb/ft^3?

1.35 The speed of light in a vacuum is 2.998×10^8 m/s. What is its speed in (a) km/h? (b) mi/min?

1.36 The volume of a certain bacterial cell is 1.72 μm^3. (a) What is its volume in cubic millimeters (mm^3)? (b) What is the volume of 10^5 cells in L?

1.37 (a) How many cubic meters of milk are in 1 qt (946.4 mL)? (b) How many liters of milk are in 835 gallons (1 gal = 4 qt)?

1.38 An empty vial weighs 55.32 g. (a) If the vial weighs 185.56 g when filled with liquid mercury (d = 13.53 g/cm^3), what is its volume? (b) How much would the vial weigh if it were filled with water (d = 0.997 g/cm^3 at 25°C)?

1.39 An Erlenmeyer flask weighs 241.3 g when empty. When filled with water (d = 1.00 g/cm^3), the flask and its contents weigh 489.1 g. (a) What is the flask's volume? (b) How much does the flask weigh when filled with chloroform (d = 1.48 g/cm^3)?

1.40 A small cube of aluminum measures 15.6 mm on a side and weighs 10.25 g. What is the density of aluminum in g/cm^3?

1.41 A steel ball-bearing with a circumference of 32.5 mm weighs 4.20 g. What is the density of the steel in g/cm³ (V of a sphere $= \frac{4}{3}\pi r^3$; circumference of a circle $= 2\pi r$)?

1.42 Perform the following conversions:
(a) 72°F (a pleasant spring day) to °C and K
(b) −164°C (the boiling point of methane, the main component of natural gas) to K and °F
(c) 0 K (absolute zero, theoretically the coldest possible temperature) to °C and °F

1.43 Perform the following conversions:
(a) 106°F (the body temperature of many birds) to K and °C
(b) 3410°C (the melting point of tungsten, the highest for any metal) to K and °F
(c) 6.1×10^3 K (the surface temperature of the sun) to °F and °C

● **Problems in Context**

1.44 Anton van Leeuwenhoek, a 17th-century pioneer in the use of the microscope, described the microorganisms he saw as "animalcules" whose length was "25 thousandths of an inch." How long were the animalcules in meters?

1.45 The distance between two adjacent peaks on a wave is called the *wavelength*. The wavelength of visible light determines its color.
(a) The wavelength of a beam of green light is 545 nanometers (nm). What is its wavelength in meters?
(b) The wavelength of a beam of red light is 683 nm. What is its wavelength in angstroms (Å)?

1.46 To the alchemists, gold was the material representation of purity. It is a very soft metal that can be hammered into extremely thin sheets. If a 1.10-g piece of gold ($d = 19.32$ g/cm³) is hammered into a sheet whose area is 40.0 ft², what is the average thickness of the sheet in centimeters?

1.47 A cylindrical tube 7.8 cm high and 0.85 cm in diameter is used to collect blood samples. How many cubic decimeters (dm³) of blood can it hold (V of a cylinder $= \pi r^2 h$)?

1.48 English and non-English units appear on grocery items. (a) If the label on a bottle of cooking oil shows "48 fl oz (1.41 L)," how many fluid ounces are in 1.00 L? (b) If the serving size of oil is exactly 1 tablespoon, which has a mass of 14 g, and there are 96 servings in the container, what is the density of the oil in grams per milliliter (g/mL)?

Uncertainty in Measurement: Significant Figures
(Sample Problem 1.7)

● **Concept Review Questions**

1.49 What is an exact number? How are exact numbers treated differently from other numbers in a calculation?

1.50 All digits other than zero are significant. State a rule that tells which zeros are significant.

1.51 A newspaper reported that the attendance at Slippery Rock's home football game was 5209. (a) How many significant figures does this number contain? (b) Was the actual number of people counted? (c) After Slippery Rock's next home game, the newspaper reported an attendance of 5000. If you assume that this number contains two significant figures, how many people could actually have been at the game?

● **Skill-Building Exercises (paired)**

1.52 Underline the significant zeros in the following numbers:
(a) 0.39   (b) 0.039   (c) 0.0390   (d) 3.0900×10^4

1.53 Underline the significant zeros in the following numbers:
(a) 5.08   (b) 508   (c) 5.080×10^3 (d) 0.05080

1.54 Round off each number to the indicated number of significant figures (sf): (a) 0.0003554 (to 2 sf); (b) 35.8348 (to 4 sf); (c) 22.4555 (to 3 sf).

1.55 Round off each number to the indicated number of significant figures (sf): (a) 231.554 (to 4 sf); (b) 0.00845 (to 2 sf); (c) 144,000 (to 1 sf).

1.56 Round off each number in the following calculation to one fewer significant figure, and find the answer:
$$\frac{19 \times 155 \times 8.3}{3.2 \times 2.9 \times 4.7}$$

1.57 Round off each number in the following calculation to one fewer significant figure, and find the answer:
$$\frac{9.8 \times 6.18 \times 2.381}{24.3 \times 1.8 \times 18.5}$$

1.58 Carry out the following calculations, making sure that your answer has the correct number of significant figures:
(a) $\dfrac{2.795 \text{ m} \times 3.10 \text{ m}}{6.48 \text{ m}}$
(b) $V = \frac{4}{3}\pi r^3$, where $r = 9.282$ cm
(c) 1.110 cm + 17.3 cm + 108.2 cm + 316 cm

1.59 Carry out the following calculations, making sure that your answer has the correct number of significant figures:
(a) $\dfrac{2.420 \text{ g} + 15.6 \text{ g}}{4.8 \text{ g}}$ (b) $\dfrac{7.87 \text{ mL}}{16.1 \text{ mL} - 8.44 \text{ mL}}$
(c) $V = \pi r^2 h$, where $r = 6.23$ cm and $h = 4.630$ cm

1.60 Write the following numbers in scientific notation:
(a) 131,000.0   (b) 0.00047   (c) 210,006   (d) 2160.5

1.61 Write the following numbers in scientific notation:
(a) 281.0   (b) 0.00380   (c) 4270.8   (d) 58,200.9

1.62 Write the following numbers in standard notation. Use a terminal decimal point when needed:
(a) 5.55×10^3 (b) 1.0070×10^4
(c) 8.85×10^{-7} (d) 3.004×10^{-3}

1.63 Write the following numbers in standard notation. Use a terminal decimal point when needed:
(a) 6.500×10^3 (b) 3.46×10^{-5} (c) 7.5×10^2 (d) 1.8856×10^2

1.64 Convert the following into correct scientific notation:
(a) 802.5×10^2 (b) 1009.8×10^{-6} (c) 0.077×10^{-9}

1.65 Convert the following into correct scientific notation:
(a) 14.3×10^1 (b) 851×10^{-2} (c) 7500×10^{-3}

1.66 Carry out each of the following calculations, paying special attention to significant figures, rounding, and units (J = joule, the SI unit of energy; mol = mole, the SI unit for amount of substance):
(a) $\dfrac{(6.626\times10^{-34} \text{ J·s}) (2.9979\times10^8 \text{ m/s})}{489\times10^{-9} \text{ m}}$
(b) $\dfrac{(6.022\times10^{23} \text{ molecules/mol}) (1.19\times10^2 \text{ g})}{46.07 \text{ g/mol}}$
(c) $(6.022\times10^{23} \text{ atoms/mol})(2.18\times10^{-18} \text{ J/atom})\left(\dfrac{1}{2^2} - \dfrac{1}{3^2}\right)$,

where the numbers 2 and 3 in the last term are exact.

1.67 Carry out each of the following calculations, paying special attention to significant figures, rounding, and units:

(a) $\dfrac{8.32\times10^7 \text{ g}}{\frac{4}{3}(3.1416)\,(1.95\times10^2 \text{ cm})^3}$ (The term $\frac{4}{3}$ is exact.)

(b) $\dfrac{(1.84\times10^2 \text{ g})\,(44.7 \text{ m/s})^2}{2}$ (The term 2 is exact.)

(c) $\dfrac{(1.07\times10^{-4} \text{ mol/L})^2\,(2.6\times10^{-3} \text{ mol/L})}{(8.35\times10^{-5} \text{ mol/L})\,(1.48\times10^{-2} \text{ mol/L})^3}$

1.68 Which statements include exact numbers?
(a) Angel Falls is 3212 ft high.
(b) There are nine known planets in the solar system.
(c) There are 453.59 g in 1 lb.
(d) There are 1000 mm in 1 m.

1.69 Which of the following include exact numbers?
(a) The speed of light in a vacuum is a physical constant; to six significant figures, it is 2.99792×10^8 m/s.
(b) The density of mercury at 25°C is 13.53 g/mL.
(c) There are 3600 s in 1 h.
(d) The English alphabet consists of 26 letters.

1.70 How long is the metal strip shown below? Be sure to answer with the correct number of significant figures.

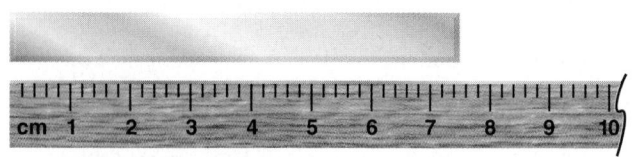

1.71 What volume of liquid is shown in the beaker at right? Be sure to report your answer with the correct number of significant figures.

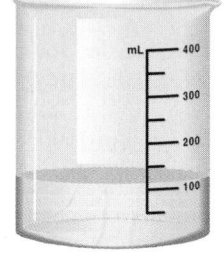

● **Problems in Context**

1.72 A gathering of 300 families attends a Founder's Day picnic in a small town, and the local Chamber of Commerce provides 3 bagels, 2 donuts, and 4 cans of iced tea to each family. After the picnic, 18 kg of garbage is collected. Which of the numbers—300, 3, 2, 4, 18—are exact numbers? Explain.

1.73 A laboratory instructor gives a sample of powdered metal to each of four students, I, II, III, and IV, and they weigh the samples. Their results for three trials follow. The true value is 8.72 g.
I: 8.72 g, 8.74 g, 8.70 g   II: 8.56 g, 8.77 g, 8.83 g
III: 8.50 g, 8.48 g, 8.51 g   IV: 8.41 g, 8.72 g, 8.55 g
(a) Calculate the average mass from each set of data, and tell which set is the most accurate.
(b) Precision is a measure of the average of the deviations of each piece of data from the average value. Which set of data is the most precise? Is this set also the most accurate?
(c) Which set of data has the best combination of accuracy and precision?
(d) Which set of data has the worst combination of accuracy and precision?

1.74 The following dartboards illustrate the types of errors often seen in measurements. The bull's-eye represents the actual value, and the darts represent the data.

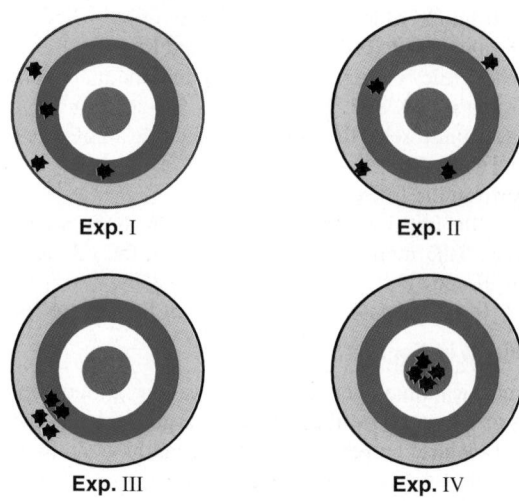

Exp. I   **Exp.** II

Exp. III   **Exp.** IV

(a) Which experiments yield the same average result?
(b) Which experiment(s) display(s) high precision?
(c) Which experiment(s) display(s) high accuracy?
(d) Which experiment(s) show(s) a systematic error?

Comprehensive Problems
Problems marked with an asterisk (*) are more challenging.

1.75 To make 2.000 gal of a powdered sports drink, a group of students measure out 2.000 gal of water with 500.-mL, 50.-mL, and 5-mL graduated cylinders. Show how they could get closest to 2.000 gal of water, using these cylinders the fewest times.

1.76 Two blank potential energy diagrams, similar to those in Figure 1.3, appear below. Beneath each diagram are objects to be placed in the diagram. Draw the objects on the dashed lines to indicate higher or lower potential energy and label each case as more or less stable:

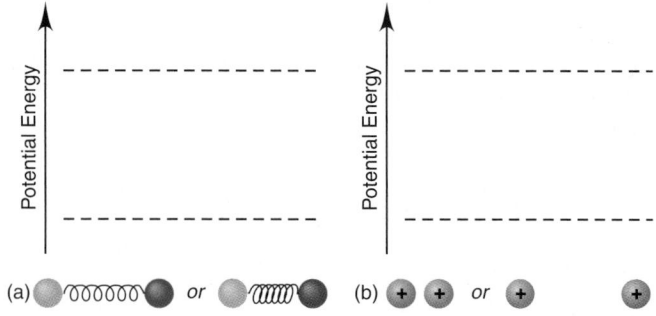

(a) Two balls attached to a relaxed spring *or* to a compressed spring.
(b) Two positive charges near each other *or* apart from each other.

1.77 A hot-air balloon rises to a height of 1.2 km. The passenger reaches down 0.81 m below the balloon basket holding a 1.4-m stick. At the end of the stick is a ball with a diameter of 32.1 cm, and a 2.7-m string hanging from the ball is connected to a small lead weight. How high (in m) is the weight above the ground?

1.78 It is possible to estimate the mass of an object from its volume and density. Most soft drinks have a density close to that of water (1.0 g/cm^3); many common metals, including iron, copper, and silver, have densities around 9.5 g/cm^3. (a) What is the mass of the liquid in a standard 12-oz bottle of diet cola? (b) What is the mass of a dime? (*Hint:* A stack of five dimes has a volume of about 1 cm^3.)

1.79 Suppose your dorm room is 11 ft wide by 12 ft long by 8.5 ft high and has an air conditioner that exchanges air at a rate of 1200 L/min. How long would it take the air conditioner to exchange the air in your room once?

1.80 The specifications for a 100-milliliter volumetric flask state that it is guaranteed to contain the specified volume to within 0.1 mL. How many significant figures should you use to record the volume contained in the flask? Explain.

1.81 An Olympic-size pool is 50.0 m long and 25.0 m wide. (a) How many gallons of water (d = 1.0 g/mL) are needed to fill the pool to an average depth of 4.8 ft? (b) What is the mass (in kg) of water in the pool?

*1.82 At room temperature (20°C) and pressure, the density of air is 1.189 g/L. An object will float in air if its density is less than that of air. In a buoyancy experiment with a new plastic, a chemist creates a rigid, thin-walled ball that weighs 0.12 g and has a volume of 560 cm^3.
(a) Will the ball float if it is evacuated?
(b) Will it float if filled with carbon dioxide (d = 1.830 g/L)?
(c) Will it float if filled with hydrogen (d = 0.0899 g/L)?
(d) Will it float if filled with oxygen (d = 1.330 g/L)?
(e) Will it float if filled with nitrogen (d = 1.165 g/L)?
(f) For any case that will float, how much weight must be added to make the ball sink?

1.83 Asbestos is a fibrous silicate mineral with remarkably high tensile strength (the mass that a bar of the material can hold without being pulled apart). But it is no longer used because airborne asbestos particles can cause lung cancer. Grunerite, a type of asbestos, has a tensile strength of 3.5×10^4 kg/cm^2 (thus, a bar of grunerite with a 1-cm^2 cross-sectional area can hold up to 3.5×10^4 kg). The tensile strengths of aluminum and Steel No. 5137 are 2.5×10^4 lb/in^2 and 5.0×10^4 lb/in^2, respectively. Calculate the cross-sectional area (in cm^2) of bars of aluminum and Steel No. 5137 that have the same tensile strength as a bar of grunerite with a cross-sectional area of 25 mm^2.

1.84 According to the lore of ancient Greece, Archimedes discovered the displacement method of density determination while bathing and used it to find the composition of the king's crown. If a crown weighing 4 lb 13 oz displaces 186 mL of water when placed in a bathtub, is the crown made of pure gold (d = 19.3 g/cm^3)?

*1.85 Brass is an alloy of copper and zinc. By varying the mass percentages of the two metals, brasses with different properties are produced. A brass called *yellow zinc* has high ductility and strength and is 34%–37% zinc by mass.
(a) Find the mass range (in g) of copper in 174 g of yellow zinc.
(b) What is the mass range (in g) of zinc in a sample of yellow zinc that contains 46.5 g of copper?

1.86 A graduated cylinder weighs 45.80 g when empty and contains 42.0 mL of water (d = 1.00 g/cm^3). A piece of lead submerged in the water brings the total volume to 67.4 mL and the mass of cylinder and contents to 396.5 g. What is the density of lead?

1.87 Liquid nitrogen is obtained from liquefied air and is used in the food industry to prepare frozen foods. It boils at 77.36 K. (a) What is this temperature in °C? (b) What is this temperature in °F? (c) At the boiling point, the density of the liquid is 809 g/L and that of the gas is 4.566 g/L. How many liters of liquid nitrogen are produced when 895.0 L of nitrogen gas is liquefied at this temperature?

1.88 The speed of sound varies according to the material through which it travels. Sound travels at 5.4×10^3 cm/s through rubber and at 1.97×10^4 ft/s through granite. Calculate each of these speeds in m/s.

1.89 A jogger runs at an average speed of 5.9 mi/h. (a) How fast is she running in m/s? (b) How many kilometers does she run in 98 min? (c) If she starts a run at 11:15 am, what time is it after she covers 4.75×10^4 ft?

*1.90 The Environmental Protection Agency (EPA) has proposed a new standard for microparticulates in air: for particles up to 2.5 μm in diameter, the maximum allowable amount is 50. μg/m^3. If your 10.0 ft × 8.25 ft × 12.5 ft dorm room just meets the new EPA standard, how many of these particles are in your room? How many of these particles are in each 0.500-L breath you take? (Assume the particles are spheres of diameter 2.5 μm and made primarily of soot, a form of carbon with a density of 2.5 g/cm^3.)

*1.91 In the figure (*right*), the cylinder contains liquid that will be added to the liquid in the flask. What volume of liquid is in the cylinder? What volume of liquid is in the flask? What will be the total volume of liquid when the cylinder is emptied into the flask?

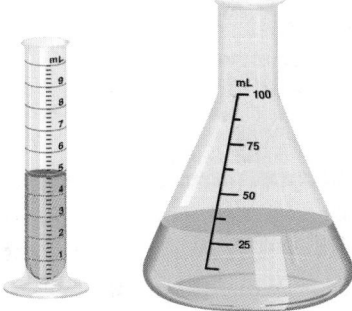

*1.92 Asphalt, a thick liquid used in roofing and road paving, is usually obtained as a residue from petroleum distillation. Yet it is sometimes found in nature. An asphalt lake in Trinidad is 113 feet deep and has an area of 115 acres. (a) What is the volume (in km^3) of the Trinidad asphalt lake (1 acre = 4047 m^2)? (b) A 21-ft wide roadway is to be paved with a 7.5-cm thick layer of asphalt from the Trinidad asphalt lake. How long (in km) can the road be?

1.93 Nutritional tables give the potassium content of a standard apple (about 3 apples/lb) as 159 mg. How many grams of potassium are in 4.25 kg of apples?

*1.94 "Copper" pennies actually contain very little copper. If a new penny is 97.3% zinc and 2.7% copper by mass, what is its density (d of copper = 8.95 g/cm^3; d of zinc = 7.14 g/cm^3)?

*1.95 The three states of matter differ greatly in their viscosity, a measure of their resistance to flow. Rank the three states from highest to lowest viscosity. Explain your ranking in submicroscopic terms.

*1.96 Temperature scales can be based on many pure liquids. If a temperature scale were based on the freezing point (5.5°C) and boiling point (80.1°C) of benzene and the temperature difference between these points was divided into 50 units (called °X), what would be the freezing and boiling points of water in °X? (See Figure 1.12, p. 25.)

CHAPTER 2

THE COMPONENTS OF MATTER

CHAPTER OUTLINE

2.1 Elements, Compounds, and Mixtures: An Atomic Overview

2.2 The Observations That Led to an Atomic View of Matter
Mass Conservation
Definite Composition
Multiple Proportions

2.3 Dalton's Atomic Theory
Postulates of the Theory
Explanation of Mass Laws
Relative Masses of Atoms

2.4 The Observations That Led to the Nuclear Atom Model
Discovery of the Electron
Discovery of the Nucleus

2.5 The Atomic Theory Today
Structure of the Atom
Atomic Number, Mass Number, and Atomic Symbol
Isotopes and Atomic Masses
Reassessing the Atomic Theory

2.6 Elements: A First Look at the Periodic Table

2.7 Compounds: Introduction to Bonding
Formation of Ionic Compounds
Formation of Covalent Compounds
Polyatomic Ions

2.8 Compounds: Formulas, Names, and Masses
Types of Chemical Formulas
Learning Names and Formulas
Ionic Compounds
Binary Covalent Compounds
Molecular Masses

2.9 Mixtures: Classification and Separation

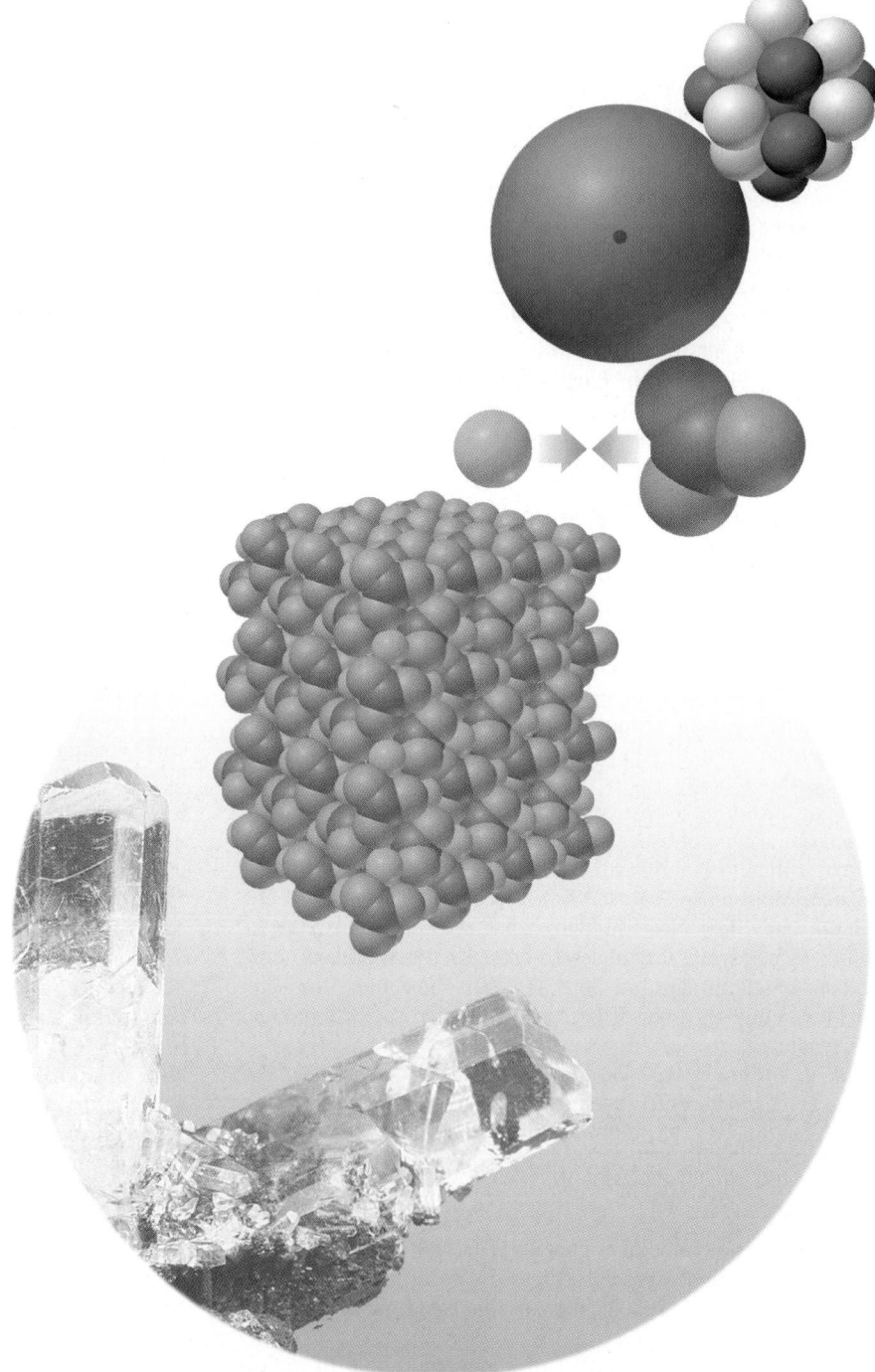

Figure: Matter and Its Parts. Magnifying our view of a sample of matter shows its components, which in turn have components, and so on to the subatomic scale. This rock specimen has embedded in it a crystalline compound (calcite), which consists of calcium and carbonate ions attracted to each other in a regular array. Each carbonate ion consists of carbon and oxygen atoms bonded to each other, and all the atoms are made up of subatomic particles. We examine each of these component parts in this chapter.

Whenever you look at an unusual stone, a beautiful shell, or even a manufactured object like a compact disk, it isn't surprising that one of your first questions may be "What is it made of?" After all, wondering about the ultimate makeup of things is an age-old practice, even though today we approach the question very differently than did many of the philosophers of ancient Greece. They believed that everything was made of one or, at most, a few elemental substances. Seeing how extensive rivers and oceans were, some believed that elemental substance to be water. Others thought it was air, which could be "thinned" into fire and "thickened" into clouds, rain, and rock. Still others believed there were four elemental substances—fire, air, water, and earth—with properties that accounted for hotness, wetness, sweetness, and many other characteristics of things.

Democritus (c. 460–370 BC), the father of atomism, took a different approach. Rather than considering a specific substance, he focused on the ultimate components of *all* substances. His reasoning went something like this: if you cut a piece of, say, copper smaller and smaller, you must eventually reach a particle of copper so small that it can no longer be cut. Therefore, matter must ultimately be composed of indivisible particles, with nothing between them but empty space. He called the particles *atoms* (Greek *atomos,* "uncuttable") and proclaimed: "According to convention, there is a sweet and a bitter, a hot and a cold, and according to convention, there is order. In truth, there are atoms and a void." However, Aristotle (384–322 BC), who elaborated the idea of four elemental substances, held that it was impossible for "nothing" to exist. Because of Aristotle's powerful influence, the concept of atoms was suppressed for 2000 years.

Finally, in the 17th century, the great English scientist Robert Boyle argued that an element is composed of "simple Bodies, not made of any other Bodies, of which all mixed Bodies are compounded, and into which they are ultimately resolved," a description that is remarkably similar to the idea of an element we hold today, in which the "simple Bodies" are atoms. This began the wonderful process of discovery, debate, and rediscovery that marks scientific inquiry, as exemplified by Lavoisier's work. Further studies in the 18th century gave rise to laws concerning the relative masses of substances that react with each other. Then, at the beginning of the 19th century, John Dalton proposed an atomic model that explained these laws and soon led to rapid progress in chemistry. By that century's close, however, further observation exposed the need to revise Dalton's model. A burst of creativity in the early 20th century gave rise to a picture of the atom with a complex internal structure, which led to our current model.

In this chapter, we examine the properties and composition of the three types of matter—elements, compounds, and mixtures—on the macroscopic and atomic scales. We discuss the mass laws and the theory Dalton proposed to explain them. To gain insight into how science works, we look at some key experiments that gave rise to our current model of the atom. The main portion of the chapter considers atomic structure, how the elements are classified, how they combine to form compounds, how we derive compound names, formulas, and masses, and how mixtures are classified and separated.

CONCEPTS & SKILLS

to review before you study this chapter
- physical and chemical change (Section 1.1)
- states of matter (Section 1.1)
- attraction and repulsion between charged particles (Section 1.1)
- meaning of a scientific model (Section 1.3)
- SI units and conversion factors (Section 1.5)
- significant figures in calculations (Section 1.6)

2.1 ELEMENTS, COMPOUNDS, AND MIXTURES: AN ATOMIC OVERVIEW

Matter can be broadly classified into three types—elements, compounds, and mixtures. An **element** is the simplest type of matter with unique physical and chemical properties. *An element consists of only one kind of atom.* Therefore, it cannot be broken down into a simpler type of matter by any physical or chemical methods. An element is one kind of **pure substance** (or just **substance**), matter whose composition is fixed. Each element has a name, such as silicon, oxygen, or copper. A sample of silicon contains only silicon atoms. A key point to remember is that the *macroscopic* properties of a piece of silicon, such as color, density, and combustibility, are different from those of a piece of copper because silicon atoms are different from copper atoms; in other words, *each element is unique because its atomic properties are unique.*

Most elements exist in nature as populations of individual atoms. Figure 2.1A shows atoms of a gaseous element such as neon. However, several elements occur naturally in molecular form: a **molecule** is an independent structural unit consisting of two or more atoms that are chemically bound together (Figure 2.1B). Oxygen, for example, occurs in air as *diatomic* (two-atom) molecules.

A **compound** is a type of matter composed of *two or more different elements that are chemically bound together.* Be sure you understand that the elements in a compound are not just mixed together; rather, as Figure 2.1C shows, their atoms have joined chemically. Ammonia, water, and carbon dioxide are some common compounds. One defining feature of a compound is that *the elements are present in fixed parts by mass* (fixed mass ratio). Because of this fixed composition, *a compound is also considered a substance.* A molecule of the compound has the same fixed parts by mass because it consists of *fixed numbers* of atoms of the component elements. For example, any sample of ammonia is 14 parts nitrogen by mass plus 3 parts hydrogen by mass. Since 1 nitrogen atom has 14 times the mass of 1 hydrogen atom, 1 molecule of ammonia always consists of 1 nitrogen atom and 3 hydrogen atoms:

> Ammonia is 14 parts N and 3 parts H by mass
> 1 N atom has 14 times the mass of 1 H atom
> *Therefore, ammonia has 1 N atom and 3 H atoms.*

Another defining feature of a compound is that *its properties are different from those of its component elements.* Table 2.1 shows a striking example. Soft, silvery sodium metal and yellow-green, poisonous chlorine gas have very different properties from the compound they form—white, crystalline sodium chloride, or common table salt! Unlike an element, a compound *can* be broken down into simpler substances—its component elements. For example, an electric current breaks down molten sodium chloride into metallic sodium and chlorine gas. Note that this breakdown is a *chemical change,* not a physical one.

Figure 2.1D depicts a **mixture,** a group of two or more substances (elements and/or compounds) that are physically intermingled. In contrast to a compound, *the components of a mixture **can** vary in their parts by mass.* Since its composition is not fixed, a mixture is *not* a substance. A mixture of the two compounds sodium chloride and water, for example, can have many different parts by mass of salt to water. Since the components are physically mixed, not chemically combined, a mixture at the atomic scale is merely a group of the individual units that

Figure 2.1 **Elements, compounds, and mixtures on the atomic scale.**
A, Most elements consist of a large collection of identical atoms. **B,** Some elements occur as molecules.
C, A molecule of a compound consists of characteristic numbers of atoms of two or more elements chemically bound together. **D,** A mixture contains the individual units of two or more elements and/or compounds that are physically intermingled. The samples shown here are gases, but elements, compounds, and mixtures occur as liquids and solids also.

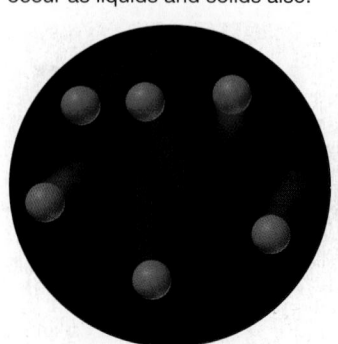

A Atoms of an element

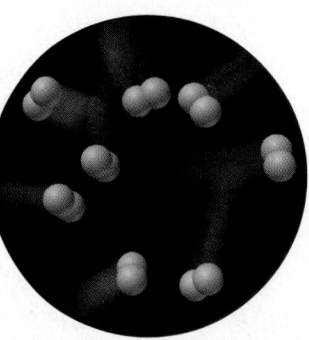

B Molecules of an element

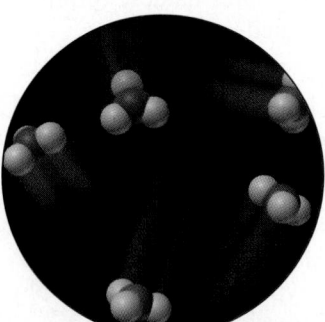

C Molecules of a compound

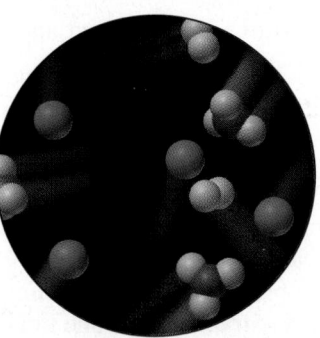

D Mixture of two elements and a compound

Table 2.1 Some Properties of Sodium, Chlorine, and Sodium Chloride

Property	Sodium	+	Chlorine	⟶	Sodium Chloride
Melting point	97.8°C		−101°C		801°C
Boiling point	881.4°C		−34°C		1413°C
Color	Silvery		Yellow-green		Colorless (white)
Density	0.97 g/cm^3		0.0032 g/cm^3		2.16 g/cm^3
Behavior in water	Reacts		Dissolves slightly		Dissolves freely

make up its component elements and compounds. Therefore, *a mixture retains many of the properties of its components.* Salt water, for instance, is colorless like water and tastes salty like sodium chloride. Unlike compounds, mixtures can be separated into their components by *physical changes;* chemical changes are not needed. In this case, the water in salt water can be boiled off, a physical process that leaves behind the sodium chloride.

SECTION SUMMARY

All matter exists as either elements, compounds, or mixtures. Elements consist of only one type of atom. A compound contains two or more elements in chemical combination; it exhibits different properties from its component elements. The elements of a compound occur in fixed parts by mass because each unit of the compound has fixed numbers of each type of atom. Elements and compounds are referred to as substances because their compositions are fixed. A mixture consists of two or more substances mixed together, not chemically combined. The components retain their individual properties and can be present in any proportion.

2.2 THE OBSERVATIONS THAT LED TO AN ATOMIC VIEW OF MATTER

Any model of the composition of matter had to explain two extremely important chemical observations that were well established by the end of the 18th century: the *law of mass conservation* and the *law of definite (or constant) composition.* As you'll see, John Dalton's atomic theory explained these laws and another observation now known as the *law of multiple proportions.*

Mass Conservation

The most fundamental chemical observation of the 18th century was the **law of mass conservation:** *the total mass of substances does not change during a chemical reaction.* The *number* of substances may change, and by definition their properties must, but the *total amount* of matter remains constant. Lavoisier had first stated this law on the basis of his combustion experiments, in which he found the mass of oxygen plus the mass of mercury equal to the mass of mercuric oxide they formed. Figure 2.2 illustrates mass conservation in a reaction that occurs in water.

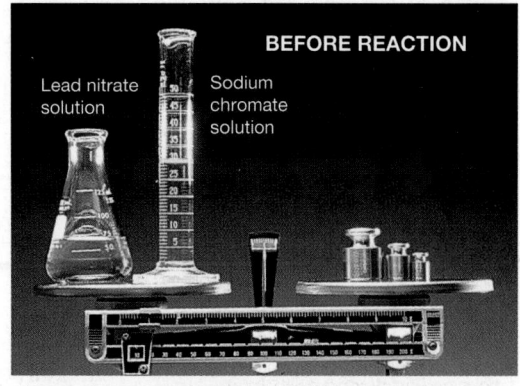

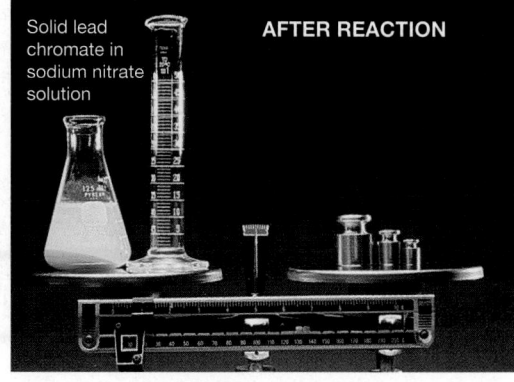

Figure 2.2 The law of mass conservation: mass remains constant during a chemical reaction. The total mass of lead nitrate solution and sodium chromate solution before they react *(top)* is the same as the total mass after they have reacted *(bottom)* to form lead chromate (yellow solid) and sodium nitrate solution.

⬤ **Immeasurable Changes in Mass**
From the work of Albert Einstein (1879–1955), we now know that mass and energy are alternate aspects of a single entity called *mass-energy.* We also know that some mass *does* change into energy during any chemical reaction, but the amount is far below the limit of detection of the best modern balance. For example, when 100 g of carbon burns in oxygen, only 0.000000036 g (3.6×10^{-8} g) of mass is converted to energy and, thus, not accounted for in the product, carbon dioxide. The energy yields of ordinary chemical reactions are relatively small; so, for all practical purposes, mass *is* conserved. (As you'll see later, however, energy changes in nuclear reactions are so large that mass changes are measured easily.)

Even in a complex biochemical change within an organism, such as the metabolism of the sugar glucose, which involves many reactions, mass is conserved:

180 g glucose + 192 g oxygen gas $\longrightarrow$ 264 g carbon dioxide + 108 g water
372 g material before change $\longrightarrow$ 372 g material after change

Mass conservation means that, based on all chemical experience, *matter cannot be created or destroyed.* ⬤

Definite Composition

Another fundamental chemical observation is summarized as the **law of definite (or constant) composition:** *no matter what its source, a particular chemical compound is composed of the same elements in the same parts (fractions) by mass.* The **fraction by mass (mass fraction)** is that part of the compound's mass contributed by the element. It is obtained by dividing the mass of each element by the total mass of compound. The **percent by mass (mass percent, mass %)** is the fraction by mass expressed as a percentage.

Let's examine the meanings of these ideas in terms of a box of marbles *(right).* The box contains three types of marbles: yellow marbles weigh 1.0 g each, purple marbles 2.0 g each, and red marbles 3.0 g each. Each type makes up a fraction of the total mass of marbles, 16.0 g. The *mass fraction* of the yellow marbles is their number times their mass divided by the total mass: $(3 \times 1.0 \text{ g})/16.0 \text{ g} = 0.19$. The *mass percent* (parts per 100 parts) of the yellow marbles is $0.19 \times 100 = 19\%$ by mass. The purple marbles have a mass fraction of

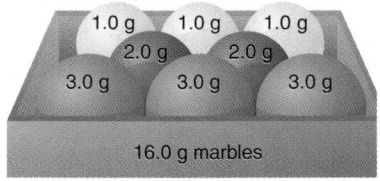

0.25 and are 25% by mass, and the red marbles have a mass fraction of 0.56 and are 56% by mass. Similarly, in a compound, each element has a *fixed* mass fraction (and mass percent).

Consider calcium carbonate, the major compound in a piece of chalk. The following results are obtained for the elemental mass composition of 20.0 g of calcium carbonate (for example, 8.0 g of calcium/20.0 g = 0.40 parts of calcium):

Analysis by Mass (grams/20.0 g)	Mass Fraction (parts/1.00 part)	Percent by Mass (parts/100 parts)
8.0 g calcium	0.40 calcium	40% calcium
2.4 g carbon	0.12 carbon	12% carbon
9.6 g oxygen	0.48 oxygen	48% oxygen
20.0 g	1.00 part by mass	100% by mass

As you can see, the sum of the mass fractions (or mass percents) equals 1.00 part (or 100%) by mass. The law of definite composition tells us that pure samples of calcium carbonate always contain the same elements in the same percents by mass (Figure 2.3).

Since a given element always constitutes the same mass fraction of a given compound, we can use the mass fraction to find the actual mass of the element in any sample of the compound:

$$\text{Mass of element} = \text{mass of compound} \times \frac{\text{part by mass of element}}{\text{one part by mass of compound}}$$

Or, more simply, since mass analysis tells us parts by mass, we can use that directly with *any* mass unit and skip the need to find the mass fraction first:

Mass of element in sample

$$= \text{mass of compound in sample} \times \frac{\text{mass of element in compound}}{\text{mass of compound}} \quad (2.1)$$

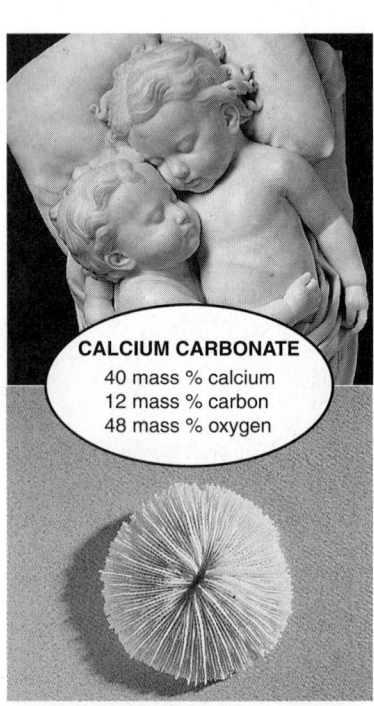

Figure 2.3 The law of definite composition. Calcium carbonate is found naturally in many forms, including marble *(top),* coral *(bottom),* chalk, and seashells. The mass percents of its component elements do not change regardless of the compound's source.

CALCIUM CARBONATE
40 mass % calcium
12 mass % carbon
48 mass % oxygen

SAMPLE PROBLEM 2.1 Calculating the Mass of an Element in a Compound

Problem Pitchblende is the most commercially important compound of uranium. Analysis shows that 84.2 g of pitchblende contains 71.4 g of uranium, with oxygen as the only other element. How many grams of uranium can be obtained from 102 kg of pitchblende?

Plan We have to find the mass of uranium in a known mass of pitchblende, given the mass of uranium in a different mass of pitchblende. The mass ratio of uranium/pitchblende is the same for any sample of pitchblende. Therefore, as shown by Equation 2.1, we multiply the mass (in kg) of pitchblende by the ratio of uranium to pitchblende we construct from the mass analysis. This gives the mass (in kg) of uranium, and we just convert kilograms to grams.

Solution Finding the mass (kg) of uranium in 102 kg of pitchblende:

$$\text{Mass (kg) of uranium} = \text{mass (kg) of pitchblende} \times \frac{\text{mass (kg) of uranium in pitchblende}}{\text{mass (kg) of pitchblende}}$$

$$\text{Mass (kg) of uranium} = 102 \text{ kg pitchblende} \times \frac{71.4 \text{ kg uranium}}{84.2 \text{ kg pitchblende}} = 86.5 \text{ kg uranium}$$

Converting the mass from kg to g:

$$\text{Mass (g) of uranium} = 86.5 \text{ kg uranium} \times \frac{1000 \text{ g}}{1 \text{ kg}} = \boxed{8.65 \times 10^4 \text{ g uranium}}$$

Check The analysis showed that most of the mass of pitchblende is due to uranium, so the large mass of uranium makes sense. Rounding off to check the math gives:

$$\sim 100 \text{ kg pitchblende} \times \frac{70}{85} = 82 \text{ kg uranium}$$

FOLLOW-UP PROBLEM 2.1 How many metric tons (t) of oxygen are combined in a sample of pitchblende that contains 2.3 t of uranium? (*Hint:* Remember that oxygen is the only other element present.)

Mass (kg) of pitchblende
multiply by mass ratio of uranium to pitchblende from analysis
Mass (kg) of uranium
1 kg = 1000 g
Mass (g) of uranium

Multiple Proportions

Dalton described a phenomenon that occurs when two elements form more than one compound. His observation is now called the **law of multiple proportions:** *if elements A and B react to form two compounds, the different masses of B that combine with a fixed mass of A can be expressed as a ratio of small whole numbers.* Consider, for example, two compounds that form from carbon and oxygen; for now, let's call them carbon oxides I and II. They have very different properties. For example, measured at the same temperature and pressure, the density of carbon oxide I is 1.25 g/L, whereas that of II is 1.98 g/L. Moreover, I is poisonous and flammable, but II is not. Analysis shows that their compositions by mass are

    Carbon oxide I: 57.1 mass % oxygen and 42.9 mass % carbon
    Carbon oxide II: 72.7 mass % oxygen and 27.3 mass % carbon

To see the phenomenon of multiple proportions, we use the mass percents of oxygen and of carbon in each compound to find the masses of these elements in a given mass, for example, 100 g, of each compound. Then we divide the mass of oxygen by the mass of carbon in each compound to obtain the mass of oxygen that combines with a fixed mass of carbon:

	Carbon Oxide I	**Carbon Oxide II**
g oxygen/100 g compound	57.1	72.7
g carbon/100 g compound	42.9	27.3
g oxygen/g carbon	$\dfrac{57.1}{42.9} = 1.33$	$\dfrac{72.7}{27.3} = 2.66$

If we then divide the grams of oxygen per gram of carbon in II by that in I, we obtain a ratio of small whole numbers:

$$\frac{2.66 \text{ g oxygen/g carbon in II}}{1.33 \text{ g oxygen/g carbon in I}} = \frac{2}{1}$$

The law of multiple proportions tells us that in two compounds of the same elements, the mass fraction of one element relative to the other element changes in *increments based on ratios of small whole numbers*. In this case, the ratio is 2:1—for a given mass of carbon, II contains *2 times* as much oxygen as I, not 1.583 times, 1.716 times, or any other intermediate amount. As you'll see next, Dalton's theory allows us to explain the composition of carbon oxides I and II on the atomic scale.

SECTION SUMMARY

Three fundamental observations known as mass laws state that (1) the total mass remains constant during a chemical reaction; (2) any sample of a given compound has the same elements present in the same parts by mass; and (3) in different compounds of the same elements, the masses of one element that combine with a fixed mass of the other can be expressed as a ratio of small whole numbers.

2.3   DALTON'S ATOMIC THEORY

With almost 200 years of hindsight, it may be easy to see how the mass laws could be explained by an atomic model—matter existing in indestructible units, each with a particular mass—but it was a major breakthrough in 1808 when John Dalton (1766–1844) presented his atomic theory of matter in *A New System of Chemical Philosophy.*

Postulates of the Atomic Theory

Dalton expressed his theory in a series of postulates. Like most great thinkers, Dalton incorporated the ideas of others into his own to create the new theory. As we go through the postulates, which are presented here in modern terms, let's see which were original and which came from others. (Later, we can examine the key differences between Dalton's postulates and our present understanding.)

1. All matter consists of **atoms,** tiny indivisible particles of an element that cannot be created or destroyed. (Derives from the "eternal, indestructible atoms" of Democritus more than 2000 years earlier and conforms to mass conservation as stated by Lavoisier.)
2. Atoms of one element *cannot* be converted into atoms of another element. In chemical reactions, the original substances separate into atoms, which recombine to form different substances. (Rejects the alchemical belief in the magical transmutation of elements.)
3. Atoms of an element are identical in mass and other properties and are different from atoms of any other element. (Contains Dalton's major new ideas: unique mass and properties for all the atoms of a given element.)
4. Compounds result from the chemical combination of a specific ratio of atoms of different elements. (Follows directly from the fact of definite composition.)

How the Theory Explains the Mass Laws

Let's see how Dalton's postulates explain the mass laws:

*   *Mass conservation.* Atoms cannot be created or destroyed (postulate 1) or converted into other types of atoms (postulate 2). Since each type of atom has a fixed mass (postulate 3), a chemical reaction, in which atoms are just combined differently with each other, cannot possibly result in a mass change.

Dalton's Revival of Atomism Although John Dalton, the son of a poor weaver, had no formal education, he established one of the most powerful concepts in science. Dalton began teaching science at 12 years of age, and later studied color blindness, a personal affliction still known as *daltonism.* In 1787, he began his life's work in meteorology, recording daily weather data until his death 57 years later. His studies on humidity and dew point led to a key discovery in the behavior of gases (Section 5.6) and eventually to his atomic theory. In 1803, he stated, "I am nearly persuaded that [the mixing of gases and their solubility in water] depends upon the mass and number of the ultimate particles. ... An enquiry into the relative masses of [these] particles of bodies is a subject ... I have lately been prosecuting ... with remarkable success." The atomic theory was published 5 years later.

- *Definite composition.* A compound is a combination of a *specific* ratio of different atoms (postulate 4), each of which has a particular mass (postulate 3). Thus, each element in a compound constitutes a fixed fraction of the total mass.
- *Multiple proportions.* Atoms of an element have the same mass (postulate 3) and are indivisible (postulate 1). Because different numbers of B atoms combine with each A atom in different compounds, the masses of element B that combine with a fixed mass of element A give a small, whole-number ratio.

The *simplest* arrangement consistent with the mass data for carbon oxides I and II in our earlier example is that one atom of oxygen combines with one atom of carbon in compound I (carbon monoxide) and that two atoms of oxygen combine with one atom of carbon in compound II (carbon dioxide) (Figure 2.4).

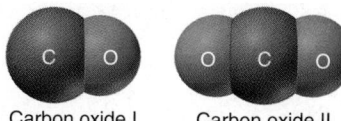

| Carbon oxide I | Carbon oxide II |
| (carbon monoxide) | (carbon dioxide) |

Figure 2.4 **The atomic basis of the law of multiple proportions.** Carbon and oxygen combine to form carbon oxide I (carbon monoxide) and carbon oxide II (carbon dioxide). The masses of oxygen in the two compounds relative to a fixed mass of carbon are in a ratio of small whole numbers.

The Relative Masses of Atoms

After publication of the atomic theory, investigators tried to determine the masses of atoms from the mass fractions of elements in compounds. But an individual atom is so small that the mass of all the atoms of one element can be determined only *relative* to the mass of all the atoms of another element. As a basis for these relative masses, Dalton assigned a mass of 1 to the hydrogen atom, the lightest known substance. Then, based on the work of Lavoisier, who had shown that water contains 8 g of oxygen for every 1 g of hydrogen, Dalton assigned a relative mass of 8 to the oxygen atom. But this relative mass would be correct *only* if a water molecule has one oxygen atom for every hydrogen atom:

$$\frac{1 \text{ oxygen (8)}}{1 \text{ hydrogen (1)}} \text{ is equivalent to } \frac{8}{1} \text{ (mass ratio)}$$

However, some other researchers, measuring the volumes of gases that react with each other, found that 2 L of hydrogen gas react with 1 L of oxygen gas. This result implied that a water molecule had one oxygen atom for every two hydrogen atoms and, therefore, that oxygen had a relative mass of 16:

$$\frac{1 \text{ oxygen (16)}}{2 \text{ hydrogen (1 each)}} = \frac{16}{2} \text{ is equivalent to } \frac{8}{1} \text{ (mass ratio)}$$

Which relative mass was correct? Many similar conflicts arose with other elements, and it took decades to resolve them. Finally, shortly after a conference in 1860 attended by over 140 of the world's most prominent chemists, a list of elements with accurate relative atomic masses was accepted—16 for oxygen, 12 for carbon, and so forth—and chemical formulas such as H_2O for water, NH_3 for ammonia, and CH_4 for methane were put into common use. Dalton's atomic model was crucial to this developing understanding because it originated the idea that the masses of reacting elements could be explained in terms of atoms.

Very soon thereafter, however, the model proved to be too limited. It could not explain *why* elements combine as they do: why, for example, do two, and not three, H atoms combine with one O atom in a water molecule? Also, Dalton's "billiard ball" view of the atom could not account for the electrically charged particles that would soon be seen in many experiments. A more complex model of the atom would be needed to understand those observations. ●

SECTION SUMMARY

Dalton's atomic theory explained the mass laws by proposing that all matter consists of indivisible, unchangeable atoms of fixed, unique mass. Mass is constant during a reaction because atoms form new combinations; each compound has a fixed mass fraction of each of its elements because it is composed of a fixed number of each type of atom; and different compounds of the same elements exhibit multiple proportions because they each consist of whole atoms. Studies of the masses and volumes of elements that combine eventually led to consistent values for relative masses of atoms.

● Atoms? Humbug! Rarely does a major new concept receive unanimous acceptance. Despite the atomic theory's impact, several major scientists denied the existence of atoms for another century. In 1877, Adolf Kolbe, an eminent organic chemist, said, "[Dalton's atoms are] . . . no more than stupid hallucinations . . . mere table-tapping and supernatural explanations." The influential physicist Ernst Mach believed that scientists should look at facts, not hypothetical entities such as atoms. It was not until 1908 that the famous chemist and outspoken opponent of atomism Wilhelm Ostwald wrote, "I am now convinced [by recent] experimental evidence of the discrete or grained nature of matter, which the atomic hypothesis sought in vain for hundreds and thousands of years." He was referring to the discovery of the electron.

The Familiar Glow of Colliding Particles The electric and magnetic properties of charged particles that collide with gas particles or hit a phosphor-coated screen have familiar applications. A "neon" sign glows because electrons collide with the gas particles in the tube, causing them to give off light. An aurora display occurs when the Earth's magnetic field bends streams of charged particles coming from the Sun, which then collide with gases in the atmosphere. In a television tube or computer monitor, the cathode ray passes back and forth over the coated screen, creating a pattern that the eye sees as a picture.

2.4   THE OBSERVATIONS THAT LED TO THE NUCLEAR ATOM MODEL

The path of discovery is often winding and unpredictable. Basic research into the nature of electricity eventually led to the discovery of *electrons,* negatively charged particles that are part of all atoms. Soon thereafter, other experiments revealed that the atom has a *nucleus*—a tiny, central core of mass and positive charge. In this section, we examine some key experiments that led to our current model of the atom.

Discovery of the Electron and Its Properties

Nineteenth-century investigators of electricity knew that matter and electric charge were somehow related. When amber is rubbed with fur, or glass with silk, positive and negative charges form—the same charges that make your hair crackle and cling to your comb on a dry day. They also knew that an electric current could decompose certain compounds into their elements.

What they did not know, however, was how electricity could be studied in the *absence* of matter. A powerful vacuum pump had been invented recently, so some investigators tried passing an electric current through nearly evacuated glass tubes. The tubes, fitted with metal electrodes that were sealed in place, were connected to an external source of electricity. When the power was turned on, a "ray" could be seen striking the phosphor-coated end of the tube and emitting a glowing spot of light. The rays were called **cathode rays** because they originated at the negative electrode (cathode) and moved to the positive electrode (anode). They were shown to travel in straight lines, to be deflected by magnetic or electric fields, and to be identical, no matter what metal was used as the cathode. Figure 2.5 summarizes the investigations of cathode ray properties. It was concluded that cathode rays consist of negatively charged particles that are found in all matter and that the rays appear when these particles collide with the few remaining gas molecules in the evacuated tubes. ⬤   Cathode ray particles were later named *electrons.*

Figure 2.5 Experiments to determine the properties of cathode rays. A cathode ray forms when high voltage is applied across the electrodes in a partially evacuated tube. A hole in the anode allows the ray to pass through and hit the coated end of the tube to produce a glow. **A,** In the absence of an external field, the ray travels in a straight path. **B,** The ray bends in an external magnetic field, so it must consist of charged particles. **C,** In an external electric field, the ray bends toward the positive plate, so the particles' charge must be negative. **D,** Any electrode material produces an identical ray, so the particles must be part of all matter—universal negatively charged particles.

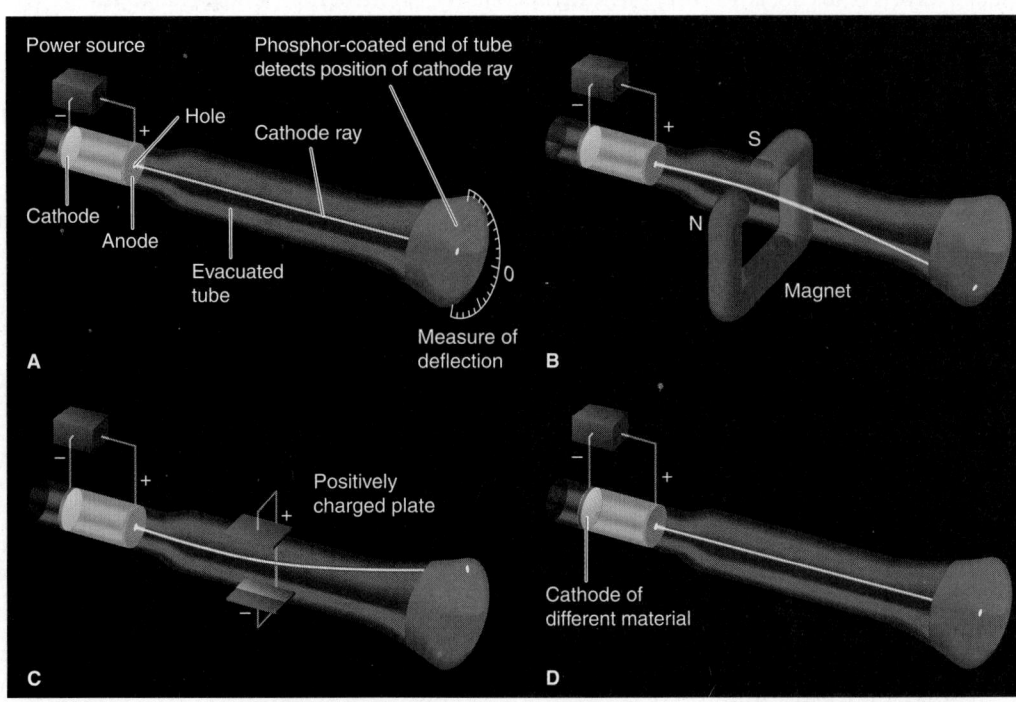

Just over a century ago, in 1897, J. J. Thomson (1856–1940) used magnetic and electric fields to measure the ratio of the cathode ray particle's mass to its charge (Figure 2.6). By comparing this value with the mass/charge ratio for the lightest charged particle in solution, Thomson estimated that the cathode ray particle weighed less than $\frac{1}{1000}$ as much as hydrogen, the lightest atom! He was shocked because this implied that, contrary to Dalton's atomic theory, *atoms are divisible into even smaller particles.* Thomson concluded, "We have in the cathode rays matter in a new state, . . . in which the subdivision of matter is carried much further . . . ; this matter being the substance from which the chemical elements are built up." Fellow scientists reacted at first with disbelief, and some even thought Thomson was joking.

In 1909, the American physicist Robert Millikan (1868–1953) measured the *charge* of the electron. He did so by observing the movement of tiny droplets of the "highest grade clock oil" in an apparatus that contained electrically charged plates and an x-ray source, as depicted in Figure 2.7. Here is a description of the basis of the experiment: X-rays knocked electrons from gas molecules in the air, and as an oil droplet fell through a hole in the positive (upper) plate, the electrons stuck to the drop, giving it a negative charge. With the electric field off, Millikan measured the mass of the droplet from its rate of fall. By turning on the field and varying its strength, he could make the drop fall more slowly, rise, or pause suspended. From these data, Millikan calculated the total charge of the droplet.

After studying many droplets, Millikan confirmed his expectation that their various charges were always some *whole-number multiple of a minimum charge.* He reasoned that different oil droplets picked up different numbers of electrons, so this minimum charge must be that of the electron itself. The value, which he calculated over 90 years ago, is within 1% of the modern value of the electron's charge, -1.602×10^{-19} C (C stands for *coulomb,* the SI unit of charge). Using the electron's mass-to-charge ratio from work by Thomson and others and this value for the electron's charge, let's calculate the electron's *extremely* small mass the way Millikan did:

$$\text{Mass of electron} = \frac{\text{mass}}{\text{charge}} \times \text{charge} = \left(-5.686 \times 10^{-12} \frac{\text{kg}}{\cancel{\text{C}}}\right)(-1.602 \times 10^{-19} \cancel{\text{C}})$$
$$= 9.109 \times 10^{-31} \text{ kg} = 9.109 \times 10^{-28} \text{ g}$$

Figure 2.6 **Joseph John Thomson.** J. J. Thomson is shown at work on his experiments to determine the mass-to-charge ratio of cathode ray particles. He received the Nobel Prize in physics in 1906 for this work.

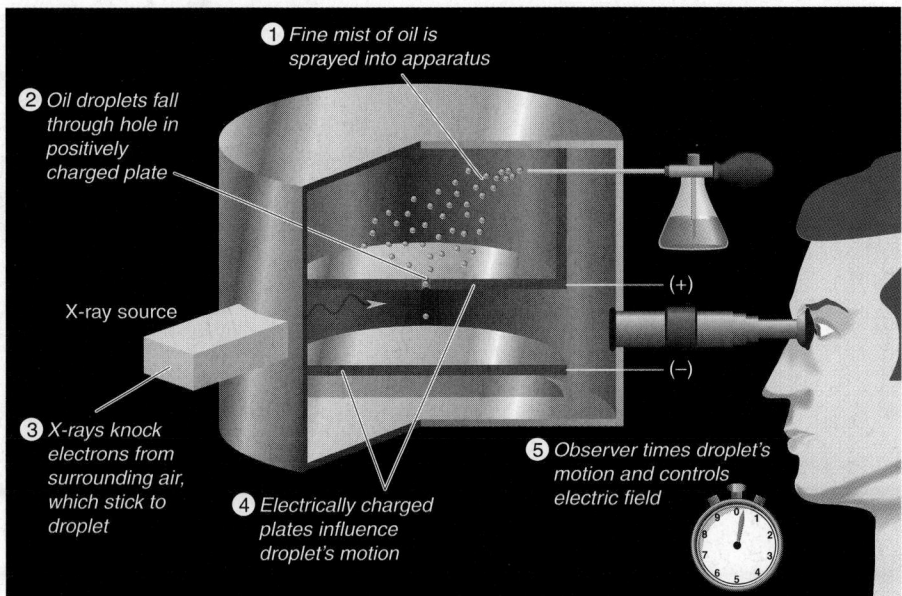

① Fine mist of oil is sprayed into apparatus

② Oil droplets fall through hole in positively charged plate

X-ray source

(+)

(−)

③ X-rays knock electrons from surrounding air, which stick to droplet

④ Electrically charged plates influence droplet's motion

⑤ Observer times droplet's motion and controls electric field

Figure 2.7 **Millikan's oil-drop experiment for measuring an electron's charge.** The motion of a given oil droplet depends on the variation in electric field and the total charge on the droplet, which depends in turn on the number of attached electrons. Millikan reasoned that the total charge must be some whole-number multiple of the charge of the electron.

Discovery of the Atomic Nucleus

Clearly, the properties of the electron posed problems about the inner structure of atoms. Since macroscopic matter is electrically neutral, the atoms that make it up must be neutral also. But if atoms contain negatively charged electrons, what positive charges balance them? And if an electron has such an incredibly tiny mass, what accounts for an atom's much larger mass? To address these issues, Thomson proposed a model of a spherical atom composed of diffuse, positively charged matter, in which electrons were embedded like "raisins in a plum pudding."

Near the turn of the 20th century, French scientists discovered radioactivity, the emission of particles and/or radiation from atoms of certain elements. A few years later, in 1910, the New Zealand-born physicist Ernest Rutherford (1871–1937) used one type of radioactive particle in a series of experiments that solved this dilemma of atomic structure. Figure 2.8 is a three-part representation of Rutherford's experiment. Tiny, dense, positively charged alpha (α) particles emitted from radium were aimed, like minute projectiles, at thin gold foil. The figure depicts (A) the "plum-pudding" hypothesis, (B) the apparatus used to measure the deflection (scattering) of the α particles from the light flashes created when the particles struck a circular, coated screen, and (C) the actual result.

With Thomson's model in mind, Rutherford expected only minor, if any, deflections of the α particles because they should act as tiny, dense, positively charged "bullets" and go right through the gold atoms. According to the model, the embedded electrons could not deflect the α particles any more than a Ping-Pong ball could deflect a speeding baseball. Initial results confirmed this, but soon the unexpected happened. As Rutherford recalled: "Then I remember two or three days later Geiger [one of his coworkers] coming to me in great excitement and saying, 'We have been able to get some of the α particles coming backwards . . . ' It was quite the most incredible event that has ever happened to me in my life. It was almost as incredible as if you fired a 15-inch shell at a piece of tissue paper and it came back and hit you."

The data showed that very few α particles were deflected at all, and that only 1 in 20,000 were deflected by more than 90° ("coming backwards"). It seemed that these few α particles were being repelled by something small, dense, and

Figure 2.8 Rutherford's α-scattering experiment and discovery of the atomic nucleus. A, The hypothesis stated that atoms consist of electrons embedded in diffuse, positively charged matter, so the speeding α particles should pass through the gold foil with, at most, minor deflections. **B,** In the experiment, α particles aimed at gold foil emit a flash of light when they pass through the gold atoms and hit a phosphor-coated screen. **C,** The actual results show occasional minor deflections and very infrequent major deflections. This could happen only if very high mass and positive charge are concentrated in a small region within the atom, the nucleus.

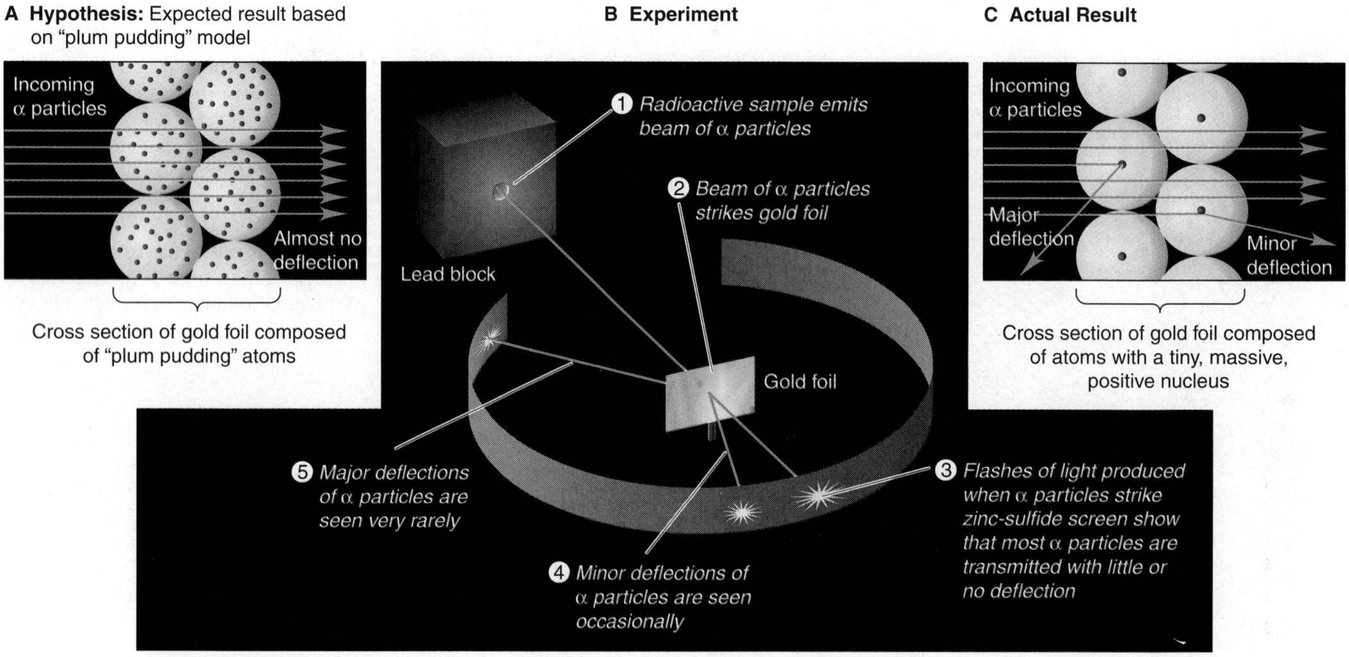

A Hypothesis: Expected result based on "plum pudding" model

Incoming α particles

Almost no deflection

Cross section of gold foil composed of "plum pudding" atoms

B Experiment

❶ Radioactive sample emits beam of α particles

❷ Beam of α particles strikes gold foil

Lead block

Gold foil

❺ Major deflections of α particles are seen very rarely

❹ Minor deflections of α particles are seen occasionally

❸ Flashes of light produced when α particles strike zinc-sulfide screen show that most α particles are transmitted with little or no deflection

C Actual Result

Incoming α particles

Major deflection

Minor deflection

Cross section of gold foil composed of atoms with a tiny, massive, positive nucleus positive within the gold atoms. From the mass, charge, and velocity of the α particles, the frequency of these large-angle deflections, and the properties of electrons, Rutherford calculated that *an atom is mostly space occupied by electrons,* but centrally located within that space lies a tiny region, which he called the **nucleus,** that contains *all the positive charge and essentially all the mass of the atom.* He proposed that positive particles lay within the nucleus and called them *protons,* and then calculated the magnitude of the nuclear charge with remarkable accuracy. Rutherford's model explained the charged nature of matter, but it could not account for all the atom's mass. It took more than 20 years to resolve this issue when, in 1932, James Chadwick discovered the *neutron,* an uncharged dense particle that also resides in the nucleus.

The **"Big Three" Subatomic Particles** Of all the subatomic particles discovered (at latest count, well over 40), the electron, proton, and neutron are the most important in chemistry because they are so long-lived. Theory postulates that, while bound in the nucleus, the neutron and proton are stable for at least 10^{30} years. The electron is considered eternal.

SECTION SUMMARY
Several major discoveries at the turn of the 20th century led to our current model of atomic structure. Cathode rays were shown to consist of negative particles (electrons) that exist in all matter. J. J. Thomson measured their mass/charge ratio and concluded that they are much smaller and lighter than atoms. Robert Millikan determined the charge of the electron, which he combined with other data to calculate its mass. Ernest Rutherford proposed that atoms consist of a tiny, massive, positive nucleus surrounded by electrons.

RUTHERFORD'S EXPERIMENT

2.5 THE ATOMIC THEORY TODAY

For nearly 200 years, scientists have known that all matter consists of atoms, and they have learned astonishing things about them. Dalton's hard, impenetrable spheres have given way to atoms with "fuzzy," indistinct boundaries and an elaborate internal architecture of subatomic particles. In this section, we examine our current model and begin to see how the properties of these particles affect the properties of atoms. Then we reassess the atomic theory in the light of our present knowledge.

Structure of the Atom

An *atom* is an electrically neutral, spherical entity composed of a positively charged central nucleus surrounded by one or more negatively charged electrons (Figure 2.9). The electrons move rapidly through the available atomic volume, held there by the attraction of the nucleus. The nucleus is incredibly dense: it contributes 99.97% of the atom's mass but occupies only about 1 ten-trillionth of its volume. (A nucleus the size of a period on this page would weigh about 100 tons, as much as 50 cars!) An atom's diameter ($\sim 10^{-10}$ m) is about 10,000 times the diameter of its nucleus ($\sim 10^{-14}$ m).

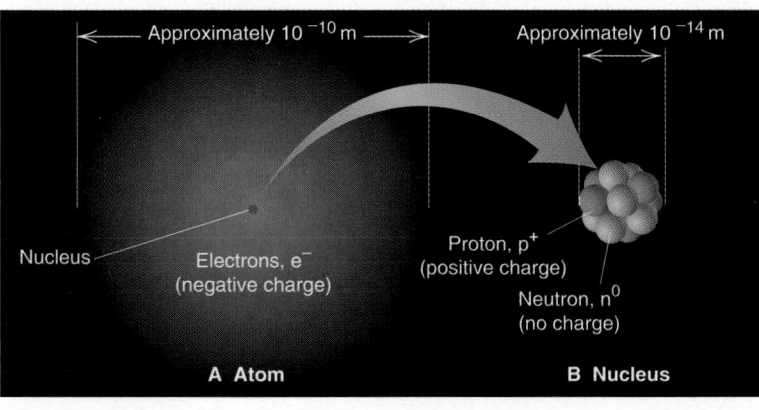

Figure 2.9 **General features of the atom. A,** A "cloud" of rapidly moving, negatively charged electrons occupies virtually all the atomic volume and surrounds the tiny, central nucleus. **B,** The nucleus contains virtually all the mass of the atom and consists of positively charged protons and uncharged neutrons. If the nucleus were actually the size in the figure ($\sim$1 cm across), the atom would be about 100 m across—slightly more than the length of a football field!

Table 2.2 **Properties of the Three Key Subatomic Particles**

Name (Symbol)	Charge		Mass		Location in Atom
	Relative	**Absolute (C)***	**Relative (amu)†**	**Absolute (g)**	
Proton (p^+)	1+	$+1.60218 \times 10^{-19}$	1.00727	1.67262×10^{-24}	Nucleus
Neutron (n^0)	0	0	1.00866	1.67493×10^{-24}	Nucleus
Electron (e^-)	1−	-1.60218×10^{-19}	0.00054858	9.10939×10^{-28}	Outside nucleus

*The coulomb (C) is the SI unit of charge.
†The atomic mass unit (amu) equals 1.66054×10^{-24} g; discussed later in this section.

An atomic nucleus consists of protons and neutrons, except for the simplest hydrogen nucleus, which is a single proton. The **proton (p^+)** has a positive charge, and the **neutron (n^0)** has no charge; thus, the positive charge of the nucleus results from the combined charges of its protons. The *magnitude* of charge possessed by a proton is equal to that of an **electron (e^-)**, but the *signs* of the charges are opposite. *An atom is neutral because the number of protons in the nucleus equals the number of electrons surrounding the nucleus.* Some properties of these three subatomic particles are listed in Table 2.2.

Atomic Number, Mass Number, and Atomic Symbol

The **atomic number (Z)** of an element equals the number of protons in the nucleus of each of its atoms. *All atoms of a particular element have the same atomic number, and each element has a different atomic number from that of any other element.* All carbon atoms ($Z = 6$) have 6 protons, all oxygen atoms ($Z = 8$) have 8 protons, and all uranium atoms ($Z = 92$) have 92 protons. There are currently 115 known elements, of which 90 occur in nature; the remaining 25 have been synthesized by nuclear processes.

The total number of protons and neutrons in the nucleus of an atom is its **mass number (A)**. Each proton and each neutron contributes one unit to the mass number. Thus, a carbon atom with 6 protons and 6 neutrons in its nucleus has a mass number of 12, and a uranium atom with 92 protons and 146 neutrons in its nucleus has a mass number of 238.

Information about the nuclear mass and charge is often included with the **atomic symbol** (or element symbol). Every element has a symbol based on its English, Latin, or Greek name, such as C for carbon, O for oxygen, S for sulfur, and Na for sodium (Latin *natrium*). The atomic number (Z) is written as a left *sub*script and the mass number (A) as a left *super*script to the symbol, so element X would be $^A_Z X$. Since the mass number is the sum of protons and neutrons, the number of neutrons (N) equals the mass number minus the atomic number:

$$\text{Number of neutrons} = \text{mass number} - \text{atomic number,} \quad \text{or} \quad N = A - Z \quad \textbf{(2.2)}$$

Thus, a chlorine atom symbolized $^{35}_{17}Cl$ has $A = 35$, $Z = 17$, and $N = 35 - 17 = 18$. Since each element has its own atomic number, we know the atomic number from the symbol. For example, every carbon atom has 6 protons, so we can write ^{12}C (spoken "carbon twelve"), with $Z = 6$ understood, for carbon with mass number 12. Another way to write this atom is carbon-12.

Isotopes and Atomic Masses of the Elements

All atoms of an element are identical in atomic number but not in mass number. All carbon atoms have 6 protons in the nucleus ($Z = 6$), but only 98.89% of naturally occurring carbon atoms have 6 neutrons in the nucleus ($A = 12$). A small percentage (1.11%) have 7 neutrons in the nucleus ($A = 13$), and even fewer (less than 0.01%) have 8 ($A = 14$). **Isotopes** of an element are atoms that have *dif-*

Naming an Element Element names have a variety of origins. Carbon comes from coal (Latin *carbo*, "ember"). Mercury (Hg) is named for the planet, but its symbol reveals its earlier name *hydragyrum* (Latin, "liquid silver"). Some names are based on a chemical action, such as hydrogen (Greek *hydros*, "water," and *genes*, "producer"), or on an obvious property, such as chlorine (Greek *chloros*, "yellow-green"). Sometimes, countries are used in the name, as in germanium for Germany and americium for America. Ytterbium, yttrium, erbium, and terbium are all named for Ytterby, the Swedish town where these elements were discovered. Paying homage to great scientists was the impetus for the names of some of the synthetic elements, such as einsteinium for Albert Einstein and curium for Marie and Pierre Curie.

ferent numbers of neutrons and therefore different mass numbers. Carbon has three naturally occurring isotopes, ^{12}C, ^{13}C, and ^{14}C. Five other isotopes of carbon, ^{9}C, ^{10}C, ^{11}C, ^{15}C, and ^{16}C, have been observed in the laboratory. All of these carbon isotopes have 6 protons and 6 electrons. Figure 2.10 depicts the atomic number, mass number, and symbol for four atoms, two of which are isotopes of the same element. A key point that we will note many times is that the chemical properties of an element are primarily determined by the number of electrons, so *all isotopes of an element have nearly identical chemical behavior,* even though they have different masses.

SAMPLE PROBLEM 2.2 Determining the Number of Subatomic Particles in the Isotopes of an Element

Problem Silicon (Si) is essential to the computer industry as a major component of semi-conductor chips. It has three naturally occurring isotopes: ^{28}Si, ^{29}Si, and ^{30}Si. Determine the numbers of protons, neutrons, and electrons in each silicon isotope.

Plan The mass number (A) of each of the three isotopes is given, so we know the sum of protons and neutrons. From the elements list on the text's inside front cover, we find the atomic number (Z, number of protons), which equals the number of electrons. We obtain the number of neutrons from Equation 2.2.

Solution From the elements list, the atomic number of silicon is 14. Therefore,

> ^{28}Si has $14p^+$, $14e^-$, and $14n^0$ $(28 - 14)$
> ^{29}Si has $14p^+$, $14e^-$, and $15n^0$ $(29 - 14)$
> ^{30}Si has $14p^+$, $14e^-$, and $16n^0$ $(30 - 14)$

FOLLOW-UP PROBLEM 2.2 How many protons, neutrons, and electrons are in **(a)** $^{11}_{5}Q$? **(b)** $^{41}_{20}X$? **(c)** $^{131}_{53}Y$? What element symbols do Q, X, and Y represent?

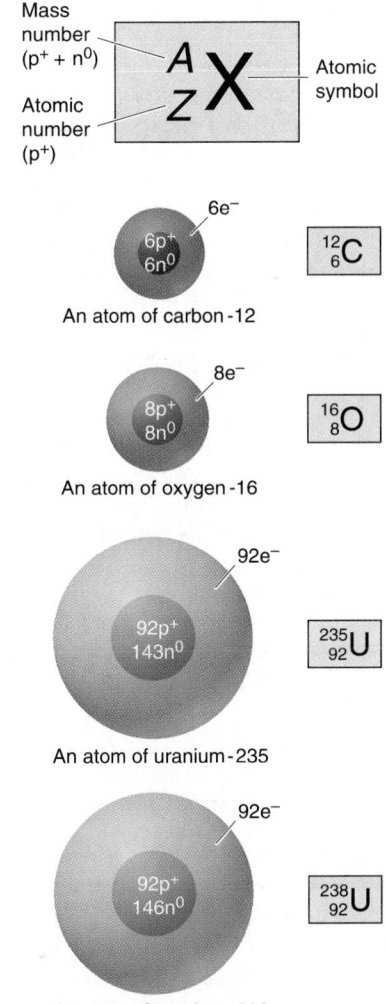

Mass number ($p^+ + n^0$) — A — Atomic symbol
Atomic number (p^+) — ZX

An atom of carbon-12 $6p^+$ $6n^0$ $6e^-$ $^{12}_{6}C$

An atom of oxygen-16 $8p^+$ $8n^0$ $8e^-$ $^{16}_{8}O$

An atom of uranium-235 $92p^+$ $143n^0$ $92e^-$ $^{235}_{92}U$

An atom of uranium-238 $92p^+$ $146n^0$ $92e^-$ $^{238}_{92}U$

The mass of an atom is measured most easily *relative* to the mass of a chosen atomic standard. The modern atomic mass standard is the carbon-12 atom. Its mass is defined as *exactly* 12 atomic mass units. Thus, the **atomic mass unit (amu)** is $\frac{1}{12}$ the mass of a carbon-12 atom. Based on this standard, the ^{1}H atom has a mass of 1.008 amu; in other words, a ^{12}C atom has almost 12 times the mass of an ^{1}H atom. We will continue to use the term *atomic mass unit* in the text, even though the name has recently been changed to the **dalton (D)**; thus, one ^{12}C atom has a mass of 12 daltons (12 D, or 12 amu). The atomic mass unit, which is a unit of relative mass, has an absolute mass of 1.66054×10^{-24} g.

The isotopic makeup of an element is determined by **mass spectrometry,** a method for measuring the relative masses and abundances of atomic-scale particles very precisely (see the Tools of the Laboratory essay on the next page). For example, using a mass spectrometer, we measure the mass ratio of ^{28}Si to ^{12}C as

$$\frac{\text{Mass of } ^{28}Si \text{ atom}}{\text{Mass of } ^{12}C \text{ standard}} = 2.331411$$

From this mass ratio, we find the **isotopic mass** of the ^{28}Si atom, the mass of the isotope relative to the mass of the standard carbon-12 isotope:

> Isotopic mass of ^{28}Si = measured mass ratio × mass of ^{12}C
> = 2.331411 × 12 amu = 27.97693 amu

Along with the isotopic mass, the mass spectrometer gives the relative abundance (fraction) of each isotope in a sample of the element. For example, the percent abundance of ^{28}Si is 92.23%. Such measurements provide data for obtaining the **atomic mass** (also called *atomic weight*) of an element, the *average* of the masses of its naturally occurring isotopes weighted according to their abundances.

Each naturally occurring isotope of an element contributes a certain portion to the atomic mass. For instance, we said that 92.23% of Si atoms are ^{28}Si. Using

Figure 2.10 Depicting the atom. Atoms of carbon-12, oxygen-16, uranium-235, and uranium-238 are shown (nuclei not drawn to scale) with their symbolic representations. The sum of the number of protons (Z) and the number of neutrons (N) equals the mass number (A). An atom is neutral, so the number of protons in the nucleus equals the number of electrons around the nucleus. The two uranium atoms are isotopes of the element.

Mass Spectrometry

Mass spectrometry, the most powerful technique for measuring the mass and abundance of charged particles, emerged from electric and magnetic deflection studies on particles formed in cathode ray experiments. When a high-energy electron collides with an atom of neon-20, for example, one of the atom's electrons is knocked away and the resulting particle has one positive charge, Ne^+ (Figure B2.1). Thus, its mass/charge ratio (m/e) equals the mass divided by $1+$. The m/e values are measured to identify the masses of different isotopes of an element.

Figure B2.2, parts A–C, depicts the core of one type of mass spectrometer and the data it provides. The sample is introduced and vaporized (if liquid or solid), then bombarded by high-energy electrons to form positively charged particles. These are attracted toward a series of negatively charged plates with slits in them, and some particles pass through into an evacuated tube exposed to a magnetic field. As the particles zoom through this region, they are deflected (their paths are bent) according to their m/e: the lightest particles are deflected most and the heaviest particles least. At the end of the magnetic

region, the particles strike a detector, which records their relative positions and abundances. For very precise work, such as determining isotopic masses and abundances, the instrument is calibrated with a substance of known amount and mass.

Mass spectrometry is also used in structural chemistry and separations science to measure the mass of virtually any atom, molecule, or fragment of a molecule. Among its many applications, mass spectrometry is employed by biochemists determining protein structures (Figure B2.2, parts D and E), materials scientists examining catalyst surfaces, forensic chemists analyzing criminal evidence, organic chemists designing new drugs, and industrial chemists investigating petroleum components.

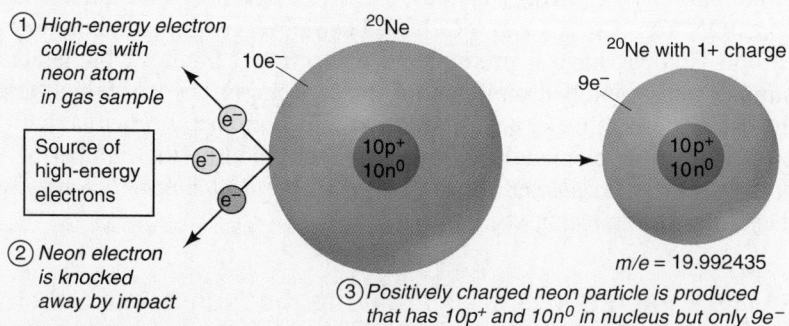

① High-energy electron collides with neon atom in gas sample

② Neon electron is knocked away by impact

③ Positively charged neon particle is produced that has $10p^+$ and $10n^0$ in nucleus but only $9e^-$

$m/e = 19.992435$

Figure B2.1 Formation of a positively charged neon (Ne) particle.

Figure B2.2 **The mass spectrometer and its data. A,** Charged particles are separated on the basis of their m/e values. Ne is the sample here. **B,** The data show the abundance of each particle. The three peaks are three Ne isotopes. **C,** The percent abundance of each particle. **D,** The mass spectrum of a protein molecule. Each peak represents a fragment of the molecule. **E,** Some modern instruments measure masses of large molecules to an accuracy of $10^5 \pm 1$ amu. Even single molecules in the 10^8 amu range have been measured.

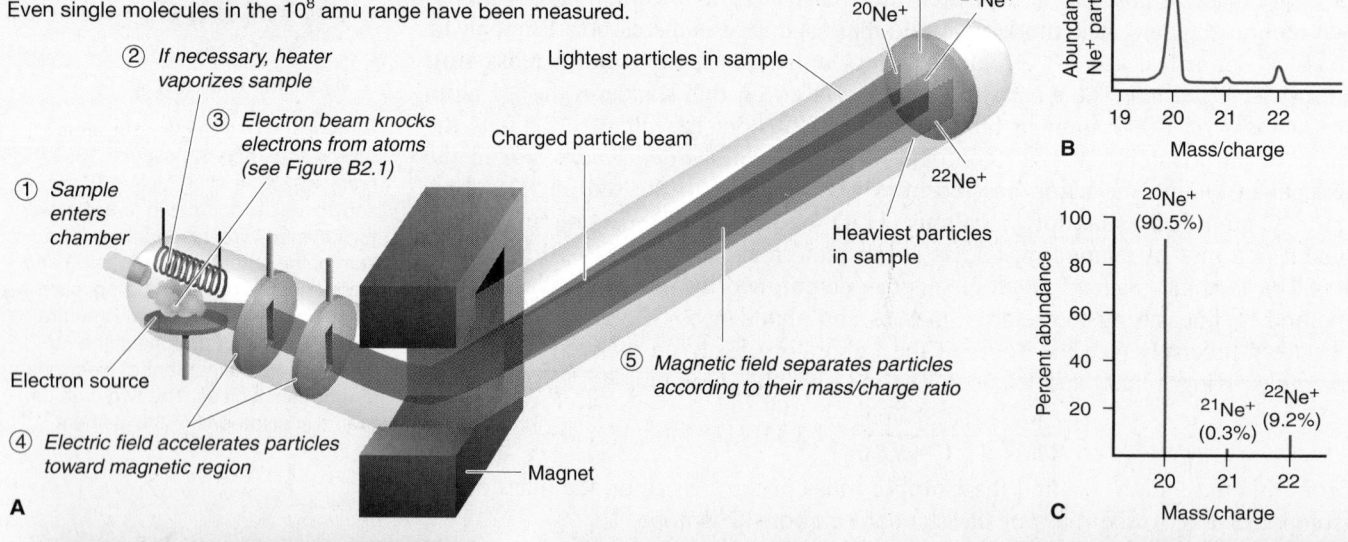

① Sample enters chamber

② If necessary, heater vaporizes sample

③ Electron beam knocks electrons from atoms (see Figure B2.1)

④ Electric field accelerates particles toward magnetic region

⑤ Magnetic field separates particles according to their mass/charge ratio

Electron source

Charged particle beam

Lightest particles in sample

Heaviest particles in sample

Magnet

Detector

A

B

C

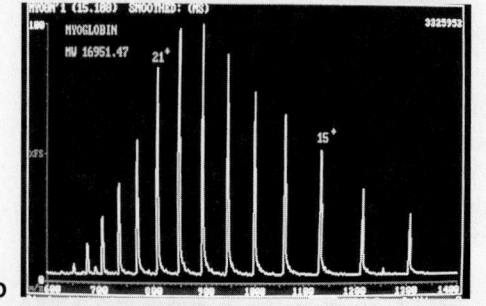

D

E

this percent abundance as a fraction and multiplying by its isotopic mass gives the portion contributed by ^{28}Si:

Portion of Si atomic mass from ^{28}Si = 27.97693 amu × 0.9223 = 25.8031 amu

(retaining two additional significant figures)

Similar calculations give the portions contributed by ^{29}Si (28.976495 amu × 0.0467 = 1.3532 amu) and by ^{30}Si (29.973770 amu × 0.0310 = 0.9292 amu), and adding the three portions together (rounding to two decimal places at the end) gives the atomic mass of silicon:

Atomic mass of Si = 25.8031 amu + 1.3532 amu + 0.9292 amu
= 28.0855 amu = 28.09 amu

SAMPLE PROBLEM 2.3 Calculating the Atomic Mass of an Element

Problem Silver (Ag; Z = 47) has 46 known isotopes, but only two occur naturally, ^{107}Ag and ^{109}Ag. Given the following mass spectrometric data, calculate the atomic mass of Ag:

Isotope	Mass (amu)	Abundance (%)
^{107}Ag	106.90509	51.84
^{109}Ag	108.90476	48.16

Plan From the mass and abundance of the two Ag isotopes, we have to find the atomic mass of Ag (weighted average of the isotopic masses). We multiply each isotopic mass by its fractional abundance to find the portion of the atomic mass contributed by each isotope. The sum of the isotopic portions is the atomic mass.

Solution Finding the portion of the atomic mass from each isotope:

Portion of atomic mass from ^{107}Ag: = isotopic mass × fractional abundance
= 106.90509 amu × 0.5184 = 55.42 amu
Portion of atomic mass from ^{109}Ag: = 108.90476 amu × 0.4816 = 52.45 amu

Finding the atomic mass of silver:

Atomic mass of Ag = 55.42 amu + 52.45 amu = 107.87 amu

Check The individual portions seem right: ~100 amu × 0.50 = 50 amu. The portions are almost the same because the two isotopic abundances are almost the same. We rounded each portion to four significant figures because that is the number of significant figures in the abundance values. This is the correct atomic mass (to two decimal places), as shown in the list of elements (*inside front cover*).

Comment Averages must be interpreted carefully. The average number of children in an American family in 1985 was 2.4. You know that no family actually has 2.4 children; you should also know that no individual silver atom has a mass of 107.87 amu. But for most laboratory purposes, we consider a sample of silver to consist of atoms with this average mass.

FOLLOW-UP PROBLEM 2.3 Boron (B; Z = 5) has two naturally occurring isotopes. Calculate the percent abundances of ^{10}B and ^{11}B from the following: atomic mass of B = 10.81 amu; isotopic mass of ^{10}B = 10.0129 amu; isotopic mass of ^{11}B = 11.0093 amu. (*Hint:* The sum of the fractional abundances is 1. If x = abundance of ^{10}B, then $1 - x$ = abundance of ^{11}B.)

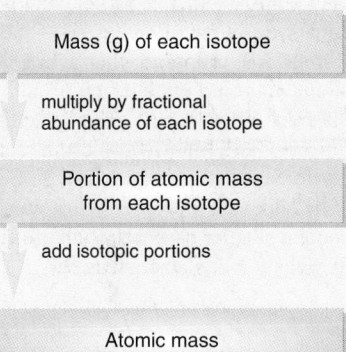

Mass (g) of each isotope

multiply by fractional abundance of each isotope

Portion of atomic mass from each isotope

add isotopic portions

Atomic mass

A Modern Reassessment of the Atomic Theory

We began discussing the atomic basis of matter with Dalton's model, which proved inaccurate in several respects. What happens to a model whose postulates are found by later experiment to be incorrect? No model can predict every possible future observation, but a powerful model evolves to retain its usefulness. Let's reexamine the atomic theory in light of what we know now:

1. *All matter is composed of atoms.* We now know that atoms *are* divisible and composed of smaller, subatomic particles (electrons, protons, and neutrons), but the atom is still the smallest body that *retains the unique identity* of an element.

The Heresy of Radioactive "Transmutation" In 1902, Rutherford performed a series of experiments with radioactive elements that shocked the scientific world. When a radioactive atom of thorium ($Z = 90$) emits an α particle ($Z = 2$), it becomes an atom of radium ($Z = 88$), which then emits another α particle and becomes an atom of radon ($Z = 86$). Rutherford proposed that when an atom emits an α particle, it turns into a different atom—one element changes into another! Many viewed this conclusion as a return to alchemy, and, as with Thomson's discovery that atoms contain smaller particles, Rutherford's findings fell on disbelieving ears.

2. *Atoms of one element cannot be converted into atoms of another element in a chemical reaction.* We now know that, in *nuclear* reactions, atoms of one element often change into atoms of another, but this *never* happens in a *chemical* reaction. ●

3. *All atoms of an element have the same number of protons and electrons, which determines the chemical behavior of the element.* We now know that isotopes of an element differ in the number of neutrons, and thus in mass number, but a sample of the element is treated as though its atoms have an *average* mass.

4. *Compounds are formed by the chemical combination of two or more elements in specific ratios.* We now know that a few compounds can have slight variations in their atom ratios, but this postulate remains essentially unchanged.

Even today, our picture of the atom is being revised. Although we are confident about the distribution of electrons within the atom (Chapters 7 and 8), the interactions among protons and neutrons within the nucleus are still on the frontier of discovery (Chapter 24).

SECTION SUMMARY

An atom has a central nucleus, which contains positively charged protons and uncharged neutrons and is surrounded by negatively charged electrons. An atom is neutral because the number of electrons equals the number of protons. An atom is represented by the notation $^{A}_{Z}X$, in which Z is the atomic number (number of protons), A the mass number (sum of protons and neutrons), and X the atomic symbol. An element occurs naturally as a mixture of isotopes, atoms with the same number of protons but different numbers of neutrons. Each isotope has a mass relative to the ^{12}C mass standard. The atomic mass of an element is the average of its isotopic masses weighted according to their natural abundances and is determined by modern instruments, such as the mass spectrometer.

2.6   ELEMENTS: A FIRST LOOK AT THE PERIODIC TABLE

At the end of the 18th century, Lavoisier compiled a list of the 23 elements known at that time; by 1870, 65 were known; by 1925, 88; today, there are 113 and still counting! These elements combine to form millions of compounds, so we clearly need some way to organize what we know about their behavior. By the mid-19th century, enormous amounts of information concerning reactions, properties, and atomic masses of the elements had been accumulated. Several researchers noted recurring, or *periodic,* patterns of behavior and proposed schemes to organize the elements according to some fundamental property.

In 1871, the Russian chemist Dmitri Mendeleev published the most successful of these organizing schemes, a table that listed the elements by increasing atomic mass, arranged so that elements with similar chemical properties fell in the same column. The modern **periodic table of the elements,** based on Mendeleev's earlier version, is one of the great classifying schemes in science and has become an indispensable tool to chemists. Throughout your study of chemistry, the periodic table will guide you through an otherwise dizzying amount of chemical and physical behavior.

A modern version of the periodic table appears in Figure 2.11 on the next page and inside the front cover. The table is formatted as follows:

1. Each element has a box that contains its atomic number, atomic symbol, and atomic mass. The boxes lie in order of *increasing atomic number* as you move from left to right.

2. The boxes are arranged into a grid of **periods** (horizontal rows) and **groups** (vertical columns). Each period has a number from 1 to 7. Each group has a number from 1 to 8 *and* either the letter A or B. A new system, with group numbers from 1 to 18 but no letters, appears in parentheses under the number-letter designations. (Most chemists still use the number-letter system, so the text retains it, but shows the new numbering system in parentheses.)

3. The eight A groups (two on the left and six on the right) contain the *main-group,* or *representative, elements*. The ten B groups, located between Groups 2A(2) and 3A(13), contain the *transition elements*. Two horizontal series of *inner transition elements,* the lanthanides and the actinides, fit *between* the elements in Group 3B(3) and Group 4B(4) and are usually placed below the main body of the table.

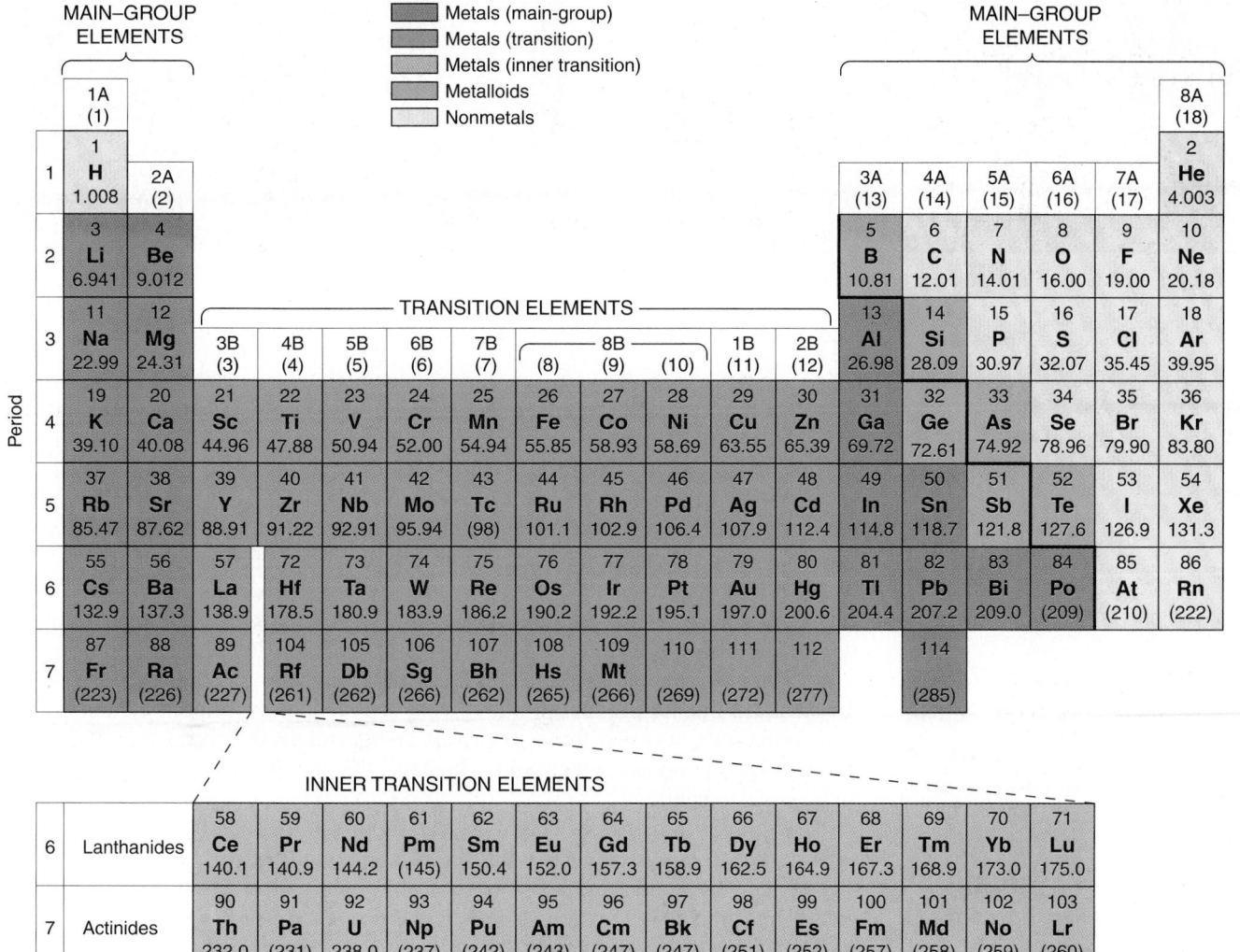

Figure 2.11 **The modern periodic table.** The table consists of element boxes arranged by *increasing* atomic number into groups (vertical columns) and periods (horizontal rows). Each box contains the atomic number, atomic symbol, and atomic mass. (A mass in parentheses is the mass number of the most stable isotope of that element.) The periods are numbered 1 to 7. The groups (sometimes called *families*) have a number-letter designation and a new group number in parentheses. The A groups are the main-group elements; the B groups are the transition elements. Two series of inner transi-tion elements are placed below the main body of the table but actually fit between the elements indicated. Metals lie below and to the left of the thick "staircase" line [top of 3A(13) to bottom of 6A(16)] and include main-group metals *(purple-blue),* transition elements *(blue),* and inner transition elements *(gray-blue).* Nonmetals *(yellow)* lie to the right of the line. Metalloids *(green)* lie along the line. We discuss the placement of hydrogen in Chapter 14. As of late 2001, elements 110 to 112, and 114 have not yet been named.

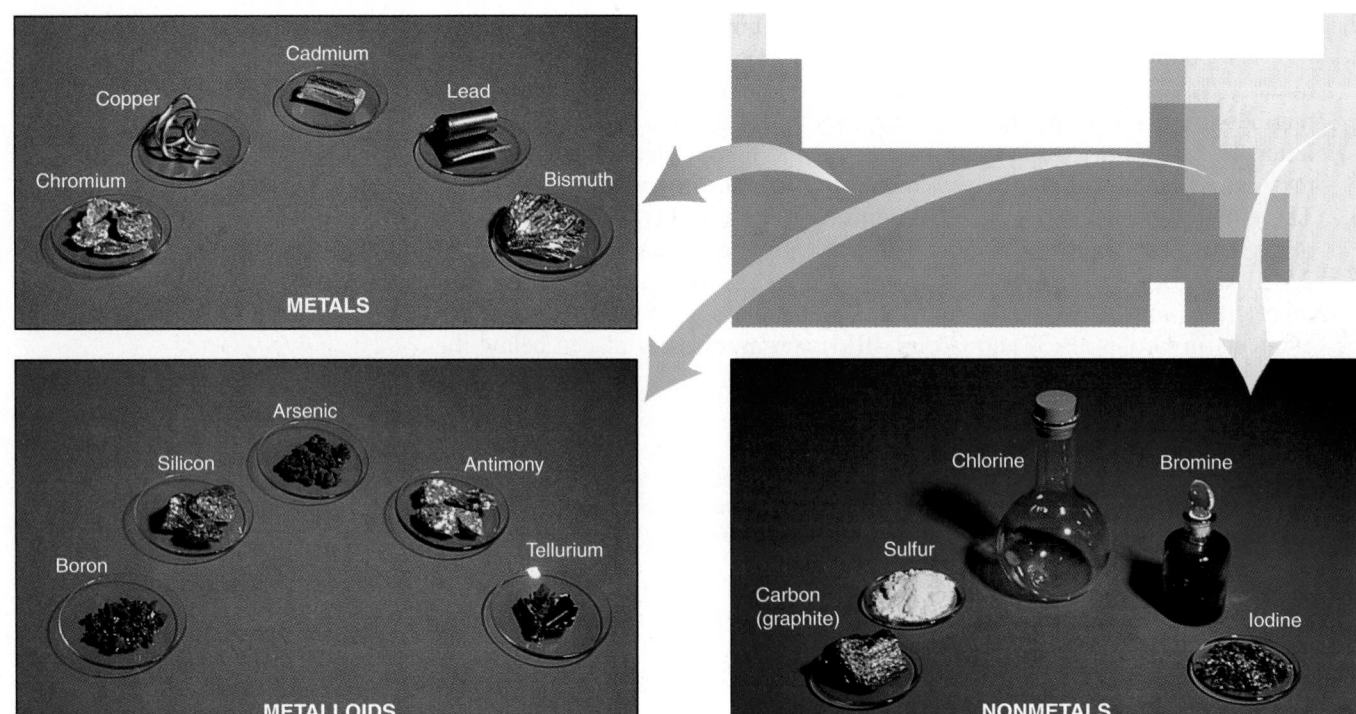

Figure 2.12 Metals, metalloids, and nonmetals.

At this point in the text, the clearest distinction among the elements is their classification as metals, nonmetals, or metalloids. The "staircase" line that runs from the top of Group 3A(13) to the bottom of Group 6A(16) is a dividing line for this classification. The **metals** (three shades of blue) appear in the large lower-left portion of the table. About three-quarters of the elements are metals, including many main-group elements and all the transition and inner transition elements. They are generally shiny solids at room temperature (mercury is the only liquid) that conduct heat and electricity well and can be tooled into sheets (malleable) and wires (ductile). The **nonmetals** (yellow) appear in the small upper-right portion of the table. They are generally gases or dull, brittle solids at room temperature (bromine is the only liquid) that conduct heat and electricity poorly. Along the staircase line lie the **metalloids** (green; also called **semimetals**), elements that have properties between those of metals and nonmetals. Several metalloids play major roles in modern electronics. Figure 2.12 shows examples of these three classes of elements.

Two major branches of chemistry can almost be defined by the elements that each studies. *Organic chemistry* studies the compounds of carbon, especially those that contain hydrogen and often oxygen, nitrogen, and a few other elements as well. This branch is concerned with fuels, drugs, dyes, polymers, and the like. *Inorganic chemistry,* on the other hand, focuses mainly on the compounds of all the other elements. It is concerned with catalysts, electronic materials, metal alloys, mineral salts, and the like. With the explosive growth in biomedical and materials research, the distinction between these traditional branches is disappearing rapidly.

It is important to learn some of the group (family) names. Group 1A(1), except for hydrogen, consists of the *alkali metals,* and Group 2A(2) consists of the *alkaline earth metals.* Both groups of metals are highly reactive elements. The *halogens,* Group 7A(17), are highly reactive nonmetals, whereas the *noble gases,* Group 8A(18), are relatively unreactive nonmetals. Other main groups [3A(13) to 6A(16)] are often named by the first element in the group; for example, Group 6A is the *oxygen family.*

A key point that we return to many times is that, in general, *elements in a group have **similar** chemical properties and elements in a period have **different** chemical properties.* We begin applying the organizing power of the periodic table in the next section, where we discuss how elements combine to form compounds.

SECTION SUMMARY

In the periodic table, the elements are arranged by atomic number into horizontal periods and vertical groups. Because of the periodic recurrence of certain key properties, elements within a group have similar behavior, whereas elements in a period have dissimilar behavior. Nonmetals appear in the upper right portion of the table, metalloids lie along a staircase line, and metals fill the rest of the table.

2.7 COMPOUNDS: INTRODUCTION TO BONDING

The overwhelming majority of elements occur in chemical combination with other elements. In fact, only a few elements occur free in nature. The noble gases—helium (He), neon (Ne), argon (Ar), krypton (Kr), xenon (Xe), and radon (Rn)—occur as separate atoms. In addition to occurring in compounds, oxygen (O), nitrogen (N), and sulfur (S) also occur commonly as uncombined molecules, such as O_2, N_2, and S_8, and carbon (C) occurs in vast, nearly pure deposits of coal. Some of the metals, such as copper (Cu), silver (Ag), gold (Au), and platinum (Pt), may also occur uncombined with other elements. But these few exceptions reinforce the general rule that elements occur combined in compounds.

The *electrons* of the atoms of interacting elements are involved in compound formation. Elements combine in two general ways:

1. *Transferring electrons* from the atoms of one element to those of another to form **ionic compounds**
2. *Sharing electrons* between atoms of different elements to form **covalent compounds**

These processes generate **chemical bonds,** the forces that hold the atoms of elements together in a compound.

The Formation of Ionic Compounds

Ionic compounds are composed of **ions,** charged particles that form when an atom (or small group of atoms) gains or loses one or more electrons. The simplest type of ionic compound is a **binary ionic compound,** one composed of just two elements. It typically forms *when a metal reacts with a nonmetal.* Each metal atom loses a certain number of its electrons and becomes a **cation,** a positively charged ion. The nonmetal atoms gain the electrons lost by the metal atoms and become **anions,** negatively charged ions. In effect, the metal atoms *transfer electrons* to the nonmetal atoms. The resulting cations and anions attract each other through electrostatic forces and form the ionic compound. A cation or anion derived from a single atom is called a **monatomic ion;** we'll discuss polyatomic ions, those derived from a small group of atoms, later.

The formation of the binary ionic compound sodium chloride, common table salt, is depicted in Figure 2.13, from the elements through the atomic-scale electron transfer to the compound. In the electron transfer, a sodium atom, which is neutral because it has the same number of protons as electrons, *loses* 1 electron and forms a sodium cation, Na^+. (The charge on the ion is written as a *right superscript*.) A chlorine atom *gains* the electron and becomes a chloride anion, Cl^-. (The name change from the nonmetal atom to the ion is discussed in the next section.) Even the tiniest visible grain of table salt contains an *enormous* number of sodium and chloride ions. The oppositely charged ions (Na^+ and Cl^-) attract each other, and the similarly charged ions (Na^+ and Na^+, or Cl^- and Cl^-) repel each other. The resulting aggregation is a regular array of alternating Na^+ and Cl^- ions that extends in all three dimensions.

The strength of the ionic bonding depends to a great extent on the net strength of these attractions and repulsions and is described by *Coulomb's law,* which can be expressed as follows: *the energy of attraction (or repulsion) between two particles is directly proportional to the product of the charges and inversely proportional to the distance between them.*

$$\text{Energy} \propto \frac{\text{charge 1} \times \text{charge 2}}{\text{distance}}$$

In other words, ions with higher charges attract (or repel) each other more strongly than ions with lower charges. Likewise, smaller ions attract (or repel) each other more strongly than larger ions, because their charges are closer together. These effects are summarized in Figure 2.14.

Ionic compounds are neutral; that is, they possess no net charge. For this to occur, they must contain equal numbers of positive and negative *charges.* This does *not* mean, however, that they must contain equal numbers of positive and negative *ions.* Because Na^+ and Cl^- each bear a unit charge ($1+$ or $1-$), equal numbers of the ions are present in sodium chloride. However, in sodium oxide, for example, twice as many Na^+ ions balance the charge of the oxide ions, O^{2-}.

FORMATION OF AN IONIC COMPOUND

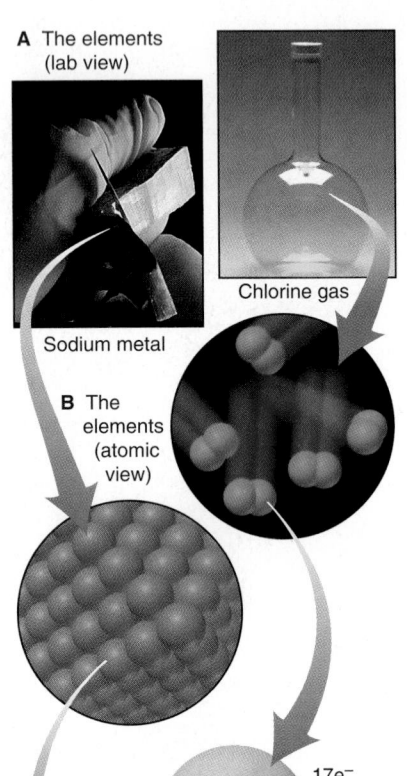

A The elements (lab view)

Chlorine gas

Sodium metal

B The elements (atomic view)

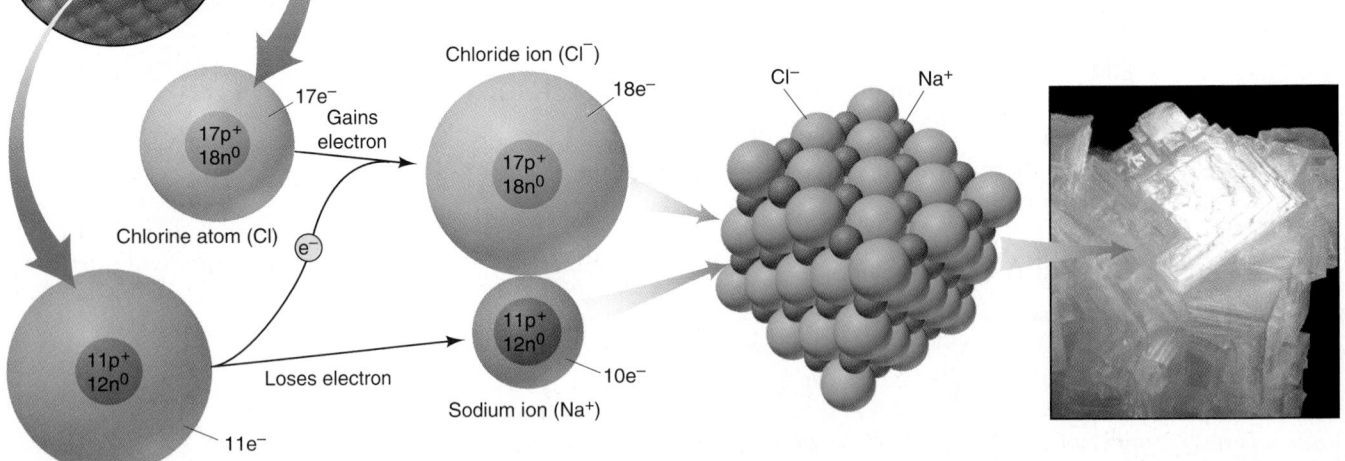

Chloride ion (Cl^-)

$17e^-$
Gains electron

$17p^+$
$18n^0$

$18e^-$

$17p^+$
$18n^0$

Cl^- Na^+

Chlorine atom (Cl)

e^-

$11p^+$
$12n^0$

$10e^-$

Sodium ion (Na^+)

$11p^+$
$12n^0$

Loses electron

$11e^-$

Sodium atom (Na)

C Electron transfer

D The compound (atomic view): Na^+ and Cl^- in the crystal

E The compound (lab view): sodium chloride crystal

Figure 2.13 The formation of an ionic compound. A, The two elements as seen in the laboratory. **B,** The elements as they might appear on the atomic scale. **C,** The neutral sodium atom loses one electron to become a sodium cation (Na^+), and the chlorine atom gains one electron to become a chloride anion (Cl^-). (Note that when atoms lose electrons, they become ions that are smaller, and when they gain electrons, they become ions that are larger.) **D,** Na^+ and Cl^- ions attract each other and lie in a three-dimensional crystalline array. **E,** This cubic array is reflected in the structure of crystalline NaCl, which occurs naturally as the mineral halite, hence the name *halogens* for the Group 7A(17) elements.

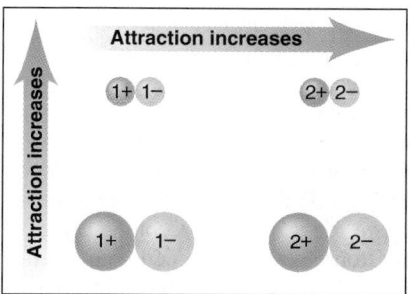

Figure 2.14 **Factors that influence the strength of ionic bonding.** For ions of a given size, strength of attraction *(arrows)* increases with higher ionic charge *(left to right)*. For ions of a given charge, strength of attraction increases with smaller ionic size *(bottom to top)*.

Can we predict the number of electrons a given atom will lose or gain when it forms an ion? In the formation of sodium chloride, for example, why does each sodium atom give up only 1 of its 11 electrons? Why doesn't each chlorine atom gain two electrons, instead of just one? For A-group elements, the periodic table provides an answer. We generally find that metals lose electrons and nonmetals gain electrons to *form ions with the same number of electrons as in the nearest noble gas* [Group 8A(18)]. Noble gases have a stability (low reactivity) that is related to their number (and arrangement) of electrons. A sodium atom ($11e^-$) can attain the stability of neon ($10e^-$), the nearest noble gas, by losing one electron. Similarly, by gaining one electron, a chlorine atom ($17e^-$) attains the stability of argon ($18e^-$), its nearest noble gas. Thus, when an element located near a noble gas forms a monatomic ion, *it gains or loses enough electrons to attain the same number as that noble gas*. Specifically, the elements in Group 1A(1) lose one electron, those in Group 2A(2) lose two, and aluminum in Group 3A(13) loses three; the elements in Group 7A(17) gain one electron, oxygen and sulfur in Group 6A(16) gain two, and nitrogen in Group 5A(15) gains three.

With the periodic table printed on a two-dimensional surface, as in Figure 2.11, it is easy to get the false impression that the elements in Group 7A(17) are "closer" to the noble gases than the elements in Group 1A(1). Actually, both groups are only one electron away from having the same number of electrons as the noble gases. To make this point, Figure 2.15 shows a modified periodic table that is cut and rejoined, with the noble gases in the center. Now you can see that fluorine (F; $Z = 9$) has one electron fewer and sodium (Na; $Z = 11$) has one electron more than the noble gas neon (Ne; $Z = 10$); thus, they form the F^- and Na^+ ions. Similarly, oxygen (O; $Z = 8$) gains two electrons and magnesium (Mg; $Z = 12$) loses two to form the O^{2-} and Mg^{2+} ions and attain the same number of electrons as neon.

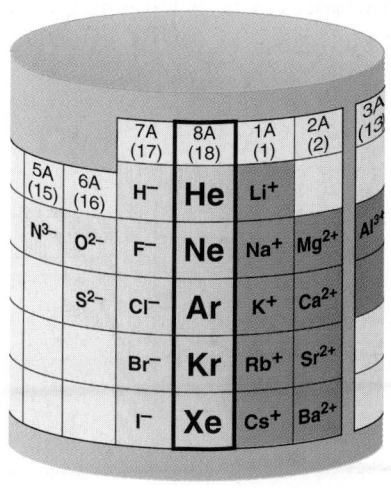

Figure 2.15 **The relationship between ions formed and the nearest noble gas.** This periodic table was redrawn to show the positions of nonmetals *(yellow)* and metals *(blue)* relative to the noble gases and to show the ions these elements form. The ionic charge equals the number of electrons lost (+) or gained (−) to attain the same number of electrons as the nearest noble gas. Species in the same row have the same number of electrons. For example, H^-, He, and Li^+ all have two electrons. [Note that H is shown here in Group 7A(17).]

A No interaction

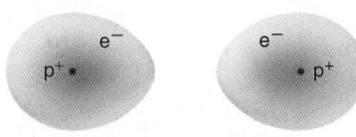

B Attraction begins

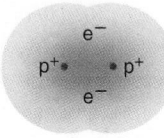

C Covalent bond

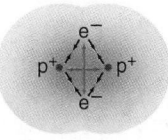

D Combination of forces

Figure 2.16 **Formation of a covalent bond between two H atoms.**
A, The distance is too great for the atoms to affect each other. **B,** As the distance decreases, the nucleus of each atom begins to attract the electron of the other. **C,** The covalent bond forms when the two nuclei mutually attract the pair of electrons at some optimum distance. **D,** The H_2 molecule is more stable than the separate atoms because the attractive forces (black arrows) between each nucleus and the two electrons are greater than the repulsive forces (red arrows) between the electrons and between the nuclei.

SAMPLE PROBLEM 2.4 Predicting the Ion an Element Forms

Problem What monatomic ions do the following elements form?
(a) Iodine ($Z = 53$)   **(b)** Calcium ($Z = 20$)   **(c)** Aluminum ($Z = 13$)
Plan We use the given Z value to find the element in the periodic table and see where its group lies relative to the noble gases. Elements in groups that lie *after* the noble gases *lose* electrons to attain the same number as the nearest noble gas and become positive ions; those in groups that lie *before* the noble gases *gain* electrons and become negative ions.
Solution (a) I^- Iodine ($_{53}I$) is a nonmetal in Group 7A(17), one of the halogens. Like any member of this group, it gains 1 electron to have the same number as the nearest Group 8A(18) member, in this case $_{54}Xe$.
(b) Ca^{2+} Calcium ($_{20}Ca$) is a member of Group 2A(2), the alkaline earth metals. Like any Group 2A member, it loses 2 electrons to attain the same number as the nearest noble gas, in this case, $_{18}Ar$.
(c) Al^{3+} Aluminum ($_{13}Al$) is a metal in the boron family [Group 3A(13)] and thus loses 3 electrons to attain the same number as its nearest noble gas, $_{10}Ne$.

FOLLOW-UP PROBLEM 2.4 What monatomic ion does each of the following elements form?
(a) $_{16}S$ **(b)** $_{37}Rb$ **(c)** $_{56}Ba$

The Formation of Covalent Compounds

Covalent compounds form when elements share electrons, which usually occurs between nonmetals. Even though relatively few nonmetals exist, they interact in many combinations to form a very large number of covalent compounds.

The simplest case of electron sharing occurs not in a compound but between two hydrogen atoms (H; $Z = 1$). Imagine two separated H atoms approaching each other, as in Figure 2.16. As they get closer, the nucleus of each atom attracts the electron of the other atom more and more strongly, and the separated atoms begin to interpenetrate each other. At some optimum distance between the nuclei, the two atoms form a **covalent bond,** a pair of electrons mutually attracted by the two nuclei. The result is a hydrogen molecule, in which each electron no longer "belongs" to a particular H atom: the two electrons are *shared* by the two nuclei. Repulsions between the nuclei and between the electrons also occur, but the net attraction is greater than the net repulsion. (We discuss the properties of covalent bonds in great detail in Chapter 9.)

A sample of hydrogen gas consists of these diatomic molecules (H_2)—pairs of atoms that are chemically bound and behave as an independent unit—*not* separate H atoms. Other nonmetals that exist as diatomic molecules at room temperature are nitrogen (N_2), oxygen (O_2), and the halogens [fluorine (F_2), chlorine (Cl_2), bromine (Br_2), and iodine (I_2)]. Phosphorus exists as tetratomic molecules (P_4), and sulfur and selenium as octatomic molecules (S_8 and Se_8) (Figure 2.17).

Figure 2.17 **Elements that form polyatomic molecules.**

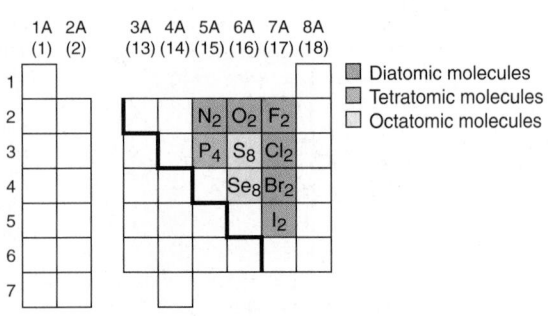

In a similar way, *atoms of different elements share electrons to form the molecules of a covalent compound.* A sample of hydrogen fluoride, for example, consists of molecules in which one H atom forms a covalent bond with one F atom; water consists of molecules in which one O atom forms covalent bonds with two H atoms:

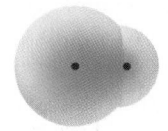 Hydrogen fluoride, HF

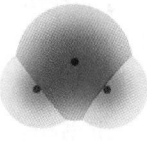 Water, H$_2$O

(As you'll see in Chapter 9, covalent bonding provides another way for atoms to attain the same number of electrons as the nearest noble gas.)

Distinguishing the Entities Involved in Covalent and Ionic Bonding A key distinction exists between the chemical entities in covalent and ionic substances. *Most covalent substances consist of molecules.* A cup of water, for example, contains a collection of individual water molecules, each surrounded by other water molecules. Each water molecule consists of an O atom connected by covalent bonds to two H atoms. In contrast, under ordinary conditions, *no molecules exist in a sample of an ionic compound.* A piece of sodium chloride, for example, is a continuous array of oppositely charged sodium and chloride ions, *not* a collection of individual "sodium chloride molecules."

Another key distinction exists between the entities involved in the attractions themselves. Covalent bonding is based on the mutual attraction between two (positively charged) nuclei and the two (negatively charged) electrons that reside between them. Ionic bonding is based on the mutual attraction between positive and negative ions.

Polyatomic Ions: Covalent Bonds Within Ions

Many ionic compounds contain **polyatomic ions,** which consist of two or more atoms bonded *covalently* and have a net positive or negative charge. For example, the ionic compound calcium carbonate is an array of polyatomic carbonate anions and monatomic calcium cations attracted to each other. The carbonate ion consists of a carbon atom covalently bonded to three oxygen atoms, and two additional electrons give the ion its 2− charge (Figure 2.18). In many reactions, a polyatomic ion stays together as a unit during interactions with other ions.

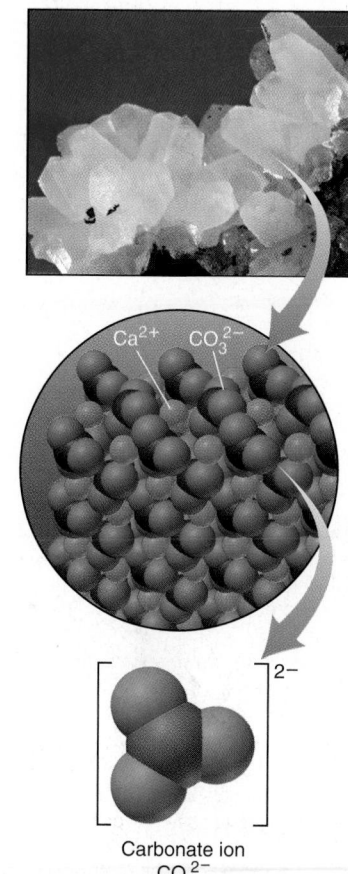

Figure 2.18 **A polyatomic ion.** Calcium carbonate is a three-dimensional array of monatomic calcium cations (*purple spheres*) and polyatomic carbonate anions. As the bottom structure shows, each carbonate ion consists of four covalently bonded atoms.

SECTION SUMMARY
Although a few elements occur uncombined in nature, the great majority exist in compounds. Ionic compounds form when a metal *transfers electrons* to a nonmetal, and the resulting positive and negative ions attract each other to form a three-dimensional array. In many cases, metal atoms lose and nonmetal atoms gain enough electrons to attain the same number of electrons as in atoms of the nearest noble gas. Covalent compounds form when elements, usually nonmetals, *share electrons.* Each covalent bond is an electron pair mutually attracted by two atomic nuclei. Monatomic ions are derived from one atom. Polyatomic ions consist of two or more covalently bonded atoms that have a net positive or negative charge due to a deficit or excess of electrons.

2.8 COMPOUNDS: FORMULAS, NAMES, AND MASSES

Names and formulas of compounds form the vocabulary of the chemical language; now it's time for you to begin speaking and writing this language. In this discussion, you'll learn the names and formulas of ionic and simple covalent compounds and how to calculate the mass of a unit of a compound from its formula.

Types of Chemical Formulas

In a **chemical formula**, element symbols and numerical subscripts show the type and number of each atom present in the smallest unit of the substance. There are several types of chemical formulas for a compound:

1. The **empirical formula** shows the *relative* number of atoms of each element in the compound. It is the simplest type of formula and is derived from the masses of the component elements. For example, in hydrogen peroxide, there is 1 part by mass of hydrogen for every 16 parts by mass of oxygen. Therefore, the empirical formula of hydrogen peroxide is HO: one H atom for every O atom.

2. The **molecular formula** shows the *actual* number of atoms of each element in a molecule of the compound. The molecular formula of hydrogen peroxide is H_2O_2; there are *actually* two H atoms and two O atoms in each molecule.

3. A **structural formula** shows the number of atoms and *the bonds between them;* that is, the relative placement and connections of atoms in the molecule. The structural formula of hydrogen peroxide is H—O—O—H; each H is bonded to an O, and the O's are bonded to each other.

Some Advice about Learning Names and Formulas

Perhaps in the future, systematic names for compounds will be used by everyone. However, many reference books, chemical supply catalogs, and practicing chemists still use many common (trivial) names, so you should learn them as well.

Here are some points to note about ion formulas:

*   Members of a periodic table group have the same ionic charge; for example, Li, Na, and K are all in Group 1A and all have a 1+ charge.

*   For A-group cations, ion charge = group number: for example, Na^+ is in Group 1A, Ba^{2+} in Group 2A. (Exceptions in Figure 2.19 are Sn^{2+} and Pb^{2+}.)

*   For anions, ion charge = group number minus 8: for example, S is in Group 6A ($6 - 8 = -2$), so the ion is S^{2-}.

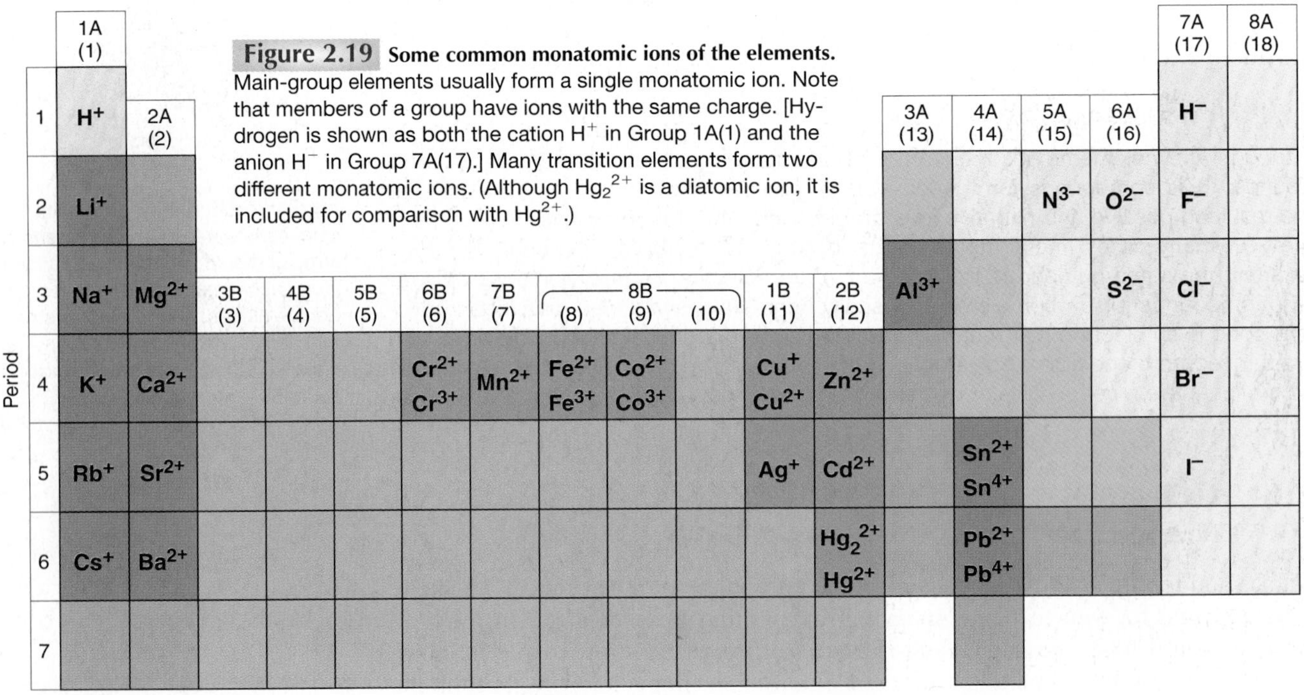

Figure 2.19 **Some common monatomic ions of the elements.** Main-group elements usually form a single monatomic ion. Note that members of a group have ions with the same charge. [Hydrogen is shown as both the cation H^+ in Group 1A(1) and the anion H^- in Group 7A(17).] Many transition elements form two different monatomic ions. (Although Hg_2^{2+} is a diatomic ion, it is included for comparison with Hg^{2+}.)

Here are some suggestions about how to learn names and formulas:

1. Memorize the A-group monatomic ions of Table 2.3 (all except Ag^+, Zn^{2+}, and Cd^{2+}) according to their positions in the periodic table of Figure 2.19. These ions have the same number of electrons as an atom of the nearest noble gas.
2. Consult Table 2.4 (page 66) and Figure 2.19 for some metals that form two different monatomic ions.
3. Divide the tables of names and charges into smaller batches, and learn a batch each day. Try "flash" cards, with the name on one side and the ion formula on the other. The most common ions are shown in **boldface** in Tables 2.3, 2.4, and 2.5, so you can focus on learning them first.

Names and Formulas of Ionic Compounds

All ionic compound names give the positive ion (cation) first and the negative ion (anion) second.

Compounds Formed from Monatomic Ions Let's first consider binary ionic compounds, those composed of ions of two elements.

- *The name of the cation is the same as the name of the metal.* Many metal names end in *-ium*.
- *The name of the anion takes the root of the nonmetal name and adds the suffix "-ide."*

For example, the anion formed from brom*ine* is named brom*ide* (brom+ide). Therefore, the compound formed from the metal calcium and the nonmetal bromine is "calcium bromide."

SAMPLE PROBLEM 2.5 Naming Binary Ionic Compounds

Problem Name the ionic compound formed from the following pairs of elements:
(a) Magnesium and nitrogen   **(b)** Iodine and cadmium
(c) Strontium and fluorine   **(d)** Sulfur and cesium
Plan The key to naming a binary ionic compound is to recognize which element is the metal and which is the nonmetal. When in doubt, check the periodic table. We place the cation name first, add the suffix *-ide* to the nonmetal root, and place the anion name last.
Solution (a) Magnesium is the metal; "nitr-" is the nonmetal root: magnesium nitride

(b) Cadmium is the metal; "iod-" is the nonmetal root: cadmium iodide

(c) Strontium is the metal; "fluor-" is the nonmetal root: strontium fluoride (Note the spelling is fluoride, not flouride.)

(d) Cesium is the metal; "sulf-" is the nonmetal root: cesium sulfide

FOLLOW-UP PROBLEM 2.5 For the following ionic compounds, give the name and periodic table group number of each of the elements present: **(a)** zinc oxide; **(b)** silver bromide; **(c)** lithium chloride; **(d)** aluminum sulfide.

Because ionic compounds are arrays of oppositely charged ions rather than separate molecular units, we write a formula for the **formula unit,** which gives the *relative* numbers of cations and anions in the compound. Thus, ionic compounds generally have only empirical formulas.* The compound has zero net charge, so the positive charges of the cations must balance the negative charges of the anions. For example, calcium bromide is composed of Ca^{2+} ions and Br^-

---

*Compounds of the mercury(I) ion, such as Hg_2Cl_2, and peroxides of the alkali metals, such as Na_2O_2, are the only two common exceptions. Their empirical formulas are HgCl and NaO, respectively.

Table 2.3 Common Monatomic Ions*

Charge	Formula	Name
Cations		
1+	H^+	hydrogen
	Li^+	**lithium**
	Na^+	**sodium**
	K^+	**potassium**
	Cs^+	cesium
	Ag^+	**silver**
2+	**Mg^{2+}**	**magnesium**
	Ca^{2+}	**calcium**
	Sr^{2+}	strontium
	Ba^{2+}	**barium**
	Zn^{2+}	**zinc**
	Cd^{2+}	cadmium
3+	**Al^{3+}**	**aluminum**
Anions		
1−	H^-	hydride
	F^-	**fluoride**
	Cl^-	**chloride**
	Br^-	**bromide**
	I^-	**iodide**
2−	**O^{2-}**	**oxide**
	S^{2-}	**sulfide**
3−	N^{3-}	nitride

*Listed by charge; those in **boldface** are most common.

ions; therefore, two Br^- balance each Ca^{2+}. The formula is $CaBr_2$, not Ca_2Br. In this and all other formulas,

- The subscript refers to the element *preceding* it.
- The *subscript 1 is understood* from the presence of the element symbol alone (that is, we do not write Ca_1Br_2).
- The charge (without the sign) of one ion becomes the subscript of the other:

$$Ca^{2+} \quad Br^{1-} \qquad \text{gives} \qquad Ca_1Br_2 \quad \text{or} \quad CaBr_2$$

Reduce the subscripts to the smallest whole numbers that retain the ratio of ions. Thus, for example, from the ions Ca^{2+} and O^{2-} we have Ca_2O_2, which we reduce to the formula CaO (but see the footnote on page 65).

SAMPLE PROBLEM 2.6 Determining Formulas of Binary Ionic Compounds

Problem Write empirical formulas for the compounds named in Sample Problem 2.5.
Plan We write the empirical formula by finding the smallest number of each ion that gives the neutral compound. These numbers appear as *right subscripts* to the element symbol.
Solution
(a) Mg^{2+} and N^{3-}; three Mg^{2+} ions (6+) balance two N^{3-} ions (6−): Mg_3N_2
(b) Cd^{2+} and I^-; one Cd^{2+} ion (2+) balances two I^- ions (2−): CdI_2
(c) Sr^{2+} and F^-; one Sr^{2+} ion (2+) balances two F^- ions (2−): SrF_2
(d) Cs^+ and S^{2-}; two Cs^+ ions (2+) balance one S^{2-} ion (2−): Cs_2S
Comment Note that ion charges do *not* appear in the compound formula. That is, for cadmium iodide, we do *not* write $Cd^{2+}I_2^-$.

FOLLOW-UP PROBLEM 2.6 Write the formulas of the compounds named in Follow-up Problem 2.5.

Compounds with Metals That Can Form More Than One Ion Many metals, particularly the transition elements (B groups), can form more than one ion, each with a particular charge. Table 2.4 shows some examples. Names of compounds

Table 2.4 Some Metals That Form More Than One Monatomic Ion*

Element	Ion Formula	Systematic Name	Common (Trivial) Name
Chromium	Cr^{2+}	chromium(II)	chromous
	Cr^{3+}	**chromium(III)**	chromic
Cobalt	Co^{2+}	cobalt(II)	
	Co^{3+}	cobalt(III)	
Copper	**Cu^+**	**copper(I)**	cuprous
	Cu^{2+}	**copper(II)**	cupric
Iron	**Fe^{2+}**	**iron(II)**	ferrous
	Fe^{3+}	**iron(III)**	ferric
Lead	**Pb^{2+}**	**lead(II)**	
	Pb^{4+}	lead(IV)	
Mercury	**Hg_2^{2+}**	mercury(I)	mercurous
	Hg^{2+}	**mercury(II)**	mercuric
Tin	**Sn^{2+}**	**tin(II)**	stannous
	Sn^{4+}	tin(IV)	stannic

*Listed alphabetically by metal name; those in **boldface** are most common.

containing these elements include a *Roman numeral within parentheses* immediately after the metal ion's name to indicate its ionic charge. For example, iron can form Fe^{2+} and Fe^{3+} ions. The two compounds that iron forms with chlorine are $FeCl_2$, named iron(II) chloride (spoken "iron two chloride"), and $FeCl_3$, named iron(III) chloride.

In common names, the Latin root of the metal is followed by either of two suffixes:

- The suffix *-ous* for the ion with the lower charge
- The suffix *-ic* for the ion with the higher charge

Thus, iron(II) chloride is also called ferr*ous* chloride and iron(III) chloride is ferr*ic* chloride. (You can easily remember this naming relationship because there is an *o* in *ous* and *lower,* and an *i* in *ic* and *higher.*)

SAMPLE PROBLEM 2.7 Determining Names and Formulas of Ionic Compounds of Elements That Form More Than One Ion

Problem Give the systematic names for the formulas or the formulas for the names of the following compounds:
(a) Tin(II) fluoride   (b) CrI_3
(c) Ferric oxide   (d) CoS
Solution (a) Tin(II) is Sn^{2+}; fluoride is F^-. Two F^- ions balance one Sn^{2+} ion: tin(II) fluoride is SnF_2. (The common name is stannous fluoride.)
(b) The anion is I^-, iodide, and the formula shows three I^-. Therefore, the cation must be Cr^{3+}, chromium(III): CrI_3 is chromium(III) iodide. (The common name is chromic iodide.)
(c) Ferric is the common name for iron(III), Fe^{3+}; oxide ion is O^{2-}. To balance the ionic charges, the formula of ferric oxide is Fe_2O_3. (The systematic name is iron(III) oxide.)
(d) The anion is sulfide, S^{2-}, which requires that the cation be Co^{2+}. The name is cobalt(II) sulfide.

FOLLOW-UP PROBLEM 2.7 Give the systematic names for the formulas or the formulas for the names of the following compounds: (a) lead(IV) oxide; (b) Cu_2S; (c) $FeBr_2$; (d) mercuric chloride.

Compounds Formed from Polyatomic Ions Ionic compounds in which one or both of the ions are polyatomic are very common. Table 2.5 gives the formulas and the names of some common polyatomic ions. Remember that *the polyatomic ion stays together as a charged unit.* The formula for potassium nitrate is KNO_3: each K^+ balances one NO_3^-. The formula for sodium carbonate is Na_2CO_3: two Na^+ balance one CO_3^{2-}. *When two or more of the same polyatomic ion are present in the formula unit, that ion appears in parentheses with the subscript written outside.* For example, calcium nitrate, which contains one Ca^{2+} and two NO_3^- ions, has the formula $Ca(NO_3)_2$. Parentheses and a subscript are *not* used unless *more than one* of the polyatomic ions is present; thus, sodium nitrate is $NaNO_3$, *not* $Na(NO_3)$.

Families of Oxoanions As Table 2.5 shows, most polyatomic ions are **oxoanions,** those in which an element, usually a nonmetal, is bonded to one or more oxygen atoms. In several cases, there exist families of two or four oxoanions that differ only in the number of oxygen atoms. A simple naming convention is used with these ions.

With two oxoanions in the family:

- The ion with *more* O atoms takes the nonmetal root and the suffix *-ate*.
- The ion with *fewer* O atoms takes the nonmetal root and the suffix *-ite*.

Table 2.5 Common Polyatomic Ions*

Formula	Name
Cations	
NH_4^+	**ammonium**
H_3O^+	**hydronium**
Anions	
CH_3COO^- (or $C_2H_3O_2^-$)	**acetate**
CN^-	cyanide
OH^-	**hydroxide**
ClO^-	hypochlorite
ClO_2^-	chlorite
ClO_3^-	**chlorate**
ClO_4^-	**perchlorate**
NO_2^-	nitrite
NO_3^-	**nitrate**
MnO_4^-	**permanganate**
CO_3^{2-}	**carbonate**
HCO_3^-	**hydrogen carbonate** (or **bicarbonate**)
CrO_4^{2-}	chromate
$Cr_2O_7^{2-}$	**dichromate**
O_2^{2-}	peroxide
PO_4^{3-}	**phosphate**
HPO_4^{2-}	hydrogen phosphate
$H_2PO_4^-$	dihydrogen phosphate
SO_3^{2-}	sulfite
SO_4^{2-}	**sulfate**
HSO_4^-	hydrog~~

*Boldface ions ar~

	Prefix	Root	Suffix
	per	*root*	ate
		root	ate
		root	ite
	hypo	*root*	ite

No. of O atoms ↑

Figure 2.20 **Naming oxoanions.** Prefixes and suffixes indicate the number of O atoms in the anion.

For example, SO_4^{2-} is the sulf*ate* ion; SO_3^{2-} is the sulf*ite* ion; similarly, NO_3^- is nitr*ate,* and NO_2^- is nitr*ite.*

With four oxoanions in the family (usually a halogen bonded to O), as Figure 2.20 shows:

- The ion with *most* O atoms has the prefix *per-,* the nonmetal root, and the suffix *-ate.*
- The ion with *one fewer* O atom has just the root and the suffix *-ate.*
- The ion with *two fewer* O atoms has just the root and the suffix *-ite.*
- The ion with *least (three fewer)* O atoms has the prefix *hypo-,* the root, and the suffix *-ite.*

For example, for the four chlorine oxoanions,

ClO_4^- is *per*chlor*ate,* ClO_3^- is chlor*ate,* ClO_2^- is chlor*ite,* ClO^- is *hypo*chlor*ite*

Hydrated Ionic Compounds Ionic compounds called **hydrates** have a specific number of water molecules associated with each formula unit. In their formulas, this number is shown after a centered dot. It is indicated in the systematic name by a Greek numerical prefix before the word *hydrate.* Table 2.6 shows these prefixes. For example, Epsom salt has the formula $MgSO_4 \cdot 7H_2O$ and the name magnesium sulfate *hepta*hydrate. Similarly, the mineral gypsum has the formula $CaSO_4 \cdot 2H_2O$ and the name calcium sulfate *di*hydrate. The water molecules, referred to as "waters of hydration," are part of the hydrate's structure. Heating can remove some or all of them, leading to a different substance. For example, the photos below show blue copper(II) sulfate pentahydrate ($CuSO_4 \cdot 5H_2O$) *(left)* being converted to white copper(II) sulfate ($CuSO_4$) *(right).*

Table 2.6 Numerical Prefixes for Hydrates and Binary Covalent Compounds

Number	Prefix
1	mono-
2	di-
3	tri-
4	tetra-
5	penta-
6	hexa-
7	hepta-
8	octa-
9	nona-
10	deca-

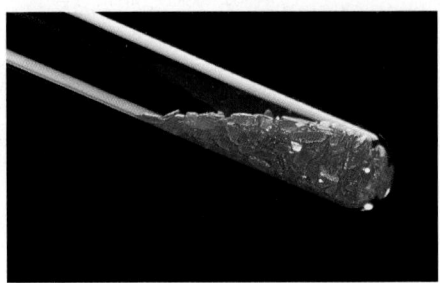

SAMPLE PROBLEM 2.8 Determining Names and Formulas of Ionic Compounds Containing Polyatomic Ions

Problem Give the systematic names for the formulas or the formulas for the names of the following compounds:
(a) $Fe(ClO_4)_2$ (b) Sodium sulfite   (c) $Ba(OH)_2 \cdot 8H_2O$
Solution (a) ClO_4^- is perchlorate; since it has a 1− charge, the cation must be Fe^{2+}. The name is iron(II) perchlorate. (The common name is ferrous perchlorate.)
(b) Sodium is Na^+; sulfite is SO_3^{2-}. Therefore, two Na^+ ions balance one SO_3^{2-} ion. The formula is Na_2SO_3. (c) Ba^{2+} is barium; OH^- is hydroxide. There are eight (octa-) water molecules in each formula unit. The name is barium hydroxide octahydrate.

FOLLOW-UP PROBLEM 2.8 Give the systematic names for the formulas or the formulas for the names of the following compounds:
(a) Cupric nitrate trihydrate   (b) Zinc hydroxide   (c) LiCN

Sometimes it's good practice to correct names or formulas that you know are wrong.

SAMPLE PROBLEM 2.9 Recognizing Incorrect Names and Formulas of Ionic Compounds

Problem Something is wrong with the second part of each statement. Provide the correct name or formula.
(a) $Ba(C_2H_3O_2)_2$ is called barium diacetate.
(b) Sodium sulfide has the formula $(Na)_2SO_3$.
(c) Iron(II) sulfate has the formula $Fe_2(SO_4)_3$.
(d) Cesium carbonate has the formula $Cs_2(CO_3)$.
Solution (a) The charge of the Ba^{2+} ion *must* be balanced by *two* $C_2H_3O_2^-$ ions, so the prefix *di-* is unnecessary. For ionic compounds, we do not indicate the number of ions with numerical prefixes. The correct name is barium acetate.
(b) Two mistakes occur here. The sodium ion is monatomic, so it does *not* require parentheses. The sulfide ion is S^{2-}, *not* SO_3^{2-} (called "sulfite"). The correct formula is Na_2S.
(c) The Roman numeral refers to the charge of the ion, *not* the number of ions in the formula. Fe^{2+} is the cation, so it requires one SO_4^{2-} to balance its charge. The correct formula is $FeSO_4$.
(d) Parentheses are *not* required when only one polyatomic ion of a kind is present. The correct formula is Cs_2CO_3.

FOLLOW-UP PROBLEM 2.9 State why the second part of each statement is incorrect, and correct it:
(a) Ammonium phosphate is $(NH_3)_4PO_4$.
(b) Aluminum hydroxide is $AlOH_3$.
(c) $Mg(HCO_3)_2$ is manganese(II) carbonate.
(d) $Cr(NO_3)_3$ is chromic(III) nitride.
(e) $Ca(NO_2)_2$ is cadmium nitrate.

Naming Acids Acids are an important group of hydrogen-containing compounds that have been used in chemical reactions since before alchemical times. In the laboratory, acids are typically used in water solution. When naming them and writing their formulas, we can consider them as anions connected to the number of hydrogen ions (H^+) needed for charge neutrality. The two common types of acids are binary acids and oxoacids:

1. *Binary acid* solutions form when certain gaseous compounds dissolve in water. For example, when gaseous hydrogen chloride (HCl) dissolves in water, it forms a solution called hydrochloric acid. The name consists of the following parts:

   Prefix *hydro-* + nonmetal *root* + suffix *-ic* + separate word *acid*
         hydro +   chlor   +   ic +   acid

   or hydrochloric acid. This naming pattern holds for many compounds in which hydrogen combines with an anion that has an *-ide* suffix.
2. *Oxoacid* names are similar to those of the oxoanions, except for two suffix changes:
   - *-ate* in the anion becomes *-ic* in the acid
   - *-ite* in the anion becomes *-ous* in the acid

   The oxoanion prefixes *hypo-* and *per-* are kept. Thus,

       BrO_4^- is *per*brom*ate*, and $HBrO_4$ is *per*brom*ic* acid
       IO_2^- is iod*ite*, and HIO_2 is iod*ous* acid

SAMPLE PROBLEM 2.10 Determining Names and Formulas of Anions and Acids

Problem Name the following anions and give the names and formulas of the acids derived from them:
(a) Br^- **(b)** IO_3^- **(c)** CN^- **(d)** SO_4^{2-} **(e)** NO_2^-
Solution **(a)** The anion is bromide; the acid is hydrobromic acid, HBr.
(b) The anion is iodate; the acid is iodic acid, HIO_3.
(c) The anion is cyanide; the acid is hydrocyanic acid, HCN.
(d) The anion is sulfate; the acid is sulfuric acid, H_2SO_4.   (In this case, the suffix is added to the element name *sulfur,* not to the root, *sulf-.*)
(e) The anion is nitrite;   the acid is nitrous acid, HNO_2.

FOLLOW-UP PROBLEM 2.10 Write formulas for the names or names for the formulas of the following acids: **(a)** chloric acid; **(b)** HF; **(c)** acetic acid; **(d)** sulfurous acid; **(e)** HBrO.

Names and Formulas of Binary Covalent Compounds

Binary covalent compounds are formed by the combination of two elements, usually nonmetals. Several are so familiar, such as ammonia (NH_3), methane (CH_4), and water (H_2O), that we use their common names, but most are named in a systematic way:

1. The element with the lower group number in the periodic table is the first word in the name; the element with the higher group number is the second word. [*Important exception:* When the compound contains oxygen and a halogen, Group 7A(17), the halogen is named first.]
2. If both elements are in the same group, the one with the higher period number is named first.
3. The second element is named with its root and the suffix *-ide*.
4. Covalent compounds have Greek numerical prefixes (see Table 2.6) to indicate the number of atoms of each element in the compound. The first word has a prefix *only* when more than one atom of the element is present; the second word *usually* has a numerical prefix.

SAMPLE PROBLEM 2.11 Determining Names and Formulas of Binary Covalent Compounds

Problem **(a)** What is the formula of carbon disulfide?
(b) What is the name of PCl_5?
(c) Give the name and formula of the compound whose molecules each consist of two N atoms and four O atoms.
Solution **(a)** The prefix *di-* means "two." The formula is CS_2.
(b) P is the symbol for phosphorus; there are five chlorine atoms, which is indicated by the prefix *penta-*. The name is phosphorus pentachloride.
(c) Nitrogen (N) comes first in the name (lower group number). The compound is dinitrogen tetraoxide, N_2O_4.

FOLLOW-UP PROBLEM 2.11 Give the name or formula for **(a)** SO_3; **(b)** SiO_2; **(c)** dinitrogen monoxide; **(d)** selenium hexafluoride.

SAMPLE PROBLEM 2.12 Recognizing Incorrect Names and Formulas of Binary Covalent Compounds

Problem Explain what is wrong with the name or formula in the second part of each statement and correct it:

(a) SF_4 is monosulfur pentafluoride.

(b) Dichlorine heptaoxide is Cl_2O_6.

(c) N_2O_3 is dinitrotrioxide.

Solution (a) There are two mistakes. *Mono-* is not needed if there is only one atom of the first element, and the prefix for four is *tetra-*, not *penta-*. The correct name is sulfur tetrafluoride.

(b) The prefix *hepta-* indicates seven, not six. The correct formula is Cl_2O_7.

(c) The full name of the first element is needed, and a space separates the two element names. The correct name is dinitrogen trioxide.

FOLLOW-UP PROBLEM 2.12 Explain what is wrong with the second part of each statement and correct it:

(a) S_2Cl_2 is disulfurous dichloride.

(b) Nitrogen monoxide is N_2O.

(c) $BrCl_3$ is trichlorine bromide.

Naming Alkanes Many organic compounds have complex structural formulas that consist of chains and/or rings of carbon atoms with branches of carbon and other atoms. Hydrocarbons, the simplest type of organic compound, contain *only* carbon and hydrogen. *Alkanes* are the simplest type of hydrocarbon, and the simplest alkanes are called *straight-chain alkanes* because they consist of chains of carbon atoms with no branches. In all alkanes, each carbon atom forms four covalent bonds. Alkanes are highly combustible, and several are important fuels, such as methane, propane, butane, and the mixture of alkanes in gasoline.

Hydrocarbons are binary covalent compounds, but they are not named in the typical way. Alkanes are named with a prefix plus the suffix *-ane*. Table 2.7 gives the names and formulas of the first 10 straight-chain alkanes. Note that only the prefixes for the four smallest are new; those for the larger alkanes are the same as the Greek prefixes in Table 2.6. You'll see organic compounds throughout the book and we'll discuss the naming of more structurally complex organic compounds in Chapter 15.

Table 2.7 The First 10 Straight-Chain Alkanes

Name	Formula
Methane	CH_4
Ethane	C_2H_6
Propane	C_3H_8
Butane	C_4H_{10}
Pentane	C_5H_{12}
Hexane	C_6H_{14}
Heptane	C_7H_{16}
Octane	C_8H_{18}
Nonane	C_9H_{20}
Decane	$C_{10}H_{22}$

Molecular Masses from Chemical Formulas

In Section 2.5, we calculated the atomic mass of an element. Using the formula of a compound to see the number of atoms of each element and the periodic table, we calculate the **molecular mass** (also called *molecular weight*) of a formula unit of the compound as the sum of the atomic masses:

$$\text{Molecular mass} = \text{sum of atomic masses} \qquad (2.3)$$

The molecular mass of a water molecule (using atomic masses to four significant figures from the periodic table) is

$$\text{Molecular mass of } H_2O = (2 \times \text{atomic mass of H}) + (1 \times \text{atomic mass of O})$$
$$= (2 \times 1.008 \text{ amu}) + 16.00 \text{ amu}$$
$$= 18.02 \text{ amu}$$

Ionic compounds are treated the same, but because they do not consist of molecules, we use the term *formula mass* for an ionic compound. Consider barium nitrate, $Ba(NO_3)_2$. To calculate its formula mass, *the number of atoms of each element inside the parentheses is multiplied by the subscript outside the parentheses:*

Formula mass of $Ba(NO_3)_2$

$= (1 \times$ atomic mass of Ba$) + (2 \times$ atomic mass of N$) + (6 \times$ atomic mass of O$)$

$= 137.3$ amu $+ (2 \times 14.01$ amu$) + (6 \times 16.00$ amu$) = 261.3$ amu

Note that atomic, not ionic, masses are used. Although masses of ions differ from those of their atoms by the masses of the electrons, electron loss equals electron gain in the compound, so electron mass is balanced.

SAMPLE PROBLEM 2.13 Calculating the Molecular Mass of a Compound

---

Problem Using the data in the periodic table, calculate the molecular (or formula) mass of the following compounds:

(a) Tetraphosphorus trisulfide   **(b)** Ammonium nitrate

Plan We first write the formula, then multiply the number of atoms (or ions) of each element by its atomic mass, and find the sum.

Solution (a) The formula is P_4S_3.

Molecular mass $= (4 \times$ atomic mass of P$) + (3 \times$ atomic mass of S$)$

$= (4 \times 30.97$ amu$) + (3 \times 32.07$ amu$)$

$=$ 220.09 amu

(b) The formula is NH_4NO_3. We count the total number of N atoms even though they belong to different ions:

Formula mass

$= (2 \times$ atomic mass of N$) + (4 \times$ atomic mass of H$) + (3 \times$ atomic mass of O$)$

$= (2 \times 14.01$ amu$) + (4 \times 1.008$ amu$) + (3 \times 16.00$ amu$)$

$=$ 80.05 amu

Check You can often find large errors by rounding atomic masses to the nearest 5 and adding: **(a)** $(4 \times 30) + (3 \times 30) = 210 \approx 220.09$. The sum has two decimal places because the atomic masses have two. **(b)** $(2 \times 15) + 4 + (3 \times 15) = 79 \approx 80.05$.

FOLLOW-UP PROBLEM 2.13 What is the formula and molecular (or formula) mass of each of the following compounds: **(a)** hydrogen peroxide; **(b)** cesium chloride; **(c)** sulfuric acid; **(d)** potassium sulfate?

The upcoming Gallery shows some of the ways that chemists picture molecules and the enormous range of molecular sizes.

SECTION SUMMARY

Chemical formulas describe the simplest atom ratio (empirical formula), actual atom number (molecular formula), and atom arrangement (structural formula) of one unit of a compound. An ionic compound is named with cation first and anion last. For metals that can form more than one ion, the charge is shown with a Roman numeral. Oxoanions have suffixes, and sometimes prefixes, attached to the element root name to indicate the number of oxygen atoms. Names of hydrates give the number of associated water molecules with numerical prefixes. Acid names are based on anion names. Names of covalent compounds have the element that is leftmost or lower down in the periodic table first, and prefixes show the number of each atom. The molecular (or formula) mass of a compound is the sum of the atomic masses in the formula. Molecules are three-dimensional objects that range in size from H_2 to biological and synthetic macromolecules.

Picturing Molecules

The most exciting thing about learning chemistry is training your mind to imagine a molecular world, one filled with tiny objects of various shapes. Molecules are depicted in a variety of useful ways, as shown below for the water molecule:

Chemical formulas show only the relative numbers of atoms.

Electron-dot and *bond-line formulas* show a bond between atoms as either a pair of dots or a line.

Ball-and-stick models show atoms as spheres and bonds as sticks, with accurate angles and relative sizes, but distances are exaggerated.

Space-filling models are accurately scaled-up versions of molecules, but they do not show bonds.

Electron-density models show the ball-and-stick model within the space-filling shape and color the regions of high (*red*) and low (*blue*) electron charge.

H₂O

H:O:H

H–O–H

Ozone (O₃, 48.00 amu) contributes to smog; natural component of stratosphere that absorbs harmful solar radiation.

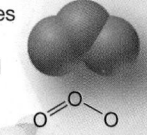

O−O−O

Carbon monoxide (CO, 28.01 amu), toxic component of car exhaust and cigarette smoke.

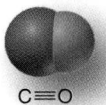

C≡O

All molecules are minute, with their relative sizes depending on composition. A water molecule is small because it consists of only three atoms. Most air pollutants, such as ozone, carbon monoxide, sulfur dioxide, and nitrogen dioxide, also consist of small molecules.

Nitrogen dioxide (NO₂, 46.01 amu) forms from car exhaust and contributes to smog and acid rain.

O−S−O

Sulfur dioxide (SO₂, 64.07 amu) forms from burning coal in power plants; contributes to acid rain.

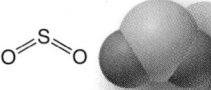

O−N−O

Many household items, such as butane, acetic acid, and aspirin, consist of somewhat larger molecules. The biologically essential molecule heme is larger still.

Butane (C₄H₁₀, 58.12 amu), fuel for cigarette lighters and camping stoves.

H H H H
| | | |
H–C–C–C–C–H
| | | |
H H H H

Acetic acid (CH₃COOH, 60.05 amu), component of vinegar.

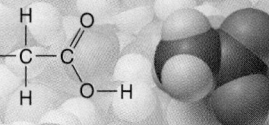

Aspirin (C₉H₈O₄, 180.15 amu), most common pain reliever in the world.

Heme (C₃₄H₃₂FeN₄O₄, 616.49 amu), part of the blood protein hemoglobin, which carries oxygen through the body.

Very large molecules, called macromolecules, can be synthetic, like nylon, or natural, like DNA, and typically consist of thousands of atoms.

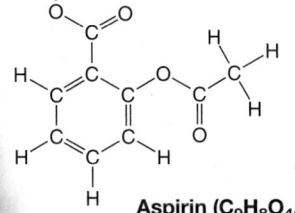

Nylon-66 (~15,000 amu), relatively small, synthetic macromolecule used to make textiles.

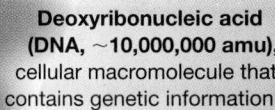

Deoxyribonucleic acid (DNA, ~10,000,000 amu), cellular macromolecule that contains genetic information.

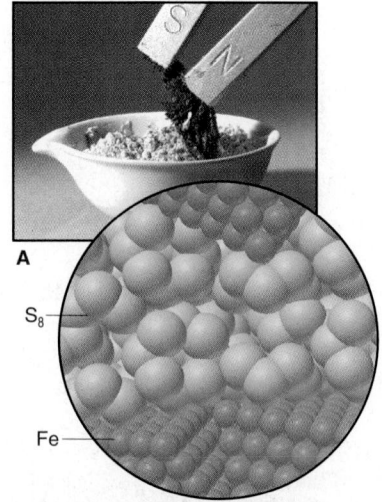

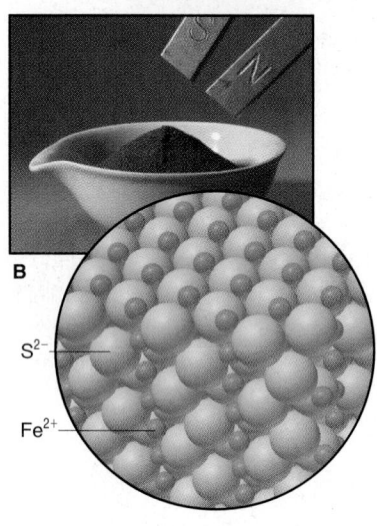

Figure 2.21 The distinction between mixtures and compounds. A, A *mixture* of iron and sulfur can be separated with a magnet because only the iron is magnetic. The blow-up shows separate regions of the two elements. **B,** After strong heating, the *compound* iron(II) sulfide forms, which is no longer magnetic. The blow-up shows the structure of the compound, in which there are no separate regions of the elements.

2.9   MIXTURES: CLASSIFICATION AND SEPARATION

Although we pay a great deal of attention to pure substances, they almost never occur around us. In the natural world, *matter usually occurs as mixtures.* A sample of clean air, for example, consists of many elements and compounds physically mixed together, including oxygen (O_2), nitrogen (N_2), carbon dioxide (CO_2), the six noble gases [Group 8A(18)], and water vapor (H_2O). The oceans are complex mixtures of dissolved ions and covalent substances, including Na^+, Mg^{2+}, Cl^-, SO_4^{2-}, O_2, CO_2, and of course H_2O. Rocks and soils are mixtures of numerous compounds—calcium carbonate ($CaCO_3$), silicon dioxide (SiO_2), aluminum oxide (Al_2O_3), iron(III) oxide (Fe_2O_3)—perhaps a few elements (gold, silver, and carbon in the form of diamond), and petroleum and coal, which are complex mixtures themselves. Living things contain thousands of substances: carbohydrates, lipids, proteins, nucleic acids, and many simpler ionic and covalent compounds.

There are two broad classes of mixtures. A **heterogeneous mixture** has one or more visible boundaries between the components. Thus, its composition is *not* uniform. Many rocks are heterogeneous, showing individual grains and flecks of different minerals. In some cases, as in milk and blood, the boundaries can be seen only with a microscope. A **homogeneous mixture** has no visible boundaries because the components are mixed as individual atoms, ions, and molecules. Thus, its composition *is* uniform. A mixture of sugar dissolved in water is homogeneous, for example, because the sugar molecules and water molecules are uniformly intermingled on the molecular level. We have no way to tell visually whether an object is a substance (element or compound) or a homogeneous mixture.

A homogeneous mixture is also called a **solution.** Although we usually think of solutions as liquid, they can exist in all three physical states. For example, air is a gaseous solution of mostly oxygen and nitrogen molecules, and wax is a solid solution of several fatty substances. Solutions in water, called **aqueous solutions,** are especially important in chemistry and comprise a major portion of the environment and of all organisms.

Recall that mixtures differ fundamentally from compounds in three ways: (1) the proportions of the components can vary; (2) the individual properties of the components are observable; and (3) the components can be separated by physical means. In some cases, if we apply enough energy to the components of the mixture, they react with each other chemically and form a compound, after which their individual properties are no longer observable. Figure 2.21 shows such a case with a mixture of iron and sulfur.

In order to investigate the properties of substances, chemists have devised many procedures for separating a mixture into its component elements and compounds. Indeed, the laws and models of chemistry could never have been formulated without this ability. Many of Dalton's critics, who thought they had found compounds with varying composition, were unknowingly studying mixtures! The upcoming Tools of the Laboratory essay describes some of the more common laboratory separation methods.

SECTION SUMMARY

Heterogeneous mixtures have visible boundaries between the components. Homogeneous mixtures have no visible boundaries because mixing occurs at the molecular level. A solution is a homogeneous mixture and can occur in any physical state. Mixtures (not compounds) can have variable proportions, can be separated physically, and retain their components' properties. Common physical separation processes include filtration, crystallization, extraction, chromatography, and distillation.

Tools of the Laboratory

Basic Separation Techniques

Some of the most challenging and time-consuming laboratory procedures involve separating mixtures and purifying the components. Several common separation techniques are described here. Note that all these methods depend on the *physical properties* of the substances in the mixture; no chemical changes occur.

Filtration separates the components of a mixture on the basis of *differences in particle size*. It is used most often to separate a liquid (smaller particles) from a solid (larger particles). Figure B2.3 shows simple filtration of a solid reaction product. In vacuum filtration, reduced pressure within the flask speeds the flow of the liquid through the filter. Filtration is a key step in the purification of the tap water you drink.

Figure B2.3 Filtration.

Crystallization is based on *differences in solubility*. The *solubility* of a substance is the amount that dissolves in a fixed volume of solvent at a given temperature. The procedure shown in Figure B2.4 applies the fact that many substances are more soluble in hot solvent than in cold. Purified compound is shown crystallizing out of the solution as it is cooled. Essential substances in computer chips and other modern electronic devices are purified by a type of crystallization.

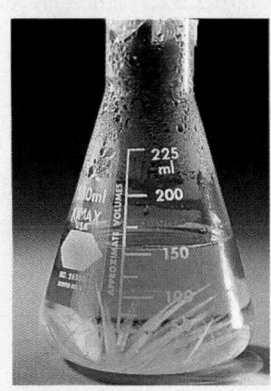

Figure B2.4 Crystallization.

Distillation separates components through *differences in volatility,* the tendency of a substance to become a gas. Ether, for example, is more volatile than water, which is much more volatile than sodium chloride. As the mixture boils, the vapor is richer in the more volatile component, which can be condensed and collected separately. The simple distillation apparatus shown in Figure B2.5 is used to separate components with *large* differences in volatility, such as water from dissolved ionic compounds. Separating components with small volatility differences requires many vaporization-condensation steps (as discussed in Chapter 13).

Extraction is also based on *differences in solubility*. In a typical procedure, a natural (often plant or animal) material is ground in a blender with a solvent that extracts (dissolves) soluble compound(s) embedded in insoluble material. This extract is separated further by the addition of a second solvent that does not dissolve in the first. After shaking in a separatory funnel, some components are extracted into the new solvent. Figure B2.6 shows the extraction of plant pigments from water into hexane, an organic solvent. *(continued)*

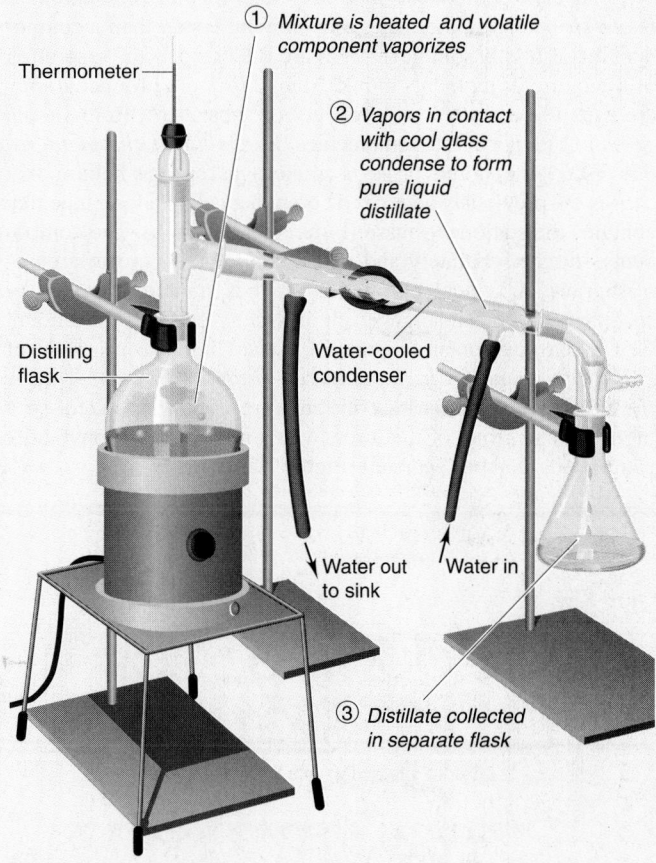

① *Mixture is heated and volatile component vaporizes*

② *Vapors in contact with cool glass condense to form pure liquid distillate*

Thermometer

Distilling flask

Water-cooled condenser

Water out to sink

Water in

③ *Distillate collected in separate flask*

Figure B2.5 Distillation.

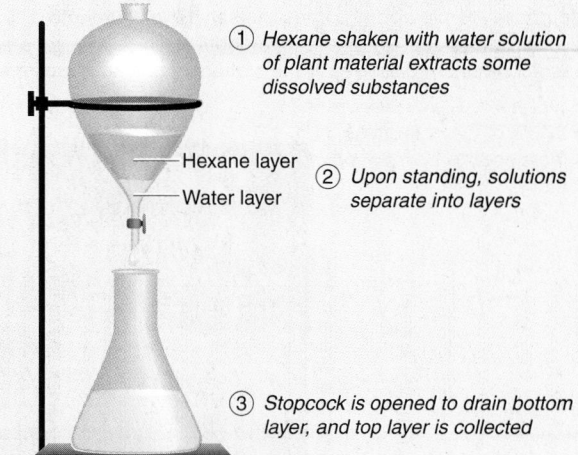

① *Hexane shaken with water solution of plant material extracts some dissolved substances*

Hexane layer

Water layer

② *Upon standing, solutions separate into layers*

③ *Stopcock is opened to drain bottom layer, and top layer is collected*

Figure B2.6 Extraction.

Chromatography is a third technique based on *differences in solubility.* The mixture is dissolved in a gas or liquid called the *mobile phase,* and the components are separated as this phase moves over a solid (or viscous liquid) surface called the *stationary phase.* A component with low solubility in the stationary phase spends less time there, thus moving faster, than a component that is highly soluble in it. Figure B2.7 depicts the separation of a mixture of pigments in ink. Many types of chromatography are used to separate a wide variety of substances, from simple gases to biological macromolecules. In *gas-liquid chromatography (GLC),* the mobile phase is an inert gas, such as helium, that carries the previously vaporized components into a long tube that contains the stationary phase (Figure B2.8, part A). The components emerge separately and reach a detector to create a chromatogram. A typical chromatogram has numerous peaks of specific position and height, each of which represents the amount of a given component (Figure B2.8, part B). The principle of *high-performance (high-pressure) liquid chromatography (HPLC)* is very similar, but the mixture is not vaporized, so a more diverse group of mixtures, which may include nonvolatile compounds, can be separated (Figure B2.9).

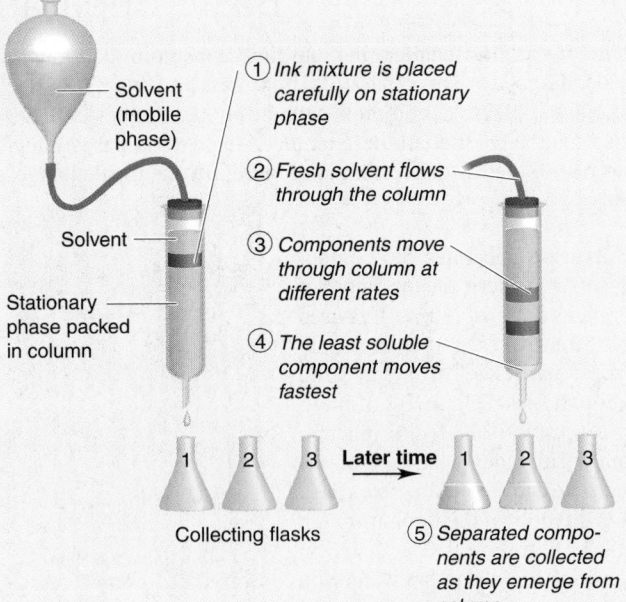

① Ink mixture is placed carefully on stationary phase

② Fresh solvent flows through the column

③ Components move through column at different rates

④ The least soluble component moves fastest

⑤ Separated components are collected as they emerge from column

Figure B2.7 Procedure for column chromatography.

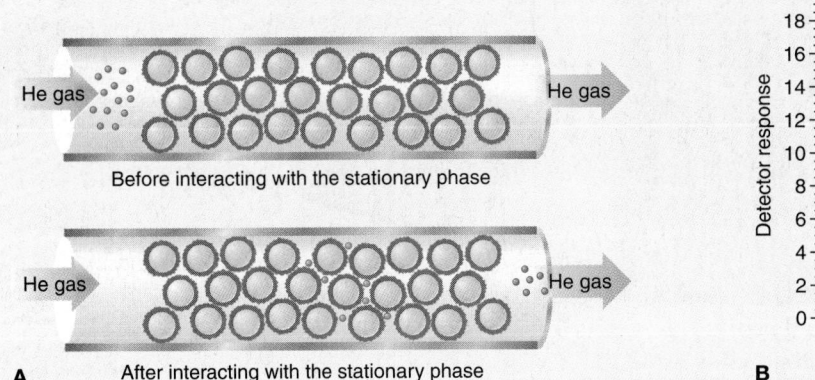

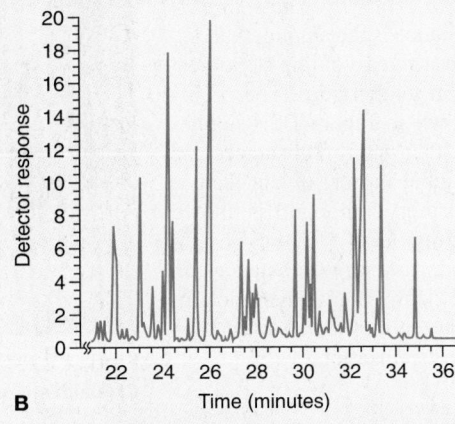

A

B

Figure B2.8 Principle of gas-liquid chromatography (GLC). **A,** The moblile phase *(purple arrow)* carries the sample mixture into a tube packed with the stationary phase *(gray outline on yellow spheres),* and each component dissolves in the stationary phase to a different extent. A component *(red)* that dissolves less readily than another *(blue)* emerges from the tube sooner. **B,** A typical gas-liquid chromatogram of a complex mixture displays each component as a peak.

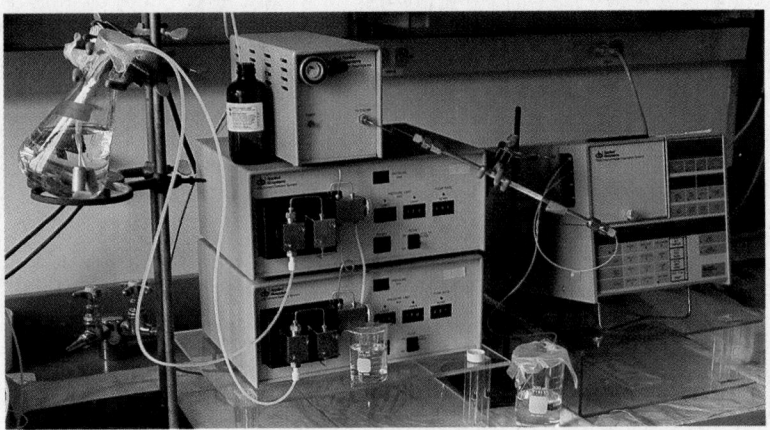

Figure B2.9 A high-performance liquid chromatograph.

Chapter Perspective

An understanding of matter at the observable and atomic levels is the essence of chemistry. In this chapter, you have learned how matter is classified in terms of its composition and how it is named in words and formulas, which are major steps toward that understanding. Figure 2.22 provides a visual review of many key terms and ideas in this chapter. In Chapter 3, we explore one of the central quantitative ideas in chemistry: how the observable amount of a substance relates to the number of atoms, molecules, or ions that make it up.

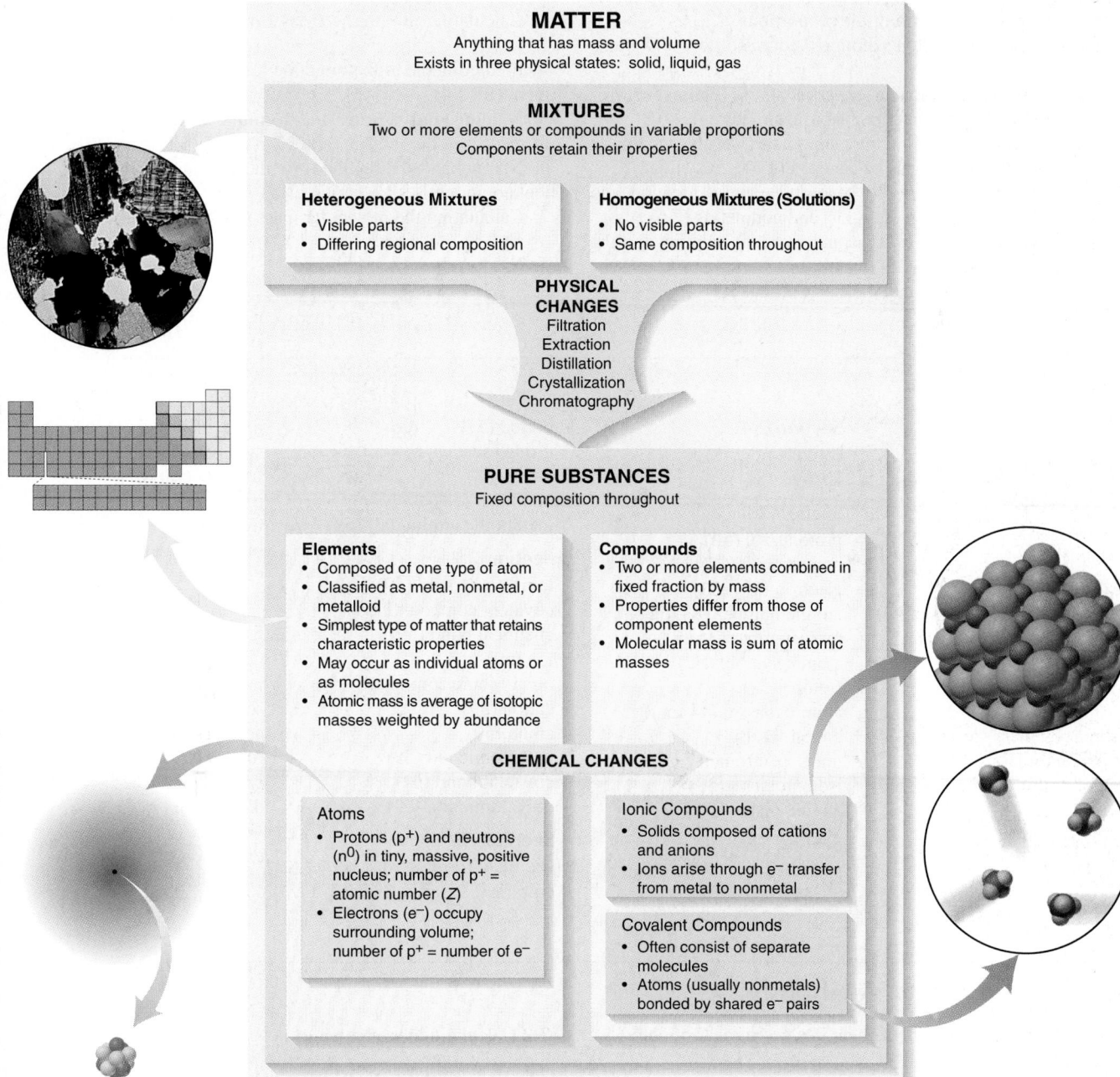

Figure 2.22 **The classification of matter from a chemical point of view.** Mixtures are separated by physical changes into elements and compounds. Chemical changes are required to convert elements into compounds, and vice versa.

For Review and Reference (Numbers in parentheses refer to pages, unless noted otherwise.)

Learning Objectives

Relevant section and/or sample problem (SP) numbers appear in parentheses.

Understand These Concepts

1. The defining characteristics of the three types of matter—element, compound, and mixture—on the macroscopic and atomic levels (2.1)
2. The significance of the three mass laws—mass conservation, definite composition, and multiple proportions (2.2)
3. The postulates of Dalton's atomic theory and how it explains the mass laws (2.3)
4. The major contribution of experiments by Thomson, Millikan, and Rutherford concerning atomic structure (2.4)
5. The structure of the atom, the main features of the subatomic particles, and the importance of isotopes (2.5)
6. The format of the periodic table and general location and characteristics of metals, metalloids, and nonmetals (2.6)
7. The essential features of ionic and covalent bonding and the distinction between them (2.7)
8. The types of mixtures and their properties (2.9)

Master These Skills

1. Using the mass ratio of element to compound to find the mass of an element in a compound (SP 2.1)
2. Using atomic notation to express the subatomic makeup of an isotope (SP 2.2)
3. Calculating an atomic mass from isotopic composition (SP 2.3)
4. Predicting the monatomic ion formed from a main-group element (SP 2.4)
5. Naming and writing the formula of an ionic compound formed from the ions in Tables 2.3 to 2.5 (SP 2.5 to 2.10)
6. Naming and writing the formula of an inorganic binary covalent compound (SP 2.11 and 2.12)
7. Calculating the molecular mass of a compound (SP 2.13)

Key Terms

Section 2.1
element (41)
pure substance (41)
molecule (42)
compound (42)
mixture (42)

Section 2.2
law of mass conservation (43)
law of definite (or constant) composition (44)
fraction by mass (mass fraction) (44)
percent by mass (mass percent, mass %) (44)
law of multiple proportions (45)

Section 2.3
atom (46)

Section 2.4
cathode ray (48)
nucleus (51)

Section 2.5
proton (p^+) (52)
neutron (n^0) (52)
electron (e^-) (52)
atomic number (Z) (52)
mass number (A) (52)
atomic symbol (52)
isotope (52)
atomic mass unit (amu) (53)
dalton (D) (53)
mass spectrometry (53)
isotopic mass (53)
atomic mass (53)

Section 2.6
periodic table of the elements (56)

period (57)
group (57)
metal (58)
nonmetal (58)
metalloid (semimetal) (58)

Section 2.7
ionic compound (59)
covalent compound (59)
chemical bond (59)
ion (59)
binary ionic compound (59)
cation (59)
anion (59)
monatomic ion (59)
covalent bond (62)
polyatomic ion (63)

Section 2.8
chemical formula (64)
empirical formula (64)

molecular formula (64)
structural formula (64)
formula unit (65)
oxoanion (67)
hydrate (68)
binary covalent compound (70)
molecular mass (71)

Section 2.9
heterogeneous mixture (74)
homogeneous mixture (74)
solution (74)
aqueous solution (74)
filtration (75)
crystallization (75)
distillation (75)
volatility (75)
extraction (75)
chromatography (76)

Key Equations and Relationships

2.1 Finding the mass of an element in a given mass of compound (44):

Mass of element in sample
$$= \text{mass of compound in sample} \times \frac{\text{mass of element}}{\text{mass of compound}}$$

2.2 Calculating the number of neutrons in an atom (52):

Number of neutrons = mass number − atomic number

or $N = A - Z$

2.3 Determining the molecular mass of a formula unit of a compound (71):

Molecular mass = sum of atomic masses

Highlighted Figures and Tables

These figures (F) and tables (T) provide a quick review of key ideas. Entries in color contain frequently used data.

F2.1 Elements, compounds, and mixtures on atomic scale (42)
F2.9 General features of the atom (51)
T2.2 Properties of the three key subatomic particles (52)
F2.11 The modern periodic table (57)
F2.14 Factors that influence the strength of ionic bonding (61)
F2.15 The relationship between ions formed and the nearest noble gas (61)
F2.16 Formation of a covalent bond between two H atoms (62)

F2.19 Some common monatomic ions of the elements (64)
T2.3 Common monatomic ions (65)
T2.4 Some metals that form more than one monatomic ion (66)
T2.5 Common polyatomic ions (67)
T2.6 Greek numerical prefixes (68)
F2.22 The classification of matter from a chemical point of view (77)

Brief Solutions to Follow-up Problems

2.1 Mass (t) of pitchblende

$$= 2.3 \text{ t uranium} \times \frac{84.2 \text{ t pitchblende}}{71.4 \text{ t uranium}}$$

$$= 2.7 \text{ t pitchblende}$$

Mass (t) of oxygen

$$= 2.7 \text{ t pitchblende} \times \frac{(84.2 - 71.4 \text{ t oxygen})}{84.2 \text{ t pitchblende}}$$

$$= 0.41 \text{ t oxygen}$$

2.2 (a) Q = B; $5p^+$, $6n^0$, $5e^-$

(b) X = Ca; $20p^+$, $21n^0$, $20e^-$

(c) Y = I; $53p^+$, $78n^0$, $53e^-$

2.3 $10.0129x + [11.0093(1 - x)] = 10.81$; $0.9964x = 0.1993$; $x = 0.2000$ and $1 - x = 0.8000$; % abundance of ^{10}B = 20.00%; % abundance of ^{11}B = 80.00%

2.4 (a) S^{2-}; (b) Rb^+; (c) Ba^{2+}

2.5 (a) Zinc [Group 2B(12)] and oxygen [Group 6A(16)]

(b) Silver [Group 1B(11)] and bromine [Group 7A(17)]

(c) Lithium [Group 1A(1)] and chlorine [Group 7A(17)]

(d) Aluminum [Group 3A(13)] and sulfur [Group 6A(16)]

2.6 (a) ZnO; (b) AgBr; (c) LiCl; (d) Al_2S_3

2.7 (a) PbO_2; (b) copper(I) sulfide (cuprous sulfide); (c) iron(II) bromide (ferrous bromide); (d) $HgCl_2$

2.8 (a) $Cu(NO_3)_2 \cdot 3H_2O$; (b) $Zn(OH)_2$; (c) lithium cyanide

2.9 (a) $(NH_4)_3PO_4$; ammonium is NH_4^+ and phosphate is PO_4^{3-}.

(b) $Al(OH)_3$; parentheses are needed around the polyatomic ion OH^-.

(c) Magnesium hydrogen carbonate; Mg^{2+} is magnesium and can have only a 2+ charge, so it does not need (II); HCO_3^- is hydrogen carbonate (or bicarbonate).

(d) Chromium(III) nitrate; the -*ic* ending is not used with Roman numerals; NO_3^- is nit*rate*.

(e) Calcium nitrite; Ca^{2+} is calcium and NO_2^- is nit*rite*.

2.10 (a) $HClO_3$; (b) hydrofluoric acid; (c) CH_3COOH (or $HC_2H_3O_2$); (d) H_2SO_3; (e) hypobromous acid

2.11 (a) Sulfur trioxide; (b) silicon dioxide; (c) N_2O; (d) SeF_6

2.12 (a) Disulfur dichloride; the -*ous* suffix is not used.

(b) NO; the name indicates one nitrogen.

(c) Bromine trichloride; Br is in a higher period in Group 7A(17), so it is named first.

2.13 (a) H_2O_2, 34.02 amu; (b) CsCl, 168.4 amu; (c) H_2SO_4, 98.09 amu; (d) K_2SO_4, 174.27 amu

Problems

Problems with **colored** numbers are answered at the back of the text. Sections match the text and provide the number(s) of relevant sample problems. Most offer Concept Review Questions, Skill-Building Exercises (in similar pairs), and Problems in Context. Then Comprehensive Problems, based on material from any section or previous chapter, follow.

Elements, Compounds, and Mixtures: An Atomic Overview

● **Concept Review Questions**

2.1 What is the key difference between an element and a compound?

2.2 List two differences between a compound and a mixture.

2.3 Which of the following are pure substances? Explain.

(a) Calcium chloride, used to melt ice on roads, consists of two elements, calcium and chlorine, in fixed mass ratios.

(b) Sulfur, which was known to the ancients, consists entirely of sulfur atoms chemically combined into molecules.

(c) Baking powder, a leavening agent, contains 26% to 30% sodium hydrogen carbonate and 30% to 35% calcium dihydrogen phosphate by mass.

(d) Sand.

2.4 Classify each substance in Problem 2.3 as an element, compound, or mixture, and explain your answers.

2.5 Explain the following statement: The smallest particles unique to an element may be atoms or molecules.

2.6 Explain the following statement: The smallest particles unique to a compound cannot be atoms.

2.7 Can the relative amounts of the components of a mixture vary? Can the relative amounts of the components of a compound vary? Explain.

● **Problems in Context**

2.8 The tap water found in many areas of the United States leaves white deposits when it evaporates. Is this tap water a mixture or a compound? Explain.

2.9 To determine whether a mineral sample is a substance or a mixture, a student grinds a portion of it to a powder, adds 150 mL of pure water, and finds that much of the powder does not dissolve. She carefully removes some of the liquid, allows the water in it to evaporate, and a white, powdery deposit remains. She weighs the deposit and dissolves it completely in a small volume of water. An identical mass of the original, powdered mineral sample does not dissolve in the same volume of water. Was the original sample a substance or a mixture? Explain.

The Observations That Led to an Atomic View of Matter
(Sample Problem 2.1)

● **Concept Review Questions**

2.10 Why was it necessary for separation techniques and methods of chemical analysis to be developed before the laws of definite composition and multiple proportions could be formulated?

2.11 To what classes of matter—element, compound, and/or mixture—do the following apply: (a) law of mass conservation; (b) law of definite composition; (c) law of multiple proportions?

2.12 In our modern view of matter and energy, is the law of mass conservation still relevant to chemical reactions? Explain.

2.13 Identify the mass law that each of the following observations demonstrates, and explain your reasoning:
(a) A sample of potassium chloride from Chile contains the same percent by mass of potassium as one from Poland.
(b) A flashbulb contains magnesium and oxygen before use and magnesium oxide afterward, but its mass does not change.
(c) Arsenic and oxygen form one compound that is 65.2 mass % arsenic and another that is 75.8 mass % arsenic.

2.14 (a) Does the percent by mass of each element in a compound depend on the amount of compound? Explain. (b) Does the mass of each element in a compound depend on the amount of compound? Explain.

2.15 Does the percent by mass of the elements in a compound depend on the amounts of the elements used to make the compound? Explain.

● **Skill-Building Exercises (paired)**

2.16 State the mass law(s) demonstrated by the following experimental results, and explain your reasoning:
Experiment 1: A student heats 1.00 g of a blue compound and obtains 0.64 g of a white compound and 0.36 g of a colorless gas.
Experiment 2: A second student heats 3.25 g of the same blue compound and obtains 2.08 g of a white compound and 1.17 g of a colorless gas.

2.17 State the mass law(s) demonstrated by the following experimental results, and explain your reasoning:
Experiment 1: A student heats 1.27 g of copper and 3.50 g of iodine to produce 3.81 g of a white compound, and 0.96 g of iodine remains.
Experiment 2: A second student heats 2.55 g of copper and 3.50 g of iodine to form 5.25 g of a white compound, and 0.80 g of copper remains.

2.18 Fluorite, a mineral of calcium, is a compound of the metal with fluorine. Analysis shows that a 2.76-g sample of fluorite contains 1.42 g of calcium. Calculate the (a) mass of fluorine in the sample; (b) mass fractions of calcium and fluorine in fluorite; (c) mass percents of calcium and fluorine in fluorite.

2.19 Galena, a mineral of lead, is a compound of the metal with sulfur. Analysis shows that a 2.34-g sample of galena contains 2.03 g of lead. Calculate the (a) mass of sulfur in the sample; (b) mass fractions of lead and sulfur in galena; (c) mass percents of lead and sulfur in galena.

2.20 Magnesium oxide (MgO) forms when the metal burns in air. (a) If 1.25 g of MgO contains 0.754 g of Mg, what is the mass ratio of magnesium to oxide? (b) How many grams of Mg are in 435 g of MgO?

2.21 Zinc sulfide (ZnS) occurs in the zincblende and wurtzite crystal structures. (a) If 2.54 g of ZnS contains 1.70 g of Zn, what is the mass ratio of zinc to sulfide? (b) How many kilograms of Zn are in 2.45 kg of ZnS?

2.22 A compound of copper and sulfur contains 88.39 g of metal and 44.61 g of nonmetal. How many grams of copper are in 5264 kg of compound? How many grams of sulfur?

2.23 A compound of iodine and cesium contains 63.94 g of metal and 61.06 g of nonmetal. How many grams of cesium are in 38.77 g of compound? How many grams of iodine?

2.24 Show, with calculations, how the following data illustrate the law of multiple proportions:
Compound 1: 47.5 mass % sulfur and 52.5 mass % chlorine
Compound 2: 31.1 mass % sulfur and 68.9 mass % chlorine

2.25 Show, with calculations, how the following data illustrate the law of multiple proportions:
Compound 1: 77.6 mass % xenon and 22.4 mass % fluorine
Compound 2: 63.3 mass % xenon and 36.7 mass % fluorine

● **Problems in Context**

2.26 Dolomite is a carbonate of magnesium and calcium. Analysis shows that 7.81 g of dolomite contains 1.70 g of Ca. Calculate the mass percent of Ca in dolomite. On the basis of the mass percent of Ca, and neglecting all other factors, which is the richer source of Ca, dolomite or fluorite (see Problem 2.18)?

2.27 The mass percent of sulfur in a sample of coal is a key factor in the environmental impact of the coal because the sulfur combines with oxygen when the coal is burned and the oxide can then be incorporated into acid rain. Which of the following coals would have the smallest environmental impact?

	Mass (g) of Sample	Mass (g) of Sulfur in Sample
Coal A	378	11.3
Coal B	495	19.0
Coal C	675	20.6

Dalton's Atomic Theory

● **Concept Review Questions**

2.28 Which of Dalton's postulates about atoms are inconsistent with later observations? Do these inconsistencies mean that Dalton was wrong? Is Dalton's model still useful? Explain.

2.29 Use Dalton's theory to explain why potassium nitrate from India or Italy has the same mass percents of K, N, and O.

The Observations That Led to the Nuclear Atom Model

● **Concept Review Questions**

2.30 Thomson was able to determine the mass/charge ratio of the electron but not its mass. How did Millikan's experiment allow determination of the electron's mass?

2.31 The following charges on individual oil droplets were obtained during an experiment similar to Millikan's. Determine a charge for the electron (in C, coulomb), and explain your answer: -3.204×10^{-19} C; -4.806×10^{-19} C; -8.010×10^{-19} C; -1.442×10^{-18} C.

2.32 Describe Thomson's model of the atom. How might it account for the production of cathode rays?

2.33 When Rutherford's coworkers bombarded gold foil with α particles, they obtained results that overturned the existing (Thomson) model of the atom. Explain.

The Atomic Theory Today
(Sample Problems 2.2 and 2.3)

● **Concept Review Questions**

2.34 Define atomic number and mass number. Which can vary without changing the identity of the element?

2.35 Choose the correct answer. The difference between the mass number of an isotope and its atomic number is (a) directly related to the identity of the element; (b) the number of electrons; (c) the number of neutrons; (d) the number of isotopes.

2.36 Even though several elements have only one naturally occurring isotope and all atomic nuclei have whole numbers of protons and neutrons, no atomic mass is a whole number. Use the data from Table 2.2 to explain this fact.

● **Skill-Building Exercises (paired)**

2.37 Argon has three naturally occurring isotopes, ^{36}Ar, ^{38}Ar, and ^{40}Ar. What is the mass number of each? How many protons, neutrons, and electrons are present in each?

2.38 Chlorine has two naturally occurring isotopes, ^{35}Cl and ^{37}Cl. What is the mass number of each isotope? How many protons, neutrons, and electrons are present in each?

2.39 Do both members of the following pairs have the same number of protons? Neutrons? Electrons?
(a) $^{16}_{8}O$ and $^{17}_{8}O$ (b) $^{40}_{18}Ar$ and $^{41}_{19}K$ (c) $^{60}_{27}Co$ and $^{60}_{28}Ni$
Which pair(s) consist(s) of atoms with the same Z value? N value? A value?

2.40 Do both members of the following pairs have the same number of protons? Neutrons? Electrons?
(a) $^{3}_{1}H$ and $^{3}_{2}He$ (b) $^{14}_{6}C$ and $^{15}_{7}N$ (c) $^{19}_{9}F$ and $^{18}_{9}F$
Which pair(s) consist(s) of atoms with the same Z value? N value? A value?

2.41 Write the $^{A}_{Z}X$ notation for each atomic depiction:

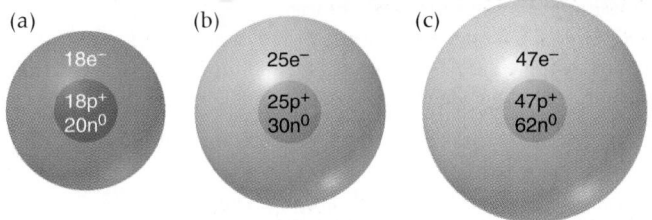

(a) 18e⁻ 18p⁺ 20n⁰
(b) 25e⁻ 25p⁺ 30n⁰
(c) 47e⁻ 47p⁺ 62n⁰

2.42 Write the $^{A}_{Z}X$ notation for each atomic depiction:

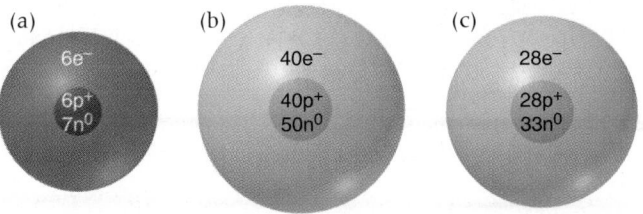

(a) 6e⁻ 6p⁺ 7n⁰
(b) 40e⁻ 40p⁺ 50n⁰
(c) 28e⁻ 28p⁺ 33n⁰

2.43 Draw atomic depictions similar to those in Problem 2.41 for (a) $^{48}_{22}Ti$; (b) $^{79}_{34}Se$; (c) $^{11}_{5}B$.

2.44 Draw atomic depictions similar to those in Problem 2.41 for (a) $^{207}_{82}Pb$; (b) $^{9}_{4}Be$; (c) $^{75}_{33}As$.

2.45 Gallium has two naturally occurring isotopes, ^{69}Ga (isotopic mass 68.9256 amu, abundance 60.11%) and ^{71}Ga (isotopic mass 70.9247 amu, abundance 39.89%). Calculate the atomic mass of gallium.

2.46 Magnesium has three naturally occurring isotopes, ^{24}Mg (isotopic mass 23.9850 amu, abundance 78.99%), ^{25}Mg (isotopic mass 24.9858 amu, abundance 10.00%), and ^{26}Mg (isotopic mass 25.9826 amu, abundance 11.01%). Calculate the atomic mass of magnesium.

2.47 Chlorine has two naturally occurring isotopes, ^{35}Cl (isotopic mass 34.9689 amu) and ^{37}Cl (isotopic mass 36.9659 amu). If chlorine has an atomic mass of 35.4527 amu, what is the percent abundance of each isotope?

2.48 Copper has two naturally occurring isotopes, ^{63}Cu (isotopic mass 62.9396 amu) and ^{65}Cu (isotopic mass 64.9278 amu). If copper has an atomic mass of 63.546 amu, what is the percent abundance of each isotope?

Elements: A First Look at the Periodic Table

● **Concept Review Questions**

2.49 Why is it useful to organize the elements in the form of the periodic table?

2.50 Correct each of the following statements:
(a) In the modern periodic table, the elements are arranged in order of increasing atomic mass.
(b) Elements in a period have similar chemical properties.
(c) Elements can be classified as either metalloids or nonmetals.

2.51 What class of elements lies along the "staircase" line in the periodic table? How do their properties compare with those of metals and nonmetals?

2.52 What are some characteristic properties of elements to the left of the elements along the "staircase"? To the right?

2.53 The elements in Groups 1A(1) and 7A(17) are all quite reactive. What is a major difference between them?

● **Skill-Building Exercises (paired)**

2.54 Give the name, atomic symbol, and group number of the element with the following Z value, and classify it as a metal, metalloid, or nonmetal:
(a) $Z = 32$ (b) $Z = 16$ (c) $Z = 2$ (d) $Z = 3$ (e) $Z = 42$

2.55 Give the name, atomic symbol, and group number of the element with the following Z value, and classify it as a metal, metalloid, or nonmetal:
(a) $Z = 33$ (b) $Z = 20$ (c) $Z = 35$ (d) $Z = 19$ (e) $Z = 12$

2.56 Fill in the blanks:
(a) The symbol and atomic number of the heaviest alkaline earth metal are _____ and _____.
(b) The symbol and atomic number of the lightest metalloid in Group 5A(15) are _____ and _____.
(c) Group 1B(11) consists of the *coinage metals*. The symbol and atomic mass of the coinage metal whose atoms have the fewest electrons are _____ and _____.
(d) The symbol and atomic mass of the halogen in Period 4 are _____ and _____.

2.57 Fill in the blanks:
(a) The symbol and atomic number of the heaviest noble gas are _____ and _____.
(b) The symbol and group number of the Period 5 transition element whose atoms have the fewest protons are _____ and _____.
(c) The elements in Group 6A(16) are sometimes called the *chalcogens*. The symbol and atomic number of the only metallic chalcogen are _____ and _____.
(d) The symbol and number of protons of the Period 4 alkali metal atom are _____ and _____.

Compounds: Introduction to Bonding
(Sample Problem 2.4)

● **Concept Review Questions**

2.58 Describe the type and nature of the bonding that occurs between reactive metals and nonmetals.
2.59 Describe the type and nature of the bonding that often occurs between two nonmetals.
2.60 How can ionic compounds be neutral if they consist of positive and negative ions?
2.61 Given that the ions in LiF and in MgO are of similar size, which compound has stronger ionic bonding? Use Coulomb's law in your explanation.
2.62 Are molecules present in a sample of KBr? Explain.
2.63 Are ions present in a sample of NH_3? Explain.
2.64 The monatomic ions of Groups 1A(1) and 7A(17) are all singly charged. In what major way do they differ? Why?
2.65 Describe the formation of solid magnesium chloride ($MgCl_2$) from large numbers of magnesium and chlorine atoms.
2.66 Describe the formation of solid potassium sulfide (K_2S) from large numbers of potassium and sulfur atoms.
2.67 Does sodium carbonate (Na_2CO_3) incorporate ionic bonding, covalent bonding, or both? Explain.

● **Skill-Building Exercises (paired)**

2.68 What monatomic ions do potassium ($Z = 19$) and iodine ($Z = 53$) form?
2.69 What monatomic ions do barium ($Z = 56$) and selenium ($Z = 34$) form?

2.70 For each ionic depiction, give the name of the parent atom, its mass number, and its group and period numbers:

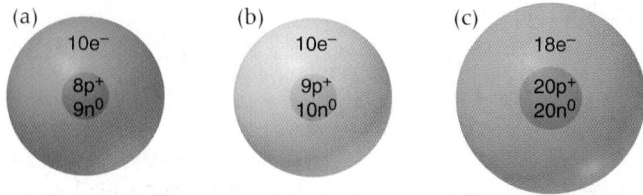

2.71 For each ionic depiction, give the name of the parent atom, its mass number, and its group and period numbers:

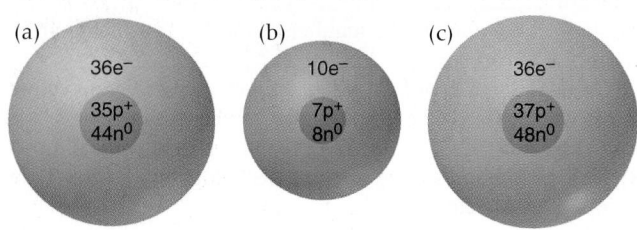

2.72 An ionic compound forms when lithium ($Z = 3$) reacts with oxygen ($Z = 8$). If a sample of the compound contains 5.3×10^{20} lithium ions, how many oxide ions does it contain?
2.73 An ionic compound forms when calcium ($Z = 20$) reacts with iodine ($Z = 53$). If a sample of the compound contains 7.4×10^{21} calcium ions, how many iodide ions does it contain?

2.74 The radii of the sodium and potassium ions are 102 pm and 138 pm, respectively. Which compound has stronger ionic attractions, sodium chloride or potassium chloride?
2.75 The radii of the lithium and magnesium ions are 76 pm and 72 pm, respectively. Which compound has stronger ionic attractions, lithium oxide or magnesium oxide?

Compounds: Formulas, Names, and Masses
(Sample Problems 2.5 to 2.13)

● **Concept Review Questions**

2.76 What is the difference between an empirical formula and a molecular formula? Can they ever be the same?
2.77 How is a structural formula similar to a molecular formula? How is it different?
2.78 Consider a mixture of 10 billion O_2 molecules and 10 billion H_2 molecules. In what way is this mixture similar to a sample containing 10 billion hydrogen peroxide (H_2O_2) molecules? In what way is it different?
2.79 For what type(s) of compound do we use Roman numerals in the name?
2.80 For what type(s) of compound do we use Greek numerical prefixes in the name?
2.81 For what type of compound are we unable to write a molecular formula?

● **Skill-Building Exercises (paired)**

2.82 Write an empirical formula for each of the following:
(a) Hydrazine, a rocket fuel, molecular formula N_2H_4
(b) Glucose, a sugar, molecular formula $C_6H_{12}O_6$
2.83 Write an empirical formula for each of the following:
(a) Ethylene glycol, car antifreeze, molecular formula $C_2H_6O_2$
(b) Peroxodisulfuric acid, a compound used to make bleaching agents, molecular formula $H_2S_2O_8$

2.84 Give the name and formula of the compound formed from the following elements: (a) lithium and nitrogen; (b) oxygen and strontium; (c) aluminum and chlorine.
2.85 Give the name and formula of the compound formed from the following elements: (a) rubidium and bromine; (b) sulfur and barium; (c) calcium and fluorine.

2.86 Give the name and formula of the compound formed from the following elements:
(a) $_{12}L$ and $_9M$ (b) $_{11}L$ and $_{16}M$ (c) $_{17}L$ and $_{38}M$
2.87 Give the name and formula of the compound formed from the following elements:
(a) $_3Q$ and $_{35}R$ (b) $_8Q$ and $_{13}R$ (c) $_{19}Q$ and $_{53}R$

2.88 Give the systematic names for the formulas or the formulas for the names: (a) tin(IV) chloride; (b) $FeBr_3$; (c) cuprous bromide; (d) Mn_2O_3.
2.89 Give the systematic names for the formulas or the formulas for the names: (a) Na_2HPO_4; (b) potassium carbonate dihydrate; (c) $NaNO_2$; (d) ammonium perchlorate.

2.90 Give the systematic names for the formulas or the formulas for the names: (a) CoO; (b) mercury(I) chloride; (c) $Pb(C_2H_3O_2)_2 \cdot 3H_2O$; (d) chromic oxide.

2.91 Give the systematic names for the formulas or the formulas for the names: (a) $Sn(SO_3)_2$; (b) potassium dichromate; (c) $FeCO_3$; (d) copper(II) nitrate.

2.92 Correct each of the following formulas:
(a) Barium oxide is BaO_2.
(b) Iron(II) nitrate is $Fe(NO_3)_3$.
(c) Magnesium sulfide is $MnSO_3$.

2.93 Correct each of the following names:
(a) CuI is cobalt(II) iodide.
(b) $Fe(HSO_4)_3$ is iron(II) sulfate.
(c) $MgCr_2O_7$ is magnesium dichromium heptaoxide.

2.94 Give the name and formula for the acid derived from each of the following anions:
(a) hydrogen sulfate   (b) IO_3^- (c) cyanide   (d) HS^-

2.95 Give the name and formula for the acid derived from each of the following anions:
(a) perchlorate   (b) NO_3^- (c) bromite   (d) F^-

2.96 Write names for the formulas or formulas for the names of the following compounds:
(a) N_2O_5 (b) chlorine trifluoride   (c) SiS_2

2.97 Write names for the formulas or formulas for the names of the following compounds:
(a) N_2O_3 (b) sulfur hexafluoride   (c) ICl_3

2.98 Give the name and formula of the compound whose molecules consist of two sulfur atoms and four fluorine atoms.

2.99 Give the name and formula of the compound whose molecules consist of two iodine atoms and seven oxygen atoms.

2.100 Correct the names of the following compounds:
(a) CO is carbon oxide.
(b) SO_2 is sodium dioxide.
(c) Cl_2O is chlorine dioxide.

2.101 Correct the formulas of the following compounds:
(a) Chlorine monoxide is ClO_1.
(b) Disulfur decafluoride is SF_{10}.
(c) Phosphorus pentafluoride is KF_5.

2.102 Give the number of atoms of the specified element in a formula unit of each of the following compounds, and calculate the molecular (formula) mass:
(a) Oxygen in aluminum sulfate, $Al_2(SO_4)_3$
(b) Hydrogen in ammonium hydrogen phosphate, $(NH_4)_2HPO_4$
(c) Oxygen in the mineral azurite, $Cu_3(OH)_2(CO_3)_2$

2.103 Give the number of atoms of the specified element in a formula unit of each of the following compounds, and calculate the molecular (formula) mass:
(a) Hydrogen in ammonium benzoate, $C_6H_5COONH_4$
(b) Nitrogen in hydrazinium sulfate, $N_2H_6SO_4$
(c) Oxygen atoms in the mineral leadhillite, $Pb_4SO_4(CO_3)_2(OH)_2$

2.104 Write the formula of each compound, and determine its molecular (formula) mass: (a) ammonium sulfate; (b) sodium dihydrogen phosphate; (c) potassium bicarbonate.

2.105 Write the formula of each compound, and determine its molecular (formula) mass: (a) sodium dichromate; (b) ammonium perchlorate; (c) magnesium nitrite trihydrate.

2.106 Calculate the molecular (formula) mass of each compound: (a) dinitrogen pentaoxide; (b) lead(II) nitrate; (c) calcium peroxide.

2.107 Calculate the molecular (formula) mass of each compound: (a) iron(II) acetate tetrahydrate; (b) sulfur tetrachloride; (c) potassium permanganate.

2.108 Give the formula, name, and molecular mass of the following molecules:

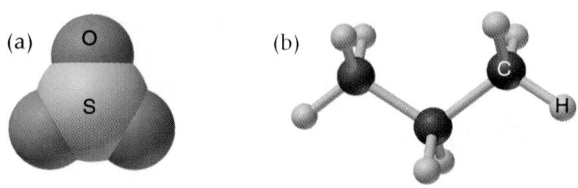

2.109 Give the formula, name, and molecular mass of the following molecules:

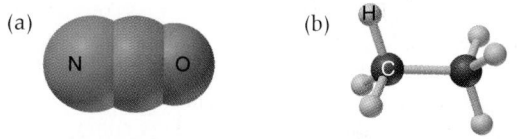

2.110 Give the name, empirical formula, and molecular mass of the molecule depicted in Figure P2.110.

2.111 Give the name, empirical formula, and molecular mass of the molecule depicted in Figure P2.111.

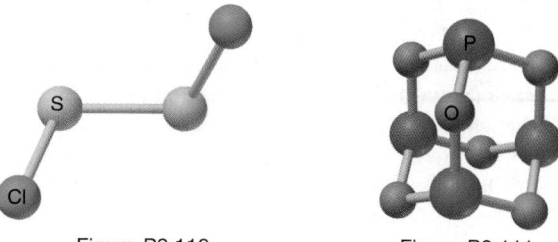

Figure P2.110   Figure P2.111

Mixtures: Classification and Separation

● **Concept Review Questions**

2.112 In what main way is separating the components of a mixture different from separating the components of a compound?

2.113 What is the difference between a homogeneous and a heterogeneous mixture?

2.114 Is a solution a homogeneous or a heterogeneous mixture? Give an example of an aqueous solution.

● **Skill-Building Exercises (paired)**

2.115 Classify each of the following as a compound, a homogeneous mixture, or a heterogeneous mixture: (a) distilled water; (b) gasoline; (c) beach sand; (d) wine; (e) air.

2.116 Classify each of the following as a compound, a homogeneous mixture, or a heterogeneous mixture: (a) orange juice; (b) vegetable soup; (c) cement; (d) calcium sulfate; (e) tea.

2.117 Name the technique(s) and briefly describe the procedure you would use to separate each of the following mixtures into two components: (a) table salt and pepper; (b) table sugar and sand; (c) drinking water contaminated with fuel oil; (d) vegetable oil and vinegar.

2.118 Name the technique(s) and briefly describe the procedure you would use to separate each of the following mixtures into two components: (a) crushed ice and crushed glass; (b) table sugar dissolved in ethanol; (c) iron and sulfur; (d) two pigments (chlorophyll *a* and chlorophyll *b*) from spinach leaves.

● **Problems in Context**

2.119 Which separation method is operating in each of the following procedures?
(a) Pouring a mixture of cooked pasta and boiling water into a colander
(b) Removing colored impurities from raw sugar to make refined sugar
(c) Preparing coffee by pouring hot water through ground coffee beans

2.120 Illicit "street" drugs are often mixed with an inactive substance, such as ascorbic acid (vitamin C). After obtaining a sample of cocaine, government chemists separate the mixture into its component substances. By calculating the mass of vitamin C per gram of drug sample, they can track the drug's distribution. For example, if different samples of cocaine obtained on the streets of New York, Los Angeles, and Paris all contain 0.6384 g of vitamin C per gram of sample, they very likely come from a common source. Is this street sample a compound, element, or mixture? In this case, does the constant mass ratio of the components exemplify the law of definite composition? Explain.

Comprehensive Problems

Problems with an asterisk (*) are more challenging.

2.121 Helium is the lightest noble gas and the second most abundant element (after hydrogen) in the universe.
(a) The radius of a helium atom is 3.1×10^{-11} m, and the radius of its nucleus is 2.5×10^{-15} m. What fraction of the spherical atomic volume is occupied by the nucleus (V of a sphere $= \frac{4}{3}\pi r^3$)?
(b) The mass of a helium-4 atom is 6.64648×10^{-24} g, and each of its two electrons has a mass of 9.10939×10^{-28} g. What fraction of this atom's mass is contributed by its nucleus?

2.122 From the following ions and their radii (in pm), choose a pair that gives the strongest ionic bonding and a pair that gives the weakest: Mg^{2+} 72; K^+ 138; Rb^+ 152; Ba^{2+} 135; Cl^- 181; O^{2-} 140; I^- 220.

2.123 Why is the atomic mass unit very close to the mass of a proton or of a neutron?

2.124 Transition metals, located in the center of the periodic table, have many essential uses as elements and form many important compounds as well. Calculate the molecular mass of the following transition-metal compounds:
(a) $[Co(NH_3)_6]Cl_3$ (b) $[Pt(NH_3)_4BrCl]Cl_2$
(c) $K_4[V(CN)_6]$ (d) $[Ce(NH_3)_6][FeCl_4]_3$

2.125 Boron has two naturally occurring isotopes, ^{10}B (19.9%) and ^{11}B (80.1%). Although the B_2 molecule does not exist naturally on Earth, it has been produced in the laboratory and has been observed in stars. (a) How many different B_2 molecules are possible? (b) What are the masses and percent abundances of each?

2.126 A rock is 5.0% by mass fayalite (Fe_2SiO_4), 7.0% by mass forsterite (Mg_2SiO_4), and the remainder silicon dioxide. What is the mass percent of each element in the rock?

2.127 Scenes A–I depict various types of matter on the atomic scale. Choose the correct scene(s) for each of the following:
(a) A mixture that fills its container
(b) A substance that cannot be broken down into simpler ones
(c) An element with a very high resistance to flow
(d) A homogeneous mixture
(e) An element that fills its container but displays a surface
(f) A gas consisting of diatomic particles
(g) A gas that can be broken down into simpler substances
(h) A substance with a 2:1 number ratio of its component atoms
(i) Matter that can be separated into its component substances by physical means
(j) A heterogeneous mixture
(k) Matter that obeys the law of definite composition

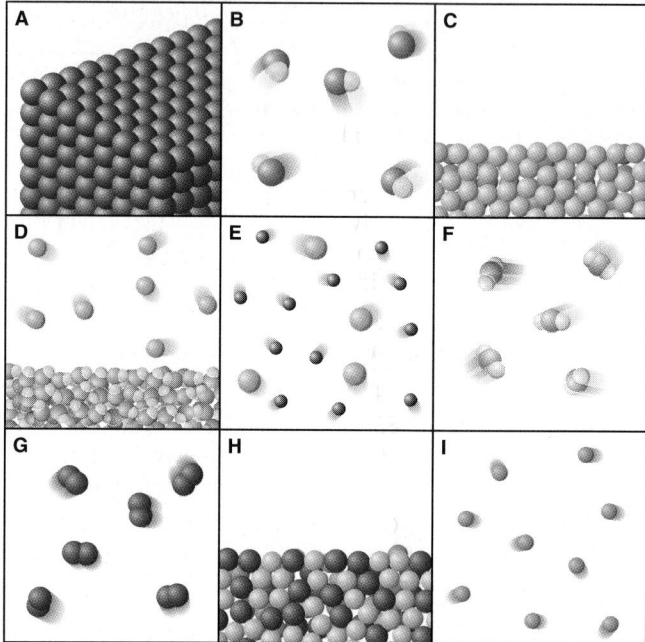

2.128 How can iodine ($Z = 53$) have a higher atomic number yet lower atomic mass than tellurium ($Z = 52$)?

2.129 The number of atoms in 1 dm^3 of aluminum is nearly the same as the number of atoms in 1 dm^3 of lead, but the densities of these metals are very different (see Table 1.5). Explain.

2.130 You are working in the laboratory preparing sodium chloride. Consider the following results for three preparations of the compound:

Case 1: 39.34 g Na + 60.66 g Cl_2 $\longrightarrow$ 100.00 g NaCl
Case 2: 39.34 g Na + 70.00 g Cl_2 $\longrightarrow$
$\qquad\qquad\qquad\qquad$ 100.00 g NaCl + 9.34 g Cl_2
Case 3: 50.00 g Na + 50.00 g Cl_2 $\longrightarrow$
$\qquad\qquad\qquad\qquad$ 82.43 g NaCl + 17.57 g Na

Explain these results in terms of the laws of conservation of mass and definite composition.

2.131 The seven most abundant ions in seawater make up more than 99% by mass of the dissolved compounds. They are listed in units of mg ion/kg seawater: chloride 18,980; sodium 10,560; sulfate 2,650; magnesium 1,270; calcium 400; potassium 380; hydrogen carbonate 140.
(a) What is the mass % of each ion in seawater?
(b) What percent of the total mass of ions is sodium ion?

(c) How does the total mass % of alkaline earth metal ions compare with the total mass % of alkali metal ions?

(d) Which makes up the larger mass fraction of dissolved components, anions or cations?

2.132 Succinic acid *(below)* is an important metabolite in biological energy production. Give the molecular formula, empirical formula, and molecular mass of succinic acid.

2.133 Fluoride ion is poisonous in relatively low amounts: 0.2 g of F^- per 70 kg of body weight can cause death. Nevertheless, in order to prevent tooth decay, F^- ions are added to drinking water at a concentration of 1 mg of F^- ion per L of water. How many liters of fluoridated drinking water would a 70-kg person have to consume in one day to reach this toxic level? How many kilograms of sodium fluoride would be needed to treat a 7.00×10^7 -gal reservoir?

2.134 Chromium is a component of many useful alloys, including stainless steel used in knives and kitchenware. Given the following data, calculate chromium's atomic mass to the appropriate number of significant figures:

Isotope	Natural Abundance (%)	Isotopic Mass (amu)
^{50}Cr	4.35	49.946046
^{52}Cr	83.79	51.940509
^{53}Cr	9.50	52.940651
^{54}Cr	2.36	53.938882

2.135 Nitrous oxide (N_2O) is a potent greenhouse gas. It enters the atmosphere from fertilizer breakdown, car exhaust, and many other sources. Some studies have shown that the isotope ratios of ^{15}N to ^{14}N and of ^{18}O to ^{16}O in N_2O depend on the source. Thus, measuring the relative abundance of molecular masses in a sample of N_2O can help determine the source. (a) What different molecular masses are possible for N_2O? (b) The percent abundance of ^{14}N is 99.6%, and that of ^{16}O is 99.8%. Which molecular mass of N_2O is least common, and which is most common?

2.136 The two isotopes of potassium with significant abundance in nature are ^{39}K (isotopic mass 38.9637 amu, 93.258%) and ^{41}K (isotopic mass 40.9618 amu, 6.730%). Fluorine has only one naturally occurring isotope, ^{19}F (isotopic mass 18.9984 amu). Calculate the formula mass of potassium fluoride.

2.137 What is the empirical formula of each of the following compounds: (a) $C_6H_{12}O_6$; (b) CH_2O; (c) $C_2H_4O_2$; (d) $C_4H_8O_4$? How can different compounds have the same empirical formula?

*2.138 The 19th-century determination of relative atomic masses was frustrated until the Italian physicist Amedeo Avogadro proposed that, at a given set of conditions, equal volumes of a gas contain equal numbers of gas particles. At fixed temperature and pressure, 1 L of hydrogen gas combines with 1 L of chlorine gas to form 2 L of hydrogen chloride gas.

(a) Does chlorine gas exist as single atoms or as diatomic molecules? Explain.

(b) How many chlorine atoms are present for every hydrogen atom in a hydrogen chloride molecule?

(c) Mass analysis of hydrogen chloride shows 0.0284 g of hydrogen for every gram of chlorine. Suggest a relative atomic mass for chlorine.

*2.139 Knowing the atomic mass and the number of atoms of an element in one molecule in a compound, you can find the mass percent of that element in the compound. Esters are organic compounds found in all plants and animals. Some are responsible for the flavors and odors of fruits and flowers. What is the percent by mass of carbon in each of the following esters?

Name	Formula	Odor
Isoamyl isovalerate	$C_4H_9COOC_5H_{11}$	apple
Amyl butyrate	$C_3H_7COOC_5H_{11}$	apricot
Isoamyl acetate	$CH_3COOC_5H_{11}$	banana
Ethyl butyrate	$C_3H_7COOC_2H_5$	pineapple

2.140 The anticancer drug Platinol (Cisplatin), $Pt(NH_3)_2Cl_2$, reacts with the cancer cell's DNA and interferes with its growth. (a) What is the mass % of platinum (Pt) in Platinol? (b) If Pt costs $19/g, how many grams of Platinol can be made for $1.00 million (assume that the cost of Pt determines the cost of the drug)?

2.141 Vanadium (V) is the 19th most abundant element on Earth; however, it is difficult to mine because its sources are widely scattered. The most abundant sources of vanadium are carnotite [$K_2(UO_2)_2(VO_4)_2 \cdot 3H_2O$] and tyuyamunite [$Ca(UO_2)_2(VO_4)_2 \cdot 6H_2O$], both of which are sources of uranium as well. An easily mined source is patronite (VS_4). Other minor sources are vanadinite [$Pb_5(VO_4)_3Cl$] and roscoelite [$KV_2(AlSi_3O_{10})(OH)_2$]. Which mineral source has the most vanadium per gram, and which has the least?

2.142 Dimercaprol ($HSCH_2CHSHCH_2OH$) is a complexing agent developed during World War I as an antidote to arsenic-based poison gas and is used today to treat heavy-metal poisoning. These agents bind and remove the toxic element from the body. (a) If each molecule binds one arsenic (As) atom, how many grams of As could be removed by 250. g of dimercaprol? (b) If one molecule binds one metal atom, calculate the mass % of each of the following metals in a metal-dimercaprol complex: mercury, thallium, chromium.

*2.143 TNT (trinitrotoluene; *below*) is used as a high explosive in construction. Calculate the mass % of each element in TNT.

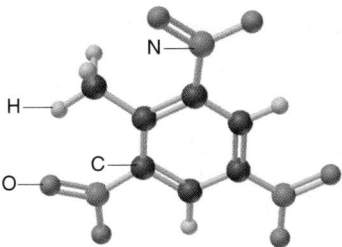

2.144 The scene represents a chemical reaction in a container. Which of the mass laws—mass conservation, definite composition, or multiple proportions—is(are) illustrated?

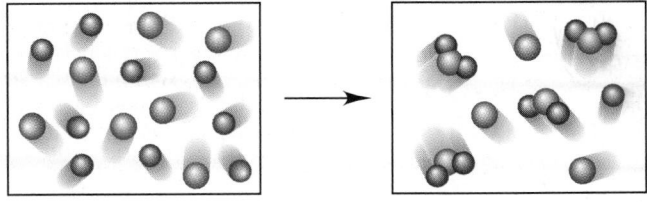

CHAPTER 3

STOICHIOMETRY: MOLE-MASS-NUMBER RELATIONSHIPS IN CHEMICAL SYSTEMS

CHAPTER OUTLINE

3.1 The Mole
Defining the Mole
Molar Mass
Mole-Mass-Number Conversions
Mass Percent

3.2 Determining the Formula of an Unknown Compound
Empirical Formulas
Molecular Formulas
Combustion Analysis
Formulas and Structures

3.3 Writing and Balancing Chemical Equations

3.4 Calculating Amounts of Reactant and Product
Molar Ratios from Balanced Equations
Reaction Sequences
Limiting Reactants
Reaction Yields

3.5 Fundamentals of Solution Stoichiometry
Molarity
Solution Mole-Mass-Number Conversions
Preparation of Molar Solutions
Reactions in Solution

Figure: Counting by weighing. The county fair challenge to guess the correct number of jellybeans in the bowl is easy if you know the weight of one jellybean and can weigh the whole bowl of them. In this chapter, you'll see that chemists take a similar approach with elements and compounds to determine the amounts of substances *and* the number of particles involved in a chemical reaction.

Chemistry is a practical science. Just imagine how useful it could be to determine the formula of a compound from the masses of its elements or to predict the amounts of substances consumed and produced in a reaction. Suppose you are a polymer chemist preparing a new plastic: how much of this new material will a given polymerization reaction yield? Or suppose you're a chemical engineer studying rocket engine thrust: what amount of exhaust gases will a test of this fuel mixture produce? Perhaps you are on a team of environmental chemists examining coal samples: what quantity of air pollutants will this sample release when burned? Or, maybe you're a biomedical researcher who has extracted a new cancer-preventing substance from a tropical plant: what is its formula, and what quantity of metabolic products will establish a safe dosage level? You can answer such questions and countless similar ones with a knowledge of **stoichiometry** (pronounced "stoy-key-AHM-uh-tree"; Greek *stoicheion,* "element or part," + *metron,* "measure"), the study of the quantitative aspects of chemical formulas and reactions.

In this chapter, you'll develop some essential skills of the chemist's trade: relating the mass of a substance to the number of chemical entities (atoms, molecules, or formula units); converting the results of mass analysis into a chemical formula; reading, writing, and thinking in the language of chemical equations; and applying the quantitative information held within them. All these abilities depend on an understanding of the *mole* concept, a central idea in all of chemistry, so that's where we'll begin.

3.1 THE MOLE

In daily life, we typically measure things out by counting or by weighing, with the choice based on convenience. It is more convenient to weigh beans or rice than to count individual pieces, and it is more convenient to count eggs or pencils than to weigh them. To measure out such things, we use mass units (a kilogram of rice) or counting units (a dozen pencils). Similarly, part of daily life in the laboratory involves measuring out substances to prepare a solution or "run" a reaction. However, an obvious problem arises when we try to do this. The atoms, molecules, or formula units are the entities that react with one another, so we would like to know the numbers of them that we mix together. But, how can we possibly count entities that are so small? To do this, chemists have devised a unit called the mole to *count chemical entities by weighing them.*

Defining the Mole

The **mole** (abbreviated **mol**) is the SI unit for amount of substance. It is defined as *the amount of a substance that contains the same number of entities as there are atoms in exactly 12 g of carbon-12.* This number is called **Avogadro's number,** in honor of the 19th-century Italian physicist Amedeo Avogadro, and as you can tell from the definition, it is enormous:

One mole (1 mol) contains 6.022×10^{23} entities (to four significant figures)	(3.1)

Thus,

1 mol of carbon-12	contains	6.022×10^{23} carbon-12 atoms
1 mol of H_2O	contains	6.022×10^{23} H_2O molecules
1 mol of NaCl	contains	6.022×10^{23} NaCl formula units

However, the mole is not just a counting unit, like the dozen, which specifies only the *number* of objects. The definition of the mole specifies the number of objects in a fixed *mass* of substance. Therefore, *1 mole of a substance represents a fixed number of chemical entities* **and** *has a fixed mass.* To see why this

to review before you study this chapter
- isotopes and atomic mass (Section 2.5)
- names and formulas of compounds (Section 2.8)
- molecular mass of a compound (Section 2.8)
- empirical and molecular formulas (Section 2.8)
- mass laws in chemical reactions (Section 2.2)

Imagine a Mole of ... A mole of any ordinary object is a staggering amount: a mole of periods (.) lined up side by side would equal the radius of our galaxy; a mole of marbles stacked tightly together would cover the United States 70 miles deep. However, atoms and molecules are not ordinary objects: a mole of water molecules (about 18 mL) can be swallowed in one gulp!

Figure 3.1 Counting objects of fixed relative mass. A, If marbles had a fixed mass, we could count them by weighing them. Each red marble weighs 7 g, and each yellow marble weighs 4 g, so 84 g of red marbles and 48 g of yellow marbles each consists of 12 marbles. Equal numbers of the two types of marbles always have a 7:4 mass ratio of red:yellow marbles. **B,** Because atoms of a substance have a fixed mass, we can weigh the substance to count the atoms. As shown here, 55.85 g of Fe (left pan) and 32.07 g of S (right pan) each consists of 6.022×10^{23} atoms (1 mol of atoms). Any two samples of iron and sulfur that contain equal numbers of atoms have a 55.85:32.07 mass ratio of iron:sulfur.

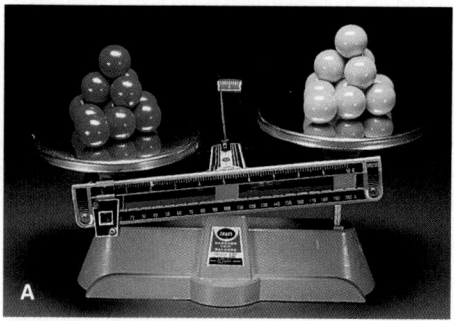

is important, consider the marbles in Figure 3.1A, which we'll use as an analogy for atoms. Suppose you have large groups of red marbles and yellow marbles; each red marble weighs 7 g and each yellow marble weighs 4 g. Right away you know that there are 12 marbles in 84 g of red marbles or in 48 g of yellow marbles. Moreover, because one red marble weighs $\frac{7}{4}$ as much as one yellow marble, any given *number* of red and of yellow marbles always has this 7:4 *mass* ratio. By the same token, any given *mass* of red and of yellow marbles always has a 4:7 *number* ratio. For example, 280 g of red marbles contains 40 marbles, and 280 g of yellow marbles contains 70 marbles. As you can see, the fixed masses of the marbles allow you to count marbles by weighing them.

Atoms have fixed masses also, and the mole gives us a practical way to determine the number of atoms, molecules, or formula units in a sample by weighing it. Let's focus on elements first and recall a key point from Chapter 2: the atomic mass of an element (which appears on the periodic table) is the weighted average of the masses of its naturally occurring isotopes. For purposes of weighing, all atoms of an element are considered to have this atomic mass. That is, all iron (Fe) atoms have an atomic mass of 55.85 amu, all sulfur (S) atoms have an atomic mass of 32.07 amu, and so forth.

The central relationship between the mass of one atom and the mass of 1 mole of those atoms is that *the atomic mass of an element expressed in* **amu** *is numerically the same as the mass of 1 mole of atoms of the element expressed in* **grams.** You can see this from the definition of the mole, which referred to the number of atoms in "**12** g of carbon-**12**." In the same way,

1 Fe atom has a mass of 55.85 amu	and	1 mol of Fe atoms has a mass of 55.85 g
1 S atom has a mass of 32.07 amu	and	1 mol of S atoms has a mass of 32.07 g
1 O atom has a mass of 16.00 amu	and	1 mol of O atoms has a mass of 16.00 g
1 O_2 molecule has a mass of 32.00 amu	and	1 mol of O_2 molecules has a mass of 32.00 g

Moreover, because of their fixed atomic masses, we know that 55.85 g of Fe atoms and 32.07 g of S atoms each contains 6.022×10^{23} atoms. As with marbles of fixed mass, one Fe atom weighs $\frac{55.85}{32.07}$ as much as one S atom, and 1 mol of Fe atoms weighs $\frac{55.85}{32.07}$ as much as 1 mol of S atoms (Figure 3.1B).

A similar relationship holds for compounds: *the molecular mass (or formula mass) of a compound expressed in* **amu** *is numerically the same as the mass of 1 mole of the compound expressed in* **grams.** Thus, for example,

1 molecule of H_2O has a mass of 18.02 amu	and	1 mol of H_2O (6.022×10^{23} molecules) has a mass of 18.02 g
1 formula unit of NaCl has a mass of 58.44 amu	and	1 mol of NaCl (6.022×10^{23} formula units) has a mass of 58.44 g

To summarize the two key points about the usefulness of the mole concept:

- The mole maintains the *same mass relationship* between macroscopic samples as exists between individual chemical entities.
- The mole relates the *number* of chemical entities to the *mass* of a sample of those entities.

A grocer cannot obtain 1 dozen eggs by weighing them, and a printer cannot obtain 1 ream (500 sheets) of paper by weighing it, because eggs and sheets of paper vary in mass. But a chemist *can* obtain 1 mol of copper atoms (6.022×10^{23} atoms) simply by weighing out 63.55 g of copper. Figure 3.2 shows 1 mol of some familiar elements and compounds.

Molar Mass

The **molar mass ($\mathcal{M}$)** of a substance is the mass per mole of its entities (atoms, molecules, or formula units). Thus, molar mass has units of grams per mole (g/mol). Table 3.1 summarizes the meanings of mass units used in this text.

Figure 3.2 One mole of some familiar substances. One mole of a substance is the amount that contains 6.022×10^{23} atoms, molecules, or formula units. From left to right: 1 mol (100.09 g) of calcium carbonate, 1 mol (32.00 g) of gaseous O_2, 1 mol (63.55 g) of copper, and 1 mol (18.02 g) of liquid H_2O.

Table 3.1 Summary of Mass Terminology*

Term	Definition	Unit
Isotopic mass	Mass of an isotope of an element	amu
Atomic mass (also called atomic weight)	Average of the masses of the naturally occurring isotopes of an element weighted according to their abundance	amu
Molecular (or formula) mass (also called molecular weight)	Sum of the atomic masses of the atoms (or ions) in a molecule (or formula unit)	amu
Molar mass ($\mathcal{M}$) (also called gram-molecular weight)	Mass of 1 mole of chemical entities (atoms, ions, molecules, formula units)	g/mol

*All terms based on the ^{12}C standard: 1 atomic mass unit = $\frac{1}{12}$ mass of one ^{12}C atom.

The periodic table is indispensable for calculating the molar mass of a substance. Here's how the calculations are done:

1. **Elements.** You find the molar mass of an element simply by looking up its atomic mass in the periodic table and then noting whether the element occurs naturally as individual atoms or as molecules.

- *Monatomic elements.* For elements that occur as collections of individual atoms, the molar mass is the numerical value from the periodic table expressed in units of grams/mole.* Thus, the molar mass of neon is 20.18 g/mol, the molar mass of iron is 55.85 g/mol, and the molar mass of gold is 197.0 g/mol.
- *Molecular elements.* For elements that occur naturally as molecules, you must know the molecular formula to determine the molar mass. For example, oxygen exists normally in air as diatomic molecules, O_2 (called *dioxygen*), so the molar mass of O_2 molecules is twice that of O atoms:

    Molar mass ($\mathcal{M}$) of O_2 = 2 × $\mathcal{M}$ of O = 2 × 16.00 g/mol = 32.00 g/mol

    The most common form of sulfur consists of octatomic molecules, S_8:

    $\mathcal{M}$ of S_8 = 8 × $\mathcal{M}$ of S = 8 × 32.07 g/mol = 256.6 g/mol

*The mass value in the periodic table is unitless because it is a *relative* atomic mass, given by the atomic mass (in amu) divided by 1 amu ($\frac{1}{12}$ mass of one ^{12}C atom in amu):

$$\text{Relative atomic mass} = \frac{\text{atomic mass (amu)}}{\frac{1}{12} \text{ mass of } ^{12}C \text{ (amu)}}$$

Therefore, you use the same number for the atomic mass (weighted average mass of one atom in amu) and the molar mass (mass of 1 mole of atoms in grams).

Table 3.2 Information Contained in the Chemical Formula of Glucose, $C_6H_{12}O_6$ ($\mathcal{M}$ = 180.16 g/mol)

	Carbon (C)	Hydrogen (H)	Oxygen (O)
Atoms/molecule of compound	6 atoms	12 atoms	6 atoms
Moles of atoms/mole of compound	6 mol of atoms	12 mol of atoms	6 mol of atoms
Atoms/mole of compound	$6(6.022\times10^{23})$ atoms	$12(6.022\times10^{23})$ atoms	$6(6.022\times10^{23})$ atoms
Mass/molecule of compound	6(12.01 amu) = 72.06 amu	12(1.008 amu) = 12.10 amu	6(16.00 amu) = 96.00 amu
Mass/mole of compound	72.06 g	12.10 g	96.00 g

2. **Compounds.** *The molar mass of a compound is the sum of the molar masses of the atoms of the elements in the formula.* For example, the formula of sulfur dioxide (SO_2) tells us that 1 mol of SO_2 molecules contains 1 mol of S atoms and 2 mol of O atoms:

$$\mathcal{M} \text{ of } SO_2 = \mathcal{M} \text{ of } S + (2 \times \mathcal{M} \text{ of } O)$$
$$= 32.07 \text{ g/mol} + (2 \times 16.00 \text{ g/mol})$$
$$= 64.07 \text{ g/mol}$$

Similarly, for ionic compounds, such as potassium sulfide (K_2S), we have

$$\mathcal{M} \text{ of } K_2S = (2 \times \mathcal{M} \text{ of } K) + \mathcal{M} \text{ of } S$$
$$= (2 \times 39.10 \text{ g/mol}) + 32.07 \text{ g/mol}$$
$$= 110.27 \text{ g/mol}$$

A key point to note is that *the subscripts in a formula refer to individual atoms (or ions), as well as to moles of atoms (or ions).* Table 3.2 presents this idea for glucose ($C_6H_{12}O_6$).

Interconverting Moles, Mass, and Number of Chemical Entities

One of the reasons the mole is such a convenient unit for laboratory work is that it allows you to calculate the mass (or the number of chemical entities) of a substance in a sample if you know the amount (number of moles) of the substance. Conversely, if you know the mass (or number of entities) of a substance, you can calculate the number of moles.

The molar mass, which expresses the equivalent relationship between 1 mole of a substance and its mass in grams, can be used as a conversion factor. We multiply by the molar mass of an element or compound ($\mathcal{M}$, in g/mol) to convert a given amount (in moles) to mass (in grams):

$$\text{Mass (g)} = \text{no. of moles} \times \frac{\text{no. of grams}}{1 \text{ mol}} \qquad (3.2)$$

Or, we divide by the molar mass (multiply by $1/\mathcal{M}$) to convert a given mass (in grams) to amount (in moles):

$$\text{No. of moles} = \text{mass (g)} \times \frac{1 \text{ mol}}{\text{no. of grams}} \qquad (3.3)$$

In a similar way, we use Avogadro's number, which expresses the equivalent relationship between 1 mole of a substance and the number of entities it contains, as a conversion factor. We multiply by Avogadro's number to convert amount of substance (in moles) to the number of entities (atoms, molecules, or formula units):

$$\text{No. of entities} = \text{no. of moles} \times \frac{6.022\times10^{23} \text{ entities}}{1 \text{ mol}} \qquad (3.4)$$

Or, we divide by Avogadro's number to do the reverse:

$$\text{No. of moles} = \text{no. of entities} \times \frac{1 \text{ mol}}{6.022\times10^{23} \text{ entities}} \qquad (3.5)$$

Converting Moles of Elements For problems involving mass-mole-number relationships of elements, keep these points in mind:

- To convert between amount (mol) and mass (g), use the molar mass ($\mathcal{M}$ in g/mol).
- To convert between amount (mol) and number of entities, use Avogadro's number (6.022×10^{23} entities/mol). For elements that occur as molecules, use the molecular formula to find atoms/mol.
- Mass and number of entities relate directly to number of moles, *not* to each other. Therefore, to convert between number of entities and mass, *first convert to number of moles*. For example, to find the number of atoms in a given mass,

$$\text{No. of atoms} = \text{mass (g)} \times \frac{1 \text{ mol}}{\text{no. of grams}} \times \frac{6.022 \times 10^{23} \text{ atoms}}{1 \text{ mol}}$$

These relationships are summarized in Figure 3.3A on the next page and demonstrated in Sample Problem 3.1.

SAMPLE PROBLEM 3.1 Calculating the Mass and Number of Atoms in a Given Number of Moles of an Element

Problem (a) Silver (Ag) is used in jewelry and tableware but no longer in U.S. coins. How many grams of Ag are in 0.0342 mol of Ag?
(b) Iron (Fe), the main component of steel, is the most important metal in industrial society. How many Fe atoms are in 95.8 g of Fe?

(a) Determining the mass (g) of Ag
Plan We know the number of moles of Ag (0.0342 mol) and have to find the mass (in g). To convert *moles* of Ag to *grams* of Ag, we multiply by the *molar mass* of Ag, which we find in the periodic table (see the top roadmap).
Solution Converting from moles of Ag to grams:

$$\text{Mass (g) of Ag} = 0.0342 \text{ mol Ag} \times \frac{107.9 \text{ g Ag}}{1 \text{ mol Ag}} = \boxed{3.69 \text{ g Ag}}$$

> Amount (mol) of Ag
>
> multiply by $\mathcal{M}$ of Ag (107.9 g/mol)
>
> Mass (g) of Ag

Check We rounded the mass to three significant figures because the number of moles has three. The units are correct. About 0.03 mol × 100 g/mol gives 3 g; the small mass makes sense because 0.0342 is a small fraction of a mole.

(b) Determining the number of Fe atoms
Plan We know the grams of Fe (95.8 g) and need the number of Fe atoms. We cannot convert directly from grams to atoms, so we first convert to moles by dividing grams of Fe by its molar mass. [This is the reverse of the step in part (a).] Then, we multiply number of moles by Avogadro's number to find number of atoms (see the bottom roadmap).
Solution Converting from grams of Fe to moles:

$$\text{Moles of Fe} = 95.8 \text{ g Fe} \times \frac{1 \text{ mol Fe}}{55.85 \text{ g Fe}} = 1.72 \text{ mol Fe}$$

> Mass (g) of Fe
>
> divide by $\mathcal{M}$ of Fe (55.85 g/mol)
>
> Amount (mol) of Fe
>
> multiply by 6.022×10^{23} atoms/mol
>
> Number of Fe atoms

Converting from moles of Fe to number of atoms:

$$\text{No. of Fe atoms} = 1.72 \text{ mol Fe} \times \frac{6.022 \times 10^{23} \text{ atoms Fe}}{1 \text{ mol Fe}}$$
$$= 10.4 \times 10^{23} \text{ atoms Fe} = \boxed{1.04 \times 10^{24} \text{ atoms Fe}}$$

Check When we approximate the mass of Fe and the molar mass of Fe, we have ~100 g/~50 g/mol = 2 mol. Therefore, the number of atoms should be about twice Avogadro's number: $2(6 \times 10^{23}) = 1.2 \times 10^{24}$.

FOLLOW-UP PROBLEM 3.1 (a) Graphite is the crystalline form of carbon used in "lead" pencils. How many moles of carbon are in 315 mg of graphite?
(b) Manganese (Mn) is a transition element essential for the growth of bones. What is the mass in grams of 3.22×10^{20} Mn atoms, the number found in 1 kg of bone?

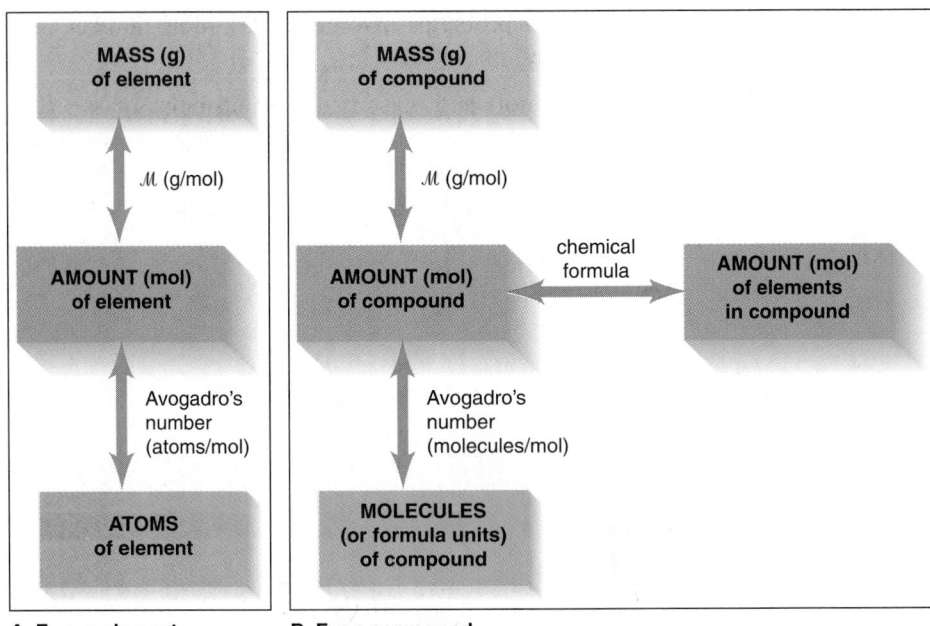

A For an element   **B For a compound**

Figure 3.3 **Summary of the mass-mole-number relationships for elements and compounds.**
A, The amount (in mol) of an element is related to its mass (in g) through the molar mass ($\mathcal{M}$ in g/mol) and to the number of atoms through Avogadro's number (6.022×10^{23} atoms/mol). For elements that occur as molecules, Avogadro's number gives the number of *molecules* per mole.
B, Similarly, moles of a compound are related to grams of the compound through the molar mass ($\mathcal{M}$ in g/mol) and to the number of molecules (or formula units) through Avogadro's number (6.022×10^{23} molecules/mol). To find the number of atoms (molecules or formula units) in a given mass, or vice versa, convert the information to moles first. With the chemical formula, you can calculate mass-mole-number information about each component element.

Converting Moles of Compounds Solving mass-mole-number problems involving compounds requires a very similar approach to the one for elements. We need the chemical formula to find the molar mass and to determine the moles of a given element in the compound. These relationships are shown in Figure 3.3B, and an example is worked through in Sample Problem 3.2.

SAMPLE PROBLEM 3.2 Calculating the Moles and Number of Formula Units in a Given Mass of a Compound

Mass (g) of $(NH_4)_2CO_3$

divide by $\mathcal{M}$ (g/mol)

Amount (mol) of $(NH_4)_2CO_3$

multiply by 6.022×10^{23}
formula units/mol

Number of $(NH_4)_2CO_3$ formula units

Problem Ammonium carbonate is a white solid that decomposes with warming. Among its many uses, it is a component of baking powder, fire extinguishers, and smelling salts. How many formula units are in 41.6 g of ammonium carbonate?
Plan We know the mass of compound (41.6 g) and need to find the number of formula units. As we saw in Sample Problem 3.1(b), to convert grams to number of entities, we have to find number of moles first, so we must divide the grams by the molar mass ($\mathcal{M}$). For this, we need $\mathcal{M}$, so we determine the formula (see Table 2.5) and take the sum of the elements' molar masses. Once we have the number of moles, we multiply by Avogadro's number to find the number of formula units.
Solution The formula is $(NH_4)_2CO_3$. Calculating molar mass:

$$\mathcal{M} = (2 \times \mathcal{M} \text{ of N}) + (8 \times \mathcal{M} \text{ of H}) + (1 \times \mathcal{M} \text{ of C}) + (3 \times \mathcal{M} \text{ of O})$$
$$= (2 \times 14.01 \text{ g/mol}) + (8 \times 1.008 \text{ g/mol}) + 12.01 \text{ g/mol} + (3 \times 16.00 \text{ g/mol})$$
$$= 96.09 \text{ g/mol}$$

Converting from grams to moles:

$$\text{Moles of } (NH_4)_2CO_3 = 41.6 \text{ g } (NH_4)_2CO_3 \times \frac{1 \text{ mol } (NH_4)_2CO_3}{96.09 \text{ g } (NH_4)_2CO_3}$$

$$= 0.433 \text{ mol } (NH_4)_2CO_3$$

Converting from moles to formula units:

$$\text{Formula units of } (NH_4)_2CO_3 = 0.433 \text{ mol } (NH_4)_2CO_3$$

$$\times \frac{6.022 \times 10^{23} \text{ formula units } (NH_4)_2CO_3}{1 \text{ mol } (NH_4)_2CO_3}$$

$$= 2.61 \times 10^{23} \text{ formula units } (NH_4)_2CO_3$$

Check The units are correct. Since the mass is less than half the molar mass ($\sim$42/96 < 0.5), the number of formula units should be less than half Avogadro's number ($\sim 2.6 \times 10^{23}/6.0 \times 10^{23}$ < 0.5).

Comment A *common mistake* is to forget the subscript 2 outside the parentheses in $(NH_4)_2CO_3$, which would give a much lower molar mass.

FOLLOW-UP PROBLEM 3.2 Tetraphosphorus decaoxide reacts with water to form phosphoric acid, a major industrial acid. In the laboratory, the oxide is used as a drying agent.
(a) What is the mass (in g) of 4.65×10^{22} molecules of tetraphosphorus decaoxide?
(b) How many P atoms are present in this sample?

Mass Percent from the Chemical Formula

Each element in a compound constitutes its own particular portion of the compound's mass. For an individual molecule (or formula unit), we use the molecular (or formula) mass and chemical formula to find the mass percent of any element X in the compound:

$$\text{Mass \% of element X} = \frac{\text{atoms of X in formula} \times \text{atomic mass of X (amu)}}{\text{molecular (or formula) mass of compound (amu)}} \times 100$$

Since the formula also tells the number of *moles* of each element in the compound, we use the molar mass to find the mass percent of each element on a mole basis:

$$\text{Mass \% of element X} = \frac{\text{moles of X in formula} \times \text{molar mass of X (g/mol)}}{\text{mass (g) of 1 mol of compound}} \times 100 \quad \textbf{(3.6)}$$

As always, the individual mass percents of the elements in the compound must add up to 100% (within rounding). As Sample Problem 3.3 demonstrates, an important practical use of mass percent is to determine the amount of an element in any size sample of a compound.

SAMPLE PROBLEM 3.3 Calculating Mass Percents and Masses of Elements in a Sample of Compound

Problem Glucose ($C_6H_{12}O_6$) is the most important nutrient in the living cell for generating chemical potential energy.
(a) What is the mass percent of each element in glucose?
(b) How many grams of carbon are in 16.55 g of glucose?

(a) Determining the mass percent of each element
Plan We know the relative number of moles of each element in glucose from the formula (6 C/12 H/6 O). We multiply the number of moles of each element by its molar mass to find grams. Dividing each element's mass by the mass of 1 mol of glucose gives the mass fraction of each element, and multiplying each fraction by 100 gives the mass percent. The calculation steps for any element X are shown in the roadmap on the next page.

Amount (mol) of element X in
1 mol of compound

multiply by $\mathcal{M}$ (g/mol) of X

Mass (g) of X in 1 mol of compound

divide by mass (g) of
1 mol of compound

Mass fraction of X

multiply by 100

Mass % of X

Solution Calculating the mass of 1 mol of $C_6H_{12}O_6$:

$$\mathcal{M} = (6 \times \mathcal{M} \text{ of C}) + (12 \times \mathcal{M} \text{ of H}) + (6 \times \mathcal{M} \text{ of O})$$
$$= (6 \times 12.01 \text{ g/mol}) + (12 \times 1.008 \text{ g/mol}) + (6 \times 16.00 \text{ g/mol})$$
$$= 180.16 \text{ g/mol}$$

Converting moles of C to grams: There are 6 mol C/1 mol glucose, so

$$\text{Mass (g) of C} = 6 \text{ mol C} \times \frac{12.01 \text{ g C}}{1 \text{ mol C}} = 72.06 \text{ g C}$$

Finding the mass fraction of C in glucose:

$$\text{Mass fraction of C} = \frac{\text{total mass C}}{\text{mass of 1 mol glucose}} = \frac{72.06 \text{ g}}{180.16 \text{ g}} = 0.4000 \text{ g C/g glucose}$$

Finding the mass percent of C:

$$\text{Mass \% of C} = \text{mass fraction of C} \times 100 = 0.4000 \times 100 = \boxed{40.00 \text{ mass \% C}}$$

Combining the steps for each of the other two elements:

$$\text{Mass \% of H} = \frac{\text{mol H} \times \mathcal{M} \text{ of H}}{\text{mass of 1 mol glucose}} \times 100 = \frac{12 \text{ mol H} \times \dfrac{1.008 \text{ g H}}{1 \text{ mol H}}}{180.16 \text{ g}} \times 100$$
$$= \boxed{6.714 \text{ mass \% H}}$$

$$\text{Mass \% of O} = \frac{\text{mol O} \times \mathcal{M} \text{ of O}}{\text{mass of 1 mol glucose}} \times 100 = \frac{6 \text{ mol O} \times \dfrac{16.00 \text{ g O}}{1 \text{ mol O}}}{180.16 \text{ g}} \times 100$$
$$= \boxed{53.29 \text{ mass \% O}}$$

Check The answers make sense: the mass % of O is greater than the mass % of C because, even though there are equal numbers of moles of each in the compound, the molar mass of O is greater than the molar mass of C. The mass % of H is small because the molar mass of H is small. The total is 100.00%.

(b) Determining the mass (g) of carbon
Plan To find the mass of C in the glucose sample, we multiply the mass of the sample by the mass fraction of C from part (a).
Solution Finding the mass of C in a given mass of glucose (showing units for mass fraction):

$$\text{Mass (g) of C} = \text{mass of glucose} \times \text{mass fraction of C}$$
$$= 16.55 \text{ g glucose} \times \frac{0.4000 \text{ g C}}{1 \text{ g glucose}}$$
$$= \boxed{6.620 \text{ g C}}$$

Check Rounding shows the answer is in the right "ballpark": 16 g times less than 0.5 parts by mass should be less than 8 g.
Comment 1. A *more direct approach* to finding the mass of element in any mass of compound is similar to the approach we used in Sample Problem 2.1 (p. 45) and eliminates the need to calculate the mass fraction. Just multiply the given mass of compound by the ratio of the total mass of element to the mass of 1 mol of compound:

$$\text{Mass (g) of C} = 16.55 \text{ g glucose} \times \frac{72.06 \text{ g C}}{180.16 \text{ g glucose}} = 6.620 \text{ g C}$$

2. From here on, you should be able to determine the molar mass of a compound, so that calculation will no longer be shown.

FOLLOW-UP PROBLEM 3.3 Ammonium nitrate is used as a fertilizer and to manufacture explosives. Agronomists base the effectiveness of fertilizers on their nitrogen content.
(a) Calculate the mass percent of N in ammonium nitrate.
(b) How many grams of N are in 35.8 kg of ammonium nitrate?

SECTION SUMMARY

A mole of substance is the amount that contains Avogadro's number (6.022×10^{23}) of chemical entities (atoms, molecules, ions, or formula units). The mass (in grams) of a mole has the same numerical value as the mass (in amu) of the entity. Thus, the mole allows us to count entities by weighing them. Using the molar mass ($\mathcal{M}$, g/mol) of an element (or compound) and Avogadro's number as conversion factors, we can convert among amount (mol), mass (g), and number of entities. The mass fraction of element X in a compound is used to calculate the mass of X in any amount of the compound.

3.2 DETERMINING THE FORMULA OF AN UNKNOWN COMPOUND

In Sample Problem 3.3, we knew the chemical formula and used it to find the mass percent (or mass fraction) of an element in a compound *and* the mass of the element in a given mass of compound. In this section, we do the reverse: use the masses of elements in a compound to find its formula.

Empirical Formulas

An analytical chemist investigating a compound decomposes it into simpler substances, finds the mass of each component element, converts these masses to numbers of moles, and then mathematically converts the moles to whole-number subscripts. ● This procedure yields the empirical formula, the *simplest whole-number ratio* of moles of each element in the compound (see Section 2.8, p. 64). Here's an example to show how the subscripts are obtained from the moles of each element.

Analysis of an unknown compound shows that the sample contains 0.21 mol of zinc, 0.14 mol of phosphorus, and 0.56 mol of oxygen. Since the subscripts in a formula represent individual atoms or moles of atoms, we write a preliminary formula that contains fractional subscripts: $Zn_{0.21}P_{0.14}O_{0.56}$. Next, we convert these fractional subscripts to whole numbers using one or two simple arithmetic steps (rounding when needed):

1. Divide each subscript by the smallest subscript:

    $$Zn_{\frac{0.21}{0.14}}P_{\frac{0.14}{0.14}}O_{\frac{0.56}{0.14}} \longrightarrow Zn_{1.5}P_{1.0}O_{4.0}$$

    This step alone often gives integer subscripts.
2. If any of the subscripts are still not integers, multiply through by the *smallest integer* that will turn all the subscripts into integers. Here, we multiply by 2, the smallest integer that will make 1.5 (the subscript for Zn) into an integer:

    $$Zn_{(1.5 \times 2)}P_{(1.0 \times 2)}O_{(4.0 \times 2)} \longrightarrow Zn_{3.0}P_{2.0}O_{8.0}, \text{ or } Zn_3P_2O_8$$

Notice that since we multiplied *all* the subscripts by 2, the *relative* number of moles has not changed. Always check that the subscripts are the smallest set of integers with the same ratio as the original numbers of moles; that is, 3:2:8 are *in the same ratio* as 0.21:0.14:0.56. A more conventional way to write this formula is $Zn_3(PO_4)_2$; the compound is zinc phosphate, a dental cement.

The following three sample problems (3.4, 3.5, and 3.6) demonstrate other common approaches to determining chemical formulas. In the first problem, the empirical formula is found from analytical data given as grams of each element rather than as moles.

● **A Rose by Any Other Name ...**
Chemists studying natural substances from animals and plants isolate compounds and determine their formulas. Geraniol ($C_{10}H_{18}O$) is the main compound that gives a rose its odor. It is used in many perfumes and cosmetics. Geraniol is also found in citronella and lemongrass oils and as part of a larger compound in geranium leaves, from which its name is derived.

SAMPLE PROBLEM 3.4 Determining an Empirical Formula from Masses of Elements

Problem Elemental analysis of a sample of an ionic compound gave the following results: 2.82 g of Na, 4.35 g of Cl, and 7.83 g of O. What is the empirical formula and name of the compound?

Plan This problem is similar to the one we just discussed, except that we are given element *masses,* so we must convert the masses into integer subscripts. We first divide each mass by the element's molar mass to find *number of moles.* Then we construct a preliminary formula and convert the numbers of moles to integers.

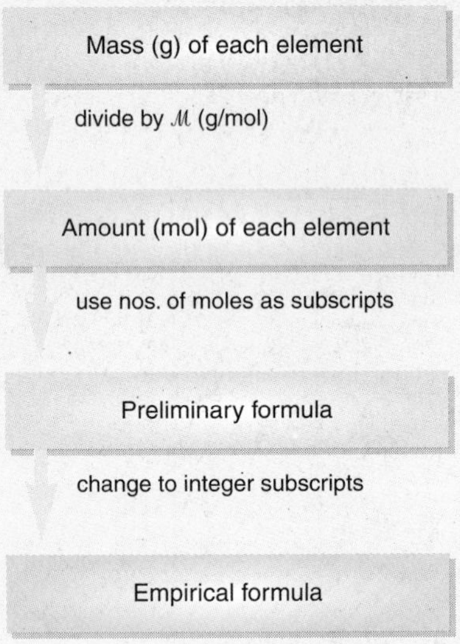

Solution Finding moles of elements:

$$\text{Moles of Na} = 2.82 \text{ g Na} \times \frac{1 \text{ mol Na}}{22.99 \text{ g Na}} = 0.123 \text{ mol Na}$$

$$\text{Moles of Cl} = 4.35 \text{ g Cl} \times \frac{1 \text{ mol Cl}}{35.45 \text{ g Cl}} = 0.123 \text{ mol Cl}$$

$$\text{Moles of O} = 7.83 \text{ g O} \times \frac{1 \text{ mol O}}{16.00 \text{ g O}} = 0.489 \text{ mol O}$$

Constructing a preliminary formula:

$$Na_{0.123}Cl_{0.123}O_{0.489}$$

Converting to integer subscripts (dividing all by the smallest subscript):

$$Na_{\frac{0.123}{0.123}}Cl_{\frac{0.123}{0.123}}O_{\frac{0.489}{0.123}} \longrightarrow Na_{1.00}Cl_{1.00}O_{3.98} \approx Na_1Cl_1O_4, \quad \text{or} \quad NaClO_4$$

We rounded the subscript of O from 3.98 to 4. The empirical formula is $NaClO_4$; the name is sodium perchlorate.

Check The moles seem correct because the masses of Na and Cl are slightly more than 0.1 of their molar masses. The mass of O is greatest and its molar mass is smallest, so it should have the greatest number of moles. The ratio of subscripts, 1:1:4, is the same as the ratio of moles, 0.123:0.123:0.489 (within rounding).

FOLLOW-UP PROBLEM 3.4 An unknown metal M reacts with sulfur to form a compound with formula M_2S_3. If 3.12 g of M reacts with 2.88 g of S, what are the names of M and M_2S_3? (*Hint:* Determine number of moles of S and use the formula to find number of moles of M.)

Sidebar flowchart:

Mass (g) of each element

divide by $\mathcal{M}$ (g/mol)

Amount (mol) of each element

use nos. of moles as subscripts

Preliminary formula

change to integer subscripts

Empirical formula

Molecular Formulas

If we know the molar mass of a compound, we can use the empirical formula to obtain the molecular formula, the *actual* number of moles of each element in 1 mol of compound. In some cases, such as water (H_2O), ammonia (NH_3), and methane (CH_4), the empirical and molecular formulas are identical, but in many others the molecular formula is a *whole-number multiple* of the empirical formula. Hydrogen peroxide, for example, has the empirical formula HO and the molecular formula H_2O_2. Dividing the molar mass of H_2O_2 (34.02 g/mol) by the empirical formula mass (17.01 g/mol) gives the whole-number multiple:

$$\text{Whole-number multiple} = \frac{\text{molar mass (g/mol)}}{\text{empirical formula mass (g/mol)}}$$

$$= \frac{34.02 \text{ g/mol}}{17.01 \text{ g/mol}} = 2.000 = 2$$

Commercial analytical laboratories often provide composition data as mass percent of each element. From this, we can determine the empirical formula by (1) assuming 100.0 g of compound, which allows us to express mass percent directly as mass, (2) converting the mass to number of moles, and (3) constructing the empirical formula. With the molar mass available to find the whole-number multiple, we can then determine the molecular formula.

SAMPLE PROBLEM 3.5 Determining a Molecular Formula from Elemental Analysis and Molar Mass

Problem During physical activity, lactic acid ($\mathcal{M}$ = 90.08 g/mol) forms in muscle tissue and is responsible for muscle soreness. Elemental analysis shows that it contains 40.0 mass % C, 6.71 mass % H, and 53.3 mass % O.
(a) Determine the empirical formula of lactic acid.
(b) Determine the molecular formula.

(a) Determining the empirical formula

Plan We know the mass % of each element and must convert each to an integer subscript. The mass of lactic acid is not given but, since mass % is the same for any mass of compound, we assume 100.0 g of lactic acid and express each mass % directly as grams. Then, we convert grams to moles and construct the empirical formula as we did in Sample Problem 3.4.

Solution Expressing mass % as grams, assuming 100.0 g of lactic acid:

$$\text{Mass (g) of C} = \frac{40.0 \text{ parts C by mass}}{100 \text{ parts by mass}} \times 100.0 \text{ g}$$
$$= 40.0 \text{ g C}$$

Similarly, we have 6.71 g of H and 53.3 g of O.
Converting from grams of each element to moles:

$$\text{Moles of C} = \text{mass of C} \times \frac{1}{\mathcal{M} \text{ of C}} = 40.0 \text{ g C} \times \frac{1 \text{ mol C}}{12.01 \text{ g C}}$$
$$= 3.33 \text{ mol C}$$

Similarly, we have 6.66 mol of H and 3.33 mol of O.
Constructing the preliminary formula: $C_{3.33}H_{6.66}O_{3.33}$
Converting to integer subscripts:

$$C_{\frac{3.33}{3.33}}H_{\frac{6.66}{3.33}}O_{\frac{3.33}{3.33}} \longrightarrow C_{1.00}H_{2.00}O_{1.00} \approx C_1H_2O_1; \text{ the empirical formula is } \boxed{CH_2O}$$

Check The numbers of moles seem correct: the masses of C and O are each slightly more than 3 times their molar masses (e.g., for C, 40 g/12 g/mol > 3 mol), and the mass of H is over 6 times its molar mass.

(b) Determining the molecular formula

Plan The molecular formula subscripts are whole-number multiples of the empirical formula subscripts. To find this whole number, we divide the given molar mass (90.08 g/mol) by the empirical formula mass, which we find from the sum of the elements' molar masses. Then we multiply the whole number by each subscript in the empirical formula.

Solution The empirical-formula molar mass is 30.03 g/mol. Finding the whole-number multiple:

$$\text{Whole-number multiple} = \frac{\mathcal{M} \text{ of lactic acid}}{\mathcal{M} \text{ of empirical formula}} = \frac{90.08 \text{ g/mol}}{30.03 \text{ g/mol}}$$
$$= 3.000 = 3$$

Determining the molecular formula:

$$C_{(1\times3)}H_{(2\times3)}O_{(1\times3)} = \boxed{C_3H_6O_3}$$

Check The calculated molecular formula has the same ratio of moles of elements (3:6:3) as the empirical formula (1:2:1) and corresponds to the given molar mass:

$$\mathcal{M} \text{ of lactic acid} = (3 \times \mathcal{M} \text{ of C}) + (6 \times \mathcal{M} \text{ of H}) + (3 \times \mathcal{M} \text{ of O})$$
$$= (3 \times 12.01) + (6 \times 1.008) + (3 \times 16.00) = 90.08 \text{ g/mol}$$

FOLLOW-UP PROBLEM 3.5 One of the most widespread environmental carcinogens (cancer-causing agents) is benzo[*a*]pyrene ($\mathcal{M}$ = 252.30 g/mol). It is found in coal dust, in cigarette smoke, and even in charcoal-grilled meat. Analysis of this hydrocarbon shows 95.21 mass % C and 4.79 mass % H. What is the molecular formula of benzo[*a*]pyrene?

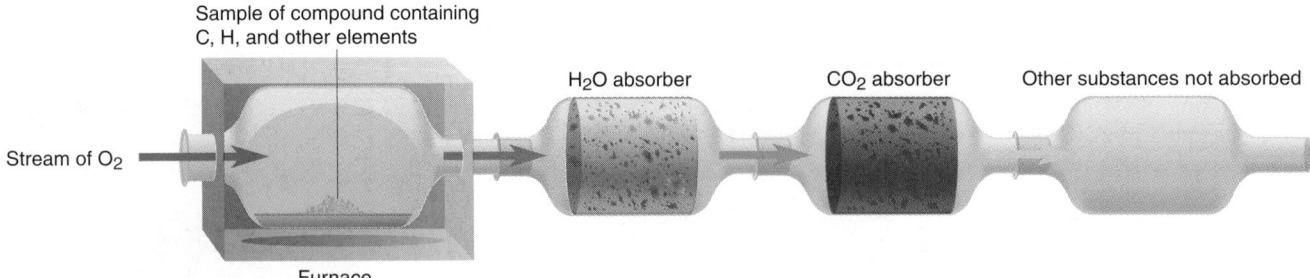

Stream of O_2

Sample of compound containing C, H, and other elements

H_2O absorber

CO_2 absorber

Other substances not absorbed

Furnace

Figure 3.4 Combustion apparatus for determining formulas of organic compounds. A sample of compound that contains C and H (and perhaps other elements) is burned in a stream of O_2 gas. The CO_2 and H_2O formed from the C and H in the sample are absorbed separately, while any other element oxides are carried through by the O_2 gas stream. H_2O is absorbed by $Mg(ClO_4)_2$; CO_2 is absorbed by NaOH on asbestos. The increases in mass of the CO_2 and H_2O absorbers due to the combustion products are used to calculate the amounts (mol) of C and H present in the sample.

Combustion Analysis

A third approach to determining a chemical formula is **combustion analysis,** a method used to measure the amounts of carbon and hydrogen in a combustible organic compound. The unknown is burned in pure O_2 in an apparatus that consists of a combustion chamber and chambers containing compounds that absorb either H_2O or CO_2 (Figure 3.4). All the H in the compound is converted to H_2O, which is absorbed in the first chamber, and all the C is converted to CO_2, which is absorbed in the second. By weighing the contents of the chambers before and after combustion, we find the masses of CO_2 and H_2O and use them to calculate the masses of C and H in the compound, from which we find the empirical formula. Many organic compounds also contain oxygen, nitrogen, or halogen. As long as this third element doesn't interfere with the absorption of CO_2 and H_2O, we can calculate its mass by subtracting the masses of C and H from the compound's original mass.

SAMPLE PROBLEM 3.6 Determining a Molecular Formula from Combustion Analysis

Problem Vitamin C ($\mathcal{M} = 176.12$ g/mol) is a compound of C, H, and O found in many natural sources, especially citrus fruits. When a 1.000-g sample of vitamin C is placed in a combustion chamber and burned, the following data are obtained:

Mass of CO_2 absorber after combustion = 85.35 g
Mass of CO_2 absorber before combustion = 83.85 g
Mass of H_2O absorber after combustion = 37.96 g
Mass of H_2O absorber before combustion = 37.55 g

What is the molecular formula of vitamin C?

Plan We find the masses of CO_2 and H_2O by subtracting the masses of the absorbers before the reaction from the masses after. From the mass of CO_2, we use the mass fraction of C in CO_2 to find the mass of C (see Comment in Sample Problem 3.3). Similarly, we find the mass of H from the mass of H_2O. The mass of vitamin C (1.000 g) minus the sum of the C and H masses gives the mass of O, the third element present. Then, we proceed as in Sample Problem 3.5: calculate numbers of moles using the elements' molar masses, construct the empirical formula, determine the whole-number multiple from the given molar mass, and construct the molecular formula.

Solution Finding the masses of combustion products:

Mass (g) of CO_2 = mass of CO_2 absorber after − mass before
= 85.35 g − 83.85 g
= 1.50 g CO_2
Mass (g) of H_2O = mass of H_2O absorber after − mass before
= 37.96 g − 37.55 g
= 0.41 g H_2O

Calculating masses of C and H using their mass fractions:

$$\text{Mass of element} = \text{mass of compound} \times \frac{\text{mass of element in compound}}{\text{mass of 1 mol of compound}}$$

$$\text{Mass (g) of C} = \text{mass of CO}_2 \times \frac{1 \text{ mol C} \times \mathcal{M} \text{ of C}}{\text{mass of 1 mol CO}_2}$$

$$= 1.50 \text{ g } \cancel{CO_2} \times \frac{12.01 \text{ g C}}{44.01 \text{ g } \cancel{CO_2}} = 0.409 \text{ g C}$$

$$\text{Mass (g) of H} = \text{mass of H}_2\text{O} \times \frac{2 \text{ mol H} \times \mathcal{M} \text{ of H}}{\text{mass of 1 mol H}_2\text{O}}$$

$$= 0.41 \text{ g } \cancel{H_2O} \times \frac{2.016 \text{ g H}}{18.02 \text{ g } \cancel{H_2O}} = 0.046 \text{ g H}$$

Calculating the mass of O:

$$\text{Mass (g) of O} = \text{mass of vitamin C sample} - (\text{mass of C} + \text{mass of H})$$
$$= 1.000 \text{ g} - (0.409 \text{ g} + 0.046 \text{ g}) = 0.545 \text{ g O}$$

Finding the amounts (mol) of elements: Dividing the grams of each element by its molar mass gives 0.0341 mol of C, 0.046 mol of H, and 0.0341 mol of O.

Constructing the preliminary formula: $C_{0.0341}H_{0.046}O_{0.0341}$

Determining the empirical formula: Dividing through by the smallest subscript gives

$$C_{\frac{0.0341}{0.0341}}H_{\frac{0.046}{0.0341}}O_{\frac{0.0341}{0.0341}} = C_{1.00}H_{1.3}O_{1.00}$$

By trial and error, we find that 3 is the smallest integer that will make all subscripts approximately into integers:

$$C_{(1.00\times3)}H_{(1.3\times3)}O_{(1.00\times3)} = C_{3.00}H_{3.9}O_{3.00} \approx C_3H_4O_3$$

Determining the molecular formula:

$$\text{Whole-number multiple} = \frac{\mathcal{M} \text{ of vitamin C}}{\mathcal{M} \text{ of empirical formula}} = \frac{176.12 \text{ g/mol}}{88.06 \text{ g/mol}} = 2.000 = 2$$

$$C_{(3\times2)}H_{(4\times2)}O_{(3\times2)} = \boxed{C_6H_8O_6}$$

Check The element masses seem correct: carbon makes up slightly more than 0.25 of the mass of CO_2 (12 g/44 g > 0.25), as do the masses in the problem (0.409 g/1.50 g > 0.25). Hydrogen makes up slightly more than 0.10 of the mass of H_2O (2 g/18 g > 0.10), as do the masses in the problem (0.046 g/0.41 g > 0.10). The molecular formula has the same ratio of subscripts (6:8:6) as the empirical formula (3:4:3) and adds up to the given molar mass:

$$(6 \times \mathcal{M} \text{ of C}) + (8 \times \mathcal{M} \text{ of H}) + (6 \times \mathcal{M} \text{ of O}) = \mathcal{M} \text{ of vitamin C}$$
$$(6 \times 12.01) + (8 \times 1.008) + (6 \times 16.00) = 176.12 \text{ g/mol}$$

Comment In determining the subscript for H, if we string the calculation steps together, we obtain the subscript 4.0, rather than 3.9, and don't need to round:

$$\text{Subscript of H} = 0.41 \text{ g } H_2O \times \frac{2.016 \text{ g H}}{18.02 \text{ g } H_2O} \times \frac{1 \text{ mol H}}{1.008 \text{ g H}} \times \frac{1}{0.0341 \text{ mol}} \times 3 = 4.0$$

FOLLOW-UP PROBLEM 3.6 A dry-cleaning solvent ($\mathcal{M}$ = 146.99 g/mol) that contains C, H, and Cl is suspected to be a cancer-causing agent. When a 0.250-g sample was studied by combustion analysis, 0.451 g of CO_2 and 0.0617 g of H_2O formed. Calculate the molecular formula.

Chemical Formulas and the Structures of Molecules

Let's take a short break from calculations to recall a key point: *a formula represents a real three-dimensional object.* How much information about the structure of the object is contained in its formula? The empirical formula tells the *relative* number of each type of atom. However, as you can see in Table 3.3 on the next page, *different compounds can have the same **empirical** formula.*

Table 3.3 Some Compounds with Empirical Formula CH$_2$O (Composition by Mass: 40.0% C, 6.71% H, 53.3% O)

Name	Molecular Formula	Whole-Number Multiple	$\mathcal{M}$ (g/mol)	Use or Function
Formaldehyde	CH$_2$O	1	30.03	Disinfectant; biological preservative
Acetic acid	C$_2$H$_4$O$_2$	2	60.05	Acetate polymers; vinegar (5% solution)
Lactic acid	C$_3$H$_6$O$_3$	3	90.08	Causes milk to sour; forms in muscle during exercise
Erythrose	C$_4$H$_8$O$_4$	4	120.10	Forms during sugar metabolism
Ribose	C$_5$H$_{10}$O$_5$	5	150.13	Component of many nucleic acids and vitamin B$_2$
Glucose	C$_6$H$_{12}$O$_6$	6	180.16	Major nutrient for energy in cells

A molecular formula, which tells the *actual* number of each type of atom, provides as much information as it is possible to obtain from mass analysis. Yet *different compounds can also have the same **molecular** formula* because the same types and numbers of atoms can bond to each other in more than one arrangement, that is, in more than one *structural formula*. (Such compounds are called *constitutional, or structural, isomers* of each other, and we'll discuss them fully later in the text.) Only by knowing a molecule's structure—the relative placement of atoms and the distances and angles separating them—can we begin to predict its behavior. For example, as Table 3.4 shows, two very different compounds have the molecular formula C$_2$H$_6$O. Their distinct physical and chemical behaviors result from their different molecular structures.

As molecular complexity (the number and types of atoms) increases, the number of structural formulas that can be written for a given molecular formula also increases: C$_2$H$_6$O has two possible structural formulas, C$_3$H$_8$O three, and C$_4$H$_{10}$O, seven. Imagine how many there are for C$_{16}$H$_{19}$N$_3$O$_4$S! Of all the possible structural formulas for this molecular formula, only one is the antibiotic ampicillin (Figure 3.5). Whenever you write or think about a formula, try to remember that it represents a real object.

Table 3.4 Two Compounds with Molecular Formula C$_2$H$_6$O

Property	Ethanol	Dimethyl Ether
$\mathcal{M}$ (g/mol)	46.07	46.07
Color	Colorless	Colorless
Melting point	$-117°C$	$-138.5°C$
Boiling point	$78.5°C$	$-25°C$
Density (at 20°C)	0.789 g/mL (liquid)	0.00195 g/mL (gas)
Use	Intoxicant in alcoholic beverages	In refrigeration
Structural formula and space-filling model		

SECTION SUMMARY

From the masses of elements in an unknown compound, the relative amounts (in moles) can be found and the empirical formula determined. If the molar mass is known, the molecular formula can also be determined. Methods such as combustion analysis provide data on the masses of elements in a compound, which can be used to obtain the formula. Because atoms can bond in different arrangements, a single molecular formula may correspond to more than one compound.

3.3 WRITING AND BALANCING CHEMICAL EQUATIONS

Perhaps the most important reason for thinking in terms of moles is because it greatly clarifies the amounts of substances taking part in a reaction. Comparing masses doesn't tell the ratio of substances reacting but comparing numbers of moles does. It allows us to view substances as large populations of interacting particles rather than as grams of material. To clarify this idea, consider the formation of hydrogen fluoride gas from H_2 and F_2, a reaction that occurs explosively at room temperature. If we weigh the gases, we find that

2.016 g of H_2 and 38.00 g of F_2 react to form 40.02 g of HF

This information tells us little except that mass is conserved. However, if we convert these masses (in grams) to amounts (in moles), we find that

1 mol of H_2 and 1 mol of F_2 react to form 2 mol of HF

This information reveals that equal-size populations of H_2 and F_2 molecules combine to form twice as large a population of HF molecules. Dividing through by Avogadro's number shows us the chemical event between individual molecules:

1 H_2 molecule and 1 F_2 molecule react to form 2 HF molecules

Figure 3.6 shows that when we express the reaction in terms of moles, *the macroscopic (molar) change corresponds to the submicroscopic (molecular) change.* As you'll see, a balanced chemical equation shows both changes.

A **chemical equation** is a statement in formulas that expresses the identities and quantities of the substances involved in a chemical or physical change. Equations are the "sentences" of chemistry, just as chemical formulas are the "words" and atomic symbols are the "letters." The left side of an equation shows the amount of each substance present before the change, and the right side shows the amounts present afterward. *For an equation to depict these amounts accurately, it must be balanced; that is, the same number of each type of atom must appear*

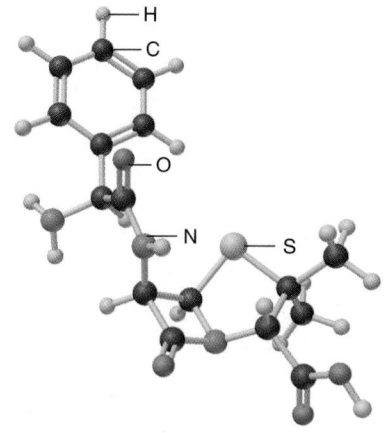

Figure 3.5 Ampicillin. Of the many possible structural formulas corresponding to the molecular formula $C_{16}H_{19}N_3O_4S$, only this particular arrangement of the atoms is the widely used antibiotic ampicillin.

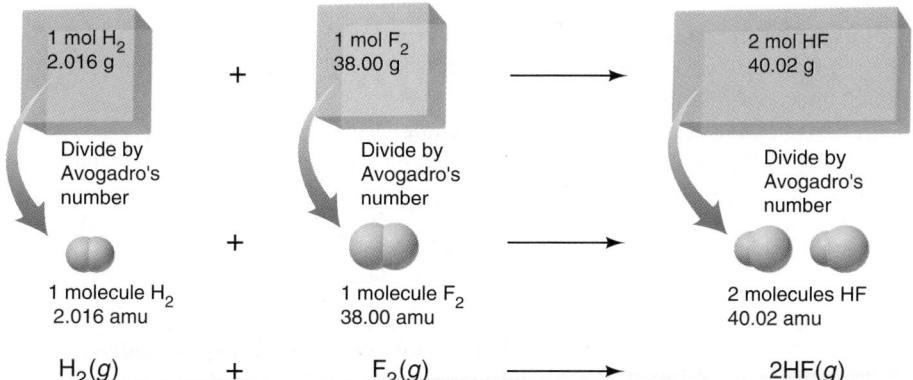

Figure 3.6 The formation of HF gas on the macroscopic and molecular levels. When 1 mol of H_2 (2.016 g) and 1 mol of F_2 (38.00 g) react, 2 mol of HF (40.02 g) forms. Dividing by Avogadro's number shows the change at the molecular level.

on both sides of the equation. This requirement follows directly from the mass laws and the atomic theory:

- In a chemical process, atoms cannot be created, destroyed, or changed, only rearranged into different combinations.
- A formula represents a fixed ratio of the elements in a compound, so a different ratio represents a different compound.

Consider the chemical change that occurs in an old-fashioned photographic flashbulb: magnesium wire and oxygen gas yield powdery magnesium oxide. (Light and heat are produced as well, but here we're concerned only with the substances involved.) Let's convert this chemical statement into a balanced equation through the following steps:

1. *Translating the statement.* We first translate the chemical statement into a "skeleton" equation: chemical formulas arranged in an equation format. All the substances that react during the change, called **reactants,** are placed to the left of a "yield" arrow, which points to all the substances produced, called **products:**

$$\overbrace{\_\_Mg \quad + \quad \_\_O_2}^{\text{reactants}} \quad \xrightarrow{\text{yield}} \quad \overbrace{\_\_MgO}^{\text{product}}$$

$$\underset{\text{magnesium and oxygen}}{} \qquad \underset{\text{yield}}{} \quad \underset{\text{magnesium oxide}}{}$$

At the beginning of the balancing process, we put a blank in front of each substance to remind us that we have to account for its atoms.

2. *Balancing the atoms.* The next step involves shifting your attention back and forth from right to left in order to *match the number of each type of atom on each side.* At the end of this step, each blank will contain a **balancing (stoichiometric) coefficient,** a numerical multiplier of *all the atoms* in the formula that follows it. In general, balancing is easiest when you

- Start with the most complex substance, the one with the largest number of atoms or different types of atoms.
- End with the least complex substance, such as an element by itself.

In this case, MgO is the most complex, so we place a coefficient 1 *in front of* the compound:

$$\_\_Mg + \_\_O_2 \longrightarrow \underline{1}\,MgO$$

To balance the Mg in MgO on the right, we place a 1 in front of Mg on the left:

$$\underline{1}\,Mg + \_\_O_2 \longrightarrow \underline{1}\,MgO$$

The O atom on the right must be balanced by one O atom on the left. One-half an O_2 molecule provides one O atom:

$$\underline{1}\,Mg + \tfrac{1}{2}\,O_2 \longrightarrow \underline{1}\,MgO$$

In terms of number and type of atom, the equation is balanced.

3. *Adjusting the coefficients.* There are several conventions about the final form of the coefficients:

- In most cases, *the smallest whole-number coefficients are preferred.* Whole numbers allow entities such as O_2 molecules to be treated as intact particles. One-half of an O_2 molecule cannot exist, so we multiply the equation by 2:

$$2Mg + 1O_2 \longrightarrow 2MgO$$

- We used the coefficient 1 to remind us to balance each substance. In the final form, a coefficient of 1 is implied by the formula of the substance, so we don't need to write it:

$$2Mg + O_2 \longrightarrow 2MgO$$

(This convention is similar to that of not writing a subscript 1 in a formula.)

4. *Checking*. After balancing and adjusting the coefficients, always check that the equation is balanced:

$$\text{Reactants (2 Mg, 2 O)} \longrightarrow \text{products (2 Mg, 2 O)}$$

5. *Specifying the states of matter*. The final equation also indicates the physical state of each substance or whether it is dissolved in water. The abbreviations used for these states are solid (*s*), liquid (*l*), gas (*g*), and aqueous solution (*aq*). From the original statement, Mg "wire" is solid, O_2 is a gas, and "powdery" MgO is also solid. The balanced equation is

$$2Mg(s) + O_2(g) \longrightarrow 2MgO(s)$$

Of course, the key point to realize is, as was pointed out in Figure 3.6, *the balancing coefficients refer to both individual chemical entities and moles of chemical entities*. Thus, 2 mol of Mg and 1 mol of O_2 yield 2 mol of MgO. Figure 3.7 shows this reaction from three points of view—as you see it on the macroscopic level, as chemists (and you!) can imagine it on the atomic level (darker colored atoms represent the stoichiometry), and on the symbolic level of the chemical equation.

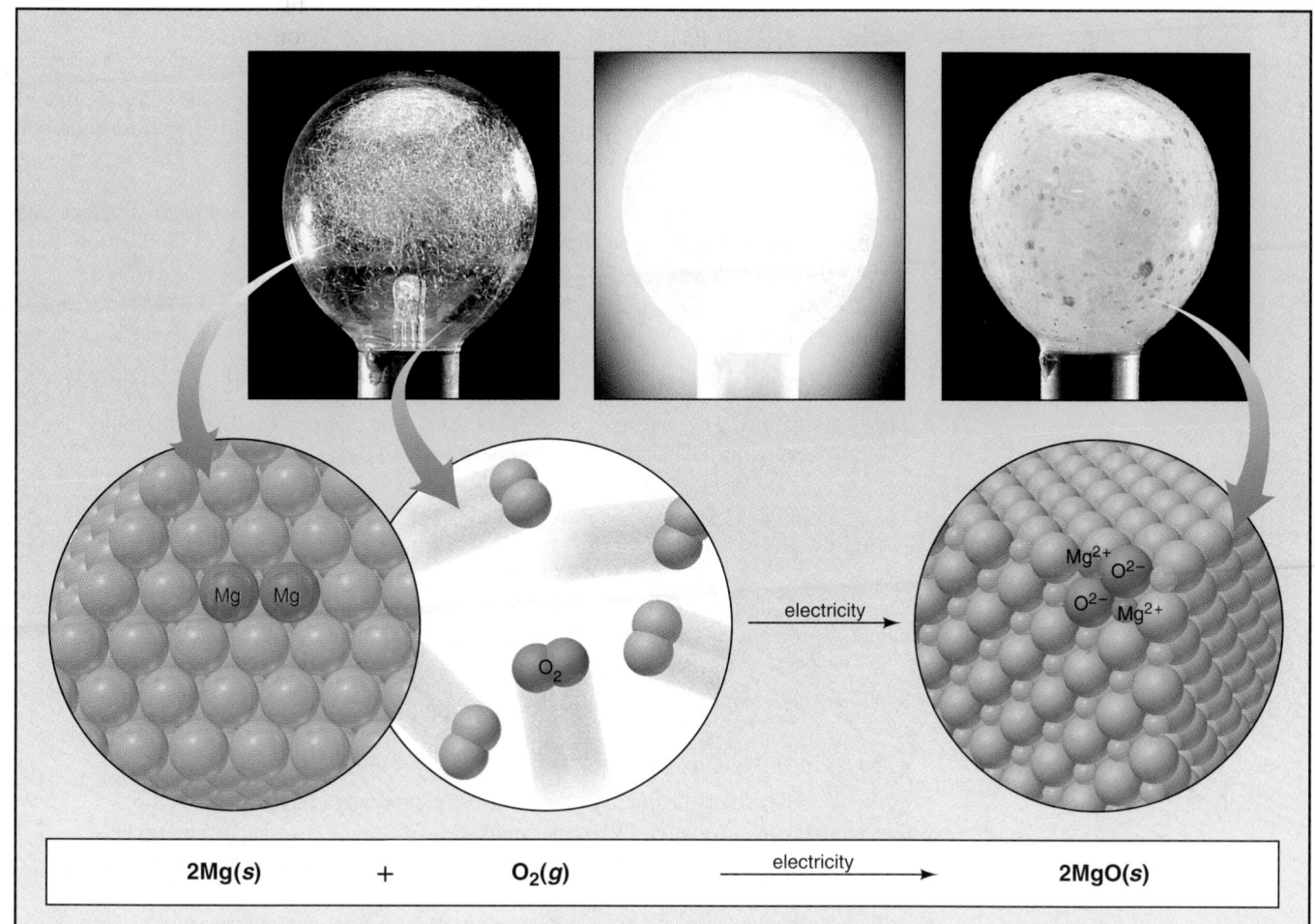

Figure 3.7 A three-level view of the chemical reaction in a flashbulb. The photos present the macroscopic view that you see. Before the reaction occurs, a fine magnesium filament is surrounded by oxygen (*left*). After the reaction, white, powdery magnesium oxide coats the bulb's inner surface (*right*). The blow-up arrows lead to an atomic-scale view, a representation of the chemist's mental picture of the reaction. The darker colored spheres show the stoichiometry. By knowing the substances before and after a reaction, we can write a balanced equation (*bottom*), the chemist's symbolic shorthand for the change.

Keep in mind these other key points about the balancing process:

- A coefficient operates on *all the atoms in the formula* that follows it: 2MgO means 2 × (MgO), or 2 Mg atoms and 2 O atoms; $2Ca(NO_3)_2$ means 2 × $[Ca(NO_3)_2]$, or 2 Ca atoms, 4 N atoms, and 12 O atoms.
- In balancing an equation, *chemical formulas cannot be altered*. In step 2 of the example, we *cannot* balance the O atoms by changing MgO to MgO_2 because MgO_2 has a different elemental composition and thus is a different compound.
- We *cannot add other reactants or products* to balance the equation because this would represent a different chemical reaction. For example, we *cannot* balance the O atoms by changing O_2 to O or by adding one O atom to the products, because the chemical statement does not say that the reaction involves O atoms.
- A balanced equation remains balanced even if you multiply all the coefficients by the same number. For example,

$$4Mg(s) + 2O_2(g) \longrightarrow 4MgO(s)$$

is also balanced: it is just the original balanced equation multiplied by 2. But, we balance an equation with the *smallest* whole-number coefficients.

SAMPLE PROBLEM 3.7 Balancing Chemical Equations

Problem Within the cylinders of a car's engine, the hydrocarbon octane (C_8H_{18}), one of many components of gasoline, mixes with oxygen from the air and burns to form carbon dioxide and water vapor. Write a balanced equation for this reaction.

Solution

1. *Translate* the statement into a skeleton equation (with coefficient blanks). Octane and oxygen are reactants; "oxygen from the air" implies molecular oxygen, O_2. Carbon dioxide and water vapor are products:

$$\underline{}C_8H_{18} + \underline{}O_2 \longrightarrow \underline{}CO_2 + \underline{}H_2O$$

2. *Balance the atoms.* We start with the most complex substance, C_8H_{18}, and balance O_2 last:

$$\underline{1}\,C_8H_{18} + \underline{}O_2 \longrightarrow \underline{}CO_2 + \underline{}H_2O$$

The C atoms in C_8H_{18} end up in CO_2. Each CO_2 contains one C atom, so 8 molecules of CO_2 are needed to balance the 8 C atoms in each C_8H_{18}:

$$\underline{1}\,C_8H_{18} + \underline{}O_2 \longrightarrow \underline{8}\,CO_2 + \underline{}H_2O$$

The H atoms in C_8H_{18} end up in H_2O. The 18 H atoms in C_8H_{18} require a coefficient 9 in front of H_2O:

$$\underline{1}\,C_8H_{18} + \underline{}O_2 \longrightarrow \underline{8}\,CO_2 + \underline{9}\,H_2O$$

There are 25 atoms of O on the right (16 in $8CO_2$ plus 9 in $9H_2O$), so we place the coefficient $\frac{25}{2}$ in front of O_2:

$$\underline{1}\,C_8H_{18} + \frac{25}{2}O_2 \longrightarrow \underline{8}\,CO_2 + \underline{9}\,H_2O$$

3. *Adjust the coefficients.* Multiply through by 2 to obtain whole numbers:

$$2C_8H_{18} + 25O_2 \longrightarrow 16CO_2 + 18H_2O$$

4. *Check* that the equation is balanced:

$$\text{Reactants (16 C, 36 H, 50 O)} \longrightarrow \text{products (16 C, 36 H, 50 O)}$$

5. *Specify* states of matter. C_8H_{18} is liquid; O_2, CO_2, and H_2O vapor are gases:

$$2C_8H_{18}(l) \quad + \quad 25O_2(g) \quad \longrightarrow \quad 16CO_2(g) \quad + \quad 18H_2O(g)$$

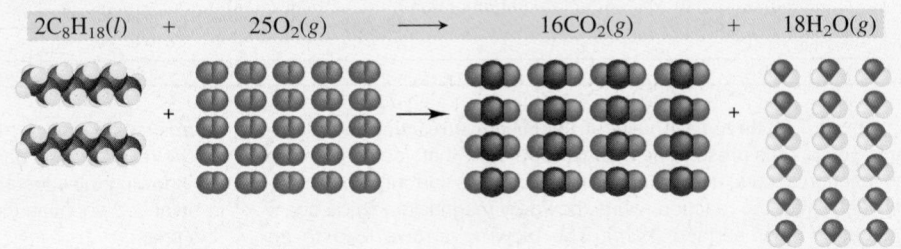

Comment This is an example of a combustion reaction. *Any* compound containing C and H that burns in an excess of air produces CO_2 and H_2O.

FOLLOW-UP PROBLEM 3.7 Write balanced equations for each of the following chemical statements:
(a) A characteristic reaction of Group 1A(1) elements: chunks of sodium react violently with water to form hydrogen gas and sodium hydroxide solution.
(b) The destruction of marble statuary by acid rain: aqueous nitric acid reacts with calcium carbonate to form carbon dioxide, water, and aqueous calcium nitrate.
(c) Halogen compounds exchanging bonding partners: phosphorus trifluoride is prepared by the reaction of phosphorus trichloride and hydrogen fluoride; hydrogen chloride is the other product. The reaction involves gases only.
(d) Explosive decomposition of dynamite: liquid nitroglycerine ($C_3H_5N_3O_9$) explodes to produce a mixture of gases—carbon dioxide, water vapor, nitrogen, and oxygen.

SECTION SUMMARY

To conserve mass and maintain the fixed composition of compounds, a chemical equation must be balanced in terms of number and type of each atom. A balanced equation has reactant formulas on the left of a yield arrow and product formulas on the right. Balancing coefficients are integer multipliers for *all* the atoms in a formula and apply to the individual entity or to moles of entities.

3.4 CALCULATING AMOUNTS OF REACTANT AND PRODUCT

A balanced equation contains a wealth of quantitative information relating individual chemical entities, amounts of chemical entities, and masses of substances. Table 3.5 presents this information for the combustion of propane, a hydrocarbon fuel used in cooking and water heating. A balanced equation is essential for all calculations involving amounts of reactants and products: *if you know the number of moles of one substance, the balanced equation tells you the number of moles of all the others in the reaction.*

Table 3.5 Information Contained in a Balanced Equation

Viewed in Terms of	Reactants $C_3H_8(g) + 5O_2(g)$	$\longrightarrow$	Products $3CO_2(g) + 4H_2O(g)$
Molecules	1 molecule C_3H_8 + 5 molecules O_2	$\longrightarrow$	3 molecules CO_2 + 4 molecules H_2O
Amount (mol)	1 mol C_3H_8 + 5 mol O_2	$\longrightarrow$	3 mol CO_2 + 4 mol H_2O
Mass (amu)	44.09 amu C_3H_8 + 160.00 amu O_2	$\longrightarrow$	132.03 amu CO_2 + 72.06 amu H_2O
Mass (g)	44.09 g C_3H_8 + 160.00 g O_2	$\longrightarrow$	132.03 g CO_2 + 72.06 g H_2O
Total mass (g)	204.09 g	$\longrightarrow$	204.09 g

Stoichiometrically Equivalent Molar Ratios from the Balanced Equation

In a balanced equation, *the number of moles of one substance is stoichiometrically equivalent to the number of moles of any other substance*. The term *stoichiometrically equivalent* means that a definite amount of one substance is formed from, produces, or reacts with a definite amount of the other. These quantitative relationships are expressed as *stoichiometrically equivalent molar ratios* that we use as conversion factors to calculate these amounts. The propane combustion reaction (Table 3.5) serves as our example:

$$C_3H_8(g) + 5O_2(g) \longrightarrow 3CO_2(g) + 4H_2O(g)$$

If we view the reaction quantitatively in terms of C_3H_8, we see that

> 1 mol of C_3H_8 reacts with 5 mol of O_2
> 1 mol of C_3H_8 produces   3 mol of CO_2
> 1 mol of C_3H_8 produces   4 mol of H_2O

Therefore, in this reaction,

> 1 mol of C_3H_8 is stoichiometrically equivalent to 5 mol of O_2
> 1 mol of C_3H_8 is stoichiometrically equivalent to 3 mol of CO_2
> 1 mol of C_3H_8 is stoichiometrically equivalent to 4 mol of H_2O

We chose to look at C_3H_8, but any two of the substances are stoichiometrically equivalent to each other. Thus,

> 3 mol of CO_2 is stoichiometrically equivalent to 4 mol of H_2O
> 5 mol of O_2 is stoichiometrically equivalent to 3 mol of CO_2

and so on.

Here's a typical problem that shows how stoichiometric equivalence is used to create conversion factors: In the combustion of propane, how many moles of O_2 are consumed when 10.0 mol of H_2O are produced? In order to solve this problem, we have to find the molar ratio between O_2 and H_2O. From the balanced equation, we see that for every 5 mol of O_2 consumed, 4 mol of H_2O are formed:

> 5 mol of O_2 is stoichiometrically equivalent to 4 mol of H_2O

We can construct two conversion factors from this equivalence, depending on the quantity we want to find:

$$\frac{5 \text{ mol } O_2}{4 \text{ mol } H_2O} \quad \text{or} \quad \frac{4 \text{ mol } H_2O}{5 \text{ mol } O_2}$$

Since we want to find moles of O_2 and we know moles of H_2O, we choose "5 mol O_2/4 mol H_2O" to cancel "mol H_2O":

$$\text{Moles of } O_2 \text{ consumed} = 10.0 \text{ mol } H_2O \times \frac{5 \text{ mol } O_2}{4 \text{ mol } H_2O} = 12.5 \text{ mol } O_2$$

$$\text{mol } H_2O \quad \xrightarrow[\substack{\text{molar ratio as} \\ \text{conversion factor}}]{} \quad \text{mol } O_2$$

Obviously, we could not have solved this problem without the balanced equation. Here is a general approach for solving *any* stoichiometry problem that involves a chemical reaction:

1. Write a balanced equation for the reaction.
2. Convert the given mass (or number of entities) of the first substance to amount (mol).
3. Use the appropriate molar ratio from the balanced equation to calculate the amount (mol) of the second substance.
4. Convert the amount of the second substance to the desired mass (or number of entities).

This approach is shown in Figure 3.8 and demonstrated in the following sample problems.

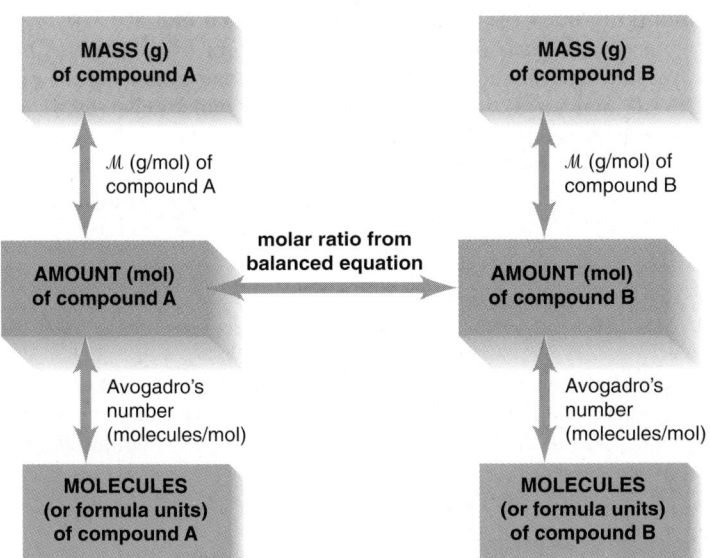

Figure 3.8 **Summary of the mass-mole-number relationships in a chemical reaction.** The amount of one substance in a reaction is related to that of any other. Quantities are expressed in terms of grams, moles, or number of entities (atoms, molecules, or formula units). Start at any box in the diagram (known) and move to any other box (unknown) by using the information on the arrows as conversion factors. As an example, if you know the mass (in g) of A and want to know the number of molecules of B, the path involves three calculation steps:
1. Grams of A to moles of A, using the molar mass ($\mathcal{M}$) of A
2. Moles of A to moles of B, using the molar ratio from the balanced equation
3. Moles of B to molecules of B, using Avogadro's number
Steps 1 and 3 refer to calculations discussed in Section 3.1 (see Figure 3.3).

SAMPLE PROBLEM 3.8 Calculating Amounts of Reactants and Products

Problem In a lifetime, the average American uses 1750 lb (794 kg) of copper in coins, plumbing, and wiring. Copper is obtained from sulfide ores, such as chalcocite, or copper(I) sulfide, by a multistage process. After an initial grinding step, the first stage is to "roast" the ore (heat it strongly with oxygen gas) to form powdered copper(I) oxide and gaseous sulfur dioxide.
(a) How many moles of oxygen are required to roast 10.0 mol of copper(I) sulfide?
(b) How many grams of sulfur dioxide are formed when 10.0 mol of copper(I) sulfide is roasted?
(c) How many kilograms of oxygen are required to form 2.86 kg of copper(I) oxide?

(a) Determining the moles of O_2 needed to roast 10.0 mol of Cu_2S
Plan We *always* write the balanced equation first. The formulas of the reactants are Cu_2S and O_2, and the formulas of the products are Cu_2O and SO_2:

$$2Cu_2S(s) + 3O_2(g) \longrightarrow 2Cu_2O(s) + 2SO_2(g)$$

We are given the *moles* of Cu_2S and need to find the *moles* of O_2. The balanced equation shows that 3 mol of O_2 is needed for every 2 mol of Cu_2S consumed, so the conversion factor is "3 mol O_2/2 mol Cu_2S" (see the roadmap).
Solution Calculating number of moles of O_2:

$$\text{Moles of } O_2 = 10.0 \text{ mol } Cu_2S \times \frac{3 \text{ mol } O_2}{2 \text{ mol } Cu_2S} = \boxed{15.0 \text{ mol } O_2}$$

Check The units are correct, and the answer is reasonable because this O_2/Cu_2S molar ratio (15:10) is equivalent to the ratio in the balanced equation (3:2).
Comment A *common mistake* is to use the incorrect conversion factor; the calculation would then be

$$\text{Moles of } O_2 = 10.0 \text{ mol } Cu_2S \times \frac{2 \text{ mol } Cu_2S}{3 \text{ mol } O_2} = \frac{6.67 \text{ mol}^2 \, Cu_2S}{1 \text{ mol } O_2}$$

Such strange units should signal that you made an error in setting up the conversion factor. In addition, the size of the answer, 6.67, is *less* than 10.0, whereas the balanced equation shows that *more* moles of O_2 than of Cu_2S are needed. Be sure to think through the calculation when setting up the conversion factor and cancel units.

Amount (mol) of Cu₂S

molar ratio

Amount (mol) of O₂

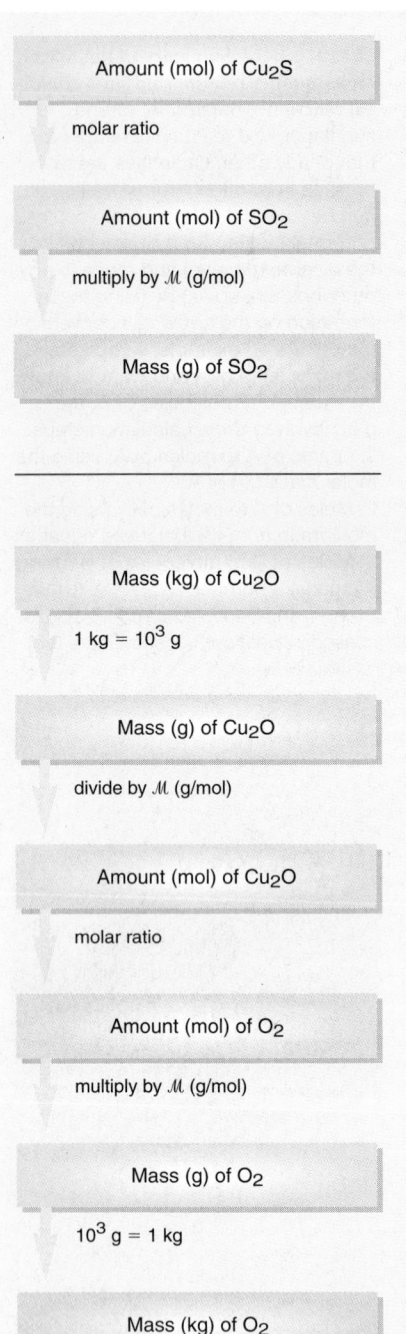

(b) Determining the mass (g) of SO_2 formed from 10.0 mol of Cu_2S

Plan Here we need the *grams* of product (SO_2) that form from the given *moles* of reactant (Cu_2S). We first find the moles of SO_2 using the molar ratio from the balanced equation (2 mol SO_2/2 mol Cu_2S) and then multiply by its molar mass (64.07 g/mol) to find grams of SO_2. The steps appear in the top roadmap.

Solution Combining the two conversion steps into one calculation, we have

$$\text{Mass (g) of } SO_2 = 10.0 \text{ mol } Cu_2S \times \frac{2 \text{ mol } SO_2}{2 \text{ mol } Cu_2S} \times \frac{64.07 \text{ g } SO_2}{1 \text{ mol } SO_2}$$
$$= \boxed{641 \text{ g } SO_2}$$

Check The answer makes sense, since the molar ratio shows that 10.0 mol of SO_2 are formed and each weighs about 64 g. We rounded to three significant figures.

(c) Determining the mass (kg) of O_2 that yields 2.86 kg of Cu_2O

Plan Here the mass of product (Cu_2O) is known, and we need the mass of reactant (O_2) that reacts to form it. We first convert the quantity of Cu_2O from *kilograms* to *moles* (in two steps, as shown on the bottom roadmap). Then, we use the molar ratio (3 mol O_2/2 mol Cu_2O) to find the *moles* of O_2 required. Finally, we convert *moles* of O_2 to *kilograms* (in two steps).

Solution Converting from kilograms of Cu_2O to moles of Cu_2O: Combining the mass unit conversion with the mass-to-mole conversion gives

$$\text{Moles of } Cu_2O = 2.86 \text{ kg } Cu_2O \times \frac{10^3 \text{ g}}{1 \text{ kg}} \times \frac{1 \text{ mol } Cu_2O}{143.10 \text{ g } Cu_2O} = 20.0 \text{ mol } Cu_2O$$

Converting from moles of Cu_2O to moles of O_2:

$$\text{Moles of } O_2 = 20.0 \text{ mol } Cu_2O \times \frac{3 \text{ mol } O_2}{2 \text{ mol } Cu_2O} = 30.0 \text{ mol } O_2$$

Converting from moles of O_2 to kilograms of O_2: Combining the mole-to-mass conversion with the mass unit conversion gives

$$\text{Mass (kg) of } O_2 = 30.0 \text{ mol } O_2 \times \frac{32.00 \text{ g } O_2}{1 \text{ mol } O_2} \times \frac{1 \text{ kg}}{10^3 \text{ g}}$$
$$= \boxed{0.960 \text{ kg } O_2}$$

Check The units are correct. Round off to check the math: for example, in the final step, ~30 mol × 30 g/mol × 1 kg/10^3 g = 0.90 kg. The answer seems reasonable: even though the amount (mol) of O_2 is greater than the amount (mol) of Cu_2O, the mass of O_2 is less than the mass of Cu_2O because $\mathcal{M}$ of O_2 is less than $\mathcal{M}$ of Cu_2O.

Comment This problem highlights a key point for solving stoichiometry problems: *convert the information given into moles*. Then, use the appropriate molar ratio and any other conversion factors to complete the problem.

FOLLOW-UP PROBLEM 3.8 Thermite is a mixture of iron(III) oxide and aluminum powders that was once used to weld railroad tracks. It undergoes a spectacular reaction to yield solid aluminum oxide and molten iron.
(a) How many grams of iron form when 135 g of aluminum reacts?
(b) How many atoms of aluminum react for every 1.00 g of aluminum oxide that is formed?

Chemical Reactions That Occur in a Sequence

In many situations, a product of one reaction becomes a reactant for the next reaction in a sequence of reactions. For stoichiometric purposes, when the same substance forms in one reaction and is used up in the next, we eliminate this common substance in an **overall (net) equation:**

1. Write the sequence of balanced equations.
2. Adjust the equations arithmetically to cancel out the common substance.
3. Add the adjusted equations together to obtain the overall balanced equation.

Sample Problem 3.9 shows the approach by continuing the copper recovery process that was begun in Sample Problem 3.8.

SAMPLE PROBLEM 3.9 Calculating Amounts of Reactants and Products in a Reaction Sequence

Problem Roasting is the first step in extracting copper from chalcocite, the ore used in the previous problem. In the next step, copper(I) oxide reacts with powdered carbon to yield copper metal and carbon monoxide gas. Write a balanced overall equation for the two-step sequence.

Plan To obtain the overall equation, we write the individual equations in sequence, adjust coefficients to cancel the common substance (or substances), and add the equations together. In this case, only Cu_2O appears as a product in one equation and as a reactant in the other, so it is the common substance.

Solution Writing the individual balanced equations:

$2Cu_2S(s) + 3O_2(g) \longrightarrow 2Cu_2O(s) + 2SO_2(g)$ [equation 1; see Sample Problem 3.8(a)]
$Cu_2O(s) + C(s) \longrightarrow 2Cu(s) + CO(g)$ [equation 2]

Adjusting the coefficients: Since 2 mol of Cu_2O are produced in equation 1 but only 1 mol of Cu_2O reacts in equation 2, we double *all* the coefficients in equation 2. Thus, the amount of Cu_2O formed in equation 1 is used up in equation 2:

$2Cu_2S(s) + 3O_2(g) \longrightarrow 2Cu_2O(s) + 2SO_2(g)$ [equation 1]
$2Cu_2O(s) + 2C(s) \longrightarrow 4Cu(s) + 2CO(g)$ [equation 2, doubled]

Adding the two equations and canceling the common substance: We keep the reactants of both equations on the left and the products of both equations on the right:

$2Cu_2S(s) + 3O_2(g) + \cancel{2Cu_2O(s)} + 2C(s) \longrightarrow \cancel{2Cu_2O(s)} + 2SO_2(g) + 4Cu(s) + 2CO(g)$

Or, $\qquad 2Cu_2S(s) + 3O_2(g) + 2C(s) \longrightarrow 2SO_2(g) + 4Cu(s) + 2CO(g)$

Check Reactants (4 Cu, 2 S, 6 O, 2 C) $\longrightarrow$ products (4 Cu, 2 S, 6 O, 2 C)
Comment 1. Even though Cu_2O *does* participate in the chemical change, it is not involved in the reaction stoichiometry. An overall equation *may not* show which substances actually react; for example, $C(s)$ and $Cu_2S(s)$ do not interact directly here, even though both are shown as reactants.
2. The SO_2 formed in metal extraction contributes to acid rain (see Follow-up Problem). To help control the problem, chemists have devised microbial and electrochemical methods to extract metals without roasting sulfide ores. Such methods are among many examples of *green chemistry,* university-government-industry initiatives designed to reduce hazardous substances in the environment.
3. These reactions were shown to explain how to obtain an overall equation. The actual extraction of copper is more complex and will be discussed in Chapter 22.

FOLLOW-UP PROBLEM 3.9 The SO_2 produced in copper recovery reacts in air with oxygen and forms sulfur trioxide. This gas, in turn, reacts with water to form a sulfuric acid solution that falls as rain or snow. Write a balanced overall equation for this process.

Multistep reaction sequences called *metabolic pathways* are common in biological systems. In most cells, the chemical energy in glucose is released through a sequence of about 30 individual reactions. The product of each reaction is the reactant of the next, so that all the common substances cancel. The overall equation is

$$C_6H_{12}O_6(aq) + 6O_2(g) \longrightarrow 6CO_2(g) + 6H_2O(l)$$

We eat food that contains glucose, inhale O_2, and excrete CO_2 and H_2O. In our cells, these reactants and products are many steps apart: O_2 never reacts *directly* with glucose. However, the molar ratios are the same as if the glucose burned in a combustion apparatus filled with pure O_2 and formed CO_2 and H_2O directly.

Chemical Reactions That Involve a Limiting Reactant

In problems up to now, the amount of *one* reactant was given, and we assumed there was enough of any other reactant for the first reactant to be completely used up. For example, to find the amount of SO_2 that forms when 100 g of Cu_2S reacts, we convert the grams of Cu_2S to moles and assume that the Cu_2S reacts with as much O_2 as needed. Because all the Cu_2S is used up, its initial amount determines, or limits, how much SO_2 can form. We call Cu_2S the **limiting reactant** (or *limiting reagent*) because the reaction stops once the Cu_2S is gone, no matter how much O_2 is present. Suppose, however, that the amounts of both Cu_2S *and* O_2 are given and we want to know how much SO_2 forms. We first have to determine whether Cu_2S *or* O_2 is the limiting reactant (that is, which one is completely used up) because the amount of that reactant limits how much SO_2 can form. The other reactant is present *in excess,* and however much of it is not used is left over.

To clarify the idea of limiting reactant, let's consider a much more appetizing situation. Suppose you have a job making sundaes in an ice cream parlor. Each sundae requires two scoops (12 oz) of ice cream, one cherry, and 50 mL of chocolate syrup:

$$\text{2 scoops (12 oz)} + \text{1 cherry} + \text{50 mL syrup} \longrightarrow \text{1 sundae}$$

A mob of 25 ravenous school kids enters, and every one wants a vanilla sundae. Can you feed them all? You have 300 oz of vanilla ice cream, 30 cherries, and 1 L of syrup, so a quick calculation shows the number of sundaes you can make from each ingredient:

$$\text{Ice cream: No. of sundaes} = 300 \; \cancel{\text{oz}} \times \frac{2 \; \cancel{\text{scoops}}}{12 \; \cancel{\text{oz}}} \times \frac{1 \; \text{sundae}}{2 \; \cancel{\text{scoops}}} = 25 \; \text{sundaes}$$

$$\text{Cherries: No. of sundaes} = 30 \; \cancel{\text{cherries}} \times \frac{1 \; \text{sundae}}{1 \; \cancel{\text{cherry}}} = 30 \; \text{sundaes}$$

$$\text{Syrup: No. of sundaes} = 1000 \; \cancel{\text{mL syrup}} \times \frac{1 \; \text{sundae}}{50 \; \cancel{\text{mL syrup}}} = 20 \; \text{sundaes}$$

The syrup is the limiting "reactant" here because it limits the total amount of "product" that can "form": of the three ingredients, the syrup allows the *fewest* sundaes to be made (Figure 3.9).

Figure 3.9 **An ice cream sundae analogy for limiting reactants. A,** The "components" combine in specific amounts to form a sundae. **B,** In this example, the number of sundaes possible is limited by the amount of syrup, the limiting "reactant." Here, only two sundaes can be made. Four scoops of ice cream and four cherries remain "in excess." See text for another situation involving these components.

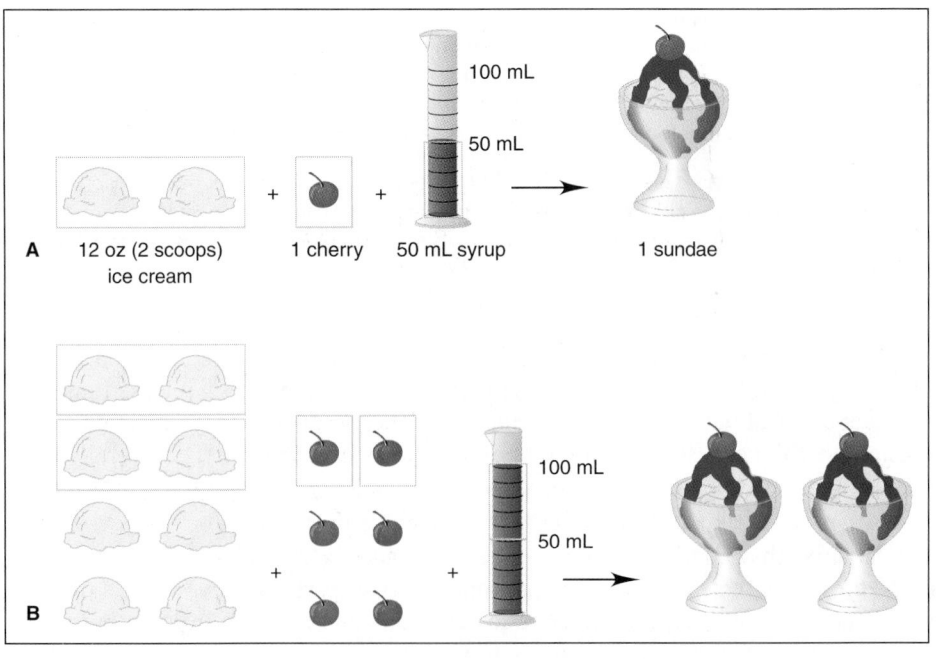

Some ice cream and cherries are left "unreacted" when all the syrup is gone, so they are present in excess:

300 oz (50 scoops) + 30 cherries + 1 L syrup $\longrightarrow$

20 sundaes + 60 oz (10 scoops) + 10 cherries

In limiting-reactant problems, the amounts of two (or more) reactants are given, and we must first determine which is limiting. To do so, *we choose the reactant that yields the **lower** amount of product.* One approach, illustrated in Sample Problem 3.10, is to perform two calculations, each of which assumes an excess of one or the other reactant. ●

SAMPLE PROBLEM 3.10 Calculating Amounts of Reactant and Product in Reactions Involving a Limiting Reactant

Problem A fuel mixture used in the early days of rocketry is composed of two liquids, hydrazine (N_2H_4) and dinitrogen tetraoxide (N_2O_4), which ignite on contact to form nitrogen gas and water vapor. How many grams of nitrogen gas form when 1.00×10^2 g of N_2H_4 and 2.00×10^2 g of N_2O_4 are mixed?

Plan As always, we first write the balanced equation. *The fact that the amounts of two reactants are given tells us that this is a limiting-reactant problem.* To determine which reactant is limiting, we calculate the mass of N_2 formed from each reactant *assuming an excess of the other.* We convert the mass of each reactant to number of moles and use the appropriate molar ratio to find the number of moles of N_2 each forms. Whichever yields *less* N_2 is the limiting reactant. Then, we convert this lower number of moles of N_2 to mass. The roadmap shows the steps.

Solution Writing the balanced equation:

$$2N_2H_4(l) + N_2O_4(l) \longrightarrow 3N_2(g) + 4H_2O(g)$$

Finding the moles of N_2 from the moles of N_2H_4 (if N_2H_4 is limiting):

$$\text{Moles of } N_2H_4 = 1.00\times10^2 \text{ g } N_2H_4 \times \frac{1 \text{ mol } N_2H_4}{32.05 \text{ g } N_2H_4} = 3.12 \text{ mol } N_2H_4$$

$$\text{Moles of } N_2 = 3.12 \text{ mol } N_2H_4 \times \frac{3 \text{ mol } N_2}{2 \text{ mol } N_2H_4} = \textbf{4.68 mol } N_2$$

Finding the moles of N_2 from the moles of N_2O_4 (if N_2O_4 is limiting):

$$\text{Moles of } N_2O_4 = 2.00\times10^2 \text{ g } N_2O_4 \times \frac{1 \text{ mol } N_2O_4}{92.02 \text{ g } N_2O_4} = 2.17 \text{ mol } N_2O_4$$

$$\text{Moles of } N_2 = 2.17 \text{ mol } N_2O_4 \times \frac{3 \text{ mol } N_2}{1 \text{ mol } N_2O_4} = \textbf{6.51 mol } N_2$$

Thus, N_2H_4 is the limiting reactant because it yields fewer moles of N_2. Converting from moles of N_2 to grams:

$$\text{Mass (g) of } N_2 = 4.68 \text{ mol } N_2 \times \frac{28.02 \text{ g } N_2}{1 \text{ mol } N_2}$$

$$= \boxed{131 \text{ g } N_2}$$

Check Even though the mass of N_2O_4 is greater than that of N_2H_4, there are fewer moles because $\mathcal{M}$ of N_2O_4 is much higher. Round off to check the math; for example, for N_2H_4, 100 g $N_2H_4 \times 1$ mol/32 g ≈ 3 mol; ~3 mol $\times \frac{3}{2} \approx 4.5$ mol N_2; ~4.5 mol $\times 30$ g/mol ≈ 135 g N_2

Comment 1. Here are two *common mistakes* that can arise in solving limiting-reactant problems:
- The limiting reactant is not the *reactant* present in fewer moles (2.17 mol of N_2O_4 vs. 3.12 mol of N_2H_4). Rather, it is the reactant that forms fewer moles of *product.*
- Similarly, the limiting reactant is not the *reactant* with lower mass. Rather, it is the reactant that forms the lower mass of *product.*

LIMITING REACTANT

Limiting Reactants in Everyday Life Limiting-"reactant" situations arise in business all the time. A car assembly plant manager must order more tires if there are 1500 car bodies and only 4000 tires, and a clothes manufacturer must cut more sleeves if there are 320 sleeves for 170 shirt bodies. You've probably faced such situations in daily life as well. A muffin recipe calls for 2 cups of flour and 1 cup of sugar, but you have 3 cups of flour and only $\frac{3}{4}$ cup of sugar. Clearly, the flour is in excess and the sugar limits the number of muffins you can make. Or, you're in charge of making cheeseburgers for a picnic, and you have 10 buns, 12 meat patties, and 15 slices of cheese. Here, the number of buns limits the cheeseburgers you can make. Or, there are 26 students and only 23 microscopes in a cell biology lab. You'll find that the number of times limiting-reactant situations arise is almost limitless.

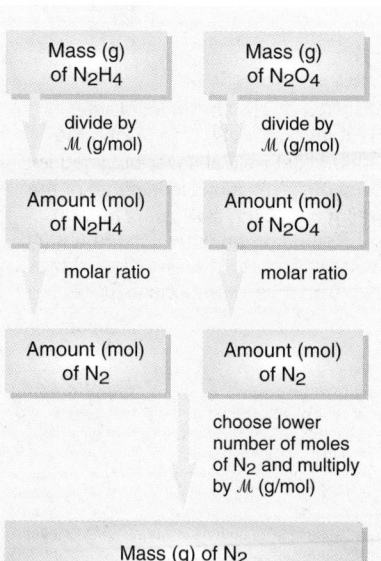

Mass (g) of N_2H_4		Mass (g) of N_2O_4
divide by $\mathcal{M}$ (g/mol)		divide by $\mathcal{M}$ (g/mol)
Amount (mol) of N_2H_4		Amount (mol) of N_2O_4
molar ratio		molar ratio
Amount (mol) of N_2		Amount (mol) of N_2
	choose lower number of moles of N_2 and multiply by $\mathcal{M}$ (g/mol)	
	Mass (g) of N_2	

2. Here is an *alternative approach* to determining which reactant is limiting. Find the moles of each reactant that would be needed to react with the other reactant. Then see which of the amounts actually given in the problem is sufficient. That substance is in excess and the other substance is limiting. For example, the balanced equation shows that 2 mol of N_2H_4 reacts with 1 mol of N_2O_4. The moles of N_2O_4 needed to react with the given moles of N_2H_4 are

$$\text{Moles of } N_2O_4 \text{ needed} = 3.12 \text{ mol } N_2H_4 \times \frac{1 \text{ mol } N_2O_4}{2 \text{ mol } N_2H_4} = 1.56 \text{ mol } N_2O_4$$

The moles of N_2H_4 needed to react with the given moles of N_2O_4 are

$$\text{Moles of } N_2H_4 \text{ needed} = 2.17 \text{ mol } N_2O_4 \times \frac{2 \text{ mol } N_2H_4}{1 \text{ mol } N_2O_4} = 4.34 \text{ mol } N_2H_4$$

We are given 2.17 mol of N_2O_4, which is *more* than the amount of N_2O_4 that is needed (1.56 mol) to react with the given amount of N_2H_4, and we are given 3.12 mol of N_2H_4, which is *less* than the amount of N_2H_4 needed (4.34 mol) to react with the given amount of N_2O_4. Therefore, N_2H_4 is limiting, and N_2O_4 is in excess. Once we determine this, we continue with the final calculation to find the amount of N_2.

FOLLOW-UP PROBLEM 3.10 How many grams of solid aluminum sulfide can be prepared by the reaction of 10.0 g of aluminum and 15.0 g of sulfur? How much of the nonlimiting reactant is in excess?

Chemical Reactions in Practice: Theoretical, Actual, and Percent Yields

Up until now, we've been optimistic about the amount of product obtained from a reaction. We have assumed that 100% of the limiting reactant becomes product, that ideal separation and purification methods exist for isolating the product, and that we use perfect lab technique to collect all the product formed. In other words, we have assumed that we obtain the **theoretical yield,** the amount indicated by the stoichiometrically equivalent molar ratio in the balanced equation.

It's time to face reality. The theoretical yield is *never* obtained, for reasons that are largely uncontrollable. For one thing, although the major reaction predominates, many reactant mixtures also proceed through one or more **side reactions** that form smaller amounts of different products, as shown in Figure 3.10. In the previous rocket fuel reaction, for example, the reactants might form some NO in the following side reaction:

$$2N_2O_4(l) + N_2H_4(l) \longrightarrow 6NO(g) + 2H_2O(g)$$

This reaction decreases the amounts of reactants available for N_2 production (see Problem 3.116 at the end of the chapter). Even more important, as we'll discuss in Chapter 4, many reactions seem to stop before they are complete, which leaves some limiting reactant unused. But, even when a reaction does go completely to product, losses occur in virtually every step of a separation procedure (see Tools of the Laboratory, Section 2.9): a tiny amount of product clings to filter paper, some distillate evaporates, a small amount of extract remains in the separatory funnel, and so forth. With careful technique, you can minimize these losses but never eliminate them. The amount of product that you actually obtain is the **actual yield.** The **percent yield (% yield)** is the actual yield expressed as a percent of the theoretical yield:

$$\% \text{ yield} = \frac{\text{actual yield}}{\text{theoretical yield}} \times 100 \tag{3.7}$$

Since the actual yield must be less than the theoretical yield, the percent yield is always less than 100%. Theoretical and actual yields are expressed in units of amount (moles) or mass (grams).

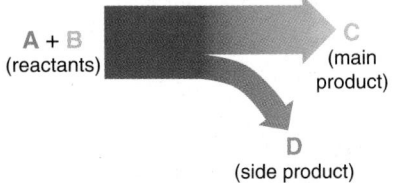

A + B
(reactants)

C
(main product)

D
(side product)

Figure 3.10 The effect of side reactions on yield. One reason the theoretical yield is never obtained is that other reactions lead some of the reactants along side paths to form undesired products.

SAMPLE PROBLEM 3.11 Calculating Percent Yield

Problem Silicon carbide (SiC) is an important ceramic material that is made by allowing sand (silicon dioxide, SiO_2) to react with powdered carbon at high temperature. Carbon monoxide is also formed. When 100.0 kg of sand are processed, 51.4 kg of SiC are recovered. What is the percent yield of SiC in this process?

Plan We are given the actual yield of SiC (51.4 kg), so we need the theoretical yield to calculate the percent yield. After writing the balanced equation, we convert the given mass of SiO_2 (100.0 kg) to amount (mol). We use the molar ratio to find the amount of SiC formed and convert that amount to mass (kg) to obtain the theoretical yield [see Sample Problem 3.8(c)]. Then, we use Equation 3.7 to find the percent yield.

Solution Writing the balanced equation:

$$SiO_2(s) + 3C(s) \longrightarrow SiC(s) + 2CO(g)$$

Converting from kilograms of SiO_2 to moles:

$$\text{Moles of } SiO_2 = 100.0 \text{ kg } SiO_2 \times \frac{1000 \text{ g}}{1 \text{ kg}} \times \frac{1 \text{ mol } SiO_2}{60.09 \text{ g } SiO_2} = 1664 \text{ mol } SiO_2$$

Converting from moles of SiO_2 to moles of SiC: The molar ratio is 1 mol SiC/1 mol SiO_2, so

$$\text{Moles of } SiO_2 = \text{moles of SiC} = 1664 \text{ mol SiC}$$

Converting from moles of SiC to kilograms:

$$\text{Mass (kg) of SiC} = 1664 \text{ mol SiC} \times \frac{40.10 \text{ g SiC}}{1 \text{ mol SiC}} \times \frac{1 \text{ kg}}{1000 \text{ g}} = 66.73 \text{ kg SiC}$$

Calculating the percent yield:

$$\% \text{ yield of SiC} = \frac{\text{actual yield}}{\text{theoretical yield}} \times 100 = \frac{51.4 \text{ kg}}{66.73 \text{ kg}} \times 100 = \boxed{77.0\%}$$

Check Rounding shows that the mass of SiC seems correct: ~1500 mol × 40 g/mol × 1 kg/1000 g = 60 kg. The molar ratio of SiC:SiO_2 is 1:1, and the $\mathcal{M}$ of SiC is about two-thirds ($\frac{40}{60}$) the $\mathcal{M}$ of SiO_2, so 100 kg of SiO_2 should form about 66 kg of SiC.

FOLLOW-UP PROBLEM 3.11 Marble (calcium carbonate) reacts with hydrochloric acid solution to form calcium chloride solution, water, and carbon dioxide. What is the % yield of carbon dioxide if 3.65 g of the gas is collected when 10.0 g of marble reacts?

In multistep reaction sequences, the percent yields of the steps are expressed as fractions and multiplied together to find the overall percent yield. Even when the yield of each step is high, the final result can be surprisingly low. For example, suppose a six-step reaction sequence has a 90.0% yield for each step; that is, you are able to recover 90.0% of the theoretical yield of product in each step. The overall recovery is only slightly more than 50%:

Overall % yield = $(0.900 \times 0.900 \times 0.900 \times 0.900 \times 0.900 \times 0.900) \times 100 = 53.1\%$

Such multistep sequences are common in the synthesis of medicines, dyes, pesticides, and many other organic compounds. In a typical synthesis, large amounts of inexpensive, simple reactants are converted to small amounts of expensive, complex product; thus, overall percent yield greatly influences the commercial potential of a product.

SECTION SUMMARY

The substances in a balanced equation are related to each other by stoichiometrically equivalent molar ratios, which can be used as conversion factors to find the moles of one substance given the moles of another. In limiting-reactant problems, the amounts of two (or more) reactants are given, and one of them limits the amount of product that forms. The limiting reactant is the one that forms the lower amount of product. In practice, side reactions, incomplete reactions, and physical losses result in an actual yield of product that is less than the theoretical yield, the amount based solely on the molar ratio.

3.5 FUNDAMENTALS OF SOLUTION STOICHIOMETRY

In the popular media, you may have seen a chemist portrayed as a person in a white lab coat, surrounded by odd-shaped glassware, pouring one colored solution into another, which produces frothing bubbles and billowing fumes. Although most reactions in solution are not this dramatic and good technique offers safer mixing procedures, the image is true to the extent that aqueous solution chemistry is a central part of laboratory activity. Liquid solutions are more convenient to store and mix than solids or gases, and the amounts of substances in solution can be measured very precisely. Since many environmental reactions and almost all biochemical reactions occur in solution, an understanding of reactions in solution is extremely important in chemistry and related sciences.

We'll discuss solution chemistry at many places in the text, but here we focus on solution stoichiometry. Only one aspect of the stoichiometry of dissolved substances is different from what we've seen so far. We know the amounts of pure substances by converting their masses directly into moles. For dissolved substances, we must know the *concentration*—the number of moles present in a certain volume of solution—to find the volume that contains a given number of moles. Of the many ways to express concentration, the most important is *molarity,* so we discuss it here (and wait until Chapter 13 to discuss the other ways). Then, we see how to prepare a solution of a specific molarity and how to use solutions in stoichiometric calculations.

Expressing Concentration in Terms of Molarity

A typical solution consists of a smaller amount of one substance, the **solute,** dissolved in a larger amount of another substance, the **solvent.** When a solution forms, the solute's individual chemical entities become evenly dispersed throughout the available volume and surrounded by solvent molecules. The **concentration** of a solution is usually expressed as *the amount of solute dissolved in a given amount of solution.* Concentration is an *intensive* quantity (like density or temperature) and thus independent of the volume of solution: a 50-L tank of a given solution has the *same concentration* (solute amount/solution amount) as a 50-mL beaker of the solution. **Molarity (*M*)** expresses the concentration in units of *moles of solute per liter of solution:*

$$\text{Molarity} = \frac{\text{moles of solute}}{\text{liters of solution}} \quad \text{or} \quad M = \frac{\text{mol solute}}{\text{L soln}} \tag{3.8}$$

SAMPLE PROBLEM 3.12 Calculating the Molarity of a Solution

Problem Hydrobromic acid (HBr) is a solution of hydrogen bromide gas in water. Calculate the molarity of hydrobromic acid solution if 455 mL contains 1.80 mol of hydrogen bromide.

Plan The molarity is the number of moles of solute in each liter of solution. We are given the number of moles (1.80 mol) and the volume (455 mL), so we divide moles by volume and convert the volume to liters to find the molarity.

Solution

Amount (mol) of HBr

divide by volume (mL)

Concentration (mol/mL) of HBr

10^3 mL = 1 L

Molarity (mol/L) of HBr

$$\text{Molarity} = \frac{1.80 \text{ mol HBr}}{455 \text{ mL soln}} \times \frac{1000 \text{ mL}}{1 \text{ L}}$$
$$= 3.96 \ M \text{ HBr}$$

Check A quick look at the math shows that there are almost 2 mol of HBr and about 0.5 L of solution, so the concentration should be about 4 mol/L, or 4 *M*.

FOLLOW-UP PROBLEM 3.12 How many moles of KI are in 84 mL of 0.50 *M* KI?

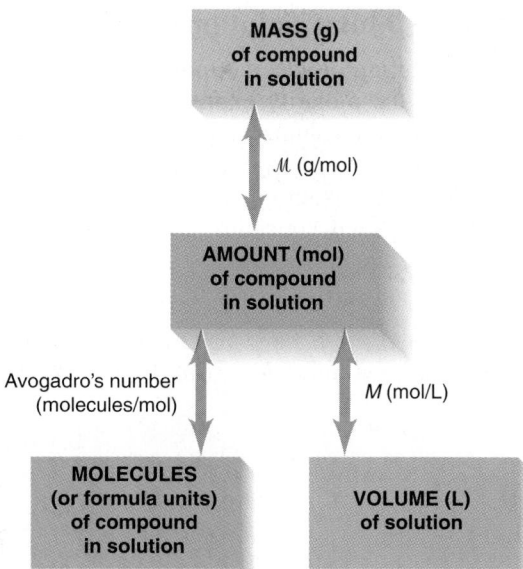

Figure 3.11 **Summary of mass-mole-number-volume relationships in solution.** The amount (in moles) of a compound in solution is related to the volume of solution in liters through the molarity (*M*) in moles per liter. The other relationships shown are identical to those in Figure 3.3, except that here they refer to the quantities *in solution*. As in previous cases, to find the quantity of substance expressed in one form or another, convert the given information to moles first.

Mole-Mass-Number Conversions Involving Solutions

Molarity can be thought of as a conversion factor used to convert between volume of solution and amount (mol) of solute, from which we find the mass or the number of entities of solute. Figure 3.11 shows this new stoichiometric relationship, and Sample Problem 3.13 applies it.

SAMPLE PROBLEM 3.13 Calculating Mass of Solute in a Given Volume of Solution

Problem How many grams of solute are in 1.75 L of 0.460 *M* sodium monohydrogen phosphate?

Plan We know the solution volume (1.75 L) and molarity (0.460 *M*), and we need the mass of solute. We use the known quantities to find the amount (mol) of solute and then convert moles to grams with the solute molar mass.

Solution Calculating moles of solute in solution:

$$\text{Moles of } Na_2HPO_4 = 1.75 \text{ L-soln} \times \frac{0.460 \text{ mol } Na_2HPO_4}{1 \text{ L-soln}}$$

$$= 0.805 \text{ mol } Na_2HPO_4$$

Converting from moles of solute to grams:

$$\text{Mass (g) } Na_2HPO_4 = 0.805 \text{ mol } Na_2HPO_4 \times \frac{141.96 \text{ g } Na_2HPO_4}{1 \text{ mol } Na_2HPO_4}$$

$$= \boxed{114 \text{ g } Na_2HPO_4}$$

Check The answer seems to be correct: ∼ 1.8 L of 0.5 mol/L contains 0.9 mol, and 150 g/mol × 0.9 mol = 135 g, which is close to 114 g of solute.

FOLLOW-UP PROBLEM 3.13 In biochemistry laboratories, solutions of sucrose (table sugar, $C_{12}H_{22}O_{11}$) are used in high-speed centrifuges to separate the parts of a biological cell. How many liters of 3.30 *M* sucrose contain 135 g of solute?

Volume (L) of solution

↓ multiply by *M* (mol/L)

Amount (mol) of solute

↓ multiply by *M* (g/mol)

Mass (g) of solute

Preparing and Diluting Molar Solutions

Whenever you prepare a solution of specific molarity, remember that the volume term in the denominator of the molarity expression is the *solution* volume, **not** the *solvent* volume. The solution volume includes contributions from solute *and* solvent, so you cannot simply dissolve 1 mol of solute in 1 L of solvent and expect a 1 *M* solution. The solute would increase the solution volume above 1 L, resulting in a lower-than-expected concentration. The correct preparation of a solution containing a solid solute consists of four steps. Let's go through them to prepare 0.500 L of 0.350 *M* nickel(II) nitrate hexahydrate [$Ni(NO_3)_2 \cdot 6H_2O$]:

1. *Weigh the solid needed.* First, calculate the mass of solid needed by converting from liters to moles and from moles to grams:

$$\text{Mass (g) of solute} = 0.500 \text{ L soln} \times \frac{0.350 \text{ mol } Ni(NO_3)_2 \cdot 6H_2O}{1 \text{ L soln}}$$

$$\times \frac{290.82 \text{ g } Ni(NO_3)_2 \cdot 6H_2O}{1 \text{ mol } Ni(NO_3)_2 \cdot 6H_2O}$$

$$= 50.9 \text{ g } Ni(NO_3)_2 \cdot 6H_2O$$

2. *Carefully transfer the solid to a volumetric flask that contains about half the final volume of solvent.* Since we need 0.500 L of solution, we choose a 500-mL volumetric flask. Add about 250 mL of distilled water and then transfer the solid. Wash down any solid clinging to the neck with a small amount of solvent.

3. *Dissolve the solid thoroughly by swirling.* If some solute remains undissolved, the solution will be less concentrated than expected, so be sure the solute is dissolved. If necessary, wait for the solution to reach room temperature. (As we'll discuss in Chapter 13, the solution process is often accompanied by heating or cooling.)

4. *Add solvent until the solution reaches its final volume.* Add distilled water to bring the volume exactly to the line on the flask neck, then cover and mix thoroughly again. Figure 3.12 shows the last three steps.

Figure 3.12 Laboratory preparation of molar solutions. After the desired mass of solid has been weighed out, the solution is prepared by **A,** carefully adding the solid to a volumetric flask about half full of solvent; **B,** swirling to dissolve the solid completely; and **C,** adding solvent to the mark on the flask neck, shown in **D.** It would *not* be correct to add the solid to the entire volume of solvent, because the total volume would exceed the desired volume and result in a lower-than-calculated concentration.

As Figure 3.13 shows, *only solvent is added when a solution is diluted,* so the solute is dispersed in a larger final volume. Thus, a given volume of the final solution contains fewer solute particles and has a lower concentration. If various low concentrations of a solution are used frequently, it is common practice to prepare a more concentrated solution (called a *stock solution*), which is stored and diluted as needed.

MAKING A SOLUTION

SAMPLE PROBLEM 3.14 Preparing a Dilute Solution from a Concentrated Solution

Problem "Isotonic saline" is a 0.15 *M* aqueous solution of NaCl that simulates the total concentration of ions found in many cellular fluids. Its uses range from a cleansing rinse for contact lenses to a washing medium for red blood cells. How would you prepare 0.80 L of isotonic saline from a 6.0 *M* stock solution?

Plan To dilute a concentrated solution, we add only solvent, so the *moles of solute are the same in both solutions.* We know the volume (0.80 L) and molarity (0.15 *M*) of the dilute (dil) NaCl we need, so we find the moles of NaCl it contains and then find the unknown volume of concentrated (conc; 6.0 *M*) NaCl that contains the same number of moles. Then, we dilute this volume with pure solvent *up to* the final volume.

Solution Finding moles of solute in dilute solution:

$$\text{Moles of NaCl in dil soln} = 0.80 \text{ L soln} \times \frac{0.15 \text{ mol NaCl}}{1 \text{ L soln}} = 0.12 \text{ mol NaCl}$$

Finding moles of solute in concentrated solution: Since we add only solvent to dilute the solution,

$$\text{Moles of NaCl in dil soln} = \text{moles of NaCl in conc soln} = 0.12 \text{ mol NaCl}$$

Finding the volume of concentrated solution that contains 0.12 mol of NaCl:

$$\text{Volume (L) of conc NaCl soln} = 0.12 \text{ mol NaCl} \times \frac{1 \text{ L soln}}{6.0 \text{ mol NaCl}} = 0.020 \text{ L soln}$$

To prepare 0.80 L of dilute solution, place 0.020 L of 6.0 M NaCl in a 1.0-L cylinder, add distilled water (~780 mL) to the 0.80-L mark, and stir thoroughly.

Check The answer seems reasonable because a small volume of concentrated solution is used to prepare a large volume of dilute solution. Also, the ratio of volumes (0.020 L:0.80 L) is the same as the ratio of concentrations (0.15 M:6.0 M).

Comment An *alternative approach* to solving dilution problems makes use of the formula

$$M_{dil} \times V_{dil} = \text{number of moles} = M_{conc} \times V_{conc} \qquad \textbf{(3.9)}$$

where the M and V terms are the molarity and volume of the *dil*ute and *conc*entrated solutions. We need the volume of concentrated solution to use, so we solve for V_{conc}:

$$V_{conc} = \frac{M_{dil} \times V_{dil}}{M_{conc}} = \frac{0.15 \, M \times 0.80 \text{ L}}{6.0 \, M} = 0.020 \text{ L}$$

The method worked out in the Solution (above) is actually the same calculation broken into two parts to emphasize the thinking process:

$$V_{conc} = 0.80 \text{ L} \times \frac{0.15 \text{ mol NaCl}}{1 \text{ L}} \times \frac{1 \text{ L}}{6.0 \text{ mol NaCl}} = 0.020 \text{ L}$$

FOLLOW-UP PROBLEM 3.14 If 25.0 mL of 7.50 M sulfuric acid are diluted to exactly 500. mL, what is the mass of sulfuric acid per milliliter?

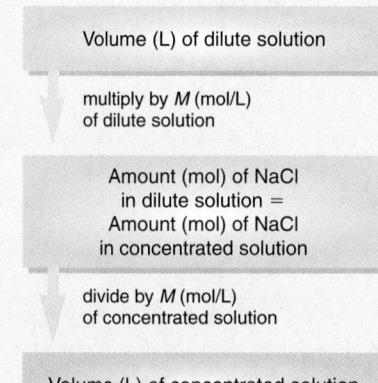

Volume (L) of dilute solution

multiply by M (mol/L) of dilute solution

Amount (mol) of NaCl in dilute solution = Amount (mol) of NaCl in concentrated solution

divide by M (mol/L) of concentrated solution

Volume (L) of concentrated solution

Figure 3.13 **Converting a concentrated solution to a dilute solution.** When a solution is diluted, only solvent is added. The solution volume increases while the total number of moles of solute remains the same. Therefore, as shown in the blow-up views, a unit volume of concentrated solution contains more solute particles than the same unit volume of dilute solution.

Concentrated solution: More solute particles per unit volume

Dilute solution: Fewer solute particles per unit volume

Stoichiometry of Chemical Reactions in Solution

Solving stoichiometry problems for reactions in solution requires the same approach as before, with the additional step of converting the volume of reactant or product to moles: (1) balance the equation, (2) find the number of moles of one substance, (3) relate it to the stoichiometrically equivalent number of moles of another substance, and (4) convert to the desired units.

SAMPLE PROBLEM 3.15 Calculating Amounts of Reactants and Products for a Reaction in Solution

Problem Specialized cells in the stomach release HCl to aid digestion. If they release too much, the excess can be neutralized with antacids. A common antacid contains magnesium hydroxide, which reacts with the acid to form water and magnesium chloride solution. As a government chemist testing commercial antacids, you use 0.10 M HCl to simulate the acid concentration in the stomach. How many liters of "stomach acid" react with a tablet containing 0.10 g of magnesium hydroxide?

Plan We know the mass of $Mg(OH)_2$ (0.10 g) that reacts and the acid concentration (0.10 M), and we must find the acid volume. After writing the balanced equation, we convert the grams of $Mg(OH)_2$ to moles, use the molar ratio to find the moles of HCl that react with these moles of $Mg(OH)_2$, and then use the molarity of HCl to find the volume that contains this number of moles. The steps appear in the roadmap.

Solution Writing the balanced equation:

$$Mg(OH)_2(s) + 2HCl(aq) \longrightarrow MgCl_2(aq) + 2H_2O(l)$$

Converting from grams of $Mg(OH)_2$ to moles:

$$\text{Moles of } Mg(OH)_2 = 0.10 \text{ g } Mg(OH)_2 \times \frac{1 \text{ mol } Mg(OH)_2}{58.33 \text{ g } Mg(OH)_2}$$

$$= 1.7 \times 10^{-3} \text{ mol } Mg(OH)_2$$

Converting from moles of $Mg(OH)_2$ to moles of HCl:

$$\text{Moles of HCl} = 1.7 \times 10^{-3} \text{ mol } Mg(OH)_2 \times \frac{2 \text{ mol HCl}}{1 \text{ mol } Mg(OH)_2}$$

$$= 3.4 \times 10^{-3} \text{ mol HCl}$$

Converting from moles of HCl to liters:

$$\text{Volume (L) of HCl} = 3.4 \times 10^{-3} \text{ mol HCl} \times \frac{1 \text{ L}}{0.10 \text{ mol HCl}}$$

$$= 3.4 \times 10^{-2} \text{ L}$$

Check The size of the answer seems reasonable: a small volume of dilute acid (0.034 L of 0.10 M) reacts with a small amount of antacid (0.0017 mol).

Comment The reaction as written is an oversimplification in that HCl and $MgCl_2$ exist as separated ions in solution. These points will be covered in great detail in Chapters 4 and 18.

FOLLOW-UP PROBLEM 3.15 Another active ingredient found in some antacids is aluminum hydroxide. Which is more effective at neutralizing stomach acid, magnesium hydroxide or aluminum hydroxide? [*Hint:* Effectiveness refers to the amount of acid that reacts with a given mass of antacid. You already know the effectiveness of 0.10 g of $Mg(OH)_2$.]

Mass (g) of $Mg(OH)_2$

divide by $\mathcal{M}$ (g/mol)

Amount (mol) of $Mg(OH)_2$

molar ratio

Amount (mol) of HCl

divide by M (mol/L)

Volume (L) of HCl

In limiting-reactant problems for reactions in solution, we first determine which reactant is limiting and then determine the yield, as demonstrated in the next sample problem.

SAMPLE PROBLEM 3.16 Solving Limiting-Reactant Problems for Reactions in Solution

Problem Mercury and its compounds have many uses, from filling teeth (as an alloy with silver, copper, and tin) to the industrial production of chlorine. Because of their toxicity, however, soluble mercury compounds, such as mercury(II) nitrate, must be removed from industrial wastewater. One removal method reacts the wastewater with sodium sulfide solution to produce solid mercury(II) sulfide and sodium nitrate solution. In a laboratory simulation, 0.050 L of 0.010 M mercury(II) nitrate reacts with 0.020 L of 0.10 M sodium sulfide. How many grams of mercury(II) sulfide form?

Plan This is a limiting-reactant problem because *the amounts of two reactants are given.* After balancing the equation, we must determine the limiting reactant. The molarity (0.010 M) and volume (0.050 L) of the mercury(II) nitrate solution tell us the moles of one reactant, and the molarity (0.10 M) and volume (0.020 L) of the sodium sulfide solution tell us the moles of the other. Then, we use the molar ratio to find the moles of HgS that form from each reactant, *assuming the other reactant is present in excess.* The limiting reactant is the one that forms fewer moles of HgS, which we convert to mass using the HgS molar mass. The roadmap shows the process.

Solution Writing the balanced equation:

$$Hg(NO_3)_2(aq) + Na_2S(aq) \longrightarrow HgS(s) + 2NaNO_3(aq)$$

Finding moles of HgS assuming $Hg(NO_3)_2$ is limiting: Combining the steps gives

$$\text{Moles of HgS} = 0.050 \text{ L soln} \times \frac{0.010 \text{ mol Hg(NO}_3)_2}{1 \text{ L soln}} \times \frac{1 \text{ mol HgS}}{1 \text{ mol Hg(NO}_3)_2}$$

$$= \mathbf{5.0 \times 10^{-4} \text{ mol HgS}}$$

Finding moles of HgS assuming Na_2S is limiting: Combining the steps gives

$$\text{Moles of HgS} = 0.020 \text{ L soln} \times \frac{0.10 \text{ mol Na}_2S}{1 \text{ L soln}} \times \frac{1 \text{ mol HgS}}{1 \text{ mol Na}_2S}$$

$$= \mathbf{2.0 \times 10^{-3} \text{ mol HgS}}$$

$Hg(NO_3)_2$ is the limiting reactant because it forms fewer moles of HgS. Converting the moles of HgS formed from $Hg(NO_3)_2$ to grams:

$$\text{Mass (g) of HgS} = 5.0 \times 10^{-4} \text{ mol HgS} \times \frac{232.7 \text{ g HgS}}{1 \text{ mol HgS}}$$

$$= \boxed{0.12 \text{ g HgS}}$$

Check Let's use the alternative method for finding the limiting reactant as a check (see Comment in Sample Problem 3.10, p. 112). Finding moles of reactants available:

$$\text{Moles of Hg(NO}_3)_2 = 0.050 \text{ L soln} \times \frac{0.010 \text{ mol Hg(NO}_3)_2}{1 \text{ L soln}}$$

$$= 5.0 \times 10^{-4} \text{ mol Hg(NO}_3)_2$$

$$\text{Moles of Na}_2S = 0.020 \text{ L soln} \times \frac{0.10 \text{ mol Na}_2S}{1 \text{ L soln}}$$

$$= 2.0 \times 10^{-3} \text{ mol Na}_2S$$

The molar ratio of the reactants is 1 $Hg(NO_3)_2$/1 Na_2S. Therefore, $Hg(NO_3)_2$ is limiting because there are fewer moles of it than are needed to react with the moles of Na_2S. Finding grams of product from moles of limiting reactant and the molar ratio:

$$\text{Mass (g) of HgS} = 5.0 \times 10^{-4} \text{ mol Hg(NO}_3)_2 \times \frac{1 \text{ mol HgS}}{1 \text{ mol Hg(NO}_3)_2} \times \frac{232.7 \text{ g HgS}}{1 \text{ mol HgS}}$$

$$= 0.12 \text{ g HgS}$$

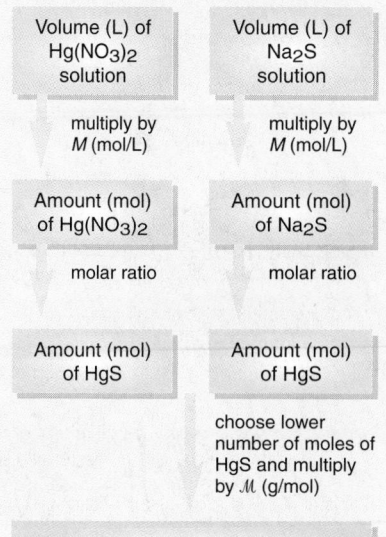

FOLLOW-UP PROBLEM 3.16 Despite their toxicity, many compounds of lead are used as pigments.
(a) What volume of 1.50 M lead(II) acetate contains 0.400 mol of Pb^{2+} ions?
(b) When this volume reacts with 125 mL of 3.40 M sodium chloride, how many grams of solid lead(II) chloride can form? (Sodium acetate solution also forms.)

SECTION SUMMARY

When reactions occur in solution, reactant and product amounts are given in terms of their concentration and volume. Molarity is the number of moles of solute dissolved in one liter of solution. Using molarity as a conversion factor, we apply the principles of stoichiometry to all aspects of reactions in solution.

Chapter Perspective

You apply the mole concept every time you measure a substance or think about how much of it will react. Figure 3.14 combines the individual stoichiometry summary diagrams into one overall review diagram. Use it for homework, to study for exams, or to obtain an overview of the various ways that the amounts involved in a reaction are interrelated.

We apply stoichiometry next to some of the most important types of chemical reactions (Chapter 4), to systems of reacting gases (Chapter 5), and to the heat involved in a reaction (Chapter 6). Stoichiometry appears at many places later in the text as well.

Figure 3.14 **An overview of the key mole-mass-number stoichiometric relationships.**

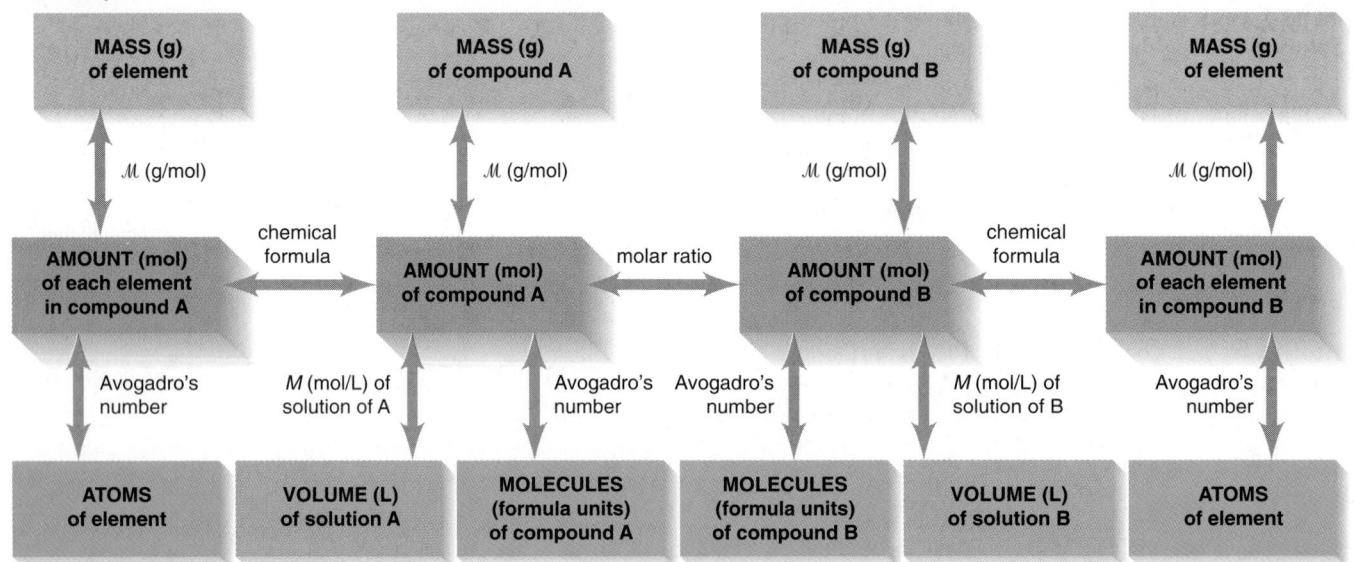

For Review and Reference (Numbers in parentheses refer to pages, unless noted otherwise.)

Learning Objectives

Relevant section and/or sample problem (SP) numbers appear in parentheses.

Understand These Concepts

1. The meaning and usefulness of the mole (3.1)
2. The relation between molecular (or formula) mass and molar mass (3.1)
3. The relations among amount of substance (in moles), mass (in grams), and number of chemical entities (3.1)
4. The information in a chemical formula (3.1)
5. The procedure for finding the empirical and molecular formulas of a compound (3.2)
6. The importance of balancing equations for the quantitative study of chemical reactions (3.3)
7. The mole-mass-number information contained in a balanced equation (3.4)
8. The relation between amounts of reactants and products (3.4)
9. Why one reactant limits the yield of product (3.4)

10. The causes of lower-than-expected yields and the distinction between theoretical and actual yields (3.4)
11. The meanings of concentration and molarity (3.5)
12. The effect of dilution on the concentration of solute (3.5)
13. How reactions in solution differ from those of pure reactants (3.5)

Master These Skills

1. Calculating the molar mass of any substance (3.1; also SPs 3.2 and 3.3)
2. Converting between amount of substance (in moles), mass (in grams), and number of chemical entities (SPs 3.1, 3.2)
3. Using mass percent to find the mass of element in a given mass of compound (SP 3.3)
4. Determining the empirical and molecular formulas of a compound from mass % and molar mass of elements (SPs 3.4, 3.5)

5. Determining a molecular formula from combustion analysis (SP 3.6)
6. Converting a chemical statement into a balanced equation (SP 3.7)
7. Using stoichiometrically equivalent molar ratios to calculate amounts of reactants and products in reactions of pure and dissolved substances (SPs 3.8 and 3.15)
8. Writing an overall equation from a series of equations (SP 3.9)

9. Recognizing limiting-reactant problems and choosing the limiting reactant in reactions of pure and dissolved substances (SPs 3.10, 3.16)
10. Calculating % yield (SP 3.11)
11. Calculating molarity and the mass of substance in solution (SPs 3.12, 3.13)
12. Preparing a dilute solution from a concentrated one (SP 3.14)

Key Terms

stoichiometry (87)

Section 3.1
mole (mol) (87)
Avogadro's number (87)
molar mass ($\mathcal{M}$) (89)

Section 3.2
combustion analysis (98)

Section 3.3
chemical equation (101)
reactant (102)
product (102)
balancing (stoichiometric)
   coefficient (102)

Section 3.4
overall (net) equation (108)
limiting reactant (110)
theoretical yield (112)
side reaction (112)
actual yield (112)
percent yield (% yield) (112)

Section 3.5
solute (114)
solvent (114)
concentration (114)
molarity (M) (114)

Key Equations and Relationships

3.1 Number of entities in one mole (87):
   1 mole contains 6.022×10^{23} entities (to 4 sf)

3.2 Converting amount (mol) to mass using $\mathcal{M}$ (90):
$$\text{Mass (g)} = \text{no. of moles} \times \frac{\text{no. of grams}}{1 \text{ mol}}$$

3.3 Converting mass to amount (mol) using (90):
$$\text{No. of moles} = \text{mass (g)} \times \frac{1 \text{ mol}}{\text{no. of grams}}$$

3.4 Converting amount (mol) to number of entities (90):
$$\text{No. of entities} = \text{no. of moles} \times \frac{6.022 \times 10^{23} \text{ entities}}{1 \text{ mol}}$$

3.5 Converting number of entities to amount (mol) (90):
$$\text{No. of moles} = \text{no. of entities} \times \frac{1 \text{ mol}}{6.022 \times 10^{23} \text{ entities}}$$

3.6 Calculating mass % (93):
Mass % of element X
$$= \frac{\text{moles of X in formula} \times \text{molar mass of X (g/mol)}}{\text{mass (g) of 1 mol of compound}} \times 100$$

3.7 Calculating percent yield (112):
$$\% \text{ yield} = \frac{\text{actual yield}}{\text{theoretical yield}} \times 100$$

3.8 Defining molarity (114):
$$\text{Molarity} = \frac{\text{moles of solute}}{\text{liters of solution}} \quad \text{or} \quad M = \frac{\text{mol solute}}{\text{L soln}}$$

3.9 Diluting a concentrated solution (117):
$$M_{\text{dil}} \times V_{\text{dil}} = \text{number of moles} = M_{\text{conc}} \times V_{\text{conc}}$$

Highlighted Figures and Tables

These figures (F) and tables (T) provide a quick review of key ideas.

T3.1 Summary of mass terminology (89)
T3.2 Information contained in a formula (90)
F3.3 Summary of mass-mole-number relationships for elements and compounds (92)
T3.5 Information contained in a balanced equation (105)

F3.8 Summary of mass-mole-number relationships in a chemical reaction (107)
F3.11 Summary of mass-mole-number-volume relationships in solution (115)
F3.14 Overview of mole-mass-number relationships (120)

Brief Solutions to Follow-up Problems

3.1 (a) Moles of C = $315 \text{ mg C} \times \dfrac{1 \text{ g}}{10^3 \text{ mg}} \times \dfrac{1 \text{ mol C}}{12.01 \text{ g C}}$
$$= 2.62 \times 10^{-2} \text{ mol C}$$
(b) Mass (g) of Mn = $3.22 \times 10^{20} \text{ Mn atoms}$
$$\times \frac{1 \text{ mol Mn}}{6.022 \times 10^{23} \text{ Mn atoms}} \times \frac{54.94 \text{ g Mn}}{1 \text{ mol Mn}}$$
$$= 2.94 \times 10^{-2} \text{ g Mn}$$

3.2 (a) Mass (g) of P_4O_{10} = $4.65 \times 10^{22} \text{ molecules } P_4O_{10}$
$$\times \frac{1 \text{ mol } P_4O_{10}}{6.022 \times 10^{23} \text{ molecules } P_4O_{10}} \times \frac{283.88 \text{ g } P_4O_{10}}{1 \text{ mol } P_4O_{10}}$$
$$= 21.9 \text{ g } P_4O_{10}$$
(b) No. of P atoms = $4.65 \times 10^{22} \text{ molecules } P_4O_{10}$
$$\times \frac{4 \text{ atoms P}}{1 \text{ molecule } P_4O_{10}} = 1.86 \times 10^{23} \text{ P atoms}$$

Brief Solutions to Follow-up Problems *(continued)*

3.3 (a) Mass % of N $= \dfrac{2 \text{ mol N} \times \dfrac{14.01 \text{ g N}}{1 \text{ mol N}}}{80.05 \text{ g NH}_4\text{NO}_3} \times 100$

$= 35.00$ mass % N

(b) Mass (g) of N $= 35.8 \text{ kg NH}_4\text{NO}_3 \times \dfrac{10^3 \text{ g}}{1 \text{ kg}} \times \dfrac{0.3500 \text{ g N}}{1 \text{ g NH}_4\text{NO}_3}$

$= 1.25 \times 10^4$ g N

3.4 Moles of S $= 2.88 \text{ g S} \times \dfrac{1 \text{ mol S}}{32.07 \text{ g S}} = 0.0898$ mol S

Moles of M $= 0.0898 \text{ mol S} \times \dfrac{2 \text{ mol M}}{3 \text{ mol S}} = 0.0599$ mol M

Molar mass of M $= \dfrac{3.12 \text{ g M}}{0.0599 \text{ mol M}} = 52.1$ g/mol

M is chromium, and M_2S_3 is chromium(III) sulfide.

3.5 Assuming 100.00 g of compound, we have 95.21 g of C and 4.79 g of H:

Moles of C $= 95.21 \text{ g C} \times \dfrac{1 \text{ mol C}}{12.01 \text{ g C}}$

$= 7.928$ mol C; also, 4.75 mol H

Preliminary formula $= C_{7.928}H_{4.75} \approx C_{1.67}H_{1.00}$

Empirical formula $= C_5H_3$

Whole-number multiple $= \dfrac{252.30 \text{ g/mol}}{63.07 \text{ g/mol}} = 4$

Molecular formula $= C_{20}H_{12}$

3.6 Mass (g) of C $= 0.451 \text{ g CO}_2 \times \dfrac{12.01 \text{ g C}}{44.01 \text{ g CO}_2}$

$= 0.123$ g C; also, 0.00690 g H

Mass (g) of Cl $= 0.250 \text{ g} - (0.123 \text{ g} + 0.00690 \text{ g})$

$= 0.120$ g Cl

Moles of elements

$= 0.0102$ mol C; 0.00685 mol H; 0.00339 mol Cl

Empirical formula $= C_3H_2Cl$; multiple $= 2$

Molecular formula $= C_6H_4Cl_2$

3.7 (a) $2Na(s) + 2H_2O(l) \longrightarrow H_2(g) + 2NaOH(aq)$

(b) $2HNO_3(aq) + CaCO_3(s) \longrightarrow$

$H_2O(l) + CO_2(g) + Ca(NO_3)_2(aq)$

(c) $PCl_3(g) + 3HF(g) \longrightarrow PF_3(g) + 3HCl(g)$

(d) $4C_3H_5N_3O_9(l) \longrightarrow$

$12CO_2(g) + 10H_2O(g) + 6N_2(g) + O_2(g)$

3.8 $Fe_2O_3(s) + 2Al(s) \longrightarrow Al_2O_3(s) + 2Fe(l)$

(a) Mass (g) of Fe

$= 135 \text{ g Al} \times \dfrac{1 \text{ mol Al}}{26.98 \text{ g Al}} \times \dfrac{2 \text{ mol Fe}}{2 \text{ mol Al}} \times \dfrac{55.85 \text{ g Fe}}{1 \text{ mol Fe}}$

$= 279$ g Fe

(b) No. of Al atoms

$= 1.00 \text{ g Al}_2\text{O}_3 \times \dfrac{1 \text{ mol Al}_2\text{O}_3}{101.96 \text{ g Al}_2\text{O}_3} \times \dfrac{2 \text{ mol Al}}{1 \text{ mol Al}_2\text{O}_3}$

$\times \dfrac{6.022 \times 10^{23} \text{ Al atoms}}{1 \text{ mol Al}} = 1.18 \times 10^{22}$ Al atoms

3.9

$$2SO_2(g) + O_2(g) \longrightarrow 2SO_3(g)$$
$$2SO_3(g) + 2H_2O(l) \longrightarrow 2H_2SO_4(aq)$$
$$\overline{2SO_2(g) + O_2(g) + 2H_2O(l) \longrightarrow 2H_2SO_4(aq)}$$

3.10 $2Al(s) + 3S(s) \longrightarrow Al_2S_3(s)$

Mass (g) of Al_2S_3 formed from Al

$= 10.0 \text{ g Al} \times \dfrac{1 \text{ mol Al}}{26.98 \text{ g Al}} \times \dfrac{1 \text{ mol Al}_2\text{S}_3}{2 \text{ mol Al}} \times \dfrac{150.17 \text{ g Al}_2\text{S}_3}{1 \text{ mol Al}_2\text{S}_3}$

$= 27.8$ g Al_2S_3

Similarly, mass (g) of Al_2S_3 formed from S $= 23.4$ g Al_2S_3.

Therefore, S is limiting reactant and 23.4 g of Al_2S_3 can form.

Mass (g) of Al in excess

$=$ total mass of Al $-$ mass of Al used

$= 10.0$ g Al

$- \left(15.0 \text{ g S} \times \dfrac{1 \text{ mol S}}{32.07 \text{ g S}} \times \dfrac{2 \text{ mol Al}}{3 \text{ mol S}} \times \dfrac{26.98 \text{ Al}}{1 \text{ mol Al}} \right)$

$= 1.6$ g Al

(We would obtain the same answer if sulfur were shown more correctly as S_8.)

3.11 $2HCl(aq) + CaCO_3(s) \longrightarrow$

$CaCl_2(aq) + H_2O(l) + CO_2(g)$

Theoretical yield (g) of CO_2

$= 10.0 \text{ g CaCO}_3 \times \dfrac{1 \text{ mol CaCO}_3}{100.09 \text{ g CaCO}_3} \times \dfrac{1 \text{ mol CO}_2}{1 \text{ mol CaCO}_3}$

$\times \dfrac{44.01 \text{ g CO}_2}{1 \text{ mol CO}_2} = 4.40$ g CO_2

% yield $= \dfrac{3.65 \text{ g CO}_2}{4.40 \text{ g CO}_2} \times 100 = 83.0\%$

3.12 Moles of KI $= 84 \text{ mL soln} \times \dfrac{1 \text{ L}}{10^3 \text{ mL}} \times \dfrac{0.50 \text{ mol KI}}{1 \text{ L soln}}$

$= 0.042$ mol KI

3.13 Volume (L) of soln

$= 135 \text{ g sucrose} \times \dfrac{1 \text{ mol sucrose}}{342.30 \text{ g sucrose}} \times \dfrac{1 \text{ L soln}}{3.30 \text{ mol sucrose}}$

$= 0.120$ L soln

3.14 M_{dil} of $H_2SO_4 = \dfrac{7.50 \, M \times 25.0 \text{ mL}}{500. \text{ mL}} = 0.375 \, M \, H_2SO_4$

Mass (g) of H_2SO_4/mL soln

$= \dfrac{0.375 \text{ mol H}_2\text{SO}_4}{1 \text{ L soln}} \times \dfrac{1 \text{ L}}{10^3 \text{ mL}} \times \dfrac{98.09 \text{ g H}_2\text{SO}_4}{1 \text{ mol H}_2\text{SO}_4}$

$= 3.68 \times 10^{-2}$ g/mL soln

3.15 $Al(OH)_3(s) + 3HCl(aq) \longrightarrow AlCl_3(aq) + 3H_2O(l)$

Volume (L) of HCl consumed

$= 0.10 \text{ g Al(OH)}_3 \times \dfrac{1 \text{ mol Al(OH)}_3}{78.00 \text{ g Al(OH)}_3} \times \dfrac{3 \text{ mol HCl}}{1 \text{ mol Al(OH)}_3}$

$\times \dfrac{1 \text{ L soln}}{0.10 \text{ mol HCl}} = 3.8 \times 10^{-2}$ L soln

Therefore, $Al(OH)_3$ is more effective than $Mg(OH)_2$.

3.16 (a) Volume (L) of soln

$= 0.400 \text{ mol Pb}^{2+} \times \dfrac{1 \text{ mol Pb(C}_2\text{H}_3\text{O}_2)_2}{1 \text{ mol Pb}^{2+}}$

$\times \dfrac{1 \text{ L soln}}{1.50 \text{ mol Pb(C}_2\text{H}_3\text{O}_2)_2} = 0.267$ L soln

(b) $Pb(C_2H_3O_2)_2(aq) + 2NaCl(aq) \longrightarrow$

$PbCl_2(s) + 2NaC_2H_3O_2(aq)$

Mass (g) of $PbCl_2$ from $Pb(C_2H_3O_2)_2$ soln $= 111$ g $PbCl_2$

Mass (g) of $PbCl_2$ from NaCl soln $= 59.1$ g $PbCl_2$

Thus, NaCl is the limiting reactant and 59.1 g of $PbCl_2$ can form.

Problems

Problems with **colored** numbers are answered at the back of the text. Sections match the text and provide the number(s) of relevant sample problems. Most offer Concept Review Questions, Skill-Building Exercises (in similar pairs), and Problems in Context. Then Comprehensive Problems, based on material from any section or previous chapters, follow.

The Mole
(Sample Problems 3.1 to 3.3)

● **Concept Review Questions**

3.1 The atomic mass of Cl is 35.45 amu and that of Al is 26.98 amu. What are the masses in grams of 2 mol of Al atoms and of 3 mol of Cl atoms?

3.2 (a) How many moles of C atoms are in 1 mol of sucrose ($C_{12}H_{22}O_{11}$)? (b) How many C atoms are in 1 mol of sucrose?

3.3 Why might the expression "1 mol of nitrogen" be confusing? What change would remove any uncertainty? For what other elements might a similar confusion exist? Why?

3.4 How is the molecular mass of a compound the same as the molar mass, and how is it different?

3.5 What advantage is there to using a counting unit (the mole) in chemistry rather than a mass unit?

3.6 Each of the following balances weighs the indicated numbers of atoms of two elements:

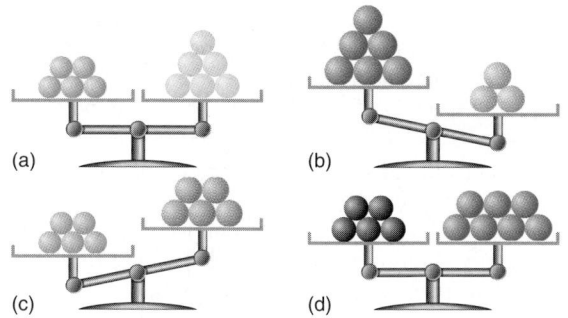

(a)   (b)
(c)   (d)

Which element—left, right, or neither,
(a) Has the higher molar mass?
(b) Has more atoms per gram?
(c) Has fewer atoms per gram?
(d) Has more atoms per mole?

3.7 You need to calculate the number of P_4 molecules that can form from 2.5 g of $Ca_3(PO_4)_2$. Explain how you would proceed (that is, write a solution "plan" without doing any calculations).

● **Skill-Building Exercises (paired)**

3.8 Calculate the molar mass of each of the following:
(a) $Sr(OH)_2$ (b) N_2O (c) $NaClO_3$ (d) Cr_2O_3

3.9 Calculate the molar mass of each of the following:
(a) $(NH_4)_3PO_4$ (b) CH_2Cl_2 (c) $CuSO_4 \cdot 5H_2O$ (d) BrF_5

3.10 Calculate the molar mass of each of the following:
(a) SnO_2 (b) BaF_2 (c) $Al_2(SO_4)_3$ (d) $MnCl_2$

3.11 Calculate the molar mass of each of the following:
(a) N_2O_4 (b) C_8H_{10} (c) $MgSO_4 \cdot 7H_2O$ (d) $Ca(C_2H_3O_2)_2$

3.12 Calculate each of the following quantities:
(a) Mass in grams of 0.57 mol of $KMnO_4$
(b) Moles of O atoms in 8.18 g of $Mg(NO_3)_2$
(c) Number of O atoms in 8.1×10^{-3} g of $CuSO_4 \cdot 5H_2O$

3.13 Calculate each of the following quantities:
(a) Mass in kilograms of 3.8×10^{20} molecules of NO_2
(b) Moles of Cl atoms in 0.0425 g of $C_2H_4Cl_2$
(c) Number of H^- ions in 4.92 g of SrH_2

3.14 Calculate each of the following quantities:
(a) Mass in grams of 0.64 mol of $MnSO_4$
(b) Moles of compound in 15.8 g of $Fe(ClO_4)_3$
(c) Number of N atoms in 92.6 g of NH_4NO_2

3.15 Calculate each of the following quantities:
(a) Total number of ions in 38.1 g of CaF_2
(b) Mass in milligrams of 3.58 mol of $CuCl_2 \cdot 2H_2O$
(c) Mass in kilograms of 2.88×10^{22} formula units of $Bi(NO_3)_3 \cdot 5H_2O$

3.16 Calculate each of the following quantities:
(a) Mass in grams of 8.41 mol of copper(I) carbonate
(b) Mass in grams of 2.04×10^{21} molecules of dinitrogen pentaoxide
(c) Number of moles and formula units in 57.9 g of sodium perchlorate
(d) Number of sodium ions, perchlorate ions, Cl atoms, and O atoms in the mass in part (c)

3.17 Calculate each of the following quantities:
(a) Mass in grams of 3.52 mol of chromium(III) sulfate decahydrate
(b) Mass in grams of 9.64×10^{24} molecules of dichlorine heptaoxide
(c) Number of moles and formula units in 56.2 g of lithium sulfate
(d) Number of lithium ions, sulfate ions, S atoms, and O atoms in the mass in part (c)

3.18 Calculate each of the following:
(a) Mass % of H in ammonium bicarbonate
(b) Mass % of O in sodium dihydrogen phosphate heptahydrate

3.19 Calculate each of the following:
(a) Mass % of I in strontium periodate
(b) Mass % of Mn in potassium permanganate

3.20 Calculate each of the following:
(a) Mass fraction of C in cesium acetate
(b) Mass fraction of O in uranyl sulfate trihydrate (the uranyl ion is UO_2^{2+})

3.21 Calculate each of the following:
(a) Mass fraction of Cl in calcium chlorate
(b) Mass fraction of P in tetraphosphorus hexaoxide

● **Problems in Context**

3.22 Oxygen is required for metabolic combustion of foods. Calculate the number of atoms in 38.0 g of oxygen gas, the amount absorbed from the lungs at rest in about 15 minutes.

3.23 Cisplatin (below), or Platinol, is a powerful drug used in the treatment of certain cancers. Calculate (a) the moles of compound in 285.3 g of cisplatin; (b) the number of hydrogen atoms in 0.98 mol of cisplatin.

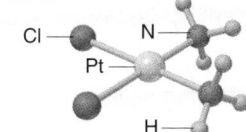

3.24 Allyl sulfide gives garlic its characteristic odor.

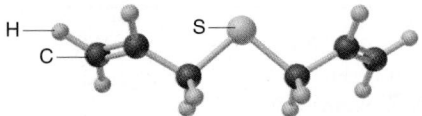

Calculate (a) the mass in grams of 1.63 mol of allyl sulfide;
(b) the number of carbon atoms in 4.77 g of allyl sulfide.
3.25 Iron reacts slowly with oxygen and water to form a compound commonly called rust ($Fe_2O_3 \cdot 4H_2O$). For 65.2 kg of rust, calculate (a) the moles of compound; (b) the moles of Fe_2O_3; (c) the grams of iron.
3.26 Propane is widely used in liquid form as a fuel for barbecue grills and camp stoves. For 75.3 g of propane, calculate (a) the moles of compound; (b) the grams of carbon.
3.27 The effectiveness of a nitrogen fertilizer is determined mainly by its mass % N. Rank the following fertilizers in terms of their effectiveness: potassium nitrate; ammonium nitrate; ammonium sulfate; urea, $CO(NH_2)_2$.
3.28 One useful method for recycling scrap aluminum converts some of it to common alum [potassium aluminum sulfate dodecahydrate, $KAl(SO_4)_2 \cdot 12H_2O$], an important dye fixative and food preservative.
(a) How many moles of potassium ions are in 352 g of alum?
(b) What is the mass % S in alum?
(c) What is the maximum amount of alum (in metric tons) that can be prepared from 18.5 metric tons of scrap aluminum (1 metric ton = 1000 kg)?
3.29 Hemoglobin, a protein found in red blood cells, carries O_2 from the lungs to the body's cells. Iron (as ferrous ion, Fe^{2+}) makes up 0.33 mass % of hemoglobin. If the molar mass of hemoglobin is 6.8×10^4 g/mol, how many Fe^{2+} ions are present in one molecule?

Determining the Formula of an Unknown Compound
(Sample Problems 3.4 to 3.6)

● **Concept Review Questions**

3.30 List three ways compositional data may be given in a problem that involves finding an empirical formula.
3.31 Which of the following sets of information allows you to obtain the molecular formula of a covalent compound? In each case that allows it, explain how you would proceed (write a solution "plan").
(a) Number of moles of each type of atom in a given sample of the compound
(b) Mass % of each element and the total number of atoms in a molecule of the compound
(c) Mass % of each element and the number of atoms of one element in a molecule of the compound
(d) Empirical formula and the mass % of each element in the compound
(e) Structural formula of the compound
3.32 Is $MgCl_2$ an empirical or a molecular formula for magnesium chloride? Explain.

● **Skill-Building Exercises (paired)**

3.33 What is the empirical formula and empirical formula mass for each of the following compounds?
(a) C_2H_4 (b) $C_2H_6O_2$ (c) N_2O_5 (d) $Ba_3(PO_4)_2$ (e) Te_4I_{16}

3.34 What is the empirical formula and empirical formula mass for each of the following compounds?
(a) C_4H_8 (b) $C_3H_6O_3$ (c) P_4O_{10} (d) $Ga_2(SO_4)_3$ (e) Al_2Br_6
3.35 What is the molecular formula of each compound?
(a) Empirical formula CH_2 ($M = 42.08$ g/mol)
(b) Empirical formula NH_2 ($M = 32.05$ g/mol)
(c) Empirical formula NO_2 ($M = 92.02$ g/mol)
(d) Empirical formula CHN ($M = 135.14$ g/mol)
3.36 What is the molecular formula of each compound?
(a) Empirical formula CH ($M = 78.11$ g/mol)
(b) Empirical formula $C_3H_6O_2$ ($M = 74.08$ g/mol)
(c) Empirical formula $HgCl$ ($M = 472.1$ g/mol)
(d) Empirical formula $C_7H_4O_2$ ($M = 240.20$ g/mol)
3.37 Determine the empirical formula of each of the following compounds:
(a) 0.063 mol of chlorine atoms combined with 0.22 mol of oxygen atoms
(b) 2.45 g of silicon combined with 12.4 g of chlorine
(c) 27.3 mass % carbon and 72.7 mass % oxygen
3.38 Determine the empirical formula of each of the following compounds:
(a) 0.039 mol of iron atoms combined with 0.052 mol of oxygen atoms
(b) 0.903 g of phosphorus combined with 6.99 g of bromine
(c) A hydrocarbon with 79.9 mass % carbon
3.39 An oxide of nitrogen contains 30.45 mass % N. (a) What is the empirical formula of the oxide? (b) If the molar mass is 90 ± 5 g/mol, what is the molecular formula?
3.40 A chloride of silicon contains 79.1 mass % Cl. (a) What is the empirical formula of the chloride? (b) If the molar mass is 269 g/mol, what is the molecular formula?
3.41 A sample of 0.600 mol of a metal M reacts completely with excess fluorine to form 46.8 g of MF_2.
(a) How many moles of F are in the sample of MF_2 that forms?
(b) How many grams of M are in this sample of MF_2?
(c) What element is represented by the symbol M?
3.42 A sample of 0.370 mol of a metal oxide (M_2O_3) weighs 55.4 g.
(a) How many moles of O are in the sample?
(b) How many grams of M are in the sample?
(c) What element is represented by the symbol M?

● **Problems in Context**

3.43 Nicotine is a poisonous, addictive compound found in tobacco. A sample of nicotine contains 6.16 mmol of C, 8.56 mmol of H, and 1.23 mmol of N [1 mmol (1 millimole) = 10^{-3} mol]. What is the empirical formula?
3.44 Cortisol ($M = 362.47$ g/mol), one of the major steroid hormones, is a key factor in the synthesis of protein. Its profound effect on the reduction of inflammation explains its use in the treatment of rheumatoid arthritis. Cortisol is 69.6% C, 8.34% H, and 22.1% O by mass. What is its molecular formula?
3.45 Acetaminophen *(right)* is one of the most popular nonaspirin, "over-the-counter" pain relievers. What is the mass % of each element in acetaminophen?

3.46 Menthol ($\mathcal{M}$ = 156.3 g/mol), a strong-smelling substance used in cough drops, is a compound of carbon, hydrogen, and oxygen. When 0.1595 g of menthol was subjected to combustion analysis, it produced 0.449 g of CO_2 and 0.184 g of H_2O. What is its molecular formula?

Writing and Balancing Chemical Equations
(Sample Problem 3.7)

● **Concept Review Questions**

3.47 What three types of information does a balanced chemical equation provide? How?

3.48 How does a balanced chemical equation apply the law of conservation of mass?

3.49 In the process of balancing the equation

$$Al + Cl_2 \longrightarrow AlCl_3$$

Student I writes: $Al + Cl_2 \longrightarrow AlCl_2$
Student II writes: $Al + Cl_2 + Cl \longrightarrow AlCl_3$
Student III writes: $2Al + 3Cl_2 \longrightarrow 2AlCl_3$
Is the approach of Student I valid? That of Student II? Of Student III? Explain.

3.50 The following boxes represent a chemical reaction between elements A (red) and B (green):

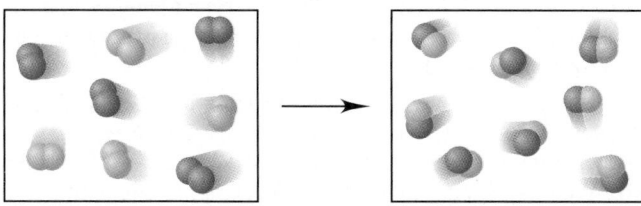

Which of the following best represents the balanced equation for the reaction?
(a) $2A + 2B \longrightarrow A_2 + B_2$ (b) $A_2 + B_2 \longrightarrow 2AB$
(c) $B_2 + 2AB \longrightarrow 2B_2 + A_2$ (d) $4A_2 + 4B_2 \longrightarrow 8AB$

● **Skill-Building Exercises (paired)**

3.51 Write balanced equations for each of the following by inserting the correct coefficients in the blanks:
(a) __Cu(s) + __S_8(s) ⟶ __Cu_2S(s)
(b) __P_4O_{10}(s) + __H_2O(l) ⟶ __H_3PO_4(l)
(c) __B_2O_3(s) + __NaOH(aq) ⟶
     __Na_3BO_3(aq) + __H_2O(l)
(d) __CH_3NH_2(g) + __O_2(g) ⟶
     __CO_2(g) + __H_2O(g) + __N_2(g)

3.52 Write balanced equations for each of the following by inserting the correct coefficients in the blanks:
(a) __$Cu(NO_3)_2$(aq) + __KOH(aq) ⟶
     __$Cu(OH)_2$(s) + __KNO_3(aq)
(b) __BCl_3(g) + __H_2O(l) ⟶ __H_3BO_3(s) + __HCl(g)
(c) __$CaSiO_3$(s) + __HF(g) ⟶
     __SiF_4(g) + __CaF_2(s) + __H_2O(l)
(d) __$(CN)_2$(g) + __H_2O(l) ⟶ __$H_2C_2O_4$(aq) + __NH_3(g)

3.53 Write balanced equations for each of the following by inserting the correct coefficients in the blanks:
(a) __SO_2(g) + __O_2(g) ⟶ __SO_3(g)
(b) __Sc_2O_3(s) + __H_2O(l) ⟶ __$Sc(OH)_3$(s)
(c) __H_3PO_4(aq) + __NaOH(aq) ⟶
     __Na_2HPO_4(aq) + __H_2O(l)
(d) __$C_6H_{10}O_5$(s) + __O_2(g) ⟶ __CO_2(g) + __H_2O(g)

3.54 Write balanced equations for each of the following by inserting the correct coefficients in the blanks:
(a) __As_4S_6(s) + __O_2(g) ⟶ __As_4O_6(s) + __SO_2(g)
(b) __$Ca_3(PO_4)_2$(s) + __SiO_2(s) + __C(s) ⟶
     __P_4(g) + __$CaSiO_3$(l) + __CO(g)
(c) __Fe(s) + __H_2O(g) ⟶ __Fe_3O_4(s) + __H_2(g)
(d) __S_2Cl_2(l) + __NH_3(g) ⟶
     __S_4N_4(s) + __S_8(s) + __NH_4Cl(s)

3.55 Convert the following into balanced equations:
(a) When gallium metal is heated in oxygen gas, it melts and forms solid gallium(III) oxide.
(b) Liquid hexane burns in oxygen gas to form carbon dioxide gas and water vapor.
(c) When solutions of calcium chloride and sodium phosphate are mixed, solid calcium phosphate forms and sodium chloride remains in solution.

3.56 Convert the following into balanced equations:
(a) When lead(II) nitrate solution is added to potassium iodide solution, solid lead(II) iodide forms and potassium nitrate solution remains.
(b) Liquid disilicon hexachloride reacts with water to form solid silicon dioxide, hydrogen chloride gas, and hydrogen gas.
(c) When nitrogen dioxide is bubbled into water, a solution of nitric acid forms and gaseous nitrogen monoxide is released.

Calculating Amounts of Reactant and Product
(Sample Problems 3.8 to 3.11)

● **Concept Review Questions**

3.57 What does the term *stoichiometrically equivalent molar ratio* mean, and how is it applied in solving problems?

3.58 Reactants A and B form product C. Write a detailed "plan" to find the mass of C when 5 g of A reacts with excess B.

3.59 Reactants D and E form product F. Write a detailed "plan" to find the mass of F when 17 g of D reacts with 11 g of E.

3.60 Percent yields are generally calculated from mass quantities. Would the result be the same if mole quantities were used instead? Why?

● **Skill-Building Exercises (paired)**

3.61 Chlorine gas can be made in the laboratory by the reaction of hydrochloric acid and manganese(IV) oxide:

$$4HCl(aq) + MnO_2(s) \longrightarrow MnCl_2(aq) + 2H_2O(g) + Cl_2(g)$$

When 1.82 mol of HCl reacts with excess MnO_2, (a) how many moles of Cl_2 form? (b) How many grams of Cl_2 form?

3.62 Bismuth oxide reacts with carbon to form bismuth metal:

$$Bi_2O_3(s) + 3C(s) \longrightarrow 2Bi(s) + 3CO(g)$$

When 352 g of Bi_2O_3 reacts with excess carbon, (a) how many moles of Bi_2O_3 react? (b) How many moles of Bi form?

3.63 Potassium nitrate decomposes on heating, producing potassium oxide and gaseous nitrogen and oxygen:

$$4KNO_3(s) \longrightarrow 2K_2O(s) + 2N_2(g) + 5O_2(g)$$

To produce 88.6 kg of oxygen, how many (a) moles of KNO_3 must be heated? (b) Grams of KNO_3 must be heated?

3.64 Chromium(III) oxide reacts with hydrogen sulfide (H_2S) gas to form chromium(III) sulfide and water:

$$Cr_2O_3(s) + 3H_2S(g) \longrightarrow Cr_2S_3(s) + 3H_2O(l)$$

To produce 421 g of Cr_2S_3, (a) how many moles of Cr_2O_3 are required? (b) How many grams of Cr_2O_3 are required?

3.65 Calculate the mass of each product formed when 33.61 g of diborane (B_2H_6) reacts with excess water:

$$B_2H_6(g) + H_2O(l) \longrightarrow H_3BO_3(s) + H_2(g) \quad \text{[unbalanced]}$$

3.66 Calculate the mass of each product formed when 174 g of silver sulfide reacts with excess hydrochloric acid:

$$Ag_2S(s) + HCl(aq) \longrightarrow AgCl(s) + H_2S(g) \quad \text{[unbalanced]}$$

3.67 Elemental phosphorus occurs as tetratomic molecules, P_4. What mass of chlorine gas is needed for complete reaction with 355 g of phosphorus to form phosphorus pentachloride?

3.68 Elemental sulfur occurs as octatomic molecules, S_8. What mass of fluorine gas is needed for complete reaction with 17.8 g of sulfur to form sulfur hexafluoride?

3.69 Solid iodine trichloride is prepared by reaction between solid iodine and gaseous chlorine to form iodine monochloride crystals, followed by treatment with additional chlorine.
(a) Write a balanced equation for each step.
(b) Write an overall balanced equation for the formation of iodine trichloride.
(c) How many grams of iodine are needed to prepare 31.4 kg of final product?

3.70 Lead can be prepared from galena [lead(II) sulfide] by first roasting the galena in oxygen gas to form lead(II) oxide and sulfur dioxide. Heating the metal oxide with more galena forms the molten metal and more sulfur dioxide.
(a) Write a balanced equation for each step.
(b) Write an overall balanced equation for the process.
(c) How many metric tons of sulfur dioxide form for every metric ton of lead obtained?

3.71 Many metals react with oxygen gas to form the metal oxide. For example, calcium reacts as follows:

$$2Ca(s) + O_2(g) \longrightarrow 2CaO(s)$$

You wish to calculate the mass of calcium oxide that can be prepared from 4.20 g of Ca and 2.80 g of O_2.
(a) How many moles of CaO can be produced from the given mass of Ca?
(b) How many moles of CaO can be produced from the given mass of O_2?
(c) Which is the limiting reactant?
(d) How many grams of CaO can be produced?

3.72 Metal hydrides react with water to form hydrogen gas and the metal hydroxide. For example,

$$SrH_2(s) + 2H_2O(l) \longrightarrow Sr(OH)_2(s) + 2H_2(g)$$

You wish to calculate the mass of hydrogen gas that can be prepared from 5.63 g of SrH_2 and 4.80 g of H_2O.
(a) How many moles of H_2 can be produced from the given mass of SrH_2?
(b) How many moles of H_2 can be produced from the given mass of H_2O?
(c) Which is the limiting reactant?
(d) How many grams of H_2 can be produced?

3.73 Calculate the maximum numbers of moles and grams of iodic acid (HIO_3) that can form when 685 g of iodine trichloride reacts with 117.4 g of water:

$$ICl_3 + H_2O \longrightarrow ICl + HIO_3 + HCl \quad \text{[unbalanced]}$$

What mass of the excess reactant remains?

3.74 Calculate the maximum numbers of moles and grams of H_2S that can form when 158 g of aluminum sulfide reacts with

131 g of water:

$$Al_2S_3 + H_2O \longrightarrow Al(OH)_3 + H_2S \quad \text{[unbalanced]}$$

What mass of the excess reactant remains?

3.75 When 0.100 mol of carbon is burned in a closed vessel with 8.00 g of oxygen, how many grams of carbon dioxide can form? Which reactant is in excess, and how many grams of it remain after the reaction?

3.76 A mixture of 0.0359 g of hydrogen and 0.0175 mol of oxygen in a closed container is sparked to initiate a reaction. How many grams of water can form? Which reactant is in excess, and how many grams of it remain after the reaction?

3.77 Aluminum nitrite and ammonium chloride react to form aluminum chloride, nitrogen, and water. What mass of each substance is present after 62.5 g of aluminum nitrite and 54.6 g of ammonium chloride react completely?

3.78 Calcium nitrate and ammonium fluoride react to form calcium fluoride, dinitrogen monoxide, and water vapor. What mass of each substance is present after 16.8 g of calcium nitrate and 17.50 g of ammonium fluoride react completely?

3.79 Two successive reactions, A $\longrightarrow$ B and B $\longrightarrow$ C, have yields of 82% and 65%, respectively. What is the overall percent yield for conversion of A to C?

3.80 Two successive reactions, D $\longrightarrow$ E and E $\longrightarrow$ F, have yields of 48% and 73%, respectively. What is the overall percent yield for conversion of D to F?

3.81 What is the percent yield of a reaction in which 41.5 g of tungsten(VI) oxide (WO_3) reacts with excess hydrogen gas to produce metallic tungsten and 9.50 mL of water ($d = 1.00$ g/mL)?

3.82 What is the percent yield of a reaction in which 200. g of phosphorus trichloride reacts with excess water to form 128 g of HCl and aqueous phosphorous acid (H_3PO_3)?

3.83 When 18.5 g of methane and 43.0 g of chlorine gas undergo a reaction that has an 80.0% yield, what mass of chloromethane (CH_3Cl) forms? Hydrogen chloride also forms.

3.84 When 56.6 g of calcium and 30.5 g of nitrogen gas undergo a reaction that has a 93.0% yield, what mass of calcium nitride forms?

● **Problems in Context**

3.85 Cyanogen, $(CN)_2$, has been observed in the atmosphere of Titan, Saturn's largest moon, and in the gases of interstellar nebulas. On Earth, it is used as a welding gas and a fumigant. In its reaction with fluorine gas, carbon tetrafluoride and nitrogen trifluoride gases are produced. What mass of carbon tetrafluoride forms when 80.0 g of each reactant is used?

3.86 An intermediate step in the industrial production of nitric acid involves the reaction of ammonia with oxygen gas to form nitrogen monoxide and water. How many grams of nitrogen monoxide can form by the reaction of 466 g of ammonia with 812 g of oxygen?

3.87 Gaseous butane is compressed and used as a liquid fuel in disposable cigarette lighters and lightweight camping stoves. Suppose a lighter contains 6.50 mL of butane ($d = 0.579$ g/mL).
(a) How many grams of oxygen are needed to burn the butane completely?
(b) How many moles of CO_2 form when all the butane burns?
(c) How many total molecules of gas form when the butane burns completely?

3.88 Diborane (B_2H_6), a useful reactant in organic synthesis, may be prepared by the following reaction, which occurs in a nonaqueous solvent:

$$NaBH_4(s) + BF_3(g) \longrightarrow B_2H_6(g) + NaBF_4(s) \text{ [unbalanced]}$$

If the reaction has a 75.0% yield of diborane, how many grams of $NaBH_4$ are needed to make 25.0 g of B_2H_6?

Fundamentals of Solution Stoichiometry
(Sample Problems 3.12 to 3.16)

● **Concept Review Questions**

3.89 Box A represents a unit volume of a solution. Choose from boxes B and C the one representing the same unit volume of solution that has (a) more solute added; (b) more solvent added; (c) higher molarity; (d) lower concentration.

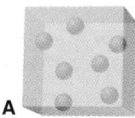
A

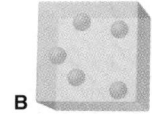
B

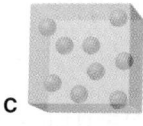
C

3.90 Given the volume and molarity of a $CaCl_2$ solution, how do you determine the number of moles and the mass of solute?

3.91 Are the following instructions for diluting a 10.0 M solution to a 1.00 M solution correct: "Take 100.0 mL of the 10.0 M solution and add 900.0 mL water"? Explain.

● **Skill-Building Exercises** *(paired)*

3.92 Calculate each of the following quantities:
(a) Grams of solute in 175.8 mL of 0.207 M calcium acetate
(b) Molarity of 500. mL of solution containing 21.1 g of potassium iodide
(c) Moles of solute in 145.6 L of 0.850 M sodium cyanide

3.93 Calculate each of the following quantities:
(a) Volume in liters of 2.26 M potassium hydroxide that contains 8.42 g of solute
(b) Number of Cu^{2+} ions in 52 L of 2.3 M copper(II) chloride
(c) Molarity of 275 mL of solution containing 135 mmol of glucose

3.94 Calculate each of the following quantities:
(a) Grams of solute needed to make 475 mL of 5.62×10^{-2} M potassium sulfate
(b) Molarity of a solution that contains 6.55 mg of calcium chloride in each milliliter
(c) Number of Mg^{2+} ions in each milliliter of 0.184 M magnesium bromide

3.95 Calculate each of the following quantities:
(a) Molarity of the solution resulting from dissolving 46.0 g of silver nitrate in enough water to give a final volume of 335 mL
(b) Volume in liters of 0.385 M manganese(II) sulfate that contains 57.0 g of solute
(c) Volume in milliliters of 6.44×10^{-2} M adenosine triphosphate (ATP) that contains 1.68 mmol of ATP

3.96 Calculate each of the following quantities:
(a) Molarity of a solution prepared by diluting 37.00 mL of 0.250 M potassium chloride to 150.00 mL
(b) Molarity of a solution prepared by diluting 25.71 mL of 0.0706 M ammonium sulfate to 500.00 mL
(c) Molarity of sodium ion in a solution made by mixing 3.58 mL of 0.288 M sodium chloride with 500. mL of 6.51×10^{-3} M sodium sulfate (assume volumes are additive)

3.97 Calculate each of the following quantities:
(a) Volume of 2.050 M copper(II) nitrate that must be diluted with water to prepare 750.0 mL of a 0.8543 M solution
(b) Volume of 1.03 M calcium chloride that must be diluted with water to prepare 350. mL of a 2.66×10^{-2} M chloride ion solution
(c) Final volume of a 0.0700 M solution prepared by diluting 18.0 mL of 0.155 M lithium carbonate with water

3.98 A sample of concentrated nitric acid has a density of 1.41 g/mL and contains 70.0% HNO_3 by mass.
(a) What mass of HNO_3 is present per liter of solution?
(b) What is the molarity of the solution?

3.99 Concentrated sulfuric acid (18.3 M) has a density of 1.84 g/mL.
(a) How many moles of sulfuric acid are present per milliliter of solution?
(b) What is the mass % of H_2SO_4 in the solution?

3.100 How many milliliters of 0.383 M HCl are needed to react with 16.2 g of $CaCO_3$?

$$2HCl(aq) + CaCO_3(s) \longrightarrow CaCl_2(aq) + CO_2(g) + H_2O(l)$$

3.101 How many grams of NaH_2PO_4 are needed to react with 38.74 mL of 0.275 M NaOH?

$$NaH_2PO_4(s) + 2NaOH(aq) \longrightarrow Na_3PO_4(aq) + 2H_2O(l)$$

3.102 How many grams of solid barium sulfate form when 25.0 mL of 0.160 M barium chloride reacts with 68.0 mL of 0.055 M sodium sulfate? Aqueous sodium chloride is the other product.

3.103 How many moles of which reactant are in excess when 350.0 mL of 0.210 M sulfuric acid reacts with 0.500 L of 0.196 M sodium hydroxide to form water and aqueous sodium sulfate?

● **Problems in Context**

3.104 Ordinary household bleach is an aqueous solution of sodium hypochlorite. What is the molarity of a bleach solution that contains 20.5 g of sodium hypochlorite in a total volume of 375 mL?

3.105 Muriatic acid, an industrial grade of concentrated HCl, is used to clean masonry and etch cement for painting. Its concentration is 11.7 M.
(a) Write instructions for diluting the concentrated acid to make 5.0 gallons of 3.5 M acid for routine use (1 gal = 4 qt; 1 qt = 0.946 L).
(b) How many milliliters of the muriatic acid solution contain 9.55 g of HCl?

3.106 A sample of impure magnesium was analyzed by allowing it to react with excess HCl solution:

$$Mg(s) + 2HCl(aq) \longrightarrow MgCl_2(aq) + H_2(g)$$

After 1.32 g of the impure metal was treated with 0.100 L of 0.750 M HCl, 0.0125 mol of HCl remained. Assuming the impurities do not react with the acid, what is the mass % Mg in the sample?

Comprehensive Problems

3.107 The mole is defined in terms of the carbon-12 atom. Use the definition to find (a) the mass in grams equal to 1 atomic mass unit; (b) the ratio of the gram to the atomic mass unit.

3.108 The study of sulfur-nitrogen compounds is an active area of chemical research, made more so by the discovery in the early 1980s of one such compound that conducts electricity like a metal. The first sulfur-nitrogen compound was prepared in 1835 and serves today as a reactant for preparing many of the others. Mass spectrometry of the compound shows a molar mass of 184.27 g/mol, and analysis shows it to contain 2.288 g of S for every 1.000 g of N. What is its molecular formula?

3.109 Hydrogen-containing fuels have a "fuel value" based on their mass % H. Rank the following compounds from highest mass % H to lowest: ethane, propane, benzene, ethanol, cetyl palmitate (whale oil, $C_{32}H_{64}O_2$).

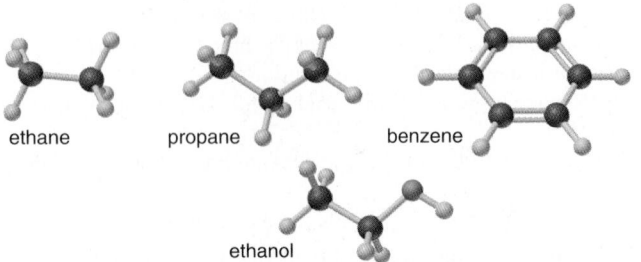

ethane   propane   benzene

ethanol

3.110 Narceine is a narcotic in opium. It crystallizes from water solution as a hydrate that contains 10.8 mass % water. If the molar mass of narceine hydrate is 499.52 g/mol, determine x in narceine·xH$_2$O.

3.111 Serotonin ($\mathcal{M} = 176$ g/mol) is a compound that conducts nerve impulses in brain and muscle. It contains 68.2 mass % C, 6.86 mass % H, 15.9 mass % N, and 9.08 mass % O. What is its molecular formula?

3.112 Convert the following descriptions of reactions into balanced equations:
(a) In a gaseous reaction, hydrogen sulfide burns in oxygen to form sulfur dioxide and water vapor.
(b) When crystalline potassium chlorate is heated to just above its melting point, it reacts to form two different crystalline compounds, potassium chloride and potassium perchlorate.
(c) When hydrogen gas is passed over powdered iron(III) oxide, iron metal and water vapor form.
(d) The combustion of gaseous ethane in air forms carbon dioxide and water vapor.
(e) Iron(II) chloride can be converted to iron(III) fluoride by treatment with chlorine trifluoride gas. Chlorine gas is also formed.

3.113 Isobutylene is a hydrocarbon used in the manufacture of synthetic rubber. When 0.847 g of isobutylene was analyzed by combustion (using an apparatus similar to that of Figure 3.4), the gain in mass of the CO_2 absorber was 2.657 g and that of the H_2O absorber was 1.089 g. What is the empirical formula of isobutylene?

3.114 One of the compounds used to increase the octane rating of gasoline is toluene (*right*). Suppose 15.0 mL of toluene ($d = 0.867$ g/mL) is consumed when a sample of gasoline burns in air.

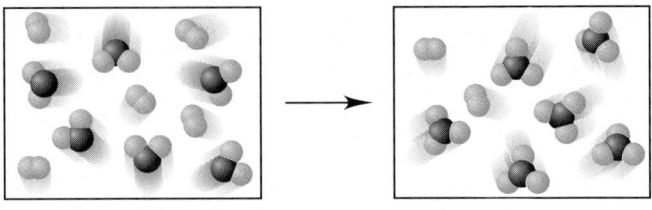

(a) How many grams of oxygen are needed for complete combustion of the toluene?
(b) How many total moles of gaseous products form?
(c) How many molecules of water vapor form?

3.115 The smelting of ferric oxide to form elemental iron occurs at high temperatures in a blast furnace through a reaction sequence with carbon monoxide. In the first step, ferric oxide re-

acts with carbon monoxide to form Fe$_3$O$_4$. This substance reacts with more carbon monoxide to form iron(II) oxide, which reacts with still more carbon monoxide to form molten iron. Carbon dioxide is also produced in each step. (a) Write an overall balanced equation for the iron-smelting process. (b) How many grams of carbon monoxide are required to form 40.0 metric tons of iron from ferric oxide?

3.116 During studies of the reaction in Sample Problem 3.10,
$$N_2O_4(l) + 2N_2H_4(l) \longrightarrow 3N_2(g) + 4H_2O(g)$$
a chemical engineer measured a less-than-expected yield of N_2 and discovered that the following side reaction occurs:
$$2N_2O_4(l) + N_2H_4(l) \longrightarrow 6NO(g) + 2H_2O(g)$$
In one experiment, 10.0 g of NO formed when 100.0 g of each reactant was used. What is the highest percent yield of N_2 that can be expected?

3.117 A mathematical equation useful for dilution calculations is $M_{dil} \times V_{dil} = M_{conc} \times V_{conc}$. What does each symbol mean, and why does the equation work?

3.118 The following boxes represent a chemical reaction between AB$_2$ and B$_2$:

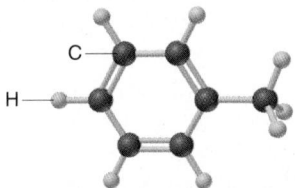

(a) Write a balanced equation for the reaction.
(b) What is the limiting reactant in this reaction?
(c) How many moles of product can be made from 3.0 mol of B$_2$ and 5.0 mol of AB$_2$?
(d) How many moles of excess reactant remain after the reaction in part (c)?

3.119 Calculate each of the following quantities:
(a) Volume of 18.0 M sulfuric acid that must be added to water to prepare 2.00 L of a 0.309 M solution
(b) Molarity of the solution obtained by diluting 80.6 mL of 0.225 M ammonium chloride to 0.250 L
(c) Volume of water added to 0.150 L of 0.0262 M sodium hydroxide to obtain a 0.0100 M solution (assume the volumes are additive at these low concentrations)
(d) Mass of calcium nitrate in each milliliter of a solution prepared by diluting 64.0 mL of 0.745 M calcium nitrate to a final volume of 0.100 L

3.120 Agricultural biochemists use 6-benzylaminopurine ($C_{12}H_{11}N_5$) in trace amounts as a plant growth regulator. In a typical application, 150. mL of a solution contains 0.030 mg of the compound. What is the molarity of the solution?

3.121 One of Germany's largest chemical manufacturers has built a plant to produce 3.74×10^8 lb of vinyl acetate per year. This compound (*right*) is used to make some of the polymers in adhesives, paints, fabric coatings, floppy disks, plastic films, and so on. Assuming production goals are met, how many moles of vinyl acetate will the plant produce each month?

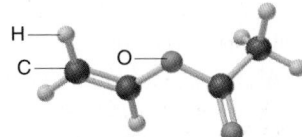

*3.122** Government and industry analysts routinely measure the quantity of ascorbic acid (vitamin C, $C_6H_8O_6$) in commercial products such as fruit juices and vitamin tablets. The ascorbic acid reacts with excess iodine, and the amount of I_2 remaining is determined with sodium thiosulfate ($Na_2S_2O_3$). In one analysis, a vitamin tablet is mixed in water and treated with 53.20 mL of 0.1030 M I_2 according to the equation

$$C_6H_8O_6(s) + I_2(aq) \longrightarrow C_6H_6O_6(aq) + 2HI(aq)$$

After all the ascorbic acid reacts, the excess I_2 reacts with 27.54 mL of 0.1153 M $Na_2S_2O_3$ according to the equation

$$I_2(aq) + 2Na_2S_2O_3(aq) \longrightarrow 2NaI(aq) + Na_2S_4O_6(aq)$$

How many grams of ascorbic acid are in the vitamin tablet?

3.123 Is each of the following statements true or false? Correct any that are false:

(a) A mole of one substance has the same number of atoms as a mole of any other substance.

(b) The theoretical yield for a reaction is based on the balanced chemical equation.

(c) A limiting-reactant problem is presented when the quantity of available material is given in moles for one of the reactants.

(d) To prepare 1.00 L of 3.00 M NaCl, weigh 175.5 g of NaCl and dissolve it in 1.00 L of distilled water.

(e) The concentration of a solution is an intensive property, but the amount of solute in a solution is an extensive property.

3.124 Box A represents one unit volume of solution A. Which box—B, C, or D—represents one unit volume after adding enough solvent to solution A to (a) triple its volume; (b) double its volume; (c) quadruple its volume?

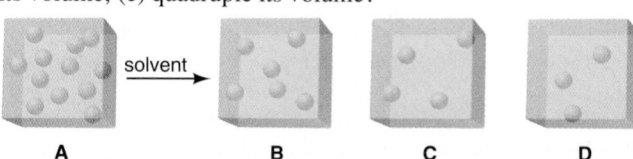

3.125 In each pair, choose the larger of the indicated quantities or state that the samples are equal:

(a) Entities: 0.4 mol of O_3 molecules or 0.4 mol of O atoms

(b) Grams: 0.4 mol of O_3 molecules or 0.4 mol of O atoms

(c) Moles: 4.0 g of N_2O_4 or 3.3 g of SO_2

(d) Grams: 0.6 mol of C_2H_4 or 0.6 mol of F_2

(e) Total ions: 2.3 mol of sodium chlorate or 2.2 mol of magnesium chloride

(f) Molecules: 1.0 g of H_2O or 1.0 g of H_2O_2

(g) Na^+ ions: 0.500 L of 0.500 M NaBr or 0.0146 kg of NaCl

(h) Grams: 6.022×10^{23} atoms of ^{235}U or 6.022×10^{23} atoms of ^{238}U

3.126 A balanced equation may be interpreted in several ways (see Table 3.5). Write a balanced equation for the reaction between solid tetraphosphorus trisulfide and oxygen gas to form solid tetraphosphorus decaoxide and gaseous sulfur trioxide. Tabulate the equation in terms of (a) molecules, (b) moles, and (c) grams.

3.127 Hydrogen gas has been suggested as a clean fuel because it produces only water vapor when it burns. If the reaction has a 98.8% yield, what mass of hydrogen forms 85.0 kg of water?

3.128 Assuming that the volumes are additive, what is the concentration of KBr in a solution prepared by mixing 0.200 L of 0.053 M KBr with 0.550 L of 0.078 M KBr?

3.129 Many reactive boron compounds are analyzed by reaction with water, which converts the boron to boric acid (H_3BO_3). The boric acid content is then determined with NaOH, which reacts with boric acid on a 1:1 mole basis. If 0.1352 g of a boron compound reacts with water, and the resulting boric acid requires 23.47 mL of 0.205 M NaOH for reaction, what was the mass % of boron in the compound?

3.130 Calculate each of the following quantities:

(a) Moles of compound in 0.588 g of ammonium bromide

(b) Number of potassium ions in 68.5 g of potassium nitrate

(c) Mass in grams of 5.85 mol of glycerol ($C_3H_8O_3$)

(d) Volume of 2.55 mol of chloroform ($CHCl_3$; $d = 1.48$ g/mL)

(e) Number of sodium ions in 2.11 mol of sodium carbonate

(f) Number of atoms in 10.0 μg of cadmium

(g) Number of atoms in 0.0015 mol of fluorine gas

3.131 Elements X (green) and Y (purple) react according to the following equation: $X_2 + 3Y_2 \longrightarrow 2XY_3$. Which molecular scene represents the product of the reaction?

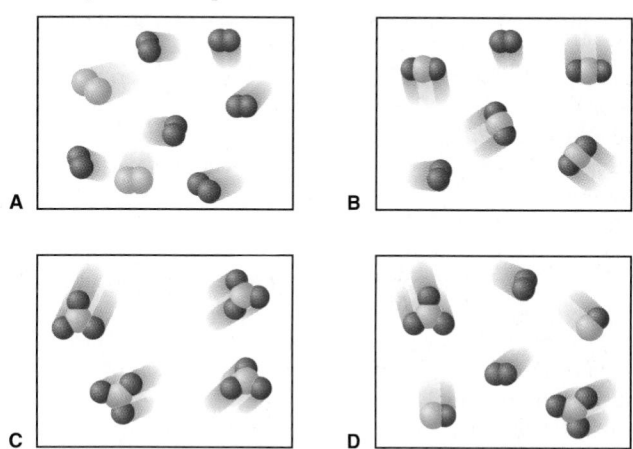

3.132 Hydrocarbon mixtures are used as fuels. How many grams of $CO_2(g)$ are produced by the combustion of 200. g of a mixture that is 25.0% CH_4 and 75.0% C_3H_8 by mass?

3.133 To 1.20 L of 0.325 M HCl, you add 3.37 L of a second HCl solution of unknown concentration. The resulting solution is 0.893 M HCl. Assuming the volumes are additive, calculate the molarity of the second HCl solution.

*3.134 Nitrogen (N), phosphorus (P), and potassium (K) are the main nutrients in plant fertilizers. According to an industry convention, the numbers on the label refer to the mass percents of N, P_2O_5, and K_2O, in that order. Calculate the N:P:K ratio of a 30:10:10 fertilizer in terms of moles of each element, and express it as $x:y:1.0$.

3.135 A 0.652-g sample of a pure strontium halide reacts with excess sulfuric acid, and the solid strontium sulfate formed is separated, dried, and found to weigh 0.755 g. What is the formula of the original halide?

3.136 Methane and ethane are the two simplest hydrocarbons. What is the mass % C in a mixture that is 40.0% methane and 60.0% ethane by mass?

*3.137 When carbon-containing compounds are burned in a limited amount of air, some $CO(g)$ as well as $CO_2(g)$ is produced. A gaseous product mixture is 35.0 mass % CO and 65.0 mass % CO_2. What is the mass % C in the mixture?

3.138 Ferrocene, first synthesized in 1951, was the first organic iron compound with direct Fe—C bonds. An understanding of ferrocene's structure gave rise to new ideas about chemical bonding and led to the preparation of many useful compounds. In the combustion analysis of ferrocene, which contains only Fe, C, and H, a 0.9437-g sample produced 2.233 g of CO_2 and 0.457 g of H_2O. What is the empirical formula of ferrocene?

*3.139 Citric acid *(below)* is concentrated in citrus fruits and plays a central metabolic role in nearly every animal and plant cell. (a) What are the molar mass and formula of citric acid? (b) How many moles of citric acid are in 1.50 qt of lemon juice ($d = 1.09$ g/mL) that is 6.82% citric acid by mass?

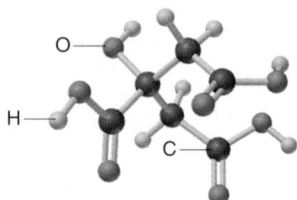

*3.140 Various nitrogen oxides, as well as oxides of sulfur, contribute to acidic rainfall through complex reaction sequences. Atmospheric nitrogen and oxygen combine to form nitrogen monoxide gas, which reacts with more oxygen to form nitrogen dioxide gas. In contact with water vapor, nitrogen dioxide forms aqueous nitric acid and more nitrogen monoxide. (a) Write balanced equations for these reactions. (b) Use the three equations to write one overall balanced equation that does *not* include nitrogen monoxide and nitrogen dioxide. (c) How many metric tons (t) of nitric acid form when 1.25×10^3 t of atmospheric nitrogen is consumed (1 t = 1000 kg)?

*3.141 In the chemical analysis of an unknown, chemists often add an excess of a reactant, determine the amount of that reactant remaining after the reaction with the unknown, and use those amounts to calculate the amount of the unknown. For the analysis of an unknown NaOH solution, you add 50.0 mL of the solution to 0.150 L of an acid solution prepared by dissolving 0.588 g of solid oxalic acid dihydrate ($H_2C_2O_4 \cdot 2H_2O$) in water:

$$2NaOH(aq) + H_2C_2O_4(aq) \longrightarrow 2H_2O(l) + Na_2C_2O_4(aq)$$

The unreacted NaOH then reacts completely with 9.65 mL of 0.116 *M* HCl:

$$NaOH(aq) + HCl(aq) \longrightarrow H_2O(l) + NaCl(aq)$$

What is the molarity of the original NaOH solution?

*3.142 When 1.5173 g of an organic iron compound containing Fe, C, H, and O was burned in O_2, 2.838 g of CO_2 and 0.8122 g of H_2O were produced. In a separate experiment to determine the mass % of iron, 0.3355 g of the compound yielded 0.0758 g of Fe_2O_3. What is the empirical formula of the compound?

*3.143 Fluorides of the relatively unreactive noble gas xenon can be formed by direct reaction of the elements at high pressure and temperature. Depending on the temperature and the reactant amounts, the product mixture may include XeF_2, XeF_4, and XeF_6. Under conditions that produce only XeF_4 and XeF_6, 1.85×10^{-4} mol of Xe reacted with 5.00×10^{-4} mol of F_2, and 9.00×10^{-6} mol of Xe was found to be in excess. What are the mass percents of XeF_4 and XeF_6 in the product?

3.144 Hemoglobin is 6.0% heme ($C_{34}H_{32}FeN_4O_4$) by mass. To remove the heme, hemoglobin is treated with acetic acid and NaCl to form hemin ($C_{34}H_{32}N_4O_4FeCl$). At a crime scene, a blood sample contains 0.45 g of hemoglobin. (a) How many grams of heme are in the sample? (b) How many moles of heme? (c) How many grams of Fe? (d) How many grams of hemin could be formed for a forensic chemist to measure?

*3.145 Manganese is a key component of extremely hard steel. The element occurs naturally in many oxides. A 542.3-g sample of a manganese oxide has an Mn:O ratio of 1.00:1.42 and con-

sists of braunite (Mn_2O_3) and manganosite (MnO).
(a) What masses of braunite and manganosite are in the ore?
(b) What is the ratio $Mn^{3+}:Mn^{2+}$ in the ore?

3.146 The human body excretes nitrogen in the form of urea, NH_2CONH_2. The key biochemical step in urea formation is the reaction of water with arginine to produce urea and ornithine:

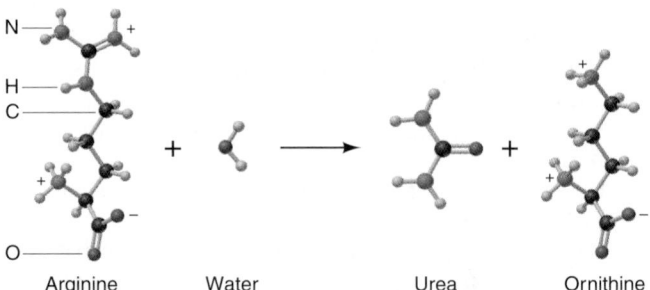

Arginine   Water   Urea   Ornithine

(a) What is the mass percent of nitrogen in urea, arginine, and ornithine. (b) How many grams of nitrogen can be excreted as urea when 143.2 g of ornithine is produced.

3.147 Lead(II) chromate ($PbCrO_4$) is used as a yellow pigment to designate traffic lanes, but has been banned from house paint because of the potential for lead poisoning. The compound is produced from chromite ($FeCr_2O_4$), an ore of chromium:

$$4FeCr_2O_4(s) + 8K_2CO_3(aq) + 7O_2(g)$$
$$\longrightarrow 2Fe_2O_3(s) + 8K_2CrO_4(aq) + 8CO_2(g)$$

Lead(II) ion then replaces the K^+ ion. If a yellow paint is 0.511% $PbCrO_4$ by mass, how many grams of chromite are needed per kilogram of paint?

*3.148 Ethanol (CH_3CH_2OH), the intoxicant in alcoholic beverages, is also used as a precursor of many other organic compounds. In concentrated sulfuric acid, ethanol forms diethyl ether:

$$2CH_3CH_2OH(l) \longrightarrow CH_3CH_2OCH_2CH_3(l) + H_2O(g)$$

In a side reaction, some ethanol forms ethylene:

$$CH_3CH_2OH(l) \longrightarrow CH_2{=}CH_2(g) + H_2O(g)$$

(a) If 50.0 g of ethanol yields 33.9 g of diethyl ether, what is the percent yield of diethyl ether? (b) During the process, 50.0% of the ethanol that did not produce diethyl ether reacts by the side reaction. What mass of ethylene is produced?

*3.149 When powdered zinc is heated with sulfur, a violent reaction occurs, and zinc sulfide forms:

$$Zn(s) + S_8(s) \longrightarrow ZnS(s) \text{ [unbalanced]}$$

Some of the reactants also combine with oxygen in air to form zinc oxide and sulfur dioxide. When 85.2 g of Zn reacts with 52.4 g of S_8, 105.4 g of ZnS forms. What is the percent yield of ZnS? (b) If all the remaining reactants combine with oxygen, how many grams of each of the two oxides form?

3.150 Cocaine ($C_{17}H_{21}O_4N$) is a natural substance found in coca leaves, which have been used for centuries as a local anesthetic and stimulant. Illegal cocaine arrives in the United States either as the pure compound or as the hydrochloride salt ($C_{17}H_{21}O_4NHCl$). At 25°C, the salt is very soluble in water (2.50 kg/L), but cocaine is much less so (1.70 g/L). (a) What is the maximum amount (in g) of the salt that can dissolve in 50.0 mL of water? (b) If the solution in part (a) is treated with NaOH, the salt is converted to cocaine. How much additional water (in L) is needed to dissolve it?

CHAPTER 4

THE MAJOR CLASSES OF CHEMICAL REACTIONS

CHAPTER OUTLINE

4.1 The Role of Water as a Solvent
Solubility of Ionic Compounds
Polar Nature of Water

4.2 Writing Equations for Aqueous Ionic Reactions

4.3 Precipitation Reactions
Driving Precipitation Reactions
Predicting Precipitation Reactions

4.4 Acid-Base Reactions
Driving Force and Net Change
Acid-Base Titrations
Proton Transfer in Acid-Base Reactions

4.5 Oxidation-Reduction (Redox) Reactions
Driving Redox Processes
Redox Terminology
Oxidation Numbers
Balancing Redox Equations
Redox Titrations

4.6 Elemental Substances in Redox Reactions

4.7 Reversible Reactions: An Introduction to Chemical Equilibrium

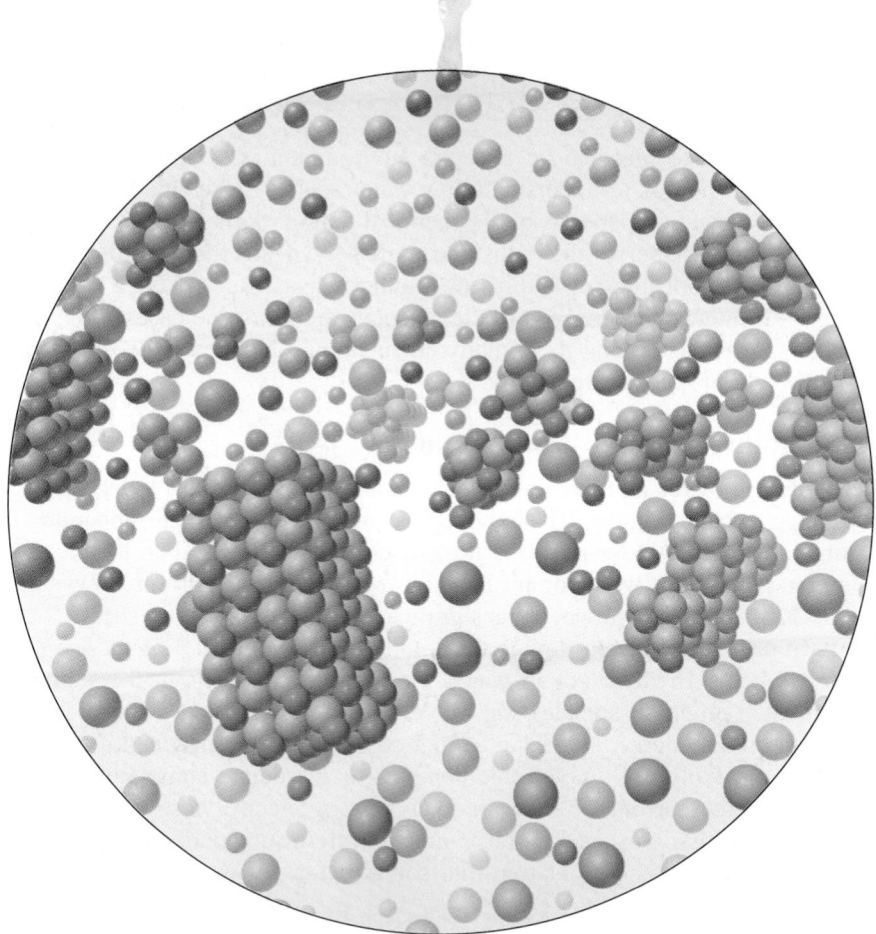

Figure: Precipitation up close. Where aqueous sodium fluoride solution from the cylinder meets aqueous calcium chloride solution, Ca^{2+} *(blue)* and F^- *(yellow)* ions attract each other into particles of solid calcium fluoride. In this chapter, you'll see that precipitation is one of a remarkably small number of reaction classes that describes the great majority of chemical reactions.

CONCEPTS & SKILLS

to review before you study this chapter

- names and formulas of compounds (Section 2.8)
- nature of ionic and covalent bonding (Section 2.7)
- mole-mass-number conversions (Section 3.1)
- molarity and mole-volume conversions (Section 3.5)
- balancing chemical equations (Section 3.3)
- calculating the amounts of reactants and products (Section 3.4)

The amazing variety that we see in nature is largely a consequence of the amazing variety of chemical reactions. Rapid chemical changes occur among gas molecules as sunlight bathes the atmosphere or lightning rips through a stormy sky. Oceans are gigantic containers in which aqueous reaction chemistry goes on unceasingly. In every cell of your body, thousands of reactions are taking place right now.

Of the millions of chemical reactions occurring in and around you, we have examined only a tiny fraction so far, and it would be impossible to examine them all. Fortunately, it isn't necessary to catalog every reaction, because when we survey even a small percentage of them, a few major reaction patterns emerge.

In this chapter, we examine the underlying nature of the three most common reaction processes. Since one of our main themes is aqueous reaction chemistry, we first investigate the crucial role water plays as a solvent for many reactions. With that as background, we focus on two of the major reaction processes—precipitation and acid-base—examining why they occur and describing the use of ionic equations to depict them. Next, we discuss the nature of the third process—oxidation-reduction—perhaps the most important of all. We classify several important types of oxidation-reduction reactions that include elemental substances as reactants or products. The chapter ends with an introductory look at the reversible nature of all chemical reactions.

4.1 THE ROLE OF WATER AS A SOLVENT

Many reactions take place in an aqueous environment, so our first step toward comprehending them is to understand how water acts as a solvent. The role a solvent plays in a reaction depends on its chemical nature. Some solvents play a passive role. They disperse the substances dissolved in them into individual molecules but do not interact with them in other ways. Water plays a much more active role. It interacts strongly with the reactants and, in some cases, even affects their bonds. Water is crucial to so many chemical and physical processes that we will discuss its properties frequently.

The Solubility of Ionic Compounds

When water dissolves an ionic solid, such as potassium bromide (KBr), an important change occurs. Figure 4.1 shows this change with a simple apparatus that measures *electrical conductivity*, the flow of electric current. When the electrodes are immersed in pure water or pushed into solid KBr, no current flows. In the aqueous KBr solution, however, a significant current flows, as shown by the brightly lit bulb. The current flow in the solution implies the *movement of charged particles:* when KBr dissolves in water, the K^+ and Br^- ions in the solid separate from each other (dissociate) and move toward the electrode whose charge is opposite the ion's charge. A substance that conducts a current when dissolved in water is an **electrolyte.** Soluble ionic compounds dissociate completely into ions and create a large current, so they are called *strong* electrolytes.

As the KBr dissolves, each ion becomes **solvated,** surrounded by solvent molecules. We express this dissociation into solvated ions in water as follows:

$$KBr(s) \xrightarrow{\text{H}_2\text{O}} K^+(aq) + Br^-(aq)$$

The "H_2O" above the arrow indicates that water is required but is not a reactant in the usual sense. When any water-soluble ionic compound dissolves, *the oppositely charged ions separate from each other, become surrounded by water molecules, and spread randomly throughout the solution.*

The formula of the compound tells us the number of moles of different ions that result when the compound dissolves. Thus, 1 mol of KBr dissociates into 2 mol of ions—1 mol of K^+ and 1 mol of Br^-.

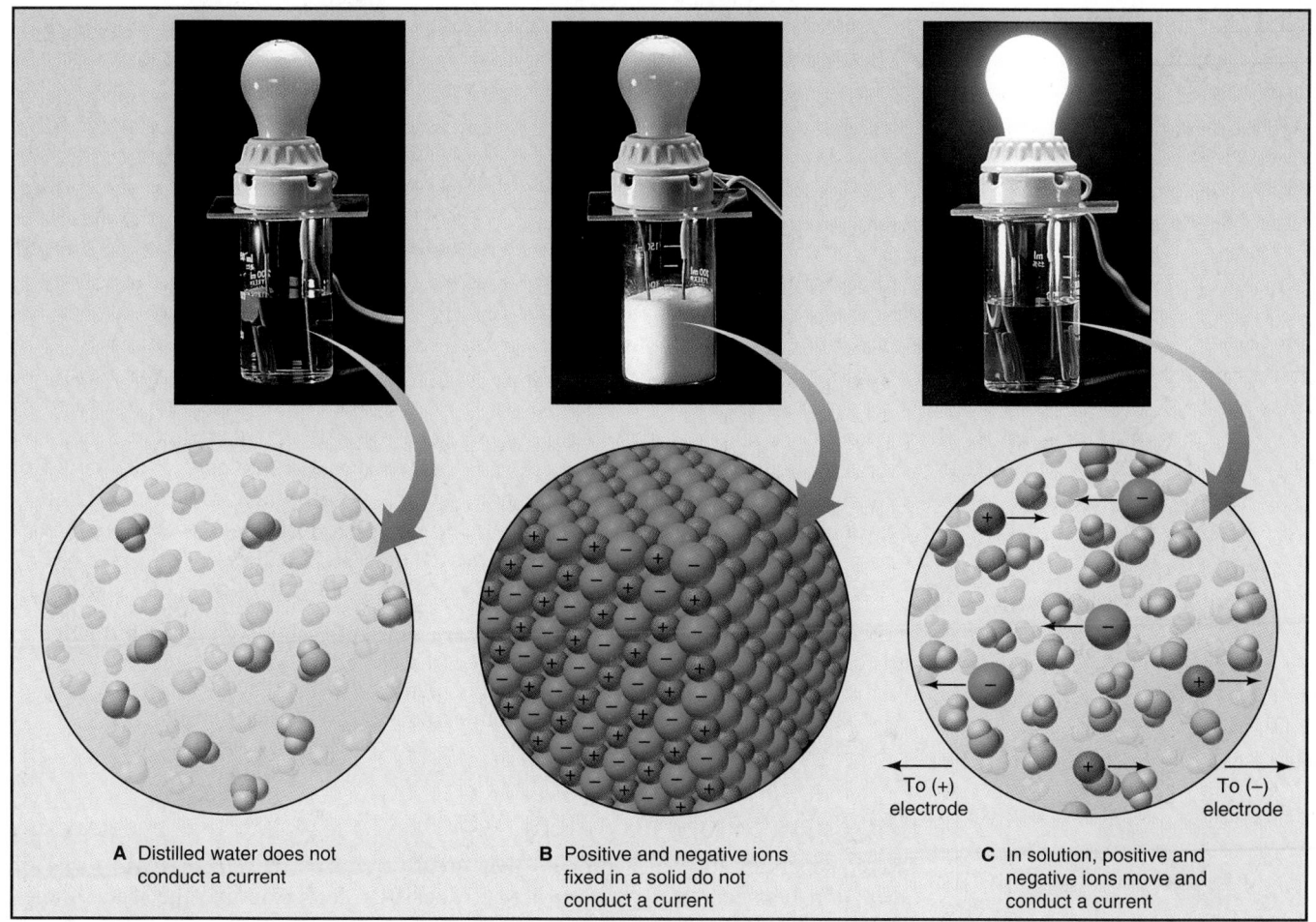

A Distilled water does not conduct a current

B Positive and negative ions fixed in a solid do not conduct a current

C In solution, positive and negative ions move and conduct a current

Figure 4.1 The electrical conductivity of ionic solutions. A, When electrodes connected to a power source are placed in distilled water, no current flows and the bulb is unlit. **B,** A solid ionic compound, such as KBr, conducts no current because the ions are bound tightly together. **C,** When KBr dissolves in H_2O, the ions separate and move through the solution toward the oppositely charged electrodes, thereby conducting a current.

SAMPLE PROBLEM 4.1 Determining Moles of Ions in Aqueous Ionic Solutions

Problem How many moles of each ion are in the following solutions?
(a) 5.0 mol of ammonium sulfate dissolved in water
(b) 78.5 g of cesium bromide dissolved in water
(c) 7.42×10^{22} formula units of copper(II) nitrate dissolved in water
(d) 35 mL of 0.84 M zinc chloride

Plan We write an equation that shows the numbers of moles of ions released when 1 mol of compound dissolves. In **(a)**, we multiply the moles of ions released by 5.0. In **(b)**, we first convert grams to moles. In **(c)**, we first convert formula units to moles. In **(d)**, we first convert molarity and volume to moles.

Solution **(a)** $(NH_4)_2SO_4(s) \xrightarrow{H_2O} 2NH_4^{+}(aq) + SO_4^{2-}(aq)$

Remember that, in general, *polyatomic ions remain as intact units in solution.*
Calculating moles of NH_4^{+} ions:

$$\text{Moles of } NH_4^{+} = 5.0 \text{ mol } (NH_4)_2SO_4 \times \frac{2 \text{ mol } NH_4^{+}}{1 \text{ mol } (NH_4)_2SO_4} = \boxed{10. \text{ mol } NH_4^{+}}$$

$\boxed{5.0 \text{ mol of } SO_4^{2-}}$ is also present.

(b) $CsBr(s) \xrightarrow{H_2O} Cs^{+}(aq) + Br^{-}(aq)$

Converting from grams to moles:

$$\text{Moles of } CsBr = 78.5 \text{ g } CsBr \times \frac{1 \text{ mol } CsBr}{212.8 \text{ g } CsBr} = 0.369 \text{ mol } CsBr$$

Thus, $\boxed{0.369 \text{ mol of } Cs^{+} \text{ and } 0.369 \text{ mol of } Br^{-}}$ are present.

(c) $Cu(NO_3)_2(s) \xrightarrow{H_2O} Cu^{2+}(aq) + 2NO_3^-(aq)$

Converting from formula units to moles:

$$\text{Moles of } Cu(NO_3)_2 = 7.42 \times 10^{22} \text{ formula units } Cu(NO_3)_2$$
$$\times \frac{1 \text{ mol } Cu(NO_3)_2}{6.022 \times 10^{23} \text{ formula units } Cu(NO_3)_2}$$
$$= 0.123 \text{ mol } Cu(NO_3)_2$$

$$\text{Moles of } NO_3^- = 0.123 \text{ mol } Cu(NO_3)_2 \times \frac{2 \text{ mol } NO_3^-}{1 \text{ mol } Cu(NO_3)_2} = \boxed{0.246 \text{ mol } NO_3^-}$$

$\boxed{0.123 \text{ mol of } Cu^{2+}}$ is also present.

(d) $ZnCl_2(aq) \longrightarrow Zn^{2+}(aq) + 2Cl^-(aq)$

Converting from liters to moles:

$$\text{Moles of } ZnCl_2 = 35 \text{ mL} \times \frac{1 \text{ L}}{10^3 \text{ mL}} \times \frac{0.84 \text{ mol } ZnCl_2}{1 \text{ L}} = 2.9 \times 10^{-2} \text{ mol } ZnCl_2$$

$$\text{Moles of } Cl^- = 2.9 \times 10^{-2} \text{ mol } ZnCl_2 \times \frac{2 \text{ mol } Cl^-}{1 \text{ mol } ZnCl_2} = \boxed{5.8 \times 10^{-2} \text{ mol } Cl^-}$$

$\boxed{2.9 \times 10^{-2} \text{ mol of } Zn^{2+}}$ is also present.

Check After you round off to check the math, see if the relative moles of ions are consistent with the formula. For instance, in (a), 10 mol NH_4^+/5.0 mol SO_4^{2-} = 2 NH_4^+/ 1 SO_4^{2-}, or $(NH_4)_2SO_4$. In (d), 0.029 mol Zn^{2+}/0.058 mol Cl^- = 1 Zn^{2+}/2 Cl^-, or $ZnCl_2$.

FOLLOW-UP PROBLEM 4.1 How many moles of each ion are in each solution?
(a) 2 mol of potassium perchlorate dissolved in water
(b) 354 g of magnesium acetate dissolved in water
(c) 1.88×10^{24} formula units of ammonium chromate dissolved in water
(d) 1.32 L of 0.55 M sodium bisulfate

The Polar Nature of Water

Water separates ions in a process that greatly reduces the *electrostatic force* of attraction between them. To see how it does this, let's examine the water molecule closely. Water's power as an ionizing solvent results from two features of the water molecule: *the distribution of its bonding electrons* and *its overall shape*.

Recall from Section 2.7 that the electrons in a covalent bond are shared between the bonded atoms. In a covalent bond that exists between identical atoms (as in H_2, Cl_2, O_2, etc.), the sharing is equal. As Figure 4.2A shows, the shared electrons in H_2 are distributed equally, so no imbalance of charge appears. On the other hand, in covalent bonds between nonidentical atoms, the sharing is unequal: one atom attracts the electron pair more strongly than the other. For reasons discussed in Chapter 9, an O atom attracts electrons more strongly than an H atom. Therefore, in each of the O—H bonds of water, the electrons spend more time closer to the O (Figure 4.2B). This unequal distribution of the electron pair's negative charge creates partially charged "poles" at the ends of each O—H bond. The O end acts as a slightly negative pole (represented by the red shading and the δ−), and the H end acts as a slightly positive pole (represented by the blue shading and the δ+). In Figure 4.2C, the bond's polarity is also indicated by a *polar arrow* (the arrowhead points to the negative pole and the tail is crossed to make a "plus"). Partial charges, like the ones on the O and H atoms in water, are much less than full ionic charges. For instance, in an ionic compound like KBr, the electron has been *transferred* from the K atom to the Br atom and two ions exist. In a covalent compound like water, no ions exist; in each polar O—H bond, the electrons have just *shifted* their average position nearer to the O atom.

The water molecule also has a bent shape: the atoms in H—O—H form an angle, not a straight line. The combined effects of its bent shape and its polar bonds make water a **polar molecule.** As you can see in Figure 4.2D, the O portion of the molecule is the partially negative pole, and the region midway between the H atoms is the partially positive pole.

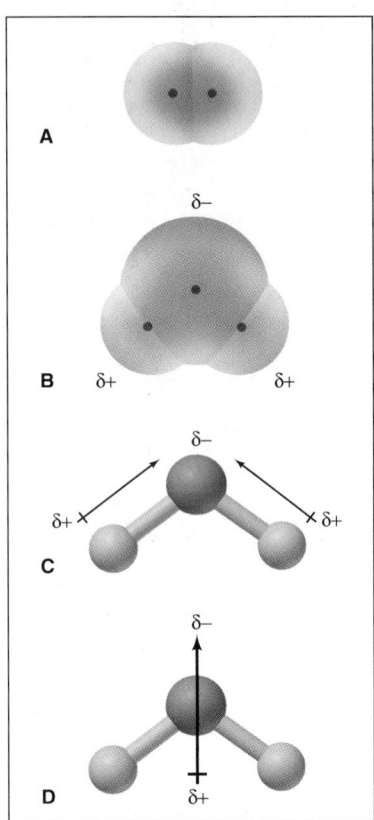

Figure 4.2 **Electron distribution in molecules of H_2 and H_2O. A,** In H_2, the nuclei are identical, so they attract the electrons equally. Note how the central region of higher electron density (red) is balanced by the two outer regions of lower electron density (blue). **B,** In H_2O, the O nucleus attracts the shared electrons more strongly than the H nucleus, creating an uneven charge distribution in each bond. The partially negative O end is designated δ− and the partially positive H end is designated δ+. **C,** In this ball-and-stick model of H_2O, a polar arrow points to the negative end of each O—H bond. **D,** The two polar O—H bonds and the bent molecular shape give rise to the polar H_2O molecule, with the partially positive end located *between* the two H atoms.

Ionic Compounds in Water Now imagine a granule of an ionic compound surrounded by bent, polar water molecules. They congregate near the orderly array of ions at the granule's surface; the negative ends of some water molecules are attracted to the cations, and the positive ends of others are attracted to the anions (Figure 4.3). An electrostatic "tug of war" occurs as the ions become partially solvated. The attraction between each ion and the water molecules gradually outweighs the attraction of the oppositely charged ions for each other. The ions become solvated as they separate from each other and move randomly throughout the solution. A similar scene occurs whenever an ionic compound dissolves in water. ⬤

Although many ionic compounds dissolve in water, many others do not. In such cases, the electrostatic attraction among ions in the compound is greater than the attraction between ions and water molecules, so the substance remains intact. Actually, these so-called insoluble substances *do* dissolve to a very small extent, usually several orders of magnitude less than so-called soluble substances. Compare, for example, the solubilities of NaCl (a "soluble" compound) and AgCl (an "insoluble" compound):

Solubility of NaCl in H_2O at 20°C = 365 g/L
Solubility of AgCl in H_2O at 20°C =   0.009 g/L

Actually, the process of dissolving is more complex than just a contest between the relative energies of attraction of the particles for each other or of the particles for the solvent. In Chapter 13, we'll see that it also involves the natural tendency of the particles to disperse randomly through the solution.

⬤ **Solid Solvents for Ions** Because of its partial charges, water is an excellent liquid solvent for ionic species, but some solids behave in a similar way. For example, poly(ethylene oxide) is a polymer with a repeating structure written as

where the *n* indicates many identical groups linked together covalently. The partial negative charges on the oxygen atoms can surround metal cations, such as Li^+, and solvate them while remaining in the solid state. Poly(ethylene oxide) and related polymers are parts of some lithium-ion batteries used in laptop computers and other portable electronic devices.

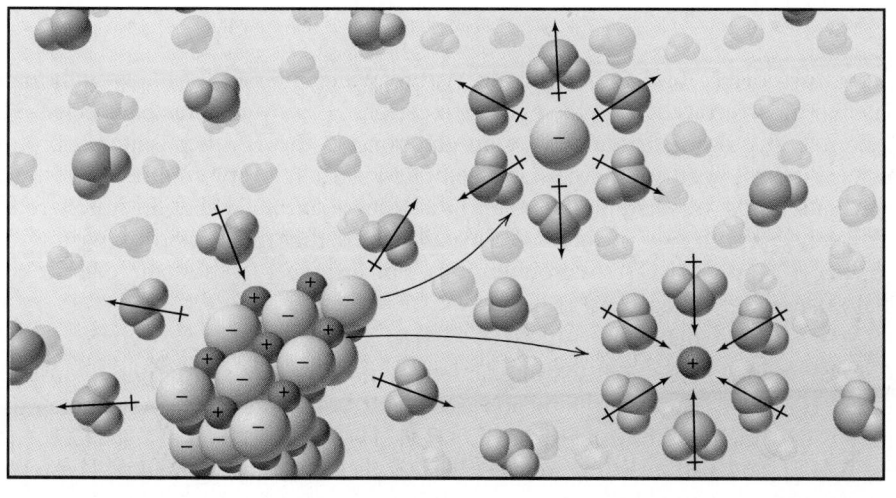

Figure 4.3 **The dissolution of an ionic compound.** When an ionic compound dissolves in water, H_2O molecules separate, surround, and disperse the ions into the liquid. Note that the negative ends of the H_2O molecules face the positive ions and the positive ends face the negative ions.

Covalent Compounds in Water Water dissolves many covalent compounds also. Table sugar (sucrose, $C_{12}H_{22}O_{11}$), beverage (grain) alcohol (ethanol, C_2H_6O), and automobile antifreeze (ethylene glycol, $C_2H_6O_2$) are some familiar examples. All contain their own polar O—H bonds, which interact with those of water. However, even though these substances dissolve, they do not dissociate into ions but remain as intact molecules. Since their aqueous solutions do not conduct an electric current, these substances are called **nonelectrolytes.** Many other covalent substances, such as benzene (C_6H_6) and octane (C_8H_{18}), do not contain polar bonds, and these substances do not dissolve appreciably in water.

A small, but very important, group of H-containing covalent compounds interacts so strongly with water that their molecules *do* dissociate into ions.

DISSOLUTION OF AN IONIC
COMPOUND AND A
COVALENT COMPOUND

In aqueous solution, these substances are all *acids,* as you'll see shortly. The molecules contain polar bonds to hydrogen, in which the atom bonded to H pulls more strongly on the shared electron pair. A good example is hydrogen chloride gas. The Cl end of the HCl molecule is partially negative, and the H end is partially positive. When HCl dissolves in water, the partially charged poles of H_2O molecules are attracted to the oppositely charged poles of HCl. The H—Cl bond breaks, with the H becoming the solvated cation $H^+(aq)$ (but see the discussion following the sample problem) and the Cl becoming the solvated anion $Cl^-(aq)$. Hydrogen bromide behaves similarly when it dissolves in water:

$$HBr(g) \xrightarrow{H_2O} H^+(aq) + Br^-(aq)$$

SAMPLE PROBLEM 4.2 Determining the Molarity of H^+ Ions in Aqueous Solutions of Acids

Problem Nitric acid is a major chemical in the fertilizer and explosives industries. In aqueous solution, each molecule dissociates and the H becomes a solvated H^+ ion. What is the molarity of $H^+(aq)$ in 1.4 *M* nitric acid?
Plan We know the molarity of acid (1.4 *M*), so we just need the formula to find the number of moles of $H^+(aq)$ present in 1 L of solution.
Solution Nitrate ion is NO_3^-, so nitric acid is HNO_3. Thus, 1 mol of $H^+(aq)$ is released per mole of acid:

$$HNO_3(l) \xrightarrow{H_2O} H^+(aq) + NO_3^-(aq)$$

Therefore, 1.4 *M* HNO_3 contains 1.4 mol of $H^+(aq)$/L and is ⟨1.4 *M* $H^+(aq)$.⟩

FOLLOW-UP PROBLEM 4.2 How many moles of $H^+(aq)$ are present in 451 mL of 3.20 *M* hydrobromic acid?

The Nature of H^+ and Other Ions in Water Water interacts strongly with many ions, but most strongly with the hydrogen cation, H^+, a very unusual species. The H atom is a proton surrounded by an electron, so the H^+ ion is just a proton. Because its full positive charge is concentrated in such a tiny volume, H^+ attracts the negative pole of surrounding water molecules so strongly that it actually forms a covalent bond to one of them. We usually show this interaction by writing the aqueous H^+ ion as H_3O^+ (hydronium ion). For instance, to show more accurately what takes place when $HBr(g)$ dissolves, we write

$$HBr(g) + H_2O(l) \longrightarrow H_3O^+(aq) + Br^-(aq)$$

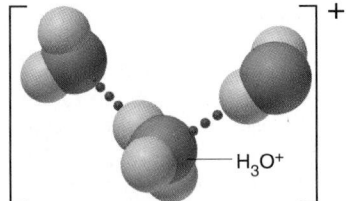

Figure 4.4 The hydrated proton. The charge of the H^+ ion is highly concentrated because the ion is so small. In aqueous solution, it forms a covalent bond to a water molecule and exists as an H_3O^+ ion associated tightly with other H_2O molecules. Here, the $H_7O_3^+$ ion is shown.

To make a point here about the interactions with water, let's write the hydronium ion as $(H_2O)H^+$. The hydronium ion associates with other water molecules in a mixture that includes $H_5O_2^+$ [or $(H_2O)_2H^+$], $H_7O_3^+$ [or $(H_2O)_3H^+$], $H_9O_4^+$ [or $(H_2O)_4H^+$], and still larger aggregates; $H_7O_3^+$ is shown in Figure 4.4. These various species exist together, but we use $H^+(aq)$ as a general, simplified notation. Later in this chapter and much of the rest of the text, we show the solvated proton as $H_3O^+(aq)$ to emphasize water's role. Water interacts covalently with many metal ions as well. For example, Fe^{3+} exists in water as $Fe(H_2O)_6^{3+}$, an Fe^{3+} ion bound to six H_2O molecules. We discuss these species fully in later chapters.

SECTION SUMMARY
When an ionic compound dissolves in water, the ions dissociate from each other and become solvated by water molecules. Because the ions are free to move, their solutions conduct electricity. Water plays an active role in dissolving ionic compounds because it consists of polar molecules that are attracted to the ions. Water also dissolves many covalent substances, and it interacts with some H-containing molecules so strongly that it breaks their covalent bonds and dissociates them into $H^+(aq)$ ions and anions. In water, the H^+ ion is bound to an H_2O and exists as H_3O^+.

4.2 WRITING EQUATIONS FOR AQUEOUS IONIC REACTIONS

Of the many thousands of reactions that occur in the environment and in organisms, the overwhelming majority take place in aqueous solution, and many of those involve ions. Chemists use three types of equations to represent aqueous ionic reactions: molecular, total ionic, and net ionic equations. In the two types of ionic equations, atoms and charges must balance; as you'll see, by balancing the atoms, we balance the charges also.

Let's examine a reaction to see what each of these equations shows. When solutions of silver nitrate and sodium chromate are mixed, the brick-red solid silver chromate (Ag_2CrO_4) forms. Figure 4.5 depicts three views of this reaction: the change you would see if you mixed these solutions in the lab, how you might imagine the change at the atomic level among the ions, and how you can symbolize the change with the three types of equations.

The **molecular equation** (*top*) reveals the least about the species in solution, and is actually somewhat misleading, because *it shows all the reactants and products as if they were intact, undissociated compounds*:

$$2AgNO_3(aq) + Na_2CrO_4(aq) \longrightarrow Ag_2CrO_4(s) + 2NaNO_3(aq)$$

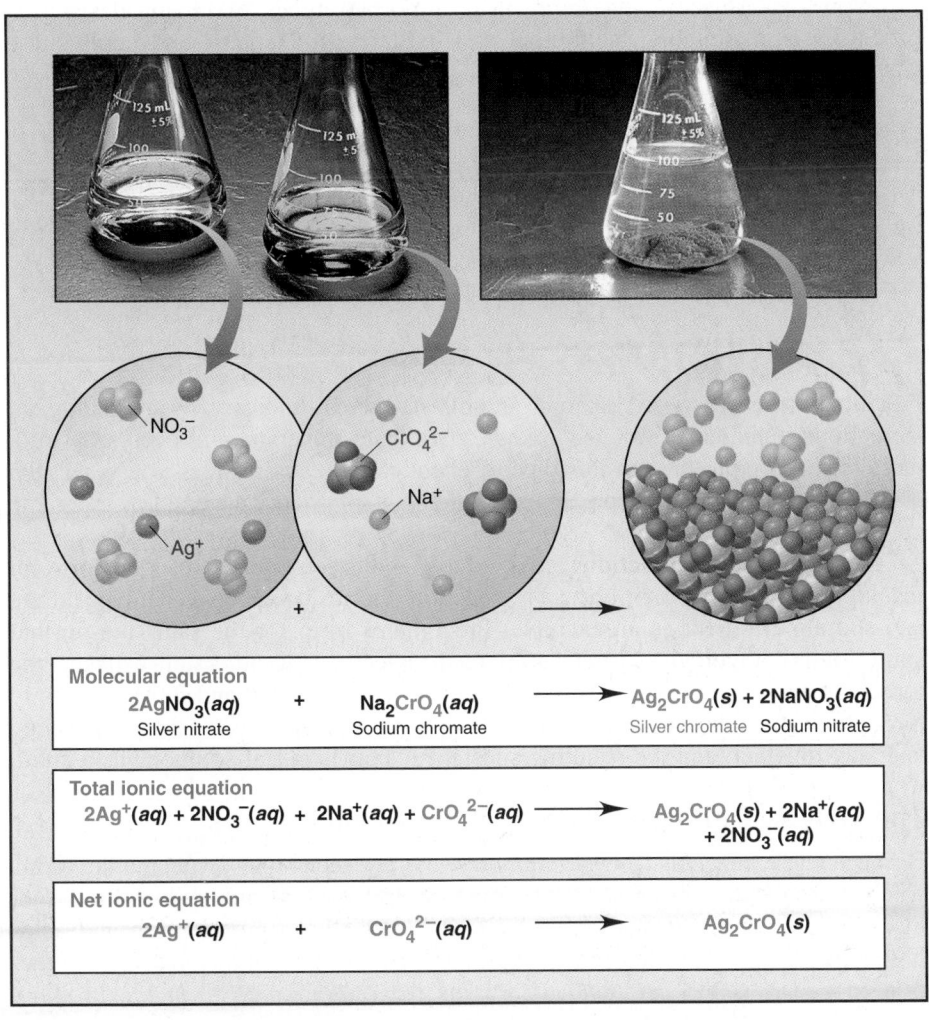

Molecular equation

$$2AgNO_3(aq) \quad + \quad Na_2CrO_4(aq) \quad \longrightarrow \quad Ag_2CrO_4(s) + 2NaNO_3(aq)$$

Silver nitrate   Sodium chromate   Silver chromate   Sodium nitrate

Total ionic equation

$$2Ag^+(aq) + 2NO_3^-(aq) + 2Na^+(aq) + CrO_4^{2-}(aq) \longrightarrow Ag_2CrO_4(s) + 2Na^+(aq) + 2NO_3^-(aq)$$

Net ionic equation

$$2Ag^+(aq) \quad + \quad CrO_4^{2-}(aq) \quad \longrightarrow \quad Ag_2CrO_4(s)$$

Figure 4.5 **A precipitation reaction and its equations.** When silver nitrate and sodium chromate solutions are mixed, a reaction occurs that forms solid silver chromate and a solution of sodium nitrate. The photos present the macroscopic view of the reaction, the view the chemist sees in the lab. The blow-up arrows lead to an atomic-scale view, a representation of the chemist's mental picture of the reactants and products. (The pale ions are spectator ions, present for electrical neutrality, but not involved in the reaction.) Three equations represent the reaction in symbols. The molecular equation shows all substances intact. The total ionic equation shows all *soluble* substances as separate, solvated ions. The net ionic equation eliminates the spectator ions to show only the reacting species.

The **total ionic equation** (*middle*) is a much more accurate representation of the reaction because *it shows all the soluble ionic substances dissociated into ions.* Now the $Ag_2CrO_4(s)$ stands out as the only undissociated substance:

$$2Ag^+(aq) + 2NO_3^-(aq) + 2Na^+(aq) + CrO_4^{2-}(aq) \longrightarrow$$
$$Ag_2CrO_4(s) + 2Na^+(aq) + 2NO_3^-(aq)$$

Notice that the $Na^+(aq)$ and $NO_3^-(aq)$ ions appear in the same form on both sides of the equation. They are called **spectator ions** because they are not involved in the actual chemical change. These ions are present as part of the reactants to balance the charge. That is, we can't add Ag^+ ions without also adding an anion, in this case, NO_3^- ion. Also notice that charges balance: there are four positive and four negative charges on the left for a net zero charge, and there are two positive and two negative charges on the right for a net zero charge.

The **net ionic equation** (*bottom*) is the most useful because *it eliminates the spectator ions and shows the actual chemical change taking place:*

$$2Ag^+(aq) + CrO_4^{2-}(aq) \longrightarrow Ag_2CrO_4(s)$$

The formation of solid silver chromate from silver ions and chromate ions *is* the only change. In fact, if we had originally mixed solutions of potassium chromate, $K_2CrO_4(aq)$, and silver acetate, $AgC_2H_3O_2(aq)$, instead of sodium chromate and silver nitrate, the same change would have occurred. Only the spectator ions would differ—$K^+(aq)$ and $C_2H_3O_2^-(aq)$ instead of $Na^+(aq)$ and $NO_3^-(aq)$. Thus, writing the net ionic equation is an excellent way to isolate the key chemical event.

Now, let's discuss the three most important types of chemical reaction processes—precipitation, acid-base, and oxidation-reduction—and apply these ways of writing equations.

SECTION SUMMARY

A molecular equation for an aqueous ionic reaction shows undissociated substances. A total ionic equation shows all soluble ionic compounds as separate, solvated ions. Spectator ions appear unchanged on both sides of the equation. By eliminating them, you see the actual chemical change in a net ionic equation.

4.3   PRECIPITATION REACTIONS

Precipitation reactions are common in both nature and industry. Many geological formations, including coral reefs, some gems and minerals, and deep-sea structures form, in part, through this type of chemical process. And, as you'll see, the chemical industry employs precipitation methods to produce several key inorganic compounds.

In **precipitation reactions,** two soluble ionic compounds react to form an insoluble product, a **precipitate.** The reaction you just saw between silver nitrate and sodium chromate is an example. Precipitates form for the same reason that some ionic compounds do not dissolve: the electrostatic attraction between the ions outweighs the tendency of the ions to become solvated and move randomly throughout the solution. When solutions of such ions are mixed, the ions collide and stay together, and the resulting substance "comes out of solution" as a solid.

The Driving Force for a Precipitation Reaction

Because these reactions are so common, it may seem that a precipitate forms whenever aqueous solutions of two ionic compounds are mixed, but that is not the case. In many aqueous ionic reactions, especially in precipitation and acid-base reactions, the "driving force" for the reaction, the event that causes the reaction to occur, is the *net removal of ions from solution to form the product.* Consider this example. When solid sodium iodide and potassium nitrate are each dissolved in water, each solution consists of separated ions dispersed throughout the solution:

$$NaI(s) \xrightarrow{H_2O} Na^+(aq) + I^-(aq)$$
$$KNO_3(s) \xrightarrow{H_2O} K^+(aq) + NO_3^-(aq)$$

Does a reaction occur if you mix these solutions? To answer this, *you must examine the possible ion combinations to see whether some of the ions are removed from solution*; that is, whether any of the possible products are insoluble. The reactant ions are

$$Na^+(aq) + I^-(aq) + K^+(aq) + NO_3^-(aq) \longrightarrow ?$$

In addition to the two original reactants, NaI and KNO_3, which you know are soluble, the other possible cation-anion combinations are $NaNO_3$ and KI. A reaction does *not* occur when you mix these starting solutions because $NaNO_3$ and KI are also soluble ionic compounds. Therefore, all the ions just remain in solution. (You'll see shortly how to tell if a product is soluble or not.)

Now substitute a solution of lead(II) nitrate, $Pb(NO_3)_2$, for the KNO_3; when you mix $Pb(NO_3)_2$ solution with NaI solution, a yellow solid forms (Figure 4.6). In addition to the two soluble reactants, the other two possible ion combinations are $NaNO_3$ and PbI_2, so the solid must be lead(II) iodide. Therefore, in this case, a reaction *does* occur because ions are removed from the solution to form PbI_2:

$$2Na^+(aq) + 2I^-(aq) + Pb^{2+}(aq) + 2NO_3^-(aq) \longrightarrow$$
$$2Na^+(aq) + 2NO_3^-(aq) + PbI_2(s)$$

If we use color and look closely at the molecular equation for this reaction, we see that *the ions are exchanging partners*:

$$2NaI(aq) + Pb(NO_3)_2(aq) \longrightarrow PbI_2(s) + 2NaNO_3(aq)$$

Such reactions are called double-displacement, or **metathesis** (pronounced *meh-TA-thuh-sis*) **reactions**. Several are important in industry, such as the preparation of silver bromide for the manufacture of black-and-white film:

$$AgNO_3(aq) + KBr(aq) \longrightarrow AgBr(s) + KNO_3(aq)$$

Figure 4.6 The reaction of Pb(NO₃)₂ and NaI. When aqueous solutions of these ionic compounds are mixed, the yellow solid PbI₂ forms.

Predicting Whether a Precipitation Reaction Will Occur

As you just saw, three steps allow you to predict whether a precipitate will form:

1. Note the ions present in the reactants.
2. Consider the possible cation-anion combinations.
3. Decide whether any of the combinations is insoluble.

A difficulty arises with the last step because there is no simple way to decide whether a given combination is insoluble. Rather, you must memorize a short list of solubility rules (Table 4.1). These rules don't cover every possibility, but learning them allows you to predict the outcome of many precipitation reactions.

PRECIPITATION REACTIONS

Table 4.1 Solubility Rules for Ionic Compounds in Water

Soluble Ionic Compounds	Insoluble Ionic Compounds
1. All common compounds of Group 1A(1) ions (Li^+, Na^+, K^+, etc.) and ammonium ion (NH_4^+) are soluble.	1. All common metal hydroxides are insoluble, *except* those of Group 1A(1) and the larger members of Group 2A(2) (beginning with Ca^{2+}).
2. All common nitrates (NO_3^-), acetates (CH_3COO^- or $C_2H_3O_2^-$), and most perchlorates (ClO_4^-) are soluble.	2. All common carbonates (CO_3^{2-}) and phosphates (PO_4^{3-}) are insoluble, *except* those of Group 1A(1) and NH_4^+.
3. All common chlorides (Cl^-), bromides (Br^-), and iodides (I^-) are soluble, *except* those of Ag^+, Pb^{2+}, Cu^+, and Hg_2^{2+}.	3. All common sulfides are insoluble *except* those of Group 1A(1), Group 2A(2), and NH_4^+.
4. All common sulfates (SO_4^{2-}) are soluble, *except* those of Ca^{2+}, Sr^{2+}, Ba^{2+}, and Pb^{2+}.	

SAMPLE PROBLEM 4.3 Predicting Whether a Precipitation Reaction Occurs; Writing Ionic Equations

Problem Predict whether a reaction occurs when each of the following pairs of solutions are mixed. If a reaction does occur, write balanced molecular, total ionic, and net ionic equations, and identify the spectator ions.
(a) Sodium sulfate(*aq*) + strontium nitrate(*aq*) $\longrightarrow$
(b) Ammonium perchlorate(*aq*) + sodium bromide(*aq*) $\longrightarrow$
Plan For each pair of solutions, we note the ions present in the reactants, write the cation-anion combinations, and refer to Table 4.1 to see if any are insoluble. For the molecular equation, we predict the products. For the total ionic equation, we write the soluble compounds as separate ions. For the net ionic equation, we eliminate the spectator ions.
Solution **(a)** In addition to the reactants, the two other ion combinations are strontium sulfate and sodium nitrate. Table 4.1 shows that strontium sulfate is insoluble, so a reaction *does* occur. Writing the molecular equation:

$$Na_2SO_4(aq) + Sr(NO_3)_2(aq) \longrightarrow SrSO_4(s) + 2NaNO_3(aq)$$

Writing the total ionic equation:

$$2Na^+(aq) + SO_4^{2-}(aq) + Sr^{2+}(aq) + 2NO_3^-(aq) \longrightarrow$$
$$SrSO_4(s) + 2Na^+(aq) + 2NO_3^-(aq)$$

Writing the net ionic equation:

$$Sr^{2+}(aq) + SO_4^{2-}(aq) \longrightarrow SrSO_4(s)$$

The spectator ions are Na^+ and NO_3^-.
(b) The other ion combinations are ammonium bromide and sodium perchlorate. Table 4.1 shows that all ammonium, sodium, and most perchlorate compounds are soluble, and all bromides are soluble except those of Ag^+, Pb^{2+}, Cu^+, and Hg_2^{2+}. Therefore, **no** reaction occurs. The compounds remain dissociated in solution as solvated ions.

FOLLOW-UP PROBLEM 4.3 Predict whether a reaction occurs, and write balanced total and net ionic equations:
(a) Iron(III) chloride(*aq*) + cesium phosphate(*aq*) $\longrightarrow$
(b) Sodium hydroxide(*aq*) + cadmium nitrate(*aq*) $\longrightarrow$
(c) Magnesium bromide(*aq*) + potassium acetate(*aq*) $\longrightarrow$
(d) Silver sulfate(*aq*) + barium chloride(*aq*) $\longrightarrow$

SECTION SUMMARY
Precipitation reactions involve the formation of an insoluble ionic compound from two soluble ones. They occur because electrostatic attractions among certain pairs of solvated ions are strong enough to cause their removal from solution. Such reactions can be predicted by noting whether any possible ion combinations are insoluble, based on a set of solubility rules.

4.4 ACID-BASE REACTIONS

Aqueous acid-base reactions involve water not only as solvent but also in the more active roles of reactant and product. These reactions are the essential chemical events in processes as diverse as the biochemical synthesis of proteins, the industrial production of several fertilizers, and some of the proposed methods for revitalizing lakes damaged by acid rain.

Obviously, an **acid-base reaction** (also called a **neutralization reaction**) occurs when an acid reacts with a base, but the definitions of these terms and the scope of this reaction process have changed considerably over the years. For our purposes at this point, we'll use definitions that apply to the chemicals you'll commonly encounter in the lab:

- An **acid** is a substance that produces H^+ ions when dissolved in water.

$$HX \xrightarrow{H_2O} H^+(aq) + X^-(aq)$$

- A **base** is a substance that produces OH^- ions when dissolved in water.

$$MOH \xrightarrow{H_2O} M^+(aq) + OH^-(aq)$$

(Other definitions of acid and base are presented later in this section and again in Chapter 18, along with a fuller meaning of neutralization.) Acids and bases are the active ingredients in many everyday products: most drain, window, and oven cleaners contain bases; vinegar and lemon juice contain acids.

Acids and bases are electrolytes and are often categorized in terms of "strength," which refers to their degree of dissociation into ions in aqueous solution. *Strong acids and strong bases dissociate completely* into ions when they dissolve in water. Therefore, like soluble ionic compounds, they are *strong* electrolytes and conduct a current well (see left photo). In contrast, *weak acids and weak bases dissociate so little that most of their molecules remain intact.* As a result, they conduct only a small current (see right photo), and are *weak* electrolytes. Table 4.2 lists some acids and bases in terms of their strength.

Both strong and weak acids have one or more H atoms as part of their structure. Strong bases have either the OH^- or the O^{2-} ion as part of their structure. Soluble ionic oxides, such as K_2O, act as strong bases because the oxide ion is not stable in water and reacts immediately to form hydroxide ion:

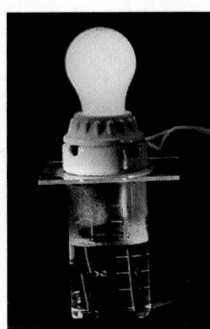
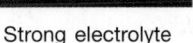

Strong electrolyte   Weak electrolyte

$$K_2O(s) + H_2O(l) \longrightarrow 2K^+(aq) + 2OH^-(aq)$$

Weak bases, such as ammonia, do not contain OH^- ions, but they produce them in a reaction with water that occurs to a small extent:

$$NH_3(g) + H_2O(l) \rightleftharpoons NH_4^+(aq) + OH^-(aq)$$

(Note the reaction arrow in the preceding equation. This type of arrow indicates that the reaction proceeds in both directions; we'll discuss this important idea further in Section 4.7.)

The Driving Force and Net Change: Formation of H_2O from H^+ and OH^-

Let's use ionic equations to see why acid-base reactions occur. We begin with the molecular equation for the reaction between the strong acid HCl and the strong base $Ba(OH)_2$:

$$2HCl(aq) + Ba(OH)_2(aq) \longrightarrow BaCl_2(aq) + 2H_2O(l)$$

Since HCl and $Ba(OH)_2$ dissociate completely and H_2O remains undissociated, the total ionic equation is

$$2H^+(aq) + 2Cl^-(aq) + Ba^{2+}(aq) + 2OH^-(aq) \longrightarrow Ba^{2+}(aq) + 2Cl^-(aq) + 2H_2O(l)$$

In the net ionic equation, we eliminate the spectator ions $Ba^{2+}(aq)$ and $Cl^-(aq)$ and see the actual reaction:

$$2H^+(aq) + 2OH^-(aq) \longrightarrow 2H_2O(l) \quad \text{or} \quad H^+(aq) + OH^-(aq) \longrightarrow H_2O(l)$$

Thus, *the essential change in all aqueous reactions between strong acids and strong bases is that an H^+ ion from the acid and an OH^- ion from the base form a water molecule;* only the spectator ions differ from one reaction to another.

Now it's easy to understand the driving force for these reactions: like precipitation reactions, acid-base reactions are driven by *the electrostatic attraction of ions and their removal from solution* in the formation of the product. In this case, the ions are H^+ and OH^- and the product is H_2O, which consists almost entirely of undissociated molecules. (Actually, water molecules dissociate *very*

Table 4.2 Selected Acids and Bases

Acids

Strong

Hydrochloric acid, HCl
Hydrobromic acid, HBr
Hydriodic acid, HI
Nitric acid, HNO_3
Sulfuric acid, H_2SO_4
Perchloric acid, $HClO_4$

Weak

Hydrofluoric acid, HF
Phosphoric acid, H_3PO_4
Acetic acid, CH_3COOH
  (or $HC_2H_3O_2$)

Bases

Strong

Sodium hydroxide, NaOH
Potassium hydroxide, KOH
Calcium hydroxide, $Ca(OH)_2$
Strontium hydroxide, $Sr(OH)_2$
Barium hydroxide, $Ba(OH)_2$

Weak

Ammonia, NH_3

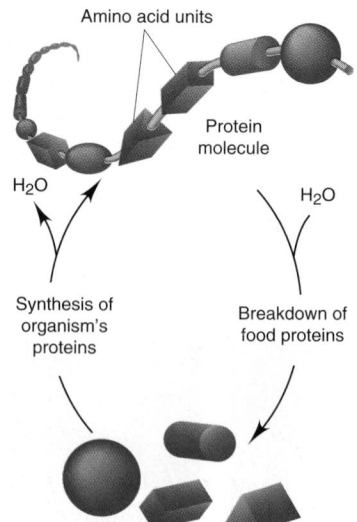

Amino acid units

Protein molecule

H_2O

H_2O

Synthesis of organism's proteins

Breakdown of food proteins

Amino acid molecules

Displacement Reactions Inside You The digestion of food proteins and the formation of an organism's own proteins form a continuous cycle of displacement reactions. A protein consists of hundreds or thousands of smaller molecules, called *amino acids*, linked in a long chain. When you eat proteins, your digestive processes use H_2O to displace one amino acid at a time. These are transported by the blood to your cells, where other metabolic processes link them together, displacing H_2O, to make your own proteins.

slightly, as you'll see in Chapter 18, but the formation of water in a neutralization reaction represents an enormous net removal of H^+ and OH^- ions.)

If we evaporate the water from the above reaction mixture, the ionic solid barium chloride remains. An ionic compound that results from the reaction of an acid and a base is called a **salt.** Thus, in a typical aqueous neutralization reaction, *the reactants are an acid and a base, and the products are a salt solution and water:*

$$HX(aq) + MOH(aq) \longrightarrow MX(aq) + H_2O(l)$$
$$\text{acid} \qquad \text{base} \qquad \text{salt} \qquad \text{water}$$

The color shows that *the cation of the salt comes from the base and the anion comes from the acid.*

As this general equation indicates, acid-base reactions, like precipitation reactions, are metathesis (double-displacement) reactions. The molecular equation for the reaction of aluminum hydroxide, the active ingredient in some antacid tablets, with HCl, the major component of stomach acid, shows this clearly:

$$3HCl(aq) + Al(OH)_3(s) \longrightarrow AlCl_3(aq) + 3H_2O(l)$$

Acid-base reactions occur frequently in the synthesis and breakdown of biological macromolecules.

SAMPLE PROBLEM 4.4 Writing Ionic Equations for Acid-Base Reactions

Problem Write balanced molecular, total ionic, and net ionic equations for each of the following acid-base reactions and identify the spectator ions:
(a) Strontium hydroxide(*aq*) + perchloric acid(*aq*) $\longrightarrow$
(b) Barium hydroxide(*aq*) + sulfuric acid(*aq*) $\longrightarrow$
Plan All are strong acids and bases (see Table 4.2), so the essential reaction is between H^+ and OH^-. The products are H_2O and a salt solution consisting of the spectator ions. Note that in **(b)**, the salt ($BaSO_4$) is insoluble (see Table 4.1), so virtually all ions are removed from solution.
Solution (a) Writing the molecular equation:

$$Sr(OH)_2(aq) + 2HClO_4(aq) \longrightarrow Sr(ClO_4)_2(aq) + 2H_2O(l)$$

Writing the total ionic equation:

$$Sr^{2+}(aq) + 2OH^-(aq) + 2H^+(aq) + 2ClO_4^-(aq) \longrightarrow$$
$$Sr^{2+}(aq) + 2ClO_4^-(aq) + 2H_2O(l)$$

Writing the net ionic equation:

$$2OH^-(aq) + 2H^+(aq) \longrightarrow 2H_2O(l) \quad \text{or} \quad OH^-(aq) + H^+(aq) \longrightarrow H_2O(l)$$

$Sr^{2+}(aq)$ and $ClO_4^-(aq)$ are the spectator ions.
(b) Writing the molecular equation:

$$Ba(OH)_2(aq) + H_2SO_4(aq) \longrightarrow BaSO_4(s) + 2H_2O(l)$$

Writing the total ionic equation:

$$Ba^{2+}(aq) + 2OH^-(aq) + 2H^+(aq) + SO_4^{2-}(aq) \longrightarrow BaSO_4(s) + 2H_2O(l)$$

The net ionic equation is the same as the total ionic equation. This is a precipitation *and* a neutralization reaction. There are no spectator ions because all the ions are used to form the two products.

FOLLOW-UP PROBLEM 4.4 Write balanced molecular, total ionic, and net ionic equations for the reaction between aqueous solutions of calcium hydroxide and nitric acid.

Acid-Base Titrations

Chemists study acid-base reactions quantitatively through titrations. In any **titration,** *one solution of known concentration is used to determine the concentration of another solution through a monitored reaction.* In a typical acid-base titration, a *standardized* solution of base, one whose concentration is *known,* is added slowly to an acid solution of *unknown* concentration.

The laboratory procedure is straightforward but requires careful technique (Figure 4.7). A known volume of the acid solution is placed in a flask, and a few drops of indicator solution are added. An *acid-base indicator* is a substance whose color is different in acid than in base. (We examine indicators in Chapters 18 and 19.) The standardized solution of base is added slowly to the flask from a buret. As the titration nears its end, indicator molecules change color near a drop of added base due to the temporary excess of OH^- ions there. As soon as the solution is swirled, however, the indicator's acidic color returns. The **equivalence point** in the titration occurs when *all the moles of H^+ ions present in the original volume of acid solution have reacted with an equivalent number of moles of OH^- ions added from the buret:*

Moles of H^+ (originally in flask) = moles of OH^- (added from buret)

The **end point** of the titration occurs when a tiny excess of OH^- ions changes the indicator permanently to its color in base. In calculations, we assume this tiny excess is insignificant, and therefore *the amount of base needed to reach the end point is the same as the amount needed to reach the equivalence point.*

$$H^+(aq) + X^-(aq) + M^+(aq) + OH^-(aq) \longrightarrow H_2O(l) + M^+(aq) + X^-(aq)$$

Figure 4.7 An acid-base titration. A, In this procedure, a measured volume of the unknown acid solution is placed in a flask beneath a buret containing the known (standardized) base solution. A few drops of indicator are added to the flask; the indicator used here is phenolphthalein, which is colorless in acid and pink in base. After an initial buret reading, base (OH^- ions) is added slowly to the acid (H^+ ions). **B,** Near the end of the titration, the indicator momentarily changes to its base color but reverts to its acid color with swirling. **C,** When the end point is reached, a tiny excess of OH^- is present, shown by the permanent change in color of the indicator. The difference between the final buret reading and the initial buret reading gives the volume of base used.

SAMPLE PROBLEM 4.5 Finding the Concentration of Acid from an Acid-Base Titration

Problem You perform an acid-base titration to standardize an HCl solution by placing 50.00 mL of HCl in a flask with a few drops of indicator solution. You put 0.1524 M NaOH into the buret, and the initial reading is 0.55 mL. At the end point, the buret reading is 33.87 mL. What is the concentration of the HCl solution?

Plan We must find the molarity of acid from the volume of acid (50.00 mL), the initial (0.55 mL) and final (33.87 mL) volumes of base, and the molarity of base (0.1524 M). First, we balance the equation. We find the volume of base added from the difference in buret readings and use the base's molarity to calculate the amount (mol) of base added. Then, we use the molar ratio from the balanced equation to find the amount (mol) of acid originally present and divide by the acid's original volume to find the molarity.

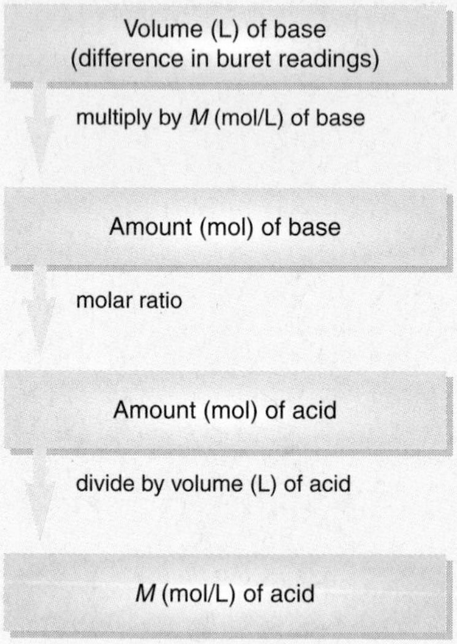

Solution Writing the balanced equation:

$$NaOH(aq) + HCl(aq) \longrightarrow NaCl(aq) + H_2O(l)$$

Finding volume (L) of NaOH solution added:

$$\text{Volume (L) of solution} = (33.87 \text{ mL soln} - 0.55 \text{ mL soln}) \times \frac{1 \text{ L}}{1000 \text{ mL}}$$
$$= 0.03332 \text{ L soln}$$

Finding amount (mol) of NaOH added:

$$\text{Moles of NaOH} = 0.03332 \text{ L soln} \times \frac{0.1524 \text{ mol NaOH}}{1 \text{ L soln}}$$
$$= 5.078 \times 10^{-3} \text{ mol NaOH}$$

Finding amount (mol) of HCl originally present: Since the molar ratio is 1:1,

$$\text{Moles of HCl} = 5.078 \times 10^{-3} \text{ mol NaOH} \times \frac{1 \text{ mol HCl}}{1 \text{ mol NaOH}} = 5.078 \times 10^{-3} \text{ mol HCl}$$

Calculating molarity of HCl:

$$\text{Molarity of HCl} = \frac{5.078 \times 10^{-3} \text{ mol HCl}}{50.00 \text{ mL}} \times \frac{1000 \text{ mL}}{1 \text{ L}} = \boxed{0.1016 \ M \text{ HCl}}$$

Check The answer makes sense: a larger volume of less concentrated acid neutralized a smaller volume of more concentrated base. Rounding shows that the moles of H^+ and OH^- are about equal: 50 mL $\times$ 0.1 M H^+ = 0.005 mol = 33 mL $\times$ 0.15 M OH^-.

FOLLOW-UP PROBLEM 4.5 What volume of 0.1292 M $Ba(OH)_2$ would neutralize 50.00 mL of the HCl solution standardized in the sample problem above?

Sidebar flowchart:

> Volume (L) of base
> (difference in buret readings)
>
> *multiply by M (mol/L) of base*
>
> Amount (mol) of base
>
> *molar ratio*
>
> Amount (mol) of acid
>
> *divide by volume (L) of acid*
>
> M (mol/L) of acid

Acid-Base Reactions as Proton-Transfer Processes

We gain deeper insight into acid-base reactions if we look more closely at the actual species present in solution. Let's see what takes place when HCl gas dissolves in water. As we discussed earlier, the polar water molecules pull apart the HCl molecule and the H^+ ion bonds to a water molecule. In essence, we can say that HCl *transfers its proton* to H_2O:

$$\overset{\frown \ H^+ \text{ transfer} \ \searrow}{HCl(g) \ + \ H_2O(l)} \longrightarrow H_3O^+(aq) + Cl^-(aq)$$

Thus, hydrochloric acid (an aqueous solution of HCl gas) actually consists of solvated H_3O^+ and solvated Cl^- ions.

When sodium hydroxide solution is added, the H_3O^+ ion transfers a proton to the OH^- ion of the base (with the product water shown here as HOH):

$$[H_3O^+(aq) + Cl^-(aq)] + [Na^+(aq) + OH^-(aq)] \longrightarrow$$
$$H_2O(l) + Na^+(aq) + Cl^-(aq) + HOH(l)$$

Without the spectator ions, the transfer of a proton from H_3O^+ to OH^- is obvious:

$$\overset{\frown \ H^+ \text{ transfer} \ \searrow}{H_3O^+(aq) \ + \ OH^-(aq)} \longrightarrow H_2O(l) + HOH(l) \quad [\text{or } 2H_2O(l)]$$

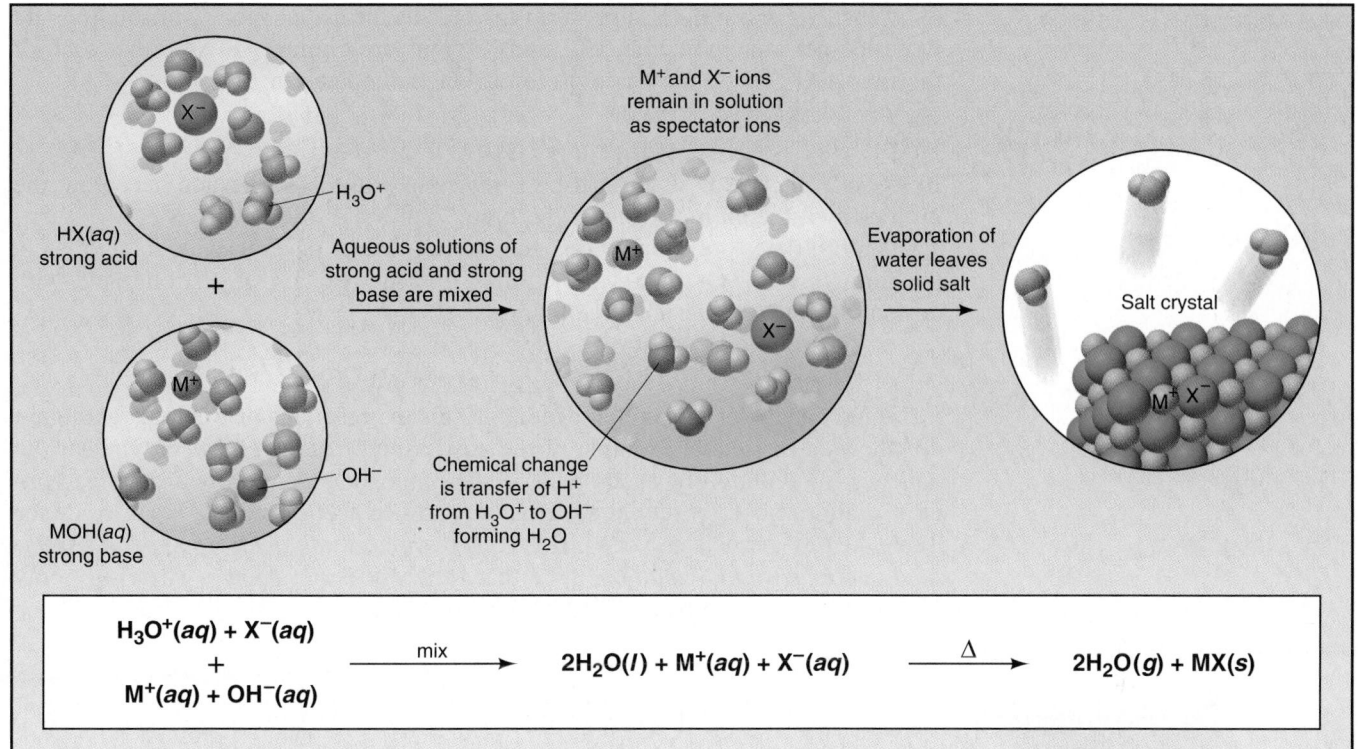

Figure 4.8 **An aqueous strong acid–strong base reaction on the atomic scale.** When solutions of a strong acid (HX) and a strong base (MOH) are mixed, the H_3O^+ from the acid transfers a proton to the OH$^-$ from the base to form an H_2O molecule. Evaporation of the water leaves the spectator ions, X$^-$ and M$^+$, as a solid ionic compound called a *salt*.

Compare this net ionic reaction with the one we saw earlier (see p. 141):

$$H^+(aq) + OH^-(aq) \longrightarrow H_2O(l)$$

and you'll see they are identical, with the additional H_2O molecule coming from the H_3O^+. Clearly, *an acid-base reaction is a proton-transfer process.* The Na$^+$ and Cl$^-$ ions remain in solution, and if the water is evaporated, they crystallize as the salt NaCl. Figure 4.8 shows this process on the atomic level.

In the early 20th century, the chemists Johannes Brønsted and Thomas Lowry realized the proton-transfer nature of acid-base reactions. They defined *an acid as a molecule (or ion) that donates a proton, and a base as a molecule (or ion) that accepts a proton.* Therefore, in the aqueous reaction between strong acid and strong base, *H_3O^+ ion acts as the acid and OH$^-$ ion acts as the base.* Because they ionize completely, a given amount of strong acid (or strong base) creates an equivalent amount of H_3O^+ (or OH$^-$) when it dissolves in water. (We discuss the Brønsted-Lowry concept thoroughly in Chapter 18.)

Thinking of acid-base reactions as proton-transfer processes helps us understand another common type of aqueous ionic reaction, those that form a gaseous product. For example, when an ionic carbonate, such as K_2CO_3, is treated with an acid, such as HCl, one of the products is carbon dioxide. *The driving force for this and similar reactions is formation of a gas **and** water because both products remove reactant ions from solution:*

$$2HCl(aq) + K_2CO_3(aq) \longrightarrow 2KCl(aq) + [H_2CO_3(aq)]$$
$$[H_2CO_3(aq)] \longrightarrow H_2O(l) + CO_2(g)$$

The product H_2CO_3 is shown in brackets to indicate that it is very unstable. It decomposes immediately into water and carbon dioxide. Combining these two equations gives the overall equation:

$$2HCl(aq) + K_2CO_3(aq) \longrightarrow 2KCl(aq) + H_2O(l) + CO_2(g)$$

When we show H_3O^+ ions from the HCl as the actual species in solution and write the net ionic equation, Cl^- and K^+ ions are eliminated. Note that each of the two H_3O^+ ions transfers a proton to the carbonate ion:

$$2H_3O^+(aq) \; + \; \overset{\overbrace{\quad 2H^+ \text{ transfer} \quad}}{CO_3{}^{2-}(aq)} \longrightarrow 2H_2O(l) + [H_2CO_3(aq)] \longrightarrow 3H_2O(l) + CO_2(g)$$

In essence, then, this is an acid-base reaction, with carbonate ion accepting the protons and, therefore, acting as the base.

Several other polyatomic ions act similarly when reacting with an acid, as in the formation of SO_2 from ionic sulfites. For the reaction of a strong acid with an ionic sulfite, the net ionic equation is

$$2H_3O^+(aq) \; + \; \overset{\overbrace{\quad 2H^+ \text{ transfer} \quad}}{SO_3{}^{2-}(aq)} \longrightarrow 3H_2O(l) + SO_2(g)$$

Ionic equations are written differently for the reactions of weak acids. Figure 4.9 shows a household example of the gas-forming reaction between vinegar (an aqueous 5% solution of acetic acid) and baking soda (sodium hydrogen carbonate) solution. Look closely at the equations. Since acetic acid is a weak acid (see Table 4.2), it dissociates very little. To show this, *weak acids appear undissociated in the net ionic equation;* note that H_3O^+ does not appear. Therefore, only $Na^+(aq)$ is a spectator ion; $CH_3COO^-(aq)$ is not.

Molecular equation
$$NaHCO_3(aq) + CH_3COOH(aq) \longrightarrow CH_3COONa(aq) + CO_2(g) + H_2O(l)$$

Total ionic equation
$$Na^+(aq) + HCO_3{}^-(aq) + CH_3COOH(aq) \longrightarrow CH_3COO^-(aq) + Na^+(aq) + CO_2(g) + H_2O(l)$$

Net ionic equation
$$HCO_3{}^-(aq) + CH_3COOH(aq) \longrightarrow CH_3COO^-(aq) + CO_2(g) + H_2O(l)$$

Figure 4.9 **An acid-base reaction that forms a gaseous product.** Carbonates and hydrogen carbonates react with acids to form gaseous CO_2 and H_2O. Here, dilute acetic acid solution (vinegar) is added to sodium hydrogen carbonate (baking soda) solution, and bubbles of CO_2 gas form. (Note that the net ionic equation includes acetic acid because it is a weak acid and does *not* dissociate into ions to an appreciable extent.)

SECTION SUMMARY

Acid-base (neutralization) reactions occur when ions are removed as an acid (an H^+-yielding substance) and a base (an OH^--yielding substance) form a water molecule. Strong acids and bases dissociate completely; weak acids and bases dissociate slightly. In a titration, a known concentration of one reactant is used to determine the concentration of the other. An acid-base reaction can also be viewed as the transfer of a proton from an acid to a base. An ionic gas-forming reaction is an acid-base reaction in which an acid transfers a proton to a polyatomic ion (carbonate or sulfite), forming a gas that leaves the reaction mixture.

4.5   OXIDATION-REDUCTION (REDOX) REACTIONS

Redox reactions are the third and, perhaps, most important type of chemical processes. They include the formation of a compound from its elements (and vice versa), all combustion reactions, the reactions that generate electricity in batteries, the reactions that produce cellular energy, and many others. In this section, we examine the redox process in detail and learn some essential terminology. We see one way to balance redox equations and one way to apply them quantitatively.

The Driving Force for Redox Processes

In **oxidation-reduction** (or **redox**) **reactions,** the key chemical event is the *net movement of electrons* from one reactant to the other. The driving force for these reactions is that the movement of electrons occurs in a specific direction—*from the reactant (or atom in the reactant) with less attraction for electrons to the reactant with more attraction for electrons.*

Such movement of electron charge occurs in the formation of both ionic and covalent compounds. As an example, let's reconsider the flashbulb reaction (see Figure 3.7, p. 103), in which an ionic compound, MgO, forms from its elements:

$$2Mg(s) + O_2(g) \longrightarrow 2MgO(s)$$

Figure 4.10A shows that during the reaction, each Mg atom loses two electrons and each O atom gains them; that is, two electrons move from each Mg atom to each O atom. This change represents a ***transfer*** *of electron charge* away from each Mg atom and toward each O atom, resulting in the formation of Mg^{2+} and O^{2-} ions. The ions aggregate and form an ionic solid.

During the formation of a covalent compound from its elements, there is again a net movement of electrons; but it is more of a "shift" in electron charge than a transfer and, thus, *not* enough to form ions. Consider the formation of HCl gas:

$$H_2(g) + Cl_2(g) \longrightarrow 2HCl(g)$$

To see the electron movement here, compare the electron charge distributions in the reactant bonds and in the product bonds. As Figure 4.10B shows, H_2 and Cl_2 molecules are each held together by pure covalent bonds; that is, the electrons are shared equally between the bonded atoms. In the HCl molecule, the electrons are shared unequally because the Cl atom attracts them more strongly than the H atom does. Thus, in HCl, the H has less electron charge *(blue shading)* than it had in H_2, whereas the Cl has more charge *(red shading)* than it had in Cl_2. In other words, in the formation of HCl, there has been a relative ***shift*** *of electron charge* away from the H atom and toward the Cl atom. This electron *shift* is not nearly as extreme as the electron *transfer* during MgO formation. In fact, in some cases, the net movement of electrons may be very slight, but the reaction is still a redox process.

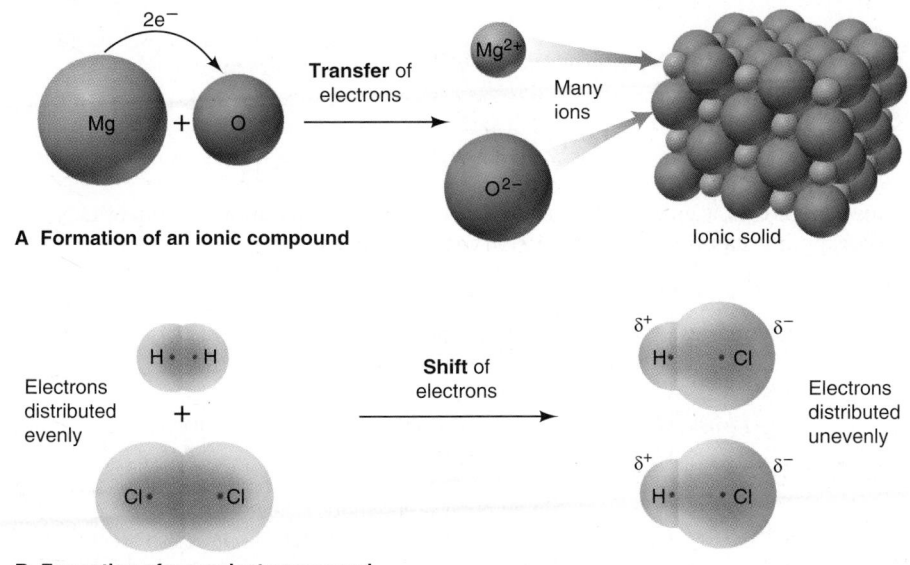

A Formation of an ionic compound

B Formation of a covalent compound

Figure 4.10 The redox process in compound formation. A, In forming the ionic compound MgO, each Mg atom transfers two electrons to each O atom. (Note that atoms become smaller when they lose electrons and larger when they gain electrons.) The resulting Mg^{2+} and O^{2-} ions aggregate with many others to form an ionic solid. **B,** In the reactants H_2 and Cl_2, the electron pairs are shown centrally located to indicate that they are shared equally. In the covalent compound HCl, the Cl attracts the shared electrons more strongly than the H does. In effect, the H electron shifts toward Cl. Note the polar nature of the HCl molecule, as shown by higher electron density *(red)* near the Cl end and lower electron density *(blue)* near the H end.

Some Essential Redox Terminology

Chemists use some important terminology to describe the movement of electrons that occurs in oxidation-reduction reactions. **Oxidation** is the *loss* of electrons, and **reduction** is the *gain* of electrons. (The original meaning of *reduction* comes from the process of reducing large amounts of metal ore to smaller amounts of metal, but you'll see shortly why we use the term for the act of gaining.) For example, during the formation of magnesium oxide, Mg undergoes oxidation (electron loss) and O_2 undergoes reduction (electron gain). The loss and gain are simultaneous, but we can imagine them occurring in separate steps:

$$\text{Oxidation (electron loss by Mg):} \qquad Mg \longrightarrow Mg^{2+} + 2e^-$$
$$\text{Reduction (electron gain by } O_2\text{):} \quad \tfrac{1}{2}O_2 + 2e^- \longrightarrow O^{2-}$$

Since O_2 gained the electrons that Mg lost when it was oxidized, we say that *O_2 oxidized Mg.* Thus, O_2 is the **oxidizing agent,** the species doing the oxidizing. Similarly, since Mg gave up the electrons that O_2 gained when it was reduced, we say that *Mg reduced O_2.* Thus, Mg is the **reducing agent,** the species doing the reducing.

Once this relationship is clear, you will realize that *the oxidizing agent becomes reduced* because it removes the electrons (and thus gains them), whereas *the reducing agent becomes oxidized* because it gives up the electrons (and thus loses them). In the formation of HCl, Cl_2 oxidizes H_2 (H loses some electron charge and Cl gains it), which is the same as saying that H_2 reduces Cl_2. The reducing agent, H_2, is oxidized and the oxidizing agent, Cl_2, is reduced.

Using Oxidation Numbers to Monitor the Movement of Electron Charge

Chemists have devised a useful "bookkeeping" system to monitor which atom loses electron charge and which atom gains it. Each atom in a molecule (or ionic compound) is assigned an **oxidation number (O.N.),** or *oxidation state,* the charge the atom would have *if* electrons were not shared but were transferred completely. Oxidation numbers are determined by the set of rules in Table 4.3. [Note that an oxidation number has the sign *before* the number ($+2$), whereas an ionic charge has the sign *after* the number ($2+$).]

Table 4.3 Rules for Assigning an Oxidation Number (O.N.)

General rules

1. For an atom in its elemental form (Na, O_2, Cl_2, etc.):   O.N. = 0
2. For a monatomic ion: O.N. = ion charge
3. The sum of O.N. values for the atoms in a compound equals zero. The sum of O.N. values for the atoms in a polyatomic ion equals the ion's charge.

Rules for specific atoms or periodic table groups

1. For Group 1A(1):	O.N. = +1 in all compounds
2. For Group 2A(2):	O.N. = +2 in all compounds
3. For hydrogen:	O.N. = +1 in combination with nonmetals
	O.N. = −1 in combination with metals and boron
4. For fluorine:	O.N. = −1 in all compounds
5. For oxygen:	O.N. = −1 in peroxides
	O.N. = −2 in all other compounds (except with F)
6. For Group 7A(17):	O.N. = −1 in combination with metals, nonmetals (except O), and other halogens lower in the group

SAMPLE PROBLEM 4.6 Determining the Oxidation Number of an Element

Problem Determine the oxidation number (O.N.) of each element in these compounds:
(a) Zinc chloride   **(b)** Sulfur trioxide   **(c)** Nitric acid
Plan We apply Table 4.3, noting the general rules that the O.N. values in a compound add up to zero, and the O.N. values in a polyatomic ion add up to the ion's charge.
Solution (a) $ZnCl_2$. The sum of O.N.s for the monatomic ions in the compound must equal zero. The O.N. of the Zn^{2+} ion is +2. The O.N. of each Cl^- ion is −1, for a total of −2. The sum of O.N.s is +2 + (−2), or 0.
(b) SO_3. The O.N. of each oxygen is −2, for a total of −6. Since the O.N.s must add up to zero, the O.N. of S is +6.
(c) HNO_3. The O.N. of H is +1, so the O.N.s of the NO_3 group must add up to −1 to give zero for the compound. The O.N. of each O is −2 for a total of −6. Therefore, the O.N. of N is +5.

FOLLOW-UP PROBLEM 4.6 Determine the O.N. of each element in the following:
(a) Scandium oxide (Sc_2O_3)   **(b)** Gallium chloride ($GaCl_3$)
(c) Hydrogen phosphate ion   **(d)** Iodine trifluoride

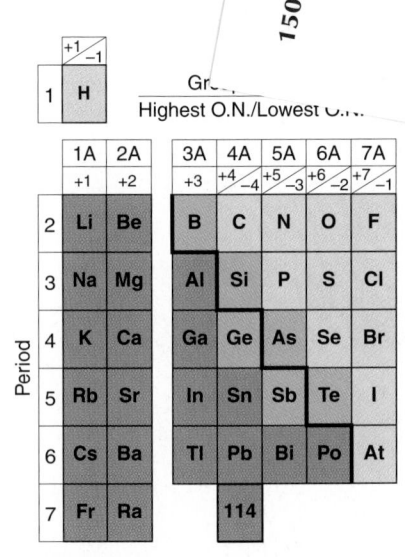

Figure 4.11 **Highest and lowest oxidation numbers of reactive main-group elements.** The A-group number shows the highest possible oxidation number (O.N.) for a main-group element. (Two important exceptions are O, which never has an O.N. of +6, and F, which never has an O.N. of +7.) For nonmetals *(yellow)* and metalloids *(green)*, the A-group number minus 8 gives the lowest possible oxidation number.

The periodic table is a great help in learning the highest and lowest oxidation numbers of most main-group elements, as Figure 4.11 shows:

- For most main-group elements, the A-group number (1A, 2A, and so on) is the *highest* oxidation number (always positive) of any element in the group. The exceptions are O and F (see Table 4.3).
- For main-group nonmetals and some metalloids, the A-group number minus 8 is the *lowest* oxidation number (always negative) of any element in the group.

For example, the highest oxidation number of S (Group 6A) is +6, as in SF_6, and the lowest is (6 − 8), or −2, as in FeS and other metal sulfides.

As you can see, the oxidation number for an element in a binary *ionic* compound has a value based in reality, because it usually equals the ionic charge. On the other hand, the oxidation number has a very unrealistic value for an element in a *covalent* compound (or polyatomic ion) because whole charges don't exist on the atoms in those species.

Another way to define a redox reaction is one in which *the oxidation numbers of the species change,* and the most important use of oxidation numbers is to monitor these changes:

- If a given atom has a higher (more positive or less negative) oxidation number in the product than it had in the reactant, the reactant molecule or ion that contained the atom was oxidized (lost electrons). Thus, *oxidation is represented by an increase in oxidation number.*
- If an atom has a lower (more negative or less positive) oxidation number in the product than it had in the reactant, the reactant molecule or ion that contained the atom was reduced (gained electrons). Thus, *the gain of electrons is represented by a decrease (a "reduction") in oxidation number. (Reduction,* as mentioned earlier, refers to an ore being "reduced" to the metal. The reducing agents used provide electrons that convert the metal ion to its elemental form.)

Figure 4.12 summarizes redox terminology. Since oxidation numbers are assigned according to the relative attraction of an atom for electrons, they are ultimately based on atomic properties, as you'll see in Chapters 8 and 9. (For the remainder of this section and the next, blue oxidation numbers represent oxidation, and red oxidation numbers indicate reduction.)

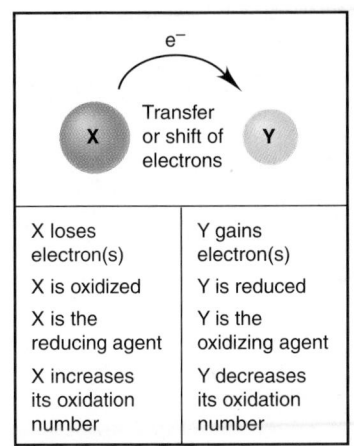

e^-	
X loses electron(s)	Y gains electron(s)
X is oxidized	Y is reduced
X is the reducing agent	Y is the oxidizing agent
X increases its oxidation number	Y decreases its oxidation number

Figure 4.12 **A summary of terminology for oxidation-reduction (redox) reactions.**

SAMPLE PROBLEM 4.7 Recognizing Oxidizing and Reducing Agents

Problem Identify the oxidizing agent and reducing agent in each of the following:
(a) $2Al(s) + 3H_2SO_4(aq) \longrightarrow Al_2(SO_4)_3(aq) + 3H_2(g)$
(b) $PbO(s) + CO(g) \longrightarrow Pb(s) + CO_2(g)$
(c) $2H_2(g) + O_2(g) \longrightarrow 2H_2O(g)$
Plan We first assign an oxidation number (O.N.) to each atom (or ion) based on the rules in Table 4.3. The reactant is the reducing agent if it contains an atom that is oxidized (O.N. increased from left side to right side of the equation). The reactant is the oxidizing agent if it contains an atom that is reduced (O.N. decreased).
Solution (a) Assigning oxidation numbers:

$$\overset{0}{2Al}(s) + \overset{+1\ \ -2\ \ +6}{3H_2SO_4}(aq) \longrightarrow \overset{+3\ \ -2\ \ +6}{Al_2(SO_4)_3}(aq) + \overset{0}{3H_2}(g)$$

The O.N. of Al increased from 0 to +3 (Al lost electrons), so Al was oxidized; Al is the reducing agent.
The O.N. of H decreased from +1 to 0 (H gained electrons), so H^+ was reduced; H_2SO_4 is the oxidizing agent.
(b) Assigning oxidation numbers:

$$\overset{+2\ \ -2}{PbO}(s) + \overset{+2\ \ -2}{CO}(g) \longrightarrow \overset{0}{Pb}(s) + \overset{+4\ \ -2}{CO_2}(g)$$

Pb decreased its O.N. from +2 to 0, so PbO was reduced; PbO is the oxidizing agent.
C increased its O.N. from +2 to +4, so CO was oxidized; CO is the reducing agent.
In general, when a substance (such as CO) bonds to more O atoms (as in CO_2), it is oxidized; and when a substance (such as PbO) bonds to fewer O atoms (as in Pb), it is reduced.
(c) Assigning oxidation numbers:

$$\overset{0}{2H_2}(g) + \overset{0}{O_2}(g) \longrightarrow \overset{+1\ -2}{2H_2O}(g)$$

O_2 was reduced (O.N. of O decreased from 0 to −2); O_2 is the oxidizing agent.
H_2 was oxidized (O.N. of H increased from 0 to +1); H_2 is the reducing agent.
Oxygen is always the oxidizing agent in a combustion reaction.
Comment 1. Compare the O.N. values in (c) with those in another common reaction that forms water—the net ionic equation for an acid-base reaction:

$$\overset{+1}{H^+}(aq) + \overset{-2\ \ +1}{OH^-}(aq) \longrightarrow \overset{+1\ -2}{H_2O}(l)$$

Note that the O.N. values remain the same on both sides of the acid-base equation. Therefore, *an acid-base reaction is* **not** *a redox reaction.*
2. If a substance occurs in its elemental form on one side of an equation, it can't possibly be in its elemental form on the other side, so the reaction must be a redox process. Notice that elements appear in all three cases above.

FOLLOW-UP PROBLEM 4.7 Identify each oxidizing agent and each reducing agent:
(a) $2Fe(s) + 3Cl_2(g) \longrightarrow 2FeCl_3(s)$
(b) $2C_2H_6(g) + 7O_2(g) \longrightarrow 4CO_2(g) + 6H_2O(g)$
(c) $5CO(g) + I_2O_5(s) \longrightarrow I_2(s) + 5CO_2(g)$

Balancing Redox Equations

It is essential to realize that *the reducing agent loses electrons and the oxidizing agent gains them* **simultaneously.** A chemical change can be an "oxidation-reduction reaction" but *not* an "oxidation reaction" *or* a "reduction reaction." The transferred electrons are never free, which means that we can balance a redox reaction by making sure that *the number of electrons lost by the reducing agent equals the number of electrons gained by the oxidizing agent.*

Two methods used to balance redox equations are the *oxidation number* method and the *half-reaction* method. This section describes the oxidation number method in detail; the half-reaction method is covered in Chapter 21. (If your professor chooses to cover that section here, it is completely transferable with no loss in continuity.)

The **oxidation number method** for balancing redox equations consists of five steps that use the changes in oxidation numbers to generate balancing coefficients. The first two steps are identical to those in Sample Problem 4.7:

Step 1. Assign oxidation numbers to all elements in the reaction.
Step 2. From the changes in oxidation numbers, identify the oxidized and reduced species.
Step 3. Compute the number of electrons lost in the oxidation and gained in the reduction from the oxidation number changes. (Draw tie-lines between these atoms to show the changes.)
Step 4. Multiply one or both of these numbers by appropriate factors to make the electrons lost equal the electrons gained, and use the factors as balancing coefficients.
Step 5. Complete the balancing by inspection, adding states of matter.

SAMPLE PROBLEM 4.8 Balancing Redox Equations by the Oxidation Number Method

Problem Use the oxidation number method to balance the following equations:
(a) $Cu(s) + HNO_3(aq) \longrightarrow Cu(NO_3)_2(aq) + NO_2(g) + H_2O(l)$
(b) $PbS(s) + O_2(g) \longrightarrow PbO(s) + SO_2(g)$
Solution **(a)** *Step 1.* Assign oxidation numbers to all elements:

$$\overset{0}{Cu} + \overset{+1}{H}\overset{+5}{N}\overset{-2}{O_3} \longrightarrow \overset{+2}{Cu}(\overset{+5}{N}\overset{-2}{O_3})_2 + \overset{+4}{N}\overset{-2}{O_2} + \overset{+1}{H_2}\overset{-2}{O}$$

Step 2. Identify oxidized and reduced species. The O.N. of Cu increased from 0 (in Cu metal) to +2 (in Cu^{2+}); Cu was oxidized. The O.N. of N decreased from +5 (in HNO_3) to +4 (in NO_2); HNO_3 was reduced. Note that some NO_3^- also acts as a spectator ion, appearing unchanged in the $Cu(NO_3)_2$; this is common in redox reactions.
Step 3. Compute e^- lost and e^- gained and draw tie-lines between the atoms. In the oxidation, $2e^-$ were lost from Cu. In the reduction, $1e^-$ was gained by N:

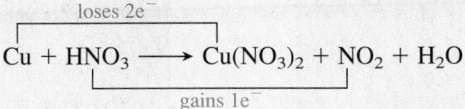

$$\overset{\text{loses } 2e^-}{Cu + HNO_3 \longrightarrow Cu(NO_3)_2 + NO_2 + H_2O}$$
$$\underset{\text{gains } 1e^-}{}$$

Step 4. Multiply by factors to make e^- lost equal e^- gained, and use the factors as coefficients. Cu lost $2e^-$, so the $1e^-$ gained by N should be multiplied by 2. We put the coefficient 2 before NO_2 and HNO_3:

$$Cu + 2HNO_3 \longrightarrow Cu(NO_3)_2 + 2NO_2 + H_2O$$

Step 5. Complete the balancing by inspection. Balancing N atoms requires a 4 in front of HNO_3 because two additional N atoms are in the NO_3^- ions in $Cu(NO_3)_2$:

$$Cu + 4HNO_3 \longrightarrow Cu(NO_3)_2 + 2NO_2 + H_2O$$

Then, balancing H atoms requires a 2 in front of H_2O, and we add states of matter:

$$Cu(s) + 4HNO_3(aq) \longrightarrow Cu(NO_3)_2(aq) + 2NO_2(g) + 2H_2O(l)$$

Check

Reactants (1 Cu, 4 H, 4 N, 12 O) $\longrightarrow$ products [1 Cu, 4 H, (2 + 2) N, (6 + 4 + 2) O]

Copper in nitric acid

(b) *Step 1.* Assign oxidation numbers:

$$\overset{+2}{P}\overset{-2}{b}\overset{}{S} + \overset{0}{O_2} \longrightarrow \overset{+2}{P}\overset{-2}{b}\overset{}{O} + \overset{+4}{S}\overset{-2}{O_2}$$

Step 2. Identify species that are oxidized and reduced. PbS was oxidized: the O.N. of S increased from −2 in PbS to +4 in SO_2. O_2 was reduced: the O.N. of O decreased from 0 in O_2 to −2 in PbO and in SO_2.

Step 3. Compute e^- lost and e^- gained and draw tie-lines. The S lost $6e^-$ and each O gained $2e^-$:

$$\begin{array}{c} \overbrace{}^{\text{loses } 6e^-} \\ PbS + O_2 \longrightarrow PbO + SO_2 \\ \underbrace{}_{\text{gains } 4e^- \; (2e^- \text{ per O})} \end{array}$$

Step 4. Multiply by factors to make e^- lost equal e^- gained. The S atom loses $6e^-$, and each O in O_2 gains $2e^-$, for a total gain of $4e^-$. Thus, placing the coefficient $\frac{3}{2}$ before O_2 gives 3 O atoms that each gain $2e^-$, for a total gain of $6e^-$:

$$PbS + \tfrac{3}{2}O_2 \longrightarrow PbO + SO_2$$

Step 5. Complete the balancing by inspection. The atoms are balanced, but all coefficients must be multiplied by 2 to obtain integers, and we add states of matter:

$$2PbS(s) + 3O_2(g) \longrightarrow 2PbO(s) + 2SO_2(g)$$

Check Reactants (2 Pb, 2 S, 6 O) $\longrightarrow$ products [2 Pb, 2 S, (2 + 4) O]

FOLLOW-UP PROBLEM 4.8 Use the oxidation number method to balance the following: $K_2Cr_2O_7(aq) + HI(aq) \longrightarrow KI(aq) + CrI_3(aq) + I_2(s) + H_2O(l)$

Redox Titrations

In an acid-base titration, a known concentration of base is used to find an unknown concentration of an acid (or vice versa). Similarly, in a redox titration, a known concentration of oxidizing agent is used to find an unknown concentration of reducing agent (or vice versa). This application of stoichiometry is used in a wide range of situations, including measuring the iron content in drinking water and the vitamin C content in fruits and vegetables.

The permanganate ion, MnO_4^-, is a common oxidizing agent in these titrations because it is strongly colored and, thus, also serves as an indicator. In Figure 4.13, MnO_4^- is used to oxidize the oxalate ion, $C_2O_4^{2-}$, to determine its

Figure 4.13 A redox titration. The oxidizing agent in the buret, $KMnO_4$, is strongly colored, so it also serves as the indicator. When it reacts with the reducing agent $C_2O_4^{2-}$ in the flask, its color changes from deep purple to almost colorless *(left).* When all the $C_2O_4^{2-}$ is oxidized, the next drop of $KMnO_4$ remains unreacted and turns the solution light purple *(right),* signaling the end point of the titration.

Net ionic equation:

$$2\overset{+7}{M}nO_4^-(aq) + 5\overset{+3}{C_2}O_4^{2-}(aq) + 16H^+(aq) \longrightarrow 2\overset{+2}{M}n^{2+}(aq) + 10\overset{+4}{C}O_2(g) + 8H_2O(l)$$

concentration. As long as any $C_2O_4^{2-}$ is present, it reduces the deep purple MnO_4^- to the very faint pink (nearly colorless) Mn^{2+} ion (Figure 4.13, *left*). As soon as all the available $C_2O_4^{2-}$ has been oxidized, the next drop of MnO_4^- turns the solution light purple (Figure 4.13, *right*). This color change indicates the end point, the point at which the electrons lost by the oxidized species ($C_2O_4^{2-}$) equal the electrons gained by the reduced species (MnO_4^-). We then calculate the concentration of the $C_2O_4^{2-}$ solution from its known volume, the known volume and concentration of the MnO_4^- solution, and the balanced equation.

Preparing a sample for a redox titration sometimes requires several laboratory steps. In Sample Problem 4.9, for instance, the Ca^{2+} ion concentration of blood is determined. The Ca^{2+} is first precipitated as calcium oxalate (CaC_2O_4), using excess sodium oxalate ($Na_2C_2O_4$). The precipitate is filtered and dilute H_2SO_4 dissolves it and releases the $C_2O_4^{2-}$ ion, which is then titrated with standardized $KMnO_4$ solution. After the $C_2O_4^{2-}$ concentration has been determined, it is used to find the original blood Ca^{2+} concentration.

SAMPLE PROBLEM 4.9 Finding an Unknown Concentration by a Redox Titration

Problem Calcium ion (Ca^{2+}) is required for blood to clot and for many other cell processes. An abnormal Ca^{2+} concentration is indicative of disease. To measure the Ca^{2+} concentration, 1.00 mL of human blood was treated with $Na_2C_2O_4$ solution. The resulting CaC_2O_4 precipitate was filtered and dissolved in dilute H_2SO_4. This solution required 2.05 mL of 4.88×10^{-4} M $KMnO_4$ to reach the end point. The unbalanced equation is

$$KMnO_4(aq) + CaC_2O_4(s) + H_2SO_4(aq) \longrightarrow$$
$$MnSO_4(aq) + K_2SO_4(aq) + CaSO_4(s) + CO_2(g) + H_2O(l)$$

(a) Calculate the amount (mol) of Ca^{2+}.
(b) Calculate the Ca^{2+} ion concentration expressed in units of mg Ca^{2+}/100 mL blood.

(a) Calculating the moles of Ca^{2+}
Plan As always, we first balance the equation. All the Ca^{2+} ion in the 1.00-mL blood sample is precipitated and then dissolved in the H_2SO_4. We find the number of moles of $KMnO_4$ needed to reach the end point from the volume (2.05 mL) and molarity (4.88×10^{-4} M) and use the molar ratio to calculate the number of moles of CaC_2O_4 dissolved in the H_2SO_4. Then, from the chemical formula, we find moles of Ca^{2+} ions.
Solution Balancing the equation:

$$2KMnO_4(aq) + 5CaC_2O_4(s) + 8H_2SO_4(aq) \longrightarrow$$
$$2MnSO_4(aq) + K_2SO_4(aq) + 5CaSO_4(s) + 10CO_2(g) + 8H_2O(l)$$

Converting from milliliters and molarity to moles of $KMnO_4$ to reach the end point:

$$\text{Moles of } KMnO_4 = 2.05 \text{ mL soln} \times \frac{1 \text{ L}}{1000 \text{ mL}} \times \frac{4.88 \times 10^{-4} \text{ mol } KMnO_4}{1 \text{ L soln}}$$
$$= 1.00 \times 10^{-6} \text{ mol } KMnO_4$$

Converting from moles of $KMnO_4$ to moles of CaC_2O_4 titrated:

$$\text{Moles of } CaC_2O_4 = 1.00 \times 10^{-6} \text{ mol } KMnO_4 \times \frac{5 \text{ mol } CaC_2O_4}{2 \text{ mol } KMnO_4}$$
$$= 2.50 \times 10^{-6} \text{ mol } CaC_2O_4$$

Finding moles of Ca^{2+} present:

$$\text{Moles of } Ca^{2+} = 2.50 \times 10^{-6} \text{ mol } CaC_2O_4 \times \frac{1 \text{ mol } Ca^{2+}}{1 \text{ mol } CaC_2O_4}$$
$$= \boxed{2.50 \times 10^{-6} \text{ mol } Ca^{2+}}$$

Check A very small volume of dilute $KMnO_4$ is needed, so 10^{-6} mol of $KMnO_4$ seems reasonable. The molar ratio of $5CaC_2O_4/2KMnO_4$ gives 2.5×10^{-6} mol of CaC_2O_4 and thus 2.5×10^{-6} mol of Ca^{2+}.

(b) Expressing the Ca^{2+} concentration as mg/100 mL blood
Plan The amount in part (a) is the moles of Ca^{2+} ion present in 1.00 mL of blood. We multiply by 100 to obtain the moles of Ca^{2+} ion in 100 mL of blood and then use the atomic mass of Ca to convert to grams and then milligrams of Ca^{2+}/100 mL blood.

Volume (L) of $KMnO_4$ solution

↓ multiply by M (mol/L)

Amount (mol) of $KMnO_4$

↓ molar ratio

Amount (mol) of CaC_2O_4

↓ ratio of elements in chemical formula

Amount (mol) of Ca^{2+}

Solution Finding moles of Ca^{2+}/100 mL blood:

$$\text{Moles of } Ca^{2+}/100 \text{ mL blood} = \frac{2.50 \times 10^{-6} \text{ mol } Ca^{2+}}{1.00 \text{ mL blood}} \times 100$$

$$= 2.50 \times 10^{-4} \text{ mol } Ca^{2+}/100 \text{ mL blood}$$

Converting from moles of Ca^{2+} to milligrams:

$$\text{Mass (mg) } Ca^{2+}/100 \text{ mL blood} = \frac{2.50 \times 10^{-4} \text{ mol } Ca^{2+}}{100 \text{ mL blood}} \times \frac{40.08 \text{ g } Ca^{2+}}{1 \text{ mol } Ca^{2+}} \times \frac{1000 \text{ mg}}{1 \text{ g}}$$

$$= 10.0 \text{ mg } Ca^{2+}/100 \text{ mL blood}$$

Check The relative amounts of Ca^{2+} make sense. If there is 2.5×10^{-6} mol/mL blood, there is 2.5×10^{-4} mol/100 mL blood. A molar mass of about 40 g/mol for Ca^{2+} gives 100×10^{-4} g, or 10×10^{-3} g/100 mL blood. It is easy to make an order-of-magnitude (power of 10) error in this type of calculation, so be sure to include all units.

Comment 1. The normal range of Ca^{2+} concentration in a human adult is 9.0 to 11.5 mg Ca^{2+}/100 mL blood, so our value seems reasonable.

2. When blood is donated, the receiving bag contains $Na_2C_2O_4$ solution, which precipitates the Ca^{2+} ion and, thus, prevents clotting.

3. A redox titration is analogous to an acid-base titration: in redox processes, electrons are lost and gained, whereas in acid-base processes, H^+ ions are lost and gained.

FOLLOW-UP PROBLEM 4.9 A 2.50-mL sample of low-fat milk was treated with sodium oxalate, and the precipitate was filtered and dissolved in H_2SO_4. This solution required 6.53 mL of 4.56×10^{-3} M $KMnO_4$ to reach the end point.
(a) Calculate the molarity of Ca^{2+} in the milk.
(b) What is the concentration of Ca^{2+} in g/L? Is this value consistent with the typical value in milk of about 1.2 g Ca^{2+}/L?

SECTION SUMMARY

When there is a net movement of electron charge from one reactant to another, a redox process takes place. The greater attraction of the electrons by one reactant over the other drives the reaction. Electron gain (reduction) and electron loss (oxidation) occur simultaneously. The redox process is tracked by assigning oxidation numbers to each atom in a reaction. The species that is oxidized (contains an atom that increases in oxidation number) is the reducing agent; the species that is reduced (contains an atom that decreases in oxidation number) is the oxidizing agent. Redox reactions can be balanced by keeping track of the changes in oxidation number. A redox titration is used to determine the concentration of either the oxidizing or the reducing agent from the known concentration of the other.

4.6 ELEMENTAL SUBSTANCES IN REDOX REACTIONS

As you've seen in this chapter, the modern approach to classifying reactions is based on the underlying chemical process—precipitation, acid-base, or redox. An older method that was once very common classified reactions by comparing the number of reactants with the number of products. By that approach, three types of reactions were recognized:

- *Combination reactions* were those in which two or more reactants formed one product:

$$X + Y \longrightarrow Z$$

- *Decomposition reactions* were those in which one reactant formed two or more products:

$$Z \longrightarrow X + Y$$

- *Displacement reactions* were those in which the same number of substances were on both sides of the equation, but the atoms (or ions) exchanged places:

$$X + YZ \longrightarrow XZ + Y$$

One feature of this older classification that is especially relevant to oxidation-reduction processes is that *whenever an elemental substance appears as either reactant or product, the change is a redox reaction.*

There are many redox reactions that do *not* involve free elements, such as the one between MnO_4^- and $C_2O_4^{2-}$ that we saw earlier, but we'll focus here on those types that *do* involve elements as free, uncombined substances.

Combining Two Elements Two elements may react to form binary ionic or covalent compounds. In every case, there is a net change in the distribution of electron charge, and so the elements change their oxidation numbers. Here are some important examples:

1. *Metal and nonmetal form an ionic compound.* Figure 4.14 shows the reaction between an alkali metal and a halogen on the observable and atomic scales. Note the change in oxidation numbers. As you can see, K is oxidized, so it is the reducing agent; Cl_2 is reduced, so it is the oxidizing agent.

Aluminum reacts with O_2, as does nearly every metal, to form ionic oxides:

$$\overset{0}{4Al(s)} + \overset{0}{3O_2(g)} \longrightarrow \overset{+3\,-2}{2Al_2O_3(s)}$$

$$\overset{0}{2K(s)} + \overset{0}{Cl_2(g)} \longrightarrow \overset{+1\ -1}{2KCl(s)}$$
Potassium   Chlorine   Potassium chloride

Figure 4.14 Combining elements to form an ionic compound. When the metal potassium and the nonmetal chlorine react, they form the solid ionic compound potassium chloride. The photos *(top)* present the view the chemist sees in the laboratory. The blow-up arrows lead to an atomic-scale view *(middle)*; the stoichiometry is indicated by the more darkly colored spheres. The balanced redox equation is shown with oxidation numbers *(bottom)*.

2. *Two nonmetals form a covalent compound.* In one of thousands of examples, ammonia forms from nitrogen and hydrogen in a reaction that occurs in industry on an enormous scale:

$$\overset{0}{N_2}(g) + 3\overset{0}{H_2}(g) \longrightarrow 2\overset{-3 \; +1}{NH_3}(g)$$

Halogens form many compounds with other nonmetals, as in the formation of phosphorus trichloride, a major reactant in the production of pesticides and other organic compounds:

$$\overset{0}{P_4}(s) + 6\overset{0}{Cl_2}(g) \longrightarrow 4\overset{+3 \; -1}{PCl_3}(l)$$

Nearly every nonmetal reacts with O_2 to form a covalent oxide, as when nitrogen monoxide forms at the very high temperatures created in air by lightning:

$$\overset{0}{N_2}(g) + \overset{0}{O_2}(g) \longrightarrow 2\overset{+2 \; -2}{NO}(g)$$

Combining Compound and Element Many binary covalent compounds react with nonmetals to form larger compounds. Many nonmetal oxides react with additional O_2 to form "higher" oxides (those with more O atoms in each molecule). For example, a key step in the generation of urban smog is

$$2\overset{+2 \; -2}{NO}(g) + \overset{0}{O_2}(g) \longrightarrow 2\overset{+4 \; -2}{NO_2}(g)$$

Similarly, many nonmetal halides combine with additional halogen:

$$\overset{+3 \; -1}{PCl_3}(l) + \overset{0}{Cl_2}(g) \longrightarrow \overset{+5 \; -1}{PCl_5}(s)$$

Decomposing Compounds into Elements A decomposition reaction occurs when a reactant absorbs enough energy for one or more of its bonds to break. The energy can take many forms—heat, electricity, light, mechanical, and so forth—but we'll focus in this discussion on heat and electricity. The products are either elements or elements and smaller compounds. Here are some common examples:

1. *Thermal decomposition.* When the energy absorbed is heat, the reaction is a thermal decomposition. (A Greek *delta*, Δ, above a reaction arrow indicates that heat is required for the reaction.) Many metal oxides, chlorates, and perchlorates release oxygen when strongly heated. Figure 4.15 shows the decomposition, on the macroscopic and atomic scales, of mercury(II) oxide, used by Lavoisier and Priestley in their classic experiments. Heating potassium chlorate is a modern method for forming small amounts of oxygen in the laboratory:

$$2\overset{+1 \; +5 \; -2}{KClO_3}(s) \overset{\Delta}{\longrightarrow} 2\overset{+1 \; -1}{KCl}(s) + 3\overset{0}{O_2}(g)$$

Notice that, in these cases, the lone reactant is the oxidizing *and* the reducing agent. For example, with HgO, O^{2-} reduces Hg^{2+} (and Hg^{2+} oxidizes O^{2-}).

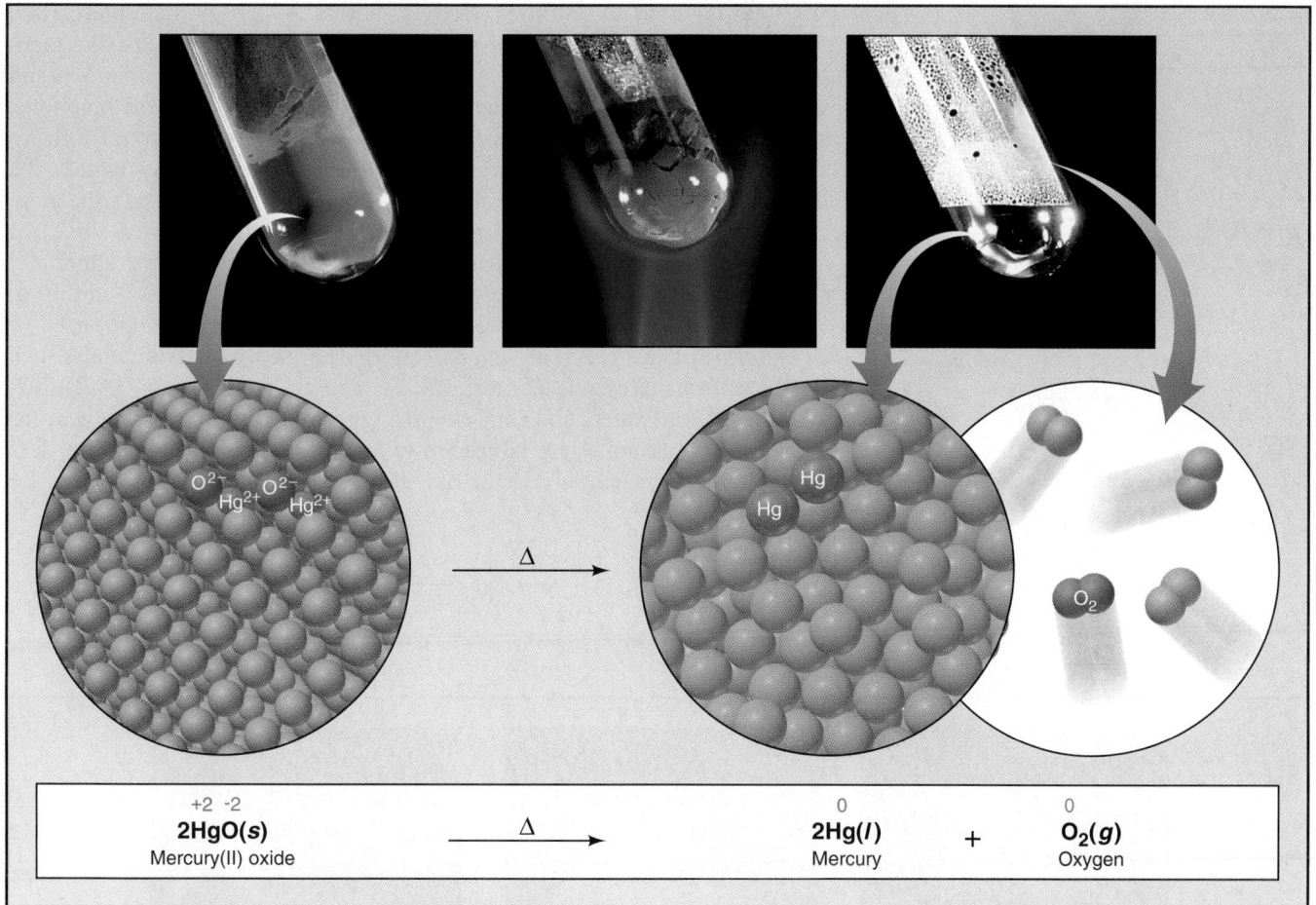

Figure 4.15 Decomposing a compound to its elements. Heating solid mercury(II) oxide decomposes it to liquid mercury and gaseous oxygen: the macroscopic (laboratory) view *(top)*; the atomic-scale view, with the more darkly colored spheres showing the stoichiometry *(middle)*; and the balanced redox equation *(bottom)*.

2. *Electrolytic decomposition*. In many cases, a compound can absorb electrical energy and decompose into its elements, in a process known as *electrolysis*. Observation of the electrolysis of water was crucial in the establishment of atomic masses:

$$\overset{+1-2}{2H_2O(l)} \xrightarrow{\text{electricity}} \overset{0}{2H_2(g)} + \overset{0}{O_2(g)}$$

Many active metals, such as sodium, magnesium, and calcium, are produced industrially by electrolysis of their molten halides:

$$\overset{+2-1}{MgCl_2(l)} \xrightarrow{\text{electricity}} \overset{0}{Mg(l)} + \overset{0}{Cl_2(g)}$$

(We'll examine the details of electrolysis in Chapters 21 and 22.)

Displacing One Element by Another; Activity Series As was pointed out earlier, displacement reactions have the same number of reactants as products. We mentioned double-displacement (metathesis) reactions in discussing precipitation and acid-base reactions. Here we consider so-called *single-displacement* reactions, which are all oxidation-reduction processes. They occur when one atom displaces the ion of a different atom from solution. When the reaction involves metals, the atom reduces the ion; when it involves nonmetals (specifically halogens), the atom oxidizes the ion. Chemists rank various elements into activity series—one for metals and one for halogens—in order of their ability to displace one another.

1. *The activity series of the metals.* Metals can be ranked by their ability to displace H_2 from various sources or to displace one another from solution.

(a) *A metal displaces H_2 from water or acid.* The most reactive metals, such as those from Group 1A(1) and Ca, Sr, and Ba from Group 2A(2), displace H_2 from water, and they do so vigorously. Figure 4.16 shows this reaction for lithium. Heat is needed to speed the reaction of slightly less reactive metals, such as Al and Zn, so these displace H_2 from steam:

$$\overset{0}{2Al(s)} + \overset{+1\,-2}{6H_2O(g)} \longrightarrow \overset{+3\,-2\,+1}{2Al(OH)_3(s)} + \overset{0}{3H_2(g)}$$

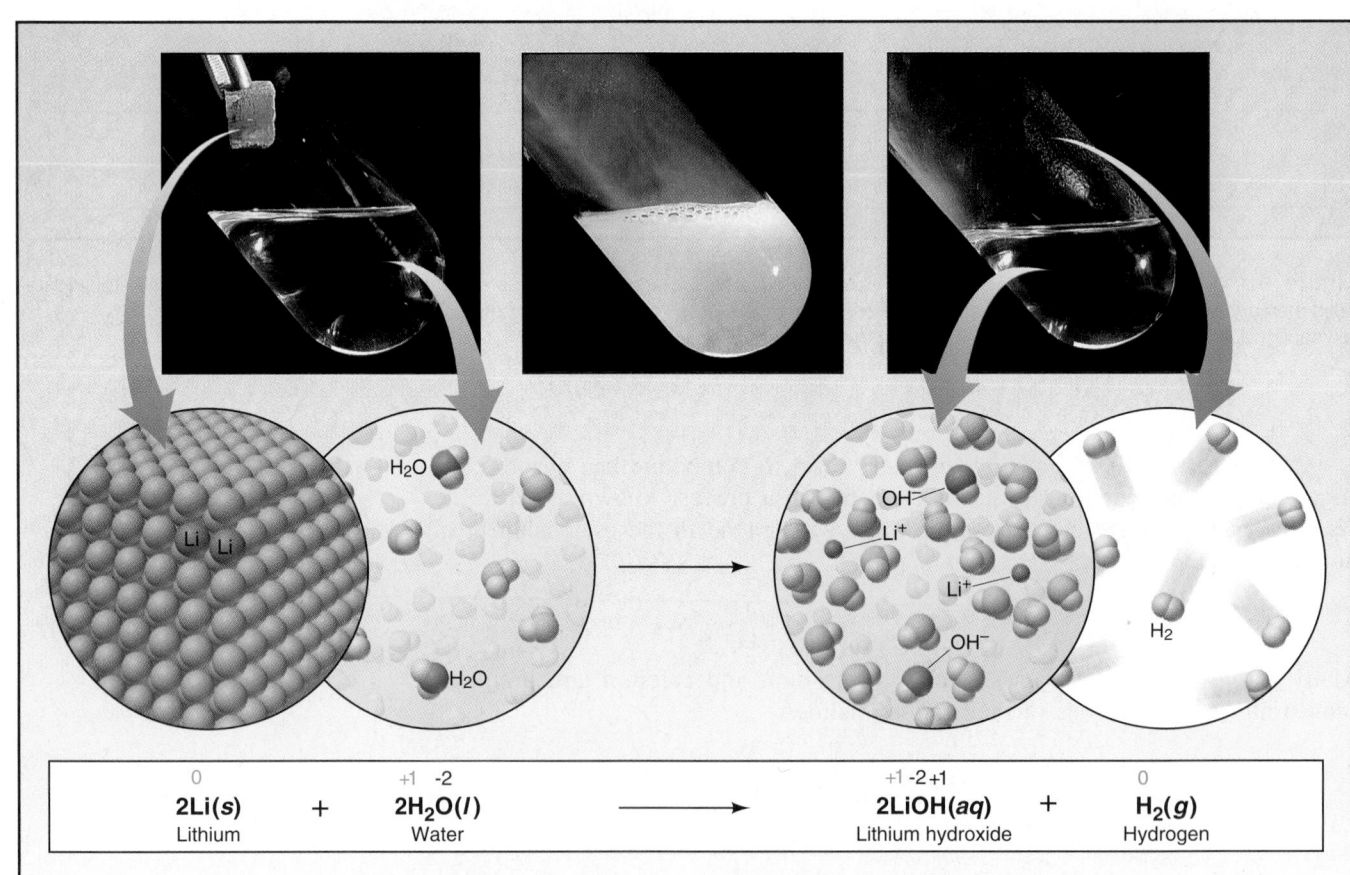

$$\overset{0}{\underset{\text{Lithium}}{2Li(s)}} + \overset{+1\ -2}{\underset{\text{Water}}{2H_2O(l)}} \longrightarrow \overset{+1\ -2+1}{\underset{\text{Lithium hydroxide}}{2LiOH(aq)}} + \overset{0}{\underset{\text{Hydrogen}}{H_2(g)}}$$

Figure 4.16 An active metal displacing hydrogen from water. Lithium displaces hydrogen from water in a vigorous reaction that yields an aqueous solution of lithium hydroxide and hydrogen gas, as shown on the macroscopic scale *(top)*, at the atomic scale *(middle)*, and as a balanced equation *(bottom)*. (For clarity, the atomic-scale view of water has been greatly simplified, and only water molecules involved in the reaction are colored red and blue.)

Still less reactive metals, such as nickel and tin, do not react with water but *do* react with acids, from which H_2 is displaced more easily (Figure 4.17). Here is the net ionic equation:

$$\overset{0}{Ni}(s) + 2\overset{+1}{H}^{+}(aq) \longrightarrow \overset{+2}{Ni}^{2+}(aq) + \overset{0}{H_2}(g)$$

Notice that in all such reactions, the metal is the reducing agent (O.N. of metal increases), and water or acid is the oxidizing agent (O.N. of H decreases). The least reactive metals, such as silver and gold, cannot displace H_2 from any source.

(b) *A metal displaces another metal ion from solution.* Direct comparisons of metal reactivity are clearest in these reactions. For example, zinc metal displaces copper(II) ion from copper(II) sulfate solution, as the total ionic equation shows:

$$\overset{+2}{Cu}^{2+}(aq) + \overset{+6}{\overset{-2}{SO_4}}{}^{2-}(aq) + \overset{0}{Zn}(s) \longrightarrow \overset{0}{Cu}(s) + \overset{+2}{Zn}^{2+}(aq) + \overset{+6}{\overset{-2}{SO_4}}{}^{2-}(aq)$$

Figure 4.18 demonstrates in atomic detail that copper metal can displace silver ion from solution. Thus, zinc is more reactive than copper, which is more reactive than silver.

Figure 4.17 **The displacement of H_2 from acid by nickel.**

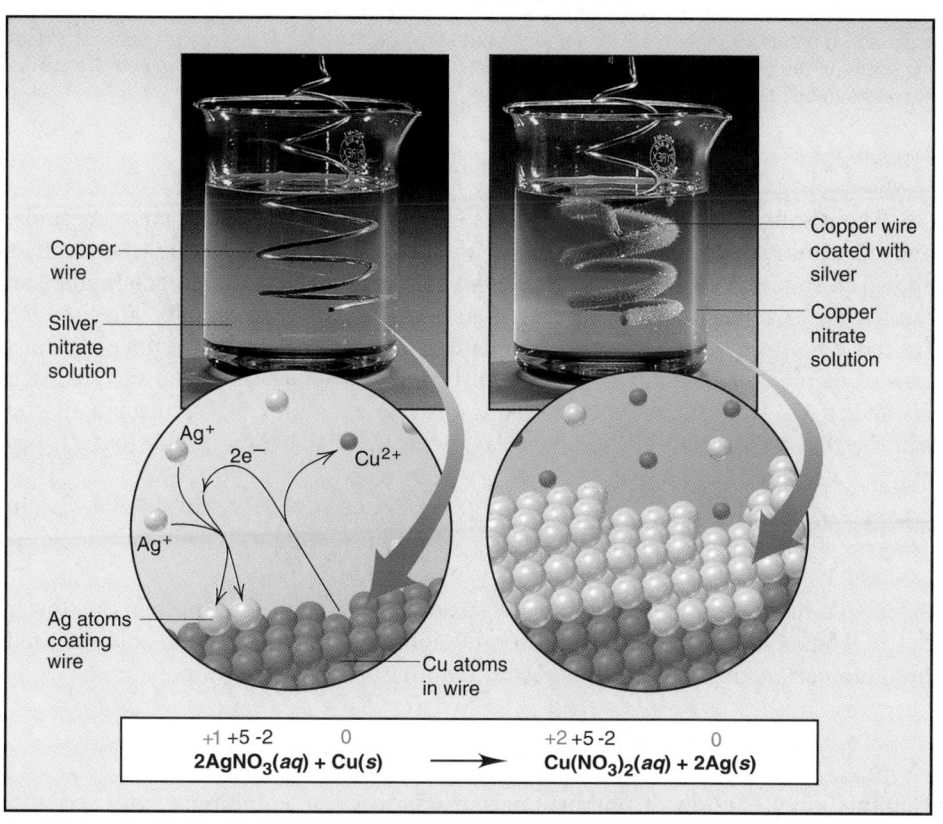

$$\overset{+1\ +5\ -2}{2AgNO_3}(aq) + \overset{0}{Cu}(s) \longrightarrow \overset{+2\ +5\ -2}{Cu(NO_3)_2}(aq) + \overset{0}{2Ag}(s)$$

Figure 4.18 **Displacing one metal with another.** More reactive metals can displace less reactive metals from solution. In this reaction, Cu atoms become Cu^{2+} ions and leave the wire, as they transfer electrons to two Ag^+ ions that become Ag atoms and coat the wire. The reaction is depicted as the laboratory view *(top),* the atomic-scale view *(middle),* and the balanced redox equation *(bottom).*

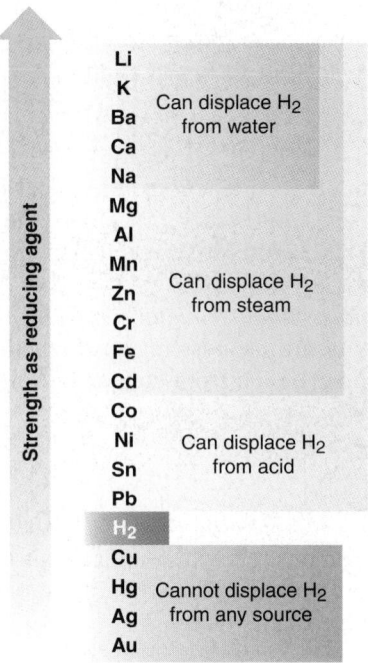

Figure 4.19 **The activity series of the metals.** This list of metals (and H_2) is arranged with the most active metal (strongest reducing agent) at the top and the least active metal (weakest reducing agent) at the bottom. The four metals below H_2 cannot displace it from any source. (The ranking refers to behavior in aqueous solution.)

The results of many such reactions between metals and water, aqueous acids, and metal-ion solutions form the basis of the **activity series of the metals.** It is shown in Figure 4.19 as a list of elements arranged such that elements higher on the list are stronger reducing agents than elements lower down. In other words, for those that are stable in water, elements higher on the list can reduce aqueous ions of elements lower down. The list also shows whether the metal can displace H_2 and, if so, from which source. Using just the examples we've discussed, you can see that Li, Al, and Ni lie above H_2, while Ag lies below it, and that Zn lies above Cu, which lies above Ag.

2. *The activity series of the halogens.* Reactivity decreases down Group 7A(17), so we can arrange the halogens into their own activity series:

$$F_2 > Cl_2 > Br_2 > I_2$$

A halogen higher in the periodic table is a stronger oxidizing agent than one lower down. Thus, chlorine can oxidize bromide ions or iodide ions from solution, and bromine can oxidize iodide ions. Here, chlorine displaces bromine:

$$\overset{-1}{2Br^-}(aq) + \overset{0}{Cl_2}(aq) \longrightarrow \overset{0}{Br_2}(aq) + \overset{-1}{2Cl^-}(aq)$$

Combustion Reactions Combustion is the process of combining with oxygen, usually with the release of large amounts of heat and light, often as a flame. Burning in air is a common example. Even though they do not always fall neatly into classes based on the number of reactants and products, *all combustion reactions are redox processes* because elemental oxygen is a reactant:

$$S_8(s) + 8O_2(g) \longrightarrow 8SO_2(g)$$

The combustion reactions we use to produce energy involve organic mixtures such as coal, gasoline, and natural gas as reactants. These mixtures consist of sub-

Space-Age Combustion Without a Flame Combustion reactions are used to generate large amounts of energy. In most common applications, the fuel is burned and the energy is released as heat (in a furnace) or as a combination of work and heat (in a combustion engine). Aboard the space shuttle, devices called *fuel cells* generate electrical energy from the flameless combustion of hydrogen gas. The H_2 is the reducing agent, and O_2 is the oxidizing agent in a complex, controlled reaction process that yields water—which the astronauts use for drinking. On Earth, fuel cells based on the reaction of either H_2 or of methanol (CH_3OH) with O_2 are being investigated for use in automobile engines.

stances with many carbon-carbon and carbon-hydrogen bonds. During the reaction, these bonds break, and each C and H atom combines with oxygen. Therefore, the products typically consist of CO_2 and H_2O. The combustion of the hydrocarbon butane, which is used in cigarette lighters, is typical:

$$2C_4H_{10}(g) + 13O_2(g) \longrightarrow 8CO_2(g) + 10H_2O(g)$$

Cellular *respiration* can be thought of as a combustion process that occurs within our bodies' cells—fortunately without flame—when we "burn" organic foodstuffs, such as glucose, for energy:

$$C_6H_{12}O_6(s) + 6O_2(g) \longrightarrow 6CO_2(g) + 6H_2O(g) + energy$$

SAMPLE PROBLEM 4.10 Identifying the Type of Redox Reaction

Problem Classify each of the following redox reactions as a combination, decomposition, or displacement reaction, write a balanced molecular equation for each, as well as total and net ionic equations for part (c), and identify the oxidizing and reducing agents:
(a) Magnesium(s) + nitrogen(g) $\longrightarrow$ magnesium nitride(s)
(b) Hydrogen peroxide(l) $\longrightarrow$ water + oxygen gas
(c) Aluminum(s) + lead(II) nitrate(aq) $\longrightarrow$ aluminum nitrate(aq) + lead(s)
Plan To decide on reaction type, recall that combination reactions produce fewer products than reactants, decomposition reactions produce more products, and displacement reactions have the same number of reactants and products. The oxidizing number (O.N.) becomes more positive for the reducing agent and less positive for the oxidizing agent.
Solution (a) Combination: two substances form one. This reaction occurs, along with formation of magnesium oxide, when magnesium burns in air:

$$\overset{0}{3Mg}(s) + \overset{0}{N_2}(g) \longrightarrow \overset{+2\ -3}{Mg_3N_2}(s)$$

Mg is the reducing agent; N_2 is the oxidizing agent.
(b) Decomposition: one substance forms two. This reaction occurs within every bottle of this common household antiseptic. Hydrogen peroxide is very unstable and breaks down from heat, light, or just shaking:

$$\overset{+1\ -1}{2H_2O_2}(l) \longrightarrow \overset{+1\ -2}{2H_2O}(l) + \overset{0}{O_2}(g)$$

H_2O_2 is both the oxidizing *and* the reducing agent. The O.N. of O in peroxides is -1. It increases to 0 in O_2 and decreases to -2 in H_2O.

(c) Displacement: two substances form two. As Figure 4.19 shows, Al is more active than Pb and, thus, displaces it from aqueous solution:

$$\overset{0}{2Al}(s) + \overset{+2\ +5\ -2}{3Pb(NO_3)_2}(aq) \longrightarrow \overset{+3\ +5\ -2}{2Al(NO_3)_3}(aq) + \overset{0}{3Pb}(s)$$

Al is the reducing agent; $Pb(NO_3)_2$ is the oxidizing agent.
The total ionic equation is

$$2Al(s) + 3Pb^{2+}(aq) + 6NO_3^-(aq) \longrightarrow 2Al^{3+}(aq) + 6NO_3^-(aq) + 3Pb(s)$$

The net ionic equation is

$$2Al(s) + 3Pb^{2+}(aq) \longrightarrow 2Al^{3+}(aq) + 3Pb(s)$$

FOLLOW-UP PROBLEM 4.10 Classify each of the following redox reactions as a combination, decomposition, or displacement reaction, write a balanced molecular equation for each, as well as total and net ionic equations for parts (b) and (c), and identify the oxidizing and reducing agents:
(a) $S_8(s) + F_2(g) \longrightarrow SF_4(g)$
(b) $CsI(aq) + Cl_2(aq) \longrightarrow CsCl(aq) + I_2(aq)$
(c) $Ni(NO_3)_2(aq) + Cr(s) \longrightarrow Cr(NO_3)_3(aq) + Ni(s)$

SECTION SUMMARY

Any reaction that includes a free element as reactant or product is a redox reaction. In combination reactions, elements combine to form a compound, or a compound and an element combine. Decomposition of compounds by absorption of heat or electricity can form elements or a compound and an element. In displacement reactions, one element displaces another from solution. Activity series rank elements in order of reactivity. The activity series of the metals ranks metals by their ability to displace H_2 from water, steam, acid or to displace one another from solution. Combustion typically releases heat and light energy through reaction of a substance with O_2.

4.7 REVERSIBLE REACTIONS: AN INTRODUCTION TO CHEMICAL EQUILIBRIUM

So far, we have viewed reactions as occurring from "left to right," from reactants to products and continuing until they are complete, that is, until the limiting reactant is used up. However, many reactions seem to stop before this happens. The reason is that another reaction, the reverse of the first one, is also taking place. The forward (left-to-right) reaction has not stopped, but the reverse (right-to-left) reaction is taking place at the same rate. Therefore, *no further changes occur in the amounts of reactants or products.* At this point, the reaction mixture has reached **dynamic equilibrium.** On the macroscopic scale, the reaction is *static,* but it is *dynamic* on the molecular scale. In principle, all reactions are reversible and will eventually reach dynamic equilibrium as long as all products remain available for the reverse reaction.

Let's examine equilibrium with a particular set of substances. Calcium carbonate breaks down when heated to calcium oxide and carbon dioxide:

$$CaCO_3(s) \longrightarrow CaO(s) + CO_2(g) \quad \text{[breakdown]}$$

It can also form when calcium oxide and carbon dioxide react:

$$CaO(s) + CO_2(g) \longrightarrow CaCO_3(s) \quad \text{[formation]}$$

The formation is just the reverse of the breakdown. Suppose we place 10 g of $CaCO_3$ in an *open* steel reaction flask and heat it to around 900°C, as shown in Figure 4.20A. The $CaCO_3$ starts breaking down to CaO and CO_2, and the CO_2 escapes from the open flask. The reaction goes to completion because the reverse reaction (formation) can occur only if CO_2 is present.

In Figure 4.20B, we perform the same experiment in a *closed* steel flask, so that the CO_2 remains in contact with the CaO. The breakdown (forward reaction) begins, but at first, when very little $CaCO_3$ has broken down, very little CO_2 and CaO are present; thus, the formation (reverse reaction) just barely begins. As the $CaCO_3$ continues to break down, the amounts of CO_2 and CaO in the flask increase. They react with each other more frequently, and the formation occurs a bit faster. As the amounts of CaO and CO_2 increase, the formation reaction gradually speeds up. Eventually, the reverse reaction (formation) happens just as fast as the forward reaction (breakdown), and the amounts of $CaCO_3$, CaO, and CO_2 no longer change: the system has

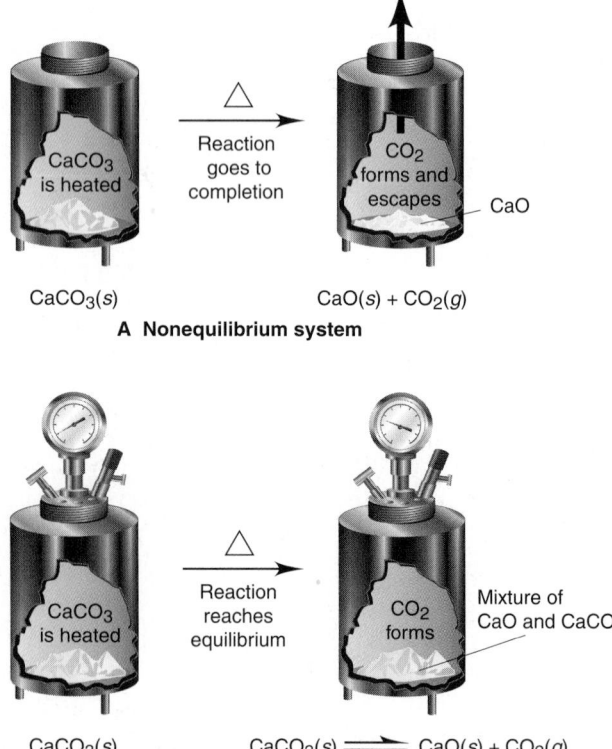

$CaCO_3(s)$ $CaO(s) + CO_2(g)$

A Nonequilibrium system

$CaCO_3(s)$ $CaCO_3(s) \rightleftharpoons CaO(s) + CO_2(g)$

B Equilibrium system

Figure 4.20 The equilibrium state. A, In an *open* steel reaction chamber, strong heating breaks down $CaCO_3$ completely because the product CO_2 escapes and is not present to react with the other product, CaO. **B,** When $CaCO_3$ breaks down in a *closed* chamber, the CO_2 *is* present to react with CaO and re-form $CaCO_3$ in a reaction that is the reverse of the breakdown. At a given temperature, no further change in the amounts of products and reactants means that the reaction has reached equilibrium.

reached equilibrium. We indicate this with a pair of arrows pointing in opposite directions:

$$CaCO_3(s) \rightleftharpoons CaO(s) + CO_2(g)$$

Bear in mind that equilibrium can be established only when *all the substances involved are kept in contact with each other*. The breakdown of $CaCO_3$ goes to completion in the open flask because the CO_2 escapes.

Aqueous acid-base reactions that form a gaseous product go to completion in an open flask for the same reason: the gas escapes, so the reverse reaction cannot take place. Precipitation and other acid-base reactions also go largely to completion, even though all the products remain in the reaction vessel. In those cases, the *ions* are not available to any appreciable extent to participate in the reverse process, because they are tied up either as an insoluble solid (precipitation) or as water molecules (acid-base).

The concept of reaction reversibility also relates to the behavior of acids and bases that are weak electrolytes, that is, those that dissociate into ions only to a small extent. The dissociation quickly becomes balanced by a reassociation, such that equilibrium is reached with very few ions present. For example, when acetic acid dissolves in water, some of the CH_3COOH molecules transfer a proton to H_2O and form H_3O^+ and CH_3COO^- ions. As more of these ions form, they react with each other more often to re-form acetic acid and water:

$$CH_3COOH(aq) + H_2O(l) \rightleftharpoons H_3O^+(aq) + CH_3COO^-(aq)$$

In fact, in 0.1 *M* CH_3COOH at 25°C, only about 1.3% of the acid molecules are dissociated at any given moment. Similarly, the weak base ammonia reacts with water to form ions. As the product ions interact, they re-form ammonia and water again, and the rates of the reverse and forward reactions soon balance:

$$NH_3(aq) + H_2O(l) \rightleftharpoons NH_4^+(aq) + OH^-(aq)$$

Thus, as you've seen, some reactions proceed very little before they reach equilibrium, while others proceed essentially to completion. And even two reactions that proceed very little, such as those we've just discussed between acetic acid or ammonia and water, proceed to different extents. So, a fundamental question arises: why does each process, even under the same conditions, reach equilibrium with its own particular ratio of product concentrations to reactant concentrations? The full answer will have to wait until later chapters, but we can hint at the factors here. The energy available for a reaction to occur is called its *free energy*. The point at which a process reaches equilibrium occurs when the reaction mixture has its lowest free energy. Two components of this free energy are the heat that a reaction releases (or absorbs) and the change in orderliness of the particles—solids are more orderly than gases, pure substances are more orderly than solutions, and so forth. A particular combination of these factors, discussed quantitatively in Chapter 20, determines the free energy and, thus, the equilibrium point for a given mixture of reactants and products at a given temperature.

Many aspects of dynamic equilibrium are relevant to natural systems, from the cycling of water in the environment to the balance of lion and antelope on the plains of Africa to the nuclear processes occurring in stars. We examine equilibrium in chemical and physical systems in Chapters 12, 13, and 17 through 21.

SECTION SUMMARY

Every reaction is reversible if all the substances are kept in contact with one another. As the amounts of products increase, the reactants begin to re-form. When the reverse reaction happens as rapidly as the forward reaction, the amounts of the substances no longer change, and the reaction mixture has reached dynamic equilibrium. A reaction goes to completion if a product is removed from the system (as a gas) or exists in a form that prevents it from reacting (precipitate or undissociated molecule). Weak acids and bases reach equilibrium in water with a very small proportion of their molecules dissociated.

Chapter Perspective

Classifying facts is the first step toward understanding them, and this chapter classified many of the most important facts of reaction chemistry into three major processes—precipitation, acid-base, and oxidation-reduction. We also examined the great influence that water has on reaction chemistry and introduced the state of dynamic equilibrium, which is related to the central question of why physical and chemical changes occur. All these topics appear again at many places in the text.

In the next chapter, our focus changes to the physical behavior of gases. You'll find that your growing appreciation of events on the molecular level has become indispensable for understanding the nature of the three physical states.

For Review and Reference (Numbers in parentheses refer to pages, unless noted otherwise.)

Learning Objectives

Relevant section and/or sample problem (SP) numbers appear in parentheses.

Understand These Concepts

1. How water dissolves ionic compounds and dissociates them into ions (4.1)
2. Distinguish between the species present when ionic and covalent compounds dissolve in water; distinguish between strong and weak electrolytes (4.1)
3. The use of ionic equations to specify the essential nature of an aqueous reaction (4.2)
4. The driving force for aqueous ionic reactions (4.3, 4.4, 4.5)
5. How to decide whether a precipitation reaction occurs (4.3)
6. The main distinction between strong and weak aqueous acids and bases (4.4)
7. The essential character of aqueous acid-base reactions as proton-transfer processes (4.4)
8. The importance of net movement of electrons in the redox process (4.5)
9. The relation between change in oxidation number and identity of oxidizing and reducing agents (4.5)
10. The presence of elements in some important types of redox reactions: combination, decomposition, displacement (4.6)

11. The balance between forward and reverse rates of a chemical reaction that leads to dynamic equilibrium; why some acids and bases are weak (4.7)

Master These Skills

1. Using the formula of a compound to find the number of moles of ions in solution (SP 4.1)
2. Determining the concentration of H^+ ion in an aqueous acid solution (SP 4.2)
3. Predicting whether a precipitation reaction occurs (SP 4.3)
4. Writing ionic equations to describe precipitation and acid-base reactions (SPs 4.3 and 4.4)
5. Calculating an unknown concentration from an acid-base or redox titration (SPs 4.5 and 4.9)
6. Determining the oxidation number of any element in a compound (SP 4.6)
7. Identifying the oxidizing and reducing agents in a redox reaction (SP 4.7)
8. Balancing redox equations (SP 4.8)
9. Identifying combination, decomposition, and displacement redox reactions (SP 4.10)

Key Terms

Section 4.1
electrolyte (132)
solvated (132)
polar molecule (134)
nonelectrolyte (135)

Section 4.2
molecular equation (137)
total ionic equation (138)
spectator ion (138)
net ionic equation (138)

Section 4.3
precipitation reaction (138)
precipitate (138)
metathesis reaction (139)

Section 4.4
acid-base reaction (140)
neutralization reaction (140)
acid (141)
base (141)
salt (142)

titration (143)
equivalence point (143)
end point (143)

Section 4.5
oxidation-reduction
 (redox) reaction (147)
oxidation (148)
reduction (148)
oxidizing agent (148)
reducing agent (148)

oxidation number (O.N.)
 (or oxidation state) (148)
oxidation number
 method (151)

Section 4.6
activity series of the
 metals (160)

Section 4.7
dynamic equilibrium (162)

Highlighted   Figures and Tables

These figures (F) and tables (T) provide a quick review of key ideas. Entries in color contain frequently used data.

F4.2 Electron distribution in H_2 and H_2O (134)
F4.3 Dissolution of an ionic compound (135)
F4.5 Depicting a precipitation reaction with ionic equations (137)
T4.1 Solubility rules for ionic compounds in water (139)
T4.2 Selected acids and bases (141)
F4.8 An aqueous strong acid–strong base reaction on the atomic scale (145)

F4.10 The redox process in compound formation (147)
T4.3 Rules for assigning an oxidation number (148)
F4.11 Highest and lowest oxidation numbers of reactive main-group elements (149)
F4.12 A summary of terminology for redox reactions (149)
F4.19 The activity series of the metals (160)

Brief Solutions to Follow-up Problems

4.1 (a) $KClO_4(s) \xrightarrow{H_2O} K^+(aq) + ClO_4^-(aq)$;
2 mol of K^+ and 2 mol of ClO_4^-

(b) $Mg(C_2H_3O_2)_2(s) \xrightarrow{H_2O} Mg^{2+}(aq) + 2C_2H_3O_2^-(aq)$;
2.49 mol of Mg^{2+} and 4.97 mol of $C_2H_3O_2^-$

(c) $(NH_4)_2CrO_4(s) \xrightarrow{H_2O} 2NH_4^+(aq) + CrO_4^{2-}(aq)$;
6.24 mol of NH_4^+ and 3.12 mol of CrO_4^{2-}

(d) $NaHSO_4(s) \xrightarrow{H_2O} Na^+(aq) + HSO_4^-(aq)$;
0.73 mol of Na^+ and 0.73 mol of HSO_4^-

4.2 Moles of H^+ = 451 mL $\times \dfrac{1\ L}{10^3\ mL}$

$\times \dfrac{3.20\ mol\ HBr}{1\ L\ soln} \times \dfrac{1\ mol\ H^+}{1\ mol\ HBr}$

= 1.44 mol H^+

4.3 (a) $Fe^{3+}(aq) + 3Cl^-(aq) + 3Cs^+(aq) + PO_4^{3-}(aq) \longrightarrow$
$FePO_4(s) + 3Cl^-(aq) + 3Cs^+(aq)$
$Fe^{3+}(aq) + PO_4^{3-}(aq) \longrightarrow FePO_4(s)$

(b) $2Na^+(aq) + 2OH^-(aq) + Cd^{2+}(aq) + 2NO_3^-(aq) \longrightarrow$
$2Na^+(aq) + 2NO_3^-(aq) + Cd(OH)_2(s)$
$2OH^-(aq) + Cd^{2+}(aq) \longrightarrow Cd(OH)_2(s)$

(c) No reaction occurs

(d) $2Ag^+(aq) + SO_4^{2-}(aq) + Ba^{2+}(aq) + 2Cl^-(aq) \longrightarrow$
$2AgCl(s) + BaSO_4(s)$

Total and net ionic equations are identical.

4.4 $Ca(OH)_2(aq) + 2HNO_3(aq) \longrightarrow$
$Ca(NO_3)_2(aq) + 2H_2O(l)$
$Ca^{2+}(aq) + 2OH^-(aq) + 2H^+(aq) + 2NO_3^-(aq) \longrightarrow$
$Ca^{2+}(aq) + 2NO_3^-(aq) + 2H_2O(l)$
$H^+(aq) + OH^-(aq) \longrightarrow H_2O(l)$

4.5 $Ba(OH)_2(aq) + 2HCl(aq) \longrightarrow BaCl_2(aq) + 2H_2O(l)$
Volume (L) of soln

$= 50.00\ mL\ HCl\ soln \times \dfrac{1\ L}{10^3\ mL} \times \dfrac{0.1016\ mol\ HCl}{1\ L\ soln}$

$\times \dfrac{1\ mol\ Ba(OH)_2}{2\ mol\ HCl} \times \dfrac{1\ L\ soln}{0.1292\ mol\ Ba(OH)_2}$

= 0.01966 L

4.6 (a) O.N. of Sc = +3; O.N. of O = -2
(b) O.N. of Ga = +3; O.N. of Cl = -1

(c) O.N. of H = +1; O.N. of P = +5; O.N. of O = -2
(d) O.N. of I = +3; O.N. of F = -1

4.7 (a) Fe is reducing agent; Cl_2 is oxidizing agent.
(b) C_2H_6 is reducing agent; O_2 is oxidizing agent.
(c) CO is reducing agent; I_2O_5 is oxidizing agent.

4.8 $K_2Cr_2O_7(aq) + 14HI(aq) \longrightarrow$
$2KI(aq) + 2CrI_3(aq) + 3I_2(s) + 7H_2O(l)$

4.9 (a) Moles of Ca^{2+} = 6.53 mL soln $\times \dfrac{1\ L}{10^3\ mL}$

$\times \dfrac{4.56\times10^{-3}\ mol\ KMnO_4}{1\ L\ soln}$

$\times \dfrac{5\ mol\ CaC_2O_4}{2\ mol\ KMnO_4} \times \dfrac{1\ mol\ Ca^{2+}}{1\ mol\ CaC_2O_4}$

= 7.44×10^{-5} mol Ca^{2+}

Molarity of Ca^{2+} = $\dfrac{7.44\times10^{-2}\ mol\ Ca^{2+}}{2.50\ mL\ milk} \times \dfrac{10^3\ mL}{1\ L}$

= $2.98\times10^{-2}\ M\ Ca^{2+}$

(b) Conc. of Ca^{2+} (g/L)

$= \dfrac{2.98\times10^{-2}\ mol\ Ca^{2+}}{1\ L} \times \dfrac{40.08\ g\ Ca^{2+}}{1\ mol\ Ca^{2+}}$

$= \dfrac{1.19\ g\ Ca^{2+}}{1\ L}$

4.10 (a) Combination:
$S_8(s) + 16F_2(g) \longrightarrow 8SF_4(g)$
S_8 is the reducing agent; F_2 is the oxidizing agent.
(b) Displacement:
$2CsI(aq) + Cl_2(aq) \longrightarrow 2CsCl(aq) + I_2(aq)$
Cl_2 is the oxidizing agent; CsI is the reducing agent
$2Cs^+(aq) + 2I^-(aq) + Cl_2(aq) \longrightarrow$
$2Cs^+(aq) + 2Cl^-(aq) + I_2(aq)$
$2I^-(aq) + Cl_2(aq) \longrightarrow 2Cl^-(aq) + I_2(aq)$
(c) Displacement:
$3Ni(NO_3)_2(aq) + 2Cr(s) \longrightarrow 3Ni(s) + 2Cr(NO_3)_3(aq)$
$3Ni^{2+}(aq) + 6NO_3^-(aq) + 2Cr(s) \longrightarrow$
$3Ni(s) + 2Cr^{3+}(aq) + 6NO_3^-(aq)$
$3Ni^{2+}(aq) + 2Cr(s) \longrightarrow 3Ni(s) + 2Cr^{3+}(aq)$
Cr is the reducing agent; $Ni(NO_3)_2$ is the oxidizing agent.

Problems

Problems with **colored** numbers are answered at the back of the text. Sections match the text and provide the number(s) of relevant sample problems. Most offer Concept Review Questions, Skill-Building Exercises (in similar pairs), and Problems in Context. Then Comprehensive Problems, based on material from any section or previous chapter, follow.

The Role of Water as a Solvent
(Sample Problems 4.1 and 4.2)

● **Concept Review Questions**

4.1 What two factors cause water to be polar?

4.2 What types of substances are most likely to be soluble in water?

4.3 What must be present in an aqueous solution for it to conduct an electric current? What general classes of compounds form solutions that conduct?

4.4 What occurs on the molecular level when an ionic compound dissolves in water?

4.5 Examine each of the following aqueous solutions and determine which represents: (a) $CaCl_2$; (b) Li_2SO_4; (c) NH_4Br.

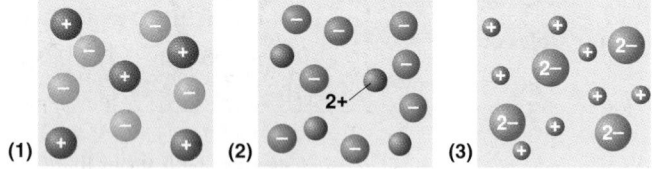

(1)   (2)   (3)

4.6 Which of the following best represents a volume from a solution of magnesium nitrate?

● = magnesium ion   ● = nitrate ion

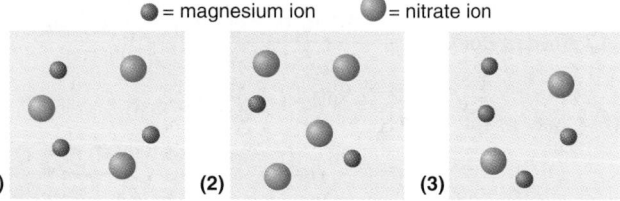

(1)   (2)   (3)

4.7 Why are some ionic compounds soluble in water and others are not?

4.8 Why are some covalent compounds soluble in water and others are not?

4.9 Some covalent compounds dissociate into ions when they dissolve in water. What atom do these compounds have in their structures? What type of aqueous solution do they form? Name three examples of such an aqueous solution.

● **Skill-Building Exercises** *(paired)*

4.10 State whether each of the following substances is likely to be very soluble in water. Explain.
(a) Benzene, C_6H_6 (b) Sodium hydroxide
(c) Ethanol, CH_3CH_2OH (d) Potassium acetate

4.11 State whether each of the following substances is likely to be very soluble in water. Explain.
(a) Lithium nitrate (b) Glycine, H_2NCH_2COOH
(c) Pentane (d) Ethylene glycol, $HOCH_2CH_2OH$

4.12 State whether an aqueous solution of each of the following substances conducts an electric current. Explain your reasoning.
(a) Cesium iodide (b) Hydrogen bromide

4.13 State whether an aqueous solution of each of the following substances conducts an electric current. Explain your reasoning.
(a) Sodium hydroxide (b) Glucose, $C_6H_{12}O_6$

4.14 How many total moles of ions are released when each of the following samples dissolves completely in water?
(a) 0.25 mol of NH_4Cl (b) 26.4 g of $Ba(OH)_2 \cdot 8H_2O$
(c) 1.78×10^{20} formula units of LiCl

4.15 How many total moles of ions are released when each of the following samples dissolves completely in water?
(a) 0.805 mol of Cs_2SO_4 (b) 1.55×10^{-3} g of $Ca(NO_3)_2$
(c) 3.85×10^{27} formula units of $Sr(HCO_3)_2$

4.16 How many total moles of ions are released when each of the following samples dissolves completely in water?
(a) 0.15 mol of Na_3PO_4 (b) 47.9 g of $NiBr_2 \cdot 3H_2O$
(c) 4.23×10^{22} formula units of $FeCl_3$

4.17 How many total moles of ions are released when each of the following samples dissolves completely in water?
(a) 0.382 mol of K_2HPO_4 (b) 6.80 g of $MgSO_4 \cdot 7H_2O$
(c) 6.188×10^{21} formula units of $NiCl_2$

4.18 How many moles and numbers of ions of each type are present in the following aqueous solutions?
(a) 95.5 mL of 2.45 *M* aluminum chloride
(b) 2.50 L of a solution containing 4.59 g/L sodium sulfate
(c) 80.5 mL of a solution containing 2.68×10^{22} formula units of magnesium bromide per liter

4.19 How many moles and numbers of ions of each type are present in the following aqueous solutions?
(a) 3.8 mL of 1.88 *M* magnesium chloride
(b) 345 mL of a solution containing 4.22 g/L aluminum sulfate
(c) 2.66 L of a solution containing 6.63×10^{21} formula units of lithium nitrate per liter

4.20 How many moles of H^+ ions are present in the following aqueous solutions?
(a) 0.140 L of 2.5 *M* perchloric acid
(b) 6.8 mL of 0.52 *M* nitric acid
(c) 2.5 L of 0.056 *M* hydrochloric acid

4.21 How many moles of H^+ ions are present in the following aqueous solutions?
(a) 1.4 L of 0.48 *M* hydrobromic acid
(b) 47 mL of 1.8 *M* hydriodic acid
(c) 425 mL of 0.27 *M* nitric acid

● **Problems in Context**

4.22 In laboratory studies of ocean-dwelling organisms, marine biologists use salt mixtures that simulate the ion concentrations in seawater. A 1.00-kg sample of simulated seawater is prepared by mixing 26.5 g of NaCl, 2.40 g of $MgCl_2$, 3.35 g of $MgSO_4$, 1.20 g of $CaCl_2$, 1.05 g of KCl, 0.315 g of $NaHCO_3$, and 0.098 g of NaBr in distilled water.
(a) If the density of this solution is 1.04 g/cm³, what is the molarity of each ion?
(b) What is the total molarity of alkali metal ions?
(c) What is the total molarity of alkaline earth metal ions?
(d) What is the total molarity of anions?

4.23 Water "softeners" remove metal ions such as Fe^{2+}, Fe^{3+}, Ca^{2+}, and Mg^{2+} (which make water "hard") by replacing them with enough Na^+ ions to maintain the same number of positive charges in the solution. If 1.0×10^3 L of hard water is 0.015 M Ca^{2+} and 0.0010 M Fe^{3+}, how many moles of Na^+ are needed to replace these ions?

Writing Equations for Aqueous Ionic Reactions

● Concept Review Questions

4.24 Which ions do not appear in a net ionic equation? Why?

4.25 Write two equations (both molecular and total ionic) with different reactants to obtain the same net ionic equation as the following equation:

$$Ba(NO_3)_2(aq) + Na_2CO_3(aq) \longrightarrow BaCO_3(s) + 2NaNO_3(aq)$$

Precipitation Reactions
(Sample Problem 4.3)

● Concept Review Questions

4.26 Why do some pairs of ions precipitate and others do not?

4.27 Use Table 4.1 to determine which of the following combinations leads to a reaction. How can you identify the spectator ions in the reaction?
(a) Calcium nitrate(aq) + potassium chloride(aq) $\longrightarrow$
(b) Sodium chloride(aq) + lead(II) nitrate(aq) $\longrightarrow$

4.28 The beakers represent the aqueous reaction of $AgNO_3$ and NaCl. Silver ions are gray. What colors are used to represent NO_3^-, Na^+, and Cl^-? Write molecular, total ionic, and net ionic equations for the reaction.

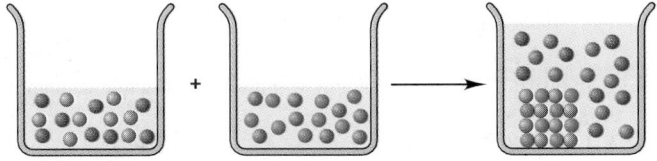

● Skill-Building Exercises *(paired)*

4.29 When each of the following pairs of aqueous solutions is mixed, does a precipitation reaction occur? If so, write the formula and name of the precipitate:
(a) Sodium nitrate + copper(II) sulfate
(b) Ammonium iodide + silver nitrate

4.30 When each of the following pairs of aqueous solutions is mixed, does a precipitation reaction occur? If so, write the formula and name of the precipitate:
(a) Potassium carbonate + barium hydroxide
(b) Aluminum nitrate + sodium phosphate

4.31 When each of the following pairs of aqueous solutions is mixed, does a precipitation reaction occur? If so, write the formula and name of the precipitate:
(a) Potassium chloride + iron(II) nitrate
(b) Ammonium sulfate + barium chloride

4.32 When each of the following pairs of aqueous solutions is mixed, does a precipitation reaction occur? If so, write the formula and name of the precipitate:
(a) Sodium sulfide + nickel(II) sulfate
(b) Lead(II) nitrate + potassium bromide

4.33 Complete the following precipitation reactions with balanced molecular, total ionic, and net ionic equations, and identify the spectator ions:

(a) $Hg_2(NO_3)_2(aq) + KI(aq) \longrightarrow$
(b) $FeSO_4(aq) + Ba(OH)_2(aq) \longrightarrow$

4.34 Complete the following precipitation reactions with balanced molecular, total ionic, and net ionic equations, and identify the spectator ions:
(a) $CaCl_2(aq) + Cs_3PO_4(aq) \longrightarrow$
(b) $Na_2S(aq) + ZnSO_4(aq) \longrightarrow$

4.35 If 35.0 mL of lead(II) nitrate solution reacts completely with excess sodium iodide solution to yield 0.628 g of precipitate, what is the molarity of lead(II) ion in the original solution?

4.36 If 25.0 mL of silver nitrate solution reacts with excess potassium chloride solution to yield 0.842 g of precipitate, what is the molarity of silver ion in the original solution?

● Problems in Context

4.37 The mass percent of Cl^- in a seawater sample is determined by titrating 25.00 mL of seawater with $AgNO_3$ solution, causing a precipitation reaction. An indicator is used to detect the end point, which occurs when free Ag^+ ion is present in solution after all the Cl^- has reacted. If 43.63 mL of 0.3020 M $AgNO_3$ is required to reach the end point, what is the mass percent of Cl^- in the seawater (d of seawater = 1.04 g/mL)?

4.38 Aluminum sulfate, known as cake alum, has a remarkably wide range of uses, from dyeing leather and cloth to purifying sewage. In aqueous solution, it reacts with base to form a white precipitate. (a) Write balanced total and net ionic equations for its reaction with aqueous NaOH. (b) What mass of precipitate forms when 135.5 mL of 0.633 M NaOH is added to 517 mL of a solution that contains 12.8 g of aluminum sulfate per liter?

Acid-Base Reactions
(Sample Problems 4.4 and 4.5)

● Concept Review Questions

4.39 Is the total ionic equation the same as the net ionic equation when $Sr(OH)_2(aq)$ and $H_2SO_4(aq)$ react? Explain.

4.40 State a general equation for a neutralization reaction.

4.41 (a) Name three common strong acids. (b) Name three common strong bases. (c) What is a characteristic behavior of a strong acid or a strong base?

4.42 (a) Name three common weak acids. (b) Name one common weak base. (c) What is the major difference between a weak acid and a strong acid or between a weak base and a strong base, and what experiment would you perform to observe it?

4.43 Do either of the following reactions go to completion? If so, what factor(s) drives each to completion?
(a) $MgSO_3(s) + 2HCl(aq) \longrightarrow$
$$MgCl_2(aq) + SO_2(g) + H_2O(l)$$
(b) $3Ba(OH)_2(aq) + 2H_3PO_4(aq) \longrightarrow Ba_3(PO_4)_2(s) + 6H_2O(l)$

4.44 The net ionic equation for the aqueous neutralization reaction between acetic acid and sodium hydroxide is different from that for the reaction between hydrochloric acid and sodium hydroxide. Explain by writing balanced net ionic equations.

● Skill-Building Exercises *(paired)*

4.45 Complete the following acid-base reactions with balanced molecular, total ionic, and net ionic equations, and identify the spectator ions:
(a) Potassium hydroxide(aq) + hydriodic acid(aq) $\longrightarrow$
(b) Ammonia(aq) + hydrochloric acid(aq) $\longrightarrow$

4.46 Complete the following acid-base reactions with balanced molecular, total ionic, and net ionic equations, and identify the spectator ions:
(a) Cesium hydroxide(aq) + nitric acid(aq) $\longrightarrow$
(b) Calcium hydroxide(aq) + acetic acid(aq) $\longrightarrow$

4.47 Limestone (calcium carbonate) is insoluble in water but dissolves when a hydrochloric acid solution is added. Why? Write balanced total ionic and net ionic equations, showing hydrochloric acid as it actually exists in water and the reaction as a proton-transfer process.

4.48 Zinc hydroxide is insoluble in water but dissolves when a nitric acid solution is added. Why? Write balanced total ionic and net ionic equations, showing nitric acid as it actually exists in water and the reaction as a proton-transfer process.

4.49 If 15.98 mL of a standard 0.1080 M KOH solution reacts with 52.00 mL of CH_3COOH solution, what is the molarity of the acid solution?

4.50 If 26.35 mL of a standard 0.1650 M NaOH solution is required to neutralize 35.00 mL of H_2SO_4, what is the molarity of the acid solution?

● **Problems in Context**

4.51 An auto mechanic spills 85 mL of 2.6 M H_2SO_4 solution from a rebuilt auto battery. How many milliliters of 2.5 M $NaHCO_3$ must be poured on the spill to react completely with the sulfuric acid?

4.52 Sodium hydroxide is used extensively in acid-base titrations because it is a strong, inexpensive base. A sodium hydroxide solution was standardized by titrating 25.00 mL of 0.1528 M standard hydrochloric acid. The initial buret reading of the sodium hydroxide was 2.24 mL, and the final reading was 39.21 mL. What was the molarity of the base solution?

4.53 One of the first steps in the enrichment of uranium for use in nuclear power plants involves a displacement reaction between UO_2 and aqueous HF:

$$UO_2(s) + 4HF(aq) \longrightarrow UF_4(s) + 2H_2O(l)$$

How many liters of 6.50 M HF are needed to react with 3.25 kg of UO_2?

Oxidation-Reduction (Redox) Reactions
(Sample Problems 4.6 to 4.9)

● **Concept Review Questions**

4.54 Describe how to determine the oxidation number of sulfur in (a) H_2S and (b) SO_3.

4.55 Is the following a redox reaction? Explain.
$$NH_3(aq) + HCl(aq) \longrightarrow NH_4Cl(aq)$$

4.56 Explain why an oxidizing agent undergoes reduction.

4.57 Why must every redox reaction involve an oxidizing agent and a reducing agent?

4.58 In which of the following equations does sulfuric acid act as an oxidizing agent? In which does it act as an acid? Explain.
(a) $4H^+(aq) + SO_4^{2-}(aq) + 2NaI(s) \longrightarrow$
$$2Na^+(aq) + I_2(s) + SO_2(g) +> 2H_2O(l)$$
(b) $BaF_2(s) + 2H^+(aq) + SO_4^{2-}(aq) \longrightarrow$
$$2HF(aq) + BaSO_4(s)$$

4.59 Identify the oxidizing agent and the reducing agent in the following reaction, and explain your answer:
$$8NH_3(g) + 6NO_2(g) \longrightarrow 7N_2(g) + 12H_2O(l)$$

● **Skill-Building Exercises** *(paired)*

4.60 Give the oxidation number of carbon in the following:
(a) CF_2Cl_2 (b) $Na_2C_2O_4$ (c) HCO_3^- (d) C_2H_6

4.61 Give the oxidation number of bromine in the following:
(a) KBr (b) BrF_3 (c) $HBrO_3$ (d) CBr_4

4.62 Give the oxidation number of nitrogen in the following:
(a) NH_2OH (b) N_2H_4 (c) NH_4^+ (d) HNO_2

4.63 Give the oxidation number of sulfur in the following:
(a) $SOCl_2$ (b) H_2S_2 (c) H_2SO_3 (d) Na_2S

4.64 Give the oxidation number of arsenic in the following:
(a) AsH_3 (b) H_3AsO_4 (c) $AsCl_3$

4.65 Give the oxidation number of phosphorus in the following:
(a) $H_2P_2O_7^{2-}$ (b) PH_4^+ (c) PCl_5

4.66 Give the oxidation number of manganese in the following:
(a) MnO_4^{2-} (b) Mn_2O_3 (c) $KMnO_4$

4.67 Give the oxidation number of chromium the following:
(a) CrO_3 (b) $Cr_2O_7^{2-}$ (c) $Cr_2(SO_4)_3$

4.68 Identify the oxidizing agent and the reducing agent in each of the following:
(a) $5H_2C_2O_4(aq) + 2MnO_4^-(aq) + 6H^+(aq) \longrightarrow$
$$2Mn^{2+}(aq) + 10CO_2(g) + 8H_2O(l)$$
(b) $3Cu(s) + 8H^+(aq) + 2NO_3^-(aq) \longrightarrow$
$$3Cu^{2+}(aq) + 2NO(g) + 4H_2O(l)$$

4.69 Identify the oxidizing agent and the reducing agent in each of the following:
(a) $Sn(s) + 2H^+(aq) \longrightarrow Sn^{2+}(aq) + H_2(g)$
(b) $2H^+(aq) + H_2O_2(aq) + 2Fe^{2+}(aq) \longrightarrow$
$$2Fe^{3+}(aq) + 2H_2O(l)$$

4.70 Identify the oxidizing agent and the reducing agent in each of the following:
(a) $8H^+(aq) + 6Cl^-(aq) + Sn(s) + 4NO_3^-(aq) \longrightarrow$
$$SnCl_6^{2-}(aq) + 4NO_2(g) + 4H_2O(l)$$
(b) $2MnO_4^-(aq) + 10Cl^-(aq) + 16H^+(aq) \longrightarrow$
$$5Cl_2(g) + 2Mn^{2+}(aq) + 8H_2O(l)$$

4.71 Identify the oxidizing agent and the reducing agent in each of the following:
(a) $8H^+(aq) + Cr_2O_7^{2-}(aq) + 3SO_3^{2-}(aq) \longrightarrow$
$$2Cr^{3+}(aq) + 3SO_4^{2-}(aq) + 4H_2O(l)$$
(b) $NO_3^-(aq) + 4Zn(s) + 7OH^-(aq) + 6H_2O(l) \longrightarrow$
$$4Zn(OH)_4^{2-}(aq) + NH_3(aq)$$

4.72 Discuss each conclusion from a study of redox reactions:
(a) The sulfide ion functions only as a reducing agent.
(b) The sulfate ion functions only as an oxidizing agent.
(c) Sulfur dioxide functions as an oxidizing or a reducing agent.

4.73 Discuss each conclusion from a study of redox reactions:
(a) The nitride ion functions only as a reducing agent.
(b) The nitrate ion functions only as an oxidizing agent.
(c) The nitrite ion functions as an oxidizing or a reducing agent.

4.74 Use the oxidation number method to balance the following equations by placing coefficients in the blanks. Identify the reducing and oxidizing agents:
(a) __$HNO_3(aq)$ + __$K_2CrO_4(aq)$ + __$Fe(NO_3)_2(aq)$ $\longrightarrow$
__$KNO_3(aq)$ + __$Fe(NO_3)_3(aq)$ + __$Cr(NO_3)_3(aq)$ + __$H_2O(l)$
(b) __$HNO_3(aq)$ + __$C_2H_6O(l)$ + __$K_2Cr_2O_7(aq)$ $\longrightarrow$
__$KNO_3(aq)$ + __$C_2H_4O(l)$ + __$H_2O(l)$ + __$Cr(NO_3)_3(aq)$

(c) $\_\_HCl(aq) + \_\_NH_4Cl(aq) + \_\_K_2Cr_2O_7(aq) \longrightarrow$
$\_\_KCl(aq) + \_\_CrCl_3(aq) + \_\_N_2(g) + \_\_H_2O(l)$

(d) $\_\_KClO_3(aq) + \_\_HBr(aq) \longrightarrow$
$\_\_Br_2(l) + \_\_H_2O(l) + \_\_KCl(aq)$

4.75 Use the oxidation number method to balance the following equations by placing coefficients in the blanks. Identify the reducing and oxidizing agents:

(a) $\_\_HCl(aq) + \_\_FeCl_2(aq) + \_\_H_2O_2(aq) \longrightarrow$
$\_\_FeCl_3(aq) + \_\_H_2O(l)$

(b) $\_\_I_2(s) + \_\_Na_2S_2O_3(aq) \longrightarrow$
$\_\_Na_2S_4O_6(aq) + \_\_NaI(aq)$

(c) $\_\_HNO_3(aq) + \_\_KI(aq) \longrightarrow$
$\_\_NO(g) + \_\_I_2(s) + \_\_H_2O(l) + \_\_KNO_3(aq)$

(d) $\_\_PbO(s) + \_\_NH_3(aq) \longrightarrow$
$\_\_N_2(g) + \_\_H_2O(l) + \_\_Pb(s)$

● **Problems in Context**

4.76 The active agent in many hair bleaches is hydrogen peroxide. The amount of hydrogen peroxide in 13.8 g of hair bleach was determined by titration with a standard potassium permanganate solution:

$$2MnO_4^-(aq) + 5H_2O_2(aq) + 6H^+(aq) \longrightarrow$$
$$5O_2(g) + 2Mn^{2+}(aq) + 8H_2O(l)$$

(a) How many moles of MnO_4^- were required for the titration if 43.2 mL of 0.105 M $KMnO_4$ was needed to reach the end point?
(b) How many moles of H_2O_2 were present in the 13.8-g sample of bleach?
(c) How many grams of H_2O_2 were in the sample?
(d) What is the mass percent of H_2O_2 in the sample?
(e) What is the reducing agent in the redox reaction?

4.77 A person's blood alcohol (C_2H_5OH) level can be determined by titrating a sample of blood plasma with a potassium dichromate solution. The balanced equation is

$$16H^+(aq) + 2Cr_2O_7^{2-}(aq) + C_2H_5OH(aq) \longrightarrow$$
$$4Cr^{3+}(aq) + 2CO_2(g) + 11H_2O(l)$$

If 35.46 mL of 0.05961 M $Cr_2O_7^{2-}$ is required to titrate 28.00 g of plasma, what is the mass percent of alcohol in the blood?

Elemental Substances in Redox Reactions
(Sample Problem 4.10)

● **Concept Review Questions**

4.78 What is the name of the type of reaction that leads to the following?
(a) An increase in the number of substances
(b) A decrease in the number of substances
(c) No change in the number of substances

4.79 Why do decomposition reactions typically have compounds as reactants, whereas combination and displacement reactions have one or more elements?

4.80 Which of the three types of reactions discussed in this section commonly produce one or more compounds?

4.81 Give an example of a combination reaction that is a redox reaction. Give an example of a combination reaction that is not a redox reaction.

4.82 Are all combustion reactions redox reactions? Explain.

● **Skill-Building Exercises** *(paired)*

4.83 Balance each of the following redox reactions and classify it as a combination, decomposition, or displacement reaction:
(a) $Ca(s) + H_2O(l) \longrightarrow Ca(OH)_2(aq) + H_2(g)$

(b) $NaNO_3(s) \longrightarrow NaNO_2(s) + O_2(g)$
(c) $C_2H_2(g) + H_2(g) \longrightarrow C_2H_6(g)$

4.84 Balance each of the following redox reactions and classify it as a combination, decomposition, or displacement reaction:
(a) $HI(g) \longrightarrow H_2(g) + I_2(g)$
(b) $Zn(s) + AgNO_3(ag) \longrightarrow Zn(NO_3)_2(aq) + Ag(s)$
(c) $NO(g) + O_2(g) \longrightarrow N_2O_4(l)$

4.85 Balance each of the following redox reactions and classify it as a combination, decomposition, or displacement reaction:
(a) $Sb(s) + Cl_2(g) \longrightarrow SbCl_3(s)$
(b) $AsH_3(g) \longrightarrow As(s) + H_2(g)$
(c) $Mn(s) + Fe(NO_3)_3(aq) \longrightarrow Mn(NO_3)_2(aq) + Fe(s)$

4.86 Balance each of the following redox reactions and classify it as a combination, decomposition, or displacement reaction:
(a) $Mg(s) + H_2O(g) \longrightarrow Mg(OH)_2(s) + H_2(g)$
(b) $Cr(NO_3)_3(aq) + Al(s) \longrightarrow Al(NO_3)_3(aq) + Cr(s)$
(c) $PF_3(g) + F_2(g) \longrightarrow PF_5(g)$

4.87 Predict the product(s) and write a balanced equation for each of the following redox reactions:
(a) $Ca(s) + Br_2(l) \longrightarrow$
(b) $Ag_2O(s) \xrightarrow{\Delta}$
(c) $Mn(s) + Cu(NO_3)_2(aq) \longrightarrow$

4.88 Predict the product(s) and write a balanced equation for each of the following redox reactions:
(a) $Mg(s) + HCl(aq) \longrightarrow$
(b) $LiCl(l) \xrightarrow{electricity}$
(c) $SnCl_2(aq) + Co(s) \longrightarrow$

4.89 Predict the product(s) and write a balanced equation for each of the following redox reactions:
(a) $N_2(g) + H_2(g) \longrightarrow$
(b) $NaClO_3(s) \xrightarrow{\Delta}$
(c) $Ba(s) + H_2O(l) \longrightarrow$

4.90 Predict the product(s) and write a balanced equation for each of the following redox reactions:
(a) $Fe(s) + HClO_4(aq) \longrightarrow$
(b) $S_8(s) + O_2(g) e \longrightarrow$
(c) $BaCl_2(aq) \xrightarrow{electricity}$

4.91 Predict the product(s) and write a balanced equation for each of the following redox reactions:
(a) Cesium + iodine $\longrightarrow$
(b) Aluminum + aqueous manganese(II) sulfate $\longrightarrow$
(c) Sulfur dioxide + oxygen $\longrightarrow$
(d) Propane and oxygen $\longrightarrow$
(e) Write a balanced net ionic equation for (b).

4.92 Predict the product(s) and write a balanced equation for each of the following redox reactions:
(a) Pentane and oxygen $\longrightarrow$
(b) Phosphorus trichloride + chlorine $\longrightarrow$
(c) Zinc + hydrobromic acid $\longrightarrow$
(d) Aqueous potassium iodide + bromine $\longrightarrow$
(e) Write a balanced net ionic equation for (d).

4.93 How many grams of O_2 can be prepared from the complete decomposition of 4.27 kg of HgO? Name and calculate the mass (in kg) of the other product.

4.94 How many grams of lime (CaO) can be produced from the complete decomposition of 114.52 g of limestone ($CaCO_3$)? Name and calculate the mass (in grams) of the other product.

4.95 In a combination reaction, 1.62 g of lithium is mixed with 6.00 g of oxygen.
(a) Which reactant is present in excess?
(b) How many moles of product are formed?
(c) After reaction, how many grams of each reactant and product are present?

4.96 In a combination reaction, 2.22 g of magnesium is heated with 3.75 g of nitrogen.
(a) Which reactant is present in excess?
(b) How many moles of product are formed?
(c) After reaction, how many grams of each reactant and product are present?

4.97 A mixture of $KClO_3$ and KCl with a mass of 0.900 g was heated to produce O_2. After heating, the mass of residue was 0.700 g. Assuming all the $KClO_3$ decomposed to KCl and O_2, calculate the mass percent of $KClO_3$ in the original mixture.

4.98 A mixture of $CaCO_3$ and CaO weighing 0.693 g was heated to produce CO_2. After heating, the remaining solid weighed 0.508 g. Assuming all the $CaCO_3$ decomposed to CaO and CO_2, calculate the mass percent of $CaCO_3$ in the original mixture.

● **Problems in Context**

4.99 Before arc welding was developed, a displacement reaction involving aluminum and iron(III) oxide was commonly used to produce molten iron (the thermite process; see photo). This reaction was used, for example, to connect sections of iron railroad track. Calculate the mass of molten iron produced when 1.00 kg of aluminum reacts with 2.00 mol of iron(III) oxide.

4.100 Iron reacts rapidly with chlorine gas to form the reddish-brown ionic Compound A, which contains iron in the higher of its two common oxidation states. Strong heating decomposes Compound A to Compound B, another ionic compound, which contains iron in the lower of its two oxidation states. When Compound A is formed by the reaction of 50.6 g of Fe and 83.8 g of Cl_2 and then heated, how much Compound B forms?

Reversible Reactions: An Introduction to Chemical Equilibrium

● **Concept Review Questions**

4.101 Why is the equilibrium state called "dynamic"?

4.102 In a decomposition reaction involving a gaseous product, what must be done for the reaction to reach equilibrium?

4.103 Describe what happens on the molecular level when acetic acid dissolves in water.

4.104 When either a mixture of NO and Br_2 or pure nitrosyl bromide (NOBr) is placed in a reaction vessel, the product mixture contains NO, Br_2, and NOBr. Explain.

● **Problems in Context**

4.105 Ammonia is produced by the millions of tons annually for use as a fertilizer. It is commonly made from N_2 and H_2 by the Haber process. Because the reaction reaches equilibrium before going completely to product, the stoichiometric amount of ammonia is not obtained. At a particular temperature and pressure, 10.0 g of H_2 reacts with 20.0 g of N_2 to form ammonia. When equilibrium is reached, 15.0 g of NH_3 has formed.
(a) Calculate the percent yield.
(b) How many moles of N_2 and H_2 are present at equilibrium?

Comprehensive Problems

Problems with an asterisk (*) are more challenging.

4.106 Nutritional biochemists have known for decades that acidic foods cooked in cast-iron cookware can supply significant amounts of dietary (ferrous) iron.
(a) Write a balanced net ionic equation, with oxidation numbers, that supports this fact.
(b) Measurements show an increase from 3.3 mg of iron to 49 mg of iron per $\frac{1}{2}$-cup (125-g) serving during the slow preparation of tomato sauce in a cast-iron pot. How many ferrous ions are present in a 26-oz (737-g) jar of the tomato sauce?

4.107 Limestone ($CaCO_3$) is used to remove acidic pollutants from smokestack flue gases in a sequence of decomposition-combination reactions. It is heated to form lime (CaO), which reacts with sulfur dioxide to form calcium sulfite. Assuming a 70.% yield in the overall reaction, what mass of limestone is required to remove all the sulfur dioxide formed by the combustion of 8.5×10^4 kg of coal that is 0.33 mass % sulfur?

4.108 The brewing industry uses yeast microorganisms to convert glucose to ethanol for wine and beer. The baking industry uses the carbon dioxide they produce to make bread rise:

$$C_6H_{12}O_6(s) \xrightarrow{\text{yeast}} 2C_2H_5OH(l) + 2CO_2(g)$$

How many grams of ethanol can be produced from the decomposition of 10.0 g of glucose? What volume of CO_2 is produced? (Assume 1 mol of gas occupies 22.4 L at the conditions used.)

4.109 A chemical engineer determines the mass percent of iron in an ore sample by converting the Fe to Fe^{2+} in acid and then titrating the Fe^{2+} with MnO_4^-. A 1.1081-g sample was dissolved in acid and then titrated with 39.32 mL of 0.03190 M $KMnO_4$. The balanced equation is

$$8H^+(aq) + 5Fe^{2+}(aq) + MnO_4^-(aq) \longrightarrow$$
$$5Fe^{3+}(aq) + Mn^{2+}(aq) + 4H_2O(l)$$

Calculate the mass percent of iron in the ore.

4.110 Mixtures of $CaCl_2$ and $NaCl$ are used for salting roads to prevent ice formation. A dissolved 1.9348-g sample of such a mixture was analyzed by using excess $Na_2C_2O_4$ to precipitate the Ca^{2+} as CaC_2O_4. The CaC_2O_4 was separated from the solution and then dissolved with sulfuric acid. The resulting $H_2C_2O_4$ was titrated with 37.68 mL of 0.1019 M $KMnO_4$ solution.
(a) Write the balanced net ionic equation for the precipitation reaction.
(b) Write the balanced net ionic equation for the titration reaction. (See Sample Problem 4.9.)
(c) What is the oxidizing agent?
(d) What is the reducing agent?
(e) Calculate the mass percent of $CaCl_2$ in the original sample.

***4.111** A student has three beakers that contain the same volume of solution. The first contains 0.1 M $AgNO_3$, the second 0.02 M CaS, and the third 0.05 M Na_2SO_4. While out of the room, her lab partner accidentally mixes the solutions. Assuming complete precipitation of any solids that form, determine which solids form and the concentration of each ion remaining in solution.

4.112 Precipitation reactions are often used to prepare useful ionic compounds. For example, thousands of tons of silver bromide are prepared annually for use in making black-and-white photographic film. (a) What mass (in kilograms) of silver bromide forms when 5.85 m^3 of 1.68 M potassium bromide reacts with 3.51 m^3 of 2.04 M silver nitrate? (b) After the solid silver bromide is removed, what ions are present in the remaining solution? Determine the molarity of each ion. (Assume the total volume is the sum of the reactant volumes.)

4.113 The flask (right) depicts the products of the titration of 25 mL of sulfuric acid with 25 mL of sodium hydroxide.
(a) Write balanced molecular, total ionic, and net ionic equations for the reaction.
(b) If each orange sphere represents 0.010 mol of sulfate ion, how many moles of acid and of base reacted?
(c) What are the molarities of the acid and the base?

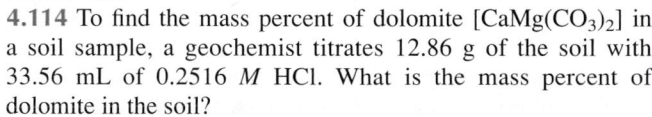

4.114 To find the mass percent of dolomite [$CaMg(CO_3)_2$] in a soil sample, a geochemist titrates 12.86 g of the soil with 33.56 mL of 0.2516 M HCl. What is the mass percent of dolomite in the soil?

4.115 The calcium carbonate impurity in a sample of phosphate rock is removed by treatment with hydrochloric acid; the products are carbon dioxide, water, and aqueous calcium chloride. When 15.5 g of the rock is treated with excess hydrochloric acid, 1.81 g of carbon dioxide is formed. Calculate the mass percent of calcium carbonate in the rock.

4.116 Nitric acid is used to make explosives and fertilizers and is produced by the Ostwald process:

Step 1. $4NH_3(g) + 5O_2(g) \longrightarrow 4NO(g) + 6H_2O(l)$
Step 2. $2NO(g) + O_2(g) \longrightarrow 2NO_2(g)$
Step 3. $3NO_2(g) + H_2O(l) \longrightarrow 2HNO_3(l) + NO(g)$

If you start with 72.5 kg of NH_3 and excess O_2: (a) How many grams of HNO_3 can form? (b) How many kilograms of O_2 are used? (c) How many kilograms of H_2O are produced? (d) For each step, determine the substance being oxidized, that being reduced, the oxidizing agent, and the reducing agent.

4.117 Complete each of the following reactions and write the net ionic equation:
1. $NaOH(aq) + HCl(aq) \longrightarrow$
2. $KOH(aq) + HNO_3(aq) \longrightarrow$
3. $Ba(OH)_2(aq) + 2HBr(aq) \longrightarrow$
(a) What do you conclude from these equations about the nature of the reactants?
(b) What are the spectator ions in each reaction?

4.118 Use the oxidation number method to balance the following equations by placing coefficients in the blanks. Identify the reducing and oxidizing agents:
(a) __$KOH(aq)$ + __$H_2O_2(aq)$ + __$Cr(OH)_3(s)$ $\longrightarrow$
__$K_2CrO_4(aq)$ + __$H_2O(l)$
(b) __$MnO_4^-(aq)$ + __$ClO_2^-(aq)$ + __$H_2O(l)$ $\longrightarrow$
__$MnO_2(s)$ + __$ClO_4^-(aq)$ + __$OH^-(aq)$
(c) __$KMnO_4(aq)$ + __$Na_2SO_3(aq)$ + __$H_2O(l)$ $\longrightarrow$
__$MnO_2(s)$ + __$Na_2SO_4(aq)$ + __$KOH(aq)$

4.119 When two solutions of ionic compounds are mixed, a reaction occurs only if ions are removed from solution to form product. (a) What are three ways in which this removal can oc-

cur? (b) What type of reaction is involved in each case? (c) Write a balanced total ionic equation to illustrate each case.

4.120 Use the oxidation number method to balance the following reactions by placing coefficients in the blanks. Identify the reducing and oxidizing agents:
(a) __$CrO_4^{2-}(aq)$ + __$HSnO_2^-(aq)$ + __$H_2O(l)$ $\longrightarrow$
__$CrO_2^-(aq)$ + __$HSnO_3^-(aq)$ + __$OH^-(aq)$
(b) __$KMnO_4(aq)$ + __$NaNO_2(aq)$ + __$H_2O(l)$ $\longrightarrow$
__$MnO_2(s)$ + __$NaNO_3(aq)$ + __$KOH(aq)$
(c) __$I^-(aq)$ + __$O_2(g)$ + __$H_2O(l)$ $\longrightarrow$
__$I_2(s)$ + __$OH^-(aq)$

4.121 Sodium peroxide (Na_2O_2) is often used in self-contained breathing devices, such as those used in fire emergencies, because it reacts with exhaled CO_2 to form Na_2CO_3 and O_2. How many liters of respired air can react with 80.0 g of Na_2O_2 if each liter of respired air contains 0.0720 g of CO_2?

4.122 Magnesium is used in airplane bodies and other lightweight alloys. The metal is obtained from seawater in a process that includes precipitation, neutralization, evaporation, and electrolysis. How many kilograms of magnesium can be obtained from 1.00 km^3 of seawater if the initial Mg^{2+} concentration is 0.13% by mass (d of seawater = 1.04 g/mL)?

*__**4.123** A typical formulation for window glass is 75% SiO_2, 15% Na_2O, and 10.% CaO by mass. What masses of sand (SiO_2), sodium carbonate, and calcium carbonate must be combined to produce 1.00 kg of glass after carbon dioxide is driven off by thermal decomposition of the carbonates?

4.124 The field of sports medicine has become an important specialty, with physicians routinely treating athletes and dancers. Ethyl chloride, a local anesthetic commonly used for simple injuries, is the product of the combination of ethylene with hydrogen chloride:

$$C_2H_4(g) + HCl(g) \longrightarrow C_2H_5Cl(g)$$

If 0.100 kg of C_2H_4 and 0.100 kg of HCl react:
(a) How many molecules of gas (reactants plus products) are present when the reaction is complete?
(b) How many moles of gas are present when half the product forms?

4.125 The *salinity* of a solution is defined as the grams of total salts per kilogram of solution. An agricultural chemist uses a solution whose salinity is 35.0 g/kg to test the effect of irrigating farmland with high-salinity river water. The two solutes are NaCl and $MgSO_4$, and there are twice as many moles of NaCl as $MgSO_4$. What masses of NaCl and $MgSO_4$ are contained in 1.00 kg of the solution?

*__**4.126** Thyroxine ($C_{15}H_{11}I_4NO_4$) is a hormone synthesized by the thyroid gland and used to control many metabolic functions in the body. A physiologist determines the mass % of thyroxine in a thyroid extract by igniting 0.4332 g of extract with sodium carbonate, which converts the iodine to iodide. The iodide is dissolved in water, and bromine and hydrochloric acid are added, which convert the iodide to iodate.
(a) How many moles of iodate form per mole of thyroxine?
(b) Excess bromine is boiled off and more iodide is added, which reacts as in the following *unbalanced* equation:

$$IO_3^-(aq) + H^+(aq) + I^-(aq) \longrightarrow I_2(aq) + H_2O(l)$$

How many moles of iodine are produced per mole of thyroxine? (*Hint:* Be sure to balance the charges as well as the atoms.) What are the oxidizing and reducing agents in the reaction?

(c) The iodine reacts completely with 17.23 mL of 0.1000 M thiosulfate as in the following *unbalanced* equation:

$$I_2(aq) + S_2O_3^{2-}(aq) \longrightarrow I^-(aq) + S_4O_6^{2-}(aq)$$

What is the mass % of thyroxine in the thyroid extract?

4.127 Carbon dioxide is removed from the atmosphere of space capsules by reaction with a solid metal hydroxide. The products are water and the metal carbonate.
(a) Calculate the mass of CO_2 that can be removed by reaction with 3.50 kg of lithium hydroxide.
(b) How many grams of CO_2 can be removed by 1.00 g of each of the following: lithium hydroxide, magnesium hydroxide, and aluminum hydroxide?

*4.128 Calcium dihydrogen phosphate, $Ca(H_2PO_4)_2$, and sodium hydrogen carbonate, $NaHCO_3$, are ingredients of baking powder that react with each other to produce CO_2, which causes dough or batter to rise:

$$Ca(H_2PO_4)_2(s) + NaHCO_3(s) \longrightarrow$$
$$CO_2(g) + H_2O(g) + CaHPO_4(s) + Na_2HPO_4(s) \text{ [unbalanced]}$$

If the baking powder contains 31% $NaHCO_3$ and 35% $Ca(H_2PO_4)_2$ by mass:
(a) How many moles of CO_2 are produced from 1.00 g of baking powder?
(b) If 1 mol of CO_2 occupies 37.0 L at 350°F (a typical baking temperature), what volume of CO_2 is produced from 1.00 g of baking powder?

4.129 During the process of developing black-and-white film, unexposed silver bromide is removed in a displacement reaction with sodium thiosulfate solution:

$$AgBr(s) + 2Na_2S_2O_3(aq) \longrightarrow Na_3Ag(S_2O_3)_2(aq) + NaBr(aq)$$

What volume of 0.105 M $Na_2S_2O_3$ solution is needed to remove 2.66 g of AgBr from a roll of film?

4.130 Ionic hydrates lose their "waters of hydration" when thermally decomposed. When 25.36 g of hydrated copper(II) sulfate is heated, it forms 16.21 g of anhydrous copper(II) sulfate. What is the formula of the hydrate?

*4.131 In 1997, at the United Nations Conference on Climate Change, the major industrial nations agreed to expand their research efforts to develop renewable sources of carbon-based fuels. For more than a decade, Brazil has been engaged in a program to replace gasoline with ethanol derived from the root crop manioc (cassava).
(a) Write separate balanced equations for the complete combustion of ethanol (C_2H_5OH) and of gasoline (represented by the formula C_8H_{18}).
(b) What mass of oxygen is required to burn completely 1.00 L of a mixture that is 90.0% gasoline ($d = 0.742$ g/mL) and 10.0% ethanol ($d = 0.789$ g/mL) by volume?
(c) If 1.00 mol of O_2 occupies 22.4 L, what volume of O_2 is needed to burn 1.00 L of the mixture?
(d) Air is 20.9% O_2 by volume. What volume of air is needed to burn 1.00 L of the mixture?

*4.132 In a car engine, gasoline (represented by C_8H_{18}) does not burn completely, and some CO, a toxic pollutant, forms along with CO_2 and H_2O. If 5.0% of the gasoline forms CO:
(a) What is the ratio of CO_2 to CO molecules in the exhaust?
(b) What is the mass ratio of CO_2 to CO?

(c) What percent of the gasoline must form CO for the mass ratio of CO_2 to CO to be exactly 1:1?

4.133 One of the molecules responsible for atmospheric ozone depletion is the refrigerant and aerosol propellant Freon-12 (CF_2Cl_2). It can be prepared in a sequence of two reactions:
1. A combination reaction between hydrogen and fluorine gases
2. A displacement reaction between the product of Reaction 1 and liquid carbon tetrachloride. Hydrogen chloride gas also forms.
(a) Write balanced equations for the two reactions.
(b) What is the maximum mass (in kg) of Freon-12 that can be produced from 0.760 kg of fluorine?

4.134 In a blast furnace, iron ore is reduced to the free metal, and CO is the reducing agent. For 156.8 g of each of the following iron ores, calculate the mass of CO required: (a) wuestite (FeO); (b) hematite (Fe_2O_3); (c) siderite ($FeCO_3$).

4.135 Interhalogens are covalent compounds of one halogen with another. When 3.299 g of I_2 reacts with F_2, 5.768 g of IF_x is formed. What is the value of x in the product formula?

4.136 In the process of *salting-in*, protein solubility in a dilute salt solution is increased by adding more salt. Because the protein solubility depends on the total ion concentration as well as the ion charge, salts yielding divalent ions are often more effective than those yielding monovalent ions. (a) How many grams of $MgCl_2$ must dissolve to equal the ion concentration of 12.4 g of NaCl? (b) How many grams of CaS must dissolve? (c) Which of the three salt solutions would dissolve the most protein?

4.137 Elemental fluorine is such a strong oxidizing agent that it cannot be produced from the abundant minerals fluorspar (CaF_2) and fluorapatite [$Ca_5(PO_4)_3F$]. One method uses potassium hexafluoromanganate(IV) with antimony(V) fluoride to produce MnF_4, which spontaneously decomposes into MnF_3 and F_2:

$$K_2MnF_6(s) + SbF_5(l) \longrightarrow KSbF_6(s) + MnF_3(s) + F_2(g)$$

Balance this equation by the oxidation number method, and identify the oxidizing and reducing agents.

*4.138 In the process of *pickling*, rust is removed from newly produced steel by washing in hydrochloric acid:

1. $6HCl(aq) + Fe_2O_3(s) \longrightarrow 2FeCl_3(aq) + 3H_2O(l)$

During the process, some iron is lost as well:

2. $2HCl(aq) + Fe(s) \longrightarrow FeCl_2(aq) + H_2(g)$

(a) Which reaction, if either, is a redox process? (b) If Reaction 2 did not occur and all the HCl were used, how many grams of Fe_2O_3 could be removed and $FeCl_3$ produced in a 2.50×10^3-L bath of 3.00 M HCl? (c) If Reaction 1 did not occur and all the HCl were used, how many grams of Fe could be lost and $FeCl_2$ produced in a 2.50×10^3-L bath of 3.00 M HCl? (d) If 0.280 g of Fe is lost per gram of Fe_2O_3 removed, what is the mass ratio of $FeCl_2$ to $FeCl_3$?

4.139 At liftoff, the space shuttle uses a solid mixture of ammonium perchlorate and aluminum powder to obtain great thrust from the volume change of solid to gas. In the presence of a catalyst, the mixture forms solid aluminum oxide and aluminum trichloride and gaseous water and nitric oxide. (a) Write a balanced equation for the reaction, and identify the reducing and oxidizing agents. (b) How many total moles of gas (water vapor and nitric oxide) are produced when 50.0 kg of ammonium perchlorate reacts with a stoichiometric amount of Al? (c) What is the volume change from this reaction? (d of $NH_4ClO_4 = 1.95$ g/cc, Al = 2.70 g/cc, $Al_2O_3 = 3.97$ g/cc, and $AlCl_3 = 2.44$ g/cc; assume 1 mol of gas occupies 22.4 L.)

GASES AND THE KINETIC-MOLECULAR THEORY

CHAPTER OUTLINE

5.1 An Overview of the Physical States of Matter

5.2 Gas Pressure and Its Measurement
Laboratory Devices
Units of Pressure

5.3 The Gas Laws and Their Experimental Foundations
Boyle's Law
Charles's Law
Avogadro's Law
Standard Conditions
The Ideal Gas Law
Solving Gas Law Problems

5.4 Further Applications of the Ideal Gas Law
Density of a Gas
Molar Mass of a Gas
Partial Pressure of a Gas

5.5 The Ideal Gas Law and Reaction Stoichiometry

5.6 The Kinetic-Molecular Theory: A Model for Gas Behavior
Explaining the Gas Laws
Effusion and Diffusion
Mean Free Path and Collision Frequency

5.7 Real Gases: Deviations from Ideal Behavior
Effects of Extreme Conditions
The van der Waals Equation

Figure: Expanding with heat. One of several fundamental behaviors of gases, the change in volume with a change in temperature, applies to countless everyday phenomena, from baking to breathing. In this chapter, we examine these behaviors to learn the essential nature of one of the three states of matter.

CONCEPTS & SKILLS

to review before you study this chapter
• physical states of matter (Section 1.1)
• SI unit conversions (Section 1.5)
• mole-mass-number conversions
  (Section 3.1)

Gases are everywhere. People have been observing their behavior, and that of matter in other states, throughout history—three of the four "elements" of the ancients were air (gas), water (liquid), and earth (solid)—but many questions remain. In this chapter and its companion, Chapter 12, we examine these states and their interrelations. Here, we highlight the gaseous state, the one we understand best.

The Earth's atmosphere—the gaseous envelope that surrounds the planet—is a colorless, odorless mixture of nearly 20 elements and compounds that extends from the surface upward more than 500 km until it merges with outer space. Some of its components—O_2, N_2, H_2O vapor, and CO_2—are essential for life because they take part in complex cycles of redox reactions throughout the environment, and you participate in those cycles with every breath you take. ● Gases also have essential roles in industry (Table 5.1).

Table 5.1 Some Important Industrial Gases

Name (Formula)	Origin and Use
Methane (CH_4)	Natural deposits; domestic fuel
Ammonia (NH_3)	From $N_2 + H_2$; fertilizers, explosives
Chlorine (Cl_2)	Electrolysis of seawater; bleaching and disinfecting
Oxygen (O_2)	Liquefied air; steelmaking
Ethylene (C_2H_4)	High-temperature decomposition of natural gas; plastics

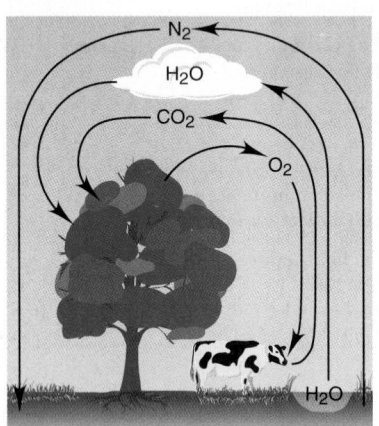

● **Atmosphere-Biosphere Redox Interconnections** The diverse organisms that make up the biosphere interact intimately with the gases of the atmosphere. Powered by solar energy, green plants reduce atmospheric CO_2 and incorporate the C atoms into their own substance. In the process, O atoms in H_2O are oxidized and released to the air as O_2. Certain microbes that live on plant roots reduce N_2 to NH_3 and form compounds that the plant uses to make its proteins. Other microbes that feed on dead plants (and animals) oxidize the proteins and release N_2 again. Animals eat plants and other animals, use O_2 to oxidize their food, and return CO_2 and H_2O to the air.

Although the *chemical* behavior of a gas depends on its composition, all gases have remarkably similar *physical* behavior, which is the focus of this chapter. For instance, although the particular gases differ, the same physical behavior is at work in the operation of a car and in the baking of bread, in the thrust of a rocket engine and in the explosion of a kernel of popcorn. The process of breathing involves the same physical principles as the creation of thunder.

In this chapter, we first contrast gases with liquids and solids and then discuss gas pressure. We consider several natural laws that describe gas behavior and then examine the ideal gas law, which encompasses the others, and apply it to reaction stoichiometry. We explain the observable behavior of gases with a simple molecular model, but then find that studies of real gases, especially under extreme conditions, require refinements of the ideal gas law and the model. Finally, we apply these principles to the properties of planetary atmospheres.

5.1 AN OVERVIEW OF THE PHYSICAL STATES OF MATTER

Under appropriate conditions of pressure and temperature, most substances can exist as a solid, a liquid, or a gas. In Chapters 1 and 2, we described these physical states in terms of how each fills a container and began to develop a molecular view that explains this macroscopic behavior: a solid has a fixed shape regardless of the container shape because its particles are held rigidly in place; a liquid conforms to the container shape but has a definite volume and a surface because its particles are close together but free to move around each other; and a gas fills the container because its particles are far apart and moving randomly. Several other aspects of their behavior distinguish gases from liquids and solids:

1. *Gas volume changes greatly with pressure.* When a sample of gas is confined to a container of variable volume, such as the piston-cylinder assembly of a car engine, an external force can compress the gas. Removing the external force allows the gas volume to increase again. In contrast, a liquid or solid resists significant changes in volume.

2. *Gas volume changes greatly with temperature.* When a gas sample at constant pressure is heated, its volume increases; when it is cooled, its volume decreases. This volume change is 50 to 100 times greater for gases than for liquids or solids.

3. *Gases have relatively low viscosity.* Gases flow much more freely than liquids and solids. Low viscosity allows gases to flow through pipes over long distances and to leak rapidly out of small holes.

4. *Most gases have relatively low densities* under normal conditions. Gas density is usually tabulated in units of grams per *liter,* whereas liquid and solid densities are in grams per *milliliter,* about 1000 times as dense (see Table 1.5). For example, at 20°C and normal atmospheric pressure, the density of $O_2(g)$ is 1.3 g/**L**, whereas the density of $H_2O(l)$ is 1.0 g/**mL** and that of NaCl(s) is 2.2 g/**mL**. When a gas is cooled, its density increases because its volume decreases: at 0°C, the density of $O_2(g)$ increases to 1.4 g/L.

5. *Gases are miscible.* *Miscible* substances mix with one another in any proportion to form a solution. Clean dry air, for example, is a solution of about 18 gases. Two liquids, however, may or may not be miscible: water and ethanol are, but water and gasoline are not. Two solids generally do not form a solution unless they are mixed as molten liquids and then allowed to solidify.

Each of these observable properties offers a clue to the molecular properties of gases. For example, consider these density data. At 20°C and normal atmospheric pressure, gaseous N_2 has a density of 1.25 g/L. If cooled below −196°C, it condenses to liquid N_2 and its density becomes 0.808 g/mL. (Note the change in units.) The same amount of nitrogen occupies less than $\frac{1}{600}$ as much space! Further cooling to below −210°C yields solid N_2 ($d = 1.03$ g/mL), which is only somewhat more dense than the liquid. These values show again that *the molecules are much farther apart in the gas than in either the liquid or the solid.* Moreover, a large amount of space between molecules is consistent with gases' miscibility, low viscosity, and compressibility. Figure 5.1 compares macroscopic and molecular views of the physical states of a real substance.

POW! P-s-s-s-t! POP! A jackhammer uses the force of rapidly expanding compressed air to break through rock and cement. When the nozzle on a can of spray paint is pressed, the pressurized propellant gases expand into the lower pressure of the surroundings and expel droplets of paint. The rapid expansion of heated gases results in such phenomena as the destruction caused by a bomb, the liftoff of a rocket, and the popping of kernels of corn.

A Gas: Molecules are far apart and fill the available space

B Liquid: Molecules are close together but move relative to each other

C Solid: Molecules are tightly packed in a regular array and move very little relative to each other

Figure 5.1 **The three states of matter.** Many pure substances, such as bromine (Br_2), can exist under appropriate conditions of pressure and temperature as **A,** a gas; **B,** a liquid; or **C,** a solid. In **A,** the liquid form of bromine has been vaporized, whereas in **C,** it has been frozen. The molecular views show that molecules are much farther apart in a gas than in a liquid or solid.

5.2   GAS PRESSURE AND ITS MEASUREMENT

Snowshoes and the Meaning of Pressure Snowshoes allow you to walk on powdery snow without sinking because they distribute your weight over a much larger area than a boot does, thereby greatly decreasing your weight per square inch. The area of a snowshoe is typically about 10 times as large as that of a boot sole, so the snowshoe exerts only about one-tenth as much pressure as the boot. The wide, padded paws of snow leopards accomplish this, too. For the same reason, high-heeled shoes exert much more pressure than flat shoes.

Blowing up a balloon provides clear evidence that a gas exerts pressure on the walls of its container. **Pressure (*P*)** is defined as the force exerted per unit of surface area:

$$\text{Pressure} = \frac{\text{force}}{\text{area}}$$

The Earth's gravitational attraction pulls the atmospheric gases toward its surface, where they exert a force on all objects. The force, or weight, of these gases creates a pressure of about 14.7 pounds per square inch (lb/in^2; psi) of surface.

As we'll discuss later, the molecules in a gas are moving in every direction, so the pressure of the atmosphere is exerted uniformly on the floor, walls, ceiling, and every object in a room. The pressure on the outside of your body is equalized by the pressure on the inside, so there is no net pressure on your body's outer surface. What would happen if this were not the case? As an analogy, consider the empty metal can attached to a vacuum pump in Figure 5.2. With the pump off, the can maintains its shape because the pressure on the outside is equal to the pressure on the inside. With the pump on, the internal pressure decreases greatly, and the ever-present external pressure easily crushes the can. A vacuum-filtration flask (and tubing), which you may have used in the lab, has thick walls that can withstand the external pressure when the flask is evacuated.

Figure 5.2 **Effect of atmospheric pressure on objects at the Earth's surface. A,** A metal can filled with air has equal pressure on the inside and outside. **B,** When the air inside the can is removed, the atmospheric pressure crushes the can.

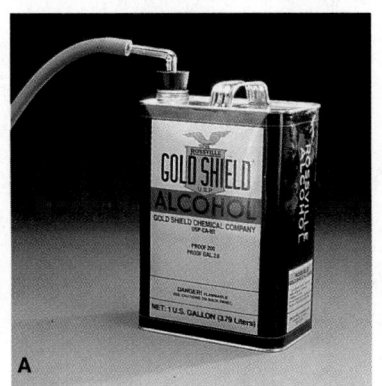
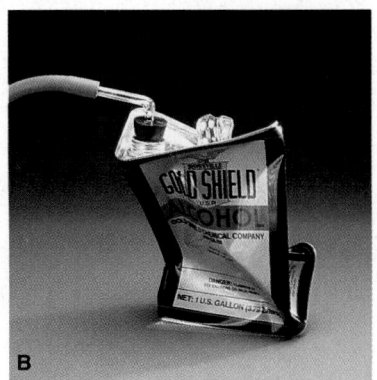

Laboratory Devices for Measuring Gas Pressure

A **barometer** is a device used to measure atmospheric pressure. Invented in 1643 by Evangelista Torricelli, the barometer is still basically just a tube about 1 m long, closed at one end, filled with mercury, and inverted into a dish containing more mercury (Figure 5.3). When the tube is inverted, some mercury flows out, forming a vacuum above the mercury remaining in the tube. At sea level under

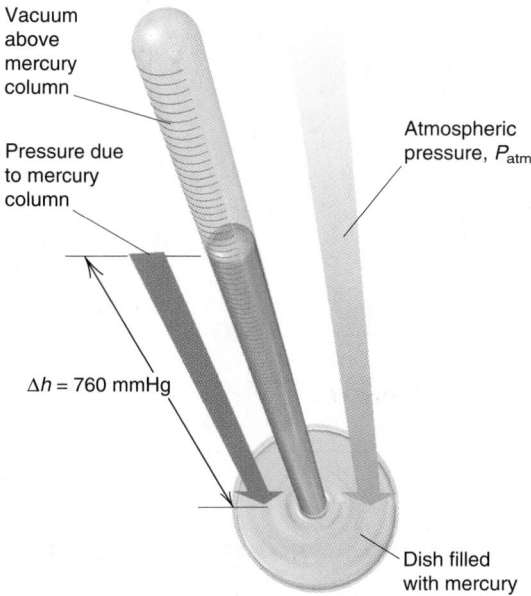

Figure 5.3 **A mercury barometer.**

ordinary atmospheric conditions, the outward flow of mercury stops when the surface of the mercury in the tube is about 760 mm above the surface of the mercury in the dish. It stops at 760 mm because at that point the column of mercury in the tube exerts the same pressure (weight/area) on the mercury surface in the dish as does the column of air that extends from the dish to the outer reaches of the atmosphere. The air pushing down keeps the mercury in the tube from flowing out any further. Likewise, if you place an evacuated tube into a dish filled with mercury, the mercury rises about 760 mm into the tube because the atmosphere pushes the mercury up to that height.

Several centuries ago, people ascribed mysterious "suction" forces to a vacuum. We know now that a vacuum does not suck up mercury into the barometer tube any more than it "sucks" in the walls of the crushed can in Figure 5.2. Only matter—in this case, the atmospheric gases—can exert a force.

Notice that we did not specify the diameter of the barometer tube. If the mercury in a 1-cm diameter tube rises to a height of 760 mm, the mercury in a 2-cm diameter tube will rise to that height also. The *weight* of mercury is greater in the wider tube, but the area is larger also; thus the *pressure,* the *ratio* of weight to area, is the same.

Since the pressure of the mercury column is directly proportional to its height, a unit commonly used for pressure is the height of the mercury (atomic symbol Hg) column in millimeters (mmHg). We discuss this and other units of pressure shortly. At sea level and 0°C, normal atmospheric pressure is 760 mmHg, but at the top of Mt. Everest (29,028 ft, or 8848 m), the atmospheric pressure is only about 270 mmHg. Thus, *pressure decreases with altitude:* the column of air above sea level is taller and weighs more than the column of air above Mt. Everest.

Laboratory barometers contain mercury rather than some other liquid because its high density allows the barometer to be a convenient size. For example, the pressure of the atmosphere would equal the pressure of a column of water about 10,300 mm, almost 34 ft, high. ● Note that, for a given pressure, the ratio of heights (h) of the liquid columns is inversely related to the ratio of the densities (d) of the liquids:

$$\frac{h_{H_2O}}{h_{Hg}} = \frac{d_{Hg}}{d_{H_2O}}$$

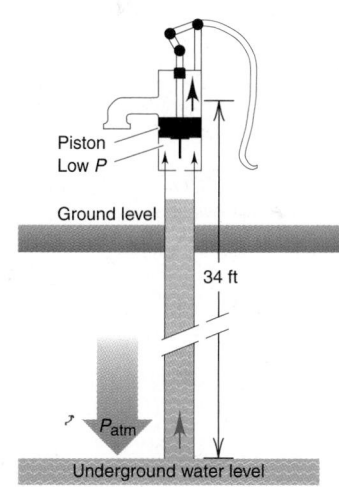

● **The Mystery of the Suction Pump** When you drink through a straw, you create lower pressure above the liquid, and the atmosphere pushes the liquid up. Similarly, a "suction" pump is a tube dipping into a water source, with a piston and handle that lower the air pressure above the water level. The pump can raise water from a well no deeper than 34 ft. This 34-ft limit was a mystery until the great 17th-century Italian scientist Galileo showed that the atmosphere pushes the water up into the tube and that its pressure can support only a 34-ft column of water. Modern pumps that draw water from deeper sources use compressed air to increase the pressure exerted on the water.

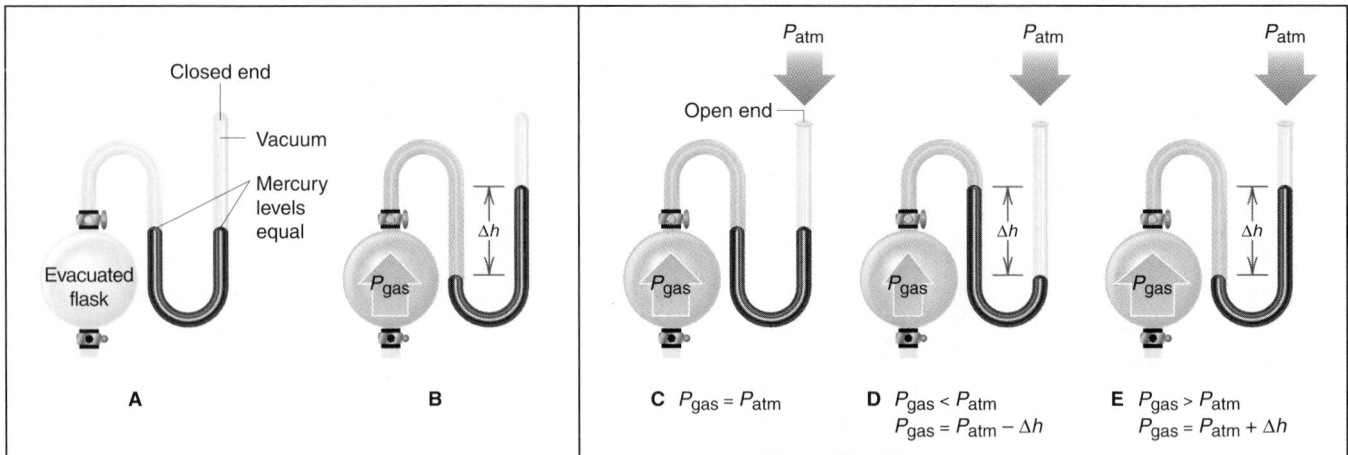

Figure 5.4 **Two types of manometer.** **A,** A closed-end manometer with an evacuated flask attached has the mercury levels equal. **B,** A gas exerts pressure on the mercury in the arm attached to the flask. The difference in heights (Δh) equals the gas pressure. **C–E,** An open-end manometer is shown with gas pressure equal to atmospheric pressure, **(C),** gas pressure lower than atmospheric pressure **(D),** and gas pressure higher than atmospheric pressure **(E).**

Manometers are devices used to measure the pressure of a gas in an experiment. Figure 5.4 shows two types of manometer. Part A shows a *closed-end manometer,* a mercury-filled, curved tube, *closed* at one end and attached to a flask at the other. When the flask is evacuated, the mercury levels in the two arms of the tube are the same because no gas exerts pressure on either mercury surface. When a gas is in the flask (part B), it pushes down the mercury level in the near arm, so the level rises in the far arm. The *difference* in column heights (Δh) equals the gas pressure. Note that if we open the lower stopcock of the evacuated flask in part A, air rushes in, Δh equals atmospheric pressure, and the closed-end manometer becomes a barometer.

The *open-end manometer,* shown in parts C–E, also consists of a curved tube filled with mercury, but one end of the tube is *open* to the atmosphere and the other is connected to the gas sample. The atmosphere pushes on one mercury level and the gas pushes on the other. Since Δh equals the difference between two pressures, to calculate the gas pressure with an open-end manometer, we must measure the atmospheric pressure separately with a barometer.

Units of Pressure

Pressure results from a force exerted on an area. The SI unit of force is the newton (N): $1\ N = 1\ kg \cdot m/s^2$. The SI unit of pressure is the **pascal (Pa),** which equals a force of one newton exerted on an area of one square meter:

$$1\ Pa = 1\ N/m^2$$

A much larger unit is the **standard atmosphere (atm),** the average atmospheric pressure measured at sea level and 0°C. It is defined in terms of the pascal:

$$1\ atm = 101.325\ kilopascals\ (kPa) = 1.01325 \times 10^5\ Pa$$

Another common pressure unit is the **millimeter of mercury (mmHg),** which is based on measurement with a barometer or manometer. In honor of Torricelli, this unit has been named the **torr:**

$$1\ torr = 1\ mmHg = \frac{1}{760}\ atm = \frac{101.325}{760}\ kPa = 133,322\ Pa$$

The *bar* is coming into more common use in chemistry:

$$1\ bar = 1 \times 10^2\ kPa = 1 \times 10^5\ Pa$$

Despite a gradual change to SI units, many chemists still express pressure in torrs and atmospheres, so they are used in this text, with frequent reference to pascals. Table 5.2 lists some important pressure units used in various scientific fields.

Table 5.2 Common Units of Pressure

Unit	Atmospheric Pressure	Scientific Field
pascal (Pa); kilopascal (kPa)	1.01325×10^5 Pa; 101.325 kPa	SI unit; physics, chemistry
atmosphere (atm)	1 atm*	Chemistry
millimeters of mercury (mmHg)	760 mmHg*	Chemistry, medicine, biology
torr	760 torr*	Chemistry
pounds per square inch (psi or lb/in^2)	14.7 lb/in^2	Engineering
bar	1.01325 bar	Meteorology, chemistry, physics

*This is an exact quantity; in calculations, we use as many significant figures as necessary.

SAMPLE PROBLEM 5.1 Converting Units of Pressure

Problem A geochemist heats a limestone ($CaCO_3$) sample and collects the CO_2 released in an evacuated flask attached to a closed-end manometer (see Figure 5.4B). After the system comes to room temperature, $\Delta h = 291.4$ mmHg. Calculate the CO_2 pressure in torrs, atmospheres, and kilopascals.

Plan The CO_2 pressure is given in units of mmHg, so we construct conversion factors from Table 5.2 to find the pressure in the other units.

Solution Converting from mmHg to torr:

$$P_{CO_2}(\text{torr}) = 291.4 \text{ mmHg} \times \frac{1 \text{ torr}}{1 \text{ mmHg}} = \boxed{291.4 \text{ torr}}$$

Converting from torr to atm:

$$P_{CO_2}(\text{atm}) = 291.4 \text{ torr} \times \frac{1 \text{ atm}}{760 \text{ torr}} = \boxed{0.3834 \text{ atm}}$$

Converting from atm to kPa:

$$P_{CO_2}(\text{kPa}) = 0.3834 \text{ atm} \times \frac{101.325 \text{ kPa}}{1 \text{ atm}} = \boxed{38.85 \text{ kPa}}$$

Check There are 760 torr in 1 atm, so ~300 torr should be <0.5 atm. There are ~100 kPa in 1 atm, so <0.5 atm should be <50 kPa.

Comment 1. In the conversion from torr to atm, we retained four significant figures because this unit conversion factor involves *exact* numbers; that is, 760 torr has as many significant figures as the calculation requires.

2. From here on, except in particularly complex situations, *the canceling of units in calculations is no longer shown.*

FOLLOW-UP PROBLEM 5.1 The CO_2 released from another mineral sample was collected in an evacuated flask connected to an open-end manometer (see Figure 5.4D). If the barometer reading is 753.6 mmHg and Δh is 174.0 mmHg, calculate P_{CO_2} in torrs, pascals, and lb/in^2.

SECTION SUMMARY

Gases exert pressure (force/area) on all surfaces with which they make contact. A barometer measures atmospheric pressure in terms of the height of the mercury column that the atmosphere can support (760 mmHg at sea level and 0°C). Both closed-end and open-end manometers are used to measure the pressure of a gas sample. Chemists measure pressure in units of atmospheres (atm), torr (equivalent to mmHg), or pascals (Pa, the SI unit).

5.3 THE GAS LAWS AND THEIR EXPERIMENTAL FOUNDATIONS

The physical behavior of a sample of gas can be described completely by four variables: pressure (P), volume (V), temperature (T), and amount (number of moles, n). The variables are interdependent: *any one of them can be determined by measuring the other three*. We know now that this quantitatively predictable behavior is a direct outcome of the structure of gases on the molecular level. Yet, it was discovered, for the most part, before Dalton's atomic theory was published!

Three key relationships exist among the four gas variables—Boyle's, Charles's, and Avogadro's laws. Each of these gas laws expresses the effect of one variable on another, with the remaining two variables held constant. Since the volume of a gas is so easy to measure, the laws are traditionally expressed as the effect on gas volume of changing the pressure, temperature, or amount of gas.

These three laws are special cases of an all-encompassing relationship among gas variables called the *ideal gas law*. This unifying observation quantitatively describes the state of a so-called **ideal gas,** one that exhibits simple linear relationships among volume, pressure, temperature, and amount. Although no ideal gas actually exists, most simple gases, such as N_2, O_2, H_2, and the noble gases, show nearly ideal behavior at ordinary temperatures and pressures. We discuss the ideal gas law after the three special cases.

The Relationship Between Volume and Pressure: Boyle's Law

Following Torricelli's invention of the barometer, the great English chemist Robert Boyle performed a series of experiments that led him to conclude that at a given temperature, *the volume occupied by a gas is inversely related to its pressure.* Figure 5.5 shows Boyle's experiment and some typical data he might have collected. Boyle fashioned a J-shaped glass tube, sealed the shorter end, and

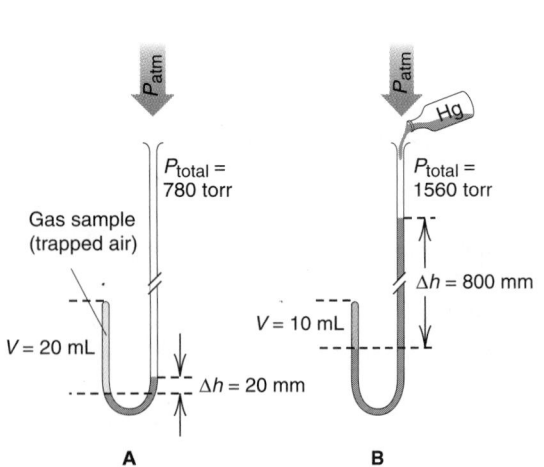

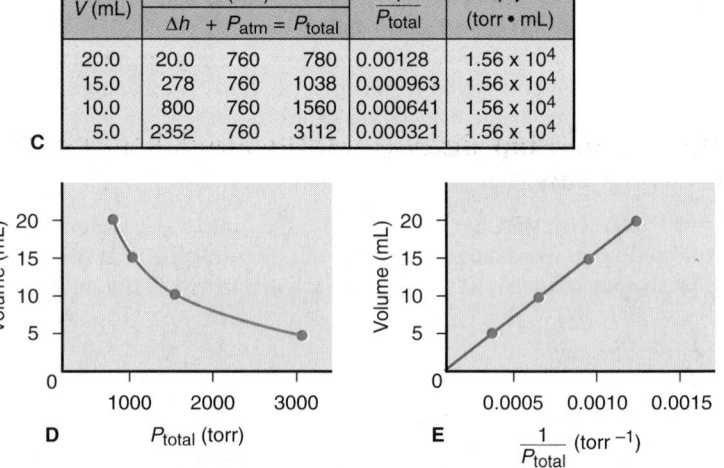

V (mL)	P (torr)			$\dfrac{1}{P_{total}}$	PV (torr • mL)
	Δh	$+ P_{atm}$	$= P_{total}$		
20.0	20.0	760	780	0.00128	1.56×10^4
15.0	278	760	1038	0.000963	1.56×10^4
10.0	800	760	1560	0.000641	1.56×10^4
5.0	2352	760	3112	0.000321	1.56×10^4

Figure 5.5 **The relationship between the volume and pressure of a gas. A,** A small amount of air (the gas) is trapped in the short arm of a J tube; n and T are fixed. The total pressure on the gas (P_{total}) is the sum of the pressure due to the difference in heights of the mercury columns (Δh) and the pressure of the atmosphere (P_{atm}). If $P_{atm} = 760$ torr, $P_{total} = 780$ torr. **B,** As mercury is added, the total pressure

on the gas increases and its volume (V) decreases. Note that if P_{total} is doubled (to 1560 torr), V is halved. (The mercury heights are not drawn to scale.) **C,** Some typical pressure-volume data from the experiment. **D,** A plot of V vs. P_{total} shows that V is inversely proportional to P. **E,** A plot of V vs. $1/P_{total}$ is a straight line whose slope is a constant characteristic of *any* gas that behaves ideally.

poured mercury into the longer end, thereby trapping some air, the gas in the experiment. From the height of the trapped air column and the diameter of the tube, he calculated the air volume. The total pressure applied to the trapped air was the pressure of the atmosphere (measured with a barometer) plus that of the mercury column (part A). By adding mercury, Boyle increased the total pressure exerted on the air, and the air volume decreased (part B). Since the temperature and amount of air were constant, Boyle could directly measure the effect of the applied pressure on the volume of air.

Note the following results (Figure 5.5):

- The product of corresponding P and V values is a constant (part C, rightmost column)
- V is *inversely* proportional to P (part D)
- V is *directly* proportional to $1/P$ (part E) and generates a linear plot of V against $1/P$. This *linear relationship between two gas variables* is a hallmark of ideal gas behavior.

The generalization of Boyle's observations is known as **Boyle's law:** *at constant temperature, the volume occupied by a fixed amount of gas is inversely proportional to the applied (external) pressure,* or

$$V \propto \frac{1}{P} \qquad [T \text{ and } n \text{ fixed}] \qquad\qquad (5.1)$$

This relationship can also be expressed as

$$PV = \text{constant} \qquad \text{or} \qquad V = \frac{\text{constant}}{P} \qquad [T \text{ and } n \text{ fixed}]$$

The constant is the same for the great majority of gases. Thus, tripling the external pressure reduces the volume to one-third its initial value; halving the external pressure doubles the volume; and so forth.

The wording of Boyle's law focuses on *external* pressure. In his experiment, however, adding more mercury causes the mercury level to rise until the pressure of the trapped air stops the rise at some new level. At that point, the pressure exerted *on* the gas equals the pressure exerted *by* the gas. In other words, by measuring the applied pressure, Boyle was also measuring the gas pressure. Thus, when gas volume doubles, gas pressure is halved. In general, if V_{gas} increases, P_{gas} decreases, and vice versa.

The Relationship Between Volume and Temperature: Charles's Law

One question raised by Boyle's work was why the pressure-volume relationship holds only at constant temperature. Experience had shown that gas volume depends on temperature, but it was not until the early 19th century, through the separate work of the French scientists J. A. C. Charles and J. L. Gay-Lussac, that the relationship was clearly understood.

Let's examine this relationship by measuring the volume of a fixed amount of a gas under constant pressure but at different temperatures. A straight tube, closed at one end, traps a fixed amount of air under a small mercury plug. The tube is immersed in a water bath that can be warmed with a heater or cooled with ice. After each change of water temperature, we measure the length of the air column, which is proportional to its volume. The pressure exerted on the gas is constant because the mercury plug and the atmospheric pressure do not change (see Figure 5.6A and B on the next page).

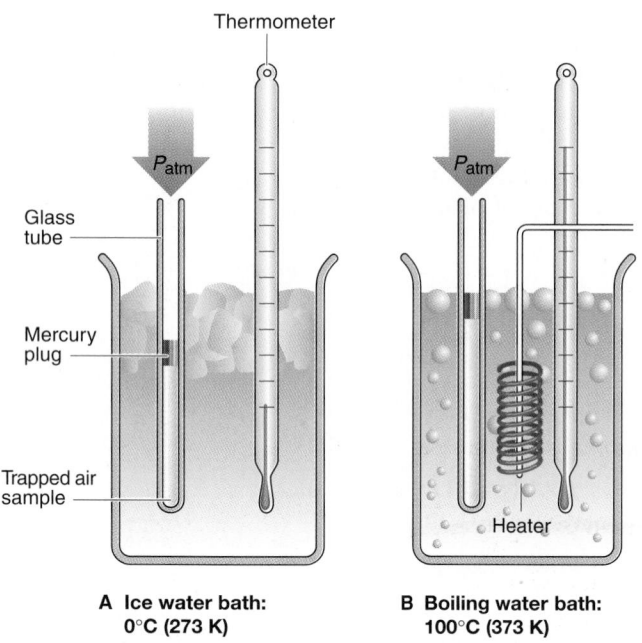

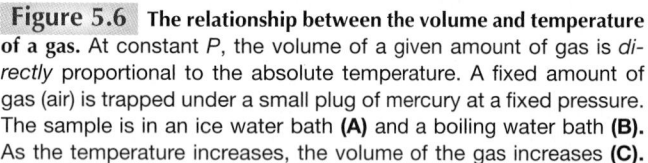

A   **Ice water bath:**
    0°C (273 K)

B   **Boiling water bath:**
    100°C (373 K)

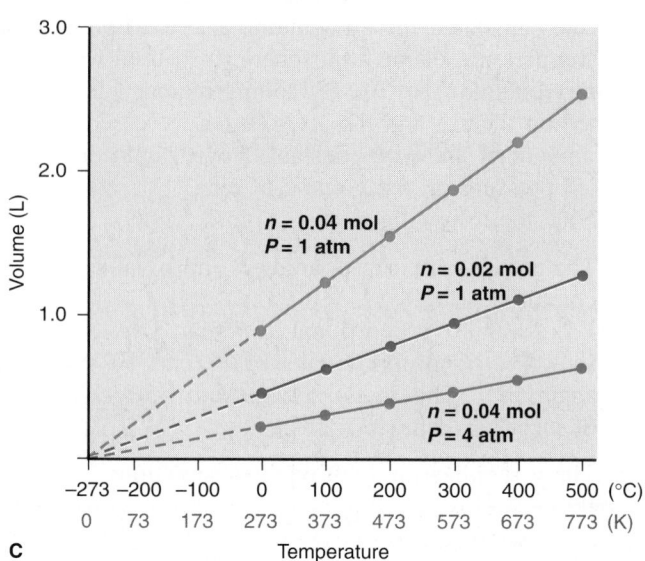

C   Temperature

Figure 5.6 **The relationship between the volume and temperature of a gas.** At constant *P*, the volume of a given amount of gas is *directly* proportional to the absolute temperature. A fixed amount of gas (air) is trapped under a small plug of mercury at a fixed pressure. The sample is in an ice water bath **(A)** and a boiling water bath **(B)**. As the temperature increases, the volume of the gas increases **(C)**.

The three lines show the effect of amount (*n*) of gas (compare red and green) and pressure (*P*) of gas (compare red and blue). The dashed lines extrapolate the data to lower temperatures. For any amount of an ideal gas at any pressure, the volume is theoretically zero at −273.15°C (0 K).

Some typical data are shown for different amounts and pressures of gas in Figure 5.6C. Again, note the linear relationships, but this time the variables are *directly* proportional: for a given amount of gas at a given pressure, *volume increases as temperature increases.* For example, the red line shows how the volume of 0.04 mol of gas at 1 atm pressure changes as the temperature changes. Extending (*extrapolating*) the line to lower temperatures (dashed portion) shows that the volume shrinks until the gas occupies a theoretical zero volume at −273°C (the intercept of the temperature axis). Similar plots for a different amount of gas (green) and a different gas pressure (blue) show lines with different slopes, but they all converge at this temperature.

One-half century after Charles's and Gay-Lussac's work, William Thomson (Lord Kelvin) used this linear relation between gas volume and temperature to devise the absolute temperature scale (Section 1.5). In this scale, absolute zero (0 K or −273.15°C) is the temperature at which an ideal gas would have zero volume. (Absolute zero has never been reached, but physicists have attained temperatures as low as 10^{-9} K.) Of course, no sample of matter can have zero volume, and every real gas condenses to a liquid at some temperature higher than 0 K. Nevertheless, *this linear dependence of volume on absolute temperature holds for most common gases over a wide temperature range.*

The modern statement of the volume-temperature relationship is known as **Charles's law:** *at constant pressure, the volume occupied by a fixed amount of gas is directly proportional to its absolute temperature,* or

$$V \propto T \qquad [P \text{ and } n \text{ fixed}] \qquad (5.2)$$

This relationship can also be expressed as

$$\frac{V}{T} = \text{constant} \qquad \text{or} \qquad V = \text{constant} \times T \qquad [P \text{ and } n \text{ fixed}]$$

If T increases, V increases, and vice versa. Once again, for any given P and n, the constant is the same for the great majority of gases.

The dependence of gas volume on *absolute* temperature means that you must *use the Kelvin scale in gas law calculations.* For instance, if the temperature changes from 200 K to 400 K, the volume of 1 mol of gas doubles. But, if the temperature changes from 200°C to 400°C, the volume increases by a factor of 1.42; that is, $\left(\dfrac{400°C + 273.15}{200°C + 273.15}\right) = \dfrac{673}{473} = 1.42$.

Other Relationships Based on Boyle's and Charles's Laws Two other important relationships emerge from an understanding of Boyle's and Charles's laws:

1. *The pressure-temperature relationship.* Charles's law is expressed as the effect of a temperature change on gas *volume.* However, volume and pressure are interdependent, so the effect of temperature on volume is closely related to its effect on pressure (sometimes referred to as *Amontons's law*). Measure the pressure in your car's tires before and after a long drive, and you will find that it has increased. Frictional heating between the tire and the road increases the air temperature inside the tire, but since the tire volume doesn't change appreciably, the air exerts more pressure. Thus, *at constant volume, the pressure exerted by a fixed amount of gas is directly proportional to the absolute temperature:*

$$P \propto T \qquad [V \text{ and } n \text{ fixed}] \qquad (5.3)$$

or

$$\frac{P}{T} = \text{constant} \qquad \text{or} \qquad P = \text{constant} \times T$$

2. *The combined gas law.* A simple combination of Boyle's and Charles's laws gives the *combined gas law,* which applies to situations when two of the three variables (V, P, T) change and you must find the effect on the third:

$$V \propto \frac{T}{P} \qquad \text{or} \qquad V = \text{constant} \times \frac{T}{P} \qquad \text{or} \qquad \frac{PV}{T} = \text{constant}$$

The Relationship Between Volume and Amount: Avogadro's Law

Boyle's and Charles's laws both specify a fixed amount of gas. Let's see why. Figure 5.7 shows an experiment involving two small test tubes, each fitted with a piston-cylinder assembly. We add 0.10 mol (4.4 g) of dry ice (frozen CO_2) to tube A and 0.20 mol (8.8 g) to tube B. As the solid warms, it changes directly to gaseous CO_2, which expands into the cylinder and pushes up the piston. When all the solid has changed to gas and the temperature is constant, we find that cylinder B has twice the volume of cylinder A. (We can neglect the volume of the tube because it is so much smaller than the volume of the cylinder.)

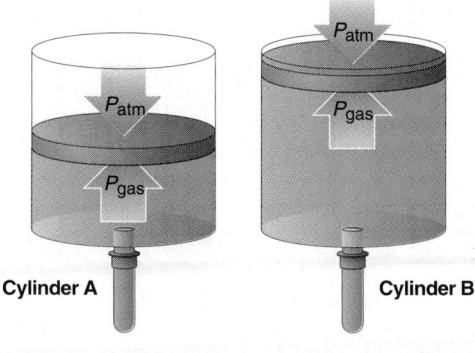

Cylinder A   **Cylinder B**

Figure 5.7 An experiment to study the relationship between the volume and amount of a gas. At a given external P and T, twice the mass of $CO_2(s)$ is put in tube B as in tube A. After the solid has changed to gas, the volume of cylinder B is twice that of cylinder A. Thus, at fixed P and T, the volume (V) of a gas is directly proportional to the amount of gas (n).

This result shows that twice the amount (mol) of gas occupies twice the volume. Note, for both cylinders, T of the gas equals room temperature and P of the gas equals atmospheric pressure. Thus, *at fixed temperature and pressure, the volume occupied by a gas is directly proportional to the amount (mol) of gas:*

$$V \propto n \qquad [P \text{ and } T \text{ fixed}] \qquad (5.4)$$

As n increases, V increases, and vice versa. This relationship is also expressed as

$$\frac{V}{n} = \text{constant} \qquad \text{or} \qquad V = \text{constant} \times n$$

PROPERTIES OF GASES

The constant is the same for all gases at a given temperature and pressure. This relationship is another way of expressing **Avogadro's law,** which states that *at fixed temperature and pressure, equal volumes of **any** ideal gas contain equal numbers of particles (or moles).*

Many familiar phenomena are based on the relationships among volume, temperature, and amount of gas. For example, in a car engine, a reaction occurs in which fewer moles of gasoline and O_2 form more moles of CO_2 and H_2O vapor, which expand as a result of the released heat and push back the piston. Dynamite explodes because a solid decomposes rapidly to form hot gases. Dough rises in a warm room because yeast forms CO_2 bubbles in the dough, which expand during baking to give the bread a still larger volume.

Breathing and the Gas Laws Taking a deep breath is a combined application of the gas laws. When you inhale, muscles are coordinated such that your diaphragm moves down and your rib cage moves out. This movement increases the volume of the lungs, which decreases the pressure of the air inside them relative to that outside, so air rushes in (Boyle's). The greater amount of air stretches the elastic tissue of the lungs and expands the volume further (Avogadro's). The air also expands slightly as it warms to body temperature (Charles's). When you exhale, the diaphragm relaxes and moves up, the rib cage moves in, and the lung volume decreases. The inside air pressure becomes greater than the outside pressure, and air rushes out.

Gas Behavior at Standard Conditions

To better understand the factors that influence gas behavior, chemists use a set of *standard conditions* called **standard temperature and pressure (STP):**

$$\text{STP:} \qquad 0°C \ (273.15 \ K) \text{ and } 1 \text{ atm } (760 \text{ torr}) \qquad (5.5)$$

Under these conditions, the volume of 1 mol of an ideal gas is called the **standard molar volume:**

$$\text{Standard molar volume} = 22.4141 \ L \quad \text{or} \quad 22.4 \ L \quad [\text{to 3 sf}] \qquad (5.6)$$

Figure 5.8 compares the properties of three simple gases at STP.

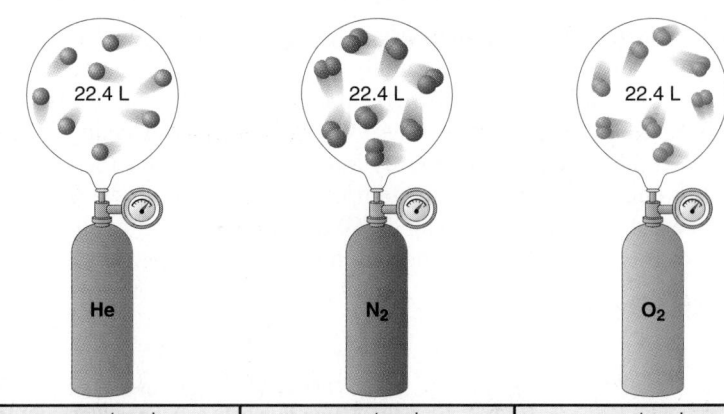

$n = 1$ mol	$n = 1$ mol	$n = 1$ mol
$P = 1$ atm (760 torr)	$P = 1$ atm (760 torr)	$P = 1$ atm (760 torr)
$T = 0°C$ (273 K)	$T = 0°C$ (273 K)	$T = 0°C$ (273 K)
$V = 22.4$ L	$V = 22.4$ L	$V = 22.4$ L
Number of gas particles = 6.022×10^{23}	Number of gas particles = 6.022×10^{23}	Number of gas particles = 6.022×10^{23}
Mass = 4.003 g	Mass = 28.02 g	Mass = 32.00 g
$d = 0.179$ g/L	$d = 1.25$ g/L	$d = 1.43$ g/L

Figure 5.8 **Standard molar volume.** One mole of an ideal gas occupies 22.4 L at STP (0°C and 1 atm). At STP, helium, nitrogen, oxygen, and most other simple gases behave ideally. Note that the *mass* of a gas, and thus its density (*d*), depends on its molar mass.

Figure 5.9 compares the volumes of some familiar objects with the standard molar volume of an ideal gas.

The Ideal Gas Law

Each of the gas laws focuses on the effect of changes in one variable on gas volume:

- Boyle's law focuses on pressure ($V \propto 1/P$).
- Charles's law focuses on temperature ($V \propto T$).
- Avogadro's law focuses on amount (mol) of gas ($V \propto n$).

We can combine these individual effects into one relationship, called the **ideal gas law** (or *ideal gas equation*):

$$V \propto \frac{nT}{P} \qquad \text{or} \qquad PV \propto nT \qquad \text{or} \qquad \frac{PV}{nT} = R$$

Figure 5.9 **The volume of 1 mol of an ideal gas compared with some familiar objects.** A basketball (7.5 L), 5-gal fish tank (18.9 L), 13-in television (21.6 L), and 22.4 L of He gas in a balloon.

where R is a proportionality constant known as the **universal gas constant.** Rearranging gives the most common form of the ideal gas law:

$$PV = nRT \tag{5.7}$$

We can obtain a value of R by measuring the volume, temperature, and pressure of a given amount of gas and substituting the values into the ideal gas law. For example, using standard conditions for the gas variables, we have

$$R = \frac{PV}{nT} = \frac{1 \text{ atm} \times 22.4141 \text{ L}}{1 \text{ mol} \times 273.15 \text{ K}} = 0.082058 \; \frac{\text{atm·L}}{\text{mol·K}} = 0.0821 \; \frac{\text{atm·L}}{\text{mol·K}} \quad [\text{3 sf}] \tag{5.8}$$

This numerical value of R corresponds to the gas variables P, V, and T expressed *in these units*. R has a different numerical value when different units are used. For example, later in this chapter, R has the value 8.314 J/mol·K (J stands for joule, the SI unit of energy).

Figure 5.10 makes a central point: the ideal gas law *becomes* one of the individual gas laws when two of the four variables are kept constant. When initial conditions (subscript $_1$) change to final conditions (subscript $_2$), we have

$$P_1V_1 = n_1RT_1 \qquad \text{and} \qquad P_2V_2 = n_2RT_2$$

Thus, $\qquad \dfrac{P_1V_1}{n_1T_1} = R \qquad$ and $\qquad \dfrac{P_2V_2}{n_2T_2} = R, \qquad$ so $\qquad \dfrac{P_1V_1}{n_1T_1} = \dfrac{P_2V_2}{n_2T_2}$

Notice that if two of the variables remain constant, say P and T, then $P_1 = P_2$ and $T_1 = T_2$, and we obtain an expression for Avogadro's law:

$$\frac{P_1V_1}{n_1T_1} = \frac{P_2V_2}{n_2T_2} \qquad \text{or} \qquad \frac{V_1}{n_1} = \frac{V_2}{n_2}$$

We use rearrangements of the ideal gas law such as this one to solve gas law problems, as you'll see next. The point to remember is that there is no need to memorize the individual gas laws.

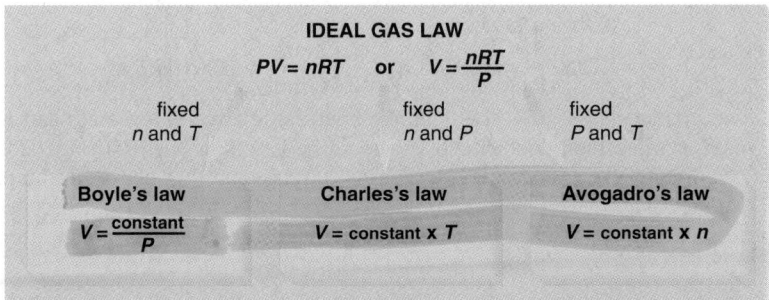

IDEAL GAS LAW

$$PV = nRT \qquad \text{or} \qquad V = \frac{nRT}{P}$$

fixed n and T	fixed n and P	fixed P and T
Boyle's law	**Charles's law**	**Avogadro's law**
$V = \dfrac{\text{constant}}{P}$	$V = \text{constant} \times T$	$V = \text{constant} \times n$

Figure 5.10 **Relationship between the ideal gas law and the individual gas laws.** Boyle's, Charles's, and Avogadro's laws are contained within the ideal gas law.

Solving Gas Law Problems

Gas law problems are phrased in many ways, but they can usually be grouped into two main types:

  1. *A change in one of the four variables causes a change in another, while the two remaining variables are held constant.* In this type, the ideal gas law reduces to one of the individual gas laws, and you solve for the new value of the variable. Units must be consistent, T must always be in kelvins, but R is not involved. Sample Problems 5.2 to 5.4 are of this type. [A variation on this type involves the combined gas law (p. 183) for simultaneous changes in two of the variables that cause a change in a third.]

  2. *One variable is unknown, but the other three are known and no change occurs.* In this type, which is exemplified by Sample Problem 5.5, the ideal gas law is applied directly to find the unknown, and the units must conform to those in R.

These problems are far easier to solve if you follow a systematic approach:

  • Summarize the information: identify the changing gas variables—knowns and unknown—and those held constant.
  • Predict the direction of the change, and later check your answer against the prediction.
  • Perform any necessary unit conversions.
  • Rearrange the ideal gas law to obtain the appropriate relationship of gas variables, and solve for the unknown variable.

The following series of sample problems applies each of these gas behaviors.

SAMPLE PROBLEM 5.2 Applying the Volume-Pressure Relationship

Problem Boyle's apprentice finds that the air trapped in a J tube occupies 24.8 cm³ at 1.12 atm. By adding mercury to the tube, he increases the pressure on the trapped air to 2.64 atm. Assuming constant temperature, what is the new volume of air (in L)?

Plan We must find the final volume (V_2) in liters, given the initial volume (V_1), initial pressure (P_1), and final pressure (P_2). The temperature and amount of gas are fixed. We convert the units of V_1 from cm³ to mL and then to L, rearrange the ideal gas law to the appropriate form, and solve for V_2. We can predict the direction of the change: since P increases, V will decrease; thus, $V_2 < V_1$. (Note that the roadmap has two parts.)

Solution Summarizing the gas variables:

$$P_1 = 1.12 \text{ atm} \qquad\qquad P_2 = 2.64 \text{ atm}$$
$$V_1 = 24.8 \text{ cm}^3 \text{ (convert to L)} \qquad V_2 = \text{unknown} \qquad T \text{ and } n \text{ remain constant}$$

Converting V_1 from cm³ to L:

$$V_1 = 24.8 \text{ cm}^3 \times \frac{1 \text{ mL}}{1 \text{ cm}^3} \times \frac{1 \text{ L}}{1000 \text{ mL}} = 0.0248 \text{ L}$$

Arranging the ideal gas law and solving for V_2: At fixed n and T, we have

$$\frac{P_1 V_1}{n_1 T_1} = \frac{P_2 V_2}{n_2 T_2} \qquad \text{or} \qquad P_1 V_1 = P_2 V_2$$

$$V_2 = V_1 \times \frac{P_1}{P_2} = 0.0248 \text{ L} \times \frac{1.12 \text{ atm}}{2.64 \text{ atm}} = \boxed{0.0105 \text{ L}}$$

Check As we predicted, $V_2 < V_1$. Let's think about the relative values of P and V as we check the math. P more than doubled, so V_2 should be less than $\frac{1}{2}V_1$ (0.0105/0.0248 $< \frac{1}{2}$).
Comment Predicting the direction of the change provides another check on the problem setup: To make $V_2 < V_1$, we must multiply V_1 by a number *less than* 1. This means the ratio of pressures must be *less than* 1, so the larger pressure (P_2) must be in the denominator, P_1/P_2.

FOLLOW-UP PROBLEM 5.2 A sample of argon gas occupies 105 mL at 0.871 atm. If the temperature remains constant, what is the volume (in L) at 26.3 kPa?

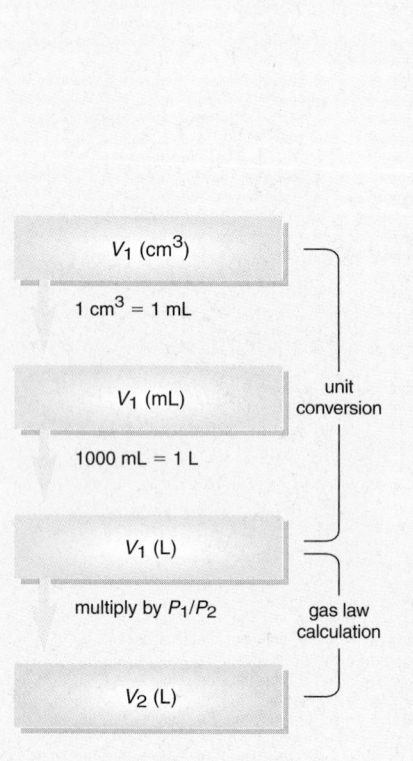

V_1 (cm³)

1 cm³ = 1 mL

V_1 (mL)

unit conversion

1000 mL = 1 L

V_1 (L)

multiply by P_1/P_2 gas law calculation

V_2 (L)

SAMPLE PROBLEM 5.3 Applying the Pressure-Temperature Relationship

Problem A 1-L steel tank is fitted with a safety valve that opens if the internal pressure exceeds 1.00×10^3 torr. It is filled with helium at 23°C and 0.991 atm and placed in boiling water at exactly 100°C. Will the safety valve open?

Plan The question "Will the safety valve open?" translates into "Is P_2 greater than 1.00×10^3 torr at T_2?" Thus, P_2 is the unknown, and T_1, T_2, and P_1 are given, with V (steel container) and n fixed. We convert both T values to kelvins and P_1 to torrs in order to compare P_2 with the safety-limit pressure. We rearrange the ideal gas law to the appropriate form and solve for P_2. Since $T_2 > T_1$, we predict that $P_2 > P_1$.

Solution Summary of gas variables:

$P_1 = 0.991$ atm (convert to torr) $P_2 =$ unknown
$T_1 = 23$°C (convert to K) $T_2 = 100$°C (convert to K)
V and n remain constant

Converting T from °C to K:

$$T_1 \text{ (K)} = 23\text{°C} + 273.15 = 296 \text{ K}$$
$$T_2 \text{ (K)} = 100\text{°C} + 273.15 = 373 \text{ K}$$

Converting P from atm to torr:

$$P_1 \text{ (torr)} = 0.991 \text{ atm} \times \frac{760 \text{ torr}}{1 \text{ atm}} = 753 \text{ torr}$$

Arranging the ideal gas law and solving for P_2: At fixed n and V, we have

$$\frac{P_1 V_1}{n_1 T_1} = \frac{P_2 V_2}{n_2 T_2} \qquad \text{or} \qquad \frac{P_1}{T_1} = \frac{P_2}{T_2}$$

$$P_2 = P_1 \times \frac{T_2}{T_1} = 753 \text{ torr} \times \frac{373 \text{ K}}{296 \text{ K}} = 949 \text{ torr}$$

P_2 is less than 1.00×10^3 torr, so the valve will *not* open.

Check Our prediction is correct: because $T_2 > T_1$, we have $P_2 > P_1$. Thus, the temperature ratio should be >1 (T_2 in the numerator). The T ratio is about 1.25 (373/296), so the P ratio should also be about 1.25 (950/750 ≈ 1.25).

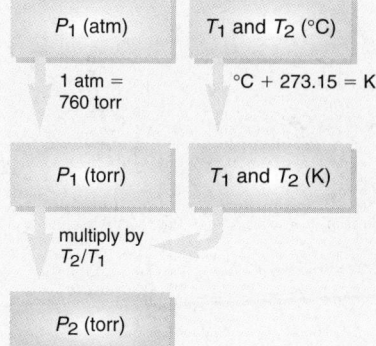

FOLLOW-UP PROBLEM 5.3 An engineer pumps air at 0°C into a newly designed piston-cylinder assembly. The volume measures 6.83 cm^3. At what temperature (in K) would the volume be 9.75 cm^3?

SAMPLE PROBLEM 5.4 Applying the Volume-Amount Relationship

Problem A scale model of a blimp rises when it is filled with helium to a volume of 55.0 dm^3. When 1.10 mol of He is added to the blimp, the volume is 26.2 dm^3. How many more grams of He must be added to make it rise? Assume constant T and P.

Plan We are given the initial amount of helium (n_1), the initial volume of the blimp (V_1), and the volume needed for it to rise (V_2), and we need the additional mass of helium to make it rise. So we first need to find n_2. We rearrange the ideal gas law to the appropriate form, solve for n_2, subtract n_1 to find the additional amount ($n_{\text{add'l}}$), and then convert moles to grams. We predict that $n_2 > n_1$ because $V_2 > V_1$.

Solution Summary of gas variables:

$n_1 = 1.10$ mol $n_2 =$ unknown (find, and then subtract n_1)
$V_1 = 26.2 \text{ dm}^3$ $V_2 = 55.0 \text{ dm}^3$
P and T remain constant

Arranging the ideal gas law and solving for n_2: At fixed P and T, we have

$$\frac{P_1 V_1}{n_1 T_1} = \frac{P_2 V_2}{n_2 T_2} \qquad \text{or} \qquad \frac{V_1}{n_1} = \frac{V_2}{n_2}$$

$$n_2 = n_1 \times \frac{V_2}{V_1} = 1.10 \text{ mol He} \times \frac{55.0 \text{ dm}^3}{26.2 \text{ dm}^3} = 2.31 \text{ mol He}$$

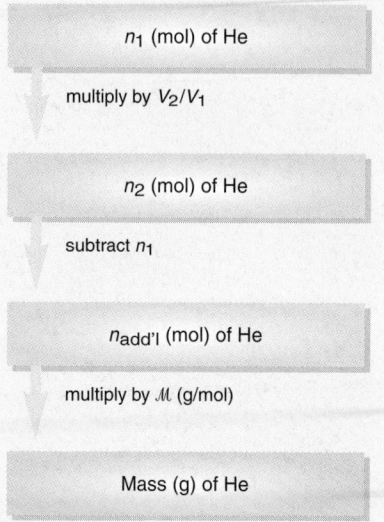

Finding the additional amount of He:

$$n_{add'l} = n_2 - n_1 = 2.31 \text{ mol He} - 1.10 \text{ mol He} = 1.21 \text{ mol He}$$

Converting moles of He to grams:

$$\text{Mass (g) of He} = 1.21 \text{ mol He} \times \frac{4.003 \text{ g He}}{1 \text{ mol He}}$$

$$= \boxed{4.84 \text{ g He}}$$

Check Since V_2 is about twice V_1 ($55/26 \approx 2$), n_2 should be about twice n_1 ($2.3/1.1 \approx 2$). Since $n_2 > n_1$, we were right to multiply n_1 by a number >1 (that is, V_2/V_1). About 1.2 mol $\times$ 4 g/mol $\approx$ 4.8 g.

Comment 1. A different sequence of steps will give you the same answer: first find the additional volume ($V_{add'l} = V_2 - V_1$), and then solve directly for $n_{add'l}$. Try it for yourself.

2. You saw that Charles's law ($V \propto T$ at fixed P and n) translates into a similar relationship between P and T at fixed V and n. The follow-up problem demonstrates that Avogadro's law ($V \propto n$ at fixed P and T) translates into an analogous relationship at fixed V and T.

FOLLOW-UP PROBLEM 5.4 A rigid plastic container holds 35.0 g of ethylene gas (C_2H_4) at a pressure of 793 torr. What is the pressure if 5.0 g of ethylene is removed at constant temperature?

SAMPLE PROBLEM 5.5 Solving for an Unknown Gas Variable at Fixed Conditions

---

Problem A steel tank has a volume of 438 L and is filled with 0.885 kg of O_2. Calculate the pressure of O_2 at 21°C.

Plan We are given V, T, and the mass of O_2, and we must find P. Since conditions are not changing, we apply the ideal gas law without rearranging it. We use the given V in liters, convert T to kelvins and mass of O_2 to moles, and solve for P.

Solution Summary of gas variables:

$$V = 438 \text{ L} \qquad\qquad T = 21°C \text{ (convert to K)}$$
$$n = 0.885 \text{ kg } O_2 \text{ (convert to mol)} \qquad P = \text{unknown}$$

Converting T from °C to K:

$$T \text{ (K)} = 21°C + 273.15 = 294 \text{ K}$$

Converting from mass of O_2 to moles:

$$n = \text{mol of } O_2 = 0.885 \text{ kg } O_2 \times \frac{1000 \text{ g}}{1 \text{ kg}} \times \frac{1 \text{ mol } O_2}{32.00 \text{ g } O_2} = 27.7 \text{ mol } O_2$$

Solving for P (note the unit canceling here):

$$P = \frac{nRT}{V} = \frac{27.7 \text{ mol} \times 0.0821 \dfrac{\text{atm·L}}{\text{mol·K}} \times 294 \text{ K}}{438 \text{ L}}$$

$$= \boxed{1.53 \text{ atm}}$$

Check The amount of O_2 seems correct: ~900 g/(30 g/mol) = 30 mol. To check the approximate size of the final calculation, round off the values, including that for R:

$$P = \frac{30 \text{ mol } O_2 \times 0.1 \dfrac{\text{atm·L}}{\text{mol·K}} \times 300 \text{ K}}{450 \text{ L}} = 2 \text{ atm}$$

which is reasonably close to 1.53 atm.

FOLLOW-UP PROBLEM 5.5 The tank in the sample problem develops a slow leak that is discovered and sealed. The new measured pressure is 1.37 atm. How many grams of O_2 remain?

SECTION SUMMARY

Four variables define the physical behavior of an ideal gas: volume (V), pressure (P), temperature (T), and amount (number of moles, n). Most simple gases display nearly ideal behavior at ordinary temperature and pressure. Boyle's, Charles's, and Avogadro's laws relate volume to pressure, to temperature, and to amount of gas, respectively. At STP (0°C and 1 atm), 1 mol of an ideal gas occupies 22.4 L. The ideal gas law incorporates the individual gas laws into one equation: $PV = nRT$, where R is the universal gas constant.

5.4 FURTHER APPLICATIONS OF THE IDEAL GAS LAW

The ideal gas law can be recast in additional ways to determine other properties of gases. In this section, we use it to find gas density, molar mass, and the partial pressure of each gas in a mixture.

The Density of a Gas

One mole of any gas occupies nearly the same volume at a given temperature and pressure, so differences in gas density ($d = m/V$) depend on differences in molar mass (see Figure 5.8). For example, at STP, 1 mol of O_2 occupies the same volume as 1 mol of N_2, but since each O_2 molecule has a greater mass than each N_2 molecule, O_2 is denser.

All gases are miscible when thoroughly mixed, but in the absence of mixing, a less dense gas will lie above a more dense one.  There are many familiar examples of this phenomenon. Some types of fire extinguishers release CO_2 because it is denser than air and will sink onto the fire, preventing more O_2 from reaching the burning material. Enormous air masses of different densities and temperatures moving past each other around the globe give rise to much of our weather.

We can rearrange the ideal gas law to calculate the density of a gas from its molar mass. Recall that the number of moles (n) is the mass (m) divided by the molar mass ($\mathcal{M}$), $n = m/\mathcal{M}$. Substituting for n in the ideal gas law gives

$$PV = \frac{m}{\mathcal{M}}RT$$

Rearranging to isolate m/V gives

$$\frac{m}{V} = d = \frac{\mathcal{M} \times P}{RT} \qquad\qquad (5.9)$$

Two important ideas are expressed by Equation 5.9:

- *The density of a gas is directly proportional to its molar mass* because a given amount of a heavier gas occupies the same volume as that amount of a lighter gas (Avogadro's law).
- *The density of a gas is inversely proportional to the temperature.* As the volume of a gas increases with temperature (Charles's law), the same mass occupies more space; thus, the density is lower.

Architectural designers and heating engineers apply the second idea when they place heating ducts near the floor of a room: the less dense warm air from the ducts rises and heats the room air. Safety experts recommend staying near the floor when escaping from a fire to avoid the hot, and therefore less dense, noxious gases. We use Equation 5.9 to find the density of a gas at any temperature and pressure near standard conditions.

Gas Density and Human Disasters Many gases that are denser than air have been involved in natural and human-caused disasters. The dense gases in smog that blanket urban centers, such as Los Angeles (see photo), contribute greatly to respiratory illness. On a far more horrific scale, in World War I, phosgene ($COCl_2$) was used against ground troops as they lay in trenches. More recently, the unintentional release of methylisocyanate from a Union Carbide India Ltd. chemical plant in Bhopal, India, killed thousands of people as vapors spread from the outskirts into the city. In 1986 in Cameroon, CO_2 released naturally from Lake Nyos suffocated thousands as it flowed down valleys into villages. Some paleontologists suggest that a similar process in volcanic lakes may have contributed to dinosaur kills.

SAMPLE PROBLEM 5.6 Calculating Gas Density

Problem Calculate the density (in g/L) of carbon dioxide and the number of molecules per liter **(a)** at STP (0°C and 1 atm) and **(b)** at ordinary room conditions (20.°C and 1.00 atm).

Plan We must find the density (d) and number of molecules of gas, given the name of the substance and two sets of P and T data. The name gives us the formula, so we can find $\mathcal{M}$; we convert T to kelvins and calculate d with Equation 5.9. Then we convert the mass per liter to molecules per liter with Avogadro's number.

Solution **(a)** Density and molecules per liter of carbon dioxide at STP. Summary of gas properties:

$$T = 0°C + 273.15 = 273 \text{ K} \qquad P = 1 \text{ atm} \qquad \mathcal{M} \text{ of } CO_2 = 44.01 \text{ g/mol}$$

Calculating density (note the unit canceling here):

$$d = \frac{\mathcal{M} \times P}{RT} = \frac{44.01 \text{ g/}\cancel{mol} \times 1.00 \cancel{atm}}{0.0821 \frac{\cancel{atm}\cdot L}{\cancel{mol}\cdot \cancel{K}} \times 273 \cancel{K}}$$

$$= \boxed{1.96 \text{ g/L}}$$

Converting from mass/L to molecules/L:

$$\text{Molecules } CO_2/L = \frac{1.96 \text{ g } CO_2}{1 \text{ L}} \times \frac{1 \text{ mol } CO_2}{44.01 \text{ g } CO_2} \times \frac{6.022 \times 10^{23} \text{ molecules } CO_2}{1 \text{ mol } CO_2}$$

$$= \boxed{2.68 \times 10^{22} \text{ molecules } CO_2/L}$$

(b) Density and molecules of carbon dioxide per liter at room conditions. Summary of gas properties:

$$T = 20.°C + 273.15 = 293 \text{ K} \qquad P = 1.00 \text{ atm} \qquad \mathcal{M} \text{ of } CO_2 = 44.01 \text{ g/mol}$$

Calculating density:

$$d = \frac{\mathcal{M} \times P}{RT} = \frac{44.01 \text{ g/mol} \times 1.00 \text{ atm}}{0.0821 \frac{atm\cdot L}{mol\cdot K} \times 293 \text{ K}}$$

$$= \boxed{1.83 \text{ g/L}}$$

Converting from mass/L to molecules/L:

$$\text{Molecules } CO_2/L = \frac{1.83 \text{ g } CO_2}{1 \text{ L}} \times \frac{1 \text{ mol } CO_2}{44.01 \text{ g } CO_2} \times \frac{6.022 \times 10^{23} \text{ molecules } CO_2}{1 \text{ mol } CO_2}$$

$$= \boxed{2.50 \times 10^{22} \text{ molecules } CO_2/L}$$

Check Round off to check the density values; for example, in (a), at STP:

$$\frac{50 \text{ g/mol} \times 1 \text{ atm}}{0.1 \frac{atm\cdot L}{mol\cdot K} \times 250 \text{ K}} = 2 \text{ g/L} \approx 1.96 \text{ g/L}$$

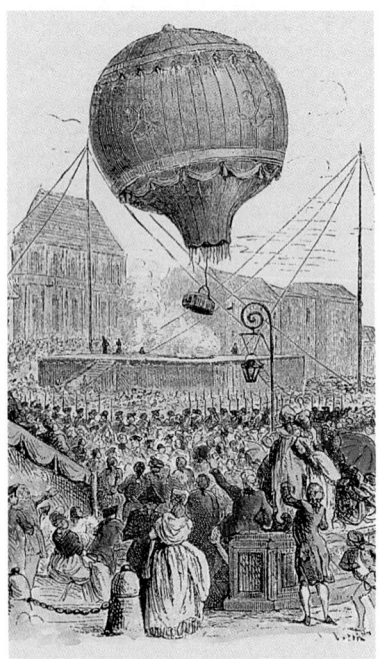

Up, Up, and Away! When the gas in a hot-air balloon is heated, its volume increases and the balloon inflates. Further heating causes some of the gas to escape. By these means, the gas density decreases and the balloon rises. Two pioneering hot-air balloonists used their knowledge of gas behavior to excel at their hobby. Jacques Charles (of Charles's law) made one of the first balloon flights in 1783. Twenty years later, Joseph Gay-Lussac (who studied the pressure-temperature relationship) set a solo altitude record that held for 50 years.

At the higher temperature in (b), the density should decrease, which can happen only if there are fewer molecules per liter, so the answer is reasonable.

Comment 1. An *alternative approach* for finding the density of most simple gases, but *at STP only,* is to divide the molar mass by the standard molar volume, 22.4 L:

$$d = \frac{\mathcal{M}}{V} = \frac{44.01 \text{ g/mol}}{22.4 \text{ L/mol}} = 1.96 \text{ g/L}$$

Then, since you know the density at one temperature (0°C), you can find it at any other temperature with the following relationship: $d_1/d_2 = T_2/T_1$.

2. Note that we have different numbers of significant figures for the pressure values. In (a), "1 atm" is part of the definition of STP, so it is an exact number. In (b), we specified "1.00 atm" to allow three significant figures in the answer.

3. Hot-air balloonists have always applied the change in density with temperature. ●

FOLLOW-UP PROBLEM 5.6 Compare the density of CO_2 at 0°C and 380 torr with its density at STP.

The Molar Mass of a Gas

Through another simple rearrangement of the ideal gas law, we can determine the molar mass of an unknown gas or volatile liquid (one that is easily vaporized):

$$n = \frac{m}{\mathcal{M}} = \frac{PV}{RT} \quad \text{so} \quad \mathcal{M} = \frac{mRT}{PV} \quad \text{or} \quad \mathcal{M} = \frac{dRT}{P} \quad (5.10)$$

Notice that this equation is just a rearrangement of Equation 5.9.

The French chemist J. B. A. Dumas (1800–1884) pioneered an ingenious method for finding the molar mass of a volatile liquid. Figure 5.11 shows the apparatus. Place a small volume of the liquid in a preweighed flask of known volume. Close the flask with a stopper that contains a narrow tube and immerse it in a water bath whose fixed temperature exceeds the liquid's boiling point. As the liquid vaporizes, the gas fills the flask and some flows out the tube. When the liquid is gone, the pressure of the gas filling the flask equals the atmospheric pressure. Remove the flask from the water bath and cool it, and the gas condenses to a liquid. Reweigh the flask to obtain the mass of the liquid, *which equals the mass of gas* that remained in the flask.

By this procedure, you have directly measured all the variables needed to calculate the molar mass of the gas: the mass of gas (*m*) occupies the flask volume (*V*) at a pressure (*P*) equal to the barometric pressure and at the temperature (*T*) of the water bath.

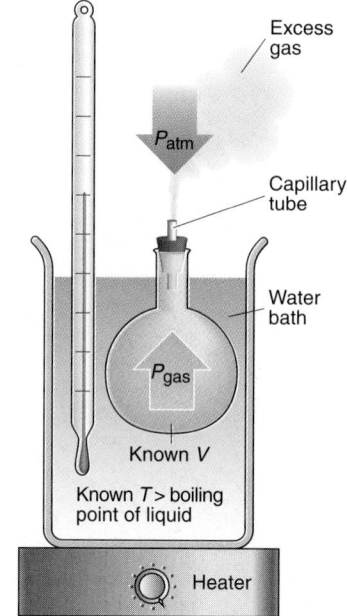

Figure 5.11 Determining the molar mass of an unknown volatile liquid. A small amount of unknown liquid is vaporized, and the gas fills the flask of known volume at the known temperature of the bath. Excess gas escapes through the capillary tube until $P_{gas} = P_{atm}$. When the flask is cooled, the gas condenses, the liquid is weighed, and the ideal gas law is used to calculate $\mathcal{M}$ (see text).

SAMPLE PROBLEM 5.7 Finding the Molar Mass of a Volatile Liquid

Problem An organic chemist isolates from a petroleum sample a colorless liquid with the properties of cyclohexane (C_6H_{12}). She uses the Dumas method and obtains the following data to determine its molar mass:

Volume (*V*) of flask = 213 mL	*T* = 100.0°C *P* = 754 torr
Mass of flask + gas = 78.416 g	Mass of flask = 77.834 g

Is the calculated molar mass consistent with the liquid being cyclohexane?

Plan We are given *V*, *T*, *P*, and mass data and must find the molar mass ($\mathcal{M}$) of the liquid to see if it is close to the $\mathcal{M}$ of cyclohexane. We convert *V* to liters, *T* to kelvins, and *P* to atmospheres, find the mass of gas by subtracting the mass of the empty flask, and use Equation 5.10 to solve for $\mathcal{M}$. Then we compare it with $\mathcal{M}$ based on the formula.

Solution Summary of gas variables:

$$m = 78.416 \text{ g} - 77.834 \text{ g} = 0.582 \text{ g} \qquad P \text{ (atm)} = 754 \text{ torr} \times \frac{1 \text{ atm}}{760 \text{ torr}} = 0.992 \text{ atm}$$

$$V \text{ (L)} = 213 \text{ mL} \times \frac{1 \text{ L}}{1000 \text{ mL}} = 0.213 \text{ L} \qquad T \text{ (K)} = 100.0°C + 273.15 = 373.2 \text{ K}$$

Solving for $\mathcal{M}$:

$$\mathcal{M} = \frac{mRT}{PV} = \frac{0.582 \text{ g} \times 0.0821 \frac{\text{atm·L}}{\text{mol·K}} \times 373.2 \text{ K}}{0.992 \text{ atm} \times 0.213 \text{ L}} = 84.4 \text{ g/mol}$$

Finding $\mathcal{M}$ from the formula:

$$\mathcal{M} \text{ of } C_6H_{12} = (6 \times 12.01 \text{ g/mol}) + (12 \times 1.008 \text{ g/mol}) = 84.16 \text{ g/mol}$$

Within experimental error, the molar mass *is* consistent with the liquid being cyclohexane.

Check Rounding to check the arithmetic, we have

$$\frac{0.6 \text{ g} \times 0.08 \frac{\text{atm·L}}{\text{mol·K}} \times 375 \text{ K}}{1 \text{ atm} \times 0.2 \text{ L}} = 90 \text{ g/mol} \qquad \text{which is close to 84.4 g/mol.}$$

FOLLOW-UP PROBLEM 5.7 At 10.0°C and 102.5 kPa, the density of dry air is 1.26 g/L. What is the average "molar mass" of dry air at these conditions?

The Partial Pressure of a Gas in a Mixture of Gases

Notice that all of the behaviors we've discussed so far were observed from experiments with air, which is a complex mixture of gases. The ideal gas law holds for virtually any gas, whether pure or a mixture, at ordinary conditions for two reasons:

- Gases mix homogeneously (form a solution) in any proportions.
- Each gas in a mixture behaves as if it were the only gas present (assuming no chemical interactions).

Dalton's Law of Partial Pressures The second point above was discovered by John Dalton in his studies of humidity. He observed that when water vapor is added to dry air, the total air pressure increases by an increment equal to the pressure of the water vapor:

$$P_{\text{humid air}} = P_{\text{dry air}} + P_{\text{added water vapor}}$$

In other words, each gas in the mixture exerts a **partial pressure,** a portion of the total pressure of the mixture, that is the same as the pressure it exerts *by itself.* This observation is formulated as **Dalton's law of partial pressures:** *in a mixture of unreacting gases, the total pressure is the sum of the partial pressures of the individual gases:*

$$P_{\text{total}} = P_1 + P_2 + P_3 + \cdots \tag{5.11}$$

As an example, suppose you have a tank of fixed volume that contains nitrogen gas at a certain pressure, and you introduce a sample of hydrogen gas into the tank. Each gas behaves independently, so we can write an ideal gas law expression for each:

$$P_{N_2} = \frac{n_{N_2}RT}{V} \qquad \text{and} \qquad P_{H_2} = \frac{n_{H_2}RT}{V}$$

Because each gas occupies the same total volume and is at the same temperature, the pressure of each gas depends only on its amount, n. Therefore, the total pressure is

$$\begin{aligned}
P_{\text{total}} &= P_{N_2} + P_{H_2} \\
&= \frac{n_{N_2}RT}{V} + \frac{n_{H_2}RT}{V} = \frac{(n_{N_2} + n_{H_2})RT}{V} \\
&= \frac{n_{\text{total}}RT}{V}
\end{aligned}$$

where $n_{\text{total}} = n_{N_2} + n_{H_2}$.

Each component in a mixture contributes a fraction of the total number of moles in the mixture, which is the **mole fraction** (X) of that component. Multiplying X by 100 gives the mole percent. Keep in mind that the sum of the mole fractions of all components in any mixture must be 1, and the sum of the mole percents must be 100%. For N_2, the mole fraction is

$$X_{N_2} = \frac{n_{N_2}}{n_{\text{total}}} = \frac{n_{N_2}}{n_{N_2} + n_{H_2}}$$

Since the total pressure is due to the total number of moles, the partial pressure of gas A is the total pressure multiplied by the mole fraction of A, X_A:

$$P_A = X_A \times P_{\text{total}} \tag{5.12}$$

Equation 5.12 is a very important result. To see that it is valid for the mixture of N_2 and H_2, we recall that $X_{N_2} + X_{H_2} = 1$ and obtain

$$\begin{aligned}
P_{\text{total}} &= P_{N_2} + P_{H_2} \\
&= (X_{N_2} \times P_{\text{total}}) + (X_{H_2} \times P_{\text{total}}) = (X_{N_2} + X_{H_2})P_{\text{total}} \\
&= 1 \times P_{\text{total}}
\end{aligned}$$

SAMPLE PROBLEM 5.8 Applying Dalton's Law of Partial Pressures

Problem In a study of O_2 uptake by muscle at high altitude, a physiologist prepares an atmosphere consisting of 79 mole % N_2, 17 mole % $^{16}O_2$, and 4.0 mole % $^{18}O_2$. (The isotope ^{18}O will be measured to determine the O_2 uptake.) The pressure of the mixture is 0.75 atm to simulate high altitude. Calculate the mole fraction and partial pressure of $^{18}O_2$ in the mixture.

Plan We must find $X_{^{18}O_2}$ and $P_{^{18}O_2}$ from P_{total} (0.75 atm) and the mole % of $^{18}O_2$ (4.0). Dividing the mole % by 100 gives the mole fraction, $X_{^{18}O_2}$. Then, using Equation 5.12, we multiply $X_{^{18}O_2}$ by P_{total} to find $P_{^{18}O_2}$.

Solution Calculating the mole fraction of $^{18}O_2$:

$$X_{^{18}O_2} = \frac{4.0 \text{ mol \% } ^{18}O_2}{100} = \boxed{0.040}$$

Solving for the partial pressure of $^{18}O_2$:

$$P_{^{18}O_2} = X_{^{18}O_2} \times P_{total} = 0.040 \times 0.75 \text{ atm} = \boxed{0.030 \text{ atm}}$$

Check $X_{^{18}O_2}$ is small because the mole % is small, so $P_{^{18}O_2}$ should be small also.

Comment At high altitudes, specialized brain cells that are sensitive to O_2 and CO_2 levels in the blood trigger an increase in rate and depth of breathing for several days, until a person becomes acclimated.

FOLLOW-UP PROBLEM 5.8 A mixture of noble gases consisting of 5.50 g of He, 15.0 g of Ne, and 35.0 g of Kr is placed in a piston-cylinder assembly at STP. Calculate the partial pressure of each gas.

Mole % of $^{18}O_2$
divide by 100
Mole fraction, $X_{^{18}O_2}$
multiply by P_{total}
Partial pressure, $P_{^{18}O_2}$

Collecting a Gas over Water The law of partial pressures is frequently used to determine the yield of a water-insoluble gas formed in a reaction. The gaseous product bubbles through water and is collected into an inverted container, as shown in Figure 5.12. The water vapor that mixes with the gas contributes a portion of the total pressure, called the *vapor pressure*, which depends only on the water temperature.

In order to determine the yield of gaseous product, we find the appropriate vapor pressure value from a list, such as the one in Table 5.3, and subtract it from the total gas pressure (corrected to barometric pressure) to get the partial pressure of the gaseous product. With V and T known, we can calculate the amount of product.

Table 5.3 Vapor Pressure of Water (P_{H_2O}) at Different T

T (°C)	P (torr)
0	4.6
5	6.5
10	9.2
11	9.8
12	10.5
13	11.2
14	12.0
15	12.8
16	13.6
18	15.5
20	17.5
22	19.8
24	22.4
26	25.2
28	28.3
30	31.8
35	42.2
40	55.3
45	71.9
50	92.5
55	118.0
60	149.4
65	187.5
70	233.7
75	289.1
80	355.1
85	433.6
90	525.8
95	633.9
100	760.0

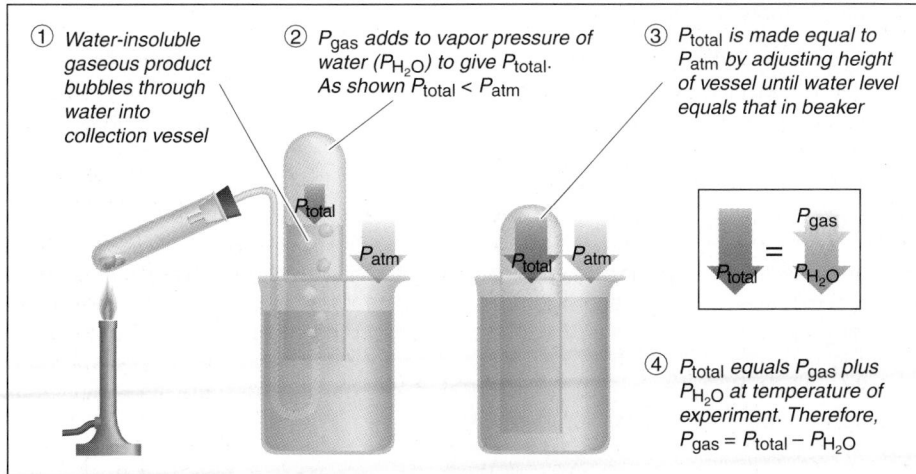

① Water-insoluble gaseous product bubbles through water into collection vessel

② P_{gas} adds to vapor pressure of water (P_{H_2O}) to give P_{total}. As shown $P_{total} < P_{atm}$

③ P_{total} is made equal to P_{atm} by adjusting height of vessel until water level equals that in beaker

④ P_{total} equals P_{gas} plus P_{H_2O} at temperature of experiment. Therefore, $P_{gas} = P_{total} - P_{H_2O}$

$P_{total} = P_{gas}$, P_{H_2O}

Figure 5.12 Collecting a water-insoluble gaseous reaction product and determining its pressure.

SAMPLE PROBLEM 5.9 Calculating the Amount of Gas Collected over Water

Problem Acetylene (C_2H_2), an important fuel in welding, is produced in the laboratory when calcium carbide (CaC_2) reacts with water:

$$CaC_2(s) + 2H_2O(l) \longrightarrow C_2H_2(g) + Ca(OH)_2(aq)$$

For a sample of acetylene that is collected over water, the total gas pressure (adjusted to barometric pressure) is 738 torr and the volume is 523 mL. At the temperature of the gas (23°C), the vapor pressure of water is 21 torr. How many grams of acetylene are collected?

Plan In order to find the mass of C_2H_2, we first need to find the number of moles of C_2H_2, $n_{C_2H_2}$, which we can obtain from the ideal gas law by calculating $P_{C_2H_2}$. The barometer reading gives us P_{total}, which is the sum of $P_{C_2H_2}$ and P_{H_2O}, and we are given P_{H_2O}, so we subtract to find $P_{C_2H_2}$. We are also given V and T, so we convert to consistent units, and find $n_{C_2H_2}$ from the ideal gas law. Then we convert moles to grams using the molar mass from the formula.

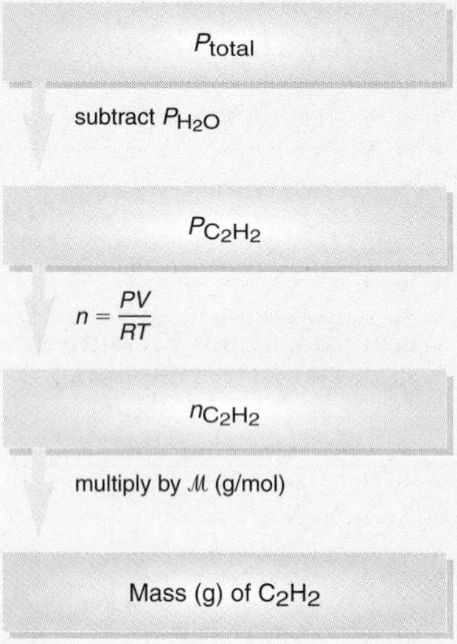

Solution Summary of gas variables:

$$P_{C_2H_2} \text{ (torr)} = P_{total} - P_{H_2O} = 738 \text{ torr} - 21 \text{ torr} = 717 \text{ torr}$$

$$P_{C_2H_2} \text{ (atm)} = 717 \text{ torr} \times \frac{1 \text{ atm}}{760 \text{ torr}} = 0.943 \text{ atm}$$

$$V \text{ (L)} = 523 \text{ mL} \times \frac{1 \text{ L}}{1000 \text{ mL}} = 0.523 \text{ L}$$

$$T \text{ (K)} = 23°C + 273.15 = 296 \text{ K}$$

$$n_{C_2H_2} = \text{unknown}$$

Solving for $n_{C_2H_2}$:

$$n_{C_2H_2} = \frac{PV}{RT} = \frac{0.943 \text{ atm} \times 0.523 \text{ L}}{0.0821 \dfrac{\text{atm·L}}{\text{mol·K}} \times 296 \text{ K}} = 0.0203 \text{ mol}$$

Converting $n_{C_2H_2}$ to mass:

$$\text{Mass (g) of } C_2H_2 = 0.0203 \text{ mol } C_2H_2 \times \frac{26.04 \text{ g } C_2H_2}{1 \text{ mol } C_2H_2}$$

$$= \boxed{0.529 \text{ g } C_2H_2}$$

Check Rounding to one significant figure, a quick arithmetic check for n gives

$$n \approx \frac{1 \text{ atm} \times 0.5 \text{ L}}{0.08 \dfrac{\text{atm·L}}{\text{mol·K}} \times 300 \text{ K}} = 0.02 \text{ mol} \approx 0.0203 \text{ mol}$$

Comment The C_2^{2-} ion (called the *carbide*, or *acetylide, ion*) is an interesting anion. It is simply $^-C{\equiv}C^-$, which acts as a base in water, removing an H^+ ion from two H_2O molecules to form acetylene, $H-C{\equiv}C-H$.

FOLLOW-UP PROBLEM 5.9 A small piece of zinc reacts with dilute HCl to form H_2, which is collected over water at 16°C into a large flask. The total pressure is adjusted to barometric pressure (752 torr), and the volume is 1495 mL. Use Table 5.3 to help calculate the partial pressure and mass of H_2.

Sidebar flowchart

P_{total}

subtract P_{H_2O}

$P_{C_2H_2}$

$n = \dfrac{PV}{RT}$

$n_{C_2H_2}$

multiply by $\mathcal{M}$ (g/mol)

Mass (g) of C_2H_2

SECTION SUMMARY

The ideal gas law can be rearranged to calculate the density and molar mass of a gas. In a mixture of gases, each component contributes its own partial pressure to the total pressure (Dalton's law of partial pressures). The mole fraction of each component is the ratio of its partial pressure to the total pressure. When a gas is in contact with water, the total pressure is the sum of the gas pressure and the vapor pressure of water at the given temperature.

5.5 THE IDEAL GAS LAW AND REACTION STOICHIOMETRY

In Chapters 3 and 4, we encountered many reactions that involved gases as reactants (e.g., combustion with O_2) or as products (e.g., acid treatment of a carbonate). From the balanced equation, we used stoichiometrically equivalent molar ratios to calculate the amounts (moles) of reactants and products and converted these quantities into masses, numbers of molecules, or solution volumes (see Figures 3.11 and 3.14). Figure 5.13 shows how you can expand your problem-solving repertoire by using the ideal gas law to convert between gas variables (P, T, and V) and amounts (moles) of gaseous reactants and products. In effect, you combine a gas law problem with a stoichiometry problem. This approach reflects more realistic situations because we usually measure gas volume, temperature, and pressure rather than mass.

Figure 5.13 **Summary of the stoichiometric relationships among the amount (mol, n) of gaseous reactant or product and the gas variables pressure (P), volume (V), and temperature (T).**

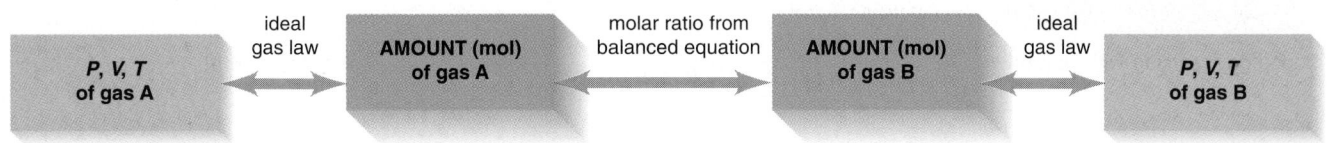

SAMPLE PROBLEM 5.10 Using Gas Variables to Find Amounts of Reactants or Products

Problem A laboratory-scale method for reducing a metal oxide is to heat it with H_2. The pure metal and H_2O are products. What volume of H_2 at 765 torr and 225°C is needed to form 35.5 g of Cu from copper(II) oxide?

Plan This is a stoichiometry *and* gas law problem. To find V_{H_2}, we first need n_{H_2}. We write and balance the equation. Next, we convert the given mass of Cu (35.5 g) to amount (mol) and use the molar ratio to find moles of H_2 needed (stoichiometry portion). Then, we use the ideal gas law to convert moles of H_2 to liters (gas law portion). A roadmap is shown, but you are familiar with all the steps.

Solution Writing the balanced equation:

$$CuO(s) + H_2(g) \longrightarrow Cu(s) + H_2O(g)$$

Calculating n_{H_2}:

$$n_{H_2} = 35.5 \text{ g Cu} \times \frac{1 \text{ mol Cu}}{63.55 \text{ g Cu}} \times \frac{1 \text{ mol } H_2}{1 \text{ mol Cu}} = 0.559 \text{ mol } H_2$$

Summary of other gas variables:

$$V = \text{unknown} \qquad P \text{ (atm)} = 765 \text{ torr} \times \frac{1 \text{ atm}}{760 \text{ torr}} = 1.01 \text{ atm}$$

$$T \text{ (K)} = 225°C + 273.15 = 498 \text{ K}$$

Solving for V_{H_2}:

$$V = \frac{nRT}{P} = \frac{0.559 \text{ mol} \times 0.0821 \frac{\text{atm·L}}{\text{mol·K}} \times 498 \text{ K}}{1.01 \text{ atm}} = \boxed{22.6 \text{ L}}$$

Check One way to check the answer is to compare it with the molar volume of an ideal gas at STP (22.4 L at 273.15 K and 1 atm). Since 1 mol of H_2 at STP occupies about 22 L, about 0.5 mol would occupy about 11 L. T is about twice 273 K, so V should be about twice 11 L. In this case, the decrease in n and increase in T nearly cancel each other.

Comment The main point here is that the stoichiometry provides one gas variable (n), two more are given, and the ideal gas law is used to find the fourth.

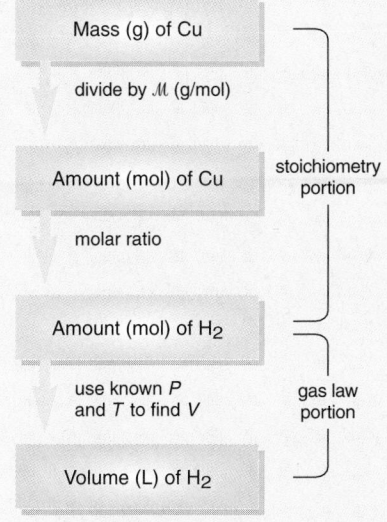

FOLLOW-UP PROBLEM 5.10 Sulfuric acid reacts with sodium chloride to form aqueous sodium sulfate and hydrogen chloride gas. How many milliliters of gas form at STP when 0.117 kg of sodium chloride reacts with excess sulfuric acid?

Chlorine gas reacting with potassium

SAMPLE PROBLEM 5.11 Using the Ideal Gas Law in a Limiting-Reactant Problem

Problem The alkali metals [Group 1A(1)] react with the halogens [Group 7A(17)] to form ionic metal halides. What mass of potassium chloride forms when 5.25 L of chlorine gas at 0.950 atm and 293 K reacts with 17.0 g of potassium?

Plan The only difference between this and previous limiting-reactant problems (see Sample Problem 3.10, p. 111) is that here we use the ideal gas law to find the amount (n) of gaseous reactant from the known V, P, and T. We first write the balanced equation and then use it to find the limiting reactant and the amount of product.

Solution Writing the balanced equation:

$$2K(s) + Cl_2(g) \longrightarrow 2KCl(s)$$

Summary of gas variables:

$$P = 0.950 \text{ atm} \qquad V = 5.25 \text{ L}$$
$$T = 293 \text{ K} \qquad n = \text{unknown}$$

Solving for n_{Cl_2}:

$$n_{Cl_2} = \frac{PV}{RT} = \frac{0.950 \text{ atm} \times 5.25 \text{ L}}{0.0821 \dfrac{\text{atm·L}}{\text{mol·K}} \times 293 \text{ K}} = 0.207 \text{ mol}$$

Converting from grams of potassium (K) to moles:

$$\text{Moles of K} = 17.0 \text{ g K} \times \frac{1 \text{ mol K}}{39.10 \text{ g K}} = 0.435 \text{ mol K}$$

Determining the limiting reactant: If Cl_2 is limiting,

$$\text{Moles of KCl} = 0.207 \text{ mol Cl}_2 \times \frac{2 \text{ mol KCl}}{1 \text{ mol Cl}_2} = 0.414 \text{ mol KCl}$$

If K is limiting,

$$\text{Moles of KCl} = 0.435 \text{ mol K} \times \frac{2 \text{ mol KCl}}{2 \text{ mol K}} = 0.435 \text{ mol KCl}$$

Cl_2 is the limiting reactant because it forms less KCl.
Converting from moles of KCl to grams:

$$\text{Mass (g) of KCl} = 0.414 \text{ mol KCl} \times \frac{74.55 \text{ g KCl}}{1 \text{ mol KCl}}$$
$$= \boxed{30.9 \text{ g KCl}}$$

Check The gas law calculation seems correct. At STP, 22 L of Cl_2 gas contains about 1 mol, so a 5-L volume would contain a bit less than 0.25 mol of Cl_2. Moreover, since P (in numerator) is slightly lower than STP and T (in denominator) is slightly higher, these should lower the calculated n further below the ideal value. The mass of KCl seems correct: less than 0.5 mol of KCl gives $< 0.5 \times \mathcal{M}$ (30.9 g $< 0.5 \times 75$ g).

FOLLOW-UP PROBLEM 5.11 Ammonia and hydrogen chloride gases react to form solid ammonium chloride. A 10.0-L reaction flask contains ammonia at 0.452 atm and 22°C, and 155 mL of hydrogen chloride gas at 7.50 atm and 271 K is introduced. After the reaction occurs and the temperature returns to 22°C, what is the pressure inside the flask? (Neglect the volume of the solid product.)

SECTION SUMMARY

By expressing the amount (mol) of gaseous reactant or product in terms of the other gas variables, we can solve stoichiometry problems for reactions involving gases.

5.6 THE KINETIC-MOLECULAR THEORY: A MODEL FOR GAS BEHAVIOR

So far we have discussed observations of macroscopic samples of gas: decreasing cylinder volume, increasing tank pressure, and so forth. This section presents the central model that explains macroscopic gas behavior at the level of individual particles: the **kinetic-molecular theory.** The theory draws conclusions through rigorous mathematical derivations, but our discussion here will be largely qualitative.

How the Kinetic-Molecular Theory Explains the Gas Laws

Developed by some of the great scientists of the 19th century, most notably James Clerk Maxwell and Ludwig Boltzmann, the kinetic-molecular theory was able to explain the gas laws that some of the great scientists of the century before had arrived at empirically.

Questions Concerning Gas Behavior To model gas behavior, we must rationalize certain questions at the molecular level:

1. *Origin of pressure.* Pressure is a measure of the force a gas exerts on a surface. How do individual gas particles create this force?
2. *Boyle's law.* A change in gas pressure in one direction causes a change in gas volume in the other. What happens to the particles when external pressure compresses the gas volume? And why aren't liquids and solids compressible?
3. *Dalton's law.* The pressure of a gas mixture is the sum of the pressures of the individual gases. Why does each gas contribute to the total pressure in proportion to its mole fraction?
4. *Charles's law.* A change in temperature is accompanied by a corresponding change in volume. What effect does higher temperature have on gas particles that increases the volume—or increases the pressure if volume is fixed? This question raises a more fundamental one: what does temperature measure on the molecular scale?
5. *Avogadro's law.* Gas volume (or pressure) depends on the number of moles present, not on the nature of the particular gas. But shouldn't 1 mol of larger molecules occupy more space than 1 mol of smaller molecules? And why doesn't 1 mol of heavier molecules exert more pressure than 1 mol of lighter molecules?

Postulates of the Kinetic-Molecular Theory The theory is based on three postulates (assumptions):

Postulate 1. *Particle volume.* A gas consists of a large collection of individual particles. The volume of an individual particle is *extremely* small compared with the volume of the container. In essence, the model pictures gas particles as having mass but no volume.

Postulate 2. *Particle motion.* Gas particles are in constant, random, straight-line motion, except when they collide with the container walls or with each other.

Postulate 3. *Particle collisions.* Collisions are *elastic,* which means that, somewhat like minute billiard balls, the colliding molecules exchange energy but do not lose any through friction. Thus, their total kinetic energy (E_k) is constant. Between collisions, the molecules do not influence each other at all.

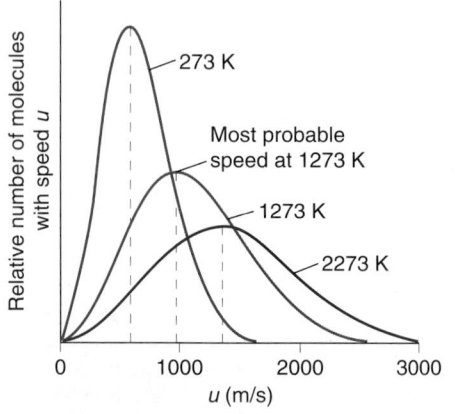

Figure 5.14 **Distribution of molecular speeds at three temperatures.** At a given temperature, a plot of the relative number of N_2 molecules vs. molecular speed (u) results in a skewed bell-shaped curve, with the most probable speed at the peak. Note that the curves spread at higher temperatures and the most probable speed is directly proportional to the temperature.

Picture the scene envisioned by the postulates: countless particles moving in every direction, smashing into the container walls and one another. Any given particle changes its speed with each collision, perhaps one instant standing nearly still from a head-on crash and the next instant zooming away from a smash on the side. Thus, *the particles have an average speed*, with most moving near the average speed, some moving faster, and some slower.

Figure 5.14 depicts the distribution of molecular speeds (u) for N_2 gas at three temperatures. The curves flatten and spread at higher temperatures. Note especially that the *most probable speed (the peak of each curve) increases as the temperature increases*. This increase occurs because the average kinetic energy of the molecules ($\overline{E_k}$; the overbar indicates the average value of a quantity), which incorporates the most probable speed, is proportional to the absolute temperature: $\overline{E_k} \propto T$, or $\overline{E_k} = c \times T$, where c is a constant that is the same for any gas. (We'll return to this equation shortly.) Thus, a major conclusion based on the distribution of speeds, which arises directly from postulate 3, is that *at a given temperature, all gases have the same average kinetic energy.*

A Molecular View of the Gas Laws Let's continue visualizing the particles to see how the theory explains the macroscopic behavior of gases and answers the questions posed above:

1. *Origin of pressure.* When a moving object collides with a surface, it exerts a force. We conclude from postulate 2, which describes particle motion, that when a particle collides with the container wall, it too exerts a force. Many such collisions result in the observed pressure. The greater the number of molecules in a given container, the more frequently they collide with the walls, and the greater the pressure is.

2. *Boyle's law* ($V \propto 1/P$). Gas molecules are points of mass with empty space between them (postulate 1), so as the pressure exerted *on* the sample increases at constant temperature, the distance between molecules decreases, and the sample volume decreases. The pressure exerted *by* the gas increases simultaneously because in a smaller volume of gas, there are shorter distances between gas molecules and the walls and between the walls themselves; thus, collisions are more frequent (Figure 5.15). The fact that liquids and solids cannot be compressed very much means that there is very little free space between the molecules.

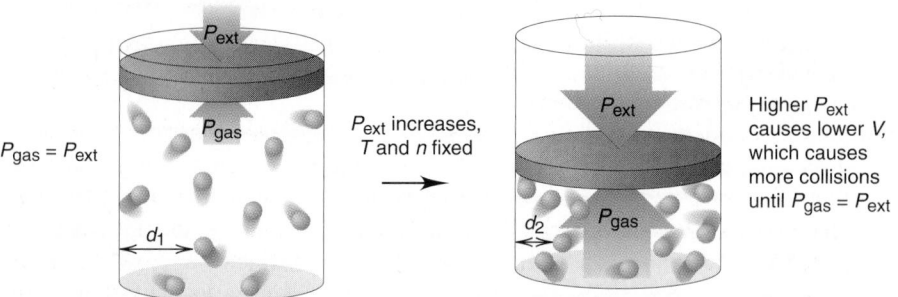

Figure 5.15 **A molecular description of Boyle's law.** At a given T, gas molecules collide with the walls across an average distance (d_1) and give rise to a pressure (P_{gas}) that equals the external pressure (P_{ext}). If P_{ext} increases, V decreases, so the average distance between a molecule and the walls is shorter ($d_2 < d_1$). Molecules strike the walls more often, and P_{gas} increases until it again equals P_{ext}. Thus, V decreases when P increases.

3. *Dalton's law of partial pressures* ($P_{\text{total}} = P_A + P_B$). Adding a given amount of gas A to a given amount of gas B causes an increase in total number of molecules in proportion to the amount of A that was added. This increase causes a corresponding increase in the number of collisions per second with the walls (postulate 2), which causes a corresponding increase in the pressure (Figure 5.16). Thus, each gas exerts a fraction of the total pressure based on the fraction of molecules (or fraction of moles; that is, the mole fraction) of that gas in the mixture.

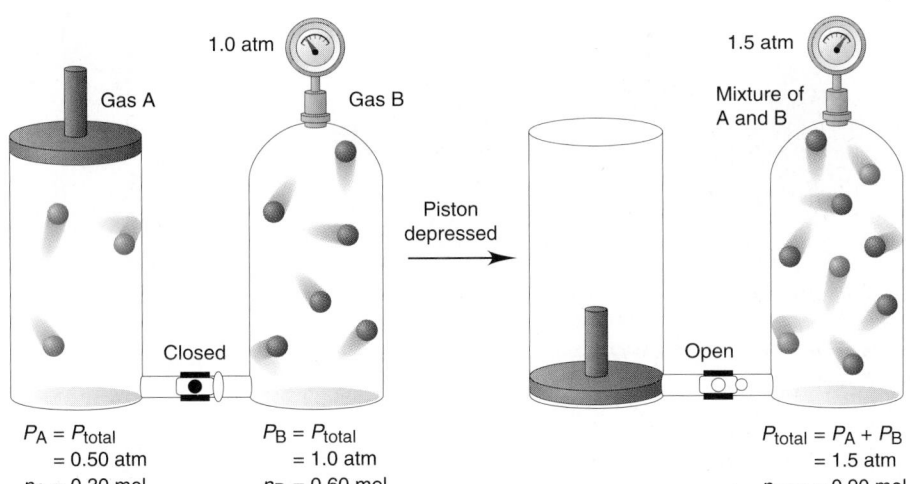

$P_A = P_{\text{total}}$
$= 0.50$ atm
$n_A = 0.30$ mol

$P_B = P_{\text{total}}$
$= 1.0$ atm
$n_B = 0.60$ mol

$P_{\text{total}} = P_A + P_B$
$= 1.5$ atm
$n_{\text{total}} = 0.90$ mol
$X_A = 0.33$ mol
$X_B = 0.67$ mol

Figure 5.16 A molecular description of Dalton's law of partial pressures. A piston-cylinder assembly containing 0.30 mol of gas A at 0.50 atm is connected to a tank of fixed volume containing 0.60 mol of gas B at 1.0 atm. When the piston is depressed at fixed temperature, gas A is forced into the tank of gas B, the gases mix, and the new total pressure, 1.5 atm, equals the sum of the partial pressures, which is related to the new total amount of gas, 0.90 mol. Thus, each gas undergoes a fraction of the total collisions related to its fraction of the total number of molecules (moles), which is equal to its mole fraction.

4. *Charles's law* ($V \propto T$). As the temperature increases, the most probable molecular speed and average kinetic energy increase (postulate 3). Thus, the molecules hit the walls more frequently *and* more energetically. A higher frequency of collisions causes higher internal pressure. As a result, the walls move outward, which increases the volume and restores the starting pressure (Figure 5.17).

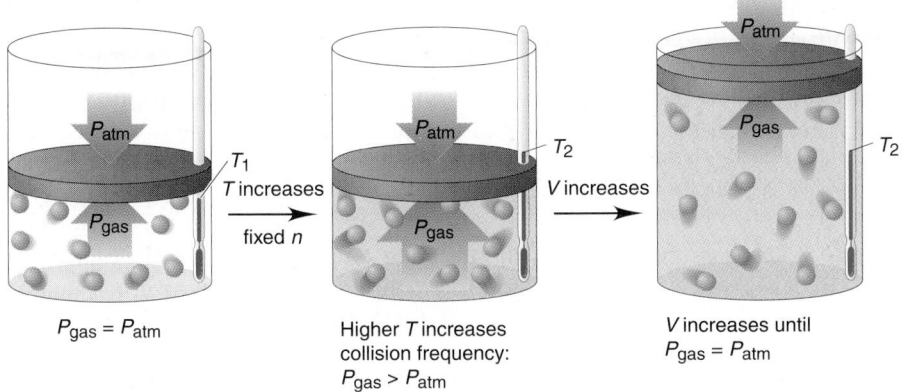

$P_{\text{gas}} = P_{\text{atm}}$

Higher T increases collision frequency: $P_{\text{gas}} > P_{\text{atm}}$

V increases until $P_{\text{gas}} = P_{\text{atm}}$

Figure 5.17 A molecular description of Charles's law. At a given temperature (T_1), $P_{\text{gas}} = P_{\text{atm}}$. When the gas is heated to T_2, the molecules move faster and collide with the walls more often, which increases P_{gas}. This increases V, so the molecules collide less often until P_{gas} again equals P_{atm}. Thus, V increases when T increases.

5. *Avogadro's law* ($V \propto n$). Adding more molecules to a container increases the total number of collisions with the walls and, therefore, the internal pressure. As a result, the volume expands until the number of collisions per unit of wall area is the same as it was before the addition (Figure 5.18).

Figure 5.18 A molecular description of Avogadro's law. At a given T, a certain amount (n) of gas gives rise to a pressure (P_{gas}) equal to P_{atm}. When more gas is added, n increases, so collisions with the walls are more frequent and P_{gas} increases. This leads to an increase in V until $P_{gas} = P_{atm}$ again. Thus, V increases when n increases.

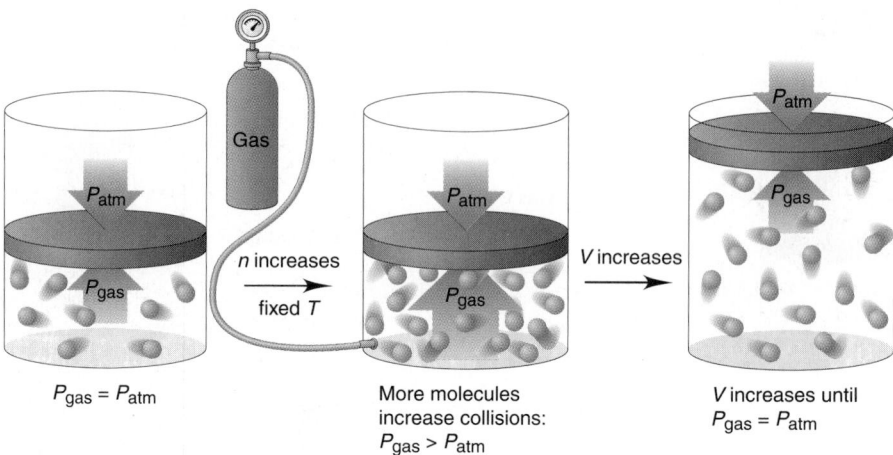

$P_{gas} = P_{atm}$

More molecules increase collisions: $P_{gas} > P_{atm}$

V increases until $P_{gas} = P_{atm}$

We still need to explain why equal numbers of molecules of two different gases, such as O_2 and H_2, occupy the same volume. To do so, we must first understand why the heavier O_2 particles *do not* hit the container walls with more energy than the lighter H_2 particles. The reason involves the meaning of kinetic energy. Recall from Chapter 1 that the kinetic energy of an object is the energy associated with its motion. This energy is related to the mass *and* the speed of the object:

$$E_k = \tfrac{1}{2} \text{ mass} \times \text{speed}^2$$

This equation shows that if a heavy object and a light object have the same kinetic energy, *the heavy object must be moving more slowly.* For a large population of molecules, the average kinetic energy is

$$\overline{E_k} = \tfrac{1}{2}m\overline{u^2}$$

where m is the molecular mass and $\overline{u^2}$ is the average of the squares of the molecular speeds. The square root of $\overline{u^2}$ is called the root-mean-square speed, or **rms speed** (u_{rms}). *A molecule moving at this speed has the average kinetic energy.* The rms speed is somewhat higher than the most probable speed, but the speeds are proportional to each other and we will use them interchangeably. The rms speed is related to the temperature and the molar mass as follows:

$$u_{rms} = \sqrt{\frac{3RT}{\mathcal{M}}} \tag{5.13}$$

where R is the gas constant, T is the absolute temperature, and $\mathcal{M}$ is the molar mass. (Because we want u in m/s, and R includes the joule, which has units of kg·m^2/s^2, we use the value 8.314 J/mol·K for R and express $\mathcal{M}$ in kg/mol.)

Postulate 3 leads to the conclusion that different gases at the same temperature have the same average kinetic energy. Therefore, Avogadro's law requires that, on average, *molecules with a higher mass have a lower speed.* In other words, at the same temperature, O_2 molecules move more slowly, on average, than H_2 molecules. Figure 5.19 shows that, in general, at the same temperature, lighter gases have higher speeds. This means that H_2 molecules collide with the

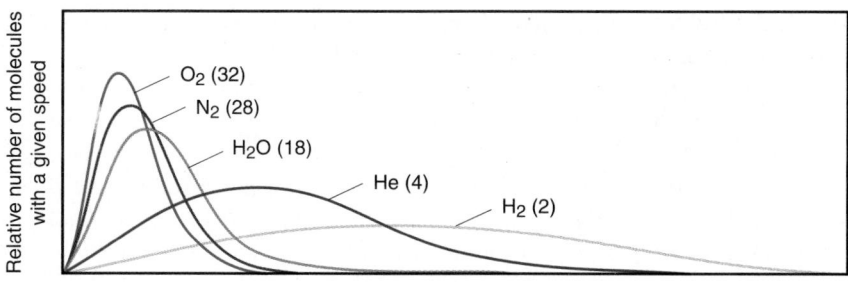

Figure 5.19 **Relationship between molar mass and molecular speed.** At a given temperature, gases with lower molar masses (numbers in parentheses) have higher most probable speeds (peak of each curve).

wall more often than do O_2 molecules, but each collision has less force. Because, at the same T, both light and heavy molecules hit the walls with the same average kinetic energy, they have the same pressure and, thus, the same volume.

The Meaning of Temperature Earlier we said that the average kinetic energy of a particle was equal to the absolute temperature times a constant, that is, $\overline{E_k}$ = constant $\times$ T. A derivation of the full relationship gives the following equation:

$$\overline{E_k} = \tfrac{3}{2}\left(\frac{R}{N_A}\right)T$$

where R is the gas constant and N_A is Avogadro's number. This equation expresses the important point that *temperature is related to the average energy of molecular motion.* Note that it is not related to the *total* energy, which depends on the size of the sample, but to the *average* energy: as T increases, $\overline{E_k}$ increases. What occurs when, say, a beaker of water containing a thermometer is heated over a flame? The rapidly moving gas particles in the flame collide more energetically with the atoms of the beaker, transferring some of their kinetic energy. The atoms vibrate faster and transfer kinetic energy to the molecules of water, which move faster and transfer kinetic energy to the atoms in the thermometer bulb. These, in turn, transfer kinetic energy to the atoms of mercury. They collide with each other more frequently and more energetically, which causes the mercury volume to expand and the level to rise. In the macroscopic world, we see this as a temperature increase; in the molecular world, it is a sequence of kinetic energy transfers from higher energy particles to lower energy particles, such that each group of particles increases its average kinetic energy.

Effusion and Diffusion

The movement of gases, either through one another or into regions of very low pressure, has many important applications.

The Process of Effusion One of the early triumphs of the kinetic-molecular theory was an explanation of **effusion,** the process by which a gas escapes from its container through a tiny hole into an evacuated space. In 1846, Thomas Graham studied this process and concluded that the effusion rate was inversely proportional to the square root of the gas density. The effusion rate is the number of moles (or molecules) of gas effusing per unit time. Since density is directly proportional to molar mass, we state **Graham's law of effusion** as follows: *the rate of effusion of a gas is inversely proportional to the square root of its molar mass,*

$$\text{Rate of effusion} \propto \frac{1}{\sqrt{\mathcal{M}}}$$

Preparing Nuclear Fuel One of the most important applications of Graham's law is the enrichment of nuclear reactor fuel: separating nonfissionable, more abundant ^{238}U from fissionable ^{235}U to increase the proportion of ^{235}U in the mixture. Since the two isotopes have identical chemical properties, they are separated by differences in the effusion rates of their gaseous compounds. Uranium ore is converted to gaseous UF_6 (a mixture of $^{238}UF_6$ and $^{235}UF_6$), which is pumped through a series of chambers with porous barriers. Because they move very slightly faster, molecules of $^{235}UF_6$ ($\mathcal{M} = 349.03$) effuse through each barrier at a rate 1.0043 times greater than that of molecules of $^{238}UF_6$ ($\mathcal{M} = 352.04$). Many passes are made, each increasing the fraction of $^{235}UF_6$ until a mixture is obtained that contains enough $^{235}UF_6$. This isotope-enrichment process was developed during the latter years of World War II and produced enough ^{235}U for two of the world's first three atomic bombs.

Argon (Ar) is lighter than krypton (Kr), so it effuses faster, assuming equal pressures of the two gases. Thus, the ratio of the rates is

$$\frac{\text{Rate}_{Ar}}{\text{Rate}_{Kr}} = \frac{\sqrt{\mathcal{M}_{Kr}}}{\sqrt{\mathcal{M}_{Ar}}} \qquad \text{or, in general,} \qquad \frac{\text{rate}_A}{\text{rate}_B} = \frac{\sqrt{\mathcal{M}_B}}{\sqrt{\mathcal{M}_A}} \qquad \textbf{(5.14)}$$

The kinetic-molecular theory explains that, at a given temperature and pressure, *the gas with the lower molar mass effuses faster because the most probable speed of its molecules is higher; therefore, more molecules escape per unit time.*

SAMPLE PROBLEM 5.12 Applying Graham's Law of Effusion

Problem Calculate the ratio of the effusion rates of helium and methane (CH_4).
Plan The effusion rate is inversely proportional to $\sqrt{\mathcal{M}}$, so we find the molar mass of each substance from the formula and take its square root. The inverse of the ratio of the square roots is the ratio of the effusion rates.
Solution

$$\mathcal{M} \text{ of } CH_4 = 16.04 \text{ g/mol} \qquad \mathcal{M} \text{ of } He = 4.003 \text{ g/mol}$$

Calculating the ratio of the effusion rates:

$$\frac{\text{Rate}_{He}}{\text{Rate}_{CH_4}} = \sqrt{\frac{\mathcal{M}_{CH_4}}{\mathcal{M}_{He}}} = \sqrt{\frac{16.04 \text{ g/mol}}{4.003 \text{ g/mol}}} = \sqrt{4.007}$$
$$= 2.002$$

Check A ratio >1 makes sense because the lighter He should effuse faster than the heavier CH_4. Because the molar mass of He is about one-fourth that of CH_4, He should effuse about twice as fast (the inverse of $\sqrt{\frac{1}{4}}$).

FOLLOW-UP PROBLEM 5.12 If it takes 1.25 min for 0.010 mol of He to effuse, how long will it take for the same amount of ethane (C_2H_6) to effuse?

Graham's law is also used to determine the molar mass of an unknown gas, X, by comparing its effusion rate with that of a known gas, such as He:

$$\frac{\text{Rate}_X}{\text{Rate}_{He}} = \frac{\sqrt{\mathcal{M}_{He}}}{\sqrt{\mathcal{M}_X}}$$

Squaring both sides and solving for the molar mass of X gives

$$\mathcal{M}_X = \mathcal{M}_{He} \times \left(\frac{\text{rate}_{He}}{\text{rate}_X}\right)^2$$

The Process of Diffusion Closely related to effusion is the process of gaseous **diffusion,** the movement of one gas through another. Diffusion rates are also described generally by Graham's law:

$$\text{Rate of diffusion} \propto \frac{1}{\sqrt{\mathcal{M}}}$$

For two gases at equal pressures, such as NH_3 and HCl, moving through another gas or a mixture of gases, such as air, we find

$$\frac{\text{Rate}_{NH_3}}{\text{Rate}_{HCl}} = \sqrt{\frac{\mathcal{M}_{HCl}}{\mathcal{M}_{NH_3}}}$$

The reason for this dependence on molar mass is the same as for effusion rates: lighter molecules have higher molecular speeds than heavier molecules, so they move farther in a given amount of time.

If gas molecules move at hundreds of meters per second at ordinary temperatures (see Figure 5.14), why does it take a second or two after you open a bottle of perfume to smell the fragrance? Although convection plays an important role, a molecule moving by diffusion does not travel very far before it collides with a molecule in the air. As you can see from Figure 5.20, the path of each molecule is tortuous. Imagine walking through an empty room, and then imagine walking through a room crowded with other moving people.

Diffusion also occurs in liquids (and even to a small extent in solids). However, because the distances between molecules are much shorter in a liquid than in a gas, collisions are much more frequent; thus diffusion is *much* slower. Diffusion of a gas through a liquid is a vital process in biological systems. For example, it plays a key part in the movement of O_2 from lungs to blood. Many organisms have evolved elaborate ways to speed the diffusion of nutrients (for example, sugar and metal ions) through their cell membranes and to slow, or even stop, the diffusion of toxins.

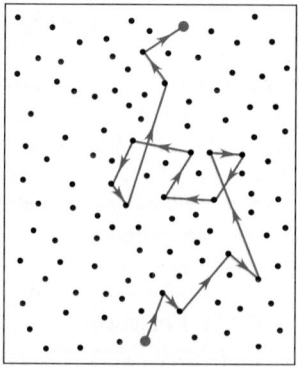

Figure 5.20 Diffusion of a gas particle through a space filled with other particles. In traversing a space, a gas molecule collides with many other molecules, which gives it a tortuous path. For clarity, the path of only one particle *(red dot)* is shown *(red lines)*.

The Chaotic World of Gases: Mean Free Path and Collision Frequency

Refinements of the basic kinetic-molecular theory provide a view into the amazing, chaotic world of gas molecules. Let's follow an "average" N_2 molecule in a room at 20°C and 1 atm pressure—perhaps the room you are in now.

Distribution of Molecular Speeds Our molecule is hurtling through the room at an average speed of 0.47 km/s, or nearly 1100 mi/h (rms speed = 0.51 km/s), and continually changing speed as it collides with other molecules. At any instant, it may be traveling at 2500 mi/h or standing still as it collides head on, but these extreme speeds are *much* less likely than the most probable one (see Figure 5.19).

Mean Free Path From a molecule's diameter, we can use the kinetic-molecular theory to obtain the **mean free path**, the average distance the molecule travels between collisions at a given temperature and pressure. Our average N_2 molecule $(3.7 \times 10^{-10}$ m in diameter) travels 6.6×10^{-8} m before smashing into a fellow traveler, which means it travels about 180 molecular diameters between collisions. (An analogy in the macroscopic world would be an N_2 molecule the size of a billiard ball traveling about 30 ft before colliding with another.) Therefore, even though gas molecules are *not* points of mass, a gas sample *is* mostly empty space. Mean free path is a key factor in the rate of diffusion and the rate of heat flow through a gas.

Collision Frequency Divide the most probable speed (distance per second) by the mean free path (distance per collision) and you obtain the **collision frequency,** the average number of collisions per second that each molecule undergoes: ●

$$\text{Collision frequency} = \frac{4.7 \times 10^2 \text{ m/s}}{6.6 \times 10^{-8} \text{ m/collision}} = 7.1 \times 10^9 \text{ collisions/s}$$

Distribution of speed (and kinetic energy) and collision frequency are essential ideas for understanding the speed of a chemical reaction, as you'll see in Chapter 16. As the upcoming Chemical Connections essay shows, the kinetic-molecular theory applies directly to the behavior of our planet's atmosphere.

● **Danger on Molecular Highways** To give you some idea of how astounding events are in the molecular world, we can express the collision frequency of a molecule in terms of a common experience in the macroscopic world: driving your compact car on the highway. Since a car is much larger than an N_2 molecule, to match the collision frequency of the molecule you would have to travel 2.8 billion mi/s (an impossibility, given that it is much faster than the speed of light) and would smash into another car every 700 yd!

SECTION SUMMARY

The kinetic-molecular theory postulates that gas molecules take up a negligible portion of the gas volume, move in straight-line paths between elastic collisions, and have average kinetic energies proportional to the absolute temperature. This theory explains the gas laws in terms of changes in distances between molecules and the container walls and changes in molecular speed. Temperature is a measure of the average kinetic energy of molecules. Effusion and diffusion rates are inversely proportional to the square root of the molar mass (Graham's law) because they are directly proportional to molecular speed. Molecular motion is characterized by a temperature-dependent most probable speed within a range of speeds, a mean free path, and a collision frequency. The atmosphere is a complex mixture of gases that exhibits variations in pressure, temperature, and composition with altitude.

Chemical Connections

Chemistry in Planetary Science

Structure and Composition of the Earth's Atmosphere

An **atmosphere** is the envelope of gases that extends continuously from a planet's surface outward, eventually thinning to a point at which it is indistinguishable from interplanetary space. On Earth, complex changes in pressure, temperature, and composition occur within this mixture of gases, and the present atmosphere is very different from the one that existed during our planet's early history.

Variation in Pressure

Since gases are compressible (Boyle's law), the pressure of the atmosphere *decreases* smoothly with distance from the surface, with a more rapid decrease at lower altitudes (Figure B5.1). Although no specific boundary delineates the outermost fringe of the atmosphere, the density and composition at around 10,000 km from the surface are identical with those of outer space. About 99% of the atmosphere's mass lies within 30 km of the surface, and 75% lies within the lowest 11 km.

Variation in Temperature

Unlike the change in pressure, temperature does *not* decrease smoothly with altitude, and the atmosphere is usually classified into regions based on the direction of temperature change (Figure B5.1). In the *troposphere,* which includes the region from the surface to around 11 km, temperatures *drop* 7°C per kilometer to −55°C (218 K). All our weather occurs in the troposphere, and all but a few aircraft fly there. Temperatures then *rise* through the *stratosphere* from −55°C to about 7°C (280 K) at 50 km; we'll discuss the reason shortly.

In the *mesosphere,* temperatures *drop* smoothly again to −93°C (180 K) at around 80 km. Within the *thermosphere,* which extends to around 500 km, temperatures *rise* again, but vary between 700 and 2000 K, depending on the intensity of solar radiation and sunspot activity.

The *exosphere,* the outermost region, maintains these temperatures and merges with outer space.

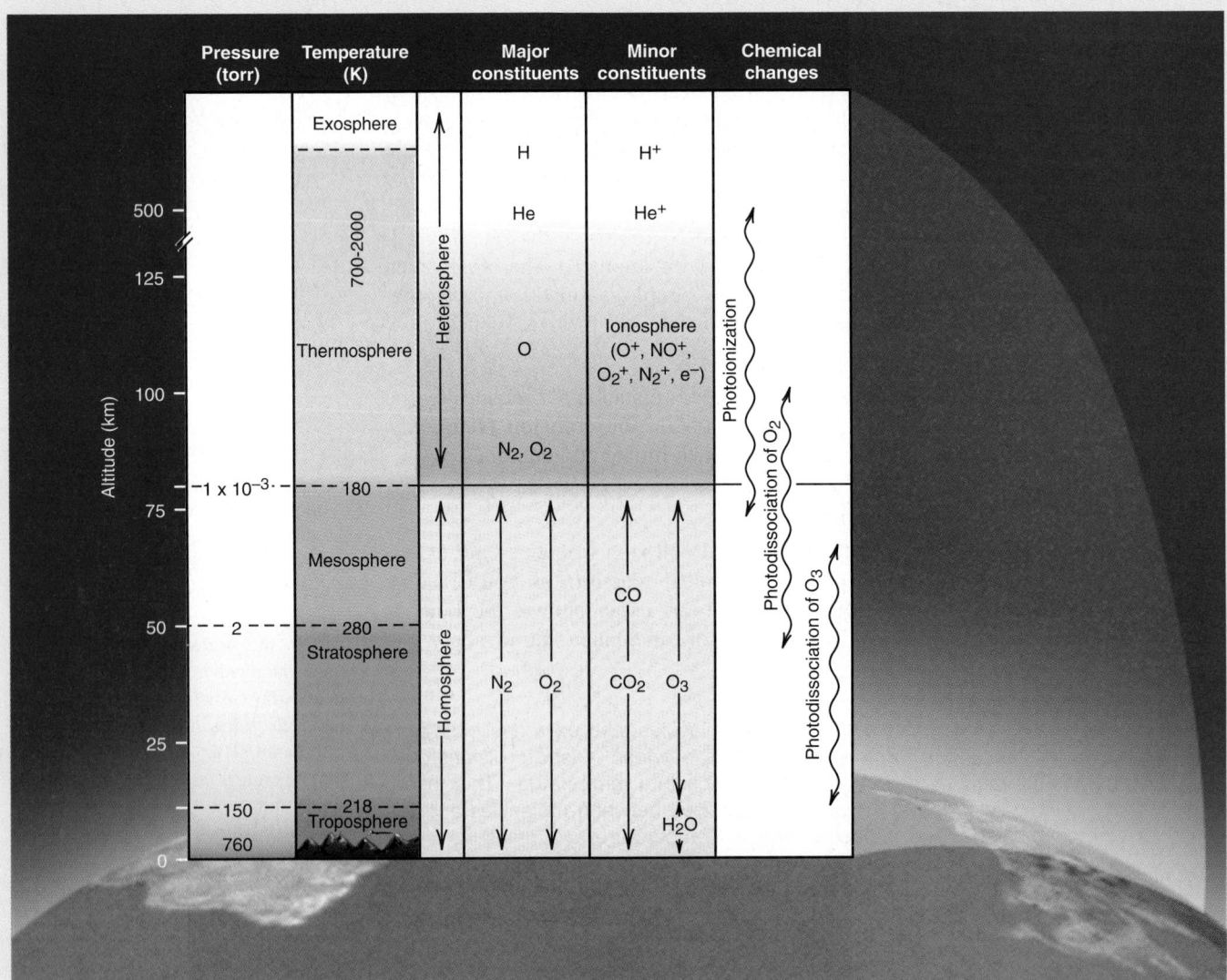

Figure B5.1 Variations in pressure, temperature, and composition of the Earth's atmosphere.

What does it actually mean to have a temperature of 2000 K at 500 km (300 mi) above the Earth's surface? Would a piece of iron (melting point ≈ 1700 K) glow red-hot and melt in the thermosphere? Our everyday use of the words "hot" and "cold" refers to measurements near the surface, where the density of the atmosphere is 10^6 times greater than in the thermosphere. At an altitude of 500 km, where collision frequency is extremely low, a thermometer, or any other object, experiences *very little transfer of kinetic energy.* Thus, the object does not become "hot" in the usual sense; in fact, it is very "cold." Recall that absolute temperature is proportional to the average kinetic energy of the particles. The high-energy solar radiation reaching these outer regions is transferred to relatively few particles, so their average kinetic energy becomes extremely high, as indicated by the high temperature. For this reason, supersonic aircraft, such as the SST, do not reach maximum speed until they reach maximum altitude, where the air is less dense, so that collisions with gas molecules are less frequent and the aircraft material becomes less hot.

Variation in Composition

In terms of chemical composition, the atmosphere is usually classified into two major regions, *homosphere* and *heterosphere.* Superimposing the regions defined by temperature on these shows that the homosphere includes the troposphere, stratosphere, and mesosphere, and the heterosphere includes the thermosphere and exosphere (Figure B5.1).

The Homosphere The homosphere has a relatively constant composition, containing, by volume, approximately 78% N_2, 21% O_2, and 1% a mixture of other gases (mostly argon). Under the conditions that occur in the homosphere, the atmospheric gases behave ideally, so volume percent is equal to mole percent (Avogadro's law), and the mole fraction of a component is directly related to its partial pressure (Dalton's law). Table B5.1 shows the components of a sample of clean, dry air at sea level.

The composition of the homosphere is uniform because of *convective mixing.* Air directly in contact with land is warmer than the air above it. The warmer air expands (Charles's law), its density decreases, and it rises through the cooler, denser air, thereby mixing the components. The cooler air sinks, becomes warmer by contact with the land, and the convection continues. The warm air currents that rise from the ground, called *thermals,* are used by soaring birds and glider pilots to stay aloft.

An important effect of convection is that the air above industrialized areas becomes cleaner as the rising air near the surface carries up ground-level pollutants, which are dispersed by winds. However, under certain weather and geographical conditions, a warm air mass remains stationary over a cool one. The resulting *temperature inversion* blocks normal convection, and harmful pollutants build up, causing severe health problems.

The Heterosphere The heterosphere has variable composition, consisting of regions dominated by a few atomic or molecular species. Convective heating does not reach these heights, so the gas particles become layered according to molar mass: nitrogen and oxygen molecules in the lower levels, oxygen atoms (O) in the next, then helium atoms (He), and free hydrogen atoms (H) in the highest level.

Embedded within the lower heterosphere is the ionosphere, containing ionic species such as O^+, NO^+, O_2^+, N_2^+, and free electrons (Figure B5.1). Ionospheric chemistry involves numer-

Table B5.1 Composition of Clean, Dry Air at Sea Level

Component	Mole Fraction
Nitrogen (N_2)	0.78084
Oxygen (O_2)	0.20946
Argon (Ar)	0.00934
Carbon dioxide (CO_2)	0.00033
Neon (Ne)	1.818×10^{-5}
Helium (He)	5.24×10^{-6}
Methane (CH_4)	2×10^{-6}
Krypton (Kr)	1.14×10^{-6}
Hydrogen (H_2)	5×10^{-7}
Dinitrogen monoxide (N_2O)	5×10^{-7}
Carbon monoxide (CO)	1×10^{-7}
Xenon (Xe)	8×10^{-8}
Ozone (O_3)	2×10^{-8}
Ammonia (NH_3)	6×10^{-9}
Nitrogen dioxide (NO_2)	6×10^{-9}
Nitrogen monoxide (NO)	6×10^{-10}
Sulfur dioxide (SO_2)	2×10^{-10}
Hydrogen sulfide (H_2S)	2×10^{-10}

ous light-induced bond breaking (*photodissociation*) and light-induced electron removal (*photoionization*) processes. One of the simpler ways that O atoms form, for instance, involves a four-step sequence that absorbs energy:

$$N_2 \longrightarrow N_2^+ + e^- \text{ [photoionization]}$$
$$N_2^+ + e^- \longrightarrow N + N$$
$$N + O_2 \longrightarrow NO + O$$
$$N + NO \longrightarrow N_2 + O$$
$$\overline{O_2 \longrightarrow O + O \text{ [overall photodissociation]}}$$

When the resulting high-energy O atoms collide with other neutral or ionic components, the average kinetic energy of thermospheric particles increases.

The Importance of Stratospheric Ozone Although most high-energy radiation is absorbed by the thermosphere, a small amount reaches the stratosphere and breaks O_2 into O atoms. The energetic O atoms collide with more O_2 to form ozone (O_3), another molecular form of oxygen:

$$O_2(g) \xrightarrow{\text{high-energy radiation}} 2O(g)$$
$$M + O(g) + O_2(g) \longrightarrow O_3(g) + M$$

where M is any particle that can carry away excess energy. This reaction releases heat, which is the reason stratospheric temperatures increase with altitude.

Stratospheric ozone is vital to life on the surface because it absorbs a great proportion of solar ultraviolet (UV) radiation, which results in decomposition of the ozone:

$$O_3(g) \xrightarrow{\text{UV light}} O_2(g) + O(g)$$

UV radiation is extremely harmful because it is strong enough to break chemical bonds and, thus, interrupt normal biological processes. Without the presence of stratospheric ozone, much more of this radiation would reach the surface, resulting in increased mutation and cancer rates. The depletion of the ozone layer as a result of industrial gases is discussed in Chapter 16.

(continued)

Earth's Primitive Atmosphere

The composition of the present atmosphere bears little resemblance to that covering the young Earth, but scientists disagree about what that primitive composition actually was: Did the carbon and nitrogen have low oxidation numbers, as in CH_4 (O.N. of C = -4) and NH_3 (O.N. of N = -3)? Or did these atoms have higher oxidation numbers, as in CO_2 (O.N. of C = +4) and N_2 (O.N. of N = 0)? One point generally accepted is that the primitive mixture did not contain free O_2.

Origin-of-life models propose that about 1 billion years after the earliest organisms appeared, blue-green algae evolved. These one-celled plants used solar energy to produce glucose by photosynthesis:

$$6CO_2(g) + 6H_2O(l) \xrightarrow{\text{light}} C_6H_{12}O_6(\text{glucose}) + 6O_2(g)$$

As a result of this reaction, the O_2 content of the atmosphere increased and the CO_2 content decreased. More O_2 allowed more oxidation to occur, which changed the geological and biological makeup of the early Earth. Iron(II) minerals changed to iron(III) minerals, sulfites changed to sulfates, and eventually organisms evolved that could use O_2 to oxidize other organisms to obtain energy. For these organisms to have survived exposure to the more energetic forms of solar radiation (particularly UV light), enough O_2 must have formed to create a protective ozone layer. Estimates indicate that the level of O_2 increased to the current level of about 20 mol % approximately 1.5 billion years ago.

A Survey of Planetary Atmospheres

Earth's combination of pressure and temperature and its oxygen-rich atmosphere and watery surface are unique in the Solar System. (Indeed, if similar conditions and composition were discovered on a planet circling any other star, excitement about the possibility of life there would be enormous.) Atmospheres on the Sun's other planets are strikingly different from Earth's.

Some, especially those on the outer planets, exist under conditions that cause extreme deviations from ideal gas behavior (Section 5.7). Based on current data from NASA spacecraft and Earth-based observations, Table B5.2 lists conditions and composition of the atmospheres on the planets within the Solar System and on some of their moons.

Table B5.2 Planetary Atmospheres

Planet (Satellite)	Pressure* (atm)	Temperature[†] (K)	Composition (mol %)
Mercury	$<10^{-12}$	~700 (day) ~100 (night)	He, H_2, O_2, Ar, Ne (Na and K from solar wind)
Venus	~90	~730	CO_2 (96), N_2 (3), He, SO_2, H_2O, Ar, Ne
Earth	1.0	avg. range 250–310	N_2 (78), O_2 (21), Ar (0.9), H_2O, CO_2, Ne, He, CH_4, Kr
(Moon)	$\sim2\times10^{-14}$	370 (day) 120 (night)	Ne, Ar, He
Mars	7×10^{-3}	300 (summer day) 140 (pole in winter) 218 average	CO_2 (95), N_2(3), Ar (1.6), O_2, H_2O, Ne, CO, Kr
Jupiter	($\sim4\times10^6$)	(~140)	H_2 (89), He (11), CH_4, NH_3, C_2H_6, C_2H_2, PH_3
(Io)	$\sim10^{-10}$	~110	SO_2, S vapor
Saturn	($\sim4\times10^6$)	(~130)	H_2 (93), He (7), CH_4, NH_3, H_2O, C_2H_6, PH_3
(Titan)	1.6	~94	N_2 (90), Ar (<6), CH_4 (3?), C_2H_6, C_2H_2, C_2H_4, HCN, H_2
Uranus	($>10^6$)	(~60)	H_2 (83), He (15), CH_4 (2)
Neptune	($>10^6$)	(~60)	H_2 (<90), He (~10), CH_4
Pluto	$\sim10^{-6}$	~50	N_2, CO, CH_4

*Values in parentheses refer to interior pressures.
[†]Values in parentheses refer to cloud-top temperatures.

5.7 REAL GASES: DEVIATIONS FROM IDEAL BEHAVIOR

A fundamental principle of science is that simpler models are more useful than complex ones—as long as they explain the data. You can certainly appreciate the usefulness of the kinetic-molecular theory. With simple postulates, it explains ideal gas behavior in terms of particles acting like infinitesimal "billiard balls," moving at speeds governed by the absolute temperature, and experiencing only perfectly elastic collisions.

In reality, however, you know that *molecules are not points of mass*. They have volumes determined by the sizes of their atoms and the lengths of their bonds. You also know that atoms contain charged particles, which give rise to *attractive and repulsive forces among molecules*. (In fact, such forces cause substances to undergo changes of states; we'll discuss these forces in great detail in Chapter 12.) Therefore, we expect these real properties of molecules to cause deviations from ideal behavior under some conditions, and this is indeed the case. We must alter the simple model and the ideal gas law to predict gas behavior at low temperatures and very high pressures.

Effects of Extreme Conditions on Gas Behavior

At ordinary conditions—relatively high temperatures and low pressures—most simple gases exhibit nearly ideal behavior. Even at STP (0°C and 1 atm), however, gases deviate *slightly* from ideal behavior. Table 5.4 shows the standard molar volumes of several gases to five significant figures. Note that they do not quite equal the ideal value. The phenomena that cause these slight deviations under standard conditions exert more influence as the temperature decreases toward the condensation point of the gas, the temperature at which it liquefies. As you can see, the largest deviations in the table are for Cl_2 and NH_3, which are already close to their condensation points at the standard temperature of 0°C.

At pressures greater than 10 atm, we begin to see significant deviations from ideal behavior in many gases. Figure 5.21 shows a plot of PV/RT versus P_{ext} for *1 mol* of several real gases and an ideal gas. For 1 mol of an *ideal* gas, the ratio PV/RT is equal to 1 at any pressure. The values on the x axis are the external

Table 5.4 Molar Volume of Some Common Gases at STP (0°C and 1 atm)

Gas	Molar Volume (L/mol)	Condensation Point (°C)
He	22.435	−268.9
H_2	22.432	−252.8
Ne	22.422	−246.1
Ideal gas	**22.414**	—
Ar	22.397	−185.9
N_2	22.396	−195.8
O_2	22.390	−183.0
CO	22.388	−191.5
Cl_2	22.184	−34.0
NH_3	22.079	−33.4

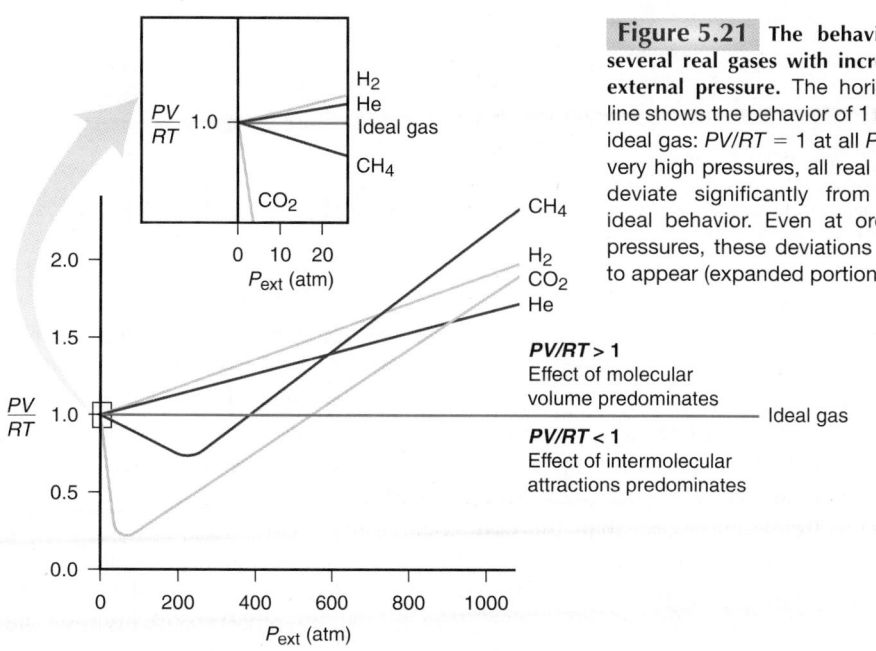

Figure 5.21 The behavior of several real gases with increasing external pressure. The horizontal line shows the behavior of 1 mol of ideal gas: $PV/RT = 1$ at all P_{ext}. At very high pressures, all real gases deviate significantly from such ideal behavior. Even at ordinary pressures, these deviations begin to appear (expanded portion).

PV/RT > 1
Effect of molecular volume predominates

PV/RT < 1
Effect of intermolecular attractions predominates

pressures at which the *PV/RT* ratios are calculated, and they range from normal (*PV/RT* = 1 at 1 atm) to very high (*PV/RT* ≈ 1.6 to 2.3 at ~1000 atm).

The *PV/RT* curve for 1 mol of methane (CH₄) is typical of that for most real gases: it decreases below the ideal value at moderately high pressures and then rises above it as pressure increases further. This shape arises from two overlapping effects of the two characteristics of real molecules just mentioned:

1. At moderately high pressure, values of *PV/RT* lower than ideal (less than 1) are due predominantly to *intermolecular attractions*.
2. At very high pressure, values of *PV/RT* greater than ideal (more than 1) are due predominantly to *molecular volume*.

Let's examine these effects on the molecular level:

1. *Intermolecular attractions.* Attractive forces between molecules are *much* weaker than the covalent bonding forces that hold a molecule together. Most intermolecular attractions are caused by slight imbalances in electron distributions and are important only over relatively short distances. At normal pressures, the spaces between gas molecules are so large that attractions are negligible, and the gas behaves nearly ideally. As the pressure rises and the volume of the sample decreases, however, the average intermolecular distance becomes smaller and attractions have a greater effect.

Picture a molecule at these higher pressures (Figure 5.22). As it approaches the container wall, nearby molecules attract it, which lessens the force of its impact. *This effect repeated throughout the sample results in decreased gas pressure and, thus, a smaller numerator in the PV/RT ratio.* Lowering the temperature significantly has the same effect because it slows the molecules, so attractive forces exert an influence for a longer time. At a low enough temperature, the attractions among molecules become overwhelming, and the gas condenses to a liquid.

Figure 5.22 **The effect of intermolecular attractions on measured gas pressure.** At ordinary pressures, the volume is large and gas molecules are too far apart to experience significant attractions. At moderately high external pressures, the volume decreases enough for the molecules to influence each other. As the close-up shows, a gas molecule approaching the container wall experiences intermolecular attractions from neighboring molecules that reduce the force of its impact. As a result, gases exert *less* pressure than the ideal gas law predicts.

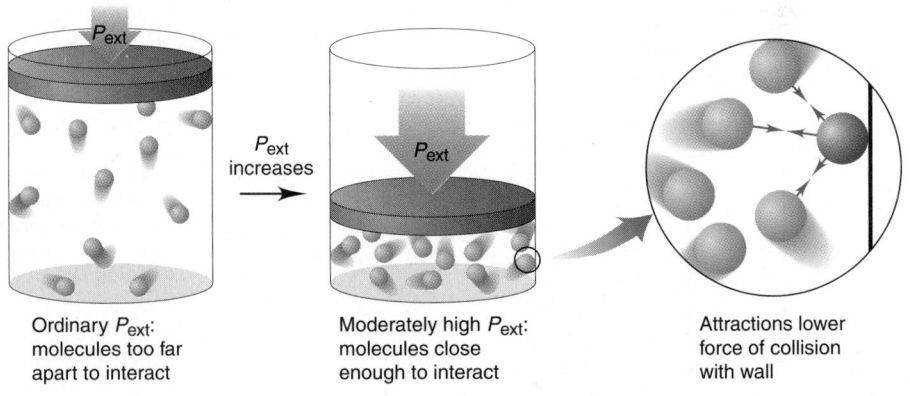

Ordinary P_{ext}: molecules too far apart to interact

Moderately high P_{ext}: molecules close enough to interact

Attractions lower force of collision with wall

2. *Molecular volume.* At normal pressures, the space between molecules (free volume) is enormous compared with the volume of the molecules *themselves* (molecular volume), so the free volume is essentially equal to the container volume. As the applied pressure increases, however, and the free volume decreases, the molecular volume makes up a greater proportion of the container volume, which you can see in Figure 5.23. Thus, at very high pressures, the free volume becomes significantly *less* than the container volume. However, we continue to use the container volume as the *V* in the *PV/RT* ratio, so the ratio is artificially high. This makes the numerator artificially high. The molecular volume effect becomes more important as the pressure increases, eventually outweighing the

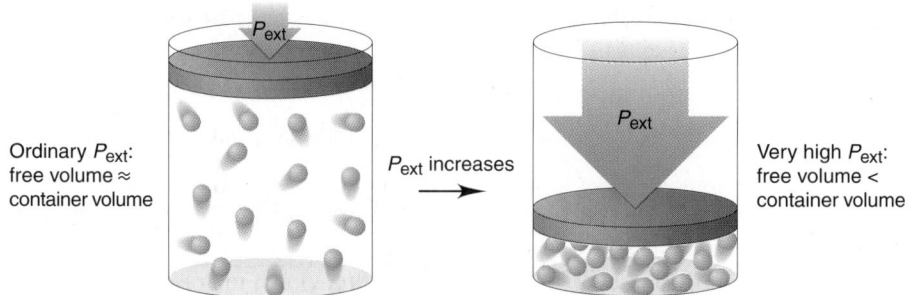

Figure 5.23 The effect of molecular volume on measured gas volume. At ordinary pressures, the volume *between* molecules (free volume) is essentially equal to the container volume because the molecules occupy only a tiny fraction of the available space. At very high external pressures, however, the free volume is significantly *less* than the container volume because of the volume of the molecules themselves.

importance of the intermolecular attractions and causing PV/RT to rise above the ideal value.

In Figure 5.21, the H_2 and He curves do not show the typical dip at moderate pressures. These gases consist of particles with such weak intermolecular attractions that the molecular-volume effect predominates at all pressures.

The van der Waals Equation: The Ideal Gas Law Redesigned

To describe real gas behavior more accurately, we need to "redesign" the ideal gas equation to do two things:

1. Adjust the measured pressure *up* by adding a factor that accounts for intermolecular attractions, and
2. Adjust the measured volume *down* by subtracting a factor from the entire container volume that accounts for the molecular volume.

In 1873, Johannes van der Waals realized the limitations of the ideal gas law and proposed an equation that accounts for the behavior of real gases. The **van der Waals equation** for n moles of a real gas is

$$\left(P + \frac{n^2 a}{V^2}\right)(V - nb) = nRT$$

$$\underset{\substack{\text{adjusts}\\P\text{ up}}}{} \qquad \underset{\substack{\text{adjusts}\\V\text{ down}}}{}$$

where P is the measured pressure, V is the container volume, n and T have their usual meanings, and a and b are **van der Waals constants,** experimentally determined positive numbers specific for a given gas. Values of these constants for several gases are given in Table 5.5. The constant a relates to the number of electrons, which in turn relates to the complexity of a molecule and the strength of its intermolecular attractions. The constant b relates to molecular volume.

Consider this typical application of the van der Waals equation to calculate a gas variable. A 1.98-L vessel contains 215 g (4.89 mol) of dry ice. After standing at 26°C (299 K), the $CO_2(s)$ changes to $CO_2(g)$. The pressure is measured (P_{real}) and calculated by the ideal gas law (P_{IGL}) and, using the appropriate values of a and b, by the van der Waals equation (P_{VDW}). The results are revealing:

$$P_{real} = 44.8 \text{ atm} \qquad P_{IGL} = 60.6 \text{ atm} \qquad P_{VDW} = 45.9 \text{ atm}$$

Comparing the real with each calculated value shows that P_{IGL} is 35.3% greater than P_{real}, but P_{VDW} is only 2.5% greater than P_{real}. At these conditions, CO_2 deviates so much from ideal behavior that the ideal gas law is not very useful.

Here is one final point to realize: According to kinetic-molecular theory, the constants a and b are zero for an ideal gas because the particles do not attract

Table 5.5 Van der Waals Constants for Some Common Gases		
Gas	$a\left(\dfrac{\text{atm·L}^2}{\text{mol}^2}\right)$	$b\left(\dfrac{\text{L}}{\text{mol}}\right)$
He	0.034	0.0237
Ne	0.211	0.0171
Ar	1.35	0.0322
Kr	2.32	0.0398
Xe	4.19	0.0511
H_2	0.244	0.0266
N_2	1.39	0.0391
O_2	1.36	0.0318
Cl_2	6.49	0.0562
CO_2	3.59	0.0427
CH_4	2.25	0.0428
NH_3	4.17	0.0371
H_2O	5.46	0.0305

each other and have no volume. Yet, even for a real gas at ordinary pressures, the molecules are very far apart. Thus,

- Attractive forces are miniscule, so $P + \dfrac{n^2a}{V^2} \approx P$

- Molecular volume is a miniscule fraction of the container volume, so $V - nb \approx V$

Therefore, *at ordinary conditions, the van der Waals equation becomes the ideal gas equation.*

SECTION SUMMARY

At very high pressures or low temperatures, all gases deviate greatly from ideal behavior. As pressure increases, most real gases exhibit first a lower and then a higher *PV/RT* ratio than the value for 1 mol of an ideal gas. These deviations are due to attractions between molecules, which lower the pressure (and the ratio), and to the larger fraction of the container volume occupied by the molecules, which increases the ratio. By including parameters characteristic of each gas, the van der Waals equation corrects for these deviations.

Chapter Perspective

As with the atomic model (Chapter 2), we have seen in this chapter how a simple molecular-scale model can explain macroscopic observations and how it often must be revised to predict a wider range of chemical behavior. Gas behavior is relatively easy to understand because gas structure is so randomized—very different from the structures of liquids and solids with their complex molecular interactions, as you'll see in Chapter 12. In Chapter 6, we return to chemical reactions, but from the standpoint of the heat involved in such changes. The meaning of kinetic energy, which we discussed in this chapter, bears directly on this central topic.

For Review and Reference (Numbers in parentheses refer to pages, unless noted otherwise.)

Learning Objectives

Relevant section and/or sample problem (SP) numbers appear in parentheses.

Understand These Concepts

1. How gases differ in their macroscopic properties from liquids and solids (5.1)
2. The meaning of pressure and the operation of a barometer and a manometer (5.2)
3. The relations among gas variables expressed by Boyle's, Charles's, and Avogadro's laws (5.3)
4. How the individual gas laws are incorporated into the ideal gas law (5.3)
5. How the ideal gas law can be used to study gas density and molar mass (5.4)
6. The relation between gas density and temperature (5.4)
7. The meaning of Dalton's law and the relation between partial pressure and mole fraction of a gas; how Dalton's law applies to collecting a gas over water (5.4)
8. How the postulates of the kinetic-molecular theory can be applied to explain the origin of pressure and the empirical gas laws (5.6)
9. The relations among molecular speed, average kinetic energy, and temperature (5.6)
10. The meanings of effusion and diffusion and how their rates are related to molar mass (5.6)

11. The relations among mean free path, molecular speed, and collision frequency (5.6)
12. Why intermolecular attractions and molecular volume cause gases to deviate from ideal behavior at low temperatures and high pressures (5.7)
13. How the van der Waals equation corrects the ideal gas law for extreme conditions (5.7)

Master These Skills

1. Interconverting among the units of pressure (atm, mmHg, torr, pascal, psi) (SP 5.1)
2. Reducing the ideal gas law to each of the individual gas laws (SPs 5.2–5.5)
3. Rearranging the ideal gas law to calculate gas density (SP 5.6) and molar mass of a volatile liquid (SP 5.7)
4. Calculating the mole fraction and partial pressure of a gas (SP 5.8)
5. Using the vapor pressure of water to find the amount of a gas collected over water (SP 5.9)
6. Applying stoichiometry and gas laws to calculate amounts of reactants and products (SPs 5.10 and 5.11)
7. Using Graham's law to solve problems of gaseous effusion (SP 5.12)

Key Terms

Section 5.2
pressure (*P*) (176)
barometer (176)
manometer (178)
pascal (Pa) (178)
standard atmosphere (atm) (178)
millimeter of mercury (mmHg) (178)
torr (178)

Section 5.3
ideal gas (180)
Boyle's law (181)
Charles's law (182)
Avogadro's law (184)
standard temperature and pressure (STP) (184)
standard molar volume (184)
ideal gas law (185)
universal gas constant (*R*) (185)

Section 5.4
partial pressure (192)
Dalton's law of partial pressures (192)
mole fraction (*X*) (192)

Section 5.6
kinetic-molecular theory (197)
rms speed (u_{rms}) (200)
effusion (201)

Graham's law of effusion (201)
diffusion (202)
mean free path (203)
collision frequency (203)
atmosphere (204)

Section 5.7
van der Waals equation (209)
van der Waals constants (209)

Key Equations and Relationships

5.1 Expressing the volume-pressure relationship (Boyle's law) (181):

$$V \propto \frac{1}{P} \quad \text{or} \quad PV = \text{constant} \quad [T \text{ and } n \text{ fixed}]$$

5.2 Expressing the volume-temperature relationship (Charles's law) (182):

$$V \propto T \quad \text{or} \quad \frac{V}{T} = \text{constant} \quad [P \text{ and } n \text{ fixed}]$$

5.3 Expressing the pressure-temperature relationship (Amontons's law) (183):

$$P \propto T \quad \text{or} \quad \frac{P}{T} = \text{constant} \quad [V \text{ and } n \text{ fixed}]$$

5.4 Expressing the volume-amount relationship (Avogadro's law) (184):

$$V \propto n \quad \text{or} \quad \frac{V}{n} = \text{constant} \quad [P \text{ and } T \text{ fixed}]$$

5.5 Defining standard temperature and pressure (184):
STP: 0°C (273.15 K) and 1 atm (760 torr)

5.6 Defining the volume of 1 mol of a gas at STP (184):
Standard molar volume = 22.414 L = 22.4 L [3 sf]

5.7 Relating volume to pressure, temperature, and amount (ideal gas law) (185):

$$PV = nRT \quad \text{and} \quad \frac{P_1 V_1}{n_1 T_1} = \frac{P_2 V_2}{n_2 T_2}$$

5.8 Calculating the value of *R* (185):

$$R = \frac{PV}{nT} = \frac{1 \text{ atm} \times 22.4 \text{ L}}{1 \text{ mol} \times 273.15 \text{ K}} = 0.0821 \frac{\text{atm·L}}{\text{mol·K}} \quad [3 \text{ sf}]$$

5.9 Rearranging the ideal gas law to find gas density (189):

$$PV = \frac{m}{M} RT \quad \text{so} \quad \frac{m}{V} = d = \frac{M \times P}{RT}$$

5.10 Rearranging the ideal gas law to find molar mass (191):

$$n = \frac{m}{M} = \frac{PV}{RT} \quad \text{so} \quad M = \frac{mRT}{PV} \quad \text{or} \quad M = \frac{dRT}{P}$$

5.11 Relating the total pressure of a gas mixture to the partial pressures of the components (Dalton's law of partial pressures) (192):

$$P_{total} = P_1 + P_2 + P_3 + \cdots$$

5.12 Relating partial pressure to mole fraction (192):

$$P_A = X_A \times P_{total}$$

5.13 Defining rms speed as a function of molar mass and temperature (200):

$$u_{rms} = \sqrt{\frac{3RT}{M}}$$

5.14 Applying Graham's law of effusion (202):

$$\frac{\text{Rate}_A}{\text{Rate}_B} = \frac{\sqrt{M_B}}{\sqrt{M_A}}$$

Highlighted Figures and Tables

These figures (F) and tables (T) provide a quick review of key ideas. Entries in color contain frequently used data.

F5.1 The three states of matter (175)
T5.2 Common units of pressure (179)
F5.5 The relationship between the volume and pressure of a gas (180)
F5.6 The relationship between the volume and temperature of a gas (182)
F5.8 Standard molar volume (184)
F5.10 Relationship between the ideal gas law and the individual gas laws (185)
T5.3 Vapor pressure of water at different *T* (193)

F5.13 Stoichiometric relationships for gases (195)
F5.14 Distribution of molecular speeds at three *T* (198)
F5.15 Molecular description of Boyle's law (198)
F5.16 Molecular description of Dalton's law (199)
F5.17 Molecular description of Charles's law (199)
F5.18 Molecular description of Avogadro's law (200)
F5.19 Relation between molar mass and molecular speed (201)
F5.21 The behavior of several real gases with increasing external pressure (207)
T5.5 Van der Waals constants for some gases (209)

Brief Solutions to Follow-up Problems

5.1 P_{CO_2} (torr) = (753.6 mmHg − 174.0 mmHg) × $\dfrac{1\ torr}{1\ mmHg}$

$= 579.6\ torr$

P_{CO_2} (Pa) = 579.6 torr × $\dfrac{1\ atm}{760\ torr}$ × $\dfrac{1.01325 \times 10^5\ Pa}{1\ atm}$

$= 7.727 \times 10^4\ Pa$

P_{CO_2} (lb/in^2) = 579.6 torr × $\dfrac{1\ atm}{760\ torr}$ × $\dfrac{14.7\ lb/in^2}{1\ atm}$

$= 11.2\ lb/in^2$

5.2 P_2 (atm) = 26.3 kPa × $\dfrac{1\ atm}{101.325\ kPa}$ = 0.260 atm

V_2 (L) = 105 mL × $\dfrac{1\ L}{1000\ mL}$ × $\dfrac{0.871\ atm}{0.260\ atm}$ = 0.352 L

5.3 T_2 (K) = 273 K × $\dfrac{9.75\ cm^3}{6.83\ cm^3}$ = 390. K

5.4 P_2 (torr) = 793 torr × $\dfrac{35.0\ g - 5.0\ g}{35.0\ g}$ = 680. torr

(There is no need to convert mass to moles because the ratio of masses equals the ratio of moles.)

5.5 $n = \dfrac{PV}{RT} = \dfrac{1.37\ atm \times 438\ L}{0.0821\ \frac{atm \cdot L}{mol \cdot K} \times 294\ K}$ = 24.9 mol O$_2$

Mass (g) of O$_2$ = 24.9 mol O$_2$ × $\dfrac{32.00\ g\ O_2}{1\ mol\ O_2}$

$= 7.97 \times 10^2\ g\ O_2$

5.6 $d = \dfrac{44.01\ g/mol \times \dfrac{380\ torr}{760\ torr/atm}}{0.0821\ \frac{atm \cdot L}{mol \cdot K} \times 273\ K}$ = 0.982 g/L

The density is lower at the smaller P because V is larger. In this case, d is lowered by one-half because P is one-half as much.

5.7 $\mathcal{M} = \dfrac{1.26\ g \times 0.0821\ \frac{atm \cdot L}{mol \cdot K} \times 283.2\ K}{\dfrac{102.5\ kPa}{101.325\ kPa/1\ atm} \times 1.00\ L}$ = 29.0 g/mol

5.8 $n_{total} = \left(5.50\ g\ He \times \dfrac{1\ mol\ He}{4.003\ g\ He}\right)$

$+ \left(15.0\ g\ Ne \times \dfrac{1\ mol\ Ne}{20.18\ g\ Ne}\right)$

$+ \left(35.0\ g\ Kr \times \dfrac{1\ mol\ Kr}{83.80\ g\ Kr}\right)$

$= 2.53\ mol$

$P_{He} = \left(\dfrac{5.50\ g\ He \times \dfrac{1\ mol\ He}{4.003\ g\ He}}{2.53\ mol}\right) \times 1\ atm = 0.543\ atm$

$P_{Ne} = 0.294\ atm \qquad P_{Kr} = 0.165\ atm$

5.9 P_{H_2} = 752 torr − 13.6 torr = 738 torr

Mass (g) of H$_2$ = $\left(\dfrac{\dfrac{738\ torr}{760\ torr/atm} \times 1.495\ L}{0.0821\ \frac{atm \cdot L}{mol \cdot K} \times 289\ K}\right) \times \dfrac{2.016\ g\ H_2}{1\ mol\ H_2}$

$= 0.123\ g\ H_2$

5.10 $H_2SO_4(aq) + 2NaCl(s) \longrightarrow Na_2SO_4(aq) + 2HCl(g)$

n_{HCl} = 0.117 kg NaCl × $\dfrac{10^3\ g}{1\ kg}$ × $\dfrac{1\ mol\ NaCl}{58.44\ g\ NaCl}$ × $\dfrac{2\ mol\ HCl}{2\ mol\ NaCl}$

$= 2.00\ mol\ HCl$

At STP, V (mL) = 2.00 mol × $\dfrac{22.4\ L}{1\ mol}$ × $\dfrac{10^3\ mL}{1\ L}$

$= 4.48 \times 10^4\ mL$

5.11 $NH_3(g) + HCl(g) \longrightarrow NH_4Cl(s)$

n_{NH_3} = 0.187 mol $\qquad n_{HCl}$ = 0.0522 mol

n_{NH_3} after reaction

$= 0.187\ mol\ NH_3 - \left(0.0522\ mol\ HCl \times \dfrac{1\ mol\ NH_3}{1\ mol\ HCl}\right)$

$= 0.135\ mol\ NH_3$

$P = \dfrac{0.135\ mol \times 0.0821\ \frac{atm \cdot L}{mol \cdot K} \times 295\ K}{10.0\ L}$ = 0.327 atm

5.12 $\dfrac{Rate\ of\ He}{Rate\ of\ C_2H_6} = \sqrt{\dfrac{30.07\ g/mol}{4.003\ g/mol}}$ = 2.741

Time for C$_2$H$_6$ to effuse = 1.25 min × 2.741 = 3.43 min

Problems

Problems with **colored** numbers are answered at the back of the text. Sections match the text and provide the number(s) of relevant sample problems. Most offer Concept Review Questions, Skill-Building Exercises (in similar pairs), and Problems in Context. Then Comprehensive Problems, based on material from any section or previous chapter, follow.

An Overview of the Physical States of Matter

● **Concept Review Questions**

5.1 How does a sample of gas differ in its behavior from a sample of liquid in each of the following situations?
(a) The sample is transferred from one container to a larger one.

(b) The sample is heated in an expandable container, but no change of state occurs.
(c) The sample is placed in a cylinder with a piston, and an external force is applied.

5.2 Are the particles in a gas farther apart or closer together than the particles in a liquid? Use your answer to this question in order to explain each of the following general observations:
(a) Gases are more compressible than liquids.
(b) Gases have lower viscosities than liquids.
(c) After thorough stirring, all gas mixtures are solutions.
(d) The density of a substance in the gas state is lower than in the liquid state.

Gas Pressure and Its Measurement
(Sample Problem 5.1)

● **Concept Review Questions**

5.3 How does a barometer work? Is the column of mercury in a barometer shorter when it is on a mountaintop or at sea level? Explain.

5.4 How can a unit of length such as millimeter of mercury (mmHg) be used as a unit of pressure, which has the dimensions of force per unit area?

5.5 In a closed-end manometer, the mercury level in the arm attached to the flask can never be higher than the mercury level in the other arm, whereas in an open-end manometer, it *can* be higher. Explain.

● **Skill-Building Exercises** *(paired)*

5.6 On a cool, rainy day, the barometric pressure is 725 mmHg. Calculate the barometric pressure in centimeters of water (cmH_2O) (d of Hg = 13.5 g/mL; d of H_2O = 1.00 g/mL).

5.7 A long glass tube, sealed at one end, has an inner diameter of 10.0 mm. The tube is filled with water and inverted into a pail of water. If the atmospheric pressure is 755 mmHg, how high (in mmH_2O) is the column of water in the tube (d of Hg = 13.5 g/mL; d of H_2O = 1.00 g/mL)?

5.8 If the barometer in Figure P5.8 reads 738.5 torr, what is the pressure of the gas in the flask in atmospheres?

5.9 If the barometer in Figure P5.9 reads 765.2 mmHg, what is the pressure of the gas in the flask in kilopascals?

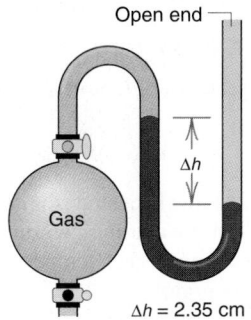

Figure P5.8

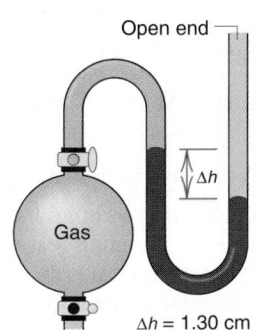

Figure P5.9

5.10 If the sample flask in Figure P5.10 is open to the air, what is the atmospheric pressure in atmospheres?

5.11 What is the pressure in pascals of the gas in the flask in Figure P5.11?

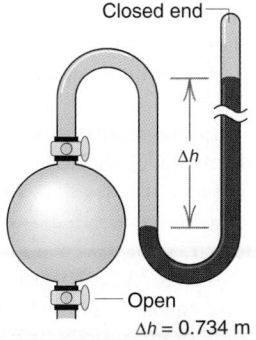

Figure P5.10

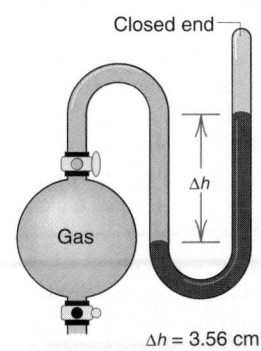

Figure P5.11

5.12 Convert the following:
(a) 0.745 atm to mmHg   (b) 992 torr to bar
(c) 365 kPa to atm   (d) 804 mmHg to kPa

5.13 Convert the following:
(a) 74.8 cmHg to atm   (b) 27.0 atm to kPa
(c) 8.50 atm to bar   (d) 0.907 kPa to torr

● **Problems in Context**

5.14 Convert the following pressures into atmospheres:
(a) At the peak of Mt. Everest, atmospheric pressure is only 2.75×10^2 mmHg.
(b) A cyclist fills her bike tires to 91 psi.
(c) The surface of Venus has an atmospheric pressure of 9.15×10^6 Pa.
(d) At 100 ft below sea level, a scuba diver experiences a pressure of 2.44×10^4 torr.

5.15 The gravitational force exerted by an object is given by $F = mg$, where F is the force in newtons, m is the mass in kilograms, and g is the acceleration due to gravity (9.81 m/s^2).
(a) Use the definition of the pascal to calculate the mass (in kg) of the atmosphere on 1 m^2 of ocean.
(b) Osmium ($Z = 76$) has the highest density of any element (22.6 g/mL). If an osmium column is 1 m^2 in area, how high must it be for its pressure to equal atmospheric pressure? [Use the answer from part (a) in your calculation.]

The Gas Laws and Their Experimental Foundations
(Sample Problems 5.2 to 5.5)

● **Concept Review Questions**

5.16 When asked to state Boyle's law, a student replies, "The volume of a gas is inversely proportional to its pressure." How is this statement incomplete? Give a correct statement of Boyle's law.

5.17 Which quantities are variables and which are fixed in each of the following: (a) Charles's law; (b) Avogadro's law; (c) Amontons's law?

5.18 Boyle's law relates gas volume to pressure, and Avogadro's law relates gas volume to number of moles. State a relationship between gas pressure and number of moles.

5.19 The volume of the cylinder in A can change to the volume in B in many ways. Choose three ways this can happen and state the variable(s) that change(s) and remain(s) fixed.

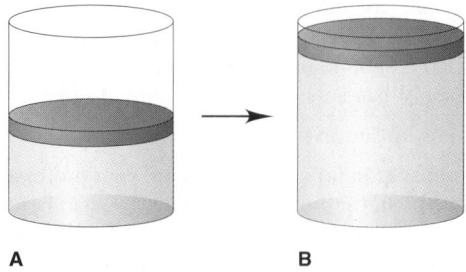

A   B

● **Skill-Building Exercises** *(paired)*

5.20 What is the effect of the following on the volume of 1 mol of an ideal gas?
(a) The pressure is tripled (at constant T).
(b) The absolute temperature is increased by a factor of 2.5 (at constant P).
(c) Two more moles of the gas are added (at constant P and T).

5.21 What is the effect of the following on the volume of 1 mol of an ideal gas?
(a) The pressure is reduced by a factor of 4 (at constant T).
(b) The pressure changes from 760 torr to 202 kPa, and the temperature changes from 37°C to 155 K.
(c) The temperature changes from 305 K to 32°C, and the pressure changes from 2 atm to 101 kPa.

5.22 What is the effect of the following on the volume of 1 mol of an ideal gas?
(a) The temperature is decreased from 700 K to 350 K (at constant P).
(b) The temperature is increased from 350°C to 700°C (at constant P).
(c) The pressure is increased from 2 atm to 8 atm (at constant T).

5.23 What is the effect of the following on the volume of 1 mol of an ideal gas?
(a) Half the gas escapes through a stopcock (at constant P and T).
(b) The initial pressure is 722 torr, and the final pressure is 0.950 atm; the initial temperature is 32°F, and the final temperature is 273 K.
(c) Both the pressure and temperature are reduced by one-fourth of their initial values.

5.24 A sample of sulfur hexafluoride gas occupies a volume of 5.10 L at 198°C. Assuming that the pressure remains constant, what temperature (in °C) is needed to reduce the volume to 2.50 L?

5.25 A 93-L sample of dry air is cooled from 145°C to −22°C while the pressure is maintained at 2.85 atm. What is the final volume?

5.26 A sample of Freon-12 (CF_2Cl_2) occupies 25.5 L at 298 K and 153.3 kPa. Find its volume at STP.

5.27 Calculate the volume of a sample of carbon monoxide that is at −14°C and 367 torr if it occupies 3.65 L at 298 K and 745 torr.

5.28 A sample of chlorine gas is confined in a 5.0-L container at 228 torr and 27°C. How many moles of gas are present in the sample?

5.29 If 1.47×10^{-3} mol of argon occupies a 75.0-mL container at 26°C, what is the pressure (in torr)?

5.30 You have 207 mL of chlorine trifluoride gas at 699 mmHg and 45°C. What is the mass (in g) of the sample?

5.31 A 75.0-g sample of dinitrogen monoxide is confined in a 3.1-L vessel. What is the pressure (in atm) at 115°C?

● **Problems in Context**

5.32 In preparation for a demonstration, your professor brings a 1.5-L bottle of sulfur dioxide into the lecture hall before class to allow the gas to reach room temperature. If the pressure gauge reads 85 psi and the lecture hall is 23°C, how many moles of sulfur dioxide are in the bottle? (*Hint:* The gauge reads zero when 14.7 psi of gas remains.)

5.33 A gas-filled weather balloon with a volume of 55.0 L is released at sea-level conditions of 755 torr and 23°C. The balloon can expand to a maximum volume of 835 L. When the balloon rises to an altitude at which the temperature is −5°C and the pressure is 0.066 atm, will it reach its maximum volume?

Further Applications of the Ideal Gas Law
(Sample Problems 5.6 to 5.9)

● **Concept Review Questions**

5.34 Why is moist air less dense than dry air?
5.35 To collect a beaker of H_2 gas by displacing the air already in the beaker, would you hold the beaker upright or inverted? Why? How would you hold the beaker to collect CO_2?
5.36 Why can we use a gas mixture, such as air, to study the general behavior of an ideal gas under ordinary conditions?
5.37 How does the partial pressure of gas A in a mixture compare to its mole fraction in the mixture? Explain.

● **Skill-Building Exercises** *(paired)*

5.38 What is the density of Xe gas at STP?
5.39 What is the density of Freon-11 ($CFCl_3$) at 120°C and 1.5 atm?

5.40 How many moles of gaseous arsine (AsH_3) will occupy 0.0400 L at STP? What is the density of gaseous arsine?
5.41 The density of a noble gas is 2.71 g/L at 3.00 atm and 0°C. Identify the gas.

5.42 Calculate the molar mass of a gas at 388 torr and 45°C if 206 ng occupies 0.206 μL.
5.43 When an evacuated 63.8-mL glass bulb is filled with a gas at 22°C and 747 mmHg, the bulb gains 0.103 g in mass. Is the gas N_2, Ne, or Ar?

5.44 When 0.600 L of Ar at 1.20 atm and 227°C is mixed with 0.200 L of O_2 at 501 torr and 127°C in a 400-mL flask at 27°C, what is the pressure in the flask?
5.45 A 355-mL container holds 0.146 g of Ne and an unknown amount of Ar at 35°C and a total pressure of 626 mmHg. Calculate the moles of Ar present.

● **Problems in Context**

5.46 The air in a hot-air balloon at 744 torr is heated from 17°C to 60.0°C. Assuming that the moles of air and the pressure remain constant, what is the density of the air at each temperature? (The average molar mass of air is 28.8 g/mol.)
5.47 On a certain winter day in Utah, the average atmospheric pressure is 650. torr. What is the molar density (in mol/L) of air when the temperature is −25°C?
5.48 A sample of a liquid hydrocarbon known to consist of molecules with five carbon atoms is vaporized in a 0.204-L flask by immersion in a water bath at 101°C. The barometric pressure is 767 torr, and the remaining gas condenses to 0.482 g of liquid. What is the molecular formula of the hydrocarbon?
5.49 A sample of air contains 78.08% nitrogen, 20.94% oxygen, 0.05% carbon dioxide, and 0.93% argon, by volume. How many molecules of each gas are present in 1.00 L of the sample at 25°C and 1.00 atm?
5.50 An environmental chemist sampling industrial exhaust gases from a coal-burning plant collects a CO_2-SO_2-H_2O mixture in a 21-L steel tank until the pressure reaches 850. torr at 45°C.
(a) How many moles of gas are collected?
(b) If the SO_2 concentration in the mixture is 7.95×10^3 parts per million by volume (ppmv), what is its partial pressure? [*Hint:* ppmv = (volume of component/volume of mixture) × 10^6.]

The Ideal Gas Law and Reaction Stoichiometry
(Sample Problems 5.10 and 5.11)

● **Skill-Building Exercises** *(paired)*

5.51 How many grams of phosphorus react with 35.5 L of O_2 at STP to form tetraphosphorus decaoxide?

$$P_4(s) + 5O_2(g) \longrightarrow P_4O_{10}(s)$$

5.52 How many grams of potassium chlorate decompose to potassium chloride and 638 mL of O_2 at 128°C and 752 torr?

$$2KClO_3(s) \longrightarrow 2KCl(s) + 3O_2(g)$$

5.53 How many grams of phosphine (PH_3) can form when 37.5 g of phosphorus and 83.0 L of hydrogen gas react at STP?

$$P_4(s) + H_2(g) \longrightarrow PH_3(g) \quad \text{[unbalanced]}$$

5.54 When 35.6 L of ammonia and 40.5 L of oxygen gas at STP burn, nitrogen monoxide and water are produced. After the products return to STP, how many grams of nitrogen monoxide are present?

$$NH_3(g) + O_2(g) \longrightarrow NO(g) + H_2O(l) \quad \text{[unbalanced]}$$

5.55 Aluminum reacts with excess hydrochloric acid to form aqueous aluminum chloride and 35.8 mL of hydrogen gas over water at 27°C and 751 mmHg. How many grams of aluminum reacted?

5.56 How many liters of hydrogen gas are collected over water at 18°C and 725 mmHg when 0.84 g of lithium reacts with water? Aqueous lithium hydroxide also forms.

● **Problems in Context**

5.57 "Strike anywhere" matches contain the compound tetraphosphorus trisulfide, which burns to form tetraphosphorus decaoxide and sulfur dioxide gas. How many milliliters of sulfur dioxide, measured at 725 torr and 32°C, can be produced from burning 0.800 g of tetraphosphorus trisulfide?

5.58 Freon-12 (CF_2Cl_2), a refrigerant and aerosol propellant, is a dangerous air pollutant. In the troposphere, it traps heat 25 times as effectively as CO_2, and in the stratosphere, it participates in the breakdown of ozone. It is prepared industrially by reaction of gaseous carbon tetrachloride with hydrogen fluoride. Hydrogen chloride gas also forms. How many grams of carbon tetrachloride are required for the production of 16.0 dm^3 of Freon-12 at 27°C and 1.20 atm?

5.59 Xenon hexafluoride was one of the first noble gas compounds synthesized. The solid reacts rapidly with the silicon dioxide in glass or quartz containers to form liquid $XeOF_4$ and gaseous silicon tetrafluoride. What is the pressure in a 1.00-L container at 25°C after 2.00 g of xenon hexafluoride reacts? (Assume that silicon tetrafluoride is the only gas present and that it occupies the entire volume.)

5.60 Roasting galena [lead(II) sulfide] is an early step in the industrial isolation of lead. How many liters of sulfur dioxide, measured at STP, are produced by the reaction of 3.75 of kg galena with 228 L of oxygen gas at 220°C and 2.0 atm? Lead(II) oxide also forms.

5.61 In one of his most critical studies into the nature of combustion, Lavoisier heated mercury(II) oxide and isolated elemental mercury and oxygen gas. If 40.0 g of mercury(II) oxide is heated in a 502-mL vessel and 20.0% (by mass) decomposes, what is the pressure (in atm) of the oxygen that forms at 25.0°C? (Assume the gas occupies the entire volume.)

The Kinetic-Molecular Theory: A Model for Gas Behavior
(Sample Problem 5.12)

● **Concept Review Questions**

5.62 Use the kinetic-molecular theory to explain the change in gas pressure that results from warming a sample of gas.

5.63 How does the kinetic-molecular theory explain why 1 mol of krypton and 1 mol of helium have the same volume at STP?

5.64 Is the rate of effusion of a gas higher than, lower than, or equal to its rate of diffusion? Explain. For two gases with molecules of approximately the same size, is the ratio of their effusion rates higher than, lower than, or equal to the ratio of their diffusion rates? Explain.

5.65 Consider two 1-L samples of gas, one H_2 and one O_2, both at 1 atm and 25°C. How do the samples compare in terms of (a) mass, (b) density, (c) mean free path, (d) average molecular kinetic energy, (e) average molecular speed, and (f) time for a given fraction of molecules to effuse?

5.66 Three 5-L flasks, fixed with pressure gauges and small valves, each contain 4 g of gas at 273 K. Flask A contains H_2, flask B contains He, and flask C contains CH_4. Rank flask contents in terms of (a) pressure, (b) average molecular kinetic energy, (c) diffusion rate after the valve is opened, (d) total kinetic energy of the molecules, (e) density, and (f) collision frequency.

● **Skill-Building Exercises** *(paired)*

5.67 What is the ratio of effusion rates for the lightest gas, H_2, to the heaviest known gas, UF_6?

5.68 What is the ratio of effusion rates for O_2 to Kr?

5.69 The graph below shows the distribution of molecular speeds of argon and helium at the same temperature.

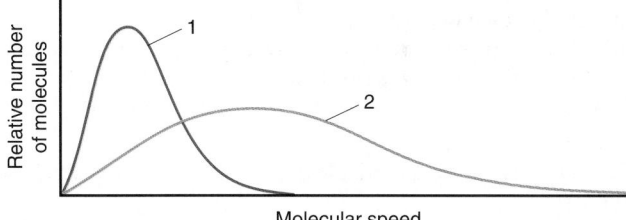

(a) Does curve 1 or 2 better represent the behavior of argon?
(b) Which curve represents the gas that effuses more slowly?
(c) Which curve more closely represents the behavior of fluorine gas? Explain.

5.70 The graph below shows the distribution of molecular speeds of a gas at two different temperatures.

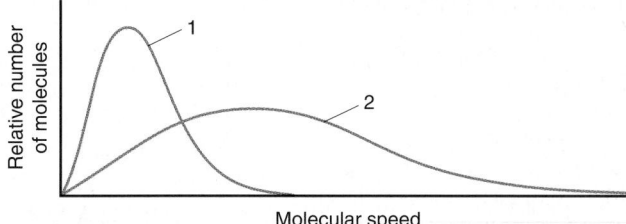

(a) Does curve 1 or 2 better represent the behavior of the gas at the lower temperature?
(b) Which curve represents the sample with the higher $\overline{E_k}$?
(c) Which curve represents the sample that diffuses more quickly?

5.71 At a particular pressure and temperature, it takes 4.55 min for a 1.5-L sample of He to effuse through a porous membrane. How long does it take for 1.5 L of F_2 to effuse under the same conditions?

5.72 A sample of an unknown gas effuses in 11.1 min. An equal volume of H_2 in the same apparatus at the same temperature and pressure effuses in 2.42 min. What is the molar mass of the unknown gas?

● **Problems in Context**

5.73 Solid white phosphorus melts and then vaporizes at high temperature. Gaseous white phosphorus effuses at a rate that is 0.404 times that of neon in the same apparatus under the same conditions. How many atoms are in a molecule of gaseous white phosphorus?

5.74 Helium is the lightest noble gas component of air, and xenon is the heaviest. [For this problem, use $R = 8.314$ J/(mol·K) and $\mathcal{M}$ in kg/mol.]
(a) Calculate the rms speed of helium in winter (0.°C) and in summer (30.°C).
(b) Compare the rms speed of helium with that of xenon at 30.°C.
(c) Calculate the average kinetic energy per mole of helium and of xenon at 30.°C.
(d) Calculate the average kinetic energy per molecule of helium at 30.°C.

Real Gases: Deviations from Ideal Behavior

● **Skill-Building Exercises** *(paired)*

5.75 Do intermolecular attractions cause negative or positive deviations from the PV/RT ratio of an ideal gas? Use data from Table 5.5 to rank Kr, CO_2, and N_2 in order of increasing magnitude of these deviations.

5.76 Does molecular size cause negative or positive deviations from the PV/RT ratio of an ideal gas? Use data from Table 5.5 to rank Cl_2, H_2, and O_2 in order of increasing magnitude of these deviations.

5.77 Does N_2 behave more ideally at 1 atm or at 500 atm? Explain.

5.78 Does SF_6 (boiling point 16°C at 1 atm) behave more ideally at 150°C or at 20°C? Explain.

Comprehensive Problems

Problems with an asterisk (*) are more challenging.

5.79 An "empty" gasoline can with dimensions 15.0 cm by 40.0 cm by 12.5 cm is attached to a vacuum pump and evacuated. If the atmospheric pressure is 14.7 lb/in², what is the total force (in pounds) on the outside of the can?

5.80 Hemoglobin is the protein that transports O_2 through the blood from the lungs to the rest of the body. In doing so, each molecule of hemoglobin combines with four molecules of O_2. If 1.00 g of hemoglobin combines with 1.53 mL of O_2 at 37°C and 743 torr, what is the molar mass of hemoglobin?

5.81 A baker uses sodium hydrogen carbonate (baking soda) as the leavening agent in a banana-nut quickbread. The baking soda decomposes according to two possible reactions:
Reaction 1: $2NaHCO_3(s) \longrightarrow Na_2CO_3(s) + H_2O(l) + CO_2(g)$
Reaction 2: $NaHCO_3(s) + H^+(aq) \longrightarrow$
$$H_2O(l) + CO_2(g) + Na^+(aq)$$

Calculate the mL of CO_2 that forms at 200.°C and 0.975 atm per gram of $NaHCO_3$ by each of the reaction processes.

5.82 A weather balloon containing 600. L of He is released near the equator at 1.01 atm and 305 K. It rises to a point where conditions are 0.489 atm and 218 K and eventually lands in the northern hemisphere under conditions of 1.01 atm and 250 K. If one-fourth of the helium leaked out during this journey, what is the volume (in L) of the balloon at landing?

*5.83 Chlorine is produced from concentrated seawater by the electrochemical chlor-alkali process. During the process, the chlorine is collected in a container that is isolated from the other products to prevent unwanted (and explosive) reactions. If a 15.00-L container holds 0.5850 kg of Cl_2 gas at 225°C, calculate
(a) P_{IGL}

(b) P_{VDW} $\left(\text{Use } R = 0.08206 \dfrac{\text{atm·L}}{\text{mol·K}}. \right)$

*5.84 Hydrogen can also be produced industrially by the chlor-alkali process. When 1 mol of $H_2(g)$ at 0°C is compressed until its volume is 0.4483 L, the pressure is observed to be 51.60 atm.
(a) If the hydrogen is behaving ideally, what is its calculated pressure?
(b) What is its real pressure?
(c) Compare the percentage errors in the pressures predicted by the two calculations.

5.85 Will the volume of a gas increase, decrease, or remain unchanged for each of the following sets of changes?
(a) The pressure is decreased from 2 atm to 1 atm, while the temperature is decreased from 200°C to 100°C.
(b) The pressure is increased from 1 atm to 3 atm, while the temperature is increased from 100°C to 300°C.
(c) The pressure is increased from 3 atm to 6 atm, while the temperature is increased from −73°C to 127°C.
(d) The pressure is increased from 0.2 atm to 0.4 atm, while the temperature is decreased from 300°C to 150°C.

5.86 When air is inhaled, it enters the alveoli of the lungs, and varying amounts of the component gases exchange with dissolved gases in the blood. As a result, the alveolar gas mixture is quite different from the atmospheric mixture. The following table presents selected data on the composition and partial pressure of four gases in the atmosphere and in the alveoli:

| | Atmosphere (sea level) | | Alveoli | |
| | | Partial | | Partial |
Gas	Mol %	Pressure (torr)	Mol %	Pressure (torr)
N_2	78.6	—	—	569
O_2	20.9	—	—	104
CO_2	0.04	—	—	40
H_2O	0.46	—	—	47

If the total pressure of each gas mixture is 1.00 atm, calculate:
(a) The partial pressure (in torr) of each gas in the atmosphere
(b) The mol % of each gas in the alveoli
(c) The number of O_2 molecules in 0.50 L of alveolar air (volume of an average breath at rest) at 37°C

5.87 Radon (Rn) is the heaviest, and only radioactive, member of Group 8A(18) (noble gases). It is a product of the disintegration of heavier radioactive nuclei found in minute concentrations in many common rocks used for building and construction. In recent years, health concerns about the cancers caused from inhaled residential radon have grown. If 1.0×10^{15} atoms of radium (Ra) produce an average of 1.373×10^4 atoms of Rn per second, how many liters of Rn, measured at STP, are produced per day by 1.0 g of Ra?

5.88 At 1400. mmHg and 286 K, a skin diver exhales a 208-mL bubble of air that is 77% N_2, 17% O_2, and 6.0% CO_2 by volume.
(a) How many milliliters would the volume of the bubble be if it were exhaled at the surface at 1 atm and 298 K?
(b) How many moles of N_2 are in the bubble?

5.89 The mass of the Earth's atmosphere is estimated as 5.14×10^{15} t (1 t = 1000 kg).
(a) If the average molar mass of air is 28.8 g/mol, how many moles of gas are in the atmosphere?
(b) How many liters would the atmosphere occupy at 25°C and 1 atm?
(c) If the surface area of the Earth is 5.100×10^8 km^2, how high should this volume of air extend? Why does the atmosphere actually extend so much higher?

5.90 Nitrogen dioxide has opposing roles in our world. It is essential in the industrial production of nitric acid, but is also a contributor to acid rain and photochemical smog. Calculate the volume of nitrogen dioxide produced at 735 torr and 28.2°C by the reaction of 4.95 cm^3 of copper (d = 8.95 g/cm^3) with 230.0 mL of nitric acid (d = 1.42 g/cm^3, 68.0% HNO_3 by mass):

$$Cu(s) + 4HNO_3(aq) \longrightarrow Cu(NO_3)_2(aq) + 2NO_2(g) + 2H_2O(l)$$

5.91 In the average adult male, the residual volume (RV) of the lungs, the volume of air remaining after a forced exhalation, is 1200 mL.
(a) How many moles of air are present in the RV at 1.0 atm and 37°C?
(b) How many molecules of gas are present under these conditions?

5.92 In a bromine-producing plant, how many liters of gaseous elemental bromine at 300°C and 0.855 atm are formed by the reaction of 275 g of sodium bromide and 175.6 g of sodium bromate in aqueous acid solution? (Assume no Br_2 dissolves.)

$$5NaBr(aq) + NaBrO_3(aq) + 3H_2SO_4(aq) \longrightarrow$$
$$3Br_2(g) + 3Na_2SO_4(aq) + 3H_2O(g)$$

5.93 Automobile air bags respond to a collision of a preset strength by electrically triggering the explosive decomposition of sodium azide (NaN_3) to its elements. In an industrial lab simulation, 15.3 mL of nitrogen gas was collected over water at 25°C and 755 torr. How many grams of azide decomposed?

5.94 An anesthetic gas contains 64.81% carbon, 13.60% hydrogen, and 21.59% oxygen, by mass. If 2.00 L of the gas at 25°C

and 0.420 atm weighs 2.57 g, what is the molecular formula of the anesthetic?

5.95 Aluminum chloride is easily vaporized at temperatures above 180°C. The gas escapes through a pinhole 0.122 times as fast as helium at the same conditions of temperature and pressure in the same apparatus. What is the molecular formula of gaseous aluminum chloride?

*5.96 (a) What is the total volume of gaseous *products,* measured at 350°C and 735 torr, when an automobile engine burns 100. g of C_8H_{18} (a typical component of gasoline)? (b) For part (a), the source of O_2 is air, which is about 78% N_2, 21% O_2, and 1.0% Ar by volume. Assuming all the O_2 reacts, but none of the N_2 or Ar does, what is the total volume of gaseous *exhaust*?

*5.97 An atmospheric chemist studying the reactions of the pollutant SO_2 places a mixture of SO_2 and O_2 in a 2.00-L container at 900. K and an initial pressure of 1.95 atm. When the reaction occurs, gaseous SO_3 forms, and the pressure eventually falls to 1.65 atm. How many moles of SO_3 form?

5.98 A sample of liquid nitrogen trichloride was heated in a 2.50-L closed reaction vessel until it decomposed completely to gaseous elements. The resulting mixture exerted a pressure of 754 mmHg at 95°C.
(a) What was the partial pressure of each gas in the container?
(b) What was the mass of the original sample?

5.99 Ammonium nitrate, a common fertilizer, is used as an explosive in fireworks and by terrorists. It was the material used in the devastating and tragic explosion of the Oklahoma City federal building in 1995. How many liters of gas at 307°C and 1.00 atm are formed by the explosive decomposition of 15.0 kg of ammonium nitrate to nitrogen, oxygen, and water vapor?

5.100 An environmental engineer submits a sample of air contaminated with sulfur dioxide to the lab for analysis. To a 500.-mL sample at 700. torr and 38°C, she adds 20.00 mL of 0.01017 M aqueous iodine, which reacts as follows:

$$SO_2(aq) + I_2(aq) + H_2O(l) \longrightarrow$$
$$HSO_4^-(aq) + I^-(aq) + H^+(aq) \quad \text{[unbalanced]}$$

The unreacted I_2 is titrated with 11.37 mL of 0.0105 M sodium thiosulfate:

$$I_2(aq) + S_2O_3^{2-}(aq) \longrightarrow I^-(aq) + S_4O_6^{2-}(aq) \quad \text{[unbalanced]}$$

What is the volume % of SO_2 in the air sample?

5.101 Canadian chemists have developed a modern variation of the 1899 Mond process for preparing extremely pure metallic nickel. A sample of impure nickel reacts with carbon monoxide at 50°C to form gaseous nickel carbonyl, $Ni(CO)_4$.
(a) How many grams of nickel can be converted to the carbonyl with 3.55 m^3 of CO at 100.7 kPa?
(b) The carbonyl is then decomposed at 21 atm and 155°C to pure (>99.95%) nickel. How many grams of nickel are obtained per cubic meter of the carbonyl?
(c) The released carbon monoxide is cooled and collected for reuse by passing it through water at 35°C. If the barometric pressure is 769 torr, what volume (in m^3) of CO is formed per cubic meter of carbonyl?

5.102 Analysis of a newly discovered gaseous silicon-fluorine compound shows that it contains 33.01 mass % silicon. At 27°C, 2.60 g of the compound exerts a pressure of 1.50 atm in a 0.250-L vessel. What is the molecular formula of the compound?

5.103 A gaseous organic compound containing only carbon, hydrogen, and nitrogen is burned in oxygen gas, and the individual

volume of each reactant and product is measured under the same conditions of temperature and pressure. Reaction of four volumes of the compound produces four volumes of CO_2, two volumes of N_2, and ten volumes of water vapor.
(a) What volume of oxygen gas was required?
(b) What is the empirical formula of the compound?

5.104 A piece of dry ice (solid CO_2, $d = 0.900$ g/mL) weighing 10.0 g is placed in a 0.800-L bottle filled with air at 0.980 atm and 550.0°C. The bottle is capped, and the dry ice changes to gas. What is the final pressure inside the bottle?

5.105 Containers A, B, and C are attached by closed stopcocks of negligible volume.

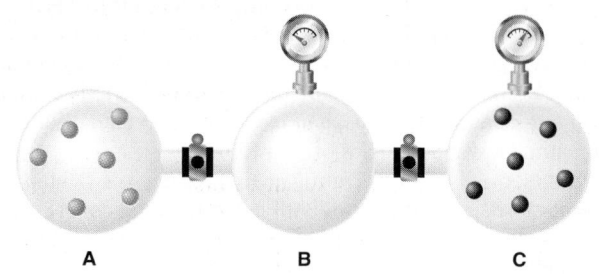

If each particle in the figure represents 10^6 particles,
(a) How many blue particles and black particles are in B after the stopcocks are opened and the system reaches equilibrium?
(b) How many blue particles and black particles are in A after the stopcocks are opened and the system reaches equilibrium?
(c) If the pressure in C, P_C, is 750 torr before the stopcocks are opened, what is P_C afterward?
(d) What is P_B afterward?

5.106 At the temperatures existing in the thermosphere (see Figure B5.1), instruments would break down and astronauts would be killed. Yet satellites function in orbit there for many years and astronauts routinely repair equipment on space walks. Explain.

*5.107 By what factor would a scuba diver's lungs expand if she ascended rapidly to the surface without inhaling or exhaling from a depth of 125 ft? If an expansion factor greater than 1.5 causes lung rupture, how far could the diver safely ascend from 125 ft without breathing? Assume constant temperature (d of seawater = 1.04 g/mL; d of Hg = 13.5 g/mL).

5.108 When 15.0 g of fluorite (CaF_2) reacts with excess sulfuric acid, hydrogen fluoride gas is collected at 744 torr and 25.5°C. Solid calcium sulfate is the other product. What gas temperature is required to store the gas in an 8.63-L container at 875 torr?

5.109 Dilute aqueous hydrogen peroxide is used as a bleaching agent and for disinfecting surfaces and small cuts. Its concentration is sometimes given as a certain number of "volumes hydrogen peroxide," which refers to the number of volumes of O_2 gas, measured at STP, that a given volume of hydrogen peroxide solution will release when it decomposes to O_2 and liquid H_2O. How many grams of hydrogen peroxide are in 0.100 L of "20 volumes hydrogen peroxide" solution?

5.110 At a height of 300 km above the Earth's surface, an astronaut finds that the atmospheric pressure is about 10^{-8} mmHg and the temperature is 500 K. How many molecules of gas are there per milliliter at this altitude?

5.111 (a) What is the rms speed of O_2 molecules at STP? (b) If the mean free path of O_2 molecules at STP is 6.33×10^{-8} m,

what is their collision frequency? [Use $R = 8.314$ J/(mol·K) and $\mathcal{M}$ in kg/mol.]

5.112 A barometer tube is 1.00×10^2 cm long and has a cross-sectional area of 1.20 cm². The height of the mercury column is 74.0 cm, and the temperature is 24°C. A small amount of N_2 is introduced into the evacuated space above the mercury, which causes the mercury level to drop to a height of 66.0 cm. How many grams of N_2 were introduced?

5.113 What is the molar concentration of the cleaning solution formed when 10.0 L of ammonia gas at 31°C and 735 torr dissolves in enough water to give a final volume of 0.750 L?

5.114 For each of the following, which shows the greater deviation from ideal behavior at the same set of conditions? Explain your choice.
(a) Argon or xenon
(b) Water vapor or neon
(c) Mercury vapor or radon
(d) Water vapor or methane

5.115 How many liters of gaseous hydrogen bromide at 27°C and 0.975 atm will a chemist need if she wishes to prepare 3.50 L of 1.20 M hydrobromic acid?

5.116 A mixture consisting of 7.0 g of CO and 10.0 g of SO_2, two atmospheric pollutants, has a pressure of 0.33 atm when placed in a sealed container. What is the partial pressure of CO?

*5.117 A mixture of CO_2 and Kr weighs 35.0 g and exerts a pressure of 0.708 atm in its container. Since Kr is expensive, you wish to recover it from the mixture. After the CO_2 is completely removed by absorption with NaOH(s), the pressure in the container is 0.250 atm. How many grams of CO_2 were originally present? How many grams of Kr can you recover?

5.118 When a car accelerates quickly, the passengers feel a force that presses them back into their seats, but a balloon filled with helium floats forward. Why?

*5.119 Aqueous sulfurous acid (H_2SO_3) was made by dissolving 0.200 L of sulfur dioxide gas at 20.°C and 740. mmHg in water to yield 500.0 mL of solution. The acid solution required 10.0 mL of sodium hydroxide solution to reach the titration end point. What was the molarity of the sodium hydroxide solution?

5.120 The lunar surface reaches 370 K at midday. The atmosphere consists of neon, argon, and traces of helium at a total pressure of only 2×10^{-14} atm. Calculate the rms speed of each component in the lunar atmosphere. [Use $R = 8.314$ J/(mol·K) and $\mathcal{M}$ in kg/mol.]

*5.121 A person inhales air richer in O_2 and exhales air richer in CO_2 and water vapor. During each hour of sleep, a person exhales a total of about 300 L of this CO_2-enriched and H_2O-enriched air.
(a) If the partial pressures of CO_2 and H_2O in exhaled air are each 30.0 torr at 37.0°C, calculate the mass of CO_2 and of H_2O exhaled in 1 h of sleep.
(b) How many grams of body mass does the person lose in an 8-h sleep if all the CO_2 and H_2O exhaled come from the metabolism of glucose?

$$C_6H_{12}O_6(s) + 6O_2(g) \longrightarrow 6CO_2(g) + 6H_2O(g)$$

5.122 Popcorn pops because the horny endosperm, a tough, elastic material, resists gas pressure within the heated kernel until it reaches explosive force. A 0.25-mL kernel has a water content of 1.5% by mass, and the water vapor reaches 175°C and 9.0 atm before the kernel ruptures. Assume the water vapor can occupy 75% of the kernel's volume. (a) What is the mass of the kernel? (b) How many milliliters would this amount of water vapor occupy at 25°C and 1.00 atm?

*5.123 Given these relationships for average kinetic energy,

$$\overline{E_k} = \tfrac{1}{2}m\overline{u^2} \qquad \text{and} \qquad \overline{E_k} = \tfrac{3}{2}\left(\frac{R}{N_A}\right)T$$

where m is molecular mass, u is rms speed, R is the gas constant [in J/(mol·K)], N_A is Avogadro's number, and T is absolute temperature:
(a) Derive Equation 5.13.
(b) Derive Equation 5.14.

5.124 What would you observe if you tilted a barometer 30° from the vertical? Explain.

5.125 The cylinder in part A of the figure below contains 0.1 mol of a gas that behaves ideally. Choose the cylinder (B, C, or D) that correctly represents the volume of the gas after each of the following changes. If none of the cylinders is correct, specify "none":
(a) P is doubled at fixed n and T.
(b) T is reduced from 400 K to 200 K at fixed n and P.
(c) T is increased from 100°C to 200°C at fixed n and P.
(d) 0.1 mol of gas is added at fixed P and T.
(e) 0.1 mol of gas is added and P is doubled at fixed T.

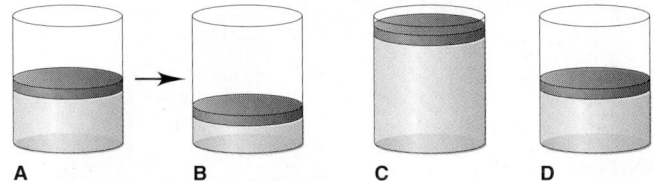

A   **B**   **C**   **D**

5.126 A 6.0-L flask contains a mixture of methane (CH_4), argon, and helium at 45°C and 1.75 atm. If the mole fractions of helium and argon are 0.25 and 0.35, respectively, how many molecules of methane are present?

*5.127 A large portion of metabolic energy arises from the biological combustion of glucose:

$$C_6H_{12}O_6(s) + 6O_2(g) \longrightarrow 6CO_2(g) + 6H_2O(g)$$

(a) If this reaction is carried out in an expandable container at 35°C and 780. torr, what volume of CO_2 is produced from 18.0 g of glucose and excess O_2?
(b) If the reaction is carried out at the same conditions with the stoichiometric amount of O_2, what is the partial pressure of each gas when the reaction is 50% complete (9.0 g of glucose remains)?

5.128 What is the average kinetic energy and rms speed of N_2 molecules at STP? Compare these values with those of H_2 molecules under the same conditions. [Use $R = 8.314$ J/(mol·K) and $\mathcal{M}$ in kg/mol.]

5.129 Gases are kept in pressurized tanks to reduce their volume for ease in shipping. (a) If an empty 10.0-L tank weighs 2.0 kg, how many moles of an ideal gas are in the tank at 298 K if the pressure is 2.50×10^3 psi? (b) What is the total mass if the tank is filled with N_2 (assume N_2 behaves ideally)? (c) What volume (in L) would the N_2 occupy at 1.00 atm?

*5.130 An equimolar mixture of Ne and Xe is accidentally placed in a container that has a tiny leak. After a short while, a very small proportion of the mixture has escaped. What is the mole fraction of Ne in the effusing gas?

*5.131 One way to utilize naturally occurring uranium (0.72% ^{235}U and 99.27% ^{238}U) as a nuclear fuel is to enrich it (increase its ^{235}U content) by allowing gaseous UF_6 to effuse through a porous membrane (see margin note, p. 202). From the relative rates of effusion of $^{235}UF_6$ and $^{238}UF_6$, find the number of steps needed to produce uranium that is 3.0 mole % ^{235}U, the enriched fuel used in many nuclear reactors.

*5.132 A slight deviation from ideal behavior exists even at normal conditions. If it behaved ideally, 1 mol of CO would occupy 22.414 L and exert 1 atm pressure at 273.15 K. Calculate P_{VDW} for 1.000 mol of CO at 273.15 K. $\left(\text{Use } R = 0.08206 \ \dfrac{\text{atm·L}}{\text{mol·K}}; \text{ for } \right.$

$\left. \text{CO}, a = 1.45 \ \dfrac{\text{atm·L}^2}{\text{mol}^2} \text{ and } b = 0.0395 \text{ L/mol} \right)$

*5.133 In preparation for a combustion demonstration, a professor's assistant fills a balloon with equal molar amounts of H_2 and O_2, but the demonstration has to be postponed until the next day. During the night, both gases leak through pores in the balloon. If 45% of the H_2 leaks, what is the O_2/H_2 ratio in the balloon the next day?

5.134 A truck tire has a volume of 208 L and is filled with air to 35.0 psi at 295 K. After a drive, the air heats up to 319 K. (a) If the tire volume is constant, what is the pressure? (b) If the tire volume increases 2.0%, what is the pressure? (c) If the tire leaks 1.5 g of air per minute and the temperature is constant, how many minutes will it take for the tire to reach the original pressure of 35.0 psi? ($\mathcal{M}$ of air = 28.8 g/mol)

*5.135 When fluorine and solid iodine are heated to high temperatures, the iodine sublimes and gaseous iodine heptafluoride forms. If 350. torr of fluorine gas and 2.50 g of solid iodine are put into a 2.50-L container at 250. K and the container is heated to 550. K, what is the final pressure? What is the partial pressure of iodine gas?

5.136 Many water treatment plants use chlorine gas to kill microorganisms before the water is released for residential use. A plant engineer has to maintain the chlorine pressure in a tank below the 85.0-atm rating and, to be safe, decides to fill the tank to 80.0% of this maximum pressure. (a) How many moles of Cl_2 gas can be kept in the 850.-L tank at 298 K if she uses the ideal gas law in the calculation? (b) What is the tank pressure if she uses the van der Waals equation for this amount of gas? (c) Did the engineer fill the tank to the desired pressure?

CHAPTER 6

THERMOCHEMISTRY: ENERGY FLOW AND CHEMICAL CHANGE

CHAPTER OUTLINE

6.1 Forms of Energy and Their Interconversion
System and Surroundings
Energy Flow to and from a System
Heat and Work
Energy Conservation
Units of Energy
State Functions

6.2 Enthalpy: Heats of Reaction and Chemical Change
The Meaning of Enthalpy
Comparing ΔE and ΔH
Exothermic and Endothermic Processes
Types of Enthalpy Change
Bond Strength and Heat of Reaction

6.3 Calorimetry: Laboratory Measurement of Heats of Reaction
Specific Heat Capacity
The Practice of Calorimetry

6.4 Stoichiometry of Thermochemical Equations

6.5 Hess's Law of Heat Summation

6.6 Standard Heats of Reaction (ΔH^0_{rxn})
Formation Equations
Determining ΔH^0_{rxn} from ΔH^0_f

Figure: The wonder of a burning match. When a match burns, the chemical reaction that occurs releases energy, as do many other processes. Still others absorb energy. Our ability to understand and measure changes in matter and energy is crucial to the ultimate fate of modern society, and in this chapter, we begin an exploration of the factors that govern these changes.

All changes in matter, whether chemical or physical, are accompanied by changes in the energy content of the matter. In the inferno of a forest fire, as wood is converted to ash and gases, its energy content changes, and that difference in energy before and after the change is *released* as heat and light. In contrast, some of the energy in a flash of lightning is *absorbed* when atmospheric N_2 and O_2 react to form NO. Energy is *absorbed* when snow melts and is *released* when water vapor condenses to rain.

The production and usage of energy in its many forms have an enormous impact on society. Some of the largest industries manufacture products that release, absorb, or limit the flow of energy. Common fuels—oil, wood, coal, and natural gas—release chemical energy for heating and to power combustion engines and steam turbines. Fertilizers help crops absorb solar energy and convert it to the chemical energy of food, which our bodies convert into other forms. Many plastic, fiberglass, and ceramic materials serve as insulators that limit energy flow.

In this chapter, we investigate the heat, or *thermal energy*, associated with changes in matter. We first examine some basic ideas of **thermodynamics,** the study of heat and its transformations. Then, we discuss **thermochemistry,** the branch of thermodynamics that deals with the heat involved in chemical reactions. We explore changes at the molecular level to find out where the heat change of a reaction comes from. Then we learn how heat is measured in order to focus on the key value: the quantity of heat released or absorbed in a reaction. The chapter ends with an overview of current and future energy sources and the conflicts between energy demand and environmental quality.

Thermodynamics is a fascinating field, a rigorously logical, highly mathematical branch of science that is as relevant in everyday life as it is in the laboratory or the environment. Our discussion of thermodynamics here, and later in Chapter 20, is confined mostly to chemical applications, but also tries to show how widely these ideas can be applied.

6.1 FORMS OF ENERGY AND THEIR INTERCONVERSION

As we discussed in Chapter 1, all energy is either potential or kinetic, and these forms are convertible from one to the other. An object has potential energy by virtue of its position and kinetic energy by virtue of its motion. The potential energy of a weight raised above the ground is converted to kinetic energy as it falls (see Figure 1.3). When the weight hits the ground, it transfers some kinetic energy to the soil and pebbles, causing them to move, thereby doing *work*. In addition, some of the transferred kinetic energy appears as *heat*, as it slightly warms the soil and pebbles. Thus, the potential energy of the weight is converted to kinetic energy, which is transferred to the ground as work and as heat.

Modern atomic theory allows us to consider other forms of energy—solar, electrical, nuclear, and chemical—as examples of potential and kinetic energy on the atomic and molecular scales. No matter what the details of the situation, *when energy is transferred from one object to another, it appears as work and/or as heat.* In this section, we examine this idea in terms of the loss or gain of energy that takes place during a chemical or physical change.

The System and Its Surroundings

To make a meaningful observation and measurement of a change in energy, we must first define the **system,** that part of the universe whose change we are going to observe. The moment we define the system, everything else relevant to the change is defined as the **surroundings.**

Wherever You Look, There Is a System In the example of the weight hitting the ground, if we define the falling weight as the system, the soil and pebbles that are moved and warmed are the surroundings. An astronomer may define a galaxy as the system and nearby galaxies as the surroundings. An ecologist studying African wildlife can define a zebra herd as the system and other animals, plants, and water supplies as the surroundings. A microbiologist may define a certain cell as the system and the extracellular solution as the surroundings. Thus, in general, it is the experiment and the experimenter that define the system and the surroundings.

Figure 6.1 A chemical system and its surroundings. Once the contents of the flask (the orange solution) are defined as the system, the flask and the laboratory become defined as the surroundings.

Figure 6.1 shows a typical chemical system and its surroundings: the system is the contents of the flask; the flask itself, the other equipment, and perhaps the rest of the laboratory are the surroundings. In principle, the rest of the universe is the surroundings, but in practice, we need to consider only the portions of the universe relevant to the system. That is, it's not likely that a thunderstorm in central Asia or a methane blizzard on Neptune will affect the contents of the flask, but the temperature, pressure, and humidity of the lab might.

Energy Flow to and from a System

Each particle in a system has potential and kinetic energy, and the sum of these energies for all the particles in the system is the **internal energy, E** (some texts use the symbol U). When a chemical system, such as the contents of the flask in Figure 6.1, changes from reactants to products and the products return to the starting temperature, the internal energy has changed. To determine this change, ΔE, we measure the difference between the system's internal energy *after* the change (E_{final}) and *before* the change ($E_{initial}$):

$$\Delta E = E_{final} - E_{initial} = E_{products} - E_{reactants} \qquad \textbf{(6.1)}$$

where Δ (Greek *delta*) means "change (or difference) in." Note especially that Δ refers to the *final state of the system **minus** the initial state*.

A change in the energy of the system is always accompanied by an **opposite** change in the energy of the surroundings. We often represent this change with an *energy diagram* in which the final and initial states are horizontal lines on a vertical energy axis. The change in internal energy, ΔE, is the difference between the heights of the two lines. A reacting chemical system can change its internal energy in either of two ways:

1. By losing some energy *to* the surroundings, as shown in Figure 6.2A:

$$E_{final} < E_{initial} \qquad \Delta E < 0$$

2. By gaining some energy *from* the surroundings, as shown in Figure 6.2B:

$$E_{final} > E_{initial} \qquad \Delta E > 0$$

Note that the change is a *transfer* of energy from system to surroundings, and vice versa.

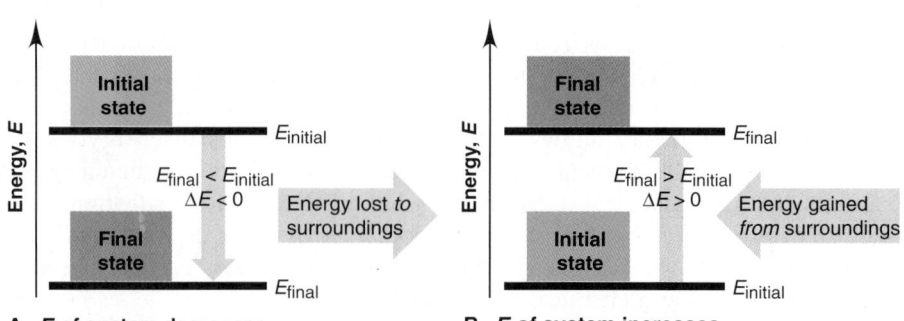

A *E* of system decreases   **B** *E* of system increases

Figure 6.2 Energy diagrams for the transfer of internal energy (*E*) between a system and its surroundings. **A,** When the internal energy of a system *decreases,* the change in energy (ΔE) is lost *to* the surroundings; therefore, ΔE of the system ($E_{final} - E_{initial}$) is negative. **B,** When the system's internal energy *increases,* ΔE is gained *from* the surroundings and is positive. Note that the vertical yellow arrow, which signifies the direction of the change in energy, *always* has its tail at the initial state and its head at the final state.

Heat and Work: Two Forms of Energy Transfer

Just as we saw when a weight hits the ground, energy transfer outward from the system or inward from the surroundings can appear in two forms, heat and work. **Heat** (or *thermal energy,* symbol *q*) is the energy transferred between a system and its surroundings as a result of a difference in their temperatures only. Energy in the form of heat is transferred from hot soup (system) to the bowl, air, and table (surroundings) because the surroundings have a lower temperature. All other forms of energy transfer (mechanical, electrical, and so on) involve some type of **work (*w*),** the energy transferred when an object is moved by a force. When you (system) kick a football (surroundings), energy is transferred as work to move the ball. If it needs to be inflated, the inside air (system) exerts a force on the inner wall of the ball and nearby air (surroundings) and does work to move it outward.

Since energy can be transferred as heat and/or work, the total change in a system's internal energy is

$$\Delta E = q + w \qquad\qquad (6.2)$$

The numerical values of *q* and *w* can be either positive or negative, depending on the change the *system* undergoes. In other words, *we define the sign of the energy transfer from the system's perspective.* Energy coming *into* the system is *positive.* Energy going *out from* the system is *negative.* Of the innumerable changes possible in the system's internal energy, let's examine four simple cases—two that involve only heat and two that involve only work:

Energy Transfer as Heat Only For a system that does no work but transfers energy only as heat (*q*), we know that *w* = 0. Therefore, from Equation 6.2, we have $\Delta E = q + 0 = q$.

  1. *Heat flowing **out from** a system.* Suppose a sample of hot water is the system; then, the beaker containing it and the rest of the lab are the surroundings. The water transfers energy as heat to the surroundings until the temperature of the water equals that of the surroundings. The system's energy decreases as heat flows *out from* the system, so the final energy of the system is less than its initial energy. Heat was lost by the system, so *q is negative,* and therefore ΔE *is negative.* Figure 6.3A shows this situation.

  2. *Heat flowing **into** a system.* On the other hand, if the system consists of ice water, it gains energy as heat from the surroundings until the temperature of the water equals that of the surroundings. In this case, energy is transferred *into* the system, so the final energy of the system is higher than its initial energy. Heat was gained by the system, so *q is positive,* and therefore ΔE *is positive.* Figure 6.3B shows this situation.

Thermodynamics in the Kitchen The air in a refrigerator (surroundings) has a lower temperature than a newly added piece of food (system), so the food loses energy as heat to the refrigerator air, $q < 0$. The air in a hot oven (surroundings) has a higher temperature than a newly added piece of food (system), so the food gains energy as heat from the oven air, $q > 0$.

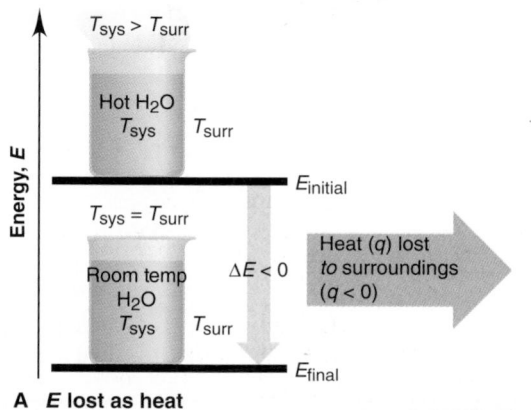

A *E lost as heat*

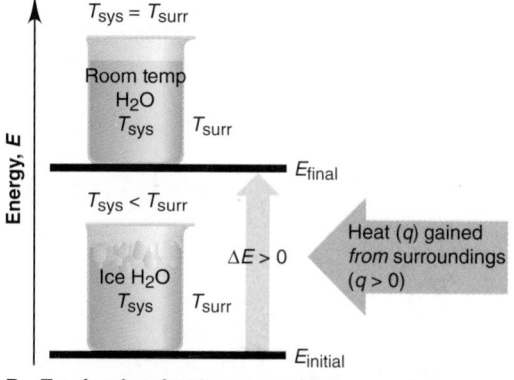

B *E gained as heat*

Figure 6.3 A system transferring energy as heat only. A, Hot water (the system, sys) transfers energy as heat (*q*) *to* the surroundings (surr) until $T_{sys} = T_{surr}$. Since $E_{initial} > E_{final}$ and $w = 0$, $\Delta E < 0$ and the sign of *q* is negative. **B,** Ice water gains energy as heat (*q*) *from* the surroundings until $T_{sys} = T_{surr}$. Since $E_{initial} < E_{final}$ and $w = 0$, $\Delta E > 0$ and the sign of *q* is positive.

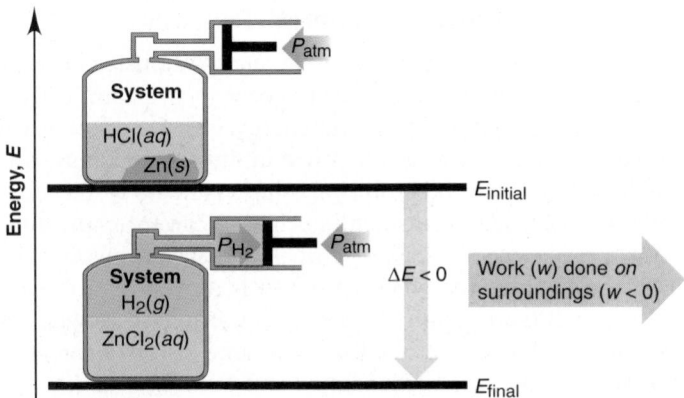

Figure 6.4 A system losing energy as work only. The internal energy of the system decreases as the reactants form products because the $H_2(g)$ does work (w) *on* the surroundings by pushing back the piston. The reaction vessel is insulated, so $q = 0$. Since $E_{initial} > E_{final}$, $\Delta E < 0$ and the sign of w is negative.

Energy Transfer as Work Only For a system that transfers energy only as work (w) and not as heat, we know that $q = 0$; therefore, $\Delta E = 0 + w = w$.

1. *Work done by a system.* Consider the reaction between zinc and hydrochloric acid as it takes place in an insulated container attached to a piston-cylinder assembly. We define the system as the atoms that make up the substances. In the initial state, the system's internal energy is that of the atoms in the form of the reactants, metallic Zn and aqueous H^+ and Cl^- ions. In the final state, the system's internal energy is that of the same atoms in the form of the products, H_2 gas and aqueous Zn^{2+} and Cl^- ions:

$$Zn(s) + 2H^+(aq) + 2Cl^-(aq) \longrightarrow H_2(g) + Zn^{2+}(aq) + 2Cl^-(aq)$$

As the H_2 gas forms, some of the internal energy is used *by* the system to do work *on* the surroundings and push the piston outward. Energy is lost by the system as work, so *w is negative* and *ΔE is negative,* as you see in Figure 6.4. The H_2 gas is doing **pressure-volume work (*PV* work),** the type of work in which a volume changes against an external pressure. The work done here is not very useful because it simply pushes back the piston and outside air. But, if the system is a ton of burning coal and O_2, and the surroundings are a locomotive engine, much of the internal energy lost from the system does the work of moving a train.

2. *Work done on a system.* If we increase the external pressure on the piston in Figure 6.4, the system gains energy because work is done *on* the system *by* the surroundings: *w is positive,* so ΔE *is positive.*

Table 6.1 summarizes the sign conventions of q and w and their effect on the sign of ΔE.

Table 6.1 The Sign Conventions* for q, w, and ΔE

q	$+$	w	$=$	ΔE
$+$		$+$		$+$
$+$		$-$		Depends on *sizes* of q and w
$-$		$+$		Depends on *sizes* of q and w
$-$		$-$		$-$

*For q: $+$ means system *gains* heat; $-$ means system *loses* heat.
For w: $+$ means work done *on* system; $-$ means work done *by* system.

The Law of Energy Conservation

As you've seen, when a system gains energy, the surroundings supply it, and when a system loses energy, the surroundings gain it. Energy can be converted from one form to another as these transfers take place, but it cannot simply appear or disappear—it cannot be created or destroyed. The **law of conservation of energy** restates this basic observation as follows: *the total energy of the universe is constant*. This law is also known as the **first law of thermodynamics.**

Conservation of energy applies everywhere. As gasoline burns in a car engine, the released energy appears as an equivalent amount of heat and work. The heat warms the car parts, passenger compartment, and surrounding air. The work appears as mechanical energy to turn the car's wheels and belts. That energy is converted further into the electrical energy of the clock and radio, the radiant energy of the headlights, the chemical energy of the battery, the heat due to friction, and so forth. If you took the sum of all these energy forms, you would find that it equals the change in energy between the reactants and products as the gasoline is burned. Complex biological processes also obey energy conservation. Through photosynthesis, green plants convert radiant energy from the Sun into chemical energy, as low-energy CO_2 and H_2O are used to make high-energy carbohydrates (such as wood) and O_2. When the wood is burned in air, those low-energy compounds form again, and the energy difference is released to the surroundings.

Thus, energy transfers between system and surroundings can be in the forms of heat and/or various types of work—mechanical, electrical, radiant, chemical—but *the energy of the system plus the energy of the surroundings remains constant: energy is conserved*. A mathematical expression of the law of conservation of energy (first law of thermodynamics) is

$$\Delta E_{\text{universe}} = \Delta E_{\text{system}} + \Delta E_{\text{surroundings}} = 0 \qquad (6.3)$$

This profound idea pertains to all systems, from a burning match to the movement of continents, from the inner workings of your heart to the formation of the Solar System.

Units of Energy

The SI unit of energy is the **joule (J),** a derived unit composed of three base units:
$$1 \text{ J} = 1 \text{ kg·m}^2/\text{s}^2$$
Both heat and work are expressed in joules. Let's see how these units arise in the case of work. The work (w) done on a mass is the force (F) times the distance (d) that the mass moves: $w = F \times d$. A *force* changes the velocity of (accelerates) a mass. Velocity has units of meters per second (m/s), so acceleration (a) has units of m/s^2. Force, therefore, has units of mass (m, in kilograms) times acceleration:
$$F = m \times a \quad \text{in units of} \quad \text{kg·m/s}^2$$
Therefore, $w = F \times d$ has units of $(\text{kg·m/s}^2) \times \text{m} = \text{kg·m}^2/\text{s}^2 = \text{J}$
Potential energy, kinetic energy, and PV work are combinations of the same physical quantities and are also expressed in joules.

The **calorie (cal)** is an older unit that was defined originally as the quantity of energy needed to raise the temperature of 1 g of water by 1°C (from 14.5°C to 15.5°C). The calorie is now defined in terms of the joule:

$$1 \text{ cal} \equiv 4.184 \text{ J} \quad \text{or} \quad 1 \text{ J} = \frac{1}{4.184} \text{ cal} = 0.2390 \text{ cal}$$

Since the quantities of energy involved in chemical reactions are usually quite large, chemists use the kilojoule (kJ), or sometimes the kilocalorie (kcal):
$$1 \text{ kJ} = 1000 \text{ J} = 0.2390 \text{ kcal} = 239.0 \text{ cal}$$

The Tragic Life of the First Law's Discoverer After studying the work habits, the food intake, and the blood of sailors in the tropics and in northern Europe, the young German doctor J. R. von Mayer concluded that the energy of food is used to heat the body *and* do work, and thus heat and work are different forms of energy. When colleagues ridiculed this idea, von Mayer became despondent. Soon thereafter, James Joule, an English brewer and amateur scientist, demonstrated this idea experimentally but gave von Mayer no credit for the concept. Around this time, von Mayer's children died unexpectedly, and he became severely depressed. He attempted suicide and was seriously injured and later sent to an insane asylum. Tainted by his conduct, his family declared him legally dead. Many years later, von Mayer was released and finally received recognition for his great insight.

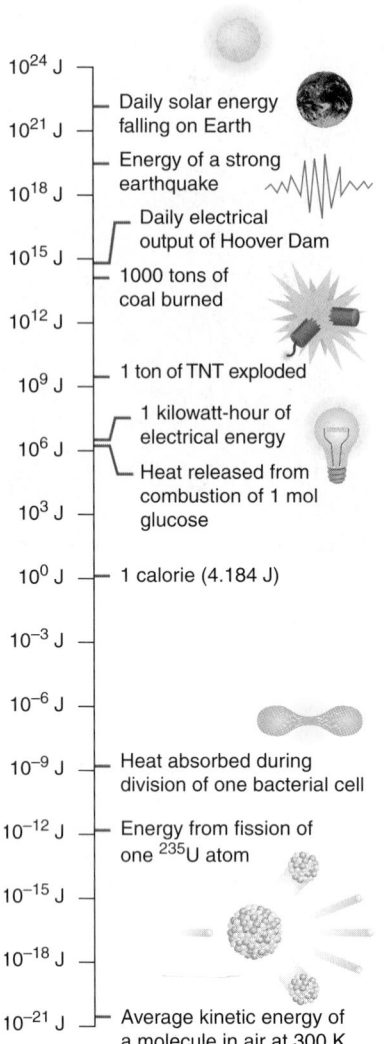

Figure 6.5 **Some interesting quantities of energy.** Note that the vertical scale is exponential.

The nutritional Calorie (note the capital C), the unit that diet tables use to show the energy available from food, is actually a kilocalorie. The *British thermal unit (Btu),* a unit in engineering that you may have seen used to indicate energy output of appliances, is the quantity of energy required to raise the temperature of 1 lb of water by 1°F and is equivalent to 1055 J. In general, the SI unit (J or kJ) is used throughout this text. Some interesting quantities of energy appear in Figure 6.5.

SAMPLE PROBLEM 6.1 Determining the Change in Internal Energy of a System

---

Problem When gasoline burns in a car engine, the heat released causes the products CO_2 and H_2O to expand, which pushes the pistons outward. Excess heat is removed by the car's cooling system. If the expanding gases do 451 J of work on the pistons and the system loses 325 J to the surroundings as heat, calculate the change in energy (ΔE) in J, kJ, and kcal.

Plan We must define system and surroundings, assign signs to q and w, and then calculate ΔE with Equation 6.2. The system is the reactants and products, and the surroundings are the pistons, the cooling system, and the rest of the car. Heat is released by the system, so q is negative. Work is done by the system to push the pistons outward, so w is also negative. We obtain the answer in J and then convert it to kJ and kcal.

Solution Calculating ΔE (from Equation 6.2) in J:

$$q = -325 \text{ J}$$
$$w = -451 \text{ J}$$
$$\Delta E = q + w = -325 \text{ J} + (-451 \text{ J})$$
$$= \boxed{-776 \text{ J}}$$

Converting from J to kJ:

$$\Delta E = -776 \text{ J} \times \frac{1 \text{ kJ}}{1000 \text{ J}}$$

$$= \boxed{-0.776 \text{ kJ}}$$

Converting from kJ to kcal:

$$\Delta E = -0.776 \text{ kJ} \times \frac{1 \text{ kcal}}{4.184 \text{ kJ}}$$

$$= \boxed{-0.185 \text{ kcal}}$$

Check The answer is reasonable: combustion releases energy from the system, so $E_{\text{final}} < E_{\text{initial}}$ and ΔE should be negative. Rounding shows that, since 4 kJ ≈ 1 kcal, nearly 0.8 kJ should be nearly 0.2 kcal.

FOLLOW-UP PROBLEM 6.1 In a reaction, gaseous reactants form a liquid product. The heat absorbed by the surroundings is 26.0 kcal, and the work done on the system is 15.0 Btu. Calculate ΔE (in kJ).

State Functions and the Path Independence of the Energy Change

An important point to understand is that there is no particular sequence by which the internal energy (E) of a system must change. This is because E is a **state function,** a property dependent only on the *current* state of the system (its composition, volume, pressure, and temperature), *not* on the path the system took to reach that state.

In fact, the energy change of a system can occur by countless combinations of heat (q) and work (w). No matter what the combination, however, the same

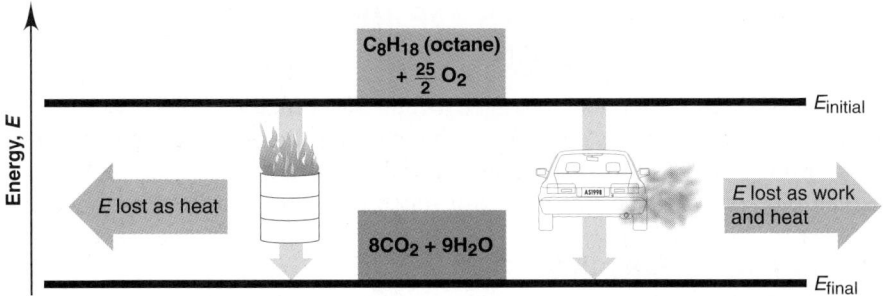

Figure 6.6 Two different paths for the energy change of a system. The change in internal energy when a given amount of octane burns in air is the same no matter how the energy is transferred. On the left, the fuel is burned in an open can, and the energy is lost almost entirely as heat. On the right, it is burned in a car engine, so a portion of the energy is lost as work to move the car, and thus less is lost as heat.

overall energy change occurs, because ΔE *does **not** depend on how the change takes place*. As an example, let's define a system in its initial state as 1 mol of octane (a component of gasoline) together with enough O_2 to burn it. In its final state, the system is the CO_2 and H_2O that form (a fractional coefficient is needed for O_2 because we specified 1 mol of octane):

$$C_8H_{18}(l) + \tfrac{25}{2}O_2(g) \longrightarrow 8CO_2(g) + 9H_2O(g)$$

initial state (E_{initial}) final state (E_{final})

Energy is released to warm the surroundings and/or do work on them, so ΔE is negative. Two of the ways the change can occur are shown in Figure 6.6. If we burn the octane in an open container, ΔE appears almost completely as heat (with a small amount of work done to push back the atmosphere). If we burn it in a car engine, a much larger portion ($\sim$30%) of ΔE appears as work that moves the car, with the rest used to heat the car, air, and exhaust gases. If we burn the octane in a lawn mower or a plane, ΔE appears as other combinations of work and heat.

Thus, even though the separate quantities of work and heat available from the change *do* depend on how the change occurs, the change in internal energy (the *sum* of the heat and work) *does not*. In other words, for a given change, ΔE *(sum of q and w) is constant, even though q and w can vary.* Thus, heat and work are *not* state functions because their values *do* depend on the path the system takes in undergoing the energy change.

The pressure (P) of an ideal gas or the volume (V) of water in a beaker are other examples of state functions. This path independence means that *changes in state functions—ΔE, ΔP, and ΔV—depend only on their initial and final states.* (Note that symbols for state functions, such as E, P, and V, are capitalized.)

Your Personal Financial State Function The *balance* in your checkbook is a state function of your personal financial system. You can open a new account with a birthday gift of $50, or you can open a new account with a deposit of a $100 paycheck and then write two $25 checks. The two paths to the balance are different, but the balance (current state) is the same.

SECTION SUMMARY

Energy is transferred as heat (q) when the system and surroundings are at different temperatures; energy is transferred as work (w) when an object is moved by a force. Heat or work gained by a system ($q > 0$; $w > 0$) increases its internal energy (E); heat or work lost by the system ($q < 0$; $w < 0$) decreases E. The total change in the system's internal energy is the sum of the heat and work: $\Delta E = q + w$. Heat and work are measured in joules (J). Energy is always conserved: it changes from one form into another, moving into or out of the system, but the total quantity of energy in the universe (system *plus* surroundings) is constant. Energy is a state function; therefore, the same ΔE can occur through any combination of q and w.

6.2 ENTHALPY: HEATS OF REACTION AND CHEMICAL CHANGE

Most physical and chemical changes occur at virtually constant atmospheric pressure—a reaction in an open flask, the freezing of a lake, a drug response in an organism. Defining a thermodynamic variable called *enthalpy,* which takes such constant pressure into account, makes it much easier to measure energy changes.

The Meaning of Enthalpy

To determine ΔE, we must measure both heat and work. The two most important types of chemical work are electrical work, the work done by moving charged particles (Chapter 21), and *PV* work, the work done by an expanding gas. We find the quantity of *PV* work done by multiplying the external pressure (P) by the change in volume of the gas (ΔV, or $V_{final} - V_{initial}$). In an open flask (or a cylinder with a weightless, frictionless piston), a gas does work by pushing back the atmosphere. As Figure 6.7 shows, this work done *on* the surroundings is defined as negative because the system loses energy:

$$w = -P\Delta V \tag{6.4}$$

For reactions that occur at constant pressure, a thermodynamic variable called **enthalpy (H)** eliminates the need to consider *PV* work. The enthalpy of a system is defined as the internal energy *plus* the product of the pressure and volume:

$$H = E + PV$$

The **change in enthalpy (ΔH)** is the change in internal energy *plus* the product of the constant pressure and the change in volume:

$$\Delta H = \Delta E + P\Delta V \tag{6.5}$$

Combining Equations 6.2 ($\Delta E = q + w$) and 6.4 leads to a key point about ΔH:

$$\Delta E = q + w = q + (-P\Delta V) = q - P\Delta V$$

At constant pressure, we denote q as q_P and solve for it:

$$q_P = \Delta E + P\Delta V$$

Notice that the right side of this equation is identical to the right side of Equation 6.5; that is,

$$q_P = \Delta E + P\Delta V = \Delta H \tag{6.6}$$

Thus, *the change in enthalpy equals the heat gained or lost at constant pressure.* Since most changes occur at constant pressure, ΔH is more relevant than ΔE *and* easier to find: *to find ΔH, measure q_P.*

We discuss the laboratory method for measuring the heat involved in a chemical or physical change in Section 6.3.

Comparing ΔE and ΔH

Knowing the *enthalpy* change of a system tells us a lot about its *energy* change as well. In fact, since many reactions involve little (if any) *PV* work, most (or all) of the energy change occurs as a transfer of heat. Here are three cases:

1. *Reactions that do not involve gases.* Gases do not appear in many reactions (precipitation, many acid-base, and many redox reactions). For example,

$$2KOH(aq) + H_2SO_4(aq) \longrightarrow K_2SO_4(aq) + 2H_2O(l)$$

Because liquids and solids undergo *very* small volume changes, $\Delta V \approx 0$; thus $P\Delta V \approx 0$ and $\Delta H \approx \Delta E$.

2. *Reactions in which the amount (mol) of gas does **not** change.* When the total amount of gaseous reactants equals the total amount of gaseous products, $\Delta V = 0$, so $P\Delta V = 0$ and $\Delta H = \Delta E$. For example,

$$N_2(g) + O_2(g) \longrightarrow 2NO(g)$$

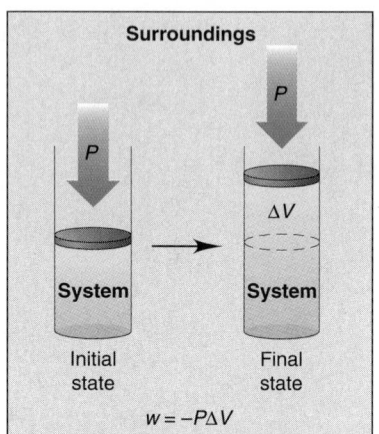

Surroundings

P

P

ΔV

System → **System**

Initial state   Final state

$w = -P\Delta V$

Figure 6.7 Pressure-volume work. When the volume (V) of a system increases by an amount ΔV against an external pressure (P), the system pushes back, and thus does *PV* work *on* the surroundings ($w = -P\Delta V$).

3. *Reactions in which the amount (mol) of gas **does** change.* In these cases, $P\Delta V \neq 0$. However, q_P is usually so much larger than $P\Delta V$ that ΔH is very close to ΔE. For instance, in the combustion of H_2, 3 mol of gas yields 2 mol of gas:

$$2H_2(g) + O_2(g) \longrightarrow 2H_2O(g)$$

In this reaction, $\Delta H = -483.6$ kJ and $P\Delta V = -2.5$ kJ, so (from Equation 6.5), we get

$$\Delta E = \Delta H - P\Delta V = -483.6 \text{ kJ} - (-2.5 \text{ kJ}) = -481.1 \text{ kJ}$$

Obviously, most of ΔE occurs as heat transfer, so $\Delta H \approx \Delta E$.

The key point to realize from these three cases is that *for many reactions, ΔH equals, or is very close to, ΔE.*

Exothermic and Endothermic Processes

Because E, P, and V are state functions, H is also a state function, which means that ΔH depends only on the *difference* between H_{final} and $H_{initial}$. The enthalpy change of a reaction, also called the **heat of reaction, ΔH_{rxn},** *always refers to H_{final} **minus** $H_{initial}$:*

$$\Delta H = H_{final} - H_{initial} = H_{products} - H_{reactants}$$

Therefore, because $H_{products}$ can be either more or less than $H_{reactants}$, the sign of ΔH indicates whether heat is absorbed or released in the change. We determine the sign of ΔH by imagining *the heat as a "reactant" or "product."* When methane burns in air, for example, we know that heat is produced, so we show it as a product (on the right):

$$CH_4(g) + 2O_2(g) \longrightarrow CO_2(g) + 2H_2O(g) + heat$$

Because heat is released to the surroundings, the products (1 mol of CO_2 and 2 mol of H_2O) must have less enthalpy than the reactants (1 mol of CH_4 and 2 mol of O_2). Therefore, ΔH ($H_{final} - H_{initial}$) is negative, as the **enthalpy diagram** in Figure 6.8A shows. An **exothermic** ("heat out") **process** *releases* heat and results in a *decrease* in the enthalpy of the system:

$$\text{Exothermic:} \quad H_{final} < H_{initial} \quad \Delta H < 0$$

An **endothermic** ("heat in") **process** *absorbs* heat and results in an *increase* in the enthalpy of the system. When ice melts, for instance, heat flows *into* the ice from the surroundings, so we show the heat as a reactant (on the left):

$$heat + H_2O(s) \longrightarrow H_2O(l)$$

Because heat is absorbed, the enthalpy of the liquid water is higher than that of the solid water, as Figure 6.8B shows. Therefore, ΔH ($H_{water} - H_{ice}$) is positive:

$$\text{Endothermic:} \quad H_{final} > H_{initial} \quad \Delta H > 0$$

In general, the value of an enthalpy change refers to reactants and products at the same temperature.

ENERGY FLOW

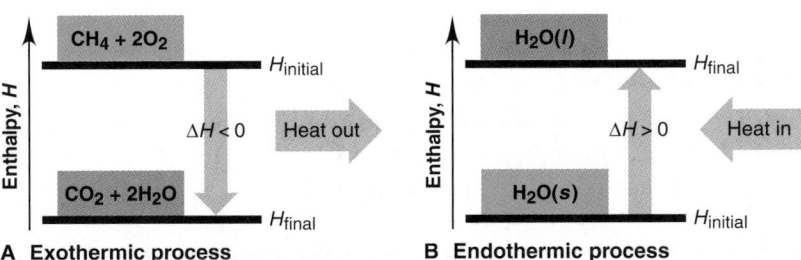

A Exothermic process   **B** Endothermic process

Figure 6.8 Enthalpy diagrams for exothermic and endothermic processes. **A,** Methane burns with a decrease in enthalpy because heat *leaves* the system. Therefore, $H_{final} < H_{initial}$, and the process is exothermic: $\Delta H < 0$. **B,** Ice melts with an increase in enthalpy because heat *enters* the system. Since $H_{final} > H_{initial}$, the process is endcthermic: $\Delta H > 0$.

SAMPLE PROBLEM 6.2 Drawing Enthalpy Diagrams and Determining the Sign of ΔH

Problem In each of the following cases, determine the sign of ΔH, state whether the reaction is exothermic or endothermic, and draw an enthalpy diagram:

(a) $H_2(g) + \frac{1}{2}O_2(g) \longrightarrow H_2O(l) + 285.8$ kJ

(b) 40.7 kJ $+ H_2O(l) \longrightarrow H_2O(g)$

Plan We inspect each equation to see whether heat is a "product" (exothermic; $\Delta H < 0$) or a "reactant" (endothermic; $\Delta H > 0$). For exothermic reactions, reactants are above products on the enthalpy diagram; for endothermic reactions, reactants are below products. The ΔH arrow *always* points from reactants to products.

Solution (a) Heat is a product (on the right), so $\Delta H < 0$ and the reaction is exothermic. The enthalpy diagram appears in the margin (*top*).

(b) Heat is a reactant (on the left), so $\Delta H > 0$ and the reaction is endothermic. The enthalpy diagram appears in the margin (*bottom*).

Check Substances on the same side of the equation as the heat have less enthalpy than substances on the other side, so make sure they are placed on the lower line of the diagram.

Comment ΔH values depend on conditions. In (b), for instance, $\Delta H = 40.7$ kJ at 1 atm and 100°C; at 1 atm and 25°C, $\Delta H = 44.0$ kJ.

FOLLOW-UP PROBLEM 6.2 When 1 mol of nitroglycerine decomposes, it causes a violent explosion and releases 5.72×10^3 kJ of heat:

$$C_3H_5(NO_3)_3(l) \longrightarrow 3CO_2(g) + \tfrac{5}{2}H_2O(g) + \tfrac{1}{4}O_2(g) + \tfrac{3}{2}N_2(g)$$

Is the reaction exothermic or endothermic? Draw an enthalpy diagram for it.

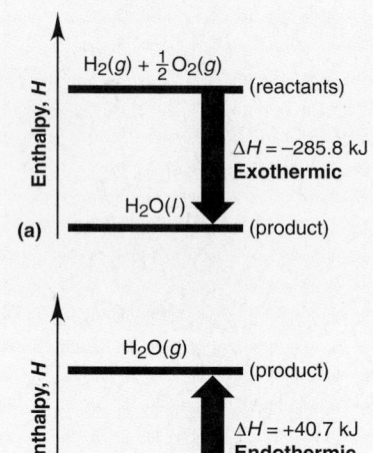

Some Important Types of Enthalpy Change

Certain enthalpy changes are very frequently studied and have special names:

- When 1 mol of a substance combines with O_2 in a combustion reaction as written, the heat of reaction is the **heat of combustion (ΔH_{comb}):**

$$C_4H_{10}(l) + \tfrac{13}{2}O_2(g) \longrightarrow 4CO_2(g) + 5H_2O(l) \quad \Delta H = \Delta H_{comb}$$

- When 1 mol of a compound is produced from its elements, the heat of reaction is the **heat of formation (ΔH_f):**

$$K(s) + \tfrac{1}{2}Br_2(l) \longrightarrow KBr(s) \quad \Delta H = \Delta H_f$$

- When 1 mol of a substance melts, the enthalpy change is the **heat of fusion (ΔH_{fus}):**

$$NaCl(s) \longrightarrow NaCl(l) \quad \Delta H = \Delta H_{fus}$$

- When 1 mol of a substance vaporizes, the enthalpy change is the **heat of vaporization (ΔH_{vap}):**

$$C_6H_6(l) \longrightarrow C_6H_6(g) \quad \Delta H = \Delta H_{vap}$$

We encounter heats of combustion and formation shortly; other special enthalpy changes are discussed in later chapters.

Changes in Bond Strengths, or Where Does the Heat of Reaction Come From?

Let's stop for a moment to consider a major question about the heat of reaction. When 2 g of H_2 (1 mol) and 38 g of F_2 (1 mol) react at 298 K (25°C), 40 g of HF (2 mol) forms and 546 kJ of heat is released:

$$H_2(g) + F_2(g) \longrightarrow 2HF(g) + 546 \text{ kJ}$$

The amount (mol) of gas does not change, so the change in enthalpy equals the change in internal energy. Where does this heat come from?

Components of Internal Energy Let's break down the system's internal energy into kinetic energy (E_k) and potential energy (E_p) components and examine the contributions to these components to see which one changes during the reaction. We'll focus on the HF molecule but a similar analysis can be made of H_2 and F_2.

There are several contributions to E_k and E_p, some major and some minor, but the most important are listed below and depicted schematically in Figure 6.9:

1. *Contributions to the kinetic energy:*
   - The molecule moving through space, $E_{k(translation)}$
   - The molecule rotating, $E_{k(rotation)}$
   - The bound atoms vibrating, $E_{k(vibration)}$
   - The electrons moving within each atom, $E_{k(electron)}$
2. *Contributions to the potential energy:*
   - Forces between the bound atoms vibrating, $E_{p(vibration)}$
   - Forces between nucleus and electrons and between electrons in each atom, $E_{p(atom)}$
   - Forces between the protons and neutrons in each nucleus, $E_{p(nuclei)}$
   - Forces between nuclei and shared electron pair in each bond, $E_{p(bond)}$

Now let's see which contributions we can eliminate because they don't change significantly during the reaction:

1. Among the *kinetic energy* contributions, the first three—$E_{k(translation)}$, $E_{k(rotation)}$, and $E_{k(vibration)}$—are proportional to the absolute temperature, which is constant at 298 K (25°C), so they don't change. Electron motion is not affected by the reaction, so $E_{k(electron)}$ doesn't change either. Therefore, the total E_k is virtually unchanged during the reaction.

2. Among the *potential energy* contributions, we can eliminate $E_{p(atom)}$ and $E_{p(nuclei)}$ because the atoms and nuclei don't change. The potential energy of vibration, $E_{p(vibration)}$, changes slightly because the bonded atoms change, but this change is small. The only energy contribution that changes significantly in a chemical reaction is $E_{p(bond)}$, the strength of attraction of the nuclei for the shared electron pair that constitutes the covalent bond.

Breaking and Forming Bonds We can think of a reaction as a process in which *the bonds of the reactants absorb energy when they break, and the bonds of the products release energy when they form.* When 1 mol of H—H bonds and 1 mol of F—F bonds absorb energy and break, the atoms form 2 mol of H—F bonds, which releases energy. This reaction has a net release of energy as heat (exothermic), which means the energy released when the bonds in HF form is more than the energy absorbed when the bonds in H_2 and F_2 break. Chemists speak in terms of the relative strengths of bonds. *Weaker bonds are easier to break than stronger bonds because they are higher in energy (less stable, more reactive).* As shown in the diagram, heat is released when HF forms from its elements because the bonds in the reactants H_2 and F_2 are weaker (less stable) than the bonds in the product HF (more stable). If, on the other hand, a reaction absorbs more heat to break reactant bonds than it releases to form product bonds, there is a net absorption of energy as heat (endothermic). In either case, this difference in energy appears as the heat of reaction, ΔH_{rxn}.

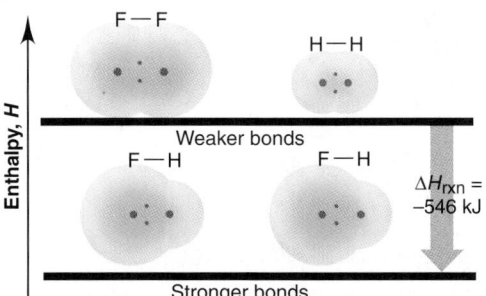

In other words, the answer to our opening question is that the heat doesn't really "come from" anywhere: *the energy released or absorbed during a chemical change is due to differences between the strengths of reactant bonds and product bonds.* We'll discuss these ideas quantitatively in Chapter 9.

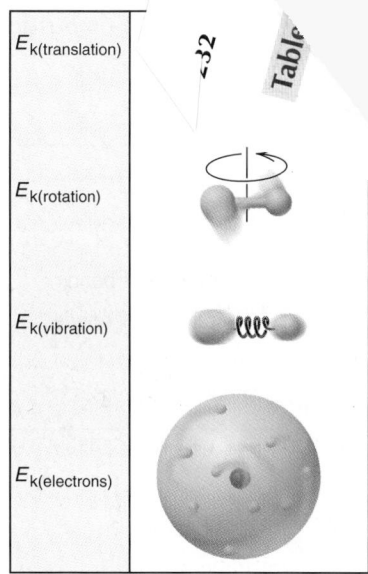

A Contributions to kinetic energy (E_k)

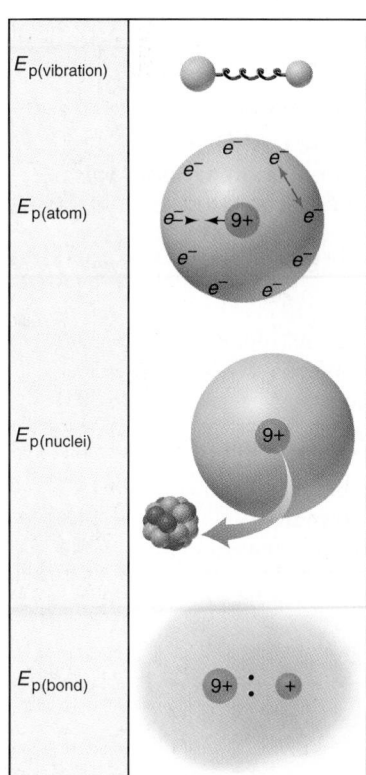

B Contributions to potential energy (E_p)

Figure 6.9 Components of internal energy (E). The total energy of a system is the sum of the E_k **(A)** and E_p **(B)** of its particles. The contributions to E_k and E_p are shown for a hydrogen fluoride molecule (or, where appropriate, just for the fluorine atom). Only $E_{p(bond)}$ changes significantly during a chemical reaction.

6.2 Heats of Combustion (ΔH_{comb}) of Some Carbon Compounds

	Two-Carbon Compounds		One-Carbon Compounds	
	Ethane (C_2H_6)	Ethanol (C_2H_5OH)	Methane (CH_4)	Methanol (CH_3OH)
Structural Formula	H H | | H—C—C—H | | H H	H H | | H—C—C—O—H | | H H	H | H—C—H | H	H | H—C—O—H | H
Sum of C—C and C—H bonds	7	6	4	3
Sum of C—O and O—H bonds	0	2	0	2
ΔH_{comb} (kJ/mol)	−1560	−1367	−890	−727
ΔH_{comb} (kJ/g)	−51.88	−29.67	−55.5	−22.7

Bonds in Fuels and Foods The most common fuels for machines are hydrocarbons and coal, and the most common ones for organisms are fats and carbohydrates. Both types of fuel are composed of large organic molecules with mostly C—C and C—H bonds. When the fuel reacts with O_2, the bonds break, and the C, H, and O atoms form C—O and O—H bonds in the products CO_2 and H_2O. Since burning these fuels releases energy, we know that the total strength of the bonds in the products is greater than the total strength of the bonds in the fuel and O_2.

Fuels with more weak (less stable, higher energy) bonds yield more energy than fuels with fewer weak bonds. Table 6.2 demonstrates this point for some small organic compounds. As the number of C—C and C—H bonds decreases and/or the number of C—O and O—H bonds (shown in red) increases, less energy is released from combustion; that is, ΔH_{comb} is less negative. In other words, the fewer bonds to O in a fuel, the more energy it releases when burned.

Both fats and carbohydrates are organic compounds that serve as high-energy foods. Fats consist of chains of carbon atoms (C—C bonds) attached to hydrogen atoms (C—H bonds), with very few, if any, C—O and O—H bonds (shown in red below). Carbohydrates have many more C—O and O—H bonds. Both foods are metabolized in the body to CO_2 and H_2O. Why do fats "contain more Calories" per gram than carbohydrates? Fats have fewer bonds to O (red bonds below), so they release more energy than do carbohydrates, as Table 6.3 confirms.

Table 6.3 Heats of Combustion of Some Fats and Carbohydrates

Substance	ΔH_{comb} (kJ/g)
Fats	
Vegetable oil	−37.0
Margarine	−30.1
Butter	−30.0
Carbohydrates	
Table sugar (sucrose)	−16.2
Brown rice	−14.9
Maple syrup	−10.4

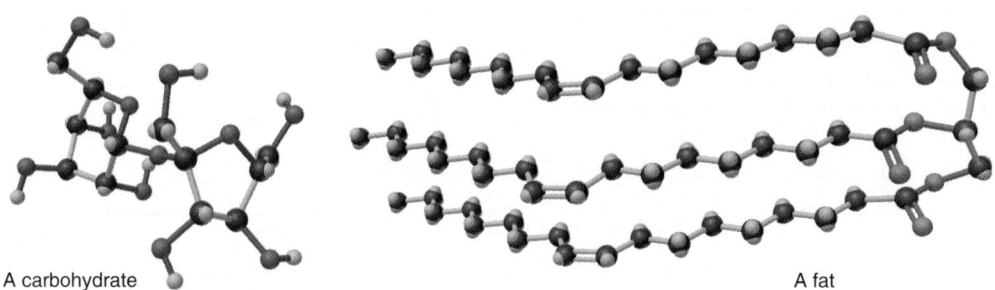

A carbohydrate   A fat

SECTION SUMMARY

The change in enthalpy, ΔH, is the heat lost or gained at constant pressure, q_P. In most cases, ΔH is equal, or very close, to ΔE. A chemical (or physical) change that releases heat is exothermic ($\Delta H < 0$); one that absorbs heat is endothermic ($\Delta H > 0$). On the molecular level, the heat lost or gained represents a difference in the strengths of the bonds in the products and the reactants. A reaction involves breaking reactant bonds and forming product bonds. Bonds in fuels and foods are weaker (less stable, higher energy) than bonds in waste products.

6.3 CALORIMETRY: LABORATORY MEASUREMENT OF HEATS OF REACTION

It is clear that the energy values of fuels and foods are important to measure. Table 6.3 shows, for example, that when 1 g of table sugar (sucrose, $C_{12}H_{22}O_{11}$) burns, 16.2 kJ of heat is released. But how do we make that measurement? It might seem that we can simply measure the enthalpies of the reactants (sucrose and O_2) and subtract them from the enthalpies of the products (CO_2 and H_2O). The problem is that the enthalpy (*H*) of a system (in this case, certain amounts of substances) in a given state cannot be measured because we have no starting point with which to compare it, no zero enthalpy. However, we *can* measure the *change* in enthalpy (ΔH) of a system. In this section, we'll see how ΔH values are determined.

Recall that ΔH is the heat released or absorbed by the system at constant pressure (q_P). To measure q_P accurately, we construct "surroundings" that retain the heat and measure the temperature change on a thermometer immersed in these surroundings. Then, we have to relate the quantity of heat released (or absorbed) to that temperature change. This relation involves a physical property of a substance called the *specific heat capacity*.

Specific Heat Capacity

You know from everyday experience that the more heat an object absorbs, the hotter it gets; that is, the quantity of heat (*q*) absorbed by an object is proportional to its temperature change:

$$q \propto \Delta T \quad \text{or} \quad q = \text{constant} \times \Delta T \quad \text{or} \quad \frac{q}{\Delta T} = \text{constant}$$

Every object has its own particular capacity for absorbing heat, its own **heat capacity,** the quantity of heat required to change its temperature by 1 K. Heat capacity is the proportionality constant in the preceding equation:

$$\text{Heat capacity} = \frac{q}{\Delta T} \quad \text{[in units of J/K]}$$

A related property is **specific heat capacity (*c*),** the quantity of heat required to change the temperature of 1 *gram* of a substance by 1 K:*

$$\text{Specific heat capacity } (c) = \frac{q}{\text{mass} \times \Delta T} \quad \text{[in units of J/g·K]}$$

If we know *c* of the substance being heated (or cooled), we can measure its mass and temperature change and calculate the heat absorbed or released:

$$q = c \times \text{mass} \times \Delta T \tag{6.7}$$

Notice that when an object gets hotter, ΔT (that is, $T_{final} - T_{initial}$) is positive. The object gains heat, so $q > 0$ as we expect. Similarly, when an object gets cooler, ΔT is negative; so $q < 0$ because heat is lost. Table 6.4 lists the specific heat capacities of several common substances.

Closely related to the specific heat capacity is the **molar heat capacity (*C*;** note capital letter), the quantity of heat required to change the temperature of 1 *mole* of a substance by 1 K:

$$\text{Molar heat capacity } (C) = \frac{q}{\text{moles} \times \Delta T} \quad \text{[in units of J/mol·K]}$$

The specific heat capacity of liquid water is 4.184 J/g·K, so

$$C \text{ of } H_2O(l) = 4.184 \frac{J}{g \cdot K} \times \frac{18.02 \text{ g}}{1 \text{ mol}} = 75.40 \frac{J}{\text{mol} \cdot K}$$

---

*Some texts use the term *specific heat* in place of *specific heat capacity*. This usage is very common but somewhat incorrect. *Specific heat* is the ratio of the heat capacity of 1 g of a substance to the heat capacity of 1 g of H_2O and therefore has no units.

Table 6.4 Specific Heat Capacities of Some Elements, Compounds, and Materials

Substance	Specific Heat Capacity (J/g·K)*
Elements	
Aluminum, Al	0.900
Graphite, C	0.711
Iron, Fe	0.450
Copper, Cu	0.387
Gold, Au	0.129
Compounds	
Water, $H_2O(l)$	4.184
Ethyl alcohol, $C_2H_5OH(l)$	2.46
Ethylene glycol, $(CH_2OH)_2(l)$	2.42
Carbon tetrachloride, $CCl_4(l)$	0.862
Solid materials	
Wood	1.76
Cement	0.88
Glass	0.84
Granite	0.79
Steel	0.45

*At 298 K (25°C).

SAMPLE PROBLEM 6.3 Calculating the Quantity of Heat from the Specific Heat Capacity

Problem A layer of copper welded to the bottom of a skillet weighs 125 g. How much heat is needed to raise the temperature of the copper layer from 25°C to 300.°C? The specific heat capacity (c) of Cu is 0.387 J/g·K.

Plan We know the mass and c of Cu and can find ΔT in °C, which equals ΔT in K. We use this ΔT and Equation 6.7 to solve for the heat.

Solution Calculating ΔT and q:

$$\Delta T = T_{\text{final}} - T_{\text{initial}} = 300.°C - 25°C = 275°C = 275 \text{ K}$$

$$q = c \times \text{mass (g)} \times \Delta T = 0.387 \text{ J/g·K} \times 125 \text{ g} \times 275 \text{ K} = \boxed{1.33 \times 10^4 \text{ J}}$$

Check Heat is absorbed by the copper bottom (system), so q is positive. Rounding shows that the arithmetic seems reasonable: $q \approx 0.4 \text{ J/g·K} \times 100 \text{ g} \times 300 \text{ K} = 1.2 \times 10^4 \text{ J}$.

FOLLOW-UP PROBLEM 6.3 Calculate the heat transferred when a 5.5-g iron nail is cooled from 37°C to 25°C. (See Table 6.4.)

The Practice of Calorimetry

The **calorimeter** is a device used to measure the heat released (or absorbed) by a physical or chemical process. We know that heat is transferred from system to surroundings (and vice versa). The calorimeter is the "surroundings" we construct to measure the temperature change that results from this transferred heat. Two common types are the constant-pressure and constant-volume calorimeters.

Constant-Pressure Calorimetry The "coffee-cup" calorimeter shown in Figure 6.10 is used to measure the heat transferred (q_P) for many processes that are open to the laboratory atmosphere. One common use is to determine the specific heat capacity of a solid, as long as it does not react with or dissolve in water. The solid (system) is weighed, heated to some known temperature, and added to a sample of water (surroundings) of known temperature and mass in the calorimeter. Stirring distributes the released heat, and the final water temperature, which is also the final temperature of the solid, is measured.

The heat lost by the system ($-q$) is equal in magnitude but opposite in sign to the heat gained ($+q$) by the surroundings:

$$-q_{\text{solid}} = q_{\text{H}_2\text{O}}$$

Substituting Equation 6.7 for each side of this equality gives

$$-(c_{\text{solid}} \times \text{mass}_{\text{solid}} \times \Delta T_{\text{solid}}) = c_{\text{H}_2\text{O}} \times \text{mass}_{\text{H}_2\text{O}} \times \Delta T_{\text{H}_2\text{O}}$$

All the quantities are known or measured except c_{solid}:

$$c_{\text{solid}} = -\frac{c_{\text{H}_2\text{O}} \times \text{mass}_{\text{H}_2\text{O}} \times \Delta T_{\text{H}_2\text{O}}}{\text{mass}_{\text{solid}} \times \Delta T_{\text{solid}}}$$

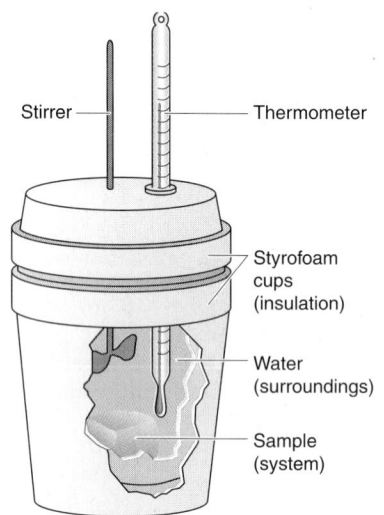

Stirrer

Thermometer

Styrofoam cups (insulation)

Water (surroundings)

Sample (system)

Figure 6.10 Coffee-cup calorimeter. This simple apparatus is used to measure the heat at constant pressure (q_P). It consists of a known mass of water (or solution) in an insulated container equipped with a thermometer and stirrer. T_{initial} of the water is measured, the process takes place (addition of a heated object, soluble salt, solution, etc.), the contents are stirred, and T_{final} of the water is measured.

SAMPLE PROBLEM 6.4 Determining the Specific Heat Capacity of a Solid

Problem A 25.64-g sample of a solid was heated in a test tube to 100.00°C in boiling water and carefully added to a coffee-cup calorimeter containing 50.00 g of water. The water temperature increased from 25.10°C to 28.49°C. What is the specific heat capacity of the solid? (Assume all the heat is gained by the water.)

Plan We know the masses of water (50.00 g) and solid (25.64 g) and the temperature before and after. We subtract to find ΔT of the water and the solid and change ΔT (°C) directly to ΔT (K). We know $c_{\text{H}_2\text{O}}$, so we can find c_{solid}.

Solution It is helpful to summarize the information given:

	Mass (g)	c (J/g·K)	T_{initial} (°C)	T_{final} (°C)	ΔT (K)
Solid	25.64	?	100.00	28.49	−71.51
H₂O	50.00	4.184	25.10	28.49	3.39

Calculating c_{solid}:

$$c_{solid} = -\frac{c_{H_2O} \times mass_{H_2O} \times \Delta T_{H_2O}}{mass_{solid} \times \Delta T_{solid}} = -\frac{4.184 \text{ J/g·K} \times 50.00 \text{ g} \times 3.39 \text{ K}}{25.64 \text{ g} \times (-71.51 \text{ K})}$$

$$= \boxed{0.387 \text{ J/g·K}}$$

Check Since $-q_{solid} = q_{H_2O}$, we can check to see if the numerical values are equal:

$$q_{H_2O} = 4.184 \text{ J/g·K} \times 50.00 \text{ g} \times 3.39 \text{ K} = 709 \text{ J}$$
$$q_{solid} = 0.387 \text{ J/g·K} \times 25.64 \text{ g} \times (-71.51 \text{ K}) = -710. \text{ J}$$

The slight difference is caused by rounding.

Comment 1. A *common mistake* is to write the wrong sign for ΔT: remember that Δ means *final − initial*.

2. The problem statement said to assume that all heat is absorbed by the water, but that is not valid for precise work because some heat must be gained by the other parts of the calorimeter (stirrer, thermometer, stopper, coffee-cup walls) as well. For more accurate measurements, discussed next, the *heat capacity of the entire calorimeter* must be known.

FOLLOW-UP PROBLEM 6.4 As a purity check for industrial diamonds, a 10.25-carat (1 carat = 0.2000 g) diamond is heated to 74.21°C and immersed in 26.05 g of water in a constant-pressure calorimeter. The initial temperature of the water is 27.20°C. Calculate ΔT of the water and of the diamond ($c_{diamond} = 0.519$ J/g·K).

Constant-Volume Calorimetry A common type of constant-volume apparatus is the *bomb calorimeter,* designed to measure very precisely the heat released in a combustion reaction. As Sample Problem 6.5 will show, this need for greater precision requires that we know (or determine) the heat capacity of the calorimeter.

Figure 6.11 depicts the preweighed combustible sample in a metal-walled chamber (bomb), which is filled with oxygen gas and immersed in an insulated water bath fitted with motorized stirrer and thermometer. A heating coil connected to an electrical source ignites the sample, and the heat evolved raises the temperature of the bomb, water, and other calorimeter parts. Knowing the mass of the sample and the heat capacity of the calorimeter, we use the measured ΔT to calculate the heat released.

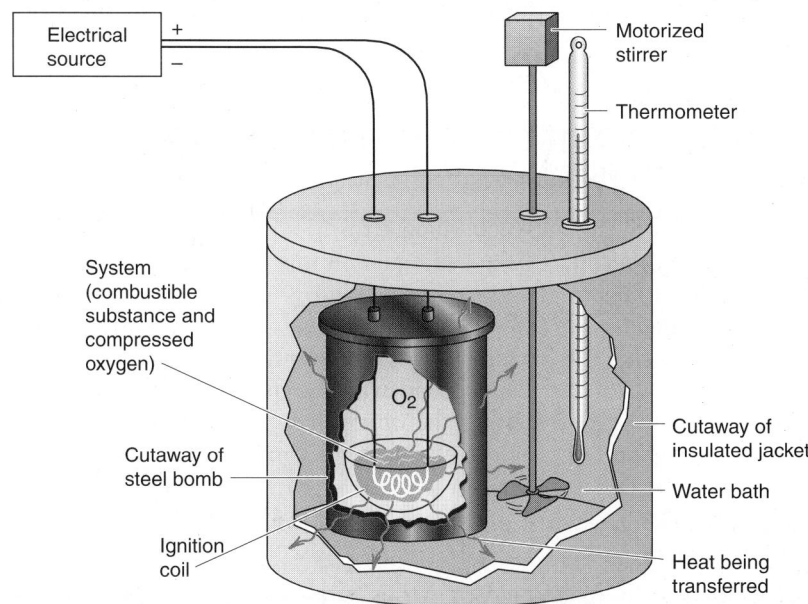

Figure 6.11 A bomb calorimeter. This device (not drawn to scale) is used to measure heat of combustion at constant volume (q_V). With continual stirring, the initial temperature of the preweighed water bath is noted. The electrically heated coil ignites the system (combustible substance in O_2) within the steel bomb. Heat released from the combustion reaction is transferred to the rest of the calorimeter, and the maximum temperature is measured.

SAMPLE PROBLEM 6.5 Calculating the Heat of Combustion

Problem A manufacturer claims that its new dietetic dessert has "fewer than 10 Calories per serving." To test the claim, a chemist at the Department of Consumer Affairs places one serving in a bomb calorimeter and burns it in O_2 (heat capacity of the calorimeter = 8.151 kJ/K). The temperature increases 4.937°C. Is the manufacturer's claim correct?

Plan When the dessert burns, the heat released is gained by the calorimeter:

$$-q_{sample} = q_{calorimeter}$$

To find the heat, we multiply the given heat capacity of the calorimeter (8.151 kJ/K) by ΔT (4.937°C).

Solution Calculating the heat gained by the calorimeter:

$$q_{calorimeter} = \text{heat capacity} \times \Delta T = 8.151 \text{ kJ/K} \times 4.937 \text{ K} = 40.24 \text{ kJ}$$

Recall that 1 Calorie = 1 kcal = 4.184 kJ. Therefore, 10 Calories = 41.84 kJ, so the claim is correct.

Check A quick math check shows the answer is reasonable: 8 kJ/K × 5 K = 40 kJ.

Comment Since the volume of the steel bomb is fixed, ΔV and thus $P\Delta V = 0$. Thus, the energy change measured is the *heat at constant volume* (q_V), which equals ΔE, not ΔH:

$$\Delta E = q + w = q_V + 0 = q_V$$

Recall from Section 6.2, however, that even though the number of moles of gas may change, ΔH is usually very close to ΔE. For example, ΔH is only 0.5% larger than ΔE for the combustion of H_2 and only 0.2% smaller for the combustion of octane (see Problem 6.109 at the end of the chapter).

FOLLOW-UP PROBLEM 6.5 A chemist burns 0.8650 g of graphite (a form of carbon) in a new bomb calorimeter, and CO_2 forms. If 393.5 kJ of heat is released per mole of graphite and T increases 2.613 K, what is the heat capacity of the bomb calorimeter?

SECTION SUMMARY

We calculate ΔH of a process by measuring the heat at constant pressure (q_P). To do this, we determine ΔT of the surroundings and relate it to q_P through the mass of the substance and its specific heat capacity (c), the quantity of energy needed to raise the temperature of 1 g of the substance by 1 K.  Calorimeters measure the heat released from a system either at constant pressure $(q_P = \Delta H)$ or at constant volume $(q_V = \Delta E)$.

6.4 STOICHIOMETRY OF THERMOCHEMICAL EQUATIONS

A good way to record the enthalpy change for a particular reaction is through a **thermochemical equation,** a balanced equation that also states the heat of reaction (ΔH_{rxn}). One thing to always keep in mind is that the ΔH_{rxn} value shown refers to the *amounts (moles) of substances and their states of matter in that specific equation.* The enthalpy change of any process has two aspects:

1. *Sign.* The sign of ΔH depends on whether the reaction is exothermic or endothermic. A forward reaction has the *opposite* sign of the reverse reaction. Decomposition of 2 mol of water to its elements (endothermic):

$$2H_2O(l) \longrightarrow 2H_2(g) + O_2(g) \quad \Delta H_{rxn} = 572 \text{ kJ}$$

Formation of 2 mol of water from its elements (exothermic):

$$2H_2(g) + O_2(g) \longrightarrow 2H_2O(l) \quad \Delta H_{rxn} = -572 \text{ kJ}$$

2. *Magnitude.* The magnitude of ΔH is *proportional to the amount of substance* reacting.

Formation of 1 mol of water from its elements (half the amount in the preceding equation):

$$H_2(g) + \tfrac{1}{2}O_2(g) \longrightarrow H_2O(l) \quad \Delta H_{rxn} = -286 \text{ kJ}$$

Imagine an Earth Without Water Liquid water has an unusually high specific heat capacity of nearly 4.2 J/g·K, about six times that of rock (~0.7 J/g·K). If the Earth were devoid of oceans, the Sun's energy would heat a planet composed of rock. It would take only 0.7 J of energy to increase the temperature of each gram of rock by 1 K. Daytime temperatures would soar. The oceans also limit the temperature drop when the Sun sets, because the energy absorbed during the day is released at night. If the Earth had a rocky surface, temperatures would be frigid every night.

Note that, in thermochemical equations, we use fractional coefficients when necessary to specify the magnitude of ΔH_{rxn} for a *particular amount of substance*. Moreover, we find that, *in a particular reaction,* a certain amount of substance is thermochemically equivalent to a certain quantity of energy. *In this reaction,*

286 kJ is thermochemically equivalent to 1 mol of $H_2(g)$
286 kJ is thermochemically equivalent to $\frac{1}{2}$ mol of $O_2(g)$
286 kJ is thermochemically equivalent to 1 mol of $H_2O(l)$

Just as we use stoichiometrically equivalent molar ratios to find amounts of substances, we use thermochemically equivalent quantities to find the heat of reaction for a given amount of substance. Also, just as we use molar mass (in g/mol of substance) to convert moles of a substance to grams, we use the heat of reaction (in kJ/mol of substance) to convert moles of a substance to an equivalent quantity of heat (in kJ). Figure 6.12 shows this new relationship, and the next sample problem applies it.

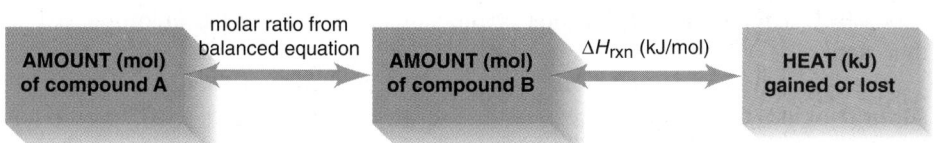

Figure 6.12 Summary of the relationship between amount (mol) of substance and the heat (kJ) transferred during a reaction.

SAMPLE PROBLEM 6.6 Using the Heat of Reaction (ΔH_{rxn}) to Find Amounts

Problem The major source of aluminum in the world is bauxite (mostly aluminum oxide). Its thermal decomposition can be represented by

$$Al_2O_3(s) \xrightarrow{\Delta} 2Al(s) + \tfrac{3}{2}O_2(g) \quad \Delta H_{rxn} = 1676 \text{ kJ}$$

If aluminum is produced this way (see Comment), how many grams of aluminum can form when 1.000×10^3 kJ of heat is transferred?

Plan From the balanced equation and enthalpy change, we see that 2 mol of Al forms when 1676 kJ of heat is absorbed. With this equivalent quantity, we convert the given kJ transferred to moles formed and then convert moles to grams.

Solution Combining steps to convert from heat transferred to mass of Al:

$$\text{Mass (g) of Al} = (1.000 \times 10^3 \text{ kJ}) \times \frac{2 \text{ mol Al}}{1676 \text{ kJ}} \times \frac{26.98 \text{ g Al}}{1 \text{ mol Al}} = \boxed{32.20 \text{ g Al}}$$

Check The mass of aluminum seems correct: ~1700 kJ forms about 2 mol of Al (54 g), so 1000 kJ should form a bit more than half that amount (27 g).

Comment In practice, aluminum is not obtained by heating but by supplying electrical energy. Because ΔH is a state function, however, the total energy required for this chemical change is the same no matter how it occurs. (We examine the industrial method in Chapter 22.)

FOLLOW-UP PROBLEM 6.6 As Lavoisier, Priestley, and others knew, mercury(II) oxide decomposes readily with heat. If it takes 90.8 kJ to decompose 1 mol of HgO, how much heat is required to produce 1 short ton (2000 lb; 907 kg) of Hg?

Heat (kJ)

↓

1676 kJ = 2 mol Al

Amount (mol) of Al

↓

multiply by $\mathcal{M}$ (g/mol)

Mass (g) of Al

SECTION SUMMARY

A thermochemical equation shows the balanced equation *and* its ΔH_{rxn}. The sign of ΔH for a forward reaction is opposite that for the reverse reaction. The magnitude of ΔH depends on the amount and physical state of the substance reacting and the ΔH per mole of substance. We use the thermochemically equivalent amounts of substance and heat from the balanced equation as conversion factors to find the quantity of heat when a given amount of substance reacts.

6.5 HESS'S LAW OF HEAT SUMMATION

Many reactions are difficult, even impossible, to perform as an individual change. A reaction may be part of a complex biochemical process; or it may take place only under extreme environmental conditions; or it may even require a change in conditions while it is occurring. If we can't run a reaction in the lab, is it possible to find its enthalpy change? One of the most powerful applications of the state-function property of enthalpy (H) allows us to do just that, to find the ΔH of *any* reaction for which we can write an equation, even if it is impossible to carry out.

This application is based on **Hess's law of heat summation:** *the enthalpy change of an overall process is the sum of the enthalpy changes of its individual steps.* To use Hess's law, we imagine an overall reaction as the sum of a series of reaction steps, whether or not it really occurs that way. Each step is chosen because its ΔH is known. Because the overall ΔH depends only on the initial and final states, Hess's law says that we add together the known ΔH values for the steps to get ΔH of the overall reaction. Similarly, if we know the ΔH values for the overall reaction and all but one of the steps, we can find the unknown ΔH of that step.

Let's see how we apply Hess's law in the case of the oxidation of sulfur to sulfur trioxide, the central process in the industrial production of sulfuric acid and in the formation of acid rain. (To introduce the approach, we'll simplify the equations by using S as the formula for sulfur, rather than the more correct S_8.) When we burn S in an excess of O_2, sulfur dioxide (SO_2) forms, *not* sulfur trioxide (SO_3). Equation 1 shows this step and its ΔH. If we change conditions and then add more O_2, we can oxidize SO_2 to SO_3 (Equation 2). In other words, we cannot put S and O_2 in a calorimeter and find ΔH for the overall reaction of S to SO_3 (Equation 3). But, we can find it with Hess's law. The three equations are

$$\text{Equation 1:} \qquad S(s) + O_2(g) \longrightarrow SO_2(g) \qquad \Delta H_1 = -296.8 \text{ kJ}$$
$$\text{Equation 2:} \qquad 2SO_2(g) + O_2(g) \longrightarrow 2SO_3(g) \qquad \Delta H_2 = -198.4 \text{ kJ}$$
$$\text{Equation 3:} \qquad S(s) + \tfrac{3}{2}O_2(g) \longrightarrow SO_3(g) \qquad \Delta H_3 = ?$$

Hess's law tells us that if we can manipulate Equations 1 and/or 2 so that they add up to Equation 3, then ΔH_3 will be the sum of the manipulated ΔH values of Equations 1 and 2.

First, we identify Equation 3 as our "target" equation, the one whose ΔH we want to find, and we carefully note the number of moles of reactants and products in it. We also note that ΔH_1 and ΔH_2 are the values for Equations 1 and 2 *as written.* Now we manipulate Equations 1 and/or 2 as follows to make them add up to Equation 3:

* Equations 1 and 3 contain the same amount of S, so we leave Equation 1 unchanged.
* Equation 2 has twice as much SO_3 as Equation 3, so we multiply it by $\tfrac{1}{2}$, being sure to halve ΔH_2 as well.
* With the targeted amounts of reactants and products present, we add Equation 1 to the halved Equation 2 and cancel terms that appear on both sides:

Equation 1:	$S(s) + O_2(g) \longrightarrow SO_2(g)$	$\Delta H_1 = -296.8$ kJ
$\tfrac{1}{2}$(Equation 2):	$SO_2(g) + \tfrac{1}{2}O_2(g) \longrightarrow SO_3(g)$	$\tfrac{1}{2}(\Delta H_2) = -99.2$ kJ

Equation 3: $S(s) + O_2(g) + SO_2(g) + \tfrac{1}{2}O_2(g) \longrightarrow SO_2(g) + SO_3(g) \quad \Delta H_3 = ?$

or, $\qquad\qquad\qquad\qquad\qquad S(s) + \tfrac{3}{2}O_2(g) \longrightarrow SO_3(g)$

Since Equation 1 plus the halved Equation 2 equals Equation 3, we have

$$\Delta H_3 = \Delta H_1 + \tfrac{1}{2}(\Delta H_2) = -296.8 \text{ kJ} + (-99.2 \text{ kJ}) = -396.0 \text{ kJ}$$

Once again, the key point is that H is a state function, so the overall ΔH depends on the difference between the initial and final enthalpies only. Hess's law tells us that the difference between the enthalpies of the reactants (1 mol of S and

$\frac{3}{2}$ mol of O_2) and that of the product (1 mol of SO_3) has the same value, whether S is oxidized directly to SO_3 (impossible) or through the formation of SO_2 first (actual).

To summarize, calculating an unknown ΔH involves three steps:

1. Identify the target equation, the step whose ΔH is unknown, and note the number of moles of reactants and products.
2. Manipulate the equations for which ΔH is known so that the target numbers of moles of reactants and products are on the correct sides. Remember to:
   • Change the sign of ΔH when you reverse an equation.
   • Multiply number of moles and ΔH by the same factor.
3. Add the manipulated equations to obtain the target equation. All substances except those in the target equation must cancel. Add their ΔH values to obtain the unknown ΔH.

SAMPLE PROBLEM 6.7 Using Hess's Law to Calculate an Unknown ΔH

Problem Two gaseous pollutants that form in auto exhaust are CO and NO. An environmental chemist is studying ways to convert them to less harmful gases through the following equation:

$$CO(g) + NO(g) \longrightarrow CO_2(g) + \tfrac{1}{2}N_2(g) \quad \Delta H = ?$$

Given the following information, calculate the unknown ΔH:

Equation A: $CO(g) + \tfrac{1}{2}O_2(g) \longrightarrow CO_2(g) \quad \Delta H_A = -283.0 \text{ kJ}$

Equation B: $N_2(g) + O_2(g) \longrightarrow 2NO(g) \quad \Delta H_B = 180.6 \text{ kJ}$

Plan We note the numbers of moles of substances in the target equation, manipulate Equations A and/or B *and* their ΔH values, and then add them together to obtain the target equation and the unknown ΔH.

Solution Noting moles of substances in the target equation: There is 1 mol each of reactants CO and NO, 1 mol of product CO_2, and $\tfrac{1}{2}$ mol of product N_2.

Manipulating the given equations: Equation A has the same number of moles of CO and CO_2 as the target, so we leave it as written. Equation B has twice the needed amounts of N_2 and NO, and they are on the opposite sides from the target; therefore, we reverse Equation B, change the sign of ΔH_B, and multiply both by $\tfrac{1}{2}$:

$$\tfrac{1}{2}[2NO(g) \longrightarrow N_2(g) + O_2(g)] \quad \Delta H = -\tfrac{1}{2}(\Delta H_B) = -\tfrac{1}{2}(180.6 \text{ kJ})$$

$$NO(g) \longrightarrow \tfrac{1}{2}N_2(g) + \tfrac{1}{2}O_2(g) \quad \Delta H = -90.3 \text{ kJ}$$

Adding the manipulated equations to obtain the target equation:

Equation A:	$CO(g) + \tfrac{1}{2}O_2(g) \longrightarrow CO_2(g)$	$\Delta H = -283.0 \text{ kJ}$
$\tfrac{1}{2}$(Equation B reversed):	$NO(g) \longrightarrow \tfrac{1}{2}N_2(g) + \tfrac{1}{2}O_2(g)$	$\Delta H = -90.3 \text{ kJ}$
Target:	$CO(g) + NO(g) \longrightarrow CO_2(g) + \tfrac{1}{2}N_2(g)$	$\Delta H = -373.3 \text{ kJ}$

Check Obtaining the desired target equation is its own check. Be sure to remember to change the sign of any equation you reversed.

FOLLOW-UP PROBLEM 6.7 Nitrogen oxides undergo many interesting reactions. Calculate ΔH for the overall equation $2NO_2(g) + \tfrac{1}{2}O_2(g) \longrightarrow N_2O_5(s)$ from the following information:

$$N_2O_5(s) \longrightarrow 2NO(g) + \tfrac{3}{2}O_2(g) \quad \Delta H = 223.7 \text{ kJ}$$

$$NO(g) + \tfrac{1}{2}O_2(g) \longrightarrow NO_2(g) \quad \Delta H = -57.1 \text{ kJ}$$

SECTION SUMMARY

Since H is a state function, $\Delta H = H_{final} - H_{initial}$ and does not depend on how the reaction takes place. Using Hess's law ($\Delta H_{total} = \Delta H_1 + \Delta H_2 + \cdots + \Delta H_n$), we can determine ΔH of any overall equation by manipulating the coefficients of appropriate individual equations and their known ΔH values.

Table 6.5 Selected Standard Heats of Formation at 25°C (298 K)

Formula	ΔH_f^0 (kJ/mol)
Calcium	
Ca(s)	0
CaO(s)	−635.1
CaCO₃(s)	−1206.9
Carbon	
C(graphite)	0
C(diamond)	1.9
CO(g)	−110.5
CO₂(g)	−393.5
CH₄(g)	−74.9
CH₃OH(l)	−238.6
HCN(g)	135
CS₂(l)	87.9
Chlorine	
Cl(g)	121.0
Cl₂(g)	0
HCl(g)	−92.3
Hydrogen	
H(g)	218.0
H₂(g)	0
Nitrogen	
N₂(g)	0
NH₃(g)	−45.9
NO(g)	90.3
Oxygen	
O₂(g)	0
O₃(g)	143
H₂O(g)	−241.8
H₂O(l)	−285.8
Silver	
Ag(s)	0
AgCl(s)	−127.0
Sodium	
Na(s)	0
Na(g)	107.8
NaCl(s)	−411.1
Sulfur	
S₈(rhombic)	0
S₈(monoclinic)	0.3
SO₂(g)	−296.8
SO₃(g)	−396.0

6.6 STANDARD HEATS OF REACTION (ΔH_{rxn}^0)

In this section, we see how Hess's law is used to determine the ΔH values of an enormous number of reactions. To begin we must take into account that thermodynamic variables, such as ΔH, vary somewhat with conditions. Therefore, to use heats of reaction as well as other thermodynamic data, chemists have established **standard states,** a set of specified conditions and concentrations:

- For a *gas,* the standard state is 1 atm* and the gas behaving ideally.
- For a substance in *aqueous solution,* the standard state is 1 M concentration (1 mol/L solution).
- For a *pure substance* (element or compound), the standard state is usually the most stable form of the substance at 1 atm and the temperature of interest. In this text, that temperature is usually 25°C (298 K).[†]

A right superscript zero indicates that the thermodynamic variable has been determined with all substances in their standard states. For example, when ΔH_{rxn} has been measured with all substances in their standard states, it is the **standard heat of reaction, ΔH_{rxn}^0.**

Formation Equations and Their Standard Enthalpy Changes

In a **formation equation,** 1 mol of a compound forms from its elements. The **standard heat of formation (ΔH_f^0)** is the enthalpy change for the formation equation when all the substances are in their standard states. For instance, the formation equation for methane (CH_4) is

$$C(graphite) + 2H_2(g) \longrightarrow CH_4(g) \quad \Delta H_f^0 = -74.9 \text{ kJ}$$

Thus, the standard heat of formation of methane is −74.9 kJ/mol. Some other examples are

$$Na(s) + \tfrac{1}{2}Cl_2(g) \longrightarrow NaCl(s) \quad \Delta H_f^0 = -411.1 \text{ kJ}$$
$$2C(graphite) + 3H_2(g) + \tfrac{1}{2}O_2(g) \longrightarrow C_2H_5OH(l) \quad \Delta H_f^0 = -277.6 \text{ kJ}$$

Standard heats of formation have been tabulated for many compounds. Table 6.5 shows ΔH_f^0 values for several substances, and a much more extensive table appears in Appendix B.

The values in Table 6.5 were selected to make two points:

1. *An element in its standard state is assigned a ΔH_f^0 of zero.* For example, note that $\Delta H_f^0 = 0$ for Na(s), but $\Delta H_f^0 = 107.8$ kJ/mol for Na(g). This means that the gaseous state is not the most stable state of sodium at 1 atm and 298.15 K, and that heat is required to form Na(g). Note also that the standard state of chlorine is Cl_2 molecules, not Cl atoms. Several elements exist in different forms, only one of which is the standard state. Thus, the standard state of carbon is graphite, not diamond, so ΔH_f^0 of C(graphite) = 0. Similarly, the standard state of oxygen is dioxygen (O_2), not ozone (O_3), and the standard state of sulfur is S_8 arranged in its rhombic crystal form, rather than its monoclinic form.

2. *Most compounds have a negative ΔH_f^0.* That is, most compounds have exothermic formation reactions under standard conditions. This means that, in most cases, *the compound is more stable than its component elements.*

---

*The definition of the standard state for gases has been changed to 1 bar, a slightly lower pressure than the 1 atm standard on which the data in this book are based (1 atm = 101.3 kPa = 1.013 bar). For most purposes, this makes *very* little difference in the standard enthalpy values.

[†]In the case of phosphorus, the most *common* form, white phosphorus (P_4), is chosen as the standard state, even though red phosphorus is more stable at 1 atm and 298 K.

SAMPLE PROBLEM 6.8 Writing Formation Equations

Problem Write balanced equations for the formation of 1 mol of the following compounds from their elements in their standard states and include ΔH_f^0.
(a) Silver chloride, AgCl, a solid at standard conditions
(b) Calcium carbonate, $CaCO_3$, a solid at standard conditions
(c) Hydrogen cyanide, HCN, a gas at standard conditions
Plan We write the elements as the reactants and 1 mol of the compound as the product, being sure all substances are in their standard states. Then, we balance the atoms and obtain the ΔH_f^0 values from Table 6.5.
Solution (a) $Ag(s) + \frac{1}{2}Cl_2(g) \longrightarrow AgCl(s)$ $\Delta H_f^0 = -127.0 \text{ kJ}$

(b) $Ca(s) + C(graphite) + \frac{3}{2}O_2(g) \longrightarrow CaCO_3(s)$ $\Delta H_f^0 = -1206.9 \text{ kJ}$

(c) $\frac{1}{2}H_2(g) + C(graphite) + \frac{1}{2}N_2(g) \longrightarrow HCN(g)$ $\Delta H_f^0 = 135 \text{ kJ}$

FOLLOW-UP PROBLEM 6.8 Write balanced equations for the formation of 1 mol of **(a)** $CH_3OH(l)$, **(b)** $CaO(s)$, and **(c)** $CS_2(l)$ from their elements in their standard states. Include ΔH_f^0 for each reaction.

Determining ΔH_{rxn}^0 from ΔH_f^0 Values of Reactants and Products

By applying Hess's law, we can use ΔH_f^0 values to determine ΔH_{rxn}^0 for any reaction. All we have to do is view the reaction as an imaginary two-step process.
Step 1. Each reactant decomposes to its elements. This step is the *reverse* of the formation reaction for each reactant, so each standard enthalpy change is $-\Delta H_f^0$.
Step 2. Each product forms from its elements. This step *is* the formation reaction for each product, so each standard enthalpy change is $+\Delta H_f^0$.
According to Hess's law, we add the enthalpy changes for these steps to obtain the overall enthalpy change for the reaction (ΔH_{rxn}^0). Figure 6.13 depicts the conceptual process. Suppose we want ΔH_{rxn}^0 for

$$TiCl_4(l) + 2H_2O(g) \longrightarrow TiO_2(s) + 4HCl(g)$$

We write this equation as though it were the sum of four individual equations, one for each compound. The first two of these equations show the decomposition of the reactants to their elements (*reverse* of their formation), and the second two show the formation of the products from their elements:

$$TiCl_4(l) \longrightarrow Ti(s) + 2Cl_2(g) \qquad -\Delta H_f^0[TiCl_4(l)]$$
$$2H_2O(g) \longrightarrow 2H_2(g) + O_2(g) \qquad -2\Delta H_f^0[H_2O(g)]$$
$$Ti(s) + O_2(g) \longrightarrow TiO_2(s) \qquad \Delta H_f^0[TiO_2(s)]$$
$$2H_2(g) + 2Cl_2(g) \longrightarrow 4HCl(g) \qquad 4\Delta H_f^0[HCl(g)]$$

$TiCl_4(l) + 2H_2O(g) + \cancel{Ti(s)} + \cancel{O_2(g)} + \cancel{2H_2(g)} + \cancel{2Cl_2(g)} \longrightarrow$
$\qquad\qquad \cancel{Ti(s)} + \cancel{2Cl_2(g)} + \cancel{2H_2(g)} + \cancel{O_2(g)} + TiO_2(s) + 4HCl(g)$

Or, $TiCl_4(l) + 2H_2O(g) \longrightarrow TiO_2(s) + 4HCl(g)$

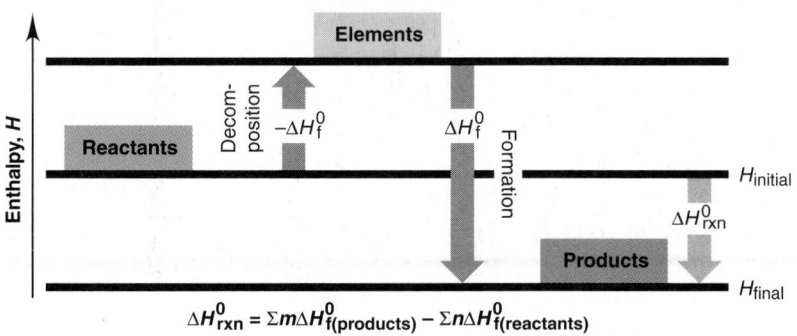

$$\Delta H_{rxn}^0 = \Sigma m \Delta H_{f(products)}^0 - \Sigma n \Delta H_{f(reactants)}^0$$

Figure 6.13 **The general process for determining ΔH_{rxn}^0 from ΔH_f^0 values.** For any reaction, ΔH_{rxn}^0 can be considered as the sum of the enthalpy changes for the decomposition of reactants to their elements $[-\Sigma n \Delta H_{f(reactants)}^0]$ and the formation of products from their elements $[+\Sigma m \Delta H_{f(products)}^0]$. [The factors m and n are the amounts (mol) of the products and reactants and equal the coefficients in the balanced equation, and Σ is the symbol for "sum of."]

It's important to realize that when titanium(IV) chloride and water react, the reactants don't *actually* decompose to their elements, which then recombine to form the products. But that is the great usefulness of Hess's law and the state-function concept. Since ΔH^0_{rxn} is the difference between two state functions, $H^0_{products}$ minus $H^0_{reactants}$, it doesn't matter *how* the change actually occurs. We simply add the individual enthalpy changes to find ΔH^0_{rxn}:

$$\Delta H^0_{rxn} = \Delta H^0_f[TiO_2(s)] + 4\Delta H^0_f[HCl(g)] + \{-\Delta H^0_f[TiCl_4(l)]\} + \{-2\Delta H^0_f[H_2O(g)]\}$$

$$= \underbrace{\{\Delta H^0_f[TiO_2(s)] + 4\Delta H^0_f[HCl(g)]\}}_{\text{Products}} - \underbrace{\{\Delta H^0_f[TiCl_4(l)] + 2\Delta H^0_f[H_2O(g)]\}}_{\text{Reactants}}$$

Notice the pattern here. By generalizing it, we state that *the standard heat of reaction is the sum of the standard heats of formation of the **products** minus the sum of the standard heats of formation of the **reactants*** (see Figure 6.13):

$$\Delta H^0_{rxn} = \Sigma m \Delta H^0_{f\text{(products)}} - \Sigma n \Delta H^0_{f\text{(reactants)}} \qquad \text{(6.8)}$$

where the symbol Σ means "sum of," and m and n are the amounts (mol) of the products and reactants indicated by the coefficients from the balanced equation.

SAMPLE PROBLEM 6.9 Calculating the Heat of Reaction from Heats of Formation

Problem Nitric acid, whose worldwide annual production is about 8 billion kg, is used to make many products, including fertilizers, dyes, and explosives. The first step in the industrial production process is the oxidation of ammonia:

$$4NH_3(g) + 5O_2(g) \longrightarrow 4NO(g) + 6H_2O(g)$$

Calculate ΔH^0_{rxn} from ΔH^0_f values.

Plan We look up ΔH^0_f values from Table 6.5 and apply Equation 6.8 to find ΔH^0_{rxn}.

Solution Calculating ΔH^0_{rxn}:

$$\Delta H^0_{rxn} = \Sigma m \Delta H^0_{f\text{(products)}} - \Sigma n \Delta H^0_{f\text{(reactants)}}$$

$$= \{4\Delta H^0_f[NO(g)] + 6\Delta H^0_f[H_2O(g)]\} - \{4\Delta H^0_f[NH_3(g)] + 5\Delta H^0_f[O_2(g)]\}$$

$$= (4 \text{ mol})(90.3 \text{ kJ/mol}) + (6 \text{ mol})(-241.8 \text{ kJ/mol})$$

$$-[(4 \text{ mol})(-45.9 \text{ kJ/mol}) + (5 \text{ mol})(0 \text{ kJ/mol})]$$

$$= 361 \text{ kJ} - 1451 \text{ kJ} + 184 \text{ kJ} - 0 \text{ kJ} = \boxed{-906 \text{ kJ}}$$

Check One way to check is to write formation equations for the amounts of individual compounds in the correct direction and take their sum:

$4NH_3(g) \longrightarrow 2N_2(g) + 6H_2(g)$	$-4(-45.9 \text{ kJ}) = 184 \text{ kJ}$
$2N_2(g) + 2O_2(g) \longrightarrow 4NO(g)$	$4(90.3 \text{ kJ}) = 361 \text{ kJ}$
$6H_2(g) + 3O_2(g) \longrightarrow 6H_2O(g)$	$6(-241.8 \text{ kJ}) = -1451 \text{ kJ}$
$4NH_3(g) + 5O_2(g) \longrightarrow 4NO(g) + 6H_2O(g)$	-906 kJ

Comment In this problem, we know the individual ΔH^0_f values and find the sum, ΔH^0_{rxn}. In the follow-up problem, we know the sum and want to find an individual value.

FOLLOW-UP PROBLEM 6.9 Use the following information to find ΔH^0_f of methanol [$CH_3OH(l)$]:

$$CH_3OH(l) + \tfrac{3}{2}O_2(g) \longrightarrow CO_2(g) + 2H_2O(g) \qquad \Delta H^0_{comb} = -638.5 \text{ kJ}$$

$$\Delta H^0_f \text{ of } CO_2(g) = -393.5 \text{ kJ/mol} \qquad \Delta H^0_f \text{ of } H_2O(g) = -241.8 \text{ kJ/mol}$$

In the following Chemical Connections essay, we apply key ideas from this chapter to new approaches for energy utilization.

The Future of Energy Use

Out of necessity, we are seeing a global rethinking of the issue of energy use. The dwindling supply of the world's fuel and the environmental impact of its combustion threaten our standard of living and have become questions of survival for hundreds of millions of people in less developed countries. Chemical aspects of energy production and utilization provide scientists and engineers with some of the greatest challenges of our time.

Energy consumption increased enormously during the Industrial Revolution and again after World War II, and it continues to increase today. A century-long changeover from the use of wood to the use of coal and then petroleum took place. The **fossil fuels**—coal, petroleum, and natural gas—remain our major sources of energy, but their polluting combustion products and eventual depletion are major drawbacks to their use.

Natural processes form fossil fuels *much* more slowly than we consume them, so these resources are *nonrenewable:* once we use them up, they are gone. Estimates indicate that known oil reserves could be 90% depleted by the year 2040. Coal reserves are much more plentiful, so chemical research seeks new uses for coal. The most important *renewable* combustible fuels are wood, other fuels derived from *biomass* (plant and animal matter), and hydrogen gas. The following discussion highlights some approaches to more rational energy use and incorporates some aspects of *green chemistry,* the government-sponsored effort with industry and academia to devise processes that avoid the release of harmful products into the environment: converting coal to cleaner fuels, producing fuels from biomass, developing a hydrogen-fueled economy, understanding the effect of carbon-based fuels on climate, utilizing energy sources that do not involve combustion, and, most important, conserving the fuels we have.

Chemical Approaches to Cleaner Fuels

Coal Current U.S. reserves of coal are enormous, but coal is a highly polluting fuel because it produces SO_2 when it burns. Two processes—desulfurization and gasification—are designed to reduce this noxious pollutant.

Flue-gas *desulfurization* devices (*scrubbers*) heat powdered limestone ($CaCO_3$) or spray lime-water slurries [$Ca(OH)_2$] to remove SO_2 from the gaseous product mixture of coal combustion:

$$CaCO_3(s) + SO_2(g) \xrightarrow{\Delta} CaSO_3(s) + CO_2(g)$$
$$2Ca(OH)_2(aq) + 2SO_2(g) + O_2(g) \longrightarrow 2CaSO_4(s) + 2H_2O(l)$$

Currently, however, the cost of operation plus disposal of the solids (nearly 1 ton per power-plant customer each year) presents serious problems for this approach.

In **coal gasification,** bonds in the large molecules in coal are broken to form much smaller molecules of sulfur-free gaseous fuels. First, high temperatures (600°C to 800°C) release volatile compounds that decompose coal to methane and a char:

$$\text{Coal}(s) \xrightarrow{\Delta} \text{coal volatiles}(g) \xrightarrow{\Delta} CH_4(g) + \text{char}(s)$$

The char (mostly carbon) reacts with steam in the endothermic *water-gas reaction* (also known as the steam-carbon reaction) to form a fuel mixture of CO and H_2 called synthesis gas (**syngas**):

$$C(s) + H_2O(g) \longrightarrow CO(g) + H_2(g) \quad \Delta H^0 = 131 \text{ kJ}$$

However, syngas has a much lower fuel value than methane. For example, a mixture containing 0.5 mol of CO and 0.5 mol of H_2 (that is, 1.0 mol of syngas) releases about one-third as much energy as 1.0 mol of methane ($\Delta H^0_{comb} = -802$ kJ/mol):

$$\tfrac{1}{2}H_2(g) + \tfrac{1}{4}O_2(g) \longrightarrow \tfrac{1}{2}H_2O(g) \qquad \Delta H^0 = -121 \text{ kJ}$$
$$\tfrac{1}{2}CO(g) + \tfrac{1}{4}O_2(g) \longrightarrow \tfrac{1}{2}CO_2(g) \qquad \Delta H^0 = -142 \text{ kJ}$$
$$\overline{\tfrac{1}{2}H_2(g) + \tfrac{1}{2}CO(g) + \tfrac{1}{2}O_2(g) \longrightarrow \tfrac{1}{2}H_2O(g) + \tfrac{1}{2}CO_2(g)}$$
$$\Delta H^0 = -263 \text{ kJ}$$

To upgrade its fuel value, syngas can be converted to other fuels, such as methane. In the *CO-shift reaction* (also known as the water-gas shift reaction), some CO reacts with more steam to form CO_2 and additional H_2:

$$CO(g) + H_2O(g) \longrightarrow CO_2(g) + H_2(g) \quad \Delta H^0 = -41 \text{ kJ}$$

The CO_2 is removed, and the H_2-enriched mixture reacts to form methane and water vapor:

$$CO(g) + 3H_2(g) \longrightarrow CH_4(g) + H_2O(g) \quad \Delta H^0 = -206 \text{ kJ}$$

Drying the product mixture gives **synthetic natural gas (SNG).** Thus, through a process that involves removal of volatiles and three reaction steps, coal is converted to methane.

Wood and Other Types of Biomass Nearly half the world's people rely on wood for their energy. In principle, wood and other forms of biomass are renewable, but worldwide deforestation for fuel, and for lumber and paper in richer countries, has resulted in wood being depleted much faster than it can grow back.

A better use of vegetable (sugarcane, corn, beets) and tree waste is **biomass conversion,** which combines chemical and microbial methods to convert sugar, cellulose, and other plant matter into combustible fuels. One such fuel is ethanol (C_2H_5OH), which is mixed with gasoline to form *gasohol*. Another is methane, which is produced in biogas generators through the microbial breakdown of plant and animal waste. China has active facilities applying these techniques. The U.S. Department of Energy and the Environmental Protection Agency are funding long-range programs that employ similar technologies to produce gaseous fuels from garbage and sewage.

Hydrogen Although combustion of H_2 ($\Delta H^0_{comb} = -242$ kJ/mol) produces only about one-third as much energy per mole as CH_4 ($\Delta H^0_{comb} = -802$ kJ/mol), it yields nonpolluting water vapor. Formation of H_2 from the decomposition of water, however, is endothermic (ΔH^0_{rxn} with liquid $H_2O = 286$ kJ/mol), and *direct* methods are still very costly. For example, at present, decomposing water with electricity is too expensive, but if future methods for generating electricity prove economical, the oceans provide an inexhaustible source of starting material.

Promising alternative methods for producing hydrogen apply Hess's law to decompose water *indirectly* to its components (ΔH^0_{rxn} with gaseous $H_2O = 242$ kJ/mol). Here is one example:

$$3FeCl_2(s) + 4H_2O(g) \xrightarrow{500°C} Fe_3O_4(s) + 6HCl(g) + H_2(g) \qquad \Delta H^0 = 318 \text{ kJ}$$
$$Fe_3O_4(s) + \tfrac{3}{2}Cl_2(g) + 6HCl(g) \xrightarrow{100°C} 3FeCl_3(s) + 3H_2O(g) + \tfrac{1}{2}O_2(g) \qquad \Delta H^0 = -249 \text{ kJ}$$
$$3FeCl_3(s) \xrightarrow{300°C} 3FeCl_2(s) + \tfrac{3}{2}Cl_2(g) \qquad \Delta H^0 = 173 \text{ kJ}$$
$$H_2O(g) \longrightarrow H_2(g) + \tfrac{1}{2}O_2(g) \qquad \Delta H^0 = 242 \text{ kJ}$$

(continued)

Transporting and storing the H_2 fuel could be accomplished with technologies used currently by utility companies. However, the actual method by which the H_2 will fuel, say, the family car, is an open question. Earlier research showed that many metals, such as palladium and niobium, absorb large amounts of H_2 into the spaces between their atoms, and it was hoped that they might be used as fuel cartridges. But more recent studies show that the metals release the H_2 too slowly to be practical. The use of H_2 in fuel cells, however, is an active area of electrochemical research.

Carbon-Based Fuels, the Greenhouse Effect, and Global Warming

Many natural processes produce CO_2, which plays a key temperature-regulating role in the atmosphere. Much of the sunlight that shines on the Earth is absorbed by the land and oceans and converted to heat (Figure B6.1). Atmospheric CO_2 does not absorb visible light from the Sun, but it does absorb heat. Thus, some of the heat radiating from the surface is trapped by the CO_2 and transferred to the atmosphere.

Over several billion years, due largely to the spread of plant life that uses CO_2 in photosynthesis, the large amount of CO_2 originally present in the Earth's atmosphere decreased to a relatively constant 0.028% by volume. However, for about 150 years, this amount has been increasing as a result of human use of fossil fuels, and it slightly exceeds 0.036% today. Thus, although the same amount of solar energy passes through the atmosphere, more is being trapped as heat in a process called the *greenhouse effect*, which has begun to cause *global warming* (Figure B6.2).

Based on current fossil fuel use, CO_2 concentrations are predicted to reach 0.070% to 0.075% by 2100. But, even if these projected increases in CO_2 occur, two closely related questions remain: (1) How much will the temperature of the atmosphere rise? (2) How will a temperature rise affect life on Earth?

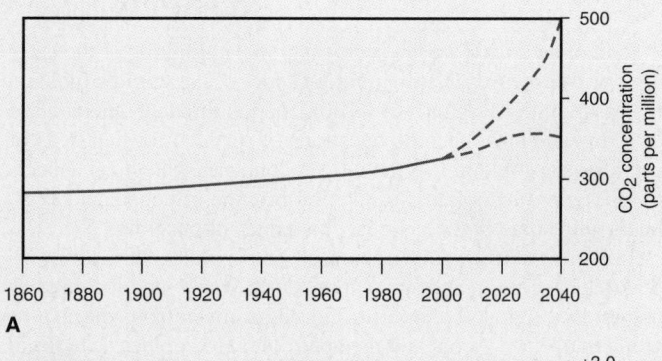

A

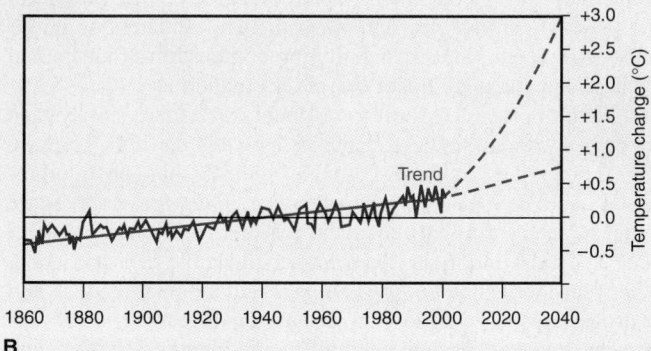

B

Figure B6.2 **The accumulating evidence for the greenhouse effect.** **A,** As a result of increased fossil fuel use since the mid-19th century, atmospheric CO_2 concentrations have increased. **B,** Coincident with this CO_2 increase, the average global temperature has risen about 0.6°C. (Zero equals the average global temperature for the period 1957 to 1970.) The projections in both graphs *(dashed lines)* are based on current fossil fuel consumption and deforestation either continuing *(upper line)* or being curtailed *(lower line)*.

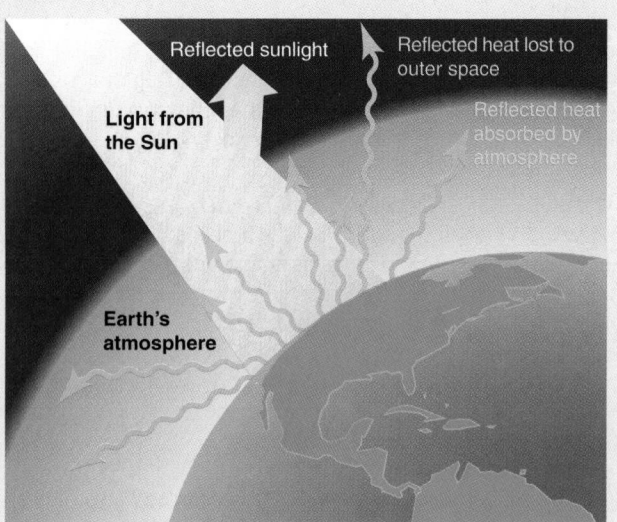

Figure B6.1 **The trapping of heat by the atmosphere.** About 25% of sunlight is reflected by the atmosphere. The remainder enters the atmosphere, where much of it is converted into heat by the air and the Earth's surface. Some of the heat emanating from the surface is trapped by atmospheric gases, mostly CO_2, and some is lost to outer space.

For over a decade now, many academic, government, and private laboratories have been devising computer models to simulate the effects of increasing amounts of CO_2 and other greenhouse gases (most importantly, CH_4) on the temperature of the atmosphere. However, the system is so complex that a definitive answer is difficult to obtain. Natural fluctuations in temperature must be taken into account, as well as cyclic changes in solar activity, which are expected to provide some additional warming for the first decade or so of the 21st century. Moreover, as the amount of CO_2 increases from fossil-fuel burning, so does the amount of soot, which may block sunlight and have a cooling effect. Water vapor also traps heat, and as temperatures rise, more water vapor forms. The increased amounts of water vapor may thicken the cloud cover and lead to cooling. Recent data indicate that decreased volcanic activity coupled with power-plant innovations may reduce the SO_2-based particulates that also cool the atmosphere.

Nevertheless, despite these opposing factors, most models predict a net warming of the atmosphere, and scientists are now observing it and documenting the predicted climate disruptions. The average temperature has increased by 0.6°C over the past 50 years. When extra heat is trapped by elevated levels of greenhouse gases, 20% of the heat warms the air and 80% evaporates water. Globally, the 10 warmest years on record have occurred since 1980, with 1997–1999 among the hottest. The decade from

1987 to 1997 included 10 times as many catastrophic floods from storms as did the previous decade, a trend that continued in 1998 and 1999. Other recently documented evidence includes a 10% decrease in global snow cover, a 40% thinning of Arctic Sea ice, a 20% decrease in glacier extent, and nearly a 10% increase in rainfall in the northern hemisphere.

Five years ago, the most accepted models predicted a temperature rise of 1.0–3.5°C. Today, the best predictions are 1.4–5.8°C (2.5–10.4°F), about 40% greater. An average increase of 1°C in the U.S. Midwest could lower the grain yield by 20%; an increase of 3°C could alter rainfall patterns and crop yields throughout the world; a 5°C increase could melt large parts of the polar ice caps, increasing sea level by 3.5–34 inches, thus flooding low-lying regions such as the Netherlands, half of Florida, and much of India, and inundating some Pacific island nations. To make matters worse, as we burn fossil fuels and wood that *release* CO_2, we cut down the forests that naturally *absorb* CO_2.

International conferences, such as the 1992 Earth Summit in Rio de Janeiro, Brazil, and the 1997 Conference on Climate Change in Kyoto, Japan, provided a forum for politicians, business interests, and scientists to agree on ways to cut emissions of greenhouse gases. Lack of progress at the 2000 International Panel on Climate Change at The Hague, Netherlands, disappointed all sides. As a result, many expect the pace of global warming and its associated effects on Earth's climate to quicken.

Solar Energy and Other Noncombustible Energy Sources

The quantity of solar energy striking the United States annually is *600 times greater than all our energy needs*. Despite this astounding abundance, solar energy is difficult to collect, store, and convert to other forms of energy. The Sun's total output is enormous but not concentrated, so vast surface areas must be devoted to collecting it.

Storage of the solar energy is necessary because intense sunlight is available over most regions for only 6 to 8 hours a day, and it is greatly reduced in rainy or cloudy weather. One storage approach makes use of ionic hydrates, such as $Na_2SO_4 \cdot 10H_2O$. When warmed by sunlight to over 32°C, the 3 mol of ions dissolve in the 10 mol of water in an endothermic process:

$$Na_2SO_4 \cdot 10H_2O(s) \xrightarrow{>32°C} Na_2SO_4(aq) \quad \Delta H^0 = 354 \text{ kJ}$$

When cooled below 32°C after sunset, the solution recrystallizes, releasing the absorbed energy for heating:

$$Na_2SO_4(aq) \xrightarrow{<32°C} Na_2SO_4 \cdot 10H_2O(s) \quad \Delta H^0 = -354 \text{ kJ}$$

Photovoltaic cells convert light directly into electricity, but the electrical energy produced is only about 10% of the radiant energy supplied. However, novel combinations of elements from Groups 3A(13) and 5A(15), such as gallium arsenide (GaAs), result in higher yields and are approaching economic parity with certain types of fossil-fuel use.

Energy Conservation: More from Less

All systems on Earth, whether living organisms or industrial power plants, waste some energy, which in effect is a waste of fuel. In terms of the energy used to produce it, for example, every discarded aluminum can is equivalent to 0.25 L of gasoline. Energy conservation lowers production costs, extends our supply of fossil fuels, and reduces pollution.

One example of an energy-conserving device is a high-efficiency gas-burning furnace for domestic heating. Its design channels hot waste gases through the furnace to transfer more heat to the room and to an attached domestic hot water system. Moreover, the channeling path cools the gases to <100°C, so the water vapor condenses, thus releasing about 10% more heat:

$$\begin{aligned} CH_4(g) + 2O_2(g) &\longrightarrow CO_2(g) + 2H_2O(g) & \Delta H^0 &= -802 \text{ kJ} \\ 2H_2O(g) &\longrightarrow 2H_2O(l) & \Delta H^0 &= -88 \text{ kJ} \\ \hline CH_4(g) + 2O_2(g) &\longrightarrow CO_2(g) + 2H_2O(l) & \Delta H^0 &= -890 \text{ kJ} \end{aligned}$$

Although engineers and chemists will be in the forefront in exploring new energy directions, a more hopeful energy future ultimately depends on our wisdom in obtaining and conserving planetary resources.

SECTION SUMMARY

Standard states are chosen conditions for substances. When 1 mol of a compound forms from its elements with all substances in their standard states, the enthalpy change is ΔH_f^0. Hess's law allows us to picture a reaction as the decomposition of reactants to their elements, followed by the formation of products from their elements. We use tabulated ΔH_f^0 values to find ΔH_{rxn}^0 or use known ΔH_{rxn}^0 and ΔH_f^0 values to find an unknown ΔH_f^0. Because of dwindling resources and environmental concerns, such as global warming, chemists are developing new energy alternatives, including coal and biomass conversion, hydrogen fuel, and noncombustible energy sources.

Chapter Perspective

Our investigation of energy in this chapter helps explain many chemical and physical phenomena: how gases do work, how reactions release or absorb heat, how the chemical energy in fuels and foods changes to other forms, and how we can use energy more wisely. We reexamine several of these ideas later in the text to address the question of why physical and chemical changes occur. In Chapter 7, you'll see how the energy released or absorbed by the atom created a revolutionary view of matter and energy that led to our current model of atomic energy states.

For Review and Reference (Numbers in parentheses refer to pages, unless noted otherwise.)

Learning Objectives

Relevant section and/or sample problem (SP) numbers appear in parentheses.

Understand These Concepts

1. The distinction between a system and its surroundings (6.1)
2. How energy is transferred to or from a system as heat and/or work (6.1)
3. The relation among the internal energy change, heat, and work (6.1)
4. The meaning of energy conservation (6.1)
5. The meaning of a state function and why ΔE is constant even though q and w vary (6.1)
6. The meaning of enthalpy and the relation between ΔE and ΔH (6.2)
7. The meaning of ΔH and the distinction between exothermic and endothermic reactions (6.2)
8. How changes in bond strength account for the heat of reaction (6.2)
9. The relation between specific heat capacity and heat transferred (6.3)
10. How constant-pressure (coffee-cup) and constant-volume (bomb) calorimeters work (6.3)
11. The relation between ΔH^0_{rxn} and amount of substance (6.4)
12. The importance of Hess's law and the manipulation of ΔH values (6.5)
13. The meaning of a formation equation and the standard heat of formation (6.6)
14. How a reaction can be viewed as the decomposition of reactants followed by the formation of products (6.6)

Master These Skills

1. Interconverting energy units (J, kJ, kcal) (SP 6.1)
2. Drawing enthalpy diagrams for chemical and physical changes (SP 6.2)
3. Solving problems involving specific heat capacity, change in temperature, and heat of reaction (SPs 6.3 to 6.5)
4. Relating the heat of reaction and the amounts of substances changing (SP 6.6)
5. Using Hess's law to find an unknown ΔH (SP 6.7)
6. Writing formation equations and using ΔH^0_f values to find ΔH^0_{rxn} (SPs 6.8 and 6.9)

Key Terms

thermodynamics (221)
thermochemistry (221)

Section 6.1
system (221)
surroundings (221)
internal energy (E) (222)
heat (q) (223)
work (w) (223)
pressure-volume work (PV work) (224)
law of conservation of energy (first law of thermodynamics) (225)
joule (J) (225)

calorie (cal) (225)
state function (226)

Section 6.2
enthalphy (H) (228)
change in enthalpy (ΔH) (228)
heat of reaction (ΔH_rxn) (229)
enthalpy diagram (229)
exothermic process (229)
endothermic process (229)
heat of combustion (ΔH_comb) (230)
heat of formation (ΔH_f) (230)
heat of fusion (ΔH_fus) (230)

heat of vaporization (ΔH_vap) (230)

Section 6.3
heat capacity (233)
specific heat capacity (c) (233)
molar heat capacity (C) (233)
calorimeter (234)

Section 6.4
thermochemical equation (236)

Section 6.5
Hess's law of heat summation (238)

Section 6.6
standard states (240)
standard heat of reaction (ΔH_rxn^0) (240)
formation equation (240)
standard heat of formation (ΔH_f^0) (240)
fossil fuel (243)
coal gasification (243)
syngas (243)
synthetic natural gas (SNG) (243)
biomass conversion (243)
photovoltaic cell (245)

Key Equations and Relationships

6.1 Defining the change in internal energy (222):
$$\Delta E = E_{final} - E_{initial} = E_{products} - E_{reactants}$$

6.2 Expressing the change in internal energy in terms of heat and work (223):
$$\Delta E = q + w$$

6.3 Stating the first law of thermodynamics (law of energy conservation) (225):
$$\Delta E_{universe} = \Delta E_{system} + \Delta E_{surroundings} = 0$$

6.4 Determining the work due to a change in volume at constant pressure (PV work) (228):
$$w = -P\Delta V$$

6.5 Relating the enthalpy change to the internal energy change (228):
$$\Delta H = \Delta E + P\Delta V$$

6.6 Identifying the enthalpy change with the heat gained or lost at constant pressure (228):
$$q_P = \Delta E + P\Delta V = \Delta H$$

6.7 Calculating the heat when a substance undergoes a temperature change (233):
$$q = c \times mass \times \Delta T$$

6.8 Calculating the standard heat of reaction (242):
$$\Delta H^0_{rxn} = \Sigma m\Delta H^0_{f\,(products)} - \Sigma n\Delta H^0_{f\,(reactants)}$$

Highlighted Figures and Tables

These figures (F) and tables (T) provide a quick review of key ideas.

F6.2 Energy diagrams for the transfer of internal energy (E) between a system and its surroundings (222)
T6.1 Sign conventions for q, w, and ΔE (224)
F6.8 Enthalpy diagrams for exothermic and endothermic processes (229)

T6.4 Specific heat capacities of some elements, compounds, and materials (233)
F6.12 Summary of the relationship between amount (mol) of substance and heat (kJ) transferred during a reaction (237)
F6.13 The general process for determining ΔH^0_{rxn} from ΔH^0_f values (241)

Brief Solutions to Follow-up Problems

6.1 $\Delta E = q + w = \left(-26.0 \text{ kcal} \times \dfrac{4.184 \text{ kJ}}{1 \text{ kcal}}\right)$
$\qquad\qquad + \left(+15.0 \text{ Btu} \times \dfrac{1.055 \text{ kJ}}{1 \text{ Btu}}\right) = -93 \text{ kJ}$

6.2 The reaction is exothermic.

6.3 $\Delta T = 25°C - 37°C = -12°C = -12 \text{ K}$
$\qquad q = 0.450 \text{ J/g·K} \times 5.5 \text{ g} \times (-12 \text{ K}) = -30. \text{ J}$

6.4 $\qquad\qquad -q_{solid} = q_{water}$
$-\,[(0.519 \text{ J/g·K}) (2.050 \text{ g}) (x - 74.21)]$
$\qquad\qquad = [(4.184 \text{ J/g·K}) (26.05 \text{ g}) (x - 27.20)]$
$\qquad\qquad\qquad x = 27.65 \text{ K}$
$\qquad\qquad \Delta T_{diamond} = -46.56 \text{ K} \ \text{ and } \ \Delta T_{water} = 0.45 \text{ K}$

6.5 $\qquad\qquad\qquad\qquad -q_{sample} = q_{calorimeter}$
$-(0.8650 \text{ g C})\left(\dfrac{1 \text{ mol C}}{12.01 \text{ g C}}\right)(-393.5 \text{ kJ/mol C}) = (2.613 \text{ K})x$
$\qquad\qquad\qquad\qquad x = 10.85 \text{ kJ/K}$

6.6 $HgO(s) \xrightarrow{\Delta} Hg(l) + \tfrac{1}{2}O_2(g) \qquad \Delta H = 90.8 \text{ kJ}$

Heat (kJ) $= 907 \text{ kg Hg} \times \dfrac{10^3 \text{ g}}{1 \text{ kg}} \times \dfrac{1 \text{ mol Hg}}{200.6 \text{ g Hg}} \times \dfrac{90.8 \text{ kJ}}{1 \text{ mol Hg}}$
$\qquad = 4.11 \times 10^5 \text{ kJ}$

6.7 $2NO(g) + \tfrac{3}{2}O_2(g) \longrightarrow N_2O_5(s) \qquad \Delta H = -223.7 \text{ kJ}$
$\qquad 2NO_2(g) \longrightarrow 2NO(g) + O_2(g)$

$\qquad\qquad\qquad\qquad\qquad\qquad\qquad\qquad \Delta H = \quad 114.2 \text{ kJ}$

$2\cancel{NO(g)} + \tfrac{1}{2}\cancel{\tfrac{3}{2}}O_2(g) + 2NO_2(g) \longrightarrow$
$\qquad\qquad\qquad\qquad N_2O_5(s) + \cancel{2NO(g)} + \cancel{O_2(g)}$
$2NO_2(g) + \tfrac{1}{2}O_2(g) \longrightarrow N_2O_5(s) \qquad \Delta H = -109.5 \text{ kJ}$

6.8 (a) $C(\text{graphite}) + 2H_2(g) + \tfrac{1}{2}O_2(g) \longrightarrow CH_3OH(l)$
$\qquad\qquad\qquad\qquad\qquad\qquad\qquad \Delta H^0_f = -238.6 \text{ kJ}$
(b) $Ca(s) + \tfrac{1}{2}O_2(g) \longrightarrow CaO(s) \quad \Delta H^0_f = -635.1 \text{ kJ}$
(c) $C(\text{graphite}) + \tfrac{1}{4}S_8(\text{rhombic}) \longrightarrow CS_2(l) \qquad \Delta H^0_f = 87.9 \text{ kJ}$

6.9 ΔH^0_f of $CH_3OH(l)$
$\qquad = -\Delta H^0_{comb} + 2\Delta H^0_f[H_2O(g)] + \Delta H^0_f[CO_2(g)]$
$\qquad = 638.5 \text{ kJ} + (2 \text{ mol})(-241.8 \text{ kJ/mol})$
$\qquad\qquad + (1 \text{ mol}) (-393.5 \text{ kJ/mol})$
$\qquad = -238.6 \text{ kJ}$

Problems

Problems with **colored** numbers are answered at the back of the text. Sections match the text and provide the number(s) of relevant sample problems. Most offer Concept Review Questions, Skill-Building Exercises (in similar pairs), and Problems in Context. Then Comprehensive Problems, based on material from any section or previous chapter, follow.

Forms of Energy and Their Interconversion
(Sample Problem 6.1)

● **Concept Review Questions**

6.1 Why do heat (q) and work (w) have positive values when entering a system and negative values when leaving?
6.2 If you feel warm after exercising, have you increased the internal energy of your body? Explain.

6.3 An *adiabatic* process is one that involves no heat transfer. What is the relationship between work and the change in internal energy in an adiabatic process?
6.4 State two ways that you increase the internal energy of your body and two ways that you decrease it.
6.5 Name a common device for each energy change:
(a) Electrical energy to thermal energy
(b) Electrical energy to sound energy
(c) Electrical energy to light energy
(d) Mechanical energy to electrical energy
(e) Chemical energy to electrical energy
6.6 In winter, an electric heater uses a certain amount of electrical energy to heat a room to 20°C. In summer, an air conditioner uses the same amount of electrical energy to cool the room to

20°C. Is the change in internal energy of the heater larger, smaller, or the same as that of the air conditioner? Explain.

6.7 Imagine lifting your textbook into the air and dropping it onto a desktop. Describe all the energy transformations (from one form to another) that occur, moving backward in time from a moment after impact.

● **Skill-Building Exercises** *(paired)*

6.8 A system receives 425 J of heat and delivers 425 J of work to its surroundings. What is the change in internal energy of the system (in J)?

6.9 A system conducts 255 cal of heat to the surroundings while delivering 428 cal of work. What is the change in internal energy of the system (in cal)?

6.10 What is the change in internal energy (in J) of a system that releases 675 J of thermal energy to its surroundings and has 525 cal of work done on it?

6.11 What is the change in internal energy (in J) of a system that absorbs 0.615 kJ of heat from its surroundings and has 0.247 kcal of work done on it?

6.12 Complete combustion of 1.0 metric ton of coal (assuming pure carbon) to gaseous carbon dioxide releases 3.3×10^{10} J of heat. Convert this energy to (a) kilojoules; (b) kilocalories; (c) British thermal units.

6.13 Thermal decomposition of 5.0 metric tons of limestone to lime and carbon dioxide requires 9.0×10^6 kJ of heat. Convert this energy to (a) joules; (b) calories; (c) British thermal units.

● **Problems in Context**

6.14 The nutritional calorie (Calorie) is equivalent to 1 kcal. One pound of body fat is equivalent to about 4.1×10^3 Calories. Express this energy equivalence in joules and kilojoules.

6.15 If an athlete expends 1850 kJ/h, how long does she have to play to work off 1.0 lb of body fat? (See Problem 6.14.)

Enthalpy: Heats of Reaction and Chemical Change
(Sample Problem 6.2)

● **Concept Review Questions**

6.16 Why is the work done when a system expands against a constant external pressure assigned a negative sign?

6.17 Why is it usually more convenient to measure ΔH than ΔE?

6.18 "Hot packs" used by skiers, climbers, and others for warmth are based on the crystallization of sodium acetate from a highly concentrated solution. What is the sign of ΔH for this crystallization? Is the reaction exothermic or endothermic?

6.19 Classify the following processes as exothermic or endothermic: (a) freezing of water; (b) boiling of water; (c) digestion of food; (d) a person running; (e) a person growing; (f) wood being chopped; (g) heating with a furnace.

6.20 What are the two main components of the internal energy of a substance? On what are they based?

6.21 For each of the following processes, state whether ΔH is less than (more negative), equal to, or greater than ΔE of the system. Explain.
(a) An ideal gas is cooled at constant pressure.
(b) A mixture of gases undergoes an exothermic reaction in a container of fixed volume.
(c) A solid yields a mixture of gases in an exothermic reaction that takes place in a container of variable volume.

● **Skill-Building Exercises** *(paired)*

6.22 Draw an enthalpy diagram for a general exothermic reaction; label axis, reactants, products, and ΔH with its sign.

6.23 Draw an enthalpy diagram for a general endothermic reaction; label axis, reactants, products, and ΔH with its sign.

6.24 Write a balanced equation and draw an approximate enthalpy diagram for each of the following: (a) the combustion of 1 mol of methane in oxygen; (b) the freezing of liquid water.

6.25 Write a balanced equation and draw an approximate enthalpy diagram for each of the following: (a) the formation of 1 mol of sodium chloride from its elements (heat is released); (b) the vaporization of liquid benzene.

6.26 Write a balanced equation and draw an approximate enthalpy diagram for each of the following changes: (a) the combustion of 1 mol of liquid ethanol (C_2H_5OH); (b) the formation of 1 mol nitrogen dioxide from its elements (heat is absorbed).

6.27 Write a balanced equation and draw an approximate enthalpy diagram for each of the following changes: (a) the sublimation of dry ice [conversion of $CO_2(s)$ directly to $CO_2(g)$]; (b) the reaction of 1 mol of sulfur dioxide with oxygen.

6.28 Which of the following gases would you expect to have the greater heat of combustion per mole? Why?

methane   or   formaldehyde

6.29 Which of the following two fuel candidates would you expect to have the greater heat of combustion per mole? Why?

ethanol   or   methanol

Calorimetry: Laboratory Measurement of Heats of Reaction
(Sample Problems 6.3 to 6.5)

● **Concept Review Questions**

6.30 Why can we measure only *changes* in enthalpy, not absolute enthalpy values?

6.31 What data do you need to determine the specific heat capacity of a substance?

6.32 Is the specific heat capacity of a substance an intensive or extensive property? Explain.

6.33 Distinguish between "specific heat capacity" and "heat capacity." Which parameter would you more likely use if you were calculating heat changes in (a) a chrome-plated, brass bathroom fixture; (b) a sample of high-purity copper wire; (c) a sample of pure water? Explain.

6.34 Both a coffee-cup calorimeter and a bomb calorimeter can be used to measure the heat involved in a reaction. Which measures ΔE and which measures ΔH? Explain.

● **Skill-Building Exercises** *(paired)*

6.35 Calculate q when 12.0 g of water is heated from 20.°C to 100.°C.

6.36 Calculate q when 0.10 g of ice is cooled from 10.°C to -75°C ($c_{ice} = 2.087$ J/g·K).

6.37 A 295-g aluminum engine part at an initial temperature of 3.00°C absorbs 85.0 kJ of heat. What is the final temperature of the part (c of Al = 0.900 J/g·K)?

6.38 A 27.7-g sample of ethylene glycol, a car radiator coolant, loses 688 J of heat. What was the initial temperature of the ethylene glycol if the final temperature is 32.5°C (c of ethylene glycol = 2.42 J/g·K)?

6.39 Two iron bolts of equal mass—one at 100.°C, the other at 55°C—are placed in an insulated container. Assuming the heat capacity of the container is negligible, what is the final temperature inside the container (c of iron = 0.450 J/g·K)?

6.40 One piece of copper jewelry at 105°C has exactly twice the mass of another piece, which is at 45°C. Both pieces are placed inside a calorimeter whose heat capacity is negligible. What is the final temperature inside the calorimeter (c of copper = 0.387 J/g·K)?

6.41 When 165 mL of water at 22°C is mixed with 85 mL of water at 82°C, what is the final temperature? (Assume that no heat is lost to the surroundings; d of water is 1.00 g/mL.)

6.42 An unknown volume of water at 18.2°C is added to 24.4 mL of water at 35.0°C. If the final temperature is 23.5°C, what was the unknown volume? (Assume that no heat is lost to the surroundings; d of water is 1.00 g/mL.)

6.43 A 505-g piece of copper tubing is heated to 99.9°C and placed in an insulated vessel containing 59.8 g of water at 24.8°C. Assuming no loss of water and a heat capacity for the vessel of 10.0 J/K, what is the final temperature of the system (c of copper = 0.387 J/g·K)?

6.44 A 30.5-g sample of an alloy at 93.0°C is placed into 50.0 g of water at 22.0°C in an insulated coffee cup with a heat capacity of 9.2 J/K. If the final temperature of the system is 31.1°C, what is the specific heat capacity of the alloy?

● **Problems in Context**

6.45 Two aircraft rivets, one of iron and the other of copper, are placed in a calorimeter that has an initial temperature of 20.°C. The data for the metals are as follows:

	Iron	**Copper**
Mass (g)	30.0	20.0
Initial T (°C)	0.0	100.0
c (J/g·K)	0.450	0.387

(a) Will heat flow from Fe to Cu or from Cu to Fe?
(b) What other information is needed in order to correct any of the measurements that would be made in an actual experiment?
(c) What is the maximum final temperature of the system (assuming the heat capacity of the calorimeter is negligible)?

6.46 A chemical engineer studying the properties of fuels placed 1.500 g of a hydrocarbon in the bomb of a calorimeter and filled it with O_2 gas (see Figure 6.11). The bomb was immersed in 2.500 L of water and the reaction initiated. The water temperature rose from 20.00°C to 23.55°C. If the calorimeter (excluding the water) had a heat capacity of 403 J/K, what was the heat of combustion (q_V) per gram of the fuel?

Stoichiometry of Thermochemical Equations
(Sample Problem 6.6)

● **Concept Review Questions**

6.47 Does a negative ΔH_{rxn} mean that the heat of reaction can be thought of as a reactant or as a product?

6.48 Would you expect $O_2(g) \longrightarrow 2O(g)$ to have a positive or a negative ΔH_{rxn}? Explain.

6.49 Is ΔH positive or negative when 1 mol of water vapor condenses to liquid water? Why? How does this value compare with the value from the vaporization of 2 mol of liquid water to water vapor?

● **Skill-Building Exercises (paired)**

6.50 Consider the following balanced thermochemical equation for a reaction sometimes used for H_2S production:

$$\tfrac{1}{8}S_8(s) + H_2(g) \longrightarrow H_2S(g) \quad \Delta H_{rxn} = -20.2 \text{ kJ}$$

(a) Is this an exothermic or endothermic reaction?
(b) What is ΔH_{rxn} for the reverse reaction?
(c) What is ΔH when 3.2 mol of S_8 reacts?
(d) What is ΔH when 20.0 g of S_8 reacts?

6.51 Consider the following balanced thermochemical equation for the decomposition of the mineral magnesite:

$$MgCO_3(s) \longrightarrow MgO(s) + CO_2(g) \quad \Delta H_{rxn} = 117.3 \text{ kJ}$$

(a) Is heat absorbed or released in the reaction?
(b) What is ΔH_{rxn} for the reverse reaction?
(c) What is ΔH when 5.35 mol of CO_2 reacts with excess MgO?
(d) What is ΔH when 35.5 g of CO_2 reacts with excess MgO?

6.52 When 1 mol of NO(g) forms from its elements, 90.29 kJ of heat is absorbed.
(a) Write a balanced thermochemical equation for this reaction.
(b) How much heat is involved when 1.50 g of NO decomposes to its elements?

6.53 When 1 mol of KBr(s) decomposes to its elements, 394 kJ of heat is absorbed.
(a) Write a balanced thermochemical equation for this reaction.
(b) How much heat is released when 10.0 kg of KBr forms from its elements?

● **Problems in Context**

6.54 Liquid hydrogen peroxide is an oxidizing agent in many rocket fuel mixtures because it releases oxygen gas on decomposition:

$$2H_2O_2(l) \longrightarrow 2H_2O(l) + O_2(g) \quad \Delta H_{rxn} = -196.1 \text{ kJ}$$

How much heat is released when 732 kg of H_2O_2 decomposes?

6.55 Compounds of boron and hydrogen are remarkable for their unusual bonding (described in Section 14.5) and also for their reactivity. With the more reactive halogens, for example, diborane (B_2H_6) forms trihalides even at low temperatures:

$$B_2H_6(g) + 6Cl_2(g) \longrightarrow 2BCl_3(g) + 6HCl(g)$$
$$\Delta H_{rxn} = -755.4 \text{ kJ}$$

How much heat is released per kilogram of diborane that reacts?

6.56 Deterioration of buildings, bridges, and other structures through the rusting of iron costs millions of dollars every day. Although the actual process also requires water, a simplified equation (with rust shown as Fe_2O_3) is

$$4Fe(s) + 3O_2(g) \longrightarrow 2Fe_2O_3(s) \quad \Delta H_{rxn} = -1.65 \times 10^3 \text{ kJ}$$

(a) How much heat is evolved when 0.100 kg of iron rusts?
(b) How much rust forms when 4.93×10^3 kJ of heat is released?

6.57 A mercury mirror forms inside a test tube by the thermal decomposition of mercury(II) oxide:

$$2HgO(s) \longrightarrow 2Hg(l) + O_2(g) \quad \Delta H_{rxn} = 181.6 \text{ kJ}$$

(a) How much heat is needed to decompose 555 g of the oxide?
(b) If 275 kJ of heat is absorbed, how many grams of mercury form?

6.58 Ethylene (C_2H_4) is the starting material for the preparation of polyethylene. Although typically made during the processing of petroleum, ethylene occurs naturally as a fruit-ripening hormone and as a component of natural gas.
(a) If the heat of combustion of C_2H_4 is -1411 kJ/mol, write a balanced thermochemical equation for the combustion of C_2H_4.
(b) How many grams of C_2H_4 must burn to give 70.0 kJ of heat?

6.59 Sucrose ($C_{12}H_{22}O_{11}$, table sugar) is oxidized in the body by O_2 via a complex set of reactions that ultimately produces $CO_2(g)$ and $H_2O(g)$ and releases 5.64×10^3 kJ/mol sucrose.
(a) Write a balanced thermochemical equation for this reaction.
(b) How much heat is released per gram of sucrose oxidized?

Hess's Law of Heat Summation

(Sample Problem 6.7)

● **Concept Review Questions**

6.60 Express Hess's law in your own words.
6.61 What is the main use of Hess's law?
6.62 It is very difficult to burn carbon in a deficiency of O_2 and produce only CO; some CO_2 forms as well. However, carbon burns in excess O_2 to form only CO_2, and CO burns in excess O_2 to form only CO_2. Use the heats of the latter two reactions (from Appendix B) to calculate ΔH_{rxn} for the following reaction:

$$C(graphite) + \tfrac{1}{2}O_2(g) \longrightarrow CO(g)$$

● **Skill-Building Exercises (paired)**

6.63 Calculate ΔH_{rxn} for

$$Ca(s) + \tfrac{1}{2}O_2(g) + CO_2(g) \longrightarrow CaCO_3(s)$$

given the following set of reactions:

$$Ca(s) + \tfrac{1}{2}O_2(g) \longrightarrow CaO(s) \qquad \Delta H = -635.1 \text{ kJ}$$
$$CaCO_3(s) \longrightarrow CaO(s) + CO_2(g) \quad \Delta H = 178.3 \text{ kJ}$$

6.64 Calculate ΔH_{rxn} for

$$2NOCl(g) \longrightarrow N_2(g) + O_2(g) + Cl_2(g)$$

given the following set of reactions:

$$\tfrac{1}{2}N_2(g) + \tfrac{1}{2}O_2(g) \longrightarrow NO(g) \qquad \Delta H = 90.3 \text{ kJ}$$
$$NO(g) + \tfrac{1}{2}Cl_2(g) \longrightarrow NOCl(g) \quad \Delta H = -38.6 \text{ kJ}$$

6.65 At a specific set of conditions, 241.8 kJ is given off when 1 mol of $H_2O(g)$ forms from its elements. Under the same conditions, 285.8 kJ is given off when 1 mol of $H_2O(l)$ forms from its elements. Calculate the heat of vaporization of water at these conditions.

6.66 When 1 mol of $CS_2(l)$ forms from its elements at 1 atm and 25°C, 89.7 kJ is absorbed, and it takes 27.7 kJ to vaporize 1 mol of the liquid. How much heat is absorbed when 1 mol of $CS_2(g)$ forms from its elements at these conditions?

6.67 Write the balanced overall equation for the following process (equation 3), calculate $\Delta H_{overall}$, and match the number of the equation with the letter of the arrow in Figure P6.67:

(1) $N_2(g) + O_2(g) \longrightarrow 2NO(g)$ $\qquad \Delta H = 180.6 \text{ kJ}$
(2) $2NO(g) + O_2(g) \longrightarrow 2NO_2(g)$ $\qquad \Delta H = -114.2 \text{ kJ}$
(3) $\qquad\qquad\qquad\qquad\qquad\qquad \Delta H_{overall} = ?$

6.68 Write the balanced overall equation for the following process (equation 3), calculate $\Delta H_{overall}$, and match the number of the equation with the letter of the arrow in Figure P6.68:

(1) $P_4(s) + 6Cl_2(g \longrightarrow 4PCl_3(g)$ $\qquad \Delta H = -1148 \text{ kJ}$
(2) $4PCl_3(g) + 4Cl_2(g) \longrightarrow 4PCl_5(g)$ $\quad \Delta H = -460 \text{ kJ}$
(3) $\qquad\qquad\qquad\qquad\qquad\qquad \Delta H_{overall} = ?$

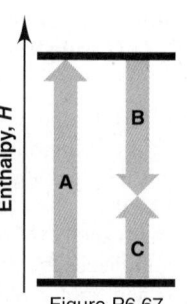

Figure P6.67

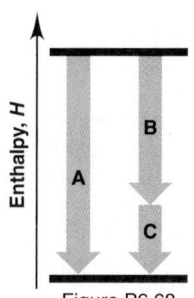

Figure P6.68

● **Problems in Context**

6.69 Diamond and graphite are two crystalline forms of carbon. At 1 atm and 25°C, diamond changes to graphite so slowly that the enthalpy change of the process must be obtained indirectly. Determine ΔH_{rxn} for

$$C(diamond) \longrightarrow C(graphite)$$

with equations from the following list:

(1) $C(diamond) + O_2(g) \longrightarrow CO_2(g)$ $\qquad \Delta H = -395.4 \text{ kJ}$
(2) $2CO_2(g) \longrightarrow 2CO(g) + O_2(g)$ $\qquad \Delta H = 566.0 \text{ kJ}$
(3) $C(graphite) + O_2(g) \longrightarrow CO_2(g)$ $\qquad \Delta H = -393.5 \text{ kJ}$
(4) $2CO(g) \longrightarrow C(graphite) + CO_2(g)$ $\qquad \Delta H = -172.5 \text{ kJ}$

Standard Heats of Reaction (ΔH^0_{rxn})

(Sample Problems 6.8 and 6.9)

● **Concept Review Questions**

6.70 What is the difference between the standard heat of formation and the standard heat of reaction?
6.71 How are ΔH^0_f values used to calculate ΔH^0_{rxn}?
6.72 Make any changes needed in each of the following equations to make ΔH^0_{rxn} equal to ΔH^0_f for the compound present:
(a) $Cl(g) + Na(s) \longrightarrow NaCl(s)$
(b) $H_2O(g) \longrightarrow 2H(g) + \tfrac{1}{2}O_2(g)$
(c) $\tfrac{1}{2}N_2(g) + \tfrac{3}{2}H_2(g) \longrightarrow NH_3(g)$

● **Skill-Building Exercises (paired)**

6.73 Use Table 6.5 or Appendix B to write balanced formation equations at standard conditions for each of the following compounds: (a) $CaCl_2$; (b) $NaHCO_3$; (c) CCl_4; (d) HNO_3.
6.74 Use Table 6.5 or Appendix B to write balanced formation equations at standard conditions for each of the following compounds: (a) HI; (b) SiF_4; (c) O_3; (d) $Ca_3(PO_4)_2$.

6.75 Calculate ΔH^0_{rxn} for each of the following:
(a) $2H_2S(g) + 3O_2(g) \longrightarrow 2SO_2(g) + 2H_2O(g)$
(b) $CH_4(g) + Cl_2(g) \longrightarrow CCl_4(l) + HCl(g)$ [unbalanced]
6.76 Calculate ΔH^0_{rxn} for each of the following:
(a) $SiO_2(s) + 4HF(g) \longrightarrow SiF_4(g) + 2H_2O(l)$
(b) $C_2H_6(g) + O_2(g) \longrightarrow CO_2(g) + H_2O(g)$ [unbalanced]

6.77 Copper(I) oxide can be oxidized to copper(II) oxide:

$$Cu_2O(s) + \tfrac{1}{2}O_2(g) \longrightarrow 2CuO(s) \quad \Delta H^0_{rxn} = -146.0 \text{ kJ}$$

Given that ΔH^0_f of $Cu_2O(s) = -168.6$ kJ/mol, what is ΔH^0_f of $CuO(s)$?

6.78 Acetylene burns in air according to the following equation:

$$C_2H_2(g) + \tfrac{5}{2}O_2(g) \longrightarrow 2CO_2(g) + H_2O(g)$$

$$\Delta H^0_{rxn} = -1255.8 \text{ kJ}$$

Given that ΔH^0_f of $CO_2(g)$ = -393.5 kJ/mol and that ΔH^0_f of $H_2O(g)$ = -241.8 kJ/mol, what is ΔH^0_f of $C_2H_2(g)$?

● **Problems in Context**

6.79 Nitroglycerine, $C_3H_5(NO_3)_3(l)$, a powerful explosive used in mining, detonates to produce a hot gaseous mixture of nitrogen, water, carbon dioxide, and oxygen.
(a) Write a balanced equation for this reaction using the smallest whole-number coefficients.
(b) If $\Delta H^0_{rxn} = -2.29 \times 10^4$ kJ for the equation as written in part (a), calculate ΔH^0_f of nitroglycerine.

6.80 The common lead-acid car battery produces a large burst of current, even at low temperatures, and is rechargeable. The reaction that occurs while recharging a "dead" battery is

$$2PbSO_4(s) + 2H_2O(l) \longrightarrow Pb(s) + PbO_2(s) + 2H_2SO_4(l)$$

(a) Use ΔH^0_f values from Appendix B to calculate ΔH^0_{rxn}.
(b) Use the following equations to check your answer in part (a):
(1) $Pb(s) + PbO_2(s) + 2SO_3(g) \longrightarrow 2PbSO_4(s)$

$$\Delta H^0 = -768 \text{ kJ}$$

(2) $SO_3(g) + H_2O(l) \longrightarrow H_2SO_4(l)$ $\quad \Delta H^0 = -132$ kJ

Comprehensive Problems

Problems with an asterisk (*) are more challenging.
6.81 Stearic acid ($C_{18}H_{36}O_2$) is a typical fatty acid, a molecule with a long hydrocarbon chain and an organic acid group (COOH) at the end. It is used to make cosmetics, ointments, soaps, and candles and is found in animal tissue as part of many saturated fats. In fact, when you eat meat, chances are that you are ingesting some fats that contain stearic acid.
(a) Write a balanced equation for the complete combustion of stearic acid to gaseous products.
(b) Calculate its heat of combustion ($\Delta H^0_f = -948$ kJ/mol).
(c) Calculate the heat (q) in kJ and kcal when 1.00 g of stearic acid is burned completely.
(d) The nutritional information on a candy bar states that 1 serving contains 11.0 g of fat and 100. Cal from fat (1 Cal = 1 kcal). Is this information consistent with your answer in part (c)?
***6.82** When you dilute sulfuric acid with water, you must be very careful because the dilution process is highly exothermic:

$$H_2SO_4(l) \xrightarrow{H_2O} H_2SO_4(aq) + heat$$

(a) Use Appendix B to calculate ΔH^0 for diluting 1.00 mol of $H_2SO_4(l)$ to 1 L of 1.00 M $H_2SO_4(aq)$.
(b) Suppose you carry out the dilution in a calorimeter. The initial T is 25.0°C, the density of the final solution is 1.060 g/mL, and its specific heat capacity is 3.50 J/g·K. What is the final T?
(c) Use the idea of heat capacity to explain why you should carry out the dilution by adding acid to water rather than water to acid.
6.83 A would-be around-the-world balloonist is preparing to make the trip in a helium-filled balloon. The trip begins in early morning at a temperature of 15°C. By midafternoon, the temperature has increased to 30.°C. Assuming the pressure remains constant at 1.00 atm, for each mole of helium, calculate:
(a) The initial and final volumes
(b) The change in internal energy, ΔE [*Hint:* Helium behaves like an ideal gas, so $E = \tfrac{3}{2}nRT$. Be sure the units of R are consistent with those of E.]

(c) The work (w) done by the helium (in J)
(d) The heat (q) transferred (in J)
(e) ΔH for the process (in J)
(f) Explain the relationship between the answers to (d) and (e).
6.84 In his studies of acid-base reactions, Arrhenius discovered an important fact demonstrated by the following equations:

$$KOH(aq) + HNO_3(aq) \longrightarrow KNO_3(aq) + H_2O(l)$$
$$NaOH(aq) + HCl(aq) \longrightarrow NaCl(aq) + H_2O(l)$$

(a) Use the data for the individual ions in Appendix B to calculate ΔH^0 for these reactions.
(b) Explain your results and use them to predict ΔH^0 for

$$KOH(aq) + HCl(aq) \longrightarrow KCl(aq) + H_2O(l)$$

6.85 Nitrogen gas is one of the most heavily produced substances in the chemical industry. Nearly all of it comes from distilling liquefied air, but small amounts of very pure N_2 are prepared by heating aqueous ammonium nitrite:

$$NH_4^+(aq) + NO_2^-(aq) \longrightarrow 2H_2O(l) + N_2(g) \; \Delta H^0_{rxn} = -305 \text{ kJ}$$

The aqueous ammonium nitrite, in turn, is prepared by adding ammonium chloride to aqueous sodium nitrite:

$$Na^+(aq) + NO_2^-(aq) + NH_4Cl(s) \longrightarrow$$
$$Na^+(aq) + Cl^-(aq) + NH_4^+(aq) + NO_2^-(aq)$$

(a) What is the heat of reaction for this change [ΔH^0_f of $NH_4^+(aq) = -132.8$ kJ/mol]?
(b) How many kilojoules are released when 50.5 g of solid NH_4Cl are used to make N_2?
6.86 Silicon carbide (SiC) is used as an abrasive. After the blue-black crystals are produced, they are crushed and glued on paper to make sandpaper or set in clay to make grinding wheels. Silicon carbide is formed by heating sand (SiO_2) with coke, a form of carbon. Carbon monoxide gas is also formed in this endothermic process, which absorbs 624.7 kJ per mole of SiC formed.
(a) What is ΔH^0_f of SiC? (Assume coke is the same as graphite.)
(b) How many kilojoules are needed per kilogram of SiC produced?
6.87 Iron metal is produced in a blast furnace through a complex series of reactions that involve reduction of iron(III) oxide with carbon monoxide.
(a) Write a balanced overall equation for the process, including the other product.
(b) Use the equations below to calculate ΔH^0_{rxn} for the overall equation:
(1) $3Fe_2O_3(s) + CO(g) \longrightarrow 2Fe_3O_4(s) + CO_2(g)$

$$\Delta H^0 = -48.5 \text{ kJ}$$

(2) $Fe(s) + CO_2(g) \longrightarrow FeO(s) + CO(g)$ $\quad \Delta H^0 = -11.0$ kJ
(3) $Fe_3O_4(s) + CO(g) \longrightarrow 3FeO(s) + CO_2(g)$ $\quad \Delta H^0 = 22$ kJ
6.88 Pure liquid octane (C_8H_{18}; $d = 0.702$ g/mL) is used as the fuel in a test of a new automobile drive train.
(a) How much energy (in kJ) is released by complete combustion of the octane in a 20.4-gal fuel tank to gases ($\Delta H^0_{comb} = -5.45 \times 10^3$ kJ/mol)?
(b) The energy delivered to the wheels at 65 mph is 5.5×10^4 kJ/h. Assuming all the energy is transferred to the wheels, what is the cruising range (in km) of the car on a full tank?
(c) If the actual cruising range is 455 miles, explain your answer to part (b).
6.89 When simple sugars, called *monosaccharides,* link together, they form a variety of complex sugars and, ultimately, *polysaccharides,* such as starch. Glucose and fructose are simple

sugars with the same formula, $C_6H_{12}O_6$, but different arrangements of atoms. The two molecules link together to form a molecule of sucrose (common table sugar) and a molecule of liquid water. The standard heats of formation of glucose, fructose, and sucrose are −1273 kJ/mol, −1266 kJ/mol, and −2226 kJ/mol, respectively. Write a balanced equation for this reaction and calculate the heat of reaction.

6.90 Physicians and nutritional biochemists recommend eating vegetable oils rather than animal fats to lower risks of heart disease. In olive oil, one of the healthier choices, the main fatty acid is oleic acid ($C_{18}H_{34}O_2$), whose $\Delta H^0_{comb} = -1.11 \times 10^4$ kJ/mol. Calculate ΔH^0_f of oleic acid. [Assume $H_2O(g)$.]

6.91 Oxidation of ClF by F_2 yields ClF_3, an important fluorinating agent:

$$ClF(g) + F_2(g) \longrightarrow ClF_3(l)$$

Use the following thermochemical equations to calculate ΔH^0_{rxn} for the production of ClF_3:

(1) $2ClF(g) + O_2(g) \longrightarrow Cl_2O(g) + OF_2(g)$ $\Delta H^0 = 167.5$ kJ
(2) $2F_2(g) + O_2(g) \longrightarrow 2OF_2(g)$ $\Delta H^0 = -43.5$ kJ
(3) $2ClF_3(l) + 2O_2(g) \longrightarrow Cl_2O(g) + 3OF_2(g)$
$\Delta H^0 = 394.1$ kJ

6.92 Silver halides are essential components of black-and-white photographic films. Ordinary film is coated with the bromide, and extremely high-speed film uses the iodide.
(a) When 50.0 mL of 5.0 g/L $AgNO_3$ is mixed with 50.0 mL of 5.0 g/L NaI at 25°C, what mass of AgI forms?
(b) Use Appendix B to find ΔH^0_{rxn}.

6.93 The calorie (4.184 J) was originally defined as the quantity of energy required to raise the temperature of 1.00 g of liquid water 1.00°C. The British thermal unit (Btu) is defined as the quantity of energy required to raise the temperature of 1.00 lb of liquid water 1.00°F.
(a) How many joules are in 1.00 British thermal unit (1 lb = 453.6 g; a change of 1.0°C = 1.8°F)?
(b) The "therm" is a unit of energy consumption that is used by natural gas companies in the United States and is defined as 100,000 Btu. How many joules are in 1.00 therm?
(c) How many moles of methane must be burned to give 1.00 therm of energy? (Assume water forms as a gas.)
(d) If natural gas costs $0.46 per therm, what is the cost per mole of methane? (Assume natural gas is pure methane.)
(e) How much would it cost to warm 308 gal of water in a hot tub from 15.0°C to 40.0°C (1 gal = 3.78 L)?

6.94 Diborane (B_2H_6) is a very reactive gas that forms gaseous boron trichloride and hydrogen chloride when it reacts with chlorine. When boron trichloride is bubbled through water, it forms solid boric acid [H_3BO_3 or $B(OH)_3$] and aqueous hydrochloric acid. Write a balanced equation for the overall reaction from B_2H_6 to H_3BO_3, and calculate the standard heat of reaction in kJ/mol of H_3BO_3 formed.

6.95 The heat of atomization (ΔH^0_{atom}) is the heat needed to form separated gaseous atoms from a substance in its standard state. The equation for the atomization of graphite is

$$C(\text{graphite}) \longrightarrow C(g)$$

Use Hess's law to calculate ΔH^0_{atom} of graphite from these data:
(1) ΔH^0_f of $CH_4 = -74.9$ kJ/mol
(2) ΔH^0_{atom} of $CH_4 = 1660$ kJ/mol
(3) ΔH^0_{atom} of $H_2 = 432$ kJ/mol

6.96 A reaction is carried out in a steel vessel within a chamber filled with argon gas. Below are molecular views of the argon adjacent to the surface of the reaction vessel before and after the reaction. Was the reaction exothermic or endothermic? Explain.

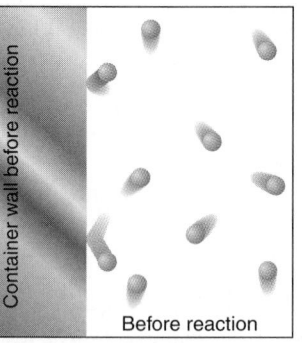 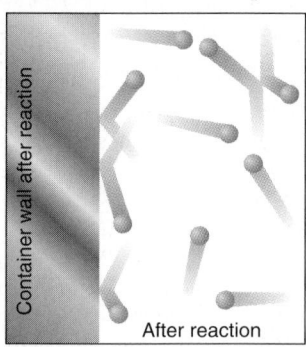

Before reaction   After reaction

6.97 For each of the following events, the system is in *italics*. State whether heat, work, or both is (are) transferred and specify the direction of each transfer.
(a) You pump air into an automobile *tire*.
(b) A *tree* rots in a forest.
(c) You strike a *match*.
(d) You cool juice in an *ice chest*.
(e) You cook *food* on a kitchen range.

6.98 Ammonia, which has N in its lowest oxidation state, is sometimes used to reduce metal oxides. At high temperature, ammonia reduces copper(II) oxide to the free metal:
Step 1: $2NH_3(g) + 6CuO(s) \longrightarrow N_2(g) + 3Cu_2O(s) + 3H_2O(g)$
Step 2: $2NH_3(g) + 3Cu_2O(s) \longrightarrow N_2(g) + 6Cu(s) + 3H_2O(g)$
(a) Use Appendix B to calculate the heat of reaction for each step.
(b) Write the overall equation, calculate its heat of reaction, and show how the results are consistent with Hess's law.

6.99 Triglycerides are the main form in which fats are stored in the body. During periods of starvation, a person's fat stores are used for energy. Tristearin ($C_{57}H_{110}O_6$) is a typical animal fat that is oxidized according to the following equation:

$$2C_{57}H_{110}O_6(s) + 163O_2(g) \longrightarrow 114CO_2(g) + 110H_2O(l)$$

If $\Delta H^0_{rxn} = -7.0 \times 10^4$ kJ, how much heat is released
(a) Per mole of O_2 consumed?
(b) Per mole of CO_2 formed?
(c) Per gram of tristearin oxidized?
(d) When 325 L of O_2 at 37°C and 755 torr is used, how many grams of tristearin can be oxidized?

6.100 Show that the SI units for pressure times volume (*PV*) equal those for energy (*E*).

6.101 Kerosene, a common space-heater fuel, is a mixture of hydrocarbons whose "average" formula is $C_{12}H_{26}$.
(a) Write a balanced equation, using the simplest whole-number coefficients, for the complete combustion of kerosene to gases.
(b) If $\Delta H^0_{comb} = -1.50 \times 10^4$ kJ for the equation as written in part (a), determine ΔH^0_f of kerosene.
(c) Calculate the heat produced by combustion of 0.50 gal of kerosene (*d* of kerosene = 0.749 g/mL).
(d) How many gallons of kerosene must be burned for a kerosene furnace to produce 1250. Btu (1 Btu = 1.055 kJ)?

6.102 Pioneers and cowboys sometimes heated their coffee by placing an iron poker from the fire directly into the coffee. If a cup held 0.50 L of coffee at 18°C, what would be the final temperature of the coffee when a 502-g iron poker at 800.°C was placed in it? (Assume no heat is lost to the surroundings and no water vaporizes.)

6.103 Phosphorus pentachloride is used in the industrial preparation of organic phosphorus compounds. Equation 1 shows its preparation from PCl_3 and Cl_2:
(1) $PCl_3(l) + Cl_2(g) \longrightarrow PCl_5(s)$
Use equations 2 and 3 to calculate ΔH_{rxn} of equation 1:
(2) $P_4(s) + 6Cl_2(g) \longrightarrow 4PCl_3(l)$ $\Delta H = -1280$ kJ
(3) $P_4(s) + 10Cl_2(g) \longrightarrow 4PCl_5(s)$ $\Delta H = -1774$ kJ

6.104 Consider the following hydrocarbon fuels:
(I) $CH_4(g)$ (II) $C_2H_4(g)$ (III) $C_2H_6(g)$
(a) Rank them in terms of heat released per mole. [Assume $H_2O(g)$ forms and combustion is complete.]
(b) Rank them in terms of heat released per gram.

6.105 A typical candy bar weighs about 2 oz (1.00 oz = 28.4 g).
(a) Assuming that a candy bar is 100% sugar and that 1.0 g of sugar is equivalent to about 4.0 Calories of energy, calculate the energy (in kJ) contained in a typical candy bar.
(b) Assuming that your mass is 58 kg and you convert chemical potential energy to work with 100% efficiency, how high would you have to climb to work off the energy in a candy bar? (Potential energy = mass × g × height, where g = 9.8 m/s².)
(c) Why is your actual conversion of potential energy to work less than 100% efficient?

6.106 Silicon tetrachloride is produced annually on the multi-kiloton scale for making transistor-grade silicon. It can be made directly from the elements (reaction 1) or, more cheaply, by heating sand and graphite with chlorine gas (reaction 2). If water is present in reaction 2, some tetrachloride may be lost in an unwanted side reaction (reaction 3):
(1) $Si(s) + 2Cl_2(g) \longrightarrow SiCl_4(g)$
(2) $SiO_2(s) + 2C(graphite) + 2Cl_2(g) \longrightarrow SiCl_4(g) + 2CO(g)$
(3) $SiCl_4(g) + 2H_2O(g) \longrightarrow SiO_2(s) + 4HCl(g)$
$\Delta H_{rxn}^0 = -139.5$ kJ
(a) Use reaction 3 to calculate the heats of reaction of reactions 1 and 2.
(b) What is the heat of reaction for the new reaction that is the sum of reactions 2 and 3?

*__6.107__ You want to determine ΔH^0 for the reaction
$Zn(s) + 2HCl(aq) \longrightarrow ZnCl_2(aq) + H_2(g)$
(a) To do so, you first determine the heat capacity of a calorimeter using the following reaction, whose ΔH is known:
$NaOH(aq) + HCl(aq) \longrightarrow NaCl(aq) + H_2O(l)$
$\Delta H^0 = -57.32$ kJ
Calculate the heat capacity of the calorimeter from these data:
Amounts used: 50.0 mL of 2.00 M HCl and 50.0 mL of 2.00 M NaOH
Initial T of both solutions: 16.9°C
Maximum T recorded during reaction: 30.4°C
Density of resulting NaCl solution: 1.04 g/mL
c of 1.00 M NaCl(aq) = 3.93 J/g·K
(b) Use the result from part (a) and the following data to determine ΔH_{rxn}^0 for the reaction between zinc and HCl(aq):
Amounts used: 100.0 mL of 1.00 M HCl and 1.3078 g of Zn
Initial T of HCl solution and Zn: 16.8°C
Maximum T recorded during reaction: 24.1°C
Density of 1.0 M HCl solution = 1.015 g/mL
c of resulting ZnCl₂(aq) = 3.95 J/g·K

(c) Given the values below, what is the error in your experiment?
ΔH_f^0 of HCl(aq) = -1.652×10^2 kJ/mol
ΔH_f^0 of ZnCl₂(aq) = -4.822×10^2 kJ/mol

*__6.108__ One mole of nitrogen gas confined within a cylinder by a piston is heated from 0°C to 819°C at 1.00 atm.
(a) Calculate the work of expansion of the gas in joules (1 J = 9.87×10^{-3} atm·L). Assume all the energy is used to do work.
(b) What would be the temperature change if the gas were heated with the same amount of energy in a container of fixed volume? (Assume the specific heat capacity of N₂ is 1.00 J/g·K.)

*__6.109__ The combustion of eight-carbon hydrocarbons (C_8H_{18}), such as those found in gasoline, is one of the most common redox reactions in modern society.
(a) Calculate ΔH_{comb}^0 of 1 mol of C_8H_{18} in excess O₂, assuming water forms as a gas. $\Delta H_f^0 = -208.45$ kJ/mol.
(b) Calculate ΔE for this reaction (1 J = 9.87×10^{-3} atm·L). (Hint: Use $P\Delta V = \Delta n_{gas}RT$ to convert between ΔH and ΔE.)

6.110 The chemistry of nitrogen oxides is very versatile. Given the following reactions and their standard enthalpy changes,
(1) $NO(g) + NO_2(g) \longrightarrow N_2O_3(g)$ $\Delta H_{rxn}^0 = -39.8$ kJ
(2) $NO(g) + NO_2(g) + O_2(g) \longrightarrow N_2O_5(g)$ $\Delta H_{rxn}^0 = -112.5$ kJ
(3) $2NO_2(g) \longrightarrow N_2O_4(g)$ $\Delta H_{rxn}^0 = -57.2$ kJ
(4) $2NO(g) + O_2(g) \longrightarrow 2NO_2(g)$ $\Delta H_{rxn}^0 = -114.2$ kJ
(5) $N_2O_5(g) \longrightarrow N_2O_5(s)$ $\Delta H_{rxn}^0 = -54.1$ kJ
calculate the heat of reaction for
$$N_2O_3(g) + N_2O_5(s) \longrightarrow 2N_2O_4(g)$$

6.111 High-purity benzoic acid (C_6H_5COOH; $\Delta H_{comb} = -3227$ kJ/mol) is a combustion standard for calibrating bomb calorimeters. A 1.221-g sample burns in a calorimeter (heat capacity = 1365 J/°C) that contains exactly 1.200 kg of water. What temperature change is observed?

*__6.112__ Electric generating plants transport large amounts of hot water through metal pipes, and oxygen dissolved in the water can cause a major corrosion problem. Hydrazine (N_2H_4) added to the water avoids the problem by reacting with the oxygen:
$$N_2H_4(aq) + O_2(g) \longrightarrow N_2(g) + 2H_2O(l)$$
About 4×10^7 kg of hydrazine is produced every year by reacting ammonia with sodium hypochlorite in the Raschig process:
$2NH_3(aq) + NaOCl(aq) \longrightarrow N_2H_4(aq) + NaCl(aq) + H_2O(l)$
$\Delta H_{rxn}^0 = -151$ kJ
(a) If ΔH_f^0 of NaOCl(aq) = -346 kJ/mol, calculate ΔH_f^0 of $N_2H_4(aq)$.
(b) What is the heat released when aqueous N₂H₄ is added to 5.00×10^3 L of plant water that is 2.50×10^{-4} M O₂?

*__6.113__ Liquid methanol (CH_3OH) is being used as an alternative fuel in cars and trucks. An industrial method for preparing methanol, sometimes called the Fischer-Tropsch process, uses the catalytic hydrogenation of carbon monoxide:
$$CO(g) + 2H_2(g) \xrightarrow{catalyst} CH_3OH(l)$$
How much heat (in kJ) is released when 15.0 L of CO at 85°C and 112 kPa reacts with 18.5 L of H₂ at 75°C and 744 torr?

*__6.114__ (a) How much heat is released when 25.0 g of methane burns in excess O₂ to form gaseous products?
(b) Calculate the temperature of the product mixture if the methane and air are both at an initial temperature of 0.0°C. Assume a stoichiometric ratio of methane to oxygen from the air, and air is 21% O₂ by volume (c of CO₂ = 57.2 J/mol·K; c of H₂O(g) = 36.0 J/mol·K; c of N₂ = 30.5 J/mol·K).

CHAPTER 7

QUANTUM THEORY AND ATOMIC STRUCTURE

CHAPTER OUTLINE

7.1 The Nature of Light
Wave Nature of Light
Particle Nature of Light

7.2 Atomic Spectra
Bohr Model of the Hydrogen Atom
Limitations of the Bohr Model
Energy States of the Hydrogen Atom

7.3 The Wave-Particle Duality of Matter and Energy
Wave Nature of Electrons and Particle Nature of Photons
Heisenberg Uncertainty Principle

7.4 The Quantum-Mechanical Model of the Atom
The Atomic Orbital
Quantum Numbers
Shapes of Atomic Orbitals
Energy Levels of the Hydrogen Atom

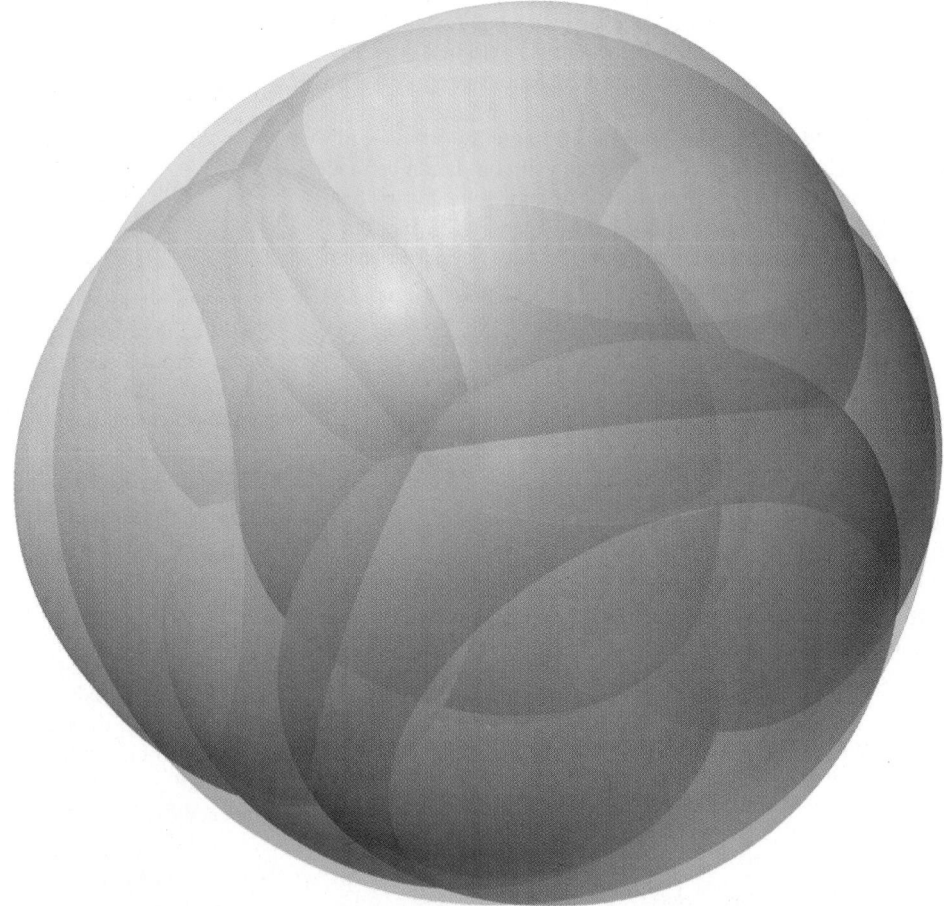

Figure: What does an atom "look" like? This mostly spherical object represents an image of the hydrogen atom. Depending on the atom's energy, differently shaped regions of space (shown in red, blue, and yellow) house the electron—at least 90% of the time! In this chapter, we follow the astounding scientific theories and discoveries of the young 20th century that led to our current atomic model.

Over a few remarkable decades—from around 1890 to 1930—a revolution took place in how we view the makeup of the universe. ● But revolutions in science are not the violent upheavals of political overthrow. Rather, flaws appear in an established model as conflicting evidence mounts, a startling discovery or two widens the flaws into cracks, and the conceptual structure crumbles gradually from its inconsistencies. New insight, verified by experiment, then guides the building of a model more consistent with reality. So it was when Lavoisier's theory of combustion overthrew the phlogiston model, when Dalton's atomic theory established the idea of individual units of matter, and when Rutherford's nuclear model substituted atoms with rich internal structure for "billiard balls" or "plum puddings." In this chapter, you will see this process unfold again with the development of modern atomic theory.

Almost as soon as Rutherford proposed his nuclear model, a major problem arose. A nucleus and an electron attract each other, so if they are to remain apart, the energy of the electron's motion (kinetic energy) must balance the energy of attraction (potential energy). However, the laws of classical physics had established that a negative particle moving in a curved path around a positive one *must* emit radiation and thus lose energy. If this requirement applied to atoms, why didn't the orbiting electron lose energy continuously and spiral into the nucleus? Clearly, if electrons behaved the way classical physics predicted, all atoms would have collapsed eons ago! The behavior of subatomic matter seemed to violate real-world experience and accepted principles.

The breakthroughs that soon followed forced a complete rethinking of the classical picture of matter and energy. In the macroscopic world, the two are distinct. Matter occurs in chunks you can hold and weigh, and you can change the amount of matter in a sample piece by piece. In contrast, energy is "massless," and its quantity changes in a continuous manner. Matter moves in specific paths, whereas light and other types of energy travel in diffuse waves. As soon as 20th-century scientists probed the subatomic world, however, these clear distinctions between particulate matter and wavelike energy began to fade.

In this chapter and the next, we discuss *quantum mechanics,* the theory that explains our current picture of atomic structure. After considering the wave properties of energy, we examine the theories and experiments that led to a *quantized,* or particulate, model of light. We see why the light emitted by an excited hydrogen (H) atom—its *atomic spectrum*—suggests an atom with distinct energy levels, and we look briefly at how atomic spectra are applied to chemical analysis today. Then we examine wave-particle duality, the central idea of quantum mechanics. The last section describes the current model of the H atom and presents the quantum numbers we use to identify the regions of space an electron occupies in an atom. In Chapter 8, we consider atoms with more than one electron and relate electron number and distribution to chemical behavior.

7.1 THE NATURE OF LIGHT

Visible light is one type of **electromagnetic radiation** (also called *electromagnetic energy* or *radiant energy*). Other familiar types include x-rays, microwaves, and radio waves. All electromagnetic radiation consists of energy propagated by means of electric and magnetic fields that alternately increase and decrease in intensity as they move through space. This classical wave model distinguishes clearly between waves and particles, and is essential for understanding why rainbows form, how magnifying glasses work, why objects look distorted under water, and many other everyday observations. But, it cannot explain observations on the atomic scale because, in those cases, energy behaves as though it consists of particles!

CONCEPTS & SKILLS

to review before you study this chapter
- discovery of the electron and atomic nucleus (Section 2.4)
- major features of atomic structure (Section 2.5)
- changes in energy state of a system (Section 6.1)

● Hooray for the Human Mind
The invention of the car, radio, and airplane fostered a feeling of unlimited human ability, and the discovery of x-rays, radioactivity, the electron, and the atomic nucleus led to the sense that the human mind would soon unravel all of nature's mysteries. Indeed, some people were convinced that few, if any, mysteries remained.

1895 Röntgen discovers x-rays.
1896 Becquerel discovers radioactivity.
1897 Thomson discovers the electron.
1898 Curie discovers radium.
1900 Freud proposes theory of the unconscious mind.
1900 Planck develops quantum theory.
1901 Marconi invents the radio.
1903 Wright brothers fly an airplane.
1905 Ford uses assembly line to build cars.
1905 Rutherford explains radioactivity.
1905 Einstein publishes relativity and photon theories.
1906 St. Denis develops modern dance.
1908 Matisse and Picasso develop modern art.
1909 Schoenberg and Berg develop modern music.
1911 Rutherford presents nuclear model.
1913 Bohr proposes atomic model.
1914 to 1918 World War I is fought.
1923 Compton demonstrates photon momentum.
1924 De Broglie publishes wave theory of matter.
1926 Schroedinger develops wave equation.
1927 Heisenberg presents uncertainty principle.
1932 Chadwick discovers the neutron.

The Wave Nature of Light

The wave properties of electromagnetic radiation are described by two interdependent variables, as Figure 7.1 shows:

- **Frequency** (v, Greek *nu*) is the number of cycles the wave undergoes per second and is expressed in units of 1/second [s^{-1}; also called *hertz* (Hz)].
- **Wavelength** (λ, Greek *lambda*) is the distance between any point on a wave and the corresponding point on the next wave; that is, the distance the wave travels during one cycle. Wavelength is expressed in meters and often, for very short wavelengths, in nanometers (nm, 10^{-9} m), picometers (pm, 10^{-12} m), or the non-SI unit angstroms (Å, 10^{-10} m).

The speed of the wave, the distance traveled per unit time (in units of meters per second), is the product of its frequency (cycles per second) and its wavelength (meters per cycle):

$$\text{Units for speed of wave:} \quad \frac{\cancel{\text{cycles}}}{\text{s}} \times \frac{\text{m}}{\cancel{\text{cycle}}} = \frac{\text{m}}{\text{s}}$$

In a vacuum, all types of electromagnetic radiation travel at 2.99792458×10^8 m/s (3.00×10^8 m/s to three significant figures), a constant called the **speed of light** (*c*):

$$c = v \times \lambda \tag{7.1}$$

Since, as Equation 7.1 shows, the product of v and λ is a constant, the individual terms have a reciprocal relationship to each other: *radiation with a high frequency has a short wavelength, and vice versa.*

Another characteristic of a wave is its **amplitude.** As Figure 7.2 shows, amplitude is the height of the crest (or depth of the trough) of each wave. The amplitude of an electromagnetic wave is a measure of the strength of its electric and magnetic fields. Thus, amplitude is related to the *intensity* of the radiation, which we perceive as brightness in the case of visible light. Light of a particular color, fire-engine red for instance, has a specific frequency and wavelength, but it can be dimmer (lower amplitude) or brighter (higher amplitude).

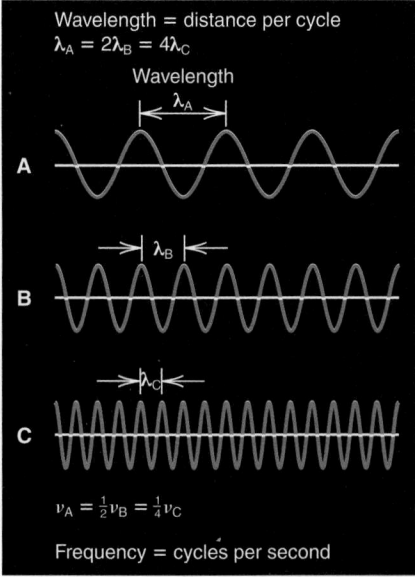

Figure 7.1 Frequency and wavelength. Three waves with different wavelengths (λ) and thus different frequencies (v) are shown. Note that as the wavelength decreases, the frequency increases, and vice versa. The wavelength of the top wave is twice that of the middle wave and four times that of the bottom wave. Thus, the frequency of the top wave is one-half that of the middle wave and one-fourth that of the bottom wave.

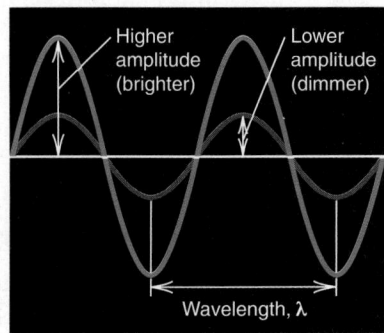

Figure 7.2 Amplitude (intensity) of a wave. Amplitude is represented by the height of the crest (or depth of the trough) of the wave. The two waves shown have the same wavelength (color) but different amplitudes and, therefore, different brightnesses (intensities).

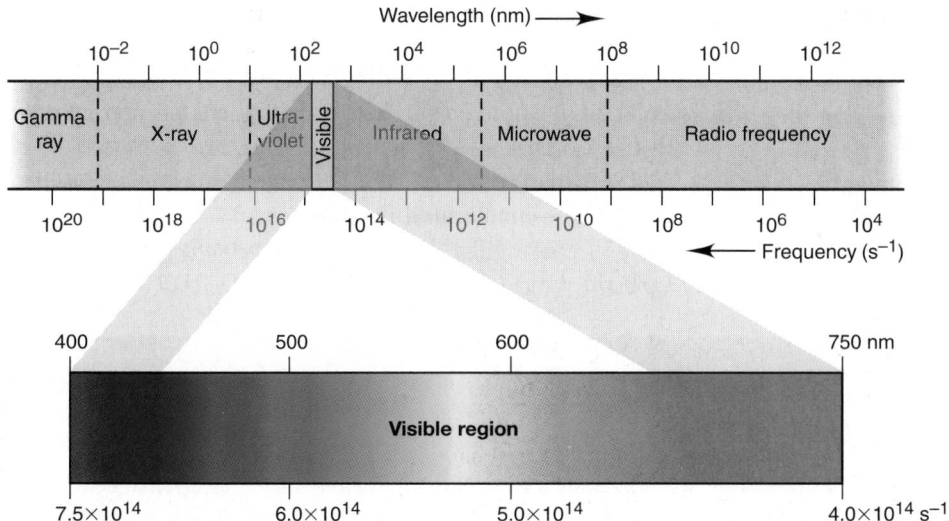

Figure 7.3 **Regions of the electromagnetic spectrum.** The electromagnetic spectrum extends from the very short wavelengths (very high frequencies) of gamma rays through the very long wavelengths (very low frequencies) of radio waves. The relatively narrow visible region is expanded (and the scale made linear) to show the component colors.

The Electromagnetic Spectrum Visible light represents a small portion of the continuum of radiant energy known as the **electromagnetic spectrum** (Figure 7.3). *The waves in the spectrum all travel at the same speed through a vacuum but differ in frequency and, therefore, wavelength.* Some regions of the spectrum are utilized by particular devices; for example, the long-wavelength, low-frequency portion makes up the microwave and radio regions. Note that each region overlaps the next. For instance, the **infrared (IR)** region overlaps the microwave region on one end and the visible region on the other.

We perceive different wavelengths (or frequencies) of *visible* light as different colors, from red ($\lambda \approx 750$ nm) to violet ($\lambda \approx 400$ nm). Light of a single wavelength is called *monochromatic* (Greek, "one color"), whereas light of many wavelengths is *polychromatic*. White light is polychromatic. The region adjacent to visible light on the short-wavelength end consists of **ultraviolet (UV)** radiation (also called *ultraviolet light*). Still shorter wavelengths (higher frequencies) make up the x-ray and gamma (γ) ray regions. Thus, a TV signal, the green light from a traffic signal, and a gamma ray emitted by a radioactive element all travel at the same speed but differ in their frequency (and wavelength).

SAMPLE PROBLEM 7.1 Interconverting Wavelength and Frequency

Problem A dental hygienist uses x-rays ($\lambda = 1.00$ Å) to take a series of dental radiographs while the patient listens to a radio station ($\lambda = 325$ cm) and looks out the window at the blue sky ($\lambda = 473$ nm). What is the frequency (in s^{-1}) of the electromagnetic radiation from each source? (Assume that the radiation travels at the speed of light, 3.00×10^8 m/s.)
Plan We are given the wavelengths, so we use Equation 7.1 to find the frequencies. However, we must first convert the wavelengths to meters because c has units of m/s.

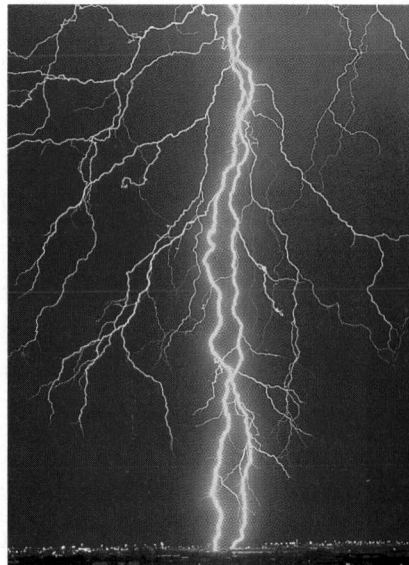

Electromagnetic Emissions Everywhere We are bathed in electromagnetic radiation from the Sun. Radiation from human activities bombards us as well: radio and TV signals; microwave radiation from traffic radar systems and telephone relay stations; and radiation given off by lightbulbs, medical x-ray equipment, car motors, and so forth. Natural sources on Earth bombard us also: lightning, radioactive decay, and even the light from fireflies! Virtually all our knowledge of the distant universe comes from radiation entering our light, x-ray, and radio telescopes.

Wavelength (given units)

$1\ \text{Å} = 10^{-10}\ \text{m}$
$1\ \text{cm} = 10^{-2}\ \text{m}$
$1\ \text{nm} = 10^{-9}\ \text{m}$

Wavelength (m)

$\nu = \dfrac{c}{\lambda}$

Frequency (s^{-1} or Hz)

Solution For x-rays: Converting from angstroms to meters,

$$\lambda = 1.00\ \text{Å} \times \frac{10^{-10}\ \text{m}}{1\ \text{Å}} = 1.00 \times 10^{-10}\ \text{m}$$

Calculating the frequency:

$$\nu = \frac{c}{\lambda} = \frac{3.00 \times 10^8\ \text{m/s}}{1.00 \times 10^{-10}\ \text{m}} = \boxed{3.00 \times 10^{18}\ \text{s}^{-1}}$$

For the radio station: Combining steps to calculate the frequency,

$$\nu = \frac{c}{\lambda} = \frac{3.00 \times 10^8\ \text{m/s}}{325\ \text{cm} \times \dfrac{10^{-2}\ \text{m}}{1\ \text{cm}}} = \boxed{9.23 \times 10^7\ \text{s}^{-1}}$$

For the blue sky: Combining steps to calculate the frequency,

$$\nu = \frac{c}{\lambda} = \frac{3.00 \times 10^8\ \text{m/s}}{473\ \text{nm} \times \dfrac{10^{-9}\ \text{m}}{1\ \text{nm}}} = \boxed{6.34 \times 10^{14}\ \text{s}^{-1}}$$

Check The orders of magnitude are correct for the regions of the electromagnetic spectrum (see Figure 7.3): x-rays (10^{19} to $10^{16}\ \text{s}^{-1}$), radio waves (10^9 to $10^4\ \text{s}^{-1}$), and visible light (7.5×10^{14} to $4.0 \times 10^{14}\ \text{s}^{-1}$).

Comment The radio station here is broadcasting at $92.3 \times 10^6\ \text{s}^{-1}$, or 92.3 million Hz (92.3 MHz), about midway in the FM range.

FOLLOW-UP PROBLEM 7.1 Some diamonds appear yellow because they contain nitrogen compounds that absorb purple light of frequency 7.23×10^{14} Hz. Calculate the wavelength (in nm and Å) of the absorbed light.

The Distinction Between Energy and Matter In the everyday world around us, energy and matter behave very differently. Let's examine some important observations about light and see how they contrast with the behavior of particles. Light of a given wavelength travels at different speeds through different transparent media—vacuum, air, water, quartz, and so forth. Therefore, when a light wave passes from one medium into another, say from air to water, the speed of the wave changes. Figure 7.4A shows the phenomenon known as **refraction.** If the

Figure 7.4 Different behaviors of waves and particles. A, A wave passing from air into water is *refracted* (bent at an angle). **B,** In contrast, a particle of matter (such as a pebble) entering a pond moves in a curved path, because gravity and the greater resistance (drag) of the water slow it down gradually. **C,** A wave is *diffracted* through a small opening, which gives rise to a circular wave on the other side. (The lines represent the crests of water waves as seen from above.) **D,** In contrast, when a collection of moving particles encounters a small opening, such as when a handful of sand is thrown at a hole in a fence, some particles move through and continue along their individual paths.

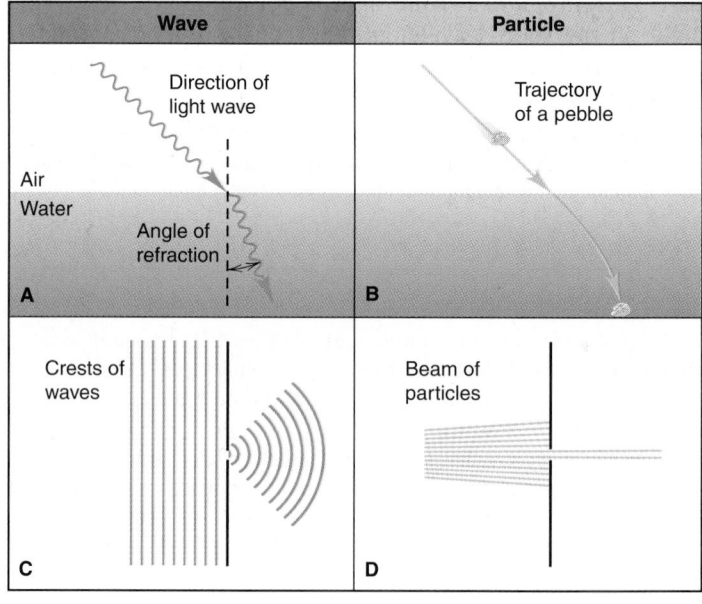

wave strikes the boundary, say between air and water, at an angle other than 90°, the change in speed causes a change in direction, and the wave continues at a different angle. The new angle (angle of refraction) depends on the materials on either side of the boundary and the wavelength of the light. In the process of *dispersion,* white light separates (disperses) into its component colors, as when it passes through a prism, because each incoming wavelength is refracted at a slightly different angle.  In contrast, particles do not undergo refraction when passing from one medium to another. Figure 7.4B shows that when a pebble thrown through the air enters a lake, its speed changes abruptly and then it continues to slow down gradually in a curved path.

When a wave strikes the edge of an object, it bends around it in a phenomenon called **diffraction.** If the wave passes through a slit about as wide as its wavelength, it bends around both edges of the slit and forms a semicircular wave on the other side of the opening, as shown in Figure 7.4C. Once again, particles act very differently. Figure 7.4D shows that if a collection of particles is aimed at a small opening, some particles hit the edge, while others go through and continue linearly in a narrower group.

If waves of light pass through two adjacent slits, the emerging circular waves interact through the process of *interference* to create a diffraction pattern of brighter and darker regions (Figure 7.5). Streams of particles, however, moving through adjacent openings continue in straight paths, with some possibly colliding and moving at different angles. At the end of the 19th century, all experience seemed to confirm these classical distinctions between the wave nature of energy and the particle nature of matter.

Rainbows and Diamonds
You can see a rainbow only when the Sun is at your back. Light entering the near surface of a water droplet is dispersed and reflected off the far surface. Since red light is bent least, it reaches your eye from droplets higher in the sky, whereas violet appears from droplets that are lower. The colors in a diamond's sparkle are due to its facets, which are cleaved at angles that disperse and reflect the incoming light, lengthening its path enough for the different wavelengths to separate.

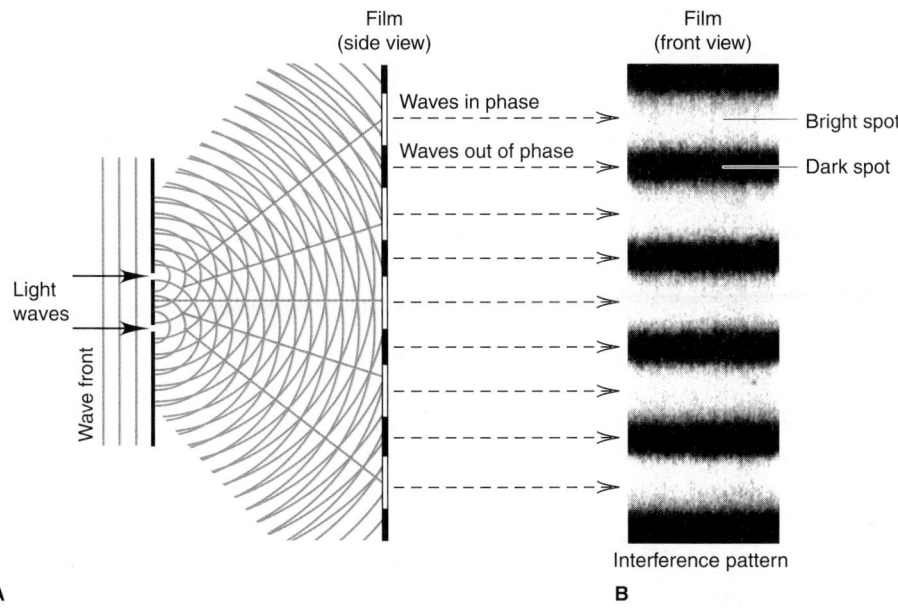

A

B

Figure 7.5 The diffraction pattern caused by light passing through two adjacent slits. A, After passing through two closely spaced slits, emerging circular waves interfere with each other. They create a diffraction (interference) pattern on a sheet of film. Bright regions appear where crests coincide and enhance each other (in phase), and dark regions appear where crests meet troughs and cancel each other (out of phase). **B,** The interference pattern obtained on the film.

The Particle Nature of Light

Three phenomena involving matter and light were especially confounding to physicists of the early 20th century: (1) blackbody radiation, (2) the photoelectric effect, and (3) atomic spectra. Explaining them required a radically new way to picture energy. We discuss the first two here and the third in the next section.

Blackbody Radiation and the Quantization of Energy When a solid object is heated to about 1000 K, it begins to emit visible light, as you can see in the soft red glow of smoldering coal (Figure 7.6, *left*). At about 1500 K, the light is brighter and more orange, like that from an electric heating coil (Figure 7.6, *center*). At temperatures greater than 2000 K, the light is still brighter and whiter, as seen in the filament of a lightbulb (Figure 7.6, *right*). These changes in intensity and wavelength of emitted light as an object is heated are characteristic of *blackbody radiation,* light given off by a hot *blackbody.* All attempts to account for the observed changes by applying classical electromagnetic theory failed.

Figure 7.6 Blackbody radiation. Familiar examples of the change in intensity and wavelength of heated objects.

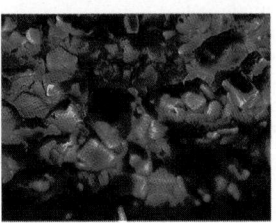  

In 1900, the German physicist Max Planck (1858–1947) developed a formula that fit the data perfectly; to find a physical explanation for his formula, however, Planck was forced to make a radical assumption, which eventually led to an entirely new view of energy. He proposed that the hot, glowing object could emit (or absorb) only *certain quantities* of energy:

$$E = nh\nu$$

where E is the energy of the radiation, ν is its frequency, n is a positive integer (1, 2, 3, and so on) called a **quantum number,** and h is a proportionality constant now called **Planck's constant.** With energy in joules (J) and frequency in s^{-1}, h has units of J·s:

$$h = 6.626 \times 10^{-34} \text{ J·s}$$

Later interpretations of Planck's proposal stated that the hot object's radiation is emitted by the atoms contained within it. If an atom can *emit* only certain quantities of energy, it follows that *the atom itself can have only certain quantities of energy.* Thus, the energy of an atom is *quantized:* it exists only in certain fixed quantities, rather than being continuous. Each change in the atom's energy results from the gain or loss of one or more "packets" of energy. Each energy packet is called a **quantum** ("fixed quantity"; plural, *quanta*) and has energy equal to $h\nu$. To restate this idea, *an atom changes its energy state by emitting (or absorbing) one or more quanta of energy.* Thus, the energy of the emitted (or absorbed) radiation is equal to the *difference in the atom's energy states:*

$$\Delta E_{\text{atom}} = E_{\text{emitted (or absorbed) radiation}} = \Delta nh\nu$$

Because n is a positive integer, the atom can change its energy only by integer multiples of $h\nu$. Therefore, the smallest energy change for an atom in a given energy state occurs when it changes to an adjacent energy state, that is, when $\Delta n = 1$:

$$\Delta E = h\nu \tag{7.2}$$

The Photoelectric Effect and the Photon Theory of Light Despite the idea that energy is quantized, Planck and other physicists continued to picture the emitted energy as traveling in waves. However, the wave model could not explain the

photoelectric effect, the flow of current when monochromatic light of sufficient energy shines on a metal plate (Figure 7.7). The existence of the current was not puzzling: it could be understood as arising when the light transfers energy to the electrons at the metal surface, which break free and are collected by the positive electrode. However, the photoelectric effect had certain confusing features, in particular, the *presence of a threshold frequency* and the *absence of a time lag:*

1. *Presence of a threshold frequency.* Light shining on the metal must have a minimum *frequency,* or no current flows. (Different metals have different minimum frequencies.) The wave theory, however, associates the energy of the light with the *amplitude* (intensity) of the wave, not with its frequency (color). Thus, the wave theory predicts that an electron would break free when it absorbed enough energy from light of *any* color.

2. *Absence of a time lag.* Current flows the moment light of this minimum frequency shines on the metal, regardless of the intensity of the light. The wave theory, however, predicts that in dim light there would be a time lag before the current would flow, because the electrons would have to absorb enough energy to break free.

Carrying Planck's idea of packeted energy further, Einstein proposed that light itself is particulate, occurring as quanta of electromagnetic energy, later called **photons.** In terms of Planck's work, we can say that each atom changes its energy whenever it absorbs or emits one photon, one "particle" of light, whose energy is fixed by its *frequency:*

$$E_{photon} = h\nu = \Delta E_{atom}$$

Let's see how Einstein's photon theory explains the photoelectric effect:

1. *Presence of a threshold frequency.* According to the photon theory, a beam of light consists of an enormous number of photons. Light intensity (brightness) is related to the number of photons striking the surface per unit time, but *not* to their energy. Therefore, a photon of a certain minimum *energy* must be absorbed for an electron to be freed. Since energy depends on frequency ($h\nu$), the theory predicts a threshold frequency.

2. *Absence of a time lag.* An electron cannot "save up" energy from several photons below the minimum energy until it has enough to break free. Rather, one electron breaks free the moment it absorbs one photon of *enough* energy. The current is weaker in dim light than in bright light because fewer photons of enough energy are present, so fewer electrons break free per unit time. But some current flows the moment photons reach the metal plate.

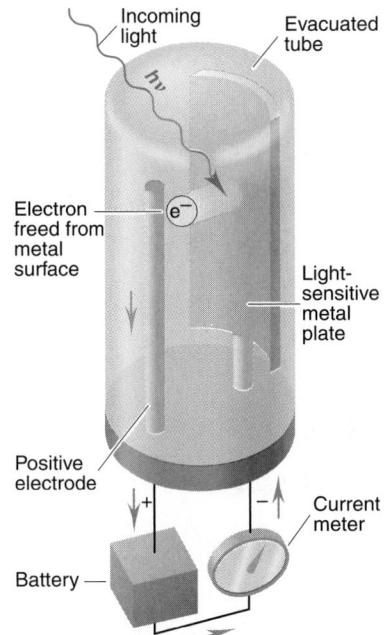

Figure 7.7 Demonstration of the photoelectric effect. When monochromatic light of high enough frequency strikes the metal plate, electrons are freed from the plate and travel to the positive electrode, which creates a current.

Ping-Pong Photons Consider this analogy for the fact that light of insufficient energy cannot free an electron from the metal surface. If one Ping-Pong ball does not have enough energy to knock a book off a shelf, neither does a series of Ping-Pong balls, because the book cannot save up the energy from the individual impacts. But one baseball traveling at the same speed does have enough energy. Whereas the energy of a ball is related to its mass and velocity, the energy of a photon is related to its frequency.

SAMPLE PROBLEM 7.2 Calculating the Energy of Radiation from Its Wavelength

Problem A cook uses a microwave oven to heat a meal. The wavelength of the radiation is 1.20 cm. What is the energy of one photon of this microwave radiation?

Plan We know λ in centimeters (1.20 cm) so we convert to meters, find the frequency with Equation 7.1, and then find the energy of one photon with Equation 7.2.

Solution Combining steps to find the energy:

$$E = h\nu = \frac{hc}{\lambda} = \frac{(6.626\times10^{-34}\ \text{J·s})(3.00\times10^{8}\ \text{m/s})}{(1.20\ \text{cm})\left(\dfrac{10^{-2}\ \text{m}}{1\ \text{cm}}\right)} = \boxed{1.66\times10^{-23}\ \text{J}}$$

Check Checking the order of magnitude gives $\dfrac{10^{-33}\ \text{J·s}\times10^{8}\ \text{m/s}}{10^{-2}\ \text{m}} = 10^{-23}\ \text{J}.$

FOLLOW-UP PROBLEM 7.2 Calculate the energies of one photon of ultraviolet ($\lambda = 1\times10^{-8}$ m), visible ($\lambda = 5\times10^{-7}$ m), and infrared ($\lambda = 1\times10^{-4}$ m) light. What do the answers indicate about the relationship between the wavelength and energy of light?

Planck's quantum theory and Einstein's photon theory assigned properties to energy that, until then, had always been reserved for matter: fixed quantity and discrete particles. These properties have since proved essential to explaining the interactions of matter and energy at the atomic level. But how can a particulate model of energy be made to fit the facts of diffraction and refraction, phenomena explained only in terms of waves? As you'll see shortly, the photon model does not *replace* the wave model. Rather, we have to accept *both* to understand reality. Before we discuss this astonishing notion, however, let's see how the new idea of quantized energy led to a key understanding about atomic behavior.

SECTION SUMMARY

Electromagnetic radiation travels in waves of specific wavelength (λ) and frequency (ν). All electromagnetic waves travel through a vacuum at the speed of light, c (3.00×10^8 m/s), which is equal to $\nu \times \lambda$. The intensity (brightness) of a light wave is related to its amplitude. The electromagnetic spectrum ranges from very long radio waves to very short gamma rays and includes the visible region [750 nm (red) to 400 nm (violet)]. Refraction and diffraction indicate that electromagnetic radiation is wavelike, but blackbody radiation and the photoelectric effect indicate that it is particle-like. Light exists as photons (quanta) that have an energy proportional to the frequency. According to quantum theory, an atom has only certain quantities of energy ($E = nh\nu$), which it can change only by absorbing or emitting a photon.

7.2 ATOMIC SPECTRA

The third key observation about matter and energy that late 19th-century physicists could not explain involved the light emitted when an element is vaporized and then thermally or electrically excited, as you see in a neon sign. Figure 7.8A shows the result when light from excited hydrogen atoms passes through a nar-

Figure 7.8 The line spectra of several elements. A, A sample of gaseous H_2 is dissociated into atoms and excited by an electric discharge. The emitted light passes through a slit and a prism, which disperses the light into individual wavelengths. The line spectrum of atomic H is shown *(top)*. **B,** The continuous spectrum of white light is compared with the line spectra of mercury and strontium. Note that each line spectrum is different from the others.

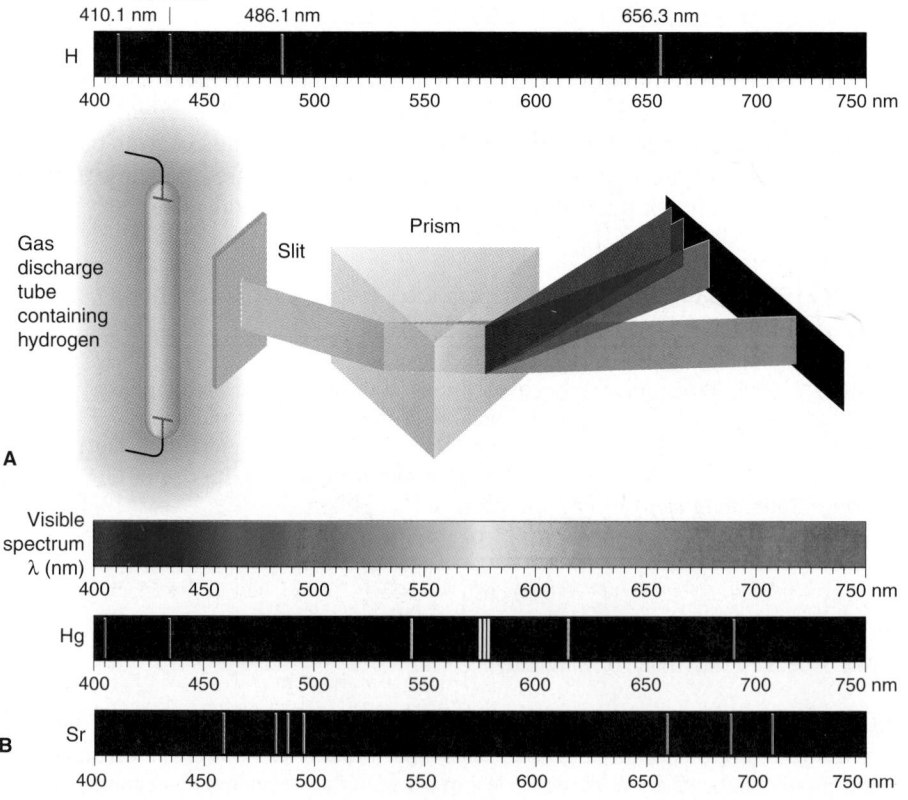

row slit and is then refracted by a prism. Note that this light does not create a *continuous spectrum,* or rainbow, as sunlight does. Instead, it creates a **line spectrum,** a series of fine lines of individual colors separated by colorless (black) spaces.* The wavelengths of these spectral lines are characteristic of the element producing them (Figure 7.8B).

Spectroscopists studying the spectrum of atomic hydrogen had identified several series of such lines in different regions of the electromagnetic spectrum. Figure 7.9 shows three of these series of lines. Equations of the following general form, called the *Rydberg equation,* were found to predict the position and wavelength of any line in a given series:

$$\frac{1}{\lambda} = R\left(\frac{1}{n_1^2} - \frac{1}{n_2^2}\right) \tag{7.3}$$

where λ is the wavelength of a spectral line, n_1 and n_2 are positive integers with $n_2 > n_1$, and R is the Rydberg constant (1.096776×10^7 m^{-1}). For the visible series of lines, $n_1 = 2$:

$$\frac{1}{\lambda} = R\left(\frac{1}{2^2} - \frac{1}{n_2^2}\right), \qquad \text{with } n_2 = 3, 4, 5, \dots$$

The Rydberg equation and the value of the constant are based on data rather than theory. No one knew *why* the spectral lines of hydrogen appear in this pattern. (Problems 7.23 and 7.24 are two of several that apply the Rydberg equation.)

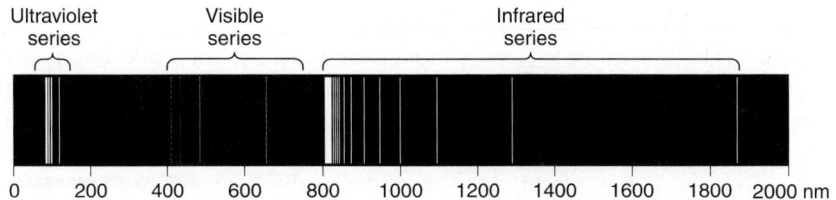

Figure 7.9 Three series of spectral lines of atomic hydrogen. These series appear in different regions of the electromagnetic spectrum. The hydrogen spectrum shown in Figure 7.8A is the visible series.

The observation of line spectra did not correlate with classical theory for one major reason. As was mentioned in the chapter introduction, as the electron spiraled into the nucleus, it should emit radiation. Moreover, the frequency of the radiation is related to the time of revolution. On the spiral path inward, the time should decrease smoothly, so the frequency of the radiation should change smoothly and create a continuous spectrum. Rutherford's nuclear model seemed totally at odds with these atomic spectra.

The Bohr Model of the Hydrogen Atom

Soon after the nuclear model was proposed, Niels Bohr (1885–1962), a young Danish physicist working in Rutherford's laboratory, suggested a model for the H atom that predicted the existence of line spectra. In his model, Bohr used Planck's and Einstein's ideas about quantized energy and proposed three postulates:

1. *The H atom has only certain allowable energy levels,* which Bohr called **stationary states.** Each of these states is associated with a fixed circular orbit of the electron around the nucleus.

2. *The atom does **not** radiate energy while in one of its stationary states.* That is, even though it violates the ideas of classical physics, the atom does not change energy while the electron moves *within* an orbit.

---

*The appearance of the spectrum as a series of lines results from the construction of the apparatus. If the light passed through a small hole, rather than a narrow slit, the spectrum would appear as a circular field of dots rather than a horizontal series of lines. The key point is that *the spectrum is discrete, rather than continuous.*

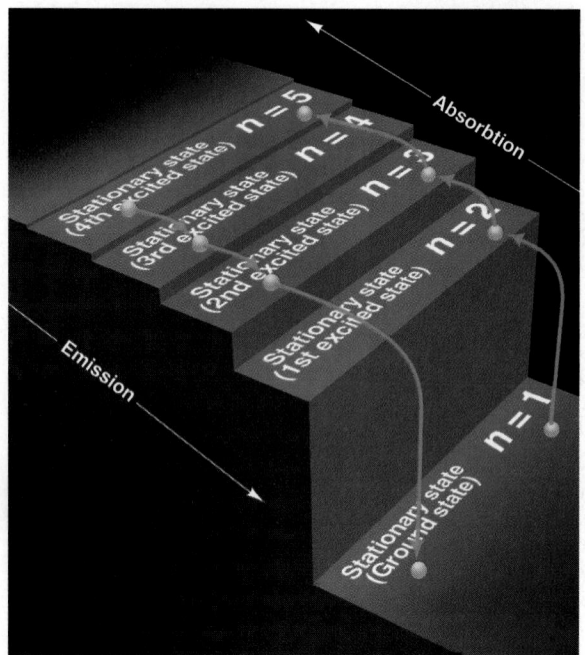

Figure 7.10 Quantum staircase. In this analogy for the energy levels of the hydrogen atom, an electron can absorb a photon and jump up to a higher "step" (stationary state) or emit a photon and jump down to a lower one. But the electron cannot lie between two steps.

EMISSION SPECTRA

3. *The atom changes to another stationary state* (the electron moves to another orbit) *only by absorbing or emitting a photon whose energy equals the difference in energy between the two states:*

$$E_{\text{photon}} = E_{\text{state A}} - E_{\text{state B}} = h\nu$$

where the energy of state A is higher than that of state B. A spectral line results when a photon of specific energy (and thus specific frequency) is *emitted* as the electron moves from a higher energy state to a lower one. Therefore, Bohr's model explains that an atomic spectrum is not continuous because *the atom's energy has only certain discrete levels, or states.*

In Bohr's model, the quantum number n (1, 2, 3, . . .) is associated with the radius of an electron orbit, which is directly related to the electron's energy: *the lower the n value, the smaller the radius of the orbit, and the lower the energy level.* When the electron is in the first orbit ($n = 1$), the orbit closest to the nucleus, the H atom is in its lowest (first) energy level, called the **ground state.** If the H atom absorbs a photon whose energy equals the *difference* between the first and second energy levels, the electron moves to the second orbit ($n = 2$), the next orbit out from the nucleus. When the electron is in the second or any higher orbit, the atom is said to be in an **excited state.** If the H atom in the first excited state (the electron in the second orbit) emits a photon of that same energy, it returns to the ground state. Figure 7.10 shows an analogy for this behavior.

Figure 7.11A shows how Bohr's model accounts for the three line spectra of hydrogen. When a sample of gaseous H atoms is excited, different atoms absorb different quantities of energy. Each atom has one electron, but so many atoms are present that all the energy levels (orbits) are populated by electrons. When the electrons drop from outer orbits to the $n = 3$ orbit (second excited state), the emitted photons create the infrared series of lines. The visible series arises when electrons drop to the $n = 2$ orbit (first excited state). Figure 7.11B shows that the ultraviolet series arises when electrons drop to the $n = 1$ orbit (ground state).

Limitations of the Bohr Model

Despite its great success in accounting for the spectral lines of the H atom, the Bohr model failed to predict the spectrum of any other atom, even that of helium, the next simplest element. In essence, the Bohr model is a one-electron model. It works beautifully for the H atom and for other one-electron species, such as several created in the lab or seen in the spectra of stars: He^+ ($Z = 2$), Li^{2+} ($Z = 3$), Be^{3+} ($Z = 4$), B^{4+} ($Z = 5$), C^{5+} ($Z = 6$), N^{6+} ($Z = 7$), and O^{7+} ($Z = 8$). But, it does not work for atoms with more than one electron because in these systems, additional nucleus-electron attractions and electron-electron repulsions are present. However, there is a more fundamental reason for the model's limitations: electrons do *not* travel in fixed orbits. As you'll see, electron movement is far less clearly defined. As a picture of the atom, the Bohr model is incorrect, but we still use the terms "ground state" and "excited state" and retain one of its central ideas in our current model: *the energy of an atom occurs in discrete levels.*

The Energy States of the Hydrogen Atom

A very useful result from Bohr's work is an equation for calculating the energy levels of an atom, which he derived from the classical principles of electrostatic attraction and circular motion:

$$E = -2.18 \times 10^{-18} \text{ J} \left(\frac{Z^2}{n^2} \right)$$

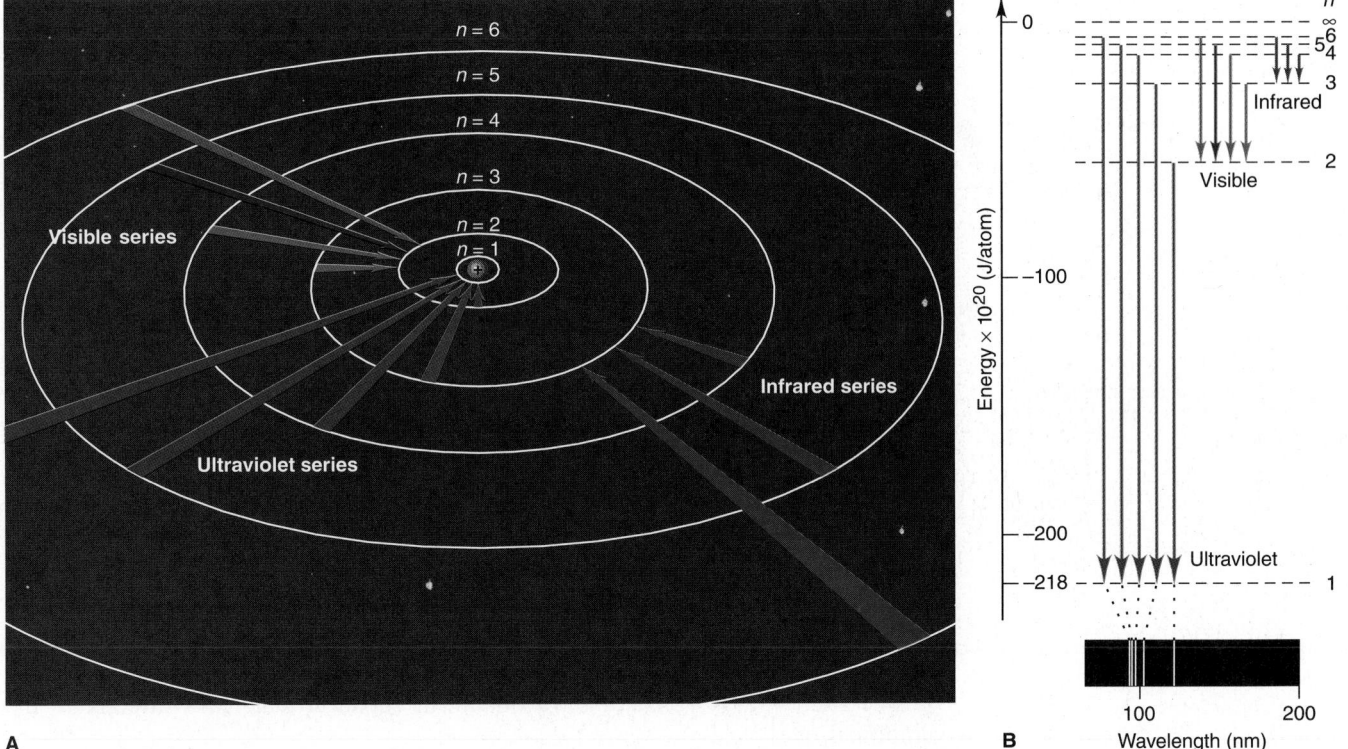

A   **B**

Figure 7.11 The Bohr explanation of the three series of spectral lines. A, According to the Bohr model, when an electron drops from an outer orbit (one farther from the nucleus) to an inner orbit (one closer to the nucleus), it emits a photon of a specific energy. Note that each of the three series is characterized by a particular inner orbit (value of n_1 in the Rydberg equation). The orbit radius is proportional to n^2. Only the first five orbits are shown. **B,** An energy diagram shows how the ultraviolet series arises. When an electron drops from a specific outer orbit to a specific inner orbit, the energy difference (depicted as a downward arrow) appears as a photon of specific wavelength and gives rise to one of the spectral lines in the series. Within each series, the greater the *difference* in orbit radii, the greater the difference in energy levels, and the higher the energy of the photon emitted. For example, in the ultraviolet series, in which $n_1 = 1$, a drop from $n = 5$ to $n = 1$ emits a photon with more energy (shorter λ, higher ν) than a drop from $n = 2$ to $n = 1$. [The axis shows negative values because $n = \infty$ (the electron completely separated from the nucleus) is *defined* as the atom with zero energy.]

where Z is the charge of the nucleus. For the H atom, $Z = 1$, so we have

$$E = -2.18\times10^{-18} \text{ J}\left(\frac{1^2}{n^2}\right) = -2.18\times10^{-18} \text{ J}\left(\frac{1}{n^2}\right)$$

Therefore, the energy of the ground state ($n = 1$) is

$$E = -2.18\times10^{-18} \text{ J}\left(\frac{1}{1^2}\right) = -2.18\times10^{-18} \text{ J}$$

Don't be confused by the negative sign for the energy values (see the axis in Figure 7.11B). It appears because we *define* the zero point of the atom's energy when *the electron is completely removed from the nucleus*. Thus, $E = 0$ when $n = \infty$, so $E < 0$ for any smaller n. As an analogy, consider a book resting on the floor. You can define the zero point of the book's potential energy in many ways. If you define zero when the book is on the floor, the energy is positive when the book is on your desk. But, if you define zero when the book is on your desk, the energy is negative when the book lies on the floor; the latter case is analogous to the energy of the H atom (Figure 7.12).

Since n is in the denominator of the energy equation, as the electron moves closer to the nucleus (n decreases), the atom becomes more stable (less energetic) and its energy becomes a *larger negative number*. As the electron moves away from the nucleus (n increases), the atom's energy increases (becomes a smaller negative number).

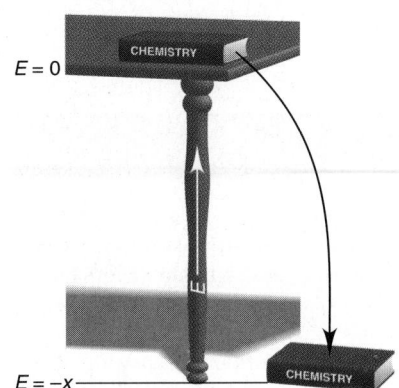

Figure 7.12 A desktop analogy for the H atom's energy. If you define the potential energy of the system as zero when a book rests on a desk, the system has negative energy when the book lies on the floor. Similarly, the H atom is defined as having zero energy when its electron is completely separated from the nucleus, so its energy is negative when the electron is attracted by the nucleus.

This equation is easily adapted to find the energy difference between any two levels:

$$\Delta E = E_{\text{final}} - E_{\text{initial}} = -2.18 \times 10^{-18} \text{ J} \left(\frac{1}{n_{\text{final}}^2} - \frac{1}{n_{\text{initial}}^2} \right) \qquad (7.4)$$

With it, we can predict the wavelengths of the spectral lines of the H atom. (In fact, Bohr obtained a value for the Rydberg constant that differed from the spectroscopists' value by only 0.05%!) Note that if we combine Equation 7.4 with Planck's expression for the change in an atom's energy (Equation 7.2), we obtain the Rydberg equation (Equation 7.3):

$$\Delta E = h\nu = \frac{hc}{\lambda} = -2.18 \times 10^{-18} \text{ J} \left(\frac{1}{n_{\text{final}}^2} - \frac{1}{n_{\text{initial}}^2} \right)$$

Therefore, $\dfrac{1}{\lambda} = -\dfrac{2.18 \times 10^{-18} \text{ J}}{hc} \left(\dfrac{1}{n_{\text{final}}^2} - \dfrac{1}{n_{\text{initial}}^2} \right)$

$$= -\frac{2.18 \times 10^{-18} \text{ J}}{(6.626 \times 10^{-34} \text{ J·s})(3.00 \times 10^8 \text{ m/s})} \left(\frac{1}{n_{\text{final}}^2} - \frac{1}{n_{\text{initial}}^2} \right)$$

$$= 1.10 \times 10^7 \text{ m}^{-1} \left(\frac{1}{n_{\text{final}}^2} - \frac{1}{n_{\text{initial}}^2} \right)$$

where $n_{\text{final}} = n_2$, $n_{\text{initial}} = n_1$, and $1.10 \times 10^7 \text{ m}^{-1}$ is the Rydberg constant ($1.096776 \times 10^7 \text{ m}^{-1}$) to three significant figures. Thus, from classical relationships of charge and of motion combined with the idea that the H atom can have only certain values of energy, we obtain an equation that leads directly to the empirical one!

We can use Equation 7.4 to find the quantity of energy needed to completely remove the electron from an H atom. In other words, what is ΔE for the following change?

$$\text{H}(g) \longrightarrow \text{H}^+(g) + \text{e}^-$$

We substitute $n_{\text{final}} = \infty$ and $n_{\text{initial}} = 1$ and obtain

$$\Delta E = E_{\text{final}} - E_{\text{initial}} = -2.18 \times 10^{-18} \text{ J} \left(\frac{1}{\infty^2} - \frac{1}{1^2} \right)$$

$$= -2.18 \times 10^{-18} \text{ J}(0 - 1) = 2.18 \times 10^{-18} \text{ J}$$

ΔE is positive because energy is *absorbed* to remove the electron from the vicinity of the nucleus. For 1 mol of H atoms,

$$\Delta E = \left(2.18 \times 10^{-18} \frac{\text{J}}{\text{atom}} \right) \left(6.022 \times 10^{23} \frac{\text{atoms}}{\text{mol}} \right) \left(\frac{1 \text{ kJ}}{10^3 \text{ J}} \right) = 1.31 \times 10^3 \text{ kJ/mol}$$

This is the *ionization energy* of the H atom, the quantity of energy required to form 1 mol of gaseous H^+ ions from 1 mol of gaseous H atoms. We return to this idea in Chapter 8.

Spectroscopic analysis of the H atom led to the Bohr model, the first step toward our current model of the atom. From its use by 19th-century chemists as a means of identifying elements and compounds, spectrometry has developed into a major tool of modern chemistry (see the Tools of the Laboratory essay on the next two pages). ●

What Are Stars Made Of? In 1868, the French astronomer Pierre Janssen noted a bright yellow line in the solar emission spectrum. After he and other scientists could not reproduce the line from any known element, they considered it to be due to an element unique to the Sun and named it helium (Greek *helios,* "sun"). In 1888, the British chemist William Ramsay examined the spectrum of an inert gas obtained by heating uranium-containing minerals, and it showed the same bright yellow line. Analysis of light from stars has shown many of the elements already known on Earth, but helium is the only element that was first discovered on a star.

SECTION SUMMARY

To explain the line spectrum of atomic hydrogen, Bohr proposed that the atom's energy is quantized because the electron's motion is restricted to fixed orbits. The electron can move from one orbit to another only if the atom absorbs or emits a photon whose energy equals the difference in energy levels (orbits). Line spectra are produced because these energy changes correspond to photons of specific wavelength. Bohr's model predicted the hydrogen atomic spectrum but could not predict that of any other atom because electrons do not have fixed orbits. Despite this, Bohr's idea that atoms have quantized energy levels is a cornerstone of our current atomic model. Spectrophotometry is an instrumental technique in which emission and absorption spectra are used to identify and measure concentrations of substances.

Spectrophotometry in Chemical Analysis

The use of spectral data to identify and quantify substances is essential to modern chemical analysis. The terms *spectroscopy, spectrometry,* and **spectrophotometry** denote a large group of instrumental techniques that obtain spectra corresponding to a substance's atomic and molecular energy levels.

The two types of spectra most often obtained are emission and absorption spectra. An **emission spectrum,** such as the H atom line spectrum, is produced when atoms in an excited state *emit* photons characteristic of the element as they return to lower energy states. Some elements produce a very intense spectral line (or several closely spaced ones) that serves as a marker of their presence. Such an intense line is the basis of **flame tests,** rapid qualitative procedures performed by placing a granule of an ionic compound or a drop of its solution in a flame (Figure B7.1, A). Some of the colors of fireworks and flares are due to emissions

from the same elements shown in the flame tests: crimson from strontium salts and blue-green from copper salts (Figure B7.1, B). The characteristic colors of sodium-vapor and mercury-vapor streetlamps, seen in many towns and cities, are due to one or a few prominent lines in their emission spectra.

An **absorption spectrum** is produced when atoms *absorb* photons of certain wavelengths and become excited from lower to higher energy states. Therefore, the absorption spectrum of an element appears as dark lines against a bright background. When white light passes through sodium vapor, for example, it gives rise to a sodium absorption spectrum, and the dark lines appear at the same wavelengths as those for the yellow-orange lines in the sodium emission spectrum (Figure B7.2).

(continued)

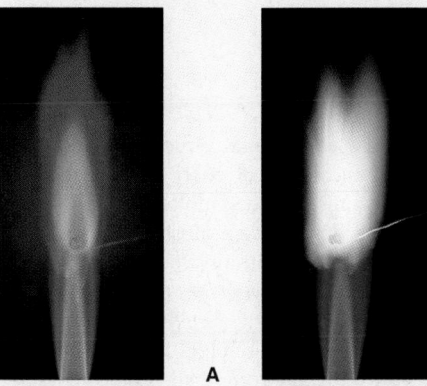

A

Figure B7.1 **Flame tests and fireworks. A,** In general, the color of the flame is created by a strong emission in the line spectrum of the element and therefore is often taken as preliminary evidence of the presence of the element in a sample. Shown here are the crimson of strontium and the blue-green of copper. **B,** The same emissions from compounds that contain these elements often appear in the brilliant displays of fireworks.

B

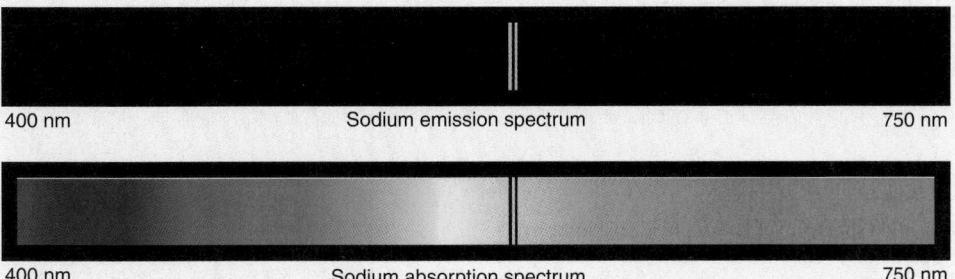

400 nm   Sodium emission spectrum   750 nm

400 nm   Sodium absorption spectrum   750 nm

Figure B7.2 **Emission and absorption spectra of sodium atoms.** The wavelengths of the bright emission lines correspond to those of the dark absorption lines because both are created by the same energy change: $\Delta E_{emission} = -\Delta E_{absorption}$. (Only the two most intense lines in the sodium atomic spectra are shown here.)

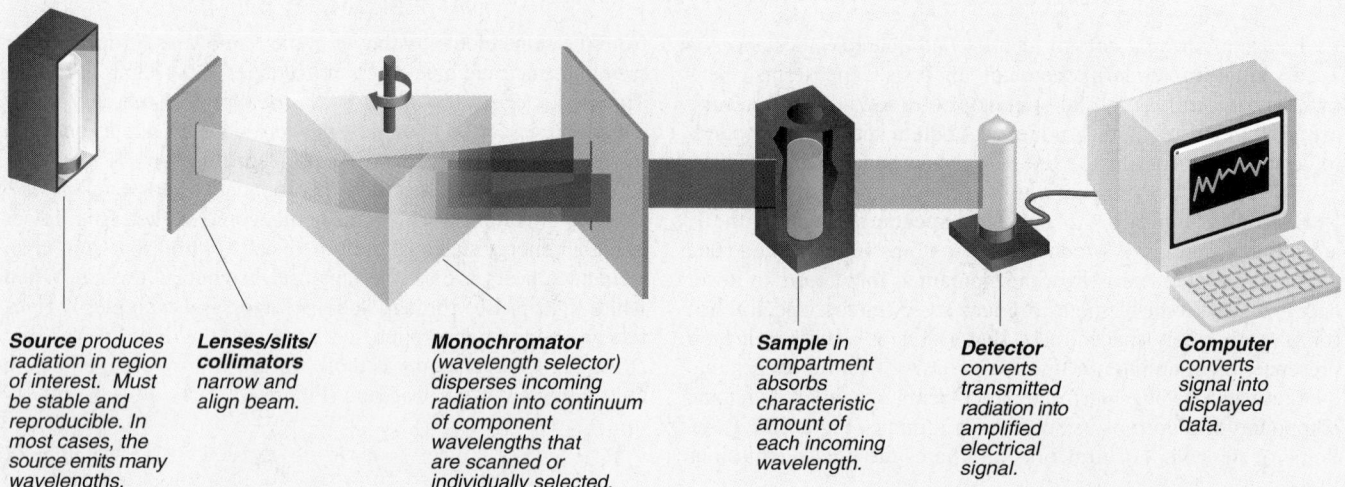

Source produces radiation in region of interest. Must be stable and reproducible. In most cases, the source emits many wavelengths.

Lenses/slits/collimators narrow and align beam.

Monochromator (wavelength selector) disperses incoming radiation into continuum of component wavelengths that are scanned or individually selected.

Sample in compartment absorbs characteristic amount of each incoming wavelength.

Detector converts transmitted radiation into amplified electrical signal.

Computer converts signal into displayed data.

Figure B7.3 The main components of a typical spectrometer.

Instruments based on absorption spectra are much more common than those based on emission spectra, for several reasons. When a solid, liquid, or dense gas is excited, it *emits* so many lines that the spectrum is a continuum (recall the continuum of colors in sunlight). Absorption is also less destructive of fragile organic and biological molecules.

Despite differences that depend on the region of the electromagnetic spectrum used to irradiate the sample, all modern spectrometers have components that perform the same basic functions (Figure B7.3). (We discuss infrared spectroscopy and nuclear magnetic resonance spectroscopy in later chapters.)

Visible light is often used to study colored substances, which absorb only some of the wavelengths from white light. A leaf looks green, for example, because its chlorophyll absorbs red and blue wavelengths strongly and green weakly, so most of the green light is reflected. The absorption spectrum of chlorophyll *a* in ether solution appears in Figure B7.4.

The overall shape of the curve and the wavelengths of the major peaks are characteristic of chlorophyll *a*, so its spectrum serves as a means of identifying it from an unknown source. The curve varies in height because chlorophyll *a* absorbs incoming wavelengths to different extents. The absorptions appear as broad bands, rather than as the distinct lines we saw earlier for individual gaseous atoms, because dissolved substances, as well as pure

solids and liquids, absorb many more wavelengths due to the greater numbers and types of energy levels within a molecule, among molecules, and between molecules and solvent.

In addition to identifying a substance, a spectrometer can be used to measure its concentration because *the absorbance, the amount of light of a given wavelength absorbed by a substance, is proportional to the number of molecules*. Suppose you want to determine the concentration of chlorophyll in an ether solution of leaf extract. You select a strongly absorbed wavelength from the chlorophyll spectrum (such as 663 nm in Figure B7.4), measure the absorbance of the leaf-extract solution, and compare it with the absorbances of a series of ether solutions with known chlorophyll concentrations.

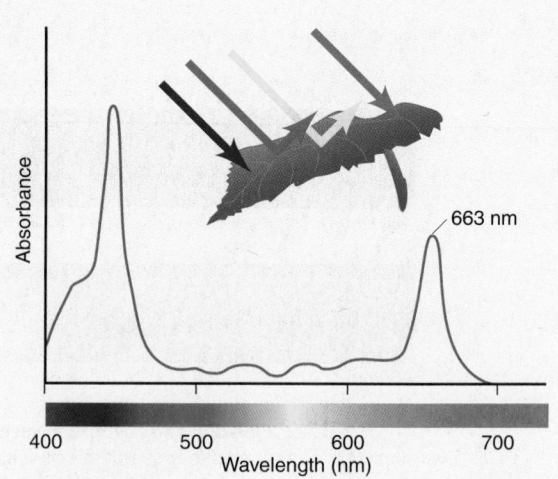

Figure B7.4 The absorption spectrum of chlorophyll *a*. Chlorophyll *a* is one of several leaf pigments. It absorbs red and blue wavelengths strongly but almost no green or yellow wavelengths. Thus, leaves containing large amounts of chlorophyll *a* appear green. The strong absorption at 663 nm can be used to quantify the amount of chlorophyll *a* present in a plant extract.

7.3 THE WAVE-PARTICLE DUALITY OF MATTER AND ENERGY

The year 1905 was a busy one for Albert Einstein. In addition to presenting the photon theory of light and explaining the photoelectric effect, he found time to explain Brownian motion (Chapter 13), which helped establish the molecular view of matter, and to introduce a new branch of physics with his theory of relativity. One of its many startling revelations was that matter and energy are alternate forms of the same entity. This idea is embodied in his famous equation $E = mc^2$, which relates the quantity of energy equivalent to a given mass, and vice versa. Relativity theory does not depend on quantum theory, but together they have completely blurred the sharp divisions we normally perceive between matter (chunky and massive) and energy (diffuse and massless).

The early proponents of quantum theory demonstrated that *energy is particle-like*. Physicists who developed the theory turned this proposition upside down and showed that *matter is wavelike*. Strange as this idea may seem, it is the key to our modern atomic model.

The Wave Nature of Electrons and the Particle Nature of Photons

Bohr's efforts were a perfect case of fitting theory to data: he *assumed* that an atom has only certain allowable energy levels in order to *explain* the observed line spectrum. However, his assumption had no basis in physical theory. Then, in the early 1920s, a young French physics student named Louis de Broglie proposed a startling reason for fixed energy levels: *if energy is particle-like, perhaps matter is wavelike*. De Broglie had been thinking of other systems that display only certain allowed motions, such as the wave of a plucked guitar string. Figure 7.13 shows that, because the ends of the string are fixed, only certain vibrational frequencies (and wavelengths) are possible. De Broglie reasoned that *if electrons have wavelike motion* and are restricted to orbits of fixed radii, that would explain why they have only certain possible frequencies and energies.

"He'll Never Make a Success of Anything" This comment, attributed to the principal of young Albert Einstein's primary school, remains a classic of misperception. Contrary to myth, the greatest physicist of the 20th century (some say of all time) was not a poor student but an independent one, preferring his own path to that prescribed by authority—a trait that gave him the intense focus characteristic of all his work. A friend recalls finding him in his small apartment, rocking his baby in its carriage with one hand while holding a pencil stub and scribbling on a pad with the other. At age 26, he was working on one of the four papers he published in 1905 that would revolutionize the way the universe is perceived and lead to his 1921 Nobel Prize.

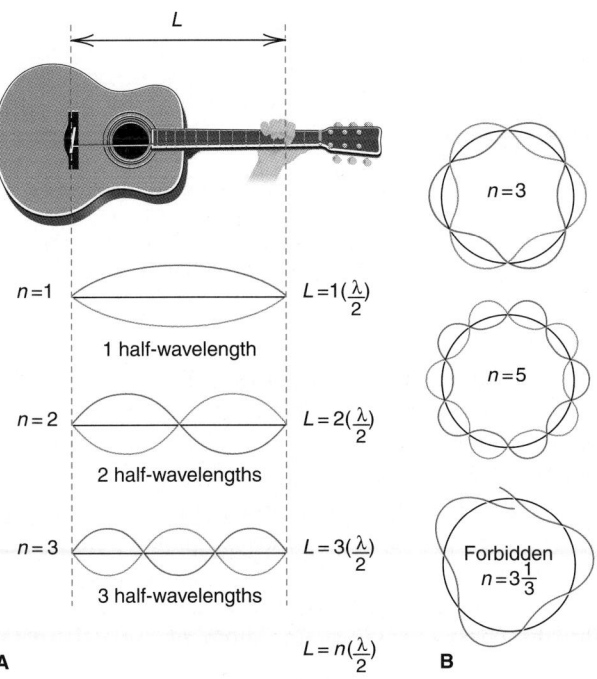

A

$n=1$ $L=1(\frac{\lambda}{2})$
1 half-wavelength

$n=2$ $L=2(\frac{\lambda}{2})$
2 half-wavelengths

$n=3$ $L=3(\frac{\lambda}{2})$
3 half-wavelengths

$L=n(\frac{\lambda}{2})$

B

$n=3$

$n=5$

Forbidden $n=3\frac{1}{3}$

Figure 7.13 Wave motion in restricted systems. A, In a musical analogy to electron waves, one half-wavelength (λ/2) is the "quantum" of the guitar string's vibration. The string length L is fixed, so the only allowed vibrations occur when L is a whole-number multiple (n) of λ/2. **B,** If an electron occupies a circular orbit, only whole numbers of wavelengths are allowed ($n = 3$ and $n = 5$ are shown). A wave with a fractional number of wavelengths (such as $n = 3\frac{1}{3}$) is "forbidden" because it rapidly dies out through overlap of crests and troughs.

Combining the equation for mass-energy equivalence ($E = mc^2$) with that for the energy of a photon ($E = h\nu = hc/\lambda$), de Broglie derived an equation for the wavelength of any particle of mass m—whether planet, baseball, or electron—moving at speed u:

$$\lambda = \frac{h}{mu} \tag{7.5}$$

According to this equation for the **de Broglie wavelength,** *matter behaves as though it moves in a wave.* Note also that an object's wavelength is *inversely* proportional to its mass, so heavy objects such as planets and baseballs have wavelengths that are *many* orders of magnitude smaller than the object itself, as you can see in Table 7.1.

Table 7.1 The de Broglie Wavelengths of Several Objects

Substance	Mass (g)	Speed (m/s)	λ (m)
Slow electron	9×10^{-28}	1.0	7×10^{-4}
Fast electron	9×10^{-28}	5.9×10^6	1×10^{-10}
Alpha particle	6.6×10^{-24}	1.5×10^7	7×10^{-15}
One-gram mass	1.0	0.01	7×10^{-29}
Baseball	142	25.0	2×10^{-34}
Earth	6.0×10^{27}	3.0×10^4	4×10^{-63}

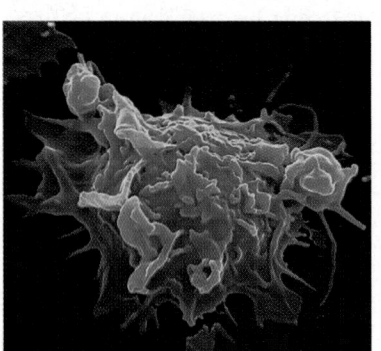

● **The Electron Microscope** In a transmission electron microscope, a beam is focused by a lens and passes through a thin section of the specimen to a second lens. The resulting image is then magnified by a third lens to a final image. The differences between this and a light microscope are that the "beam" consists of high-speed electrons and the "lenses" are electromagnetic fields, which can be adjusted to give up to 200,000-fold magnification and 0.5-nm resolution. In a scanning electron microscope, the electron beam scans the specimen, knocking electrons from it, which creates a current that varies with surface irregularities. The current generates an image that looks like the object's surface, as in this micrograph of a type of lymphocyte (11,000×). The great advantage of electron microscopes is that high-speed electrons have wavelengths much smaller than those of visible light and, thus, allow much higher image resolution.

SAMPLE PROBLEM 7.3 Calculating the de Broglie Wavelength of an Electron

Problem Find the de Broglie wavelength of an electron with a speed of 1.00×10^6 m/s (electron mass = 9.11×10^{-31} kg; $h = 6.626 \times 10^{-34}$ kg·m²/s).
Plan We know the speed (1.00×10^6 m/s) and mass (9.11×10^{-31} kg) of the electron, so we substitute these into Equation 7.5 to find λ.
Solution

$$\lambda = \frac{h}{mu} = \frac{6.626 \times 10^{-34} \text{ kg·m}^2/\text{s}}{(9.11 \times 10^{-31} \text{ kg})(1.00 \times 10^6 \text{ m/s})} = \boxed{7.27 \times 10^{-10} \text{ m}}$$

Check The order of magnitude and units seem correct:

$$\lambda \approx \frac{10^{-33} \text{ kg·m}^2/\text{s}}{(10^{-30} \text{ kg})(10^6 \text{ m/s})} = 10^{-9} \text{ m}$$

Comment As you'll see in the upcoming discussion, such fast-moving electrons, with wavelengths in the range of atomic sizes, exhibit remarkable properties.

FOLLOW-UP PROBLEM 7.3 What is the speed of an electron that has a de Broglie wavelength of 100. nm?

If particles travel in waves, electrons should exhibit diffraction and interference (see Section 7.1). ● Since a fast-moving electron has a wavelength of about 10^{-10} m, the spaces between atoms in a crystal serve as perfect "adjacent slits." In 1927, C. Davisson and L. Germer guided a beam of electrons at a nickel crystal and obtained a diffraction pattern. Figure 7.14 shows the diffraction patterns obtained when either x-rays or electrons impinge on aluminum foil. Apparently, electrons—particles with mass and charge—create diffraction patterns, just as electromagnetic waves do! Even though electrons do not have orbits of fixed radius, as de Broglie thought, the energy levels of the atom *are* related to the wave nature of the electron.

If electrons have properties of energy, do photons have properties of matter? The de Broglie equation suggests that we can calculate the momentum (p), the product of mass and speed, for a photon of a given wavelength. Substituting the speed of light (c) for speed u in Equation 7.5 and solving for p gives

$$\lambda = \frac{h}{mc} = \frac{h}{p} \quad \text{and} \quad p = \frac{h}{\lambda}$$

Notice the inverse relationship between p and λ. This means that shorter wavelength (higher energy) photons have greater momentum. Thus, a decrease in a photon's momentum should appear as an increase in its wavelength. In 1923, Arthur Compton directed a beam of x-ray photons at a sample of graphite and observed that the wavelength of the reflected photons increased. This result means that the photons transferred some of their momentum to the electrons in the carbon atoms of the graphite, just as colliding billiard balls transfer momentum to one another. In this experiment, photons behave as particles with momentum!

To scientists of the time, these results were very unsettling. Classical experiments had shown matter to be particulate and energy to be wavelike, but these new studies showed just the opposite. The understanding of matter and energy had come full circle: every characteristic trait used to define the one now also defined the other. Figure 7.15 summarizes the conceptual and experimental breakthroughs that led to this juncture.

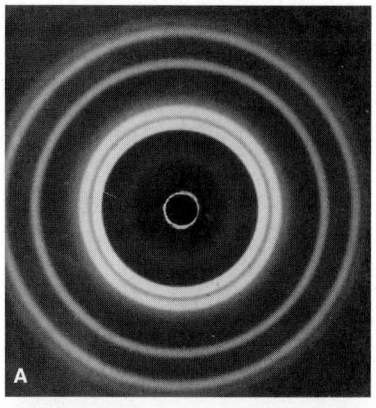

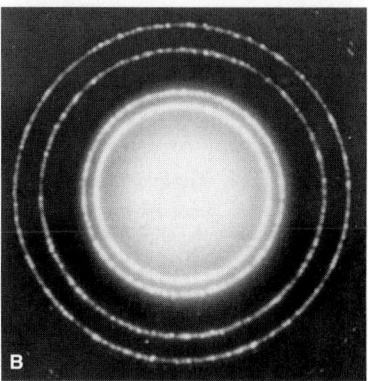

Figure 7.14 **Comparing diffraction patterns of x-rays and electrons.** **A,** X-ray diffraction pattern of Al foil. **B,** Electron diffraction pattern of Al foil. This behavior implies that both x-rays, which are electromagnetic radiation, and electrons, which are particles, travel in waves.

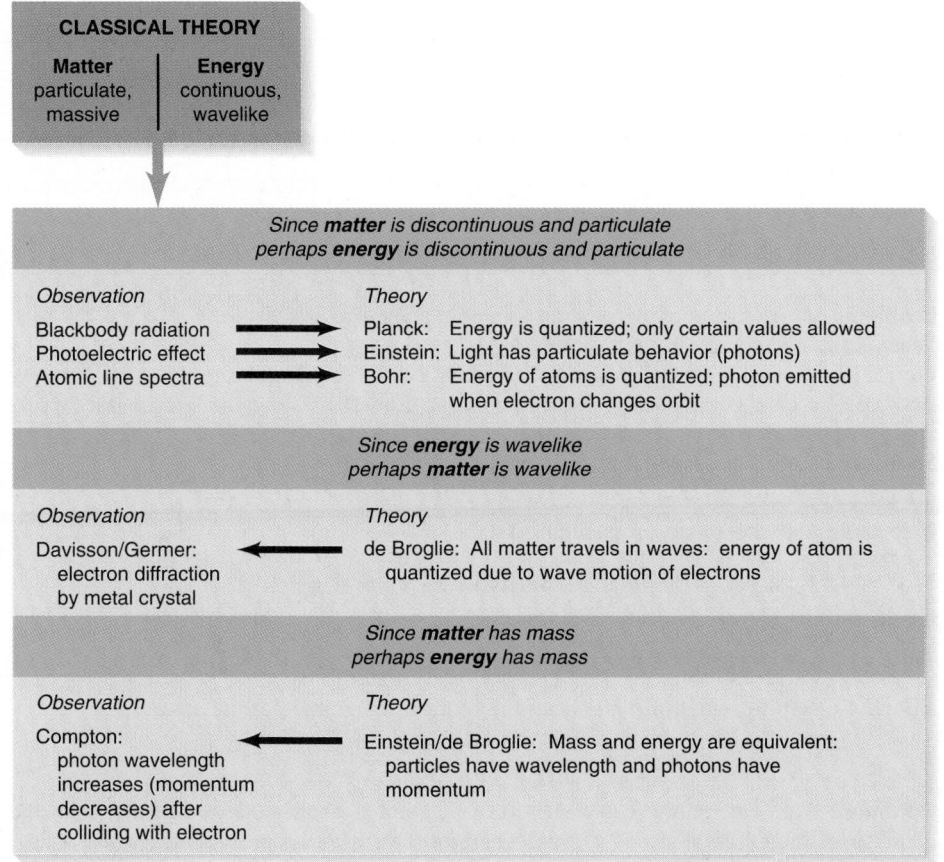

Figure 7.15 Summary of the major observations and theories leading from classical theory to quantum theory. As often happens in science, an observation (experiment) stimulates the need for an explanation (theory), and/or a theoretical insight provides the impetus for an experimental test.

CLASSICAL THEORY

Matter particulate, massive	**Energy** continuous, wavelike

Since **matter** is discontinuous and particulate perhaps **energy** is discontinuous and particulate

Observation	Theory
Blackbody radiation	Planck: Energy is quantized; only certain values allowed
Photoelectric effect	Einstein: Light has particulate behavior (photons)
Atomic line spectra	Bohr: Energy of atoms is quantized; photon emitted when electron changes orbit

Since **energy** is wavelike perhaps **matter** is wavelike

Observation	Theory
Davisson/Germer: electron diffraction by metal crystal	de Broglie: All matter travels in waves: energy of atom is quantized due to wave motion of electrons

Since **matter** has mass perhaps **energy** has mass

Observation	Theory
Compton: photon wavelength increases (momentum decreases) after colliding with electron	Einstein/de Broglie: Mass and energy are equivalent: particles have wavelength and photons have momentum

QUANTUM THEORY

Energy *same as* **Matter** particulate, massive, wavelike

The truth is that *both* matter and energy show *both* behaviors: each possesses both "faces." In some experiments, we observe one face; in other experiments, we observe the other face. The distinction between a particle and a wave is meaningful only in the macroscopic world, *not* in the atomic world. The distinction between matter and energy is in our minds and our limiting definitions, not in nature. This dual character of matter and energy is known as the **wave-particle duality.**

The Heisenberg Uncertainty Principle

In the macroscopic world, a moving particle has a definite location at any instant, whereas a wave is spread out in space. If an electron has the properties of both a particle and a wave, what can we determine about its position in the atom? In 1927, the German physicist Werner Heisenberg postulated the **uncertainty principle,** which states that it is impossible to know simultaneously the exact position *and* momentum of a particle. For a particle with constant mass m, the principle is expressed mathematically as

$$\Delta x \cdot m\Delta u \geq \frac{h}{4\pi} \qquad (7.6)$$

where Δx is the uncertainty in position and Δu is the uncertainty in speed. The more accurately we know the position of the particle (smaller Δx), the less accurately we know its speed (larger Δu), and vice versa.

By knowing the position and speed of a pitched baseball and using the classical laws of motion, we can predict its trajectory and whether it will be a strike or a ball. For a baseball, Δx and Δu are insignificant because its mass is enormous compared with $h/4\pi$. Knowing the position and speed of an electron, and from them its trajectory, is another matter entirely, as Sample Problem 7.4 demonstrates.

SAMPLE PROBLEM 7.4 Applying the Uncertainty Principle

Problem An electron moving near an atomic nucleus has a speed of $6\times10^6 \pm 1\%$ m/s. What is the uncertainty in its position (Δx)?

Plan The uncertainty in the speed (Δu) is given as 1%, so we multiply u (6×10^6 m/s) by Δu (0.01) to calculate its value, substitute it into Equation 7.6, and solve for the uncertainty in position (Δx).

Solution Finding the uncertainty in speed, Δu:

$$\Delta u = 1\% \text{ of } u = 0.01(6\times10^6 \text{ m/s}) = 6\times10^4 \text{ m/s}$$

Calculating the uncertainty in position, Δx:

$$\Delta x \cdot m\Delta u \geq \frac{h}{4\pi}$$

Thus, $\quad \Delta x \geq \dfrac{h}{4\pi m\Delta u} \geq \dfrac{6.626\times10^{-34} \text{ kg·m}^2\text{/s}}{4\pi(9.11\times10^{-31} \text{ kg})(6\times10^4 \text{ m/s})} \geq \boxed{1\times10^{-9} \text{ m}}$

Check Be sure to round off and check the order of magnitude of the answer:

$$\Delta x \geq \frac{10^{-33} \text{ kg·m}^2\text{/s}}{(10^1)(10^{-30} \text{ kg})(10^5 \text{ m/s})} = 10^{-9} \text{ m}$$

Comment The uncertainty in the electron's position is about 10 times greater than the diameter of the entire atom (10^{-10} m)! Therefore, we have no precise idea where in the atom the electron is located. In the follow-up problem, see if an umpire has any better idea where a baseball is located when calling balls and strikes.

FOLLOW-UP PROBLEM 7.4 How accurately can an umpire know the position of a baseball (mass = 0.142 kg) moving at $100.0 \pm 1.00\%$ mi/h ($44.7 \pm 1.00\%$ m/s)?

As the results of Sample Problem 7.4 show, the uncertainty principle has profound implications for an atomic model. It means that *we cannot assign fixed paths for electrons*, such as the circular orbits of Bohr's model. As you'll see in the next section, the most we can ever hope to know is the *probability*—the odds—of finding an electron in a given region of space. However, we are not *sure* it is there any more than a gambler is sure of the next roll of the dice. ●

SECTION SUMMARY

As a result of quantum theory and relativity theory, we can no longer view matter and energy as distinct entities. The de Broglie wavelength proposes that electrons (and all matter) have wavelike motion. Allowed atomic energy levels are related to allowed wavelengths of the electron's motion. Electrons exhibit diffraction patterns, as do waves of energy, and photons exhibit transfer of momentum, as do particles of mass. The wave-particle duality of matter and energy is observable only on the atomic scale. According to the uncertainty principle, we cannot know simultaneously the exact position and speed of an electron.

7.4 THE QUANTUM-MECHANICAL MODEL OF THE ATOM

Acceptance of the dual nature of matter and energy and of the uncertainty principle culminated in the field of **quantum mechanics,** which examines the wave motion of objects on the atomic scale. In 1926, Erwin Schrödinger derived an equation that is the basis for the quantum-mechanical model of the hydrogen atom. The model describes an atom that has certain allowed quantities of energy due to the allowed wavelike motion of an electron whose exact location is impossible to know.

The Atomic Orbital and the Probable Location of the Electron

The electron's matter-wave moves in three-dimensional space near the nucleus and experiences a continuous, but varying, influence from the nuclear charge. As a result, the **Schrödinger equation** is quite complex, but a simplified form is

$$\mathscr{H}\psi = E\psi$$

where E is the energy of the atom. The symbol ψ (Greek *psi,* pronounced "sigh") is called a **wave function,** a mathematical description of the motion of the electron's matter-wave in terms of time and position. The symbol $\mathscr{H}$, called the Hamiltonian operator, represents a set of mathematical operations that, when carried out on a particular ψ, yields an allowed energy state.*

Each solution to the equation (that is, each energy state of the atom) is associated with a given wave function, also called an **atomic orbital.** It's important to keep in mind that an "orbital" in the quantum-mechanical model *bears no resemblance* to an "orbit" in the Bohr model: the *orbit* was a path supposedly followed by the electron, whereas the *orbital* is a mathematical function with no physical meaning.

We cannot know precisely where the electron is at any moment, but we can describe where it *probably* is, that is, where it is most likely to be found, or where

●**Uncertainty Is Unacceptable?** Although the uncertainty principle is universally accepted by today's physicists, Einstein found some aspects of it difficult to accept, as reflected in his famous statement that "God does not play dice with the universe." Rutherford was also skeptical. When Niels Bohr *(right),* who had become a champion of the new physics, delivered a lecture in Rutherford's laboratory on the principle, Rutherford *(left)* said, "You know, Bohr, your conclusions seem to me as uncertain as the premises on which they are built." Acceptance of radical ideas does not come easily, even among fellow geniuses.

---

*The complete form of the Schrödinger equation in terms of the three linear axes is

$$\frac{d^2\psi}{dx^2} + \frac{d^2\psi}{dy^2} + \frac{d^2\psi}{dz^2} + \frac{8\pi^2 m_e}{h^2}[E - V(x,y,z)]\psi(x,y,z) = 0$$

where ψ is the wave function; the first three terms describe how ψ changes in space; m_e is the electron's mass; E is the total quantized energy of the atomic system; and V is the potential energy at point (x,y,z). Solving the equation for almost any practical application requires a significant amount of computer time.

it spends most of its time. Although the wave function (atomic orbital) has no physical meaning, we do ascribe a physical meaning to the square of the wave function, ψ^2. The quantity ψ^2 expresses the *probability* that the electron is at a particular point (or, more precisely, in a particular very small volume) within the atom. (Whereas ψ can have positive or negative values, ψ^2 is always positive, which makes sense for a value that expresses a probability.) For a given energy level, we can depict this probability with an *electron probability density diagram,* or more simply, an **electron density diagram.** In Figure 7.16A, the value of ψ^2 for a given volume is represented by a certain density of dots (the electron density): the greater the density of dots, the higher the probability of finding the electron within that volume.

Electron density diagrams are sometimes called **electron cloud** representations. If we *could* take a time-exposure snapshot of the electron whirling around the nucleus in wavelike motion, it would appear as a "cloud" of electron positions. The electron cloud is an *imaginary* picture of the electron changing its position rapidly over time; it does *not* mean that an electron is a diffuse cloud of charge. Note that *the electron density decreases with distance from the nucleus along a line, r.* The same concept is shown graphically in the plot of ψ^2 vs. r in Figure 7.16B. Although the thickness of the printed line shows the curve touching the axis, *the probability of the electron being far from the nucleus is very small, but not zero.*

The *total* probability of finding the electron at any distance r from the nucleus is also important. To find this, we mentally divide the volume around the nucleus into thin, concentric, spherical layers, like the layers of an onion (shown in cross section in Figure 7.16C), and ask in which *spherical layer* we are most likely to find the electron. This is the same as asking for the *sum of ψ^2 values* within each spherical layer. The steep falloff in electron density with distance (see Figure 7.16B) has an important effect. Near the nucleus, the volume of each layer increases faster than its electron density decreases. As a result, the *total* probability of finding the electron in the second layer is higher than in the first. Electron density drops off so quickly, however, that this effect soon diminishes with greater distance. Thus, even though the volume of each layer continues to

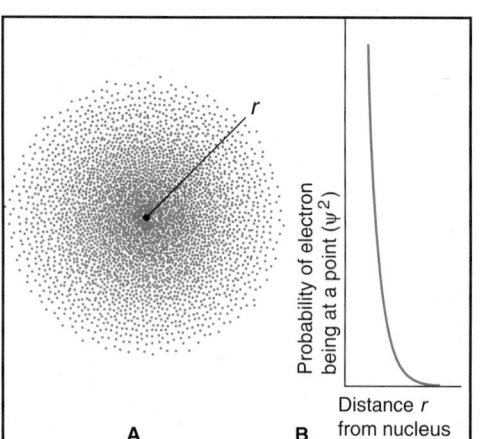

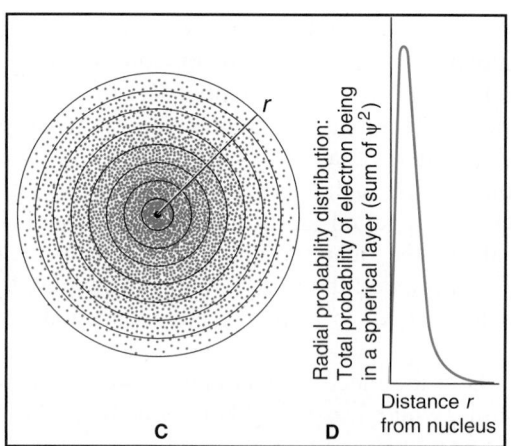

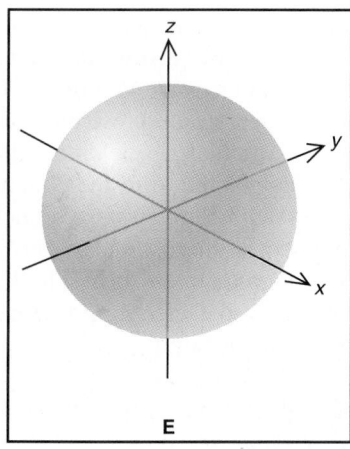

Figure 7.16 **Electron probability in the ground-state H atom.** **A,** An electron density diagram shows a cross section of the H atom. The dots, each representing the probability of the electron being at a point (actually, a tiny volume), decrease along a line outward from the nucleus. **B,** A plot of the data in **A** shows that the probability (ψ^2) at any point along the line decreases with distance from the nucleus but does not reach zero (the thickness of the line makes it appear to do so). **C,** Dividing the atom's volume into thin, concentric, spherical lay-

ers (shown in cross section) and counting the dots within each layer gives the total probability of finding the electron within that layer. **D,** A radial probability distribution plot shows total electron density in each spherical layer vs. r. Since electron density decreases more slowly than the volume of each concentric layer increases, the plot shows a peak. **E,** A 90% probability contour shows the ground state of the H atom (orbital of lowest energy) and represents the volume in which the electron spends 90% of its time.

increase, the total probability in a given layer eventually decreases. Because of these opposing effects of decreasing electron density and increasing layer volume, the total probability peaks in some layer near the nucleus but beyond the first one. Figure 7.16D shows this as a **radial probability distribution plot.**

The peak of the radial probability distribution for the ground-state H atom appears at the same distance from the nucleus (0.529Å, or 5.29×10^{-11} m) as the closest Bohr orbit. Thus, at least for the ground state, the Schrödinger model predicts that the electron spends *most* of its time at the same distance that the Bohr model predicted it spent *all* of its time. The difference between "most" and "all" reflects the uncertainty of the electron's location in the Schrödinger model.

How far away from the nucleus can we find the electron? This is the same as asking "How large is the atom?" Recall from Figure 7.16B that the probability of finding the electron far from the nucleus is not zero. Therefore, we *cannot* assign a definite volume to an atom. However, we often visualize atoms with a 90% **probability contour,** such as in Figure 7.16E, which shows the volume within which the electron of the hydrogen atom spends 90% of its time.

Quantum Numbers of an Atomic Orbital

So far we have discussed the electron density for the *ground* state of the H atom. When the atom absorbs energy, it exists in an *excited* state and the wave motion of the electron is described by a different atomic orbital (wave function). As you'll see, each atomic orbital has a distinctive radial probability distribution and probability contour.

An atomic orbital is specified by three quantum numbers. One is related to the orbital's size, another to its shape, and the third to its orientation in space.* The quantum numbers have a hierarchical relationship: the size-related number limits the shape-related number, which limits the orientation-related number. Let's examine this hierarchy and then look at the shapes and orientations.

1. The **principal quantum number (n)** *is a positive integer* (1, 2, 3, and so forth). It indicates the relative *size* of the orbital and therefore the relative *distance from the nucleus* of the peak in the radial probability distribution plot. The principal quantum number specifies the *energy level* of the H atom: *the higher the n value, the higher the energy level.* When the electron occupies an orbital with $n = 1$, the H atom is in its ground state and has lower energy than when the electron occupies the $n = 2$ orbital (first excited state).

2. The **angular momentum quantum number (l)** *is an integer from 0 to $n - 1$.* It is related to the *shape* of the orbital and is sometimes called the *orbital-shape quantum number.* Note that the principal quantum number sets a limit on the values for the angular momentum quantum number; that is, n limits l. For an orbital with $n = 1$, l can have a value of only 0. For orbitals with $n = 2$, l can have a value of 0 or 1; for those with $n = 3$, l can be 0, 1, or 2; and so forth. Note that the number of possible l values equals the value of n.

3. The **magnetic quantum number (m_l)** *is an integer from $-l$ through 0 to $+l$.* It prescribes the *orientation* of the orbital in the space around the nucleus and is sometimes called the *orbital-orientation quantum number.* The possible values of an orbital's magnetic quantum number are set by its angular momentum quantum number (that is, l determines m_l). An orbital with $l = 0$ can have only $m_l = 0$. However, an orbital with $l = 1$ can have any one of three m_l values, -1, 0, or $+1$; thus, there are three possible orbitals with $l = 1$, each with its own orientation. Note that the number of possible m_l values *equals* the number of orbitals, which is $2l + 1$ for a given l value.

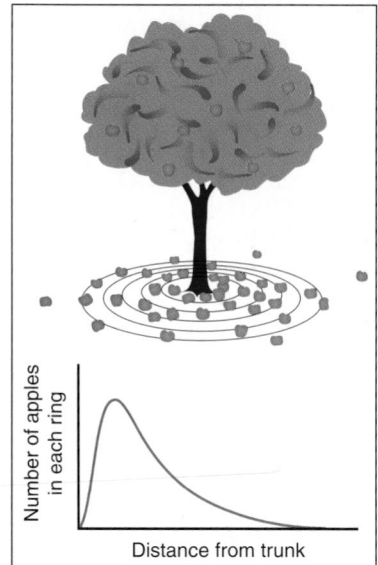

Number of apples in each ring — Distance from trunk

A Radial Probability Distribution of Apples An analogy might clarify why the curve in the radial probability distribution plot peaks and then falls off. Picture the fallen apples around the base of an apple tree: the density of apples is greatest near the trunk and decreases with distance. Divide the ground under the tree into foot-wide concentric rings and collect the apples within each ring. Apple density is greatest in the first ring, but the area of the second ring is larger, so it contains a greater *total* number of apples. Farther out near the edge of the tree, rings have more area but lower apple "density," so the total number of apples decreases. A plot of "number of apples within a ring" vs. "distance of ring from trunk" shows a peak at some close distance from the trunk, as in Figure 7.16D.

---

*For ease in discussion, we refer to the size, shape, and orientation of an "atomic orbital," although we really mean the size, shape, and orientation of an "atomic orbital's radial probability distribution." This usage is common in both introductory and advanced texts.

Table 7.2 **The Hierarchy of Quantum Numbers for Atomic Orbitals**

Name, Symbol (Property)	Allowed Values	Quantum Numbers
Principal, n (size, energy)	Positive integer $(1, 2, 3, \ldots)$	
Angular momentum, l (shape)	0 to $n - 1$	
Magnetic, m_l (orientation)	$-l, \ldots, 0, \ldots, +l$	

Table 7.2 summarizes the relationships among the three quantum numbers. The total number of orbitals for a given n value is n^2.

SAMPLE PROBLEM 7.5 Determining Quantum Numbers for an Energy Level

Problem What values of the angular momentum (l) and magnetic (m_l) quantum numbers are allowed for a principal quantum number (n) of 3? How many orbitals are allowed for $n = 3$?

Plan We determine allowable quantum numbers with the rules from the text: l values are integers from 0 to $n - 1$, and m_l values are integers from $-l$ to 0 to $+l$. One m_l value is assigned to each orbital, so the number of m_l values gives the number of orbitals.

Solution Determining l values: for $n = 3$, $l = 0, 1, 2$

Determining m_l for each l value:

For $l = 0$, $m_l = 0$

For $l = 1$, $m_l = -1, 0, +1$

For $l = 2$, $m_l = -2, -1, 0, +1, +2$

There are nine m_l values, so there are nine orbitals with $n = 3$.

Check Table 7.2 shows that we are correct. The total number of orbitals for a given n value is n^2, and for $n = 3$, $n^2 = 9$.

FOLLOW-UP PROBLEM 7.5 Specify the l and m_l values for $n = 4$.

The energy states and orbitals of the atom are described with specific terms and associated with one or more quantum numbers:

1. *Level.* The atom's energy **levels,** or *shells,* are given by the n value: the smaller the n value, the lower the energy level and the greater the probability of the electron being closer to the nucleus.

2. *Sublevel.* The atom's levels contain **sublevels,** or *subshells,* which designate the orbital shape. Each sublevel has a letter designation:

$l = 0$ is an s sublevel.

$l = 1$ is a p sublevel.

$l = 2$ is a d sublevel.

$l = 3$ is an f sublevel.

(The letters derive from the names of spectroscopic lines: *s*harp, *p*rincipal, *dif*fuse, and *f*undamental.) Sublevels are named by joining the n value and the letter designation. For example, the sublevel (subshell) with $n = 2$ and $l = 0$ is called the $2s$ sublevel.

3. *Orbital*. Each allowed combination of n, l, and m_l values specifies one of the atom's *orbitals*. Thus, the three quantum numbers that describe an orbital express its size (energy), shape, and spatial orientation. You can easily give the quantum numbers of the orbitals in any sublevel if you know the sublevel letter designation and the quantum number hierarchy. For example, the hierarchy prescribes that the $2s$ sublevel has only one orbital, and its quantum numbers are $n = 2$, $l = 0$, and $m_l = 0$. The $3p$ sublevel has three orbitals: one with $n = 3$, $l = 1$, and $m_l = -1$; another with $n = 3$, $l = 1$, and $m_l = 0$; and a third with $n = 3$, $l = 1$, and $m_l = +1$.

SAMPLE PROBLEM 7.6 Determining Sublevel Names and Orbital Quantum Numbers

---

Problem Give the name, magnetic quantum numbers, and number of orbitals for each sublevel with the following quantum numbers:
(a) $n = 3, l = 2$ (b) $n = 2, l = 0$ (c) $n = 5, l = 1$ (d) $n = 4, l = 3$
Plan To name the sublevel (subshell), we combine the n value and l letter designation. Since we know l, we can find the possible m_l values, whose total number equals the number of orbitals.
Solution

n	l	Sublevel Name	Possible m_l Values	No. of Orbitals
(a) 3	2	$3d$	$-2, 1, 0, +1, +2$	5
(b) 2	0	$2s$	0	1
(c) 5	1	$5p$	$-1, 0, +1$	3
(d) 4	3	$4f$	$-3, -2, -1, 0, +1, +2, +3$	7

Check Check the number of orbitals in each sublevel using
$$\text{No. of orbitals} = \text{no. of } m_l \text{ values} = 2l + 1$$

FOLLOW-UP PROBLEM 7.6 What are the n, l, and possible m_l values for the $2p$ and $5f$ sublevels?

SAMPLE PROBLEM 7.7 Identifying Incorrect Quantum Numbers

Problem What is wrong with each of the following quantum number designations and/or sublevel names?

n	l	m_l	Name
(a) 1	1	0	$1p$
(b) 4	3	+1	$4d$
(c) 3	1	−2	$3p$

Solution (a) A sublevel with $n = 1$ can have only $l = 0$, not $l = 1$. The only possible sublevel name is $1s$.
(b) A sublevel with $l = 3$ is an f sublevel, not a d sublevel. The name should be $4f$.
(c) A sublevel with $l = 1$ can have only m_l of $-1, 0, +1$, not -2.
Check Check that l is always less than n, and m_l is always $\geq -l$ and $\leq +l$.

FOLLOW-UP PROBLEM 7.7 Supply the missing quantum numbers and sublevel names.

n	l	m_l	Name
(a) ?	?	0	$4p$
(b) 2	1	0	?
(c) 3	2	−2	?
(d) ?	?	?	$2s$

Shapes of Atomic Orbitals

Each sublevel of the H atom corresponds to orbitals with characteristic shapes. As you'll see in Chapter 8, orbitals for the other atoms have similar shapes.

The *s* Orbital An orbital with $l = 0$ has a *spherical* shape with the nucleus at its center and is called an ***s* orbital.** The H atom's ground state, for example, has the electron in the $1s$ orbital. Figure 7.17A shows that the electron density is highest at the nucleus, whether it is depicted graphically *(top)* or as a section of a three-dimensional electron cloud *(middle)*. On the other hand, the radial probability distribution *(bottom)*, which represents the *total* probability of finding the electron (that is, where the electron spends most of its time), is highest slightly out from the nucleus. Both plots fall off smoothly with distance.

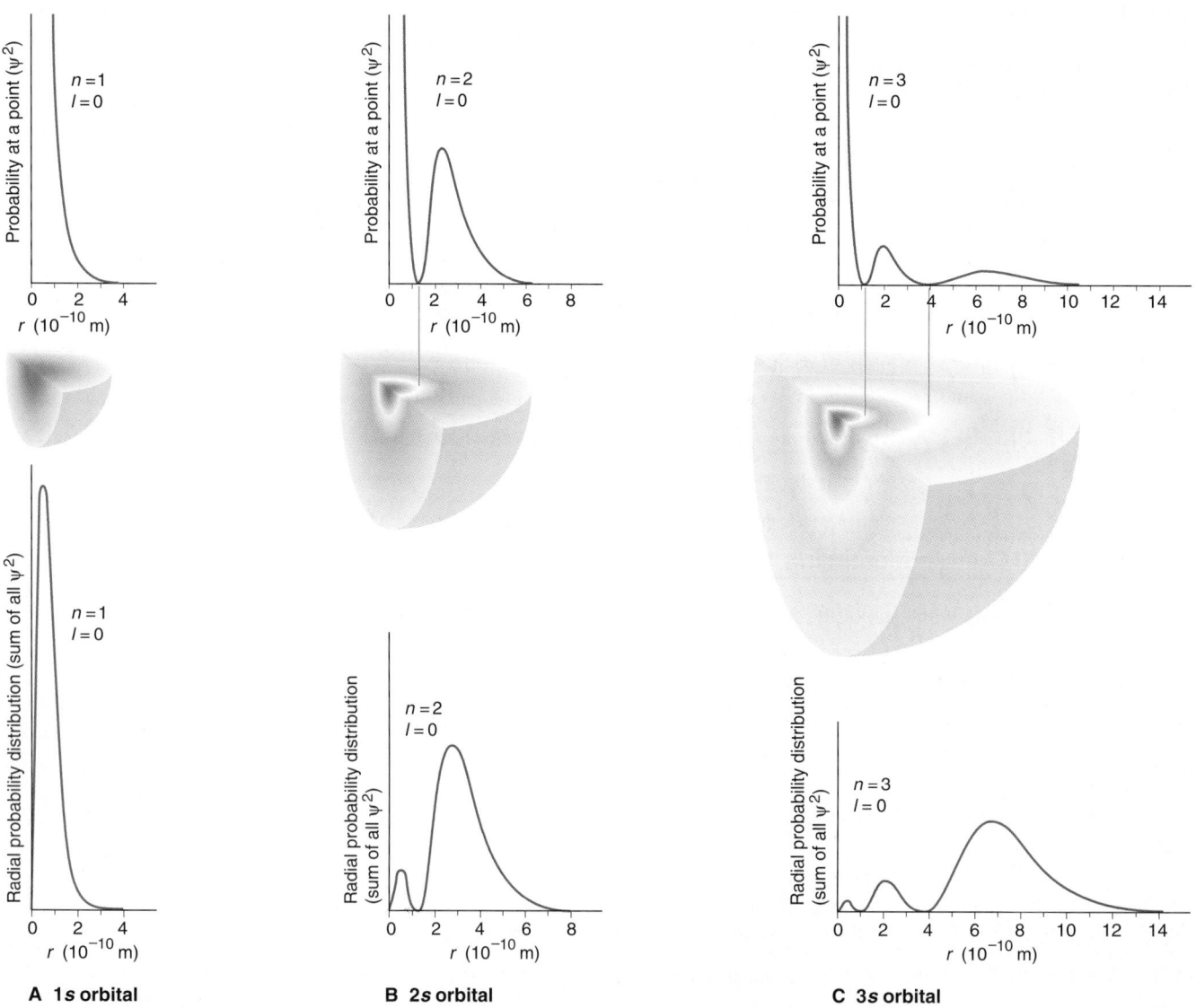

A 1s orbital   **B 2s orbital**   **C 3s orbital**

Figure 7.17 **The 1s, 2s, and 3s orbitals.** Information for each of the *s* orbitals is shown in three ways: as a plot of electron density *(top)*; as an electron cloud representation *(middle)*, in which shading coincides with peaks in the plot above; and as a radial probability distribution *(bottom)* that shows where the electron spends its time. **A,** The 1s orbital. **B,** The 2s orbital. **C,** The 3s orbital. Note the nodes (regions of zero probability) in the 2s and 3s orbitals.

The 2s orbital, shown in Figure 7.17B, has two regions of higher electron density. The radial probability distribution (*total* probability; *bottom*) of the more distant region is *higher* than that of the closer one because the sum of its ψ^2 is taken over a much larger volume. Between the two regions is a spherical **node,** a shell-like region where the probability drops to zero ($\psi^2 = 0$ at the node, analogous to zero amplitude of a wave). Because the 2s orbital is larger than the 1s, an electron in this orbital spends more time *farther* from the nucleus than when it occupies the 1s.

The 3s orbital, shown in Figure 7.17C, has three regions of higher electron density and two nodes. Here again, the highest radial probability is at the greatest distance from the nucleus because the sum of all ψ^2 is taken over a larger volume. This pattern of more nodes and higher probability with distance continues for s orbitals of higher n value. Since an s orbital has a spherical shape, it can have only one orientation and, thus, only one value for the magnetic quantum number: for any s orbital, $m_l = 0$.

The *p* Orbital An orbital with $l = 1$ has two regions (lobes) of high probability, one on *either side* of the nucleus, and is called a ***p* orbital.** Thus, as you can see in Figure 7.18, the *nucleus lies at the nodal plane* of this "dumbbell-shaped" orbital. Since the maximum value of l is $n - 1$, only levels with $n = 2$ or higher can have a p orbital. Therefore, the lowest energy p orbital (the one closest to the nucleus) is the 2p. Keep in mind that *one p orbital consists of both lobes* and that the electron spends *equal* time in both. As we would expect from the pattern of s orbitals, a 3p orbital is larger than a 2p orbital, a 4p orbital is larger than a 3p orbital, and so forth.

Unlike an s orbital, each p orbital has a specific orientation in space. The $l = 1$ value has three possible m_l values: -1, 0, and $+1$, which refer to three *mutually perpendicular p orbitals*. They are identical in size, shape, and energy, differing only in orientation. For convenience, we associate p orbitals with the x, y, and z axes (but there is no necessary relation between a spatial axis and a given m_l value): the p_x orbital lies along the x axis, the p_y along the y axis, and the p_z along the z axis.

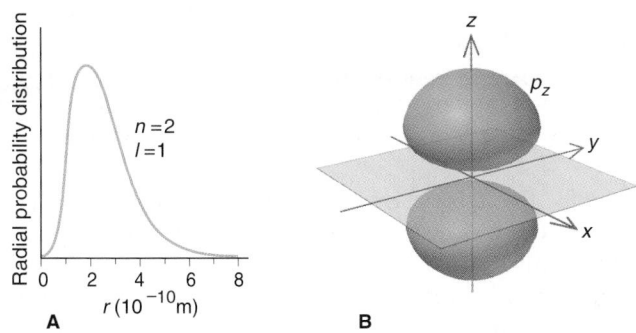
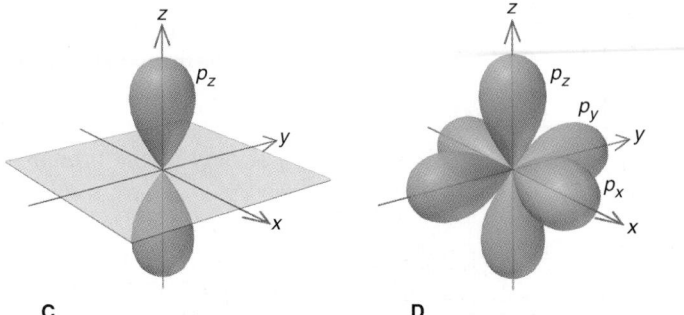

Figure 7.18 **The 2p orbitals. A,** A radial probability distribution plot of the 2p orbital shows a single peak. It lies at nearly the same distance from the nucleus as the larger peak in the 2s plot (shown in Figure 7.17B). **B,** An accurate representation of the probability contour of the 2p_z orbital. Note the nodal plane at the nucleus. An electron occupies both regions of a 2p orbital equally and spends 90% of its time within this volume. The 2p_x orbital and 2p_y orbital have identical shapes but lie along the x and y axes, respectively. **C,** The stylized depiction of the 2p probability contour used throughout the text. **D,** In the atom, the three 2p orbitals occupy mutually perpendicular regions of space, contributing to an atom's overall spherical shape.

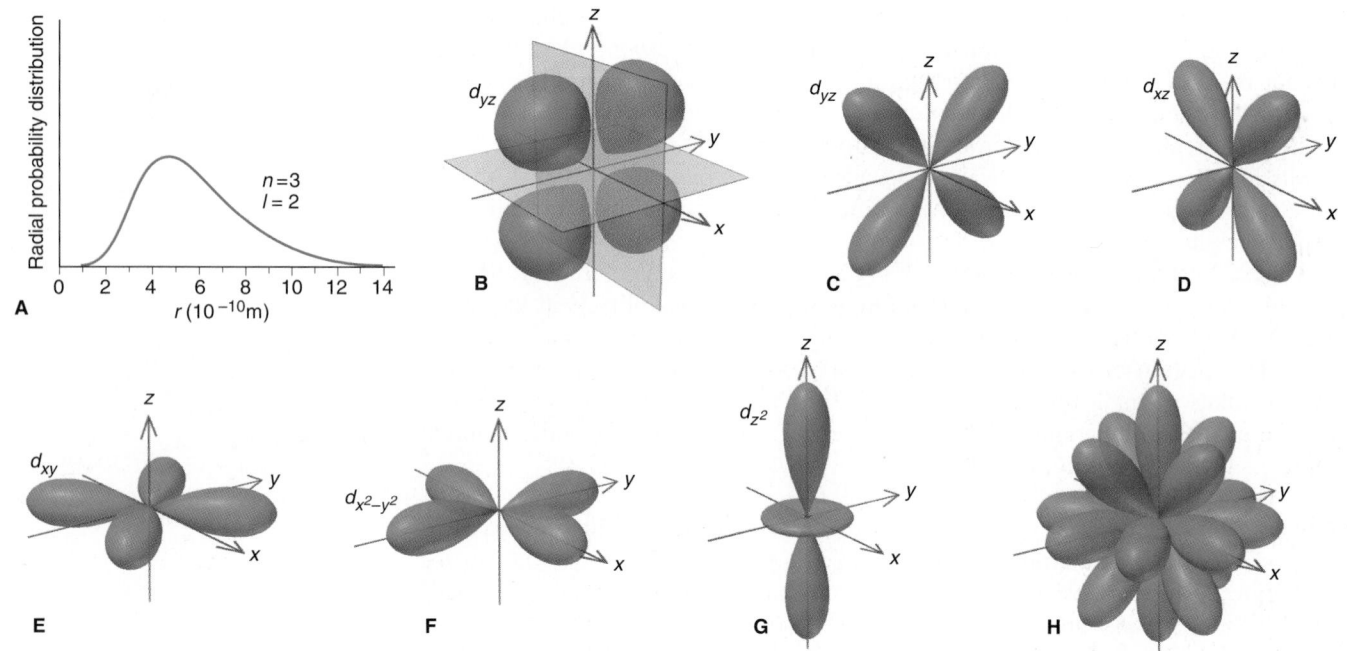

Figure 7.19 The 3*d* orbitals. **A,** A radial probability distribution plot. **B,** An accurate representation of the 3d_{yz} orbital probability contour. Note the mutually perpendicular nodal planes and the lobes lying *between* the axes. **C,** The stylized depiction of the 3d_{yz} orbital used throughout the text. **D,** The 3d_{xz} orbital. **E,** The 3d_{xy} orbital. **F,** The lobes of the 3$d_{x^2-y^2}$ orbital lie *on* the *x* axis and *y* axis. **G,** The 3d_{z^2} orbital has two lobes and a central, donut-shaped region. **H,** A composite of the five 3*d* orbitals, which again contributes to an atom's overall spherical shape.

The *d* Orbital An orbital with $l = 2$ is called a ***d* orbital.** There are five possible m_l values for the $l = 2$ value: -2, -1, 0, $+1$, and $+2$. Thus, a *d* orbital can have any one of five different orientations, as shown in Figure 7.19. Four of the five *d* orbitals have four lobes ("cloverleaf shape") prescribed by two mutually perpendicular nodal planes, with the nucleus lying at the junction of the lobes. Three of these orbitals lie in the mutually perpendicular *xy*, *xz*, and *yz* planes, with their lobes *between* the axes, and are called the d_{xy}, d_{xz}, and d_{yz} orbitals. A fourth, the $d_{x^2-y^2}$ orbital, also lies in the *xy* plane, but its lobes are directed *along* the axes. The fifth *d* orbital, the d_{z^2}, has a different shape: two major lobes lie along the *z* axis, and a donut-shaped region of electron density girdles the center. An electron associated with a given *d* orbital has equal probability of being in any of the orbital's lobes.

As we said for the *p* orbitals, the axis designations are *not* associated with a given m_l value. In keeping with the quantum number rules, a *d* orbital ($l = 2$) must have a principal quantum number of $n = 3$ or greater. The 4*d* orbitals extend farther from the nucleus than the 3*d* orbitals, and the 5*d* orbitals extend still farther.

Orbitals with Higher *l* Values Orbitals with $l = 3$ are *f* orbitals and must have a principal quantum number of at least $n = 4$. There are seven *f* orbitals ($2l + 1 = 7$), each with a complex, multilobed shape; Figure 7.20 shows one of them. Orbitals with $l = 4$ are *g* orbitals, but we will not discuss them further because they play no known role in chemical bonding.

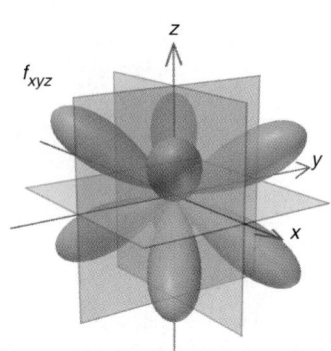

Figure 7.20 One of the seven possible 4*f* orbitals. The 4f_{xyz} orbital has eight lobes and three nodal planes. The other six 4*f* orbitals have multilobed contours also.

Energy Levels of the Hydrogen Atom

The energy state of the H atom depends on the principal quantum number n only. An electron in an orbital with a higher n value spends its time (on average) farther from the nucleus, so it is higher in energy. As you'll see in Chapter 8, the energy state of an atom with more than one electron depends on both the n and l values of the occupied orbitals. Thus, in the case of the H atom *only*, all four $n = 2$ orbitals (one $2s$ and three $2p$) have the same energy, and all nine $n = 3$ orbitals (one $3s$, three $3p$, and five $3d$) have the same energy (Figure 7.21).

SECTION SUMMARY

The electron's wave function (ψ, atomic orbital) is a mathematical description of the electron's wavelike motion in an atom. Each wave function is associated with one of the atom's allowed energy states. The probability of finding the electron at a particular location is represented by ψ^2. An electron density diagram and a radial probability distribution plot show how the electron occupies the space near the nucleus for a particular energy level. Three features of the atomic orbital are described by quantum numbers: size (n), shape (l), and orientation (m_l). Orbitals with the same n and l values constitute a sublevel; sublevels with the same n value constitute an energy level. A sublevel with $l = 0$ has a spherical (s) orbital; a sublevel with $l = 1$ has three, two-lobed (p) orbitals; and a sublevel with $l = 2$ has five, four-lobed (d) orbitals. In the case of the H atom, the energy levels depend on the n value only.

Figure 7.21 **The energy levels in the H atom.** In the H atom, the energy level depends only on the n value of the orbital. For example, the $2s$ and three $2p$ orbitals (shown as short lines) all have the same energy.

ATOMIC LINE SPECTRA

Chapter Perspective

In this brief exploration of the origins of quantum physics, we see the everyday distinctions between matter and energy disappear and a new picture of the hydrogen atom gradually come into focus. By the atom's very nature, however, this focus cannot be perfectly sharp. The sequel to these remarkable discoveries unfolds in Chapter 8, where we begin our discussion of how the periodic behavior of the elements emerges from the properties of the atom, which in turn arise from the electron occupancy of its orbitals.

For Review and Reference (Numbers in parentheses refer to pages, unless noted otherwise.)

Learning Objectives

Relevant section and/or sample problem (SP) numbers appear in parentheses.

Understand These Concepts

1. The wave characteristics of light (the interrelations of frequency, wavelength, and speed; the meaning of amplitude) and a general overview of the electromagnetic spectrum (7.1)
2. How particles and waves differ in terms of the phenomena of refraction, diffraction, and interference (7.1)
3. The quantization of energy and the fact that an atom changes its energy by emitting or absorbing quanta of radiation (7.1)
4. How the photon theory explains the photoelectric effect and the relation between photon absorbed and electron released (7.1)
5. How the Bohr theory explained line spectra of the H atom; the importance of discrete atomic energy levels (7.2)
6. The wave-particle duality of matter and energy and the relevant theories and experiments that led to it (de Broglie wavelength, electron diffraction, photon momentum) (7.3)
7. The meaning of the uncertainty principle and how it limits our knowledge of electron properties (7.3)
8. The distinction between ψ (atomic orbital) and ψ^2 (probability of finding the electron at some distance from the nucleus) (7.4)

9. How electron density diagrams and radial probability distribution plots depict electron location within the atom (7.4)
10. The hierarchy of the quantum numbers that describe the size (n, energy), shape (l), and orientation (m_l) of an orbital (7.4)
11. The distinction between level (shell), sublevel (subshell), and orbital (7.4)
12. The shapes and nodes of s, p, and d orbitals (7.4)

Master These Skills

1. Interconverting wavelength and frequency (SP 7.1)
2. Calculating the energy of a photon from its frequency and/or wavelength (SP 7.2)
3. Applying de Broglie's equation to find the wavelength of a particle (SP 7.3)
4. Applying the uncertainty principle to see that location and speed cannot be determined simultaneously for particles of extremely small mass (SP 7.4)
5. Determining quantum numbers and sublevel designations (SPs 7.5 to 7.7)

Key Terms

Section 7.1
electromagnetic radiation (255)
frequency (ν) (256)
wavelength (λ) (256)
speed of light (c) (256)
amplitude (256)
electromagnetic spectrum (257)
infrared (IR) (257)
ultraviolet (UV) (257)
refraction (258)
diffraction (259)
quantum number (260)

Planck's constant (h) (260)
quantum (260)
photoelectric effect (261)
photon (261)

Section 7.2
line spectrum (263)
stationary state (263)
ground state (264)
excited state (264)
spectrophotometry (267)
emission spectrum (267)
flame test (267)
absorption spectrum (267)

Section 7.3
de Broglie wavelength (270)
wave-particle duality (272)
uncertainty principle (272)

Section 7.4
quantum mechanics (273)
Schrödinger equation (273)
wave function (atomic orbital) (273)
electron density diagram (274)
electron cloud (274)
radial probability distribution plot (275)

probability contour (275)
principal quantum number (n) (275)
angular momentum quantum number (l) (275)
magnetic quantum number (m_l) (275)
level (shell) (276)
sublevel (subshell) (276)
s orbital (278)
node (279)
p orbital (279)
d orbital (280)

Key Equations and Relationships

7.1 Relating the speed of light to its frequency and wavelength (256):

$$c = \nu \times \lambda$$

7.2 Determining the smallest change in an atom's energy (260):

$$\Delta E = h\nu$$

7.3 Calculating the wavelength of any line in the H atom spectrum (Rydberg equation) (263):

$$\frac{1}{\lambda} = R\left(\frac{1}{n_1^2} - \frac{1}{n_2^2}\right)$$

7.4 Finding the difference between two energy levels in the H atom (266):

$$\Delta E = E_{\text{final}} - E_{\text{initial}} = -2.18 \times 10^{-18}\,\text{J}\left(\frac{1}{n_{\text{final}}^2} - \frac{1}{n_{\text{initial}}^2}\right)$$

7.5 Calculating the wavelength of any moving particle (de Broglie wavelength) (270):

$$\lambda = \frac{h}{mu}$$

7.6 Finding the uncertainty in position or speed of a particle (Heisenberg uncertainty principle) (272):

$$\Delta x \cdot m\Delta u \geq \frac{h}{4\pi}$$

Highlighted Figures and Tables

These figures (F) and tables (T) provide a quick review of key ideas.

F7.3 The electromagnetic spectrum (257)
F7.15 From classical to quantum theory (271)
F7.16 Electron probability in the ground-state H atom (274)
T7.2 The hierarchy of quantum numbers (276)

F7.17 The 1s, 2s, and 3s orbitals (278)
F7.18 The 2p orbitals (279)
F7.19 The 3d orbitals (280)

Brief Solutions to Follow-up Problems

7.1 λ (nm) $= \dfrac{3.00 \times 10^8\,\text{m/s}}{7.23 \times 10^{14}\,\text{s}^{-1}} \times \dfrac{10^9\,\text{nm}}{1\,\text{m}} = 415\,\text{nm}$

λ (Å) $= 415\,\text{nm} \times \dfrac{10\,\text{Å}}{1\,\text{nm}} = 4150\,\text{Å}$

7.2 UV: $E = hc/\lambda$

$= \dfrac{(6.626 \times 10^{-34}\,\text{J·s})(3.00 \times 10^8\,\text{m/s})}{1 \times 10^{-8}\,\text{m}} = 2 \times 10^{-17}\,\text{J}$

Visible: $E = 4 \times 10^{-19}$ J; IR: $E = 2 \times 10^{-21}$ J
As λ increases, E decreases.

7.3 $u = \dfrac{h}{m\lambda} = \dfrac{6.626 \times 10^{-34}\,\text{kg·m}^2/\text{s}}{(9.11 \times 10^{-31}\,\text{kg})\left(100\,\text{nm} \times \dfrac{1\,\text{m}}{10^9\,\text{nm}}\right)}$

$= 7.27 \times 10^3$ m/s

7.4 $\Delta x \geq \dfrac{6.626 \times 10^{-34}\,\text{kg·m}^2/\text{s}}{4\pi(0.142\,\text{kg})(0.447\,\text{m/s})} \geq 8.31 \times 10^{-34}$ m

7.5 $n = 4$, so $l = 0, 1, 2, 3$. In addition to the m_l values in Sample Problem 7.5, we have those for $l = 3$:

$$m_l = -3, -2, -1, 0, +1, +2, +3$$

7.6 For 2p: $n = 2$, $l = 1$, $m_l = -1, 0, +1$
For 5f: $n = 5$, $l = 3$, $m_l = -3, -2, -1, 0, +1, +2, +3$

7.7 (a) $n = 4$, $l = 1$; (b) name is 2p; (c) name is 3d;
(d) $n = 2$, $l = 0$, $m_l = 0$

Problems

Problems with **colored** numbers are answered at the back of the text. Sections match the text and provide the number(s) of relevant sample problems. Most offer Concept Review Questions, Skill-Building Exercises (in similar pairs), and Problems in Context. Then Comprehensive Problems, based on material from any section or previous chapter, follow.

The Nature of Light
(Sample Problems 7.1 and 7.2)

● **Concept Review Questions**

7.1 In what ways are microwave and ultraviolet radiation the same? In what ways are they different?

7.2 Consider the following types of electromagnetic radiation:
(1) microwave   (2) ultraviolet   (3) radio waves
(4) infrared   (5) x-ray   (6) visible
(a) Arrange them in order of increasing wavelength.
(b) Arrange them in order of increasing frequency.
(c) Arrange them in order of increasing energy.

7.3 Define each of the following wave phenomena, and give an example of where each occurs: (a) refraction; (b) diffraction; (c) dispersion; (d) interference.

7.4 In the mid-17th century, Isaac Newton proposed that light existed as a stream of particles, and the wave-particle debate continued for over 250 years until the revolutionary ideas of Planck and Einstein. Give two pieces of evidence for the wave model and two for the particle model.

7.5 What new idea about energy did Planck use to explain black-body radiation?

7.6 What new idea about light did Einstein use to explain the photoelectric effect? Why does the photoelectric effect exhibit a threshold frequency? Why does it *not* exhibit a time lag?

● **Skill-Building Exercises *(paired)***

7.7 An AM station broadcasts rock music at "960 on your radio dial." Units for AM frequencies are given in kilohertz (kHz). Find the wavelength of these radio waves in meters (m), nanometers (nm), and angstroms (Å).

7.8 An FM station broadcasts classical music at 93.5 MHz (megahertz, or 10^6 Hz). Find the wavelength (in m, nm, and Å) of these radio waves.

7.9 A radio wave has a frequency of 3.6×10^{10} Hz. What is the energy (in J) of one photon of this radiation?

7.10 An x-ray has a wavelength of 1.3 Å. Calculate the energy (in J) of one photon of this radiation.

7.11 Rank the following photons in terms of increasing energy: (a) blue ($\lambda = 453$ nm); (b) red ($\lambda = 660$ nm); (c) yellow ($\lambda = 595$ nm).

7.12 Rank the following photons in terms of decreasing energy: (a) IR ($\nu = 6.5 \times 10^{13}$ s^{-1}); (b) microwave ($\nu = 9.8 \times 10^{11}$ s^{-1}); (c) UV ($\nu = 8.0 \times 10^{15}$ s^{-1}).

● **Problems in Context**

7.13 Police often monitor traffic with "K-band" radar guns, which operate in the microwave region at 22.235 GHz (1 GHz = 10^9 Hz). Find the wavelength (in nm and Å) of this radiation.

7.14 Covalent bonds in a molecule absorb radiation in the IR region and vibrate at characteristic frequencies.

(a) The C—O bond in an organic compound absorbs radiation of wavelength 9.6 μm. What frequency (in s^{-1}) corresponds to that wavelength?
(b) The H—Cl bond has a frequency of vibration of 8.652×10^{13} Hz. What wavelength (in μm) corresponds to that frequency?

7.15 Cobalt-60 is a radioactive isotope used to treat cancers of the brain and other tissues. A gamma ray emitted by an atom of this isotope has an energy of 1.33 MeV (million electron volts; 1 eV = 1.602×10^{-19} J). What is the frequency (in Hz) and wavelength (in m) of this gamma ray?

7.16 (a) The first step in the formation of ozone in the upper atmosphere occurs when oxygen molecules absorb UV radiation of wavelengths ≤ 242 nm. Calculate the frequency and energy of the least energetic of these photons.
(b) Ozone absorbs light having wavelengths of 2200 to 2900 Å, thus protecting organisms on the Earth's surface from this high-energy UV radiation. What are the frequency and energy of the most energetic of these photons?

Atomic Spectra

● **Concept Review Questions**

7.17 How is n_1 in the Rydberg equation (Equation 7.3) related to the quantum number n in the Bohr model?

7.18 How would a planetary (Solar System) model of the atom differ from a key assumption in Bohr's model? What was the theoretical basis from which Bohr made this assumption?

7.19 Distinguish between an absorption spectrum and an emission spectrum. With which did Bohr work?

7.20 Which of these electron transitions correspond to absorption of energy and which to emission?
(a) $n = 2$ to $n = 4$ (b) $n = 3$ to $n = 1$
(c) $n = 5$ to $n = 2$ (d) $n = 3$ to $n = 4$

7.21 Why could the Bohr model not predict line spectra for atoms other than hydrogen?

7.22 The H atom and Be^{3+} ion each have one electron. Does the Bohr model predict their spectra accurately? Would you expect their line spectra to be identical? Explain.

● **Skill-Building Exercises *(paired)***

7.23 Use the Rydberg equation (Equation 7.3) to calculate the wavelength (in nm) of the photon emitted when a hydrogen atom undergoes a transition from $n = 5$ to $n = 2$.

7.24 Use the Rydberg equation to calculate the wavelength (in Å) of the photon absorbed when a hydrogen atom undergoes a transition from $n = 1$ to $n = 3$.

7.25 What is the wavelength (in nm) of the least energetic spectral line in the infrared series of the H atom?

7.26 What is the wavelength (in nm) of the least energetic spectral line in the visible series of the H atom?

7.27 Calculate the energy difference (ΔE) for the transition in Problem 7.23 for 1 mol of H atoms.

7.28 Calculate the energy difference (ΔE) for the transition in Problem 7.24 for 1 mol of H atoms.

7.29 Arrange the following H atom electron transitions in order of *increasing* frequency of the photon absorbed or emitted:
(a) $n = 2$ to $n = 4$ (b) $n = 2$ to $n = 1$
(c) $n = 2$ to $n = 5$ (d) $n = 4$ to $n = 3$

7.30 Arrange the following H atom electron transitions in order of *decreasing* wavelength of the photon absorbed or emitted:
(a) $n = 2$ to $n = \infty$ (b) $n = 4$ to $n = 20$
(c) $n = 3$ to $n = 10$ (d) $n = 2$ to $n = 1$

7.31 The electron in a ground-state H atom absorbs a photon of wavelength 97.20 nm. To what energy level does the electron move?

7.32 An electron in the $n = 5$ level of an H atom emits a photon of wavelength 1281 nm. To what energy level does the electron move?

● **Problems in Context**

7.33 In addition to continuous radiation, fluorescent lamps emit sharp lines in the visible region from a mercury discharge within the tube. Much of this light has a wavelength of 436 nm. What is the energy (in J) of one photon of this light?

7.34 The oxidizing agents used in most fireworks consist of potassium salts, such as $KClO_4$ or $KClO_3$, rather than the corresponding sodium salts. One of the problems with using sodium salts is their extremely intense yellow-orange emission at 589 nm, which obscures other colors in the display. What is the energy (in J) of one photon of this light? What is the energy (in kJ) of 1 einstein of this light (1 einstein = 1 mol of photons)?

The Wave-Particle Duality of Matter and Energy
(Sample Problems 7.3 and 7.4)

● **Concept Review Questions**

7.35 In what sense is the wave motion of a guitar string analogous to the motion of an electron in an atom?

7.36 What experimental support did de Broglie's concept receive?

7.37 If particles have wavelike motion, why don't we observe that motion in the macroscopic world?

7.38 Why can't we overcome the uncertainty predicted by Heisenberg's principle by building more precise devices to reduce the error in measurements below the $h/4\pi$ limit?

● **Skill-Building Exercises (paired)**

7.39 A 220-lb fullback runs the 40-yd dash at a speed of 19.6 ± 0.1 mi/h.
(a) What is his de Broglie wavelength (in meters)?
(b) What is the uncertainty in his position?

7.40 An alpha particle (mass = 6.6×10^{-24} g) emitted by radium travels at $3.4\times10^7 \pm 0.1\times10^7$ mi/h.
(a) What is its de Broglie wavelength (in meters)?
(b) What is the uncertainty in its position?

7.41 How fast must a 56.5-g tennis ball travel in order to have a de Broglie wavelength that is equal to that of a photon of green light (5400 Å)?

7.42 How fast must a 142-g baseball travel in order to have a de Broglie wavelength that is equal to that of an x-ray photon with $\lambda = 100.$ pm?

7.43 The sodium flame test has a characteristic yellow color due to emissions of wavelength 589 nm. What is the mass equivalence of one photon of this wavelength (1 J = 1 $kg \cdot m^2/s^2$)?

7.44 The lithium flame test has a characteristic red color due to emissions of wavelength 671 nm. What is the mass equivalence of 1 mol of photons of this wavelength (1 J = 1 $kg \cdot m^2/s^2$)?

The Quantum-Mechanical Model of the Atom
(Sample Problems 7.5 to 7.7)

● **Concept Review Questions**

7.45 What physical meaning is attributed to the square of the wave function, ψ^2?

7.46 Explain in your own words what the "electron density" in a particular tiny volume of space means.

7.47 Explain in your own words what it means for the peak in the radial probability distribution plot for the $n = 1$ level of a hydrogen atom to be at 0.529 Å. Is the probability of finding an electron at 0.529 Å from the nucleus greater for the $1s$ or the $2s$ orbital?

7.48 What feature of an orbital is related to each of the following?
(a) Principal quantum number (n)
(b) Angular momentum quantum number (l)
(c) Magnetic quantum number (m_l)

● **Skill-Building Exercises (paired)**

7.49 How many orbitals in an atom can have each of the following designations: (a) $1s$; (b) $4d$; (c) $3p$; (d) $n = 3$?

7.50 How many orbitals in an atom can have each of the following designations: (a) $5f$; (b) $4p$; (c) $5d$; (d) $n = 2$?

7.51 Give all possible m_l values for orbitals that have each of the following: (a) $l = 2$; (b) $n = 1$; (c) $n = 4, l = 3$.

7.52 Give all possible m_l values for orbitals that have each of the following: (a) $l = 3$; (b) $n = 2$; (c) $n = 6, l = 1$.

7.53 Draw probability contours (with axes) for each of the following orbitals: (a) s; (b) p_x.

7.54 Draw probability contours (with axes) for each of the following orbitals: (a) p_z; (b) d_{xy}.

7.55 For each of the following, give the sublevel designation, the allowable m_l values, and the number of orbitals:
(a) $n = 4, l = 2$ (b) $n = 5, l = 1$ (c) $n = 6, l = 3$

7.56 For each of the following, give the sublevel designation, the allowable m_l values, and the number of orbitals:
(a) $n = 2, l = 0$ (b) $n = 3, l = 2$ (c) $n = 5, l = 1$

7.57 For each of the following sublevels, give the n and l values and the number of orbitals: (a) $5s$; (b) $3p$; (c) $4f$.

7.58 For each of the following sublevels, give the n and l values and the number of orbitals: (a) $6g$; (b) $4s$; (c) $3d$.

7.59 Are the following quantum number combinations allowed? If not, show two ways to correct them:
(a) $n = 2; l = 0; m_l = -1$ (b) $n = 4; l = 3; m_l = -1$
(c) $n = 3; l = 1; m_l = 0$ (d) $n = 5; l = 2; m_l = +3$

7.60 Are the following quantum number combinations allowed? If not, show two ways to correct them:
(a) $n = 1; l = 0; m_l = 0$ (b) $n = 2; l = 2; m_l = +1$
(c) $n = 7; l = 1; m_l = +2$ (d) $n = 3; l = 1; m_l = -2$

Comprehensive Problems
Problems with an asterisk (*) are more challenging.

7.61 The orange pigment in carrots and orange peel is mostly β-carotene, an organic compound that is insoluble in water but very soluble in benzene and chloroform. Describe an experimental method for determining the concentration of β-carotene in the oil expressed from orange peel.

7.62 The quantum-mechanical treatment of the hydrogen atom gives the energy, E, of the electron as a function of the principal quantum number, n:

$$E = -\frac{h^2}{8\pi^2 m_e a_0^2 n^2} \qquad (n = 1, 2, 3, \ldots)$$

where h is Planck's constant, m_e is the electron mass, and a_0 is 52.92×10^{-12} m.
(a) Write the expression in the form $E = -(\text{constant})\frac{1}{n^2}$, evaluate the constant (in J), and compare it with the corresponding expression from Bohr's theory.
(b) Use the expression to find the energy change between $n = 2$ and $n = 3$.
(c) Calculate the wavelength of the photon that corresponds to this energy change. Is this photon seen in the hydrogen spectrum obtained from experiment (see Figure 7.8)?

7.63 The following figure illustrates the photoelectric effect graphically. It is a plot of the kinetic energies of electrons ejected from the surface of potassium metal or silver metal at different frequencies of incident light.

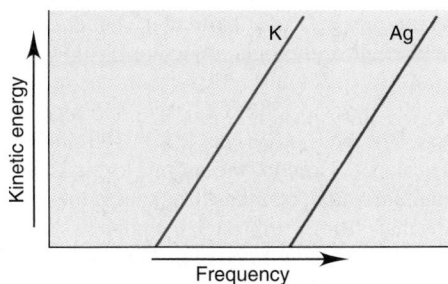

(a) Why don't the lines begin at the origin?
(b) Why don't the lines begin at the same point?
(c) Which metal requires light of shorter wavelength to eject an electron?
(d) Why are the slopes of the lines equal?

7.64 The human eye is a complex sensing device for visible light. The optic nerve needs a minimum of 2.0×10^{-17} J of energy to trigger a series of impulses that eventually reaches the brain.
(a) How many photons of red light (700. nm) are needed?
(b) How many photons of blue light (475 nm)?

7.65 One reason carbon monoxide (CO) is toxic is that it binds to the blood protein hemoglobin more strongly than oxygen does. The hemoglobin—CO bond absorbs radiation of 1953 cm^{-1}. (The units are the reciprocal of the wavelength in centimeters.) Calculate the wavelength (in nm and Å) and the frequency (in Hz) of the absorbed radiation.

7.66 A metal ion M^{n+} has a single electron. The highest energy line in its emission spectrum occurs at a frequency of 2.961×10^{16} Hz. Identify the ion.

7.67 TV and radio stations transmit in specific frequency bands of the radio region of the electromagnetic spectrum.
(a) TV channels 2 to 13 (VHF) broadcast signals between the frequencies of 59.5 and 215.8 MHz, whereas FM radio broadcasts signals with wavelengths between 2.78 and 3.41 m. Do these bands of signals overlap?
(b) AM radio broadcasts signals with frequencies between 550 and 1600 kHz. Which has a broader transmission band, AM or FM?

7.68 Compare the wavelengths of an electron (mass = 9.11×10^{-31} kg) and a proton (mass = 1.67×10^{-27} kg), each having (a) a speed of 3.0×10^6 m/s; (b) a kinetic energy of 2.5×10^{-15} J.

*7.69 Five lines in the H atom spectrum have the following wavelengths (in Å): (a) 1212.7; (b) 4340.5; (c) 4861.3; (d) 6562.8; (e) 10938. Three lines result from transitions to $n_{\text{final}} = 2$ (visible series). The other two result from transitions in different series, one with $n_{\text{final}} = 1$ and the other with $n_{\text{final}} = 3$. Identify n_{initial} for each line.

7.70 In his explanation of the threshold frequency in the photoelectric effect, Einstein reasoned that the absorbed photon must have the minimum energy required to dislodge an electron from the metal surface. This energy is called the *work function* (ϕ) of that metal. What is the longest wavelength of radiation (in nm) that could cause the photoelectric effect in each of these metals?
(a) Calcium, $\phi = 4.60 \times 10^{-19}$ J
(b) Titanium, $\phi = 6.94 \times 10^{-19}$ J
(c) Sodium, $\phi = 4.41 \times 10^{-19}$ J

7.71 You have three metal samples—A, B, and C—that are tantalum (Ta), barium (Ba), and tungsten (W), but you don't know which is which. Metal A emits electrons in response to visible light; metals B and C require UV light. (a) Identify metal A, and find the longest wavelength that removes an electron.
(b) What range of wavelengths would distinguish B and C? [The work functions are Ta (6.81×10^{-19} J), Ba (4.30×10^{-19} J), and W (7.16×10^{-19} J); work function is explained in Problem 7.70.]

7.72 Refractometry is an analytical method based on the difference between the speed of light as it travels through a substance (v) and its speed in a vacuum (c). In the procedure, light of known wavelength passes through a fixed thickness of the substance at a known temperature. The index of refraction equals c/v. Using yellow light ($\lambda = 589$ nm) at 20°C, for example, the index of refraction of water is 1.33 and that of diamond is 2.42. Calculate the speed of light in (a) water and (b) diamond.

7.73 Carbon-carbon bonds form the "backbone" of nearly every organic and biological molecule. The average bond energy of the C—C bond is 347 kJ/mol. Calculate the frequency and wavelength of the least energetic photon that can break this bond. In what region of the electromagnetic spectrum is this radiation?

7.74 Imagining a world in which some physical constants have radically different values can often be fascinating. For example, suppose Planck's constant were 1.13×10^{-9} J·s, and a single photon of light had enough energy to vaporize a cup of coffee (6.00×10^5 J). What is the maximum wavelength of this photon, and what color would it be?

7.75 The following quantum number combinations are not allowed. Assuming the n and m_l values are correct, change the l value to create an allowable combination:
(a) $n = 3$; $l = 0$; $m_l = -1$ (b) $n = 3$; $l = 3$; $m_l = +1$
(c) $n = 7$; $l = 2$; $m_l = +3$ (d) $n = 4$; $l = 1$; $m_l = -2$

*7.76 A ground-state H atom absorbs a photon of wavelength 94.91 nm, and its electron attains a higher energy level. The atom then emits two photons: one of wavelength 1281 nm to reach an intermediate level, and a second to return to the ground state.
(a) What higher level did the electron reach?
(b) What intermediate level did the electron reach?
(c) What was the wavelength of the second photon emitted?

*7.77 Consider these ground-state ionization energies of one-electron species:

H $= 1.31\times10^3$ kJ/mol
$He^+ = 5.24\times10^3$ kJ/mol
$Li^{2+} = 1.18\times10^4$ kJ/mol

(a) Write a general expression for the ionization energy of any one-electron species. (b) Use your expression to calculate the ionization energy of B^{4+}. (c) What is the minimum wavelength required to remove the electron from the $n = 3$ level of He^+? (d) What is the minimum wavelength required to remove the electron from the $n = 2$ level of Be^{3+}?

*7.78 Why do the spaces between spectral lines within a series decrease as the wavelength becomes shorter?

7.79 In the course of developing his model, Bohr arrived at the following formula for the radius of the electron's orbit: $r_n = n^2h^2\epsilon_0/\pi m_e e^2$, where m_e is the electron mass, e is its charge, and ϵ_0 is a constant related to charge attraction in a vacuum. Given that $m_e = 9.109\times10^{-31}$ kg, $e = 1.602\times10^{-19}$ C, and $\epsilon_0 = 8.854\times10^{-12}$ C^2/J·m, calculate the following:
(a) The radius of the 1st ($n = 1$) orbit in the H atom
(b) The radius of the 10th ($n = 10$) orbit in the H atom

*7.80 (a) Calculate the Bohr radius of an electron in the $n = 3$ orbit of a hydrogen atom. (See Problem 7.79.)
(b) What is the energy (in J) of the atom in part (a)?
(c) What is the energy of an Li^{2+} ion when its electron is in the $n = 3$ orbit?
(d) Why are the answers to parts (b) and (c) different?

7.81 Enormous numbers of microwave photons are needed to warm macroscopic samples of matter. A portion of soup containing 252 g of water is heated in a microwave oven from 20.°C to 98°C, with radiation of wavelength 1.55×10^{-2} m. How many photons are absorbed by the water in the soup?

*7.82 The quantum-mechanical treatment of the hydrogen atom gives an expression for the wave function, ψ, of the 1s orbital:

$$\psi = \frac{1}{\sqrt{\pi}}\left(\frac{1}{a_0}\right)^{\frac{3}{2}}e^{-r/a_0}$$

where r is the distance from the nucleus and a_0 is 52.92 pm. The electron density is the probability of finding the electron at a particular point (or tiny volume) at distance r from the nucleus and is proportional to ψ^2. The radial probability is the total probability of finding the electron at all points at distance r from the nucleus and is proportional to $4\pi r^2\psi^2$. Calculate the values (to three significant figures) of ψ, ψ^2, and $4\pi r^2\psi^2$ to fill in the following table, and sketch plots of these quantities versus r. Compare the latter two plots with those in Figure 7.17A:

r (pm)	ψ ($pm^{-\frac{3}{2}}$)	ψ^2 (pm^{-3})	$4\pi r^2\psi^2$ (pm^{-1})
0			
50			
100			
200			

*7.83 Lines in one spectral series can overlap lines in another.
(a) Use the Rydberg equation to show whether the range of wavelengths in the $n_1 = 1$ series overlaps with the range in the $n_1 = 2$ series.

(b) Use the Rydberg equation to show whether the range of wavelengths in the $n_1 = 3$ series overlaps with the range in the $n_1 = 4$ series.
(c) How many lines in the $n_1 = 4$ series lie in the range of the $n_1 = 5$ series?
(d) What does this overlap imply about the hydrogen spectrum at longer wavelengths?

*7.84 The following values are the only allowable energy levels of a hypothetical one-electron atom:

$E_6 = -2\times10^{-19}$ J	$E_5 = -7\times10^{-19}$ J
$E_4 = -11\times10^{-19}$ J	$E_3 = -15\times10^{-19}$ J
$E_2 = -17\times10^{-19}$ J	$E_1 = -20\times10^{-19}$ J

(a) If the electron were in the $n = 3$ level, what would be the highest frequency (and minimum wavelength) of radiation that could be emitted?
(b) What is the ionization energy (in kJ/mol) of the atom in its ground state?
(c) If the electron were in the $n = 4$ level, what would be the shortest wavelength (in nm) of radiation that could be absorbed without causing ionization?

7.85 In fireworks displays, light of a characteristic wavelength is related to the presence of a particular element. What are the frequency and color of the light associated with each of the following?
(a) Li^+, $\lambda = 671$ nm
(b) Cs^+, $\lambda = 456$ nm
(c) Ca^{2+}, $\lambda = 649$ nm
(d) Na^+, $\lambda = 589$ nm

7.86 On the planet Switcho, the atomic orbitals have slightly different quantum numbers: n has the same values as on Earth, l has integral values from $-n$ through n, and m has integral values from 0 through the absolute value of l. (a) How many orbitals are there for $n = 3$ on Switcho? (b) What are the possible l values for $n = 4$? (c) What are the possible m values for $n = 5$, $l = -5$?

7.87 Visible light provides green plants with the energy needed to drive photosynthesis. Horticulturists know that, for many plants, leaf color depends on how brightly lit the growing area is: dark green leaves are associated with low light levels, and pale green with high levels. (a) Use the photon theory of light to explain why a plant adapts this way. (b) What change in leaf composition might account for this behavior?

*7.88 In compliance with conservation of energy, Einstein explained that in the photoelectric effect, the energy of a photon ($h\nu$) absorbed by a metal is the sum of the work function (ϕ), the minimum energy needed to dislodge an electron from the metal's surface, and the kinetic energy (E_k) of the electron: $h\nu = \phi + E_k$. When light of wavelength 358.1 nm falls on the surface of potassium metal, the speed (u) of the dislodged electron is 6.40×10^5 m/s.
(a) What is E_k ($\frac{1}{2}mu^2$) of the dislodged electron?
(b) What is ϕ (in J) of potassium?

7.89 An electron microscope focuses electrons through magnetic lenses to observe objects at higher magnification than is possible with a light microscope. For any microscope, the smallest object that can be observed is one-half the wavelength of the light used. Thus, for example, the smallest object that can be observed with light of 400 nm is 2×10^{-7} m. (a) What is the smallest object observable with an electron microscope using electrons moving at 5.5×10^4 m/s? (b) At 3.0×10^7 m/s?

7.90 In a typical fireworks device, the heat of the reaction between a strong oxidizing agent, such as $KClO_4$, and an organic compound excites certain salts, which emit specific colors. Strontium salts have an intense emission at 641 nm, and barium salts have one at 493 nm. (a) What colors do these emissions produce? (b) What is the energy (in kJ) of these emissions for 1.00 g each of the chloride salts of Sr and Ba? (Assume that all the heat released is converted to light emitted.)

*7.91 Atomic hydrogen produces well-known series of spectral lines in several regions of the electromagnetic spectrum. Each series fits the Rydberg equation with its own particular n_1 value. Calculate the value of n_1 (by trial and error if necessary) that would produce a series of lines in which:
(a) The *highest* energy line has a wavelength of 3282 nm.
(b) The *lowest* energy line has a wavelength of 7460 nm.

7.92 Fish-liver oil is an excellent animal source of vitamin A. Its concentration is measured spectrophotometrically at a wavelength of 329 nm.
(a) Suggest a reason for using this wavelength.
(b) In what region of the spectrum does this wavelength lie?
(c) When 0.1232 g of fish-liver oil is dissolved in 500. mL of solvent, the absorbance is 0.724 units. When 1.67×10^{-3} g of vitamin A is dissolved in 250. mL of solvent, the absorbance is 1.018 units. Calculate the vitamin A concentration in the fish-liver oil.

7.93 Many calculators use photocells to provide their energy. Find the maximum wavelength needed to remove an electron from silver ($\phi = 7.59 \times 10^{-19}$ J). Is silver is a good choice for a photocell that uses visible light?

7.94 In a game of "Clue," Ms. White is killed in the conservatory with a lead pipe. You have rigged up a device in each room to help you find the murderer—a spectrometer that emits the entire visible spectrum to indicate who is in that room. For example, if someone wearing yellow is in a room, light at 580 nm is reflected. The suspects are Col. Mustard, Prof. Plum, Mr. Green, Ms. Peacock (blue), and Ms. Scarlet. At the time of the murder, the spectrometer in the dining room recorded a reflection at 520 nm, those in the lounge and study recorded reflections of lower frequencies, and the one in the library recorded a reflection of the shortest possible wavelength. Who killed Ms. White? Explain.

7.95 In the 1998 Winter Olympics, Claudia Pechstein of Germany won the 5000-m speed skating competition with a time of 6 min 59.61 s. (a) Assuming the rightmost digit in the distance and time values has an uncertainty of one unit, what is her speed in mi/h? (b) Assuming her racing weight is 134 lbs, what is her de Broglie wavelength? (c) What is the uncertainty in her position as she crossed the finish line?

7.96 Electric power is typically stated in units of watts (1 W = 1 J/s). About 95% of the power output of an incandescent bulb is converted to heat and 5% to light. If 10% of that light shines on your chemistry text, how many photons per second shine on the book from a 75-W bulb? (Assume the photons have a wavelength of 550 nm.)

7.97 The flame test for sodium is based on its very intense emission at 589 nm, and the test for potassium is based on its emission at 404 nm. When both elements are present, the Na^+ emission is so strong that the K^+ emission can't be seen, except by looking through a cobalt-glass filter. (a) What are the colors of these Na^+ and K^+ emissions? (b) What does the cobalt-glass filter do?

*7.98 Only certain electron transitions are allowed from one energy level to another. In one-electron species, the change in the angular-momentum quantum number, l, of an allowed transition must be ± 1. For example, a $3p$ electron can drop directly to a $2s$ orbital but not to a $2p$. Thus, in the UV series, where $n_{final} = 1$, allowed electron transitions can start in a p orbital ($l = 1$) of $n = 2$ or higher, not in an s ($l = 0$) or d ($l = 2$) orbital of $n = 2$ or higher. From what orbital do each of the allowed electron transitions start for the first four emission lines in the visible series ($n_{final} = 2$)?

7.99 The use of phosphate compounds in detergents and their subsequent environmental discharge has led to serious imbalances in the natural life cycle of freshwater lakes. A chemist studying water pollution used a spectrophotometric method to measure total phosphate and obtained the following data for known standards:

Absorbance (400 nm)	Concentration (mol/L)
0	0.0
0.10	2.5×10^{-5}
0.16	3.2×10^{-5}
0.20	4.4×10^{-5}
0.25	5.6×10^{-5}
0.38	8.4×10^{-5}
0.48	10.5×10^{-5}
0.62	13.8×10^{-5}
0.76	17.0×10^{-5}
0.88	19.4×10^{-5}

(a) Draw a curve plotting absorbance at 400 nm vs. phosphate concentration.
(b) If a sample of lake water has an absorbance of 0.55, what is its phosphate concentration?

CHAPTER 8

ELECTRON CONFIGURATION AND CHEMICAL PERIODICITY

CHAPTER OUTLINE

8.1 Development of the Periodic Table

8.2 Characteristics of Many-Electron Atoms
- The Electron-Spin Quantum Number
- The Exclusion Principle
- Electrostatic Effects and the Splitting of Energy Levels

8.3 The Quantum-Mechanical Model and the Periodic Table
- Building Up Periods 1 and 2
- Building Up Period 3
- Electron Configurations Within Groups
- Building Up Period 4
- General Principles of Electron Configurations
- Complex Patterns

8.4 Trends in Some Key Periodic Atomic Properties
- Trends in Atomic Size
- Trends in Ionization Energy
- Trends in Electron Affinity

8.5 The Connection Between Atomic Structure and Chemical Reactivity
- Trends in Metallic Behavior
- Properties of Monatomic Ions

Figure: Patterns in nature. With the regularity of a heartbeat, a solar eclipse, or, as shown here, the arrangement of seeds in the head of a sunflower (*Helianthus annuus*), nature exhibits periodic patterns. As you'll see in this chapter, the arrangement of electrons in atoms recurs periodically, too, which allows us to explain why many properties of the elements recur periodically and to predict physical and chemical behavior.

In Chapter 7, you saw how an outpouring of scientific creativity by early 20th-century physicists led to a new understanding of matter and energy, which in turn led to the quantum-mechanical model of the atom. But you can be sure that late 19th-century chemists were not sitting idly by, waiting for their colleagues in physics to develop that model. They were exploring the nature of electrolytes, establishing the kinetic-molecular theory, and developing chemical thermodynamics. The fields of organic chemistry and biochemistry were born, as were the fertilizer, explosives, glassmaking, soapmaking, bleaching, and dyestuff industries. And, for the first time, chemistry became a university subject in Europe and America. Superimposed on this activity was the accumulation of an enormous body of facts about the elements, which became organized into the periodic table.

The goal of this chapter is to show how the organization of the table, condensed from countless hours of laboratory work, was explained perfectly by the new quantum-mechanical atomic model. This model answers one of the central questions in chemistry: why do the elements behave as they do? Or, rephrasing the question to fit the main topic of this chapter: how does the **electron configuration** of an element—*the distribution of electrons within the orbitals of its atoms*—relate to its chemical and physical properties?

First, we briefly discuss the historical development of the periodic table. Then, we take up where Chapter 7 left off. There, we focused on applying the quantum-mechanical model to the H atom. Here, we extend it to *many-electron atoms,* those with more than one electron. We'll see how the model defines a unique set of quantum numbers for every electron in the atoms of every element. We consider the order of filling orbitals with electrons and see how it correlates with the order of elements in the periodic table. All this leads to an understanding of how electron configuration and nuclear charge give rise to trends in atomic properties and how these trends account for the periodic patterns of chemical reactivity. Finally, we apply this understanding to a discussion of metals, nonmetals, and their ions.

8.1 DEVELOPMENT OF THE PERIODIC TABLE

An essential requirement for the amazing growth in theoretical and practical chemistry in the second half of the 19th century was the ability to organize the facts known about element behavior. The earliest organizing attempt was made by Johann Döbereiner, who placed groups of three elements with similar properties, such as calcium, strontium, and barium, into "triads." Later, John Newlands noted similarities between every eighth element (arranged by atomic mass), like the similarity between every eighth note in the musical scale, and placed elements into "octaves." As more elements were discovered, however, these early numerical schemes lost much of their validity.

In Chapter 2, you saw that the most successful organizing scheme was made by the Russian chemist Dmitri Mendeleev. In 1870, he arranged the 65 elements then known into a *periodic table* and summarized their behavior in the **periodic law:** when arranged by atomic mass, the elements exhibit a periodic recurrence of similar properties. It is a curious quirk of history that Mendeleev and the German chemist Julius Lothar Meyer arrived at virtually the same organization simultaneously, yet independently. Mendeleev focused on chemical properties and Meyer on physical properties. The greater credit has gone to Mendeleev because he was able to *predict* the properties of several as-yet-undiscovered elements, for which he had left blank spaces in his table. Table 8.1 (on the next page) compares the actual properties of germanium, which Mendeleev gave the provisional name "eka silicon" ("first under silicon"), with his predictions for it.

Mendeleev's Great Contribution Born in a small Siberian town, Dmitri Ivanovich Mendeleev was the youngest of 17 children of the local schoolteacher. He showed early talent for mathematics and science, so his mother took him to St. Petersburg, where he remained to study and work for much of his life. Early research interests centered on the physical properties of gases and liquids, and later he was consulted frequently on the industrial processing of petroleum. In developing his periodic table, he prepared a note card on the properties of each element and arranged and rearranged them until he realized that properties repeated when the elements were placed in order of increasing atomic mass.

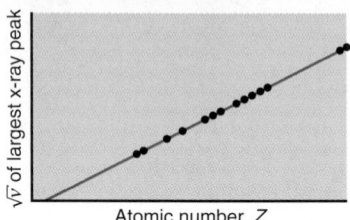

Table 8.1 Mendeleev's Predicted Properties of Germanium ("eka Silicon") and Its Actual Properties

Property	Predicted Properties of eka Silicon (E)	Actual Properties of Germanium (Ge)
Atomic mass	72 amu	72.61 amu
Appearance	Gray metal	Gray metal
Density	5.5 g/cm^3	5.32 g/cm^3
Molar volume	13 cm^3/mol	13.65 cm^3/mol
Specific heat capacity	0.31 J/g·K	0.32 J/g·K
Oxide formula	EO$_2$	GeO$_2$
Oxide density	4.7 g/cm^3	4.23 g/cm^3
Sulfide formula and solubility	ES$_2$; insoluble in H$_2$O; soluble in aqueous (NH$_4$)$_2$S	GeS$_2$; insoluble in H$_2$O; soluble in aqueous (NH$_4$)$_2$S
Chloride formula (boiling point)	ECl$_4$ ($<$100°C)	GeCl$_4$ (84°C)
Chloride density	1.9 g/cm^3	1.844 g/cm^3
Element preparation	Reduction of K$_2$EF$_6$ with sodium	Reduction of K$_2$GeF$_6$ with sodium

Moseley and Atomic Number

When a metal is bombarded with high-energy electrons, an inner electron is knocked from the atom, an outer electron moves down to fill in the space, and x-rays are emitted. Bohr had proposed that the x-ray spectrum of an element had wavelengths proportional to the nuclear charge. In 1913, Moseley studied the x-ray spectra of a series of metals and was able to correlate the largest peak in the x-ray spectrum of each metal with its order in the periodic table (its *atomic number*). He showed the nuclear charge increased by one unit for each element (see the graph above). Among other results, the findings confirmed the placement of Co ($Z = 27$) *before* Ni ($Z = 28$), despite cobalt's higher atomic mass, and also confirmed that the gap between Cl ($Z = 17$) and K ($Z = 19$) is the place for Ar ($Z = 18$). Tragically, Moseley died the next year at the age of 26 as a pilot in World War I.

Today's periodic table, which appears on the inside front cover of the text, resembles Mendeleev's in most details, although it includes 49 elements that were unknown in 1870. The only substantive change is that the elements are now arranged in order of *atomic number* (number of protons) rather than atomic mass. This change was based on the work of the British physicist Henry G. J. Moseley, who found a direct dependence between an element's nuclear charge and its position in the periodic table.

8.2 CHARACTERISTICS OF MANY-ELECTRON ATOMS

Like the Bohr model, the Schrödinger equation does not give *exact* solutions for many-electron atoms. However, unlike the Bohr model, the Schrödinger equation gives very good *approximate* solutions. These solutions show that the atomic orbitals of many-electron atoms are *hydrogen-like;* that is, they resemble those of the H atom. This conclusion means we can use the same quantum numbers that we used for the H atom to describe the orbitals of other atoms.

Nevertheless, the existence of more than one electron in an atom requires us to consider three features that were not relevant in the case of hydrogen: (1) the need for a fourth quantum number, (2) a limit on the number of electrons allowed in a given orbital, and (3) a more complex set of orbital energy levels. Let's examine these new features and then go on to determine the electron configuration for each element.

The Electron-Spin Quantum Number

Recall from Chapter 7 that the three quantum numbers n, l, and m_l describe the size (energy), shape, and orientation, respectively, of an atomic orbital. However, an additional quantum number is needed to describe a property of the electron itself, called *spin*, which is not a property of the orbital. Electron spin becomes important when more than one electron is present.

When a beam of H atoms passes through a nonuniform magnetic field, as shown in Figure 8.1, it splits into two beams that bend away from each other.

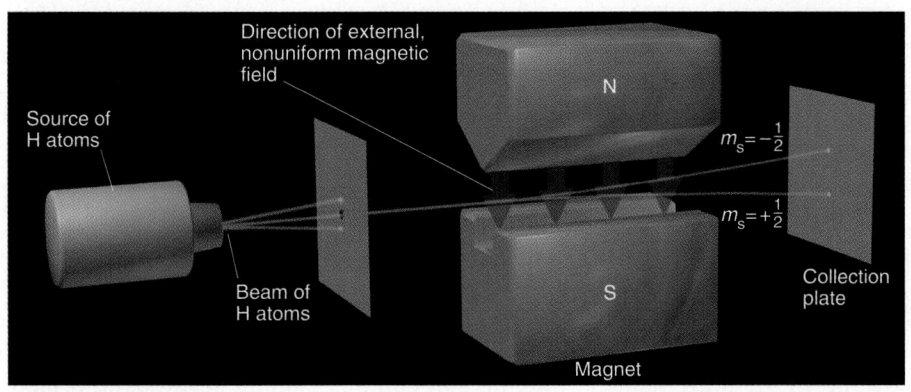

Source of
H atoms

Direction of external,
nonuniform magnetic
field

N

$m_s = -\frac{1}{2}$

$m_s = +\frac{1}{2}$

Beam of
H atoms

S

Collection
plate

Magnet

Figure 8.1 Observing the effect of electron spin. A nonuniform magnetic field, created by magnet faces with different shapes, splits a beam of hydrogen atoms in two. The split beam results from the two possible directions for the electron's spin within each atom.

The explanation of the split beam is that the electron generates a tiny magnetic field, as though it were a spinning charge. The single electron in each H atom can spin in one of two opposite directions, each of which generates a magnetic field. The opposing directions of these fields cause half of the electrons to be attracted into the region of the large external magnetic field and the other half to be repelled; so the beam of H atoms splits.

The **spin quantum number (m_s)** indicates the direction of the electron spin about its own axis and can have one of two possible values, $+\frac{1}{2}$ or $-\frac{1}{2}$. As a result, *each electron in an atom is described completely by a set of **four** quantum numbers: the first three describe its orbital, and the fourth describes its spin.* The quantum numbers are summarized in Table 8.2.

Table 8.2 Summary of Quantum Numbers of Electrons in Atoms

Name	Symbol	Permitted Values	Property
Principal	n	Positive integers (1, 2, 3, etc.)	Orbital energy (size)
Angular momentum	l	Integers from 0 to $n-1$	Orbital shape (The l values 0, 1, 2, and 3 correspond to s, p, d, and f orbitals, respectively.)
Magnetic	m_l	Integers from $-l$ to 0 to $+l$	Orbital orientation
Spin	m_s	$+\frac{1}{2}$ or $-\frac{1}{2}$	Direction of e^- spin

Now we can write a set of four quantum numbers for any electron in the ground state of any atom. For example, the set of quantum numbers for the lone electron in hydrogen (H; $Z = 1$) is $n = 1$, $l = 0$, $m_l = 0$, and $m_s = +\frac{1}{2}$. (The spin quantum number for this electron could just as well have been $-\frac{1}{2}$, but by convention, we assign $+\frac{1}{2}$ for the first electron in an orbital.)

The Exclusion Principle

The element after hydrogen is helium (He; $Z = 2$), the first with atoms having more than one electron. The first electron in the He ground state has the same set of quantum numbers as the electron in the H atom, but the second He electron does not. Based on observations of the excited states of atoms, the Austrian physicist Wolfgang Pauli formulated the **exclusion principle:** *no two electrons in the same atom can have the same four quantum numbers.* That is, each electron must have a unique "identity" as expressed by its set of quantum numbers. Therefore, the second He electron occupies the same orbital as the first but has an opposite spin: $n = 1$, $l = 0$, $m_l = 0$, and $m_s = -\frac{1}{2}$.

Baseball Quantum Numbers The unique set of quantum numbers that describes an electron is analogous to the unique location of a box seat at a baseball game. The stadium (atom) is divided into section (n, level), box (l, sublevel), row (m_l, orbital), and seat (m_s, spin). Only one person (electron) can have this particular set of stadium "quantum numbers."

Because the spin quantum number (m_s) can have only two values, the major consequence of the exclusion principle is that *an atomic orbital can hold a maximum of two electrons and they must have opposing spins.* We say that the 1*s* orbital in He is *filled* and that the electrons have *paired spins.* Thus, a beam of He atoms is not split in an experiment like that in Figure 8.1.

Electrostatic Effects and the Splitting of Energy Levels

Electrostatic effects play a major role in determining the energy states of many-electron atoms. These effects lead to a more complex set of energy states than exists in the H atom and to a splitting of energy levels. To see this result, let's move on to the third element, lithium, and continue adding electrons to orbitals.

The first two electrons in the ground state of lithium (Li; $Z = 3$) have the same two sets of quantum numbers as the electrons in He, so the 1*s* orbital of Li is filled. By definition, *the electrons of an atom in its ground state occupy the orbitals of lowest energy.* Therefore, with the lowest energy (1*s*) orbital filled, the third Li electron must go into the orbital of next-lowest energy—the 2*s* orbital. The set of quantum numbers for this electron are $n = 2$, $l = 0$, $m_l = 0$, and $m_s = +\frac{1}{2}$. But the $n = 2$ energy level has 2*s and* 2*p* orbitals, so why is the 2*s* orbital lower in energy than the 2*p*?

Recall the end of Chapter 7, where we noted that the energy state of the H atom is determined *only* by the *n* value of the occupied orbital. All sublevels of a given level, such as the 2*s* and 2*p*, have the same energy because the only electrostatic interaction is the nucleus-electron attraction. On the other hand, the energy states of many-electron atoms arise from nucleus-electron attractions *and* electron-electron repulsions. One major consequence of these additional electrostatic interactions is *the splitting of energy levels into sublevels of differing energies: the energy of an orbital in a many-electron atom depends mostly on its n value (size) and somewhat on its l value (shape).*

Evidence for the splitting of energy levels is seen in the complex line spectra of many-electron atoms. Figure 8.2 shows that even though helium has only one more electron than hydrogen, it displays many more spectral lines, indicating more available orbital energies in its excited states.

Two factors essential to understanding energy-level splitting are described by Coulomb's law (Section 2.7):

1. The farther apart opposite charges are, the weaker the attraction is. When nucleus and electron are far apart, the energy is higher (the system is less stable) than when they are close together. (See also Figure 1.3C, page 7.)
2. The higher opposite charges are, the stronger the attraction is. When a nucleus of higher charge attracts an electron, the energy is lower (the system is more stable) than when a nucleus of lower charge attracts an electron.

Let's apply these factors, together with the basic rule that opposite charges attract and like charges repel, to see why energy levels split and, thus, why the 2*s* orbital is lower in energy than the 2*p*.

Figure 8.2 Spectral evidence of energy-level splitting in many-electron atoms. More spectral lines for He than for H indicate that an He atom has more available orbital energies. This evidence is consistent with energy levels splitting into sublevels in many-electron atoms.

Chemists measure an orbital's energy in terms of the energy needed to remove an electron from that orbital in an atom (expressed as the energy, in kJ/mol, needed to remove 1 mol of electrons from 1 mol of atoms). It takes more energy to remove an electron in a more stable (lower energy) orbital than it does to remove one in a less stable (higher energy) orbital. Recall that, by definition, an atom's energy has a negative value. The more stable an orbital, the lower (more negative) its energy is.

Now we'll compare several atomic systems to isolate the effect on orbital energy of *nuclear charge, electron repulsions,* and *orbital shape.*

The Effect of Nuclear Charge (Z) on Orbital Energy To isolate the effect of nuclear charge, let's compare the H atom and the He$^+$ ion. Both of these atomic systems have one electron in the $1s$ orbital, but they have different nuclear charges—the H atom has a 1+ charge, and the He$^+$ ion has a 2+ charge (Figure 8.3). The orbital energies are

E of $1s$ in He$^+$ = −5250 kJ/mol and E of $1s$ in H = −1311 kJ/mol

As the more negative orbital energy shows, the He$^+$ electron is harder to remove because it lies in a more stable (lower energy) orbital, as a result of its stronger attraction to the nucleus. Thus, *higher nuclear charge lowers orbital energy (stabilizes the system) by increasing nucleus-electron attractions.*

Shielding: The Effect of Electron Repulsions on Orbital Energy Now let's consider the effect of additional electrons. We'll look separately at the effects of other electrons in the same orbital and in inner orbitals.

1. *Additional electron in the same orbital.* To isolate the effect of another electron in the same orbital, let's compare the He atom and the He$^+$ ion. Both atomic systems have a 2+ nuclear charge, but He has two electrons in the $1s$ orbital and He$^+$ has only one (Figure 8.4). The orbital energies are

E of $1s$ in He$^+$ = −5250 kJ/mol and E of $1s$ in He = −2372 kJ/mol

It takes about half as much energy to remove an electron from He as from He$^+$. Why does the second electron in the He atom make an electron so much easier to remove? Each electron "feels" not only the nuclear attraction but also the other electron's repulsion. This repulsion counteracts the attraction somewhat and makes each electron easier to remove, so the He $1s$ orbital is less stable (higher energy). You might think of it as one electron pushing away the other. Thus, *an additional electron raises the orbital energy through electron-electron repulsions.* It is almost as if each electron "shields" the other electron somewhat from the full nuclear charge. This **shielding** (or screening) reduces the full nuclear charge to an **effective nuclear charge (Z_{eff}),** *the nuclear charge an electron actually experiences.* (Other electrons in the same sublevel have a similar effect.) Thus, *shielding by other electrons makes an electron easier to remove.*

2. *Additional electrons in inner orbitals.* To isolate the effect of additional electrons in inner orbitals, the atomic systems we'll compare are a bit unusual—the ground-state Li atom and the first excited state of the Li^{2+} ion. Both have a 3+ nuclear charge. The Li atom has two inner ($1s$) electrons and one outer ($2s$) electron; the Li^{2+} ion has only one electron, which, in the first excited state, occupies the $2s$ orbital (Figure 8.5). Orbital energies are

E of $2s$ in Li^{2+} = −2954 kJ/mol and E of $2s$ in Li = −520 kJ/mol

It takes only about one-sixth as much energy to remove the $2s$ electron from the Li atom. Because the inner ($1s$) electrons spend nearly all their time *between* the outer ($2s$) electron and the nucleus, they shield the $2s$ electron from the nuclear attraction very effectively; this makes the $2s$ electron in the Li atom much easier to remove. Clearly, *inner electrons shield outer electrons more effectively than do electrons in the same sublevel.*

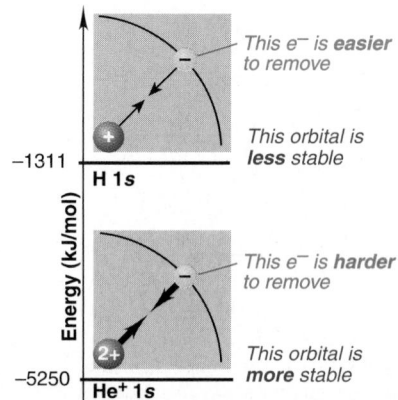

Figure 8.3 **The effect of nuclear charge.** Greater nuclear charge lowers orbital energy.

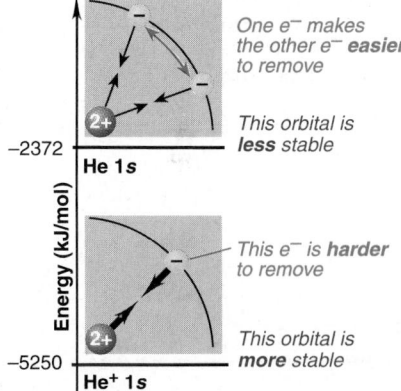

Figure 8.4 **The effect of another electron in the same orbital.** Each electron *shields* the other somewhat from the full nuclear charge and raises orbital energy.

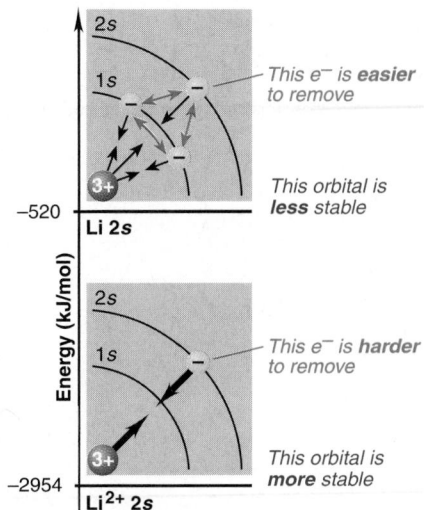

Figure 8.5 **The effect of other electrons in inner orbitals.** Inner electrons shield outer electrons very well and raise orbital energy greatly.

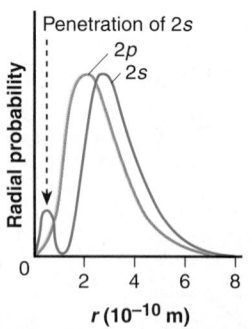

Figure 8.6 **The effect of orbital shape.** Radial probability distributions show that a 2s electron spends most of its time slightly farther from the nucleus than does a 2p electron but penetrates near the nucleus for a small fraction of the time. This penetration by the 2s electron increases its attraction to the nucleus and, thus, lowers the orbital energy.

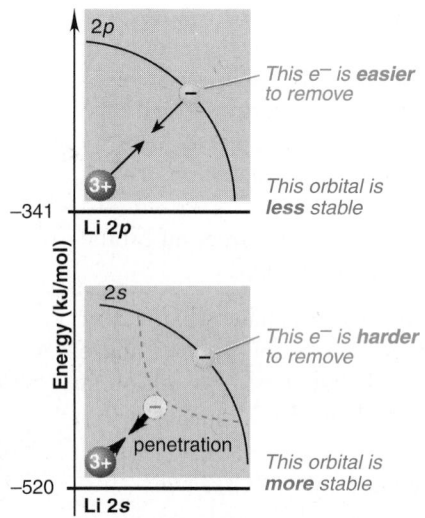

Penetration: The Effect of Orbital Shape on Orbital Energy To isolate the effect of orbital shape, we'll compare the ground state and first excited state of the Li atom. Both atomic systems are identical except that the outer electron occupies orbitals with different *l* values. In addition to the filled 1s orbital, the ground-state Li atom has an electron in the 2s orbital, whereas the Li atom in its first excited state has its third electron in a 2p orbital. The orbital energies are

$$E \text{ of } 2s \text{ in Li} = -520 \text{ kJ/mol}$$

$$E \text{ of } 2p \text{ in Li} = -341 \text{ kJ/mol}$$

Why is the 2s orbital more stable (lower energy)? At first, we might expect the 2s electron to be easier to remove because, as the graph in Figure 8.6 shows, a 2s electron (*blue curve*) is, on average, slightly farther from the nucleus than a 2p electron (*orange curve*). But note that a small portion of the 2s orbital's radial probability distribution appears very near the nucleus. In effect, the 2s electron spends part of its time "penetrating" very close to the nucleus. This **penetration** increases the overall attraction of the 2s electron to the nucleus and allows it to shield the 2p electron better. At the same time, the 2p electron is less able to shield the 2s. As a result, the 2p electron is easier to remove than the 2s.

In general, *penetration and the resulting effect on shielding cause an energy level to split into sublevels.* The lower the *l* value of an orbital, the more its electrons penetrate, and so the greater their attraction to the nucleus. Therefore, *for a given n value, the lower the l value is, the lower the sublevel energy:*

Order of sublevel energies: $s < p < d < f$

Thus, the 2s ($l = 0$) is lower in energy than the 2p ($l = 1$), the 3p ($l = 1$) is lower than the 3d ($l = 2$), and so forth.

Figure 8.7 shows the general energy order in which levels and sublevels are filled with electrons. Note how a given energy level (*n* value) is split into several sublevels (*l* values) of differing energies. Next, we use this energy order to fill orbitals with electrons and construct a periodic table of ground-state atoms.

SECTION SUMMARY

Differentiating electrons in many-electron atoms requires four quantum numbers: three (n, l, m_l) describe the orbitals, and a fourth (m_s) describes electron spin. The Pauli exclusion principle requires that each electron have a unique set of four quantum numbers; therefore, an orbital can hold no more than two electrons, and their spins must be paired (opposite). Electrostatic interactions determine orbital energies as follows:

1. Greater nuclear charge lowers orbital energy and makes electrons harder to remove.

2. Electron-electron repulsions raise orbital energy and make electrons easier to remove. Repulsions have the effect of *shielding* electrons from the full nuclear charge, reducing that charge to an effective nuclear charge, Z_{eff}. Inner electrons shield outer electrons most effectively.

3. Greater radial probability distribution near the nucleus (greater penetration) makes the electron harder to remove because it is attracted more strongly and shielded less effectively. As a result, an energy level (shell) is split into sublevels (subshells) with the energy order $s < p < d < f$.

8.3 THE QUANTUM-MECHANICAL MODEL AND THE PERIODIC TABLE

Quantum mechanics provides the theoretical foundation for the experimentally based periodic table. In this section, we fill the table with elements and determine their electron configurations—the distributions of electrons within their atoms' orbitals. Note especially the *recurring pattern in electron configurations, which is the basis for the recurring pattern in element behavior.*

Building Up Periods 1 and 2

A useful way to determine the electron configurations of the elements is to start at the beginning of the periodic table and add one electron per element to the *lowest energy orbital available.* (Of course, one proton and one or more neutrons are also added to the nucleus.) This approach is called the **aufbau principle** (German *aufbauen,* "to build up"), and it results in *ground-state* electron configurations. Let's assign sets of quantum numbers to the electrons in the ground state of the first 10 elements, those in the first two periods (horizontal rows).

For the electron in H, as you've seen, the set is

$$\text{H } (Z = 1): \quad n = 1, \; l = 0, \; m_l = 0, \; m_s = +\tfrac{1}{2}$$

You also saw that the first electron in He has the same set as the electron in H, but the second He electron has opposing spin (exclusion principle):

$$\text{He } (Z = 2): \quad n = 1, \; l = 0, \; m_l = 0, \; m_s = -\tfrac{1}{2}$$

When we present an element and its atomic number (Z) in this discussion, the quantum numbers that follow refer to the element's *last added* electron. [Of course, an atom is neutral, so the number of protons (Z) equals the *total* number of electrons.]

Before proceeding, here are two common ways to show the orbital occupancy:

1. *The electron configuration.* This shorthand notation consists of the principal energy level (*n* value), the letter designation of the sublevel (*l* value), and the number of electrons (#) in the sublevel, written as a superscript: $nl^{\#}$. The electron configuration of H is $1s^1$ (spoken "one-ess-one"); that of He is $1s^2$ (spoken "one-ess-two," *not* "one-ess-squared"). This notation does *not* indicate electron spin but assumes you know that the two $1s$ electrons have paired (opposite) spins.

2. *The orbital diagram.* An **orbital diagram** consists of a box (or circle, or just a line) for each orbital in a given energy level, grouped by sublevel, with an arrow indicating an electron's presence *and* its direction of spin. (Traditionally, ↑ is $+\tfrac{1}{2}$ and ↓ is $-\tfrac{1}{2}$, but these are arbitrary; it is necessary only to be consistent. Throughout the text, orbital occupancy is also indicated by color intensity: an orbital with no color is empty, pale color means half-filled, and full color indicates filled.) The electron configurations and orbital diagrams for the first two elements are

$$\text{H } (Z = 1) \; 1s^1 \; \boxed{\uparrow} \qquad \text{He } (Z = 2) \; 1s^2 \; \boxed{\uparrow\downarrow}$$
$$\qquad\qquad\qquad\;\; 1s \qquad\qquad\qquad\qquad\qquad 1s$$

The exclusion principle tells us that an orbital can hold only two electrons, so the $1s$ orbital in He is filled, and the $n = 1$ level is also filled. The $n = 2$ level is filled next, beginning with the $2s$ orbital, the next lowest in energy. As we said earlier, the first two electrons in Li fill the $1s$ orbital, and the last added Li electron is $n = 2, \; l = 0, \; m_l = 0, \; m_s = +\tfrac{1}{2}$. The electron configuration for Li is $1s^22s^1$. Note that the orbital diagram shows all the orbitals for $n = 2$, whether or not they are occupied:

$$\text{Li } (Z = 3) \; 1s^22s^1 \; \boxed{\uparrow\downarrow} \; \boxed{\uparrow} \; \boxed{\,|\,\,|\,}$$
$$\qquad\qquad\qquad\qquad\;\; 1s \quad\;\; 2s \qquad\;\; 2p$$

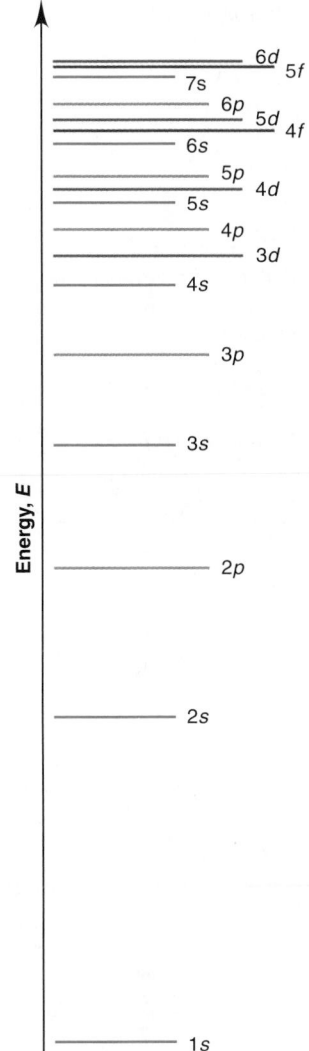

Figure 8.7 **Order for filling energy sublevels with electrons.** For filling many-electron atoms, the relative energies of sublevels increase with principal quantum number *n* ($1 < 2 < 3$, etc.) and angular momentum quantum number *l* ($s < p < d < f$). Note also that as *n* increases, the sublevel energies become closer together. The penetration effect, together with this narrowing of energy differences, results in the overlap of some sublevels; for example, the $4s$ sublevel is slightly lower in energy than the $3d$, so it is filled first. (The lines are color coded by sublevel type and have different lengths only for ease in labeling.)

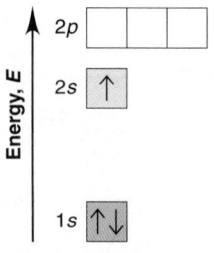

Figure 8.8 A vertical orbital diagram for the Li ground state. Sublevel energy increases from bottom to top.

Keep in mind that the energy of the sublevels (each group of boxes is a sublevel) increases from left to right. Figure 8.8 emphasizes this energy order by arranging the sublevels vertically.

With the $2s$ orbital only half-filled in Li, the fourth electron of beryllium fills it with the electron's spin paired: $n = 2$, $l = 0$, $m_l = 0$, $m_s = -\frac{1}{2}$.

$$\text{Be } (Z = 4) \quad 1s^2 2s^2$$

The next lowest energy sublevel is the $2p$. A p sublevel has $l = 1$, so the m_l (orientation) values can be -1, 0, or $+1$. The three orbitals in the $2p$ sublevel have *equal energy* (same n and l values), which means that the fifth electron of boron can go into *any one of the $2p$ orbitals*. For convenience, let's label the boxes from left to right, -1, 0, $+1$. By convention, we place the electron in the $m_l = -1$ orbital: $n = 2$, $l = 1$, $m_l = -1$, $m_s = +\frac{1}{2}$.

$$\text{B } (Z = 5) \quad 1s^2 2s^2 2p^1$$

To minimize electron-electron repulsions, the last added (sixth) electron of carbon enters one of the *unoccupied* $2p$ orbitals; by convention, we place it in the $m_l = 0$ orbital. Experiment shows that the spin of this electron is *parallel* to (in the same direction as) the spin of the other $2p$ electron: $n = 2$, $l = 1$, $m_l = 0$, $m_s = +\frac{1}{2}$.

$$\text{C } (Z = 6) \quad 1s^2 2s^2 2p^2$$

The placement of electrons for carbon exemplifies **Hund's rule:** *when orbitals of equal energy are available, the electron configuration of lowest energy has the maximum number of unpaired electrons with parallel spins.* Based on Hund's rule, nitrogen's seventh electron enters the last empty $2p$ orbital, with its spin parallel to the two other $2p$ electrons: $n = 2$, $l = 1$, $m_l = +1$, $m_s = +\frac{1}{2}$.

$$\text{N } (Z = 7) \quad 1s^2 2s^2 2p^3$$

The eighth electron in oxygen must enter one of these three half-filled $2p$ orbitals and "pair up" with (have opposing spin to) the electron already present. Since the $2p$ orbitals all have the same energy, we proceed as before and place the electron in the orbital previously designated $m_l = -1$. The quantum numbers are $n = 2$, $l = 1$, $m_l = -1$, $m_s = -\frac{1}{2}$.

$$\text{O } (Z = 8) \quad 1s^2 2s^2 2p^4$$

Fluorine's ninth electron enters either of the two remaining half-filled $2p$ orbitals: $n = 2$, $l = 1$, $m_l = 0$, $m_s = -\frac{1}{2}$.

$$\text{F } (Z = 9) \quad 1s^2 2s^2 2p^5$$

Only one unfilled orbital remains in the $2p$ sublevel, so the tenth electron of neon occupies it: $n = 2$, $l = 1$, $m_l = +1$, $m_s = -\frac{1}{2}$. With neon, the $n = 2$ level is filled.

$$\text{Ne } (Z = 10) \quad 1s^2 2s^2 2p^6$$

SAMPLE PROBLEM 8.1 Determining Quantum Numbers from Orbital Diagrams

Problem Write a set of quantum numbers for the third electron and a set for the eighth electron of the F atom.

Plan Based on the orbital diagram, we count to the electron of interest and note its level (n), sublevel (l), orbital (m_l), and direction of spin (m_s).

Solution The third electron is in the $2s$ orbital. The upward arrow indicates a spin of $+\frac{1}{2}$:

$$n = 2, l = 0, m_l = 0, m_s = +\tfrac{1}{2}$$

The eighth electron is in the first $2p$ orbital, which is designated $m_l = -1$, and has a downward arrow:

$$n = 2, l = 1, m_l = -1, m_s = -\tfrac{1}{2}$$

FOLLOW-UP PROBLEM 8.1 Use the periodic table to identify the element with the electron configuration $1s^2 2s^2 2p^4$. Write its orbital diagram, and give the quantum numbers of its sixth electron.

With so much attention paid to these notations, it's easy to forget that atoms are real spherical objects and that the electrons occupy volumes with specific shapes and orientations. Figure 8.9 shows the first 10 elements arranged in the periodic table format, with orbital contours depicted.

Even at this early stage of filling the table, we can make an important correlation between chemical behavior and electron configuration: elements in the same group have similar outer electron configurations. As an example, helium (He) and neon (Ne) in Group 8A(18) both have filled outer shells—$n = 1$ for helium and $n = 2$ for neon—and neither element forms compounds. Apparently, *filled outer shells make these elements unreactive.*

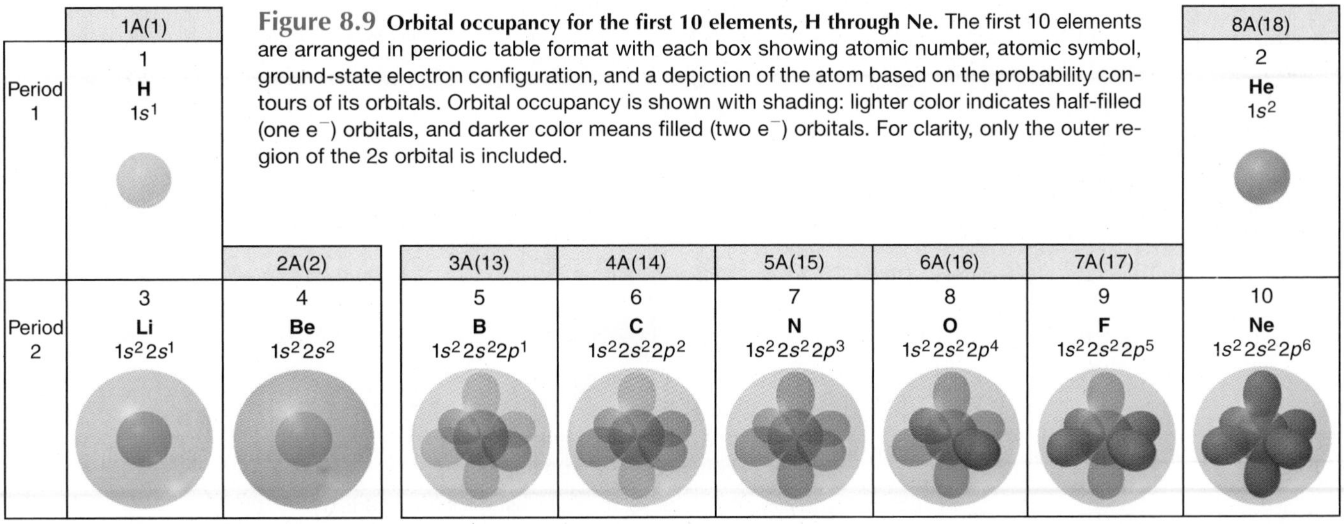

Figure 8.9 Orbital occupancy for the first 10 elements, H through Ne. The first 10 elements are arranged in periodic table format with each box showing atomic number, atomic symbol, ground-state electron configuration, and a depiction of the atom based on the probability contours of its orbitals. Orbital occupancy is shown with shading: lighter color indicates half-filled (one e⁻) orbitals, and darker color means filled (two e⁻) orbitals. For clarity, only the outer region of the $2s$ orbital is included.

1A(1)							8A(18)
1 **H** $1s^1$							2 **He** $1s^2$
	2A(2)	3A(13)	4A(14)	5A(15)	6A(16)	7A(17)	
3 **Li** $1s^2 2s^1$	4 **Be** $1s^2 2s^2$	5 **B** $1s^2 2s^2 2p^1$	6 **C** $1s^2 2s^2 2p^2$	7 **N** $1s^2 2s^2 2p^3$	8 **O** $1s^2 2s^2 2p^4$	9 **F** $1s^2 2s^2 2p^5$	10 **Ne** $1s^2 2s^2 2p^6$

Period 1 / Period 2

Building Up Period 3

The Period 3 elements, sodium through argon, lie directly under the Period 2 elements, lithium through neon. As the orbitals become filled, note the *group similarities in outer electron configuration.* The sublevels of the $n = 3$ level are filled in the order 3s, 3p, 3d. Table 8.3 presents *partial* orbital diagrams (3s and 3p sublevels only) and electron configurations for the eight elements in Period 3 (with *filled inner levels* in brackets and the sublevel to which the last electron is added in colored type).

In sodium (the second alkali metal) and magnesium (the second alkaline earth metal), electrons are added to the 3s sublevel, which contains the 3s orbital only, just as they filled the 2s sublevel in lithium and beryllium above. Then, just as for boron, carbon, and nitrogen in Period 2, the last electron added to aluminum, silicon, and phosphorus in Period 3 half-fills each of the 3p orbitals with spins parallel (Hund's rule). The last electrons added to sulfur, chlorine, and argon enter each of the half-filled 3p orbitals, thereby filling the 3p sublevel. With argon, the next noble gas after helium and neon, we arrive at the end of Period 3. (As you'll see shortly, the 3d orbitals are filled in Period 4.)

The rightmost column of Table 8.3 shows the *condensed electron configuration.* In this simplified notation, the electron configuration of the previous noble gas is represented by its element symbol in brackets, and it is followed by the electron configuration of the energy level being filled. The condensed electron configuration of oxygen, for example, is [He] $2s^2 2p^4$, where [He] represents $1s^2$; for sodium, the condensed electron configuration is [Ne] $3s^1$, where [Ne] stands for $1s^2 2s^2 2p^6$ (as Table 8.3 shows); and so forth.

Table 8.3 Partial Orbital Diagrams and Electron Configurations* for the Elements in Period 3

Atomic Number/ Element	Partial Orbital Diagram (3s and 3p Sublevels Only)		Full Electron Configuration	Condensed Electron Configuration
	3s	**3p**		
11/Na	↑	☐ ☐ ☐	$[1s^2 2s^2 2p^6]\, 3s^1$	[Ne] $3s^1$
12/Mg	↑↓	☐ ☐ ☐	$[1s^2 2s^2 2p^6]\, 3s^2$	[Ne] $3s^2$
13/Al	↑↓	↑ ☐ ☐	$[1s^2 2s^2 2p^6]\, 3s^2 3p^1$	[Ne] $3s^2 3p^1$
14/Si	↑↓	↑ ↑ ☐	$[1s^2 2s^2 2p^6]\, 3s^2 3p^2$	[Ne] $3s^2 3p^2$
15/P	↑↓	↑ ↑ ↑	$[1s^2 2s^2 2p^6]\, 3s^2 3p^3$	[Ne] $3s^2 3p^3$
16/S	↑↓	↑↓ ↑ ↑	$[1s^2 2s^2 2p^6]\, 3s^2 3p^4$	[Ne] $3s^2 3p^4$
17/Cl	↑↓	↑↓ ↑↓ ↑	$[1s^2 2s^2 2p^6]\, 3s^2 3p^5$	[Ne] $3s^2 3p^5$
18/Ar	↑↓	↑↓ ↑↓ ↑↓	$[1s^2 2s^2 2p^6]\, 3s^2 3p^6$	[Ne] $3s^2 3p^6$

*Colored type indicates sublevel to which last electron is added.

Period	1A (1)	2A (2)	3A (13)	4A (14)	5A (15)	6A (16)	7A (17)	8A (18)
1	1 **H** $1s^1$							2 **He** $1s^2$
2	3 **Li** [He] $2s^1$	4 **Be** [He] $2s^2$	5 **B** [He] $2s^22p^1$	6 **C** [He] $2s^22p^2$	7 **N** [He] $2s^22p^3$	8 **O** [He] $2s^22p^4$	9 **F** [He] $2s^22p^5$	10 **Ne** [He] $2s^22p^6$
3	11 **Na** [Ne] $3s^1$	12 **Mg** [Ne] $3s^2$	13 **Al** [Ne] $3s^23p^1$	14 **Si** [Ne] $3s^23p^2$	15 **P** [Ne] $3s^23p^3$	16 **S** [Ne] $3s^23p^4$	17 **Cl** [Ne] $3s^23p^5$	18 **Ar** [Ne] $3s^23p^6$

Figure 8.10 Condensed ground-state electron configurations in the first three periods. The first 18 elements, H through Ar, are arranged in three periods containing two, eight, and eight elements. Each box shows the atomic number, atomic symbol, and condensed ground-state electron configuration. Note that elements in a group have similar outer electron configurations *(color)*.

Electron Configurations Within Groups

One of the central points in all chemistry is that *similar outer electron configurations correlate with similar chemical behavior.* Figure 8.10 shows the condensed electron configurations of the first 18 elements. Note the similarities within each group. Here are some examples from just two groups:

- In Group 1A(1), lithium and sodium have the condensed electron configuration [noble gas] ns^1 (where n is the quantum number of the outermost energy level), as do all the other alkali metals (K, Rb, Cs, Fr). All are highly reactive metals that form ionic compounds with nonmetals with formulas such as MCl, M_2O, and M_2S (where M represents the alkali metal), and all react vigorously with water to displace H_2 (Figure 8.11A).
- In Group 7A(17), fluorine and chlorine have the condensed electron configuration [noble gas] ns^2np^5, as do the other halogens (Br, I, At). Little is known about rare, radioactive astatine (At), but all the others are reactive nonmetals that occur as diatomic molecules, X_2 (where X represents the halogen). All form ionic compounds with metals (KX, MgX_2) (Figure 8.11B), covalent compounds with hydrogen (HX) that yield acidic solutions in water, and covalent compounds with carbon (CX_4).

To summarize, here is the major connection between quantum mechanics and chemical periodicity: *orbitals are filled in order of increasing energy, which leads to outer electron configurations that recur periodically, which leads to chemical properties that recur periodically.*

The First *d*-Orbital Transition Series: Building Up Period 4

The 3*d* orbitals are filled in Period 4. Note, however, that *the 4s orbital is filled before the 3d.* This switch in filling order is due to the shielding and penetration effects that we discussed in Section 8.2. The radial probability distribution of the 3*d* orbital is greater outside the filled, inner $n = 1$ and $n = 2$ levels, so a 3*d* electron is shielded very effectively from the nuclear charge. In contrast, penetration by the 4*s* electron means that it spends a significant part of its time near the nucleus and feels a greater nuclear attraction. Thus, the 4*s* orbital is slightly *lower* in energy than the 3*d*, so it fills first. Similarly, the 5*s* orbital fills before the 4*d*, and the 6*s* fills before the 5*d*. In general, then, *the ns sublevel fills before the* $(n - 1)d$ *sublevel.* As we proceed through the transition series, however, you'll see several exceptions to this pattern because the energies of the *ns* and $(n - 1)d$ sublevels become extremely close with higher values of *n*.

1A(1)
ns^1
$_3$Li
$_{11}$Na
$_{19}$K
$_{37}$Rb
$_{55}$Cs
$_{87}$Fr

A

7A(17)
ns^2np^5
$_9$F
$_{17}$Cl
$_{35}$Br
$_{53}$I
$_{85}$At

B

Figure 8.11 Similar reactivities within a group. A, Potassium metal reacting. All alkali metals [Group 1A(1)] react vigorously with water and displace H_2. **B,** Chlorine and potassium metal reacting. All halogens [Group 7A(17)] react with metals to form ionic halides.

Table 8.4 shows the partial orbital diagrams and ground-state electron configurations for the 18 elements in Period 4 (again with filled inner levels in brackets and the sublevel to which the last electron has been added in colored type). The first two elements of the period, potassium and calcium, are the next alkali and alkaline earth metals, respectively, and their electrons fill the $4s$ sublevel. The third element, scandium ($Z = 21$), is the first of the **transition elements,** those in which d orbitals are being filled. The last electron in scandium occupies any one of the five $3d$ orbitals because they are equal in energy. Scandium has the electron configuration [Ar] $4s^23d^1$.

The filling of $3d$ orbitals proceeds one at a time, as in p orbitals, except in two cases: chromium ($Z = 24$) and copper ($Z = 29$). Vanadium ($Z = 23$), the element before chromium, has three half-filled d orbitals ([Ar] $4s^23d^3$). Rather than the last electron in chromium entering a fourth empty d orbital to give [Ar] $4s^23d^4$, chromium has one electron in the $4s$ sublevel and five in the $3d$ sublevel. Thus, both the $4s$ and the $3d$ sublevels are *half-filled* (see margin).

The other anomalous filling pattern occurs with copper. Following nickel ([Ar] $4s^23d^8$), we would expect copper to have the [Ar] $4s^23d^9$ configuration.

Cr ($Z = 24$)   [Ar] $4s^13d^5$

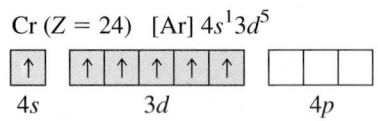

Table 8.4 Partial Orbital Diagrams and Electron Configurations* for the Elements in Period 4

Atomic Number	Element	Partial Orbital Diagram ($4s$, $3d$, and $4p$ Sublevels Only)			Full Electron Configuration	Condensed Electron Configuration
		$4s$	$3d$	$4p$		
19	K	↑			$[1s^22s^22p^63s^23p^6]\,4s^1$	[Ar] $4s^1$
20	Ca	↑↓			$[1s^22s^22p^63s^23p^6]\,4s^2$	[Ar] $4s^2$
21	Sc	↑↓	↑		$[1s^22s^22p^63s^23p^6]\,4s^23d^1$	[Ar] $4s^23d^1$
22	Ti	↑↓	↑ ↑		$[1s^22s^22p^63s^23p^6]\,4s^23d^2$	[Ar] $4s^23d^2$
23	V	↑↓	↑ ↑ ↑		$[1s^22s^22p^63s^23p^6]\,4s^23d^3$	[Ar] $4s^23d^3$
24	Cr	↑	↑ ↑ ↑ ↑ ↑		$[1s^22s^22p^63s^23p^6]\,4s^13d^5$	[Ar] $4s^13d^5$
25	Mn	↑↓	↑ ↑ ↑ ↑ ↑		$[1s^22s^22p^63s^23p^6]\,4s^23d^5$	[Ar] $4s^23d^5$
26	Fe	↑↓	↑↓ ↑ ↑ ↑ ↑		$[1s^22s^22p^63s^23p^6]\,4s^23d^6$	[Ar] $4s^23d^6$
27	Co	↑↓	↑↓ ↑↓ ↑ ↑ ↑		$[1s^22s^22p^63s^23p^6]\,4s^23d^7$	[Ar] $4s^23d^7$
28	Ni	↑↓	↑↓ ↑↓ ↑↓ ↑ ↑		$[1s^22s^22p^63s^23p^6]\,4s^23d^8$	[Ar] $4s^23d^8$
29	Cu	↑	↑↓ ↑↓ ↑↓ ↑↓ ↑↓		$[1s^22s^22p^63s^23p^6]\,4s^13d^{10}$	[Ar] $4s^13d^{10}$
30	Zn	↑↓	↑↓ ↑↓ ↑↓ ↑↓ ↑↓		$[1s^22s^22p^63s^23p^6]\,4s^23d^{10}$	[Ar] $4s^23d^{10}$
31	Ga	↑↓	↑↓ ↑↓ ↑↓ ↑↓ ↑↓	↑	$[1s^22s^22p^63s^23p^6]\,4s^23d^{10}4p^1$	[Ar] $4s^23d^{10}4p^1$
32	Ge	↑↓	↑↓ ↑↓ ↑↓ ↑↓ ↑↓	↑ ↑	$[1s^22s^22p^63s^23p^6]\,4s^23d^{10}4p^2$	[Ar] $4s^23d^{10}4p^2$
33	As	↑↓	↑↓ ↑↓ ↑↓ ↑↓ ↑↓	↑ ↑ ↑	$[1s^22s^22p^63s^23p^6]\,4s^23d^{10}4p^3$	[Ar] $4s^23d^{10}4p^3$
34	Se	↑↓	↑↓ ↑↓ ↑↓ ↑↓ ↑↓	↑↓ ↑ ↑	$[1s^22s^22p^63s^23p^6]\,4s^23d^{10}4p^4$	[Ar] $4s^23d^{10}4p^4$
35	Br	↑↓	↑↓ ↑↓ ↑↓ ↑↓ ↑↓	↑↓ ↑↓ ↑	$[1s^22s^22p^63s^23p^6]\,4s^23d^{10}4p^5$	[Ar] $4s^23d^{10}4p^5$
36	Kr	↑↓	↑↓ ↑↓ ↑↓ ↑↓ ↑↓	↑↓ ↑↓ ↑↓	$[1s^22s^22p^63s^23p^6]\,4s^23d^{10}4p^6$	[Ar] $4s^23d^{10}4p^6$

*Colored type indicates sublevel(s) whose occupancy changes when last electron is added.

Instead, the 4*s* orbital of copper is half-filled (1 electron), and the 3*d* orbitals are *filled* with 10 electrons (see margin). The anomalous filling patterns in Cr and Cu lead us to conclude that *half-filled and filled sublevels are unexpectedly stable.* These are the first two cases of a pattern seen with many other elements.

In zinc, both the 4*s* and 3*d* sublevels are completely filled, and the first transition series ends. As Table 8.4 shows, the 4*p* sublevel is then filled by the next six elements. Period 4 ends with krypton, the next noble gas.

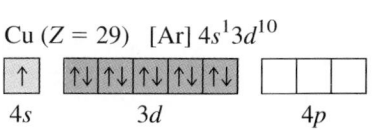

General Principles of Electron Configurations

There are 77 known elements beyond the 36 we have considered. Let's examine the ground-state electron configurations to highlight some key ideas.

Similar Outer Electron Configurations Within a Group To repeat one of chemistry's central themes and the key to the usefulness of the periodic table, *elements in a group have similar chemical properties because they have similar outer electron configurations* (Figure 8.12). Among the main-group elements (A groups)—the *s*-block and *p*-block elements—outer electron configurations within a group are essentially identical, as shown by the group headings in Figure 8.12. Some variations in the transition elements (B groups, *d* block) and inner transition elements (*f* block) occur, as you'll see.

Figure 8.12 **A periodic table of partial ground-state electron configurations.** These ground-state electron configurations show the electrons beyond the previous noble gas in the sublevel block being filled (excluding filled inner sublevels). For main-group elements, the group heading identifies the general outer configuration. Anomalous electron configurations occur often among the *d*-block and *f*-block elements, the first two appearing for Cr (*Z* = 24) and Cu (*Z* = 29). Helium is colored as an *s*-block element but placed with the other members of Group 8A(18). Configurations for elements 110 to 112, and 114 have not yet been confirmed.

Main-Group Elements (*s* block)

Transition Elements (*d* block)

Main-Group Elements (*p* block)

Inner Transition Elements (*f* block)

Period number: highest occupied energy level

	1A (1) ns^1	2A (2) ns^2	3B (3)	4B (4)	5B (5)	6B (6)	7B (7)	(8)	8B (9)	(10)	1B (11)	2B (12)	3A (13) ns^2np^1	4A (14) ns^2np^2	5A (15) ns^2np^3	6A (16) ns^2np^4	7A (17) ns^2np^5	8A (18) ns^2np^6
1	1 **H** $1s^1$																	2 **He** $1s^2$
2	3 **Li** $2s^1$	4 **Be** $2s^2$											5 **B** $2s^22p^1$	6 **C** $2s^22p^2$	7 **N** $2s^22p^3$	8 **O** $2s^22p^4$	9 **F** $2s^22p^5$	10 **Ne** $2s^22p^6$
3	11 **Na** $3s^1$	12 **Mg** $3s^2$											13 **Al** $3s^23p^1$	14 **Si** $3s^23p^2$	15 **P** $3s^23p^3$	16 **S** $3s^23p^4$	17 **Cl** $3s^23p^5$	18 **Ar** $3s^23p^6$
4	19 **K** $4s^1$	20 **Ca** $4s^2$	21 **Sc** $4s^23d^1$	22 **Ti** $4s^23d^2$	23 **V** $4s^23d^3$	24 **Cr** $4s^13d^5$	25 **Mn** $4s^23d^5$	26 **Fe** $4s^23d^6$	27 **Co** $4s^23d^7$	28 **Ni** $4s^23d^8$	29 **Cu** $4s^13d^{10}$	30 **Zn** $4s^23d^{10}$	31 **Ga** $4s^24p^1$	32 **Ge** $4s^24p^2$	33 **As** $4s^24p^3$	34 **Se** $4s^24p^4$	35 **Br** $4s^24p^5$	36 **Kr** $4s^24p^6$
5	37 **Rb** $5s^1$	38 **Sr** $5s^2$	39 **Y** $5s^24d^1$	40 **Zr** $5s^24d^2$	41 **Nb** $5s^14d^4$	42 **Mo** $5s^14d^5$	43 **Tc** $5s^24d^5$	44 **Ru** $5s^14d^7$	45 **Rh** $5s^14d^8$	46 **Pd** $4d^{10}$	47 **Ag** $5s^14d^{10}$	48 **Cd** $5s^24d^{10}$	49 **In** $5s^25p^1$	50 **Sn** $5s^25p^2$	51 **Sb** $5s^25p^3$	52 **Te** $5s^25p^4$	53 **I** $5s^25p^5$	54 **Xe** $5s^25p^6$
6	55 **Cs** $6s^1$	56 **Ba** $6s^2$	57 **La*** $6s^25d^1$	72 **Hf** $6s^25d^2$	73 **Ta** $6s^25d^3$	74 **W** $6s^25d^4$	75 **Re** $6s^25d^5$	76 **Os** $6s^25d^6$	77 **Ir** $6s^25d^7$	78 **Pt** $6s^15d^9$	79 **Au** $6s^15d^{10}$	80 **Hg** $6s^25d^{10}$	81 **Tl** $6s^26p^1$	82 **Pb** $6s^26p^2$	83 **Bi** $6s^26p^3$	84 **Po** $6s^26p^4$	85 **At** $6s^26p^5$	86 **Rn** $6s^26p^6$
7	87 **Fr** $7s^1$	88 **Ra** $7s^2$	89 **Ac**** $7s^26d^1$	104 **Rf** $7s^26d^2$	105 **Db** $7s^26d^3$	106 **Sg** $7s^26d^4$	107 **Bh** $7s^26d^5$	108 **Hs** $7s^26d^6$	109 **Mt** $7s^26d^7$	110 $7s^26d^8$	111 $7s^26d^9$	112 $7s^26d^{10}$	114 $7s^27p^2$					

		58 **Ce** $6s^24f^15d^1$	59 **Pr** $6s^24f^3$	60 **Nd** $6s^24f^4$	61 **Pm** $6s^24f^5$	62 **Sm** $6s^24f^6$	63 **Eu** $6s^24f^7$	64 **Gd** $6s^24f^75d^1$	65 **Tb** $6s^24f^9$	66 **Dy** $6s^24f^{10}$	67 **Ho** $6s^24f^{11}$	68 **Er** $6s^24f^{12}$	69 **Tm** $6s^24f^{13}$	70 **Yb** $6s^24f^{14}$	71 **Lu** $6s^24f^{14}5d^1$
6	*Lanthanides														
7	**Actinides	90 **Th** $7s^26d^2$	91 **Pa** $7s^25f^26d^1$	92 **U** $7s^25f^36d^1$	93 **Np** $7s^25f^46d^1$	94 **Pu** $7s^25f^6$	95 **Am** $7s^25f^7$	96 **Cm** $7s^25f^76d^1$	97 **Bk** $7s^25f^9$	98 **Cf** $7s^25f^{10}$	99 **Es** $7s^25f^{11}$	100 **Fm** $7s^25f^{12}$	101 **Md** $7s^25f^{13}$	102 **No** $7s^25f^{14}$	103 **Lr** $7s^25f^{14}6d^1$

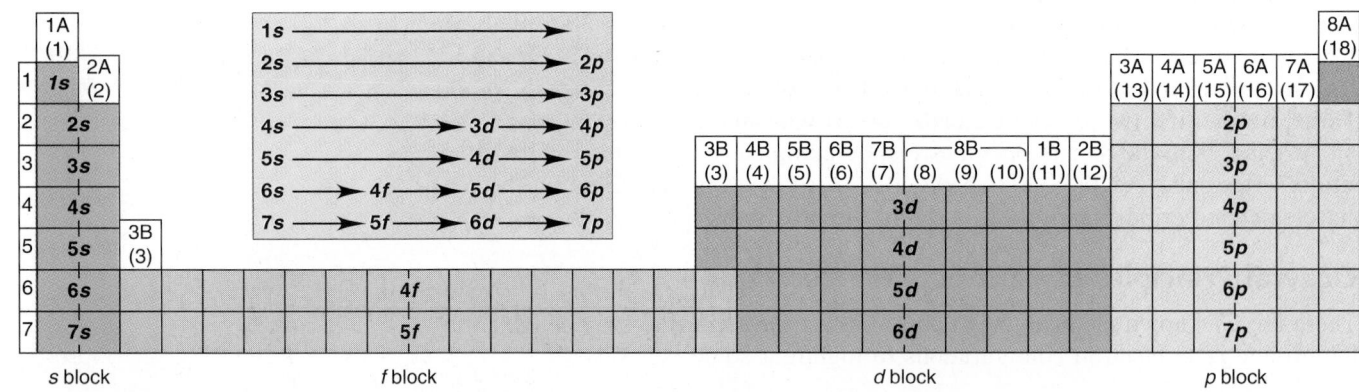

Figure 8.13 **The relation between orbital filling and the periodic table.** If we "read" the periods like the words on a page, the elements are arranged into sublevel blocks that occur in the order of increasing energy. This form of the periodic table shows the sublevel blocks. (The *f* blocks fit between the first and second elements of the *d* blocks in Periods 6 and 7.) *Inset:* A simple version of sublevel order.

Orbital Filling Order When the elements are "built up" by filling their levels and sublevels in order of increasing energy, we obtain the actual sequence of elements in the periodic table. Thus, reading the table from left to right, as you do words on a page, gives the energy order of levels and sublevels, which is shown in Figure 8.13. The arrangement of the periodic table is certainly the best way to learn the orbital filling order of the elements.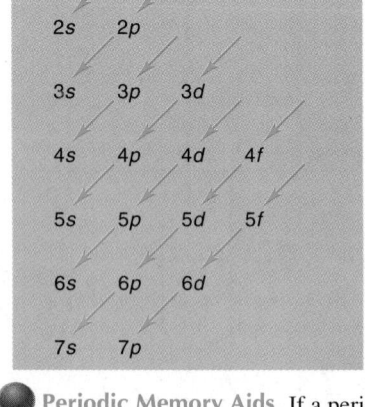

Categories of Electrons A survey of the periodic table distinguishes three categories of electrons:

1. **Inner (core) electrons** are those in the previous noble gas and any completed transition series. They fill all the *lower energy levels* of an atom.
2. **Outer electrons** are those in the *highest energy level* (highest *n* value). They spend most of their time farthest from the nucleus.
3. **Valence electrons** are those involved in forming compounds. *Among the main-group elements, the valence electrons **are** the outer electrons.* Among the transition elements, some inner *d* electrons are also often involved in bonding and are counted among the valence electrons.

Group and Period Numbers Several key pieces of information are embedded in the periodic table:

1. Among the main-group elements (A groups), *the group number equals the number of outer electrons* (those with the highest *n*). For example, chlorine (Cl; Period 3, Group **7**A) has 7 outer electrons, tellurium (Te; Period 5, Group **6**A) has 6, and so forth.
2. *The period number is the n value of the highest energy level.* Thus, in Period 2, the $n = 2$ level has the highest energy; in Period 5, it is the $n = 5$ level.
3. The *n* value squared (n^2) gives the total number of *orbitals* in that energy level. Because an orbital can hold no more than two electrons (exclusion principle), $2n^2$ gives the maximum number of *electrons* (or elements) in the energy level. For example, for the $n = 3$ level, the number of orbitals is $n^2 = 9$: one 3*s*, three 3*p*, and five 3*d*. The number of electrons is $2n^2$, or 18: two 3*s* and six 3*p* electrons occur in the eight elements of Period 3, and ten 3*d* electrons are added in the ten transition elements of Period 4.

Complex Patterns: The Transition and Inner Transition Elements

Periods 4, 5, 6, and 7 incorporate the *d*-block transition elements. The general pattern, as you've seen, is that the $(n - 1)d$ orbitals are filled between the *ns* and *np* orbitals. Thus, Period 5 follows the same general pattern as Period 4. In Period 6, the 6*s* sublevel is filled in cesium (Cs) and barium (Ba), and then lanthanum

Periodic Memory Aids If a periodic table is unavailable, this arrangement of sublevels can remind you of the filling order. List the sublevels as shown, and read from 1*s*, following the direction of the arrows. Note that:
• values of *n* are constant horizontally
• values of *l* are constant vertically
• combined values of $n + l$ are constant diagonally

(La; $Z = 57$), the first member of the $5d$ transition series, occurs. At this point, the first series of **inner transition elements,** those in which f orbitals are being filled, intervenes (Figure 8.13). The f orbitals have $l = 3$, so the possible m_l values are -3, -2, -1, 0, $+1$, $+2$, and $+3$; that is, there are seven f orbitals, for a total of 14 elements in *each* of the two inner transition series.

The Period 6 inner transition series fills the $4f$ orbitals and consists of the **lanthanides** (or **rare earths**), so called because they occur after and are similar to lanthanum. The other inner transition series holds the **actinides,** which fill the $5f$ orbitals that appear in Period 7 after actinium (Ac; $Z = 89$). In both series, the $(n - 2)f$ orbitals are filled, after which filling of the $(n - 1)d$ orbitals proceeds. Period 6 ends by filling the $6p$ orbitals as in other p-block elements. Period 7 is still incomplete because only two elements with $7p$ electrons are known at this time.

Several irregularities in filling pattern occur in both the d and f blocks. Two already mentioned occur in chromium (Cr) and copper (Cu) in Period 4. Silver (Ag) and gold (Au), the two elements under Cu in Group 1B(11), follow copper's pattern. Molybdenum (Mo) follows the pattern of Cr in Group 6B(6), but tungsten (W) does not. Other anomalous configurations appear among the transition elements in Periods 5 and 6. Note, however, that even though minor variations from the expected configurations occur, the sum of ns electrons and $(n - 1)d$ electrons always equals the *new* group number. For instance, despite variations in the electron configurations in Group 6B(**6**)—Cr, Mo, W, and Sg—the sum of ns and $(n - 1)d$ electrons is 6; in Group 8B(**10**)—Ni, Pd, and Pt—the sum is 10.

Whenever our observations differ from our expectations, remember that the fact always takes precedence over the model; in other words, the electrons don't "care" what orbitals *we* think they should occupy. As the atomic orbitals in larger atoms fill with electrons, sublevel energies differ very little, which results in these variations from the expected pattern.

SAMPLE PROBLEM 8.2 Determining Electron Configurations

Problem Using the periodic table on the inside cover of the text (not Figure 8.12 or Table 8.4), give the full and condensed electron configurations, partial orbital diagrams showing valence electrons, and number of inner electrons for the following elements:
(a) Potassium (K; $Z = 19$)   **(b)** Molybdenum (Mo; $Z = 42$)   **(c)** Lead (Pb; $Z = 82$)
Plan The atomic number tells us the number of electrons, and the periodic table shows the order for filling sublevels. In the partial orbital diagrams, we include all electrons after those of the previous noble gas *except* those in *filled* inner sublevels. The number of inner electrons is the sum of those in the previous noble gas and in filled d and f sublevels.
Solution (a) For K ($Z = 19$), the full electron configuration is $1s^2 2s^2 2p^6 3s^2 3p^6 4s^1$.
The condensed configuration is $[\text{Ar}]\, 4s^1$.
The partial orbital diagram for valence electrons is

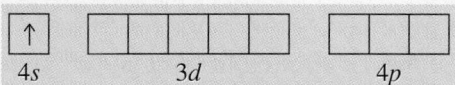

K is a main-group element in Group 1A(1) of Period 4, so there are 18 inner electrons.
(b) For Mo ($Z = 42$), we would expect the full electron configuration to be $1s^2 2s^2 2p^6 3s^2 3p^6 4s^2 3d^{10} 4p^6 5s^2 4d^4$. However, Mo lies under Cr in Group 6B(6) and exhibits the same variation in filling pattern in the ns and $(n - 1)d$ sublevels:
$1s^2 2s^2 2p^6 3s^2 3p^6 4s^2 3d^{10} 4p^6 5s^1 4d^5$.
The condensed electron configuration is $[\text{Kr}]\, 5s^1 4d^5$.
The partial orbital diagram for valence electrons is

Mo is a transition element in Group 6B(6) of Period 5, so there are 36 inner electrons.

(c) For Pb ($Z = 82$), the full electron configuration is

$$1s^2 2s^2 2p^6 3s^2 3p^6 4s^2 3d^{10} 4p^6 5s^2 4d^{10} 5p^6 6s^2 4f^{14} 5d^{10} 6p^2.$$

The condensed electron configuration is $[Xe]\, 6s^2 4f^{14} 5d^{10} 6p^2$. The partial orbital diagram for valence electrons (no filled inner sublevels) is

↑↓		↑	↑	
6s		6p		

Pb is a main-group element in Group 4A(14) of Period 6, so there are 54 (in Xe) + 14 (in 4f series) + 10 (in 5d series) = 78 inner electrons.

Check Be sure the sum of the superscripts (electrons) in the full electron configuration equals the atomic number, and that the number of *valence* electrons in the condensed configuration equals the number of electrons in the partial orbital diagram.

FOLLOW-UP PROBLEM 8.2 Without referring to Table 8.4 or Figure 8.12, give full and condensed electron configurations, partial orbital diagrams showing valence electrons, and the number of inner electrons for the following elements:
(a) Ni ($Z = 28$)   **(b)** Sr ($Z = 38$)   **(c)** Po ($Z = 84$)

SECTION SUMMARY

In the aufbau method, one electron is added to an atom of each successive element in accord with Pauli's exclusion principle (no two electrons can have the same set of quantum numbers) and Hund's rule (orbitals of equal energy become half-filled, with electron spins parallel, before any pairing occurs). The elements of a group have similar outer electron configurations and similar chemical behavior. For the main-group elements, valence electrons (those involved in reactions) are in the outer (highest energy) level only. For transition elements, inner *d* electrons are often involved in reactions also. In general, $(n - 1)d$ orbitals fill after *ns* and before *np* orbitals. In Periods 6 and 7, $(n - 2)f$ orbitals fill between the first and second $(n - 1)d$ orbital elements.

8.4 TRENDS IN SOME KEY PERIODIC ATOMIC PROPERTIES

All physical and chemical behavior of the elements is based ultimately on the electron configurations of their atoms. In this section, we focus on three properties of atoms that are directly influenced by electron configuration: atomic size, ionization energy (the energy required to remove an electron from a gaseous atom), and electron affinity (the energy change involved in adding an electron to a gaseous atom). These properties are *periodic:* they generally increase and decrease in a recurring manner throughout the periodic table. As a result, their relative magnitudes can often be predicted, and they often exhibit consistent changes, or *trends,* within a group or period that correlate with element behavior.

Trends in Atomic Size

In Chapter 7, we noted that an electron in an atom can lie relatively far from the nucleus, so we commonly represent atoms as spheres in which the electrons spend 90% of their time. However, we *define* atomic size in terms of how closely one atom lies next to another. In practice, we measure the distance between identical, adjacent atomic nuclei in a sample of an element and divide that distance in half. (See the experimental method in Chapter 12.) Because atoms do not have hard surfaces, the size of an atom in a compound depends somewhat on the atoms near it. In other words, *atomic size varies slightly from substance to substance.*

Figure 8.14 shows two common definitions of atomic size. The **metallic radius** is one-half the distance between nuclei of adjacent atoms in a crystal of the element; we typically use this definition for metals. For elements commonly occurring as molecules, mostly nonmetals, we define atomic size by the **covalent radius,** one-half the distance between nuclei of identical covalently bonded atoms.

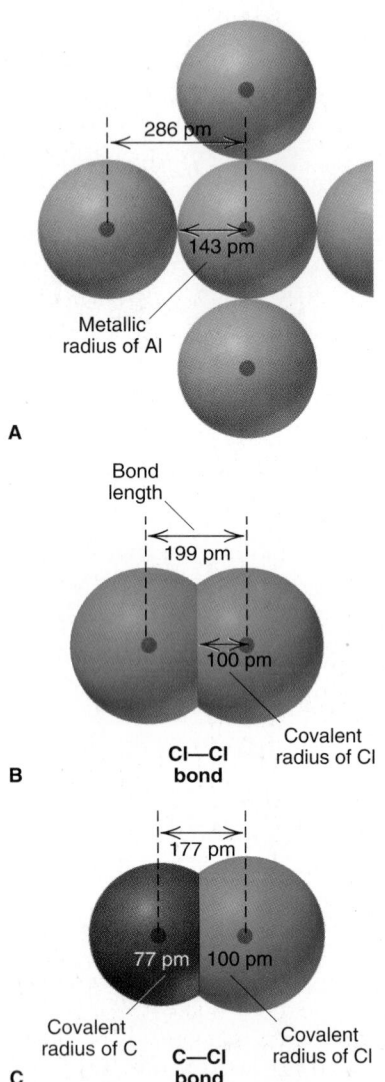

A

B

C

Figure 8.14 Defining metallic and covalent radii. A, The metallic radius is one-half the distance between nuclei of adjacent atoms in a crystal of the element, as shown here for aluminum. **B,** The covalent radius is one-half the distance between bonded nuclei in a molecule of the element, as shown here for chlorine. In effect, it is one-half the bond length. **C,** In a covalent compound, the bond length and known covalent radii are used to determine other radii. Here the C—Cl bond length (177 pm) and the covalent radius of Cl (100 pm) are used to find a value for the covalent radius of C (177 pm − 100 pm = 77 pm).

Trends Among the Main-Group Elements Atomic size greatly influences other atomic properties and is critical to understanding element behavior. Figure 8.15 shows the atomic radii of the main-group elements and most of the transition elements. Among the main-group elements, note that atomic size varies within both a group and a period. These variations in atomic size are the result of two opposing influences:

1. *Changes in n.* As the principal quantum number (n) increases, the outer electrons spend more time farther from the nucleus, so the atoms are larger.

2. *Changes in Z_{eff}.* As the effective nuclear charge (Z_{eff})—the positive charge "felt" by an electron—increases, outer electrons are pulled closer to the nucleus, so the atoms are smaller.

The net effect of these influences depends on shielding of the increasing nuclear charge by inner electrons:

1. *Down a group, n dominates.* As we move down a main group, each member has one more level of *inner* electrons; these shield the *outer* electrons very effectively. Calculations show Z_{eff} changes little, rising only slightly for each element, so the atomic radius increases as a result of increasing n value. *Atomic radius generally **increases** in a group from top to bottom.*

2. *Across a period, Z_{eff} dominates.* As we move across a period of main-group elements, electrons are added to the same outer level, so the shielding by inner electrons does not change. Because outer electrons shield each other poorly, Z_{eff} rises significantly, and so the outer electrons are pulled closer. *Atomic radius generally **decreases** in a period from left to right.*

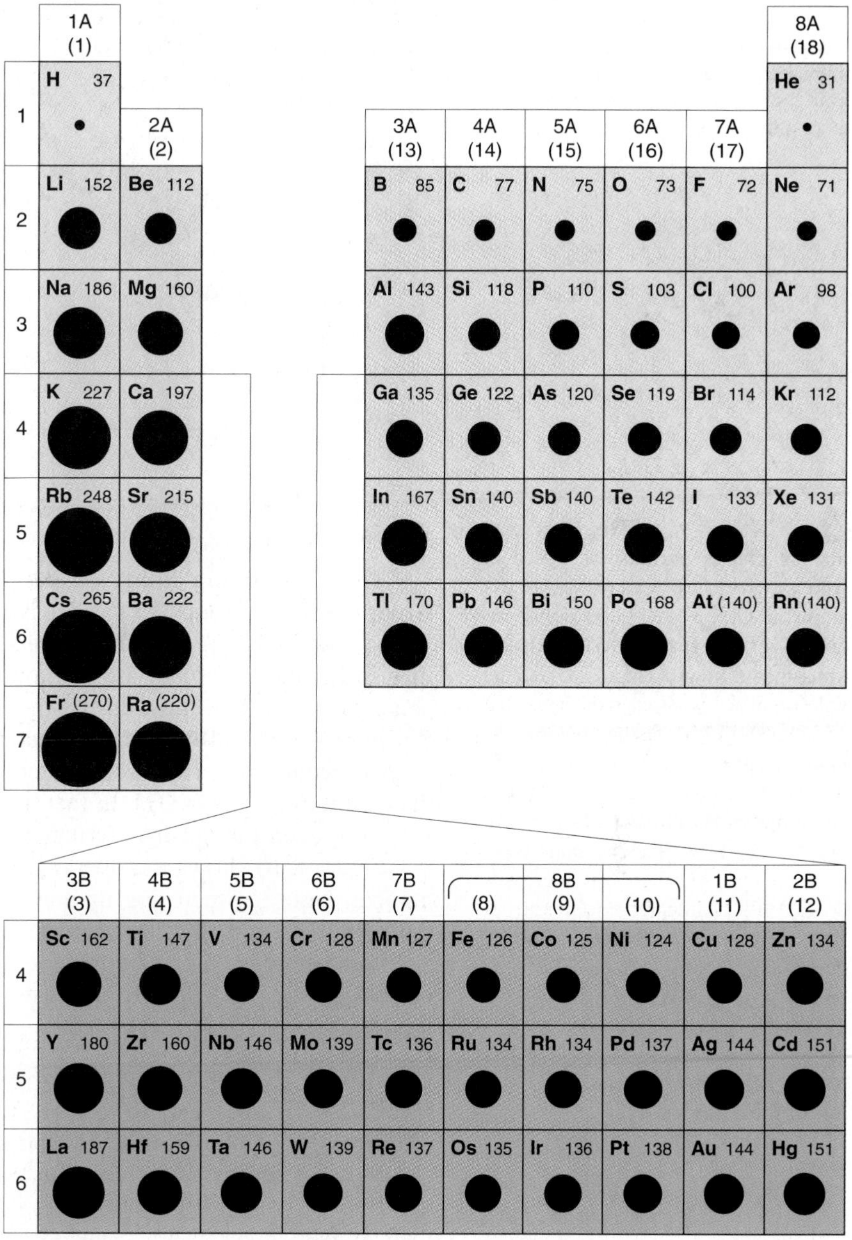

Figure 8.15 — Atomic radii of the main-group and transition elements (in picometers):

Main-group elements:

Period	1A (1)	2A (2)	3A (13)	4A (14)	5A (15)	6A (16)	7A (17)	8A (18)
1	H 37							He 31
2	Li 152	Be 112	B 85	C 77	N 75	O 73	F 72	Ne 71
3	Na 186	Mg 160	Al 143	Si 118	P 110	S 103	Cl 100	Ar 98
4	K 227	Ca 197	Ga 135	Ge 122	As 120	Se 119	Br 114	Kr 112
5	Rb 248	Sr 215	In 167	Sn 140	Sb 140	Te 142	I 133	Xe 131
6	Cs 265	Ba 222	Tl 170	Pb 146	Bi 150	Po 168	At (140)	Rn (140)
7	Fr (270)	Ra (220)						

Transition elements:

Period	3B (3)	4B (4)	5B (5)	6B (6)	7B (7)	8B (8)	8B (9)	8B (10)	1B (11)	2B (12)
4	Sc 162	Ti 147	V 134	Cr 128	Mn 127	Fe 126	Co 125	Ni 124	Cu 128	Zn 134
5	Y 180	Zr 160	Nb 146	Mo 139	Tc 136	Ru 134	Rh 134	Pd 137	Ag 144	Cd 151
6	La 187	Hf 159	Ta 146	W 139	Re 137	Os 135	Ir 136	Pt 138	Au 144	Hg 151

Figure 8.15 Atomic radii of the main-group and transition elements. Atomic radii (in picometers) are shown as circles of proportional size for the main-group elements (*tan*) and the transition elements (*blue*). Among the main-group elements, atomic radius generally increases from top to bottom and decreases from left to right. The transition elements do not exhibit this size trend. (Values in parentheses have only two significant figures.)

Trends Among the Transition Elements

As you can see from Figure 8.15, these general trends hold quite well for the main-group elements but *not* for the transition elements. Remember that inner *d* orbitals are filled within a transition series. As we move from left to right, size shrinks through the first two or three elements because of the increasing nuclear charge. But, from then on, *the size remains relatively constant* because shielding by these inner *d* electrons counteracts the usual increase in Z_{eff}. For instance, vanadium (V; $Z = 23$), the third Period 4 transition metal, has the same radius as zinc (Zn; $Z = 30$), the last. This pattern also appears in Periods 5 and 6 in the *d*-block transition series and in both series of inner transition elements.

Figure 8.16 **Periodicity of atomic radius.** A plot of atomic radius vs. atomic number for the elements in Periods 1 through 6 shows a periodic change: the radius generally decreases through a period to the noble gas [Group 8A(18); *purple*] and then increases suddenly to the next alkali metal [Group 1A(1); *brown*]. Deviation from the general decrease occurs among the transition elements.

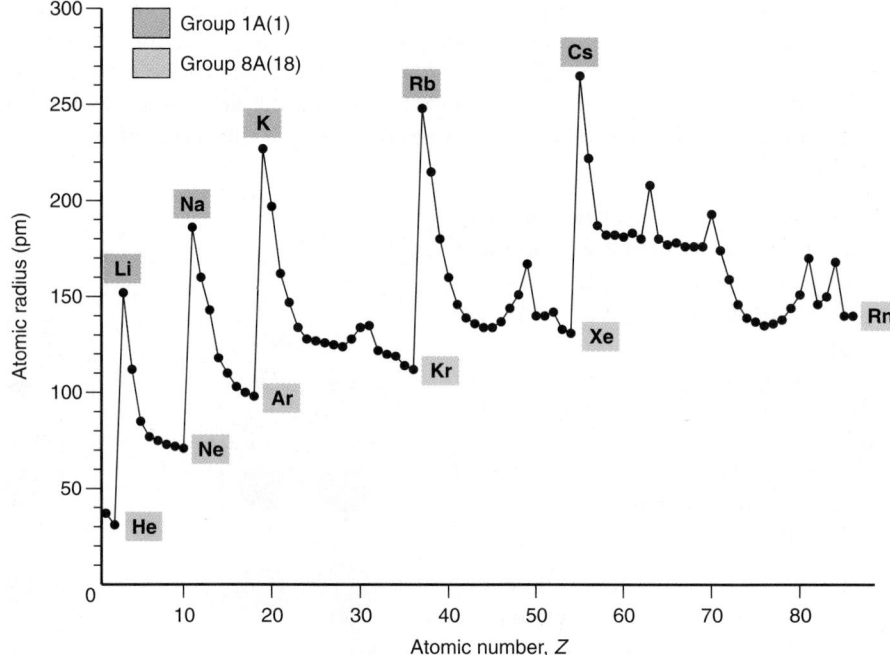

Packing 'Em In The increasing nuclear charge shrinks the space in which each electron can move. For example, in Group 1A(1), the atomic radius of cesium (Cs; Z = 55) is only 1.7 times that of lithium (Li; Z = 3); so the volume of Cs is about five times that of Li, even though Cs has 18 times as many electrons. At the opposite ends of Period 2, neon (Ne; Z = 10) has about the same volume as Li (Z = 3), but Ne has three times as many electrons. Thus, whether size increases down a group or decreases across a period, the attraction caused by the larger number of protons in the nucleus greatly crowds the electrons.

This intervening filling of *d* electrons causes a major size decrease from Group 2A(2) to Group 3A(13), the two main groups that flank the transition series. The size decrease in Periods 4, 5, and 6 (*with* a transition series) is much greater than in Period 3 (*without* a transition series). Because electrons in the *np* orbitals penetrate more than those in the $(n − 1)d$ orbitals, the first *np* electron "feels" a Z_{eff} increased by all the protons added during the transition series. The greatest change occurs in Period 4, in which calcium (Ca; Z = 20) is nearly 50% larger than gallium (Ga; Z = 31). In fact, gallium is slightly *smaller* than aluminum (Al; Z = 13), even though it is farther down the same group!

Figure 8.16 shows the overall variation in atomic size with increasing atomic number. Note the recurring up-and-down pattern as size drops across a period to the noble gas and then leaps up to the alkali metal that begins the next period. Also note how each transition series, beginning with that in Period 4 (K to Kr), throws off the smooth size decrease.

SAMPLE PROBLEM 8.3 Ranking Elements by Atomic Size

Problem Using only the periodic table (not Figure 8.15), rank each set of main-group elements in order of *decreasing* atomic size:
(a) Ca, Mg, Sr   **(b)** K, Ga, Ca   **(c)** Br, Rb, Kr   **(d)** Sr, Ca, Rb
Plan To rank the elements by atomic size, we find them in the periodic table. They are main-group elements, so size increases down a group and decreases across a period.
Solution (a) Sr > Ca > Mg. These three elements are in Group 2A(2), and size decreases up the group.
(b) K > Ca > Ga. These three elements are in Period 4, and size decreases across a period.
(c) Rb > Br > Kr. Rb is largest because it has one more energy level and is farthest to the left. Kr is smaller than Br because Kr is farther to the right in Period 4.
(d) Rb > Sr > Ca. Ca is smallest because it has one fewer energy level. Sr is smaller than Rb because it is farther to the right.
Check From Figure 8.15, we see that the rankings are correct.

FOLLOW-UP PROBLEM 8.3 Using only the periodic table, rank the elements in each set in order of *increasing* size: **(a)** Se, Br, Cl; **(b)** I, Xe, Ba.

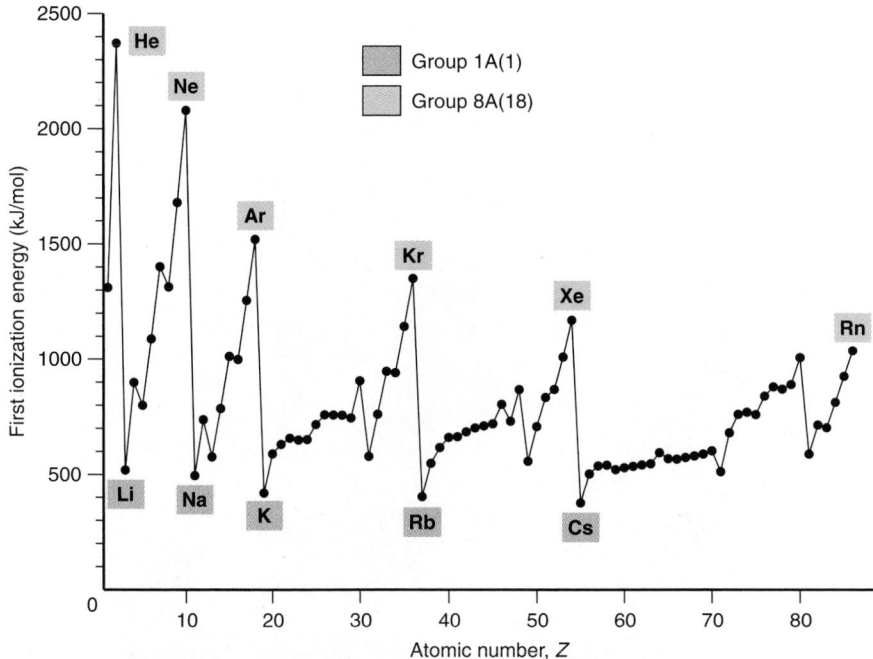

Figure 8.17 **Periodicity of first ionization energy (IE$_1$).** A plot of IE$_1$ vs. atomic number for the elements in Periods 1 through 6 shows a periodic pattern: the lowest values occur for the alkali metals (*brown*) and the highest for the noble gases (*purple*). This is the *inverse* of the trend in atomic size (see Figure 8.16).

Trends in Ionization Energy

The **ionization energy (IE)** is the energy required for the *complete removal* of 1 mol of electrons from 1 mol of gaseous atoms or ions. Pulling an electron away from a nucleus *requires* energy to overcome the attraction. Because energy flows *into* the system, the ionization energy is always positive (like ΔH of an endothermic reaction).

In Chapter 7, you saw that the ionization energy of the H atom is the energy difference between $n = 1$ and $n = \infty$, the point at which the electron is completely removed. Many-electron atoms can lose more than one electron. The first ionization energy (IE$_1$) removes an outermost electron (highest energy sublevel) from the gaseous atom:

$$\text{Atom}(g) \longrightarrow \text{ion}^+(g) + \text{e}^- \qquad \Delta E = \text{IE}_1 > 0$$

The second ionization energy (IE$_2$) removes a second electron. This electron is pulled away from a positively charged ion, so IE$_2$ is always larger than IE$_1$:

$$\text{Ion}^+(g) \longrightarrow \text{ion}^{2+}(g) + \text{e}^- \qquad \Delta E = \text{IE}_2 \text{ (always > IE}_1)$$

The first ionization energy is a key factor in an element's chemical reactivity because, as you'll see, *atoms with a low IE$_1$ tend to form cations during reactions, whereas those with a high IE$_1$ (except the noble gases) often form anions.*

Variations in First Ionization Energy The elements exhibit a periodic pattern in first ionization energy, as shown in Figure 8.17. By comparing this figure with Figure 8.16, you can see a roughly *inverse* relationship between IE$_1$ and atomic size: *as size decreases, it takes more energy to remove an electron.* This inverse relationship appears throughout the groups and periods of the table. Let's examine the two trends and their exceptions.

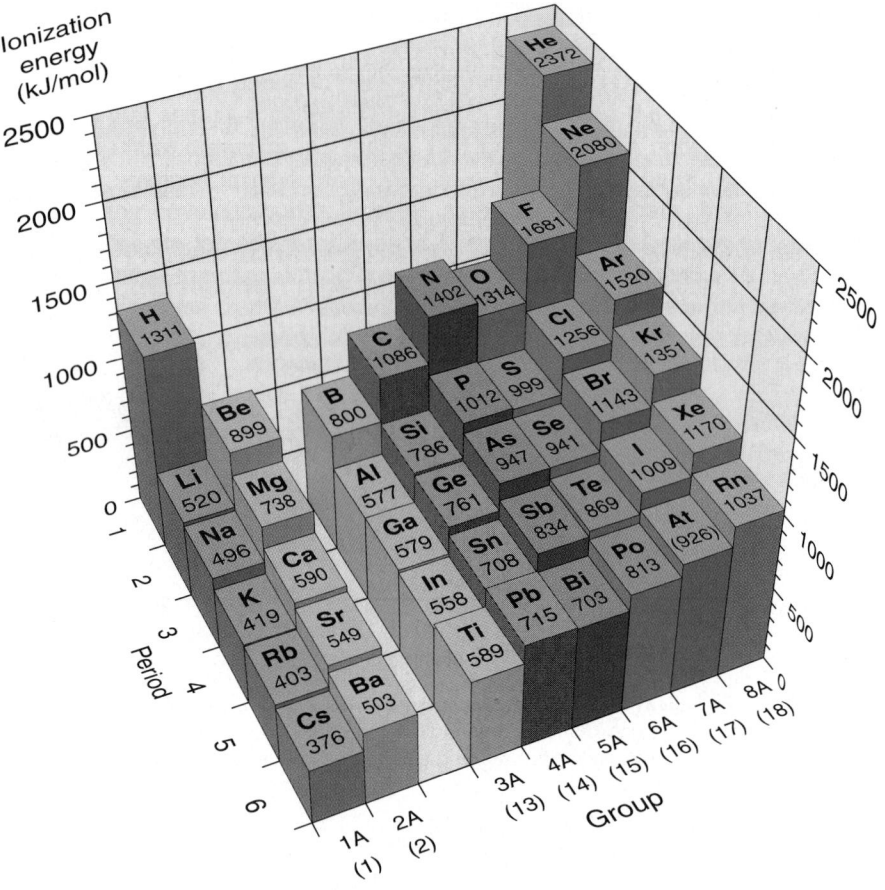

1. *Down a group.* As we move *down* a main group, atomic size increases. As the distance from nucleus to outermost electron increases, the attraction between them lessens, so the electron is easier to remove. Figure 8.18 shows that *ionization energy generally **decreases** down a group.*

The only significant exception to this pattern occurs in Group 3A(13), right after the transition series, and is due to the effect of the series on atomic size: IE₁ decreases from boron (B) to aluminum (Al), but not for the rest of the group. Filling the *d* sublevels in Periods 4, 5, and 6 causes a greater-than-expected Z_{eff}, which holds the outer electrons more tightly in the larger Group 3A members.

2. *Across a period.* As we move across a period, Z_{eff} increases, so atomic size decreases. As a result, the attraction between nucleus and outer electrons increases, so an electron is harder to remove. *Ionization energy generally **increases** across a period.* Clearly, it is easier to remove an electron from an alkali metal than from a noble gas.

There are several small "dips" in the otherwise smooth increase in ionization energy. They occur in Group 3A(13) for B and Al and in Group 6A(16) for O and S. The dips in Group 3A occur because these electrons are the first in the *np* sublevel. This sublevel is higher in energy than the *ns*, so the electron is pulled off more easily, leaving a stable, filled *ns* sublevel. The dips in Group 6A occur because the np^4 electron is the first to pair up with another *np* electron, and electron-electron repulsions raise the orbital energy. Removing this electron relieves the repulsions and leaves a stable, half-filled *np* sublevel; so the fourth *p* electron comes off more easily than the third one does.

SAMPLE PROBLEM 8.4 Ranking Elements by First Ionization Energy

Problem Using the periodic table only, rank the elements in each of the following sets in order of *decreasing* IE$_1$:
(a) Kr, He, Ar   **(b)** Sb, Te, Sn   **(c)** K, Ca, Rb   **(d)** I, Xe, Cs
Plan As in Sample Problem 8.3, we first find the elements in the periodic table and then apply the general trends of decreasing IE$_1$ down a group and increasing IE$_1$ across a period.
Solution (a) He > Ar > Kr. These three are all in Group 8A(18), and IE$_1$ decreases down a group.
(b) Te > Sb > Sn. These three are all in Period 5, and IE$_1$ increases across a period.
(c) Ca > K > Rb. IE$_1$ of K is larger than IE$_1$ of Rb because K is higher in Group 1A(1). IE$_1$ of Ca is larger than IE$_1$ of K because Ca is farther to the right in Period 4.
(d) Xe > I > Cs. IE$_1$ of I is smaller than IE$_1$ of Xe because I is farther to the left. IE$_1$ of I is larger than IE$_1$ of Cs because I is farther to the right and in the previous period.
Check Since trends in IE$_1$ are generally the opposite of the trends in size, you can rank the elements by size and check that you obtain the reverse order.

FOLLOW-UP PROBLEM 8.4 Rank the elements in each of the following sets in order of *increasing* IE$_1$: **(a)** Sb, Sn, I; **(b)** Sr, Ca, Ba.

Variations in Successive Ionization Energies Successive ionization energies (IE$_1$, IE$_2$, and so on) of a given element increase because each electron is pulled away from an ion with a higher and higher positive charge. Note from Figure 8.19, however, that this increase is not smooth, but includes an enormous jump.

A more complete picture is presented in Table 8.5, which shows successive ionization energies for the elements in Period 2 and the first element in Period 3. Move horizontally through the values for a given element, and you reach a point that separates relatively low from relatively high IE values (shaded area to right of line). This jump appears *after* the outer (valence) electrons have been removed and, thus, reflects the much greater energy needed to remove an inner (core) electron. For example, follow the values for boron (B): IE$_1$ is lower than IE$_2$, which is lower than IE$_3$, which is *much* lower than IE$_4$. Thus, boron has three electrons in the highest energy level ($1s^2 2s^2 2p^1$). Because of the significantly greater energy needed to remove core electrons, they are *not* involved in chemical reactions.

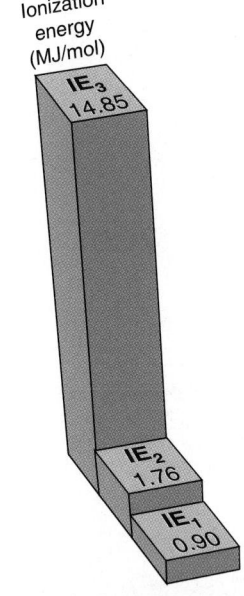

Ionization energy (MJ/mol)

IE$_3$ 14.85

IE$_2$ 1.76

IE$_1$ 0.90

Figure 8.19 **The first three ionization energies of beryllium (in MJ/mol).** Successive ionization energies always increase, but an exceptionally large increase occurs when the first core electron is removed. For Be, this occurs with the third electron (IE$_3$). (Also see Table 8.5.)

Table 8.5 Successive Ionization Energies of the Elements Lithium Through Sodium

Z	Element	Number of Valence Electrons	Ionization Energy (MJ/mol)*									
			IE$_1$	IE$_2$	IE$_3$	IE$_4$	IE$_5$	IE$_6$	IE$_7$	IE$_8$	IE$_9$	IE$_{10}$
3	Li	1	0.52	7.30	11.81							
4	Be	2	0.90	1.76	14.85	21.01			Core electrons			
5	B	3	0.80	2.43	3.66	25.02	32.82					
6	C	4	1.09	2.35	4.62	6.22	37.83	47.28				
7	N	5	1.40	2.86	4.58	7.48	9.44	53.27	64.36			
8	O	6	1.31	3.39	5.30	7.47	10.98	13.33	71.33	84.08		
9	F	7	1.68	3.37	6.05	8.41	11.02	15.16	17.87	92.04	106.43	
10	Ne	8	2.08	3.95	6.12	9.37	12.18	15.24	20.00	23.07	115.38	131.43
11	Na	1	0.50	4.56	6.91	9.54	13.35	16.61	20.11	25.49	28.93	141.37

*MJ/mol, or megajoules per mole = 10^3 kJ/mol.

SAMPLE PROBLEM 8.5 Identifying an Element from Successive Ionization Energies

Problem Name the Period 3 element with the following ionization energies (in kJ/mol) and write its electron configuration:

IE_1	IE_2	IE_3	IE_4	IE_5	IE_6
1012	1903	2910	4956	6278	22,230

Plan We look for a large jump in the IE values, which occurs after all valence electrons have been removed. Then we refer to the periodic table to find the Period 3 element with this number of valence electrons and write its electron configuration.

Solution The exceptionally large jump occurs after IE_5, so the element has five valence electrons and, thus, is in Group 5A(15). This Period 3 element is phosphorus (P; Z = 15). Its electron configuration is $1s^22s^22p^63s^23p^3$.

FOLLOW-UP PROBLEM 8.5 Element Q is in Period 3 and has the following ionization energies (in kJ/mol):

IE_1	IE_2	IE_3	IE_4	IE_5	IE_6
577	1816	2744	11,576	14,829	18,375

Name element Q and write its electron configuration.

Trends in Electron Affinity

The **electron affinity (EA)** is the energy change accompanying the *addition* of 1 mol of electrons to 1 mol of gaseous atoms or ions. As with ionization energy, there is a first electron affinity, a second, and so forth. The *first electron affinity* (EA_1) accompanies the formation of 1 mol of monovalent (1−) gaseous anions:

$$\text{Atom}(g) + e^- \longrightarrow \text{ion}^-(g) \qquad \Delta E = EA_1$$

In most cases, *energy is released when the first electron is added* because it is attracted to the atom's nuclear charge. Thus, EA_1 is usually negative (just as ΔH for an exothermic reaction is negative).* The second electron affinity (EA_2), on the other hand, is always positive because energy must be *absorbed* in order to overcome electrostatic repulsions and add another electron to a negative ion.

Factors other than Z_{eff} and atomic size affect electron affinities, so trends are not as regular as those for the previous two properties. For instance, we might expect electron affinities to decrease smoothly down a group (smaller negative number) because the nucleus is farther away from an electron being added. But, as Figure 8.20 shows, only Group 1A(1) exhibits this behavior. We might also expect a regular increase in electron affinities across a period (larger negative number) because size decreases and the increasing Z_{eff} should attract the electron being added more strongly. An overall left-to-right increase in magnitude is there, but we certainly would not say that it is a regular increase. These exceptions arise from changes in sublevel energy and in electron-electron repulsion.

1A (1)	2A (2)	3A (13)	4A (14)	5A (15)	6A (16)	7A (17)	8A (18)
H −72.8							He (0.0)
Li −59.6	Be (+18)	B −26.7	C −122	N +7	O −141	F −328	Ne (+29)
Na −52.9	Mg (+21)	Al −42.5	Si −134	P −72.0	S −200	Cl −349	Ar (+35)
K −48.4	Ca (+186)	Ga −28.9	Ge −119	As −78.2	Se −195	Br −325	Kr (+39)
Rb −46.9	Sr (+146)	In −28.9	Sn −107	Sb −103	Te −190	I −295	Xe (+41)
Cs −45.5	Ba (+46)	Tl −19.3	Pb −35.1	Bi −91.3	Po −183	At −270	Rn (+41)

Figure 8.20 Electron affinities of the main-group elements. The electron affinities (in kJ/mol) of the main-group elements are shown. Negative values indicate that energy is released when the anion forms. Positive values, which occur in Groups 2A(2) and 8A(18), indicate that energy is absorbed to form the anion; in fact, these anions are unstable and the values are estimated.

*Tables of first electron affinities often list them as positive values, showing the *quantity* of energy released, rather than the difference between the final and initial energy values. Keep this convention in mind when researching these values in reference texts. Electron affinities are more difficult to measure than ionization energies, so values are frequently being corrected as more accurate methods become available.

Despite irregularities, three key points emerge when we examine ionization energy and electron affinity values:

1. *Reactive nonmetals.* The elements in Groups 6A(16) and especially 7A(17) (halogens) have high ionization energies and highly negative (exothermic) electron affinities. These elements lose electrons with difficulty but attract them strongly. Therefore, *in their ionic compounds, they form negative ions.*

2. *Reactive metals.* The elements in Group 1A(1) have low ionization energies and slightly negative (exothermic) electron affinities. Those in Group 2A(2) have low first *and* second ionization energies and positive (endothermic) electron affinities. Both groups lose electrons readily but attract them only weakly, if at all. Therefore, *in their ionic compounds, they form positive ions.*

3. *Noble gases.* The elements in Group 8A(18) have very high ionization energies and slightly positive (endothermic) electron affinities. Therefore, *these elements do **not** tend to lose or gain electrons.* In fact, only the larger members of the group (Kr, Xe, Rn) form any compounds at all.

In the next section, we'll continue to focus on the relation between atomic properties and chemical behavior.

SECTION SUMMARY

Trends in three atomic properties are summarized in Figure 8.21. Atomic size increases down a main group and decreases across a period. Across a transition series, size remains relatively constant. First ionization energy (the energy required to remove the outermost electron from a mole of gaseous atoms) is inversely related to atomic size: IE_1 decreases down a main group and increases across a period. Successive ionization energies show an exceptionally large increase when the first inner (core) electron is removed. Electron affinity (the energy involved in adding an electron to a mole of gaseous atoms) shows many variations from expected trends. As a result of the relative sizes of IEs and EAs, in their ionic compounds, Group 1A(1) and 2A(2) elements tend to form cations, and Group 6A(16) and 7A(17) elements tend to form anions.

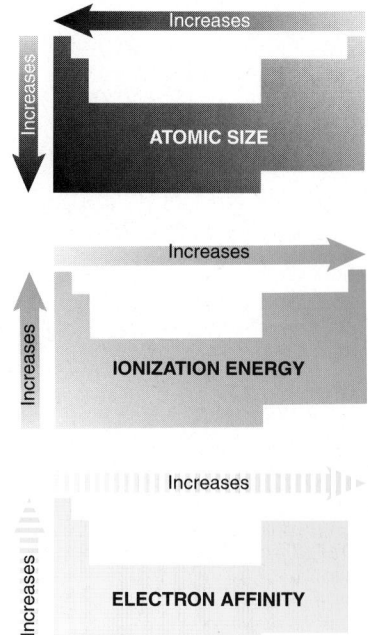

Figure 8.21 **Trends in three atomic properties.** Periodic trends are depicted as gradations in shading on miniature periodic tables, with arrows indicating the direction of general increase in a group or period. (For electron affinity, Group 8A(18) is not shown, and the dashed arrows imply the numerous exceptions to expected trends.)

8.5 THE CONNECTION BETWEEN ATOMIC STRUCTURE AND CHEMICAL REACTIVITY

Our main purpose for discussing atomic properties is, of course, to see how they affect element behavior. In this section, you'll see how the properties we just examined influence metallic behavior and determine the type of ion an element can form, as well as how electron configuration relates to magnetic properties.

Trends in Metallic Behavior

Metals are located in the left and lower three-quarters of the periodic table. They are typically shiny solids that have moderate to high melting points, are good thermal and electrical conductors, and tend to lose electrons in reactions with nonmetals. *Nonmetals* are located in the upper right quarter of the table. They are typically not shiny, have relatively low melting points, are poor thermal and electrical conductors, and tend to gain electrons in reactions with metals. *Metalloids* are located in the region between the other two classes and have properties between them as well. Thus, *metallic behavior decreases left to right and increases top to bottom* in the periodic table (Figure 8.22).

It's important to keep in mind, however, that all the elements may not fall neatly into our classifications. For instance, carbon is a nonmetal, but in the form of graphite, it is a good electrical conductor. Iodine, another nonmetal, is a shiny solid. Gallium and cesium are both metals, but they melt at temperatures below body temperature, and mercury is a liquid at room temperature. Although such exceptions exist, we can make several generalizations about metallic behavior.

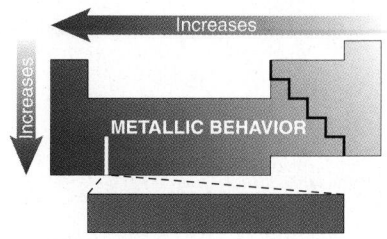

Figure 8.22 **Trends in metallic behavior.** The gradation in metallic behavior among the elements is depicted as a gradation in shading from bottom left to top right, with arrows showing the direction of increase. Elements that behave as metals appear in the left and lower three-quarters of the table. (Because hydrogen is a nonmetal, it appears next to helium in this periodic table.)

Relative Tendency to Lose Electrons Metals tend to lose electrons during chemical reactions because they have low ionization energies compared to nonmetals. The increase in metallic behavior down a group is most obvious in the physical and chemical behavior of the elements in Groups 3A(13) through 6A(16), which contain more than one class of element. For example, consider the elements in Group 5A(15), which appear vertically in Figure 8.23. Here, the change is so great that, with regard to monatomic ions, *elements at the top tend to form anions and those at the bottom tend to form cations.* Nitrogen (N) is a gaseous nonmetal, and phosphorus (P) a solid nonmetal. Both occur occasionally as 3− anions in their compounds. Arsenic (As) and antimony (Sb) are metalloids, with Sb the more metallic of the two; neither forms ions readily. Bismuth (Bi), the largest member, is a typical metal, forming mostly ionic compounds in which it appears as a 3+ cation. Even in Group 2A(2), which consists entirely of metals, the tendency to form cations increases down the group. Beryllium (Be), for example, forms covalent compounds with nonmetals, whereas the compounds of barium (Ba) are ionic.

As we move across a period, it becomes more difficult to lose an electron (IE increases) and easier to gain one (EA becomes more negative). Therefore, with regard to monatomic ions, *elements at the left tend to form cations and those at the right tend to form anions.* The typical decrease in metallic behavior across a period is clear among the elements in Period 3, which appear horizontally in Figure 8.23. Sodium and magnesium are metals. Sodium is shiny when freshly cut under mineral oil, but it loses an electron so readily to O_2 that, if cut in air, its surface is coated immediately with a dull oxide. These metals exist naturally as Na^+ and Mg^{2+} ions in oceans, minerals, and organisms. Aluminum is metallic in its physical properties and forms the Al^{3+} ion in some compounds, but it bonds covalently in most others. Silicon (Si) is a shiny metalloid that does not occur as a monatomic ion. The most common form of phosphorus is a white, waxy nonmetal that, as noted above, forms the P^{3-} ion in a few compounds. Sulfur is a crumbly yellow nonmetal that forms the sulfide ion (S^{2-}) in many compounds. Diatomic chlorine (Cl_2) is a yellowish, gaseous nonmetal that attracts electrons avidly and exists in nature as the Cl^- ion.

Acid-Base Behavior of the Element Oxides Metals are also distinguished from nonmetals by the acid-base behavior of their oxides in water:

- Most main-group metals *transfer* electrons to oxygen, so their *oxides are ionic. In water, these oxides act as bases,* producing OH^- ions and reacting with acids. Calcium oxide is an example (turns indicator pink in photo below, *left*).
- Nonmetals *share* electrons with oxygen, so *nonmetal oxides are covalent. In water, they act as acids,* producing H^+ ions and reacting with bases. Tetraphosphorus decaoxide is an example (turns indicator yellow in photo below, *right*).

Figure 8.23 **The change in metallic behavior in Group 5A(15) and Period 3.**
Moving down from nitrogen to bismuth shows an *increase* in metallic behavior
(and decrease in ionization energy). Moving left to right from sodium to chlorine
shows a *decrease* in metallic behavior (and general increase in ionization energy).
Each box shows the element and its atomic number, symbol, and first ionization
energy (in kJ/mol).

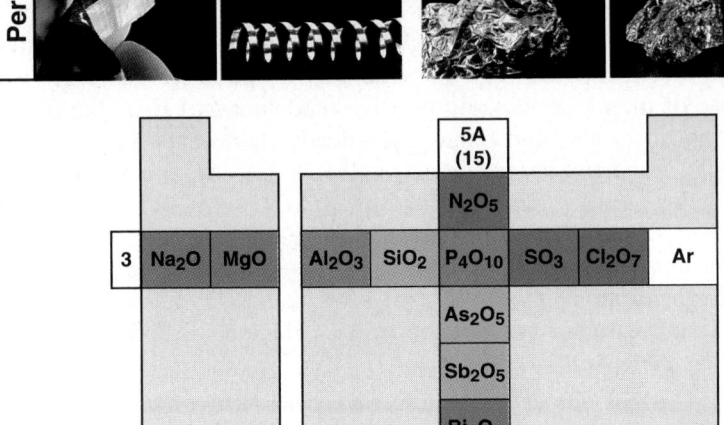

Figure 8.24 **The trend in acid-base behavior of element oxides.** The trend in
acid-base behavior for common oxides of Group 5A(15) and Period 3 elements is
shown as a gradation in color (*red* = acidic; *blue* = basic). Note that the metals
form basic oxides and the nonmetals form acidic oxides. Aluminum forms an oxide (*purple*) that can act as an acid and as a base. Thus, as atomic size increases,
oxide basicity increases.

Some metals and many metalloids form oxides that are **amphoteric:** they can act
as acids *and* as bases in water.

Figure 8.24 classifies the acid-base behavior of some common oxides, focusing once again on the elements in Group 5A(15) and Period 3. Note that *as the
elements become more metallic down a group, their oxides become more basic.*
In Group 5A, dinitrogen pentaoxide, N_2O_5, forms nitric acid:

$$N_2O_5(s) + H_2O(l) \longrightarrow 2HNO_3(aq)$$

Tetraphosphorus decaoxide, P_4O_{10}, forms the weaker acid H_3PO_4:

$$P_4O_{10}(s) + 6H_2O(l) \longrightarrow 4H_3PO_4(aq)$$

The oxide of the metalloid arsenic is weakly acidic, whereas that of the metalloid antimony is weakly basic. Bismuth, the most metallic of the group, forms a
basic oxide that is insoluble in water but that forms a salt and water with acid:

$$Bi_2O_3(s) + 6HNO_3(aq) \longrightarrow 2Bi(NO_3)_3(aq) + 3H_2O(l)$$

Note that *as the elements become less metallic across a period, their oxides
become more acidic.* In Period 3, sodium and magnesium form the strongly basic
oxides Na_2O and MgO. Metallic aluminum forms amphoteric aluminum oxide
(Al_2O_3), which reacts with acid and with base:

$$Al_2O_3(s) + 6HCl(aq) \longrightarrow 2AlCl_3(aq) + 3H_2O(l)$$
$$Al_2O_3(s) + 2NaOH(aq) + 3H_2O(l) \longrightarrow 2NaAl(OH)_4(aq)$$

Silicon dioxide is weakly acidic, forming a salt and water with base:

$$SiO_2(s) + 2NaOH(aq) \longrightarrow Na_2SiO_3(aq) + H_2O(l)$$

The common oxides of phosphorus, sulfur, and chlorine form acids of increasing
strength: H_3PO_4, H_2SO_4, and $HClO_4$.

Properties of Monatomic Ions

So far we've focused on the reactants—the atoms—in the process of electron loss and gain. Now we focus on the products—the ions. In particular, we examine electron configurations, magnetic properties, and sizes of ions relative to those of their parent atoms.

Electron Configurations of Main-Group Ions In Chapter 2, you learned the symbols and charges of many monatomic ions. But *why* does an ion have that charge in its compounds? Why is a sodium ion Na^+ and not Na^{2+}, and why is a fluoride ion F^- and not F^{2-}? For elements at the left and right ends of the periodic table, the explanation concerns the very low reactivity of the noble gases. As we said earlier, because of their high IEs and positive (endothermic) EAs, the noble gases typically do not form ions and remain chemically stable with a *filled* outer energy level (ns^2np^6). *Elements in Groups 1A(1), 2A(2), 6A(16), and 7A(17) that readily form ions either lose or gain electrons to attain a filled outer level and thus a noble gas configuration.* Their ions are said to be **isoelectronic** (Greek *iso*, "same") with the nearest noble gas (Figure 8.25; also see Figure 2.15, p. 61).

When an alkali metal atom [Group 1A(1)] loses its single valence electron, it becomes isoelectronic with the *previous* noble gas. The Na^+ ion, for example, is isoelectronic with neon (Ne):

$$Na\ (1s^22s^22p^63s^1) \longrightarrow Na^+\ (1s^22s^22p^6)\ [\text{isoelectronic with Ne } (1s^22s^22p^6)] + e^-$$

When a halogen atom [Group 7A(17)] adds a single electron to the five in its *np* sublevel, it becomes isoelectronic with the *next* noble gas. Bromide ion, for example, is isoelectronic with krypton (Kr):

$$Br\ ([Ar]\ 4s^23d^{10}4p^5) + e^- \longrightarrow Br^-\ ([Ar]\ 4s^23d^{10}4p^6)\ [\text{isoelectronic with Kr } ([Ar]\ 4s^23d^{10}4p^6)]$$

The energy needed to remove the electrons from metals to attain the previous noble gas configuration is supplied during their exothermic reactions with nonmetals. Removing more than one electron from Na or more than two from Mg means removing core electrons, which requires 5 to 10 times as much energy as is available in a reaction. This is the reason Na^{2+} or Mg^{3+} ions do *not* form. Similarly, adding more than one electron to F to form F^{2-} or more than two to O to form O^{3-} means placing these electrons in a higher outer energy level, which also requires much more energy than is available from a typical reaction.

The larger metals of Groups 3A(13), 4A(14), and 5A(15) form cations through a different process, because it would be energetically impossible for them to lose enough electrons to attain a noble gas configuration. For example, tin (Sn; $Z = 50$) would have to lose 14 electrons—two $5p$, ten $4d$, and two $5s$—to be isoelectronic with krypton (Kr; $Z = 36$), the previous noble gas. Instead, tin loses far fewer electrons and attains two different stable configurations. In the tin(IV) ion (Sn^{4+}), the metal atom empties its outer energy level and attains the stability of empty $5s$ and $5p$ sublevels and a filled inner $4d$ sublevel. This $(n - 1)d^{10}$ configuration is called a **pseudo–noble gas configuration:**

$$Sn\ ([Kr]\ 5s^24d^{10}5p^2) \longrightarrow Sn^{4+}\ ([Kr]\ 4d^{10}) + 4e^-$$

Alternatively, in the more common tin(II) ion (Sn^{2+}), the atom loses the two $5p$ electrons only and attains the stability of filled $5s$ and $4d$ sublevels:

$$Sn\ ([Kr]\ 5s^24d^{10}5p^2) \longrightarrow Sn^{2+}\ ([Kr]\ 5s^24d^{10}) + 2e^-$$

The retained ns^2 electrons are sometimes called an *inert pair* because they seem difficult to remove. Thallium, lead, and bismuth, the largest and most metallic members of Groups 3A(13) to 5A(15), commonly form ions that retain the ns^2 pair of electrons: Tl^+, Pb^{2+}, and Bi^{3+}.

Excessively high energy cost is also the reason that some elements do not form monatomic ions at all in their reactions. For instance, carbon would have to lose four electrons to form C^{4+} and attain the He configuration, or gain four to

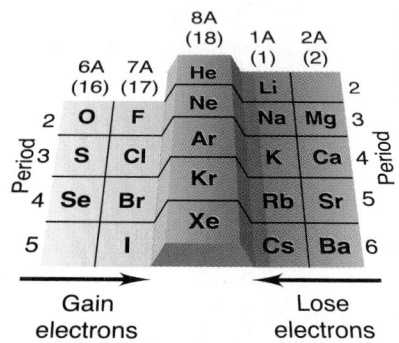

Figure 8.25 **Main-group ions and the noble gas electron configurations.** Most of the elements that form monatomic ions that are isoelectronic with a noble gas lie in the four groups that flank Group 8A(18), two on either side.

form C^{4-} and attain the Ne configuration, but neither ion forms. (Such multivalent ions *are* observed in the spectra of stars, however, where temperatures exceed 10^6 K.) As you'll see in Chapter 9, carbon and other atoms that do not form ions attain a filled shell by *sharing* electrons through covalent bonding.

SAMPLE PROBLEM 8.6 Writing Electron Configurations of Main-Group Ions

Problem Using condensed electron configurations, write reactions for the formation of the common ions of the following elements:
(a) Iodine ($Z = 53$)   (b) Potassium ($Z = 19$)   (c) Indium ($Z = 49$)
Plan We identify the element's position in the periodic table and recall two general points:
- Ions of elements in Groups 1A(1), 2A(2), 6A(16), and 7A(17) are typically isoelectronic with the nearest noble gas.
- Metals in Groups 3A(13) to 5A(15) can lose the *ns* and *np* electrons or just the *np* electrons.
Solution (a) Iodine is in Group 7A(17), so it gains one electron and is isoelectronic with xenon:
$$I\ ([Kr]\ 5s^2 4d^{10} 5p^5) + e^- \longrightarrow I^-\ ([Kr]\ 5s^2 4d^{10} 5p^6)\ (\text{same as Xe})$$

(b) Potassium is in Group 1A(1), so it loses one electron and is isoelectronic with argon:
$$K\ ([Ar]\ 4s^1) \longrightarrow K^+\ ([Ar]) + e^-$$

(c) Indium is in Group 3A(13), so it loses either three electrons to form In^{3+} (pseudo–noble gas) or one to form In^+ (inert pair):
$$In\ ([Kr]\ 5s^2 4d^{10} 5p^1) \longrightarrow In^{3+}\ ([Kr]\ 4d^{10}) + 3e^-$$
$$In\ ([Kr]\ 5s^2 4d^{10} 5p^1) \longrightarrow In^+\ ([Kr]\ 5s^2 4d^{10}) + e^-$$

Check Be sure that the number of electrons in the ion's electron configuration, plus those gained or lost to form the ion, equals Z.

FOLLOW-UP PROBLEM 8.6 Using condensed electron configurations, write reactions showing formation of the common ions of the following elements:
(a) Ba ($Z = 56$)   (b) O ($Z = 8$)   (c) Pb ($Z = 82$)

Electron Configurations of Transition Metal Ions In contrast to most main-group ions, *transition metal ions rarely attain a noble gas configuration*, and the reason, once again, is that energy costs are too high. The exceptions in Period 4 are scandium, which forms Sc^{3+}, and titanium, which occasionally forms Ti^{4+} in some compounds. The typical behavior of a transition element is to *form more than one cation by losing all of its ns and some of its (n − 1)d electrons.* (We focus here on the Period 4 transition series, but these points hold for the Periods 5 and 6 series as well.)

In the aufbau process of building up the ground-state atoms, Period 3 ends with the noble gas argon. At the beginning of Period 4, the radial probability distribution of the 4s orbital near the nucleus makes it more stable than the empty 3d. Therefore, the first and second electrons added in the period enter the 4s in K and Ca. But, as soon as we reach the transition elements and the 3d orbitals begin to fill, the increasing nuclear charge attracts their electrons more and more strongly. Moreover, the added 3d electrons fill inner orbitals, so they are not very well shielded from the increasing nuclear charge by the 4s electrons. As a result, the 3d orbital becomes *more stable* than the 4s. In effect, a *crossover in orbital energy* occurs as we enter the transition series (Figure 8.26). The effect on ion formation is critical: because the 3d orbitals are more stable, *the 4s electrons are lost before the 3d electrons to form the Period 4 transition metal ions*. Thus, the 4s electrons are added before the 3d to form the atom and lost before the 3d to form the ion: "first-in, first-out."

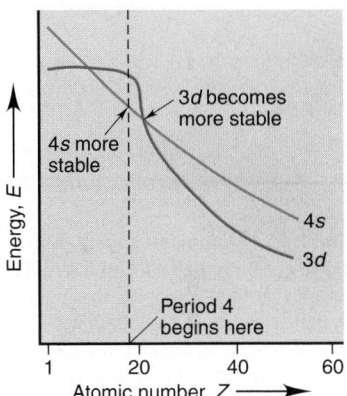

Figure 8.26 The Period 4 crossover in sublevel energies. The 3d orbitals are empty in elements at the beginning of Period 4. Because the 4s electron penetrates closer to the nucleus, the energy of the 4s orbital is lower in K and Ca; thus, the 4s fills before the 3d. But as the 3d orbitals fill, beginning with $Z > 20$, these inner electrons are attracted by the increasing nuclear charge, and they also shield the 4s electrons. As a result, there is an energy crossover such that the 3d sublevel becomes lower in energy than the 4s. As a result, the 4s electrons are removed first when the transition metal ion forms. In other words,
- For a main-group metal ion, the highest *n* level of electrons is "last-in, first-out."
- For a transition metal ion, the highest *n* level of electrons is "first-in, first out."

To summarize, *electrons with the highest n value are removed first.* Here are a few simple rules for forming the ion of any main-group or transition element:

- For main-group, *s*-block metals, remove all electrons with the highest *n* value.
- For main-group, *p*-block metals, remove *np* electrons before *ns* electrons.
- For transition (*d*-block) metals, remove *ns* electrons before $(n − 1)d$ electrons.
- For nonmetals, add electrons to the *p* orbitals of highest *n* value.

Magnetic Properties of Transition Metal Ions If we can't see electrons in orbitals, how do we know that a particular electron configuration is correct? Although analysis of atomic spectra is the most important method for determining configuration, the magnetic properties of an element and its compounds can support or refute conclusions from spectra. Recall that an electron's spin generates a tiny magnetic field, which causes a beam of H atoms to split in an external magnetic field (see Figure 8.1). Only chemical species (atoms, ions, or molecules) with one or more *unpaired* electrons are affected by the external field. The species used in the original 1921 split-beam experiment was the silver atom:

$$\text{Ag } (Z = 47) \quad [\text{Kr}] \; 5s^1 4d^{10}$$

$$\boxed{\uparrow}_{5s} \quad \boxed{\uparrow\downarrow \; \uparrow\downarrow \; \uparrow\downarrow \; \uparrow\downarrow \; \uparrow\downarrow}_{4d} \quad \boxed{\;\;}_{5p}$$

Note the unpaired 5*s* electron. A beam of cadmium atoms, the element after silver, is not split because its 5*s* electrons are *paired* (Cd: $[\text{Kr}] \; 5s^2 4d^{10}$).

A species with unpaired electrons exhibits **paramagnetism:** it is attracted by an external magnetic field. A species with all electrons paired exhibits **diamagnetism:** it is not attracted (and, in fact, is slightly repelled) by a magnetic field. Figure 8.27 shows how this magnetic behavior is studied. Many transition metals and their compounds are paramagnetic because their atoms and ions have unpaired electrons.

Figure 8.27 Apparatus for measuring the magnetic behavior of a sample. The substance is weighed on a very sensitive balance in the absence of an external magnetic field. **A,** If the substance is diamagnetic (has all *paired* electrons), its apparent mass is unaffected (or slightly reduced) when the magnetic field is "on." **B,** If the substance is paramagnetic (has *unpaired* electrons), its apparent mass increases when the field is "on" because the balance arm feels an additional force. This method is used to estimate the number of unpaired electrons in transition metal compounds.

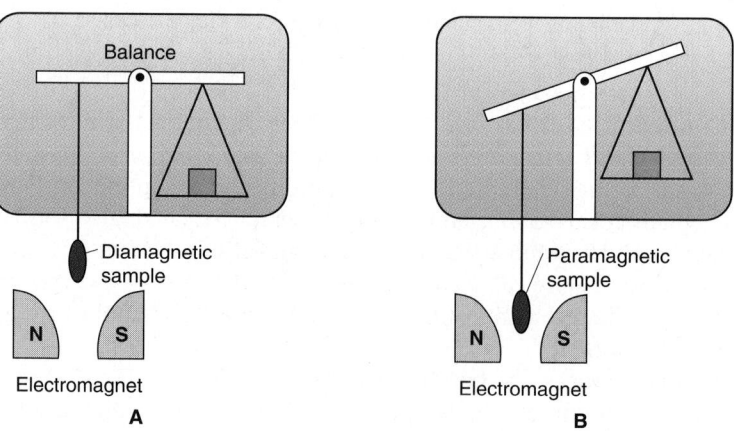

Let's see how studies of paramagnetism might be used to provide additional evidence for a proposed electron configuration. Spectral analysis of the titanium atom yields the configuration $[\text{Ar}] \; 4s^2 3d^2$. Experiment shows that Ti metal is paramagnetic, which is consistent with the presence of unpaired electrons in its atoms. Spectral analysis of the Ti^{2+} ion yields the configuration $[\text{Ar}] \; 3d^2$, indicating loss of the two 4*s* electrons. Once again, experiment supports these findings by showing that Ti^{2+} compounds are paramagnetic. If Ti had lost its two 3*d* electrons during ion formation, its compounds would be diamagnetic because the 4s electrons are paired. Thus, the [Ar] $3d^2$ configuration supports the conclusion that electrons of highest n value are lost first:

$$\text{Ti ([Ar] } 4s^23d^2) \longrightarrow \text{Ti}^{2+} \text{ ([Ar] } 3d^2) + 2e^-$$

The partial orbital diagrams are

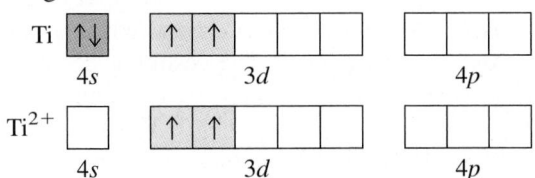

An increase in paramagnetism occurs when iron metal (Fe) forms Fe^{3+} compounds. This fact is consistent with Fe losing its 4s electrons and one of its paired 3d electrons:

$$\text{Fe ([Ar] } 4s^23d^6) \longrightarrow \text{Fe}^{3+} \text{ ([Ar] } 3d^5) + 3e^-$$

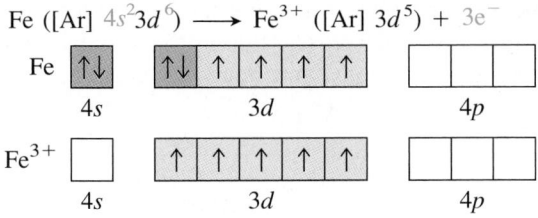

Copper (Cu) is paramagnetic, but zinc (Zn) is diamagnetic, as are the Cu^+ and Zn^{2+} ions. These observations support the electron configurations proposed by spectral analysis. The two ions are isoelectronic:

$$\text{Cu ([Ar] } 4s^13d^{10}) \longrightarrow \text{Cu}^+ \text{ ([Ar] } 3d^{10}) + e^-$$
$$\text{Zn ([Ar] } 4s^23d^{10}) \longrightarrow \text{Zn}^{2+} \text{ ([Ar] } 3d^{10}) + 2e^-$$

$$\text{Cu}^+ \text{ or Zn}^{2+}$$

SAMPLE PROBLEM 8.7 Writing Electron Configurations and Predicting Magnetic Behavior of Transition Metal Ions

Problem Use condensed electron configurations to write the reaction for the formation of each transition metal ion, and predict whether the ion is paramagnetic:
(a) Mn^{2+} (Z = 25)   **(b)** Cr^{3+} (Z = 24)   **(c)** Hg^{2+} (Z = 80)

Plan We first write the condensed electron configuration of the atom, noting the irregularity for Cr in (b). Then we remove electrons, beginning with ns electrons, to attain the ion charge. If unpaired electrons are present, the ion is paramagnetic.

Solution (a) Mn ([Ar] $4s^23d^5$) $\longrightarrow$ Mn^{2+} ([Ar] $3d^5$) + 2e$^-$

There are five unpaired e$^-$, so Mn^{2+} is paramagnetic.

(b) Cr ([Ar] $4s^13d^5$) $\longrightarrow$ Cr^{3+} ([Ar] $3d^3$) + 3e$^-$

There are three unpaired e$^-$, so Cr^{3+} is paramagnetic.

(c) Hg ([Xe] $6s^24f^{14}5d^{10}$) $\longrightarrow$ Hg^{2+} ([Xe] $4f^{14}5d^{10}$) + 2e$^-$

There are no unpaired e$^-$, so Hg^{2+} is *not* paramagnetic (is diamagnetic).

Check We removed the ns electrons first, and the sum of the lost electrons and those in the electron configuration of the ion equals Z.

FOLLOW-UP PROBLEM 8.7 Write the condensed electron configuration of each of the following transition metal ions, and predict whether it is paramagnetic:
(a) V^{3+} (Z = 23)   **(b)** Ni^{2+} (Z = 28)   **(c)** La^{3+} (Z = 57)

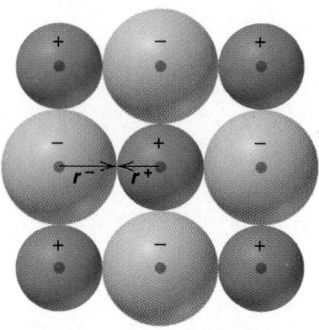

Figure 8.28 **Depicting ionic radius.**
The cation radius (r^+) and the anion radius (r^-) each make up a portion of the total distance between the nuclei of adjacent ions in a crystalline ionic compound.

Ionic Size vs. Atomic Size The **ionic radius** is an estimate of the size of an ion in a crystalline ionic compound. You can picture it as one ion's portion of the distance between the nuclei of neighboring ions in the solid (Figure 8.28). From the relation between effective nuclear charge and atomic size, we can predict the size of an ion relative to its parent atom:

- *Cations are smaller than their parent atoms.* When a cation forms, electrons are *removed from* the outer level. The resulting decrease in electron repulsions allows the nuclear charge to pull the remaining electrons closer.
- *Anions are larger than their parent atoms.* When an anion forms, electrons are *added to* the outer level. The increase in repulsions causes the electrons to occupy more space.

Figure 8.29 shows the radii of some common main-group monatomic ions relative to their parent atoms. As you can see, *ionic size increases down a group* because the number of energy levels increases. Across a period, however, the pattern is more complex. Size decreases among the cations, then increases tremendously when we reach the anions, and finally decreases again among the anions.

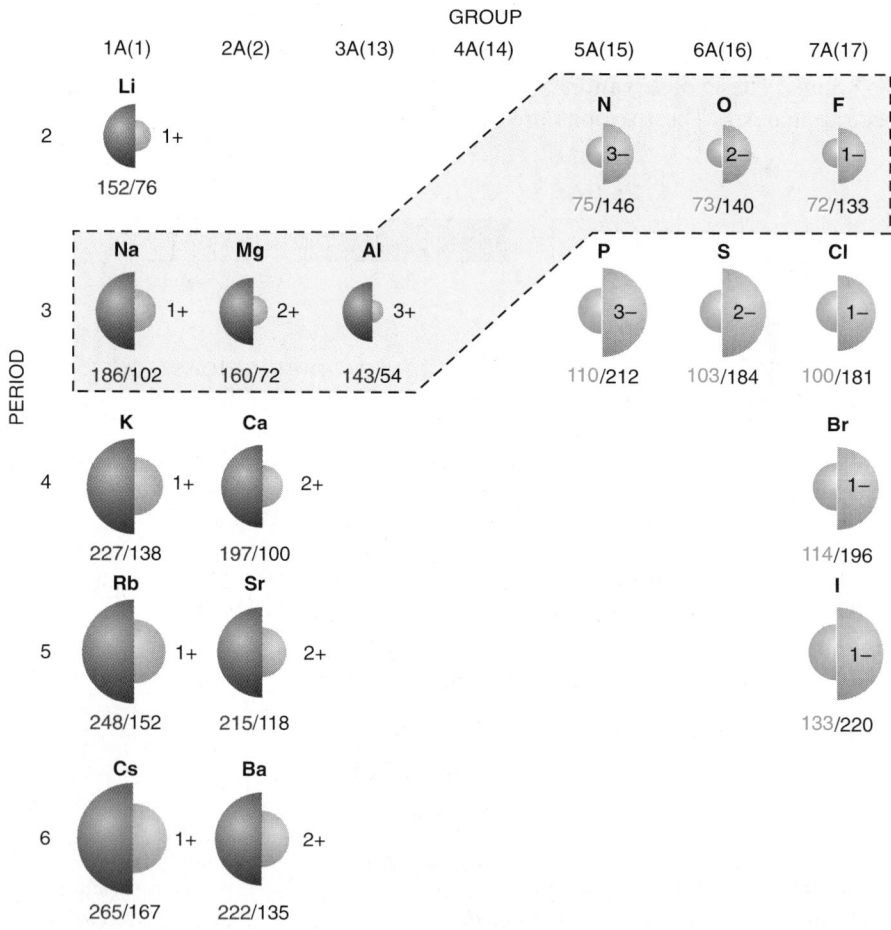

Figure 8.29 **Ionic vs. atomic radii.** The atomic radii (*colored half-spheres*) and ionic radii (*gray half-spheres*) of some main-group elements are arranged in periodic table format (with all radii values in picometers). Note that metals (*blue*) form positive ions that are smaller than the atoms, whereas nonmetals (*red*) form negative ions that are larger than the atoms. The dashed outline sets off ions of Period 2 nonmetals and Period 3 metals that are *isoelectronic* with each other. Note the size decrease from anion to cation.

This pattern results from changes in effective nuclear charge and electron-electron repulsions. In Period 3 (Na through Cl), for example, increasing Z_{eff} from left to right makes Na^+ larger than Mg^{2+}, which in turn is larger than Al^{3+}. The great jump in size from cations to anions occurs because we are *adding* electrons rather than removing them, so repulsions increase sharply. For instance, P^{3-} has eight more electrons than Al^{3+}. Then, the ongoing rise in Z_{eff} makes P^{3-} larger than S^{2-}, which is larger than Cl^-. These factors lead to some striking effects even among ions with the same number of electrons. Look at the ions within the dashed outline in Figure 8.29, which are all isoelectronic with neon. Even though the cations form from elements in the next period, the anions are still much larger. The pattern is

$$3- > 2- > 1- > 1+ > 2+ > 3+$$

ISOELECTRONIC SERIES

When an element forms more than one cation, *the greater the ionic charge, the smaller the ionic radius.* Consider Fe^{2+} and Fe^{3+}. The number of protons is the same, but Fe^{3+} has one fewer electron, so electron repulsions are reduced somewhat. As a result, Z_{eff} increases, which pulls all the electrons closer, so Fe^{3+} is smaller than Fe^{2+}.

To summarize the main points,

- Ionic size increases down a group.
- Ionic size decreases across a period but increases from cation to anion.
- Ionic size decreases with increasing positive (or decreasing negative) charge in an isoelectronic series.
- Ionic size decreases as charge increases for different cations of a given element.

SAMPLE PROBLEM 8.8 Ranking Ions by Size

Problem Rank each set of ions in order of *decreasing* size, and explain your ranking:
(a) Ca^{2+}, Sr^{2+}, Mg^{2+} (b) K^+, S^{2-}, Cl^- (c) Au^+, Au^{3+}
Plan We find the position of each element in the periodic table and apply the ideas presented in the text.
Solution (a) Since Mg^{2+}, Ca^{2+}, and Sr^{2+} are all from Group 2A(2), they decrease in size up the group: $Sr^{2+} > Ca^{2+} > Mg^{2+}$.
(b) The ions K^+, S^{2-}, and Cl^- are isoelectronic. S^{2-} has a lower Z_{eff} than Cl^-, so it is larger. K^+ is a cation, and has the highest Z_{eff}, so it is smallest: $S^{2-} > Cl^- > K^+$.
(c) Au^+ has a lower charge than Au^{3+}, so it is larger: $Au^+ > Au^{3+}$.

FOLLOW-UP PROBLEM 8.8 Rank the ions in each set in order of *increasing* size:
(a) Cl^-, Br^-, F^- (b) Na^+, Mg^{2+}, F^- (c) Cr^{2+}, Cr^{3+}

SECTION SUMMARY

Highly metallic behavior correlates with large atomic size and low ionization energy. Thus, metallic behavior increases down a group and decreases across a period. Within the main groups, metal oxides are basic and nonmetal oxides acidic. Thus, oxides become more acidic across a period and more basic down a group. Many main-group elements form ions that are isoelectronic with the nearest noble gas. Removing (or adding) more electrons than needed to give the noble gas configuration involves a prohibitive amount of energy. Metals in Groups 3A(13) to 5A(15) lose either their np electrons or both their ns and np electrons. Transition metals lose ns electrons before $(n - 1)d$ electrons and commonly form more than one ion. Many transition metals and their compounds are paramagnetic because their atoms (or ions) have unpaired electrons. Cations are smaller and anions larger than their parent atoms. Ionic radius increases down a group. Across a period, cationic and anionic radii decrease, but a large increase occurs from cation to anion.

Chapter Perspective

This chapter is our springboard to understanding the chemistry of the elements. We have begun to see that recurring electron configurations lead to trends in atomic properties, which in turn lead to trends in chemical behavior. With this insight, we can go on to investigate how atoms bond (Chapter 9), how molecular shapes arise (Chapter 10), how molecular shapes and other properties can be explained using certain models (Chapter 11), how the physical properties of liquids and solids emerge from atomic properties (Chapter 12), and how those properties influence the solution process (Chapter 13). We will briefly review these ideas and gain a perspective on where they lead (Interchapter) before we can survey elemental behavior in greater detail (Chapter 14) and see how it applies to the remarkable diversity of organic compounds (Chapter 15). Within this group of chapters, you will see chemical models come alive in chemical facts.

For Review and Reference (Numbers in parentheses refer to pages, unless noted otherwise.)

Learning Objectives

Relevant section and/or sample problem (SP) numbers appear in parentheses.

Understand These Concepts

1. The meaning of the periodic law and the arrangement of elements by atomic number (8.1)
2. The reason for the spin quantum number and its two possible values (8.2)
3. How the exclusion principle applies to orbital filling (8.2)
4. The effects of nuclear charge, shielding, and penetration on the splitting of orbital energies; the meaning of effective nuclear charge (8.2)
5. How the order in the periodic table is based on the order of orbital energies (8.3)
6. How orbitals are filled in main-group and transition elements; the importance of Hund's rule (8.3)
7. How outer electron configuration within a group is related to chemical behavior (8.3)
8. The distinction among inner, outer, and valence electrons (8.3)
9. The meaning of atomic radius, ionization energy, and electron affinity (8.4)
10. How n value and effective nuclear charge give rise to the periodic trends of atomic size and ionization energy (8.4)
11. The importance of core electrons to the pattern of successive ionization energies (8.4)

12. How atomic properties are related to the tendency to form ions (8.4)
13. The general properties of metals and nonmetals (8.5)
14. How the vertical and horizontal trends in metallic behavior are related to ion formation and oxide acidity (8.5)
15. Why main-group ions are either isoelectronic with the nearest noble gas or have a pseudo–noble gas electron configuration (8.5)
16. The origin of paramagnetic and diamagnetic behavior (8.5)
17. The relation between ionic and atomic size and the trends in ionic size (8.5)

Master These Skills

1. Using orbital diagrams and electron configurations to learn the set of quantum numbers for any electron in an atom (SP 8.1)
2. Writing full and condensed electron configurations of an element (SP 8.2)
3. Using periodic trends to rank elements by atomic size and first ionization energy (SPs 8.3 and 8.4)
4. Identifying an element from its successive ionization energies (SP 8.5)
5. Writing electron configurations of main-group and transition metal ions (SPs 8.6 and 8.7)
6. Using periodic trends to rank ions by relative size (SP 8.8)

Key Terms

electron configuration (289)

Section 8.1
periodic law (289)

Section 8.2
spin quantum number (m_s) (291)
exclusion principle (291)
shielding (293)
effective nuclear charge (Z_{eff}) (293)

penetration (294)

Section 8.3
aufbau principle (295)
orbital diagram (295)
Hund's rule (296)
transition elements (300)
inner (core) electrons (302)
outer electrons (302)
valence electrons (302)

inner transition elements (303)
lanthanides (rare earths) (303)
actinides (303)

Section 8.4
metallic radius (304)
covalent radius (304)
ionization energy (IE) (307)
electron affinity (EA) (310)

Section 8.5
amphoteric (313)
isoelectronic (314)
pseudo–noble gas configuration (314)
paramagnetism (316)
diamagnetism (316)
ionic radius (318)

Highlighted Figures and Tables

These figures (F) and tables (T) provide a quick review of key ideas. Entries in color contain frequently used data.

T8.2 Summary of quantum numbers of electrons in atoms (291)

F8.3 Lower nuclear charge makes H less stable than He^+ (293)

F8.4 Shielding by another electron in the orbital makes He less stable than He^+ (293)

F8.5 Shielding by inner electrons makes Li less stable than Li^{2+} (293)

F8.6 Penetration by the $2s$ electron in Li makes the $2s$ orbital more stable than the $2p$ (294)

F8.7 Order for filling energy sublevels with electrons (295)

F8.12 Periodic table of partial ground-state electron configurations (301)

F8.13 Relation between orbital filling and the periodic table (302)

F8.15 Atomic radii of the main-group and transition elements (305)

F8.16 Periodicity of atomic radius (306)

F8.17 Periodicity of first ionization energy (IE_1) (307)

F8.18 First ionization energies of the main-group elements (308)

F8.21 Trends in three atomic properties (311)

F8.22 Trends in metallic behavior (311)

F8.24 Trend in acid-base behavior of element oxides (313)

F8.25 Main-group ions and the noble gas electron configurations (314)

F8.29 Ionic vs. atomic radii (318)

Brief Solutions to Follow-up Problems

8.1 The element has eight electrons, so $Z = 8$: oxygen.

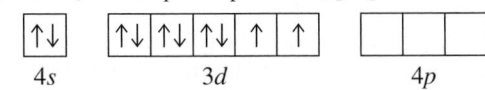

1s   2s   2p

Sixth electron: $n = 2$, $l = 1$, $m_l = 0$, $m_s = +\frac{1}{2}$

8.2 (a) For Ni, $1s^22s^22p^63s^23p^64s^23d^8$; [Ar] $4s^23d^8$

4s   3d   4p

Ni has 18 inner electrons.

(b) For Sr, $1s^22s^22p^63s^23p^64s^23d^{10}4p^65s^2$; [Kr] $5s^2$

5s   4d   5p

Sr has 36 inner electrons.

(c) For Po, $1s^22s^22p^63s^23p^64s^23d^{10}4p^65s^24d^{10}5p^66s^24f^{14}5d^{10}6p^4$

[Xe] $6s^24f^{14}5d^{10}6p^4$

6s   6p

Po has 78 inner electrons.

8.3 (a) Cl < Br < Se; (b) Xe < I < Ba

8.4 (a) Sn < Sb < I; (b) Ba < Sr < Ca

8.5 Q is aluminum: $1s^22s^22p^63s^23p^1$

8.6 (a) Ba ([Xe] $6s^2$) $\longrightarrow$ Ba^{2+} ([Xe]) + $2e^-$

(b) O ([He] $2s^22p^4$) + $2e^- \longrightarrow$
O^{2-} ([He] $2s^22p^6$) (same as Ne)

(c) Pb ([Xe] $6s^24f^{14}5d^{10}6p^2$) $\longrightarrow$
Pb^{2+} ([Xe] $6s^24f^{14}5d^{10}$) + $2e^-$

Pb ([Xe] $6s^24f^{14}5d^{10}6p^2$) $\longrightarrow$ Pb^{4+} ([Xe] $4f^{14}5d^{10}$) + $4e^-$

8.7 (a) V^{3+}: [Ar] $3d^2$; paramagnetic

(b) Ni^{2+}: [Ar] $3d^8$; paramagnetic

(c) La^{3+}: [Xe]; not paramagnetic (diamagnetic)

8.8 (a) $F^- < Cl^- < Br^-$; (b) $Mg^{2+} < Na^+ < F^-$;
(c) $Cr^{3+} < Cr^{2+}$

Problems

Problems with **colored** numbers are answered at the back of the text. Sections match the text and provide the number(s) of relevant sample problems. Most offer Concept Review Questions, Skill-Building Exercises (in similar pairs), and Problems in Context. Then Comprehensive Problems, based on material from any section or previous chapter, follow.

Development of the Periodic Table

● **Concept Review Questions**

8.1 What would be your reaction to a claim that a new element had been discovered and it fit between tin (Sn) and antimony (Sb) in the periodic table?

8.2 Based on results of his study of atomic x-ray spectra, the English physicist Henry Moseley made a discovery that replaced atomic mass as the criterion for ordering the elements. By what criterion are the elements now ordered in the periodic table? Give an example of a sequence of element order that was confirmed by Moseley's findings.

● **Skill-Building Exercises (paired)**

8.3 Before Mendeleev published his periodic table, Johann Döbereiner grouped similar elements into triads, in which the unknown properties of one member could be predicted by averaging known values of the properties of the others. To test this idea, predict the values of the following quantities:
(a) The atomic mass of K from the atomic masses of Na and Rb
(b) The melting point of Br_2 from the melting points of Cl_2 ($-101.0°C$) and I_2 ($113.6°C$) (actual value = $-7.2°C$)

8.4 To test Döbereiner's idea (Problem 8.3), predict:
(a) The boiling point of HBr from the boiling points of HCl ($-84.9°C$) and HI ($-35.4°C$) (actual value $= -67.0°C$)
(b) The boiling point of AsH_3 from the boiling points of PH_3 ($-87.4°C$), and SbH_3 ($-17.1°C$) (actual value $= -55°C$)

Characteristics of Many-Electron Atoms

● **Concept Review Questions**

8.5 Summarize the rules for the allowable values of the four quantum numbers of an electron in an atom.
8.6 Which of the quantum numbers relate(s) to the electron only? Which relate(s) to the orbital?
8.7 State the exclusion principle in your own words. What does it imply about the number and spin of electrons in an atomic orbital?
8.8 What is the key distinction between sublevel energies in one-electron species, such as the H atom, and those in many-electron species, such as the C atom? What factors lead to this distinction? Would you expect the pattern of sublevel energies in Be^{3+} to be more like that in H or that in C? Explain.
8.9 Define *shielding* and *effective nuclear charge*. What is the connection between the two?
8.10 What is the penetration effect? How is it related to shielding? Use the penetration effect to explain the difference in relative orbital energies of a $3p$ electron and a $3d$ electron in the same atom.

● **Skill-Building Exercises** *(paired)*

8.11 How many electrons in an atom can have each of the following quantum number or sublevel designations?
(a) $n = 2, l = 1$ (b) $3d$ (c) $4s$
8.12 How many electrons in an atom can have each of the following quantum number or sublevel designations?
(a) $n = 2, l = 1, m_l = 0$ (b) $5p$ (c) $n = 4, l = 3$

8.13 How many electrons in an atom can have each of the following quantum number or sublevel designations?
(a) $4p$ (b) $n = 3, l = 1, m_l = +1$ (c) $n = 5, l = 3$
8.14 How many electrons in an atom can have each of the following quantum number or sublevel designations?
(a) $2s$ (b) $n = 3, l = 2$ (c) $6d$

The Quantum-Mechanical Model and the Periodic Table
(Sample Problems 8.1 and 8.2)

● **Concept Review Questions**

8.15 State the periodic law in your own words, and explain its relation to electron configuration. (Use Na and K in your explanation.)
8.16 State Hund's rule in your own words, and show its application in the orbital box diagram of the nitrogen atom.
8.17 How does the aufbau principle, in connection with the periodic law, lead to the format of the periodic table?
8.18 For main-group elements, are outer electron configurations similar or different within a group? Within a period? Explain.
8.19 For which blocks of elements are outer electrons the same as valence electrons? For which are d electrons often included among valence electrons?
8.20 What is the general rule for the electron capacity of an energy level? Apply it to determine the capacity of the fourth level.

● **Skill-Building Exercises** *(paired)*

8.21 Write a full set of quantum numbers for the following:
(a) The outermost electron in an Rb atom
(b) The electron gained when an S^- ion becomes an S^{2-} ion
(c) The electron lost when an Ag atom ionizes
(d) The electron gained when an F^- ion forms from an F atom
8.22 Write a full set of quantum numbers for the following:
(a) The outermost electron in an Li atom
(b) The electron gained when a Br atom becomes a Br^- ion
(c) The electron lost when a Cs atom ionizes
(d) The highest energy electron in the ground-state B atom

8.23 Write the full ground-state electron configuration for each:
(a) Rb   (b) Ge   (c) Ar
8.24 Write the full ground-state electron configuration for each:
(a) Br   (b) Mg   (c) Se

8.25 Write the full ground-state electron configuration for each:
(a) Cl   (b) Si   (c) Sr
8.26 Write the full ground-state electron configuration for each:
(a) S   (b) Kr   (c) Cs

8.27 Draw an orbital diagram showing valence electrons, and write the condensed ground-state electron configuration for each:
(a) Ti   (b) Cl   (c) V
8.28 Draw an orbital diagram showing valence electrons, and write the condensed ground-state electron configuration for each:
(a) Ba   (b) Co   (c) Ag

8.29 Draw an orbital diagram showing valence electrons, and write the condensed ground-state electron configuration for each:
(a) Mn   (b) P   (c) Fe
8.30 Draw an orbital diagram showing valence electrons, and write the condensed ground-state electron configuration for each:
(a) Ga   (b) Zn   (c) Sc

8.31 Draw the partial (valence-level) orbital diagram, and write the symbol, group number, and period number of the element:
(a) $[He] 2s^2 2p^4$ (b) $[Ne] 3s^2 3p^3$
8.32 Draw the partial (valence-level) orbital diagram, and write the symbol, group number, and period number of the element:
(a) $[Kr] 5s^2 4d^{10}$ (b) $[Ar] 4s^2 3d^8$

8.33 Draw the partial (valence-level) orbital diagram, and write the symbol, group number, and period number of the element:
(a) $[Ne] 3s^2 3p^5$ (b) $[Ar] 4s^2 3d^{10} 4p^3$
8.34 Draw the partial (valence-level) orbital diagram, and write the symbol, group number, and period number of the element:
(a) $[Ar] 4s^2 3d^5$ (b) $[Kr] 5s^2 4d^2$

8.35 From each partial (valence-level) orbital diagram, write the ground-state electron configuration and group number:

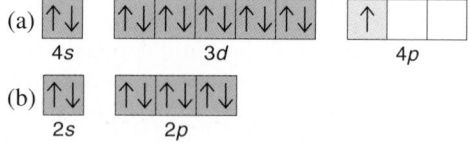

8.36 From each partial (valence-level) orbital diagram, write the ground-state electron configuration and group number:

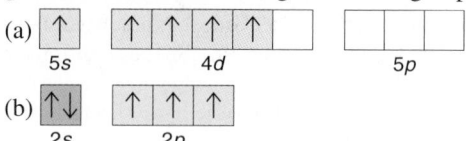

8.37 How many inner, outer, and valence electrons are present in an atom of the following elements?
(a) O   (b) Sn   (c) Ca   (d) Fe   (e) Se

8.38 How many inner, outer, and valence electrons are present in an atom of the following elements?
(a) Br   (b) Cs   (c) Cr   (d) Sr   (e) F

8.39 Identify each element below, and give the symbols of the other elements in its group:
(a) [He] $2s^2 2p^1$ (b) [Ne] $3s^2 3p^4$ (c) [Xe] $6s^2 5d^1$

8.40 Identify each element below, and give the symbols of the other elements in its group:
(a) [Ar] $4s^2 3d^{10} 4p^4$ (b) [Xe] $6s^2 4f^{14} 5d^2$ (c) [Ar] $4s^2 3d^5$

8.41 Identify each element below, and give the symbols of the other elements in its group:
(a) [He] $2s^2 2p^2$ (b) [Ar] $4s^2 3d^3$ (c) [Ne] $3s^2 3p^3$

8.42 Identify each element below, and give the symbols of the other elements in its group:
(a) [Ar] $4s^2 3d^{10} 4p^2$ (b) [Ar] $4s^2 3d^7$ (c) [Kr] $5s^2 4d^5$

● **Problems in Context**

8.43 After an atom in its ground state absorbs energy, it exists in an excited state. Spectral lines are produced when the atom returns to its ground state. The yellow-orange line in the sodium spectrum, for example, is produced by the emission of energy when excited sodium atoms return to their ground state. Write the electron configuration and the orbital diagram of the first excited state of sodium. (*Hint:* The outermost electron is excited.)

8.44 One reason spectroscopists study excited states is to gain information about the energies of orbitals that are unoccupied in an atom's ground state. Each of the following electron configurations represents an atom in an excited state. Identify the element, and write its condensed ground-state configuration:
(a) $1s^2 2s^2 2p^6 3s^1 3p^1$ (b) $1s^2 2s^2 2p^6 3s^2 3p^4 4s^1$
(c) $1s^2 2s^2 2p^6 3s^2 3p^6 4s^2 3d^4 4p^1$ (d) $1s^2 2s^2 2p^5 3s^1$

Trends in Some Key Periodic Atomic Properties
(Sample Problems 8.3 to 8.5)

● **Concept Review Questions**

8.45 If the exact outer limit of an isolated atom cannot be measured, what criterion can we use to determine atomic radii? What is the difference between a covalent radius and a metallic radius?

8.46 Explain the relationship between the trends in atomic size and in ionization energy within the main groups.

8.47 In what region of the periodic table will you find elements with relatively high IEs? With relatively low IEs?

8.48 Why do successive IEs of a given element always increase? When the difference between successive IEs of a given element is exceptionally large (for example, between IE_1 and IE_2 of K), what do we learn about its electron configuration?

8.49 In a plot of IE_1 for the Period 3 elements, why do the values for elements in Groups 3A(13) and 6A(16) drop slightly below the generally increasing trend?

8.50 Which group in the periodic table has elements with high (endothermic) IE_1 and very negative (exothermic) first electron affinities (EA_1)? Give the charge on the ions these atoms form.

8.51 The EA_2 of an oxygen atom is positive, even though its EA_1 is negative. Why does this change of sign occur? Which other elements exhibit a positive EA_2? Explain.

8.52 How does d-electron shielding influence atomic size among the Period 4 transition elements?

● **Skill-Building Exercises (paired)**

8.53 Arrange each set in order of *increasing* atomic size:
(a) Rb, K, Cs   (b) C, O, Be   (c) Cl, K, S   (d) Mg, K, Ca

8.54 Arrange each set in order of *decreasing* atomic size:
(a) Ge, Pb, Sn   (b) Sn, Te, Sr   (c) F, Ne, Na   (d) Be, Mg, Na

8.55 Arrange each set of atoms in order of *increasing* IE_1:
(a) Sr, Ca, Ba   (b) N, B, Ne   (c) Br, Rb, Se   (d) As, Sb, Sn

8.56 Arrange each set of atoms in order of *decreasing* IE_1:
(a) Na, Li, K   (b) Be, F, C   (c) Cl, Ar, Na   (d) Cl, Br, Se

8.57 Write the full electron configuration of the Period 2 element with the following successive IEs (in kJ/mol):
$IE_1 = 801$ $IE_2 = 2427$ $IE_3 = 3659$
$IE_4 = 25{,}022$ $IE_5 = 32{,}822$

8.58 Write the full electron configuration of the Period 3 element with the following successive IEs (in kJ/mol):
$IE_1 = 738$ $IE_2 = 1450$ $IE_3 = 7732$
$IE_4 = 10{,}539$ $IE_5 = 13{,}628$

8.59 Which element in each of the following sets would you expect to have the *highest* IE_2?
(a) Na, Mg, Al   (b) Na, K, Fe   (c) Sc, Be, Mg

8.60 Which element in each of the following sets would you expect to have the *lowest* IE_3?
(a) Na, Mg, Al   (b) K, Ca, Sc   (c) Li, Al, B

The Connection Between Atomic Structure and Chemical Reactivity
(Sample Problems 8.6 to 8.8)

● **Concept Review Questions**

8.61 List three ways in which metals and nonmetals differ.

8.62 Summarize the trend in metallic character as a function of position in the periodic table. Is it the same as the trend in atomic size? Ionization energy?

8.63 Summarize the acid-base behavior of main-group metal and nonmetal oxides in water. How does oxide acidity in water change down a group and across a period?

8.64 What ions are possible for the two largest elements in Group 4A(14)? How does each arise?

8.65 What is a pseudo–noble gas configuration? Give an example of one from Group 3A(13).

8.66 How are measurements of paramagnetism used to support electron configurations derived spectroscopically? Use Cu(I) and Cu(II) chlorides as examples.

8.67 The charges of a set of isoelectronic ions vary from 3+ to 3−. Place ions in order of increasing size.

● **Skill-Building Exercises (paired)**

8.68 Which element would you expect to be *more* metallic?
(a) Ca or Rb   (b) Mg or Ra   (c) Br or I

8.69 Which element would you expect to be *more* metallic?
(a) S or Cl   (b) In or Al   (c) As or Br

8.70 Which element would you expect to be *less* metallic?
(a) Sb or As   (b) Si or P   (c) Be or Na

8.71 Which element would you expect to be *less* metallic?
(a) Cs or Rn   (b) Sn or Te   (c) Se or Ge

8.72 Does the reaction of a main-group nonmetal oxide in water produce an acidic solution or a basic solution? Write a balanced equation for the reaction of a Group 6A(16) nonmetal oxide with water.

8.73 Does the reaction of a main-group metal oxide in water produce an acidic solution or a basic solution? Write a balanced equation for the reaction of a Group 2A(2) oxide with water.

8.74 Write the charge and full ground-state electron configuration of the monatomic ion most likely to be formed by each:
(a) Cl   (b) Na   (c) Ca

8.75 Write the charge and full ground-state electron configuration of the monatomic ion most likely to be formed by each:
(a) Rb   (b) N   (c) Br

8.76 Write the charge and full ground-state electron configuration of the monatomic ion most likely to be formed by each:
(a) Al   (b) S   (c) Sr

8.77 Write the charge and full ground-state electron configuration of the monatomic ion most likely to be formed by each:
(a) P   (b) Mg   (c) Se

8.78 How many unpaired electrons are present in the ground state of an atom from each of the following groups?
(a) 2A(2)   (b) 5A(15)   (c) 8A(18)   (d) 3A(13)

8.79 How many unpaired electrons are present in the ground state of an atom from each of the following groups?
(a) 4A(14)   (b) 7A(17)   (c) 1A(1)   (d) 6A(16)

8.80 Which of these are paramagnetic in their ground state?
(a) Ga   (b) Si   (c) Be   (d) Te

8.81 Are compounds of these ground-state ions paramagnetic?
(a) Ti^{2+} (b) Zn^{2+} (c) Ca^{2+} (d) Sn^{2+}

8.82 Write the condensed ground-state electron configurations of these transition metal ions, and state which are paramagnetic:
(a) V^{3+} (b) Cd^{2+} (c) Co^{3+} (d) Ag^{+}

8.83 Write the condensed ground-state electron configurations of these transition metal ions, and state which are paramagnetic:
(a) Mo^{3+} (b) Au^{+} (c) Mn^{2+} (d) Hf^{2+}

8.84 Palladium (Pd; $Z = 46$) is diamagnetic. Draw partial orbital diagrams to show which of the following electron configurations is consistent with this fact:
(a) $[Kr] 5s^2 4d^8$ (b) $[Kr] 4d^{10}$ (c) $[Kr] 5s^1 4d^9$

8.85 Niobium (Nb; $Z = 41$) has an anomalous ground-state electron configuration for a Group 5B(5) element: $[Kr] 5s^1 4d^4$. What is the expected electron configuration for elements in this group? Draw partial orbital diagrams to show how paramagnetic measurements could be used to support niobium's actual configuration.

8.86 Rank the ions in each set in order of *increasing* size, and explain your ranking:
(a) Li^+, K^+, Na^+ (b) Se^{2-}, Rb^+, Br^- (c) O^{2-}, F^-, N^{3-}

8.87 Rank the ions in each set in order of *decreasing* size, and explain your ranking:
(a) Se^{2-}, S^{2-}, O^{2-} (b) Te^{2-}, Cs^+, I^- (c) Sr^{2+}, Ba^{2+}, Cs^+

Comprehensive Problems

Problems with an asterisk (*) are more challenging.

8.88 Some versions of the periodic table show hydrogen at the top of Group 1A(1) *and* at the top of Group 7A(17). What properties of hydrogen justify each of these placements?

8.89 Test your familiarity with the layout of the periodic table. Which element is described by each of the following? Place each part letter (a, b, c, etc.) in the correct element box.
(a) Smallest atomic radius in Group 6A
(b) Largest atomic radius in Period 6
(c) Smallest atomic radius in Period 3
(d) Highest IE_1 in Group 14
(e) Lowest IE_1 in Period 5
(f) Most metallic in Group 15
(g) Group 3A element that forms most basic oxide
(h) Period 4 element with filled outer level
(i) Condensed ground-state electron configuration is $[Ne] 3s^2 3p^2$
(j) Condensed ground-state electron configuration is $[Kr] 5s^2 4d^6$
(k) Forms 2+ ion with electron configuration $[Ar] 3d^3$
(l) Period 5 element that forms 3+ ion with pseudo–noble gas configuration
(m) Period 4 transition element that forms 3+ diamagnetic ion
(n) Period 4 transition element that forms 2+ ion with a half-filled *d* sublevel

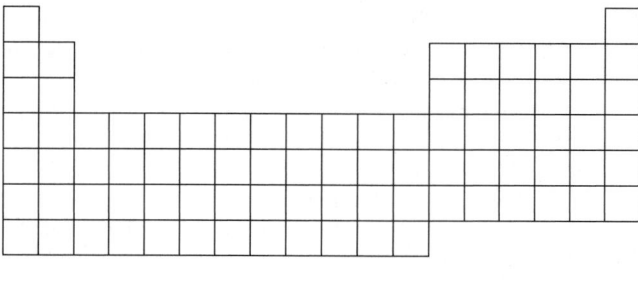

8.90 Based on the general vertical and horizontal trends in periodic properties, explain why it might be difficult to rank the following pairs:
(a) K and Sr in terms of atomic size
(b) Mn and Fe in terms of IE_1
(c) Na and Ca in terms of metallic character
(d) P and Se in terms of oxide acidity

8.91 The NaCl crystal structure consists of alternating Na^+ and Cl^- ions lying next to each other in three dimensions. If the Na^+ radius is 56.4% of the Cl^- radius and the distance between Na^+ nuclei is 566 pm, what are the radii of the two ions?

8.92 Name each element, and write its full ground-state electron configuration:
(a) Smallest metal in Period 3
(b) Heaviest lanthanide
(c) Lightest transition metal in Period 5
(d) Period 3 member whose 2− ion is isoelectronic with Ar
(e) Alkaline earth metal whose cation is isoelectronic with Kr
(f) Group 5A(15) metalloid with the most acidic oxide

*8.93 A fundamental relationship of electrostatics states that the energy required to separate opposite charges of magnitudes Q_1 and Q_2 that are the distance d apart is proportional to $\dfrac{Q_1 \times Q_2}{d}$. Use this relationship and any other factors to explain the following observations: (a) the IE_2 of He ($Z = 2$) is *more* than twice the IE_1 of H ($Z = 1$); (b) the IE_1 of He is *less* than twice the IE_1 of H.

*8.94 You are trapped in an alternative universe in which the quantum number assignments are the same as in ours, except that the quantum number l is an integer from 0 to n. What are the atomic numbers for the two smallest noble gases in the alternative universe?

8.95 Write the formula and name of the compound formed from the following ionic interactions:
(a) The 2+ ion and the 1− ion are both isoelectronic with the atoms of a chemically unreactive Period 4 element.
(b) The 2+ ion and the 2− ion are both isoelectronic with Period 3 noble gas.
(c) The 2+ ion is the smallest with a filled d subshell; the anion forms from the smallest halogen.
(d) The ions form from the largest and smallest ionizable atoms in Period 2.

*8.96 For over a century, chemists have noted several "diagonal relationships" in the periodic table: exceptional similarities between an element in Period 2 and another in Period 3 in the next higher group. Suggest a reason based on atomic and ionic properties for each of the following: (a) the chemical behavior of Li is very similar to that of Mg; (b) the acidity of $Be(OH)_2$ is very similar to that of $Al(OH)_3$.

8.97 Predict the sign of ΔH for each of the following reactions and explain your prediction:
(a) $Li(g) + Na^-(g) \longrightarrow Li^-(g) + Na(g)$
(b) $Na^+(g) + K(g) \longrightarrow Na(g) + K^+(g)$

8.98 Calculate the longest wavelength of electromagnetic radiation that could ionize an atom of each of the following elements:
(a) Li; $IE_1 = 520.1$ kJ/mol   (b) Au; $IE_1 = 889.9$ kJ/mol
(c) Cl; $IE_1 = 1255.7$ kJ/mol

8.99 The hot glowing gases around the Sun, the *corona*, can reach millions of degrees Celsius, high enough to remove many electrons from gaseous atoms. Iron ions with charges as high as 14+ have been observed in the corona. Which ions from Fe^+ to Fe^{14+} are paramagnetic? Which would be most attracted to a magnetic field?

*8.100 There are some exceptions to the trends of first and successive ionization energies. For each of the following pairs, explain which ionization energy would be higher:
(a) IE_1 of Ga or IE_1 of Ge   (b) IE_2 of Ga or IE_2 of Ge
(c) IE_3 of Ga or IE_3 of Ge   (d) IE_4 of Ga or IE_4 of Ge

8.101 An object has the color of the mixture of wavelengths of visible light that it reflects. White clothing reflects the most visible light energy, and black clothing absorbs the most. To stay cool on a hot, sunny day, should you wear orange/red or blue/purple clothing? Explain.

8.102 Half of the first 18 elements have an odd number of electrons, and half have an even number. Show why these elements aren't half paramagnetic and half diamagnetic.

8.103 Scientists have speculated that as yet unknown superheavy elements might be moderately stable. In fact, in 1976, it was mistakenly believed that Element 126 had been discovered in a sample of mica. Predict the n and l quantum numbers for the outermost electron in an atom of this element. What is the sublevel designation of the outermost orbital? How many orbitals would be in this sublevel?

*8.104 A substance is colored because it absorbs only some of the wavelengths of incoming visible light. The absorbed light excites electrons into unfilled orbitals.

(a) Why are many salts of transition metals colored, whereas nearly all main-group salts are colorless?
(b) The energy difference between the $5d$ and $6s$ sublevels in gold accounts for its color. Assuming this energy difference is about 2.7 eV (electron volt; 1 eV = 1.602×10^{-19} J), explain why gold has a warm yellow color.

8.105 Because they have been so carefully measured, emission lines from mercury discharge tubes are often used to calibrate spectrometers. In one of many series of transitions, mercury goes from a higher to a lower excited state with a line at 436 nm and then returns to the ground state with a line at 254 nm. (a) What is the energy of a photon that corresponds to each emission line? (b) What is the energy difference between each excited state and the ground state? (c) What is the wavelength of a single photon emitted when an electron drops from the higher excited state to the ground state?

8.106 Draw the partial (valence-level) orbital diagram and write the electron configuration of the atom and monatomic ion of the element with the following ionization energies (kJ/mol):

IE_1	IE_2	IE_3	IE_4	IE_5	IE_6	IE_7	IE_8
999	2251	3361	4564	7013	8495	27,106	31,669

*8.107 Use electron configurations to account for the stability of the lanthanide ions Ce^{4+} and Eu^{2+}.

8.108 As a science major, you are assigned to the "Atomic Dorm" at college. The first floor contains one dorm room with one bedroom that has a bunk bed for two students. The second floor contains two dorm rooms, one like the room on the first floor, and the other with three bedrooms, each with a bunk bed for two students. The third floor contains three dorm rooms, two like the rooms on the second floor, and a larger third room that contains five bedrooms, each with a bunk bed for two students. Entering students choose room and bed on a first-come, first-serve basis by criteria in the following order of importance:
(1) They want to be on the lowest available floor.
(2) They want to be in a smallest available dorm room.
(3) They want to be in a lower bunk bed if available.
(a) Which bed does the first student choose? (b) How many students are already in top bunks when the 17th student chooses? (c) Which bed does the 21st student choose? (d) How many students are already in bottom bunks when the 25th student chooses?

*8.109 On the planet Zog in the Andromeda galaxy, all of the stable elements have been studied. Data for some main-group elements are shown below (Zoggian units are unknown on Earth and, therefore, not shown). Limited communications with the Zoggians have indicated that balloonium is a monatomic gas with two positive charges in its nucleus. Use the data to deduce the names that Earthlings give to these elements:

Name	Atomic Radius	IE_1	EA_1
Balloonium	10	339	+3.0
Inertium	24	297	+4.1
Allotropium	34	143	−28.6
Brinium	63	70.9	−7.6
Canium	47	101	−15.3
Fertilium	25	200	0
Liquidium	38	163	−46.4
Utilium	48	82.4	−6.1
Crimsonium	72	78.4	+23.9

CHAPTER 9

MODELS OF CHEMICAL BONDING

CHAPTER OUTLINE

9.1 Atomic Properties and Chemical Bonds
Types of Chemical Bonding
Lewis Electron-Dot Symbols

9.2 The Ionic Bonding Model
Importance of Lattice Energy
Periodic Trends in Lattice Energy
How the Model Explains the Properties
of Ionic Compounds

9.3 The Covalent Bonding Model
Formation of a Covalent Bond
Bond Energy and Bond Length
How the Model Explains the Properties
of Covalent Compounds

**9.4 Between the Extremes: Electronegativity
and Bond Polarity**
Electronegativity
Polar Covalent Bonds and Bond Polarity
Partial Ionic Character
Continuum of Bonding Across a Period

9.5 An Introduction to Metallic Bonding
The Electron-Sea Model
How the Model Explains the Properties
of Metals

Figure: Why are diamonds so hard? Consisting entirely of carbon, the properties of a diamond, including its hardness, reflect the strength and arrangement of the bonds holding its atoms together. In this chapter, we examine the three types of chemical bonds and see how their characteristics can account for the macroscopic behavior of elements and compounds.

Why is table salt (or any other ionic compound) a hard, brittle, high-melting solid that conducts a current only when molten or dissolved in water? Why are many covalent substances, such as wax, low melting, soft, and nonconducting, while others, like diamond, are high melting and extremely hard? And why are metals shiny and bendable and able to conduct a current whether molten or solid? The answers lie in the *type of bonding within the substance.* In Chapter 8, we examined the properties of individual atoms and ions. Yet, in virtually all the substances in and around you, these particles occur bound together. As you'll see in this chapter, much of the excitement of chemistry comes in discovering how the properties of atoms influence the types of chemical bonds they form, because these are ultimately responsible for the behavior of the substance itself.

This chapter begins with an overview of how atomic properties give rise to the three major bonding models—ionic, covalent, and metallic. As each model is described, you'll see how it explains the properties of each type of substance. First, we discuss ionic bonding and the energy involved in the formation of an ionic solid from its elements. Covalent bonding, which occurs in the vast majority of compounds, is treated in the next two sections. The first describes the formation and characteristics of a covalent bond. The second describes the range of bonding, from pure covalent to ionic, and shows that most bonds fall somewhere in between. The chapter ends with an introduction to metallic bonding.

9.1 ATOMIC PROPERTIES AND CHEMICAL BONDS

Before we examine the types of chemical bonding, we should ask why atoms bond at all. In general terms, they do so for one overriding reason: *bonding lowers the potential energy between positive and negative particles,* whether those particles are oppositely charged ions or atomic nuclei and the electrons between them. Just as the electron configuration and the strength of the nucleus-electron attraction determine the properties of an atom, the type and strength of chemical bonds determine the properties of a substance.

Types of Chemical Bonding

On the atomic level, we distinguish a metal from a nonmetal on the basis of several properties that correlate with position in the periodic table (Figure 9.1 and inside the front cover). Recall from Chapter 8 that metallic properties generally

CONCEPTS & SKILLS

to review before you study this chapter

- characteristics of ionic and covalent bonding (Section 2.7)
- polar covalent bonds and the polarity of water (Section 4.1)
- Hess's law, ΔH^0_{rxn}, and ΔH^0_f (Section 6.5)
- atomic and ionic electron configurations (Sections 8.3 and 8.5)
- trends in atomic properties and metallic behavior (Sections 8.4 and 8.5)

Figure 9.1 **A general comparison of metals and nonmetals. A,** The key indicates the position of metals, nonmetals, and metalloids within the periodic table. **B,** The relative magnitudes of some key atomic properties vary from left to right within a period and correlate with an element being metallic or nonmetallic. For example, within a given period, metal atoms are larger than nonmetal atoms.

PROPERTY	METAL ATOM	NONMETAL ATOM
Atomic size	Larger	Smaller
Z_{eff}	Lower	Higher
IE	Lower	Higher
EA	Less negative	More negative

B Relative magnitudes of atomic properties within a period

Figure 9.2 **The three models of chemical bonding. A,** In ionic bonding, metal atoms transfer electron(s) to nonmetal atoms, forming oppositely charged ions that attract each other to form a solid. **B,** In covalent bonding, two atoms share an electron pair localized between their nuclei (shown here as a bond line). Most covalent substances consist of individual molecules, each consisting of two or more atoms. **C,** In metallic bonding, many metal atoms pool their valence electron(s) to form a delocalized "electron sea" that holds the metal-ion cores together.

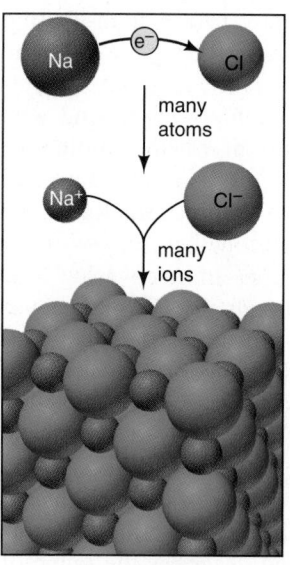

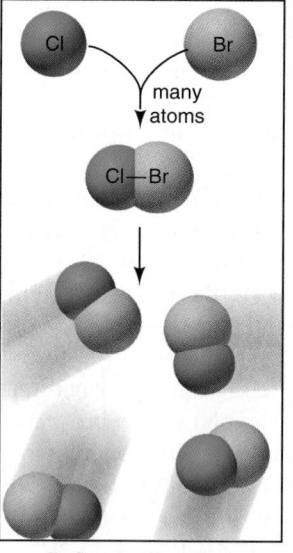

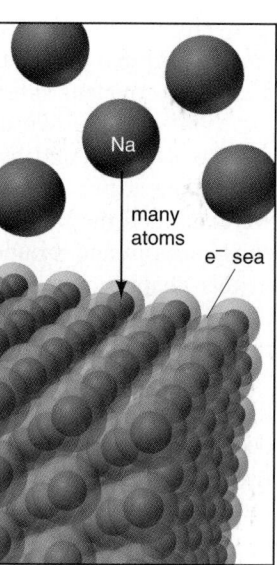

A Ionic bonding   **B** Covalent bonding   **C** Metallic bonding

FORMATION OF AN IONIC COMPOUND

decrease across a period and increase down a group. As a first step toward appreciating the importance of bonding, let's identify the three types of bonding that result from combining the two types of atoms—metal with nonmetal, nonmetal with nonmetal, and metal with metal:

1. *Metal with nonmetal: electron transfer and ionic bonding* (Figure 9.2A). We typically observe **ionic bonding** between atoms with large differences in their tendency to lose or gain electrons. Such differences occur between reactive metals [Groups 1A(1) and 2A(2)] and nonmetals [Groups 7A(17) and the top of 6A(16)]. The metal atom (low IE) loses its one or two valence electrons, whereas the nonmetal atom (highly negative EA) gains the electrons. *Electron transfer* from metal to nonmetal occurs, and each atom forms an ion with a noble gas configuration. The electrostatic attraction between these positive and negative ions draws them into the three-dimensional array of an ionic solid, whose chemical formula represents the cation-to-anion ratio (empirical formula).

2. *Nonmetal with nonmetal: electron sharing and covalent bonding* (Figure 9.2B). When two atoms have a small difference in their tendency to lose or gain electrons, we observe *electron sharing* and **covalent bonding.** This type of bonding most commonly occurs between nonmetal atoms (although a pair of metal atoms can form a covalent bond also). Each nonmetal atom holds onto its own electrons tightly (high IE) and tends to attract other electrons as well (highly negative EA). The attraction of each nucleus for the valence electrons of the other draws the atoms together. A shared electron pair is considered to be *localized* between the two atoms because it spends most of its time there, linking them in a covalent bond of a given length and strength. In most cases, separate molecules form when covalent bonding occurs, and the chemical formula reflects the actual numbers of atoms in the molecule (molecular formula).

3. *Metal with metal: electron pooling and metallic bonding* (Figure 9.2C). In general, metal atoms are relatively large, and their few outer electrons are well shielded by filled inner levels. Thus, they lose outer electrons comparatively easily (low IE) but do not gain them very readily (small or positive EA). These properties lead large numbers of metal atoms to share their valence electrons, but in a different way from covalent bonding. In the simplest model of **metallic bonding,** all the metal atoms in a sample *pool* their valence electrons into an evenly distributed "sea" of electrons that "flows" between and around the metal-ion cores (nucleus plus inner electrons) and attracts them, thereby holding them together.

Unlike the localized electrons in covalent bonding, electrons in metallic bonding are *delocalized,* moving freely throughout the piece of metal.

It's important to remember that there are exceptions to these idealized bonding models in the world of real substances. You cannot always predict bond type solely from the elements' positions in the periodic table. For instance, all binary ionic compounds contain a metal and a nonmetal, but all metals do not form binary ionic compounds with all nonmetals. As just one example, when the metal beryllium [Group 2A(2)] combines with the nonmetal chlorine [Group 7A(17)], the bonding fits the covalent model better than the ionic model. In other words, just as we see a gradation in metallic behavior within groups and periods, we also see a gradation in bonding from one type to another.

Lewis Electron-Dot Symbols: Depicting Atoms in Chemical Bonding

Before turning to the individual models, let's discuss a method for depicting the valence electrons of interacting atoms. In the **Lewis electron-dot symbol** (named for the American chemist G. N. Lewis), the element symbol represents the nucleus *and* inner electrons, and the surrounding dots represent the valence electrons, as shown in Figure 9.3. Note the recurring pattern of dots for elements within a group.

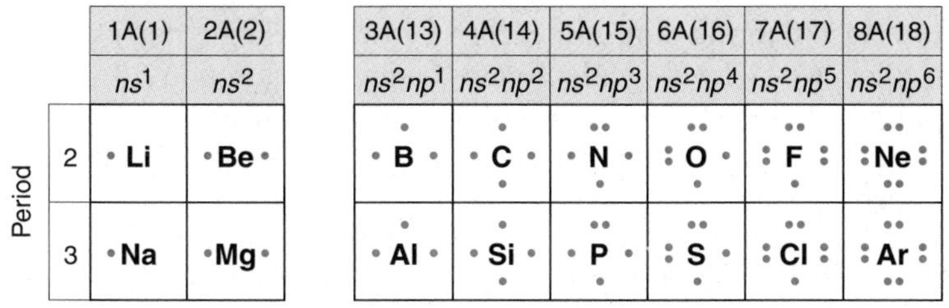

	1A(1)	2A(2)		3A(13)	4A(14)	5A(15)	6A(16)	7A(17)	8A(18)
	ns^1	ns^2		ns^2np^1	ns^2np^2	ns^2np^3	ns^2np^4	ns^2np^5	ns^2np^6
2	·Li	·Be·		· B ·	· C ·	· N ·	: O ·	: F :	:Ne:
3	·Na	·Mg·		· Al ·	· Si ·	· P ·	: S ·	: Cl :	: Ar :

Period

Figure 9.3 **Lewis electron-dot symbols for elements in Periods 2 and 3.** The element symbol represents the nucleus and inner electrons, and the dots around it represent valence electrons, either paired or unpaired. The number of unpaired dots indicates the number of electrons a metal atom loses, the number a nonmetal gains, or the number of covalent bonds a nonmetal atom usually forms.

It's easy to write the Lewis symbol for any main-group element:

1. Note its A-group number (1A to 8A), which gives the number of valence electrons.
2. Place one dot at a time on the four sides (top, right, bottom, left) of the element symbol.
3. Pair up the dots until all are used.

The specific placement of dots is not important; that is, in addition to the one shown in Figure 9.3, the Lewis symbol for nitrogen can *also* be written

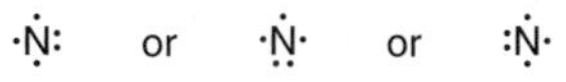

$$·\overset{..}{N}: \quad \text{or} \quad ·\overset{.}{\underset{.}{N}}· \quad \text{or} \quad :\overset{.}{N}·$$

The number and pairing of dots provide information about an element's bonding behavior:

- For a metal, the *total* number of dots is the maximum number of electrons it loses to form a cation.
- For a nonmetal, the number of *unpaired* dots is the number of electrons that become paired either through electron gain or through electron sharing. Thus, the number of unpaired dots equals either the negative charge of the anion an atom forms or the number of covalent bonds it forms.

To illustrate the last point, look at the Lewis symbol for carbon. Rather than having one pair of dots and two unpaired dots, as its electron configuration ([He] $2s^2 2p^2$) would indicate, carbon has four unpaired dots because it forms four bonds. (In Chapter 10, we use some other dot arrangements for the larger nonmetals.)

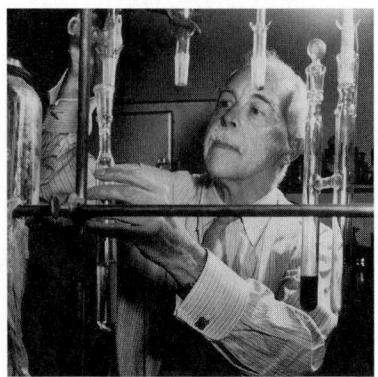

The Remarkable Insights of G. N. Lewis Many of the ideas in this text emerged from the mind of Gilbert Newton Lewis (1875–1946). As early as 1902, nearly a decade *before* Rutherford proposed the nuclear model of the atom, Lewis's notebooks show a scheme that involves the filling of outer electron "shells" to explain the way elements combine. His electron-dot symbols and associated structural formulas (discussed in Chapter 10) have become standards for representing bonding. Among his many other contributions are his insights into a general understanding of the behavior of acids and bases (which we discuss in Chapter 18).

SECTION SUMMARY

Nearly all naturally occurring substances consist of atoms or ions bound to others. Chemical bonding allows atoms to lower their energy. Ionic bonding occurs when a metal atom transfers electrons to a nonmetal atom, and the resulting ions attract each other to form an ionic solid. Covalent bonding occurs most often between nonmetal atoms and commonly results in molecules. The bonded atoms share a pair of electrons, which remain localized between them. Metallic bonding occurs when many metal atoms pool their valence electrons in a delocalized electron sea that holds all the atoms together. The Lewis electron-dot symbol of an atom depicts the number of valence electrons for a main-group element.

9.2 THE IONIC BONDING MODEL

The central idea of the ionic bonding model is the *transfer of electrons from metal to nonmetal to form ions that come together in a solid ionic compound.* For nearly every main-group element that forms a monatomic ion, the ionic electron configuration has a filled outer level: either two or eight electrons, the same number as in the nearest noble gas. Lewis incorporated this fact into the **octet rule:** *when atoms bond, they lose, gain, or share electrons to attain a filled outer shell of eight (or two) electrons.* The octet rule holds for the great majority of the compounds of Period 2 elements and a large number of others as well.

Figure 9.4 shows three ways to depict the transfer of an electron from a lithium atom to a fluorine atom. In each, Li loses its single outer electron and is left with a filled $n = 1$ level, while F gains a single electron to fill its $n = 2$ level. In this case, each atom is one electron away from its nearest noble gas—He for Li and Ne for F, so the number of electrons lost by each Li equals the number gained by each F. Therefore, equal numbers of Li^+ and F^- ions form, as the formula LiF indicates. In all cases of ionic bonding, *the total number of electrons lost by the metal atoms equals the total number of electrons gained by the nonmetal atoms.*

SAMPLE PROBLEM 9.1 Depicting Ion Formation

Problem Use partial orbital diagrams and Lewis symbols to depict the formation of Na^+ and O^{2-} ions from the atoms, and determine the formula of the compound.
Plan First we draw the orbital diagrams and Lewis symbols for the Na and O atoms. To attain filled outer levels, Na loses one electron and O gains two. Therefore, to make the number of electrons lost equal the number gained, two Na atoms are needed for each O atom.
Solution

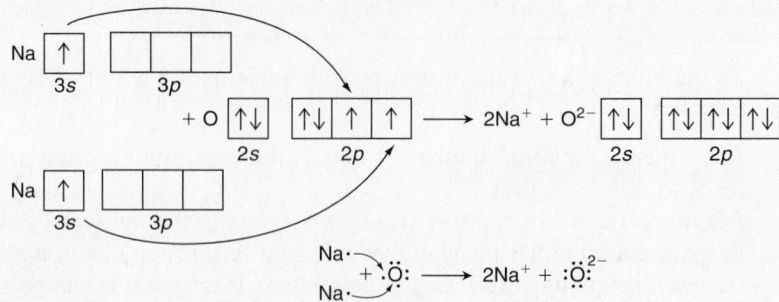

The formula is Na_2O.

FOLLOW-UP PROBLEM 9.1 Use condensed electron configurations and Lewis electron-dot symbols to depict the formation of Mg^{2+} and Cl^- ions from the atoms, and write the formula of the compound.

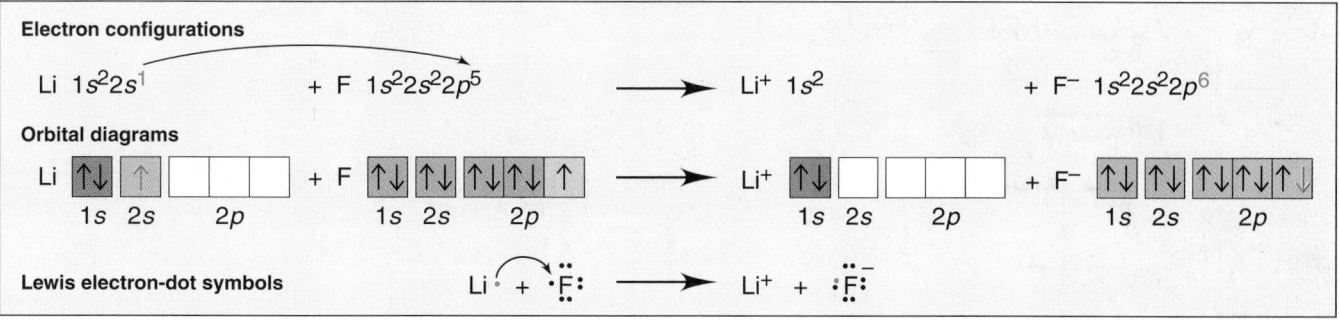

Figure 9.4 **Three ways to represent the formation of Li^+ and F^- through electron transfer.** The electron being transferred is indicated in red.

Energy Considerations in Ionic Bonding: The Importance of Lattice Energy

You may be surprised to learn that the electron-transfer process by itself actually *absorbs* energy! In fact, the reason ionic compounds form is because of the enormous *release* of energy that occurs when the ions come together and form the solid. Consider just the electron-transfer process for the formation of lithium fluoride, which involves two steps—a gaseous Li atom losing an electron and a gaseous F atom gaining it:

- The first ionization energy (IE_1) of Li is the energy required for 1 mol of gaseous Li atoms to lose 1 mol of outer electrons:
$$Li(g) \longrightarrow Li^+(g) + e^- \qquad IE_1 = 520 \text{ kJ}$$

- The electron affinity (EA) of F is the energy change that occurs when 1 mol of F atoms gains 1 mol of electrons:
$$F(g) + e^- \longrightarrow F^-(g) \qquad EA = -328 \text{ kJ}$$

Adding these values, we find that the two-step electron-transfer process *by itself* requires energy:
$$Li(g) + F(g) \longrightarrow Li^+(g) + F^-(g) \qquad IE_1 + EA = 192 \text{ kJ}$$

The total energy cost of ion formation is even greater than this because metallic lithium and diatomic fluorine must first be converted to the separate gaseous atoms, which also requires energy. Despite this, the standard heat of formation (ΔH_f^0) of solid LiF is -617 kJ/mol; that is, 617 kJ is *released* per mole of LiF(s) formed. The case of LiF is typical: despite an endothermic electron transfer, ionic solids form readily, often vigorously. Figure 9.5 shows the formation of NaBr.

Clearly, if the overall reaction of Li(s) and $F_2(g)$ to form LiF(s) releases energy, there must be some other energy component(s) exothermic enough to overcome the endothermic steps. The exothermic component(s) result(s) from the strong *attraction between oppositely charged ions*. When 1 mol of $Li^+(g)$ and 1 mol of $F^-(g)$ form 1 mol of gaseous LiF molecules, a large quantity of heat is released:
$$Li^+(g) + F^-(g) \longrightarrow LiF(g) \qquad \Delta H^0 = -755 \text{ kJ}$$

Of course, LiF does not consist of gaseous molecules under ordinary conditions because *much more energy is released when the gaseous ions coalesce into a crystalline solid*. When that occurs, each ion attracts several others of opposite charge. The **lattice energy** is the enthalpy change that accompanies gaseous ions coalescing into an ionic solid:
$$Li^+(g) + F^-(g) \longrightarrow LiF(s) \qquad \Delta H_{lattice}^0 \text{ of LiF} = \text{lattice energy} = -1050 \text{ kJ}$$

(You'll see how we calculate this value in a moment.) The magnitude of the lattice energy indicates the strength of ionic interactions and influences the melting point, hardness, and solubility of ionic compounds.

A

B

Figure 9.5 **The reaction between sodium and bromine. A,** Despite the endothermic electron-transfer process, all the Group 1A(1) metals react exothermically with any of the Group 7A(17) nonmetals to form solid alkalimetal halides. The example shown is sodium (in beaker under mineral oil) and bromine (dark orange-brown liquid in flask). **B,** The reaction is usually rapid and vigorous.

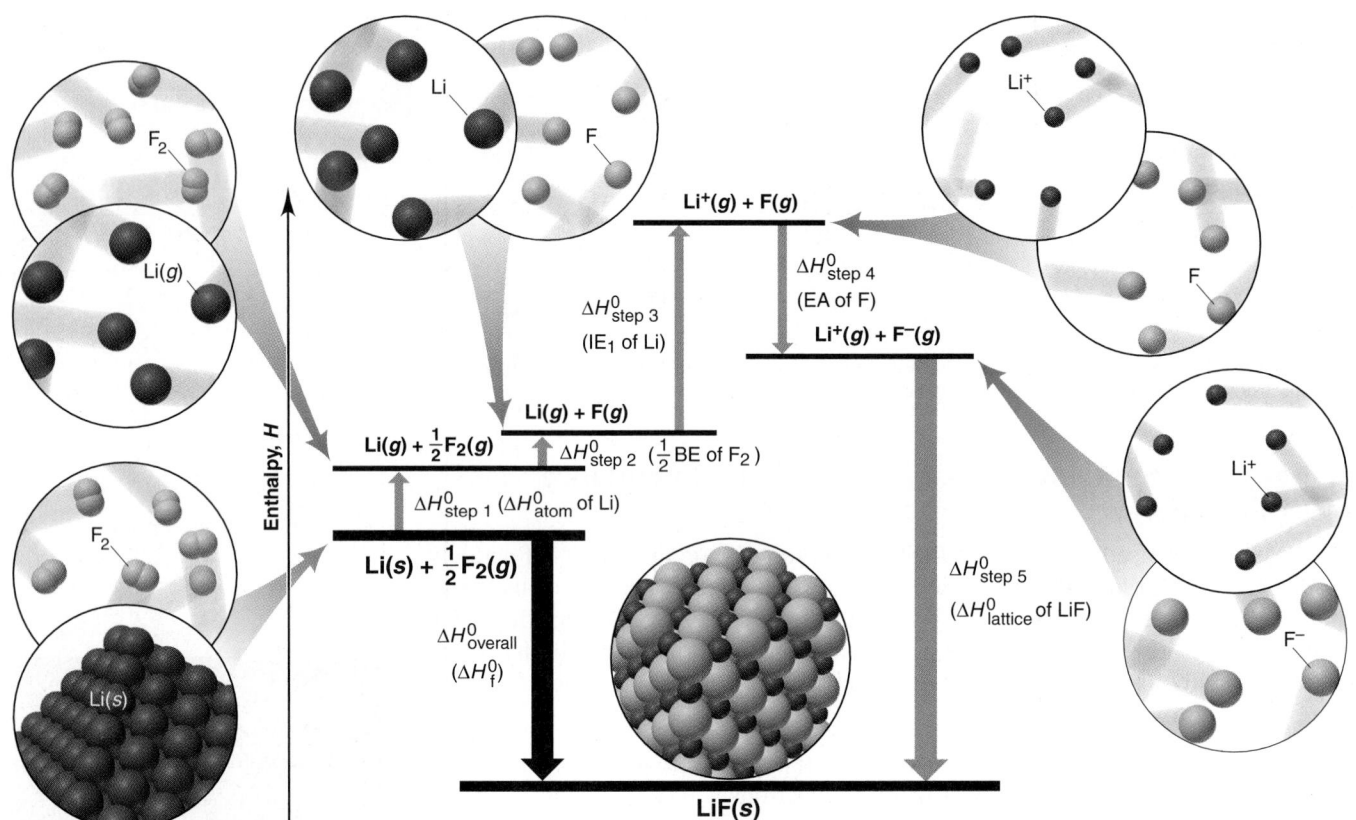

Figure 9.6 **The Born-Haber cycle for lithium fluoride.** The formation of LiF(s) from its elements is shown happening either in one overall reaction *(black arrow)* or in five hypothetical steps, each with its own enthalpy change *(orange arrows)*. The overall enthalpy change for the process (ΔH_f^0) is known, as are $\Delta H_{\text{step 1}}^0$ through $\Delta H_{\text{step 4}}^0$. Therefore, $\Delta H_{\text{step 5}}^0$, the lattice energy ($\Delta H_{\text{lattice}}^0$ of LiF), can be calculated. (ΔH_{atom}^0 is the heat of atomization; BE is the bond energy.)

Lattice energy plays the critical role in formation of ionic compounds, but it cannot be measured directly. One way to determine lattice energy applies Hess's law (see Section 6.5), which states that the enthalpy change of an overall reaction is the sum of the enthalpy changes for the individual reactions that make it up: $\Delta H_{\text{total}} = \Delta H_1 + \Delta H_2 + \cdots$. Lattice energies are calculated by means of a **Born-Haber cycle,** a series of chosen steps from elements to ionic compound for which all the enthalpies* are known except the lattice energy.

Let's go through the Born-Haber cycle for lithium fluoride. Figure 9.6 shows two possible paths, either the direct combination reaction (black arrow) or the multistep path (orange arrows), one step of which releases the unknown lattice energy. Hess's law tells us both paths involve the same overall enthalpy change:

$$\Delta H_f^0 \text{ of LiF}(s) = \text{sum of } \Delta H^0 \text{ for multistep path}$$

It's important to realize that we *choose hypothetical steps whose enthalpy changes we can measure* to depict the energy components of LiF formation, even though these are **not** *the actual steps that occur when lithium reacts with fluorine.*

We begin with the elements in their standard states, metallic lithium and gaseous diatomic fluorine. In the multistep process, the elements are converted to individual gaseous atoms (steps 1 and 2), the electron-transfer steps form gaseous ions (steps 3 and 4), and the ions form a solid (step 5). We identify each ΔH^0 by its step number:

Step 1. Converting 1 mol of solid Li to separate gaseous Li atoms involves breaking metallic bonds, so it requires energy:

$$\text{Li}(s) \longrightarrow \text{Li}(g) \qquad \Delta H_{\text{step 1}}^0 = 161 \text{ kJ}$$

(This process is called *atomization,* and the enthalpy change is ΔH_{atom}^0.)

---

*Strictly speaking, ionization energy (IE) and electron affinity (EA) are internal energy changes (ΔE), not enthalpy changes (ΔH), but in these cases, $\Delta H = \Delta E$ because $\Delta V = 0$ (see Section 6.2).

Step 2. Converting an F_2 molecule to F atoms involves breaking a covalent bond, so it requires energy, which, as we discuss later, is called the *bond energy* (BE). We need 1 mol of F atoms for 1 mol of LiF, so we start with $\frac{1}{2}$ mol of F_2:

$$\tfrac{1}{2}F_2(g) \longrightarrow F(g) \qquad \Delta H^0_{\text{step 2}} = \tfrac{1}{2}(\text{BE of } F_2)$$
$$= \tfrac{1}{2}(159 \text{ kJ}) = 79.5 \text{ kJ}$$

Step 3. Removing the $2s$ electron from 1 mol of Li to form 1 mol of Li^+ requires energy:

$$Li(g) \longrightarrow Li^+(g) + e^- \qquad \Delta H^0_{\text{step 3}} = IE_1 = 520 \text{ kJ}$$

Step 4. Adding an electron to 1 mol of F to form 1 mol of F^- releases energy:

$$F(g) + e^- \longrightarrow F^-(g) \qquad \Delta H^0_{\text{step 4}} = EA = -328 \text{ kJ}$$

Step 5. Forming 1 mol of the crystalline ionic solid from the gaseous ions is the step whose enthalpy change (the lattice energy) is unknown:

$$Li^+(g) + F^-(g) \longrightarrow LiF(s) \qquad \Delta H^0_{\text{step 5}} = \Delta H^0_{\text{lattice}} \text{ of LiF} = ?$$

We know the enthalpy change of the formation reaction,

$$Li(s) + \tfrac{1}{2}F_2(g) \longrightarrow LiF(s) \qquad \Delta H^0_{\text{overall}} = \Delta H^0_f = -617 \text{ kJ}$$

Using Hess's law, we set the known ΔH^0_f equal to the sum of the steps and calculate the lattice energy:

$$\Delta H^0_f = \Delta H^0_{\text{step 1}} + \Delta H^0_{\text{step 2}} + \Delta H^0_{\text{step 3}} + \Delta H^0_{\text{step 4}} + \Delta H^0_{\text{lattice}} \text{ of LiF}$$

Solving for $\Delta H^0_{\text{lattice}}$ of LiF gives

$$\Delta H^0_{\text{lattice}} \text{ of LiF} = \Delta H^0_f - (\Delta H^0_{\text{step 1}} + \Delta H^0_{\text{step 2}} + \Delta H^0_{\text{step 3}} + \Delta H^0_{\text{step 4}})$$
$$= -617 \text{ kJ} - [161 \text{ kJ} + 79.5 \text{ kJ} + 520 \text{ kJ} + (-328 \text{ kJ})] = -1050 \text{ kJ}$$

Note that *the magnitude of the lattice energy dominates the multistep process.*

The Born-Haber cycle reveals a central point: *ionic solids exist **only** because the lattice energy drives the energetically unfavorable electron transfer.* In other words, the energy *required* for elements to lose or gain electrons is *supplied* by the attraction between the ions they form: energy is expended to form the ions, but it is more than regained when they attract each other and form a solid.

Periodic Trends in Lattice Energy

Because the lattice energy is the result of electrostatic interactions among ions, we expect its magnitude to depend on several factors, including ionic size, ionic charge, and ionic arrangement in the solid. In Chapter 2, you were introduced to **Coulomb's law** as a fundamental aspect of ionic bonding, and we've applied it several times since to other situations in which charges attract each other. Coulomb's law states that the electrostatic *force* associated with two charges (A and B) is directly proportional to the product of their magnitudes and inversely proportional to the square of the distance between them:

$$\text{Electrostatic force} \propto \frac{\text{charge A} \times \text{charge B}}{\text{distance}^2}$$

Given that energy = force × distance, the electrostatic *energy* is inversely proportional to the *first* power of the distance between the charges:

$$\text{Electrostatic energy} \propto \frac{\text{charge A} \times \text{charge B}}{\text{distance}} \qquad (9.1)$$

In an ionic solid, there are charged particles at various distances, so the calculation requires more than just a simple extension of Coulomb's law. Nevertheless, the lattice energy ($\Delta H^0_{\text{lattice}}$) is directly proportional to the electrostatic energy. Cations and anions lie as close to each other as possible, and the distance between them is the distance between their centers, which equals the sum of their radii (see Figure 8.28):

$$\text{Electrostatic energy} \propto \frac{\text{cation charge} \times \text{anion charge}}{\text{cation radius} + \text{anion radius}} \propto \Delta H^0_{\text{lattice}} \qquad (9.2)$$

This relationship helps us predict trends in lattice energy and explain the effects of ionic size and charge:

1. *Effect of ionic size.* As we move down a group of metals or nonmetals, ionic radii increase. Therefore, the energy of attraction between cations and anions should decrease because the interionic distance is greater; thus, the lattice energies of their compounds should decrease as well. This prediction is borne out in the lattice energies of the alkali-metal halides shown in Figure 9.7. A smooth decrease occurs in lattice energy down a group whether we hold the cation constant (LiF to LiI) or the anion constant (LiF to RbF).

2. *Effect of ionic charge.* When we compare lithium fluoride with magnesium oxide, we find cations of about equal radii (Li^+ = 76 pm and Mg^{2+} = 72 pm), and anions of about equal radii (F^- = 133 pm and O^{2-} = 140 pm). Thus, the only significant difference is the ionic charge: LiF contains the univalent Li^+ and F^- ions, whereas MgO contains the divalent Mg^{2+} and O^{2-} ions. The difference in their lattice energies is striking:

$$\Delta H^0_{lattice} \text{ of LiF} = -1050 \text{ kJ/mol} \qquad \text{and} \qquad \Delta H^0_{lattice} \text{ of MgO} = -3923 \text{ kJ/mol}$$

This nearly fourfold increase in $\Delta H^0_{lattice}$ reflects the fourfold increase in the product of the charges (1×1 vs. 2×2) in the numerator of Equation 9.2.

The high lattice energy of MgO explains why ionic solids with 2+ ions even exist. After all, much more energy is needed to form 2+ ions than 1+ ions. Form-

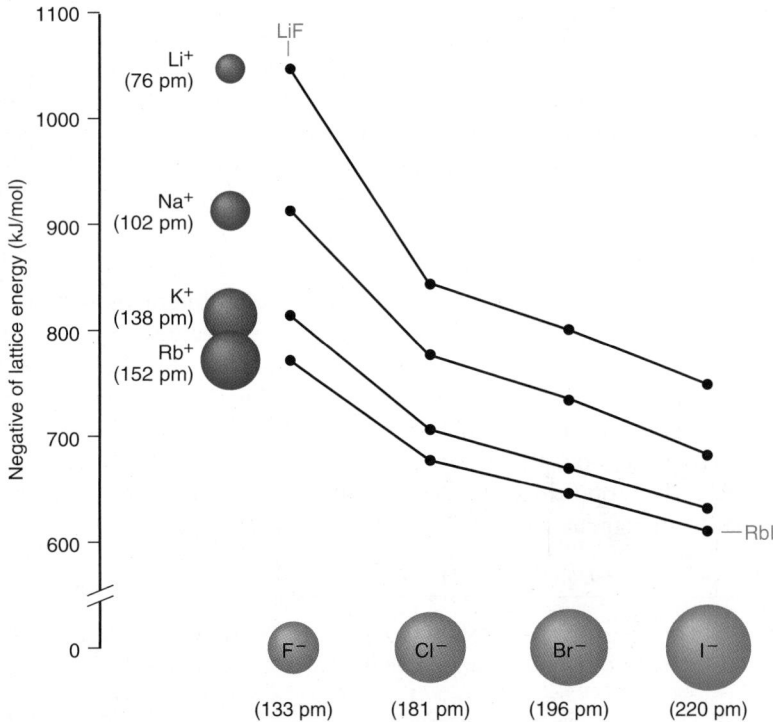

Figure 9.7 **Trends in lattice energy.** The lattice energies for many of the alkali-metal halides are shown. Each series of four points represents a given Group 1A(1) cation *(left side)* combining with the various Group 7A(17) anions *(bottom)*. As ionic radii increase, the electrostatic attractions decrease, so the lattice energies of the compounds decrease as well. Thus, LiF (smallest ions shown) has the highest lattice energy and RbI (largest ions) has the lowest.

ing 1 mol of Mg^{2+} involves loss of the first *and* second electrons (and an enthalpy change equal to the sum of the first and second ionization energies):

$$Mg(g) \longrightarrow Mg^{2+}(g) + 2e^- \qquad \Delta H^0 = IE_1 + IE_2 = 738 \text{ kJ} + 1450 \text{ kJ} = 2188 \text{ kJ}$$

Adding 1 mol of electrons to 1 mol of O atoms (first electron affinity, EA_1) is exothermic, but adding a second mole of electrons (second electron affinity, EA_2) is endothermic because an electron is being added to the negative O^- ion. The overall formation of 1 mol of O^{2-} ions is endothermic:

$$
\begin{array}{llll}
O(g) + e^- \longrightarrow O^-(g) & \Delta H^0 = EA_1 & = & -141 \text{ kJ} \\
O^-(g) + e^- \longrightarrow O^{2-}(g) & \Delta H^0 = EA_2 & = & 878 \text{ kJ} \\
\hline
O(g) + 2e^- \longrightarrow O^{2-}(g) & \Delta H^0 = EA_1 + EA_2 = & & 737 \text{ kJ}
\end{array}
$$

In addition, there are the endothermic steps for converting 1 mol of $Mg(s)$ to $Mg(g)$ (148 kJ) and breaking $\frac{1}{2}$ mol of O_2 molecules into 1 mol of O atoms (249 kJ). Nevertheless, as a result of the ions' 2+ charges, solid MgO readily forms whenever Mg burns in air ($\Delta H_f^0 = -601$ kJ/mol). Clearly, the enormous lattice energy ($\Delta H_{lattice}^0$ of MgO $= -3923$ kJ/mol) more than compensates for these endothermic steps.

How the Model Explains the Properties of Ionic Compounds

The first and most important job of any model is to explain the facts. By magnifying our view, we can see how the ionic bonding model accounts for the properties of ionic solids. You may have seen a piece of rock salt (NaCl). It is *hard* (does not dent), *rigid* (does not bend), and *brittle* (cracks without deforming). These properties are due to the powerful attractive forces that hold the ions *in specific positions* throughout the crystal. Moving the ions out of position requires overcoming these forces, so the sample resists denting and bending. If *enough* pressure is applied, ions of like charge are brought next to each other, and their repulsions crack the sample suddenly (Figure 9.8).

Most ionic compounds *do not* conduct electricity in the solid state but *do* conduct it when melted or when dissolved in water. (Notable exceptions include so-called superionic conductors, such as AgI, and the superconducting ceramics, which have remarkable conductivity in the solid state.) According to the ionic

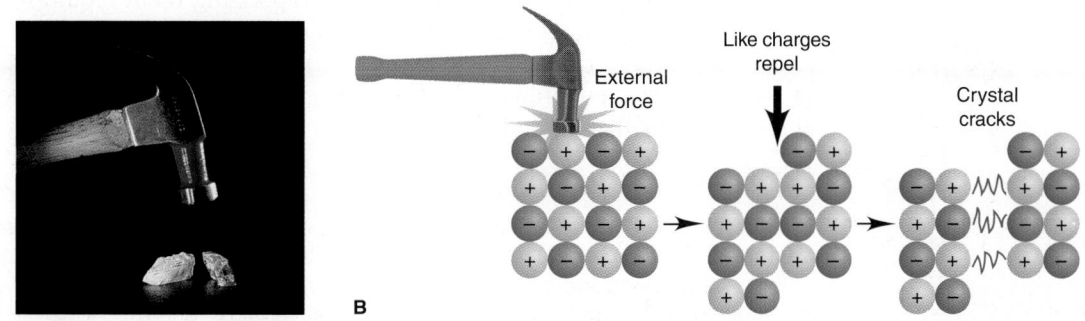

Figure 9.8 Electrostatic forces and the reason ionic compounds crack. A, Ionic compounds are hard and will crack, rather than bend, when struck with enough force. **B,** The positive and negative ions in the crystal are arranged to maximize their attraction. When an external force is applied, like charges move near each other, and the repulsions crack the piece apart.

Table 9.1 Melting and Boiling Points of Some Ionic Compounds		
Compound	mp (°C)	bp (°C)
CsBr	636	1300
NaI	661	1304
$MgCl_2$	714	1412
KBr	734	1435
$CaCl_2$	782	>1600
NaCl	801	1413
LiF	845	1676
KF	858	1505
MgO	2852	3600

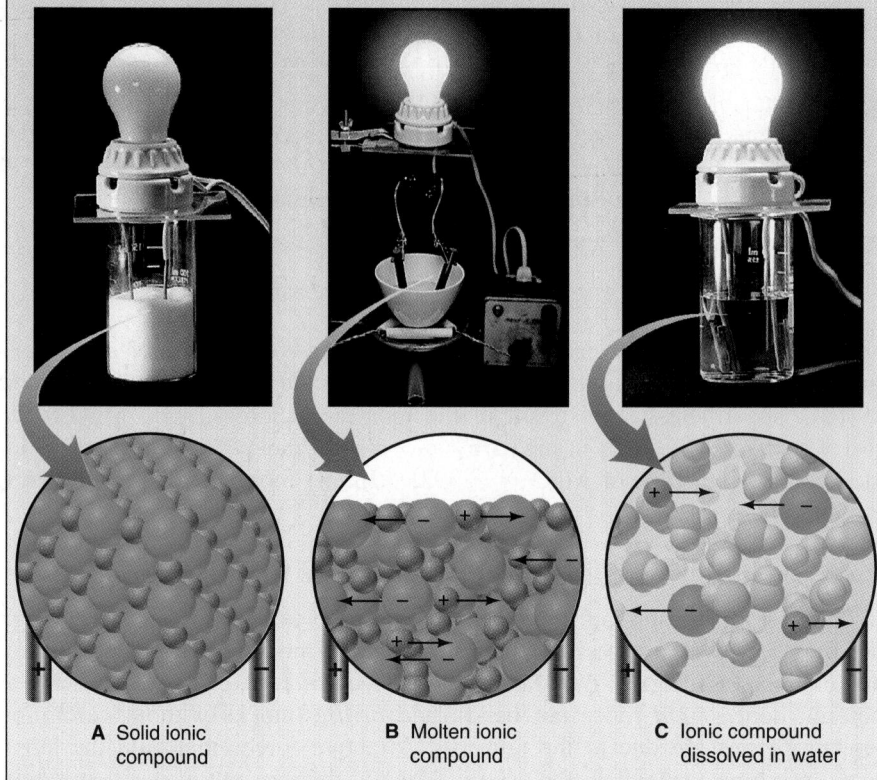

Figure 9.9 **Electrical conductance and ion mobility. A,** No current flows in the ionic solid because ions are immobile. **B,** In the molten compound, mobile ions flow toward the oppositely charged electrodes and carry a current. **C,** In an aqueous solution of the compound, mobile solvated ions carry a current.

bonding model, the solid consists of immobilized ions. When it melts or dissolves, however, the ions are free to move and carry an electric current, as shown in Figure 9.9.

The model also explains that high temperatures are needed to melt and boil an ionic compound (Table 9.1) because freeing the ions from their positions (melting) requires large amounts of energy, and vaporizing them requires even more. In fact, the ionic attraction is so strong that, as Figure 9.10 shows, the vapor consists of **ion pairs,** gaseous ionic molecules rather than individual ions. But keep in mind that in their ordinary (solid) state, ionic compounds consist of rows of alternating ions that extend in all directions, and *no separate molecules exist.*

SECTION SUMMARY

In ionic bonding, a metal transfers electrons to a nonmetal, and the resulting ions attract each other strongly to form a solid. Main-group elements often attain a filled outer level of electrons (either eight or two) by forming ions with the configuration of the nearest noble gas. Ion formation by itself *requires* energy. However, the lattice energy, the energy *released* when the gaseous ions form a solid, is large, and it is the major reason ionic solids exist. The magnitude of the lattice energy, which depends on ionic size and charge, can be determined by applying Hess's law in a Born-Haber cycle.

The ionic bonding model pictures oppositely charged ions held rigidly in position by strong electrostatic attractions and explains why ionic solids crack rather than bend and why they conduct electric current only when melted or dissolved. The ions occur as gaseous ion pairs when the compound vaporizes, which requires very high temperatures.

Figure 9.10 **Vaporizing an ionic compound.** Ionic compounds generally have very high boiling points because the ions must have high enough kinetic energies to break free from surrounding ions. In fact, ionic compounds usually vaporize as ion pairs.

9.3 THE COVALENT BONDING MODEL

Look through any large reference source of chemical compounds, such as the *Handbook of Chemistry and Physics,* and you'll quickly find that the number of known covalent compounds dwarfs the number of known ionic compounds. Molecules held together by covalent bonds range from diatomic hydrogen to biological and synthetic macromolecules with molar masses of several million. We also find covalent bonds in the polyatomic ions of many ionic compounds. Without doubt, sharing electrons is the principal way that atoms interact chemically.

The Formation of a Covalent Bond

A sample of hydrogen gas consists of H_2 molecules. But why *do* the atoms exist bound together in pairs? Look at Figure 9.11 as you imagine what would happen to two isolated H atoms that approach each other from a distance. When the atoms are far apart, each behaves as though the other were not present (point 1). As the distance between the nuclei decreases, each nucleus starts to attract the other atom's electron, which lowers the potential energy of the system. The attractions continue to draw the atoms closer, and the system becomes progressively lower in energy (point 2). As attractions increase, however, so do repulsions between the nuclei and between the electrons. At some internuclear distance, the maximum attraction is achieved in the face of the increasing repulsion, and the system has its minimum energy (point 3). Any shorter distance would increase the repulsions

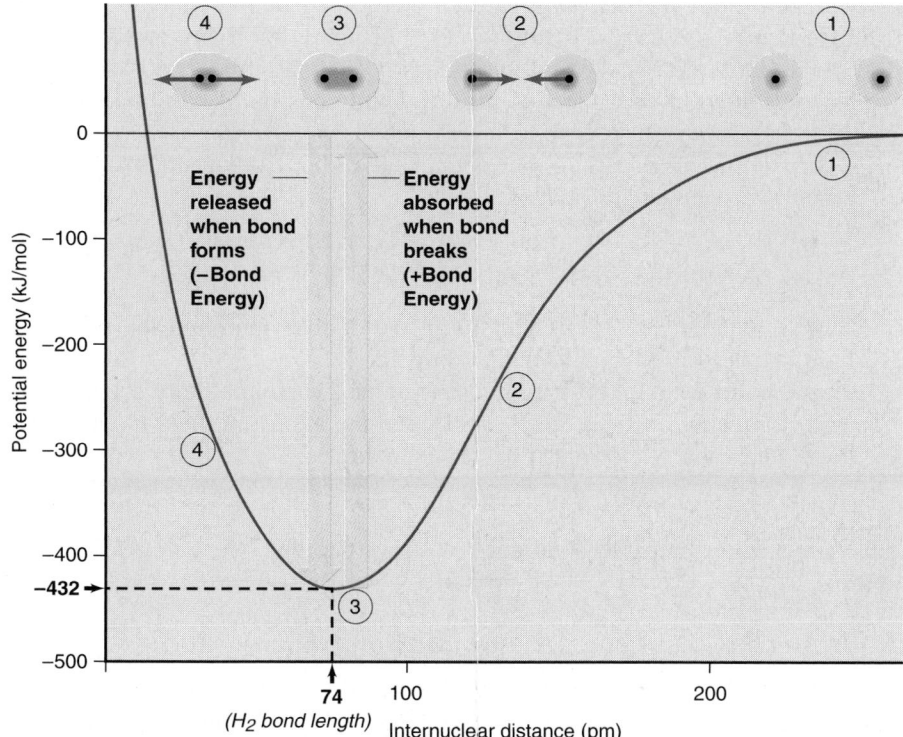

Figure 9.11 Covalent bond formation in H_2. The potential energy of a system of two H atoms plotted against the distance between the nuclei is shown, with a depiction of the atomic systems above. At point 1, the atoms are too far apart to attract each other. At 2, each nucleus attracts the other atom's electron. At 3, the combination of nucleus-electron attractions and electron-electron and nucleus-nucleus repulsions gives the minimum energy of the system. The energy difference between points 1 and 3 is the H_2 bond energy (432 kJ/mol). It is released when the bond forms and absorbed when the bond breaks. The internuclear distance at point 3 is the H_2 bond length (74 pm). If the atoms move closer, as at point 4, repulsions increase the system's energy and force the atoms apart to point 3 again.

and cause the potential energy to rise (point 4). The mutual attraction between nuclei and electron pair constitutes the **covalent bond** that holds the atoms together as an H_2 molecule.

Bonding Pairs and Lone Pairs In covalent bonding, as in ionic bonding, each atom achieves a full outer level of electrons, but this is accomplished by differ-ent means. *Each atom in a covalent bond "counts" the shared electrons as belonging entirely to itself.* Thus, the two electrons in the shared electron pair simultaneously fill the outer level of *both* H atoms. The **shared pair,** or **bonding pair,** is represented by either a pair of dots or a line, H:H or H—H.

An electron pair that is part of an atom's valence shell but *not* involved in bonding is called a **lone pair,** or **unshared pair.** The bonding pair in HF fills the outer shell of the H atom *and,* together with three lone pairs, fills the outer shell of the F atom as well:

Similarly, in F_2, the bonding pair and three lone pairs fill the outer shell of *each* F atom:

(This text generally shows bonding pairs as lines and lone pairs as dots.)

Types of Bonds and Bond Order The **bond order** is the number of electron pairs being shared between any two bonded atoms. The covalent bond in H_2, HF, and F_2 is a **single bond,** one that consists of a single bonding pair of electrons. *A sin-gle bond has a bond order of one.*

Single bonds are the most common type of bond, but many molecules (and ions) contain multiple bonds. Multiple bonds occur most frequently when C, O, N, or S atoms bond to each other. A **double bond** consists of two bonding pairs, four electrons shared between two atoms, so *the bond order is two.* Ethylene (C_2H_4) is a simple hydrocarbon that contains a carbon-carbon double bond and four carbon-hydrogen single bonds:

Each carbon "counts" the four electrons in the double bond and the four in its two single bonds to hydrogen to obtain an octet.

A **triple bond** consists of three bonding pairs; two atoms share six electrons, so *the bond order is three.* In the N_2 molecule, the atoms are held together by a triple bond, and each N atom also has a lone pair:

Six electrons from the triple bond and two from the lone pair give *each* N atom an octet.

The Properties of a Covalent Bond: Bond Energy and Bond Length

As we noted earlier, and Figure 9.12 depicts, the covalent bond results from the balance between the nucleus-electron attractions and the electron-electron and nucleus-nucleus repulsions. The strength of the bond depends on the magnitude of the mutual attraction between the bonded nuclei and the shared electrons. The **bond energy (BE)** (also called *bond enthalpy* or *bond strength*) is the energy required to overcome this attraction. It is defined as the standard enthalpy change for breaking the bond in 1 mol of gaseous molecules. *Bond breakage is an endothermic process, so the bond energy is always positive:*

$$A—B(g) \longrightarrow A(g) + B(g) \qquad \Delta H^0_{bond\ breaking} = BE_{A—B} \ (always > 0)$$

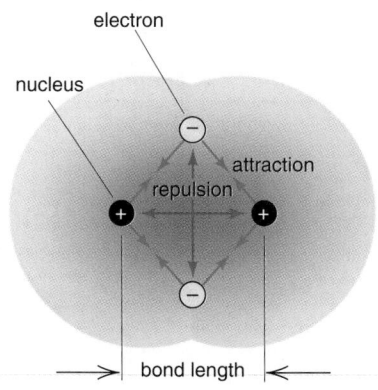

electron

nucleus

attraction

repulsion

bond length

Figure 9.12 The attractive and repulsive forces in covalent bonding. Nucleus-electron attractions and nucleus-nucleus and electron-electron repulsions occur simultaneously. At some optimum distance (the bond length), the attractive forces balance the repulsive forces. The attraction of nuclei for shared electrons determines the bond energy. (Although shown at specific locations in the figure, the electrons are actually distributed throughout the blue shaded volume.)

Put another way, the bond energy is the difference in energy between the separated atoms and the bonded atoms (the potential energy difference between points 1 and 3 in Figure 9.11). The same amount of energy that is absorbed to break the bond is released when it forms. *Bond formation is an exothermic process, so the sign of the enthalpy change is negative:*

$$A(g) + B(g) \longrightarrow A{-}B(g) \qquad \Delta H^0_{\text{bond forming}} = -BE_{A-B} \text{ (always} < 0)$$

Because bond energies depend on the bonded atoms—their electron configurations, nuclear charges, and atomic radii—each type of bond has its own bond energy. Table 9.2 lists bond energies for some common bonds. Moreover, the energy of a given bond varies slightly from molecule to molecule, and even within the same molecule, so *the tabulated value is an average bond energy.*

Table 9.2 Average Bond Energies (kJ/mol)

Bond	Energy	Bond	Energy	Bond	Energy	Bond	Energy
Single Bonds							
H—H	432	N—H	391	Si—H	323	S—H	347
H—F	565	N—N	160	Si—Si	226	S—S	266
H—Cl	427	N—P	209	Si—O	368	S—F	327
H—Br	363	N—O	201	Si—S	226	S—Cl	271
H—I	295	N—F	272	Si—F	565	S—Br	218
		N—Cl	200	Si—Cl	381	S—I	~170
C—H	413	N—Br	243	Si—Br	310		
C—C	347	N—I	159	Si—I	234	F—F	159
C—Si	301					F—Cl	193
C—N	305	O—H	467	P—H	320	F—Br	212
C—O	358	O—P	351	P—Si	213	F—I	263
C—P	264	O—O	204	P—P	200	Cl—Cl	243
C—S	259	O—S	265	P—F	490	Cl—Br	215
C—F	453	O—F	190	P—Cl	331	Cl—I	208
C—Cl	339	O—Cl	203	P—Br	272	Br—Br	193
C—Br	276	O—Br	234	P—I	184	Br—I	175
C—I	216	O—I	234			I—I	151
Multiple Bonds							
C=C	614	N=N	418	C≡C	839	N≡N	945
C=N	615	N=O	607	C≡N	891		
C=O	745	O_2	498	C≡O	1070		
	(799 in CO_2)						

Table 9.3 Average Bond Lengths (pm)

Bond	Length	Bond	Length	Bond	Length	Bond	Length
Single Bonds							
H—H	74	N—H	101	Si—H	148	S—H	134
H—F	92	N—N	146	Si—Si	234	S—P	210
H—Cl	127	N—P	177	Si—O	161	S—S	204
H—Br	141	N—O	144	Si—S	210	S—F	158
H—I	161	N—S	168	Si—N	172	S—Cl	201
		N—F	139	Si—F	156	S—Br	225
		N—Cl	191	Si—Cl	204	S—I	234
C—H	109	N—Br	214	Si—Br	216		
C—C	154	N—I	222	Si—I	240	F—F	143
C—Si	186					F—Cl	166
C—N	147	O—H	96	P—H	142	F—Br	178
C—O	143	O—P	160	P—Si	227	F—I	187
C—P	187	O—O	148	P—P	221	Cl—Cl	199
C—S	181	O—S	151	P—F	156	Cl—Br	214
C—F	133	O—F	142	P—Cl	204	Cl—I	243
C—Cl	177	O—Cl	164	P—Br	222	Br—Br	228
C—Br	194	O—Br	172	P—I	243	Br—I	248
C—I	213	O—I	194			I—I	266
Multiple Bonds							
C=C	134	N=N	122	C≡C	121	N≡N	110
C=N	127	N=O	120	C≡N	115	N≡O	106
C=O	123	O_2	121	C≡O	113		

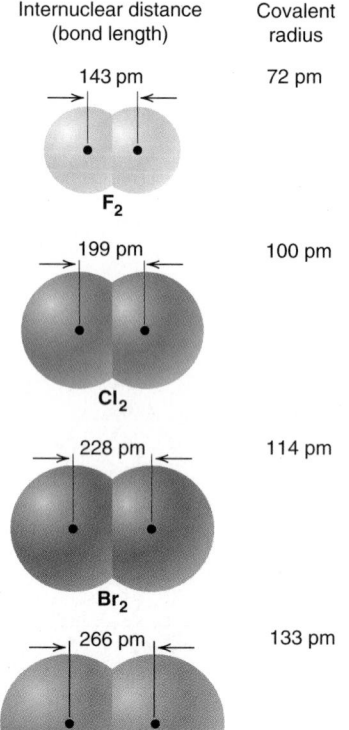

Internuclear distance (bond length)	Covalent radius
143 pm	72 pm
F_2	
199 pm	100 pm
Cl_2	
228 pm	114 pm
Br_2	
266 pm	133 pm
I_2	

A covalent bond has a **bond length,** the distance between the nuclei of two bonded atoms. In Figure 9.11, bond length is shown as the distance between the nuclei at the point of minimum energy. Table 9.3 shows the lengths of some covalent bonds. Here, too, the values represent *average* bond lengths for the given bond in different substances. Bond length is related to the sum of the radii of the bonded atoms. In fact, most atomic radii are calculated from measured bond lengths (see Figure 8.14C). Bond lengths for a series of similar bonds increase with atomic size, as shown in Figure 9.13 for the halogens.

A close relationship exists among bond order, bond length, and bond energy. Two nuclei are more strongly attracted to two shared electron pairs than to one: the atoms are drawn closer together *and* are more difficult to pull apart. Therefore, *for a given pair of atoms, a higher bond order results in a shorter bond length and a higher bond energy.* So, as Table 9.4 shows, for a given pair of atoms, *a shorter bond is a stronger bond.*

In some cases, we can extend this relationship among atomic size, bond length, and bond strength by holding one atom in the bond constant and varying the other atom within a group or period. For example, the trend in carbon-halogen single bond lengths, C—I > C—Br > C—Cl, parallels the trend in atomic size, I > Br > Cl, and is opposite to the trend in bond energy, C—Cl > C—Br > C—I. Thus, for *single bonds,* longer bonds are usually weaker.

Figure 9.13 Bond length and covalent radius. Within a series of similar substances, such as the diatomic halogen molecules, bond length increases as covalent radius increases.

Table 9.4 The Relation of Bond Order, Bond Length, and Bond Energy

Bond	Bond Order	Average Bond Length (pm)	Average Bond Energy (kJ/mol)
C—O	1	143	358
C=O	2	123	745
C≡O	3	113	1070
C—C	1	154	347
C=C	2	134	614
C≡C	3	121	839
N—N	1	146	160
N=N	2	122	418
N≡N	3	110	945

FORMATION OF A COVALENT BOND

SAMPLE PROBLEM 9.2 Comparing Bond Length and Bond Strength

Problem Using the periodic table, but not Tables 9.2 and 9.3, rank the bonds in each set in order of *decreasing* bond length and bond strength:
(a) S—F, S—Br, S—Cl   (b) C=O, C—O, C≡O
Plan In part (a), S is singly bonded to a halogen atom, so all members of the set have a bond order of one. Bond length increases and bond strength decreases as the halogen's atomic radius increases, which we can see from its position in the periodic table. In all the bonds in part (b), the same two atoms are bonded, but the bond orders differ. In this case, bond strength increases and bond length decreases as bond order increases.
Solution (a) Atomic size increases down a group, so F < Cl < Br.

Bond length: S—Br > S—Cl > S—F

Bond strength: S—F > S—Cl > S—Br

(b) By ranking the bond orders, C≡O > C=O > C—O, we obtain

Bond length: C—O > C=O > C≡O

Bond strength: C≡O > C=O > C—O

Check From Tables 9.2 and 9.3, we see that the rankings are correct.
Comment Remember that for different atoms, as in part (a), *the relationship between length and strength holds* only *for single bonds* and not in every case, so apply it carefully.

FOLLOW-UP PROBLEM 9.2 Rank the bonds in each set in order of *increasing* bond length and bond strength: (a) Si—F, Si—C, Si—O; (b) N=N, N—N, N≡N.

How the Model Explains the Properties of Covalent Compounds

The covalent bonding model proposes that electron sharing between pairs of atoms leads to strong, localized bonds, usually within individual molecules. At first glance, however, it seems that the model is inconsistent with some of the familiar physical properties of covalent substances. After all, most are gases (such as methane and ammonia), liquids (such as benzene and water), or low-melting solids (such as sulfur and paraffin wax). Covalent bonds are strong (~200 to 500 kJ/mol), so why do covalent substances melt and boil at such low temperatures?

To answer this question, we must distinguish between two different sets of forces: (1) the *strong covalent bonding forces* holding the atoms together within the molecule (those we have been discussing), and (2) the *weak intermolecular forces* holding the molecules near each other in the macroscopic sample. It is these weak forces *between* the molecules, not the strong covalent bonds *within* each molecule, that are responsible for the physical properties of covalent substances.

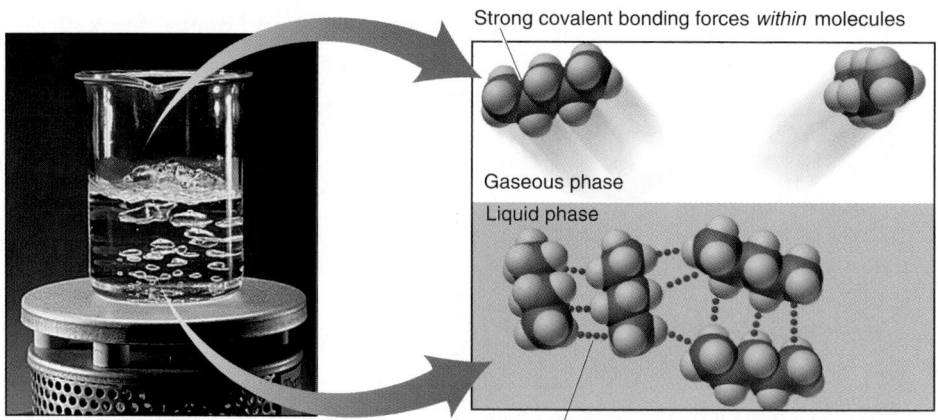

Strong covalent bonding forces *within* molecules

Gaseous phase

Liquid phase

Weak intermolecular forces *between* molecules

Figure 9.14 Strong forces within molecules and weak forces between them. When pentane boils, weak forces *between* molecules (intermolecular forces) are overcome, but the strong covalent bonds holding the atoms together within each molecule remain unaffected. Thus, the pentane molecules leave the liquid phase as intact units.

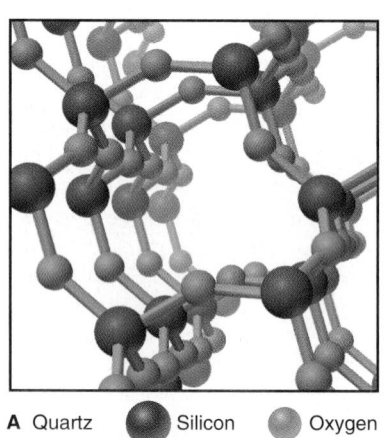

A Quartz ● Silicon ● Oxygen

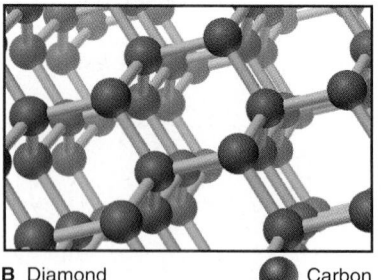

B Diamond ● Carbon

Figure 9.15 Covalent bonds of network covalent solids. A, In quartz (SiO$_2$), each Si atom is bonded covalently to four O atoms and each O atom is bonded to two Si atoms in a pattern that extends throughout the sample. Because no separate SiO$_2$ molecules are present, the melting point and the hardness are very high. **B,** In diamond, each C atom is covalently bonded to four other C atoms throughout the crystal. Diamond is the hardest natural substance known and has an extremely high melting point.

Consider, for example, what happens when pentane (C$_5$H$_{12}$) boils. As Figure 9.14 shows, the weak interactions *between* the pentane molecules are affected, not the strong C—C and C—H covalent bonds *within* each molecule.

Some covalent substances, called *network covalent solids,* do not consist of separate molecules. Rather, they are held together by covalent bonds that extend in three dimensions *throughout* the sample. If the model is correct, the properties of these substances *should* reflect the strength of their covalent bonds, and this is indeed the case. Two examples, quartz and diamond, are shown in Figure 9.15. Quartz (SiO$_2$) is very hard and melts at 1550°C. It is composed of silicon and oxygen atoms connected by covalent bonds that extend throughout the sample; no separate SiO$_2$ molecules exist. Diamond consists of covalent bonds connecting each carbon atom to four others throughout the sample. It is the hardest substance known and melts at around 3550°C. Clearly, covalent bonds *are* strong, but because most covalent substances consist of separate molecules with weak forces between them, their physical properties do not reflect this strength. (We discuss intermolecular forces in detail in Chapter 12.)

Unlike ionic compounds, most covalent substances are poor electrical conductors, even when melted or when dissolved in water. An electric current is carried by either mobile electrons or mobile ions. In covalent substances, the electrons are localized as either shared or unshared pairs, so they are not free to move, and no ions are present. The Tools of the Laboratory essay describes an important laboratory tool for studying covalent substances.

SECTION SUMMARY

A shared pair of valence electrons attracts the nuclei of two atoms and holds them together in a covalent bond while filling each atom's outer shell. The number of shared pairs between the two atoms is the bond order. For a given type of bond, the bond energy is the average energy required to completely separate the bonded atoms; the bond length is the average distance between their nuclei. For a given pair of bonded atoms, bond order is directly related to bond energy and inversely related to bond length. Substances that consist of separate molecules are generally soft and low melting because of the weak forces between molecules. Those held together throughout by covalent bonds are extremely hard and high melting. Most covalent substances have low electrical conductivity because electrons are localized and ions are absent. The atoms in a covalent bond vibrate, and the energy of these vibrations can be studied with IR spectroscopy.

Infrared Spectroscopy

Infrared (IR) spectroscopy is an instrumental technique used primarily for measuring the absorption of IR radiation by covalently bonded molecules. It plays a major role in our understanding of molecular structure and bonding. IR spectrometers are found in most research laboratories, particularly where organic molecules are studied, and are essential aids in identifying unknown substances. Their key components are the same as those of any spectrometer (see pages 267–268). The source emits radiation of many wavelengths, and those in the IR region are selected and directed at the sample. Certain wavelengths are absorbed more than others, and the IR spectrum of the compound is generated.

What property of a molecule is displayed in its IR spectrum? All molecules, whether in a gas, liquid, or solid, undergo continual rotations and vibrations. Consider, for instance, a sample of ethane gas. The H_3C-CH_3 molecules zoom throughout the container, colliding with the walls and each other. If we could look closely at one molecule, however, and disregard its motion through space, we would see the molecule rotating and its two CH_3 groups rotating relative to each other about the C—C bond. The bonded atoms also vibrate, that is, move back and forth as though their bonds were flexible springs: stretching and compressing, twisting, bending, rocking, and wagging (Figure B9.1). (Thus, the length of a given bond even within a given substance is actually the *average* distance between nuclei, analogous to the average length of a spring stretching and compressing.)

Each vibrational motion has its own natural frequency, which is based on the type of motion, the masses of the atoms, and the strengths of the bonds between them. These frequencies correspond to wavelengths between 2.5 and 25 μm, which make up a part of the IR region of the electromagnetic spectrum (see Figure 7.3). The energy of each of these vibrations is quantized. Just as an atom can absorb a photon whose energy corresponds to the difference between two quantized *electron* energy levels, a molecule can absorb an IR photon whose energy corresponds to

Diatomic Molecule
Stretch

Nonlinear Triatomic Molecule
Wagging, twisting, and rocking

Linear Triatomic Molecule
Stretch asymmetrical

Stretch symmetrical

Bending

Figure B9.1 Some vibrational motions in general diatomic and triatomic molecules.

the difference between two of its quantized *vibrational* energy levels.

The IR spectrum is important in compound identification because of two related factors. First, *each kind of bond has a characteristic range of IR wavelengths it can absorb.* For example, a C—C bond absorbs IR photons in a different wavelength range from those absorbed by a C=C bond, a C—H bond, a C=O bond, and so forth. Second, the exact wavelength and quantity of IR radiation absorbed by a given bond depend on the *overall structure* of the molecule. This means that *each compound has a characteristic IR spectrum* that can be used to identify it, much as a fingerprint identifies a person. The spectrum appears as a series of downward pointing peaks, varying in depth and sharpness, that correspond to absorption frequencies across the IR region scanned. Figure B9.2 shows the IR spectrum of acrylonitrile, a compound that is used to manufacture synthetic rubber and plastics; no other compound has exactly the same IR spectrum.

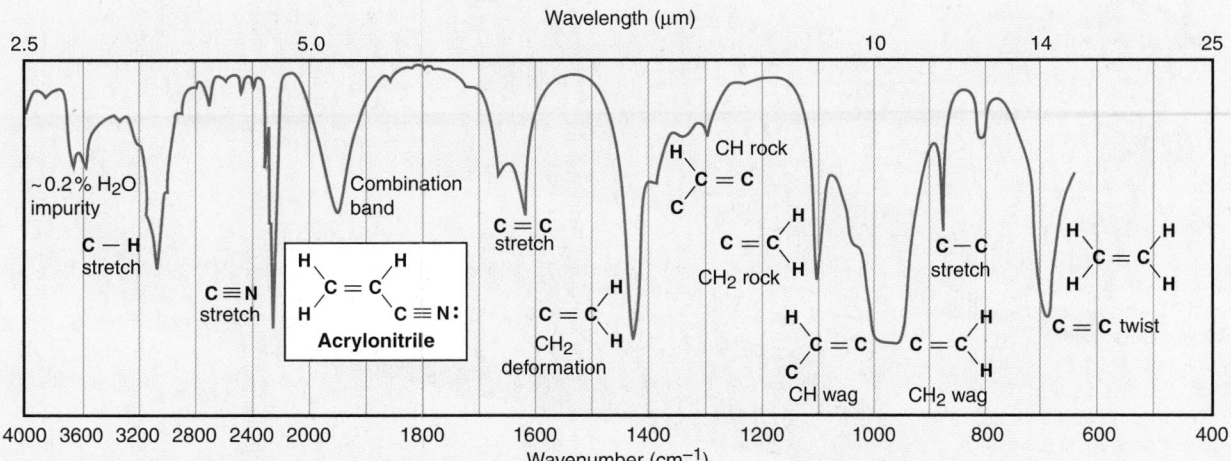

Figure B9.2 The infrared (IR) spectrum of acrylonitrile. The IR spectrum of acrylonitrile is typical of a molecule with several types of covalent bonds. There are many absorption bands (peaks) of differing depth and sharpness. Most peaks correspond to a particular type of vibration in a particular group of bonded atoms. Some broad peaks (for example, "combination band") represent several overlapping types of vibrations. The spectrum is reproducible and unique for acrylonitrile. (The bottom axis shows wavenumbers, the inverse of wavelength, so its units are those of length^{-1}. The scale expands to the right of 2000 cm^{-1}.)

9.4 BETWEEN THE EXTREMES: ELECTRONEGATIVITY AND BOND POLARITY

Scientific models are idealized descriptions of reality. As we've discussed them so far, the ionic and covalent bonding models portray compounds as being formed by *either* complete electron transfer *or* complete electron sharing. However, the great majority of compounds have bonds that are more accurately thought of as partially ionic and partially covalent. Now that you're familiar with the ideal models, we explore the actual continuum in bonding that lies between these extremes.

Electronegativity

One of the most important concepts in chemical bonding is **electronegativity (EN),** the relative ability of a bonded atom to attract the shared electrons.* More than 50 years ago, the American chemist Linus Pauling developed the most common scale of relative EN values for the elements. Here is an example to show the basis of Pauling's approach. We might expect the bond energy of the HF bond to be the average of the energies of an H—H bond (432 kJ/mol) and an F—F bond (159 kJ/mol), or 296 kJ/mol. However, the actual bond energy of H—F is 565 kJ/mol, or 269 kJ/mol *higher* than the average. Pauling reasoned that this difference is due to an *electrostatic (charge) contribution* to the H—F bond energy.

---

*Electronegativity is not the same as electron affinity (EA), although many elements with a high EN also have a highly negative EA. Electronegativity refers to a bonded atom attracting the shared electron pair; electron affinity refers to a separate atom in the gas phase gaining an electron to form a gaseous anion.

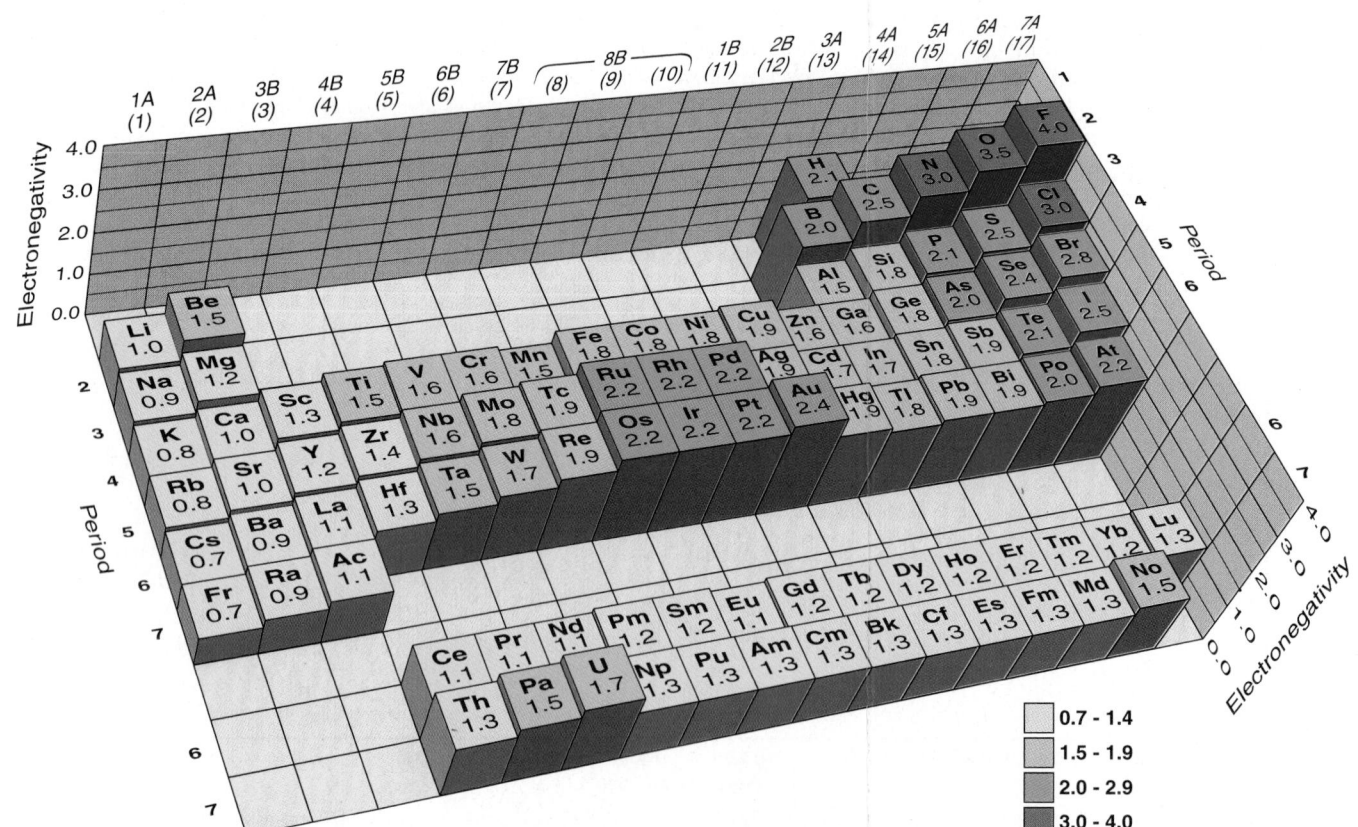

Figure 9.16 **The Pauling electronegativity (EN) scale.** The EN is shown by the height of the post with the value on top. The key indicates arbitrary EN cutoffs. In the main groups, EN generally *increases* from left to right and bottom to top. The noble gases are not shown. The transition and inner transition elements show relatively little change in EN. Hydrogen is shown near elements of similar EN.

If F attracts the shared electron pair more strongly than H, that is, if F is more electronegative than H, the electrons will spend more time closer to F. This unequal sharing of electrons makes the F end of the bond partially negative and the H end partially positive, and the attraction between these partial charges *increases* the energy required to break the bond.

From similar studies with the remaining hydrogen halides and many other compounds, Pauling arrived at the scale of *relative EN values* shown in Figure 9.16. The values themselves are not measured quantities but are based on Pauling's assignment of the highest EN value, 4.0, to fluorine.

Trends in Electronegativity Because the nucleus of a smaller atom is closer to the shared pair than that of a larger atom, it attracts bonding electrons more strongly (Figure 9.17A). So, in general, electronegativity is inversely related to atomic size. As Figure 9.17B makes clear for the main-group elements, *electronegativity generally increases up a group and across a period*. The figure includes the atomic size (on top of each post) and shows this size decreasing with the electronegativity increase. In Pauling's electronegativity scale, and in any of the several others available, *nonmetals are more electronegative than metals*. The most electronegative element is fluorine, with oxygen a close second. Thus, except when it bonds with fluorine, oxygen always pulls bonding electrons toward itself. The least electronegative element (also referred to as the most *electropositive*) is francium, in the lower left corner of the periodic table, but it is radioactive and extremely rare, so for all practical purposes, cesium is the most electropositive.*

---

*In 1934, the American physicist Robert S. Mulliken developed an approach to electronegativity based solely on atomic properties: EN = (IE − EA)/2. By this approach as well, fluorine, with a high ionization energy (IE) and a large negative electron affinity (EA), has a high EN; and cesium, with a low IE and small EA, has a low EN.

Figure 9.17 **Electronegativity and atomic size. A,** In general, an element with a smaller atomic size *(top)* has a higher electronegativity than one with larger atomic size *(bottom)* because the nucleus of the smaller atom is closer to the bonding pair, and thus it attracts the pair more strongly. **B,** The electronegativities of the main-group elements from Periods 2 to 6 (excluding noble gases) are shown as posts of different heights. On top of each post is a hemisphere showing the relative atomic size.

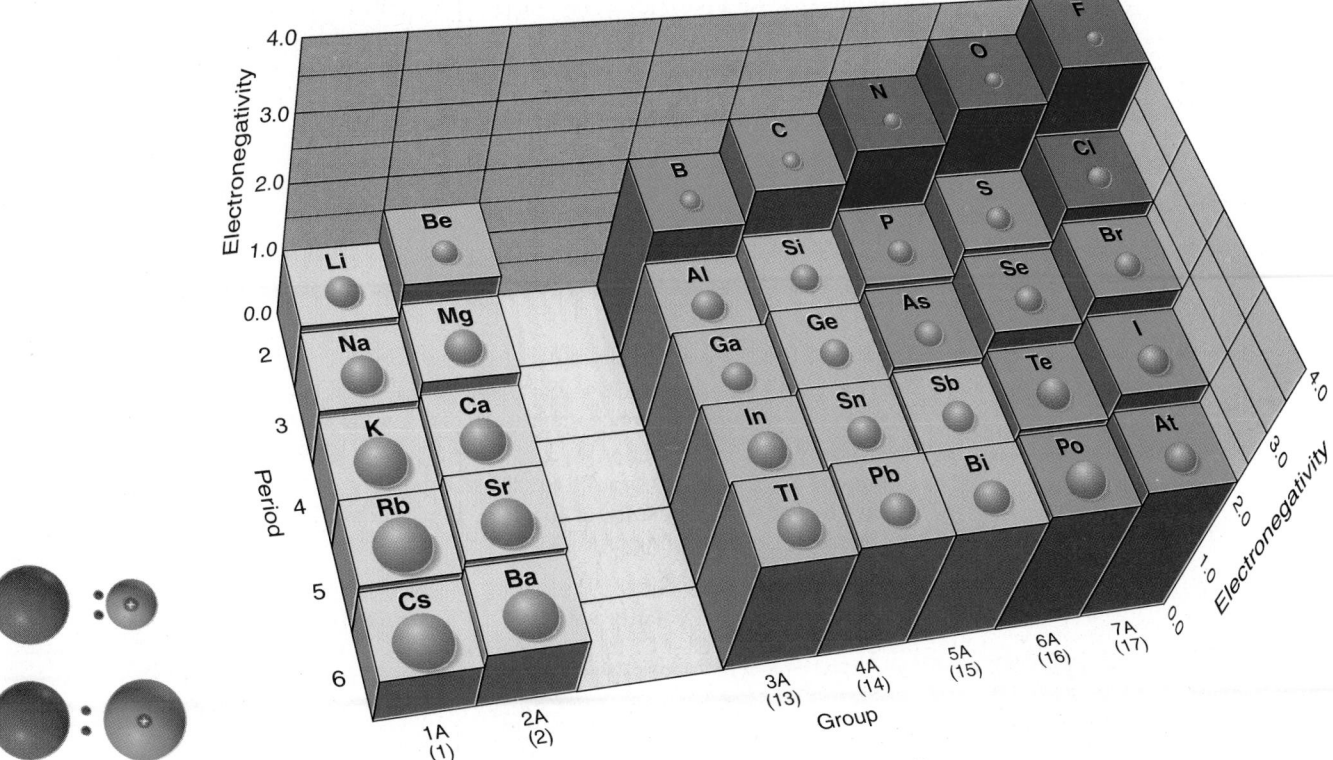

Electronegativity and Oxidation Number One important use of electronegativity is in determining an atom's oxidation number (O.N.; see Section 4.5):

1. The more electronegative atom in a bond is assigned *all* the *shared* electrons; the less electronegative atom is assigned *none*.
2. Each atom in a bond is assigned *all* of its *unshared* electrons.
3. The oxidation number is given by

$$\text{O.N.} = \text{no. of valence } e^- - (\text{no. of shared } e^- + \text{no. of unshared } e^-)$$

In HCl, for example, Cl is more electronegative than H. It has 7 valence electrons but is assigned 8 (2 shared + 6 unshared), so its oxidation number is $7 - 8 = -1$. The H atom has 1 valence electron and is assigned none, so its oxidation number is $1 - 0 = +1$.

Polar Covalent Bonds and Bond Polarity

From our discussion of HF, you can see that when atoms with different electronegativities form a bond, the bonding pair is shared *un*equally, so the bond has partially negative and positive poles. This type of bond is called a **polar covalent bond** and is depicted by a polar arrow ($\longmapsto$) pointing toward the negative pole, or by $\delta+$ and $\delta-$ symbols, where the lowercase Greek letter delta, δ, represents a partial charge:

$$\overset{\longmapsto}{\text{H—}\ddot{\text{F}}\text{:}} \quad \text{or} \quad \overset{\delta+\ \ \delta-}{\text{H—}\ddot{\text{F}}\text{:}}$$

In the H—H and F—F bonds, the atoms are identical, so the bonding pair is shared equally, and the bond is called a **nonpolar covalent bond.** Thus, by knowing the EN values of the atoms in a bond, you can determine the direction of the bond polarity.

IONIC VS. COVALENT BONDING

SAMPLE PROBLEM 9.3 Determining Bond Polarity from EN Values

Problem **(a)** Use a polar arrow to indicate the polarity of each bond: N—H, F—N, I—Cl.
(b) Rank the following bonds in order of increasing polarity: H—N, H—O, H—C.
Plan **(a)** We use Figure 9.16 to find the EN values of the bonded atoms and point the polar arrow toward the negative end. **(b)** Each choice has H bonded to an atom from Period 2. Since EN increases across a period, the polarity is greatest for the bond whose Period 2 atom is farthest to the right.
Solution **(a)** The EN of N = 3.0 and the EN of H = 2.1, so N is more electronegative than H: $\overset{\longleftarrow}{\text{N—H}}$

The EN of F = 4.0 and the EN of N = 3.0, so F is more electronegative: $\overset{\longleftarrow}{\text{F—N}}$

The EN of I = 2.5 and the EN of Cl = 3.0, so I is less electronegative: $\overset{\longmapsto}{\text{I—Cl}}$

(b) The order of increasing EN is C < N < O, and each has a higher EN than H. Therefore, O pulls most on the bonded pair to H, and C pulls least; so the order of bond polarity is $\boxed{\text{H—C} < \text{H—N} < \text{H—O.}}$

Comment In Chapter 10, you'll see that the polarity of the bonds in a molecule contributes to the overall polarity of the molecule, which is a major factor determining the magnitudes of physical properties.

FOLLOW-UP PROBLEM 9.3 Arrange each set of bonds in order of increasing polarity, and indicate bond polarity with $\delta+$ and $\delta-$ symbols:
(a) Cl—F, Br—Cl, Cl—Cl   **(b)** Si—Cl, P—Cl, S—Cl, Si—Si

The Partial Ionic Character of Polar Covalent Bonds

If you ask "Is an X—Y bond ionic or covalent?" the answer in almost every case is "Both, partially!" A better question is *"To what extent* is the bond ionic or covalent?" The existence of partial charges means that a polar covalent bond behaves as if it were partially ionic. The **partial ionic character** of a bond is related directly to the **electronegativity difference (ΔEN),** the difference between the EN values of the bonded atoms: *a greater ΔEN results in larger partial charges and a higher partial ionic character.* For example, ΔEN for LiF(*g*) is 4.0 − 1.0 = 3.0; for HF(*g*), it is 4.0 − 2.1 = 1.9; and for $F_2(g)$, it is 4.0 − 4.0 = 0. Thus, the bond in LiF has more ionic character than the H—F bond, which has more than the F—F bond.

Various attempts have been made to classify the ionic character of bonds, but they all use arbitrary cutoff values, which is inconsistent with the gradation of ionic character observed experimentally. One approach uses ΔEN values to divide bonds into ionic, polar covalent, and nonpolar covalent. Based on a ΔEN range from 0 (completely nonpolar) to 3.3 (highly ionic), some approximate guidelines appear in Figure 9.18.

Another approach calculates the *percent ionic character* of a bond by comparing the actual behavior of a polar molecule in an electric field with the behavior it would have if the electron were transferred completely (pure ionic). A value of 50% ionic character is often chosen to divide substances we recognize as "ionic" from those we recognize as "covalent." Such methods show 43% ionic character for the H—F bond and expected decreases for the other hydrogen halides: H—Cl is 19% ionic, H—Br 11%, and H—I 4%. A plot of percent ionic character vs. ΔEN for a variety of gaseous diatomic molecules is shown in Figure 9.19. The specific values are not important, but note that *percent ionic character generally increases with ΔEN.* Another point to note is that whereas some molecules, such as $Cl_2(g)$, have 0% ionic character, none has 100% ionic

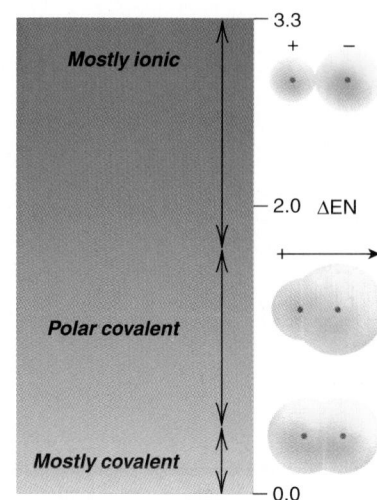

ΔEN	IONIC
>1.7	Mos...
0.4-1.7	Polar covale...
<0.4	Mostly covalent
0	Nonpolar covalent

A

Figure 9.18 Boundary ranges for classifying ionic character of chemical bonds. A, The electronegativity difference (ΔEN) between bonded atoms shows cutoff values that act as a general guide to a bond's relative ionic character. **B,** The gradation in ionic character is shown with shading across the entire bonding range from ionic (*green*) to covalent (*yellow*).

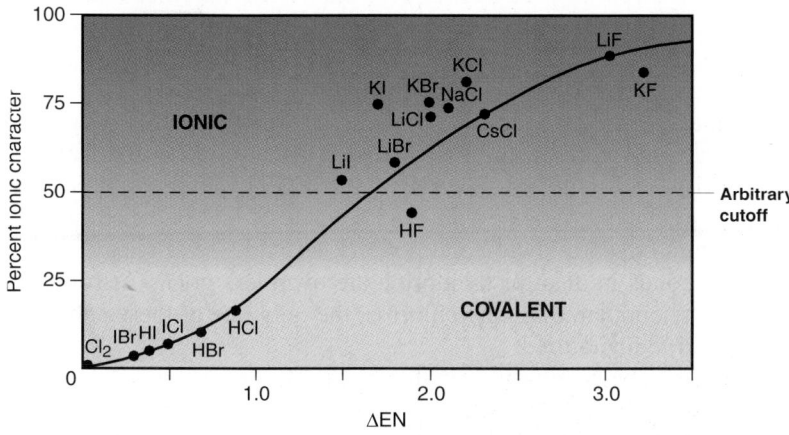

Figure 9.19 Percent ionic character as a function of electronegativity difference (ΔEN). The percent ionic character is plotted against ΔEN for some simple gaseous diatomic molecules. Note that, in general, ΔEN correlates with ionic character. (The arbitrary cutoff for an ionic compound is >50% ionic character.)

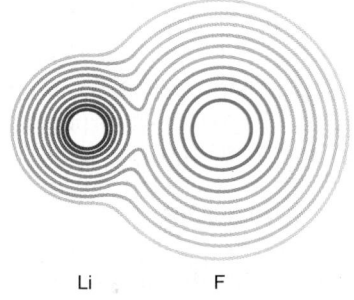

Figure 9.20 The charge density of LiF. A cross section of a diatomic LiF molecule showing contours of electron charge density *(color intensity)* that decrease from the nuclei outward. Note that even in this ionic interaction, there is significant overlap of electron density between the atoms, indicating some covalent character.

character. Thus, *electron sharing occurs to some extent in every bond,* even one between an alkali metal and a halogen (Figure 9.20).

The Continuum of Bonding Across a Period

A metal and a nonmetal—elements from opposite sides of the periodic table—have a relatively large ΔEN and typically interact by electron transfer to form an ionic compound. Two nonmetals—elements from the same side of the table—have a small ΔEN and interact by electron sharing to form a covalent compound. If we combine a nonmetal such as Cl_2 with each of the other elements in Period 3, we expect to observe a steady decrease in ΔEN and a gradation in type of bonding from ionic to polar covalent to nonpolar covalent.

Figure 9.21 shows samples of the most common Period 3 chlorides—NaCl, $MgCl_2$, $AlCl_3$, $SiCl_4$, PCl_3, S_2Cl_2, and Cl_2—along with structural formulas (discussed at the beginning of the next chapter) and some properties. The first compound is sodium chloride, a white (colorless) crystalline solid with typical ionic properties: high melting point and high conductivity when molten or dissolved. With these properties and a ΔEN of 2.1, NaCl is ionic by any criterion. Magnesium chloride is still considered ionic, with a ΔEN of 1.8, but it has a lower melting point and lower conductivity.

Aluminum chloride is still less ionic. Its ΔEN value of 1.5 indicates a highly polar covalent Al—Cl bond. Rather than a lattice of Al^{3+} and Cl^- ions, it consists of extended layers of linked Al and Cl atoms. Strong polar interactions hold the particles together in each layer, but weak forces between layers result in a much lower melting point. The low electrical conductivity of molten $AlCl_3$ is consistent with a scarcity of ions.

Silicon tetrachloride's melting point is very low because of weak forces *between* separate molecules, and it has no measurable electrical conductivity. Each molecule consists of strong Si—Cl bonds with a ΔEN value of 1.2, near the middle of the polar covalent region. With a ΔEN value of 0.9, phosphorus trichloride continues the trend toward lower bond polarity, as evidenced by its very low melting point. The bonds in disulfur dichloride are even less polar (ΔEN = 0.5). The series ends with nonpolar, diatomic chlorine, the only one of these substances that is a gas at room temperature.

Thus, *as ΔEN becomes smaller, the bond becomes more covalent,* and the properties of the Period 3 chlorides change from those of a solid consisting of ions to those of a gas consisting of individual molecules.

SECTION SUMMARY

An atom's electronegativity is related to its ability to pull bonded electrons toward it, which generates partial charges at the ends of the bond. Electronegativity increases across a period and decreases down a group, the reverse of the trend in size. The greater the ΔEN for the two atoms in a bond, the more polar the bond is and the greater its ionic character. There is a gradation of bond type, from ionic to polar covalent to nonpolar covalent; this can be seen by combining a halogen with the other elements in its period.

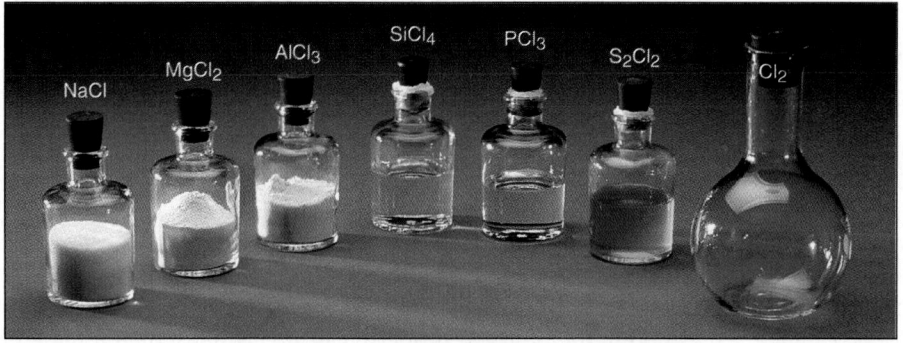

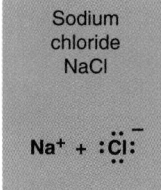 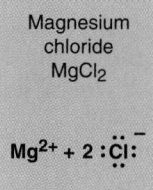 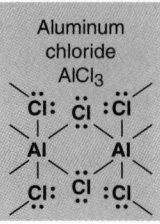 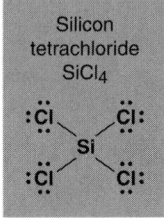 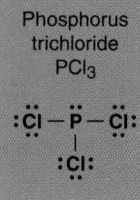 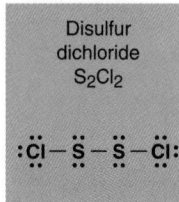 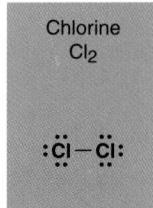

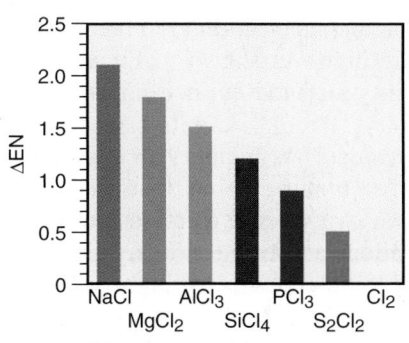 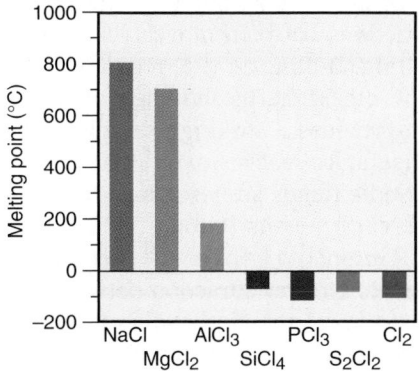 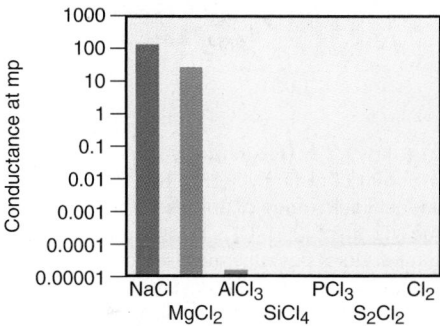

Figure 9.21 **Properties of the Period 3 chlorides.** Samples of the compounds formed from each of the Period 3 elements with chlorine are shown in periodic table sequence in the photo, with their structural formulas below. Note the trend in properties displayed in the bar graphs: as ΔEN decreases, both melting point and electrical conductivity (at the melting point) decrease as well. These trends are consistent with a change in bond type from ionic through polar covalent to nonpolar covalent.

9.5 AN INTRODUCTION TO METALLIC BONDING

The third type of chemical bonding we consider is metallic bonding, which occurs among large numbers of metal atoms. In this section, you'll see a qualitative model; a more quantitative one is presented in Chapter 12.

The Electron-Sea Model

In reactions with nonmetals, reactive metals (such as Na) transfer their outer electrons and form ionic solids (such as NaCl). Two metal atoms can also share their valence electrons in a covalent bond and form gaseous, diatomic molecules (such as Na₂). But what holds the atoms together in a piece of sodium metal? The **electron-sea model** of metallic bonding proposes that all the metal atoms in the sample contribute their valence electrons to form an "electron sea" that is delocalized throughout the substance. The metal ions (the nuclei with their core electrons) are submerged within this electron sea in an orderly array (see Figure 9.2C).

In contrast to ionic bonding, the metal ions are not held in place as rigidly as in an ionic solid. In contrast to covalent bonding, no particular pair of metal

Table 9.5 Melting and Boiling Points of Some Metals		
Element	mp (°C)	bp (°C)
Lithium (Li)	180	1347
Tin (Sn)	232	2623
Aluminum (Al)	660	2467
Barium (Ba)	727	1850
Silver (Ag)	961	2155
Copper (Cu)	1083	2570
Uranium (U)	1130	3930

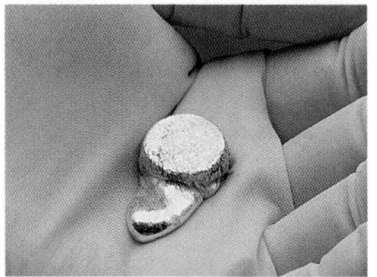

Figure 9.22 The unusually low melting point of gallium. Gallium has the widest liquid range of any element. Its melting point (29.8°C) is below body temperature, because metallic bonding forces need not be broken when the metal ions move from their lattice positions. But gallium boils at 2403°C, because these forces are strong enough to resist separation of the atoms.

atoms is bonded through any localized pair of electrons. Rather, *the valence electrons are shared among all the atoms in the substance.* The piece of metal is held together by the mutual attraction of the metal cations for the mobile, highly delocalized electrons.

Although there are metallic compounds, metals typically form **alloys,** solid mixtures with variable composition. Many familiar metallic materials are alloys, such as those used for car parts, airplane bodies, building and bridge supports, coins, jewelry, and dental work.

How the Model Explains the Properties of Metals

Although the physical properties of metals vary over a wide range, most are solids with moderate to high melting points and much higher boiling points (Table 9.5). Metals typically bend or dent rather than crack or shatter. Many can be flattened into sheets (malleable) and pulled into wires (ductile). Unlike typical ionic and covalent substances, metals conduct heat and electricity well in *both* the solid and liquid states.

Two features of the electron-sea model that account for these properties are the *regularity,* but not rigidity, of the metal-ion array and the *mobility* of the valence electrons. The melting and boiling points of metals are related to the energy of the metallic bonding. Melting points are only moderately high because the attractions between moveable cations and electrons need not be broken during melting. Boiling a metal requires each cation and its electron(s) to break away from the others, so the boiling points are quite high. As shown in Figure 9.22, gallium provides a striking example: it melts in your hand but doesn't boil until the temperature reaches over 2400°C.

Periodic trends are also consistent with the model. As Figure 9.23 shows, the alkaline earth metals [Group 2A(2)] have higher melting points than the alkali metals [Group 1A(1)]. The 2A metal atoms have two valence electrons and form 2+ cations. Greater attraction between these cations and twice as many electrons means stronger metallic bonding than for the 1A metal atoms, so higher temperatures are needed to melt the 2A solids.

Mechanical and conducting properties are also explained by the model. When a piece of metal is deformed by a hammer, the metal ions slide past each other

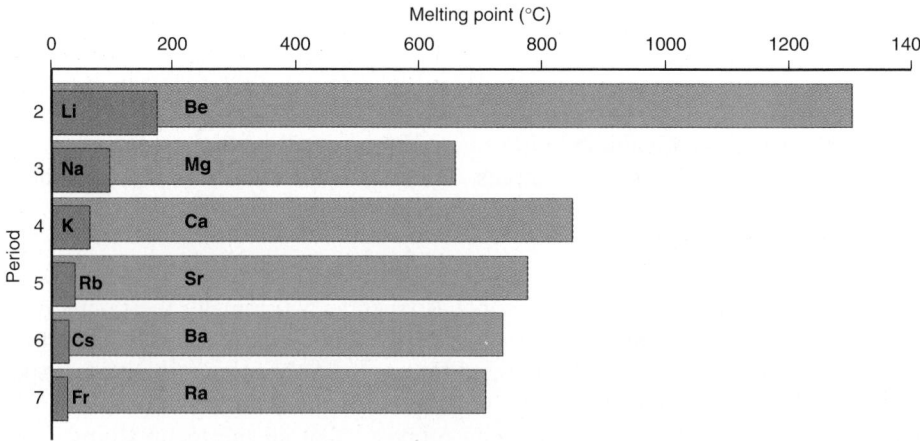

Figure 9.23 Melting points of the Group 1A(1) and Group 2A(2) elements. The alkaline earth metals [Group 2A(2), *blue*] have higher melting points than the alkali metals [Group 1A(1), *brown*] because their ions have 2+ charges and the electron sea has twice as many valence electrons, which results in stronger attractions.

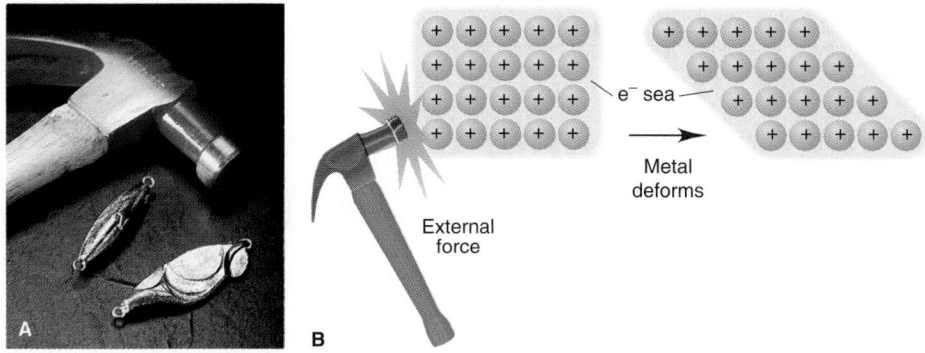

Figure 9.24 **The reason metals deform. A,** An external force applied to a piece of metal deforms the piece without breaking it. **B,** At the atomic level, the force simply moves metal ions past each other through the surrounding electron sea.

through the electron sea to new lattice positions. Thus, the metal-ion cores do not repel each other (Figure 9.24). Compare this behavior with the repulsions that occur when an ionic solid is struck (see Figure 9.8).

Metals are good conductors of electricity because they have mobile electrons. When a piece of metal is attached to a battery, electrons flow from one terminal into the metal and replace electrons flowing from the metal into the other terminal. Irregularities in the array of metal atoms reduce this conductivity. Ordinary copper wire used to carry an electric current, for example, is more than 99.99% pure because traces of other atoms can drastically restrict the flow of electrons.

The mobile electrons also make metals good conductors of heat. If you place your hand on a piece of metal and a piece of wood that are both at room temperature, the metal feels colder because it conducts body heat away from your hand much faster than the wood. The delocalized electrons in the metal disperse the heat from your hand more quickly than the localized electron pairs in the covalent bonds of wood.

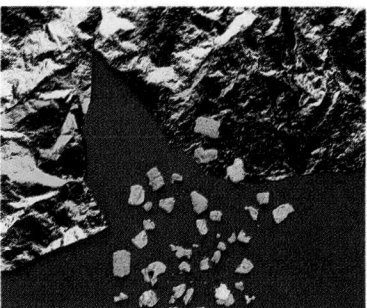

The Amazing Malleability of Gold All the Group 1B(11) metals—copper, silver, and gold—are soft enough to be machined easily, but gold is in a class by itself. One gram of gold forms a cube 0.37 cm on a side or a sphere the size of a small ball-bearing. It is so ductile that it can be drawn into a wire 20 μm thick and 165 m long, and so malleable that it can be hammered into a 1.0-m² sheet that is only 230 atoms (about 70 nm) thick!

SECTION SUMMARY

According to the electron-sea model, the valence electrons of the metal atoms in a sample are highly delocalized and attract the metal cations together. Metals have only moderate melting points because the metal ions remain attracted to the electron sea even if their relative positions change. Yet they have high boiling points because boiling requires completely overcoming these bonding attractions. Metals can be deformed because the electron sea prevents repulsions among the cations. Metals conduct electricity and heat because their electrons are mobile.

Chapter Perspective

Our theme throughout this chapter has been that the particular type of chemical bonding—whether ionic, covalent, metallic, or some blend of these—is governed by the properties of the bonding atoms. This fundamental idea reappears many times as we investigate the forces that give rise to the properties of liquids, solids, and solutions (Chapters 12 and 13), the behavior of the main-group elements (Chapter 14), and the organic chemistry of carbon (Chapter 15). Before we can develop these topics, however, you'll see in Chapter 10 how the relative placement of atoms and the arrangement of bonding and lone pairs gives a molecule its characteristic shape and how that shape influences many of the compound's properties. Then, in Chapter 11, you'll see how covalent bonding theory explains the nature of the bond itself and the properties of compounds.

For Review and Reference (Numbers in parentheses refer to pages, unless noted otherwise.)

Learning Objectives

Relevant section and/or sample problem (SP) numbers appear in parentheses.

Understand These Concepts

1. How differences in atomic properties lead to differences in bond type; the basic distinctions among the three types of bonding (9.1)
2. The essential features of ionic bonding: electron transfer to form ions, and their electrostatic attraction to form a solid (9.2)
3. How lattice energy is ultimately responsible for formation of ionic compounds (9.2)
4. How ionic compound formation is conceptualized as occurring in hypothetical steps (Born-Haber cycle) to calculate lattice energy (9.2)
5. How Coulomb's law explains the periodic trends in lattice energy (9.2)
6. Why ionic compounds are brittle, high melting, and conduct electricity only when molten or dissolved in water (9.2)
7. How nonmetal atoms form a covalent bond (9.3)
8. How bonding and lone pairs fill the valence shell of each atom in a molecule (9.3)
9. The interrelationships among bond order, bond length, and bond energy (9.3)
10. How the distinction between bonding and nonbonding forces explains the properties of covalent molecules and network covalent solids (9.3)

11. The periodic trends in electronegativity and the inverse relation of EN values to atomic sizes (9.4)
12. How bond polarity arises from differences in electronegativity of bonded atoms; the direction of polarity (9.4)
13. The change in partial ionic character with ΔEN and the change in bonding from ionic to polar covalent to nonpolar covalent across a period (9.4)
14. The role of delocalized electrons in metallic bonding (9.5)
15. How the electron-sea model explains why metals bend, have very high boiling points, and conduct electricity in solid or molten forms (9.5)

Master These Skills

1. Using Lewis electron-dot symbols to depict main-group atoms (9.1)
2. Depicting the formation of binary ionic compounds with electron configurations, orbital diagrams, and Lewis electron-dot symbols (SP 9.1)
3. Calculating lattice energy from the energy components of ionic compound formation (9.2)
4. Ranking similar covalent bonds according to their length and strength (SP 9.2)
5. Ranking bonds by polarity and predicting their direction (SP 9.3)

Key Terms

Section 9.1
ionic bonding (328)
covalent bonding (328)
metallic bonding (328)
Lewis electron-dot symbol (329)

Section 9.2
octet rule (330)
lattice energy (331)

Born-Haber cycle (332)
Coulomb's law (333)
ion pair (336)

Section 9.3
covalent bond (338)
bonding (shared) pair (338)
lone (unshared) pair (338)
bond order (338)
single bond (338)

double bond (338)
triple bond (338)
bond energy (BE) (338)
bond length (340)
infrared (IR) spectroscopy (343)

Section 9.4
electronegativity (EN) (344)
polar covalent bond (346)

nonpolar covalent bond (346)
partial ionic character (347)
electronegativity difference (ΔEN) (347)

Section 9.5
electron-sea model (349)
alloy (350)

Key Equations and Relationships

9.1 Expressing the energy of attraction (or repulsion) between charges (333):

$$\text{Electrostatic energy} \propto \frac{\text{charge A} \times \text{charge B}}{\text{distance}}$$

9.2 Relating the energy of attraction to the lattice energy (333):

$$\text{Electrostatic energy} \propto \frac{\text{cation charge} \times \text{anion charge}}{\text{cation radius} + \text{anion radius}}$$
$$\propto \Delta H^0_{\text{lattice}}$$

Highlighted Figures and Tables

These figures (F) and tables (T) provide a quick review of key ideas. Entries in color contain frequently used data.

F9.1 A general comparison of metals and nonmetals (327)
F9.2 The three models of chemical bonding (328)
F9.3 Lewis electron-dot symbols for elements in Periods 2 and 3 (329)
F9.6 The Born-Haber cycle for LiF (332)
F9.7 Trends in lattice energy (334)

F9.11 Covalent bond formation in H_2 (337)
T9.2 Average bond energies (339)
T9.3 Average bond lengths (340)
F9.16 The Pauling electronegativity scale (344)
F9.17 Electronegativity and atomic size (345)
F9.21 Properties of the Period 3 chlorides (349)

Brief Solutions to Follow-up Problems

9.1 Mg ([Ne] $3s^2$) + 2 Cl ([Ne] $3s^2 3p^5$) $\longrightarrow$
$\qquad\qquad$ Mg^{2+} ([Ne]) + $2Cl^-$ ([Ne] $3s^2 3p^6$)

$\cdot Mg\cdot$ + $\vdots\ddot{\underset{\cdot\cdot}{Cl}}\vdots$ $\longrightarrow$ Mg^{2+} + 2 $:\ddot{Cl}:^-$ Formula is $MgCl_2$

9.2 (a) Bond length: Si—F < Si—O < Si—C
Bond strength: Si—C < Si—O < Si—F

(b) Bond length: N≡N < N=N < N—N
Bond strength: N—N < N=N < N≡N
9.3 (a) $\overset{\delta+}{Cl}$—$\overset{\delta-}{Cl}$ < $\overset{\delta+}{Br}$—$\overset{\delta-}{Cl}$ < $\overset{\delta+}{Cl}$—$\overset{\delta-}{F}$;
(b) Si—Si < $\overset{\delta+}{S}$—$\overset{\delta-}{Cl}$ < $\overset{\delta+}{P}$—$\overset{\delta-}{Cl}$ < $\overset{\delta+}{Si}$—$\overset{\delta-}{Cl}$

Problems

Problems with colored numbers are answered at the back of the text. Sections match the text and provide the number(s) of relevant sample problems. Most offer Concept Review Questions, Skill-Building Exercises (in similar pairs), and Problems in Context. Then Comprehensive Problems, based on material from any section or previous chapter, follow.

Atomic Properties and Chemical Bonds

● **Concept Review Questions**

9.1 In general terms, how does each of the following atomic properties influence the metallic character of the main-group elements in a period?
(a) Ionization energy $\qquad$ (b) Atomic radius
(c) Number of outer electrons $\qquad$ (d) Effective nuclear charge
9.2 Both nitrogen and bismuth are members of Group 5A(15). Which is more metallic? Explain your answer in terms of atomic properties.
9.3 What is the relationship between the tendency of a main-group element to form a monatomic ion and its position in the periodic table? In what part of the table are the main-group elements that typically form cations? Anions?

● **Skill-Building Exercises (paired)**

9.4 Which member of each pair is *more* metallic?
(a) Na or Cs $\qquad$ (b) Mg or Rb $\qquad$ (c) As or N
9.5 Which member of each pair is *less* metallic?
(a) I or O $\qquad$ (b) Be or Ba $\qquad$ (c) Se or Ge

9.6 State the type of bonding—ionic, covalent, or metallic—you would expect in (a) CsF(s); (b) $N_2(g)$; (c) Na(s).
9.7 State the type of bonding—ionic, covalent, or metallic—you would expect in (a) $ICl_3(g)$; (b) $N_2O(g)$; (c) LiCl(s).

9.8 State the type of bonding—ionic, covalent, or metallic—you would expect in (a) $O_3(g)$; (b) $MgCl_2(s)$; (c) $BrO_2(g)$.
9.9 State the type of bonding—ionic, covalent, or metallic—you would expect in (a) Cr(s); (b) $H_2S(s)$; (c) CaO(s).

9.10 Draw a Lewis electron-dot symbol for each atom:
(a) Rb $\qquad$ (b) Si $\qquad$ (c) I
9.11 Draw a Lewis electron-dot symbol for each atom:
(a) Ba $\qquad$ (b) Kr $\qquad$ (c) Br
9.12 Draw a Lewis electron-dot symbol for each atom:
(a) Sr $\qquad$ (b) P $\qquad$ (c) S
9.13 Draw a Lewis electron-dot symbol for each atom:
(a) As $\qquad$ (b) Se $\qquad$ (c) Ga

9.14 Give the group number and general electron configuration of an element with the following electron-dot symbol:
(a) $\cdot\ddot{X}:$ $\qquad$ (b) $\dot{X}\cdot$
9.15 Give the group number and general electron configuration of an element with the following electron-dot symbol:
(a) $\cdot\ddot{X}:$ $\qquad$ (b) $\cdot\dot{X}\cdot$

The Ionic Bonding Model
(Sample Problem 9.1)

● **Concept Review Questions**

9.16 If energy is required to form monatomic ions from metals and nonmetals, why do ionic compounds exist?
9.17 In general, how does the lattice energy of an ionic compound depend on the charges and sizes of the ions?
9.18 When gaseous Na^+ and Cl^- ions form gaseous NaCl ion pairs, 548 kJ/mol of energy is released. Why, then, does NaCl occur as a solid under ordinary conditions?
9.19 To form S^{2-} ions from gaseous sulfur atoms requires 214 kJ/mol, but these ions exist in solids such as K_2S. Explain.

● **Skill-Building Exercises (paired)**

9.20 Use condensed electron configurations and Lewis electron-dot symbols to depict the monatomic ions formed from each of the following reactants, and predict the formula of the compound the ions produce:
(a) Ba and Cl   (b) Sr and O   (c) Al and F   (d) Rb and O

9.21 Use condensed electron configurations and Lewis electron-dot symbols to depict the monatomic ions formed from each of the following reactants, and predict the formula of the compound the ions produce:
(a) Cs and S   (b) O and Ga   (c) N and Mg   (d) Br and Li

9.22 Identify the main group to which X belongs in each ionic compound formula: (a) XF_2; (b) MgX; (c) X_2SO_4.

9.23 Identify the main group to which X belongs in each ionic compound formula: (a) X_3PO_4; (b) $X_2(SO_4)_3$; (c) $X(NO_3)_2$.

9.24 Identify the main group to which X belongs in each ionic compound formula: (a) X_2O_3; (b) XCO_3; (c) Na_2X.

9.25 Identify the main group to which X belongs in each ionic compound formula: (a) CaX_2; (b) Al_2X_3; (c) XPO_4.

9.26 For each pair, choose the compound with the higher (more negative) lattice energy, and explain your choice:
(a) BaS or CsCl   (b) LiCl or CsCl

9.27 For each pair, choose the compound with the higher (more negative) lattice energy, and explain your choice:
(a) CaO or CaS   (b) BaO or SrO

9.28 For each pair, choose the compound with the lower (less negative) lattice energy, and explain your choice:
(a) CaS or BaS   (b) NaF or MgO

9.29 For each pair, choose the compound with the lower (less negative) lattice energy, and explain your choice:
(a) NaF or NaCl   (b) K_2O or K_2S

9.30 Use the following to calculate the $\Delta H^0_{lattice}$ of NaCl:

$Na(s) \longrightarrow Na(g)$ $\Delta H^0 = 109$ kJ
$Cl_2(g) \longrightarrow 2Cl(g)$ $\Delta H^0 = 243$ kJ
$Na(g) \longrightarrow Na^+(g) + e^-$ $\Delta H^0 = 496$ kJ
$Cl(g) + e^- \longrightarrow Cl^-(g)$ $\Delta H^0 = -349$ kJ
$Na(s) + \frac{1}{2}Cl_2(g) \longrightarrow NaCl(s)$ $\Delta H^0_f = -411$ kJ

Compared with the lattice energy of LiF (-1050 kJ/mol), is the magnitude of the NaCl value what you expected? Explain.

9.31 Use the following to calculate the $\Delta H^0_{lattice}$ of MgF_2:

$Mg(s) \longrightarrow Mg(g)$ $\Delta H^0 = 148$ kJ
$F_2(g) \longrightarrow 2F(g)$ $\Delta H^0 = 159$ kJ
$Mg(g) \longrightarrow Mg^+(g) + e^-$ $\Delta H^0 = 738$ kJ
$Mg^+(g) \longrightarrow Mg^{2+}(g) + e^-$ $\Delta H^0 = 1450$ kJ
$F(g) + e^- \longrightarrow F^-(g)$ $\Delta H^0 = -328$ kJ
$Mg(s) + F_2(g) \longrightarrow MgF_2(s)$ $\Delta H^0_f = -1123$ kJ

Compared with the lattice energy of LiF (-1050 kJ/mol) or the lattice energy you calculated for NaCl in Problem 9.30, does the relative magnitude of the value for MgF_2 surprise you? Explain.

● **Problems in Context**

9.32 Aluminum oxide (Al_2O_3) is a widely used industrial abrasive (emery, corundum), with its specific application depending on the hardness of the crystal. What does this hardness imply about the magnitude of the lattice energy? Would you have predicted from the chemical formula that Al_2O_3 is hard? Explain.

9.33 Born-Haber cycles were used to obtain the first reliable values for electron affinity by considering the EA value as the unknown and using a theoretically calculated value for the lattice energy. Use a Born-Haber cycle for KF and the following values to calculate a value for the electron affinity of fluorine:

$K(s) \longrightarrow K(g)$ $\Delta H^0 = 90$ kJ
$K(g) \longrightarrow K^+(g) + e^-$ $\Delta H^0 = 419$ kJ
$F_2(g) \longrightarrow 2F(g)$ $\Delta H^0 = 159$ kJ
$K(s) + \frac{1}{2}F_2(g) \longrightarrow KF(s)$ $\Delta H^0_f = -569$ kJ
$K^+(g) + F^-(g) \longrightarrow KF(s)$ $\Delta H^0_{lattice} = -821$ kJ

The Covalent Bonding Model
(Sample Problem 9.2)

● **Concept Review Questions**

9.34 Describe the interactions that occur between individual chlorine atoms as they approach each other and form Cl_2. What combination of forces gives rise to the energy holding the atoms together and to the final internuclear distance?

9.35 Define bond energy using the H—Cl bond as an example. When this bond breaks, is energy absorbed or released? Is the accompanying ΔH value positive or negative? How do the magnitude and sign of this ΔH value relate to the value that accompanies H—Cl bond formation?

9.36 For single bonds between similar types of atoms, how does the strength of the bond relate to the sizes of the atoms? Explain.

9.37 How does the energy of the bond between a given pair of atoms relate to the bond order? Why?

9.38 When liquid benzene (C_6H_6) boils, does the gas consist of molecules, ions, or separate atoms? Explain.

● **Skill-Building Exercises (paired)**

9.39 Using the periodic table only, arrange the members of each of the following sets in order of increasing bond *strength:*
(a) Br—Br, Cl—Cl, I—I   (b) S—H, S—Br, S—Cl
(c) C=N, C—N, C≡N

9.40 Using the periodic table only, arrange the members of each of the following sets in order of increasing bond *length:*
(a) H—F, H—I, H—Cl   (b) C—S, C=O, C—O
(c) N—H, N—S, N—O

● **Problems in Context**

9.41 Formic acid (HCOOH) is secreted by certain species of ants when they bite. It has this structural formula:

$$:\!\overset{\displaystyle :O:}{\underset{\displaystyle H-C-\ddot{O}-H}{\|}}$$

Rank the relative strengths of the C—O bond, C=O bond, and attractive force between two HCOOH molecules.

9.42 In Figure B9.2, the peak labeled "C=C stretch" occurs at a shorter wavelength than that labeled "C—C stretch," as it does in the IR spectrum of any substance with those bonds. Explain the relative positions of these peaks. In what relative position along the wavelength scale of Figure B9.2 would you expect to find a peak for a C≡C stretch? Explain.

Between the Extremes: Electronegativity and Bond Polarity
(Sample Problem 9.3)

● **Concept Review Questions**

9.43 Describe the vertical and horizontal trends in electronegativity (EN) among the main-group elements. According to Pauling's scale, what are the two most electronegative elements? The two least electronegative elements?

9.44 What is the general relationship between IE_1 and EN for the elements? Why?

9.45 Is the H—O bond in water nonpolar covalent, polar covalent, or ionic? Define each term, and explain your choice.

9.46 How does electronegativity differ from electron affinity?

9.47 How is the partial ionic character of a diatomic molecule related to the ΔEN of the bonded atoms? Why?

● **Skill-Building Exercises (paired)**

9.48 Using the periodic table only, arrange the elements in each set in order of *increasing* EN: (a) S, O, Si; (b) Mg, P, As.

9.49 Using the periodic table only, arrange the elements in each set in order of *increasing* EN: (a) I, Br, N; (b) Ca, H, F.

9.50 Using the periodic table only, arrange the elements in each set in order of *decreasing* EN: (a) N, P, Si; (b) Ca, Ga, As.

9.51 Using the periodic table only, arrange the elements in each set in order of *decreasing* EN: (a) Br, Cl, P; (b) I, F, O.

9.52 Use Figure 9.16 to indicate the polarity of each bond with *polar arrows:* (a) N—B; (b) N—O; (c) C—S; (d) S—O; (e) N—H; (f) Cl—O.

9.53 Use Figure 9.16 to indicate the polarity of each bond with *polar arrows:* (a) Br—Cl; (b) F—Cl; (c) H—O; (d) Se—H; (e) As—H; (f) S—N.

9.54 Which is the more polar bond in each of the following pairs in Problem 9.52: (a) or (b); (c) or (d); (e) or (f)?

9.55 Which is the more polar bond in each of the following pairs in Problem 9.53: (a) or (b); (c) or (d); (e) or (f)?

9.56 Are the bonds in each of the following substances ionic, nonpolar covalent, or polar covalent? Arrange the substances with polar covalent bonds in order of increasing bond polarity: (a) S_8 (b) RbCl (c) PF_3 (d) SCl_2 (e) F_2 (f) SF_2

9.57 Are the bonds in each of the following substances ionic, nonpolar covalent, or polar covalent? Arrange the substances with polar covalent bonds in order of increasing bond polarity: (a) KCl (b) P_4 (c) BF_3 (d) SO_2 (e) Br_2 (f) NO_2

9.58 Rank the members of each set of compounds in order of *increasing* ionic character of their bonds. Use *polar arrows* to indicate the bond polarity of each: (a) HBr, HCl, HI (b) H_2O, CH_4, HF (c) SCl_2, PCl_3, $SiCl_4$

9.59 Rank the members of each set of compounds in order of *decreasing* ionic character of their bonds. Use *partial charges* to indicate the bond polarity of each: (a) PCl_3, PBr_3, PF_3 (b) BF_3, NF_3, CF_4 (c) SeF_4, TeF_4, BrF_3

● **Problems in Context**

9.60 The energy of the C—C bond is 347 kJ/mol, and that of the Cl—Cl bond is 243 kJ/mol. Which of the following values might you expect for the C—Cl bond energy? Explain.
(a) 590 kJ/mol (sum of the values given)
(b) 104 kJ/mol (difference of the values given)
(c) 295 kJ/mol (average of the values given)
(d) 339 kJ/mol (greater than the average of the values given)

An Introduction to Metallic Bonding

● **Concept Review Questions**

9.61 (a) List four physical characteristics of a solid metal.
(b) List two chemical characteristics that lead to an element being classified as a metal.

9.62 Briefly account for the following relative values:
(a) The melting point of sodium is 89°C, whereas that of potassium is 63°C.
(b) The melting points of Li and Be are 180°C and 1287°C, respectively.
(c) Lithium boils more than 1100°C higher than it melts.

9.63 Magnesium metal is easily deformed by an applied force, whereas magnesium fluoride is shattered. Why do these two solids behave so differently?

Comprehensive Problems

Problems with an asterisk (*) are more challenging.

9.64 Geologists have a rule of thumb: when molten rock cools and solidifies, crystals of compounds with the smallest (least negative) lattice energies appear at the bottom of the mass. Suggest a reason for this.

9.65 During welding, a highly exothermic reaction, such as the combustion of acetylene (ethyne; HC≡CH) in pure oxygen, heats the metal pieces and fuses them. (a) Use Table 9.2 to find the heat of reaction per mole of acetylene (with water formed as a gas). (b) When 500.0 g of acetylene burns, how many kilojoules of heat are given off? (c) How many grams of CO_2 are produced? (d) How many liters of O_2 at 298 K and 18.0 atm are consumed?

9.66 Use Lewis electron-dot symbols to represent the formation of (a) BrF_3 from bromine and fluorine atoms; (b) AlF_3 from aluminum and fluorine atoms.

*9.67 Even though so much energy is required to form a divalent metal cation, the alkaline earth metals form halides with general formula MX_2, rather than MX. Let's see why.
(a) Use the following data to calculate the ΔH_f^0 of MgCl:

$$\begin{aligned} Mg(s) &\longrightarrow Mg(g) & \Delta H^0 &= 148 \text{ kJ} \\ Cl_2(g) &\longrightarrow 2Cl(g) & \Delta H^0 &= 243 \text{ kJ} \\ Mg(g) &\longrightarrow Mg^+(g) + e^- & \Delta H^0 &= 738 \text{ kJ} \\ Cl(g) + e^- &\longrightarrow Cl^-(g) & \Delta H^0 &= -349 \text{ kJ} \end{aligned}$$

$\Delta H_{\text{lattice}}^0$ of MgCl = -783.5 kJ/mol

(b) Is MgCl stable relative to its elements? Explain.
(c) Use Hess's law to calculate ΔH^0 for the conversion of MgCl to $MgCl_2$ and Mg (ΔH_f^0 of $MgCl_2 = -641.6$ kJ/mol).
(d) Is the formation of MgCl favored relative to the formation of $MgCl_2$? Explain.

*9.68 The first 21 ionization energies (IEs) for zinc (in MJ/mol) are 0.939, 1.80, 3.97, 5.94, 8.26, 10.8, 13.4, 17.4, 20.3, 23.8, 27.4, 31.1, 42.0, 45.4, 49.0, 54.2, 57.9, 61.9, 69.8, 73.8, and 186.
(a) Draw a graph of IE vs. number of electrons removed.
(b) Explain the general trend in the curve and the reason for any significant deviations from this trend.

9.69 Some gases react explosively with one another. If their reaction is sufficiently exothermic, the heat released as some product molecules form is transferred to other reactant molecules, which release more heat as they react, and so on. The rapid heat release causes a rapid expansion of the gases, and a thermal explosion results. Use bond energies to calculate ΔH^0 of the following reactions and predict which will occur explosively:
(a) $H_2(g) + Cl_2(g) \longrightarrow 2HCl(g)$
(b) $H_2(g) + I_2(g) \longrightarrow 2HI(g)$
(c) $2H_2(g) + O_2(g) \longrightarrow 2H_2O(g)$

*9.70 By using photons of specific wavelengths, chemists can dissociate gaseous HI to produce H atoms with accurately

known speeds. When HI dissociates, the H atoms move away rapidly, whereas the relatively heavy I atoms move little.
(a) What is the longest wavelength (in nm) that can dissociate a molecule of HI?
(b) If a photon of 254 nm is used, what is the excess energy (in J) over that needed for the dissociation?
(c) If all this excess energy is carried away by the H atom as kinetic energy, what is its speed (in m/s)?

9.71 Carbon dioxide is a linear molecule. The vibrations of the three atoms include symmetrical stretching, bending, and asymmetrical stretching, (see Figure B9.1, p. 343) and their frequencies are 4.02×10^{13} s^{-1}, 2.00×10^{13} s^{-1}, and 7.05×10^{13} s^{-1}, respectively. (a) What region of the electromagnetic spectrum corresponds to these frequencies? (b) Calculate the energy (in J) of each vibration. Which occurs most readily (takes the least energy)?

9.72 In developing the concept of electronegativity, Pauling used the term *excess bond energy* for the difference between the actual bond energy of X—Y and the average bond energies of X—X and Y—Y (see text discussion for the case of HF). Based on the values in Figure 9.16, which of the following substances contains bonds with *no* excess bond energy?
(a) PH$_3$ (b) CS$_2$ (c) BrCl   (d) BH$_3$ (e) Se$_8$

9.73 Use electron configurations to predict the relative hardnesses and melting points of rubidium ($Z = 37$), vanadium ($Z = 23$), and cadmium ($Z = 48$).

9.74 Chemists employ lasers to emit light of a given energy to initiate bond breakage. (a) What is the minimum energy and frequency of a photon that can cause the dissociation of a Cl$_2$ molecule? (b) The first key step in the destruction of stratospheric ozone by industrial chlorofluorocarbons is thought to be photodissociation of a C—Cl bond. What is the longest wavelength of a photon that can cause this dissociation?

*9.75 Bond energies can be combined with values for other atomic properties to obtain ΔH values that cannot be measured directly. Use bond energy, ionization energy, and electron affinity values to calculate the ΔH^0_{rxn} for the ionic dissociation of Cl$_2$:

$$Cl_2(g) \longrightarrow Cl^+(g) + Cl^-(g)$$

*9.76 The HF bond length is 92 pm, 16% shorter than the sum of the covalent radii of H (37 pm) and F (72 pm). Why is the bond length in HF less than the sum of the covalent radii? Similar calculations for the other hydrogen halides show that the difference between actual and calculated bond lengths becomes smaller down the group from HF to HI. Explain.

*9.77 Chemists distinguish between two types of covalent bond breakage. Homolytic breakage occurs when each atom in the bond gets one of the shared electrons. Heterolytic breakage occurs when one atom gets both electrons and the other gets none.

In some cases, the electronegativity of neighboring atoms affects the energy required for the breakage.
(a) Which type of bond breakage results in bond energies such as those in Table 9.2? Explain.
(b) Which type results in the formation of ions? Explain.
(c) Suggest a reason why the C—C bond energy in H$_3$C—CCl$_3$ (BE = 372 kJ/mol) deviates so far from the average C—C bond energy of 347 kJ/mol.
(d) Would you expect it to take more than 372 kJ/mol to break the C—C bond in H$_3$C—CH$_3$ heterolytically? Explain.

*9.78 Homolytic cleavage of O$_2$ forms two O atoms and requires 249.2 kJ/mol of O, whereas heterolytic cleavage forms O$^-$ and O$^+$ ions (see Problem 9.77). Calculate the heat of reaction for the heterolytic cleavage of O$_2$.

9.79 The work function (ϕ) of a metal is the minimum energy needed to remove an electron from its surface. The ionization energy (IE) of an element is the energy needed to remove 1 mol of electrons from 1 mol of gaseous atoms. (a) Is it easier to remove an electron from a gaseous silver atom or from the surface of solid silver ($\phi = 7.59\times10^{-19}$ J; IE = 731 kJ/mol)? (b) Explain the results in terms of the metallic bonding model.

9.80 Lattice energies can also be calculated for covalent solids using a Born-Haber cycle, and the network solid silicon dioxide has one of the highest. Silicon dioxide is found in pure crystalline form as transparent rock quartz. Much harder than glass, this material was once prized for making lenses for optical devices and expensive spectacles. Use Appendix B and the following data to calculate $\Delta H^0_{lattice}$ of SiO$_2$:

Si$(s) \longrightarrow$ Si(g)	$\Delta H^0 = 454$ kJ
Si$(g) \longrightarrow$ Si$^{4+}(g) + 4e^-$	$\Delta H^0 = 9949$ kJ
O$_2(g) \longrightarrow$ 2O(g)	$\Delta H^0 = 498$ kJ
O$(g) + 2e^- \longrightarrow$ O$^{2-}(g)$	$\Delta H^0 = 737$ kJ

9.81 We can write equations for the formation of methane from ethane (C$_2$H$_6$) with its C—C bond, from ethene (C$_2$H$_4$) with its C=C bond, and from ethyne (C$_2$H$_2$) with its C≡C bond:

C$_2$H$_6(g)$ + H$_2(g) \longrightarrow$ 2CH$_4(g)$	$\Delta H^0_{rxn} = -65.07$ kJ/mol
C$_2$H$_4(g)$ + 2H$_2(g) \longrightarrow$ 2CH$_4(g)$	$\Delta H^0_{rxn} = -202.21$ kJ/mol
C$_2$H$_2(g)$ + 3H$_2(g) \longrightarrow$ 2CH$_4(g)$	$\Delta H^0_{rxn} = -376.74$ kJ/mol

Given that the average C—H bond energy in CH$_4$ is 415 kJ/mol, use Table 9.2 to calculate the average C—H bond energy in ethane, ethene, and ethyne.

9.82 The Sun's emissions include infrared, visible, and ultraviolet radiation. Some of this radiation can damage skin tissue, and sunscreens are made to block it. (a) What is the ratio of the energies of yellow light of 575 nm to UV light of 300 nm? (b) What is the ratio of the energies of yellow light of 575 nm to IR radiation of 1000 nm? (c) Can any of these three types of radiation break a C—C bond? (d) A C—H bond?

CHAPTER 10

THE SHAPES OF MOLECULES

CHAPTER OUTLINE

10.1 Depicting Molecules and Ions with Lewis Structures
Using the Octet Rule
Resonance
Formal Charge
Exceptions to the Octet Rule

10.2 Using Lewis Structures and Bond Energies to Calculate Heats of Reaction

10.3 Valence-Shell Electron-Pair Repulsion (VSEPR) Theory and Molecular Shape
Electron-Group Arrangements and Molecular Shapes
Molecular Shape with Two Electron Groups
Shapes with Three Electron Groups
Shapes with Four Electron Groups
Shapes with Five Electron Groups
Shapes with Six Electron Groups
Using VSEPR Theory to Determine Molecular Shape
Shapes with More Than One Central Atom

10.4 Molecular Shape and Molecular Polarity
Bond Polarity, Bond Angle, and Dipole Moment
Molecular Polarity and Behavior

Figure: Interconnecting shapes. In everyday life, the shapes of many objects fit together, like this puzzle piece in its slot, to perform a function—key and lock, mortise and tenon, hand and glove. Similarly, an organism's molecules have shapes that fit together to trigger the processes of life. In this chapter, we learn to depict molecules as two-dimensional structural formulas and then, more importantly, to visualize them as three-dimensional objects.

CONCEPTS & SKILLS

to review before you study this chapter
- electron configurations of main-group elements (Section 8.3)
- electron-dot symbols (Section 9.1)
- the octet rule (Section 9.2)
- bond order, bond length, and bond energy (Section 9.3)
- polar covalent bonds and bond polarity (Section 9.4)

Armed with your understanding of how atoms bond, you can now explore an idea essential to chemistry—*and* biology! The printed page, covered with atomic symbols, lines, and pairs of dots, makes it easy to forget the amazing, three-dimensional reality of molecular shape. In any molecule, each atom, bonding pair, and lone pair has its own position in space relative to the others, determined by the attractive and repulsive forces that govern all matter. With definite angles and distances between the nuclei, a molecule is an independent, minute architecture, extending throughout its tiny volume of space. Whether we consider the details of simple reactions, the properties of synthetic materials, or the intricate life-sustaining processes of living cells, molecular shape is a crucial factor.

Although our focus in this chapter is on the shapes of molecules, we begin by seeing how to convert the *molecular* formula of a compound into a flat *structural* formula, called a *Lewis structure,* which shows how the atoms are attached to each other within the molecule but does *not* reveal the overall shape. We use Lewis structures to calculate heats of reaction from bond energies. Then we examine VSEPR (valence-shell electron-pair repulsion) theory, with which we convert these two-dimensional structural formulas into three-dimensional shapes. We'll see the five basic shape classes that many simple molecules adopt and the ways these can combine to form the shapes of more complex molecules. You'll learn how shape and bond polarity create a polarity for the entire molecule and glimpse the influence of shape on biological function.

10.1 DEPICTING MOLECULES AND IONS WITH LEWIS STRUCTURES

The first step toward visualizing what a molecule looks like is to convert its molecular formula to its **Lewis structure** (or **Lewis formula**). This two-dimensional structural formula consists of electron-dot symbols that depict each atom and its neighbors, the bonding pairs that hold them together, and the lone pairs that fill each atom's outer shell.* In many cases, the octet rule (Section 9.2) guides us in allotting electrons to the atoms in a Lewis structure; in many other cases, however, we set the rule aside.

Using the Octet Rule to Write Lewis Structures

To write a Lewis structure from the molecular formula, we decide on the relative placement of the atoms in the molecule (or ion)—that is, which atoms are adjacent and become bonded to each other—and distribute the total number of valence electrons as bonding and lone pairs. Let's begin by examining Lewis structures for species that "obey" the octet rule—those in which each atom fills its outer level with eight electrons (or two for hydrogen).

Lewis Structures for Molecules with Single Bonds First, we discuss the steps for writing Lewis structures for molecules that have only single bonds, using nitrogen trifluoride, NF_3, as an example. You may want to refer to Figure 10.1 as we go through the steps.

---

*A Lewis *structure* may be more correctly called a Lewis *formula* because it provides information about the relative placement of atoms in a molecule or ion and shows which atoms are bonded to each other, but it does **not** indicate the three-dimensional shape. Nevertheless, use of the term Lewis "structure" is a convention that we follow.

Figure 10.1 **The steps in converting a molecular formula into a Lewis structure.**

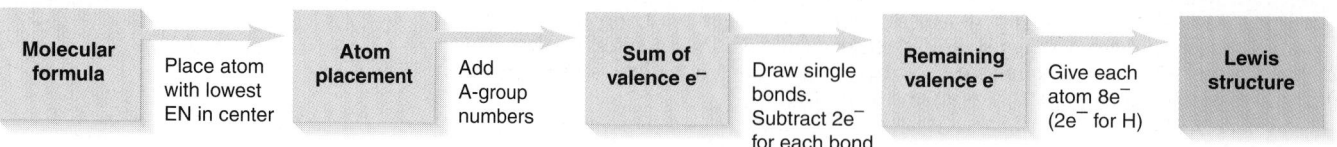

Molecular formula		Atom placement		Sum of valence e⁻		Remaining valence e⁻		Lewis structure
	Place atom with lowest EN in center →		Add A-group numbers →		Draw single bonds. Subtract 2e⁻ for each bond →		Give each atom 8e⁻ (2e⁻ for H) →	

Step 1. Place the atoms relative to each other. For compounds of molecular formula AB_n, place the atom with *lower group number* in the center because it needs more electrons to attain an octet. In NF_3, the N (Group 5A) has five electrons so it needs three, whereas each F (Group 7A) has seven and so needs only one; thus, N goes in the center with the three F atoms around it:

$$F$$
$$_F\ ^N\ _F$$

If the atoms in a compound have the same group number, such as those in SO_3 or ClF_3, place the atom with the *higher period number* in the center. This type of placement usually results in the *less electronegative atom* in the center (EN of N = 3.0 and EN of F = 4.0). H can form only one bond, so it is *never* a central atom.

Step 2. Determine the total number of valence electrons available. For molecules, add up the valence electrons of all the atoms. (Recall that the number of valence electrons equals the A-group number.) In NF_3, N has five valence electrons, and each F has seven:

$$[1 \times N(5e^-)] + [3 \times F(7e^-)] = 5e^- + 21e^- = 26 \text{ valence } e^-$$

For polyatomic ions, *add* one e^- for each negative charge of the ion, or *subtract* one e^- for each positive charge.

Step 3. Draw a single bond from each surrounding atom to the central atom, and subtract two valence electrons for each bond. There must be at least a single bond between bonded atoms:

$$F$$
$$|$$
$$F{-}N{-}F$$

Subtract $2e^-$ for each single bond from the total number of valence electrons available (from step 2) to find the number remaining:

$$3 \text{ N—F bonds} \times 2e^- = 6e^- \quad \text{so} \quad 26e^- - 6e^- = 20e^- \text{ remaining}$$

Step 4. Distribute the remaining electrons in pairs so that each atom obtains eight electrons (or two for H). First, place lone pairs on the *surrounding (more electronegative) atoms* to give each an octet. If any electrons remain, place them around the central atom. Then check that each atom has $8e^-$:

$$:\!\ddot{F}\!:$$
$$|$$
$$:\!\ddot{F}\!-\!N\!-\!\ddot{F}\!:$$

This is the Lewis structure for NF_3. It's always a good idea to check that the total number of electrons in the Lewis structure (bonds plus lone pairs) equals the sum of the valence electrons: $6e^-$ in three bonds plus $20e^-$ in ten lone pairs equals 26 valence electrons.

This particular arrangement of F atoms around the N atom was chosen because it resembles the molecular shape of NF_3, as you'll see in Section 10.3. Lewis structures do not indicate shape, however, so an equally correct depiction of NF_3 would be

$$:\!\ddot{F}\!:$$
$$|$$
$$:\!\ddot{F}\!-\!N\!-\!\ddot{F}\!:$$

or any other that retains the *same connectivity* of atoms, that is, the same connections among the atoms—a central N atom connected by single bonds to three surrounding F atoms.

Using these four steps, you can write a Lewis structure for any singly bonded molecule whose central atom is C, N, or O, as well as for some molecules with central atoms from higher periods.

SAMPLE PROBLEM 10.1 Writing Lewis Structures for Molecules with One Central Atom

Problem Write a Lewis structure for CCl_2F_2, one of the compounds responsible for the depletion of stratospheric ozone.

Solution *Step 1.* Place the atoms relative to each other. In CCl_2F_2, carbon has the lowest group number and EN, so it is the central atom. The other atoms surround it, but their specific positions are not important:

$$\begin{array}{ccc} & Cl & \\ F & C & F \\ & Cl & \end{array}$$

Step 2. Determine the total number of valence electrons (from A-group numbers): C is in Group 4A, F is in Group 7A, and Cl is in Group 7A, too. Therefore, we have

$$[1 \times C(4e^-)] + [2 \times F(7e^-)] + [2 \times Cl(7e^-)] = 32 \text{ valence } e^-$$

Step 3. Draw single bonds to the central atom and subtract $2e^-$ for each bond:

$$\begin{array}{c} Cl \\ | \\ F-C-F \\ | \\ Cl \end{array}$$

Four single bonds use $8e^-$, so $32e^- - 8e^-$ leaves $24e^-$ remaining.

Step 4. Distribute the remaining electrons in pairs, beginning with the surrounding atoms, so that each atom has an octet:

$$\begin{array}{c} :\ddot{C}l: \\ | \\ :\ddot{F}-C-\ddot{F}: \\ | \\ :\ddot{C}l: \end{array}$$

Check Counting the electrons shows that each atom has an octet. Remember that bonding electrons are counted as belonging to each atom in the bond. The total number of electrons in bonds (8) and lone pairs (24) equals 32 valence electrons.

FOLLOW-UP PROBLEM 10.1 Write Lewis structures for **(a)** H_2S; **(b)** OF_2; **(c)** $SOCl_2$.

A slightly more complex situation occurs when molecules have two or more central atoms bonded to each other, with the other atoms around them.

SAMPLE PROBLEM 10.2 Writing Lewis Structures for Molecules with More than One Central Atom

Problem Write the Lewis structure for methanol (molecular formula CH_4O), an important industrial alcohol that is being used as a gasoline alternative in car engines.

Solution *Step 1.* Place the atoms relative to each other. The H atoms can have only one bond, so C and O must be adjacent to each other. In nearly all their compounds, C has four bonds and O has two, so we arrange the H atoms to show this:

$$\begin{array}{cccc} & H & & \\ H & C & O & H \\ & H & & \end{array}$$

Step 2. Find the sum of valence electrons:

$$[1 \times C(4e^-)] + [1 \times O(6e^-)] + [4 \times H(1e^-)] = 14e^-$$

Step 3. Add single bonds and subtract $2e^-$ for each bond:

$$\begin{array}{c} H \\ | \\ H-C-O-H \\ | \\ H \end{array}$$

Five bonds use $10e^-$, so $14e^- - 10e^-$ leaves $4e^-$ remaining.

Step 4. Add the remaining electrons in pairs:

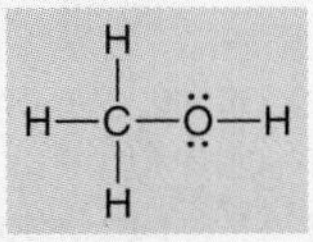

Carbon already has an octet, so the four remaining valence e⁻ form two lone pairs on O. We now have the Lewis structure for methanol.

Check Each H atom has 2e⁻, and the C and O each have 8e⁻. The total number of valence electrons is 14e⁻, which equals 10e⁻ in bonds plus 4e⁻ in lone pairs.

FOLLOW-UP PROBLEM 10.2 Write Lewis structures for **(a)** hydroxylamine (NH_3O) and **(b)** dimethyl ether (C_2H_6O; no O—H bonds).

Lewis Structures for Molecules with Multiple Bonds Sometimes, you'll find that, after steps 1 to 4, there are not enough electrons for the central atom (or one of the central atoms) to attain an octet. This usually means that a multiple bond is present, and the following additional step is needed:

Step 5. Cases involving multiple bonds. If, after step 4, a central atom still does not have an octet, make a multiple bond by changing a lone pair from one of the surrounding atoms into a bonding pair to the central atom.

SAMPLE PROBLEM 10.3 Writing Lewis Structures for Molecules with Multiple Bonds

---

Problem Write Lewis structures for the following:
(a) Ethylene (C_2H_4), the most important reactant in the manufacture of polymers
(b) Nitrogen (N_2), the most abundant atmospheric gas
Plan We begin the solution after steps 1 to 4: placing the atoms, counting the total valence electrons, making single bonds, and distributing the remaining valence electrons in pairs to attain octets. Then we continue with step 5, if needed.
Solution (a) For C_2H_4. After steps 1 to 4, we have

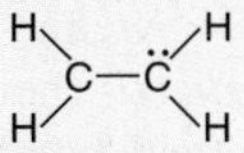

Step 5. Change a lone pair to a bonding pair. The C on the right has an octet, but the C on the left has only 6e⁻, so we convert the lone pair to another bonding pair between the two C atoms:

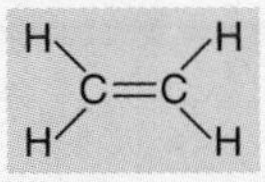

(b) For N_2. After steps 1 to 4, we have :N̈—N̈:
Step 5. Neither N has an octet, so we change a lone pair to a bonding pair:

:N̈=N:

In this case, moving one lone pair to make a double bond still does not give the N on the right an octet, so we move a lone pair from the left N to make a triple bond:

:N≡N:

Check In part (a), each C counts the 4e⁻ in the double bond as part of its own octet. The valence electron total is 12e⁻, all in six bonds. In part (b), each N counts the 6e⁻ in the triple bond as part of its own octet. The valence electron total is 10e⁻, which equals the electrons in three bonds and two lone pairs.

FOLLOW-UP PROBLEM 10.3 Write Lewis structures for **(a)** CO (the only common molecule in which C has only three bonds); **(b)** HCN; **(c)** CO_2.

Resonance: Delocalized Electron-Pair Bonding

We can often write more than one Lewis structure, each with the same relative placement of atoms, for a molecule or ion with *double bonds next to single bonds*. Consider ozone (O_3), a serious air pollutant at ground level but a life-sustaining absorber of harmful ultraviolet (UV) radiation in the stratosphere. Two valid Lewis structures (with lettered O atoms for clarity) are

In structure I, oxygen B has a double bond to oxygen A and a single bond to oxygen C. In structure II, the single and double bonds are reversed. These are *not* two different types of O_3 molecules, just two different Lewis structures for the same molecule.

In fact, *neither* Lewis structure depicts O_3 accurately. Bond length and bond energy measurements indicate that the two bonds in O_3 are identical, with properties that lie between those of an O—O bond and an O==O bond, something like a "one-and-a-half" bond. The molecule is shown more correctly with two Lewis structures, called **resonance structures** (or **resonance forms**), and a two-headed resonance arrow (⟷) between them. Resonance structures *have the same relative placement of atoms but different locations of bonding and lone electron pairs*. You can convert one resonance form to another by moving lone pairs to bonding positions, and vice versa:

Resonance structures are not real bonding depictions: O_3 does *not* change back and forth from structure I this instant to structure II the next. The actual molecule is a **resonance hybrid,** an average of the resonance forms.

Our need for more than one Lewis structure to depict the ozone molecule is the result of **electron-pair delocalization.** In a single, double, or triple bond, each electron pair is attracted by the nuclei of the two bonded atoms, and the electron density is greatest in the region between the nuclei: each electron pair is *localized*. In the resonance hybrid for O_3, however, two of the electron pairs (one bonding and one nonbonding) are *delocalized:* their density is "spread" over the entire molecule.* This results in two identical bonds, each consisting of a single bond (the localized electron pair) and a *partial bond* (the contribution from one of the delocalized electron pairs). We draw the resonance hybrid with a curved dashed line to show the delocalized pairs:

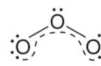

Electron delocalization diffuses electron density over a greater volume, which reduces electron-electron repulsions and thus stabilizes the molecule. Resonance is very common, and many molecules (and ions) are best depicted as resonance hybrids. Benzene (C_6H_6), for example, has two important resonance forms in which alternating single and double bonds have different positions. The actual molecule has six identical carbon-carbon bonds because there are six C—C bonds

A Purple Mule, Not a Blue Horse and a Red Donkey A mule is a genetic mix, a hybrid, of a horse and a donkey; it is not a horse one instant and a donkey the next. Similarly, the color purple is a mix of two other colors, red and blue, not red one instant and blue the next. In the same sense, a resonance hybrid is one molecular species, not this resonance form one instant and that resonance form the next. The problem lies in our inability to depict the hybrid accurately with a single Lewis structure.

Blue horse   Red donkey

Purple mule

---

*Keep in mind that electron delocalization in a metal (Section 9.5), which extends over all the atoms in the entire sample, is *much* greater than in a covalent resonance hybrid, where it extends over a few atoms only.

and three electron pairs delocalized over all six C atoms, often shown as a dashed circle (or simply a circle):

resonance forms   resonance hybrid

Partial bonding, such as occurs in resonance hybrids, often leads to fractional bond orders. For O_3, we have

$$\text{Bond order} = \frac{3 \text{ electron pairs}}{2 \text{ atom-to-atom linkages}} = 1\tfrac{1}{2}$$

The carbon-to-carbon bond order in benzene is 9 pairs/6 linkages, or $1\tfrac{1}{2}$ also. For the carbonate ion, CO_3^{2-}, three resonance structures can be drawn. Each has 4 electron pairs shared among 3 linkages, so the bond order is $1\tfrac{1}{3}$. One of the three resonance structures for CO_3^{2-} is

Note that *the Lewis structure of a polyatomic ion is shown in square brackets, with its charge as a right superscript outside the brackets.*

SAMPLE PROBLEM 10.4 Writing Resonance Structures

Problem Write resonance structures for the nitrate ion, NO_3^{-}.
Plan We write a Lewis structure using the steps outlined earlier, remembering to add $1e^{-}$ to the total number of valence electrons because of the $1-$ ionic charge. Then we move lone and bonding pairs to write other resonance forms and connect them with the resonance arrow.
Solution After steps 1 to 4, we have

Step 5. Since N has only $6e^{-}$, we change one lone pair on an O atom to a bonding pair and form a double bond, which gives each atom an octet. All the O atoms are equivalent, however, so we can move a lone pair from any of the three O atoms and obtain three resonance structures:

Check Each structure has the same relative placement of atoms, an octet around each atom, and $24e^{-}$ (the sum of the valence electron total and $1e^{-}$ from the ionic charge distributed in four bonds and eight lone pairs).
Comment Remember that no double bond actually occurs in the NO_3^{-} ion. The ion is a resonance hybrid of these three structures with a bond order of $1\tfrac{1}{3}$.

FOLLOW-UP PROBLEM 10.4 One of the three resonance structures for CO_3^{2-} was shown just before Sample Problem 10.4. Draw the other two.

Formal Charge: Selecting the Best Resonance Structure

In the previous examples, the resonance forms were equally mixed to form the resonance hybrid because the molecules (or ions) were symmetrical: the surrounding atoms were all the same. When this is not the case, one resonance form may look more like the hybrid than the others. In other words, because the resonance hybrid is an average of the resonance forms, one form may contribute more and "weight" the average in its favor. One way to select the most important resonance form is to determine each atom's **formal charge,** the charge it would have *if the bonding electrons were shared equally.* An atom's formal charge is its total number of valence electrons *minus* the number of valence electrons it "owns" in the molecule: it owns *all* of its unshared valence electrons and *half* of its shared valence electrons. Thus,

> Formal charge of atom =
> no. of valence e$^-$ − (no. of unshared valence e$^-$ + $\frac{1}{2}$ no. of shared valence e$^-$)   **(10.1)**

For example, in O_3, the formal charge of oxygen A (O_A) in resonance form I is

$$6 \text{ valence e}^- - (4 \text{ unshared e}^- + \tfrac{1}{2} \text{ of 4 shared e}^-) = 6 - 4 - 2 = 0$$

The formal charges of all the atoms in the two O_3 resonance forms are

$$O_A[6 - 4 - \tfrac{1}{2}(4)] = 0$$
$$O_B[6 - 2 - \tfrac{1}{2}(6)] = +1$$
$$O_C[6 - 6 - \tfrac{1}{2}(2)] = -1$$

$$O_A[6 - 6 - \tfrac{1}{2}(2)] = -1$$
$$O_B[6 - 2 - \tfrac{1}{2}(6)] = +1$$
$$O_C[6 - 4 - \tfrac{1}{2}(4)] = 0$$

Forms I and II are symmetrical—each has the same formal charges but on different atoms—so they contribute equally to the resonance hybrid. Note that *the formal charges must sum to the actual charge on the species:* zero for a molecule and the ionic charge for an ion.

Three criteria help us choose the more important resonance structures:

- Smaller formal charges (whether positive or negative) are preferable to larger ones.
- Like charges on adjacent atoms are not desirable.
- A more negative formal charge should reside on a more electronegative atom.

Let's apply these criteria to the cyanate ion, NCO^-, which is not symmetrical. Three resonance forms with formal charges are

Formal charges:
Resonance forms:

We eliminate form A because it has larger formal charges than the others and a positive formal charge on the (more electronegative) O. Forms B and C have the same magnitude of formal charges, but form C has a −1 charge on O, which is more electronegative than N. Therefore, B and C are significant contributors to the resonance hybrid of the cyanate ion, with C the more important.

Formal charge (used to examine resonance structures) is *not* the same as oxidation number (used to monitor redox reactions):

- In determining formal charge, the bonding electrons are assigned *equally* to the bonded atoms (as if the bonding were *nonpolar covalent*), so each atom receives half of them:

$$\text{Formal charge} = \text{valence e}^- - (\text{lone pair e}^- + \tfrac{1}{2} \text{ bonding e}^-)$$

- In determining oxidation number, the bonding electrons are assigned *completely* to the more electronegative atom (as if the bonding were *ionic*):

$$\text{Oxidation number} = \text{valence e}^- - (\text{lone pair e}^- + \text{ bonding e}^-)$$

Here are the formal charges and oxidation numbers of the three cyanate ion resonance structures:

Formal charges:

$$\underset{-3 \quad +4 \quad -2}{\overset{(-2) \quad (0) \quad (+1)}{[:\ddot{N}-C\equiv O:]^-}} \longleftrightarrow \underset{-3 \quad +4 \quad -2}{\overset{(-1) \quad (0) \quad (0)}{[\ddot{N}=C=\ddot{O}:]^-}} \longleftrightarrow \underset{-3 \quad +4 \quad -2}{\overset{(0) \quad (0) \quad (-1)}{[:N\equiv C-\ddot{O}:]^-}}$$

Oxidation numbers:

Note that the oxidation numbers *do not* change from one resonance form to another (because the electronegativities do not change), but the formal charges *do* change (because the numbers of bonding and lone pairs change).

Lewis Structures for Exceptions to the Octet Rule

The octet rule is a useful guide for most molecules with Period 2 central atoms, but it doesn't hold for every one. Also, many molecules have central atoms from higher periods. As you'll see, some central atoms have fewer than eight electrons around them, and others have more. The most significant octet rule exceptions are electron-deficient molecules, odd-electron molecules, and especially molecules with expanded valence shells.

Electron-Deficient Molecules Gaseous molecules containing either beryllium or boron as the central atom are often **electron deficient;** that is, they have *fewer* than eight electrons around the Be or B atom. The Lewis structures of gaseous beryllium chloride* and boron trifluoride are

$$:\ddot{C}l-Be-\ddot{C}l: \qquad \overset{\displaystyle :\ddot{F}:}{\underset{:\ddot{F}:}{\overset{|}{B}}}\ddot{F}:$$

There are only four electrons around beryllium and six around boron. Why don't nonbonding pairs from the surrounding halogen atoms form multiple bonds to the central atoms, thereby satisfying the octet rule? Halogens are much more electronegative than beryllium or boron, and formal charges show that the following are unlikely structures:

$$\underset{(+1) \quad (-2) \quad (+1)}{:\ddot{C}l=Be=\ddot{C}l:} \qquad \overset{(0)}{\underset{(0)}{:\ddot{F}}}\overset{:\ddot{F}:}{\underset{(-1)}{\overset{|}{B}}}\ddot{F}:_{(+1)}$$

(Some data for BF_3 show a shorter than expected B—F bond. Shorter bonds indicate double-bond character, so the structure with the B=F bond may be a minor contributor to a resonance hybrid.) The main way electron-deficient molecules attain an octet is by forming additional bonds in reactions. When BF_3 reacts with ammonia, for instance, a compound forms in which boron attains its octet:[†]

$$\overset{:\ddot{F}:}{\underset{:\ddot{F}:}{\overset{|}{B}}}\ddot{F}: \quad \overset{H}{\underset{H}{\overset{|}{N}}} \quad \longrightarrow \quad :\ddot{F}-\overset{:\ddot{F}:}{\underset{:\ddot{F}:}{\overset{|}{B}}}-\overset{H}{\underset{H}{\overset{|}{N}}}-H$$

Odd-Electron Molecules A few molecules contain an odd number of valence electrons, so they cannot possibly have all their electrons in pairs. Such species, called **free radicals,** contain a lone (unpaired) electron, which makes them paramagnetic (Section 8.5) and extremely reactive. 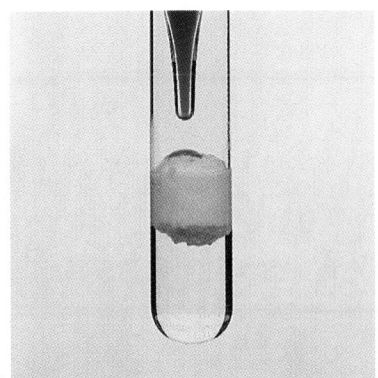 Strictly speaking, Lewis structures are based on an electron-pair model, so they don't apply directly to species with a lone electron, but we'll use formal charges to decide where the

---

*Despite beryllium being a member of the alkaline earth metals [Group 2A(2)], most of its compounds have properties consistent with covalent, rather than ionic, bonding. For example, molten $BeCl_2$ does not conduct electricity, indicating the absence of ions (Chapter 14).

[†]Reactions of the sort shown here—in which one species "donates" an electron pair to another to form a covalent bond—are examples of Lewis acid-base reactions, an extremely important and widely encountered type of reaction that we discuss in Chapter 18.

Deadly Free-Radical Activity
Free radicals can be extremely dangerous to biological systems because they rupture bonds in cells' biomolecules. If a free radical reacts with a biomolecule, it typically forms a covalent bond to one of the H atoms and removes it, and the biomolecule is left with an unpaired electron, thereby becoming a new free radical. That species repeats the process and creates other species with lone electrons that proliferate to rupture genes and cell membranes. We benefit from this ability when we apply the disinfectant hydrogen peroxide to a cut because it forms free radicals that destroy bacterial membranes. (The photo shows H_2O_2 reacting vigorously with a drop of blood.) On the other hand, recent studies have suggested that several disease states, including certain forms of cancer, may be attributed to free radicals. Also, vitamin E is believed to interrupt free-radical proliferation.

lone electron resides. Most odd-electron molecules have a central atom from an odd-numbered group, such as N [Group 5A(15)] or Cl [Group 7A(17)]. Consider NO_2 as an example. Whichever atom, N or doubly bonded O, has the lone electron will have a formal charge of +1. The structure with the lone electron, and thus the +1 charge, on N is preferred because N is less electronegative than O:

Formal charges:

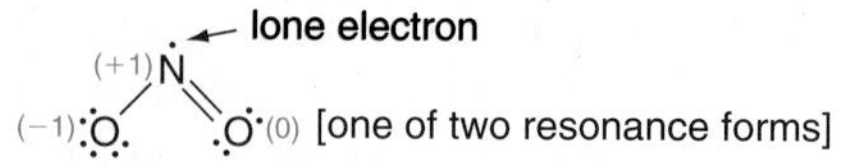

[structure with lone electron label] [one of two resonance forms]

Nitrogen dioxide is formed when the NO in auto exhaust reacts with O_2 in sunlight; it is a major contributor to urban smog. Free radicals can react with each other to pair up their lone electrons. When two NO_2 molecules collide, for example, they form dinitrogen tetraoxide, N_2O_4, and each N attains an octet:

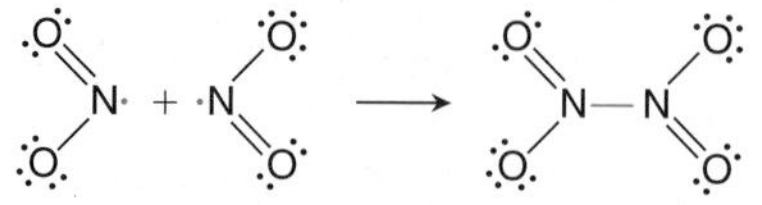

Expanded Valence Shells Many molecules and ions have more than eight valence electrons around the central atom. *An atom expands its valence shell to form more bonds,* a process that releases energy. The only way a central atom can accommodate additional pairs is to use empty *outer d* orbitals in addition to occupied *s* and *p* orbitals. Therefore, **expanded valence shells** occur only with a *central nonmetal from Period 3 or higher,* those in which *d* orbitals are available.

One example is sulfur hexafluoride, SF_6, a remarkably dense and inert gas used as an insulator in electrical equipment. The central sulfur is surrounded by six covalent bonds, one to each fluorine, for a total of 12 electrons:

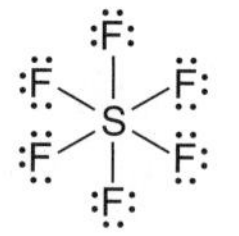

Another example is phosphorus pentachloride, PCl_5, a fuming yellow-white solid used in the manufacture of lacquers and films. PCl_5 is formed when phosphorus trichloride, PCl_3, reacts with chlorine gas. The P in PCl_3 has an octet, but it uses the lone pair to form two more bonds to chlorine and expands its valence shell in PCl_5 to a total of 10 electrons. Note that when PCl_5 forms, *one* Cl—Cl bond breaks (left side of the equation), and *two* P—Cl bonds form (right side), for a net increase of one bond:

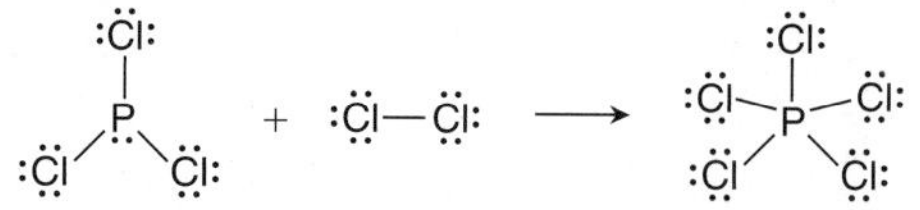

In the cases we've looked at so far, the central atom forms bonds to *more than four* atoms. But there are many cases of expanded valence shells in which the central atom bonds to *four or fewer* atoms. Consider sulfuric acid, the industrial chemical produced in the greatest quantity. Two of the resonance forms for H_2SO_4, with formal charges, are

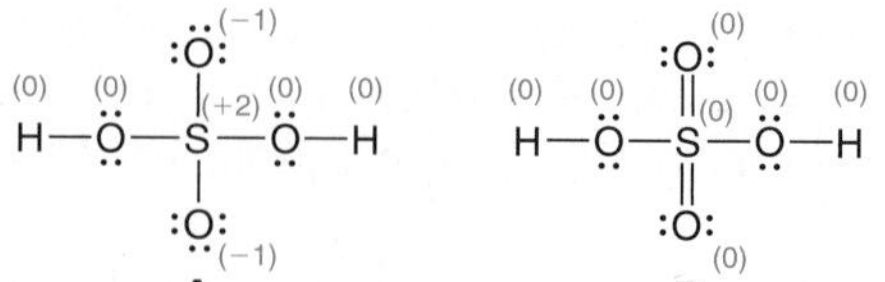

In form B, sulfur has an expanded valence shell of 12 electrons. Note that form B is a more significant contributor than A to the resonance hybrid because it has lower formal charges. Most importantly, form B is consistent with observed bond lengths. In gaseous H_2SO_4, the two sulfur-oxygen bonds *with* H atoms attached to O are 157 pm long, whereas the two sulfur-oxygen bonds *without* H atoms attached to O are 142 pm long. This shorter bond indicates double-bond character, which is shown in form B.

When sulfuric acid loses two H^+ ions, it forms the sulfate ion, SO_4^{2-}. All sulfur-oxygen bonds in SO_4^{2-} are 149 pm long, which is intermediate in length between the two $S=O$ bonds (~142 pm) and the two $S-O$ bonds (~157 pm) in the parent acid. Two of six resonance forms consistent with these data are

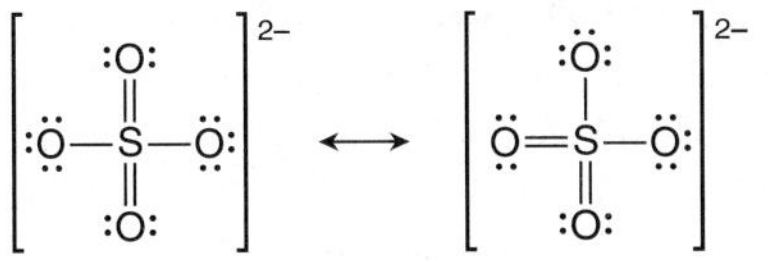

Thus, the SO_4^{2-} ion is a resonance hybrid with four $S-O$ bonds and two more bonding pairs delocalized over the structure; so each sulfur-oxygen bond has a bond order of $1\frac{1}{2}$.

Measurements show that the sulfur-oxygen bonds in SO_2 and SO_3 are all approximately the $S=O$ bond length (142 pm), so Lewis structures for these molecules, with formal charges, are

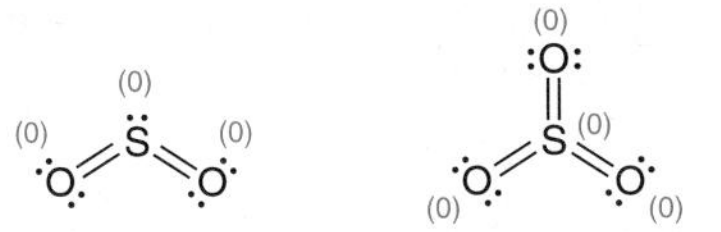

Sulfur and phosphorus often accommodate 12 electrons in their compounds, and iodine as many as 14. Keep in mind that these *atoms expand their valence shells to form more bonds and minimize formal charge.*

SAMPLE PROBLEM 10.5 Writing Lewis Structures for Exceptions to the Octet Rule

Problem Write Lewis structures for **(a)** H_3PO_4 and **(b)** $BFCl_2$. In **(a),** decide on the most likely structure.
Plan We write each Lewis structure and examine it for exceptions to the octet rule. In (a), the central atom is P, which is in Period 3, so it can use *d* orbitals to have more than an octet. Therefore, we can write more than one Lewis structure. We use formal charge to decide if one resonance form is more important. In (b), the central atom is B, which can have fewer than an octet of electrons.
Solution (a) For H_3PO_4, two possible Lewis structures, with formal charges, are

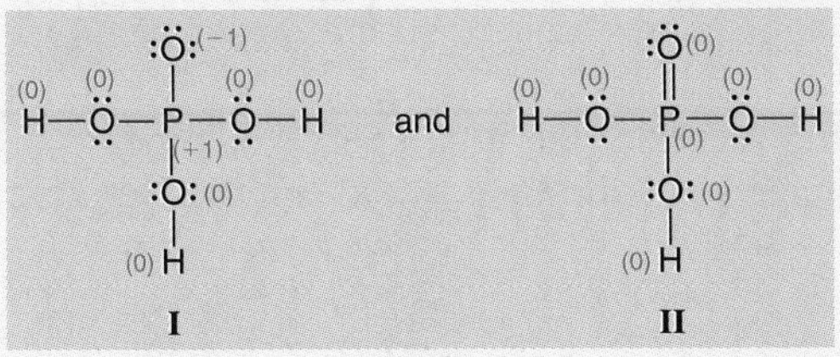

Structure **II** has lower formal charges, so it is the more important resonance form.
(b) For $BFCl_2$, the Lewis structure leaves B with only six electrons surrounding it:

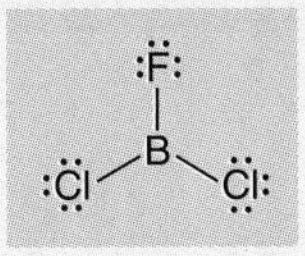

Comment In (a), structure **II** is also consistent with bond length measurements, which show one shorter (152 pm) phosphorus-oxygen bond and three longer (157 pm) ones.

FOLLOW-UP PROBLEM 10.5 Write the most likely Lewis structure for **(a)** $POCl_3$; **(b)** ClO_2; **(c)** XeF_4.

SECTION SUMMARY
A stepwise process converts a molecular formula into a Lewis structure, a *two-dimensional* representation of a molecule (or ion) that shows the relative placement of atoms and distribution of valence electrons among bonding and lone pairs. When two or more Lewis structures can be drawn for the same relative placement of atoms, the actual structure is a hybrid of those resonance forms. Formal charge can be used to determine the most important contributor to the hybrid. Electron-deficient molecules (central Be or B) and odd-electron species (free radicals) have less than an octet around the central atom but often attain an octet in reactions. It is common for a central nonmetal atom from Period 3 or higher to form more bonds in a molecule (or ion) by using *d* orbitals to expand its valence shell and hold more than eight electrons.

10.2 USING LEWIS STRUCTURES AND BOND ENERGIES TO CALCULATE HEATS OF REACTION

We can view a chemical reaction as a process in which certain bonds break in reacting species to give molecular fragments that recombine into product species by forming new bonds. We use Lewis structures and bond energies (bond enthalpies) to calculate the heat of reaction (ΔH^0_{rxn}) by noting which reactant bonds break and which product bonds form. But it is often simpler to assume that *all the reactant bonds break to give individual atoms, from which all the product bonds form* (Figure 10.2). Even though the actual reaction may not occur this way, Hess's law allows us to sum the bond energies (with their appropriate signs) to arrive at the overall heat of reaction. (This method assumes that the heat of reaction is due entirely to changes in bond energy, which requires that all reactants and products be gases. When liquids or solids are present, the additional heat involved in changing physical state must be taken into account. We address this topic in Chapter 12.)

Heat is absorbed to break bonds in the reactants (ΔH^0 is positive) and is released when bonds form and make products (ΔH^0 is negative). The sum of these enthalpy changes is the heat of reaction:

$$\Delta H^0_{rxn} = \Delta H^0_{\text{reactant bonds broken}} + \Delta H^0_{\text{product bonds formed}} \qquad \textbf{(10.2)}$$

As you can see from Equation 10.2,

- In an exothermic reaction, the total energy of product bonds formed is *more* than that of reactant bonds broken, so the *larger* negative sum makes ΔH^0_{rxn} *negative*.
- In an endothermic reaction, the total energy of product bonds formed is *less* than that of reactant bonds broken, so the *smaller* negative sum makes ΔH^0_{rxn} *positive*.

Figure 10.2 **Using bond energies to calculate ΔH^0_{rxn}.** Any chemical reaction can be divided conceptually into two hypothetical steps: (1) reactant bonds break to yield separate atoms in a step that absorbs heat (+ sum of BE), and (2) the atoms combine to form product bonds in a step that releases heat (− sum of BE). When the total bond energy of the products is greater than that of the reactants, more energy is released than is absorbed, so the reaction is exothermic (as shown); ΔH^0_{rxn} is negative. When the total bond energy of the products is less than that of the reactants, the reaction is endothermic; ΔH^0_{rxn} is positive.

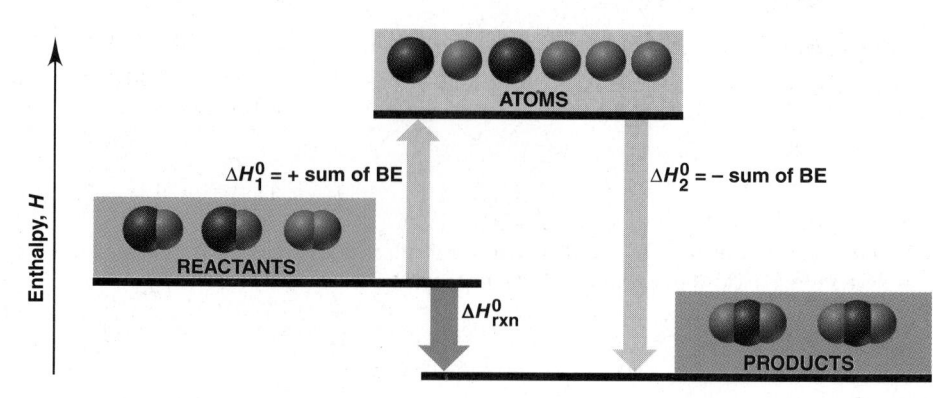

Let's apply this method to calculate ΔH^0_{rxn} for the combustion of methane and compare it with the value obtained from calorimetry, which is

$$CH_4(g) + 2O_2(g) \longrightarrow CO_2(g) + 2H_2O(g) \quad \Delta H^0_{rxn} = -802 \text{ kJ}$$

From the Lewis structures in Figure 10.3, we see that the reaction involves breaking all the bonds in CH_4 and O_2 and forming all the bonds in CO_2 and H_2O. We look up the bond energy values (see Table 9.2, p. 339), using a positive sign for bonds broken and a negative sign for bonds formed:

Bonds broken

$$
\begin{aligned}
4 \times \text{C—H} &= (4 \text{ mol})(413 \text{ kJ/mol}) = 1652 \text{ kJ} \\
2 \times \text{O}_2 &= (2 \text{ mol})(498 \text{ kJ/mol}) = 996 \text{ kJ} \\
\hline
\text{Total } \Delta H^0_{\text{reactant bonds broken}} &= 2648 \text{ kJ}
\end{aligned}
$$

Bonds formed

$$
\begin{aligned}
2 \times \text{C=O} &= (2 \text{ mol})(-799 \text{ kJ/mol}) = -1598 \text{ kJ} \\
4 \times \text{O—H} &= (4 \text{ mol})(-467 \text{ kJ/mol}) = -1868 \text{ kJ} \\
\hline
\text{Total } \Delta H^0_{\text{product bonds formed}} &= -3466 \text{ kJ}
\end{aligned}
$$

Adding these totals together gives

$$
\begin{aligned}
\Delta H^0_{rxn} &= \Delta H^0_{\text{reactant bonds broken}} + \Delta H^0_{\text{product bonds formed}} \\
&= 2648 \text{ kJ} + (-3466 \text{ kJ}) = -818 \text{ kJ}
\end{aligned}
$$

Why is there a discrepancy between the bond-energy value (-818 kJ) and the calorimetric value (-802 kJ)? Variations in experimental method always introduce small discrepancies, but there is a more basic reason. As noted earlier, bond energies are *average* values obtained from many different compounds in which the bond occurs. The energy of the bond *in a particular substance* is usually close, but not equal, to this average. For example, the tabulated C—H bond energy of 413 kJ/mol is the average value of C—H bonds in many different molecules. In fact, in methane, 1660 kJ is actually required to break 4 mol of C—H bonds, or 415 kJ/mol C—H bonds, which gives a ΔH^0_{rxn} slightly closer to the calorimetric value. Thus, it isn't surprising to find discrepancies between the two ΔH^0_{rxn} values. What is surprising—and satisfying in its confirmation of bond theory—is that the values are so close.

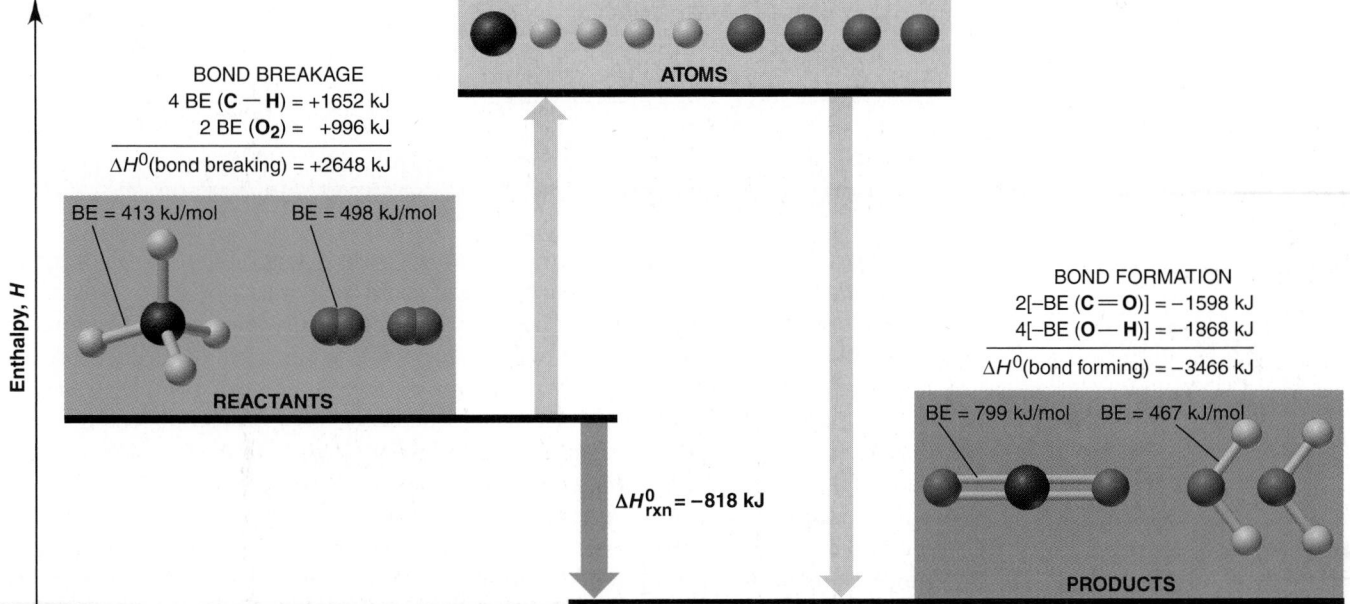

Figure 10.3 Using bond energies to calculate ΔH^0_{rxn} of methane. Treating the combustion of methane as a hypothetical two-step process (see Figure 10.2) means breaking all the bonds in the reactants and forming all the bonds in the products.

SAMPLE PROBLEM 10.6 Calculating Enthalpy Changes from Bond Energies

Problem Use Table 9.2 (p. 339) to calculate ΔH_{rxn}^{0} for the following reaction:

$$CH_4(g) + 3Cl_2(g) \longrightarrow CHCl_3(g) + 3HCl(g)$$

Plan First, we write the Lewis structures of all the substances. We assume that all reactant bonds break and all product bonds form, and we find their energies in Table 9.2. Then, we substitute the two sums with correct signs into Equation 10.2.

Solution Writing the Lewis structures:

Calculating ΔH_{rxn}^{0}: For bonds broken, the values are

$$
\begin{aligned}
4 \times \text{C—H} &= (4 \text{ mol})(413 \text{ kJ/mol}) = 1652 \text{ kJ} \\
3 \times \text{Cl—Cl} &= (3 \text{ mol})(243 \text{ kJ/mol}) = 729 \text{ kJ} \\
\hline
\Delta H_{\text{bonds broken}}^{0} &= 2381 \text{ kJ}
\end{aligned}
$$

For bonds formed, the values are

$$
\begin{aligned}
3 \times \text{C—Cl} &= (3 \text{ mol})(-339 \text{ kJ/mol}) = -1017 \text{ kJ} \\
1 \times \text{C—H} &= (1 \text{ mol})(-413 \text{ kJ/mol}) = -413 \text{ kJ} \\
3 \times \text{H—Cl} &= (3 \text{ mol})(-427 \text{ kJ/mol}) = -1281 \text{ kJ} \\
\hline
\Delta H_{\text{bonds formed}}^{0} &= -2711 \text{ kJ}
\end{aligned}
$$

$$\Delta H_{rxn}^{0} = \Delta H_{\text{bonds broken}}^{0} + \Delta H_{\text{bonds formed}}^{0} = 2381 \text{ kJ} + (-2711 \text{ kJ}) = \boxed{-330 \text{ kJ}}$$

Check The signs of the enthalpy changes are correct: $\Delta H_{\text{bonds broken}}^{0}$ should be >0, and $\Delta H_{\text{bonds formed}}^{0} < 0$. More energy is released than absorbed, so ΔH_{rxn}^{0} is negative:

$$\sim 2400 \text{ kJ} + [\sim(-2700 \text{ kJ})] = -300 \text{ kJ}$$

FOLLOW-UP PROBLEM 10.6 Use bond energies to calculate the enthalpy changes for the following reactions:
(a) $N_2(g) + 3H_2(g) \longrightarrow 2NH_3(g)$ **(b)** $C_2H_4(g) + HBr(g) \longrightarrow C_2H_5Br(g)$

SECTION SUMMARY

Lewis structures of reactants and products can be used to determine the bonds broken and the bonds formed during a reaction. Applying Hess's law, we use tabulated bond energies to calculate the heat of reaction.

10.3 VALENCE-SHELL ELECTRON-PAIR REPULSION (VSEPR) THEORY AND MOLECULAR SHAPE

Virtually every biochemical process hinges to a great extent on the shapes of interacting molecules. Every medicine you take, odor you smell, or flavor you taste depends on part or all of one molecule fitting physically together with another. This universal importance of molecular shape in the functioning of each organism carries over to the ecosystem. Biologists have learned of complex interactions regulating behaviors, such as mating, defense, navigation, and feeding, that depend on one molecule recognizing the shape of another. In this section, we discuss a model for understanding and predicting molecular shape.

The Lewis structure of a molecule is something like the blueprint of a building: a flat drawing showing the relative placement of parts (atom cores), the structural connections (groups of bonding valence electrons), and the various attachments (nonbonding lone pairs of valence electrons). To construct the molecular shape from the Lewis structure, chemists employ **valence-shell electron-pair repulsion (VSEPR) theory.** Its basic principle is that *each group of valence electrons around a central atom is located as far away as possible from the others*

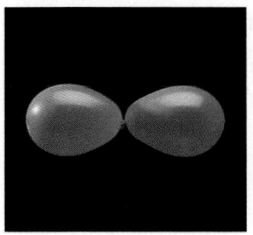 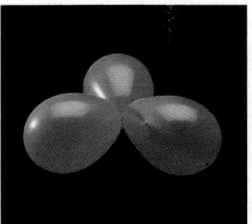   

Figure 10.4 A balloon analogy for the mutual repulsion of electron groups. Attached balloons move apart so that each can occupy as much space as possible. Five geometric arrangements arise by attaching two, three, four, five, or six balloons. Electron groups repel each other and become arranged in a similar way around a central atom, as Figure 10.5 at the bottom of the page shows.

in order to minimize repulsions. We define a "group" of electrons as any number of electrons that occupy a localized region around an atom. Thus, an electron group may consist of a single bond, a double bond, a triple bond, a lone pair, or even a lone electron.* Each of these groups of valence electrons repels the other groups to maximize the angles between them. It is the three-dimensional arrangement of nuclei joined by these groups that gives rise to the molecular shape.

Electron-Group Arrangements and Molecular Shapes

When two, three, four, five, or six objects attached to a central point maximize the space that each can occupy around that point, five geometric patterns result. Figure 10.4 depicts these patterns with balloons. If the objects are the valence-electron groups of a central atom, their repulsions maximize the space each occupies and give rise to the five *electron-group arrangements* of minimum energy seen in the great majority of molecules and polyatomic ions.

The electron-group arrangement is defined by the valence-electron groups, both bonding and nonbonding, around the central atom. On the other hand, the **molecular shape** is defined by the relative positions of the atomic nuclei. Figure 10.5 shows the molecular shapes that occur when *all* the surrounding electron groups are *bonding* groups. (Note the correspondence with the five balloon arrangements in Figure 10.4 above.) When some are *nonbonding* groups, different molecular shapes occur. Thus, *the same electron-group arrangement can give rise to different molecular shapes:* some with all bonding groups (as in Figure 10.5) and others with bonding and nonbonding groups. To classify molecular shapes, we assign each a specific AX_mE_n designation, where m and n are integers, A is the central atom, X is a surrounding atom, and E is a nonbonding valence-electron group (usually a lone pair).

The **bond angle** is the angle formed by the nuclei of two surrounding atoms with the nucleus of the central atom at the vertex. The angles shown for the shapes in Figure 10.5 are *ideal* bond angles, those predicted by simple geometry alone. These are observed when all the bonding electron groups around a central atom are identical and are connected to atoms of the same element. When this is not the case, the bond angles deviate from the ideal angles, as you'll see shortly.

*The two electron pairs in a double bond (or the three pairs in a triple bond) occupy separate orbitals, so they remain near each other and act as one electron group (see Chapter 11).

Figure 10.5 **Electron-group repulsions and the five basic molecular shapes.** When a given number of electron groups attached to a central atom (*red*) repel each other, they maximize the angle between themselves and become oriented as far apart as possible in space. If each electron group is a bonding group to a surrounding atom (*dark gray*), the molecular shapes and bond angles shown here are observed, and the name of the molecular shape is the same as that of the electron-group arrangement. When one or more of the electron groups is a lone pair, other molecular shapes are observed, as you'll see in upcoming figures.

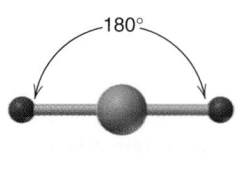

Linear

Trigonal planar

Tetrahedral

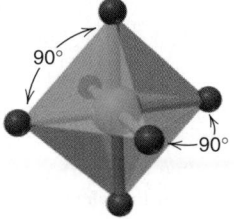

Trigonal bipyramidal

Octahedral

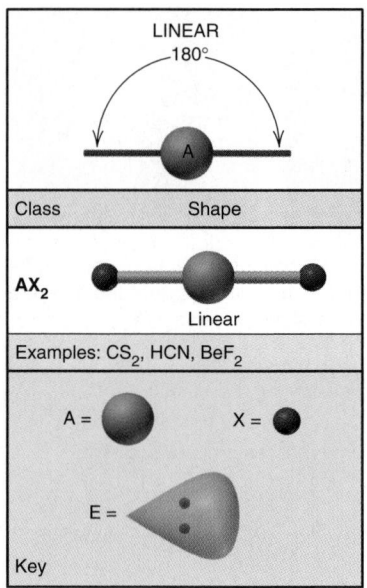

Figure 10.6 **The single molecular shape of the linear electron-group arrangement.** The key *(bottom)* for A, X, and E also refers to Figures 10.7, 10.8, 10.10, and 10.11.

It's important to realize that we use the VSEPR model to account for the molecular shapes observed by means of various types of spectroscopy. In almost every case, VSEPR predictions are in accord with actual observations. (We discuss some of these observational methods in Chapter 12.)

The Molecular Shape with Two Electron Groups (Linear Arrangement)

When two electron groups attached to a central atom are oriented as far apart as possible, they point in opposite directions. The **linear arrangement** of electron groups results in a **linear molecular shape** and a bond angle of 180°. Figure 10.6 shows the general arrangement (top), shape (middle) with VSEPR shape class (AX_2), and the formulas of some linear molecules.

Gaseous beryllium chloride ($BeCl_2$) is a linear molecule (AX_2). Gaseous Be compounds are electron deficient, with only two electron pairs around the central Be atom:

In carbon dioxide, the central C atom forms two double bonds with the O atoms:

Each double bond acts as a separate electron group and is oriented 180° away from the other, so CO_2 is linear. Notice that the lone pairs on the O atoms of CO_2 or on the Cl atoms of $BeCl_2$ are not involved in the molecular shape: only electron groups around the *central* atom influence shape.

Molecular Shapes with Three Electron Groups (Trigonal Planar Arrangement)

Three electron groups around the central atom repel each other to the corners of an equilateral triangle, which gives the **trigonal planar arrangement,** shown in Figure 10.7, and an ideal bond angle of 120°. This arrangement has two possible molecular shapes, one with three surrounding atoms and the other with two atoms and one lone pair. It provides our first opportunity to see the effects of double bonds and lone pairs on bond angles.

When the three electron groups are bonding groups, the molecular shape is *trigonal planar* (AX_3). Boron trifluoride (BF_3), another electron-deficient molecule, is an example. It has six electrons around the central B atom in three single bonds to F atoms. The nuclei lie in a plane, and each F—B—F angle is 120°:

The nitrate ion (NO_3^-) is one of several polyatomic ions with the trigonal planar shape. One of three resonance forms of the nitrate ion is shown below (see Sample Problem 10.4). The resonance hybrid has three identical bonds of bond order $1\frac{1}{3}$, so the ideal bond angle is observed:

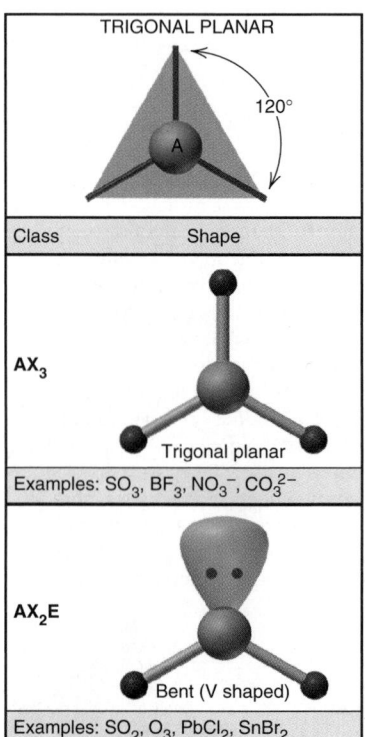

Figure 10.7 **The two molecular shapes of the trigonal planar electron-group arrangement.**

Effect of Double Bonds. How do bond angles deviate from the ideal angles when the surrounding atoms and electron groups are not identical? Consider formaldehyde (CH_2O), a substance with many uses, including the manufacture of Formica countertops, the production of methanol, and the preservation of cadavers. Its trigonal planar shape is due to two types of surrounding atoms (O and H) and two types of electron groups (single and double bonds):

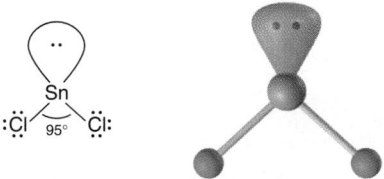

ideal $\qquad$ actual

The actual bond angles deviate from the ideal because *the double bond, with its greater electron density, repels the two single bonds more strongly than they repel each other.*

Effect of Lone Pairs. The molecular shape is defined only by the positions of the nuclei, so when one of the three electron groups is a lone pair (AX_2E), the shape is **bent,** or **V shaped,** not trigonal planar. Gaseous tin(II) chloride is an example, with the three electron groups in a trigonal plane and the lone pair at one of the triangle's corners. A lone pair can have a major effect on bond angle. Because a lone pair is held by only one nucleus, it is less confined and exerts stronger repulsions than a bonding pair. Thus, *a lone pair repels bonding pairs more strongly than bonding pairs repel each other.* This stronger repulsion *decreases* the angle between bonding pairs. Note the decrease from the ideal 120° angle in $SnCl_2$:

Molecular Shapes with Four Electron Groups (Tetrahedral Arrangement)

The shapes described so far have all been easy to depict in two dimensions, but four electron groups must use three dimensions to achieve maximal separation. This is a good time for you to recall that *Lewis structures do **not** depict shape.* Consider the shape of methane. The Lewis structure (shown below) indicates four bonds pointing to the corners of a square, which suggests a 90° bond angle. However, in three dimensions, the four electron groups can move farther apart than 90° and point to the vertices of a tetrahedron, a polyhedron with four faces made of identical equilateral triangles. Methane has a bond angle of 109.5°. Perspective drawings, such as these for methane, indicate depth by using wedges (or wedges and dashed lines) for some of the bonds:

The normal lines represent shared electron groups in the plane of the page, one wedge is the bond between the central atom and a group lying toward you above the page, and the other wedge is the bond between the central atom and a group lying away from you below the page. Another convention is to show the bond below the page as a dashed line. The ball-and-stick model shows the tetrahedral shape clearly.

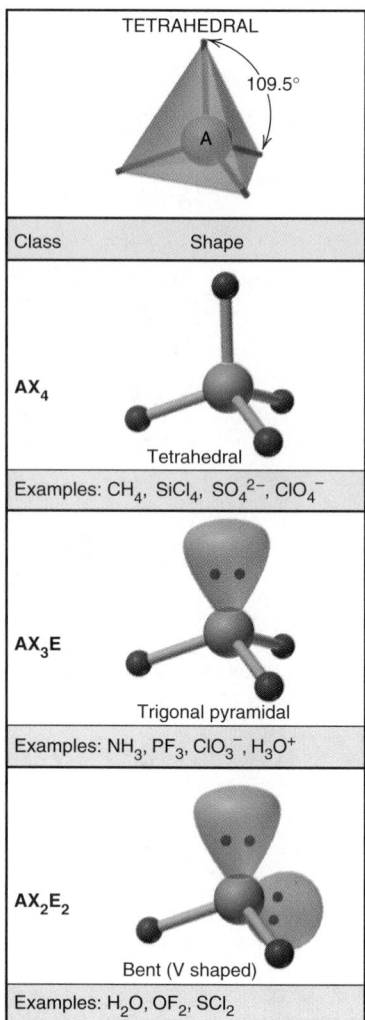

TETRAHEDRAL

109.5°

A

Class	Shape

AX₄

Tetrahedral

Examples: CH₄, SiCl₄, SO₄²⁻, ClO₄⁻

AX₃E

Trigonal pyramidal

Examples: NH₃, PF₃, ClO₃⁻, H₃O⁺

AX₂E₂

Bent (V shaped)

Examples: H₂O, OF₂, SCl₂

Figure 10.8 **The three molecular shapes of the tetrahedral electron-group arrangement.**

*All molecules or ions with four electron groups around a central atom adopt the **tetrahedral arrangement*** (Figure 10.8). When all four electron groups are bonding groups, as in the case of methane, the molecular shape is also *tetrahedral* (AX₄), a very common geometry in organic molecules. In Sample Problem 10.1, we drew the Lewis structure for the tetrahedral molecule dichlorodifluoromethane (CCl₂F₂), without regard to how the halogen atoms surround the C atom. Because Lewis structures are flat, it seems that we can write two different structures for CCl₂F₂, but these actually represent the same molecule, as Figure 10.9 makes clear.

When one of the four electron groups in the tetrahedral arrangement is a lone pair, the molecular shape is that of a **trigonal pyramid** (AX₃E), a tetrahedron with one vertex "missing." As we would expect from the stronger repulsions due to the lone pair, the measured bond angle is slightly less than the ideal 109.5°. In ammonia (NH₃), for example, the lone pair forces the N—H bonding pairs closer, and the H—N—H bond angle is 107.3°.

Picturing molecular shapes is a great way to visualize what happens during a reaction. For instance, when ammonia reacts with the proton from an acid, the lone pair on the N atom of trigonal pyramidal NH₃ forms a covalent bond to the H⁺ and yields the ammonium ion (NH₄⁺), one of many tetrahedral polyatomic ions. Note how the H—N—H bond angle expands from 107.3° in NH₃ to 109.5° in NH₄⁺, as the lone pair becomes another bonding pair:

When the four electron groups around the central atom include two bonding and two nonbonding groups, the molecular shape is *bent, or V shaped* (AX₂E₂). [In the trigonal planar arrangement, the shape with two bonding groups and one lone pair is also called bent (AX₂E), but its ideal bond angle is 120°, not 109.5°.] Water is the most important V-shaped molecule with the tetrahedral arrangement. We might expect the repulsions from its two lone pairs to have a *greater* effect

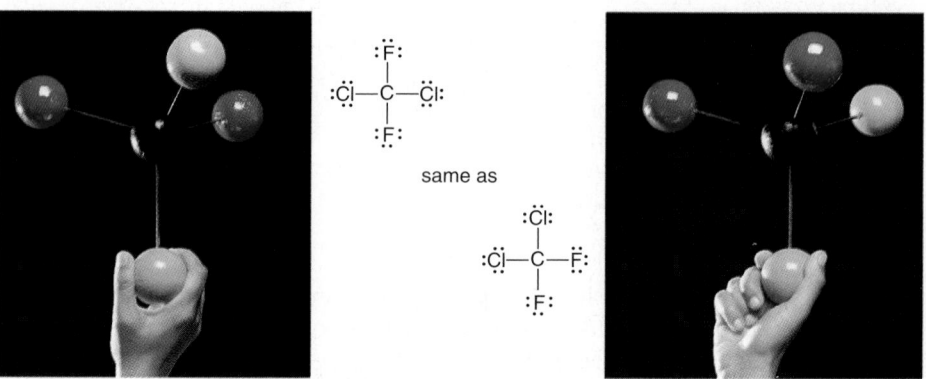

Figure 10.9 **Lewis structures and molecular shapes.** Lewis structures do not indicate geometry. For example, it may seem as if two different Lewis structures can be written for CCl₂F₂, but a twist of the model (Cl, *green;* F, *yellow*) shows that they represent the same molecule.

on the bond angle than the repulsions from the single lone pair in NH_3. Indeed, the H—O—H bond angle is reduced to 104.5°:

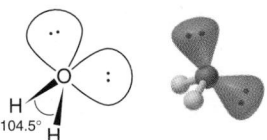

Thus, for similar molecules with a given electron-group arrangement, electron-pair repulsions cause deviations from ideal bond angles in the following order:

> Lone pair–lone pair > lone pair–bonding pair > bonding pair–bonding pair   **(10.3)**

Molecular Shapes with Five Electron Groups (Trigonal Bipyramidal Arrangement)

All molecules with five or six electron groups have a central atom from Period 3 or higher because only these atoms have the *d* orbitals available to expand the valence shell beyond eight electrons.

When five electron groups maximize their separation, they form the **trigonal bipyramidal arrangement.** In a trigonal bipyramid, two trigonal pyramids share a common base, as shown in Figure 10.10. Note that, in a molecule with this arrangement, *there are two types of positions for surrounding electron groups and two ideal bond angles.* Three **equatorial groups** lie in a trigonal plane that includes the central atom, and two **axial groups** lie above and below this plane. Therefore, a 120° bond angle separates equatorial groups, and a 90° angle separates axial from equatorial groups. In general, the greater the bond angle, the weaker the repulsions, so *equatorial-equatorial (120°) repulsions are weaker than axial-equatorial (90°) repulsions.* The tendency of the electron groups to occupy *equatorial* positions, and thus minimize the stronger 90° repulsions, governs the four shapes of the trigonal bipyramidal arrangement.

With all five positions occupied by bonded atoms, the molecule has the *trigonal bipyramidal* shape (AX_5), as in phosphorus pentachloride (PCl_5):

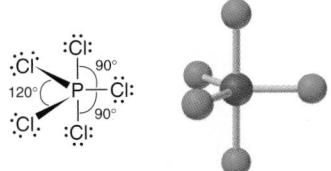

Three other shapes arise for molecules with lone pairs. Since lone pairs exert stronger repulsions than bonding pairs, we find that *lone pairs occupy equatorial positions.* With one lone pair present at an equatorial position, the molecule has a **seesaw shape** (AX_4E). Sulfur tetrafluoride (SF_4), a powerful fluorinating agent, has this shape, shown here and in Figure 10.10 with the "seesaw" tipped up on an end. Note how the equatorial lone pair repels all four bonding pairs to reduce the bond angles:

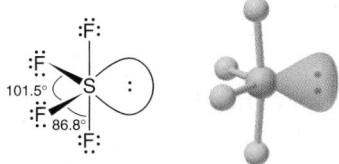

The tendency of lone pairs to occupy equatorial positions causes molecules with three bonding groups and two lone pairs to have a **T shape** (AX_3E_2). Bromine trifluoride (BrF_3), one of many compounds with fluorine bound to a

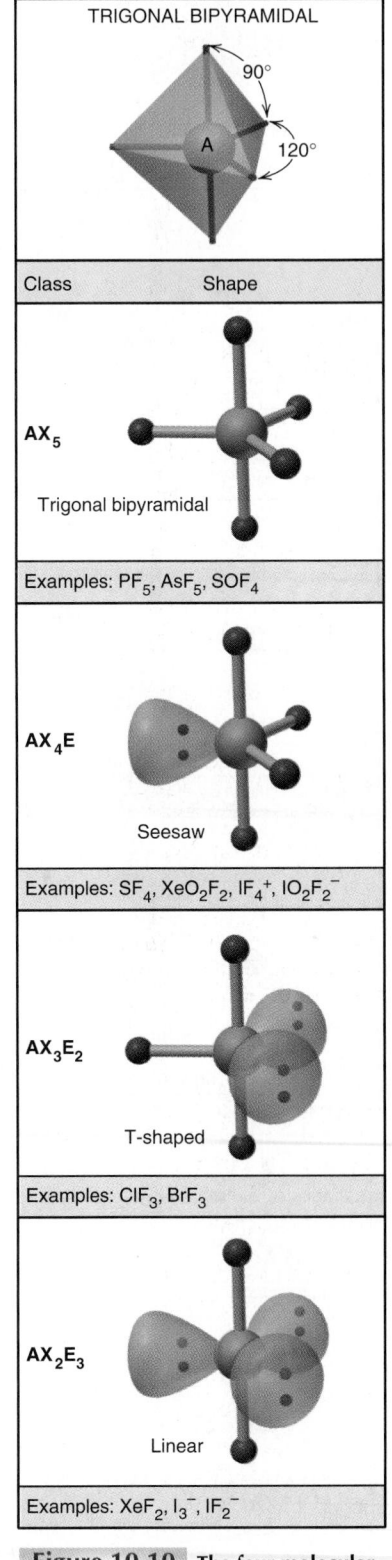

Class	Shape

TRIGONAL BIPYRAMIDAL

AX_5

Trigonal bipyramidal

Examples: PF_5, AsF_5, SOF_4

AX_4E

Seesaw

Examples: SF_4, XeO_2F_2, IF_4^+, $IO_2F_2^-$

AX_3E_2

T-shaped

Examples: ClF_3, BrF_3

AX_2E_3

Linear

Examples: XeF_2, I_3^-, IF_2^-

Figure 10.10 The four molecular shapes of the trigonal bipyramidal electron-group arrangement.

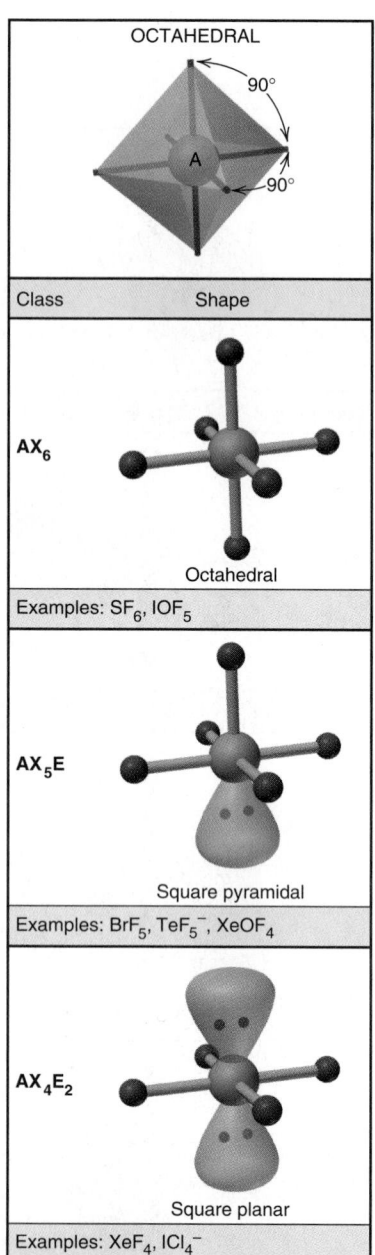

Class	Shape
AX₆	Octahedral

Examples: SF₆, IOF₅

AX₅E	Square pyramidal

Examples: BrF₅, TeF₅⁻, XeOF₄

AX₄E₂	Square planar

Examples: XeF₄, ICl₄⁻

Figure 10.11 The three molecular shapes of the octahedral electron-group arrangement.

VSEPR THEORY AND THE SHAPES
OF MOLECULES

VSEPR

larger halogen, has this shape. Note the predicted decrease from the ideal 90° F—Br—F bond angle:

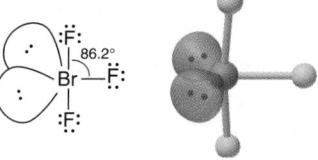

Molecules with three lone pairs in equatorial positions must have the two bonding groups in axial positions, which gives the molecule a *linear* shape (AX_2E_3) and a 180° axial-to-central-to-axial (X—A—X) bond angle. For example, the triiodide ion (I_3^-), which forms when I_2 dissolves in aqueous I^- solution, is linear:

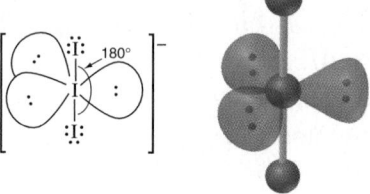

Molecular Shapes with Six Electron Groups (Octahedral Arrangement)

The last of the five major electron-group arrangements is the **octahedral arrangement.** An octahedron is a polyhedron with eight faces made of identical equilateral triangles and six identical vertices, as shown in Figure 10.11. In a molecule (or ion) with this arrangement, six electron groups surround the central atom and each points to one of the six vertices, which gives all the groups a 90° ideal bond angle. Three important molecular shapes occur with this arrangement.

With six bonding groups, the molecular shape is *octahedral* (AX_6). When the seesaw-shaped SF_4 reacts with additional F_2, the central S atom expands its valence shell further to form octahedral sulfur hexafluoride (SF_6):

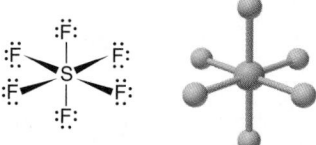

Because there is only one ideal bond angle, it makes no difference which position one lone pair occupies. Five bonded atoms and one lone pair define the **square pyramidal shape** (AX_5E), as shown for iodine pentafluoride (IF_5):

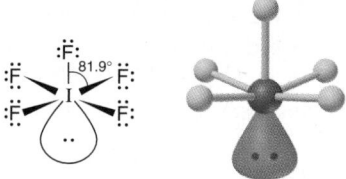

When a molecule has two lone pairs, however, they always lie at *opposite vertices* to avoid the stronger lone pair–lone pair repulsions at 90°. This positioning gives the **square planar shape** (AX_4E_2), as in xenon tetrafluoride (XeF_4):

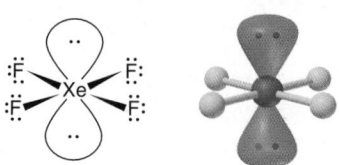

Using VSEPR Theory to Determine Molecular Shape

Let's apply a stepwise method for using the VSEPR theory to determine a molecular shape from a molecular formula:

Step 1. Write the Lewis structure from the molecular formula to see the relative placement of atoms and the number of electron groups (Figure 10.1, p. 358).

Step 2. Assign an electron-group arrangement by counting *all* electron groups around the central atom, bonding *plus* nonbonding.

Step 3. Predict the ideal bond angle from the electron-group arrangement and *the direction of any deviation* caused by lone pairs or double bonds.

Step 4. Draw and name the molecular shape by counting bonding groups and nonbonding groups separately.

Figure 10.12 summarizes these steps, and the next two sample problems apply them.

Figure 10.12 **The steps in determining a molecular shape.** Four steps are needed to convert a molecular formula to a molecular shape. For the first step, writing the Lewis structure, see Figure 10.1, p. 358.

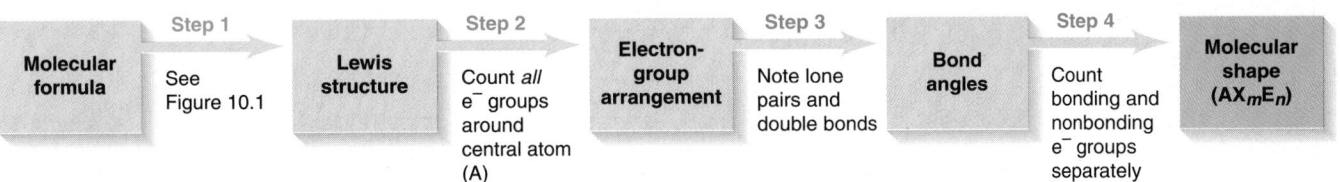

SAMPLE PROBLEM 10.7 Predicting Molecular Shapes with Two, Three, or Four Electron Groups

Problem Draw the molecular shape and predict the bond angles (relative to the ideal angles) of **(a)** PF_3 and **(b)** $COCl_2$.

Solution (a) For PF_3.

Step 1. Write the Lewis structure from the formula (see below left).

Step 2. Assign the electron-group arrangement: Three bonding groups plus one lone pair give four electron groups around P and the *tetrahedral arrangement*.

Step 3. Predict the bond angle: For the tetrahedral electron-group arrangement, the ideal angle is 109.5°. There is one lone pair, so the actual bond angle should be less than 109.5°.

Step 4. Draw and name the molecular shape: With four electron groups, one of them a lone pair, PF_3 has a trigonal pyramidal shape (AX_3E):

$$:\ddot{F}-\ddot{P}-\ddot{F}:\ \xrightarrow[\text{groups}]{4\ e^-}\ \text{Tetrahedral}\ \xrightarrow[\text{pair}]{1\ \text{lone}}\ <109.5°\ \xrightarrow[\text{groups}]{3\ \text{bonding}}$$

arrangement

AX_3E

(b) For $COCl_2$.

Step 1. Write the Lewis structure from the formula (see below left).

Step 2. Assign the electron-group arrangement: Two single bonds plus one double bond give three electron groups around C and the *trigonal planar arrangement*.

Step 3. Predict the bond angles: The ideal angle is 120°, but the double bond between C and O should compress the Cl—C—Cl angle to less than 120°.

Step 4. Draw and name the molecular shape: With three electron groups and no lone pairs, $COCl_2$ has a trigonal planar shape (AX_3):

$$\xrightarrow[\text{groups}]{3\ e^-}\ \text{Trigonal planar}\ \xrightarrow[\text{bond}]{1\ \text{double}}\ \begin{array}{l}\text{Cl—C—O} >120°\\ \text{Cl—C—Cl} <120°\end{array}\ \xrightarrow[\text{groups}]{3\ \text{bonding}}$$

arrangement

AX_3

Check We compare the answers with the general information in Figures 10.7 and 10.8.

Comment Be sure the Lewis structure is correct because it determines the other steps.

FOLLOW-UP PROBLEM 10.7 Draw the shape and predict the bond angles (relative to the ideal angles) of **(a)** CS_2; **(b)** $PbCl_2$; **(c)** CBr_4; **(d)** SF_2.

SAMPLE PROBLEM 10.8 Predicting Molecular Shapes with Five or Six Electron Groups

Problem Determine the molecular shape and predict the bond angles (relative to the ideal angles) of (a) SbF_5 and (b) BrF_5.

Plan We proceed as in Sample Problem 10.7, keeping in mind the need to minimize the number of 90° repulsions.

Solution (a) For SbF_5.

Step 1. Lewis structure (see below left)

Step 2. Electron-group arrangement: With five electron groups, this is the *trigonal bipyramidal* arrangement.

Step 3. Bond angles: All the groups and surrounding atoms are identical, so the bond angles are ideal: 120° between equatorial groups and 90° between axial and equatorial groups.

Step 4. Molecular shape: Five electron groups and no lone pairs give the trigonal bipyramidal shape (AX_5):

(b) For BrF_5.

Step 1. Lewis structure (see below left).

Step 2. Electron-group arrangement: Six electron groups give the *octahedral* arrangement.

Step 3. Bond angles: The lone pair should make all bond angles less than the ideal 90°.

Step 4. Molecular shape: With six electron groups and one of them a lone pair, BrF_5 has the square pyramidal shape (AX_5E):

Check We compare our answers with Figures 10.10 and 10.11.

Comment The linear, tetrahedral, square planar, and octahedral shapes also arise in an important group of substances called *coordination compounds,* which we discuss in Chapter 23.

FOLLOW-UP PROBLEM 10.8 Draw the molecular shapes and predict the bond angles (relative to the ideal angles) of (a) ICl_2^-; (b) ClF_3; (c) SOF_4.

Molecular Shapes with More Than One Central Atom

As you know, many molecules, especially those in living systems, have more than one central atom. The shapes of these molecules are combinations of the molecular shapes with a single central atom. Consider ethane (C_2H_6), a component of natural gas (Figure 10.13A). With four bonding groups and no lone pairs around each of the two central carbons, ethane is shaped like two overlapping tetrahedra.

Ethanol (CH_3CH_2OH; molecular formula C_2H_6O), the intoxicating substance in beer and wine, has three atom centers (Figure 10.13B). The CH_3— group is tetrahedrally shaped, and the —CH_2— group has four bonding groups around its central C atom, so it is tetrahedrally shaped also. The O atom has four electron groups and two lone pairs around it, which gives the V shape (AX_2E_2). For these more complex molecules, we find the molecular shape around one central atom at a time.

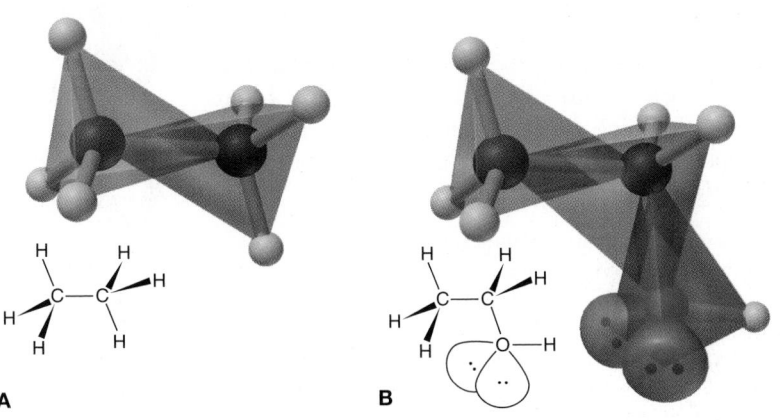

Figure 10.13 The tetrahedral centers of ethane and of ethanol. When a molecule has more than one central atom, the shape is a composite of the shape around each center. **A,** Ethane's shape can be viewed as two overlapping tetrahedra. **B,** Ethanol's shape can be viewed as three overlapping tetrahedral arrangements, with the shape around the O atom bent (V shaped) because of its two lone pairs.

A　　　　　　　**B**

SAMPLE PROBLEM 10.9 Predicting Molecular Shapes with More Than One Central Atom

Problem Determine the shape around each of the central atoms in acetone, $(CH_3)_2C{=}O$.
Plan There are three central atoms, all C, two of which are in $CH_3{-}$ groups. We determine the shape around one central atom at a time.
Solution
Step 1. Lewis structure (see below left).
Step 2. Electron-group arrangement: Each $CH_3{-}$ group has four electron groups around its central C, so its electron-group arrangement is *tetrahedral.* The third C atom has three electron groups around it, so it has the *trigonal planar* arrangement.
Step 3. Bond angles: The H—C—H angle in the $CH_3{-}$ groups should be near the ideal tetrahedral angle of 109.5°. The C=O double bond should compress the C—C—C angle to less than the ideal 120°.
Step 4. Shapes around central atoms: With four electron groups and no lone pairs, the shapes around the two C atoms in the $CH_3{-}$ groups are ☐ tetrahedral ☐ (AX_4). With three electron groups and no lone pairs, the shape around the middle C atom is ☐ trigonal planar ☐ (AX_3):

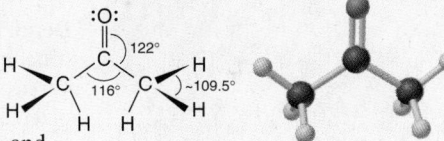

FOLLOW-UP PROBLEM 10.9 Determine the shape around each central atom and predict any deviations from ideal bond angles in the following: **(a)** H_2SO_4; **(b)** propyne (C_3H_4; there is one C≡C bond); **(c)** S_2F_2.

In recent years, chemists have synthesized some organic molecules with remarkable—some would say beautiful—shapes. Some of their uses may still be speculative, but there is no question about their geometric elegance. A few of these molecules appear in the upcoming Gallery.

SECTION SUMMARY

The VSEPR theory proposes that each group of electrons (single bond, multiple bond, lone pair, or lone electron) around a central atom remains as far away from the others as possible. Five electron-group arrangements result when two, three, four, five, or six electron groups surround a central atom. Each arrangement is associated with one or more molecular shapes. Ideal bond angles are prescribed by the regular geometric shapes; deviations from these angles occur when the surrounding atoms or electron groups are not identical. Lone pairs and double bonds exert greater repulsions than single bonds. Larger molecules have shapes that are composites of the shapes around each central atom.

Molecular Beauty:
Odd Shapes with Useful Functions

Chemists see a strange beauty in the geometric intricacy of nature's most minute objects. The simplicity of spherical atoms disappears when they combine to form pentagons, helices, and countless other shapes. Moreover, many "beautiful" molecules have marvelous practical uses.

Fullerenes Buckminsterfullerene ("bucky ball," named for R. Buckminster Fuller, the architect and philosopher who designed domes based on this shape) is a truncated icosahedron of C atoms, a molecular "soccer ball" with 60 vertices and 32 faces (12 pentagons and 20 hexagons). This C_{60} structure, discovered in soot in 1985 and prepared in bulk in 1990, represents a third form of crystalline carbon (graphite and diamond are the other two). It is the parent of a new family of structures called *fullerenes* and has spawned a new field of synthesis. By inserting atoms, such as potassium, inside the ball (see models) and bonding countless types of chemical groups to the outside, researchers have discovered a whole new range of possible applications, including lubricants, superconductors, rocket fuels, lasers, batteries, magnetic films, cancer and AIDS drugs—the list grows daily!

Nanotubes
These younger cousins of fullerenes consist of extremely long, thin, graphite-like cylinders with fullerene ends. They are often nested within one another (as in photo behind model). Despite thicknesses of a few nanometers, these structures are highly conductive along their length and about 40 times stronger than steel! Dreams abound of nanoscale electronic components made with nanotubes or atom-thick wires formed by inserting metal atoms within their interiors.

Dendrimers Named for their structural similarity to the branching of trees, dendrimers form when one molecule with several bonding groups reacts with itself. They have excellent surface properties and are being studied for applications as films, fibers, and paint binders. (The model shows one-third of the circular schematic.)

Cubanes
Bending carbon's normal tetrahedral bond angle to 90° requires energy, which becomes stored in the bond. Thus, cyclobutane, a square of C atoms with two H atoms at each corner, is unstable above 500 K. Much more energy is stored in the bonds of cubane, a cube of C atoms with one H at each corner, synthesized in 1964. The bond-strain energy of cubane can yield enormous explosive power. Tetranitrocubane (shown here) is a very powerful explosive, and octanitrocubane, which was synthesized in 2000, is considered the most powerful non-nuclear explosive known. Other properties of cubanes, which are apparently unrelated to the bond energy, include attacking cancer cells and inactivating an enzyme involved in Parkinson's disease.

10.4 MOLECULAR SHAPE AND MOLECULAR POLARITY

Knowing the shape of a substance's molecules is a key to understanding its physical and chemical behavior. One of the most important and far-reaching effects of molecular shape is molecular polarity, which can influence melting and boiling points, solubility, chemical reactivity, and even biological function.

In Chapter 9, you learned that a covalent bond is *polar* when it joins atoms of different electronegativities because the atoms share the electrons unequally. In diatomic molecules, such as HF, where there is only one bond, the bond polarity causes the molecule itself to be polar. Molecules with a net imbalance of charge have a **molecular polarity.** In molecules with more than two atoms, *both shape and bond polarity determine molecular polarity.* In an electric field, polar molecules become oriented, on average, with their partial charges pointing toward the oppositely charged electric plates, as shown for HF in Figure 10.14. The **dipole moment (μ)** is the product of these partial charges and the distance between them. It is typically measured in *debye* (D) units; using the SI units of charge (coulomb, C) and length (meter, m), $1\ D = 3.34 \times 10^{-30}\ C \cdot m$. [The unit is named for Peter Debye (1884–1966), the Dutch American chemist and physicist who won the Nobel Prize in 1936 for his major contributions to our understanding of molecular structure and solution behavior.]

Bond Polarity, Bond Angle, and Dipole Moment

When determining molecular polarity, we must take shape into account because the presence of polar bonds does not *always* lead to a polar molecule. In carbon dioxide, for example, the large electronegativity difference between C (EN = 2.5) and O (EN = 3.5) makes each C=O bond quite polar. However, CO_2 is linear, so its bonds point 180° from each other. As a result, the two identical bond polarities are counterbalanced and give the molecule *no net dipole moment* (μ = 0 D). Note that the electron density model shows regions of high negative charge (*red*) distributed equally on either side of the central region of high positive charge (*blue*):

Water also has identical atoms bonded to the central atom, but it *does* have a significant dipole moment (μ = 1.85 D). In each O—H bond, electron density is pulled toward the more electronegative O atom. Here, the bond polarities are *not* counterbalanced, because the water molecule is V shaped (also see Figure 4.2, p. 134). Instead, the bond polarities are partially reinforced, and the O end of the molecule is more negative than the other end (the region between the H atoms), which the electron density model shows clearly:

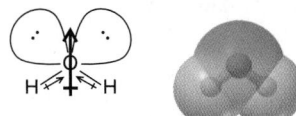

(The molecular polarity of water has some amazing effects, from determining the composition of the oceans to supporting life itself, as you'll see in Chapter 12.)

In the two previous examples, molecular shape influences polarity. When different molecules have the same shape, the nature of the atoms surrounding the central atom can have a major effect on polarity. Consider carbon tetrachloride (CCl_4) and chloroform ($CHCl_3$), two tetrahedral molecules with very different

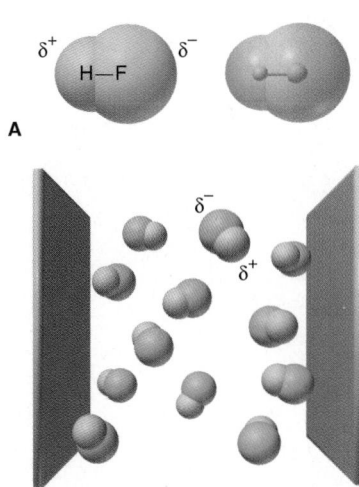

A

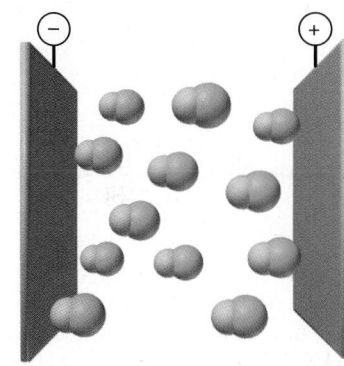

B Electric field off

C Electric field on

Figure 10.14 The orientation of polar molecules in an electric field. A, A space-filling model of HF (*left*) shows the partial charges of this polar molecule. The electron density model (*right*) shows high electron density (*red*) associated with the F end and low electron density (*blue*) with the H end. **B,** In the absence of an external electric field, HF molecules are oriented randomly. **C,** In the presence of the field, the molecules, on average, become oriented with their partial charges pointing toward the oppositely charged plates.

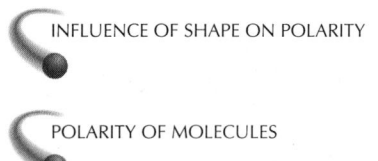

INFLUENCE OF SHAPE ON POLARITY

POLARITY OF MOLECULES

polarities. In CCl_4, the surrounding atoms are all Cl atoms. Although each C—Cl bond is polar ($\Delta EN = 0.5$), the molecule is nonpolar ($\mu = 0$ D) because the individual bond polarities counterbalance each other. In $CHCl_3$, H substitutes for one Cl atom, disrupting the balance and giving chloroform a significant dipole moment ($\mu = 1.01$ D):

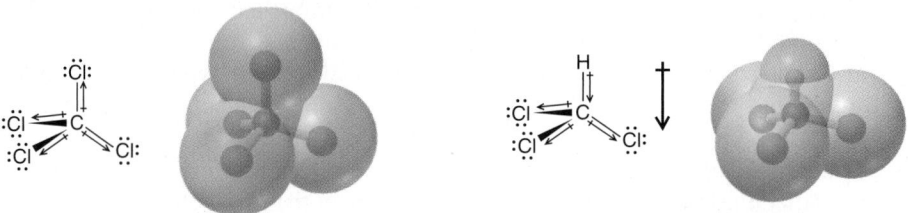

SAMPLE PROBLEM 10.10 Predicting the Polarity of Molecules

Problem From electronegativity (EN) values and their periodic trends (see Figure 9.16), predict whether each of the following molecules is polar and show the direction of bond dipoles and the overall molecular dipole when applicable:
(a) Ammonia, NH_3 **(b)** Boron trifluoride, BF_3
(c) Carbonyl sulfide, COS (atom sequence SCO)
Plan First, we draw and name the molecular shape. Then, using relative EN values, we decide on the direction of each bond dipole. Finally, we see if the bond dipoles balance or reinforce each other in the molecule as a whole.
Solution **(a)** For NH_3. The molecular shape is trigonal pyramidal. From Figure 9.16, we see that N (EN = 3.0) is more electronegative than H (EN = 2.1), so the bond dipoles point toward N. The bond dipoles partially reinforce each other, and thus the molecular dipole points toward N:

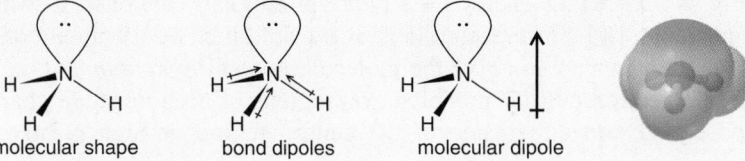

Therefore, ammonia is polar.
(b) For BF_3. The molecular shape is trigonal planar. Because F (EN = 4.0) is farther to the right in Period 2 than B (EN = 2.0), it is more electronegative; thus, each bond dipole points toward F. However, the bond angle is 120°, so the three bond dipoles counterbalance each other, and BF_3 has no molecular dipole:

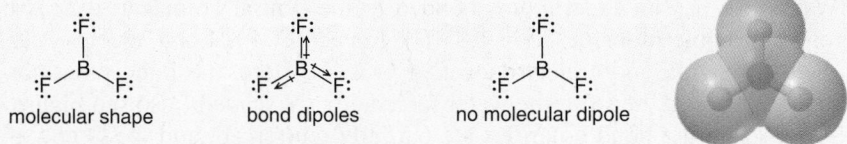

Therefore, boron trifluoride is nonpolar.

(c) For COS. The molecular shape is linear. With C and S having the same EN, the C=S bond is nonpolar, but the C=O bond is quite polar ($\Delta EN = 1.0$), so there is a net molecular dipole toward the O:

$$\ddot{S}=C=\ddot{O}$$
molecular shape

$$\ddot{S}=C=\ddot{O}$$
bond dipoles

$$\ddot{S}=C=\ddot{O}$$
molecular dipole

Therefore, carbonyl sulfide is polar.

FOLLOW-UP PROBLEM 10.10 Show the bond dipoles and molecular dipole, if any, for **(a)** dichloromethane (CH_2Cl_2); **(b)** iodine oxide pentafluoride (IOF_5); **(c)** nitrogen tribromide (NBr_3).

The Effect of Molecular Polarity on Behavior

To get a sense of the influence of molecular polarity on physical behavior, consider what effect a molecular dipole might have when many polar molecules lie near each other, as they do in a liquid. How does a molecular property such as dipole moment affect a macroscopic property such as boiling point? A liquid boils when its molecules have enough energy to form bubbles of gas. To enter the bubble, the molecules in the liquid must overcome the weak attractive forces *between* them. A molecular dipole influences the strength of these attractions.

Consider the two dichloroethylenes shown below. These compounds have the *same* molecular formula ($C_2H_2Cl_2$) and therefore the *same* molar mass, but they have *different* physical and chemical properties. In particular, *cis*-1,2-dichloroethylene boils 13°C higher than *trans*-1,2-dichloroethylene. VSEPR theory predicts, and spectroscopic analysis confirms, that all the nuclei lie in the same plane, with a trigonal planar shape around each C atom. The *trans* compound has no dipole moment ($\mu = 0$ D) because the C—Cl bond polarities balance each other. In contrast, the *cis* compound *is* polar ($\mu = 1.90$ D) because the bond dipoles partially reinforce each other, with the molecular dipole pointing between the Cl atoms. In the liquid state, the polar *cis* molecules attract each other more strongly than the nonpolar *trans* molecules, so more energy is needed to overcome these stronger forces. Therefore, the *cis* compound has a higher boiling point. We extend these ideas in Chapter 12. The Chemical Connections essay on the next two pages discusses some of the many fundamental effects of molecular shape and polarity on biological behavior.

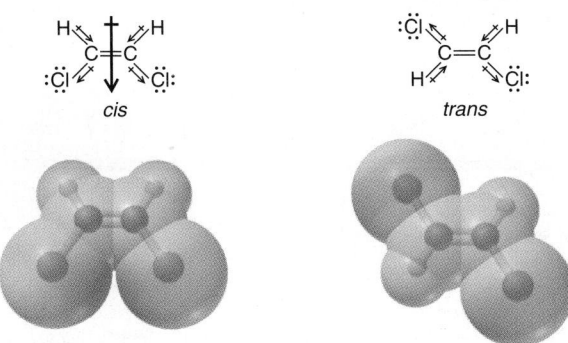

cis   *trans*

SECTION SUMMARY

Bond polarity and molecular shape determine molecular polarity, which is measured as a dipole moment. When bond polarities counterbalance each other, the molecule is nonpolar; when they reinforce each other, the molecule is polar. Molecular shape and polarity can affect physical properties, such as boiling point, and they play a central role in many aspects of biological function.

Chapter Perspective

In this chapter, you saw how chemists use the simple idea of electrostatic repulsion between shared and/or unshared electron groups to account for the shapes of molecules and, ultimately, their properties. In the next chapter, you'll learn how chemists explain molecular shape and behavior through the nature of the orbitals that electrons occupy. The landscape of a molecule—its geometric contours and regions of charge—gives rise to all its other properties. In upcoming chapters, we'll explore the physical behavior of liquids and solids (Chapter 12), the properties of solutions (Chapter 13), and the behavior of the main-group elements (Chapter 14) and organic molecules (Chapter 15), all of which emerge from atomic properties and their resulting molecular properties.

Chemical Connections Chemistry in Sensory Physiology
Molecular Shape, Biological Receptors, and the Sense of Smell

In a very simple sense, a biological cell is a membrane-bound sack filled with molecular shapes interacting in an aqueous fluid. As a result of the magnificent internal organization of cells, many complex processes in an organism begin when a molecular "key" fits into a correspondingly shaped molecular "lock." The key can be a relatively small molecule circulating in a body fluid, whereas the lock is usually a large molecule, known as a *biological receptor,* that is often found embedded in a cell membrane. The receptor contains a precisely shaped cavity, or *receptor site,* that is exposed to the passing fluid. Thousands of molecules collide with this site, but when one with the correct shape (that is, the molecular key) lands on it, the receptor "grabs" it through intermolecular attractions, and the biological response begins.

Let's see how this fitting together of molecular shapes operates in the sense of smell (olfaction). A substance must have certain properties to have an odor. An odorous molecule travels through the air, so it must come from a gas or a volatile liquid or solid. To reach the receptor, it must be soluble, at least to a small extent, in the thin film of aqueous solution that lines the nasal passages. Most important, the odorous molecule, or a portion of it, must have a shape that fits into one of the olfactory receptor sites that cover the nerve endings deep within the nasal passage (Figure B10.1). When this happens, the resulting nerve impulses travel from these endings to the brain, which interprets the impulses as a specific odor.

In the 1950s, the stereochemical theory of odor (*stereo* means "three-dimensional") was introduced to explain the relationship between odor and molecular shape. Its basic premise is that a molecule's shape (and sometimes its polarity), but *not* its composition, is the primary determinant of its odor. According to this theory, there are seven primary odors, each corresponding to one of seven different types of olfactory receptor sites. The seven odors are camphor-like, musky, floral, minty, ethereal, pungent, and putrid. (The last two odors depend on molecular polarity more than shape, and their receptors have partial charges opposite to those of the molecules landing there.) Figure B10.2 shows the proposed shapes of three of the seven receptor sites, each occupied by a molecule having the associated odor.

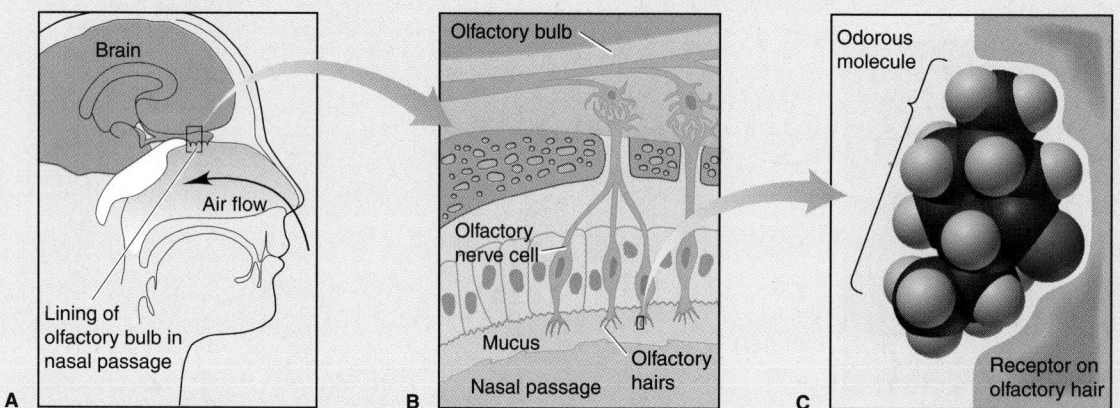

Figure B10.1 **The location of olfactory receptors within the nose. A,** The olfactory area lies at the top of the nasal passage very close to the brain. Air containing the odorous molecules is sniffed in, warmed, moistened, and channeled toward this region. **B,** A blow-up of the region shows olfactory nerve cells and their hairlike endings protruding into the liquid-coated nasal passage. **C,** A further blow-up shows a receptor site on one of the endings containing an odorous molecule that matches its shape. This particular molecule has a minty odor.

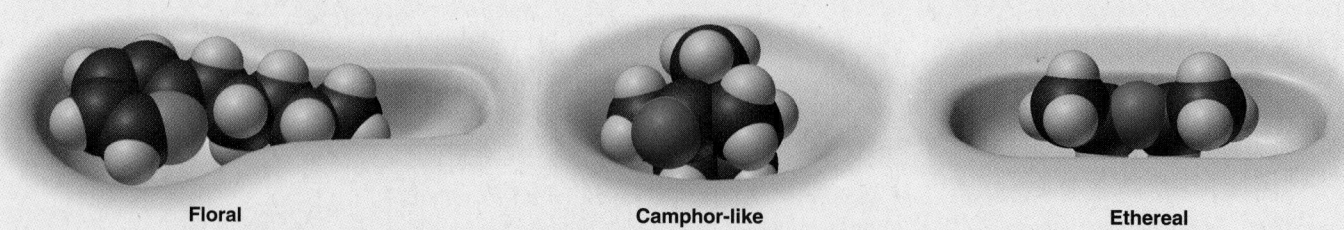

Floral   Camphor-like   Ethereal

Figure B10.2 **Shapes of some olfactory receptor sites.** Three of the seven proposed olfactory receptors are shown with a molecule having that odor occupying the site.

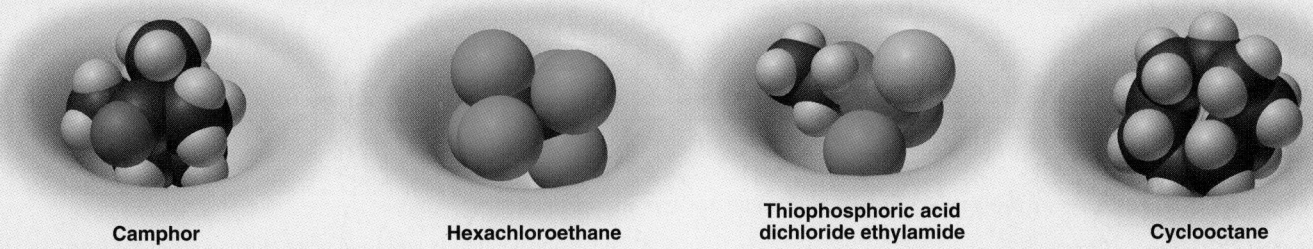

Camphor	Hexachloroethane	Thiophosphoric acid dichloride ethylamide	Cyclooctane

Figure B10.3 Different molecules with the same odor. The theory stating that odor is based on shape, not composition, is supported by these four different substances: all have a camphor-like (moth repellent) odor.

Several predictions of this original version of the theory have been verified by experiment. If two substances fit the same receptors, they should have the same odor, even if their compositions vary. The four molecules in Figure B10.3 all fit the "bowl-shaped" camphor-like receptor and all smell like moth repellent, despite their different formulas. If different portions of one molecule fit different receptors, the molecule should have a mixed odor. Portions of the benzaldehyde molecule fit the camphor-like, floral, and minty receptors, which gives it an almond odor; other molecules that smell like almonds fit the same three receptors.

Despite these and other results consistent with the theory, recent evidence suggests that the original model is too simple. In many cases, a prediction of odor based on molecular shape turns out to be incorrect when the substance is actually smelled. One reason for this discrepancy is that a molecule in the gas phase may have a very different shape in solution at the receptor. By the early 1990s, evidence for the presence of some 1000 receptor sites had been obtained, and it is thought that various combinations of stimulation produce the more than 10,000 odors humans can distinguish. So, although a key premise of the theory is still accepted—odor depends on molecular shape—the nature of the dependence is complex and is an active area of research in the food, cosmetics, and insecticide industries. The sense of smell is so vital for survival that these topics are important areas of study for biologists as well.

Many other biochemical processes are controlled by one molecule fitting into a receptor site on another. Enzymes are proteins that bind cellular reactants and speed their reaction (Figure B10.4). Nerve impulses are transmitted when small molecules released from one nerve fit into receptors on the next. Mind-altering drugs act by chemically disrupting the molecular fit at such nerve receptors in the brain. One type of immune response is triggered when a molecule on a bacterial surface binds to the receptors on "killer" cells in the bloodstream. Hormones regulate energy production and growth by fitting into and activating key receptors. Genes function when certain nucleic acid molecules fit into specific regions of other nucleic acids. Indeed, *no molecular property is more crucial to living systems than shape*.

Figure B10.4 Molecular shape and enzyme action. A small sugar molecule (*below*) is about to move up and fit into the receptor site of an enzyme molecule and undergo reaction.

For Review and Reference (Numbers in parentheses refer to pages, unless noted otherwise.)

Learning Objectives

Relevant section and/or sample problem (SP) numbers appear in parentheses.

Understand These Concepts

1. How Lewis structures depict the atoms and their bonding and nonbonding electron pairs in a molecule or ion (10.1)
2. How resonance and electron delocalization explain bond properties in many compounds with double bonds adjacent to single bonds (10.1)
3. The meaning of formal charge and how it is used to select the most important resonance structure; the difference between formal charge and oxidation number (10.1)
4. The octet rule and its three major exceptions—molecules with an electron deficiency, an odd number of electrons, or an expanded valence shell (10.1)
5. How a reaction can be divided conceptually into bond-breaking and bond-forming steps (10.2)
6. How electron-group repulsions lead to molecular shapes (10.3)
7. The five electron-group arrangements and their associated molecular shapes (10.3)

8. Why double bonds and lone pairs cause deviations from ideal bond angles (10.3)
9. How bond polarities and molecular shape combine to give a molecular polarity (10.4)

Master These Skills

1. Using a stepwise method for writing a Lewis structure from a molecular formula (SPs 10.1 to 10.3 and 10.5)
2. Writing resonance structures for molecules and ions (SP 10.4)
3. Calculating the formal charge of any atom in a molecule or ion (10.1)
4. Using Lewis structures to calculate ΔH^0_{rxn} from bond energies (SP 10.6)
5. Predicting molecular shapes from Lewis structures (SPs 10.7 to 10.9)
6. Using molecular shape and electronegativity values to predict the direction of a molecular dipole (SP 10.10)

Key Terms

Section 10.1

Lewis structure (Lewis formula) (358)
resonance structure (resonance form) (362)
resonance hybrid (362)
electron-pair delocalization (362)
formal charge (364)
electron deficient (365)
free radical (365)

expanded valence shell (366)

Section 10.3

valence-shell electron-pair repulsion (VSEPR) theory (370)
molecular shape (371)
bond angle (371)
linear arrangement (372)
linear shape (372)

trigonal planar arrangement (372)
bent shape (V shape) (373)
tetrahedral arrangement (374)
trigonal pyramidal shape (374)
trigonal bipyramidal arrangement (375)
equatorial group (375)
axial group (375)

seesaw shape (375)
T shape (375)
octahedral arrangement (376)
square pyramidal shape (376)
square planar shape (376)

Section 10.4

molecular polarity (381)
dipole moment (μ) (381)

Key Equations and Relationships

10.1 Calculating the formal charge on an atom (364):

Formal charge of atom

$$= \text{no. of valence } e^- - (\text{no. of unshared valence } e^- + \tfrac{1}{2} \text{ no. of shared valence } e^-)$$

10.2 Calculating heat of reaction from bond energies (368):

$$\Delta H^0_{rxn} = \Delta H^0_{\text{reactant bonds broken}} + \Delta H^0_{\text{product bonds formed}}$$

10.3 Ranking the effect of electron-pair repulsions on bond angle (375):

Lone pair–lone pair > lone pair–bonding pair
> bonding pair–bonding pair

Highlighted Figures and Tables

These figures (F) and tables (T) provide a quick review of key ideas.

F10.1 The steps in writing a Lewis structure (358)
F10.2 Using bond energies to calculate the enthalpy change of a reaction (ΔH^0_{rxn}) (368)
F10.5 Electron-group repulsions and the five basic molecular shapes (371)
F10.6 The single molecular shape of the linear electron-group arrangement (372)
F10.7 The two molecular shapes of the trigonal planar electron-group arrangement (372)

F10.8 The three molecular shapes of the tetrahedral electron-group arrangement (374)
F10.10 The four molecular shapes of the trigonal bipyramidal electron-group arrangement (375)
F10.11 The three molecular shapes of the octahedral electron-group arrangement (376)
F10.12 The steps in determining a molecular shape (377)

Brief Solutions to Follow-up Problems

10.1 (a)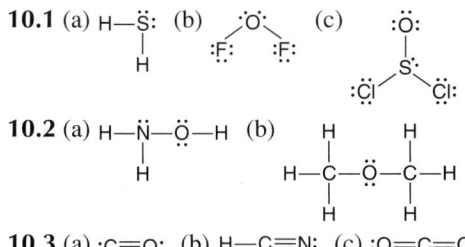

10.2 (a) H—N̈—Ö—H (b)

10.3 (a) :C≡O: (b) H—C≡N: (c) :Ö=C=Ö:

10.4

10.5 (a) (b) (c)

10.6 (a)

:N≡N: + 3 H—H ⟶ 2 :N—H

$\Delta H^0_{\text{bonds broken}} = 1\ N{\equiv}N + 3\ H{-}H$
$= 945\ \text{kJ} + 1296\ \text{kJ} = 2241\ \text{kJ}$
$\Delta H^0_{\text{bonds formed}} = 6\ N{-}H = -2346\ \text{kJ}$
$\Delta H^0_{\text{rxn}} = -105\ \text{kJ}$

(b)

$\Delta H^0_{\text{bonds broken}} = 1\ C{=}C + 4\ C{-}H + 1\ H{-}Br$
$= 614\ \text{kJ} + 1652\ \text{kJ} + 363\ \text{kJ} = 2629\ \text{kJ}$
$\Delta H^0_{\text{bonds formed}} = 1\ C{-}C + 5\ C{-}H + 1\ C{-}Br$
$= -(347\ \text{kJ} + 2065\ \text{kJ} + 276\ \text{kJ}) = -2688\ \text{kJ}$
$\Delta H^0_{\text{rxn}} = -59\ \text{kJ}$

10.7 (a) S̈=C=S̈   (b) 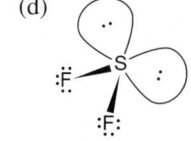 V shaped, <120°

Linear, 180°

(c) 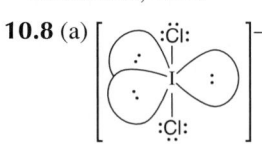 Tetrahedral, 109.5°

(d) 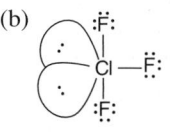 V shaped, <109.5°

10.8 (a) 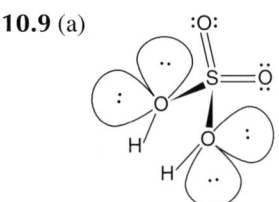 Linear, 180°   (b) T shaped, <90°

(c) 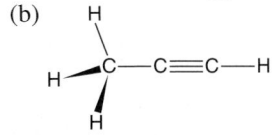   Trigonal bipyramidal, F_{eq}—S—F_{eq} angle <120° and F_{ax}—S—F_{eq} angle <90°.

10.9 (a) 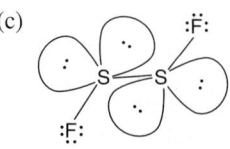   S is tetrahedral; double bonds compress O—S—O angle to <109.5°. Each central O is V shaped; lone pairs compress H—O—S angle to <109.5°.

(b)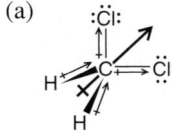

(c)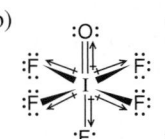

Methyl C is tetrahedral ~109.5°; other C atoms are linear, 180°.

Each S is V shaped; F—S—S angle <109.5°.

10.10

(a) (b) (c)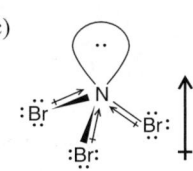

Problems

Problems with **colored** numbers are answered at the back of the text. Sections match the text and provide the number(s) of relevant sample problems. Most offer Concept Review Questions, Skill-Building Exercises (in similar pairs), and Problems in Context. Then Comprehensive Problems, based on material from any section or previous chapter, follow.

Depicting Molecules and Ions with Lewis Structures
(Sample Problems 10.1 to 10.5)

● **Concept Review Questions**

10.1 Which of these atoms *cannot* serve as a central atom in a Lewis structure: (a) O; (b) He; (c) F; (d) H; (e) P? Explain.
10.2 When is a resonance hybrid needed to adequately depict the bonding in a molecule? Using NO_2 as an example, explain

how a resonance hybrid is consistent with the actual bond length, bond strength, and bond order.

10.3 In which of the following bonding patterns does X obey the octet rule?

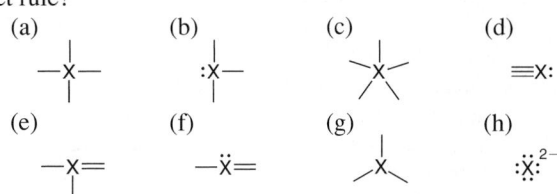

10.4 What is required for an atom to expand its valence shell? Which of the following atoms can expand its valence shell: F, S, H, Al, Se, Cl?

● **Skill-Building Exercises** *(paired)*

10.5 Draw a Lewis structure for (a) SiF_4; (b) $SeCl_2$; (c) COF_2 (C is central).

10.6 Draw a Lewis structure for (a) PH_4^+; (b) C_2F_4; (c) SbH_3.

10.7 Draw a Lewis structure for (a) PF_3; (b) H_2CO_3 (both H atoms are attached to O atoms); (c) CS_2.

10.8 Draw a Lewis structure for (a) CH_4S; (b) S_2Cl_2; (c) $CHCl_3$.

10.9 Draw Lewis structures of all the important resonance forms of (a) NO_2; (b) NO_2F (N is central).

10.10 Draw Lewis structures of all the important resonance forms of (a) HNO_3 ($HONO_2$); (b) $HAsO_4^{2-}$ ($HOAsO_3^{2-}$).

10.11 Draw Lewis structures of all the important resonance forms of (a) N_3^-; (b) NO_2^-.

10.12 Draw Lewis structures of all the important resonance forms of (a) HCO_2^- (H is attached to C); (b) $HBrO_4$ ($HOBrO_3$).

10.13 Draw a Lewis structure and calculate the formal charge of each atom in (a) IF_5; (b) AlH_4^-.

10.14 Draw a Lewis structure and calculate the formal charge of each atom in (a) COS (C is central); (b) NO.

10.15 Draw a Lewis structure and calculate the formal charge of each atom in (a) CN^-; (b) ClO^-.

10.16 Draw a Lewis structure and calculate the formal charge of each atom in (a) BF_4^-; (b) ClNO.

10.17 Draw a Lewis structure for the most important resonance form of each ion, showing formal charges and oxidation numbers of the atoms: (a) BrO_3^-; (b) SO_3^{2-}.

10.18 Draw a Lewis structure for the most important resonance form of each ion, showing formal charges and oxidation numbers of the atoms: (a) AsO_4^{3-}; (b) ClO_2^-.

10.19 The following species do not obey the octet rule. Draw a Lewis structure for each, and state the type of octet rule exception: (a) BH_3; (b) AsF_4^-; (c) $SeCl_4$.

10.20 The following species do not obey the octet rule. Draw a Lewis structure for each, and state the type of octet rule exception: (a) PF_6^-; (b) ClO_3; (c) H_3PO_3 (there is one P—H bond).

10.21 The following species do not obey the octet rule. Draw a Lewis structure for each, and state the type of octet rule exception: (a) BrF_3; (b) ICl_2^-; (c) BeF_2.

10.22 The following species do not obey the octet rule. Draw a Lewis structure for each, and state the type of octet rule exception: (a) O_3^-; (b) XeF_2; (c) SbF_4^-.

● **Problems in Context**

10.23 Molten beryllium chloride reacts with chloride ion from molten NaCl to form the $BeCl_4^{2-}$ ion, in which the Be atom attains an octet. Show the net ionic reaction with Lewis structures.

10.24 Despite many attempts, the perbromate ion (BrO_4^-) was not prepared in the laboratory until about 1970. (Indeed, articles were published explaining theoretically why it could never be prepared!) Draw a Lewis structure for BrO_4^- in which all atoms have lowest formal charges.

10.25 Cryolite (Na_3AlF_6) is an indispensable component in the electrochemical manufacture of aluminum. Draw a Lewis structure for the AlF_6^{3-} ion.

10.26 Phosgene is a colorless, highly toxic gas employed against troops in World War I and used today as a key reactant in organic synthesis. Use formal charges to select the most important of the following resonance structures:

A　　　**B**　　　**C**

Using Lewis Structures and Bond Energies to Calculate Heats of Reaction
(Sample Problem 10.6)

● **Concept Review Questions**

10.27 Write a solution plan (without actual numbers, but including the bond energies you would use and how you would combine them algebraically) for calculating the enthalpy change of the following reaction:

$$H_2(g) + O_2(g) \longrightarrow H_2O_2(g)$$

10.28 The text points out that, for similar types of substances, one with weaker bonds is usually more reactive than one with stronger bonds. Why is this generally true?

10.29 Why is there a discrepancy between a heat of reaction obtained from calorimetry and one obtained from bond energies?

● **Skill-Building Exercises** *(paired)*

10.30 Use bond energies from Table 9.2 (p. 339) to calculate the heat of reaction:

10.31 Use bond energies from Table 9.2 (p. 339) to calculate the heat of reaction:

10.32 Use Lewis structures and bond energies to calculate the enthalpy change of the following reaction:

$$C_2H_4(g) + H_2O(g) \longrightarrow CH_3CH_2OH(g)$$

10.33 Use Lewis structures and bond energies to calculate the enthalpy change of the following reaction:

$$HCN(g) + 2H_2(g) \longrightarrow CH_3NH_2(g)$$

● **Problems in Context**

10.34 An important industrial route to extremely pure acetic acid is the reaction of methanol with carbon monoxide. Calculate the heat of reaction for this process.

10.35 Acetylene gas (C_2H_2) burns with oxygen in an oxyacetylene torch to produce carbon dioxide, water vapor, and the great heat needed for welding metals. The heat of combustion of acetylene is -1259 kJ/mol. Estimate the bond energy of the $C\equiv C$ bond, and compare your value with that in Table 9.2 (p. 339).

Valence-Shell Electron-Pair Repulsion (VSEPR) Theory and Molecular Shape
(Sample Problems 10.7 to 10.9)

● **Concept Review Questions**

10.36 If you know the formula of a molecule or ion, what is the first step in predicting its shape?

10.37 In what situation is the name of the molecular shape the same as the name of the electron-group arrangement?

10.38 Which of the following numbers of electron groups can give rise to a bent (V-shaped) molecule: two, three, four, five, six? Draw an example for each case, showing the shape classification (AX_mE_n) and the ideal bond angle.

10.39 Name all the molecular shapes that have a tetrahedral electron-group arrangement.

10.40 Why aren't lone pairs considered along with surrounding bonded groups when describing the molecular shape?

10.41 Use wedge-line perspective drawings (if necessary) to sketch the atom positions in a general molecule of formula (not shape class) AX_n that has each of the following shapes: (a) V shaped; (b) trigonal planar; (c) trigonal bipyramidal; (d) T shaped; (e) trigonal pyramidal; (f) square pyramidal.

10.42 What would you expect to be the electron-group arrangement around atom A in each of the following cases? For each arrangement, give the ideal bond angle and the direction of any expected deviation:

(a)
```
     X
     |
  X—A:
     |
     X
```
(b) X—A≡X

(c)
```
     X
     ‖
  X=A—X
```
(d) X—Ä—X

(e) X=A=X

(f)
```
      ⟋X
  :A
   |⟍X
   X
```

● **Skill-Building Exercises** *(paired)*

10.43 Determine the electron-group arrangement, molecular shape, and ideal bond angle(s) for each of the following: (a) O_3; (b) H_3O^+; (c) NF_3.

10.44 Determine the electron-group arrangement, molecular shape, and ideal bond angle(s) for each of the following: (a) SO_4^{2-}; (b) NO_2^-; (c) PH_3.

10.45 Determine the electron-group arrangement, molecular shape, and ideal bond angle(s) for each of the following: (a) CO_3^{2-}; (b) SO_2; (c) CF_4.

10.46 Determine the electron-group arrangement, molecular shape, and ideal bond angle(s) for each of the following: (a) SO_3; (b) N_2O (N is central); (c) CH_2Cl_2.

10.47 Name the shape and give the AX_mE_n classification and ideal bond angle(s) for each of the following general molecules:

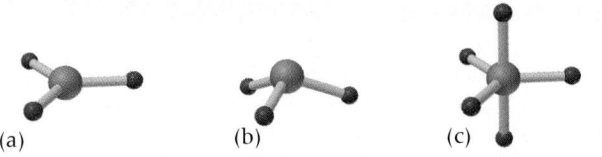

(a)   (b)   (c)

10.48 Name the shape and give the AX_mE_n classification and ideal bond angle(s) for each of the following general molecules:

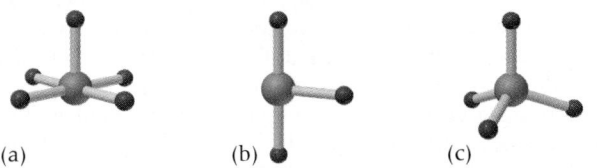

(a)   (b)   (c)

10.49 Determine the shape, ideal bond angle(s), and the direction of any deviation from these angles for each of the following: (a) ClO_2^-; (b) PF_5; (c) SeF_4; (d) KrF_2.

10.50 Determine the shape, ideal bond angle(s), and the direction of any deviation from these angles for each of the following: (a) ClO_3^-; (b) IF_4^-; (c) $SeOF_2$; (d) TeF_5^-.

10.51 Determine the shape around each central atom in each molecule, and explain any deviation from ideal bond angles: (a) CH_3OH; (b) N_2O_4 (O_2NNO_2).

10.52 Determine the shape around each central atom in each molecule, and explain any deviation from ideal bond angles: (a) H_3PO_4 (no H—P bond); (b) CH_3—O—CH_2CH_3.

10.53 Determine the shape around each central atom in each molecule, and explain any deviation from ideal bond angles: (a) CH_3COOH; (b) H_2O_2.

10.54 Determine the shape around each central atom in each molecule, and explain any deviation from ideal bond angles: (a) H_2SO_3 (no H—S bond); (b) N_2O_3 ($ONNO_2$).

10.55 Arrange the following AF_n species in order of *increasing* F—A—F bond angles: BF_3, BeF_2, CF_4, NF_3, OF_2.

10.56 Arrange the following ACl_n species in order of *decreasing* Cl—A—Cl bond angles: SCl_2, OCl_2, PCl_3, $SiCl_4$, $SiCl_6^{2-}$.

10.57 State ideal values for each of the bond angles in each molecule, and note where you expect deviations:

(a)
```
   H
   |
H—C=N—Ö—H
```

(b)
```
   H   H
   |   |
H—C—Ö—C—H
   |   |
   H   H
```

(c)
```
      H
      |
     :O:
      |
H—Ö—B—Ö—H
```

10.58 State ideal values for each of the bond angles in each molecule, and note where you expect deviations:

(a)
```
Ö=N—Ö—H
   |
  :O:
```

(b)
```
   H
   |
H—C—C=Ö
   | |
   H H
```

(c)
```
  :O:
   ‖
H—C—Ö—H
```

● **Problems in Context**

10.59 Because both tin and carbon are members of Group 4A(14), they form structurally similar compounds. However, tin exhibits a greater variety of structures because it forms several ionic species. Predict the shapes and ideal bond angles, including any deviations, for the following: (a) $Sn(CH_3)_2$; (b) $SnCl_3^-$; (c) $Sn(CH_3)_4$; (d) SnF_5^-; (e) SnF_6^{2-}.

10.60 In the gas phase, phosphorus pentachloride exists as separate molecules. In the solid phase, however, the compound is composed of alternating PCl_4^+ and PCl_6^- ions. What change(s) in molecular shape occur(s) as PCl_5 solidifies? How does the Cl—P—Cl angle change?

Molecular Shape and Molecular Polarity
(Sample Problem 10.10)

● **Concept Review Questions**

10.61 For molecules of general formula AX_n (where $n > 2$), how do you determine if a molecule is polar?

10.62 How can a molecule with polar covalent bonds fail to be a polar molecule? Give an example.

10.63 Explain in general why the shape of a biomolecule is important to its function.

● **Skill-Building Exercises** *(paired)*

10.64 Consider the molecules SCl_2, F_2, CS_2, CF_4, and $BrCl$.
(a) Which has bonds that are the most polar?
(b) Which have a molecular dipole moment?
10.65 Consider the molecules BF_3, PF_3, BrF_3, SF_4, and SF_6.
(a) Which has bonds that are the most polar?
(b) Which have a molecular dipole moment?

10.66 Which molecule in each pair has the greater dipole moment: (a) SO_2 or SO_3; (b) ICl or IF; (c) SiF_4 or SF_4; (d) H_2O or H_2S? Give the reason for your choice.
10.67 Which molecule in each pair has the greater dipole moment: (a) ClO_2 or SO_2; (b) HBr or HCl; (c) $BeCl_2$ or SCl_2; (d) AsF_3 or AsF_5? Give the reason for your choice.

● **Problems in Context**

10.68 There are three different dichloroethylenes (molecular formula $C_2H_2Cl_2$), which we can designate X, Y, and Z. Compound X has no dipole moment, but compound Z does. Compounds X and Z each combine with hydrogen to give the same product:

$$C_2H_2Cl_2 \text{ (X or Z)} + H_2 \longrightarrow ClCH_2-CH_2Cl$$

What are the structures of X, Y, and Z? Would you expect compound Y to have a dipole moment?
10.69 Dinitrogen difluoride, N_2F_2, is the only stable, simple inorganic molecule with an $N=N$ bond. The compound occurs in *cis* and *trans* forms.
(a) Draw the molecular shapes of the two forms of N_2F_2.
(b) Predict the polarity, if any, of each form.

Comprehensive Problems

Problems with an asterisk (*) are more challenging.
10.70 In addition to ammonia, nitrogen forms three other hydrides: hydrazine (N_2H_4), diazene (N_2H_2), and tetrazene (N_4H_4).
(a) Use Lewis structures to compare the strength, length, and order of nitrogen-nitrogen bonds in hydrazine, diazene, and N_2.
(b) Tetrazene (atom sequence H_2NNNNH_2) decomposes above 0°C to hydrazine and nitrogen gas. Draw a Lewis structure for tetrazene, and calculate ΔH_{rxn}^0 for this decomposition.
10.71 Draw a Lewis structure for each species: (a) PF_5; (b) CCl_4; (c) H_3O^+; (d) ICl_3; (e) BeH_2; (f) PH_2^-; (g) $GeBr_4$; (h) CH_3^-; (i) BCl_3; (j) BrF_4^+; (k) XeO_3; (l) TeF_4.
10.72 Determine the molecular shape of each of the species in Problem 10.71.
10.73 Both aluminum and iodine form chlorides, Al_2Cl_6 and I_2Cl_6, with "bridging" Cl atoms. Lewis structures, showing lone pairs on the Cl atoms only, are

(a) What is the formal charge on each atom? (b) Which of these molecules has a planar shape? Explain.
10.74 Nitrosyl fluoride (NOF) has an atom sequence in which all atoms have formal charges of zero. Write the Lewis structure consistent with this fact.
10.75 The VSEPR model was developed in the 1950s, before any xenon compounds had been prepared. Thus, in the early 1960s, these compounds provided an excellent test of the model's predictive power. What would you have predicted for the shapes of XeF_2, XeF_4, and XeF_6?

10.76 Boron trifluoride is an important reactant in organic syntheses but is a very reactive gas at room temperature and thus difficult to handle. Boron's ability to accept another pair of electrons and attain an octet is used to prepare an easily stored liquid form: a compound of BF_3 and diethyl ether ($CH_3-CH_2-O-CH_2-CH_3$). Draw Lewis structures for BF_3, diethyl ether, and the product, and indicate how the molecular shapes around B and O change during the reaction.
10.77 The actual bond angle in NO_2 is 134.3°, and in NO_2^- it is 115.4°, although the ideal bond angle is 120° in both. Explain.
10.78 "Inert" xenon actually forms several compounds, especially with the highly electronegative elements oxygen and fluorine. The simple fluorides XeF_2, XeF_4, and XeF_6 are all formed by direct reaction of the elements. As you might expect from the size of the xenon atom, the Xe—F bond is not a strong one. Calculate the Xe—F bond energy in XeF_6, given that the heat of formation is -402 kJ/mol.
10.79 Propylene oxide is used to make many products, including plastics such as polyurethane. One method for synthesizing it involves oxidizing propene with hydrogen peroxide:

(a) What is the molecular shape and ideal bond angle around each carbon atom in propylene oxide? (b) Predict any deviation from the ideal for the actual bond angle (assume the three atoms in the ring form an equilateral triangle).
10.80 Chloral, $Cl_3C-CH=O$, reacts with water to form the sedative and hypnotic agent chloral hydrate, $Cl_3C-CH(OH)_2$. Draw Lewis structures for these substances, and describe the change in molecular shape, if any, that occurs around each of the carbon atoms during the reaction.
10.81 Dichlorine heptaoxide, Cl_2O_7, can be viewed as two ClO_4 groups sharing an O atom. Draw a Lewis structure for Cl_2O_7 with the lowest formal charges, and predict any deviation from the ideal for the Cl—O—Cl bond angle.
*10.82 Like several other bonds, carbon-oxygen bonds have lengths and strengths that depend on the bond order. Draw Lewis structures for the following species, and arrange them in order of increasing carbon-oxygen bond length and then by increasing carbon-oxygen bond strength: (a) CO; (b) CO_3^{2-}; (c) H_2CO; (d) CH_4O; (e) HCO_3^- (H attached to O).
10.83 In the 1980s, there was an international agreement to destroy all stockpiles of mustard gas, $ClCH_2CH_2SCH_2CH_2Cl$. When this substance contacts the moisture in eyes, nasal passages, and skin, the OH groups in water replace the Cl atoms and create high local concentrations of hydrochloric acid, which cause severe blistering and tissue destruction. Write a balanced equation for this reaction, and calculate ΔH_{rxn}^0.
10.84 The four bonds of carbon tetrachloride (CCl_4) are polar, but the molecule is nonpolar because the bond polarity is canceled by the symmetric tetrahedral shape. When other atoms substitute for some of the Cl atoms, the symmetry is broken and the molecule becomes polar. Use Figure 9.16 to rank the following molecules from the least polar to the most polar: CH_2Br_2, CF_2Cl_2, CH_2F_2, CH_2Cl_2, CBr_4, CF_2Br_2.
*10.85 Ethanol (CH_3CH_2OH) is being used as a gasoline additive or alternative in many parts of the world.
(a) Use bond energies to calculate the heat of combustion of gaseous ethanol.

(b) In its standard state at 25°C, ethanol is a liquid. Its vaporization requires 40.5 kJ/mol. Correct the value from part (a) to find the heat of combustion of liquid ethanol.

(c) How does the value from part (b) compare with the value you calculate from standard heats of formation (Appendix B)?

*10.86 In the following compounds, the C atoms form a single ring. Draw a Lewis structure for each, identify cases for which resonance exists, and determine the carbon-carbon bond order(s): (a) C_3H_4; (b) C_3H_6; (c) C_4H_6; (d) C_4H_4; (e) C_6H_6.

10.87 An oxide of nitrogen is 25.9% N by mass, has a molar mass of 108 g/mol, and contains no nitrogen-nitrogen nor oxygen-oxygen bonds. Draw its Lewis structure, and name it.

*10.88 An experiment requires 50.0 mL of 0.040 M NaOH for the titration of 1.00 mmol of acid. Mass analysis of the acid shows 2.24% hydrogen, 26.7% carbon, and 71.1% oxygen. Draw the Lewis structure of the acid.

*10.89 A gaseous compound has a composition by mass of 24.8% carbon, 2.08% hydrogen, and 73.1% chlorine. At STP, the gas has a density of 4.3 g/L. Draw a Lewis structure that satisfies these facts. Would another structure also satisfy them? Explain.

10.90 Perchlorates are powerful oxidizing agents used in fireworks, flares, and the booster rockets of the space shuttle. Lewis structures for the perchlorate ion (ClO_4^-) can be drawn with all single bonds or with one, two, or three double bonds. Draw each of these possible resonance forms, use formal charges to determine the most important, and calculate its average bond order.

10.91 Use Lewis structures to determine which *two* of the following are unstable: (a) SF_2; (b) SF_3; (c) SF_4; (d) SF_5; (e) SF_6.

*10.92 A major short-lived, neutral species in flames is OH.

(a) What is unusual about the electronic structure of OH?

(b) Use the standard heat of formation of OH(g) and bond energies to calculate the O—H bond energy in OH(g) (ΔH_f^0 of OH(g) = 39.0 kJ/mol).

(c) From the average value for the O—H bond energy in Table 9.2 (p. 339) and your value for the O—H bond energy in OH(g), find the energy needed to break the first O—H bond in water.

10.93 Pure HN_3 (atom sequence HNNN) is a very explosive compound. In aqueous solution, it is a weak acid (comparable to acetic acid) that yields the azide ion, N_3^-. Draw resonance structures to explain why the nitrogen-nitrogen bond lengths are equal in N_3^- but unequal in HN_3.

10.94 Except for nitrogen, the elements of Group 5A(15) all form pentafluorides, and most form pentachlorides. The chlorine atoms of PCl_5 can be replaced with fluorine atoms one at a time to give, successively, PCl_4F, PCl_3F_2, . . . , PF_5.

(a) Given the sizes of F and Cl, would you expect the first two F substitutions to be at axial or equatorial positions? Explain.

(b) Which of the five fluorine-containing molecules have no dipole moment?

10.95 Dinitrogen monoxide (N_2O), used as the anesthetic "laughing gas" in dental surgery, supports combustion in a manner similar to oxygen, with the nitrogen atoms forming N_2. Draw three resonance structures for N_2O (one N is central) and use formal charges to decide the relative importance of each. What correlation can you suggest between the most important structure and the observation that N_2O supports combustion?

10.96 Oxalic acid ($H_2C_2O_4$) is found in toxic concentrations in rhubarb leaves. The acid forms two ions, $HC_2O_4^-$ and $C_2O_4^{2-}$, by the sequential loss of H^+ ions. Draw Lewis structures for the three species, and comment on the relative lengths and strengths

of their carbon-oxygen bonds. The connections among the atoms are shown below with single bonds only.

10.97 Some scientists speculate that many organic molecules required for life on the young Earth arrived on meteorites. The Murchison meteorite that landed in Australia in 1969 contained 92 different amino acids, including 21 found in Earth organisms. A skeleton structure (single bonds only) of one of these extraterrestrial amino acids is

Draw a Lewis structure, and identify any atoms with a nonzero formal charge.

10.98 Hydrazine (N_2H_4) is used as a rocket fuel because it reacts very exothermically with oxygen to form nitrogen gas and water vapor. The heat released and the increase in number of moles of gas provide thrust. Calculate the heat of reaction.

10.99 In addition to propyne (see Follow-up Problem 10.9), there are two other structures with molecular formula C_3H_4. Draw a Lewis structure for each, determine the shape around each carbon, and predict any deviations from ideal bond angles.

10.100 A molecule of formula AY_3 is found experimentally to be polar. Which molecular shapes are possible and which impossible for AY_3?

*10.101 In contrast to the cyanate ion (NCO^-), which is stable and found in many compounds, the fulminate ion (CNO^-), with its different atom sequence, is unstable and forms compounds with heavy metal ions, such as Ag^+ and Hg^{2+}, that are explosive. Like the cyanate ion, the fulminate ion has three resonance structures. Which is the most important contributor to the resonance hybrid? Suggest a reason for the instability of fulminate.

10.102 Ethylene, C_2H_4, and tetrafluoroethylene, C_2F_4, are used to make the polymers polyethylene and polytetrafluoroethylene (Teflon), respectively.

(a) Draw the Lewis structures for C_2H_4 and C_2F_4, and give the ideal H—C—H and F—C—F bond angles.

(b) The actual H—C—H and F—C—F bond angles are 117.4° and 112.4°, respectively. Explain these deviations.

10.103 Lewis structures of mescaline, a hallucinogenic compound in peyote cactus, and dopamine, a neurotransmitter in mammalian brain, appear below. Suggest a reason for mescaline's ability to disrupt nerve impulses.

mescaline   dopamine

CHAPTER 11

THEORIES OF COVALENT BONDING

CHAPTER OUTLINE

11.1 Valence Bond (VB) Theory and Orbital Hybridization
The Central Themes of VB Theory
Types of Hybrid Orbitals

11.2 The Mode of Orbital Overlap and the Types of Covalent Bonds
VB Treatment of Single and Multiple Bonds
Orbital Overlap and Molecular Rotation

11.3 Molecular Orbital (MO) Theory and Electron Delocalization
The Central Themes of MO Theory
Homonuclear Diatomic Molecules of the Period 2 Elements
Some Heteronuclear Diatomic Molecules
MO Descriptions of Ozone and Benzene

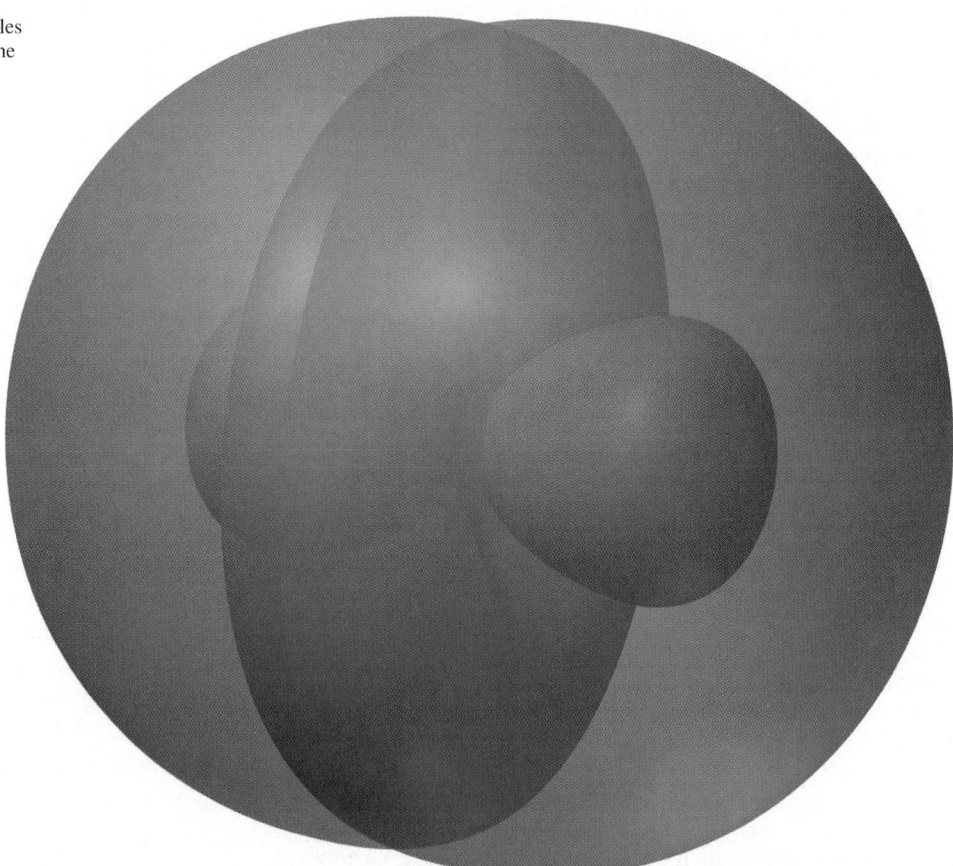

Figure: Bonding theory visualized. How do molecular shapes emerge from interacting atoms? One theory proposes that new types of orbitals arise during bonding, such as these *sp* hybrid orbitals of the central beryllium in $BeCl_2$. In this chapter, we explore two models that rationalize the properties of molecules and their covalent bonds.

Ill scientific models have limitations because they are simplifications of reality. The VSEPR model accounts for molecular shapes by assuming that electron groups tend to minimize their repulsions, which results in their occupying as
much space as possible around a central atom. But it does *not* tell us how those
shapes, which we observe experimentally, can be explained from the interactions
of atomic orbitals. After all, none of the orbitals we examined in Chapter 7 are
oriented toward the vertices of a tetrahedron or a trigonal bipyramid, to mention
just two of the common molecular shapes. Moreover, even though the shape of a
molecule is crucial to many aspects of its behavior, the shape does not help us
interpret its magnetic or spectral properties; only an understanding of its orbitals
and their energy levels can do that.

To understand these factors, we apply a quantum-mechanical model, which
has been so successful in explaining atomic properties, to the properties of molecules as well. In this chapter, we discuss two theories of covalent bonding that
are based on quantum mechanics. Valence bond (VB) theory explains the interactions of atomic orbitals that create covalent bonds and shows how observed
molecular shapes are rationalized by these interactions. We also apply VB theory
to a description of the two types of covalent bonds. Molecular orbital (MO) theory explains molecular energy levels and associated properties by proposing the
existence of orbitals that extend over the whole molecule. Each theory comple-
ments the other and is indispensable to a full understanding of covalent bonding.

Don't be discouraged by our need for more than one model to explain the
observations in a topic as wide-ranging as chemical bonding. In every science,
one model accounts for a particular aspect of a topic better than another, and several models are called into service to explain a broader range of phenomena.

11.1 VALENCE BOND (VB) THEORY AND ORBITAL HYBRIDIZATION

What *is* a covalent bond, and what characteristic gives it strength? Most importantly, how can we explain the *molecular* shapes we see based on the interactions
of *atomic* orbitals? The most useful approach for answering these questions is
valence bond (VB) theory.

The Central Themes of VB Theory

The basic principle of VB theory is that *a covalent bond forms when the orbitals
of two atoms overlap and are occupied by a pair of electrons that have the highest probability of being located between the nuclei.* Three central themes of VB
theory derive from this principle:

1. *Opposing spins of the electron pair.* As the exclusion principle (Section
8.2) prescribes, the space formed by the overlapping orbitals *has a maximum
capacity of two electrons that must have opposite spins.* When a molecule of H_2
forms, for instance, the two 1s electrons of two H atoms occupy the overlapping
1s orbitals and have opposite spins (Figure 11.1A).

2. *Maximum overlap of bonding orbitals.* The bond strength depends on the
attraction of the nuclei for the shared electrons, so *the greater the orbital overlap, the stronger (more stable) the bond.* The extent of overlap depends on the
shape and direction of the orbitals involved. An s orbital is spherical, but p and
d orbitals have more electron density in one direction than in another, so a bond
involving p or d orbitals tends to be oriented in the direction that maximizes overlap. In the HF bond, for example, the 1s orbital of H overlaps the half-filled 2p
orbital of F *along the long axis* of that orbital (Figure 11.1B). Any other direction would result in less overlap and, thus, a weaker bond. Similarly, in the F—F
bond of F_2, the two 2p orbitals interact end to end, that is, *along the long axes*
of the orbitals, to maximize overlap (Figure 11.1C).

CONCEPTS & SKILLS

to review before you study this chapter
- atomic orbital shapes (Section 7.4)
- the exclusion principle (Section 8.2)
- Hund's rule (Section 8.3)
- writing Lewis structures (Section 10.1)
- resonance in covalent bonding (Section 10.1)
- molecular shapes (Section 10.3)
- molecular polarity (Section 10.4)

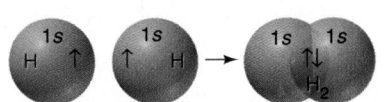

A Hydrogen, H_2

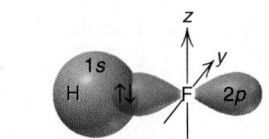

B Hydrogen fluoride, HF

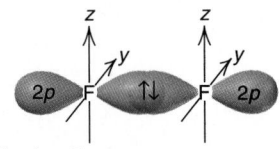

C Fluorine, F_2

**Figure 11.1 Orbital overlap and
spin pairing in three diatomic molecules. A,** In the H_2 molecule, the two
overlapping 1s orbitals are occupied
by the two 1s electrons with opposite
spins. (The electrons, shown as arrows, spend most time between the
nuclei but move throughout the overlapping orbitals.) **B,** To maximize overlap in HF, half-filled H 1s and F 2p
orbitals overlap along the long axis of
the 2p orbital involved in bonding.
(The $2p_x$ orbital is shown bonding; the
other two 2p orbitals of F are not
shown.) **C,** In F_2, the half-filled $2p_x$ orbital on one F points end to end to-
ward the similar orbital on the other F
to maximize overlap.

3. *Hybridization of atomic orbitals*. To account for the bonding in simple diatomic molecules such as HF, we picture the direct overlap of *s* and *p* orbitals of isolated atoms. But how can we account for the shapes observed in so many molecules and larger polyatomic ions through the overlap of spherical *s* orbitals, dumbbell-shaped perpendicular *p* orbitals, and cloverleaf-shaped *d* orbitals?

Consider a methane molecule. It is composed of four H atoms bonded to a central C atom. An isolated ground-state C atom ([He] $2s^2 2p^2$) has four valence electrons: two in the $2s$ orbital and one each in two of the three $2p$ orbitals. We might easily see how the two half-filled *p* orbitals of C could overlap with the $1s$ orbitals of two H atoms to form *two* C—H bonds with a 90° H—C—H bond angle. But how can these orbitals overlap to form the *four* identical C—H bonds with the 109.5° bond angle that occurs in methane?

Linus Pauling answered such questions by proposing that *the valence atomic orbitals in the molecule are **different** from those in the isolated atoms*. Quantum-mechanical calculations show that mathematical mixing of certain combinations of orbitals in a given atom gives rise to *new atomic orbitals*. The spatial orientations of these new orbitals lead to more stable bonds and are consistent with the observed molecular shapes. The process of orbital mixing is called **hybridization,** and the new atomic orbitals are called **hybrid orbitals.** Two key points about the number and type of hybrid orbitals are that

- The *number* of hybrid orbitals obtained *equals* the number of atomic orbitals mixed.
- The *type* of hybrid orbitals obtained *varies* with the types of atomic orbitals mixed.

You can imagine hybridization as a process in which atomic orbitals mix, hybrid orbitals form, and electrons enter them with spins parallel (Hund's rule) to create stable bonds. In truth, though, hybridization is a mathematically derived result from quantum mechanics that accounts for the molecular shapes we observe.

Types of Hybrid Orbitals

We postulate the presence of a certain type of hybrid orbital *after* we observe the molecular shape. As we discuss the five common types of hybridization, notice that the spatial orientation of each type of hybrid orbital corresponds with one of the five common electron-group arrangements predicted by VSEPR theory.

***sp* Hybridization** When two electron groups surround the central atom, we observe a linear shape, which means that the bonding orbitals must have a linear orientation. VB theory explains this by proposing that mixing two *nonequivalent* orbitals of a central atom, one *s* and one *p*, gives rise to two *equivalent* ***sp* hybrid orbitals** that lie 180° apart, as shown in Figure 11.2A. Note the shape of the hybrid orbital: with one large and one small lobe, it differs markedly from the shapes of the atomic orbitals that were mixed. The orientations of hybrid orbitals extend electron density in the bonding direction and minimize repulsions between the electrons that occupy them. Thus, *both shape and orientation maximize overlap with the orbital of the other atom in the bond*.

In gaseous $BeCl_2$, for example, the Be atom is said to be *sp* hybridized. The rest of Figure 11.2 depicts the hybridization of Be in $BeCl_2$ in terms of a traditional orbital box diagram of the valence level (part B), a box diagram with orbital contours (C), and the bond formation with Cl (D). The $2s$ and one of the three $2p$ orbitals of Be mix and form two *sp* orbitals. These overlap $3p$ orbitals of two Cl atoms, and the four valence electrons—two from Be and one from each Cl—occupy the overlapped orbitals in pairs with opposite spins. The two unhybridized $2p$ orbitals of Be lie perpendicular to each other and to the bond axes. Thus, through hybridization, the paired $2s$ electrons in the isolated Be atom are distributed into two *sp* orbitals, which form two Be—Cl bonds.

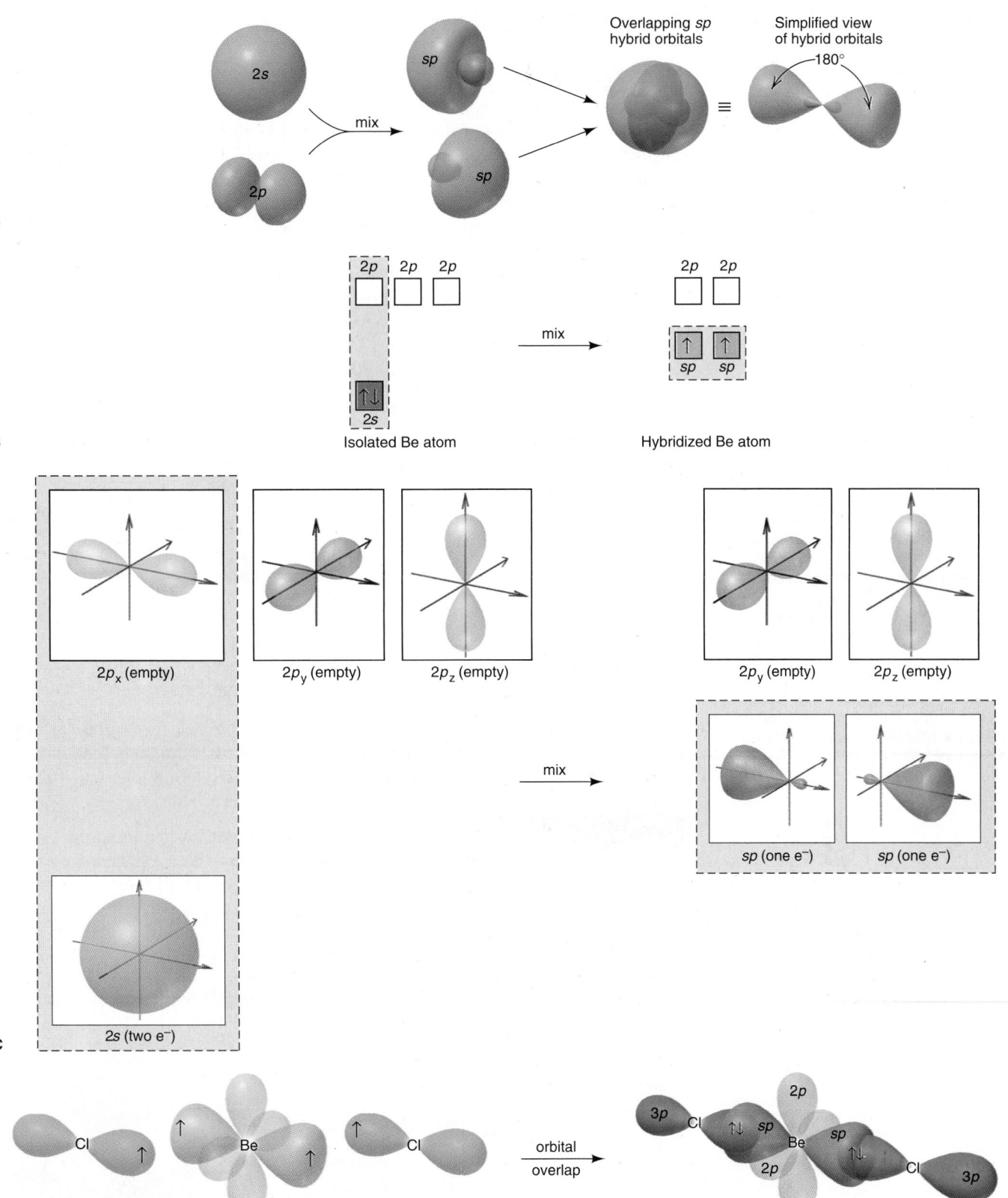

Figure 11.2 **The *sp* hybrid orbitals in gaseous BeCl₂. A,** One 2*s* and one 2*p* atomic orbital mix to form two *sp* hybrid orbitals (shown above and below one another and slightly to the side for ease of viewing). Note the large and small lobes of the hybrid orbitals. In the molecule, the two *sp* orbitals of Be are oriented in opposite directions. For clarity, the simplified hybrid orbitals will be used throughout the text, usually without the small lobe. **B,** The orbital box diagram for hybridization in Be is drawn vertically and shows that the 2*s* and one

of the three 2*p* orbitals form two *sp* hybrid orbitals, and the two other 2*p* orbitals remain unhybridized. Electrons half-fill the *sp* hybrid orbitals. During bonding, each *sp* orbital fills by sharing an electron from Cl (not shown). **C,** The orbital box diagram is shown with orbital contours instead of electron arrows. **D,** BeCl₂ forms by overlap of the two *sp* hybrids with the 3*p* orbitals of two Cl atoms; the two unhybridized Be 2*p* orbitals lie perpendicular to the *sp* hybrids. (For clarity, only the 3*p* orbital involved in bonding is shown for each Cl.)

sp² **Hybridization** In order to rationalize the trigonal planar electron-group arrangement and the shapes of molecules based on it, we introduce the mixing of one *s* and two *p* orbitals of the central atom to give three hybrid orbitals that point toward the vertices of an equilateral triangle, their axes 120° apart. These are called *sp²* **hybrid orbitals.** Note that, unlike electron configuration notation, hybrid orbital notation uses superscripts for the *number* of atomic orbitals of a given type that are mixed, *not* for the number of electrons in the orbital: here, one *s* and two *p* orbitals were mixed, so we have s^1p^2, or sp^2.

VB theory proposes that the central B atom in the BF_3 molecule is sp^2 hybridized. Figure 11.3 shows the three sp^2 orbitals in the trigonal plane, with the third 2*p* orbital unhybridized and perpendicular to this plane. Each sp^2 orbital overlaps the 2*p* orbital of an F atom, and the six valence electrons—three from B and one from each of the three F atoms—form three bonding pairs.

Figure 11.3 **The *sp²* hybrid orbitals in BF₃. A,** The orbital box diagram shows that the 2*s* and two of the three 2*p* orbitals of the B atom mix to make three *sp²* hybrid orbitals. Three electrons (shown as upward arrows) half-fill the *sp²* hybrids. The third 2*p* orbital remains empty and unhybridized. **B,** BF₃ forms through overlap of 2*p* orbitals on three F atoms with the *sp²* hybrids. During bonding, each *sp²* orbital fills by addition of an electron from F (shown as a downward arrow). The three *sp²* hybrids of B lie 120° apart, and the unhybridized 2*p* orbital is perpendicular to the trigonal bonding plane.

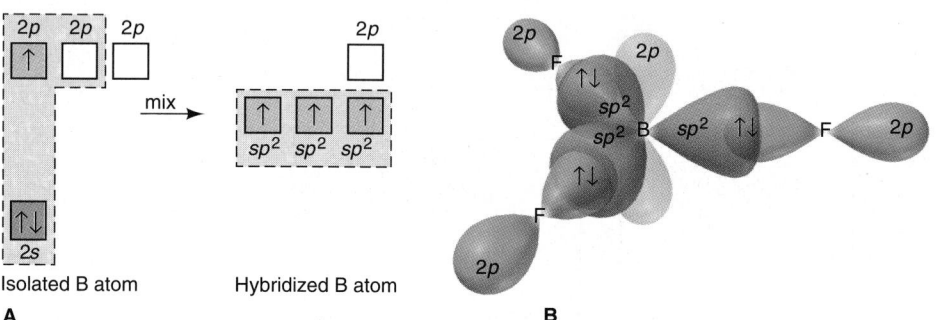

To account for other molecular shapes within a given electron-group arrangement, we postulate that one or more of the hybrid orbitals contains lone pairs. In ozone (O_3), for example, the central O is sp^2 hybridized and a lone pair fills one of its three sp^2 orbitals, so ozone has a bent molecular shape.

sp³ **Hybridization** Now let's return to the question posed earlier about the orbitals in methane, the same question that arises for any species with a tetrahedral electron-group arrangement. VB theory proposes that the one *s* and all three *p* orbitals of the central atom mix and form four *sp³* **hybrid orbitals,** which point toward the vertices of a tetrahedron. As shown in Figure 11.4, the C atom in methane is sp^3 hybridized. Its four valence electrons half-fill the four sp^3 hybrids, which overlap the half-filled 1*s* orbitals of four H atoms and form four C—H bonds.

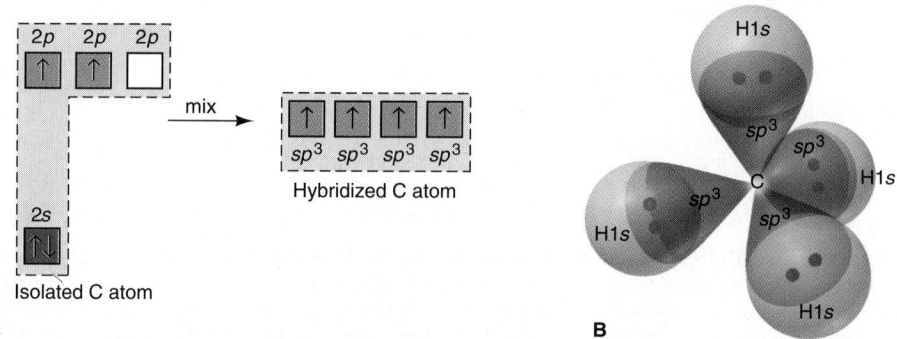

Figure 11.4 **The *sp³* hybrid orbitals in CH₄. A,** The 2*s* and all three 2*p* orbitals of C are mixed to form four *sp³* hybrids. Carbon's four valence electrons half-fill the *sp³* hybrids. **B,** In methane, the four *sp³* orbitals of C point toward the corners of a tetrahedron and overlap the 1*s* orbitals of four H atoms. Each *sp³* orbital fills by addition of an electron from H (shown as dots).

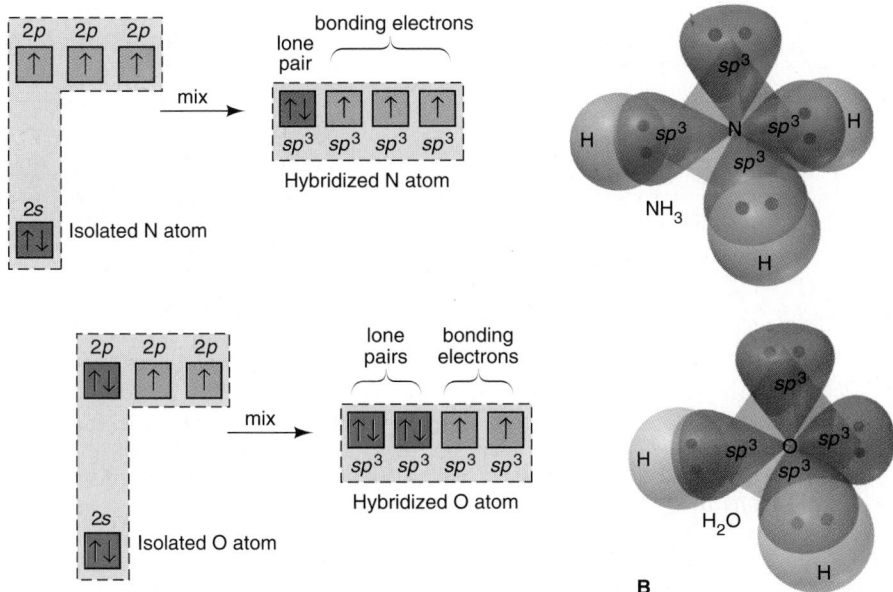

Figure 11.5 The sp^3 hybrid orbitals in NH₃ and H₂O. A, The orbital box diagrams show sp^3 hybridization, as in CH₄. In NH₃ *(top)*, one of the sp^3 orbitals is filled with a lone pair. In H₂O *(bottom)*, two of the sp^3 orbitals are filled with lone pairs. **B,** Contour diagrams show the tetrahedral orientation of the sp^3 orbitals and the overlap of the bonded H atoms. Each half-filled sp^3 orbital fills by addition of an electron from H. (Shared pairs and lone pairs are shown as dots.)

Figure 11.5 shows the bonding in molecules of the other shapes with the tetrahedral arrangement. The trigonal pyramidal shape of NH₃ arises when a lone pair fills one of the four sp^3 orbitals of N, and the bent shape of H₂O arises when lone pairs fill two of the sp^3 orbitals of O.

sp^3d Hybridization The shapes of molecules with trigonal bipyramidal or octahedral electron-group arrangements are rationalized with VB theory through similar arguments. The only new point is that such molecules have central atoms from Period 3 or higher, so atomic *d* orbitals, as well as *s* and *p* orbitals, are mixed to form the hybrid orbitals.

To rationalize the trigonal bipyramidal shape of the PCl₅ molecule, for example, the VB model proposes that the one 3*s*, the three 3*p*, and one of the five 3*d* orbitals of the central P atom mix and form five **sp^3d hybrid orbitals,** which point to the vertices of a trigonal bipyramid. Each hybrid orbital overlaps a 3*p* orbital of a Cl atom, and the five valence electrons of P, together with one from each of the five Cl atoms, pair up to form five P—Cl bonds, as shown in Figure 11.6. Other shapes in this electron-group arrangement have lone pairs in one or more of the central atom's sp^3d orbitals.

Figure 11.6 The sp^3d hybrid orbitals in PCl₅. A, The orbital box diagram shows that one 3*s*, three 3*p*, and one of the five 3*d* orbitals of P mix to form five sp^3d orbitals that are half-filled. Four 3*d* orbitals are unhybridized and empty. **B,** The trigonal bipyramidal PCl₅ molecule forms by the overlap of a 3*p* orbital from each of the five Cl atoms with the sp^3d hybrid orbitals of P (unhybridized, empty 3*d* orbitals not shown). During bonding, each sp^3d orbital fills by addition of an electron from Cl. (The five bonding pairs are not shown.)

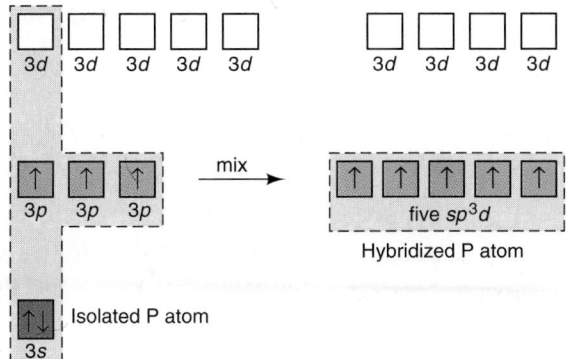

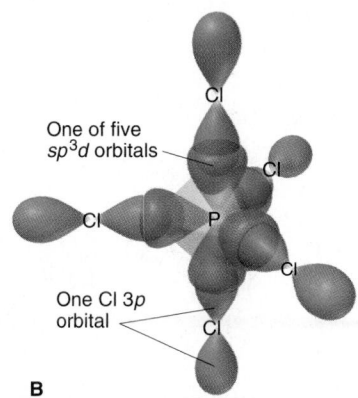

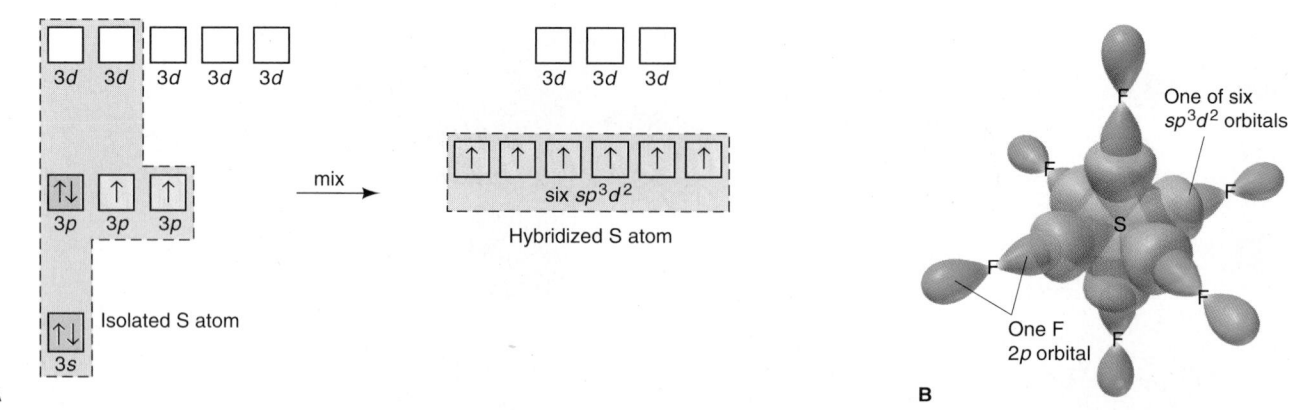

A

B

Figure 11.7 **The sp^3d^2 hybrid orbitals in SF_6. A,** The orbital box diagram shows that one $3s$, three $3p$, and two $3d$ orbitals of S mix to form six sp^3d^2 orbitals that are half-filled. Three $3d$ orbitals are unhybridized and empty. **B,** The octahedral SF_6 molecule forms from overlap of a $2p$ orbital from each of six F atoms with the sp^3d^2 orbitals of S (unhybridized, empty $3d$ orbitals not shown). During bonding, each sp^3d^2 orbital fills by addition of an electron from F. (The six bonding pairs are not shown.)

MOLECULAR SHAPES AND
ORBITAL HYBRIDIZATION

sp^3d^2 **Hybridization** To rationalize the shape of SF_6, the VB model proposes that the one $3s$, the three $3p$, and two of the five $3d$ orbitals of the central S atom mix and form six sp^3d^2 **hybrid orbitals,** which point to the vertices of an octahedron. Each hybrid orbital overlaps a $2p$ orbital of an F atom, and the six valence electrons of S, together with one from each of the six F atoms, pair up to form six S—F bonds (Figure 11.7). Molecules with square pyramidal and square planar shapes have lone pairs in one or two of the central atom's sp^3d^2 orbitals, respectively.

Table 11.1 summarizes the number and types of atomic orbitals that are mixed to obtain the five types of hybrid orbitals. Once again, note the similarities between the orientations of the hybrid orbitals proposed by VB theory and the shapes predicted by VSEPR theory (see Figure 10.5, p. 371). Figure 11.8 shows the three conceptual steps from molecular formula to postulating the hybrid orbitals in the molecule, and Sample Problem 11.1 details the end of that process.

Table 11.1 **Composition and Orientation of Hybrid Orbitals**

	Linear	Trigonal Planar	Tetrahedral	Trigonal Bipyramidal	Octahedral
Atomic orbitals mixed	one s one p	one s two p	one s three p	one s three p one d	one s three p two d
Hybrid orbitals formed	two sp	three sp^2	four sp^3	five sp^3d	six sp^3d^2
Unhybridized orbitals remaining	two p	one p	none	four d	three d
Orientation					

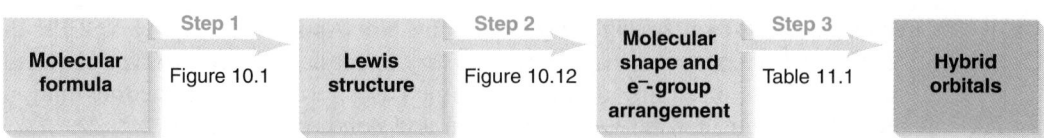

Figure 11.8 The conceptual steps from molecular formula to the hybrid orbitals used in bonding. (See Figures 10.1, page 358, and 10.12, page 377.)

SAMPLE PROBLEM 11.1 Postulating Hybrid Orbitals in a Molecule

Problem Use partial orbital diagrams to describe how mixing of atomic orbitals on the central atoms leads to the hybrid orbitals in each of the following:
(a) Methanol, CH_3OH **(b)** Sulfur tetrafluoride, SF_4

Plan From the Lewis structure, we determine the number and arrangement of electron groups around the central atoms along with the molecular shape. From that, we postulate the type of hybrid orbitals involved. Then, we write the partial orbital diagram for each central atom before and after the orbitals are hybridized.

Solution **(a)** For CH_3OH. The electron-group arrangement is tetrahedral around both C and O atoms. Therefore, each central atom is sp^3 hybridized. The C atom has four half-filled sp^3 orbitals:

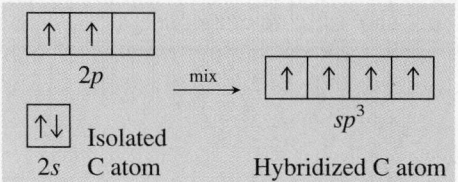

The O atom has two half-filled sp^3 orbitals and two filled with lone pairs:

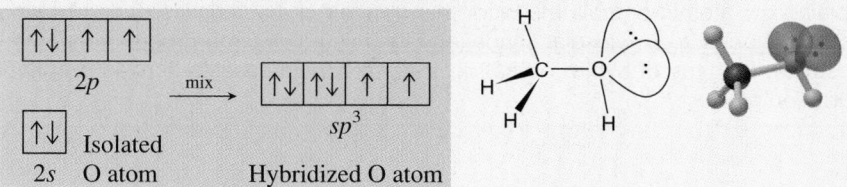

During bonding, each half-filled C or O orbital becomes filled, with the second electron in each case coming from H.

(b) For SF_4. The molecular shape is seesaw, which is based on the trigonal bipyramidal electron-group arrangement. Thus, the central S atom is surrounded by five electron groups, which implies sp^3d hybridization. One $3s$ orbital, three $3p$ orbitals, and one $3d$ orbital are mixed. One hybrid orbital is filled with a lone pair, and four are half-filled. Four unhybridized $3d$ orbitals remain empty:

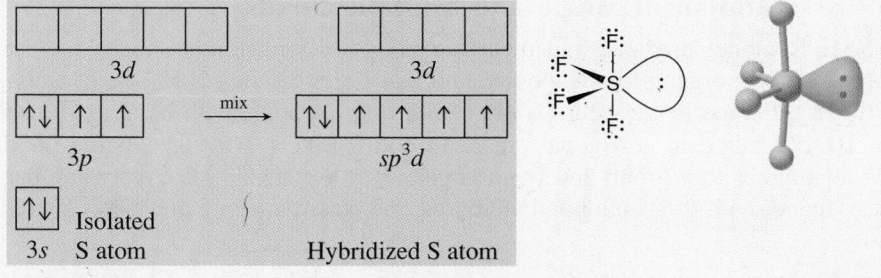

During bonding, each half-filled S orbital becomes filled, with the second electrons coming from F atoms.

FOLLOW-UP PROBLEM 11.1 Use partial orbital diagrams to show how atomic orbitals on the central atoms mix to form hybrid orbitals in **(a)** beryllium fluoride, BeF_2; **(b)** silicon tetrachloride, $SiCl_4$; **(c)** xenon tetrafluoride, XeF_4.

When the Concept of Hybridization May Not Apply We employ VSEPR theory and VB theory *whenever it is necessary* to rationalize an observed molecular shape in terms of atomic orbitals. In some cases, however, the theories may not be needed. Consider the Lewis structure and bond angle of H_2S:

$$H \underset{92°}{\overset{\cdot\cdot}{\diagup}} \overset{:\ddot{S}:}{\diagdown} H$$

Based on VSEPR theory, we would predict that, as in H_2O, the four electron groups around H_2S point to the vertices of a tetrahedron, and the two lone pairs compress the H—S—H bond angle somewhat below the ideal 109.5°. Based on VB theory, we would propose that the $3s$ and $3p$ orbitals of the central S atom mix and form four sp^3 hybrids, two of which are filled with lone pairs, while the other two overlap $1s$ orbitals of two H atoms and are filled with bonding pairs.

The problem is that observation does *not* support these arguments. In fact, the H_2S molecule has a bond angle of 92°, close to the 90° angle expected between *unhybridized,* perpendicular, atomic p orbitals. Similar angles occur in the other Group 6A(16) hydrides and also in the larger hydrides of Group 5A(15). It makes no sense to apply a theory when the facts don't warrant it. In the case of H_2S and these other nonmetal hydrides, neither VSEPR theory nor the concept of hybridization applies. It's important to remember that real factors, such as bond length, atomic size, and electron-electron repulsions, influence molecular shape. Apparently, with these larger central atoms and their longer bonds to H, crowding and electron-electron repulsions decrease, and the simple overlap of unhybridized atomic orbitals rationalizes the observed shapes perfectly well.

SECTION SUMMARY

VB theory explains that a covalent bond forms when two atomic orbitals overlap. The bond holds two electrons with paired (opposite) spins. Orbital hybridization allows us to explain how atomic orbitals mix and change their characteristics during bonding. Based on the observed molecular shape (and the related electron-group arrangement), we postulate the type of hybrid orbital needed. In certain cases, we need not invoke hybridization at all.

11.2 THE MODE OF ORBITAL OVERLAP AND THE TYPES OF COVALENT BONDS

In this section, we employ VB theory to focus on the *mode* by which orbitals overlap—end to end or side to side—to understand the types of covalent bonds and the detailed makeup of multiple bonds.

The VB Treatment of Single and Multiple Bonds

The VSEPR model predicts, and measurements verify, different shapes for the three two-carbon organic molecules ethane (C_2H_6), ethylene (C_2H_4), and acetylene (C_2H_2). Ethane is tetrahedrally shaped at both carbons, with bond angles of about 109.5°. Ethylene is trigonal planar at both carbons, with the double bond acting as one electron group and bond angles near the ideal 120°. Acetylene has a linear shape, with the triple bond acting as one electron group and bond angles of 180°:

ethane   ethylene   acetylene

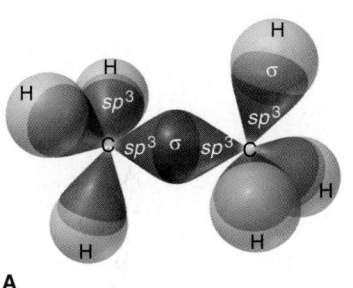

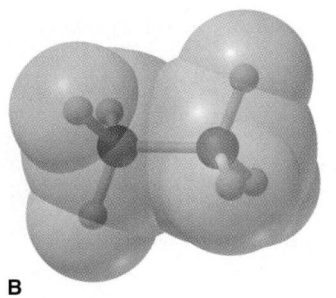

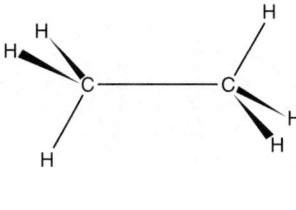

A **B** **C**

Figure 11.9 The σ bonds in ethane (C_2H_6). A, Both C atoms in ethane are sp^3 hybridized. One sp^3 orbital of each C overlaps to form a C—C σ bond. The other sp^3 orbitals overlap the 1s orbitals of six H atoms to form six C—H σ bonds. **B,** Electron density is distributed relatively evenly among the σ bonds of the ethane molecule. **C,** Bond-line drawing.

A close look at the bonds shows two types of orbital overlap here:

1. *End-to-end overlap and sigma (σ) bonding.* Both C atoms of ethane are sp^3 hybridized (Figure 11.9). The C—C bond involves overlap of one sp^3 orbital from each C, and each of the six C—H bonds involves overlap of a C sp^3 orbital with an H 1s orbital. The bonds in ethane are like all the others described so far in this section. Look closely at the C—C bond, for example. It involves the overlap of the end of one orbital with the end of the other. The bond resulting from such *end-to-end* overlap is called a **sigma (σ) bond.** It has its *highest electron density along the bond axis* (an imaginary line joining the nuclei) and is shaped like an ellipse rotated about its long axis (the shape resembles a football). All single bonds, formed by any combination of overlapping hybrid, s, or p orbitals, have their electron density concentrated along the bond axis, and thus are σ bonds.

2. *Side-to-side overlap and pi (π) bonding.* A close look at Figure 11.10 reveals the double nature of the carbon-carbon bond in ethylene. Here, each C atom is sp^2 hybridized. Each C atom's four valence electrons half-fill its three sp^2 orbitals and its unhybridized 2p orbital, which lies perpendicular to the sp^2 plane. Two sp^2 orbitals of each C form C—H σ bonds by overlapping the 1s orbitals of two H atoms. The third sp^2 orbital forms a C—C σ bond with an sp^2 orbital of the other C because their orientation allows end-to-end overlap. With the σ-bonded C atoms near each other, their half-filled unhybridized 2p orbitals are close enough to overlap *side to side.* Such overlap forms another type of covalent bond called a **pi (π) bond.** It has *two regions of electron density,* one above and one below the σ-bond axis. *One π bond holds two electrons that move through both regions of the bond. A double bond always consists of one σ bond and one π bond.* As Figure 11.10C shows, the double bond increases electron density between the C atoms.

Now we can answer a question brought up in our discussion of VSEPR theory in Chapter 10: why do the electron pairs in a double bond act as one electron group; that is, why don't the two electron pairs push each other apart? The answer is that each electron pair occupies a distinct orbital, a specific region of electron density, so repulsions are reduced.

Figure 11.10 The σ and π bonds in ethylene (C_2H_4). A, The two C atoms are sp^2 hybridized and form one C—C σ bond and four C—H σ bonds. Their half-filled unhybridized 2p orbitals lie perpendicular to the σ-bond axis and are shown overlapping side to side. **B,** The two overlapping regions comprise *one* π bond, which is occupied by two electrons. The σ bonds are shown as a line and wedges. The two lobes of the π bond lie above and below the C—C σ-bond line. **C,** With four electrons (one σ bond and one π bond) between the C atoms, the electron density *(red shading)* is high there. **D,** Bond-line drawing.

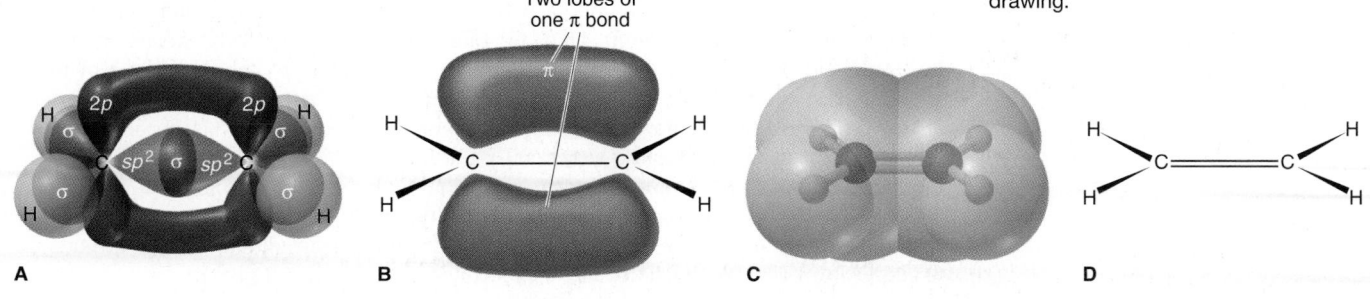

A **B** **C** **D**

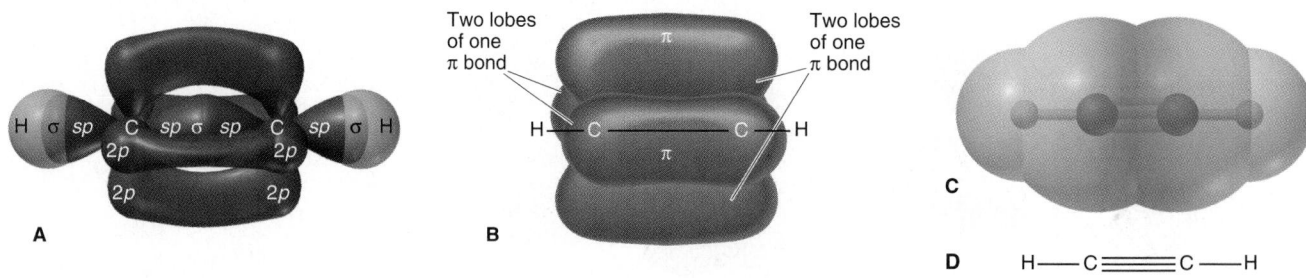

Figure 11.11 The σ bonds and π bonds in acetylene (C$_2$H$_2$).
A, The σ bonds in acetylene are formed when an *sp* orbital of each C overlaps to form a C—C σ bond and the other *sp* orbital of each C overlaps an *s* orbital of an H atom to form two C—H σ bonds. Two unhybridized 2*p* orbitals on each C remain and are shown overlapping. **B,** Two π bonds result from side-to-side overlap. Here, one π bond is shown with its lobes above and below the C—C σ bond (shown as a line), and the other is shown with its lobes in front of and behind this σ bond. In the actual molecule, however, the electron clouds of the two π bonds partially coincide, which gives the molecule cylindrical symmetry. **C,** The six electrons—one σ bond and two π bonds—create even greater electron density between the C atoms than in ethylene. **D,** Bond-line drawing.

A triple bond, such as the C≡C bond in acetylene, *consists of one σ and two π bonds* (Figure 11.11). To maximize overlap in a linear shape, one *s* and one *p* orbital in *each* C atom form two *sp* hybrids, and two 2*p* orbitals remain unhybridized. The four valence electrons half-fill all four orbitals. Each C uses one of its *sp* orbitals to form a σ bond with an H atom and uses the other to form the C—C σ bond. Side-to-side overlap of one pair of 2*p* orbitals gives one π bond, with electron density above and below the σ bond. Side-to-side overlap of the other pair of 2*p* orbitals gives the other π bond, 90° away from the first, with electron density in front of and behind the σ bond. The result is a *cylindrically symmetrical* H—C≡C—H molecule. Note the still greater electron density between the C atoms created by the six bonding electrons.

The extent of overlap influences bond strength. Because side-to-side overlap is not as extensive as end-to-end overlap, we expect a π bond to be weaker than a σ bond, and thus a double bond should be less than twice as strong as a single bond. Indeed, this expectation is borne out for carbon-carbon bonds. However, many factors, such as lone-pair repulsions, bond polarities, and other electrostatic contributions, affect overlap and the relative strength of σ and π bonds between other elements. Thus, as a rough approximation, a double bond is about twice as strong as a single bond, and a triple bond is about three times as strong.

SAMPLE PROBLEM 11.2 Describing the Bonding in Molecules with Multiple Bonds

Problem Describe the types of bonds and orbitals in acetone, (CH$_3$)$_2$CO.
Plan Referring to Sample Problem 11.1, we note the shapes around each atom to postulate the hybrid orbitals used and pay special attention to the multiple bonding of the C=O bond.
Solution In Sample Problem 10.9, we determined the shapes of the three central atoms of acetone to be tetrahedral around each C of the two CH$_3$ (methyl) groups and trigonal planar around the middle C atom. Thus, the middle C has three *sp*^{2} orbitals and one unhybridized *p* orbital. Each of the two methyl C atoms has four *sp*^{3} orbitals. Three of these *sp*^{3} orbitals overlap the 1*s* orbitals of the H atoms to form σ bonds; the fourth overlaps an *sp*^{2} orbital of the middle C atom. Thus, two of the three *sp*^{2} orbitals of the middle C form σ bonds to the terminal C atoms.

The O atom is *sp*^{2} hybridized also and has an unhybridized *p* orbital that can form a π bond. Two of the O atom's *sp*^{2} orbitals hold lone pairs, and the third forms a σ bond with the third *sp*^{2} orbital of the middle C atom. The unhybridized, half-filled *p* orbitals on C and O form a π bond. The σ and π bonds constitute the C=O bond:

σ bonds   π bond

Comment The molecular shape is our guide to the orbitals used in bonding. Therefore, we cannot know directly whether a terminal atom, such as the O atom in acetone, is hybridized. After all, it could use two perpendicular p orbitals for the σ and π bonds with C and leave the other p and the s orbital to hold the two lone pairs. However, a lone pair in an s orbital would increase repulsions by locating additional electrons partly in the bonding region; having the lone pair in an sp^2 orbital away from the C=O bond avoids this. Furthermore, some compounds are known in which the O of a C=O group is bonded to another atom, and the bond angles in those cases suggest that the O is sp^2 hybridized.

FOLLOW-UP PROBLEM 11.2 Describe the types of bonds and the orbitals in **(a)** hydrogen cyanide, HCN, and **(b)** carbon dioxide, CO_2.

Orbital Overlap and Molecular Rotation

The type of bond influences the ability of one part of a molecule to rotate relative to another part. A σ *bond allows free rotation* of the parts of the molecule with respect to each other because the extent of overlap is not affected. If you could hold one CH_3 group of the ethane molecule, the other CH_3 group could spin like a pinwheel without affecting the C—C σ-bond overlap (see Figure 11.9).

However, p orbitals must be parallel to engage in side-to-side overlap, so *a* π *bond restricts rotation* around it. Rotating one CH_2 group in ethylene with respect to the other must decrease the side-to-side overlap and break the π bond. Now you can see why distinct *cis* and *trans* structures can exist for double-bonded molecules, such as 1,2-dichloroethylene (Section 10.4). As Figure 11.12 shows, the π bond allows two *different* arrangements of atoms around the two C atoms, which can have a major effect on molecular polarity (see p. 383). (Rotation around a triple bond is not meaningful: because each triple-bonded C atom is attached to only one other group, there can be no difference in their relative positions.)

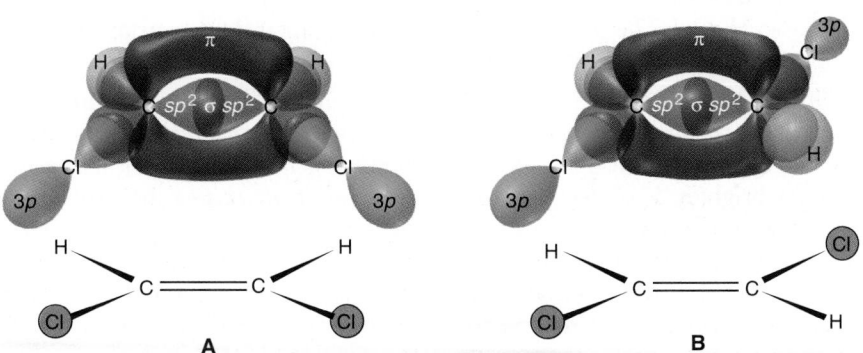

Figure 11.12 Restricted rotation of π-bonded molecules. A, *Cis-* and **B,** *trans-*1,2-dichloroethylene occur as distinct molecules because the π bond between the C atoms restricts rotation and maintains two different relative positions of the H and Cl atoms.

End-to-end overlap of atomic orbitals forms a σ bond and allows free rotation of the parts of the molecule. Side-to-side overlap forms a π bond, which restricts rotation. A multiple bond consists of a σ bond and either one π bond (double bond) or two π bonds (triple bond). Multiple bonds have greater electron density than single bonds.

11.3 MOLECULAR ORBITAL (MO) THEORY AND ELECTRON DELOCALIZATION

Scientists choose the model that helps them answer a particular question best. If the question concerns shape, chemists choose the VSEPR model, followed by hybrid-orbital analysis with VB theory. However, VB theory does not adequately explain the magnetic or spectral properties of molecules, and it understates the importance of electron delocalization. In order to deal with these phenomena, which involve a molecule's energy levels, chemists choose **molecular orbital (MO) theory.**

In VB theory, a molecule is pictured as a group of atoms bound together through *localized* overlap of valence-shell atomic orbitals. In MO theory, a molecule is pictured as a collection of nuclei with the electron orbitals *delocalized* over the entire molecule. The MO model is a quantum-mechanical treatment for molecules similar to the one for individual atoms that we discussed in Chapter 8. Just as an atom has atomic orbitals (AOs) that have a given energy and shape and are occupied by the atom's electrons, a molecule has **molecular orbitals (MOs)** that have a given energy and shape and are occupied by the molecule's electrons. Despite the great usefulness of MO theory, it too has a drawback: molecular orbitals are more difficult to visualize than the easily depicted shapes of VSEPR theory or the hybrid orbitals of VB theory.

The Central Themes of MO Theory

Several key ideas of MO theory appear in its description of the hydrogen molecule and other simple species. These ideas include the formation of MOs, their energy and shape, and how they fill with electrons.

Formation of Molecular Orbitals Because electron motion is so complex, approximations are required to solve the Schrödinger equation for any atom with more than one electron. Similar complications arise even with H_2, the simplest molecule, so approximations are required to solve for the properties of MOs. The most common approximation mathematically *combines* (adds or subtracts) the atomic orbitals (atomic wave functions) of nearby atoms to form molecular orbitals (molecular wave functions).

When two H nuclei lie near each other, as in H_2, their AOs overlap. The two modes of combining the AOs are as follows:

- *Adding the wave functions together.* This combination forms a **bonding MO,** which has a *region of high electron density between the nuclei*. Additive overlap is analogous to two light waves reinforcing each other, making the resulting light brighter. For electron waves, the overlap *increases* the probability that the electrons are between the nuclei (Figure 11.13A).
- *Subtracting the wave functions from each other.* This combination forms an **antibonding MO,** which has *a node between the nuclei, a region of zero electron density* (Figure 11.13B). Subtractive overlap is analogous to two light waves that cancel each other, so the light disappears. With electron waves, the probability that the electrons occupy the space between the nuclei *decreases* to zero.

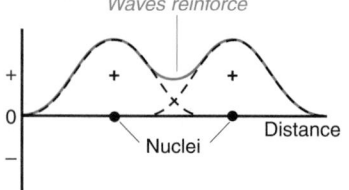

A Amplitudes of wave functions added

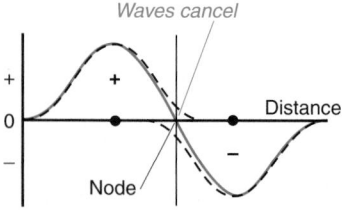

B Amplitudes of wave functions subtracted

Figure 11.13 **An analogy between light waves and atomic wave functions.** When light waves undergo interference, their amplitudes either add together or subtract. **A,** When the amplitudes of atomic wave functions *(dashed lines)* are added, a bonding molecular orbital (MO) results and electron density *(red line)* increases between the nuclei. **B,** Conversely, when the amplitudes of the wave functions are subtracted, an antibonding MO results, which has a node (region of zero electron density) between the nuclei.

The two possible combinations for hydrogen atoms H_A and H_B are

AO of H_A + AO of H_B = bonding MO of H_2 (more e^- density between nuclei)

AO of H_A − AO of H_B = antibonding MO of H_2 (less e^- density between nuclei)

Notice that *the number of AOs combined always equals the number of MOs formed:* two H atomic orbitals combine to form two H_2 molecular orbitals.

Energy and Shape of H_2 Molecular Orbitals *The bonding MO is lower in energy and the antibonding MO higher in energy than the AOs that combined to form them.* Let's examine Figure 11.14 to see why this is so.

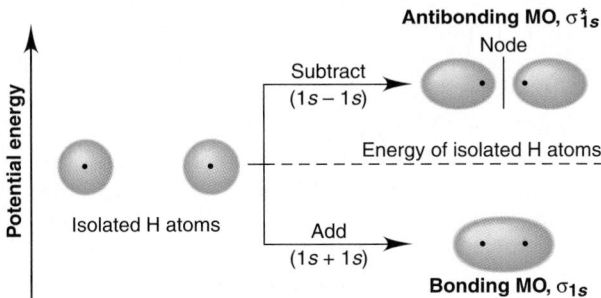

Figure 11.14 **Contours and energies of the bonding and antibonding molecular orbitals (MOs) in H_2.** When two H 1s atomic orbitals (AOs) combine, they form two H_2 MOs. The bonding MO (σ_{1s}) forms from addition of the AOs and is lower in energy than the AOs because most of its electron density lies *between* the nuclei (shown as dots). The antibonding MO (σ_{1s}^*) forms from subtraction of the AOs and is higher in energy because there is a *node* between the nuclei and most of the electron density lies outside the internuclear region.

The bonding MO in H_2 is spread mostly *between* the nuclei, with the nuclei attracted to the intervening electrons. An electron in this MO can delocalize its charge over a much larger volume than is possible in its individual AO. Because the electron-electron repulsions are reduced, the bonding MO is *lower* in energy than the isolated AOs. Therefore, when electrons occupy this orbital, the molecule is *more stable* than the separate atoms. In contrast, the antibonding MO has a node between the nuclei and most of its electron density *outside* the internuclear region. The electrons do not shield one nucleus from the other, which increases nuclear repulsions and makes the antibonding MO *higher* in energy than the isolated AOs. Therefore, when the antibonding orbital is occupied, the molecule is *less stable* than when this orbital is empty.

Both the bonding and antibonding MOs of H_2 are **sigma (σ) MOs** because they are cylindrically symmetrical about an imaginary line that runs through the nuclei. The bonding MO is denoted by σ_{1s}, that is, a σ MO formed by combination of 1s AOs. Antibonding orbitals are denoted with a superscript star, so the one derived from the 1s AOs is σ_{1s}^* (spoken "sigma, one ess, star").

To interact effectively and form MOs, *atomic orbitals must have similar energy and orientation.* The 1s orbitals on two H atoms have identical energy and orientation, so they interact strongly. This point will be important when we consider molecules composed of atoms with many sublevels.

Filling Molecular Orbitals with Electrons Electrons fill MOs just as they fill AOs:

- Orbitals are filled in order of increasing energy (aufbau principle).
- An orbital has a maximum capacity of two electrons with opposite spins (exclusion principle).
- Orbitals of equal energy are half-filled, with spins parallel, before any is filled (Hund's rule).

Molecular orbital (MO) diagrams show the relative energy and number of electrons in each MO, as well as the AOs from which they formed. Figure 11.15 is the MO diagram for H_2.

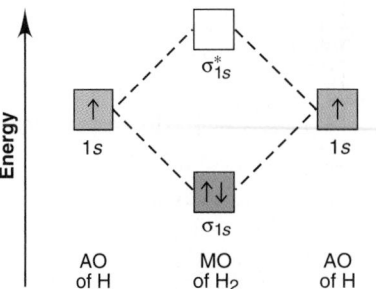

H_2 bond order = $\frac{1}{2}(2 - 0) = 1$

Figure 11.15 **The MO diagram for H_2.** The orbital boxes indicate the relative energies and electron occupancy of the MOs and the AOs from which they formed. Two electrons, one from each H atom, fill the lower energy σ_{1s} MO, while the higher energy σ_{1s}^* MO remains empty. Orbital occupancy is also shown by color (darker = full, paler = half-filled, no color = empty).

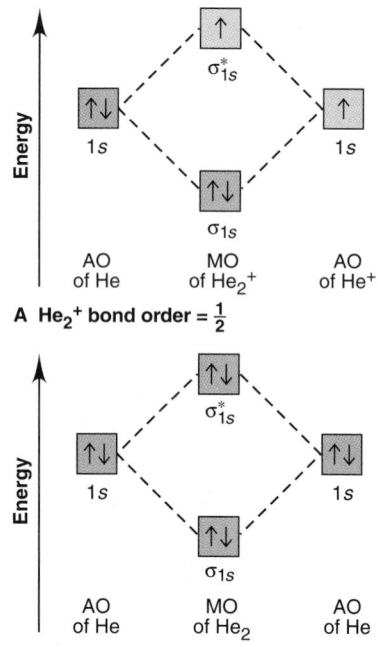

A He_2^+ bond order = $\frac{1}{2}$

B He_2 bond order = 0

Figure 11.16 **MO diagrams for** He_2^+ **and** He_2. **A,** In He_2^+, three electrons enter MOs in order of increasing energy to give a filled σ_{1s} MO and a half-filled σ_{1s}^* MO. The bond order of $\frac{1}{2}$ implies that He_2^+ exists. **B,** In He_2, the four valence electrons fill both the σ_{1s} and the σ_{1s}^* MOs, so there is no net stabilization (bond order = 0).

MO theory redefines bond order. In a Lewis structure, bond order is the number of electron pairs per linkage. The **MO bond order** is the number of electrons in bonding MOs minus the number in antibonding MOs, divided by two:

Bond order = $\frac{1}{2}$[(no. of e⁻ in bonding MO) − (no. of e⁻ in antibonding MO)] **(11.1)**

Thus, for H_2, the bond order is $\frac{1}{2}(2 - 0) = 1$. A bond order greater than zero indicates that the molecular species is stable relative to the separate atoms, whereas a bond order of zero implies no net stability and, thus, no likelihood that the species will form. In general, *the higher the bond order, the stronger the bond.*

Another similarity between models using MOs and AOs is that we can write electron configurations for a molecule. The symbol of each occupied MO is shown in parentheses, and the number of electrons in it is written outside as a superscript. Thus, the electron configuration of H_2 is $(\sigma_{1s})^2$.

One of the early triumphs of MO theory was its ability to *predict* the existence of He_2^+, the dihelium molecule-ion, which is composed of two He nuclei and three electrons. Let's use MO theory to see why He_2^+ exists and, at the same time, why He_2 does not. In He_2^+, the $1s$ atomic orbitals form the molecular orbitals, so the MO diagram, shown in Figure 11.16A, is similar to that for H_2. The three electrons enter the MOs to give a pair in the σ_{1s} MO and a lone electron in the σ_{1s}^* MO. The bond order is $\frac{1}{2}(2 - 1) = \frac{1}{2}$. Thus, He_2^+ has a relatively weak bond, but it should exist. Indeed, this molecular ionic species has been observed frequently when He atoms collide with He⁺ ions. Its electron configuration is $(\sigma_{1s})^2(\sigma_{1s}^*)^1$.

On the other hand, He_2 has four electrons to place in its σ_{1s} and σ_{1s}^* MOs. As Figure 11.16B shows, both the bonding and antibonding orbitals are filled. The stabilization gained by the electron pair in the bonding MO is canceled by the destabilization from the electron pair in the antibonding MO. From its zero bond order $[\frac{1}{2}(2 - 2) = 0]$, we predict, and experiment has so far confirmed, that a covalent He_2 molecule does not exist.

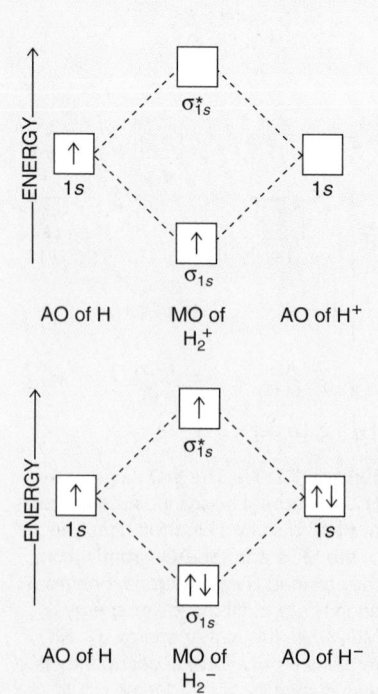

AO of H   MO of H_2^+ AO of H⁺

AO of H   MO of H_2^- AO of H⁻

SAMPLE PROBLEM 11.3 Predicting Species Stability Using MO Diagrams

Problem Use MO diagrams to predict whether H_2^+ and H_2^- exist. Determine their bond orders and electron configurations.

Plan In these species, the $1s$ orbitals form MOs, so the MO diagrams are similar to that for H_2. We determine the number of electrons in each species and distribute the electrons in pairs to the bonding and antibonding MOs in order of increasing energy. We obtain the bond order with Equation 11.1 and write the electron configuration as described in the text.

Solution For H_2^+. H_2 has two e⁻, so H_2^+ has only one, as shown in the margin (top diagram). The bond order is $\frac{1}{2}(1 - 0) = \frac{1}{2}$, so we predict that H_2^+ does exist. The electron configuration is $(\sigma_{1s})^1$.

For H_2^-. Since H_2 has two e⁻, H_2^- has three, as shown in the margin (bottom diagram). The bond order is $\frac{1}{2}(2 - 1) = \frac{1}{2}$, so we predict that H_2^- does exist. The electron configuration is $(\sigma_{1s})^2(\sigma_{1s}^*)^1$.

Check The number of electrons in the MOs equals the number of electrons in the AOs, as it should.

Comment Both these species have been detected spectroscopically: H_2^+ occurs in the hydrogen-containing material around stars; H_2^- has been formed in the laboratory.

FOLLOW-UP PROBLEM 11.3 Use an MO diagram to predict whether two hydride ions (H⁻) will form H_2^{2-}. Calculate the bond order of H_2^{2-} and write its electron configuration.

Homonuclear Diatomic Molecules of the Period 2 Elements

Homonuclear diatomic molecules are those composed of two identical atoms. You're already familiar with several from Period 2—N_2, O_2, and F_2—as the elemental forms under standard conditions. Others in Period 2—Li_2, Be_2, B_2, C_2, and Ne_2—are observed, if at all, only in high-temperature gas-phase experiments. An MO description of these species provides some interesting tests of the model. Let's look first at those from the *s* block, Groups 1A(1) and 2A(2), and then at those from the *p* block, Groups 3A(13) through 8A(18).

Bonding in the *s*-Block Diatomic Molecules Both Li and Be occur as metals under normal conditions, but let's see what MO theory predicts for their stability as the diatomic gases dilithium (Li_2) and diberyllium (Be_2).

These atoms have both 1*s* and 2*s* electrons. Like the MOs formed from 1*s* AOs, those formed from 2*s* AOs are σ orbitals, so they are cylindrically symmetrical around the internuclear axis. The relative energies of the MOs follow the order of the AOs that form them: a 2*s* AO is higher in energy than a 1*s*, so the σ_{2s} and σ_{2s}^* MOs are higher in energy than the σ_{1s} and σ_{1s}^* MOs. The six electrons from two Li atoms enter the MOs in order of increasing energy, two electrons with opposing spins per orbital, as in Figure 11.17A. Dilithium has four electrons in bonding MOs and two in the lower energy antibonding MO, so its bond order is $\frac{1}{2}(4 - 2) = 1$. Li_2 *has* been observed. Its complete MO electron configuration is $(\sigma_{1s})^2(\sigma_{1s}^*)^2(\sigma_{2s})^2$, but we usually show only the configuration involving valence electrons, in this case, the 2*s*; thus, we have $(\sigma_{2s})^2$.

The MO diagram for Be_2 (Figure 11.17B) has filled σ_{2s} and σ_{2s}^* MOs. This is similar to the case of He_2. The bond order is $\frac{1}{2}(4 - 4) = 0$, and thus the ground state of the Be_2 molecule has never been observed. Note that, in the MO diagrams for Li_2 and Be_2, the σ_{1s} and σ_{1s}^* MOs are filled. Because the stabilizing effect of the filled σ_{1s} MO cancels the destabilizing effect of the σ_{1s}^* MO, these MOs provide no *net* contribution to the bonding. From here on, the MO diagrams show only the MOs created by combinations of the *valence*-electron AOs.

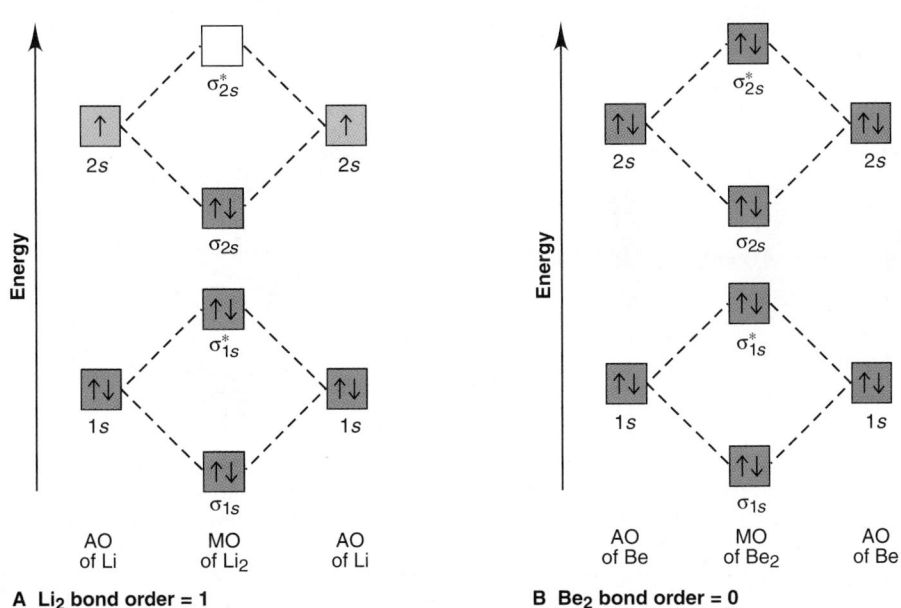

A Li₂ bond order = 1   **B Be₂ bond order = 0**

Figure 11.17 Bonding in *s*-block homonuclear diatomic molecules. A, Li_2. MOs obtained from combinations of 1*s* AOs are lower in energy than those obtained from 2*s* AOs. The six electrons from two Li atoms fill the lowest three MOs, and the σ_{2s}^* remains empty. With a bond order of 1, Li_2 does form. **B,** Be_2. The eight electrons from two Be atoms fill all the available MOs to give no net stabilization. Ground-state Be_2 has a zero bond order and has never been observed.

Molecular Orbitals from Atomic *p*-Orbital Combinations As we move to the *p* block, atomic 2*p* orbitals become involved, so we first consider the shapes and energies of the MOs that result from their combinations. Recall that *p* orbitals can interact with each other from two different directions, as shown in Figure 11.18. End-to-end combination gives a pair of σ MOs, the σ_{2p} and σ^*_{2p}. Side-to-side combination gives a pair of **pi (π) MOs**, π_{2p} and π^*_{2p}. Similar to MOs formed from *s* orbitals, bonding MOs from *p*-orbital combinations have their greatest electron density *between* the nuclei; whereas antibonding MOs have a node between the nuclei and most of their electron density *outside* the internuclear region.

The order of MO energy levels, whether bonding or antibonding, is based on the order of AO energy levels *and* on the mode of the *p*-orbital combination:

- MOs formed from 2*s* orbitals are lower in energy than MOs formed from 2*p* orbitals because 2*s* AOs are lower in energy than 2*p* AOs.
- Bonding MOs are lower in energy than antibonding MOs, so σ_{2p} is lower in energy than σ^*_{2p}, and π_{2p} is lower than π^*_{2p}.
- Atomic *p* orbitals can interact more extensively end to end than they can side to side. Thus, the σ_{2p} MO is lower in energy than the π_{2p} MO. Similarly, the destabilizing effect of the σ^*_{2p} MO is greater than that of the π^*_{2p} MO.

Thus, the energy order for MOs derived from 2*p* orbitals is

$$\sigma_{2p} < \pi_{2p} < \pi^*_{2p} < \sigma^*_{2p}$$

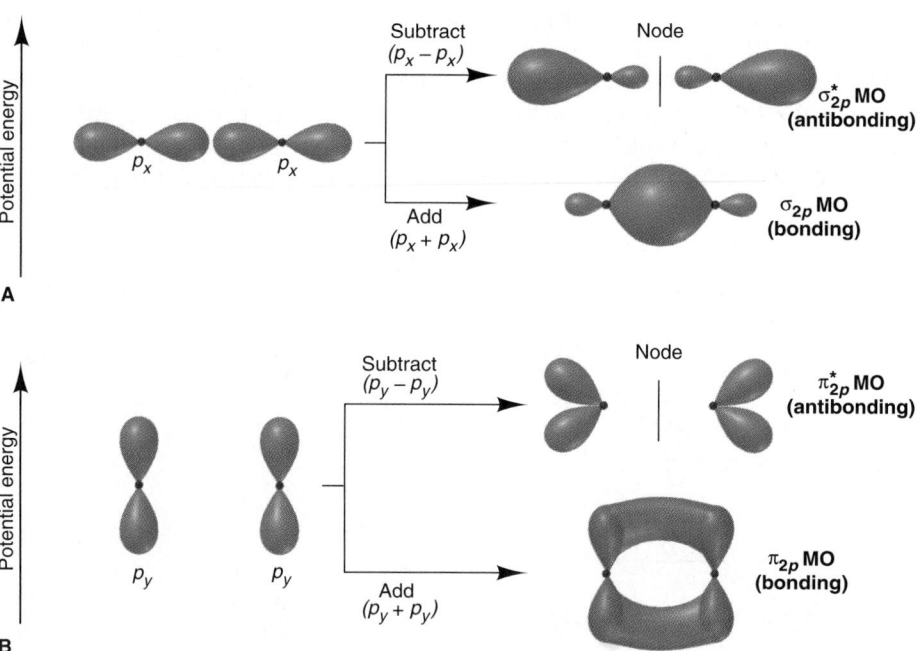

Figure 11.18 **Contours and energies of σ and π MOs through combinations of 2*p* atomic orbitals. A,** The *p* orbitals lying along the line between the atoms (usually designated p_x) undergo end-to-end overlap and form σ_{2p} and σ^*_{2p} MOs. Note the greater electron density between the nuclei for the bonding orbital and the node between the nuclei for the antibonding orbital. **B,** The *p* orbitals that lie perpendicular to the internuclear axis (p_y and p_z) undergo side-to-side overlap to form two π MOs. The p_z interactions are the same as those shown here for the p_y orbitals, so a total of four π MOs form. A π_{2p} is a bonding MO with its greatest density above and below the internuclear axis; a π^*_{2p} is an antibonding MO with a node between the nuclei and its electron density outside the internuclear region.

There are three perpendicular $2p$ orbitals in each atom, so when six p orbitals combine, the two facing each other end to end form a σ and a σ^* MO, and the two pairs that interact side to side form two π MOs of the same energy and two π^* MOs of the same energy. Combining these orientations with the energy order gives the *expected* MO diagram for the p-block Period 2 homonuclear diatomic molecules, which is shown in Figure 11.19A.

One other factor influences the MO energy order. Recall that only AOs of similar energy interact to form MOs. The order in Figure 11.19A assumes that the s and p AOs are so different in energy that they do not interact with each other: in MO terminology, the orbitals do not *mix*. This is true for O, F, and Ne. In these atoms, repulsions occur as the $2p$ electrons pair up; these repulsions raise the energy of the $2p$ orbitals high enough above the energy of the $2s$ orbitals to minimize orbital mixing. However, when the $2p$ AOs are half-filled, as in B, C, and N atoms, repulsions are small, so the $2p$ energies are much closer to the $2s$ energy. As a result, some mixing occurs between the $2s$ orbital of one atom and the end-on $2p$ orbital of the other. This orbital mixing *lowers* the energy of the σ_{2s} and σ_{2s}^* MOs and *raises* the energy of the σ_{2p} and σ_{2p}^* MOs; the π MOs are not affected. The MO diagram for B_2 through N_2 (Figure 11.19B) reflects this AO mixing. The only change that affects this discussion is the *reverse in order* of the σ_{2p} and π_{2p} MOs.

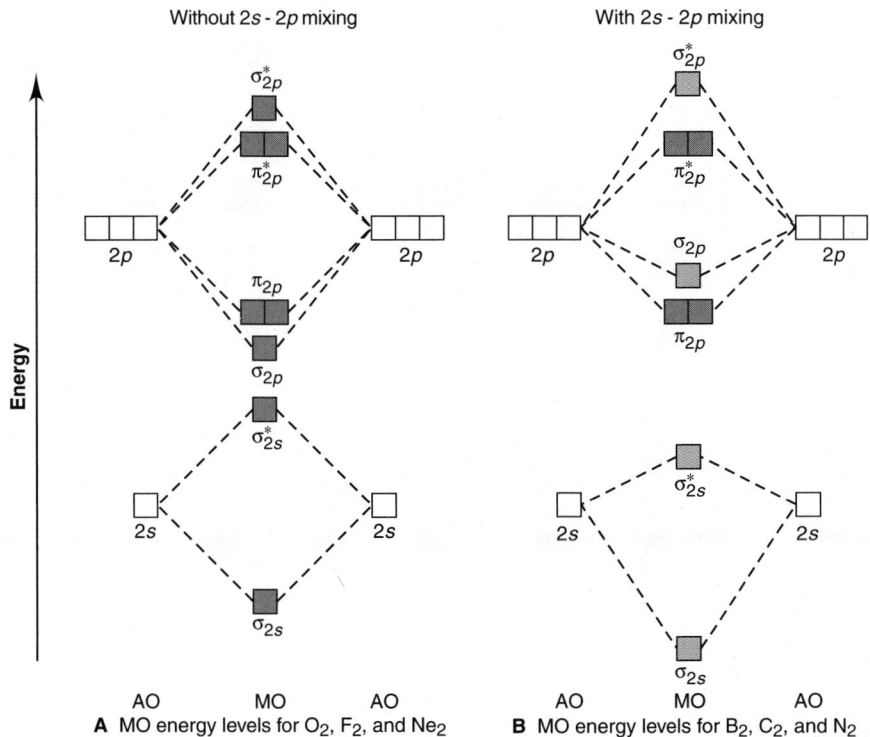

Figure 11.19 Relative MO energy levels for Period 2 homonuclear diatomic molecules. A, MO energy levels for O_2, F_2, and Ne_2. The six $2p$ orbitals of the two atoms form six MOs that are higher in energy than the two MOs formed from the two $2s$ orbitals. The AOs forming π orbitals give rise to two bonding MOs (π_{2p}) of equal energy and two antibonding MOs (π_{2p}^*) of equal energy. This sequence of energy levels arises from minimal $2s$-$2p$ orbital mixing. **B,** MO energy levels for B_2, C_2, and N_2. Because of significant $2s$-$2p$ orbital mixing, the energies of σ MOs formed from $2p$ orbitals increase and those formed from $2s$ orbitals decrease. The major effect of this orbital mixing on the MO sequence is that the σ_{2p} is higher in energy than the π_{2p}. (For clarity, those MOs affected by mixing are shown in purple.)

Bonding in the *p*-Block Diatomic Molecules Figure 11.20 shows the sequence of MOs, their electron occupancy, and some properties of B_2 through Ne_2. Note how *a higher bond order correlates with a greater bond energy and shorter bond length.* Also note how orbital occupancy correlates with magnetic properties. Recall from Chapter 8 that the spins of unpaired electrons in an atom (or ion) cause the substance to be *paramagnetic,* attracted to an external magnetic field. If all the electron spins are paired, the substance is *diamagnetic,* unaffected (or weakly repelled) by the magnetic field. The same observations apply to molecules. These properties are not addressed directly in VSEPR or VB theory.

The B_2 molecule has six outer electrons to place in its MOs. Four of these fill the σ_{2s} and σ_{2s}^{*} MOs. The remaining two electrons enter the two π_{2p} MOs, one in each orbital, in keeping with Hund's rule. With four electrons in bonding MOs and two electrons in antibonding MOs, the bond order of B_2 is $\frac{1}{2}(4 - 2) = 1$. As expected from its MO diagram, B_2 is paramagnetic.

The two additional electrons present in C_2 fill the two π_{2p} MOs. Since C_2 has two more bonding electrons than B_2, it has a bond order of 2 and the expected

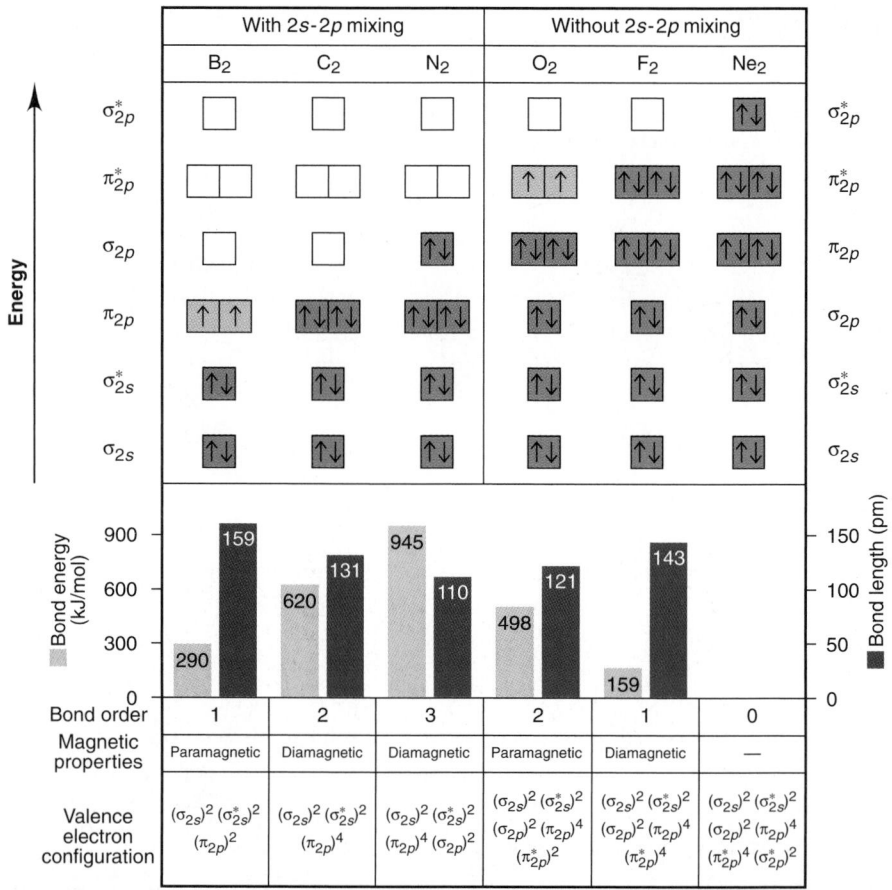

Figure 11.20 **MO occupancy and molecular properties for B_2 through Ne_2.** The sequence of MOs and their electron populations are shown for the homonuclear diatomic molecules in the *p* block of Period 2 [Groups 3A(13) to 8A(18)]. The bond energy, bond length, bond order, magnetic properties, and valence electron configuration appear below the orbital diagrams. Note the correlation between bond order and bond energy, both of which are inversely related to bond length.

stronger, shorter bond. All the electrons are paired, and, as the model predicts, C_2 is diamagnetic.

In N_2, the two additional electrons enter and fill the σ_{2p} MO. The resulting bond order is 3, which is consistent with the triple bond in the Lewis structure. As the model predicts, the bond energy is higher and the bond length shorter, and N_2 is diamagnetic.

With O_2, we really see the power of MO theory compared to theories based on localized orbitals. For years, it seemed impossible to reconcile bonding theories with the bond strength and magnetic behavior of O_2. On the one hand, the data show a double-bonded molecule that is paramagnetic. On the other hand, we can write two possible Lewis structures for O_2, but neither gives such a molecule. One has a double bond and all electrons paired, the other a single bond and two electrons unpaired:

$$\ddot{\text{O}}{=}\ddot{\text{O}} \quad \text{or} \quad {:}\dot{\text{O}}{-}\dot{\text{O}}{:}$$

MO theory resolves this paradox beautifully. As Figure 11.20 shows, the bond order of O_2 is 2: eight electrons occupy bonding MOs and four occupy antibonding MOs $[\frac{1}{2}(8 - 4) = 2]$. Note the lower bond energy and greater bond length relative to N_2. The *two* electrons with highest energy occupy the *two* π^*_{2p} MOs with unpaired (parallel) spins, making the molecule paramagnetic. Figure 11.21 shows liquid O_2 suspended between the poles of a powerful magnet.

The two additional electrons in F_2 fill the π^*_{2p} orbitals, which decreases the bond order to 1, and the absence of unpaired electrons makes F_2 diamagnetic. As expected, the bond energy is lower and the bond distance longer than in O_2. Note that the bond energy for F_2 is only about half that for B_2, even though they have the same bond order. F is smaller than B, so we might expect a stronger bond. But the 18 electrons in the smaller volume of F_2 compared with the 10 electrons in B_2 cause greater repulsions and make the F_2 single bond easier to break.

The final member of the series, Ne_2, does not exist for the same reason that He_2 does not: all the MOs are filled, so the stabilization from bonding electrons cancels the destabilization from antibonding electrons, and the bond order is zero.

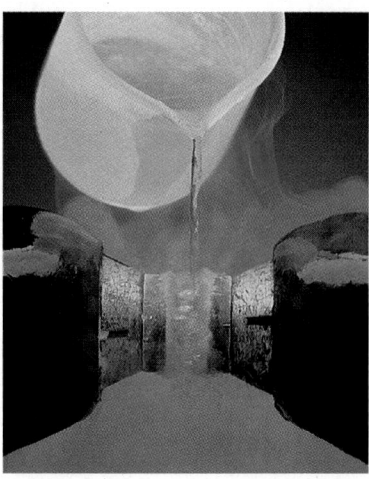

Figure 11.21 The paramagnetic properties of O_2. Liquid O_2 is attracted to the poles of a magnet because it is paramagnetic, as MO theory predicts. A diamagnetic substance would fall between the poles.

SAMPLE PROBLEM 11.4 Using MO Theory to Explain Bond Properties

Problem As the following data show, removing an electron from N_2 forms an ion with a weaker, longer bond than in the parent molecule, whereas the ion formed from O_2 has a stronger, shorter bond:

	N_2	N_2^+	O_2	O_2^+
Bond energy (kJ/mol)	945	841	498	623
Bond length (pm)	110	112	121	112

Explain these facts with diagrams that show the sequence and occupancy of MOs.
Plan We first determine the number of valence electrons in each species. Then, we draw the sequence of MO energy levels for the four species, recalling that they differ for N_2 and O_2 (see Figures 11.19 and 11.20), and fill them with electrons. Finally, we calculate bond orders and compare them with the data. Recall that bond order is related directly to bond energy and inversely to bond length.
Solution Determining the valence electrons:

N has 5 valence e^-, so N_2 has 10 and N_2^+ has 9
O has 6 valence e^-, so O_2 has 12 and O_2^+ has 11

Drawing and filling the MO diagrams:

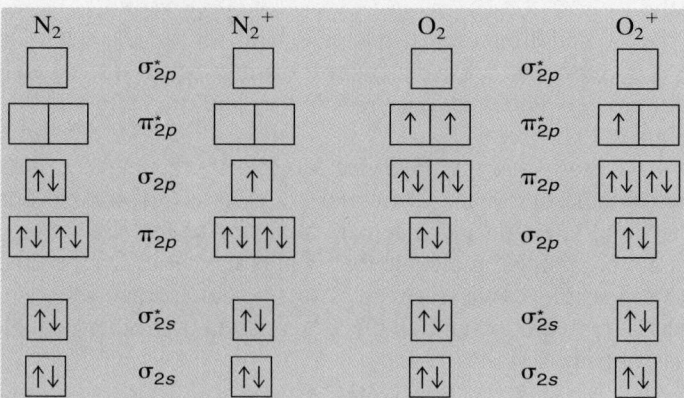

Calculating bond orders:

$$\tfrac{1}{2}(8-2)=3 \qquad \tfrac{1}{2}(7-2)=2.5 \qquad \tfrac{1}{2}(8-4)=2 \qquad \tfrac{1}{2}(8-3)=2.5$$

When N_2 forms N_2^+, a *bonding* electron is removed, so the bond order decreases. Thus, N_2^+ has a weaker, longer bond than N_2. When O_2 forms O_2^+, an *antibonding* electron is removed, so the bond order increases. Thus, O_2^+ has a stronger, shorter bond than O_2.
Check The answer makes sense in terms of the relationships among bond order, bond energy, and bond length. Check that the total number of bonding and antibonding electrons equals the number of valence electrons calculated.

FOLLOW-UP PROBLEM 11.4 Determine the bond orders for the following species: F_2^{2-}, F_2^-, F_2, F_2^+, F_2^{2+}. List the species in order of increasing bond energy and in order of increasing bond length.

MO Description of Some Heteronuclear Diatomic Molecules

Heteronuclear diatomic molecules, those composed of two *different* atoms, have asymmetric MO diagrams because the atomic orbitals of the two atoms have unequal energies. Atoms with greater effective nuclear charge (Z_{eff}) draw their electrons closer to the nucleus and thus have atomic orbitals of lower energy (Section 8.2). Greater Z_{eff} also gives these atoms higher electronegativity values. Let's apply MO theory to the bonding in HF and NO.

Bonding in HF To form the MOs in HF, we combine appropriate AOs from isolated H and F atoms. The high effective nuclear charge of F holds all its electrons more tightly than the H nucleus holds its electron. As a result, *all* occupied atomic orbitals of F have lower energy than the 1s orbital of H. Therefore, the H 1s orbital interacts only with the F 2p orbitals, and only one of the three 2p orbitals, say the $2p_z$, leads to end-on overlap. This results in a σ MO and a σ^* MO. The two other p orbitals of F ($2p_x$ and $2p_y$) are not involved in bonding and are called **nonbonding MOs;** they have the same energy as the isolated AOs (Figure 11.22).

Because the occupied bonding MO is closer in energy to the AOs of F, we say that the F orbital contributes more to the bonding in HF than the H orbital does. Generally, in polar covalent molecules, *bonding MOs are closer in energy to the AOs of the more electronegative atom.* In effect, fluorine's greater electronegativity lowers the energy of the bonding MO and draws the bonding electrons closer to its nucleus.

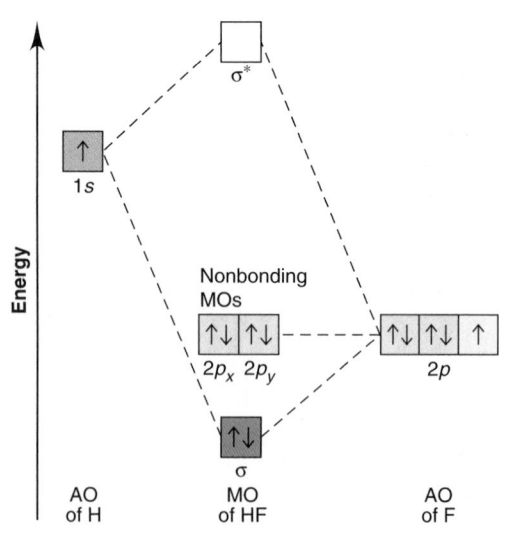

Figure 11.22 The MO diagram for HF. For a polar covalent molecule, the MO diagram is asymmetric because the more electronegative atom has AOs of lower energy. In HF, the bonding MO is closer in energy to the 2p orbital of F. Electrons that are not involved in bonding occupy nonbonding MOs. (The 2s AO of F is not shown.)

Bonding in NO Nitrogen monoxide (nitric oxide) is a highly reactive molecule because it has a lone electron. Two possible Lewis structures for NO, with formal charges (Section 10.1), are

$$:\!N\!=\!\dot{O}\!:\quad \text{or}\quad :\!\dot{N}\!=\!\dot{O}\!:$$
$$\mathbf{I}\qquad\qquad\mathbf{II}$$

Both structures show a double bond, but the *measured* bond energy suggests a bond order *higher* than 2. Furthermore, it is not clear where the lone electron resides, although the lower formal charges for structure I suggest that it is on the N atom.

MO theory predicts the bond order and indicates lone-electron placement with no difficulty. The MO diagram in Figure 11.23 is asymmetric, with the AOs of the more electronegative O lower in energy. The 11 valence electrons of NO enter MOs in order of increasing energy, leaving the lone electron in one of the π^*_{2p} orbitals. The eight bonding electrons and three antibonding electrons give a bond order of $\frac{1}{2}(8-3)=2.5$, more in keeping with experiment than either Lewis structure. The bonding electrons lie in MOs closer in energy to the AOs of the O atom. The lone electron occupies an antibonding orbital. Because this orbital receives a greater contribution from the $2p$ orbitals of N, the lone electron resides closer to the N atom.

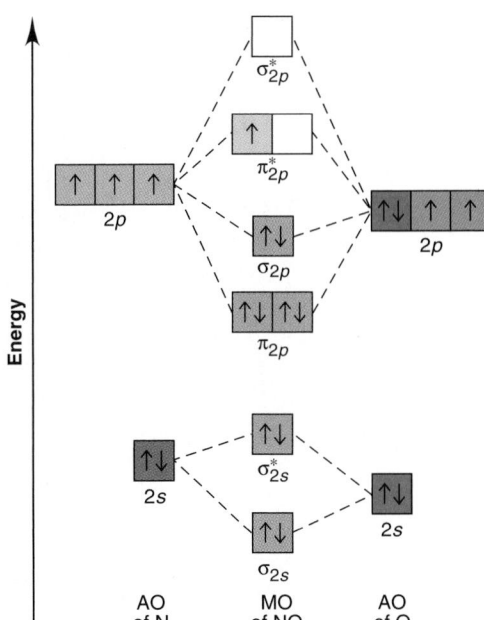

Figure 11.23 The MO diagram for NO. Eleven electrons occupy the MOs in nitric oxide. Note that the lone electron occupies an antibonding MO whose energy is closer to that of the AO of the (less electronegative) N atom.

MO Descriptions of Ozone and Benzene

MO theory extends to molecules with more than two atoms, but the orbital shapes and MO diagrams are too complex for a detailed treatment here. However, let's briefly discuss how the model eliminates the need for resonance forms and helps explain the effects of the absorption of energy.

Recall that we cannot draw a single Lewis structure that adequately depicts bonding in molecules such as ozone and benzene because the adjacent single and double bonds actually have identical properties. Instead, we draw more than one structure and mentally combine them into a resonance hybrid. The VB model also relies on resonance because it depicts *localized* electron-pair bonds.

In contrast, MO theory pictures a structure of delocalized σ and π bonding and antibonding MOs. Figure 11.24 shows the shapes of the π-bonding MOs of lowest energy in benzene and in ozone. Each holds one pair of electrons. The extended electron densities allow delocalization of this π-electron pair over the entire molecule, thus eliminating the need for separate resonance forms. In benzene, the upper and lower hexagonal lobes of this π-bonding MO lie above and below the σ plane of all six carbon nuclei. In ozone, the two lobes of the lowest energy π-bonding MO extend over and under all three oxygen nuclei. Another

A Benzene, C$_6$H$_6$

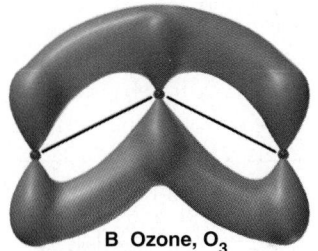

B Ozone, O$_3$

Figure 11.24 The lowest energy π-bonding MOs in benzene and ozone. A, The most stable π-bonding MO in C$_6$H$_6$ has hexagonal lobes of electron density above and below the σ plane of the six C atoms. **B,** The most stable π-bonding MO in O$_3$ extends above and below the σ plane (shown as bond lines) of the three O atoms.

advantage of MO theory is it can explain excited states and spectra of molecules: for instance, why O_3 decomposes when it absorbs ultraviolet radiation in the stratosphere (bonding electrons are excited and enter empty antibonding orbitals), and why the ultraviolet spectrum of benzene has its characteristic absorption bands.

SECTION SUMMARY

MO theory treats a molecule as a collection of nuclei with molecular orbitals delocalized over the entire structure. Atomic orbitals of comparable energy can be added and subtracted to obtain bonding and antibonding MOs, respectively. Bonding MOs have most of their electron density between the nuclei and are lower in energy than the atomic orbitals; most of the electron density of antibonding MOs does not lie between the nuclei, so these MOs are higher in energy. MOs are filled in order of their energy with paired electrons having opposing spins. MO diagrams show energy levels and orbital occupancy. Those for homonuclear diatomic molecules of Period 2 explain observed bond energy, bond length, and magnetic behavior. In heteronuclear diatomic molecules, the more electronegative atom contributes more to the bonding MOs. MO theory eliminates the need for resonance forms to depict larger molecules.

Chapter Perspective

In this chapter, you've seen that the two most important orbital-based models of covalent bonding, each with its own benefits and limitations, complement each other. Next, you'll see how the electron distribution in the bonds of molecules gives rise to the physical behavior of liquids and solids. MO theory will come into play for a deeper understanding of metals. And, following Chapter 12 come two chapters that apply the concepts you've learned from Chapter 7 onward to the behavior of inorganic, organic, and biochemical substances.

For Review and Reference (Numbers in parentheses refer to pages, unless noted otherwise.)

Learning Objectives

Relevant section and/or sample problem (SP) numbers appear in parentheses.

Understand These Concepts

1. The main ideas of valence bond theory—orbital overlap, opposing electron spins, and hybridization as a means of rationalizing molecular shapes (11.1)
2. How orbitals mix to form hybrid orbitals with different spatial orientations (11.1)
3. The distinction between end-to-end and side-to-side overlap and the origin of sigma (σ) and pi (π) bonds in simple molecules (11.2)
4. How two modes of orbital overlap lead to single, double, and triple bonds (11.2)
5. Why π bonding restricts rotation around double bonds (11.2)
6. The distinction between the localized bonding of VB theory and the delocalized bonding of MO theory (11.3)
7. How addition or subtraction of AOs forms bonding or antibonding MOs (11.3)

8. Shapes of MOs formed from combinations of two *s* orbitals and combinations of two *p* orbitals (11.3)
9. How MO bond order predicts the stability of molecular species (11.3)
10. How MO theory explains the bonding and properties of homonuclear and heteronuclear diatomic molecules of Period 2 (11.3)

Master These Skills

1. Using molecular shape to postulate the hybrid orbitals used by a central atom (SP 11.1)
2. Describing the types of orbitals and bonds in a molecule (SP 11.2)
3. Drawing MO diagrams, writing electron configurations, and calculating bond orders of molecular species (SP 11.3)
4. Explaining bond properties with MO diagrams (SP 11.4)

Key Terms

Section 11.1

valence bond (VB) theory (393)
hybridization (394)
hybrid orbital (394)
sp hybrid orbital (394)
sp^2 hybrid orbital (396)

sp^3 hybrid orbital (396)
sp^3d hybrid orbital (397)
sp^3d^2 hybrid orbital (398)

Section 11.2

sigma (σ) bond (401)
pi (π) bond (401)

Section 11.3

molecular orbital (MO) theory (404)
molecular orbital (MO) (404)
bonding MO (404)
antibonding MO (404)
sigma (σ) MO (405)

molecular orbital (MO) diagram (405)
MO bond order (406)
homonuclear diatomic molecule (407)
pi (π) MO (408)
nonbonding MO (412)

Key Equations and Relationships

11.1 Calculating the MO bond order (406):

$$\text{Bond order} = \tfrac{1}{2}[(\text{no. of } e^- \text{ in bonding MO}) - (\text{no. of } e^- \text{ in antibonding MO})]$$

Highlighted Figures and Tables

These figures (F) and tables (T) provide a quick review of key ideas.

F11.2 The sp hybrid orbitals in gaseous $BeCl_2$ (395)
T11.1 Composition and orientation of hybrid orbitals (398)
F11.8 The conceptual steps from molecular formula to hybrid orbitals used in bonding (399)
F11.10 The σ and π bonds in ethylene (C_2H_4) (401)
F11.14 Contours and energies of the bonding and antibonding MOs in H_2 (405)

F11.18 Contours and energies of σ and π MOs through combinations of $2p$ atomic orbitals (408)
F11.20 MO occupancy and molecular properties for B_2 through Ne_2 (410)

Brief Solutions to Follow-up Problems

11.1 (a) Shape is linear, so Be is sp hybridized.

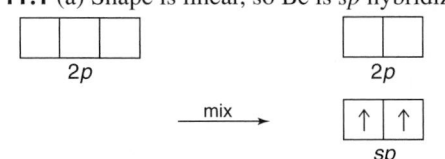

Isolated Be atom   Hybridized Be atom

(b) Shape is tetrahedral, so Si is sp^3 hybridized.

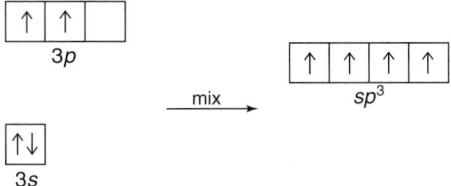

Isolated Si atom   Hybridized Si atom

(c) Shape is square planar, so Xe is sp^3d^2 hybridized.

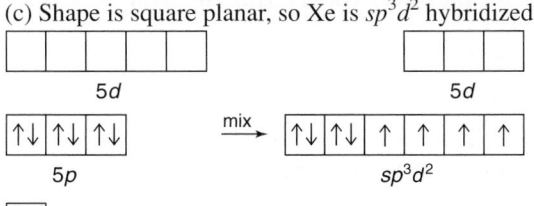

Isolated Xe atom   Hybridized Xe atom

11.2 (a) H—C≡N:

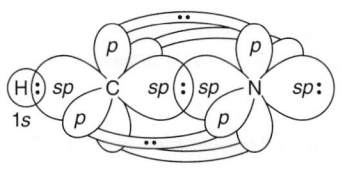

HCN is linear, so C is sp hybridized. N is also sp hybridized. One sp of C overlaps the $1s$ of H to form a σ bond. The other sp of C overlaps one sp of N to form a σ bond. The other sp of N holds a lone pair. Two unhybridized p orbitals of N and two of C overlap to form two π bonds.

(b) Ö=C=Ö

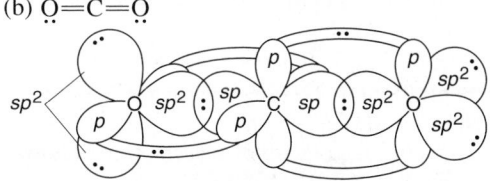

CO_2 is linear, so C is sp hybridized. Both O atoms are sp^2 hybridized. Each sp of C overlaps one sp^2 of an O to form two σ bonds. Each of the two unhybridized p orbitals of C forms a π bond with the unhybridized p of one of the two O atoms. Two sp^2 of each O hold lone pairs.

11.3

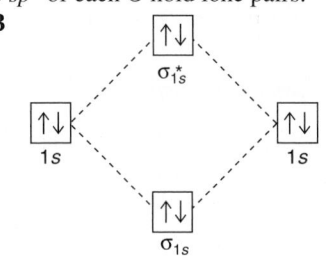

AO of H^- MO of H_2^{2-} AO of H^-

Not likely to exist: BO $= \tfrac{1}{2}(2-2) = 0$; $(\sigma_{1s})^2(\sigma_{1s}^*)^2$

11.4 Bond orders: $F_2^{2-} = 0$; $F_2^- = \tfrac{1}{2}$; $F_2 = 1$; $F_2^+ = 1\tfrac{1}{2}$
$F_2^{2+} = 2$

Bond energy: $F_2^{2-} < F_2^- < F_2 < F_2^+ < F_2^{2+}$
Bond length: $F_2^{2+} < F_2^+ < F_2 < F_2^-$; F_2^{2-} has no bond

Problems

Problems with **colored** numbers are answered at the back of the text. Sections match the text and provide the number(s) of relevant sample problems. Most offer Concept Review Questions, Skill-Building Exercises (in similar pairs), and Problems in Context. Then Comprehensive Problems, based on material from any section or previous chapter, follow.

Valence Bond (VB) Theory and Orbital Hybridization
(Sample Problem 11.1)

● **Concept Review Questions**

11.1 What type of central-atom orbital hybridization corresponds to each electron-group arrangement: (a) trigonal planar; (b) octahedral; (c) linear; (d) tetrahedral; (e) trigonal bipyramidal?

11.2 What is the orbital hybridization of a central atom that has one lone pair and bonds to: (a) two other atoms; (b) three other atoms; (c) four other atoms; (d) five other atoms?

11.3 How do carbon and silicon differ with regard to the *types* of orbitals available for hybridization? Explain.

11.4 How many hybrid orbitals form when four orbitals on a central atom mix during a reaction? Explain.

● **Skill-Building Exercises (paired)**

11.5 Give the number and type of hybrid orbital that forms when each set of atomic orbitals mixes:
(a) two d, one s, and three p (b) three p and one s

11.6 Give the number and type of hybrid orbital that forms when each set of atomic orbitals mixes:
(a) one p and one s (b) three p, one d, and one s

11.7 What is the hybridization of nitrogen in each of the following: (a) NO; (b) NO_2; (c) NO_2^-?

11.8 What is the hybridization of carbon in each of the following: (a) CO_3^{2-}; (b) $C_2O_4^{2-}$; (c) NCO^-?

11.9 What is the hybridization of chlorine in each of the following: (a) ClO_2; (b) ClO_3^-; (c) ClO_4^-?

11.10 What is the hybridization of bromine in each of the following: (a) BrF_3; (b) BrO_2^-; (c) BrF_5?

11.11 Which types of atomic orbitals of the central atom mix to form hybrid orbitals in (a) $SiClH_3$; (b) CS_2?

11.12 Which types of atomic orbitals of the central atom mix to form hybrid orbitals in (a) Cl_2O; (b) $BrCl_3$?

11.13 Which types of atomic orbitals of the central atom mix to form hybrid orbitals in (a) SCl_3F; (b) NF_3?

11.14 Which types of atomic orbitals of the central atom mix to form hybrid orbitals in (a) PF_5; (b) SO_3^{2-}?

11.15 Use partial orbital diagrams to show how the atomic orbitals of the central atom lead to hybrid orbitals in (a) $GeCl_4$; (b) BCl_3; (c) CH_3^+.

11.16 Use partial orbital diagrams to show how the atomic orbitals of the central atom lead to hybrid orbitals in (a) BF_4^-; (b) PO_4^{3-}; (c) SO_3.

11.17 Use partial orbital diagrams to show how the atomic orbitals of the central atom lead to hybrid orbitals in (a) $SeCl_2$; (b) H_3O^+; (c) IF_4^-.

11.18 Use partial orbital diagrams to show how the atomic orbitals of the central atom lead to hybrid orbitals in (a) $AsCl_3$; (b) $SnCl_2$; (c) PF_6^-.

● **Problems in Context**

11.19 Methyl isocyanate, $CH_3—\ddot{N}\text{=}C\text{=}\ddot{O}:$, is an intermediate in the manufacture of many pesticides. It received notoriety in 1984 when a leak from a manufacturing plant resulted in the death of more than 2000 people in Bhopal, India. What is the hybridization of the N atom and the two C atoms in methyl isocyanate? Sketch the molecular shape.

The Mode of Orbital Overlap and the Types of Covalent Bonds
(Sample Problem 11.2)

● **Concept Review Questions**

11.20 Are the statements true or false? Correct any that are false.
(a) Two σ bonds comprise a double bond.
(b) A triple bond consists of one π bond and two σ bonds.
(c) Bonds formed from atomic s orbitals are always σ bonds.
(d) A π bond restricts rotation about the σ-bond axis.
(e) A π bond consists of two pairs of electrons.
(f) End-to-end overlap results in a bond with electron density above and below the bond axis.

● **Skill-Building Exercises (paired)**

11.21 Describe the hybrid orbitals used by the central atom and the type(s) of bonds formed in (a) NO_3^-; (b) CS_2; (c) CH_2O.

11.22 Describe the hybrid orbitals used by the central atom and the type(s) of bonds formed in (a) O_3; (b) I_3^-; (c) $COCl_2$ (C is central).

11.23 Describe the hybrid orbitals used by the central atom(s) and the type(s) of bonds formed in (a) FNO; (b) C_2F_4; (c) $(CN)_2$.

11.24 Describe the hybrid orbitals used by the central atom(s) and the type(s) of bonds formed in (a) BrF_3; (b) $CH_3C\text{≡}CH$; (c) SO_2.

● **Problems in Context**

11.25 2-Butene ($CH_3CH\text{=}CHCH_3$) is a starting material in the manufacture of lubricating oils and many other compounds. Draw two different structures for 2-butene, indicating the σ and π bonds in each.

Molecular Orbital (MO) Theory and Electron Delocalization
(Sample Problems 11.3 and 11.4)

● **Concept Review Questions**

11.26 Two p orbitals from one atom and two p orbitals from another atom are combined to form molecular orbitals for the joined atoms. How many MOs will result from this combination? Explain.

11.27 Certain atomic orbitals on two atoms were combined to form the following MOs. Name the atomic orbitals used and the MOs formed, and explain which MO has higher energy:

11.28 How do the bonding and antibonding MOs formed from a given pair of AOs compare to each other with respect to (a) energy; (b) presence of nodes; (c) internuclear electron density?

11.29 Antibonding MOs always have at least one node. Can a bonding MO have a node? If so, draw an example.

● **Skill-Building Exercises (paired)**

11.30 How many electrons does it take to fill (a) a σ bonding MO; (b) a π antibonding MO; (c) the MOs formed from combination of the $1s$ orbitals of two atoms?

11.31 How many electrons does it take to fill (a) the MOs formed from combination of the $2p$ orbitals of two atoms; (b) a σ^*_{2p} MO; (c) the MOs formed from combination of the $2s$ orbitals of two atoms?

11.32 Show the shapes of bonding and antibonding MOs formed by combination of (a) an s orbital and a p orbital; (b) two p orbitals (end to end).

11.33 Show the shapes of bonding and antibonding MOs formed by combination of (a) two s orbitals; (b) two p orbitals (side to side).

11.34 Use MO diagrams and the bond orders you obtain from them to answer: (a) Is Be_2^+ stable? (b) Is Be_2^+ diamagnetic? (c) What is the valence electron configuration of Be_2^+?

11.35 Use MO diagrams and the bond orders you obtain from them to answer: (a) Is O_2^- stable? (b) Is O_2^- paramagnetic? (c) What is the valence electron configuration of O_2^-?

11.36 Use MO diagrams to place C_2^-, C_2, and C_2^+ in order of (a) increasing bond energy; (b) increasing bond length.

11.37 Use MO diagrams to place B_2^+, B_2, and B_2^- in order of (a) decreasing bond energy; (b) decreasing bond length.

Comprehensive Problems

Problems with an asterisk (*) are more challenging.

11.38 Predict the shape, state the hybridization of the central atom, and give the ideal bond angles and any expected deviations for (a) BrO_3^-; (b) $AsCl_4^-$; (c) SeO_4^{2-}; (d) BiF_5^{2-}; (e) SbF_4^+; (f) AlF_6^{3-}; (g) IF_4^+.

11.39 Butadiene (shown below) is a colorless gas used to make synthetic rubber and many other compounds:

(a) How many σ bonds and π bonds does the molecule have?
(b) Are *cis-trans* structural arrangements about the double bonds possible? Explain.

11.40 Epinephrine (or adrenaline) is a naturally occurring hormone that is also manufactured commercially for use as a heart stimulant, a nasal decongestant, and a glaucoma treatment. A valid Lewis structure is

(a) What is the hybridization of each C, O, and N atom?
(b) How many σ bonds does the molecule have?
(c) How many π electrons are delocalized in the ring?

11.41 Use partial orbital diagrams to show how the atomic orbitals on the central atom lead to the hybrid orbitals in (a) IF_2^-; (b) ICl_3; (c) $XeOF_4$; (d) BHF_2.

11.42 Isoniazid is an antibacterial agent that is particularly useful against many common strains of tuberculosis. A valid Lewis structure is

(a) How many σ bonds are in the molecule?
(b) What is the hybridization of each C and N atom?

11.43 Hydrazine, N_2H_4, and carbon disulfide, CS_2, form a cyclic molecule with the following Lewis structure:

(a) Draw Lewis structures for N_2H_4 and CS_2.
(b) How do electron-group arrangement, molecular shape, and hybridization of N change when hydrazine reacts to form the product? (c) How do electron-group arrangement, molecular shape, and hybridization of C change when carbon disulfide reacts to form the product?

11.44 In each of the following equations, what hybridization change, if any, occurs at the underlined atom?
(a) $\underline{B}F_3 + NaF \longrightarrow Na^+BF_4^-$
(b) $\underline{P}Cl_3 + Cl_2 \longrightarrow PCl_5$
(c) $H\underline{C}\equiv CH + H_2 \longrightarrow H_2C=CH_2$
(d) $\underline{Si}F_4 + 2F^- \longrightarrow SiF_6^{2-}$
(e) $\underline{S}O_2 + \frac{1}{2}O_2 \longrightarrow SO_3$

***11.45** Like atoms, molecules also have ionization energies,
$$XY(g) \longrightarrow XY^+(g) + e^- \qquad \Delta H = IE_1 \text{ of } XY$$
and electron affinities:
$$XY(g) + e^- \longrightarrow XY^-(g) \qquad \Delta H = EA_1 \text{ of } XY$$
And, just as for atoms, the magnitudes of IE_1 and EA_1 for molecules depend on factors such as nuclear charge, atomic size, orbital occupancy, and shielding. Thus, for example, just as IE_1 increases from C to N to O in Period 2, IE_1 increases from C_2 to CN to N_2. (a) Use an MO diagram to explain why nitric oxide, NO, does not follow this trend but has a lower IE_1 than C_2. (b) How do bond order, bond length, bond energy, and magnetic behavior change as NO loses an electron to form NO^+? (c) How would these properties change if NO gained an electron to form NO^-?

11.46 Nitroglycerine ($C_3H_5N_3O_9$) decomposes explosively to produce a gaseous mixture of N_2, CO_2, H_2O, and O_2. A Lewis structure for nitroglycerine is

(a) Describe the hybridization of C, N, and O in the C—O—N groupings.
(b) Use Table 9.2 (p. 339) to calculate the heat of reaction for the decomposition of 1 mol of nitroglycerine.

***11.47** The sulfate ion can be represented with four S—O bonds or with two S—O and two S=O bonds. (a) Which representation is better from the standpoint of formal charges? (b) What is the shape of the sulfate ion, and what hybrid orbitals of S are postulated for the σ bonding? (c) In view of the answer to part (b), what orbitals of S must be used for the π bonds? What orbitals of O? (d) Draw a diagram to show how one atomic orbital from S and one from O overlap to form a π bond.

11.48 Tryptophan is one of the amino acids typically found in proteins:

(a) What is the hybridization of the numbered C, N, and O atoms?
(b) How many σ bonds are present in tryptophan?
(c) Predict the bond angles at points a, b, and c.

11.49 Sulfur forms oxides, oxoanions, and halides. What is the hybridization of the central S in SO_2, SO_3, SO_3^{2-}, SCl_4, SCl_6, and S_2Cl_2 (atom sequence Cl—S—S—Cl)?

*11.50 The hydrocarbon allene, $H_2C=C=CH_2$, is obtained indirectly from petroleum and used as a precursor for several types of plastics. What is the hybridization of each C atom in allene? Draw a bonding picture for allene with lines for σ bonds, and show the arrangement of the π bonds. Be sure to represent the geometry of the molecule in three dimensions.

11.51 Many unusual heteronuclear diatomic molecules form in hot gases. Draw an MO diagram, write the valence electron configuration, and predict the bond order for: (a) BC; (b) OF.

*11.52 Linoleic acid is an essential fatty acid found in many vegetable oils, such as soy, peanut, and cottonseed. A key structural feature of the molecule is the *cis* orientation around its two double bonds, where R_1 and R_2 represent two different groups that form the rest of the molecule.

(a) How many different compounds are possible, changing only the *cis-trans* arrangements around these two double bonds?
(b) How many are possible for a similar compound with three double bonds?

11.53 The molecule B_2 is paramagnetic (see Figure 11.20). How does this behavior support the order of MO energy levels arising from $2s$-$2p$ mixing? Would the magnetic behavior of B_2 be different if there were no $2s$-$2p$ mixing? Explain.

11.54 Although LiF consists of ions in the solid state, it exists as highly polar LiF molecules in the gaseous state. Based on the relative magnitudes of properties such as ionization energy, electron affinity, and electronegativity, explain why the molecules are polar. Draw an MO diagram for LiF, similar to that for HF, assuming that the Li $2s$ orbital interacts with the F $2p$ orbital.

11.55 There is unanimous concern in health-related government agencies that the American diet contains too much meat, and nu-

merous recommendations have been made to include more fruit and vegetables in the diet. One of the richest sources of vegetable protein is soy, available in many forms. Among these is soybean curd, or tofu, which is a staple of many Asian diets. Chemists have isolated an anticancer agent called *genistein* from tofu, which may explain the much lower incidence of cancer among people in the Far East. A valid Lewis structure for genistein is

(a) Is the hybridization of each C in the right-hand ring the same? Explain.
(b) Is the hybridization of the O atom in the center ring the same as that of the O atoms in OH groups? Explain.
(c) How many carbon-oxygen σ bonds are there? How many carbon-oxygen π bonds?
(d) Do all the lone pairs on oxygens occupy the same type of hybrid orbital? Explain.

*11.56 Simple proteins consist of amino acids linked together in a long chain, a small portion of which is

Experiment shows that rotation about the C—N bond (indicated by the arrow) is somewhat restricted. Explain with resonance structures, and show the types of bonding involved.

11.57 The compound 2,6-dimethylpyrazine gives chocolate its odor and is used in flavorings. A valid Lewis structure is

(a) Which atomic orbitals mix to form the hybrid orbitals of N?
(b) In what type of hybrid orbital do the lone pairs of N reside?
(c) Is the hybridization of C in each CH_3 group the same as that of each C in the ring? Explain.

11.58 Acetylsalicylic acid (aspirin), the most widely used medicine in the world, has the following Lewis structure:

(a) What is the hybridization of each C and each O atom?
(b) How many localized π bonds are present?
(c) How many C atoms have a trigonal planar shape around them? A tetrahedral shape?

INTERMOLECULAR FORCES: LIQUIDS, SOLIDS, AND PHASE CHANGES

CHAPTER OUTLINE

12.1 An Overview of Physical States and Phase Changes

12.2 Quantitative Aspects of Phase Changes
Heat Involved in Phase Changes
Equilibrium Nature of Phase Changes
Phase Diagrams

12.3 Types of Intermolecular Forces
Ion-Dipole Forces
Dipole-Dipole Forces
The Hydrogen Bond
Charge-Induced Dipole Forces
Dispersion (London) Forces

12.4 Properties of the Liquid State
Surface Tension
Capillarity
Viscosity

12.5 The Uniqueness of Water
Solvent Properties
Thermal Properties
Surface Properties
Density of Solid and Liquid Water

12.6 The Solid State: Structure, Properties, and Bonding
Structural Features of Solids
Crystalline Solids
Amorphous Solids
Bonding in Solids

12.7 Advanced Materials
Electronic Materials
Liquid Crystals
Ceramic Materials
Polymeric Materials
Nanotechnology

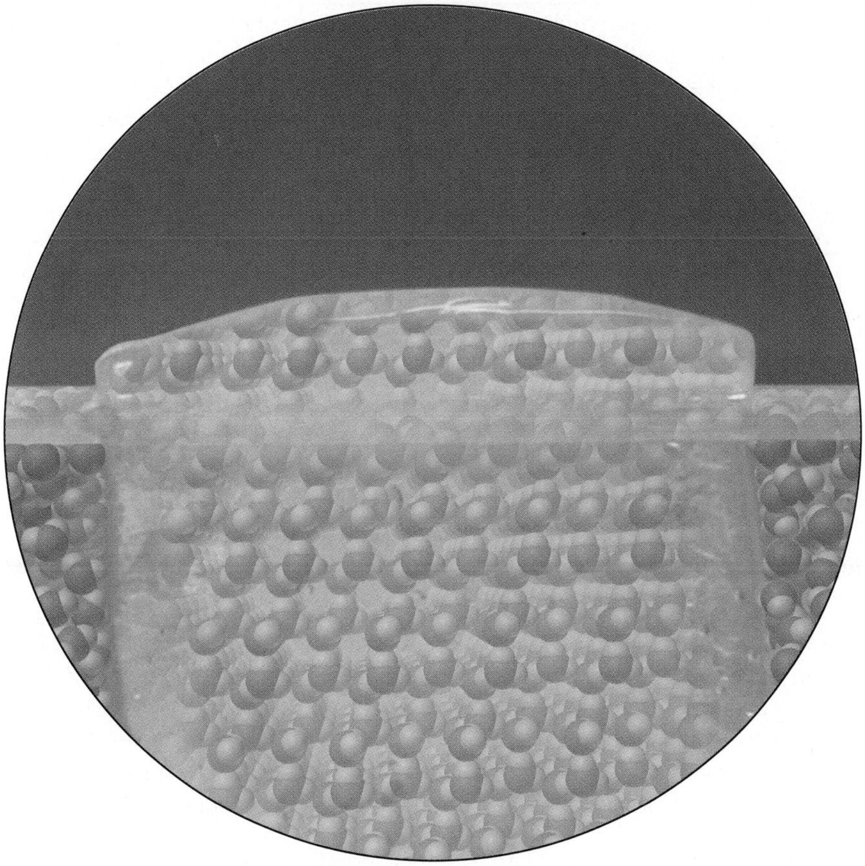

Order floating on chaos. A molecular-scale view of an ice cube in a glass of water reveals the order of the crystalline solid state and the disorder of the liquid state. It also reveals the essence of these two states—different physical properties created by specific interactions among identical particles under different conditions. We'll examine these interactions and their wide-ranging effects thoroughly in this chapter and, along the way, discover why this ordinary scence is so extraordinary.

CONCEPTS & SKILLS

to review before you study this chapter
- properties of gases, liquids, and solids (Section 5.1)
- kinetic-molecular theory of gases (Section 5.6)
- kinetic and potential energy (Section 6.1)
- heat capacity, enthalpy change, and Hess's law (Sections 6.2, 6.3, and 6.5)
- Coulomb's law (Section 9.2)
- chemical bonding models (Chapter 9)
- molecular polarity (Section 10.4)
- molecular orbital treatment of diatomic molecules (Section 11.3)

All the matter around us occurs in one of three physical states—gas, liquid, or solid. In fact, under specific conditions, many pure substances can exist in any of the states. We are all very familiar with the states of water: we inhale and exhale gaseous water; drink, excrete, and wash with liquid water; and slide on or cool our drinks with solid water. Look around and you'll find other examples of solids (jewelry, table salt), liquids (gasoline, antifreeze), and gases (air, CO_2 bubbles) that can exist in one or both of the other states, even though the conditions needed for the change in state may be unusual.

The three states were introduced in Chapter 1 and their properties compared when we examined gases in Chapter 5. In this chapter, we turn our attention to liquids and solids. A physical state is one type of **phase,** any physically distinct, homogeneous part of a system. The water in a glass constitutes a single phase; add some ice and you have two phases. Liquids and solids are called *condensed* phases (or condensed states) because their particles are extremely close together.

The potential energy between the particles—molecules, atoms, or ions—in a sample of matter results from attractive and repulsive forces called *interparticle forces* or, more commonly, **intermolecular forces.** The interplay between these forces and the kinetic energy of the particles gives rise to the properties of each state (or phase) and to **phase changes,** the changes from one phase to another.

This chapter begins with an overview of the states and their phase changes. Then we focus on phase changes at the molecular level, calculate the energy changes involved, and highlight the effects of temperature and pressure with phase diagrams. We next consider the various types of intermolecular forces and then discuss the properties of liquids. The next section combines ideas from several chapters to show how the unique properties of water arise from the electron configurations of its atoms. Then we discuss the properties of solids, emphasizing the relationship between type of bonding and intermolecular force. You will learn how the atomic-scale structures of solids are determined experimentally and, in the final section, how some exciting modern materials are produced and used.

12.1 AN OVERVIEW OF PHYSICAL STATES AND PHASE CHANGES

Imagine yourself among the particles of a molecular substance and you'll discover two types of electrostatic forces at work:

- *Intra*molecular forces (bonding forces) exist *within* each molecule (or polyatomic ion) and influence the *chemical* properties of the substance.
- *Inter*molecular forces exist *between* the molecules (or ions) and influence the *physical* properties of the substance.

Now imagine a molecular view of three states of the same substance. Take water as an example and focus on one molecule from gaseous water, one from liquid water, and one from ice. They appear identical—bent, polar H—O—H molecules. In fact, in their *chemical* behavior, the three states of water *are* identical because their molecules are held together by the same *intra*molecular covalent bonding forces. However, the *physical* behavior of these states differs greatly.

A Kinetic-Molecular View of the Three States Whether a substance is a gas, liquid, or solid depends on the interplay between the potential energy of the intermolecular attractions, which tends to draw the molecules together, and the kinetic energy of the molecules, which tends to disperse them. According to Coulomb's law, the potential energy depends on the charges of the particles and the distances between them (see Section 9.2). The average kinetic energy, which is related to the particles' average speed, is proportional to the absolute temperature.

Table 12.1 A Macroscopic Comparison of Gases, Liquids, and Solids

State	Shape and Volume	Compressibility	Ability to Flow
Gas	Conforms to shape and volume of container	High	High
Liquid	Conforms to shape of container; volume limited by surface	Very low	Moderate
Solid	Maintains its own shape and volume	Almost none	Almost none

This interplay between potential and kinetic energies directly affects the properties that define the three states. In Table 12.1, we focus on three properties—shape, compressibility, and ability to flow—that we discussed in earlier chapters. The following kinetic-molecular view of the three states is an extension of the model we used to understand gases (see Section 5.6):

- *In a gas,* the energy of attraction is small relative to the energy of motion; so, on average, the particles are far apart. This large interparticle distance has several macroscopic consequences. A gas moves randomly throughout its container and fills it. Gases are highly compressible, and they flow and diffuse through one another easily.
- *In a liquid,* the attractions are stronger because the particles are in virtual contact. But their kinetic energy still allows them to tumble randomly over and around each other. Therefore, a liquid conforms to the shape of its container but has a surface. With very little free space between the particles, liquids resist an applied external force and thus compress only very slightly. They flow and diffuse but *much* more slowly than gases.
- *In a solid,* the attractions dominate the motion to such an extent that the particles remain in position relative to one another, jiggling in place. With the particles usually slightly closer together than in a liquid and their positions fixed, a solid has a specific shape. Consequently, solids compress even less than liquids, and their particles do not flow significantly.

Types of Phase Changes Phase changes are also determined by the interplay between kinetic energy and intermolecular forces. As the temperature increases, the average kinetic energy increases as well, so the faster moving particles can overcome attractions more easily; conversely, lower temperatures allow the forces to draw the slower moving particles together.

What happens when gaseous water is cooled? First, a mist appears as the particles form tiny microdroplets that then collect into a bulk sample of liquid with a single surface. The process by which a gas changes into a liquid is called **condensation;** the opposite process, changing from a liquid into a gas, is called **vaporization.** With further cooling, the particles move even more slowly and become fixed in position as the liquid solidifies in the process of **freezing;** the opposite change is called **melting,** or **fusion.** In common speech, the term *freezing* implies low temperature because we typically think of water, but many substances freeze at much higher temperatures; gold, for example, freezes (solidifies) at 1064°C.

Enthalpy changes accompany these phase changes. As the molecules of a gas attract each other and come closer together in the liquid, and then become fixed in the solid, the system of particles loses energy, which is released as heat. Thus, *condensing and freezing are exothermic changes.* On the other hand, energy must be absorbed to overcome the attractive forces that keep the particles in a liquid

Environmental Flow The environment demonstrates beautifully the differences in the abilities of the three states to flow and diffuse. Atmospheric gases mix so well that the 80 km of air closest to Earth has a uniform composition. Much less mixing occurs in the oceans, and differences in composition at various depths support different species. Rocky solids (see photo) intermingle so little that adjacent strata remain separated for millions of years.

Frozen Gold The phase change that occurs when liquids freeze is essential for casting applications. In the preparation of a dental crown, the niches and crevices of the tooth impression are filled with a molten metal, often a gold alloy. After cooling, the metal solidifies and withstands years of high pressure from chewing. The casting of bronze statues, glass decorations, candles, and numerous plastic objects employs the same principle.

A Cooling Phase Change The evaporation of sweat has a cooling effect because heat from your body is used to vaporize the water. Many fur-covered animals achieve this cooling effect, too. Cats (and many rodents) lick themselves, transferring water from inside their bodies to the surface where it can evaporate, and dogs just open their mouths, hang out their tongues, and pant!

together and that keep them fixed in place in a solid. Thus, *melting and vaporizing are endothermic changes.*

For a pure substance, each phase change has a specific enthalpy change *per mole* (measured at 1 atm and the temperature of the change). For vaporization, it is called the **heat of vaporization (ΔH^0_{vap}),** and for fusion, it is the **heat of fusion (ΔH^0_{fus}).** In the case of water, we have

$$H_2O(l) \longrightarrow H_2O(g) \qquad \Delta H = \Delta H^0_{vap} = 40.7 \text{ kJ/mol (at 100°C)}$$

$$H_2O(s) \longrightarrow H_2O(l) \qquad \Delta H = \Delta H^0_{fus} = 6.02 \text{ kJ/mol (at 0°C)}$$

The reverse processes, condensing and freezing, have enthalpy changes of the *same magnitude but opposite sign:*

$$H_2O(g) \longrightarrow H_2O(l) \qquad \Delta H = -\Delta H^0_{vap} = -40.7 \text{ kJ/mol}$$

$$H_2O(l) \longrightarrow H_2O(s) \qquad \Delta H = -\Delta H^0_{fus} = -6.02 \text{ kJ/mol}$$

Water is typical of most pure substances in that it takes less energy to melt 1 mol of solid than to vaporize 1 mol of liquid: $\Delta H^0_{fus} < \Delta H^0_{vap}$. Figure 12.1 shows several examples. This difference occurs because a phase change is essentially a change in intermolecular distance and freedom of motion. Thus, less energy is needed to overcome the forces holding the molecules in their fixed positions (to melt a solid) than to separate them completely from each other (to vaporize a liquid).

The three states of water are so common because they all are stable under ordinary conditions. Carbon dioxide, on the other hand, is familiar as a gas and a solid (dry ice), but liquid CO_2 occurs only at external pressures greater than 5 atm. Thus, at ordinary conditions, solid CO_2 becomes a gas without first becoming a liquid. This process is called **sublimation.** On a clear wintry day, you can dry clothes outside on a line, even though it may be too cold for ice to melt, because the ice sublimes. Freeze-dried foods are prepared by sublimation. The opposite process, changing from a gas directly into a solid, is called **deposition**—you may have seen ice crystals form on a cold window from the deposition of water vapor. The **heat of sublimation (ΔH^0_{subl})** is the enthalpy change when 1 mol of the substance sublimes. Hess's law says that it equals the sum of the heats of fusion and vaporization:

$$
\begin{array}{ll}
\text{Solid} \longrightarrow \text{liquid} & \Delta H^0_{fus} \\
\text{Liquid} \longrightarrow \text{gas} & \Delta H^0_{vap} \\
\hline
\text{Solid} \longrightarrow \text{gas} & \Delta H^0_{subl}
\end{array}
$$

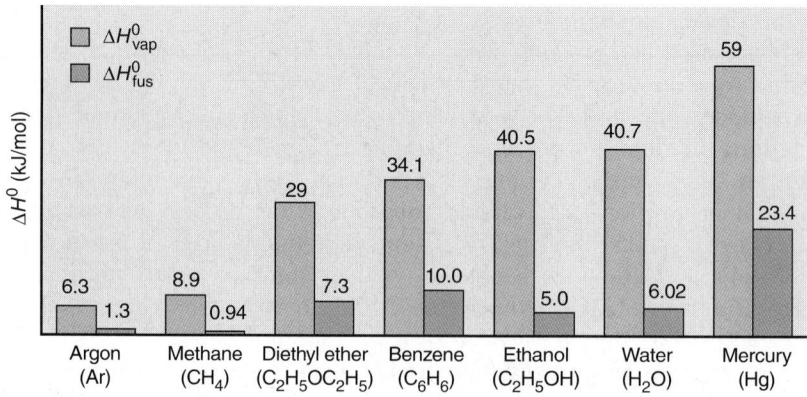

Figure 12.1 **Heats of vaporization and fusion for several common substances.** ΔH^0_{vap} is always larger than ΔH^0_{fus} because it takes more energy to separate particles completely than just to free them from their fixed positions in the solid.

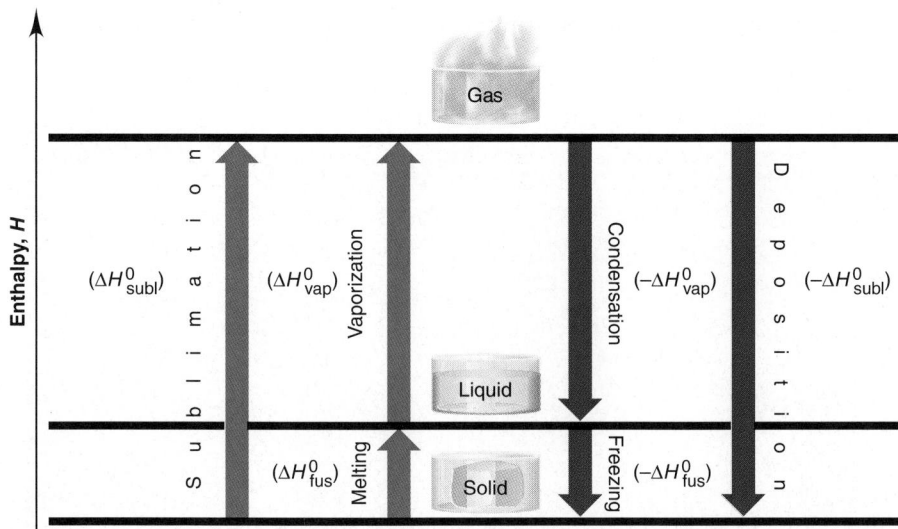

Figure 12.2 **Phase changes and their enthalpy changes.** Each type of phase change, with its associated enthalpy change, appears on a general enthalpy diagram. Note that fusion (or melting), vaporization, and sublimation are endothermic changes (positive ΔH^0), whereas freezing, condensation, and deposition are exothermic changes (negative ΔH^0).

Figure 12.2 summarizes the terminology of the various phase changes and shows the enthalpy changes associated with them.

SECTION SUMMARY

Because of the relative magnitudes of intermolecular forces and kinetic energy, the particles in a gas are far apart and moving randomly, those in a liquid are in contact but still moving relative to each other, and those in a solid are in contact and fixed relative to one another in a rigid structure. These molecular-level differences in the states of matter account for macroscopic differences in shape, compressibility, and ability to flow. When a solid becomes a liquid (melting, or fusion) or a liquid becomes a gas (vaporization), energy is absorbed to overcome intermolecular forces and increase the average distance between particles. When particles come closer together in the reverse changes (freezing and condensation), energy is released. Sublimation is the changing of a solid directly into a gas. Each phase change is associated with a given enthalpy change under specified conditions.

12.2 QUANTITATIVE ASPECTS OF PHASE CHANGES

Some of the most spectacular, and familiar, phase changes occur in the daily weather. When it rains, water vapor condenses to a liquid, which changes back to a gas as puddles dry up. In the spring, snow melts and streams fill; in winter, water freezes and falls to earth. These remarkable transformations also take place whenever you make a pot of tea or a tray of ice cubes. In this section, we examine the heat absorbed or released in a phase change and the equilibrium nature of the process.

Heat Involved in Phase Changes: A Kinetic-Molecular Approach

We can apply the kinetic-molecular theory quantitatively to phase changes by means of a **heating-cooling curve,** which shows the changes that occur when heat is added or removed at a constant rate from a particular sample of matter. As an

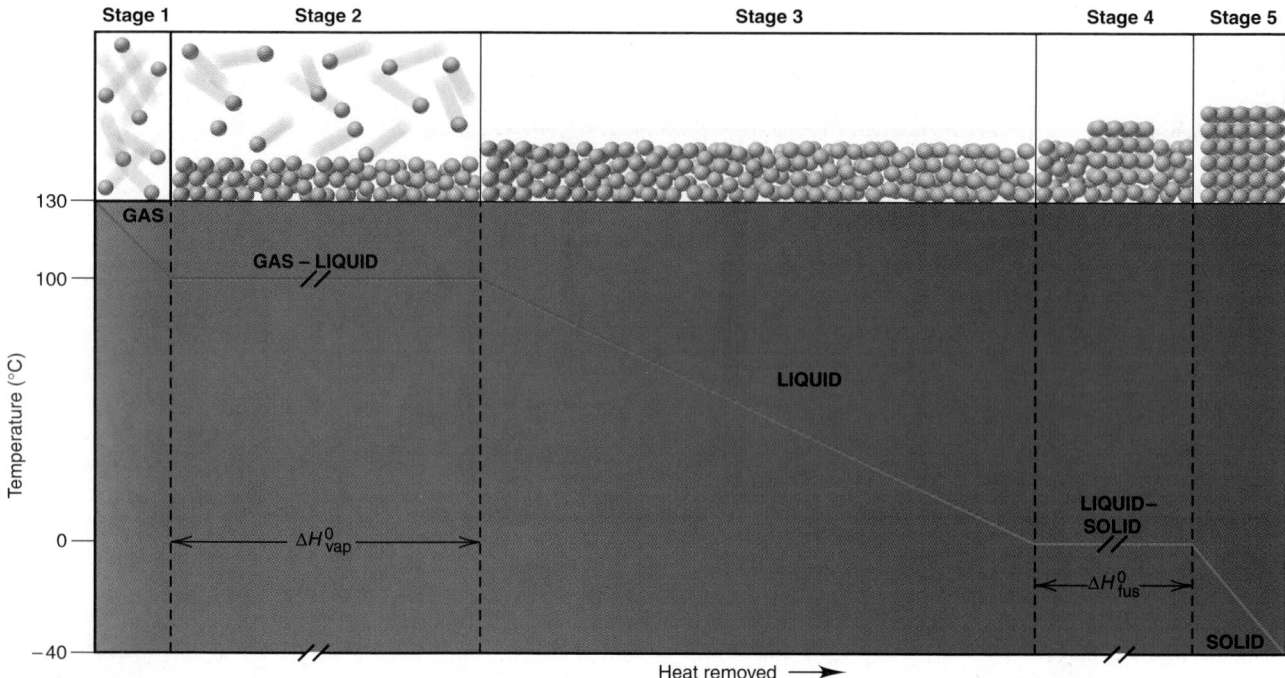

Figure 12.3 **A cooling curve for the conversion of gaseous water to ice.** A plot is shown of temperature vs. heat removed as gaseous water at 130°C changes to ice at −40°C. This process occurs in five stages, with a molecular-level depiction shown for each stage. Stage 1: Gaseous water cools. Stage 2: Gaseous water condenses. Stage 3: Liquid water cools. Stage 4: Liquid water freezes. Stage 5: Solid water cools. The slopes of the lines in stages 1, 3, and 5 reflect the magnitudes of the molar heat capacities of the phases. Although not drawn to scale, the line in stage 2 is longer than the line in stage 4 because ΔH^0_{vap} of water is greater than ΔH^0_{fus}. A plot of temperature vs. heat added would have the same steps but in reverse order.

example, the cooling process is depicted in Figure 12.3 for a 2.50-mol sample of gaseous water in a piston-cylinder assembly, with the pressure kept at 1 atm and the temperature changing from 130°C to −40°C. To an observer, the process is continuous, but we can divide it into five heat-releasing (exothermic) stages that correspond to the five portions of the curve—the gas cools, it condenses to a liquid, the liquid cools, it freezes to a solid, and the solid cools a bit further:

Stage 1. Gaseous water cools. Picture a collection of water molecules behaving like a typical gas: zooming around chaotically at a range of speeds, smashing into each other and the container walls. At a high enough temperature, the most probable speed, and thus the average kinetic energy (E_k), of the molecules is high enough to overcome the potential energy (E_p) of attractions among them. As the temperature falls, the average E_k decreases, so the attractions become increasingly important.

The change is $H_2O(g)$ [130°C] $\longrightarrow$ $H_2O(g)$ [100°C]. The heat (q) is the product of the amount (number of moles, n), the molar heat capacity of *gaseous water*, $C_{water(g)}$ and the temperature change, ΔT ($T_{final} - T_{initial}$):

$$q = n \times C_{water(g)} \times \Delta T = (2.50 \text{ mol}) (33.1 \text{ J/mol·°C}) (100°C - 130°C)$$
$$= -2482 \text{ J} = -2.48 \text{ kJ}$$

The minus sign indicates that heat is released. (For purposes of canceling, the units for molar heat capacity, C, include °C, rather than K. The magnitude of C is not affected because the kelvin and the Celsius degree represent the same temperature increment.)

Stage 2. Gaseous water condenses. At the condensation point, the slowest of the molecules are near each other long enough for intermolecular attractions to form groups of molecules, which aggregate into microdroplets and then a bulk liquid. Note that while the state is changing from gas to liquid, *the temperature remains constant*, so the average E_k is constant. This means that, even though molecules in a gas certainly move *farther* between collisions than those in a liquid, their average *speed* is the same at a given temperature. Thus, removing heat from the system involves a decrease in the average E_p, as the molecules approach and attract each other more strongly, but not a decrease in the average E_k. In other words, at $100°C$, gaseous water and liquid water have the same average E_k, but the liquid has lower E_p.

The change is $H_2O(g)$ $[100°C]$ $\longrightarrow$ $H_2O(l)$ $[100°C]$. The heat released is the amount (n) times the negative of the heat of vaporization ($-\Delta H^0_{vap}$):

$$q = n(-\Delta H^0_{vap}) = (2.50 \text{ mol}) (-40.7 \text{ kJ/mol}) = -102 \text{ kJ}$$

This step contributes the *greatest portion of the total heat released* because of the decrease in potential energy that occurs with the enormous decrease in distance between molecules in a gas and those in a liquid.

Stage 3. Liquid water cools. The molecules have now condensed to the liquid state. The continued loss of heat appears as a decrease in temperature, that is, as a decrease in the most probable molecular speed and, thus, the average E_k. The temperature decreases as long as the sample remains liquid.

The change is $H_2O(l)$ $[100°C]$ $\longrightarrow$ $H_2O(l)$ $[0°C]$. The heat depends on amount (n), the molar heat capacity of *liquid water,* and ΔT:

$$q = n \times C_{water(l)} \times \Delta T = (2.50 \text{ mol}) (75.4 \text{ J/mol·°C}) (0°C - 100°C)$$
$$= -18850 \text{ J} = -18.8 \text{ kJ}$$

Stage 4. Liquid water freezes. At the freezing temperature of water, $0°C$, intermolecular attractions overcome the motion of the molecules around one another. Beginning with the slowest, the molecules lose E_p and align themselves into the crystalline structure of ice. Molecular motion continues, but only as vibration of atoms about their fixed positions. As during condensation, the temperature and average E_k remain constant during freezing.

The change is $H_2O(l)$ $[0°C]$ $\longrightarrow$ $H_2O(s)$ $[0°C]$. The heat released is n times the negative of the heat of fusion ($-\Delta H^0_{fus}$):

$$q = n(-\Delta H^0_{fus}) = (2.50 \text{ mol}) (-6.02 \text{ kJ/mol}) = -15.0 \text{ kJ}$$

Stage 5. Solid water cools. With motion restricted to jiggling in place, further cooling merely reduces the average speed of this jiggling.

The change is $H_2O(s)$ $[0°C]$ $\longrightarrow$ $H_2O(s)$ $[-40°C]$. The heat released depends on n, the molar heat capacity of *solid water,* and ΔT:

$$q = n \times C_{water(s)} \times \Delta T = (2.50 \text{ mol}) (37.6 \text{ J/mol·°C}) (-40°C - 0°C)$$
$$= -3760 \text{ J} = -3.76 \text{ kJ}$$

According to Hess's law, the total heat released is the sum of the heats released for the individual stages. The sum of q for stages 1 to 5 is -142 kJ.

Two key points stand out in this or any similar process (at constant pressure), whether exothermic or endothermic:

- *Within a phase,* a change in heat is accompanied by *a change in temperature,* which is associated with a change in average E_k as *the most probable speed of the molecules changes.* The heat lost or gained depends on the amount of substance, the molar heat capacity for that phase, and the change in temperature.
- *During a phase change,* a change in heat occurs at a *constant temperature,* which is associated with a change in E_p, as *the average distance between molecules changes.* Both physical states are present during a phase change. The heat lost or gained depends on the amount of substance and the enthalpy of the phase change (ΔH^0_{vap} or ΔH^0_{fus}).

The Equilibrium Nature of Phase Changes

In everyday experience, phase changes take place in open containers—the outdoors, a pot on a stove, the freezer compartment of a refrigerator—so such a change is not reversible. In a closed container under controlled conditions, however, *phase changes of many substances are reversible and reach equilibrium,* just as chemical changes do.

Liquid-Gas Equilibria Picture an *open* flask containing a pure liquid at constant temperature and focus on the molecules at the surface. Within their range of molecular speeds, some are moving fast enough and in the right direction to overcome attractions, so they vaporize. Nearby molecules immediately fill the gap, and with energy supplied by the constant-temperature surroundings, the process continues until the entire liquid phase is gone.

Now picture starting with a *closed* flask at constant temperature, as in Figure 12.4A, and assume that a vacuum exists above the liquid. As before, some of the molecules at the surface have a high enough E_k to vaporize. As the number of molecules in the vapor phase increases, the pressure of the vapor increases. At the same time, some of the molecules in the vapor that collide with the surface have a low enough E_k to become attracted too strongly to leave the liquid and they condense. For a given surface area, the number of molecules that make up the surface is constant; therefore, the rate of vaporization—the number of molecules leaving the surface per unit time—is also constant. On the other hand, as the vapor becomes more populated, molecules collide with the surface more often, so the rate of condensation slowly increases. As condensation continues to offset vaporization, the increase in the pressure of the vapor slows. Eventually, the rate of condensation equals the rate of vaporization, as depicted in Figure 12.4B. From this time onward, *the pressure of the vapor is constant at that temperature.* Macroscopically, the situation seems static, but at the molecular level, molecules are entering and leaving the liquid surface at equal rates. The system has reached a state of *dynamic equilibrium:*

VAPOR PRESSURE

$$\text{Liquid} \rightleftharpoons \text{gas}$$

Figure 12.4C depicts the entire process graphically.

The pressure exerted by the vapor at equilibrium is called the *equilibrium vapor pressure*, or just the **vapor pressure,** of the liquid at that temperature. If we use a larger flask, more molecules are present in the gas phase at equilibrium; as long as some liquid remains, however, the vapor pressure does not change. Suppose we disturb this equilibrium system by pumping out some of the vapor at constant temperature, thereby lowering the pressure. (If this were a cylinder with piston, we could lower the pressure by moving the piston outward to increase the volume.) The rate of condensation temporarily falls below the rate of vaporization (the forward process is faster). Fewer molecules re-enter the liquid than leave it, so the pressure of the vapor rises until, after a short period, the conden-

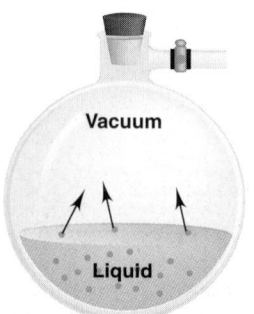

A Molecules in liquid vaporize.

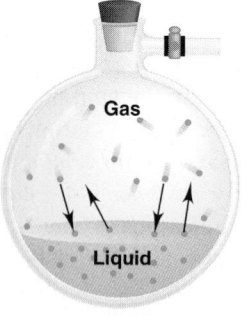

B Molecules enter and leave liquid at same rate.

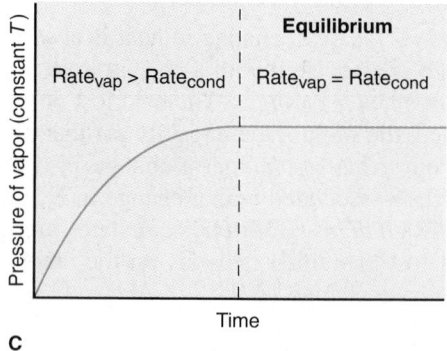

C

Figure 12.4 **Liquid-gas equilibrium. A,** In a closed flask at constant temperature with the air removed, the initial pressure is zero. As molecules leave the surface and enter the space above the liquid, the pressure of the vapor rises. **B,** At equilibrium, the same number of molecules leave as enter the liquid within a given time, so the pressure of the vapor reaches a constant value. **C,** A graph of pressure vs. time shows that the pressure of the vapor increases as long as the rate of vaporization is greater than the rate of condensation. At equilibrium, the pressure is constant because the rates are equal. The pressure at this point is the *vapor pressure* of the liquid at that temperature.

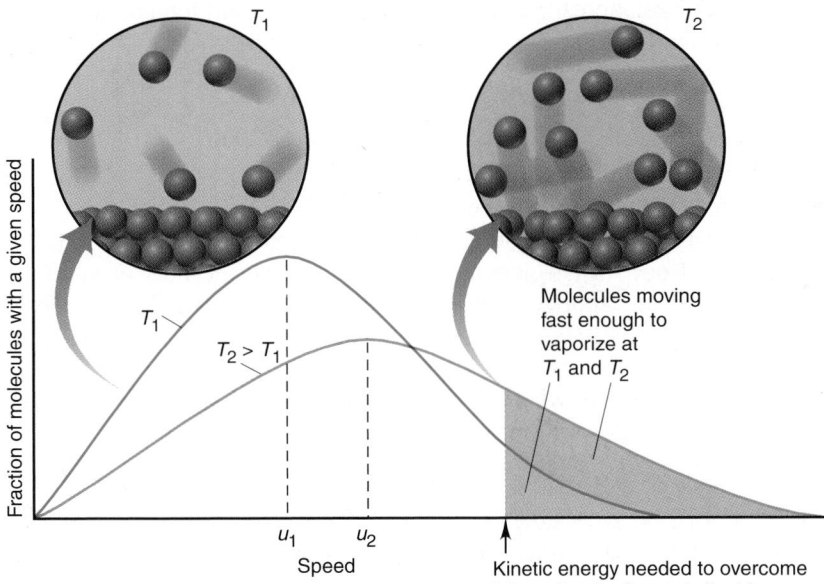

Figure 12.5 **The effect of temperature on the distribution of molecular speeds in a liquid.** With the temperature T_1 lower than T_2, the most probable molecular speed u_1 is less than u_2. (Note the similarity to Figure 5.14, p. 198.) The fraction of molecules with enough energy to escape the liquid *(shaded area)* is *greater at the higher temperature*. The molecular views show that at the higher T, equilibrium is reached with more gas molecules in the same volume and thus at a higher vapor pressure.

sation rate increases enough for equilibrium to be reached again. Similarly, if we disturb the system by pumping in additional vapor (or moving the piston inward to decrease the volume) at constant temperature, thereby raising the pressure, the rate of condensation temporarily exceeds the rate of vaporization. More molecules enter the liquid than leave it (the backward process is faster), but soon the condensation rate decreases, and the pressure again reaches the equilibrium value. This behavior of a pure liquid in contact with its vapor is a general one for any system: *when a system at equilibrium is disturbed, it counteracts the disturbance and eventually reestablishes a state of equilibrium.* We'll return to this key idea often in later chapters.

The Effects of Temperature and Intermolecular Forces on Vapor Pressure The vapor pressure of a substance depends on the temperature. Raising the temperature of a liquid *increases* the fraction of molecules moving fast enough to escape the liquid and *decreases* the fraction moving slowly enough to be recaptured. This important idea is shown in Figure 12.5. In general, *the higher the temperature is, the higher the vapor pressure.*

The vapor pressure also depends on the intermolecular forces present. The average E_k is the same for different substances at a given temperature. Therefore, molecules with weaker intermolecular forces vaporize more easily. In general, *the weaker the intermolecular forces are, the higher the vapor pressure.*

Figure 12.6 shows the vapor pressure of three liquids as a function of temperature. There are two points to notice. First, each curve rises more steeply as the temperature increases. Second, at a given temperature, the substance with the weakest intermolecular forces has the highest vapor pressure: the intermolecular forces in diethyl ether are weaker than those in ethanol, which are weaker than those in water.

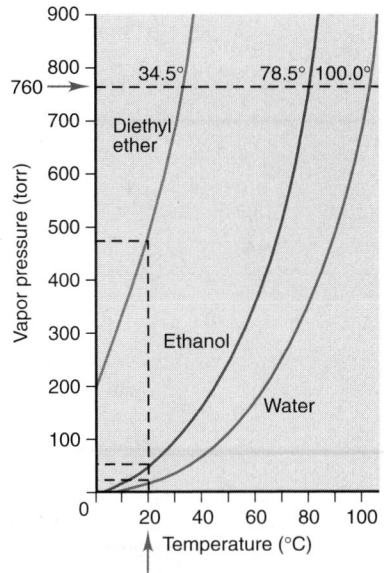

Figure 12.6 **Vapor pressure as a function of temperature and intermolecular forces.** The vapor pressures of three liquids are plotted against temperature. At any given temperature (see the vertical dashed line at 20°C), diethyl ether has the highest vapor pressure and water the lowest, because diethyl ether has the weakest intermolecular forces and water the strongest. The horizontal dashed line at 760 torr shows the normal boiling points of the liquids, the temperature at which the vapor pressure equals atmospheric pressure at sea level.

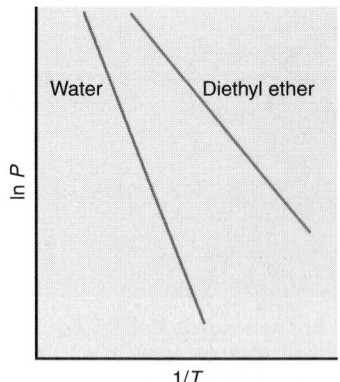

Figure 12.7 A linear plot of the vapor pressure–temperature relationship. The Clausius-Clapeyron equation gives a straight line when the natural logarithm of the vapor pressure (ln P) is plotted against the inverse of the absolute temperature (1/T). The slopes ($-\Delta H_{vap}/R$) allow determination of the heats of vaporization of the two liquids. Note that the slope is steeper for water because its ΔH_{vap} is greater.

The nonlinear relationship between vapor pressure and temperature shown in Figure 12.6 can be expressed as a linear relationship between ln P and 1/T:

$$\ln P = \frac{-\Delta H_{vap}}{R}\left(\frac{1}{T}\right) + C$$
$$y \quad = \quad m \quad \; x \; + \; b$$

where ln P is the natural logarithm of the vapor pressure, ΔH_{vap} is the heat of vaporization, R is the universal gas constant (8.31 J/mol·K), T is the absolute temperature, and C is a constant (not related to heat capacity). This is the **Clausius-Clapeyron equation,** which gives us a way of finding the heat of vaporization, the energy needed to vaporize 1 mol of molecules in the liquid state. The blue equation beneath the Clausius-Clapeyron equation is the equation for a straight line, where $y = \ln P$, $x = 1/T$, m (the slope) $= -\Delta H_{vap}/R$, and b (the y-axis intercept) $= C$. Figure 12.7 shows plots for diethyl ether and water. A two-point version of the equation allows a nongraphical determination of ΔH_{vap}:

$$\ln \frac{P_2}{P_1} = \frac{-\Delta H_{vap}}{R}\left(\frac{1}{T_2} - \frac{1}{T_1}\right) \tag{12.1}$$

If ΔH_{vap} and P_1 at T_1 are known, we can calculate the vapor pressure (P_2) at any other temperature (T_2) or the temperature at any other pressure.

SAMPLE PROBLEM 12.1 Using the Clausius-Clapeyron Equation

Problem The vapor pressure of ethanol is 115 torr at 34.9°C. If ΔH_{vap} of ethanol is 40.5 kJ/mol, calculate the temperature (in °C) when the vapor pressure is 760 torr.
Plan We are given ΔH_{vap}, P_1, P_2, and T_1 and substitute them into Equation 12.1 to solve for T_2. The value of R here is 8.31 J/mol·K, so we must convert T_1 to K to obtain T_2, and then convert T_2 to °C.
Solution Substituting the values into Equation 12.1 and solving for T_2:

$$\ln \frac{P_2}{P_1} = \frac{-\Delta H_{vap}}{R}\left(\frac{1}{T_2} - \frac{1}{T_1}\right)$$

$$T_1 = 34.9°C + 273.15 = 308.0 \text{ K}$$

$$\ln \frac{760 \text{ torr}}{115 \text{ torr}} = \left(-\frac{40.5 \times 10^3 \text{ J/mol}}{8.314 \text{ J/mol·K}}\right)\left(\frac{1}{T_2} - \frac{1}{308.0 \text{ K}}\right)$$

$$1.888 = (-4.87 \times 10^3)\left[\frac{1}{T_2} - (3.247 \times 10^{-3})\right]$$

$$T_2 = 350. \text{ K}$$

Converting T_2 from K to °C:

$$T_2 = 350. \text{ K} - 273.15 = \boxed{77°C}$$

Check Round off to check the math. The change is in the right direction: higher P should occur at higher T. As we discuss next, a substance has a vapor pressure of 760 torr at its normal boiling point. Checking the *CRC Handbook of Chemistry and Physics* shows that the boiling point of ethanol is 78.5°C, very close to our answer.

FOLLOW-UP PROBLEM 12.1 At 34.1°C, the vapor pressure of water is 40.1 torr. What is the vapor pressure at 85.5°C? The ΔH_{vap} of water is 40.7 kJ/mol.

Vapor Pressure and Boiling Point In an *open* container, the atmosphere bears down on the liquid surface. As the temperature rises, molecules leave the surface more often, and they also move more quickly throughout the liquid. At some temperature, the average E_k of the molecules in the liquid is great enough for bubbles of vapor to form *in the interior,* and the liquid boils. At any lower temperature, the bubbles collapse as soon as they start to form because the external pressure is greater than the vapor pressure inside the bubbles. Thus, the **boiling point** is *the temperature at which the vapor pressure equals the external pressure,* usually that of the atmosphere. Once boiling begins, the temperature remains constant until the liquid phase is gone because the applied heat is used by the molecules to overcome attractions and enter the gas phase.

The boiling point varies with elevation, because the atmospheric pressure does. At high elevations, a lower pressure is exerted on the liquid surface, so molecules in the interior need less kinetic energy to form bubbles. At low elevations, the opposite is true. Thus, *the boiling point depends on the applied pressure.* The *normal boiling point* is observed at standard atmospheric pressure (760 torr, or 101.3 kPa; see the horizontal dashed line in Figure 12.6).

Since the boiling point is the temperature at which the vapor pressure equals the external pressure, we can also interpret the curves in Figure 12.6 as a plot of external pressure vs. boiling point. For instance, the H_2O curve shows that water boils at 100°C at 760 torr (sea level), at 94°C at 610 torr (Boulder, Colorado), and at about 72°C at 270 torr (top of Mt. Everest).

Solid-Liquid Equilibria At the molecular level, the particles in a crystal are continually vibrating about their fixed positions. As the temperature rises, the particles vibrate more violently, until some have enough kinetic energy to break free of their positions, and melting begins. As more molecules enter the liquid (molten) phase, some collide with the solid and become fixed again. Because the phases remain in contact, a dynamic equilibrium is established when the melting rate equals the freezing rate. The temperature at which this occurs is the **melting point;** it is the same temperature as the freezing point, differing only in the direction of the energy flow. As with the boiling point, the temperature remains at the melting point as long as both phases are present.

Because liquids and solids are nearly incompressible, a change in pressure has little effect on the rate of movement to or from the solid. Therefore, in contrast to the boiling point, the melting point is affected by pressure only very slightly, and a plot of pressure (*y* axis) vs. temperature (*x* axis) for a solid-liquid phase change is typically a straight, *nearly* vertical line.

Solid-Gas Equilibria Solids have much lower vapor pressures than liquids. Sublimation, the process of a solid changing directly into a gas, is much less familiar than vaporization because the necessary conditions of pressure and temperature are uncommon for most substances. Some solids *do* have high enough vapor pressures to sublime at ordinary conditions, including dry ice (carbon dioxide), iodine (Figure 12.8), moth repellents, and solid room deodorizers. A substance sublimes rather than melts because the combination of intermolecular attractions and atmospheric pressure is not great enough to keep the particles near one another when they leave the solid state. The pressure vs. temperature plot for the solid-gas transition shows a large effect of temperature on the pressure of the vapor; thus, it resembles the liquid-gas line in curving upward at higher temperatures.

Figure 12.8 Iodine subliming. At ordinary atmospheric pressure, solid iodine sublimes (changes directly from a solid into a gas). When the I_2 vapor comes in contact with a cold surface, such as the water-filled inner tube, it deposits I_2 crystals. Sublimation is a means of purification that, along with distillation, was probably discovered by the alchemists.

Cooking Under Low or High Pressure People who live or hike in mountainous regions cook their meals under lower atmospheric pressure. The resulting lower boiling point of the liquid means the food takes *more* time to cook. On the other hand, in a pressure cooker, the pressure exceeds that of the atmosphere, so the temperature rises above the normal boiling point. The food becomes hotter and, thus, takes *less* time to cook.

Phase Diagrams: The Effect of Pressure and Temperature on Physical State

To describe the phase changes of a substance at various conditions of temperature and pressure, we construct a **phase diagram,** which combines the liquid-gas, solid-liquid, and solid-gas curves. The shape of the phase diagram for CO_2, shown in Figure 12.9A, is typical for most substances. A phase diagram has these three features:

1. *Regions of the diagram.* Each region corresponds to one phase of the substance. A particular phase is stable for any combination of pressure and temperature within its region. If any of the other phases is placed under those conditions, it will change to the stable phase. In general, the solid is stable at low temperature and high pressure, the gas at high temperature and low pressure, and the liquid at intermediate conditions.

2. *Lines between regions.* The lines separating the regions represent the phase-transition curves discussed earlier. Any point along a line shows the pressure and temperature at which the two phases exist in equilibrium. Note that the solid-liquid line has a *positive* slope (slants to the *right* with increasing pressure) because, for most substances, the solid is more dense than the liquid. Because the liquid occupies slightly more space than the solid, an increase in pressure favors the solid phase, in most cases. (Water is the major exception, as you'll see later in the chapter.)

3. *The critical point.* The liquid-gas line ends at the **critical point.** Picture a liquid in a closed container. As it is heated, it expands, so its density decreases. At the same time, more liquid vaporizes, so the density of the vapor increases. The liquid and vapor densities become closer and closer to each other until, at the *critical temperature* (T_c), the two densities are equal and the phase boundary disappears. The pressure at this temperature is the *critical pressure* (P_c). At this point, the average E_k of the molecules is so high that the vapor cannot be condensed no matter what pressure is applied. The two most common gases in air have values far below room temperature: no matter what the pressure, O_2 will not condense above $-119°C$, and N_2 will not condense above $-147°C$. Beyond the critical temperature, a *supercritical fluid* exists rather than separate liquid and gaseous phases. (See margin note on facing page.) ●

Figure 12.9 **Phase diagrams for CO_2 and H_2O.** Each region depicts the temperatures and pressures under which the phase is stable. The lines between any two regions show the conditions at which the two phases exist in equilibrium. The critical point shows the conditions beyond which separate liquid and gas phases no longer exist. Above the critical point, the liquid phase cannot exist, regardless of the pressure. At the triple point, the three phases exist in equilibrium. (The axes are not linear.) **A,** The phase diagram for CO_2 is typical of most substances in that the solid-liquid line slopes to the right with increasing pressure: the solid is *more* dense than the liquid. **B,** Water is one of the few substances whose solid-liquid line slopes to the left with increasing pressure: the solid is *less* dense than the liquid. (The slopes of the solid-liquid lines in both diagrams are exaggerated.)

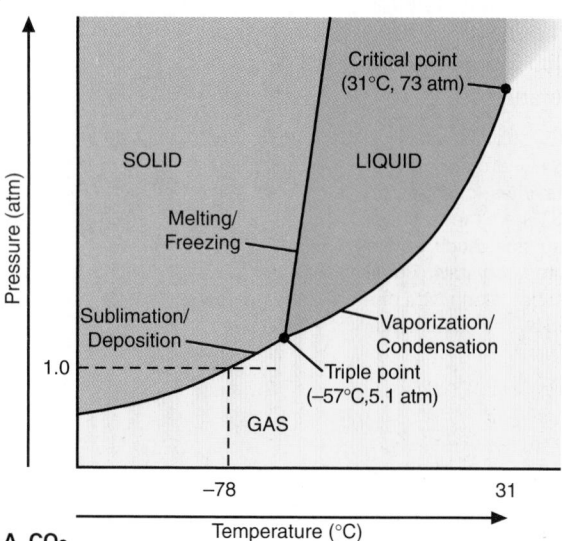

A CO_2

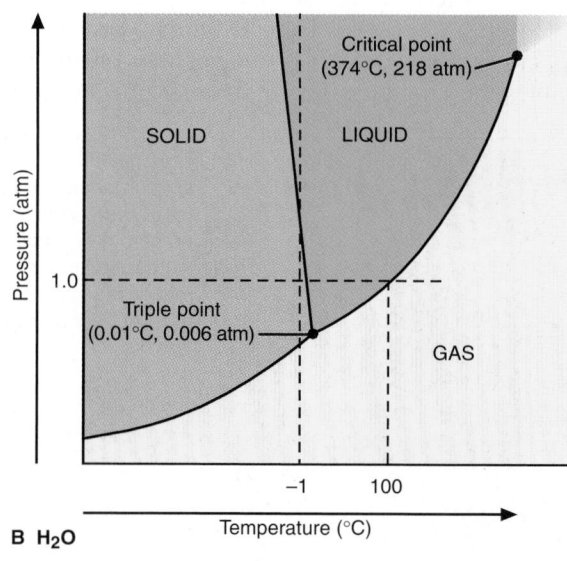

B H_2O

4. *The triple point.* The three phase-transition curves meet at the **triple point:** the pressure and temperature at which three phases are in equilibrium. Phase diagrams for substances with several solid forms, such as sulfur, have more than one triple point. As strange as it sounds, at the triple point in Figure 12.9A, CO_2 is subliming and depositing, melting and freezing, and vaporizing and condensing simultaneously!

The CO_2 phase diagram explains why dry ice (solid CO_2) doesn't melt under ordinary conditions. The triple-point pressure for CO_2 is 5.1 atm; therefore, at around 1 atm, liquid CO_2 does not occur. By following the horizontal dashed line in Figure 12.9A, you can see that when solid CO_2 is heated at 1.0 atm, it sublimes at $-78°C$ to gaseous CO_2 rather than melting. If normal atmospheric pressure were 5.2 atm, liquid CO_2 would be common.

The phase diagram for water differs in one key respect from the general case and reveals an extremely important property (Figure 12.9B). Unlike almost every other substance, solid water is *less dense* than liquid water. Because the solid occupies more space than the liquid, *water expands on freezing.* This behavior results from the unique open crystal structure of ice, which we discuss in a later section. As always, an increase in pressure favors the phase that occupies less space, but in the case of water, this is the *liquid* phase. Therefore, the solid-liquid line for water has a *negative* slope (slants to the *left* with increasing pressure): the higher the pressure, the lower the temperature at which water freezes. For example, the vertical dashed line at $-1°C$ crosses the solid-liquid line, which means that ice melts with only an increase in pressure.

The triple point of water occurs at low pressure (0.006 atm). Therefore, when solid water is heated at 1.0 atm, the horizontal dashed line crosses the solid-liquid line (at 0°C, the normal melting point) and enters the liquid region. Thus, at ordinary pressures, ice melts rather than sublimes. As the temperature rises, the horizontal line crosses the liquid-gas curve (at 100°C, the normal boiling point) and enters the gas region.

SECTION SUMMARY

A heating-cooling curve depicts the change in temperature with heat loss or gain. Within a phase, temperature (and average E_k) change as heat is added or removed. During a phase change, temperature (and average E_k) are constant, but E_p changes. The total heat change for the curve is calculated using Hess's law. In a closed container, equilibrium is established between the liquid and gas phases. Vapor pressure, the pressure of the gas at equilibrium, is related directly to temperature and inversely to the strength of the intermolecular forces. The Clausius-Clapeyron equation uses ΔH_{vap} to relate the vapor pressure to the temperature. A liquid in an open container boils when its vapor pressure equals the external pressure. Solid-liquid equilibrium occurs at the melting point. Many solids sublime at low pressures and high temperatures. A phase diagram shows the phase that exists at a given pressure and temperature and the conditions at the critical point and the triple point of a substance. Water differs from most substances in that its solid phase is less dense than its liquid phase, so its solid-liquid line has a negative slope.

The Remarkable Behavior of a Supercritical Fluid (SCF) What sort of material lies beyond the familiar regions of liquid and gas? An SCF expands and contracts like a gas but has the solvent properties of a liquid, properties that chemists can alter by controlling the density. Supercritical CO_2 has received the most attention so far from government and industry. It extracts nonpolar ingredients from complex mixtures, such as caffeine from coffee beans, nicotine from tobacco, and fats from potato and corn chips, while leaving taste and aroma ingredients behind, to produce healthier consumer products. Lower the pressure and the SCF disperses immediately as a harmless gas. Supercritical CO_2 dissolves the fat from meat, along with pesticide and drug residues, which can then be quantified and monitored. It is currently being studied as an environmentally friendly dry cleaning agent. In an unexpected finding of great potential use, supercritical H_2O was shown to dissolve nonpolar substances even though liquid water does not! Studies are under way to carry out the large-scale removal of nonpolar organic toxins, such as PCBs, from industrial waste by extracting them into supercritical H_2O; after this step, O_2 gas is added to oxidize the toxins to small, harmless molecules.

PHASE DIAGRAMS AND THE
STATES OF MATTER

12.3 TYPES OF INTERMOLECULAR FORCES

As we said, the nature of the phases and their changes are due primarily to forces among the molecules. Both bonding (intramolecular) forces and intermolecular forces arise from electrostatic attractions between opposite charges. Bonding forces are due to the attraction between cations and anions (ionic bonding), nuclei and electron pairs (covalent bonding), or metal cations and delocalized valence electrons (metallic bonding). Intermolecular forces, on the other hand, are due to the attraction between molecules as a result of partial charges, or the attraction

Figure 12.10 Covalent and van der Waals radii. As shown here for solid chlorine, the van der Waals (VDW) radius is one-half the distance between adjacent *nonbonded* atoms ($\frac{1}{2}$ × VDW distance), and the covalent radius is one-half the distance between *bonded* atoms ($\frac{1}{2}$ × bond length).

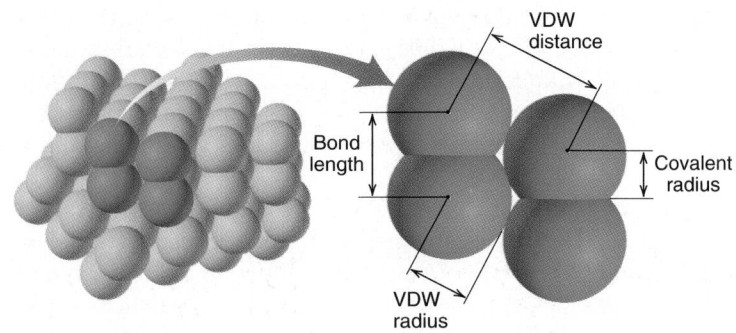

between ions and molecules. The two types of forces differ in magnitude, and Coulomb's law explains why:

- *Bonding forces are relatively strong* because they involve larger charges that are closer together.
- *Intermolecular forces are relatively weak* because they typically involve smaller charges that are farther apart.

How far apart are the charges between molecules that give rise to intermolecular forces? Consider Cl_2 as an example. When we measure the distances between two Cl nuclei in a sample of solid Cl_2, we obtain two different values, as shown in Figure 12.10. The shorter distance is between *two **bonded** Cl atoms in the same molecule.* It is, as you know, called the *bond length,* and one-half this distance is the *covalent radius.* The longer distance is between *two **nonbonded** Cl atoms in adjacent molecules.* It is called the *van der Waals distance* (named after the Dutch physicist Johannes van der Waals, who studied the effects of intermolecular forces on the behavior of real gases). This distance is the closest one Cl_2 molecule can approach another, the point at which intermolecular attractions balance electron-cloud repulsions. One-half this distance is the **van der Waals radius,** one-half the closest distance between the nuclei of identical *nonbonded* Cl atoms. *The van der Waals radius of an atom is always larger than its covalent radius,* but van der Waals radii decrease across a period and increase down a group, just as covalent radii do. Figure 12.11 shows these relationships for many of the nonmetals.

There are several types of intermolecular forces: ion-dipole, dipole-dipole, hydrogen bonding, dipole–induced dipole, and dispersion forces. As we discuss these *intermolecular* forces (also called *van der Waals forces*), look at Table 12.2, which compares them with the stronger *intramolecular* (bonding) forces and lists them in decreasing order of their relative strengths.

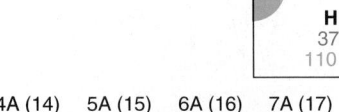

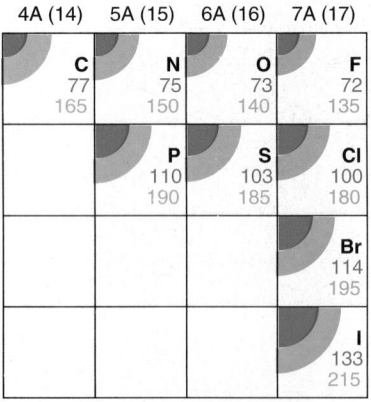

Figure 12.11 Periodic trends in covalent and van der Waals radii (in pm). Like covalent radii *(blue quarter-circles and top numbers)*, van der Waals radii *(green quarter-circles and bottom numbers)* increase down a group and decrease across a period. The covalent radius of an element is *always* less than its van der Waals radius.

Ion-Dipole Forces

When an ion and a nearby polar molecule (dipole) attract each other, an **ion-dipole force** results. The most important example takes place when an ionic compound dissolves in water. The ions become separated because the attractions between the ions and the oppositely charged poles of the H_2O molecules overcome the attractions between the ions themselves. Ion-dipole forces in solutions and their associated energy are discussed fully in Chapter 13.

Dipole-Dipole Forces

In Figure 10.14 (p. 381), you saw how gaseous polar molecules are oriented by a large, external electric field. When polar molecules lie near one another, as in liquids and solids, their partial charges act as tiny electric fields that orient them and give rise to **dipole-dipole forces:** the positive pole of one molecule attracts the negative pole of another. Figure 12.12 depicts these orientations.

Table 12.2	Comparison of Bonding and Nonbonding (Intermolecular) Forces			
Force	**Model**	**Basis of Attraction**	**Energy (kJ/mol)**	**Example**
Bonding				
Ionic		Cation—anion	400–4000	NaCl
Covalent		Nuclei—shared e^- pair	150–1100	H—H
Metallic		Cations—delocalized electrons	75–1000	Fe
Nonbonding (Intermolecular)				
Ion-dipole		Ion charge—dipole charge	40–600	$Na^+ \cdots O\!\!<^H_H$

H bond	$\delta^- \;\; \delta^+ \qquad \delta^-$
	 —A—H········:B—
		Polar bond to H—dipole charge (high EN of N, O, F)	10–40	:Ö—H····:Ö—H
				   | |
				   H H

Dipole-dipole		Dipole charges	5–25	I—Cl····I—Cl
Ion–induced dipole		Ion charge—polarizable e^- cloud	3–15	$Fe^{2+} \cdots O_2$
Dipole–induced dipole		Dipole charge—polarizable e^- cloud	2–10	H—Cl····Cl—Cl
Dispersion (London)		Polarizable e^- clouds	0.05–40	F—F····F—F

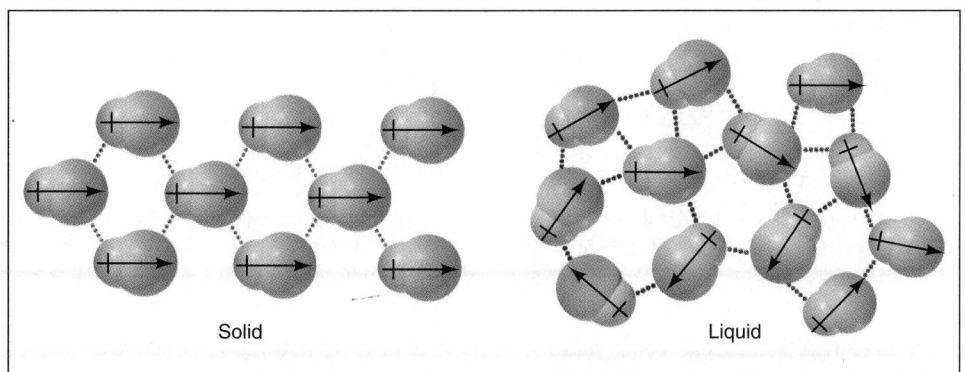

Figure 12.12 Orientation of polar molecules because of dipole-dipole forces. In solid and liquid states, the polar molecules are close enough for partial charges of one molecule to attract nearby opposite charges, and the molecules become oriented. The arrangement is more orderly in the solid phase (*left*) of a substance than in the liquid (*right*) because, at the lower temperatures required for freezing, the average kinetic energy of the particles is lower. (Interparticle spaces are increased for clarity.)

Solid

Liquid

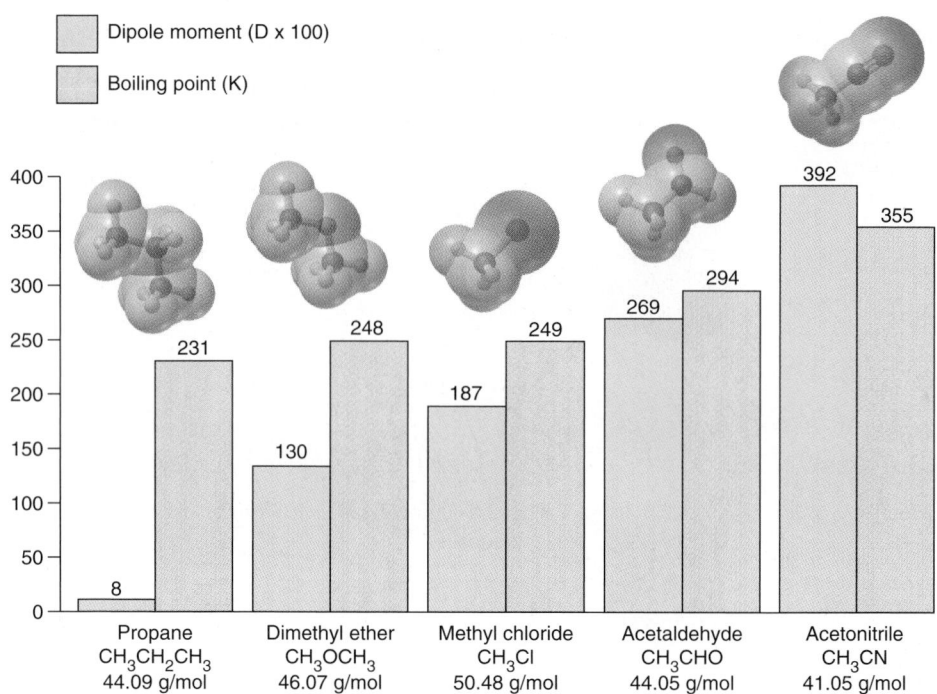

Figure 12.13 Dipole moment and boiling point. For compounds of similar molar mass, the boiling point increases with increasing dipole moment. (Note the increasing color intensities in the electron density models.) The greater dipole moment creates stronger dipole-dipole forces, which require higher temperatures to overcome.

These are the forces that give polar *cis*-1,2-dichloroethylene a higher boiling point than the nonpolar *trans* compound (see p. 383). In fact, for molecular compounds of approximately the same size and molar mass, the greater the dipole moment, the greater the dipole-dipole forces between the molecules are, and so the more energy it takes to separate them. Consider the boiling points of the compounds in Figure 12.13. Methyl chloride, for instance, has a smaller dipole moment than acetaldehyde, so less energy is needed to overcome the dipole-dipole forces between its molecules and it boils at a lower temperature.

The Hydrogen Bond

A special type of dipole-dipole force arises between molecules that have *an H atom bound to a small, highly electronegative atom with lone electron pairs*. The most important atoms that fit this description are N, O, and F. The H—N, H—O, and H—F bonds are very polar, so electron density is withdrawn from H. As a result, the partially positive H of one molecule is attracted to the partially negative lone pair on the N, O, or F of another molecule, and a **hydrogen bond (H bond)** forms. Thus, the atom sequence that leads to an H bond (dotted line) is —B:⋯⋯H—A— , where *both* A and B are N, O, or F. Three such examples are

$$—\ddot{F}:\cdots\cdots H—\ddot{O}— \qquad —\ddot{O}:\cdots\cdots H—\ddot{N}— \qquad —\ddot{N}:\cdots\cdots H—\ddot{F}:$$

The small sizes of N, O, and F* are essential to H-bonding for two reasons:

1. It makes these atoms so electronegative that their covalently bonded H is highly positive.
2. It allows the lone pair on the other N, O, or F to come close to the H.

---

*H-bond–type interactions occur with the larger P, S, and Cl, but those are so much weaker than the interactions with N, O, and F that we will not consider them.

SAMPLE PROBLEM 12.2 Drawing Hydrogen Bonds Between Molecules of a Substance

Problem Which of the following substances exhibits H bonding? For those that do, draw two molecules of the substance with the H bonds between them.

$$(a)\ C_2H_6 \qquad (b)\ CH_3OH \qquad (c)\ CH_3\overset{\overset{\displaystyle O}{\|}}{C}{-}NH_2$$

Plan We draw each structure to see if it contains N, O, or F covalently bonded to H. If it does, we draw two molecules of the substance in the —B:····H—A— pattern.

Solution (a) For C_2H_6. No H bonds are formed.

(b) For CH_3OH. The H covalently bonded to the O in one molecule forms an H bond to the lone pair on the O of an adjacent molecule:

H
|
H—C—H   H
|
H—Ö:····H—Ö—C—H
|
H

(c) For $CH_3\overset{\overset{\displaystyle O}{\|}}{C}{-}NH_2$. Two of these molecules can form one H bond from the H of N to O, or they can form two such H bonds:

H :O: H   H
| || | |
H—C—C—N—H····:Ö=C—C—H   or   H—C—C ...

A third possibility (not shown) could be between an H attached to N on one molecule and the N lone pair of another molecule.

Check The —B:····H—A— sequence (with A and B either N, O, or F) is present.

Comment Note that *H covalently bonded to C does not form H bonds* because carbon is not electronegative enough to make the C—H bond very polar.

FOLLOW-UP PROBLEM 12.2 Which of these substances exhibits H bonding? Draw the H bonds between two molecules of the substance where appropriate.

$$(a)\ CH_3\overset{\overset{\displaystyle O}{\|}}{C}{-}OH \qquad (b)\ CH_3CH_2OH \qquad (c)\ CH_3\overset{\overset{\displaystyle O}{\|}}{C}CH_3$$

The Significance of H Bonding H bonding has a profound impact in many systems. Figure 12.14 shows its effect on the boiling points of the binary hydrides of Groups 4A(14) through 7A(17). For reasons that we'll discuss shortly, boiling points typically rise as molar mass increases, as you can see in the Group 4A hydrides, CH_4 through SnH_4. In the other groups, however, the first member in each series—NH_3, H_2O, and HF—deviates enormously from this expected increase because H bonds keep these molecules together. Therefore, additional energy is required to break the H bonds before the molecules can separate and enter the gas phase. For example, on the basis of molar mass alone, we would expect water to boil about 200°C lower than it does (dashed line in figure). (Water's H bonds have many major effects in nature; see Section 12.5.)

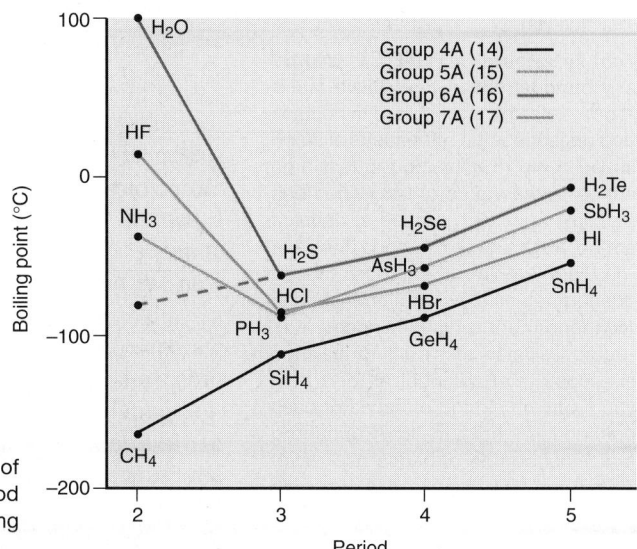

Figure 12.14 **Hydrogen bonding and boiling point.** The boiling points of the binary hydrides of Groups 4A(14) to 7A(17) are plotted against period number. The H bonds in NH_3, H_2O, and HF give them much higher boiling points than if the trend were based on molar mass, as it is for Group 4A.

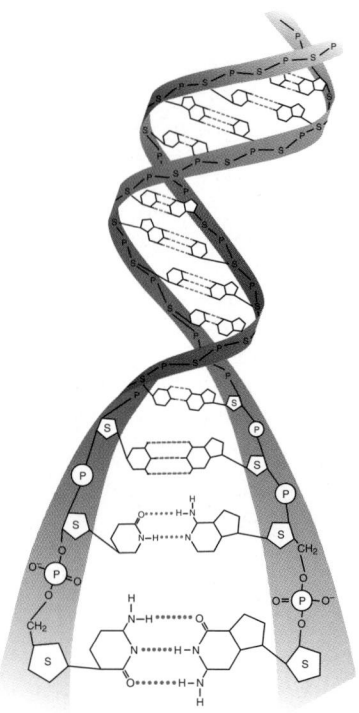

Figure 12.15 **Covalent bonding and H bonding in the structure of deoxyribonucleic acid (DNA).** The DNA molecule consists of two chains. Each is held together by strong covalent bonds, but millions of H bonds link one chain to the other to form a double helix.

Even though the strength of one H bond is relatively small (~5% of a typical covalent single bond energy), the combined strength of many H bonds can be large. Consider the H bonds in deoxyribonucleic acid (DNA), the giant molecule in all cells that acts as a genetic "blueprint," governing the function and appearance of the entire organism. As Figure 12.15 shows, DNA consists of two molecular chains wrapped around each other in a long double helix (see also Gallery, p. 73). Strong covalent bonds link the atoms in each chain, but the two intertwining chains are joined by millions of H bonds. The total energy of the H bonds keeps the chains together during many processes; each H bond is so weak, however, that a few at a time break so that the chains can separate during the crucial processes of protein synthesis and cell reproduction. (We discuss these processes in Chapter 15.)

Polarizability and Charge-Induced Dipole Forces

Even though electrons are localized in bonding or lone pairs, they are in constant motion, so that we often picture them as "clouds" of negative charge. A nearby electric field can distort a cloud, pulling electron density toward a positive charge or pushing it away from a negative charge. In effect, the field *induces* a distortion in the electron cloud. For a nonpolar molecule, this distortion results in a temporary, induced dipole moment; for a polar molecule, it enhances the dipole moment that is already present. The source of the electric field can be the electrodes of a battery, the charge of a nearby ion, or even the partial charges of a nearby polar molecule.

The ease with which a particle's electron cloud can be distorted is called its **polarizability.** Smaller atoms (or ions) are less polarizable than larger ones because their electrons are closer to the nucleus and held more tightly. Thus,

- *Polarizability increases down a group* of atoms (or ions) because size increases and larger electron clouds are more easily distorted.
- *Polarizability decreases from left to right across a period* because the increasing effective nuclear charge holds the electrons more tightly.
- Cations are *less* polarizable than their parent atom because they are smaller, whereas anions are *more* polarizable because they are larger.

Ion–induced dipole and dipole–induced dipole forces are the two types of charge-induced dipole forces; they are most important in solution, so we'll focus on them in Chapter 13. Nevertheless, the concept of polarizability is a general one that applies to all intermolecular forces.

Dispersion (London) Forces

Up to this point, we've discussed intermolecular forces that depend on an existing charge, either in an ion or a polar molecule. But what forces cause nonpolar substances like octane, chlorine, and the noble gases to condense and solidify? An attractive force must be acting between the particles, or these substances would be gases under any conditions. The intermolecular force primarily responsible for the condensed states of nonpolar substances is the **dispersion force** (or **London force,** named for Fritz London, the German physicist who first explained the quantum-mechanical basis of the attraction).

Dispersion forces are caused by *momentary oscillations of electron charge.* Over time, the electron charge within an atom is distributed uniformly around the nucleus, and the atom is nonpolar. But at any instant, this may not be the case, and the atom has an *instantaneous dipole* that can influence nearby atoms. Picture two argon atoms. When far apart, the atoms do not influence each other. When near each other, however, *the instantaneous dipole in each atom induces a dipole in its neighbor.* The result is a synchronized motion of the electrons in the

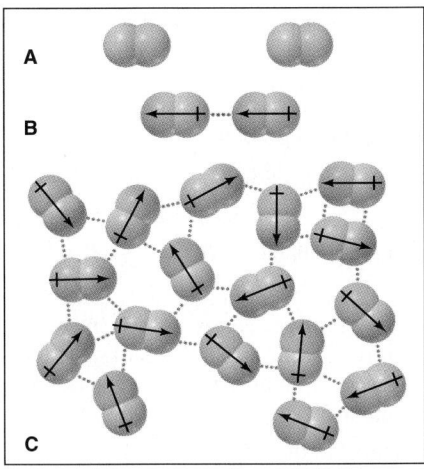

Figure 12.16 Dispersion forces among nonpolar molecules. The dispersion force is responsible for the condensed states of noble gases and nonpolar molecules. **A,** Separated Cl_2 molecules are nonpolar. **B,** An instantaneous dipole in each molecule induces a dipole in its neighbor. These partial charges attract the molecules together. **C,** This process takes place among molecules throughout the sample.

two atoms, which causes an attraction between them. This process occurs with other nearby atoms and, thus, throughout the sample. At low enough temperatures, the attractions among the dipoles keep all the atoms together. Thus, dispersion forces are *instantaneous dipole–induced dipole forces.* Figure 12.16 depicts the dispersion forces among nonpolar molecules.

*Dispersion forces are weak, but they exist between **any** particles.* Therefore, except for small, polar molecules with large dipole moments or those able to form H bonds, *the dispersion force is the dominant intermolecular force between identical molecules.* The relative strength of the dispersion force depends on the polarizability of the particle, which in turn depends on its size. In general terms, therefore, *dispersion forces increase with number of electrons, which correlates closely with molar mass* because particles with greater mass have more atoms and/or larger (heavier) atoms and thus more electrons. For example, Figure 12.17 shows that as molar mass increases down the halogens or noble gases, dispersion forces become greater, which is reflected in higher boiling points.

For nonpolar substances with the same molar mass, the strength of the dispersion forces may be influenced by molecular shape. Shapes that allow more points of contact have more area over which electron clouds can be distorted, so stronger attractions result. Consider the hydrocarbons *n*-pentane and neopentane, which have the same formula (C_5H_{12}) but different shapes. As shown in Figure 12.18, *n*-pentane is a chain of five carbon atoms and is somewhat cylindrical. Neopentane has two CH_3 branches off a three-carbon chain and, thus, a more spherical shape. As a result, two *n*-pentane molecules make more contact with each other than do two neopentane molecules. Greater contact allows the dispersion forces to act at more points, so *n*-pentane has the higher boiling point.

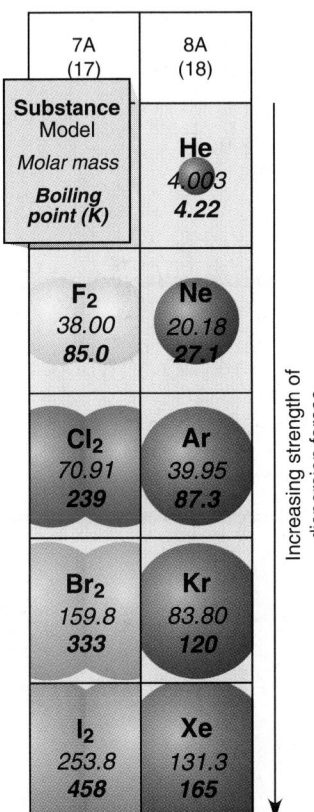

	7A (17)	8A (18)
Substance Model *Molar mass* **Boiling point (K)**		**He** 4.003 4.22
	F_2 38.00 85.0	**Ne** 20.18 27.1
	Cl_2 70.91 239	**Ar** 39.95 87.3
	Br_2 159.8 333	**Kr** 83.80 120
	I_2 253.8 458	**Xe** 131.3 165

Increasing strength of dispersion forces →

Figure 12.17 Molar mass and boiling point. The strength of dispersion forces increases with number of electrons, which usually correlates with molar mass. One effect of this is the higher boiling points going down the groups for halogens and noble gases.

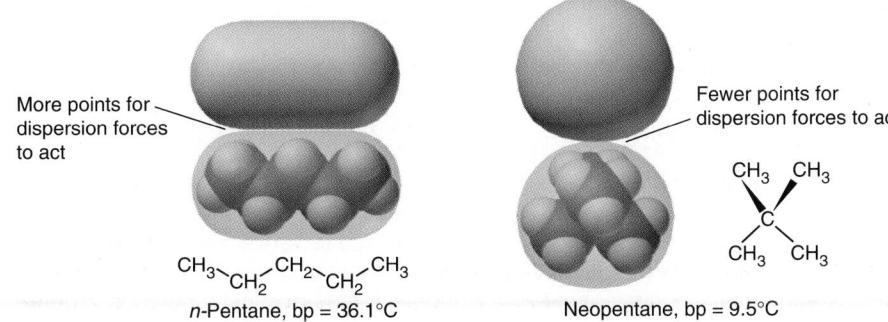

More points for dispersion forces to act

Fewer points for dispersion forces to act

CH_3 CH_3
\C/
CH_3 CH_3

CH₃～CH₂～CH₂～CH₂～CH₃
n-Pentane, bp = 36.1°C

Neopentane, bp = 9.5°C

Figure 12.18 Molecular shape and boiling point. The spherical neopentane molecules make less contact with each other than do the cylindrical *n*-pentane molecules, so neopentane has a lower boiling point.

SAMPLE PROBLEM 12.3 Predicting the Type and Relative Strength of Intermolecular Forces

Problem For each pair of substances, identify the dominant intermolecular forces in each substance, and select the substance with the higher boiling point:
(a) $MgCl_2$ or PCl_3
(b) CH_3NH_2 or CH_3F
(c) CH_3OH or CH_3CH_2OH
(d) Hexane ($CH_3CH_2CH_2CH_2CH_2CH_3$) or 2,2-dimethylbutane $CH_3\overset{\overset{\displaystyle CH_3}{|}}{\underset{\underset{\displaystyle CH_3}{|}}{C}}CH_2CH_3$

Plan We examine the formulas and picture (or draw) the structures to identify key differences between members of each pair: Are ions present? Are molecules polar or nonpolar? Is N, O, or F bonded to H? Do the molecules have different masses or shapes? To rank their strengths, we consult Table 12.2 and remember that

• Bonding forces are stronger than nonbonding (intermolecular) forces.
• Hydrogen bonding is a strong type of dipole-dipole force.
• Dispersion forces are decisive when the difference is molar mass or molecular shape.

Solution (a) $MgCl_2$ consists of Mg^{2+} and Cl^- ions held together by ionic bonding forces; PCl_3 consists of polar molecules, so intermolecular dipole-dipole forces are present. The forces in $MgCl_2$ are stronger, so it has a higher boiling point.
(b) CH_3NH_2 and CH_3F both consist of polar molecules of about the same molar mass. CH_3NH_2 has N—H bonds, so it can form H bonds (see margin). CH_3F contains a C—F bond but no H—F bond, so dipole-dipole forces occur but not H bonds. Therefore, CH_3NH_2 has the higher boiling point.
(c) CH_3OH and CH_3CH_2OH molecules both contain an O—H bond, so they can form H bonds (see margin). CH_3CH_2OH has an additional —CH_2— group and thus a larger molar mass, which correlates with stronger dispersion forces; therefore, it has a higher boiling point.
(d) Hexane and 2,2-dimethylbutane are nonpolar molecules of the same molar mass but different molecular shapes (see margin). Cylindrical hexane molecules can make more extensive intermolecular contact than the more spherical 2,2-dimethylbutane molecules can, so hexane should have greater dispersion forces and a higher boiling point.
Check The actual boiling points show that our predictions are correct:
(a) $MgCl_2$ (1412°C) and PCl_3 (76°C)
(b) CH_3NH_2 (−6.3°C) and CH_3F (−78.4°C)
(c) CH_3OH (64.7°C) and CH_3CH_2OH (78.5°C)
(d) Hexane (69°C) and 2,2-dimethylbutane (49.7°C)
Comment Dispersion forces are *always* present, but in parts (a) and (b), they are much less significant than the other forces involved.

(b)
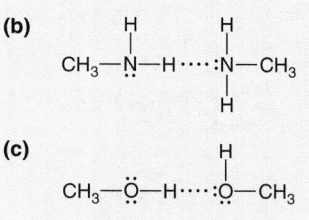

(c)
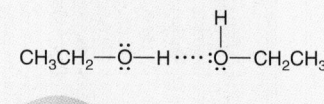

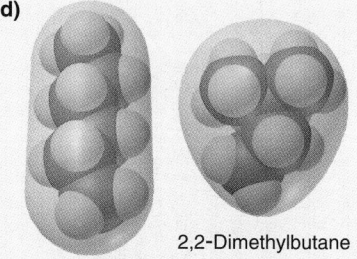

(d)

2,2-Dimethylbutane

Hexane

FOLLOW-UP PROBLEM 12.3 In each pair, identify all the intermolecular forces present for each substance, and select the substance with the higher boiling point:
(a) CH_3Br or CH_3F (b) $CH_3CH_2CH_2OH$ or $CH_3CH_2OCH_3$ (c) C_2H_6 or C_3H_8

SECTION SUMMARY

The van der Waals radius determines the shortest distance over which intermolecular forces operate; it is always larger than the covalent radius. Intermolecular forces are much weaker than bonding (intramolecular) forces. Ion-dipole forces occur between ions and polar molecules. Dipole-dipole forces occur between oppositely charged poles on polar molecules. Hydrogen bonding, a special type of dipole-dipole force, occurs when H bonded to N, O, or F is attracted to the lone pair of N, O, or F on another molecule. Electron clouds can be distorted (polarized) in an electric field. Ion– and dipole–induced dipole forces arise between a charge and the dipole it induces in another molecule. Dispersion (London) forces are instantaneous dipole–induced dipole forces that occur among all particles and increase with number of electrons (molar mass). Molecular shape determines the extent of contact between molecules and thus can be a factor in the strength of dispersion forces.

12.4 PROPERTIES OF THE LIQUID STATE

Of the three states of matter, the liquid is the least understood at the molecular level. Because of the *randomness* of the particles in a gas, any region of the sample is virtually identical to any other. As you'll see in Section 12.6, different regions of a crystalline solid are identical because of the *orderliness* of the particles. Liquids, however, have a combination of these attributes that changes continually: a region that is orderly one moment becomes random the next, and vice versa. Despite this complexity at the molecular level, the macroscopic properties of liquids are well understood. In this section, we discuss three liquid properties —surface tension, capillarity, and viscosity.

Surface Tension

In a liquid sample, intermolecular forces exert different effects on a molecule at the surface than on one in the interior (Figure 12.19). Interior molecules are attracted by others on all sides, whereas molecules at the surface have others only below and to the sides. As a result, molecules at the surface experience a *net attraction downward* and move toward the interior to increase attractions and become more stable. Therefore, *a liquid surface tends to have the smallest possible area,* that of a sphere, and behaves like a "taut skin" covering the interior.

To increase the surface area, molecules must move to the surface by breaking some attractions in the interior, which requires energy. The **surface tension** is the energy required to increase the surface area by a unit amount. Some representative values, in units of J/m^2, are presented in Table 12.3. By comparing these values with those in Table 12.2, you can see that, in general, *the stronger the forces are between the particles in a liquid, the greater the surface tension.* Water has a high surface tension because its molecules form multiple H bonds. *Surfactants* (*surf*ace-*act*ive age*nts*), such as soaps, detergents, petroleum recovery agents, and biological fat emulsifiers, decrease the surface tension of water by congregating at the surface and disrupting the H bonds.

Figure 12.19 The molecular basis of surface tension. Molecules in the interior of a liquid experience intermolecular attractions in all directions. Molecules at the surface experience a net attraction downward (red arrow) and move toward the interior. Thus, a liquid tends to minimize the number of molecules at the surface, which results in surface tension.

Table 12.3 Surface Tension and Forces Between Particles

Substance	Formula	Surface Tension (J/m^2) at 20°C	Major Force(s)
Diethyl ether	$CH_3CH_2OCH_2CH_3$	1.7×10^{-2}	Dipole-dipole; dispersion
Ethanol	CH_3CH_2OH	2.3×10^{-2}	H bonding
Butanol	$CH_3CH_2CH_2CH_2OH$	2.5×10^{-2}	H bonding; dispersion
Water	H_2O	7.3×10^{-2}	H bonding
Mercury	Hg	48×10^{-2}	Metallic bonding

Capillarity

The rising of a liquid through a narrow space against the pull of gravity is called *capillary action,* or **capillarity.** This phenomenon is applied during simple blood-screening tests, by pricking a finger and holding a narrow tube, called a *capillary tube,* against the skin opening (see photo). Capillarity results from a competition between the intermolecular forces within the liquid (cohesive forces) and those between the liquid and the tube walls (adhesive forces).

Picture what occurs at the molecular level when you place a glass capillary tube in water. Glass is mostly silicon dioxide (SiO_2), so the water molecules form H bonds to the oxygen atoms of the tube's inner wall. Because the adhesive forces (H bonding) between the water and the wall are stronger than the cohesive forces (H bonding) within the water, a thin film of water creeps up the wall. At the same

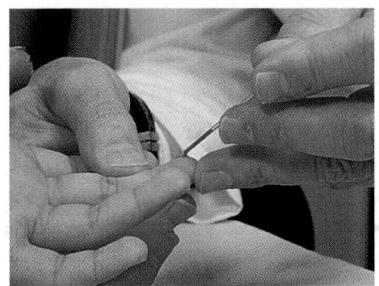

Blood being drawn into a capillary tube for medical tests.

Figure 12.20 Shape of water or mercury meniscus in glass. A, Water displays a concave meniscus in a glass tube because the adhesive (H-bond) forces between the H₂O molecules and the O—Si—O groups of the glass are *stronger* than the cohesive (H-bond) forces within the water. **B,** Mercury displays a convex meniscus in a glass tube because the cohesive (metallic bonding) forces within the mercury are *stronger* than the adhesive (dispersion) forces between the mercury and the glass.

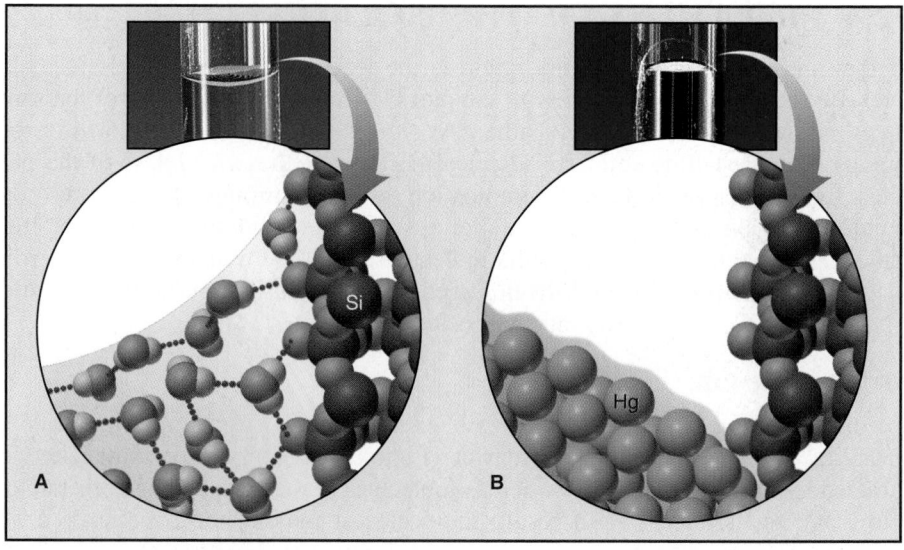

time, the cohesive forces that give rise to surface tension pull the liquid surface taut. These adhesive and cohesive forces combine to raise the water level and produce the familiar concave meniscus (Figure 12.20A). The liquid rises until gravity pulling down is balanced by the adhesive forces pulling up.

On the other hand, if you place a glass capillary tube in a dish of mercury, the mercury level in the tube drops below that in the dish. Mercury has a higher surface tension than water (see Table 12.3), which means it has stronger cohesive forces (metallic bonding). The cohesive forces among the mercury atoms are much stronger than the adhesive forces (mostly dispersion) between mercury and glass, so the liquid tends to pull away from the walls. At the same time, the surface atoms are being pulled toward the interior of the mercury by its high surface tension, so the level drops. These combined forces produce a convex meniscus (Figure 12.20B, seen in a laboratory barometer or manometer).

Viscosity

When a liquid flows, the molecules slide around and past each other. A liquid's **viscosity,** its resistance to flow, results from intermolecular attractions resisting this movement. Both gases and liquids flow, but liquid viscosities are higher because the intermolecular forces operate over much shorter distances.

Viscosity decreases with heating, as Table 12.4 shows for water. When molecules move faster at higher temperatures, they can overcome intermolecular forces more easily, so the resistance to flow decreases. Next time you put cooking oil in a pan and heat it, watch the oil flow more easily and spread out in a thin layer.

Molecular shape plays a role in a liquid's viscosity. Long molecules make more contact with each other than spherical ones do, so for the same types of forces, liquids containing longer molecules have higher viscosities. A striking example of a change in viscosity occurs during the making of syrup. Even at room temperature, a concentrated aqueous sugar solution has a higher viscosity than water because of H bonding among the many hydroxyl (—OH) groups on the ring-shaped sugar molecules. When the solution is slowly heated to boiling, the sugar molecules react with each other and link covalently, gradually forming long chains. Hydrogen bonds and dispersion forces occur at many points along the chains, and the resulting syrup is a viscous liquid that pours slowly and clings to a spoon. When a viscous syrup is cooled, it may become stiff enough to be picked up and stretched—into taffy candy. The Gallery shows other familiar examples of these three liquid properties.

Table 12.4 Viscosity of Water at Several Temperatures

Temperature (°C)	Viscosity (N·s/m²)*
20	1.00×10^{-3}
40	0.65×10^{-3}
60	0.47×10^{-3}
80	0.35×10^{-3}

*The units of viscosity are newton-seconds per square meter.

Properties of Liquids

Of the three states of matter, only liquids combine the ability to flow with the strength that comes from intermolecular contact, and this combination appears in numerous applications.

Beaded droplets on waxy surfaces

The adhesive (dipole–induced dipole) forces between water and a nonpolar surface are much weaker than the cohesive (H-bond) forces within water. As a result, water pulls away from a nonpolar surface and forms beaded droplets. You have seen this effect when water beads on a flower petal or a freshly waxed car after a rainfall.

Minimizing a surface

In the low-gravity environment of an orbiting space shuttle, the tendency of a liquid to minimize its surface creates perfectly spherical droplets, unlike the flattened drops we see on Earth. For the same reason, bubbles in a soft drink are spherical because the liquid uses the minimum number of molecules needed to surround the gas. A water strider flits across a pond on widespread legs that do not exert enough pressure to exceed the surface tension.

Maintaining motor oil viscosity

To protect engine parts during long drives or in hot weather, when an oil would ordinarily become too thin, motor oils contain additives, called *polymeric viscosity index improvers,* that act as thickeners. As the oil heats up, the additive molecules change shape from compact spheres to spaghetti-like strands and become tangled with the hydrocarbon oil molecules. As a result of the greater dispersion forces, there is an increase in viscosity that compensates for the decrease due to heating.

Capillary action after a shower

Paper and cotton consist of fibers of cellulose, long carbon-containing molecules with many attached hydroxyl (—OH) groups. A towel dries you in two ways:
First, capillary action draws the water molecules away from your body between the closely spaced cellulose molecules. Second, the water molecules themselves form adhesive H bonds to the cellulose —OH groups.

How a ballpoint pen works

The essential parts of a ballpoint pen are the moving ball and the viscous ink. The material of the ball is chosen for the strong adhesive forces between it and the ink. Cohesive forces within the ink are replaced by those adhesive forces when the ink "wets" the ball. As the ball rolls along the paper, the adhesive forces between ball and ink are overcome by those between ink and paper. The rest of the ink stays in the pen because of its high viscosity.

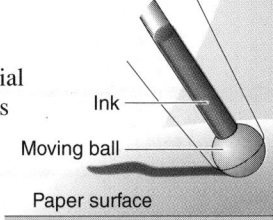

Ink

Moving ball

Paper surface

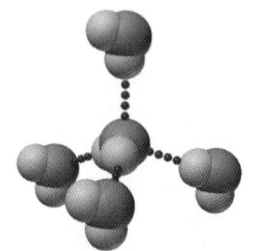

Figure 12.21 **The H-bonding ability of the water molecule.** Because it has two O—H bonds and two lone pairs, one H_2O molecule can engage in as many as four H bonds to surrounding H_2O molecules, which are arranged tetrahedrally.

12.5 THE UNIQUENESS OF WATER

Water is absolutely amazing stuff, but it is so familiar that we take it for granted. In fact, water has some of the most unusual properties of any substance. These properties, which are vital to our very existence, arise inevitably from the nature of the H and O atoms that make up the molecule. Each atom attains a filled outer level by sharing electrons in single covalent bonds. With two bonding pairs and two lone pairs around the O atom and a large electronegativity difference in each O—H bond, the H_2O molecule is bent and highly polar. This arrangement is crucial because it allows each water molecule to engage in four H bonds with its neighbors (Figure 12.21). From these fundamental atomic and molecular properties emerges unique and remarkable macroscopic behavior.

Solvent Properties of Water

The *great solvent power* of water is the result of its polarity and its exceptional H-bonding ability. Water dissolves ionic compounds through ion-dipole forces that separate the ions from the solid and keep them in solution (see Figure 4.3, p. 135). Water dissolves many polar nonionic substances, such as ethanol (CH_3CH_2OH) and glucose ($C_6H_{12}O_6$), by forming H bonds to them. Water even dissolves nonpolar gases, such as those in the atmosphere, to a limited extent, through dipole–induced dipole and dispersion forces. Chapter 13 highlights these solvent properties.

Because it can dissolve so many substances, water is the environmental and biological solvent, forming the complex solutions we know as oceans, rivers, lakes, and cellular fluids. Aquatic animals could not survive without dissolved O_2, nor could aquatic plants survive without dissolved CO_2. The coral reefs that girdle the sea bottom in tropical latitudes are composed of carbonates made from dissolved CO_2 and HCO_3^- by tiny marine animals. Life is thought to have begun in a "primordial soup," an aqueous mixture of simple biomolecules from which emerged the larger molecules whose self-sustaining reactions are characteristic of living things. From a chemical point of view, all organisms, from bacteria to humans, can be thought of as highly organized systems of membranes enclosing and compartmentalizing complex aqueous solutions.

Thermal Properties of Water

Water has an exceptionally *high specific heat capacity,* higher than almost any other liquid. Recall from Section 6.3 that heat capacity is a measure of the heat absorbed by a substance for a given temperature rise. When a substance is heated, some of the energy increases the average molecular speed, some increases molecular vibration and rotation, and some is used to overcome intermolecular forces. Because water has so many strong H bonds, it has a high specific heat capacity. With oceans and seas covering 70% of the Earth's surface, daytime heat from the Sun causes relatively small changes in temperature, allowing our planet to support life. On the waterless, airless Moon, temperatures can range from 100°C (212°F) to −150°C (−238°F) in the same lunar day. Even in Earth's deserts, where a dense atmosphere tempers these extremes, day-night temperature differences of 40°C (72°F) are common.

Numerous strong H bonds also give water an exceptionally *high heat of vaporization*. A quick calculation shows how essential this property is to our existence. When 1000 g of water absorbs about 4 kJ of heat, its temperature rises 1°C. With a ΔH^0_{vap} of 2.3 kJ/g, however, less than 2 g of water must evaporate to keep the temperature constant for the remaining 998 g. The average human adult has 40 kg of body water and generates about 10,000 kJ of heat each day from metabolism. If this heat were used only to increase the average E_k of body water, the temperature rise would mean immediate death. Instead, the heat is converted to E_p as it breaks H bonds and evaporates sweat, resulting in a stable body temperature and minimal loss of body fluid. The Sun's energy supplies the heat of vaporization for ocean water. Water vapor, formed in warm latitudes, moves through the atmosphere, and its potential energy is released as heat to warm cooler regions when the vapor condenses to rain. The enormous energy involved in this cycling of water powers the storms and winds all over the planet.

Surface Properties of Water

The H bonds that give water its remarkable thermal properties are also responsible for its *high surface tension* and *high capillarity.* Except for some metals and molten salts, water has the highest surface tension of any liquid. This property is vital for surface aquatic life because it keeps plant debris resting on a pond surface, which provides shelter and nutrients for many fish, microorganisms, and insects. Water's high capillarity, a result of its high surface tension, is crucial to land plants. During dry periods, plant roots absorb deep groundwater, which rises by capillary action through the tiny spaces between soil particles.

The Density of Solid and Liquid Water

As you saw in Figure 12.21, through the H bonds of water, each O atom becomes connected to as many as four other O atoms via four H atoms. Continuing this pattern through many molecules in a fixed array gives ice the hexagonal, *open structure* shown in Figure 12.22A. The symmetrical beauty of snowflakes, as shown in Figure 12.22B, reflects this hexagonal organization.

This organization explains the negative slope of the solid-liquid curve in the phase diagram for water (see Figure 12.9B, p. 430): As pressure is applied, some H bonds break; as a result, some water molecules enter the spaces. The crystal structure breaks down, and the sample liquefies.

The large spaces within ice give *the solid state a lower density than the liquid state.* When the surface of a lake freezes in winter, the ice floats on the liquid water below. If the solid were denser than the liquid, as is true for nearly

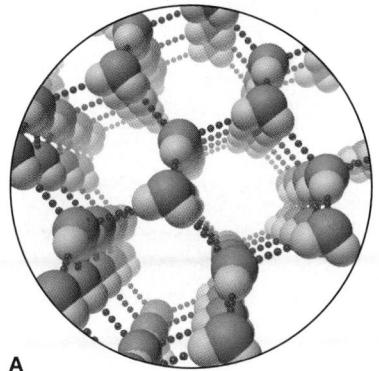

A

B

Figure 12.22 The hexagonal structure of ice. A, As a consequence of the geometric arrangement of the H bonds in H_2O, ice has an open, hexagonally shaped crystal structure. Thus, when liquid water freezes, the volume *increases.* (See also the opening figure, p. 419.) **B,** The delicate six-pointed beauty of a snowflake reflects the hexagonal crystal structure of ice.

Figure 12.23 The expansion and contraction of water. As water freezes and thaws repeatedly, the stress due to expanding and contracting can break rocks. Over eons of geologic time, this process creates sand and soil.

every other substance, the surface of a lake would freeze and sink repeatedly until the entire lake was solid. Aquatic life would not survive from year to year.

The density of water changes in a complex way. When ice melts at 0°C, the tetrahedral arrangement around each O atom breaks down, and the loosened molecules pack much more closely, filling spaces in the collapsing solid structure. As a result, water is most dense (1.000 g/mL) at around 4°C (3.98°C). With more heating, the density decreases through normal thermal expansion.

This change in density is vital for freshwater life. As lake water becomes colder in the fall and early winter, it becomes more dense *before* it freezes. Similarly, in spring, less dense ice thaws to form more dense water *before* the water expands. During both these seasonal density changes, the top layer of water reaches the high-density point first and sinks. The next layer of water rises because it is slightly less dense, reaches 4°C, and likewise sinks. This sinking and rising distribute nutrients and dissolved oxygen.

The expansion and contraction of water dramatically affect the land as well. When rain fills the crevices in rocks and then freezes, great outward stress is applied, to be relieved only when the ice melts. In time, this repeated freeze-thaw stress cracks the rocks apart (Figure 12.23). Over eons of geologic change, this effect has helped to produce the sand and soil of the planet.

The properties of water and their far-reaching consequences illustrate the theme that runs throughout this book: the world we know is the stepwise, macroscopic outgrowth of the atomic world we seek to know. Figure 12.24 offers a fanciful summary of water's unique properties in the form of a tree whose "roots" (atomic properties) give foundation to the base of the "trunk" (molecular proper-

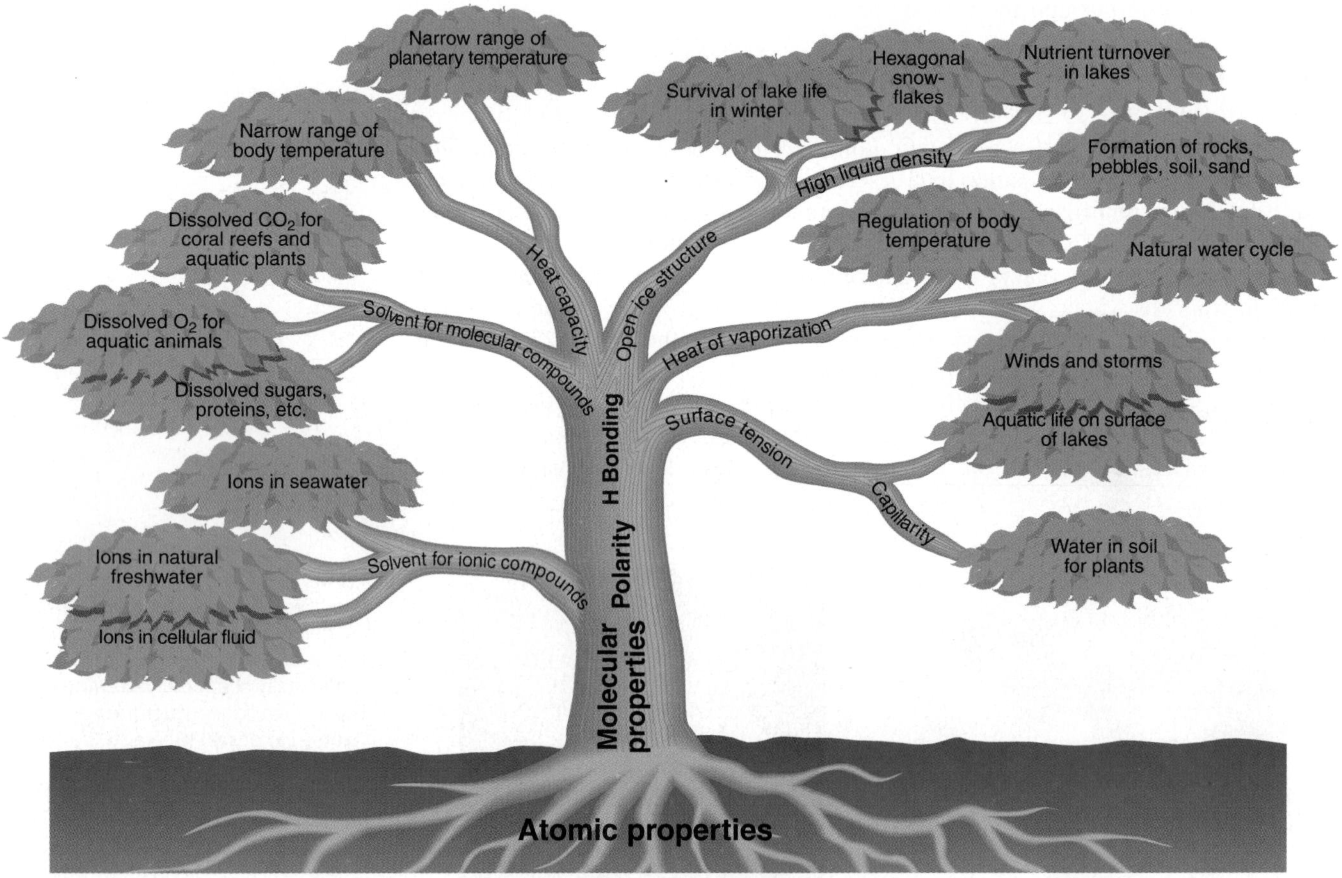

Figure 12.24 The macroscopic properties of water and their atomic and molecular "roots."

ties), which strengthens the next portion (polarity), from which grows another portion (H bonding). These two portions of the trunk "branch" out to the other properties of water and their global effects.

SECTION SUMMARY

The atomic properties of hydrogen and oxygen atoms result in the water molecule's bent shape, polarity, and H-bonding ability. These properties of water enable it to dissolve many ionic and polar compounds. Water's hydrogen bonding results in a high specific heat capacity and a high heat of vaporization, which combine to give the Earth and its organisms a narrow temperature range. H bonds also confer high surface tension and capillarity, which are essential to plant and animal life. Water expands on freezing because of its H bonds, which lead to an open structure in ice. Seasonal density changes foster nutrient mixing in lakes. Seasonal freeze-thaw stress on rocks results eventually in soil formation.

12.6 THE SOLID STATE: STRUCTURE, PROPERTIES, AND BONDING

Stroll through the mineral collection of any school or museum and you'll be struck by the extraordinary variety and beauty of these solids. In this section, we first discuss the general structural features of crystalline solids and then examine a laboratory method for studying them. We survey the properties of the major types of solids and find the whole range of intermolecular forces at work. We then present a model for bonding in solids that explains many of their properties.

Structural Features of Solids

We can divide solids into two broad categories based on the orderliness of their shapes, which in turn is based on the orderliness of their particles. **Crystalline solids** generally have a well-defined shape, as shown in Figure 12.25, because their particles—atoms, molecules, or ions—occur in an orderly arrangement. **Amorphous solids** have poorly defined shapes because they lack extensive molecular-level ordering of their particles. We focus here, for the most part, on crystalline solids.

Figure 12.25 The striking beauty of crystalline solids. The symmetry and regularity of crystals are due to the orderly arrangement of their particles. *Left to right:* celestite, pyrite, amethyst, and halite.

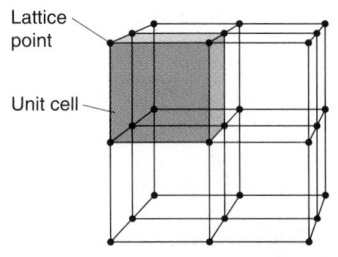

A Portion of 3-D lattice

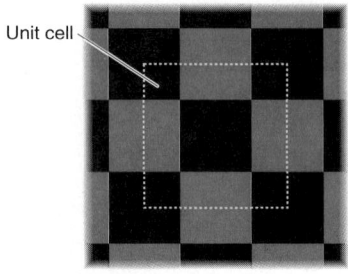

B Portion of 2-D lattice

Figure 12.26 **The crystal lattice and the unit cell. A,** The lattice is an array of points that defines the positions of the particles in a crystal structure. It is shown here with the points connected by lines. A unit cell *(colored)* is the simplest array of points that produces the lattice, when repeated in all directions. A simple cubic unit cell, one of the 14 types of unit cell that occur in nature, is shown. **B,** A section of checkerboard is a two-dimensional analogy for a lattice.

The Crystal Lattice and the Unit Cell If you could see the particles within a crystal, you would find them packed tightly together in an orderly, three-dimensional array. Consider the simplest case, in which all the particles are identical spheres, and imagine a point at the same location within each particle in this array, say at the center. The points form a regular pattern throughout the crystal that is called the crystal **lattice.** Thus, *the lattice consists of all points with identical surroundings.* Put another way, if you were to remove the rest of each particle, leaving only the lattice point, and were transported from one point to another, you would not be able to tell that you had moved. Keep in mind that there is no preexisting array of lattice points; rather, *the arrangement of the particle points defines the lattice.*

Figure 12.26A shows a portion of a lattice and the **unit cell**, the *smallest* portion of the crystal that, if repeated in all three directions, gives the crystal. A two-dimensional analogy for a unit cell and a crystal can be seen in a checkerboard (as shown in Figure 12.26B), a section of tiled floor, a piece of wallpaper, or any other pattern that is constructed from a repeating unit. The **coordination number** of a particle in a crystal is the number of nearest neighbors surrounding it.

There are 7 crystal systems and 14 types of unit cells that occur in nature, but we will be concerned primarily with the *cubic system,* which gives rise to the cubic lattice. The solid states of a majority of metallic elements, some covalent compounds, and many ionic compounds occur as cubic lattices. (We also describe the hexagonal unit cell a bit later.) There are three types of cubic unit cells within the cubic system:

1. In the **simple cubic unit cell,** shown in Figure 12.27A, the centers of eight particles define the corners of a cube. Attractions pull the particles together, so they touch along the cube's edges; but they do not touch diagonally along the cube's faces or through its center. The coordination number of each particle is 6: four in its own layer, one in the layer above, and one in the layer below.
2. In the **body-centered cubic unit cell,** shown in Figure 12.27B, a particle lies at each corner *and* in the center of the cube. Those at the corners do not touch each other, but they all touch the one in the center. Each particle is surrounded by eight nearest neighbors, four above and four below, so the coordination number is 8.
3. In the **face-centered cubic unit cell,** shown in Figure 12.27C, a particle lies at each corner *and* in the center of each face but not in the center of the cube. Those at the corners touch those in the faces but not each other. The coordination number is 12.

One unit cell lies adjacent to another throughout the crystal, with no gaps, so a particle at a corner or face is *shared* by adjacent unit cells. As you can see from Figure 12.27 (third row from the top), in the three cubic unit cells, the particle at each corner is part of eight adjacent cells, so one-eighth of each of these particles belongs to each unit cell (bottom row). There are eight corners in a cube, so each simple cubic cell contains $8 \times \frac{1}{8}$ particle = 1 particle. The body-centered cubic unit cell contains one particle from the eight corners and one in the center, for a total of two particles; and the face-centered cubic unit cell contains four particles, one from the eight corners and three from the half-particles in each of the six faces.

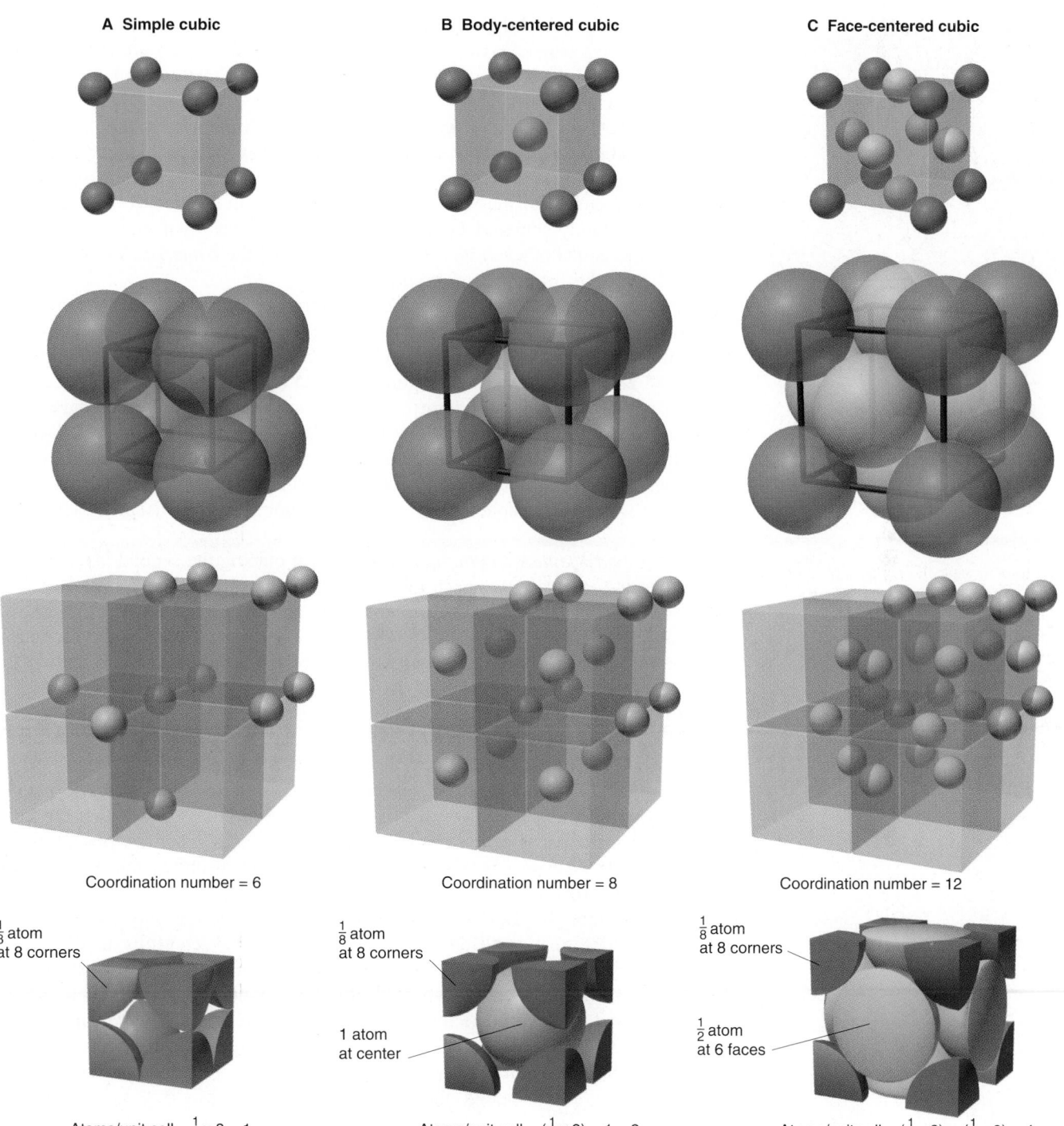

A Simple cubic

B Body-centered cubic

C Face-centered cubic

Coordination number = 6

Coordination number = 8

Coordination number = 12

$\frac{1}{8}$ atom at 8 corners

$\frac{1}{8}$ atom at 8 corners

1 atom at center

$\frac{1}{8}$ atom at 8 corners

$\frac{1}{2}$ atom at 6 faces

Atoms/unit cell = $\frac{1}{8} \times 8 = 1$

Atoms/unit cell = $(\frac{1}{8} \times 8) + 1 = 2$

Atoms/unit cell = $(\frac{1}{8} \times 8) + (\frac{1}{2} \times 6) = 4$

Figure 12.27 **The three cubic unit cells. A,** Simple; **B,** body-centered; and **C,** face-centered cubic unit cells. *Top row:* Cubic arrangements of atoms in expanded view. *Second row:* Space-filling view of these cubic arrangements. For clarity, corner atoms are blue, body-centered atoms pink, and face-centered atoms yellow. *Third row:* A unit cell *(shaded blue, upper right rear)* in a portion of the crystal. One particle *(dark blue in center)* surrounded by the given number of nearest neighbors shows the coordination number. *Bottom row:* Total numbers of atoms in the actual unit cells. The simple cubic has one atom; the body-centered, two; and the face-centered, four.

The efficient packing of fruit.

Packing Efficiency and the Creation of Unit Cells The unit cells found in nature result from the various ways atoms are packed together, which are similar to the ways macroscopic spheres—marbles, golf balls, fruit—are packed for shipping or displayed (see photo).

For particles of the same size, *the higher the coordination number of the crystal is, the greater the number of particles in a given volume.* Therefore, as the coordination numbers in Figure 12.27 indicate, a crystal structure based on the face-centered cubic cell has more particles packed into a given volume than one based on the body-centered cubic cell, which has more than one based on the simple cubic cell. Let's see how to pack spheres of the *same size* to create these unit cells and the hexagonal unit cell as well:

1. *The simple cubic cell.* Suppose we arrange the first layer of spheres as shown in Figure 12.28A. Note the large diamond-shaped spaces (cutaway portion). If we place the next layer of spheres *directly above* the first, as shown in Figure 12.28B, we obtain an arrangement based on the *simple* cubic unit cell. By calculating the **packing efficiency** of this arrangement—the percentage of the total volume occupied by the spheres themselves—we find that only 52% of the available unit-cell volume is occupied by spheres and 48% consists of the empty space between them (see Problem 12.129 at the end of the chapter). This is a very inefficient way to pack spheres, so neither fruit nor atoms are usually packed this way.

2. *The body-centered cubic cell.* Rather than placing the second layer directly above the first, we can use space more efficiently by placing the spheres (colored differently for clarity) on the diamond-shaped spaces in the first layer, as shown in Figure 12.28C. Then we pack the third layer onto the spaces in the second so that the first and third layers line up vertically. This arrangement is based on the *body-centered* cubic unit cell, and its packing efficiency is 68%—much higher than for the simple cubic. Several metallic elements, including chromium, iron, and all the Group 1A(1) elements, have a crystal structure based on the body-centered cubic unit cell.

3. *The hexagonal and face-centered cubic unit cells.* Spheres can be packed even more efficiently. First, we shift rows in the bottom layer such that the large diamond-shaped spaces become smaller triangular spaces. Then we place the second layer over these spaces. Figure 12.28D shows this arrangement, with the first layer labeled *a* (*orange*) and the second layer *b* (*green*).

We can place the third layer of spheres in two different ways, and how we do so gives rise to two different unit cells. If you look carefully at the spaces formed in layer *b* of Figure 12.28D, you'll see that some are orange because they lie above *spheres* in layer *a*, whereas others are white because they lie above *spaces* in layer *a*. If we place the third layer of spheres (*orange*) over the orange spaces (down and left to Figure 12.28E), they lie directly over spheres in layer *a*, and we obtain an *abab. . .* layering pattern because every other layer is placed identically. This gives **hexagonal closest packing,** which is based on the *hexagonal unit cell.*

On the other hand, if we place the third layer of spheres (*blue*) over the white spaces (down and right to Figure 12.28F), the spheres lie over spaces in layer *a*. This placement is different from both layers *a* and *b*, so we obtain an *abcabc. . .* pattern. This gives **cubic closest packing,** which is based on the *face-centered* cubic unit cell.

The packing efficiency of both hexagonal and cubic closest packing is 74%, and the coordination number of both is 12. There is no way to pack spheres of equal size more efficiently. Most metallic elements crystallize in either of these

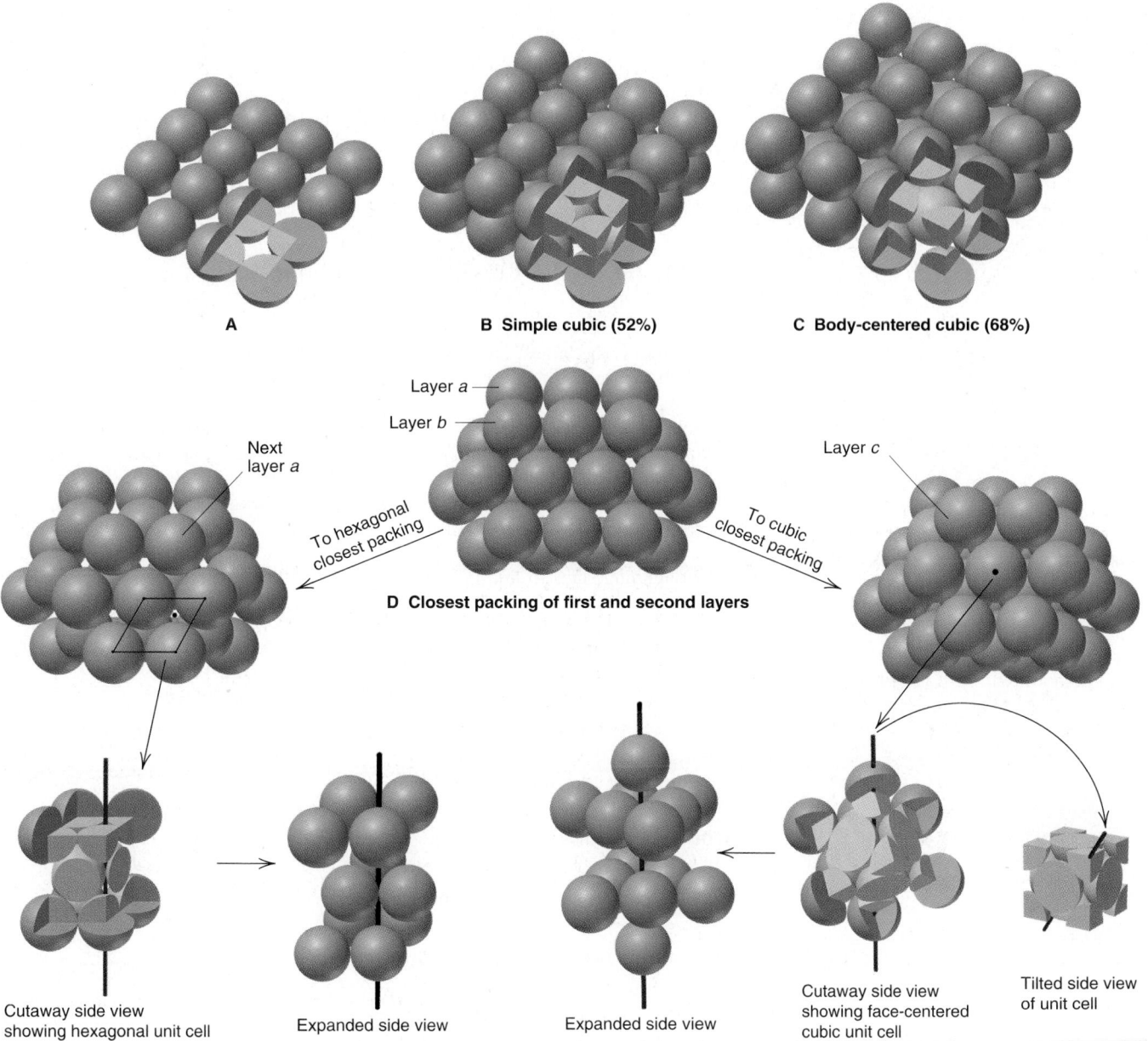

A

B Simple cubic (52%)

C Body-centered cubic (68%)

Layer *a*

Layer *b*

Next layer *a*

Layer *c*

To hexagonal closest packing

To cubic closest packing

D Closest packing of first and second layers

Cutaway side view showing hexagonal unit cell

Expanded side view

Expanded side view

Cutaway side view showing face-centered cubic unit cell

Tilted side view of unit cell

E Hexagonal closest packing (*abab*...) (74%)

F Cubic closest packing (*abcabc*...) (74%)

Figure 12.28 **Packing of spheres. A,** In the first layer, each sphere lies next to another horizontally and vertically; note the large diamond-shaped spaces *(cutaway)*. **B,** If the spheres in the next layer lie *directly* over those in the first, the packing is based on the *simple cubic* unit cell *(pale orange cube, lower right corner)*. **C,** If the spheres in the next layer lie in the diamond-shaped spaces of the preceding layer, the packing is based on the *body-centered cubic* unit cell *(lower right corner)*. **D,** The closest possible packing of the first layer *(a; orange)* is obtained by shifting every other row in part A, thus reducing the diamond-shaped spaces to smaller triangular spaces. The spheres of the second layer *(b; green)* are placed above these spaces; note the orange and white spaces that result. **E,** Follow the left arrow from part D to obtain hexagonal closest packing. When the third layer *(a; orange)* is placed directly over the first, that is, over the orange spaces, we obtain an *abab*. . . pattern. Rotating the layers 90° produces the side view, with the hexagonal unit cell shown as a cutaway segment, and the expanded side view. **F,** Follow the right arrow from part D to obtain cubic closest packing. When the third layer *(c; blue)* covers the white spaces, it lies in a different position from the first *and* second layers to give an *abcabc*. . . pattern. Rotating the layers 90° shows the side view, with the face-centered cubic unit cell as a cutaway, and a further tilt shows the unit cell clearly; finally, we see the expanded view. The packing efficiency is given in parentheses for each unit cell.

arrangements. Magnesium, titanium, and zinc are some elements that adopt the hexagonal structure; nickel, copper, and lead adopt the cubic structure, as do many ionic compounds and other substances, such as frozen carbon dioxide, methane, and most noble gases.

In Sample Problem 12.4, we use the density of an element and the packing efficiency of its crystal structure to calculate its atomic radius. Variations of this approach are used to find the molar mass and as one of the ways to determine Avogadro's number.

SAMPLE PROBLEM 12.4 Determining Atomic Radius from Crystal Structure

---

Problem Barium is the largest nonradioactive alkaline earth metal. It has a body-centered cubic unit cell and a density of 3.62 g/cm^3. What is the atomic radius of barium? (Volume of a sphere: $V = \frac{4}{3}\pi r^3$.)

Plan Because an atom is spherical, we can find its radius from its volume. If we multiply the reciprocal of density (volume/mass) by the molar mass (mass/mole), we find the volume of 1 mol of Ba metal. The metal crystallizes in the body-centered cubic structure, so 68% of this volume is occupied by 1 mol of the atoms themselves (see Figure 12.28C). Dividing by Avogadro's number gives the volume of one Ba atom, from which we find the radius.

Solution Combining steps to find the volume of 1 mol of Ba metal:

$$\text{Volume/mole of Ba metal} = \frac{1}{\text{density}} \times \mathcal{M}$$

$$= \frac{1 \text{ cm}^3}{3.62 \text{ g Ba}} \times \frac{137.3 \text{ g Ba}}{1 \text{ mol Ba}}$$

$$= 37.9 \text{ cm}^3/\text{mol Ba}$$

Finding the volume of 1 mol of Ba *atoms:*

$$\text{Volume/mole of Ba atoms} = \text{volume/mol Ba} \times \text{packing efficiency}$$
$$= 37.9 \text{ cm}^3/\text{mol Ba} \times 0.68 = 26 \text{ cm}^3/\text{mol Ba atoms}$$

Finding the volume of one Ba atom:

$$\text{Volume of Ba atom} = \frac{26 \text{ cm}^3}{1 \text{ mol Ba atoms}} \times \frac{1 \text{ mol Ba atoms}}{6.022 \times 10^{23} \text{ Ba atoms}}$$
$$= 4.3 \times 10^{-23} \text{ cm}^3/\text{Ba atom}$$

Finding the atomic radius of Ba from the volume of a sphere:

$$V \text{ of Ba atom} = \frac{4}{3}\pi r^3$$

So

$$r^3 = \frac{3V}{4\pi}$$

Thus

$$r = \sqrt[3]{\frac{3V}{4\pi}} = \sqrt[3]{\frac{3(4.3 \times 10^{-23} \text{ cm}^3)}{4 \times 3.14}}$$

$$= 2.2 \times 10^{-8} \text{ cm}$$

Check The order of magnitude is correct for an atom ($\sim 10^{-8}$ cm $\approx 10^{-10}$ m). The actual value is 2.22×10^{-8} cm (see Figure 8.15), so our answer seems correct.

FOLLOW-UP PROBLEM 12.4 Iron crystallizes in a body-centered cubic structure. The volume of one Fe atom is 8.38×10^{-24} cm^3, and the density of Fe is 7.874 g/cm^3. Calculate an approximate value for Avogadro's number.

Density (g/cm^3) of Ba metal

find reciprocal and multiply by $\mathcal{M}$ (g/mol)

Volume (cm^3) per mole of Ba metal

multiply by packing efficiency

Volume (cm^3) per mole of Ba atoms

divide by Avogadro's number

Volume (cm^3) of Ba atom

$V = \frac{4}{3}\pi r^3$

Radius (cm) of Ba atom

Our understanding of solids is based on the ability to "see" their crystal structures. Two techniques for doing this are described in the Tools of the Laboratory essay.

Tools of the Laboratory

X-Ray Diffraction Analysis and Scanning Tunneling Microscopy

In this chapter and in Chapter 10, we have discussed crystal structures and molecular shapes as if they have actually been seen. You may have been wondering how chemists know atomic radii, bond lengths, and bond angles when the objects exhibiting them are so incredibly minute. Various tools exist for peering into the molecular world and measuring its dimensions, but two of the most powerful are **x-ray diffraction analysis** and **scanning tunneling microscopy.**

X-Ray Diffraction Analysis

This technique has been used for decades to determine crystal structures. In Chapter 7, we discussed wave diffraction and saw how interference patterns of bright and dark regions appear when light passes through slits that are spaced at the distance of the light's wavelength (see Figure 7.5). In 1912, the Swiss physicist Max von Laue suggested that, since x-ray wavelengths are about the same size as the spaces between layers of particles in many solids, the layers might diffract x-rays. (Actually, the suggestion was made to test whether x-rays were particulate or wavelike.) X-ray diffraction was soon recognized as a powerful tool for determining the structure of a solid.

Let's see how this technique is used to measure a key parameter in a crystal structure: the distance (d) between layers of atoms. Figure B12.1 depicts a side view of two layers in a simplified lattice. Two waves impinge on the crystal at an angle θ and are diffracted at the same angle by adjacent layers. When the first wave strikes the top layer and the second strikes the next layer, the waves are *in phase* (peaks aligned with peaks and troughs with troughs). If they are still in phase after being diffracted, a spot appears on a nearby photographic plate. Note that this will occur only if the additional distance traveled by the second wave ($DE + EF$ in the figure) is a whole number of wavelengths, $n\lambda$, where n is an integer (1, 2, 3, and so on). From trigonometry, we find that

$$n\lambda = 2d \sin \theta$$

where θ is the known angle of incoming light, λ is its known wavelength, and d is the unknown distance between the layers in the crystal. This relationship is the *Bragg equation,* named for

W. H. Bragg and his son W. L. Bragg, who shared the Nobel Prize in 1915 for their work on crystal structure analysis.

Rotating the crystal changes the angle of incoming radiation and produces a different set of spots, eventually yielding a complete diffraction pattern that is used to determine the distances and angles within the lattice (Figure B12.2). The diffraction pattern is not an actual picture of the structure; the pattern must be analyzed mathematically to obtain the dimensions of the crystal. Modern x-ray diffraction equipment automatically rotates the crystal and measures thousands of diffractions, and a computer calculates the parameters of interest.   *(continued)*

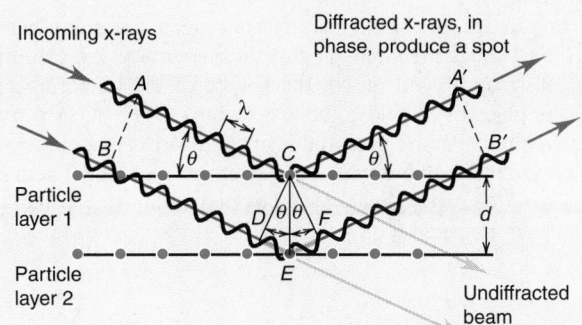

Figure B12.1 Diffraction of x-rays by crystal planes. As in-phase x-ray beams *A* and *B* pass into a crystal at angle θ, they are diffracted by interaction with the particles. Beam *B* travels the distance *DE + EF* farther than beam *A*. If this additional distance is equal to a whole number of wavelengths, the beams remain in phase and create a spot on a screen or photographic plate. From the pattern of spots and the Bragg equation, $n\lambda = 2d \sin \theta$, the distance *d* between layers of particles can be calculated.

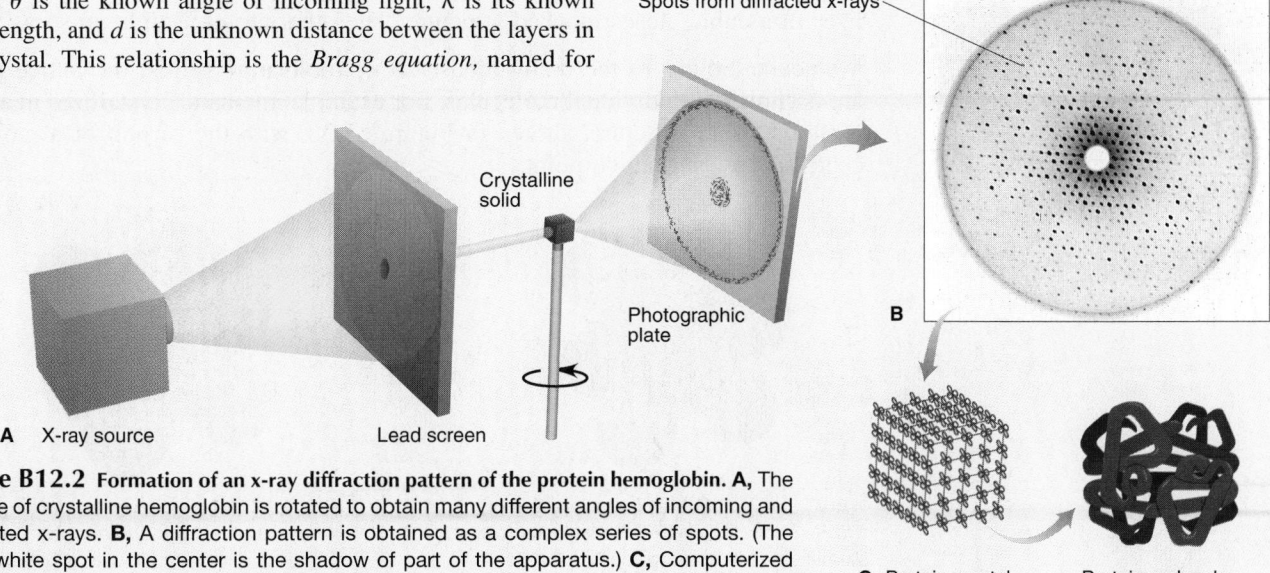

Figure B12.2 Formation of an x-ray diffraction pattern of the protein hemoglobin. A, The sample of crystalline hemoglobin is rotated to obtain many different angles of incoming and diffracted x-rays. **B,** A diffraction pattern is obtained as a complex series of spots. (The large white spot in the center is the shadow of part of the apparatus.) **C,** Computerized analysis relates the pattern to distances and angles within the crystal, providing data used to generate a picture of the hemoglobin molecule.

X-ray analysis is used to answer questions in many branches of chemistry, but its greatest impact has been in biochemistry. It has shown that DNA exists as a double helix, and it is currently helping biochemists learn how a protein's three-dimensional structure is related to its function.

Scanning Tunneling Microscopy

This technique, a much newer method than x-ray analysis, is used to observe surfaces on the atomic scale. It was invented in the early 1980s by Gerd Binnig and Heinrich Rohrer, two Swiss physicists who won the Nobel Prize for their work in 1986. The technique is based on the idea that an electron in an atom has a small probability of existing far from the nucleus, so given the right conditions, it can move ("tunnel") to end up closer to another atom.

In practice, the tunneling electrons create a current that can be used to image the atoms of an adjacent surface. An extremely sharp tungsten-tipped probe, the source of the tunneling electrons, is placed very close (about 0.5 nm) to the surface under study. A small electric potential is applied across this minute gap to increase the probability that the electrons will tunnel across it. The size of the gap is kept constant by maintaining a constant tunneling current generated by the moving electrons. For this to oc-

cur, the probe must move tiny distances up and down, thus following the atomic contour of the surface. This movement is electronically monitored, and after many scans, a three-dimensional map of the surface is obtained. The method has revealed magnificent images of atoms and molecules coated on surfaces and is being used to study many aspects of surfaces, such as the nature of defects and the adhesion of films (Figure B12.3).

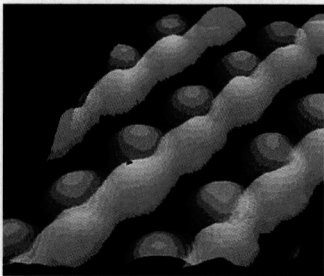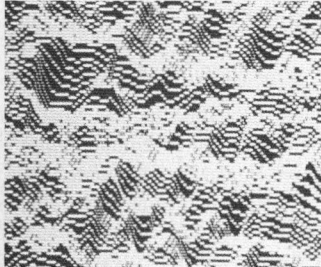

Figure B12.3 Scanning tunneling micrographs. The light-sensitive semiconducting substance gallium arsenide (*left*) exhibits remarkable order (Ga, *blue*; As, *pink*). High-purity gold that has been rolled and annealed (*right*) shows a chaotic surface due to many lattice defects. (Magnification ×6,000,000.)

Types and Properties of Crystalline Solids

Now we can turn to the five most important types of solids, which are summarized in Table 12.5. Each is defined by the type of particle in the crystal, which determines the forces between them. You may want to review the bonding models (Chapter 9) to clarify how they relate to the properties of different solids.

Atomic Solids Individual atoms held together by dispersion forces form an **atomic solid.** The noble gases [Group 8A(18)] are the only examples, and their physical properties reflect the very weak forces among the atoms. Melting and boiling points and heats of vaporization and fusion are all very low, rising smoothly with increasing molar mass. As shown in Figure 12.29, argon crystallizes in a cubic closest packed structure. The other atomic solids do so as well.

Molecular Solids In the many thousands of **molecular solids,** the lattice points are occupied by individual molecules. For example, methane crystallizes in a face-centered cubic structure, shown in Figure 12.30, with the carbon of a molecule centered on each lattice point.

Figure 12.29 Cubic closest packing of frozen argon. Group 8A(18) elements form atomic solids that adopt cubic closest packing (face-centered cubic cell).

Figure 12.30 Cubic closest packing of frozen methane. Methane adopts a face-centered cubic unit cell, with a C atom centered on each lattice point. Only one CH_4 molecule is shown at a site position.

Table 12.5 Characteristics of the Major Types of Crystalline Solids

	Particle(s)	Interparticle Forces	Physical Behavior	Examples [mp, °C]
Atomic	Atoms	Dispersion	Soft, very low mp, poor thermal and electrical conductors	Group 8A(18) [Ne −249 to Rn −71]
Molecular	Molecules	Dispersion, dipole-dipole, H bonds	Fairly soft, low to moderate mp, poor thermal and electrical conductors	*Nonpolar** O_2 [−219], C_4H_{10} [−138] Cl_2 [−101], C_6H_{14} [−95] P_4 [44.1] *Polar* SO_2 [−73], $CHCl_3$ [−64] HNO_3 [−42], H_2O [0.0] CH_3COOH [17]
Ionic	Positive and negative ions	Ion-ion attraction	Hard and brittle, high mp, good thermal and electrical conductors when molten	NaCl [801] CaF_2 [1423] MgO [2852]
Metallic	Atoms	Metallic bond	Soft to hard, low to very high mp, excellent thermal and electrical conductors, malleable and ductile	Na [97.8] Zn [420] Fe [1535]
Network	Atoms	Covalent bond	Very hard, very high mp, usually poor thermal and electrical conductors	SiO_2 (quartz) [1610] C(diamond) [~4000]

*Nonpolar molecular solids are arranged in order of increasing molar mass. Note the correlation with increasing melting point (mp).

Various combinations of dipole-dipole, dispersion, and H bonding forces are at work in molecular solids, which accounts for their wide range of physical properties. Dispersion forces are the principal force acting in nonpolar substances, so melting points generally increase with molar mass (Table 12.5). Among polar molecules, dipole-dipole forces and, where possible, H bonding dominate. Except for those substances consisting of the simplest molecules, molecular solids have higher melting points than the atomic solids (noble gases). Nevertheless, intermolecular forces are still relatively weak, so the melting points are much lower than those of ionic, network covalent, and metallic solids.

Ionic Solids In crystalline **ionic solids,** the unit cell contains particles with whole, rather than partial, charges. As a result, the interparticle forces (ionic bonding forces) are *much* stronger than the van der Waals forces in atomic or molecular solids. To maximize attractions, cations are surrounded by as many anions as possible, and vice versa, with *the smaller of the two ions lying in the spaces (holes) formed by the packing of the larger.* Since the unit cell is the smallest portion of the crystal that maintains the overall spatial arrangement, it is also the smallest portion that maintains the overall chemical composition. In other words, *the unit cell has the same cation:anion ratio as the empirical formula.*

Ionic compounds adopt several different crystal structures, but many use cubic closest packing. Let's first consider two structures that have a 1:1 ratio of ions. The *sodium chloride structure* is found in many compounds, including most of the alkali metal [Group 1A(1)] halides and hydrides, the alkaline earth metal [Group 2A(2)] oxides and sulfides, several transition-metal oxides and sulfides, and most of the silver halides. To visualize this structure, first imagine Cl^- anions

NaCl unit cell

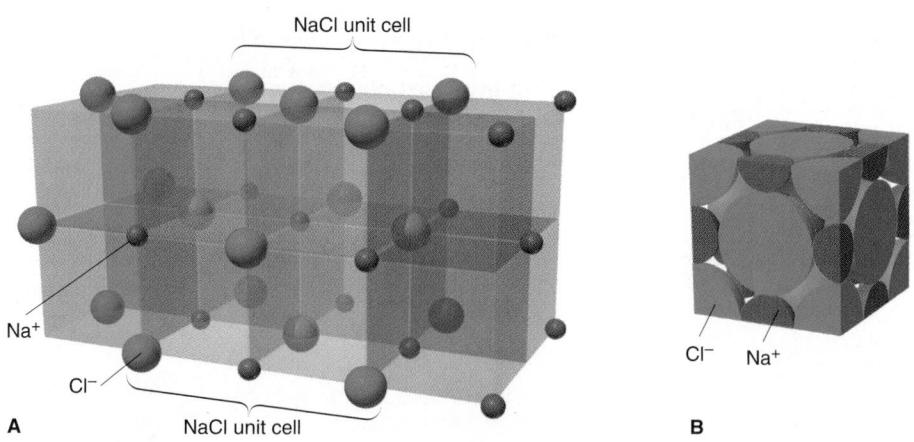

A   NaCl unit cell   B

Figure 12.31   **The sodium chloride structure. A,** In an expanded view, the sodium chloride structure can be pictured as resulting from the interpenetration of two face-centered cubic arrangements, one of Na^+ ions *(brown)* and the other of Cl^- ions *(green)*. **B,** A space-filling view of the NaCl unit cell (central overlapped portion in part A), which consists of four Cl^- ions and four Na^+ ions.

and Na^+ cations organized separately in face-centered cubic (cubic closest packing) arrays. The crystal structure arises when these two arrays penetrate each other such that the smaller Na^+ ions end up in the holes between the larger Cl^- ions, as shown in Figure 12.31A. Thus, each Na^+ is surrounded by six Cl^-, and vice versa (coordination number = 6). Figure 12.31B is a space-filling depiction of the unit cell showing a face-centered cube of Cl^- ions with Na^+ ions between them. Note the four Cl^- $[(8 \times \frac{1}{8}) + (6 \times \frac{1}{2}) = 4 \ Cl^-]$ and four Na^+ $[(12 \times \frac{1}{4}) + 1$ in the center = 4 $Na^+]$, giving a 1:1 ion ratio.

Another structure with a 1:1 ion ratio is the *zinc blende (ZnS) structure*. It can be pictured as two face-centered cubic arrays, one of Zn^{2+} ions and the other of S^{2-} ions, penetrating each other such that each ion is tetrahedrally surrounded by four ions of opposite charge (coordination number = 4). Note the 1:1 ratio of ions in the unit cell shown in Figure 12.32. Many other compounds, including AgI, CdS, and the Cu(I) halides adopt the zinc blende structure.

The *fluorite (CaF_2) structure* is common among salts with a 1:2 cation:anion ratio, especially those having relatively large cations and relatively small anions.

CUBIC UNIT CELLS AND THEIR ORIGINS

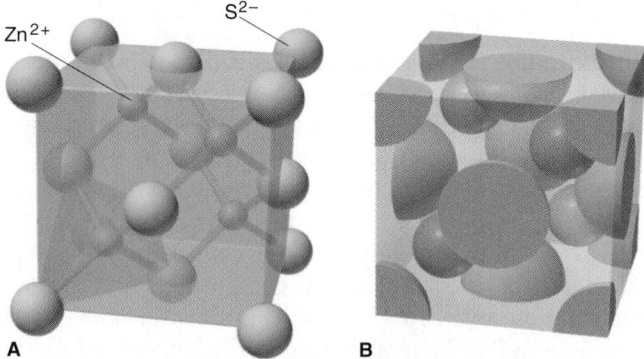

A   B

Figure 12.32   The zinc blende structure. Zinc sulfide adopts the zinc blende structure. **A,** The translucent cube shows a face-centered cubic array of four $[(8 \times \frac{1}{8}) + (6 \times \frac{1}{2}) = 4] \ S^{2-}$ ions *(yellow)* tetrahedrally surrounding each of four Zn^{2+} ions *(gray)* to give the 1:1 empirical formula. (Bonds are shown only for clarity.) **B,** The actual unit cell slightly expanded to show the interior ions.

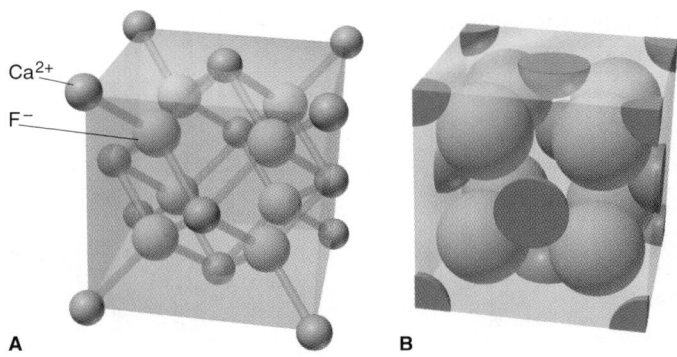

Figure 12.33 **The fluorite structure.** Calcium fluoride adopts the fluorite structure. **A,** The translucent cube shows a face-centered cubic array of four Ca^{2+} ions *(blue)* tetrahedrally surrounding each of eight F^- ions *(yellow)* to give the 4:8, or 1:2, ratio. **B,** The actual unit cell (slightly expanded).

In the case of CaF_2, the unit cell is a face-centered cubic array of Ca^{2+} ions with F^- ions occupying *all* eight available holes (Figure 12.33). This results in a $Ca^{2+} : F^-$ ratio of 4:8, or 1:2. SrF_2 and $BaCl_2$ have the fluorite structure also. The *antifluorite structure* is often seen in compounds having a cation:anion ratio of 2:1 and a relatively large anion (for example, K_2S). In this structure, the cations occupy all eight holes formed by the cubic closest packing of the anions, just the opposite of the fluorite structure.

The properties of ionic solids are a direct consequence of the *fixed ion positions* and *very strong interionic forces,* which create a high lattice energy. Thus, ionic solids typically have high melting points and low electrical conductivities. When a large amount of heat is supplied and the ions gain enough kinetic energy to break free of their positions, the solid melts and the mobile ions conduct a current. Ionic compounds are hard because only a strong external force can change the relative positions of many trillions of interacting ions. If enough force *is* applied to move them, ions of like charge are brought near each other, and their repulsions crack the crystal (see Figure 9.8, p. 335).

Metallic Solids In contrast to the weak dispersion forces between the atoms in atomic solids, powerful metallic bonding forces hold individual atoms together in **metallic solids.** Most metallic elements crystallize in one of the two closest packed structures (Figure 12.34).

The properties of metals—high electrical and thermal conductivity, luster, and malleability—result from the presence of delocalized electrons, the essential feature of metallic bonding (introduced in Section 9.5). Metals have a wide range of melting points and hardnesses, which are related to the packing efficiency of the crystal structure and the number of valence electrons available for bonding. For example, Group 2A metals are harder and higher melting than Group 1A metals (see Figure 9.23, p. 350), because the 2A metals have closest packed structures (except Ba) and twice as many delocalized valence electrons.

Network Covalent Solids In the final type of crystalline solid, separate particles are not present. Instead, strong covalent bonds link the atoms together throughout a **network covalent solid.** As a consequence of the strong bonding, all these substances have extremely high melting and boiling points, but their conductivity and hardness depend on the details of their bonding.

The two common crystalline forms of elemental carbon are examples of network covalent solids. Although graphite and diamond have the same composition,

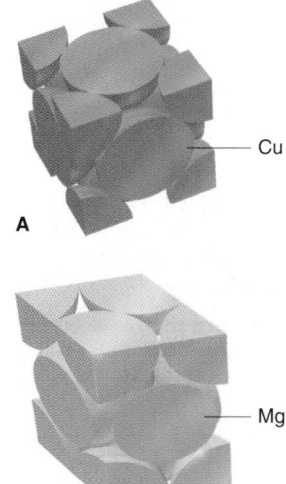

Figure 12.34 **Crystal structures of metals.** Most metallic elements crystallize in one of the closest packed arrangements. **A,** Copper adopts cubic closest packing. **B,** Magnesium adopts hexagonal closest packing.

Table 12.6 Comparison of the Properties of Diamond and Graphite

Property	Graphite		Diamond	
Density (g/cm^3)	2.27		3.51	
Hardness	<1 (very soft)		10 (hardest)	
Melting point (K)	4100		4100	
Color	Shiny black		Colorless transparent	
Electrical conductivity	High (along sheet)		None	
ΔH^0_{comb} (kJ/mol)	−393.5		−395.4	
ΔH^0_f (kJ/mol)	0 (standard state)		1.90	

A Diamond Film on Every Pot
In the 1950s, synthetic diamonds were made slowly and expensively by exposing graphite to extreme temperatures (~1400°C) and pressures (50,000 atm). In the 1960s, a much cheaper method was discovered; by the 1990s, it was being used to deposit thin films of diamond onto virtually any surface. In the process of chemical vapor deposition (CVD), a stream of methane molecules break down at moderate temperatures (~600°C) and low pressures (~0.001 atm), and the carbon atoms deposit on the surface at rates build up a thickness greater than 100 μm per hour. Because diamond is incredibly hard and has high thermal conductivity, applications for diamond films are myriad: scratch-proof cookware, watch crystals, hard discs, and eyeglasses; lifetime drill bits, ball bearings, and razor blades; high-temperature semiconductors. . . . The list goes on and on.

their properties are strikingly different, as Table 12.6 shows. Graphite occurs as stacked flat sheets of hexagonal carbon rings with a strong σ-bond framework and delocalized π bonds, reminiscent of benzene. The arrangement of hexagons looks like chicken wire or honeycomb. Whereas the π-bonding electrons of benzene are delocalized over one ring, those of graphite are delocalized over the entire sheet. These mobile electrons allow graphite to conduct electricity, but only in the plane of the sheets. Graphite is a common electrode material and was once used for lightbulb filaments. The sheets interact via dispersion forces. Common impurities, such as O_2, that lodge between the sheets allow them to slide past each other easily, which explains why graphite is so soft. Diamond crystallizes in a face-centered cubic unit cell, with each carbon atom tetrahedrally surrounded by four others in one virtually endless array. Strong, single bonds throughout the crystal make diamond the hardest substance known.   Because of its localized bonding electrons, diamond (like most network covalent solids) is unable to conduct electricity.

By far the most important network covalent solids are the *silicates*. They utilize a variety of bonding patterns, but nearly all consist of extended arrays of covalently bonded silicon and oxygen atoms. Quartz (SiO_2) is a common example. We'll discuss silicates, which form the structure of clays, rocks, and many minerals, when we consider the chemistry of silicon in Chapter 14.

Amorphous Solids

Amorphous solids are noncrystalline. Many have small, somewhat ordered regions connected by large disordered regions. Charcoal, rubber, and glass are some familiar examples of amorphous solids.

The process that forms quartz glass is typical of that for many amorphous solids. Crystalline quartz (SiO_2) has a cubic closest packed structure. The crystalline form is melted, and the viscous liquid is cooled rapidly to prevent it from recrystallizing. The chains of silicon and oxygen atoms cannot orient themselves quickly enough into an orderly structure, so they solidify in a distorted jumble containing many gaps and misaligned rows (Figure 12.35). The absence of regularity in the structure confers some properties of a liquid; in fact, glasses are sometimes referred to as *supercooled liquids*.

Bonding in Solids: Molecular Orbital Band Theory

Chapter 9 introduced a qualitative model of metallic bonding that pictures metal ions submerged in a "sea" of mobile, delocalized valence electrons. Quantum mechanics offers another model, an extension of molecular orbital (MO) theory, called **band theory.** It is more quantitative than the electron-sea model, and therefore more useful. We'll pay special attention to bonding in metals and differences in electrical conductivity of metals, metalloids, and nonmetals.

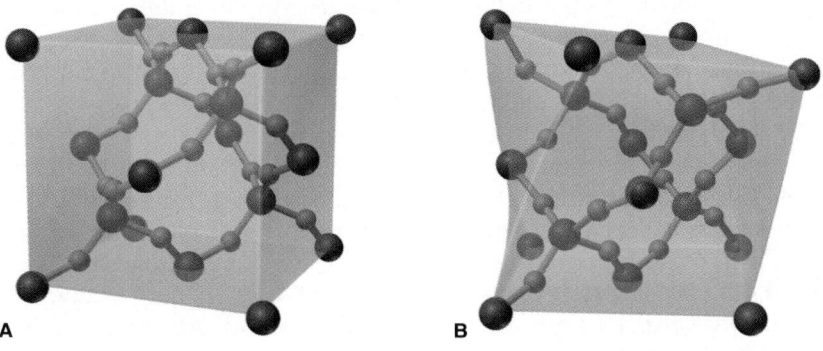

Figure 12.35 **Crystalline and amorphous silicon dioxide. A,** The atomic arrangement of cristobalite, one of the many crystalline forms of silica (SiO_2), shows the regularity of cubic closest packing. **B,** The atomic arrangement of a quartz glass is amorphous with a generally disordered structure.

Recall that when two atoms form a diatomic molecule, their atomic orbitals (AOs) combine to form an equal number of molecular orbitals (MOs). Let's consider lithium as an example. Figure 12.36 shows the formation of MOs in lithium. In diatomic lithium, Li_2, each atom has four valence orbitals (one $2s$ and three $2p$). (Recall that in Section 11.3, we focused primarily on the $2s$ orbitals.) They combine to form eight MOs, four bonding and four antibonding, spread over both atoms. If two more Li atoms combine, they form Li_4, a slightly larger aggregate, with 16 delocalized MOs. As more Li atoms join the cluster, more MOs are created, their energy levels lying closer and closer together. Extending this process to a 7-g sample of lithium metal (the molar mass) results in 1 mol of Li atoms (Li_{N_A}) combining to form an extremely large number ($4 \times$ Avogadro's number) of delocalized MOs, with *energies so closely spaced that they form a continuum, or band, of MOs*. It is almost as though the entire piece of metal were one enormous Li molecule.

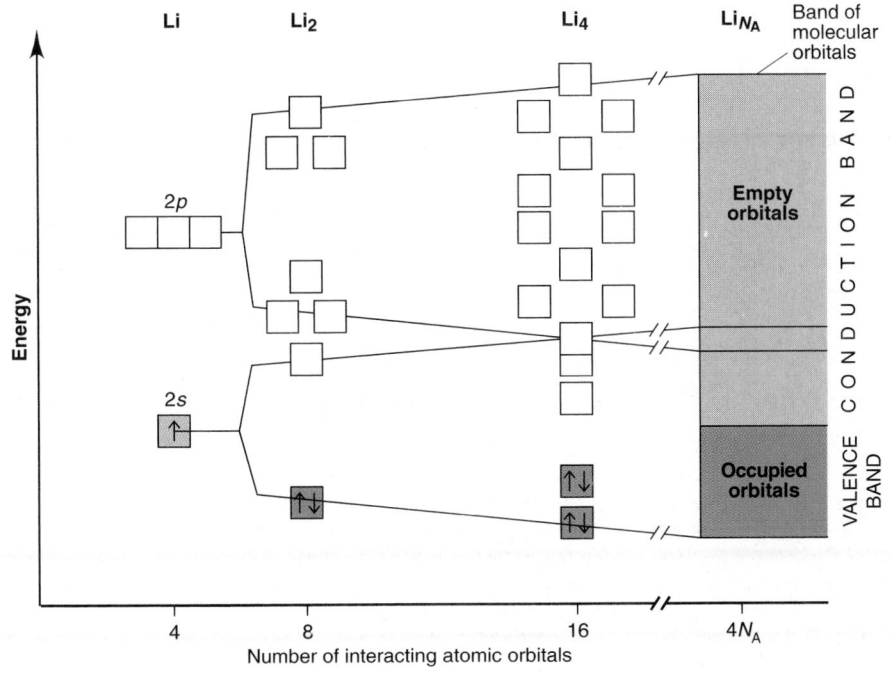

Figure 12.36 **The band of molecular orbitals in lithium metal.** Lithium atoms contain four valence orbitals, one $2s$ and three $2p$ *(left).* When two lithium atoms combine (Li_2), their AOs form eight MOs within a certain range of energy. Four Li atoms (Li_4) form 16 MOs. A mole of Li atoms forms $4N_A$ MOs (N_A = Avogadro's number). The orbital energies are so close together that they form a continuous band. The valence electrons enter the lower energy portion (valence band), while the higher energy portion (conduction band) remains empty. In lithium (and other metals), the valence and conduction bands have no gap between them.

The band model proposes that the lower energy MOs are occupied by the valence electrons and make up the **valence band.** The empty MOs that are higher in energy make up the **conduction band.** In Li metal, the valence band is derived from the $2s$ AOs, and the conduction band is derived mostly from an intermingling of the $2s$ and $2p$ AOs. In Li_2, two valence electrons fill the lowest energy MO and leave the antibonding MO empty. Similarly, in Li metal, 1 mol of valence electrons fills the valence band and leaves the conduction band empty.

The key to understanding metallic properties is that *in metals, the valence and conduction bands are contiguous,* which means that electrons can jump from the filled valence band to the unfilled conduction band if they receive even an infinitesimally small quantity of energy. In other words, the electrons are completely delocalized: *they are free to move throughout the piece of metal.* Thus, metals conduct electricity so well because an applied electric field easily excites the highest energy electrons into empty orbitals, and they move through the sample.

Metallic luster (shininess) is another effect of the continuous band of MO energy levels. With so many closely spaced levels available, electrons can absorb and release photons of many frequencies as they move between the valence and conduction bands. Malleability and thermal conductivity also result from the completely delocalized electrons. Under an externally applied force, layers of positive metal ions simply move past each other, always protected from mutual repulsions by the presence of the delocalized electrons (see Figure 9.24B, p. 351). When a metal wire is heated, the highest energy electrons are excited and their extra energy is transferred as kinetic energy along the wire's length.

Large numbers of nonmetal or metalloid atoms can also combine to form bands of MOs. Metals conduct a current well (conductors), whereas most nonmetals do not (insulators), and the conductivity of metalloids lies somewhere in between (semiconductors). Band theory explains these differences in terms of the size of the energy gaps between the valence and conduction bands, as shown in Figure 12.37:

1. *Conductors (metals).* The valence and conduction bands of a **conductor** have no gap between them, so electrons flow when even a tiny electrical potential difference is applied. When the temperature is raised, greater random motion of the atoms hinders electron movement, which *decreases* the conductivity of a metal.

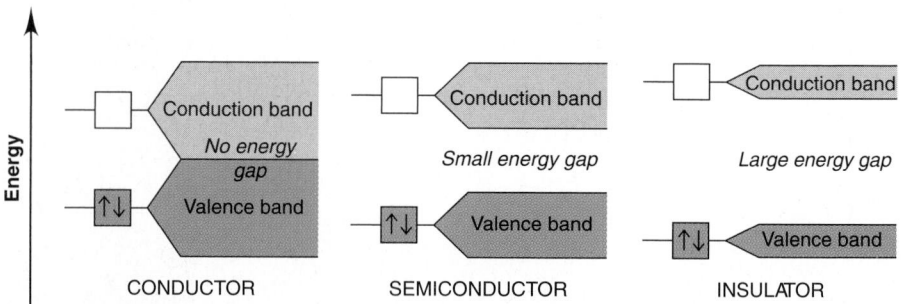

Figure 12.37 **Electrical conductivity in a conductor, semiconductor, and insulator.** Band theory explains differences in electrical conductivity in terms of the size of the energy gap between the material's valence and conduction bands. In conductors (metals), there is no gap. In semiconductors (many metalloids), electrons can jump the small gap if they are given energy, as when the sample is heated. In insulators (most nonmetals), electrons cannot jump the large energy gap.

2. *Semiconductors (metalloids).* In a **semiconductor,** a relatively small energy gap exists between the valence and conduction bands. Thermally excited electrons can cross the gap, allowing a small current to flow. Thus, in contrast to a conductor, the conductivity of a semiconductor *increases* when it is heated.
3. *Insulators (nonmetals).* In an **insulator,** the gap between the bands is too large for electrons to jump even when the substance is heated, so no current is observed.

Another type of electrical conductivity, called **superconductivity,** has been generating intense interest for more than a decade now. When metals conduct at ordinary temperatures, electron flow is restricted by collisions with atoms vibrating in their lattice sites. Such restricted flow appears as resistive heating and represents a loss of energy. To conduct with no energy loss—to superconduct—requires extreme cooling to minimize atom movement. This remarkable phenomenon had been observed in metals only by cooling them to near absolute zero, which can be done only with liquid helium (bp = 4 K; price = $11/L).

In 1986, all this changed with the synthesis of certain ionic oxides that superconduct near the boiling point of liquid nitrogen (bp = 77 K; price = $0.25/L). Like metal conductors, oxide superconductors have no band gap. In the case of $YBa_2Cu_3O_7$, x-ray analysis shows that the Cu ions in the oxide lattice are aligned, which may be related to the superconducting property. In 1989, oxides with Bi and Tl instead of Y and Ba were synthesized and found to superconduct at 125 K; recently, an oxide with Hg, Ba, and Ca, in addition to Cu and O, was shown to superconduct at 133 K. Engineering dreams for these materials include storage and transmission of electricity with no loss of energy, electric power plants operating far from cities, ultrasmall microchips for ultrafast computers, electromagnets to levitate superfast railway trains (Figure 12.38), and inexpensive medical diagnostic equipment with remarkable image clarity.

However, manufacturing problems must be overcome. The oxides are very brittle and fragile, but some recent developments may solve this drawback. A more fundamental concern is that when the oxide is warmed or placed in a strong magnetic field, the superconductivity may disappear and not return again on cooling. Clearly, research into superconductivity will involve chemists, physicists, and engineers for many years to come. We consider other remarkable materials in the next section.

Figure 12.38 The levitating power of a superconducting oxide. A magnet remains suspended above a cooled high-temperature superconductor. Someday, this phenomenon may be used to levitate trains above their tracks for quiet, fast travel.

SECTION SUMMARY
The particles in crystalline solids lie at points that form a structure of repeating unit cells. The three types of unit cells in the cubic system are simple, body-centered, and face-centered. The most efficient packing arrangements are cubic closest packing and hexagonal closest packing. Bond angles and distances in a crystal structure can be determined with x-ray diffraction analysis and scanning tunneling microscopy. Atomic solids have a closest packed structure, with atoms held together by very weak dispersion forces. Molecular solids have molecules at the lattice points, often in a cubic closest packed structure. Their intermolecular forces (dispersion, dipole-dipole, H bonding) and resulting physical properties vary greatly. Ionic solids often crystallize with one type of ion filling holes in a cubic closest packed structure of the other. The high melting points, hardness, and low conductivity of these solids arise from strong ionic attractions. Most metals have a closest packed structure. The atoms of network covalent solids are covalently bonded throughout the sample. Amorphous solids have very little regularity among their particles. Band theory proposes that orbitals in the atoms of solids combine to form a continuum, or band, of molecular orbitals. Metals are electrical conductors because electrons move freely from the filled (valence band) to the empty (conduction band) portions of this energy continuum. Insulators have a large energy gap between the two portions; semiconductors have a small gap.

12.7 ADVANCED MATERIALS

In the last few decades, the exciting field of materials science has grown from solid-state chemistry, physics, and engineering, and it is changing our lives in astonishing ways. Objects that were once considered futuristic fantasies of science-fiction writers are realities or soon will be: powerful, ultrafast computers no bigger than this book, connected electronically to millions of others throughout the world; cars powered by sunlight and made of nonmetallic parts stronger than steel and lighter than aluminum; sporting goods made of the materials of space vehicles; ultrasmall machines constructed by manipulating individual atoms and molecules. In this section, we briefly discuss some of these remarkable materials.

Electronic Materials

The ideal of a perfectly ordered crystal is attainable only if it is grown very slowly under carefully controlled conditions. When crystals form more rapidly, **crystal defects** inevitably form. Planes of particles are misaligned, particles are out of place or missing entirely, and foreign particles are lodged in the lattice.

Although crystal defects usually weaken a substance, they are sometimes introduced intentionally to create materials with improved properties, such as increased strength or hardness or, as you'll see in a moment, to increase a material's conductivity for use in electronic devices. In the process of welding two metals together, for example, *vacancies* form near the surface when atoms vaporize, and then these vacancies move deeper as atoms from lower rows rise to fill the gaps. Welding causes the two types of metal atoms to intermingle and fill each other's vacancies. Metal alloying introduces several kinds of defects, as when some atoms of a second metal occupy lattice sites of the first. Often, the alloy is harder than the pure metal; an example is brass, an alloy of copper with zinc. One reason the welded metals are stronger and the alloy is harder is that the second metal contributes additional valence electrons for metallic bonding.

Doped Semiconductors The manufacture and miniaturization of doped semiconductors have revolutionized the communications, home-entertainment, and information industries. By controlling the number of valence electrons through the creation of specific types of defects, chemists and engineers can greatly increase the conductivity of a semiconductor.

Pure silicon (Si), which lies below carbon in Group 4A(14), conducts poorly at room temperature because an energy gap separates its filled valence band from its conduction band (Figure 12.39A). Its conductivity can be greatly enhanced by **doping,** adding small amounts of other elements to increase or decrease the number of valence electrons in the bands. When Si is doped with phosphorus [or another Group 5A(15) element], P atoms occupy some of the lattice sites. Since P has one more valence electron than Si, this additional electron must enter an empty orbital in the conduction band, thus bridging the energy gap and increasing conductivity. Such doping creates an *n-type semiconductor,* so called because extra *n*egative charges (electrons) are present (Figure 12.39B).

When Si is doped with gallium [or another Group 3A(13) element], Ga atoms occupy some sites (Figure 12.39C). Since Ga has one fewer valence electron than Si, some of the orbitals in the valence band are empty, which creates a positive site. Si electrons can migrate to these empty orbitals, thereby increasing conductivity. Such doping creates a *p-type semiconductor,* so called because the empty orbitals act as *p*ositive holes.

In contact with each other, an n-type and a p-type semiconductor form a *p-n junction.*  When the negative terminal of a battery is connected to the

Solar Cells One of the more common types of solar cell is, in essence, a p-n junction with an n-type surface exposed to a light source. The light provides energy to free electrons from the n-type region and accelerate them through an external circuit into the p-type region, thus providing a current flow to power a calculator, light a bulb, and so on. Arrays of solar cells supply electric power to many residences and businesses, as well as to the space shuttle and most communications satellites (see illustration).

A Pure silicon crystal   **B** n-Type doping with phosphorus   **C** p-Type doping with gallium

Figure 12.39 Crystal structures and band representations of doped semiconductors. A, Pure silicon has the same crystal structure as diamond but acts as a semiconductor, the energy gap between its valence and conduction bands keeping conductivity low at room temperature. **B,** Doping silicon with phosphorus *(purple)* adds additional valence electrons, which are free to move through the crystal. They enter the lower portion of the conduction band, which is adjacent to higher energy empty orbitals, thereby increasing conductivity. **C,** Doping silicon with gallium *(orange)* removes electrons from the valence band and introduces positive "holes." Nearby Si electrons can enter these empty orbitals, thereby increasing conductivity. The orbitals from which the Si electrons moved become vacant; in effect, the holes moved.

n-type portion and the positive terminal to the p-type portion, electrons flow freely in the n-to-p direction, which has the simultaneous effect of moving holes in the p-to-n direction (Figure 12.40). No current flows if the terminals are reversed. Such unidirectional current flow makes a p-n junction act as a *rectifier,* a device that converts alternating current into direct current. A p-n junction in a modern integrated circuit can be made smaller than a square 10 μm on a side. Before the p-n junction was created, rectifiers were bulky, expensive vacuum tubes.

A modern computer chip the size of a nickel may incorporate millions of p-n junctions in the form of *transistors*. One of the most common types, an n-p-n transistor, is made by sandwiching a p-type portion between two n-type portions to form adjacent p-n junctions. The current flowing through one junction controls the current flowing through the other and results in an amplified signal. Once again, to accomplish signal amplification before the advent of doped semiconductors required large, and often unreliable, vacuum tubes. Today, minute transistors are found in every radio, TV, and computer and have made possible the multibillion-dollar electronics industry.

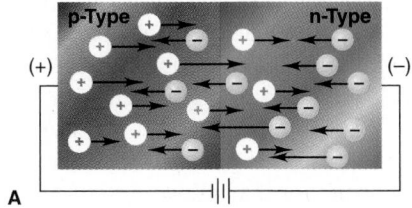
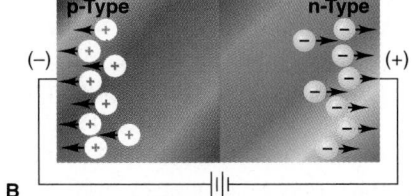

Figure 12.40 The p-n junction. Placing a p-type semiconductor adjacent to an n-type creates a p-n junction. **A,** If the negative battery terminal is on the n-type portion, electrons *(yellow sphere with minus sign)* flow toward the p-type portion, which, in effect, moves holes *(white circle with plus sign)* toward the n-type portion. The small arrows indicate the net direction of movement. Note that any hole in the n-type portion lies opposite an electron that moved to the p-type portion. **B,** If the positive battery terminal is on the n-type portion, electrons are attracted to it, so no flow takes place across the junction.

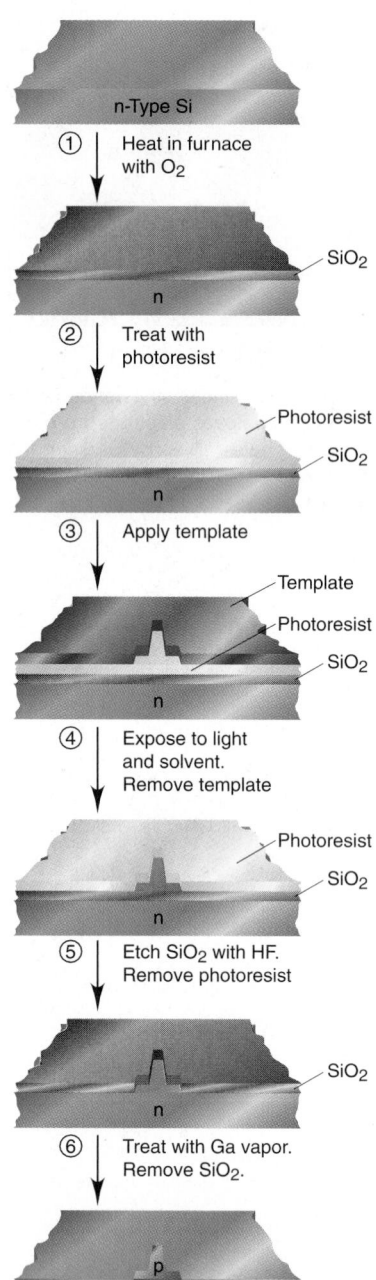

Manufacturing a p-n Junction Several steps are required to manufacture a simple p-n junction. The process, as shown in Figure 12.41, begins with a thin wafer made from a single crystal of n-type silicon:

Step 1. Forming an oxide surface. In a furnace, the wafer is heated in O_2 to form a surface layer of SiO_2.

Step 2. Covering with a photoresist. The oxide-coated wafer is covered by a light-sensitive wax or polymer film called a *photoresist.*

Step 3. Applying the template. A template with spaces having the desired shape of the p-type region is applied. Thus, all other areas are "masked" from the subsequent treatments.

Step 4. Exposing the photoresist and removing the template. Light shining on the template alters the exposed areas of photoresist so that it can be dissolved away in specific solvents, revealing the oxide surface. The template is removed.

Step 5. Etching the oxide surface and removing the photoresist. Treatment with hydrofluoric acid etches the template-shaped area of oxide coating to expose the n-type Si surface. Following this, the remaining photoresist is also removed.

Step 6. Creating the p-n junction. The wafer is exposed at high temperature to vapor of a Group 3A(13) element, which diffuses into the bare areas to create a template-shaped region of p-type Si adjacent to the n-type Si. The remaining SiO_2 is removed at this point also.

Typically, these steps are repeated to create an n-type region adjacent to the newly formed p-type region, thus forming an n-p-n transistor.

Liquid Crystals

Incorporated in the membrane of every cell in your body and in the display of just about every digital watch, pocket calculator, and laptop computer are unique substances known as **liquid crystals.** These materials flow like liquids but, like crystalline solids, pack at the molecular level with a high degree of order.

Properties, Preparation, and Types of Liquid Crystals To understand the properties of liquid crystals, let's first examine how particles are ordered in the three common physical states and how this affects their properties. The extent of order among particles distinguishes crystalline solids clearly from gases and liquids. Gases have no order and liquids have little more. Both are considered *isotropic,* which means that their physical properties are the same in every direction within the phase. For example, the viscosity of a gas or a liquid is the same regardless of direction. Glasses and other amorphous solids are also isotropic because they have no regular lattice structure.

In contrast, crystalline solids have a high degree of order among their particles. The properties of a crystal *do* depend on direction, so a crystal is *anisotropic.* The facets in a cut diamond, for instance, arise because the crystal cracks in one direction more easily than in another. Liquid crystals are also anisotropic in that several physical properties, including the electrical and optical properties that lead to their most important applications, differ with direction through the phase.

Like crystalline molecular solids, liquid crystal phases consist of individual molecules. In most cases, the molecules that form liquid crystal phases have two characteristics: a long, cylindrical shape and a structure that allows intermolecular attractions through dispersion and dipole-dipole or H-bonding forces, but that inhibits perfect crystalline packing. Figure 12.42 shows the structures of two molecules that form liquid crystal phases. Note the rodlike shapes and the presence of certain groups—in these cases, flat, benzene-like ring systems—that keep the molecules extended. Many of these types of molecules also have a molecular dipole associated with the long molecular axis. A sufficiently strong electric field

Figure 12.41 Steps in manufacturing a p-n junction. Steps 1 to 5 prepare the desired shape of the p-type portion on the n-type wafer. Step 6 dopes that shape with the Group 3A(13) element.

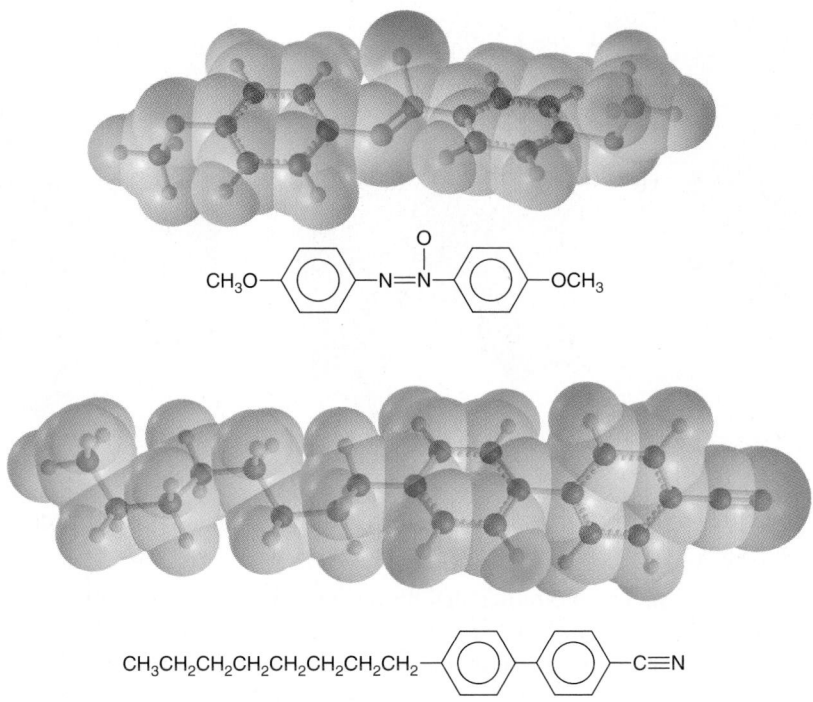

Figure 12.42 Structures of two typical liquid crystal molecules. Note the long, extended shapes and the regions of high (*red*) and low (*blue*) electron density.

can orient large numbers of these polar molecules in approximately the same direction, like compass needles in a magnetic field.

The viscosity of a liquid crystal phase is lowest in the direction parallel to the long axis. Like moistened microscope slides, it is easier for the molecules to slide along each other (because the total attractive force remains the same), than it is for them to pull apart from each other sideways. As a result, the molecules tend to align while the phase flows.

Liquid crystal phases can arise in two general ways, and, sometimes, either way can occur in the same substance. A *thermotropic* phase develops as a result of a change in temperature. As a crystalline solid is heated, the molecules leave their lattice sites, but the intermolecular interactions are still strong enough to keep the molecules aligned with each other along their long axes. Like any other phase, the liquid crystal phase has sharp transition temperatures; however, it exists over a relatively small temperature range. Further heating provides enough kinetic energy for the molecules to become disordered, as in a normal liquid. The typical range for liquid crystal phases of pure substances is from <1°C to around 10°C, but mixing phases of two or more substances can greatly extend this range. For this reason, the liquid crystal phases used within display devices, as well as those within cell membranes, consist of mixtures of molecules.

A *lyotropic* phase occurs in solution as the result of changes in concentration, but the conditions for forming such a phase vary for different substances. For example, when purified, some biomolecules that exist naturally in mammalian cell membranes form lyotropic phases in water at the mild temperature that occurs within the organism. At the other extreme, Kevlar, a fiber used in bulletproof vests and high-performance sports equipment, forms a lyotropic phase at high temperatures in concentrated H_2SO_4 solution.

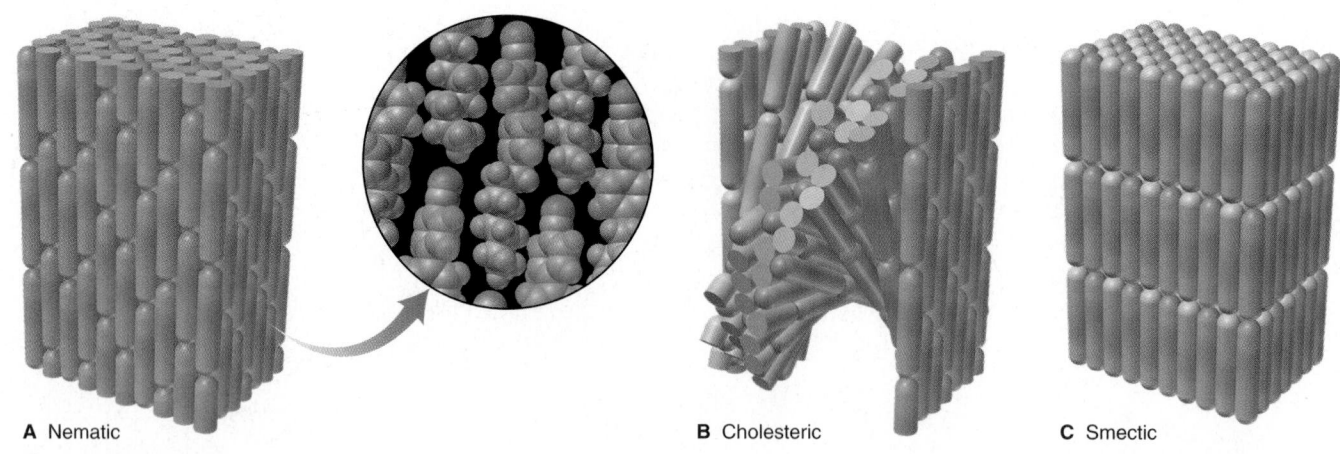

A Nematic

B Cholesteric

C Smectic

Figure 12.43 **The three common types of liquid crystal phases.** **A,** Nematic phase. A rectangular volume of the phase, with expanded view, shows a closeup of the arrangement of molecules. **B,** Cholesteric phase. Note the corkscrew-like arrangement of the layers. **C,** Smectic phase. A rectangular volume of the phase shows the more orderly stacking of layers.

Molecules that form liquid crystals can exhibit various types of order. Three common types are nematic, cholesteric, and smectic:

- In a *nematic* phase, the molecules lie in the same direction but their ends are not aligned, much like a school of fish swimming in synchrony (Figure 12.43A). The nematic phase is the least ordered type of liquid crystal phases.

- In a *cholesteric* phase, which is somewhat more ordered, the molecules lie in layers that each exhibit nematic-type ordering. Rather than lying in parallel fashion, however, each layer is rotated by a fixed angle with respect to the next layer. The result is a helical (corkscrew) arrangement, so a cholesteric phase is often called a *twisted nematic phase* (Figure 12.43B).

- In a *smectic* phase, which is the most ordered, the molecules lie parallel to each other, *with* their ends aligned, in layers that are stacked directly over each other, much like a supermarket display of shelves filled with identical bottles (Figure 12.43C). The long molecular axis has a well-defined angle (shown in the figure as 90°) with respect to the plane of the layer. The molecules in Figure 12.42 are typical of those that form nematic or smectic phases. Liquid crystalline patterns appear in many biological systems (Figure 12.44).

In some cases, a substance that forms a given liquid crystal phase under one set of conditions forms other phases under different conditions. Thus, a given thermotropic liquid crystal substance can pass from disordered liquid through a series of distinct liquid crystal phases to an ordered crystal through a decrease in temperature. A lyotropic substance can undergo similar changes through an increase in concentration.

Applications of Liquid Crystals The ability to control the orientation of the molecules in a liquid crystal allows us to produce materials with high strength or unique optical properties.

Figure 12.44 **Liquid crystals in biological systems** **A,** Nematic arrays of tobacco mosaic virus particles within the fluid of a tobacco leaf. **B,** The orderly arrangement of actin and myosin protein filaments in voluntary muscle cells.

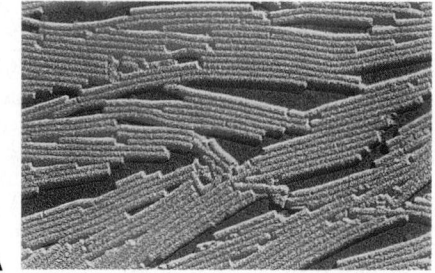

A

B

High-strength applications involve the use of extremely long molecules called *polymers*. While in a thermotropic liquid crystal phase and during their flow through the processing equipment, these molecules become highly aligned, like the fibers in wood. Cooling solidifies them into fibers, rods, and sheets that can be shaped into materials with superior mechanical properties in the direction of the long molecular axis. Sporting equipment, supersonic aircraft parts, and the sails used in the America's Cup races are fabricated from these polymeric materials. (We discuss the structure and physical behavior of polymers later in this section and their synthesis in Chapter 15.)

Far more important in today's consumer market are the liquid crystal displays (LCDs) used in watches, calculators, and computers. All depend on *changes in molecular orientation in an electric field*. The most common type, called a *twisted nematic* display, is shown schematically in Figure 12.45 as a small portion of a wristwatch LCD. The liquid crystal phase consists of layers of nematic phases sandwiched between thin glass plates that incorporate transparent electrodes. Through a special coating process, the long molecular axis lies parallel to the plane of the glass plates. The distance between the plates (6–8 μm) is chosen so that the molecular axis within each succeeding layer twists just enough for the molecular orientation at the bottom plate to be 90° from that at the top plate. Above and below this "sandwich" are thin polarizing filters (like those used in Polaroid lenses or camera sunlight filters), which allow light waves oriented in only one direction to pass through. The filters are placed in a "crossed" arrangement, so that light passing through the top filter must twist 90° to pass through the bottom filter. The orientation and optical properties of the molecules twist the orientation of the light by exactly this amount. This whole grouping of filters, plates, and liquid crystal phase lies on a mirror.

A current generated by the watch battery controls the orientation of the molecules within the phase. With the current on in one region of the display, the molecules become oriented *toward* the field, and thus block the light from passing through to the bottom filter, so that region appears dark. With the current off in another region, light passes through the molecules and bottom filter to the mirror and back again, so that region appears bright.

Cholesteric liquid crystals are used in applications that involve color changes with temperature. The helical twist in these phases "unwinds" with heating, and the extent of the unwinding determines the color. Liquid crystal thermometers that include a mixture of substances to widen their range of temperatures use this effect, as do novelty items such as "mood" rings. New and more important uses include "mapping" the area of a tumor, detecting faulty connections in electronic circuit boards, and nondestructive testing of materials under stress.

Ceramic Materials

First developed by Stone Age people, **ceramics** are defined as nonmetallic, non-polymeric solids that are hardened by heating to high temperatures. Clay ceramics consist of silicate microcrystals suspended in a glassy cementing medium. In "firing" a ceramic pot, for example, a kiln heats the object made of an aluminosilicate clay, such as kaolinite, to 1500°C and the clay loses water:

$$Si_2Al_2O_5(OH)_4(s) \longrightarrow Si_2Al_2O_7(s) + 2H_2O(g)$$

During the heating process, the structure rearranges to an extended network of Si-centered and Al-centered tetrahedra of O atoms (Section 14.6).

Bricks, porcelain, glazes, and other clay ceramics are useful because of their hardness and resistance to heat and chemicals. Today's high-tech ceramics incorporate these traditional characteristics in addition to superior electrical and magnetic properties (Table 12.7, on the next page). As just one example, consider the unusual electrical behavior of certain zinc oxide (ZnO) composites. Ordinarily a semiconductor, ZnO can be doped so that it becomes a conductor. Imbedding

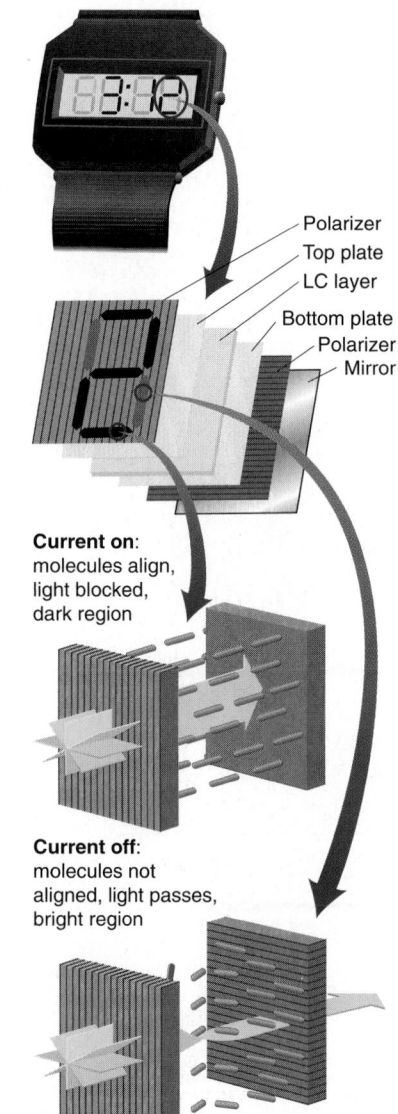

Figure 12.45 Schematic of a liquid crystal display (LCD). A close-up of the "2" on a wristwatch LCD reveals two polarizers sandwiching two glass plates, which sandwich an LC layer, all lying on a mirror. When light waves oriented in all directions enter the first polarizer, only waves oriented in one direction emerge to enter the LC layer. Enlarging a dark region of the numeral *(top blow-up)* shows that with the current on, the LC molecules are aligned and the light cannot pass through the other polarizer to the mirror, so the viewer sees no light. Enlarging a bright region *(bottom blow-up)* shows that with the current off, the LC molecules lie in a twisted nematic arrangement, which rotates the plane of the light waves and allows them to pass through the other polarizer to the mirror. Reflecting and retracing this path (not shown), the light reaches the viewer.

Labels on figure:
Polarizer
Top plate
LC layer
Bottom plate
Polarizer
Mirror

Current on: molecules align, light blocked, dark region

Current off: molecules not aligned, light passes, bright region

Table 12.7 Some Uses of New Ceramics and Ceramic Mixtures

Ceramic	Applications
SiC, Si_3N_4, TiB_2, Al_2O_3	Whiskers (fibers) to strengthen Al and other ceramics
Si_3N_4	Car engine parts; turbine rotors for "turbo" cars; electronic sensor units
Si_3N_4, BN, Al_2O_3	Supports or layering materials (as insulators) in electronic microchips
SiC, Si_3N_4, TiB_2, ZrO_2, Al_2O_3, BN	Cutting tools, edge sharpeners (as coatings and whole devices), scissors, surgical tools, industrial "diamond"
BN, SiC	Armor-plating reinforcement fibers (as in Kevlar composites)
ZrO_2, Al_2O_3	Surgical implants (hip and knee joints)

particles of the doped oxide into an insulating ceramic produces a variable resistor: at low voltage, the material conducts poorly, but at high voltage, it conducts well. Best of all, the changeover voltage can be "preset" by controlling the size of ZnO particles and the thickness of the insulating medium.

Preparing Modern Ceramics Among the important modern ceramics are silicon carbide (SiC) and nitride (Si_3N_4), boron nitride (BN), and the superconducting oxides. They are prepared by standard chemical methods that involve driving off a volatile component during the reaction.

The SiC ceramics are made from compounds used in silicone polymer manufacture (we'll discuss their structures and uses in Section 14.6):

$$n(CH_3)_2SiCl_2(l) + 2nNa(s) \longrightarrow 2nNaCl(s) + [(CH_3)_2Si]_n(s)$$

This product is heated to 800°C to form the ceramic:

$$[(CH_3)_2Si]_n \longrightarrow nCH_4(g) + nH_2(g) + nSiC(s)$$

SiC can also be prepared by direct reaction of Si and graphite under vacuum:

$$Si(s) + C(graphite) \xrightarrow{\sim 1500°C} SiC(s)$$

The nitride is also prepared by reaction of the elements:

$$3Si(s) + 2N_2(g) \xrightarrow{>1300°C} Si_3N_4(s)$$

The formation of a BN ceramic begins with the reaction of boron trichloride or boric acid with ammonia:

$$B(OH)_3(s) + 3NH_3(g) \longrightarrow B(NH_2)_3(s) + 3H_2O(g)$$

Heat drives off some of the bound nitrogen as NH_3 to yield the ceramic:

$$B(NH_2)_3(s) \xrightarrow{\Delta} 2NH_3(g) + BN(s)$$

One of the common high-temperature superconducting oxides is a ceramic made by heating a mixture of barium carbonate and copper and yttrium oxides, followed by further heating in the presence of O_2:

$$4BaCO_3(s) + 6CuO(s) + Y_2O_3(s) \xrightarrow{\Delta} 2YBa_2Cu_3O_{6.5}(s) + 4CO_2(g)$$
$$YBa_2Cu_3O_{6.5}(s) + \tfrac{1}{4}O_2(g) \xrightarrow{\Delta} YBa_2Cu_3O_7(s)$$

Ceramic Structures and Uses Structures of several ceramic materials are shown in Figure 12.46. Note the diamond-like structure of silicon carbide. Network covalent bonding gives this material great strength. SiC is being made into thin fibers, called *whiskers,* to reinforce other ceramics in a composite structure and prevent cracking, much like steel rods reinforcing concrete. Silicon nitride is virtually inert chemically, retains its strength and wear resistance for extended periods above

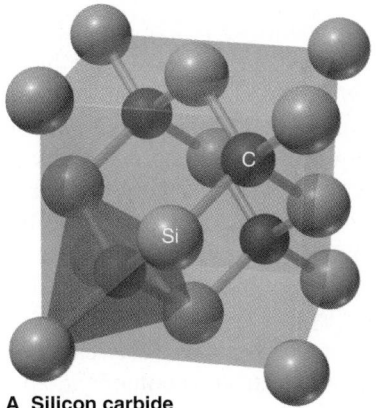

A Silicon carbide

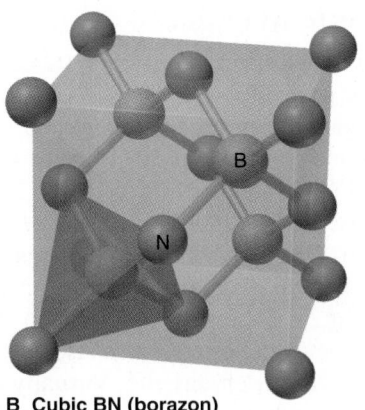

B Cubic BN (borazon)

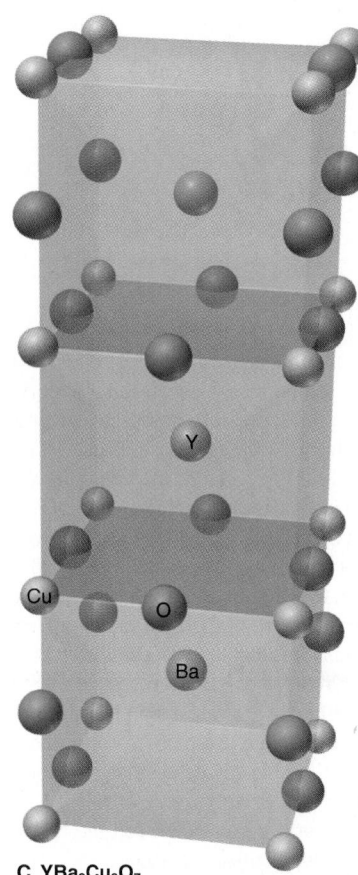

C YBa$_2$Cu$_3$O$_7$

Figure 12.46 **Unit cells of some modern ceramic materials.** SiC **(A)** and the high-pressure form of BN **(B)** both have a crystal structure similar to that of diamond and are extremely hard. YBa$_2$Cu$_3$O$_7$ **(C)** is one of the high-temperature superconducting oxides.

1000°C, is dense and hard, and acts as an electrical insulator. Japanese and American automakers are testing it in high-efficiency car and truck engines because it allows an ideal combination of low weight, high operating temperature, and little need for lubrication.

The BN ceramics exist in two structures (Section 14.5), analogous to the common crystalline forms of carbon. In the graphite-like form, BN has extraordinary properties as an electrical insulator. At high temperature and very high pressure (1800°C and 8.5×10^4 atm), it converts to a diamond-like structure, which is extremely hard and durable. Both forms are virtually invisible to radar.

Earlier, we mentioned some potential uses of the superconducting oxides. In nearly every one of these ceramic materials, copper occurs in an unusual oxidation state. In YBa$_2$Cu$_3$O$_7$, for instance, assuming oxidation states of +3 for Y, +2 for Ba, and −2 for O, the three Cu atoms have a total oxidation state of +7. This is allocated as Cu(II)$_2$Cu(III), with one Cu in the unusual +3 state. X-ray crystallography indicates that a distortion in the structure makes four of the oxide ions unusually close to the Y^{3+} ion, which aligns the Cu ions into chains within the crystal. It is suspected that a specific half-filled $3d$ orbital in Cu oriented toward a neighboring O^{2-} ion may be associated with superconductivity, although the process is still poorly understood. Because of their brittleness, it has been difficult to fashion these ceramics into wires, but methods for making superconducting films and ribbons have recently been developed.

Research in ceramic processing is beginning to overcome the inherent brittleness of this entire class of materials. This brittleness arises from the strength of the ionic-covalent bonding in these solids and their resulting inability to deform. Under stress, a microfine crystal defect widens and lengthens until the material cracks. One new method forms defect-free ceramics using controlled packing and heat-treating of extremely small, uniform oxide particles coated with organic polymers. Another method is aimed at arresting a widening crack. These ceramics are embedded with zirconia (ZrO$_2$), whose crystal structure expands up to 5% under the mechanical stress of a crack tip: the moment the advancing crack reaches them, the zirconia particles effectively pinch it shut. Very recently, a third method was reported. Japanese researchers prepared a ceramic material made of precisely grown single crystals of Al$_2$O$_3$ and GdAlO$_3$, which become entangled during the solidification process. The material bends without cracking at temperatures above 1800 K.

Despite remaining technical difficulties, chemical ingenuity will continue to develop new ceramic materials and apply their amazing and useful properties well into the 21st century.

Polymeric Materials

In its simplest form, a **polymer** (Greek, "many parts") is an extremely large molecule, or **macromolecule**, consisting of a covalently linked chain of smaller molecules, called **monomers** (Greek, "one part"). The monomer is the *repeat unit* of the polymer, and a typical polymer may have anywhere from hundreds to hundreds of thousands of repeat units. *Synthetic* polymers are created by chemical reactions in the laboratory; *natural* polymers (or *biopolymers*) are created by chemical reactions within organisms. There are many types of monomers, and their chemical structures allow for the complete repertoire of intermolecular forces.

Synthetic polymers, such as plastics, rubbers, and crosslinked glasses have revolutionized everyday life. Virtually every home, car, electronic component, and processed food contains synthetic polymers in its structure or packaging. You interact with dozens of these materials each day—from paints to floor coverings to clothing to the additives and adhesives in this textbook. Some of these materials, like those used in food containers, are very long-lived in the environment and have created a serious waste-disposal problem. Others are being actively recycled into the same or other useful products, such as garbage bags, outdoor furniture, roofing tiles, and even marine pilings and roadside curbs. Still others, such as artificial skin, heart valve components, and hip joints, are designed to have as long a life as possible.

In this section, we'll examine the physical nature of synthetic polymers and explore the role intermolecular forces play in their properties and uses. In Chapter 15, we'll examine the types of monomers, look at the preparation of synthetic polymers, and then focus on the structures and vital functions of the biopolymers.

One Strand or Many Pieces? By the mid-19th century, entrepreneurs had transformed cheap natural polymers into valuable materials, such as rubber and the cellulose nitrate ("celluloid") film used in the young movie industry. Studies had shown the presence of repeat units, and most believed that polymers were small molecules held together by intermolecular forces. But, the young German chemist Hermann Staudinger was convinced that polymers were large molecules held together by covalent bonds, and, despite ridicule from many prominent chemists, his covalent-linkage hypothesis was eventually confirmed. For this work, he was awarded the Nobel Prize in chemistry in 1953.

Dimensions of a Polymer Chain: Mass, Size, and Shape Because of their great lengths, polymers are unlike smaller molecules in several important ways. Let's see how chemists describe the mass, size, and shape of a polymer chain and how the chains exist in a sample. We'll focus throughout on polyethylene, by far the most common synthetic polymer.

1. *Polymer mass.* The molar mass of a polymer chain ($\mathcal{M}_{polymer}$, in g/mol, often referred to as the *molecular weight*) depends on two parameters—the molar mass of the repeat unit ($\mathcal{M}_{repeat}$) and the **degree of polymerization (n),** the number of repeat units in the chain:

$$\mathcal{M}_{polymer} = \mathcal{M}_{repeat} \times n$$

For example, the molar mass of the ethylene repeat unit is 28 g/mol. If an individual polyethylene chain in a plastic grocery bag has a degree of polymerization of 7100, the molar mass of that particular chain is

$$\mathcal{M}_{polymer} = \mathcal{M}_{repeat} \times n = (28 \text{ g/mol}) (7.1 \times 10^3) = 2.0 \times 10^5 \text{ g/mol}$$

Table 12.8 shows some other examples.

Table 12.8 Molar Masses of Some Common Polymers

Name	$\mathcal{M}_{polymer}$ (g/mol)	n	Uses
Acrylates	2×10^5	2×10^3	Rugs, carpets
Polyamide (nylons)	1.5×10^4	1.2×10^2	Tires, fishing line
Polycarbonate	1×10^5	4×10^2	Compact disks
Polyethylene	3×10^5	1×10^4	Grocery bags
Polyethylene (ultra-high molecular weight)	5×10^6	2×10^5	Hip joints
Poly(ethylene terephthalate)	2×10^4	1×10^2	Soda bottles
Polystyrene	3×10^5	3×10^3	Packing; coffee cups
Poly(vinyl chloride)	1×10^5	1.5×10^3	Plumbing

However, even though any *given* chain within a sample of a polymer has a fixed molar mass, the degree of polymerization often varies considerably from chain to chain. As a result, *all samples of synthetic polymers have a distribution of chain lengths*. For this reason, polymer chemists use various definitions of *average* molar mass, and a common one is the *number-average molar mass*, $\mathcal{M}_n$:

$$\mathcal{M}_n = \frac{\text{total mass of all chains}}{\text{number of moles of chains}}$$

Thus, even though the number-average molar mass of the polyethylene in grocery bags is, say, 1.6×10^5 g/mol, the chains may vary in molar mass from about 7.0×10^4 to 3.0×10^5 g/mol.

2. *Polymer size and shape.* The long axis of a polymer chain is called its *backbone*. The length of an *extended* backbone is simply the number of repeat units (degree of polymerization, *n*) times the length of each repeat unit (l_0). For instance, the length of an ethylene repeat unit is about 250 pm, so the extended length of our particular grocery-bag polyethylene chain is

Length of extended chain $= n \times l_0 = (7.1 \times 10^3)(2.5 \times 10^2 \text{ pm}) = 1.8 \times 10^6$ pm

Comparing this length with the thickness of the chain, which is only about 40 pm, gives a good picture of the threadlike dimensions of the extended chain.

It's very important to understand, however, that a polymer molecule, whether pure or in solution, doesn't exist as an extended chain but, in fact, is far more compact. To picture the actual shape, polymer chemists assume, as a first approximation, that the shape of the chain arises as a result of free rotation around all of its single bonds. Thus, as each repeat unit rotates randomly, the chain continuously changes direction, turning back on itself many times and eventually arriving at the **random coil** shape that most polymers adopt (Figure 12.47). In reality, of course, rotation is not completely free because, as one portion of a chain bends and twists near other portions of the same chain or of other nearby chains, they attract each other. Thus, the nature of intermolecular forces between chain portions, between other chains, and/or between chain and solvent becomes a key factor in establishing the actual shape of a polymer chain.

Figure 12.47 The random-coil shape of a polymer chain. A short ball-and-stick section *(left)*, showing bond lengths and angles, leads to a space-filling depiction of a polyethylene chain. Note the random coiling of the chain's carbon atoms *(black)*. Sections of several nearby chains *(red, green, and yellow)* are entangled with this chain, kept near one another by dispersion forces. In reality, entangling chains fill the gaps seen here. The radius of gyration (*R*g) represents the average distance from the center of the coiled molecule to its outer edge.

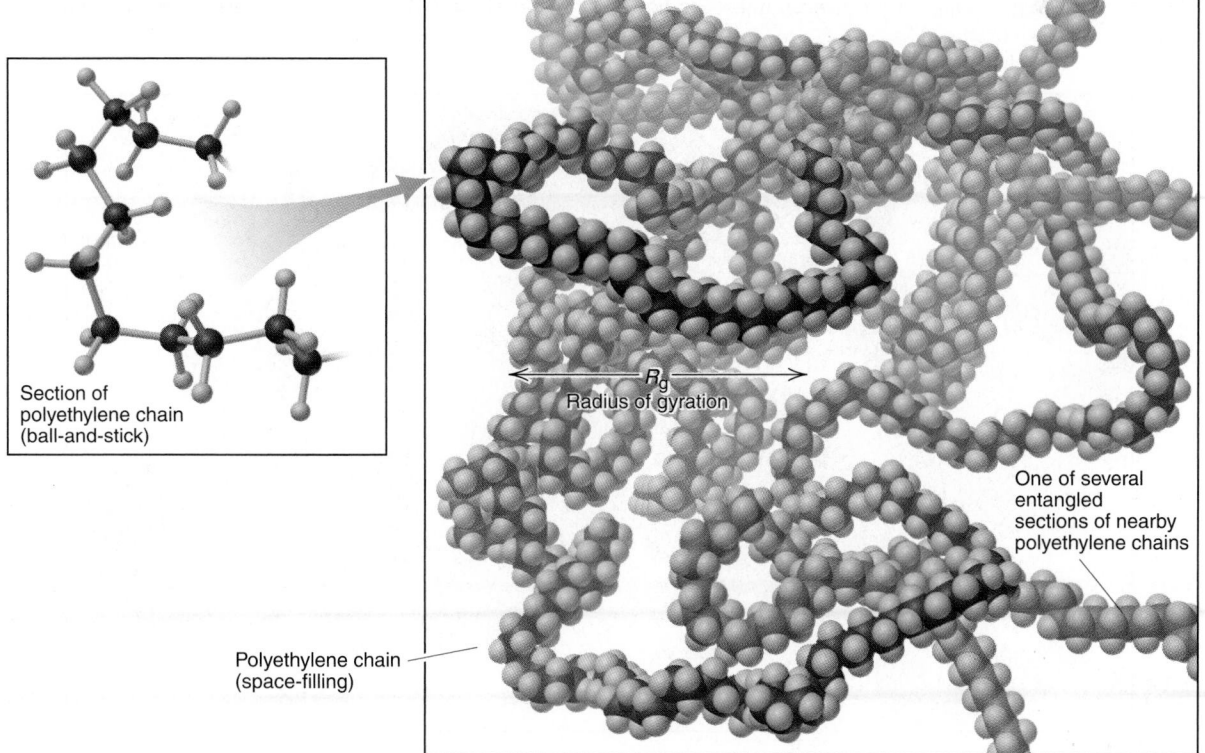

Section of polyethylene chain (ball-and-stick)

R_g
Radius of gyration

One of several entangled sections of nearby polyethylene chains

Polyethylene chain (space-filling)

The size of the coiled polymer chain is expressed by its **radius of gyration, R_g**, the average distance from the center of mass to the outside edge of the coil (see Figure 12.47). Even though Rg is reported as a single value for a given polymer, it represents an average value of many chains. The mathematical expression for the radius of gyration considers the length of each repeat unit and its randomized direction in space, as well as the bond angles between atoms in a unit and between adjacent units:*

$$R_g = \sqrt{\frac{nl_0^2}{6}}$$

As we would expect, the radius of gyration increases with the degree of polymerization, and thus with the molar mass as well. Most importantly, light-scattering experiments and other laboratory measurements correlate with the calculated results, so for many polymers the radius of gyration can be determined experimentally.

For our grocery-bag polyethylene chain, we have

$$R_g = \sqrt{\frac{nl_0^2}{6}} = \sqrt{\frac{(7.1 \times 10^3)(2.5 \times 10^2 \text{ pm})^2}{6}} = 8.6 \times 10^3 \text{ pm}$$

Doubling the radius gives a diameter of 1.7×10^4 pm, less than one-hundredth the length of the extended chain!

3. *Polymer crystallinity*. You may get the impression from the discussion so far that a sample of a given polymer is just a disorderly jumble of chains, but this is often not the case. If the molecular structure allows neighboring chains to pack together and if the chemical groups lead to favorable dipole-dipole, H-bonding, or dispersion forces, portions of the chains can align regularly and exhibit crystallinity.

However, the crystallinity of a polymer is very different from the crystallinity of the simple compounds we discussed earlier. There, the orderly array extends over many molecules, and the unit cell includes at least one molecule. In contrast, the orderly regions of a polymer rarely, if ever, involve even one whole molecule (Figure 12.48). At best, polymers are *semicrystalline*, because only parts of

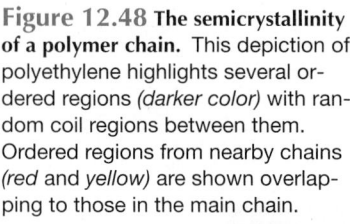

Figure 12.48 The semicrystallinity of a polymer chain. This depiction of polyethylene highlights several ordered regions *(darker color)* with random coil regions between them. Ordered regions from nearby chains *(red and yellow)* are shown overlapping to those in the main chain.

*The mathematical derivation of R_g is beyond the scope of this text, but it is analogous to the two-dimensional "walk of the drunken sailor." With each step, the sailor stumbles in random directions and, given enough time, ends up very close to the starting position. The radius of gyration quantifies how far the end of the polymer chain (the sailor) has gone from the origin.

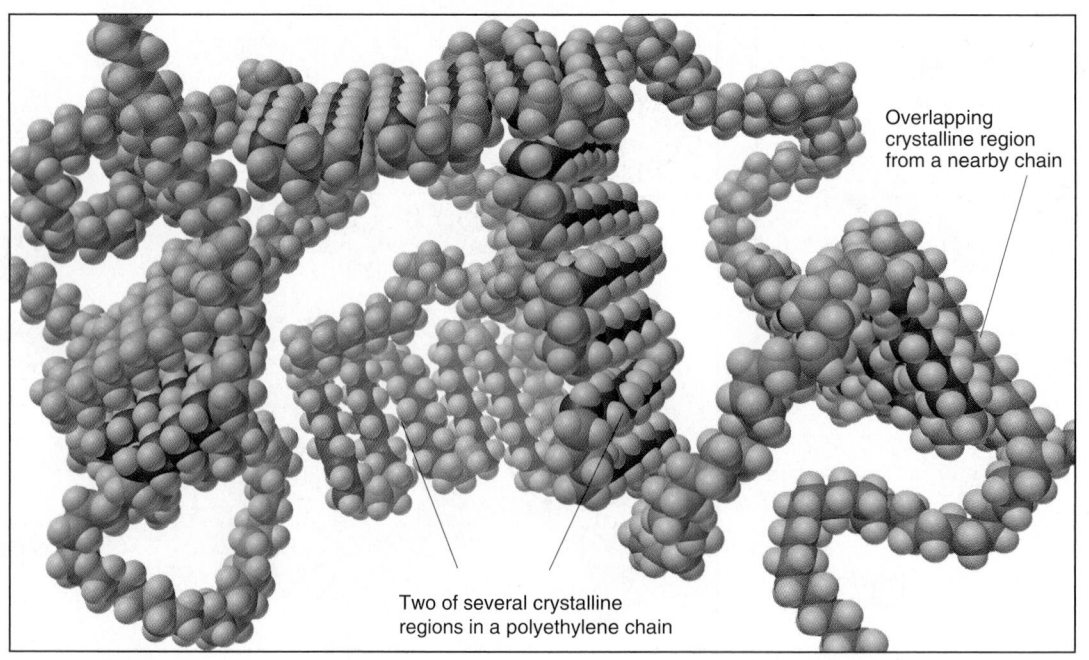

Overlapping crystalline region from a nearby chain

Two of several crystalline regions in a polyethylene chain

the molecule align with parts of neighboring molecules (or with other parts of the same chain), while most of the chain remains as a random coil. Thus, the unit cell of a polymer includes only a small part of the molecule.

Flow Behavior of Polymers. In Section 12.4, we defined the viscosity of a fluid as its resistance to flow. Some of the most important uses of polymers arise from their ability to change the viscosity of a solvent in which they have been dissolved and to undergo temperature-dependent changes in their own viscosity.

When an appreciable amount of polymer (about 5–15 mass %) dissolves, the viscosity of the solution is much higher than that of the pure solvent. This behavior is put to use by adding polymers to increase the viscosity of many common materials, such as motor oil, paint, and salad dressing. A dissolved polymer increases the viscosity of the solution by interacting with the solvent. As the random coil of a polymer moves through a solution, solvent molecules are attracted to its exterior and interior through intermolecular forces (Figure 12.49). Thus, the polymer coil drags along many solvent molecules that are attracted to other solvent molecules and other coils, and flow is lessened. Increasing the polymer concentration increases the viscosity because the coils are more likely to become entangled in one another. In order for each coil to flow, it must disentangle from its neighbors or drag them along.

Viscosity is a basic property that characterizes the behavior of a particular polymer-solvent pair at a given temperature. Just as it does for a pure polymer, the size (radius of gyration) of a random coil of a polymer in solution increases with molar mass, and so does the viscosity. To study basic interactions and for use in polymer manufacture, chemists and other industrial scientists have

Figure 12.49 The viscosity of a polymer in solution. A ball-and-stick section *(left panel)* of a poly(ethylene oxide) chain in water shows the H bonds that form between the lone pairs of the chain O atoms and the H atoms of solvent molecules. The polymer chain, depicted as a blue and red coiled rod *(center panel)*, forms many H bonds with solvent. Note the H bonds to water that allow one chain to interact with others nearby. As the concentration of polymer increases *(three right panels)*, the viscosity of the solution increases because the movement of each chain is restricted through its interactions with solvent and with other chains.

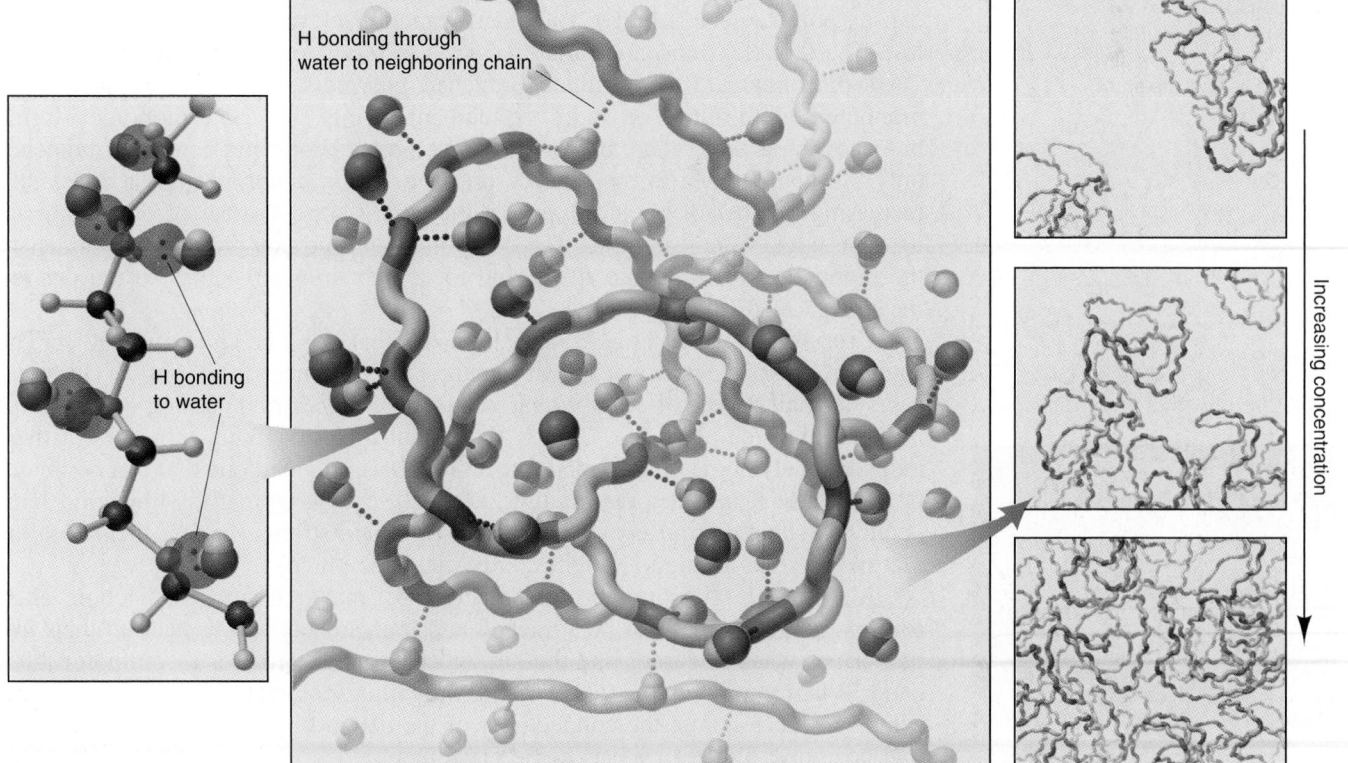

H bonding through water to neighboring chain

H bonding to water

Increasing concentration

developed essential quantitative equations to predict the viscosity of polymers of different molar masses in a variety of solvents (see Problem 12.150).

Intermolecular forces also play a major role in the flow of a pure polymer sample. At temperatures high enough to melt them, many polymers exist as viscous liquids, flowing more like honey than water. The forces between the chains, as well as chain entanglement, hinder the molecules from flowing past each other. As the temperature decreases, the intermolecular attractions exert a greater effect, and the eventual result is a rigid solid. If the chains don't crystallize, the resulting material is called a *polymer glass*. The transition from a liquid to a glass occurs over a narrow (10–20°C) temperature range for a given polymer, but chemists define a single temperature at the midpoint of the range as the *glass transition temperature, T_g*. Like window glass, many polymer glasses are transparent, such as polystyrene in drinking cups and polycarbonate in compact discs.

The flow-related effects of polymers give rise to their familiar **plastic** mechanical behavior. The word "plastic" refers to a material that, when deformed, retains its new shape; in contrast, when an "elastic" object is deformed, it returns to its original shape. Many polymers can be deformed (stretched, bent, twisted) when warm and retain their deformed shape when cooled. In this way, they are made into countless everyday objects—milk bottles, car parts, refrigerator interiors, and so forth.

Molecular Architecture of Polymers The polymer's architecture—its overall spatial layout and molecular structure—is crucial to its properties. In addition to the linear chains we've discussed so far, chemists create polymers with more complex architectures through the processes of branching and crosslinking.

Branches are smaller chains appended to a polymer backbone. As branching increases, the chains cannot pack together as well, so the degree of crystallinity decreases; as a result, the polymer is less rigid. A small amount of branching occurs as a side reaction in the preparation of high-density polyethylene (HDPE). Because it is still largely linear, though, it is rigid enough for use in milk containers. In contrast, much more branching is intentionally induced to prepare low-density polyethylene (LDPE). The chains cannot pack well, so crystallinity is very low. The flexible, transparent material used in food storage bags results.

Dendrimers are the ultimate branched polymers. They are prepared from monomers with three or more attachment points, so each monomer forms branches. In essence, then, dendrimers have no backbone and consist of branches only. As you can see in the Gallery on page 380, a dendrimer has a constantly increasing number of branches and an incredibly large number of end groups at its outer edge. Chemists have used dendrimers to bind one polymer to another in the production of films and fibers and to deliver drug molecules to the desired location in medical applications.

Crosslinks can be thought of as branches that link one chain to another. The extent of crosslinking can result in remarkable differences in properties. In many cases, a small degree of crosslinking yields a *thermoplastic* polymer, one that still flows at high temperatures. But, as the extent of crosslinking increases, a thermoplastic polymer is transformed into a *thermoset* polymer, one that can no longer flow because it has become a single network. Below their glass transition temperatures, some thermosets are extremely rigid and strong, making them ideal as matrix materials in high-strength composites (see photo).

Above their glass transition temperatures, many thermosets become **elastomers**, polymers that can be stretched and immediately spring back to their initial shapes when released, like the net under a trapeze artist or a common rubber band. When you stretch a rubber band, individual polymer chains flow for only a short distance before the connectivity of the network returns them to their original positions. Table 12.9 lists some elastomers.

Thermoset polymer in tennis racket.

Table 12.9 Some Common Elastomers

Name	T_g(°C)	Uses
Poly(dimethyl siloxane)	−123	Breast implants
Polybutadiene	−106	Rubber bands
Polyisoprene	−65	Surgical gloves
Polychloroprene (neoprene)	−43	Footwear; medical tubing

Differences in monomer sequences often influence polymer properties as well. A *homopolymer* consists of one type of monomer, such as A—A—A—A—A—. . . , whereas a **copolymer** consists of two or more types. The simplest copolymer is a *block copolymer*, often called an *AB block copolymer* because a chain of monomer A and a chain of monomer B are linked at one point:

. . . A—A—A—A—A—B—B—B—B—B . . .

If the intermolecular forces between the A and B portions of the chain are weaker than those between different regions within each portion, the A and B portions form their own random coils. This ability makes AB block copolymers ideal adhesives for joining two polymer surfaces covalently. An ABA block copolymer has A chains linked at each end of a B chain:

. . . —A—A—A—B—(B)$_n$—B—A—A—A— . . .

Some of these block copolymers act as *thermoplastic elastomers*, materials shaped at high temperature that become elastomers at room temperature; not surprisingly, some of these materials have revolutionized the footware industry.

Polymer chemists are continually tailoring polymers to control properties such as viscosity, strength, toughness, and flexibility. The importance of intermolecular forces on their physical properties is only part of the story of these fascinating and remarkably useful materials. Silicone polymers are described in a Gallery in Chapter 14, and the organic reactions that form polymer chains from their monomers are examined in Chapter 15.

Nanotechnology: Designing Materials Atom by Atom

At the frontier of interdisciplinary science and expanding at an incredible pace, the exciting new field of *nanotechnology* is joining researchers from physics, materials science, chemistry, biology, environmental science, medicine, and many branches of engineering. International conferences, science journals, and an ever-growing list of university and industrial web sites herald the enormous potential impact of nanotechnology on society.

Nanotechnology is the science and engineering of nanoscale systems—those in the size range from 1 to 50 nm. Until recently, physical scientists had focused primarily on the structures and properties of atoms, which are smaller (around 1×10^{-1} nm), or of crystals, which are larger (around 1×10^5 nm and up). Between these size extremes, nanotechnologists examine the chemical and physical properties of nanostructures, manipulating atoms one at a time to synthesize particles, clusters, and layers with properties very different from either individual molecules or their bulk phases. The scanning tunneling microscope (see Tools of the Laboratory, p. 452) and the similar scanning probe and atomic force microscopes are among the precise tools required to place individual atoms, molecules, and clusters into the positions required to build a structure or cause a desired reaction.

The key to this futuristic technology lies in two features of nanoscale construction that occur routinely in nature. The first is *self-assembly*, the ability of smaller, simpler parts to organize themselves into a larger, more complex whole.

On the molecular scale, it refers to atoms or small molecules aggregating through intermolecular forces, especially dipole-dipole, H-bonding, and dispersion forces, which can act as "glue." Oppositely charged regions on two such particles make contact to form a larger particle, which in turn forms a still larger one. The other feature is *controlled orientation*, the positioning of two molecules near each other long enough for the intermolecular forces to take effect. Some industrial catalysts and all biological catalysts (enzymes) act this way. (Catalysts, discussed in Chapter 16, are substances that speed reactions.) Common synthesis and assembly strategies for nanoscale construction include approaches that mimic biological self-assembly, sophisticated precipitation methods, and a variety of physical and chemical aerosol techniques for making nanoclusters and then manipulating them into more consolidated structures.

With change occurring so rapidly, it is possible to provide only a very general description of the current research directions in nanotechnology. A recent report sponsored by the National Science and Technology Council describes worldwide efforts in four key areas—dispersions and coatings, high-surface-area materials, functional devices, and consolidated materials.

Dispersions and Coatings A wide range of optical, thermal, and electrical applications of dispersions and coatings are becoming available through nanostructuring, including products related to printing, sunscreens, photography, and pharmaceuticals. Some examples are thermal and optical barriers, image enhancement, ink-jet materials, coated abrasive slurries, and information-recording layers. Highly ordered, iron/platinum nanoparticles of extremely uniform size and exceptional magnetic properties have been prepared and may be adapted as coatings for high-density information storage, allowing one million times as much data per unit of surface area as current materials. Another exciting application involves highly ordered, one-molecule-thick films—in effect, two-dimensional crystals—that can be coated onto a variety of surfaces. One approach is to layer light-sensitive molecules that change reversibly from one form to another. A finely focused laser would change the molecular form to create specific patterns, thus storing information at the density of one bit of data per molecule!

High-Surface-Area Materials These applications take advantage of the incredibly large surface areas of nanoscale building blocks. For example, a particle 5 nm in diameter has about half of its atoms on its surface. When such nanoparticles are assembled, the resulting surface area is enormous. Current projects include porous membranes for water purification and batteries, drug-delivery systems, and multilayer films that incorporate photosynthetic molecules for high-efficiency solar cells. There is great potential for molecule-specific sensors. Biosensors now allow detection of as little as 10^{-14} mol of DNA using color changes in gold nanoparticles. In a recent development, 50-nm gold particles sensed slight differences in two DNA portions, which may allow detection of subtle genetic mutations. An eventual goal is to develop biosensors that could circulate freely in the bloodstream, measure levels of specific disease-related molecules, and deliver drugs to individual cells, even individual genes.

Functional Devices The need for ever smaller machines is the driving force in this area, and the nanoscale computer is the dream. The ongoing race for faster, smaller, traditional computers will soon hit a fundamental "brick wall." Since the 1960s, computing speeds have doubled about every two years as the light-etched rules between transistors and diodes on the silicon chip were made closer and closer. Current UV photolithography makes rules about 180 nm apart. As this distance approaches 50 nm, however, doping inconsistencies, heat generation, and other inherent limitations arise, and the silicon-based chip reaches its lower size limit. Imagine the potential impact of ongoing work to develop supercomputers consisting of molecule-sized diodes and transistors bound to an organic surface—the ultimate reduction in size and, thus, increase in speed.

To realize this dream, one of the major research efforts focuses on developing the single-electron transistor (SET). SETs will be made into arrays using methods similar to those of biological self-assembly. Such nanoscale devices must have nanoscale connections, and a related area involves fabricating nanowires. In one method, a metal is electroplated from solution [such as cobalt from $Co(NO_3)_2$] to fill uniform nanopores created by controlled oxidation of aluminum surfaces. But the greatest research activity is focused on carbon nanotubes (see Gallery, p. 380). These can be made single- or multiwalled, with insulating, semiconducting, or metallic properties. Whether produced by high-temperature (800°–1000°C) reduction of a hydrocarbon or by electron-beam irradiation of fullerenes, nanotubes are then physically separated according to size and properties using atomic force or scanning tunneling microscopes. Adding specific chemical groups to the nanotubes expands their applications further (Figure 12.50).

Consolidated Materials It is known that mechanical, magnetic, and optical properties change dramatically when bulk materials are consolidated from nanoscale building blocks. Nanostructuring will greatly increase the hardness and strength of metals and the ductility and plasticity of ceramics. Nanoparticle fillers can yield nanocomposites with unique properties—soft magnets, tough cutting tools, ultrastrong ductile cements, magnetic refrigerants, and a wide range of nanoparticle-filled elastomers, thermoplastics, and thermosets.

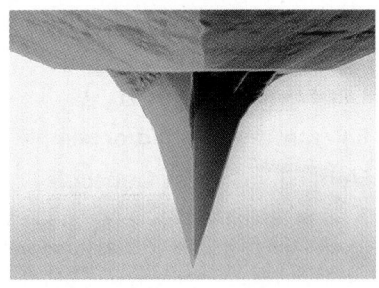

A

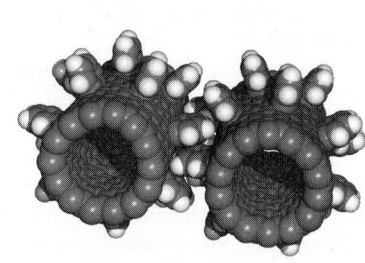

B

Figure 12.50 Manipulating atoms. A, The tip of an atomic force microscope, one of the key tools used to build nanodevices. **B,** This computer simulation of a nanogear shows carbon nanotubes with attached benzene rings acting as gear teeth.

SECTION SUMMARY

Doping increases the conductivity of semiconductors and is essential to modern electronic materials. Doping silicon with Group 5A atoms introduces negative sites (n-type) by adding valence electrons to the conduction band, whereas doping with Group 3A atoms adds positive holes (p-type) by emptying some orbitals in the valence band. Placing these two types of doped Si next to one another forms a p-n junction. Sandwiching one type between two pieces of the other forms a transistor.

Liquid crystal phases flow like liquids but have molecules ordered like crystalline solids. Typically, the molecules have rodlike shapes, and their intermolecular forces keep them aligned. Thermotropic phases are prepared by heating the solid; lyotropic phases form when the solvent concentration is varied. The nematic, cholesteric, and smectic types of liquid crystals differ in their molecular order. Liquid crystal applications depend on controlling the orientation of the molecules.

Ceramics are very resistant to heat and chemicals. Most are network covalent solids formed at high temperature from simple reactants. They add lightweight strength to other materials.

Polymers are extremely large molecules that adopt random coil shapes as a result of intermolecular forces. A polymer sample has an average molar mass because it consists of a range of chain lengths. The high viscosity of the sample arises from attractions between chains and, in the case of a dissolved polymer, between the chain and the solvent molecules. By varying the degrees of branching, cross-linking, and ordering (crystallinity), chemists tailor polymers with remarkable properties.

Nanoscale materials can be made one atom or molecule at a time through mechanical processes involving self-assembly and controlled orientation of molecules through intermolecular forces.

Chapter Perspective

Our central focus—macroscopic behavior resulting from molecular behavior—became still clearer in this chapter, as we built on earlier ideas of bonding, molecular shape, and polarity to understand the intermolecular forces that create the properties of liquids and solids. In Chapter 13, we'll find many parallels to the key ideas in this chapter: the same intermolecular forces create solutions, temperature influences solubility just as it does vapor pressure, and the equilibrium state also underlies the properties of solutions.

For Review and Reference (Numbers in parentheses refer to pages, unless noted otherwise.)

Learning Objectives

Relevant section and/or sample problem (SP) numbers appear in parentheses.

Understand These Concepts

1. How the interplay between kinetic and potential energy underlies the properties of the three states of matter and their phase changes (12.1)

2. The processes involved, both within a phase and during a phase change, when heat is added or removed from a pure substance (12.2)

3. The meaning of vapor pressure and how phase changes are dynamic equilibrium processes (12.2)

4. How temperature and intermolecular forces influence vapor pressure (12.2)

5. The relation between vapor pressure and boiling point (12.2)

6. How a phase diagram shows the phases of a substance at differing conditions of pressure and temperature (12.2)

7. The distinction between bonding and intermolecular forces on the basis of Coulomb's law and the meaning of the van der Waals radius of an atom (12.3)

8. The types and relative strengths of intermolecular forces acting in a substance (dipole-dipole, H-bonding, dispersion), the impact of H bonding on physical properties, and the meaning of polarizability (12.3)

9. The meanings of surface tension, capillarity, and viscosity and how intermolecular forces influence their magnitudes (12.4)

10. How the important macroscopic properties of water arise from atomic and molecular properties (12.5)

11. The meaning of crystal lattice and the characteristics of the three types of cubic unit cells (12.6)

12. How packing of spheres gives rise to the hexagonal and cubic unit cells (12.6)

13. Types of crystalline solids and how their intermolecular forces give rise to their properties (12.6)

14. How band theory accounts for the properties of metals and the relative conductivities of metals, nonmetals, and metalloids (12.6)

15. The structures, properties, and functions of modern materials (doped semiconductors, liquid crystals, ceramics, polymers, and nanostructures) on the atomic scale (12.7)

Master These Skills

1. Calculating the overall enthalpy change when heat is added to or removed from a pure substance (12.2)

2. Using the Clausius-Clapeyron equation to examine the relationship between vapor pressure and temperature (SP 12.1)

3. Using a phase diagram to predict the physical state and/or phase change of a substance (12.2)

4. Determining whether a substance can form H bonds and drawing the H-bonded structures (SP 12.2)

5. Predicting the types and relative strength of the intermolecular forces acting within a substance from its structure (SP 12.3)

6. Finding the number of particles in a unit cell (12.6)

7. Calculating atomic radius from the density and crystal structure of an element (SP 12.4)

Key Terms

phase (420)
intermolecular forces (420)
phase change (420)

Section 12.1
condensation (421)
vaporization (421)
freezing (421)
melting (fusion) (421)
heat of vaporization (ΔH^0_{vap}) (422)
heat of fusion (ΔH^0_{fus}) (422)
sublimation (422)
deposition (422)
heat of sublimation (ΔH^0_{subl}) (422)

Section 12.2
heating-cooling curve (423)
vapor pressure (426)
Clausius-Clapeyron equation (428)
boiling point (428)

melting point (429)
phase diagram (430)
critical point (430)
triple point (431)

Section 12.3
van der Waals radius (432)
ion-dipole force (432)
dipole-dipole force (432)
hydrogen bond (H bond) (434)
polarizability (436)
dispersion (London) force (436)

Section 12.4
surface tension (439)
capillarity (439)
viscosity (440)

Section 12.6
crystalline solid (445)
amorphous solid (445)
lattice (446)
unit cell (446)

coordination number (446)
simple cubic unit cell (446)
body-centered cubic unit cell (446)
face-centered cubic unit cell (446)
packing efficiency (448)
hexagonal closest packing (448)
cubic closest packing (448)
x-ray diffraction analysis (451)
scanning tunneling microscopy (451)
atomic solid (452)
molecular solid (452)
ionic solid (453)
metallic solid (455)
network covalent solid (455)
band theory (456)
valence band (458)
conduction band (458)
conductor (458)

semiconductor (459)
insulator (459)
superconductivity (459)

Section 12.7
crystal defect (460)
doping (460)
liquid crystal (462)
ceramic (465)
polymer (468)
macromolecule (468)
monomer (468)
degree of polymerization (n) (468)
random coil (469)
radius of gyration (R_g) (470)
plastic (472)
branch (472)
crosslink (472)
elastomer (472)
copolymer (473)
nanotechnology (473)

Key Equations and Relationships

12.1 Using the vapor pressure at one temperature to find the vapor pressure at another temperature (two-point form of the Clausius-Clapeyron equation) (428):

$$\ln \frac{P_2}{P_1} = \frac{-\Delta H_{\text{vap}}}{R}\left(\frac{1}{T_2} - \frac{1}{T_1}\right)$$

Highlighted Figures and Tables

These figures (F) and tables (T) provide a quick review of key ideas.

T12.1 Macroscopic comparison of gases, liquids, and solids (421)
F12.2 Phase changes and their enthalpy changes (423)
F12.3 A cooling curve for the conversion of gaseous water to ice (424)
F12.4 Liquid-gas equilibrium (426)
F12.5 The effect of temperature on the distribution of molecular speeds in a liquid (427)
F12.6 Vapor pressure as a function of temperature and intermolecular forces (427)
F12.9 Phase diagrams for CO_2 and H_2O (430)
F12.10 Covalent and van der Waals radii (432)
F12.11 Periodic trends in covalent and van der Waals radii (432)
T12.2 Bonding and nonbonding (intermolecular) forces (433)

F12.14 Hydrogen bonding and boiling point (435)
F12.17 Molar mass and boiling point (437)
F12.21 The H-bonding ability of the water molecule (442)
F12.24 The macroscopic properties of water and their atomic and molecular "roots" (444)
F12.26 The crystal lattice and the unit cell (446)
F12.27 The three cubic unit cells (447)
F12.28 Packing of spheres (449)
T12.5 The major types of crystalline solids (453)
F12.31 The sodium chloride structure (454)
F12.36 The band of molecular orbitals in lithium metal (457)
F12.37 Electrical conductivity in a conductor, semiconductor, and insulator (458)

Brief Solutions to Follow-up Problems

12.1 $\ln \dfrac{P_2}{P_1} = \left(\dfrac{-40.7\times10^3 \text{ J/mol}}{8.314 \text{ J/mol·K}}\right)$

$\times \left(\dfrac{1}{273.15 + 85.5 \text{ K}} - \dfrac{1}{273.15 + 34.1 \text{ K}}\right)$

$= (-4.90\times10^3 \text{ K})(-4.66\times10^{-4} \text{ K}^{-1}) = 2.28$

$\dfrac{P_2}{P_1} = 9.8$; thus, $P_2 = 40.1 \text{ torr} \times 9.8 = 3.9\times10^2 \text{ torr}$

12.2 (a) (b) (c) No H bonding

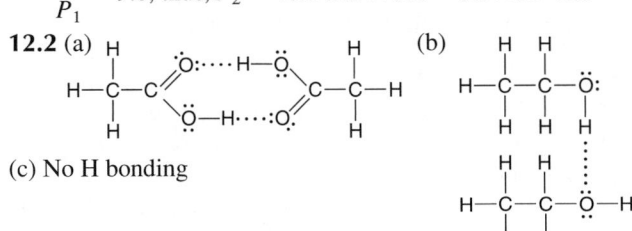

12.3 (a) Dipole-dipole, dispersion; CH_3Br
(b) H bonds, dipole-dipole, dispersion; $CH_3CH_2CH_2OH$
(c) Dispersion; C_3H_8

12.4 Avogadro's no. $= \dfrac{1 \text{ cm}^3}{7.874 \text{ g Fe}} \times \dfrac{55.85 \text{ g Fe}}{1 \text{ mol Fe}} \times 0.68$

$\times \dfrac{1 \text{ Fe atom}}{8.38\times10^{-24} \text{ cm}^3}$

$= 5.8\times10^{23}$ Fe atoms/mol Fe

Problems

Problems with **colored** numbers are answered at the back of the text. Sections match the text and provide the number(s) of relevant sample problems. Most offer Concept Review Questions, Skill-Building Exercises (in similar pairs), and Problems in Context. Then Comprehensive Problems, based on material from any section or previous chapter, follow.

An Overview of Physical States and Phase Changes

● **Concept Review Questions**

12.1 How does the energy of attraction between particles compare with their energy of motion in a gas and in a solid? As part of your answer, identify two macroscopic properties that differ between a gas and a solid.

12.2 What types of forces, intramolecular or intermolecular,
(a) prevent ice cubes from adopting the shape of their container?
(b) are overcome when ice melts?
(c) are overcome when liquid water is vaporized?
(d) are overcome when gaseous water is converted to hydrogen gas and oxygen gas?
12.3 (a) Why are gases more easily compressed than liquids?
(b) Why do liquids have a greater ability to flow than solids?
12.4 (a) Why is the heat of fusion (ΔH_{fus}) of a substance smaller than its heat of vaporization (ΔH_{vap})?
(b) Why is the heat of sublimation (ΔH_{subl}) of a substance greater than its ΔH_{vap}?
(c) At a given temperature and pressure, how does the magnitude of the heat of vaporization of a substance compare with that of its heat of condensation?

● **Skill-Building Exercises** *(paired)*

12.5 Which forces are intramolecular and which intermolecular?
(a) Those preventing oil from evaporating at room temperature
(b) Those preventing butter from melting in a refrigerator
(c) Those allowing silver to tarnish
(d) Those preventing O_2 in air from forming O atoms
12.6 Which forces are intramolecular and which intermolecular?
(a) Those allowing fog to form on a cool, humid evening
(b) Those allowing water to form when H_2 is sparked
(c) Those allowing liquid benzene to crystallize when cooled
(d) Those responsible for the low boiling point of hexane

12.7 Name the phase change in each of these events:
(a) Dew appears on a lawn in the morning.
(b) Icicles change into liquid water.
(c) Wet clothes dry on a summer day.
12.8 Name the phase change in each of these events:
(a) A diamond film forms on a surface from gaseous carbon atoms in a vacuum.
(b) Mothballs in a bureau drawer disappear over time.
(c) Molten iron from a blast furnace is cast into ingots ("pigs").

● **Problems in Context**

12.9 Liquid propane, a widely used fuel, is produced by compressing gaseous propane at 20°C. During the process, approximately 15 kJ of energy is released for each mole of gas liquefied. Where does this energy come from?
12.10 Many heat-sensitive and oxygen-sensitive solids, such as camphor, are purified by warming under vacuum. The solid vaporizes directly, and the vapor crystallizes on a cool surface. What phase changes are involved in this method?

Quantitative Aspects of Phase Changes
(Sample Problem 12.1)

● **Concept Review Questions**

12.11 Describe the changes (if any) in potential energy and in kinetic energy among the molecules when gaseous PCl_3 condenses to a liquid at a fixed temperature.
12.12 When benzene is at its melting point, two processes occur simultaneously and balance each other. Describe these processes on the macroscopic and molecular levels.
12.13 Liquid hexane (bp = 69°C) is placed in a closed container at room temperature. At first, the pressure of the vapor phase increases, but after a short time, it stops changing. Why?
12.14 Explain the effect of strong intermolecular forces on each of these parameters: (a) critical temperature; (b) boiling point; (c) vapor pressure; (d) heat of vaporization.
12.15 At 1.1 atm, will water boil at 100.°C? Explain.
12.16 A liquid is in equilibrium with its vapor in a closed vessel at a fixed temperature. The vessel is connected by a stopcock to an evacuated vessel. When the stopcock is opened, will the final pressure of the vapor be different from the original value if (a) some liquid remains; (b) all the liquid is first removed? Explain.
12.17 The phase diagram for substance A has a solid-liquid curve with a positive slope, and the one for substance B has a solid-liquid line with a negative slope. What macroscopic property can distinguish A from B?
12.18 Why does water vapor at 100°C cause more severe burns than liquid water at 100°C?

● **Skill-Building Exercises** *(paired)*

12.19 From the data below, calculate the total heat (in J) needed to convert 12.00 g of ice at −5.00°C to liquid water at 0.500°C:

| mp at 1 atm | 0.0°C | ΔH^0_{fus} | 6.02 kJ/mol |
| c_{liquid} | 4.21 J/g·°C | c_{solid} | 2.09 J/g·°C |

12.20 From the data below, calculate the total heat (in J) needed to convert 0.333 mol of ethanol gas at 300°C and 1 atm to liquid ethanol at 25.0°C and 1 atm:

| bp at 1 atm | 78.5°C | ΔH^0_{vap} | 40.5 kJ/mol |
| c_{gas} | 1.43 J/g·°C | c_{liquid} | 2.45 J/g·°C |

12.21 A liquid has a ΔH^0_{vap} of 35.5 kJ/mol and a boiling point of 122°C at 1.00 atm. What is its vapor pressure at 109°C?
12.22 Diethyl ether has a ΔH^0_{vap} of 29.1 kJ/mol and a vapor pressure of 0.703 atm at 25.0°C. What is its vapor pressure at 95.0°C?
12.23 What is the ΔH^0_{vap} of a liquid that has a vapor pressure of 641 torr at 85.2°C and a boiling point of 95.6°C at 1 atm?
12.24 Methane (CH_4) has a boiling point of −164°C at 1 atm and a vapor pressure of 42.8 atm at −100°C. What is the heat of vaporization of CH_4?

12.25 Use these data to draw a qualitative phase diagram for ethylene (C_2H_4). Is $C_2H_4(s)$ more or less dense than $C_2H_4(l)$?

bp at 1 atm	−103.7°C
mp at 1 atm	−169.16°C
Critical point	9.9°C and 50.5 atm
Triple point	−169.17°C and 1.20×10^{-3} atm

12.26 Use these data to draw a qualitative phase diagram for H_2. Does H_2 sublime at 0.05 atm? Explain.

mp at 1 atm	13.96 K
bp at 1 atm	20.39 K
Triple point	13.95 K and 0.07 atm
Critical point	33.2 K and 13.0 atm
Vapor pressure of solid at 10 K	0.001 atm

● **Problems in Context**

12.27 In the process of freeze-drying, ice sublimes and the water vapor is removed by vacuum. In which region of Figure 12.9B does sublimation take place? What is the highest temperature at which water can sublime? Why is freeze-drying difficult (or uneconomical) at very low temperatures?
12.28 Butane is a common fuel used in cigarette lighters and camping stoves. Normally supplied in metal containers under pressure, the fuel exists as a mixture of liquid and gas, so high temperatures may cause the container to explode. At 25.0°C, the vapor pressure of butane is 2.3 atm. What is the pressure in the container at 150.°C (ΔH_{vap} = 24.3 kJ/mol)?
12.29 Use Figure 12.9A to answer the following questions:
(a) Carbon dioxide is sold in steel cylinders under pressures of approximately 20 atm. Is there liquid CO_2 in the cylinder at room temperature (~20°C)? At 40°C? At −40°C? At −120°C?
(b) Carbon dioxide is also sold as solid chunks, called *dry ice*, in insulated containers. If the chunks are warmed by leaving them in an open container at room temperature, will they melt?
(c) If a container is nearly filled with dry ice and then sealed and warmed to room temperature, will the dry ice melt?
(d) If dry ice is compressed at a temperature below its triple point, will it melt?
(e) Will liquid CO_2 placed in a beaker at room temperature boil?

Types of Intermolecular Forces
(Sample Problems 12.2 and 12.3)

● Concept Review Questions

12.30 Why are covalent bonds typically much stronger than intermolecular forces?

12.31 Even though molecules are neutral, the dipole-dipole force is one of the most important interparticle forces that exists among them. Explain.

12.32 Oxygen and selenium are members of Group 6A(16). Water forms H bonds, but H_2Se does not. Explain.

12.33 In solid I_2, is the distance between the two I nuclei of one I_2 molecule longer or shorter than the distance between two I nuclei of adjacent I_2 molecules? Explain.

12.34 Polar molecules exhibit dipole-dipole forces. Do they also exhibit dispersion forces? Explain.

12.35 Distinguish between *polarizability* and *polarity*. How does each influence intermolecular forces?

12.36 How can one nonpolar molecule induce a dipole in a nearby nonpolar molecule?

● Skill-Building Exercises (paired)

12.37 What is the strongest interparticle force in a sample of (a) CH_3OH; (b) CCl_4; (c) Cl_2?

12.38 What is the strongest interparticle force in a sample of (a) H_3PO_4; (b) SO_2; (c) $MgCl_2$?

12.39 What is the strongest interparticle force in a sample of (a) CH_3Br; (b) CH_3CH_3; (c) NH_3?

12.40 What is the strongest interparticle force in a sample of (a) Kr; (b) BrF; (c) H_2SO_4?

12.41 Which member of each pair of compounds forms intermolecular H bonds? Draw the H-bonded structures in each case:
(a) CH₃CHCH₃ or CH_3SCH_3 (b) HF or HBr
     |
    OH

12.42 Which member of each pair of compounds forms intermolecular H bonds? Draw the H-bonded structures in each case:
(a) $(CH_3)_2NH$ or $(CH_3)_3N$ (b) $HOCH_2CH_2OH$ or FCH_2CH_2F

12.43 Which forces oppose vaporization of each substance?
(a) hexane   (b) water   (c) $SiCl_4$

12.44 Which forces oppose vaporization of each substance?
(a) Br_2 (b) SbH_3 (c) CH_3NH_2

12.45 Which has the greater polarizability? Explain.
(a) Br^- or I^- (b) $CH_2{=}CH_2$ or $CH_3{-}CH_3$ (c) H_2O or H_2Se

12.46 Which has the greater polarizability? Explain.
(a) Ca^{2+} or Ca   (b) CH_3CH_3 or $CH_3CH_2CH_3$ (c) CCl_4 or CF_4

12.47 Which member in each pair of liquids has the *higher* vapor pressure at a given temperature? Explain.
(a) C_2H_6 or C_4H_{10} (b) CH_3CH_2OH or CH_3CH_2F
(c) NH_3 or PH_3

12.48 Which member in each pair of liquids has the *lower* vapor pressure at a given temperature? Explain.
(a) $HOCH_2CH_2OH$ or $CH_3CH_2CH_2OH$
(b) CH_3COOH or $(CH_3)_2C{=}O$ (c) HF or HCl

12.49 Which substance has the *higher* boiling point? Explain.
(a) LiCl or HCl   (b) NH_3 or PH_3 (c) Xe or I_2

12.50 Which substance has the *higher* boiling point? Explain.
(a) CH_3CH_2OH or $CH_3CH_2CH_3$ (b) NO or N_2 (c) H_2S or H_2Te

12.51 Which substance has the *lower* boiling point? Explain.
(a) $CH_3CH_2CH_2CH_3$ or CH₂—CH₂   (b) NaBr or PBr_3
            | |
           CH₂—CH₂

(c) H_2O or HBr

12.52 Which substance has the *lower* boiling point? Explain.
(a) CH_3OH or CH_3CH_3 (b) FNO or ClNO
(c)

● Problems in Context

12.53 For pairs of molecules in the gas phase, average H-bond dissociation energies are 17 kJ/mol for NH_3, 22 kJ/mol for H_2O, and 29 kJ/mol for HF. Explain this increase in H-bond strength.

12.54 Dispersion forces are the only intermolecular forces present in motor oil, yet it has a high boiling point. Explain.

12.55 Why does the antifreeze ingredient ethylene glycol ($HOCH_2CH_2OH$; $\mathcal{M}$ = 62.07 g/mol) have a boiling point of 197.6°C, whereas propanol ($CH_3CH_2CH_2OH$; $\mathcal{M}$ = 60.09 g/mol), a compound with a similar molar mass, has a boiling point of only 97.4°C?

Properties of the Liquid State

● Concept Review Questions

12.56 Before the phenomenon of surface tension was understood, physicists described the surface of water as being covered with a "skin." What causes this skinlike phenomenon?

12.57 Small, equal-sized drops of oil, water, and mercury lie on a waxed floor. How does each liquid behave? Explain.

12.58 Why does an aqueous solution of ethanol (CH_3CH_2OH) have a lower surface tension than water?

12.59 Why are units of energy per area (J/m^2) used for surface tension values?

12.60 Does the *strength* of the intermolecular forces in a liquid change as the liquid is heated? Explain. Why does liquid viscosity decrease with rising temperature?

● Skill-Building Exercises (paired)

12.61 Rank the following in order of *increasing* surface tension at a given temperature, and explain your ranking:
(a) $CH_3CH_2CH_2OH$ (b) $HOCH_2CH(OH)CH_2OH$
(c) $HOCH_2CH_2OH$

12.62 Rank the following in order of *decreasing* surface tension at a given temperature, and explain your ranking:
(a) CH_3OH (b) CH_3CH_3 (c) $H_2C{=}O$

12.63 Rank the compounds in Problem 12.61 in order of *increasing* viscosity at a given temperature; explain your ranking.

12.64 Rank the compounds in Problem 12.62 in order of *decreasing* viscosity at a given temperature; explain your ranking.

● Problems in Context

12.65 Are the same cohesive and adhesive forces involved when a paper towel absorbs apple juice as when it absorbs cooking oil? Explain.

12.66 Viscosity-enhancing additives increase the viscosity of a motor oil as the temperature increases. How do they work?

12.67 Pentanol ($C_5H_{11}OH$; $\mathcal{M}$ = 88.15 g/mol) has nearly the same molar mass as hexane (C_6H_{14}; $\mathcal{M}$ = 86.17 g/mol) but is more than 12 times as viscous at 20°C. Explain.

The Uniqueness of Water

● **Concept Review Questions**

12.68 For what types of substances is water a good solvent? For what types is it a poor solvent? Explain.

12.69 A water molecule can engage in as many as four H bonds. Explain.

12.70 Warm-blooded animals have a narrow range of body temperature because their bodies have a high water content. Explain.

12.71 What property of water keeps plant debris on the surface of lakes and ponds? What is the ecological significance of this?

12.72 A drooping plant can be made upright by watering the ground around it. Explain.

12.73 Describe the molecular basis of the property of water responsible for the presence of ice on the surface of a frozen lake.

12.74 Describe in molecular terms what occurs when ice melts.

The Solid State: Structure, Properties, and Bonding
(Sample Problem 12.4)

● **Concept Review Questions**

12.75 What is the difference between an amorphous solid and a crystalline solid on the macroscopic and molecular levels? Give an example of each.

12.76 How are a solid's unit cell and crystal structure related?

12.77 For structures consisting of identical atoms, how many atoms are contained in the simple, body-centered, and face-centered cubic unit cells? Explain how you obtained the values.

12.78 An element has a crystal structure in which the width of the cubic unit cell equals the diameter of an atom. What type of unit cell does it have?

12.79 What specific difference in the positioning of spheres gives a crystal structure based on the face-centered cubic unit cell less empty space than one based on the body-centered cubic unit cell?

12.80 Both solid Kr and solid Cu consist of individual atoms. Why do their physical properties differ so much?

12.81 What is the energy gap in band theory? Compare its size in superconductors, conductors, semiconductors, and insulators.

12.82 Predict the effect (if any) of an increase in temperature on the electrical conductivity of (a) a conductor; (b) a semiconductor; (c) an insulator.

12.83 Besides the type of unit cell, what atomic properties are needed to find the density of a solid consisting of identical atoms?

● **Skill-Building Exercises** *(paired)*

12.84 What type of crystal lattice does each metal form? (The number of atoms per unit cell is given in parentheses.)
(a) Ni (4)   (b) Cr (2)   (c) Ca (4)

12.85 What is the number of atoms per unit cell for each metal?
(a) Polonium, Po   (b) Iron, Fe   (c) Silver, Ag

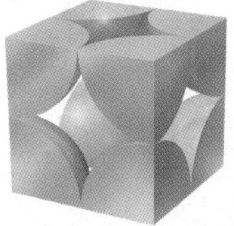  

12.86 Of the five major types of crystalline solid, which does each of the following form: (a) Sn; (b) Si; (c) Xe?

12.87 Of the five major types of crystalline solid, which does each of the following form: (a) cholesterol ($C_{27}H_{45}OH$); (b) KCl; (c) BN?

12.88 Of the five major types of crystalline solid, which does each of the following form, and why: (a) Ni; (b) F_2; (c) CH_3OH?

12.89 Of the five major types of crystalline solid, which does each of the following form, and why: (a) SiC; (b) Na_2SO_4; (c) SF_6?

12.90 Zinc oxide adopts the zinc blende crystal structure (Figure P12.90). How many Zn^{2+} ions are in the ZnO unit cell?

12.91 Calcium sulfide adopts the sodium chloride crystal structure (Figure P12.91). How many S^{2-} ions are in the CaS unit cell?

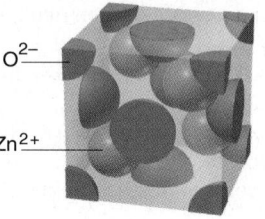 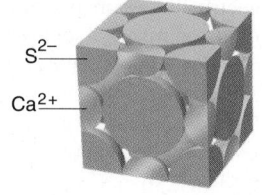

Figure P12.90   Figure P12.91

12.92 Zinc selenide (ZnSe) crystallizes in the zinc blende structure and has a density of 5.42 g/cm^3.
(a) How many Zn and Se particles are in each unit cell?
(b) What is the mass of a unit cell?
(c) What is the volume of a unit cell?
(d) What is the edge length of a unit cell?

12.93 An element crystallizes in a face-centered cubic lattice and has a density of 1.45 g/cm^3. The edge of its unit cell is 4.52×10^{-8} cm.
(a) How many atoms are in each unit cell?
(b) What is the volume of a unit cell?
(c) What is the mass of a unit cell?
(d) Calculate an approximate atomic mass for the element.

12.94 Classify each of the following as a conductor, insulator, or semiconductor: (a) phosphorus; (b) mercury; (c) germanium.

12.95 Classify each of the following as a conductor, insulator, or semiconductor: (a) carbon (graphite); (b) sulfur; (c) platinum.

12.96 Predict the effect (if any) of an increase in temperature on the electrical conductivity of (a) antimony, Sb; (b) tellurium, Te; (c) bismuth, Bi.

12.97 Predict the effect (if any) of a decrease in temperature on the electrical conductivity of (a) silicon, Si; (b) lead, Pb; (c) germanium, Ge.

● **Problems in Context**

12.98 Polonium, the largest member of Group 6A(16), is a rare radioactive metal that is the only element with a crystal structure based on the simple cubic unit cell. If its density is 9.142 g/cm^3, calculate an approximate atomic radius for polonium.

12.99 The coinage metals—copper, silver, and gold—crystallize in a cubic closest packed structure. Use the density of copper (8.95 g/cm^3) and its molar mass (63.55 g/mol) to calculate an approximate atomic radius for copper.

12.100 Many common rocks contain large regions of silicate networks held together by ionic interactions. What properties of rocks can be attributed to this general composition?

12.101 Semiconductors are used in many devices that measure temperature. Use band theory to explain the nature of the property that makes semiconductors suitable for this purpose.

Advanced Materials

● **Concept Review Questions**

12.102 When tin is added to copper, the resulting alloy (bronze) is much harder than copper. Explain.

12.103 In the process of doping a semiconductor, certain impurities are added to increase its electrical conductivity. Explain this process for an n-type and a p-type semiconductor.

12.104 State two molecular characteristics of substances that typically form liquid crystals. How is each of them related to function?

12.105 Distinguish between isotropic and anisotropic properties. Which describes the behavior of a liquid crystal?

12.106 How are the properties of modern ceramics the same as those of traditional clays, and how are they different? Refer to specific substances in your answer.

12.107 Why is the average molar mass of a polymer sample different from the molar mass of an individual chain?

12.108 How does the random coil shape relate to the radius of gyration of a polymer chain?

12.109 What factor(s) influence the viscosity of a polymer solution? What factor(s) influence the viscosity of a molten polymer? What is a polymer glass?

12.110 Use an example to show how branching and crosslinking can affect the physical behavior of a polymer.

● **Skill-Building Exercises** *(paired)*

12.111 Silicon and germanium are both semiconducting elements from Group 4A(14) that can be doped to improve their conductivity. Would the following form an n-type or a p-type semiconductor: (a) Ge doped with P; (b) Si doped with In?

12.112 Would the following form an n-type or a p-type semiconductor: (a) Ge doped with As; (b) Si doped with B?

12.113 The repeat unit in a polystyrene coffee cup has the formula $C_6H_5CHCH_2$. If the molar mass of the polymer is 3.5×10^5 g/mol, what is the degree of polymerization?

12.114 The monomer of poly(vinyl chloride) has the formula C_2H_3Cl. If there are 1565 repeat units in a single chain of the polymer, what is the molecular mass (in amu) of that chain?

12.115 The polypropylene (repeat unit CH_3CHCH_2) in a plastic toy has a molar mass of 2.5×10^5 g/mol and a repeat unit length of 0.252 pm. Calculate the radius of gyration.

12.116 The polymer that is used to make 2-L soda bottles [poly(ethylene terephthalate)] has a repeat unit with molecular formula $C_{10}H_8O_4$ and a length of 1.075 nm. Calculate the radius of gyration of a chain with molar mass of 2.30×10^4 g/mol.

Comprehensive Problems

Problems with an asterisk (*) are more challenging.

12.117 A 0.75-L bottle is cleaned, dried, and closed in a room where the air is 22°C and 44% relative humidity (that is, the water vapor in the air is 0.44 of the equilibrium vapor pressure at 22°C). The bottle is brought outside and stored at 0.0°C.
(a) What mass of liquid water condenses inside the bottle?
(b) Would liquid water condense at 10°C? (See Table 5.3.)

***12.118** In an experiment, 4.00 L of N_2 is saturated with water vapor at 22°C and then compressed to half its volume.
(a) What is the partial pressure of H_2O in the compressed gas mixture?
(b) What mass of water vapor condenses to liquid?

12.119 Which forces are overcome when the following events occur: (a) NaCl dissolves in water; (b) krypton boils; (c) water boils; (d) CO_2 sublimes?

12.120 Changes in pressure cause phase changes analogous to the changes with temperature shown in Figure 12.3.
(a) Refer to Figure 12.9B and draw a curve of pressure vs. time for water similar to Figure 12.3; label the phase changes as the pressure is continuously *increased* at 2°C.
(b) Refer to Figure 12.9A and draw a curve for carbon dioxide as the pressure is continuously *increased* at −50°C.

12.121 Because bismuth has several well-characterized solid, crystalline phases, it is used to calibrate instruments employed in high-pressure studies. The following phase diagram for bismuth shows the liquid phase and five different solid phases stable above 1 katm (1000 atm) and up to 300°C. (a) Which solid phases are stable at 25°C? (b) Which phase is stable at 50 katm and 175°C? (c) Identify the phase transitions that bismuth undergoes at 200°C as the pressure is reduced from 100 to 1 katm. (d) What phases are present at each of the triple points?

12.122 Does it take more or less heat to make a sample of ice sublime than to make the same amount of liquid water vaporize? Would the same be true for dry ice?

12.123 In making computer chips, a 5.00-kg cylindrical ingot of ultrapure n-type doped silicon that is 5.25 inches in diameter is sliced into wafers 1.02×10^{-4} m thick.
(a) Assuming no waste, how many wafers can be made?
(b) What is the mass of a wafer (d of Si = 2.34 g/cm³; V of a cylinder = $\pi r^2 h$)?
(c) A key step in making p-n junctions for the chip is chemical removal of the oxide layer on the wafer through treatment with gaseous HF. Write a balanced equation for this reaction.
(d) If 0.750% of the Si atoms are removed during the treatment in part (c), how many moles of HF are required per wafer, assuming 100% reaction yield?

12.124 Why do the densities of most liquids increase as they are cooled and solidified? How does water differ in this regard?

12.125 Methyl salicylate, $C_8H_8O_3$, the odorous constituent of oil of wintergreen and of the flavor of birch beer, has a vapor pressure of 1.00 torr at 54.3°C and 10.0 torr at 95.3°C.
(a) What is its vapor pressure at 25°C?
(b) How many liters of air must pass over the compound at 25 °C to vaporize 1.0 mg?

12.126 Mercury vapor is very toxic and is readily absorbed through the lungs. At 20.°C, mercury (ΔH_{vap} = 59.1 kJ/mol) has a vapor pressure of 1.20×10^{-3} torr, which is high enough to be hazardous. To reduce the danger to workers who may be exposed in processing plants, mercury is cooled to lower its vapor pressure. At what temperature would the vapor pressure of mercury be at the relatively safe level of 5.0×10^{-5} torr?

*12.127 Polytetrafluoroethylene (Teflon) has a repeat unit with the formula F_2C—CF_2. A sample of the polymer consists of fractions with the following distribution of chains:

Fraction	Average no. of repeat units	Amount (mol) of polymer
1	273	0.10
2	330	0.40
3	368	1.00
4	483	0.70
5	525	0.30
6	575	0.10

(a) Determine the molar mass of each fraction.
(b) Determine the number-average molar mass of the sample.
(c) Another type of average molar mass of a polymer sample is called the *weight-average molar mass, M_w*:

$$M_w = \frac{\Sigma (M \text{ of fraction} \times \text{mass of fraction})}{\text{total mass of all fractions}}$$

Calculate the weight-average molar mass of the above sample.

12.128 A greenhouse contains 216 m^3 of air at a temperature of 26°C, and a humidifier in it vaporizes 4.00 L of water.
(a) What is the pressure of water vapor in the greenhouse, assuming that none escapes and that the air was originally completely dry (d of H_2O = 1.00 g/mL)?
(b) What total volume of liquid water would have to be vaporized to saturate the air (i.e., achieve 100% relative humidity)? (See Table 5.4, p. 207)

*12.129 The packing efficiency of spheres is the percentage of the total space (volume) of the unit cell occupied by the spheres:

$$\text{Packing efficiency (\%)} = \frac{\text{volume of spheres in unit cell}}{\text{volume of unit cell}} \times 100$$

Using spheres of radius r and unit-cell edge length A, calculate the packing efficiency of (a) a simple cubic unit cell; (b) a face-centered cubic unit cell; (c) a body-centered cubic unit cell (volume of a sphere = $\frac{4}{3}\pi r^3$). (d) A solid consisting of identical atoms has a density of 9.0 g/cm^3 and crystallizes in a cubic closest packed structure. What is its density if the atoms are rearranged to a structure with a body-centered cubic unit cell? A simple cubic unit cell?

12.130 Consider the phase diagram for substance X:
(a) What phase(s) is(are) present at point A? E? F? H? B? C?
(b) Which point corresponds to the critical point? Which point corresponds to the triple point?
(c) What curve corresponds to conditions at which the solid and gas are in equilibrium?

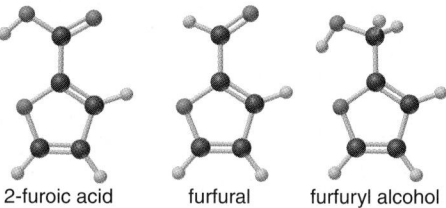

(d) Describe what happens when you start at point A and increase the temperature at constant pressure.
(e) Describe what happens when you start at point H and decrease the pressure at constant temperature.
(f) Is liquid X more or less dense than solid X?

12.131 When you step on the brake pedal of a car, the pressure of your foot is transferred through the brake lines to the brake pad. Are the brake lines filled with gas, liquid, or solid? How is this phase best suited for the function?

12.132 Why do uninsulated water pipes burst in cold weather?
*12.133 The only alkali halides that do not adopt the NaCl structure are CsCl, CsBr, and CsI, formed from the largest alkali cation and the three largest halide ions. These crystallize in the *cesium chloride structure* (shown here for CsCl). This structure has been used as an example of how dispersion forces can dominate in the presence of ionic forces. Use the ideas of coordination number and polarizability to explain why the CsCl structure exists.

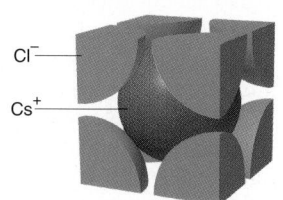

12.134 H bonding often causes individual molecules to join with others and remain together under varying conditions. Each of the following has been identified in either the gas or liquid phase. Draw structures for each showing the H bonds: (a) $(HF)_2$, dimer; (b) $(CH_3OH)_4$, cyclic; (c) $(HF)_6$, cyclic.
*12.135 Corn is a valuable source of industrial chemicals. For example, furfural is prepared from corncobs. It is an important reactant in plastics manufacture and a key solvent for the production of cellulose acetate, which is used to make everything from videotape to waterproof fabric. It can be reduced to furfuryl alcohol or oxidized to 2-furoic acid.

2-furoic acid   furfural   furfuryl alcohol

(a) Which of these compounds can form H bonds? Draw structures in each case.
(b) The molecules of some substances can form an "internal" H bond, that is, an H bond within a single molecule. This takes the form of a polygon with atoms as corners and bonds as sides and an H bond as one of the sides. Which of these molecules is (are) likely to form a stable internal H bond? Draw the structure. (*Hint:* Structures with 5 or 6 atoms as corners are most stable.)
12.136 The density of solid gallium at its melting point is 5.9 g/cm^3, whereas that of liquid gallium is 6.1 g/cm^3. Is the temperature at the triple point higher or lower than the normal melting point? Is the slope of the solid-liquid curve for gallium positive or negative?
12.137 When supercooling occurs, the temperature of the liquid phase drops smoothly to a point slightly below the normal freezing point. When the first crystals of solid appear, the temperature rises sharply to the freezing point, and from then on the cooling curve appears normal. Sketch a general cooling curve, with labeled axes and regions, for a substance that changes from gas to liquid to solid and exhibits supercooling.
*12.138 A 4.7-L sealed bottle containing 0.33 g of ethanol, C_2H_6O, is placed in a refrigerator and reaches equilibrium with its vapor at -11°C. (a) What mass of ethanol is present in the vapor? (b) When the container is removed and warmed to room temperature, 20.°C, will all the ethanol vaporize? (c) How much liquid ethanol would be present at 0.0°C? The vapor pressure of ethanol is 10. torr at -2.3°C and 40. torr at 19°C.
12.139 A cubic unit cell contains atoms of element A at each corner and atoms of element Z on each face. What is the empirical formula of the compound?

12.140 The boiling point of amphetamine, $C_9H_{13}N$, is 201°C at 760 torr and 83°C at 13 torr. What is the concentration (in g/m^3) of amphetamine when it is in contact with 20.°C air?

*12.141 (a) The sodium chloride, or "rock salt," crystal structure (see Figure 12.31) forms at room temperature and 1 atm pressure. Find the volume of one unit cell. (Radius of Na^+ is 102 pm, and that of Cl^- is 181 pm.) Use this value to calculate the density of NaCl at room temperature and 1 atm pressure. (b) As the pressure increases at fixed temperature, the ions become smaller, the volume of the unit cell decreases, and the density of the sample increases. At 300 katm, the volume has been reduced by 35%. What is the density of NaCl at this pressure? Assuming that both ionic radii are reduced by the same amount, what are the effective radii of the ions at this pressure? (c) At high temperature and pressure, NaCl undergoes a phase transition to the cesium chloride structure (see Problem 12.133). Unlike Cs^+, the Na^+ ion is small enough to fit in the center of a simple cubic unit cell of Cl^- ions. Use the Cl^- radius at 300 katm to find the volume of one unit cell in this crystal structure. What is the density of NaCl with this crystal structure?

12.142 Is it possible for a salt of formula AB_3 to have a face-centered cubic unit cell of anions with cations in all the eight available holes? Explain.

*12.143 Substance A has the following properties:

mp at 1 atm	−20.°C	bp at 1 atm	85°C
ΔH_{fus}	180. J/g	ΔH_{vap}	500. J/g
c_{solid}	1.0 J/g·°C	c_{liquid}	2.5 J/g·°C
c_{gas}	0.5 J/g·°C		

At 1 atm, a 25-g sample of A is heated from −40.°C to 100.°C at a constant rate of 450. J/min. (a) How many minutes does it take to heat the sample to its melting point? (b) How many minutes does it take to melt the sample? (c) Perform any other necessary calculations, and draw a curve of temperature vs. time for the entire heating process.

*12.144 An aerospace manufacturer is building a prototype experimental aircraft that cannot be detected by radar. Boron nitride is chosen for incorporation into the body parts, and the boric acid/ammonia method is used to prepare the ceramic material. Given 85.5% and 86.8% yields for the two reaction steps, how much boron nitride can be prepared from 1.00 metric ton of boric acid and 12.5 m^3 of ammonia at 275 K and 3.07×10^3 kPa? Assume that ammonia behaves ideally under these conditions and is recycled completely in the reaction process.

*12.145 A sample of polystyrene (repeat unit $C_6H_5CHCH_2$) has a molar mass of 104,160 g/mol. Here is a section of the backbone, with average C—C bond length and angle:
(a) Calculate the extended length of the polymer chain.
(b) Calculate the radius of gyration of the polymer chain.

*12.146 In a body-centered cubic unit cell, the central atom lies on an internal diagonal of the cell and touches the corner atoms. (a) Find the length of the diagonal in terms of r, the atomic radius. (b) If the edge length of the cube is a, what is the length of a *face* diagonal? (c) Derive an expression for a in terms of r. (d) How many atoms are in this unit cell? (e) What fraction of the unit cell volume is filled with spheres?

12.147 Ice is held in a cylinder by means of a piston at −10°C. What happens to the contents of the cylinder as the piston is slowly withdrawn at a constant temperature?

*12.148 Sodium has a crystal structure based on the body-centered cubic unit cell. What is the mass of one unit cell of sodium?

*12.149 KF has the same type of crystal structure as NaCl. The unit cell of KF has an edge length of 5.39 Å. Calculate the density of KF.

*12.150 The *intrinsic viscosity* of a polymer chain in a given solvent, $[\eta]_{solvent}$, is a measure of the intermolecular interactions between solvent and polymer. Generally, a larger $[\eta]_{solvent}$ indicates a stronger interaction. Intrinsic viscosities of polymer chains in solution are given by the Mark-Houwink equation, $[\eta]_{solvent} = K\mathcal{M}^a$, where $\mathcal{M}$ is the molar mass of the polymer, and K and a are constants specific for polymer and solvent. Use the following information for 25°C in the remainder of the problem:

Polymer	Solvent	K (mL/g)	a
Polystyrene	Benzene	9.5×10^{-3}	0.74
	Cyclohexane	8.1×10^{-2}	0.50
Polyisobutylene	Benzene	8.3×10^{-2}	0.50
	Cyclohexane	2.6×10^{-1}	0.70

(a) A polystyrene sample has a molar mass of 104,160 g/mol. Calculate the intrinsic viscosity in benzene and in cyclohexane. Which solvent has stronger interactions with the polymer?
(b) A different polystyrene sample has a molar mass of 52,000 g/mol. Calculate its $[\eta]_{benzene}$. Given a polymer standard of known $\mathcal{M}$, how could you use its measured $[\eta]$ in a given solvent to determine the molar mass of any sample of that polymer?
(c) Compare $[\eta]$ of a polyisobutylene sample [repeat unit $(CH_3)_2CCH_2$] with a molar mass of 104,160 g/mol with that of the polystyrene in part (a). What does this suggest about the solvent-polymer interactions of the two samples?

*12.151 One way of purifying gaseous H_2 is to pass it under high pressure through the holes of a metal's crystal structure. Palladium, which adopts a cubic closest packed structure, absorbs more H_2 than any other element and is one of the metals currently used for this purpose. Although the metal-hydrogen interaction is unclear, it is estimated that the density of absorbed H_2 approaches that of liquid hydrogen (70.8 g/L). What volume (in L) of gaseous H_2, measured at STP, can be packed into the spaces of 1 dm^3 of palladium metal?

*12.152 On a humid day in New Orleans, the temperature is 22.0°C, and the partial pressure of water vapor in the air is 31.0 torr. The 9000-ton air-conditioning system in the Louisiana Superdome maintains an inside air temperature of 22.0°C also, but a partial pressure of water vapor of 10.0 torr. The volume of air in the dome is 2.4×10^6 m^3, and the total pressure inside and outside the dome are both 1.0 atm. (a) What mass of water (in metric tons) must be removed every time the inside air is completely replaced with outside air? (*Hint:* How many moles of gas are in the dome? How many moles of water vapor? How many moles of dry air? How many moles of outside air must be added to the air in the dome to simulate the composition of outside air?) (b) Find the heat released when this mass of water condenses.

CHAPTER 13

THE PROPERTIES OF MIXTURES: SOLUTIONS AND COLLOIDS

CHAPTER OUTLINE

13.1 Types of Solutions: Intermolecular Forces and Predicting Solubility
Intermolecular Forces in Solution
Liquid Solutions
Gas Solutions and Solid Solutions

13.2 Energy Changes in the Solution Process
Heats of Solution and Solution Cycles
Heats of Hydration
The Tendency Toward Disorder

13.3 Solubility as an Equilibrium Process
Effect of Temperature
Effect of Pressure

13.4 Quantitative Ways of Expressing Concentration
Molarity and Molality
Parts of Solute by Parts of Solution
Converting Units of Concentration

13.5 Colligative Properties of Solutions
Nonvolatile Nonelectrolyte Solutions
Solute Molar Mass
Volatile Nonelectrolyte Solutions
Electrolyte Solutions

13.6 The Structure and Properties of Colloids

Figure: Swirls of sweetness. Changes in density are visible near a dissolving sugar cube, as molecules of sucrose leave the solid to interact with water molecules and form a homogeneous mixture (solution). In this chapter, you'll learn why mixtures form, how chemists quantify them, and how we make use of the difference in properties between solutions and pure solvents.

Nearly all the gases, liquids, and solids that make up our world are *mixtures—* two or more substances physically mixed together but not chemically combined. Synthetic mixtures, such as glass and soap, usually contain relatively few components, whereas natural mixtures, such as seawater and soil, are more complex, often containing more than 50 different substances. Living mixtures, such as trees and students, are the most complex—even a simple bacterial cell contains well over 5000 different compounds (Table 13.1).

Table 13.1 Approximate Composition of a Bacterium

Substance	Mass % of Cell	Number of Types	Number of Molecules
Water	~70	1	5×10^{10}
Ions	1	20	?
Sugars*	3	200	3×10^{8}
Amino acids*	0.4	100	5×10^{7}
Lipids*	2	50	3×10^{7}
Nucleotides*	0.4	200	1×10^{7}
Other small molecules	0.2	~200	?
Macromolecules (proteins, nucleic acids, polysaccharides)	23	~5000	6×10^{6}

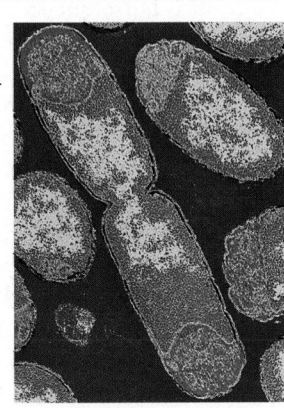

*Includes precursors and metabolites.

Recall from Chapter 2 that a mixture has two defining characteristics: *its composition is variable,* and *it retains some properties of its components.* In this chapter, we focus on two common types of mixtures: solutions and colloids. *A solution is a homogeneous mixture,* one with no boundaries separating its components; thus, a solution exists as one phase. A *heterogeneous mixture* has two or more phases. The pebbles in concrete or the bubbles in champagne are visible indications that these are heterogeneous mixtures. A *colloid* is a heterogeneous mixture in which one component is dispersed as very fine particles in another component, so distinct phases are not easy to see. Smoke and milk are colloids. The essential difference between a solution and a colloid is one of particle *size:*

- In a solution, the particles are individual atoms, ions, or small molecules.
- In a colloid, the particles are typically either macromolecules or aggregations of small molecules that are not large enough to settle out.

As you'll see later in the chapter, these molecular-scale differences result in many observable differences.

This chapter opens with a survey of the types of solutions and the role of intermolecular forces in their formation. Predicting solubility is a major focus of the first section, in which we look at systems ranging from salts in water to antibiotics in bacterial cells. Next, we investigate why a substance dissolves, in terms of two key factors: the enthalpy change and the disordering that occur when a solution forms. To understand the second factor, the concept of entropy changes in the solution process is introduced. Then, we examine the equilibrium nature of solubility and see how temperature and pressure affect it. We discuss ways of expressing concentration and then use them to explore the differences between the physical properties of solutions and those of pure substances. An investigation of the properties of colloids follows, and the chapter closes with the application of solution and colloid chemistry to the purification of water.

13.1 TYPES OF SOLUTIONS: INTERMOLECULAR FORCES AND PREDICTING SOLUBILITY

We often describe solutions in terms of one substance dissolving in another: the **solute** dissolves in the **solvent.** Usually, *the solvent is the most abundant component* of a given solution. In some cases, however, the substances are **miscible,** that is, soluble in each other in any proportion; in such cases, it may not be meaningful to call one the solute and the other the solvent.

The **solubility (S)** of a solute is the maximum amount that dissolves in a fixed quantity of a particular solvent at a specified temperature, given that excess solute is present. Different solutes have different solubilities. For example, for sodium chloride (NaCl), $S = 39.12$ g/100. mL water at 100.°C, whereas for silver chloride (AgCl), $S = 0.0021$ g/100. mL water at 100.°C. Obviously, NaCl is much more soluble in water than AgCl is. (Solubility is also expressed in other units, as you'll see later.) Although solubility has a quantitative meaning, *dilute* and *concentrated* are qualitative terms that refer to the *relative amounts of solute:* a dilute solution contains much less dissolved solute than a concentrated one.

From everyday experience, you know that some solvents can dissolve a given solute, whereas others cannot. For example, butter does not dissolve in water, but it does in cooking oil. A major factor determining whether a solution forms is the *relative strength of the intermolecular forces within and between solute and solvent.* From a knowledge of these forces, we can often predict which solutes will dissolve in which solvents.

Intermolecular Forces in Solution

All the intermolecular forces we discussed in Chapter 12 for pure substances also occur within mixtures. Figure 13.1 summarizes these forces in order of decreasing strength.

Ion-dipole forces are a principal factor in the solubility of ionic compounds in water. When a salt dissolves, each ion on the crystal's surface attracts the oppositely charged end of the water dipole. These attractive forces overcome those between the ions and break down the crystal structure. As each ion becomes separated, more water molecules cluster around it in **hydration shells** (Figure 13.2). Water molecules in the closest hydration shell are H bonded to others slightly farther away; those form a less structured hydration shell and are in turn H bonded to other molecules in the bulk solvent. For monatomic ions, the number of water molecules in the closest hydration shell depends on the ion's size. Four water molecules can fit tetrahedrally around small ions, such as Li^+, while larger ions, such as Na^+ and F^-, usually have six water molecules surrounding them octahedrally. It's important to see, and we'll return to this point later, that this process does not merely lead to a random jumble of ions and water molecules; rather, there is some order in the orientation of the molecules around the ions.

Dipole-dipole forces and the special type of dipole-dipole force known as the *H bond* are both very important in solutions. The H bond is a primary factor in water's ability to dissolve numerous oxygen- and nitrogen-containing organic and biological compounds, such as alcohols, sugars, amines, and amino acids. (Recall that O and N are small, so their lone pairs can get very close to the partially positive H of H_2O.)

The two types of *charge-induced dipole forces* rely on the polarizability of the components. When an ion's charge distorts the electron cloud of a nearby nonpolar particle, an **ion–induced dipole force** results. To cite one essential biological example, this intermolecular force plays a role in the binding between the Fe^{2+} ion in hemoglobin and an O_2 molecule in the bloodstream. An ion can also increase the magnitude of an existing dipole in a nearby molecule. Thus, this force contributes to the formation of any solution that contains ions—salts dissolved in

Ion-dipole
(40–600)

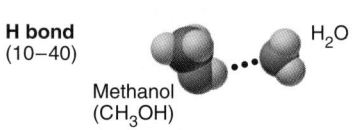

H bond
(10–40)

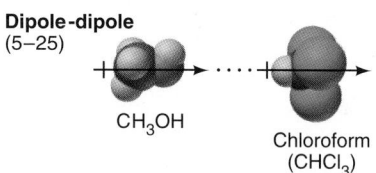

Methanol
(CH_3OH) H_2O

Dipole-dipole
(5–25)

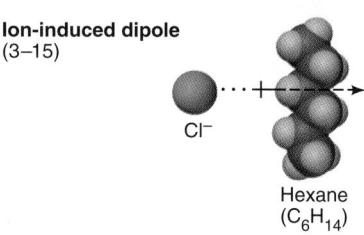

CH_3OH

Chloroform
($CHCl_3$)

Ion-induced dipole
(3–15)

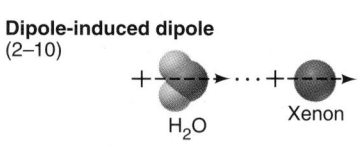

Cl^-

Hexane
(C_6H_{14})

Dipole-induced dipole
(2–10)

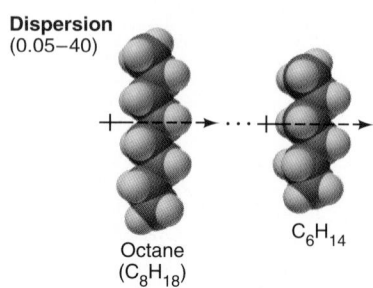

H_2O Xenon

Dispersion
(0.05–40)

Octane
(C_8H_{18}) C_6H_{14}

Figure 13.1 **The major types of intermolecular forces in solutions.** Forces are listed in decreasing order of strength (with values in kJ/mol), and an example of each is shown with space-filling models.

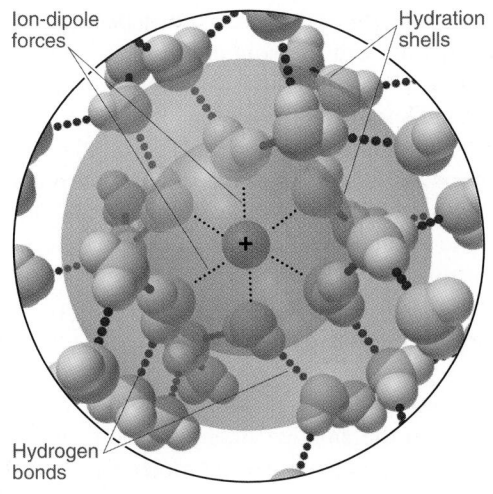

Ion-dipole forces

Hydration shells

Hydrogen bonds

Figure 13.2 Hydration shells around an aqueous ion. When an ionic compound dissolves in water, ion-dipole forces orient water molecules around the separated ions to form hydration shells. The cation shown here is octahedrally surrounded by six water molecules, which form H bonds with water molecules in the next hydration shell, and those form H bonds with others farther away.

water or in less polar solvents. Coulomb's law says that a greater charge results in a stronger energy of attraction. Therefore, the ion–induced dipole force is stronger than the **dipole–induced dipole force,** which arises when the partial charges of a polar molecule distort the electron cloud of a nearby nonpolar molecule. The solubility in water, limited though it is, of the nonpolar atmospheric gases O_2, N_2, and the noble gases, is due in part to dipole–induced dipole forces. Paint thinners and grease solvents also function through these forces.

We must always appreciate the importance of ever-present *dispersion forces* because they contribute to the solubility of all solutes in all solvents. In fact, they are the *principal* attractive force in solutions of nonpolar substances; for example, petroleum exists as a homogeneous mixture because of dispersion forces.

With these intermolecular forces in mind, let's examine the types of solutions. Solutions can be gaseous, liquid, or solid. In most cases, *the physical state of the solvent determines that of the solution.* We highlight solutions in which the solvent is a liquid because they are by far the most common and important.

Liquid Solutions and the Role of Molecular Polarity

From cytoplasm to tree sap, gasoline to cleaning fluid, iced tea to urine, solutions in which a liquid is the solvent are commonplace in everyday life. Water is the most prominent liquid solvent because it is so abundant in the environment and can dissolve both ionic compounds and polar molecules. It dissolves many substances through the formation of H bonds. Nonpolar substances, such as cooking oil and car wax, do not dissolve appreciably in water but dissolve freely in less polar or nonpolar solvents.

Experience shows *substances with similar types of intermolecular forces dissolve in each other.* This fact is summarized in the old rule-of-thumb **like dissolves like,** which often provides a sound qualitative way to predict solubility.

Liquid-Liquid and Solid-Liquid Solutions Many salts dissolve in water because the strong ion-dipole attractions that water forms with the ions are very similar to the strong attractions between the ions themselves and, therefore, can *substitute* for them. The same salts are insoluble in hexane (C_6H_{14}) because the weak ion–induced dipole forces their ions could form with this nonpolar solvent *cannot* substitute for attractions between the ions. Similarly, oil does not dissolve in water because the weak dipole–induced dipole forces between oil and water molecules cannot substitute for the strong H bonds between water molecules. Oil *does* dissolve in hexane, however, because the dispersion forces in one substitute readily for the dispersion forces in the other. Thus, for a solution to form, "like dissolves like" means that the forces created between solute and solvent must be *comparable in strength* to the forces destroyed within both the solute and the solvent.

To examine this idea further, let's compare the solubilities of a series of alcohols in two solvents that act through very different intermolecular forces—water and hexane. Alcohols are organic molecules with a hydroxyl (—OH) group bound to a hydrocarbon group. The simplest type of alcohol has the general formula $CH_3(CH_2)_nOH$; we'll consider alcohols with $n = 0$ to 5. We can view an alcohol molecule as consisting of two portions: the polar —OH group and the nonpolar hydrocarbon chain. The —OH portion forms strong H bonds with water and weak dipole–induced dipole forces with hexane. The hydrocarbon portion interacts through dispersion forces with hexane and through dipole–induced dipole forces with water.

In Table 13.2, the models show the relative change in size of the polar and nonpolar portions of the alcohol molecules. In the smaller alcohols (one to three carbons), the hydroxyl group is a relatively large portion, so the molecules interact with each other through H bonding, just as water molecules do. When they mix with water, H bonding within solute and within solvent is replaced by H bonding *between* solute and solvent (Figure 13.3). As a result, these smaller alcohols are miscible with water.

Water solubility decreases dramatically for alcohols larger than three carbons, and those with chains longer than six carbons are insoluble in water. For these larger alcohols to dissolve, the nonpolar chains have to move between the water molecules, substituting their weak attractions with those water molecules for strong H bonds among the water molecules themselves. The —OH portion of the alcohol does form H bonds to water, but these don't outweigh the H bonds between water molecules that have to break to make room for the hydrocarbon portion.

Table 13.2 Solubility* of a Series of Alcohols in Water and Hexane

Alcohol	Model	Solubility in Water	Solubility in Hexane
CH_3OH (methanol)		∞	1.2
CH_3CH_2OH (ethanol)		∞	∞
$CH_3(CH_2)_2OH$ (propanol)		∞	∞
$CH_3(CH_2)_3OH$ (butanol)		1.1	∞
$CH_3(CH_2)_4OH$ (pentanol)		0.30	∞
$CH_3(CH_2)_5OH$ (hexanol)		0.058	∞

*Expressed in mol alcohol/1000 g solvent at 20°C.

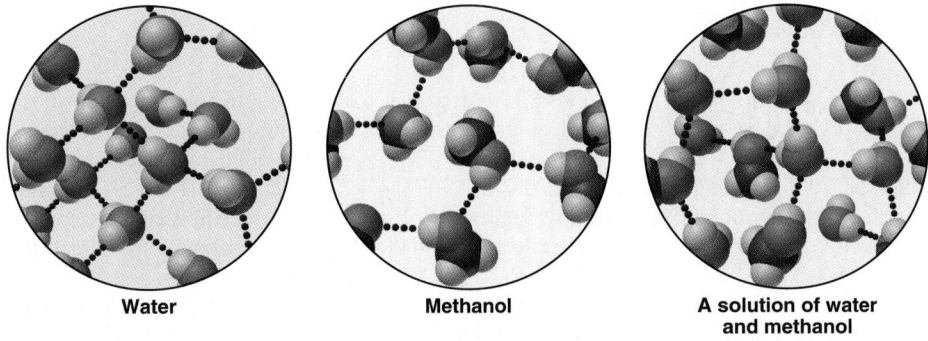

Water   **Methanol**   **A solution of water and methanol**

Figure 13.3 **Like dissolves like: solubility of methanol in water.** The H bonds in water and in methanol are similar in type and strength, so they can substitute for one another. Thus, methanol is soluble in water; in fact, they are miscible.

The table shows that the opposite trend occurs with hexane. Now the major solute-solvent *and* solvent-solvent interactions are dispersion forces. The weak forces between the —OH group of methanol (CH_3OH) and hexane cannot substitute for the strong H bonding among CH_3OH molecules, so the solubility of methanol in hexane is relatively low. In any larger alcohol, however, dispersion forces become increasingly more important, and these *can* substitute for dispersion forces in pure hexane, so solubility increases. With no strong solvent-solvent forces to be replaced by weak solute-solvent forces, even the two-carbon chain of ethanol has strong enough solute-solvent attractions to be miscible in hexane.

Many other organic molecules have polar and nonpolar portions, and the predominance of one portion or the other determines their solubility. The dual polarity of some substances, such as soaps, detergents, and many antibiotics, even determines their function, as the next Chemical Connections essay describes.

SAMPLE PROBLEM 13.1 Predicting Relative Solubilities of Substances

Problem Predict which solvent will dissolve more of the given solute:
(a) Sodium chloride in methanol (CH_3OH) or in propanol ($CH_3CH_2CH_2OH$)
(b) Ethylene glycol ($HOCH_2CH_2OH$) in hexane ($CH_3CH_2CH_2CH_2CH_2CH_3$) or in water
(c) Diethyl ether ($CH_3CH_2OCH_2CH_3$) in water or in ethanol (CH_3CH_2OH)
Plan We examine the formulas of the solute and each solvent to determine the types of forces that could occur. A solute tends to be more soluble in a solvent whose intermolecular forces are similar to, and therefore can substitute for, its own.
Solution (a) Methanol. NaCl is an ionic solid that dissolves through ion-dipole forces. Both methanol and propanol contain a polar —OH group, but propanol's longer hydrocarbon chain can form only weak forces with the ions, so it is less effective at substituting for the ionic attractions in the solute.
(b) Water. Ethylene glycol molecules have two —OH groups, so the molecules interact with each other through H bonding. They are more soluble in H_2O, whose H bonds can substitute for their own H bonds better than the dispersion forces in hexane can.
(c) Ethanol. Diethyl ether molecules interact with each other through dipole and dispersion forces and can form H bonds to both H_2O and ethanol. The ether is more soluble in ethanol because that solvent can form H bonds *and* substitute for the ether's dispersion forces. Water, on the other hand, can form H bonds with the ether, but it lacks any hydrocarbon portion, so it forms much weaker dispersion forces with that solute.

FOLLOW-UP PROBLEM 13.1 Which solute is more soluble in the given solvent?
(a) Butanol ($CH_3CH_2CH_2CH_2OH$) or 1,4-butanediol ($HOCH_2CH_2CH_2CH_2OH$) in water
(b) Chloroform ($CHCl_3$) or carbon tetrachloride (CCl_4) in water

The Mode of Action of Soaps and Antibiotics

Strange as it may seem, soaps function by the same principle as many antibiotics do—the influence on solubility of dual polarity in one molecule.

A **soap** is the salt of an acid-base reaction between a strong base (metal hydroxide) and a fatty acid, a molecule with a long hydrocarbon chain bonded to an organic acid group (—COOH). A typical soap molecule has an even number of carbon atoms, and is usually made up of a nonpolar "tail" 15–19 carbons long and a polar-ionic "head" consisting of the —COO$^-$ group and a cation. (Actually, the cation of a soap influences its properties greatly. Lithium soaps are hard and high melting and are used in car lubricants. Softer, more water-soluble sodium soaps are used as common bar soap. Potassium soaps are low melting and used as liquid soaps. The insolubility of calcium soaps explains why it is difficult to clean clothes in hard water, which typically contains Ca^{2+} ions.)

Sodium stearate [$CH_3(CH_2)_{16}COONa$; inset of Figure B13.1] is a major component of many bar soaps. The tail is soluble in greasy materials, and the head is soluble in water. When a greasy cloth is immersed in soapy water, the nonpolar tails of the soap molecules interact with the nonpolar grease molecules through dispersion forces, while the polar-ionic heads interact with the water through ion-dipole forces and H bonds. Agitation creates tiny aggregates of grease molecules in which are embedded soap molecules whose polar-ionic heads stick into the water. These aggregates are flushed away by added water (Figure B13.1). Detergents have a similar structure of nonpolar tail and polar-ionic head (see Figure 15.17, p. 632). Soaps and detergents are called *surfactants* (a combination of the words *surf*ace-*act*ive age*nts*) because they interact with the surfaces of two phases, such as oil and water.

Certain antibiotics function via dual polarities as well. All cells have membranes that maintain their internal concentrations of ions in a delicate balance with the concentrations in the external aqueous fluid: some ions, such as Na^+, are excluded from the cell, whereas others, such as K^+, are kept inside the cell. If this ionic segregation of Na^+ and K^+ is disrupted, the organism cannot survive.

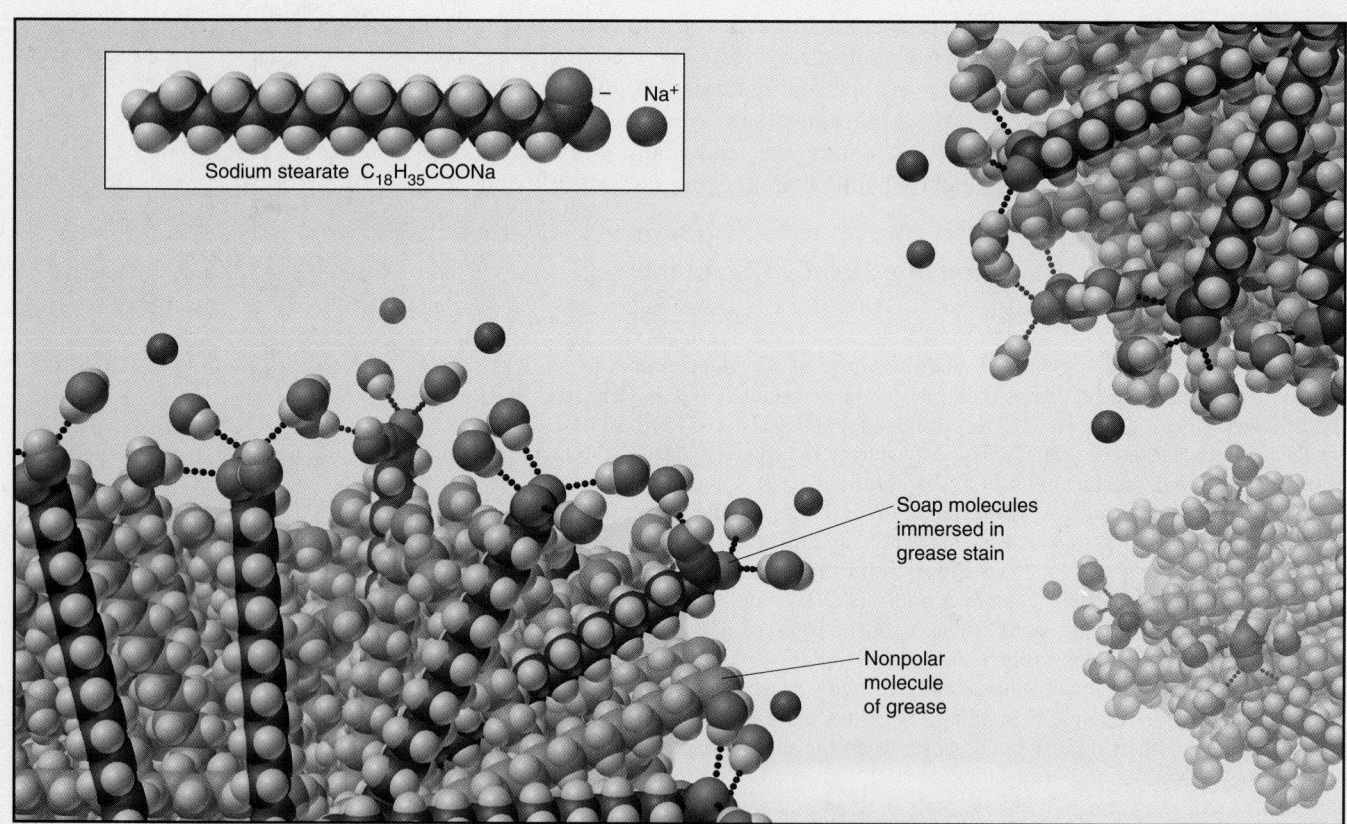

Sodium stearate $C_{18}H_{35}COONa$

Soap molecules immersed in grease stain

Nonpolar molecule of grease

Figure B13.1 The structure and function of a soap. The inset shows the structure of sodium stearate, a typical soap. Note the nonpolar tail and polar-ionic head. The tails of soap molecules interact with nonpolar molecules in grease through dispersion forces, while the heads interact with water through H bonds and ion-dipole forces. Agitation breaks up the grease stain into small soap-grease aggregates that rinsing flushes away.

Cell membranes are thin envelopes that consist mostly of a double layer of molecules. Each molecule combines two fatty acids, an alcohol, and a phosphate group. The nonpolar tails of the fatty-acid portions interact through dispersion forces within the membrane, while the charged, polar heads of the phosphate portion interact through ion-dipole forces with the watery interior and exterior fluids.

Antibiotics are natural substances produced by one microorganism to chemically destroy another. Many have been adopted by medical science as potent weapons against bacterial infection. Several classes of antibiotics function by using their molecular polarities to disrupt an organism's aqueous ionic makeup. Some antibiotics disrupt the ion balance by forming a channel in the bacterial membrane through which aqueous ions can flow. Figure B13.2 shows the action of the antibiotic gramicidin A. This antibiotic has a dual polarity on one tube-shaped molecule: nonpolar groups on the outside of the tube and polar groups on the inside. The nonpolar outside helps the molecule dissolve into the nonpolar membrane of the target bacterium through dispersion forces, and the polar inside allows hydrated ions to enter through H bonds and ion-dipole forces. Two gramicidin A molecules lying end to end span the bacterial membrane and remain there, providing a polar passageway through which ions can diffuse in and out. The cell's interior-exterior ionic segregation is disrupted, and the organism dies.

Valinomycin is a donut-shaped antibiotic molecule that works in a different way. The polar hole of the donut binds an ion, while the nonpolar rim allows the molecule to dissolve into the bacterial membrane. Rather than remaining lodged there as gramicidin A does, valinomycin moves through to the other side of the membrane and releases the ion. Millions of such transits by many valinomycin molecules soon disrupt the ionic segregation, and the organism dies.

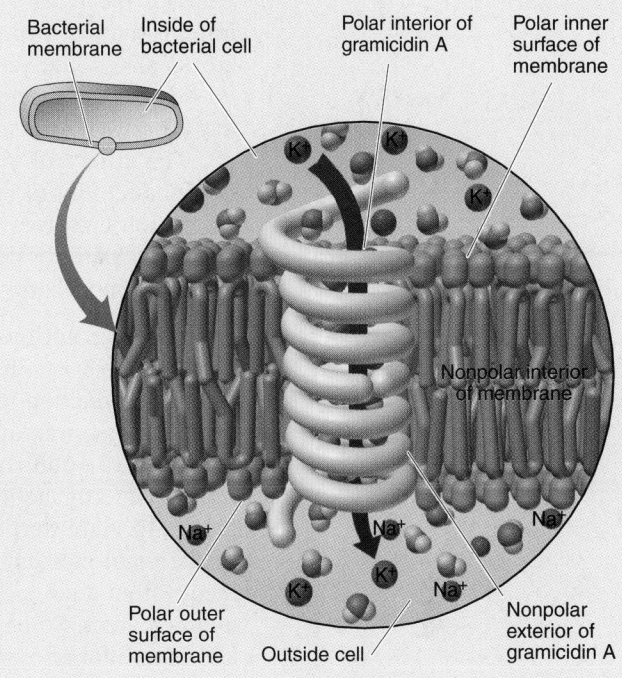

Figure B13.2 The mode of action of an antibiotic. Gramicidin A is a tubelike molecule, nonpolar on the outside and polar on the inside. After dissolving into the nonpolar interior of a bacterial cell membrane via dispersion forces, pairs of gramicidin molecules aligned end to end span the membrane's thickness. Ions diffuse through the polar interior of these molecules, which destroys the cell's ionic balance and its segregation of Na^+ outside and K^+ inside.

Gas-Liquid Solutions Gases that are nonpolar, such as N_2, or are nearly so, such as NO, have low boiling points because their intermolecular attractions are weak. Likewise, they are not very soluble in water because solute-solvent forces are weak. In fact, as Table 13.3 shows, for nonpolar gases, boiling point generally correlates with solubility in water.

In some cases, the small amount of a nonpolar gas that does dissolve is essential to a process. The most important environmental example is the solubility of O_2 in water. At 25°C and 1 atm, the solubility of O_2 is only 3.2 mL/100. mL of water, but aquatic animal life would die without this small amount. In other cases, the solubility of a gas may *seem* high, but the gas is actually reacting with the solvent or another component. Oxygen seems much more soluble in blood than in water because O_2 molecules are continually bonding with hemoglobin molecules in red blood cells. Similarly, carbon dioxide, which is essential for aquatic plants and coral-reef growth, seems very soluble in water (~81 mL of CO_2/100. mL of H_2O at 25°C and 1 atm) because it is reacting, in addition to simply dissolving:

$$CO_2(g) + H_2O(l) \longrightarrow H^+(aq) + HCO_3^-(aq)$$

Table 13.3 Correlation Between Boiling Point and Solubility in Water

Gas	Solubility (M)*	bp (K)
He	4.2×10^{-4}	4.2
Ne	6.6×10^{-4}	27.1
N_2	10.4×10^{-4}	77.4
CO	15.6×10^{-4}	81.6
O_2	21.8×10^{-4}	90.2
NO	32.7×10^{-4}	121.4

*At 273 K and 1 atm.

Gas Solutions and Solid Solutions

Despite the central place of liquid solutions in chemistry, gaseous solutions and solid solutions have vital importance and numerous applications.

Gas-Gas Solutions *All gases are infinitely soluble in one another.* Air is the classic example of a gaseous solution, consisting of about 18 gases in widely differing proportions. Anesthetic gas proportions are finely adjusted to the needs of the patient and the length of the surgical procedure. The ratios of components in many industrial gas mixtures, such as $CO:H_2$ in syngas production or $N_2:H_2$ in ammonia production, are controlled to optimize product yield under varying conditions of temperature and pressure.

Gas-Solid Solutions *When a gas dissolves in a solid, it occupies the spaces between the closely packed particles.* Hydrogen can be purified by passing an impure sample through a solid metal such as palladium. Only the H_2 molecules are small enough to enter the spaces between the Pd atoms, where they form Pd—H covalent bonds. Under high H_2 pressure, the H atoms are passed along the Pd crystal structure and emerge from the solid as H_2 molecules.

The ability of gases to penetrate a solid also has disadvantages. The electrical conductivity of copper, for example, is drastically reduced by the presence of O_2, which dissolves into the crystal structure and reacts to form copper(I) oxide. High-conductivity copper is prepared by melting and recasting the copper in an O_2-free atmosphere.

Solid-Solid Solutions Because solids diffuse so little, their mixtures are usually heterogeneous, as in gravel mixed with sand. Some solid-solid solutions can be formed by melting the solids and then mixing them and allowing them to freeze. Many **alloys,** mixtures of elements that have an overall metallic character, are examples of solid-solid solutions (although several common alloys have microscopic heterogeneous regions). In *substitutional* alloys, such as brass (Figure 13.4A) and sterling silver, atoms of another element substitute for some atoms of the main element in the structure. In *interstitial* alloys, atoms of another element (often a nonmetal) fill some of the spaces *(interstices)* between atoms of the main element. Some forms of carbon steel, shown in Figure 13.4B, for example, are interstitial alloys of iron with a small amount of carbon.

Figure 13.4 The arrangement of atoms in two types of alloys. A, Brass is a substitutional alloy in which zinc atoms substitute for copper atoms at many of copper's face-centered cubic packing sites. **B,** Some forms of carbon steel are interstitial alloys in which carbon atoms lie in the holes (interstices) of the body-centered cubic crystal structure of iron.

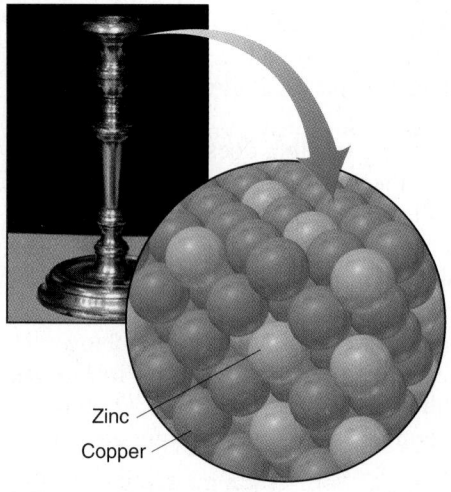

Zinc

Copper

A Brass, a substitutional alloy

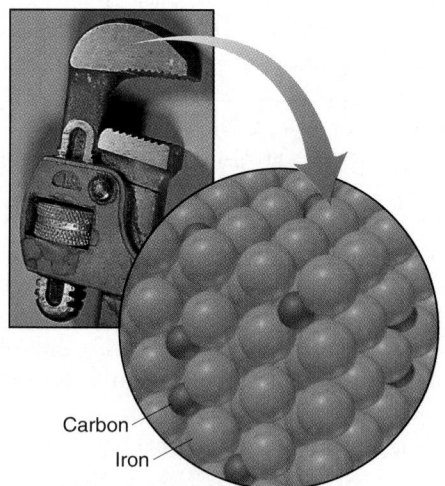

Carbon

Iron

B Carbon steel, an interstitial alloy

Waxes are another familiar type of solid-solid solution. Most waxes are amorphous solids that may contain small regions of crystalline regularity. A natural *wax* is a solid of biological origin that is insoluble in water but dissolves in nonpolar solvents. Many plants have waxy coatings on their leaves and fruit to prevent evaporation of interior water. ●

SECTION SUMMARY

Solutions are homogeneous mixtures consisting of a solute dissolved in a solvent through the action of intermolecular forces. The solubility of a solute in a given amount of solvent is the maximum amount, with excess solute present, that can dissolve at a specified temperature. (For gaseous solutes, the pressure must be specified also.) In addition to the intermolecular forces that exist in pure substances, ion-dipole, ion–induced dipole, and dipole–induced dipole forces occur in solution. If similar intermolecular forces occur in solute and solvent, a solution will likely form. When ionic compounds dissolve in water, the ions become surrounded by hydration shells of H-bonded water molecules. Solubility of organic molecules in water depends on their polarity and the extent of their polar and nonpolar portions. Soaps and some antibiotics act through the dual polarities of their molecules. The solubility of nonpolar gases in water is low because of weak intermolecular forces. Gases are miscible with one another and dissolve in solids by fitting into spaces in the crystal structure. Solid-solid solutions, such as alloys and waxes, form when the components are mixed while molten.

● Waxes for Home and Auto
Beeswax, the remarkable structural material that bees secrete to build their combs, is a complex mixture of fatty acids and hydrocarbons in which some of the molecules contain chains more than 40 carbon atoms long. Carnauba wax, from a South American palm, is a mixture of compounds, each consisting of a fatty acid bound to a long-chain alcohol. It is hard but forms a thick gel in nonpolar solvents, perfect for car waxes.

13.2 ENERGY CHANGES IN THE SOLUTION PROCESS

As a qualitative predictive tool, "like dissolves like" is helpful in many cases. As you might expect, this handy *macroscopic* rule is based on the *molecular* interactions that occur between solute and solvent particles. To see *why* like dissolves like, we can break down the solution process conceptually into steps and examine them in terms of changes in *enthalpy* and *entropy* of the system. We discussed enthalpy in Chapter 6 and focus on it first here. The concept of entropy is introduced at the end of this section and treated quantitatively in Chapter 20.

Heats of Solution and Solution Cycles

Picture a general solute and solvent about to form a solution. Both consist of particles attracting each other. For one substance to dissolve in another, three events must occur: (1) solute particles must separate from each other, (2) some solvent particles must separate to make room for the solute particles, and (3) solute and solvent particles must mix together. No matter what the nature of the attractions within the solute and within the solvent, some energy is *absorbed* when particles separate, and some energy is *released* when they mix and attract each other. As a result of these changes in attractions, the solution process is typically accompanied by a change in enthalpy. We can divide the process into these three steps, each with its own enthalpy term:

Step 1. Solute particles separate from each other. This step involves overcoming intermolecular attractions, so it is *endothermic:*

Solute (aggregated) + *heat* $\longrightarrow$ solute (separated) $\Delta H_{\text{solute}} > 0$

Step 2. Solvent particles separate from each other. This step also involves overcoming attractions, so it is *endothermic*, too:

Solvent (aggregated) + *heat* $\longrightarrow$ solvent (separated) $\Delta H_{\text{solvent}} > 0$

Step 3. Solute and solvent particles mix. The particles attract each other, so this step is *exothermic:*

Solute (separated) + solvent (separated) $\longrightarrow$ solution + *heat* $\Delta H_{\text{mix}} < 0$

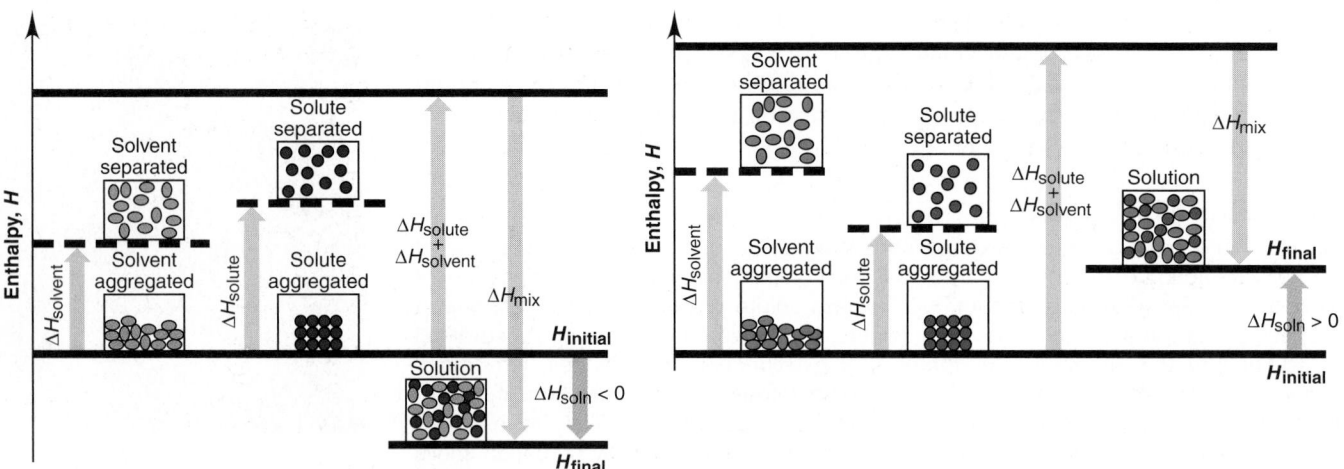

A Exothermic solution process

B Endothermic solution process

Figure 13.5 **Solution cycles and the enthalpy components of the heat of solution.** ΔH_{soln} can be thought of as the sum of three enthalpy changes: $\Delta H_{solvent}$ (separating the solvent; always >0), ΔH_{solute} (separating the solute; always >0), and ΔH_{mix} (mixing solute and solvent; always <0). In **A**, ΔH_{mix} is larger than the sum of ΔH_{solute} and $\Delta H_{solvent}$, so ΔH_{soln} is negative (exothermic process). In **B**, ΔH_{mix} is smaller than the sum of the others, so ΔH_{soln} is positive (endothermic process).

The total enthalpy change that occurs when a solution forms from solute and solvent is the **heat of solution (ΔH_{soln}),** and we combine the three individual enthalpy changes to find it. The overall process is called a *thermochemical solution cycle* and is yet another application of Hess's law:

$$\Delta H_{soln} = \Delta H_{solute} + \Delta H_{solvent} + \Delta H_{mix} \qquad \textbf{(13.1)}$$

(In Section 9.2 we conceptually broke down the heat of formation of an ionic compound into component enthalpies in a similar way using a Born-Haber cycle.) If the sum of the endothermic terms ($\Delta H_{solute} + \Delta H_{solvent}$) is smaller than the exothermic term (ΔH_{mix}), ΔH_{soln} is negative; that is, the process is exothermic. Figure 13.5A is an enthalpy diagram for the formation of such a solution. If the sum of the endothermic terms is larger than the exothermic term, ΔH_{soln} is positive; that is, the process is endothermic (Figure 13.5B). However, if ΔH_{soln} is highly positive, the solute may not dissolve to any significant extent in that solvent.

Heats of Hydration: Ionic Solids in Water

Before we discuss the entropy factor in the solution process, we can simplify the solution cycle for aqueous systems. The $\Delta H_{solvent}$ and ΔH_{mix} components of the heat of solution are difficult to measure individually. Combined, these terms represent the enthalpy change during **solvation,** the process of surrounding a solute particle with solvent particles. Solvation in water is called **hydration.** Thus, the enthalpy changes for separating the water molecules ($\Delta H_{solvent}$) and mixing the solute with them (ΔH_{mix}) are combined into the **heat of hydration (ΔH_{hydr}).** In water, Equation 13.1 becomes

$$\Delta H_{soln} = \Delta H_{solute} + \Delta H_{hydr}$$

The heat of hydration is a crucial factor in dissolving an ionic solid. The breaking of several H bonds in water is more than compensated for when several strong ion-dipole forces form, so hydration of an ion is *always* exothermic. The ΔH_{hydr} of an ion is defined as the enthalpy change for the hydration of 1 mol of separated (gaseous) ions:

$$M^+(g) \text{ [or } X^-(g)] \xrightarrow{H_2O} M^+(aq) \text{ [or } X^-(aq)] \qquad \Delta H_{hydr \text{ of the ion}} \text{ (always } <0)$$

Heats of hydration exhibit trends based on the **charge density** of the ion, the ratio of the ion's charge to its volume. In general, the higher the charge density is, the more negative ΔH_{hydr} is. According to Coulomb's law, the greater the ion's charge is and the closer the ion can approach the oppositely charged end of the water molecule's dipole, the stronger the attraction. Therefore,

- A 2+ ion attracts H_2O molecules more strongly than a 1+ ion of similar size.
- A small 1+ ion attracts H_2O molecules more strongly than a large 1+ ion.

Down a group of ions, from Li^+ to Cs^+ in Group 1A(1), for example, the charge stays the same and the size increases; thus, the charge densities decrease, as do the heats of hydration. Going across the periodic table from Group 1A(1) to Group 2A(2), the 2A ion has a smaller radius *and* a greater charge, so its charge density and ΔH_{hydr} are greater. Table 13.4 shows these relationships for some ions.

The energy required to separate an ionic solute (ΔH_{solute}) into gaseous ions is the negative of its lattice energy ($-\Delta H_{lattice}$), so ΔH_{solute} is highly positive:

$$M^+X^-(s) \longrightarrow M^+(g) + X^-(g) \qquad \Delta H_{solute} \text{ (always >0)} = -\Delta H_{lattice}$$

Thus, the heat of solution for ionic compounds in water combines the negative of the lattice energy (always positive) and the combined heats of hydration of cation and anion (always negative),

$$\Delta H_{soln} = -\Delta H_{lattice} + \Delta H_{hydr \text{ of the ions}} \qquad (13.2)$$

The sizes of the individual terms determine the sign of the heat of solution.

Figure 13.6 shows enthalpy diagrams for dissolving three ionic solutes in water. The first, NaCl, has a slightly positive heat of solution ($\Delta H_{soln} = 3.9$ kJ/mol). Its lattice energy term is only slightly greater than the combined ionic heats of hydration, so if you dissolve NaCl in water in a flask, you do not notice any temperature change. On the other hand, if you dissolve NaOH in water, the flask feels hot. The lattice energy term for NaOH is much smaller than the combined ionic heats of hydration, so dissolving NaOH is highly exothermic ($\Delta H_{soln} = -44.5$ kJ/mol). Finally, if you dissolve NH_4NO_3 in water, the flask feels cold.

Table 13.4 Trends in Ionic Heats of Hydration

Ion	Ionic Radius (pm)	ΔH_{hydr} (kJ/mol)
Group 1A(1)		
Li^+	76	−510
Na^+	102	−410
K^+	138	−336
Rb^+	152	−315
Cs^+	167	−282
Group 2A(2)		
Mg^{2+}	72	−1903
Ca^{2+}	100	−1591
Sr^{2+}	118	−1424
Ba^{2+}	135	−1317
Group 7A(17)		
F^-	133	−431
Cl^-	181	−313
Br^-	196	−284
I^-	220	−247

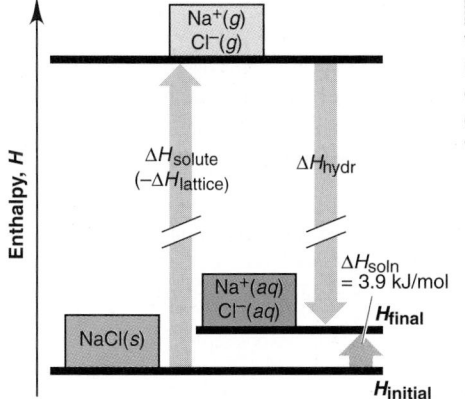

A NaCl. $\Delta H_{lattice}$ is slightly larger than ΔH_{hydr}: ΔH_{soln} is small and positive.

Figure 13.6 **Dissolving ionic compounds in water.** The enthalpy diagram for dissolving an ionic compound in water includes the negative of $\Delta H_{lattice}$ (ΔH_{solute}; always positive) and the combined ionic heats of hydration (ΔH_{hydr}; always negative).

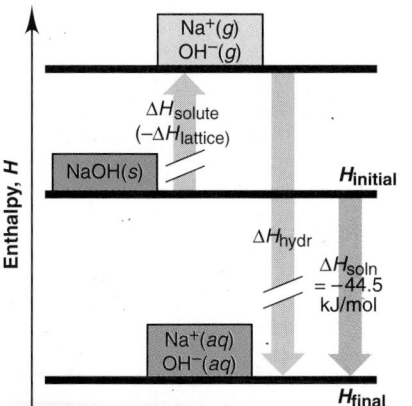

B NaOH. ΔH_{hydr} dominates: ΔH_{soln} is large and negative.

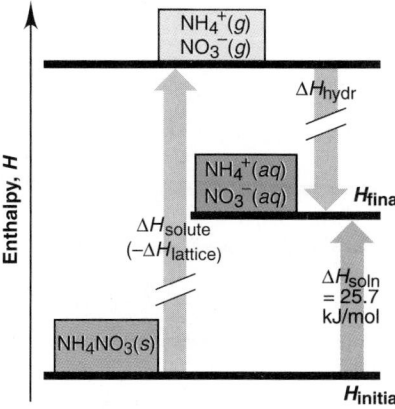

C NH_4NO_3. $\Delta H_{lattice}$ dominates: ΔH_{soln} is large and positive.

Hot Packs, Cold Packs, and Self-Heating Soup Strain a muscle or sprain a joint and you may obtain temporary relief by applying the appropriate heat of solution. Hot packs and cold packs consist of a thick outer pouch containing water and a thin inner pouch containing a salt. A squeeze on the outer pouch breaks the inner pouch, and the salt dissolves. Most hot packs use anhydrous $CaCl_2$ (ΔH_{soln} = −82.8 kJ/mol), whereas cold packs use NH_4NO_3 (ΔH_{soln} = 25.7 kJ/mol). The change in temperature can be quite large—a cold pack, for instance, can bring the solution from room temperature down to 0°C—but their usable time is limited to around half an hour. In Japan, some cans of soup have double walls, with an ionic compound in a packet and water between the walls. When you open the can, the packet breaks, and the exothermic solution process, which can quickly reach about 90°C, warms the soup.

In this case, the lattice energy term is much larger than the combined ionic heats of hydration, so the process is highly endothermic (ΔH_{soln} = 25.7 kJ/mol).

The Solution Process and the Tendency Toward Disorder

The other major factor that determines whether a substance dissolves is somewhat more abstract than the change in heat, but it must be considered along with it. This factor concerns the *natural tendency of most systems to become more disordered*. In thermodynamic terms, we say that the **entropy** of the system tends to increase. *Entropy is a measure of a system's disorder.*

Here is an obvious example of this tendency. Stack piles of nickels, dimes, and quarters heads up in a closed box. This system is very ordered, so it has low entropy. Now shake the box and look inside. The coins lie randomly in a pile, some heads up and some heads down. The disorder of the system has increased, and so has its entropy. Common sense tells you that if you shook the box for another century, the coins would not wind up in the stacks you started with. To get them that way, you must expend energy sorting and restacking them to counteract their natural tendency to become mixed up. A solution is more disordered, has greater entropy, than the pure solute and solvent. Thus, *solutions form naturally; pure solutes and solvents do not.* Water treatment plants, oil refineries, metal foundries, and many other industrial plants expend enormous quantities of energy reversing this natural tendency and separating mixtures into pure components.

The solution process involves two factors—the change in enthalpy ***and*** *the change in entropy.* In many cases, their relative magnitudes determine whether a solution forms. *The increase in entropy that accompanies dissolving always favors solution formation.* However, the change in enthalpy can work either way: a decrease in enthalpy of solution (ΔH_{soln} < 0) favors solution formation, but an increase in enthalpy hinders it.

To see this interplay of enthalpy and entropy, let's consider two solute-solvent pairs. The first does *not* form a solution because its enthalpy factor outweighs its entropy factor; the second *does* form a solution because its entropy factor dominates. Sodium chloride does not dissolve in hexane, as you would predict from the dissimilar intermolecular forces. As Figure 13.7A shows, separating the solvent is relatively easy because of the weak dispersion forces, but separating the ionic solute requires supplying energy equal in magnitude to $\Delta H_{lattice}$. Mixing releases very little energy because the ion–induced dipole attractions between Na^+ (or Cl^-) ions and hexane are weak. The sum of the endothermic terms is much larger than the exothermic term, so ΔH_{soln} is highly positive. In this case, disso-

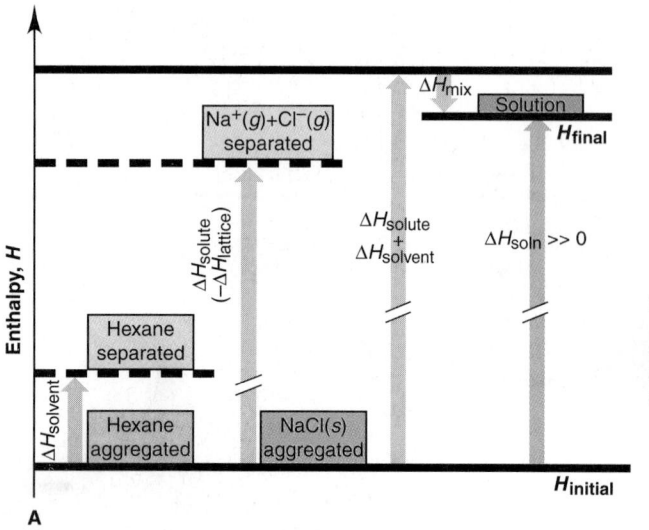

Figure 13.7 Enthalpy diagrams for dissolving NaCl and octane in hexane. A, Since attractions between ions and hexane molecules are weak, ΔH_{mix} is *much* smaller than ΔH_{solute}. Thus, ΔH_{soln} is so positive that NaCl does *not* dissolve in hexane. **B,** Intermolecular forces in octane and in hexane are so similar that ΔH_{soln} is very small. Octane dissolves in hexane because the solution has greater entropy (more disorder) than the pure components.

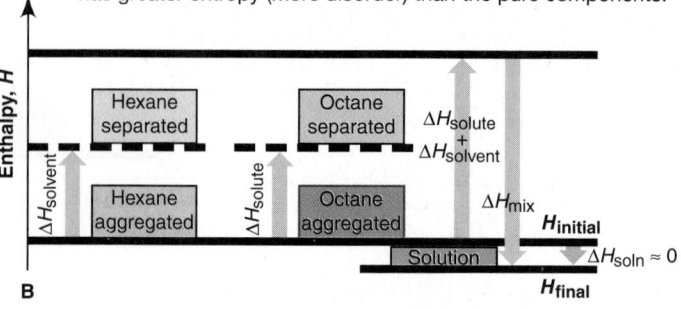

lution does not occur because the entropy increase that would occur when solute and solvent mix cannot outweigh the large enthalpy increase.

Now consider the second solute-solvent pair, octane (C_8H_{18}) and hexane (C_6H_{14}). Both consist of nonpolar molecules held together by dispersion forces of comparable strength. We would therefore predict that octane is soluble in hexane; in fact, they are infinitely soluble (miscible). Miscibility does not necessarily mean a great deal of heat is released, however, and in this case, the heat of solution is around zero (Figure 13.7B). With no enthalpy change driving the process, why do these substances form a solution? Octane dissolves in hexane because *the entropy increases greatly when pure substances mix and become disordered.* Thus, the large increase in entropy drives formation of this solution.

In some cases, a large enough increase in entropy can cause a solution to form even when the enthalpy increases significantly ($\Delta H_{soln} > 0$). As Figure 13.6C showed earlier, when NH_4NO_3 dissolves in water, the process is very endothermic. Nevertheless, in that case, the increase in entropy that occurs when the orderly ionic crystal breaks down and the ions mix with water molecules more than compensates for the heat absorbed. The importance of entropy in physical and chemical systems is covered in depth in Chapter 20.

SECTION SUMMARY

An overall heat of solution can be obtained from a thermochemical solution cycle as the sum of two endothermic steps (solute separation and solvent separation) and one exothermic step (solute-solvent mixing). In aqueous solutions, the combination of solvent separation and mixing is called *hydration.* Ionic heats of hydration are always negative because of strong ion-dipole forces. Most systems have a natural tendency to increase their entropy (become disordered), and a solution has greater entropy (more disorder) than the pure solute and solvent. The combination of enthalpy and entropy changes determines whether a solution forms. A substance with a positive ΔH_{soln} dissolves *only* if the entropy increase is large enough to outweigh it.

13.3 SOLUBILITY AS AN EQUILIBRIUM PROCESS

Imagine what takes place on the molecular level when an ionic solid dissolves. Ions leave the solid and become dispersed in the solvent. Some dissolved ions collide occasionally with the undissolved solute and recrystallize. As long as the rate of dissolving is greater than the rate of recrystallizing, the concentration of ions rises. Eventually, ions from the solid are dissolving at the same rate as ions in the solution are recrystallizing (Figure 13.8). At this point, even though dissolving and recrystallizing continue, there is no further change in the concentration with time. The system has reached equilibrium; that is, *excess undissolved solute is in equilibrium with the dissolved solute:*

$$\text{Solute (undissolved)} \rightleftharpoons \text{solute (dissolved)}$$

Figure 13.8 **Equilibrium in a saturated solution.** In a saturated solution, equilibrium exists between excess solid solute and dissolved solute. At a particular temperature, the number of solute particles dissolving per unit time equals the number recrystallizing.

Seed crystal

A

B

C

Figure 13.9 Sodium acetate crystallizing from a supersaturated solution. When a seed crystal of sodium acetate is added to a supersaturated solution of the compound (**A**), solute begins to crystallize out of solution (**B**) and continues to do so until the solution remaining above is saturated (**C**).

A Saturated Solution Is Like a Pure Liquid and Its Vapor Equilibrium processes in a saturated solution are analogous to those for a pure liquid and its vapor in a closed flask (Section 12.2). For the liquid, rates of vaporizing and condensing are equal; in the solution, rates of dissolving and recrystallizing are equal. In the liquid-vapor system, particles leave the liquid to enter the vapor, and their concentration (pressure) increases until, at equilibrium, the available space above the liquid is "saturated" with vapor at a given temperature. In the solution, particles leave the solid solute to enter the solvent, and their concentration increases until, at equilibrium, the available solvent is saturated with solute at a given temperature.

This solution is called **saturated:** it contains the maximum amount of dissolved solute at a given temperature in the presence of undissolved solute. Filter off the saturated solution and add more solute to it, and the added solute will not dissolve. A solution that contains less than this amount of dissolved solute is called **unsaturated:** add more solute, and more will dissolve until the solution becomes saturated.

In some cases, a solution may be prepared that contains more than the equilibrium amount of dissolved solute. Such a solution is called **supersaturated.** It is unstable relative to the saturated solution, which means that if you add a "seed" crystal of solute, or just tap the container, the excess solute crystallizes immediately, leaving a saturated solution (Figure 13.9). You can often prepare a supersaturated solution of a solute that has greater solubility at a higher temperature. While warming the contents of the flask, you dissolve more than the amount of solute required to prepare a saturated solution at some lower temperature and then slowly cool the solution. If the excess solute remains dissolved, the solution is supersaturated.

Effect of Temperature on Solubility

Temperature affects the solubility of most substances. You may have noticed, for example, that not only does sugar dissolve more quickly in hot tea than in iced tea, but *more* sugar dissolves; in other words, the solubility of sugar in tea is greater at higher temperatures. Let's examine the effects of temperature on the solubility of solids and of gases.

Temperature and the Solubility of Solids Like sugar, *most solids are more soluble at higher temperatures.* Figure 13.10 shows several examples of ionic compounds in water. Note that most have upward-curving lines. Cerium sulfate is the only exception shown, but several other salts, mostly sulfates, behave similarly to it. Some salts exhibit increasing solubility up to a certain temperature and then decreasing solubility at still higher temperatures.

We might think that the sign of ΔH_{soln} would indicate the effect of temperature. Most ionic solids have a positive ΔH_{soln} because for most the lattice energy is greater than the heat of hydration. Thus, heat is *absorbed* to form the solution from solute and solvent, and if we think of heat as a reactant, a rise in temperature should increase the rate of the forward process:

$$\text{Solute} + \text{solvent} + heat \rightleftharpoons \text{saturated solution}$$

Tabulated ΔH_{soln} values refer to the enthalpy change for a solution to reach the standard state of 1 M, but in order to understand the effect of temperature, we need the sign of the enthalpy change very close to the point of saturation, and

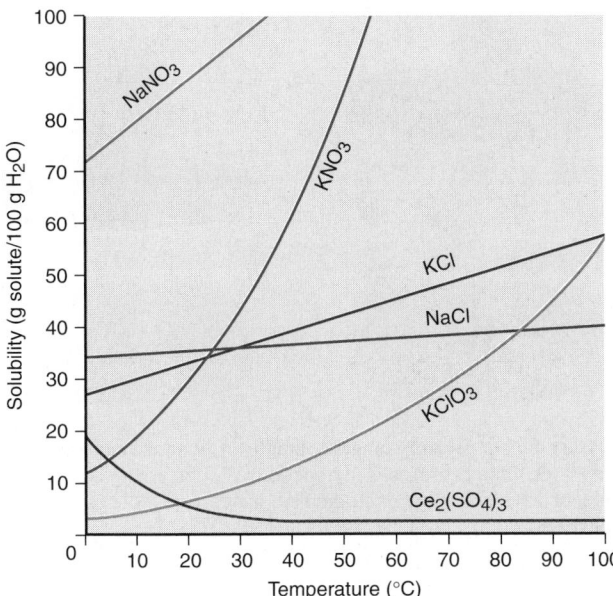

Figure 13.10 The relation between solubility and temperature for several ionic compounds. Most ionic compounds have higher solubilities at higher temperatures. Cerium sulfate is one of several exceptions.

this sign may not be the same. To use earlier examples, tabulated values give a negative ΔH_{soln} for NaOH and a positive one for NH_4NO_3, yet both compounds are more soluble at higher temperatures. The point is that solubility is a complex behavior, and, although the effect of temperature ultimately reflects the equilibrium nature of solubility, no single measure can help us predict it for a given solute.

Temperature and Gas Solubility in Water The effect of temperature on gas solubility is much more predictable. When a solid dissolves in a liquid, the solute particles must separate, so energy must be added; thus, for a solid, $\Delta H_{solute} > 0$. In contrast, gas particles are already separated, so $\Delta H_{solute} \approx 0$. Because the hydration step is exothermic ($\Delta H_{hydr} < 0$), the sum of these two terms must be negative. Thus, for all gases in water, $\Delta H_{soln} < 0$:

$$\text{Solute}(g) + \text{water}(l) \rightleftharpoons \text{saturated solution}(aq) + \text{heat}$$

This equation means that *gas solubility in water **decreases** with rising temperature*. Gases have weak intermolecular forces, so there are relatively weak intermolecular forces between a gas and water. When the temperature rises, the average kinetic energy of the particles in solution increases, allowing the gas particles to easily overcome these weak forces and re-enter the gas phase.

This behavior can lead to an environmental problem known as *thermal pollution*. During many industrial processes, large amounts of water are taken from a nearby river or lake, pumped through the system to cool liquids, gases, and equipment, and then returned to the body of water at a higher temperature. The metabolic rates of fish and other aquatic animals increase in the warmer water released near the plant outlet; thus, their need for O_2 increases, but the concentration of dissolved O_2 is lower at the higher temperature. This oxygen depletion can harm these aquatic populations. Moreover, the warmer water is less dense, so it floats on the cooler water below and prevents O_2 from reaching it. Thus, creatures living at lower levels become oxygen deprived as well. Farther from the plant, the water temperature returns to ambient levels, the O_2 solubility increases, and the temperature layering disappears. One way to lessen the problem is with cooling towers, which cool the water before it exits the plant (Figure 13.11).

Figure 13.11 Fighting thermal pollution. Steam billowing from cooling towers is common at large electric power plants. Hot water resulting from the cooling of gases and equipment enters near the top of the tower and cools in contact with air. The cooler water is then released back into the water source.

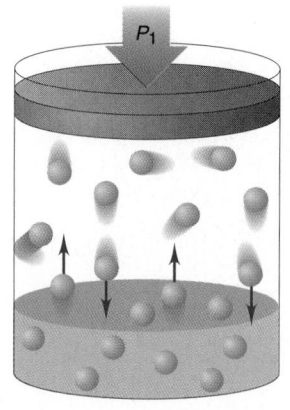
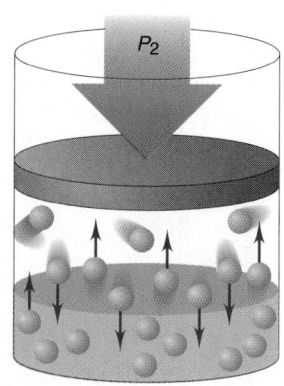

Figure 13.12 The effect of pressure on gas solubility. **A,** A saturated solution of a gas is in equilibrium at pressure P_1. **B,** If the pressure is increased to P_2, the volume of the gas decreases. As a result, the frequency of collisions with the surface increases, and more gas is in solution when equilibrium is re-established.

Scuba Diving and Soda Pop Although N_2 is barely soluble in water or blood at sea level, high external pressures increase its solubility significantly. Scuba divers who breathe compressed air and dive below about 50 ft have much more N_2 dissolved in their blood. If they ascend quickly to lower pressures, they may be afflicted with decompression sickness ("the bends"), as the dissolved N_2 bubbles out of their blood and causes painful, sometimes fatal, blockage of capillaries. This condition can be avoided by using less soluble gases, such as He, in the breathing mixtures. In principle, the same thing happens when you open a can of a carbonated drink. In the closed can, dissolved CO_2 is at equilibrium with 4 atm of CO_2 gas in the space above it. In the open can, the dissolved CO_2 is under 1 atm of air ($P_{CO_2} = 3\times10^{-4}$ atm), so CO_2 bubbles out of solution. With time, the system reaches equilibrium with the P_{CO_2} in air: the drink goes "flat."

Effect of Pressure on Solubility

Because liquids and solids are almost incompressible, pressure has little effect on their solubility, but it has a major effect on gas solubility. Consider the piston-cylinder assembly in Figure 13.12, with a gas above a saturated aqueous solution of the gas. At a given pressure, the same number of gas molecules enter and leave the solution per unit time; that is, the system is at equilibrium:

$$\text{Gas} + \text{solvent} \rightleftharpoons \text{saturated solution}$$

If you push down on the piston, the gas volume decreases, its pressure increases, and gas particles collide with the liquid surface more often. Therefore, more gas particles enter than leave the solution per unit time. Higher gas pressure disturbs the balance at equilibrium, so more gas dissolves to reduce this disturbance (the previous equation shifts to the right) until the system re-establishes equilibrium.

Henry's law expresses the quantitative relationship between gas pressure and solubility: *the solubility of a gas (S_{gas}) is directly proportional to the partial pressure of the gas (P_{gas}) above the solution:*

$$S_{gas} = k_H \times P_{gas} \qquad \text{(13.3)}$$

where k_H is the Henry's law constant and is specific for a given gas-solvent combination at a given temperature. With S_{gas} in mol/L and P_{gas} in atm, the units of k_H are mol/L·atm.

SAMPLE PROBLEM 13.2 Using Henry's Law to Calculate Gas Solubility

Problem The partial pressure of carbon dioxide gas inside a bottle of cola is 4 atm at 25°C. What is the solubility of CO_2? The Henry's law constant for CO_2 dissolved in water is 3.3×10^{-2} mol/L·atm at 25°C.

Plan We know P_{CO_2} (4 atm) and the value of k_H (3.3×10^{-2} mol/L·atm), so we substitute them into Equation 13.3 to find S_{CO_2}.

Solution $S_{CO_2} = k_H \times P_{CO_2} = (3.3\times10^{-2}$ mol/L·atm$)(4$ atm$) = $ 0.1 mol/L

Check The units are correct for solubility. We rounded the answer to one significant figure because there is only one in the pressure value.

FOLLOW-UP PROBLEM 13.2 If air contains 78% N_2 by volume, what is the solubility of N_2 in water at 25°C and 1 atm (k_H for N_2 in H_2O at 25°C = 7×10^{-4} mol/L·atm)?

A solution that contains the maximum amount of dissolved solute in the presence of excess solute is saturated. A state of equilibrium exists when a saturated solution is in contact with excess solute, because solute particles are entering and leaving the solution at the same rate. Most solids are more soluble at higher temperatures. All gases have a negative ΔH_{soln} in water, so heating lowers gas solubility in water. Henry's law says that the solubility of a gas is directly proportional to its partial pressure above the solution.

13.4 QUANTITATIVE WAYS OF EXPRESSING CONCENTRATION

Concentration is the *proportion* of a substance in a mixture, so it is an intensive property, one that does not depend on the quantity of mixture present: 1.0 L of 0.1 M NaCl has the same concentration as 1.0 mL of 0.1 M NaCl. Concentration is most often expressed as the ratio of the quantity of solute to the quantity of *solution,* but sometimes it is the ratio of solute to *solvent.* Because both parts of the ratio can be given in terms of mass, volume, or amount (mol), chemists employ several concentration terms, including molarity, molality, and various expressions of "parts of solute per part of solution" (Table 13.5).

Table 13.5 Concentration Definitions

Concentration Term	Ratio
Molarity (M)	$\dfrac{\text{amount (mol) of solute}}{\text{volume (L) of solution}}$
Molality (m)	$\dfrac{\text{amount (mol) of solute}}{\text{mass (kg) of solvent}}$
Parts by mass	$\dfrac{\text{mass of solute}}{\text{mass of solution}}$
Parts by volume	$\dfrac{\text{volume of solute}}{\text{volume of solution}}$
Mole fraction (X)	$\dfrac{\text{amount (mol) of solute}}{\text{amount (mol) of solute} + \text{amount (mol) of solvent}}$

Molarity and Molality

Molarity (M) is the *number of moles of solute dissolved in 1 L of solution:*

$$\text{Molarity } (M) = \frac{\text{amount (mol) of solute}}{\text{volume (L) of solution}} \tag{13.4}$$

In Chapter 3, you used molarity to convert liters of solution into moles of dissolved solute. Expressing concentration in terms of molarity may have drawbacks, however. Because volume is affected by temperature, so is molarity. A solution expands when heated, so a unit volume of hot solution contains slightly less solute than a unit volume of cold solution. This can be a source of error in very precise work. More importantly, because of solute-solvent interactions that are difficult to predict, *solution volumes may not be additive;* that is, adding 500. mL of one solution to 500. mL of another may not give 1000. mL. Therefore, in precise work, a solution with a desired molarity may not be easy to prepare.

A concentration term that does not contain volume in its ratio is **molality (*m*),** *the number of moles of solute dissolved in 1000 g (1 kg) of solvent:*

$$\text{Molality } (m) = \frac{\text{amount (mol) of solute}}{\text{mass (kg) of solvent}} \quad \text{(13.5)}$$

Note that molality is expressed in terms of the quantity of *solvent,* not solution. Molal solutions are prepared by measuring *masses* of solute and solvent, not solvent or solution volume. Mass does not change with temperature, so neither does molality. Moreover, unlike volumes, masses *are* additive: adding 500. g of one solution to 500. g of another *does* give 1000. g of final solution. For these reasons, molality is a preferred unit when temperature, and hence density, may change, as in the examination of solutions' physical properties. It's important to realize that, because 1 L of water has a mass of 1 kg, *molality and molarity are nearly the same for dilute aqueous solutions.*

SAMPLE PROBLEM 13.3 Calculating Molality

Problem What is the molality of a solution prepared by dissolving 32.0 g of $CaCl_2$ in 271 g of water?
Plan To use Equation 13.5, we convert mass of $CaCl_2$ (32.0 g) to amount (mol) with the molar mass (g/mol) and then divide by the mass of water (271 g), being sure to convert from grams to kilograms.
Solution Converting from grams of solute to moles:

$$\text{Moles of } CaCl_2 = 32.0 \text{ g } CaCl_2 \times \frac{1 \text{ mol } CaCl_2}{110.98 \text{ g } CaCl_2} = 0.288 \text{ mol } CaCl_2$$

Finding molality:

$$\text{Molality} = \frac{\text{mol solute}}{\text{kg solvent}} = \frac{0.288 \text{ mol } CaCl_2}{271 \text{ g} \times \dfrac{1 \text{ kg}}{10^3 \text{ g}}} = \boxed{1.06 \ m \ CaCl_2}$$

Check The answer seems reasonable: the given moles of $CaCl_2$ and kilograms of H_2O are about the same numerically, so their ratio is about 1.

FOLLOW-UP PROBLEM 13.3 How many grams of glucose ($C_6H_{12}O_6$) must be dissolved in 563 g of ethanol (C_2H_5OH) to prepare a $2.40 \times 10^{-2} \ m$ solution?

Mass (g) of $CaCl_2$

divide by $\mathcal{M}$ (g/mol)

Amount (mol) of $CaCl_2$

divide by kg of water

Molality (*m*) of $CaCl_2$ solution

Parts of Solute by Parts of Solution

Several concentration terms are based on the number of solute (or solvent) parts present in a specific number of *solution* parts. The solution parts can be expressed in terms of mass, volume, or amount (mol).

Parts by Mass The most common of the parts-by-mass terms is **mass percent,** which you encountered in Chapter 3. The word *percent* means "per hundred," so mass percent of solute means the mass of solute dissolved in every 100. parts by mass of solution, or the mass fraction times 100:

$$\text{Mass percent} = \frac{\text{mass of solute}}{\text{mass of solute} + \text{mass of solvent}} \times 100$$
$$= \frac{\text{mass of solute}}{\text{mass of solution}} \times 100 \quad \text{(13.6)}$$

Sometimes mass percent is written as **% (w/w),** the "w/w" indicating that the percent is a ratio of weights (more accurately, masses). You may have seen mass percent values on bottles of solid chemicals to indicate the amounts of impurities present. Two very similar units are parts per million (ppm) by mass and parts per billion (ppb) by mass: grams of solute per million or per billion grams of solution. (In Equation 13.6, you multiply by 10^6 or by 10^9, instead of by 100.)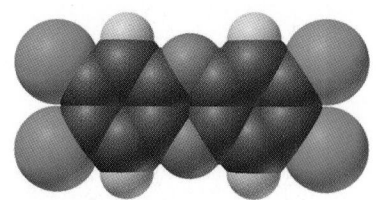

Parts by Volume The most common of the parts-by-volume terms is **volume percent,** the volume of solute in 100. volumes of solution:

$$\text{Volume percent} = \frac{\text{volume of solute}}{\text{volume of solution}} \times 100 \qquad \textbf{(13.7)}$$

A common symbol for this term is **% (v/v).** Commercial rubbing alcohol, for example, is an aqueous solution of isopropyl alcohol (a three-carbon alcohol) that contains 70 volumes of isopropyl alcohol per 100. volumes of solution, and the label indicates this as "70% (v/v)." Parts-by-volume concentrations are most often used for liquids and gases. Minor atmospheric components occur in parts per million by volume (ppmv). For example, there are about 0.05 ppmv (50 ppbv) of the toxic gas carbon monoxide (CO) in clean air, 1000 times as much (about 50 ppmv of CO) in air over urban traffic, and 10,000 times as much (about 500 ppmv of CO) in the smoke of one cigarette. Some extremely tiny concentrations are found in nature. *Pheromones* are organic compounds secreted by one member of a species to signal other members about food, danger, sexual readiness, and so forth. Many organisms, including dogs and monkeys, release pheromones, and researchers suspect that humans do as well. Most studies of pheromones focus on social insects. Some insect pheromones are active at only a few hundred molecules per milliliter of air, about 100 parts per quadrillion by volume (Figure 13.13).

A symbol frequently used for aqueous solutions is **% (w/v),** a ratio of solute *weight* (actually mass) to solution *volume.* Thus, a 1.5% (w/v) NaCl solution contains 1.5 g of NaCl per 100. mL of *solution.* This way of expressing concentrations is particularly common in medical labs and other health-related facilities.

Mole Fraction The **mole fraction (*X*)** of a solute is the ratio of number of solute moles to the total number of moles (solute plus solvent), that is, parts by mole. The *mole percent* is the mole fraction expressed as a percentage:

$$\text{Mole fraction } (X) = \frac{\text{amount (mol) of solute}}{\text{amount (mol) of solute} + \text{amount (mol) of solvent}} \qquad \textbf{(13.8)}$$

$$\text{Mole percent (mol \%)} = \text{mole fraction} \times 100$$

We discussed these terms in Chapter 5 in relation to Dalton's law of partial pressures for mixtures of gases, but they apply to liquids and solids as well. Concentrations given as mole fractions provide the clearest picture of the actual proportion of solute (or solvent) particles among all the particles in the solution.

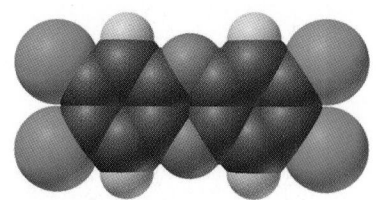**Unhealthy Ultralow Concentrations** Environmental toxicologists often measure extremely low concentrations of pollutants. The Centers for Disease Control and Prevention consider TCDD (*t*etra*chlorod*ibenzo*di*oxin, *above*) one of the most toxic substances known. It is a member of the dioxin family of chlorinated hydrocarbons, a by-product of bleaching paper. TCDD is considered unsafe at soil levels above 1 ppb. From contact with air, water, and soil, most North Americans have an average of 0.01 ppb TCDD in their fatty tissue. The Environmental Protection Agency recently assessed TCDD as unsafe at *all* levels measured.

 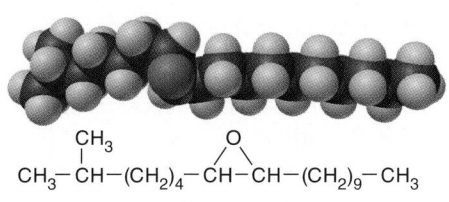

CH₃−CH−(CH₂)₄−CH−CH−(CH₂)₉−CH₃

Figure 13.13 The sex attractant of the gypsy moth. The pheromone secreted by the female gypsy moth (called *disparlure*) can attract a male over 0.5 mi away. Chemists are studying ways to synthesize pheromones as a means of insect pest control.

SAMPLE PROBLEM 13.4 Expressing Concentration in Parts by Mass, Parts by Volume, and Mole Fraction

Problem (a) Find the concentration of calcium (in ppm) in a 3.50-g pill that contains 40.5 mg of Ca.

(b) The label on a 0.750-L bottle of Italian chianti indicates "11.5% alcohol by volume." How many liters of alcohol does the wine contain?

(c) A sample of rubbing alcohol contains 142 g of isopropyl alcohol (C_3H_7OH) and 58.0 g of water. What are the mole fractions of alcohol and water?

Plan (a) We are given the masses of Ca (40.5 mg) and the pill (3.50 g). We convert the mass of Ca from mg to g, find the ratio of mass of Ca to mass of pill, and multiply by 10^6 to obtain ppm. (b) We know the volume % (11.5%, or 11.5 parts by volume of alcohol to 100 parts of chianti) and the total volume (0.750 mL), so we use Equation 13.7 to find the volume of alcohol. (c) We know the mass and formula of each component, so we convert them to amount (mol) and apply Equation 13.8 to find mole fraction.

Solution (a) Finding parts per million by mass of Ca. Combining the steps, we have

$$\text{ppm Ca} = \frac{\text{mass of Ca}}{\text{mass of pill}} \times 10^6 = \frac{40.5 \text{ mg Ca} \times \dfrac{1 \text{ g}}{10^3 \text{ mg}}}{3.50 \text{ g}} \times 10^6 = \boxed{1.16 \times 10^4 \text{ ppm Ca}}$$

(b) Finding volume (L) of alcohol:

$$\text{Volume (L) of alcohol} = 0.750 \text{ L chianti} \times \frac{11.5 \text{ L alcohol}}{100. \text{ L chianti}} = \boxed{0.0862 \text{ L}}$$

(c) Finding mole fractions. Converting from grams to moles:

$$\text{Moles of } C_3H_7OH = 142 \text{ g } C_3H_7OH \times \frac{1 \text{ mol } C_3H_7OH}{60.09 \text{ g } C_3H_7OH} = 2.36 \text{ mol } C_3H_7OH$$

$$\text{Moles of } H_2O = 58.0 \text{ g } H_2O \times \frac{1 \text{ mol } H_2O}{18.02 \text{ g } H_2O} = 3.22 \text{ mol } H_2O$$

Calculating mole fractions:

$$X_{C_3H_7OH} = \frac{\text{moles of } C_3H_7OH}{\text{total moles}} = \frac{2.36 \text{ mol}}{2.36 \text{ mol} + 3.22 \text{ mol}} = \boxed{0.423}$$

$$X_{H_2O} = \frac{\text{moles of } H_2O}{\text{total moles}} = \frac{3.22 \text{ mol}}{2.36 \text{ mol} + 3.22 \text{ mol}} = \boxed{0.577}$$

Check (a) The mass ratio is about 0.04 g/4 g $= 10^{-2}$, and $10^{-2} \times 10^6 = 10^4$ ppm, so it seems correct. (b) The volume % is a bit more than 10%, so the volume of alcohol should be a bit more than 75 mL (0.075 L). (c) Always check that the *mole fractions add up to one:* 0.423 + 0.577 = 1.000.

FOLLOW-UP PROBLEM 13.4 An alcohol solution contains 35.0 g of propanol (C_3H_7OH) and 150. g of ethanol (C_2H_5OH). Calculate the mass percent and the mole fraction of each alcohol.

Converting Units of Concentration

All the units we just discussed represent different ways of expressing concentration, so they are interconvertible. Keep these points in mind:

- To convert a unit based on amount (mol) to one based on mass, you need the molar mass. These conversions are similar to mass-mole conversions you've done earlier.
- To convert a unit based on mass to one based on volume, you need the solution *density*. Given the mass of a solution, the density (mass/volume) gives you the volume, or vice versa.
- Molality involves quantity of *solvent,* whereas the others involve quantity of *solution.*

SAMPLE PROBLEM 13.5 Converting Concentration Units

Problem Hydrogen peroxide is a powerful oxidizing agent used in concentrated solution in rocket fuels and in dilute solution as a hair bleach. An aqueous solution of H_2O_2 is 30.0% by mass and has a density of 1.11 g/mL. Calculate its
(a) Molality   **(b)** Mole fraction of H_2O_2 **(c)** Molarity
Plan We know the mass % (30.0) and the density (1.11 g/mL). **(a)** For molality, we need the amount (mol) of solute and the mass (kg) of *solvent*. Assuming 100.0 g of H_2O_2 solution allows us to express the mass % directly as grams of substance. We subtract the grams of H_2O_2 to obtain the grams of solvent. To find molality, we convert grams of H_2O_2 to moles and divide by mass of solvent (converting g to kg). **(b)** To find the mole fraction, we use the number of moles of H_2O_2 (from part a) and convert the grams of H_2O to moles. Then we divide the moles of H_2O_2 by the total moles. **(c)** To find molarity, we assume 100.0 g of solution and use the given solution density to find the volume. Then we divide the amount (mol) of H_2O_2 (from part a) by *solution* volume (in L).
Solution (a) From mass % to molality. Finding mass of solvent (assuming 100.0 g of solution):

$$\text{Mass (g) of } H_2O = 100.0 \text{ g solution} - 30.0 \text{ g } H_2O_2 = 70.0 \text{ g } H_2O$$

Converting from grams of H_2O_2 to moles:

$$\text{Moles of } H_2O_2 = 30.0 \text{ g } H_2O_2 \times \frac{1 \text{ mol } H_2O_2}{34.02 \text{ g } H_2O_2} = 0.882 \text{ mol } H_2O_2$$

Calculating molality:

$$\text{Molality of } H_2O_2 = \frac{0.882 \text{ mol } H_2O_2}{70.0 \text{ g } \times \frac{1 \text{ kg}}{10^3 \text{ g}}} = 12.6 \text{ } m \text{ } H_2O_2$$

(b) From mass % to mole fraction:

$$\text{Moles of } H_2O_2 = 0.882 \text{ mol } H_2O_2 \quad [\text{from part a}]$$

$$\text{Moles of } H_2O = 70.0 \text{ g } H_2O \times \frac{1 \text{ mol } H_2O}{18.02 \text{ g } H_2O} = 3.88 \text{ mol } H_2O$$

$$X_{H_2O_2} = \frac{0.882 \text{ mol}}{0.882 \text{ mol} + 3.88 \text{ mol}} = 0.185$$

(c) From mass % and density to molarity. Converting from solution mass to volume:

$$\text{Volume (mL) of solution} = 100.0 \text{ g} \times \frac{1 \text{ mL}}{1.11 \text{ g}} = 90.1 \text{ mL}$$

Calculating molarity:

$$\text{Molarity} = \frac{\text{mol } H_2O_2}{\text{L soln}} = \frac{0.882 \text{ mol } H_2O_2}{90.1 \text{ mL} \times \frac{1 \text{ L soln}}{10^3 \text{ mL}}} = 9.79 \text{ } M \text{ } H_2O_2$$

Check Rounding shows the answers seem reasonable: **(a)** The ratio of ~0.9 mol/0.07 kg is greater than 10. **(b)** ~0.9 mol H_2O_2/(1 mol + 4 mol) ≈ 0.2. **(c)** The ratio of moles to liters (0.9/0.09) is around 10.

FOLLOW-UP PROBLEM 13.5 A sample of commercial concentrated hydrochloric acid is 11.8 *M* HCl and has a density of 1.190 g/mL. Calculate the mass %, molality, and mole fraction of HCl.

SECTION SUMMARY

The concentration of a solution is independent of the amount of solution and can be expressed by molarity (mol solute/L solution) and molality (mol solute/kg solvent), as well as parts by mass (mass solute/mass solution), parts by volume (volume solute/volume solution), and mole fraction [mol solute/(mol solute + mol solvent)]. The choice of units depends on convenience or the nature of the solution. If, in addition to the quantities of solute and solution, the solution density is also known, all concentration units are interconvertible.

13.5 COLLIGATIVE PROPERTIES OF SOLUTIONS

We might expect the presence of solute particles to make the physical properties of a solution different from those of the pure solvent. However, what we might *not* expect is that, in the case of four important solution properties, the *number* of solute particles makes the difference, not their chemical identity. These properties, known as **colligative properties** (*colligative* means "collective"), are vapor pressure lowering, boiling point elevation, freezing point depression, and osmotic pressure. Even though most of these effects are small, they have many practical applications, including some that are vital to biological systems.

Historically, colligative properties were measured to explore the nature of a solute in aqueous solution and its extent of dissociation into ions. In Chapter 4, we classified solutes by their ability to conduct an electric current, which requires moving ions to be present. Recall that an aqueous solution of an **electrolyte** conducts a current because the solute separates into ions as it dissolves. Soluble salts, strong acids, and strong bases dissociate completely. Their solutions conduct a large current, so these solutes are *strong electrolytes*. Weak acids and bases dissociate very little. They are *weak electrolytes* because their solutions conduct little current. Many compounds, such as sugar and alcohol, do not dissociate into ions at all. They are **nonelectrolytes** because their solutions do not conduct a current. Figure 13.14 shows these behaviors.

Figure 13.14 The three types of electrolytes. A, Strong electrolytes conduct a large current because they dissociate completely into ions. **B,** Weak electrolytes conduct a small current because they dissociate very little. **C,** Nonelectrolytes do not conduct a current because they do not dissociate.

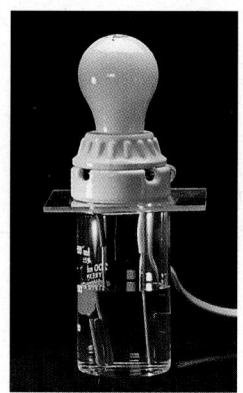

A   B   C

To predict the magnitude of a colligative property, we refer to the solute formula to find the number of particles in solution. Each mole of nonelectrolyte yields 1 mol of particles in the solution. For example, 0.35 M glucose contains 0.35 mol of solute particles per liter. In principle, each mole of strong electrolyte dissociates into the number of moles of ions in the formula unit: 0.4 M Na_2SO_4 contains 0.8 mol of Na^+ ions and 0.4 mol of SO_4^{2-} ions, or 1.2 mol of particles, per liter (see Sample Problem 4.1, p. 133). (We examine the equilibrium nature of the dissociation of weak electrolytes in Chapters 18 and 19.)

Colligative Properties of Nonvolatile Nonelectrolyte Solutions

In this section, we focus most of our attention on the simplest case, the colligative properties of solutes that do not dissociate into ions and have negligible vapor pressure even at the boiling point of the solvent. Such solutes are called *nonvolatile nonelectrolytes;* sucrose (table sugar) is an example. Later, we briefly explore the properties of volatile nonelectrolytes and of strong electrolytes.

Vapor Pressure Lowering The vapor pressure of a solution of a nonvolatile non-electrolyte is always *lower* than the vapor pressure of the pure solvent. We can understand this **vapor pressure lowering (ΔP)** if we recall that natural processes tend to occur in a direction that increases disorder. A pure solvent vaporizes because particles moving randomly in a vapor phase are more disordered (have higher entropy) than when they are lying near each other in the liquid. However, a solution is already more disordered than a pure liquid, so the solvent has less tendency to vaporize in order to attain the same degree of disorder. Thus, equilibrium is reached—the rates of vaporization and condensation become equal—at a lower vapor pressure for the solution. The molecular views in Figure 13.15 illustrate this point.

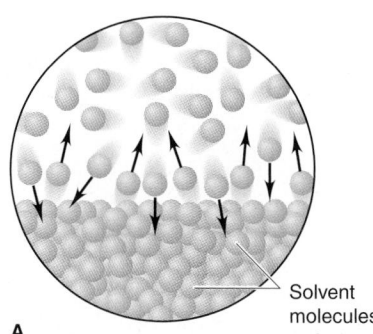

Solvent molecules

A

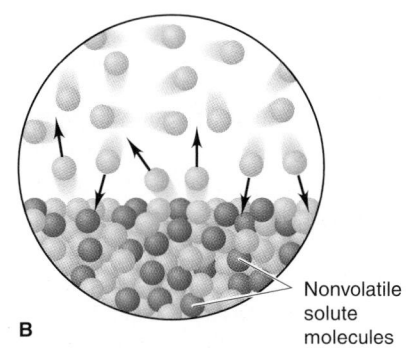

Nonvolatile solute molecules

B

Figure 13.15 **The effect of the solute on the vapor pressure of a solution.** Vaporization occurs because of the tendency of a system to become more disordered (increase its entropy). **A,** Equilibrium is established between a pure liquid and its vapor when the numbers of molecules vaporizing and condensing in a given time are equal. **B,** The presence of a dissolved solute already increases the disorder of the system, so fewer solvent molecules vaporize in a given time. Therefore, fewer molecules need to condense to balance them, so equilibrium is established at a lower vapor pressure.

In quantitative terms, we find that the vapor pressure of solvent above the solution ($P_{solvent}$) equals the mole fraction of solvent in the solution ($X_{solvent}$) times the vapor pressure of the pure solvent ($P^0_{solvent}$). This relationship is expressed by **Raoult's law:**

$$P_{solvent} = X_{solvent} \times P^0_{solvent} \qquad \textbf{(13.9)}$$

In a solution, $X_{solvent}$ is always less than 1, so $P_{solvent}$ is always less than $P^0_{solvent}$. An **ideal solution** is one that follows Raoult's law at any concentration. However, just as most gases deviate from ideality, so do most solutions. In practice, Raoult's law gives a good approximation of the behavior of *dilute* solutions only and is exact at infinite dilution.

How does the *amount* of solute affect the *magnitude* of the vapor pressure lowering, ΔP? The solution consists of solvent and solute, so the sum of their mole fractions equals 1:

$$X_{solvent} + X_{solute} = 1; \quad \text{thus,} \quad X_{solvent} = 1 - X_{solute}$$

From Raoult's law, we have

$$P_{solvent} = X_{solvent} \times P^0_{solvent} = (1 - X_{solute}) \times P^0_{solvent}$$

Multiplying through on the right side gives

$$P_{solvent} = P^0_{solvent} - (X_{solute} \times P^0_{solvent})$$

Rearranging and introducing ΔP gives

$$P^0_{solvent} - P_{solvent} = \Delta P = X_{solute} \times P^0_{solvent} \qquad \textbf{(13.10)}$$

Thus, the *magnitude* of ΔP equals the mole fraction of solute times the vapor pressure of the pure solvent. This relationship is applied in the next Sample Problem.

SAMPLE PROBLEM 13.6 Using Raoult's Law to Find the Vapor Pressure Lowering

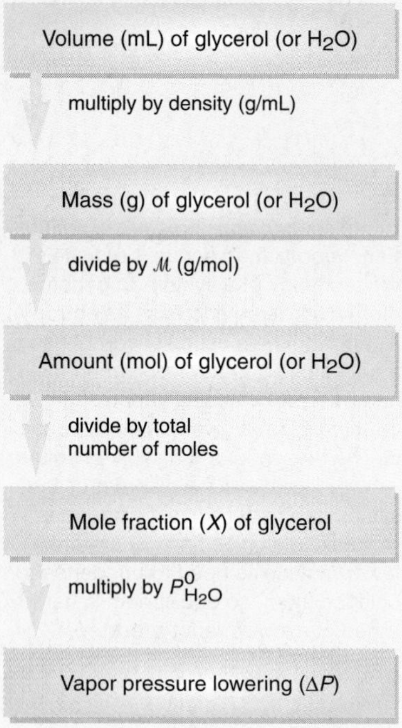

Problem Calculate the vapor pressure lowering, ΔP, when 10.0 mL of glycerol ($C_3H_8O_3$) is added to 500. mL of water at 50.°C. At this temperature, the vapor pressure of pure water is 92.5 torr and its density is 0.988 g/mL. The density of glycerol is 1.26 g/mL.

Plan To calculate ΔP, we use Equation 13.10. We are given the vapor pressure of pure water ($P^0_{H_2O} = 92.5$ torr), so we just need the mole fraction of glycerol, $X_{glycerol}$. We convert the given volume of glycerol (10.0 mL) to mass using the given density (1.26 g/L), find the molar mass from the formula, and convert mass (g) to amount (mol). The same procedure gives amount of H_2O. From these, we find $X_{glycerol}$ and ΔP.

Solution Calculating the amount (mol) of glycerol and of water:

$$\text{Moles of glycerol} = 10.0 \text{ mL glycerol} \times \frac{1.26 \text{ g glycerol}}{1 \text{ mL glycerol}} \times \frac{1 \text{ mol glycerol}}{92.09 \text{ g glycerol}}$$

$$= 0.137 \text{ mol glycerol}$$

$$\text{Moles of } H_2O = 500. \text{ mL } H_2O \times \frac{0.988 \text{ g } H_2O}{1 \text{ mL } H_2O} \times \frac{1 \text{ mol } H_2O}{18.02 \text{ g } H_2O} = 27.4 \text{ mol } H_2O$$

Calculating the mole fraction of glycerol:

$$X_{glycerol} = \frac{0.137 \text{ mol}}{0.137 \text{ mol} + 27.4 \text{ mol}} = 0.00498$$

Finding the vapor pressure lowering:

$$\Delta P = X_{glycerol} \times P^0_{H_2O} = 0.00498 \times 92.5 \text{ torr} = \boxed{0.461 \text{ torr}}$$

Check The amounts of each component seem correct: for glycerol, $\sim$10 mL $\times$ 1.25 g/mL $\div$ 100 g/mol = 0.125 mol; for H_2O, $\sim$500 mL $\times$ 1 g/mL $\div$ 20 g/mol = 25 mol. The small ΔP is reasonable because the mole fraction of solute is small.

Comment The calculation assumes that glycerol is nonvolatile. At 1 atm, glycerol boils at 290.0°C, so the vapor pressure of glycerol at 50°C is so low it can be neglected.

FOLLOW-UP PROBLEM 13.6 Calculate the vapor pressure lowering of a solution of 2.00 g of aspirin ($M = 180.15$ g/mol) in 50.0 g of methanol (CH_3OH) at 21.2°C. Pure methanol has a vapor pressure of 101 torr at this temperature.

Sidebar flowchart (left margin):

Volume (mL) of glycerol (or H_2O)

multiply by density (g/mL)

Mass (g) of glycerol (or H_2O)

divide by M (g/mol)

Amount (mol) of glycerol (or H_2O)

divide by total number of moles

Mole fraction (X) of glycerol

multiply by $P^0_{H_2O}$

Vapor pressure lowering (ΔP)

Boiling Point Elevation *A solution boils at a higher temperature than the pure solvent.* Let's see why. The boiling point (boiling temperature, T_b) of a liquid is the temperature at which its vapor pressure equals the external pressure. The vapor pressure of a solution is lower than the external pressure at the solvent's boiling point because the vapor pressure of a solution is lower than that of the pure solvent at any temperature. Therefore, the solution does not yet boil. A higher temperature is needed to raise the solution's vapor pressure to equal the external pressure. We can see this **boiling point elevation (ΔT_b)** by superimposing a phase diagram for the solution on one for the pure solvent (Figure 13.16). Note that the gas-liquid curve for the solution lies *below* that for the pure solvent at any temperature and to the right of it at any pressure.

Like the vapor pressure lowering, the magnitude of the boiling point elevation is proportional to the concentration of solute particles:

$$\Delta T_b \propto m \quad \text{or} \quad \Delta T_b = K_b m \qquad (13.11)$$

where m is the solution molality, K_b is the *molal boiling point elevation constant,* and ΔT_b is the boiling point elevation. We typically speak of ΔT_b as a positive value, so we subtract the lower temperature from the higher; that is, we subtract the solvent T_b from the solution T_b:

$$\Delta T_b = T_{b(solution)} - T_{b(solvent)}$$

Molality is used as the concentration unit because it is directly related to mole fraction and thus to particles of solute. It also involves mass rather than volume of solvent, so it is not affected by temperature changes. The constant K_b has units

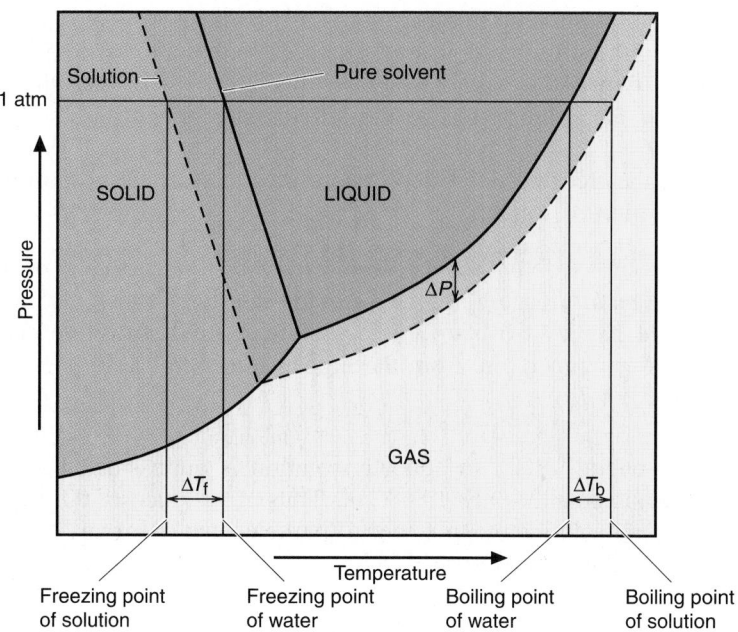

Figure 13.16 **Phase diagrams of solvent and solution.** Phase diagrams of an aqueous solution (*dashed lines*) and of pure water (*solid lines*) show that, by lowering the vapor pressure (ΔP), a dissolved solute elevates the boiling point (ΔT_b) and depresses the freezing point (ΔT_f).

of degrees Celsius per molal unit (°C/*m*) and is specific for a given solvent; Table 13.6 lists some examples.

Notice that the K_b for water is only 0.512°C/*m*, so the changes in boiling point are quite small: if you dissolved 1.00 mol of glucose (180. g; 1.00 mol of particles) in 1.00 kg of water, or 0.500 mol of NaCl (29.2 g; a strong electrolyte, so also 1.00 mol of particles) in 1.00 kg of water, the boiling points of the resulting solutions at 1 atm would be only 100.512°C instead of 100.000°C.

Freezing Point Depression As you just saw, only solvent molecules can vaporize from the solution, so molecules of the nonvolatile solute are left behind. Similarly, in many cases, *only solvent molecules can solidify,* again leaving solute molecules behind to form a slightly more concentrated solution. The freezing point of a solution is that temperature at which its vapor pressure equals that of the pure solvent. At this temperature, the two phases—solid solvent and liquid solution—are in equilibrium. Because the vapor pressure of the solution is lower than that of the solvent at any temperature, the solution freezes at a lower

Table 13.6 Molal Boiling Point Elevation and Freezing Point Depression Constants of Several Solvents

Solvent	Boiling Point (°C)*	K_b (°C/*m*)	Melting Point (°C)	K_f (°C/*m*)
Acetic acid	117.9	3.07	16.6	3.90
Benzene	80.1	2.53	5.5	4.90
Carbon disulfide	46.2	2.34	−111.5	3.83
Carbon tetrachloride	76.5	5.03	−23	30.
Chloroform	61.7	3.63	−63.5	4.70
Diethyl ether	34.5	2.02	−116.2	1.79
Ethanol	78.5	1.22	−117.3	1.99
Water	100.0	0.512	0.0	1.86

*At 1 atm.

temperature than the solvent. In other words, the number of solvent particles leaving and entering the solid per unit time become equal at a lower temperature. The **freezing point depression (ΔT_f)** is shown in Figure 13.16; note that the solid-liquid curve for the solution lies to the left of that for the pure solvent at any pressure.

Like ΔT_b, the freezing point depression has a magnitude proportional to the molal concentration of solute:

$$\Delta T_f \propto m \quad \text{or} \quad \Delta T_f = K_f m \qquad \textbf{(13.12)}$$

where K_f is the *molal freezing point depression constant,* which also has units of °C/m (see Table 13.6). Also like ΔT_b, ΔT_f is considered a positive value, so we subtract the lower temperature from the higher, but in this case it is the solution T_f from the solvent T_f:

$$\Delta T_f = T_{f(\text{solvent})} - T_{f(\text{solution})}$$

Here, too, the overall effect in aqueous solution is quite small because the K_f value for water is small—only 1.86°C/m. Thus, 1 m glucose, 0.5 m NaCl, and 0.33 m K$_2$SO$_4$, all solutions with 1 mol of particles per kilogram of water, freeze at −1.86°C at 1 atm instead of at 0.00°C.

SAMPLE PROBLEM 13.7 Determining the Boiling Point Elevation and Freezing Point Depression of a Solution

Problem You add 1.00 kg of ethylene glycol antifreeze (C$_2$H$_6$O$_2$) to your car radiator, which contains 4450 g of water. What are the boiling and freezing points of the solution?
Plan To find the boiling and freezing points of the solution, we first find the molality by converting the given mass of solute (1.00 kg) to amount (mol) and dividing by mass of solvent (4450 g). Then we calculate ΔT_b and ΔT_f from Equations 13.11 and 13.12 (using constants from Table 13.6). We add ΔT_b to the solvent boiling point and subtract ΔT_f from the solvent freezing point. The roadmap shows the steps.
Solution Calculating the molality:

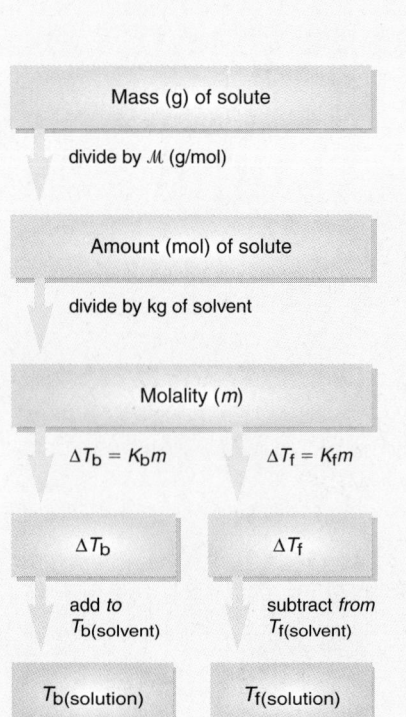

$$\text{Moles of C}_2\text{H}_6\text{O}_2 = 1.00 \text{ kg C}_2\text{H}_6\text{O}_2 \times \frac{10^3 \text{ g}}{1 \text{ kg}} \times \frac{1 \text{ mol C}_2\text{H}_6\text{O}_2}{62.07 \text{ g C}_2\text{H}_6\text{O}_2} = 16.1 \text{ mol C}_2\text{H}_6\text{O}_2$$

$$\text{Molality} = \frac{\text{mol solute}}{\text{kg solvent}} = \frac{16.1 \text{ mol C}_2\text{H}_6\text{O}_2}{4450 \text{ g H}_2\text{O} \times \frac{1 \text{ kg}}{10^3 \text{ g}}} = 3.62 \ m \text{ C}_2\text{H}_6\text{O}_2$$

Finding the boiling point elevation and $T_{b(\text{solution})}$, with $K_b = 0.512$°C/m:

$$\Delta T_b = \frac{0.512°\text{C}}{m} \times 3.62 \ m = 1.85°\text{C}$$

$$T_{b(\text{solution})} = T_{b(\text{solvent})} + \Delta T_b = 100.00°\text{C} + 1.85°\text{C} = \boxed{101.85°\text{C}}$$

Finding the freezing point depression and $T_{f(\text{solution})}$, with $K_f = 1.86$°C/m:

$$\Delta T_f = \frac{1.86°\text{C}}{m} \times 3.62 \ m = 6.73°\text{C}$$

$$T_{f(\text{solution})} = T_{f(\text{solvent})} - \Delta T_f = 0.00°\text{C} - 6.73°\text{C} = \boxed{-6.73°\text{C}}$$

Check The changes in boiling and freezing points should be in the same proportion as the constants used. That is, $\Delta T_b/\Delta T_f$ should equal K_b/K_f: 1.85/6.73 = 0.275 = 0.512/1.86.
Comment These answers are only approximate because the concentration far exceeds that of a *dilute* solution, for which Raoult's law is most useful.

FOLLOW-UP PROBLEM 13.7 What is the minimum concentration (molality) of ethylene glycol solution that will protect the car's cooling system from freezing at 0.00°F? (Assume the solution is ideal.)

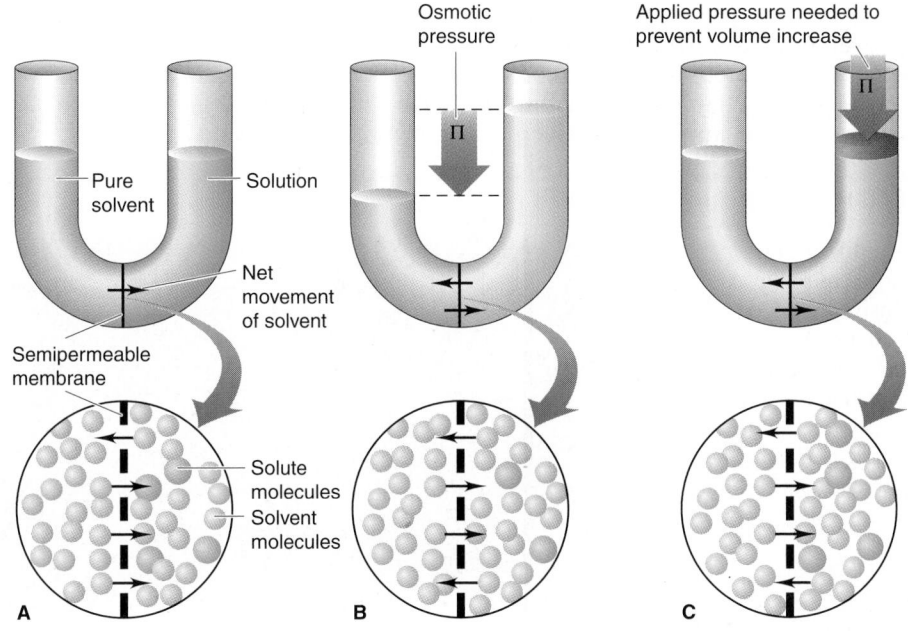

Figure 13.17 **The development of osmotic pressure. A,** In the process of osmosis, a solution and a solvent (or solutions of different concentrations) are separated by a semipermeable membrane. Pores in the membrane allow only solvent to pass through. The molecular-scale view (*below*) shows that more solvent molecules enter the solution than leave it in a given time. **B,** As a result, the solution volume increases, so its concentration decreases. The difference in heights of liquid in the two compartments at equilibrium creates a pressure called the *osmotic pressure* (Π). The greater height in the solution compartment exerts a backward pressure that eventually equalizes the flow of solvent in both directions. **C,** The osmotic pressure is also defined as the applied pressure required to *prevent* this volume change.

Osmotic Pressure The fourth colligative property applies only to aqueous solutions. It arises when two solutions of different concentrations are separated by a **semipermeable membrane,** one that allows water, but *not* solute, to pass through. This process is called **osmosis.** Many parts of organisms have semipermeable membranes that regulate internal concentrations by osmosis.

Consider a simple apparatus in which a membrane lies at the curve of a U tube and separates an aqueous sugar solution from pure water. The tiny pores in the membrane allow water molecules to pass in *either* direction, but not the larger sugar molecules. Because the solute is present, fewer water molecules touch the membrane on the solution side, so fewer water molecules leave the solution in a given time than enter it (Figure 13.17A). This *net flow of water into the solution* increases the volume of the solution and thus decreases its concentration.

As the height of the solution rises and that of the solvent falls, the resulting pressure difference pushes some water molecules *from* the solution back through the membrane. Equilibrium is reached when water molecules are pushed out of the solution at the same rate they enter it (Figure 13.17B). The pressure difference at equilibrium is the **osmotic pressure (Π),** which is defined as the applied pressure required to *prevent* the net movement of water from solvent to solution (Figure 13.17C).

The osmotic pressure is proportional to the number of solute particles in a given *volume* of solution, that is, to the molarity (*M*):

$$\Pi \propto \frac{n_{solute}}{V_{soln}} \quad \text{or} \quad \Pi \propto M$$

The proportionality constant is *R* times the absolute temperature *T*. Thus,

$$\Pi = \frac{n_{solute}}{V_{soln}} RT = MRT \qquad (13.13)$$

The similarity of Equation 13.13 to the ideal gas law ($P = nRT/V$) is not surprising, because both relate the pressure of a system to its concentration and temperature. The Gallery shows some important applications of colligative properties.

Colligative Properties in Industry and Biology

Colligative properties—especially freezing point depression and osmotic pressure—have much practical relevance to everyday life. Some common applications of freezing point depression make life safer during cold winter months. Others are essential to the electronics industry. Applications of osmotic pressure are found throughout nature and in the health and biological sciences because, without question, the most important semipermeable membranes surround living cells.

Applications of Freezing Point Depression

Plane de-icing and car antifreezing Ethylene glycol ($C_2H_6O_2$) is miscible with water through extensive hydrogen bonding and has a high enough boiling point for it to be essentially nonvolatile at 100°C. It is the major ingredient in airplane "de-icers." In automobile "year-round" anti-freeze, it lowers the freezing point of water in the radiator in the winter and raises its boiling point in the summer.

Biological antifreeze To survive in the Arctic and in northern winters, many fish and insects, including the common housefly, produce large amounts of glycerol ($C_3H_8O_3$)—a substance with a structure very similar to that of ethylene glycol and also miscible with water—which lowers the freezing point of their blood.

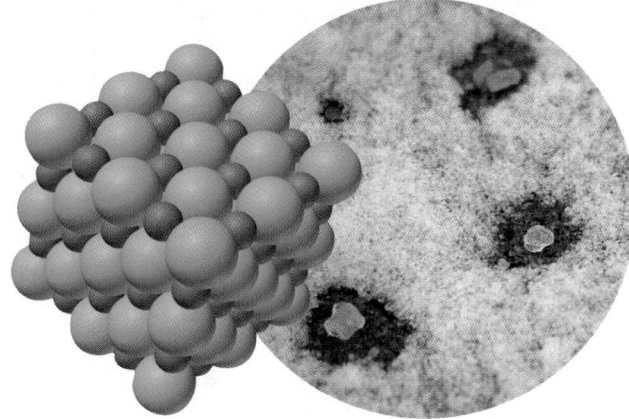

Salts for slippery streets Highway crews use salts, such as mixtures of NaCl and $CaCl_2$, to melt ice on streets. A small amount of salt dissolves in the ice by lowering its freezing point and melting it. More salt dissolves, more ice melts, and so forth. An advantage of $CaCl_2$ is that it has a highly negative ΔH_{soln}, so heat is released when it dissolves, which melts more ice.

Silicon rod with impurities   Zone of molten silicon   Inert gas

Heating coil   Purer silicon

Zone refining Manufacturing computer chips requires extremely pure starting materials. In the process of zone refining, a rod of impure silicon (or other metal or metalloid) is passed slowly through a heating coil in an inert atmosphere, and the first narrow zone of impure solid melts. As the next zone melts, the dissolved impurities from the first zone lower the freezing point of the solution, while the purer solvent refreezes. The process continues, zone by zone, throughout the entire rod. After several passes through the coil, in which each zone's impurities combine with those from other zones, the refrozen silicon becomes extremely pure—more than 99.999999%. The ultrapure silicon is then sliced into wafers for the production of computer chips (Section 12.7).

Applications of Osmotic Pressure

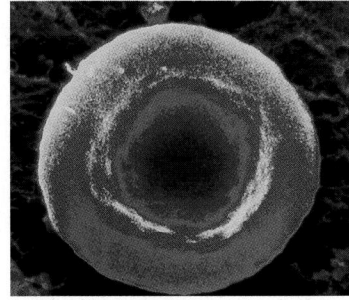

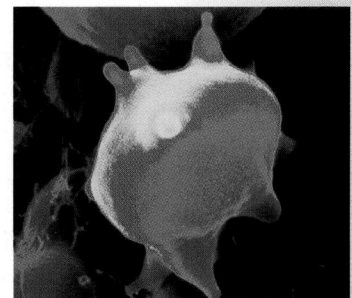

Controlling cell shape The word *tonicity* refers to the tone, or firmness, of a biological cell. An *isotonic* solution has the same concentration of particles as in the cell fluid, so water enters the cell at the same rate that it leaves, thereby maintaining the cell's normal shape.

To study cell contents, biochemists rupture membranes by placing the cell in a *hypotonic* solution, one that has a lower concentration of solute particles than the cell fluid. As a result, water enters the cell faster than it leaves, causing the cell to expand and burst.

In a *hypertonic* solution, one that contains a higher concentration of solute particles than the cell fluid, the cell shrinks from the net outward flow of water.

Isotonic daily care Contact-lens rinses consist of isotonic saline (0.15 *M* NaCl) to prevent any changes in the volume of corneal cells. Solutions for the intravenous delivery of nutrients or drugs are always isotonic.

Hypotonic watering of trees The dissolved substances in tree sap create a more concentrated solution than the surrounding groundwater. Water enters membranes in the roots and rises into the tree, creating an osmotic pressure that can exceed 20 atm in the tallest trees!

Sodium ion: the extracellular osmoregulator Of the four major biological cations—Na^+, K^+, Mg^{2+}, and Ca^{2+}—Na^+ is essential for all animals to regulate their fluid volume (which includes you!). The Na^+ ion accounts for more than 90 mol % of all cations *outside* a cell. A high Na^+ concentration draws water out of the cell by osmosis; a low concentration leaves more inside. The primary role of Na^+ is to regulate the water volume of the body, and the primary role of the kidneys is to regulate the concentration of Na^+. Changes in blood pressure (volume) activate nerves and hormones to adjust blood flow and alter kidney function.

Hypertonic food preservation Before refrigeration was common, salt was used as a preservative. The salt causes microbes on the food's surface to shrivel and die from loss of water. (Salt was so highly prized for this purpose that Roman soldiers were paid in salt, from which practice comes the word *salary*.) In 1772, Captain James Cook wisely brought on board his ship large amounts of cabbage as food stores and preserved them with salt, thus converting the perishable vegetable into long-lived sauerkraut. The high vitamin-C content of the pickled cabbage prevented the debilitating effects of scurvy and allowed Cook's crew to continue explorations and partake in experiments that would revolutionize longitude measurements and, thus, oceangoing navigation.

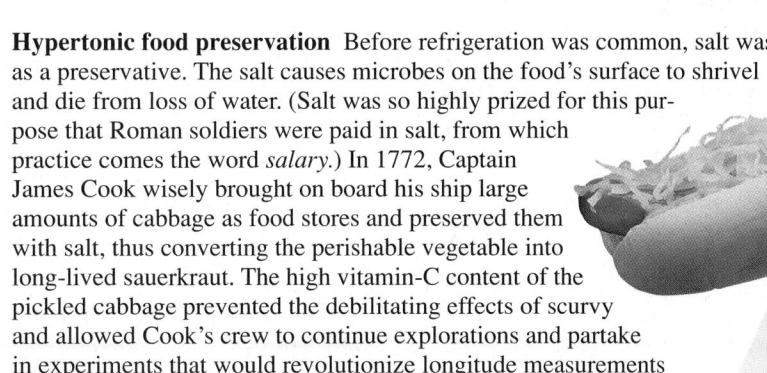

The Underlying Theme of Colligative Properties A common thread runs through our explanations of the four colligative properties of nonvolatile solutes. Each property rests on the inability of solute particles to cross between two phases. They cannot enter the gas phase, which leads to vapor pressure lowering and boiling point elevation. They cannot enter the solid phase, which leads to freezing point depression. They cannot cross a semipermeable membrane, which leads to the development of osmotic pressure. The presence of solute decreases the mole fraction of solvent, which lowers the number of solvent particles leaving the solution per unit time; this lowering requires an adjustment to reach equilibrium again. This adjustment to reach the new balance in numbers of particles crossing between two phases per unit time results in the measured colligative property.

Using Colligative Properties to Find Solute Molar Mass

Each colligative property relates concentration to some measurable quantity—the number of degrees the freezing point is lowered, the magnitude of osmotic pressure created, and so forth. From these measurements, we can determine the amount (mol) of solute particles and, for a known mass of solute, the molar mass of the solute as well.

In principle, any of the colligative properties can be used to find the solute's molar mass, but in practice, some systems provide more accurate data than others. For example, to determine the molar mass of an unknown solute by freezing point depression, you would select a solvent with as large a molal freezing point depression constant as possible (see Table 13.6). If the solute is soluble in acetic acid, for instance, a 1 m concentration of it depresses the freezing point of acetic acid by 3.90°C, more than twice the change in water (1.86°C).

Of the four colligative properties, osmotic pressure creates the largest changes and therefore the most precise measurements. Biological and polymer chemists estimate molar masses as great as 10^5 g/mol by measuring osmotic pressure. Because only a tiny fraction of a mole of a macromolecular solute dissolves, it would create too small a change in the other colligative properties.

SAMPLE PROBLEM 13.8 Determining Molar Mass from Osmotic Pressure

Problem Biochemists have discovered more than 400 mutant varieties of hemoglobin, the blood protein that carries oxygen throughout the body. A physician studying a variety associated with a fatal disease first finds its molar mass ($\mathcal{M}$). She dissolves 21.5 mg of the protein in water at 5.0°C to make 1.50 mL of solution and measures an osmotic pressure of 3.61 torr. What is the molar mass of this hemoglobin variety?

Plan We know the osmotic pressure ($\Pi = 3.61$ torr), R, and T (5.0°C). We convert Π from torr to atm, and T from °C to K, and then use Equation 13.13 to solve for molarity (M). Then we calculate the amount (mol) of hemoglobin from the known volume (1.50 mL) and use the known mass (21.5 mg) to find $\mathcal{M}$.

Solution Combining unit conversion steps and solving for molarity from Equation 13.13:

$$M = \frac{\Pi}{RT} = \frac{\dfrac{3.61 \text{ torr}}{760 \text{ torr}/1 \text{ atm}}}{\left(0.0821 \dfrac{\text{atm·L}}{\text{mol·K}}\right)(273.15 \text{ K} + 5.0)} = 2.08 \times 10^{-4} \, M$$

Finding amount (mol) of solute (after changing mL to L):

$$\text{Moles of solute} = M \times V = \frac{2.08 \times 10^{-4} \text{ mol}}{1 \text{ L soln}} \times 0.00150 \text{ L soln} = 3.12 \times 10^{-7} \text{ mol}$$

Calculating molar mass of hemoglobin (after changing mg to g):

$$\mathcal{M} = \frac{0.0215 \text{ g}}{3.12 \times 10^{-7} \text{ mol}} = \boxed{6.89 \times 10^4 \text{ g/mol}}$$

Sidebar flowchart:

Π (atm)

$\downarrow$ $M = \Pi/RT$

M (mol/L)

$\downarrow$ multiply by volume (L) of solution

Amount (mol) of solute

$\downarrow$ divide *into* mass (g) of solute

$\mathcal{M}$ (g/mol)

Check The answers seem reasonable: The small osmotic pressure implies a very low molarity. Hemoglobin is a protein, a biological macromolecule, so we expect a small number of moles [($\sim$2×10^{-4} mol/L) (1.5×10^{-3} L) = 3×10^{-7} mol] and a high molar mass ($\sim$21×10^{-3} g/3×10^{-7} mol = 7×10^{4} g/mol).

FOLLOW-UP PROBLEM 13.8 A 0.30 *M* solution of sucrose that is at 37°C has approximately the same osmotic pressure as blood does. What is the osmotic pressure of blood?

Colligative Properties of Volatile Nonelectrolyte Solutions

What is the effect on vapor pressure when the solute *is* volatile, that is, when the vapor consists of solute *and* solvent molecules? From Raoult's law (Equation 13.9), we know that

$$P_{solvent} = X_{solvent} \times P^0_{solvent} \quad \text{and} \quad P_{solute} = X_{solute} \times P^0_{solute}$$

where $X_{solvent}$ and X_{solute} refer to the mole fractions in the liquid phase. According to Dalton's law of partial pressures, the total vapor pressure is the sum of the partial vapor pressures:

$$P_{total} = P_{solvent} + P_{solute} = (X_{solvent} \times P^0_{solvent}) + (X_{solute} \times P^0_{solute})$$

Just as a nonvolatile solute lowers the vapor pressure of the solvent by making the mole fraction of the solvent less than 1, *the presence of each volatile component lowers the vapor pressure of the other* by making each mole fraction less than 1.

Let's examine this effect in a solution that contains equal amounts (mol) of benzene (C_6H_6) and toluene (C_7H_8): $X_{ben} = X_{tol} = 0.500$. At 25°C, the vapor pressure of pure benzene (P^0_{ben}) is 95.1 torr and that of pure toluene (P^0_{tol}) is 28.4 torr; note that benzene is more volatile than toluene. We find the partial pressures from Raoult's law:

$$P_{ben} = X_{ben} \times P^0_{ben} = 0.500 \times 95.1 \text{ torr} = 47.6 \text{ torr}$$
$$P_{tol} = X_{tol} \times P^0_{tol} = 0.500 \times 28.4 \text{ torr} = 14.2 \text{ torr}$$

As you can see, the presence of benzene lowers the vapor pressure of toluene, and vice versa.

Does the composition of the vapor differ from that of the solution? To see, let's calculate the mole fraction of each substance *in the vapor* by applying Dalton's law. Recall from Section 5.4 that $X_A = P_A/P_{total}$. Therefore, for benzene and toluene in the vapor,

$$X_{ben} = \frac{P_{ben}}{P_{total}} = \frac{47.6 \text{ torr}}{47.6 \text{ torr} + 14.2 \text{ torr}} = 0.770$$

$$X_{tol} = \frac{P_{tol}}{P_{total}} = \frac{14.2 \text{ torr}}{47.6 \text{ torr} + 14.2 \text{ torr}} = 0.230$$

The vapor composition is very different from the solution composition. The essential point to notice is that *the vapor has a higher mole fraction of the **more** volatile solution component.* The 50:50 ratio of benzene:toluene in the liquid created a 77:23 ratio of benzene:toluene in the vapor. Condense this vapor into a separate container, and that new *solution* would have this 77:23 composition, and the new *vapor* above it would be enriched still further in benzene.

In the process of **fractional distillation**, this phenomenon is used to separate a mixture of volatile components. Numerous vaporization-condensation steps continually enrich the vapor, until the vapor reaching the top of the fractionating column consists solely of the most volatile component. In the industrial process of petroleum refining (Figure 13.18), fractional distillation is used to separate the hundreds of individual compounds in crude oil into a small number of "fractions" based on boiling point range.

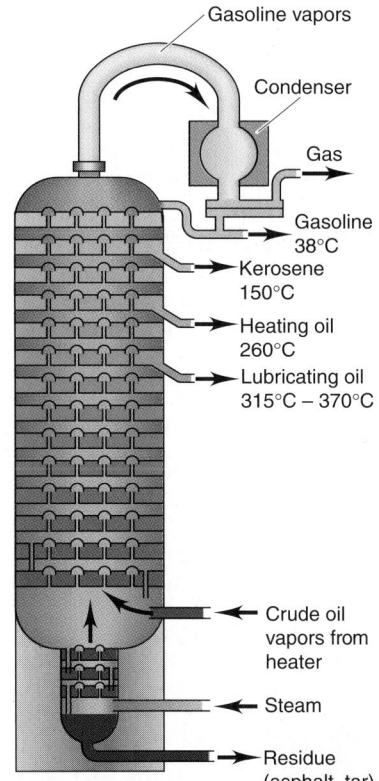

Figure 13.18 The process of fractional distillation. In the laboratory, a solution of two or more volatile components is attached to a *fractionating column* that is packed with glass beads and connected to a condenser. As the solution is heated, the vapor mixture rises and condenses repeatedly on the beads, each time forming a liquid enriched in the more volatile component. The vapor reaching the condenser consists of the more volatile component only. In industry, this process is used to separate petroleum into many products. A 30-m high fractionating tower can separate components that differ by a few tenths of a degree in their boiling points. (The illustration is a simplified version of a multipart process.)

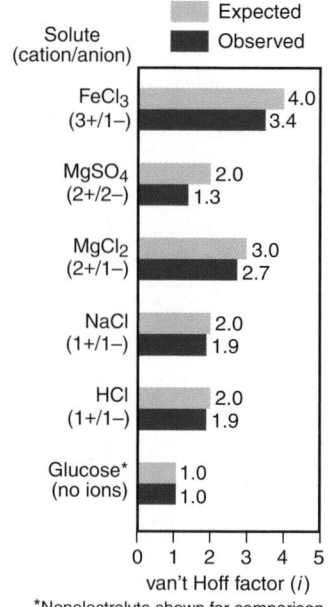

Solute (cation/anion) — Expected / Observed

Solute	value
FeCl₃ (3+/1−)	4.0 / 3.4
MgSO₄ (2+/2−)	2.0 / 1.3
MgCl₂ (2+/1−)	3.0 / 2.7
NaCl (1+/1−)	2.0 / 1.9
HCl (1+/1−)	2.0 / 1.9
Glucose* (no ions)	1.0 / 1.0

van't Hoff factor (*i*)
*Nonelectrolyte shown for comparison

Figure 13.19 **Nonideal behavior of electrolyte solutions.** The van't Hoff factors (*i*) for various ionic solutes in dilute (0.05 *m*) aqueous solution show that the observed value (*dark blue*) is always *lower* than the expected value (*light blue*). This deviation is due to ionic interactions that, in effect, reduce the number of free ions in solution. The deviation is greatest for multivalent ions.

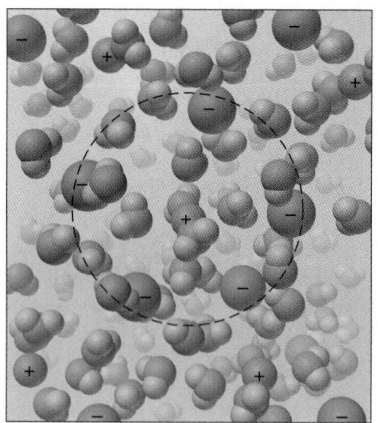

Figure 13.20 **An ionic atmosphere model for nonideal behavior of electrolyte solutions.** Hydrated anions cluster near cations, and vice versa, to form ionic atmospheres of net opposite charge. Because the ions do not act independently, their concentrations are effectively *less* than expected. Such interactions cause deviations from the ideal behavior expressed by Raoult's law.

Colligative Properties of Electrolyte Solutions

When we consider colligative properties of electrolyte solutions, the solute formula tells us the number of particles. For instance, the boiling point elevation (ΔT_b) of 0.050 *m* NaCl should be twice that of 0.050 *m* glucose ($C_6H_{12}O_6$), because NaCl dissociates into two particles per formula unit. Thus, we include a multiplying factor in the equations for the colligative properties of electrolyte solutions. The *van't Hoff factor (i)*, named after the Dutch chemist Jacobus van't Hoff (1852–1911), is the ratio of the *measured* value of the colligative property in the electrolyte solution to the *expected* value for a nonelectrolyte solution:

$$i = \frac{\text{measured value for electrolyte solution}}{\text{expected value for nonelectrolyte solution}}$$

To calculate the colligative properties of electrolyte solutions, we incorporate the van't Hoff factor into the equation:

For vapor pressure lowering:	$\Delta P = i(X_{\text{solute}} \times P^0_{\text{solvent}})$
For boiling point elevation:	$\Delta T_b = i(K_b m)$
For freezing point depression:	$\Delta T_f = i(K_f m)$
For osmotic pressure:	$\Pi = i(MRT)$

If electrolyte solutions behaved ideally, the factor *i* would be the amount (mol) of particles in solution divided by the amount (mol) of dissolved solute; that is, *i* would be 2 for NaCl, 3 for $Mg(NO_3)_2$, and so forth. Careful experiment shows, however, that *most electrolyte solutions are **not** ideal*. For example, comparing the boiling point elevation for 0.050 *m* NaCl solution with that for 0.050 *m* glucose solution gives a factor *i* of 1.9, not 2.0:

$$i = \frac{\Delta T_b \text{ of } 0.050 \, m \text{ NaCl}}{\Delta T_b \text{ of } 0.050 \, m \text{ glucose}} = \frac{0.049°C}{0.026°C} = 1.9$$

The measured value of the van't Hoff factor is typically *lower* than that expected from the formula. This deviation implies that the ions are not behaving as independent particles. However, we know from other evidence that soluble salts dissociate completely into ions. The fact that the deviation is greater with divalent and trivalent ions is a strong indication that the ionic charge is somehow involved (Figure 13.19).

To explain this nonideal behavior, we picture ions as separate but near each other. Clustered near a positive ion are, on average, more negative ions, and vice versa. Figure 13.20 shows each ion surrounded by an **ionic atmosphere** of net opposite charge. Through these electrostatic associations, each type of ion behaves as if it were "tied up," so its concentration seems *lower* than it actually is. Thus, we often speak of an *effective* concentration, obtained by multiplying *i* by the *stoichiometric* concentration based on the formula. The greater the charge, the stronger the electrostatic associations, so the deviation from ideal behavior is greater for compounds that dissociate into multivalent ions.

At ordinary conditions and concentrations, nonideal behavior of solutions is much more common, and the deviations much larger, than nonideal behavior of gases, because the particles in solutions are so much closer together. Nevertheless, the two systems exhibit some interesting similarities. Gases display nearly ideal behavior at low pressures because the distances between particles are large. Similarly, van't Hoff factors (*i*) approach their ideal values as the solution becomes more dilute, that is, as the distance between ions increases. In gases, attractions between particles cause deviations from the expected pressure. In solutions, attractions between particles cause deviations from the expected size of a colligative property. Finally, for both real gases and real solutions, we use empirically determined numbers (van der Waals constants or van't Hoff factors) to transform theories (the ideal gas law or Raoult's law) into more useful relations.

SECTION SUMMARY
Colligative properties are related to the number of dissolved solute particles, not their chemical nature. Compared with the pure solvent, a solution of a nonvolatile non-electrolyte has a lower vapor pressure (Raoult's law), an elevated boiling point, a depressed freezing point, and an osmotic pressure. Colligative properties can be used to determine the solute molar mass. When solute *and* solvent are volatile, the vapor pressure of each is lowered by the presence of the other. The vapor pressure of the more volatile component is always higher. Electrolyte solutions exhibit nonideal behavior because ionic interactions reduce the effective concentration of the ions.

13.6 THE STRUCTURE AND PROPERTIES OF COLLOIDS

Stir a handful of fine sand into a glass of water and watch what happens. The sand particles are suspended at first but then gradually settle to the bottom. Sand in water is an example of a **suspension,** a *heterogeneous* mixture containing particles large enough to be seen with the naked eye and clearly distinct from the surrounding fluid. In contrast, stirring sugar into water forms a solution, a *homogeneous* mixture in which the particles are individual molecules distributed evenly throughout the surrounding fluid.

Between the extremes of suspensions and solutions is a large group of mixtures called *colloidal dispersions,* or simply **colloids,** in which a dispersed (solute-like) substance is distributed throughout a dispersing (solvent-like) substance. Colloidal particles are larger than simple molecules but small enough to remain distributed and not settle out. They have a range of diameters between 1 and 1000 nm (10^{-9} to 10^{-6} m). A colloidal particle may consist of a single macromolecule (such as a protein or synthetic polymer) or an aggregate of many atoms, ions, or molecules. Whatever their composition, colloidal particles have an enormous total surface area as a result of their small size. Consider a cube with 1-cm sides. It has a total surface area of 6 cm^2. If it were divided equally into 10^{12} cubes, the cubes would be the size of large colloidal particles and have a total surface area of 60,000 cm^2, or 6 m^2. This enormous surface area allows many more interactions to exert a great total adhesive force, which attracts other particles and gives rise to some of the practical uses of colloids.

Colloids are classified in Table 13.7 according to whether the dispersed and dispersing substances are gases, liquids, or solids. Many familiar commercial products and natural objects are colloids. For example, whipped cream is a *foam,* a gas dispersed in a liquid. Firefighting foams, such as those used at emergency

Table 13.7 **Types of Colloids**

Colloid Type	Dispersed Substance	Dispersing Medium	Example
Aerosol	Liquid	Gas	Fog
Aerosol	Solid	Gas	Smoke
Foam	Gas	Liquid	Whipped cream
Solid foam	Gas	Solid	Marshmallow
Emulsion	Liquid	Liquid	Milk
Solid emulsion	Liquid	Solid	Butter
Sol	Solid	Liquid	Paint; cell fluid
Solid sol	Solid	Solid	Opal

In the small intestine, fats are digested by *bile salts,* soaplike molecules with smaller polar-ionic and larger nonpolar portions. Secreted by the liver, stored in the gallbladder, and released in the intestine, bile salts emulsify fats just as soap emulsifies grease: fatty aggregates are broken down into particles of colloidal size and dispersed in the watery fluid. In this form, the fats are broken down further and transported into the blood for cellular metabolism.

airplane landings, are liquid mixtures of proteins in water that are made frothy with fine jets of air. Most biological fluids are aqueous *sols,* solids dispersed in water. Within a typical cell, proteins and nucleic acids are colloidal-size particles dispersed in an aqueous solution of ions and small molecules. The action of soaps and detergents occurs by the formation of an *emulsion,* a liquid (soap dissolved in grease) dispersed in another liquid (water). ⬤

Most colloids are cloudy or opaque, but some are transparent to the naked eye. When light passes through a colloid, it is scattered randomly by the dispersed particles because their sizes are similar to the wavelengths of visible light (400 to 750 nm). Viewed from the side, the scattered light beam is visibly broader than one passing through a solution. This light-scattering phenomenon is called the **Tyndall effect** (Figure 13.21). Dust in air displays this effect when sunlight shines through it, as does mist pierced by headlights at night.

Under low magnification, you can see colloidal particles exhibit *Brownian motion,* a characteristic movement in which the particles change speed and direction erratically. This motion results because the colloidal particles are being pushed this way and that by molecules of the dispersing medium. These collisions are primarily responsible for keeping colloidal particles from settling out. (Einstein's explanation of Brownian motion in 1905 was a principal factor in the acceptance of the molecular nature of matter.)

When colloidal particles collide, why don't they aggregate into larger particles and settle out? Interparticle forces provide the explanation. Water-dispersed colloids remain dispersed because the particles have charged surfaces that interact strongly with the water through ion-dipole forces. Soap molecules form spherical *micelles,* with the charged heads forming the micelle exterior and the nonpolar tails interacting via dispersion forces in the interior. Aqueous proteins are typically spherical and mimic this micellar arrangement, with charged amino acid groups facing the water and uncharged groups buried within the molecule. Nonpolar oily particles can be dispersed in water by introducing ions, which are adsorbed onto their surfaces by dispersion forces. Charge repulsions between adsorbed ions prevent the particles from aggregating.

Despite these forces, various methods can coagulate the particles and "destroy" the colloid. Heating a colloid makes the particles move faster and collide more often and with enough force to coalesce into heavier particles that settle out. Adding an electrolyte solution introduces oppositely charged ions that

A

B

Figure 13.21 Light scattering and the Tyndall effect. A, When a beam of light passes through a solution (*left jar*), its path remains narrow and barely visible. When it passes through a colloid (*right jar*), it is scattered and broadened by the particles and thus easily visible. **B,** Sunlight is scattered as it shines through misty air in a forest.

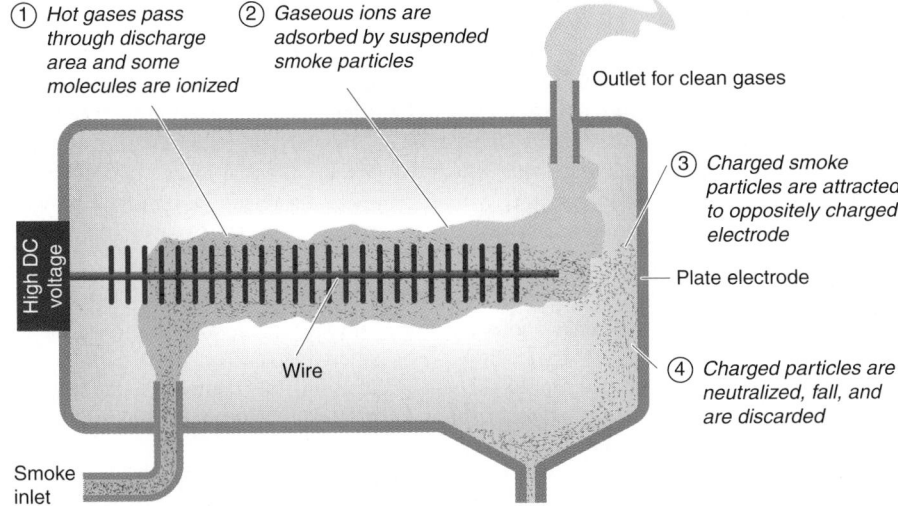

① Hot gases pass through discharge area and some molecules are ionized

② Gaseous ions are adsorbed by suspended smoke particles

Outlet for clean gases

③ Charged smoke particles are attracted to oppositely charged electrode

Plate electrode

High DC voltage

Wire

④ Charged particles are neutralized, fall, and are discarded

Smoke inlet

Figure 13.22 A Cottrell precipitator for removing particulates from industrial smokestack gases.

neutralize the particles' surface charges, which allows the particles to coagulate and settle. ● Uncharged colloidal particles in smokestack gases are removed by creating ions that become adsorbed on the particles, which are then attracted to the charged plates of a Cottrell precipitator. Such precipitators are installed in the smokestacks of coal-burning power plants (Figure 13.22) and collect about 90% of colloidal smoke particulates, thus preventing their release into the air.

The upcoming Chemical Connections essay applies solution and colloid chemistry to the purification of water for residential and industrial use.

SECTION SUMMARY

Colloidal particles are smaller than those in a suspension and larger than those in a solution. Colloids are classified by the physical states of the dispersed and dispersing substances and are formed from many combinations of gas, liquid, and solid. Colloids have extremely large surface areas, exhibit random (Brownian) motion, and scatter incoming light (Tyndall effect). Colloidal particles in water have charged surfaces that keep them dispersed, but they can be coagulated by heating or by the addition of ions.

Chapter Perspective

In Chapter 12, you saw how pure liquids and solids behave, and here you've seen how their behaviors change when they are mixed together. Two features of these systems reappear in later chapters: their equilibrium nature in Chapters 17 to 19, and their tendency toward disorder in Chapter 20.

This chapter completes our general discussion of atomic and molecular properties and their influence on the properties of matter. Immediately following is an Interchapter feature that reviews these properties in a pictorial format and helps preview their application in the next two chapters. In Chapter 14, we travel through the main groups of the periodic table, applying the ideas you've learned so far to the chemical and physical behavior of the elements. Then, in Chapter 15, we apply them to carbon and its nearest neighbors to see how their properties are responsible for the fascinating world of organic substances, including synthetic polymers and biomolecules.

● **From Colloid to Civilization** At times, civilizations have been born where colloids were being coagulated by electrolyte solutions. At the mouths of rivers, where salt concentrations increase near an ocean or sea, the clay particles dispersed in the river water come together to form muddy deltas, such as those of the Nile (*reddish area in photo*) and the Mississippi. The ancient Egyptian empire and the city of New Orleans are the results of this global colloid chemistry.

Chemistry in Sanitary Engineering

Solutions and Colloids in Water Purification

C lean water is a priceless and limited resource that we've begun to treasure only recently, after decades of pollution and waste. Because of the natural tendency of systems to become disordered, it requires energy to remove dissolved, dispersed, and suspended particles from water to make it clean enough for humans to use.

Most water destined for human use comes from lakes, rivers, or reservoirs that may serve also as the final sink after the water is used. Many mineral ions, such as NO_3^- and Fe^{3+}, may be present in high concentrations. Dissolved organic compounds, some of them toxic, may be present as well. Fine clay particles and a whole spectrum of microorganisms are dispersed in colloidal form. Larger particles and debris of every variety may be present in suspension.

Water Treatment Plants

As a sample of water moves from the natural source into a water treatment facility, the largest particles are physically removed at the intake site by screens (Figure B13.3, step 1). Finer particles, including microorganisms, are removed in large settling tanks by treatment with lime (CaO) and cake alum [$Al_2(SO_4)_3$] (step 2), which react to form a fluffy, gel-like mass of $Al(OH)_3$:

$$3CaO(s) + 3H_2O(l) + Al_2(SO_4)_3(s) \longrightarrow$$
$$2Al(OH)_3(colloidal\ gel) + 3CaSO_4(aq)$$

The fine particles are trapped within or adsorbed onto the enormous surface area of the gel, which coagulates, settles out, and is filtered through a sand bed (step 3).

With suspended and colloidal particles removed, the water is aerated in large sprayers to saturate it with oxygen, which speeds the oxidation of dissolved organic compounds (step 4). The water is then disinfected, usually by treatment with Cl_2 gas and/or aqueous solutions of hypochlorite ion, ClO^- (step 5), which may give the water an unpleasant odor. Chlorine can also form toxic chlo-

rinated hydrocarbons, but these can be removed by adsorption onto activated charcoal particles. These steps through the treatment facility dispose of debris and grit, colloidal clay, microorganisms, and much of the oxidizable organic matter, but dissolved ions remain. Many of these can be removed by water softening and reverse osmosis.

Water Softening via Ion Exchange

Water that contains large amounts of divalent cations, such as Ca^{2+}, Mg^{2+}, and Fe^{2+}, is called **hard water.** These cations cause several problems. During cleaning, they combine with the anions of fatty acids in soaps to produce insoluble deposits on clothes, washing machine parts, and sinks:

$$Ca^{2+}(aq) + 2C_{17}H_{35}COONa(aq) \longrightarrow$$
$$\underset{soap}{} \qquad \underset{deposit}{(C_{17}H_{35}COO)_2Ca(s)} + 2Na^+(aq)$$

When a large amount of bicarbonate (HCO_3^-) is present in the water, the hard-water cations cause a buildup of *scale*, insoluble carbonate deposits within boilers and hot-water pipes that interfere with the transfer of heat and damage plumbing:

$$Ca^{2+}(aq) + 2HCO_3^-(aq) \xrightarrow{\Delta} CaCO_3(s) + CO_2(g) + H_2O(l)$$

The removal of hard-water ions, called **water softening,** solves these problems. It is accomplished by exchanging "soft-water" Na^+ ions for the hard-water cations. A typical domestic **ion-exchange** system contains an *ion-exchange resin,* an insoluble polymer that has covalently bound anion groups, such as $-SO_3^-$ or $-COO^-$, to which Na^+ ions are attached to balance the charges (Figure B13.4). The divalent cations in hard water are attracted to the resin's anionic groups and displace the Na^+ ions into the water: one type of ion is exchanged for another. The resin is replaced when all the resin sites are occupied, or it can be "regenerated" by treating it with a very concentrated Na^+ solution, which exchanges Na^+ ions for the bound Ca^{2+}.

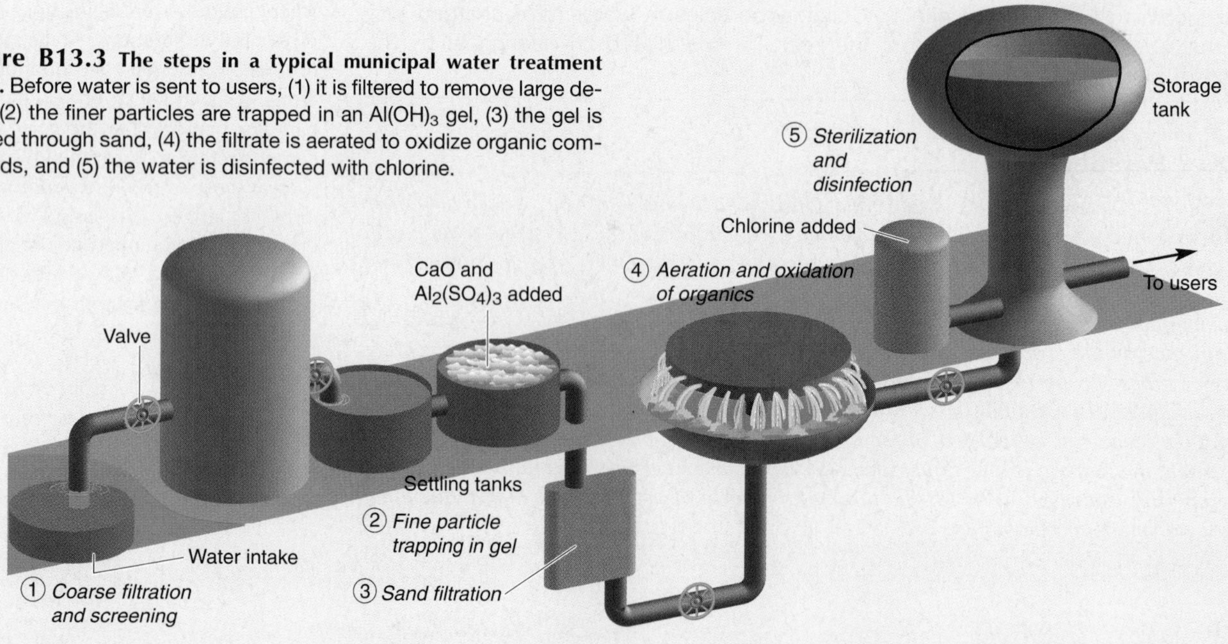

Figure B13.3 The steps in a typical municipal water treatment plant. Before water is sent to users, (1) it is filtered to remove large debris, (2) the finer particles are trapped in an $Al(OH)_3$ gel, (3) the gel is filtered through sand, (4) the filtrate is aerated to oxidize organic compounds, and (5) the water is disinfected with chlorine.

⑤ *Sterilization and disinfection*

Storage tank

Chlorine added

④ *Aeration and oxidation of organics*

To users

CaO and $Al_2(SO_4)_3$ added

Valve

Settling tanks

② *Fine particle trapping in gel*

Water intake

① *Coarse filtration and screening*

③ *Sand filtration*

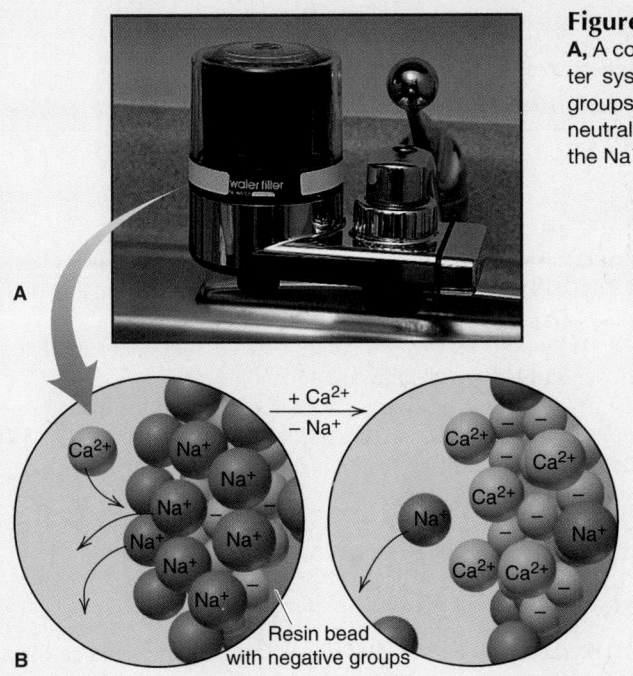

Figure B13.4 Ion exchange for removal of hard-water cations.
A, A commercial ion-exchange column is installed in a household water system. **B,** In a typical ion-exchange resin, negatively charged groups are covalently bound to resin beads, with Na^+ ions present to neutralize the charges. Hard-water ions, such as Ca^{2+}, exchange with the Na^+ ions, which are displaced into the flowing water.

$+ Ca^{2+}$
$- Na^+$

Resin bead
with negative groups

Reverse Osmosis

Another way to remove ions and other dissolved substances from water is by **reverse osmosis.** In osmosis, water moves from a dilute to a concentrated solution through a semipermeable membrane. The resulting difference in water volumes creates an osmotic pressure. In reverse osmosis, water moves out of the concentrated solution when a pressure *greater* than the osmotic pressure is *applied* to the solution, forcing the water back through the membrane and leaving the ions behind—in a sense, filtering out the ions at the molecular level.

In domestic water systems, reverse osmosis is used to remove toxic ions, such as the *heavy-metal ions* Pb^{2+}, Cd^{2+}, and Hg^{2+}, present at concentrations too low for removal by ion exchange. On a much larger scale, reverse osmosis is used in **desalination** plants, which remove large amounts of ions from seawater (Figure B13.5). Reverse osmosis plants find increasing use in arid regions, such as the Middle East. Seawater is pumped under high pressure into tubes containing millions of hollow fiber membranes, each the thickness of a human hair. Water molecules, but not ions, pass through the membranes into the fiber to be collected. Seawater containing about 40,000 ppm of total dissolved solids can be purified to a level around 400 ppm (suitable for drinking) in one pass through such a system.

Water Treatment after Use

Used water is called **wastewater,** or *sewage,* and it must be treated before being returned to the groundwater, river, or lake. Sewage treatment is especially important for industrial wastewater, which may contain toxic components. In *primary* sewage treatment, wastewater undergoes the same steps as are used to treat water coming into the system. Most municipalities now also include *secondary* sewage treatment. In this stage, bacteria biologically degrade the organic compounds and some microorganisms still present in solution or in the solids from the settling tanks. In certain cases, secondary treatment may be supplemented by *tertiary* treatment in a process tailored to the specific pollutant involved. Heavy-metal ions, for instance, can be eliminated by a precipitation step before primary and secondary treatment. Tertiary methods also exist for phosphate, nitrate, and toxic organic substances.

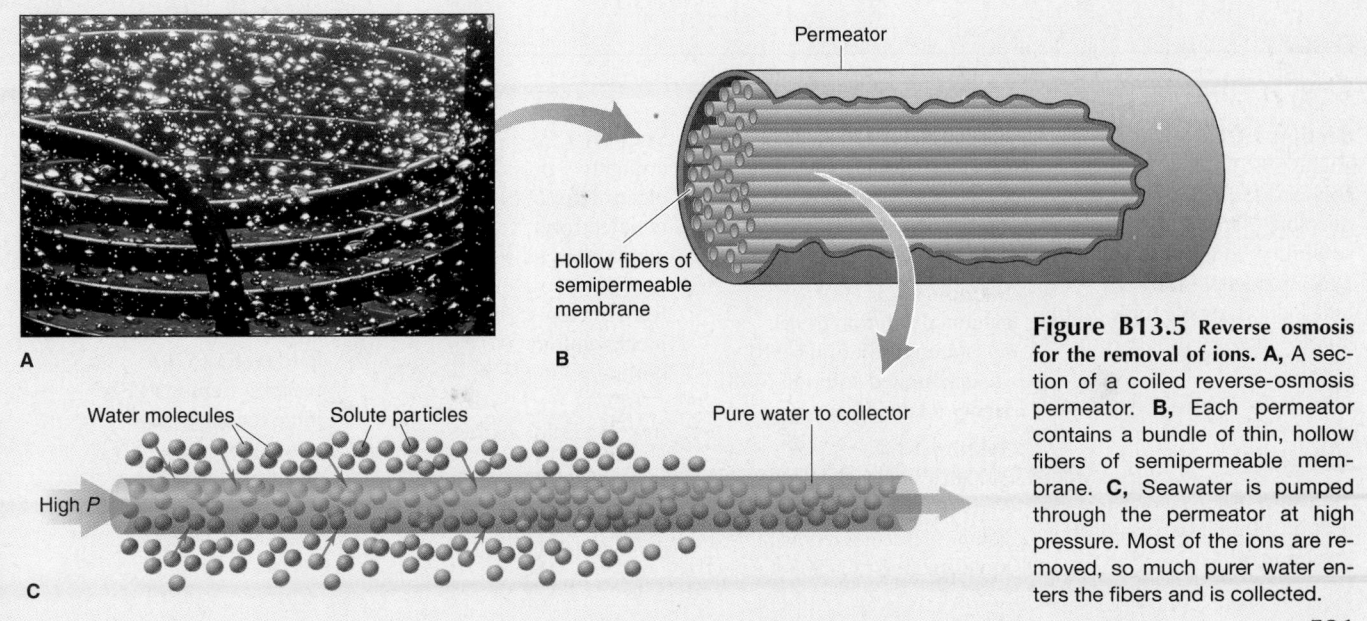

Figure B13.5 Reverse osmosis for the removal of ions. A, A section of a coiled reverse-osmosis permeator. **B,** Each permeator contains a bundle of thin, hollow fibers of semipermeable membrane. **C,** Seawater is pumped through the permeator at high pressure. Most of the ions are removed, so much purer water enters the fibers and is collected.

Permeator

Hollow fibers of
semipermeable
membrane

Water molecules   Solute particles   Pure water to collector

High *P*

521

For Review and Reference (Numbers in parentheses refer to pages, unless noted otherwise.)

Learning Objectives

Section and/or sample problem (SP) numbers appear in parentheses.

Understand These Concepts

1. The quantitative meaning of solubility (13.1)
2. The major types of intermolecular forces in solution and their relative strengths (13.1)
3. How the "like-dissolves-like" rule depends on intermolecular forces (13.1)
4. Why gases have relatively low solubilities in water (13.1)
5. General characteristics of solutions formed by various combinations of gases, liquids, and solids (13.1)
6. The enthalpy components of a solution cycle and their effect on ΔH_{soln} (13.2)
7. The dependence of ΔH_{hydr} on ionic charge density and the factors that determine whether ionic solution processes are exothermic or endothermic (13.2)
8. The meaning of entropy and how the balance between the change in enthalpy and the change in entropy governs the solution process (13.2)
9. The distinctions among saturated, unsaturated, and supersaturated solutions and the equilibrium nature of a saturated solution (13.3)
10. The relation between temperature and the solubility of solids (13.3)
11. Why the solubility of gases in water decreases with a rise in temperature (13.3)
12. The effect of gas pressure on solubility and its quantitative expression as Henry's law (13.3)
13. The meaning of molarity, molality, mole fraction, and parts by mass or by volume of a solution (13.4)
14. The distinction between electrolytes and nonelectrolytes in solution (13.5)
15. The four colligative properties and their dependence on number of dissolved particles (13.5)
16. Ideal solutions and the importance of Raoult's law (13.5)
17. How the phase diagram of a solution differs from that of the pure solvent (13.5)
18. Why the vapor over a solution of volatile nonelectrolyte is richer in the more volatile component (13.5)
19. Why electrolyte solutions are not ideal and the meanings of the van't Hoff factor and the ionic-atmosphere concept (13.5)
20. How particle size distinguishes suspensions, colloids, and solutions (13.6)
21. How colloidal behavior is demonstrated by Brownian motion and the Tyndall effect (13.6)

Master These Skills

1. Predicting relative solubilities from intermolecular forces (SP 13.1)
2. Using Henry's law to calculate the solubility of a gas (SP 13.2)
3. Expressing concentration in terms of molality, parts by mass, parts by volume, and mole fraction (SPs 13.3 and 13.4)
4. Interconverting among the various units for expressing concentration (SP 13.5)
5. Using Raoult's law to calculate the vapor pressure lowering of a solution (SP 13.6)
6. Determining boiling point elevation and freezing point depression (SP 13.7)
7. Using a colligative property to calculate the molar mass of a solute (SP 13.8)
8. Calculating the composition of vapor over a solution of volatile nonelectrolyte (13.5)
9. Calculating the van't Hoff factor i from the magnitude of a colligative property (13.5)

Key Terms

Section 13.1
solute (486)
solvent (486)
miscible (486)
solubility (S) (486)
hydration shell (486)
ion–induced dipole force (486)
dipole–induced dipole force (487)
like-dissolves-like rule (487)
soap (490)
alloy (492)

Section 13.2
heat of solution (ΔH_{soln}) (494)
solvation (494)

hydration (494)
heat of hydration (ΔH_{hydr}) (494)
charge density (495)
entropy (496)

Section 13.3
saturated solution (498)
unsaturated solution (498)
supersaturated solution (498)
Henry's law (500)

Section 13.4
molality (m) (502)
mass percent [% (w/w)] (502)
volume percent [% (v/v)] (503)
mole fraction (X) (503)

Section 13.5
colligative property (506)
electrolyte (506)
nonelectrolyte (506)
vapor pressure lowering (ΔP) (507)
Raoult's law (507)
ideal solution (507)
boiling point elevation (ΔT_b) (508)
freezing point depression (ΔT_f) (510)
semipermeable membrane (511)
osmosis (511)

osmotic pressure (Π) (511)
fractional distillation (515)
ionic atmosphere (516)

Section 13.6
suspension (517)
colloid (517)
Tyndall effect (518)
hard water (520)
water softening (520)
ion exchange (520)
reverse osmosis (521)
desalination (521)
wastewater (521)

Key Equations and Relationships

13.1 Dividing the general heat of solution into component enthalpies (494):

$$\Delta H_{soln} = \Delta H_{solute} + \Delta H_{solvent} + \Delta H_{mix}$$

13.2 Dividing the heat of solution of an ionic compound in water into component enthalpies (495):

$$\Delta H_{soln} = -\Delta H_{lattice} + \Delta H_{hydr\ of\ the\ ions}$$

13.3 Relating gas solubility to its partial pressure (Henry's law) (500):

$$S_{gas} = k_H \times P_{gas}$$

13.4 Defining concentration in terms of molarity (501):

$$\text{Molarity } (M) = \frac{\text{amount (mol) of solute}}{\text{volume (L) of solution}}$$

13.5 Defining concentration in terms of molality (502):

$$\text{Molality } (m) = \frac{\text{amount (mol) of solute}}{\text{mass (kg) of solvent}}$$

13.6 Defining concentration in terms of mass percent (502):

$$\text{Mass percent } [\% \text{ (w/w)}] = \frac{\text{mass of solute}}{\text{mass of solution}} \times 100$$

13.7 Defining concentration in terms of volume percent (503):

$$\text{Volume percent } [\% \text{ (v/v)}] = \frac{\text{volume of solute}}{\text{volume of solution}} \times 100$$

13.8 Defining concentration in terms of mole fraction (503):

Mole fraction (X)

$$= \frac{\text{amount (mol) of solute}}{\text{amount (mol) of solute} + \text{amount (mol) of solvent}}$$

13.9 Expressing the relationship between the vapor pressure of solvent above a solution and its mole fraction in the solution (Raoult's law) (507):

$$P_{solvent} = X_{solvent} \times P^0_{solvent}$$

13.10 Calculating the vapor pressure lowering due to solute (507):

$$\Delta P = X_{solute} \times P^0_{solvent}$$

13.11 Calculating the boiling point elevation of a solution (508):

$$\Delta T_b = K_b m$$

13.12 Calculating the freezing point depression of a solution (510):

$$\Delta T_f = K_f m$$

13.13 Calculating the osmotic pressure of a solution (511):

$$\Pi = \frac{n_{solute}}{V_{soln}} RT = MRT$$

Highlighted Figures and Tables

These figures (F) and tables (T) provide a quick review of key ideas.

F13.1 Major types of intermolecular forces in solutions (486)
F13.2 Hydration shells around an aqueous ion (487)
F13.5 Solution cycles and the enthalpy components of ΔH_{soln} (494)
T13.4 Trends in ionic heats of hydration (495)
F13.6 Enthalpy diagrams for dissolving three ionic compounds in water (495)
F13.8 Equilibrium in a saturated solution (497)

F13.12 Effect of pressure on gas solubility (500)
T13.5 Concentration definitions (501)
F13.15 Effect of the solute on vapor pressure of a solution (507)
F13.16 Phase diagrams of solvent and solution (509)
F13.17 Development of osmotic pressure (511)
F13.19 Nonideal behavior of electrolyte solutions (516)
T13.7 Types of colloids (517)

Brief Solutions to Follow-up Problems

13.1 (a) 1,4-Butanediol is more soluble in water because it can form more H bonds.
(b) Chloroform is more soluble in water because it can form dipole-dipole forces.
13.2 $S_{N_2} = (7 \times 10^{-4} \text{ mol/L·atm})(0.78 \text{ atm})$
$= 5 \times 10^{-4} \text{ mol/L}$

13.3 Mass (g) of glucose $= 563 \text{ g ethanol} \times \dfrac{1 \text{ kg}}{10^3 \text{ g}}$

$\times \dfrac{2.40 \times 10^{-2} \text{ mol glucose}}{1 \text{ kg ethanol}} \times \dfrac{180.16 \text{ g glucose}}{1 \text{ mol glucose}} = 2.43 \text{ g glucose}$

13.4 Mass % $C_3H_7OH = \dfrac{35.0 \text{ g}}{35.0 \text{ g} + 150. \text{ g}} \times 100 = 18.9 \text{ mass \%}$

Mass % $C_2H_5OH = 100.0 - 18.9 = 81.1 \text{ mass \%}$

$$X_{C_3H_7OH} = \frac{35.0 \text{ g } C_3H_7OH \times \dfrac{1 \text{ mol } C_3H_7OH}{60.09 \text{ g } C_3H_7OH}}{\left(35.0 \text{ g } C_3H_7OH \times \dfrac{1 \text{ mol } C_3H_7OH}{60.09 \text{ g } C_3H_7OH}\right) + \left(150. \text{ g } C_2H_5OH \times \dfrac{1 \text{ mol } C_2H_5OH}{46.07 \text{ g } C_2H_5OH}\right)} = 0.152$$

$$X_{C_2H_5OH} = 1.000 - 0.152 = 0.848$$

Brief Solutions to Follow-up Problems (continued)

13.5 Mass % HCl $= \dfrac{\text{mass of HCl}}{\text{mass of soln}} \times 100$

$$= \dfrac{\dfrac{11.8 \text{ mol HCl}}{1 \text{ L soln}} \times \dfrac{36.46 \text{ g HCl}}{1 \text{ mol HCl}}}{\dfrac{1.190 \text{ g}}{1 \text{ mL soln}} \times \dfrac{10^3 \text{ mL}}{1 \text{ L}}} \times 100$$

$= 36.2 \text{ mass \% HCl}$

Mass (kg) of soln $= 1 \text{ L soln} \times \dfrac{1.190 \times 10^{-3} \text{ kg soln}}{1 \times 10^{-3} \text{ L soln}}$

$= 1.190 \text{ kg soln}$

Mass (kg) of HCl $= 11.8 \text{ mol HCl} \times \dfrac{36.46 \text{ g HCl}}{1 \text{ mol HCl}} \times \dfrac{1 \text{ kg}}{10^3 \text{ g}}$

$= 0.430 \text{ kg HCl}$

Molality of HCl $= \dfrac{\text{mol HCl}}{\text{kg water}} = \dfrac{\text{mol HCl}}{\text{kg soln} - \text{kg HCl}}$

$= \dfrac{11.8 \text{ mol HCl}}{0.760 \text{ kg H}_2\text{O}} = 15.5 \text{ } m \text{ HCl}$

$X_{HCl} = \dfrac{\text{mol HCl}}{\text{mol HCl} + \text{mol H}_2\text{O}}$

$= \dfrac{11.8 \text{ mol}}{11.8 \text{ mol} + \left(760 \text{ g H}_2\text{O} \times \dfrac{1 \text{ mol}}{18.02 \text{ g H}_2\text{O}}\right)} = 0.219$

13.6 $\Delta P = X_{aspirin} \times P^0_{methanol}$

$= \dfrac{\dfrac{2.00 \text{ g}}{180.15 \text{ g/mol}}}{\dfrac{2.00 \text{ g}}{180.15 \text{ g/mol}} + \dfrac{50.0 \text{ g}}{32.04 \text{ g/mol}}} \times 101 \text{ torr}$

$= 0.713 \text{ torr}$

13.7 $m = \dfrac{(0.00°\text{F} - 32°\text{F})\left(\dfrac{5°\text{C}}{9°\text{F}}\right)}{1.86°\text{C}/m} = 9.56 \text{ } m$

13.8 $\Pi = MRT$

$= (0.30 \text{ mol/L})\left(0.0821 \dfrac{\text{atm·L}}{\text{mol·K}}\right)(37°\text{C} + 273.15)$

$= 7.6 \text{ atm}$

Problems

Problems with colored numbers are answered at the back of the text. Sections match the text and provide the number(s) of relevant sample problems. Most offer Concept Review Questions, Skill-Building Exercises (in similar pairs), and Problems in Context. Then Comprehensive Problems, based on material from any section or previous chapter, follow.

Types of Solutions: Intermolecular Forces and Predicting Solubility
(Sample Problem 13.1)

● **Concept Review Questions**

13.1 Describe how properties of seawater illustrate the two characteristics that define mixtures.

13.2 What types of intermolecular forces give rise to hydration shells in an aqueous solution of sodium chloride?

13.3 Acetic acid is miscible with water. Would you expect carboxylic acids, general formula $CH_3(CH_2)_n COOH$, to become more or less water soluble as n increases? Explain.

13.4 Which would you expect to be more effective as a soap, sodium acetate or sodium stearate? Explain.

13.5 Hexane and methanol are miscible as gases but only slightly soluble in each other as liquids. Explain.

13.6 Hydrogen chloride (HCl) gas is much more soluble than propane gas (C_3H_8) in water, even though HCl has a lower boiling point. Explain.

● **Skill-Building Exercises (paired)**

13.7 Which gives the more concentrated solution, (a) KNO_3 in H_2O or (b) KNO_3 in carbon tetrachloride (CCl_4)? Explain.

13.8 Which gives the more concentrated solution, (a) stearic acid $[CH_3(CH_2)_{16}COOH]$ in H_2O or (b) stearic acid in CCl_4? Explain.

13.9 What is the strongest type of intermolecular force between solute and solvent in each solution?
(a) $CsCl(s)$ in $H_2O(l)$
(b) $CH_3\overset{\displaystyle O}{\overset{\|}{C}}CH_3(l)$ in $H_2O(l)$
(c) $CH_3OH(l)$ in $CCl_4(l)$

13.10 What is the strongest type of intermolecular force between solute and solvent in each solution?
(a) $Cu(s)$ in $Ag(s)$
(b) $CH_3Cl(g)$ in $CH_3OCH_3(g)$
(c) $CH_3CH_3(g)$ in $CH_3CH_2CH_2NH_2(l)$

13.11 What is the strongest type of intermolecular force between solute and solvent in each solution?
(a) $CH_3OCH_3(g)$ in $H_2O(l)$
(b) $Ne(g)$ in $H_2O(l)$
(c) $N_2(g)$ in $C_4H_{10}(g)$

13.12 What is the strongest type of intermolecular force between solute and solvent in each solution?
(a) $C_6H_{14}(l)$ in $C_8H_{18}(l)$
(b) $H_2C{=}O(g)$ in $CH_3OH(l)$
(c) $Br_2(l)$ in $CCl_4(l)$

13.13 Which member of each pair is more soluble in diethyl ether? Why?
(a) $NaCl(s)$ or $HCl(g)$
(b) $H_2O(l)$ or $CH_3\overset{\displaystyle O}{\overset{\|}{C}}H(l)$
(c) $MgBr_2(s)$ or $CH_3CH_2MgBr(s)$

13.14 Which member of each pair is more soluble in water? Why?
(a) $CH_3CH_2OCH_2CH_3(l)$ or $CH_3CH_2OCH_3(g)$

(b) $CH_2Cl_2(l)$ or $CCl_4(l)$

(c)

cyclohexane   or   tetrahydropyran

Problems in Context

13.15 The dictionary defines *homogeneous* as "uniform in composition throughout." River water is a mixture of dissolved compounds, such as calcium bicarbonate, and suspended soil particles. Is river water homogeneous? Explain.

13.16 Gluconic acid is a derivative of glucose used in cleaners and in the dairy and brewing industries. Caproic acid is a short-chain fatty acid used in the flavoring industry. Although both are six-carbon acids (see structures below), gluconic acid is soluble in water and nearly insoluble in hexane, whereas caproic acid has the opposite solubility behavior. Explain.

$$\underset{\text{gluconic acid}}{\overset{\displaystyle OH \quad OH \quad OH \quad OH \quad OH}{CH_2-CH-CH-CH-CH-COOH}}$$

$$\underset{\text{caproic acid}}{CH_3-CH_2-CH_2-CH_2-CH_2-COOH}$$

Energy Changes in the Solution Process

● Concept Review Questions

13.17 What is the relationship between solvation and hydration?

13.18 For a general solvent, which enthalpy terms in the thermochemical solution cycle are combined to obtain $\Delta H_{solvation}$?

13.19 (a) What is the charge density of an ion, and what two properties of an ion affect it?
(b) Arrange the following in order of increasing charge density:

(c) How do the two properties in part (a) affect the ionic heat of hydration, ΔH_{hydr}?

13.20 For ΔH_{soln} to be zero (or very small), what quantities must be nearly equal in magnitude? Will their signs be the same or opposite?

13.21 Aside from stacks of coins being shaken in a box, mentioned in the text, describe three common systems undergoing an increase in entropy.

13.22 A flask containing solid NH_4Cl becomes colder as water is added and the salt dissolves.
(a) Is the dissolving of NH_4Cl in water exothermic or endothermic?
(b) Is the magnitude of $\Delta H_{lattice}$ of NH_4Cl larger or smaller than the combined ΔH_{hydr} of the ions? Explain.
(c) Given the answer to part (a), why does NH_4Cl dissolve in water?

13.23 An ionic compound has a highly negative heat of solution in water. Would you expect it to be very soluble or nearly insoluble in water? Explain in terms of enthalpy and entropy changes.

● Skill-Building Exercises *(paired)*

13.24 Sketch a qualitative enthalpy diagram for the process of dissolving $KCl(s)$ in H_2O (endothermic).
13.25 Sketch a qualitative enthalpy diagram for the process of dissolving $NaI(s)$ in H_2O (exothermic).

13.26 Which ion in each pair has the greater charge density? Explain.
(a) Na^+ or Cs^+ (b) Sr^{2+} or Rb^+ (c) Na^+ or Cl^-
(d) O^{2-} or F^- (e) OH^- or SH^-
13.27 Which ion has the lower ratio of charge to ionic volume? Explain.
(a) Br^- or I^- (b) Sc^{3+} or Ca^{2+} (c) Br^- or K^+
(d) S^{2-} or Cl^- (e) Sc^{3+} or Al^{3+}

13.28 Which ion in each pair of Problem 13.26 has the *larger* ΔH_{hydr}?
13.29 Which ion in each pair of Problem 13.27 has the *smaller* ΔH_{hydr}?

13.30 (a) Use the following data to calculate the combined heats of hydration for the ions in potassium bromate ($KBrO_3$):
$$\Delta H_{lattice} = -745 \text{ kJ/mol} \qquad \Delta H_{soln} = 41.1 \text{ kJ/mol}$$
(b) Which ion do you think contributes more to the answer to part (a)? Why?
13.31 (a) Use the following data to calculate the combined heats of hydration for the ions in sodium acetate ($NaC_2H_3O_2$):
$$\Delta H_{lattice} = -763 \text{ kJ/mol} \qquad \Delta H_{soln} = 17.3 \text{ kJ/mol}$$
(b) Which ion do you think contributes more to the answer to part (a)? Why?

13.32 State whether the entropy of the system increases or decreases in each of the following processes:
(a) A glass vase is shattered.
(b) Gold is extracted and purified from its ore.
(c) Ethanol (CH_3CH_2OH) dissolves in propanol ($CH_3CH_2CH_2OH$).
(d) "Bugs" in a computer program are eliminated.
13.33 State whether the entropy of the system increases or decreases in each of the following processes:
(a) Pure gases are mixed to prepare an anesthetic.
(b) Electronic-grade silicon is prepared from sand.
(c) Dry ice (solid CO_2) sublimes.
(d) Gasoline is burned in a car engine.

● Problems in Context

13.34 Silver nitrate is used industrially to produce silver halides for photographic film and in forensic science to accomplish a similar task. The sodium chloride left behind in the sweat of a fingerprint is treated with silver nitrate solution to form silver chloride. This precipitate is then developed to show the black-and-white fingerprint. Given that the lattice energy of silver nitrate is -822 kJ/mol and its heat of hydration is -799 kJ/mol, calculate its heat of solution.

Solubility as an Equilibrium Process
(Sample Problem 13.2)

● Concept Review Questions

13.35 You are given a bottle of solid X and three aqueous solutions of X, one saturated, one unsaturated, and one supersaturated. How would you determine which solution is which?

13.36 Potassium permanganate ($KMnO_4$) has a solubility of 6.4 g/100 g of H_2O at 20°C and a curve of solubility vs. temperature that slopes upward to the right. How would you prepare a supersaturated solution of $KMnO_4$?

13.37 Why does the solubility of any gas in water decrease with rising temperature?

● **Skill-Building Exercises (paired)**

13.38 For a saturated aqueous solution of each of the following at 20°C and 1 atm, will the solubility increase, decrease, or stay the same when the indicated change occurs?
(a) $O_2(g)$, increase P (b) $N_2(g)$, increase V

13.39 For a saturated aqueous solution of each of the following at 20°C and 1 atm, will the solubility increase, decrease, or stay the same when the indicated change occurs?
(a) $He(g)$, decrease T (b) $RbI(s)$, increase P

13.40 The Henry's law constant (k_H) for O_2 in water at 20°C is 1.28×10^{-3} mol/L·atm.
(a) How many grams of O_2 will dissolve in 2.00 L of H_2O that is in contact with pure O_2 at 1.00 atm?
(b) How many grams of O_2 will dissolve in 2.00 L of H_2O that is in contact with air, where the partial pressure of O_2 is 0.209 atm?

13.41 Argon makes up 0.93% by volume of the atmosphere. Calculate its solubility (mol/L) in water at 20°C and 1.0 atm. The Henry's law constant for Ar under these conditions is 1.5×10^{-3} mol/L·atm.

● **Problems in Context**

13.42 Caffeine is about 10 times as soluble in hot water as in cold water. A chemist puts a hot-water extract of caffeine into an ice bath, and some caffeine crystallizes. Is the remaining solution saturated, unsaturated, or supersaturated?

13.43 The partial pressure of CO_2 gas above the liquid in a bottle of champagne at 20°C is 5.5 atm. What is the solubility of CO_2 in champagne? Assume Henry's law constant is the same for champagne as for water: at 20°C, $k_H = 3.7 \times 10^{-2}$ mol/L·atm.

13.44 Individuals with respiratory problems are often treated with devices that deliver air with a higher partial pressure of O_2 than normal air. Why?

Quantitative Ways of Expressing Concentration
(Sample Problems 13.3 to 13.5)

● **Concept Review Questions**

13.45 Explain the difference between molarity and molality. Under what circumstances would molality be a more accurate measure of the prepared concentration of a solution than molarity? Why?

13.46 Which of the different ways of expressing concentration include the following: (a) volume of solution; (b) mass of solution; (c) mass of solvent?

13.47 A solute has a solubility in water of 21 g/kg solvent. Is this value the same as 21 g/kg solution? Explain.

13.48 You want to convert among molarity, molality, and mole fraction of a solution. You know the masses of solute and solvent and the volume of solution. Is this enough information to carry out all the conversions? Explain.

13.49 When a solution is heated, which of the different ways of expressing concentration change in value? Which remain unchanged? Explain.

● **Skill-Building Exercises (paired)**

13.50 Calculate the molarity of each aqueous solution:
(a) 42.3 g of table sugar ($C_{12}H_{22}O_{11}$) in 100. mL of solution
(b) 5.50 g of $LiNO_3$ in 505 mL of solution

13.51 Calculate the molarity of each aqueous solution:
(a) 0.82 g of ethanol (C_2H_5OH) in 10.5 mL of solution
(b) 1.22 g of gaseous NH_3 in water to make 33.5 mL of solution

13.52 Calculate the molarity of each aqueous solution:
(a) 75.0 mL of 0.250 M NaOH diluted to 0.250 L with water
(b) 35.5 mL of 1.3 M HNO_3 diluted to 0.150 L with water

13.53 Calculate the molarity of each aqueous solution:
(a) 25.0 mL of 6.15 M HCl diluted to 0.500 L with water
(b) 8.55 mL of 2.00×10^{-2} M KI diluted to 10.0 mL with water

13.54 How would you prepare the following aqueous solutions?
(a) 355 mL of 8.74×10^{-2} M KH_2PO_4 from solid KH_2PO_4
(b) 425 mL of 0.315 M NaOH from 1.25 M NaOH

13.55 How would you prepare the following aqueous solutions?
(a) 3.5 L of 0.55 M NaCl from solid NaCl
(b) 17.5 L of 0.3 M urea [$(NH_2)_2C=O$] from 2.2 M urea

13.56 How would you prepare the following aqueous solutions?
(a) 1.50 L of 0.257 M KBr from solid KBr
(b) 355 mL of 0.0956 M $LiNO_3$ from 0.244 M $LiNO_3$

13.57 How would you prepare the following aqueous solutions?
(a) 67.5 mL of 1.33×10^{-3} M $Cr(NO_3)_3$ from solid $Cr(NO_3)_3$
(b) 6.8×10^3 m³ of 1.55 M NH_4NO_3 from 3.00 M NH_4NO_3

13.58 Calculate the molality of the following:
(a) A solution containing 88.4 g of glycine (NH_2CH_2COOH) dissolved in 1.250 kg of H_2O
(b) A solution containing 8.89 g of glycerol ($C_3H_8O_3$) in 75.0 g of ethanol (C_2H_6O)

13.59 Calculate the molality of the following:
(a) A solution containing 164 g of HCl in 753 g of H_2O
(b) A solution containing 16.5 g of naphthalene ($C_{10}H_8$) in 53.3 g of benzene (C_6H_6)

13.60 What is the molality of a solution consisting of 34.0 mL of benzene (C_6H_6; $d = 0.877$ g/mL) in 187 mL of hexane (C_6H_{14}; $d = 0.660$ g/mL)?

13.61 What is the molality of a solution consisting of 2.77 mL of carbon tetrachloride (CCl_4; $d = 1.59$ g/mL) in 79.5 mL of methylene chloride (CH_2Cl_2; $d = 1.33$ g/mL)?

13.62 How would you prepare the following aqueous solutions?
(a) 3.00×10^2 g of 0.115 m ethylene glycol ($C_2H_6O_2$) from ethylene glycol and water
(b) 1.00 kg of 2.00 mass % HNO_3 from 62.0 mass % HNO_3

13.63 How would you prepare the following aqueous solutions?
(a) 1.00 kg of 0.0555 m ethanol (C_2H_5OH) from ethanol and water
(b) 475 g of 15.0 mass % HCl from 37.1 mass % HCl

13.64 A solution of isopropanol (C_3H_7OH) is made by dissolving 0.30 mol of isopropanol in 0.80 mol of water.
(a) What is the mole fraction of isopropanol?
(b) What is the mass percent of isopropanol?
(c) What is the molality of isopropanol?

13.65 A solution is made by dissolving 0.100 mol of NaCl in 8.60 mol of water.
(a) What is the mole fraction of NaCl?
(b) What is the mass percent of NaCl?
(c) What is the molality of NaCl?

13.66 What mass of cesium chloride must be added to 0.500 L of water (d = 1.00 g/mL) to produce a 0.400 m solution? What are the mole fraction and the mass percent of CsCl?

13.67 What are the mole fraction and the mass percent of a solution made by dissolving 0.30 g of KBr in 0.400 L of water (d = 1.00 g/mL)?

13.68 An 8.00 mass % aqueous solution of ammonia has a density of 0.9651 g/mL. Calculate the molality, molarity, and mole fraction of NH_3.

13.69 A 28.8 mass % aqueous solution of iron(III) chloride has a density of 1.280 g/mL. Calculate the molality, molarity, and mole fraction of $FeCl_3$.

● **Problems in Context**

13.70 Wastewater from a cement factory contains 0.22 g of Ca^{2+} ion and 0.066 g of Mg^{2+} ion per 100.0 L of solution. The solution density is 1.001 g/mL. Calculate the Ca^{2+} and Mg^{2+} concentrations in ppm (by mass).

13.71 An automobile antifreeze mixture is made by mixing equal volumes of ethylene glycol (d = 1.114 g/mL; $\mathcal{M}$ = 62.07 g/mol) and water (d = 1.00 g/mL) at 20°C. The density of the mixture is 1.070 g/mL. Express the concentration of ethylene glycol as each of the following:
(a) volume percent (b) mass percent (c) molarity
(d) molality (e) mole fraction

Colligative Properties of Solutions
(Sample Problems 13.6 to 13.8)

● **Concept Review Questions**

13.72 The composition (chemical formula) of a solute does *not* affect the extent of the solution's colligative properties. What characteristic of a solute *does* affect these properties? Name a physical property of a solution that *is* affected by the composition of the solute.

13.73 What is a nonvolatile nonelectrolyte? Why does this type of solute provide the simplest case for examining colligative properties?

13.74 In what sense is a strong electrolyte "strong"? What property of the substance makes it a strong electrolyte?

13.75 Express Raoult's law in words. Is Raoult's law valid for a solution of a volatile solute? Explain.

13.76 What are the most important differences between the phase diagram of a pure solvent and the phase diagram of a solution of that solvent?

13.77 Is the composition of the vapor at the top of a fractionating column different from the composition at the bottom? Explain.

13.78 Is the boiling point of 0.01 m KF(aq) higher or lower than that of 0.01 m glucose(aq)? Explain.

13.79 Which aqueous solution has a boiling point closer to its predicted value, 0.050 m NaF or 0.50 m KCl? Explain.

13.80 Which aqueous solution has a freezing point closer to its predicted value, 0.01 m NaBr or 0.01 m $MgCl_2$? Explain.

13.81 The freezing point depression constants of the solvents cyclohexane and naphthalene are 20.1°C/m and 6.94°C/m, respectively. Which would give a more accurate determination by freezing point depression of the molar mass of a substance that is soluble in either solvent? Why?

● **Skill-Building Exercises** *(paired)*

13.82 Classify the following substances as strong electrolytes, weak electrolytes, or nonelectrolytes:
(a) hydrogen chloride (HCl)
(b) potassium nitrate (KNO_3)
(c) glucose ($C_6H_{12}O_6$)
(d) ammonia (NH_3)

13.83 Classify the following substances as strong electrolytes, weak electrolytes, or nonelectrolytes:
(a) sodium permanganate ($NaMnO_4$)
(b) acetic acid (CH_3COOH)
(c) methanol (CH_3OH)
(d) calcium acetate [$Ca(C_2H_3O_2)_2$]

13.84 How many moles of solute particles are present in 1 L of each of the following aqueous solutions?
(a) 0.2 M KI   (b) 0.070 M HNO_3
(c) 10^{-4} M K_2SO_4 (d) 0.07 M ethanol (C_2H_5OH)

13.85 How many moles of solute particles are present in 1 mL of each of the following aqueous solutions?
(a) 0.01 M $CuSO_4$ (b) 0.005 M $Ba(OH)_2$
(c) 0.06 M pyridine (C_5H_5N)   (d) 0.05 M $(NH_4)_2CO_3$

13.86 Which solution has the lower freezing point?
(a) 10.0 g of CH_3OH in 100. g of H_2O *or*
20.0 g of CH_3CH_2OH in 200. g of H_2O
(b) 10.0 g of H_2O in 1.00 kg of CH_3OH *or*
10.0 g of CH_3CH_2OH in 1.00 kg of CH_3OH

13.87 Which solution has the higher boiling point?
(a) 35.0 g of $C_3H_8O_3$ in 250. g of ethanol *or*
35.0 g of $C_2H_6O_2$ in 250. g of ethanol
(b) 20. g of $C_2H_6O_2$ in 0.50 kg of H_2O *or*
20. g of NaCl in 0.50 kg of H_2O

13.88 Rank the following aqueous solutions
(I)   0.100 m $NaNO_3$
(II)   0.200 m glucose
(III) 0.100 m $CaCl_2$
in order of increasing (a) osmotic pressure; (b) boiling point; (c) freezing point; (d) vapor pressure at 50°C.

13.89 Rank the following aqueous solutions
(I)   0.04 m urea [$(NH_2)_2C{=}O$]
(II)   0.02 m $AgNO_3$
(III) 0.02 m $CuSO_4$
in order of decreasing (a) osmotic pressure; (b) boiling point; (c) freezing point; (d) vapor pressure at 298 K.

13.90 Calculate the vapor pressure of a solution of 44.0 g of glycerol ($C_3H_8O_3$) in 500.0 g of water at 25°C. The vapor pressure of water at 25°C is 23.76 torr. (Assume ideal behavior.)

13.91 Calculate the vapor pressure of a solution of 0.39 mol of cholesterol in 5.4 mol of toluene at 32°C. Pure toluene has a vapor pressure of 41 torr at this temperature. (Assume ideal behavior.)

13.92 What is the freezing point of 0.111 m urea in water?

13.93 What is the boiling point of 0.200 m lactose in water?

13.94 The boiling point of ethanol (C_2H_5OH) is 78.5°C. What is the boiling point of a solution of 3.4 g of vanillin ($\mathcal{M}$ = 152.14 g/mol) in 50.0 g of ethanol (K_b of ethanol = 1.22°C/m)?

13.95 The freezing point of benzene is 5.5°C. What is the freezing point of a solution of 5.00 g of naphthalene ($C_{10}H_8$) in 444 g of benzene (K_f of benzene = 4.90°C/m)?

13.96 What is the minimum mass of ethylene glycol ($C_2H_6O_2$) that must be dissolved in 14.5 kg of water to prevent the solution from freezing at $-10.0°F$? (Assume ideal behavior.)

13.97 What is the minimum mass of glycerol ($C_3H_8O_3$) that must be dissolved in 11.0 mg of water to prevent the solution from freezing at $-25°C$? (Assume ideal behavior.)

13.98 Calculate the molality and van't Hoff factor (i) for the following aqueous solutions:
(a) 1.00 mass % NaCl, freezing point = $-0.593°C$
(b) 0.500 mass % CH_3COOH, freezing point = $-0.159°C$

13.99 Calculate the molality and van't Hoff factor i for the following aqueous solutions:
(a) 0.500 mass % KCl, freezing point = $-0.234°C$
(b) 1.00 mass % H_2SO_4, freezing point = $-0.423°C$

● **Problems in Context**

13.100 Wastewater discharged into a stream by a sugar refinery contains sucrose ($C_{12}H_{22}O_{11}$) as its main impurity. The solution contains 3.42 g of sucrose/L. A government-industry project is designed to test the feasibility of removing the sugar by reverse osmosis. What pressure must be applied to the apparatus at 20.°C to produce pure water?

13.101 In a study designed to prepare new gasoline-resistant coatings, a polymer chemist dissolves 6.053 g of poly(vinyl alcohol) in enough water to make 100.0 mL of solution. At 25°C, the osmotic pressure of this solution is 0.272 atm. What is the molar mass of the polymer sample?

13.102 The U.S. Food and Drug Administration lists dichloromethane (CH_2Cl_2) and carbon tetrachloride (CCl_4) among the many chlorinated organic compounds that are carcinogenic. What are the partial pressures of these substances in the vapor above a solution of 1.50 mol of CH_2Cl_2 and 1.00 mol of CCl_4 at 23.5°C? The vapor pressures of pure CH_2Cl_2 and CCl_4 at this temperature are 352 torr and 118 torr, respectively. (Assume ideal behavior.)

The Structure and Properties of Colloids

● **Concept Review Questions**

13.103 Is the fluid inside a bacterial cell considered a solution, a colloid, or both? Explain.

13.104 What type of colloid is each of the following?
(a) milk   (b) fog   (c) shaving cream

13.105 What is Brownian motion, and what causes it?

13.106 In a movie theater, you can often see the beam of projected light. What phenomenon does this exemplify? Why does it occur?

13.107 Why don't soap micelles coagulate and form large globules? Is soap a more effective cleaner in freshwater or in seawater? Why?

Comprehensive Problems

Problems with an asterisk (*) are more challenging.

13.108 Nitrous oxide (N_2O) is used in whipped cream containers as the gas that makes the cream foam. Some experiments to use carbon dioxide (CO_2) instead have proven unsuccessful. What does this suggest about the relative sizes of the Henry's law constant for N_2O and CO_2 in cream? Explain.

13.109 An aqueous solution is 10.% glucose by mass ($d = 1.039$ g/mL at 20°C). Calculate its freezing point, boiling point at 1 atm, and osmotic pressure.

13.110 Gramicidin A is a common antibiotic. If the molecule were polar on the outside and nonpolar on the inside, would its function be affected? Explain.

13.111 Which of the following simple depictions best represents a molecular-scale view of an ionic compound in aqueous solution? Explain.

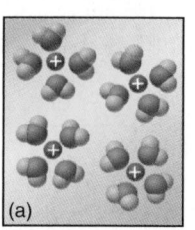 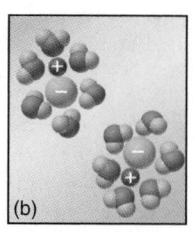 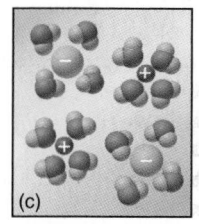
(a)   (b)   (c)

13.112 Solutes elevate the boiling point of a solvent but depress the freezing point. Explain in molecular terms.

13.113 Gold occurs in seawater at an average concentration of 1.1×10^{-2} ppb. How many liters of seawater must be processed to recover 1 troy ounce of gold, assuming 79.5% efficiency (*d* of seawater = 1.025 g/mL; 1 troy ounce = 31.1 g)?

13.114 Use atomic properties to explain why xenon is more than 25 times as soluble as helium in water at 0°C.

13.115 The text points out that molality and molarity are nearly the same for dilute aqueous solutions. Is this true for nonaqueous solvents as well? Explain.

13.116 Pyridine (see structure below) is an essential portion of many biologically active compounds, such as nicotine and vitamin B_6. Like ammonia, it has a lone pair on N, which makes it act as a weak base. Because it is miscible in a wide range of solvents, from water to benzene, pyridine is one of the most important bases and solvents in organic syntheses. Account for its solubility behavior in terms of intermolecular forces.

pyridine

13.117 "De-icing salt" is used to melt snow and ice on streets. The highway department of a small town is deciding whether to buy NaCl or $CaCl_2$ for the job. The town can obtain NaCl for 22 cents per kg. What is the maximum the town should pay for $CaCl_2$ to be cost effective?

13.118 A mixture of solid NaCl and saturated NaCl solution at 2°C is warmed slowly with constant stirring. The solid is completely dissolved at 10°C. The solution is cooled slowly to 5°C, and no solid appears. What will happen if a crystal of NaCl is added to this solution? What will happen to the temperature?

13.119 Is a solution that is 50% by mass of methanol in ethanol different from one that is 50% by mass of ethanol in methanol? Explain.

13.120 An industrial chemist is studying small organic compounds for their potential use as an automobile antifreeze. When 0.243 g of a compound is dissolved in 25.0 mL of water, the freezing point of the solution is $-0.201°C$.

(a) Calculate the molar mass of the compound (d of water = 1.00 g/mL at the temperature of the experiment).

(b) Compositional analysis of the compound gives 53.31 mass % C and 11.18 mass % H, the remainder being O. Calculate the empirical and molecular formulas of the compound.

(c) Draw two possible Lewis structures for a compound with this formula, one that forms H bonds and one that does not.

13.121 Use the concept of osmotic pressure to explain why drinking seawater does not quench your thirst.

13.122 Give brief answers for each of the following:

(a) Why are lime (CaO) and cake alum [$Al_2(SO_4)_3$] added during water purification?

(b) Why is water that contains large amounts of Ca^{2+}, Mg^{2+}, or Fe^{2+} difficult to use for cleaning?

(c) What is the meaning of "reverse" in reverse osmosis?

(d) Why might a water treatment plant use ozone as the final disinfectant instead of chlorine, even though ozone is more expensive?

(e) How does passing a saturated NaCl solution through a "spent" ion-exchange resin regenerate the resin?

13.123 Which ion in each pair has the *larger* ΔH_{hydr}?

(a) Mg^{2+} or Ba^{2+} (b) Mg^{2+} or Na^+ (c) NO_3^- or CO_3^{2-}
(d) SO_4^{2-} or ClO_4^- (e) Fe^{3+} or Fe^{2+} (f) Ca^{2+} or K^+

13.124 β-Pinene ($C_{10}H_{16}$) and α-terpineol ($C_{10}H_{18}O$) are two of the many compounds used in perfumes and cosmetics to provide a "fresh pine" scent. At 367 K, the pure substances have vapor pressures of 100.3 torr and 9.8 torr, respectively. What is the composition of the vapor (in terms of mole fractions) above a solution containing equal masses of these compounds at 367 K? (Assume ideal behavior.)

*13.125 A solution is prepared by dissolving 1.50 g of a compound in 25.0 mL of H_2O at 25°C. The boiling point of the solution is 100.45°C.

(a) What is the molar mass of the compound if it is a nonvolatile nonelectrolyte and the solution behaves ideally (d of H_2O at 25°C = 0.997 g/mL)?

(b) Independent conductivity measurements indicate that the compound is actually ionic with general formula AB_2 or A_2B. What is the molar mass of the compound *if* the solution behaves ideally?

(c) Compositional analysis indicates an empirical formula of CaN_2O_6. Explain the difference between the actual formula mass and that calculated from the boiling point elevation experiment.

(d) Calculate the van't Hoff factor (i) for this solution.

*13.126 A pharmaceutical preparation made with ethanol (C_2H_5OH) is contaminated with methanol (CH_3OH). A sample of vapor above the liquid mixture is found to contain a 97:1 mass ratio of C_2H_5OH:CH_3OH. What is the mass ratio of these alcohols in the liquid? At the temperature of the liquid, the vapor pressures of C_2H_5OH and CH_3OH are 60.5 torr and 126.0 torr, respectively.

13.127 A science student decides to make a drink very precisely by dissolving 1.00 cup of colored, granulated sucrose in water to prepare 2.00 L of drink.

(a) Assuming the coloring has negligible mass, calculate the molarity, mass percent, and molality of sucrose in the drink ($\mathcal{M}$ of sucrose = 342.3 g/mol; mass of sucrose = 190. g/cup; d of water = 0.998 g/mL).

(b) If the sucrose were in one piece, rather than granulated, what volume would be needed (d of sucrose = 1.58 g/cm^3; 1 cup = 237 mL)?

13.128 A saturated Na_2CO_3 solution is prepared, and a small excess of solid is present. A seed crystal of $Na_2^{14}CO_3$ (^{14}C is a radioactive isotope of ^{12}C) is introduced (see the figure), and the radioactivity is measured over time.

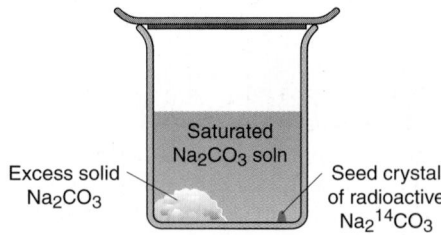

(a) Would you expect to find radioactivity in the solution? Explain.

(b) Would you expect to find radioactivity in all the solid or just in the seed crystal? Explain.

13.129 A biochemical engineer isolates a bacterial gene fragment and dissolves a 10.0-mg sample of the material in enough water to make 30.0 mL of solution. The osmotic pressure of the solution is 0.340 torr at 25°C.

(a) What is the molar mass of the gene fragment?

(b) If the solution density is 0.997 g/mL, how large is the freezing point depression for this solution (K_f of water = 1.86°C/m)?

13.130 A river is contaminated with 0.75 mg/L of dichloroethylene ($C_2H_2Cl_2$). What is the concentration of dichloroethylene (in ng/L) at 21°C in the air breathed by the people sitting along the riverbank? The Henry's law constant for dichloroethylene in water is 0.033 mol/L·atm.

13.131 A simple device used for estimating the concentration of total dissolved solids in an aqueous solution works by measuring the electrical conductivity of the solution. The method assumes that equal concentrations of different solids give approximately the same conductivity, and that the conductivity is proportional to concentration. The table below gives some actual electrical conductivities (in arbitrary units) for solutions of selected solids at the indicated concentrations (in ppm by mass):

Sample	**Conductivity**		
	0 ppm	5.00×10^3 ppm	10.00×10^3 ppm
$CaCl_2$	0.0	8.0	16.0
K_2CO_3	0.0	7.0	14.0
Na_2SO_4	0.0	6.0	11.0
Seawater (dil)	0.0	8.0	15.0
Sucrose ($C_{12}H_{22}O_{11}$)	0.0	0.0	0.0
Urea [$(NH_2)_2C{=}O$]	0.0	0.0	0.0

(a) Comment on the reliability of these measurements for estimating concentrations of dissolved solids.

(b) For what types of substances is this method likely to be seriously in error? Why?

(c) Based on this method, an aqueous $CaCl_2$ solution has a conductivity of 14.0 units. Calculate its mole fraction and molality.

*13.132 Two beakers are placed in a closed container (see figure, *left*). One beaker contains water, the other a concentrated aqueous sugar solution. With time, the solution volume increases and the water volume decreases (*right*). Explain on the molecular level.

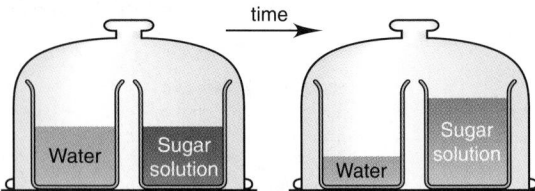

13.133 Glyphosate is the active ingredient in a common weed and grass killer. It is sold as an 18.0% by mass solution with a density of 8.94 lb/gal. (a) How many grams of Glyphosate are in a 16.0 fl oz container (1 gal = 128 fl oz)? (b) To treat a patio area of 300. ft^2, it is recommended that 3.00 fl oz be diluted with water to 1.00 gal. What is the mass percent of Glyphosate in the diluted solution (1 gal = 3.785 L)?

13.134 How would you prepare 250. g of 0.150 *m* aqueous $NaHCO_3$?

13.135 Tartaric acid can be produced from crystalline residues found in wine vats. It is used in baking powders and as an additive in foods. Analysis shows that it contains 32.3% by mass carbon and 3.97% by mass hydrogen; the balance is oxygen. When 0.981 g of tartaric acid is dissolved in 11.23 g of water, the solution freezes at −1.26°C. Use these data to find the empirical and molecular formulas of tartaric acid.

*13.136 Methanol (CH_3OH) and ethanol (C_2H_5OH) are miscible because the strongest intermolecular force for both compounds is hydrogen bonding. In some methanol-ethanol solutions, the mole fraction of methanol is higher, but the mass percent of ethanol is higher. What is the range of mole fraction of methanol for these solutions?

13.137 Carbon dioxide is less soluble in dilute $HCl(aq)$ than in dilute $NaOH(aq)$. Explain.

*13.138 Derive a general equation that expresses the relationship between the molarity and the molality of a solution, and use it to explain why the numerical values of these two terms are approximately equal for very dilute aqueous solutions.

*13.139 A florist prepares a solution of nitrogen-phosphorus fertilizer by dissolving 5.66 g of NH_4NO_3 and 4.42 g of $(NH_4)_3PO_4$ in enough water to make 20.0 L of solution. What are the molarities of NH_4^+ and of PO_4^{3-} in the solution?

13.140 Urea is a white crystalline solid that has uses as a fertilizer, in the pharmaceutical industry, and in the manufacture of certain polymer resins. Analysis of urea reveals that, by mass, it is 20.1% carbon, 6.7% hydrogen, 46.5% nitrogen and the balance oxygen.
(a) Calculate the empirical formula of urea.
(b) A 5.0 g/L solution of urea in water has an osmotic pressure of 2.04 atm, measured at 25°C. What is the molar mass and molecular formula of urea?

13.141 The total concentration of dissolved particles in blood is 0.30 *M*. An intravenous (IV) solution must be isotonic with blood, which means it has the same concentration.
(a) To relieve dehydration, a patient has 100. mL/h of IV glucose ($C_6H_{12}O_6$) for 2.5 h. How many grams of glucose did she receive?

(b) If isotonic saline (NaCl) were used, what is the molarity of the solution?
(c) If the patient has 150. mL/h of IV saline for 1.5 h, how many grams of NaCl did she receive?

13.142 The survival of fish in natural waters depends on the solubility of air in the water, which is 0.147 cm^3 of air (measured at STP) per gram of water at 20.°C and 0.10 MPa (megapascal) pressure.
(a) What is the solubility in a mountain stream where the pressure is 0.060 MPa?
(b) Calculate k_H of air in water at 20.°C.
(c) At 20.°C and 0.50 MPa, the solubility is 0.825 cm^3 of air (at STP) per gram of water. What solubility would be calculated from Henry's law? Calculate the error, as a percent of the measured solubility, from relying on Henry's law at this high pressure.

13.143 When making ice cream, the temperature of the ingredients is kept below the freezing point of water with a surrounding ice bath that contains a large amount of salt.
(a) Assuming that NaCl dissolves completely and forms an ideal solution, what mass of it is needed to lower the melting point of 5.5 kg of ice to −5.0°C?
(b) Making the same assumptions, calculate the mass of $CaCl_2$ needed.

13.144 Several different ionic compounds are each being recrystallized by the following procedure:
Step 1. A saturated aqueous solution of the compound is prepared at 50°C.
Step 2. The mixture is filtered to remove undissolved compound.
Step 3. The filtrate is cooled to 0°C.
Step 4. The crystals that form are filtered, dried, and weighed.
(a) Using Figure 13.11, which of the following compounds would have the highest % recovery and which the lowest: KNO_3, $KClO_3$, KCl, NaCl? Explain.
(b) Starting with 100. g of each compound in your answer to part (a), how many grams of each can be recovered?

13.145 An automobile radiator antifreeze must prevent the radiator fluid from freezing in winter and boiling in summer. Ethylene glycol is currently used as an antifreeze, but it is poisonous.
(a) What properties are necessary for a useful antifreeze?
(b) Is ethanol (C_2H_6O) a suitable substitute? Explain.
(c) If glycerol ($C_3H_8O_3$) were used under the same conditions as in Sample Problem 13.7, what would be the freezing and boiling points of the solution? Based on this result, would glycerol be a suitable substitute?

*13.146 The phase diagram of water in Figure 12.9B (p. 430) describes the phase changes when only water is present. Consider how the diagram would change if air were also present at normal atmospheric pressure (about 1 atm) and dissolved in the water.
(a) Would the three phases of water still attain equilibrium at some temperature? Explain.
(b) Would that temperature be higher, lower, or the same as the triple point for pure water? Explain.
(c) Would ice sublime at a few degrees below the freezing point under this pressure? Explain.
(d) Would the liquid have the same vapor pressure at 100°C as that shown in Figure 12.9B? At 120°C?

A PERSPECTIVE ON THE PROPERTIES
OF THE ELEMENTS

TOPICS

Topic 1 The Key Atomic Properties
Topic 2 Characteristics of Chemical Bonding
Topic 3 Metallic Behavior
Topic 4 Acid-Base Behavior of the Element
Oxides
Topic 5 Redox Behavior of the Elements
Topic 6 Physical States and Phase Changes

Figure: The dramatic chemical change that occurs when sodium reacts with water is just one example of the countless reactions of the main-group elements. As you have seen, these changes are based on atomic and molecular properties, which we revisit in the upcoming pages.

Chemistry has a central, underlying principle: *The behavior of a sample of matter emerges from the properties of its component atoms.* This illustrated Interchapter reviews many ideas about these properties from earlier chapters so that you can see in upcoming chapters how they lead to the behavior of the main-group elements. Keep in mind that, despite our categories, clear dividing lines rarely appear in nature: instead, the periodic properties of matter display gradual changes from one substance to another.

INTERCHAPTER INTERCHAPTER INTERCHAPTER INTERCHAPTER

The Key Atomic Properties

Four atomic properties are critical to the behavior of an element: *electron configuration, atomic size, ionization energy,* and *electronegativity.*

s Block

n	1A (1) ns^1	2A (2) ns^2
1	1 **H** $1s^1$	2 **He** $1s^2$
2	3 **Li** $2s^1$	4 **Be** $2s^2$
3	11 **Na** $3s^1$	12 **Mg** $3s^2$
4	19 **K** $4s^1$	20 **Ca** $4s^2$
5	37 **Rb** $5s^1$	38 **Sr** $5s^2$
6	55 **Cs** $6s^1$	56 **Ba** $6s^2$
7	87 **Fr** $7s^1$	88 **Ra** $7s^2$

p Block

3A (13) ns^2np^1	4A (14) ns^2np^2	5A (15) ns^2np^3	6A (16) ns^2np^4	7A (17) ns^2np^5	8A (18) ns^2np^6
					He
5 **B** $2s^2p^1$	6 **C** $2s^2p^2$	7 **N** $2s^2p^3$	8 **O** $2s^2p^4$	9 **F** $2s^2p^5$	10 **Ne** $2s^2p^6$
13 **Al** $3s^23p^1$	14 **Si** $3s^23p^2$	15 **P** $3s^23p^3$	16 **S** $3s^23p^4$	17 **Cl** $3s^23p^5$	18 **Ar** $3s^23p^6$
31 **Ga** $4s^24p^1$	32 **Ge** $4s^24p^2$	33 **As** $4s^24p^3$	34 **Se** $4s^24p^4$	35 **Br** $4s^24p^5$	36 **Kr** $4s^24p^6$
49 **In** $5s^25p^1$	50 **Sn** $5s^25p^2$	51 **Sb** $5s^25p^3$	52 **Te** $5s^25p^4$	53 **I** $5s^25p^5$	54 **Xe** $5s^25p^6$
81 **Tl** $6s^26p^1$	82 **Pb** $6s^26p^2$	83 **Bi** $6s^26p^3$	84 **Po** $6s^26p^4$	85 **At** $6s^26p^5$	86 **Rn** $6s^26p^6$
	114 $7s^27p^2$				

Transition Elements (d block): 21, 22, 23, 24, 25, 26, 27, 28, 29, 30; 39, 40, 41, 42, 43, 44, 45, 46, 47, 48; 57, 72, 73, 74, 75, 76, 77, 78, 79, 80; 89, 104, 105, 106, 107, 108, 109, 110, 111, 112

Inner Transition Elements (f block): 58, 59, 60, 61, 62, 63, 64, 65, 66, 67, 68, 69, 70, 71; 90, 91, 92, 93, 94, 95, 96, 97, 98, 99, 100, 101, 102, 103

Electron configuration ($nl^{\#}$) is the distribution of electrons in the energy levels and sublevels of an atom *(Sections 7.4 and 8.3):*
- The *n* value (positive integer) indicates the *energy* and *relative distance* from the nucleus of orbitals in a level.
- The *l* value (and its more commonly used letter designation *s, p, d, f*) indicates the *shape* of orbitals in the sublevel *(right).*
- The superscript ($^{\#}$) tells the *number* of electrons in the sublevel.

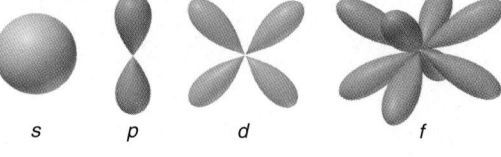

s   *p*   *d*   *f*

The periodic table above shows the sublevel blocks of elements and the ground-state (lowest energy), outer (valence-level) electron configuration of the main-group (*s* and *p* block) elements. Note that
- The **s block** lies to the left of the transition elements and the **p block** lies to the right. [Although H is not an alkali metal, like the elements in Group 1A(1), and He belongs in Group 8A(18), both are part of the *s* block.]
- Outer electron configurations are *similar within a group.*
- Outer electron configurations are *different within a period.*
- Outer electrons occupy the *ns* and *np* sublevels (*n* = period number).
- Four valence-level orbitals (one *ns* + three *np*) occur among the main-group elements.
- Outer electrons *are* the valence electrons of the main-group elements.
- The A-group number (**1A** to **8A**) equals the number of valence electrons.

Outer electrons are *shielded* from the full nuclear charge by electrons in the same level and, especially, by those in inner levels. Shielding reduces the attraction they experience to a much lower *effective nuclear charge* (Z_{eff}) *(Section 8.2).* Within a level, electrons that *penetrate* more (spend more time near the nucleus) shield others more effectively. The extent of penetration is in the order $s > p > d > f$. The radial probability distribution curves *(right)* show that
- Inner (*n* = 1) electrons shield outer (*n* = 2) electrons very effectively.
- 2*s* electrons shield 2*p* electrons somewhat because 2*s* electrons spend more time near the nucleus.

Z_{eff} greatly influences atomic properties. In general,
- Z_{eff} increases significantly left to right *across* a period.
- Z_{eff} increases slightly *down* a group.

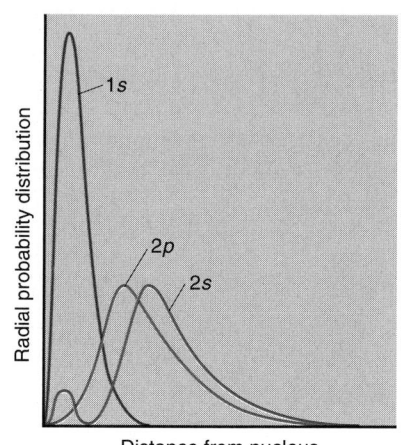

Radial probability distribution — 1s, 2p, 2s

Distance from nucleus

Atomic size is based on atomic radius, one-half the distance between nuclei of identical, bonded atoms *(Section 8.4)*. The small red periodic table shows the trends in atomic size among the main-group elements. Note that

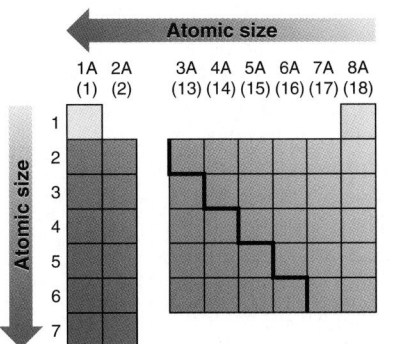

- Atomic size generally *decreases left to right across a period:* increasing Z_{eff} pulls outer electrons closer.
- Atomic size generally *increases down a group:* outer electrons in higher periods lie farther from the nucleus.

Ionization energy (IE) is the energy required to remove the highest energy electron from 1 mol of gaseous atoms *(Section 8.4)*. The relative magnitude of the IE influences the types of bonds an atom forms: an element with a *low IE* is more likely to *lose* electrons, and one with a *high IE* is more likely to *share (or gain)* electrons (excluding the noble gases). From the small yellow periodic table, note that

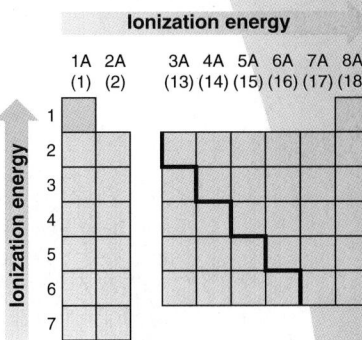

- IE generally *increases left to right across a period:* higher Z_{eff} holds electrons tighter.
- IE generally *decreases down a group:* greater distance from the nucleus lowers the attraction for electrons.

Thus, the trends in IE are *opposite* those in atomic size: it is easier to remove an electron (lower IE) that is farther from the nucleus (larger atomic size).

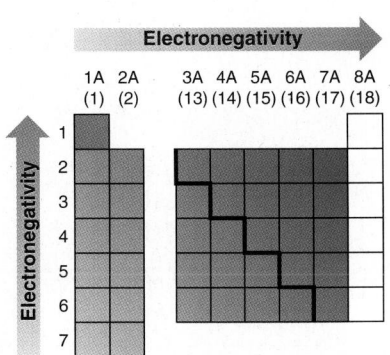

Electronegativity (EN) is a number that refers to the relative ability of an atom in a covalent bond to attract shared electrons *(Section 9.4)*. From the small green periodic table, note that

- EN generally *increases left to right across a period:* higher Z_{eff} and shorter distance from the nucleus strengthen the attraction for the shared pair.
- EN generally *decreases down a group:* greater distance from the nucleus weakens the attraction for the shared pair. (Group 8A is not shaded because the noble gases form few compounds.)

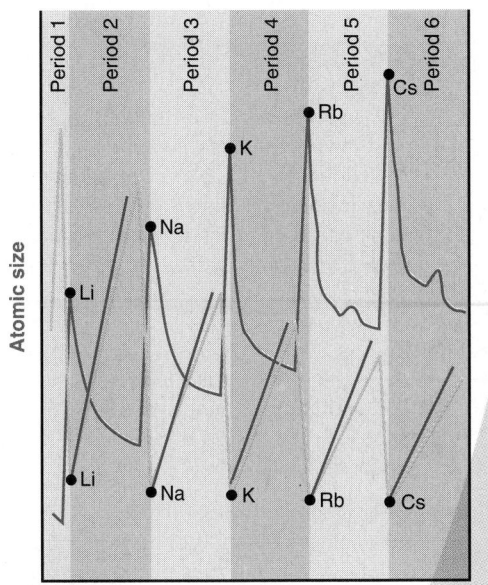

Thus, the trends in EN are *opposite* those in atomic size and the *same* as those in IE.

The graph shows simplified plots of atomic size *(red),* ionization energy *(yellow),* and electronegativity *(green)* versus atomic number. Note that

- Atomic size decreases gradually left to right within a period *(vertical gray band)* and then increases suddenly at the beginning of the next period.
- IE and EN display the opposite pattern, increasing left to right within a period and decreasing at the beginning of the next period.

The difference in electronegativity (ΔEN) between the atoms in a bond greatly influences physical and chemical behavior of the compound, as the block diagram shows.

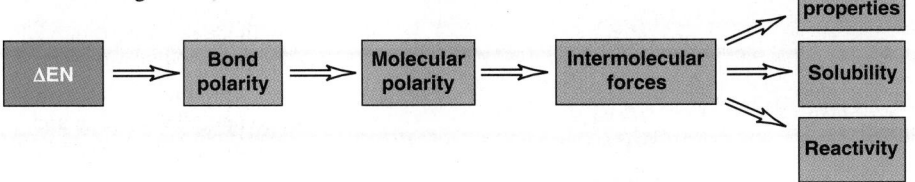

533

Characteristics of Chemical Bonding

Chemical bonds are the forces that hold atoms (or ions) together in an element or compound. The type of bonding, bond properties, nature of orbital overlap, and number of bonds determine physical and chemical behavior.

TOPIC 2

Types of Bonding

There are three idealized bonding models: *ionic, covalent,* and *metallic.*

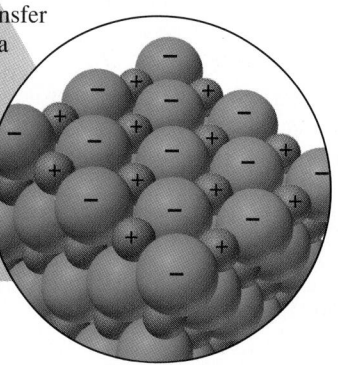

Covalent bonding results from the attraction between two nuclei and a localized electron pair. The bond arises through electron sharing between atoms with a small ΔEN (usually two nonmetals) and leads to discrete molecules with specific shapes or to extended networks *(Section 9.3).*

Ionic bonding results from the attraction between positive and negative ions. The ions arise through electron transfer between atoms with a large ΔEN (from metal to nonmetal). This bonding leads to crystalline solids with ions packed tightly in regular arrays *(Section 9.2).*

Metallic bonding results from the attraction between the cores of metal atoms (metal cations) and their delocalized valence electrons. This bonding arises through the shared pooling of valence electrons from many atoms and leads to crystalline solids *(Sections 9.5 and 12.6).*

The actual bonding in real substances usually lies between these distinct models *(Section 9.4).* The electron density depictions show that electron clouds overlap slightly even in ionic bonding (NaCl). They overlap significantly in polar covalent bonding (an SiCl bond from SiCl₄) and completely in nonpolar covalent bonding (Cl₂).

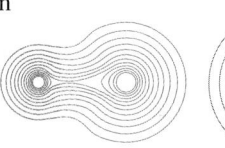
NaCl

SiCl bond in SiCl₄

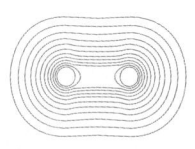
Cl₂

The triangular diagram shows the continuum of bond types among all the Period 3 main-group elements:

• Along the left side of the triangle, compounds of each element with chlorine display a gradual change from ionic to covalent bonding and a *decrease* in bond polarity from bottom to top.

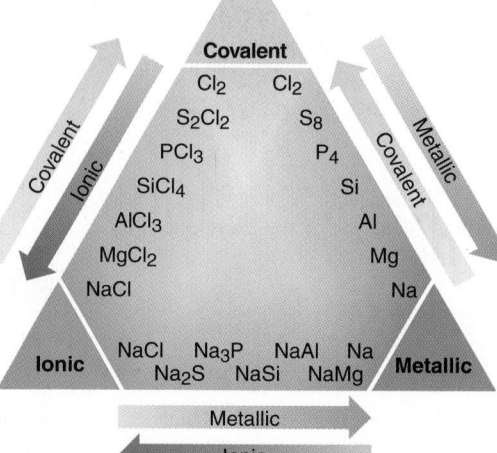

• Along the right side, the elements themselves display a gradual change from covalent to metallic bonding.

• Along the base, compounds of each element with sodium display a gradual change from ionic to metallic bonding and, once again, a *decrease* in bond polarity from left to right.

Bond Properties

There are two important properties of a covalent bond *(Section 9.3):*
Bond length is the distance between the nuclei of bonded atoms.
Bond energy (bond strength) is the enthalpy change required to break a given bond in 1 mol of gaseous molecules.

Among similar compounds, these bond properties are related to each other and to reactivity, as shown in the graph for the carbon tetrahalides (CX_4). Note that

- *As bond length increases, bond energy decreases:* shorter bonds are stronger bonds.
- *As bond energy decreases, reactivity increases.*

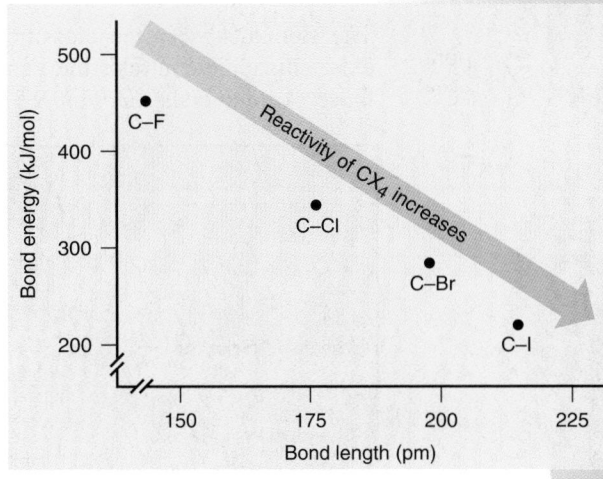

Nature of Orbital Overlap

In a covalent bond, the shared electrons reside in the entire region composed of the overlapping orbitals of the two atoms. The diagram depicts the bonding in ethylene (C_2H_4).

Bond order is one-half the number of electrons shared. Bond orders of 1 (single bond) and 2 (double bond) are common; a bond order of 3 (triple bond) is much less common. Fractional bond orders occur when there are resonance structures for species with adjacent single and double bonds *(Sections 9.3 and 10.1).*

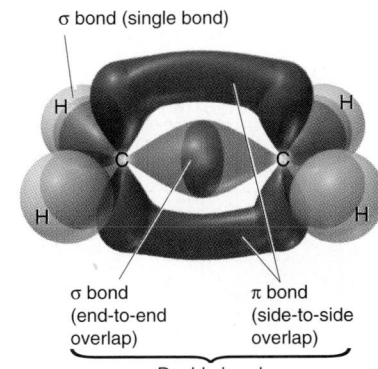

σ bond (single bond)

σ bond (end-to-end overlap)   π bond (side-to-side overlap)

Double bond

Orbitals overlap in two ways, which leads to two types of bonds *(Section 11.2):*

- **End-to-end overlap** (of *s*, *p*, and hybrid atomic orbitals) leads to a sigma (σ) bond, one with electron density distributed symmetrically along the bond axis. A single bond is a σ bond.
- **Side-to-side overlap** (of *p* with *p*, or sometimes *d*, orbitals) leads to a pi (π) bond, one with electron density distributed above and below the bond axis. A double bond consists of one σ bond and one π bond. A π bond restricts rotation around the bond axis, allowing for different spatial arrangements of the atoms and, therefore, different molecules. Pi bonds are often sites of reactivity; for example,

$$CH_2{=}CH_2(g) + H{-}Cl(g) \longrightarrow CH_3{-}CH_2{-}Cl(g)$$
$$CH_3{-}CH_3(g) + H{-}Cl(g) \longrightarrow \text{no reaction}$$

Number of Bonds and Molecular Shape

The shape of a molecule is defined by the positions of the nuclei of the bonded atoms. According to VSEPR theory *(Section 10.3),* the number of electron groups in the valence level of a central atom, which is based on the number of bonding and lone pairs, is the key factor that determines molecular shape. The small periodic table shows that

- *The elements in Period 2 cannot form more than four bonds* because they have a maximum of four (one *s* and three *p*) valence orbitals. (Only carbon forms four bonds routinely.) Molecular shapes *(small circle)* are based on linear, trigonal planar, and tetrahedral electron-group arrangements.
- *Many elements in Period 3 or higher can form more than four bonds* by using empty *d* orbitals and, thus, expanding their valence levels. Shapes include those above and others based on trigonal bipyramidal and octahedral electron-group arrangements *(large circle).*

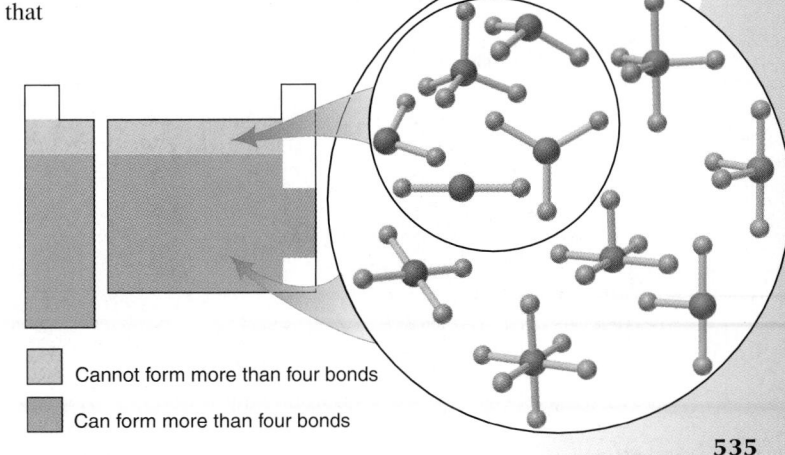

Cannot form more than four bonds

Can form more than four bonds

535

Metallic Behavior

The elements are often classified as metals, metalloids, or nonmetals. Although exceptions exist, this table contrasts the general atomic, physical, and chemical properties of metals with those of nonmetals *(Section 8.5)*.

	Metals	**Nonmetals**
Atomic Properties	Have fewer valence electrons (in period) Have larger atomic size Have lower ionization energies Have lower electronegativities	Have more valence electrons (in period) Have smaller atomic size Have higher ionization energies Have higher electronegativities
Physical Properties	Occur as solids at room temperature Conduct electricity and heat well Are malleable and ductile	Occur in all three physical states Conduct electricity and heat poorly Are not malleable or ductile
Chemical Properties	Lose electron(s) to become cations React with nonmetals to form ionic compounds Mix with other metals to form solid solutions (alloys)	Gain electron(s) to become anions React with metals to form ionic compounds React with other nonmetals to form covalent compounds

This small periodic table shows the location of metals, metalloids, and nonmetals among the main-group elements. Note that

- Metals lie in the lower-left portion of the table.
- Nonmetals lie in the upper-right portion of the table.
- Metalloids lie between the metals and nonmetals. These elements have intermediate values of atomic size, IE, and EN and display an intermediate metallic character: shiny solids with low conductivity; cation-like with nonmetals (e.g., AsF_3) and anion-like with metals (e.g., Na_3As).

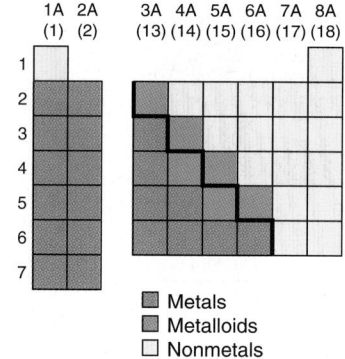

From the blue, shaded periodic table, note that

- *Metallic behavior* changes gradually among the elements.
- *Metallic behavior parallels atomic size:* Larger members of a group *(bottom)* or period *(left)* are more metallic; smaller members are less metallic.

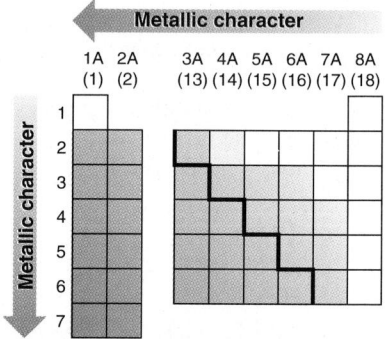

Metals and nonmetals typically form crystalline ionic compounds when they react with each other, and ionic size and charge determine the packing in these solids. Monatomic ions exhibit clear size trends:

- Cations are smaller than their parent atoms, and anions are larger.
- Ionic size *increases down* a group.
- Ionic size *decreases left to right across* a period, but anions are much larger than cations.

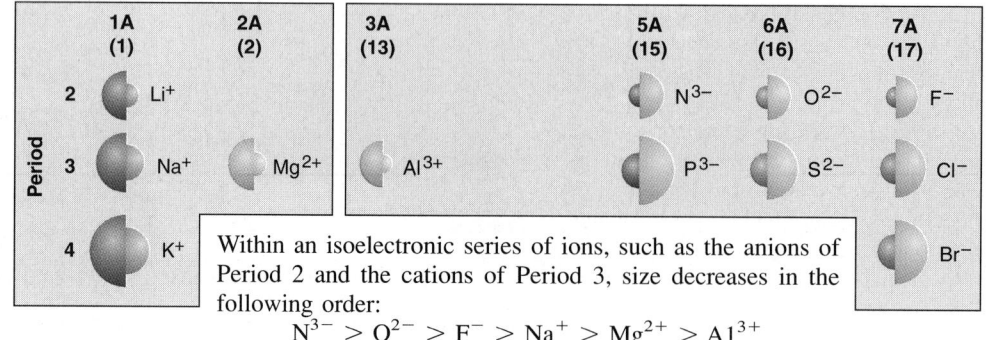

Within an isoelectronic series of ions, such as the anions of Period 2 and the cations of Period 3, size decreases in the following order:

$$N^{3-} > O^{2-} > F^- > Na^+ > Mg^{2+} > Al^{3+}$$

Acid-Base Behavior of the Element Oxides

Oxides are known for almost every element. The metallic behavior of an element correlates with the acid-base behavior of its oxide in water *(Section 8.5)*. Recall that

- An acid produces H^+ ions when dissolved in water and reacts with a base to form a salt
  and water.
- A base produces OH^- ions when dissolved in water and reacts with an acid to form a salt
  and water.

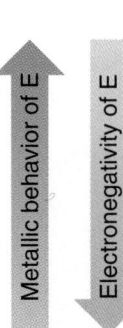

This chart shows that the
electronegativity and metallic
behavior of an element (E)
determine the type of bonding
(E-to-O) in its oxide and, therefore, the acid-base behavior of
the oxide dissolved in water.

Basic Oxide (Ionic E-to-O)	$+ H_2O \longrightarrow$	more OH^-
	$+$ acid $\longrightarrow$	salt (E cation) $+ H_2O$
Amphoteric Oxide	$+$ acid $\longrightarrow$	salt (E cation) $+ H_2O$
	$+$ base $\longrightarrow$	salt (E oxoanion) $+ H_2O$
Acidic Oxide (Covalent E-to-O)	$+$ base $\longrightarrow$	salt (E oxoanion) $+ H_2O$
	$+ H_2O \longrightarrow$	more H^+

Note that, among the main groups,

- *Elements with low EN (metals) form basic oxides.* The E-to-O bonding is ionic.
  The O^{2-} ion is the basic species:
  $$O^{2-}(s) + H_2O(l) \longrightarrow 2OH^-(aq)$$
  For example, barium forms the basic oxide BaO:
  Reacts with water: $BaO(s) + H_2O(l) \longrightarrow Ba^{2+}(aq) + 2OH^-(aq)$
  Reacts with acid: $BaO(s) + 2H^+(aq) \longrightarrow Ba^{2+}(aq) + H_2O(l)$

- *Elements with high EN (nonmetals) form acidic oxides.* The E-to-O bonding is covalent.
  Water bonds to E of the oxide to form an acid, which releases H^+. For example, sulfur forms the
  acidic oxide SO_2:
  Reacts with water: $SO_2(g) + H_2O(l) \rightleftharpoons H_2SO_3(aq) \rightleftharpoons H^+(aq) + HSO_3^-(aq)$
  Reacts with base: $SO_2(g) + 2OH^-(aq) \longrightarrow SO_3^{2-}(aq) + H_2O(l)$

- *Elements with intermediate EN (some metalloids and metals) form amphoteric oxides, which react
  with acid and base.* For example, Al_2O_3 is an amphoteric oxide:
  Reacts with acid: $Al_2O_3(s) + 6H^+(aq) \longrightarrow 2Al^{3+}(aq) + 3H_2O(l)$
  Reacts with base: $Al_2O_3(s) + 2OH^-(aq) + 3H_2O(l) \longrightarrow 2Al(OH)_4^-(aq)$

1A (1)	2A (2)		3A (13)	4A (14)	5A (15)	6A (16)	7A (17)	8A (18)
Li_2O	BeO		B_2O_3	CO_2	N_2O_5 N_2O_3			
Na_2O	MgO		Al_2O_3	SiO_2	P_4O_{10} P_4O_6	SO_3 SO_2	Cl_2O_7 Cl_2O	
K_2O	CaO		Ga_2O_3	GeO_2	As_2O_5 As_4O_6	SeO_3 SeO_2	Br_2O	
Rb_2O	SrO		In_2O_3 In_2O	SnO_2 SnO	Sb_2O_5 Sb_4O_6	TeO_3 TeO_2	I_2O_5	
Cs_2O	BaO		Tl_2O	PbO_2 PbO	Bi_2O_3	PoO_2 PoO		
Fr_2O	RaO							

- ▨ Strongly basic
- ▢ Weakly basic
- ▢ Amphoteric
- ▨ Strongly acidic
- ▨ Moderately acidic
- ▢ Weakly acidic

The periodic table shows the acid-base
behavior of many main-group oxides in
water. Note that

- *Oxide acidity increases left to right
  across a period and decreases down a
  group,* which is *opposite* the trend in
  metallic behavior (and atomic size).
- When an element forms two oxides, *the
  more acidic oxide has the element in a
  higher oxidation state* (see also Topic 5).
  Thus, for example, SO_2 forms the weak
  acid H_2SO_3, whereas SO_3 forms the
  strong acid H_2SO_4.

TOPIC 4

537

Redox Behavior of the Elements

The relative ability of an element to lose or gain electrons (or electron charge) when reacting with other elements defines its redox (oxidation-reduction) behavior *(Section 4.5)*:

- *The oxidation number, O.N. (or oxidation state)* of an atom in an element is zero. In a compound, the O.N. is the number of electrons that have shifted away from the atom (positive O.N.) or toward it (negative O.N.). O.N. values are determined by a set of rules *(Section 4.5)*.
- *An oxidation-reduction (redox) reaction* occurs when the O.N. values of any atoms in the reactants are different from those in the products.
- Among the countless redox reactions are all those that involve *an elemental substance as reactant or product*. These include all combustion reactions and all formation reactions, such as this one between potassium and chlorine: $2K + Cl_2 \longrightarrow 2KCl$

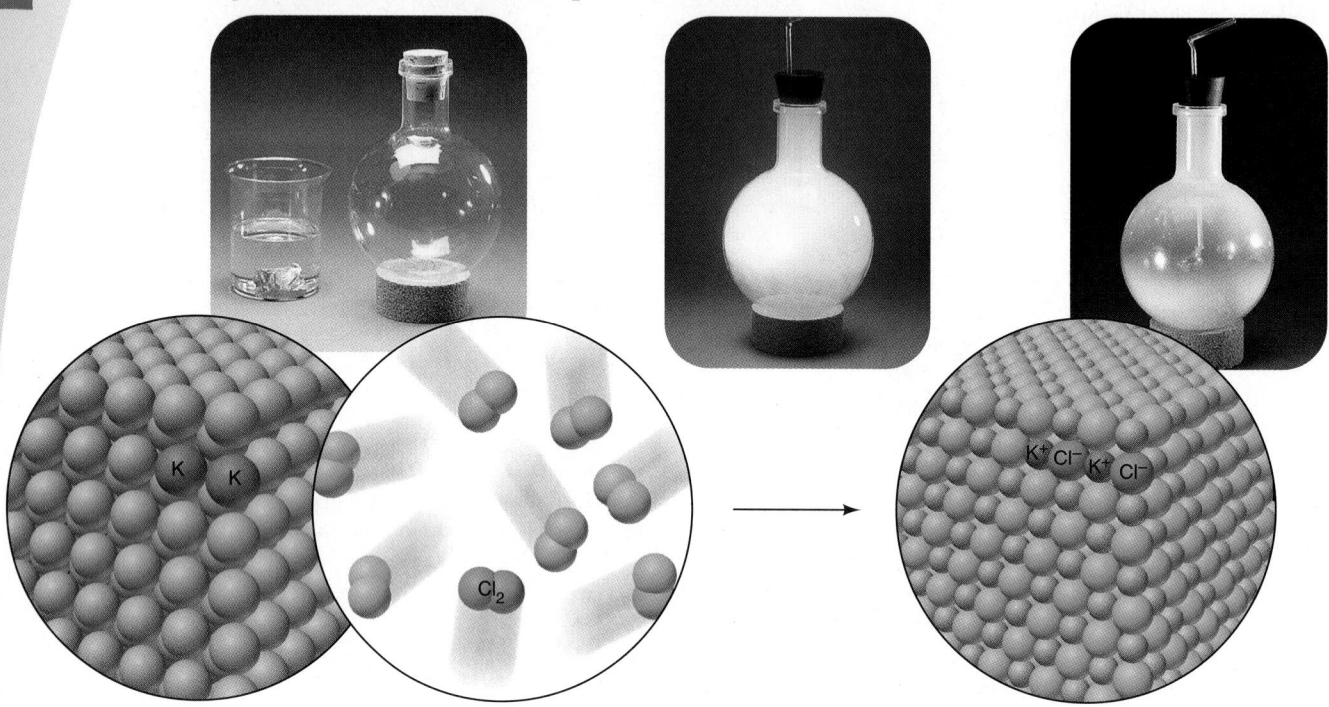

Reducing and Oxidizing Agents

All redox reactions include a reducing *and* an oxidizing agent among the reactants. This chart shows that

- The reducing agent gives electrons *to* (reduces) the oxidizing agent and becomes oxidized (more positive O.N.).
- The oxidizing agent removes electrons *from* (oxidizes) the reducing agent and becomes reduced (more negative O.N.).

Reducing agent	Oxidizing agent
X loses electrons	Y gains electrons
X is oxidized by Y	Y is reduced by X
X oxidation number increases	Y oxidation number decreases

The small periodic table shows that oxidizing and reducing ability are related to atomic properties:

- Elements with low IE and low EN [Groups 1A(1) and 2A(2)] are *strong reducing agents*.
- Elements with high IE and high EN [Group 7A(17) and oxygen in Group 6A(16)] are *strong oxidizing agents*.

Strong reducing agent
Strong oxidizing agent

Oxidation States of the Main-Group Elements

The periodic table below shows some oxidation states (most common states in **bold** type) of the elements in their compounds. Note that

- *The highest (most positive) state in a group equals the A-group number.* It occurs when all of an atom's outer (valence) electrons shift toward a *more* electronegative atom.
- *Among nonmetals, the lowest (most negative) state equals the A-group number minus eight.* It occurs when a nonmetal atom fills its outer level with electrons that shift away from a *less* electronegative atom.
- *Nonmetals have more oxidation states than metals* in the same group or period. (Oxygen and fluorine are exceptions.)
- Odd-numbered groups generally have odd-numbered states and even-numbered groups have even-numbered states. *Oxidation states differ by units of two* because electrons are lost (or gained) in pairs *(Section 14.9).*
- For many metals and metalloids with more than one oxidation state [Groups 3A(13) to 5A(15)], *the lower state becomes more common down the group.* This state results when *np* electrons only are lost *(Section 8.5).*
- An element with more than one oxidation state exhibits *greater metallic behavior in its lower state.* For example, arsenic(III) oxide is more basic, more like a metal oxide, than is arsenic(V) oxide (see Topic 4).

Most common oxidation states shown in **bold.**

Physical States and Phase Changes

Physical state and the heat of a phase change reflect the relative strengths of the bonding and/or intermolecular forces between the atoms, ions, or molecules that make up an element or compound *(Sections 12.2, 12.3, and 12.6).*

Physical States of the Elements

The large periodic table shows the type of particle or solid structure *(shape in box),* physical state *(color of shape),* and dominant interparticle force *(background color of box)* in the most common form of each of the main-group elements at room temperature. Note that

- *Metals (left and bottom) are solids:* strong metallic bonding holds the atoms in their crystal structures.
- *Metalloids (along staircase line) and carbon are solids:* strong covalent bonding holds the atoms together in extensive networks.
- *Lighter nonmetals and Group 8A(18) (right and top) are gases:* dispersion forces are weak between molecules (H_2, N_2, O_2, F_2, Cl_2) or atoms with smaller, less polarizable electron clouds.
- *Heavier nonmetals are liquid (Br_2) or soft solids (P_4, S_8, and I_2):* dispersion forces are stronger between molecules with larger, more polarizable electron clouds.

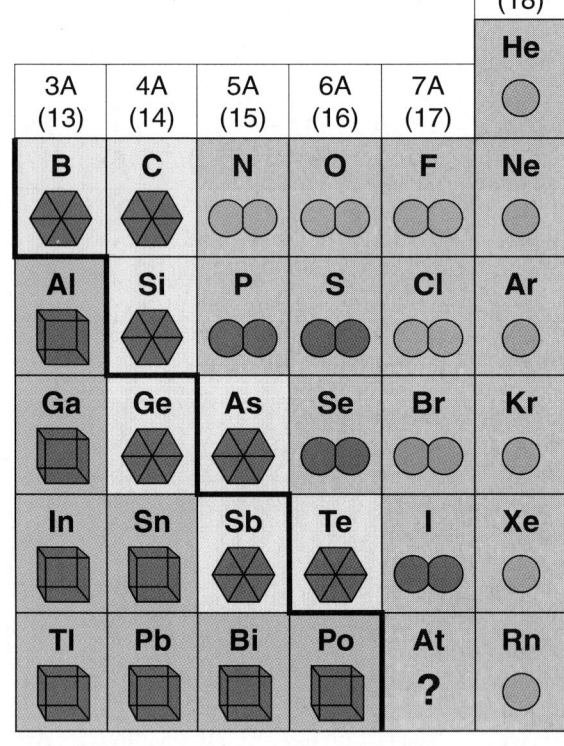

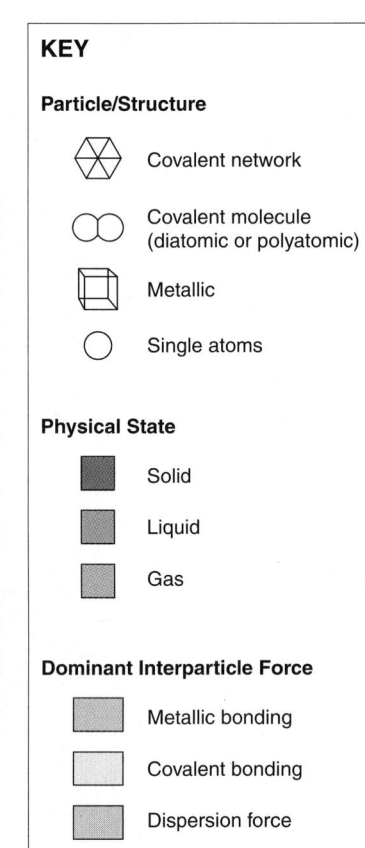

Phase Changes of the Elements

The small periodic table shows trends in melting point, boiling point, ΔH_{fus}, and ΔH_{vap} for Groups 1A(1), 7A(17), and 8A(18):

- In Group 1A(1), these properties generally *increase up* the group: smaller atom cores attract delocalized electrons more strongly, which leads to stronger metallic bonding.
- In Groups 7A(17) and 8A(18), these properties generally *increase down* the group: dispersion forces become stronger with larger, more polarizable atoms.
- In Groups 3A(13) to 6A(16), these properties reflect changes in interparticle forces down the group: lower values for molecular nonmetals, higher values for covalent networks of metalloids (and carbon), and intermediate values for metals.

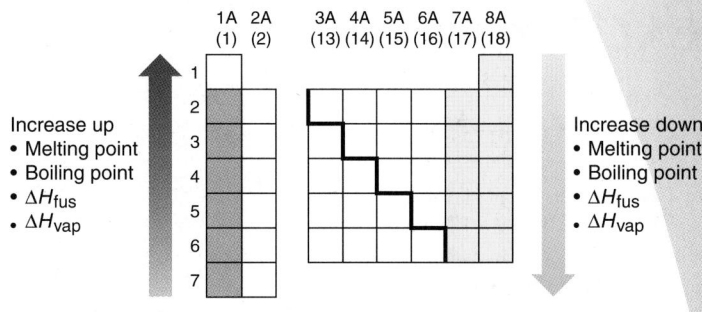

Increase up
- Melting point
- Boiling point
- ΔH_{fus}
- ΔH_{vap}

Increase down
- Melting point
- Boiling point
- ΔH_{fus}
- ΔH_{vap}

Physical Properties of Compounds

Compounds have physical properties based on the types of bonding and intermolecular forces.

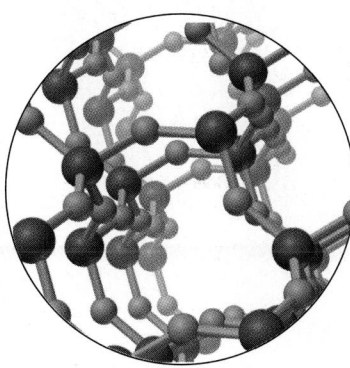

Molecular compounds, such as methane, have a physical state that depends on intermolecular forces. For polar compounds, dipole-dipole forces dominate; for nonpolar compounds, dispersion forces are most important. Most of these substances are gases, liquids, or low-melting solids at room temperature.

Hydrogen bonding arises when H is bonded to N, O, or F. It has a major effect on physical properties. For example, despite similar molar masses, H bonding in H_2O (18.02 g/mol) gives it a much higher melting point, boiling point, ΔH_{fus}, and ΔH_{vap} than CH_4 (16.04 g/mol). Water's unusually high specific heat capacity, surface tension, and viscosity also result from H bonding *(Section 12.5)*.

Network covalent compounds, such as silica, have extremely high melting points, boiling points, ΔH_{fus}, and ΔH_{vap}.

Ionic compounds, such as sodium chloride, have very high melting points, boiling points, ΔH_{fus}, and ΔH_{vap}.

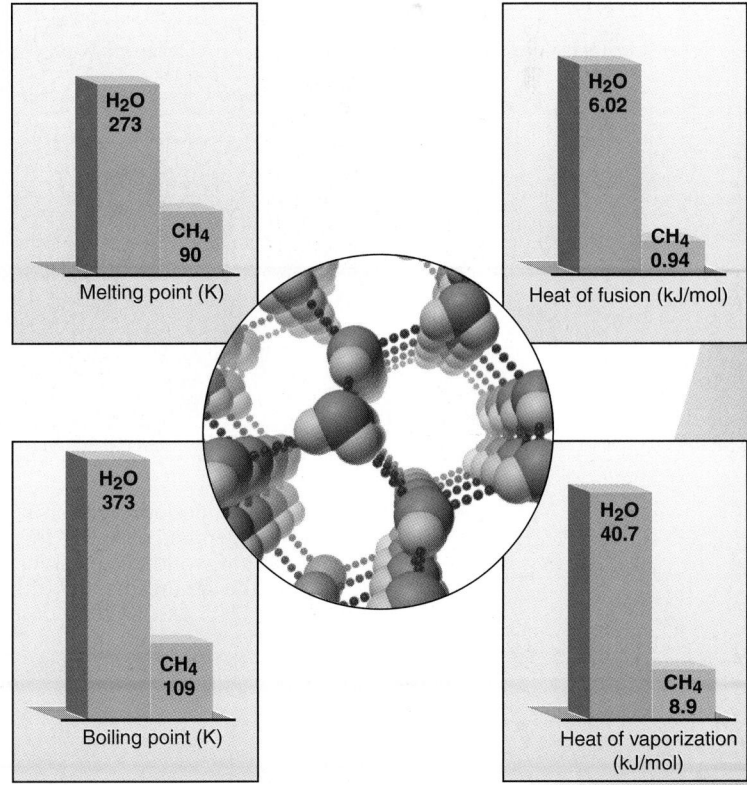

541

CHAPTER 14

PERIODIC PATTERNS IN THE MAIN-GROUP ELEMENTS: BONDING, STRUCTURE, AND REACTIVITY

CHAPTER OUTLINE

14.1 Hydrogen, the Simplest Atom
Highlights of Hydrogen Chemistry

14.2 Trends Across the Periodic Table: The Period 2 Elements

14.3 Group 1A(1): The Alkali Metals

14.4 Group 2A(2): The Alkaline Earth Metals

14.5 Group 3A(13): The Boron Family
Highlights of Boron Chemistry

14.6 Group 4A(14): The Carbon Family
Highlights of Carbon Chemistry
Highlights of Silicon Chemistry

14.7 Group 5A(15): The Nitrogen Family
Highlights of Nitrogen Chemistry
Highlights of Phosphorus Chemistry

14.8 Group 6A(16): The Oxygen Family
Highlights of Oxygen Chemistry
Highlights of Sulfur Chemistry

14.9 Group 7A(17): The Halogens
Highlights of Halogen Chemistry

14.10 Group 8A(18): The Noble Gases

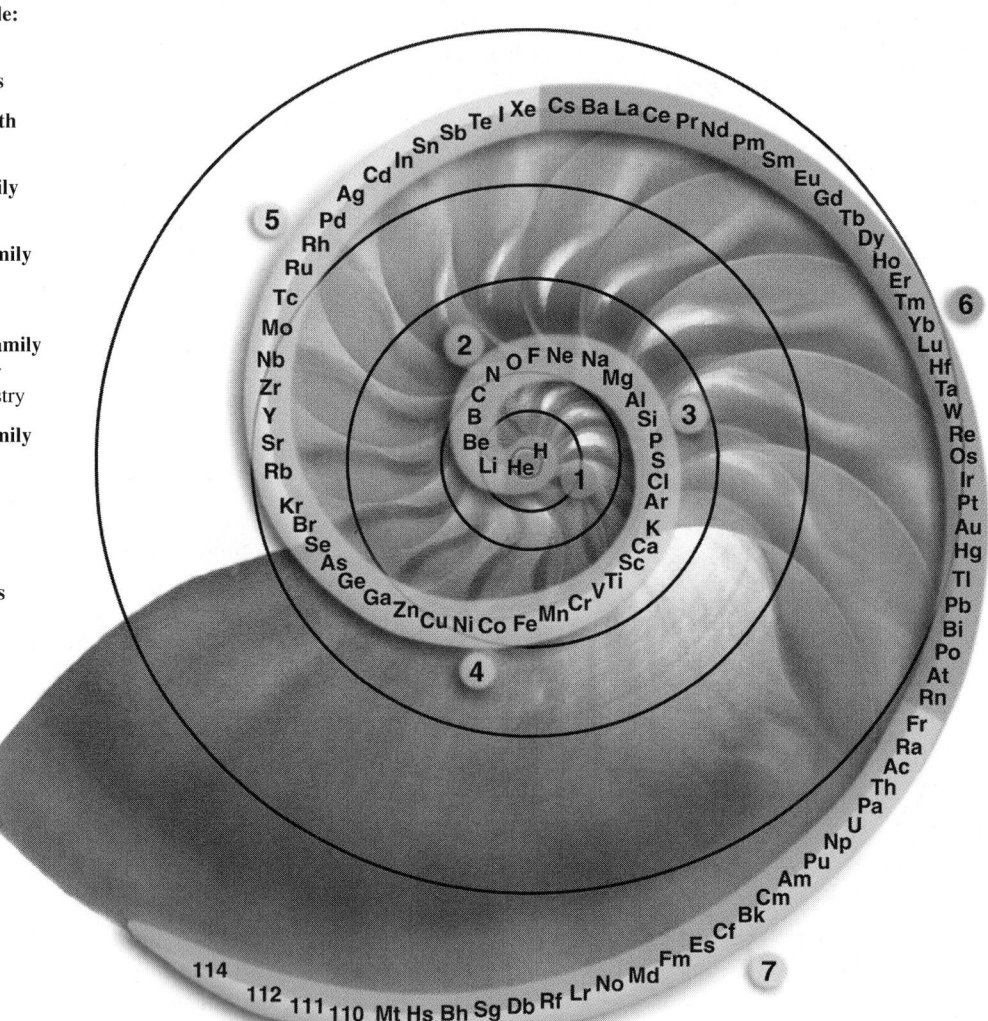

Figure: Periodic patterns in nature. From the movement of the planets to the arrangement of seeds in a sunflower, recurring patterns appear throughout the natural world. The regular increase in the spiral of the *Nautilus* shell is mimicked by the arrangement of this periodic table (colored by period and redrawn from the 1905 text *Outline of Inorganic Chemistry* by Frank Austin Gooch and Claude Frederic Walker). In this chapter, we examine the patterns in physical and chemical behavior that emerge from the recurring atomic properties of the main-group elements.

In your study of chemistry so far, you've learned how to name compounds, balance equations, and calculate reaction yields. You've seen how heat is related to chemical and physical change, how electron configuration influences atomic properties, how elements bond to form compounds, and how the arrangement of bonding and lone pairs accounts for molecular shapes. Most recently, you've seen how atomic and molecular properties give rise to the macroscopic properties of gases, liquids, solids, and solutions.

The purpose of this knowledge, of course, is to make sense of the magnificent diversity of chemical and physical behavior around you. Therefore, in this chapter, we apply these ideas to understand the bonding, structure, and reactivity of the main-group elements and to see how their behavior correlates with their position in the periodic table. We begin with hydrogen, the simplest and, in some ways, most important of all the elements, and then use Period 2 to survey the general changes *across* the periodic table. Each of the remaining sections deals with one of the eight families of main-group elements. (If you haven't already read the illustrated Interchapter feature preceding this chapter, now is the perfect time to do so. It reviews the major trends seen throughout the periodic table in preparation for the upcoming discussion.)

The periodic table was derived from chemical facts observed in countless hours of 18th- and 19th-century research. One of the greatest achievements in science is 20th-century quantum theory, which provides a theoretical foundation for the periodic table's arrangement. But bear in mind that one theory can rarely account for *all* the facts. Therefore, be amazed by the predictive power of the patterns in the table, but do not be concerned if occasional exceptions to these patterns occur. After all, our models are simple, but nature is complex.

CONCEPTS & SKILLS
to review before you study this chapter
• see Interchapter feature

14.1 HYDROGEN, THE SIMPLEST ATOM

A hydrogen atom consists of a nucleus with a single positive charge, surrounded by a single electron. Despite this simple structure, or perhaps because of it, hydrogen may be the most important element of all. In the Sun, hydrogen (H) nuclei combine to form helium (He) nuclei in a process that provides nearly all the energy on Earth. About 90% of all the atoms in the universe are H atoms, making it the most abundant element by far. On Earth, only tiny amounts of the free element occur naturally (as H_2), but hydrogen is abundant in combination with oxygen in water. Because of its simple structure and low molar mass, nonpolar gaseous H_2 is colorless and odorless, and its extremely weak dispersion forces result in very low melting ($-259°C$) and boiling points ($-253°C$).

Where Does Hydrogen Fit in the Periodic Table?

Hydrogen has no perfectly suitable position in the periodic table (Figure 14.1). Because of its single valence electron and common $+1$ oxidation state, it might almost fit in Group 1A(1). However, unlike the alkali metals, hydrogen *shares* its electron with nonmetals rather than losing it to them. Like a nonmetal, it has a relatively high ionization energy (IE = 1311 kJ/mol)—much higher than that of lithium (IE = 520 kJ/mol), whose IE is the highest in Group 1A—and a much higher electronegativity as well (EN of H = 2.1; EN of Li = 1.0).

On the other hand, hydrogen might almost fit in Group 7A(17). Like the halogens, it occurs as diatomic molecules and fills its outer shell either by electron sharing or by gaining one electron from a metal to form a monatomic anion, the hydride ion (H^-; O.N. = -1). However, hydrogen has a lower EN than any of the halogens and lacks their three valence electron pairs. Moreover, the H^- ion is rare and reactive, whereas halide ions are common and stable.

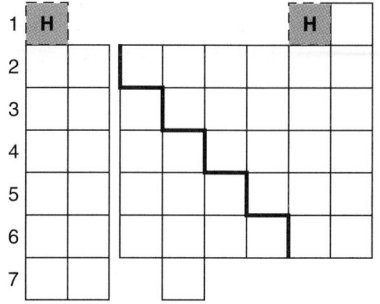

Figure 14.1 Where does hydrogen belong? Hydrogen's small size and single electron confer properties that are not fully consistent with the members of any one periodic group. Depending on the property, hydrogen may fit better in Group 1A(1) or Group 7A(17).

Hydrogen's unique behavior is due to its tiny size. It has a high IE because its electron is very close to the nucleus, with no inner electrons to shield it from the positive charge. It has a low EN (for a nonmetal) because it has only one proton to attract bonded electrons. In this chapter, hydrogen will appear in either Group 1A(1) or 7A(17) depending on the property being considered.

Highlights of Hydrogen Chemistry

In Chapters 12 and 13, we discussed hydrogen bonding and its impact on physical properties (melting and boiling points, heats of fusion and vaporization, and specific heat capacities) and solubilities. You learned that H bonding plays a critical part in stabilizing the Earth's climate as well as your body temperature, and you glimpsed its role in the functioning of biomolecules. In Chapter 6, we considered the possible future use of hydrogen as a fuel, and later in the text, we'll see how hydrogen is used to recover some metals from their ores. Hydrogen is very reactive, combining with nearly every element, so we focus here on the three types of hydrides.

Ionic (Saltlike) Hydrides With very reactive metals, such as those in Groups 1A(1) and the larger members of 2A(2) (Ca, Sr, and Ba), hydrogen forms *saltlike hydrides:* white, crystalline solids composed of the metal cation and the hydride ion:

$$2Li(s) + H_2(g) \longrightarrow 2LiH(s)$$
$$Ca(s) + H_2(g) \longrightarrow CaH_2(s)$$

In water, H^- is a strong base that pulls H^+ from surrounding H_2O molecules to form H_2 and OH^-:

$$NaH(s) + H_2O(l) \longrightarrow Na^+(aq) + OH^-(aq) + H_2(g)$$

The hydride ion is also a powerful reducing agent; for example, it reduces Ti(IV) to the free metal:

$$TiCl_4(l) + 4LiH(s) \longrightarrow Ti(s) + 4LiCl(s) + 2H_2(g)$$

Covalent (Molecular) Hydrides Hydrogen reacts with nonmetals to form many *covalent hydrides,* such as CH_4, NH_3, H_2O, and HF. Most are gases consisting of small molecules, but many hydrides of boron and carbon are liquids or solids that consist of much larger molecules. In most covalent hydrides, hydrogen has an oxidation number of $+1$ because the other nonmetal has a higher electronegativity than hydrogen.

Conditions for preparing the covalent hydrides depend on the reactivity of the other nonmetal. For example, with stable, triple-bonded N_2, hydrogen reacts at high temperatures ($\sim$400°C) and pressures ($\sim$250 atm), and the reaction needs a catalyst to proceed at any practical speed:

$$N_2(g) + 3H_2(g) \xrightarrow{\text{catalyst}} 2NH_3(g) \quad \Delta H^0_{rxn} = -91.8 \text{ kJ}$$

Industrial facilities throughout the world use this reaction to produce millions of tons of ammonia each year for fertilizers, explosives, and synthetic fibers. On the other hand, hydrogen combines rapidly with reactive, single-bonded F_2, even at extremely low temperatures (-196°C):

$$F_2(g) + H_2(g) \longrightarrow 2HF(g) \quad \Delta H^0_{rxn} = -546 \text{ kJ}$$

Metallic (Interstitial) Hydrides Many transition elements form *metallic (interstitial) hydrides,* in which H_2 molecules (and H atoms) occupy the holes in the metal's crystal structure. Thus, such hydrides are *not* compounds but gas-solid solutions. Also, unlike ionic and covalent hydrides, interstitial hydrides, such as $TiH_{1.7}$, typically lack a single, stoichiometric formula because the metal can incorporate variable amounts of hydrogen, depending on the pressure and temperature of the gas. ●

14.2 TRENDS ACROSS THE PERIODIC TABLE: THE PERIOD 2 ELEMENTS

Table 14.1 on the next two pages presents the trends in atomic properties of the Period 2 elements, lithium through neon, and the physical and chemical properties that emerge from them. In general, these trends apply to the other periods as well. Note the following points:

- Electrons fill the one *ns* and the three *np* orbitals according to the Pauli exclusion principle and Hund's rule.
- As a result of increasing nuclear charge and the addition of electrons to orbitals of the same energy level (same *n* value), atomic size generally decreases, whereas first ionization energy and electronegativity generally increase (see bar graphs on p. 547).
- Metallic character decreases with increasing nuclear charge as elements change from metals to metalloids to nonmetals.
- Reactivity is highest at the left and right ends of the period, except for the inert noble gas.
- Bonding between atoms of an element changes from metallic, to covalent in networks, to covalent in individual molecules, to none (noble gases exist as separate atoms). As expected, physical properties change abruptly at the network/molecule boundary, which occurs in Period 2 between carbon (solid) and nitrogen (gas).
- Bonding between each element and an active nonmetal changes from ionic, to polar covalent, to covalent. Bonding between each element and an active metal changes from metallic to polar covalent to ionic.
- The acid-base behavior of the common oxides in water changes from basic to amphoteric to acidic as the bond between the element and oxygen becomes more covalent.
- Reducing strength decreases through the metals, and oxidizing strength increases through the nonmetals. In Period 2, common oxidation numbers (O.N.s) equal the A-group number for Li and Be and the A-group number minus eight for O and F. Boron has several O.N.s, Ne has none, and C and N show all possible O.N.s for their groups.

One point that will be coming up often but is not shown in Table 14.1 is the anomalous behavior of the Period 2 elements within their groups. The Period 2 elements display some behavior that is *not* representative of their groups because they have a relatively *small size* and a *small number of orbitals* within the energy level.

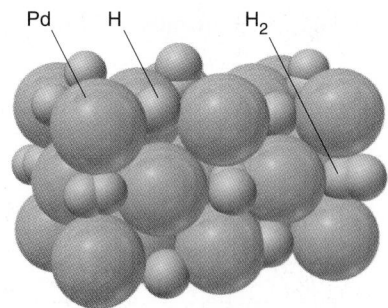

Pd   H H_2

● **Fill 'Er Up with Hydrogen? Not Likely** Dreams of a future hydrogen economy had included using metallic hydrides as "storage containers" for hydrogen fuel in cars. Just drive up to your local refueling station, insert a cylinder of H-packed metal in your car, and zip away again! It is true that some metals, such as palladium (see figure) and niobium, and some alloys, such as $LaNi_5$, store large amounts of hydrogen. However, research over the past 15 years has uncovered some serious problems, especially the facts that the metals that store the most hydrogen are expensive and heavy and they hold the gas tightest, releasing it only at very high temperatures. As a result, in 1996, U.S. government funding for the research stopped. Some groups are keeping the dream alive, however, through their studies of carbon nanotubes, which store large amounts of H_2, and of an Na-Al alloy that is lightweight, stores 5% H_2 by mass, and releases it below 200°C. Only time and careful testing will tell!

Table 14.1 Trends in Atomic, Physical, and Chemical Properties of the Period 2 Elements

Group: Element/At. No.:	1A(1) Lithium (Li) $Z = 3$	2A(2) Beryllium (Be) $Z = 4$	3A(13) Boron (B) $Z = 5$	4A(14) Carbon (C) $Z = 6$
Atomic Properties				
Condensed electron configuration; partial orbital box diagram	[He] $2s^1$ 2s 2p	[He] $2s^2$ 2s 2p	[He] $2s^2 2p^1$ 2s 2p	[He] $2s^2 2p^2$ 2s 2p
Physical Properties				
Appearance				
Metallic character	Metal	Metal	Metalloid	Nonmetal
Hardness	Soft	Hard	Very hard	Graphite: soft Diamond: extremely hard
Melting point/boiling point	Low mp for a metal	High mp	Extremely high mp	Extremely high mp
Chemical Properties				
General reactivity	Reactive	Low reactivity at room temperature	Low reactivity at room temperature	Low reactivity at room temperature; graphite more reactive
Bonding among atoms of element	Metallic	Metallic	Network covalent	Network covalent
Bonding with nonmetals	Ionic	Polar covalent	Polar covalent	Covalent (π bonds common)
Bonding with metals	Metallic	Metallic	Polar covalent	Polar covalent
Acid/base behavior of common oxide	Strongly basic	Amphoteric	Very weakly acidic	Very weakly acidic
Redox behavior (O.N.)	Strong reducing agent (+1)	Moderately strong reducing agent (+2)	Complex hydrides good reducing agents (+3, −3)	Every oxidation state from +4 to −4
Relevance/Uses of Element and Compounds				
	Li soaps for auto grease; thermonuclear bombs; high-voltage, low-weight batteries; manic-depressive treatment (Li_2CO_3)	Rocket nose cones; alloys for springs and gears; nuclear reactor parts; x-ray tubes	Cleaning agent (borax); eyewash, antiseptic (boric acid); armor (B_4C); borosilicate glass; plant nutrient	Graphite: lubricant, structural fiber Diamond: jewelry, cutting tools, protective films Limestone ($CaCO_3$) Organic compounds: drugs, fuels, textiles, etc.

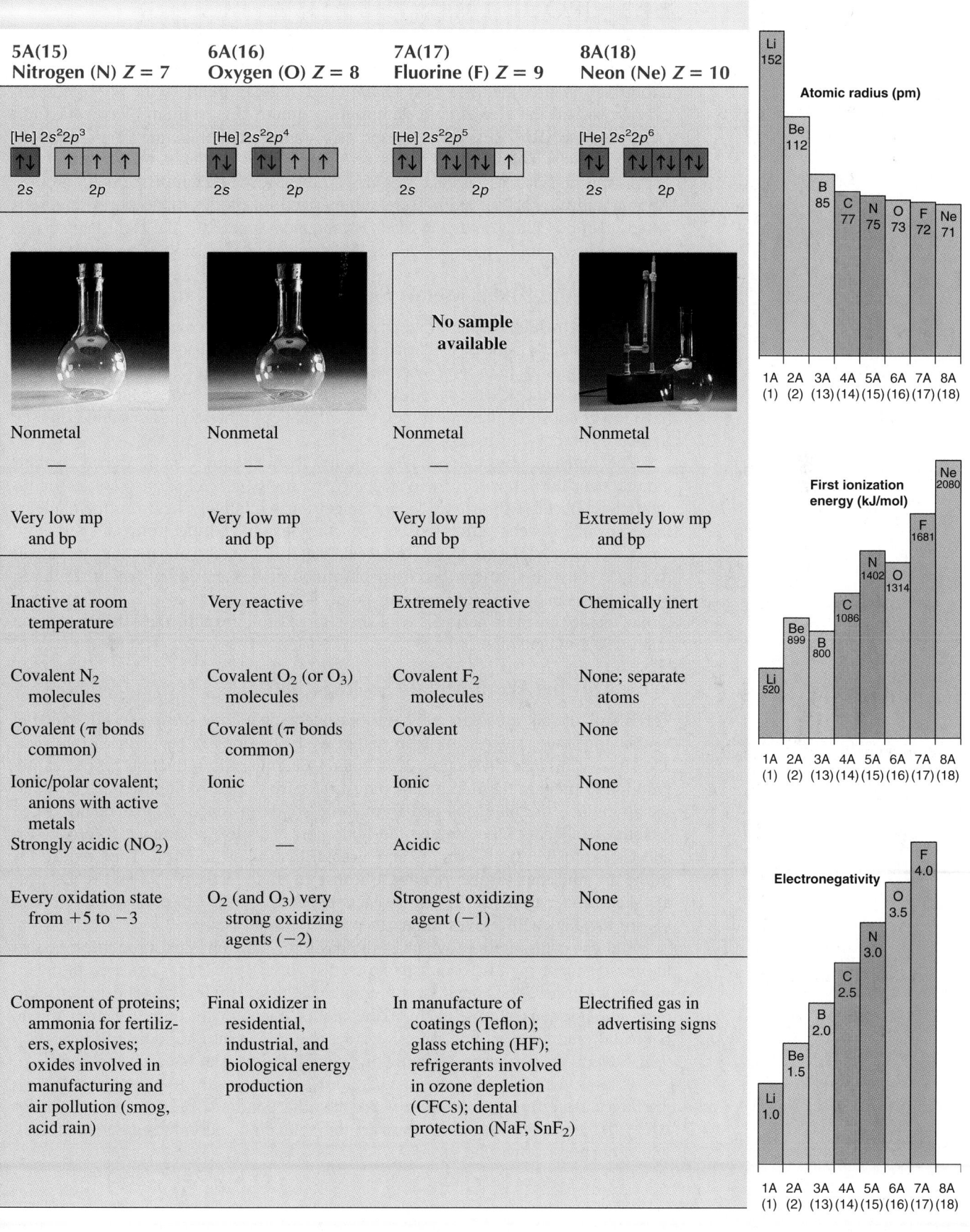

5A(15) Nitrogen (N) $Z = 7$	6A(16) Oxygen (O) $Z = 8$	7A(17) Fluorine (F) $Z = 9$	8A(18) Neon (Ne) $Z = 10$
[He] $2s^2 2p^3$	[He] $2s^2 2p^4$	[He] $2s^2 2p^5$	[He] $2s^2 2p^6$
Nonmetal	Nonmetal	Nonmetal	Nonmetal
—	—	—	—
Very low mp and bp	Very low mp and bp	Very low mp and bp	Extremely low mp and bp
Inactive at room temperature	Very reactive	Extremely reactive	Chemically inert
Covalent N_2 molecules	Covalent O_2 (or O_3) molecules	Covalent F_2 molecules	None; separate atoms
Covalent (π bonds common)	Covalent (π bonds common)	Covalent	None
Ionic/polar covalent; anions with active metals	Ionic	Ionic	None
Strongly acidic (NO_2)	—	Acidic	None
Every oxidation state from +5 to −3	O_2 (and O_3) very strong oxidizing agents (−2)	Strongest oxidizing agent (−1)	None
Component of proteins; ammonia for fertilizers, explosives; oxides involved in manufacturing and air pollution (smog, acid rain)	Final oxidizer in residential, industrial, and biological energy production	In manufacture of coatings (Teflon); glass etching (HF); refrigerants involved in ozone depletion (CFCs); dental protection (NaF, SnF$_2$)	Electrified gas in advertising signs

No sample available

Atomic radius (pm)

Li 152, Be 112, B 85, C 77, N 75, O 73, F 72, Ne 71

1A(1) 2A(2) 3A(13) 4A(14) 5A(15) 6A(16) 7A(17) 8A(18)

First ionization energy (kJ/mol)

Li 520, Be 899, B 800, C 1086, N 1402, O 1314, F 1681, Ne 2080

1A(1) 2A(2) 3A(13) 4A(14) 5A(15) 6A(16) 7A(17) 8A(18)

Electronegativity

Li 1.0, Be 1.5, B 2.0, C 2.5, N 3.0, O 3.5, F 4.0

1A(1) 2A(2) 3A(13) 4A(14) 5A(15) 6A(16) 7A(17) 8A(18)

14.3 GROUP 1A(1): THE ALKALI METALS

The first group of elements in the periodic table is named for the alkaline (basic) nature of their oxides and for the basic solutions the elements form in water. Group 1A(1) provides the best example of regular trends with no significant exceptions. All the elements in the group—lithium (Li), sodium (Na), potassium (K), rubidium (Rb), cesium (Cs), and rare, radioactive francium (Fr)—are very reactive metals. The Family Portrait of Group 1A(1) elements, on pages 550 and 551, is the first in a series that provides an overview of each of the main groups. Atomic and physical properties are summarized on the left-hand page; representative reactions and some important compounds appear on the right-hand page. Refer to these family portraits during the discussions for supporting information.

Why Are the Alkali Metals Soft, Low Melting, and Lightweight?

Unlike most metals, the alkali metals are soft: Na has the consistency of cold butter, and K can be squeezed like clay. The alkali metals also have lower melting and boiling points than any other group of metals. Except for Li, they all melt below 100°C, and Cs melts a few degrees above room temperature. They also have lower densities than most metals: Li floats on lightweight household oil (see Sample Problem 1.5, p. 23).

The unusual physical behavior of the alkali metals can be traced to their atomic size, the largest in their respective periods, and to the ns^1 electron configuration. Because the single valence electron is relatively far from the nucleus, there is only a weak attraction between the delocalized electrons and the atom cores in the crystal structure. This weak metallic bonding means that the alkali metal crystal structure can be easily deformed or broken down, which results in a soft consistency and low melting point. These elements also have the lowest molar masses in their periods; with their large atomic radii, they therefore have relatively low densities.

Why Are the Alkali Metals So Reactive?

The alkali metals are extremely reactive elements. They are *powerful reducing agents* and, thus, always occur in nature as 1+ cations rather than as free metals. (As we discuss in Section 22.4, highly endothermic reduction processes are required to prepare the free metals industrially from their molten salts.) The alkali metals reduce the halogens and form ionic solids in reactions that release large quantities of heat. They reduce the hydrogen in water, reacting vigorously (Rb and Cs explosively) to form H_2 and a metal hydroxide solution. They reduce O_2 in the air, and thus tarnish rapidly. Because of this reactivity, Na and K are usually kept under mineral oil (an unreactive liquid) in the laboratory, and Rb and Cs are handled with gloves under an inert argon atmosphere.

The ns^1 configuration, which is the basis for their physical properties, is also the reason these metals form salts so readily. In a Born-Haber cycle of the reaction between an alkali metal and a nonmetal, for example, the solid metal separates into gaseous atoms, each of the atoms transfers its outer electron to the nonmetal, and the resulting cations attract the anions into an ionic solid (Section 9.2). Several properties based on the ns^1 configuration relate to each of these steps.

1. *Low heat of atomization* (ΔH^0_{atom}). Consistent with the alkali metals' low melting and boiling points, their weak metallic bonding leads to low values for ΔH^0_{atom} (the energy needed to convert the solid into individual gaseous atoms), which decrease down the group:*

$$E(s) \longrightarrow E(g) \qquad \Delta H_{atom} \; (Li > Na > K > Rb > Cs)$$

---

*Throughout the chapter, we use E to represent any element in a group.

2. *Low IE and small ionic radius.* Each alkali metal has the largest size and the lowest IE in its period. A great *decrease* in size occurs when the outer electron is lost: the volume of the Li^+ ion is less than 13% that of the Li atom! Thus, Group 1A(1) ions are small spheres with considerable charge density.

3. *High lattice energy.* When the salts crystallize, large amounts of energy are released because the small cations lie close to the anions. Thus, the endothermic atomization and ionization steps are easily outweighed by the highly exothermic formation of the solid. For a given anion, the trend in lattice energy is the inverse of the trend in cation size: *as the cation becomes larger, the lattice energy has a smaller magnitude.* This steady decrease in the magnitude of the lattice energy within the Group 1A(1) and 2A(2) chlorides is shown in Figure 14.2.

Despite the strong ionic attractions in the solid, *nearly all Group 1A salts are water soluble.* The ions attract water molecules to create a highly exothermic heat of hydration (ΔH_{hydr}), and a large increase in disorder arises when ions in the organized crystal become randomized and hydrated in solution; together, these factors outweigh the high lattice energy.

The magnitude of the hydration energy *decreases* as ionic size increases:

$$E^+(g) \longrightarrow E^+(aq) \qquad \Delta H = -\Delta H_{hydr} \; (Li^+ > Na^+ > K^+ > Rb^+ > Cs^+)$$

Interestingly, the *smaller* ions attract water molecules strongly enough to form *larger hydrated ions.* This size trend has a major effect on the function of nerves, kidneys, and cell membranes because the *sizes* of $Na^+(aq)$ and $K^+(aq)$, the most common cations in cell fluids, influence their movement in and out of cells.

The Anomalous Behavior of Lithium

As we noted in discussing Table 14.1, all the Period 2 elements display some anomalous (unrepresentative) behavior within their groups. Even within the regular, predictable trends of Group 1A(1), Li has some atypical properties. It is the only member that forms a simple oxide and nitride, Li_2O and Li_3N, on reaction with O_2 and N_2 in air. Only Li forms molecular compounds with hydrocarbon groups from organic halides:

$$2Li(s) + CH_3CH_2Cl(g) \longrightarrow CH_3CH_2Li(s) + LiCl(s)$$

Organolithium compounds, such as CH_3CH_2Li, are liquids or low-melting solids that dissolve in nonpolar solvents and contain polar covalent $^{\delta-}C-Li^{\delta+}$ bonds. They are important reactants in the synthesis of organic compounds.

Because of its small size, Li^+ has a relatively high charge density. Therefore, it can deform nearby electron clouds to a much greater extent than the other 1A ions can, which gives many lithium salts significant covalent character (Figure 14.3). Thus, LiCl, LiBr, and LiI are much more soluble in polar organic solvents, such as ethanol and acetone, than are the halides of Na and K, because the lithium halide dipole interacts with these solvents through dipole-dipole forces. The small, highly positive Li^+ makes dissociation of Li salts into ions more difficult in water; thus, the fluoride, carbonate, hydroxide, and phosphate of Li are much less soluble in water than those of Na and K.

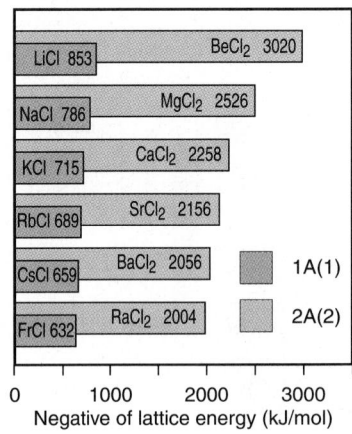

Figure 14.2 Lattice energies of the Group 1A(1) and 2A(2) chlorides. The magnitude of the lattice energy decreases regularly in both groups of metal chlorides as the cations become larger. Lattice energies for the 2A chlorides are greater because the 2A cations have higher charge and smaller size.

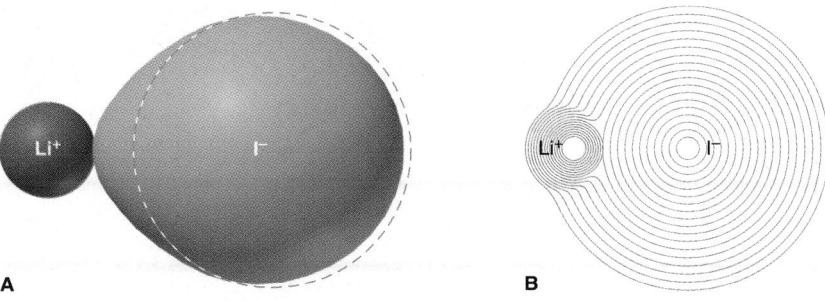

Figure 14.3 The effect of Li$^+$ charge density on a nearby electron cloud. The ability of the Li$^+$ ion to deform nearby electron clouds gives rise to many anomalous properties. In this case, Li$^+$ polarizes an I$^-$ ion, which gives some covalent character to lithium iodide, LiI, as depicted by models **(A)** and by electron density contours **(B).**

Group 1A(1): The Alkali Metals

Key Atomic and Physical Properties

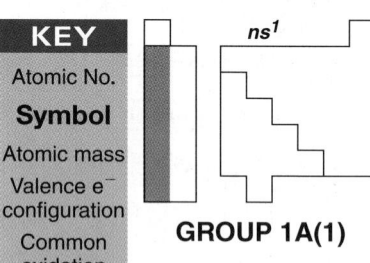

KEY
Atomic No.
Symbol
Atomic mass
Valence e⁻ configuration
Common oxidation states

GROUP 1A(1)

ns^1

Atomic Properties

Group electron configuration is ns^1. All members have the +1 oxidation state and form an E^+ ion.

Atoms have the largest size and lowest IE and EN in their periods.

Down the group, atomic and ionic size increase, while IE and EN decrease.

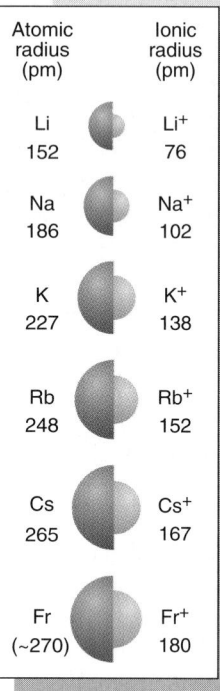

Atomic radius (pm)		Ionic radius (pm)	
Li	152	Li⁺	76
Na	186	Na⁺	102
K	227	K⁺	138
Rb	248	Rb⁺	152
Cs	265	Cs⁺	167
Fr	(~270)	Fr⁺	180

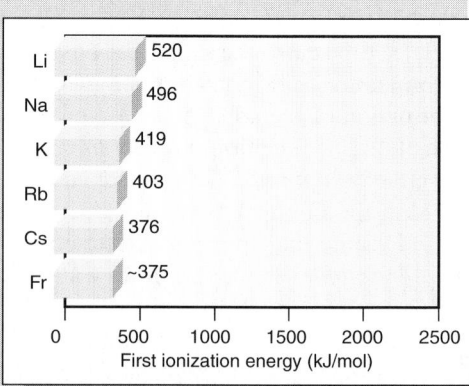

First ionization energy (kJ/mol): Li 520, Na 496, K 419, Rb 403, Cs 376, Fr ~375

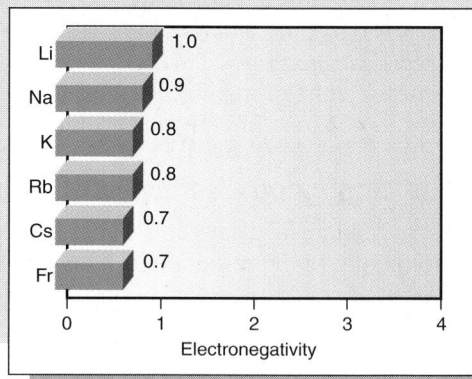

Electronegativity: Li 1.0, Na 0.9, K 0.8, Rb 0.8, Cs 0.7, Fr 0.7

Physical Properties

Metallic bonding is relatively weak because there is only one valence electron. Therefore, these metals are soft with relatively low melting and boiling points. These values decrease down the group because larger atom cores attract delocalized electrons less strongly.

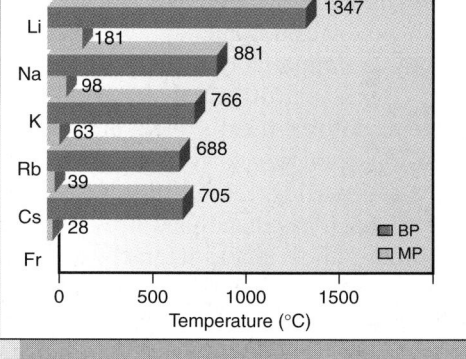

Temperature (°C) — BP / MP:
Li 181 / 1347, Na 98 / 881, K 63 / 766, Rb 39 / 688, Cs 28 / 705, Fr

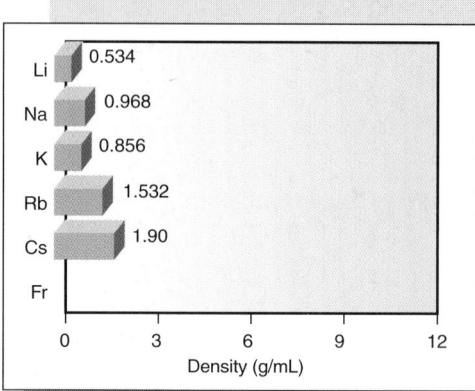

Density (g/mL): Li 0.534, Na 0.968, K 0.856, Rb 1.532, Cs 1.90, Fr

Large atomic size and low atomic mass result in low density; thus density generally increases down the group because mass increases more than size.

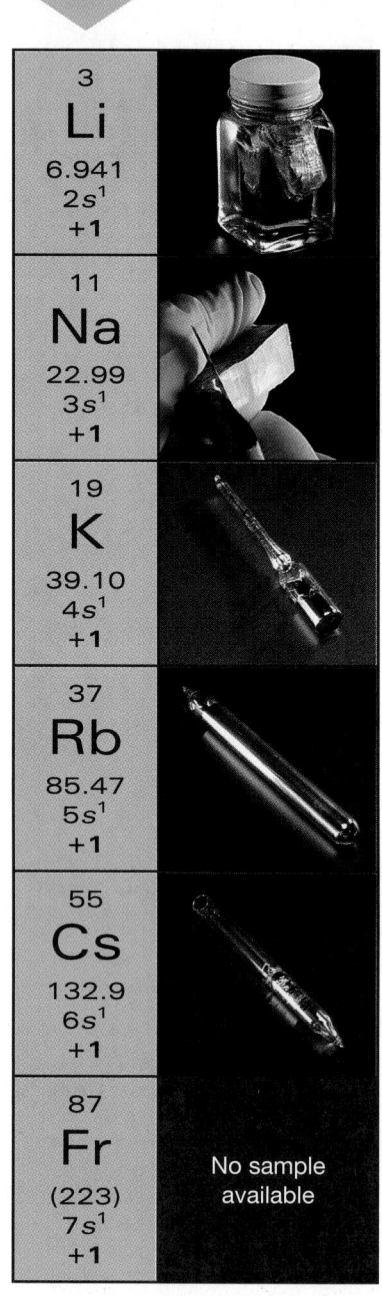

3	**Li**	6.941 · 2s¹ · +1
11	**Na**	22.99 · 3s¹ · +1
19	**K**	39.10 · 4s¹ · +1
37	**Rb**	85.47 · 5s¹ · +1
55	**Cs**	132.9 · 6s¹ · +1
87	**Fr**	(223) · 7s¹ · +1

No sample available

Some Reactions and Compounds

Important Reactions

The reducing power of the alkali metals (E) is shown in reactions 1 to 4. Some industrial applications of Group 1A(1) compounds are shown in reactions 5 to 7.

1. The alkali metals reduce H in H_2O from the $+1$ to the 0 oxidation state:

$$2E(s) + 2H_2O(l) \longrightarrow 2E^+(aq) + 2OH^-(aq) + H_2(g)$$

The reaction becomes more vigorous down the group (*see photo*).

Potassium reacting with water

2. The alkali metals reduce oxygen, but the product depends on the metal. Li forms the oxide, Li_2O; Na forms the peroxide, Na_2O_2; K, Rb, and Cs form the superoxide, EO_2:

$$4Li(s) + O_2(g) \longrightarrow 2Li_2O(s)$$
$$K(s) + O_2(g) \longrightarrow KO_2(s)$$

In emergency breathing units, KO_2 reacts with H_2O and CO_2 in exhaled air to release O_2.

3. The alkali metals reduce hydrogen to form ionic (saltlike) hydrides:

$$2E(s) + H_2(g) \longrightarrow 2EH(s)$$

NaH is an industrial base and reducing agent that is used to prepare other reducing agents, such as $NaBH_4$.

4. The alkali metals reduce halogens to form ionic halides:

$$2E(s) + X_2 \longrightarrow 2EX(s) \quad (X = F, Cl, Br, I)$$

5. Sodium chloride is the most important alkali metal halide.
(a) In the Downs process for the production of sodium metal (*Section 22.4*), reaction 4 is reversed by supplying electricity to molten NaCl:

$$2NaCl(l) \xrightarrow{\text{electricity}} 2Na(l) + Cl_2(g)$$

(b) In the chlor-alkali process (*Section 22.5*), NaCl(aq) is electrolyzed to form several key industrial chemicals:

$$2NaCl(aq) + 2H_2O(l) \longrightarrow 2NaOH(aq) + H_2(g) + Cl_2(g)$$

(c) In its reaction with sulfuric acid, NaCl forms two major products:

$$2NaCl(s) + H_2SO_4(aq) \longrightarrow Na_2SO_4(aq) + 2HCl(g)$$

Sodium sulfate is important in the paper industry; HCl is essential in steel, plastics, textiles, and food production.

6. Sodium hydroxide is used in the formation of bleaching solutions:

$$2NaOH(aq) + Cl_2(g) \longrightarrow NaClO(aq) + NaCl(aq) + H_2O(l)$$

7. In an ion-exchange process (*Chemical Connections, Section 13.6*), water is "softened" by removal of dissolved hard-water cations, which displace Na^+ from a resin:

$$M^{2+}(aq) + Na_2Z(s) \longrightarrow MZ(s) + 2Na^+(aq)$$
$$(M = Mg, Ca; Z = resin)$$

Important Compounds

1. Lithium chloride and lithium bromide, LiCl and LiBr. Because the Li^+ ion is so small, Li salts have a high affinity for H_2O and yet a positive heat of solution. Thus, they are used in dehumidifiers and air-cooling units.

2. Lithium carbonate, Li_2CO_3. Used to make porcelain enamels and toughened glasses and as a drug in the treatment of manic-depressive disorders.

3. Sodium chloride, NaCl. Millions of tons used in the industrial production of Na, NaOH, $Na_2CO_3/NaHCO_3$, Na_2SO_4, and HCl or purified for use as table salt.

4. Sodium carbonate and sodium hydrogen carbonate, Na_2CO_3 and $NaHCO_3$. Carbonate used as an industrial base and to make glass. Hydrogen carbonate, which releases CO_2 at low temperatures ($50°$ to $100°C$), used in baking powder and in fire extinguishers.

5. Sodium hydroxide, NaOH. Most important industrial base; used to make bleach, sodium phosphates, and alcohols.

6. Potassium nitrate, KNO_3. Powerful oxidizing agent used in gunpowder and fireworks (*see photo*).

Versatile Magnesium Magnesium, second of the alkaline earths, forms a tough, adherent oxide coating that prevents further reaction of the metal in air and confers stability for many uses. The metal can be rolled, forged, welded, and riveted into virtually any shape, and it forms strong, low-density alloys: some weigh 25% as much as steel but are just as strong. Alloyed with aluminum and zinc, magnesium is used for everyday objects, such as camera bodies, luggage, and the "mag" wheels of bicycles and sports cars. Alloyed with the lanthanides (first series of inner transition elements), it is used in objects that require great strength at high temperature, such as auto engine blocks and missile parts.

Lime: The Most Useful Metal Oxide Calcium oxide (lime) is a slightly soluble, basic oxide that is among the five most heavily produced industrial compounds in the world; it is made by roasting limestone to high temperatures [Group 2A(2) Family Portrait, reaction 7]. It has essential roles in steelmaking, water treatment, and smokestack "scrubbing." Lime reacts with carbon to form calcium carbide, which is used to make acetylene, and with arsenic acid (H_3AsO_4) to make insecticides for treating cotton, tobacco, and potato plants. Most glass contains about 12% lime by mass. The paper industry uses lime to prepare the bleach [$Ca(ClO)_2$] that whitens paper. Farmers "dust" fields with lime to "sweeten" acidic soil. The food industry uses it to neutralize compounds in milk to make cream and butter and to neutralize impurities in crude sugar.

14.4 GROUP 2A(2): THE ALKALINE EARTH METALS

The Group 2A(2) elements are called *alkaline earth metals* because their oxides give basic (alkaline) solutions and melt at such high temperatures that they remained as solids ("earths") in the alchemists' fires. The group includes a fascinating collection of elements: rare beryllium (Be), common magnesium (Mg) and calcium (Ca), less familiar strontium (Sr) and barium (Ba), and radioactive radium (Ra). The Group 2A(2) Family Portrait (pp. 554 and 555) presents an overview of these elements.

How Do the Physical Properties of the Alkaline Earth and Alkali Metals Compare?

In general terms, the elements in Groups 1A(1) and 2A(2) behave as close cousins. Whatever differences occur between the groups are those of degree, not kind, and are due to the change in electron configuration: ns^2 vs. ns^1. Two electrons are available for metallic bonding, and the nucleus contains one additional positive charge. These factors make the attraction between delocalized electrons and atom cores greater. Consequently, 2A melting and boiling points are much higher than for the corresponding 1A metals; in fact, the 2A elements melt at around the same temperatures as the 1A elements boil! Compared with transition metals, such as iron and chromium, the alkaline earths are soft and lightweight, but they are much harder and more dense than the alkali metals.

How Do the Chemical Properties of the Alkaline Earth and Alkali Metals Compare?

The alkaline earth metals display a wider range of chemical behavior than the alkali metals, largely because of the behavior of beryllium, as you'll see shortly. Because the second valence electron lies in the same sublevel as the first, it is not shielded from the additional nuclear charge very well, so Z_{eff} is greater. Therefore, Group 2A(2) elements have smaller atomic radii and higher ionization energies than Group 1A(1) elements. Despite the higher IEs, *all the alkaline earths (except Be) form ionic compounds* as 2+ cations. Beryllium behaves differently because so much energy is needed to remove two electrons from this tiny atom that it never forms discrete Be^{2+} ions, and its bonds are polar covalent.

Like the alkali metals, the alkaline earth metals are *strong reducing agents*. The elements reduce O_2 in air to form the oxide (Ba also forms the peroxide, BaO_2). Except for Be and Mg, which form adherent oxide coatings, the alkaline earths reduce H_2O at room temperature to form H_2. And, except for Be, they reduce the halogens, N_2, and H_2 to form ionic compounds. The Group 2A oxides are strongly basic (except for amphoteric BeO) and react with acidic oxides to form salts, such as sulfites and carbonates; for example,

$$SrO(s) + CO_2(g) \longrightarrow SrCO_3(s)$$

Natural carbonates, such as limestone and marble, are major structural materials and the commercial sources for most 2A compounds.

The alkaline earth metals are reactive because the high lattice energies of their compounds more than compensate for the large total IE needed to form the 2+ cation (Section 9.2). Group 2A salts have much higher lattice energies than Group 1A salts (see Figure 14.2) because the 2A cations are smaller and doubly charged.

One of the main differences between the two groups is the lower solubility of 2A salts in water. Their ions are smaller and more highly charged than 1A ions, resulting in much higher charge densities. Even though this increases heats of hydration, it increases lattice energies even more. In fact, most 2A fluorides, carbonates, phosphates, and sulfates are considered insoluble, unlike the corresponding 1A compounds. Nevertheless, the ion-dipole attraction of 2+ ions for water is so strong that many slightly soluble 2A salts crystallize as hydrates; two examples are Epsom salt, $MgSO_4 \cdot 7H_2O$, used as a soak for inflammations, and gypsum, $CaSO_4 \cdot 2H_2O$, used as the bonding material between the paper sheets in wallboard and as the cement in surgical casts.

The Anomalous Behavior of Beryllium

The Period 2 element Be displays far more anomalous behavior than Li in Group 1A(1). If the Be^{2+} ion did exist, its high charge density would polarize nearby electron clouds very strongly and cause extensive orbital overlap. In fact, *all Be compounds exhibit covalent bonding.* Even BeF_2, the most ionic Be compound, has a relatively low melting point and, when melted, a low electrical conductivity.

With only two valence electrons, Be does not attain an octet in its simple gaseous compounds (Section 10.1). When it bonds to an electron-rich atom, however, this electron deficiency is overcome as the gas condenses. Consider beryllium chloride ($BeCl_2$) (Figure 14.4). At temperatures greater than 900°C, it consists of linear molecules in which two *sp* hybrid orbitals hold four electrons around the central Be. As it cools, the molecules bond together, solidifying in long chains, with each Be sp^3 hybridized to finally attain an octet.

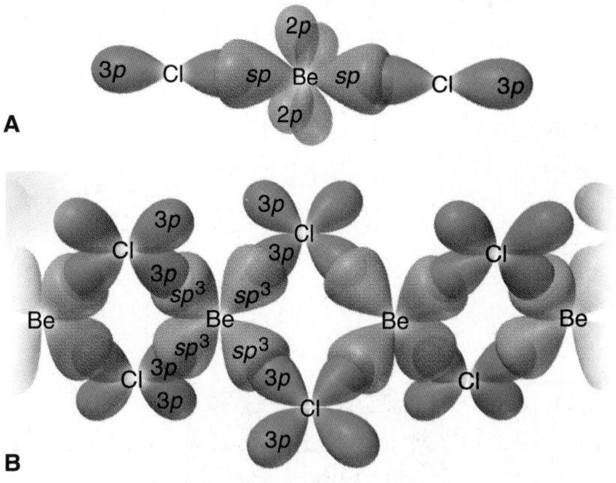

Diagonal Relationships: Lithium and Magnesium

One of the clearest ways in which atomic properties influence chemical behavior appears in the **diagonal relationships**, similarities between a Period 2 element and one diagonally down and to the right in Period 3 (Figure 14.5).

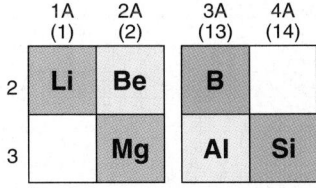

Figure 14.5 Three diagonal relationships in the periodic table. Certain Period 2 elements exhibit behaviors that are very similar to those of the Period 3 elements immediately below and to the right. Three such diagonal relationships exist: Li and Mg, Be and Al, and B and Si.

The first of three such relationships occurs between Li and Mg and reflects similarities in atomic and ionic size. Note that *one period down increases atomic (or ionic) size and one group to the right decreases it.* Thus, Li has a radius of 152 pm and Mg has a radius of 160 pm; the Li^+ radius is 76 pm and that of Mg^{2+} is 72 pm. From similar atomic properties emerge similar chemical properties. Both elements form nitrides with N_2, hydroxides and carbonates that decompose easily with heat, organic compounds with a polar covalent metal-carbon bond, and salts with similar solubilities. We'll discuss the diagonal relationships between Be and Al and between B and Si in later sections.

Looking Backward and Forward: Groups 1A(1), 2A(2), and 3A(13)

Throughout this chapter, comparing the previous, current, and upcoming groups (Figure 14.6) will help you to keep horizontal trends in mind while examining vertical groups. Not much changes from 1A to 2A, and the elements behave as metals both physically and chemically. With smaller atomic sizes and stronger metallic bonding, 2A elements are harder, higher melting, and denser than those in 1A. Nearly all 1A and most 2A compounds are ionic. The higher ionic charge in Group 2A (2+ vs. 1+) leads to higher lattice energies and less soluble salts. The range of behavior in 2A is wider than that in 1A because of Be, and the range widens much further in Group 3A from metalloid boron to metallic thallium.

Figure 14.4 Overcoming electron deficiency in beryllium chloride. **A,** At high temperatures, $BeCl_2$ is a gas with only four electrons around each Be. **B,** Solid $BeCl_2$ occurs in long chains with each Cl bridging two Be atoms, which gives each Be an octet.

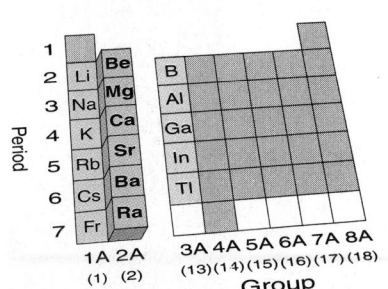

Figure 14.6 Standing in Group 2A(2), looking backward to 1A(1) and forward to 3A(13).

Group 2A(2): The Alkaline Earth Metals

Key Atomic and Physical Properties

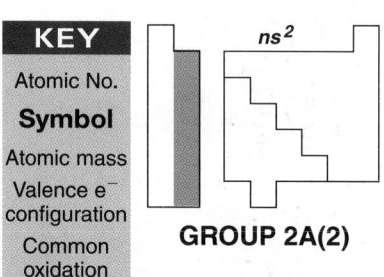

KEY

Atomic No.
Symbol
Atomic mass
Valence e⁻ configuration
Common oxidation states

ns^2

GROUP 2A(2)

4 **Be** 9.012 $2s^2$ +2	
12 **Mg** 24.30 $3s^2$ +2	
20 **Ca** 40.08 $4s^2$ +2	
38 **Sr** 87.62 $5s^2$ +2	
56 **Ba** 137.3 $6s^2$ +2	
88 **Ra** (226) $7s^2$ +2	No sample available

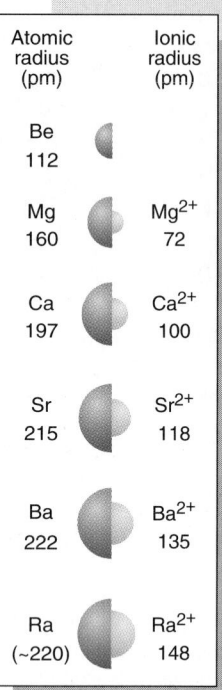

	Atomic radius (pm)	Ionic radius (pm)
Be	112	
Mg	160	Mg²⁺ 72
Ca	197	Ca²⁺ 100
Sr	215	Sr²⁺ 118
Ba	222	Ba²⁺ 135
Ra	(~220)	Ra²⁺ 148

Atomic Properties

Group electron configuration is ns^2 (filled ns sublevel). All members have the +2 oxidation state and, except for Be, form compounds with an E^{2+} ion.

Atomic and ionic sizes increase down the group but are smaller than for the corresponding 1A(1) element.

IE and EN decrease down the group but are higher than for the corresponding 1A(1) element.

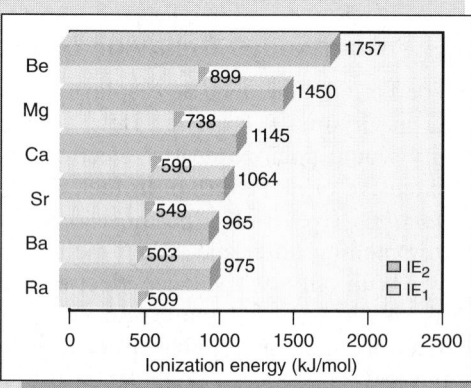

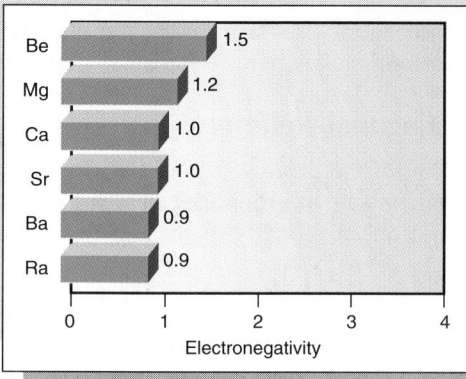

Physical Properties

Metallic bonding involves two valence e⁻. These metals are still relatively soft but are much harder than the 1A(1) metals.

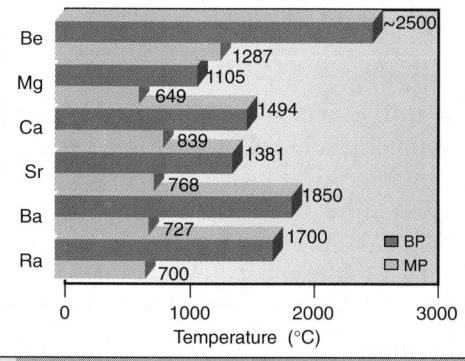

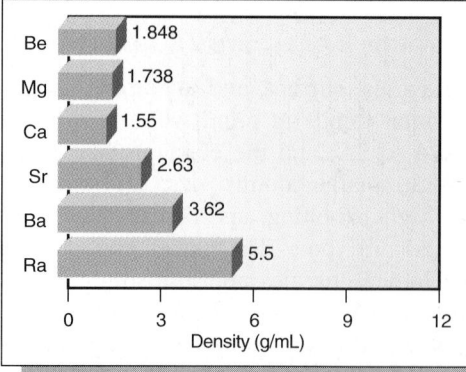

Melting and boiling points generally decrease, and densities generally increase down the group. These values are much higher than for 1A(1), and the trend is not as regular.

Some Reactions and Compounds

Important Reactions

The elements (E) act as reducing agents in reactions 1 to 5; note the similarity to reactions of Group 1A(1). Reaction 6 shows the general basicity of the 2A(2) oxides; reaction 7 shows the general instability of their carbonates at high temperature.

1. The metals reduce O_2 to form the oxides:
$$2E(s) + O_2(g) \longrightarrow 2EO(s)$$
Ba also forms the peroxide, BaO_2.

Magnesium ribbon burning

2. The larger metals reduce water to form hydrogen gas:
$$E(s) + 2H_2O(l) \longrightarrow E^{2+}(aq) + 2OH^-(aq) + H_2(g)$$
$$(E = Ca, Sr, Ba)$$
Be and Mg form an adherent oxide coating that allows only slight reaction.

3. The metals reduce halogens to form ionic halides:
$$E(s) + X_2 \longrightarrow EX_2(s) \quad (X = F, Cl, Br, I)$$

4. Most of the elements reduce hydrogen to form ionic hydrides:
$$E(s) + H_2(g) \longrightarrow EH_2(s) \quad (E = \text{all except Be})$$

5. Most of the elements reduce nitrogen to form ionic nitrides:
$$3E(s) + N_2(g) \longrightarrow E_3N_2(s) \quad (E = \text{all except Be})$$

6. Except for amphoteric BeO, the oxides are basic:
$$EO(s) + H_2O(l) \longrightarrow E^{2+}(aq) + 2OH^-(aq)$$
$Ca(OH)_2$ is a component of cement and mortar.

7. All carbonates undergo thermal decomposition to the oxide:
$$ECO_3(s) \xrightarrow{\Delta} EO(s) + CO_2(g)$$
This reaction is used to produce CaO (lime) in huge amounts from naturally occurring limestone (*see Margin Note, p. 552*).

Important Compounds

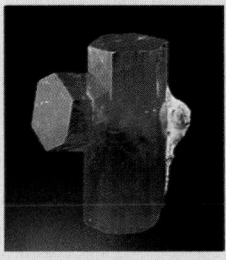

1. Beryl, $Be_3Al_2Si_6O_{18}$. Beryl, the industrial source of Be metal, also occurs as a gemstone with a variety of colors (*see photo*). It is chemically identical to emerald, except for the trace of Cr^{3+} that gives emerald its green color.

2. Magnesium oxide, MgO. Because of its high melting point (2852°C), MgO is used as a refractory material for furnace brick (*see photo*) and wire insulation.

Industrial kiln

3. Alkylmagnesium halides, RMgX (R = hydrocarbon group; X = halogen). These compounds are used to synthesize many organic compounds. Organotin agricultural fungicides are made by treating RMgX with $SnCl_4$:
$$3RMgCl + SnCl_4 \longrightarrow 3MgCl_2 + R_3SnCl$$

4. Calcium carbonate, $CaCO_3$. Occurs as enormous natural deposits of limestone, marble, chalk, and coral. Used as a building material, to make lime, and, in high purity, as a toothpaste abrasive and an antacid (*see photo*).

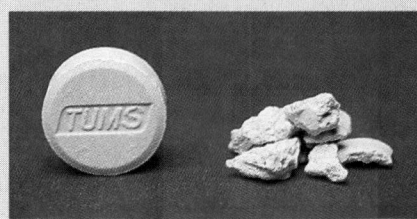

Antacid tablet   Limestone

14.5 GROUP 3A(13): THE BORON FAMILY

The third family of main-group elements contains some unusual members and some familiar ones, some exotic bonding, and some strange physical properties. Boron (B) heads the family, but, as you'll see, its properties certainly do not represent the other members. Metallic aluminum (Al) has properties that are more typical of the group, but its great abundance and importance contrast with the rareness of gallium (Ga), indium (In), and thallium (Tl). The atomic, physical, and chemical properties of these elements are summarized in the Group 3A(13) Family Portrait (pp. 558 and 559).

How Do the Transition Elements Influence Group 3A(13) Properties?

Group 3A(13) is the first of the *p* block. If you look at the main groups only, the elements of this group seem to be just one group away from those of Group 2A(2). In Period 4 and higher, however, a large gap separates the two groups (see Figure 8.12). The gap holds 10 transition elements (*d* block) each in Periods 4, 5, and 6 and an additional 14 inner transition elements (*f* block) in Period 6. Recall from Section 8.2 that *d* and *f* electrons penetrate very little, and so spend very little time near the nucleus. Thus, the heavier 3A members—Ga, In, and Tl—have nuclei with many more protons, but their outer (*s* and *p*) electrons are only partially shielded from the much higher positive charge; as a result, these elements have greater Z_{eff} values than the two lighter members of the group.

Many properties of these heavier 3A elements are influenced by this stronger nuclear attraction. Figure 14.7 compares several properties of Group 3A(13) with those of Group 3B(3) (the first group of transition elements), in which the additional protons of the *d* block and *f* block have *not* been added. Note the regular changes exhibited by the 3B elements in contrast to the irregular patterns for those in 3A. The deviations for Ga reflect the *d*-block *contraction in size* and can be explained by limited shielding by the *d* electrons of the 10 additional protons of the first transition series. Similarly, the deviations for Tl reflect the *f*-block (lanthanide) contraction and can be explained by limited shielding by the *f* electrons of the 14 additional protons of the first inner transition series.

Physical properties are influenced by the type of bonding that occurs in the element. Boron is a network covalent metalloid—black, hard, and very high melting. The other group members are metals—shiny and relatively soft and low melting. Aluminum's low density and three valence electrons make it an exceptional conductor: for a given mass, aluminum conducts a current twice as effectively as copper. Gallium has the largest liquid temperature range of any element: it melts in your hand but does not boil until 2403°C (see Figure 9.22). Its metallic bonding is too weak to keep the Ga atoms fixed when the solid is warmed, but it is strong enough to keep them from escaping the molten metal until it is very hot. ●

What New Features Appear in the Chemical Properties of Group 3A(13)?

Looking down Group 3A(13), we see a wide range of chemical behavior. Boron, the anomalous member from Period 2, is the first metalloid we've encountered so far and the only one in the group. It is much less reactive at room temperature than the other members and forms covalent bonds exclusively. Although aluminum acts like a metal physically, its halides exist in the gas phase as covalent *dimers*—molecules formed by joining two identical smaller molecules (Figure 14.8)—and its oxide is amphoteric rather than basic. Most of the other 3A compounds are ionic, but they have more covalent character than similar 2A com-

Gallium Arsenide: The Next Wave of Semiconductors An important modern use of gallium is in the production of gallium arsenide (GaAs) semiconductors. One of the factors limiting the efficiency of computer chips is the speed with which electrons can move, and they move 10 times faster through GaAs than through Si-based chips. GaAs chips also have novel optical properties: a current is created when they absorb light, and, conversely, they emit light when a current is supplied. GaAs devices are already used in light-powered calculators, wristwatches, and solar panels. In addition, GaAs-based lasers are much smaller and more powerful than other types.

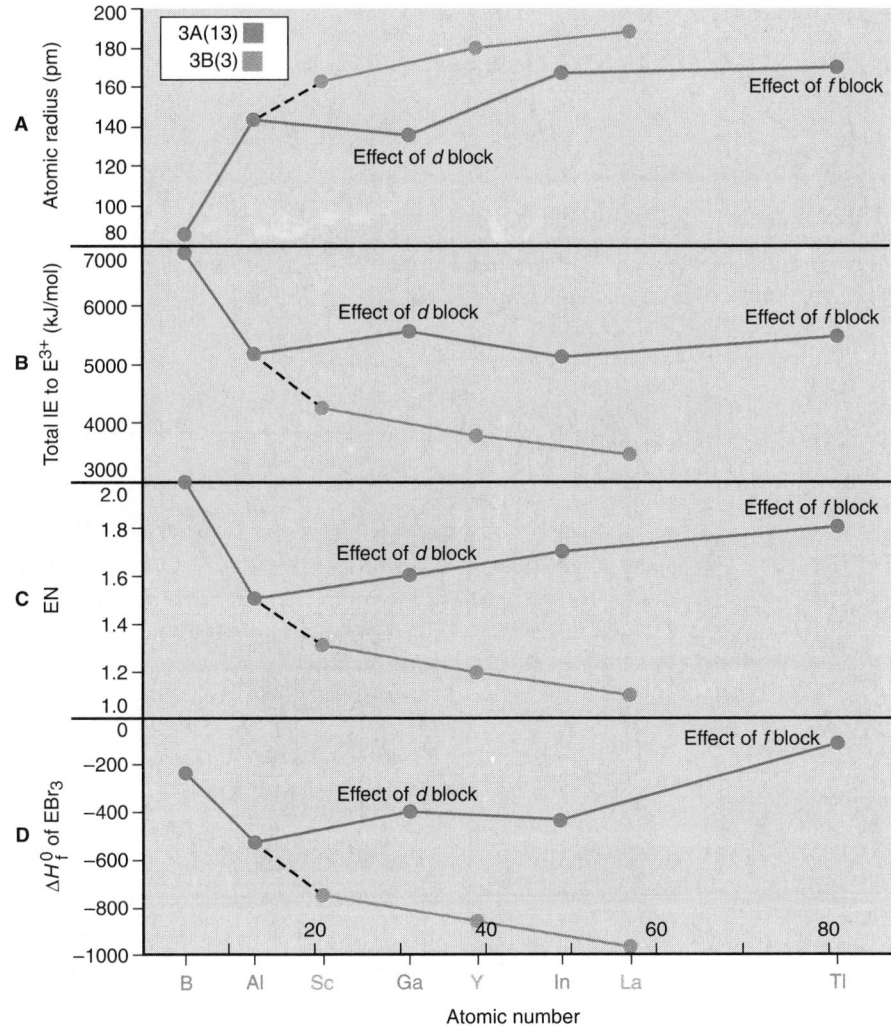

Figure 14.7 **The effect of transition elements on properties: Group 3B(3) vs. Group 3A(13).** The additional protons in the nuclei of transition elements exert an exceptionally strong attraction because d and f electrons shield the nuclear charge poorly. This greater effective nuclear charge affects the p-block elements in Periods 4 to 6, as can be seen by comparing properties of Group 3B(3), the first group after the s block, with those of Group 3A(13), the first group after the d block. **A,** Atomic size. In Group 3B, size increases smoothly, whereas in 3A, Ga and Tl are smaller than expected. **B,** Total ionization energy for E^{3+}. The deviations in size lead to deviations in total IE ($IE_1 + IE_2 + IE_3$). Note the regular decrease for Group 3B and the unexpectedly higher values for Ga and Tl in 3A. **C,** Electronegativity. The deviations in size also make Ga and Tl more electronegative than expected. **D,** Heat of formation of EBr_3. The higher IE values for Ga and Tl mean that less heat is released upon formation of the ionic compound, thus the magnitudes of the ΔH_f^0 values for $GaBr_3$ and $TlBr_3$ are smaller than expected.

pounds. Because the 3A ions are smaller and more highly charged than the 2A ions, they polarize the anion electron cloud more effectively.

The redox behavior of the elements in this group provides a chance to note three general principles that appear in Groups 3A(13) to 6A(16):

1. *Presence of multiple oxidation states.* Many of the larger elements in these groups also have an important oxidation state *two lower than the A-group number.* The lower state occurs when the atoms lose their np electrons only, not their two ns electrons. This fact is often called the *inert-pair effect* (Section 8.5), but it has nothing to do with any inertness of ns electrons; for example, the $6s$ electrons in Tl are lost *more* easily than the $4s$ electrons in Ga. The reason for the common appearance of the lower state involves bond energies of the compounds. Bond energy decreases as atomic size, and therefore bond length, increases. Consider the Tl—Cl bond. It is relatively long and weak, and it takes more energy to remove the two $6s$ electrons from Tl^+ to make Tl^{3+} than is released when two more weak Tl—Cl bonds form in $TlCl_3$. In fact, $TlCl_3$ is so unstable that it decomposes readily to Cl_2 gas and TlCl.

2. *Increasing prominence of the lower oxidation state.* When a group exhibits more than one oxidation state, *the lower state becomes more prominent going down the group.* In Group 3A(13), for instance, all members exhibit the $+3$ state, but the $+1$ state first appears with some compounds of gallium and becomes the only important state of thallium.

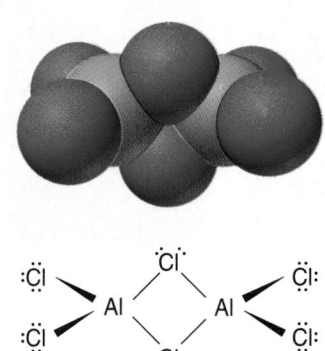

Figure 14.8 **The dimeric structure of gaseous aluminum chloride.** Despite its name, aluminum trichloride exists in the gas phase as the dimer, Al_2Cl_6.

Key Atomic and Physical Properties

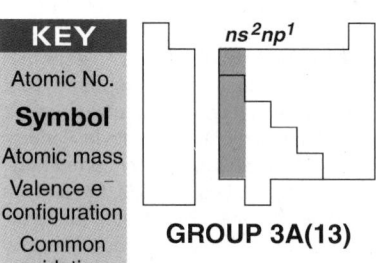

GROUP 3A(13)

ns^2np^1

	5
	B
	10.81
	$2s^2 2p^1$
	+3

	13
	Al
	26.98
	$3s^2 3p^1$
	+3

	31
	Ga
	69.72
	$4s^2 4p^1$
	+3, +1

	49
	In
	114.8
	$5s^2 5p^1$
	+3, +1

	81
	Tl
	204.4
	$6s^2 6p^1$
	+1

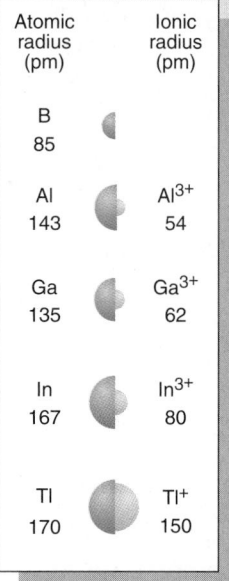

Atomic radius (pm)	Ionic radius (pm)
B 85	
Al 143	Al^{3+} 54
Ga 135	Ga^{3+} 62
In 167	In^{3+} 80
Tl 170	Tl^+ 150

Atomic Properties

Group electron configuration is ns^2np^1. All except Tl commonly display the +3 oxidation state. The +1 state becomes more common down the group.

Atomic size is smaller and EN is higher than for 2A(2) elements; IE is lower, however, because it is easier to remove e⁻ from the higher energy p sublevel.

Atomic size, IE, and EN do not change as expected down the group because there are intervening transition and inner transition elements.

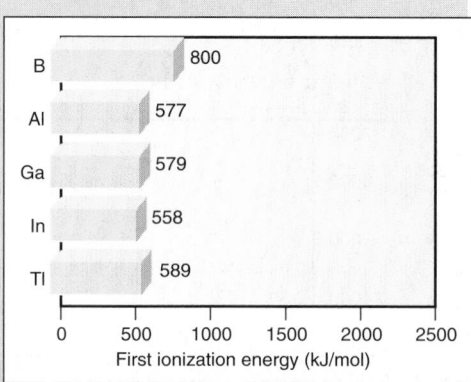

First ionization energy (kJ/mol): B 800, Al 577, Ga 579, In 558, Tl 589

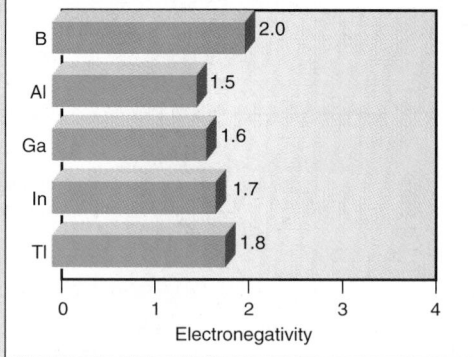

Electronegativity: B 2.0, Al 1.5, Ga 1.6, In 1.7, Tl 1.8

Physical Properties

Bonding changes from network covalent in B to metallic in the rest of the group. Thus, B has a much higher melting point than the others, but there is no overall trend. Boiling points decrease down the group.

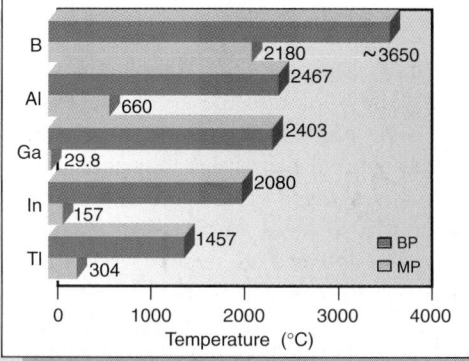

Temperature (°C) — BP/MP:
B: 2180 / ~3650
Al: 660 / 2467
Ga: 29.8 / 2403
In: 157 / 2080
Tl: 304 / 1457

Densities increase down the group.

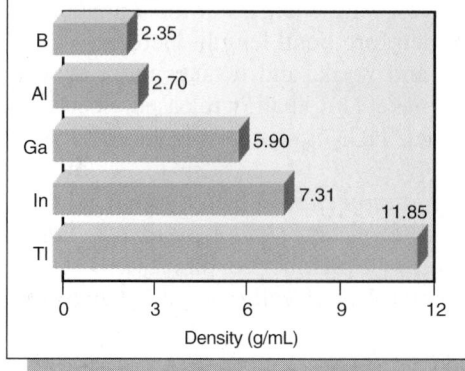

Density (g/mL): B 2.35, Al 2.70, Ga 5.90, In 7.31, Tl 11.85

Some Reactions and Compounds

Important Reactions

In reactions 1 to 3, note that the elements (E) usually require higher temperatures to react than those of Groups 1A(1) and 2A(2); note the lower oxidation state of Tl. Reactions 4 to 6 show some compounds in key industrial processes.

1. The elements react sluggishly, if at all, with water:

$$2Ga(s) + 6H_2O(hot) \longrightarrow 2Ga^{3+}(aq) + 6OH^-(aq) + 3H_2(g)$$
$$2Tl(s) + 2H_2O(steam) \longrightarrow 2Tl^+(aq) + 2OH^-(aq) + H_2(g)$$

Al becomes covered with a layer of Al_2O_3 that prevents further reaction (*see photo*).

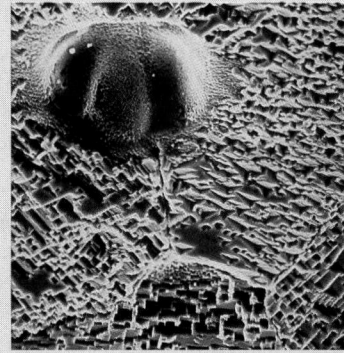

Electron micrograph of oxide coating on Al surface

2. When strongly heated in pure O_2, all members form oxides:

$$4E(s) + 3O_2(g) \xrightarrow{\Delta} 2E_2O_3(s) \qquad (E = B, Al, Ga, In)$$
$$4Tl(s) + O_2(g) \xrightarrow{\Delta} 2Tl_2O(s)$$

Oxide acidity decreases down the group: B_2O_3 (weakly acidic) > Al_2O_3 > Ga_2O_3 > In_2O_3 > Tl_2O (strongly basic), and the +1 oxide is more basic than the +3 oxide.

3. All members reduce halogens (X_2):

$$2E(s) + 3X_2 \longrightarrow 2EX_3 \qquad (E = B, Al, Ga, In)$$
$$2Tl(s) + X_2 \longrightarrow 2TlX(s)$$

BX_3 are volatile covalent compounds. Trihalides of Al, Ga, and In are (mostly) ionic solids but occur as covalent dimers in the gas phase; in this way, the 3A atom attains a filled outer level.

4. Acid treatment of Al_2O_3 is important in water purification:

$$Al_2O_3(s) + 3H_2SO_4(l) \longrightarrow Al_2(SO_4)_3(s) + 3H_2O(l)$$

In water, $Al_2(SO_4)_3$ and CaO form a colloid that aids in removing suspended particles.

5. The overall reaction in the production of aluminum metal is a redox process:

$$2Al_2O_3(s) + 3C(s) \longrightarrow 4Al(s) + 3CO_2(g)$$

This process is carried out electrochemically in the presence of cryolite (Na_3AlF_6), which lowers the melting point of the reactant mixture and takes part in the change (*Section 22.4*).

6. A displacement reaction produces gallium arsenide, GaAs (*see margin note, p. 556*):

$$(CH_3)_3Ga(g) + AsH_3(g) \longrightarrow 3CH_4(g) + GaAs(s)$$

Important Compounds

1. Boron oxide, B_2O_3. Used in the production of borosilicate glass (*see margin note, p. 560*).

2. Borax, $Na_2[B_4O_5(OH)_4] \cdot 8H_2O$. Major mineral source of boron compounds and B_2O_3. Used as a fireproof insulation material and as a washing powder (20-Mule Team Borax).

3. Boric acid, H_3BO_3 [or $B(OH)_3$]. Used as external disinfectant, eyewash, and insecticide.

4. Diborane, B_2H_6. A powerful reducing agent for possible use as a rocket fuel. Used to synthesize higher boranes, compounds that led to new theories of chemical bonding.

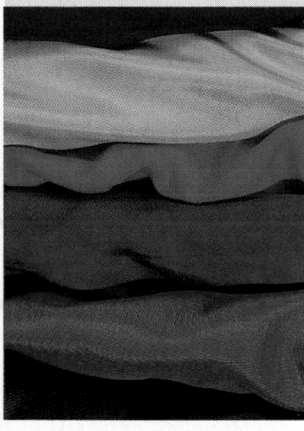

5. Aluminum sulfate (cake alum), $Al_2(SO_4)_3 \cdot 18H_2O$. Used in purifying water, tanning leather, and sizing paper; as a fixative for dyeing cloth (*see photo*); and as an antiperspirant.

6. Aluminum oxide, Al_2O_3. Major compound in natural source (bauxite) of Al metal. Used as abrasive in sandpaper, sanding and cutting tools, and toothpaste. Large crystals with metal ion impurities often of gemstone quality (*see photo*). Inert support for chromatography. In fibrous forms, woven into heat-resistant fabrics; also used to strengthen ceramics and metals.

Ruby

7. $Tl_2Ba_2Ca_2Cu_3O_{10}$. Becomes a high-temperature superconductor at 125 K, which is readily attained with liquid N_2 (*see photo*).

Magnet levitated by superconductor

3. *Relative basicity of oxides*. In general, *oxides with the element in a lower oxidation state are more basic than oxides with the element in a higher oxidation state*. A good example in Group 3A is that In_2O is more basic than In_2O_3. In general, when an element has more than one oxidation state, *it acts more like a metal in its lower state*, and this, too, is related to ionic charge density. In this example, the lower charge of In^+ does not polarize the O^{2-} ion as much as the higher charge of In^{3+} does. Thus, in compounds of general formula E_2O, the E-to-O bonding is more ionic than it is in E_2O_3 compounds, so the O^{2-} ion is more available to act as a base (Interchapter, Topic 4).

Highlights of Boron Chemistry

Like the other Period 2 elements, the chemical behavior of boron is strikingly different from that of the other members of its group. As pointed out earlier, *all boron compounds are covalent,* and unlike the other Group 3A(13) members, boron forms network covalent compounds or large molecules with metals, H, O, N, and C. The unifying feature of many boron compounds is their *electron deficiency,* but boron adopts two strategies to fill its outer level: accepting a bonding pair from an electron-rich atom and forming bridge bonds with electron-poor atoms.

Accepting a Bonding Pair from an Electron-Rich Atom In gaseous boron trihalides (BX_3), the B atom is electron deficient, with only six electrons around it (Section 10.1). To attain an octet, the B atom accepts a lone pair of electrons from an electron-rich atom and forms a covalent bond:

$$BF_3(g) + :NH_3(g) \longrightarrow F_3B—NH_3(g)$$

(Such reactions, in which one reactant accepts an electron pair from another to form a covalent bond, are very widespread in inorganic, organic, and biochemical processes. They are known as *Lewis acid-base reactions,* and we'll discuss them in Chapters 18 and 23 and see examples of them throughout the second half of the text.)

Similarly, B has only six electrons in boric acid, $B(OH)_3$ (sometimes written as H_3BO_3). In water, the acid itself does not release a proton. Rather, it accepts an electron pair from the O in H_2O, forming a fourth bond and releasing an H^+ ion:

$$B(OH)_3(s) + H_2O(l) \rightleftharpoons B(OH)_4^-(aq) + H^+(aq)$$

Boron's outer shell is filled in the wide variety of borate salts, such as the mineral borax (sodium borate), $Na_2[B_4O_5(OH)_4]\cdot 8H_2O$, used for decades as a household cleaning agent.

Boron also attains an octet in several boron-nitrogen compounds whose structures are amazingly similar to that of elemental carbon and some of its organic compounds (Figure 14.9). These similarities are due to the fact that the size, IE, and EN of C are between those of B and N (see Table 14.1). Moreover, C has four valence electrons, whereas B has three and N has five, so the $\cdot\dot{C}—\dot{C}\cdot$ and $\cdot\dot{B}—\dot{N}\cdot$ groupings have the same number of valence electrons. Therefore, $H_3C—CH_3$ (ethane) and $H_3B—NH_3$ (amine-borane) are isoelectronic, as are benzene and borazine (Figure 14.9A and B).

Boron nitride (Figure 14.9C) has a structure consisting of hexagons fused into sheets, very similar to that of graphite. Moreover, like graphite, its π electrons are highly delocalized through resonance. However, molecular-orbital calculations show that there is a large energy gap between the filled valence band and the empty conductance band in boron nitride, but no such gap in graphite (see Figure 12.37). As a result, boron nitride is a white electrical insulator, whereas graphite is a black electrical conductor. At high pressure and temperature, boron nitride forms borazon (Figure 14.9D), which has a crystal structure like that of diamond and is extremely hard and abrasive.

Borates in Your Labware Strong heating of boric acid (or borate salts) drives off water molecules and gives molten boron oxide: $2B(OH)_3(s) \xrightarrow{\Delta} B_2O_3(l) + 3H_2O(g)$. The molten oxide dissolves metal oxides to form borate glasses. When mixed with silica (SiO_2), it forms borosilicate glass. Its transparency and small change in size when heated or cooled make borosilicate glass useful in cookware and in the glassware you use in the lab.

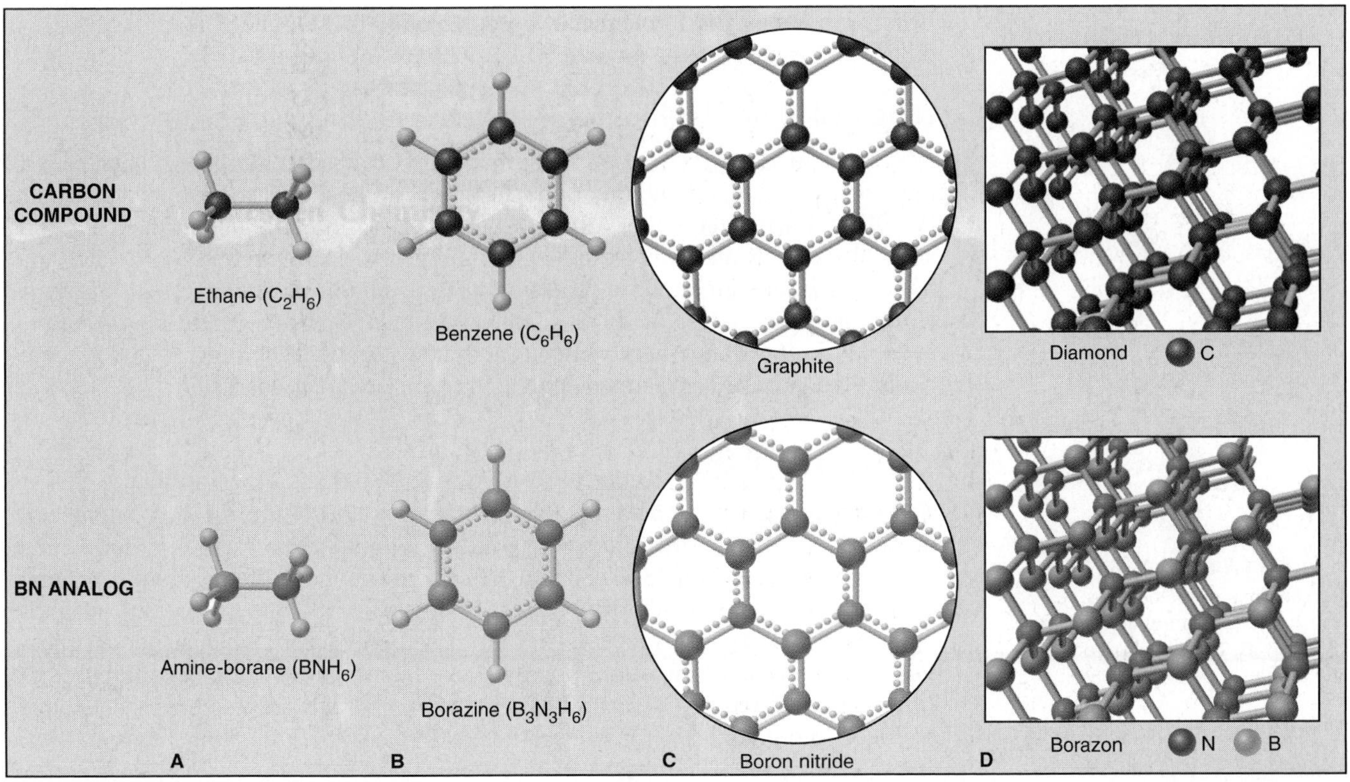

CARBON COMPOUND

Ethane (C_2H_6)

Benzene (C_6H_6)

Graphite

Diamond   ● C

BN ANALOG

Amine-borane (BNH_6)

Borazine ($B_3N_3H_6$)

Boron nitride

Borazon   ● N   ● B

A   **B**   **C**   **D**

Figure 14.9 **Similarities between substances with C—C bonds and those with B—N bonds.** Note that in each case, B attains an octet of electrons by bonding with electron-rich N. **A,** Ethane and its BN analog. **B,** Benzene and borazine, which is often referred to as "inor-ganic" benzene. **C,** Graphite and the similar extended hexagonal structure of boron nitride. **D,** Diamond and borazon have the same crystal structure and are among the hardest substances known.

Forming Bridge Bonds with Electron-Poor Atoms In elemental boron and its many hydrides (boranes), there is no electron-rich atom to supply boron with electrons. In these substances, boron attains an octet through some unusual bonding. In diborane (B_2H_6) and many larger boranes, for example, two types of B—H bonds exist. The first type is a typical electron-pair bond. The valence bond picture in Figure 14.10 shows an sp^3 orbital of B overlapping the H $1s$ orbital in each of the four terminal B—H bonds, using two of the three electrons in the valence shell of each B atom.

The other type of bond is a hydride **bridge bond** (or three-center, two-electron bond), in which *each B—H—B grouping is held together by only two electrons.* Two sp^3 orbitals, one from *each* B, overlap an H $1s$ orbital between them. Two electrons move through this extended bonding orbital—one from one of the B atoms and the other from the H atom—and join the two B atoms via the H atom bridge. Notice that *each B atom is surrounded by eight electrons:* four from the two normal B—H bonds and four from the two B—H—B bridge bonds.

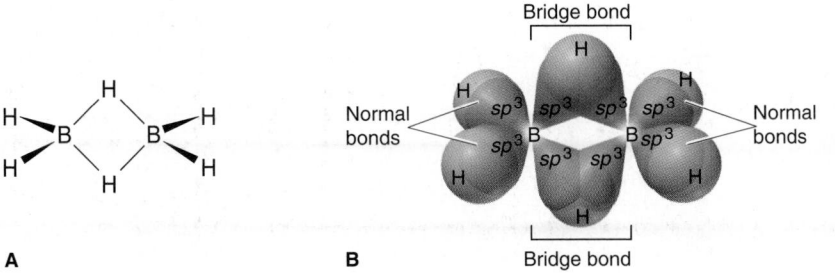

A   **B**

Figure 14.10 **The two types of covalent bonding in diborane. A,** A perspective diagram of B_2H_6 shows the unusual B—H—B bridge bond and the tetrahedral arrangement around each B atom. **B,** A valence bond depiction shows each sp^3-hybridized B forming normal covalent bonds with two hydrogens and two bridge bonds, in which two electrons bind three atoms, at the two central B—H—B groupings.

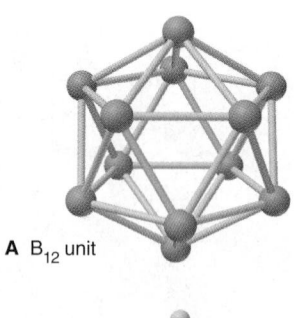

A B_{12} unit

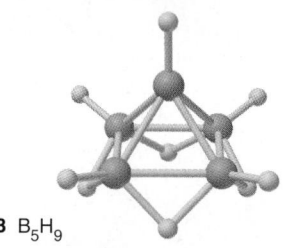

B B_5H_9

Figure 14.11 The boron icosahedron and one of the boranes. A, The icosahedral structural unit of elemental boron. **B,** The structure of B_5H_9, one of many boranes.

In many boranes and in elemental boron (Figure 14.11), one B atom bridges two others in a three-center, two-electron B—B—B bond.

Diagonal Relationships: Beryllium and Aluminum

Beryllium in Group 2A(2) and aluminum in Group 3A(13) are another pair of diagonally related elements. Both form oxoanions in strong base: beryllate, $Be(OH)_4{}^{2-}$, and aluminate, $Al(OH)_4{}^-$. Both have bridge bonds in their hydrides and chlorides. Both form oxide coatings impervious to reaction with water, and both oxides are amphoteric, extremely hard, and high melting. Although the atomic and ionic sizes of these elements differ, the small, highly charged Be^{2+} and Al^{3+} ions polarize nearby electron clouds strongly. Therefore, some Al compounds and all Be compounds have significant covalent character.

14.6 GROUP 4A(14): THE CARBON FAMILY

The whole range of elemental behavior occurs within Group 4A(14): nonmetallic carbon (C) leads off, followed by the metalloids silicon (Si) and germanium (Ge), with metallic tin (Sn) and lead (Pb) at the bottom of the group. Information about the compounds of C and of Si fills libraries: organic chemistry, most polymer chemistry, and biochemistry are based on carbon, whereas geochemistry and some extremely important polymer and electronic technologies are based on silicon. The Group 4A(14) Family Portrait (pp. 564 and 565) summarizes atomic, physical, and chemical properties.

How Does the Bonding in an Element Affect Physical Properties?

The elements of Group 4A(14) and their neighbors in Groups 3A(13) and 5A(15) illustrate how physical properties, such as melting point and heat of fusion (ΔH_{fus}), depend on the type of bonding in an element (Table 14.2). Within Group 4A, the large decrease in melting point between the covalent network solids C and Si is due to longer, weaker bonds in the Si structure; the large decrease between Ge and Sn is due to the change from covalent network to metallic bonding. Similarly, looking horizontally, the large increases in melting point and ΔH_{fus} across a period between Al and Si and between Ga and Ge reflect the change from metallic to covalent network bonding. Note the abrupt rises in the values for these properties from metallic Al, Ga, and Sn to the network-covalent metalloids Si, Ge, and Sb, and note the abrupt drops from the covalent networks of C and Si to the individual molecules of N and P in Group 5A.

Table 14.2 Bond Type and the Melting Process in Groups 3A(13) to 5A(15)

Period	Group 3A(13)				Group 4A(14)				Group 5A(15)				Key:	
	Element	Bond Type	Melting Point (°C)	ΔH_{fus} (kJ/mol)	Element	Bond Type	Melting Point (°C)	ΔH_{fus} (kJ/mol)	Element	Bond Type	Melting Point (°C)	ΔH_{fus} (kJ/mol)	⬡	Metallic
2	B	⬡	2180	23.6	C	⬡	4100	Very high	N	∞	−210	0.7	⬡	Covalent network
3	Al	☐	660	10.5	Si	⬡	1420	50.6	P	∞	44.1	2.5	∞	Covalent molecule
4	Ga	☐	30	5.6	Ge	⬡	945	36.8	As	⬡	816	27.7	▨	Metal
5	In	☐	157	3.3	Sn	☐	232	7.1	Sb	⬡	631	20.0	▨	Metalloid
6	Tl	☐	304	4.3	Pb	☐	327	4.8	Bi	☐	271	10.5	☐	Nonmetal

Allotropism: Different Forms of an Element Striking variations in physical properties often appear among **allotropes,** different crystalline or molecular forms of a substance. One allotrope is usually more stable than another at a particular pressure and temperature. Group 4A(14) provides the first dramatic example of allotropism, in the forms of carbon. It is difficult to imagine two substances made entirely of the same atom that are more different than graphite and diamond. Graphite is a black electrical conductor that is soft and "greasy," whereas diamond is a colorless electrical insulator that is extremely hard. Graphite is the standard state of carbon, the more stable form at ordinary temperature and pressure, as the phase diagram in Figure 14.12 shows. Fortunately for jewelry owners, diamond changes to graphite at a negligible rate under normal conditions.

In the mid-1980s, a newly discovered allotrope of carbon generated great interest. Mass spectrometric analysis of soot had shown evidence for a soccer ball–shaped molecule of formula C_{60} (Figure 14.13A; see also the Gallery, p. 380). More recent findings reveal the molecule in geological samples formed by meteorite impact, even the one that occurred around the time the dinosaurs became extinct. The molecule has been dubbed *buckminsterfullerene* (informally called a "buckyball") after the architect-engineer R. Buckminster Fuller, who designed structures with similar shapes. Excitement rose in 1990, when scientists learned how to prepare multigram quantities of C_{60} and related fullerenes, enough to study macroscopic behavior and possible applications. Since then, metal atoms have been incorporated into the structure and many different groups (fluorine, hydroxyl, sugars, etc.) have been attached to prepare compounds with a range of useful properties.

Then, in 1991, scientists passed an electric discharge through graphite rods sealed with helium gas in a container and obtained extremely thin (~1 nm in diameter) graphite-like tubes with fullerene ends (Figure 14.13B). These *nanotubes* are rigid and, on a mass basis, much stronger than steel along their long axis. They also conduct electricity along this axis because of the delocalized electrons. Reports about fullerenes and nanotubes appear almost daily in scientific journals. With potential applications in nanoscale electronics, catalysis, polymers, and medicine, fullerene and nanotube chemistry will receive increasing attention well into the 21st century.

Tin has two allotropes. White β-tin is stable at room temperature and above, whereas gray α-tin is the more stable form below 13°C (56°F). When white tin is kept for long periods at a low temperature, some converts to microcrystals of gray tin. The random formation and growth of these regions of gray tin, which has a different crystal structure, weaken the metal and make it crumble. In the unheated cathedrals of medieval northern Europe, tin pipes of magnificent organs sometimes crumbled as a result of "tin disease" caused by this allotropic transition from white to gray tin.

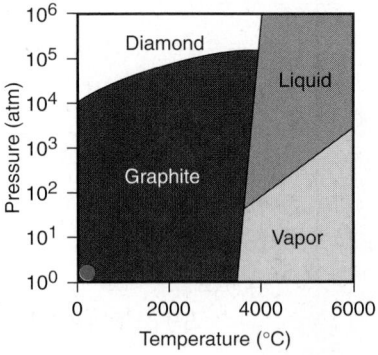

Figure 14.12 Phase diagram of carbon. Graphite is the more stable form at ordinary conditions (*small red circle at extreme lower left*). Diamond is more stable at very high pressure.

Figure 14.13 Buckyballs and nanotubes. A, Crystals of buckminsterfullerene (C_{60}) are shown leading to a ball-and-stick model. The parent of the fullerenes, the "buckyball," is a soccer ball–shaped molecule of 60 carbon atoms. **B,** Nanotubes are single or, as shown in this colorized transmission electron micrograph, concentric graphite-like tubes with fullerene ends (see the Gallery, p. 380).

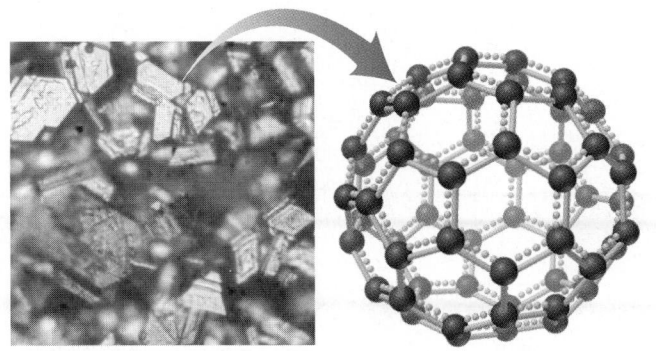

A

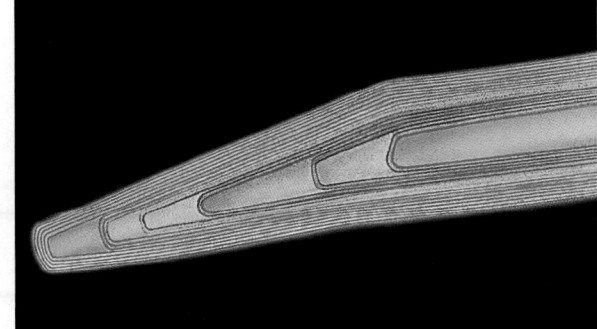

B

Key Atomic and Physical Properties

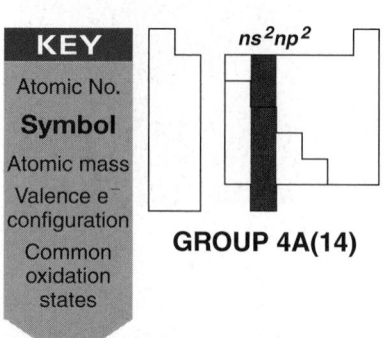

ns^2np^2

GROUP 4A(14)

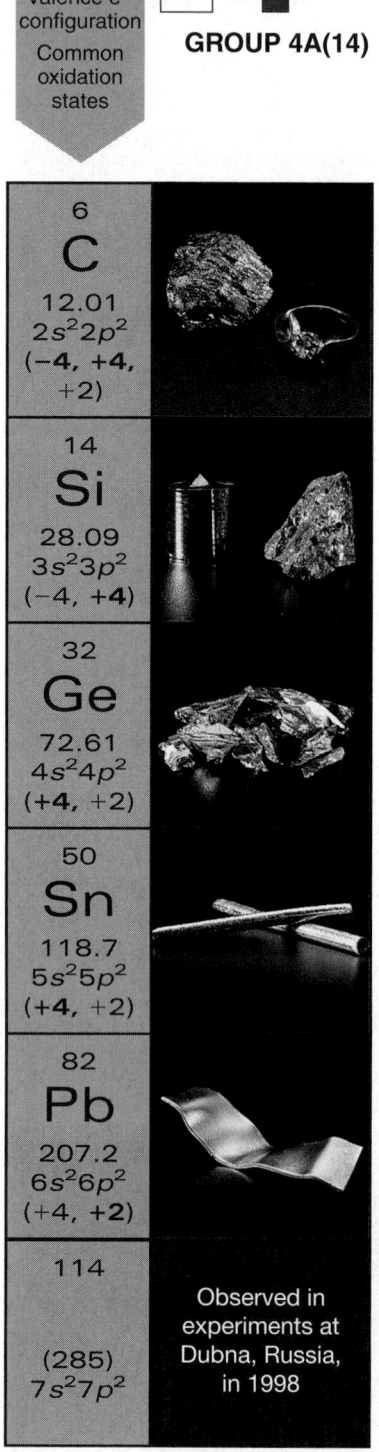

| 6 |
| C |
| 12.01 |
| $2s^22p^2$ |
| (−4, +4, +2) |

| 14 |
| Si |
| 28.09 |
| $3s^23p^2$ |
| (−4, +4) |

| 32 |
| Ge |
| 72.61 |
| $4s^24p^2$ |
| (+4, +2) |

| 50 |
| Sn |
| 118.7 |
| $5s^25p^2$ |
| (+4, +2) |

| 82 |
| Pb |
| 207.2 |
| $6s^26p^2$ |
| (+4, +2) |

| 114 |
| |
| (285) |
| $7s^27p^2$ |

Observed in experiments at Dubna, Russia, in 1998

Atomic Properties

Group electron configuration is ns^2np^2. Down the group, the number of oxidation states decreases, and the lower (+2) state becomes more common.

Down the group, size increases. Because transition and inner transition elements intervene, IE and EN do not decrease smoothly.

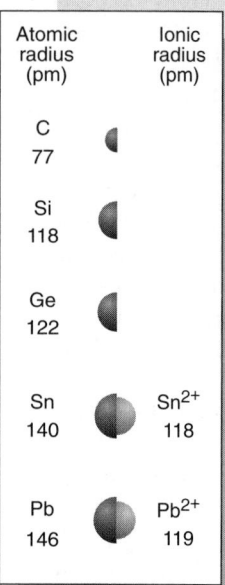

Atomic radius (pm)	Ionic radius (pm)
C 77	
Si 118	
Ge 122	
Sn 140	Sn²⁺ 118
Pb 146	Pb²⁺ 119

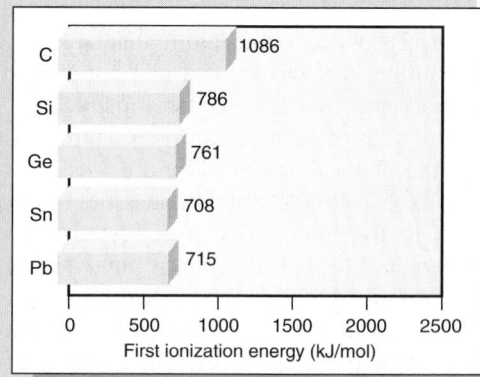

First ionization energy (kJ/mol)

C	1086
Si	786
Ge	761
Sn	708
Pb	715

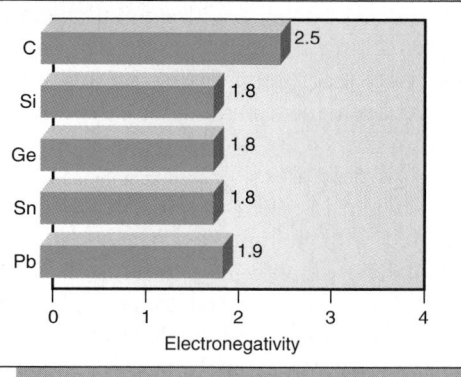

Electronegativity

C	2.5
Si	1.8
Ge	1.8
Sn	1.8
Pb	1.9

Physical Properties

Trends in properties, such as decreasing hardness and melting point, are due to changes in types of bonding within the solid: covalent network in C, Si, and Ge; metallic in Sn and Pb (*see text*).

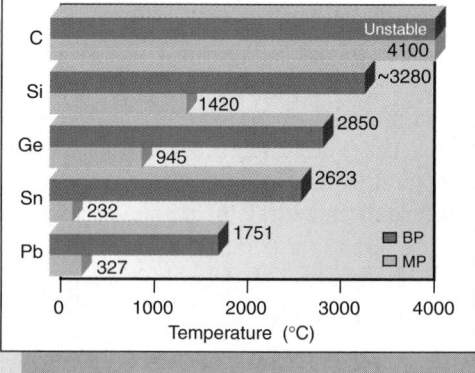

Temperature (°C)

	BP	MP
C	Unstable	4100
Si	~3280	1420
Ge	2850	945
Sn	2623	232
Pb	1751	327

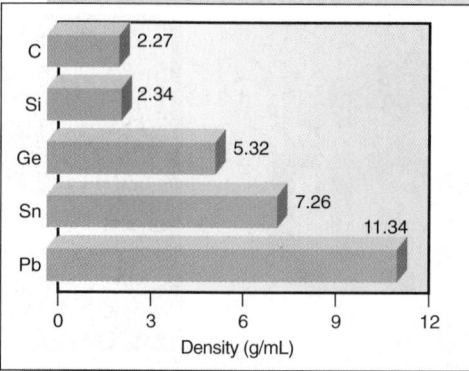

Density (g/mL)

C	2.27
Si	2.34
Ge	5.32
Sn	7.26
Pb	11.34

Down the group, density increases because of several factors, including differences in crystal packing.

Some Reactions and Compounds

Important Reactions

Reactions 1 and 2 pertain to all the elements (E); reactions 3 to 7 concern industrial uses for compounds of C and Si.

1. The elements are oxidized by halogens:

$$E(s) + 2X_2 \longrightarrow EX_4 \quad (E = C, Si, Ge)$$

The +2 halides are more stable for tin and lead, SnX_2 and PbX_2.

2. The elements are oxidized by O_2:

$$E(s) + O_2(g) \longrightarrow EO_2 \quad (E = C, Si, Ge, Sn)$$

Pb forms the +2 oxide, PbO. Oxides become more basic down the group. The reaction of CO_2 and H_2O provides the weak acidity of natural unpolluted waters:

$$CO_2(g) + H_2O(l) \rightleftharpoons [H_2CO_3(aq)]$$
$$\rightleftharpoons H^+(aq) + HCO_3^-(aq)$$

3. Air and steam passed over hot coke produce gaseous fuel mixtures (producer gas and water gas):

$$C(s) + air(g) + H_2O(g) \longrightarrow CO(g) + CO_2(g) + N_2(g) + H_2(g)$$
[not balanced]

4. Hydrocarbons react with O_2 to form CO_2 and H_2O. The reaction for methane is adapted to yield heat or electricity:

$$CH_4(g) + 2O_2(g) \longrightarrow CO_2(g) + 2H_2O(g)$$

Oxyacetylene torch

5. Certain metal carbides react with water to produce acetylene:

$$CaC_2(s) + 2H_2O(l) \longrightarrow$$
$$Ca(OH)_2(aq) + C_2H_2(g)$$

The gas is used to make other organic compounds and as a fuel in welding (*see photo*).

6. Freons (chlorofluorocarbons) are formed by fluorinating carbon tetrachloride:

$$CCl_4(l) + HF(g) \longrightarrow CFCl_3(g) + HCl(g)$$

Production of trichlorofluoromethane (Freon-11), the major refrigerant in the world, is being eliminated because of its severe effects on the environment (*see margin note, p. 567*).

7. Silica is reduced to form elemental silicon:

$$SiO_2(s) + 2C(s) \longrightarrow Si(s) + 2CO(g)$$

This crude silicon is made ultrapure through zone refining (*see Gallery, p. 512*) for manufacture of computer chips (*see photo*).

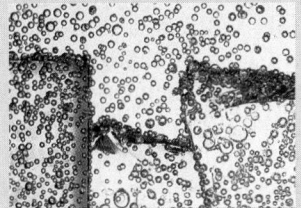
Computer chip

Important Compounds

1. Carbon monoxide, CO. Used as a gaseous fuel, as a precursor for organic compounds, and as a reactant in the purification of nickel. Formed in internal combustion engines and released as a toxic air pollutant.

2. Carbon dioxide, CO_2. Atmospheric component used by photosynthetic plants to make carbohydrates and O_2. The final oxidation product of all carbon-based fuels; its increase in the atmosphere is causing global warming. Used industrially as a refrigerant gas, a blanketing gas in fire extinguishers, and an effervescent gas in beverages (*see photo*). Combined with NH_3 to form urea for fertilizers and plastics manufacture.

3. Methane, CH_4. Used as a fuel and in the production of many organic compounds. Major component of natural gas. Formed by anaerobic decomposition of plants (swamp gas) and by microbes in termites and certain mammals. May contribute to global warming.

4. Silicon dioxide, SiO_2. Occurs in many amorphous (glassy) and crystalline forms, quartz being the most common. Used to make glass and as an inert chromatography support material.

5. Silicon carbide, SiC. Known as *carborundum*, a major industrial abrasive and a highly refractory ceramic for tough, high-temperature uses. Can be doped to form a high-temperature semiconductor.

6. Organotin compounds, R_4Sn. Used to stabilize PVC, poly(vinyl chloride), plastics (*see photo*), and to cure silicone rubbers. Agricultural biocide for insects, fungi, and weeds.

7. Tetraethyl lead, $(C_2H_5)_4Pb$. Once used as a gasoline additive to improve fuel efficiency, but now banned because of its inactivation of auto catalytic converters. Major source of lead as a toxic air pollutant.

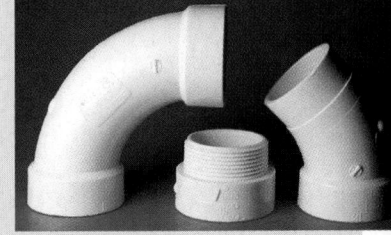

How Does the Type of Bonding Change in Group 4A(14) Compounds?

The Group 4A(14) elements display a wide range of chemical behavior, from the covalent compounds of carbon to the ionic compounds of lead. Carbon's intermediate EN of 2.5 ensures that it virtually always forms covalent bonds, but the larger members of the group form bonds with increasing ionic character. With nonmetals, Si and Ge form strong polar covalent bonds, such as the Si—O bond, one of the strongest of any Period 3 element (BE = 368 kJ/mol). This bond is responsible for the physical and chemical stability of the Earth's solid surface, as we discuss later in this section. Although individual Sn or Pb ions rarely exist, the bonding of either element with a nonmetal has considerable ionic character.

The pattern of elements with more than one oxidation state that we first observed in Group 3A(13) appears here as well. After silicon, the shift to greater importance of the lower oxidation state occurs: Si compounds of the +4 state are much more stable than those of the +2 state, whereas Pb compounds of the +2 state are more stable than those of the +4 state. The elements also behave more like metals in the lower oxidation state. Consider the chlorides and oxides of Sn and Pb. The +2 chlorides $SnCl_2$ and $PbCl_2$ are white, relatively high-melting, water-soluble crystals—typical properties of a salt (Figure 14.14). In contrast, $SnCl_4$ is a volatile, benzene-soluble liquid, and $PbCl_4$ is a thermally unstable oil. Both likely consist of individual, tetrahedral molecules. The +2 oxides SnO and PbO are more basic than the +4 oxides SnO_2 and PbO_2 because the +2 metals are less able to polarize the O^{2-} ion, so the E-to-O bonding is more ionic.

Figure 14.14 The greater metallic character of tin and lead in the lower oxidation state. Metals with more than one oxidation state exhibit more metallic behavior in the lower state. Tin(II) and lead(II) chlorides are white, crystalline, saltlike solids. In contrast, tin(IV) and lead(IV) chlorides are volatile liquids, indicating the presence of individual molecules.

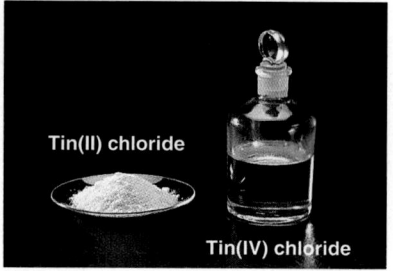

Tin(II) chloride

Tin(IV) chloride

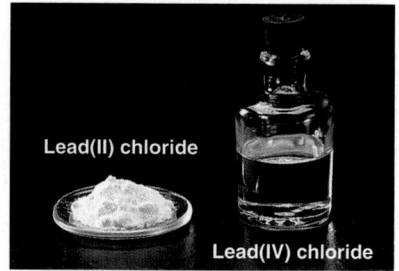

Lead(II) chloride

Lead(IV) chloride

Highlights of Carbon Chemistry

Like the other Period 2 elements, carbon is an anomaly in its group; indeed, it may be an anomaly in the entire periodic table. Carbon forms bonds with the smaller Group 1A(1) and 2A(2) metals, many transition metals, the halogens, and its neighbors B, Si, N, O, P, and S. It also exhibits every oxidation state possible for its group, from +4 in CO_2 and halides like CCl_4 through −4 in CH_4.

Two features stand out in the chemistry of carbon: its ability to bond to itself, a process known as *catenation*, and its ability to form multiple bonds. As a result of its small size and its capacity for four bonds, carbon can form chains, branches, and rings that lead to myriad structures. Add a lot of H, some O and N, a bit of S, P, halogens, and a few metals, and you have the whole organic world! Figure 14.15 shows three of the several million organic compounds known. Multiple bonding is common in carbon structures because the C—C bond is short enough for side-to-side overlap of two half-filled $2p$ orbitals to form π bonds. (In Chapter 15, we discuss in detail how the atomic properties of carbon give rise to the diverse structure and reactivity of organic compounds.)

Because the other 4A members are larger, E—E bonds become longer and weaker down the group; thus, in terms of bond strength, C—C > Si—Si > Ge—Ge. The empty *d* orbitals of these larger atoms make the chains much more

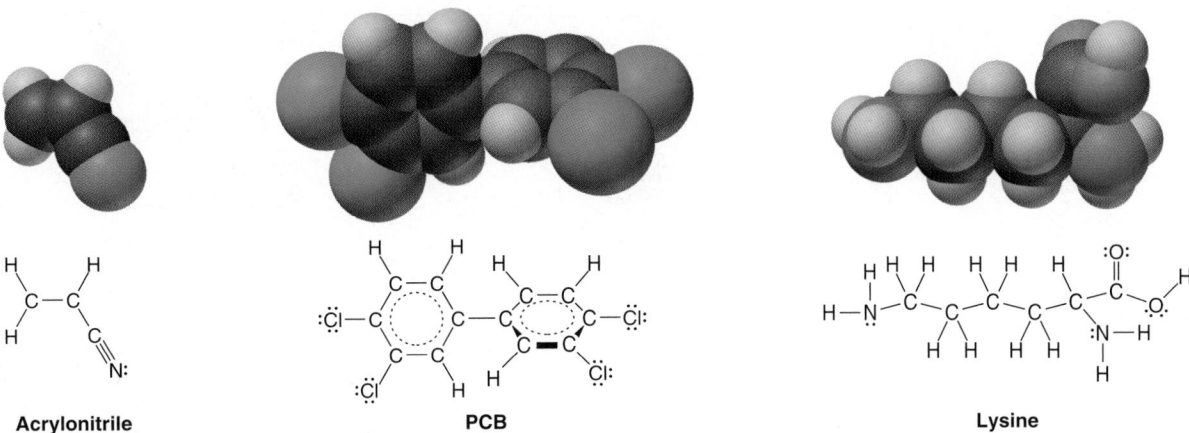

Figure 14.15 Three of the several million known organic compounds of carbon. Acrylonitrile, a precursor of acrylic fibers. PCB (one of the polychlorinated biphenyls). Lysine, one of about 20 amino acids that occur in proteins.

susceptible to chemical attack. Thus, some compounds with long Si chains are known but they are very reactive, and the longest Ge chain has only eight atoms. Moreover, longer bonds do not typically allow sufficient overlap of *p* orbitals for π bonding. Only very recently have compounds with π bonds between Si atoms and between Ge atoms been prepared, but they too are extremely reactive.

In contrast to its organic compounds, carbon's inorganic compounds are simple. Metal carbonates are the main mineral form. Marble, limestone, chalk, coral, and several other types are found in enormous deposits throughout the world. Many of these compounds are remnants of fossilized marine organisms. Carbonates are used in several common antacids because they react with the HCl in stomach acid [see Group 2A(2) Family Portrait, compound 4]:

$$CaCO_3(s) + 2HCl(aq) \longrightarrow CaCl_2(aq) + CO_2(g) + H_2O(l)$$

Identical net ionic reactions with sulfuric and nitric acids protect lakes bounded by limestone deposits from the harmful effects of acid rain.

Unlike the other 4A members, which form only solid network-covalent or ionic oxides, carbon forms two common gaseous oxides, CO_2 and CO. Carbon dioxide is essential to all life: it is the primary source of carbon in plants and animals through photosynthesis. Its aqueous solution is the cause of acidity in natural waters. However, its atmospheric buildup from deforestation and excessive use of fossil fuels may severely affect the global climate. Carbon monoxide forms when carbon or its compounds burn in an inadequate supply of O_2:

$$2C(s) + O_2(g) \longrightarrow 2CO(g)$$

Carbon monoxide is a key component of syngas fuels (see Chemical Connections, p. 243) and is widely used in the production of methanol, formaldehyde, and other major industrial compounds.

CO binds strongly to many transition metals. When inhaled in cigarette smoke or polluted air, it enters the blood and binds strongly to the Fe(II) in hemoglobin, preventing the normal binding of O_2, and to other iron-containing proteins. The cyanide ion (CN⁻) is *isoelectronic* with CO:

$$[:C{\equiv}N:]^- \quad \text{same electronic structure as} \quad :C{\equiv}O:$$

Cyanide binds to many of the same iron-containing proteins and is also toxic.

Monocarbon halides (or halomethanes) are tetrahedral molecules whose stability to heat and light decreases as the size of the halogen atom increases and the C—X bond becomes longer (weaker) (see Interchapter, Topic 2). The chlorofluorocarbons (CFCs, or Freons) have important industrial uses; their short, strong carbon-halogen bonds are exceptionally stable, however, and the environmental persistence of these compounds has created major problems. ●

CFCs: The Good, the Bad, and the Strong The strengths of the C—F and C—Cl bonds make CFCs, such as Freon-12 shown here, both useful and harmful. Chemically and thermally stable, nontoxic, and nonflammable, they are excellent cleaners for electronic parts, coolants in refrigerators and air conditioners, and propellants in aerosol cans. However, their bond strengths also mean that CFCs decompose *very* slowly near the Earth's surface. In the lower atmosphere, they contribute to climate warming, absorbing infrared radiation 16,000 times as effectively as CO_2. Once in the stratosphere, however, they are bombarded by ultraviolet (UV) radiation, which breaks the otherwise stable C—Cl bonds, releasing free Cl atoms that initiate ozone-destroying reactions (Chapter 16). Legal production of CFCs has ended in the United States, but international production and smuggling are widespread.

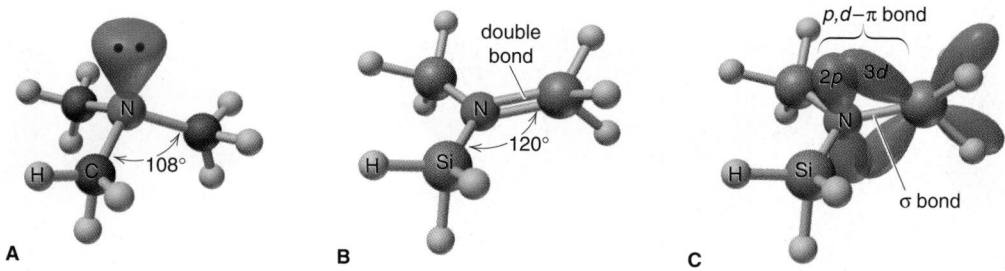

A   **B**   **C**

Figure 14.16 The impact of *p,d*-π bonding on the structure of trisilylamine. A, The molecular shape of trimethylamine is trigonal *pyramidal*, as for ammonia. **B,** Trisilylamine, the silicon analog, has a trigonal *planar* shape because of the formation of a double bond between N and Si. The ball-and-stick model shows one of three resonance forms. **C,** The lone pair in an unhybridized *p* orbital of N overlaps with an empty *d* orbital of Si to give a *p,d*-π bond. In the resonance hybrid, a *d* orbital on each Si is involved in the π bond.

Highlights of Silicon Chemistry

Silicon halides are more reactive than carbon halides because Si has empty *d* orbitals that are available for bond formation. Surprisingly, the silicon halides have longer yet *stronger* bonds than the corresponding carbon halides (see Tables 9.2 and 9.3). One explanation for this unusual strength is that the Si—X bond has some double-bond character because of the presence of a σ bond *and* a different type of π bond, called a *p,d*-π *bond*. This bond arises from side-to-side overlap of an Si *d* orbital and a halogen *p* orbital. A similar type of *p,d*-π bonding leads to the trigonal planar molecular shape of trisilylamine, in contrast to the trigonal pyramidal shape of trimethylamine (Figure 14.16).

To a great extent, the chemistry of silicon is the chemistry of the *silicon-oxygen bond*. Just as carbon forms unending C—C chains, the —Si—O— grouping repeats itself endlessly in a wide variety of **silicates,** the most important minerals on the planet, and in **silicones,** synthetic polymers that have many applications:

1. *Silicate minerals.* From common sand and clay to semiprecious amethyst and carnelian, silicate minerals are the dominant form of matter in the nonliving world. Oxygen, the most abundant element on Earth, and silicon, the next most abundant, compose these minerals and thus account for four of every five atoms on the surface of the planet!

The silicate building unit is the *orthosilicate* grouping, —SiO$_4$—, a tetrahedral arrangement of four oxygens around a central silicon. Several well-known minerals contain SiO$_4^{4-}$ ions or small groups of them linked together. The gemstone zircon (ZrSiO$_4$) contains one unit; hemimorphite [Zn$_4$(OH)$_2$Si$_2$O$_7$·H$_2$O] contains two units linked through an oxygen corner; and beryl (Be$_3$Al$_2$Si$_6$O$_{18}$), the major source of beryllium, contains six units joined into a cyclic ion (Figure 14.17). As you'll see shortly, in addition to these separate ions, SiO$_4$ units are linked more extensively to create much of the planet's mineral structure.

2. *Silicone polymers.* Unlike the naturally occurring silicates, silicone polymers are manufactured substances consisting of alternating Si and O atoms with two organic groups bound to each Si atom. A key starting material is formed by the reaction of silicon with methyl chloride:

$$Si(s) + 2CH_3Cl(g) \xrightarrow[\sim 300°C]{\text{Cu catalyst}} (CH_3)_2SiCl_2(g)$$

Although the corresponding carbon compound [(CH$_3$)$_2$CCl$_2$] is inert in water, the empty *d* orbitals in Si make this compound reactive:

$$(CH_3)_2SiCl_2(l) + 2H_2O(l) \longrightarrow (CH_3)_2Si(OH)_2(l) + 2HCl(g)$$

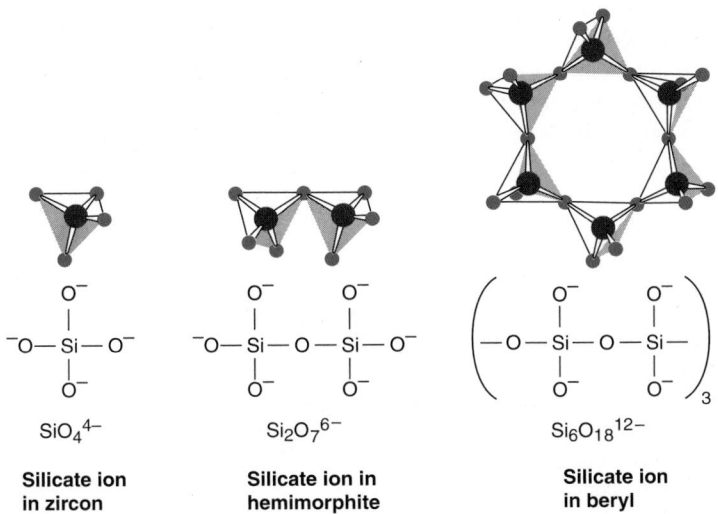

SiO_4^{4-}

Silicate ion in zircon

$Si_2O_7^{6-}$

Silicate ion in hemimorphite

$Si_6O_{18}^{12-}$

Silicate ion in beryl

Figure 14.17 Structures of the silicate anions in some minerals.

The product reacts with itself to form water and a silicone chain molecule called a *poly(dimethyl siloxane):*

$$n(CH_3)_2Si(OH)_2 \longrightarrow \left[O-\underset{\underset{CH_3}{|}}{\overset{\overset{CH_3}{|}}{Si}} \right]_n + nH_2O$$

Silicones have properties of both plastics and minerals. The organic groups give them the flexibility and weak intermolecular forces between chains that are characteristic of a plastic, while the O—Si—O backbone confers the thermal stability and nonflammability of a mineral. In the Gallery on the following pages, note the structural parallels between silicates and silicones.

Diagonal Relationships: Boron and Silicon

The final diagonal relationship that we consider occurs between the metalloids B and Si. Both exhibit the electrical properties of a semiconductor (Section 12.6). Both elements and their mineral oxoanions—borates and silicates—occur in extended covalent networks. Both boric acid $[B(OH)_3]$ and silicic acid $[Si(OH)_4]$ are weakly acidic and occur in layers with widespread H bonding. Both elements form flammable, low-melting compounds with hydrogen—the boranes and silanes—that act as reducing agents.

Looking Backward and Forward: Groups 3A(13), 4A(14), and 5A(15)

Standing in Group 4A(14), we look back at Group 3A(13) as the transition from the *s* block of metals to the *p* block of mostly metalloids and nonmetals (Figure 14.18). Major changes occur in physical behavior as we move across from metals to covalent networks. Changes in chemical behavior occur as well, as cations give way to covalent tetrahedra. Looking ahead to Group 5A(15), we find covalent bonding and (as electronegativity increases) the expanded valence shells of nonmetals occurring more often, as well as the first appearance of monatomic anions.

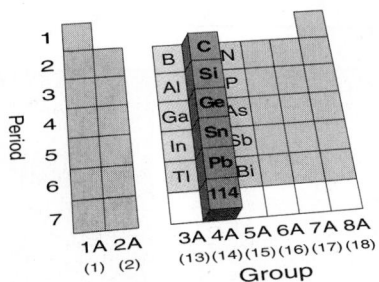

Figure 14.18 Standing in Group 4A(14), looking backward to 3A(13) and forward to 5A(15).

Silicate Minerals and Silicone Polymers

The silicates and silicones show beautifully how organization at the molecular level manifests itself in the properties of macroscopic substances. Interestingly, both of these types of materials exhibit the same three structural classes: chains, sheets, and frameworks.

Silicate Minerals

Chain Silicates

The simplest structural class of silicates occurs when each SiO_4 unit shares two of its O corners with other SiO_4 units, forming a chain. Two chains can link laterally into a ribbon; the most common ribbon has repeating $Si_4O_{11}^{6-}$ units. Metal ions bind the polyanionic ribbons together into neutral sheaths. With only weak intermolecular forces between sheaths, the material occurs in fibrous strands, as in the family of asbestos minerals.

Asbestos

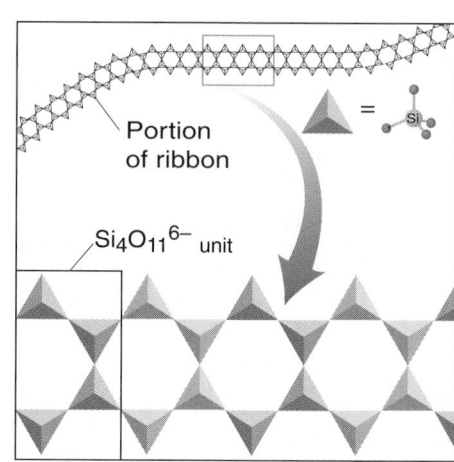

Portion of ribbon

$= \,\, \bullet Si$

$Si_4O_{11}^{6-}$ unit

Sheet (Layer) Silicates

The next structural class of silicates arises when each SiO_4 unit shares three of its four O corners with other SiO_4 units to form a sheet; double sheets arise when the fourth O is shared with another sheet. In talc, the softest mineral, the sheets interact through weak forces, so talcum powder feels slippery. If Al substitutes for some Si, or if $Al(OH)_3$ layers interleave with silicate sheets, an aluminosilicate results, such as the clay kaolinite. Different substitutions and/or interlayers of ions give the micas. In muscovite mica, ions lie between aluminosilicate double layers. Mica flakes when the ionic attractions are overcome.

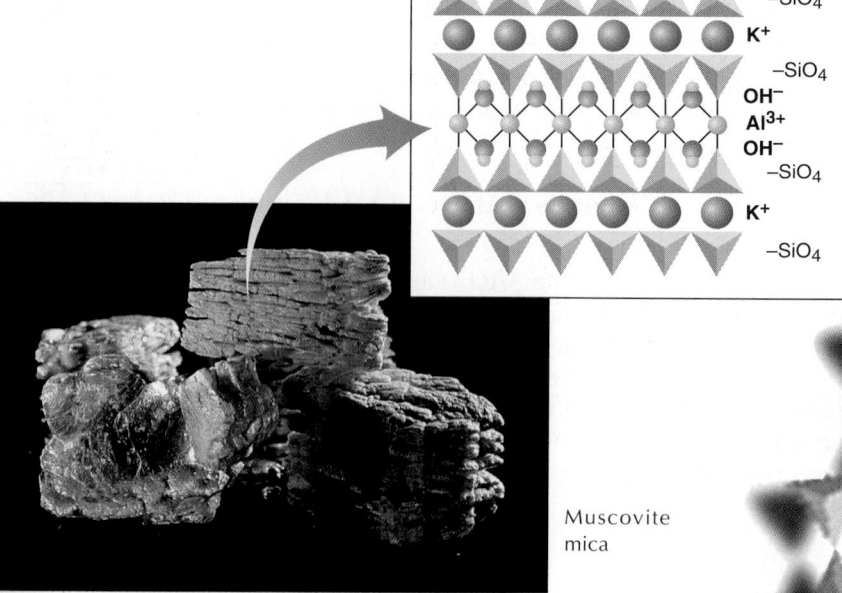

$-SiO_4$
K^+
$-SiO_4$
OH^-
Al^{3+}
OH^-
$-SiO_4$
K^+
$-SiO_4$

Muscovite mica

Framework Silicates

The final structural class of silicates arises when SiO_4 units share all four O corners to give a three-dimensional framework silicate, such as silica (SiO_2), which occurs most often as α-quartz. Some of silica's 12 crystalline forms exist as semiprecious gems. Feldspars, which comprise 60% of the Earth's crust, result when some Si is replaced by Al; granite consists of microcrystals of feldspar, mica, and quartz. Eons of weathering convert feldspars into clays. Zeolites have open frameworks of polyhedra that give rise to minute tunnels. Synthetic zeolites are made with cavities of specific sizes to trap particular molecules; they are used to dry gas mixtures, separate hydrocarbons, and prepare catalysts.

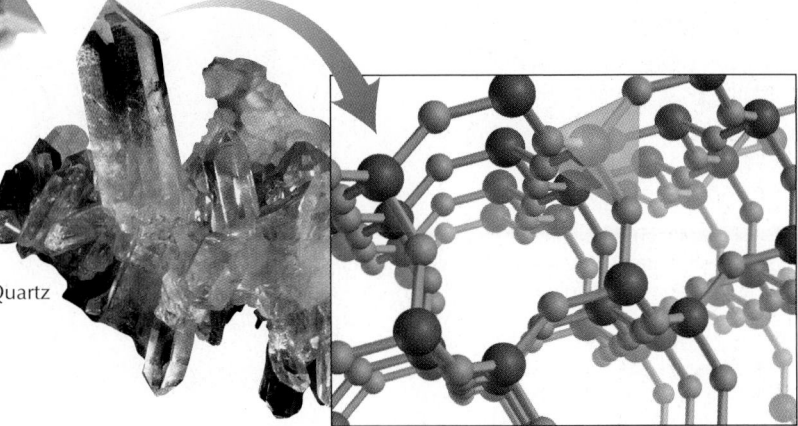

Zeolite bead

Quartz

Silicone Polymers

Polymer Chemistry

The design, synthesis, and production of polymers form one of the largest branches of modern chemical science. Nearly half of all industrial chemists and chemical engineers are involved in this pursuit. A major branch of polymer chemistry is devoted to exploring the properties and countless uses of silicones.

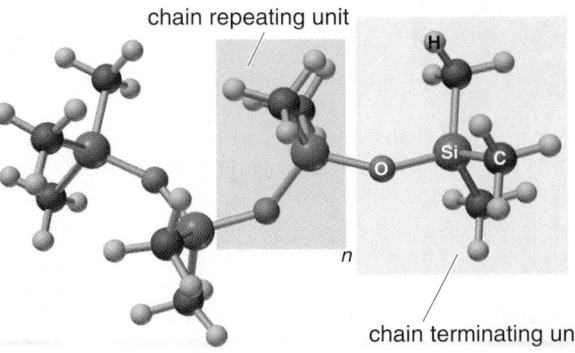

chain repeating unit

chain terminating unit

Chain Silicones: Oils and Greases

The simplest structural class of silicones consists of the poly(dimethyl siloxane) chain, in which each $(CH_3)_2Si(OH)_2$ unit uses both OH groups to link to two others. A chain-terminating compound with a third organic group, such as $(CH_3)_3SiOH$, is added to control chain length. These polymers are unreactive, oily liquids with high viscosity and low surface tension. They are used as hydraulic oils and lubricants, as antifoaming agents for frying potato chips, and as components of suntan oil, car polish, digestive aids, and makeup.

(continued)   **571**

Sheet Silicones: Elastomers

In the next class of silicones, the third OH group of an added bridging compound, such as $CH_3Si(OH)_3$, reacts to condense chains laterally into gummy sheets that are made into elastomers (rubbers), which are flexible, elastic, and stable from $-100°C$ to $250°C$. These are used in gaskets, rollers, cable insulation, space suits, contact lenses, and dentures.

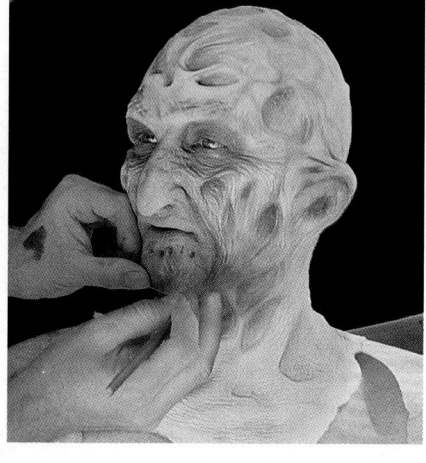

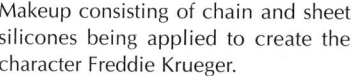

Makeup consisting of chain and sheet silicones being applied to create the character Freddie Krueger.

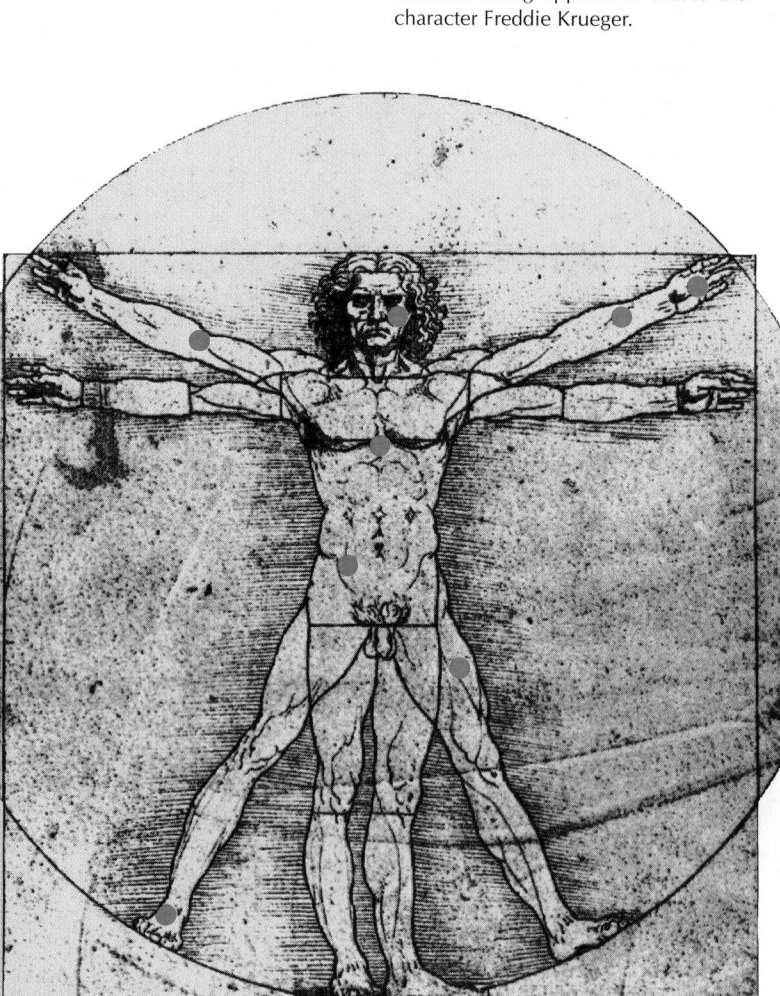

Framework Silicones: Resins

In the final structural class of silicones, reactions that free OH groups, while substituting larger organic groups for some CH_3 groups, interlink sheets to produce strong, thermally stable resins. These are used as insulating laminates on printed circuit boards, and as nonstick coatings on cookware. Sheet and framework silicones have revolutionized modern surgical practice by providing numerous parts that can be implanted permanently in a patient to replace damaged ones. Some of these (indicated by dots on Leonardo's man) are artificial skin, bone, joints, blood vessels, and organ parts.

14.7 GROUP 5A(15): THE NITROGEN FAMILY

The first two elements of Group 5A(15), gaseous nonmetallic nitrogen (N) and solid nonmetallic phosphorus (P), will occupy most of our attention. The industrial and environmental significance of their compounds is matched only by their importance in the structures and functions of biomolecules. Below them are two metalloids, arsenic (As) and antimony (Sb), followed by the sole metal, bismuth (Bi), the last nonradioactive element in the periodic table. The Group 5A(15) Family Portrait (pp. 574 and 575) provides an overview.

What Accounts for the Wide Range of Physical Behavior in Group 5A(15)?

Group 5A(15) displays the widest range of physical behavior we've seen so far because of large changes in bonding and intermolecular forces. Nitrogen occurs as a gas consisting of N_2 molecules, which interact through such weak dispersion forces that it *boils* more than 200°C below room temperature. Elemental phosphorus exists most commonly as tetrahedral P_4 molecules. However, because P is heavier and more polarizable than N, stronger dispersion forces are present and the element melts about 25°C above room temperature. Arsenic consists of extended, puckered sheets in which each As atom is covalently bonded to three others and forms nonbonding interactions with three nearest neighbors in adjacent sheets. This arrangement gives As the highest melting point in the group. A similar covalent network for Sb gives it a much higher melting point than the next member of the group, Bi, which has metallic bonding.

Phosphorus has several allotropes. The white and red forms have very different properties because of the way the atoms are linked. The highly reactive white form is prepared as a gas and condensed to a whitish, waxy solid under cold water to prevent it from igniting in air. It consists of tetrahedral molecules of four P atoms bonded *to each other,* with no atom in the center (Figure 14.19A). Each P atom uses its half-filled $3p$ orbitals to bond to the other three. Whereas the $3p$ orbitals of an isolated P atom lie 90° apart, the bond angles in P_4 are 60°. With the smaller angle comes poor orbital overlap and a P—P bond strength only about 80% that of a normal P—P bond (Figure 14.19B). Weaker bonds break more easily, so they contribute to the white allotrope's high reactivity. Heating the white form in the absence of air breaks one of the P—P bonds in each tetrahedron, and those orbitals overlap with others to form the chains of P_4 units that make up the red form (Figure 14.19C). Individual molecules in white P make it highly reactive, low melting (44.1°C), and soluble in nonpolar solvents; the chains in red P make it much less reactive, high melting (~600°C), and insoluble.

Figure 14.19 Two allotropes of phosphorus. A, White phosphorus exists as individual P_4 molecules, with the P—P bonds forming the edges of a tetrahedron. **B,** The reactivity of P_4 is due in part to the bond strain that arises from the 60° bond angle. Note how overlap of the $3p$ orbitals is decreased because they do not meet directly end to end (overlap is shown here for only three of the P—P bonds), which makes the bonds easier to break. **C,** In red phosphorus, one of the P—P bonds of the white form has broken and links the tetrahedra together into long chains. Lone pairs (not shown) reside in s orbitals in both allotropes.

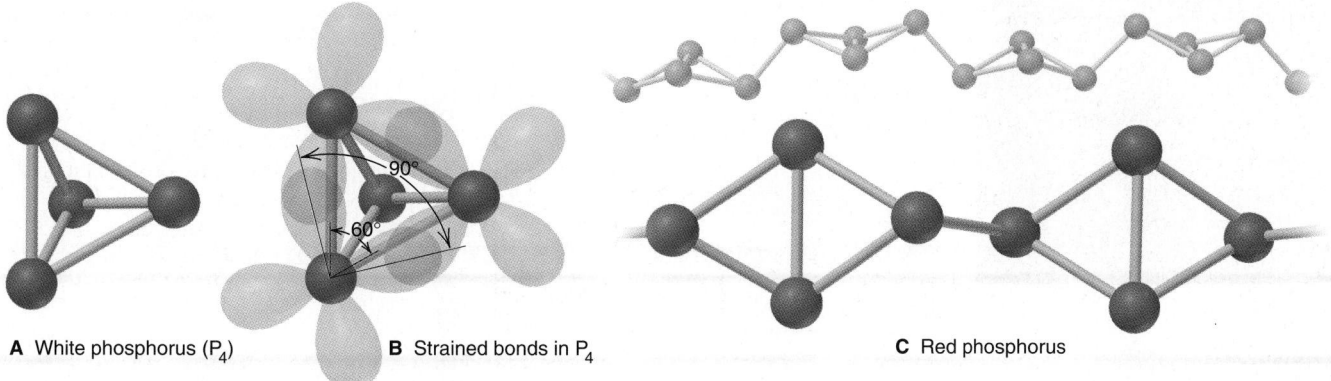

A White phosphorus (P_4) **B** Strained bonds in P_4 **C** Red phosphorus

Key Atomic and Physical Properties

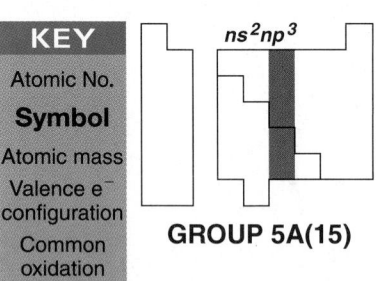

ns^2np^3

GROUP 5A(15)

Atomic radius (pm)		Ionic radius (pm)
N 75		N³⁻ 146
P 110		P³⁻ 212
As 120		
Sb 140		
Bi 150		Bi³⁺ 103

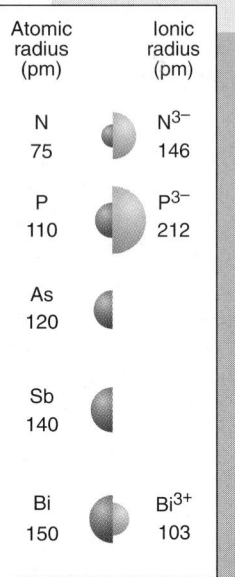

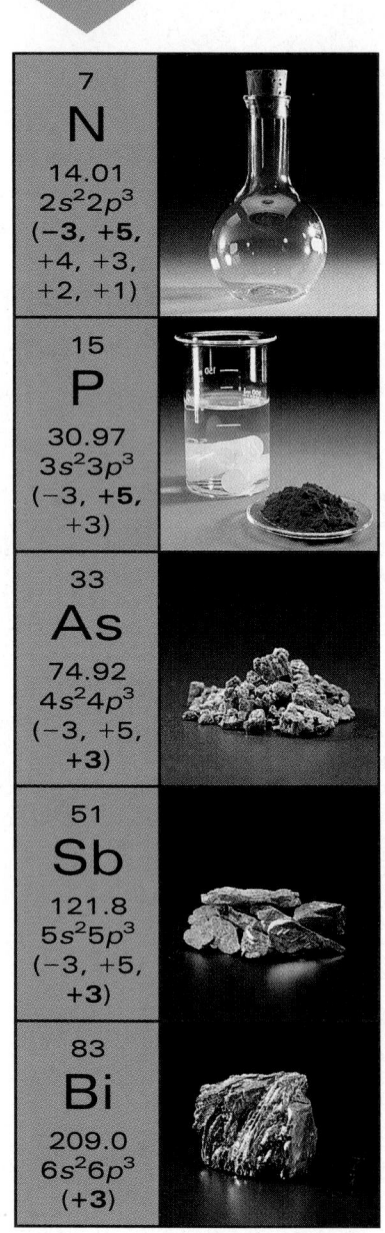

7	
N	
14.01	
$2s^22p^3$	
(−3, +5, +4, +3, +2, +1)	

15	
P	
30.97	
$3s^23p^3$	
(−3, +5, +3)	

33	
As	
74.92	
$4s^24p^3$	
(−3, +5, +3)	

51	
Sb	
121.8	
$5s^25p^3$	
(−3, +5, +3)	

83	
Bi	
209.0	
$6s^26p^3$	
(+3)	

Atomic Properties

Group electron configuration is ns^2np^3. The np sublevel is half-filled, with each p orbital containing one electron. The number of oxidation states decreases down the group, and the lower (+3) state becomes more common.

Atomic properties follow generally expected trends. The large (~50%) increase in size from N to P correlates with the much lower IE and EN of P.

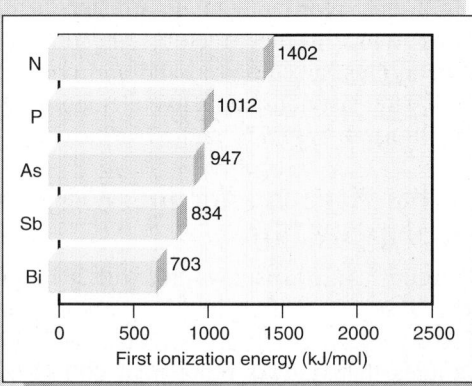

First ionization energy (kJ/mol)

- N 1402
- P 1012
- As 947
- Sb 834
- Bi 703

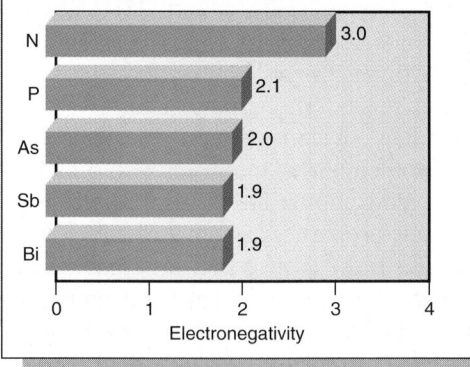

Electronegativity

- N 3.0
- P 2.1
- As 2.0
- Sb 1.9
- Bi 1.9

Physical Properties

Physical properties reflect the change from individual molecules (N, P) to covalent network (As, Sb) to metal (Bi). Thus, melting points increase and then decrease.

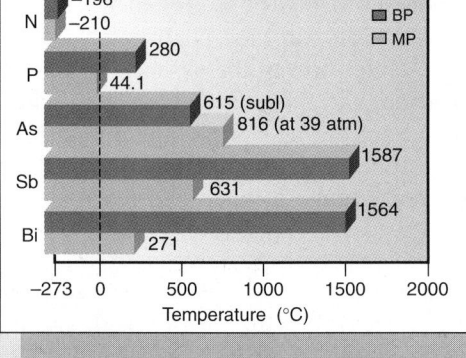

Temperature (°C)

- N: −196 (BP), −210 (MP)
- P: 280 (BP), 44.1 (MP)
- As: 615 (subl), 816 (at 39 atm)
- Sb: 1587 (BP), 631 (MP)
- Bi: 1564 (BP), 271 (MP)

Large atomic size and low atomic mass result in low density. Because mass increases more than size down the group, the density of the elements as solids increases. The dramatic increase from P to As is due to the intervening transition elements.

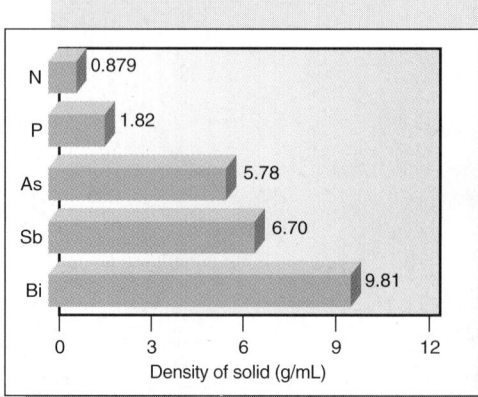

Density of solid (g/mL)

- N 0.879
- P 1.82
- As 5.78
- Sb 6.70
- Bi 9.81

Some Reactions and Compounds

Important Reactions

General group behavior is shown in reactions 1 to 3, whereas phosphorus chemistry is the theme in reactions 4 and 5.

1. Nitrogen is "fixed" industrially in the Haber process:

$$N_2(g) + 3H_2(g) \rightleftharpoons 2NH_3(g)$$

Further reactions convert NH_3 to NO, NO_2, and HNO_3 (*see Highlights of Nitrogen Chemistry*). Hydrides of some other group members are formed from reaction in water (or H_3O^+) of a metal phosphide, arsenide, and so forth:

$$Ca_3P_2(s) + 6H_2O(l) \longrightarrow 2PH_3(g) + 3Ca(OH)_2(aq)$$

2. Halides are formed by direct combination of the elements:

$$2E(s) + 3X_2 \longrightarrow 2EX_3 \quad \text{(E = all except N)}$$
$$EX_3 + X_2 \longrightarrow EX_5 \quad \text{(E = all except N with}$$
$$X = \text{F and Cl (but no } BiCl_5\text{);}$$
$$E = \text{P for } X = \text{Br)}$$

3. Oxoacids are formed from the halides in a reaction with water that is common to many nonmetal halides:

$$EX_3 + 3H_2O(l) \longrightarrow H_3EO_3(aq) + 3HX(aq)$$
$$\text{(E = all except N)}$$
$$EX_5 + 4H_2O(l) \longrightarrow H_3EO_4(aq) + 5HX(aq)$$
$$\text{(E = all except N and Bi)}$$

Note that the oxidation number of E does *not* change.

4. Phosphate ions are dehydrated to form polyphosphates:

$$3NaH_2PO_4(s) \longrightarrow Na_3P_3O_9(s) + 3H_2O(g)$$

5. When P_4 reacts in basic solution, its oxidation state both decreases *and* increases (disproportionation):

$$P_4(s) + 3OH^-(aq) + 3H_2O(l) \longrightarrow PH_3(g) + 3H_2PO_2^-(aq)$$

Analogous reactions are typical of many nonmetals, such as S_8 and X_2 (halogens).

Important Compounds

1. Ammonia, NH_3. First substance formed when atmospheric N_2 is used to make N-containing compounds. Annual multimillion-ton production for use in fertilizers, explosives, rayon, and polymers such as nylon, urea-formaldehyde resins, and acrylics.

2. Hydrazine, N_2H_4 (*see margin note, p. 576*).

3. Nitric oxide (NO), nitrogen dioxide (NO_2), and nitric acid (HNO_3). Oxides are intermediates in HNO_3 production. This acid is used in fertilizer manufacture, nylon production, metal etching, and the explosives industry (*see Highlights of Nitrogen Chemistry*).

Detonation of explosives

4. Amino acids, H_3N^+—$CH(R)$—COO^- (R = one of 20 different organic groups). Occur in every organism, both free and linked together into proteins. Essential to the growth and function of all cells. Synthetic amino acids are used as dietary supplements.

5. Phosphorus trichloride, PCl_3. Used to form many organic phosphorus compounds, including oil and fuel additives, plasticizers, flame retardants, and insecticides. Also used to make PCl_5, $POCl_3$, and other important P-containing compounds.

6. Tetraphosphorus decaoxide (P_4O_{10}) and phosphoric acid (H_3PO_4) (*see Highlights of Phosphorus Chemistry*).

7. Sodium tripolyphosphate, $Na_5P_3O_{10}$. As a water-softening agent (Calgon), it combines with hard-water Mg^{2+} and Ca^{2+}

Algal growth in polluted lake

ions, preventing them from reacting with soap anions, and thus improves cleaning action. Its use has been curtailed in the United States because it pollutes lakes and streams by causing excessive algal growth (*see photo*).

8. Adenosine triphosphate (ATP) and other biophosphates. ATP acts to transfer chemical energy in the cell; necessary for all biological processes requiring energy. Phosphate groups occur in sugars, fats, proteins, and nucleic acids.

9. Bismuth subsalicylate, $BiO(C_7H_5O_3)$. The active ingredient in Pepto-Bismol (*see photo*), a widely used remedy for diarrhea and nausea. (The pink color is not due to this white compound.)

What Patterns Appear in the Chemical Behavior of Group 5A(15)?

The same general pattern of chemical behavior that we discussed for Group 4A(14) appears again in this group, reflected in the change from nonmetallic N to metallic Bi. The overwhelming majority of Group 5A(15) compounds have *covalent bonds*. Whereas N can form no more than four bonds, the next three members can expand their valence shells by using empty *d* orbitals.

For a 5A element to form an ion with a noble gas electron configuration, it must *gain* three electrons, the last two in endothermic steps. Nevertheless, the enormous lattice energy released when such highly charged anions attract cations drives their formation, but this occurs with N only in compounds with active metals, such as Li_3N and Mg_3N_2 (and perhaps with P in Na_3P). Metallic Bi forms mostly covalent compounds but exists as a cation in a few compounds, such as BiF_3 and $Bi(NO_3)_3 \cdot 5H_2O$, by *losing* its three valence *p* electrons.

Again, we see the patterns first seen in Groups 3A and 4A as we move down the group. Fewer oxidation states occur, with the lower state becoming more prominent: N exhibits every state possible for a 5A element, from +5 to −3; only the +5 and +3 states are common for P, As, and Sb; and +3 is the only common state of Bi. The oxides change from acidic to amphoteric to basic, reflecting the increase in the metallic character of the elements. In addition, the lower oxide of an element is more basic than the higher oxide, reflecting the greater ionic character of the E-to-O bonding in the lower oxide.

All the Group 5A(15) elements form gaseous hydrides of formula EH_3. Except for NH_3, these are extremely reactive and poisonous and are synthesized by reaction of a metal phosphide, arsenide, and so forth, which acts as a strong base in water or aqueous acid. For example,

$$Ca_3As_2(s) + 6H_2O(l) \longrightarrow 2AsH_3(g) + 3Ca(OH)_2(aq)$$

Ammonia is made industrially by direct combination of the elements at high pressure and moderately high temperature:

$$N_2(g) + 3H_2(g) \rightleftharpoons 2NH_3(g)$$

Molecular properties of the Group 5A(15) hydrides reveal some interesting bonding and structural patterns:

- Despite its much lower molar mass, NH_3 melts and boils at higher temperatures than the other 5A hydrides as a result of *H bonding*.
- Bond angles decrease from 107.3° for NH_3 to around 90° for the other hydrides, which suggests that the larger atoms use unhybridized *p* orbitals.
- E—H bond lengths increase down the group, so bond strength and thermal stability decrease: AsH_3 decomposes at 250°C, SbH_3 at 20°C, and BiH_3 at −45°C.

We'll see these features—H bonding for the smallest member, change in bond angles, change in bond energies—in the hydrides of Group 6A(16) as well.

The Group 5A(15) elements form all the trihalides (EX_3) and the pentafluorides (EF_5), but few other pentahalides (PCl_5, PBr_5, $AsCl_5$, and $SbCl_5$). Nitrogen forms the trihalide only, because it cannot expand its valence shell. Most trihalides are prepared by direct combination:

$$P_4(s) + 6Cl_2(g) \longrightarrow 4PCl_3(l)$$

The pentahalides form with excess halogen:

$$PCl_3(l) + Cl_2(g) \longrightarrow PCl_5(s)$$

As with the hydrides, the thermal stability of the halides decreases as the E—X bond becomes longer. Among the nitrogen halides, for example, NF_3 is a stable, rather unreactive gas. NCl_3 is explosive and reacts rapidly with water. (The chemist who first prepared it lost three fingers and an eye!) NBr_3 can only be made below −87°C. NI_3 has never been prepared, but an ammoniated product ($NI_3 \cdot NH_3$) explodes at the slightest touch.

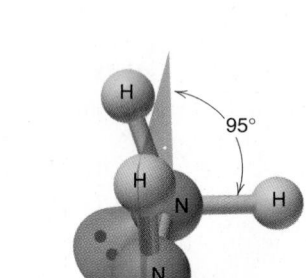

95°

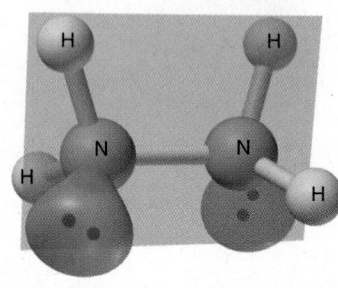

Hydrazine, Nitrogen's Other Hydride Aside from NH_3, the most important Group 5A(15) hydride is hydrazine, N_2H_4. Like NH_3, it is a weak base, forming $N_2H_5^+$ and $N_2H_6^{2+}$ ions with acids. Hydrazine is also used to make antituberculin drugs, plant growth regulators, and fungicides, and organic hydrazine derivatives are used in rocket and missile fuels. We might expect the lone pairs in the molecule to lie opposite each other and give a symmetrical structure with no dipole moment. However, repulsions between each lone pair and the bonding pairs on the other atom force the lone pairs to lie below the N—N bond (see model), giving a large dipole moment ($\mu = 1.85$ D in the gas phase). A similar arrangement occurs in hydrogen peroxide (H_2O_2), the analogous hydride of oxygen.

In an aqueous reaction pattern *typical of many nonmetal halides,* the 5A halides react with water to yield the hydrogen halide and the oxoacid, in which E has the *same* oxidation state as it had in the original halide. For example, PX_5 (O.N. of P = +5) produces phosphoric acid (O.N. of P = +5) and HX:

$$PCl_5(s) + 4H_2O(l) \longrightarrow H_3PO_4(l) + 5HCl(g)$$

Highlights of Nitrogen Chemistry

Surely the most obvious highlight of nitrogen chemistry is the inertness of N_2 itself. Nearly four-fifths of the atmosphere consists of N_2, and the other fifth is nearly all O_2, a very strong oxidizing agent. Nevertheless, the searing temperature of a lightning bolt is required for significant amounts of atmospheric nitrogen oxides to form. Thus, even though N_2 is inert at moderate temperatures, it reacts at high temperatures with H_2, Li, Group 2A(2) members, B, Al, C, Si, Ge, O_2, and many transition elements. In fact, nearly every element in the periodic table forms bonds to N. Here we focus on the oxides and the oxoacids and their salts.

Nitrogen Oxides Nitrogen is remarkable for having six stable oxides, each with a *positive* heat of formation because of the great strength of the $N{\equiv}N$ bond (BE = 945 kJ/mol). Their structures and some properties are shown in Table 14.3. Unlike the hydrides and halides of nitrogen, the oxides are planar. Nitrogen displays all its positive oxidation states in these compounds, and in N_2O and N_2O_3, the two N atoms have different states.

Table 14.3 Structures and Properties of the Nitrogen Oxides

Formula	Name	Space-filling Model	Lewis Structure	Oxidation State of N	ΔH_f^0 (kJ/mol) at 298 K	Comment
N_2O	Dinitrogen monoxide (dinitrogen oxide; nitrous oxide)		$:N{\equiv}N{-}\ddot{O}:$	+1 (0, +2)	82.0	Colorless gas; used as dental anesthetic ("laughing gas") and aerosol propellant
NO	Nitrogen monoxide (nitrogen oxide; nitric oxide)		$:\dot{N}{=}\ddot{O}:$	+2	90.3	Colorless, paramagnetic gas; biochemical messenger; air pollutant
N_2O_3	Dinitrogen trioxide			+3 (+2, +4)	83.7	Reddish brown gas (reversibly dissociates to NO and NO_2)
NO_2	Nitrogen dioxide			+4	33.2	Orange-brown, paramagnetic gas formed during HNO_3 manufacture; poisonous air pollutant
N_2O_4	Dinitrogen tetraoxide			+4	9.16	Colorless to yellow liquid (reversibly dissociates to NO_2)
N_2O_5	Dinitrogen pentaoxide			+5	11.3	Colorless, volatile solid consisting of NO_2^+ and NO_3^-; gas consists of N_2O_5 molecules

Dinitrogen monoxide (N_2O; also called dinitrogen oxide or nitrous oxide) is the dental anesthetic "laughing gas" and the propellant in canned whipped cream. It is a linear molecule with an electronic structure best described by three resonance forms (note formal charges; see Problem 10.95):

$$\overset{0 \quad +1 \quad -1}{:N\equiv N-\overset{..}{\underset{..}{O}}:} \longleftrightarrow \overset{-1 \quad +1 \quad 0}{:\underset{..}{N}=N=\overset{..}{\underset{..}{O}}:} \longleftrightarrow \overset{-2 \quad +1 \quad +1}{:\overset{..}{\underset{.}{N}}-N\equiv O:}$$

most important   least important

Nitrogen monoxide (NO; also called nitrogen oxide or nitric oxide) is an odd-electron molecule with a vital biochemical function. In Section 11.3, we used MO theory to explain its bond properties. The commercial preparation of NO through the oxidation of ammonia occurs as a first step in the production of nitric acid:

$$4NH_3(g) + 5O_2(g) \longrightarrow 4NO(g) + 6H_2O(g)$$

Nitrogen monoxide is also produced whenever air is heated to high temperatures, as in a car engine or a lightning storm:

$$N_2(g) + O_2(g) \xrightarrow{\text{high } T} 2NO(g)$$

Heating converts NO to two other oxides:

$$3NO(g) \xrightarrow{\Delta} N_2O(g) + NO_2(g)$$

This type of redox reaction is called a **disproportionation.** It occurs when a substance *acts as both an oxidizing and a reducing agent in a reaction.* In the process, an atom with an intermediate oxidation state in the reactant occurs in the products in both lower and higher states: the oxidation state of N in NO ($+2$) is intermediate between that in N_2O ($+1$) and that in NO_2 ($+4$).

Nitrogen dioxide (NO_2), a brown poisonous gas, also forms when NO reacts with additional oxygen:

$$2NO(g) + O_2(g) \longrightarrow 2NO_2(g)$$

Like NO, NO_2 is also an odd-electron molecule, but the electron is more localized on the N atom. Thus, NO_2 dimerizes reversibly to dinitrogen tetraoxide:

$$O_2N\cdot(g) + \cdot NO_2(g) \rightleftharpoons O_2N\!-\!NO_2(g) \quad \text{(or } N_2O_4)$$

Thunderstorms form NO and NO_2 and carry them down to the soil, where they act as natural fertilizers. In urban traffic, however, their formation leads to *photochemical smog* (Figure 14.20). During the morning commute, NO forms in car and truck (especially diesel) engines and is then oxidized in air to NO_2. As the Sun climbs higher, radiant energy breaks down some NO_2:

$$NO_2(g) \xrightarrow{\text{sunlight}} NO(g) + O(g)$$

The O atoms collide with O_2 molecules and form ozone, O_3, a powerful oxidizing agent that damages synthetic rubber and plastic, as well as plant and animal tissue:

$$O_2(g) + O(g) \longrightarrow O_3(g)$$

A complex series of reactions between NO_2, O_3, and unburned hydrocarbons in gasoline fumes forms *peroxyacylnitrates* (PANs), a group of potent nose and eye irritants. The outcome of this fascinating atmospheric chemistry is choking, brown smog.

Nitrogen Oxoacids and Oxoanions The two common nitrogen oxoacids are nitric acid and nitrous acid (Figure 14.21). The first two steps in the *Ostwald process* for the production of nitric acid—the oxidations of NH_3 to NO and of NO to NO_2—have been shown. The final step is a disproportionation, as the oxidation numbers show:

$$\overset{+4}{3NO_2}(g) + H_2O(l) \longrightarrow \overset{+5}{2HNO_3}(aq) + \overset{+2}{NO}(g)$$

The NO is recycled to make more NO_2.

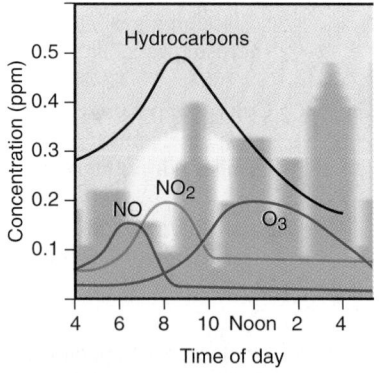

Figure 14.20 **The formation of photochemical smog.** Localized atmospheric concentrations of the precursor components of smog change during the day. In early morning traffic, NO and hydrocarbon levels increase from auto exhaust, followed by increases in NO_2 as NO reacts with air. Ozone peaks later as NO_2 breaks down in stronger sunlight and releases O atoms that react with O_2. By mid-afternoon, all the components for the production of peroxyacylnitrates (PANs) are present and photochemical smog results.

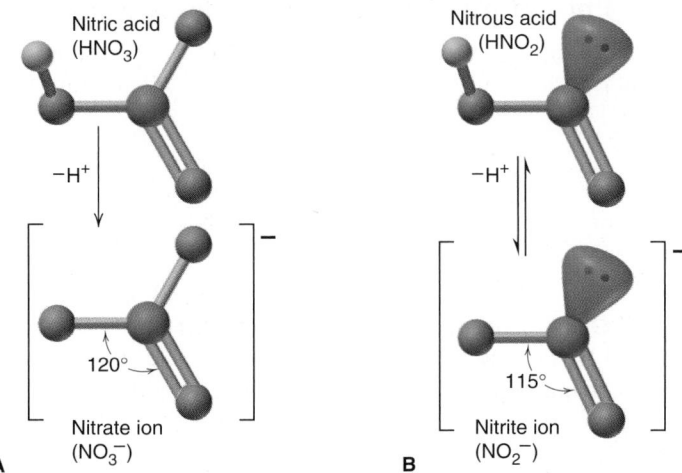

Figure 14.21 The structures of nitric and nitrous acids and their oxoanions. A, Nitric acid loses a proton (H^+) to form the trigonal planar nitrate ion (one of three resonance forms is shown). **B,** Nitrous acid, a much weaker acid, forms the planar nitrite ion. Note the effect of nitrogen's lone pair (the lone pairs on the oxygens are not shown) in reducing the ideal 120° bond angle to 115° (one of two resonance forms is shown).

In nitric acid, as in all oxoacids, *the acidic H is attached to one of the O atoms.* In the laboratory, nitric acid is used as a strong oxidizing acid. The products of its reaction with metals vary with the metal's reactivity and the acid's concentration. In the following examples, notice from the net ionic equations that *the NO_3^- ion is the oxidizing agent.* Nitrate ion that is not reduced is a spectator ion and does not appear in the net ionic equations.

- With an active metal, such as Al, and dilute acid, N is reduced from the +5 state all the way to the −3 state in the ammonium ion:

$$8Al(s) + 30HNO_3(aq;\ 1\ M) \longrightarrow 8Al(NO_3)_3(aq) + 3NH_4NO_3(aq) + 9H_2O(l)$$
$$8Al(s) + 30H^+(aq) + 3NO_3^-(aq) \longrightarrow 8Al^{3+}(aq) + 3NH_4^+(aq) + 9H_2O(l)$$

- With a less reactive metal, such as Cu, and more concentrated acid, N is reduced to the +2 state in NO:

$$3Cu(s) + 8HNO_3(aq;\ 3\ to\ 6\ M) \longrightarrow 3Cu(NO_3)_2(aq) + 4H_2O(l) + 2NO(g)$$
$$3Cu(s) + 8H^+(aq) + 2NO_3^-(aq) \longrightarrow 3Cu^{2+}(aq) + 4H_2O(l) + 2NO(g)$$

- With still more concentrated acid, N is reduced only to the +4 state in NO_2:

$$Cu(s) + 4HNO_3(aq;\ 12\ M) \longrightarrow Cu(NO_3)_2(aq) + 2H_2O(l) + 2NO_2(g)$$
$$Cu(s) + 4H^+(aq) + 2NO_3^-(aq) \longrightarrow Cu^{2+}(aq) + 2H_2O(l) + 2NO_2(g)$$

Nitrates form when HNO_3 reacts with metals and with their hydroxides, oxides, or carbonates. *All nitrates are soluble in water.*

Nitrous acid, HNO_2, a much weaker acid than HNO_3, forms when metal nitrites are treated with a strong acid:

$$NaNO_2(aq) + HCl(aq) \longrightarrow HNO_2(aq) + NaCl(aq)$$

These two acids reveal a *general pattern in relative acid strength among oxoacids:* the more O atoms bound to the central nonmetal, the stronger the acid. Thus, HNO_3 is stronger than HNO_2. The O atoms pull electron density from the N atom, which in turn pulls electron density from the O of the O—H bond, facilitating the release of the H^+ ion. The O atoms also act to stabilize the resulting oxoanion by delocalizing its negative charge. The same pattern occurs in the oxoacids of sulfur and the halogens; we'll discuss the pattern quantitatively in Chapter 18.

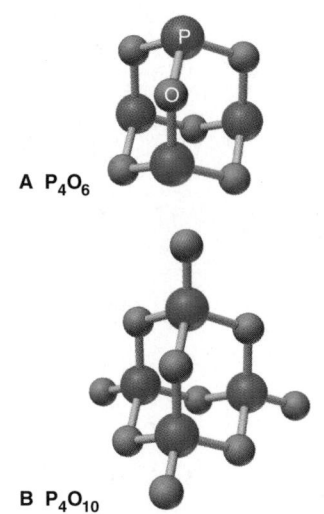

A P_4O_6

B P_4O_{10}

Figure 14.22 Important oxides of phosphorus. **A,** P_4O_6. **B,** P_4O_{10}.

Highlights of Phosphorus Chemistry: Oxides and Oxoacids

Phosphorus forms two important oxides, P_4O_6 and P_4O_{10}. Tetraphosphorus hexaoxide, P_4O_6, has P in its +3 oxidation state and forms when white P_4 reacts with limited oxygen:

$$P_4(s) + 3O_2(g) \longrightarrow P_4O_6(s)$$

P_4O_6 has the tetrahedral orientation of the P atoms in P_4, with an O atom between each pair of P atoms (Figure 14.22A). It reacts with water to form phosphor*ous* acid (note the spelling):

$$P_4O_6(s) + 6H_2O(l) \longrightarrow 4H_3PO_3(l)$$

The formula H_3PO_3 is misleading because the acid has only two acidic H atoms; the third is bound to the central P and does not dissociate. Phosphorous acid is a weak acid in water but reacts completely in two steps with excess strong base:

Salts of phosphorous acid contain the phosphite ion, $HPO_3{}^{2-}$.

In tetraphosphorus decaoxide, P_4O_{10}, P is in the +5 oxidation state. Commonly known as "phosphorus pentoxide" from the empirical formula (P_2O_5), it forms when P_4 burns in excess O_2:

$$P_4(s) + 5O_2(g) \longrightarrow P_4O_{10}(s)$$

Its structure can be viewed as that of P_4O_6 with another O atom bonded to each of the four corner P atoms (Figure 14.22B). P_4O_{10} is a powerful drying agent and, in a vigorous exothermic reaction with water, forms phosphoric acid (H_3PO_4), one of the "top-10" most important compounds in chemical manufacturing:

$$P_4O_{10}(s) + 6H_2O(l) \longrightarrow 4H_3PO_4(l)$$

The presence of many H bonds makes pure H_3PO_4 syrupy, more than 75 times as viscous as water. The laboratory-grade concentrated acid is an 85% by mass aqueous solution. H_3PO_4 is a weak triprotic acid; in water, it loses one proton in the following equilibrium reaction:

$$H_3PO_4(l) + H_2O(l) \rightleftharpoons H_2PO_4{}^-(aq) + H_3O^+(aq)$$

In excess strong base, however, the three protons dissociate completely in three steps to give the three phosphate oxoanions:

dihydrogen phosphate ion   hydrogen phosphate ion   phosphate ion

Phosphoric acid has a central role in fertilizer production, but it is also used as a polishing agent for aluminum car trim and as an additive in soft drinks to give a touch of tartness. The various phosphate salts have numerous essential applications.

Polyphosphates are formed by heating hydrogen phosphates, which lose water as they form P—O—P linkages. This type of reaction, in which an H_2O molecule is lost for every pair of OH groups that join, is called a **dehydration-condensation;** it occurs frequently in the formation of polyoxoanion chains and other polymeric structures, both synthetic and natural. For example, sodium diphosphate, $Na_4P_2O_7$, is prepared by heating sodium hydrogen phosphate:

$$2Na_2HPO_4(s) \xrightarrow{\Delta} Na_4P_2O_7(s) + H_2O(g)$$

The Countless Uses of Phosphates Phosphates have an amazing array of applications in home and industry. Na_3PO_4 is a paint stripper and grease remover and is still used in cleaning powders in other countries. Na_2HPO_4 is a laxative ingredient and is used to adjust the acidity of boiler water. The potassium salt K_3PO_4 is used to stabilize latex for synthetic rubber, and K_2HPO_4 is a radiator corrosion inhibitor. Ammonium phosphates are used as fertilizers and as flame retardants on curtains and paper costumes. Calcium phosphates are used in baking powders and toothpastes, as mineral supplements in stock feed, and (on the hundred-million-ton scale) as fertilizers throughout the world.

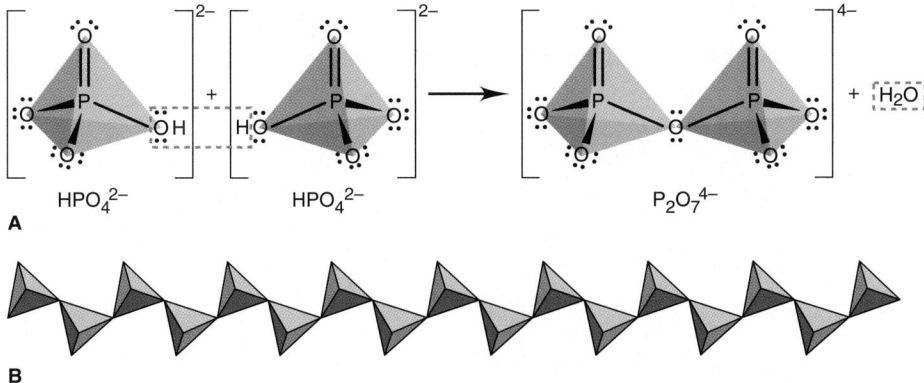

A

B

Figure 14.23 The diphosphate ion and polyphosphates. A, When two hydrogen phosphate ions undergo a dehydration-condensation reaction, they lose a water molecule and join through a shared O atom to form a diphosphate ion. **B,** Polyphosphates are chains of many such tetrahedral PO_4 units. Chain shapes depend on the orientation of the individual units. Note the similarity to silicate chains (see Gallery, Section 14.6).

The diphosphate ion, the smallest of the polyphosphates, consists of two PO_4 units linked through a common oxygen corner (Figure 14.23A). Its reaction with water, the reverse of the previous reaction, generates heat:

$$P_2O_7^{4-}(aq) + H_2O(l) \longrightarrow 2HPO_4^-(aq) + heat$$

A similar process is put to great use by organisms, when a third PO_4 unit linked to diphosphate creates the triphosphate grouping, part of the all-important high-energy biomolecule adenosine triphosphate (ATP). In Chapters 20 and 21, we discuss the central role of ATP in biological energy production. Extended polyphosphate chains consist of many tetrahedral PO_4 units (Figure 14.23B) and are structurally similar to silicate chains. As a final look at the chemical versatility of phosphorus, consider some of its numerous important compounds with sulfur and with nitrogen.

14.8 GROUP 6A(16): THE OXYGEN FAMILY

The first two members of this family—gaseous nonmetallic oxygen (O) and solid nonmetallic sulfur (S)—are among the most important elements in industry, the environment, and living things. Two metalloids, selenium (Se) and tellurium (Te), appear below them, and the lone metal, radioactive polonium (Po), ends the group. The Group 6A(16) Family Portrait (pp. 582 and 583) displays the features of these elements.

How Do the Oxygen and Nitrogen Families Compare Physically?

Group 6A(16) resembles Group 5A(15) in many respects, so let's look at some common themes. The pattern of physical properties we saw in Group 5A appears again in this group. Like nitrogen, oxygen occurs as a low-boiling diatomic gas. Like phosphorus, sulfur occurs as a polyatomic molecular solid. Like arsenic, selenium commonly occurs as a gray metalloid. Like antimony, tellurium is slightly more metallic than the preceding group member but still displays network bonding. Finally, like bismuth, polonium has a metallic crystal structure. As we expect by comparison with the 5A elements, electrical conductivities increase steadily down the group as bonding changes from individual molecules (insulators) to metalloid networks (semiconductors) to a metallic solid (conductor).

Match Heads, Bug Sprays, and O-Rings Phosphorus forms many sulfides and nitrides. P_4S_3 is used in "strike-anywhere" match heads, and P_4S_{10} is used in the manufacture of organophosphorus pesticides, such as malathion. Polyphosphazenes have properties similar to those of silicones. Indeed, the $—(R_2)P{=}N—$ unit is isoelectronic with the silicone unit, $—(R_2)Si—O—$. Sheets, films, fibers, and foams of polyphosphazene are water repellent, flame resistant, solvent resistant, and flexible at low temperatures—perfect for the gaskets and O-rings in spacecraft and polar vehicles.

Group 6A(16): The Oxygen Family

Key Atomic and Physical Properties

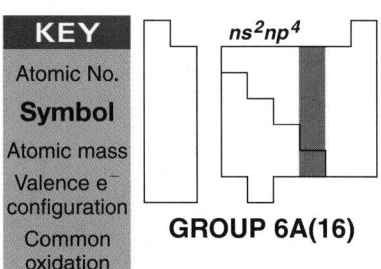

ns^2np^4

GROUP 6A(16)

Atomic Properties

Group electron configuration is ns^2np^4. As in Groups 3A(13) and 5A(15), a lower (+4) oxidation state becomes more common down the group.

Down the group, atomic and ionic size increase, and IE and EN decrease.

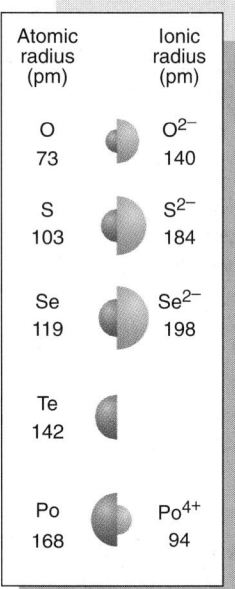

Atomic radius (pm)		Ionic radius (pm)
O 73		O²⁻ 140
S 103		S²⁻ 184
Se 119		Se²⁻ 198
Te 142		
Po 168		Po⁴⁺ 94

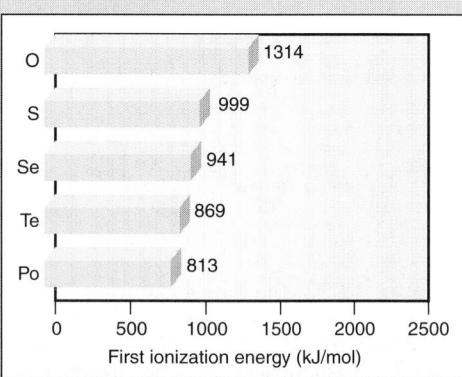

First ionization energy (kJ/mol)

- O 1314
- S 999
- Se 941
- Te 869
- Po 813

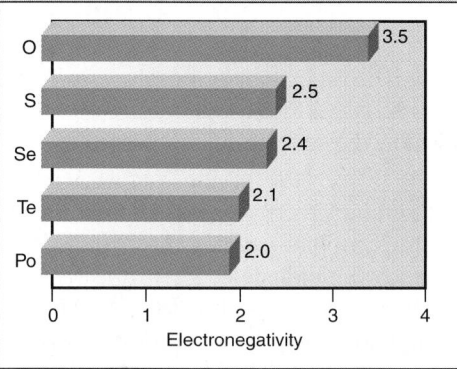

Electronegativity

- O 3.5
- S 2.5
- Se 2.4
- Te 2.1
- Po 2.0

Physical Properties

Melting points increase through Te, which has covalent bonding, and then decrease for Po, which has metallic bonding.

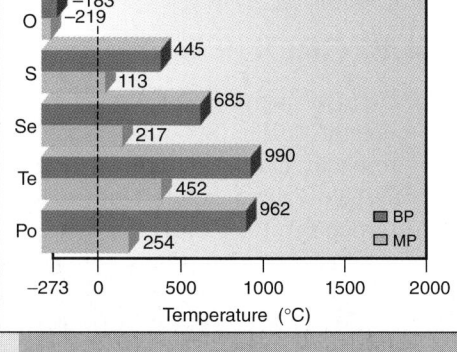

Temperature (°C)

- O −183 / −219
- S 445 / 113
- Se 685 / 217
- Te 990 / 452
- Po 962 / 254

☐ BP
☐ MP

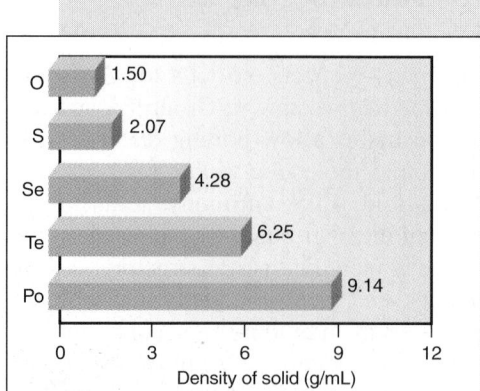

Density of solid (g/mL)

- O 1.50
- S 2.07
- Se 4.28
- Te 6.25
- Po 9.14

Densities of the elements as solids increase steadily.

8
O
16.00
$2s^22p^4$
(−1, **−2**)

16
S
32.07
$3s^23p^4$
(**−2, +6,** **+4,** +2)

34
Se
78.96
$4s^24p^4$
(**−2, +6,** **+4,** +2)

52
Te
127.6
$5s^25p^4$
(**−2, +6,** **+4,** +2)

84
Po
(209)
$6s^26p^4$
(**+4,** +2)

Some Reactions and Compounds

Important Reactions

Halogenation and oxidation of the elements (E) appear in reactions 1 and 2, and sulfur chemistry in reactions 3 and 4.

1. Halides are formed by direct combination:

$$E(s) + X_2(g) \longrightarrow \text{various halides}$$
$$(E = S, Se, Te; X = F, Cl)$$

2. The other elements in the group are oxidized by O_2:

$$E(s) + O_2(g) \longrightarrow EO_2 \quad (E = S, Se, Te, Po)$$

SO_2 is oxidized further, and the product is used in the final step of H_2SO_4 manufacture (*see Highlights of Sulfur Chemistry*):

$$2SO_2(g) + O_2(g) \longrightarrow 2SO_3(g)$$

3. Sulfur is recovered when hydrogen sulfide is oxidized:

$$8H_2S(g) + 4O_2(g) \longrightarrow S_8(s) + 8H_2O(g)$$

This reaction is used to obtain sulfur when natural deposits are not available.

4. The thiosulfate ion is formed when an alkali metal sulfite reacts with sulfur, as in the preparation of photographer's "hypo":

$$S_8(s) + 8Na_2SO_3(aq) \longrightarrow 8Na_2S_2O_3(aq)$$

Important Compounds

1. Water, H_2O. The single most important compound on Earth (*Section 12.5*).

2. Hydrogen peroxide, H_2O_2. Used as an oxidizing agent, disinfectant, and bleach, and in the production of peroxy compounds for polymerization (*see margin note, p. 585*).

3. Hydrogen sulfide, H_2S. Vile-smelling toxic gas formed during anaerobic decomposition of plant and animal matter, in volcanoes, and in deep-sea thermal vents. Used as a source of sulfur and in the 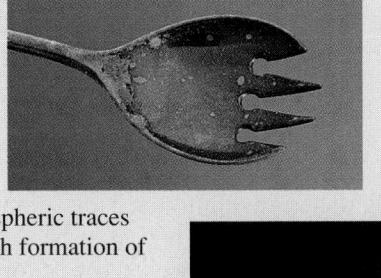 manufacture of paper. Atmospheric traces cause silver to tarnish through formation of black Ag_2S (*see photo*).

4. Sulfur dioxide, SO_2. Colorless, choking gas formed in volcanoes (*see photo*) or whenever an S-containing material (coal, oil, metal sulfide ores, and so on) is burned. More than 90% of SO_2 produced is used to make sulfuric acid. Also used as a fumigant and a preservative of fruit, syrups, and wine. As a reducing agent, removes excess Cl_2 from industrial wastewater, removes O_2 from petroleum handling tanks, and prepares ClO_2 for bleaching paper. Atmospheric pollutant in acid rain.

5. Sulfur trioxide (SO_3) and sulfuric acid (H_2SO_4). SO_3, formed from SO_2 over a V_2O_5 catalyst, is then converted to H_2SO_4. The acid is the cheapest strong acid and is so widely used in industry that its production level is an indicator of a nation's economic strength. It is a strong dehydrating agent that removes water from any organic source (*see Highlights of Sulfur Chemistry*).

6. Sulfur hexafluoride, SF_6. Extremely inert gas used as an electrical insulator.

Allotropism is more common in Group 6A(16) than in Group 5A(15). Oxygen has two allotropes: life-giving dioxygen (O_2), and poisonous triatomic ozone (O_3). Oxygen gas is colorless, odorless, paramagnetic, and thermally stable. In contrast, ozone gas is bluish, has a pungent odor, is diamagnetic, and decomposes in heat and especially in ultraviolet (UV) light:

$$2O_3(g) \xrightarrow{\text{UV}} 3O_2(g)$$

This ability to absorb high-energy photons makes stratospheric ozone vital to life. A thinning of the ozone layer, observed above the North and especially the South Poles, means that more UV light will reach the Earth's surface, with potentially hazardous effects. (We'll discuss the chemical causes of ozone depletion in Chapter 16.)

Sulfur is the allotrope "champion" of the periodic table, with more than 10 forms. The S atom's ability to bond to other S atoms creates numerous rings and chains, with S—S bond lengths that range from 180 pm to 260 pm and bond angles from 90° to 180°. At room temperature, the sulfur molecule is a crown-shaped ring of eight atoms, called *cyclo-*S_8 (Figure 14.24). The most stable allotrope is orthorhombic α-S_8, which consists entirely of these molecules; all other S allotropes eventually revert to this one.

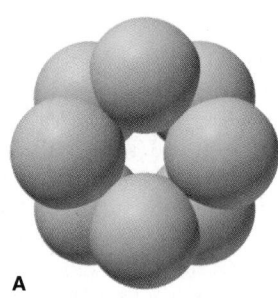

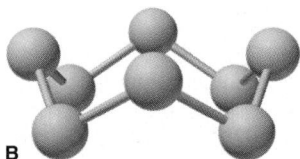

Figure 14.24 The cyclo-S_8 molecule. A, Top view of a space-filling model of the cyclo-S_8 molecule. **B,** Side view of a ball-and-stick model of the molecule; note the crownlike shape.

Selenium and Xerography
Photocopying was invented in the 1940s as a rapid, inexpensive, dry means of copying documents (*xerography,* Greek "dry writing"). The process is based on the ability of selenium to conduct a current when illuminated. A film of amorphous Se is deposited on an aluminum drum and electrostatically charged. Exposure to a document produces an "image" of low and high positive charges that correspond to the document's bright and dark areas. Negatively charged black, dry ink (toner) particles are attracted to the regions of high charge more than to those of low charge. This pattern of black particles is transferred electrostatically to paper, and the particles are fused to the paper's surface by heat or solvent. Excess toner is removed from the Se film, the charges are "erased" by exposure to light, and the film is ready for the next page.

Selenium also has several allotropes, some consisting of crown-shaped Se_8 molecules. Gray Se is composed of layers of helical chains. Its electrical conductivity in visible light has revolutionized the photocopying industry. When molten glass, cadmium sulfide, and gray Se are mixed and heated in the absence of air, a ruby-red glass forms, which you see whenever you stop at a traffic light.

How Do the Oxygen and Nitrogen Families Compare Chemically?

Changes in chemical behavior in this group are also similar to those in the previous group. Even though O and S occur as anions much more often than do N and P, like N and P, they bond covalently with almost every other nonmetal. Covalent bonds appear in the compounds of Se and Te (as in those of As and Sb), whereas Po behaves like a metal (as does Bi) in some of its saltlike compounds. In contrast to nitrogen, oxygen has few common oxidation states, but the earlier pattern returns with the other members: the +6, +4, and −2 states occur most often, with the lower positive (+4) state becoming more common in Te and Po [as the lower positive (+3) state does in Sb and Bi].

The range in atomic properties is wider in this group than in Group 5A(15) because of oxygen's high EN (3.5) and great oxidizing strength, second only to that of fluorine. As in 5A and earlier groups, the behavior of the Period 2 element stands out. In fact, aside from a similar outer electron configuration, the larger members of Group 6A behave very little like oxygen: they are much less electronegative, form anions much less often (S^{2-} occurs with active metals), and their hydrides exhibit no H bonding.

Except for O, all the 6A elements form foul-smelling, poisonous, gaseous hydrides (H_2E) by treatment with acid of the metal sulfide, selenide, and so forth. For example,

$$FeSe(s) + 2HCl(aq) \longrightarrow H_2Se(g) + FeCl_2(aq)$$

Hydrogen sulfide also forms naturally in swamps from the breakdown of organic matter. It is as toxic as HCN, and even worse, it anesthetizes your olfactory nerves, so that as its concentration increases, you smell it less! The other hydrides are about 100 times *more* toxic.

In their bonding and thermal stability, these Group 6A hydrides have several features in common with those of Group 5A:

- Only water can form H bonds, so it melts and boils much higher than the other H_2E compounds (see Figure 12.14). (Oxygen's other hydride, H_2O_2, is also extensively H bonded.)
- Bond angles drop from the nearly tetrahedral value for H_2O (104.5°) to around 90° for the larger hydrides, suggesting that the central atom uses unhybridized *p* orbitals.
- E—H bond length increases (bond energy decreases) down the group. Thus, H_2Te decomposes above 0°C, and H_2Po can be made only in extreme cold because thermal energy from the radioactive Po decomposes it. Another result of longer (weaker) bonds is that the 6A hydrides are acids in water, and their acidity increases from H_2S to H_2Po.

Except for O, the Group 6A elements form a wide range of halides, whose structure and reactivity patterns depend on the *sizes of the central atom and the surrounding halogens:*

- Sulfur forms many fluorides, a few chlorides, one bromide, but no stable iodides.
- As the central atom becomes larger, the halides become more stable. Thus, tetrachlorides and tetrabromides of Se, Te, and Po are known, as are tetraiodides of Te and Po. Hexafluorides are known only for S, Se, and Te.

The inverse relationship between bond length and bond strength that we've seen previously does not account for this pattern. Rather, it is based on the effect of electron repulsions due to crowding of lone pairs and halogen (X) atoms around the central Group 6A atom. With S, the larger X atoms become too crowded, which explains why sulfur iodides do not occur. With increasing size of E and therefore length of E—X bonds, however, lone pairs and X atoms do not crowd each other as much, and a greater number of stable halides form.

Two sulfur fluorides illustrate clearly how crowding and orbital availability affect reactivity. Sulfur tetrafluoride (SF_4) is extremely reactive. It forms SO_2 and HF when exposed to moisture and is commonly used to fluorinate many compounds:

$$3SF_4(g) + 4BCl_3(g) \longrightarrow 4BF_3(g) + 3SCl_2(l) + 3Cl_2(g)$$

In contrast, sulfur hexafluoride (SF_6) is almost as inert as a noble gas! It is odorless, tasteless, nonflammable, nontoxic, and insoluble. Hot metals, boiling HCl, molten KOH, and high-pressure steam have no effect on it. It is used as an insulating gas in high-voltage generators, withstanding over 10^6 volts across

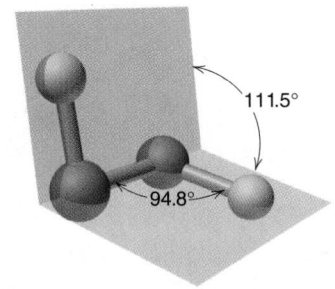

111.5°
94.8°

Hydrogen Peroxide: Hydrazine's Cousin Oxygen forms another hydride called hydrogen peroxide, H_2O_2 (HO—OH). Like hydrazine (H_2N—NH_2), the related hydride of nitrogen, it has a skewed shape. It is a colorless liquid with a high density, viscosity, and boiling point because of extensive H bonding. In peroxides, O is in the -1 oxidation state, midway between that in O_2 (zero) and that in oxides (-2); thus, H_2O_2 readily disproportionates:

$$H_2O_2(l) \longrightarrow H_2O(l) + \tfrac{1}{2}O_2(g)$$

Aside from the familiar use of H_2O_2 as a hair bleach and disinfectant, more than 70% of the half-million tons produced each year are used to bleach paper pulp, textiles, straw, and leather and to make other chemicals. H_2O_2 is also used in tertiary sewage treatment (Section 13.6) to oxidize foul-smelling effluents and restore O_2 to wastewater.

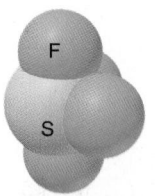 

Sulfur tetrafluoride (SF₄)   Sulfur hexafluoride (SF₆)

Figure 14.25 **Structural differences between SF₄ and SF₆.** SF₄ has a lone pair and empty *d* orbitals that can become involved in bonding. SF₆ already has the maximum number of bonds that can be formed by S, and closely packed F atoms envelop the central S atom, rendering SF₆ chemically inert.

electrodes only 50 mm apart. A look at the structures of these two fluorides provides the key to these astounding differences (Figure 14.25). Sulfur tetrafluoride can form bonds to another atom either by donating its lone electron pair or by accepting a lone pair into one of its empty *d* orbitals. On the other hand, SF₆ has no central lone pair, and the six F atoms form an octahedral sheath around the central S atom that blocks chemical attack.

Highlights of Oxygen Chemistry: Range of Oxide Properties

Oxygen is the most abundant element on the Earth's surface, occurring both as the free element and in innumerable oxides, silicates, carbonates, and phosphates, as well as in water. Virtually all free O_2 has a biological origin, having been formed for billions of years by photosynthetic algae and multicellular plants in an overall equation that looks deceptively simple:

$$n\text{H}_2\text{O}(l) + n\text{CO}_2(g) \xrightarrow{\text{light}} n\text{O}_2(g) + (\text{CH}_2\text{O})_n \text{ (carbohydrates)}$$

The reverse process occurs during combustion and respiration. Through these O_2-forming and O_2-utilizing processes, the 1.5×10^9 km³ of water on Earth is, on average, used and remade every 2 million years!

Every element (except He, Ne, and Ar) forms at least one oxide, many by direct combination. A broad spectrum of properties characterizes these compounds. Some oxides are gases that condense at very low temperatures, such as CO (bp = −192°C); others are solids that melt at extremely high temperatures, such as BeO (mp = 2530°C). They cover the full range of conductivity: insulators (MgO), semiconductors (NiO), conductors (ReO₃), and superconductors (YBa₂Cu₃O₇). Some oxides have endothermic heats of formation (ΔH_f^0 of NO = +90.3 kJ/mol), whereas other oxides have exothermic ones (ΔH_f^0 of CO_2 = −393.5 kJ/mol). They may be thermally stable (CaO) or unstable (HgO), as well as chemically reactive (Li₂O) or inert (Fe₂O₃).

Given this vast range of behavior, another useful way to classify element oxides is by their acid-base properties (see Interchapter, Topic 4). The oxides of Group 6A(16) exhibit expected trends in acidity, with SO₃ the most acidic and PoO₂ the most basic.

Highlights of Sulfur Chemistry: Oxides, Oxoacids, and Sulfides

Like phosphorus, sulfur forms two important oxides, sulfur dioxide (SO₂) and sulfur trioxide (SO₃). Sulfur is in its +4 oxidation state in SO₂, a colorless, choking gas that forms whenever S, H₂S, or a metal sulfide burns in air:

$$2\text{H}_2\text{S}(g) + 3\text{O}_2(g) \longrightarrow 2\text{H}_2\text{O}(g) + 2\text{SO}_2(g)$$
$$4\text{FeS}_2(s) + 11\text{O}_2(g) \longrightarrow 2\text{Fe}_2\text{O}_3(s) + 8\text{SO}_2(g)$$

Most SO₂ is used to produce sulfuric acid.

In water, sulfur dioxide forms sulfurous acid, which exists in equilibrium with hydrated SO₂ rather than as isolable H₂SO₃ molecules:

$$\text{SO}_2(aq) + \text{H}_2\text{O}(l) \rightleftharpoons [\text{H}_2\text{SO}_3(aq)] \rightleftharpoons \text{H}^+(aq) + \text{HSO}_3^-(aq)$$

(Similarly, carbonic acid occurs in equilibrium with hydrated CO₂ and cannot be isolated as H₂CO₃ molecules.) Sulfurous acid is weak and has two acidic protons, forming the hydrogen sulfite (bisulfite, HSO₃⁻) and sulfite (SO₃²⁻) ions with strong base. Because the S in SO₃²⁻ is in the +4 state and is easily oxidized to the +6 state, sulfites are good reducing agents and are used to preserve foods and wine by eliminating undesirable products of air oxidation.

En route to sulfuric acid, SO₂ is first oxidized to SO₃ (S in the +6 state) by heating in O₂ over a catalyst:

$$\text{SO}_2(g) + \tfrac{1}{2}\text{O}_2(g) \underset{}{\overset{\text{V}_2\text{O}_5/\text{K}_2\text{O catalyst}}{\rightleftharpoons}} \text{SO}_3(g)$$

Acid from the Sky Awareness of SO₂ as a major air pollutant is now widespread. Even though enormous amounts of SO₂ form during volcanic and other geothermal activity, they are dwarfed by the amounts emitted from human sources: coal-burning power plants, petroleum refineries, and metal-ore smelters. In the atmosphere, SO₂ reacts with O₂ and water to form sulfuric acid, which rains, snows, and dusts down on animals, plants, buildings, and lakes. The destructive effects of acid precipitation are being intensively studied, and remedial action is intensifying (Chapter 19).

Figure 14.26 The dehydration of carbohydrates by sulfuric acid. The protons of concentrated H_2SO_4 combine exothermically with water, dehydrating many organic materials. When table sugar is treated with sulfuric acid, the components of water are removed from the carbohydrate molecules $(CH_2O)_n$. Steam forms, and the remaining carbon expands to a porous mass.

(We discuss how catalysts work in Chapter 16 and how H_2SO_4 is produced in Chapter 22.) The SO_3 is absorbed into concentrated H_2SO_4 and treated with additional H_2O:

$$SO_3(\text{in concentrated } H_2SO_4) + H_2O(l) \longrightarrow H_2SO_4(l)$$

With more than 40 million tons produced each year in the United States alone, H_2SO_4 ranks first among all industrial chemicals. Fertilizer production, metal, pigment, and textile processing, and soap and detergent manufacturing are just a few of the major industries that depend on sulfuric acid.

Concentrated laboratory-grade sulfuric acid is a viscous, colorless liquid that is 98% H_2SO_4 by mass. Like other strong acids, H_2SO_4 dissociates completely in water, forming the hydrogen sulfate (or bisulfate) ion, a much weaker acid:

hydrogen sulfate ion   sulfate ion

Most common hydrogen sulfates and sulfates are water soluble, but those of Group 2A(2) (except $MgSO_4$), Pb^{2+}, and Hg_2^{2+} are not.

Concentrated sulfuric acid is an excellent dehydrating agent. Its loosely held proton transfers to water in a highly exothermic formation of hydronium (H_3O^+) ions. This process can occur even when the reacting substance contains no free water. For example, H_2SO_4 dehydrates wood, natural fibers, and many other organic substances by removing the components of water from the molecular structure, leaving behind a carbonaceous mass (Figure 14.26).

Thiosulfuric acid ($H_2S_2O_3$) is a structural analog of sulfuric acid in which a second S substitutes for one of the O atoms (*thio-* means "containing sulfur in place of oxygen"). The thiosulfate ion ($S_2O_3^{2-}$) is an important reducing agent in chemical analysis. However, its largest commercial use is in photography, where sodium thiosulfate pentahydrate ($Na_2S_2O_3\cdot5H_2O$), known as "hypo," is involved in fixing the image (Section 23.3).

Many metals combine directly with S to form *metal sulfides*. Indeed, naturally occurring sulfides are ores, which are mined for the extraction of many metals, including copper, zinc, lead, and silver. Aside from the sulfides of Groups 1A(1) and 2A(2), most metal sulfides do not have discrete S^{2-} ions. Several transition metals, such as chromium, iron, and nickel, form covalent, alloy-like, nonstoichiometric compounds with S, such as $Cr_{0.88}S$ or $Fe_{0.86}S$. Some important minerals contain S_2^{2-} ions; an example is iron pyrite, or "fool's gold" (FeS_2) (Figure 14.27). We discuss the metallurgy of ores in Chapter 22.

Figure 14.27 A common sulfide mineral. Pyrite (FeS_2), or fool's gold, is a beautiful but relatively cheap mineral that contains the S_2^{2-} ion, but no gold.

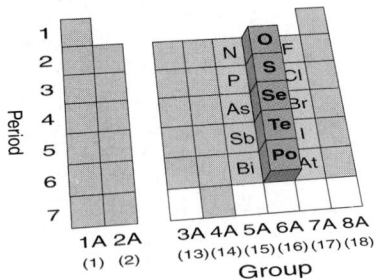

Figure 14.28 **Standing in Group 6A(16), looking backward to Group 5A(15) and forward to Group 7A(17).**

Looking Backward and Forward: Groups 5A(15), 6A(16), and 7A(17)

Groups 5A(15) and 6A(16) are very similar in their physical and chemical trends and in the versatility of phosphorus and sulfur (Figure 14.28). Their greatest difference is the sluggish behavior of N_2 compared with the striking reactivity of O_2. In both groups, metallic character appears only in the largest members. From here on, metals and even metalloids are left behind: all Group 7A(17) elements are reactive nonmetals. Anion formation, which was rare in 5A and more common in 6A, is one of the dominant features of 7A. Another feature of the 7A elements is the number of covalent compounds they form with oxygen *and* with each other.

14.9 GROUP 7A(17): THE HALOGENS

Our last chance to view elements of great reactivity occurs in Group 7A(17). The halogens begin with fluorine (F), the strongest electron "grabber" of all. Chlorine (Cl), bromine (Br), and iodine (I) also form compounds with most elements, and even extremely rare astatine (At) is thought to be reactive. The key features of the halogens are presented in the Group 7A(17) Family Portrait (pp. 590 and 591).

What Accounts for the Regular Changes in the Halogens' Physical Properties?

Like the alkali metals at the other end of the periodic table, the halogens display regular trends in their physical properties. However, whereas melting and boiling points and heats of fusion and vaporization *decrease* down Group 1A(1), these properties *increase* down Group 7A(17) (see Interchapter, Topic 6). The reason for these opposite trends is the different type of bonding in the elements. The alkali metals consist of atoms held together by metallic bonding, which *decreases* in strength as the atoms become larger. The halogens, on the other hand, exist as diatomic molecules that interact through dispersion forces, which *increase* in strength as the atoms become larger and more easily polarized. Thus, F_2 is a very pale yellow gas, Cl_2 a yellow-green gas, Br_2 a brown-orange liquid, and I_2 a purple-black solid.

Why Are the Halogens So Reactive?

The Group 7A(17) elements react with most metals and nonmetals to form many ionic and covalent compounds: metal and nonmetal halides, halogen oxides, and oxoacids. The reason for halogen reactivity is the same as that for alkali metal reactivity—an electron configuration one electron away from that of a noble gas. Whereas a 1A metal atom must *lose* one electron to attain a filled outer shell, *a 7A nonmetal atom must gain one electron to fill its outer shell.* It accomplishes this filling in either of two ways:

1. Gaining an electron from a metal atom, thus forming a negative ion as the metal forms a positive one
2. Sharing an electron pair with a nonmetal atom, thus forming a covalent bond

Down the group, reactivity reflects the decrease in electronegativity: F_2 is the most reactive and I_2 the least. The exceptional reactivity of elemental F_2 is also related to the weakness of the F—F bond. Because F is small, the bond is short; however, the lone pairs on each F atom repel those on the other, which weakens the bond (Figure 14.29). As a result of these factors, F_2 reacts with every element (except He, Ne, and Ar), in many cases, explosively.

The halogens display the largest range in electronegativity of any group, but all are electronegative enough to behave as nonmetals. They act as *oxidizing*

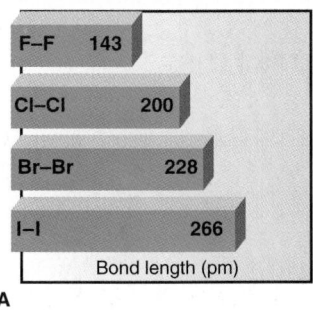

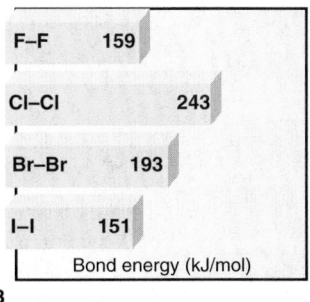

A   B

Figure 14.29 **Bond energies and bond lengths of the halogens. A,** In keeping with the increase in atomic size down the group, bond lengths increase steadily. **B,** The halogens show a general decrease in bond energy as bond length increases. However, F_2 deviates from this trend because its small, close, electron-rich atoms repel each other, thereby lowering its bond energy.

agents in the majority of their reactions, and halogens higher in the group can oxidize halide ions lower down:

$$F_2(g) + 2X^-(aq) \longrightarrow 2F^-(aq) + X_2(aq) \quad (X = Cl, Br, I)$$

Thus, the oxidizing ability of X_2 *decreases* down the group: the lower the EN, the less strongly it pulls electrons. Similarly, the reducing ability of X^- *increases* down the group: the larger the ion, the more easily it gives up its electron (Figure 14.30).

The halogens undergo some important aqueous redox chemistry. Fluorine is such a powerful oxidizing agent that it tears water apart, oxidizing the O to produce O_2, some O_3, and HFO (hypofluorous acid). The other halogens undergo disproportionations (note the oxidation numbers):

$$\overset{0}{X_2} + H_2O(l) \rightleftharpoons \overset{-1}{H}X(aq) + \overset{+1}{H}XO(aq) \quad (X = Cl, Br, I)$$

At equilibrium, very little product is present unless excess OH^- ion is added, which reacts with the HX and HXO and drives the reaction to completion:

$$X_2 + 2OH^-(aq) \longrightarrow X^-(aq) + XO^-(aq) + H_2O(l)$$

When X is Cl, the product mixture acts as a bleach: household bleach is a dilute solution of sodium hypochlorite (NaClO). Heating causes XO^- to disproportionate further, creating oxoanions with X in a higher oxidation state:

$$3\overset{+1}{X}O^-(aq) \xrightarrow{\Delta} 2\overset{-1}{X}{}^-(aq) + \overset{+5}{X}O_3{}^-(aq)$$

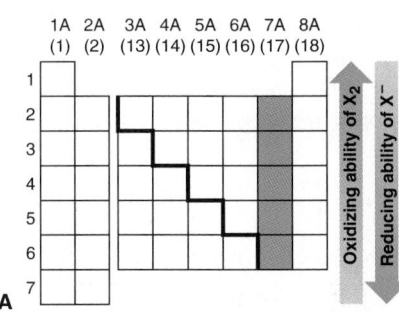

A

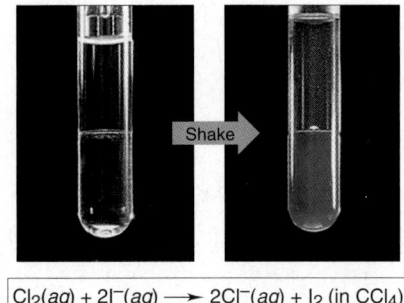

B $Cl_2(aq) + 2I^-(aq) \longrightarrow 2Cl^-(aq) + I_2 \text{ (in } CCl_4)$

Figure 14.30 **The relative oxidizing ability of the halogens. A,** Halogen redox behavior is based on atomic properties such as electron affinity, ionic charge density, and electronegativity. A halogen (X_2) higher in the group can oxidize a halide ion (X^-) lower down. **B,** As an example, when aqueous Cl_2 is added to a solution of I^- (*top layer*), it oxidizes the I^- to I_2, which dissolves in the CCl_4 solvent (*bottom layer*) to give a purple solution.

Key Atomic and Physical Properties

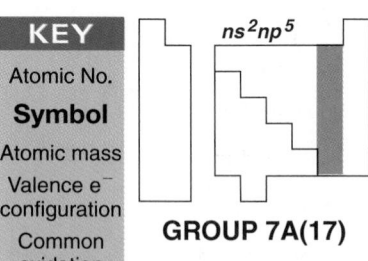

ns^2np^5

GROUP 7A(17)

Atomic Properties

Group electron configuration is ns^2np^5; elements lack one e⁻ to complete their outer shell. The −1 oxidation state is the most common for all members. Except for F, the halogens exhibit all odd-numbered states (+7 through −1).

Down the group, atomic and ionic size increase steadily, as IE and EN decrease.

	Atomic radius (pm)		Ionic radius (pm)
F	72		F⁻ 133
Cl	100		Cl⁻ 181
Br	114		Br⁻ 196
I	133		I⁻ 220
At	(140)		no data

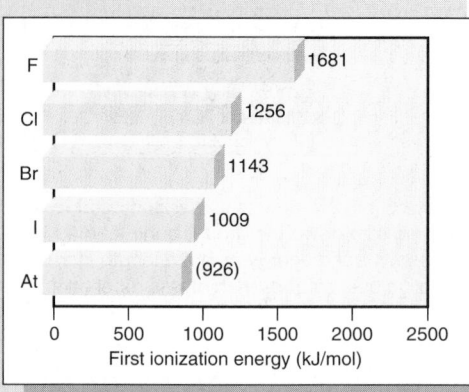

First ionization energy (kJ/mol)

- F 1681
- Cl 1256
- Br 1143
- I 1009
- At (926)

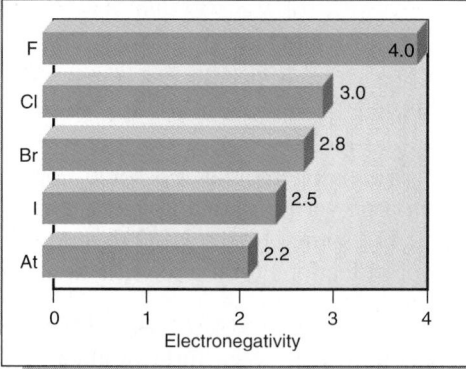

Electronegativity

- F 4.0
- Cl 3.0
- Br 2.8
- I 2.5
- At 2.2

9		
F	Photograph not available	
19.00		
$2s^22p^5$		
(−1)		
17		
Cl		
35.45		
$3s^23p^5$		
(−1, +7, +5, +3, +1)		
35		
Br		
79.90		
$4s^24p^5$		
(−1, +7, +5, +3, +1)		
53		
I		
126.9		
$5s^25p^5$		
(−1, +7, +5, +3, +1)		
85		
At	Extremely rare, no sample available	
(210)		
$6s^26p^5$		
(−1)		

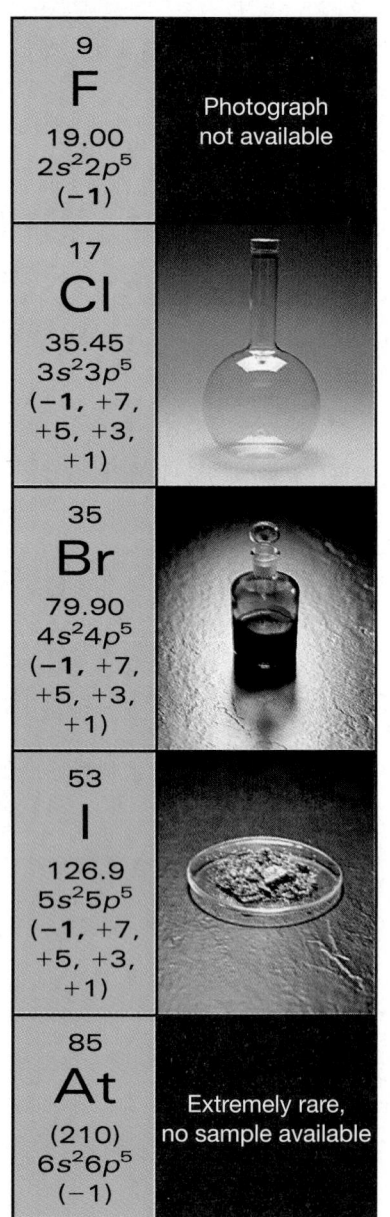

Physical Properties

Phase-change temperatures and enthalpies exhibit smooth trends. Down the group, melting and boiling points increase as a result of stronger dispersion forces between heavier molecules.

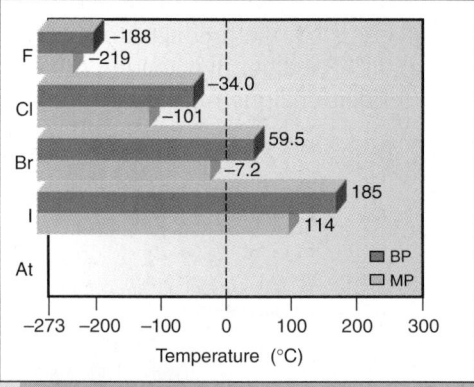

Temperature (°C)

- F −188 / −219
- Cl −34.0 / −101
- Br 59.5 / −7.2
- I 185 / 114
- At

☐ BP   ☐ MP

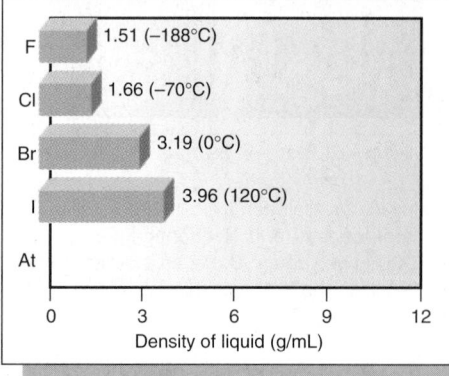

Density of liquid (g/mL)

- F 1.51 (−188°C)
- Cl 1.66 (−70°C)
- Br 3.19 (0°C)
- I 3.96 (120°C)
- At

The densities of the elements as liquids (at given T) increase steadily with molar mass.

Some Reactions and Compounds

Important Reactions

Oxidizing strength and aqueous redox chemistry of the halogens are shown in reactions 1 and 2, and industrial processes involving fluorine in reactions 3 and 4.

1. The halogens (X_2) oxidize many metals and nonmetals. The reaction with hydrogen, although not used commercially for HX production (except for high-purity HCl), is characteristic of these strong oxidizing agents:

$$X_2 + H_2(g) \longrightarrow 2HX(g)$$

2. The halogens disproportionate in water:

$$X_2 + H_2O(l) \rightleftharpoons HX(aq) + HXO(aq)$$
$$(X = Cl, Br, I)$$

In aqueous base, the reaction goes to completion to form hypohalites (*see text*) and, at higher temperatures, halates; for example:

$$3Cl_2(g) + 6OH^-(aq) \xrightarrow{\Delta} ClO_3^-(aq) + 5Cl^-(aq) + 3H_2O(l)$$

3. F_2 is produced electrolytically at moderate temperature:

$$2HF \text{ (as a soln of KF in HF)} \xrightarrow[90°C]{\text{electrolysis}} H_2(g) + F_2(g)$$

A major use of F_2 is in the preparation of UF_6 for nuclear fuel.

4. Glass (amorphous silica) is etched with HF:

$$SiO_2(s) + 6HF(g) \longrightarrow H_2SiF_6(aq) + 2H_2O(l)$$

Space shuttle

Important Compounds

Fluorite

1. Fluorspar (fluorite), CaF_2. Widely distributed mineral used as a flux in steelmaking and in the production of HF (*see photo*).

2. Hydrogen fluoride, HF. Colorless, extremely toxic gas used to make F_2, organic fluorine compounds, and polymers. Also used in aluminum manufacture and in glass etching (*see margin note, p. 592*).

3. Hydrogen chloride, HCl. Extremely water-soluble gas that forms hydrochloric acid, which occurs naturally in the stomach fluids of mammals (humans produce 1.5 L of 0.1 M HCl daily) and in volcanic gases (from reaction of H_2O and sea salt). Made by reaction of NaCl and H_2SO_4 and as a by-product of plastics (PVC) production. Used in the "pickling" of steel (removal of adhering oxides) and in the production of syrups, rayon, and plastic.

4. Sodium hypochlorite, NaClO, and calcium hypochlorite, $Ca(ClO)_2$. Oxidizing agents used to bleach wood pulp and textiles and to disinfect swimming pools, foods, and sewage (also used to disinfect the Apollo 11 on return from the moon). Household bleach is 5.25% NaClO by mass in water.

5. Ammonium perchlorate, NH_4ClO_4. Strong oxidizing agent used in the space shuttle program (*see photo*).

6. Potassium iodide, KI. Most common soluble iodide. Added to table salt to prevent thyroid disease (goiter). Used in chemical analysis because it is easily oxidized to I_2, which provides a colored end point.

7. Polychlorinated biphenyls, PCBs. Mixture of chlorinated organic compounds used as nonflammable insulating liquids in electrical transformers. Production discontinued because of persistence in the environment, where they become concentrated in fish, birds, and mammals, and cause reproductive disturbances and possibly cancer.

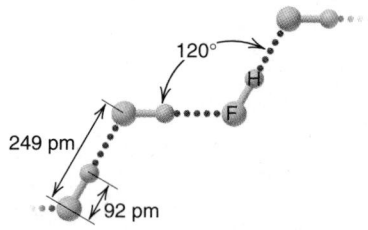

HF: Unusual Structure, Familiar Uses At room temperature, all the hydrogen halides except HF are diatomic gases. As a result of its extensive H bonding, gaseous HF exists as short chains or rings of formula $(HF)_6$; liquid HF boils at 19.5°C, more than 50 degrees higher than even the much heavier HI; and solid HF exists as an H-bonded polymer. HF has many uses, including the synthesis of cryolite (Na_3AlF_6) for aluminum production, of fluorocarbons for refrigeration, and of NaF for water fluoridation. HF is also used in nuclear fuel processing and for glass etching in the manufacture of lightbulbs, transistors, and TV tubes.

Highlights of Halogen Chemistry

In this section, we examine the compounds the halogens form with hydrogen and with each other, as well as the oxides, oxoanions, and oxoacids.

The Hydrogen Halides The halogens form gaseous hydrogen halides (HX) through direct combination with H_2 or through the action of a concentrated acid on the metal halide (a nonoxidizing acid is used for HBr and HI):

$$CaF_2(s) + H_2SO_4(l) \longrightarrow CaSO_4(s) + 2HF(g)$$
$$3NaBr(s) + H_3PO_4(l) \longrightarrow Na_3PO_4(s) + 3HBr(g)$$

Commercially, most HCl is formed as a by-product in the chlorination of hydrocarbons for plastics production:

$$CH_2{=}CH_2(g) + Cl_2(g) \longrightarrow ClCH_2CH_2Cl(l) \xrightarrow{500°C} CH_2{=}CHCl(g) + HCl(g)$$
$$\text{vinyl chloride}$$

In this case, the vinyl chloride reacts in a separate process to form poly(vinyl chloride), or PVC, a polymer used extensively in modern plumbing pipes.

In water, gaseous HX molecules form a *hydrohalic acid.* Only HF, with its relatively short, strong bond, forms a weak acid:

$$HF(g) + H_2O(l) \rightleftharpoons H_3O^+(aq) + F^-(aq)$$

The others dissociate completely to form the stoichiometric amount of H_3O^+ ions:

$$HBr(g) + H_2O(l) \longrightarrow H_3O^+(aq) + Br^-(aq)$$

(We saw reactions similar to these in Chapter 4. They involve transfer of a proton from acid to H_2O and are classified as *Brønsted-Lowry acid-base reactions.* In Chapter 18, we discuss them thoroughly and examine the relation between bond length and acidity of the larger HX molecules.)

Interhalogen Compounds: The "Halogen Halides" Halogens react exothermically with one another to form many **interhalogen compounds.** The simplest are diatomic molecules, such as ClF or BrCl. Every binary combination of the four common halogens is known. The more electronegative halogen is in the −1 oxidation state, and the less electronegative is in the +1 state. Interhalogens of general formula XY_n ($n = 3, 5, 7$) form when the larger members (X) use d orbitals to expand their valence shells. In every case, the central atom has the *lower electronegativity* and a positive oxidation state.

The commercially useful interhalogens are powerful *fluorinating agents,* some of which react with metals, nonmetals, and oxides—even wood and asbestos:

$$Sn(s) + ClF_3(l) \longrightarrow SnF_2(s) + ClF(g)$$
$$P_4(s) + 5ClF_3(l) \longrightarrow 4PF_3(g) + 3ClF(g) + Cl_2(g)$$
$$2B_2O_3(s) + 4BrF_3(l) \longrightarrow 4BF_3(g) + 2Br_2(l) + 3O_2(g)$$

Their reactions with water are nearly explosive and yield HF and *the oxoacid in which the central halogen has the same oxidation state.* For example,

$$3H_2O(l) + \overset{+5}{BrF_5}(l) \longrightarrow 5HF(g) + \overset{+5}{HBrO_3}(aq)$$

The Oddness and Evenness of Oxidation States In Topic 5 of the Interchapter, we saw that odd-numbered groups exhibit odd-numbered oxidation states and even-numbered groups exhibit even-numbered states. The reason for this general behavior is that *almost all stable molecules have paired electrons,* either as bonding or lone pairs. Therefore, *when bonds form or break, two electrons are involved, so the oxidation state changes by two.*

Consider the interhalogens. Four general formulas are XY, XY_3, XY_5, and XY_7; examples are shown in Figure 14.31. With Y in the −1 state, X must be in the +1, +3, +5, and +7 state, respectively. The −1 state arises when Y fills its valence level; the +7 state arises when the central halogen (X) is completely oxidized, that is, when all seven valence electrons have shifted away from it to the more electronegative Y atoms around it.

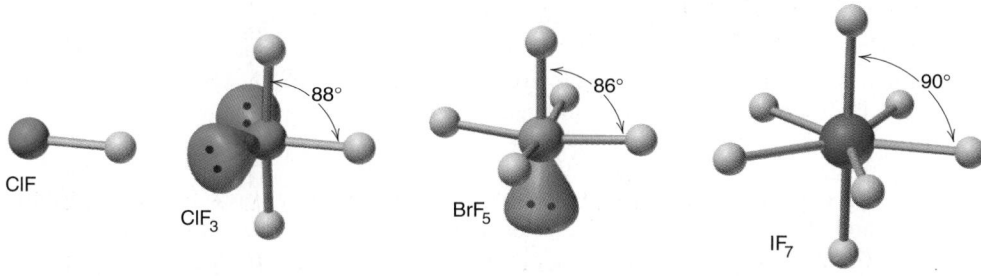

Linear, XY	T shaped, XY$_3$	Square pyramidal, XY$_5$	Pentagonal bipyramidal, XY$_7$

Figure 14.31 Molecular shapes of the main types of interhalogen compounds.

Let's examine the iodine fluorides to see why the oxidation states jump by two units. When I_2 reacts with F_2, IF forms (note the oxidation number of I):

$$I_2 + F_2 \longrightarrow 2\overset{+1}{I}F$$

In IF_3, I uses *two* more valence electrons to form *two* more bonds:

$$\overset{+1}{I}F + F_2 \longrightarrow \overset{+3}{I}F_3$$

Otherwise, an unstable lone-electron species containing two fluorines would form. With more fluorine, another jump of two units occurs and the pentafluoride forms:

$$\overset{+3}{I}F_3 + F_2 \longrightarrow \overset{+5}{I}F_5$$

With still more fluorine, the heptafluoride forms:

$$\overset{+5}{I}F_5 + F_2 \longrightarrow \overset{+7}{I}F_7$$

An element in an even-numbered group, such as sulfur in Group 6A(16), shows the same tendency for its compounds to have paired electrons. Elemental sulfur (oxidation number, O.N. = 0) gains or shares two electrons to complete its shell (O.N. = −2). It uses two electrons to react with fluorine, for example, and form SF_2 (O.N. = +2), two more electrons for SF_4 (O.N. = +4), and two more for SF_6 (O.N. = +6). Thus, an element with one even state typically has all even states, and an element with one odd state typically has all odd states. To reiterate the main point, *successive oxidation states differ by two units because stable molecules have electrons in pairs around their atoms.*

Halogen Oxides, Oxoacids, and Oxoanions The Group 7A(17) elements form many oxides that are *powerful oxidizing agents and acids in water.* Dichlorine monoxide (Cl_2O) and especially chlorine dioxide (ClO_2) are used to bleach paper (Figure 14.32). ClO_2 is unstable to heat and shock, so it is prepared on site, and more than 100,000 tons are used annually:

$$2NaClO_3(s) + SO_2(g) + H_2SO_4(aq) \longrightarrow 2ClO_2(g) + 2NaHSO_4(aq)$$

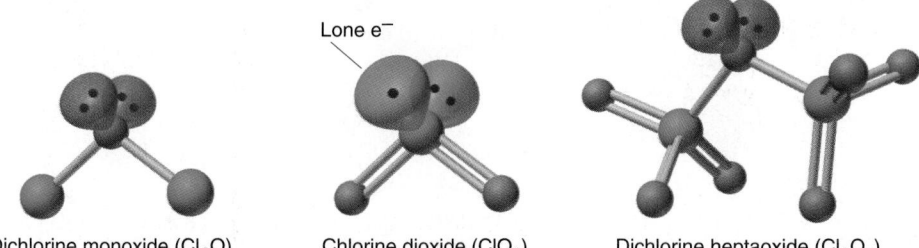

Dichlorine monoxide (Cl_2O)	Chlorine dioxide (ClO_2)	Dichlorine heptaoxide (Cl_2O_7)

Figure 14.32 Chlorine oxides. Dichlorine monoxide and chlorine dioxide are bent molecules; note the lone electron in ClO_2. Dichlorine heptaoxide can be viewed as two ClO_4 tetrahedra joined through an O corner. (Each ball-and-stick model shows the structure in which each atom has its lowest formal charge. Lone pairs are shown on the central atoms only.)

The dioxide has an unpaired electron and Cl in the unusual $+4$ oxidation state.

Chlorine is in its highest $(+7)$ oxidation state in dichlorine heptaoxide, Cl_2O_7, which is a symmetrical molecule formed when two $HClO_4$ ($HO-ClO_3$) molecules undergo a dehydration-condensation reaction:

$$O_3Cl-O\boxed{H + HO}-ClO_3 \longrightarrow O_3Cl-\ddot{O}-ClO_3(l) + H_2O(l)$$

The halogen oxoacids and oxoanions are produced by reaction of the halogens and their oxides with water. Most of the oxoacids are stable only in solution. Table 14.4 shows ball-and-stick models of the acids in which each atom has its lowest formal charge; note the formulas, which emphasize that H is bonded to O. The hypohalites (XO^-), halites (XO_2^-), and halates (XO_3^-) are oxidizing agents formed by aqueous disproportionation reactions [see Group 7A(17) Family Portrait, reaction 2]. You may have heated solid alkali chlorates in the laboratory to form small amounts of O_2:

$$2MClO_3(s) \xrightarrow{\Delta} 2MCl(s) + 3O_2(g)$$

Potassium chlorate is the oxidizer in "safety" matches.

Several perhalates are also strong oxidizing agents. Ammonium perchlorate, prepared from sodium perchlorate, is the oxidizing agent for the aluminum powder in the solid-fuel booster rocket of the space shuttle; each launch uses more than 700 tons of NH_4ClO_4:

$$10Al(s) + 6NH_4ClO_4(s) \longrightarrow 4Al_2O_3(s) + 12H_2O(g) + 3N_2(g) + 2AlCl_3(g)$$

The relative strengths of the halogen oxoacids depend on two factors:

1. *Electronegativity of the halogen.* Among oxoacids in the same oxidation state, such as the halic acids, HXO_3 (or $HOXO_2$), acid strength decreases as the halogen EN decreases:

$$HOClO_2 > HOBrO_2 > HOIO_2$$

The more electronegative the halogen, the more electron density it removes from the O—H bond, and the more easily the proton is lost.

2. *Oxidation state of the halogen.* Among oxoacids of a given halogen, such as chlorine, acid strength decreases as the oxidation state of the halogen decreases:

$$HOClO_3 > HOClO_2 > HOClO > HOCl$$

The higher the oxidation state (number of bound O atoms) of the halogen, the more electron density it pulls from the O—H bond. We consider these trends quantitatively in Chapter 18.

Pyrotechnic Perchlorates Thousands of tons of perchlorates are made each year for use in explosives and fireworks. The white flash and thundering boom of a fireworks display are caused by $KClO_4$ reacting with powdered sulfur and aluminum. At rock concerts and other theatrical productions, mixtures of $KClO_4$ and Mg are often used for special effects.

Table 14.4 The Known Halogen Oxoacids*

Central Atom	Hypohalous Acid (HOX)	Halous Acid (HOXO)	Halic Acid (HOXO$_2$)	Perhalic Acid (HOXO$_3$)
Fluorine	HOF	—	—	—
Chlorine	HOCl	HOClO	HOClO$_2$	HOClO$_3$
Bromine	HOBr	(HOBrO)?	HOBrO$_2$	HOBrO$_3$
Iodine	HOI	—	HOIO$_2$	HOIO$_3$, (HO)$_5$IO
Oxoanion	Hypohalite	Halite	Halate	Perhalate

*Lone pairs are shown only on the halogen atom.

14.10 GROUP 8A(18): THE NOBLE GASES

The last main group consists of individual atoms too "noble" to interact with others. The Group 8A(18) elements display regular trends in physical properties and very low, if any, reactivity. The group consists of helium (He), the second most abundant element in the universe, neon (Ne) and argon (Ar), krypton (Kr) and xenon (Xe), the only members for which compounds have been well studied, and finally, radioactive radon (Rn). The noble gases make up about 1% by volume of the atmosphere, primarily due to the high abundance of Ar. Their properties appear in the Group 8A(18) Family Portrait on the next page.

How Can Noble Gases Form Compounds?

Lying at the far right side of the periodic table, the Group 8A(18) elements consist of individual atoms with filled outer levels and the smallest radii in their periods: even Li, the smallest alkali metal (152 pm), is bigger than Rn, the largest noble gas (140 pm). These elements come as close to behaving as ideal gases as any other substances. Only at very low temperatures do they condense and solidify. In fact, He is the only substance that does *not* solidify by a reduction in temperature alone; it requires an increase in pressure as well. Helium has the lowest melting point known ($-272.2°C$ at 25 atm), only one degree above 0 K ($-273.15°C$), and it boils only about three degrees higher. Weak dispersion forces hold these elements in condensed states, with melting and boiling points that increase, as expected, with molar mass.

Figure 14.33 Crystals of xenon tetrafluoride (XeF_4).

Ever since their discovery in the late 19th century, these elements had been considered, and even formerly named, the "inert" gases. Atomic theory and, more important, all experiments had supported this idea. Then, in 1962, all this changed when the first noble gas compound was prepared. How, with filled outer levels and extremely high ionization energies, *can* noble gases react?

The discovery of noble gas reactivity is a classic example of clear thinking in the face of an unexpected event. At the time, a young inorganic chemist named Neil Bartlett was studying platinum fluorides, known to be strong oxidizing agents. When he accidentally exposed PtF_6 to air, its deep-red color lightened slightly, and analysis showed that the PtF_6 had oxidized O_2 to form the ionic compound $[O_2]^+[PtF_6]^-$. Knowing that the ionization energy of the oxygen molecule ($O_2 \longrightarrow O_2^+ + e^-$; IE = 1175 kJ/mol) is very close to IE_1 of xenon (1170 kJ/mol), Bartlett reasoned that PtF_6 might be able to oxidize xenon. Shortly thereafter, he prepared $XePtF_6$, an orange-yellow solid. Within a few months, the white crystalline XeF_2 and XeF_4 (Figure 14.33) were also prepared. In addition to the +2 and +4 oxidation states, Xe occurs in the +6 state in several compounds, such as XeF_6, and in the +8 state in the unstable oxide, XeO_4. A few compounds of Kr have also been made. The xenon fluorides react rapidly in water to form HF and various other products, including different xenon compounds.

Looking Backward and Forward: Groups 7A(17), 8A(18), and 1A(1)

With the disappearance of metallic behavior, the Group 7A(17) elements form a host of anions, covalent oxides, and oxoanions, which change oxidation states readily in a rich aqueous chemistry. The vigorous reactivity of the halogens is in stark contrast to the inertness of their 8A neighbors. Filled outer levels render the noble gas atoms largely inert, despite a limited ability to react with the most electronegative elements, fluorine and oxygen. The least reactive family in the periodic table stands between the two most reactive: the halogens, which need one more electron to fill their outer level, and the alkali metals, which need one fewer (Figure 14.34). As you know, atomic, physical, and chemical properties change dramatically from Group 8A(18) to Group 1A(1).

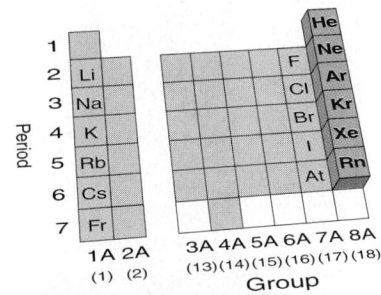

Figure 14.34 Standing in Group 8A(18), looking backward at the halogens, Group 7A(17), and ahead to the alkali metals, Group 1A(1).

Key Atomic and Physical Properties

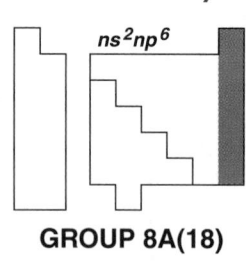

ns^2np^6

GROUP 8A(18)

Atomic Properties

Group electron configuration is $1s^2$ for He and ns^2np^6 for the others. The valence shell is filled. Only Kr and Xe (and perhaps Rn) are known to form compounds. The more reactive Xe exhibits all even oxidation states (+2 to +8).

This group contains the smallest atoms with the highest IEs in their periods. Down the group, atomic size increases and IE decreases steadily. (EN values are given only for Kr and Xe.)

Atomic radius (pm)	
He	31
Ne	71
Ar	98
Kr	112
Xe	131
Rn	(140)

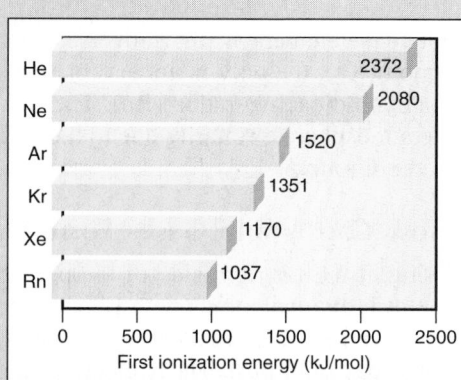

First ionization energy (kJ/mol)

He	2372
Ne	2080
Ar	1520
Kr	1351
Xe	1170
Rn	1037

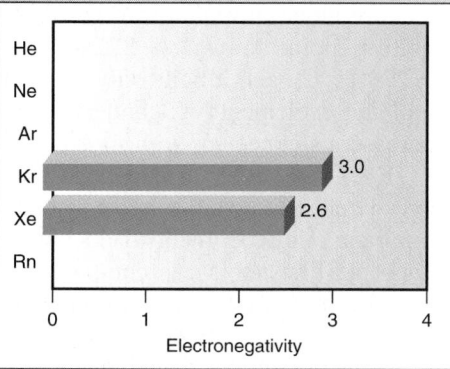

Electronegativity

Kr	3.0
Xe	2.6

Physical Properties

Melting and boiling points of these gaseous elements are extremely low but increase down the group because of stronger dispersion forces. Note the extremely small liquid ranges.

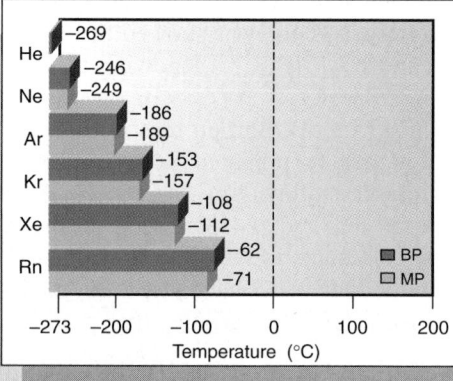

Temperature (°C) — BP / MP

	BP	MP
He	−269	
Ne	−246	−249
Ar	−186	−189
Kr	−153	−157
Xe	−108	−112
Rn	−62	−71

Densities (at STP) increase steadily, as expected.

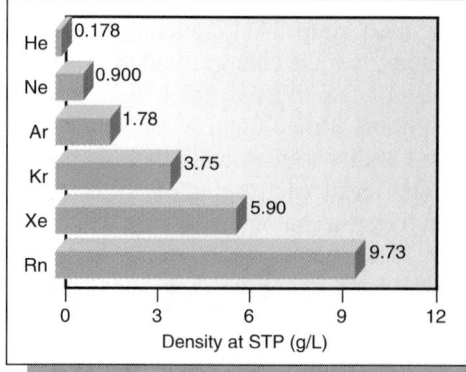

Density at STP (g/L)

He	0.178
Ne	0.900
Ar	1.78
Kr	3.75
Xe	5.90
Rn	9.73

2	
He	
4.003	
$1s^2$	
(none)	

10	
Ne	
20.18	
$2s^22p^6$	
(none)	

18	
Ar	
39.95	
$3s^23p^6$	
(none)	

36	
Kr	
83.80	
$4s^24p^6$	
(+2)	

54	
Xe	
131.3	
$5s^25p^6$	
(+8, +6, +4, +2)	

86	
Rn	
(222)	
$6s^26p^6$	
(+2)	

Mass spectral peak

Chapter Perspective

Our excursion through the bonding and reactivity patterns of the elements has come full circle, but we've only been able to touch on some of the most important features. Clearly, the properties of the atoms and the resultant physical and chemical behaviors of the elements are magnificent testimony to nature's diversity. In Chapter 15, we continue with our theme of macroscopic behavior emerging from atomic properties in an investigation of the marvelously complex organic compounds of carbon. In Chapter 22, we revisit the most important main-group elements to see how they occur in nature and how we isolate and use them.

For Review and Reference (Numbers in parentheses refer to pages, unless noted otherwise.)

Learning Objectives

Relevant section numbers appear in parentheses.

Understand These Concepts

Note: Many characteristic reactions appear in the "Important Reactions" section within each group's Family Portrait.

1. How hydrogen is similar to, yet different from, alkali metals and halogens; the differences between saltlike and covalent hydrides (Section 14.1)
2. Key horizontal trends in atomic properties, types of bonding, oxide acid-base properties, and redox behavior of the elements as they change from metal to nonmetal (Section 14.2)
3. How the ns^1 configuration accounts for the physical and chemical properties of the alkali metals (Section 14.3)
4. How small atomic size and limited number of valence orbitals account for the anomalous behavior of the Period 2 member of each group (see each section for details)
5. How the ns^2 configuration accounts for the key differences between Groups 1A(1) and 2A(2) (Section 14.4)
6. The basis of the three important diagonal relationships (Li/Mg, Be/Al, B/Si) (Sections 14.4 to 14.6)
7. How the presence of inner $(n-1)d$ electrons affects properties in Group 3A(13) (Section 14.5)
8. Patterns among larger members of Groups 3A(13) to 6A(16): two common oxidation states (inert-pair effect), lower state more important down the group, and more basic lower oxide (Sections 14.5 to 14.8)
9. How boron attains an octet of electrons (Section 14.5)
10. The effect of bonding on the physical behavior of Groups 4A(14) to 6A(16) (Section 14.6)
11. Allotropism in carbon, phosphorus, and sulfur (Sections 14.6 to 14.8)
12. How atomic properties lead to catenation and multiple bonding in organic compounds (Section 14.6)
13. Structures and properties of the silicates and silicones (Section 14.6)
14. Patterns of behavior among hydrides and halides of Groups 5A(15) and 6A(16); reactions of nonmetal halides with water (Sections 14.7 to 14.9)
15. Structure and chemistry of the nitrogen oxides and oxoacids (Section 14.7)
16. Structure and chemistry of the phosphorus oxides and oxoacids (Section 14.7)
17. Dehydration-condensation reactions and polymeric structures (Section 14.7)
18. Structure and chemistry of the sulfur oxides and oxoacids (Section 14.8)
19. How the ns^2np^5 configuration accounts for halogen reactivity with metals (Section 14.9)
20. The meaning of disproportionation and its importance in halogen aqueous chemistry (Section 14.9)
21. Why the oxidation states of an element change by two units (Section 14.9)
22. Structure and chemistry of the halogen oxides and oxoacids (Section 14.9)
23. How the ns^2np^6 configuration accounts for the relative inertness of noble gases (Section 14.10)

Key Terms

Section 14.4
diagonal relationship (553)
Section 14.5
bridge bond (561)

Section 14.6
allotrope (563)
silicate (568)
silicone (568)

Section 14.7
disproportionation reaction (578)
dehydration-condensation reaction (580)

Section 14.9
interhalogen compound (592)

Highlighted Figures and Tables

These figures (F) and tables (T) provide a quick review of key ideas.

T14.1 The Period 2 elements (546–547)
Family Portrait Group 1A(1) (550–551)
F14.5 Three diagonal relationships (553)
Family Portrait Group 2A(2) (554–555)
Family Portrait Group 3A(13) (558–559)
T14.2 Bond type and the melting process (562)

Family Portrait Group 4A(14) (564–565)
Family Portrait Group 5A(15) (574–575)
T14.3 The nitrogen oxides (577)
Family Portrait Group 6A(16) (582–583)
Family Portrait Group 7A(17) (590–591)
Family Portrait Group 8A(18) (596)

Problems

Problems with **colored** numbers are answered at the back of the text. Sections match the text. Most offer Concept Review Questions, Skill-Building Exercises (in similar pairs), and Problems in Context. Then Comprehensive Problems, based on material from any section or previous chapter, follow.
Note: This set begins with problems based on material from the Interchapter.

Interchapter Review

● **Concept Review Questions**

14.1 (a) Define the four atomic properties that are reviewed in the Interchapter.
(b) To what do the three parts of the electron configuration ($nl^{\#}$) correlate?
(c) Which part of the electron configuration is primarily associated with the size of the atom?
(d) What is the major distinction between outer electron configurations in a group compared with those in a period?
(e) What correlation, if any, exists between the group number and the number of valence electrons?

14.2 (a) What trends, if any, exist for Z_{eff} across a period and down a group?
(b) How does Z_{eff} influence atomic size, IE_1, and EN across a period?

14.3 Iodine monochloride and elemental bromine have nearly the same molar mass and liquid density but very different boiling points.
(a) What molecular property is primarily responsible for this difference in boiling point? What atomic property gives rise to it? Explain.
(b) Which substance has a higher boiling point? Why?

14.4 How does the trend in atomic size differ from the trend in ionization energy? Explain.

14.5 How are bond energy, bond length, and reactivity related for similar compounds?

14.6 How are covalent and metallic bonding similar? How are they different?

14.7 If the leftmost element in a period combined with each of the others in the period, how would the type of bonding change from left to right? Explain in terms of atomic properties.

14.8 Why is rotation about the bond axis possible for single-bonded atoms but not double-bonded atoms?

14.9 Explain the horizontal irregularity in size of the most common ions of Period 3 elements.

14.10 Would you expect S^{2-} to be larger or smaller than Cl^-? Would you expect Mg^{2+} to be larger or smaller than Na^+? Explain.

14.11 (a) How does the type of bonding in element oxides correlate with the electronegativity of the elements?
(b) How does the acid-base behavior of element oxides correlate with the electronegativity of the elements?

14.12 (a) How does the metallic character of an element correlate with the *acidity* of its oxide?
(b) What trends, if any, exist in oxide *basicity* across a period and down a group?

14.13 How are atomic size, IE_1, and EN related to redox behavior of the elements in Groups 1A(1), 2A(2), 6A(16), and 7A(17)?

14.14 How do the physical properties of a network covalent solid and a molecular covalent solid differ? Why?

● **Skill-Building Exercises (paired)**

14.15 Rank the following elements in order of *increasing*
(a) atomic size: Ba, Mg, Sr
(b) IE_1: P, Na, Al
(c) EN: Br, Cl, Se
(d) number of valence electrons: Bi, Ga, Sn

14.16 Rank the following elements in order of *decreasing*
(a) atomic size: N, Si, P
(b) IE_1: Kr, K, Ar
(c) EN: In, Rb, I
(d) number of valence electrons: Sb, S, Cs

14.17 Which of the following pairs react to form *ionic* compounds: (a) Cl and Br; (b) Na and Br; (c) P and Se; (d) H and Ba?

14.18 Which of the following pairs react to form *covalent* compounds: (a) Be and C; (b) Sr and O; (c) Ca and Cl; (d) P and F?

14.19 Draw a Lewis structure for a compound with a bond order of 2 throughout the molecule.

14.20 Draw a Lewis structure for a polyatomic ion with a fractional bond order throughout the molecule.

14.21 Rank the following in order of *increasing* bond length: $SiCl_4$, CF_4, $GeBr_4$.

14.22 Rank the following in order of *decreasing* bond energy: NF_3, NI_3, NCl_3.

14.23 Rank the following in order of *increasing* radius: (a) O^{2-}, F^-, Na^+ (b) S^{2-}, P^{3-}, Cl^-

14.24 Rank the following in order of *decreasing* radius: (a) Ca^{2+}, K^+, Ga^{3+} (b) Br^-, Sr^{2+}, Rb^+

14.25 Rank O^-, O^{2-}, and O in order of *increasing* size.

14.26 Rank Tl^{3+}, Tl, and Tl^+ in order of *decreasing* size.

14.27 Which member of each pair gives the more *basic* solution in water: (a) CaO or SO_3; (b) BeO or BaO; (c) CO_2 or NO_2; (d) P_4O_{10} or K_2O? For the pair in part (a), write an equation for each oxide dissolving to support your answer.

14.28 Which member of each pair gives the more *acidic* solution in water: (a) NO_2 or SrO; (b) SnO or SnO_2; (c) Cl_2O or Na_2O; (d) SO_2 or MgO? For the pair in part (a), write an equation for each oxide dissolving to support your answer.

14.29 Which member of each pair has more *covalent* character in its bonds: (a) LiCl or KCl; (b) $AlCl_3$ or PCl_3; (c) NCl_3 or $AsCl_3$?

14.30 Which member of each pair has more *ionic* character in its bonds: (a) BeF_2 or CaF_2; (b) PbF_2 or PbF_4; (c) GeF_4 or PF_3?

14.31 Rank the following in order of *increasing* (a) melting point: Na, Si, Ar   (b) ΔH_{fus}: Rb, Cs, Li

14.32 Rank the following in order of *decreasing* (a) boiling point: O_2, Br_2, As(s)   (b) ΔH_{vap}: Cl_2, Ar, I_2

Hydrogen, the Simplest Atom

● **Concept Review Questions**

14.33 Hydrogen has only one proton, but its IE_1 is much greater than that of lithium, which has three protons. Explain.

14.34 Sketch a periodic table, and label the areas containing elements that give rise to the three types of hydrides discussed in the text.

● **Skill-Building Exercises (paired)**

14.35 Draw Lewis structures for the following compounds, and predict which member of each pair will form hydrogen bonds: (a) NF_3 or NH_3 (b) CH_3OCH_3 or CH_3CH_2OH

14.36 Draw Lewis structures for the following compounds, and predict which member of each pair will form hydrogen bonds: (a) NH_3 or AsH_3 (b) CH_4 or H_2O

14.37 Complete and balance the following equations: (a) An active metal reacting with acid,

$$Al(s) + HCl(aq) \longrightarrow$$

(b) A saltlike (alkali metal) hydride reacting with water,

$$LiH(s) + H_2O(l) \longrightarrow$$

14.38 Complete and balance the following equations: (a) A saltlike (alkaline earth metal) hydride reacting with water,

$$CaH_2(s) + H_2O(l) \longrightarrow$$

(b) Reduction by hydrogen to form a metal,

$$PdCl_2(aq) + H_2(g) \longrightarrow$$

● **Problems in Context**

14.39 Compounds such as $NaBH_4$, $Al(BH_4)_3$, and $LiAlH_4$ are complex hydrides used as reducing agents in many chemical syntheses.

(a) Give the oxidation number of each element in these three compounds.

(b) Write a Lewis structure for the polyatomic anion in $NaBH_4$, and predict its shape.

14.40 Unlike the F^- ion, which has an ionic radius close to 133 pm in all alkali metal fluorides, the ionic radius of H^- varies from 137 pm in LiH to 152 pm in CsH. Suggest an explanation for the large variability in H^- but not in F^-.

Trends Across the Periodic Table

● **Concept Review Questions**

14.41 How does the maximum oxidation number vary across a period for the main group elements? Is the pattern in Period 2 different?

14.42 What correlation, if any, exists for the Period 2 elements between group number and the number of covalent bonds the element typically forms? How is the correlation different for elements in higher periods?

14.43 Each of the chemically active Period 2 elements forms stable compounds that have bonds to fluorine.

(a) What are the names and formulas of the substances formed?

(b) Does ΔEN increase or decrease from left to right?

(c) Does percent ionic character increase or decrease from left to right?

(d) Draw Lewis structures for those compounds with predominantly covalent bonding.

14.44 Period 6 is unusual in several ways.

(a) It is the longest period in the table. How many elements belong to Period 6? How many metals?

(b) It contains no metalloids. Where is the metal-nonmetal boundary in Period 6?

14.45 An element forms an oxide, E_2O_3, and a fluoride, EF_3.

(a) Of what two groups might E be a member?

(b) How does the group to which E belongs affect the properties of the oxide and the fluoride?

14.46 Fluorine lies between oxygen and neon in Period 2. Whereas atomic sizes and ionization energies of these three elements change smoothly, their electronegativities display a dramatic change. What is this change, and how do their electron configurations explain it?

Group 1A(1): The Alkali Metals

● **Concept Review Questions**

14.47 Lithium salts are often much less soluble in water than the corresponding salts of other alkali metals. For example, at 18°C, the concentration of a saturated LiF solution is 1.0×10^{-2} M, whereas that of a saturated KF solution is 1.6 M. How would you explain this behavior?

14.48 The alkali metals play virtually the same general chemical role in all their reactions. (a) What is this role? (b) How is it based on atomic properties? (c) Using sodium, write two balanced equations that illustrate this role.

14.49 How do atomic properties account for the low densities of the Group 1A(1) elements?

● **Skill-Building Exercises (paired)**

14.50 Each of the following properties shows regular trends in Group 1A(1). Predict whether each increases or decreases *down* the group: (a) density; (b) ionic size; (c) E—E bond energy; (d) IE_1; (e) magnitude of ΔH_{hydr} of E^+ ion.

14.51 Each of the following properties shows regular trends in Group 1A(1). Predict whether each increases or decreases *up* the group: (a) melting point; (b) E—E bond length; (c) hardness; (d) molar volume; (e) lattice energy of EBr.

14.52 Write a balanced equation for the formation from its elements of sodium peroxide, an industrial bleach.

14.53 Write a balanced equation for the formation of rubidium bromide through a neutralization reaction of a strong acid and a strong base.

● **Problems in Context**

14.54 Although the alkali halides can be prepared directly from the elements, the far less expensive industrial route was treatment of the carbonate or hydroxide with aqueous hydrohalic acid (HX) followed by recrystallization. Balance the reaction between potassium carbonate and aqueous hydriodic acid.

14.55 The total terrestrial abundance of francium ($Z = 87$) is estimated at 2×10^{-18} ppm, or about 15 g of Fr in the top kilometer of the Earth's crust. Nevertheless, it has been predicted that if a weighable quantity of Fr could be obtained, it would be a liquid at room temperature. What is the basis of this prediction?

Group 2A(2): The Alkaline Earth Metals

● **Concept Review Questions**

14.56 How do Groups 1A(1) and 2A(2) compare with respect to reaction of the metals with water?

14.57 Alkaline earth metals are involved in two key diagonal relationships in the periodic table. (a) Give the two pairs of elements in these diagonal relationships. (b) For each pair, cite two similarities that demonstrate the relationship. (c) Why are the members of each pair so similar in behavior?

14.58 The melting points of alkaline earth metals are many times higher than those of the alkali metals. Explain this difference on the basis of atomic properties. Name three other physical properties for which Group 2A(2) metals have higher values than the corresponding 1A(1) metals.

● **Skill-Building Exercises (paired)**

14.59 Write a balanced equation for each reaction:
(a) "Slaking" of lime (treatment with water)
(b) Combustion of calcium in air

14.60 Write a balanced equation for each reaction:
(a) Thermal decomposition of witherite (barium carbonate)
(b) Neutralization of stomach acid (HCl) by milk of magnesia (magnesium hydroxide)

● **Problems in Context**

14.61 Lime (CaO) is one of the most abundantly produced chemicals in the world. Write balanced equations for
(a) The preparation of lime from natural sources
(b) The use of slaked lime to remove SO_2 from flue gases
(c) The reaction of lime with arsenic acid (H_3AsO_4) to manufacture the insecticide calcium arsenate
(d) The regeneration of NaOH in the paper industry by reaction of lime with aqueous sodium carbonate

14.62 In some reactions, Be behaves like a typical alkaline earth metal; in others, it does not. Complete and balance the following equations:
(a) $BeO(s) + H_2O(l) \longrightarrow$

(b) $BeCl_2(l) + Cl^-(l; \text{from molten NaCl}) \longrightarrow$
In which reaction does Be behave like the other Group 2A(2) members?

Group 3A(13): The Boron Family

● **Concept Review Questions**

14.63 How do the transition metals in Period 4 affect the pattern of ionization energies in Group 3A(13)? How does this pattern compare with that in Group 3B(3)?

14.64 How do the acidities of aqueous solutions of Tl_2O and Tl_2O_3 compare with each other? Explain.

14.65 Despite the expected decrease in atomic size, there is an unexpected drop in the first ionization energy between Groups 2A(2) and 3A(13) in Periods 2 through 4 (e.g., Be to B). Explain this pattern in terms of electron configuration and orbital energies.

14.66 Many compounds of Group 3A(13) elements have chemical behavior that reflects an electron deficiency.
(a) What is the meaning of *electron deficiency?*
(b) Give two reactions that illustrate this behavior.

14.67 Boron chemistry is not typical of its group.
(a) Cite three ways in which boron and its compounds differ significantly from the other 3A(13) members and their compounds.
(b) What is the reason for these differences?

● **Skill-Building Exercises (paired)**

14.68 Rank the following oxides in order of increasing aqueous *acidity:* Ga_2O_3, Al_2O_3, In_2O_3.

14.69 Rank the following hydroxides in order of increasing aqueous *basicity:* TlOH, $B(OH)_3$, $In(OH)_3$.

14.70 Thallium forms the compound TlI_3. What is the apparent oxidation state of Tl in this compound? Given that the anion is I_3^-, what is the actual oxidation state of Tl? Draw the shape of the anion, giving its VSEPR class and bond angles. Propose a reason why the compound does not exist as $(Tl^{3+})(I^-)_3$.

14.71 Very stable dihalides of the Group 3A(13) metals are known. What is the apparent oxidation state of Ga in $GaCl_2$? Given that $GaCl_2$ consists of a Ga^+ cation and a $GaCl_4^-$ anion, what are the actual oxidation states of Ga? Draw the shape of the anion, giving its VSEPR class and bond angles.

● **Problems in Context**

14.72 Give the name and symbol or formula of a Group 3A(13) element or compound that fits each description or use:
(a) Component of heat-resistant (Pyrex-type) glass
(b) Manufacture of high-speed computer chips
(c) Largest temperature range for liquid state of an element
(d) Elementary substance with three-center, two-electron bonds
(e) Metal protected from oxidation by adherent oxide coat
(f) Mild antibacterial agent (e.g., for eye infections)
(g) Toxic metal that lies in the periodic table between two other toxic metals.

14.73 Many compounds of the Group 3A(13) and 5A(15) elements are technologically important as semiconductors. They can be prepared from the elements at high pressure and temperature, but must be kept encased to prevent decomposition in moist air to the 3A hydroxide and the 5A hydride.
(a) Write a balanced equation for the formation of indium(III) arsenide.

(b) Write a balanced equation for the decomposition of gallium antimonide in moist air.

14.74 Use VSEPR theory to draw structures, with ideal bond angles, for boric acid and the anion it forms in reaction with water.

14.75 Halides of Al, Ga, and In occur as dimers in the gas and liquid phases (those containing the larger halogens form dimeric solids also). Use VSEPR theory to draw the structure, with ideal bond angles, of the $GaBr_3$ dimer.

Group 4A(14): The Carbon Family

● **Concept Review Questions**

14.76 How does the basicity of SnO_2 in water compare with that of CO_2? Explain.

14.77 Nearly every compound of silicon has the element in the +4 oxidation state. In contrast, most compounds of lead have the element in the +2 state.
(a) What general observation do these facts illustrate?
(b) What is the explanation for these facts in terms of atomic and molecular properties?
(c) Give an analogous example from the chemistry of Group 3A(13).

14.78 The sum of IE_1 through IE_4 for Group 4A(14) elements shows a decrease from C to Si, a slight increase from Si to Ge, a decrease from Ge to Sn, and an increase from Sn to Pb. (a) What is the expected trend for ionization energy down a group? (b) Suggest an explanation for the deviations from the expected trend. (c) Which group might you expect to show even greater deviations?

14.79 Give explanations for the large drops in melting point from C to Si and from Ge to Sn.

14.80 What is an allotrope? Name two Group 4A(14) elements that exhibit allotropy and name two of their allotropes.

14.81 Even though EN values vary relatively little down Group 4A(14), the elements change from nonmetal to metal. Explain.

14.82 How do atomic properties account for the enormous number of carbon compounds? Why don't other Group 4A(14) elements behave similarly?

● **Skill-Building Exercises (paired)**

14.83 Draw a Lewis structure for
(a) The cyclic silicate ion $Si_4O_{12}{}^{8-}$
(b) A cyclic hydrocarbon with formula C_4H_8
14.84 Draw a Lewis structure for
(a) The cyclic silicate ion $Si_6O_{18}{}^{12-}$
(b) A cyclic hydrocarbon with formula C_6H_{12}

14.85 Show three units of a linear silicone polymer made from $(CH_3)_2Si(OH)_2$ with some $(CH_3)_3SiOH$ added to end the chain.
14.86 Show two chains of three units each of a sheet silicone polymer made from $(CH_3)_2Si(OH)_2$ with $CH_3Si(OH)_3$ added to crosslink the chains.

● **Problems in Context**

14.87 Although it is often stated that double bonding is limited to compounds of Period 2 elements, compounds of Period 3 elements also contain double bonds.
(a) What types of atomic orbitals are used for the π bond in $C{=}O$?

(b) Why are the same types of orbitals not satisfactory for an $Si{=}O$ π bond?
(c) What orbital type could be used for an $Si{=}O$ π bond?
(d) Which atom(s) would supply the electrons in such a π bond?

14.88 Give the name and symbol or formula of a Group 4A(14) element or compound that fits each description or use:
(a) Hardest known substance
(b) Medicinal antacid
(c) Atmospheric gas implicated in greenhouse effect
(d) Waterproofing polymer based on silicon
(e) Synthetic abrasive composed entirely of Group 4A elements
(f) Product formed when coke burns in a limited supply of air
(g) Toxic metal found in plumbing and paints

14.89 One similarity between B and Si is the explosive combustion of their hydrides in air. Write balanced equations for the combustion of B_2H_6 and of Si_4H_{10}.

Group 5A(15): The Nitrogen Family

● **Concept Review Questions**

14.90 Which Group 5A(15) elements form trihalides? Pentahalides? Explain.

14.91 As you move down Group 5A(15), the melting points of the elements increase and then decrease. Explain.

14.92 (a) What is the range of oxidation states shown by the elements of Group 5A(15) as you move down the group? (b) How does this range illustrate the general rule for the range of oxidation states in groups on the right side of the periodic table?

14.93 Bismuth(V) compounds are such powerful oxidizing agents that they have not been prepared in pure form. How is this fact consistent with the location of Bi in the periodic table?

14.94 Rank the following oxides in order of increasing acidity in water: Sb_2O_3, Bi_2O_3, P_4O_{10}, Sb_2O_5.

● **Skill-Building Exercises (paired)**

14.95 Assuming that acid strength relates directly to electronegativity of the central atom, rank H_3PO_4, HNO_3, and H_3AsO_4 in order of *increasing* acid strength.

14.96 Assuming that acid strength relates directly to number of O atoms on the central atom, rank $H_2N_2O_2$ [or $(HON)_2$]; HNO_3 (or $HONO_2$); and HNO_2 (or $HONO$) in order of *decreasing* acid strength.

14.97 Complete and balance the following:
(a) $As(s) + \text{excess } O_2(g) \longrightarrow$
(b) $Bi(s) + \text{excess } F_2(g) \longrightarrow$
(c) $Ca_3As_2(s) + H_2O(l) \longrightarrow$
14.98 Complete and balance the following:
(a) $\text{Excess } Sb(s) + Br_2(l) \longrightarrow$
(b) $HNO_3(aq) + MgCO_3(s) \longrightarrow$
(c) $K_2HPO_4(s) \xrightarrow{\Delta}$

14.99 Complete and balance the following:
(a) $N_2(g) + Al(s) \xrightarrow{\Delta}$
(b) $PF_5(g) + H_2O(l) \longrightarrow$
14.100 Complete and balance the following:
(a) $AsCl_3(l) + H_2O(l) \longrightarrow$
(b) $Sb_2O_3(s) + NaOH(aq) \longrightarrow$

14.101 Based on the relative sizes of F and Cl, predict the structure of PF_2Cl_3.

14.102 Use the VSEPR model to predict the structure of the cyclic ion $P_3O_9^{3-}$.

● **Problems in Context**

14.103 The white and red allotropes of phosphorus differ dramatically in toxicity. One is relatively unreactive in the body, whereas the other is quite poisonous. Suggest how their structures lead to a difference in toxicity, and name the nontoxic allotrope.

14.104 Give the name and symbol or formula of a Group 5A(15) element or compound that fits each of the following descriptions or uses:
(a) Hydride produced at multimillion-ton level
(b) Element(s) essential in plant nutrition
(c) Hydride used in the manufacture of rocket propellants and drugs
(d) Odd-electron molecule (two examples)
(e) Amphoteric hydroxide with 5A element in its +3 state
(f) Phosphorus-containing water softener
(g) Element that is an electrical conductor

14.105 Would you expect hydrazine (N_2H_4) to form H bonds? Explain with structures.

14.106 Nitrous oxide (N_2O), the "laughing gas" used as an anesthetic by dentists, is made by thermal decomposition of solid NH_4NO_3. Write a balanced equation for this reaction. What are the oxidation states of N in NH_4NO_3 and in N_2O?

14.107 Write balanced equations for the thermal decomposition of potassium nitrate (O_2 is also formed in both cases): (a) at low temperature to the nitrite; (b) at high temperature to the metal oxide and nitrogen.

Group 6A(16): The Oxygen Family

● **Concept Review Questions**

14.108 Rank the following in order of increasing electrical conductivity, and explain your ranking: Po, S, Se.

14.109 The oxygen and nitrogen families have some obvious similarities and differences.
(a) State two general physical similarities between Group 5A(15) and 6A(16) elements.
(b) State two general chemical similarities between Group 5A(15) and 6A(16) elements.
(c) State two chemical similarities between P and S.
(d) State two physical similarities between N and O.
(e) State two chemical differences between N and O.

14.110 A molecular property of the Group 6A(16) hydrides changes abruptly down the group. This change has been explained in terms of a change in hybridization.
(a) Between what periods does the change occur?
(b) What is the change in the molecular property?
(c) What is the change in hybridization?
(d) What other group displays a similar change?

● **Skill-Building Exercises** *(paired)*

14.111 Complete and balance the following:
(a) $NaHSO_4(aq) + NaOH(aq) \longrightarrow$
(b) $S_8(s) + \text{excess } F_2(g) \longrightarrow$
(c) $FeS(s) + HCl(aq) \longrightarrow$
(d) $Te(s) + I_2(s) \longrightarrow$

14.112 Complete and balance the following:
(a) $H_2S(g) + O_2(g) \longrightarrow$

(b) $SO_3(g) + H_2O(l) \longrightarrow$
(c) $SF_4(g) + H_2O(l) \longrightarrow$
(d) $Al_2Se_3(s) + H_2O(l) \longrightarrow$

14.113 Is each oxide basic, acidic, or amphoteric in water: (a) SeO_2; (b) N_2O_3; (c) K_2O; (d) BeO; (e) BaO?
14.114 Is each oxide basic, acidic, or amphoteric in water: (a) MgO; (b) N_2O_5; (c) CaO; (d) CO_2; (e) TeO_2?

14.115 Rank the following hydrides in order of *increasing* acid strength: H_2S, H_2O, H_2Te.

14.116 Rank the following species in order of *decreasing* acid strength: H_2SO_4, H_2SO_3, HSO_3^-.

● **Problems in Context**

14.117 Describe the physical changes observed when solid sulfur is heated from room temperature to 440°C and then poured quickly into cold water. Explain the molecular changes responsible for the macroscopic changes.

14.118 Give the name and symbol or formula of a Group 6A(16) element or compound that fits each of the following descriptions or uses:
(a) Unreactive gas used as an electrical insulator
(b) Unstable allotrope of oxygen
(c) Oxide having sulfur in the same oxidation state as in sulfuric acid
(d) Air pollutant produced by burning sulfur-containing coal
(e) Powerful dehydrating agent
(f) Compound used in solution in the photographic process
(g) Trace gas that tarnishes silver

14.119 In addition to the monatomic sulfide anion, S^{2-}, sulfur forms a variety of polyatomic sulfide anions, S_n^{2-}. The simplest, S_2^{2-}, is found in pyrite ores, such as FeS_2, and others through S_6^{2-} are known.
(a) Name and draw a Lewis structure for the oxygen anion that is analogous to S_2^{2-}.
(b) Predict the shape and bond angle of the anion in BaS_3.

14.120 Disulfur decafluoride is intermediate in reactivity between SF_4 and SF_6. It disproportionates at 150°C to these monosulfur fluorides. Write a balanced equation for this reaction, and give the oxidation state of S in each compound.

Group 7A(17): The Halogens

● **Concept Review Questions**

14.121 (a) What is the physical state and color of each of the halogens at STP?
(b) Explain the change in physical state down Group 7A(17) in terms of molecular properties.

14.122 (a) What are the common oxidation states of the halogens?
(b) Give an explanation based on electron configuration for the range and values of the oxidation states of chlorine.
(c) Why is fluorine an exception to the pattern of oxidation states found for the other group members?

14.123 How many electrons does a halogen atom need to complete its octet? Give examples of the different ways that a Cl atom can do so.

14.124 Select the stronger bond in each pair:
(a) Cl—Cl or Br—Br (b) Br—Br or I—I (c) F—F or Cl—Cl
Why doesn't the F—F bond strength follow the group trend?

14.125 In addition to interhalogen compounds, many polyatomic interhalogen ions exist. Would you expect interhalogen ions with a $1+$ or a $1-$ charge to have an even or odd number of atoms? Explain.

14.126 (a) A halogen (X_2) disproportionates in base in several steps to X^- and XO_3^-. Write the overall equation for the disproportionation of Br_2 to Br^- and BrO_3^-.
(b) Write a balanced equation for the reaction of ClF_5 with aqueous base (by analogy with the reaction of BrF_5 shown in text).

● **Skill-Building Exercises (paired)**

14.127 Complete and balance the following equations. If no reaction occurs, write NR:
(a) $Rb(s) + Br_2(l) \longrightarrow$
(b) $I_2(s) + H_2O(l) \longrightarrow$
(c) $Br_2(l) + I^-(aq) \longrightarrow$
(d) $CaF_2(s) + H_2SO_4(l) \longrightarrow$

14.128 Complete and balance the following equations. If no reaction occurs, write NR:
(a) $H_3PO_4(l) + NaI(s) \longrightarrow$
(b) $Cl_2(g) + I^-(aq) \longrightarrow$
(c) $Br_2(l) + Cl^-(aq) \longrightarrow$
(d) $ClF(g) + F_2(g) \longrightarrow$

14.129 Rank the following acids in order of *increasing* acid strength: $HClO$, $HClO_2$, $HBrO$, HIO.

14.130 Rank the following acids in order of *decreasing* acid strength: $HBrO_3$, $HBrO_4$, HIO_3, $HClO_4$.

● **Problems in Context**

14.131 Give the name and symbol or formula of a Group 7A(17) element or compound that fits each description or use:
(a) Used in etching glass
(b) Naturally occurring source (ore) of fluorine
(c) Oxide used in bleaching paper pulp and textiles
(d) Weakest hydrohalic acid
(e) Compound used as food additive to prevent goiter (thyroid disorder)
(f) Element that is produced in the largest quantity
(g) Organic chloride used to make plastics

14.132 An industrial chemist treats solid NaCl with concentrated H_2SO_4 and obtains gaseous HCl and $NaHSO_4$. When she substitutes solid NaI for NaCl, gaseous H_2S, solid I_2, and S_8 are obtained but no HI.
(a) What type of reaction did the H_2SO_4 undergo with NaI?
(b) Why does NaI, but not NaCl, cause this type of reaction?
(c) To produce $HI(g)$ by the reaction of NaI with an acid, how does the acid have to differ from sulfuric acid?

14.133 Rank the halogens Cl, Br, and I in order of increasing oxidizing strength based on their products with metallic Re: $ReCl_6$, $ReBr_5$, ReI_4. Explain your ranking.

Group 8A(18): The Noble Gases

14.134 Which noble gas is the most abundant in the universe? In the atmosphere?

14.135 What oxidation states does Xe show in its compounds?

14.136 Why do the noble gases have such low boiling points?

14.137 Explain why Xe, and to a limited extent Kr, form compounds, whereas He, Ne, and Ar do not.

14.138 (a) Why do the stable xenon fluorides have an even number of fluorine atoms?
(b) Why do the ionic species XeF_3^+ and XeF_7^- have odd numbers of F atoms?
(c) Predict the shape of XeF_3^+.

Comprehensive Problems

Problems with an asterisk (*) are more challenging.

14.139 Sodium hydride is used to prepare sodium dithionate $(Na_2S_2O_4)$, used in bleaching paper pulp:
$$NaH(s) + SO_2(l) \longrightarrow Na_2S_2O_4(s) + H_2(g)$$
(a) Balance the equation, and give the O.N. of each element.
(b) How many liters of hydrogen gas measured at 758 torr and $-45°C$ can form when exactly 1 metric ton of NaH reacts with 1 metric ton of SO_2?

14.140 Given the following information,
$$H^+(g) + H_2O(g) \longrightarrow H_3O^+(g) \qquad \Delta H = -720 \text{ kJ}$$
$$H^+(g) + H_2O(l) \longrightarrow H_3O^+(aq) \qquad \Delta H = -1090 \text{ kJ}$$
$$H_2O(l) \longrightarrow H_2O(g) \qquad \Delta H = 40.7 \text{ kJ}$$
calculate the heat of solution of the hydronium ion:
$$H_3O^+(g) \xrightarrow{H_2O} H_3O^+(aq)$$

14.141 The electronic transition in Na from $3p^1$ to $3s^1$ gives rise to a bright yellow-orange emission at 589.2 nm. What is the energy of this transition?

14.142 Unlike the other Group 2A(2) metals, beryllium acts like aluminum and zinc in reacting with concentrated aqueous base to release hydrogen gas and form oxoanions of formula $M(OH)_4^{n-}$. Write equations for the reactions of these metals with NaOH.

14.143 Just as boron nitride is isoelectronic with carbon, other compounds of Groups 3A(13) and 5A(15) are isoelectronic with elements in Group 4A(14). What element is isoelectronic with (a) aluminum phosphide; (b) gallium arsenide?

14.144 Cyclopropane (C_3H_6), the smallest cyclic hydrocarbon, is reactive for the same reason that white phosphorus is. Use bond properties and valence bond theory to explain the high reactivity of C_3H_6.

14.145 Thionyl chloride $(SOCl_2)$ is a sulfur oxohalide that is used industrially to dehydrate metal halide hydrates.
(a) Write a balanced equation for its reaction with magnesium chloride hexahydrate, in which SO_2 and HCl form along with the metal halide.
(b) Draw the Lewis structure of thionyl chloride that has the lowest formal charges.

14.146 The main reason alkali metal dihalides (MX_2) do *not* form is the high IE_2 of the metal.
(a) Why is the IE_2 so high for alkali metals?
(b) The IE_2 for Cs is 2255 kJ/mol, low enough for CsF_2 to form exothermically $(\Delta H_f^0 = -125 \text{ kJ/mol})$. This compound cannot be synthesized, however, because CsF forms with a much greater release of heat $(\Delta H_f^0 = -530 \text{ kJ/mol})$. Thus, the breakdown of CsF_2 to CsF happens readily. Write the equation for this breakdown, and calculate the heat of reaction per mole of CsF.

14.147 Semiconductors made from elements in Groups 3A(13) and 5A(15) are typically prepared by direct reaction of the elements at high temperature. An engineer treats 32.5 g of molten gallium with 20.4 L of white phosphorus vapor at 515 K and 195 kPa. If purification losses are 7.2% by mass, how many grams of gallium phosphide can be prepared?

*****14.148** Two substances that have the empirical formula HNO are hyponitrous acid $(\mathcal{M} = 62.04 \text{ g/mol})$ and nitroxyl $(\mathcal{M} = 31.02 \text{ g/mol})$.
(a) What is the molecular formula of each species?

(b) Draw the Lewis structure having the lowest formal charges for each species. (*Hint:* Hyponitrous acid has an N$=$N bond.)

(c) Predict the shape of each species around the N atoms.

(d) When hyponitrous acid loses two protons, it forms the hyponitrite ion. Draw *cis* and *trans* forms of this ion.

14.149 The species CO, CN$^-$, and C$_2{}^{2-}$ are isoelectronic.

(a) Draw their Lewis structures.

(b) Draw their MO diagrams (assume 2s-2p mixing, as in N$_2$), and give the bond order and electron configuration for each.

*14.150 The Ostwald process is a series of three reactions used for the industrial production of nitric acid from ammonia.

(a) Write out the series of balanced equations for the Ostwald process.

(b) If NO is *not* recycled, how many moles of NH$_3$ are consumed per mole of HNO$_3$ produced?

(c) In a typical industrial unit, the process is very efficient, with a 96% yield for the first step. Assuming 100% yields for the subsequent steps, what volume of nitric acid (60.% by mass; $d = 1.37$ g/mL) can be prepared per cubic meter of a gas mixture that is 90.% air and 10.% NH$_3$ by volume at the industrial conditions of 5.0 atm and 850.°C?

*14.151 Organolithium compounds such as LiCH$_3$ are important reagents in organic synthesis. The compounds are not simple substances, as the formula suggests, but exist in the solid state, and in solution with nonpolar solvents, as tetramers [e.g., methyllithium is Li$_4$(CH$_3$)$_4$].

(a) How many electrons are available for bonding among the four lithium atoms and the four methyl groups?

(b) Considering this number of available electrons and the total number of atoms bonded in the cluster, what unusual type of bonding might you expect to occur in some of the bonds in methyllithium?

14.152 Perhaps surprisingly, some pure liquid interhalogens that contain fluorine have high electrical conductivity. The explanation is that one molecule transfers a fluoride ion to another molecule. Write the equation that would explain the high electrical conductivity of bromine trifluoride.

14.153 Solid PCl$_5$ exists not as molecules but rather as two ions, PCl$_4{}^+$ and PCl$_6{}^-$. Similarly, solid PBr$_5$ exists as ions, but in this case as PBr$_4{}^+$ and Br$^-$. Suggest a reason for this difference in solid state structure.

*14.154 Producer gas is a fuel formed by passing air over red-hot coke (amorphous carbon). It consists of approximately 25% CO, 5.0% CO$_2$, and 70.% N$_2$ by mass. What mass of producer gas can be formed from 1.75 metric tons of coke, assuming an 87% yield?

14.155 Hypofluorous acid (HFO, or more correctly HOF) was first prepared in weighable amounts only in 1971. The synthesis required treatment of fluorine with ice followed by rapid removal of the oxoacid to prevent the formation of oxygen difluoride through its reaction with more F$_2$.

(a) Write balanced equations for these reactions.

(b) What is the O.N. of O in HOF and in OF$_2$?

(c) Predict the shapes of HOF and OF$_2$.

14.156 What is a disproportionation reaction, and which of the following fit the description?

(a) I$_2$(s) + KI(aq) $\longrightarrow$ KI$_3$(aq)

(b) 2ClO$_2$(g) + H$_2$O(l) $\longrightarrow$ HClO$_3$(aq) + HClO$_2$(aq)

(c) Cl$_2$(g) + 2NaOH(aq) $\longrightarrow$
$$\text{NaCl}(aq) + \text{NaClO}(aq) + \text{H}_2\text{O}(l)$$

(d) NH$_4$NO$_2$(s) $\longrightarrow$ N$_2$(g) + 2H$_2$O(g)

(e) 3MnO$_4{}^{2-}$(aq) + 2H$_2$O(l) $\longrightarrow$
$$2\text{MnO}_4{}^-(aq) + \text{MnO}_2(s) + 4\text{OH}^-(aq)$$

(f) 3AuCl(s) $\longrightarrow$ AuCl$_3$(s) + 2Au(s)

14.157 Explain the following observations:

(a) Phosphorus forms PCl$_5$ in addition to the expected PCl$_3$, but nitrogen forms only NCl$_3$.

(b) Carbon tetrachloride is unreactive toward water, but silicon tetrachloride reacts rapidly and completely. (To give what?)

(c) The oxygen-sulfur bond in SO$_4{}^{2-}$ is shorter than expected for an S—O single bond.

(d) Chlorine forms ClF$_3$ and ClF$_5$, but ClF$_4$ is unknown.

14.158 Which group(s) of the periodic table is (are) described by the following general statements?

(a) The elements form neutral compounds of VSEPR class AX$_3$E.

(b) The free elements are strong oxidizing agents and form monatomic ions and oxoanions.

(c) The valence shell allows the atoms to form compounds by combining with two atoms that donate one electron each.

(d) The free elements are strong reducing agents, show only one nonzero oxidation state, and form mainly ionic compounds.

(e) The elements can form stable compounds with only three bonds, but as a central atom, they can accept a pair of electrons from a fourth atom without expanding their valence shell.

(f) Only larger members of the group are known to be chemically active.

14.159 Diiodine pentaoxide (I$_2$O$_5$) was discovered by Gay-Lussac in 1813, but its structure was unknown until 1970! Like Cl$_2$O$_7$, it can be prepared by the dehydration-condensation of the corresponding oxoacid.

(a) Name the precursor oxoacid, write a reaction for formation of the oxide, and draw a likely Lewis structure.

(b) Data show that the bonds to the terminal O are shorter than the bonds to the bridging O. Why?

(c) I$_2$O$_5$ is one of the few chemicals that can oxidize CO rapidly and completely; elemental iodine forms in the process. Write a balanced equation for this reaction.

14.160 An important starting material for the manufacture of polyphosphazenes is the cyclic molecule (NPCl$_2$)$_3$. The molecule has a symmetrical six-membered ring of alternating N and P atoms, with the Cl atoms bound to the P atoms. The nitrogen-phosphorus bond length is significantly less than that expected for an N—P single bond.

(a) Draw a likely Lewis structure for the molecule.

(b) How many lone pairs of electrons appear on the ring atoms?

(c) What is the order of the nitrogen-phosphorus bond?

14.161 Potassium fluotitanate (K$_2$TiF$_6$), a starting material for other titanium salts, is made by dissolving titanium(IV) oxide in concentrated HF, adding aqueous KF, and then heating to dryness to drive off excess HF. The impure solid is recrystallized by dissolving in hot water and then cooling to 0°C, which leaves excess KF in solution. For the reaction of 5.00 g of TiO$_2$, 250. mL of hot water is used to recrystallize the product. How many grams of purified K$_2$TiF$_6$ are obtained? (Solubility of K$_2$TiF$_6$ in water at 0°C is 0.60 g /100 mL.)

14.162 What are the molecular formulas for the most common allotrope of oxygen and of sulfur? Why are these formulas so different?

14.163 Account for the following facts:

(a) Ca^{2+} and Na^+ have very nearly the same radii.

(b) CaF_2 is insoluble in water, but NaF is quite soluble.

(c) Molten $BeCl_2$ is a poor electrical conductor, whereas molten $CaCl_2$ is an excellent conductor.

14.164 Cake alum (aluminum sulfate) is used as a "flocculating agent" in water purification. The "floc" is a gelatinous precipitate of aluminum hydroxide that carries out of solution small suspended particles and bacteria for removal by filtration.

(a) The hydroxide is formed by the reaction of the aluminum ion with water. Write the equation for this reaction.

(b) The acidity from this reaction is partially neutralized by the sulfate ion. Write an equation for this reaction.

14.165 Carbon tetrachloride is made by passing Cl_2 in the presence of a catalyst through liquid CS_2 near its boiling point. Disulfur dichloride (S_2Cl_2) also forms, in addition to the by-products SCl_2 and CCl_3SCl. How many grams of CS_2 reacted if excess Cl_2 yields 50.0 g of CCl_4 and 0.500 g of CCl_3SCl? Is it possible to determine how much S_2Cl_2 and SCl_2 also formed? Explain.

14.166 H_2 may act as a reducing agent or an oxidizing agent, depending on the substance reacting with it. Using, in turn, sodium and chlorine as the other reactant, write balanced equations for the two reactions, and characterize the redox role of hydrogen in each.

14.167 P_4 is prepared by heating phosphate rock [principally $Ca_3(PO_4)_2$] with sand and coke:

$$Ca_3(PO_4)_2(s) + SiO_2(s) + C(s) \longrightarrow$$
$$CaSiO_3(s) + CO(g) + P_4(g) \text{ [unbalanced]}$$

How many kilograms of phosphate rock are needed to produce 315 mol of P_4, assuming that the conversion is 90.% efficient?

14.168 Assume you can use either NH_4NO_3 or N_2H_4 as fertilizer. Which contains more moles of N per mole of compound? Which contains the higher mass of N per gram of compound?

14.169 From its formula, one might expect CO to be quite polar, but its dipole moment is actually low (0.11 D).

(a) Draw the Lewis structure for CO.

(b) Calculate the formal charges.

(c) Based on your answers to parts (a) and (b), explain why the dipole moment is so low.

*14.170 In addition to Al_2Cl_6, aluminum forms other species with bridging halide ions to two aluminum atoms. One such species is the ion $Al_2Cl_7^-$. The ion is symmetrical, with a linear $Al-Cl-Al$ bond.

(a) What orbitals does Al use to bond with the Cl atoms?

(b) What is the shape around each Al?

(c) What is the hybridization of the central Cl?

(d) What do the shape and hybridization suggest about the presence of lone pairs of electrons on the central Cl?

14.171 When an alkaline earth carbonate is heated, it releases CO_2, leaving the metal oxide. The temperature at which each Group 2A carbonate yields a CO_2 partial pressure of 1 atm is

Carbonate	Temperature (°C)
$MgCO_3$	542
$CaCO_3$	882
$SrCO_3$	1155
$BaCO_3$	1360

(a) Suggest a reason for this trend. (b) Mixtures of $CaCO_3$ and MgO are used to absorb dissolved silicates from boiler water. How would you prepare a mixture of $CaCO_3$ and MgO from dolomite, which contains $CaCO_3$ and $MgCO_3$?

14.172 The bond angles in the nitrite ion, nitrogen dioxide, and the nitronium ion (NO_2^+) are 115°, 134°, and 180°, respectively. Rationalize these values using Lewis structures and VSEPR theory.

14.173 A common method for producing a gaseous hydride is to treat a salt containing the anion of the volatile hydride with a strong acid.

(a) Write an equation for each of the following examples: (1) the production of HF from CaF_2; (2) the production of HCl from $NaCl$; (3) the production of H_2S from FeS.

(b) In some cases even a weak acid such as water will suffice if the anion of the salt has a sufficiently strong attraction for protons. An example is the production of PH_3 from Ca_3P_2 and water. Write the equation for this reaction.

(c) By analogy, predict the products and write the equation for the reaction of Al_4C_3 with water.

14.174 A former use of chlorine trifluoride in the United States was in the production of uranium hexafluoride for the nuclear industry:

$$U(s) + 3ClF_3(l) \longrightarrow UF_6(l) + 3ClF(g)$$

How many grams of UF_6 can form from 1.00 metric ton of uranium ore that is 1.55% by mass uranium and 12.75 L of chlorine trifluoride ($d = 1.88$ g/mL)?

*14.175 Chlorine is used to make household bleach solutions containing 5.25% (by mass) $NaClO$. Assuming 100% yield in the reaction producing $NaClO$ from Cl_2, how many liters of $Cl_2(g)$ at STP will be needed to make 1000. L of bleach solution ($d = 1.07$ g/mL)?

14.176 The triatomic molecular ion H_3^+ was first detected and characterized by J. J. Thomson using mass spectrometry. Use the bond energy of H_2 (432 kJ/mol) and the proton affinity of H_2 ($H_2 + H^+ \longrightarrow H_3^+$; $\Delta H = -337$ kJ/mol) to calculate the heat of reaction for

$$H + H + H^+ \longrightarrow H_3^+$$

14.177 An atomic hydrogen torch is used for cutting and welding thick sheets of metal. When H_2 passes through an electric arc, the molecules decompose into atoms, which react with O_2 at the end of the torch. Temperatures in excess of 5000°C are reached, which can melt all metals. Write equations for H_2 breakdown to H atoms and for the subsequent overall reaction of the H atoms with oxygen. Use Appendix B to find the standard heat of each reaction per mole of product.

14.178 In aqueous HF solution, an important species is the ion HF_2^-, which has the bonding arrangement FHF^-. Draw the Lewis structure for this ion, and explain how it arises.

14.179 Which of the following oxygen ions are paramagnetic: O^+, O^-, O^{2-}, O^{2+}?

*14.180 Hydrogen peroxide can act as either an oxidizing agent or a reducing agent. (a) When H_2O_2 is treated with aqueous KI, I_2 forms. In which role is H_2O_2 acting? What is the oxygen-containing product formed? (b) When H_2O_2 is treated with aqueous $KMnO_4$, the purple color of MnO_4^- disappears and a gas forms. In which role is H_2O_2 acting? What is the oxygen-containing product formed?

CHAPTER 15

ORGANIC COMPOUNDS AND THE ATOMIC PROPERTIES OF CARBON

CHAPTER OUTLINE

15.1 The Special Nature of Carbon and the Characteristics of Organic Molecules
Structural Complexity
Chemical Diversity

15.2 The Structures and Classes of Hydrocarbons
Carbon Skeletons and Hydrogen Skins
Alkanes
Constitutional Isomers
Optical Isomers
Alkenes and Geometric Isomers
Alkynes
Aromatic Hydrocarbons
Catenated Inorganic Hydrides

15.3 Some Important Classes of Organic Reactions
Types of Organic Reactions
Organic Redox Reactions

15.4 Properties and Reactivities of Common Functional Groups
Single-Bonded Groups
Double-Bonded Groups
Single- and Double-Bonded Groups
Triple-Bonded Groups

15.5 The Monomer-Polymer Theme I: Synthetic Macromolecules
Addition Polymers
Condensation Polymers

15.6 The Monomer-Polymer Theme II: Biological Macromolecules
Sugars and Polysaccharides
Amino Acids and Proteins
Nucleotides and Nucleic Acids

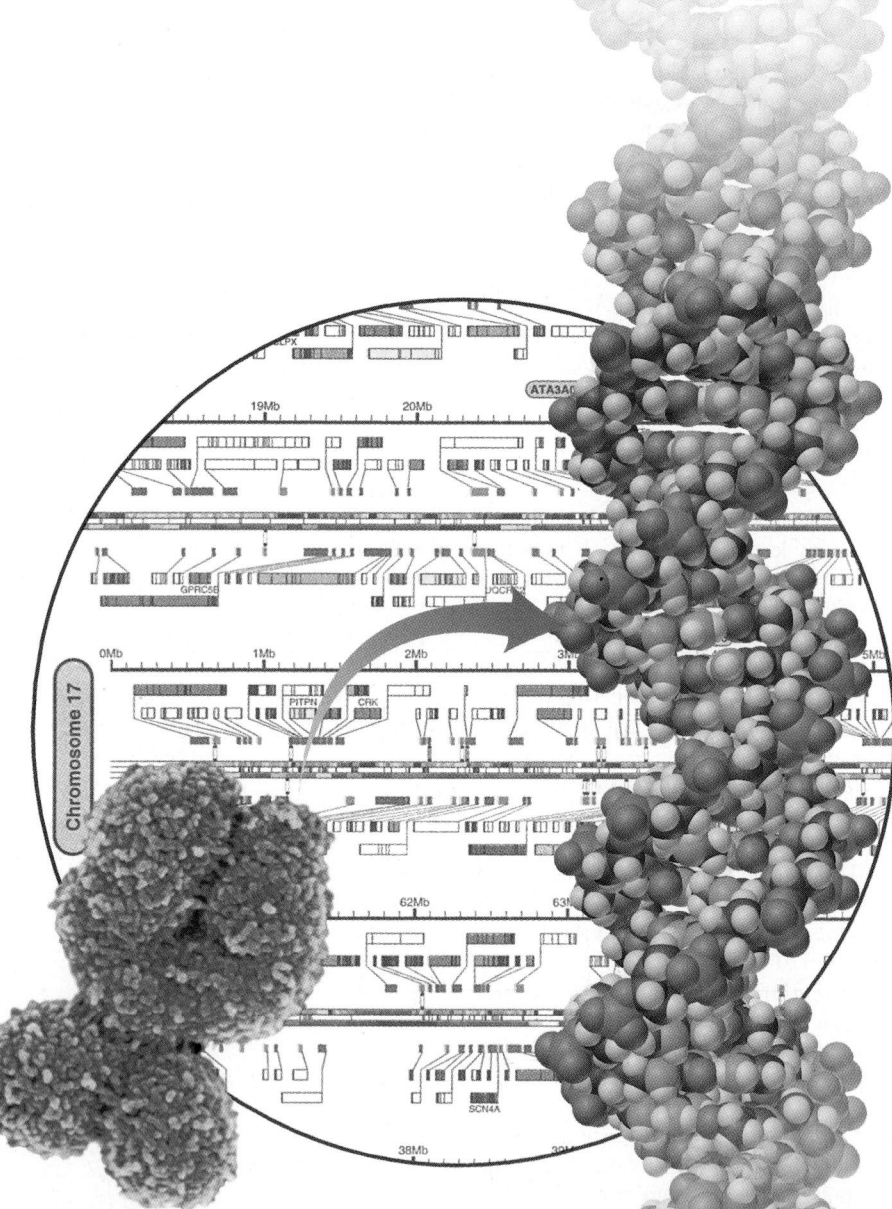

Figure: Know Thyself—Genetically! This genetic adaptation of the ancient dictum can finally come true based on the recent determination of the entire human genome, an astonishing laboratory feat that will provide understanding of our biology as never before. A small section of the genome "map" underlies one of the 23 distinct human chromosomes and a very short section of DNA. Our biological nature derives ultimately from our biochemical nature, which, as you'll see in this chapter, derives ultimately from the amazing nature of the carbon atom.

s there any chemical system more remarkable than a living cell? Through delicately controlled mechanisms, it oxidizes food for energy, maintains the concentrations of thousands of aqueous components, interacts continuously with its environment, synthesizes both simple and complex molecules, and even reproduces itself! With all our technological prowess, no human-made system even approaches the cell for sheer elegance of function.

This amazing chemical machine consumes, creates, and consists largely of *organic compounds*. Except for a few inorganic salts and ever-present water, everything you put into or on your body—food, medicine, cosmetics, and clothing—consists of organic compounds. Organic fuels warm our homes, cook our meals, and power our society. Major industries are devoted to producing organic compounds, such as polymers, pharmaceuticals, and insecticides.

What *is* an organic compound? Dictionaries define it as "a compound of carbon," but that definition includes carbonates, cyanides, carbides, cyanates, and other carbon-containing ionic compounds that most chemists classify as inorganic. Here is a compositional definition: all **organic compounds** contain carbon, nearly always bonded to itself and to hydrogen, and often to other elements as well.

The word *organic* has a biological connotation arising from a major misconception that stifled research into the chemistry of living systems for many decades. In the early 19th century, many prominent thinkers believed that an unobservable spiritual energy, a "vital force," existed *within* the compounds of living things, making them impossible to synthesize and fundamentally different from compounds of the mineral world. This idea of *vitalism* was challenged in 1828, when the young German chemist Friedrich Wöhler heated ammonium cyanate, a "mineral-world" compound, and produced urea, a "living-world" compound:

$$NH_4OCN \xrightarrow{\Delta} H_2N-\overset{\overset{\displaystyle O}{\|}}{C}-NH_2$$

Although Wöhler did not appreciate the significance of this reaction—he was more interested in the fact that two compounds can have the same molecular formula—his experiment is considered a key event in the origin of organic chemistry. Chemists soon synthesized methane, acetic acid, acetylene, and many other organic compounds from inorganic sources. ● Today, we know that *the same chemical principles govern organic and inorganic systems* because the behavior of a compound arises from the properties of its elements, no matter how marvelous that behavior may be.

In this brief introduction to an enormous and ever-growing field, we continue the text's central theme—how the structure and reactivity of organic molecules emerge naturally from the properties of the component atoms, in this case, carbon and its handful of bonding partners. First, we review the rather special atomic properties of carbon and see how they relate to organic molecules. We initially focus on hydrocarbons to give you a feel for writing and naming organic compounds. Then we classify the main types of organic reactions and, in the central portion of the chapter, apply them to the major families of organic compounds. Finally, we extend these ideas to the giant molecules of commerce and life—synthetic and natural polymers.

15.1 THE SPECIAL NATURE OF CARBON AND THE CHARACTERISTICS OF ORGANIC MOLECULES

Although there is nothing mystical about organic molecules, their indispensible role in biology and industry leads us to ask if carbon has some extraordinary attributes that give it a special chemical "personality." Of course, each element has its own specific properties, and carbon is no more unique than sodium, hafnium, or any other element. But the atomic properties of carbon do give it

urea

● **Organic Chemistry Is Enough to Drive One Mad** The vitalists did not change their beliefs overnight. Indeed, organic compounds *do* seem different from inorganic compounds because of their complex structures and compositions. Imagine the consternation of the 19th-century chemist who was accustomed to studying compounds with formulas such as $CaCO_3$, $Ba(NO_3)_2$, and $CuSO_4 \cdot 5H_2O$ and then isolated white crystals from a gallstone and found their empirical formula to be $C_{27}H_{46}O$ (cholesterol). Even Wöhler later said, "Organic chemistry . . . is enough to drive one mad. It gives me the impression of a primeval tropical forest, full of the most remarkable things, a monstrous and boundless thicket, with no way of escape, into which one may well dread to enter."

bonding capabilities beyond those of any other element, which in turn lead to the two obvious characteristics of organic molecules—structural complexity and chemical diversity.

The Structural Complexity of Organic Molecules

Most organic molecules have much more complex structures than most inorganic molecules, and a quick review of carbon's atomic properties and bonding behavior shows why:

1. *Electron configuration, electronegativity, and covalent bonding.* Carbon's ground-state electron configuration of [He] $2s^2 2p^2$—four electrons more than He and four fewer than Ne—means that the formation of carbon ions is energetically impossible under ordinary conditions: to form a C^{4+} cation requires energy equal to the sum of IE_1 through IE_4, and a C^{4-} anion requires the sum of EA_1 through EA_4, the last three steps being endothermic. Carbon's position in the periodic table (Figure 15.1) gives it an electronegativity midway between the most metallic and nonmetallic elements of Period 2: Li = 1.0, C = 2.5, F = 4.0. Therefore, *carbon shares electrons to attain a filled shell,* bonding covalently in all its elemental forms and compounds.

2. *Bond properties, catenation, and molecular shape.* The *number* and *strength* of carbon's bonds lead to its outstanding ability to *catenate* (bond to itself), which allows it to form a multitude of chemically and thermally stable chain, ring, and branched compounds. Through the process of orbital hybridization (Section 11.1), *carbon forms four bonds in virtually all its compounds,* and they point in as many as four different directions. Because it is small, *carbon forms relatively short, strong bonds,* which allows close approach to another atom and thus greater orbital overlap. The C—C bond is short enough to allow side-to-side overlap of half-filled, unhybridized *p* orbitals and the formation of *multiple bonds,* which restrict rotation of attached groups (see Figure 11.12, p. 403). These features add more possibilities for the shapes of carbon compounds.

3. *Molecular stability.* Although silicon and several other elements also catenate, none can compete with carbon. Atomic and bonding properties confer three crucial differences between C and Si chains that explain why C chains are so stable and, therefore, so common:

- Atomic size and bond strength. As atomic size increases down Group 4A(14), bonds between identical atoms become longer and weaker. Thus, a C—C bond (347 kJ/mol) is much stronger than an Si—Si bond (226 kJ/mol).
- Relative heats of reaction. A C—C bond (347 kJ/mol) and a C—O bond (358 kJ/mol) have nearly the same energy, so relatively little heat is released when a C chain reacts and one bond replaces the other. In contrast, an Si—O bond (368 kJ/mol) is much stronger than an Si—Si bond (226 kJ/mol), so a large quantity of heat is released when an Si chain reacts.
- Orbitals available for reaction. Unlike C, Si has low-energy *d* orbitals that can be attacked (occupied) by the lone pairs of incoming reactants. Thus, for example, ethane (CH_3—CH_3) is stable in water and does not react in air unless sparked, whereas disilane (SiH_3—SiH_3) breaks down in water and ignites spontaneously in air.

The Chemical Diversity of Organic Molecules

In addition to their elaborate geometries, organic compounds are noted for their sheer number and diverse chemical behavior. Look in any compendium of known compounds, such as the *CRC Handbook of Chemistry and Physics,* and you'll

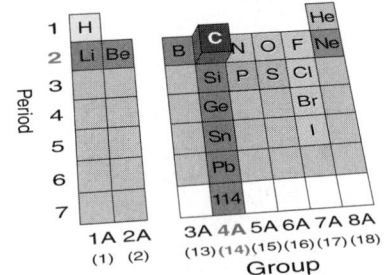

Figure 15.1 The position of carbon in the periodic table. Carbon lies at the center of Period 2, so it has an intermediate electronegativity (EN), and at the top of Group 4A(14), so it is relatively small. Other elements common in organic compounds are H, N, O, P, S, and the halogens.

find that the number of organic compounds dwarfs the number of inorganic compounds of all the other elements combined! Several million organic compounds are known, and thousands more are discovered or synthesized each year.

This incredible diversity is also founded on atomic and bonding behavior and is due to three interrelated factors:

1. *Bonding to heteroatoms.* Carbon frequently forms bonds to **heteroatoms,** atoms other than C or H. The most common heteroatoms are N and O, but S, P, and the halogens often occur, and organic compounds with other elements are known as well. To get an idea of the diversity created by a heteroatom, look at Figure 15.2, which shows that 23 different molecular structures are produced by various arrangements of four C atoms singly bonded to each other, the necessary number of H atoms, and just one O atom (either singly or doubly bonded).

2. *Electron density and reactivity.* Most reactions start—that is, a new bond begins to form—*when a region of high electron density on one molecule meets a region of low electron density on another.* These regions may be due to the presence of a multiple bond or to the partial charges that occur in carbon-heteroatom bonds. For example, consider four common bonds found in organic molecules:

- The C—C bond. When C is singly bonded to itself, as it is in portions of nearly every organic molecule, the EN values are equal and the bond is nonpolar. Therefore, in general, *C—C bonds are unreactive.*
- The C—H bond. This bond, which also occurs in nearly every organic molecule, is very nearly nonpolar because it is short (109 pm) and the EN values of H (2.1) and C (2.5) are close. Thus, *C—H bonds are largely unreactive* as well.
- The C—O bond. In contrast to the previous two bonds, this bond, which occurs in many types of organic molecules, is highly polar ($\Delta EN = 1.0$), with the O end of the bond electron rich ($\delta-$) and the C end electron poor ($\delta+$). As a result of this imbalance in electron density, *the C—O bond is reactive* and, given appropriate conditions, a reaction will occur there.
- Bonds to other heteroatoms. Even when a carbon-heteroatom bond has a small ΔEN, as for C—Br ($\Delta EN = 0.3$), or none at all, as for C—S ($\Delta EN = 0$), these heteroatoms are large, and so their bonds to carbon are long, weak, and thus reactive.

3. *Nature of functional groups.* One of the most important concepts in organic chemistry is that of the **functional group,** a specific combination of bonded atoms that reacts in a *characteristic* way, no matter what molecule it occurs in. In nearly every case, *the reaction of an organic compound takes place at the functional group.* In fact, as you'll see, we often substitute a general symbol for the rest of the molecule because it usually stays the same when the functional group undergoes a reaction. Functional groups vary from a carbon-carbon multiple bond to any combination of carbon-heteroatom bonds. In other words, any bonded grouping of atoms other than C—C and C—H constitutes a functional group, and each has its own pattern of reactivity. A particular bond may be a functional group or part of one or more functional groups. For example, the C—O bond occurs in four functional groups. We will discuss three of these groups in this chapter: the

alcohol group $-\overset{|}{\underset{|}{C}}-\overset{..}{\underset{..}{O}}-H$, the carboxylic acid group $-\overset{:O:}{\overset{\|}{C}}-\overset{..}{\underset{..}{O}}-H$, and the

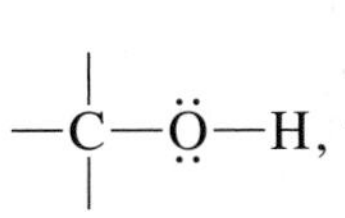

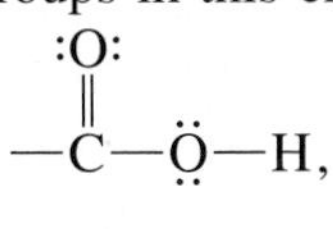

ester group $-\overset{:O:}{\overset{\|}{C}}-\overset{..}{\underset{..}{O}}-\overset{|}{\underset{|}{C}}-$.

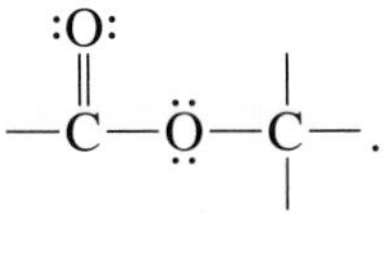

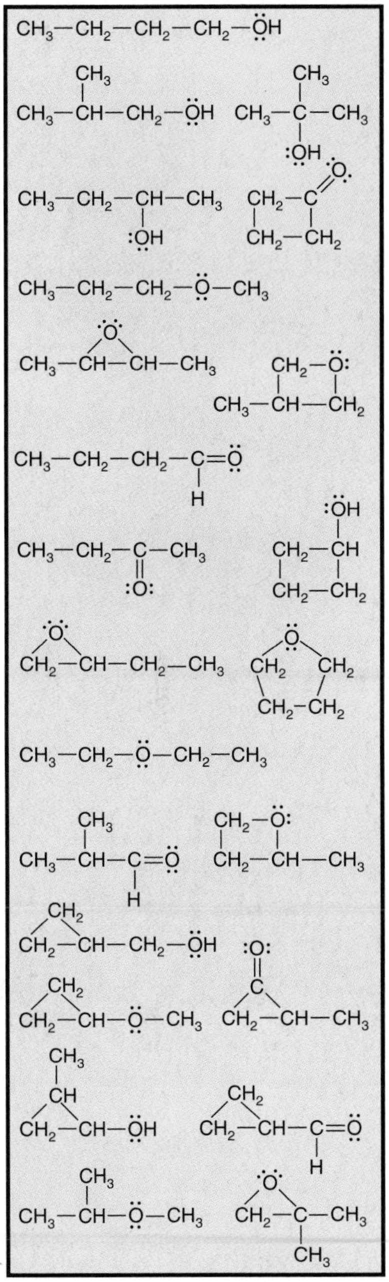

Figure 15.2 The chemical diversity of organic compounds. Different arrangements of chains, branches, rings, and heteroatoms give rise to many structures. There are 23 different compounds possible from just four C atoms joined by single bonds, one O atom, and the necessary H atoms.

SECTION SUMMARY

The structural complexity of organic compounds arises from carbon's small size, intermediate EN, four bonding orbitals, ability to form π bonds, and the absence of *d* orbitals in its valence shell. These factors lead to chains, branches, and rings of C atoms joined by strong, chemically resistant bonds that point in as many as four directions from each C. The chemical diversity of organic compounds arises from C's ability to bond to itself and to many other elements and to form multiple bonds to itself and to O and N. These factors lead to compounds that contain functional groups, specific portions of molecules that react in characteristic ways.

15.2 THE STRUCTURES AND CLASSES OF HYDROCARBONS

A fanciful, anatomical analogy can be made between an organic molecule and an animal. The carbon-carbon bonds form the skeleton, with the longest continual chain the backbone and any branches the limbs. Covering the skeleton is a skin of hydrogen atoms, with functional groups protruding at specific locations, like chemical fingers ready to grab an incoming reactant.

In this section, we "dissect" one group of compounds down to their skeletons and see how to name and draw them. **Hydrocarbons,** the simplest type of organic compound, are a large group of substances containing only H and C atoms. Some common fuels, such as natural gas and gasoline, are hydrocarbon mixtures. Hydrocarbons are also important *feedstocks,* precursor reactants used to make other compounds. Ethylene, acetylene, and benzene, for example, are feedstocks for hundreds of other substances.

Carbon Skeletons and Hydrogen Skins

Let's begin by examining the possible bonding arrangements of C atoms only (we'll leave off the H atoms at first) in simple skeletons without multiple bonds or rings. To distinguish different arrangements, focus on the *sequence* of C atoms (that is, the successive linkages of one to another) and keep in mind that *groups joined by single (sigma) bonds are relatively free to rotate* (Section 11.2).

Structures with one, two, or three carbons can be arranged in only one way. Whether you draw three C atoms in a line or with a bend, the *sequence* is the same. Four C atoms, however, have two possible arrangements—a four-C chain or a three-C chain with a one-C branch at the central C:

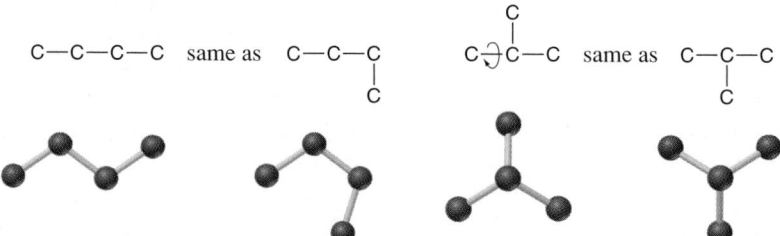

Even when we show the chain more realistically with the bends due to the tetrahedral shape around the C atoms, as in the ball-and-stick models, the situation is the same. Notice that if the branch is added to either end of the three-C chain, it is simply a bend in a four-C chain, *not* a different arrangement. Similarly, if the branch points down instead of up, it represents the same arrangement because groups joined by single bonds rotate.

As the total number of C atoms increases, the number of different arrangements increases as well. Five C atoms have 3 arrangements; 6 C atoms can be arranged in 5 ways, 7 C atoms in 9 ways, 10 C atoms in 75 ways, and 20 C atoms

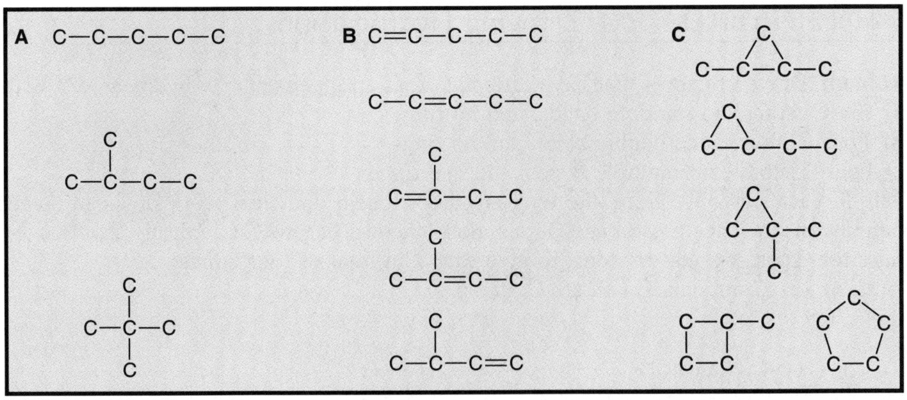

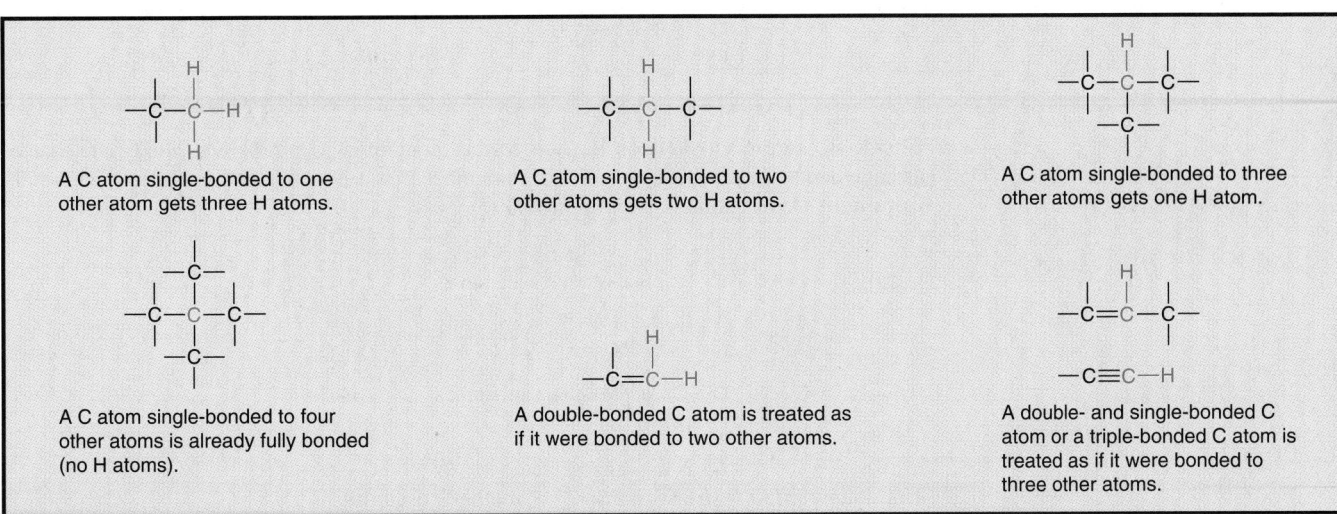

Figure 15.3 **Some five-carbon skeletons. A,** Three five-C skeletons are possible with only single bonds. **B,** Five more skeletons are possible with one C=C bond present. **C,** Five more skeletons are possible with one ring present. Even more would be possible with a ring *and* a double bond.

in more than 300,000 ways! If we include multiple bonds and rings, the number of arrangements increases further. For example, including one C=C bond in the five-C skeletons creates 5 more arrangements, and including one ring creates 5 more (Figure 15.3).

When determining the number of different skeletons, remember that

- Each C can form a *maximum* of four single bonds, or two single and one double bond, or one single and one triple bond.
- The *arrangement* of C atoms determines the skeleton, so a straight chain and a bent chain represent the same skeleton.
- Groups joined by single bonds can *rotate,* so a branch pointing down is the same as one pointing up. (Recall that a double bond restricts rotation.)

If we put a hydrogen "skin" on a carbon skeleton, we obtain a hydrocarbon. Figure 15.4 shows that the skeleton has the correct number of H atoms when each C has four bonds. Sample Problem 15.1 provides practice in drawing hydrocarbons.

A C atom single-bonded to one other atom gets three H atoms.

A C atom single-bonded to two other atoms gets two H atoms.

A C atom single-bonded to three other atoms gets one H atom.

A C atom single-bonded to four other atoms is already fully bonded (no H atoms).

A double-bonded C atom is treated as if it were bonded to two other atoms.

A double- and single-bonded C atom or a triple-bonded C atom is treated as if it were bonded to three other atoms.

Figure 15.4 **Adding the H-atom skin to the C-atom skeleton.** In a hydrocarbon molecule, each carbon atom bonds to as many hydrogen atoms as needed to give the carbon a total of four bonds.

SAMPLE PROBLEM 15.1 Drawing Hydrocarbons

Problem Draw structures that have different atom arrangements for hydrocarbons with
(a) Six C atoms, no multiple bonds, and no rings
(b) Four C atoms, one double bond, and no rings
(c) Four C atoms, no multiple bonds, and one ring
Plan In each case, we draw the longest carbon chain and then work down to smaller chains with branches at different points along them. The process typically involves trial and error. Then, we add H atoms to give each C a total of four bonds.
Solution **(a)** Compounds with six C atoms:

(b) Compounds with four C atoms and one double bond:

(c) Compounds with four C atoms and one ring:

Check Be sure each skeleton has the correct number of C atoms and bonds and that no arrangements are repeated or omitted; remember that a double bond counts as two bonds.
Comment Avoid some *common mistakes:*

In (a): C—C—C—C—C is the same skeleton as C—C—C—C—C
with C branch below second carbon   with C branch above fourth carbon

C—C—C—C is the same skeleton as C—C—C—C
with branches

In (b): C—C—C=C is the same skeleton as C=C—C—C
The double bond restricts rotation, so another possibility is

Too many bonds in

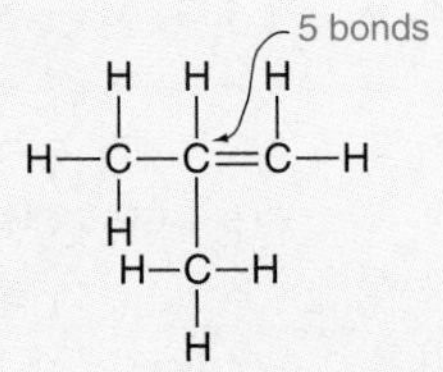

In (c): Too many bonds in

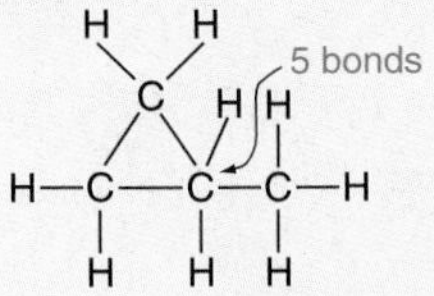

FOLLOW-UP PROBLEM 15.1 Draw all hydrocarbons that have different atom arrangements with
(a) Seven C atoms, no multiple bonds, and no rings (nine arrangements)
(b) Five C atoms, one triple bond, and no rings (three arrangements)

The enormous number of hydrocarbons can be classified into four main groups. For the remainder of this section, we examine some structural features and physical properties of each group. Later, we discuss the chemical behavior of the hydrocarbons.

Alkanes: Hydrocarbons with Only Single Bonds

A hydrocarbon that contains only single bonds is an **alkane** (general formula C_nH_{2n+2}, where n is a positive integer). For example, if $n = 5$, the formula is $C_5H_{[(2\times5)+2]}$, or C_5H_{12}. These compounds comprise a **homologous series**, one in which each member differs from the next by a $—CH_2—$ (methylene) group. In alkanes, each C is sp^3 hybridized. Since each C is bonded to the *maximum number of other atoms* (C or H), alkanes are referred to as **saturated hydrocarbons.**

Naming Alkanes You learned the names of the first ten unbranched alkanes in Chapter 2 (see Table 2.7). Here we discuss general rules for naming any alkane and, by extension, other organic compounds as well. The key point to remember is that *each chain, branch, or ring has a name based on the **number** of C atoms.* The entire name of the compound has three portions:

PREFIX + ROOT + SUFFIX

- *Root:* The root tells the number of C atoms in the longest *continuous* chain in the molecule. The roots are shown in Table 15.1. Recall that there are special roots for compounds of one to four C atoms; roots of longer chains are based on Greek numbers.
- *Prefix:* Each prefix identifies a *group attached to the main chain* and the number of the carbon to which it is attached. Prefixes identifying hydrocarbon branches are the same as root names (Table 15.1) but take a *-yl* ending. Each prefix is placed *before* the root.
- *Suffix:* The suffix tells the *type of organic compound* the molecule represents, that is, it identifies the functional group the molecule possesses. The suffix is placed *after* the root.

For example, in the name 2-methylbutane, *2-methyl-* is the prefix (a methyl group is attached to C-2 of the main chain), *-but-* is the root (the main chain has four C atoms), and *-ane* is the suffix (the compound is an alkane).

The alkanes are used here to demonstrate this general approach to naming; other organic compounds are named in the same way, with a variety of other prefixes and suffixes (see Table 15.5, p. 629). In addition to the *systematic* names

Table 15.1 Numerical Roots for Carbon Chains and Branches

Roots	Number of C Atoms
meth-	1
eth-	2
prop-	3
but-	4
pent-	5
hex-	6
hept-	7
oct-	8
non-	9
dec-	10

Table 15.2 Rules for Naming an Organic Compound

1. Naming the longest chain (root)
   (a) Find the longest *continuous* chain of C atoms.
   (b) Select the root that corresponds to the number of C atoms in this chain.

$$CH_3-\underset{\underset{CH_2-CH_3}{|}}{CH}-CH-CH_2-CH_2-CH_3$$

6 carbons $\Longrightarrow$ hex-

2. Naming the compound type (suffix)
   (a) For alkanes, add the suffix *-ane* to the chain root. (Other suffixes appear in Table 15.5 with their functional group and compound type.)
   (b) If the chain forms a ring, the name is preceded by *cyclo-*.

hex- + -ane $\Longrightarrow$ hexane

3. Naming the branches (prefix)
   (a) Each branch name consists of a subroot (number of C atoms) and the ending *-yl* to signify that it is not part of the main chain.
   (b) Branch names precede the chain name. When two or more branches are present, name them in *alphabetical* order.
   (c) To specify where the branch occurs along the chain, number the main-chain C atoms consecutively, starting at the end *closer* to a branch, to achieve the *lowest* numbers for the branches. Precede each branch name with the number of the chain C to which that branch is attached.
   (d) If the compound has no branches, the name consists of the root and suffix.

CH_3 methyl

$$CH_3-\underset{\underset{CH_2-CH_3}{|} \text{ ethyl}}{CH}-CH-CH_2-CH_2-CH_3$$

ethylmethylhexane

$$\underset{1}{CH_3}-\underset{2}{\underset{|}{CH}}-\underset{3}{CH}-\underset{4}{CH_2}-\underset{5}{CH_2}-\underset{6}{CH_3}$$

3-ethyl-2-methylhexane

introduced here, many compounds are still known by *common* names; we'll note the most important of those. To obtain the systematic name of a compound,

1. Name the longest chain (root).
2. Add the compound type (suffix).
3. Name any branches (prefix).

Table 15.2 presents the rules for naming an organic compound and applies them to an alkane component of gasoline.

Depicting Alkanes with Formulas and Models Nowhere is it more important— or useful—to visualize molecules than in discussions of organic compounds, so chemists use several ways of depicting these complex structures. On the two-dimensional surface of a page or blackboard, expanded and condensed structural formulas are the easiest to draw. With molecular model kits or computer software, three-dimensional models make the actual molecular shapes vivid.

The *expanded formula* shows each atom and bond; this type appears in Sample Problem 15.1. A common type of *condensed formula* groups the H atoms with the C atom to which they are bound. Figure 15.5 shows the expanded and con-

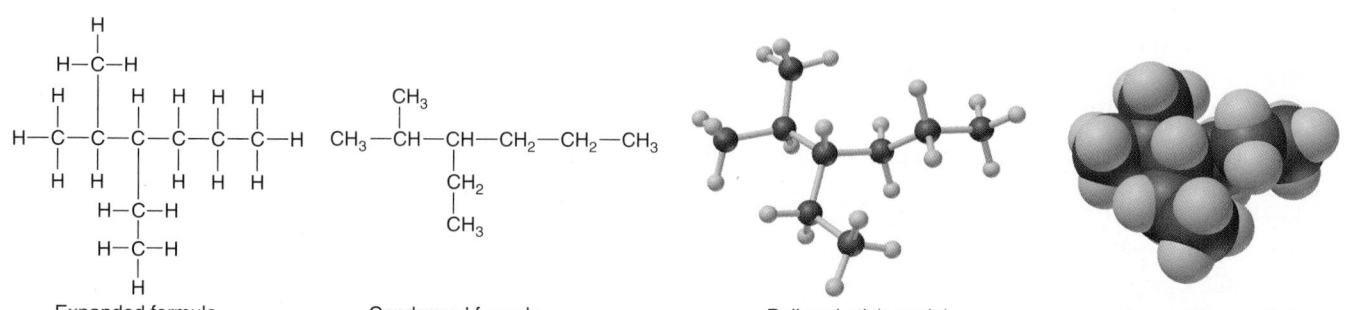

Expanded formula   Condensed formula   Ball-and-stick model   Space-filling model

Figure 15.5 Ways of depicting formulas and models of an alkane.

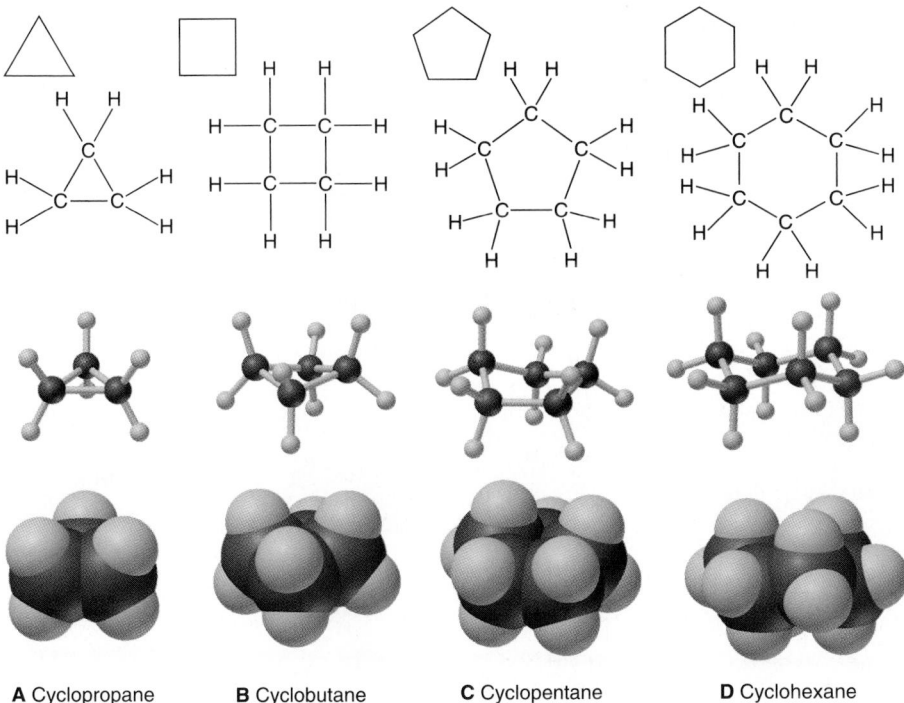

A Cyclopropane   **B** Cyclobutane   **C** Cyclopentane   **D** Cyclohexane

Figure 15.6 Depicting cycloalkanes. Cycloalkanes are usually drawn as regular polygons. Each side is a C—C bond, and each corner represents a C atom with its required number of H atoms. The expanded formulas show each bond in the molecule. The ball-and-stick and space-filling models show that, except for cyclopropane, the rings are not planar. These arrangements minimize electron repulsions between adjacent H atoms. Cyclohexane **(D)** is shown in its more stable chair conformation.

densed formulas, together with the ball-and-stick and space-filling models, of the compound named in Table 15.2.

Cyclic Hydrocarbons A **cyclic hydrocarbon** contains one or more rings in its structure. When a straight-chain alkane (C_nH_{2n+2}) forms a ring, two H atoms are lost as the C—C bond forms to join the two ends of the chain. Thus, *cycloalkanes* have the general formula C_nH_{2n}. Cyclic hydrocarbons are often drawn using a shorthand method that shows *only* C—C bond lines. Each corner of the ring is understood to be *a C atom with the correct number of H atoms* attached to give a total of four bonds; expanded and shorthand structures for some cycloalkanes, together with the three-dimensional models, are shown in Figure 15.6. Note that, except for three-C rings, *cycloalkanes are nonplanar*. This structural feature arises from the tetrahedral shape around each C atom and the need to minimize electron repulsions between adjacent H atoms. As a result, orbital overlap of adjacent C atoms is maximized. The most stable form of cyclohexane (Figure 15.6D) is called the *chair conformation*.

Constitutional Isomerism and the Physical Properties of Alkanes

Two or more compounds with the same molecular formula but different properties are called **isomers.** Those with *different arrangements of bonded atoms* are **constitutional (or structural) isomers;** alkanes with a given number of C atoms but different skeletons are examples. The smallest alkane to exhibit constitutional isomerism has four C atoms: two different compounds have the formula C_4H_{10}

Table 15.3 The Constitutional Isomers of C_4H_{10} and C_5H_{12}

Systematic Name (Common Name)	Condensed Formula	Expanded Formula	Space-filling Model	Density (g/mL)	Boiling Point (°C)
Butane (*n*-butane)	$CH_3-CH_2-CH_2-CH_3$	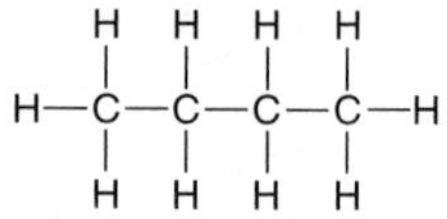	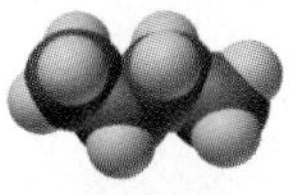	0.579	−0.5

2-Methylpropane (isobutane)	$CH_3-CH-CH_3$
	        $|$
	       CH_3
				0.549	−11.6

Pentane (*n*-pentane)	$CH_3-CH_2-CH_2-CH_2-CH_3$	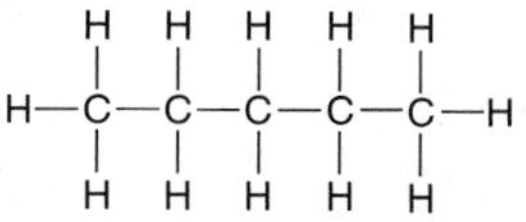		0.626	36.1

2-Methylbutane (isopentane)	$CH_3-CH-CH_2-CH_3$
	        $|$
	       CH_3
				0.620	27.8

2,2-Dimethylpropane (neopentane)	CH_3
	       $|$
	 CH_3-C-CH_3
	       $|$
	      CH_3
				0.614	9.5

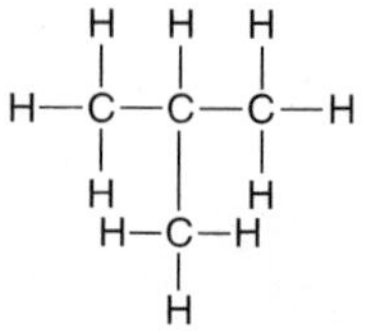

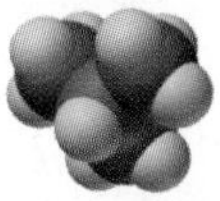

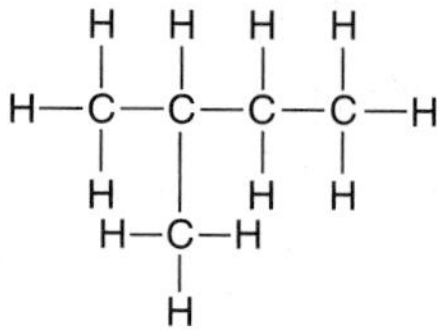

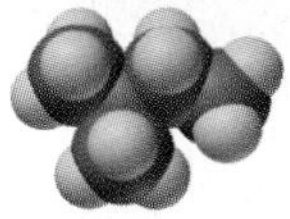

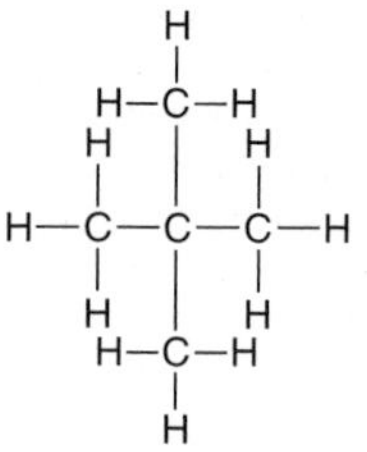

(Table 15.3). The unbranched one is butane (common name, *n*-butane; *n*- stands for "normal," or straight chain), and the other is 2-methylpropane (common name, *iso*butane). Similarly, three compounds have the formula C_5H_{12}. The unbranched isomer is pentane (common name, *n*-pentane); the one with a methyl group at C-2 of a four-C chain is 2-methylbutane (common name, *iso*pentane). The third isomer has two methyl branches on C-2 of a three-C chain, so its name is 2,2-dimethylpropane (common name, *neo*pentane).

Because alkanes are nearly nonpolar, we expect their physical properties to be determined by dispersion forces, and the boiling points in Table 15.3 certainly bear this out. The four-C alkanes boil lower than the five-C compounds. Moreover, within each group of isomers, the more spherical member (isobutane or neopentane) boils lower than the more elongated one (*n*-butane or *n*-pentane). As you saw in Chapter 12, this trend occurs because a spherical shape leads to less intermolecular contact, and thus lower total dispersion forces, than does an elongated shape.

A particularly clear example of the effect of dispersion forces on physical properties occurs among the unbranched alkanes (*n*-alkanes). Among these compounds, boiling points increase steadily with chain length: the longer the chain, the greater the contact, the stronger the dispersion forces, and the higher the boiling point (Figure 15.7). Pentane (five C atoms) is the smallest *n*-alkane that exists as a liquid at room temperature. The solubility of alkanes, and of all hydrocarbons, is easy to predict from the like-dissolves-like rule (Section 13.1). Alkanes are miscible in each other and in other nonpolar solvents, such as benzene, but

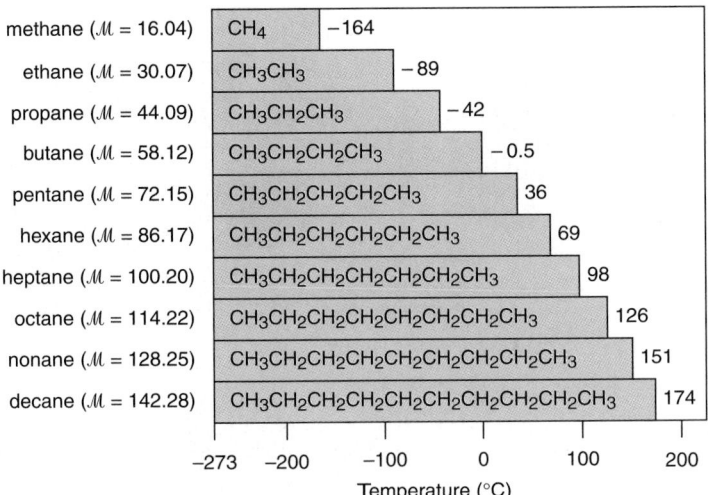

are nearly insoluble in water. The solubility of pentane in water, for example, is only 0.36 g/L at room temperature.

Chiral Molecules and Optical Isomerism

Another type of isomerism in some alkanes and many other organic (as well as some inorganic) compounds is called *stereoisomerism.* **Stereoisomers** are molecules with the same arrangement of atoms *but different orientations of groups in space.* *Optical isomerism* is one type of stereoisomerism: *when two objects are mirror images of each other and cannot be superimposed, they are* **optical isomers,** also called *enantiomers.* To use a familiar example, your right hand is an optical isomer of your left. Look at your right hand in a mirror, and you will see that the *image* is identical to your left hand (Figure 15.8). No matter how you twist your arms around, however, your hands cannot lie on top of each other with all parts superimposed. They are not superimposable because each is *asymmetric:* there is no plane of symmetry that divides your hand into two identical parts.

An asymmetric molecule is called **chiral** (Greek *cheir,* "hand"). Typically, an organic molecule *is chiral if it contains a carbon atom that is bonded to four **different** groups.* This C atom is called a *chiral center* or an asymmetric carbon. In 3-methylhexane, for example, C-3 is a chiral center because it is bonded to four different groups: H—, CH_3—, CH_3—CH_2—, and CH_3—CH_2—CH_2— (Figure 15.9A). Like your two hands, the two forms are mirror images and cannot be superimposed on each other: when two of the groups are superimposed, the other two are opposite each other. Thus, the two forms are optical isomers. The central C atom in the amino acid alanine is also a chiral center (Figure 15.9B).

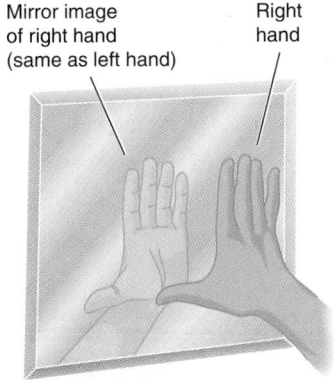

Mirror image of right hand (same as left hand)   Right hand

Figure 15.8 An analogy for optical isomers. The reflection of your right hand looks like your left hand. Each hand is asymmetric, so you cannot superimpose them with your palms facing in the same direction.

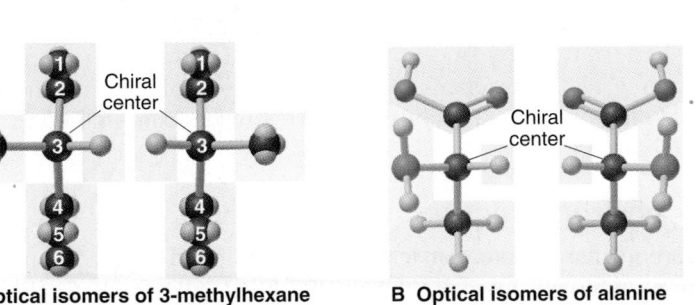

A Optical isomers of 3-methylhexane   **B Optical isomers of alanine**

Figure 15.9 Two chiral molecules. A, 3-Methylhexane is chiral because C-3 is bonded to four different groups. These two models are optical isomers (enantiomers). **B,** The central C in the amino acid alanine is also bonded to four different groups.

Figure 15.10 The rotation of plane-polarized light by an optically active substance. The source emits light oscillating in all planes. When the light passes through the first polarizing filter, light oscillating in only one plane emerges. The plane-polarized beam enters the sample compartment, which contains a known concentration of an optical isomer, and the plane rotates as it passes through the solution. For the experimenter to see the light, it must pass through a second polarizing filter (called the *analyzer*), which is attached to a movable ring that is calibrated in degrees and measures the angle of rotation.

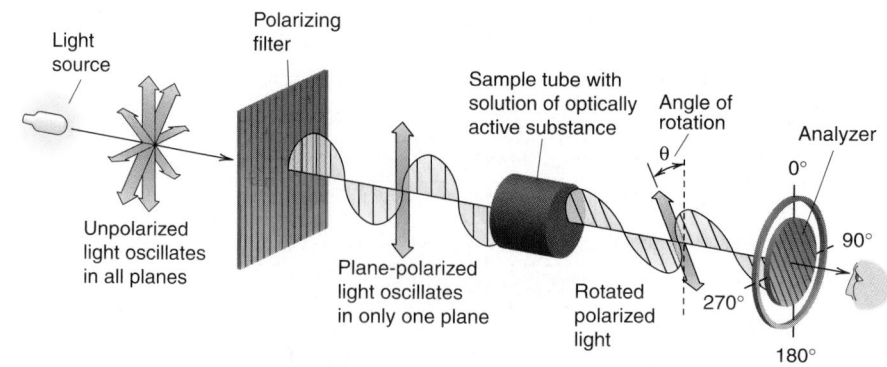

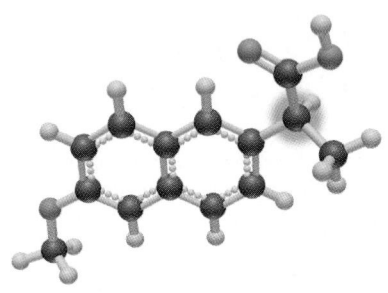

Chiral Medicines Many drugs are chiral molecules of which one optical isomer is biologically active and the other has either a different type of activity or none at all. Naproxen, the pain reliever and anti-inflammatory agent, is an example; one isomer (see the model) is active as an antiarthritic agent, and the other is a potent liver toxin that must be removed from the mixture during synthesis. The notorious drug thalidomide is another example. One optical isomer is active against depression, whereas the other causes fetal mutations and deaths. Tragically, the drug was sold in the 1950s as the racemic mixture and caused limb malformations in many children whose mothers had it prescribed to relieve "morning sickness" during pregnancy.

Unlike constitutional isomers, optical isomers are identical in all but two respects:

1. In their physical properties, *optical isomers differ only in the direction that each isomer rotates the plane of polarized light.* A **polarimeter** is used to measure the angle that the plane is rotated (Figure 15.10). A beam of light consists of waves that oscillate in all planes. A polarizing filter blocks all waves except those in one plane, so the light emerging through the filter is *plane-polarized.* An optical isomer is **optically active** because it rotates the plane of this polarized light. (Liquid-crystal displays incorporate polarizing filters and optically active compounds; see Figure 12.45, p. 465.) The *dextrorotatory* isomer (designated *d* or +) rotates the plane of light clockwise; the *levorotatory* isomer (designated *l* or −) is the mirror image of the first and rotates the plane counterclockwise. An equimolar mixture of the two isomers (called a *racemic mixture*) does not rotate the plane at all because the dextrorotation cancels the levorotation. The *specific rotation* is a characteristic, measurable property of the isomer at a certain temperature, concentration, and wavelength of light.

2. In their chemical properties, *optical isomers differ only in a chiral (asymmetric) chemical environment,* one that distinguishes "right-handed" from "left-handed" molecules. As an analogy, your right hand fits well in your right glove but not in your left glove. Typically, one isomer of an optically active reactant is added to a mixture of optical isomers of another compound. The products of the reaction have different properties and can be separated. For example, when the anion *d*-lactate is added to a mixture of *d*-alanine and *l*-alanine in their cationic forms, the *d* isomer of alanine crystallizes as the *d*-lactate/*d*-alanine salt, but the *d*-lactate/*l*-alanine salt is more soluble and remains in solution.

Optical isomerism plays a vital role in living cells. Nearly all carbohydrates and amino acids are optically active, but only one of the isomers is biologically usable. For example, *d*-glucose is metabolized for energy, but *l*-glucose is excreted unused. Similarly, *l*-alanine is incorporated naturally into proteins, but *d*-alanine is not. The organism distinguishes one optical isomer from the other through its enzymes, large molecules that speed virtually every reaction in the cell by binding to the reactants. A specific portion of the enzyme provides the asymmetric environment, so only one of the isomers can bind there and undergo reaction (Figure 15.11).

Alkenes: Hydrocarbons with Double Bonds

A hydrocarbon that contains at least one C=C bond is called an **alkene.** With two H atoms removed to make the double bond, alkenes have the general formula, C_nH_{2n}. The double-bonded C atoms are sp^2 hybridized. Because their carbon atoms bond fewer than the maximum of four atoms each, alkenes are considered **unsaturated hydrocarbons.**

Alkene names differ from those of alkanes in two respects:

1. The main chain (root) *must* contain both C atoms of the double bond, even if it is not the longest chain. The chain is numbered from the end *closer* to the C=C bond, and the position of the bond is indicated by the number of the *first* C atom in it.
2. The suffix for alkenes is *-ene*.

For example, there are three four-C alkenes (C_4H_8), two unbranched and one branched (Sample Problem 15.1, part b). The branched isomer is 2-methylpropene; the unbranched isomer with the C=C bond between C-1 and C-2 is 1-butene; the unbranched isomer with the C=C bond between C-2 and C-3 is 2-butene. As you'll see next, there are two isomers of 2-butene, but they are of a different sort than we have discussed so far.

The C=C Bond and Geometric (*cis-trans*) Isomerism There are two major structural differences between alkenes and alkanes. First, alkanes have a *tetrahedral* geometry (~109.5°) around each C atom, whereas the double-bonded C atoms in alkenes are *trigonal planar* (~120°). Second, the C—C bond *allows* rotation of bonded groups, so the atoms in an alkane continually change their relative positions. In contrast, the π bond of the C=C bond *restricts* rotation, which fixes the relative positions of the atoms bonded to it.

This rotational restriction leads to another type of stereoisomerism. **Geometric isomers** (also called ***cis-trans* isomers**) have different orientations of groups around a double bond (or similar structural feature). Table 15.4 shows the two geometric isomers of 2-butene (also see Comment, Sample Problem 15.1). One isomer, *cis*-2-butene, has the CH_3 groups on the *same* side of the C=C bond, whereas the other isomer, *trans*-2-butene, has them on *opposite* sides of the C=C bond. In general, a *cis* isomer has the *larger portions of the main chain on the same side* of the double bond, and a *trans* isomer has them on opposite sides. (This type of geometric isomerism is also called *cis-trans isomerism*. It also occurs in many transition metal compounds, which have a different structural feature that restricts rotation, as we discuss in Chapter 23.) For a molecule to have geometric isomers, *each C atom in the C=C bond must be bonded to two different groups*. Like all isomers, geometric isomers have different properties. Note in Table 15.4 that the two 2-butenes differ in molecular shape *and* physical properties. The *cis* isomer has a bend in the chain that the *trans* isomer lacks. In Chapters 10 and 12, you saw how such a difference affects molecular polarity and physical properties, which arise from differing strengths of intermolecular attractions. The upcoming Chemical Connections essay shows how this simple difference in geometry has profound effects in biological systems as well.

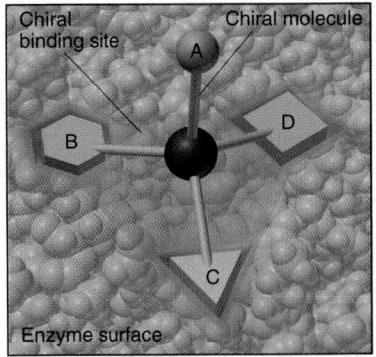

Figure 15.11 The binding site of an enzyme. Organisms can utilize only one of a pair of optical isomers because their enzymes have binding sites with shapes that are chiral (asymmetric). The shape of the mirror image of this molecule does not allow it to bind at this site.

Table 15.4 The Geometric Isomers of 2-Butene

Systematic Name	Condensed Formula	Space-filling Model	Density (g/mL)	Boiling Point (°C)
cis-2-Butene	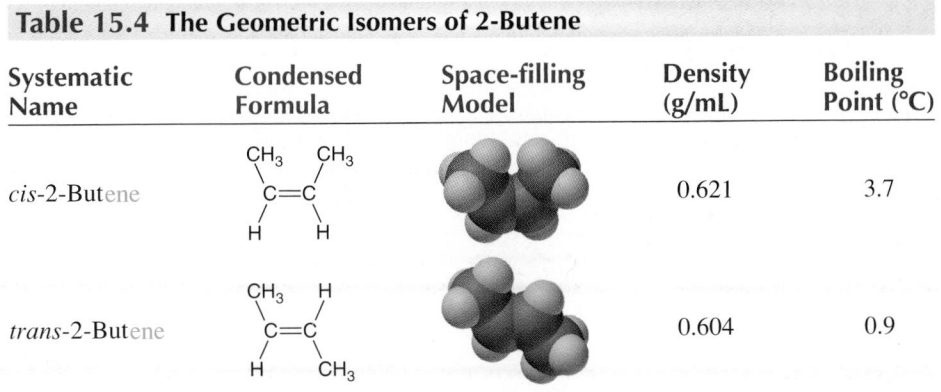		0.621	3.7
trans-2-Butene			0.604	0.9

Chemical Connections   Chemistry in Sensory Physiology
Geometric Isomers and the Chemistry of Vision

Of all our senses, sight provides the most information about the external world. Light bouncing off objects enters the lens of the eye and is focused on the retina. From there, molecular signals are converted to mechanical signals and then to electrical signals that are transmitted to the brain. This remarkable sequence is one of the few physiological processes that we understand thoroughly at the molecular level. The first step relies on the different shapes of geometric isomers.

The molecule responsible for receiving the light energy is *retinal*. It is derived from retinol (vitamin A), which we obtain mostly from β-carotene in yellow and green vegetables. Retinal is a 20-C compound consisting of a 15-C chain and five 1-C branches. As you can see from the ball-and-stick structure in Figure B15.1, the chain includes five C=C bonds, a six-C ring at one end, and a C=O bond at the other. There are two biologically occurring isomers of retinal. The all-*trans* isomer, shown on the right, has a *trans* orientation around all five double bonds. The 11-*cis* isomer, shown on the left, has a *cis* orientation around the C=C bond between C-11 and C-12. Note the significant difference in shape between the two isomers.

Certain cells of the retina are densely packed with *rhodopsin,* which consists of the relatively small 11-*cis*-retinal covalently bonded to a large protein. The initial chemical event in vision occurs when rhodopsin absorbs a photon of visible light.

The energy of visible photons (between 165 and 293 kJ/mol) lies in the range needed to break a C=C π bond (~250 kJ/mol). Retinal is bonded to the protein in such a way that the 11-*cis* π bond is the most susceptible to breakage. The photon is absorbed in a few trillionths of a second, the *cis* π bond breaks, the groups attached to the intact σ bond rotate, and a π bond re-forms to produce all-*trans*-retinal in another few millionths of a second. In effect, light energy is converted into the mechanical energy of the moving molecular chain.

This rapid and relatively large change in the shape of retinal causes the protein portion of rhodopsin to change shape as well, a process that breaks the bond to retinal. The change in protein shape triggers a flow of ions into the retina cells, initiating electrical impulses to the optic nerve, which leads to the brain. Meanwhile, the free all-*trans*-retinal diffuses away and is changed back to the *cis* form, which then binds to the protein portion again.

Retinal must be ideally suited to its function because it has been selected through evolution as a photon absorber in organisms as different as purple bacteria, mollusks, insects, and vertebrates. The features that make it the perfect choice are its strong absorption in the visible region, the efficiency with which light converts the *cis* to the *trans* form, and the large structural change it undergoes when this takes place.

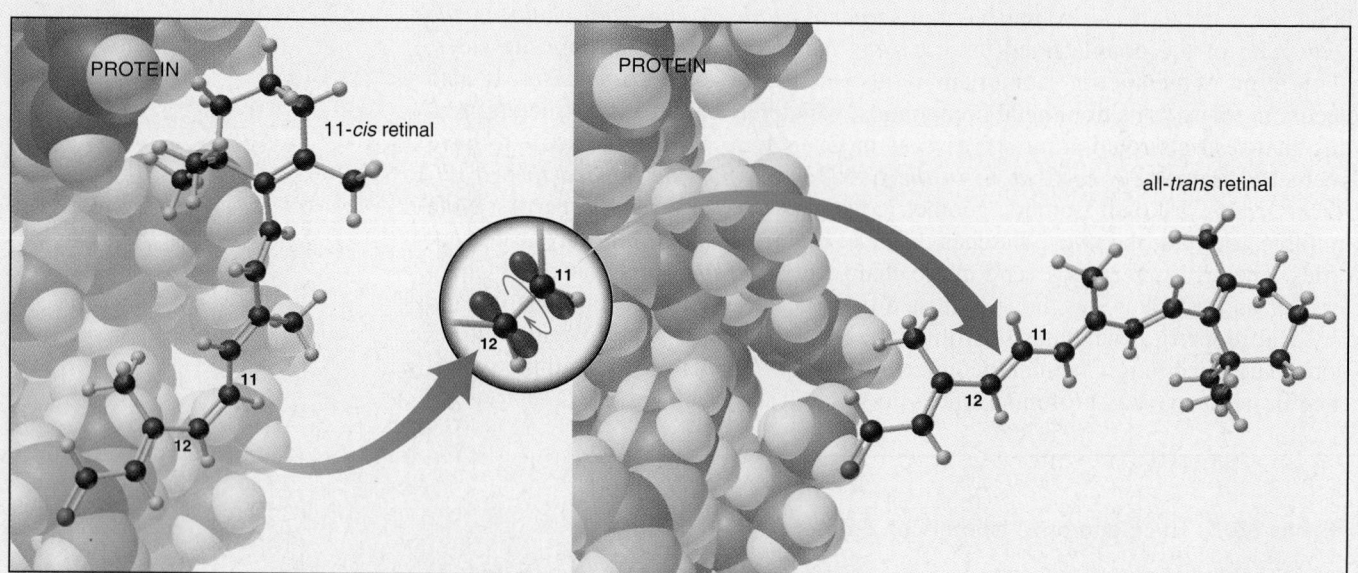

Figure B15.1 The initial chemical event in vision. In rhodopsin, the geometric isomer 11-*cis*-retinal is bonded to (*shown here as lying on*) a large protein. A photon absorbed by the retinal breaks the 11-*cis* π bond. The two parts of the retinal chain rotate around the σ bond (*shown in circle*), and the π bond forms again but with a *trans* orientation. This change in the shape of retinal changes the shape of the protein, causing it to release the all-*trans*-retinal.

Alkynes: Hydrocarbons with Triple Bonds

Hydrocarbons that contain at least one C≡C bond are called **alkynes.** Their general formula is C_nH_{2n-2} because they have two H atoms fewer than alkenes of the same length. Because a carbon in a C≡C bond can bond to only one other atom, the geometry around each C atom is linear (180°): each C is *sp* hybridized. Alkynes are named in the same way as alkenes, except that the suffix is *-yne*. Because of their localized π electrons, C=C and C≡C bonds are electron rich and act as functional groups. Thus, alkenes and alkynes are much more reactive than alkanes, as we'll discuss in Section 15.4.

SAMPLE PROBLEM 15.2 Naming Alkanes, Alkenes, and Alkynes

Problem Give the systematic name for each of the following, indicate the chiral center in part **(d)**, and draw two geometric isomers for part **(e)**:

(a)

$$CH_3\!-\!\underset{\underset{CH_3}{|}}{\overset{\overset{CH_3}{|}}{C}}\!-\!CH_2\!-\!CH_3$$

(b)

$$CH_3\!-\!CH_2\!-\!\underset{\underset{\underset{CH_3}{|}}{CH_2}}{\overset{\overset{CH_3}{|}}{CH}}\!-\!CH\!-\!CH_3$$

(c)

[cyclopentane ring with CH₃ and CH₂—CH₃ substituents]

(d)

$$CH_3\!-\!CH_2\!-\!\underset{\underset{CH_3}{|}}{CH}\!-\!CH\!=\!CH_2$$

(e)

$$CH_3\!-\!CH_2\!-\!CH\!=\!\underset{\underset{CH_3}{|}}{\overset{\overset{CH_3}{|}}{C}}\!-\!CH\!-\!CH_3$$

Plan For **(a)** to **(c)**, we refer to Table 15.2. We first name the longest chain (*root-* + *-ane*). Then we find the *lowest* branch numbers by counting C atoms from the end *closer* to a branch. Finally, we name each branch (*root-* + *-yl*) and put them alphabetically before the chain name. For **(d)** and **(e)**, the longest chain that *includes* the multiple bond is numbered from the end closer to it. For **(d)**, the chiral center is the C atom bonded to four different groups. In **(e)**, the *cis* isomer has larger groups on the same side of the double bond, and the *trans* isomer has them on opposite sides.

Solution

(a)

[structure with methyl, butane, methyl, methyl labels]

2,2-dimethylbutane

When a type of branch appears more than once, we group the chain numbers and indicate the number of branches with a numerical prefix, such as 2,2-*di*methyl.

(b)

[structure labeled hexane, methyl]

3,4-dimethylhexane

In this case, we can number the chain from either end because the branches are the same and are attached to the two central C atoms.

(c)

[cyclopentane structure with methyl and ethyl labels]

1-ethyl-2-methylcyclopentane

We number the ring C atoms so that a branch is attached to C-1.

(d)

[structure labeled methyl, 1-pentene, chiral center]

3-methyl-1-pentene

(e)

[cis structure labeled methyl, 3-hexene, methyl]

cis-2,3-dimethyl-3-hexene

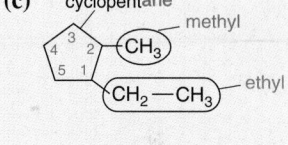

trans-2,3-dimethyl-3-hexene

Check A good check (and excellent practice) is to reverse the process by drawing structures for the names to see if you come up with the structures given in the problem.
Comment In part **(b)**, C-3 and C-4 are also chiral centers, as are C-1 and C-2 in part **(c)**. However, in **(b)** the molecule is not chiral: it has a plane of symmetry between C-3 and C-4, so one half of the molecule rotates light opposite to the other half. Avoid these common mistakes: In **(b)**, 2-ethyl-3-methylpentane is wrong: the longest chain is *hexane*. In **(c)**, 1-methyl-2-ethylcyclopentane is wrong: the branches are named *alphabetically*.

FOLLOW-UP PROBLEM 15.2 Draw condensed formulas for the compounds with the following names: **(a)** 3-ethyl-3-methyloctane; **(b)** 1-ethyl-3-propylcyclohexane; **(c)** 3,3-diethyl-l-hexyne; **(d)** *trans*-3-methyl-3-heptene.

Aromatic Hydrocarbons: Cyclic Molecules with Delocalized π Electrons

Unlike the cycloalkanes, **aromatic hydrocarbons** are planar molecules, usually with one or more rings of six C atoms, and are often drawn with alternating single and double bonds. As you learned for benzene (Section 10.1), however, all the ring bonds are identical, with values of length and strength *between* those of a C—C and a C=C bond. To indicate this, benzene is also shown as a resonance hybrid, with a circle (or dashed circle) representing the delocalized character of the π electrons (Figure 15.12A). An orbital picture shows the two lobes of the delocalized π cloud above and below the hexagonal plane of the σ-bonded C atoms (Figure 15.12B).

Figure 15.12 Representations of benzene. A, Benzene is often drawn with two resonance forms, showing alternating single and double bonds in different positions. The molecule is more correctly depicted as the resonance hybrid, with the delocalized electrons shown as an unbroken or dashed circle. **B,** The delocalized π cloud and the σ-bond plane of the benzene ring.

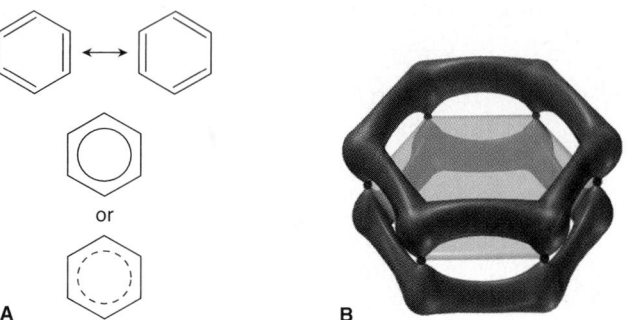

The systematic naming of simple aromatic compounds is quite straightforward. Usually, benzene is the parent compound, and attached groups, or *substituents,* are named as prefixes. However, many common names are still in use. For example, benzene with one methyl group is systematically named *methylbenzene* but is better known by its common name *toluene*. With only one substituent present in toluene, we do not number the ring C atoms; when two or more groups are attached, however, we number in such a way that one of the groups is attached to ring C-1. Thus, toluene and the three structural isomers with two methyl groups attached are

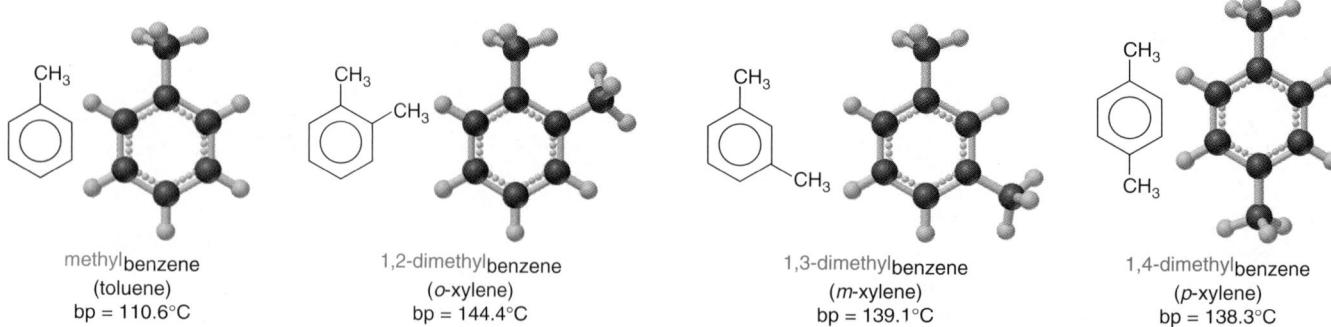

methyl benzene
(toluene)
bp = 110.6°C

1,2-dimethyl benzene
(*o*-xylene)
bp = 144.4°C

1,3-dimethyl benzene
(*m*-xylene)
bp = 139.1°C

1,4-dimethyl benzene
(*p*-xylene)
bp = 138.3°C

In common names, the positions of two groups are indicated by *o-* (ortho) for groups on adjacent ring C atoms, *m-* (meta) for groups separated by one ring C atom, and *p-* (para) for groups on opposite ring C atoms. The dimethylbenzenes (commonly known as *xylenes*) are important solvents and feedstocks for polyester fibers and dyes. (See margin note on opposite page.) ●

The number of isomers increases with more than two attached groups. For example, there are six isomers for a compound with one methyl and three nitro

(—NO₂) groups attached to a benzene ring; the explosive TNT is only one,

2,4,6-trinitromethylbenzene
(trinitrotoluene, TNT)

One of the most important methods for determining the structures of organic molecules is discussed in the upcoming Tools of the Laboratory essay.

Variations on a Theme: Catenated Inorganic Hydrides

In short discussions called *Variations on a Theme* found throughout this chapter, we examine similarities between organic and inorganic compounds. From this perspective, you'll see that the behavior of carbon is remarkable, but not unique, in the chemistry of the elements.

Although no element approaches carbon in the variety and complexity of its hydrides, catenation occurs frequently in the periodic table, and many ring, chain, and cage structures are known. Some of the most fascinating belong to the boron hydrides, or boranes (Section 14.5). Although their shapes rival even those of the hydrocarbons, the weakness of their unusual bridge bonds renders most of them thermally and chemically unstable.

An obvious structural similarity exists between alkanes and the silicon hydrides, or silanes (Section 14.6). Silanes even have an analogous general formula (Si_nH_{2n+2}). Branched silanes are also known, but no cyclic or unsaturated (Si=Si) compounds had been prepared until very recently. Unlike alkanes, silanes are unstable thermally and ignite spontaneously in air.

Sulfur's ability to catenate is second only to carbon's, and many chains and rings occur among its allotropes (Section 14.8). A large series of sulfur hydrides, or polysulfanes, is known. However, these molecules are unbranched chains with H atoms at the ends only (H—S_n—H). Like the silanes, the polysulfanes are oxidized easily and decompose readily to sulfur's only stable hydride, H_2S, and its most stable allotrope, cyclo-S_8.

SECTION SUMMARY

Hydrocarbons contain only C and H atoms, so their physical properties depend on the strength of their dispersion forces. The names of the organic compounds have a root for the longest chain, a prefix for any attached group, and a suffix for the type of compound. Alkanes (C_nH_{2n+2}) have only single bonds. Cycloalkanes (C_nH_{2n}) have ring structures that are typically nonplanar. Alkenes (C_nH_{2n}) have at least one C=C bond. Alkynes (C_nH_{2n-2}) have at least one C≡C bond. Aromatic hydrocarbons have at least one planar ring with delocalized π electrons.

Isomers are compounds with the same molecular formula but different properties. Structural isomers have different atom arrangements. Stereoisomers (optical and geometric) have the same arrangement of atoms, but their atoms are oriented differently in space. Optical isomers cannot be superimposed on each other because they are asymmetric, with four different groups bonded to the C that is the chiral center. They have identical physical and chemical properties except in their rotation of plane-polarized light and their reaction with chiral reactants. Geometric (*cis-trans*) isomers have groups oriented differently around a C=C bond, which restricts rotation. Light converts a *cis* isomer of retinal to the all-*trans* form, which initiates the visual response. ¹H-NMR spectroscopy indicates the relative numbers of H atoms in the various environments within an organic molecule.

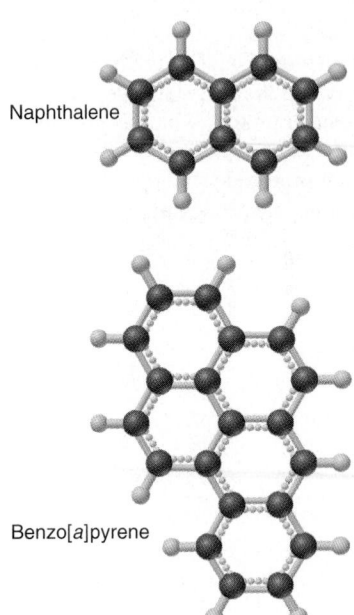

Naphthalene

Benzo[a]pyrene

Aromatic Carcinogens *Polycyclic aromatic compounds* have "fused" ring systems, in which two or more rings share one or more sides. The simplest is naphthalene, a feedstock for dyes. Many of these compounds (and benzene itself) have been shown to have carcinogenic (cancer-causing) activity. One of the most potent, benzo[*a*]pyrene, is found in soot, cigarette smoke, car exhaust, and even the smoke from barbecue grills!

Tools of the Laboratory

Nuclear Magnetic Resonance (NMR) Spectroscopy

Chemists rely on a battery of instrumental methods to determine the complex structures of organic molecules. We discussed mass spectrometry in Chapter 2 and infrared (IR) spectroscopy in Chapter 9. One of the most useful tools is **nuclear magnetic resonance (NMR) spectroscopy,** which measures environments of certain nuclei in a molecule to elucidate its structure.

In Chapter 8, we saw that an electron can spin in either of two directions, each of which creates a tiny magnetic field. Several types of nuclei behave in a similar way. The most important of these are ^1H, ^{13}C, ^{19}F, and ^{31}P. In this discussion we focus primarily on the proton, ^1H, the nucleus of the most common isotope of hydrogen, and therefore refer to ^1H-NMR spectroscopy. Generally, the magnetic fields of all the protons in a sample of compound are oriented randomly. When placed in a strong external magnetic field (H_0), however, the proton fields become aligned with it (parallel) or against it (antiparallel). The parallel orientation is slightly lower in energy, with the energy difference (ΔE) between the two energy states (spin states) lying in the radio frequency (rf) region of the spectrum.

The protons in the sample oscillate rapidly between the two spin states, and in a process known as *resonance,* the rf value is varied until it matches the frequency of this oscillation. At this point, the protons are "in resonance" with the rf radiation and they absorb and re-emit the energy, which is detected by the rf receiver of the NMR spectrometer (Figure B15.2). If all the protons in a sample required the same ΔE for resonance, an NMR spectrum would have only one peak and be useless. However, the ΔE between the two states depends on the *actual* magnetic field felt by each proton, which is affected by the tiny magnetic fields of the *electrons* on atoms adjacent to that proton. Thus, the ΔE of each proton depends on the electrons in the adjacent atoms—C atoms, electronegative atoms, multiple bonds, and aromatic rings—in other words, on the specific molecular environment. Therefore, the NMR spectrum is *unique* to that compound.

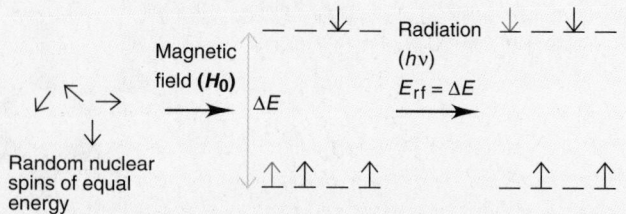

Figure B15.2 **The basis of proton spin resonance.** The randomly oriented magnetic fields of protons in a sample become aligned in a strong external field **(H_0),** with slightly more protons in the lower spin state. In a process called *resonance,* the spin flips when the proton absorbs a photon with energy equal to ΔE (radio frequency region).

The NMR spectrum of a compound is a series of peaks that represents the resonance of each proton as a function of the changing magnetic field. The *chemical shift* of the protons in a given environment is where a peak appears; it represents the ratio of the frequency at which resonance occurs for that proton to H_0. The chemical shifts are shown relative to that of an added standard, tetramethylsilane [(CH_3)$_4$Si, TMS], which has 12 protons bonded to four C atoms that are bonded to one Si atom. Because the shape of TMS is tetrahedral around the Si, the protons are in identical environments, so they produce one peak. Figure B15.3 shows the ^1H-NMR spectrum of acetone. The six protons of acetone also have identical environments—bonded to two C atoms that are each bonded to the C atom in a C$=$O bond—so they also produce one peak, but at a different position from that of TMS. (The axes need not concern us.) The spectrum of dimethoxymethane in Figure B15.4 shows *two* peaks in addition to the TMS peak. The taller one is due to the six CH_3 protons, and the shorter is due to the two CH_2 protons. The area under each peak (given here in units of chart-paper spaces) is proportional to *the number*

15.3 SOME IMPORTANT CLASSES OF ORGANIC REACTIONS

In Chapter 4, we classified chemical reactions based on the chemical process involved (precipitation, acid-base, or redox) and then briefly included a classification based on the number of reactants and products (combination, decomposition, or displacement). We take a similar approach here with organic reactions.

From here on, we use the notation of an uppercase R with a single bond, R—, to signify a general organic group attached to one of the atoms shown; you can usually picture R— as an **alkyl group,** a saturated hydrocarbon chain with one bond available. Thus, R—CH_2—Br has an alkyl group attached to a CH_2 group bearing a Br atom; R—CH$=$$CH_2$ is an alkene with an alkyl group attached to one of the carbons in the double bond; and so forth. (Often, when more than one R group is present, we write R, R', R'' to indicate that these groups may be different.)

Types of Organic Reactions

Most organic reactions are examples of three broad types that can be identified by comparing the *number of bonds to C* in the reactants and products:

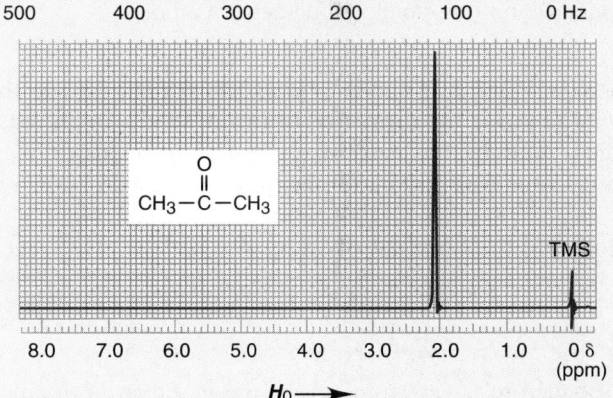

Figure B15.3 The ^{1}H-NMR spectrum of acetone. Because the six CH$_3$ protons are in identical environments, they produce one peak, which has a different chemical shift from that for the 12 protons of the TMS standard.

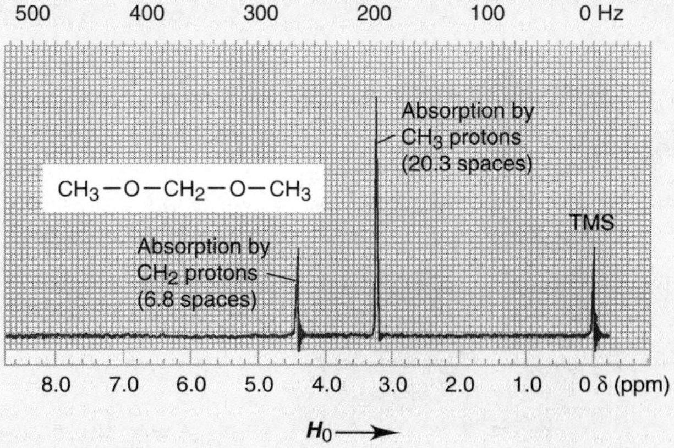

Figure B15.4 The ^{1}H-NMR spectrum of dimethoxymethane. The two peaks represent the chemical shifts of protons in two different environments. The area under each peak is proportional to the number of protons in each environment.

of protons in a given environment and is determined by the instrument. Note that the area ratio is 20.3:6.8 ≈ 3:1, the same as the ratio of six CH$_3$ protons to two CH$_2$ protons. Thus, by analyzing the chemical shifts and peak areas, the chemist learns the type and number of hydrogens in the compound.

Newer NMR methods measure the resonance of other nuclei and have many applications in biochemistry and medicine. ^{13}C-NMR is used to monitor changes in protein and nucleic acid shape and function, and ^{31}P-NMR can determine the health of various organs. For example, the extent of damage from a heart attack can be learned by using ^{31}P-NMR to measure the concentrations of specific phosphate-containing molecules involved in energy utilization by cardiac muscle tissue.

Most recently, computer-aided magnetic resonance imaging (MRI) has allowed visualization of damage to organs and greatly assists physicians in medical diagnosis. For example, an MRI scan of the head (Figure B15.5) can show various levels of metabolic activity in different regions of the brain.

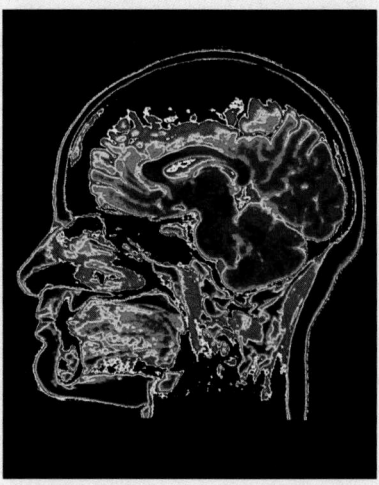

Figure B15.5 Magnetic resonance imaging (MRI) of a human head. Colors indicate a range from high (*red*) to low (*blue*) metabolic activity.

1. An **addition reaction** occurs when an unsaturated reactant becomes a saturated product:

$$R-CH{=}CH-R \ + \ X-Y \ \longrightarrow \ R-\overset{\overset{\displaystyle X}{|}}{C}H-\overset{\overset{\displaystyle Y}{|}}{C}H-R$$

Note the C atoms are bonded to *more* atoms in the product than in the reactant.

The C=C and C≡C bonds and the C=O bond commonly undergo addition reactions. In each case, the π bond breaks, leaving the σ bond intact. In the product, the C atoms (or C and O) form two additional σ bonds. Let's examine the standard heat of reaction (ΔH^0_{rxn}) for a typical addition reaction to see why these reactions occur. Consider the reaction between ethene (common name, ethylene) and HCl:

$$CH_2{=}CH_2 \ + \ H-Cl \ \longrightarrow \ H-CH_2-CH_2-Cl$$

Reactants (bonds broken)	Product (bonds formed)
1 C=C = 614 kJ	1 C—C = −347 kJ
4 C—H = 1652 kJ	5 C—H = −2065 kJ
1 H—Cl = 427 kJ	1 C—Cl = −339 kJ
Total = 2693 kJ	Total = −2751 kJ

And $\Delta H^0_{rxn} = \Delta H^0_{bonds\ broken} + \Delta H^0_{bonds\ formed} = 2693\ kJ + (-2751\ kJ) = -58\ kJ$

A   **B**

Figure 15.13 A color test for C=C bonds. A, Br_2 (in pipet) reacts with a compound that has a C=C bond (in beaker), and its orange-brown color disappears:

$$>C=C< \;+\; Br_2 \longrightarrow -\overset{\displaystyle Br}{\underset{\displaystyle |}{C}}-\overset{\displaystyle |}{\underset{\displaystyle Br}{C}}-$$

B, The compound in this beaker has no C=C bond, so the Br_2 does not react, and the orange-brown color remains.

The reaction is exothermic. By looking at the *net* change in bonds, we see that the driving force for many additions is the formation of two σ bonds (in this case, C—H and C—Cl) from one σ bond (in this case, H—Cl) and one relatively weak π bond. An addition reaction is the basis of a color test for the presence of C=C bonds (Figure 15.13).

2. **Elimination reactions** are the opposite of addition reactions. They occur when a saturated reactant becomes an unsaturated product:

$$R-\overset{\displaystyle Y}{\underset{\displaystyle |}{CH}}-\overset{\displaystyle X}{\underset{\displaystyle |}{CH_2}} \longrightarrow R-CH=CH_2 \;+\; X-Y$$

Note that the C atoms are bonded to *fewer* atoms in the product than in the reactant. Pairs of halogen atoms, an H atom and a halogen atom, or an H atom and an —OH group are typically eliminated, but C atoms are not. Thus, the driving force for many elimination reactions is the loss of a small, stable molecule, such as $HCl(g)$ or H_2O, which increases the entropy (disorder) of the system (Section 13.2):

$$CH_3-\overset{\displaystyle OH}{\underset{\displaystyle |}{CH}}-\overset{\displaystyle H}{\underset{\displaystyle |}{CH_2}} \xrightarrow{H_2SO_4} CH_3-CH=CH_2 \;+\; H-OH$$

3. A **substitution reaction** occurs when an atom (or group) from an added reagent substitutes for one in the organic reactant:

$$R-\overset{\displaystyle |}{\underset{\displaystyle |}{C}}-X \;+\; :Y \longrightarrow R-\overset{\displaystyle |}{\underset{\displaystyle |}{C}}-Y \;+\; :X$$

Note that the C atom is bonded to the *same number* of atoms in the product as in the reactant. The C atom may be saturated or unsaturated, and X and Y can be many different atoms, but generally *not* C. The main flavor ingredient in banana oil, for instance, forms through a substitution reaction; note that the O substitutes for the Cl:

$$CH_3-\overset{\displaystyle O}{\overset{\|}{C}}-Cl \;+\; H\overset{..}{O}-CH_2-CH_2-\overset{\displaystyle CH_3}{\underset{\displaystyle |}{CH}}-CH_3 \longrightarrow CH_3-\overset{\displaystyle O}{\overset{\|}{C}}-O-CH_2-CH_2-\overset{\displaystyle CH_3}{\underset{\displaystyle |}{CH}}-CH_3 \;+\; H-Cl$$

SAMPLE PROBLEM 15.3 Recognizing the Type of Organic Reaction

Problem State whether each reaction is an addition, elimination, or substitution:

(a) $CH_3-CH_2-CH_2-Br \longrightarrow CH_3-CH=CH_2 \;+\; HBr$

(b) ⬠(cyclopentene) $+ H_2 \longrightarrow$ ⬠(cyclopentane)

(c) $CH_3\overset{\displaystyle O}{\overset{\|}{C}}-\mathbf{Br} \;+\; CH_3CH_2OH \longrightarrow CH_3\overset{\displaystyle O}{\overset{\|}{C}}-OCH_2CH_3 \;+\; HBr$

Plan We determine the type of reaction by looking for any change in the number of atoms bonded to C:
• More atoms bonded to C is an *addition*.
• Fewer atoms bonded to C is an *elimination*.
• Same number of atoms bonded to C is a *substitution*.

Solution
(a) Elimination: two bonds in the reactant, C—H and C—Br, are absent in the product, so fewer atoms are bonded to C.
(b) Addition: two more C—H bonds form in the product, so more atoms are bonded to C.
(c) Substitution: the reactant C—Br bond becomes a C—O bond in the product, so the same number of atoms are bonded to C.

FOLLOW-UP PROBLEM 15.3 Write a balanced equation for the following:
(a) An addition reaction between 2-butene and Cl_2
(b) A substitution reaction between $CH_3-CH_2-CH_2-Br$ and OH^-
(c) The elimination of H_2O from $(CH_3)_3C-OH$

The Redox Process in Organic Reactions

An important process in many organic reactions is *oxidation-reduction*. But chemists do not usually monitor the change in oxidation numbers of the various C atoms in a reaction. Rather, they note the movement of electron density around a C atom by counting the number of bonds to more electronegative atoms (usually O) or to less electronegative atoms (usually H). A more electronegative atom takes some electron density from the C, whereas a less electronegative atom gives some electron density to the C. Moreover, even though a redox reaction always involves both an oxidation and a reduction, organic chemists typically *focus on the organic reactant only.* Therefore,

- When a C atom in the organic reactant forms more bonds to O or fewer bonds to H, the reactant is oxidized and the reaction is called an *oxidation.*
- When a C atom in the organic reactant forms fewer bonds to O or more bonds to H, the reactant is reduced and the reaction is called a *reduction.*

The most dramatic redox reactions are combustion reactions. All organic compounds that contain C and H atoms burn in excess O_2 to form CO_2 and H_2O. For ethane, the reaction is

$$2CH_3-CH_3 + 7O_2 \longrightarrow 4CO_2 + 6H_2O$$

Obviously, when ethane converts to CO_2 and H_2O, each of its C atoms has more bonds to O and fewer bonds to H. Thus, ethane is oxidized, and this reaction is referred to as an *oxidation,* even though O_2 is reduced as well.

Most oxidations do not involve such a total breaking apart of the molecule, however. When 2-propanol reacts with potassium dichromate in acidic solution (a common oxidizing agent in organic reactions), the organic compound that forms is 2-propanone:

$$CH_3-\underset{\underset{\text{2-propanol}}{\overset{|}{OH}}}{\overset{|}{CH}}-CH_3 \xrightarrow[H_2SO_4]{K_2Cr_2O_7} CH_3-\underset{\underset{\text{2-propanone}}{\overset{||}{O}}}{C}-CH_3$$

Note that C-2 in 2-propanone has one fewer bond to H and one more bond to O than it does in 2-propanol. Thus, the 2-propanol is oxidized, so this is an *oxidation.* Don't forget, however, that the dichromate ion is reduced at the same time:

$$Cr_2O_7^{2-} + 14H^+ + 6e^- \longrightarrow 2Cr^{3+} + 7H_2O$$

The addition of H_2 to an alkene is referred to as a *reduction:*

$$CH_2{=}CH_2 + H_2 \xrightarrow{Pd} CH_3-CH_3$$

Note that each C has more bonds to H in ethane than it has in ethene, so the ethene is reduced. (The H_2 is oxidized in the process, and the palladium shown over the arrow acts as a catalyst to speed up the reaction.)

SECTION SUMMARY

In an addition reaction, a π bond breaks and the two C atoms bond to more atoms. In an elimination, a π bond forms and the two C atoms bond to fewer atoms. In a substitution, one atom replaces another atom, but the total number of atoms bonded to C does not change. In an organic redox process, the organic reactant is oxidized if a C atom in the compound forms more bonds to O atoms (or fewer bonds to H atoms), and it is reduced if a C atom forms more bonds to H (or fewer bonds to O).

15.4 PROPERTIES AND REACTIVITIES OF COMMON FUNCTIONAL GROUPS

The central organizing principle of organic reaction chemistry is the *functional group.* To predict how an organic compound might react, we narrow our focus there because *the distribution of electron density in the functional group affects the reactivity.* The electron density can be high, as in the C=C and C≡C bonds, or it can be low at one end of a bond and high at the other, as in the C—Cl and C—O bonds. Such bond sites enhance dipoles in the other reactant. As a result, the reactants attract each other and begin a sequence of bond-forming and bond-breaking steps that lead to product. Thus, *the intermolecular forces that affect physical properties and solubility also affect reactivity.* Table 15.5 lists some of the important functional groups in organic compounds.

When we classify functional groups by bond order (single, double, and so forth), in most cases they follow certain patterns of reactivity:

- Single-bonded functional groups undergo substitution or elimination.
- Double- and triple-bonded functional groups undergo addition.
- Functional groups with both single and double bonds undergo substitution.

Functional Groups with Single Bonds

The most common functional groups with only single bonds are alcohols, haloalkanes, and amines.

Alcohols The **alcohol** functional group consists of carbon bonded to an —OH group, $-\overset{|}{\underset{|}{C}}-\ddot{O}-H$, and the general formula of an alcohol is R—OH. Alcohols are named by dropping the final *-e* from the parent hydrocarbon name and adding the suffix *-ol.* Thus, the two-carbon alcohol is ethanol (ethan- + -ol). The common name is the hydrocarbon *root-* + *-yl,* followed by "alcohol"; thus, the common name of ethanol is ethyl alcohol. (This substance, obtained from fermented grain, has been consumed by people as an intoxicant in beverages since ancient times; today, it is recognized as the most abused drug in the world.) Alcohols are important laboratory reagents, and the functional group occurs in many biomolecules, including carbohydrates, sterols, and some amino acids. Figure 15.14 shows the names, structures, and uses of some important compounds that contain the alcohol group.

You can think of an alcohol as a water molecule with an R group in place of one of the H atoms. In fact, alcohols react with very active metals, such as the alkali metals [Group 1A(1)], in a manner similar to water:

$$2Na\ +\ 2HOH\ \longrightarrow\ 2NaOH\ +\ H_2$$
$$2Na\ +\ 2CH_3-CH_2-OH\ \longrightarrow\ 2CH_3-CH_2-ONa\ +\ H_2$$

Water forms a solution of the strongly basic hydroxide ion, and an alcohol forms a solution of the strongly basic *alkoxide ion* (RO⁻): ethanol forms the ethoxide ion (CH₃—CH₂—O⁻). The physical properties of the smaller alcohols are also similar to those of water. They have high melting and boiling points as a result of hydrogen bonding, and they dissolve polar molecules and some salts. (Substitution of R groups for both H atoms of water gives *ethers,* R—O—R.)

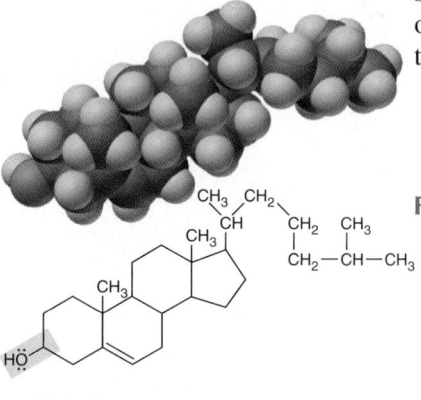

Cholesterol
Major sterol in animals; essential for cell membranes; precursor of steroid hormones

Figure 15.14 Some molecules with the alcohol functional group.

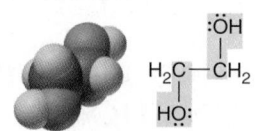

Serine
Amino acid found in most proteins

1,2-Ethanediol (ethylene glycol)
Main component of auto antifreeze

Methanol (methyl alcohol)
By-product in coal gasification; de-icing agent; gasoline substitute; precursor of organic compounds

Table 15.5 Important Functional Groups in Organic Compounds

Functional Group	Compound Type	Suffix or Prefix of Name	Example		Systematic Name (Common Name)
C=C	alkene	-ene	H₂C=CH₂ (structure)		ethene (ethylene)
—C≡C—	alkyne	-yne	H—C≡C—H		ethyne (acetylene)
—C—Ö—H	alcohol	-ol	H—C—Ö—H (structure)		methanol (methyl alcohol)
—C—Ẍ: (X=halogen)	haloalkane	halo-	H—C—Cl: (structure)		chloromethane (methyl chloride)
—C—N̈—	amine	-amine	H—C—C—N̈—H (structure)		ethylamine
—C—H (with =O)	aldehyde	-al	structure		ethanal (acetaldehyde)
—C—C—C— (with =O)	ketone	-one	structure		2-propanone (acetone)
—C—Ö—H (with =O)	carboxylic acid	-oic acid	structure		ethanoic acid (acetic acid)
—C—Ö—C— (with =O)	ester	-oate	structure		methyl ethanoate (methyl acetate)
—C—N̈— (with =O)	amide	-amide	structure		ethanamide (acetamide)
—C≡N:	nitrile	-nitrile	H—C—C≡N: (structure)		ethanenitrile (acetonitrile, methyl cyanide)

Alcohols undergo elimination and substitution reactions. Dehydration, the elimination of H and OH, requires acid and forms alkenes:

$$\underset{\text{cyclohexanol}}{\text{OH group}} \xrightarrow{\text{H}^+} \underset{\text{cyclohexene}}{\bigcirc} + \text{H}_2\text{O}$$

Elimination of two H atoms requires inorganic oxidizing agents, such as $K_2Cr_2O_7$ in aqueous H_2SO_4. As we saw in Section 15.3, the reaction is an oxidation and produces the C=O group:

$$\underset{\text{2-butanol}}{\text{CH}_3-\text{CH}_2-\overset{\overset{\text{OH}}{|}}{\text{CH}}-\text{CH}_3} \xrightarrow[\text{H}_2\text{SO}_4]{\text{K}_2\text{Cr}_2\text{O}_7} \underset{\text{2-butanone}}{\text{CH}_3-\text{CH}_2-\overset{\overset{\text{O}}{\|}}{\text{C}}-\text{CH}_3}$$

For alcohols with an OH group at the end of the chain ($R-CH_2-OH$), another oxidation occurs. Wine turns sour, for example, when the ethanol in contact with air is oxidized to acetic acid:

$$\text{CH}_3-\text{CH}_2 \xrightarrow[-\text{H}_2\text{O}]{\frac{1}{2}\text{O}_2} \text{CH}_3-\overset{\overset{\text{O}}{\|}}{\text{CH}} \xrightarrow{\frac{1}{2}\text{O}_2} \text{CH}_3-\overset{\overset{\text{O}}{\|}}{\text{C}}-\text{OH}$$

Substitution yields products with other single-bonded functional groups. With hydrohalic acids, many alcohols give haloalkanes:

$$\text{R}_2\text{CH}-\text{OH} + \text{HBr} \longrightarrow \text{R}_2\text{CH}-\text{Br} + \text{HOH}$$

As you'll see below, *the C atom undergoing the change in a substitution is bonded to a more electronegative element,* which makes it partially positive and readily attacked by an incoming negative group.

Haloalkanes A *halogen* atom (X) bonded to C gives the **haloalkane** functional group, $-\overset{|}{\underset{|}{\text{C}}}-\overset{..}{\underset{..}{\text{X}}}:$, and compounds have the general formula R—X. Haloalkanes (common name, **alkyl halides**) are named by adding the halogen as a prefix to the parent hydrocarbon name and numbering the C atom to which the halogen is attached, as in bromomethane, 2-chloropropane, or 1,3-diiodohexane.

Just as many alcohols undergo substitution to alkyl halides when treated with halide ions in acid, many halides undergo substitution to alcohols in base. For example, OH^- attacks the positive C end of the C—X bond and displaces X^-:

$$\text{CH}_3-\text{CH}_2-\text{CH}_2-\text{CH}_2-\text{Br} + \text{OH}^- \longrightarrow \text{CH}_3-\text{CH}_2-\text{CH}_2-\text{CH}_2-\text{OH} + \text{Br}^-$$
$$\underset{\text{1-bromobutane}}{} \qquad\qquad\qquad \underset{\text{1-butanol}}{}$$

Substitution by groups such as $-CN$, $-SH$, $-OR$, and $-NH_2$ allows chemists to convert alkyl halides to a host of other compounds.

Just as addition of HX *to* an alkene produces haloalkanes, elimination of HX *from* a haloalkane by reaction with a strong base, such as potassium ethoxide, produces an alkene:

$$\underset{\substack{\text{2-chloro-2-methyl-}\\\text{propane}}}{\text{CH}_3-\overset{\overset{\text{CH}_3}{|}}{\underset{\underset{\text{Cl}}{|}}{\text{C}}}-\text{CH}_3} + \underset{\substack{\text{potassium}\\\text{ethoxide}}}{\text{CH}_3-\text{CH}_2-\text{OK}} \longrightarrow \underset{\text{2-methylpropene}}{\text{CH}_3-\overset{\overset{\text{CH}_3}{|}}{\text{C}}=\text{CH}_2} + \text{KCl} + \text{CH}_3-\text{CH}_2-\text{OH}$$

Haloalkanes have many important uses, but many are carcinogenic in mammals and have severe neurological effects in humans. Their stability has made them notorious environmental contaminants. ●

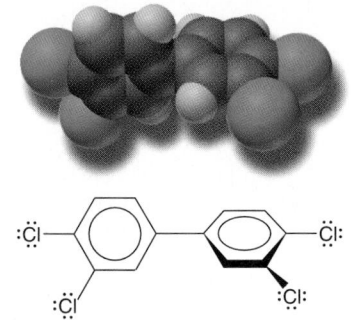

● **Pollutants in the Food Chain** Until recently, halogenated aromatics, such as the *polychlorinated biphenyls* (*PCBs;* one of 209 different compounds is shown), were used as insulating fluids in electrical transformers and then discharged in wastewater. Because of their low solubility and high stability, they accumulate for decades in river and lake sediment and are eaten by microbes and invertebrates. Fish eat the invertebrates, and birds and mammals, including humans, eat the fish. PCBs become increasingly concentrated in body fat at each stage. As a result of their health risks, PCBs in natural waters present an enormous cleanup problem.

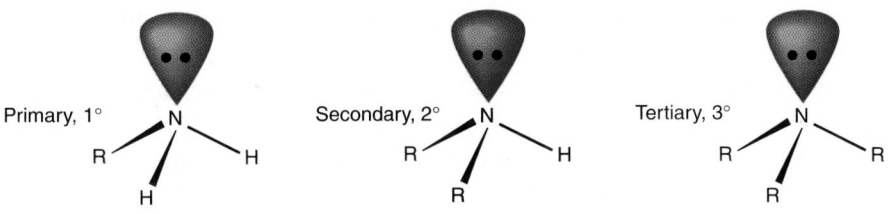

Primary, 1°

Secondary, 2°

Tertiary, 3°

Figure 15.15 General structures of amines. Amines have a trigonal pyramidal shape and are classified by the number of R groups bonded to N. The lone pair on the nitrogen atom is the key to amine reactivity.

Amines The **amine** functional group is $-\overset{|}{\underset{|}{C}}-\overset{|}{N}:$. Chemists classify amines as derivatives of ammonia, with R groups in place of one or more H atoms. *Primary* (1°) amines are RNH_2, *secondary* (2°) amines are R_2NH, and *tertiary* (3°) amines are R_3N. Like ammonia, amines have trigonal pyramidal shapes and a lone pair of electrons on a partially negative N atom (Figure 15.15). Common names are usually used, and the suffix *-amine* follows the name of the alkyl group; thus, methylamine has one methyl group attached to N, diethylamine has two ethyl groups attached, and so forth. Figure 15.16 shows that the amine functional group occurs in many biomolecules.

Primary and secondary amines can form H bonds, so they have higher melting and boiling points than hydrocarbons and alkyl halides of similar molar mass. For example, dimethylamine ($\mathcal{M} = 45.09$ g/mol) boils 45°C higher than ethyl fluoride ($\mathcal{M} = 48.06$ g/mol). Trimethylamine has a greater molar mass than dimethylamine, but it melts more than 20°C *lower* because trimethylamine molecules are not H bonded.

Amines of low molar mass are fishy smelling, water soluble, and weakly basic. The reaction with water proceeds only slightly to the right to reach equilibrium:

$$CH_3-\overset{..}{N}H_2 + H_2O \rightleftharpoons CH_3-\overset{+}{N}H_3 + OH^-$$

Amines undergo substitution reactions in which the lone pair on N attacks the partially positive C in alkyl halides to displace X^- and form a larger amine:

$$2CH_3-CH_2-\overset{..}{N}H_2 + CH_3-CH_2-Cl \longrightarrow CH_3-CH_2-\underset{\underset{CH_3-CH_2}{|}}{\overset{..}{N}H} + CH_3-CH_2-\overset{+}{N}H_3Cl^-$$

ethylamine   chloroethane   diethylamine   ethylammonium chloride

(One molecule of ethylamine participates in the substitution, while the other binds the released H^+ and prevents it from remaining on the diethylamine product.)

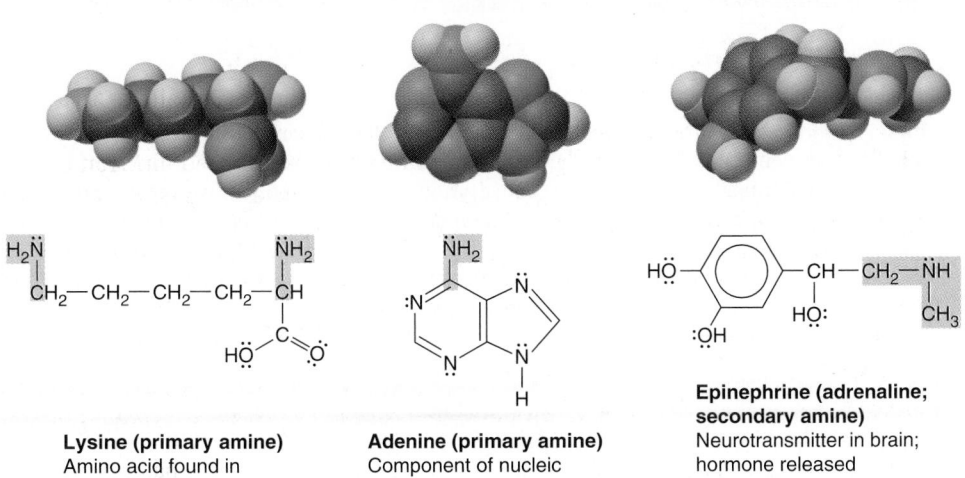

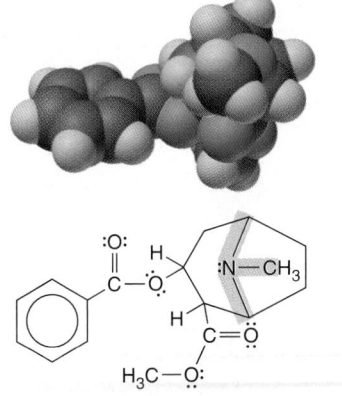

Lysine (primary amine)
Amino acid found in most proteins

Adenine (primary amine)
Component of nucleic acids

Epinephrine (adrenaline; secondary amine)
Neurotransmitter in brain; hormone released during stress

Cocaine (tertiary amine)
Brain stimulant; widely abused drug

Figure 15.16 Some biomolecules with the amine functional group.

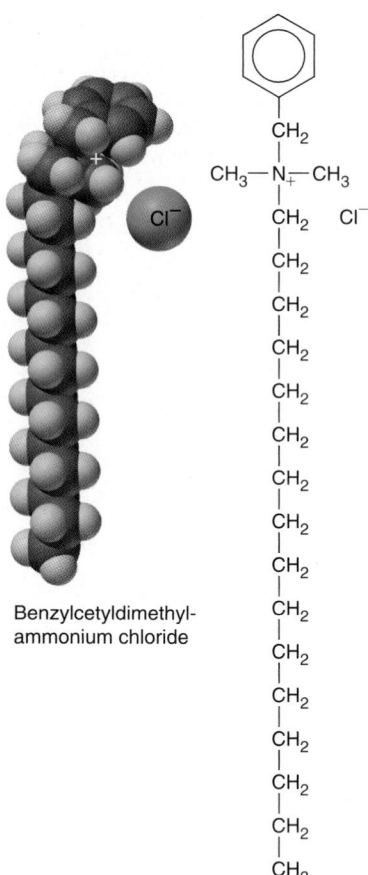

Benzylcetyldimethyl-
ammonium chloride

Figure 15.17 **Structure of a cationic detergent.** Quaternary ammonium salts are used as detergents. The charged portion dissolves in water, and the hydrocarbon portion dissolves in grease.

Four R groups bonded to an N atom give an ionic compound called a *quaternary (4°) ammonium salt:*

$$4NH_3 + 4RCl \longrightarrow 3NH_4Cl + R_4N^+Cl^- \quad \text{(a quaternary ammonium chloride)}$$

Many common liquid detergents are concentrated aqueous solutions of quaternary ammonium salts with large R groups (Figure 15.17). As with soaps (Section 13.1), the ionic portion makes such detergents water soluble, and the R groups allow them to dissolve in grease.

Variations on a Theme: Inorganic Compounds with Single Bonds to O, X, and N The —OH group occurs frequently in inorganic compounds. All oxoacids contain at least one —OH, usually bonded to a relatively electronegative nonmetal atom, which in most cases is bonded to other O atoms. Oxoacids are acidic in water because these additional O atoms pull electron density from the central nonmetal, which pulls electron density from the O—H bond, releasing an H^+ ion and stabilizing the oxoanion through resonance. Alcohols are *not* acidic in water because they lack the additional O atoms and the electronegative nonmetal.

Halides of nearly every nonmetal are known, and many undergo substitution reactions in base. As in the case of an alkyl halide, the process involves an attack on the partially positive central atom by OH^-:

$$H—O^- \quad \overset{\delta^+ \ \delta^-}{C—Cl} \longrightarrow \quad C—OH + Cl^-$$

$$H—O^- \quad \overset{\delta^+ \ \delta^-}{B—Cl} \longrightarrow \quad B—OH + Cl^-$$

Thus, alkyl halides undergo the same general reaction as other nonmetal halides, such as BCl_3, SiF_4, and PCl_5.

The bonds between nitrogen and larger nonmetals, such as Si, P, and S, have significant double-bond character, which affects structure and reactivity. For example, trisilylamine, the Si analog of trimethylamine (see Figure 14.16, p. 568), is planar, rather than trigonal pyramidal, partially as a result of p,d-π bonding. The lone pair on N is delocalized in this π bond, so trisilylamine is not basic.

SAMPLE PROBLEM 15.4 Predicting the Reactions of Alcohols, Alkyl Halides, and Amines

Problem Determine the reaction type and predict the product(s) in the following:
(a) $CH_3—CH_2—CH_2—I + NaOH \longrightarrow$
(b) $CH_3—CH_2—Br + 2CH_3—CH_2—CH_2—NH_2 \longrightarrow$
(c) $CH_3—\underset{\underset{OH}{|}}{CH}—CH_3 \xrightarrow[\text{H}_2\text{SO}_4]{\text{Cr}_2\text{O}_7{}^{2-}}$

Plan We first determine the functional group(s) of the reactant(s) and then examine any inorganic reagent(s) to decide on the possible reaction type, keeping in mind that, in general, these functional groups undergo substitution or elimination. In **(a)**, the reactant is an alkyl halide, so the OH^- of the inorganic reagent substitutes for the —I. In **(b)**, the reactants are an amine and an alkyl halide, so the N: of the amine substitutes for the —Br. In **(c)**, the reactant is an alcohol, the inorganic reagents form a strong oxidizing agent, and alcohols undergo elimination to carbonyl compounds.

Solution **(a)** Substitution: The products are $CH_3—CH_2—CH_2—OH + NaI$

(b) Substitution: The products are $CH_3—CH_2—CH_2—\underset{\underset{CH_2—CH_3}{|}}{NH} + CH_3—CH_2—CH_2—\overset{+}{N}H_3Br^-$

(c) Elimination (oxidation): The product is $CH_3—\underset{\underset{O}{\|}}{C}—CH_3$

Check The only changes should be at the functional group.

FOLLOW-UP PROBLEM 15.4 Fill in the blank in each reaction. (*Hint:* Examine any inorganic compounds and the organic product to determine the organic reactant.)

(a) _____ + CH$_3$—ONa $\longrightarrow$ CH$_3$—CH=C(CH$_3$)—CH$_3$ + NaCl + CH$_3$—OH

(b) _____ $\xrightarrow[\text{H}_2\text{SO}_4]{\text{Cr}_2\text{O}_7{}^{2-}}$ CH$_3$—CH$_2$—C(=O)—OH

Functional Groups with Double Bonds

The most important double-bonded functional groups are the C=C bond of alkenes and the C=O bond of aldehydes and ketones. Both appear in many organic and biological molecules.

Comparing the Reactivity of Alkenes and Aromatic Compounds The C=C bond is the essential portion of the alkene functional group, $\overset{\diagdown}{}$C=C$\overset{\diagup}{}$. Although they can be further unsaturated to alkynes, *alkenes typically undergo addition.* The electron-rich double bond is readily attracted to the partially positive H atoms of hydronium ions and hydrohalic acids, yielding alcohols and alkyl halides, respectively:

CH$_3$—C(CH$_3$)=CH$_2$ + H$_3$O$^+$ $\longrightarrow$ CH$_3$—C(CH$_3$)(OH)—CH$_3$ + H$^+$

2-methylpropene   2-methyl-2-propanol

CH$_3$—CH=CH$_2$ + HCl $\longrightarrow$ CH$_3$—CH(Cl)—CH$_3$

propene   2-chloropropane

The *localized* unsaturation of alkenes is very different from the *delocalized* unsaturation of benzene. Since benzene does *not* have double bonds, despite the structures of its resonance forms, it does *not* behave like an alkene. Benzene, for example, does not decolorize bromine because there are no isolated π-electron pairs to bond with Br$_2$.

In general, aromatic rings are much *less* reactive than alkenes because delocalized π electrons stabilize such a ring, making it *lower* in energy than one with localized π electrons. Data indicate that benzene is about 150 kJ/mol more stable than a six-C ring with three C=C bonds would be; in other words, benzene requires about 150 kJ/mol more energy to react.

An *addition* reaction with benzene, therefore, requires additional energy to break up the delocalized π system. Benzene does undergo many *substitution* reactions, however, in which the delocalization is retained when a ring H atom is replaced by another group:

benzene + Br$_2$ $\xrightarrow{\text{FeBr}_3}$ bromobenzene + HBr

Aldehydes and Ketones The C=O bond, or **carbonyl group,** is one of the most chemically versatile. In the **aldehyde** functional group, the carbonyl C is bonded to H (and often C), so it always occurs *at the end of a chain,* R—CH=O. Aldehyde names drop the final -*e* from the parent alkane and add -*al.* For example,

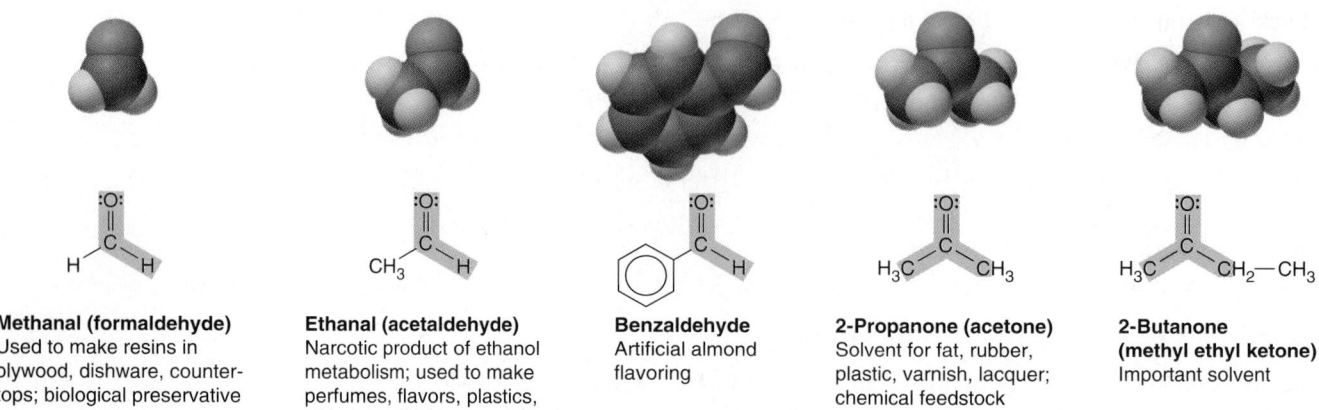

Methanal (formaldehyde)
Used to make resins in plywood, dishware, counter-tops; biological preservative

Ethanal (acetaldehyde)
Narcotic product of ethanol metabolism; used to make perfumes, flavors, plastics, other chemicals

Benzaldehyde
Artificial almond flavoring

2-Propanone (acetone)
Solvent for fat, rubber, plastic, varnish, lacquer; chemical feedstock

2-Butanone (methyl ethyl ketone)
Important solvent

Figure 15.18 **Some common aldehydes and ketones.**

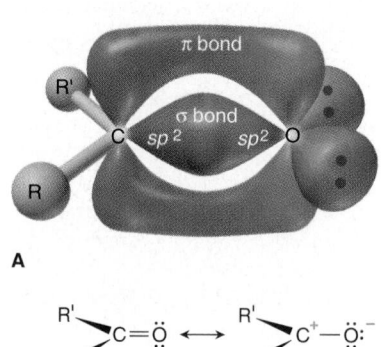

A

B

Figure 15.19 **The carbonyl group.**
A, The σ and π bonds that make up the C=O bond of the carbonyl group. **B,** The charged resonance form shows that the C=O bond is polar (ΔEN = 1.0).

the three-C aldehyde is propanal. In the **ketone** functional group, the carbonyl C is bonded to two other C atoms, $-\overset{|}{\underset{|}{C}}-\overset{:O:}{\overset{\|}{C}}-\overset{|}{\underset{|}{C}}-$, so it occurs *within the chain*.

Ketones, $R-\overset{O}{\overset{\|}{C}}-R'$, are named by numbering the carbonyl C, dropping the final -*e* from the alkane name, and adding -*one*. For example, the unbranched, five-C ketone with the carbonyl C as C-2 in the chain is named 2-pentanone. Figure 15.18 shows some common carbonyl compounds.

Like the C=C bond, the C=O bond is *electron rich;* unlike the C=C bond, it is *highly polar* (ΔEN = 1.0). Figure 15.19 emphasizes this polarity with an electron density model and a charged resonance form. Aldehydes and ketones are formed by the oxidation of alcohols:

$$CH_3-CH_2-OH \xrightarrow{\text{oxidation}} CH_3-\overset{O}{\overset{\|}{C}}-H$$
ethanol → ethanal (common name, acetaldehyde)

$$CH_3-CH_2-\overset{OH}{\underset{}{\overset{|}{CH}}}-CH_3 \xrightarrow{\text{oxidation}} CH_3-CH_2-\overset{O}{\overset{\|}{C}}-CH_3$$
2-butanol → 2-butanone

Conversely, as a result of their unsaturation, they can undergo *addition* and be reduced to alcohols:

cyclobutanone $\xrightarrow{\text{reduction}}$ cyclobutanol

As a result of bond polarity, addition often occurs with an electron-rich group bonding to the carbonyl C and an electron-poor group bonding to the carbonyl O. **Organometallic compounds,** which have a metal atom (usually Li or Mg) attached to an R group through a polar covalent bond (Sections 14.3 and 14.4), take part in this type of reaction. In a two-step sequence, they convert carbonyl compounds to alcohols with *different carbon skeletons:*

$$R-\overset{\delta+}{C}H=\overset{\delta-}{O} + \overset{\delta-}{R'}-\overset{\delta+}{Li} \longrightarrow \xrightarrow{H_2O} R-\overset{OH}{\underset{}{\overset{|}{C}H}}-R' + LiOH$$

In the following reaction steps, for example, the electron-rich C of ethyllithium, CH_3CH_2-Li, attacks the electron-poor carbonyl C of 2-propanone, adding its ethyl group; at the same time, the Li adds to the carbonyl O. Treating the mixture with water forms the C—OH group:

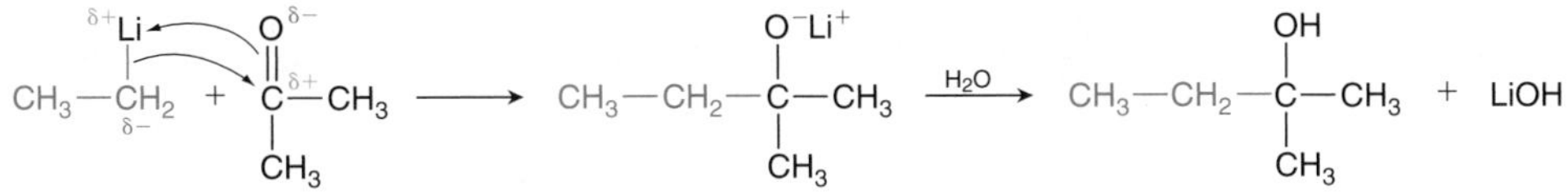

Note that the product skeleton combines the two reactant skeletons. The field of *organic synthesis* often employs organometallic compounds to create molecules with different skeletons and, thus, new compounds.

SAMPLE PROBLEM 15.5 Predicting the Steps in a Reaction Sequence

Problem Fill in the blanks in the following reaction sequence:

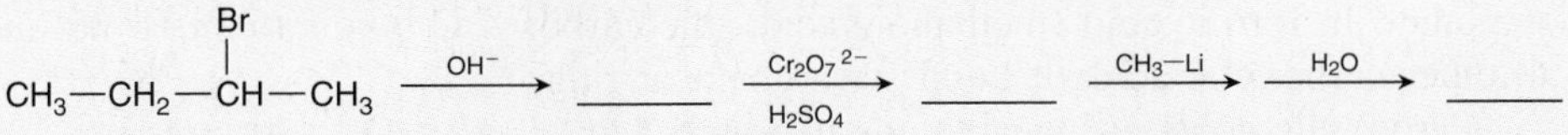

Plan For each step, we examine the functional group of the reactant and the reagent above the yield arrow to decide on the most likely product.
Solution The sequence starts with an alkyl halide reacting with OH^-. Substitution gives an alcohol. Oxidation of this alcohol with acidic dichromate gives a ketone. Finally, a two-step reaction of a ketone with CH_3—Li and then water forms an alcohol with a carbon skeleton that has the CH_3 group attached to the carbonyl C:

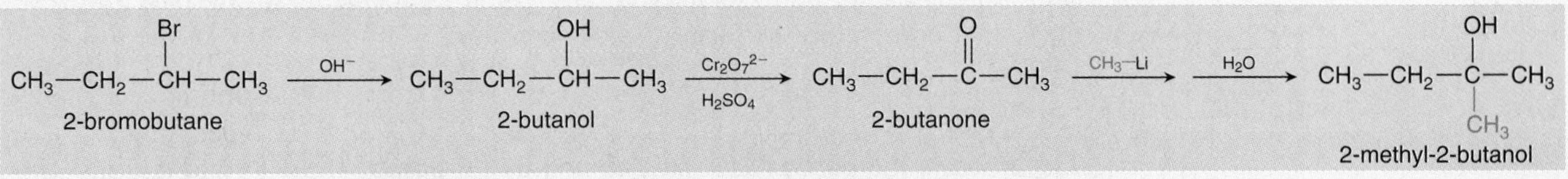

Check In this case, make sure that the first two reactions alter the functional group only and that the final steps change the C skeleton.

FOLLOW-UP PROBLEM 15.5 Choose reactants to obtain the following products:

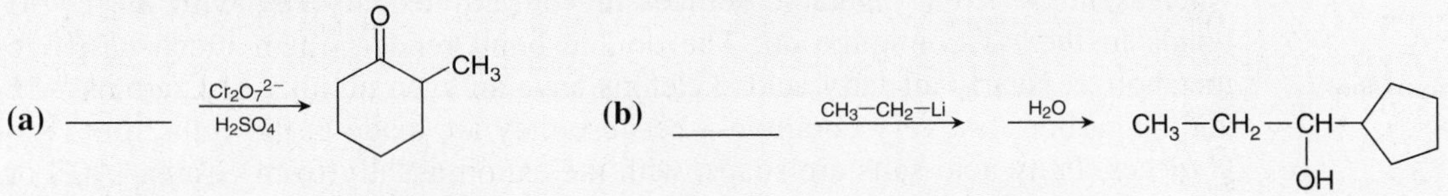

Variations on a Theme: Inorganic Compounds with Double Bonds Homonuclear (same kind of atom) double bonds are rare among atoms other than C, but we've seen many double bonds between O and other nonmetals, as in the oxides of S, N, and halogens. Like carbonyl compounds, these substances undergo addition reactions. For example, the partially negative O of water attacks the partially positive S of SO_3 to form sulfuric acid:

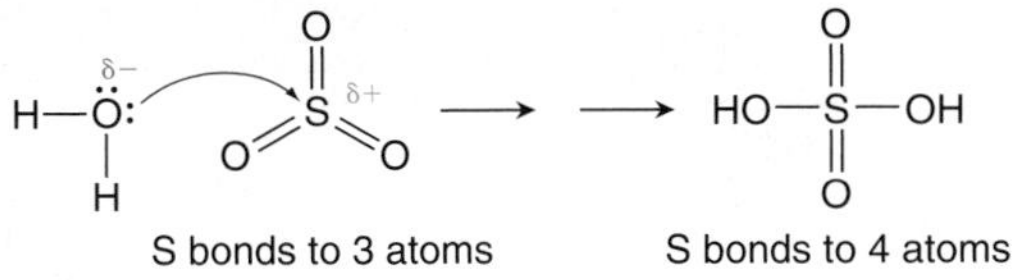

Functional Groups with Both Single and Double Bonds

A family of three functional groups contains C double bonded to O (a carbonyl group) *and* single bonded to O or N. The parent of the family is the **carboxylic acid** group, —C—ÖH, also called the *carboxyl group* and written —COOH. *The most important reaction type of the family is substitution from one member to* *another.* Substitution for the —OH by the —OR of alcohols gives the **ester** group,

$$
\begin{array}{c}
:O: \\
\parallel \\
-C-\ddot{O}-R;
\end{array}
$$

substitution by the —$\ddot{N}$— of amines gives the **amide** group,

$$
\begin{array}{c}
:O: \\
\parallel \quad | \\
-C-N-. \\
\quad\ddot{}
\end{array}
$$

Carboxylic Acids Carboxylic acids, $R-\overset{\overset{\displaystyle O}{\parallel}}{C}-OH$, are named by dropping the *-e* from the parent alkane name and adding *-oic acid;* however, many common names are used. For example, the four-C acid is butanoic acid (the carboxyl C is counted in the root); its common name is butyric acid. Figure 15.20 shows some important carboxylic acids. The carboxyl C already has three bonds, so it forms only one other. In formic acid (methanoic acid), the carboxyl C bonds to an H, but in all other carboxylic acids it bonds to a chain or ring.

Carboxylic acids are weak acids in water:

$$
CH_3-\overset{\overset{\displaystyle O}{\parallel}}{C}-OH(l) + H_2O(l) \rightleftharpoons CH_3-\overset{\overset{\displaystyle O}{\parallel}}{C}-O^-(aq) + H_3O^+(aq)
$$

ethanoic acid
(acetic acid)

At equilibrium in acid solutions of typical concentration, more than 99% of the acid molecules are undissociated at any given moment. In strong base, however, they react completely to form a salt and water:

$$
CH_3-\overset{\overset{\displaystyle O}{\parallel}}{C}-OH(l) + NaOH(aq) \longrightarrow CH_3-\overset{\overset{\displaystyle O}{\parallel}}{C}-O^-(aq) + Na^+(aq) + H_2O(l)
$$

The anion is the *carboxylate ion,* named by dropping *-oic acid* and adding *-oate;* the sodium salt of butanoic acid, for instance, is sodium butanoate.

Carboxylic acids with long hydrocarbon chains are **fatty acids,** an essential group of compounds found in all cells. Animal fatty acids have saturated chains, whereas many from vegetable sources are unsaturated, usually with the C=C bonds in the *cis* configuration. The double bond renders them much easier to metabolize. Nearly all fatty acid skeletons have an even number of C atoms—16 and 18 carbons are very common—because they are made in the cells from two-C pieces. Fatty acid salts are soaps, with the cation usually from Group 1A(1) or 2A(2) (Section 13.1).

Substitution here occurs through a two-step sequence: *addition plus elimination equals substitution.* Addition to the trigonal planar shape of the carbonyl

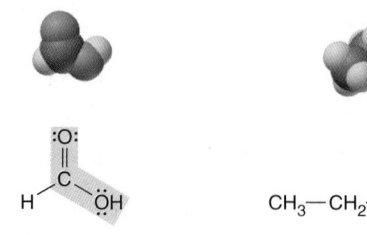

Methanoic acid (formic acid)
An irritating component
of ant and bee stings

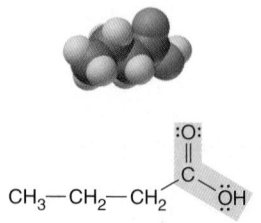

Butanoic acid (butyric acid)
Odor of rancid butter;
suspected component of
monkey sex attractant

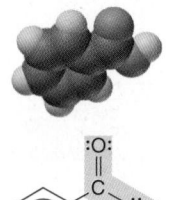

Benzoic acid
Calorimetric standard;
used in preserving food,
dyeing fabric, curing tobacco

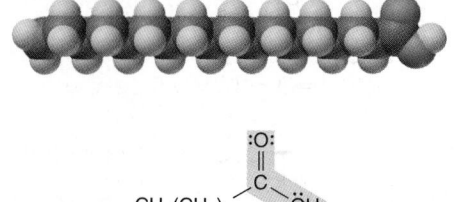

Octadecanoic acid (stearic acid)
Found in animal fats; used in
making candles and soaps

Figure 15.20 **Some molecules with the carboxylic acid functional group.**

group gives an unstable tetrahedral intermediate, which immediately undergoes elimination to revert to a trigonal planar product:

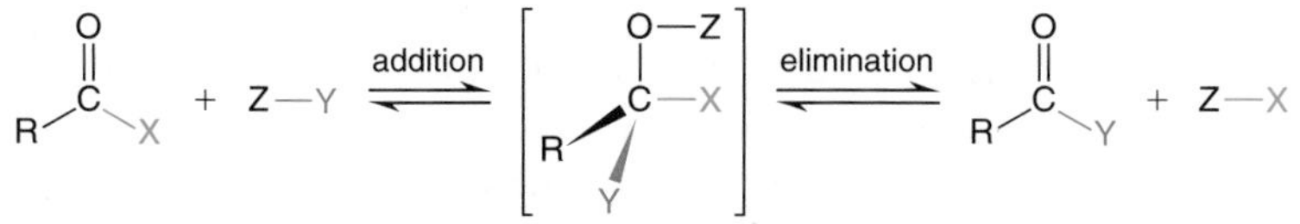

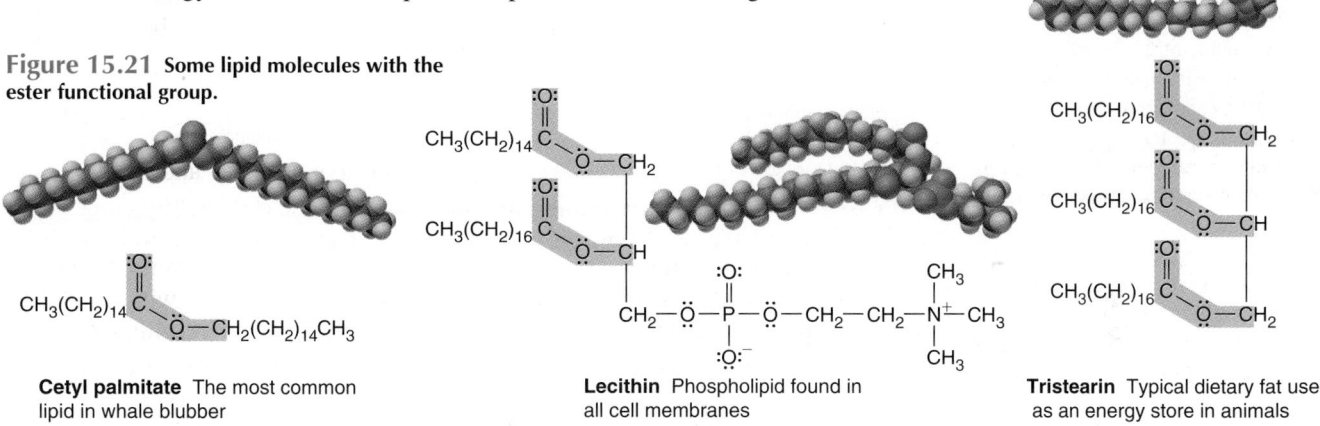

Strong heating of carboxylic acids forms an **acid anhydride** through a type of substitution called a *dehydration-condensation reaction* (Section 14.7), in which two molecules condense into one with loss of water:

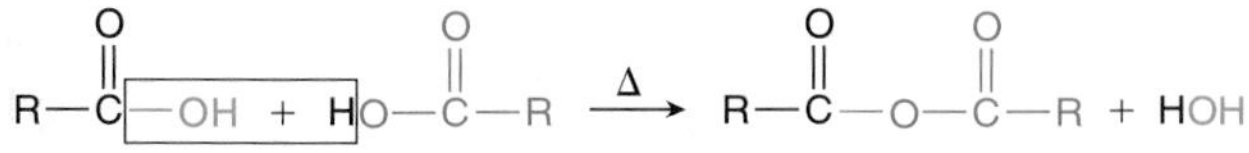

Esters *An alcohol and a carboxylic acid form an ester.* The first part of an ester name designates the alcohol portion and the second the acid portion (named in the same way as the carboxylate ion). For example, the ester formed between ethanol and ethanoic acid is ethyl ethanoate (common name, ethyl acetate), a solvent for nail polish and model glue.

The ester group occurs commonly in **lipids,** a large group of fatty biological substances. Most dietary fats are *triglycerides,* esters composed of three fatty acids linked to the alcohol 1,2,3-trihydroxypropane (common name, glycerol), which function as energy stores. Some important lipids are shown in Figure 15.21.

Figure 15.21 Some lipid molecules with the ester functional group.

Cetyl palmitate The most common lipid in whale blubber

Lecithin Phospholipid found in all cell membranes

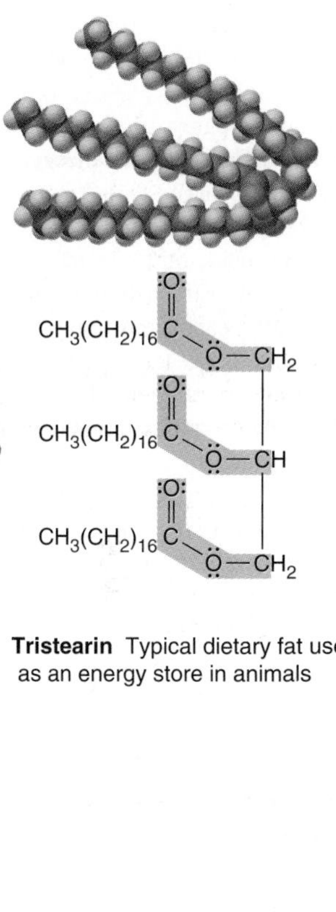

Tristearin Typical dietary fat used as an energy store in animals

Esters, like acid anhydrides, form through a dehydration-condensation reaction; in this case, it is called an *esterification*:

Isotope-labeling studies of this reaction using ^{18}O show that the alcohol O is attracted to the carboxyl C and becomes bonded to it in the ester, and the —OH of the acid becomes part of the product water (Figure 15.22). Thus, the acid supplies the $RC\!=\!O$ portion of the ester, and the alcohol supplies the —OR′ portion.

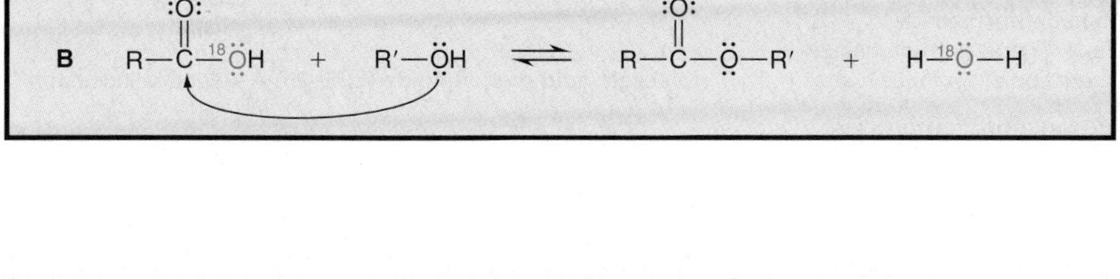

Figure 15.22 Which reactant contributes which group to the ester? An ester forms when a carboxylic acid reacts with an alcohol. To determine which reactant supplies the ester O, the acid and alcohol were labeled with the isotope ^{18}O. **A,** When $R^{18}OH$ reacts with the unlabeled acid, the ester contains ^{18}O but the water doesn't. **B,** When $RCO^{18}OH$ reacts with the unlabeled alcohol, the water contains ^{18}O. Thus, the alcohol supplies the —OR′ part of the ester, and the acid supplies the $RC\!=\!O$ part.

In Chapter 16 we will discuss the role of H⁺ in increasing the rate of this multistep reaction.

Note that the esterification reaction is reversible. The opposite of dehydration-condensation is called **hydrolysis,** in which the O atom of water is attracted to the partially positive C atom of the ester, cleaving (lysing) the molecule into two parts. One part receives water's —OH, and the other part receives water's other H. In the process of soap manufacture, or *saponification* (Latin *sapon,* "soap"), begun in ancient times, ester bonds in animal or vegetable fats are hydrolyzed with strong base:

a triglyceride + 3NaOH →Δ 3 soaps + glycerol

Amides The product of a substitution between an amine (or NH₃) and an ester is an amide. The partially negative N is attracted to the partially positive ester C, an alcohol (ROH) is lost, and an amide forms:

methyl ethanoate (methyl acetate) + ethylamine → N-ethylethanamide (N-ethylacetamide) + methanol

Amides are named by denoting the amine portion with *N-* and replacing *-oic acid* with *-amide.* In the amide from the previous reaction, the ethyl group comes from the amine, and the acid portion comes from ethanoic acid (acetic acid). Some amides are shown in Figure 15.23.

Amides are hydrolyzed in hot water (or base) to a carboxylic acid and an amine. Thus, even though amides are not normally formed in the following way, they can be viewed as the result of a reversible dehydration-condensation:

$$R-C(=O)-OH + H-N-R' \rightleftharpoons R-C(=O)-N-R' + HOH$$

The most important example of the amide group is the *peptide bond,* which links amino acids in a protein, as you'll see in Section 15.6.

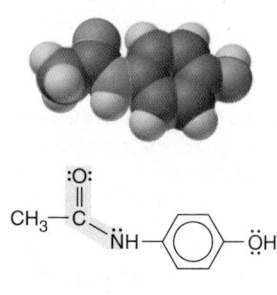

Acetaminophen
Active ingredient in nonaspirin pain relievers; used to make dyes and photographic chemicals

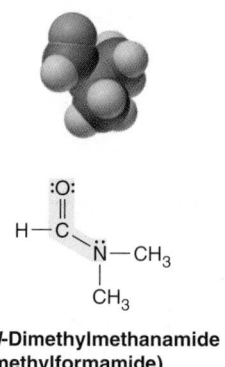

N,N-Dimethylmethanamide (dimethylformamide)
Major organic solvent; used in production of synthetic fibers

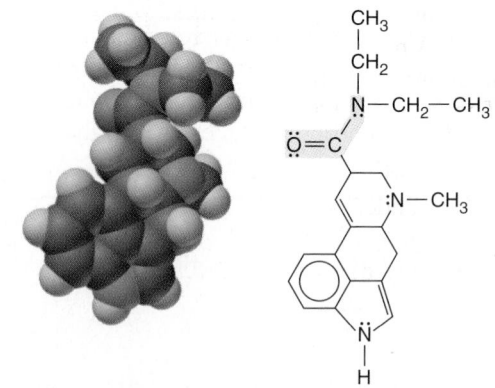

Lysergic acid diethylamide (LSD-25) A potent hallucinogen

Figure 15.23 **Some molecules with the amide functional group.**

The carboxylic acid family undergoes reactions to form other functional groups. For example, certain inorganic reducing agents convert acids and esters to alcohols, and amides to amines:

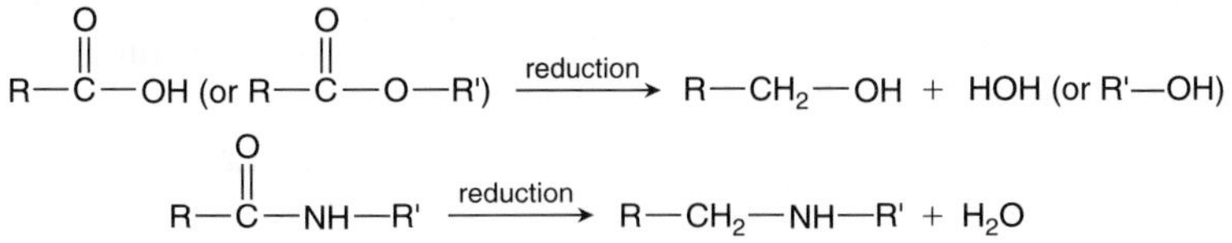

$$\underset{\substack{\\ \text{R—C—OH (or R—C—O—R')}}}{\overset{\substack{O \qquad\qquad O \\ \| \qquad\qquad \|}}{}} \xrightarrow{\text{reduction}} \text{R—CH}_2\text{—OH} + \text{HOH (or R'—OH)}$$

$$\underset{\substack{\\ \text{R—C—NH—R'}}}{\overset{\substack{O \\ \|}}{}} \xrightarrow{\text{reduction}} \text{R—CH}_2\text{—NH—R'} + \text{H}_2\text{O}$$

SAMPLE PROBLEM 15.6 Predicting the Reactions of the Carboxylic Acid Family

Problem Predict the product(s) of the following reactions:

(a) 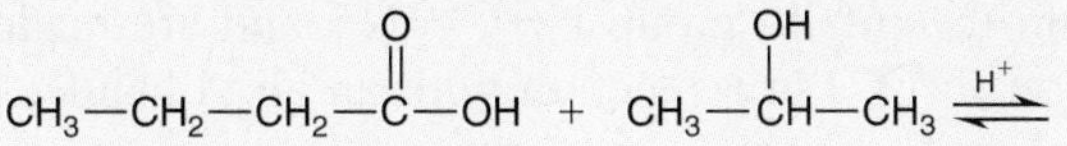 $\underset{\substack{\\ \text{CH}_3\text{—CH}_2\text{—CH}_2\text{—C—OH}}}{\overset{\substack{O \\ \|}}{}} + \underset{\substack{\\ \text{CH}_3\text{—CH—CH}_3}}{\overset{\substack{OH \\ |}}{}} \overset{H^+}{\rightleftharpoons}$

(b) 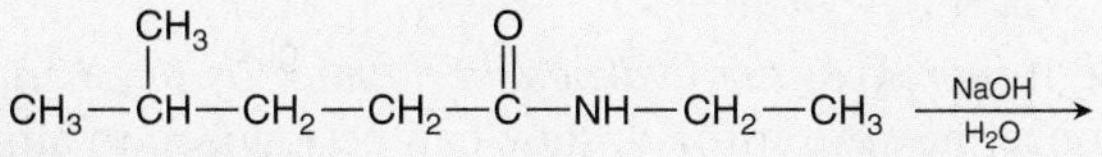 $\underset{\substack{\text{CH}_3 \\ | \\ \text{CH}_3\text{—CH—CH}_2\text{—CH}_2\text{—C—NH—CH}_2\text{—CH}_3}}{\overset{\substack{\qquad\qquad\qquad\qquad\quad O \\ \qquad\qquad\qquad\qquad\quad \|}}{}} \xrightarrow[\text{H}_2\text{O}]{\text{NaOH}}$

Plan We discussed substitution reactions (including addition-elimination and dehydration-condensation) and hydrolysis. In **(a)**, a carboxylic acid and alcohol react, so it must be a substitution to form an ester and water. In **(b)**, an amide reacts with OH⁻, so it is hydrolyzed to an amine and a sodium carboxylate.

Solution **(a)** Formation of an ester:

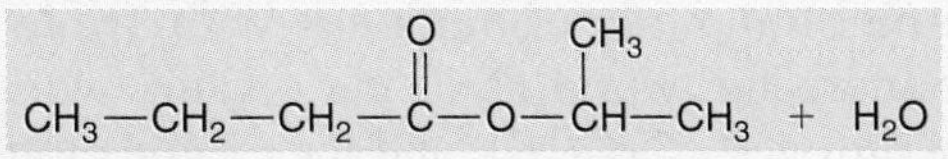

$$\underset{\substack{\\ \text{CH}_3\text{—CH}_2\text{—CH}_2\text{—C—O—CH—CH}_3}}{\overset{\substack{O \qquad\quad CH_3 \\ \| \qquad\quad |}}{}} + \text{H}_2\text{O}$$

(b) Basic hydrolysis of an amide:

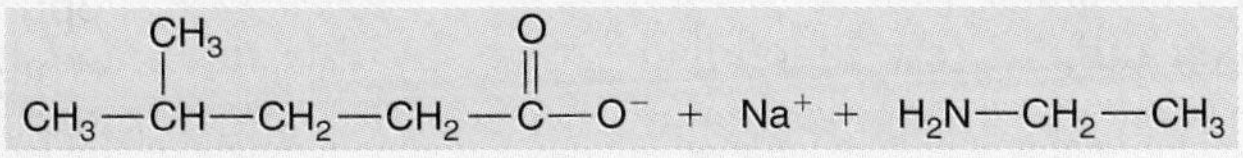

$$\underset{\substack{\text{CH}_3 \\ | \\ \text{CH}_3\text{—CH—CH}_2\text{—CH}_2\text{—C—O}^-}}{\overset{\substack{\qquad\qquad\qquad\quad O \\ \qquad\qquad\qquad\quad \|}}{}} + \text{Na}^+ + \text{H}_2\text{N—CH}_2\text{—CH}_3$$

Check Note that in part (b), the carboxylate ion forms, rather than the acid, because aqueous NaOH is present, which will react with the carboxylic acid.

FOLLOW-UP PROBLEM 15.6 Fill in the blanks in the following reactions:

(a) 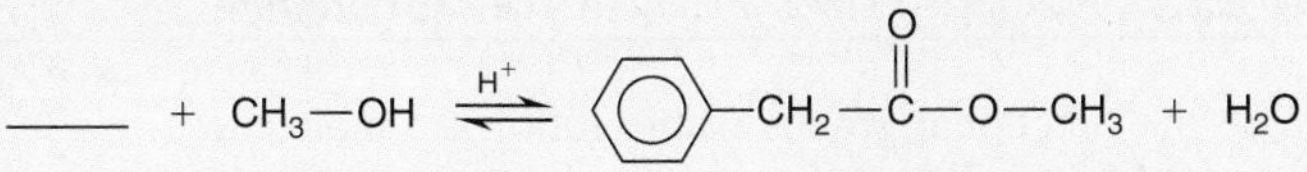 ——— + CH₃—OH $\overset{H^+}{\rightleftharpoons}$ ⬡—CH₂—$\overset{\substack{O \\ \|}}{C}$—O—CH₃ + H₂O

(b) 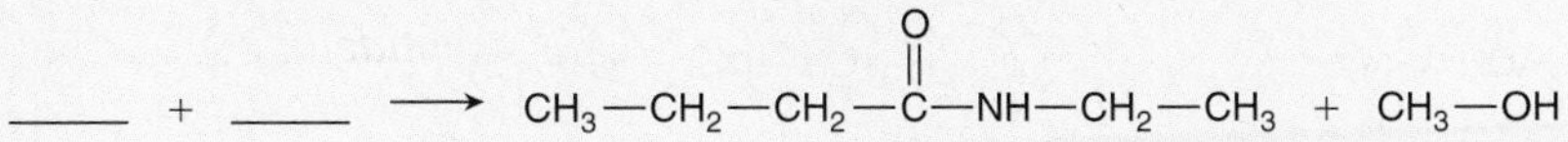 ——— + ——— ⟶ $\text{CH}_3\text{—CH}_2\text{—CH}_2\text{—}\overset{\substack{O \\ \|}}{C}\text{—NH—CH}_2\text{—CH}_3$ + CH₃—OH

Variations on a Theme: Oxoacids, Esters, and Amides of Other Nonmetals

A nonmetal that is both double and single bonded to O occurs in most inorganic oxoacids, such as phosphoric, sulfuric, and chlorous acids. Those with additional O atoms are stronger acids than carboxylic acids.

Diphosphoric and disulfuric acids are acid anhydrides formed by dehydration-condensation reactions, just as a carboxylic acid anhydride is formed (Figure 15.24). Inorganic oxoacids form esters and amides that are part of many biological molecules. We already saw that certain lipids

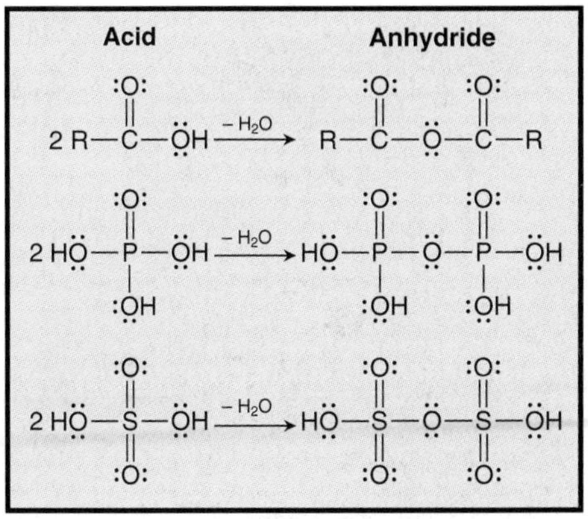

Figure 15.24 **The formation of carboxylic, phosphoric, and sulfuric acid anhydrides.**

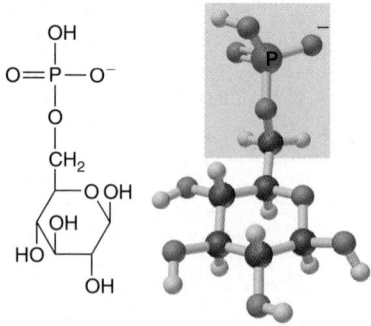

A Glucose-6-phosphate

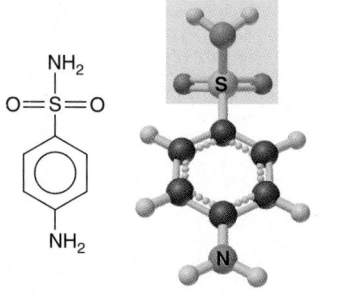

B Sulfanilamide

Figure 15.25 An ester and an amide of other nonmetals. A, Glucose-6-phosphate contains a phosphate ester group. **B,** Sulfanilamide, an important grouping in many antibiotics, contains a sulfonamide group.

are phosphate esters (see Figure 15.21). The first compound formed when glucose is digested is a phosphate ester shown in Figure 15.25A; and a similar phosphate ester is a major structural feature of nucleic acids, as you'll see shortly. Amides of organic sulfur-containing oxoacids, called *sulfonamides*, are potent antibiotics; the simplest of these is depicted in Figure 15.25B. More than 10,000 different sulfonamides have been synthesized.

Functional Groups with Triple Bonds

There are only two important triple-bonded functional groups. *Alkynes,* with their electron-rich —C≡C— groups, undergo addition (by H_2O, H_2, HX, X_2, and so forth) to form double-bonded or saturated compounds:

$$CH_3-C\equiv CH \xrightarrow{H_2} CH_3-CH=CH_2 \xrightarrow{H_2} CH_3-CH_2-CH_3$$
$$\text{propyne} \qquad\qquad \text{propene} \qquad\qquad \text{propane}$$

Nitriles (R—C≡N) contain the **nitrile** group (—C≡N:) and are made by substituting a CN^- (cyanide) ion for X^- in a reaction with an alkyl halide:

$$CH_3-CH_2-Cl + NaCN \longrightarrow CH_3-CH_2-C\equiv N + NaCl$$

This reaction is useful because it *increases the hydrocarbon chain by one C atom.* Nitriles are versatile because once they are formed, they can be reduced to amines or hydrolyzed to carboxylic acids:

$$CH_3-CH_2-CH_2-NH_2 \xleftarrow{\text{reduction}} CH_3-CH_2-C\equiv N \xrightarrow[\text{hydrolysis}]{H_3O^+,\ H_2O} CH_3-CH_2-\overset{\displaystyle O}{\overset{\|}{C}}-OH + NH_4^+$$

Variations on a Theme: Inorganic Compounds with Triple Bonds Triple bonds are as scarce in the inorganic world as in the organic world. Carbon monoxide (:C≡O:), elemental nitrogen (:N≡N:), and the cyanide ion ([:C≡N:]⁻) are the only common examples.

You've seen quite a few functional groups by this time, and it is especially important that you can recognize them in a complex organic molecule. Sample Problem 15.7 provides some practice.

SAMPLE PROBLEM 15.7 Recognizing Functional Groups

Problem Circle and name the functional groups in the following molecules:

(a)

(b)

(c)

Plan We use Table 15.5 to identify the various functional groups.

Solution

(a) carboxylic acid, ester

(b) alcohol, 2° amine

(c) ketone, haloalkane, alkene

FOLLOW-UP PROBLEM 15.7 Circle and name the functional groups:

(a)

(b)

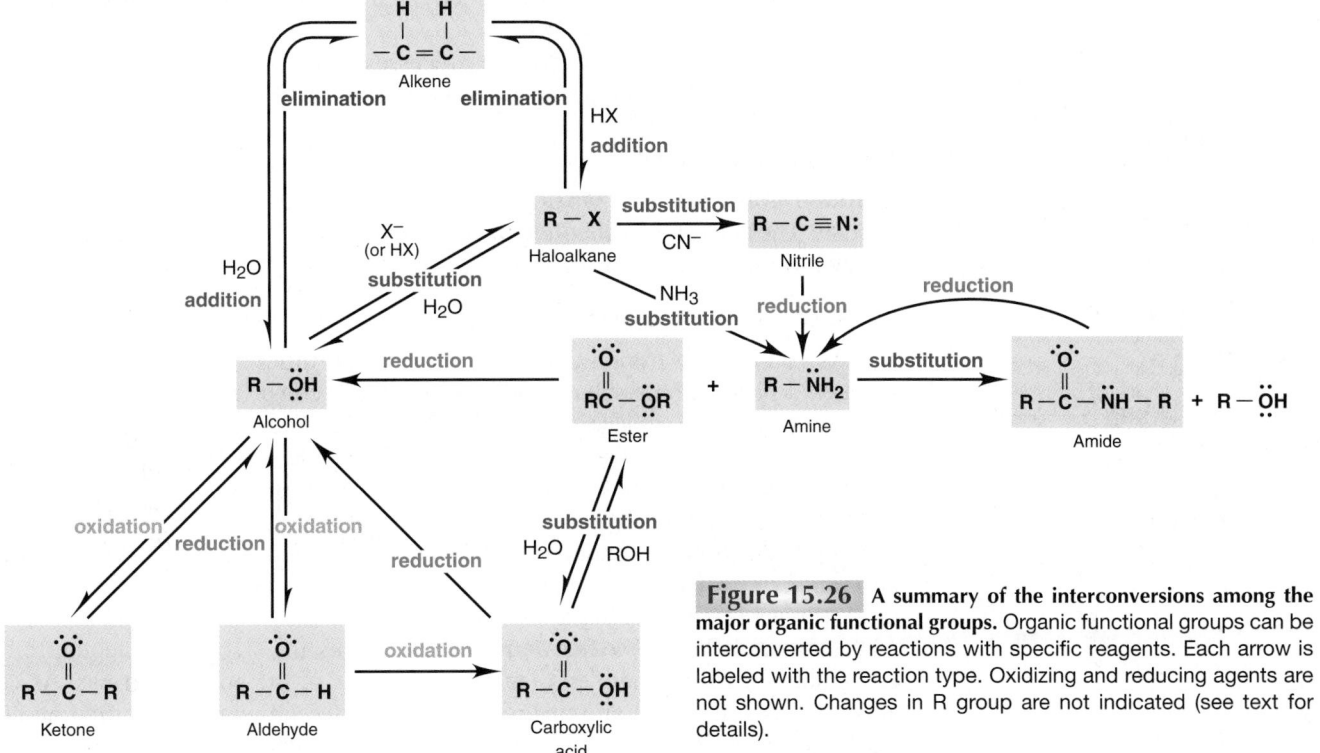

Figure 15.26 **A summary of the interconversions among the major organic functional groups.** Organic functional groups can be interconverted by reactions with specific reagents. Each arrow is labeled with the reaction type. Oxidizing and reducing agents are not shown. Changes in R group are not indicated (see text for details).

Figure 15.26 provides a summarizing overview of the names, general structures, and interconversions of the functional groups we've discussed. Note the remarkable versatility of organic functional groups.

SECTION SUMMARY

Organic reactions are initiated when regions of high and low electron density of different reactant molecules attract each other. Single-bonded groups—alcohols, amines, and alkyl halides—take part in substitution and elimination reactions. Double- and triple-bonded groups—alkenes, aldehydes, ketones, alkynes, and nitriles—generally take part in addition reactions. Aromatic compounds typically undergo substitution, rather than addition, because delocalization of the π electrons stabilizes the ring. Groups with both double and single bonds—carboxylic acids, esters, and amides—generally take part in substitution reactions. Many reactions change one functional group to another, but some, especially reactions with organometallic compounds and with the cyanide ion, change the C skeleton.

15.5 THE MONOMER-POLYMER THEME I: SYNTHETIC MACROMOLECULES

In our survey of advanced materials in Chapter 12, you saw that polymers are extremely large molecules that consist of many monomeric repeat units. In that chapter, we focused on the mass, shape, and physical properties of polymers in the context of phase changes and intermolecular forces. Now, in keeping with the focus of this chapter, we'll note how polymers are named and then discuss the two major types of organic reactions that link monomers covalently into a chain.

To name a polymer, just add the prefix *poly-* to the monomer name. Thus, the polymer made from the monomer ethylene is called *polyethylene*; the polymer made from the monomer styrene is *polystyrene*, and so forth. The only additional detail is that when the monomer has a two-word name, parentheses are used

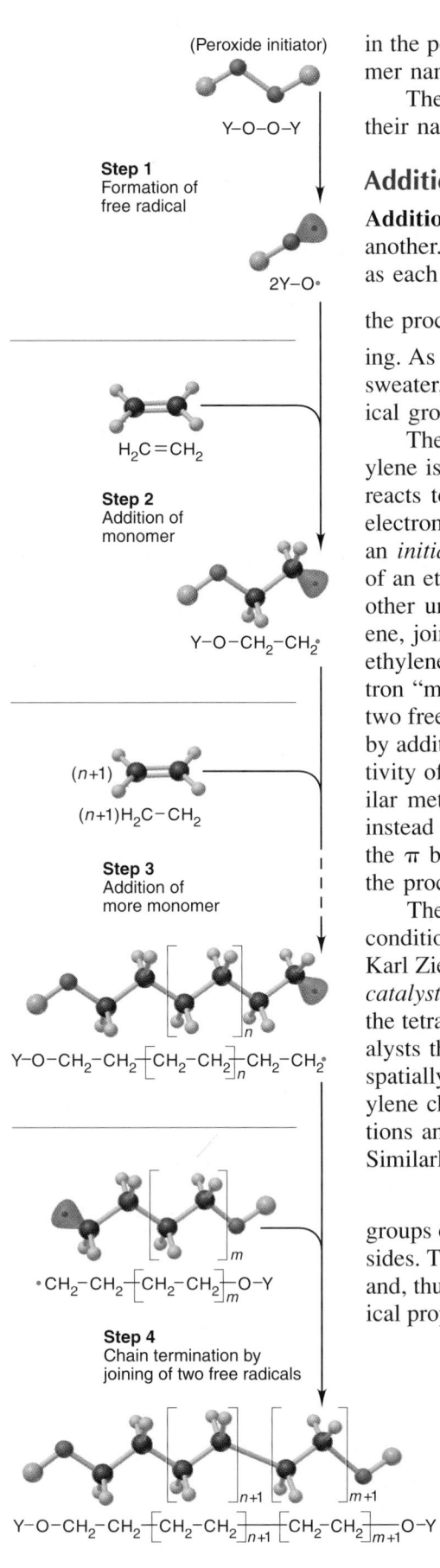

(Peroxide initiator)

Y—O—O—Y

Step 1
Formation of
free radical

2Y—O•

$H_2C=CH_2$

Step 2
Addition of
monomer

Y—O—CH$_2$—CH$_2$•

(n+1)

(n+1)H$_2$C—CH$_2$

Step 3
Addition of
more monomer

Y—O—CH$_2$—CH$_2$$\left[\text{CH}_2\text{—CH}_2\right]_nCH_2$—CH$_2$•

•CH$_2$—CH$_2$$\left[\text{CH}_2\text{—CH}_2\right]_m$O—Y

Step 4
Chain termination by
joining of two free radicals

Y—O—CH$_2$—CH$_2$$\left[\text{CH}_2\text{—CH}_2\right]_{n+1}$$\left[\text{CH}_2\text{—CH}_2\right]_{m+1}$O—Y

in the polymer name; for example, when the monomer is vinyl chloride, the polymer name is *poly(vinyl chloride)*.

The two major types of reaction processes that form synthetic polymers lend their names to the resulting classes of polymer—addition and condensation.

Addition Polymers

Addition polymers form when monomers undergo an addition reaction with one another. These are also called *chain-reaction (or chain-growth) polymers* because as each monomer adds to the chain, it forms a new reactive site to continue the process. The monomers of most addition polymers have the $\text{C}=\text{C}$ grouping. As you can see from Table 15.6, the essential differences between an acrylic sweater, a plastic grocery bag, and a bowling ball come from the different chemical groups that are attached to the double-bonded C atoms of the monomer.

The *free-radical polymerization* of ethene (ethylene, $CH_2=CH_2$) to polyethylene is a simple example of the addition process. In Figure 15.27, the monomer reacts to form a *free radical*, a species with an unpaired electron, that seeks an electron from another monomer to form a covalent bond. The process begins when an *initiator*, usually a peroxide, generates a free radical that attacks the π bond of an ethylene unit, forming a σ bond with one of the *p* electrons and leaving the other unpaired. This new free radical then attacks the π bond of another ethylene, joining it to the chain end, and the backbone of the polymer grows. As each ethylene adds, it leaves an unpaired electron on the growing end to find an electron "mate" and make the chain one repeat unit longer. This process stops when two free radicals form a covalent bond or when a very stable free radical is formed by addition of an *inhibitor* molecule. Recent progress in controlling the high reactivity of free-radical species promises an even wider range of polymers. In a similar method, polymerization is initiated by the formation of a cation (or anion) instead of a free radical. The cationic (or anionic) reactive end of the chain attacks the π bond of the next monomer to form another cationic (or anionic) end, and the process continues.

The most important polymerization reactions take place under relatively mild conditions through the use of catalysts that incorporate transition metals. In 1963, Karl Ziegler and Giulio Natta received a Nobel Prize for developing *Ziegler-Natta catalysts,* which employ an organoaluminum compound, such as $Al(C_2H_5)_3$, and the tetrachloride of titanium or vanadium. Today, chemists use organometallic catalysts that are *stereoselective* to create polymers whose repeat units have groups spatially oriented in particular ways. Through the use of these catalysts, polyethylene chains with molar masses of 10^4 to 10^5 g/mol are made by varying conditions and reagents.

Similarly, polypropylenes, $-\!\!\left[\text{CH}_2\text{—CH}\right]\!\!-_n$, can be made that have all the CH_3
$\qquad\qquad\qquad\qquad\qquad\qquad\quad |$
$\qquad\qquad\qquad\qquad\qquad\qquad CH_3$
groups of the repeat units oriented either on one side of the chain or on alternating sides. The different orientations lead to different packing efficiencies of the chains and, thus, different degrees of crystallinity, which lead to differences in such physical properties as density, rigidity, and elasticity (Section 12.7).

Figure 15.27 Steps in the free-radical polymerization of ethylene. In this polymerization method, free radicals initiate, propagate, and terminate the formation of an addition polymer. An initiator (Y—O—O—Y) is split to form two molecules of a free radical (Y—O·). The free radical attacks the π bond of a monomer and creates another free radical (Y—O—CH$_2$—CH$_2$·). The process proceeds, and the chain grows (propagates) until an inhibitor is added (not shown) or two free radicals combine.

Table 15.6 Some Major Addition Polymers

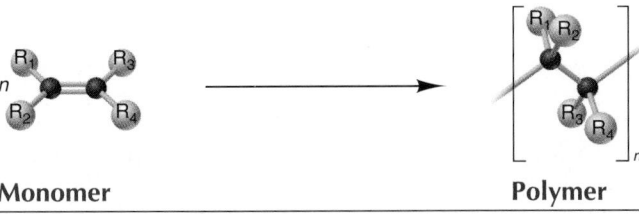

Monomer	Polymer	Applications
H₂C=CH₂	polyethylene	Plastic bags; bottles; toys
F₂C=CF₂	polytetrafluoroethylene	Cooking utensils (e.g., Teflon)
H₂C=CHCH₃	polypropylene	Carpeting (indoor-outdoor); bottles
H₂C=CHCl	poly(vinyl chloride)	Plastic wrap; garden hose; indoor plumbing
H₂C=CH(C₆H₅)	polystyrene	Insulation; furniture; packing materials
H₂C=CHCN	polyacrylonitrile	Yarns; fabrics; wigs (e.g., Orlon, Acrilon)
H₂C=CHOC(O)CH₃	poly(vinyl acetate)	Adhesives; paints; textile coatings; computer disks
H₂C=CCl₂	poly(vinylidene chloride)	Food wrap (e.g., Saran)
H₂C=C(CH₃)C(O)OCH₃	poly(methyl methacrylate)	Glass substitute (e.g., Lucite, Plexiglas); bowling balls; paint

Condensation Polymers

The monomers of **condensation polymers** must have *two functional groups*; we can designate such a monomer as A—**R**—B (where **R** is the rest of the molecule). Most commonly, the monomers link when an A group on one undergoes a *dehydration-condensation reaction* with a B group on another:

$$\tfrac{1}{2}n\text{H—A—}\mathbf{R}\text{—B—OH} + \tfrac{1}{2}n\text{H—A—}\mathbf{R}\text{—B—OH} \xrightarrow{-(n-1)\text{HOH}}$$

$$\text{H}\!-\!\!\left[\text{A—}\mathbf{R}\text{—B}\right]_{\!n}\!\!-\text{OH}$$

Many condensation polymers are *copolymers,* those consisting of two or more different repeat units (Section 12.7). For example, condensation of carboxylic acid and amine monomers forms *polyamides (nylons),* whereas carboxylic acid and alcohol monomers form *polyesters.*

One of the most common polyamides is *nylon-66* (see Gallery, p. 73), manufactured by mixing equimolar amounts of a six-C diamine (1,6-diaminohexane) and a six-C diacid (1,6-hexanedioic acid). The basic amine reacts with the acid to form a "nylon salt." Heating drives off water and forms the amide bonds:

In the laboratory, this nylon is made without heating by using a more reactive acid component (Figure 15.28). Covalent bonds in the chain give nylons great strength, and H bonds between chains give them great flexibility. About half of all nylons are made to reinforce automobile tires, the remainder being used for rugs, clothing, fishing line, and so forth.

Dacron, a popular polyester fiber, is woven from polymer strands formed when equimolar amounts of 1,4-benzenedicarboxylic acid and 1,2-ethanediol react. Blending these polyester fibers with various amounts of cotton gives fabrics that are durable, easily dyed, and crease resistant. Extremely thin Mylar films, used for recording tape and food packaging, are also made from this polymer.

Variations on a Theme: Inorganic Polymers You already know that some synthetic polymers have inorganic backbones. In Chapter 14, we discussed the silicones, polymers with the repeat unit —$(R_2)Si$—O—. Depending on the chain crosslinks and the R groups, silicones range from oily liquids to elastic sheets to rigid solids and have applications that include stopcock grease, artificial limbs, and spacesuits. We also mentioned the polyphosphazenes, whose repeating unit —$(R_2)P$=N— exists as a flexible chain even at low temperatures.

Figure 15.28 The formation of nylon-66. Nylons are formed industrially by the reaction of diamine and diacid monomers. In this laboratory demonstration, the six-C diacid chloride monomer, which is more reactive than the diacid, is used; the polyamide forms between the phases.

SECTION SUMMARY

Polymers are extremely large molecules made of many smaller monomers. Addition polymers are formed from unsaturated monomers that commonly link through free-radical reactions. Most condensation polymers are formed by linking two types of monomer through dehydration-condensation reactions. Reaction conditions, catalysts, and monomer groups can be varied to produce polymers with different properties.

15.6 THE MONOMER-POLYMER THEME II: BIOLOGICAL MACROMOLECULES

The monomer-polymer theme was being played out in nature eons before humans employed it to such great advantage. Biological macromolecules are nothing more than condensation polymers created by nature's reaction chemistry and improved through evolution. These remarkable molecules are the greatest proof of the versatility of carbon and its handful of atomic partners.

Natural polymers are the "stuff of life"—polysaccharides, proteins, and nucleic acids. Some have structures that make wood strong, hair curly, fingernails hard, and wool flexible. Others speed up the myriad reactions that occur in every cell or defend us against infection. Still others possess the genetic information needed to forge other biomolecules. Remarkable as these giant molecules are, the functional groups of their monomers and the reactions that link them are identical to those of other, smaller organic molecules.

Sugars and Polysaccharides

In essence, the same chemical change occurs when you burn a piece of wood or eat a piece of bread. Wood and bread are mixtures of carbohydrates, substances that provide energy through oxidation.

Monomer Structure and Linkage Glucose and other simple sugars, from the three-C *trioses* to the seven-C *heptoses,* are called **monosaccharides.** They consist of carbon chains with attached hydroxyl and carbonyl groups. In addition to their roles as individual molecules primarily engaged in energy metabolism, they serve as the monomer units of **polysaccharides.** Most natural polysaccharides are formed from five- and six-C units. In aqueous solution, an alcohol group and the aldehyde (or ketone) group of the *same* sugar molecule react with each other to form a cyclic molecule with either a five- or six-membered ring. The ring structure of glucose is shown in Figure 15.29A. When two of these cyclic units undergo a dehydration-condensation reaction, a **disaccharide** forms. For example, sucrose (table sugar) is a disaccharide of glucose and fructose (Figure 15.29B); lactose (milk sugar) is a disaccharide of glucose and galactose; maltose, used in brewing and as a sweetener, is a disaccharide of two glucose units.

Types of Polysaccharides A polysaccharide consists of *many* monosaccharide units linked together. The three major natural polysaccharides—cellulose, starch, and glycogen—consist entirely of glucose units, but they differ in the ring positions of the links, in bond orientation, and in the extent of crosslinking. Some polysaccharides also contain nitrogen in their attached groups.

Cellulose is the most abundant organic chemical on Earth. More than 50% of the carbon in plants occurs in the cellulose of stems and leaves; wood is largely cellulose, and cotton is more than 90% cellulose. This polymer consists of long chains of glucose monomers. The great strength of wood is due largely to the H bonds between cellulose chains. The monomers are linked in a particular way from C-1 in one unit to C-4 in the next. Humans lack the enzymes to break this link, so we cannot digest cellulose (unfortunately!); however, microorganisms in the digestive tracts of some animals, such as cows, sheep, and termites, can.

Starch is a mixture of polysaccharides of glucose that serves as an *energy store* in plants. When a plant needs energy, some starch is broken down by hydrolysis of the bonds between units, and the released glucose is oxidized through a multistep metabolic pathway. Starch occurs in plant cells as insoluble granules of amylose, a helical molecule of several thousand glucose units, and amylopectin, a highly branched, bushlike molecule of up to a million glucose units. Most of the glucose units are linked by C-1 to C-4 bonds, as in cellulose, but a different orientation around the chiral C-1 allows our digestive enzymes to break starch down into monomers. A C-6 to C-1 crosslink joins chains every 24 to 30 units.

Polysaccharide Skeletons of Lobsters and Roaches The variety of polysaccharide properties that arises from simple changes in the monomers is amazing. For example, substituting an —NH_2 group for the —OH group at C-2 in glucose produces *glucosamine*. The amide formed from glucosamine and acetic acid (*N*-acetylglucosamine) is the monomer of *chitin* (pronounced "KY-tin"), a polysaccharide that is the main component of the tough, brittle, external skeletons of insects and crustaceans.

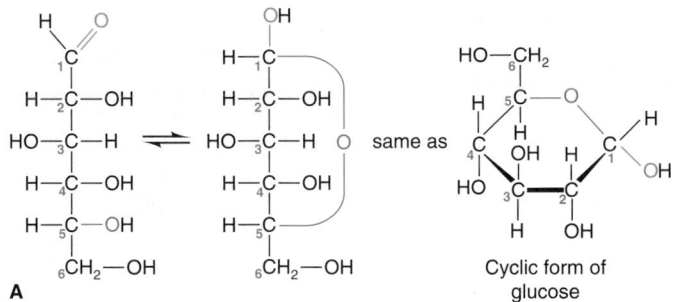

A

Cyclic form of glucose

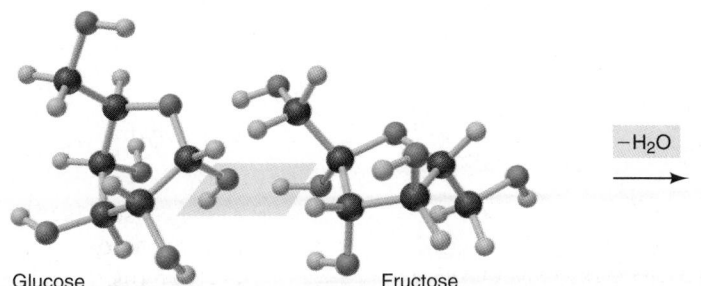

B   Glucose   Fructose

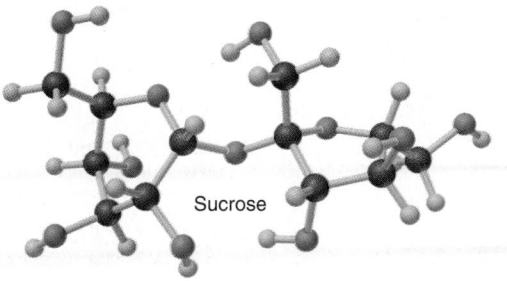

Sucrose

−H_2O

Figure 15.29 The structure of glucose in aqueous solution and the formation of a disaccharide. A, A molecule of glucose undergoes an internal addition reaction between the aldehyde group of C-1 and the alcohol group of C-5 to form a cyclic monosaccharide. **B,** Two monosaccharides, glucose and fructose, undergo a dehydration-condensation reaction to form the disaccharide sucrose (table sugar) and a water molecule.

Glycogen functions as the energy storage molecule in animals. It occurs in liver and muscle cells as large, insoluble granules consisting of glycogen molecules made from 1000 to more than 500,000 glucose units. In glycogen, the glucose units are also linked by C-1 to C-4 bonds, but the molecule is more highly crosslinked than starch, with C-6 to C-1 crosslinks every 8 to 12 units.

Amino Acids and Proteins

As you saw in Section 15.5, synthetic polyamides (such as nylon-66) are formed from two monomers, one with a carboxyl group at each end and the other with an amine group at each end. **Proteins,** the polyamides of nature, are unbranched polymers formed from monomers called **amino acids.** *Each amino acid monomer has a carboxyl end and an amine end.*

Monomer Structure and Linkage An amino acid has both a carboxyl group and an amine group attached to the α-*carbon,* the second C atom in the chain. Proteins are made up of about 20 different types of amino acids, each with its own particular R group, ranging from an H atom to a polycyclic N-containing aromatic structure, as shown in Figure 15.30.

Figure 15.30 The common amino acids. About 20 different amino acids occur in proteins. The R groups are screened yellow, and the α-carbons (*boldface*), with carboxyl and amino groups, are screened grey. Here the amino acids are shown with the charges they have under physiological conditions. They are grouped by polarity, acid-base character, and presence of an aromatic ring. The R groups play a major role in the shape and function of the protein.

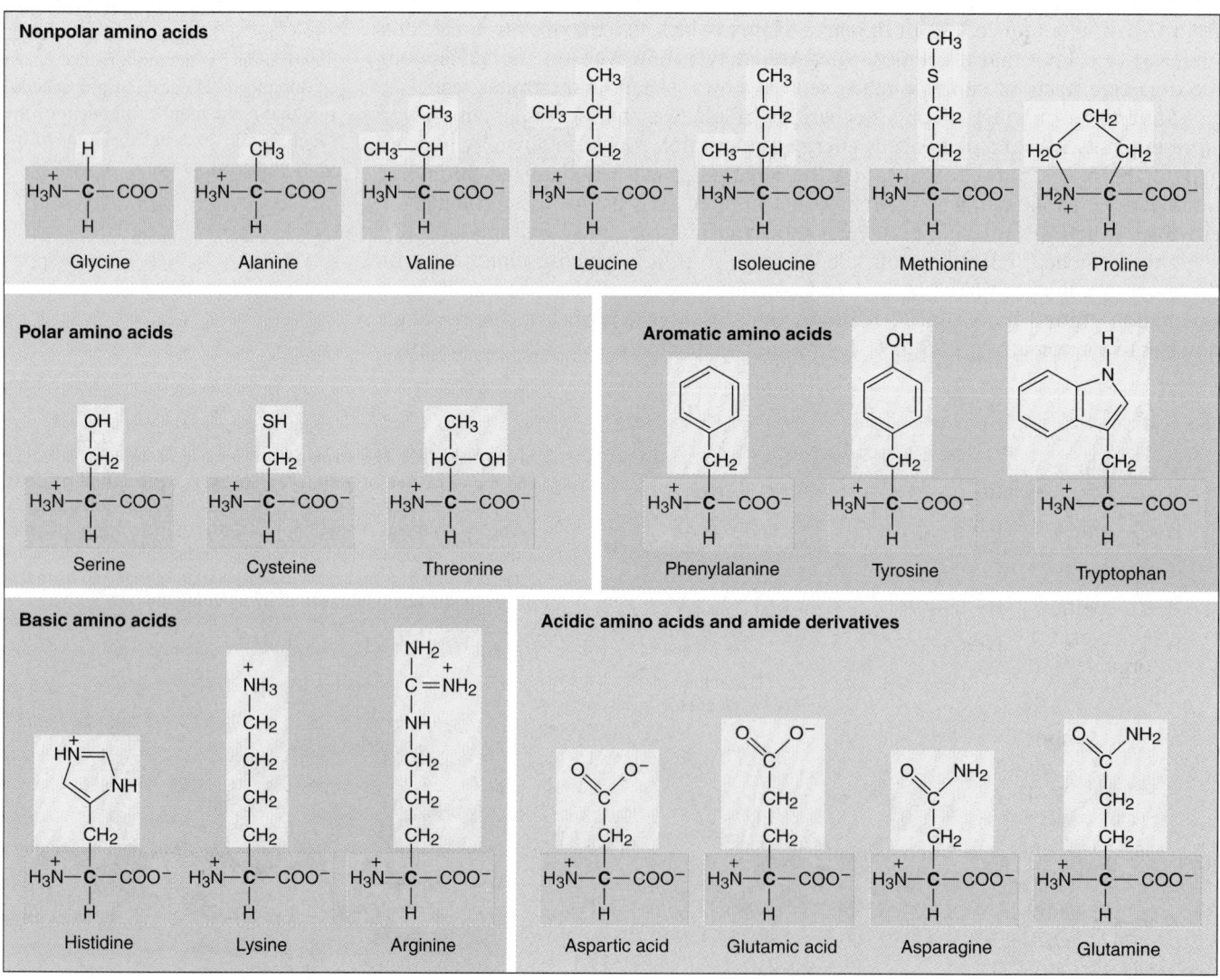

Figure 15.31 A portion of a polypeptide chain. The peptide bond holds the monomers together in a protein. Three peptide bonds (*orange screen*) joining four amino acids (*gray screen*) occur in this portion of a polypeptide chain. Note the repeating pattern of the chain: peptide bond—α-carbon—peptide bond—α-carbon—and so on. Also note that the R groups dangle off the chain.

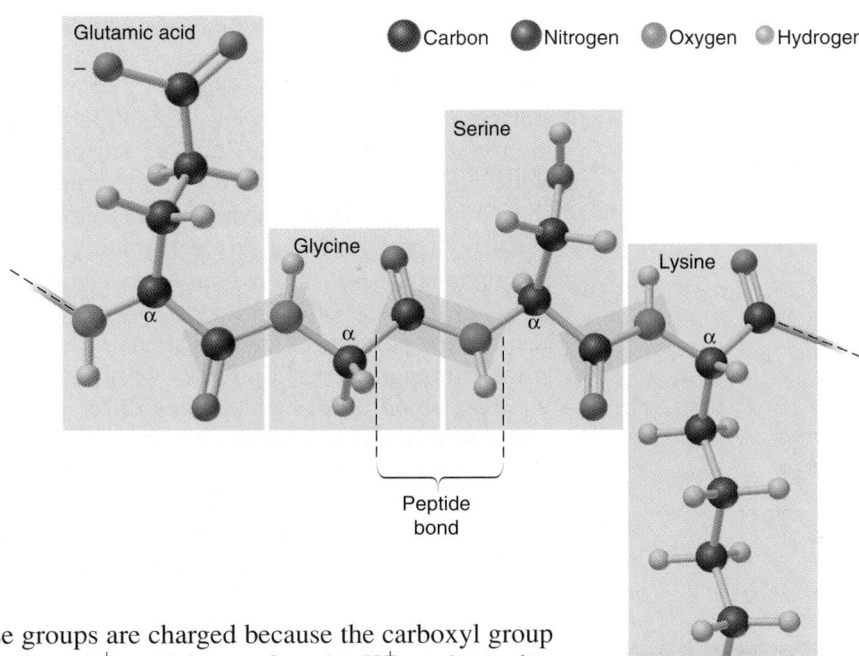

In the aqueous cell fluid, these groups are charged because the carboxyl group transfers an H^+ ion to H_2O to form H_3O^+, which transfers the H^+ to the amine group. The overall process is, in effect, an intramolecular acid-base reaction:

As you can see, an H atom is the third group bonded to the α-carbon, and the fourth is the R group (also called the *side chain*).

Each amino acid is linked to the next one by a *peptide* (amide) bond in a dehydration-condensation reaction by which the carboxyl group of one monomer reacts with the amine group of the next. Therefore, the polypeptide chain, the backbone of the protein, consists of an α-*carbon bonded to an amide group bonded to the next* α-*carbon bonded to the next amide group,* and so forth, as shown in Figure 15.31). The various R groups dangle from the α-carbons on alternate sides of the chain.

Proteins range in length from about 50 amino acids ($\mathcal{M} \approx 5 \times 10^3$ g/mol) to several thousand ($\mathcal{M} \approx 5 \times 10^5$ g/mol). Even for a relatively small protein of 100 amino acids, the number of possible sequences of the 20 types of amino acids is virtually limitless ($20^{100} \approx 10^{130}$) from a purely mathematical point of view. In fact, however, only a tiny fraction of these possibilities occur. For example, even in an organism as complex as a human being, there are about 10^5 different types of protein.

Proteins: Composition, Sequence, Shape, and Function *Each type of protein has its own amino acid composition,* a specific number and proportion of the different amino acids. However, it is not the composition that defines the protein's role in the cell; rather, *the sequence of amino acids determines the protein's shape and function.* Shapes vary from long rods to undulating sheets, from baskets with deep crevices to Y-shaped blobs; many proteins have regions of different shapes. The forces responsible for protein shapes are *the same bonding and intermolecular forces that operate for all molecules.* Because proteins are so large, however, these forces often act between distant regions of the same molecule.

Figure 15.32 is a schematic depiction of the forces operating in many proteins. The covalent peptide bonds create the chain (backbone). The —SH ends of two cysteine R groups (see Figure 15.30) often form an —S—S— bond, a covalent *disulfide bridge* that brings distant parts of the chain together. Polar and ionic R groups usually protrude into the surrounding aqueous fluid, interacting with water through ion-dipole forces and H bonds. Sometimes ionic groups secure the chain's bends: this occurs when an anionic (—COO⁻) R group lies near a cationic (—NH₃⁺) one and forms an electrostatic *salt link* (or *ion pair*). Helical and sheet-like segments arise from *H bonds between the C=O of one peptide bond and the N—H of another*. Other H bonds between peptide bond atoms or R groups act to keep distant portions of the chain near each other. Nonpolar R groups usually congregate through dispersion forces within the nonaqueous protein interior.

Two broad classes of proteins differ in the complexity of their amino acid compositions and sequences and, therefore, in their shape and function:

1. *Fibrous proteins* have relatively simple amino acid compositions, with structures shaped like extended helices or sheets. They are key components of hair, wool, skin, and connective tissue—materials that require strength and flexibility. Like synthetic polymers, these proteins have a small number of different

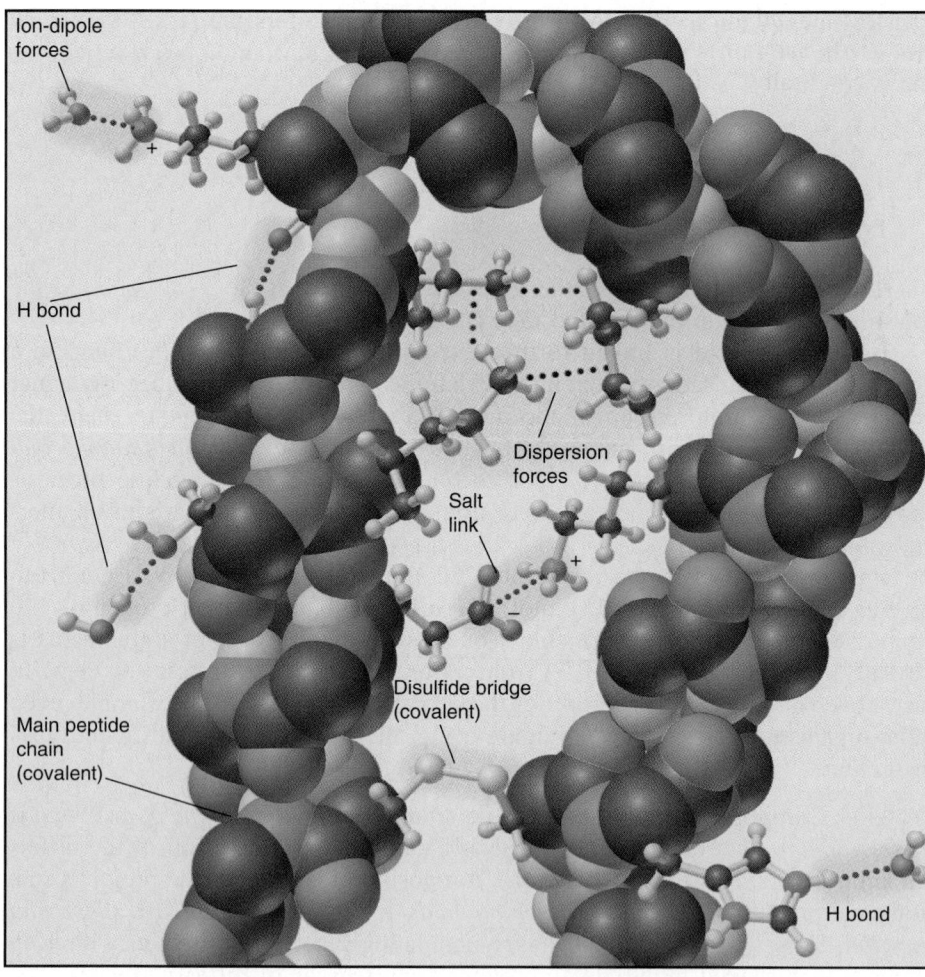

Figure 15.32 **The forces that maintain protein structure.** In this schematic of a portion of a general protein, ball-and-stick R groups are attached to a space-filling polypeptide chain. A combination of covalent, ionic, and intermolecular forces (both within the protein and between the protein and water) is responsible for the overall protein shape, which in turn determines the function.

R groups in a repeating sequence. For example, collagen (Figure 15.33A), the most common animal protein, is found in tendons and skin; more than 30% of the R groups come from glycine (G) and another 20% come from proline (P) to create a long, triple helix with the general sequence —G—X—P—G—X—P— and so on (where X is another amino acid). In silk fibroin, a protein secreted by the silk moth caterpillar, more than 85% of the R groups come from glycine, alanine, and serine (Figure 15.33B). Fibroin chain segments hydrogen bond to one another to form a *pleated sheet.* Stacks of sheets interact through dispersion forces, which make fibroin strong and flexible but not very extendable—perfect for a silkworm's cocoon.

2. *Globular proteins* have much more complex amino acid compositions, often containing varying proportions of all 20 different R groups. As the name implies, globular proteins are typically more rounded and compact, with a wide variety of shapes and a correspondingly wide range of functions: defenders against bacterial invasion, messengers that trigger cell actions, catalysts of chemical change, membrane gatekeepers that maintain aqueous concentrations, and many others. Particular R groups occur at locations crucial to the protein's function. For example, in catalytic proteins, a few R groups form a crevice that closely matches the shapes of reactant molecules. These groups hold the reactants through bonding and intermolecular forces and speed their reaction by bringing them together and twisting and stretching their bonds to form products. Experiment has shown repeatedly that a slight change in one of these critical R groups decreases function dramatically. This fact supports the essential idea that *the protein's amino acid sequence determines its structure, which in turn determines its function*:

$$\text{SEQUENCE} \Longrightarrow \text{STRUCTURE} \Longrightarrow \text{FUNCTION}$$

As you'll see next, the sequence of amino acids in every protein of every organism is prescribed by the genetic information held within the organism's nucleic acids.

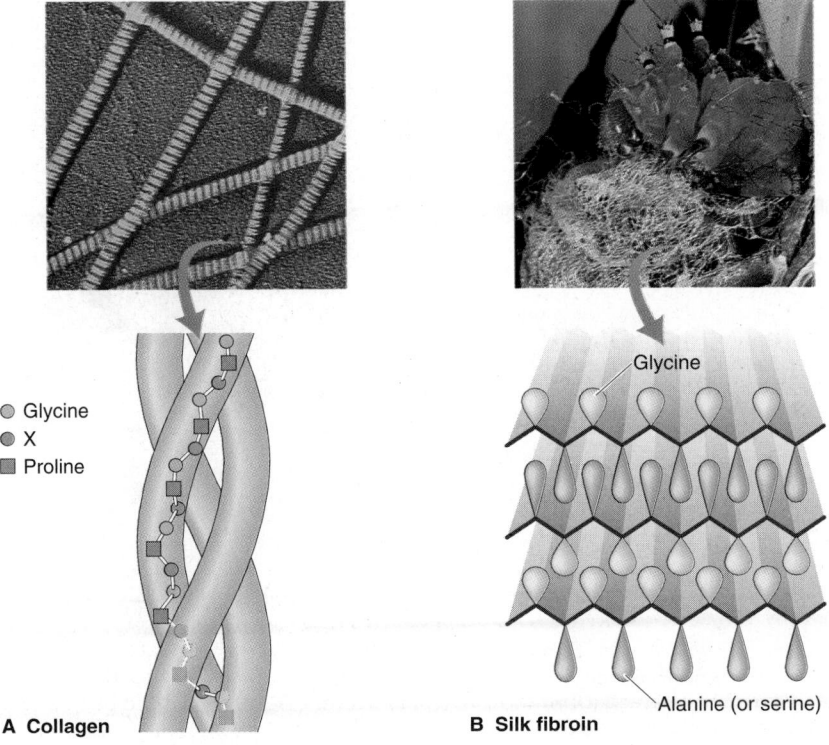

A Collagen **B Silk fibroin**

● Glycine
● X
■ Proline

Glycine

Alanine (or serine)

Figure 15.33 The shapes of fibrous proteins. Fibrous proteins have largely structural roles in an organism, and their amino acid sequence is relatively simple. **A,** The triple helix of a collagen molecule has a glycine as every third amino acid along the chain. **B,** The pleated sheet structure of silk fibroin has a repeating sequence of glycine at every other position with alanine or serine in between. The sheets interact through dispersion forces.

Nucleotides and Nucleic Acids

The chemical information that guides the design, construction, and function of all proteins is contained in **nucleic acids.**

Monomer Structure and Linkage Nucleic acids are unbranched polymers that consist of monomers called **mononucleotides;** each mononucleotide is itself a dehydration-condensation product made from three parts: an N-containing base, a sugar, and a phosphate group (Figure 15.34A). The two types of nucleic acid, *ribonucleic acid* (RNA) and *deoxyribonucleic acid* (DNA), differ in the sugar portions of their mononucleotides. RNA contains *ribose,* a five-C monosaccharide, and DNA contains *deoxyribose,* in which —H substitutes for —OH on the C-2 of ribose.

Dehydration-condensation reactions create phosphate ester links between mononucleotides to give a polynucleotide chain. Therefore, the repeating pattern of the nucleic acid backbone is *sugar linked to phosphate linked to sugar linked to phosphate,* and so on (Figure 15.34B). Attached to each sugar is one of four N-containing bases, either a pyrimidine (six-membered ring) or a purine (six- and five-membered rings sharing a side). The pyrimidines are thymine (T) and cytosine (C); the purines are guanine (G) and adenine (A); in RNA, uracil (U) substitutes for thymine. The bases dangle off the sugar-phosphate chain, similar to the way the R groups dangle off the polypeptide chain of a protein.

Figure 15.34 Mononucleotide monomers and their linkage. **A,** Each mononucleotide consists of a base (structure not shown), a sugar, and a phosphate group. In RNA, the sugar is ribose; in DNA, it is 2-deoxyribose (note the absence of an —OH group on C-2 of the ring). **B,** A tiny segment of the polynucleotide chain of DNA shows the phosphate ester link from one sugar portion to the next formed through a dehydration-condensation. The N-containing bases are dangling off the chain.

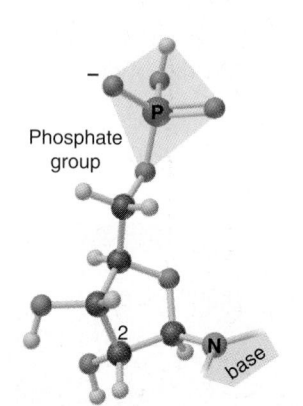

Phosphate group

Mononucleotide of ribonucleic acid (RNA)

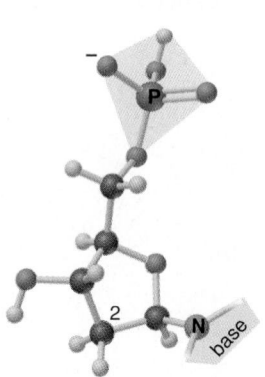

Mononucleotide of deoxyribonucleic acid (DNA)

A

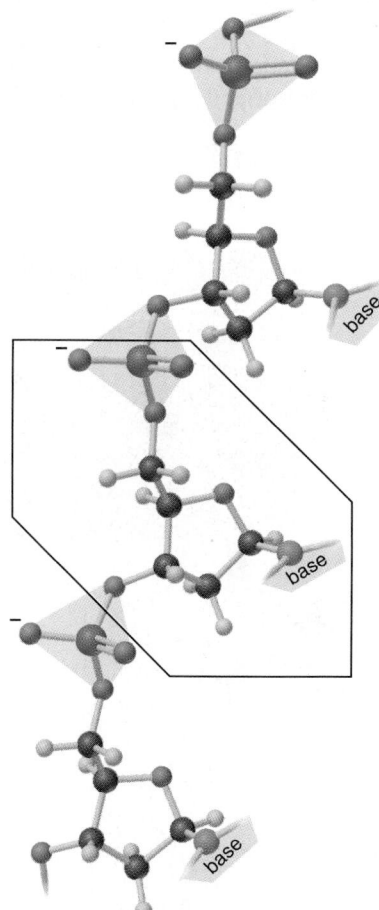

Portion of DNA polynucleotide chain

B

The Central Importance of Base Pairing In the nucleus of the cell, DNA exists as two chains wrapped around each other in a **double helix** (Figure 15.35). The negatively charged sugar-phosphate backbones face the aqueous surroundings, and each base in one chain "pairs" with a base in the other through H bonding. A double-helical DNA molecule contains hundreds of thousands of H-bonded bases in prescribed pairs. Two features of these **base pairs** are crucial to the structure and function of DNA:

- A pyrimidine and a purine are always paired, which gives the double helix a constant diameter.
- Each base is always paired with the same partner: A with T and G with C. Thus, *the base sequence on one chain is the complement of the base sequence on the other*. For example, the sequence A—C—T on one chain is *always* paired with the sequence T—G—A on the other: A with T, C with G, and T with A.

Each of the DNA molecules is folded into a tangled mass that forms one of the cell's *chromosomes*. The DNA molecule is amazingly long and thin: if the largest human chromosome were stretched out, it would be 4 cm long, although in the cell nucleus it is wound tightly into a structure only 5 nm long—8 million times shorter! Within a chromosome, linear segments of the DNA molecule act as *genes*, chemical blueprints for synthesizing the organism's proteins.

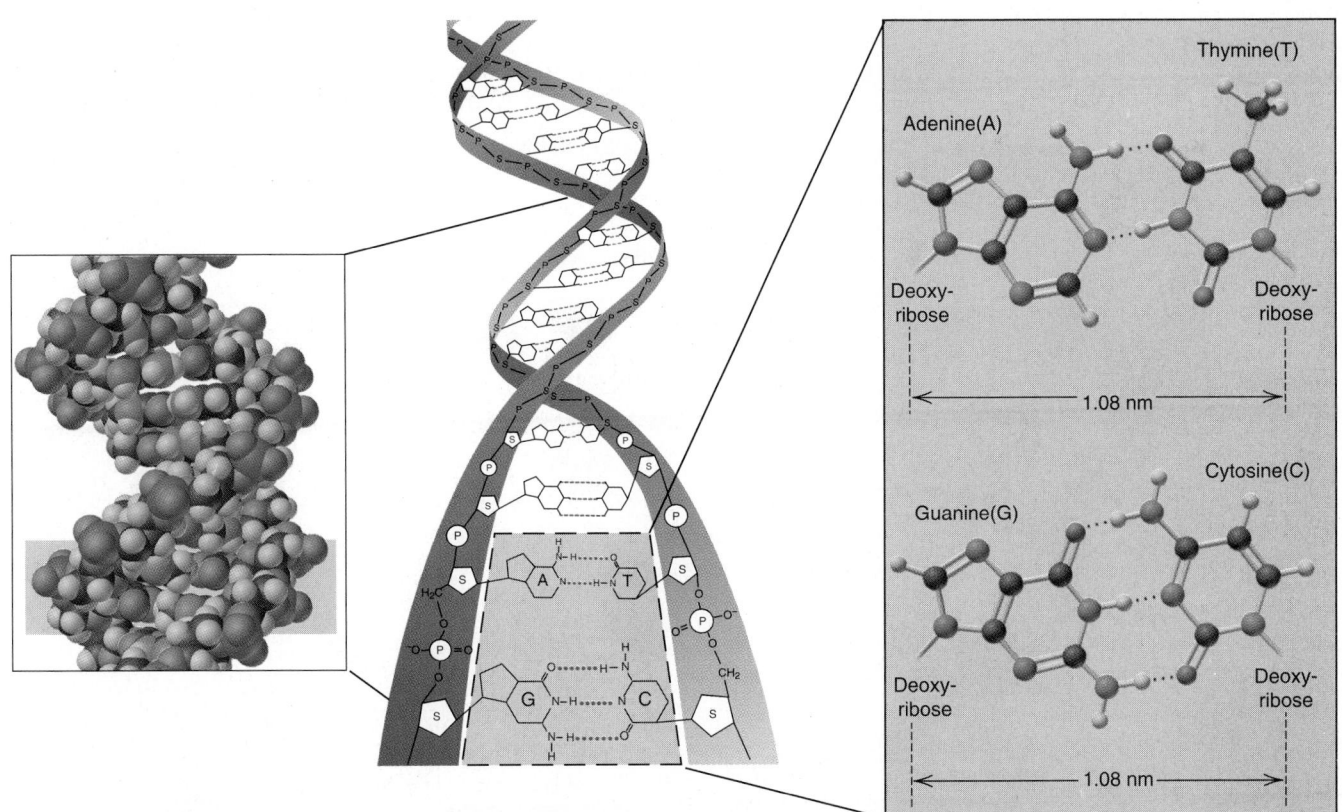

Figure 15.35 The double helix of DNA. In DNA, shown with a short segment of space-filling model (*left*) whose length matches the schematic drawing (*center*), the polar sugar(S)-phosphate(P) backbone faces the watery outside, and the nonpolar bases form H bonds to each other in the DNA core. A pyrimidine and a purine always form H-bonded base pairs to maintain the double helix width (*right*), and the members of the pairs are always the same: A pairs with T and G pairs with C.

An Outline of Protein Synthesis *The information content of a gene resides in its base sequence.* In the **genetic code,** each base acts as a "letter," each three-base sequence as a "word," and *each word codes for a specific amino acid.* For example, the sequence C—A—C codes for the amino acid histidine, A—A—G codes for lysine, and so on. Through a complex series of interactions, greatly simplified in Figure 15.36, one amino acid at a time is positioned and linked to the next in the process of protein synthesis. To fully appreciate just this aspect of the chemical basis of biology, try to keep in mind that *this amazingly complex process occurs largely through H bonding between base pairs.*

Here is an outline of the process of *protein synthesis.* DNA occurs in the cell nucleus, but the genetic message is decoded outside it, so the information must be sent to the synthesis site. RNA serves in this messenger role, as well as in several others. A portion of the DNA is temporarily unwound and acts as a *template* for the formation of a complementary chain of *messenger RNA* (mRNA) made by linking together individual mononucleotides; thus, *the DNA code words are transcribed into RNA code words* through base pairing. The messenger RNA leaves the nucleus and binds, again through base pairing, to an RNA-rich particle in the cell called a *ribosome.* The words (three-base sequences) in the messanger RNA are then decoded by molecules of *transfer RNA* (tRNA). These smaller nucleic acid "shuttles" have two key portions on opposite ends of their structures:

- A three-base sequence complementary with a word on the messenger RNA
- A binding site for the amino acid coded by that word

The ribosome moves along the bound messenger RNA, one word at a time, while transfer RNAs bind and position their amino acids near one another in preparation for peptide bond formation and synthesis of the protein.

In essence, then, protein synthesis involves the DNA message of three-base words being transcribed into the RNA message of three-base words, which is then translated into a sequence of amino acids that are linked to make a protein:

DNA BASE SEQUENCE $\Longrightarrow$ RNA BASE SEQUENCE $\Longrightarrow$ PROTEIN AMINO ACID SEQUENCE

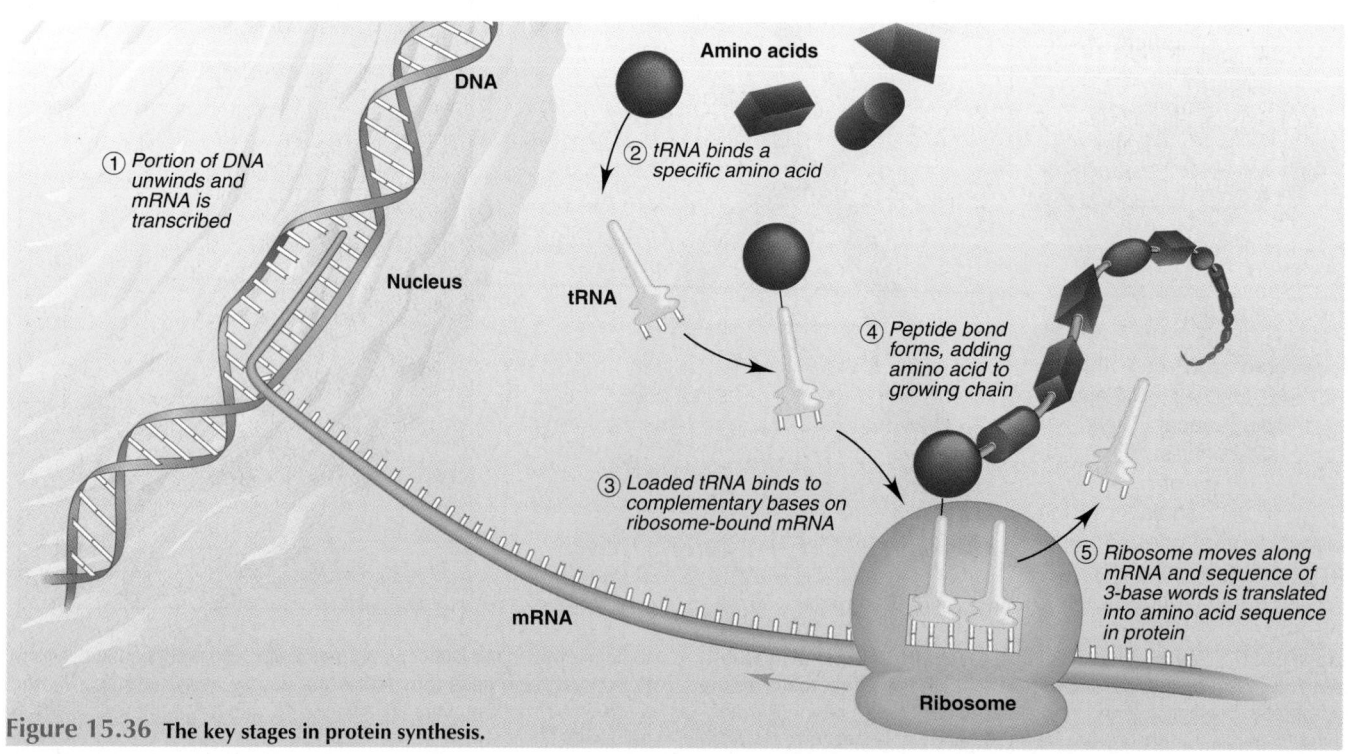

Figure 15.36 The key stages in protein synthesis.

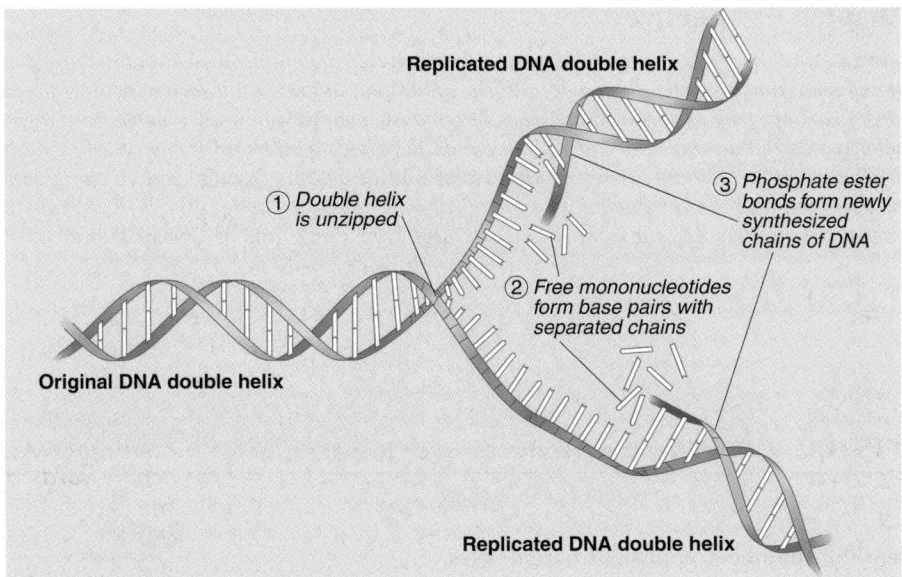

Figure 15.37 **The key stages in DNA replication.**

An Outline of DNA Replication Another complex series of interactions allows *DNA replication,* the ability of DNA to copy itself. When a cell divides, its chromosomes are replicated, or reproduced, ensuring that the new cells have the same number and types of chromosomes. In this process, a small portion of the double helix is "unzipped," and each DNA chain acts as a template for the base pairing of its mononucleotide monomers with free mononucleotides (Figure 15.37). These newly H-bonded mononucleotides are covalently linked together to make a chain. Gradually, each of the unzipped chains forms the complementary half of a double helix, leading to *two* double helices. Because a given base always pairs with its complement, the original double helix is copied and the genetic makeup of the cells is maintained.

The biopolymers provide striking evidence for the folly of vitalism, the idea with which we began the chapter, that there exists some sort of vital force that is found only in substances from living systems. No unknowable force exclusive to living things is required to explain the marvels of the living world. The same atomic properties that give rise to covalent bonds, molecular shape, and intermolecular forces provide the means for all life-forms to flourish.

SECTION SUMMARY

The three types of natural polymers—polysaccharides, proteins, and nucleic acids—are formed by dehydration-condensation reactions. Polysaccharides are formed from cyclic monosaccharides, such as glucose. Cellulose, starch, and glycogen have structural or energy-storage roles. Proteins are polyamides formed from as many as 20 different types of amino acids. Fibrous proteins have extended shapes and play structural roles. Globular proteins have compact shapes and play metabolic, immunologic, and hormonal roles. The amino acid sequence of a protein determines its shape and function. Nucleic acids (DNA and RNA) are polynucleotides formed from four different mononucleotides. The base sequence of the DNA chain determines the sequence of amino acids in an organism's proteins. Hydrogen bonding between specific base pairs is the key to protein synthesis and DNA replication.

Chapter Perspective

The amazing diversity of organic compounds highlights the importance of atomic properties in chemical and biological behavior. While gaining an appreciation for the special properties of carbon, you've seen similar types of bonding, shapes, intermolecular forces, and reactivity in organic and inorganic compounds. Upcoming chapters focus on the dynamic aspects of all reactions—their speed, extent, and direction—as we explore the kinetics, equilibrium, and thermodynamics of chemical and physical change. We'll have many opportunities to exemplify these topics with organic compounds and biochemical systems.

For Review and Reference (Numbers in parentheses refer to pages, unless noted otherwise)

Learning Objectives

Relevant section and/or sample problem (SP) numbers appear in parentheses.

Understand These Concepts

1. How carbon's atomic properties give rise to its ability to form four strong covalent bonds, multiple bonds, and bonds to itself, which results in the great structural diversity of organic compounds (Section 15.1)

2. How carbon's atomic properties give rise to its ability to bond heteroatoms, which creates regions of charge imbalance that result in functional groups (Section 15.1)

3. Structures and names of alkanes, alkenes, and alkynes (Section 15.2)

4. The distinctions among constitutional, optical, and geometric isomers (Section 15.2)

5. The importance of optical isomerism in organisms (Section 15.2)

6. The effect of restricted rotation around a π bond on alkenes' structure and properties (Section 15.2)

7. The nature of organic addition, elimination, and substitution reactions (Section 15.3)

8. The properties and reaction types of the various functional groups (Section 15.4):
- Substitution and elimination for alcohols, alkyl halides, and amines
- Addition for alkenes, alkynes, and aldehydes and ketones
- Substitution for the carboxylic acid family (acids, esters, and amides)

9. Why delocalization of electrons causes aromatic rings to have lower reactivity than alkenes (Section 15.4)

10. The polarity of the carbonyl bond and the importance of organometallic compounds in addition reactions of carbonyl compounds (Section 15.4)

11. How addition plus elimination lead to substitution in the reactions of the carboxylic acid family (Section 15.4)

12. How addition and condensation polymers form (Section 15.5)

13. The three types of biopolymers and their monomers (Section 15.6)

14. How the nature of amino acid R groups and intermolecular forces influence the shape of a protein (Section 15.6)

15. How amino acid sequence determines protein shape, which determines function (Section 15.6)

16. How complementary base pairing controls the processes of protein synthesis and DNA replication (Section 15.6)

17. How DNA base sequence determines RNA base sequence, which determines amino acid sequence (Section 15.6)

Master These Skills

1. Determining hydrocarbon structures given the number of C atoms, multiple bonds, and rings (SP 15.1)

2. Naming hydrocarbons and drawing expanded and condensed formulas (Section 15.2 and SP 15.2)

3. Drawing geometric isomers and identifying chiral centers of molecules (SP 15.2)

4. Recognizing the type of reaction from the structures of reactants and products (SP 15.3)

5. Recognizing a reaction as an oxidation or reduction from the structures of reactants and products (Section 15.3)

6. Determining the reactants and products of the reactions of alcohols, alkyl halides, and amines (SP 15.4 and Follow-up)

7. Determining the products in a stepwise reaction sequence (SP 15.5)

8. Determining the reactants of the reactions of aldehydes and ketones (Follow-up Problem 15.5)

9. Determining the reactants and products of the reactions of the carboxylic acid family (SP 15.6 and Follow-up)

10. Recognizing and naming the functional groups in an organic molecule (SP 15.7)

11. Drawing an abbreviated synthetic polymer structure based on monomer structures (SP 15.5)

12. Drawing small peptides from amino acid structures (Section 15.6)

13. Using the base-paired sequence of one DNA strand to predict the sequence of the other (Section 15.6)

Key Terms

organic compound (607)

Section 15.1
heteroatom (609)
functional group (609)

Section 15.2
hydrocarbon (610)
alkane (C_nH_{2n+2}) (613)
homologous series (613)
saturated hydrocarbon (613)
cyclic hydrocarbon (615)
isomers (615)
constitutional (structural)
    isomers (615)
stereoisomers (617)
optical isomers (617)
chiral molecule (617)

polarimeter (618)
optically active (618)
alkene (C_nH_{2n}) (618)
unsaturated hydrocarbon
    (618)
geometric (*cis-trans*) isomers
    (619)
alkyne (C_nH_{2n-2}) (621)
aromatic hydrocarbon (622)
nuclear magnetic resonance
    (NMR) spectroscopy (624)

Section 15.3
alkyl group (624)
addition reaction (625)
elimination reaction (626)
substitution reaction (626)

Section 15.4
alcohol (628)
haloalkane (alkyl halide)
    (630)
amine (631)
carbonyl group (633)
aldehyde (633)
ketone (634)
organometallic compound
    (634)
carboxylic acid (635)
ester (636)
amide (636)
fatty acid (636)
acid anhydride (637)
lipid (637)

hydrolysis (638)
nitrile (640)

Section 15.5
addition polymer (642)
condensation polymer (643)

Section 15.6
monosaccharide (645)
polysaccharide (645)
disaccharide (645)
protein (646)
amino acid (646)
nucleic acid (650)
mononucleotide (650)
double helix (651)
base pair (651)
genetic code (652)

Highlighted   Figures and Tables

These figures (F) and tables (T) provide a quick review of key ideas.

F15.4 Adding H-atom skin to the C-atom skeleton (611)
T15.1 Roots for carbon chains and branches (613)
T15.2 Naming an organic compound (614)
T15.5 Important functional groups (629)

F15.26 Interconversions of functional groups (641)
F15.31 Portion of polypeptide chain (647)
F15.32 Forces that maintain protein structure (648)
F15.34 Mononucleotide monomers of RNA and DNA (650)

Brief Solutions to Follow-up Problems

5.1 (a)

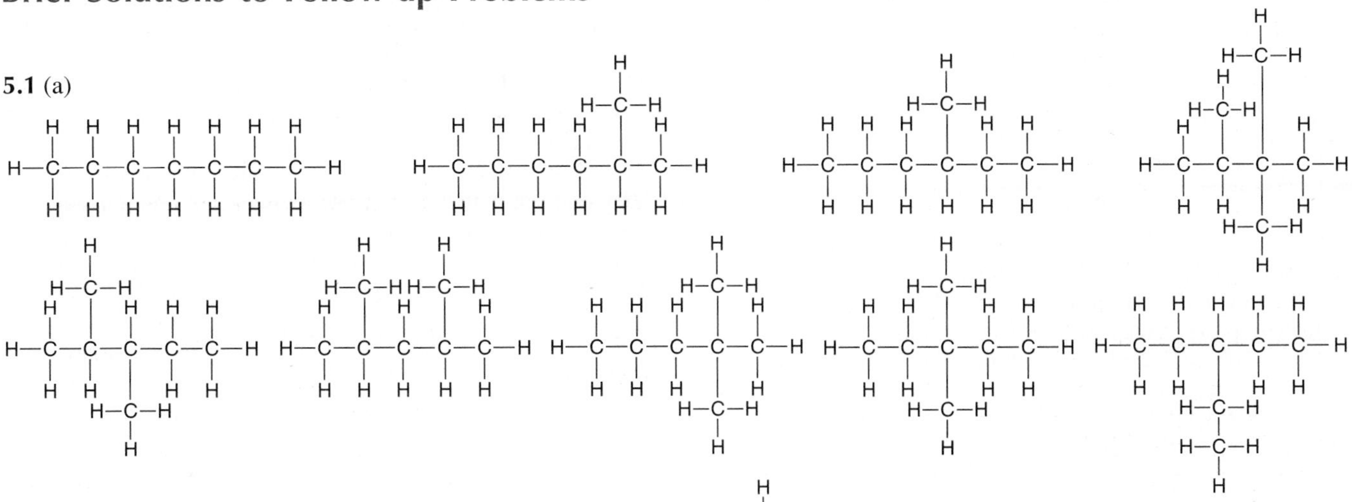

(b)

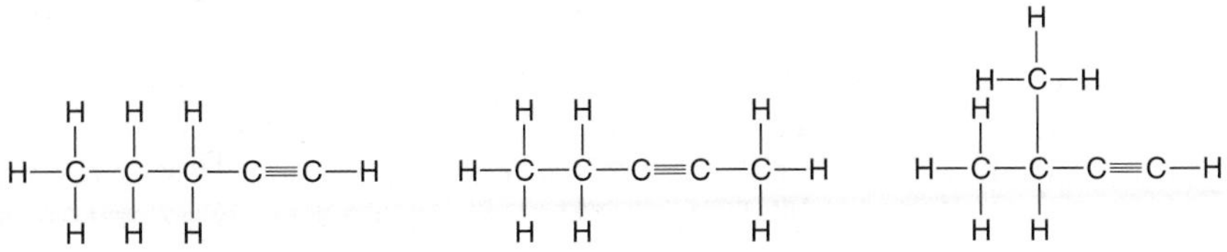

Brief Solutions to Follow-up Problems (continued)

15.2 (a) $CH_3-CH_2-CH_2-CH_2-CH_2-\underset{\underset{CH_3}{|}}{\overset{\overset{CH_2-CH_3}{|}}{C}}-CH_2-CH_3$

(b)

$CH_2-CH_2-CH_3$ ... CH_2-CH_3

(c) $HC\equiv C-\underset{\underset{CH_2-CH_3}{|}}{\overset{\overset{CH_2-CH_3}{|}}{C}}-CH_2-CH_2-CH_3$

(d) $CH_3-CH_2\diagdown \underset{CH_3}{\diagup}C=C\diagup \overset{H}{\diagdown}CH_2-CH_2-CH_3$

15.3 (a) $CH_3-CH=CH-CH_3 + Cl_2 \longrightarrow$

$CH_3-\underset{\underset{Cl}{|}}{\overset{\overset{Cl}{|}}{CH}}-CH-CH_3$

(b) $CH_3-CH_2-CH_2-Br + OH^- \longrightarrow$

$CH_3-CH_2-CH_2-OH + Br^-$

(c)

$CH_3-\underset{\underset{OH}{|}}{\overset{\overset{CH_3}{|}}{C}}-CH_3 \xrightarrow{-H_2O} CH_3-\overset{\overset{CH_3}{|}}{C}=CH_2$

15.4 (a) $CH_3-\underset{\underset{Cl}{|}}{\overset{\overset{CH_3}{|}}{CH}}-CH-CH_3$ or $CH_3-CH_2-\underset{\underset{Cl}{|}}{\overset{\overset{CH_3}{|}}{C}}-CH_3$

(b) $CH_3-CH_2-CH_2-OH$

15.5 (a) [cyclohexane ring with OH and CH_3 groups]

(b) [cyclopentane ring with $\overset{\overset{O}{||}}{C}-H$ group]

15.6 (a) [benzene ring]$-CH_2-\overset{\overset{O}{||}}{C}-OH$

(b) $CH_3-CH_2-CH_2-\overset{\overset{O}{||}}{C}-O-CH_3 + CH_3-CH_2-NH_2$

15.7 (a) [benzene ring]$-\underset{alkene}{(CH=CH)}-\underset{aldehyde}{(\overset{\overset{O}{||}}{C}-H)}$

(b) $\underset{amide}{(H_2N-\overset{\overset{O}{||}}{C})}-CH_2-\underset{haloalkane}{(\underset{\underset{Br}{|}}{CH})}-CH_3$

Problems

Problems with **colored** numbers are answered at the back of the text. Sections match the text and provide the number(s) of relevant sample problems. Most offer Concept Review Questions, Skill-Building Exercises (in similar pairs), and Problems in Context. Then Comprehensive Problems, based on material from any section or previous chapter, follow.

The Special Nature of Carbon and the Characteristics of Organic Molecules

● **Concept Review Questions**

15.1 Give the names and formulas of two carbon compounds that are organic and two that are inorganic.

15.2 Through the first quarter of the 19th century, the idea of vitalism proposed that a key difference existed between substances isolated from animate sources and those from inanimate sources. What was the central notion of vitalism? Give an example of a finding that led to its eventual downfall.

15.3 Explain each statement in terms of atomic properties:
(a) Carbon engages in covalent rather than ionic bonding.
(b) Carbon has four bonds in all its organic compounds.
(c) Carbon forms neither stable cations, like many metals, nor stable anions, like many nonmetals.
(d) Carbon bonds to itself more extensively than does any other element.
(e) Carbon forms stable multiple bonds.

15.4 Carbon bonds to many elements other than itself.
(a) Name six elements that commonly bond to carbon in organic compounds.

(b) Which of these elements are heteroatoms?
(c) Which of these elements are more electronegative than carbon? Less electronegative?
(d) How does bonding of carbon to heteroatoms increase the number of organic compounds?

15.5 Silicon lies just below carbon in Group 4A(14) and also forms four covalent bonds. Why aren't there as many silicon compounds as carbon compounds?

15.6 What is the range of oxidation states for carbon? Name a compound in which carbon has its highest oxidation state and one in which it has its lowest.

15.7 Which of these bonds to carbon would you expect to be relatively reactive: C—H, C—C, C—I, C=O, C—Li? Explain.

The Structures and Classes of Hydrocarbons
(Sample Problems 15.1 and 15.2)

● **Concept Review Questions**

15.8 (a) What structural feature is associated with each of the following hydrocarbons: an alkane; a cycloalkane; an alkene; an alkyne?
(b) Give the general formula for each.
(c) Which hydrocarbons are considered saturated?

15.9 Define each of the following types of isomers: (a) constitutional; (b) geometric; (c) optical. Which types of isomers are stereoisomers?

15.10 Among alkenes, alkynes, and aromatic hydrocarbons, only alkenes exhibit *cis-trans* isomerism. Explain why the others do not.

15.11 Which objects are asymmetric (have no plane of symmetry): (a) a circular clock face; (b) a football; (c) a dime; (d) a brick; (e) a hammer; (f) a spring?

15.12 Explain briefly how a polarimeter works and what it measures.

15.13 How does an aromatic hydrocarbon differ from a cycloalkane in terms of its bonding? How does this difference affect structure?

● Skill-Building Exercises *(paired)*

15.14 Draw all possible skeletons for a 7-C compound with
(a) A 6-C chain and 1 double bond
(b) A 5-C chain and 1 double bond
(c) A 5-C ring and no double bonds

15.15 Draw all possible skeletons for a 6-C compound with
(a) A 5-C chain and 2 double bonds
(b) A 5-C chain and 1 triple bond
(c) A 4-C ring and no double bonds

15.16 Add the correct number of hydrogens to each of the skeletons in Problem 15.14.

15.17 Add the correct number of hydrogens to each of the skeletons in Problem 15.15.

15.18 Draw correct structures, by making a single change, for any that are incorrect:

(a) 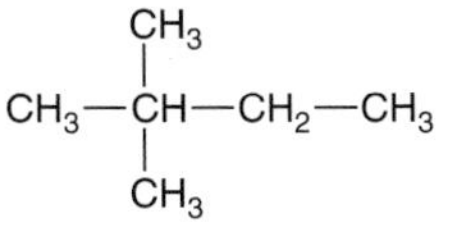
(b) $CH_3{=}CH{-}CH_2{-}CH_3$
(c) 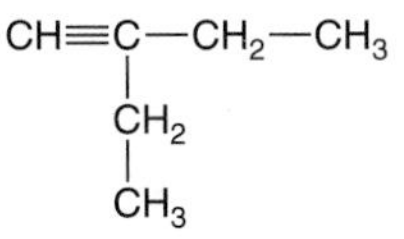
(d) 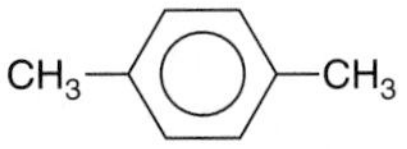

15.19 Draw correct structures, by making a single change, for any that are incorrect:

(a) $CH_3{-}CH{=}CH{-}CH_2{-}CH_3$
(b) 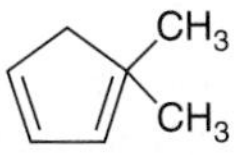
(c) $CH_3{-}C{\equiv}CH{-}CH_2{-}CH_3$
(d) 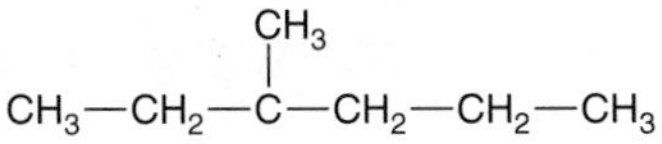

15.20 Draw the structure or give the name of each compound:
(a) 2,3-dimethyloctane
(b) 1-ethyl-3-methylcyclohexane
(c) 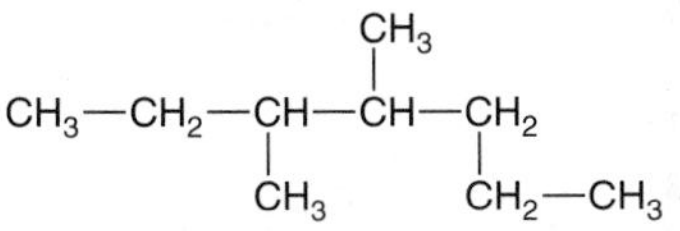
(d) 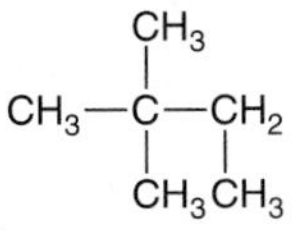

15.21 Draw the structure or give the name of each compound:
(a) 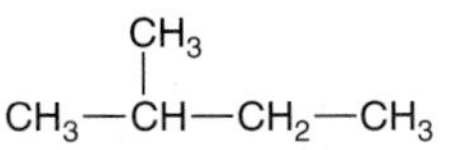
(b) 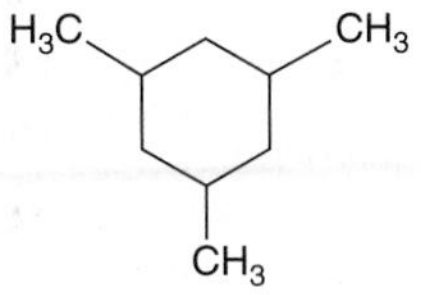
(c) 1,2-diethylcyclopentane
(d) 2,4,5-trimethylnonane

15.22 Each of the following names is wrong. Draw structures based on them, and correct the names:
(a) 4-methylhexane
(b) 2-ethylpentane
(c) 2-methylcyclohexane
(d) 3,3-methyl-4-ethyloctane

15.23 Each of the following names is wrong. Draw structures based on them, and correct the names:
(a) 3,3-dimethylbutane
(b) 1,1,1-trimethylheptane
(c) 1,4-diethylcyclopentane
(d) 1-propylcyclohexane

15.24 Each of the following compounds can exhibit optical activity. Circle the chiral center(s) in each:
(a) 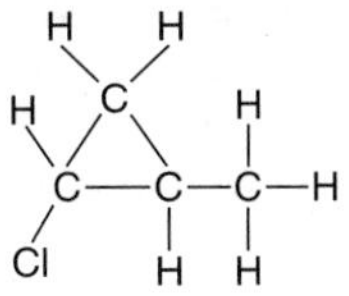
(b) 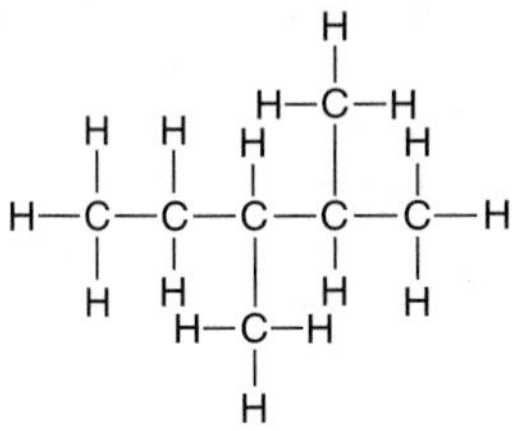

15.25 Each of the following compounds can exhibit optical activity. Circle the chiral center(s) in each:
(a) 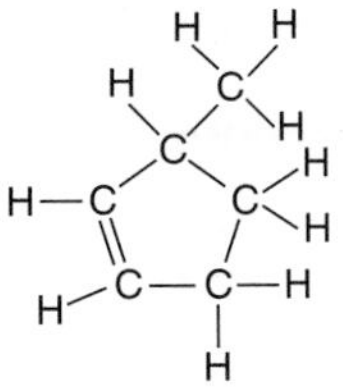
(b) 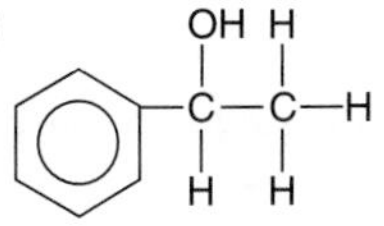

15.26 Draw structures from the following names, and determine which compounds are optically active:
(a) 3-bromohexane
(b) 3-chloro-3-methylpentane
(c) 1,2-dibromo-2-methylbutane

15.27 Draw structures from the following names, and determine which compounds are optically active:
(a) 1,3-dichloropentane
(b) 3-chloro-2,2,5-trimethylhexane
(c) 1-bromo-1-chlorobutane

15.28 Which of the following structures exhibit geometric isomerism? Draw and name the two isomers in each case:
(a) $CH_3{-}CH_2{-}CH{=}CH{-}CH_3$
(b) 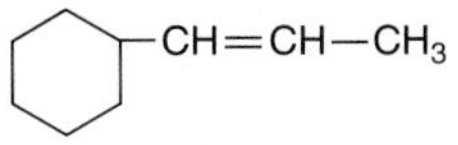
(c) 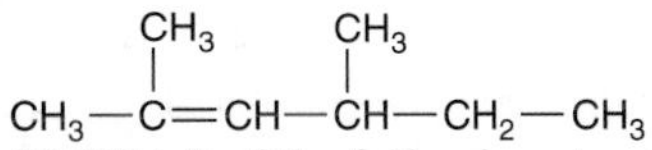

15.29 Which of the following structures exhibit geometric isomerism? Draw and name the two isomers in each case:
(a) 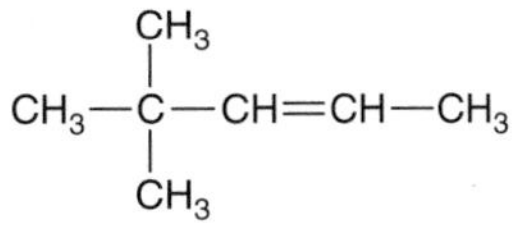
(b) 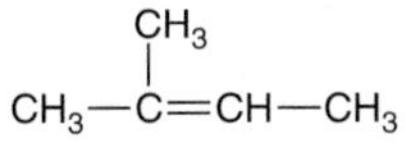
(c) 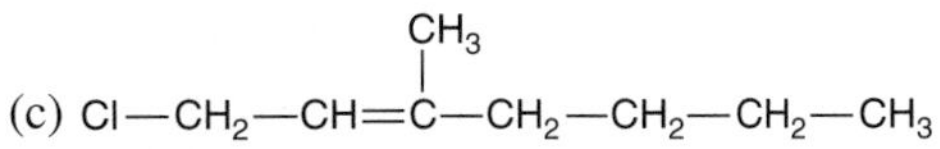

15.30 Which compounds exhibit geometric isomerism? Draw and name the two isomers in each case:
(a) propene
(b) 3-hexene
(c) 1,1-dichloroethene
(d) 1,2-dichloroethene

15.31 Which compounds exhibit geometric isomerism? Draw and name the two isomers in each case:
(a) 1-pentene
(b) 2-pentene
(c) 1-chloropropene
(d) 2-chloropropene

15.32 Draw and name all the constitutional isomers of dichlorobenzene.

15.33 Draw and name all the constitutional isomers of trimethylbenzene.

● **Problems in Context**

15.34 Butylated hydroxytoluene (BHT) is a common preservative added to cereals and other dry foods. Its systematic name is 1-hydroxy-2,6-di-*tert*-butyl-4-methylbenzene (where "*tert*-butyl" is 1,1-dimethylethyl). Draw the structure of BHT.

15.35 There are two compounds with the name 2-methyl-3-hexene, but only one with the name 2-methyl-2-hexene. Explain with structures.

15.36 The explosive TNT is one of six isomers of trinitrotoluene (see p. 623). Draw and name the other five.

15.37 Ethanol and dimethyl ether are constitutional isomers with the molecular formula C_2H_6O.
(a) Draw their structures.
(b) Aside from the TMS peak, how many peaks representing different proton environments appear in the ^1H-NMR spectrum of each isomer?

Some Important Classes of Organic Reactions
(Sample Problem 15.3)

● **Concept Review Questions**

15.38 In terms of numbers of reactant and product substances, which organic reaction type corresponds to (a) a combination reaction, (b) a decomposition reaction, (c) a displacement reaction?

15.39 The same type of bond is broken in an addition reaction and formed in an elimination reaction. Name the type.

15.40 Can a redox reaction also be an addition, elimination, or substitution reaction? Explain with examples.

● **Skill-Building Exercises** *(paired)*

15.41 Determine the type of each of the following reactions:

(a) CH_3—CH_2—$\overset{\overset{\displaystyle Br}{|}}{CH}$—$CH_3$ $\xrightarrow{\text{NaOH}}{\Delta}$
CH_3—CH=CH—CH_3 + NaBr + H_2O

(b) CH_3—CH=CH—CH_2—CH_3 + H_2 $\xrightarrow{\text{Pt}}$
CH_3—CH_2—CH_2—CH_2—CH_3

15.42 Determine the type of each of the following reactions:

(a) CH_3—$\overset{\overset{\displaystyle O}{||}}{CH}$ + HCN $\longrightarrow$ CH_3—$\overset{\overset{\displaystyle OH}{|}}{CH}$—CN

(b) CH_3—$\overset{\overset{\displaystyle O}{||}}{C}$—O—$CH_3$ + CH_3—NH_2 $\xrightarrow{H^+}$
CH_3—$\overset{\overset{\displaystyle O}{||}}{C}$—NH—$CH_3$ + CH_3—OH

15.43 Write equations for the following:
(a) An addition reaction between H_2O and 3-hexene (H^+ is a catalyst)
(b) An elimination reaction between 2-bromopropane and hot potassium ethoxide, CH_3—CH_2—OK (KBr and ethanol are also products)
(c) A light-induced substitution reaction between Cl_2 and ethane to form 1,1-dichloroethane

15.44 Write equations for the following:
(a) A substitution reaction between 2-bromopropane and KI
(b) An addition reaction between cyclohexene and Cl_2
(c) An addition reaction between 2-propanone and H_2 (Ni metal is a catalyst)

15.45 Based on the number of bonds and the nature of the bonded atoms, state whether each of the following changes is an oxidation or a reduction:
(a) =CH_2 becomes —CH_2—OH
(b) =CH— becomes —CH_2—
(c) ≡C— becomes —CH_2—

15.46 Based on the number of bonds and the nature of the bonded atoms, state whether each of the following changes is an oxidation or a reduction:

(a) —$\overset{\overset{\displaystyle |}{}}{\underset{\underset{\displaystyle |}{}}{C}}$—OH becomes —$\overset{}{\underset{\underset{\displaystyle |}{}}{C}}$=O

(b) —CH_2—OH becomes =CH_2

(c) $\overset{\overset{\displaystyle O}{||}}{-C}$—$\overset{\overset{}{}}{\underset{\underset{\displaystyle |}{}}{C}}$— becomes $\overset{\overset{\displaystyle O}{||}}{-C}$—O—

15.47 Is the organic reactant oxidized, reduced, or neither in each of the following reactions?

(a) 2-hexene $\xrightarrow[\text{cold OH}^-]{\text{KMnO}_4}$ 2,3-dihydroxyhexane

(b) cyclohexane $\xrightarrow[\text{catalyst}]{\Delta}$ benzene + $3H_2$

15.48 Is the organic reactant oxidized, reduced, or neither in each of the following reactions?

(a) 1-butyne + H_2 $\xrightarrow{\text{Pt}}$ 1-butene

(b) toluene $\xrightarrow[\text{H}_3\text{O}^+, \Delta]{\text{KMnO}_4}$ benzoic acid

● **Problems in Context**

15.49 Phenylethylamine is a natural substance that is structurally similar to amphetamine. It is found in sources as diverse as almond oil and human urine, where it occurs at elevated concentrations as a result of stress and certain forms of schizophrenia. One method of synthesizing the compound for pharmacological and psychiatric studies involves two steps:

phenylethylamine

Classify each step as an addition, elimination, or substitution.

Properties and Reactivities of Common Functional Groups
(Sample Problems 15.4 to 15.7)

● **Concept Review Questions**

15.50 Compounds with nearly identical molar masses often have very different physical properties. Choose the compound with the higher value for each of the following properties, and explain your choice.
(a) Solubility in water: chloroethane or methylethylamine

(b) Melting point: diethyl ether or 1-butanol
(c) Boiling point: trimethylamine or propylamine

15.51 Fill in each blank with a general formula for the type of compound formed:

15.52 Of the three major types of organic reactions, which do *not* occur readily with benzene? Why?

15.53 Why does the C=O group react differently from the C=C group? Show an example of the difference.

15.54 Many substitution reactions involve an initial electrostatic attraction between reactants. Show where this attraction arises in the formation of an amide from an amine and an ester.

15.55 Although both carboxylic acids and alcohols contain an —OH group, one is acidic in water and the other is not. Explain.

15.56 What reaction type is common to the formation of esters and acid anhydrides? What is the other product?

15.57 Both alcohols and carboxylic acids undergo substitution, but the processes are very different. Explain.

● **Skill-Building Exercises** (*paired*)

15.58 Name the type of organic compound from the following description of its functional group:
(a) Polar single-bonded group that does not include O or N
(b) Triple-bonded group that is polar
(c) Single- and double-bonded group that is acidic in water
(d) Double-bonded group that must be at the end of a carbon chain

15.59 Name the type of organic compound from the following description of its functional group:
(a) N-containing single- and double-bonded group
(b) Double-bonded group that is not polar
(c) Polar double-bonded group that cannot be at the end of a carbon chain
(d) Single-bonded group that is basic in water

15.60 In each of the following, circle and name the functional group(s):

(a) $CH_3-CH=CH-CH_2-OH$

(b)

(c)

(d) $N\equiv C-CH_2-\overset{\overset{\displaystyle O}{\|}}{C}-CH_3$

(e)

15.61 In each of the following, circle and name the functional group(s):

(a) $HO-\overset{\overset{\displaystyle O}{\|}}{C}-CH_2-\overset{\overset{\displaystyle O}{\|}}{CH}$

(b) $I-CH_2-CH_2-C\equiv CH$

(c) $CH_2=CH-CH_2-\overset{\overset{\displaystyle O}{\|}}{C}-O-CH_3$

(d) $CH_3-\overset{\overset{\displaystyle Br}{|}}{CH}-CH=CH-CH_2-NH-CH_3$

(e) $CH_3-NH-\overset{\overset{\displaystyle O}{\|}}{C}-\overset{\overset{\displaystyle O}{\|}}{C}-O-CH_3$

15.62 Draw all possible alcohols with the formula $C_5H_{12}O$.

15.63 Draw all possible aldehydes and ketones with the formula $C_5H_{10}O$.

15.64 Draw all possible amines with the formula $C_4H_{11}N$.

15.65 Draw all possible carboxylic acids with the formula $C_5H_{10}O_2$.

15.66 Draw the product resulting from mild oxidation of (a) 2-butanol; (b) 2-methylpropanal; (c) cyclopentanol.

15.67 Draw the alcohol you would oxidize to produce (a) 2-methylpropanal; (b) 2-pentanone; (c) 3-methylbutanoic acid.

15.68 Draw the organic product formed when the following compounds undergo a substitution reaction:
(a) Acetic acid and methylamine
(b) Butanoic acid and 2-propanol
(c) Formic acid and 2-methyl-1-propanol

15.69 Draw the organic product formed when the following compounds undergo a substitution reaction:
(a) Acetic acid and 1-hexanol
(b) Propanoic acid and dimethylamine
(c) Ethanoic acid and diethylamine

15.70 Draw structures for the carboxylic acid and alcohol portions of the following esters:

(a) $CH_3-(CH_2)_4-\overset{\overset{\displaystyle O}{\|}}{C}-O-CH_2-CH_3$

(b)

(c)

15.71 Draw structures for the carboxylic acid and amine portions of the following amides:

(a)

(b) $CH_3-\overset{\overset{\displaystyle CH_3}{|}}{CH}-\overset{\overset{\displaystyle O}{\|}}{C}-\overset{\underset{\displaystyle CH_3}{|}}{N}-CH_2-CH_3$

(c)

15.72 Fill in the expected organic products:

(a) $CH_3-CH_2-Br \xrightarrow{OH^-} \underline{\quad} \xrightarrow[H^+]{CH_3-CH_2-\overset{\overset{\displaystyle O}{\|}}{C}-OH} \underline{\quad}$

(b) $CH_3-CH_2-\overset{\overset{\displaystyle Br}{|}}{CH}-CH_3 \xrightarrow{CN^-} \underline{\quad} \xrightarrow{H_3O^+,\ H_2O} \underline{\quad}$

15.73 Fill in the expected organic products:

(a) $CH_3-CH_2-CH=CH_2 \xrightarrow{H^+, H_2O} \underline{\quad} \xrightarrow{Cr_2O_7^{2-}, H^+} \underline{\quad}$

(b) $CH_3-CH_2-\overset{\overset{\displaystyle O}{\|}}{C}-CH_3 \xrightarrow{CH_3-CH_2-Li} \xrightarrow{H_2O} \underline{\quad}$

15.74 Supply the missing organic and/or inorganic substances:

(a) $CH_3-CH_2-OH + \underline{\quad ? \quad} \xrightarrow{?}$

$CH_3-CH_2-O-\overset{\overset{\displaystyle O}{\|}}{C}-CH_2-CH_3$

(b) $CH_3-\overset{\overset{\displaystyle O}{\|}}{C}-O-CH_3 \xrightarrow{?} CH_3-CH_2-NH-\overset{\overset{\displaystyle O}{\|}}{C}-CH_3$

15.75 Supply the missing organic and/or inorganic substances:

(a) $CH_3-\overset{\overset{\displaystyle Cl}{|}}{CH}-CH_3 \xrightarrow{?} CH_3-CH=CH_2 \xrightarrow{?} CH_3-\overset{\overset{\displaystyle Br}{|}}{CH}-\overset{\overset{\displaystyle Br}{|}}{CH_2}$

(b) $CH_3-CH_2-CH_2-OH \xrightarrow{?} CH_3-CH_2-\overset{\overset{\displaystyle O}{\|}}{C}-OH + \underline{\quad ? \quad} \xrightarrow{?}$

$CH_3-CH_2-\overset{\overset{\displaystyle O}{\|}}{C}-O-CH_2-\bigcirc$

● **Problems in Context**

15.76 Which of the following compounds are highly soluble in water? Explain.

(a) $CH_3-CH_2-CH=CH_2$
(b) $CH_3-CH_2-CH_2-OH$

(c) $CH_3-CH_2-CH_2-NH_2$
(d) $CH_3-CH_2-\overset{\overset{\displaystyle O}{\|}}{C}-OH$

(e) $CH_3-(CH_2)_6-\overset{\overset{\displaystyle O}{\|}}{C}-OH$
(f) $CH_3-CH_2-\overset{\overset{\displaystyle Br}{|}}{CH}-CH_3$

15.77 Ethyl formate ($H\overset{\overset{\displaystyle O}{\|}}{C}-O-CH_2-CH_3$) is commonly added to foods to give them the flavor of rum. How would you synthesize ethyl formate from ethanol, methanol, and any inorganic reagents?

The Monomer-Polymer Theme I: Synthetic Macromolecules

● **Concept Review Questions**

15.78 Name the reaction processes that lead to the two types of synthetic polymers.
15.79 Which functional group is common to the monomers that make up addition polymers? What makes these polymers different from one another?
15.80 What is a free radical? How is it involved in polymer formation?
15.81 Which intermolecular force is primarily responsible for the different types of polyethylene? Explain.
15.82 Which of the two types of synthetic polymer is more similar chemically to biopolymers? Explain.
15.83 Which two functional groups react to form nylons? Polyesters?

● **Skill-Building Exercises (paired)**

15.84 Draw an abbreviated structure for the following polymers, with brackets around the repeat unit:

(a) Poly(vinyl chloride) (PVC) from $\overset{H}{\underset{H}{>}}C=C\overset{H}{\underset{Cl}{<}}$

(b) Polypropylene from $\overset{H}{\underset{H}{>}}C=C\overset{H}{\underset{CH_3}{<}}$

15.85 Draw an abbreviated structure for the following polymers, with brackets around the repeat unit:

(a) Teflon from $\overset{F}{\underset{F}{>}}C=C\overset{F}{\underset{F}{<}}$

(b) Polystyrene from $\overset{H}{\underset{H}{>}}C=C\overset{H}{\underset{\bigcirc}{<}}$

15.86 Write a balanced equation for the reaction between 1,4-benzenedicarboxylic acid and 1,2-dihydroxyethane to form the polyester Dacron. Draw an abbreviated structure for the polymer, with brackets around the repeat unit.
15.87 Write a balanced equation for the reaction of dihydroxydimethylsilane,

$$HO-\overset{\overset{\displaystyle CH_3}{|}}{\underset{\underset{\displaystyle CH_3}{|}}{Si}}-OH$$

to form the condensation polymer known as Silly Putty.

The Monomer-Polymer Theme II: Biological Macromolecules

● **Concept Review Questions**

15.88 Which type of polymer is formed from each of the following monomers: (a) amino acids; (b) alkenes; (c) simple sugars; (d) mononucleotides?
15.89 What is the key structural difference between fibrous and globular proteins? How is it related, in general, to the proteins' amino acid composition?
15.90 Protein shape, function, and amino acid sequence are interrelated. Which determines which?
15.91 What type of link joins the mononucleotides in each strand of DNA?
15.92 What is base pairing? How does it pertain to the structure of DNA?
15.93 RNA base sequence, protein amino acid sequence, and DNA base sequence are interrelated. Which determines which in the process of protein synthesis?

● **Skill-Building Exercises (paired)**

15.94 Draw the structure of the R group of (a) alanine; (b) histidine; (c) methionine.
15.95 Draw the structure of the R group of (a) glycine; (b) isoleucine; (c) tyrosine.

15.96 Draw the structure of each of the following tripeptides:
(a) Aspartic acid-histidine-tryptophan
(b) Glycine-cysteine-tyrosine with the charges that exist in cell fluid
15.97 Draw the structure of each of the following tripeptides:
(a) Lysine-phenylalanine-threonine
(b) Alanine-leucine-valine with the charges that exist in cell fluid

15.98 Write the sequence of the complementary DNA strand that pairs with each of the following DNA base sequences: (a) TTAGCC; (b) AGACAT.

15.99 Write the sequence of the complementary DNA strand that pairs with each of the following DNA base sequences: (a) GGTTAC; (b) CCCGAA.

15.100 Write the base sequence of DNA template from which this RNA sequence was derived: UGUUACGGA. How many amino acids are coded for in this sequence?

15.101 Write the base sequence of DNA template from which this RNA sequence was derived: GUAUCAAUGAACUUG. How many amino acids are coded for in this sequence?

● **Problems in Context**

15.102 Protein shapes are maintained by a variety of forces that arise from interactions between the amino acid R groups. Name the amino acid that possesses each R group and the force that could arise in each of the following interactions:

(a) $-CH_2-SH$ with $HS-CH_2-$

(b) $-(CH_2)_4-NH_3^+$ with $^-O-\overset{\overset{O}{\|}}{C}-CH_2-$

(c) $-CH_2-\overset{\overset{O}{\|}}{C}-NH_2$ with $HO-CH_2-$

(d) $-\overset{\overset{CH_3}{|}}{CH}-CH_3$ with 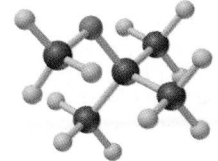$-CH_2-$

15.103 What forces are responsible for the helix and pleated sheet structures of so many proteins? From which groups do these forces arise?

Comprehensive Problems

Problems with an asterisk (*) are more challenging.

***15.104** Starting with the given organic reactant and any necessary inorganic reagents, explain how you would perform each of the following syntheses:

(a) Starting with $CH_3-CH_2-CH_2-OH$, make

$$CH_3-\overset{\overset{Br}{|}}{CH}-CH_2-Br$$

(b) Starting with CH_3-CH_2-OH, make

$$CH_3-\overset{\overset{O}{\|}}{C}-O-CH_2-CH_3$$

***15.105** Ethers (general formula $R-O-R'$) have many important uses. Until recently, methyl *tert*-butyl ether (MTBE, see structure below) was used as an octane booster and fuel additive for gasoline. It increases the oxygen content of the fuel, which reduces CO emissions during winter months. MTBE is synthesized by reacting 2-methylpropene with methanol.

(a) Write a balanced equation for the synthesis of MTBE. (*Hint:* Alcohols add to alkenes similarly to way water does.)

(b) If the government required that auto fuel mixtures contain 2.7% oxygen by mass to reduce CO emissions, how many grams of MTBE would have to be added to each 100. g of gasoline?

(c) How many liters of MTBE would be present in each liter of fuel mixture? The density of both gasoline and MTBE is 0.740 g/mL.

(d) How many liters of air, which is approximately 21% O_2 by volume, are needed at 24°C and 1.00 atm to fully combust 1.00 L of MTBE?

15.106 Compound X has the formula C_4H_8O and reacts readily with $K_2Cr_2O_7$ in H_2SO_4 to yield a carboxylic acid. X is synthesized from 2-methyl-1-propanol. Propose a structure for compound X.

***15.107** Some of the most useful compounds for organic synthesis are Grignard reagents, for the development of which Victor Grignard and Paul Sabatier were awarded the Nobel Prize in 1912. These compounds (general formula $R-MgX$, where X is a halogen) are made from $R-X$ with Mg in ether solvent and used to change the carbon skeleton of a starting carbonyl compound in a reaction similar to that with $R-Li$:

$$R'-\overset{\overset{O}{\|}}{C}-R'' + R-MgBr \longrightarrow R'-\overset{\overset{\overset{OMgBr}{|}}{C}}{\underset{R}{|}}-R'' \xrightarrow{H_2O}$$

$$R'-\overset{\overset{\overset{OH}{|}}{C}}{\underset{R}{|}}-R'' + Mg(OH)Br$$

(a) What is the product, after a final step with water, of the reaction between ethanal and the Grignard reagent of bromobenzene?

(b) What is the product, after a final step with water, of the reaction between 2-butanone and the Grignard reagent of 2-bromopropane?

(c) There are often two (or more) combinations of Grignard reagent and carbonyl compound that will give the same product. Choose another pair of reactants to give the product in (a).

(d) What carbonyl compound must react with a Grignard reagent to obtain the $-OH$ group at the *end* of the carbon chain?

(e) What Grignard reagent and carbonyl compound would you use to prepare 2-methyl-2-butanol?

15.108 A synthesis of 2-butanol was performed by treating 2-bromobutane with hot sodium hydroxide solution. The yield was 60%, indicating that a significant portion of the reactant was converted into a second product. Predict what this other product might be. What simple test would support your prediction?

***15.109** Biphenyl consists of two benzene rings connected by a single bond. When Cl atoms substitute for some of the H atoms, polychlorinated biphenyls (PCBs) are formed. There are over 200 different PCBs, but only six with one Cl on each ring. Draw these constitutional isomers.

***15.110** Cadaverine (1,5-diaminopentane) and putrescine (1,4-diaminobutane) are two compounds that are formed by bacterial action and are responsible for the odor of rotting flesh. Draw their structures. Suggest a series of reactions to synthesize putrescine from 1,2-dibromoethane and any inorganic reagents.

15.111 Pyrethrins, such as jasmolin II (shown below), are a group of natural compounds that are synthesized by flowers of the genus *Chrysanthemum* (known as pyrethrum flowers) to act as insecticides.
(a) Circle and name the functional groups in jasmolin II.
(b) What is the hybridization of the numbered carbons?
(c) Which, if any, of the numbered carbons are chiral centers?

15.112 Taste can often be affected by relatively small changes in a molecule. For example, compound A (shown below) tastes 4000 times sweeter than glucose, whereas compound B has no taste. An —NH$_2$ group attached to benzene gives *aniline*, and the carbon to which it is attached is designated 1. The CH$_3$—CH$_2$—CH$_2$—O is called a propoxy group. Name compounds A and B.

15.113 Vanillin (see structure) is a naturally occurring flavoring agent used in many food products. Name each functional group that contains oxygen. Which carbon-oxygen bond is shortest?

15.114 It has been said that "the hydrogen bond is the most important intermolecular force in nature." Give five cases to support this statement.
15.115 Which features of retinal make it so useful as a photon absorber in the visual systems of organisms?
15.116 The polypeptide chain in proteins does not exhibit free rotation because of the partial double-bond character of the peptide bond. Explain this fact with resonance structures.
*15.117 Although other nonmetals form many compounds that are structurally analogous to those of carbon, the inorganic compounds are usually more reactive. Predict any missing products and write balanced equations for each of the following:
(a) The decomposition and chlorination of diborane to boron trichloride
(b) The combustion of pentaborane (B$_5$H$_9$) in O$_2$
(c) The addition of water to the double bonds of borazine (B$_3$N$_3$H$_6$, Figure 14.9). (*Hint:* The —OH group bonds to B.)
(d) The hydrolysis of trisilane (Si$_3$H$_8$) to silica (SiO$_2$) and H$_2$
(e) The complete halogenation of disilane with Cl$_2$
(f) The thermal decomposition of H$_2$S$_5$ to hydrogen sulfide and sulfur molecules
(g) The hydrolysis of PCl$_5$

*15.118 The genetic code consists of a series of three-base words that each code for a given amino acid.
(a) Using the selections from the genetic code shown below, determine the amino acid sequence coded by the following segment of RNA:

UCCACAGCCUAUAUGGCAAACUUGAAG

AUG = methionine	CCU = proline	CAU = histidine
UGG = tryptophan	AAG = lysine	UAU = tyrosine
GCC = alanine	UUG = leucine	CGG = arginine
UGU = cysteine	AAC = asparagine	ACA = threonine
UCC = serine	GCA = alanine	UCA = serine

(b) What is the complementary DNA sequence from which this RNA sequence was made?
*15.119 Citral, the main odor constituent in oil of lemon grass, is used extensively in the flavoring and perfume industries. It occurs naturally as a mixture of the geometric isomers geranial (the *trans* isomer) and neral (the *cis* isomer). Its systematic name is 3,7-dimethyl-2,6-octadienal ("dien" indicates two C=C groups). Draw the structures of geranial and neral.
*15.120 Complete hydrolysis of a 100.00-g sample of a peptide gave the following amounts of individual amino acids (molar masses appear in parentheses): 3.00 g of glycine (75.07); 0.90 g of alanine (89.10); 3.70 g of valine (117.15); 6.90 g of proline (115.13); 7.30 g of serine (105.10); and 86.00 g of arginine (174.21).
(a) Why does the total mass of amino acids exceed the mass of peptide?
(b) What are the relative numbers of amino acids in this peptide sample?
(c) What is the minimum molar mass of the peptide?
15.121 Thiols (or mercaptans) are the sulfur analogs of alcohols; that is, S appears in the molecule in place of O. Many occur naturally, and all have very disagreeable odors. They are named by adding the suffix *-thiol* to the entire parent hydrocarbon name or by using *mercapto-* as a prefix.
(a) Draw the structure of a compound in skunk musk, 3-methyl-1-butanethiol.
(b) Write a reaction to convert 3-mercapto-1-propene (found in garlic) to propanethiol (found in onion).
*15.122 2-Butanone is reduced by hydride ion donors, such as sodium borohydride (NaBH$_4$), to the alcohol 2-butanol. Even though the alcohol has a chiral center, the product isolated from the redox reaction is not optically active. Explain.

*15.123 Sodium propanoate (CH$_3$—CH$_2$—$\overset{\text{O}}{\overset{\|}{\text{C}}}$—ONa) is a common preservative found in breads, cheeses, and pies. How would you synthesize sodium propanoate from 1-propanol and any inorganic reagents?
15.124 Because of the behavior of their R groups in water, lysine is classified as a basic amino acid and aspartic acid as an acidic amino acid. Write balanced equations that demonstrate this behavior.

CHAPTER 16

KINETICS: RATES AND MECHANISMS OF CHEMICAL REACTIONS

CHAPTER OUTLINE

16.1 Factors That Influence Reaction Rate

16.2 Expressing the Reaction Rate
Average, Instantaneous, and Initial Reaction Rates
Rate and Concentration

16.3 The Rate Law and Its Components
Initial Rate
Reaction Order Terminology
Determining Reaction Orders
Determining the Rate Constant

16.4 Integrated Rate Laws: Concentration Changes over Time
First-, Second-, and Zero-Order Reactions
Reaction Order
Reaction Half-Life

16.5 The Effect of Temperature on Reaction Rate

16.6 Explaining the Effects of Concentration and Temperature
Collision Theory
Transition State Theory

16.7 Reaction Mechanisms: Steps in the Overall Reaction
Elementary Reactions
The Rate-Determining Step
The Mechanism and the Rate Law

16.8 Catalysis: Speeding Up a Chemical Reaction
Homogeneous Catalysis
Heterogeneous Catalysis

Figure: Temperature and biological activity. The metabolic processes of cold-blooded animals like this magnificent veiled chameleon speed up as temperatures rise toward midday. In this chapter, you'll see how the speed of a reaction is influenced by several factors, including temperature, and how we can control them.

CONCEPTS & SKILLS
to review before you study this chapter
• influence of temperature on molecular
  speed and collision frequency
  (Section 5.6)

Until now we've taken a rather simple approach to chemical change: reactants mix and products form. A balanced equation is an essential quantitative tool for calculating product yields from reactant amounts, but it tells us nothing about three dynamic aspects of the reaction, which are essential to understanding chemical change and which we examine in the next several chapters:

• How fast is the reaction proceeding at a given moment?
• What will the reactant and product concentrations be when the reaction is complete?
• Will the reaction proceed by itself and release energy, or will it require energy to proceed?

In this chapter, we address the first of these questions and focus on the field of *kinetics,* which deals with the speed of a reaction and its mechanism, the stepwise changes that reactants undergo in their conversion to products. Chapters 17 through 19 are concerned with *equilibrium,* the dynamic balance between forward and reverse reactions and how external influences alter reactant and product concentrations. Chapters 20 and 21 present *thermodynamics* and its application to electrochemistry; there, we investigate why a reaction occurs and how we can put it to use. These central subdisciplines of chemistry apply to all physical and chemical change and, thus, are crucial to our understanding of modern technology, the environment, and our own biology.

Chemical kinetics is the study of *reaction rates,* the changes in concentrations of reactants (or products) as a function of time (Figure 16.1).

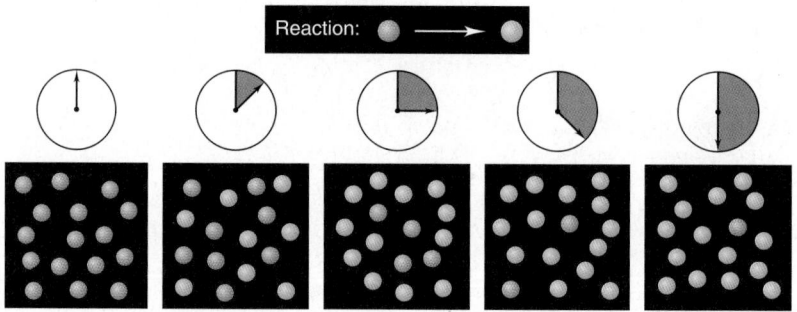

Figure 16.1 **Reaction rate: the central focus of chemical kinetics.** The rate at which reactant becomes product is the underlying theme of chemical kinetics. As time elapses, reactant *(purple) decreases* and product *(green) increases.*

Reactions occur at a wide range of rates (Figure 16.2). Some, like a neutralization, a precipitation, or an explosive redox process, seem to be over as soon as the reactants make contact—in a fraction of a second. Others, such as the reactions involved in cooking or rusting, take a moderate length of time, from minutes to months. Still others take much longer: the reactions that make up the human aging process continue for decades, and those involved in the formation of coal from dead plants take hundreds of millions of years.

Knowing how fast a chemical change occurs can be essential. How quickly a medicine acts or blood clots can make the difference between life and death. How long it takes for cement to harden, polyethylene to form, or a fabric to be dyed can make the difference between profit and loss. In general, the rates of these diverse processes depend on the same variables, most of which chemists can manipulate to maximize yields within a given time or to slow down an unwanted reaction.

We begin with a qualitative overview of the factors affecting reaction rate. Then we see how to express a rate quantitatively in the form of a *rate law,* how the components of a rate law are experimentally determined, and how concentra-

A

B

C

D

Figure 16.2 The wide range of reaction rates. Reaction processes occur at a wide range of rates. An explosion **(A)** is much faster than the process of ripening **(B),** which is much faster than the process of rusting **(C),** which is much faster than the process of human aging **(D).**

tion and temperature affect the rate. Next, we examine the models that explain those effects. Only then can we take apart the overall reaction, see the stages it may go through, and picture the structure that exists fleetingly at the moment the reactant bonds are breaking and the product bonds are forming. The chapter ends with a discussion of how catalysts increase reaction rates, highlighting the role of catalysts in two vital areas—the reactions of a living cell and the depletion of atmospheric ozone.

16.1 FACTORS THAT INFLUENCE REACTION RATE

Let's begin our study of kinetics with a qualitative look at the key factors that affect how fast a reaction proceeds. Under any given set of conditions, *each reaction has its own characteristic rate,* which is determined by the chemical nature of the reactants. At room temperature, for example, hydrogen reacts explosively with fluorine but extremely slowly with nitrogen:

$$H_2(g) + F_2(g) \longrightarrow 2HF(g) \qquad \text{[very fast]}$$
$$3H_2(g) + N_2(g) \longrightarrow 2NH_3(g) \qquad \text{[very slow]}$$

We can control four factors that affect the rate of a given reaction: the concentrations of the reactants, the physical state of the reactants, the temperature at which the reaction occurs, and the use of a catalyst. We'll consider the first three factors here and discuss the fourth later in the chapter.

1. *Concentration: molecules must collide to react.* A major factor influencing the rate of a given reaction is reactant concentration. Consider the reaction between ozone and nitric oxide that occurs in the stratosphere, where the oxide is released in the exhaust gases of supersonic aircraft:

$$NO(g) + O_3(g) \longrightarrow NO_2(g) + O_2(g)$$

Imagine what this reaction might look like at the molecular level if the reactants were confined in a reaction vessel. Nitric oxide and ozone molecules zoom every which way, crashing into each other and the vessel walls. A reaction between NO and O_3 can occur only when the molecules collide. The more molecules present in the container, the more frequently they collide, and the more often a reaction occurs. Thus, *reaction rate is proportional to the concentration of reactants:*

Rate ∝ collision frequency ∝ concentration

In this case, we're looking at a very simple reaction, one in which reactant molecules collide and form product molecules in one step, but even the rates of complex reactions depend on reactant concentration.

2. *Physical state: molecules must mix to collide.* The frequency of collisions between molecules also depends on the physical states of the reactants. When the reactants are in the same phase, as in an aqueous solution, thermal motion brings them into contact. When they are in different phases, contact occurs only at the interface, so vigorous stirring and grinding may be needed. In these cases, *the more finely divided a solid or liquid reactant, the greater its surface area per unit volume, the more contact it makes with the other reactant, and the faster the reaction occurs.* Figure 16.3A shows a steel nail heated in oxygen glowing feebly; in Figure 16.3B, the same mass of steel wool bursts into flame. For the same reason, you start a campfire with wood chips and thin branches, not logs.

Figure 16.3 The effect of surface area on reaction rate. A, A hot nail glows in O_2. **B,** The same mass of hot steel wool bursts into flame in O_2. The greater surface area per unit volume of the steel wool means that more metal makes contact with O_2, so the reaction is faster.

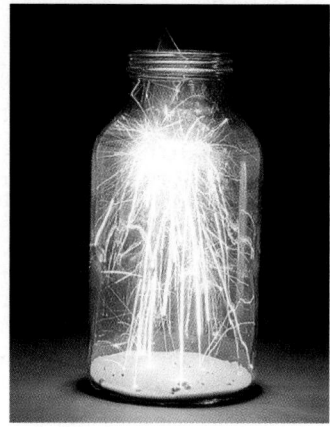

A   B

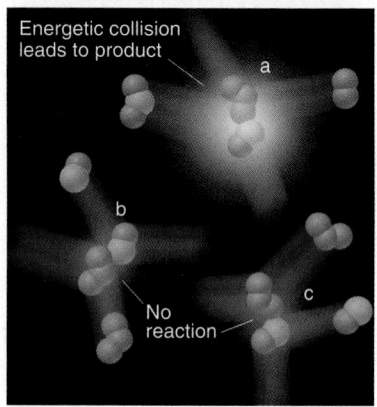

Figure 16.4 Collision energy and reaction rate. The reaction is shown in the panel. Although many NO-O_3 collisions occur, relatively few have enough energy to cause reaction. At this temperature, only collision *a* is energetic enough to lead to product; the reactant molecules in collisions *b* and *c* just bounce off each other.

3. *Temperature: molecules must collide with enough energy to react.* Temperature usually has a major effect on the speed of a reaction. Two familiar kitchen appliances employ this effect: a refrigerator slows down chemical processes that spoil food, whereas an oven speeds up other chemical processes that cook it.

Recall that molecules in a sample of gas have a range of speeds, with the most probable speed dependent on the temperature (see Figure 5.14, p. 198). Thus, *at a higher temperature, more collisions occur in a given time.* Even more important, however, is the fact that temperature affects the kinetic energy of the molecules, and thus the *energy* of the collisions. In the jumble of molecules in the NO-O_3 reaction mentioned previously, most collisions simply result in the molecules recoiling, like billiard balls, with no reaction taking place. However, some collisions occur with so much energy that the molecules react (Figure 16.4). And, at a higher temperature, more of these sufficiently energetic collisions occur. Thus, *raising the temperature increases the reaction rate by increasing the number and, especially, the energy of the collisions:*

Rate ∝ collision energy ∝ temperature

The qualitative idea that reaction rate is influenced by the frequency and energy of reactant collisions leads to several quantitative questions: How can we describe the dependence of rate on reactant concentration mathematically? Do all changes in concentration affect the rate to the same extent? Do all rates increase to the same extent with a given rise in temperature? How do reactant molecules use the energy of collision to form product molecules, and is there a way to determine this energy? What do the reactants look like as they are turning into products? We address these questions in the following sections.

SECTION SUMMARY
Chemical kinetics deals with reaction rates and the stepwise molecular events by which a reaction occurs. Under a given set of conditions, each reaction has its own rate. Concentration affects rate by influencing the frequency of collisions between reactant molecules. Physical state affects rate by determining the surface area per unit volume of reactants. Temperature affects rate by influencing the frequency but, even more importantly, the energy of the collisions.

16.2 EXPRESSING THE REACTION RATE

Before we can deal quantitatively with the effects of concentration and temperature on reaction rate, we must express the rate mathematically. A rate is a change in some variable per unit of time. The most common examples relate to the rate of motion (speed) of an object, which is the change in its position divided by the change in time. Suppose, for instance, we measure a racehorse's starting position x_1 at time t_1, and its final position x_2 at time t_2. The horse's average speed is

$$\text{Rate of motion} = \frac{\text{change in position}}{\text{change in time}} = \frac{x_2 - x_1}{t_2 - t_1} = \frac{\Delta x}{\Delta t}$$

A racehorse changing position with time.

In the case of a chemical change, we are concerned with the **reaction rate,** the changes in concentrations of reactants or products per unit time: *reactant concentrations decrease while product concentrations increase.* Consider a general reaction, A $\longrightarrow$ B. We quickly measure the starting reactant concentration (conc A$_1$) at t_1, allow the reaction to proceed, and then quickly measure the reactant concentration again (conc A$_2$) at t_2. The change in concentration divided by the change in time gives the *average* rate:

$$\text{Rate of reaction} = -\frac{\text{change in concentration of A}}{\text{change in time}}$$
$$= -\frac{\text{conc A}_2 - \text{conc A}_1}{t_2 - t_1} = -\frac{\Delta(\text{conc A})}{\Delta t}$$

Note the minus sign. By convention, reaction rate is a *positive* number, but conc A$_2$ will always be *lower* than conc A$_1$, so the *change in (final − initial) concentration of reactant A is always negative.* We use the minus sign simply to convert the negative change in reactant concentration to a positive value for the rate. Suppose the concentration of A changes from 1.2 mol/L (conc A$_1$) to 0.75 mol/L (conc A$_2$) over a 125-s period. The average rate is

$$\text{Rate} = -\frac{0.75 \text{ mol/L} - 1.2 \text{ mol/L}}{125 \text{ s} - 0 \text{ s}} = 3.6 \times 10^{-3} \text{ mol/L·s}$$

We use *square brackets [ ] to express concentration in moles per liter.* That is, [A] is the concentration of A in mol/L, so the rate expressed in terms of A is

$$\text{Rate} = -\frac{\Delta[A]}{\Delta t} \tag{16.1}$$

The rate has units of moles per liter per second (mol L^{-1} s^{-1}, or mol/L·s), or any time unit convenient for the reaction under study (minutes, years, and so on).

If instead we measure the *product* to determine the reaction rate, we find its concentration *increasing* over time. That is, conc B$_2$ is always *higher* than conc B$_1$. Thus, the *change* in product concentration, $\Delta[B]$, is *positive,* and the reaction rate for A $\longrightarrow$ B expressed in terms of B is

$$\text{Rate} = \frac{\Delta[B]}{\Delta t}$$

Average, Instantaneous, and Initial Reaction Rates

Examining the rate of a real reaction reveals an important point: *the rate itself varies with time as the reaction proceeds.* Consider the reversible gas-phase reaction between ethylene and ozone, one of many reactions that can be involved in the formation of photochemical smog:

$$C_2H_4(g) + O_3(g) \rightleftharpoons C_2H_4O(g) + O_2(g)$$

For now, we consider only reactant concentrations. You can see from the equation coefficients that for every molecule of C_2H_4 that reacts, a molecule of O_3 reacts with it. In other words, the concentrations of both reactants decrease at the same rate in this particular reaction:

$$\text{Rate} = -\frac{\Delta[C_2H_4]}{\Delta t} = -\frac{\Delta[O_3]}{\Delta t}$$

By measuring the concentration of either reactant, we can follow the reaction rate.

Suppose we have a known concentration of O_3 in a closed reaction vessel kept at 30°C (303 K). Table 16.1 shows the concentration of O_3 at various times during the first minute after we introduce C_2H_4 gas. The rate over the entire 60.0 s is the total change in concentration divided by the change in time:

$$\text{Rate} = -\frac{\Delta[O_3]}{\Delta t} = -\frac{(1.10 \times 10^{-5}\,\text{mol/L}) - (3.20 \times 10^{-5}\,\text{mol/L})}{60.0\,\text{s} - 0.0\,\text{s}}$$
$$= 3.50 \times 10^{-7}\,\text{mol/L·s}$$

This calculation gives us the **average rate** over that period; that is, during the first 60.0 s of the reaction, ozone concentration decreases an *average* of 3.50×10^{-7} mol/L each second. However, the average rate does not show that the rate is changing, and it tells us nothing about how fast the ozone concentration is decreasing *at any given instant.*

We can see the rate change during the reaction by calculating the average rate over two shorter periods—one earlier and one later. Between the starting time 0.0 s and 10.0 s, the average rate is

$$\text{Rate} = -\frac{\Delta[O_3]}{\Delta t} = -\frac{(2.42 \times 10^{-5}\,\text{mol/L}) - (3.20 \times 10^{-5}\,\text{mol/L})}{10.0\,\text{s} - 0.0\,\text{s}}$$
$$= 7.80 \times 10^{-7}\,\text{mol/L·s}$$

During the last 10.0 s, between 50.0 s and 60.0 s, the average rate is

$$\text{Rate} = -\frac{\Delta[O_3]}{\Delta t} = -\frac{(1.10 \times 10^{-5}\,\text{mol/L}) - (1.23 \times 10^{-5}\,\text{mol/L})}{60.0\,\text{s} - 50.0\,\text{s}}$$
$$= 1.30 \times 10^{-7}\,\text{mol/L·s}$$

The earlier rate is six times as fast as the later rate. Thus, *the rate decreases during the course of the reaction.* This makes perfect sense from a molecular point of view: as O_3 molecules are used up, fewer of them are present to collide with C_2H_4, so the rate decreases.

The change in rate can also be seen by plotting the concentrations vs. the times at which they were measured (Figure 16.5). A curve is obtained, which means that the rate changes. *The slope of the straight line ($\Delta y / \Delta x$, that is, $\Delta[O_3]/\Delta t$) joining any two points gives the average rate over that period.*

The shorter the time period we choose, the closer we come to the **instantaneous rate,** the rate at a particular instant during the reaction. *The slope of a line tangent to the curve at a particular point gives the instantaneous rate at that time.* For example, the rate of the reaction 35.0 s after it began is 2.50×10^{-7} mol/L·s, the slope of the line drawn tangent to the curve through the point at which $t = 35.0$ s (line *d* in Figure 16.5). In general, we use the term *reaction rate* to mean the *instantaneous* reaction rate.

As a reaction continues, the product concentrations increase, so the reverse reaction proceeds more quickly. To find the overall (net) rate, we would have to take both forward and reverse reactions into account and calculate the difference

Table 16.1 Concentration of O_3 at Various Times in Its Reaction with C_2H_4 at 303 K

Time (s)	Concentration of O_3 (mol/L)
0.0	3.20×10^{-5}
10.0	2.42×10^{-5}
20.0	1.95×10^{-5}
30.0	1.63×10^{-5}
40.0	1.40×10^{-5}
50.0	1.23×10^{-5}
60.0	1.10×10^{-5}

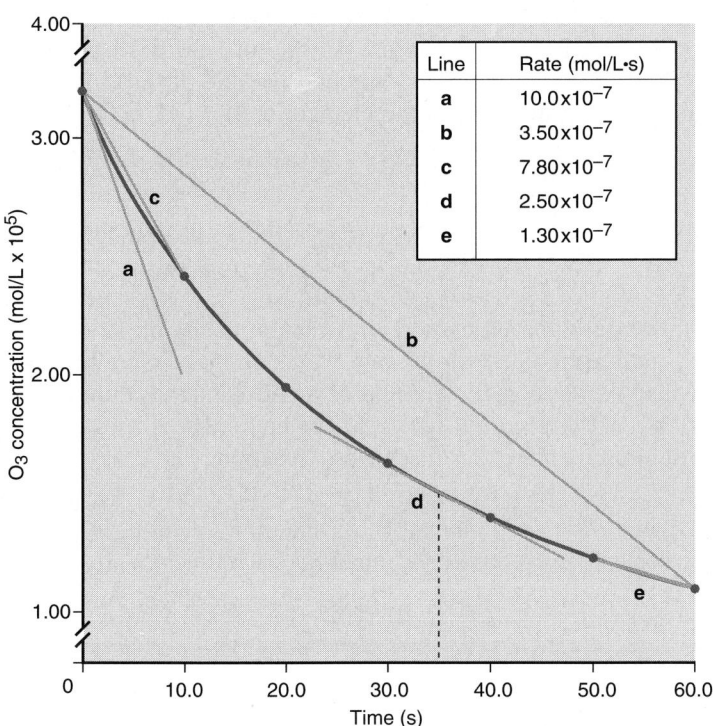

Line	Rate (mol/L·s)
a	10.0×10^{-7}
b	3.50×10^{-7}
c	7.80×10^{-7}
d	2.50×10^{-7}
e	1.30×10^{-7}

Figure 16.5 **The concentration of O_3 vs. time during its reaction with C_2H_4.** Plotting the data in Table 16.1 gives a curve because the rate changes during the reaction. The *average* rate over a given period is the slope of a line joining two points along the curve. The slope of line *b* is the average rate over the first 60.0 s of the reaction. The slopes of lines *c* and *e* give the average rate over the first and last 10.0-s intervals, respectively. Line *c* is steeper than line *e* because the average rate over the earlier period is higher. The *instantaneous* rate at 35.0 s is the slope of line *d*, the tangent to the curve at *t* = 35.0 s. The *initial* rate is the slope of line *a*, the tangent to the curve at *t* = 0.0 s.

between their rates. A common way to avoid this complication for many reactions is to measure the **initial rate,** the instantaneous rate at the moment the reactants are mixed. Under these conditions, the product concentrations are negligible, so the reverse rate is negligible. Moreover, we know the reactant concentrations from the concentrations and volumes of the solutions we mix together. The initial rate is measured by determining the slope of the line tangent to the curve at *t* = 0. In Figure 16.5, the initial rate is 10.0×10^{-7} mol/L·s (line *a*). Unless stated otherwise, we will use initial rate data to determine other kinetic parameters.

Expressing Rate in Terms of Reactant and Product Concentrations

So far, in our discussion of the O_3-C_2H_4 reaction, we've expressed the rate in terms of the decreasing concentration of O_3. The rate is the same in terms of C_2H_4, but it is exactly the opposite in terms of the products because their concentrations are *increasing*. From the balanced equation, we see that one molecule of C_2H_4O and one of O_2 appear for every molecule of C_2H_4 and of O_3 that disappear. We can express the rate in terms of any of the four substances involved:

$$\text{Rate} = -\frac{\Delta[C_2H_4]}{\Delta t} = -\frac{\Delta[O_3]}{\Delta t} = +\frac{\Delta[C_2H_4O]}{\Delta t} = +\frac{\Delta[O_2]}{\Delta t}$$

Again, note the negative values for the reactants and the positive values for the products (usually written without the plus sign). Figure 16.6 shows a plot of the simultaneous monitoring of one reactant and one product. Because, in this case, product concentration increases at the same rate that reactant concentration decreases, the curves have the same shapes but are inverted.

In the reaction between ethylene and ozone, the reactants disappear and the products appear at the same rate because all the coefficients in the balanced equation are equal. Consider next the reaction between hydrogen and iodine to form hydrogen iodide:

$$H_2(g) + I_2(g) \longrightarrow 2HI(g)$$

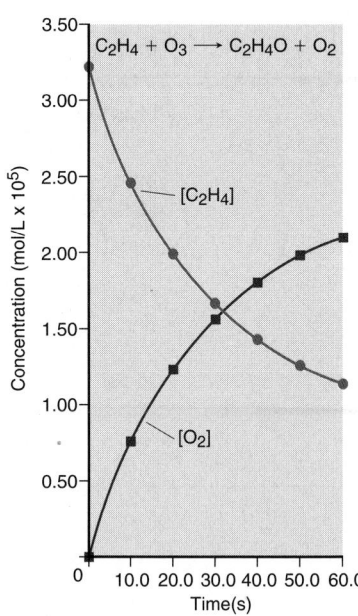

Figure 16.6 **Plots of $[C_2H_4]$ and $[O_2]$ vs. time.** Measuring reactant $[C_2H_4]$ and product $[O_2]$ gives curves of identical shape changing in opposite directions. The steep upward (positive) slope of $[O_2]$ early in the reaction mirrors the steep downward (negative) slope of $[C_2H_4]$ because the faster C_2H_4 is used up, the faster O_2 is formed. The curve shapes are identical in this case because the equation coefficients are identical.

For every molecule of H_2 that disappears, one molecule of I_2 disappears and *two* molecules of HI appear. In other words, the rate of $[H_2]$ decrease is the same as the rate of $[I_2]$ decrease, but both are only half the rate of $[HI]$ increase. By referring the change in $[I_2]$ and $[HI]$ to the change in $[H_2]$, we have

$$\text{Rate} = -\frac{\Delta[H_2]}{\Delta t} = -\frac{\Delta[I_2]}{\Delta t} = \frac{1}{2}\frac{\Delta[HI]}{\Delta t}$$

If we refer the change in $[H_2]$ and $[I_2]$ to the change in $[HI]$ instead, we obtain

$$\text{Rate} = \frac{\Delta[HI]}{\Delta t} = -2\frac{\Delta[H_2]}{\Delta t} = -2\frac{\Delta[I_2]}{\Delta t}$$

Notice that this expression is just a rearrangement of the previous one; also note that it gives a numerical value for the rate that is double the previous value. Thus, the mathematical expression for the rate of a particular reaction and *the numerical value of the rate depend on which substance serves as the reference.*

We can summarize these results for any reaction:

$$a\text{A} + b\text{B} \longrightarrow c\text{C} + d\text{D}$$

where *a*, *b*, *c*, and *d* are coefficients of the balanced equation. In general, the rate is related to reactant or product concentrations as follows:

$$\text{Rate} = -\frac{1}{a}\frac{\Delta[A]}{\Delta t} = -\frac{1}{b}\frac{\Delta[B]}{\Delta t} = \frac{1}{c}\frac{\Delta[C]}{\Delta t} = \frac{1}{d}\frac{\Delta[D]}{\Delta t} \qquad \textbf{(16.2)}$$

SAMPLE PROBLEM 16.1 Expressing Rate in Terms of Changes in Concentration with Time

Problem Because it has a nonpolluting product (water vapor), hydrogen gas is used for fuel aboard the space shuttle and may be used by Earth-bound engines in the near future:

$$2H_2(g) + O_2(g) \longrightarrow 2H_2O(g)$$

(a) Express the rate in terms of changes in $[H_2]$, $[O_2]$, and $[H_2O]$ with time.
(b) When $[O_2]$ is decreasing at 0.23 mol/L·s, at what rate is $[H_2O]$ increasing?
Plan (a) Of the three substances in the equation, let's choose O_2 as the reference because its coefficient is 1. For every molecule of O_2 that disappears, two molecules of H_2 disappear, so the rate of $[O_2]$ decrease is one-half the rate of $[H_2]$ decrease. By similar reasoning, we see that the rate of $[O_2]$ decrease is one-half the rate of $[H_2O]$ increase. **(b)** Because $[O_2]$ is decreasing, the change in its concentration must be negative. We substitute the negative value into the expression and solve for $\Delta[H_2O]/\Delta t$.
Solution (a) Expressing the rate in terms of each component:

$$\text{Rate} = -\frac{1}{2}\frac{\Delta[H_2]}{\Delta t} = -\frac{\Delta[O_2]}{\Delta t} = \frac{1}{2}\frac{\Delta[H_2O]}{\Delta t}$$

(b) Calculating the rate of change of $[H_2O]$:

$$\frac{1}{2}\frac{\Delta[H_2O]}{\Delta t} = -\frac{\Delta[O_2]}{\Delta t} = -(-0.23 \text{ mol/L·s})$$

$$\frac{\Delta[H_2O]}{\Delta t} = 2(0.23 \text{ mol/L·s}) = \boxed{0.46 \text{ mol/L·s}}$$

Check (a) A good check is to use the rate expression to obtain the balanced equation: $[H_2]$ changes twice as fast as $[O_2]$, so two H_2 molecules react for each O_2. $[H_2O]$ changes twice as fast as $[O_2]$, so two H_2O molecules form from each O_2. From this reasoning, we get $2H_2 + O_2 \longrightarrow 2H_2O$. The $[H_2]$ and $[O_2]$ decrease, so they take minus signs; $[H_2O]$ increases, so it takes a plus sign. Another check is to use Equation 16.2, with A = H_2, a = 2; B = O_2, b = 1; C = H_2O, c = 2. Thus,

$$\text{Rate} = -\frac{1}{a}\frac{\Delta[A]}{\Delta t} = -\frac{1}{b}\frac{\Delta[B]}{\Delta t} = \frac{1}{c}\frac{\Delta[C]}{\Delta t}$$

or

$$\text{Rate} = -\frac{1}{2}\frac{\Delta[H_2]}{\Delta t} = -\frac{\Delta[O_2]}{\Delta t} = \frac{1}{2}\frac{\Delta[H_2O]}{\Delta t}$$

(b) Given the rate expression, it makes sense for the numerical value of the rate of $[H_2O]$ increase to be twice that of $[O_2]$ decrease.

Comment Thinking through this type of problem at the molecular level is the best approach, but use Equation 16.2 to confirm your answer.

FOLLOW-UP PROBLEM 16.1 **(a)** Balance the following equation and express the rate in terms of the change in concentration with time for each substance:

$$NO(g) + O_2(g) \longrightarrow N_2O_3(g)$$

(b) How fast is $[O_2]$ decreasing when $[NO]$ is decreasing at a rate of 1.60×10^{-4} mol/L·s?

SECTION SUMMARY

The average reaction rate is the change in reactant (or product) concentration over a change in time, Δt. The rate slows as reactants are used up. The instantaneous rate at time t is obtained from the slope of the tangent to a concentration vs. time curve at time t. The initial rate, the instantaneous rate at $t = 0$, occurs when reactants are just mixed and before any product accumulates. The expression for a reaction rate and its value depend on which reaction component is being monitored.

16.3 THE RATE LAW AND ITS COMPONENTS

The centerpiece of any kinetic study is the **rate law** (or **rate equation**) for the reaction in question. The rate law expresses the rate as a function of reactant concentrations, product concentrations, and temperature. Any hypothesis we make about how the reaction occurs on the molecular level must conform to the rate law because it is based on experimental fact.

In this discussion, we generally consider reactions for which the products do not appear in the rate law. In these cases, *the reaction rate depends only on reactant concentrations and temperature*. First, we consider the effect of concentration on rate for reactions occurring at a fixed temperature. For a general reaction,

$$a\mathrm{A} + b\mathrm{B} + \cdots \longrightarrow c\mathrm{C} + d\mathrm{D} + \cdots$$

the rate law has the form

$$\text{Rate} = k[\mathrm{A}]^m[\mathrm{B}]^n \cdots \qquad (16.3)$$

The proportionality constant k, called the **rate constant,** is specific for a given reaction at a given temperature; it does *not* change as the reaction proceeds. (As you'll see later, k *does* change with temperature and therefore determines how temperature affects the rate.) The exponents m and n, called the **reaction orders,** define how the rate is affected by reactant concentration. Thus, if the rate doubles when $[\mathrm{A}]$ doubles, the rate depends on $[\mathrm{A}]$ raised to the first power, $[\mathrm{A}]^1$, so $m = 1$. Similarly, if the rate quadruples when $[\mathrm{B}]$ doubles, the rate depends on $[\mathrm{B}]$ raised to the second power, $[\mathrm{B}]^2$, so $n = 2$. In another reaction, the rate may not change at all when $[\mathrm{A}]$ doubles; in that case, the rate does *not* depend on $[\mathrm{A}]$ or, to put it another way, the rate depends on $[\mathrm{A}]$ raised to the zero power, $[\mathrm{A}]^0$, so $m = 0$. Keep in mind that the coefficients a and b in the general balanced equation are *not* necessarily related in any way to the reaction orders m and n.

A key point to remember is that *the components of the rate law—rate, reaction orders, and rate constant—must be found by experiment;* they cannot be deduced from the reaction stoichiometry. So let's take an experimental approach to finding the components by

1. Using concentration measurements to find the *initial rate*
2. Using initial rates from several experiments to find the *reaction orders*
3. Using these values to calculate the *rate constant*

Once we know the rate law, we can use it to predict the rate for any initial reactant concentrations.

Measuring Reaction Rates

Speculation about how a reaction occurs at the molecular level must be based on measurements of reaction rates. There are many experimental approaches, but all must obtain the results quickly and reproducibly. We consider four common methods, with specific examples.

Spectrometric Methods

These methods are used to measure the concentration of a reactant or product that absorbs (or emits) light of a narrow range of wavelengths. The reaction is typically performed *within* the sample compartment of a spectrometer set to measure a wavelength characteristic of one of the species (Figure B16.1; see also Tools of the Laboratory, pp. 267–268). For example, in the NO-O_3 reaction, only NO_2 has a color:

$$NO(g, \text{colorless}) + O_3(g, \text{colorless}) \longrightarrow$$
$$O_2(g, \text{colorless}) + NO_2(g, \text{brown})$$

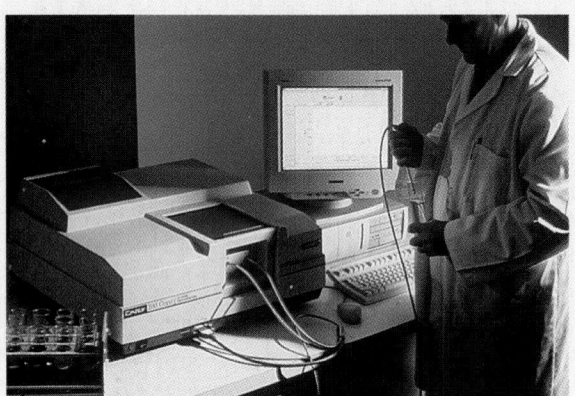

Known amounts of reactants are injected into a gas sample tube of known volume, and the rate of NO_2 formation is measured by monitoring the color over time. Reactions in aqueous solution are studied similarly.

Conductometric Methods

When nonionic reactants form ionic products, or vice versa, the change in conductivity of the solution over time can be used to measure the rate. Electrodes are immersed in the reaction mixture, and the increase (or decrease) in conductivity correlates with the formation of product (Figure B16.2). Consider the reaction between an organic halide, such as 2-bromo-2-methylpropane, and water:

$$(CH_3)_3C-Br(l) + H_2O(l) \longrightarrow$$
$$(CH_3)_3C-OH(l) + H^+(aq) + Br^-(aq)$$

The HBr that forms is a strong acid in water, so it dissociates completely into ions. As time passes, more ions form, so the conductivity of the reaction mixture increases.

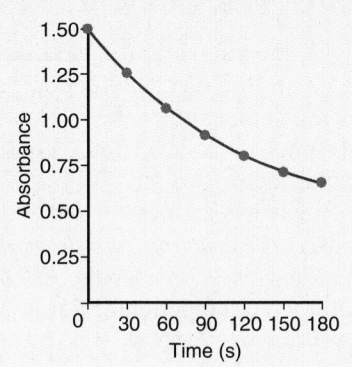

Figure B16.1 Spectrometric monitoring of a reaction. The investigator adds the reactant(s) to the sample tube and immediately places it in the spectrometer. For a reactant that is colored, rate data is determined from a plot of light absorbed vs. time.

Determining the Initial Rate

In the last section, we showed how initial rates are determined from a plot of concentration vs. time. Because we use initial rate data to determine the reaction orders and rate constant, an accurate experimental method for measuring concentration at various times during the reaction is the key to constructing the rate law. A few of the many techniques used for measuring changes in concentration over time are presented in the Tools of the Laboratory essay.

Reaction Order Terminology

Before we see how reaction orders are determined from initial rate data, let's discuss the meaning of reaction order and some important terminology. We speak of a reaction as having an *individual* order "with respect to" or "in" each reactant as well as an *overall* order, which is simply the sum of the individual orders.

In the simplest case of a reaction with a single reactant A, it is *first order* overall if the rate is directly proportional to [A]:

$$\text{Rate} = k[A]$$

It is *second order* overall if the rate is directly proportional to the square of [A]:

$$\text{Rate} = k[A]^2$$

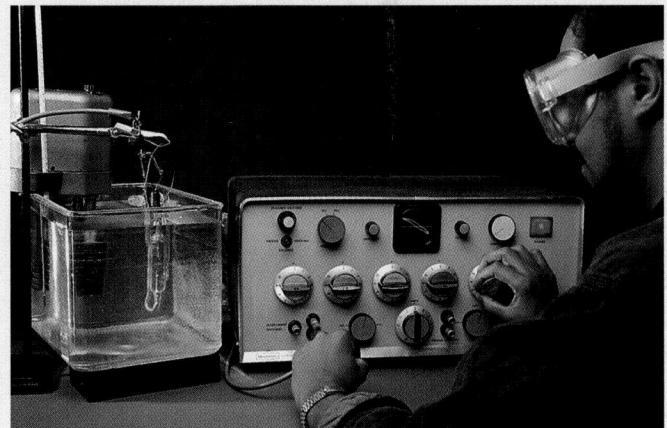

Figure B16.2 Conductometric monitoring of a reaction. When a reactant mixture differs in conductivity from the product mixture, the change in conductivity is proportional to the reaction rate. It is usually easier to monitor nonionic reactants forming ionic products.

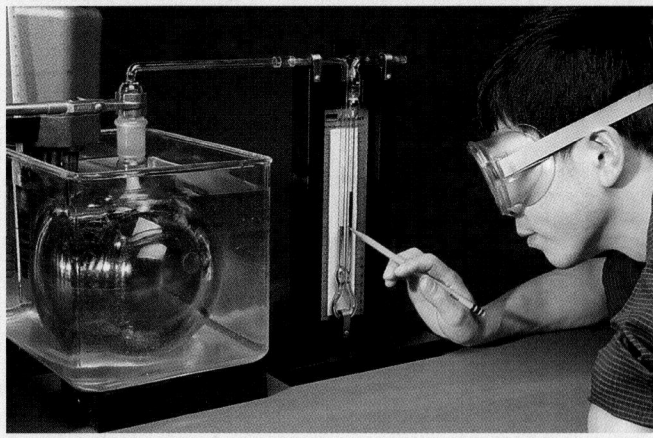

Figure B16.3 Manometric monitoring of a reaction. When a reaction results in a change in the number of moles of gas, the change in pressure with time corresponds to a change in reaction rate. The rate of formation of NO_2 from N_2O_4 is being studied here.

Manometric Methods

If a reaction involves a change in the number of moles of gas, the rate can be determined from the change in pressure (at constant volume and temperature) over time. In practice, a manometer is attached to a reaction vessel of known volume that is immersed in a constant-temperature bath. For example, the reaction between zinc and acetic acid can be monitored by this method:

$$Zn(s) + 2CH_3COOH(aq) \longrightarrow$$
$$Zn^{2+}(aq) + 2CH_3COO^-(aq) + H_2(g)$$

As H_2 forms, the pressure increases. Thus, the reaction rate is directly proportional to the rate of increase of the H_2 gas pressure. In Figure B16.3, this method is being used to study a reaction involving nitrogen dioxide.

Direct Chemical Methods

Rates of slow reactions, or of those that can be easily slowed, are often studied by direct chemical methods. A small, measured portion (called an *aliquot*) of the reaction mixture is removed, and the reaction in this portion is stopped, usually by rapid cooling. The concentration of reactant or product in the aliquot is measured, while the bulk of the reaction mixture continues to react and is sampled later. For example, the earlier reaction between an organic halide and water can also be studied by titration. The reaction rate in an aliquot is slowed by quickly transferring it to a chilled flask in an ice bath. HBr concentration in the aliquot is determined by titrating with standardized NaOH solution. To determine the change in HBr concentration with time, the procedure is repeated at regular intervals during the reaction.

And it is *zero order* overall if the rate is *not* dependent on [A] at all, a situation that is quite common in metal-catalyzed and biochemical processes:

$$\text{Rate} = k[A]^0 = k(1) = k$$

Here are some real examples. For the reaction between nitric oxide and ozone,

$$NO(g) + O_3(g) \longrightarrow NO_2(g) + O_2(g)$$

the rate law has been experimentally determined to be

$$\text{Rate} = k[NO][O_3]$$

This reaction is first order with respect to NO (or first order in NO), which means that the rate depends on NO concentration raised to the first power, that is, $[NO]^1$ (an exponent of 1 is generally omitted). It is also first order with respect to O_3, or $[O_3]^1$. This reaction is second order overall ($1 + 1 = 2$).

Now consider a different gas-phase reaction:

$$2NO(g) + 2H_2(g) \longrightarrow N_2(g) + 2H_2O(g)$$

The rate law for this reaction has been determined to be

$$\text{Rate} = k[NO]^2[H_2]$$

The reaction is second order in NO, first order in H_2, and so third order overall.

Finally, for the hydrolysis of 2-bromo-2-methylpropane, which we looked at in the Tools of the Laboratory essay,

$$(CH_3)_3C-Br(l) + H_2O(l) \longrightarrow (CH_3)_3C-OH(l) + H^+ + Br^-(aq)$$

The rate law for this reaction has been found to be

$$\text{Rate} = k[(CH_3)_3CBr]$$

This reaction is first order in 2-bromo-2-methylpropane. Note that the concentration of H_2O does not even appear in the rate law. Thus, the reaction is zero order with respect to H_2O ($[H_2O]^0$). This means that the rate does not depend on the concentration of H_2O. We can also write the rate law as

$$\text{Rate} = k[(CH_3)_3CBr][H_2O]^0$$

Overall, this is a first-order reaction.

These examples demonstrate a major point: *reaction orders* ***cannot*** *be deduced from the balanced equation*. In the reaction between NO and H_2 and in the hydrolysis of 2-bromo-2-methylpropane, the reaction orders in the rate laws do *not* correspond to the coefficients of the balanced equations. Reaction orders *must* be determined from rate data.

Reaction orders are usually positive integers or zero, but they can also be fractional or negative. For the reaction

$$CHCl_3(g) + Cl_2(g) \longrightarrow CCl_4(g) + HCl(g)$$

a fractional order appears in the rate law:

$$\text{Rate} = k[CHCl_3][Cl_2]^{1/2}$$

This reaction order means that the rate depends on the square root of the Cl_2 concentration. For example, if the initial Cl_2 concentration is increased by a factor of 4, while the initial $CHCl_3$ concentration is kept the same, the rate increases by a factor of 2, the square root of the change in $[Cl_2]$. A negative exponent means that the rate *decreases* when the concentration of that component increases. Negative orders are often seen for reactions whose rate laws include products. For example, for the atmospheric reaction

$$2O_3(g) \rightleftharpoons 3O_2(g)$$

the rate law has been shown to be

$$\text{Rate} = k[O_3]^2[O_2]^{-1} \qquad \text{or} \qquad k\frac{[O_3]^2}{[O_2]}$$

If the O_2 concentration doubles, the reaction proceeds half as fast.

SAMPLE PROBLEM 16.2 Determining Reaction Order from Rate Laws

Problem For each of the following reactions, determine the reaction order with respect to each reactant and the overall order from the given rate law:
(a) $2NO(g) + O_2(g) \longrightarrow 2NO_2(g)$; rate $= k[NO]^2[O_2]$
(b) $CH_3CHO(g) \longrightarrow CH_4(g) + CO(g)$; rate $= k[CH_3CHO]^{3/2}$
(c) $H_2O_2(aq) + 3I^-(aq) + 2H^+(aq) \longrightarrow I_3^-(aq) + 2H_2O(l)$; rate $= k[H_2O_2][I^-]$
Plan We inspect the exponents in the rate law, *not* the coefficients of the balanced equation, to find the individual orders, and then take their sum to find the overall reaction order.
Solution (a) The exponent of [NO] is 2, so the reaction is second order with respect to NO, first order with respect to O_2, and third order overall.

(b) The reaction is $\frac{3}{2}$ order in CH_3CHO and $\frac{3}{2}$ order overall.

(c) The reaction is first order in H_2O_2, first order in I^-, and second order overall. The reactant H^+ does not appear in the rate law, so the reaction is zero order in H^+.
Check Be sure that each reactant has an order and that the sum of the individual orders gives the overall order.

FOLLOW-UP PROBLEM 16.2 Experiment shows that the reaction

$$5Br^-(aq) + BrO_3^-(aq) + 6H^+(aq) \longrightarrow 3Br_2(l) + 3H_2O(l)$$

obeys this rate law: rate $= k[Br^-][BrO_3^-][H^+]^2$. What are the reaction orders in each reactant and the overall reaction order?

Determining Reaction Orders

Sample Problem 16.2 shows how to find the reaction orders from a known rate law. Now let's see how they are found from data *before* the rate law is known. Suppose we are studying the reaction between oxygen and nitric oxide, a key step in the formation of acid rain and in the industrial production of nitric acid:

$$O_2(g) + 2NO(g) \longrightarrow 2NO_2(g)$$

The rate law, expressed in general form, is

$$\text{Rate} = k[O_2]^m[NO]^n$$

To find the reaction orders, *we run a series of experiments, each of which starts with a different set of reactant concentrations, and from each we obtain an initial rate.*

Table 16.2 Initial Rates for a Series of Experiments in the Reaction Between O_2 and NO

Experiment	Initial Reactant Concentrations (mol/L)		Initial Rate (mol/L·s)
	O_2	NO	
1	1.10×10^{-2}	1.30×10^{-2}	3.21×10^{-3}
2	2.20×10^{-2}	1.30×10^{-2}	6.40×10^{-3}
3	1.10×10^{-2}	2.60×10^{-2}	12.8×10^{-3}
4	3.30×10^{-2}	1.30×10^{-2}	9.60×10^{-3}
5	1.10×10^{-2}	3.90×10^{-2}	28.8×10^{-3}

Table 16.2 shows experiments that change one reactant concentration while keeping the other constant. If we compare experiments 1 and 2, we see the effect of doubling $[O_2]$ on the rate. First, we take the ratio of their rate laws:

$$\frac{\text{Rate 2}}{\text{Rate 1}} = \frac{k[O_2]_2^m [NO]_2^n}{k[O_2]_1^m [NO]_1^n}$$

where $[O_2]_2$ is the O_2 concentration for experiment 2, $[NO]_1$ is the NO concentration for experiment 1, and so forth. Because k is a constant and $[NO]$ does not change between these two experiments, these quantities cancel:

$$\frac{\text{Rate 2}}{\text{Rate 1}} = \frac{[O_2]_2^m}{[O_2]_1^m} = \left(\frac{[O_2]_2}{[O_2]_1}\right)^m$$

Substituting the values from Table 16.2, we obtain

$$\frac{6.40 \times 10^{-3} \text{ mol/L·s}}{3.21 \times 10^{-3} \text{ mol/L·s}} = \left(\frac{2.20 \times 10^{-2} \text{ mol/L}}{1.10 \times 10^{-2} \text{ mol/L}}\right)^m$$

Dividing, we obtain

$$1.99 = (2.00)^m$$

Rounding to one significant figure gives

$$2 = 2^m; \quad \text{therefore, } m = 1$$

The reaction is first order in O_2: when $[O_2]$ doubles, the rate doubles.

To find the order with respect to NO, we compare experiments 3 and 1, in which $[O_2]$ is held constant and $[NO]$ is doubled:

$$\frac{\text{Rate 3}}{\text{Rate 1}} = \frac{k[O_2]_3^m [NO]_3^n}{k[O_2]_1^m [NO]_1^n}$$

As before, k is constant, and in this pair of experiments $[O_2]$ does not change, so these quantities cancel:

$$\frac{\text{Rate 3}}{\text{Rate 1}} = \left(\frac{[NO]_3}{[NO]_1}\right)^n$$

The actual values give

$$\frac{12.8\times10^{-3}\ \text{mol/L·s}}{3.21\times10^{-3}\ \text{mol/L·s}} = \left(\frac{2.60\times10^{-2}\ \text{mol/L}}{1.30\times10^{-2}\ \text{mol/L}}\right)^{n}$$

Dividing, we obtain

$$3.99 = (2.00)^{n}$$

Rounding gives

$$4 = 2^{n};\quad \text{therefore, } n = 2$$

The reaction is second order in NO: when [NO] doubles, the rate quadruples. Thus, the rate law is

$$\text{Rate} = k[O_2][NO]^2$$

You may want to use experiment 1 in combination with experiments 4 and 5 to check this result.

SAMPLE PROBLEM 16.3 Determining Reaction Orders from Initial Rate Data

Problem Many gaseous reactions occur in a car engine and exhaust system. One of these is

$$NO_2(g) + CO(g) \longrightarrow NO(g) + CO_2(g) \qquad \text{rate} = k[NO_2]^{m}[CO]^{n}$$

Use the following data to determine the individual and overall reaction orders:

Experiment	Initial Rate (mol/L·s)	Initial [NO$_2$] (mol/L)	Initial [CO] (mol/L)
1	0.0050	0.10	0.10
2	0.080	0.40	0.10
3	0.0050	0.10	0.20

Plan We need to solve the general rate law for the reaction orders m and n. To solve for each exponent, we proceed as in the text, taking the ratio of the rate laws for two experiments in which only the reactant in question changes.

Solution Calculating m in [NO$_2$]^{m}: We take the ratio of the rate laws for experiments 1 and 2, in which [NO$_2$] varies but [CO] is constant:

$$\frac{\text{Rate 2}}{\text{Rate 1}} = \frac{k[NO_2]_2^{m}\,[CO]_2^{n}}{k[NO_2]_1^{m}\,[CO]_1^{n}} = \left(\frac{[NO_2]_2}{[NO_2]_1}\right)^{m} \quad \text{or} \quad \frac{0.080\ \text{mol/L·s}}{0.0050\ \text{mol/L·s}} = \left(\frac{0.40\ \text{mol/L}}{0.10\ \text{mol/L}}\right)^{m}$$

This gives $16 = 4.0^{m}$, so $m = 2.0$. The reaction is second order in NO$_2$.

Calculating n in [CO]^{n}: We take the ratio of the rate laws for experiments 1 and 3, in which [CO] varies but [NO$_2$] is constant:

$$\frac{\text{Rate 3}}{\text{Rate 1}} = \frac{k[NO_2]_3^{2}[CO]_3^{n}}{k[NO_2]_1^{2}[CO]_1^{n}} = \left(\frac{[CO]_3}{[CO]_1}\right)^{n} \quad \text{or} \quad \frac{0.0050\ \text{mol/L·s}}{0.0050\ \text{mol/L·s}} = \left(\frac{0.20\ \text{mol/L}}{0.10\ \text{mol/L}}\right)^{n}$$

We have $1.0 = (2.0)^{n}$, so $n = 0$. The rate does not change when [CO] varies, so the reaction is zero order in CO.

Therefore, the rate law is

$$\text{Rate} = k[NO_2]^2[CO]^0 = k[NO_2]^2(1) = k[NO_2]^2$$

The reaction is second order overall.

Check A good check is to reason through the orders. If $m = 1$, quadrupling [NO$_2$] would quadruple the rate; but the rate *more* than quadruples, so $m > 1$. If $m = 2$, quadrupling [NO$_2$] would increase the rate by a factor of 16 (4^2). The ratio of rates is $0.080/0.005 = 16$, so $m = 2$. In contrast, increasing [CO] has no effect on the rate, which can happen only if [CO]$^{n} = 1$, so $n = 0$.

FOLLOW-UP PROBLEM 16.3 Find the rate law and the overall reaction order for the reaction H$_2$ + I$_2$ $\longrightarrow$ 2HI from the following data at 450°C:

Experiment	Initial Rate (mol/L·s)	Initial [H$_2$] (mol/L)	Initial [I$_2$] (mol/L)
1	1.9×10^{-23}	0.0113	0.0011
2	1.1×10^{-22}	0.0220	0.0033
3	9.3×10^{-23}	0.0550	0.0011
4	1.9×10^{-22}	0.0220	0.0056

Determining the Rate Constant

With the rate, reactant concentrations, and reaction orders known, the sole remaining unknown in the rate law is the rate constant, k. The rate constant is specific for a particular reaction *at a particular temperature*. These experiments in the O_2-NO series were run at the same temperature, so we can use the data from any of them to solve for k. From experiment 1 in Table 16.2, for instance, we obtain

$$k = \frac{\text{rate 1}}{[O_2]_1[NO]_1^2} = \frac{3.21 \times 10^{-3} \text{ mol/L·s}}{(1.10 \times 10^{-2} \text{ mol/L})(1.30 \times 10^{-2} \text{ mol/L})^2}$$

$$= \frac{3.21 \times 10^{-3} \text{ mol/L·s}}{1.86 \times 10^{-6} \text{ mol}^3/\text{L}^3} = 1.73 \times 10^3 \text{ L}^2/\text{mol}^2\text{·s}$$

Always check that the values of k in the same series are constant within experimental error. To three significant figures, the average value of k for the five experiments in Table 16.2 is 1.72×10^3 L^2/mol^2·s.

Note the units for the rate constant. With concentrations in mol/L and the reaction rate in units of mol/L·time, the units for k will depend on the order of the reaction and, of course, the time unit. The units for k here, L^2/mol^2·s, are required to give a rate with units of mol/L·s:

$$\frac{\text{mol}}{\text{L·s}} = \frac{\text{L}^2}{\text{mol}^2\text{·s}} \times \frac{\text{mol}}{\text{L}} \times \left(\frac{\text{mol}}{\text{L}}\right)^2$$

The rate constant will *always* have these units for an overall third-order reaction with the time unit in seconds. Table 16.3 shows the units of k for some common overall reaction orders, but you can always determine the units mathematically.

SECTION SUMMARY

An experimentally determined rate law shows how the rate of a reaction depends on concentration. If we consider only initial rates, the rate law often takes this form: rate $= k[A]^m[B]^n \cdots$. With an accurate method for obtaining initial rates, reaction orders are determined by comparing rates for different initial concentrations. To do this, using several experiments, we vary the concentration of one reactant at a time to see its effect on the rate. With rate, concentrations, and reaction orders known, the rate constant is the only remaining unknown in the rate law, so it can be calculated.

Table 16.3 Units of the Rate Constant k for Several Overall Reaction Orders

Overall Reaction Order	Units of k (t in seconds)
0	mol/L·s (or mol L^{-1} s^{-1})
1	1/s (or s^{-1})
2	L/mol·s (or L mol^{-1} s^{-1})
3	L^2/mol^2·s (or L^2 mol^{-2} s^{-1})

General formula:

$$\text{Units of } k = \frac{\left(\dfrac{\text{L}}{\text{mol}}\right)^{\text{order}-1}}{\text{unit of } t}$$

16.4 INTEGRATED RATE LAWS: CONCENTRATION CHANGES OVER TIME

Notice that the rate laws we've developed so far do not include time as a variable. They tell us the rate or concentration at a given instant, allowing us to answer the critical question, "How fast is the reaction proceeding at the moment when y moles per liter of A are reacting with z moles per liter of B?" However, by employing different forms of the rate laws, called **integrated rate laws,** we can consider the time factor and answer other questions, such as "How long will it take for x moles per liter of A to be used up?" or "What is the concentration of A after y minutes of reaction?"

Integrated Rate Laws for First-, Second-, and Zero-Order Reactions

Consider a simple first-order reaction, A $\longrightarrow$ B. (Because first- and second-order reactions are more common, we'll discuss them before zero-order reactions.) As we discussed previously, the rate can be expressed as the change in the concentration of A divided by the change in time:

$$\text{Rate} = -\frac{\Delta[A]}{\Delta t}$$

It can also be expressed in terms of the rate law:

$$\text{Rate} = k[A]$$

Setting these different expressions equal to each other gives

$$-\frac{\Delta[A]}{\Delta t} = k[A]$$

Through the methods of calculus, this expression is integrated over time to obtain the integrated rate law for a first-order reaction:

$$\ln \frac{[A]_0}{[A]_t} = kt \quad \text{(first-order reaction; rate} = k[A]) \quad \textbf{(16.4)}$$

where ln is the natural logarithm, $[A]_0$ is the concentration of A at $t = 0$, and $[A]_t$ is the concentration of A at any time t during an experiment. In mathematical terms, $\ln \frac{a}{b} = \ln a - \ln b$, so we have

$$\ln [A]_0 - \ln [A]_t = kt$$

For a general second-order reaction, the expression including time is quite complex, so let's consider the case in which the rate law contains only one reactant. Setting the rate expressions equal to each other gives

$$\text{Rate} = -\frac{\Delta[A]}{\Delta t} = k[A]^2$$

Integrating over time gives the integrated rate law for a second-order reaction involving one reactant:

$$\frac{1}{[A]_t} - \frac{1}{[A]_0} = kt \quad \text{(second-order reaction; rate} = k[A]^2) \quad \textbf{(16.5)}$$

For a zero-order reaction, we would have

$$\text{Rate} = -\frac{\Delta[A]}{\Delta t} = k[A]^0$$

Integrating over time gives the integrated rate law for a zero-order reaction:

$$[A]_t - [A]_0 = -kt \quad \text{(zero-order reaction; rate} = k[A]^0 = k) \quad \textbf{(16.6)}$$

Sample Problem 16.4 shows one way integrated rate laws are applied.

SAMPLE PROBLEM 16.4 Determining the Reactant Concentration at a Given Time

Problem At 1000°C, cyclobutane (C_4H_8) decomposes in a first-order reaction, with the very high rate constant of 87 s^{-1}, to two molecules of ethylene (C_2H_4).
(a) If the initial C_4H_8 concentration is 2.00 M, what is the concentration after 0.010 s?
(b) What fraction of C_4H_8 has decomposed in this time?
Plan (a) We must find the concentration of cyclobutane at time t, $[C_4H_8]_t$. The problem tells us this is a first-order reaction, so we use the integrated first-order rate law:

$$\ln \frac{[C_4H_8]_0}{[C_4H_8]_t} = kt$$

We know k (87 s^{-1}), t (0.010 s), and $[C_4H_8]_0$ (2.00 M), so we can solve for $[C_4H_8]_t$.
(b) The fraction decomposed is the concentration that has decomposed divided by the initial concentration:

$$\text{Fraction decomposed} = \frac{[C_4H_8]_0 - [C_4H_8]_t}{[C_4H_8]_0}$$

Solution (a) Substituting the data into the integrated rate law:

$$\ln \frac{2.00 \text{ mol/L}}{[C_4H_8]_t} = (87 \text{ s}^{-1})(0.010 \text{ s}) = 0.87$$

Taking the antilog of both sides:

$$\frac{2.00 \text{ mol/L}}{[C_4H_8]_t} = e^{0.87} = 2.4$$

Solving for $[C_4H_8]_t$:

$$[C_4H_8]_t = \frac{2.00 \text{ mol/L}}{2.4} = \boxed{0.83 \text{ mol/L}}$$

(b) Finding the fraction that has decomposed after 0.010 s:

$$\frac{[C_4H_8]_0 - [C_4H_8]_t}{[C_4H_8]_0} = \frac{2.00 \text{ mol/L} - 0.83 \text{ mol/L}}{2.00 \text{ mol/L}} = \boxed{0.58}$$

Check The concentration remaining after 0.010 s (0.83 mol/L) is less than the starting concentration (2.00 mol/L), which makes sense. Raising e to an exponent slightly less than 1 should give a number (2.4) slightly less than the value of e (2.718). Moreover, the final result makes sense: a high rate constant indicates a fast reaction, so it's not surprising that so much decomposes in such a short time.

Comment Integrated rate laws are also used to solve for the time it takes to reach a certain reactant concentration, as in Follow-up Problem 16.4.

FOLLOW-UP PROBLEM 16.4 At 25°C, hydrogen iodide breaks down very slowly to hydrogen and iodine: rate = $k[HI]^2$. The rate constant at 25°C is 2.4×10^{-21} L/mol·s. If 0.0100 mol of HI(g) is placed in a 1.0-L container, how long will it take for the concentration of HI to reach 0.00900 mol/L (10.0% reacted)?

Determining the Reaction Order from the Integrated Rate Law

Suppose you don't know the rate law for a reaction and don't have the initial rate data needed to determine the reaction orders (as we did in Sample Problem 16.3). Another method for finding reaction orders is a graphical technique that uses concentration and time data directly.

Integrated rate laws can be rearranged into the form of an equation for a straight line, $y = mx + b$, where m is the slope and b is the y-axis intercept. For a first-order reaction, we have

$$\ln [A]_0 - \ln [A]_t = kt$$

Rearranging and changing signs gives

$$\ln [A]_t = -kt + \ln [A]_0$$
$$y = mx + b$$

Thus, a plot of $\ln [A]_t$ vs. t gives a straight line with slope $= -k$ and y intercept $= \ln [A]_0$ (Figure 16.7A).

For a simple second-order reaction, we have

$$\frac{1}{[A]_t} - \frac{1}{[A]_0} = kt$$

Rearranging gives

$$\frac{1}{[A]_t} = kt + \frac{1}{[A]_0}$$
$$y = mx + b$$

In this case, a plot of $1/[A]_t$ vs. t gives a straight line with slope $= k$ and y intercept $= 1/[A]_0$ (Figure 16.7B).

For a zero-order reaction, we have

$$[A]_t - [A]_0 = -kt$$

Rearranging gives

$$[A]_t = -kt + [A]_0$$
$$y = mx + b$$

Thus, a plot of $[A]_t$ vs. t gives a straight line with slope $= -k$ and y intercept $= [A]_0$ (Figure 16.7C).

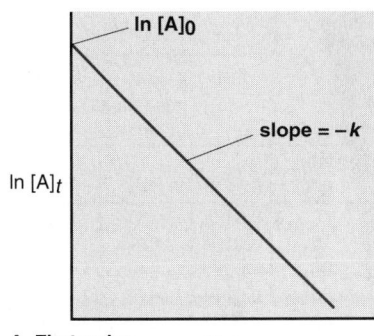

A First order

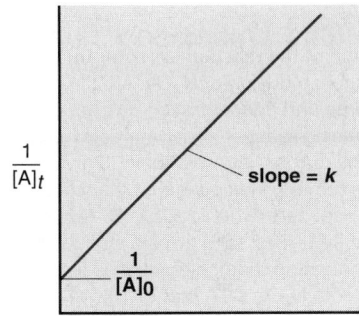

B Second order

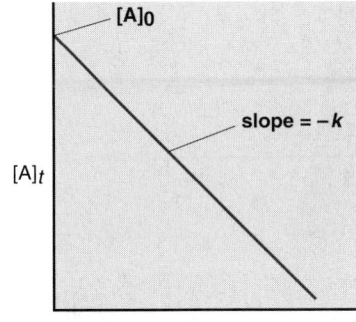

C Zero order

Figure 16.7 Integrated rate laws and reaction order. **A,** Plot of $\ln [A]_t$ vs. t gives a straight line for a reaction that is first order in A. **B,** Plot of $1/[A]_t$ vs. t gives a straight line for a reaction that is second order in A. **C,** Plot of $[A]_t$ vs. t gives a straight line for a reaction that is zero order in A.

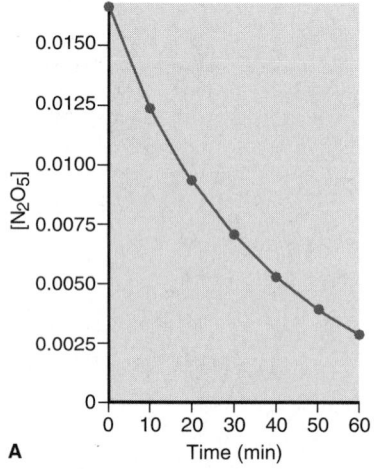

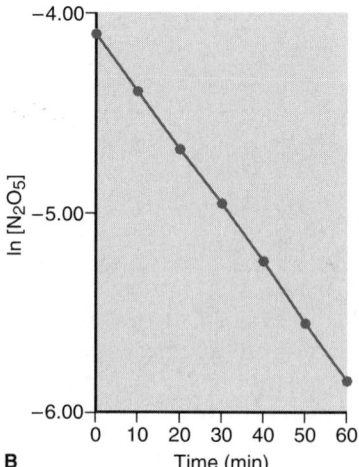

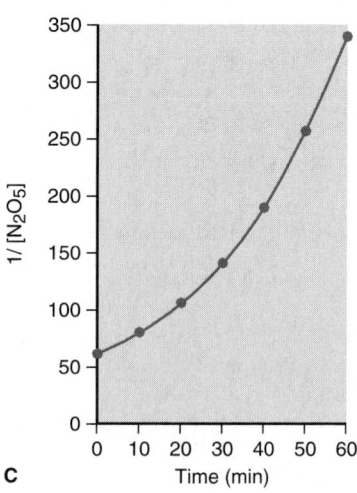

Time (min)	[N$_2$O$_5$]	ln [N$_2$O$_5$]	1/[N$_2$O$_5$]
0	0.0165	−4.104	60.6
10	0.0124	−4.390	80.6
20	0.0093	−4.68	1.1×10^2
30	0.0071	−4.95	1.4×10^2
40	0.0053	−5.24	1.9×10^2
50	0.0039	−5.55	2.6×10^2
60	0.0029	−5.84	3.4×10^2

Figure 16.8 **Graphical determination of the reaction order for the decomposition of N$_2$O$_5$.** A table of time and concentration data for determining reaction order appears below the graphs. **A,** A plot of [N$_2$O$_5$] vs. time is curved, indicating that the reaction is *not* zero order in N$_2$O$_5$. **B,** A plot of ln [N$_2$O$_5$] vs. time gives a straight line, indicating that the reaction *is* first order in N$_2$O$_5$. **C,** A plot of 1/[N$_2$O$_5$] vs. time is curved, indicating that the reaction is *not* second order in N$_2$O$_5$. Plots A and C support the conclusion from plot B.

Therefore, some trial-and-error graphical plotting is required to find the reaction order from the concentration and time data:

- If you obtain a straight line when you plot ln [reactant] vs. time, the reaction is *first order* with respect to that reactant.
- If you obtain a straight line when you plot 1/[reactant] vs. time, the reaction is *second order* with respect to that reactant.
- If you obtain a straight line when you plot [reactant] vs. time, the reaction is *zero order* with respect to that reactant.

Figure 16.8 shows how this approach is used to determine the order for the decomposition of N$_2$O$_5$. Since a plot of ln [N$_2$O$_5$] *is* linear and the plot of 1/[N$_2$O$_5$] *is not,* the decomposition of N$_2$O$_5$ must be first order in N$_2$O$_5$.

Reaction Half-Life

The **half-life** ($t_{1/2}$) of a reaction is the time required for the reactant concentration to reach half its initial value. A half-life is expressed in time units appropriate for a given reaction and is characteristic of that reaction at a given temperature.

At fixed conditions, *the half-life of a first-order reaction is a constant, independent of reactant concentration.* For example, the half-life for the first-order decomposition of N$_2$O$_5$ at 45°C is 24.0 min. The half-life means if we start with, say, 0.0600 mol/L of N$_2$O$_5$ at 45°C, after 24 min (one half-life), 0.0300 mol/L has been consumed and 0.0300 mol/L remains; after 48 min (two half-lives), 0.0150 mol/L remains; after 72 min (three half-lives), 0.0075 mol/L remains, and so forth (Figure 16.9).

We can see from the integrated rate law why the half-life of a first-order reaction is independent of concentration:

$$\ln \frac{[A]_0}{[A]_t} = kt$$

After one half-life, $t = t_{1/2}$, and $[A]_t = \frac{1}{2}[A]_0$. Substituting, we obtain

$$\ln \frac{[A]_0}{\frac{1}{2}[A]_0} = kt_{1/2} \qquad \text{or} \qquad \ln 2 = kt_{1/2}$$

Then, solving for $t_{1/2}$, we have

$$t_{1/2} = \frac{\ln 2}{k} = \frac{0.693}{k} \qquad \text{(first-order process; rate} = k[A]) \qquad \textbf{(16.7)}$$

As you can see, *the time to reach one-half the starting concentration in a first-order reaction does not depend on what that starting concentration is.*

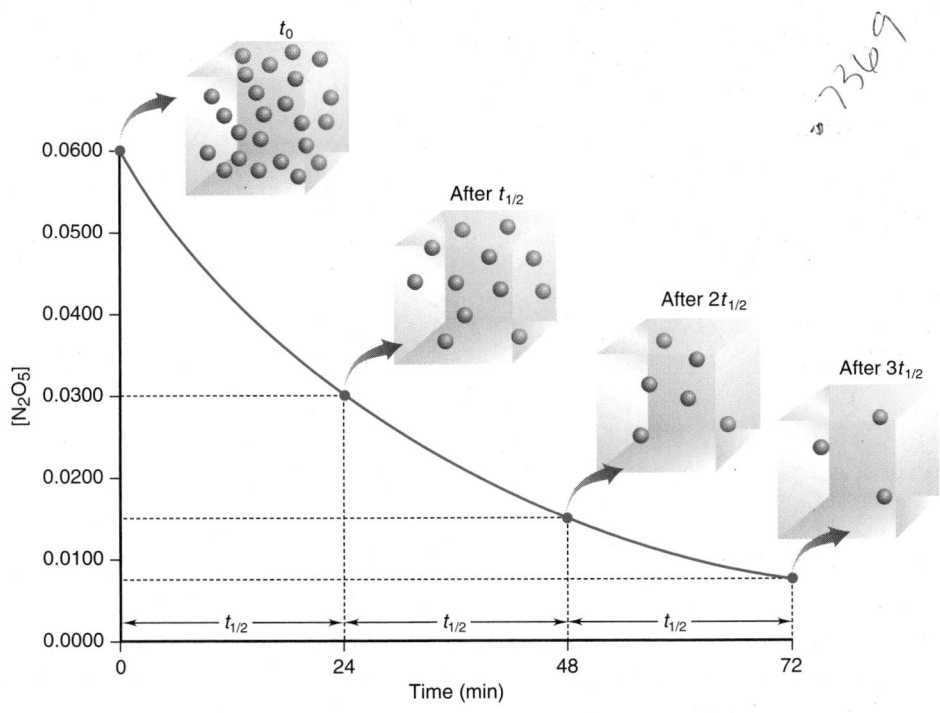

7369

Figure 16.9 A plot of [N₂O₅] vs. time for three half-lives. During each half-life, the concentration is halved. $T = 45°C$ and $[N_2O_5]_0 = 0.0600$ mol/L. The blow-up volumes, with N_2O_5 molecules as colored spheres, show that after three half-lives, $\frac{1}{2} \times \frac{1}{2} \times \frac{1}{2} = \frac{1}{8}$ of the original concentration remains.

Radioactive decay of an unstable nucleus is another example of a first-order process. For example, the half-life for the decay of uranium-235 is 7.1×10^8 yr. After 710 million years, a 1-kg sample of uranium-235 will contain 0.5 kg of uranium-235, and a 1-mg sample of uranium-235 will contain 0.5 mg. (We discuss the kinetics of radioactive decay thoroughly in Chapter 24.) Whether we consider a molecule or a radioactive nucleus, the *decomposition of each particle in a first-order process is independent of the number of other particles present.*

SAMPLE PROBLEM 16.5 Determining the Half-Life of a First-Order Reaction

Problem Cyclopropane is the smallest cyclic hydrocarbon. Because its 60° bond angles allow only poor orbital overlap, its bonds are weak. As a result, it is thermally unstable and rearranges to propene at 1000°C via the following first-order reaction:

$$H_2C\!\!-\!\!CH_2(g) \xrightarrow{\Delta} CH_3\!-\!CH\!\!=\!\!CH_2(g)$$

(with CH₂ bridging above)

The rate constant is 9.2 s^{-1}. **(a)** What is the half-life of the reaction? **(b)** How long does it take for the concentration of cyclopropane to reach one-quarter of the initial value?
Plan (a) The cyclopropane rearrangement is first order, so to find $t_{1/2}$ we use Equation 16.7 and substitute for k (9.2 s^{-1}). **(b)** Each half-life decreases the concentration to one-half its initial value, so two half-lives decrease it to one-quarter.
Solution (a) Solving for $t_{1/2}$:

$$t_{1/2} = \frac{\ln 2}{k} = \frac{0.693}{9.2 \text{ s}^{-1}} = \boxed{0.075 \text{ s}}$$

It takes 0.075 s for half the cyclopropane to form propene at this temperature.
(b) Finding the time to reach one-quarter of the initial concentration:

$$\text{Time} = 2(t_{1/2}) = 2(0.075 \text{ s}) = \boxed{0.15 \text{ s}}$$

Check (a) Rounding, we have $0.7/9$ s$^{-1} = 0.08$ s, so the answer seems correct.

FOLLOW-UP PROBLEM 16.5 Iodine-123 is used to study thyroid gland function. This radioactive isotope breaks down in a first-order process with a half-life of 13.1 h. What is the rate constant for the process?

Table 16.4 **An Overview of Zero-Order, First-Order, and Simple Second-Order Reactions**

	Zero Order	First Order	Second Order
Rate law	rate = k	rate = $k[A]$	rate = $k[A]^2$
Units for k	mol/L·s	1/s	L/mol·s
Integrated rate law in straight-line form	$[A]_t =$ $-kt + [A]_0$	$\ln [A]_t =$ $-kt + \ln [A]_0$	$1/[A]_t =$ $kt + 1/[A]_0$
Plot for straight line	$[A]_t$ vs. t	$\ln [A]_t$ vs. t	$1/[A]_t$ vs. t
Slope, y intercept	$-k, [A]_0$	$-k, \ln [A]_0$	$k, 1/[A]_0$
Half-life	$[A]_0/2k$	$(\ln 2)/k$	$1/k[A]_0$

In contrast to the half-life of a first-order reaction, the half-life of a second-order reaction *does* depend on reactant concentration:

$$t_{1/2} = \frac{1}{k[A]_0} \quad \text{(second-order process; rate = } k[A]^2\text{)}$$

Note that here *the half-life is **inversely** proportional to the initial reactant concentration*. This result means that a second-order reaction with a high initial reactant concentration has a shorter half-life, and one with a low initial reactant concentration has a longer half-life. Therefore, *as a second-order reaction proceeds, the half-life increases*.

In contrast to the half-life of a second-order reaction, *the half-life of a zero-order reaction is **directly** proportional to the initial reactant concentration*:

$$t_{1/2} = \frac{[A]_0}{2k} \quad \text{(zero-order process; rate = } k\text{)}$$

Thus, if a zero-order reaction begins with a high reactant concentration, it has a longer half-life than if it begins with a low reactant concentration. Table 16.4 summarizes the essential features of zero-, first-, and second-order reactions.

SECTION SUMMARY

Integrated rate laws are used to find the time needed to reach a certain concentration of reactant or the concentration present after a given time. Rearrangements of the integrated rate laws allow us to determine reaction orders and rate constants graphically. The half-life is the time needed to consume half the reactant; for first-order reactions, it is independent of concentration.

16.5 THE EFFECT OF TEMPERATURE ON REACTION RATE

Temperature often has a major effect on reaction rate. As Figure 16.10A shows for a common organic reaction (hydrolysis, or reaction with water, of an ester), when reactant concentrations are held constant, the rate nearly doubles with each rise in temperature of 10 K (or 10°C). In fact, for many reactions near room temperature, an increase of 10°C causes a doubling or tripling of the rate.

How does the rate law express this effect of temperature? If we collect concentration and time data for the same reaction run at *different* temperatures (T), and then solve each rate expression for k, we find that k increases as T increases. In other words, *temperature affects the rate by affecting the rate constant*. A plot of k vs. T gives a curve that increases exponentially, as shown in Figure 16.10B.

These results are consistent with studies made in 1889 by the Swedish chemist Svante Arrhenius, who discovered a key relationship between temperature and the rate constant. In its modern form, the **Arrhenius equation** is

$$k = Ae^{-E_a/RT} \tag{16.8}$$

Figure 16.10 Dependence of the rate constant on temperature. **A,** In the hydrolysis of an ester, when reactant concentrations are held constant and temperature increases, the rate and rate constant increase. Note the near doubling of k with each temperature rise of 10 K (10°C). **B,** A plot of rate constant vs. temperature for this reaction shows a smoothly increasing curve.

Exp't	[Ester]	[H_2O]	T (K)	Rate (mol/L·s)	k (L/mol·s)
1	0.100	0.200	288	1.04×10^{-3}	0.0521
2	0.100	0.200	298	2.02×10^{-3}	0.101
3	0.100	0.200	308	3.68×10^{-3}	0.184
4	0.100	0.200	318	6.64×10^{-3}	0.332

A

B

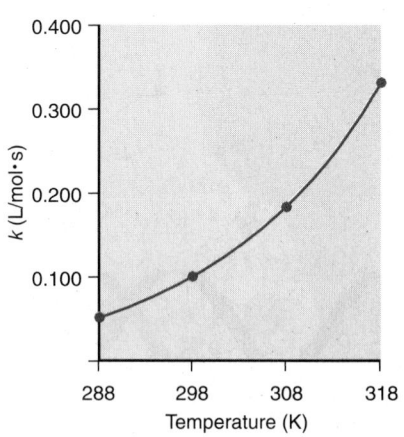

where k is the rate constant, e is the base of natural logarithms, T is the absolute temperature, and R is the universal gas constant. We'll discuss the meaning of A, which is related to the orientation of the colliding molecules, in the next section. The E_a term is the **activation energy** of the reaction, which Arrhenius considered the *minimum energy* the molecules must have to react; we discuss its meaning further in the next section as well. This negative exponential relationship between temperature and the rate constant means that *as the temperature increases, the negative exponent becomes smaller, so the value of k becomes larger, which means that the rate increases:*

$$\text{Higher } T \Longrightarrow \text{larger } k \Longrightarrow \text{increased rate}$$

We can calculate E_a from the Arrhenius equation by taking the natural logarithm of both sides and recasting the equation into the form of an equation for a straight line:

$$\ln k = \ln A - \frac{E_a}{R}\left(\frac{1}{T}\right)$$
$$y \ \ = \ \ b \ \ + \ \ mx$$

A plot of $\ln k$ vs. $1/T$ gives a straight line whose slope is $-E_a/R$ and whose y intercept is $\ln A$ (Figure 16.11). Therefore, with the constant R known, we can determine E_a graphically from a series of k values at different temperatures.

Because the relationship between $\ln k$ and $1/T$ is linear, we can use an even simpler method for finding E_a if we know the rate constant at two different temperatures, T_2 and T_1:

$$\ln k_2 = \ln A - \frac{E_a}{R}\left(\frac{1}{T_2}\right) \qquad \ln k_1 = \ln A - \frac{E_a}{R}\left(\frac{1}{T_1}\right)$$

When we subtract $\ln k_1$ from $\ln k_2$, the "$\ln A$" term drops out and the other terms can be rearranged to give

$$\ln \frac{k_2}{k_1} = -\frac{E_a}{R}\left(\frac{1}{T_2} - \frac{1}{T_1}\right) \qquad \textbf{(16.9)}$$

From this, we can solve for E_a. ●

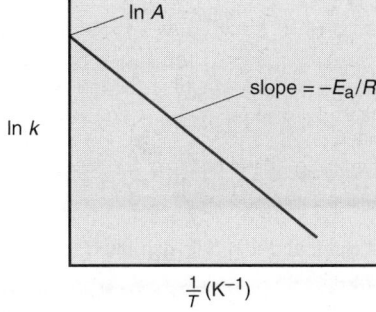

Figure 16.11 Graphical determination of the activation energy. A plot of $\ln k$ vs. $1/T$ gives a straight line with slope $-E_a/R$.

● **The Significance of R** The importance of R extends beyond the study of gases (or osmotic pressure). By expressing pressure and volume in more fundamental quantities, we obtain the dimensions for energy, E:

$$P = \frac{\text{force}}{\text{area}} = \frac{\text{force}}{(\text{length})^2}; \ V = (\text{length})^3$$

Thus,

$$PV = \frac{\text{force}}{(\text{length})^2} \times (\text{length})^3$$
$$= \text{force} \times \text{length}$$

Energy is used to move an object; that is, energy is expended when a force acts over a distance. Thus, $E =$ force × distance (or length), so

$$PV = \text{force} \times \text{length} = E$$

Solving for R in the ideal gas law and substituting for PV, we obtain

$$R = \frac{PV}{nT} = \frac{E}{\text{amount} \times T}$$

Thus, R is the proportionality constant that relates the energy, amount (mol) of substance, and temperature of any chemical system.

SAMPLE PROBLEM 16.6 Determining the Energy of Activation

Problem The decomposition of hydrogen iodide,

$$2HI(g) \longrightarrow H_2(g) + I_2(g)$$

has rate constants of 9.51×10^{-9} L/mol·s at 500. K and 1.10×10^{-5} L/mol·s at 600. K. Find E_a.

Plan We are given the rate constants, k_1 and k_2, at two temperatures, T_1 and T_2, so we substitute into Equation 16.9 and solve for E_a.

Solution Rearranging Equation 16.9 to solve for E_a:

$$\ln \frac{k_2}{k_1} = -\frac{E_a}{R}\left(\frac{1}{T_2} - \frac{1}{T_1}\right)$$

$$E_a = -R\left(\ln \frac{k_2}{k_1}\right)\left(\frac{1}{T_2} - \frac{1}{T_1}\right)^{-1}$$

$$= -(8.314 \text{ J/mol·K})\left(\ln \frac{1.10 \times 10^{-5} \text{ L/mol·s}}{9.51 \times 10^{-9} \text{ L/mol·s}}\right)\left(\frac{1}{600. \text{ K}} - \frac{1}{500. \text{ K}}\right)^{-1}$$

$$= 1.76 \times 10^5 \text{ J/mol}$$

$$= \boxed{1.76 \times 10^2 \text{ kJ/mol}}$$

Comment Be sure to retain the same number of significant figures in $1/T$ as you have in T, or a significant error could be introduced. Round to the correct number of significant figures only at the final answer. On most pocket calculators, the expression $(1/T_2 - 1/T_1)$ is entered as follows: $(T_2)(1/x) - (T_1)(1/x) =$.

FOLLOW-UP PROBLEM 16.6 The reaction $2NOCl(g) \longrightarrow 2NO(g) + Cl_2(g)$ has an E_a of 1.00×10^2 kJ/mol and a rate constant of 0.286 L/mol·s at 500. K. What is the rate constant at 490. K?

In this section and the previous two, we discussed a series of experimental and mathematical methods for the study of reaction kinetics. Figure 16.12 is a useful summary of this information. Note that the integrated rate law provides an alternative method for obtaining reaction orders and the rate constant.

SECTION SUMMARY

As the Arrhenius equation shows, rate increases with temperature because a temperature rise increases the rate constant. The activation energy, E_a, the minimum energy needed for a reaction to occur, can be determined graphically from k values at different T values.

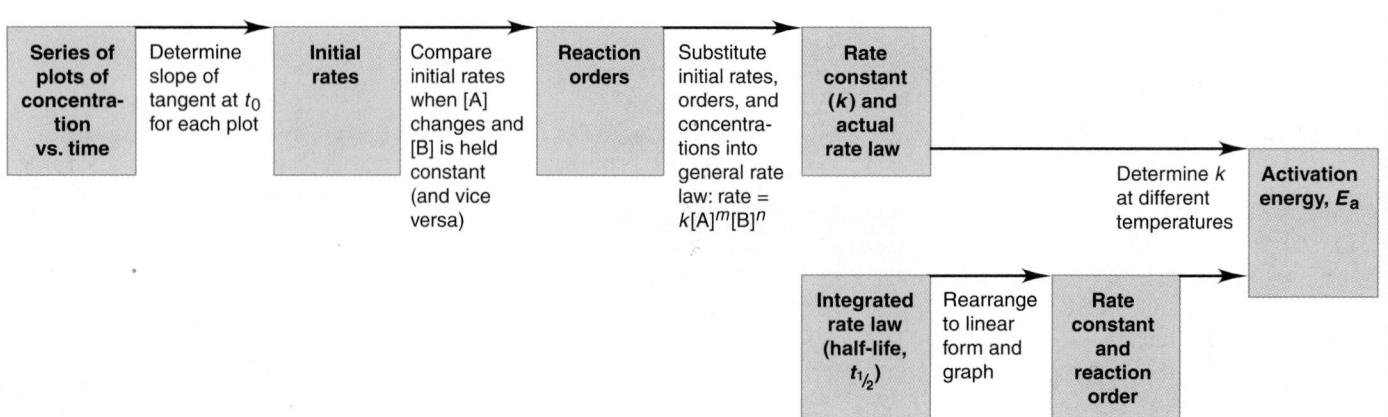

Figure 16.12 **Information sequence to determine the kinetic parameters of a reaction.** Note that the integrated rate law does not depend on the method of initial rates and that it is also used to determine reaction orders and rate constant.

16.6 EXPLAINING THE EFFECTS OF CONCENTRATION AND TEMPERATURE

The Arrhenius equation was developed empirically from the observations of many reactions. It took more than 30 years for models to appear that could explain the effects of concentration and temperature on reaction rate. The two major models we consider here are distinct but completely compatible, each highlighting different aspects of the reaction process. **Collision theory** views the reaction rate as the result of particles colliding with a certain frequency and minimum energy. **Transition state theory** offers a close-up view of how the energy of a collision converts reactant to product.

Collision Theory: Basis of the Rate Law

The basic tenet of collision theory is that reactant particles—atoms, molecules, and ions—must collide with each other to react. Therefore, the number of collisions per unit time provides an upper limit on how fast a reaction can take place. The model restricts itself to simple one-step reactions in which two particles collide and form products: A + B $\longrightarrow$ products. With its emphasis on collisions between three-dimensional particles, it elegantly explains why reactant concentrations are multiplied together in the rate law, how temperature affects the rate, and what influence molecular structure has on rate.

Why Concentrations Are Multiplied in the Rate Law If particles must collide to react, the laws of probability tell us why the rate depends on the *product* of the reactant concentrations, not their sum. Imagine that you have only two particles of A and two of B confined in a reaction vessel. Figure 16.13 shows that four A-B collisions are possible. If you add another particle of A, there can be six A-B collisions (3 × 2), not five (3 + 2); add another particle of B, and there can be nine A-B collisions (3 × 3), not six (3 + 3). Thus, the collision model is consistent with the observation that concentrations are *multiplied* in the rate law.

How Temperature Affects Rate: The Importance of Activation Energy Increasing the temperature of a reaction increases the average speed of particles and therefore their collision frequency. But collision frequency cannot be the only factor affecting rate, or every gaseous reaction would be over instantaneously: after all, at 1 atm and 20°C, the molecules in a milliliter of gas experience about 10^{27} collisions per second, or about 4×10^7 collisions/molecule, so the reaction would be over in less than one ten-millionth of a second! In fact, *in the vast majority of collisions, the molecules rebound without reacting.*

Arrhenius proposed that every reaction has an *energy threshold* that the colliding molecules must exceed in order to react. (An analogy might be an athlete who must exceed the height of the bar to accomplish a high jump.) This minimum collision energy is the *activation energy* (E_a), the energy required to activate the molecules into a state from which reactant bonds can change into product bonds. Recall that at any given temperature, molecules have a range of kinetic energies; thus, their collisions have a range of energies as well. According to collision theory, *only those collisions with enough energy to exceed E_a can lead to reaction.*

We noted earlier that many reactions near room temperature approximately double or triple their rates with a 10°C rise in temperature. Is the rate increase due to a higher number of collisions? Actually, this has only a minor effect. Calculations show that a 10°C rise increases the average molecular speed by only 2%. If an increase in speed were the only effect of temperature and if the speed of each colliding molecule increases by 2%, we would expect only a 4% increase in rate. Far more important is that *the temperature rise enlarges the fraction of*

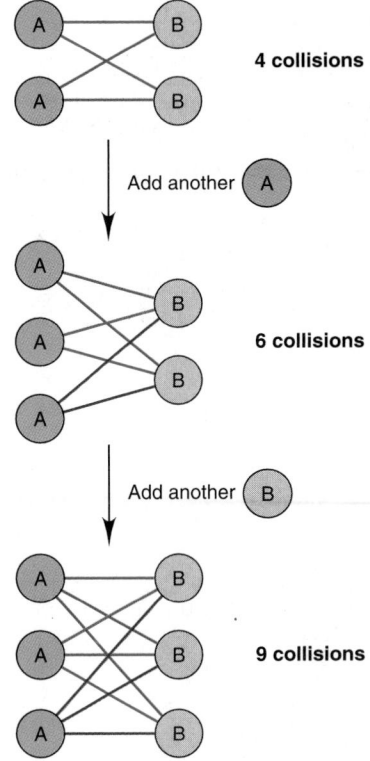

Figure 16.13 The dependence of possible collisions on the product of reactant concentrations. Concentrations are multiplied, not added, in the rate law because the number of possible collisions is the *product*, not the sum, of the numbers of particles present.

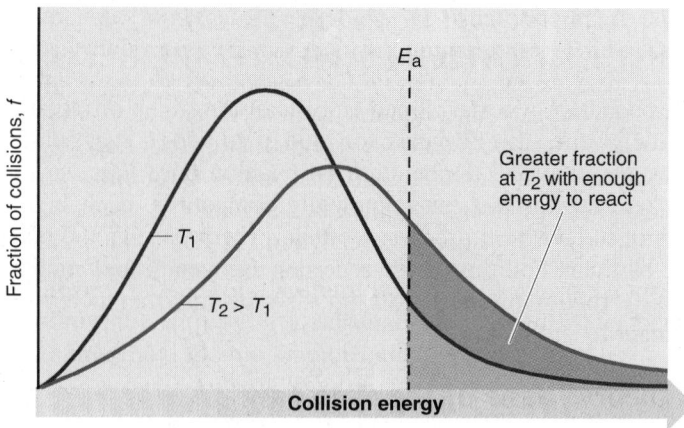

Figure 16.14 **The effect of temperature on the distribution of collision energies.** At the higher temperature, T_2, a larger fraction of collisions occurs with enough energy to exceed E_a.

Table 16.5 The Effect of E_a and T on the Fraction (f) of Collisions with Sufficient Energy to Allow Reaction	
E_a (kJ/mol)	f (at $T = 298$ K)
50	1.70×10^{-9}
75	7.03×10^{-14}
100	2.90×10^{-18}
T	f (at $E_a = 50$ kJ/mol)
25°C (298 K)	1.70×10^{-9}
35°C (308 K)	3.29×10^{-9}
45°C (318 K)	6.12×10^{-9}

collisions with enough energy to exceed the activation energy. This key point is shown in Figure 16.14.

At a given temperature, the fraction f of molecular collisions with energy greater than or equal to the activation energy E_a is given by

$$f = e^{-E_a/RT}$$

where e is the base of natural logarithms, T is the absolute temperature, and R is the universal gas constant. [Notice that the right side of this equation is the central component in the Arrhenius equation (Equation 16.8).] *The magnitudes of both E_a and T affect the fraction of sufficiently energetic collisions.* In the top portion of Table 16.5, you can see the effect of increasing E_a on this fraction of collisions at a fixed temperature. Note how much the fraction shrinks with a 25-kJ/mol increase in activation energy. (As the height of the bar is raised, fewer athletes can accomplish the jump.) In the bottom portion you can see the effect of T on the fraction for a fixed E_a of 50 kJ/mol, a typical value for many reactions. Note that the fraction nearly doubles for a 10°C increase. Doubling the fraction doubles the rate constant, which doubles the reaction rate.

A reversible reaction has two activation energies (Figure 16.15). The activation energy for the forward reaction [$E_{a(fwd)}$] is the energy difference between the activated state and the reactants; the activation energy for the reverse reaction [$E_{a(rev)}$] is the energy difference between the activated state and the products. The figure shows an energy-level diagram for an exothermic reaction, so the products are at a lower energy than the reactants, and $E_{a(fwd)}$ is less than $E_{a(rev)}$.

By turning the collision-energy distribution curve of Figure 16.14 vertically and aligning it on both sides of Figure 16.15, we obtain Figure 16.16, which shows how temperature affects the fraction of collisions that exceed the activation energy for both the forward and reverse reactions. Several ideas are illustrated in this composite figure:

- In both reaction directions, a larger fraction of collisions exceeds the activation energy at the higher temperature, T_2: higher T increases reaction rate.
- For an *exothermic* process (forward reaction here) at any temperature, the fraction of reactant collisions with energy exceeding $E_{a(fwd)}$ is *larger* than the fraction of product collisions with energy exceeding $E_{a(rev)}$; thus, the forward reaction is faster. On the other hand, in an *endothermic* process (reverse reaction here), $E_{a(fwd)}$ is greater than $E_{a(rev)}$, so the fraction of product collisions with energy exceeding $E_{a(rev)}$ is larger, and the reverse reaction is faster.

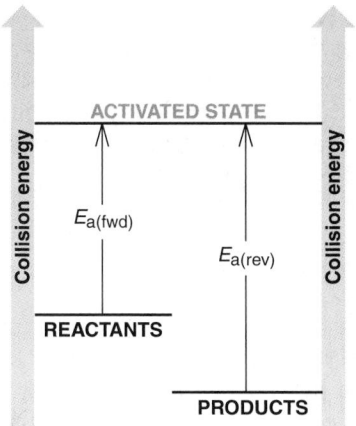

Figure 16.15 **Energy-level diagram for a reaction.** For molecules to react, they must collide with enough energy to reach an activated state. This minimum collision energy is the energy of activation, E_a. A reaction can occur in either direction, so the diagram shows two activation energies. Here, the forward reaction is exothermic because $E_{a(fwd)} < E_{a(rev)}$.

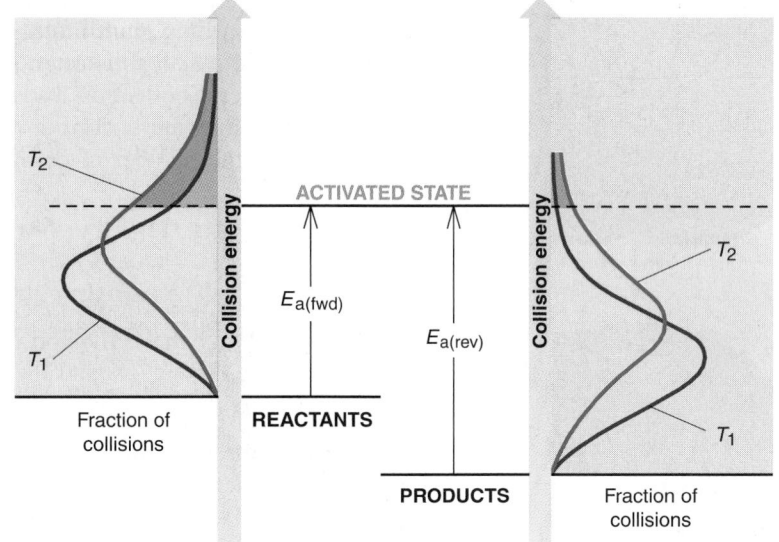

Figure 16.16 **An energy-level diagram of the fraction of collisions exceeding E_a.** When Figure 16.14 is aligned vertically on the left and right axes of Figure 16.15, we see that
- In either direction, the fraction of collisions exceeding E_a is larger at the higher T.
- In an *exothermic* reaction at any temperature, the fraction of collisions exceeding $E_{a(fwd)}$ is larger than the fraction exceeding $E_{a(rev)}$.

These conclusions are consistent with the Arrhenius equation; that is, *the larger the E_a, the smaller the value of k, and the slower the reaction:*

$$\text{Larger } E_a \Longrightarrow \text{smaller } k \Longrightarrow \text{decreased rate}$$

How Molecular Structure Affects Rate You've seen that the enormous number of collisions per second is greatly reduced when we count only those with enough energy to react. However, even this tiny fraction of the total collisions does not reveal the true number of **effective collisions,** those that actually lead to product. In addition to colliding with enough energy, *the molecules must collide such that the reacting atoms make contact.* In other words, a collision must have enough energy *and* a particular *molecular orientation* to be an effective collision.

In the Arrhenius equation, the effect of molecular orientation is contained in the factor A:

$$k = Ae^{-E_a/RT}$$

This term is called the **frequency factor,** the product of the collision frequency Z and an *orientation probability factor p,* which is specific for each reaction: $A = pZ$. The factor p is related to the structural complexity of the colliding particles. You can think of it as the ratio of effectively oriented collisions to all possible collisions. For example, Figure 16.17 shows a few of the possible collision orientations for the following simple gaseous reaction:

$$NO(g) + NO_3(g) \longrightarrow 2NO_2(g)$$

Of the five collisions shown, only one has an orientation in which the N of NO collides with an O of NO_3. Actually, the probability factor (p value) for this reaction is 0.006: only 6 collisions in every 1000 (1 in 167) have an orientation that leads to reaction.

Collisions between individual atoms have p values near 1: almost no matter how they hit, so long as the collision has enough energy, the particles react. In such cases, the rate constant depends only on the frequency and energy of the collisions. At the other extreme are biochemical reactions, in which the reactants

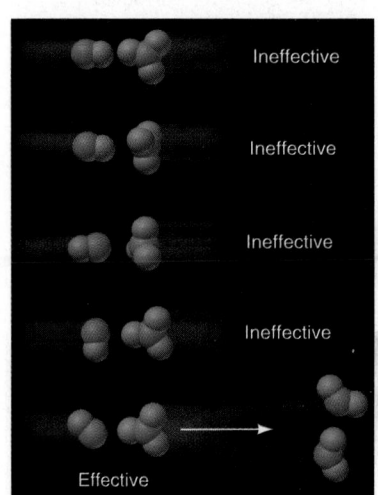

Figure 16.17 **The importance of molecular orientation to an effective collision.** Only one of the five orientations shown for the collision between NO and NO_3 has the correct orientation to lead to product. In the effective orientation, collision occurs between the atoms that will become bonded in the product.

are often two small molecules that can react only when they collide with a specific tiny region of a giant molecule—a protein or nucleic acid. The orientation factor for these reactions is often less than 10^{-6}: fewer than one in a million sufficiently energetic collisions leads to product. The fact that countless such biochemical reactions are occurring right now as you read this sentence helps make the point that the number of collisions per second is truly astounding.

Transition State Theory: Molecular Nature of the Activated State

Collision theory is a simple model that is easy to visualize, but it provides no insight about why the activation energy is needed and what the activated molecules look like. To understand these aspects of the process, we turn to transition state theory.

Visualizing the Transition State Recall from our discussion of energy changes (Chapter 6) that the internal energy of a system is the sum of its kinetic and potential energies. Two molecules that are far apart but speeding toward each other have high kinetic energy and low potential energy. As the molecules get closer, some kinetic energy is converted to potential energy as the electron clouds repel each other. At the moment of a head-on collision, the molecules stop, and their kinetic energy is converted to the potential energy of the collision. *If this potential energy is less than the activation energy, the molecules recoil,* bouncing off each other like billiard balls. Repulsions decrease, speeds increase, and the molecules zoom apart without reacting.

The tiny fraction of molecules that are oriented effectively *and* moving at the highest speed behave differently. *Their kinetic energy pushes them together with enough force to overcome repulsions and react.* Nuclei in one atom attract electrons in another; atomic orbitals overlap and electron density shifts; some bonds lengthen and weaken while others start to form. If we could watch this process in slow motion, we would see the reactant molecules gradually change their bonds and shapes as they turn into product molecules. At some point during this smooth transformation, what exists is *neither reactant nor product but a transitional species with partial bonds.* This species is extremely unstable (has very high potential energy) and exists only at the instant when the reacting system is highest in energy. It is called the **transition state,** or **activated complex,** of the reaction, and it forms only if the molecules collide in an effective orientation *and* the energy of the collision is equal to or greater than the activation energy. Thus, *the activation energy is needed to stretch and deform bonds in order to reach the transition state.* Because transition states *cannot* be isolated, our knowledge of them is indirect, coming from reasoning and studies of analogous, more stable species.

Consider the reaction between methyl bromide and hydroxide ion:

$$CH_3Br + OH^- \longrightarrow CH_3OH + Br^-$$

The electronegative bromine makes the carbon of methyl bromide partially positive. If the reactants are moving toward each other fast enough and are oriented effectively when they collide, the negative oxygen in OH^- approaches the carbon with enough energy to begin forming a C—O bond, which causes the C—Br bond to weaken. In the transition state (Figure 16.18), the carbon is surrounded by five atoms in a trigonal bipyramidal shape, which never occurs in stable carbon compounds. This high-energy species consists of three normal C—H bonds and two partial bonds, one to O and the other to Br; there are *never* five normal bonds to carbon.

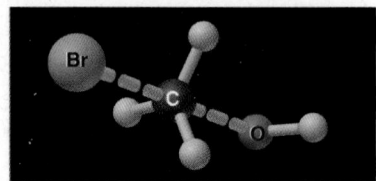

Figure 16.18 Nature of the transition state in the reaction between CH_3Br and OH^-. Note the partial (elongated) C—O and C—Br bonds and the trigonal bipyramidal shape of the transition state of this reaction.

Reaching the transition state is no guarantee that the reaction will proceed to products; a transition state can change in either direction. In this case, if the C—O bond continues to shorten and strengthen, products will form; however, if the C—Br bond becomes shorter and stronger again, the transition state will decompose back to reactants.

Depicting the Change with Reaction Energy Diagrams A useful way to depict the events we just described is with a **reaction energy diagram,** which shows the potential energy of the system during the reaction as a smooth curve. Figure 16.19 shows the reaction energy diagram for the methyl bromide–hydroxide ion reaction, with a structural (wedge-bond) formula and molecular-scale view at various points during the change. The horizontal axis, labeled "reaction progress" (or "reaction coordinate"), means that reactants change to products as we move from left to right. The reaction is exothermic, so the reactants are higher in energy than the products. This energy difference is due to differences in bond energy, which appear as the heat of reaction, ΔH_{rxn}. The diagram also shows the activation energies for the forward and reverse reactions; in this case, $E_{a(fwd)}$ is less than $E_{a(rev)}$.

Transition state theory proposes that *every reaction (and every step in an overall reaction) goes through its own transition state,* from which point it can continue in either direction. Thus, according to this model, *all reactions are reversible.* We can imagine how the transition state in many reactions might look

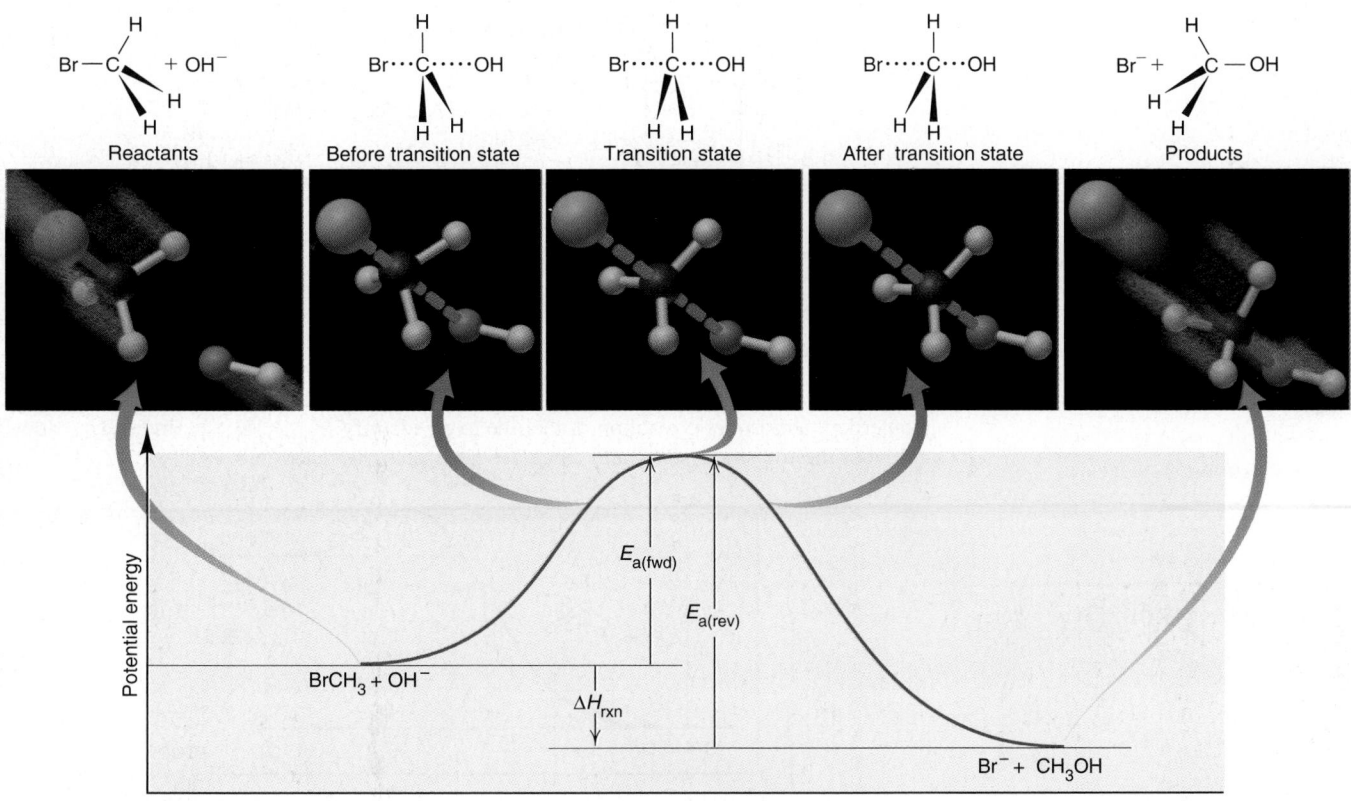

Figure 16.19 **Reaction energy diagram for the reaction between CH$_3$Br and OH$^-$.** A plot of potential energy vs. reaction progress shows the relative energy levels of reactants, products, and transition state joined by a curved line, as well as the activation energies of the forward and reverse steps and the heat of reaction. The structural formulas and molecular-scale views depict the change at five points. Note the gradual bond forming and bond breaking as the system goes through the transition state.

by examining the reactant and product bonds that undergo change. Figure 16.20 depicts reaction energy diagrams for three simple reactions. Note that the shape of the postulated transition state in each case is based on a collision between atoms that become bonded in the product.

SAMPLE PROBLEM 16.7 Drawing Reaction Energy Diagrams and Transition States

---

Problem A key reaction in the upper atmosphere is

$$O_3(g) + O(g) \longrightarrow 2O_2(g)$$

The $E_{a(fwd)}$ is 19 kJ, and the ΔH_{rxn} for the reaction as written is -392 kJ. Draw a reaction energy diagram for this reaction, postulate a transition state, and calculate $E_{a(rev)}$.

Plan The reaction is highly exothermic ($\Delta H_{rxn} = -392$ kJ), so the products are much lower in energy than the reactants. The small $E_{a(fwd)}$ (19 kJ) means that the energy of the reactants lies slightly below that of the transition state. To reach the transition state from the products would require energy equal to ΔH_{rxn} plus $E_{a(fwd)}$; thus, $E_{a(rev)} = \Delta H_{rxn} + E_{a(fwd)}$. To postulate the transition state, we sketch the species and note that one of the bonds in O_3 weakens, and this partially bonded O begins forming a bond to the separate O atom.

Solution The reaction energy diagram (not drawn to scale), with transition state, is

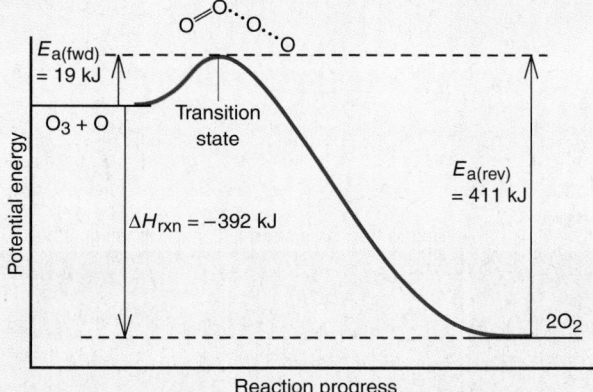

Check Rounding to find $E_{a(rev)}$ gives $\sim 390 + 20 = 410$.

FOLLOW-UP PROBLEM 16.7 The following reaction energy diagram depicts another key atmospheric reaction. Label the axes, identify $E_{a(fwd)}$, $E_{a(rev)}$, and ΔH_{rxn}, draw and label the transition state, and calculate $E_{a(rev)}$ for the reaction.

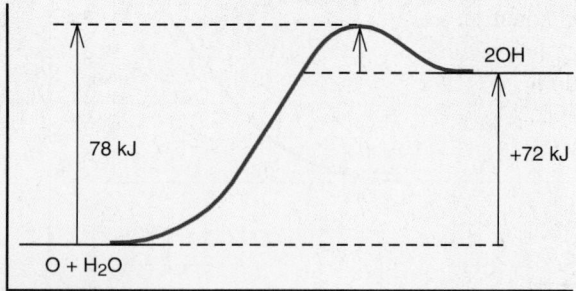

SECTION SUMMARY

According to collision theory, reactant particles must collide to react, and the number of collisions depends on the product of the reactant concentrations. At higher temperatures, more collisions have enough energy to exceed the activation energy (E_a). The relative sizes of E_a for the forward and the reverse reactions depend on whether the overall reaction is exothermic or endothermic. Molecules must collide with an effective orientation for reaction to occur, so structural complexity decreases rate.

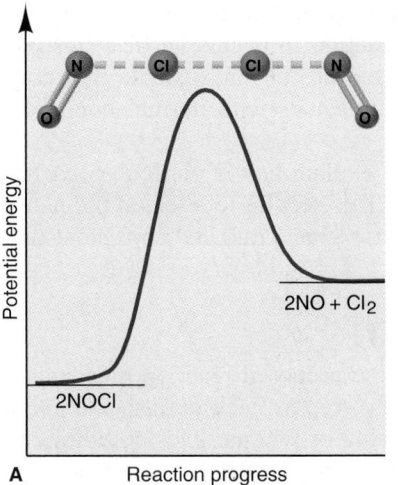
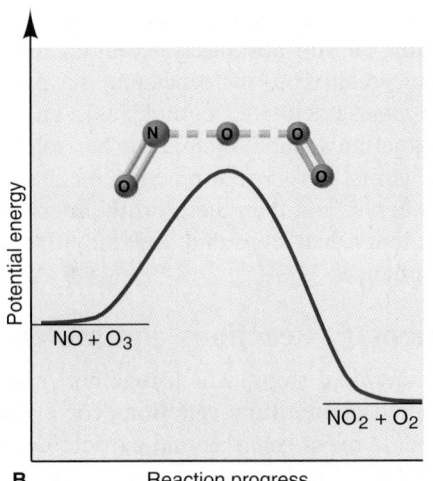
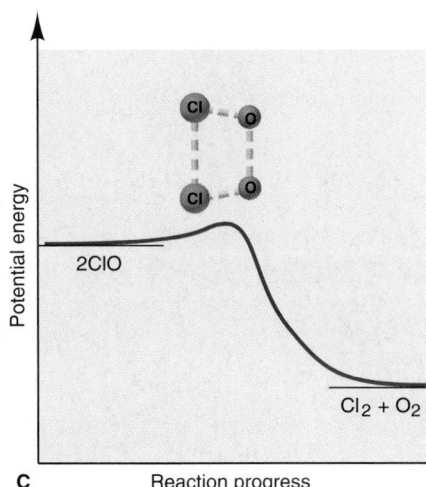

Figure 16.20 Reaction energy diagrams and possible transition states for three reactions.
A, $2NOCl(g) \longrightarrow 2NO(g) + Cl_2(g)$ (despite the formula NOCl, the atom sequence is ClNO).

B, $NO(g) + O_3(g) \longrightarrow NO_2(g) + O_2(g)$.
C, $2ClO(g) \longrightarrow Cl_2(g) + O_2(g)$.
Note that reaction **A** is endothermic, **B** and **C** are exothermic, and **C** has a very small $E_{a(fwd)}$.

Transition state theory pictures the kinetic energy of the particles changing to potential energy during a collision. Given a sufficiently energetic collision and an effective molecular orientation, the reactant species become an unstable transition state, which decomposes to either reactant or product. Reaction energy diagrams depict the changing energy of the chemical system as it progresses from reactants through transition state(s) to products.

16.7 REACTION MECHANISMS: STEPS IN THE OVERALL REACTION

Imagine trying to figure out how a car works just by examining the body, wheels, and dashboard. It can't be done—you need to look under the hood and inside the engine to see how the parts fit together and function. Similarly, because our main purpose is to know how a reaction works *at the molecular level,* examining the overall balanced equation is not much help—we must "look under the yield arrow and inside the reaction" to see how reactants change into products.

When we do so, we find that most reactions occur through a **reaction mechanism,** a sequence of single reaction steps that sum to the overall reaction. For example, a possible mechanism for the overall reaction

$$2A + B \longrightarrow E + F$$

might involve these three simpler steps:

(1) $A + B \longrightarrow C$
(2) $C + A \longrightarrow D$
(3) $D \longrightarrow E + F$

Adding them together and canceling common substances, we get the overall equation:

$$A + B + \cancel{C} + A + \cancel{D} \longrightarrow \cancel{C} + \cancel{D} + E + F \quad \text{or} \quad 2A + B \longrightarrow E + F$$

Note what happens to C and to D in this mechanism. C is a product in step 1 and a reactant in step 2, and D is a product in 2 and a reactant in 3. Each functions as a **reaction intermediate,** a substance that is formed and used up during the overall reaction. Reaction intermediates do not appear in the overall balanced equation but are absolutely essential for the reaction to occur. They are usually unstable relative to the reactants and products but are far more stable than transition states (activated complexes). They are molecules with normal bonds and are sometimes stable enough to be isolated.

Chemists *propose* a reaction mechanism to explain how a particular reaction might occur, and then they *test* the mechanism. This section focuses on the nature of the individual steps and how they fit together to give a rate law consistent with experiment.

Elementary Reactions and Molecularity

The individual steps, which together make up the proposed reaction mechanism, are called **elementary reactions** (or **elementary steps**). Each describes a *single molecular event,* such as one particle decomposing or two particles colliding and combining. *An elementary step is **not** made up of simpler steps.*

An elementary step is characterized by its **molecularity,** the number of *reactant* particles involved in the step. Consider the mechanism for the breakdown of ozone in the stratosphere. The overall reaction is

$$2O_3(g) \longrightarrow 3O_2(g)$$

A two-step mechanism has been proposed for this reaction. Notice that the two steps sum to the overall reaction. The first elementary step is a **unimolecular reaction,** one that involves the decomposition or rearrangement of a single particle:

(1) $O_3(g) \longrightarrow O_2(g) + O(g)$

The second step is a **bimolecular reaction,** one in which two particles react:

(2) $O_3(g) + O(g) \longrightarrow 2O_2(g)$

Some *termolecular* elementary steps occur, but they are extremely rare because the probability of three particles colliding simultaneously with enough energy and with an effective orientation is very small. Higher molecularities are not known. Unless evidence exists to the contrary, it makes good chemical sense to propose only unimolecular or bimolecular reactions as the elementary steps in a reaction mechanism.

The rate law for an elementary reaction, unlike that for an overall reaction, *can* be deduced from the reaction stoichiometry. An elementary reaction occurs in one step, so its rate must be proportional to the product of the reactant concentrations. Therefore, *we use the equation coefficients as the reaction orders in the rate law for an elementary step; that is, reaction order equals molecularity* (Table 16.6). Remember that this statement holds *only* when we know that the reaction is elementary; you've already seen that for an overall reaction, the reaction orders must be determined experimentally.

Table 16.6 Rate Laws for General Elementary Steps

Elementary Step	Molecularity	Rate Law
A $\longrightarrow$ product	Unimolecular	Rate = $k[A]$
2A $\longrightarrow$ product	Bimolecular	Rate = $k[A]^2$
A + B $\longrightarrow$ product	Bimolecular	Rate = $k[A][B]$
2A + B $\longrightarrow$ product	Termolecular	Rate = $k[A]^2[B]$

SAMPLE PROBLEM 16.8 Determining Molecularity and Rate Laws for Elementary Steps

Problem The following two reactions are proposed as elementary steps in the mechanism for an overall reaction:

(1) $\quad\quad\quad NO_2Cl(g) \longrightarrow NO_2(g) + Cl(g)$
(2) $NO_2Cl(g) + Cl(g) \longrightarrow NO_2(g) + Cl_2(g)$

(a) Write the overall balanced equation.
(b) Determine the molecularity of each step.
(c) Write the rate law for each step.

Plan We find the overall equation from the sum of the elementary steps. The molecularity of each step equals the total number of reactant particles. We write the rate law for each step using the molecularities as reaction orders.

Solution (a) Writing the overall equation:

$$NO_2Cl(g) \longrightarrow NO_2(g) + Cl(g)$$
$$NO_2Cl(g) + Cl(g) \longrightarrow NO_2(g) + Cl_2(g)$$
$$\overline{NO_2Cl(g) + NO_2Cl(g) + \cancel{Cl(g)} \longrightarrow NO_2(g) + \cancel{Cl(g)} + NO_2(g) + Cl_2(g)}$$
$$2NO_2Cl(g) \longrightarrow 2NO_2(g) + Cl_2(g)$$

(b) Determining the molecularity of each step: The first elementary step has only one reactant, NO_2Cl, so it is unimolecular. The second elementary step has two reactants, NO_2Cl and Cl, so it is bimolecular.

(c) Writing rate laws for the elementary reactions:

(1) $\quad Rate_1 = k_1[NO_2Cl]$
(2) $\quad Rate_2 = k_2[NO_2Cl][Cl]$

Check In part (a), be sure the equation is balanced; in part (c), be sure the substances in brackets are the reactants of each elementary step.

FOLLOW-UP PROBLEM 16.8 The following elementary steps constitute a proposed mechanism for a reaction:

(1) $\quad\quad\quad 2NO(g) \longrightarrow N_2O_2(g)$
(2) $\quad\quad\quad 2[H_2(g) \longrightarrow 2H(g)]$
(3) $N_2O_2(g) + H(g) \longrightarrow N_2O(g) + HO(g)$
(4) $2[HO(g) + H(g) \longrightarrow H_2O(g)]$
(5) $\quad H(g) + N_2O(g) \longrightarrow HO(g) + N_2(g)$

(a) Write the balanced equation for the overall reaction.
(b) Determine the molecularity of each step.
(c) Write the rate law for each step.

The Rate-Determining Step of a Reaction Mechanism

All the elementary steps in a mechanism do not have the same rate. Usually, one of the steps is much slower than the others, so it limits how fast the overall reaction proceeds. This step is called the **rate-determining step** (or **rate-limiting step**).

Because the rate-determining step limits the rate of the overall reaction, its rate law represents the rate law for the overall reaction. Consider the reaction between nitrogen dioxide and carbon monoxide:

$$NO_2(g) + CO(g) \longrightarrow NO(g) + CO_2(g)$$

If the overall reaction were an elementary reaction—that is, if the mechanism consisted of only one step—we could immediately write the overall rate law as

$$Rate = k[NO_2][CO]$$

However, as you saw in Sample Problem 16.3, experiment shows that the actual rate law is

$$Rate = k[NO_2]^2$$

From this, we know immediately that the reaction shown cannot be elementary.

Sleeping Through the Rate-Determining Step Baking provides a tasty, macroscopic example of a rate-limiting step. Five steps in making a loaf of French bread are (1) mixing the ingredients (15 min), (2) kneading the dough (10 min), (3) letting the dough rise in a refrigerator (400 min), (4) shaping the loaf (2 min), and (5) baking the loaf (25 min). Obviously, the dough-rising step limits how fast a baker can produce a loaf because it is so much slower than the other steps. Keenly aware of the baking "reaction mechanism," bakers let the dough rise overnight, which allows them to sleep through the rate-determining step.

A proposed two-step mechanism is

(1) $NO_2(g) + NO_2(g) \longrightarrow NO_3(g) + NO(g)$ [slow; rate determining]
(2) $NO_3(g) + CO(g) \longrightarrow NO_2(g) + CO_2(g)$ [fast]

Notice that NO_3 functions as a reaction intermediate in the mechanism. Rate laws for these elementary steps are

(1) $Rate_1 = k_1[NO_2][NO_2] = k_1[NO_2]^2$
(2) $Rate_2 = k_2[NO_3][CO]$

Note that if $k_1 = k$, *the rate law for the rate-determining step (step 1) is identical to the experimental rate law.* The first step is so slow compared with the second that the overall reaction takes essentially as long as the first step. Here you can see that one reason for a reactant (in this case, CO) having a zero order is that it takes part in the reaction only *after* the rate-determining step.

Correlating the Mechanism with the Rate Law

Conjuring up a reasonable reaction mechanism is one of the most exciting aspects of chemical kinetics and can be a classic example of the use of the scientific method. We use observations and data from rate experiments to hypothesize what the individual steps might be and then test our hypothesis by gathering further evidence. If the evidence supports it, we continue to apply that mechanism; if not, we propose a new one. However, *we can never prove, just from data, that a particular mechanism represents the actual chemical change.*

Regardless of the elementary steps that are proposed for a mechanism, they must meet three criteria:

1. *The elementary steps must add up to the overall equation.* We cannot wind up with more (or fewer) reactants or products than are present in the balanced equation.
2. *The elementary steps must be physically reasonable.* As we noted, most steps should involve one reactant particle (unimolecular) or two (bimolecular). Steps with three reactant particles (termolecular) are very unlikely.
3. *The mechanism must correlate with the rate law.* Most importantly, a mechanism must support the experimental facts shown by the rate law, not the other way around.

Let's see how the mechanisms of several reactions conform to these criteria and how the elementary steps fit together.

Mechanisms with a Slow Initial Step We've already seen one mechanism with a rate-determining first step in the NO_2-CO reaction shown previously. Another example is the reaction between nitrogen dioxide and fluorine gas:

$$2NO_2(g) + F_2(g) \longrightarrow 2NO_2F(g)$$

The experimental rate law is first order in NO_2 and in F_2:

$$Rate = k[NO_2][F_2]$$

The accepted mechanism for the reaction is

(1) $NO_2(g) + F_2(g) \longrightarrow NO_2F(g) + F(g)$ [slow; rate determining]
(2) $NO_2(g) + F(g) \longrightarrow NO_2F(g)$ [fast]

Molecules of reactant and product appear in both elementary steps. The free fluorine atom is a reaction intermediate.

Does this mechanism meet the three crucial criteria?

1. The elementary reactions sum to the balanced equation:

$NO_2(g) + NO_2(g) + F_2(g) + F(g) \longrightarrow NO_2F(g) + NO_2F(g) + F(g)$
or $2NO_2(g) + F_2(g) \longrightarrow 2NO_2F(g)$

2. Both steps are bimolecular, so they are chemically reasonable.

3. The mechanism gives the rate law of the overall equation. To show this, we write the rate laws for the elementary steps:

    (1) $Rate_1 = k_1[NO_2][F_2]$
    (2) $Rate_2 = k_2[NO_2][F]$

Step 1 is the rate-determining step and therefore gives the overall rate law, with $k_1 = k$. Because the second molecule of NO_2 appears in the step that follows the rate-determining step, it does not appear in the overall rate law. Thus, we see that *the overall rate law includes only species active in the reaction up to and including those in the rate-determining step*. This point was also illustrated by the NO_2-CO mechanism earlier. Carbon monoxide was absent from the overall rate law because it appeared *after* the rate-determining step.

Figure 16.21 shows a reaction energy diagram for the NO_2-F_2 reaction. Notice that

- *Each step in the mechanism has its own transition state.* (Note that only one molecule of NO_2 is involved in step 1, and only the first transition state is depicted.)
- The F atom intermediate is a reactive, unstable species (as you know from halogen chemistry), so it is higher in energy than the reactants or product.
- The first step is slower (rate limiting), so its activation energy is *larger* than that of the second.
- The overall reaction is exothermic, so the product is lower in energy than the reactants.

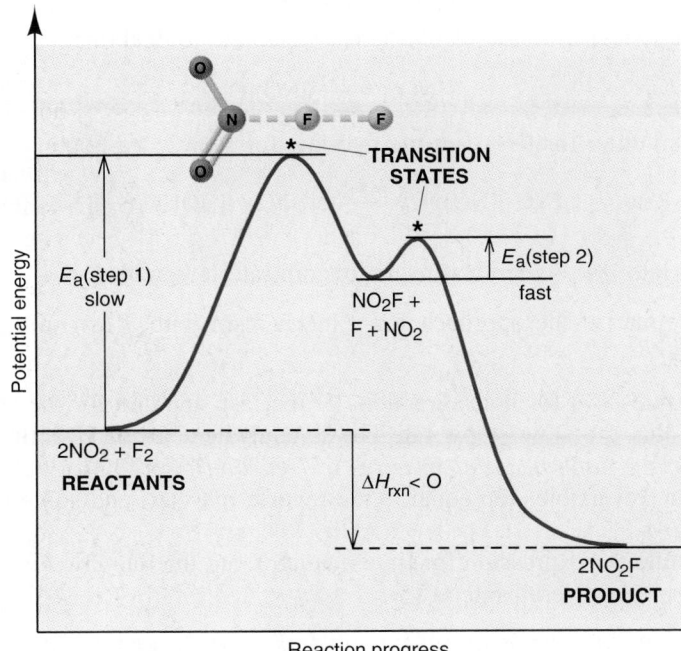

Reaction progress

Figure 16.21 Reaction energy diagram for the two-step NO_2-F_2 reaction. Each step in the mechanism has its own transition state. The proposed transition state is shown for step 1. Reactants for the second step are the F atom intermediate and the second molecule of NO_2. Note that the first step is slower (higher E_a). The overall reaction is exothermic ($\Delta H_{rxn} < 0$).

Mechanisms with a Fast Initial Step If the rate-limiting step in a mechanism is *not* the initial step, it acts as a bottleneck later in the reaction sequence. As a result, the product of a fast initial step builds up and starts reverting to reactant, while waiting for the slow step to remove it. With time, the product of the initial step is changing back to reactant as fast as it is forming. In other words, the *fast initial step reaches equilibrium*. As you'll see, this situation allows us to fit the mechanism to the overall rate law.

Consider once again the oxidation of nitric oxide:

$$2NO(g) + O_2(g) \longrightarrow 2NO_2(g)$$

The experimentally determined rate law is

$$\text{Rate} = k[NO]^2[O_2]$$

and a proposed mechanism is

(1) $NO(g) + O_2(g) \rightleftharpoons NO_3(g)$ [fast, reversible]

(2) $NO_3(g) + NO(g) \longrightarrow 2NO_2(g)$ [slow; rate determining]

Note that, with cancellation of the reaction intermediate NO_3, the first criterion is met because the sum of the steps gives the overall equation. Also note that the second criterion is met because both steps are bimolecular.

To meet the third criterion (the mechanism conforms to the overall rate law), we first write rate laws for the elementary steps:

(1) $\text{Rate}_{1(\text{fwd})} = k_1[NO][O_2]$

  $\text{Rate}_{1(\text{rev})} = k_{-1}[NO_3]$

where k_{-1} is the rate constant for the reverse reaction.

(2) $\text{Rate}_2 = k_2[NO_3][NO]$

Now we must show that the rate law for the rate-determining step (step 2) gives the overall rate law. As written, it does not, because it contains the intermediate NO_3, and *an overall rate law can include only reactants (and products)*. Therefore, we must eliminate $[NO_3]$ from the step 2 rate law. To do so, we express $[NO_3]$ in terms of reactants. Step 1 reaches equilibrium when the forward and reverse rates are equal:

$$\text{Rate}_{1(\text{fwd})} = \text{Rate}_{1(\text{rev})}$$

Or $$k_1[NO][O_2] = k_{-1}[NO_3]$$

To express $[NO_3]$ in terms of reactants, we isolate it algebraically:

$$[NO_3] = \frac{k_1}{k_{-1}}[NO][O_2]$$

Then, substituting for $[NO_3]$ in the rate law for step 2, we obtain

$$\text{Rate}_2 = k_2[NO_3][NO] = k_2\left(\frac{k_1}{k_{-1}}[NO][O_2]\right)[NO] = \frac{k_2 k_1}{k_{-1}}[NO]^2[O_2]$$

Thus, this rate law is identical to the overall rate law, with $k = \dfrac{k_2 k_1}{k_{-1}}$.

To summarize the approach for a mechanism with a fast initial, reversible step:

1.   Write rate laws for both directions of the fast step and for the slow step.
2.   Show that the slow step's rate law is equivalent to the overall rate law, by *expressing [intermediate] in terms of [reactant]:* set the forward rate law of the fast, reversible step equal to the reverse rate law, and solve for [intermediate].
3.   Substitute the expression for [intermediate] into the rate law for the slow step to obtain the overall rate law.

Several end-of-chapter problems, including 16.72 and 16.73, provide additional examples of this approach.

SECTION SUMMARY

The mechanisms of most common reactions consist of two or more elementary steps, reactions that occur in one step and depict a single chemical change. The molecularity of an elementary step equals the number of reactant particles and is the same as the reaction order of its rate law. Unimolecular and bimolecular steps are common. The rate-determining, or rate-limiting (slowest), step determines how fast the overall reaction occurs, and its rate law represents the overall rate law. Reaction intermediates are species that form in one step and react in a later one. The steps in a proposed mechanism must add up to the overall reaction, be physically reasonable, and conform to the overall rate law. If a fast step precedes the slow step, the fast step reaches equilibrium, and the concentrations of intermediates in the rate law of the slow step must be expressed in terms of reactants.

16.8 CATALYSIS: SPEEDING UP A CHEMICAL REACTION

There are many situations in which the rate of a reaction must be increased for it to be useful. In an industrial process, for example, a higher rate often determines whether a new product can be made economically. Sometimes, we can speed up a reaction sufficiently with a higher temperature, but energy is costly and many substances are heat sensitive and easily decomposed. Alternatively, we can often employ a **catalyst,** a substance that increases the rate *without* being consumed in the reaction. Because catalysts are not consumed, very small, nonstoichiometric quantities are generally all that are required. Nevertheless, these substances are employed in so many important processes that several million tons of industrial catalysts are produced annually in the United States alone! Nature is the master designer and user of catalysts. Even the simplest bacterium employs thousands of biological catalysts, known as *enzymes,* to speed up its cellular reactions. Every organism relies on enzymes to sustain life.

Each catalyst has its own specific way of functioning, but in general, *a catalyst effects a lower activation energy, which in turn makes the rate constant larger and the rate higher*. Two important points stand out in Figure 16.22:

- A catalyst speeds up the forward *and* reverse reactions. A reaction with a catalyst *does not yield more product* than one without a catalyst, but it yields the product *more quickly*.
- A catalyst effects a *lower activation energy* by providing a *different mechanism* for the reaction, a new lower energy pathway.

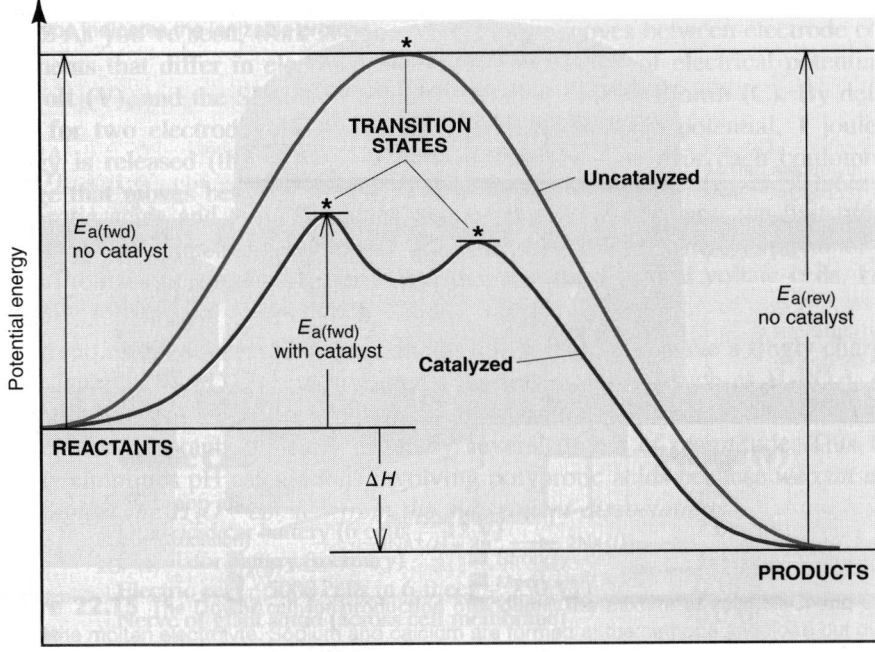

Reaction progress

Figure 16.22 Reaction energy diagram of a catalyzed and an uncatalyzed process. A catalyst speeds a reaction by providing a new lower energy pathway, in this case by replacing the one-step mechanism with a two-step mechanism. Both forward and reverse rates are increased to the same extent, so *a catalyst does not affect the overall reaction yield.* (The only activation energy shown for the catalyzed reaction is the larger one for the forward direction.)

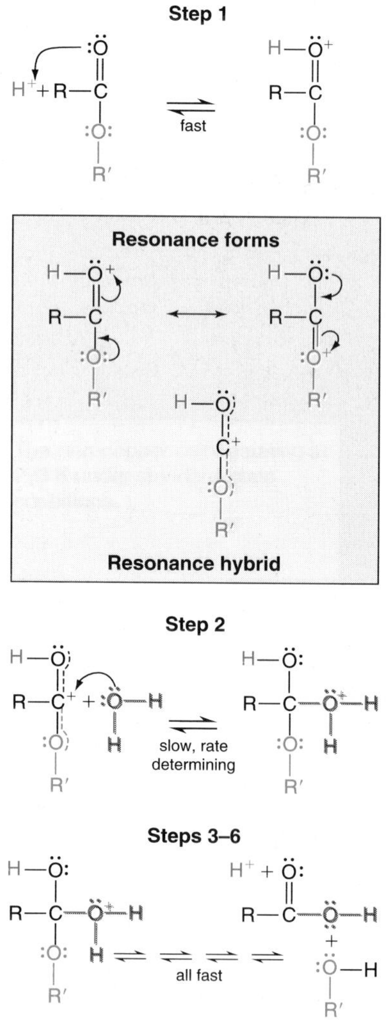

Step 1

Resonance forms

Resonance hybrid

Step 2

slow, rate
determining

Steps 3–6

all fast

Figure 16.23 Mechanism for the catalyzed hydrolysis of an organic ester. In step 1, the catalytic H^+ ion binds to the electron-rich oxygen. The resonance hybrid of this product *(see gray panel)* shows the C atom more positive than it would ordinarily be. This enhanced charge on C attracts the partially negative O of water more strongly, increasing the fraction of effective collisions and thus speeding up step 2, the rate-determining step. Loss of R′OH and release of H^+ occur in a final series of fast steps.

Consider a general *uncatalyzed* reaction that proceeds by a one-step mechanism involving a bimolecular collision:

$$A + B \longrightarrow product \qquad [slower]$$

In the *catalyzed* reaction, the reactants interact with the catalyst, so the mechanism might involve a two-step pathway:

$$A + catalyst \longrightarrow C \qquad [faster]$$
$$C + B \longrightarrow product + catalyst \qquad [faster]$$

Note that *the catalyst is not consumed,* as its definition requires. Rather, it is used and then regenerated, and the activation energies of both steps are lower than the activation energy of the uncatalyzed pathway.

There are two broad categories of catalyst—homogeneous and heterogeneous—depending on whether the catalyst is in the same phase as the reactant and product.

Homogeneous Catalysis

A **homogeneous catalyst** exists in solution with the reaction mixture. All homogeneous catalysts are gases, liquids, or soluble solids. Some industrial processes that employ these catalysts are shown in Table 16.7 *(top).*

A thoroughly studied example of homogeneous catalysis is the hydrolysis of an organic ester (RCOOR′), a reaction we examined in Section 15.4:

Here R and R′ are hydrocarbon groups, R—C—OH is an organic acid, and R′—OH is an alcohol. The reaction rate is low at room temperature but can be increased by adding a small amount of strong inorganic acid, which provides H^+ ion, the catalyst in the reaction.

In the first step of the catalyzed reaction (Figure 16.23), the H^+ ion forms a bond to the double-bonded O atom. By considering the resonance forms, we see that the bonding of H^+ then makes the C atom more positive, which *increases its attraction* for the partially negative O atom of water. In effect, H^+ increases the likelihood that the bonding of water, which is the rate-determining step, will

Table 16.7   Some Modern Processes Based on Catalysis

Reactants	Catalyst	Product	Use
Homogeneous			
Propylene, oxidizer	Mo(VI) complexes	Propylene oxide	Polyurethane foams; polyesters
Methanol, CO	$[Rh(CO)_2I_2]^-$	Acetic acid	Poly(vinyl acetate) coatings; poly(vinyl alcohol)
Butadiene, HCN	Ni/P compounds	Adiponitrile	Nylons (fibers, plastics)
α-Olefins, CO, H_2	Rh/P compounds	Aldehydes	Plasticizers, lubricants
Heterogeneous			
Ethylene, O_2	Silver, cesium chloride on alumina	Ethylene oxide	Polyesters, ethylene glycol, lubricants
Propylene, NH_3, O_2	Bismuth molybdates	Acrylonitrile	Plastics, fibers, resins
Ethylene	Organochromium and titanium halides on silica	High-density polyethylene	Molded products

take place. Several steps later, H$^+$ returns to solution. Thus, H$^+$ acts as a catalyst because it speeds up the reaction but is not itself consumed: it is used up in one step and re-formed in another.

Heterogeneous Catalysis

A **heterogeneous catalyst** speeds up a reaction that occurs in a separate phase. The catalyst is most often a solid interacting with gaseous or liquid reactants. Because reaction occurs on the solid's surface, heterogeneous catalysts usually have enormous surface areas for contact, between 1 and 500 m^2/g. Interestingly, many reactions that occur on a metal surface, such as the decomposition of HI on gold and the decomposition of N$_2$O on platinum, are zero order because the rate-determining step occurs on the surface itself. Thus, despite an enormous surface area, once the reactant gas covers the surface, increasing the reactant concentration cannot increase the rate. Table 16.7 *(bottom)* lists some polymer manufacturing processes that employ heterogeneous catalysts.

One of the most important examples of heterogeneous catalysis is the addition of H$_2$ to the C=C bonds of organic compounds to form C—C bonds. The petroleum, plastics, and food industries frequently use catalytic **hydrogenation.** The conversion of vegetable oil into margarine is one example.

The simplest hydrogenation converts ethylene to ethane:

$$H_2C=CH_2(g) + H_2(g) \longrightarrow H_3C-CH_3(g)$$

In the absence of a catalyst, the reaction occurs very slowly. At high H$_2$ pressure in the presence of finely divided nickel, palladium, or platinum, the reaction becomes rapid even at ordinary temperatures. These Group 8B(10) metals catalyze by *chemically adsorbing the reactants onto their surface* (Figure 16.24). First, H$_2$ lands and splits into separate H atoms chemically bound to the solid catalyst's metal atoms (catM):

$$H-H(g) + 2catM(s) \longrightarrow 2catM-H \text{ (H atoms bound to metal surface)}$$

Then, C$_2$H$_4$ adsorbs and reacts with two H atoms, one at a time, to form C$_2$H$_6$. The H—H bond breakage is the rate-determining step in the overall process, and interaction with the catalyst's surface provides the low-E_a step as part of an alternative reaction mechanism. ● Two Chemical Connections essays follow. The first introduces the remarkable abilities of the heterogeneous catalysts inside you, and the second discusses how both types of catalysis play a role in an atmospheric process of major concern.

Figure 16.24 The metal-catalyzed hydrogenation of ethylene.

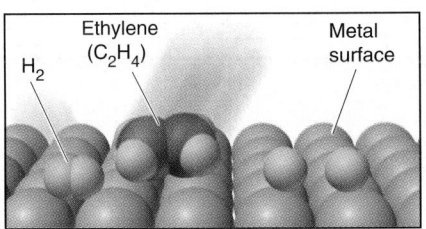

① *Before C$_2$H$_4$ lands, H$_2$ adsorbs on metal surface and, in rate-determining step, splits into H atoms.*

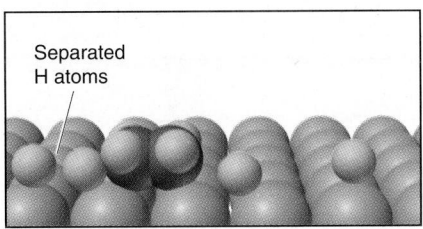

② *H atoms migrate over surface and collide.*

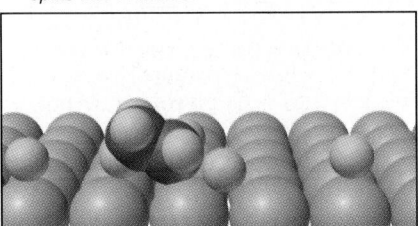

③ *One H atom bonds to adsorbed C$_2$H$_4$.*

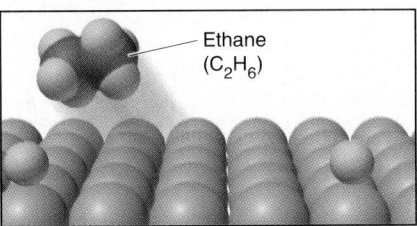

④ *Another C—H bond forms and C$_2$H$_6$ leaves the surface.*

Chemical Connections · Chemistry in Enzymology

Kinetics and Function of Biological Catalysts

Within every living cell, thousands of individual reactions occur. Many involve complex chemical changes, yet they take place in dilute solution at ordinary temperatures and pressures. The rate of each reaction responds smoothly to momentary or permanent changes in other reaction rates, various signals from other cells, and environmental stresses. Virtually every reaction in this marvelous chemical harmony is catalyzed by its own specific **enzyme,** a protein catalyst whose function has been perfected through evolution.

Enzymes have complex three-dimensional shapes and molar masses ranging from about 15,000 to 1,000,000 g/mol (Section 15.6). At a specific region on an enzyme's surface is its **active site,** a molecular crevice whose shape results from the shapes of the amino acid side chains directly involved in catalyzing the reaction (also see Figure B10.4, p. 385). When the reactant molecules, called the **substrates** of the reaction, collide with the active site, the chemical change is initiated. The active site makes up only a small part of the enzyme's surface—like a tiny hollow carved into a mountainside—and often includes amino acid groups from distant regions of the protein that exist near each other because of the folding of the protein backbone (Figure B16.4). In most cases, *substrates bind to the active site through intermolecular forces:* H bonds, dipole forces, and other weak attractions.

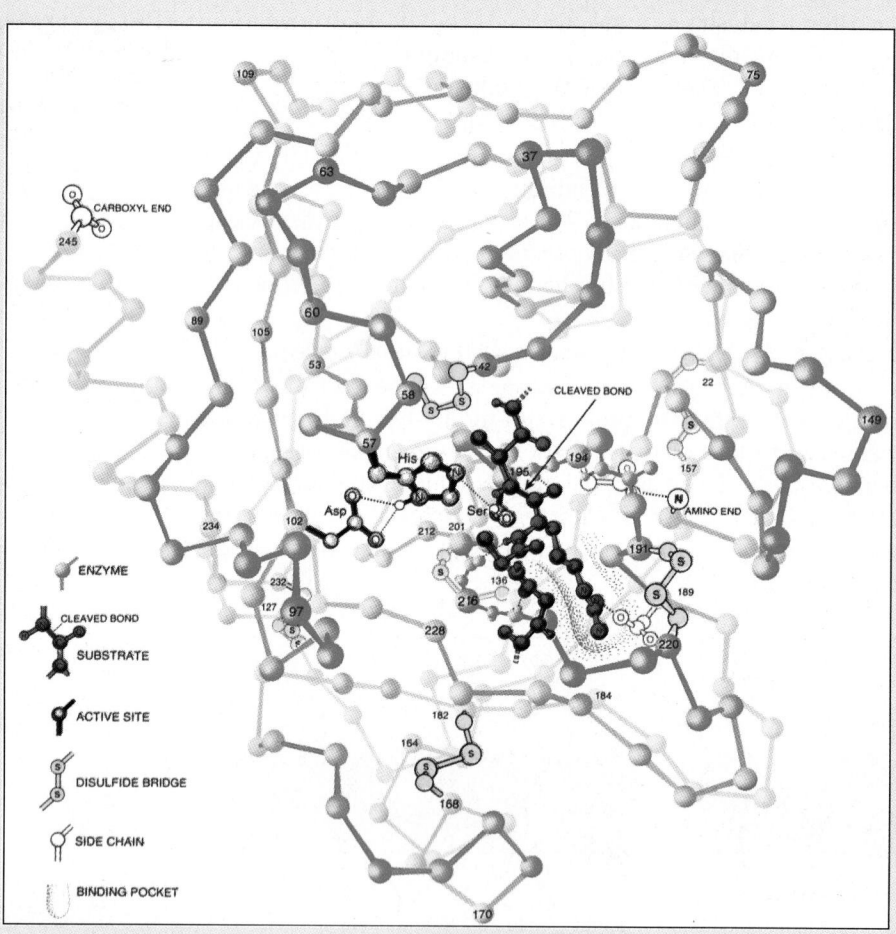

Figure B16.4 The widely separated amino acid groups that form an active site. The amino acids in chymotrypsin, a digestive enzyme, are shown as linked spheres and numbered consecutively from the beginning of the chain. The R groups of the active-site amino acids 57 (histidine, His), 102 (aspartic acid, Asp), and 195 (serine, Ser) are shown because they play a crucial role in the enzyme-catalyzed reaction. Part of the substrate lies within a binding pocket to anchor it during reaction. (Illustration by Irving Geis. Rights owned by Howard Hughes Medical Institute. Not to be used without permission.)

An enzyme has features of both a homogeneous and a heterogeneous catalyst. Most enzymes are enormous compared with their substrates, and they are often found embedded within membranes of larger cell parts. Thus, like a heterogeneous catalyst, an enzyme provides an active surface on which a reactant is immobilized temporarily, waiting for its reaction partner to land nearby. Like a homogeneous catalyst, the enzyme's amino acid groups interact actively with the substrates in multistep sequences involving intermediates.

Enzymes are incredibly *efficient* catalysts. Consider the hydrolysis of urea as an example:

$$(NH_2)_2C=O(aq) + 2H_2O(l) + H^+(aq) \longrightarrow$$
$$2NH_4^+(aq) + HCO_3^-(aq)$$

In water at room temperature, the rate constant for the uncatalyzed reaction is approximately 3×10^{-10} s^{-1}. Under the same conditions in the presence of the enzyme *urease* (pronounced "*yur*-ee-ase"), the rate constant is 3×10^4 s^{-1}, a 10^{14}-fold increase! Enzymes increase rates by 10^8 to 10^{20} times, values that industrial chemists who design catalysts can only dream of.

Enzymes are also extremely *specific:* each reaction is generally catalyzed by a particular enzyme. Urease catalyzes *only* the hydrolysis of urea, and none of the several thousand other enzymes present in the cell catalyzes that reaction. This remarkable specificity results from the particular groups that comprise the active site. Two models of enzyme action are illustrated in Figure B16.5. According to the **lock-and-key model,** when the "key" (substrate) fits the "lock" (active site), the chemical change begins. However, modern x-ray crystallographic and spectroscopic methods show that, in many cases, *the enzyme changes shape when the substrate lands at the active site.* This **induced-fit model** of enzyme action pictures the substrate inducing the active site to adopt a perfect fit. Rather than a rigidly shaped lock and key, therefore, we might picture a hand in a glove, in which the "glove" (active site) does not attain its functional shape until the "hand" (substrate) moves into place.

The kinetics of enzyme catalysis has many features in common with that of ordinary catalysis. In an uncatalyzed reaction, the rate is affected by the concentrations of the reactants; in a catalyzed reaction, the rate is affected by the *concentration of reactant bound to catalyst*. In the enzyme-catalyzed case, substrate (S) and enzyme (E) form an intermediate **enzyme-substrate complex (ES),** whose concentration determines the rate of product (P) formation. The steps common to virtually all enzyme-catalyzed reactions are

(1) E + S $\rightleftharpoons$ ES   [fast, reversible]
(2)   ES $\longrightarrow$ E + P   [slow; rate determining]

Thus, rate = k[ES]. When all the available enzyme molecules are bound to substrate and exist as ES, increasing [S] has no effect on rate and the process is zero order. The breakdown of ethanol in the body, to cite one very common example, is zero order.

Enzymes employ a variety of catalytic mechanisms. In some cases, the active site groups bring the reacting atoms of the bound substrates closer together. In other cases, the groups move apart slightly, stretching the substrate bond that is to be broken in the process. Some enzyme groups are acidic and thus are able to provide H$^+$ ions that increase the speed of a rate-determining step. *Hydrolases* are a class of enzymes that cleave bonds by such *acid catalysis*. For example, lysozyme, an enzyme that is found in tears, hydrolyzes bacterial cell walls, thus protecting the eyes from microbes; chymotrypsin, an enzyme found in the small intestine, hydrolyzes proteins into smaller molecules during digestion.

No matter what their specific mode of action, *all enzymes function by stabilizing the reaction's transition state.* For instance, in the lysozyme-catalyzed reaction, the transition state is a sugar molecule whose bonds are twisted and stretched into unusual lengths and angles, but it fits the lysozyme active site perfectly. By stabilizing the transition state through binding it effectively, lysozyme lowers the activation energy of the reaction and thus increases the rate.

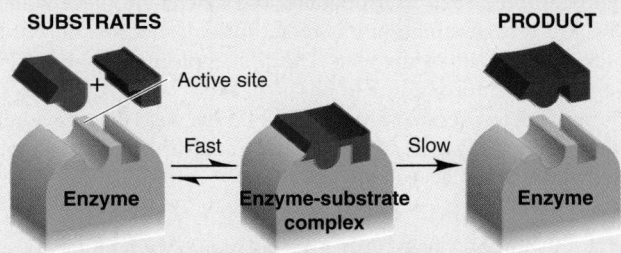

A Lock-and-key model

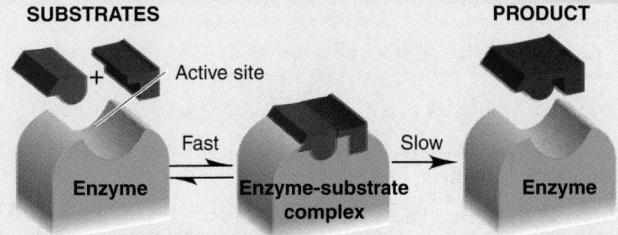

B Induced-fit model

Figure B16.5 **Two models of enzyme action. A,** In the lock-and-key model, the active site is thought to be an exact fit for the substrate shapes. **B,** In the induced-fit model, the active site is thought to change shape to fit the substrates. Most enzyme-catalyzed reactions proceed through a fast, reversible formation of an enzyme-substrate(s) complex, followed by a slow conversion to product(s) and free enzyme.

Depletion of the Earth's Ozone Layer

Both homogeneous and heterogeneous catalysts play key roles in one of the most serious environmental concerns of our time—the depletion of ozone from the stratosphere. The stratospheric ozone layer absorbs UV radiation from the Sun with wavelengths between 280 and 320 nm that would otherwise reach the Earth's surface. This radiation (called UV-B) has enough energy to break bonds in deoxyribonucleic acid (DNA) and thereby damage genes. Depletion of stratospheric ozone could significantly increase risks to human health from UV-B, particularly a higher incidence of skin cancer and cataracts (a loss of transparency of the lens of the eye). Plant life, especially the simpler forms at the base of the food chain, may also be damaged.

Historically, the stratospheric ozone concentration varied seasonally but remained nearly constant from year to year through a complex series of atmospheric reactions. Oxygen atoms are formed from dissociation of oxygen by UV radiation with wavelengths less than 280 nm. They react with O_2 to form ozone, and with ozone to regenerate O_2:

$$O_2 \xrightarrow{\text{UV}} 2O$$
$$O + O_2 \longrightarrow O_3 \quad \text{[ozone formation]}$$
$$O + O_3 \longrightarrow 2O_2 \quad \text{[ozone breakdown]}$$

However, research by Paul Crutzen, Mario Molina, and F. Sherwood Rowland, for which they received the Nobel Prize in 1995, revealed that industrially produced chlorofluorocarbons (CFCs) had shifted this balance by catalyzing the breakdown reaction.

CFCs were widely employed as aerosol propellants, plastic foam blowing agents, and air-conditioning refrigerants, leading to large quantities being released into the atmosphere. Unreactive in the troposphere, they are transported gradually into the stratosphere, where they encounter UV radiation with enough energy to split a CFC molecule and release chlorine atoms:

$$CF_2Cl_2 \xrightarrow{\text{UV}} CF_2Cl\cdot + Cl\cdot$$

(The dots are unpaired electrons resulting from bond cleavage.)

Like many species with unpaired electrons (free radicals), atomic Cl is very reactive. The Cl atoms react with stratospheric ozone to produce an intermediate, chlorine monoxide ($\cdot ClO$), which then reacts with free O atoms to regenerate Cl atoms:

$$O_3 + Cl\cdot \longrightarrow \cdot ClO + O_2$$
$$\cdot ClO + O \longrightarrow \cdot Cl + O_2$$

The sum of these steps is the ozone breakdown reaction:

$$O_3 + \cancel{\cdot Cl} + \cancel{\cdot ClO} + O \longrightarrow \cancel{\cdot ClO} + O_2 + \cancel{\cdot Cl} + O_2$$
or
$$O_3 + O \longrightarrow 2O_2$$

Note that the Cl atom acts as a homogeneous catalyst: it exists in the same phase as the reactants, speeds up the process via a different mechanism, and is regenerated. The crux of the problem is that each Cl atom has a stratospheric half-life of about 2 years, during which time it speeds the breakdown of about 100,000 ozone molecules. Bromine is a more effective catalyst, but much less ends up in the stratosphere. Halons, used as fire suppressants, and methyl bromide, an agricultural insecticide, are the main sources of stratospheric bromine.

A variety of Cl-containing species cause the Antarctic *ozone hole*, the severe reduction of stratospheric ozone over the Antarctic region when the Sun rises after the long winter darkness.

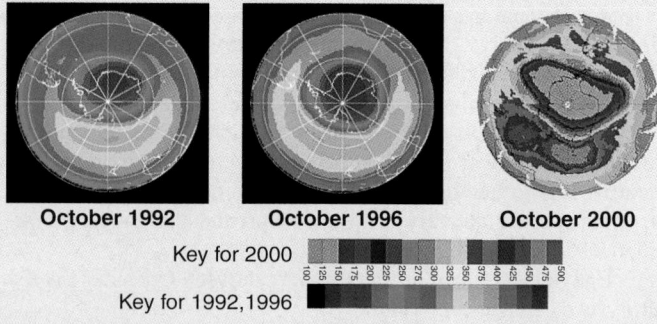

October 1992   **October 1996**   **October 2000**

Key for 2000

Key for 1992, 1996

Figure B16.6 The increasing size of the Antarctic ozone hole. Satellite images show changes in ozone concentration over the South Pole for October 1992, 1996, and 2000. Note the increasing size of the "hole" in the ozone layer (indicated by the four leftmost colors in the keys). Since 1995, data show a similar thinning, though not yet as severe, over the North Pole.

Measurements of high [$\cdot ClO$] over Antarctica are consistent with the breakdown mechanism. Atmospheric scientists have documented more than 80% ozone depletion over the South Pole (Figure B16.6). The ozone hole enlarges by heterogeneous catalysis involving polar stratospheric clouds. The clouds provide a surface for reactions that convert inactive chlorine compounds, such as HCl and chlorine nitrate ($ClONO_2$), to substances, such as Cl_2, that are cleaved by UV radiation to Cl atoms. Heterogeneous catalysis also occurs on fine particles in the stratosphere. The eruption of Mt. Pinatubo in 1991 reduced stratospheric ozone for 2 years. There is ozone thinning over the North Pole, and NASA has also documented the loss of stratospheric ozone at middle latitudes, such as over the United States. So far, both of these losses occur to a lower extent and at a lower rate.

The Montreal Protocol in 1987 and later amendments curtailed growth in production of CFCs and set dates for phasing out their production and usage, as well as that of other chlorinated and brominated compounds, such as CCl_4, CCl_3CH_3, and CH_3Br. Another change was replacement of CFC aerosol propellants by hydrocarbons, such as isobutane, $(CH_3)_3CH$. Replacement of CFCs in refrigeration units is more difficult, because it will require changes in equipment. The first replacements for CFCs were hydrochlorofluorocarbons (HCFCs), such as CHF_2Cl. These deplete ozone less than CFCs do because they have relatively short tropospheric lifetimes due to abstraction of H atoms by hydroxyl radicals ($\cdot OH$):

$$CHF_2Cl + \cdot OH \longrightarrow H_2O + \cdot CF_2Cl$$

After this initial attack, further degradation occurs rapidly.

HCFCs will be phased out by 2040 and replaced by hydrofluorocarbons (HFCs, which contain only C, H, and F) in devices such as car air conditioners. HFCs are preferable to CFCs or HCFCs because fluorine is a poor catalyst of ozone breakdown. Nevertheless, because of the long lifetimes of CFCs and the current emissions of HCFCs, until their phase out, full recovery of the ozone layer may take another century!

SECTION SUMMARY

A catalyst is a substance that increases the rate of a reaction without being consumed. It accomplishes this by providing an alternative mechanism with a lower activation energy. Homogeneous catalysts function in the same phase as the reactants. Heterogeneous catalysts act in a different phase from the reactants. The hydrogenation of carbon-carbon double bonds takes place on a solid catalyst, which speeds the breakage of the H—H bond in H_2. Enzymes are biological catalysts with spectacular efficiency and specificity. Chlorine atoms derived from CFC molecules catalyze the breakdown of stratospheric ozone.

Chapter Perspective

With this introduction to chemical kinetics, we have begun to explore the dynamic inner workings of chemical change. Variations in reaction rate are observed through concentration and temperature changes, which operate on the molecular level through the frequency and energetics of particle collisions and the details of reactant structure. Kinetics allows us to speculate about the molecular pathway of a reaction. Modern industry and biochemistry depend on its principles. However, speed and yield are very different aspects of a reaction. In Chapter 17, we'll see how opposing reaction rates give rise to the equilibrium state and examine how much product has formed once the net reaction has stopped.

For Review and Reference (Numbers in parentheses refer to pages, unless noted otherwise)

Learning Objectives

Relevant section and/or sample problem (SP) numbers appear in parentheses.

Understand These Concepts

1. How reaction rate depends on concentration, physical state, and temperature (Section 16.1)
2. The meaning of reaction rate in terms of changing concentrations over time (Section 16.2)
3. How the rate can be expressed in terms of reactant or product concentrations (Section 16.2)
4. The distinction between average and instantaneous rate and why the instantaneous rate changes during the reaction (Section 16.2)
5. The interpretation of reaction rate in terms of reactant and product concentrations (Section 16.2)
6. The experimental basis of the rate law and the information needed to determine it—initial rate data, reaction orders, and rate constant (Section 16.3)
7. The importance of reaction order in determining the rate (Section 16.3)
8. How reaction order is determined from initial rates at different concentrations (Section 16.3)
9. How integrated rate laws show the dependence of concentration on time (Section 16.4)
10. The meaning of half-life and why it is constant for a first-order reaction (Section 16.4)
11. Activation energy and the effect of temperature on the rate constant (Arrhenius equation) (Section 16.5)
12. Why concentrations are multiplied in the rate law (Section 16.6)
13. How temperature affects rate by influencing collision energy and, thus, the fraction of collisions with energy exceeding the activation energy (Section 16.6)

14. Why molecular orientation and complexity influence the number of effective collisions and the rate (Section 16.6)
15. How the transition state represents the momentary species between reactants and products and how the activation energy is needed to form it (Section 16.6)
16. How an elementary step represents a single molecular event and its molecularity equals the number of colliding particles (Section 16.7)
17. How a reaction mechanism consists of several elementary steps, with the slowest step determining the overall rate (Section 16.7)
18. The criteria for a valid reaction mechanism (Section 16.7)
19. How a catalyst speeds a reaction by lowering the activation energy (Section 16.8)
20. The distinction between homogeneous and heterogeneous catalysis (Section 16.8)

Master These Skills

1. Calculating instantaneous rate from the slope of a tangent to a concentration vs. time plot (Section 16.2)
2. Expressing rate in terms of changes in concentration over time and calculating the instantaneous rate (SP 16.1)
3. Determining reaction order from a known rate law (SP 16.2)
4. Determining reaction order from changes in initial rate with concentration (SP 16.3)
5. Calculating the rate constant and its units (Section 16.3)
6. Using an integrated rate law to find concentration at a given time or the time to reach a given concentration (SP 16.4)
7. Determining reaction order graphically with a rearranged integrated rate law (Section 16.4)

Learning Objectives *(continued)*

8. Determining the half-life of a first-order reaction (SP 16.5)
9. Using the Arrhenius equation to calculate the activation energy or the rate constant (SP 16.6)
10. Using reaction energy diagrams to depict the energy changes during a reaction (SP 16.7)

11. Postulating a transition state for a simple reaction (SP 16.7)
12. Determining the molecularity and rate law for an elementary step (SP 16.8)
13. Constructing a mechanism with either a slow or a fast initial step (Section 16.7)

Key Terms

chemical kinetics (664)

Section 16.2
reaction rate (667)
average rate (668)
instantaneous rate (668)
initial rate (669)

Section 16.3
rate law (rate equation) (671)
rate constant (671)
reaction orders (671)

Section 16.4
integrated rate law (677)

half-life ($t_{1/2}$) (680)

Section 16.5
Arrhenius equation (682)
activation energy (E_a) (683)

Section 16.6
collision theory (685)
transition state theory (685)
effective collision (687)
frequency factor (687)
transition state (activated complex) (688)

reaction energy diagram (689)

Section 16.7
reaction mechanism (691)
reaction intermediate (691)
elementary reaction (elementary step) (692)
molecularity (692)
unimolecular reaction (692)
bimolecular reaction (692)
rate-determining (rate-limiting) step (693)

Section 16.8
catalyst (697)
homogeneous catalyst (698)
heterogeneous catalyst (699)
hydrogenation (699)
enzyme (700)
active site (700)
substrate (700)
lock-and-key model (701)
induced-fit model (701)
enzyme-substrate complex (ES) (701)

Key Equations and Relationships

16.1 Expressing reaction rate in terms of reactant A (667):

$$\text{Rate} = -\frac{\Delta[A]}{\Delta t}$$

16.2 Expressing the rate of a general reaction (670):

$$aA + bB \longrightarrow cC + dD$$

$$\text{Rate} = -\frac{1}{a}\frac{\Delta[A]}{\Delta t} = -\frac{1}{b}\frac{\Delta[B]}{\Delta t} = \frac{1}{c}\frac{\Delta[C]}{\Delta t} = \frac{1}{d}\frac{\Delta[D]}{\Delta t}$$

16.3 Writing a general rate law (for a case not involving products) (671):

$$\text{Rate} = k[A]^m[B]^n \cdots$$

16.4 Calculating the time to reach a given [A] in a first-order reaction (rate = k[A]) (678):

$$\ln\frac{[A]_0}{[A]_t} = kt$$

16.5 Calculating the time to reach a given [A] in a simple second-order reaction (rate = $k[A]^2$) (678):

$$\frac{1}{[A]_t} - \frac{1}{[A]_0} = kt$$

16.6 Calculating the time to reach a given [A] in a zero-order reaction (rate = k) (678):

$$[A]_t - [A]_0 = -kt$$

16.7 Finding the half-life of a first-order process (680):

$$t_{1/2} = \frac{\ln 2}{k} = \frac{0.693}{k}$$

16.8 Relating the rate constant to the temperature (Arrhenius equation) (682):

$$k = Ae^{-E_a/RT}$$

16.9 Calculating the activation energy (rearranged Arrhenius equation in two-point form) (683):

$$\ln\frac{k_2}{k_1} = -\frac{E_a}{R}\left(\frac{1}{T_2} - \frac{1}{T_1}\right)$$

Highlighted Figures and Tables

These figures (F) and tables (T) provide a quick review of important ideas.

F16.1 Reaction rate (664)
F16.5 The concentration of O_3 vs. time (669)
T16.3 Units of k for overall reaction orders (677)
F16.7 Integrated rate laws and reaction order (679)
F16.8 Graphical determination of reaction order (680)
F16.9 [N_2O_5] vs. time for three half-lives (681)
T16.4 Zero-order, first-order, and simple second-order reactions (682)
F16.10 Dependence of the rate constant on temperature (683)
F16.11 Graphical determination of E_a (683)
F16.12 Kinetic parameters of a reaction (684)

F16.14 Effect of temperature on distribution of collision energies (686)
T16.5 Effect of E_a and T on the fraction (f) of sufficiently energetic collisions (686)
F16.15 Energy-level diagram for a reaction (687)
F16.16 Fraction of collisions exceeding E_a (687)
F16.19 Reaction energy diagram for the reaction of CH_3Br and OH^- (689)
T16.6 Rate laws for general elementary steps (692)
F16.22 Reaction energy diagram for catalyzed and uncatalyzed processes (697)

Brief Solutions to Follow-up Problems

16.1 (a) $4NO(g) + O_2(g) \longrightarrow 2N_2O_3(g)$;

$$\text{rate} = -\frac{\Delta[O_2]}{\Delta t} = -\frac{1}{4}\frac{\Delta[NO]}{\Delta t} = \frac{1}{2}\frac{\Delta[N_2O_3]}{\Delta t}$$

(b) $-\dfrac{\Delta[O_2]}{\Delta t} = -\dfrac{1}{4}\dfrac{\Delta[NO]}{\Delta t}$

$$= -\frac{1}{4}(-1.60 \times 10^{-4} \text{ mol/L·s}) = 4.00 \times 10^{-5} \text{ mol/L·s}$$

16.2 First order in Br^-, first order in BrO_3^-, second order in H^+, fourth order overall.

16.3 Rate $= k[H_2]^m[I_2]^n$. From experiments 1 and 3, $m = 1$. From experiments 2 and 4, $n = 1$.
Therefore, rate $= k[H_2][I_2]$; second order overall.

16.4 $1/[HI]_1 - 1/[HI]_0 = kt$;
$111 \text{ L/mol} - 100 \text{ L/mol} = (2.4 \times 10^{-21} \text{ L/mol·s})(t)$
$t = 4.6 \times 10^{21} \text{ s (or } 1.5 \times 10^{14} \text{ yr)}$

16.5 $t_{1/2} = (\ln 2)/k; k = 0.693/13.1 \text{ h} = 5.29 \times 10^{-2} \text{ h}^{-1}$

16.6 $\ln \dfrac{0.286 \text{ L/mol·s}}{k_1}$

$$= -\frac{1.00 \times 10^5 \text{ J/mol}}{8.314 \text{ J/mol·K}} \times \left(\frac{1}{500. \text{ K}} - \frac{1}{490. \text{ K}}\right) = 0.491$$

$k_1 = 0.175 \text{ L/mol·s}$

16.7

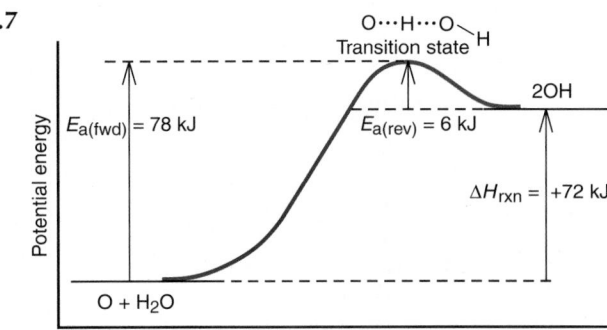

16.8 (a) Balanced equation:
$2NO(g) + 2H_2(g) \longrightarrow N_2(g) + 2H_2O(g)$
(b) Step 2 is unimolecular. All others are bimolecular.
(c) Rate$_1 = k_1[NO]^2$; rate$_2 = k_2[H_2]$; rate$_3 = k_3[N_2O_2][H]$;
rate$_4 = k_4[HO][H]$; rate$_5 = k_5[H][N_2O]$.

Problems

Problems with **colored** numbers are answered at the back of the text. Sections match the text and provide the number(s) of relevant sample problems. Most offer Concept Review Questions, Skill-Building Exercises (in similar pairs), and Problems in Context. Then Comprehensive Problems, based on material from any section or previous chapter, follow.

Factors That Influence Reaction Rate

● **Concept Review Questions**

16.1 What variable of a chemical reaction is measured over time to obtain the reaction rate?

16.2 How does an increase in pressure affect the rate of a gas-phase reaction? Explain.

16.3 A reaction is carried out with water as the solvent. How does the addition of more water to the reaction vessel affect the rate of the reaction? Explain.

16.4 How does an increase in the surface area of a solid affect the rate of its reaction with a gas? Explain.

16.5 How does an increase in temperature affect the rate of a reaction? Explain the two factors involved.

16.6 In a kinetics experiment, a chemist places crystals of iodine in a closed reaction vessel, introduces a given quantity of hydrogen gas, and obtains data to calculate the rate of hydrogen iodide formation. In a second experiment, she uses the same amounts of iodine and hydrogen but first warms the flask to 130°C; at that temperature the iodine sublimes. In which of these two experiments does the reaction proceed at a higher rate? Why?

Expressing the Reaction Rate
(Sample Problem 16.1)

● **Concept Review Questions**

16.7 Define *reaction rate*. Assuming constant temperature and a closed reaction vessel, why does rate change with time?

16.8 (a) What is the difference between an average rate and an instantaneous rate? (b) What is the difference between an initial rate and an instantaneous rate?

16.9 Give two reasons to measure initial rates in a kinetics study.

16.10 For the reaction $A(g) \longrightarrow B(g)$, sketch two curves on the same set of axes that show
(a) The formation of product as a function of time
(b) The consumption of reactant as a function of time

16.11 For the reaction $C(g) \longrightarrow D(g)$, [C] vs. time is plotted:

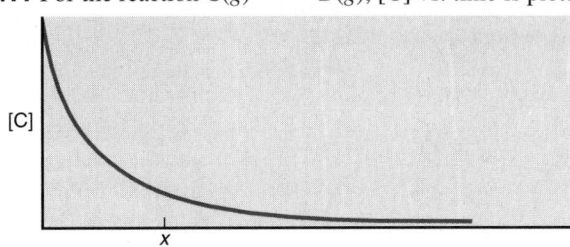

How do you determine each of the following?
(a) The average rate over the entire experiment
(b) The reaction rate at time x
(c) The initial reaction rate
(d) Would the values in parts (a), (b), and (c) be different if you plotted [D] vs. time? Explain.

● **Skill-Building Exercises** *(paired)*

16.12 The compound AX_2 decomposes according to the equation, $2AX_2(g) \longrightarrow 2AX(g) + X_2(g)$. In one experiment, $[AX_2]$ was measured at various times and these data were obtained:

Time (s)	$[AX_2]$ (mol/L)
0	0.0500
2.0	0.0448
6.0	0.0300
8.0	0.0249
10.0	0.0209
20.0	0.0088

(a) Find the average rate over the entire experiment.
(b) Is the initial rate higher or lower than the rate in part (a)? Use graphical methods to estimate the initial rate.

16.13 (a) Use the data from Problem 16.12 to calculate the average rate from 8.0 to 20.0 s.
(b) Is the rate at exactly 5.0 s higher or lower than the rate in part (a)? Use graphical methods to estimate the rate at 5.0 s.

16.14 Express the rate of reaction in terms of the change in concentration of each of the reactants and products:

$$2A(g) \longrightarrow B(g) + C(g)$$

When [C] is increasing at 2 mol/L·s, how fast is [A] decreasing?

16.15 Express the rate of reaction in terms of the change in concentration of each of the reactants and products:

$$D(g) \longrightarrow \tfrac{3}{2}E(g) + \tfrac{5}{2}F(g)$$

When [E] is increasing at 0.25 mol/L·s, how fast is [F] increasing?

16.16 Express the rate of reaction in terms of the change in concentration of each of the reactants and products:

$$A(g) + 2B(g) \longrightarrow C(g)$$

When [B] is decreasing at 0.5 mol/L·s, how fast is [A] decreasing?

16.17 Express the rate of reaction in terms of the change in concentration of each of the reactants and products:

$$2D(g) + 3E(g) + F(g) \longrightarrow 2G(g) + H(g)$$

When [D] is decreasing at 0.1 mol/L·s, how fast is [H] increasing?

16.18 The rate of a reaction is expressed in terms of changes in concentration of reactants and products. Write a balanced equation for the reaction:

$$\text{Rate} = -\frac{1}{2}\frac{\Delta[N_2O_5]}{\Delta t} = \frac{1}{4}\frac{\Delta[NO_2]}{\Delta t} = \frac{\Delta[O_2]}{\Delta t}$$

16.19 The rate of a reaction is expressed in terms of changes in concentration of reactants and products. Write a balanced equation for the reaction:

$$\text{Rate} = -\frac{\Delta[CH_4]}{\Delta t} = -\frac{1}{2}\frac{\Delta[O_2]}{\Delta t} = \frac{1}{2}\frac{\Delta[H_2O]}{\Delta t} = \frac{\Delta[CO_2]}{\Delta t}$$

● **Problems in Context**

16.20 The decomposition of nitrosyl bromide is followed manometrically because the number of moles of gas changes; it cannot be followed colorimetrically because both NOBr and Br_2 are reddish brown:

$$2NOBr(g) \longrightarrow 2NO(g) + Br_2(g)$$

Use the data below to answer the following:
(a) Determine the average rate over the entire experiment.
(b) Determine the average rate between 2.00 and 4.00 s.
(c) Use graphical methods to estimate the initial reaction rate.
(d) Use graphical methods to estimate the rate at 7.00 s.
(e) At what time does the instantaneous rate equal the average rate over the entire experiment?

Time (s)	[NOBr] (mol/L)
0.00	0.0100
2.00	0.0071
4.00	0.0055
6.00	0.0045
8.00	0.0038
10.00	0.0033

16.21 The formation of ammonia is one of the most important processes in the chemical industry:

$$N_2(g) + 3H_2(g) \longrightarrow 2NH_3(g)$$

Express the reaction rate in terms of changes in $[N_2]$, $[H_2]$, and $[NH_3]$.

16.22 Just as the depletion of stratospheric ozone today threatens life on Earth, its accumulation was one of the crucial processes that allowed life to develop in prehistoric times:

$$3O_2(g) \longrightarrow 2O_3(g)$$

(a) Express the reaction rate in terms of $[O_2]$ and $[O_3]$.
(b) At a given instant, the reaction rate in terms of $[O_2]$ is 2.17×10^{-5} mol/L·s. What is it in terms of $[O_3]$?

The Rate Law and Its Components
(Sample Problems 16.2 and 16.3)

● **Concept Review Questions**

16.23 The rate law for the general reaction

$$aA + bB + \cdots \longrightarrow cC + dD + \cdots$$

is rate $= k[A]^m[B]^n \cdots$.
(a) Explain the meaning of k.
(b) Explain the meanings of m and n. Does $m = a$ and $n = b$? Explain.
(c) If the reaction is first order in A and second order in B, and time is measured in minutes (min), what are the units for k?

16.24 You are studying the reaction

$$A_2(g) + B_2(g) \longrightarrow 2AB(g)$$

to determine its rate law. Assuming that you have a valid experimental procedure for obtaining $[A_2]$ and $[B_2]$ at various times, explain how you determine (a) the initial rate, (b) the reaction orders, and (c) the rate constant.

16.25 By what factor does the rate change in each of the following cases (assuming constant temperature)?
(a) A reaction is first order with respect to reactant A, and [A] is doubled.
(b) A reaction is second order with respect to reactant B, and [B] is halved.
(c) A reaction is second order with respect to reactant C, and [C] is tripled.

● **Skill-Building Exercises** *(paired)*

16.26 Give the individual reaction orders for each substance and the overall reaction order for the following rate law:

$$\text{Rate} = k[BrO_3^-][Br^-][H^+]^2$$

16.27 Give the individual reaction orders for each substance and the overall reaction order for the following rate law:
$$\text{Rate} = k\frac{[O_3]^2}{[O_2]}$$

16.28 By what factor does the rate in Problem 16.26 change if each of the following changes occurs: (a) $[BrO_3^-]$ is doubled; (b) $[Br^-]$ is halved; (c) $[H^+]$ is quadrupled?

16.29 By what factor does the rate in Problem 16.27 change if each of the following changes occurs: (a) $[O_3]$ is doubled; (b) $[O_2]$ is doubled; (c) $[O_2]$ is halved?

16.30 Give the individual reaction orders for each substance and the overall reaction order for this rate law:
$$\text{Rate} = k[NO_2]^2[Cl_2]$$

16.31 Give the individual reaction orders for each substance and the overall reaction order for this rate law:
$$\text{Rate} = k\frac{[HNO_2]^4}{[NO]^2}$$

16.32 By what factor does the rate in Problem 16.30 change if each of the following changes occurs: (a) $[NO_2]$ is tripled; (b) $[NO_2]$ and $[Cl_2]$ are doubled; (c) $[Cl_2]$ is halved?

16.33 By what factor does the rate in Problem 16.31 change if each of the following changes occurs: (a) $[HNO_2]$ is doubled; (b) $[NO]$ is doubled; (c) $[HNO_2]$ is halved?

16.34 For the reaction
$$4A(g) + 3B(g) \longrightarrow 2C(g)$$
the following data were obtained at constant temperature:

Experiment	Initial [A] (mol/L)	Initial [B] (mol/L)	Initial Rate (mol/L·min)
1	0.100	0.100	5.00
2	0.300	0.100	45.0
3	0.100	0.200	10.0
4	0.300	0.200	90.0

(a) What is the order with respect to each reactant?
(b) Write the rate law.
(c) Calculate k (using, for example, the data from experiment 1).

16.35 For the reaction
$$A(g) + B(g) + C(g) \longrightarrow D(g)$$
the following data were obtained at constant temperature:

Exp't	Initial [A] (mol/L)	Initial [B] (mol/L)	Initial [C] (mol/L)	Initial Rate (mol/L·s)
1	0.0500	0.0500	0.0100	6.25×10^{-3}
2	0.1000	0.0500	0.0100	1.25×10^{-2}
3	0.1000	0.1000	0.0100	5.00×10^{-2}
4	0.0500	0.0500	0.0200	6.25×10^{-3}

(a) What is the order with respect to each reactant?
(b) Write the rate law.
(c) Calculate k (using, for example, the data from experiment 1).

16.36 Without consulting Table 16.3, give the units of the rate constants for reactions with the following overall orders: (a) first order; (b) second order; (c) third order; (d) $\frac{5}{2}$ order.

16.37 Give the overall reaction order that corresponds to rate constants with the following units:
(a) mol/L·s (b) yr^{-1}
(c) $(mol/L)^{1/2}\cdot s^{-1}$ (d) $(mol/L)^{-5/2}\cdot min^{-1}$

● **Problems in Context**

16.38 Phosgene is a toxic gas prepared by the reaction of carbon monoxide with chlorine:
$$CO(g) + Cl_2(g) \longrightarrow COCl_2(g)$$
The following data were obtained in a kinetics study of its formation:

Experiment	Initial [CO] (mol/L)	Initial [Cl₂] (mol/L)	Initial Rate (mol/L·s)
1	1.00	0.100	1.29×10^{-29}
2	0.100	0.100	1.33×10^{-30}
3	0.100	1.00	1.30×10^{-29}
4	0.100	0.0100	1.32×10^{-31}

(a) Write the rate law for the formation of phosgene.
(b) Calculate the average value of the rate constant.

Integrated Rate Laws: Concentration Changes over Time
(Sample Problems 16.4 and 16.5)

● **Concept Review Questions**

16.39 How are integrated rate laws used to determine reaction order? What is the order in reactant if a plot of
(a) The natural logarithm of [reactant] vs. time is linear?
(b) The inverse of [reactant] vs. time is linear?
(c) [Reactant] vs. time is linear?

16.40 Define the *half-life* of a reaction. Explain on the molecular level why the half-life of a first-order reaction is a constant value.

● **Skill-Building Exercises (paired)**

16.41 For the simple decomposition reaction
$$AB(g) \longrightarrow A(g) + B(g)$$
rate $= k[AB]^2$ and $k = 0.2$ L/mol·s. How long will it take for [AB] to reach one-third of its initial concentration of 1.50 M?

16.42 For the reaction in Problem 16.41, what is [AB] after 10.0 s?

16.43 In a first-order decomposition reaction, 50.0% of a compound decomposes in 10.5 min.
(a) What is the rate constant of the reaction?
(b) How long does it take for 75.0% of the compound to decompose?

16.44 A decomposition reaction has a rate constant of 0.0012 yr^{-1}.
(a) What is the half-life of the reaction?
(b) How long does it take for [reactant] to reach 12.5% of its original value?

● **Problems in Context**

16.45 During a study of ammonia production, an industrial chemist discovers that the compound decomposes to its elements N_2 and H_2 in a first-order process. She collects the following data:

Time (s)	0	1.000	2.000
[NH₃] (mol/L)	4.000	3.986	3.974

(a) Use graphical methods to determine the rate constant.
(b) What is the half-life for ammonia decomposition?

The Effect of Temperature on Reaction Rate
(Sample Problem 16.6)

● **Concept Review Questions**

16.46 Use the exponential term in the Arrhenius equation to explain how temperature affects reaction rate.

16.47 How is the activation energy determined from the Arrhenius equation?

16.48 (a) Sketch a graph to show the relationship between k (y axis) and T (x axis).
(b) Sketch a graph to show the relationship between $\ln k$ (y axis) and $1/T$ (x axis). How is the activation energy determined from this graph?

● **Skill-Building Exercises (paired)**

16.49 The rate constant of a reaction is 4.7×10^{-3} s^{-1} at 25°C, and the activation energy is 33.6 kJ/mol. What is the value of k at 75°C?

16.50 The rate constant of a reaction is 4.50×10^{-5} L/mol·s at 195°C and 3.20×10^{-3} L/mol·s at 258°C. What is the activation energy of the reaction?

● **Problems in Context**

16.51 Understanding the high-temperature formation and breakdown of the nitrogen oxides is essential for controlling the pollutants generated by car engines. The second-order reaction for the breakdown of nitric oxide to its elements has rate constants of 0.0796 L/mol·s at 737°C and 0.0815 L/mol·s at 947°C. What is the activation energy of this reaction?

Explaining the Effects of Concentration and Temperature
(Sample Problem 16.7)

● **Concept Review Questions**

16.52 What is the central idea of collision theory? How does this idea explain the effect of concentration on reaction rate?

16.53 Is collision frequency the only factor affecting reaction rate? Explain.

16.54 Arrhenius proposed that each reaction has an energy threshold that must be reached for the particles to react. The kinetic theory of gases proposes that the average kinetic energy of the particles in a sample is proportional to the absolute temperature. Explain how these concepts relate to the effect of temperature on reaction rate.

16.55 (a) For a reaction with a given E_a, how does an increase in T affect the rate?
(b) For a reaction at a given T, how does a decrease in E_a affect the rate?

16.56 In the reaction AB + CD $\rightleftharpoons$ EF, 4×10^{-5} mol of AB molecules collides with 4×10^{-5} mol of CD molecules. Will 4×10^{-5} mol of EF form? Explain.

16.57 Assuming the activation energies are equal, which of the following reactions will occur at a higher rate at 50°C? Explain:

$$NH_3(g) + HCl(g) \longrightarrow NH_4Cl(s)$$
$$N(CH_3)_3(g) + HCl(g) \longrightarrow (CH_3)_3NHCl(s)$$

● **Skill-Building Exercises (paired)**

16.58 For the reaction A(g) + B(g) $\longrightarrow$ AB(g), how many unique collisions between A and B are possible if there are four particles of A and three particles of B present in the vessel?

16.59 For the reaction A(g) + B(g) $\longrightarrow$ AB(g), how many unique collisions between A and B are possible if 1.01 mol of A(g) and 2.12 mol of B(g) are present in the vessel?

16.60 At 25°C, what is the fraction of collisions with energy equal to or greater than an activation energy of 100. kJ/mol?

16.61 If the temperature in Problem 16.60 is increased to 50.°C, by what factor does the fraction of collisions with energy equal to or greater than the activation energy change?

16.62 For the reaction ABC + D $\rightleftharpoons$ AB + CD, $\Delta H^0_{rxn} = -55$ kJ/mol and $E_{a(fwd)} = 215$ kJ/mol. Assuming the reaction occurs in one step,
(a) Draw a reaction energy diagram   (b) Calculate $E_{a(rev)}$
(c) Sketch a possible transition state if A—B—C is V-shaped.

16.63 For the reaction A$_2$ + B$_2$ $\longrightarrow$ 2AB, $E_{a(fwd)} = 125$ kJ/mol and $E_{a(rev)} = 85$ kJ/mol. Assuming the reaction occurs in one step,
(a) Draw a reaction energy diagram   (b) Calculate ΔH^0_{rxn}
(c) Sketch a possible transition state

● **Problems in Context**

16.64 Aqua regia, a mixture of hydrochloric and nitric acids, has been used since alchemical times as a solvent for many metals, including gold. Its orange color is due to the presence of nitrosyl chloride. Consider this one-step gaseous reaction for the formation of this compound:

$$NO(g) + Cl_2(g) \longrightarrow NOCl(g) + Cl(g) \qquad \Delta H^0 = 83 \text{ kJ}$$

(a) Draw a reaction energy diagram for the reaction, given that $E_{a(fwd)}$ is 86 kJ/mol.
(b) Calculate $E_{a(rev)}$.
(c) Sketch a possible transition state for the reaction. (*Hint:* The atom sequence of nitrosyl chloride is Cl—N—O.)

Reaction Mechanisms: Steps in the Overall Reaction
(Sample Problem 16.8)

● **Concept Review Questions**

16.65 Is the rate of an overall reaction lower, higher, or equal to the average rate of the individual steps? Explain.

16.66 Explain why the coefficients of an elementary step equal the reaction orders of its rate law but those of an overall reaction do not.

16.67 Is it possible for more than one mechanism to be consistent with the rate law of a given reaction? Explain.

16.68 What is the difference between a reaction intermediate and a transition state?

16.69 Why is a bimolecular step more reasonable physically than a termolecular step?

16.70 If a slow step precedes a fast step in a two-step mechanism, do the substances in the fast step appear in the rate law? Explain.

16.71 If a fast step precedes a slow step in a two-step mechanism, how is the fast step affected? How is this effect used to determine the validity of the mechanism?

● **Skill-Building Exercises (paired)**

16.72 The proposed mechanism for a reaction is
(1) A(g) + B(g) $\rightleftharpoons$ X(g)   [fast]
(2) X(g) + C(g) $\longrightarrow$ Y(g)   [slow]
(3) Y(g) $\longrightarrow$ D(g)   [fast]

(a) What is the overall equation?

(b) Identify the intermediate(s), if any.

(c) What is the molecularity and rate law for each step?

(d) Is the mechanism consistent with the actual rate law: rate = k[A][B][C]?

(e) Is the following one-step mechanism equally valid: A(g) + B(g) + C(g) $\longrightarrow$ D(g)?

16.73 Consider the following mechanism:

(1) ClO$^-$(aq) + H$_2$O(l) $\rightleftharpoons$ HClO(aq) + OH$^-$(aq)   [fast]

(2) I$^-$(aq) + HClO(aq) $\longrightarrow$ HIO(aq) + Cl$^-$(aq)   [slow]

(3) OH$^-$(aq) + HIO(aq) $\longrightarrow$ H$_2$O(l) + IO$^-$(aq)   [fast]

(a) What is the overall equation?

(b) Identify the intermediate(s), if any.

(c) What is the molecularity and the rate law for each step?

(d) Is the mechanism consistent with the actual rate law: rate = k[ClO$^-$][I$^-$]?

● **Problems in Context**

16.74 In a study of nitrosyl halides, a chemist proposes the following mechanism for the synthesis of nitrosyl bromide:

$$NO(g) + Br_2(g) \rightleftharpoons NOBr_2(g) \text{ [fast]}$$
$$NOBr_2(g) + NO(g) \longrightarrow 2NOBr(g) \text{ [slow]}$$

If the rate law is rate = k[NO]^2[Br$_2$], is the proposed mechanism valid? Explain by showing that it satisfies the three criteria for validity.

16.75 The rate law for 2NO(g) + O$_2$(g) $\longrightarrow$ 2NO$_2$(g) is rate = k[NO]^2[O$_2$]. In addition to the mechanism in the text, the following ones have been proposed:

I   2NO(g) + O$_2$(g) $\longrightarrow$ 2NO$_2$(g)

II   2NO(g) $\rightleftharpoons$ N$_2$O$_2$(g)   [fast]

      N$_2$O$_2$(g) + O$_2$(g) $\longrightarrow$ 2NO$_2$(g)   [slow]

III 2NO(g) $\rightleftharpoons$ N$_2$(g) + O$_2$(g)   [fast]

      N$_2$(g) + 2O$_2$(g) $\longrightarrow$ 2NO$_2$(g)   [slow]

(a) Which of these mechanisms is consistent with the rate law?

(b) Which is most reasonable chemically? Why?

Catalysis: Speeding Up a Chemical Reaction

● **Concept Review Questions**

16.76 Consider the reaction N$_2$O(g) $\xrightarrow{\text{Au}}$ N$_2$(g) + $\frac{1}{2}$O$_2$(g).

(a) Is the gold a homogeneous or a heterogeneous catalyst?

(b) On the same set of axes, sketch the reaction energy diagrams for the catalyzed and the uncatalyzed reactions.

16.77 Does a catalyst increase reaction rate by the same means as a rise in temperature does? Explain.

16.78 In a popular classroom demonstration, hydrogen gas and oxygen gas are mixed in a balloon. Although the mixture is stable under normal conditions, when a spark is applied to the mixture or a small amount of powdered metal is dropped into it, the mixture explodes. (a) Is the spark acting as a catalyst? Explain. (b) Is the metal acting as a catalyst? Explain.

16.79 Describe three common examples of catalyzed reactions that occur around you daily.

Comprehensive Problems

Problems with an asterisk (*) are more challenging.

16.80 Experiments show that each of the following redox reactions is second order overall:

Reaction 1: NO$_2$(g) + CO(g) $\longrightarrow$ NO(g) + CO$_2$(g)

Reaction 2: NO(g) + O$_3$(g) $\longrightarrow$ NO$_2$(g) + O$_2$(g)

(a) When [NO$_2$] in reaction 1 is doubled, the rate quadruples. Write the rate law for this reaction.

(b) When [NO] in reaction 2 is doubled, the rate doubles. Write the rate law for this reaction.

(c) In each reaction, the initial concentrations of the reactants are equal. For each reaction, what is the ratio of the initial rate to the rate when the reaction is 50% complete?

(d) In reaction 1, the initial [NO$_2$] is twice the initial [CO]. What is the ratio of the initial rate to the rate when the reaction is 50% complete?

(e) In reaction 2, the initial [NO] is twice the initial [O$_3$]. What is the ratio of the initial rate to the rate when the reaction is 50% complete?

16.81 Consider the following reaction energy diagram:

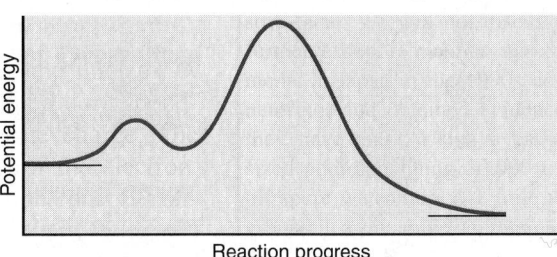

(a) How many elementary steps are in the reaction mechanism?

(b) Which step is rate limiting?

(c) Is the overall reaction exothermic or endothermic?

16.82 Reactions between certain organic (alkyl) halides and water produce alcohols. Consider the overall reaction for t-butyl bromide (2-bromo-2-methylpropane):

$$(CH_3)_3CBr(aq) + H_2O(l) \longrightarrow$$
$$(CH_3)_3COH(aq) + H^+(aq) + Br^-(aq)$$

The experimental rate law is rate = k[(CH$_3$)$_3$CBr]. The accepted mechanism for the reaction is

(1) (CH$_3$)$_3$C—Br(aq) $\longrightarrow$ (CH$_3$)$_3$C$^+$(aq) + Br$^-$(aq)   [slow]

(2) (CH$_3$)$_3$C$^+$(aq) + H$_2$O(l) $\longrightarrow$ (CH$_3$)$_3$C—OH$_2$$^+$($aq$)   [fast]

(3) (CH$_3$)$_3$C—OH$_2$$^+$($aq$) $\longrightarrow$ H$^+$(aq) + (CH$_3$)$_3$C—OH(aq)

                               [fast]

(a) Why doesn't H$_2$O appear in the rate law?

(b) Write rate laws for the elementary steps.

(c) What intermediates appear in the mechanism?

(d) Show that the mechanism is consistent with the experimental rate law.

16.83 High-flying supersonic transports release NO into the stratosphere. NO catalyzes ozone depletion by a mechanism analogous to the action of Cl. Write the reactions in the mechanism, and show why NO acts as a catalyst.

16.84 Archeologists can determine the age of artifacts made of wood or bone by measuring the concentration of the radioactive isotope ^{14}C present in the object. The amount of isotope decreases in a first-order process. If 15.5% of the original amount of ^{14}C is present in a wooden tool at the time of analysis, what is the age of the tool? The half-life of ^{14}C is 5730 yr.

16.85 A slightly bruised apple will rot extensively in about 4 days at room temperature (20°C). If it is kept in the refrigerator at 0°C, the same extent of rotting takes about 16 days. What is the activation energy for the rotting reaction?

16.86 Benzoyl peroxide, the substance most widely used against acne, has a half-life of 9.8×10^3 days when refrigerated. How long will it take to lose 5% of its potency (95% remaining)?

16.87 The rate law for the reaction

$$NO_2(g) + CO(g) \longrightarrow NO(g) + CO_2(g)$$

is rate $= k[NO_2]^2$; one possible mechanism is shown on p. 694.
(a) Draw a reaction energy diagram for that mechanism, given that $\Delta H^0_{overall} = -226$ kJ/mol.
(b) The following alternative mechanism has been proposed:

(1) $2NO_2(g) \longrightarrow N_2(g) + 2O_2(g)$	[slow]
(2) $2CO(g) + O_2(g) \longrightarrow 2CO_2(g)$	[fast]
(3) $N_2(g) + O_2(g) \longrightarrow 2NO(g)$	[fast]

Is the alternative mechanism consistent with the rate law? Is one mechanism more reasonable physically? Explain.

16.88 Consider the following general reaction and data:

$$2A + 2B + C \longrightarrow D + 3E$$

Exp't	Initial [A] (mol/L)	Initial [B] (mol/L)	Initial [C] (mol/L)	Initial Rate (mol/L·s)
1	0.024	0.085	0.032	6.0×10^{-6}
2	0.096	0.085	0.032	9.6×10^{-5}
3	0.024	0.034	0.080	1.5×10^{-5}
4	0.012	0.170	0.032	1.5×10^{-6}

(a) What is the reaction order with respect to each reactant?
(b) Calculate the rate constant.
(c) Write the rate law for this reaction.
(d) Express the rate in terms of changes in concentration with time for each of the components.

16.89 How do each of the following changes affect the reaction rate (increase, decrease, no effect)?
(a) Decreasing the pressure of a gas-liquid reaction
(b) Increasing the concentration of a reactant that occurs in a step preceding the rate-limiting step
(c) Increasing the concentration of a reactant that occurs in a step following the rate-limiting step
(d) Grinding up a solid reactant in a gas-solid reaction
(e) Rapidly stirring a reaction between immiscible liquids

16.90 Enzymes are remarkably efficient catalysts that can increase reaction rates by as many as 20 orders of magnitude.
(a) How do enzymes affect the transition state of a reaction, and how does this effect increase the reaction rate?
(b) What characteristics of enzymes give them this tremendous effectiveness as catalysts?

16.91 Biacetyl, the flavoring that makes margarine taste "just like butter," is extremely stable at room temperature, but at 200°C it undergoes a first-order breakdown with a half-life of 9.0 min. An industrial flavor-enhancing process requires that a biacetyl-flavored food be heated briefly at 200°C. How long can the food be heated and retain 85% of its buttery flavor?

16.92 Biochemists consider the citric acid cycle to be the central reaction sequence in metabolism. One of the key steps is an oxidation catalyzed by the enzyme isocitrate dehydrogenase and the oxidizing agent NAD$^+$. Under certain conditions, the reaction in yeast obeys 11th-order kinetics:

$$Rate = k[enzyme][isocitrate]^4[AMP]^2[NAD^+]^m[Mg^{2+}]^2$$

What is the order with respect to NAD$^+$?

16.93 Nitrogen oxides undergo numerous decomposition and combination reactions in both industry and the environment. One of the most thoroughly studied is the decomposition of dinitrogen pentaoxide:

$$2N_2O_5(g) \longrightarrow 4NO_2(g) + O_2(g)$$

At a given instant, $[N_2O_5]$ is decreasing at the rate of 3.15 mol/L·s. What is the rate of the reaction in terms of $[NO_2]$?

16.94 Enzymes in human liver catalyze a large number of reactions that degrade ingested toxic chemicals. By what factor is the rate of a detoxification reaction changed if a liver enzyme lowers the activation energy by 5 kJ/mol at 37°C?

16.95 Experiment shows that the rate of formation of carbon tetrachloride from chloroform,

$$CHCl_3(g) + Cl_2(g) \longrightarrow CCl_4(g) + HCl(g)$$

is first order in CHCl$_3$, $\frac{1}{2}$ order in Cl$_2$, and $\frac{3}{2}$ order overall. Show the following mechanism is consistent with the overall rate law:

(1) $Cl_2(g) \rightleftharpoons 2Cl(g)$	[fast]
(2) $Cl(g) + CHCl_3(g) \longrightarrow HCl(g) + CCl_3(g)$	[slow]
(3) $CCl_3(g) + Cl(g) \longrightarrow CCl_4(g)$	[fast]

16.96 A biochemist studying the breakdown in soil of the insecticide DDT finds that it decomposes by a first-order reaction with a half-life of 12 yr. How long does it take DDT to decompose from 275 ppbm to 10. ppbm (parts per billion by mass) in a soil sample?

16.97 Under certain conditions, the rate of a reaction is 0.192 mol/L·min.
(a) What is the rate in units of mol/L·s?
(b) What is the rate in units of mol·dm^{-3}·s^{-1}?
(c) What is the rate expressed in SI base units?

16.98 For the reaction $A(g) + B(g) \longrightarrow AB(g)$, the rate is 0.20 mol/L·s, when $[A]_0 = [B]_0 = 1.0$ mol/L. If the reaction is first order in B and second order in A, what is the rate when $[A]_0 = 2.0$ mol/L and $[B]_0 = 3.0$ mol/L?

***16.99** The hydrolysis of table sugar (sucrose) occurs by the following overall reaction:

$$\underset{sucrose}{C_{12}H_{22}O_{11}(s)} + H_2O(l) \longrightarrow \underset{glucose}{C_6H_{12}O_6(aq)} + \underset{fructose}{C_6H_{12}O_6(aq)}$$

A nutritional biochemist studies the kinetics of the process and obtains the following data:

[Sucrose] (mol/L)	Time (h)
0.501	0
0.451	0.50
0.404	1.00
0.363	1.50
0.267	3.00

(a) Use the data to determine the rate constant and the half-life of the reaction.
(b) How long does it take to hydrolyze 75% of the sucrose?
(c) Other studies have shown that this reaction is actually second order overall but appears to follow first-order kinetics. (Such a reaction is termed a *pseudo first-order reaction*.) Suggest a reason for this apparent first-order behavior.

16.100 Is each of these statements true? If not, explain why.
(a) At a given temperature, all molecules possess the same kinetic energy.
(b) Halving the pressure of a gaseous reaction doubles the reaction rate.
(c) The higher the activation energy of a reaction is, the lower the reaction rate.
(d) A temperature increase of 10°C doubles the rate of any reaction.
(e) If reactant molecules collide with greater energy than the activation energy, they change into product molecules.

(f) The activation energy of a reaction depends on the temperature.

(g) The rate of a reaction increases as the reaction proceeds.

(h) The activation energy of a reaction depends on collision frequency.

(i) A catalyst increases the rate by increasing collision frequency.

(j) Exothermic reactions have higher rates than endothermic ones.

(k) Temperature has no effect on the value of the Arrhenius factor (A).

(l) The activation energy of a reaction is lowered by a catalyst.

(m) For most common reactions, the enthalpy change is lowered by a catalyst.

(n) The probability factor (p) is near unity for reactions between single atoms.

(o) The initial rate of a reaction is its maximum rate.

(p) A bimolecular reaction is generally twice as fast as a unimolecular reaction.

(q) The molecularity of an elementary reaction is proportional to the molecular complexity of the reactant(s).

16.101 For the decomposition of gaseous dinitrogen pentaoxide, $2N_2O_5(g) \longrightarrow 4NO_2(g) + O_2(g)$, the rate constant is $k = 2.8 \times 10^{-3} \text{ s}^{-1}$ at 60°C. The initial concentration of N_2O_5 is 1.58 mol/L.

(a) What is $[N_2O_5]$ after 5.00 min?

(b) What fraction of the N_2O_5 has decomposed after 5.00 min?

16.102 Even when a mechanism is consistent with the rate law, later experimentation may show it to be incorrect or only one of several alternatives. As an example, the reaction between hydrogen and iodine has the following rate law: rate = $k[H_2][I_2]$. The long-accepted mechanism proposed a single bimolecular step; that is, the overall reaction was thought to be elementary:

$$H_2(g) + I_2(g) \longrightarrow 2HI(g)$$

In the 1960s, however, spectroscopic evidence showed the presence of free I atoms during the reaction. Kineticists have since proposed a three-step mechanism:

(1) $I_2(g) \rightleftharpoons 2I(g)$ [fast]
(2) $H_2(g) + I(g) \rightleftharpoons H_2I(g)$ [fast]
(3) $H_2I(g) + I(g) \longrightarrow 2HI(g)$ [slow]

Show that this mechanism is consistent with the rate law.

16.103 Suggest an experimental method for measuring the change in concentration with time for each of the following reactions:

(a) $CH_3CH_2Br(l) + H_2O(l) \longrightarrow CH_3CH_2OH(l) + HBr(aq)$

(b) $2NO(g) + Cl_2(g) \longrightarrow 2NOCl(g)$

*16.104 An atmospheric chemist fills a gas-reaction container with gaseous dinitrogen pentaoxide to a pressure of 125 kPa, and the gas decomposes to nitrogen dioxide and oxygen. What is the partial pressure of nitrogen dioxide, P_{NO_2} (in kPa), when the total pressure is 178 kPa?

16.105 Several studies have shown that small doses of aspirin are effective against the formation of blood clots in arteries. Therefore, elderly people with this form of cardiovascular disease are advised to take half a tablet each day as a precaution against stroke. Aspirin (and many other drugs) is broken down in the body by a first-order process. The half-life of aspirin in elderly humans is 3.7 h.

(a) How much aspirin remains in the bloodstream from a 160-mg dose after 24 hours?

(b) In young humans, the half-life is 2.4 h. How much aspirin remains in a young person 24 hours after the same dose?

*16.106 Iodide ion reacts with chloroform to displace chloride ion in a common organic substitution reaction:

$$I^- + CH_3Cl \longrightarrow CH_3I + Cl^-$$

(a) Draw a wedge-bond structural formula of chloroform and indicate the most effective direction of I^- attack.

(b) The analogous reaction with 2-chlorobutane [Figure P16.106(b)] results in a major change in specific rotation as measured by polarimetry. Explain, showing a wedge-bond structural formula of the product.

(c) Under different conditions, 2-chlorobutane loses Cl^- in a rate-determining step to form a planar intermediate [Figure P16.106(c)]. This cationic species reacts with HI and then loses H^+ to form a product that exhibits no optical activity. Explain, showing a wedge-bond structural formula.

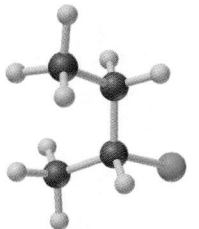
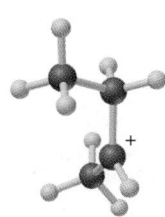

Figure P16.106(b)   Figure P16.106(c)

16.107 A pile of flour on your kitchen counter does not ignite with a match, but flour dust in a mill or silo can react explosively. Explain.

16.108 The chlorination of ethane is one of several synthetic routes to producing chloroethanes, a group of major industrial solvents. An example of this type of reaction is

$$C_2H_6(g) + 4Cl_2(g) \longrightarrow C_2H_2Cl_4(l) + 4HCl(g)$$

Express the reaction rate in terms of changes in concentration of each reactant and product.

16.109 Acetone is one of the most important solvents in organic chemistry, used to dissolve everything from fats and waxes to airplane glue and nail polish. At high temperatures, it decomposes in a first-order process to methane and ketene ($CH_2=C=O$). At 600°C, the rate constant is $8.7 \times 10^{-3} \text{ s}^{-1}$.

(a) What is the half-life of the reaction?

(b) How much time is required for 40.% of a sample of acetone to decompose?

(c) How much time is required for 90.% of a sample of acetone to decompose?

*16.110 In the lower troposphere, ozone is one of the components of photochemical smog. It is generated in air when nitrogen dioxide, formed by the oxidation of nitric oxide from car exhaust, reacts by the following mechanism:

(1) $NO_2(g) \xrightarrow{k_1} NO(g) + O(g)$

(2) $O(g) + O_2(g) \xrightarrow{k_2} O_3(g)$

Assuming the rate of formation of atomic oxygen in step 1 equals the rate of its consumption in step 2, use the data below to calculate

(a) The concentration of atomic oxygen [O]

(b) The rate of ozone formation

$k_1 = 6.0 \times 10^{-3} \text{ s}^{-1}$ $[NO_2] = 4.0 \times 10^{-9} M$
$k_2 = 1.0 \times 10^6 \text{ L/mol·s}$ $[O_2] = 1.0 \times 10^{-2} M$

16.111 The reaction and rate law for the gas-phase decomposition of dinitrogen pentaoxide are

$$2N_2O_5(g) \longrightarrow 4NO_2(g) + O_2(g) \qquad \text{rate} = k[N_2O_5]$$

Which of the following can be considered valid mechanisms for the reaction?

I   One-step collision

II $2[N_2O_5(g) \longrightarrow NO_3(g) + NO_2(g)]$ [slow]
    $2[NO_3(g) \longrightarrow NO_2(g) + O(g)]$ [fast]
    $2O(g) \longrightarrow O_2(g)$ [fast]

III $N_2O_5(g) \rightleftharpoons NO_3(g) + NO_2(g)$ [fast]
    $NO_2(g) + N_2O_5(g) \longrightarrow 3NO_2(g) + O(g)$ [slow]
    $NO_3(g) + O(g) \longrightarrow NO_2(g) + O_2(g)$ [fast]

IV $2N_2O_5(g) \rightleftharpoons 2NO_2(g) + N_2O_3(g) + 3O(g)$ [fast]
    $N_2O_3(g) + O(g) \longrightarrow 2NO_2(g)$ [slow]
    $2O(g) \longrightarrow O_2(g)$ [fast]

V $2N_2O_5(g) \longrightarrow N_4O_{10}(g)$ [slow]
    $N_4O_{10}(g) \longrightarrow 4NO_2(g) + O_2(g)$ [fast]

16.112 Carbon disulfide is a poisonous, flammable liquid that is an excellent solvent for phosphorus, sulfur, and several other nonmetals. A kinetic study of its gaseous decomposition reveals the following data:

Experiment	Initial $[CS_2]$ (mol/L)	Initial Rate (mol/L·s)
1	0.100	2.7×10^{-7}
2	0.080	2.2×10^{-7}
3	0.055	1.5×10^{-7}
4	0.044	1.2×10^{-7}

(a) Write the rate law for the decomposition of CS_2.
(b) Calculate the average value of the rate constant.

16.113 Kinetic studies of hydrogen halide formation show that the rate law for HBr formation, $H_2(g) + Br_2(g) \longrightarrow 2HBr(g)$, is rate $= k[H_2][Br_2]^{1/2}$. Which of the following mechanisms is consistent with the rate law?

I $H_2(g) + Br_2(g) \longrightarrow 2HBr(g)$

II $H_2(g) \rightleftharpoons 2H(g)$ [fast]
    $H(g) + Br_2(g) \longrightarrow HBr(g) + Br(g)$ [slow]
    $Br(g) + H(g) \longrightarrow HBr(g)$ [fast]

III $Br_2(g) \rightleftharpoons 2Br(g)$ [fast]
    $Br(g) + H_2(g) \longrightarrow HBr(g) + H(g)$ [slow]
    $H(g) + Br(g) \longrightarrow HBr(g)$ [fast]

*16.114 In a "clock" reaction, a dramatic color change occurs at a time determined by concentration and temperature. One of the most famous is the iodine clock reaction. The overall equation is

$$2I^-(aq) + S_2O_8^{2-}(aq) \longrightarrow I_2(aq) + 2SO_4^{2-}(aq)$$

As I_2 forms, it is immediately consumed by its reaction with a fixed amount of added $S_2O_3^{2-}$:

$$I_2(aq) + 2S_2O_3^{2-}(aq) \longrightarrow 2I^-(aq) + S_4O_6^{2-}(aq)$$

Once the $S_2O_3^{2-}$ is consumed, the excess I_2 forms a blue-black product with starch solution present in the mixture:

$$I_2 + \text{starch} \longrightarrow \text{starch} \cdot I_2 \text{ (blue-black)}$$

The rate of the reaction is also influenced by the total concentration of ions, so KCl and $(NH_4)_2SO_4$ are added to maintain a constant value. Use the data below to determine the following:
(a) The average rate for each trial
(b) The order with respect to each reactant
(c) The rate constant at 23°C
(d) The rate law for the overall reaction

	Exp't 1	Exp't 2	Exp't 3
0.200 M KI (mL)	10.0	20.0	20.0
0.100 M $Na_2S_2O_8$ (mL)	20.0	20.0	10.0
0.0050 M $Na_2S_2O_3$ (mL)	10.0	10.0	10.0
0.200 M KCl (mL)	10.0	0	0
0.100 M $(NH_4)_2SO_4$ (mL)	0	0	10.0
Time to color (s)	29.0	14.5	14.5

16.115 The mathematics of the first-order rate law can be applied to any situation in which a quantity decreases by a constant fraction per unit of time (or any other variable).
(a) As light moves through a solution, its intensity decreases per unit length traveled in the solution. Show that

$$\ln\left(\frac{\text{intensity of light leaving the solution}}{\text{intensity of light entering the solution}}\right)$$
$$= -\text{fraction of light removed per unit of length} \times \text{length of solution}$$

(b) The value of your savings declines under conditions of constant inflation. Show that

$$\ln(\text{value remaining})$$
$$= -\text{fraction lost per unit of time} \times \text{time of saving}$$

*16.116 Consider the following organic reaction, in which one halogen replaces another in an alkyl halide:

$$CH_3CH_2Br + KI \longrightarrow CH_3CH_2I + KBr$$

In acetone, this particular reaction goes to completion because KI is soluble in acetone but KBr is not. In the mechanism, I^- approaches the carbon *opposite* to the Br (see Figure 16.19, with I^- instead of OH^-). After Br^- has been replaced by I^- and precipitates as KBr, other I^- ions react with the ethyl iodide by the same mechanism.
(a) If we designate the carbon bonded to the halogen as C-1, what is the shape around C-1 and the hybridization of C-1 in ethyl iodide?
(b) In the transition state, one of the two lobes of the unhybridized $2p$ orbital of C-1 overlaps a p orbital of I, while the other lobe overlaps a p orbital of Br. What is the shape around C-1 and the hybridization of C-1 in the transition state?
(c) The deuterated reactant, CH_3CHDBr (where D is ^2H), has two optical isomers because C-1 is chiral. If the reaction is run with one of the isomers, the ethyl iodide is *not* optically active. Explain.

EQUILIBRIUM:
THE EXTENT OF CHEMICAL REACTIONS

CHAPTER OUTLINE

17.1 The Dynamic Nature of the Equilibrium State

17.2 The Reaction Quotient and the Equilibrium Constant
Writing the Reaction Quotient
Variations in the Form of Q

17.3 Expressing Equilibria with Pressure Terms: Relation Between K_c and K_p

17.4 Reaction Direction: Comparing Q and K

17.5 How to Solve Equilibrium Problems
Using Quantities to Determine K
Using K to Determine Quantities

17.6 Reaction Conditions and the Equilibrium State: Le Châtelier's Principle
Change in Concentration
Change in Pressure (Volume)
Change in Temperature
Lack of Effect of a Catalyst

Figure: Traffic equilibrium. Mimicking the forward and reverse reactions of a chemical system, the to-and-fro of traffic over this bridge maintains a relatively constant "concentration" of cars on either side. At equilibrium, the concentration of reactant changing to product is balanced by the concentration of product changing back again. In this chapter, the first of three dealing with the equilibrium state, you'll learn how concentrations change as a system approaches equilibrium, how we determine the extent of the overall change, and how these systems adapt to external forces.

CONCEPTS & SKILLS

to review before you study this chapter
- reversibility of reactions (Section 4.7)
- equilibrium vapor pressure (Section 12.2)
- equilibrium nature of a saturated solution (Section 13.3)
- dependence of rate on concentration (Sections 16.2 and 16.6)
- rate laws for elementary reactions (Section 16.7)
- function of a catalyst (Section 16.8)

In this chapter, we turn to the second central question in the dynamics of reaction chemistry: how much product will form under a given set of starting concentrations and conditions? Whereas chemical kinetics, the topic of Chapter 16, tells us how fast reactant and product concentrations are changing, chemical equilibrium tells us what those concentrations are once they've stopped changing. Furthermore, studies of equilibrium tell us how to affect these final concentrations by altering conditions.

Just as reactions vary greatly in their speed, they also vary in their extent. These two aspects are *not* the same. A fast reaction may go a long way or barely at all toward products. Consider the dissociation of an acid in water. In 1 *M* HCl, virtually all the hydrogen chloride molecules are dissociated into ions. In contrast, in 1 *M* CH₃COOH, fewer than 1% of the acetic acid molecules are dissociated at any given time. Yet both reactions take less than a second to reach completion. Similarly, some slow reactions eventually yield a large amount of product, whereas others yield very little. After a few years at ordinary temperatures, a steel water-storage tank will rust, and will do so completely given enough time; but no matter how long you wait, the water inside will not decompose to its elements.

The point is that the principles of kinetics and of equilibrium apply to different aspects of a reaction:

- Kinetics applies to the *speed* of a reaction, the concentration of product appearing (or of reactant disappearing) per unit time.
- Equilibrium applies to the *extent* of a reaction, the concentration of product that has appeared given unlimited time, or when no further change occurs.

Knowing how much product will form in a given reaction is crucial. How much of a new drug or new polymer can you obtain from a particular reaction? Is another reactant mixture that yields more product at a higher temperature a wiser choice? If a slow reaction has a good yield, will a catalyst speed it up enough to make the reaction useful?

This chapter opens with a description of the equilibrium state at the macroscopic and molecular levels and then focuses on the equilibrium constant and its relation to the balanced equation. You'll see how to express the equilibrium condition in terms of concentrations or pressures and how to determine whether a reaction mixture is proceeding toward products or reactants. We apply equilibrium concepts to a series of common quantitative problems and then examine how reaction conditions affect the equilibrium state. We end with a discussion of equilibrium in two areas of applied chemistry: the industrial production of ammonia and the metabolic pathways in living cells. This introduction to equilibrium principles covers mostly systems of gases and pure liquids and solids; we discuss solution equilibria in the next two chapters.

17.1 THE DYNAMIC NATURE OF THE EQUILIBRIUM STATE

Countless experiments with chemical systems have shown that, in a state of equilibrium, *the concentrations of reactants and products no longer change with time.* This apparent cessation of chemical activity occurs because *all reactions are reversible.* Let's examine a chemical system at the macroscopic and molecular levels to see how the equilibrium state arises. The system consists of two gases, colorless dinitrogen tetraoxide and brown nitrogen dioxide:

$$N_2O_4(g; \text{colorless}) \rightleftharpoons 2NO_2(g; \text{brown})$$

When we introduce some $N_2O_4(l)$ into a sealed flask kept at 100°C, a change occurs immediately. The liquid vaporizes (bp = 21°C) and the gas begins to turn

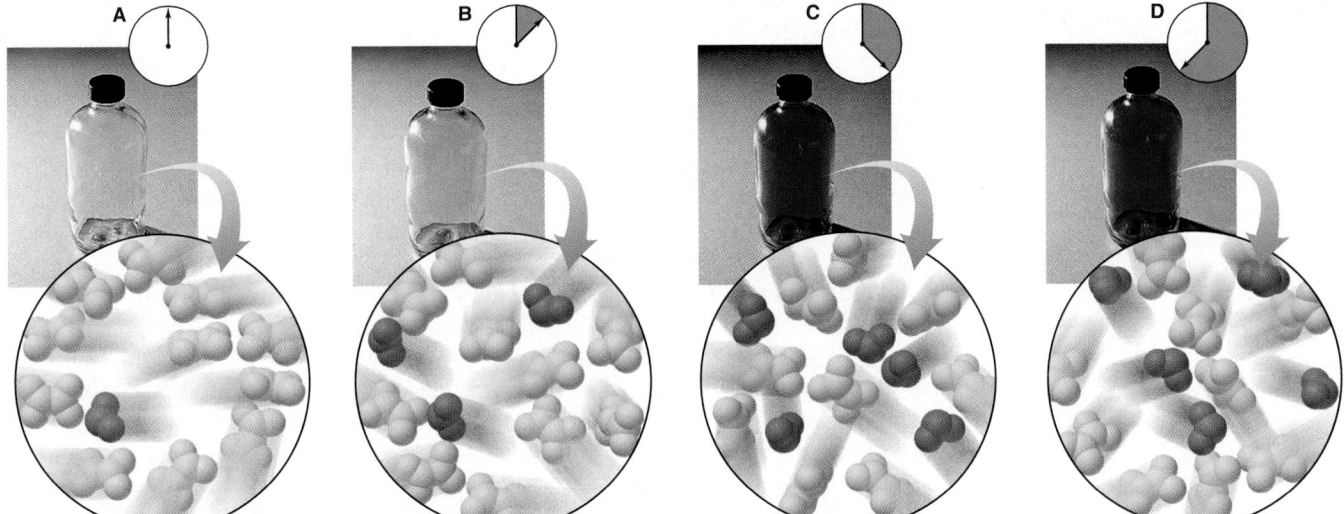

Figure 17.1 Reaching equilibrium on the macroscopic and molecular levels. A, When the experiment begins, the reaction mixture consists mostly of colorless N_2O_4. **B,** As N_2O_4 decomposes to reddish brown NO_2, the color of the mixture becomes pale brown. **C,** When equilibrium is reached, the concentrations of NO_2 and N_2O_4 are constant, and the color reaches its final color. **D,** Because the reaction continues in the forward and reverse directions at equal rates, the concentrations (and color) remain constant.

pale brown. The color slowly darkens, but after a few moments, no further color change can be seen (Figure 17.1).

As we close in on the molecular level, a much more active scene unfolds. The N_2O_4 molecules fly wildly throughout the flask, a few splitting into two NO_2 molecules. As time passes, more N_2O_4 molecules decompose and the concentration of NO_2 rises. As observers in the macroscopic world, we see the flask contents darken, because NO_2 is reddish brown. As the number of N_2O_4 molecules decreases, N_2O_4 decomposition slows. At the same time, increasing numbers of NO_2 molecules collide and combine, so re-formation of N_2O_4 speeds up. Eventually, N_2O_4 molecules decompose as fast as NO_2 molecules combine. The system has reached equilibrium: *reactant and product concentrations stop changing because the forward and reverse rates have become equal,*

$$\text{At equilibrium: } \text{rate}_{\text{fwd}} = \text{rate}_{\text{rev}} \tag{17.1}$$

Thus, a system at equilibrium continues to be dynamic at the molecular level, but we see *no further **net** change because changes in one direction are balanced by changes in the other.*

At a particular temperature, when the system reaches equilibrium, product and reactant concentrations do not change. Therefore, the ratio of these concentrations must be a constant. We'll use the rate laws for the N_2O_4-NO_2 system to derive this constant. At equilibrium, we have

$$\text{rate}_{\text{fwd}} = \text{rate}_{\text{rev}}$$

In this case, both forward and reverse reactions are elementary steps (Section 16.7), so we can write their rate laws directly from the balanced equation:

$$k_{\text{fwd}}[N_2O_4]_{\text{eq}} = k_{\text{rev}}[NO_2]^2_{\text{eq}}$$

where k_{fwd} and k_{rev} are the forward and reverse rate constants, respectively, and the subscript "eq" refers to concentrations at equilibrium. By rearranging, we set the ratio of the rate constants equal to the ratio of the concentration terms:

$$\frac{k_{\text{fwd}}}{k_{\text{rev}}} = \frac{[NO_2]^2_{\text{eq}}}{[N_2O_4]_{\text{eq}}}$$

The ratio of constants gives rise to a new overall constant called the **equilibrium constant (K):**

$$K = \frac{k_{\text{fwd}}}{k_{\text{rev}}} = \frac{[NO_2]^2_{eq}}{[N_2O_4]_{eq}} \qquad (17.2)$$

The equilibrium constant K is a number equal to a particular ratio of equilibrium product and reactant concentrations at a particular temperature. We examine this idea closely in the next section and show that it holds as well for overall reactions made up of several elementary steps.

Remember, it is the opposing *rates* that are equal at equilibrium, not necessarily the concentrations. *The magnitude of K is an indication of how far a reaction proceeds toward product at a given temperature.* Indeed, different reactions, even at the same temperature, have a wide range of concentrations at equilibrium—from almost all reactant to almost all product—and, therefore, they have a wide range of equilibrium constants (Figure 17.2). Here are three examples:

1. *Small K.* If a reaction goes very little toward product before reaching equilibrium, it has a small K, and we may even say there is "no reaction." For example, the oxidation of nitrogen barely proceeds at 1000 K:*

$$N_2(g) + O_2(g) \rightleftharpoons 2NO(g) \qquad K = 1\times10^{-30}$$

2. *Large K.* Conversely, if a reaction reaches equilibrium with very little reactant remaining, it has a large K, and we say it "goes to completion." The oxidation of carbon monoxide goes to completion at 1000 K:

$$2CO(g) + O_2(g) \rightleftharpoons 2CO_2(g) \qquad K = 2.2\times10^{22}$$

3. *Intermediate K.* When significant amounts of both reactant and product are present at equilibrium, K has an intermediate value, as when bromine monochloride breaks down to its elements at 1000 K:

$$2BrCl(g) \rightleftharpoons Br_2(g) + Cl_2(g) \qquad K = 5$$

---

*To distinguish the equilibrium constant from the Kelvin temperature unit, the equilibrium constant is written with an uppercase italic K, whereas the kelvin is an uppercase roman K. Also, since the kelvin is a unit, it always follows a number.

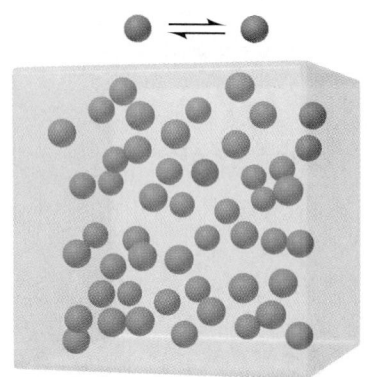

A

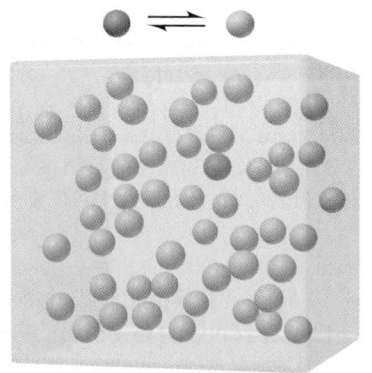

B

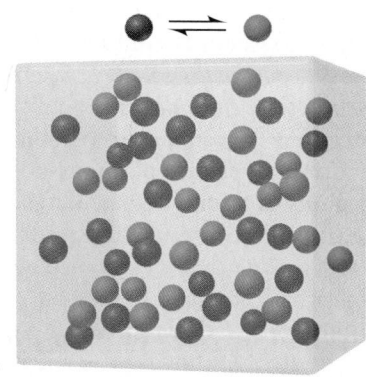

C

Figure 17.2 **The range of equilibrium constants. A,** A system that reaches equilibrium with very little product has a small K. For this reaction, $K = 1/49 = 0.020$. **B,** A system that reaches equilibrium with nearly all product has a large K. For this reaction, $K = 49/1 = 49$. **C,** A system that reaches equilibrium with significant concentrations of reactant and product has an intermediate K. For this reaction, $K = 25/25 = 1.0$.

Kinetics and equilibrium are distinct aspects of a reacting system; that is, rate and yield are not necessarily related. When the forward and reverse reactions occur at the same rate, the system has reached dynamic equilibrium and concentrations no longer change. The equilibrium constant (K) is a number based on a particular ratio of product and reactant concentrations. K is small for reactions that reach equilibrium with a high concentration of reactants and large for reactions that reach equilibrium with a low concentration of reactants.

17.2 THE REACTION QUOTIENT AND THE EQUILIBRIUM CONSTANT

Our discussion of the equilibrium constant in Section 17.1 was based on kinetics. But the fundamental observation of equilibrium studies was stated many years before the principles of kinetics were developed. In 1864, two Norwegian chemists, Cato Guldberg and Peter Waage, observed that *at a given temperature, a chemical system reaches a state in which a particular ratio of reactant and product concentrations has a constant value.* This is one way of stating the **law of chemical equilibrium,** or the **law of mass action.** Note that no mention of rates appears in this statement.

In their discovery of the law of mass action, Guldberg and Waage studied many reactions in which reactant and product concentrations varied widely. They found that, *for a particular system and temperature, the same equilibrium state is attained regardless of how the reaction is run.* For example, in the N_2O_4-NO_2 system at 100°C, we can start with pure N_2O_4, pure NO_2, or any mixture of NO_2 and N_2O_4, and given enough time, the ratio of concentrations attains the same value (within experimental error).

The particular ratio of concentration terms that we write for a given reaction is called the **reaction quotient (Q,** or **mass-action expression)**. (As you'll see shortly, it is based directly on the balanced equation for the reaction.) For the reaction of N_2O_4 to form NO_2 that we discussed previously, the reaction quotient is

$$Q = \frac{[NO_2]^2}{[N_2O_4]}$$

Although the reactant and product concentration *terms* in Q ($[NO_2]$ and $[N_2O_4]$ in this example) remain the same, the numerical *values* of those terms (the actual concentrations of these substances) change during the reaction, so the *numerical value of Q changes.* That is, as the reaction proceeds toward the equilibrium state, there is a continual, smooth change in the concentrations of reactants and products, so the ratio of values that gives Q must change also: at the beginning of the reaction, the concentrations have initial values, and Q has an initial value; a moment later, after the reaction has proceeded a bit, the concentrations have slightly different values, and so does Q; another moment into the reaction, and there is more change in the concentrations and more change in Q; and on and on, *until the reacting system reaches equilibrium.* At that point, at a given temperature, the reactant and product concentrations have reached their equilibrium concentrations and no longer change. Most importantly, the value of Q no longer changes; it has now reached its equilibrium value and equals K at that temperature:

At equilibrium: $Q = K$ (17.3)

So, monitoring Q tells whether the system has reached equilibrium, how far away it is if it has not, and, as we'll discuss later, which direction it is changing

Table 17.1 **Initial and Equilibrium Concentration Ratios for the N_2O_4-NO_2 System at 100°C**

	Initial		Ratio (Q)	Equilibrium		Ratio (K)
Exp't	$[N_2O_4]$	$[NO_2]$	$[NO_2]^2/[N_2O_4]$	$[N_2O_4]_{eq}$	$[NO_2]_{eq}$	$[NO_2]^2_{eq}/[N_2O_4]_{eq}$
1	0.1000	0.0000	0.0000	0.0491	0.1018	0.211
2	0.0000	0.1000	∞	0.0185	0.0627	0.212
3	0.0500	0.0500	0.0500	0.0332	0.0837	0.211
4	0.0750	0.0250	0.00833	0.0411	0.0930	0.210

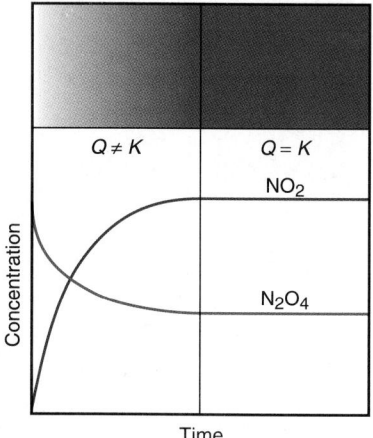

Figure 17.3 **The change in Q during the N_2O_4-NO_2 reaction.** The curved plots and the darkening brown screen above them show that $[N_2O_4]$ and $[NO_2]$, and therefore the value of Q, change with time. Before equilibrium is reached, the concentrations are changing continuously, so $Q \neq K$. Once equilibrium is reached (*vertical line*) and any time thereafter, $Q = K$.

to reach equilibrium. Table 17.1 presents four experiments, each representing a different run of the N_2O_4-NO_2 reaction. The two essential points to note are:

- The value of the ratio of *initial* concentrations varies widely but always gives the same value for the ratio of *equilibrium* concentrations.
- The *individual* equilibrium concentrations are different in each case, but this *ratio* of equilibrium concentrations is constant.

The curves in Figure 17.3 show experiment 1 in Table 17.1. Note that, as indicated by the curves, $[N_2O_4]$ and $[NO_2]$ change smoothly during the course of the reaction and, thus, so does the value of Q. Once the system reaches equilibrium, the concentrations no longer change and Q equals K. In other words, for any given chemical system, K is a special value of Q that occurs when the reactant and product terms have their equilibrium values.

Writing the Reaction Quotient

In Chapter 16, you saw that the rate law for an overall reaction cannot be written from the balanced equation, but must be determined from rate data. In contrast, the reaction quotient *can* be written directly from the balanced equation: *Q is a ratio made up of product concentration terms multiplied together and divided by reactant concentration terms multiplied together, with each term raised to the power of its stoichiometric coefficient.*

The most common form of the reaction quotient shows reactant and product terms as molar concentrations, which are designated by square brackets, [ ]. In these cases, which you've seen so far, K is the *equilibrium constant based on concentrations*, which we designate from now on as K_c. Similarly, we designate the reaction quotient based on concentrations as Q_c. For the general equation

$$aA + bB \rightleftharpoons cC + dD$$

where a, b, c, and d are the stoichiometric coefficients, the reaction quotient is

$$Q_c = \frac{[C]^c[D]^d}{[A]^a[B]^b} \tag{17.4}$$

(Another form of the reaction quotient that we discuss later shows gaseous reactant and product terms as pressures.)

To construct the reaction quotient for any reaction, write the balanced equation first. For the formation of ammonia from its elements, for example, the balanced equation is

$$N_2(g) + 3H_2(g) \rightleftharpoons 2NH_3(g)$$

To construct the reaction quotient, the product term goes in the numerator and the reactant terms go in the denominator multiplied by each other, and each term is raised to the power of its balancing coefficient:

$$Q_c = \frac{[NH_3]^2}{[N_2][H_2]^3}$$

Let's practice this essential skill.

SAMPLE PROBLEM 17.1 Writing the Reaction Quotient from the Balanced Equation

Problem Write the reaction quotient, Q_c, for each of the following reactions:
(a) The decomposition of dinitrogen pentaoxide, $N_2O_5(g) \rightleftharpoons NO_2(g) + O_2(g)$
(b) The combustion of propane gas, $C_3H_8(g) + O_2(g) \rightleftharpoons CO_2(g) + H_2O(g)$
Plan We balance the equations and then construct the reaction quotient as described by Equation 17.4.

Solution (a) $2N_2O_5(g) \rightleftharpoons 4NO_2(g) + O_2(g)$ $\quad Q_c = \dfrac{[NO_2]^4[O_2]}{[N_2O_5]^2}$

(b) $C_3H_8(g) + 5O_2(g) \rightleftharpoons 3CO_2(g) + 4H_2O(g)$ $\quad Q_c = \dfrac{[CO_2]^3[H_2O]^4}{[C_3H_8][O_2]^5}$

Check Always be sure that the exponents in Q are the same as the balancing coefficients. A good check is to reverse the process: turn the numerator into products and the denominator into reactants and change the exponents to coefficients.

FOLLOW-UP PROBLEM 17.1 Write a reaction quotient, Q_c, for each of the following reactions (unbalanced):
(a) The first step in nitric acid production, $NH_3(g) + O_2(g) \rightleftharpoons NO(g) + H_2O(g)$
(b) The disproportionation of nitric oxide, $NO(g) \rightleftharpoons N_2O(g) + NO_2(g)$

Variations in the Form of the Reaction Quotient

As you'll see in the upcoming discussion, the reaction quotient Q is a collection of terms based on the balanced equation *exactly as written* for a given reaction. Therefore, the value of Q, which varies during the reaction, and the value of K, the constant value of Q when the system has reached equilibrium, also depend on how the balanced equation is written.

A Word about Units for Q and K In this text (and most others), *the values of Q and K are shown as unitless numbers.* This is because each term in the reaction quotient represents the *ratio* of the measured quantity of the substance (molar concentration or pressure) to the thermodynamic standard-state quantity of the substance. Recall from Section 6.6 that these standard states are 1 M for a substance in solution, 1 atm for gases, and the pure substance for a liquid or solid. Thus, a concentration of 1.20 M becomes $\dfrac{1.20\ M}{1\ M} = 1.20$; similarly, a pressure of 0.53 atm becomes $\dfrac{0.53\ \text{atm}}{1\ \text{atm}} = 0.53$. (As you'll see, the molar "concentration" of a pure liquid or solid, that is, the number of moles per liter of the substance, is a constant, and the term does not appear in Q at all.) With these quantity terms unitless, the ratio of terms we use to find the value of Q (or K) is also unitless.

Form of Q for an Overall Reaction Notice that we've been writing reaction quotients without knowing whether an equation represents an individual reaction step or an overall multistep reaction. We can do this because we obtain the same expression for the overall reaction as we do when we combine the expressions for the individual steps. That is, *if an overall reaction is the **sum** of two or more reactions, the overall reaction quotient (or equilibrium constant) is the **product** of the reaction quotients (or equilibrium constants) for the steps:*

$$Q_{\text{overall}} = Q_1 \times Q_2 \times Q_3 \times \cdots$$

and

$$K_{\text{overall}} = K_1 \times K_2 \times K_3 \times \cdots \qquad (17.5)$$

Sample Problem 17.2 demonstrates this point.

SAMPLE PROBLEM 17.2 Writing the Reaction Quotient for an Overall Reaction

Smog over Los Angeles.

Problem At the high temperatures reached during the explosive combustion of gasoline within the cylinders of an auto engine, some of the N_2 and O_2 present form nitric oxide, which combines with more O_2 to form nitrogen dioxide, a toxic pollutant that contributes to photochemical smog (see photo):

(1) $N_2(g) + O_2(g) \rightleftharpoons 2NO(g)$ $K_{c1} = 4.3\times10^{-25}$
(2) $2NO(g) + O_2(g) \rightleftharpoons 2NO_2(g)$ $K_{c2} = 6.4\times10^{9}$

(a) Show that the overall Q_c for this reaction sequence is the same as the product of the Q_c's for the individual reactions.
(b) Calculate K_c for the overall reaction.

Plan In **(a)**, we first write the overall reaction by adding the individual reactions and then write the overall Q_c. Next, we write the Q_c for each reaction. Because we *add* the individual steps, we *multiply* their Q_c's and cancel common terms to obtain the overall Q_c. In **(b)**, we are given the individual K_c values (4.3×10^{-25} and 6.4×10^{9}), so we multiply them to find $K_{c(\text{overall})}$.

Solution (a) Writing the overall reaction and its reaction quotient:

$$
\begin{array}{ll}
(1) & N_2(g) + O_2(g) \rightleftharpoons \cancel{2NO(g)} \\
(2) & \cancel{2NO(g)} + O_2(g) \rightleftharpoons 2NO_2(g) \\
\hline
\text{Overall:} & N_2(g) + 2O_2(g) \rightleftharpoons 2NO_2(g)
\end{array}
$$

$$Q_{c(\text{overall})} = \frac{[NO_2]^2}{[N_2][O_2]^2}$$

Writing the reaction quotients for the individual steps:

For step 1,
$$Q_{c1} = \frac{[NO]^2}{[N_2][O_2]}$$

For step 2,
$$Q_{c2} = \frac{[NO_2]^2}{[NO]^2[O_2]}$$

Multiplying the individual reaction quotients and canceling:

$$Q_{c1} \times Q_{c2} = \frac{\cancel{[NO]^2}}{[N_2][O_2]} \times \frac{[NO_2]^2}{\cancel{[NO]^2}[O_2]} = \frac{[NO_2]^2}{[N_2][O_2]^2} = Q_{c(\text{overall})}$$

(b) Calculating the overall K_c:

$$K_{c(\text{overall})} = K_{c1} \times K_{c2} = (4.3\times10^{-25})(6.4\times10^{9}) = 2.8\times10^{-15}$$

Check Round off and check the calculation in part (b):
$$K_c \approx (4\times10^{-25})(6\times10^{9}) = 24\times10^{-16} = 2.4\times10^{-15}$$

FOLLOW-UP PROBLEM 17.2 The following sequence of individual steps has been proposed for the overall reaction between H_2 and Br_2 to form HBr:

(1) $Br_2(g) \rightleftharpoons 2Br(g)$
(2) $Br(g) + H_2(g) \rightleftharpoons HBr(g) + H(g)$
(3) $H(g) + Br(g) \rightleftharpoons HBr(g)$

Write the overall equation and show that the overall Q_c is the product of the Q_c's for the individual steps.

Form of Q for a Forward and Reverse Reaction The form of the reaction quotient depends on the *direction* in which the balanced equation is written. Consider, for example, the oxidation of sulfur dioxide to sulfur trioxide, a key step in acid rain formation and sulfuric acid production (see the photo on the opposite page):

$$2SO_2(g) + O_2(g) \rightleftharpoons 2SO_3(g)$$

The reaction quotient for this equation *as written* is

$$Q_{c(fwd)} = \frac{[SO_3]^2}{[SO_2]^2[O_2]}$$

If we had written the reverse reaction, the decomposition of sulfur trioxide,

$$2SO_3(g) \rightleftharpoons 2SO_2(g) + O_2(g)$$

the reaction quotient would be the *reciprocal* of $Q_{c(fwd)}$:

$$Q_{c(rev)} = \frac{[SO_2]^2[O_2]}{[SO_3]^2} = \frac{1}{Q_{c(fwd)}}$$

Thus, *a reaction quotient (or equilibrium constant) for a forward reaction is the* ***reciprocal*** *of the reaction quotient (or equilibrium constant) for the reverse reaction:*

$$Q_{c(fwd)} = \frac{1}{Q_{c(rev)}} \quad \text{and} \quad K_{c(fwd)} = \frac{1}{K_{c(rev)}} \tag{17.6}$$

The K_c values for the forward and reverse reactions at 1000 K are

$$K_{c(fwd)} = 261 \quad \text{and} \quad K_{c(rev)} = \frac{1}{K_{c(fwd)}} = \frac{1}{261} = 3.83 \times 10^{-3}$$

These values make sense: if the forward reaction goes far to the right (high K_c), the reverse reaction does not (low K_c).

Form of Q for a Reaction with Coefficients Multiplied by a Common Factor
Multiplying all the coefficients of the equation by some factor also changes the form of Q. For example, multiplying all the coefficients in the previous equation for the formation of SO_3 by $\frac{1}{2}$ gives

$$SO_2(g) + \tfrac{1}{2}O_2(g) \rightleftharpoons SO_3(g)$$

For this equation, the reaction quotient is

$$Q'_{c(fwd)} = \frac{[SO_3]}{[SO_2][O_2]^{1/2}}$$

Notice that Q_c for the halved equation equals Q_c for the original equation raised to the $\frac{1}{2}$ power:

$$Q'_{c(fwd)} = Q_{c(fwd)}^{1/2} = \left(\frac{[SO_3]^2}{[SO_2]^2[O_2]}\right)^{1/2} = \frac{[SO_3]}{[SO_2][O_2]^{1/2}}$$

Once again, the same property holds for the equilibrium constants. Relating the halved reaction to the original, we have

$$K'_{c(fwd)} = K_{c(fwd)}^{1/2} = (261)^{1/2} = 16.2$$

Similarly, if you double coefficients, the reaction quotient is the original expression squared; if you triple coefficients, it is the original expression cubed; and so on. It may seem that we have changed the extent of the reaction, as indicated by a change in K, merely by changing the balancing coefficients of the equation, but this clearly cannot be true. *A particular K has meaning only in relation to a particular balanced equation.* In this case, $K_{c(fwd)}$ and $K'_{c(fwd)}$ relate to different equations and thus cannot be compared directly.

In general, *if all the coefficients of the balanced equation are multiplied by some factor, that factor becomes the exponent for relating the reaction quotients and the equilibrium constants.* For a multiplying factor n, which we can write as

$$n(aA + bB \rightleftharpoons cC + dD)$$

the reaction quotient and equilibrium constant are

$$Q' = Q^n = \left(\frac{[C]^c[D]^d}{[A]^a[B]^b}\right)^n \quad \text{and} \quad K' = K^n \tag{17.7}$$

A sulfuric acid manufacturing plant.

SAMPLE PROBLEM 17.3 Determining the Equilibrium Constant for an Equation Multiplied by a Common Factor

Problem For the ammonia-formation reaction,

$$N_2(g) + 3H_2(g) \rightleftharpoons 2NH_3(g)$$

the equilibrium constant, K_c, is 2.4×10^{-3} at 1000 K. If we change the coefficients of this equation, which we'll call the reference (ref) equation, what are the values of K_c for the following balanced equations?

(a) $\frac{1}{3}N_2(g) + H_2(g) \rightleftharpoons \frac{2}{3}NH_3(g)$ **(b)** $NH_3(g) \rightleftharpoons \frac{1}{2}N_2(g) + \frac{3}{2}H_2(g)$

Plan We compare each equation with the reference equation to see how the direction and coefficients have changed. In **(a)**, the equation is the reference equation multiplied by $\frac{1}{3}$, so K_c equals $K_{c(ref)}$ (2.4×10^{-3}) raised to the $\frac{1}{3}$ power. In **(b)**, the equation is one-half the *reverse* of the reference equation, so K_c is the reciprocal of $K_{c(ref)}$ raised to the $\frac{1}{2}$ power.

Solution The reaction quotient for the reference equation is $Q_{c(ref)} = \dfrac{[NH_3]^2}{[N_2][H_2]^3}$.

(a)
$$Q_c = Q_{c(ref)}^{1/3} = \left(\frac{[NH_3]^2}{[N_2][H_2]^3}\right)^{1/3} = \frac{[NH_3]^{2/3}}{[N_2]^{1/3}[H_2]}$$

Thus,
$$K_c = K_{c(ref)}^{1/3} = (2.4\times10^{-3})^{1/3} = 0.13$$

(b)
$$Q_c = \left(\frac{1}{Q_{c(ref)}}\right)^{1/2} = \left(\frac{1}{\dfrac{[NH_3]^2}{[N_2][H_2]^3}}\right)^{1/2} = \frac{[N_2]^{1/2}[H_2]^{3/2}}{[NH_3]}$$

Thus,
$$K_c = \left(\frac{1}{K_{c(ref)}}\right)^{1/2} = \left(\frac{1}{2.4\times10^{-3}}\right)^{1/2} = 20.$$

Check A good check is to work the math backward. For part (a), $(0.13)^3 = 2.2\times10^{-3}$, within rounding of 2.4×10^{-3}. The reaction goes in the same direction, so at equilibrium, it should be mostly reactants, as the $K_c < 1$ indicates. For part (b), $1/(20.)^2 = 2.5\times10^{-3}$, again within rounding. At equilibrium, the reverse reaction should contain mostly products, as $K_c > 1$ indicates.

FOLLOW-UP PROBLEM 17.3 At 1200 K, the reaction of hydrogen and chlorine to form hydrogen chloride is

$$H_2(g) + Cl_2(g) \rightleftharpoons 2HCl(g) \qquad K_c = 7.6\times10^8$$

Calculate K_c for the following reactions:

(a) $\frac{1}{2}H_2(g) + \frac{1}{2}Cl_2(g) \rightleftharpoons HCl(g)$ **(b)** $\frac{4}{3}HCl(g) \rightleftharpoons \frac{2}{3}H_2(g) + \frac{2}{3}Cl_2(g)$

Form of Q for a Reaction Involving Pure Liquids and Solids Until now, we've looked at *homogeneous* equilibria, systems in which all the components of the reaction are in the same phase, such as a system of reacting gases. When the components are in different phases, the system reaches *heterogeneous* equilibrium.

Consider the decomposition of limestone to lime and carbon dioxide, in which a gas and two solids make up the reaction components:

$$CaCO_3(s) \rightleftharpoons CaO(s) + CO_2(g)$$

Based on the rules for writing the reaction quotient, we have

$$Q_c = \frac{[CaO][CO_2]}{[CaCO_3]}$$

A pure solid, however, such as $CaCO_3$ or CaO, always has the same *concentration* at a given temperature, that is, the same number of moles per liter of the solid, just as it has the same density at a given temperature. Moreover, since a solid's volume changes very little with temperature, its concentration also changes very little. For these reasons, the concentration of a pure solid is constant, and the same argument applies to the concentration of a pure liquid.

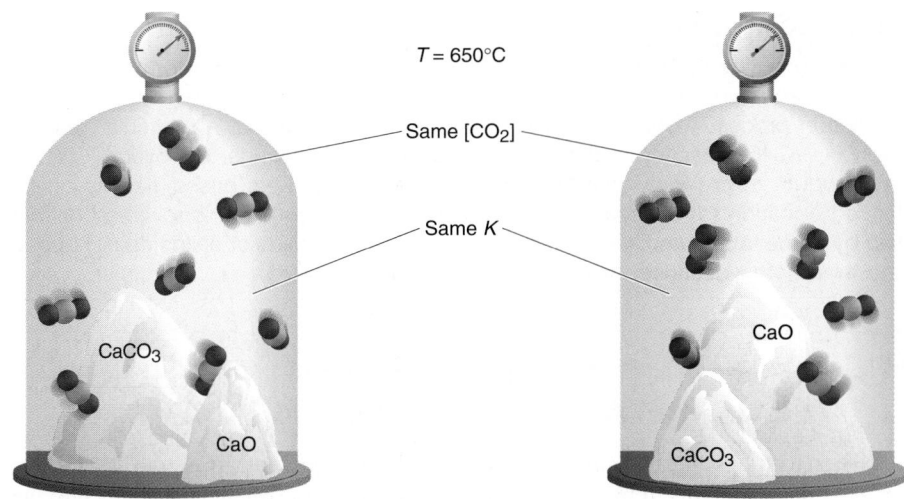

$T = 650°C$

Same $[CO_2]$

Same K

$CaCO_3$

CaO

CaO

$CaCO_3$

Figure 17.4 **The reaction quotient for a heterogeneous system.** Even though the two containers have different amounts of the two solids CaO and $CaCO_3$, as long as both solids are present, at a given temperature, the containers have the same $[CO_2]$ at equilibrium.

Because we are concerned only with concentrations that *change* as they approach equilibrium, *we eliminate the terms for pure liquids and solids from the reaction quotient.* We do this by incorporating their constant concentrations into a rearranged reaction quotient, Q_c'. We multiply both sides of the equation by $[CaCO_3]$ and divide both sides by $[CaO]$. Thus, the only substance whose concentration can change is the gaseous CO_2:

$$Q_c' = Q_c \frac{[CaCO_3]}{[CaO]} = [CO_2]$$

No matter how much CaO and $CaCO_3$ are in the reaction vessel, *so long as some of each is present,* the reaction quotient for the reaction equals the CO_2 concentration (Figure 17.4).

Table 17.2 summarizes the ways of writing Q and calculating K.

Table 17.2 Ways of Expressing Q and Calculating K

Form of Chemical Equation	Form of Q	Value of K
Reference reaction: A $\rightleftharpoons$ B	$Q_{(ref)} = \dfrac{[B]}{[A]}$	$K_{(ref)} = \dfrac{[B]_{eq}}{[A]_{eq}}$
Reverse reaction: B $\rightleftharpoons$ A	$Q = \dfrac{1}{Q_{(ref)}} = \dfrac{[A]}{[B]}$	$K = \dfrac{1}{K_{(ref)}}$
Reaction as sum of two steps: (1) A $\rightleftharpoons$ C (2) C $\rightleftharpoons$ B	$Q_1 = \dfrac{[C]}{[A]}; Q_2 = \dfrac{[B]}{[C]}$ $Q_{overall} = Q_1 \times Q_2 = Q_{(ref)}$ $= \dfrac{[C]}{[A]} \times \dfrac{[B]}{[C]} = \dfrac{[B]}{[A]}$	 $K_{overall} = K_1 \times K_2$ $= K_{(ref)}$
Coefficients multiplied by n	$Q = Q_{(ref)}^n$	$K = K_{(ref)}^n$
Reaction with pure solid or liquid component, such as A(s)	$Q = Q_{(ref)}[A] = [B]$	$K = K_{(ref)}[A] = [B]$

SECTION SUMMARY

SECTION SUMMARY
The reaction quotient, Q, is a particular ratio of product to reactant terms. Substituting experimental values into this expression gives the value of Q, which changes as the reaction proceeds. When the system reaches equilibrium at a particular temperature, $Q = K$. If a reaction is the sum of two or more reactions, the overall Q (or K) is the product of the individual Q's (or K's). The *form* of Q is based directly on the balanced equation for the reaction, so it changes if the equation is reversed or multiplied by some factor, and K changes accordingly. Pure liquids or solids do not appear in the terms of Q because their concentrations are constant.

17.3 EXPRESSING EQUILIBRIA WITH PRESSURE TERMS: RELATION BETWEEN K_c AND K_p

It is easier to measure the pressure of a gas than its concentration and, as long as the gas behaves ideally under the conditions of the experiment, the ideal gas law (Section 5.3) allows us to relate these variables to each other:

$$PV = nRT, \quad \text{so} \quad P = \frac{n}{V}RT \quad \text{or} \quad \frac{P}{RT} = \frac{n}{V}$$

where P is the pressure of a gas and n/V is its molar concentration (M). Thus, with R a constant and T kept constant, *pressure is directly proportional to molar concentration.* When the substances involved in the reaction are gases, we can express the reaction quotient and calculate its value in terms of partial pressures instead of concentrations. For example, in the reaction between gaseous NO and O_2,

$$2NO(g) + O_2(g) \rightleftharpoons 2NO_2(g)$$

the reaction quotient based on partial pressures, Q_p, is

$$Q_p = \frac{P_{NO_2}^2}{P_{NO}^2 \times P_{O_2}}$$

(In later chapters, you'll see cases where some reaction components are expressed as concentrations and others as partial pressures.) The equilibrium constant obtained when all components are present at their equilibrium partial pressures is designated K_p, the *equilibrium constant based on pressures.* In many cases, K_p has a value different from K_c, but the two constants are related; thus, if you know one, you can calculate the other by noting the *change in amount (mol) of gas,* Δn_{gas}, from the balanced equation. Let's see this relationship by converting the terms in Q_c for the NO-O_2 reaction to those in Q_p:

$$2NO(g) + O_2(g) \rightleftharpoons 2NO_2(g)$$

As the balanced equation shows,

3 mol (2 mol + 1 mol) gaseous reactants $\rightleftharpoons$ 2 mol gaseous products

With Δ meaning final minus initial (products minus reactants), we have

$$\Delta n_{gas} = \text{moles of gaseous product} - \text{moles of gaseous reactant} = 2 - 3 = -1$$

Keep this value of Δn_{gas} in mind because it appears in the algebraic conversion that follows. The reaction quotient based on concentrations is

$$Q_c = \frac{[NO_2]^2}{[NO]^2[O_2]}$$

Using the ideal gas law as $n/V = P/RT$, we first express concentrations as n/V and convert them to partial pressures, P; then we collect the RT terms and cancel:

$$Q_c = \frac{\dfrac{n_{NO_2}^2}{V^2}}{\dfrac{n_{NO}^2}{V^2} \times \dfrac{n_{O_2}}{V}} = \frac{\dfrac{P_{NO_2}^2}{(RT)^2}}{\dfrac{P_{NO}^2}{(RT)^2} \times \dfrac{P_{O_2}}{RT}} = \frac{P_{NO_2}^2}{P_{NO}^2 \times P_{O_2}} \times \frac{\dfrac{1}{(RT)^2}}{\dfrac{1}{(RT)^2} \times \dfrac{1}{RT}} = \frac{P_{NO_2}^2}{P_{NO}^2 \times P_{O_2}} \times RT$$

The far right side of the previous expression is Q_p multiplied by RT: $Q_c = Q_p(RT)$. Also, at equilibrium, $K_c = K_p(RT)$; thus, $K_p = \dfrac{K_c}{RT}$, or $K_c(RT)^{-1}$.

Notice that *the exponent of the RT term equals the change in the amount (mol) of gas* (Δn_{gas}) *from the balanced equation,* -1. Thus, in general, we have

$$K_p = K_c(RT)^{\Delta n_{gas}} \qquad (17.8)$$

The units for the partial pressure terms in K_p will be atmospheres, pascals, torr, etc., raised to some power, so the units of R must be consistent with those units. As Equation 17.8 shows, for those reactions in which the amount (mol) of gas does not change, $\Delta n_{gas} = 0$, so $K_p = K_c$.

SAMPLE PROBLEM 17.4 Converting Between K_c and K_p

Problem Calculate K_c for the following, if CO_2 pressure is in atmospheres:

$$CaCO_3(s) \rightleftharpoons CaO(s) + CO_2(g) \qquad K_p = 2.1\times10^{-4} \text{ (at 1000. K)}$$

Plan We know K_p (2.1×10^{-4}), so to convert between K_p and K_c, we must first determine Δn_{gas} from the balanced equation. Then we rearrange Equation 17.8. With gas pressure in atmospheres, R is 0.0821 atm·L/mol·K.

Solution Determining Δn_{gas}: There is 1 mol of gaseous product and none of gaseous reactant, so $\Delta n_{gas} = 1 - 0 = 1$.

Rearranging Equation 17.8 and calculating K_c:

$$K_p = K_c(RT)^1 \qquad \text{so} \qquad K_c = K_p(RT)^{-1}$$

$$K_c = (2.1\times10^{-4})(0.0821 \times 1000.)^{-1} = \boxed{2.6\times10^{-6}}$$

Check Work backward to see whether you obtain the given K_p:

$$K_p = (2.6\times10^{-6})(0.0821 \times 1000.) = 2.1\times10^{-4}$$

FOLLOW-UP PROBLEM 17.4 Calculate K_p for the following reaction:

$$PCl_3(g) + Cl_2(g) \rightleftharpoons PCl_5(g) \qquad K_c = 1.67 \text{ (at 500. K)}$$

SECTION SUMMARY

The reaction quotient and the equilibrium constant can be expressed in terms of concentrations (Q_c and K_c); for gases, they are expressed in terms of partial pressures (Q_p and K_p). The values of K_p and K_c are related by using the ideal gas law: $K_p = K_c(RT)^{\Delta n_{gas}}$.

17.4 REACTION DIRECTION: COMPARING Q AND K

Suppose you start a reaction with a mixture of reactants and products and you know the equilibrium constant at the temperature of the reaction. How do you know if the reaction has reached equilibrium? And, if it hasn't, how do you know in which direction it is progressing to reach equilibrium? The value of Q can change; thus, at any particular time during the reaction, Q can be smaller than K, larger than K, or, when the system reaches equilibrium, equal to K. By comparing the value of Q at a particular time with the known K, you can tell whether the reaction has attained equilibrium or, if not, in which direction it is progressing. With product terms in the numerator of Q and reactant terms in the denominator, *more product makes the ratio of terms larger and more reactant makes the ratio smaller.*

Figure 17.5 Reaction direction and the relative sizes of Q and K. When Q_c is smaller than K_c, the equilibrium system shifts to the right, that is, toward products. When Q_c is larger than K_c, the equilibrium system shifts to the left. Both shifts continue until $Q_c = K_c$. Note that the size of K_c remains the same throughout.

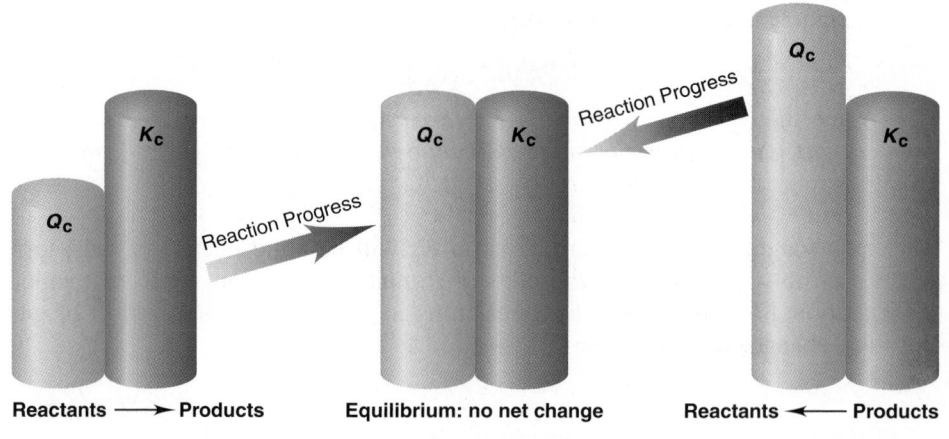

Reactants ⟶ Products   Equilibrium: no net change   Reactants ⟵ Products

The three possible relative sizes of Q and K are shown in Figure 17.5.

• $Q < K$. If the value of Q is smaller than K, the denominator (reactants) is large relative to the numerator (products). For Q to become equal to K, the denominator must decrease and the numerator increase. In other words, the reaction will progress to the right, toward products, until equilibrium is reached:

$$\text{If } Q < K, \text{ reactants} \longrightarrow \text{products}$$

• $Q > K$. If Q is larger than K, the numerator (products) will decrease and the denominator (reactants) increase until equilibrium is reached. Therefore, the reaction will progress to the left, toward reactants:

$$\text{If } Q > K, \text{ reactants} \longleftarrow \text{products}$$

• $Q = K$. This situation exists only when the reactant and product concentrations (or pressures) have attained their equilibrium values. Thus, despite the dynamic processes occurring at the molecular level, no further net change occurs:

$$\text{If } Q = K, \text{ reactants} \rightleftharpoons \text{products}$$

SAMPLE PROBLEM 17.5 Comparing Q and K to Determine Reaction Direction

Problem For the reaction $N_2O_4(g) \rightleftharpoons 2NO_2(g)$, $K_c = 0.21$ at 100°C. At a point during the reaction, $[N_2O_4] = 0.12$ M and $[NO_2] = 0.55$ M. Is the reaction at equilibrium? If not, in which direction is it progressing?
Plan We write the expression for Q_c, find its value by substituting the given concentrations, and then compare its value with the given K_c.
Solution Writing the reaction quotient and solving for Q_c:

$$Q_c = \frac{[NO_2]^2}{[N_2O_4]} = \frac{0.55^2}{0.12} = 2.5$$

With $Q_c > K_c$, the reaction is not at equilibrium and will proceed to the left until $Q_c = K_c$.
Check With $[NO_2] > [N_2O_4]$, we expect to obtain a value for Q_c that is greater than 0.21. If $Q_c > K_c$, the numerator will decrease and the denominator will increase until $Q_c = K_c$; that is, this reaction will proceed toward reactants.

FOLLOW-UP PROBLEM 17.5 Chloromethane forms by the reaction

$$CH_4(g) + Cl_2(g) \rightleftharpoons CH_3Cl(g) + HCl(g)$$

At 1500 K, $K_p = 1.6 \times 10^4$. In the reaction mixture, $P_{CH_4} = 0.13$ atm, $P_{Cl_2} = 0.035$ atm, $P_{CH_3Cl} = 0.24$ atm, and $P_{HCl} = 0.47$ atm. Is CH_3Cl or CH_4 forming?

SECTION SUMMARY
We compare the values of Q and K to determine the direction in which a reaction will proceed toward equilibrium.
- If $Q_c < K_c$, more product forms.
- If $Q_c > K_c$, more reactant forms.
- If $Q_c = K_c$, there is no net change.

As you've seen in this and the previous sections, three criteria define a system at equilibrium:
- Reactant and product concentrations are constant over time.
- The forward reaction rate equals the reverse reaction rate.
- The reaction quotient equals the equilibrium constant: $Q = K$.

17.5 HOW TO SOLVE EQUILIBRIUM PROBLEMS

Many kinds of equilibrium problems arise in the real world, as well as on chemistry exams, but we can group most of them into two types:

1. In one type, we are given equilibrium quantities (concentrations or partial pressures) and solve for K.
2. In the other type, we are given K and initial quantities and solve for the equilibrium quantities.

(From now on, the subscript "eq" is used only when it is not clear that a concentration is an equilibrium value.)

Using Quantities to Determine the Equilibrium Constant

There are two common variations on the type of equilibrium problem in which we solve for K: one involves a straightforward substitution of quantities, and the other requires first finding some of the quantities.

Substituting Given Equilibrium Quantities into Q to Find K The straightforward case is one in which we are given the equilibrium quantities and we must calculate K.

Suppose, for example, that equal amounts of gaseous hydrogen and iodine are injected into a 1.50-L reaction flask at a fixed temperature. In time, the following equilibrium is attained:

$$H_2(g) + I_2(g) \rightleftharpoons 2HI(g)$$

At equilibrium, analysis shows that the flask contains 1.80 mol of H_2, 1.80 mol of I_2, and 0.520 mol of HI. We calculate K_c by finding the concentrations and substituting them into the reaction quotient. From the balanced equation, we write the reaction quotient:

$$Q_c = \frac{[HI]^2}{[H_2][I_2]}$$

We first have to convert the amounts (mol) to concentrations (mol/L), using the flask volume of 1.50 L:

$$[H_2] = \frac{1.80 \text{ mol}}{1.50 \text{ L}} = 1.20 \, M$$

Similarly, $[I_2] = 1.20 \, M$, and $[HI] = 0.347 \, M$. Substituting these values into the expression for Q_c gives K_c:

$$K_c = \frac{(0.347)^2}{(1.20)(1.20)} = 8.36 \times 10^{-2}$$

Using a Reaction Table to Determine Equilibrium Quantities and Find K When some values are not given, we determine them first from stoichiometry and then find K. In the following example, pay close attention to a valuable device being introduced: *the reaction table*.

In a study of carbon oxidation, an evacuated vessel containing a small amount of powdered graphite is heated to 1080 K, and then CO_2 is added to a pressure of 0.458 atm. Once the CO_2 is added, the system starts to produce CO. After equilibrium is reached, the total pressure inside the vessel is 0.757 atm. Calculate K_p.

As always, we start by writing the balanced equation and the reaction quotient:

$$CO_2(g) + C(graphite) \rightleftharpoons 2CO(g)$$

The data are given in atmospheres and we must find K_p, so we write Q in terms of partial pressures (note the absence of the solid, C):

$$Q_p = \frac{P_{CO}^2}{P_{CO_2}}$$

We are given the initial P_{CO_2} and P_{total} at equilibrium. To find K_p, we must find the equilibrium pressures of CO_2 and CO, which requires solving a stoichiometry problem, and then substitute them into the expression for Q_p.

Let's stop for a moment before we begin the calculations to think through what happened in the vessel. An unknown portion of the CO_2 reacted with graphite to form an unknown amount of CO. We know the *relative* amounts of CO_2 and CO from the balanced equation: for 1 mol of CO_2 that reacts, 2 mol of CO forms, which means that when x atm of CO_2 reacts, $2x$ atm of CO forms:

$$x \text{ atm } CO_2 \longrightarrow 2x \text{ atm CO}$$

From the data given, we already know something about the size of K_p. If it were very large, almost all the CO_2 would be converted to CO, so the final (total) pressure would be twice the initial pressure [2(0.458 atm) ≈ 1 atm]. On the other hand, if K_p were very small, almost no CO would form, so the final pressure would be close to the initial pressure, 0.458 atm. However, the final pressure (0.757 atm) is between these extremes, so K_p must have an intermediate value. The pressure of CO_2 at equilibrium, $P_{CO_2(eq)}$, is the initial pressure, $P_{CO_2(init)}$, *minus* the pressure of the CO_2 that reacts, x:

$$P_{CO_2(init)} - x = P_{CO_2(eq)}$$

Similarly, the pressure of CO at equilibrium, $P_{CO(eq)}$, is the initial pressure, $P_{CO(init)}$, *plus* the pressure of the CO that forms, $2x$. Because $P_{CO(init)}$ is zero at the beginning of the reaction, we have

$$P_{CO(init)} + 2x = 0 + 2x = 2x = P_{CO(eq)}$$

A useful way to summarize this information is with a reaction table that shows the balanced equation and what we know about

- the *initial* quantities (concentrations or pressures) of reactants and products
- the *changes* in these quantities during the reaction
- the *equilibrium* quantities

Pressure (atm)	$CO_2(g)$	+	C(graphite)	$\rightleftharpoons$	$2CO(g)$
Initial	0.458		—		0
Change	$-x$		—		$+2x$
Equilibrium	$0.458 - x$		—		$2x$

Note that we *add the initial value to the change to obtain the equilibrium value* in each column. Note also that we include data *only for those substances whose concentrations change*; thus, in this case, the graphite column is blank. We use reaction tables in many of the equilibrium problems here and in later chapters.

To solve for K_p, we substitute equilibrium values into the reaction quotient, so we have to find x. To do so, we use the other piece of data given, P_{total}. According to Dalton's law of partial pressures and using the quantities from the bottom (equilibrium) row of the reaction table,

$$P_{total} = 0.757 \text{ atm} = P_{CO_2(eq)} + P_{CO(eq)} = (0.458 \text{ atm} - x) + 2x$$

Thus,

$$0.757 \text{ atm} = 0.458 \text{ atm} + x \quad \text{and} \quad x = 0.299 \text{ atm}$$

With x known, we determine the equilibrium partial pressures:

$$P_{CO_2(eq)} = 0.458 \text{ atm} - x = 0.458 \text{ atm} - 0.299 \text{ atm} = 0.159 \text{ atm}$$
$$P_{CO(eq)} = 2x = 2(0.299 \text{ atm}) = 0.598 \text{ atm}$$

Now, we substitute these values into the expression for Q_p to find K_p:

$$Q_p = \frac{P_{CO(eq)}^2}{P_{CO_2(eq)}} = \frac{0.598^2}{0.159} = 2.25 = K_p$$

As we predicted, K_p is neither very large nor very small.

SAMPLE PROBLEM 17.6 Calculating K_c from Concentration Data

Problem In a study of hydrogen halide decomposition, a researcher fills an evacuated 2.00-L flask with 0.200 mol of HI gas and allows the reaction to proceed at 453°C:

$$2HI(g) \rightleftharpoons H_2(g) + I_2(g)$$

At equilibrium, [HI] = 0.078 M. Calculate K_c.

Plan To calculate K_c, we need the equilibrium concentrations. We can find the initial [HI] from the amount (0.200 mol) and the flask volume (2.00 L), and we are given [HI] at equilibrium (0.078 M). From the balanced equation, when $2x$ mol of HI reacts, x mol of H_2 and x mol of I_2 form. We set up a reaction table, use the known [HI] at equilibrium to solve for x (the [H_2] or [I_2] that forms), and substitute the concentrations into Q_c.

Solution Calculating initial [HI]:

$$[HI] = \frac{0.200 \text{ mol}}{2.00 \text{ L}} = 0.100 \ M$$

Setting up the reaction table, with $x = $ [H_2] and [I_2] that form and $2x = $ [HI] that reacts:

Concentration (M)	2HI(g)	$\rightleftharpoons$	H₂(g)	+	I₂(g)
Initial	0.100		0		0
Change	$-2x$		$+x$		$+x$
Equilibrium	$0.100 - 2x$		x		x

Solving for x, using the known [HI] at equilibrium:

$$[HI] = 0.100 \ M - 2x = 0.078 \ M$$
$$x = 0.011 \ M$$

Therefore, the equilibrium concentrations are

$$[H_2] = [I_2] = 0.011 \ M \quad \text{and} \quad [HI] = 0.078 \ M$$

Substituting into the reaction quotient:

$$Q_c = \frac{[H_2][I_2]}{[HI]^2}$$

Thus,

$$K_c = \frac{(0.011)(0.011)}{0.078^2} = \boxed{0.020}$$

Check Rounding gives ~$0.01^2/0.08^2 = 0.02$. Because the initial [HI] of 0.100 M fell slightly at equilibrium to 0.078 M, relatively little product formed; so we expect $K_c < 1$.

FOLLOW-UP PROBLEM 17.6 The atmospheric oxidation of nitric oxide, $2NO(g) + O_2(g) \rightleftharpoons 2NO_2(g)$, was studied at 184°C with initial pressures of 1.000 atm of NO and 1.000 atm of O_2. At equilibrium, $P_{O_2} = 0.506$ atm. Calculate K_p.

Using the Equilibrium Constant to Determine Quantities

Like the type of problem that involves finding K, the type that involves finding equilibrium concentrations (or pressures) has several variations. Sample Problem 17.7 is one variation, in which we know K and some of the equilibrium concentrations and must find another equilibrium concentration.

SAMPLE PROBLEM 17.7 Determining Equilibrium Concentrations from K_c

---

Problem In a study of the conversion of methane to other fuels, a chemical engineer mixes gaseous CH_4 and H_2O in a 0.32-L flask at 1200 K. At equilibrium, the flask contains 0.26 mol of CO, 0.091 mol of H_2, and 0.041 mol of CH_4. What is $[H_2O]$ at equilibrium? $K_c = 0.26$ for the equation

$$CH_4(g) + H_2O(g) \rightleftharpoons CO(g) + 3H_2(g)$$

Plan First, we use the balanced equation to write the reaction quotient. We can calculate the equilibrium concentrations from the given numbers of moles and the flask volume (0.32 L). Substituting these into Q_c and setting it equal to the given K_c (0.26), we solve for the unknown equilibrium concentration, $[H_2O]$.

Solution Writing the reaction quotient:

$$CH_4(g) + H_2O(g) \rightleftharpoons CO(g) + 3H_2(g) \qquad Q_c = \frac{[CO][H_2]^3}{[CH_4][H_2O]}$$

Determining the equilibrium concentrations:

$$[CH_4] = \frac{0.041 \text{ mol}}{0.32 \text{ L}} = 0.13 \ M$$

Similarly, $[CO] = 0.81 \ M$ and $[H_2] = 0.28 \ M$.

Calculating $[H_2O]$ at equilibrium: Since $Q_c = K_c$, rearranging gives

$$[H_2O] = \frac{[CO][H_2]^3}{[CH_4]K_c} = \frac{(0.81)(0.28)^3}{(0.13)(0.26)} = 0.53 \ M$$

Check Always check by substituting the concentrations into Q_c to confirm K_c:

$$Q_c = \frac{[CO][H_2]^3}{[CH_4][H_2O]} = \frac{(0.81)(0.28)^3}{(0.13)(0.53)} = 0.26 = K_c$$

FOLLOW-UP PROBLEM 17.7 Nitric oxide, oxygen, and nitrogen react by the following equation: $2NO(g) \rightleftharpoons N_2(g) + O_2(g)$; $K_c = 2.3 \times 10^{30}$ at 298 K. In the atmosphere, $P_{O_2} = 0.209$ atm and $P_{N_2} = 0.781$ atm. What is the equilibrium partial pressure of NO in the air we breathe? [*Hint*: You need K_p to find the partial pressure.]

In a somewhat more involved variation, we know K and *initial* quantities and must find *equilibrium* quantities, for which we use a reaction table. In the upcoming sample problem, the amounts were chosen to simplify the math, allowing us to focus more easily on the overall approach.

SAMPLE PROBLEM 17.8 Determining Equilibrium Concentrations from Initial Concentrations and K_c

---

Problem Fuel engineers use the extent of the change from CO and H_2O to CO_2 and H_2 to regulate the proportions of synthetic fuel mixtures. If 0.250 mol of CO and 0.250 mol of H_2O are placed in a 125-mL flask at 900 K, what is the composition of the equilibrium mixture? At this temperature, K_c is 1.56 for the equation

$$CO(g) + H_2O(g) \rightleftharpoons CO_2(g) + H_2(g)$$

Plan We have to find the "composition" of the equilibrium mixture, in other words, the equilibrium concentrations. As always, we use the balanced equation to write the reaction

quotient. We find the initial [CO] and [H$_2$O] from the given amounts (0.250 mol of each) and volume (0.125 L), use the balanced equation to define x and set up a reaction table, substitute into Q_c, and solve for x, from which we calculate the concentrations.
Solution Writing the reaction quotient:

$$CO(g) + H_2O(g) \rightleftharpoons CO_2(g) + H_2(g) \qquad Q_c = \frac{[CO_2][H_2]}{[CO][H_2O]}$$

Calculating initial reactant concentrations:

$$[CO] = [H_2O] = \frac{0.250 \text{ mol}}{0.125 \text{ L}} = 2.00 \text{ } M$$

Setting up the reaction table, with $x = $ [CO] and [H$_2$O] that react:

Concentration (*M*)	CO(*g*)	+	H$_2$O(*g*)	$\rightleftharpoons$	CO$_2$(*g*)	+	H$_2$(*g*)
Initial	2.00		2.00		0		0
Change	$-x$		$-x$		$+x$		$+x$
Equilibrium	$2.00 - x$		$2.00 - x$		x		x

Substituting into the reaction quotient and solving for x:

$$Q_c = \frac{[CO_2][H_2]}{[CO][H_2O]} = \frac{(x)(x)}{(2.00 - x)(2.00 - x)} = \frac{x^2}{(2.00 - x)^2}$$

At equilibrium, we have

$$Q_c = K_c = 1.56 = \frac{x^2}{(2.00 - x)^2}$$

We can apply the following math shortcut in this case *but not in general*: Because the right side of the equation is a perfect square, we take the square root of both sides:

$$\sqrt{1.56} = \frac{x}{2.00 - x} = \pm 1.25$$

A positive number (1.56) has a positive *and* a negative square root, but *only the positive root has any chemical meaning*, so we ignore the negative root:*

$$1.25 = \frac{x}{2.00 - x} \qquad \text{or} \qquad 2.50 - 1.25x = x$$

So

$$2.50 = 2.25x; \qquad \text{therefore,} \qquad x = 1.11 \text{ } M$$

Calculating equilibrium concentrations:

$$[CO] = [H_2O] = 2.00 \text{ } M - x = 2.00 \text{ } M - 1.11 \text{ } M = \boxed{0.89 \text{ } M}$$
$$[CO_2] = [H_2] = x = \boxed{1.11 \text{ } M}$$

Check From the intermediate size of K_c, it makes sense that the changes in concentration are moderate. It's a good idea to check that the sign of x in the reaction table makes sense—only reactants were initially present, so the change had to proceed to the right: x is the change in concentration, so it has a negative sign for reactants and a positive sign for products. Also check that the equilibrium concentrations give the known K_c: $\frac{(1.11)(1.11)}{(0.89)(0.89)} = 1.56$.

FOLLOW-UP PROBLEM 17.8 The decomposition of HI at low temperature was studied by injecting 2.50 mol of HI into a 10.32-L vessel at 25°C. What is [H$_2$] at equilibrium for the reaction $2HI(g) \rightleftharpoons H_2(g) + I_2(g)$; $K_c = 1.26 \times 10^{-3}$?

---

*The negative root gives $-1.25 = \frac{x}{2.00 - x}$, or $-2.50 + 1.25x = x$.

So $\qquad -2.50 = -0.25x$, and $x = 10.$ *M*

This value has no chemical meaning because we started with 2.00 *M* of each reactant, so it is impossible for 10. *M* to react. Moreover, the square root of an equilibrium constant is another equilibrium constant, which cannot have a negative value.

Using the Quadratic Formula to Solve for the Unknown The shortcut that we used to simplify the math in Sample Problem 17.8 is a special case that occurs when the numerator and denominator of the reaction quotient are perfect squares. It worked because we started with equal concentrations of the two reactants, but that is not ordinarily the case.

Suppose, for example, we instead start the reaction in the sample problem with 2.00 M CO and 1.00 M H_2O. The reaction table is

Concentration (M)	$CO(g)$	+	$H_2O(g)$	$\rightleftharpoons$	$CO_2(g)$	+	$H_2(g)$
Initial	2.00		1.00		0		0
Change	$-x$		$-x$		$+x$		$+x$
Equilibrium	$2.00 - x$		$1.00 - x$		x		x

Substituting these values into Q_c, we obtain

$$Q_c = \frac{[CO_2][H_2]}{[CO][H_2O]} = \frac{(x)(x)}{(2.00 - x)(1.00 - x)} = \frac{x^2}{x^2 - 3.00x + 2.00}$$

At equilibrium, we have

$$1.56 = \frac{x^2}{x^2 - 3.00x + 2.00}$$

To solve for x in this case, we rearrange the previous expression into the form of a *quadratic equation: $ax^2 + bx + c = 0$.* Carrying out the arithmetic gives

$$0.56x^2 - 4.68x + 3.12 = 0$$

where $a = 0.56$, $b = -4.68$, and $c = 3.12$. Then we can find x with the quadratic formula (Appendix A):

$$x = \frac{-b \pm \sqrt{b^2 - 4ac}}{2a}$$

The $\pm$ sign means that we obtain two possible values for x:

$$x = \frac{4.68 \pm \sqrt{(-4.68)^2 - 4(0.56)(3.12)}}{2(0.56)}$$

$$x = 7.6\ M \quad \text{and} \quad x = 0.73\ M$$

Note that only one of the values for x makes sense chemically. The larger value gives negative concentrations at equilibrium (for example, 2.00 M − 7.6 M = −5.6 M), which have no meaning. Therefore, $x = 0.73\ M$, and we have

$$[CO] = 2.00\ M - x = 2.00\ M - 0.73\ M = 1.27\ M$$
$$[H_2O] = 1.00\ M - x = 0.27\ M$$
$$[CO_2] = [H_2] = x = 0.73\ M$$

Checking to see if these values give the known K_c, we have

$$K_c = \frac{(0.73)(0.73)}{(1.27)(0.27)} = 1.6 \text{ (within rounding of 1.56)}$$

Simplifying Assumptions for Finding an Unknown Quantity In many cases, we can use chemical "common sense" to make an assumption that avoids the use of the quadratic formula to find x. In general, *if a reaction has a relatively small K and a relatively large initial reactant concentration, the concentration change (x) can often be neglected* without introducing significant error. This assumption does not mean that $x = 0$, because then there would be no reaction. It means that if a reaction proceeds very little (small K) and if there is a high initial reactant concentration, very little will be used up; therefore, at equilibrium, the reactant concentration will have hardly changed:

$$[reactant]_{init} - x = [reactant]_{eq} \approx [reactant]_{init}$$

You can imagine a similar situation in everyday life. On a bathroom scale, you weigh 158 lb. Take off your wristwatch, and you still weigh 158 lb. Within the precision of the measurement, the weight of the wristwatch is so small compared with your weight that it can be neglected:

Initial body weight − weight of watch = final body weight ≈ initial body weight

Similarly, if the initial concentration of A is, for example, 0.500 M and, because of a small K_c, the concentration of A that reacts is 0.002 M, we can assume that

$$0.500\ M - 0.002\ M = 0.498\ M \approx 0.500\ M$$

that is, $\qquad$ $[A]_{init} - [A]_{reacting} = [A]_{eq} \quad \approx [A]_{init}$ $\qquad$ **(17.9)**

For the assumption that x is negligible to be justified, you must check that the error introduced is not significant. But how much is "significant"? One common, and somewhat arbitrary, criterion is the 5% rule: *if the assumption results in a change (error) in a concentration that is less than 5%, the error is not significant, and the assumption is justified.* Let's go through a sample problem and make this assumption to see how it simplifies the math, and then we'll see if the assumption is justified in the case of two different initial concentrations. We make this assumption often in Chapters 18 and 19.

SAMPLE PROBLEM 17.9 Calculating Equilibrium Concentrations with Simplifying Assumptions

Problem Phosgene is a potent chemical warfare agent that is now outlawed by international agreement. It decomposes by the reaction

$$COCl_2(g) \rightleftharpoons CO(g) + Cl_2(g) \qquad K_c = 8.3 \times 10^{-4}\ \text{(at 360°C)}$$

Calculate [CO], [Cl$_2$], and [COCl$_2$], when the following amounts of phosgene decompose and reach equilibrium in a 10.0-L flask:
(a) 5.00 mol of COCl$_2$ $\qquad$ **(b)** 0.100 mol of COCl$_2$

Plan We know from the balanced equation that when x mol of COCl$_2$ decomposes, x mol of CO and x mol of Cl$_2$ form. We convert amount (5.00 mol or 0.100 mol) to concentration, define x and set up the reaction table, and substitute the values into Q_c. Before using the quadratic formula, we simplify the calculation by assuming that x is negligibly small. After solving for x, we check the assumption and find the concentrations. If the assumption is not justified, we must use the quadratic formula to find x.

Solution (a) For 5.00 mol of COCl$_2$. Writing the reaction quotient:

$$Q_c = \frac{[CO][Cl_2]}{[COCl_2]}$$

Calculating initial [COCl$_2$]:

$$[COCl_2]_{init} = \frac{5.00\ \text{mol}}{10.0\ \text{L}} = 0.500\ M$$

Setting up the reaction table, with $x = [COCl_2]_{reacting}$:

Concentration (M)	COCl$_2$(g) $\rightleftharpoons$	CO(g) +	Cl$_2$(g)
Initial	0.500	0	0
Change	−x	+x	+x
Equilibrium	0.500 − x	x	x

If we use the equilibrium values in Q_c, we obtain

$$Q_c = \frac{[CO][Cl_2]}{[COCl_2]} = \frac{x^2}{0.500 - x} = K_c = 8.3 \times 10^{-4}$$

Because K_c is small, the reaction does not proceed very far to the right, so let's assume that x (the $[COCl_2]$ that reacts) is so much smaller than the initial concentration, 0.500 M, that the equilibrium concentration is nearly the same. Therefore,

$$0.500 \ M - x \approx 0.500 \ M$$

Using this assumption, we substitute and solve for x:

$$K_c = 8.3\times10^{-4} \approx \frac{x^2}{0.500}$$

$$x^2 \approx (8.3\times10^{-4})(0.500) \quad \text{so} \quad x \approx 2.0\times10^{-2}$$

Checking the assumption by finding the percent error:

$$\frac{2.0\times10^{-2}}{0.500} \times 100 = 4\% \text{ is less than } 5\%, \text{ so the assumption is justified}$$

Solving for the equilibrium concentrations:

$$[CO] = [Cl_2] = x = \boxed{2.0\times10^{-2} \ M}$$

$$[COCl_2] = 0.500 \ M - x = \boxed{0.480 \ M}$$

(b) For 0.100 mol of $COCl_2$. The calculation in this case is the same, except that $[COCl_2]_{init} = 0.100 \text{ mol}/10.0 \text{ L} = 0.0100 \ M$. Thus, at equilibrium, we have

$$Q_c = \frac{[CO][Cl_2]}{[COCl_2]} = \frac{x^2}{0.0100 - x} = K_c = 8.3\times10^{-4}$$

Making the assumption that $0.0100 \ M - x \approx 0.0100 \ M$ and solving for x:

$$K_c = 8.3\times10^{-4} \approx \frac{x^2}{0.0100}$$

$$x \approx 2.9\times10^{-3}$$

Checking the assumption:

$$\frac{2.9\times10^{-3}}{0.0100} \times 100 = 29\% \text{ is more than } 5\%, \text{ so the assumption is } not \text{ justified}$$

We must solve the quadratic equation, $x^2 + (8.3\times10^{-4})x - (8.3\times10^{-6}) = 0$, for which the only meaningful value of x is 2.5×10^{-3} (see Appendix A).
Solving for the equilibrium concentrations:

$$[CO] = [Cl_2] = \boxed{2.5\times10^{-3} \ M}$$

$$[COCl_2] = 1.00\times10^{-2} \ M - x = \boxed{7.5\times10^{-3} \ M}$$

Check Once again, the best check is to use the calculated values to be sure you obtain the given K_c.
Comment Note that the assumption was justified at the high initial concentration, but *not* at the low initial concentration.

FOLLOW-UP PROBLEM 17.9 In a study of halogen bond strengths, 0.50 mol of I_2 was heated in a 2.5-L vessel, and the following reaction occurred: $I_2(g) \rightleftharpoons 2I(g)$.
(a) Calculate $[I_2]$ and $[I]$ at equilibrium at 600 K; $K_c = 2.94\times10^{-10}$.
(b) Calculate $[I_2]$ and $[I]$ at equilibrium at 2000 K; $K_c = 0.209$.

Problems Involving Mixtures of Reactants and Products: Determining Reaction Direction In the problems we've worked so far, the direction of the reaction was obvious: with only reactants present, the reaction had to go toward products. Thus, in the reaction tables, we knew that the unknown change in reactant concentration had a negative sign ($-x$) and the change in product concentration had a positive sign ($+x$). Suppose, however, we start with a *mixture* of reactants and products. Whenever the reaction direction is not obvious, we first *compare the value of Q with K to find the direction* in which the reaction proceeds to reach equilibrium. This tells us the sign of x, the unknown change in concentration. (In order to focus on this idea, the next sample problem eliminates the need for the quadratic formula.)

SAMPLE PROBLEM 17.10 Predicting Reaction Direction and Calculating Equilibrium Concentrations

Problem The research and development unit of a chemical company is studying the reaction of CH_4 and H_2S, two components of natural gas:

$$CH_4(g) + 2H_2S(g) \rightleftharpoons CS_2(g) + 4H_2(g)$$

In one experiment, 1.00 mol of CH_4, 1.00 mol of CS_2, 2.00 mol of H_2S, and 2.00 mol of H_2 are mixed in a 250-mL vessel at 960°C. At this temperature, $K_c = 0.036$.
(a) In which direction will the reaction proceed to reach equilibrium?
(b) If $[CH_4] = 5.56\ M$ at equilibrium, what are the equilibrium concentrations of the other substances?
Plan (a) To find the direction, we convert the given initial amounts and volume (0.250 L) to concentrations, calculate Q_c, and compare it with K_c. (b) Based on the results from (a), we determine the sign of each concentration change for the reaction table and then use the known $[CH_4]$ at equilibrium (5.56 M) to determine x and the other equilibrium concentrations.
Solution (a) Calculating the initial concentrations:

$$[CH_4] = \frac{1.00\ \text{mol}}{0.250\ \text{L}} = 4.00\ M$$

Similarly, $[H_2S] = 8.00\ M$, $[CS_2] = 4.00\ M$, and $[H_2] = 8.00\ M$.
Calculating the value of Q_c:

$$Q_c = \frac{[CS_2][H_2]^4}{[CH_4][H_2S]^2} = \frac{(4.00)(8.00)^4}{(4.00)(8.00)^2} = 64.0$$

Comparing Q_c and K_c: $Q_c > K_c$ (64.0 > 0.036), so the reaction goes to the left. Therefore, concentrations of reactants increase and those of products decrease.
(b) Setting up a reaction table, with $x = [CS_2]$ that reacts, which equals $[CH_4]$ that forms:

Concentration (M)	$CH_4(g)$	+	$2H_2S(g)$	$\rightleftharpoons$	$CS_2(g)$	+	$4H_2(g)$
Initial	4.00		8.00		4.00		8.00
Change	$+x$		$+2x$		$-x$		$-4x$
Equilibrium	$4.00 + x$		$8.00 + 2x$		$4.00 - x$		$8.00 - 4x$

Solving for x: At equilibrium,

$$[CH_4] = 5.56\ M = 4.00\ M + x$$

So,

$$x = 1.56\ M$$

Thus,

$$[H_2S] = 8.00\ M + 2x = 8.00\ M + 2(1.56\ M) = 11.12\ M$$
$$[CS_2] = 4.00\ M - x = 2.44\ M$$
$$[H_2] = 8.00\ M - 4x = 1.76\ M$$

Check The comparison of Q_c and K_c showed the reaction proceeding to the left. The given data from part (b) confirm this because $[CH_4]$ increases from 4.00 M to 5.56 M during the reaction. Check that the concentrations give the known K_c:

$$\frac{(2.44)(1.76)^4}{(5.56)(11.12)^2} = 0.0341,\ \text{which is close to 0.036}$$

FOLLOW-UP PROBLEM 17.10 An inorganic chemist studying the reactions of phosphorus halides mixes 0.1050 mol of PCl_5 with 0.0450 mol of Cl_2 and 0.0450 mol of PCl_3 in a 0.5000-L flask at 250°C: $PCl_5(g) \rightleftharpoons PCl_3(g) + Cl_2(g)$; $K_c = 4.2 \times 10^{-2}$.
(a) In which direction will the reaction proceed?
(b) If $[PCl_5] = 0.2065\ M$ at equilibrium, what are the equilibrium concentrations of the other components?

SOLVING EQUILIBRIUM PROBLEMS

PRELIMINARY SETTING UP

1. Write the balanced equation
2. Write the reaction quotient, Q
3. Convert all amounts into the correct units (M or atm)

WORKING ON THE REACTION TABLE

4. When reaction direction is not known, compare Q with K
5. Construct a reaction table

✓ Check the sign of x, the change in the quantity

SOLVING FOR x AND EQUILIBRIUM QUANTITIES

6. Substitute the quantities into Q
7. To simplify the math, assume that x is negligible ($[A]_{init} - x = [A]_{eq} \approx [A]_{init}$)
8. Solve for x

✓ Check that assumption is justified (< 5% error). If not, solve quadratic equation for x.

9. Find the equilibrium quantities

✓ Check to see that calculated values give the known K

Figure 17.6 Steps in solving equilibrium problems. These nine steps, grouped into three tasks, provide a useful approach to calculating equilibrium quantities, given initial quantities and K.

By this time, you've seen quite a few variations on the type of equilibrium problem in which you know K and some initial quantities and must find the equilibrium quantities. Figure 17.6 presents a useful summary of the steps involved in solving these types of equilibrium problems. A good way to organize the steps is to group them into three overall parts.

SECTION SUMMARY

In most equilibrium problems, we use quantities (concentrations or pressures) of reactants and products to find K, or we use K to find quantities. We use a reaction table to summarize the initial quantities, how they change, and the equilibrium quantities. When K is small and the initial quantity of reactant is large, we assume that the unknown quantity of material that reacts (x) is so much smaller than the initial quantity that it can be neglected when added to or subtracted from the initial quantity. If this assumption is not justified (that is, if the error is greater than 5%), we use the quadratic formula to find x.

17.6 REACTION CONDITIONS AND THE EQUILIBRIUM STATE: LE CHÂTELIER'S PRINCIPLE

The most remarkable feature of a system at equilibrium is its ability to return to equilibrium after a change in conditions moves it away from that state. This drive to reattain equilibrium is stated in **Le Châtelier's principle:** when a chemical system at equilibrium is disturbed, it reattains equilibrium by undergoing a net reaction that reduces the effect of the disturbance.

Two phrases in this statement need further explanation. First, what does it mean to "disturb" a system? At equilibrium, Q equals K. When a change in conditions forces the system temporarily out of equilibrium ($Q \neq K$), we say that the system has been stressed, or disturbed. Three common disturbances are a change in concentration of a component (that appears in Q), a change in pressure (caused by a change in volume), or a change in temperature. We'll discuss each below. The other phrase, "net reaction," is often referred to as a shift in the *equilibrium position* of the system to the right or left. The equilibrium position is just the specific equilibrium concentrations (or pressures). A shift in the equilibrium position to the right means that there is a net reaction to the right (reactant to product) until equilibrium is reattained; a shift to the left means that there is a net reaction to the left (product to reactant). Thus, when a disturbance occurs, we say that the equilibrium position shifts, which means that *concentrations (or pressures) change in a way that reduces the disturbance, and the system attains a new equilibrium ($Q = K$ again).*

Le Châtelier's principle allows us to predict the direction of the shift in equilibrium position. Most importantly, it helps research and industrial chemists create conditions that maximize yields. Let's examine each of the three kinds of disturbances to see how a system at equilibrium responds; then, we'll note the effect, if any, of a catalyst. In the following discussion, we focus on the reversible gaseous reaction between phosphorus trichloride and chlorine to produce phosphorus pentachloride:

$$PCl_3(g) + Cl_2(g) \rightleftharpoons PCl_5(g)$$

However, the basis of Le Châtelier's principle holds for any system at equilibrium, whether in the natural or social sciences. (See the margin note on the opposite page.)

The Effect of a Change in Concentration

When a system at equilibrium is disturbed by a change in concentration of one of the components, the system reacts in the direction that reduces the change:

- If the concentration increases, the system reacts to consume some of it.
- If the concentration decreases, the system reacts to produce some of it.

Of course, the component must be one that appears in Q; thus, pure liquids and solids, which do not appear in Q because their concentrations are constant, are not involved.

At 523 K, the PCl_3-Cl_2-PCl_5 system reaches equilibrium when

$$Q_c = \frac{[PCl_5]}{[PCl_3][Cl_2]} = 24.0 = K_c$$

What happens if we now inject some Cl_2 gas, one of the reactants? The system will always act to reduce the disturbance, so it will reduce the increase in reactant by proceeding toward the product side, thereby consuming some additional Cl_2. In terms of the reaction quotient, when we add Cl_2, the $[Cl_2]$ term increases, so the value of Q_c immediately falls as the denominator becomes larger; thus, the system is no longer at equilibrium. As some of the added Cl_2 reacts with some of the PCl_3 present and produces more PCl_5, the denominator becomes smaller once again and the numerator larger, until eventually Q_c again equals K_c. The concentrations of the components have changed, however: the concentrations of Cl_2 and PCl_5 are higher than in the original equilibrium position, and the concentration of PCl_3 is lower. Nevertheless, the ratio of values gives the same K_c. We describe this change by saying that *the equilibrium position shifts to the right when a component on the left is added:*

$$PCl_3 + Cl_2(added) \longrightarrow PCl_5$$

What happens if, instead of adding Cl_2, we could remove some PCl_3, the other reactant? In this case, the system reduces the disturbance (the decrease in reactant), by proceeding toward the reactant side, thereby consuming some PCl_5. Once again, thinking in terms of Q_c, when we remove PCl_3, the $[PCl_3]$ term decreases, the denominator becomes smaller, and the value of Q_c rises above K_c. As some PCl_5 decomposes to PCl_3 and Cl_2, the numerator decreases and the denominator increases until Q_c equals K_c again. Here, too, the concentrations are different from those of the original equilibrium position, but K_c is not. We say that *the equilibrium position shifts to the left when a component on the left is removed:*

$$PCl_3(removed) + Cl_2 \longleftarrow PCl_5$$

The same points we just made for adding or removing a reactant also hold for adding or removing a product. If we add PCl_5, its concentration rises and the equilibrium position shifts to the left, just as it did when we removed some PCl_3; if we remove some PCl_5, the equilibrium position shifts to the right, just as it did when we added some Cl_2. In other words, no matter how the disturbance in concentration comes about, the system responds to make Q_c and K_c equal again. To summarize the effects of concentration changes:

- The equilibrium position shifts to the *right* if a reactant is added or a product is removed.
- The equilibrium position shifts to the *left* if a reactant is removed or a product is added.

In general, whenever the concentration of a component changes, *the equilibrium system reacts to consume some of the added substance or produce some of the removed substance.* In this way, the system "reduces the effect of the disturbance." The effect is not completely eliminated, however, as we will see next from a quantitative comparison of original and new equilibrium positions.

● The Universality of Le Châtelier's Principle Although Le Châtelier limited the scope of his principle to physical and chemical systems, its application is much wider. Ecologists and economists, for example, often see it at work. On the African savannah, the numbers of herbivores (antelope, wildebeest, zebra) and carnivores (lion, cheetah) are in delicate balance. Any disturbance (drought, disease) causes shifts in the relative numbers until they attain a new balance. In its purest form, the economic principle of supply-and-demand is another example of this equilibrium principle. A given demand-supply balance for a product establishes a given price. If either demand or supply is disturbed, the disturbance causes a shift in the other and, therefore, in the prevailing price until a new balance is attained.

Consider the case in which we added Cl_2 to the system at equilibrium. Suppose the original equilibrium position was established with the following concentrations: $[PCl_3] = 0.200\ M$, $[Cl_2] = 0.125\ M$, and $[PCl_5] = 0.600\ M$. Thus,

$$Q_c = \frac{[PCl_5]}{[PCl_3][Cl_2]} = \frac{0.600}{(0.200)(0.125)} = 24.0 = K_c$$

Now we add enough Cl_2 to increase its concentration by $0.075\ M$. Before any reaction occurs, this addition creates a new set of initial concentrations. Then the system reacts and comes to a new equilibrium position. From Le Châtelier's principle, we predict that adding more reactant will produce more product, that is, shift the equilibrium position to the right. Experiment shows that the new $[PCl_5]$ at equilibrium is $0.637\ M$.

Table 17.3 shows a reaction table of the entire process: the original equilibrium position, the disturbance, the (new) initial concentrations, the size and direction of the change needed to reattain equilibrium, and the new equilibrium position, and Figure 17.7 depicts the process.

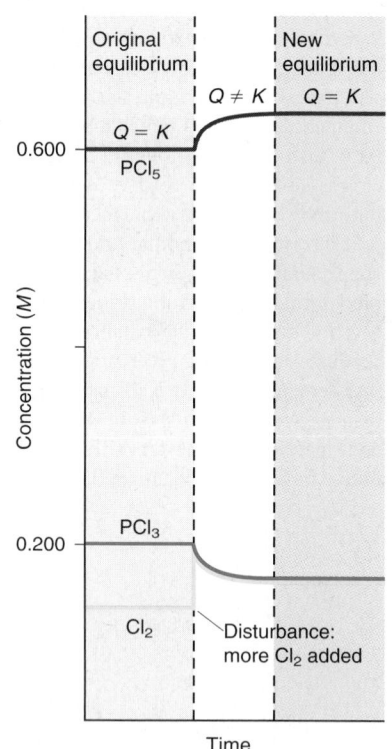

Table 17.3   The Effect of Added Cl_2 on the PCl_3-Cl_2-PCl_5 System

Concentration (M)	$PCl_3(g)$	+	$Cl_2(g)$	$\rightleftharpoons$	$PCl_5(g)$
Original equilibrium	0.200		0.125		0.600
Disturbance			+0.075		
New initial	0.200		0.200		0.600
Change	−x		−x		+x
New equilibrium	0.200 − x		0.200 − x		0.600 + x
					(0.637)*

*Experimentally determined value.

From Table 17.3,

$$[PCl_5] = 0.600\ M + x = 0.637\ M, \qquad \text{so } x = 0.037\ M$$

Also,

$$[PCl_3] = [Cl_2] = 0.200\ M - x = 0.163\ M$$

Therefore, at equilibrium,

$$K_{c(\text{original concentration})} = \frac{0.600}{(0.200)(0.125)} = 24.0$$

$$K_{c(\text{new concentration})} = \frac{0.637}{(0.163)(0.163)} = 24.0$$

There are several key points to notice about the new equilibrium concentrations that exist after Cl_2 is added:

- As we predicted, $[PCl_5]$ (0.637 M) is higher than its original concentration (0.600 M).
- $[Cl_2]$ (0.163 M) is higher than its original equilibrium concentration (0.125 M), but lower than its initial concentration just after the addition (0.200 M); thus, the disturbance (addition of Cl_2) is *reduced but not eliminated*.
- $[PCl_3]$ (0.163 M), the other left-side component, is lower than its original concentration (0.200 M) because some reacted with the added Cl_2.
- Most importantly, although the position of equilibrium shifted to the right, K_c *remains the same*.

Be sure to note that the system adjusts by changing concentrations, but the value of Q_c *at equilibrium* is the same as in the original system. In other words, *at a given temperature, K_c does **not** change with a change in concentration.*

Figure 17.7 The effect of added Cl_2 on the PCl_3-Cl_2-PCl_5 system. In the original equilibrium (*gray region*), all concentrations are constant. When Cl_2 (*yellow curve*) is added, its concentration jumps and then starts to fall as Cl_2 reacts with some PCl_3 to form more PCl_5. After a period of time, equilibrium is re-established at new concentrations (*blue region*) but with the same K.

SAMPLE PROBLEM 17.11 Predicting the Effect of a Change in Concentration on the Equilibrium Position

Problem To improve air quality and obtain a useful product, sulfur is often removed from coal and natural gas by treating the fuel contaminant hydrogen sulfide with O_2:

$$2H_2S(g) + O_2(g) \rightleftharpoons 2S(s) + 2H_2O(g)$$

What happens to
(a) $[H_2O]$ if O_2 is added?   **(b)** $[H_2S]$ if O_2 is added?
(c) $[O_2]$ if H_2S is removed?   **(d)** $[H_2S]$ if sulfur is added?

Plan We write the reaction quotient to see how Q_c is affected by each disturbance, relative to K_c. This effect tells us the direction in which the reaction proceeds for the system to reattain equilibrium and how each concentration changes.

Solution Writing the reaction quotient: $Q_c = \dfrac{[H_2O]^2}{[H_2S]^2[O_2]}$

(a) When O_2 is added, the denominator of Q_c increases, so $Q_c < K_c$. The reaction proceeds to the right until $Q_c = K_c$ again, so [H_2O] increases.

(b) As in part (a), when O_2 is added, $Q_c < K_c$. Some H_2S reacts with the added O_2 as the reaction proceeds to the right, so [H_2S] decreases.

(c) When H_2S is removed, the denominator of Q_c decreases, so $Q_c > K_c$. As the reaction proceeds to the left to re-form H_2S, more O_2 is produced as well, so [O_2] increases.

(d) The concentration of solid S is unchanged as long as some is present, so it does not appear in the reaction quotient. Adding more S has no effect, so [H_2S] is unchanged (but see Comment 2 below).

Check Apply Le Châtelier's principle to see that the reaction proceeds in the direction that lowers the increased concentration or raises the decreased concentration.

Comment 1. As you know, sulfur exists most commonly as S_8. How would this change in formula affect the answers? The balanced equation and Q_c would be

$$8H_2S(g) + 4O_2(g) \rightleftharpoons S_8(s) + 8H_2O(g) \qquad Q_c = \dfrac{[H_2O]^8}{[H_2S]^8[O_2]^4}$$

The value of K_c is different for this equation, but the changes described in the problem have the same effects. For example, in (a), if O_2 were added, the denominator of Q_c would increase, so $Q_c < K_c$. As above, the reaction would proceed to the right until $Q_c = K_c$ again. In other words, changes predicted by the Le Châtelier's principle for a given reaction are not affected by a change in the balancing coefficients.

2. In (d), you saw that adding a solid has no effect on the concentrations of other components: because *the **concentration** of the solid cannot change*, it does not appear in Q. But *the **amount** of solid can change*. Thus, adding H_2S shifts the reaction to the right, and more S forms.

FOLLOW-UP PROBLEM 17.11 In a study of the chemistry of glass etching, an inorganic chemist examines the reaction between sand (SiO_2) and hydrogen fluoride at a temperature above the boiling point of water:

$$SiO_2(s) + 4HF(g) \rightleftharpoons SiF_4(g) + 2H_2O(g)$$

Predict the effect on $[SiF_4]$ when **(a)** $H_2O(g)$ is removed; **(b)** some liquid water is added; **(c)** HF is removed; **(d)** some sand is removed.

The Effect of a Change in Pressure (Volume)

Changes in pressure have significant effects only on equilibrium systems with gaseous components. Aside from phase changes, a change in pressure has a negligible effect on liquids and solids because they are nearly incompressible. Pressure changes can occur in three ways:

- Changing the concentration of a gaseous component
- Adding an inert gas (one that does not take part in the reaction)
- Changing the volume of the reaction vessel

We just considered the effect of changing the concentration of a component, and that reasoning holds here. Next, let's see why *adding an inert gas has no effect on the equilibrium position*. Adding an inert gas does not change the volume, so all *reactant and product concentrations remain the same*. In other words, the volume and number of moles of the reactant and product gases do not change, so their *partial pressures do not change*. Because we use these (unchanged) partial pressures in the reaction quotient, the equilibrium position cannot change. Moreover, the inert gas does not appear in Q, so it cannot have an effect.

On the other hand, changing the pressure by changing the volume often causes a large shift in the equilibrium position. Suppose we let the PCl_3-Cl_2-PCl_5 system come to equilibrium in a cylinder-piston assembly. Then, we press down on the piston to halve the volume: the gas pressure immediately doubles. To reduce this increase in gas pressure, the system responds by reducing the number of gas molecules. And it does so in the only possible way—by shifting the reaction toward the side with *fewer moles of gas*, in this case, toward the product side:

$$PCl_3(g) + Cl_2(g) \longrightarrow PCl_5(g)$$
$$2 \text{ mol gas} \longrightarrow 1 \text{ mol gas}$$

Notice that *a change in volume results in a change in concentration*: a decrease in container volume raises the concentration, and an increase in volume lowers the concentration. Recall that $Q_c = \dfrac{[PCl_5]}{[PCl_3][Cl_2]}$. When the volume is halved, the concentrations double, but the denominator of Q_c is the product of two concentrations, so it quadruples while the numerator only doubles. Thus, Q_c becomes less than K_c. As a result, the system forms more PCl_5 and a new equilibrium position is reached. Because it is just another way to change the concentration, *a change in pressure due to a change in volume does **not** alter K_c*.

Thus, for a system that contains gases at equilibrium, in which the amount (mol) of gas, n_{gas}, changes during the reaction (Figure 17.8):

- If the volume becomes smaller (pressure is higher), the reaction shifts so that the total number of gas molecules decreases.
- If the volume becomes larger (pressure is lower), the reaction shifts so that the total number of gas molecules increases.

Figure 17.8 **The effect of pressure (volume) on an equilibrium system.** The system of gases *(center)* is at equilibrium. For the reaction

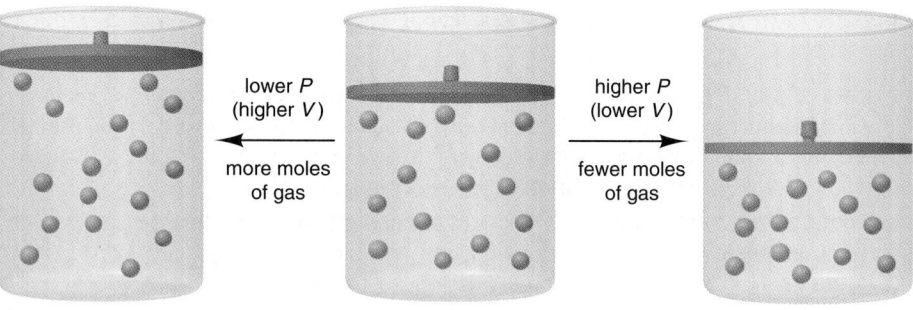

an increase in pressure *(right)* decreases the volume, so the reaction shifts to the right to make fewer molecules. A decrease in pressure (left) increases the volume, so the reaction shifts to the left to make more molecules.

In many cases, however, n_{gas} does not change ($\Delta n_{gas} = 0$). For example,

$$H_2(g) + I_2(g) \rightleftharpoons 2HI(g)$$
$$2 \text{ mol gas} \longrightarrow 2 \text{ mol gas}$$

Q_c has the same number of terms in the numerator and denominator:

$$Q_c = \frac{[HI]^2}{[H_2][I_2]} = \frac{[HI][HI]}{[H_2][I_2]}$$

Therefore, a change in volume has the same effect on the numerator and denominator. Thus, *if $\Delta n_{gas} = 0$, there is no effect on the equilibrium position.*

SAMPLE PROBLEM 17.12 Predicting the Effect of a Change in Volume (Pressure) on the Equilibrium Position

Problem How would you change the volume of each of the following reactions to *increase* the yield of the products?
(a) $CaCO_3(s) \rightleftharpoons CaO(s) + CO_2(g)$ **(b)** $S(s) + 3F_2(g) \rightleftharpoons SF_6(g)$
(c) $Cl_2(g) + I_2(g) \rightleftharpoons 2ICl(g)$
Plan Whenever gases are present, a change in volume causes a change in concentration. For reactions in which the number of moles of gas changes, if the volume decreases (pressure increases), the equilibrium position shifts to relieve the pressure by reducing the number of moles of gas. A volume increase (pressure decrease) has the opposite effect.
Solution (a) The only gas is the product CO_2. To make the system produce more CO_2, we increase the volume (decrease the pressure).
(b) With 3 mol of gas on the left and only 1 mol on the right, we decrease the volume (increase the pressure) to form more SF_6.
(c) The number of moles of gas is the same on both sides of the equation, so a change in volume (pressure) will have no effect on the yield of ICl.
Check Let's predict the relative values of Q_c and K_c. In (a), $Q_c = [CO_2]$, so increasing the volume will make $Q_c < K_c$, and the system will make more CO_2. In (b), $Q_c = [SF_6]/[F_2]^3$. Lowering the volume increases $[F_2]$ and $[SF_6]$ proportionately, but Q_c decreases because of the exponent 3 in the denominator. To make $Q_c = K_c$ again, $[SF_6]$ must increase. In (c), $Q_c = [ICl]^2/[Cl_2][I_2]$. A change in volume (pressure) affects the numerator (2 mol) and denominator (2 mol) equally, so it will have no effect.

FOLLOW-UP PROBLEM 17.12 How would you change the pressure (via a volume change) of the following reaction mixtures to *decrease* the yield of products?
(a) $2SO_2(g) + O_2(g) \rightleftharpoons 2SO_3(g)$ **(b)** $4NH_3(g) + 5O_2(g) \rightleftharpoons 4NO(g) + 6H_2O(g)$
(c) $CaC_2O_4(s) \rightleftharpoons CaCO_3(s) + CO(g)$

The Effect of a Change in Temperature

Of the three types of disturbances—change in concentration, pressure, or temperature—*only temperature changes alter K*. To see why, we must take the heat of reaction into account:

$$PCl_3(g) + Cl_2(g) \rightleftharpoons PCl_5(g) \qquad \Delta H^0_{rxn} = -111 \text{ kJ}$$

The forward reaction is exothermic (releases heat; $\Delta H^0 < 0$), so the reverse reaction is endothermic (absorbs heat; $\Delta H^0 > 0$):

$$PCl_3(g) + Cl_2(g) \longrightarrow PCl_5(g) + \textbf{\textit{heat}} \text{ (exothermic)}$$
$$PCl_3(g) + Cl_2(g) \longleftarrow PCl_5(g) + \textbf{\textit{heat}} \text{ (endothermic)}$$

If we consider *heat as a component of the equilibrium system*, a rise in temperature "adds" heat to the system and a drop in temperature "removes" heat from the system. As with a change in any other component, the system shifts to reduce the effect of the change. Therefore, *a temperature increase (adding heat) favors the endothermic (heat-absorbing) direction, and a temperature decrease (removing heat) favors the exothermic (heat-releasing) direction.*

If we start with the system at equilibrium, Q_c equals K_c. Increase the temperature, and the system responds by decomposing some PCl_5 to PCl_3 and Cl_2, which absorbs the added heat. The denominator of Q_c becomes larger and the numerator smaller, so the system reaches a new equilibrium position at a smaller ratio of concentration terms, that is, a lower K_c. Similarly, the system responds to a drop in temperature by forming more PCl_5 from some PCl_3 and Cl_2, which releases more heat. The numerator of Q_c becomes larger, the denominator smaller, and the new equilibrium position has a higher K_c. Thus,

- A temperature rise will increase K_c for a system with a positive ΔH^0_{rxn}.
- A temperature rise will decrease K_c for a system with a negative ΔH^0_{rxn}.

Let's review these ideas with a sample problem.

LE CHÂTELIER'S PRINCIPLE

SAMPLE PROBLEM 17.13 Predicting the Effect of a Change in Temperature on the Equilibrium Position

Problem How does an *increase* in temperature affect the equilibrium concentration of the underlined substance and K_c for the following reactions?

(a) $CaO(s) + H_2O(l) \rightleftharpoons \underline{Ca(OH)_2}(aq)$ $\Delta H^0 = -82$ kJ

(b) $CaCO_3(s) \rightleftharpoons CaO(s) + \underline{CO_2}(g)$ $\Delta H^0 = 178$ kJ

(c) $\underline{SO_2}(g) \rightleftharpoons S(s) + O_2(g)$ $\Delta H^0 = 297$ kJ

Plan We write each equation to show heat as a reactant or product. Increasing the temperature adds heat, so the system shifts to absorb the heat; that is, the endothermic reaction occurs. K_c will increase if the forward reaction is endothermic and decrease if it is exothermic.

Solution (a) $CaO(s) + H_2O(l) \rightleftharpoons Ca(OH)_2(aq) +$ ***heat***

Adding heat shifts the system to the left: [Ca(OH)_2] and K_c will decrease.

(b) $CaCO_3(s) +$ ***heat*** $\rightleftharpoons CaO(s) + CO_2(g)$

Adding heat shifts the system to the right: [CO_2] and K_c will increase.

(c) $SO_2(g) +$ ***heat*** $\rightleftharpoons S(s) + O_2(g)$

Adding heat shifts the system to the right: [SO_2] will decrease and K_c will increase.

Check You can check your answers by going through the reasoning for a *decrease* in temperature: heat is removed and the exothermic direction is favored. All the answers should be opposite.

FOLLOW-UP PROBLEM 17.13 How does a *decrease* in temperature affect the partial pressure of the underlined substance and the value of K_p for each of the following reactions?

(a) $C(graphite) + 2\underline{H_2}(g) \rightleftharpoons CH_4(g)$ $\Delta H^0 = -75$ kJ

(b) $\underline{N_2}(g) + O_2(g) \rightleftharpoons 2NO(g)$ $\Delta H^0 = 181$ kJ

(c) $P_4(s) + 10Cl_2(g) \rightleftharpoons 4\underline{PCl_5}(g)$ $\Delta H^0 = -1528$ kJ

The van't Hoff Equation: Changes in *K* with Changes in *T* The *van't Hoff equation* shows mathematically how the equilibrium constant is affected by changes in temperature:

$$\ln \frac{K_2}{K_1} = -\frac{\Delta H^0_{rxn}}{R}\left(\frac{1}{T_2} - \frac{1}{T_1}\right) \qquad \textbf{(17.10)}$$

where K_1 is the equilibrium constant at T_1, K_2 is the equilibrium constant at T_2, and R is the universal gas constant (8.314 J/mol·K). If we know ΔH^0_{rxn} and K at one temperature, the van't Hoff equation allows us to find K at any other temperature (or to find ΔH^0_{rxn}, given the two K's at two T's).

Here's a typical problem that requires the van't Hoff equation. Coal gasification processes usually begin with the formation of syngas from carbon and steam:

$$C(s) + H_2O(g) \rightleftharpoons CO(g) + H_2(g) \qquad \Delta H^0_{rxn} = 131 \text{ kJ/mol}$$

An engineer knows that K_p is only 9.36×10^{-17} at 25°C and therefore wants to find a temperature that allows a much higher yield. Calculate K_p at 700°C.

$$\ln \frac{K_2}{K_1} = -\frac{\Delta H^0_{rxn}}{R}\left(\frac{1}{T_2} - \frac{1}{T_1}\right)$$

The temperatures must be converted, so we have

$$\ln\left(\frac{K_{p2}}{9.36 \times 10^{-17}}\right) = -\frac{131 \times 10^3 \text{ J/mol}}{8.314 \text{ J/mol·K}}\left(\frac{1}{973 \text{ K}} - \frac{1}{298 \text{ K}}\right)$$

$$\frac{K_{p2}}{9.36 \times 10^{-17}} = 8.51 \times 10^{15}$$

$$K_{p2} = 0.797$$

Equation 17.10 confirms the qualitative prediction from Le Châtelier's principle: for a temperature rise, we have

$$T_2 > T_1 \quad \text{and} \quad 1/T_2 < 1/T_1, \quad \text{so } 1/T_2 - 1/T_1 < 0$$

Therefore,

- For an endothermic reaction ($\Delta H^0_{rxn} > 0$), the $-(\Delta H^0_{rxn}/R)$ term is < 0. With $1/T_2 - 1/T_1 < 0$, the right side of the equation is > 0. Thus, $\ln(K_2/K_1) > 0$, so $K_2 > K_1$.
- For an exothermic reaction ($\Delta H^0_{rxn} < 0$), the $-(\Delta H^0_{rxn}/R)$ term is > 0. With $1/T_2 - 1/T_1 < 0$, the right side of the equation is < 0. Thus, $\ln(K_2/K_1) < 0$, so $K_2 < K_1$.

(For further practice with the van't Hoff equation, see Problems 17.73 and 17.74.)

The Lack of Effect of a Catalyst

Let's briefly consider a final external change to the reacting system: adding a catalyst. Recall from Chapter 16 that a catalyst speeds up a reaction by providing an alternative mechanism with a lower activation energy, thereby increasing the forward *and* reverse rates to the same extent. In other words, it shortens the time needed to attain the *final* concentrations. Thus, *a catalyst shortens the time it takes to reach equilibrium but has **no** effect on the equilibrium position.*

If, for instance, we add a catalyst to a mixture of PCl_3 and Cl_2 at 523 K, the system will attain the *same* equilibrium concentrations of PCl_3, Cl_2, and PCl_5 *more quickly* than it did without the catalyst. As you'll see in a moment, however, catalysts often play key roles in optimizing reaction systems. The upcoming Chemical Connections essays highlight two areas that apply equilibrium principles: the first concerns a major industrial process, and the second examines metabolic processes in organisms.

Table 17.4 summarizes the effects of changing conditions on the position of equilibrium. Note that many changes alter the equilibrium position, but only temperature changes alter the value of the equilibrium constant.

Temperature-Dependent Systems There is a striking similarity among expressions for the temperature dependence of K, k (rate constant), and P (equilibrium vapor pressure):

$$\ln \frac{K_2}{K_1} = -\frac{\Delta H^0_{rxn}}{R}\left(\frac{1}{T_2} - \frac{1}{T_1}\right)$$

$$\ln \frac{k_2}{k_1} = -\frac{E_a}{R}\left(\frac{1}{T_2} - \frac{1}{T_1}\right)$$

$$\ln \frac{P_2}{P_1} = -\frac{\Delta H_{vap}}{R}\left(\frac{1}{T_2} - \frac{1}{T_1}\right)$$

Each concentration-related term (K, k, or P) is dependent on T through an energy term (ΔH^0_{rxn}, E_a, or ΔH_{vap}, respectively) divided by R. The similarity arises because the equations express the same relationship. K equals a ratio of rate constants, and $E_{a(fwd)} - E_{a(rev)} = \Delta H^0_{rxn}$. In the vapor pressure case, for the phase change $A(l) \rightleftharpoons A(g)$, the heat of reaction is equal to the heat of vaporization: $\Delta H_{vap} = \Delta H^0_{rxn}$. Also, the equilibrium constant equals the equilibrium vapor pressure: $K_p = P_A$.

Table 17.4 Effect of Various Disturbances on an Equilibrium System		
Disturbance	**Net Direction of Reaction**	**Effect on Value of K**
Concentration		
Increase [reactant]	Toward formation of product	None
Decrease [reactant]	Toward formation of reactant	None
Increase [product]	Toward formation of reactant	None
Decrease [product]	Toward formation of product	None
Pressure		
Increase P (decrease V)	Toward formation of fewer moles of gas	None
Decrease P (increase V)	Toward formation of more moles of gas	None
Increase P (add inert gas, no change in V)	None; concentrations unchanged	None
Temperature		
Increase T	Toward absorption of heat	Increases if $\Delta H^0_{rxn} > 0$ Decreases if $\Delta H^0_{rxn} < 0$
Decrease T	Toward release of heat	Increases if $\Delta H^0_{rxn} < 0$ Decreases if $\Delta H^0_{rxn} > 0$
Catalyst added	None; forward and reverse equilibrium attained sooner; rates increase equally	None

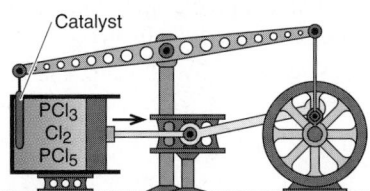

Catalyst

PCl₃
Cl₂
PCl₅

Catalyzed Perpetual Motion? The engine shown consists of a piston attached to a flywheel, whose rocker arm, holding a catalyst, moves in and out of the reaction in the cylinder. What if the catalyst *could* increase the rate of PCl_5 breakdown but not of its formation? In the cylinder, the catalyst would speed up the breakdown of PCl_5 to PCl_3 and Cl_2—1 mol of gas to 2 mol—increasing gas pressure and pushing the piston out. With the catalyst out of the cylinder, PCl_3 and Cl_2 would re-form PCl_5, lowering gas pressure and moving the piston in. The process would supply power with no external input of energy. If a catalyst *did* shift the position of equilibrium, we could design amazing machines!

The Haber Process for the Synthesis of Ammonia

Nitrogen occurs in many essential natural and synthetic compounds. By far the richest source of nitrogen is the atmosphere, where four of every five molecules are N_2. Despite this abundance, the supply of *usable* nitrogen for biological and manufacturing processes is limited because of the low chemical reactivity of N_2. Because of the strong triple bond holding the two N atoms together, the nitrogen atom is very difficult to "fix," that is, to combine with other atoms.

Natural nitrogen fixation occurs either through the elegant specificity of enzymes found in bacteria that live on plant roots or through the brute force of lightning. Nearly 13% of nitrogen fixation on Earth is accomplished industrially through the **Haber process** for the formation of ammonia from its elements:

$$N_2(g) + 3H_2(g) \rightleftharpoons 2NH_3(g) \qquad \Delta H^0_{rxn} = -91.8 \text{ kJ}$$

The process was developed by the German chemist Fritz Haber and first used in 1913. From its humble beginnings in a plant with a capacity of 12,000 tons a year, world production of ammonia has exploded to a current level of more than 110 million tons a year. On a mole basis, more ammonia is produced industrially than any other compound. Over 80% of this ammonia is used in fertilizer applications. In fact, the most common form of fertilizer is compressed anhydrous, liquid NH_3 injected directly into the soil (Figure B17.1). Other uses of NH_3 include the production of explosives, via the formation of HNO_3, and the making of nylons and other polymers. Smaller amounts are used as refrigerants, rubber stabilizers, and household cleaners and in the synthesis of pharmaceuticals and other organic chemicals.

The Haber process provides an excellent opportunity to apply equilibrium principles and see the compromises needed to make an industrial process economically worthwhile. From inspection of the balanced equation, we can see three ways to maximize the yield of ammonia:

1. *Decrease $[NH_3]$.* Ammonia is the product, so removing it as it forms will make the system produce more in a continual drive to reattain equilibrium.

Table B17.1 Effect of Temperature on K_c for Ammonia Synthesis

T (K)	K_c
200.	7.17×10^{15}
300.	2.69×10^{8}
400.	3.94×10^{4}
500.	1.72×10^{2}
600.	4.53×10^{0}
700.	2.96×10^{-1}
800.	3.96×10^{-2}

2. *Decrease volume (increase pressure).* Because 4 mol of gas reacts to form 2 mol of gas, decreasing the volume will shift the equilibrium position toward fewer moles of gas, that is, toward ammonia formation.

3. *Decrease temperature.* Because the formation of ammonia is exothermic, decreasing the temperature (removing heat) will shift the equilibrium position toward the product, thereby increasing K_c (Table B17.1).

Therefore, the ideal conditions for maximizing the yield of ammonia are continual removal of NH_3 as it forms, high pressure, and low temperature. Figure B17.2 shows the percent yield of ammonia at various conditions of pressure and temperature. Note the almost complete conversion (98.3%) to ammonia at 1000 atm and the relatively low temperature of 473 K (200.°C).

Unfortunately, a problem arises that highlights the distinction between the principles of equilibrium and kinetics. Although the *yield* is favored by low temperature, the *rate* of formation is not. In fact, ammonia forms so slowly at low temperature that the process becomes uneconomical. In practice, a compromise is achieved that optimizes yield *and* rate. High pressure and continuous removal are used to increase yield, but the temperature is

Figure B17.1 Liquid ammonia used as fertilizer.

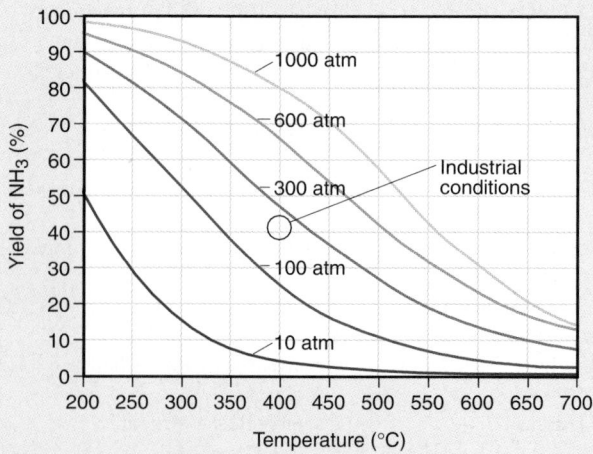

Figure B17.2 Percent yield of ammonia vs. temperature (°C) at five different operating pressures. At very high pressure and low temperature (*top left*), the yield is high, but the rate of formation is low. Industrial conditions (*circle*) are between 200 and 300 atm at about 400°C.

raised to a moderate level and a catalyst is used to increase the rate. Achieving the same rate without a catalyst requires much higher temperatures and results in a much lower yield.

Stages in the industrial production of ammonia are shown schematically in Figure B17.3. To extend equipment life and minimize cost, modern ammonia plants operate at pressures of about 200 to 300 atm and temperatures of around 673 K (400.°C). The catalyst consists of 5-mm to 10-mm chunks of iron crystals embedded in a fused mixture of MgO, Al_2O_3, and SiO_2. The stoi-chiometric ratio of compressed reactant gases (N_2:H_2 = 1:3 by volume) is injected into the heated, pressurized reaction chamber, where it flows over the catalyst beds. Some of the heat needed is supplied by the enthalpy change of the reaction. The emerging equilibrium mixture, which contains about 35% NH_3 by volume, is cooled by refrigeration coils until the NH_3 (boiling point, $-33.4°C$) condenses and is removed. The boiling points of N_2 and H_2 are much lower, they remain gaseous and are recycled by pumps back into the reaction chamber to continue the process.

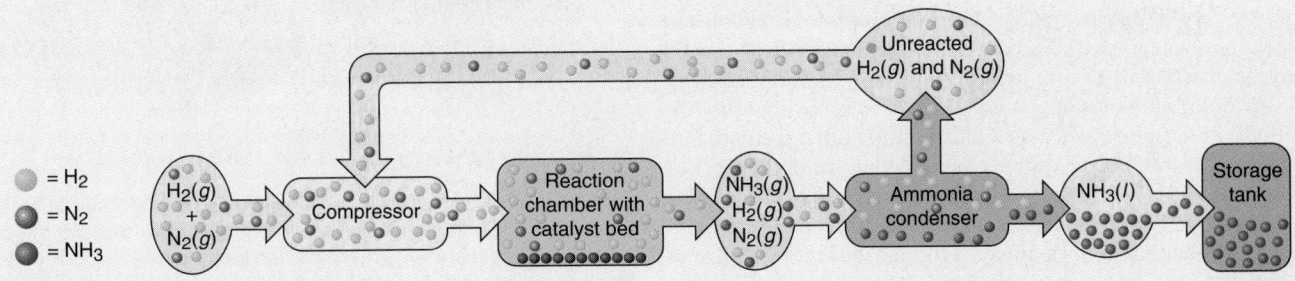

Figure B17.3 Key stages in the Haber process for synthesizing ammonia.

Chemical Connections   Chemistry in Cellular Metabolism
Design and Control of a Metabolic Pathway

Biological cells are microscopic wizards of chemical change. From the simplest bacterium to the most specialized neuron, every cell performs thousands of individual reactions that allow it to grow and reproduce, feed and excrete, move and communicate. Taken together, these chemical reactions constitute the cell's *metabolism.* These myriad feats of biochemical breakdown, synthesis, and energy flow are organized into reaction sequences called **metabolic pathways.** Some pathways disassemble the biopolymers in food into monomer building blocks: sugars, amino acids, and nucleotides. Others extract energy from these small molecules and then use much of the energy to assemble molecules with other functions, such as hormones, neurotransmitters, defensive toxins, and the cell's own biopolymers.

In principle, *each step in a metabolic pathway is a reversible reaction catalyzed by a specific enzyme* (Section 16.8). As you'll see, however, *equilibrium is never reached* in a pathway, even though the principles we discussed earlier apply. As with other multistep reaction sequences, the product of the first reaction becomes the reactant of the second, the product of the second becomes the reactant of the third, and so on. Consider, for example, the five-step pathway shown in Figure B17.4, in which one amino acid, threonine, is converted into another, isoleucine, in the cells of a bread mold. Threonine, supplied from a different region of the cell, forms ketobutyrate through the catalytic action of the first enzyme (reaction 1). The equilibrium position of reaction 1 continuously shifts to the right because the ketobutyrate is continuously removed as the reactant in reaction 2. Similarly, reaction 2 shifts to the right as its product is used in reaction 3. Each subsequent reaction shifts the equilibrium position of the previous reaction in the direction of product. The final product, isoleucine, is typically removed to make proteins elsewhere in the cell. Thus, *the entire pathway operates in one direction.*

(continued)

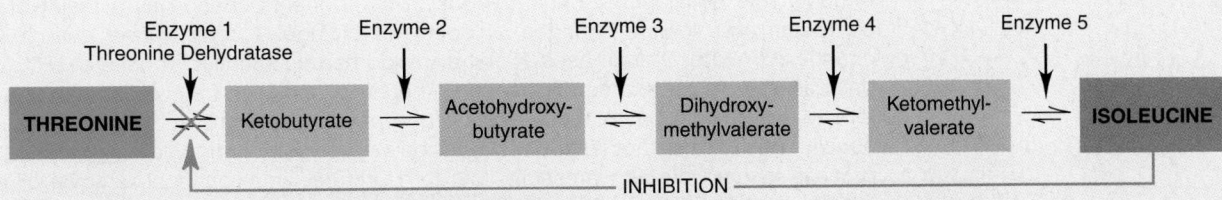

Figure B17.4 The metabolic pathway for the biosynthesis of isoleucine from threonine. Isoleucine is synthesized from threonine in a sequence of five enzyme-catalyzed reactions. As indicated by the unequal equilibrium arrows, each step is reversible but is shifted toward product because each product becomes the reactant of the subsequent step. When the cell's need for isoleucine is temporarily satisfied, this synthesis slows through end-product feedback inhibition: the concentration of isoleucine, the end product, builds up and inhibits the catalytic activity of threonine dehydratase, the first enzyme in the pathway.

This continuous shift in equilibrium position results in two major features of metabolic pathways. First, *each step proceeds with nearly 100% yield.* That is, virtually every molecule of threonine that enters this region of the cell eventually changes to ketobutyrate, every molecule of ketobutyrate to the next product, and so on.

Second, *reactant and product concentrations remain constant* or vary within extremely narrow limits, even though the system never attains equilibrium. This situation, called a *steady state,* is different in a basic way from those we have studied so far. In equilibrium systems, equal reaction rates in *opposing* directions give rise to constant concentrations of reactants and products. In steady-state systems, on the other hand, the rates of reactions in *one* direction—into, through, and out of the system—give rise to nearly constant concentrations of intermediates. Ketobutyrate, for example, is formed via reaction 1 just as fast as it is used via reaction 2, so its concentration is kept constant. (You can establish a steady-state amount of water by filling a sink and then opening the faucet and drain so that water enters just as fast as it leaves.)

It is absolutely critical that a cell regulate the concentration of the final product of a pathway. Recall from Chapter 16 that an enzyme catalyzes a reaction when the substrate (reactant) occupies the *active site.* However, substrate concentrations are so much higher than enzyme concentrations that, if no other factors were involved, the active sites on all the enzyme molecules would always be occupied and all cellular reactions would occur at their maximum rates. Although this might be ideal for an industrial process, it could be catastrophic for an organism. To regulate overall product formation, the rates of certain key steps are controlled by *regulatory enzymes,* which contain an *inhibitor site* on their surface in addition to an active site. The enzyme's three-dimensional structure is such that when the inhibitor site is occupied, the shape of the active site is deformed and the reaction cannot be catalyzed (Figure B17.5).

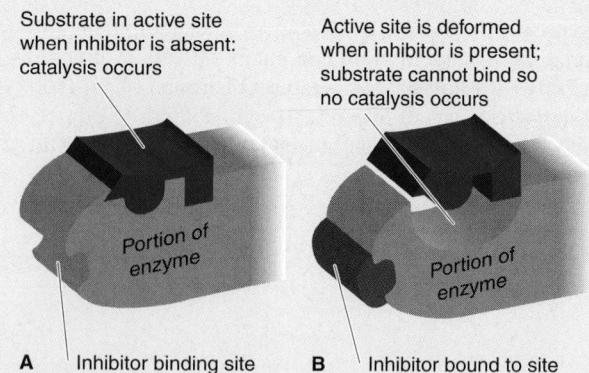

Figure B17.5 **The effect of inhibitor binding on the shape of the active site. A,** When the inhibitor site is not occupied, the enzyme catalyzes the reaction. **B,** When the inhibitor site is occupied, that enzyme molecule does not function.

In the simplest case of metabolic control (see Figure B17.4), *the final product of the pathway is also the inhibitor molecule, and the regulatory enzyme catalyzes the first step.* Suppose, for instance, that a bread mold cell is temporarily making less protein, so the isoleucine synthesized by the pathway in Figure B17.4 is not being removed as quickly. As the concentration of isoleucine rises, the chance increases that isoleucine molecules will land on the inhibitor sites of molecules of the first enzyme in the pathway, thereby inhibiting production of itself. This process is called *end-product feedback inhibition.* More complex pathways have more elaborate regulatory schemes, but many operate through similar types of inhibitory feedback (see Problem 17.111). The exquisitely detailed regulation of a cell's numerous metabolic pathways arises directly from the principles of chemical kinetics and equilibrium.

SECTION SUMMARY

Le Châtelier's principle states that if an equilibrium system is disturbed, it undergoes a net reaction that reduces the disturbance and allows it to reattain equilibrium. Changes in concentration cause a net reaction away from the added component or toward the removed component. For a reaction that involves a change in number of moles of gas, an increase in pressure (decrease in volume) causes a net reaction toward fewer moles of gas, and a decrease in pressure causes the opposite change. Although the equilibrium concentrations of components change as a result of concentration and volume changes, K does not change. A temperature change *does* change K: higher T increases K for an endothermic reaction (positive ΔH^0_{rxn}) and decreases K for an exothermic reaction (negative ΔH^0_{rxn}). A catalyst causes the system to reach equilibrium more quickly by speeding forward and reverse reactions equally, but it does not affect the equilibrium position. Ammonia is produced in a process favored by high pressure, low temperature, and continual removal of product. To make the process economical, an intermediate temperature and a catalyst are used. A metabolic pathway is a cellular reaction sequence in which each step is shifted completely toward product. Its overall yield is controlled by feedback inhibition of certain key enzymes.

Chapter Perspective

The equilibrium phenomenon is central to all natural systems. In this introduction, we discussed the nature of equilibrium on the observable and molecular levels, ways to solve relevant problems, and how conditions affect equilibrium systems. In the next two chapters, we apply these ideas to acids and bases and to other aqueous ionic systems. Then we examine the fundamental relationship between the equilibrium constant and the energy change that accompanies a reaction to understand why chemical reactions occur.

For Review and Reference (Numbers in parentheses refer to pages, unless noted otherwise.)

Learning Objectives

Relevant section and/or sample problem (SP) numbers appear in parentheses.

Understand These Concepts

1. The distinction between the rate and the extent of a reaction (Introduction)
2. Why a system attains dynamic equilibrium when forward and reverse reaction rates are equal (Section 17.1)
3. The equilibrium constant as a number that is equal to a particular ratio of rate constants and of concentration terms (Section 17.1)
4. How the magnitude of K is related to the extent of the reaction (Section 17.1)
5. Why the same equilibrium state is reached no matter what the starting concentrations of the reacting system (Section 17.2)
6. How the reaction quotient (Q) changes continuously until the system reaches equilibrium, at which point $Q = K$ (Section 17.2)
7. Why the form of Q is based exactly on the balanced equation *as written* (Section 17.2)
8. How the *sum* of reaction steps gives the overall reaction, and the *product* of Q's (or K's) gives the overall Q (or K) (Section 17.2)
9. Why pure solids and liquids do not appear in Q (Section 17.2)
10. How the interconversion of K_c and K_p is based on the ideal gas law and Δn_{gas} (Section 17.3)
11. How the reaction direction depends on the relative values of Q and K (Section 17.4)
12. How a reaction table is used to find an unknown quantity (concentration or pressure) (Section 17.5)
13. How assuming that the change in [reactant] is relatively small simplifies finding equilibrium quantities (Section 17.5)
14. How Le Châtelier's principle explains the effects of a change in concentration, pressure (volume), and temperature on a system at equilibrium and on K (Section 17.6)
15. Why a change in temperature *does* affect K (Section 17.6)
16. Why the addition of a catalyst does *not* affect K (Section 17.6)

Master These Skills

1. Writing the reaction quotient (Q) from a balanced equation (SP 17.1)
2. Writing Q and calculating K for a reaction consisting of several steps (SP 17.2)
3. Writing Q and finding K for a reaction multiplied by a common factor (SP 17.3)
4. Writing Q for heterogeneous equilibria (Section 17.2)
5. Converting between K_c and K_p (SP 17.4)
6. Comparing Q and K to determine reaction direction (SP 17.5)
7. Substituting quantities (concentrations or pressures) into Q to find K (Section 17.5)
8. Using a reaction table to determine quantities and find K (SP 17.6)
9. Finding one equilibrium quantity from other equilibrium quantities and K (SP 17.7)
10. Finding an equilibrium quantity from initial quantities and K (SP 17.8)
11. Solving a quadratic equation for an unknown equilibrium quantity (Section 17.5)
12. Assuming that the change in [reactant] is relatively small to find equilibrium quantities and checking the assumption (SP 17.9)
13. Comparing the values of Q and K to find reaction direction and sign of x, the unknown change in a quantity (SP 17.10)
14. Using the relative values of Q and K to predict the effect of a change in concentration on the equilibrium position and on K (SP 17.11)
15. Using Le Châtelier's principle and Δn_{gas} to predict the effect of a change in pressure (volume) on the equilibrium position (SP 17.12)
16. Using Le Châtelier's principle and ΔH^0 to predict the effect of a change in temperature on the equilibrium position and on K (SP 17.13)
17. Using the van't Hoff equation to calculate K at one temperature given K at another temperature (Section 17.6)

Key Terms

Section 17.1
equilibrium constant (K) (716)

Section 17.2
law of chemical equilibrium (law of mass action) (717)

reaction quotient (Q) (mass-action expression) (717)

Section 17.6
Le Châtelier's principle (736)
Haber process (744)
metabolic pathway (745)

Key Equations and Relationships

17.1 Defining equilibrium in terms of reaction rates (715):

$$\text{At equilibrium: } \text{rate}_{\text{fwd}} = \text{rate}_{\text{rev}}$$

17.2 Defining the equilibrium constant for the reaction $A \rightleftharpoons 2B$ (716):

$$K = \frac{k_{\text{fwd}}}{k_{\text{rev}}} = \frac{[B]_{\text{eq}}^2}{[A]_{\text{eq}}}$$

17.3 Defining the equilibrium constant in terms of the reaction quotient (717):

$$\text{At equilibrium: } Q = K$$

17.4 Expressing Q_c for the reaction $aA + bB \rightleftharpoons cC + dD$ (718):

$$Q_c = \frac{[C]^c[D]^d}{[A]^a[B]^b}$$

17.5 Finding the overall K for a reaction sequence (719):

$$K_{\text{overall}} = K_1 \times K_2 \times K_3 \times \cdots$$

17.6 Finding K of a process from K of the reverse process (721):

$$K_{\text{fwd}} = \frac{1}{K_{\text{rev}}}$$

17.7 Finding K of a reaction multiplied by a factor n (721):

$$K' = K^n$$

17.8 Relating K based on pressures to K based on concentrations (725):

$$K_p = K_c(RT)^{\Delta n_{\text{gas}}}$$

17.9 Assuming that ignoring the concentration reacting introduces insignificant error (733):

$$[A]_{\text{init}} - [A]_{\text{reacting}} = [A]_{\text{eq}} \approx [A]_{\text{init}}$$

17.10 Finding K at one temperature given K at another (van't Hoff equation) (742):

$$\ln \frac{K_2}{K_1} = -\frac{\Delta H_{\text{rxn}}^0}{R}\left(\frac{1}{T_2} - \frac{1}{T_1}\right)$$

Highlighted Figures and Tables

These figures (F) and tables (T) provide a quick review of key ideas.

F17.2 The range of equilibrium constants (716)
F17.3 The change in Q during a reaction (718)
T17.2 Ways of expressing Q (723)
F17.5 Reaction direction and the relative sizes of Q and K (726)

F17.6 Steps in solving equilibrium problems (736)
T17.3 Effect of added Cl_2 on the PCl_3-Cl_2-PCl_5 system (738)
F17.8 Effect of pressure (volume) on an equilibrium system (740)
T17.4 Effects of disturbances on an equilibrium system (743)

Brief Solutions to Follow-up Problems

17.1 (a) $Q_c = \dfrac{[NO]^4[H_2O]^6}{[NH_3]^4[O_2]^5}$

(b) $Q_c = \dfrac{[N_2O][NO_2]}{[NO]^3}$

17.2 $H_2(g) + Br_2(g) \rightleftharpoons 2HBr(g)$;

$Q_{c(\text{overall})} = \dfrac{[HBr]^2}{[H_2][Br_2]}$

$Q_{c(\text{overall})} = Q_{c1} \times Q_{c2} \times Q_{c3}$

$\qquad = \dfrac{[\cancel{Br}]^2}{[Br_2]} \times \dfrac{[HBr][\cancel{H}]}{[\cancel{Br}][H_2]} \times \dfrac{[HBr]}{[\cancel{H}][\cancel{Br}]}$

$\qquad = \dfrac{[HBr]^2}{[H_2][Br_2]}$

17.3 (a) $K_c = K_{c(\text{ref})}^{1/2} = 2.8 \times 10^4$

(b) $K_c = \left(\dfrac{1}{K_{c(\text{ref})}}\right)^{2/3} = 1.2 \times 10^{-6}$

17.4 $K_p = K_c(RT)^{-1} = 1.67\left(0.0821\ \dfrac{\text{atm·L}}{\text{mol·K}} \times 500.\ \text{K}\right)^{-1}$

$\qquad = 4.07 \times 10^{-2}$

17.5 $Q_p = \dfrac{(P_{CH_3Cl})(P_{HCl})}{(P_{CH_4})(P_{Cl_2})} = \dfrac{(0.24)(0.47)}{(0.13)(0.035)} = 25$;

$Q_p < K_p$, so CH_3Cl is forming.

17.6 From the reaction table for $2NO + O_2 \rightleftharpoons 2NO_2$,

$P_{O_2} = 1.000\ \text{atm} - x = 0.506\ \text{atm}; x = 0.494\ \text{atm}$

Also, $P_{NO} = 0.012\ \text{atm}$ and $P_{NO_2} = 0.988\ \text{atm}$, so

$$K_p = \frac{0.988^2}{0.012^2(0.506)} = 1.3 \times 10^4$$

17.7 Since $\Delta n_{\text{gas}} = 0$, $K_p = K_c = 2.3 \times 10^{30} = \dfrac{(0.781)(0.209)}{P_{NO}^2}$

Thus, $\qquad P_{NO} = 2.7 \times 10^{-16}\ \text{atm}$

17.8 From the reaction table,

$$[H_2] = [I_2] = x$$
$$[HI] = 0.242 - 2x$$

Thus, $\qquad K_c = 1.26 \times 10^{-3} = \dfrac{x^2}{(0.242 - 2x)^2}$

Taking the square root of both sides, neglecting the negative root, and solving gives $x = [H_2] = 8.02 \times 10^{-3}\ M$.

17.9 (a) Based on the reaction table, and assuming that $0.20\,M - x \approx 0.20\,M$,

$$K_c = 2.94\times10^{-10} \approx \frac{4x^2}{0.20} \qquad x \approx 3.8\times10^{-6}$$

Error $= 1.9\times10^{-3}\%$, so assumption is justified; therefore, at equilibrium, $[I_2] = 0.20\,M$ and $[I] = 7.6\times10^{-6}\,M$.
(b) Based on the same reaction table and assumption, $x \approx 0.10$; error is 50%, so assumption is *not* justified. Solve equation:

$$4x^2 + 0.209x - 0.042 = 0 \qquad x = 0.080\,M$$

Therefore, at equilibrium, $[I_2] = 0.12\,M$ and $[I] = 0.16\,M$.

17.10 (a) $Q_c = \dfrac{(0.0900)(0.0900)}{0.2100} = 3.86\times10^{-2}$;

$Q_c < K_c$, so reaction proceeds to the right.
(b) From the reaction table,
$\quad[PCl_5] = 0.2100\,M - x = 0.2065\,M \qquad x = 0.0035\,M$
So, $[Cl_2] = [PCl_3] = 0.0900\,M + x = 0.0935\,M$
17.11 (a) $[SiF_4]$ increases; (b) decreases; (c) decreases; (d) no effect.
17.12 (a) Decrease P; (b) increase P; (c) increase P.
17.13 (a) P_{H_2} will decrease; K_p will increase; (b) P_{N_2} will increase; K_p will decrease; (c) P_{PCl_5} will increase; K_p will increase.

Problems

Problems with **colored** numbers are answered at the back of the text. Sections match the text and provide the number(s) of relevant sample problems. Most offer Concept Review Questions, Skill-Building Exercises (in similar pairs), and Problems in Context. Then Comprehensive Problems, based on material from any section or previous chapter, follow.

The Dynamic Nature of the Equilibrium State

● **Concept Review Questions**

17.1 A change in reaction conditions increases the rate of a certain forward reaction more than that of the reverse reaction. What is the effect on the equilibrium constant and the concentrations of reactants and products at equilibrium?
17.2 When a chemical company employs a new reaction to manufacture a product, the chemists consider its rate (kinetics) and yield (equilibrium). How do each of these affect the usefulness of a manufacturing process?
17.3 If there is no change in concentrations, why is the equilibrium state considered dynamic?
17.4 Is K very large or very small for a reaction that goes essentially to completion? Explain.
17.5 Does the following reaction energy diagram depict a reaction in which K is small or large? Explain.

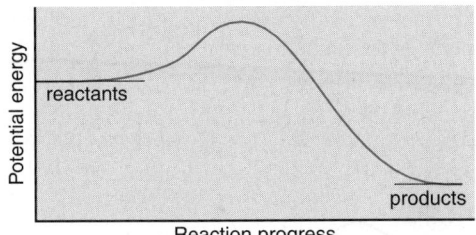

The Reaction Quotient and the Equilibrium Constant
(Sample Problems 17.1 to 17.3)

● **Concept Review Questions**

17.6 For a given reaction at a given temperature, the value of K is constant. Is the value of Q constant also? Explain.
17.7 In a series of experiments on the thermal decomposition of lithium peroxide,

$$2Li_2O_2(s) \rightleftharpoons 2Li_2O(s) + O_2(g)$$

a chemist finds that, as long as some Li_2O_2 is present at the end of the experiment, the amount of O_2 obtained in a given container at a given T reaches the same value. Explain.

17.8 In a study of the formation of HI from its elements,

$$H_2(g) + I_2(g) \rightleftharpoons 2HI(g)$$

equal amounts of H_2 and I_2 were placed in a container, which was then sealed and heated.
(a) On one set of axes, sketch concentration vs. time curves for H_2 and HI, and explain how Q changes as a function of time.
(b) Is the value of Q different if $[I_2]$ is plotted instead of $[H_2]$?
17.9 Explain the difference between a heterogeneous and a homogeneous equilibrium reaction. Give an example of each.
17.10 Does Q for the formation of 1 mol of NO from its elements differ from Q for the decomposition of 1 mol of NO to its elements? Explain and give the relationship between the two Q's.
17.11 Does Q for the formation of 1 mol of NH_3 from H_2 and N_2 differ from Q for the formation of NH_3 from H_2 and 1 mol of N_2? Explain and give the relationship between the two Q's.

● **Skill-Building Exercises** *(paired)*

17.12 Balance each reaction and write its reaction quotient, Q_c:
(a) $NO(g) + O_2(g) \rightleftharpoons N_2O_3(g)$
(b) $SF_6(g) + SO_3(g) \rightleftharpoons SO_2F_2(g)$
(c) $SClF_5(g) + H_2(g) \rightleftharpoons S_2F_{10}(g) + HCl(g)$
17.13 Balance each reaction and write its reaction quotient, Q_c:
(a) $C_2H_6(g) + O_2(g) \rightleftharpoons CO_2(g) + H_2O(g)$
(b) $CH_4(g) + F_2(g) \rightleftharpoons CF_4(g) + HF(g)$
(c) $SO_3(g) \rightleftharpoons SO_2(g) + O_2(g)$

17.14 Balance each reaction and write its reaction quotient, Q_c:
(a) $NO_2Cl(g) \rightleftharpoons NO_2(g) + Cl_2(g)$
(b) $POCl_3(g) \rightleftharpoons PCl_3(g) + O_2(g)$
(c) $NH_3(g) + O_2(g) \rightleftharpoons N_2(g) + H_2O(g)$
17.15 Balance each reaction and write its reaction quotient, Q_c:
(a) $O_2(g) \rightleftharpoons O_3(g)$
(b) $NO(g) + O_3(g) \rightleftharpoons NO_2(g) + O_2(g)$
(c) $N_2O(g) + H_2(g) \rightleftharpoons NH_3(g) + H_2O(g)$
17.16 At a particular temperature, $K_c = 1.6\times10^{-2}$ for

$$2H_2S(g) \rightleftharpoons 2H_2(g) + S_2(g)$$

Calculate K_c for each of the following reactions:
(a) $\frac{1}{2}S_2(g) + H_2(g) \rightleftharpoons H_2S(g)$
(b) $5H_2S(g) \rightleftharpoons 5H_2(g) + \frac{5}{2}S_2(g)$
17.17 At a particular temperature, $K_c = 6.5\times10^2$ for

$$2NO(g) + 2H_2(g) \rightleftharpoons N_2(g) + 2H_2O(g)$$

Calculate K_c for each of the following reactions:
(a) $NO(g) + H_2(g) \rightleftharpoons \frac{1}{2}N_2(g) + H_2O(g)$
(b) $2N_2(g) + 4H_2O(g) \rightleftharpoons 4NO(g) + 4H_2(g)$

17.18 Balance each of the following examples of heterogeneous equilibria and write its reaction quotient, Q_c:
(a) $Na_2O_2(s) + CO_2(g) \rightleftharpoons Na_2CO_3(s) + O_2(g)$
(b) $H_2O(l) \rightleftharpoons H_2O(g)$
(c) $NH_4Cl(s) \rightleftharpoons NH_3(g) + HCl(g)$

17.19 Balance each of the following examples of heterogeneous equilibria and write its reaction quotient, Q_c:
(a) $H_2O(l) + SO_3(g) \rightleftharpoons H_2SO_4(aq)$
(b) $KNO_3(s) \rightleftharpoons KNO_2(s) + O_2(g)$
(c) $S_8(s) + F_2(g) \rightleftharpoons SF_6(g)$

17.20 Balance each of the following examples of heterogeneous equilibria and write its reaction quotient, Q_c:
(a) $NaHCO_3(s) \rightleftharpoons Na_2CO_3(s) + CO_2(g) + H_2O(g)$
(b) $SnO_2(s) + H_2(g) \rightleftharpoons Sn(s) + H_2O(g)$
(c) $H_2SO_4(l) + SO_3(g) \rightleftharpoons H_2S_2O_7(l)$

17.21 Balance each of the following examples of heterogeneous equilibria and write its reaction quotient, Q_c:
(a) $Al(s) + NaOH(aq) + H_2O(l) \rightleftharpoons$
$$Na[Al(OH)_4](aq) + H_2(g)$$
(b) $CO_2(s) \rightleftharpoons CO_2(g)$
(c) $N_2O_5(s) \rightleftharpoons NO_2(g) + O_2(g)$

● **Problems in Context**

17.22 White phosphorus, P_4, is produced by the reduction of phosphate rock, $Ca_3(PO_4)_2$. If exposed to oxygen, the waxy, white solid smokes, bursts into flames, and releases a large quantity of heat. Does the reaction
$$P_4(g) + 5O_2(g) \rightleftharpoons P_4O_{10}(s)$$
have a large or small equilibrium constant? Explain.

17.23 The interhalogen ClF_3 is prepared in a two-step fluorination of chlorine gas:
$$Cl_2(g) + F_2(g) \rightleftharpoons ClF(g)$$
$$ClF(g) + F_2(g) \rightleftharpoons ClF_3(g)$$
(a) Balance each step and write the overall equation.
(b) Show that the overall Q_c equals the product of the Q_c's for the individual steps.

Expressing Equilibria with Pressure Terms: Relation Between K_c and K_p
(Sample Problem 17.4)

● **Concept Review Questions**

17.24 Guldberg and Waage proposed the original definition of the equilibrium constant as a particular ratio of *concentrations*. What relationship allows us to use a particular ratio of *partial pressures* (for a gaseous reaction) to express an equilibrium constant? Explain.

17.25 In which cases are K_c and K_p equal, and in which cases are they not equal?

17.26 A certain reaction at equilibrium has more moles of gaseous products than reactants.
(a) Is K_c larger or smaller than K_p?
(b) Write a general statement about the relative sizes of K_c and K_p for any gaseous equilibria.

● **Skill-Building Exercises (paired)**

17.27 Determine Δn_{gas} for each of the following reactions:
(a) $2KClO_3(s) \rightleftharpoons 2KCl(s) + 3O_2(g)$
(b) $2PbO(s) + O_2(g) \rightleftharpoons 2PbO_2(s)$
(c) $I_2(s) + 3XeF_2(s) \rightleftharpoons 2IF_3(s) + 3Xe(g)$

17.28 Determine Δn_{gas} for each of the following reactions:
(a) $MgCO_3(s) \rightleftharpoons MgO(s) + CO_2(g)$
(b) $2H_2(g) + O_2(g) \rightleftharpoons 2H_2O(l)$
(c) $HNO_3(l) + ClF(g) \rightleftharpoons ClONO_2(g) + HF(g)$

17.29 Calculate K_c for the following equilibria:
(a) $CO(g) + Cl_2(g) \rightleftharpoons COCl_2(g)$; $K_p = 3.9 \times 10^{-2}$ at 1000. K
(b) $S_2(g) + C(s) \rightleftharpoons CS_2(g)$; $K_p = 28.5$ at 500. K

17.30 Calculate K_c for the following equilibria:
(a) $H_2(g) + I_2(g) \rightleftharpoons 2HI(g)$; $K_p = 49$ at 730. K
(b) $2SO_2(g) + O_2(g) \rightleftharpoons 2SO_3(g)$; $K_p = 2.5 \times 10^{10}$ at 500. K

17.31 Calculate K_p for the following equilibria:
(a) $N_2O_4(g) \rightleftharpoons 2NO_2(g)$; $K_c = 6.1 \times 10^{-3}$ at 298 K
(b) $N_2(g) + 3H_2(g) \rightleftharpoons 2NH_3(g)$; $K_c = 2.4 \times 10^{-3}$ at 1000. K

17.32 Calculate K_p for the following equilibria:
(a) $H_2(g) + CO_2(g) \rightleftharpoons H_2O(g) + CO(g)$; $K_c = 0.77$ at 1020. K
(b) $3O_2(g) \rightleftharpoons 2O_3(g)$; $K_c = 1.8 \times 10^{-56}$ at 570. K

Reaction Direction: Comparing Q and K
(Sample Problem 17.5)

● **Concept Review Questions**

17.33 When the numerical value of Q is less than K, in which direction does the reaction proceed to reach equilibrium? Explain.

17.34 State three criteria that characterize a chemical system at equilibrium.

● **Skill-Building Exercises (paired)**

17.35 At 425°C, $K_p = 4.18 \times 10^{-9}$ for the reaction
$$2HBr(g) \rightleftharpoons H_2(g) + Br_2(g)$$
In one experiment, 0.20 atm of $HBr(g)$, 0.010 atm of $H_2(g)$, and 0.010 atm of $Br_2(g)$ are introduced into a container. Is the reaction at equilibrium? If not, in which direction will it proceed?

17.36 At 100°C, $K_p = 60.6$ for the reaction
$$2NOBr(g) \rightleftharpoons 2NO(g) + Br_2(g)$$
In a given experiment, 0.10 atm of each component is placed in a container. Is the system at equilibrium? If not, in which direction will it proceed?

● **Problems in Context**

17.37 The water-gas shift reaction plays a central role in the chemical methods for obtaining cleaner fuels from coal:
$$CO(g) + H_2O(g) \rightleftharpoons CO_2(g) + H_2(g)$$
At a given temperature, $K_p = 2.7$. If 0.13 mol of CO, 0.56 mol of H_2O, 0.62 mol of CO_2, and 0.43 mol of H_2 are introduced into a 2.0-L flask, in which direction must the reaction proceed to reach equilibrium?

How to Solve Equilibrium Problems
(Sample Problems 17.6 to 17.10)

● **Concept Review Questions**

17.38 In the 1980s, CFC-11 was one of the most heavily produced chlorofluorocarbons. The last step in its formation is
$$CCl_4(g) + HF(g) \rightleftharpoons CFCl_3(g) + HCl(g)$$
If you start the reaction with equal concentrations of CCl_4 and HF, you obtain equal concentrations of $CFCl_3$ and HCl at equilibrium. Are the concentrations of $CFCl_3$ and HCl equal if you start with unequal concentrations of CCl_4 and HF? Explain.

17.39 In a problem involving the catalyzed reaction of methane and steam, the following reaction table was prepared:

Pressure

(atm)	$CH_4(g)$	$+ 2H_2O(g)$	$\rightleftharpoons CO_2(g)$	$+ 4H_2(g)$
Initial	0.30	0.40	0	0
Change	$-x$	$-2x$	$+x$	$+4x$
Equilibrium	$0.30 - x$	$0.40 - 2x$	x	$4x$

Explain the entries in the "Change" and "Equilibrium" rows.

17.40 (a) What is the basis of the approximation that avoids using the quadratic formula to find an equilibrium concentration? (b) When can this approximation *not* be made?

● **Skill-Building Exercises** *(paired)*

17.41 In an experiment to study the formation of HI(g),
$$H_2(g) + I_2(g) \rightleftharpoons 2HI(g)$$
$H_2(g)$ and $I_2(g)$ were placed in a sealed container at a certain temperature. At equilibrium, $[H_2] = 6.50\times10^{-5}$ M, $[I_2] = 1.06\times10^{-3}$ M, and $[HI] = 1.87\times10^{-3}$ M. Calculate K_c for the reaction at this temperature.

17.42 Gaseous ammonia was introduced into a sealed container and heated to a certain temperature:
$$2NH_3(g) \rightleftharpoons N_2(g) + 3H_2(g)$$
At equilibrium, $[NH_3] = 0.0225$ M, $[N_2] = 0.114$ M, and $[H_2] = 0.342$ M. Calculate K_c for the reaction at this temperature.

17.43 Gaseous PCl_5 decomposes according to the reaction
$$PCl_5(g) \rightleftharpoons PCl_3(g) + Cl_2(g)$$
In one experiment, 0.15 mol of $PCl_5(g)$ was introduced into a 2.0-L container. Construct the reaction table for this process.

17.44 Hydrogen fluoride, HF, can be made from the reaction
$$H_2(g) + F_2(g) \rightleftharpoons 2HF(g)$$
In one experiment, 0.10 mol of $H_2(g)$ and 0.050 mol of $F_2(g)$ are introduced into a 0.50-L container. Construct the reaction table for this process.

17.45 For the following reaction, $K_p = 6.5\times10^4$ at 308 K:
$$2NO(g) + Cl_2(g) \rightleftharpoons 2NOCl(g)$$
At equilibrium, $P_{NO} = 0.35$ atm and $P_{Cl_2} = 0.10$ atm. What is the equilibrium partial pressure of NOCl(g)?

17.46 For the following reaction, $K_p = 0.262$ at 1000°C:
$$C(s) + 2H_2(g) \rightleftharpoons CH_4(g)$$
At equilibrium, P_{H_2} is 1.22 atm. What is the equilibrium partial pressure of $CH_4(g)$?

17.47 Ammonium hydrogen sulfide decomposes according to the following reaction, for which $K_p = 0.11$ at 250°C:
$$NH_4HS(s) \rightleftharpoons H_2S(g) + NH_3(g)$$
If 55.0 g of $NH_4HS(s)$ is placed in a sealed 5.0-L container, what is the partial pressure of $NH_3(g)$ at equilibrium?

17.48 Hydrogen sulfide decomposes according to the following reaction, for which $K_c = 9.30\times10^{-8}$ at 700°C:
$$2H_2S(g) \rightleftharpoons 2H_2(g) + S_2(g)$$
If 0.45 mol of H_2S is placed in a 3.0-L container, what is the equilibrium concentration of $H_2(g)$ at 700°C?

17.49 Even at high temperature, the formation of nitric oxide is not favored:
$$N_2(g) + O_2(g) \rightleftharpoons 2NO(g) \quad K_c = 4.10\times10^{-4} \text{ at 2000°C}$$
What is the equilibrium concentration of NO(g) when a mixture of 0.20 mol of $N_2(g)$ and 0.15 mol of $O_2(g)$ is allowed to come to equilibrium in a 1.0-L container at this temperature?

17.50 Nitrogen dioxide decomposes according to the reaction:
$$2NO_2(g) \rightleftharpoons 2NO(g) + O_2(g)$$
where $K_p = 4.48\times10^{-13}$ at a certain temperature. A pressure of 0.75 atm of NO_2 is introduced into a container and allowed to come to equilibrium. What are the equilibrium partial pressures of NO(g) and $O_2(g)$?

17.51 Hydrogen iodide decomposes according to the reaction
$$2HI(g) \rightleftharpoons H_2(g) + I_2(g)$$
A sealed 1.50-L container initially holds 0.00623 mol of H_2, 0.00414 mol of I_2, and 0.0244 mol of HI at 703 K. When equilibrium is reached, the equilibrium concentration of $H_2(g)$ is 0.00467 M. What are the equilibrium concentrations of HI(g) and $I_2(g)$?

17.52 Compound A decomposes according to the equation
$$A(g) \rightleftharpoons 2B(g) + C(g)$$
A sealed 1.00-L reaction vessel initially contains 1.75×10^{-3} mol of A(g), 1.25×10^{-3} mol of B(g), and 6.50×10^{-4} mol of C(g) at 100°C. When equilibrium is reached, the concentration of A(g) is 2.15×10^{-3} M. What are the equilibrium concentrations of B(g) and C(g)?

● **Problems in Context**

17.53 In an analysis of interhalogen reactivity, 0.500 mol of ICl was placed in a 5.00-L flask and allowed to decompose at a high temperature: $2ICl(g) \rightleftharpoons I_2(g) + Cl_2(g)$. Calculate the equilibrium concentrations of I_2, Cl_2, and ICl ($K_c = 0.110$ at this temperature).

17.54 A United Nations toxicologist studying the properties of mustard gas, $S(CH_2CH_2Cl)_2$, a blistering agent used in warfare, prepares a mixture of 0.675 M SCl_2 and 0.973 M C_2H_4 and allows it to react at room temperature (20.0°C):
$$SCl_2(g) + 2C_2H_4(g) \rightleftharpoons S(CH_2CH_2Cl)_2(g)$$
At equilibrium, $[S(CH_2CH_2Cl)_2] = 0.350$ M. Calculate K_p.

17.55 The first step in industrial nitric acid production is the catalyzed oxidation of ammonia. Without a catalyst, a different reaction predominates:
$$4NH_3(g) + 3O_2(g) \rightleftharpoons 2N_2(g) + 6H_2O(g)$$
When 0.0150 mol of $NH_3(g)$ and 0.0150 mol of $O_2(g)$ are placed in a 1.00-L container at a certain temperature, the N_2 concentration at equilibrium is 1.96×10^{-3} M. Calculate K_c for the reaction at this temperature.

17.56 A key step in the extraction of iron from its ore is
$$FeO(s) + CO(g) \rightleftharpoons Fe(s) + CO_2(g) \quad K_p = 0.403 \text{ at 1000°C}$$
This step occurs in the 700°C to 1200°C zone within a blast furnace. What are the equilibrium partial pressures of CO(g) and $CO_2(g)$ when 1.00 atm of CO(g) and excess FeO(s) react in a sealed container at 1000°C?

Reaction Conditions and the Equilibrium State: Le Châtelier's Principle
(Sample Problems 17.11 to 17.13)

● **Concept Review Questions**

17.57 What is meant by the word "disturbance" in the statement of Le Châtelier's principle?

17.58 What is the difference between the equilibrium position and the equilibrium constant of a reaction? Which changes as a result of a change in reactant concentration?

17.59 Scenes A, B, C depict this reaction at three temperatures:

$$NH_4Cl(s) \rightleftharpoons NH_3(g) + HCl(g) \qquad \Delta H^0_{rxn} = 176 \text{ kJ}$$

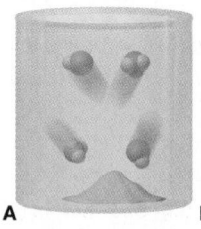 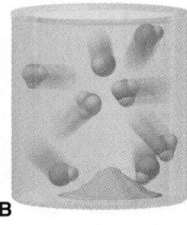 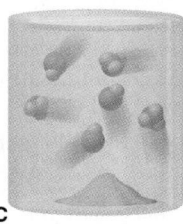

A   **B**   **C**

(a) Which best represents the reaction mixture at the highest temperature? Explain.

(b) Which best represents the reaction mixture at the lowest temperature? Explain.

17.60 What is implied by the word "constant" in the term *equilibrium constant*? Give two reaction parameters that can be changed without changing the value of an equilibrium constant.

17.61 Le Châtelier's principle is related ultimately to the rates of the forward and reverse steps in a reaction. Using this idea, explain (a) why an increase in reactant concentration shifts the equilibrium position to the right but does not change the equilibrium constant; (b) why a decrease in volume shifts the equilibrium position toward fewer moles of gas but does not change the equilibrium constant; and (c) why a rise in temperature shifts the equilibrium position of an exothermic reaction toward reactants and also changes the equilibrium constant.

17.62 Le Châtelier's principle predicts that a rise in the temperature of an endothermic reaction from T_1 to T_2 results in K_2 larger than K_1. Explain.

● **Skill-Building Exercises** *(paired)*

17.63 Consider this equilibrium system:

$$CO(g) + Fe_3O_4(s) \rightleftharpoons CO_2(g) + 3FeO(s)$$

How does the equilibrium position shift as a result of each of the following disturbances?

(a) CO is added.

(b) CO_2 is removed by adding solid NaOH.

(c) Additional $Fe_3O_4(s)$ is added to the system.

(d) Dry ice is added at constant temperature.

17.64 Sodium bicarbonate undergoes thermal decomposition according to the reaction

$$2NaHCO_3(s) \rightleftharpoons Na_2CO_3(s) + CO_2(g) + H_2O(g)$$

How does the equilibrium position shift as a result of each of the following disturbances?

(a) 0.20 atm of argon gas is added.

(b) $NaHCO_3(s)$ is added.

(c) $Mg(ClO_4)_2(s)$ is added as a drying agent to remove H_2O.

(d) Dry ice is added at constant temperature.

17.65 Predict the effect of *increasing* the container volume on the amounts of each reactant and product in the following:

(a) $F_2(g) \rightleftharpoons 2F(g)$

(b) $2CH_4(g) \rightleftharpoons C_2H_2(g) + 3H_2(g)$

17.66 Predict the effect of *increasing* the container volume on the amounts of each reactant and product in the following:

(a) $CH_3OH(l) \rightleftharpoons CH_3OH(g)$

(b) $CH_4(g) + NH_3(g) \rightleftharpoons HCN(g) + 3H_2(g)$

17.67 Predict the effect of *decreasing* the container volume on the amounts of each reactant and product in the following:

(a) $H_2(g) + Cl_2(g) \rightleftharpoons 2HCl(g)$

(b) $2H_2(g) + O_2(g) \rightleftharpoons 2H_2O(l)$

17.68 Predict the effect of *decreasing* the container volume on the amounts of each reactant and product in the following:

(a) $C_3H_8(g) + 5O_2(g) \rightleftharpoons 3CO_2(g) + 4H_2O(l)$

(b) $4NH_3(g) + 3O_2(g) \rightleftharpoons 2N_2(g) + 6H_2O(g)$

17.69 How would you adjust the *volume* of the reaction vessel in order to maximize product yield in the following reactions?

(a) $Fe_3O_4(s) + 4H_2(g) \rightleftharpoons 3Fe(s) + 4H_2O(g)$

(b) $2C(s) + O_2(g) \rightleftharpoons 2CO(g)$

17.70 How would you adjust the *volume* of the reaction vessel in order to maximize product yield in the following reactions?

(a) $Na_2O_2(s) \rightleftharpoons 2Na(l) + O_2(g)$

(b) $C_2H_2(g) + 2H_2(g) \rightleftharpoons C_2H_6(g)$

17.71 Predict the effect of *increasing* the temperature on the amounts of products in the following reactions:

(a) $CO(g) + 2H_2(g) \rightleftharpoons CH_3OH(g) \qquad \Delta H^0_{rxn} = -90.7 \text{ kJ}$

(b) $C(s) + H_2O(g) \rightleftharpoons CO(g) + H_2(g) \qquad \Delta H^0_{rxn} = 131 \text{ kJ}$

(c) $2NO_2(g) \rightleftharpoons 2NO(g) + O_2(g)$ (endothermic)

(d) $2C(s) + O_2(g) \rightleftharpoons 2CO(g)$ (exothermic)

17.72 Predict the effect of *decreasing* the temperature on the amounts of reactants in the following reactions:

(a) $C_2H_2(g) + H_2O(g) \rightleftharpoons CH_3CHO(g) \qquad \Delta H^0_{rxn} = -151 \text{ kJ}$

(b) $CH_3CH_2OH(l) + O_2(g) \rightleftharpoons CH_3CO_2H(l) + H_2O(g);$
$$\Delta H^0_{rxn} = -451 \text{ kJ}$$

(c) $2C_2H_4(g) + O_2(g) \rightleftharpoons 2CH_3CHO(g)$ (exothermic)

(d) $N_2O_4(g) \rightleftharpoons 2NO_2(g)$ (endothermic)

17.73 Deuterium (D or ^2H) is an isotope of hydrogen. The molecule D_2 undergoes an exchange reaction with ordinary H_2 that leads to isotopic equilibrium:

$$D_2(g) + H_2(g) \rightleftharpoons 2DH(g) \qquad K_p = 1.80 \text{ at } 298 \text{ K}$$

If ΔH^0_{rxn} is 0.32 kJ/mol DH, calculate K_p at 500. K.

17.74 The formation of methanol is an important industrial reaction in the processing of new fuels. At 298 K, $K_p = 2.25 \times 10^4$ for the reaction

$$CO(g) + 2H_2(g) \rightleftharpoons CH_3OH(l)$$

If $\Delta H^0_{rxn} = -128$ kJ/mol CH_3OH, calculate K_p at 0°C.

● **Problems in Context**

17.75 Lime (CaO) is used primarily in the manufacture of steel, glass, and high-quality paper. It is produced in an endothermic reaction by the thermal decomposition of limestone:

$$CaCO_3(s) \rightleftharpoons CaO(s) + CO_2(g)$$

How would you control reaction conditions to produce the maximum amount of lime?

17.76 The oxidation of SO_2 to SO_3 is an important industrial reactions because it is the key step in sulfuric acid production:

$$SO_2(g) + \tfrac{1}{2}O_2(g) \rightleftharpoons SO_3(g) \qquad \Delta H^0_{rxn} = -99.2 \text{ kJ}$$

(a) What qualitative combination of T and P maximizes SO_3 yield?

(b) How does addition of O_2 affect Q? K?

(c) Suggest a reason that catalysis is used for this reaction in the manufacture of H_2SO_4.

Comprehensive Problems

Problems with an asterisk (*) are more challenging.

17.77 The "filmstrip" represents five molecular-level scenes of a gaseous mixture as it reaches equilibrium over time:

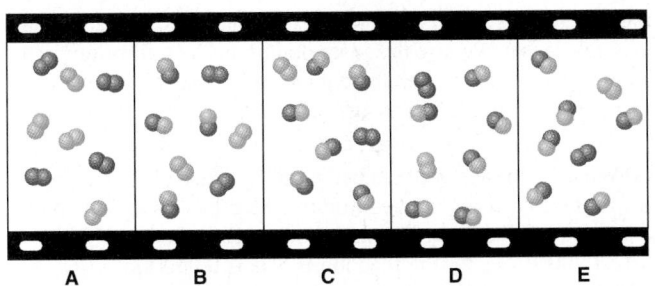

A   B   C   D   E

X is purple and Y is orange: $X_2(g) + Y_2(g) \rightleftharpoons 2XY(g)$.
(a) Write the reaction quotient, Q, for this reaction.
(b) If each particle represents 0.1 mol of particles, calculate Q for each scene.
(c) If $K > 1$, is time progressing to the right or to the left? Explain.
(d) Calculate K at this temperature.
(e) If $\Delta H^0_{rxn} < 0$, which scene, if any, best represents the mixture at a higher temperature? Explain.
(f) Which scene, if any, best represents the mixture at a higher pressure (lower volume)? Explain.

17.78 The powerful chlorinating agent sulfuryl dichloride (SO_2Cl_2) can be prepared by the following two-step sequence:
$$H_2S(g) + O_2(g) \rightleftharpoons SO_2(g) + H_2O(g)$$
$$SO_2(g) + Cl_2(g) \rightleftharpoons SO_2Cl_2(g)$$
(a) Balance each step and write the overall equation.
(b) Show that the overall Q_c equals the product of the Q_c's for the individual steps.

17.79 A mixture of 5.00 volumes of N_2 and 1.00 volume of O_2 passes through a heated furnace and reaches equilibrium at 900. K and 5.00 atm:
$$N_2(g) + O_2(g) \rightleftharpoons 2NO(g) \quad K_p = 6.70 \times 10^{-10}$$
(a) What is the partial pressure of NO?
(b) What is the concentration in micrograms per liter ($\mu g/L$) of NO in the mixture?

17.80 For the following equilibrium system, which of the changes will form more $CaCO_3$?
$$CO_2(g) + Ca(OH)_2(s) \rightleftharpoons CaCO_3(s) + H_2O(l)$$
$$\Delta H^0 = -113 \text{ kJ}$$
(a) Decrease temperature at constant pressure (no phase change)
(b) Increase volume at constant temperature
(c) Increase partial pressure of CO_2
(d) Remove one-half of the initial $CaCO_3$

*17.81 Ammonium carbamate (NH_2COONH_4) is a salt of carbamic acid that is found in the blood and urine of mammals. At 250.°C, $K_c = 1.58 \times 10^{-8}$ for the following equilibrium:
$$NH_2COONH_4(s) \rightleftharpoons 2NH_3(g) + CO_2(g)$$
If 7.80 g of NH_2COONH_4 is introduced into a 0.500-L evacuated container, what is the total pressure inside the container at equilibrium?

17.82 A study of the water-gas shift reaction (see Problem 17.37) was made in which equilibrium was reached with [CO] = [H_2O] = [H_2] = 0.10 M and [CO_2] = 0.40 M. After 0.60 mol of H_2 is added to the 2.0-L container and equilibrium is reestablished, what are the new concentrations of all the components?

17.83 The Group 8B(10) elements Ni, Pd, and Pt are used as industrial catalysts. Isolation and purification of these metals involve a series of steps. For nickel, for example, the sulfide ore is roasted in air: $Ni_3S_2(s) + O_2(g) \rightleftharpoons NiO(s) + SO_2(g)$. The metal oxide is reduced by the H_2 in water gas (CO + H_2) to impure Ni: $NiO(s) + H_2(g) \rightleftharpoons Ni(s) + H_2O(g)$. The CO in wa-

ter gas then reacts with the metal in the Mond process to form gaseous nickel carbonyl, $Ni(s) + CO(g) \rightleftharpoons Ni(CO)_4(g)$, which is subsequently decomposed to the metal.
(a) Balance each of the three steps, and obtain an overall balanced equation for the conversion of Ni_3S_2 to $Ni(CO)_4$.
(b) Show that the overall Q_c is the product of the Q_c's for the individual reactions.

17.84 One of the most important industrial sources of ethanol is the reaction of steam with ethene derived from crude oil:
$$C_2H_4(g) + H_2O(g) \rightleftharpoons C_2H_5OH(g)$$
$$\Delta H^0_{rxn} = -47.8 \text{ kJ} \quad K_c = 9 \times 10^3 \text{ at 600. K}$$
(a) At equilibrium, $P_{C_2H_5OH} = 200.$ atm and $P_{H_2O} = 400.$ atm. Calculate $P_{C_2H_4}$.
(b) Is the highest yield of ethanol obtained at high or low pressures? High or low temperatures?
(c) Calculate K_c at 450. K.
(d) In manufacturing, the yield of ammonia is increased by condensing it to a liquid and removing it from the vessel. Would condensing the C_2H_5OH work in this process? Explain.

17.85 Which of the following situations represent equilibrium?
(a) Migratory birds fly north in summer and south in winter.
(b) In a grocery store, some carts are kept inside and some outside. Customers bring carts out to their cars, while store clerks bring carts in to replace those taken out.
(c) In a tug o' war, a ribbon tied to the center of the rope moves back and forth until one side loses, after which the ribbon goes all the way to the winner's side.
(d) As a stew is cooking, water in the stew vaporizes, and the vapor condenses on the lid to droplets that drip into the stew.

*17.86 An industrial chemist introduces 2.0 atm of H_2 and 2.0 atm of CO_2 into a 1.00-L container at 25.0°C and then raises the temperature to 700.°C, at which $K_c = 0.534$:
$$H_2(g) + CO_2(g) \rightleftharpoons H_2O(g) + CO(g)$$
How many grams of H_2 are present after equilibrium is established?

17.87 As an EPA scientist studying catalytic converters and urban smog, you want to find K_c for the following reaction:
$$2NO_2(g) \rightleftharpoons N_2(g) + 2O_2(g) \quad K_c = ?$$
Use the following data to find the unknown K_c:
$$\tfrac{1}{2}N_2(g) + \tfrac{1}{2}O_2(g) \rightleftharpoons NO(g) \quad K_c = 4.8 \times 10^{-10}$$
$$2NO_2(g) \rightleftharpoons 2NO(g) + O_2(g) \quad K_c = 1.1 \times 10^{-5}$$

17.88 An inorganic chemist places 1 mol of BrCl in container A and 0.5 mol of Br_2 and 0.5 mol of Cl_2 in container B. She seals the containers and heats them to 300°C. Measurement shows that, with time, both containers hold identical mixtures of BrCl, Br_2, and Cl_2.
(a) Write a balanced equation for the reaction taking place in container A.
(b) Write the reaction quotient, Q, for this reaction.
(c) How do the values of Q in A compare with those in B over time?
(d) Explain on the molecular level how it is possible for both containers to end up with identical mixtures.

17.89 A quality control engineer examining the conversion of SO_2 to SO_3 in the manufacture of sulfuric acid determines that $K_c = 1.7 \times 10^8$ at 600. K for the reaction
$$2SO_2(g) + O_2(g) \rightleftharpoons 2SO_3(g)$$
(a) At equilibrium, $P_{SO_3} = 300.$ atm and $P_{O_2} = 100.$ atm. Calculate P_{SO_2}.

(b) The engineer places a mixture of 0.0040 mol of $SO_2(g)$ and 0.0028 mol of $O_2(g)$ in a 1.0-L container and raises the temperature to 1000 K. At equilibrium, 0.0020 mol of $SO_3(g)$ is present. Calculate K_c and P_{SO_2} for this reaction at 1000. K.

17.90 Phosgene ($COCl_2$) is a toxic substance that forms readily from carbon monoxide and chlorine at elevated temperatures:

$$CO(g) + Cl_2(g) \rightleftharpoons COCl_2(g)$$

If 0.350 mol of each reactant is placed in a 0.500-L flask at 600 K, what are the concentrations of each substance at equilibrium ($K_c = 4.95$ at this temperature)?

17.91 When 0.100 mol of $CaCO_3(s)$ and 0.100 mol of $CaO(s)$ are placed in an evacuated sealed 10.0-L container and heated to 385 K, $P_{CO_2} = 0.220$ atm after equilibrium is established:

$$CaCO_3(s) \rightleftharpoons CaO(s) + CO_2(g)$$

An additional 0.300 atm of $CO_2(g)$ is then pumped into the container. What is the total mass (in g) of $CaCO_3$ after equilibrium is re-established?

17.92 Use each of the following reaction quotients to write the balanced equation:

(a) $Q = \dfrac{[CO_2]^2[H_2O]^2}{[C_2H_4][O_2]^3}$ (b) $Q = \dfrac{[NH_3]^4[O_2]^7}{[NO_2]^4[H_2O]^6}$

17.93 Hydrogenation of carbon-carbon π bonds is important in the petroleum and food industries. The conversion of acetylene to ethylene is a simple example of the process:

$$C_2H_2(g) + H_2(g) \rightleftharpoons C_2H_4(g) \qquad K_c = 2.9\times10^8 \text{ at 2000. K}$$

The process is usually performed at much lower temperatures with the aid of a catalyst. Use ΔH_f^0 values from Appendix B to calculate K_c for this reaction at 300. K.

17.94 In combustion studies of H_2 as an alternative fuel, you find evidence that the hydroxyl radical (HO) is formed in flames by the reaction $H(g) + \frac{1}{2}O_2(g) \rightleftharpoons HO(g)$. Use the following data to calculate K_c for the reaction:

$$\tfrac{1}{2}H_2(g) + \tfrac{1}{2}O_2(g) \rightleftharpoons HO(g) \qquad K_c = 0.58$$
$$\tfrac{1}{2}H_2(g) \rightleftharpoons H(g) \qquad K_c = 1.6\times10^{-3}$$

*17.95 Highly toxic disulfur decafluoride decomposes by a free-radical process: $S_2F_{10}(g) \rightleftharpoons SF_4(g) + SF_6(g)$. In a study of the decomposition, S_2F_{10} was placed in a 2.0-L flask and heated to 100°C; $[S_2F_{10}]$ was 0.50 M at equilibrium. More S_2F_{10} was added, and, when equilibrium was reattained, $[S_2F_{10}]$ was 2.5 M. How did $[SF_4]$ and $[SF_6]$ change from the original equilibrium position to the new equilibrium position after the addition of more S_2F_{10}?

17.96 Aluminum is one of the most versatile metals. It is produced by the Hall-Heroult process, in which molten cryolite, Na_3AlF_6, is used as a solvent for the aluminum ore. Cryolite undergoes very slight decomposition with heat to produce a tiny amount of F_2, which escapes into the atmosphere above the solvent. K_c is 2×10^{-104} at 1300 K for the reaction:

$$Na_3AlF_6(l) \rightleftharpoons 3Na(l) + Al(l) + 3F_2(g)$$

What is the concentration of F_2 over a bath of molten cryolite at this temperature?

17.97 An equilibrium mixture of car exhaust gases consisting of 10.0 volumes of CO_2, 1.00 volume of unreacted O_2, and 50.0 volumes of unreacted N_2 leaves the engine at 4.0 atm and 800. K.
(a) Given this equilibrium, what is the partial pressure of CO?

$$2CO_2(g) \rightleftharpoons 2CO(g) + O_2(g) \qquad K_p = 1.4\times10^{-28} \text{ at 800. K}$$

(b) Assuming the mixture has enough time to reach equilibrium, what is the concentration in picograms per liter (pg/L) of CO in

the exhaust gas? (The actual concentration of CO in car exhaust is much higher because the gases do *not* reach equilibrium in the short transit time through the engine and exhaust system.)

*17.98 Consider the following reaction:

$$3Fe(s) + 4H_2O(g) \rightleftharpoons Fe_3O_4(s) + 4H_2(g)$$

(a) What are the apparent oxidation states of Fe and of O in Fe_3O_4?
(b) Fe_3O_4 is a compound of iron in which Fe occurs in two oxidation states. What are the oxidation states of Fe in Fe_3O_4?
(c) At 900°C, K_c for the reaction is 5.1. If 0.050 mol of $H_2O(g)$ and 0.100 mol of Fe(s) are placed in a 1.0-L container at 900°C, how many grams of Fe_3O_4 are present when equilibrium is established?

Note: The synthesis of ammonia is a major process throughout the industrialized world. Problems 17.99 to 17.107 refer to various aspects of this all-important reaction:

$$N_2(g) + 3H_2(g) \rightleftharpoons 2NH_3(g) \qquad \Delta H_{rxn}^0 = -91.8 \text{ kJ}$$

17.99 When ammonia is made industrially, the mixture of N_2, H_2, and NH_3 that emerges from the reaction chamber is far from equilibrium. Why does the plant supervisor use reaction conditions that produce less than the maximum yield of ammonia?

*17.100 The following reaction is sometimes used to produce the H_2 needed for the synthesis of ammonia:

$$CH_4(g) + CO_2(g) \rightleftharpoons 2CO(g) + 2H_2(g)$$

(a) What is the percent yield of H_2 when an equimolar mixture of CH_4 and CO_2 with a total pressure of 20.0 atm reaches equilibrium at 1200. K, at which $K_p = 3.548\times10^6$?
(b) What is the percent yield of H_2 for this system at 1300. K, at which $K_p = 2.626\times10^7$?
(c) Use the van't Hoff equation to find ΔH_{rxn}^0.

*17.101 The methane used to obtain H_2 for manufacturing NH_3 is impure and usually contains some higher hydrocarbons as well, such as propane, C_3H_8. Imagine the reaction of propane occurring in two steps:

$$C_3H_8(g) + 3H_2O(g) \rightleftharpoons 3CO(g) + 7H_2(g)$$
$$K_p = 8.175\times10^{15} \text{ at 1200. K}$$
$$CO(g) + H_2O(g) \rightleftharpoons CO_2(g) + H_2(g)$$
$$K_p = 0.6944 \text{ at 1200. K}$$

(a) Write the overall equation for the reaction of propane and steam to produce carbon dioxide and hydrogen.
(b) Calculate the equilibrium constant, K_p, for the overall process at 1200. K.
(c) When 1.00 volume of C_3H_8 and 4.00 volumes of H_2O, each at 1200. K and 5.0 atm, are mixed in a container, what is the final pressure? Assume the total volume remains constant, that the reaction is essentially complete, and that the gases behave ideally.
(d) What percentage of the C_3H_8 remains unreacted?

*17.102 Using methane and steam as a source of H_2 for ammonia synthesis requires high temperatures. Rather than burning CH_4 separately to heat the mixture, a more efficient strategy is to inject some oxygen into the reaction mixture. All of the H_2 is thus released for the synthesis, and the heat of combustion of CH_4 helps maintain the required temperature. Imagine the reaction occurring in two steps:

$$2CH_4(g) + O_2(g) \rightleftharpoons 2CO(g) + 4H_2(g)$$
$$K_p = 9.34\times10^{28} \text{ at 1000. K}$$
$$CO(g) + H_2O(g) \rightleftharpoons CO_2(g) + H_2(g)$$
$$K_p = 1.374 \text{ at 1000. K}$$

(a) Write the overall equation for the reaction of methane, steam, and oxygen to form carbon dioxide and hydrogen.
(b) What is K_p for the overall reaction?
(c) What is K_c for the overall reaction?
(d) A mixture of 2.0 mol of CH_4, 1.0 mol of O_2, and 2.0 mol of steam with a total pressure of 30. atm reacts at 1000. K in a container of constant volume. Assuming that the reaction is essentially complete and the ideal gas law is a valid approximation, what is the final pressure?

17.103 In a physical change, such as the melting of ice at 1°C, the reactant (ice) changes *completely* to the product (liquid water). In contrast, a chemical change reaches *equilibrium* in which both reactants *and* products are present. By considering how Q changes as each process occurs, explain the difference between this physical change and a chemical change.

*17.104 A mixture of 3.00 volumes of H_2 and 1.00 volume of N_2 reacts at 344°C to form ammonia. The equilibrium mixture at 110. atm contains 41.49% NH_3 by volume. Calculate K_p for the reaction, assuming that the gases behave ideally.

*17.105 One mechanism for the synthesis of ammonia proposes that N_2 and H_2 molecules catalytically dissociate into atoms:

$$N_2(g) \rightleftharpoons 2N(g) \qquad \log K_p = -43.10$$
$$H_2(g) \rightleftharpoons 2H(g) \qquad \log K_p = -17.30$$

(a) Find the partial pressure of N in N_2 at 1000. K and 200. atm.
(b) Find the partial pressure of H in H_2 at 1000. K and 600. atm.
(c) How many atoms of N and of H are present per liter of mixture?
(d) Based on these answers, which of the following is a more reasonable step to continue the mechanism after the catalytic dissociation? Explain.

$$N(g) + H(g) \longrightarrow NH(g)$$
$$N_2(g) + H(g) \longrightarrow NH(g) + N(g)$$

17.106 Consider the formation of ammonia in two experiments.
(a) To a 1.00-L container at 727°C, 1.30 mol of N_2 and 1.65 mol of H_2 are added. At equilibrium, 0.100 mol of NH_3 is present. Calculate the equilibrium concentrations of N_2 and H_2, and find K_c for the reaction: $2NH_3(g) \rightleftharpoons N_2(g) + 3H_2(g)$.
(b) In a different 1.00-L container at the same temperature, equilibrium is established with 8.34×10^{-2} mol of NH_3, 1.50 mol of N_2, and 1.25 mol of H_2 present. Calculate K_c for the reaction: $NH_3(g) \rightleftharpoons \frac{1}{2}N_2(g) + \frac{3}{2}H_2(g)$.
(c) What is the relationship between the K_c values in parts (a) and (b)? Why aren't these values the same?

*17.107 You are a member of a research team of industrial chemists who are discussing the plans to operate an ammonia processing plant:

$$N_2(g) + 3H_2(g) \rightleftharpoons 2NH_3(g)$$

(a) The plant operates at close to 700 K, at which K_p is 1.00×10^{-4}, and employs the stoichiometric 1:3 ratio of N_2:H_2. At equilibrium the partial pressure of the product NH_3 is 50. atm. Calculate the partial pressures of each reactant and the total pressure under these operating conditions.
(b) One member of the team makes the following suggestion: since the partial pressure of H_2 is cubed in the reaction quotient, the plant could produce the same amount of NH_3 if the reactants were in a 1:6 ratio of N_2:H_2 and could do so at a lower pressure, which would lower operating costs and, thus, increase profitability. Calculate the partial pressure of each reactant and the total pressure under these operating conditions, assuming an un-

changed partial pressure of 50. atm for NH_3. Is the team member's argument valid?

17.108 The two most abundant atmospheric gases react to a tiny extent at 298 K in the presence of a catalyst:

$$N_2(g) + O_2(g) \rightleftharpoons 2NO(g) \qquad K_p = 4.35 \times 10^{-31}$$

(a) What are the equilibrium pressures of the three components when the atmospheric partial pressures of O_2 (0.210 atm) and of N_2 (0.780 atm) are put into an evacuated 1.00-L flask at 298 K with catalyst?
(b) What is the total pressure in the container?
(c) Find K_c for this reaction at 298 K.

17.109 Pitchblende is the principal mineral source of uranium used for nuclear reactors. In pitchblende, uranium decays to radium, which decays to radon. Consequently, the concentration of radium in the mineral remains nearly constant. Is the constant concentration of radium in pitchblende a manifestation of an equilibrium or a steady state? Explain.

17.110 Isopentyl alcohol reacts with pure acetic acid to form isopentyl acetate, the essence of banana oil:

$$C_5H_{11}OH + CH_3COOH \rightleftharpoons CH_3COOC_5H_{11} + H_2O$$

A student adds a drying agent to remove H_2O and thus increase the yield of banana oil. Is this approach reasonable? Explain.

17.111 Many essential metabolites are products in branched pathways, such as the one shown below:

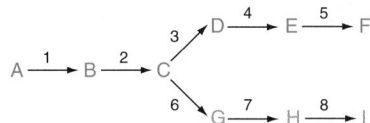

One method of control of these pathways occurs through inhibition of the first enzyme specific for a branch.
(a) Which enzyme is inhibited by F?
(b) Which enzyme is inhibited by I?
(c) What disadvantage would there be if F inhibited enzyme 1?
(d) What disadvantage would there be if F inhibited enzyme 6?

17.112 Glauber's salt, $Na_2SO_4 \cdot 10H_2O$, was used by J. R. Glauber in the 17th century as a medicinal agent. At 25°C, $K_p = 4.08 \times 10^{-25}$ for the loss of waters of hydration from Glauber's salt:

$$Na_2SO_4 \cdot 10H_2O(s) \rightleftharpoons Na_2SO_4(s) + 10H_2O(g)$$

(a) What is the vapor pressure of water at 25°C in a closed container holding a sample of $Na_2SO_4 \cdot 10H_2O(s)$?
(b) How do the following changes affect the ratio (higher, lower, same) of hydrated form/anhydrous form for the system above?
(1) Add more $Na_2SO_4(s)$ (2) Reduce the container volume
(3) Add more water vapor (4) Add N_2 gas

17.113 Synthetic diamonds are made under conditions of high temperature (2000 K) and high pressure (10^{10} Pa; 10^5 atm) in the presence of catalysts. The phase diagram for carbon is useful for understanding the conditions for the formation of both natural and synthetic diamonds. Along the diamond-graphite line, the two allotropes are in equilibrium. (a) At point A, what is the sign of ΔH for the formation of diamond from graphite? Explain.
(b) Which allotrope is denser? Explain.

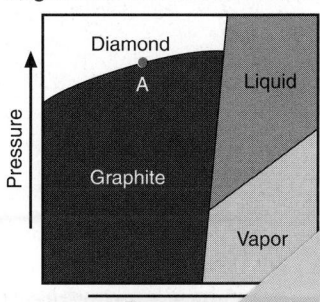

CHAPTER 18

ACID-BASE EQUILIBRIA

CHAPTER OUTLINE

18.1 Acids and Bases in Water
The Classical Acid-Base Definition
The Acid-Dissociation Constant (K_a)
Relative Strengths of Acids and Bases

18.2 Autoionization of Water and the pH Scale
Autoionization and K_w
The pH Scale

18.3 Proton Transfer and the Brønsted-Lowry Acid-Base Definition
The Conjugate Acid-Base Pair
Net Direction of Acid-Base Reactions

18.4 Solving Problems Involving Weak-Acid Equilibria
Finding K_a Given Concentrations
Finding Concentrations Given K_a
Extent of Acid Dissociation
Polyprotic Acids

18.5 Weak Bases and Their Relation to Weak Acids
Ammonia and the Amines
Anions of Weak Acids
The Relation Between K_a and K_b

18.6 Molecular Properties and Acid Strength
Nonmetal Hydrides
Oxoacids
Acidity of Hydrated Metal Ions

18.7 Acid-Base Properties of Salt Solutions
Salts That Yield Neutral Solutions
Salts That Yield Acidic Solutions
Salts That Yield Basic Solutions
Salts of Weakly Acidic Cations and Weakly Basic Anions

18.8 Generalizing the Brønsted-Lowry Concept: The Leveling Effect

18.9 Electron-Pair Donation and the Lewis Acid-Base Definition
Molecules as Lewis Acids
Metal Cations as Lewis Acids
Overview of Acid-Base Definitions

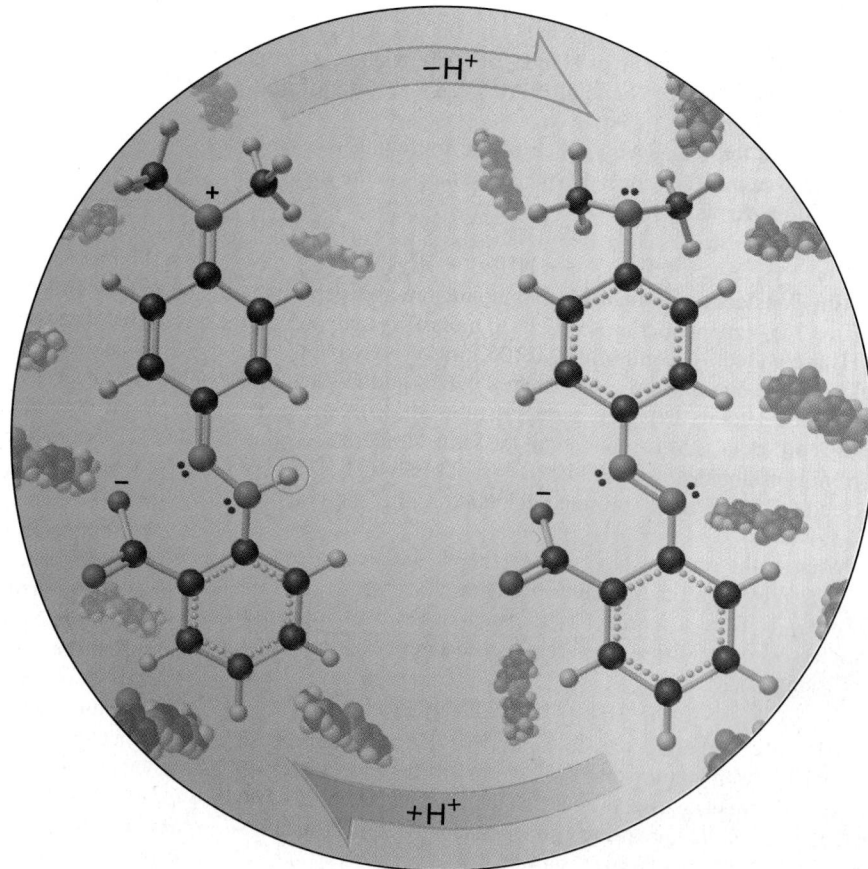

Figure: The colorful life of a proton. The reversible loss or gain of a proton can sometimes have useful applications. The acidic form *(left)* of methyl red has only one more proton (circled in green) than the basic form *(right)*. Yet its presence affects the electron distribution within the benzene ring system, which in turn affects the wavelengths of light the molecule absorbs. The result is a red acidic form and a yellow basic form—perfect for an acid-base indicator. In this chapter, you'll learn how to define acids and bases and how to apply the principles of equilibrium to their reactions.

Acids and bases have been used as laboratory chemicals since the time of the alchemists. With exotic names such as oil of vitriol (sulfuric acid), aqua fortis (nitric acid), and spirits of hartshorn (ammonia), these substances found early uses in everything from fraudulent "gold-forming" recipes to the practical metallurgy of gold, silver, and copper. In the 15th century, artists etched copper plates with nitric acid to create their works. To make glass, sand was treated with the strongly basic potash (K_2CO_3) that formed among wood ashes. In modern times, acids and bases are everyday chemicals in the home and in industry (Figure 18.1) and still play indispensable roles in the chemistry laboratory.

Table 18.1 on the next page shows some common acids and bases and their familiar uses. Notice that some of the acids (e.g., acetic and citric) have a sour taste. In fact, since the 17th century, sourness had been one of the defining properties of acids. An acid was any substance that had a sour taste; reacted with active metals, such as aluminum and zinc, to produce hydrogen gas; and turned certain organic compounds characteristic colors. (We discuss these compounds, known as *indicators,* later and in Chapter 19.) A base, on the other hand, was any substance that had a bitter taste and slippery feel and turned the same organic compounds different characteristic colors.* Moreover, it was commonly known that *when acids and bases react, each cancels the properties of the other in a process called neutralization.*

Evolving definitions are common in science. As more phenomena are understood, an early description becomes too limited and is replaced by a broader one. Even though the very early definitions of acids and bases described some distinctive properties, they inevitably gave way to definitions based on the molecular nature of acid-base behavior.

In this chapter, we develop in succession three definitions of acids and bases that allow us to understand ever-increasing numbers of reactions. In the process, you'll see how the principles of chemical equilibrium apply to this essential group of substances. After presenting the classical (Arrhenius) acid-base definition, we examine acid dissociation as an equilibrium process to see why acids vary in strength. Then we introduce the pH scale as a means of comparing the acidity or basicity of aqueous solutions. Next, we see how the Brønsted-Lowry definition greatly expands the meaning of "base" and, with it, the chemical changes we

---

*Despite what 17th-century chemists may have done, you should *NEVER* taste or touch laboratory chemicals. Many of them, including common acids and bases, are harmful and can cause severe burns. To satisfy your curiosity, note the sour taste of an acid in the vinegar or lemon juice on your next salad.

CONCEPTS & SKILLS

to review before you study this chapter
- role of water as solvent (Section 4.1)
- writing total and net ionic equations (Section 4.2)
- acids, bases, and acid-base reactions (Section 4.4)
- proton transfer in acid-base reactions (Section 4.4)
- properties of an equilibrium constant (Section 17.2)
- solving equilibrium problems (Section 17.5)

Pioneers of Acid-Base Chemistry Three 17th-century chemists laid the early foundations of acid-base chemistry. Johann Glauber (1604–1668) became renowned for his ability to prepare acids and their salts. Glauber's salt ($Na_2SO_4 \cdot 10H_2O$) is still used today for preparing dyes and printing textiles. Otto Tachenius (c. 1620–1690) is credited with first recognizing that a salt is the product of the reaction between an acid and a base. Robert Boyle (1627–1691), in addition to his studies of gas behavior, fostered the use of spot tests, flame colors, fume odors, and precipitates to analyze reactions. He was the first to associate the color change in syrup of violets (an organic dye) with the acidic or basic nature of the test solution.

A

B

Figure 18.1 Etching with acids. A, The insides of "frosted" bulbs have been etched with hydrofluoric acid. **B,** Acid is used to remove unwanted oxides of silicon and metals during the process of manufacturing computer chips.

Table 18.1 Some Common Acids and Bases and Their Household Uses

Substance	Use
Acids	
Acetic acid, CH_3COOH	Flavoring, preservative
Citric acid, $H_3C_6H_5O_7$	Flavoring
Phosphoric acid, H_3PO_4	Rust remover
Boric acid, H_3BO_3	Mild antiseptic, insecticide
Aluminum salts, $NaAl(SO_4)_2 \cdot 12H_2O$	In baking powder, with sodium hydrogen carbonate
Hydrochloric acid (muriatic acid), HCl	Brick and ceramic tile cleaner
Bases	
Sodium hydroxide (lye), NaOH	Oven cleaner, unblocking plumbing
Ammonia, NH_3	Household cleaner
Sodium carbonate, Na_2CO_3	Water softener, grease remover
Sodium hydrogen carbonate, $NaHCO_3$	Fire extinguisher, rising agent in cake mixes (baking soda), mild antacid
Sodium phosphate, Na_3PO_4	Cleaner for surfaces before painting or wallpapering

include as acid-base reactions. We investigate the molecular structures of acids and bases to rationalize variations in strength and then see that the very designations "acid" and "base" depend on the relative strengths of these substances and on the solvent in which the reaction occurs. Finally, we examine the Lewis acid-base definition, which expands the meaning of "acid" and what we consider to be acid-base behavior even further.

18.1 ACIDS AND BASES IN WATER

Although water is not an essential participant in all modern acid-base definitions, most laboratory work with acids and bases involves water, as do most important environmental, biological, and industrial applications. You know from our introductory discussion in Chapter 4 that *water is a product in all reactions between strong acids and strong bases:*

$$HCl(aq) + NaOH(aq) \longrightarrow NaCl(aq) + H_2O(l)$$

Indeed, as the net ionic equation of this reaction shows, water is *the* product:

$$H^+(aq) + OH^-(aq) \longrightarrow H_2O(l)$$

Furthermore, whenever an acid dissociates in water, solvent molecules participate in the reaction:

$$HA(g \text{ or } l) + H_2O(l) \longrightarrow A^-(aq) + H_3O^+(aq)$$

As you saw in that earlier discussion, water surrounds the proton to form H-bonded species with the general formula $H(H_2O)_n^+$. Because the proton is so small, its charge density is very high, so its attraction to water is especially strong. The proton bonds covalently to one of the lone electron pairs of a water molecule's O atom to form a **hydronium ion, H_3O^+,** which forms H bonds to several other water molecules; the $H(H_2O)_3^+$ ion is shown in Figure 18.2. To emphasize the active role of water in aqueous acid-base chemistry and the nature of the proton-water interaction, the hydrated proton is shown generally in the text as $H_3O^+(aq)$, although in some cases this hydrated species is shown more simply as $H^+(aq)$.

Proton or Hydroxide Ion Release and the Classical Acid-Base Definition

The earliest and simplest definition of acids and bases that reflects their chemical nature on the molecular scale was suggested by Svante Arrhenius, whose work on the rate constant we encountered in Chapter 16. In the **classical** (or **Arrhenius**) **acid-base definition,** acids and bases are classified in terms of their formulas and their behavior *in water:*

- An *acid* is a substance that has H in its formula and dissociates in water to yield H_3O^+.
- A *base* is a substance that has OH in its formula and dissociates in water to yield OH^-.

Some typical Arrhenius acids are HCl, HNO_3, and HCN, and some typical bases are NaOH, KOH, and $Ba(OH)_2$. Although Arrhenius bases contain discrete OH^- ions in their structures, Arrhenius acids *never* contain H^+ ions. On the contrary, these acids contain *covalently bonded H atoms that ionize in water.*

When an acid and a base react, they undergo **neutralization.** As we said, 17th-century scientists also observed this ability of an acid to cancel, or neutralize, the properties of a base, and vice versa. The meaning of acid-base reactions has changed along with the definitions of acid and base, but in the Arrhenius sense, neutralization occurs when *the H^+ ion from the acid and the OH^- ion from the base combine to form H_2O.* This description explains an observation that puzzled many of Arrhenius's colleagues. They observed that all neutralization reactions between what we now consider to be strong acids and strong bases (those that dissociate completely in water) had the same heat of reaction. No matter which strong acid and base reacted, and no matter which salt formed, ΔH_{rxn}^0 was about -56 kJ per mole of water that formed. Arrhenius suggested that the heat of reaction was always the same because the actual reaction was always the same—a hydrogen ion and a hydroxide ion formed water:

$$H^+(aq) + OH^-(aq) \longrightarrow H_2O(l) \qquad \Delta H_{rxn}^0 = -55.9 \text{ kJ}$$

The dissolved salt that formed along with the water, for example, NaCl in the reaction of sodium hydroxide with hydrochloric acid,

$$Na^+(aq) + OH^-(aq) + H^+(aq) + Cl^-(aq) \longrightarrow Na^+(aq) + Cl^-(aq) + H_2O(l)$$

did not affect the ΔH_{rxn}^0 but existed as hydrated spectator ions.

Despite its importance at the time, limitations in the classical definition soon became apparent. Arrhenius and many others realized that some substances that do *not* have OH in their formulas *do* behave as bases. For example, NH_3 and K_2CO_3 also yield OH^- in water. As you'll see shortly, broader acid-base definitions are required to include these species.

Variation in Acid Strength: The Acid-Dissociation Constant (K_a)

Acids and bases differ greatly in their *strength* in water, that is, in the amount of H_3O^+ or OH^- produced per mole of substance dissolved. We generally classify acids and bases as either strong or weak, according to the extent of their dissociation into ions in water (see Table 4.2, p. 141). It's important to remember, however, that a *gradation* in strength exists, as we'll examine quantitatively in a moment. Acids and bases are electrolytes in water, so this classification of acid

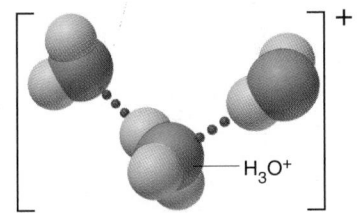

Figure 18.2 The nature of the hydrated proton. When an acid dissolves in water, the proton that is released forms a hydronium ion (H_3O^+) by bonding covalently to a water molecule. The H_3O^+ ion hydrogen bonds to other water molecules, forming a mixture of species with the general formula $H(H_2O)_n^+$. The $H(H_2O)_3^+$ (or $H_7O_3^+$) ion is shown here.

and base strength correlates with our earlier classification of electrolyte strength: *strong electrolytes dissociate completely, and weak electrolytes dissociate partially:*

- *Strong acids dissociate completely into ions in water* (Figure 18.3A):

$$HA(g \text{ or } l) + H_2O(l) \longrightarrow H_3O^+(aq) + A^-(aq)$$

  In a dilute solution of a strong acid, *virtually no HA molecules are present;* that is, $[H_3O^+] \approx [HA]_{init}$. In other words, $[HA]_{eq} \approx 0$, so the value of K_c is extremely large:

$$Q_c = \frac{[H_3O^+][A^-]}{[HA][H_2O]} \qquad \text{(at equilibrium, } Q_c = K_c \gg 1\text{)}$$

  Because the reaction is essentially complete, it is not very useful to express it as an equilibrium process. In a dilute aqueous nitric acid solution, for example, there are virtually no undissociated nitric acid molecules:

$$HNO_3(l) + H_2O(l) \longrightarrow H_3O^+(aq) + NO_3^-(aq)$$

- *Weak acids dissociate very slightly into ions in water* (Figure 18.3B):

$$HA(aq) + H_2O(l) \rightleftharpoons H_3O^+(aq) + A^-(aq)$$

  In a dilute solution of a weak acid, *the great majority of HA molecules are undissociated.* Thus, $[H_3O^+] \ll [HA]_{init}$. In other words, $[HA]_{eq} \approx [HA]_{init}$, so the value of K_c is very small. Hydrocyanic acid is an example of a weak acid:

$$HCN(aq) + H_2O(l) \rightleftharpoons H_3O^+(aq) + CN^-(aq)$$

$$Q_c = \frac{[H_3O^+][CN^-]}{[HCN][H_2O]} \qquad \text{(at equilibrium, } Q_c = K_c \ll 1\text{)}$$

Figure 18.3 **The extent of dissociation for strong and weak acids.** The bar graphs show the relative numbers of moles of species before *(left)* and after *(right)* acid dissociation occurs. **A,** When a strong acid dissolves in water, it dissociates completely, yielding $H_3O^+(aq)$ and $A^-(aq)$ ions; virtually no HA molecules are present. **B,** In contrast, when a weak acid dissolves in water, it remains mostly undissociated, yielding relatively few $H_3O^+(aq)$ and $A^-(aq)$ ions.

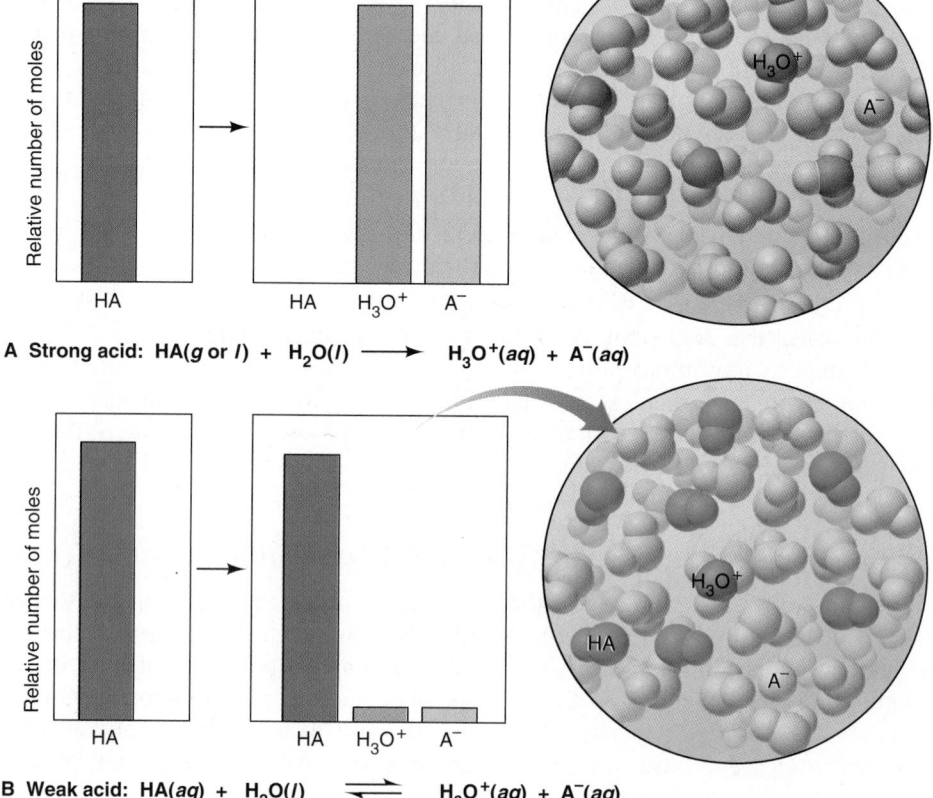

A **Strong acid:** $HA(g \text{ or } l) + H_2O(l) \longrightarrow H_3O^+(aq) + A^-(aq)$

B **Weak acid:** $HA(aq) + H_2O(l) \rightleftharpoons H_3O^+(aq) + A^-(aq)$

(From here on, square brackets without a subscript are used to mean molar concentration *at equilibrium;* that is, [X] means $[X]_{eq}$. Because we are dealing with systems *at equilibrium* in this chapter, instead of presenting Q and stating that Q equals K at equilibrium, we will express K directly as a collection of equilibrium concentration terms.)

Because of the effect of concentration on rate (Section 16.6), this dramatic difference in $[H_3O^+]$ causes a dramatic difference in the reaction rate when an equal concentration of a strong or a weak acid reacts with an active metal, such as zinc (Figure 18.4):

$$Zn(s) + 2H_3O^+(aq) \longrightarrow Zn^{2+}(aq) + 2H_2O(l) + H_2(g)$$

In a strong acid, such as 1 M HCl, zinc reacts rapidly, forming bubbles of H_2 vigorously. In a weak acid, such as 1 M CH_3COOH, zinc reacts slowly, forming bubbles of H_2 sluggishly on the metal piece. The strong acid has a high $[H_3O^+]$; therefore, much more H_3O^+ ion is available for reaction with the metal. The weak acid, on the other hand, has a low $[H_3O^+]$; therefore, much less H_3O^+ ion is available.

The Meaning of K_a There is a *specific* equilibrium constant for acid dissociation that highlights only those species whose concentrations change to any signigicant extent. The equilibrium expression for the dissociation of a general *weak acid, HA,* in water is

$$K_c = \frac{[H_3O^+][A^-]}{[HA][H_2O]}$$

The concentration of water, $[H_2O]$, is so much larger than $[HA]$ that it changes negligibly when HA dissociates, so it is treated as a constant. Therefore, as you saw in Section 17.2, we simplify the equilibrium expression by multiplying $[H_2O]$ by K_c to define a new equilibrium constant, the **acid-dissociation constant** (or **acid-ionization constant**) K_a:

$$K_c[H_2O] = K_a = \frac{[H_3O^+][A^-]}{[HA]} \qquad \textbf{(18.1)}$$

Like any equilibrium constant, K_a is a number that is temperature dependent, whose magnitude tells how far to the right the reaction has proceeded to reach equilibrium. Thus, *the stronger the acid, the higher the $[H_3O^+]$ at equilibrium, and the larger the K_a:*

$$\text{Stronger acid} \Longrightarrow \text{higher } [H_3O^+] \Longrightarrow \text{larger } K_a$$

The range of values for the acid-dissociation constants of weak acids extends over many orders of magnitude. Listed below are some benchmark K_a values for typical weak acids that will give you a general idea of the fraction of HA molecules that dissociate into ions:

- For a weak acid with a relatively high K_a ($\sim 10^{-2}$), a 1 M solution has $\sim 10\%$ of the HA molecules dissociated. The K_a of chlorous acid ($HClO_2$) is 1.12×10^{-2}, and 1 M $HClO_2$ is 10.6% dissociated.
- For a weak acid with a moderate K_a ($\sim 10^{-5}$), a 1 M solution has $\sim 0.3\%$ of the HA molecules dissociated. The K_a of acetic acid (CH_3COOH) is 1.8×10^{-5}, and 1 M CH_3COOH is 0.42% dissociated.
- For a weak acid with a relatively low K_a ($\sim 10^{-10}$), a 1 M solution has $\sim 0.001\%$ of the HA molecules dissociated. The K_a of HCN is 6.2×10^{-10}, and 1 M HCN is 0.0025% dissociated.

Thus, for solutions of the same initial HA concentration, *the smaller the K_a, the lower the percent dissociation of HA:*

$$\text{Smaller } K_a \Longrightarrow \text{lower } \% \text{ HA dissociated} \Longrightarrow \text{weaker acid}$$

Figure 18.4 Reaction of zinc with a strong and a weak acid. In the reaction of zinc with a strong acid, such as 1 M HCl (*left*), the higher concentration of H_3O^+ results in rapid formation of H_2 (bubbles). In a weak acid, such as 1 M CH_3COOH (*right*), the H_3O^+ concentration is much lower, so the formation of H_2 is much slower.

DISSOCIATION OF STRONG AND WEAK ACIDS

Table 18.2 K_a Values for Some Monoprotic Acids at 25°C

Name (Formula)	Lewis Structure*	K_a
Iodic acid (HIO$_3$)	H—O̤—I̤=O̤ with :O: below	1.6×10^{-1}
Chlorous acid (HClO$_2$)	H—O̤—C̤l=O̤	1.12×10^{-2}
Nitrous acid (HNO$_2$)	H—O̤—N̤=O̤	7.1×10^{-4}
Hydrofluoric acid (HF)	H—F̤:	6.8×10^{-4}
Formic acid (HCOOH)	H—C(=O̤)—O̤—H	1.8×10^{-4}
Benzoic acid (C$_6$H$_5$COOH)	C$_6$H$_5$—C(=O̤)—O̤—H	6.3×10^{-5}
Acetic acid (CH$_3$COOH)	H—C(H)(H)—C(=O̤)—O̤—H	1.8×10^{-5}
Propanoic acid (CH$_3$CH$_2$COOH)	H—C(H)(H)—C(H)(H)—C(=O̤)—O̤—H	1.3×10^{-5}
Hypochlorous acid (HClO)	H—O̤—C̤l:	2.9×10^{-8}
Hypobromous acid (HBrO)	H—O̤—B̤r:	2.3×10^{-9}
Hydrocyanic acid (HCN)	H—C≡N:	6.2×10^{-10}
Phenol (C$_6$H$_5$OH)	C$_6$H$_5$—O̤—H	1.0×10^{-10}
Hypoiodous acid (HIO)	H—O̤—I̤:	2.3×10^{-11}

ACID STRENGTH ↑

*Red type indicates the ionizable proton; structures have zero formal charge.

Table 18.2 lists K_a values of some weak *monoprotic* acids (those with one ionizable proton) in water. Strong acids are absent because they do not have meaningful K_a values. Note that the ionizable proton in organic acids is the H attached to oxygen in the —COOH group; the H atoms bonded to the C atoms in these acids do *not* ionize. In Section 18.4, we discuss the behavior of polyprotic acids (those with more than one ionizable proton).

Classifying the Relative Strengths of Acids and Bases

Using a table of acid-dissociation constants is the surest way to quantify relative acid strength, but you can often classify acids and bases qualitatively as strong or weak just from their formulas:

- *Strong acids.* Two types of strong acids, with examples that *you should memorize,* are
  1. The hydrohalic acids HCl, HBr, and HI
  2. Oxoacids in which the number of O atoms exceeds the number of ionizable protons by two or more, such as HNO$_3$, H$_2$SO$_4$, and HClO$_4$

- *Weak acids.* There are many *more* weak acids than strong ones. Four types, with examples, are
  1. The hydrohalic acid HF
  2. Those acids in which H is not bonded to O or to halogen, such as HCN and H_2S
  3. Oxoacids in which the number of O atoms equals or exceeds by one the number of ionizable protons, such as HClO, HNO_2, and H_3PO_4
  4. Organic acids (general formula RCOOH), such as CH_3COOH and C_6H_5COOH

- *Strong bases.* Soluble compounds containing O^{2-} or OH^- ions are strong bases. The cations are usually those of the most active metals:
  1. M_2O or MOH, where M = Group 1A(1) metal (Li, Na, K, Rb, Cs)
  2. MO or $M(OH)_2$, where M = Group 2A(2) metal (Ca, Sr, Ba)
     [MgO and $Mg(OH)_2$ are only slightly soluble, but the soluble portion dissociates completely.]

- *Weak bases.* Many compounds with an electron-rich nitrogen are weak bases (none are Arrhenius bases). The common structural feature is an N atom that has a lone electron pair in its Lewis structure (shown here in the formula):
  1. Ammonia ($\ddot{N}H_3$)
  2. Amines (general formula $R\ddot{N}H_2$, $R_2\ddot{N}H$, or $R_3\ddot{N}$), such as $CH_3CH_2\ddot{N}H_2$, $(CH_3)_2\ddot{N}H$, $(C_3H_7)_3\ddot{N}$, and $C_5H_5\ddot{N}$

SAMPLE PROBLEM 18.1 Classifying Acid and Base Strength from the Chemical Formula

Problem Classify each of the following compounds as a strong acid, weak acid, strong base, or weak base:
(a) H_2SeO_4 (b) $(CH_3)_2CHCOOH$
(c) KOH   (d) $(CH_3)_2CHNH_2$
Plan We examine the formula and classify each acid or base, using the text descriptions. Particular points to note for acids are the numbers of O atoms relative to H atoms and the presence of the —COOH group. For bases, note the nature of the cation and the presence of an N atom that has a lone pair.
Solution (a) Strong acid: H_2SeO_4 is an oxoacid in which the number of O atoms exceeds the number of ionizable protons by two.
(b) Weak acid: $(CH_3)_2CHCOOH$ is an organic acid, as indicated by the —COOH group.
(c) Strong base: KOH is one of the Group 1A(1) hydroxides.
(d) Weak base: $(CH_3)_2CHNH_2$ has a lone pair on the N and is an amine.

FOLLOW-UP PROBLEM 18.1 Which of the following is the stronger acid or base?
(a) HClO or $HClO_3$ (b) HCl or CH_3COOH (c) NaOH or CH_3NH_2

SECTION SUMMARY

Acids and bases are essential substances in home, industry, and the environment. In aqueous solution, water combines with the proton released from an acid to form the hydrated species represented by $H_3O^+(aq)$. In the classical (Arrhenius) definition, acids contain H and yield H_3O^+ in water, bases contain OH and yield OH^- in water, and an acid-base reaction (neutralization) is the reaction of H^+ and OH^- to form H_2O. Acid strength depends on $[H_3O^+]$ relative to [HA] in aqueous solution. Strong acids dissociate completely and weak acids slightly. The extent of dissociation is expressed by the acid-dissociation constant, K_a. Weak acids have K_a values ranging from about 10^{-1} to 10^{-12}. Acids and bases can be classified qualitatively as strong or weak based on their formulas.

18.2 AUTOIONIZATION OF WATER AND THE pH SCALE

Before we discuss the next major definition of acid-base behavior, let's examine a crucial property of water that enables us to quantify $[H_3O^+]$ in any aqueous system: *water is an extremely weak electrolyte*. The electrical conductivity of tap water is due almost entirely to dissolved ions, but even water that has been distilled and deionized repeatedly exhibits a tiny conductance. The reason is that water itself dissociates into ions very slightly in an equilibrium process known as **autoionization** (or self-ionization):

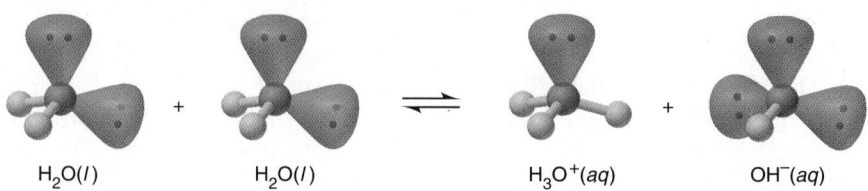

$$H_2O(l) \qquad H_2O(l) \qquad H_3O^+(aq) \qquad OH^-(aq)$$

The Equilibrium Nature of Autoionization: The Ion-Product Constant for Water (K_w)

Like any equilibrium system, the autoionization of water is described quantitatively by an equilibrium constant:

$$K_c = \frac{[H_3O^+][OH^-]}{[H_2O]^2}$$

Because the concentration of H_2O is essentially constant here, we simplify this equilibrium expression by including the constant $[H_2O]^2$ term with the value of K_c to obtain a new equilibrium constant, the **ion-product constant for water, K_w**:

$$K_c[H_2O]^2 = K_w = [H_3O^+][OH^-] = 1.0\times10^{-14} \text{ (at 25°C)} \qquad \textbf{(18.2)}$$

Notice that *one H_3O^+ ion and one OH^- ion appear for each H_2O molecule that dissociates*. Therefore, in pure water, we find that

$$[H_3O^+] = [OH^-] = \sqrt{1.0\times10^{-14}} = 1.0\times10^{-7} \text{ M (at 25°C)}$$

Pure water has a concentration of about 55.5 M $\left(\text{that is, } \dfrac{1000 \text{ g/L}}{18.02 \text{ g/mol}}\right)$, so these equilibrium concentrations are attained when only one in 555 million water molecules dissociates reversibly into ions!

Autoionization of water has two major consequences for aqueous acid-base chemistry:

1. *A change in $[H_3O^+]$ causes an inverse change in $[OH^-]$*, and vice versa:

$$\text{Higher } [H_3O^+] \Longrightarrow \text{ lower } [OH^-] \qquad \text{and} \qquad \text{Higher } [OH^-] \Longrightarrow \text{ lower } [H_3O^+]$$

Recall from our discussion of Le Châtelier's principle that a change in concentration of either ion shifts the equilibrium position, but it does *not* change the equilibrium constant. Therefore, if some acid is added, $[H_3O^+]$ increases, so $[OH^-]$ must decrease; if some base is added, $[OH^-]$ increases, so $[H_3O^+]$ must decrease. However, the addition of H_3O^+ or OH^- merely leads to the formation of H_2O, so the value of K_w is maintained.

2. *Both ions are present in all aqueous systems.* Thus, all acidic solutions contain a low concentration of OH^- ions, and all basic solutions contain a low concentration of H_3O^+ ions. The equilibrium nature of autoionization allows us to define "acidic" and "basic" solutions in terms of relative magnitudes of $[H_3O^+]$ and $[OH^-]$:

In an *acidic* solution,	$[H_3O^+] > [OH^-]$
In a *basic* solution,	$[H_3O^+] < [OH^-]$
In a *neutral* solution,	$[H_3O^+] = [OH^-]$

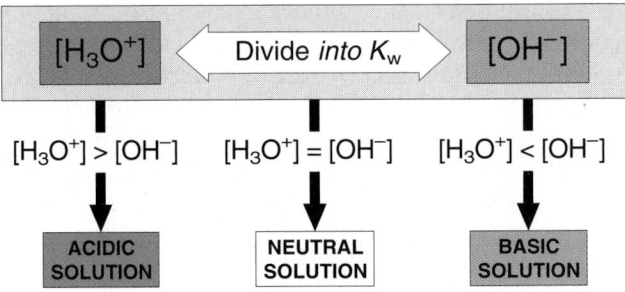

Figure 18.5 The relationship between $[H_3O^+]$ and $[OH^-]$ and the relative acidity of solutions.

Figure 18.5 summarizes these relationships and the relative solution acidity. Moreover, if you know the value of K_w at a particular temperature and the concentration of one of these ions, you can easily calculate the concentration of the other ion by solving for it in the K_w expression:

$$[H_3O^+] = \frac{K_w}{[OH^-]} \quad \text{or} \quad [OH^-] = \frac{K_w}{[H_3O^+]}$$

SAMPLE PROBLEM 18.2 Calculating $[H_3O^+]$ and $[OH^-]$ in an Aqueous Solution

Problem A research chemist adds a measured amount of HCl gas to pure water at 25°C and obtains a solution with $[H_3O^+] = 3.0 \times 10^{-4}$ M. Calculate $[OH^-]$. Is the solution neutral, acidic, or basic?
Plan We use the known value of K_w at 25°C (1.0×10^{-14}) and the given $[H_3O^+]$ (3.0×10^{-4} M) to solve for $[OH^-]$. Then, referring to Figure 18.5, we compare $[H_3O^+]$ with $[OH^-]$ to determine whether the solution is acidic, basic, or neutral.
Solution Calculating $[OH^-]$:

$$[OH^-] = \frac{K_w}{[H_3O^+]} = \frac{1.0 \times 10^{-14}}{3.0 \times 10^{-4}} = \boxed{3.3 \times 10^{-11} \ M}$$

Because $[H_3O^+] > [OH^-]$, the solution is acidic.
Check It makes sense that adding an acid to water results in an acidic solution. Moreover, since $[H_3O^+]$ is greater than 10^{-7} M, $[OH^-]$ must be less than 10^{-7} M to give a constant K_w.

FOLLOW-UP PROBLEM 18.2 Calculate $[H_3O^+]$ in a solution at 25°C whose $[OH^-] = 6.7 \times 10^{-2}$ M. Is the solution neutral, acidic, or basic?

Expressing the Hydronium Ion Concentration: The pH Scale

In aqueous solutions, $[H_3O^+]$ can vary over an enormous range: from about 10 M to 10^{-15} M. To handle numbers with negative exponents more conveniently in calculations, we convert them to positive numbers using a numerical system called a *p-scale,* the negative of the common (base-10) logarithm of the number. Applying this numerical system to $[H_3O^+]$ gives **pH,** the negative logarithm of $[H^+]$ (or $[H_3O^+]$):

$$pH = -\log [H_3O^+] \tag{18.3}$$

What is the pH of 10^{-12} M H_3O^+ solution?

$$pH = -\log [H_3O^+] = -\log 10^{-12} = (-1)(-12) = 12$$

Similarly, a 10^{-3} M H_3O^+ solution has a pH of 3, and a 5.4×10^{-4} M H_3O^+ solution has a pH of 3.27:

$$pH = -\log [H_3O^+] = (-1)(\log 5.4 + \log 10^{-4}) = 3.27$$

Logarithmic Scales in Sound and Seismology The p-scale is not the only logarithmic scale used in scientific measurements. The decibel scale measures the power of an acoustic signal, and the Richter scale measures the energy of ground movement. Urban planners study noise "pollution" by measuring decibel levels at different locations in a city at various times of the day. Seismologists record ground movement at stations around the world and try to predict the onset of earthquakes.

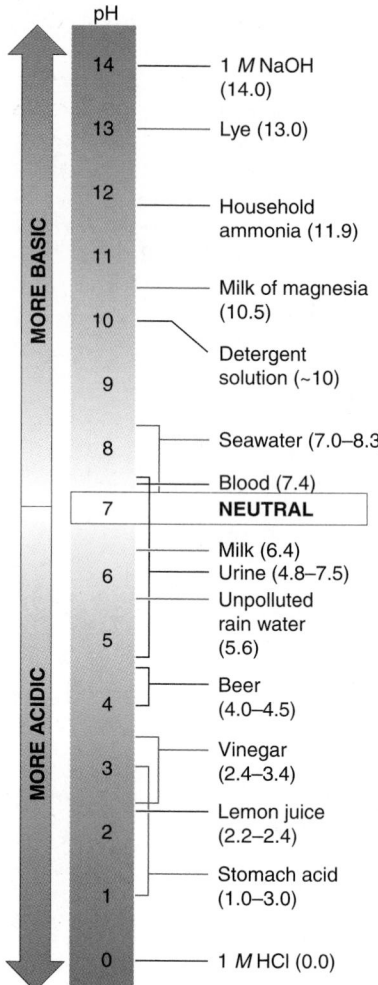

Figure 18.6 The pH values of some familiar aqueous solutions.

As with any measurement, the number of significant figures in a pH value reflects the precision with which the concentration is known. However, it is a logarithm, so the number of significant figures in the concentration equals the number of digits *to the right of the decimal point in the logarithm* (see Appendix A). In the preceding example, 5.4×10^{-4} *M* has two significant figures, so its negative logarithm, 3.27, has two digits to the right of the decimal point.

Note in particular that *the higher the pH, the lower the $[H_3O^+]$. Therefore, an acidic solution has a lower pH (higher $[H_3O^+]$) than a basic solution.* At 25°C in pure water, $[H_3O^+]$ is 1.0×10^{-7} *M*, so

pH of a neutral solution	= 7.00
pH of an acidic solution	< 7.00
pH of a basic solution	> 7.00

Figure 18.6 shows that the pH values of some familiar aqueous solutions fall within a range of 0 to 14.

Another important point arises when we compare $[H_3O^+]$ in different solutions. Because the pH scale is logarithmic, a solution of pH 1.0 has an $[H_3O^+]$ that is 10 times higher than that of a pH 2.0 solution, 100 times higher than that of a pH 3.0 solution, and so forth. To find the $[H_3O^+]$ from the pH, you perform the opposite arithmetic process; that is, you find the negative antilog of pH:

$$[H_3O^+] = 10^{-pH}$$

A p-scale is used to express other quantities as well:

- Hydroxide ion concentration can be expressed as pOH:

$$pOH = -\log [OH^-]$$

Acidic solutions have a higher pOH (lower $[OH^-]$) than basic solutions.
- Equilibrium constants can be expressed as p*K*:

$$pK = -\log K$$

A low pK corresponds to a high K. A reaction that reaches equilibrium with mostly products (proceeds far to the right) has a low p*K* (high *K*), whereas one that has mostly reactants at equilibrium has a high p*K* (low *K*). Table 18.3 shows this relationship for some weak acids.

Table 18.3 The Relationship Between K_a and pK_a

Acid Name (Formula)	K_a at 25°C	pK_a
Hydrogen sulfate ion (HSO_4^-)	1.02×10^{-2}	1.991
Nitrous acid (HNO_2)	7.1×10^{-4}	3.15
Acetic acid (CH_3COOH)	1.8×10^{-5}	4.74
Hypobromous acid (HBrO)	2.3×10^{-9}	8.64
Phenol (C_6H_5OH)	1.0×10^{-10}	10.00

The Relations Among pH, pOH, and pK_w Taking the negative log of both sides of the K_w expression gives a very useful relationship among pK_w, pH, and pOH:

$$K_w = [H_3O^+][OH^-] = 1.0 \times 10^{-14} \text{ (at 25°C)}$$
$$-\log K_w = (-\log [H_3O^+]) + (-\log [OH^-]) = -\log (1.0 \times 10^{-14})$$

$$pK_w = pH + pOH = 14.00 \quad \text{(at 25°C)} \tag{18.4}$$

Thus, the sum of pH and pOH is 14.00 in any aqueous solution at 25°C. With pH, pOH, $[H_3O^+]$, and $[OH^-]$ interrelated through K_w, knowing any one of the values allows us to determine the others (Figure 18.7).

	$[H_3O^+]$	pH	$[OH^-]$	pOH
MORE BASIC → BASIC	1.0×10^{-15}	15.00	1.0×10^{1}	-1.00
	1.0×10^{-14}	14.00	1.0×10^{0}	0.00
	1.0×10^{-13}	13.00	1.0×10^{-1}	1.00
	1.0×10^{-12}	12.00	1.0×10^{-2}	2.00
	1.0×10^{-11}	11.00	1.0×10^{-3}	3.00
	1.0×10^{-10}	10.00	1.0×10^{-4}	4.00
	1.0×10^{-9}	9.00	1.0×10^{-5}	5.00
	1.0×10^{-8}	8.00	1.0×10^{-6}	6.00
NEUTRAL	1.0×10^{-7}	7.00	1.0×10^{-7}	7.00
	1.0×10^{-6}	6.00	1.0×10^{-8}	8.00
	1.0×10^{-5}	5.00	1.0×10^{-9}	9.00
	1.0×10^{-4}	4.00	1.0×10^{-10}	10.00
MORE ACIDIC → ACIDIC	1.0×10^{-3}	3.00	1.0×10^{-11}	11.00
	1.0×10^{-2}	2.00	1.0×10^{-12}	12.00
	1.0×10^{-1}	1.00	1.0×10^{-13}	13.00
	1.0×10^{0}	0.00	1.0×10^{-14}	14.00
	1.0×10^{1}	-1.00	1.0×10^{-15}	15.00

Figure 18.7 The relations among $[H_3O^+]$, pH, $[OH^-]$, and pOH. Because K_w is constant, $[H_3O^+]$ and $[OH^-]$ are interdependent, and change in opposite directions as the acidity or basicity of the aqueous solution increases. The pH and pOH are interdependent in the same way. Note that at 25°C, the product of $[H_3O^+]$ and $[OH^-]$ is 1.0×10^{-14}, and the sum of pH and pOH is 14.00.

SAMPLE PROBLEM 18.3 Calculating $[H_3O^+]$, pH, $[OH^-]$, and pOH

Problem In an art restoration project, a conservator prepares copper-plate etching solutions by diluting concentrated HNO_3 to 2.0 M, 0.30 M, and 0.0063 M HNO_3. Calculate $[H_3O^+]$, pH, $[OH^-]$, and pOH of the three solutions at 25°C.

Plan We know from its formula that HNO_3 is a strong acid, so it dissociates completely; thus, $[H_3O^+] = [HNO_3]_{init}$. We use the given concentrations and the value of K_w at 25°C (1.0×10^{-14}) to find $[H_3O^+]$ and $[OH^-]$ and then use them to calculate pH and pOH.

Solution Calculating the values for 2.0 M HNO_3:

$$[H_3O^+] = \boxed{2.0\ M}$$

$$pH = -\log [H_3O^+] = -\log 2.0 = \boxed{-0.30}$$

$$[OH^-] = \frac{K_w}{[H_3O^+]} = \frac{1.0 \times 10^{-14}}{2.0} = \boxed{5.0 \times 10^{-15}\ M}$$

$$pOH = -\log (5.0 \times 10^{-15}) = \boxed{14.30}$$

Calculating the values for 0.30 M HNO_3:

$$[H_3O^+] = \boxed{0.30\ M}$$

$$pH = -\log [H_3O^+] = -\log 0.30 = \boxed{0.52}$$

$$[OH^-] = \frac{K_w}{[H_3O^+]} = \frac{1.0 \times 10^{-14}}{0.30} = \boxed{3.3 \times 10^{-14}\ M}$$

$$pOH = -\log (3.3 \times 10^{-14}) = \boxed{13.48}$$

Calculating the values for 0.0063 M HNO_3:

$$[H_3O^+] = \boxed{6.3 \times 10^{-3}\ M}$$

$$pH = -\log [H_3O^+] = -\log (6.3 \times 10^{-3}) = \boxed{2.20}$$

$$[OH^-] = \frac{K_w}{[H_3O^+]} = \frac{1.0 \times 10^{-14}}{6.3 \times 10^{-3}} = \boxed{1.6 \times 10^{-12}\ M}$$

$$pOH = -\log (1.6 \times 10^{-12}) = \boxed{11.80}$$

Check As the solution becomes more dilute, $[H_3O^+]$ decreases, so pH increases, as we expect. An $[H_3O^+]$ greater than 1.0 M, as in 2.0 M HNO_3, gives a positive log, so it results in a negative pH. The arithmetic seems correct because pH + pOH = 14.00 in each case.

Comment On most calculators, finding the pH requires several keystrokes. For example, to find the pH of 6.3×10^{-3} M HNO_3 solution, you enter: 6.3, EXP, 3, +/−, log, +/−.

FOLLOW-UP PROBLEM 18.3 A solution of NaOH has a pH of 9.52. What is its pOH, $[H_3O^+]$, and $[OH^-]$ at 25°C?

A

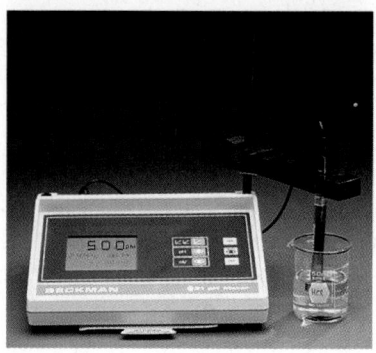

B

Figure 18.8 **Methods for measuring the pH of an aqueous solution. A,** A few drops of the solution are placed on a strip of pH paper, and the color is compared with the color chart. **B,** The electrodes of a pH meter immersed in the test solution measure [H_3O^+]. (In this instrument, the two electrodes are housed in one probe.)

Measuring pH In the laboratory, pH values are usually obtained with an acid-base indicator or, more precisely, with an instrument called a pH meter. **Acid-base indicators** are organic molecules whose colors depend on the acidity or basicity of the solution in which they are dissolved. The pH of a solution is estimated quickly with *pH paper,* a paper strip impregnated with one or a mixture of indicators. A drop of test solution is placed on the paper strip, and the color of the strip is compared with a color chart, as shown in Figure 18.8A.

The *pH meter* measures [H_3O^+] by means of two electrodes immersed in the test solution. One electrode provides a stable reference voltage; the other has an extremely thin, conducting, glass membrane that separates a known internal [H_3O^+] from the unknown external [H_3O^+]. The difference in [H_3O^+] creates a voltage difference across the membrane, which is measured and displayed in pH units (Figure 18.8B). We examine this process further in Chapter 21.

SECTION SUMMARY

Pure water has a low conductivity because it autoionizes to a small extent. This process is described by an equilibrium reaction whose equilibrium constant is the ion-product constant for water, K_w (1.0×10^{-14} at 25°C). Thus, [H_3O^+] and [OH^-] are inversely related. In acidic solution, [H_3O^+] is greater than [OH^-], the reverse is true in basic solution, and the two are equal in neutral solution. To express small values of [H_3O^+] more simply, we use the pH scale (pH $= -\log$ [H_3O^+]). A high pH represents a low [H_3O^+]. Similarly, pOH $= -\log$ [OH^-], and $pK = -\log K$. In acidic solutions, pH < 7.00; in basic solutions, pH > 7.00; and in neutral solutions, pH $= 7.00$. The sum of pH and pOH equals pK_w (14.00 at 25°C).

18.3 PROTON TRANSFER AND THE BRØNSTED-LOWRY ACID-BASE DEFINITION

Earlier we noted a major shortcoming of the classical (Arrhenius) definition: many substances that yield OH^- ions when they dissolve in water do not contain OH in their formulas. Examples include ammonia, the amines, and many salts of weak acids, such as NaF. Another limitation of the Arrhenius definition was that water had to be the solvent for acid-base reactions. In the early 20th century, J. N. Brønsted and T. M. Lowry suggested definitions that remove these limitations. (Recall that we discussed their ideas briefly in Section 4.4.) According to the **Brønsted-Lowry acid-base definition,**

- *An acid is a **proton donor,** any species that donates an H^+ ion.* An acid must contain H in its formula; HNO_3 and $H_2PO_4^-$ are two of many examples. All Arrhenius acids are Brønsted-Lowry acids.
- *A base is a **proton acceptor,** any species that accepts an H^+ ion.* A base must contain a lone pair of electrons to bind the H^+ ion; a few examples are NH_3, CO_3^{2-}, F^-, as well as OH^-. Brønsted-Lowry bases are not Arrhenius bases, but all Arrhenius bases contain the Brønsted-Lowry base OH^-.

From the Brønsted-Lowry perspective, the only requirement for an acid-base reaction is that *one species donate a proton and another species accept it: an acid-base reaction is a proton-transfer process.* Acid-base reactions can occur between gases, in nonaqueous solutions, and in heterogeneous mixtures, as well as in aqueous solutions.

An acid and a base always work together in the transfer of a proton. In other words, one species behaves as an acid only if another species *simultaneously* behaves as a base, and vice versa. Even when an acid or a base merely dissolves in water, an acid-base reaction occurs because water acts as the other partner. Consider two typical acidic and basic solutions:

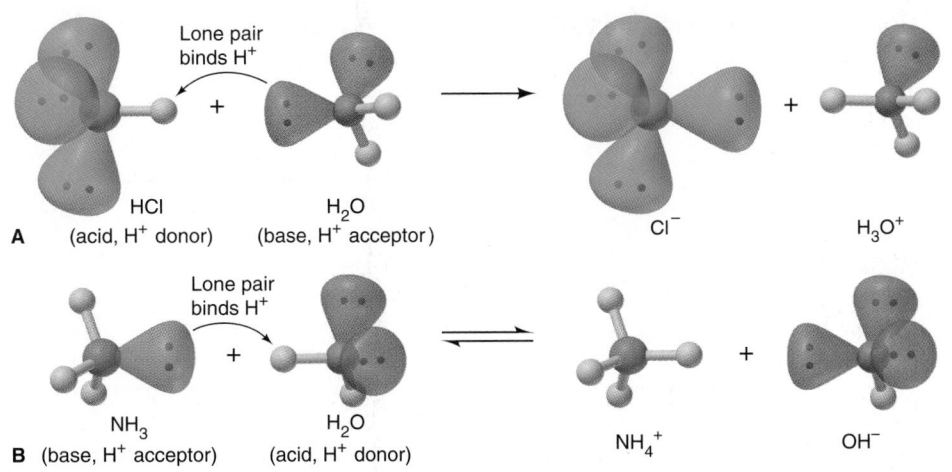

Figure 18.9 Proton transfer as the essential feature of a Brønsted-Lowry acid-base reaction. A, When HCl dissolves in water, it acts as an acid by donating a proton to water, which acts as a base by accepting it. **B,** When NH_3 dissolves in water, it acts as a base by accepting a proton from water, which acts as an acid by donating it. Thus, in the Brønsted-Lowry sense, an acid-base reaction occurs in both cases.

1. *Acid donates a proton to water* (Figure 18.9A). When HCl dissolves in water, an H^+ ion (a proton) is transferred from HCl to H_2O, where it becomes attached to a lone pair of electrons on the O atom, forming H_3O^+. In effect, HCl (the acid) has *donated* the H^+, and H_2O (the base) has *accepted* it:

$$HCl(g) + H_2\ddot{O}(l) \longrightarrow Cl^-(aq) + H_3\ddot{O}^+(aq)$$

2. *Base accepts a proton from water* (Figure 18.9B). When ammonia dissolves in water, proton transfer also occurs. An H^+ from H_2O attaches to the N atom's lone pair, forming NH_4^+. With an H^+ transferred, the water molecule becomes an OH^- ion:

$$\ddot{N}H_3(g) + H_2O(l) \rightleftharpoons NH_4^+(aq) + OH^-(aq)$$

In this case, H_2O (the acid) has *donated* the H^+, and NH_3 (the base) has *accepted* it. Thus, H_2O is *amphoteric:* it acts as a base in one case and as an acid in the other. As you'll see, many other species are amphoteric as well.

The Conjugate Acid-Base Pair

The Brønsted-Lowry definition provides a new way to look at acid-base reactions because it focuses on the reactants *and* the products. For example, let's examine the reaction between hydrogen sulfide and ammonia:

$$H_2S + NH_3 \rightleftharpoons HS^- + NH_4^+$$

In the forward reaction, H_2S acts as an acid by donating an H^+ to NH_3, which acts as a base. The reverse reaction involves another acid-base pair. The ammonium ion, NH_4^+, acts as an acid by donating an H^+ to the hydrogen sulfide ion, HS^-, which acts as a base. Notice that the acid, H_2S, becomes a base, HS^-, and the base, NH_3, becomes an acid, NH_4^+.

In Brønsted-Lowry terminology, H_2S and HS^- are a **conjugate acid-base pair:** HS^- is the conjugate base of the acid H_2S. Similarly, NH_3 and NH_4^+ form a conjugate acid-base pair: NH_4^+ is the conjugate acid of the base NH_3. *Every acid has a conjugate base, and every base has a conjugate acid.* Thus, for any conjugate acid-base pair,

- The conjugate base of the pair has one *fewer* H and one *more* negative charge than the acid.
- The conjugate acid of the pair has one *more* H and one *fewer* negative charge than the base.

A Brønsted-Lowry acid-base reaction occurs when *an acid and a base react to form their conjugate base and conjugate acid, respectively:*

$$acid_1 + base_2 \rightleftharpoons base_1 + acid_2$$

Table 18.4 The Conjugate Pairs in Some Acid-Base Reactions

	Acid	+	Base	⇌	Base	+	Acid
Reaction 1	HF	+	H_2O	⇌	F^-	+	H_3O^+
Reaction 2	HCOOH	+	CN^-	⇌	$HCOO^-$	+	HCN
Reaction 3	NH_4^+	+	CO_3^{2-}	⇌	NH_3	+	HCO_3^-
Reaction 4	$H_2PO_4^-$	+	OH^-	⇌	HPO_4^{2-}	+	H_2O
Reaction 5	H_2SO_4	+	$N_2H_5^+$	⇌	HSO_4^-	+	$N_2H_6^{2+}$
Reaction 6	HPO_4^{2-}	+	SO_3^{2-}	⇌	PO_4^{3-}	+	HSO_3^-

Table 18.4 shows some Brønsted-Lowry acid-base reactions. Notice that
- Each reaction has an acid and a base as reactants *and* as products, and these comprise two conjugate acid-base pairs.
- Acids and bases can be neutral, cationic, or anionic.
- The same species can be an acid or a base, depending on the other species reacting. Water behaves this way in reactions 1 and 4, and HPO_4^{2-} does so in reactions 4 and 6.

SAMPLE PROBLEM 18.4 Identifying Conjugate Acid-Base Pairs

Problem The following reactions are important environmental processes. Identify the conjugate acid-base pairs.
(a) $H_2PO_4^-(aq) + CO_3^{2-}(aq) \rightleftharpoons HCO_3^-(aq) + HPO_4^{2-}(aq)$
(b) $H_2O(l) + SO_3^{2-}(aq) \rightleftharpoons OH^-(aq) + HSO_3^-(aq)$
Plan To find the conjugate pairs, we find the species that donated an H^+ (acid) and the species that accepted it (base). The acid (or base) on the left becomes its conjugate base (or conjugate acid) on the right. Remember, the conjugate acid has one more H and a charge that is 1+ greater than that of its conjugate base.
Solution (a) $H_2PO_4^-$ has one more H^+ than HPO_4^{2-}; CO_3^{2-} has one fewer H^+ than HCO_3^-. Therefore, $H_2PO_4^-$ and HCO_3^- are the acids, and HPO_4^{2-} and CO_3^{2-} are the bases. The conjugate acid-base pairs are $H_2PO_4^-/HPO_4^{2-}$ and HCO_3^-/CO_3^{2-}.
(b) H_2O has one more H^+ than OH^-; SO_3^{2-} has one fewer H^+ than HSO_3^-. The acids are H_2O and HSO_3^-; the bases are OH^- and SO_3^{2-}. The conjugate acid-base pairs are H_2O/OH^- and HSO_3^-/SO_3^{2-}.

FOLLOW-UP PROBLEM 18.4 Identify the conjugate acid-base pairs:
(a) $CH_3COOH(aq) + H_2O(l) \rightleftharpoons CH_3COO^-(aq) + H_3O^+(aq)$
(b) $H_2O(l) + F^-(aq) \rightleftharpoons OH^-(aq) + HF(aq)$

Relative Acid-Base Strength and the Net Direction of Reaction

The net *direction* of an acid-base reaction depends on the relative strengths of the acids and bases involved. *A reaction proceeds to the greater extent in the direction in which a stronger acid and stronger base form a weaker acid and weaker base.* If the stronger acid and base are written on the left, the net direction is to the right, so $K_c > 1$. The net direction of the H_2S-NH_3 reaction is to the right ($K_c > 1$) because H_2S is a stronger acid than NH_4^+, the other acid present, and NH_3 is a stronger base than HS^-, the other base:

$$H_2S + NH_3 \rightleftharpoons HS^- + NH_4^+$$
stronger acid + stronger base ⟶ weaker base + weaker acid

You might think of the process as *a competition for the proton between the two bases,* NH_3 and HS^-, in which NH_3 wins.

Similarly, the extent of acid (HA) dissociation in water depends on a competition for the proton between the two bases, A^- and H_2O. When the acid HNO_3 dissolves in water, it transfers an H^+ to the base, H_2O, forming the conjugate base of HNO_3, which is NO_3^-, and the conjugate acid of H_2O, which is H_3O^+:

$$HNO_3 \quad + \quad H_2O \quad \longrightarrow \quad NO_3^- \quad + \quad H_3O^+$$

stronger acid + stronger base ⟶ weaker base + weaker acid

(In this case, the net direction is so far to the right that it would be inappropriate to show an equilibrium arrow.) HNO_3 is a stronger acid than H_3O^+, and H_2O is a stronger base than NO_3^-. Thus, with strong acids such as HNO_3, the H_2O wins the competition for the proton because A^- (NO_3^-) is a much weaker base. On the other hand, with weak acids such as HF, the A^- (F^-) wins because it is a stronger base than H_2O:

$$HF \quad + \quad H_2O \quad \rightleftharpoons \quad F^- \quad + \quad H_3O^+$$

weaker acid + weaker base ⟵ stronger base + stronger acid

From many such reactions, we can rank conjugate pairs in terms of the ability of the acid to transfer its proton (Figure 18.10). Note, especially, that *a weaker acid has a stronger conjugate base*. This makes perfect sense: the acid gives up its proton less readily because its conjugate base holds it more strongly. We can use this list to predict the direction of a reaction between any two pairs, that is, whether the equilibrium position lies predominantly to the right ($K_c > 1$) or to the left ($K_c < 1$). *An acid-base reaction proceeds to the right if the acid reacts with a base that is lower on the list* because this combination produces a weaker conjugate base and a weaker conjugate acid.

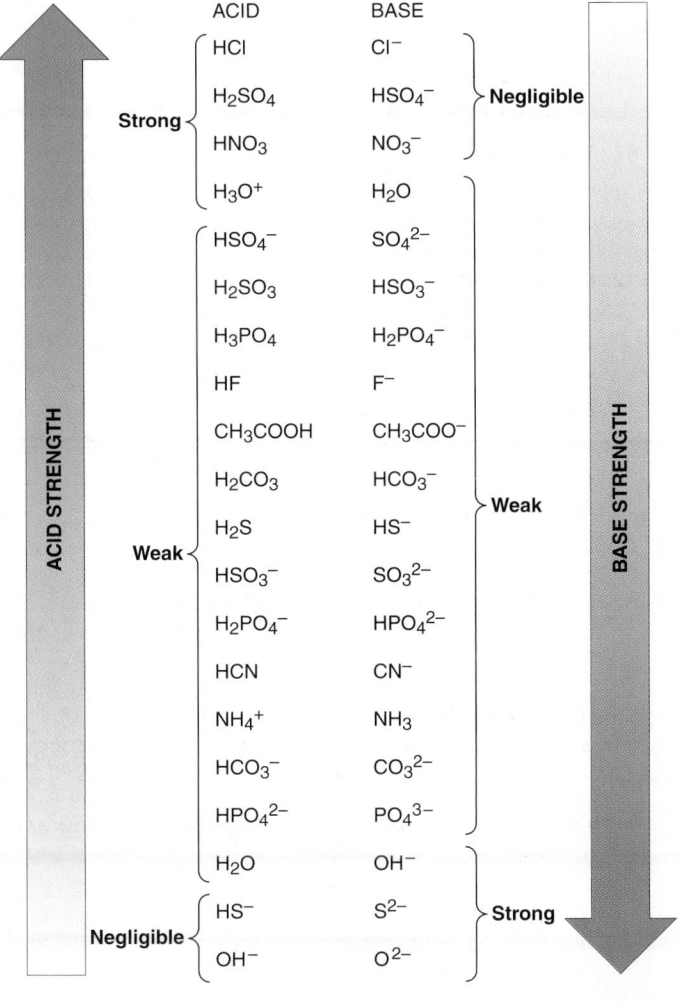

Figure 18.10 **Strengths of conjugate acid-base pairs.** The stronger the acid is, the weaker its conjugate base. The strongest acid appears on the top left and the strongest base on the bottom right. When an acid reacts with a base farther down the list, the reaction proceeds to the right ($K_c > 1$).

SAMPLE PROBLEM 18.5 Predicting the Net Direction of an Acid-Base Reaction

Problem Predict the net direction and whether K_c is greater or less than 1 for each of the following reactions (assume equal initial concentrations of all species):
(a) $H_2PO_4^-(aq) + NH_3(aq) \rightleftharpoons NH_4^+(aq) + HPO_4^{2-}(aq)$
(b) $H_2O(l) + HS^-(aq) \rightleftharpoons OH^-(aq) + H_2S(aq)$
Plan We first identify the conjugate acid-base pairs. To predict the direction, we consult Figure 18.10 to see which acid and base are stronger. The stronger acid and base form the weaker acid and base, so the reaction proceeds in that net direction. If the reaction *as written* proceeds to the right, then [products] is higher than [reactants], and $K_c > 1$.
Solution (a) The conjugate pairs are $H_2PO_4^-/HPO_4^{2-}$ and NH_4^+/NH_3. $H_2PO_4^-$ is higher on the list of acids, so it is stronger than NH_4^+; and NH_3 is lower on the list of bases, so it is stronger than HPO_4^{2-}. Therefore,

$$H_2PO_4^-(aq) + NH_3(aq) \rightleftharpoons NH_4^+(aq) + HPO_4^{2-}(aq)$$
stronger acid + stronger base $\longrightarrow$ weaker acid + weaker base

The net direction is right, so $K_c > 1$.

(b) The conjugate pairs are H_2O/OH^- and H_2S/HS^-. H_2S is higher on the list of acids, and OH^- is lower on the list of bases. Thus, we have

$$H_2O(l) + HS^-(aq) \rightleftharpoons OH^-(aq) + H_2S(aq)$$
weaker acid + weaker base $\longleftarrow$ stronger base + stronger acid

The net direction is left, so $K_c < 1$.

FOLLOW-UP PROBLEM 18.5 Explain with balanced equations, in which you show the net direction of the reaction, each of the following observations:
(a) You smell ammonia when NH_3 dissolves in water.
(b) The odor goes away when you add an excess of HCl to the solution in part (a).
(c) The odor returns when you add an excess of NaOH to the solution in part (b).

SECTION SUMMARY
The Brønsted-Lowry acid-base definition does not require that bases contain OH or that acid-base reactions occur in aqueous solution. It defines an acid as a species that donates a proton and a base as one that accepts it. Acids and bases act together in proton transfer. When an acid donates its proton, it becomes the conjugate base; when a base accepts a proton, it becomes the conjugate acid. In an acid-base reaction, acids and bases form their conjugates. A stronger acid has a weaker conjugate base, and vice versa. Thus, the reaction proceeds in the net direction in which a stronger acid and base form a weaker base and acid.

18.4 SOLVING PROBLEMS INVOLVING WEAK-ACID EQUILIBRIA

Just as you saw in Chapter 17 for equilibrium problems in general, there are two general types of equilibrium problems involving weak acids and their conjugate bases:

1. Given equilibrium concentrations, find K_a.
2. Given K_a and some concentration information, find the other equilibrium concentrations.

For all of these problems, we'll apply the same problem-solving approach, notation system, and assumptions:

- *The problem-solving approach.* As always, start with what is given in the problem statement and move logically toward what you want to find. Make a habit of applying the following steps:
  1. Write the balanced equation and K_a expression; these will tell you what to find.
  2. Define x as the unknown concentration that changes during the reaction. Frequently, $x = [HA]_{dissoc}$, the concentration of HA that dissociates, which, through the use of certain assumptions, also equals $[H_3O^+]$ and $[A^-]$ at equilibrium.
  3. Construct a reaction table that incorporates the unknown.
  4. Make assumptions that simplify the calculations, usually that x is very small relative to the initial concentration.
  5. Substitute the values into the K_a expression, and solve for x.
  6. Check that the assumptions are justified. (Apply the 5% test that was introduced in Sample Problem 17.9, p. 733.) If they are not justified, use the quadratic formula to find x.

- *The notation system.* As always, the molar concentration of each species is shown with brackets. A subscript refers to where the species comes from or when it occurs in the reaction process. For example, $[H_3O^+]_{from\ HA}$ is the molar concentration of H_3O^+ that comes from HA dissociation; $[HA]_{init}$ is the initial molar concentration of HA, that is, before the dissociation occurs; $[HA]_{dissoc}$ is the molar concentration of HA that dissociates; and so forth. Recall that *brackets with no subscript refer to the molar concentration of the species at equilibrium.*

- *The assumptions.* We make two assumptions to simplify the arithmetic:
  1. The $[H_3O^+]$ from the autoionization of water is negligible. In fact, it is so much smaller than $[H_3O^+]$ from HA dissociation that we can neglect it:

  $$[H_3O^+] = [H_3O^+]_{from\ HA} + [H_3O^+]_{from\ H_2O} \approx [H_3O^+]_{from\ HA}$$

  Indeed, Le Châtelier's principle tells us that $[H_3O^+]_{from\ HA}$ *decreases* the extent of autoionization of water, so $[H_3O^+]_{from\ H_2O}$ in the HA solution is even less than $[H_3O^+]$ in pure water. Note that each molecule of HA that dissociates forms one H_3O^+ and one A^-, so $[A^-] = [H_3O^+]$.
  2. A weak acid has a small K_a. Therefore, it dissociates to such a small extent that we can neglect the change in its concentration to find its equilibrium concentration:

  $$[HA] = [HA]_{init} - [HA]_{dissoc} \approx [HA]_{init}$$

Finding K_a Given Concentrations

This type of problem involves finding the acid-dissociation constant, K_a, of a weak acid:

$$HA(aq) + H_2O(l) \rightleftharpoons H_3O^+(aq) + A^-(aq)$$

$$K_a = \frac{[H_3O^+][A^-]}{[HA]}$$

A common laboratory approach is to prepare an aqueous solution of HA and measure its pH. You prepared the solution, so you know $[HA]_{init}$. You can calculate $[H_3O^+]$ from the measured pH and then determine $[A^-]$ and $[HA]$ at equilibrium from the assumptions you make. At this point, you substitute these values into the K_a expression and solve for K_a. We'll go through the entire approach in Sample Problem 18.6 and then simplify later problems by omitting some of the recurring steps.

SAMPLE PROBLEM 18.6 Finding the K_a of a Weak Acid from the pH of Its Solution

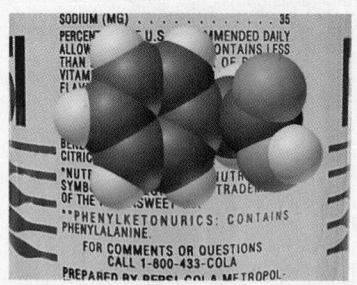

Phenylalanine, one of the amino acids that make up aspartame, is metabolized to phenylacetic acid (model).

Problem Phenylacetic acid ($C_6H_5CH_2COOH$, simplified here to HPAc; see photo) builds up in the blood of persons with phenylketonuria, an inherited disorder that, if untreated, causes mental retardation and death. A study of the acid shows that the pH of 0.12 M HPAc is 2.60. What is the K_a of phenylacetic acid?

Plan We are given $[HPAc]_{init}$ (0.12 M) and the pH (2.60) and must find K_a. We first write the equation for HPAc dissociation and the expression for K_a to see which values we need to find:

$$HPAc(aq) + H_2O(l) \rightleftharpoons H_3O^+(aq) + PAc^-(aq) \qquad K_a = \frac{[H_3O^+][PAc^-]}{[HPAc]}$$

• To find $[H_3O^+]$: We know the pH, so we can find $[H_3O^+]$. Because a pH of 2.60 is more than four pH units (10^4-fold) *lower* than the pH of pure water (pH = 7.0), we can assume that $[H_3O^+]_{from\ HPAc} \gg [H_3O^+]_{from\ H_2O}$. Therefore, $[H_3O^+]_{from\ HPAc} + [H_3O^+]_{from\ H_2O} \approx [H_3O^+]_{from\ HPAc} \approx [H_3O^+]$.

• To find $[PAc^-]$: Because each HPAc that dissociates forms one H_3O^+ and one PAc^-, $[H_3O^+] \approx [PAc^-]$.

• To find $[HPAc]$: We know $[HPAc]_{init}$. Because HPAc is a weak acid, we assume that very little dissociates, so $[HPAc]_{init} - [HPAc]_{dissoc} = [HPAc] \approx [HPAc]_{init}$.

We set up a reaction table, make the assumptions, substitute the equilibrium values, solve for K_a, and then check the assumptions.

Solution Calculating $[H_3O^+]$:

$$[H_3O^+] = 10^{-pH} = 10^{-2.60} = 2.5 \times 10^{-3}\ M$$

Setting up the reaction table, with $x = [HPAc]_{dissoc} = [H_3O^+]_{from\ HPAc} = [PAc^-] \approx [H_3O^+]$:

Concentration (M)	HPAc(aq)	+	H₂O(l)	⇌	H₃O⁺(aq)	+	PAc⁻(aq)
Initial	0.12		—		1×10^{-7}		0
Change	$-x$		—		$+x$		$+x$
Equilibrium	$0.12 - x$		—		$x + (<1 \times 10^{-7})$		x

Making the assumptions:

1. The calculated $[H_3O^+]$ ($2.5 \times 10^{-3}\ M$) $\gg [H_3O^+]_{from\ H_2O}$ ($<1 \times 10^{-7}\ M$), so we assume that $[H_3O^+] \approx [H_3O^+]_{from\ HPAc} = x$.

2. HPAc is a weak acid, so we assume that $[HPAc] = 0.12\ M - x \approx 0.12\ M$.

Solving for the equilibrium concentrations:

$$x \approx [H_3O^+] = [PAc^-] = 2.5 \times 10^{-3}\ M$$

$$[HPAc] = 0.12\ M - x = 0.12\ M - (2.5 \times 10^{-3}\ M) \approx 0.12\ M\ (\text{to 2 sf})$$

Substituting these values into K_a:

$$K_a = \frac{[H_3O^+][PAc^-]}{[HPAc]} \approx \frac{(2.5 \times 10^{-3})(2.5 \times 10^{-3})}{0.12} = \boxed{5.2 \times 10^{-5}}$$

Checking the assumptions by finding the percent error in concentration:

1. For $[H_3O^+]_{from\ H_2O}$: $\dfrac{1 \times 10^{-7}\ M}{2.5 \times 10^{-3}\ M} \times 100 = 4 \times 10^{-3}\% < 5\%$; assumption is justified.

2. For $[HPAc]_{dissoc}$: $\dfrac{2.5 \times 10^{-3}\ M}{0.12\ M} \times 100 = 2.1\% < 5\%$; assumption is justified. We had already shown above that, to two significant figures, the concentration had not changed, so this check was not really necessary.

Check The $[H_3O^+]$ makes sense: pH 2.60 should give $[H_3O^+]$ between 10^{-2} and $10^{-3}\ M$. The K_a calculation also seems in the correct range: $(10^{-3})^2/10^{-1} = 10^{-5}$, and this value seems reasonable for a weak acid.

Comment $[H_3O^+]_{from\ H_2O}$ is so small relative to $[H_3O^+]_{from\ HA}$ that, from here on, we will disregard it and enter it as zero in reaction tables.

FOLLOW-UP PROBLEM 18.6 The conjugate acid of ammonia is NH_4^+, a weak acid. If a 0.2 M NH_4Cl solution has a pH of 5.0, what is the K_a of NH_4^+?

Finding Concentrations Given K_a

The second type of equilibrium problem involving weak acids gives some concentration data and the K_a value and asks for the equilibrium concentration of some component. Such problems are very similar to those we solved in Chapter 17 in which a substance with a given initial concentration reacted to an unknown extent (see Sample Problems 17.8 to 17.10).

SAMPLE PROBLEM 18.7 Determining Concentrations from K_a and Initial [HA]

Problem Propanoic acid (CH_3CH_2COOH, which we simplify as HPr) is an organic acid whose salts are used to retard mold growth in foods. What is the $[H_3O^+]$ of 0.10 M HPr ($K_a = 1.3 \times 10^{-5}$)?

Plan We know the initial concentration (0.10 M) and K_a (1.3×10^{-5}) of HPr, and we need to find $[H_3O^+]$. First, we write the balanced equation and the expression for K_a:

$$HPr(aq) + H_2O(l) \rightleftharpoons H_3O^+(aq) + Pr^-(aq)$$

$$K_a = \frac{[H_3O^+][Pr^-]}{[HPr]} = 1.3 \times 10^{-5}$$

We know $[HPr]_{init}$ but not [HPr]. If we let $x = [HPr]_{dissoc}$, x is also $[H_3O^+]_{from\ HPr}$ and $[Pr^-]$ because each HPr that dissociates yields one H_3O^+ and one Pr^-. With this information, we can set up a reaction table. In solving for x, we assume that, because HPr has a small K_a, it dissociates very little; therefore, $[HPr]_{init} - x = [HPr] \approx [HPr]_{init}$. After we find x, we check the assumption.

Solution Setting up the reaction table, with $x = [HPr]_{dissoc} = [H_3O^+]_{from\ HPr} = [Pr^-] = [H_3O^+]$:

Concentration (M)	HPr(aq)	+	H₂O(l)	⇌	H₃O⁺(aq)	+	Pr⁻(aq)
Initial	0.10		—		0		0
Change	$-x$		—		$+x$		$+x$
Equilibrium	$0.10 - x$		—		x		x

Making the assumption: K_a is small, so x is small compared with $[HPr]_{init}$; therefore, $0.10\ M - x \approx 0.10\ M$. Substituting into the K_a expression and solving for x:

$$K_a = \frac{[H_3O^+][Pr^-]}{[HPr]} = 1.3 \times 10^{-5} \approx \frac{(x)(x)}{0.10}$$

$$x \approx \sqrt{(0.10)(1.3 \times 10^{-5})} = \boxed{1.1 \times 10^{-3}\ M = [H_3O^+]}$$

Checking the assumption:

For $[HPr]_{dissoc}$: $\dfrac{1.1 \times 10^{-3}\ M}{0.10\ M} \times 100 = 1.1\% < 5\%$; assumption is justified.

Check The $[H_3O^+]$ seems reasonable for a dilute solution of a weak acid with a moderate K_a. By reversing the calculation, we can check the math: $(1.1 \times 10^{-3})^2/0.10 = 1.2 \times 10^{-5}$, which is within rounding of the given K_a.

Comment In these problems, we assume that the concentration of HA that dissociates ($[HA]_{dissoc} = x$) can be neglected because K_a is relatively small. However, this is true only if $[HA]_{init}$ is relatively large. A simple benchmark reveals whether the assumption is justified:

• If $\dfrac{[HA]_{init}}{K_a} > 400$, the assumption is justified: neglecting x introduces an error $< 5\%$.

• If $\dfrac{[HA]_{init}}{K_a} < 400$, the assumption is *not* justified; neglecting x introduces an error $> 5\%$, so we solve a quadratic equation to find x.

The latter situation occurs in the follow-up problem.

FOLLOW-UP PROBLEM 18.7 Cyanic acid (HOCN) is an extremely acrid, unstable substance. What is the $[H_3O^+]$ and pH of 0.10 M HOCN ($K_a = 3.5 \times 10^{-4}$)?

The Effect of Concentration on the Extent of Acid Dissociation

If we repeat the calculation in Sample Problem 18.7, but start with a lower [HPr], we observe a very interesting fact about the extent of dissociation of a weak acid. Suppose the initial concentration of HPr is one-tenth as much, 0.010 *M* rather than 0.10 *M*. After filling in the reaction table and making the same assumptions, we find that

$$x = [HPr]_{dissoc} = 3.6 \times 10^{-4} \, M$$

Now, let's compare the percents of HPr molecules dissociated at the two different initial acid concentrations, using the relationship

$$\text{Percent HA dissociated} = \frac{[HA]_{dissoc}}{[HA]_{init}} \times 100 \qquad (18.5)$$

Case 1: $[HPr]_{init} = 0.10 \, M$

$$\text{Percent dissociated} = \frac{1.1 \times 10^{-3} \, M}{1.0 \times 10^{-1} \, M} \times 100 = 1.1\%$$

Case 2: $[HPr]_{init} = 0.010 \, M$

$$\text{Percent dissociated} = \frac{3.6 \times 10^{-4} \, M}{1.0 \times 10^{-2} \, M} \times 100 = 3.6\%$$

As the initial acid concentration decreases, the percent dissociation of the acid increases. Don't confuse the *concentration* of HA dissociated with the *percent* HA dissociated. The [HA]_{dissoc} is lower in the diluted HA solution because the actual *number* of dissociated HA molecules is smaller. It is the *fraction* (and thus the *percent*) of dissociated HA molecules that increases with dilution.

This phenomenon is analogous to a change in container volume (pressure) for gases at equilibrium (see the discussion in Section 17.6). In that situation, an increase in volume shifts the equilibrium position to favor more moles of gas. In the case of HA dissociation, the available volume increases as solvent is added and the solution is diluted; this increase in volume shifts the equilibrium position to favor more moles of ions.

The Behavior of Polyprotic Acids

Acids with more than one ionizable proton are **polyprotic acids.** In a solution of a polyprotic acid, one proton at a time dissociates from the acid molecule, and each dissociation step has a different K_a. For example, phosphoric acid is a triprotic acid (three ionizable protons), so it has three K_a values:

$$H_3PO_4(aq) + H_2O(l) \rightleftharpoons H_2PO_4^-(aq) + H_3O^+(aq)$$

$$K_{a1} = \frac{[H_2PO_4^-][H_3O^+]}{[H_3PO_4]} = 7.2 \times 10^{-3}$$

$$H_2PO_4^-(aq) + H_2O(l) \rightleftharpoons HPO_4^{2-}(aq) + H_3O^+(aq)$$

$$K_{a2} = \frac{[HPO_4^{2-}][H_3O^+]}{[H_2PO_4^-]} = 6.3 \times 10^{-8}$$

$$HPO_4^{2-}(aq) + H_2O(l) \rightleftharpoons PO_4^{3-}(aq) + H_3O^+(aq)$$

$$K_{a3} = \frac{[PO_4^{3-}][H_3O^+]}{[HPO_4^{2-}]} = 4.2 \times 10^{-13}$$

As you can see from the relative K_a values, H_3PO_4 is a much stronger acid than $H_2PO_4^-$, which is much stronger than HPO_4^{2-}. Table 18.5 lists some common

Table 18.5 Successive K_a Values for Some Polyprotic Acids at 25°C

Name (Formula)	Lewis Structure*	K_{a1}	K_{a2}	K_{a3}
Oxalic acid ($H_2C_2O_4$)		5.6×10^{-2}	5.4×10^{-5}	
Phosphorous acid (H_3PO_3)		3×10^{-2}	1.7×10^{-7}	
Sulfurous acid (H_2SO_3)		1.4×10^{-2}	6.5×10^{-8}	
Phosphoric acid (H_3PO_4)		7.2×10^{-3}	6.3×10^{-8}	4.2×10^{-13}
Arsenic acid (H_3AsO_4)		6×10^{-3}	1.1×10^{-7}	3×10^{-12}
Citric acid ($H_3C_6H_5O_7$)		7.5×10^{-4}	1.7×10^{-5}	4.0×10^{-7}
Carbonic acid (H_2CO_3)		4.5×10^{-7}	4.7×10^{-11}	
Hydrosulfuric acid (H_2S)		9×10^{-8}	1×10^{-17}	

ACID STRENGTH

*Red type indicates the ionizable protons.

polyprotic acids and their K_a values. Notice that, in every case, the first proton comes off to a much greater extent than the second and, where applicable, the second ionizes to a much greater extent than the third:

$$K_{a1} > K_{a2} > K_{a3}$$

This trend makes sense: it is more difficult for an H^+ ion to leave a singly charged anion (such as $H_2PO_4^-$) than to leave a neutral molecule (such as H_3PO_4), and more difficult still to leave a doubly charged anion (HPO_4^{2-}). Successive acid dissociation constants typically differ by several orders of magnitude. This fact greatly simplifies pH calculations involving polyprotic acids because *we can usually neglect the H_3O^+ coming from the subsequent dissociations.*

SAMPLE PROBLEM 18.8 Calculating Equilibrium Concentrations for a Polyprotic Acid

Problem Ascorbic acid ($H_2C_6H_6O_6$; H_2Asc for this problem), known as vitamin C, is a diprotic acid ($K_{a1} = 1.0 \times 10^{-5}$ and $K_{a2} = 5 \times 10^{-12}$) found in citrus fruit. Calculate $[H_2Asc]$, $[HAsc^-]$, $[Asc^{2-}]$, and the pH of 0.050 M H_2Asc.

Plan We know the initial concentration (0.050 M) and both K_a's for H_2Asc, and we have to calculate the equilibrium concentrations of all species and convert $[H_3O^+]$ to pH. We first write the equations and K_a expressions:

$$H_2Asc(aq) + H_2O(l) \rightleftharpoons HAsc^-(aq) + H_3O^+(aq) \qquad K_{a1} = \frac{[HAsc^-][H_3O^+]}{[H_2Asc]} = 1.0 \times 10^{-5}$$

$$HAsc^-(aq) + H_2O(l) \rightleftharpoons Asc^{2-}(aq) + H_3O^+(aq) \qquad K_{a2} = \frac{[Asc^{2-}][H_3O^+]}{[HAsc^-]} = 5 \times 10^{-12}$$

Then we assume the following:
1. Because $K_{a1} \gg K_{a2}$, the first dissociation produces almost all the H_3O^+: $[H_3O^+]_{\text{from } H_2Asc} \gg [H_3O^+]_{\text{from } HAsc^-}$.
2. Because K_{a1} is small, the amount that dissociates can be neglected relative to the initial concentration: $[H_2Asc]_{\text{init}} \approx [H_2Asc]$.

We set up a reaction table for the first dissociation, with x equal to the concentration of H_2Asc that dissociates, then solve for $[H_3O^+]$ and $[HAsc^-]$. Because the second dissociation is so much less than the first, we can substitute values from the first directly to find $[Asc^{2-}]$ of the second.

Solution Setting up a reaction table with $x = [H_2Asc]_{\text{dissoc}} = [HAsc^-] \approx [H_3O^+]$:

Concentration (M)	$H_2Asc(aq)$	$+$	$H_2O(l)$	$\rightleftharpoons$	$H_3O^+(aq)$	$+$	$HAsc^-(aq)$
Initial	0.050		—		0		0
Change	$-x$		—		$+x$		$+x$
Equilibrium	$0.050 - x$		—		x		x

Making the assumptions:
1. Because $K_{a2} \ll K_{a1}$, $[H_3O^+]_{\text{from } HAsc^-} \ll [H_3O^+]_{\text{from } H_2Asc}$. Therefore,
$$[H_3O^+]_{\text{from } H_2Asc} \approx [H_3O^+]$$
2. Because K_{a1} is small, $[H_2Asc]_{\text{init}} - x = [H_2Asc] \approx [H_2Asc]_{\text{init}}$. Thus,
$$[H_2Asc] = 0.050\ M - x \approx 0.050\ M$$

Substituting into the expression for K_{a1} and solving for x:

$$K_{a1} = \frac{[H_3O^+][HAsc^-]}{[H_2Asc]} = 1.0 \times 10^{-5} \approx \frac{x^2}{0.050}$$

$$x = [HAsc^-] \approx [H_3O^+] \approx 7.1 \times 10^{-4}\ M$$

$$\text{pH} = -\log[H_3O^+] = -\log(7.1 \times 10^{-4}) = 3.15$$

Checking the assumptions:
1. $[H_3O^+]_{\text{from } HAsc^-} \ll [H_3O^+]_{\text{from } H_2Asc}$: For any second dissociation that does occur, we have

$$[H_3O^+]_{\text{from } HAsc^-} \approx \sqrt{[HAsc^-](K_{a2})} = \sqrt{(7.1 \times 10^{-4})(5 \times 10^{-12})} = 6 \times 10^{-8}\ M$$

This is even less than $[H_3O^+]_{\text{from } H_2O}$, so the assumption is justified.

2. $[H_2Asc]_{\text{dissoc}} \ll [H_2Asc]_{\text{init}}$: $\dfrac{7.1 \times 10^{-4}\ M}{0.050\ M} \times 100 = 1.4\% < 5\%$; assumption is justified.

Also, from the Comment in Sample Problem 18.7, note that
$$\frac{[H_2Asc]_{\text{init}}}{K_{a1}} = \frac{0.050}{1.0 \times 10^{-5}} = 5000 > 400$$

Calculating $[Asc^{2-}]$:

$$K_{a2} = \frac{[H_3O^+][Asc^{2-}]}{[HAsc^-]} \qquad \text{and} \qquad [Asc^{2-}] = \frac{(K_{a2})[HAsc^-]}{[H_3O^+]}$$

$$[Asc^{2-}] = \frac{(5 \times 10^{-12})(7.1 \times 10^{-4})}{7.1 \times 10^{-4}} = 5 \times 10^{-12}\ M$$

Check $K_{a1} \gg K_{a2}$, so it makes sense that $[HAsc^-] \gg [Asc^{2-}]$ because Asc^{2-} is produced only in the second (much weaker) dissociation. Both K_a's are small, so all concentrations except $[H_2Asc]$ should be much lower than the original 0.050 M.

FOLLOW-UP PROBLEM 18.8 Oxalic acid (HOOC—COOH, or $H_2C_2O_4$) is the simplest organic diprotic acid. Its commercial uses include bleaching straw and leather and removing rust and ink stains. Calculate $[H_2C_2O_4]$, $[HC_2O_4^-]$, $[C_2O_4^{2-}]$, and the pH of a 0.150 M $H_2C_2O_4$ solution. Use K_a values from Table 18.5.

SECTION SUMMARY

Two common types of weak-acid equilibrium problems involve finding K_a from concentrations and finding concentrations from K_a. We summarize the information in a reaction table, and we simplify the arithmetic by assuming (1) $[H_3O^+]_{from\ H_2O}$ is so small relative to $[H_3O^+]_{from\ HA}$ that it can be neglected, and (2) weak acids dissociate so little that $[HA]_{init} \approx [HA]$ at equilibrium. The *fraction* of weak acid molecules that dissociates is greater in a more dilute solution, even though the total $[H_3O^+]$ is less. Polyprotic acids have more than one ionizable proton, but we assume that the first dissociation provides virtually all the H_3O^+.

18.5 WEAK BASES AND THEIR RELATION TO WEAK ACIDS

By focusing on where the proton comes from and goes to, the Brønsted-Lowry concept expands the definition of a base to encompass a host of species that the Arrhenius definition excludes: a base is any species that accepts a proton; to do so, *the base must have a lone electron pair.* (The lone electron pair also plays the central role in the Lewis acid-base definition, as you'll see later in this chapter.)

Now let's examine the equilibrium system of a weak base and focus, as we did for weak acids, on aqueous solutions. When a base (B) dissolves, it accepts a proton from H_2O, which acts here as an acid, leaving behind an OH^- ion:

$$B(aq) + H_2O(aq) \rightleftharpoons BH^+(aq) + OH^-(aq)$$

This general reaction for a base in water is described by the following equilibrium expression:

$$K_c = \frac{[BH^+][OH^-]}{[B][H_2O]}$$

Based on our earlier reasoning that $[H_2O]$ is treated as a constant in aqueous reactions, we include $[H_2O]$ in the value of K_c and obtain the **base-dissociation constant** (or **base-ionization constant**), K_b:

$$K_b = \frac{[BH^+][OH^-]}{[B]} \qquad (18.6)$$

Despite the name "base-dissociation constant," *no base dissociates in the process,* as you can see from the reaction.

As in the relation between pK_a and K_a, we know that pK_b, the negative logarithm of the base-dissociation constant, decreases with increasing K_b (that is, increasing base strength). In aqueous solution, the two large classes of weak bases are nitrogen-containing molecules, such as ammonia and the amines, and the anions of weak acids.

Molecules as Weak Bases: Ammonia and the Amines

Ammonia is the simplest nitrogen-containing compound that acts as a weak base in water:

$$NH_3(aq) + H_2O(l) \rightleftharpoons NH_4^+(aq) + OH^-(aq) \qquad K_b = 1.76 \times 10^{-5} \text{ (at 25°C)}$$

Despite labels on reagent bottles that read "ammonium hydroxide," an aqueous solution of ammonia consists largely of *intact* NH_3 molecules, as you can see

Ammonia's Picturesque Past One of today's key industrial chemicals has a long, vivid past. The name *ammonia* derives from an Egyptian god, whom the Romans called Ammon. Sacrificial wastes were piled outside temples (see photo), where they decomposed; the remaining mineral salts were called *salts of Ammon*. Later, the name *ammon* was retained for the volatile portion of these substances. In the late 15th century and for some time thereafter, ammonia was obtained from animal protein through the distillation of horns and hoofs and was called *spirits of hartshorn*.

Table 18.6 K_b Values for Some Molecular (Amine) Bases at 25°C

Name (Formula)	Lewis Structure*	K_b
Diethylamine [$(CH_3CH_2)_2NH$]		8.6×10^{-4}
Dimethylamine [$(CH_3)_2NH$]		5.9×10^{-4}
Triethylamine [$(CH_3CH_2)_3N$]		5.2×10^{-4}
Methylamine (CH_3NH_2)		4.4×10^{-4}
Ethanolamine ($HOCH_2CH_2NH_2$)		3.2×10^{-5}
Ammonia (NH_3)		1.76×10^{-5}
Pyridine (C_5H_5N)		1.7×10^{-9}
Aniline ($C_6H_5NH_2$)		4.0×10^{-10}

BASE STRENGTH

*Blue type indicates the basic nitrogen and its lone pair.

from the small K_b value here. In a 1.0 M NH$_3$ solution, for example, [OH$^-$] = [NH$_4^+$] = 4.2×10^{-3} M, so about 99.58% of the NH$_3$ is not ionized. Table 18.6 shows the K_b values for some common molecular bases.

If one or more of the H atoms in NH$_3$ is replaced by an organic group (designated as R), an *amine* results: RNH$_2$, R$_2$NH, or R$_3$N (Section 15.4; see Figure 15.15). The key structural feature of these organic compounds, as in all Brønsted-Lowry bases, is *a lone pair of electrons that can bind the proton donated by the acid.* Figure 18.11 depicts this process for methylamine, the simplest amine.

Figure 18.11 Abstraction of a proton from water by methylamine. The amines are organic derivatives of ammonia. Methylamine, the simplest amine, acts as a base in water by abstracting a proton, thereby increasing [OH$^-$].

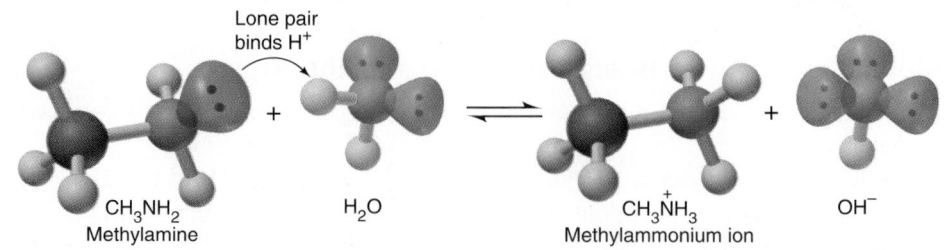

Lone pair binds H$^+$

CH$_3$NH$_2$
Methylamine

H$_2$O

CH$_3$NH$_3^+$
Methylammonium ion

OH$^-$

Finding the pH of a solution of a molecular weak base is a problem very similar to the type involving a weak acid. We write the equilibrium expression, set up a reaction table to find [base]$_{reacting}$, make the usual assumptions, and then solve for [OH$^-$]. The main difference is that we must convert [OH$^-$] to [H$_3$O$^+$] in order to calculate pH.

SAMPLE PROBLEM 18.9 Determining pH from K_b and Initial [B]

Problem Dimethylamine, (CH$_3$)$_2$NH (see margin), a key intermediate in detergent manufacture, has a K_b of 5.9×10^{-4}. What is the pH of 1.5 M (CH$_3$)$_2$NH?

Plan We know the initial concentration (1.5 M) and K_b (5.9×10^{-4}) of (CH$_3$)$_2$NH and have to find the pH. The amine reacts with water to form OH$^-$, so we have to find [OH$^-$] and then calculate [H$_3$O$^+$] and pH. The balanced equation and K_b expression are

$$(CH_3)_2NH(aq) + H_2O(l) \rightleftharpoons (CH_3)_2NH_2^+(aq) + OH^-(aq)$$

$$K_b = \frac{[(CH_3)_2NH_2^+][OH^-]}{[(CH_3)_2NH]}$$

Dimethylamine

Because $K_b \gg K_w$, the [OH$^-$] from the autoionization of water is negligible, and we disregard it. Therefore,

$$[OH^-]_{from\ base} = [(CH_3)_2NH_2^+] = [OH^-]$$

Because K_b is small, we assume that the amount of amine reacting is small, so

$$[(CH_3)_2NH]_{init} - [(CH_3)_2NH]_{reacting} = [(CH_3)_2NH] \approx [(CH_3)_2NH]_{init}$$

We proceed as usual, setting up a reaction table, making the assumption, and solving for x. Then we check the assumption and convert [OH$^-$] to [H$_3$O$^+$] using K_w; finally, we calculate pH.

Solution Setting up the reaction table, with

$$x = [(CH_3)_2NH]_{reacting} = [(CH_3)_2NH_2^+] = [OH^-]$$

Concentration (M)	(CH$_3$)$_2$NH(aq) + H$_2$O(l) $\rightleftharpoons$	(CH$_3$)$_2$NH$_2^+$(aq) +	OH$^-$(aq)
Initial	1.5	—	0	0
Change	$-x$	—	$+x$	$+x$
Equilibrium	$1.5 - x$	—	x	x

Making the assumption:
K_b is small, so $[(CH_3)_2NH]_{init} \approx [(CH_3)_2NH]$; thus, 1.5 M − $x \approx$ 1.5 M.
Substituting into the K_b expression and solving for x:

$$K_b = \frac{[(CH_3)_2NH_2^+][OH^-]}{[(CH_3)_2NH]} = 5.9\times10^{-4} \approx \frac{x^2}{1.5}$$

$$x = [OH^-] \approx 3.0\times10^{-2}\ M$$

Checking the assumption:

$$\frac{3.0\times10^{-2}\ M}{1.5\ M} \times 100 = 2.0\% < 5\%;\ \text{assumption is justified}$$

Note that the Comment from Sample Problem 18.7 applies to weak bases as well:

$$\frac{[B]_{init}}{K_b} = \frac{1.5}{5.9\times10^{-4}} = 2.5\times10^3 > 400$$

Calculating pH:

$$[H_3O^+] = \frac{K_w}{[OH^-]} = \frac{1.0\times10^{-14}}{3.0\times10^{-2}} = 3.3\times10^{-13}\ M$$

$$pH = -\log(3.3\times10^{-13}) = \boxed{12.48}$$

Check The value of x seems reasonable: $\sqrt{(\sim6\times10^{-4})(1.5)} = \sqrt{9\times10^{-4}} = 3\times10^{-2}$. Because (CH$_3$)$_2$NH is a weak base, the pH should be several pH units greater than 7.

FOLLOW-UP PROBLEM 18.9 Pyridine (C$_5$H$_5$N, see margin), a major solvent and base in organic syntheses, has a pK_b of 8.77. What is the pH of 0.10 M pyridine?

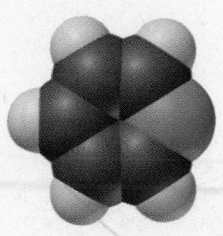
Pyridine

Anions of Weak Acids as Weak Bases

The other large group of Brønsted-Lowry bases consists of the anions of weak acids:*

$$A^-(aq) + H_2O(l) \rightleftharpoons HA(aq) + OH^-(aq) \qquad K_b = \frac{[HA][OH^-]}{[A^-]}$$

For example, F^-, the anion of the weak acid HF, acts as a weak base:

$$F^-(aq) + H_2O(l) \rightleftharpoons HF(aq) + OH^-(aq) \qquad K_b = \frac{[HF][OH^-]}{[F^-]}$$

Why is a solution of HA acidic and a solution of A^- basic? Let's approach the question by examining the relative concentrations of species present in 1 M HF and in 1 M NaF:

1. *The acidity of HA(aq).* Because HF is a weak acid, most of the HF exists in undissociated form. A small fraction of the HF molecules do dissociate, donating their protons to H_2O and yielding small concentrations of H_3O^+ and F^-. The equilibrium position of the system lies far to the left:

$$HF(aq) + H_2O(l) \overset{\longleftarrow}{\rightleftharpoons} H_3O^+(aq) + F^-(aq)$$

Water molecules also contribute minute amounts of H_3O^+ and OH^-, but their concentrations are extremely small:

$$2H_2O(l) \overset{\longleftarrow}{\rightleftharpoons} H_3O^+(aq) + OH^-(aq)$$

Of all the species present—HF, H_2O, H_3O^+, F^-, and OH^-—the two that can influence the acidity of the solution are H_3O^+, predominantly from HF, and OH^- from water. The solution is acidic because $[H_3O^+]_{from\ HF} \gg [OH^-]_{from\ H_2O}$.

2. *The basicity of A^-(aq).* Now, consider the species present in 1 M NaF. The salt dissociates completely to yield a relatively large concentration of F^-. The Na^+ ion behaves as a spectator, but some F^- reacts as a weak base with water to produce a very small amount of HF (and of OH^-):

$$F^-(aq) + H_2O(l) \overset{\longleftarrow}{\rightleftharpoons} HF(aq) + OH^-(aq)$$

As before, water dissociation contributes minute amounts of H_3O^+ and OH^-. Thus, in addition to the Na^+ ion, the species present are the same as in the HF solution: HF, H_2O, H_3O^+, F^-, and OH^-. The two species that affect the acidity are OH^-, predominantly from the F^- reaction with water, and H_3O^+ from water. In this case, $[OH^-]_{from\ F^-} \gg [H_3O^+]_{from\ H_2O}$, so the solution is basic.

To summarize, *the factor that determines the relative acidity of an HA solution or an A^- solution is the relative concentrations of HA and of A^- in each solution:*

- In an HA solution, $[HA] \gg [A^-]$ and $[H_3O^+]_{from\ HA} \gg [OH^-]_{from\ H_2O}$, so the solution is acidic.
- In an A^- solution, $[A^-] \gg [HA]$ and $[OH^-]_{from\ A^-} \gg [H_3O^+]_{from\ H_2O}$, so the solution is basic.

The Relation Between K_a and K_b of a Conjugate Acid-Base Pair

An important relationship exists between the K_a of HA and the K_b of A^-, which we can see by treating the two dissociation reactions as a reaction sequence and adding them together:

$$\begin{array}{r} \cancel{HA} + H_2O \rightleftharpoons H_3O^+ + \cancel{A^-} \\ \cancel{A^-} + H_2O \rightleftharpoons \cancel{HA} + OH^- \\ \hline 2H_2O \rightleftharpoons H_3O^+ + OH^- \end{array}$$

---

*This equation and equilibrium expression are sometimes referred to as a *hydrolysis reaction* and a *hydrolysis constant*, K_h, because water is dissociated (hydrolyzed), and its parts end up in the products. Actually, except for the charge on the base, this process is the same as the proton-abstraction process with molecular bases such as ammonia, so a new term and equilibrium constant are unnecessary. Thus, K_h is just another symbol for K_b, so we'll use K_b throughout.

The sum of the two dissociation reactions is the autoionization of water. Recall from Chapter 17 that, for a reaction that is the *sum* of two or more reactions, the overall equilibrium constant is the *product* of the individual equilibrium constants. Therefore, writing the expressions for each reaction gives

$$\frac{[H_3O^+][\cancel{A^-}]}{[\cancel{HA}]} \times \frac{[\cancel{HA}][OH^-]}{[\cancel{A^-}]} = [H_3O^+][OH^-]$$

or

$$K_a \quad \times \quad K_b \quad = \quad K_w \qquad (18.7)$$

This relationship allows us to find K_a of the acid in a conjugate pair given K_b of the base, and vice versa. Let's use this relationship to obtain a key piece of data for solving equilibrium problems. Reference tables typically have K_a and K_b values for molecular species only. The K_b for F^- or the K_a for $CH_3NH_3^+$, for example, does not appear in standard tables, but you can calculate either value simply by looking up the value of the molecular conjugate species and relating it to K_w. For the K_b value of F^-, for instance, we look up the K_a value for HF and relate it to K_w:

$$K_a \text{ of HF} = 6.8 \times 10^{-4} \text{ (from Table 18.2)}$$

So, we have

$$K_a \text{ of HF} \times K_b \text{ of } F^- = K_w$$

or,

$$K_b \text{ of } F^- = \frac{K_w}{K_a \text{ of HF}} = \frac{1.0 \times 10^{-14}}{6.8 \times 10^{-4}} = 1.5 \times 10^{-11}$$

We can use this calculated K_b value to finish solving the problem.

SAMPLE PROBLEM 18.10 Determining the pH of a Solution of A⁻

Problem Sodium acetate (CH_3COONa, or NaAc for this problem) has applications in photographic development and textile dyeing. What is the pH of 0.25 M NaAc? K_a of acetic acid (HAc) is 1.8×10^{-5}.

Plan From the formula (NaAc) and the fact that all sodium salts are water soluble, we know the initial concentration of acetate ion, Ac^-, is 0.25 M. We also know the K_a of the parent acid, HAc (1.8×10^{-5}). We have to find the pH of the solution of Ac^-, which acts as a base in water:

$$Ac^-(aq) + H_2O(l) \rightleftharpoons HAc(aq) + OH^-(aq) \qquad K_b = \frac{[HAc][OH^-]}{[Ac^-]}$$

If we calculate $[OH^-]$, we can find $[H_3O^+]$ and convert it to pH. To solve for $[OH^-]$, we need the K_b of Ac^-, which we obtain from the K_a of HAc and K_w. All sodium salts are soluble, so we know that $[Ac^-] = 0.25 M$. Our usual assumption is that $[Ac^-]_{init} \approx [Ac^-]$.

Solution Setting up the reaction table, with $x = [Ac^-]_{reacting} = [HAc] = [OH^-]$:

Concentration (M)	$Ac^-(aq)$	+ H₂O(*l*)	⇌	HAc(*aq*)	+	OH⁻(*aq*)
Initial	0.25	—		0		0
Change	−x	—		+x		+x
Equilibrium	0.25 − x	—		x		x

Solving for K_b:

$$K_b = \frac{K_w}{K_a} = \frac{1.0 \times 10^{-14}}{1.8 \times 10^{-5}} = 5.6 \times 10^{-10}$$

Making the assumption: Because K_b is small, $0.25 M - x \approx 0.25 M$.
Substituting into the expression for K_b and solving for x:

$$K_b = \frac{[HAc][OH^-]}{[Ac^-]} = 5.6 \times 10^{-10} \approx \frac{x^2}{0.25}$$

$$x = [OH^-] \approx 1.2 \times 10^{-5} M$$

Checking the assumption:

$$\frac{1.2\times10^{-5}\ M}{0.25\ M} \times 100 = 4.8\times10^{-3}\% < 5\%; \text{assumption is justified}$$

Note that

$$\frac{0.25}{5.6\times10^{-10}} = 4.5\times10^{8} > 400$$

Solving for pH:

$$[\text{H}_3\text{O}^+] = \frac{K_\text{w}}{[\text{OH}^-]} = \frac{1.0\times10^{-14}}{1.2\times10^{-5}} = 8.3\times10^{-10}\ M$$

$$\text{pH} = -\log(8.3\times10^{-10}) = \boxed{9.08}$$

Check The K_b calculation seems reasonable: $\sim10\times10^{-15}/2\times10^{-5} = 5\times10^{-10}$. Because Ac^- is a weak base, $[\text{OH}^-] > [\text{H}_3\text{O}^+]$; so the pH > 7, which makes sense.

FOLLOW-UP PROBLEM 18.10 Sodium hypochlorite (NaClO) is the active ingredient in household laundry bleach. What is the pH of 0.20 M NaClO?

SECTION SUMMARY
The extent to which a weak base abstracts a proton from water to form OH^- is expressed by a base-dissociation constant, K_b. Brønsted-Lowry bases include NH_3 and amines and the anions of weak acids. All produce basic solutions by accepting H^+ from water, which yields OH^- and thus makes $[\text{H}_3\text{O}^+] < [\text{OH}^-]$. A solution of HA is acidic because $[\text{HA}] \gg [\text{A}^-]$, so $[\text{H}_3\text{O}^+] > [\text{OH}^-]$. A solution of A^- is basic because $[\text{A}^-] \gg [\text{HA}]$, so $[\text{OH}^-] > [\text{H}_3\text{O}^+]$. By multiplying the expressions for K_a of HA and K_b of A^-, we obtain K_w. This relationship allows us to calculate either K_a of BH^+, the cationic conjugate acid of a molecular weak base B, or K_b of A^-, the anionic conjugate base of a molecular weak acid HA.

18.6 MOLECULAR PROPERTIES AND ACID STRENGTH

The strength of an acid depends on its ability to donate a proton, which in molecular terms depends in turn on the strength of the bond to the acidic proton. In this section, we apply trends in atomic and bond properties to determine the trends in acid strength of nonmetal hydrides and oxoacids and discuss the acidity of hydrated metal ions.

Trends in Acid Strength of Nonmetal Hydrides

Two factors determine how easily a proton is released from a nonmetal hydride: the electronegativity of the central nonmetal (E) and the strength of the E—H bond. Figure 18.12 displays two periodic trends:

1. *Across a period, nonmetal hydride acid strength increases.* Across a period, the electronegativity of the nonmetal E determines the trend. As E becomes more electronegative, electron density around H is withdrawn, and the E—H bond becomes more polar. As a result, an H^+ is released more easily to an O atom of a surrounding water molecule. In aqueous solution, the hydrides of Groups 3A(13) to 5A(15) do not behave as acids, but an increase in acid strength is seen in Groups 6A(16) and 7A(17). Thus, HCl is a stronger acid than H_2S because Cl is more electronegative (EN = 3.0) than S (EN = 2.5). The same relationship holds across each period.

2. *Down a group, nonmetal hydride acid strength increases.* Down a group, E—H bond strength determines the trend. As E becomes larger, the E—H bond

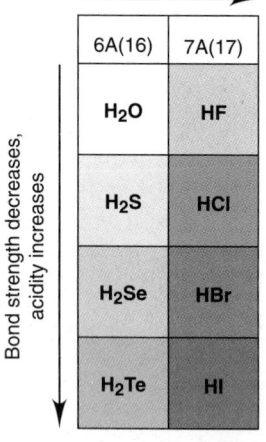

Electronegativity increases, acidity increases →

Bond strength decreases, acidity increases ↓

6A(16)	7A(17)
H_2O	HF
H_2S	HCl
H_2Se	HBr
H_2Te	HI

Figure 18.12 **The effect of atomic and molecular properties on nonmetal hydride acidity.** As the electronegativity of the nonmetal (E) bonded to the ionizable proton increases *(left to right)*, the acidity increases. As the length of the E—H bond increases *(top to bottom)*, the bond strength decreases, so the acidity increases. (In water, HCl, HBr, and HI are equally strong, for reasons discussed in Section 18.8.)

becomes longer and weaker, so H^+ comes off more easily.* Thus, the hydrohalic acids increase in strength down the group:

$$HF \ll HCl < HBr < HI$$

A similar trend in increasing acid strength is seen down Group 6A(16). (The trend in hydrohalic acid strength is not seen in aqueous solution, where HCl, HBr, and HI are all equally strong; we discuss how this trend is observed in Section 18.8.)

Trends in Acid Strength of Oxoacids

All oxoacids have the acidic H atom bound to an O atom, so bond strength (length) is not a factor in their acidity, though it is with the nonmetal hydrides. Rather, as you saw in Section 14.8, two factors determine the acid strength of oxoacids: the electronegativity of the central nonmetal (E) and the number of O atoms.

1. *For oxoacids with the **same** number of oxygens around E, acid strength increases with the electronegativity of E.* Consider the hypohalous acids (written here as HOE, where E is a halogen atom). The more electronegative E is, the more electron density it pulls from the O—H bond; the more polar the O—H bond becomes, the more easily H^+ is lost (Figure 18.13A). Because electronegativity decreases down the group, we predict that acid strength decreases: HOCl > HOBr > HOI. Our prediction is confirmed by the K_a values:

K_a of HOCl $= 2.9 \times 10^{-8}$ K_a of HOBr $= 2.3 \times 10^{-9}$ K_a of HOI $= 2.3 \times 10^{-11}$

We also predict (correctly) that in Group 6A(16), H_2SO_4 is stronger than H_2SeO_4; in Group 5A(15), H_3PO_4 is stronger than H_3AsO_4, and so forth.

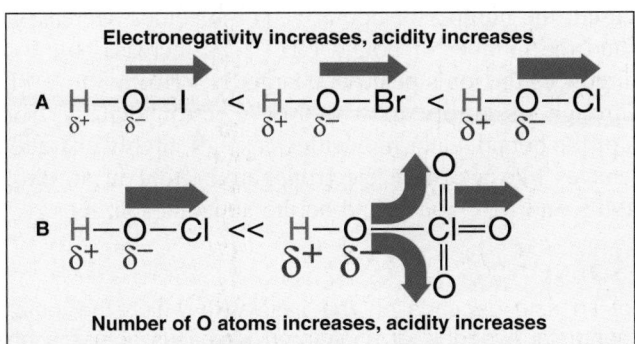

Figure 18.13 The relative strengths of oxoacids. **A,** Among the hypohalous acids (HOE, where E is a halogen), HOCl is the strongest and HOI the weakest. Because Cl is the most electronegative among the halogen atoms shown here, it withdraws electron density (indicated by thickness of green arrow) from the O—H bond most effectively, making that bond most polar in HOCl (indicated by the relative sizes of the δ symbols). **B,** Among the chlorine oxoacids, the additional O atoms in HOClO₃ pull electron density from the O—H bond, making the bond much more polar than that in HOCl (indicated once again by the sizes of the δ symbols).

2. *For oxoacids with **different** numbers of oxygens around a given E, acid strength increases with number of O atoms.* The electronegative O atoms pull electron density away from E, which makes the O—H bond more polar. The more O atoms present, the greater the shift in electron density, and the more easily the H^+ ion comes off (Figure 18.13B). Therefore, we predict, for instance, that chlorine oxoacids (written here as HOClₙ, with n from 0 to 3) increase in strength in the order HOCl < HOClO < HOClO₂ < HOClO₃. Once again, the K_a values support the prediction:

K_a of HOCl (hypochlorous acid) $= 2.9 \times 10^{-8}$
K_a of HClO₂ (chlorous acid) $= 1.12 \times 10^{-2}$
K_a of HOClO₂ (chloric acid) ≈ 1
K_a of HOClO₃ (perchloric acid) $= >10^7$

It follows from this that HNO_3 is stronger than HNO_2, that H_2SO_4 is stronger than H_2SO_3, and so forth.

---

*Actually, bond energy refers to bond breakage that forms an H atom, whereas acidity refers to bond breakage that forms an H^+ ion, so the two processes are not the same. Nevertheless, the magnitudes of the two types of bond breakage parallel each other.

Acidity of Hydrated Metal Ions

The aqueous solutions of certain metal ions are acidic because the *hydrated* metal ion transfers an H^+ ion to water. Consider a general metal nitrate, $M(NO_3)_n$, as it dissolves in water. The ions separate and become bound to a specific number of surrounding H_2O molecules. This equation shows the hydration of the cation (M^{n+}) with H_2O molecules; hydration of the anion (NO_3^-) is indicated by (*aq*):

$$M(NO_3)_n(s) + xH_2O(l) \longrightarrow M(H_2O)_x^{n+}(aq) + nNO_3^-(aq)$$

If the metal ion, M^{n+}, is *small and highly charged,* it has a high charge density and withdraws sufficient electron density from the O—H bonds of the bound water molecules for a proton to be released. That is, the hydrated cation, $M(H_2O)_x^{n+}$, acts as a typical Brønsted-Lowry acid. In the process, the bound H_2O molecule that releases the proton becomes a bound OH^- ion:

$$M(H_2O)_x^{n+}(aq) + H_2O(l) \rightleftharpoons M(H_2O)_{x-1}OH^{(n-1)+}(aq) + H_3O^+(aq)$$

Each type of hydrated metal ion that releases a proton has a characteristic K_a value. Table 18.7 shows some common examples.

Aluminum ion, for example, has the small size and high positive charge needed to produce an acidic solution. When an aluminum salt, such as $Al(NO_3)_3$, dissolves in water, the following steps occur:

$$Al(NO_3)_3(s) + 6H_2O(l) \longrightarrow Al(H_2O)_6^{3+}(aq) + 3NO_3^-(aq)$$

[dissolution and hydration]

$$Al(H_2O)_6^{3+}(aq) + H_2O(l) \rightleftharpoons Al(H_2O)_5OH^{2+}(aq) + H_3O^+(aq)$$

[dissociation of weak acid]

Note the formulas of the hydrated metal ions in the last step. When H^+ is released, the number of bound H_2O molecules decreases by 1 (from 6 to 5) and the number of bound OH^- ions increases by 1 (from 0 to 1), which reduces the ion's positive charge by 1 (from 3 to 2) (Figure 18.14).

Through its ability to withdraw electron density from the O—H bonds of the bound water molecules, a small, highly charged central metal ion behaves like a central electronegative atom in an oxoacid. Salts of most M^{2+} and M^{3+} ions yield acidic aqueous solutions.

Table 18.7 K_a Values of Some Hydrated Metal Ions at 25°C

Free Ion	Hydrated Ion	K_a
Fe^{3+}	$Fe(H_2O)_6^{3+}(aq)$	6×10^{-3}
Sn^{2+}	$Sn(H_2O)_6^{2+}(aq)$	4×10^{-4}
Cr^{3+}	$Cr(H_2O)_6^{3+}(aq)$	1×10^{-4}
Al^{3+}	$Al(H_2O)_6^{3+}(aq)$	1×10^{-5}
Be^{2+}	$Be(H_2O)_4^{2+}(aq)$	4×10^{-6}
Cu^{2+}	$Cu(H_2O)_6^{2+}(aq)$	3×10^{-8}
Pb^{2+}	$Pb(H_2O)_6^{2+}(aq)$	3×10^{-8}
Zn^{2+}	$Zn(H_2O)_6^{2+}(aq)$	1×10^{-9}
Co^{2+}	$Co(H_2O)_6^{2+}(aq)$	2×10^{-10}
Ni^{2+}	$Ni(H_2O)_6^{2+}(aq)$	1×10^{-10}

ACID STRENGTH ↑

SECTION SUMMARY

The strength of an acid depends on the ease with which the ionizable proton is released. For nonmetal hydrides, acid strength increases across a period, with the electronegativity of the nonmetal (E), and down a group, with the length of the E—H bond. For oxoacids with the same number of O atoms, acid strength increases with electronegativity of E; for oxoacids with the same E, acid strength increases with number of O atoms. Small, highly charged metal ions are acidic in water because they withdraw electron density from the O—H bonds of bound H_2O molecules, releasing an H^+ ion to the solution.

Figure 18.14 The acidic behavior of the hydrated Al^{3+} ion. When a metal ion enters water, it is hydrated as water molecules bond to it. If the ion is small and multiply charged, as is the Al^{3+} ion, it pulls sufficient electron density from the O—H bonds of the attached water molecules to make the bonds more polar, and an H^+ ion is transferred to a nearby water molecule.

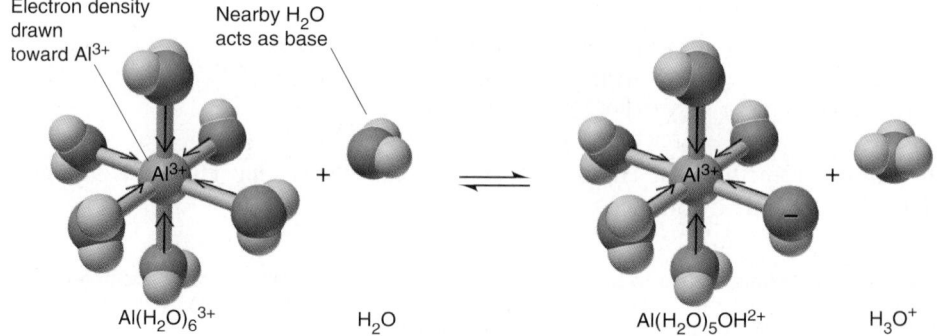

Electron density drawn toward Al^{3+} Nearby H_2O acts as base

$Al(H_2O)_6^{3+}$ H_2O $Al(H_2O)_5OH^{2+}$ H_3O^+

18.7 ACID-BASE PROPERTIES OF SALT SOLUTIONS

Up to now you've seen that cations of weak bases (such as NH_4^+) are acidic, anions of weak acids (such as CN^-) are basic, anions of polyprotic acids (such as $H_2PO_4^-$) are often acidic, and small, highly charged metal cations (such as Al^{3+}) are acidic. Therefore, when salts containing these ions dissolve in water, the pH of the solution is affected. You can predict the relative acidity of a salt solution from the relative ability of the cation and/or anion to react with water. Let's examine the ionic makeup of salts that yield neutral, acidic, or basic solutions to see how we make this prediction.

Salts That Yield Neutral Solutions

A salt consisting of the anion of a strong acid and the cation of a strong base yields a neutral solution because the ions do not react with water. In order to see why the ions don't react, let's consider the dissociation of the parent acid and base. When a strong acid such as HNO_3 dissolves, complete dissociation takes place:

$$HNO_3(l) + H_2O(l) \longrightarrow NO_3^-(aq) + H_3O^+(aq)$$

H_2O is a much stronger base than NO_3^-, so the reaction proceeds essentially to completion. The same argument can be made for any strong acid: *the anion of a strong acid is a much weaker base than water.* Therefore, a strong acid anion is hydrated, but nothing further happens.

Now consider the dissociation of a strong base, such as NaOH:

$$NaOH(s) \xrightarrow{\text{H}_2\text{O}} Na^+(aq) + OH^-(aq)$$

The Na^+ ion has a relatively large size and low charge and therefore does not bond strongly with the water molecules around it. When Na^+ enters water, it becomes hydrated but nothing further happens. *The cations of all strong bases behave this way.*

The anions of strong acids are the halide ions, except F^-, and those of strong oxoacids, such as NO_3^-, ClO_4^-, and HSO_4^-. The cations of strong bases are those from Group 1A(1) and Ca^{2+}, Sr^{2+}, and Ba^{2+} from Group 2A(2). Salts containing only these ions, such as NaCl and $Ba(NO_3)_2$, *yield neutral solutions because no reaction with water takes place.*

Salts That Yield Acidic Solutions

A salt consisting of the anion of a strong acid and the cation of a weak base yields an acidic solution because the cation acts as a weak acid, and the anion does not react. For example, NH_4Cl produces an acidic solution because the NH_4^+ ion, the cation that forms from the weak base NH_3, is a weak acid, and the Cl^- ion, the anion of a strong acid, does not react:

$$NH_4Cl(s) \xrightarrow{\text{H}_2\text{O}} NH_4^+(aq) + Cl^-(aq) \qquad \text{[dissolution and hydration]}$$
$$NH_4^+(aq) + H_2O(l) \rightleftharpoons NH_3(aq) + H_3O^+(aq) \qquad \text{[dissociation of weak acid]}$$

As you saw earlier, *small, highly charged metal ions* make up another group of cations that yield H_3O^+ in solution. For example, $Fe(NO_3)_3$ produces an acidic solution because the hydrated Fe^{3+} ion acts as a weak acid, whereas the NO_3^- ion, the anion of a strong acid, does not react:

$$Fe(NO_3)_3(s) + 6H_2O(l) \xrightarrow{\text{H}_2\text{O}} Fe(H_2O)_6^{3+}(aq) + 3NO_3^-(aq)$$

[dissolution and hydration]

$$Fe(H_2O)_6^{3+}(aq) + H_2O(l) \rightleftharpoons Fe(H_2O)_5OH^{2+}(aq) + H_3O^+(aq)$$

[dissociation of weak acid]

A third group of salts that yield H_3O^+ ions in solutions consists of *cations of strong bases and anions of polyprotic acids with another ionizable proton.* For example, NaH_2PO_4 yields an acidic solution because Na^+, the cation of a strong base, does not react, while $H_2PO_4^-$ is a weak acid:

$$NaH_2PO_4(s) \xrightarrow{H_2O} Na^+(aq) + H_2PO_4^-(aq) \quad \text{[dissolution and hydration]}$$

$$H_2PO_4^-(aq) + H_2O(l) \rightleftharpoons HPO_4^{2-}(aq) + H_3O^+(aq) \quad \text{[dissociation of weak acid]}$$

Salts That Yield Basic Solutions

A salt consisting of the anion of a weak acid and the cation of a strong base yields a basic solution in water because the anion acts as a weak base, and the cation does not react. The anion of a weak acid abstracts a proton from water to yield OH^- ion. Sodium acetate, for example, yields a basic solution because the Na^+ ion, the cation of a strong base, does not react with water, and the CH_3COO^- ion, the anion of the weak acid CH_3COOH, acts as a weak base:

$$CH_3COONa(s) \xrightarrow{H_2O} Na^+(aq) + CH_3COO^-(aq) \quad \text{[dissolution and hydration]}$$

$$CH_3COO^-(aq) + H_2O(l) \rightleftharpoons CH_3COOH(aq) + OH^-(aq) \quad \text{[reaction of weak base]}$$

Table 18.8 displays the acid-base behavior of the various types of salts in water.

Table 18.8 The Behavior of Salts in Water

Salt Solution (Examples)	pH	Nature of Ions	Ion That Reacts with Water	
Neutral [NaCl, KBr, Ba(NO$_3$)$_2$]	7.0	Cation of strong base Anion of strong acid	None	
Acidic (NH$_4$Cl, NH$_4$NO$_3$, CH$_3$NH$_3$Br)	<7.0	Cation of weak base Anion of strong acid	Cation	
Acidic [Al(NO$_3$)$_3$, CrCl$_3$, FeBr$_3$]	<7.0	Small, highly charged cation Anion of strong acid	Cation	
Acidic (NaH$_2$PO$_4$, KHSO$_4$, NaHSO$_3$)	<7.0	Cation of strong base First anion of polyprotic acid	Anion	
Basic (CH$_3$COONa, KF, Na$_2$CO$_3$)	>7.0	Cation of strong base Anion of weak acid	Anion	

SAMPLE PROBLEM 18.11 Predicting Relative Acidity of Salt Solutions

Problem Predict whether aqueous solutions of the following are acidic, basic, or neutral, and write an equation for the reaction of any ion with water:
(a) Potassium perchlorate, $KClO_4$ (b) Sodium benzoate, C_6H_5COONa
(c) Chromium trichloride, $CrCl_3$ (d) Sodium hydrogen sulfate, $NaHSO_4$
Plan We examine the formulas to determine the cations and anions. Depending on the nature of these ions, the solution will be neutral (strong-acid anion and strong-base cation), acidic (weak-base cation and strong-acid anion, highly charged metal cation, or first anion of a polyprotic acid), or basic (weak-acid anion and strong-base cation).
Solution (a) Neutral. The ions are K^+ and ClO_4^-. The K^+ ion is from the strong base KOH, and the ClO_4^- anion is from the strong acid $HClO_4$. Neither ion reacts with water.
(b) Basic. The ions are Na^+ and $C_6H_5COO^-$. Na^+ is the cation of the strong base NaOH and does not react with water. The benzoate ion, $C_6H_5COO^-$, is from the weak acid benzoic acid, so it reacts with water to produce OH^- ion:

$$C_6H_5COO^-(aq) + H_2O(l) \rightleftharpoons C_6H_5COOH(aq) + OH^-(aq)$$

(c) Acidic. The ions are Cr^{3+} and Cl^-. Cl^- is the anion of the strong acid HCl, so it does not react with water. Cr^{3+} is a small metal ion with a high positive charge, so the hydrated ion, $Cr(H_2O)_6^{3+}$, reacts with water to produce H_3O^+:

$$Cr(H_2O)_6^{3+}(aq) + H_2O(l) \rightleftharpoons Cr(H_2O)_5OH^{2+}(aq) + H_3O^+(aq)$$

(d) Acidic. The ions are Na^+ and HSO_4^-. Na^+ is the cation of the strong NaOH, so it does not react with water. HSO_4^- is the first anion of the diprotic acid H_2SO_4, and it reacts with water to produce H_3O^+:

$$HSO_4^-(aq) + H_2O(l) \rightleftharpoons SO_4^{2-}(aq) + H_3O^+(aq)$$

FOLLOW-UP PROBLEM 18.11 Write equations to predict whether solutions of the following salts are acidic, basic, or neutral: (a) $KClO_2$; (b) $CH_3NH_3NO_3$; (c) CsI.

Salts of Weakly Acidic Cations and Weakly Basic Anions

The only salts left to consider are those consisting of a cation that acts as a weak acid *and* an anion that acts as a weak base. In these cases, and there are quite a few, both ions react with water. It makes sense, then, that the overall acidity of the solution will depend on the relative acid strength or base strength of the separated ions, which can be determined by comparing their equilibrium constants.

For example, will an aqueous solution of ammonium hydrogen sulfide, NH_4HS, be acidic or basic? First, we write equations for any reactions that occur between the separated ions and water. Ammonium ion is the conjugate acid of a weak base, so it acts as a weak acid:

$$NH_4^+(aq) + H_2O(l) \rightleftharpoons NH_3(aq) + H_3O^+(aq)$$

Hydrogen sulfide ion is the anion of the weak acid H_2S, so it acts as a weak base:

$$HS^-(aq) + H_2O(l) \rightleftharpoons H_2S(aq) + OH^-(aq)$$

The reaction that goes farther to the right will have the greater influence on the pH of the solution, so we must compare the K_a of NH_4^+ with the K_b of HS^-. Recall that only molecular compounds are listed in K_a and K_b tables, so we have to calculate these values for the ions:

$$K_a \text{ of } NH_4^+ = \frac{K_w}{K_b \text{ of } NH_3} = \frac{1.0 \times 10^{-14}}{1.76 \times 10^{-5}} = 5.7 \times 10^{-10}$$

$$K_b \text{ of } HS^- = \frac{K_w}{K_{a1} \text{ of } H_2S} = \frac{1.0 \times 10^{-14}}{9 \times 10^{-8}} = 1 \times 10^{-7}$$

The difference in magnitude of the equilibrium constants ($K_b \approx 200K_a$) tells us that the abstraction of a proton from H_2O by HS^- proceeds further than the release of a proton to H_2O by NH_4^+. In other words, because K_b of $HS^- > K_a$ of NH_4^+, the NH_4HS solution is basic.

SAMPLE PROBLEM 18.12 Predicting the Relative Acidity of Salt Solutions from K_a and K_b of the Ions

Problem Determine whether an aqueous solution of zinc formate, $Zn(HCOO)_2$, is acidic, basic, or neutral.

Plan The formula consists of the small, highly charged, and therefore weakly acidic, Zn^{2+} cation and the weakly basic $HCOO^-$ anion of the weak acid HCOOH. To determine the relative acidity of the solution, we write equations that show the reactions of the ions with water, and then find K_a of Zn^{2+} (from Table 18.7) and calculate K_b of $HCOO^-$ (from K_a of HCOOH in Table 18.2) to see which ion reacts to a greater extent.

Solution Writing the reactions with water:

$$Zn(H_2O)_6^{2+}(aq) + H_2O(l) \rightleftharpoons Zn(H_2O)_5OH^+(aq) + H_3O^+(aq)$$

$$HCOO^-(aq) + H_2O(l) \rightleftharpoons HCOOH(aq) + OH^-(aq)$$

Obtaining K_a and K_b of the ions: From Table 18.7, K_a of $Zn(H_2O)_6^{2+}(aq) = 1 \times 10^{-9}$. From Table 18.2, we obtain K_a of HCOOH and solve for K_b of $HCOO^-$:

$$K_b \text{ of } HCOO^- = \frac{K_w}{K_a \text{ of HCOOH}} = \frac{1.0 \times 10^{-14}}{1.8 \times 10^{-4}} = 5.6 \times 10^{-11}$$

K_a of $Zn(H_2O)_6^{2+} > K_b$ of $HCOO^-$, so the solution is ‎ acidic.

FOLLOW-UP PROBLEM 18.12 Determine whether solutions of the following salts are acidic, basic, or neutral: **(a)** $Cu(CH_3COO)_2$; **(b)** NH_4F.

SECTION SUMMARY

Salts that yield a neutral solution consist of ions that do not react with water. Salts that yield an acidic solution contain an unreactive anion and a cation that releases a proton to water. Salts that yield a basic solution contain an unreactive cation and an anion that abstracts a proton from water. If both cation and anion react with water, the ion that reacts to the greater extent (higher K) determines the acidity or basicity of the salt solution.

18.8 GENERALIZING THE BRØNSTED-LOWRY CONCEPT: THE LEVELING EFFECT

We conclude our focus on the Brønsted-Lowry concept with an important principle that holds for acid-base behavior in any solvent. Notice that, in H_2O, all Brønsted-Lowry acids yield H_3O^+ and all Brønsted-Lowry bases yield OH^- —the ions that form when the solvent autoionizes. In general, *an acid yields the cation and a base yields the anion of solvent autoionization.*

This idea lets us examine a question you may have been wondering about: why are all strong acids and strong bases *equally* strong in water? The answer is that *in water, the strongest acid possible is H_3O^+ and the strongest base possible is OH^-.* The moment we put some gaseous HCl in water, it reacts with the base H_2O and forms H_3O^+. The same holds for HNO_3, H_2SO_4, and any strong acid. All strong acids are equally strong in water because they dissociate *completely* to form H_3O^+. Given that the strong acid is no longer present, we are actually observing the acid strength of H_3O^+.

Similarly, strong bases, such as $Ba(OH)_2$, dissociate completely in water to yield OH^-. Even those that do not contain hydroxide ions in the solid, such as K_2O, do so. The oxide ion, which is a stronger base than OH^-, immediately abstracts a proton from water to form OH^-:

$$2K^+(aq) + O^{2-}(aq) + H_2O(l) \longrightarrow 2K^+(aq) + 2OH^-(aq)$$

No matter what species we try, any acid stronger than H_3O^+ simply donates its proton to H_2O, and any base stronger than OH^- accepts a proton from H_2O. Thus, water exerts a **leveling effect** on any strong acid or base by reacting with it to form the products of water's autoionization. Acting as a base, water levels the strength of all strong acids by making them appear equally strong, and acting as an acid, it levels the strength of all strong bases as well.

To rank strong acids in terms of relative strength, we must dissolve them in a solvent that is a *weaker* base than water, one that accepts their protons less readily. For example, you saw in Figure 18.12 that the hydrohalic acids increase in strength as the halogen becomes larger, as a result of the longer, weaker H—X bond. In water, HF is weaker than the other hydrogen halides, but HCl, HBr, and HI appear equally strong because water causes them to dissociate completely. When we dissolve them in pure acetic acid, however, *the acetic acid acts as the base* and accepts a proton from the acids:

$$
\begin{array}{cccc}
\text{acid} & \text{base} & \text{base} & \text{acid} \\
HCl(g) + CH_3COOH(l) & \rightleftharpoons & Cl^-(acet) + CH_3COOH_2^+(acet) \\
HBr(g) + CH_3COOH(l) & \rightleftharpoons & Br^-(acet) + CH_3COOH_2^+(acet) \\
HI(g) + CH_3COOH(l) & \rightleftharpoons & I^-(acet) + CH_3COOH_2^+(acet)
\end{array}
$$

[The use of (*acet*) instead of (*aq*) indicates solvation by CH_3COOH.] However, because acetic acid is a *weaker base* than water, the three acids protonate it to *different* extents. Measurements show that HI protonates the solvent to a greater extent than HBr, and HBr does so more than HCl; that is, in pure acetic acid, $K_{HI} > K_{HBr} > K_{HCl}$. Therefore, HCl is a weaker acid than HBr, which is weaker than HI. Similarly, the relative strength of strong bases is determined in a solvent that is a weaker acid than H_2O, such as liquid NH_3.

SECTION SUMMARY

Strong acids (or strong bases) dissociate completely to yield H_3O^+ (or OH^-) in water; in effect, water equalizes (levels) their strengths. Acids that are equally strong in water show differences in strength when dissolved in a solvent that is a weaker base than water, such as acetic acid.

18.9 ELECTRON-PAIR DONATION AND THE LEWIS ACID-BASE DEFINITION

The final acid-base concept we consider was developed by Gilbert N. Lewis, whose contribution to understanding the importance of valence electron pairs in molecular bonding we discussed in Chapter 9. Whereas the Brønsted-Lowry concept focuses on the proton in defining a species as an acid or a base, the Lewis concept highlights the role of the *electron pair*. The **Lewis acid-base definition** holds that

- A *base* is any species that *donates* an electron pair.
- An *acid* is any species that *accepts* an electron pair.

The Lewis definition, like the Brønsted-Lowry definition, requires that a base have an electron pair to donate, so it does not expand the classes of bases. However, *it greatly expands the classes of acids*. Many species, such as CO_2 and Cu^{2+}, that do not contain H in their formula (and thus cannot be Brønsted-Lowry acids) function as Lewis acids by accepting an electron pair in their reactions. Lewis stated his objection to the proton as the defining feature of an acid this way: "To restrict the group of acids to those substances which contain hydrogen interferes as seriously with the systematic understanding of chemistry as would the restriction of the term oxidizing agent to those substances containing oxygen." Moreover, in the Lewis sense, the proton itself functions as an acid because it accepts the electron pair donated by a base:

$$B\!:\ +\ H^+ \rightleftharpoons B\!-\!H^+$$

Thus, *all Brønsted-Lowry acids donate H^+, a Lewis acid.*

The product of any Lewis acid-base reaction is called an **adduct**, *a single species that contains a **new** covalent bond:*

$$\overset{\frown}{A} + :B \rightleftharpoons A{-}B \text{ (adduct)}$$

Thus, the Lewis concept radically broadens the idea of acid-base reactions. What to Arrhenius was the formation of H_2O from H^+ and OH^- became, to Brønsted and Lowry, the transfer of a proton from a stronger acid to a stronger base to form a weaker base and weaker acid. To Lewis, the same process became *the donation and acceptance of an electron pair to form a covalent bond in an adduct.*

As we've seen, the key feature of a *Lewis base is a lone pair of electrons to donate.* The key feature of a *Lewis acid is a vacant orbital* (or the ability to rearrange its bonds to form one) to accept that lone pair and form a new bond. There are a variety of neutral molecules and positively charged ions that satisfy this requirement.

Molecules as Lewis Acids

Many neutral molecules function as Lewis acids. In every case, the atom that accepts the electron pair is low in electron density because of either an electron deficiency or a polar multiple bond.

Lewis Acids with Electron-Deficient Atoms Some molecular Lewis acids contain a central atom that is *electron deficient,* one surrounded by fewer than eight valence electrons. The most important are covalent compounds of the Group 3A(13) elements boron and aluminum. As noted in Chapters 9 and 14, these compounds react vigorously to complete their octet. For example, boron trifluoride accepts an electron pair from ammonia to form a covalent bond in a gaseous Lewis acid-base reaction:

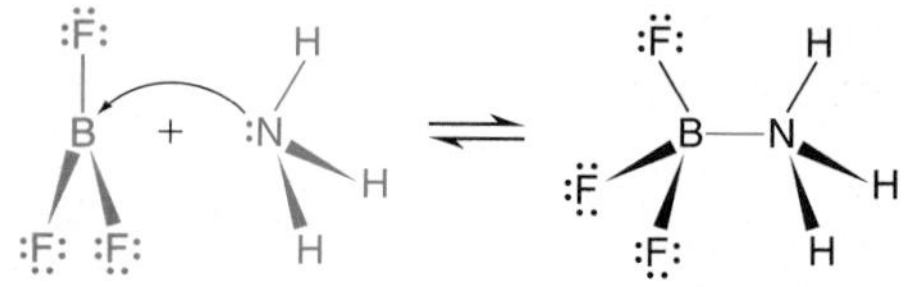

Unexpected solubility behavior is sometimes due to adduct formation. Aluminum chloride, for instance, dissolves freely in relatively nonpolar diethyl ether because of a Lewis acid-base reaction, in which the ether's O atom donates an electron pair to Al to form a covalent bond:

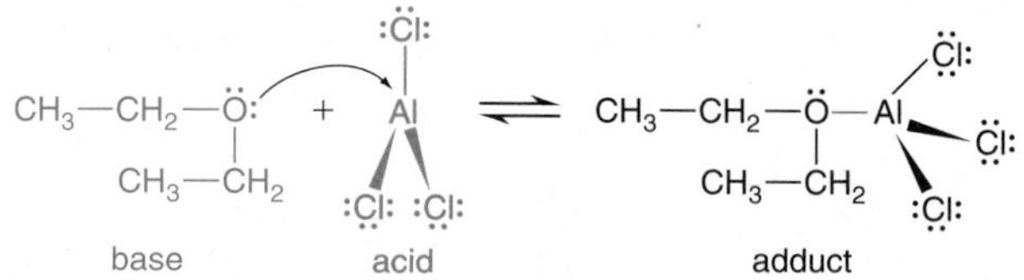

This acidic behavior of boron and aluminum halides is put to use in many organic syntheses. For example, toluene, an important solvent and organic reagent, can be made by the action of CH_3Cl on benzene in the presence of $AlCl_3$. The Lewis acid $AlCl_3$ abstracts the Lewis base Cl^- from CH_3Cl to form an adduct that has a reactive CH_3^+ group, which attacks the benzene ring:

$$\underset{\text{base}}{CH_3Cl} + \underset{\text{acid}}{AlCl_3} \rightleftharpoons \underset{\text{adduct}}{[CH_3]^+[Cl{-}AlCl_3]^-}$$

$$\underset{\text{benzene}}{C_6H_6} + [CH_3]^+[Cl{-}AlCl_3]^- \rightleftharpoons \underset{\text{toluene}}{C_6H_5CH_3} + AlCl_3 + HCl$$

Lewis Acids with Polar Multiple Bonds Molecules that contain a polar double bond also function as Lewis acids. As the electron pair on the Lewis base approaches the partially positive end of the double bond, one of the bonds breaks to form the new bond in the adduct. For example, consider the reaction that occurs when SO_2 dissolves in water. The electronegative O atoms in SO_2 withdraw electron density from the central S, so it is partially positive. The O atom of water donates a lone pair to the S, breaking one of the π bonds and forming an S—O bond, and a proton is transferred from water to that O. The resulting adduct is sulfurous acid, and the overall process is

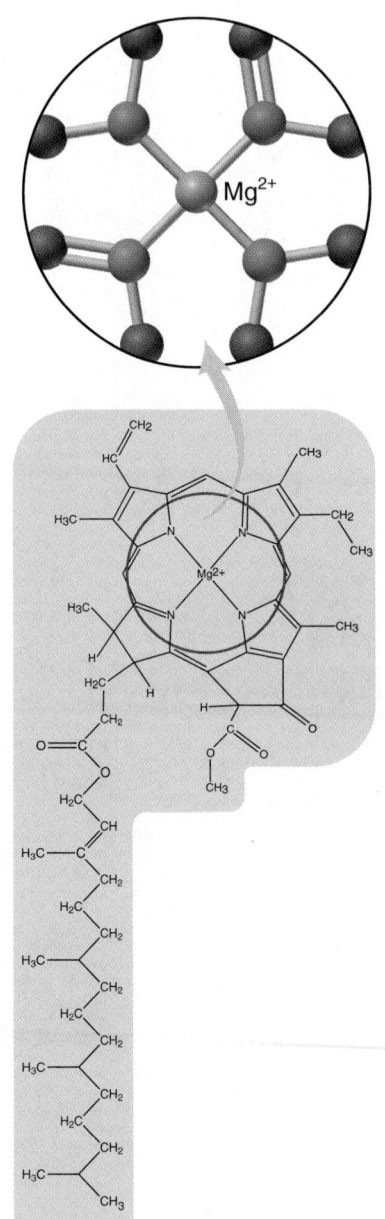

Figure 18.15 **The Mg^{2+} ion as a Lewis acid in the chlorophyll molecule.** Many biomolecules contain metal ions that act as Lewis acids. In chlorophyll, Mg^{2+} accepts electron pairs from surrounding N atoms that are part of the large organic portion of the molecule.

In the formation of carbonates from a metal oxide and carbon dioxide, an analogous reaction occurs in a nonaqueous heterogeneous system. The O^{2-} ion (shown below from CaO) donates an electron pair to the partially positive C in CO_2, a π bond breaks, and the CO_3^{2-} ion forms as the adduct:

Metal Cations as Lewis Acids

Earlier we saw that certain hydrated metal ions act as Brønsted-Lowry acids. In the Lewis sense, the hydration process itself is an acid-base reaction. The hydrated cation is the adduct, as lone electron pairs on the O atoms of water form covalent bonds to the positively charged ion; thus, *any metal ion acts as a Lewis acid when it dissolves in water:*

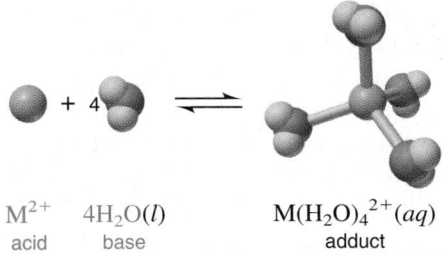

$$M^{2+} \quad 4H_2O(l) \quad \rightleftharpoons \quad M(H_2O)_4{}^{2+}(aq)$$

acid   base   adduct

Ammonia is a stronger Lewis base than water because it displaces H_2O from a hydrated ion when aqueous NH_3 is added:

$$Ni(H_2O)_6{}^{2+}(aq) + 6NH_3(aq) \rightleftharpoons Ni(NH_3)_6{}^{2+}(aq) + 6H_2O(l)$$

hydrated adduct   base   ammoniated adduct

We discuss the equilibrium nature of these acid-base reactions in greater detail in Chapter 19, and we investigate the structures of these ions in Chapter 23.

Many essential biomolecules are Lewis adducts with central metal ions. Most often, O and N atoms of organic groups, with their lone pairs, serve as the Lewis base. Chlorophyll is a Lewis adduct of a central Mg^{2+} and the four N atoms of an organic tetrapyrrole ring system (Figure 18.15). Vitamin B_{12} has a similar structure with a central Co^{3+}, and so does heme, but with a central Fe^{2+}. Several other metal ions, such as Zn^{2+}, Mo^{2+}, and Cu^{2+}, are bound at the active sites of enzymes and function as Lewis acids in the catalytic action.

SAMPLE PROBLEM 18.13 Identifying Lewis Acids and Bases

Problem Identify the Lewis acids and Lewis bases in the following reactions:
(a) $H^+ + OH^- \rightleftharpoons H_2O$
(b) $Cl^- + BCl_3 \rightleftharpoons BCl_4^-$
(c) $K^+ + 6H_2O \rightleftharpoons K(H_2O)_6^+$
Plan We examine the formulas to see which species accepts the electron pair (Lewis acid) and which donates it (Lewis base) in forming the adduct.
Solution **(a)** The H^+ ion accepts an electron pair from the OH^- ion in forming a bond. H^+ is the acid and OH^- is the base.

(b) The Cl^- ion has four lone pairs and uses one to form a new bond to the central B. Therefore, BCl_3 is the acid and Cl^- is the base.

(c) The K^+ ion does not have any valence electrons to provide, so the bond is formed when electron pairs from O atoms of water enter empty orbitals on K^+. Thus, K^+ is the acid and H_2O is the base.

Check The Lewis acids (H^+, BCl_3, and K^+) each have an unfilled valence shell that can accept an electron pair from the Lewis bases (OH^-, Cl^-, and H_2O).

FOLLOW-UP PROBLEM 18.13 Identify the Lewis acids and Lewis bases in the following reactions:
(a) $OH^- + Al(OH)_3 \rightleftharpoons Al(OH)_4^-$
(b) $SO_3 + H_2O \rightleftharpoons H_2SO_4$
(c) $Co^{3+} + 6NH_3 \rightleftharpoons Co(NH_3)_6^{3+}$

An Overview of Acid-Base Definitions

By looking closely at the essential chemical change involved, chemists can see a common theme in reactions as diverse as a standardized base being used to analyze an unknown fatty acid, or baking soda being used in breadmaking, or even oxygen binding to hemoglobin in a blood cell. From this wider perspective, the diversity of acid-base reactions takes on more unity. Let's stand back and survey the scope of the three acid-base definitions and see how they fit together.

The *classical (Arrhenius) definition,* which was the first attempt at describing acids and bases on the molecular level, is the most limited and narrow of the three definitions. It applies only to species whose structures include an H atom or OH group that is released as an ion when the species dissolves in water. Because relatively few species have these prerequisites, Arrhenius acid-base reactions are relatively few in number, and all such reactions result in the formation of H_2O.

The *Brønsted-Lowry definition* is more general, seeing acid-base reactions as proton-transfer processes and eliminating the requirement that they occur in water. Whereas a Brønsted-Lowry acid, like an Arrhenius acid, still must have an H, a Brønsted-Lowry base is defined as any species with an electron pair available to accept a transferred proton. This definition includes a great many more species as bases. Furthermore, it defines the acid-base reaction in terms of conjugate acid-base pairs, with an acid and a base on both sides of the reaction. The system reaches an equilibrium state based on the relative strengths of the acid, the base, and their conjugates.

The *Lewis definition* has the widest scope of the three. The defining event of a Lewis acid-base reaction is the donation and acceptance of an electron pair to form a new covalent bond. Lewis bases still must have an electron pair to donate, but Lewis acids—as electron-pair acceptors—include many species not encompassed by the other two definitions, such as electron-deficient compounds, compounds with polar double bonds, metal ions, and the proton itself.

The Lewis acid-base definition focuses on the donation or acceptance of an electron pair to form a new covalent bond in an adduct, the product of an acid-base reaction. Lewis bases donate the electron pair, and Lewis acids accept it. Thus, many species that do not contain H are Lewis acids. Molecules with polar double bonds act as Lewis acids, as do those with electron-deficient atoms. Metal ions act as Lewis acids when they dissolve in water, which acts as a Lewis base, to form the adduct, a hydrated cation. Many metal ions function as Lewis acids in biomolecules.

Chapter Perspective

In this chapter, we extended the principles of equilibrium to acids and bases. We also investigated one direction in which the science of chemistry matured, as narrow definitions of acids and bases progressively widened to encompass different species, physical states, solvent systems, and reaction types. Acids and bases, by whatever definition, are an extremely important group of substances. In Chapter 19, we continue our discussion of these systems and apply many of the ideas developed here to other aqueous equilibria.

For Review and Reference (Numbers in parentheses refer to pages, unless noted otherwise.)

Learning Objectives

Relevant section and/or sample problem (SP) numbers appear in parentheses.

Understand These Concepts

1. Why the proton exists bound to a water molecule, as H_3O^+, in all aqueous acid-base reactions (Section 18.1)
2. The classical (Arrhenius) definitions of an acid and a base (Section 18.1)
3. Why all reactions of a strong acid and a strong base have the same ΔH_{rxn}^0 (Section 18.1)
4. How the strength of an acid (or base) relates to the extent of its dissociation into ions in water (Section 18.1)
5. How relative acid strength is expressed by the acid-dissociation constant K_a (Section 18.1)
6. Why water is a very weak electrolyte and how its autoionization is expressed by K_w (Section 18.2)
7. Why $[H_3O^+]$ is inversely related to $[OH^-]$ in any aqueous solution (Section 18.2)
8. How the relative magnitudes of $[H_3O^+]$ and $[OH^-]$ define whether a solution is acidic, basic, or neutral (Section 18.2)
9. The Brønsted-Lowry definitions of an acid and a base and how an acid-base reaction can be viewed as a proton-transfer process (Section 18.3)
10. How water acts as a base (or as an acid) when an acid (or a base) dissolves in it (Section 18.3)
11. How a conjugate acid-base pair differs by one proton (Section 18.3)
12. How a Brønsted-Lowry acid-base reaction involves two conjugate acid-base pairs (Section 18.3)
13. Why a stronger acid and base react ($K_c > 1$) to form a weaker base and acid (Section 18.3)

14. How percent dissociation of a weak acid increases as its concentration decreases (Section 18.4)
15. How a polyprotic acid dissociates in two or more steps and why only the first step supplies significant $[H_3O^+]$ (Section 18.4)
16. How weak bases in water accept a proton rather than dissociate; the meaning of K_b and pK_b (Section 18.5)
17. How ammonia, amines, and weak-acid anions act as weak bases in water (Section 18.5)
18. Why relative concentrations of HA and A^- determine the acidity of their solution (Section 18.5)
19. The relationship of the K_a and K_b of a conjugate acid-base pair to K_w (Section 18.5)
20. The effects of electronegativity, bond polarity, and bond energy on acid strength (Section 18.6)
21. Why aqueous solutions of small, highly charged metal ions are acidic (Section 18.6)
22. The various combinations of cations and anions that lead to acidic, basic, or neutral salt solutions (Section 18.7)
23. Why the strengths of strong acids are leveled in water but differentiated in a less basic solvent (Section 18.8)
24. The Lewis definitions of an acid and a base and how a Lewis acid-base reaction involves the donation and acceptance of an electron pair to form a covalent bond (Section 18.9)
25. How Group 3A(13) halides, molecules with polar multiple bonds, and metal cations act as Lewis acids (Section 18.9)

Learning Objectives *(continued)*

Master These Skills

1. Classifying strong and weak acids and bases from their formulas (SP 18.1)
2. Using K_w to calculate $[H_3O^+]$ and $[OH^-]$ in an aqueous solution (SP 18.2)
3. Using p-scales to express $[H_3O^+]$, $[OH^-]$, and K (Section 18.2)
4. Interconverting pH, pOH, $[H_3O^+]$, and $[OH^-]$ (SP 18.3)
5. Identifying conjugate acid-base pairs (SP 18.4)
6. Using relative acid strengths to predict the net direction of an acid-base reaction (SP 18.5)
7. Calculating K_a of a weak acid from pH (SP 18.6)
8. Calculating $[H_3O^+]$ (and, thus, pH) from K_a and $[HA]_{init}$ (SP 18.7)

9. Applying the quadratic equation to find concentrations (Follow-up Problem 18.7)
10. Calculating the percent dissociation of a weak acid (Section 18.4)
11. Calculating $[H_3O^+]$ and other concentrations for a polyprotic acid (SP 18.8)
12. Calculating pH from K_b and $[B]_{init}$ (SP 18.9)
13. Finding K_b of A^- from K_a of HA and K_w (Section 18.5 and SP 18.10)
14. Calculating pH from K_b of A^- and $[A^-]_{init}$ (SP 18.10)
15. Predicting the relative acid strength of nonmetal hydrides and oxoacids (Section 18.6)
16. Predicting the relative acidity of a salt solution from the nature of the cation and anion (SPs 18.11 and 18.12)
17. Identifying Lewis acids and bases (SP 18.13)

Key Terms

Section 18.1
hydronium ion, H_3O^+ (758)
classical (Arrhenius) acid-base definition (759)
neutralization (759)
acid-dissociation (acid-ionization) constant (K_a) (761)

Section 18.2
autoionization (764)
ion-product constant for water (K_w) (764)
pH (765)
acid-base indicator (768)

Section 18.3
Brønsted-Lowry acid-base definition (768)
proton donor (768)
proton acceptor (768)
conjugate acid-base pair (769)

Section 18.4
polyprotic acid (776)

Section 18.5
base-dissociation (base-ionization) constant (K_b) (779)

Section 18.8
leveling effect (790)

Section 18.9
Lewis acid-base definition (791)
adduct (792)

Key Equations and Relationships

18.1 Defining the acid-dissociation constant (761):

$$K_a = \frac{[H_3O^+][A^-]}{[HA]}$$

18.2 Defining the ion-product constant for water (764):
$$K_w = [H_3O^+][OH^-] = 1.0 \times 10^{-14} \text{ (at 25°C)}$$

18.3 Defining pH (765):
$$pH = -\log [H_3O^+]$$

18.4 Relating pK_w to pH and pOH (766):
$$pK_w = pH + pOH = 14.00 \text{ (at 25°C)}$$

18.5 Finding the percent dissociation of HA (776):

$$\text{Percent HA dissociated} = \frac{[HA]_{dissoc}}{[HA]_{init}} \times 100$$

18.6 Defining the base-dissociation constant (779):

$$K_b = \frac{[BH^+][OH^-]}{[B]}$$

18.7 Expressing the relationship among K_a, K_b, and K_w (783):

$$K_a \times K_b = K_w$$

Highlighted Figures and Tables

These figures (F) and tables (T) provide a quick review of key ideas. Entries in color contain frequently used data.

F18.3 Extent of dissociation for strong and weak acids (760)
T18.2 K_a of some monoprotic acids at 25°C (762)
F18.5 Defining acidic, neutral, and basic solutions (765)
F18.7 Relations among $[H_3O^+]$, pH, $[OH^-]$, and pOH (767)
F18.9 Proton transfer in Brønsted-Lowry acid-base reactions (769)
T18.4 Some conjugate acid-base pairs (770)

F18.10 Strengths of conjugate acid-base pairs (771)
T18.5 K_a of some polyprotic acids at 25°C (777)
T18.6 K_b of ammonia and some amines at 25°C (780)
F18.12 Trends in nonmetal hydride acidity (784)
F18.13 Relative strengths of oxoacids (785)
T18.7 K_a of hydrated metal ions at 25°C (786)
T18.8 Behavior of salts in water (788)

Brief Solutions to Follow-up Problems

18.1 (a) $HClO_3$; (b) HCl; (c) NaOH

18.2 $[H_3O^+] = \dfrac{1.0\times10^{-14}}{6.7\times10^{-2}} = 1.5\times10^{-13}\ M$; basic

18.3 pOH $= 14.00 - 9.52 = 4.48$
$[H_3O^+] = 10^{-9.52} = 3.0\times10^{-10}\ M$
$[OH^-] = \dfrac{1.0\times10^{-14}}{3.0\times10^{-10}} = 3.3\times10^{-5}\ M$

18.4 (a) CH_3COOH/CH_3COO^- and H_3O^+/H_2O
(b) H_2O/OH^- and HF/F^-

18.5 (a) $NH_3(g) + H_2O(l) \rightleftharpoons NH_4^+(aq) + OH^-(aq)$
(b) $NH_3(g) + H_3O^+(aq; \text{ from HCl}) \longrightarrow NH_4^+(aq) + H_2O(l)$
(c) $NH_4^+(aq) + OH^-(aq; \text{ from NaOH}) \longrightarrow NH_3(g) + H_2O(l)$

18.6 $NH_4^+(aq) + H_2O(l) \rightleftharpoons NH_3(aq) + H_3O^+(aq)$
$[H_3O^+] = 10^{-pH} = 10^{-5.0} = 1\times10^{-5}\ M = [NH_3]$

From reaction table, $K_a = \dfrac{[NH_3][H_3O^+]}{[NH_4^+]} = \dfrac{(x)(x)}{0.2-x}$

$\approx \dfrac{(1\times10^{-5})^2}{0.2} = 5\times10^{-10}$

18.7 $K_a = \dfrac{[H_3O^+][OCN^-]}{[HOCN]} = \dfrac{(x)(x)}{0.10-x} = 3.5\times10^{-4}$

Since $\dfrac{[HOCN]_{init}}{K_a} = \dfrac{0.10}{3.5\times10^{-4}} = 286 < 400$, you must solve a
quadratic equation: $x^2 + (3.5\times10^{-4})x - (3.5\times10^{-5}) = 0$
$x = [H_3O^+] = 5.7\times10^{-3}\ M$; pH $= 2.24$

18.8 $K_{a1} = \dfrac{[HC_2O_4^-][H_3O^+]}{[H_2C_2O_4]} = \dfrac{x^2}{0.150-x} = 5.6\times10^{-2}$

Since $\dfrac{[H_2C_2O_4]_{init}}{K_{a1}} < 400$, you must solve a quadratic equation:
$x^2 + (5.6\times10^{-2})x - (8.4\times10^{-3}) = 0$
$x = [H_3O^+] = 0.068\ M$; pH $= 1.17$
$x = [HC_2O_4^-] = 0.068\ M$; $[H_2C_2O_4] = 0.150\ M - x = 0.082\ M$
$[C_2O_4^{2-}] = \dfrac{(K_{a2})[HC_2O_4^-]}{[H_3O^+]} = \dfrac{(5.4\times10^{-5})(0.068)}{0.068}$
$= 5.4\times10^{-5}\ M$

18.9 $K_b = \dfrac{[C_5H_5NH^+][OH^-]}{[C_5H_5N]} = 10^{-8.77} = 1.7\times10^{-9}$

Assuming $0.10\ M - x \approx 0.10\ M$, $K_b = 1.7\times10^{-9} \approx \dfrac{(x)(x)}{0.10}$;
$x = [OH^-] \approx 1.3\times10^{-5}\ M$; $[H_3O^+] = 7.7\times10^{-10}\ M$;
pH $= 9.11$

18.10 K_b of $ClO^- = \dfrac{K_w}{K_a \text{ of HClO}} = \dfrac{1.0\times10^{-14}}{2.9\times10^{-8}} = 3.4\times10^{-7}$

Assuming $0.20\ M - x \approx 0.20\ M$,
$K_b = 3.4\times10^{-7} = \dfrac{[HClO][OH^-]}{[ClO^-]} \approx \dfrac{x^2}{0.20}$;
$x = [OH^-] \approx 2.6\times10^{-4}\ M$; $[H_3O^+] = 3.8\times10^{-11}\ M$;
pH $= 10.42$

18.11 (a) Basic:
$ClO_2^-(aq) + H_2O(l) \rightleftharpoons HClO_2(aq) + OH^-(aq)$
K^+ is from strong base KOH.
(b) Acidic:
$CH_3NH_3^+(aq) + H_2O(l) \rightleftharpoons CH_3NH_2(aq) + H_3O^+(aq)$
NO_3^- is from strong acid HNO_3.
(c) Neutral: Cs^+ is from strong base CsOH; I^- is from strong acid HI.

18.12 (a) K_a of $Cu(H_2O)_6^{2+} = 3\times10^{-8}$
K_b of $CH_3COO^- = \dfrac{K_w}{K_a \text{ of } CH_3COOH} = 5.6\times10^{-10}$
Since $K_a > K_b$, $Cu(CH_3COO)_2(aq)$ is acidic.
(b) K_a of $NH_4^+ = \dfrac{K_w}{K_b \text{ of } NH_3} = 5.7\times10^{-10}$
K_b of $F^- = \dfrac{K_w}{K_a \text{ of HF}} = 1.5\times10^{-11}$
Since $K_a > K_b$, $NH_4F(aq)$ is acidic.

18.13 (a) OH^- is the Lewis base; $Al(OH)_3$ is the Lewis acid.
(b) H_2O is the Lewis base; SO_3 is the Lewis acid.
(c) NH_3 is the Lewis base; Co^{3+} is the Lewis acid.

Problems

Problems with **colored** numbers are answered at the back of the text. Sections match the text and provide the number(s) of relevant sample problems. Most offer Concept Review Questions, Skill-Building Exercises (in similar pairs), and Problems in Context. Then Comprehensive Problems, based on material from any section or previous chapter, follow.
Note: Unless stated otherwise, all problems refer to aqueous solutions at 298 K (25°C).

Acids and Bases in Water
(Sample Problem 18.1)

● Concept Review Questions

18.1 Describe the role of water according to the classical (Arrhenius) acid-base definition.

18.2 What characteristics do all Arrhenius acids have in common? What characteristics do all Arrhenius bases have in common? Explain neutralization in terms of the Arrhenius acid-base

definition. What quantitative finding led Arrhenius to propose this idea of neutralization?

18.3 Why is the Arrhenius acid-base definition considered too limited? Give an example of a case in which the Arrhenius definition does not apply.

18.4 What is meant by the words "strong" and "weak" in terms of acids and bases? Weak acids have K_a values that vary over more than 10 orders of magnitude. What do they have in common that classifies them as "weak"?

● Skill-Building Exercises (paired)

18.5 Which of the following are Arrhenius acids?
(a) H_2O (b) $Ca(OH)_2$ (c) H_3PO_3 (d) HI
18.6 Which of the following are Arrhenius acids?
(a) $NaHSO_4$ (b) CH_4 (c) NaH   (d) H_3N

18.7 Which of the following are Arrhenius bases?
(a) H_3AsO_4 (b) $Ba(OH)_2$ (c) HClO   (d) KOH

18.8 Which of the following are Arrhenius bases?
(a) CH_3COOH (b) HOH   (c) CH_3OH (d) H_2NNH_2

18.9 Write the K_a expression for each of the following:
(a) HCN   (b) HCO_3^- (c) HCOOH

18.10 Write the K_a expression for each of the following:
(a) $CH_3NH_3^+$ (b) HClO   (c) H_2S

18.11 Write the K_a expression for each of the following:
(a) HNO_2 (b) CH_3COOH (c) $HBrO_2$

18.12 Write the K_a expression for each of the following:
(a) $H_2PO_4^-$ (b) H_3PO_2 (c) HSO_4^-

18.13 Use Table 18.2 to rank the following in order of *increasing* acid strength: HIO_3, HI, CH_3COOH, HF.

18.14 Use Table 18.2 to rank the following in order of *decreasing* acid strength: HClO, HCl, HCN, HNO_2.

18.15 Classify each of the following as a strong or weak acid or base: (a) H_3AsO_4; (b) $Sr(OH)_2$; (c) HIO; (d) $HClO_4$.

18.16 Classify each of the following as a strong or weak acid or base: (a) CH_3NH_2; (b) K_2O; (c) HI; (d) HCOOH.

18.17 Classify each of the following as a strong or weak acid or base: (a) RbOH; (b) HBr; (c) H_2Te; (d) HClO.

18.18 Classify each of the following as a strong or weak acid or base: (a) $HOCH_2CH_2NH_2$; (b) H_2SeO_4; (c) HS^-; (d) $B(OH)_3$.

Autoionization of Water and the pH Scale
(Sample Problems 18.2 and 18.3)

● **Concept Review Questions**

18.19 What is an autoionization reaction? Write equations for the autoionization reactions of H_2O and of H_2SO_4.

18.20 What is the difference between K_c and K_w for the autoionization of water?

18.21 (a) What is the change in pH when $[OH^-]$ increases by a factor of 10?
(b) What is the change in $[H_3O^+]$ when the pH decreases by 2 units?

18.22 Which of these solutions has the higher pH? Explain.
(a) A 0.1 M solution of an acid with $K_a = 1\times10^{-4}$ or one with $K_a = 4\times10^{-5}$
(b) A 0.1 M solution of an acid with $pK_a = 3.0$ or one with $pK_a = 3.5$
(c) A 0.1 M solution of a weak acid or a 0.01 M solution of the same acid
(d) A 0.1 M solution of a weak acid or a 0.1 M solution of a strong acid
(e) A 0.1 M solution of an acid or a 0.1 M solution of a base
(f) A solution of pOH 6.0 or one of pOH 8.0

● **Skill-Building Exercises (paired)**

18.23 (a) What is the pH of 0.0111 M NaOH? Is the solution neutral, acidic, or basic?
(b) What is the pOH of 1.23×10^{-3} M HCl? Is the solution neutral, acidic, or basic?

18.24 (a) What is the pH of 0.0333 M HNO_3? Is the solution neutral, acidic, or basic?
(b) What is the pOH of 0.0347 M KOH? Is the solution neutral, acidic, or basic?

18.25 (a) What is the pH of 5.04×10^{-3} M HI? Is the solution neutral, acidic, or basic?

(b) What is the pOH of 2.55 M $Ba(OH)_2$? Is the solution neutral, acidic, or basic?

18.26 (a) What is the pH of 7.52×10^{-4} M CsOH? Is the solution neutral, acidic, or basic?
(b) What is the pOH of 1.59×10^{-3} M $HClO_4$? Is the solution neutral, acidic, or basic?

18.27 (a) What are $[H_3O^+]$, $[OH^-]$, and pOH in a solution with a pH of 9.78?
(b) What are $[H_3O^+]$, $[OH^-]$, and pH in a solution with a pOH of 10.43?

18.28 (a) What are $[H_3O^+]$, $[OH^-]$, and pOH in a solution with a pH of 3.47?
(b) What are $[H_3O^+]$, $[OH^-]$, and pH in a solution with a pOH of 4.33?

18.29 (a) What are $[H_3O^+]$, $[OH^-]$, and pOH in a solution with a pH of 2.77?
(b) What are $[H_3O^+]$, $[OH^-]$, and pH in a solution with a pOH of 5.18?

18.30 (a) What are $[H_3O^+]$, $[OH^-]$, and pOH in a solution with a pH of 8.97?
(b) What are $[H_3O^+]$, $[OH^-]$, and pH in a solution with a pOH of 11.27?

18.31 How many moles of H_3O^+ or OH^- must you add per liter of HA solution to adjust its pH from 3.25 to 3.65? Assume a negligible volume change.

18.32 How many moles of H_3O^+ or OH^- must you add per liter of HA solution to adjust its pH from 9.33 to 9.07? Assume a negligible volume change.

18.33 How many moles of H_3O^+ or OH^- must you add to 6.5 L of HA solution to adjust its pH from 4.82 to 5.22? Assume a negligible volume change.

18.34 How many moles of H_3O^+ or OH^- must you add to 87.5 mL of HA solution to adjust its pH from 8.92 to 6.33? Assume a negligible volume change.

● **Problems in Context**

18.35 Although the text asserts that water is an extremely weak electrolyte, parents commonly warn their children of the danger of swimming in a pool or lake during a lightning storm. Explain.

18.36 Like any equilibrium constant, K_w changes with temperature.
(a) Given that autoionization is an endothermic process, does K_w increase or decrease with rising temperature? Explain with a reaction that includes heat as reactant or product.
(b) In many medical applications, the value of K_w at 37°C (body temperature) may be more appropriate than the value at 25°C, 1.0×10^{-14}. The pH of pure water at 37°C is 6.80. Calculate K_w, pOH, and $[OH^-]$ at this temperature.

Proton Transfer and the Brønsted-Lowry Acid-Base Definition
(Sample Problems 18.4 and 18.5)

● **Concept Review Questions**

18.37 How do the Arrhenius and Brønsted-Lowry definitions of an acid and a base differ? How are they similar? Name two Brønsted-Lowry bases that are not considered Arrhenius bases. Can you do the same for acids? Explain.

18.38 What is a conjugate acid-base pair? What is the relationship between the two members of the pair?

18.39 A Brønsted-Lowry acid-base reaction proceeds in the net direction in which a stronger acid and stronger base form a weaker acid and weaker base. Explain.

18.40 What is an amphoteric species? Name one and write balanced equations that show why it is amphoteric.

● **Skill-Building Exercises (paired)**

18.41 Write balanced equations and K_a expressions for these Brønsted-Lowry acids in water:
(a) H_3PO_4 (b) C_6H_5COOH (c) HSO_4^-
18.42 Write balanced equations and K_a expressions for these Brønsted-Lowry acids in water:
(a) $HCOOH$ (b) $HClO_3$ (c) $H_2AsO_4^-$

18.43 Give the formula of the conjugate base of each of the following: (a) HCl; (b) H_2CO_3; (c) H_2O.
18.44 Give the formula of the conjugate base of each of the following: (a) HPO_4^{2-}; (b) NH_4^+; (c) HS^-.

18.45 Give the formula of the conjugate acid of each of the following: (a) NH_3; (b) NH_2^-; (c) nicotine, $C_{10}H_{14}N_2$.
18.46 Give the formula of the conjugate acid of each of the following: (a) O^{2-}; (b) SO_4^{2-}; (c) H_2O.

18.47 In each equation, label the acids, bases, and conjugate acid-base pairs:
(a) $HCl + H_2O \rightleftharpoons Cl^- + H_3O^+$
(b) $HClO_4 + H_2SO_4 \rightleftharpoons ClO_4^- + H_3SO_4^+$
(c) $HPO_4^{2-} + H_2SO_4 \rightleftharpoons H_2PO_4^- + HSO_4^-$
18.48 In each equation, label the acids, bases, and conjugate acid-base pairs:
(a) $NH_3 + HNO_3 \rightleftharpoons NH_4^+ + NO_3^-$
(b) $O^{2-} + H_2O \rightleftharpoons OH^- + OH^-$
(c) $NH_4^+ + BrO_3^- \rightleftharpoons NH_3 + HBrO_3$

18.49 In each equation, label the acids, bases, and conjugate acid-base pairs:
(a) $NH_3 + H_3PO_4 \rightleftharpoons NH_4^+ + H_2PO_4^-$
(b) $CH_3O^- + NH_3 \rightleftharpoons CH_3OH + NH_2^-$
(c) $HPO_4^{2-} + HSO_4^- \rightleftharpoons H_2PO_4^- + SO_4^{2-}$
18.50 In each equation, label the acids, bases, and conjugate acid-base pairs:
(a) $NH_4^+ + CN^- \rightleftharpoons NH_3 + HCN$
(b) $H_2O + HS^- \rightleftharpoons OH^- + H_2S$
(c) $HSO_3^- + CH_3NH_2 \rightleftharpoons SO_3^{2-} + CH_3NH_3^+$

18.51 Write balanced net ionic equations for the following reactions and label the conjugate acid-base pairs:
(a) $NaOH(aq) + NaH_2PO_4(aq) \rightleftharpoons H_2O(l) + Na_2HPO_4(aq)$
(b) $KHSO_4(aq) + K_2CO_3(aq) \rightleftharpoons K_2SO_4(aq) + KHCO_3(aq)$
18.52 Write balanced net ionic equations for the following reactions and label the conjugate acid-base pairs:
(a) $HNO_3(aq) + Li_2CO_3(aq) \rightleftharpoons LiNO_3(aq) + LiHCO_3(aq)$
(b) $2NH_4Cl(aq) + Ba(OH)_2(aq) \rightleftharpoons$
$$2H_2O(l) + BaCl_2(aq) + 2NH_3(aq)$$

18.53 The following aqueous species constitute two conjugate acid-base pairs. Use them to write one acid-base reaction with $K_c > 1$ and another with $K_c < 1$: HS^-, Cl^-, HCl, H_2S.
18.54 The following aqueous species constitute two conjugate acid-base pairs. Use them to write one acid-base reaction with $K_c > 1$ and another with $K_c < 1$: NO_3^-, F^-, HF, HNO_3.

18.55 Use Figure 18.10 to determine whether $K_c > 1$ for each reaction:
(a) $HCl + NH_3 \rightleftharpoons NH_4^+ + Cl^-$
(b) $H_2SO_3 + NH_3 \rightleftharpoons HSO_3^- + NH_4^+$
18.56 Use Figure 18.10 to determine whether $K_c > 1$ for each reaction:
(a) $OH^- + HS^- \rightleftharpoons H_2O + S^{2-}$
(b) $HCN + HCO_3^- \rightleftharpoons H_2CO_3 + CN^-$

18.57 Use Figure 18.10 to determine whether $K_c < 1$ for each reaction:
(a) $NH_4^+ + HPO_4^{2-} \rightleftharpoons NH_3 + H_2PO_4^-$
(b) $HSO_3^- + HS^- \rightleftharpoons H_2SO_3 + S^{2-}$
18.58 Use Figure 18.10 to determine whether $K_c < 1$ for each reaction:
(a) $H_2PO_4^- + F^- \rightleftharpoons HPO_4^{2-} + HF$
(b) $CH_3COO^- + HSO_4^- \rightleftharpoons CH_3COOH + SO_4^{2-}$

Solving Problems Involving Weak-Acid Equilibria
(Sample Problems 18.6 to 18.8)

● **Concept Review Questions**

18.59 In each of the following cases, would you expect the concentration of acid before and after dissociation to be nearly the same or very different? Explain your reasoning.
(a) A concentrated solution of a strong acid
(b) A concentrated solution of a weak acid
(c) A dilute solution of a weak acid
(d) A dilute solution of a strong acid
18.60 A sample of 0.0001 M HCl has $[H_3O^+]$ close to that of a sample of 0.1 M CH_3COOH. Are acetic acid and hydrochloric acid equally strong in these samples? Explain.
18.61 In which of the following solutions will $[H_3O^+]$ be approximately equal to $[CH_3COO^-]$: (a) 0.1 M CH_3COOH; (b) 1×10^{-7} M CH_3COOH; (c) a solution containing both 0.1 M CH_3COOH and 0.1 M CH_3COONa? Explain.
18.62 Why do successive acid-dissociation constants decrease for all polyprotic acids?

● **Skill-Building Exercises (paired)**

18.63 A 0.15 M solution of butanoic acid, $CH_3CH_2CH_2COOH$, contains 1.51×10^{-3} M H_3O^+. What is the K_a of butanoic acid?
18.64 A 0.035 M solution of a weak acid (HA) has a pH of 4.88. What is the K_a of the acid?

18.65 Nitrous acid, HNO_2, has a K_a of 7.1×10^{-4}. What are $[H_3O^+]$, $[NO_2^-]$, and $[OH^-]$ in 0.50 M HNO_2?
18.66 Hydrofluoric acid, HF, has a K_a of 6.8×10^{-4}. What are $[H_3O^+]$, $[F^-]$, and $[OH^-]$ in 0.75 M HF?

18.67 Chloroacetic acid, $ClCH_2COOH$, has a pK_a of 2.87. What are $[H_3O^+]$, pH, $[ClCH_2COO^-]$, and $[ClCH_2COOH]$ in 1.05 M $ClCH_2COOH$?
18.68 Hypochlorous acid, HClO, has a pK_a of 7.54. What are $[H_3O^+]$, pH, $[ClO^-]$, and $[HClO]$ in 0.115 M HClO?

18.69 A 0.25 M solution of a weak acid is 3.0% dissociated.
(a) Calculate the $[H_3O^+]$, pH, $[OH^-]$, and pOH of the solution.
(b) Calculate K_a of the acid.
18.70 A 0.735 M solution of a weak acid is 12.5% dissociated.
(a) Calculate the $[H_3O^+]$, pH, $[OH^-]$, and pOH of the solution.
(b) Calculate K_a of the acid.

18.71 A 0.250-mol sample of HX is dissolved in enough H_2O to form 655 mL of solution. If the pH of the solution is 3.44, what is the K_a of HX?

18.72 A 4.85×10^{-3} mol sample of HY is dissolved in enough H_2O to form 0.095 L of solution. If the pH of the solution is 2.68, what is the K_a of HY?

18.73 The weak acid HZ has a K_a of 1.55×10^{-4}.
(a) Calculate the pH of 0.075 M HZ.
(b) Calculate the pOH of 0.045 M HZ.

18.74 The weak acid HQ has a pK_a of 4.89.
(a) Calculate the $[H_3O^+]$ of 3.5×10^{-2} M HQ.
(b) Calculate the $[OH^-]$ of 0.65 M HQ.

18.75 (a) Calculate the pH of 0.175 M HY, if $K_a = 1.00\times10^{-4}$.
(b) Calculate the pOH of 0.175 M HX, if $K_a = 1.00\times10^{-2}$.

18.76 (a) Calculate the pH of 0.553 M $KHCO_3$; K_a of $HCO_3^- = 4.7\times10^{-11}$.
(b) Calculate the pOH of 0.044 M HIO_3; K_a of $HIO_3 = 0.16$.

18.77 Use Table 18.2 to calculate the percent dissociation of 0.25 M benzoic acid, C_6H_5COOH.

18.78 Use Table 18.2 to calculate the percent dissociation of 0.050 M CH_3COOH.

18.79 Use Table 18.5 to calculate $[H_2S]$, $[HS^-]$, $[S^{2-}]$, $[H_3O^+]$, pH, $[OH^-]$, and pOH in a 0.10 M solution of the diprotic acid hydrosulfuric acid.

18.80 Use Table 18.5 to calculate $[H_2C_2O_4]$, $[HC_2O_4^-]$, $[C_2O_4^{2-}]$, $[H_3O^+]$, pH, $[OH^-]$, and pOH in a 0.200 M solution of the diprotic acid oxalic acid.

● **Problems in Context**

18.81 Acetylsalicylic acid (aspirin), $HC_9H_7O_4$, is the most widely used pain reliever and fever reducer. Determine the pH of a 0.018 M aqueous solution of aspirin ($K_a = 3.2\times10^{-4}$).

18.82 Formic acid, HCOOH, the simplest carboxylic acid, has many uses in the textile and rubber industries. It is an extremely caustic liquid that is secreted as a defense by many species of ants (family *Formicidae*). Calculate the percent dissociation of 0.50 M HCOOH.

Weak Bases and Their Relation to Weak Acids

(Sample Problems 18.9 and 18.10)

● **Concept Review Questions**

18.83 What is the essential structural feature of all Brønsted-Lowry bases? How does this feature function in an acid-base reaction?

18.84 Why are most anions basic in H_2O? Give formulas of four anions that are not basic.

18.85 Except for the Na^+ spectator ion and neglecting H_2O, aqueous solutions of CH_3COOH and CH_3COONa contain the same species.
(a) What are the species?
(b) Why is 0.1 M CH_3COOH acidic and 0.1 M CH_3COONa basic?

● **Skill-Building Exercises** *(paired)*

18.86 Write balanced equations and K_b expressions for these Brønsted-Lowry bases in water:
(a) pyridine, C_5H_5N (b) CO_3^{2-}

18.87 Write balanced equations and K_b expressions for these Brønsted-Lowry bases in water:
(a) benzoate ion, $C_6H_5COO^-$ (b) $(CH_3)_3N$

18.88 Write balanced equations and K_b expressions for these Brønsted-Lowry bases in water:
(a) hydroxylamine, $HO-NH_2$ (b) HPO_4^{2-}

18.89 Write balanced equations and K_b expressions for these Brønsted-Lowry bases in water:
(a) guanidine, $(H_2N)_2C=NH$ (the doubly bonded N is the most basic)
(b) acetylide ion, $HC\equiv C^-$

18.90 What is the pH of 0.050 M dimethylamine?

18.91 What is the pH of 0.12 M diethylamine?

18.92 What is the pH of 0.15 M ethanolamine?

18.93 What is the pH of 0.26 M aniline?

18.94 (a) What is K_b of the acetate ion?
(b) What is K_a of the anilinium ion, $C_6H_5NH_3^+$?

18.95 (a) What is K_b of the benzoate ion, $C_6H_5COO^-$?
(b) What is K_a of the 2-hydroxyethylammonium ion, $HOCH_2CH_2NH_3^+$ (pK_b of $HOCH_2CH_2NH_2 = 4.49$)?

18.96 (a) What is pK_b of ClO_2^-?
(b) What is pK_a of the dimethylammonium ion, $(CH_3)_2NH_2^+$?

18.97 (a) What is the pK_b of NO_2^-?
(b) What is the pK_a of the hydrazinium ion, $H_2N-NH_3^+$ (K_b of hydrazine $= 8.5\times10^{-7}$)?

18.98 (a) What is the pH of 0.050 M KCN?
(b) What is the pH of 0.30 M triethylammonium chloride, $(CH_3CH_2)_3NHCl$?

18.99 (a) What is the pH of 0.100 M sodium phenolate, C_6H_5ONa, the sodium salt of phenol?
(b) What is the pH of 0.15 M methylammonium bromide, CH_3NH_3Br (K_b of $CH_3NH_2 = 4.4\times10^{-4}$)?

18.100 (a) What is the pH of 0.53 M potassium formate, HCOOK?
(b) What is the pH of 1.22 M NH_4Br?

18.101 (a) What is the pH of 0.75 M NaF?
(b) What is the pH of 0.88 M pyridinium chloride, C_5H_5NHCl?

● **Problems in Context**

18.102 Sodium hypochlorite solution is sold as "chlorine bleach" and is recognized as a potentially dangerous household solution. The dangers arise from its basicity and from ClO^-, the active bleaching ingredient. What is $[OH^-]$ in an aqueous solution that is 5.0% NaClO by mass? What is the pH of the solution? (Assume d of solution $= 1.0$ g/mL.)

18.103 Codeine ($C_{18}H_{21}NO_3$) is a narcotic pain reliever that forms a salt with HCl. What is the pH of 0.050 M codeine hydrochloride (pK_b of codeine $= 5.80$)?

Molecular Properties and Acid Strength

● **Concept Review Questions**

18.104 Across a period, how does the electronegativity of a nonmetal affect the acidity of its binary hydride?

18.105 How does the atomic size of a nonmetal affect the acidity of its binary hydride?

18.106 Does it make sense that a strong acid has a weak bond to its acidic proton, whereas a weak acid has a strong bond to its acidic proton? Explain.

18.107 Perchloric acid, $HClO_4$, is the strongest of the halogen oxoacids, and hypoiodous acid, HIO, is the weakest. What two factors govern this difference in acid strength?

● **Skill-Building Exercises (paired)**

18.108 Choose the *stronger* acid in each of the following pairs:
(a) H_2SeO_3 or H_2SeO_4 (b) H_3PO_4 or H_3AsO_4 (c) H_2S or H_2Te
18.109 Choose the *weaker* acid in each of the following pairs:
(a) HBr or H_2Se (b) $HClO_4$ or H_2SO_4 (c) H_2SO_3 or H_2SO_4

18.110 Choose the *stronger* acid in each of the following pairs:
(a) H_2Se or H_3As (b) $B(OH)_3$ or $Al(OH)_3$ (c) $HBrO_2$ or HBrO
18.111 Choose the *weaker* acid in each of the following pairs:
(a) HI or HBr   (b) H_3AsO_4 or H_2SeO_4 (c) HNO_3 or HNO_2

18.112 Use Table 18.7 to choose the solution with the *lower* pH:
(a) 0.1 M $CuSO_4$ or 0.05 M $Al_2(SO_4)_3$
(b) 0.1 M $ZnCl_2$ or 0.1 M $PbCl_2$
18.113 Use Table 18.7 to choose the solution with the *lower* pH:
(a) 0.1 M $FeCl_3$ or 0.1 M $AlCl_3$
(b) 0.1 M $BeCl_2$ or 0.1 M $CaCl_2$

18.114 Use Table 18.7 to choose the solution with the *higher* pH:
(a) 0.1 M $Ni(NO_3)_2$ or 0.1 M $Co(NO_3)_2$
(b) 0.1 M $Al(NO_3)_3$ or 0.1 M $Cr(NO_3)_3$
18.115 Use Table 18.7 to choose the solution with the *higher* pH:
(a) 0.1 M $NiCl_2$ or 0.1 M NaCl
(b) 0.1 M $Sn(NO_3)_2$ or 0.1 M $Co(NO_3)_2$

Acid-Base Properties of Salt Solutions
(Sample Problems 18.11 and 18.12)

● **Concept Review Questions**

18.116 What determines whether an aqueous solution of a salt will be acidic, basic, or neutral? Give an example of each type of salt.

18.117 Why is a solution of NaF basic, whereas a solution of NaCl is neutral?

18.118 The NH_4^+ ion forms acidic solutions and the $C_2H_3O_2^-$ ion forms basic solutions. However, a solution of ammonium acetate is almost neutral. Do all of the ammonium salts of weak acids form neutral solutions? Explain your answer.

● **Skill-Building Exercises (paired)**

18.119 Explain with equations and calculations, when necessary, whether an aqueous solution of each of the following salts is acidic, basic, or neutral:
(a) KBr   (b) NH_4I (c) KCN
18.120 Explain with equations and calculations, when necessary, whether an aqueous solution of each of the following salts is acidic, basic, or neutral:
(a) $Cr(NO_3)_3$ (b) NaHS   (c) $Zn(CH_3COO)_2$

18.121 Explain with equations and calculations, when necessary, whether an aqueous solution of each of the following salts is acidic, basic, or neutral:
(a) Na_2CO_3 (b) $CaCl_2$ (c) $Cu(NO_3)_2$

18.122 Explain with equations and calculations, when necessary, whether an aqueous solution of each of the following salts is acidic, basic, or neutral:
(a) CH_3NH_3Cl (b) $KClO_4$ (c) CoF_2

18.123 Explain with equations and calculations, when necessary, whether an aqueous solution of each of the following salts is acidic, basic, or neutral:
(a) $SrBr_2$ (b) $Ba(CH_3COO)_2$ (c) $(CH_3)_2NH_2Br$
18.124 Explain with equations and calculations, when necessary, whether an aqueous solution of each of the following salts is acidic, basic, or neutral:
(a) $Fe(HCOO)_3$ (b) $KHCO_3$ (c) K_2S

18.125 Explain with equations and calculations, when necessary, whether an aqueous solution of each of the following salts is acidic, basic, or neutral:
(a) $(NH_4)_3PO_4$ (b) Na_2SO_4 (c) LiClO
18.126 Explain with equations and calculations, when necessary, whether an aqueous solution of each of the following salts is acidic, basic, or neutral:
(a) $Pb(CH_3COO)_2$ (b) $Cr(NO_2)_3$ (c) CsI

18.127 Rank the following salts in order of *increasing* pH of their 0.1 M aqueous solutions:
(a) KNO_3, K_2SO_3, K_2S, $Fe(NO_3)_2$
(b) NH_4NO_3, $NaHSO_4$, $NaHCO_3$, Na_2CO_3
18.128 Rank the following salts in order of *decreasing* pH of their 0.1 M aqueous solutions:
(a) $FeCl_2$, $FeCl_3$, $MgCl_2$, $KClO_2$
(b) NH_4Br, $NaBrO_2$, NaBr, $NaClO_2$

Generalizing the Brønsted-Lowry Concept: The Leveling Effect

● **Concept Review Questions**

18.129 The methoxide ion, CH_3O^-, and amide ion, NH_2^-, are very strong bases that are "leveled" by water. What does this mean? Write the reactions that occur in the leveling process. What species do the two leveled solutions have in common?

18.130 Explain the differing extents of dissociation of H_2SO_4 in CH_3COOH, H_2O, and NH_3.

18.131 In H_2O, HF is weak and the other hydrohalic acids are equally strong. In NH_3, however, all the hydrohalic acids are equally strong. Explain.

Electron-Pair Donation and Lewis Acid-Base Definition
(Sample Problem 18.13)

● **Concept Review Questions**

18.132 What feature must a molecule or ion have for it to act as a Lewis base? A Lewis acid? Explain the roles of these features.

18.133 How do Lewis acids differ from Brønsted-Lowry acids? How are they similar? Do Lewis bases differ from Brønsted-Lowry bases? Explain.

18.134 (a) Is a weak Brønsted-Lowry base necessarily a weak Lewis base? Explain with an example.
(b) Identify the Lewis bases in the following reaction:
$$Cu(H_2O)_4^{2+}(aq) + 4CN^-(aq) \rightleftharpoons$$
$$Cu(CN)_4^{2-}(aq) + 4H_2O(l)$$
(c) Given that $K_c > 1$ for the reaction in part (b), which Lewis base is stronger?

18.135 In which of the three concepts of acid-base behavior discussed in the text can water be a product of an acid-base reaction? In which is it the only product?

18.136 (a) Give an example of a *substance* that is a base in two of the three acid-base definitions, but not in the third.

(b) Give an example of a *substance* that is an acid in one of the three acid-base definitions, but not in the other two.

● **Skill-Building Exercises (paired)**

18.137 Which are Lewis acids and which are Lewis bases?
(a) Cu^{2+} (b) Cl^- (c) $SnCl_2$ (d) OF_2

18.138 Which are Lewis acids and which are Lewis bases?
(a) Na^+ (b) NH_3 (c) CN^- (d) BF_3

18.139 Which are Lewis acids and which are Lewis bases?
(a) BF_3 (b) S^{2-} (c) SO_3^{2-} (d) SO_3

18.140 Which are Lewis acids and which are Lewis bases?
(a) Mg^{2+} (b) OH^- (c) SiF_4 (d) $BeCl_2$

18.141 Identify the Lewis acid and the Lewis base in each equation:
(a) $Na^+ + 6H_2O \rightleftharpoons Na(H_2O)_6^+$
(b) $CO_2 + H_2O \rightleftharpoons H_2CO_3$
(c) $F^- + BF_3 \rightleftharpoons BF_4^-$

18.142 Identify the Lewis acid and the Lewis base in each equation:
(a) $Fe^{3+} + 2H_2O \rightleftharpoons FeOH^{2+} + H_3O^+$
(b) $H_2O + H^- \rightleftharpoons OH^- + H_2$
(c) $4CO + Ni \rightleftharpoons Ni(CO)_4$

18.143 Classify the following as Arrhenius, Brønsted-Lowry, and Lewis acid-base reactions. A reaction may fit all, two, one, or none of the categories:
(a) $Ag^+ + 2NH_3 \rightleftharpoons Ag(NH_3)_2^+$
(b) $H_2SO_4 + NH_3 \rightleftharpoons HSO_4^- + NH_4^+$
(c) $2HCl \rightleftharpoons H_2 + Cl_2$
(d) $AlCl_3 + Cl^- \rightleftharpoons AlCl_4^-$

18.144 Classify the following as Arrhenius, Brønsted-Lowry, and Lewis acid-base reactions. A reaction may fit all, two, one, or none of the categories:
(a) $Cu^{2+} + 4Cl^- \rightleftharpoons CuCl_4^{2-}$
(b) $Al(OH)_3 + 3HNO_3 \rightleftharpoons Al^{3+} + 3H_2O + 3NO_3^-$
(c) $N_2 + 3H_2 \rightleftharpoons 2NH_3$
(d) $CN^- + H_2O \rightleftharpoons HCN + OH^-$

Comprehensive Problems

Problems with an asterisk (*) are more challenging.

18.145 Pantothenic acid, $C_9H_{17}NO_5$, also called *vitamin B₃*, has the structure shown below. This biologically active molecule is an optical isomer that behaves like a monoprotic Brønsted-Lowry acid in water.

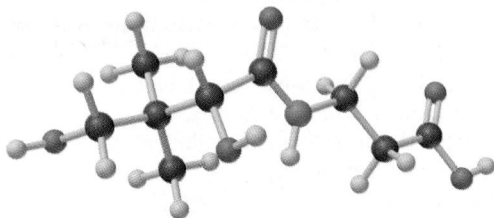

(a) Use the molecular formula to write the equation for the reaction of pantothenic acid with water and the K_a expression.

(b) Use the model to determine which C atom is the chiral center.

18.146 The pK_a of acetic acid is 4.76 in pure water and 4.53 in seawater. Is it a stronger acid in one solvent than in the other? If so, in which?

18.147 Bodily processes in humans maintain the pH of blood within a narrow range. In fact, a condition called *acidosis* occurs if the blood pH goes below 7.35, and another called *alkalosis* occurs if the pH goes above 7.45. Given that the pK_w of blood is 13.63 at 37°C (body temperature), what is the normal range of $[H_3O^+]$ and of $[OH^-]$ in blood?

18.148 One use of phenol, C_6H_5OH, is as a disinfectant. In water its pK_a is 10.0, but in methanol it is 14.4.
(a) Why are the values different?
(b) Use these two values to decide whether methanol is a stronger or weaker base than water.
(c) Write the dissociation reaction of phenol in methanol.
(d) Write an expression that defines the autoionization constant of methanol.

*18.149 When carbon dioxide dissolves in water, it undergoes a multistep equilibrium process, with $K_{overall} = 4.5 \times 10^{-7}$, which is simplifed to the following:

$$CO_2(g) + H_2O(l) \rightleftharpoons H_2CO_3(aq)$$
$$H_2CO_3(aq) + H_2O(l) \rightleftharpoons HCO_3^-(aq) + H_3O^+(aq)$$

(a) Classify each step as a Lewis or a Brønsted-Lowry acid-base reaction.
(b) What is the pH of nonpolluted rainwater in equilibrium with clean air (P_{CO_2} in clean air $= 3.2 \times 10^{-4}$ atm; Henry's law constant for CO_2 at 25°C is 0.033 mol/L·atm)?
(c) What is $[CO_3^{2-}]$ in rainwater (K_a of $HCO_3^- = 4.7 \times 10^{-11}$)?
(d) If the partial pressure of CO_2 in clean air doubles in the next few decades, what will the pH of rainwater become?

18.150 Seashells are mostly calcium carbonate, which reacts with H_3O^+ according to the equation

$$CaCO_3(s) + H_3O^+(aq) \rightleftharpoons$$
$$Ca^{2+}(aq) + HCO_3^-(aq) + H_2O(l)$$

If the extent of autoionization of water increases at higher pressure, will seashells dissolve more rapidly near the surface or at great depths? Explain.

18.151 Many molecules with central atoms from Period 3 or higher take part in Lewis acid-base reactions in which the central atom expands its valence shell. For example, $SnCl_4$ reacts with $(CH_3)_3N$ as follows:

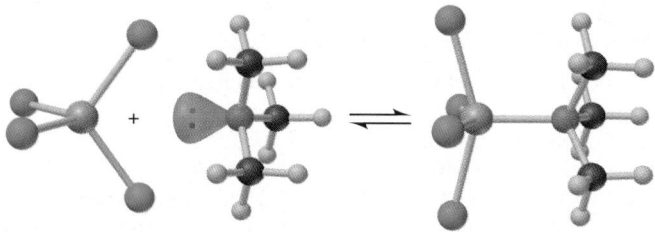

(a) Identify the Lewis acid and Lewis base in the reaction.
(b) Give the nl designation of the sublevel of the central atom in the acid that accepts the lone pair.

*18.152 A chemist makes four successive 1:10 dilutions of 1.0×10^{-5} M HCl. Calculate the pH of the original solution and of each dilution (through 1.0×10^{-9} M HCl).

18.153 At 25°C, the pK_w of seawater is 13.22; this value is lower than that for pure water as a result of the dissolved salts. If the pH of a sample of seawater is 7.54, what is the pOH of the sample?

18.154 Chlorobenzene, C_6H_5Cl, is a key intermediate in the manufacture of many aromatic compounds, including aniline dyes and chlorinated pesticides. It is made by the $FeCl_3$-catalyzed chlorination of benzene in this series of steps:

Step 1: $Cl_2 + FeCl_3 \rightleftharpoons FeCl_5$ (or $Cl^+FeCl_4^-$)
Step 2: $C_6H_6 + Cl^+FeCl_4^- \rightleftharpoons C_6H_6Cl^+ + FeCl_4^-$
Step 3: $C_6H_6Cl^+ \rightleftharpoons C_6H_5Cl + H^+$
Step 4: $H^+ + FeCl_4^- \rightleftharpoons HCl + FeCl_3$

(a) Which of the step(s) is (are) Lewis acid-base reactions?
(b) Identify the Lewis acids and Lewis bases in each of those steps.

*18.155 Hydrogen peroxide, H_2O_2 ($pK_a = 11.75$), is commonly used as a bleaching agent and an antiseptic. The product sold in stores is 3% H_2O_2 by mass and contains 0.001% phosphoric acid by mass to stabilize the solution. Which contributes more H_3O^+ to this commercial solution, the H_2O_2 or the H_3PO_4?

18.156 The strengths of acids and bases are directly related to their strengths as electrolytes (Section 4.4).
(a) Is the electrical conductivity of 0.1 M HCl higher, lower, or the same as that of 0.1 M CH_3COOH? Explain.
(b) Is the electrical conductivity of 1×10^{-7} M HCl higher, lower, or the same as that of 1×10^{-7} M CH_3COOH? Explain.

18.157 Esters, RCOOR′, are compounds formed by the reaction of carboxylic acids, RCOOH, and alcohols, R′OH, where R and R′ are hydrocarbon groups. Many esters are responsible for the odors of fruit and, thus, have important uses in the food and cosmetics industries. The mechanism of ester formation is catalyzed by H^+. The first two steps are Lewis acid-base reactions:

Step 1.

Step 2.

Identify the Lewis acids and Lewis bases in these two steps.

18.158 The metallic radius of aluminum (143 pm) is similar to that of lithium (152 pm), and the ionic radius of the aluminum ion (54 pm) is similar to that of the lithium ion (76 pm). Why is an aqueous solution of aluminum nitrate acidic but one of lithium nitrate neutral?

18.159 The corrosion of iron-containing objects has a major economic impact (see Chapter 21). In an investigation of corrosion of stainless steel by hydrochloric acid, some 0.010 M HCl is heated in a closed stainless steel container to 200°C, which increases the pressure to 10^4 atm. Under these conditions, pK_w is 9.25. What is $[OH^-]$ in the stainless steel container?

*18.160 Many substances undergo autoionization in a manner analogous to water. For example, the autoionization reaction of liquid ammonia is

$$2NH_3(l) \rightleftharpoons NH_4^+(am) + NH_2^-(am)$$

where *am* refers to solvation by ammonia.

(a) Write a K_{am} expression for the autoionization of ammonia that is analogous to the K_w expression for water.
(b) What are the strongest acid and strongest base that can exist in liquid ammonia?
(c) An aqueous solution with $[OH^-] > [H_3O^+]$ is basic and one with $[OH^-] < [H_3O^+]$ is acidic. What are the analogous relationships for solutions in liquid ammonia?
(d) The acids HNO_3 and HCOOH are leveled in liquid NH_3. Explain what this means and show equations.
(e) Pure liquid sulfuric acid also undergoes autoionization. The bases hydroxide ion, OH^-, and methoxide ion, CH_3O^-, are leveled in pure liquid H_2SO_4. Explain what this means and show equations.

*18.161 Autoionization (see Problem 18.160) occurs in methanol (CH_3OH) and in ethylenediamine ($NH_2CH_2CH_2NH_2$).
(a) The autoionization constant of methanol (K_{met}) is 2×10^{-17}. What is $[CH_3O^-]$ in pure CH_3OH?
(b) The concentration of $NH_2CH_2CH_2NH_3^+$ in pure $NH_2CH_2CH_2NH_2$ is 2×10^{-8} M. What is the autoionization constant (K_{en}) of ethylenediamine?

*18.162 Some drugs are such weak bases that titration with a strong acid to determine their concentration is incomplete in water. To resolve this situation, industrial pharmacologists titrate many of these drugs using pure acetic acid as the solvent instead of water. How does this substitution of the solvent aid in the titration?

18.163 Tris(hydroxymethyl)aminomethane, commonly known as TRIS or THAM, is a water-soluble base with industrial applications as a reactant in the synthesis of surfactants and pharmaceuticals, as an emulsifying agent in cosmetic creams and lotions, and as a component of various cleaning and polishing mixtures for textiles and leather. In biomedical research, solutions of TRIS are used to maintain nearly constant pH for the study of enzymes and other cellular components. Given that the pK_b is 5.91, calculate the pH of 0.060 M TRIS.

18.164 When iron(III) salts are dissolved in water, the solution becomes yellow and acidic due to the formation of $Fe(H_2O)_5OH^{2+}$ and H_3O^+. The overall process involves both Lewis and Brønsted-Lowry acid-base reactions. Write the equations for the process.

18.165 Vinegar is a 5.0% (w/v) solution of acetic acid in water. For culinary use, it is flavored with apples, herbs, wine, and so forth. What is the pH of vinegar?

18.166 Explain how to differentiate between a strong and a weak monoprotic acid based on the results of the following procedures:
(a) Electrical conductivity of equimolar solutions of each acid is measured.
(b) Equal moles of each are dissolved in water and tested with pH paper.
(c) Zinc metal is added to acid solutions of equal concentration.

18.167 At 50°C and 1 atm, $K_w = 5.19 \times 10^{-14}$. Calculate parts (a)–(c) under these conditions:
(a) $[H_3O^+]$ in pure water
(b) $[H_3O^+]$ in 0.010 M NaOH
(c) $[OH^-]$ in 0.0010 M $HClO_4$
(d) Calculate $[H_3O^+]$ in 0.0100 M KOH at 100°C and 1000 atm pressure ($K_w = 1.10 \times 10^{-12}$).
(e) Calculate the pH of pure water at 100°C and 1000 atm.

18.168 The catalytic efficiency of an enzyme is called its *activity* and refers to the rate at which it catalyzes the conversion of reactants into products. Most enzymes have optimum activity over a relatively narrow range of pH values, which is related to the pH of the local cellular fluid in which the reaction occurs. The pH profiles of three digestive enzymes are shown.

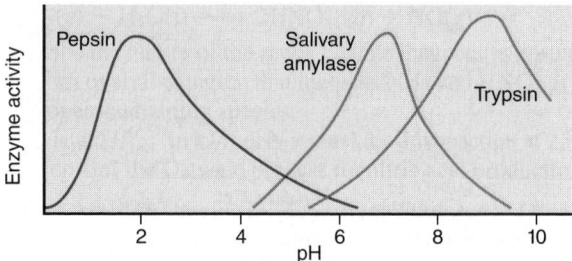

Salivary amylase begins digestion of starches in the mouth and has optimum activity at a pH of 6.8; pepsin begins protein digestion in the stomach and has optimum activity at a pH of 2.0; and trypsin, released in pancreatic juices, continues protein digestion in the small intestine and has optimum activity at a pH of 9.5. Calculate $[H_3O^+]$ in the local cellular fluid for each enzyme.

18.169 Sodium phosphate has industrial uses ranging from clarifying crude sugar to manufacturing paper. It is sold as TSP and is used in solution to remove boiler scale and to wash painted brick and concrete. What is the pH of a solution containing 33 g of Na_3PO_4 per liter? What is $[OH^-]$ of this solution?

18.170 The Group 5A(15) hydrides react with boron trihalides in a reversible Lewis acid-base reaction. When 0.15 mol of $PH_3BCl_3(s)$ is introduced into a 3.0-L container at a certain temperature, 8.4×10^{-3} mol of PH_3 is present at equilibrium: $PH_3BCl_3(s) \rightleftharpoons PH_3(g) + BCl_3(g)$.
(a) Calculate K_c for the reaction at this temperature.
(b) Draw a Lewis structure for the reactant.

*18.171 For most polyprotic acids, K_{a1} is significantly larger than K_{a2}, but for the acid $HO_2C(CH_2)_6CO_2H$, the difference in K_a values is insignificant. Explain.

18.172 The autoionization constant of HF is 1×10^{-11} at 25°C. What is the concentration of F^- in pure HF? What other ion is present with molarity equal to that of F^-?

18.173 The advent of the modern supermarket has made extended shelf life of products a major concern of consumers and scientists in the food and drug industries. Calcium propionate [calcium propanoate, $Ca(CH_3CH_2COO)_2$] is one of the more common mold inhibitors in food, tobacco, and pharmaceuticals.
(a) Use balanced equations to show whether an aqueous solution of calcium propionate is acidic, basic, or neutral. (b) Use Table 18.2 to find the pH of a solution that is made by dissolving 7.05 g of $Ca(CH_3CH_2COO)_2$ in water to give 0.500 L of solution.

18.174 What is an acidic oxide? Which is the acidic oxide in the Lewis acid-base reaction between CaO and CO_2? Are all reactions between acidic and basic oxides Lewis acid-base reactions?

18.175 An aqueous solution of beryllium nitrate is quite acidic, but aqueous solutions of other Group 2A(2) nitrates are nearly neutral. Explain.

18.176 HX ($\mathcal{M}$ = 150. g/mol) and HY ($\mathcal{M}$ = 50.0 g/mol) are weak acids. A solution containing 12.0 g/L of HX has the same pH as a solution containing 6.00 g/L of HY. Which is the stronger acid? Explain.

18.177 How are H_3O^+ and OH^- ions produced in pure water? Why do the concentrations of these ions remain constant? Why do the concentrations of these ions decrease when the temperature decreases? What happens to the ions that are "lost" at lower temperature?

18.178 Amine bases are notorious for having foul odors. Putrescine ($NH_2CH_2CH_2CH_2CH_2NH_2$), once thought to be found only in rotting animal tissue, is now known to be a component of all cells and essential for their normal and abnormal (cancerous) growth. It also plays a key role in formation of GABA, a central nervous system neurotransmitter. A 0.10 M aqueous solution of putrescine has $[OH^-] = 2.1 \times 10^{-3}$. What is the K_b?

18.179 Polymers and other large molecules are not very soluble in water, but their solubility increases if they have charged groups.
(a) Casein is a protein in milk that contains many carboxylic acid groups on its side chains. Explain how the solubility of casein in water varies with pH.
(b) Histones are proteins that are essential to the proper function of DNA. They are weakly basic due to the presence of side chains with —NH_2 and =NH groups. Explain how the solubility of histones in water varies with pH.

*18.180 Quinine ($C_{20}H_{24}N_2O_2$; see structure) is a natural product that was originally extracted from the bark of cinchona trees but is now synthesized by the pharmaceutical industry. Its antimalarial properties are responsible for saving thousands of lives during construction of the Panama Canal, and, thus, it stands as a classic example of the untold chemical wealth of the forests of Central and South America. Both N atoms are basic, but the N (colored) of the tertiary amine group is far more basic (pK_b = 5.1) than the N within the aromatic ring system (pK_b = 9.7).

(a) Quinine is not very soluble in water: a saturated solution is only 1.6×10^{-3} M. What is the pH of this solution?
(b) Show that the aromatic N contributes negligibly to the pH of the solution.
(c) Because of its low solubility as a free base, quinine is often administered as an amine salt. Quinine hydrochloride ($C_{20}H_{24}N_2O_2 \cdot HCl$), for example, is about 120 times more soluble in water than quinine. What is the pH of 0.53 M quinine hydrochloride?
(d) What is the pH of a solution (d = 1.0 g/mL) that is 1.5% quinine hydrochloride by mass, a typical antimalarial concentration?

IONIC EQUILIBRIA IN AQUEOUS SYSTEMS

CHAPTER OUTLINE

19.1 Equilibria of Acid-Base Buffer Systems
The Common-Ion Effect
The Henderson-Hasselbalch Equation
Buffer Capacity and Range
Preparing a Buffer

19.2 Acid-Base Titration Curves
Acid-Base Indicators
Strong Acid–Strong Base Titrations
Weak Acid–Strong Base Titrations
Weak Base–Strong Acid Titrations
Polyprotic Acid Titrations
Amino Acids as Polyprotic Acids

19.3 Equilibria of Slightly Soluble Ionic Compounds
The Solubility-Product Constant (K_{sp})
Calculations Involving K_{sp}
The Effect of a Common Ion
The Effect of pH
Q_{sp} vs. K_{sp}

19.4 Equilibria Involving Complex Ions
Formation of Complex Ions
Complex Ions and Solubility
Amphoteric Hydroxides

19.5 Application of Ionic Equilibria to Chemical Analysis
Selective Precipitation
Qualitative Analysis

Figure: Waterworld. Natural waters interact physically and chemically with the solid, gaseous, and living domains of the planet through intricate aqueous equilibria. In this chapter we investigate three types of ionic equilibria that occur in aqueous systems.

CONCEPTS & SKILLS

to review before you study this chapter
- solubility rules for ionic compounds (Section 4.3)
- effect of concentration on equilibrium position (Section 17.6)
- conjugate acid-base pairs (Section 18.3)
- calculations of weak-acid and weak-base equilibria (Sections 18.4 and 18.5)
- acid-base properties of salt solutions (Section 18.7)
- Lewis acids and bases (Section 18.9)

Europa, one of Jupiter's moons, has an icy surface with hints of vast oceans of liquid water beneath. Is there life on Europa? Perhaps some Europan astronomer viewing Earth is asking the same question. Every astronaut has felt awe upon viewing our "beautiful blue orb" from space. A biologist peering at the fabulous watery world of a living cell probably feels the same way. A chemist is in awe of the principles of equilibrium and their universal application to aqueous solutions wherever they occur in the external and internal worlds.

Consider just a few cases of aqueous equilibria. The magnificent formations in limestone caves and the vast expanses of oceanic coral reefs result from subtle shifts in carbonate solubility equilibria. Carbonates also influence soil pH and prevent acidification of lakes by acid rain. Equilibria involving carbon dioxide and phosphates help organisms maintain cellular pH within narrow limits. Equilibria involving clays found in soil control the availability of ionic nutrients for plants. The principles of ionic equilibrium also govern how water is softened, how substances are purified by precipitation of unwanted ions, and even how the weak acids in wine and vinegar influence the delicate taste of a fine French sauce.

In this chapter, we explore three key aqueous ionic equilibrium systems: acid-base buffers, slightly soluble salts, and complex ions. Our discussion of buffers introduces the common-ion effect, an important phenomenon in many ionic equilibria. We discuss why buffers are important, how they work, and how to prepare them. We take a closer look at acid-base titrations and how buffers are involved in that process. Then, we examine slightly soluble salts in the laboratory and in nature to see how conditions influence their solubility. Next, we investigate complex ions and how they form and change from one type to another. Finally, we see how these aqueous equilibria are employed in chemical analysis. There are few new ideas in this chapter, but there are many opportunities to apply earlier ideas in new ways.

19.1 EQUILIBRIA OF ACID-BASE BUFFER SYSTEMS

Why do some lakes become acidic when showered by acid rain, while others remain unaffected? How does blood maintain a constant pH in contact with countless cellular acid-base reactions? How can a chemist sustain a nearly constant $[H_3O^+]$ in reactions that consume or produce H_3O^+ or OH^-? The answer in each case is through the presence of a buffer.

In everyday language, a buffer is something that lessens the impact of an external force. An **acid-base buffer** is a solution that *lessens changes in $[H_3O^+]$ resulting from the addition of acid or base*. Figure 19.1 shows that when a small amount of H_3O^+ or OH^- ions is added to an unbuffered solution, *the change in pH is much larger* than the change that results from the same addition to a buffered solution.

To withstand the addition of strong acid or strong base without significantly changing its pH, a buffer must contain an acidic component that can react with added OH^- ion *and* a basic component that can react with added H_3O^+ ion. However, these buffer components cannot be just any acid and base because they would neutralize each other. Most commonly, *the components of a buffer are the conjugate acid-base pair of a weak acid*. In Figure 19.1C and D, for example, the buffer is a mixture of acetic acid (CH_3COOH) and acetate ion (CH_3COO^-).

How a Buffer Works: The Common-Ion Effect

Buffers work through a phenomenon known as the **common-ion effect.** An example of this effect occurs when acetic acid dissociates in water and some sodium acetate is added. As you know, acetic acid dissociates only slightly in water:

$$CH_3COOH(aq) + H_2O(l) \rightleftharpoons CH_3COO^-(aq) + H_3O^+(aq)$$

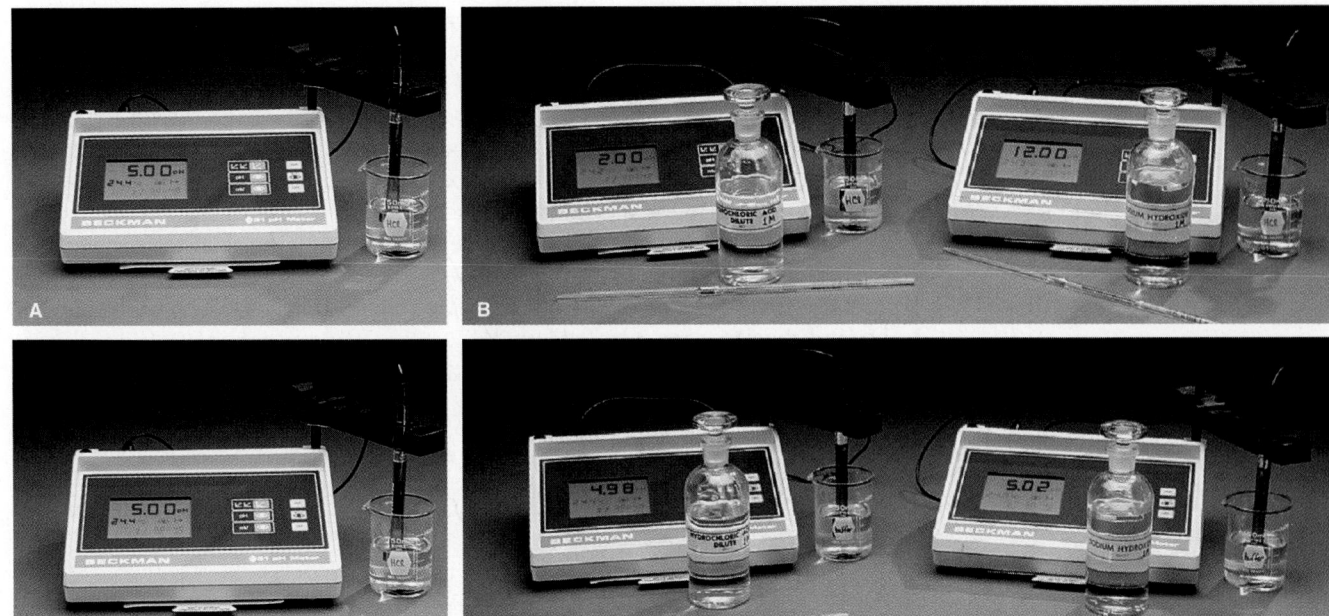

Figure 19.1 **The effect of addition of acid or base to an unbuffered or buffered solution. A,** A 100-mL sample of an unbuffered solution of dilute HCl is adjusted to pH 5.00. **B,** After the addition of 1 mL of 1 *M* HCl (*left*) or of 1 *M* NaOH (*right*), the pH change is large. **C,** A 100-mL sample of a buffered solution, made by mixing 1 *M* CH₃COOH with 1 *M* CH₃COONa, is adjusted to pH 5.00. **D,** After the addition of 1 mL of 1 *M* HCl (*left*) or of 1 *M* NaOH (*right*), the pH change is negligible.

From Le Châtelier's principle (Section 17.6), we know that if some CH_3COO^- ion is added, the equilibrium position shifts to the left; $[H_3O^+]$ decreases, in effect lowering the extent of acid dissociation:

$$CH_3COOH(aq) + H_2O(l) \rightleftharpoons CH_3COO^-(aq; \text{added}) + H_3O^+(aq)$$

Similarly, if we dissolve acetic acid in a sodium acetate solution, the acetate ion already present combines with some H_3O^+ as it dissociates from the acid, which lowers the $[H_3O^+]$. The effect again is to lower the acid dissociation. Acetate ion is called *the common ion* in this case because it is "common" to both the acetic acid and sodium acetate solutions; that is, acetate ion is added to a solution in which it is already present. *The common-ion effect occurs when a reactant containing a given ion is added to an equilibrium mixture that already contains that ion and the position of equilibrium shifts away from forming more of it.*

Table 19.1 shows the percent dissociation and the pH of an acetic acid solution containing various concentrations of acetate ion (supplied by adding solid sodium acetate). Note that the *common ion, CH_3COO^-, suppresses the percent dissociation of CH_3COOH,* which makes the solution less acidic (higher pH).

Table 19.1 The Effect of Added Acetate Ion on the Dissociation of Acetic Acid

$[CH_3COOH]_{init}$	$[CH_3COO^-]_{added}$	% Dissociation*	pH
0.10	0.00	1.3	2.89
0.10	0.050	0.036	4.44
0.10	0.10	0.018	4.74
0.10	0.15	0.012	4.92

*% Dissociation $= \dfrac{[CH_3COOH]_{dissoc}}{[CH_3COOH]_{init}} \times 100$

The Essential Feature of a Buffer In the previous case, we prepared a buffer by mixing a weak acid (CH_3COOH) and its conjugate base (CH_3COO^-). *How* does this solution resist pH changes when H_3O^+ or OH^- is added? The essential feature of a buffer is that *it consists of high concentrations of the acidic (HA) and basic (A^-) components.* When small amounts of H_3O^+ or OH^- ions are added to the buffer, they cause *a small amount of one buffer component to convert into the other,* which changes the relative concentrations of the two components. As long as the amount of H_3O^+ or OH^- added is much smaller than the amounts of HA and A^- present originally, *the added ions have little effect on the pH because they are consumed by one or the other buffer component:* the A^- consumes added H_3O^+, and the HA consumes added OH^-.

Consider what happens to a solution containing high [CH_3COOH] and [CH_3COO^-] when we add small amounts of strong acid or base. The expression for HA dissociation at equilibrium is

$$K_a = \frac{[CH_3COO^-][H_3O^+]}{[CH_3COOH]}$$

Solving for [H_3O^+], we obtain

$$[H_3O^+] = K_a \times \frac{[CH_3COOH]}{[CH_3COO^-]}$$

Note that because K_a is constant, *the [H_3O^+] of the solution depends directly on the buffer-component concentration ratio,* $\dfrac{[CH_3COOH]}{[CH_3COO^-]}$:

*   If the ratio [HA]/[A^-] goes up, [H_3O^+] goes up.
*   If the ratio [HA]/[A^-] goes down, [H_3O^+] goes down.

When we add a small amount of strong acid, the increased amount of H_3O^+ ion reacts with a nearly *stoichiometric amount* of acetate ion in the buffer to form more acetic acid:

$$H_3O^+(aq; \text{ added}) + CH_3COO^-(aq) \longrightarrow CH_3COOH(aq) + H_2O(l)$$

As a result, [CH_3COO^-] goes down by that amount and [CH_3COOH] goes up by that amount, which increases the buffer-component concentration ratio, as you can see in Figure 19.2. The [H_3O^+] increases also but only very slightly.

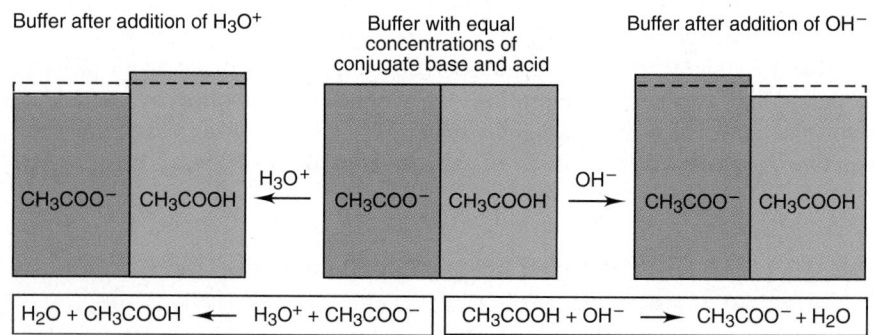

Figure 19.2 **How a buffer works.** A buffer consists of high concentrations of a conjugate acid-base pair, in this case, acetic acid (CH_3COOH) and acetate ion (CH_3COO^-). When a small amount of H_3O^+ is added (*left*), that same amount of CH_3COO^- combines with it, which increases the amount of CH_3COOH slightly. Similarly, when a small amount of OH^- is added (*right*), that amount of CH_3COOH combines with it, which increases the amount of CH_3COO^- slightly. In both cases, the relative changes in amounts of the buffer components are small, so their concentration ratio, and therefore the pH, changes very little.

Adding a small amount of strong base produces the opposite result. It supplies OH^- ions, which react with a nearly *stoichiometric amount* of CH_3COOH in the buffer, forming that much more CH_3COO^-:

$$CH_3COOH(aq) + OH^-(aq; \text{ added}) \longrightarrow CH_3COO^-(aq) + H_2O(l)$$

The buffer-component concentration ratio decreases, which decreases $[H_3O^+]$, but once again, the change is very slight.

Thus, the components consume virtually all the added H_3O^+ or OH^-. To reiterate, as long as the amount of added H_3O^+ or OH^- is small compared with the amounts of the buffer components, *the conversion of one component into the other produces a small change in the buffer-component concentration ratio and, consequently, a small change in $[H_3O^+]$ and in pH.* Sample Problem 19.1 demonstrates how small these pH changes actually are in a typical case. Note that the latter two parts of the problem combine a stoichiometry portion, like the problems in Chapter 3, and a weak-acid dissociation portion, like those in Chapter 18.

SAMPLE PROBLEM 19.1 Calculating the Effect of Added H_3O^+ and OH^- on Buffer pH

Problem Calculate the pH:
(a) Of a buffer solution consisting of 0.50 M CH_3COOH and 0.50 M CH_3COONa
(b) After adding 0.020 mol of solid NaOH to 1.0 L of the buffer solution in part (a)
(c) After adding 0.020 mol of HCl to 1.0 L of the buffer solution in part (a)
K_a of $CH_3COOH = 1.8 \times 10^{-5}$. (Assume the additions cause negligible volume changes.)
Plan In each case, we know, or can find, $[CH_3COOH]_{init}$ and $[CH_3COO^-]_{init}$ and the K_a of CH_3COOH (1.8×10^{-5}), and we need to find $[H_3O^+]$ at equilibrium and convert it to pH. In **(a)**, we use the given concentrations of buffer components (each 0.50 M) as the initial values. As in earlier problems, we assume that x, the $[CH_3COOH]$ that dissociates, and which equals $[H_3O^+]$, is so small relative to $[CH_3COOH]_{init}$ that it can be neglected. We set up a reaction table, solve for x, and check the assumption. In **(b)** and **(c)**, we assume that the added OH^- or H_3O^+ reacts completely with the buffer components to yield new $[CH_3COOH]_{init}$ and $[CH_3COO^-]_{init}$, which then dissociate to an unknown extent. We set up two reaction tables. The first summarizes the stoichiometry of adding strong base (0.020 mol) or acid (0.020 mol). The second summarizes the dissociation of the new HA concentrations, so we proceed as in **(a)** to find the new $[H_3O^+]$.
Solution (a) The original pH: $[H_3O^+]$ in the original buffer.
Setting up a reaction table with $x = [CH_3COOH]_{dissoc} = [H_3O^+]$ (as in Chapter 18, we assume that $[H_3O^+]$ from H_2O is negligible and disregard it):

Concentration (M)	$CH_3COOH(aq)$	+ $H_2O(l)$	$\rightleftharpoons$	$CH_3COO^-(aq)$	+ $H_3O^+(aq)$
Initial	0.50	—		0.50	0
Change	$-x$	—		$+x$	$+x$
Equilibrium	$0.50 - x$	—		$0.50 + x$	x

Making the assumption and finding the equilibrium $[CH_3COOH]$ and $[CH_3COO^-]$: With K_a small, x is small, so we assume

$$[CH_3COOH] = 0.50\ M - x \approx 0.50\ M \quad \text{and} \quad [CH_3COO^-] = 0.50\ M + x \approx 0.50\ M$$

Solving for x ($[H_3O^+]$ at equilibrium):

$$x = [H_3O^+] = K_a \times \frac{[CH_3COOH]}{[CH_3COO^-]} \approx (1.8 \times 10^{-5}) \times \frac{0.50}{0.50} = 1.8 \times 10^{-5}\ M$$

Checking the assumption:

$$\frac{1.8 \times 10^{-5}\ M}{0.50\ M} \times 100 = 3.6 \times 10^{-3}\% < 5\%$$

The assumption is justified, and we will use the same assumption in parts **(b)** and **(c)**.
Calculating pH:

$$pH = -\log [H_3O^+] = -\log (1.8 \times 10^{-5}) = \boxed{4.74}$$

(b) The pH after adding base (0.020 mol of NaOH to 1.0 L of buffer). Finding $[OH^-]_{added}$:

$$[OH^-]_{added} = \frac{0.020 \text{ mol } OH^-}{1.0 \text{ L soln}} = 0.020 \, M \, OH^-$$

Setting up a reaction table for the *stoichiometry* of adding OH^- to CH_3COOH:

Concentration (M)	$CH_3COOH(aq)$ +	$OH^-(aq)$	$\longrightarrow$ $CH_3COO^-(aq)$ +	$H_2O(aq)$
Before addition	0.50	—	0.50	—
Addition	—	0.020	—	—
After addition	0.48	0	0.52	—

Setting up a reaction table for the *acid dissociation,* using these new initial concentrations. As in part (a), $x = [CH_3COOH]_{dissoc} = [H_3O^+]$:

Concentration (M)	$CH_3COOH(aq)$ +	$H_2O(l)$	$\rightleftharpoons$ $CH_3COO^-(aq)$ +	$H_3O^+(aq)$
Initial	0.48	—	0.52	0
Change	$-x$	—	$+x$	$+x$
Equilibrium	$0.48 - x$	—	$0.52 + x$	x

Making the assumption that x is small, and solving for x:

$[CH_3COOH] = 0.48 \, M - x \approx 0.48 \, M$ and $[CH_3COO^-] = 0.52 \, M + x \approx 0.52 \, M$

$$[H_3O^+] = K_a \times \frac{[CH_3COOH]}{[CH_3COO^-]} \approx (1.8 \times 10^{-5}) \times \frac{0.48}{0.52} = 1.7 \times 10^{-5} \, M$$

Calculating the pH:

$$pH = -\log [H_3O^+] = -\log (1.7 \times 10^{-5}) = \boxed{4.77}$$

The addition of strong base increased the concentration of the basic buffer component at the expense of the acidic buffer component. Note especially that the pH *increased only slightly,* from 4.74 to 4.77.

(c) The pH after adding acid (0.020 mol of HCl to 1.0 L of buffer). Finding $[H_3O^+]_{added}$:

$$[H_3O^+]_{added} = \frac{0.020 \text{ mol } H_3O^+}{1.0 \text{ L soln}} = 0.020 \, M \, H_3O^+$$

Now we proceed as in part (b), by first setting up a reaction table for the *stoichiometry* of adding H_3O^+ to CH_3COO^-:

Concentration (M)	$CH_3COO^-(aq)$ +	$H_3O^+(aq)$	$\longrightarrow$ $CH_3COOH(aq)$ +	$H_2O(l)$
Before addition	0.50	—	0.50	—
Addition	—	0.020	—	—
After addition	0.48	0	0.52	—

The reaction table for the acid dissociation, with $x = [CH_3COOH]_{dissoc} = [H_3O^+]$ is

Concentration (M)	$CH_3COOH(aq)$ +	$H_2O(l)$	$\rightleftharpoons$ $CH_3COO^-(aq)$ +	$H_3O^+(aq)$
Initial	0.52	—	0.48	0
Change	$-x$	—	$+x$	$+x$
Equilibrium	$0.52 - x$	—	$0.48 + x$	x

Making the assumption that x is small, and solving for x:

$[CH_3COOH] = 0.52 \, M - x \approx 0.52 \, M$ and $[CH_3COO^-] = 0.48 \, M + x \approx 0.48 \, M$

$$x = [H_3O^+] = K_a \times \frac{[CH_3COOH]}{[CH_3COO^-]} \approx (1.8 \times 10^{-5}) \times \frac{0.52}{0.48} = 2.0 \times 10^{-5} \, M$$

Calculating the pH:

$$pH = -\log [H_3O^+] = -\log (2.0 \times 10^{-5}) = \boxed{4.70}$$

The addition of strong acid increased the concentration of the acidic buffer component at the expense of the basic buffer component and *lowered* the pH only slightly, from 4.74 to 4.70.

Check The changes in [CH₃COOH] and [CH₃COO⁻] occur in opposite directions in parts (b) and (c), which makes sense. The additions were of equal amounts, so the pH increase in (b) should equal the pH decrease in (c), within rounding.
Comment In part (a), we justified our assumption that x can be neglected. Therefore, in parts (b) and (c), we can use the "After addition" values in the last line of the stoichiometry tables directly for the ratio of buffer components and dispense with the reaction table for the dissociation. In subsequent problems in this chapter, we follow this simplified approach.

FOLLOW-UP PROBLEM 19.1 Calculate the pH of a buffer consisting of 0.50 M HF and 0.45 M F⁻ **(a)** before and **(b)** after addition of 0.40 g of NaOH to 1.0 L of the buffer (K_a of HF $= 6.8 \times 10^{-4}$).

The Henderson-Hasselbalch Equation

For any weak acid, HA, the equation and K_a expression for the acid dissociation at equilibrium are

$$HA + H_2O \rightleftharpoons H_3O^+ + A^-$$

$$K_a = \frac{[H_3O^+][A^-]}{[HA]}$$

A simple mathematical rearrangement turns this expression into a more useful form for buffer calculations. Remember that the key variable that determines $[H_3O^+]$ is the concentration *ratio* of acid species to base species. Isolating $[H_3O^+]$ gives

$$[H_3O^+] = K_a \times \frac{[HA]}{[A^-]}$$

Taking the negative logarithm (base 10) of both sides gives

$$-\log [H_3O^+] = -\log K_a - \log \left(\frac{[HA]}{[A^-]}\right)$$

from which we obtain

$$pH = pK_a + \log \left(\frac{[A^-]}{[HA]}\right)$$

(Note the inversion of the buffer-component concentration ratio when the sign of the logarithm is changed.) A key point that we'll note again later is that when $[A^-] = [HA]$, their ratio becomes 1; the log term then becomes 0, and thus $pH = pK_a$.

Generalizing the previous equation for any conjugate acid-base pair gives the **Henderson-Hasselbalch equation:**

$$pH = pK_a + \log \left(\frac{[base]}{[acid]}\right) \qquad \textbf{(19.1)}$$

This relationship allows us to solve directly for pH instead of having to calculate $[H_3O^+]$ first. For instance, by applying the Henderson-Hasselbalch equation in part (b) of Sample Problem 19.1, we could have found the pH of the buffer after the addition of NaOH as follows:

$$pH = pK_a + \log \left(\frac{[CH_3COO^-]}{[CH_3COOH]}\right)$$

$$= 4.74 + \log \left(\frac{0.52}{0.48}\right) = 4.77$$

(We just derived the Henderson-Hasselbalch equation from fundamental definitions and simple algebra. It is always a good idea to derive simple relationships this way rather than having to memorize them.)

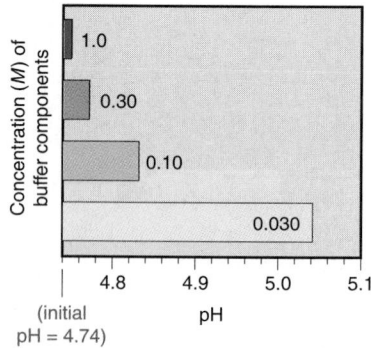

Figure 19.3 **The relation between buffer capacity and pH change.** The four bars in the graph represent CH_3COOH/CH_3COO^- buffers with the same initial pH (4.74) but different component concentrations (labeled on or near each bar). When a given amount of strong base is added to each buffer, the pH increases. The length of the bar corresponds to the pH increase. Note that the more concentrated the buffer, the greater its capacity, and the smaller the pH change.

Buffer Capacity and Buffer Range

As you've seen, buffers resist a pH change as long as the concentrations of buffer components are *large* compared with the amount of strong acid or base added. **Buffer capacity** depends on the component concentrations and is a measure of the ability to resist pH change: *the more concentrated the components of a buffer, the greater the buffer capacity.* In other words, you must add more H_3O^+ or OH^- to a high-capacity (concentrated) buffer than to a low-capacity (dilute) buffer to obtain a given pH change. Conversely, adding the same amount of H_3O^+ or OH^- to buffers of different capacities produces a smaller pH change in the higher capacity buffer, as shown in Figure 19.3. This makes sense because the more concentrated buffer exhibits a smaller change in the ratio of its component concentrations on addition of a given amount of H_3O^+ or OH^-. It's important to realize that *the pH of a buffer is distinct from its buffer capacity.* A buffer made of equal volumes of 1.0 M CH_3COOH and 1.0 M CH_3COO^- has the same pH (4.74) as a buffer that is made of equal volumes of 0.10 M CH_3COOH and 0.10 M CH_3COO^-, but it has a much larger capacity for resisting a pH change.

Buffer capacity is also affected by the *relative* concentrations of the buffer components. As a buffer functions, the concentration of one component increases relative to the other. Since the ratio of these concentrations determines the pH, the less the ratio changes, the less the pH changes. *For a given addition of acid or base, the concentration ratio changes less for similar buffer-component concentrations than it does for different concentrations.* To illustrate this point, we must keep the sum of the acidic and basic components constant. Suppose we have a buffer in which $[HA] = [A^-] = 1.000$ M. When we add 0.010 mol of OH^- to 1.00 L of buffer, $[A^-]$ becomes 1.010 M and $[HA]$ becomes 0.990 M:

$$\frac{[A^-]_{init}}{[HA]_{init}} = \frac{1.000\ M}{1.000\ M} = 1.000 \qquad \frac{[A^-]_{final}}{[HA]_{final}} = \frac{1.010\ M}{0.990\ M} = 1.02$$

$$\text{Percent change} = \frac{1.02 - 1.000}{1.000} \times 100 = 2\%$$

Now suppose that the buffer-component concentrations are $[HA] = 0.250$ M and $[A^-] = 1.750$ M. The same addition of 0.010 mol of OH^- to 1.00 L of buffer gives $[HA] = 0.240$ M and $[A^-] = 1.760$ M, so the ratios are

$$\frac{[A^-]_{init}}{[HA]_{init}} = \frac{1.750\ M}{0.250\ M} = 7.00 \qquad \frac{[A^-]_{final}}{[HA]_{final}} = \frac{1.760\ M}{0.240\ M} = 7.33$$

$$\text{Percent change} = \frac{7.33 - 7.00}{7.00} \times 100 = 4.7\%$$

As you can see, the change in the buffer-component concentration ratio is much larger when the initial component concentrations are very different.

It follows that *a buffer has the highest capacity when the component concentrations are equal,* that is, when $[A^-]/[HA] = 1$:

$$\text{pH} = \text{p}K_a + \log\left(\frac{[A^-]}{[HA]}\right) = \text{p}K_a + \log 1 = \text{p}K_a + 0 = \text{p}K_a$$

Note this important result: for a given concentration, *a buffer whose pH is equal to or near the pK_a of its acid component has the highest buffer capacity.*

The **buffer range** is the pH range over which the buffer acts effectively. The further the buffer-component concentration ratio is from 1, the less effective the buffering action (that is, the lower the buffer capacity). In practice, we find that if the $[A^-]/[HA]$ ratio is greater than 10 or less than 0.1—that is, if one component is more than 10 times as concentrated as the other—buffering action is poor. Given that $\log 10 = +1$ and $\log 0.1 = -1$, we find that *buffers have a usable range within ±1 pH unit of the pK_a of the acid component:*

$$\text{pH} = \text{p}K_a + \log\left(\frac{10}{1}\right) = \text{p}K_a + 1 \qquad \text{and} \qquad \text{pH} = \text{p}K_a + \log\left(\frac{1}{10}\right) = \text{p}K_a - 1$$

Preparing a Buffer

Any large chemical supply-house catalog lists many common buffers available in a variety of pH values and concentrations. So, you might ask, why learn how to prepare a buffer? In many cases, a common buffer is simply not available with the desired pH or concentration, and you have to make it yourself. In many modern medical or biological research applications, a buffer of unusual composition may be required to simulate a cell system or stabilize a fragile biological macromolecule. Even the most sophisticated, automated laboratory frequently relies on personnel versed in wet chemical techniques with a good knowledge of basic chemistry to prepare a buffer. Several steps are required to prepare a buffer of a desired pH, so let's go through them with an actual example:

1. *Choose the conjugate acid-base pair.* First, decide on the chemical composition of the buffer, that is, the conjugate acid-base pair. This choice is determined to a large extent by the desired pH. Remember that a buffer is most effective when the ratio of its component concentrations is close to 1, in which case the pH $\approx$ pK_a of the acid.

Suppose you need a buffer whose pH is 3.90: the pK_a of the acid component should be as close to 3.90 as possible; or $K_a = 10^{-3.90} = 1.3 \times 10^{-4}$. Scanning a table of acid-dissociation constants (see Table 18.2, p. 762) shows that formic acid ($K_a = 1.8 \times 10^{-4}$; p$K_a = 3.74$) is a good choice. Therefore, the buffer components will be formic acid, HCOOH, and formate ion, HCOO$^-$, supplied by a soluble salt, such as sodium formate, HCOONa.

2. *Calculate the ratio of buffer component concentrations.* Next, find the ratio of [A$^-$]/[HA] that gives the desired pH. With the Henderson-Hasselbalch equation, we have

$$\text{pH} = \text{p}K_a + \log\left(\frac{[\text{A}^-]}{[\text{HA}]}\right) \quad \text{or} \quad 3.90 = 3.74 + \log\left(\frac{[\text{HCOO}^-]}{[\text{HCOOH}]}\right)$$

$$\log\left(\frac{[\text{HCOO}^-]}{[\text{HCOOH}]}\right) = 0.16 \quad \text{so} \quad \left(\frac{[\text{HCOO}^-]}{[\text{HCOOH}]}\right) = 10^{0.16} = 1.4$$

Thus, for every 1.0 mol of HCOOH in a given volume of solution, you need 1.4 mol of HCOONa.

3. *Determine the buffer concentration.* Next, decide how concentrated the buffer should be. Remember that the higher the concentrations of components, the greater the buffer capacity. For most laboratory-scale applications, concentrations of about 0.50 *M* are suitable, but the decision is often based on availability of stock solutions.

Suppose you have a large stock of 0.40 *M* HCOOH and you need approximately 1.0 L of final buffer. A mole-to-gram calculation gives the amount (mol) of sodium formate you will need:

$$\text{Moles of HCOOH} = 1.0 \text{ L soln} \times \frac{0.40 \text{ mol HCOOH}}{1.0 \text{ L soln}} = 0.40 \text{ mol HCOOH}$$

$$\text{Moles of HCOONa} = 0.40 \text{ mol HCOOH} \times \frac{1.4 \text{ mol HCOONa}}{1.0 \text{ mol HCOOH}} = 0.56 \text{ mol HCOONa}$$

$$\text{Mass (g) of HCOONa} = 0.56 \text{ mol HCOONa} \times \frac{68.01 \text{ g HCOONa}}{1 \text{ mol HCOONa}} = 38 \text{ g HCOONa}$$

4. *Mix the solution and adjust the pH.* The buffer is prepared by thoroughly dissolving 38 g of solid sodium formate in 0.40 *M* HCOOH to a total volume of 1.0 L. Due to the behavior of nonideal solutions (Section 13.5) and other ionic phenomena, a buffer prepared in this way may vary from the desired pH by as much as several tenths of a pH unit. Therefore, after making up the solution, *adjust the buffer pH* to the desired value by adding strong acid or strong base, while monitoring the solution with a pH meter. The following sample problem does not refer to the Henderson-Hasselbalch equation.

SAMPLE PROBLEM 19.2 Preparing a Buffer

Problem An environmental chemist needs a carbonate buffer of pH 10.00 to study the effects of the acid rain on limestone-rich soils. How many grams of Na_2CO_3 must she add to 1.5 L of freshly prepared 0.20 M $NaHCO_3$ to make the buffer? K_a of HCO_3^- is 4.7×10^{-11}.

Plan The conjugate pair is already chosen, HCO_3^- (acid) and CO_3^{2-} (base), as are the volume (1.5 L) and concentration (0.20 M) of HCO_3^-, so we must find the buffer-component concentration ratio that gives pH 10.00 and the mass of Na_2CO_3 to dissolve. We first convert pH to $[H_3O^+]$ and use the K_a expression from the equation to solve for the required $[CO_3^{2-}]$. Multiplying by the volume of solution gives the amount (mol) of CO_3^{2-} required, and then we use the molar mass to find the mass (g) of Na_2CO_3.

Solution Calculating $[H_3O^+]$: $[H_3O^+] = 10^{-pH} = 10^{-10.00} = 1.0 \times 10^{-10}$ M

Solving for $[CO_3^{2-}]$ in the concentration ratio:

$$HCO_3^-(aq) + H_2O(l) \rightleftharpoons H_3O^+(aq) + CO_3^{2-}(aq) \qquad K_a = \frac{[H_3O^+][CO_3^{2-}]}{[HCO_3^-]}$$

So

$$[CO_3^{2-}] = K_a \frac{[HCO_3^-]}{[H_3O^+]} = \frac{(4.7 \times 10^{-11})(0.20)}{1.0 \times 10^{-10}} = 0.094 \; M$$

Calculating the amount (mol) of CO_3^{2-} needed for the given volume:

$$\text{Moles of } CO_3^{2-} = 1.5 \text{ L soln} \times \frac{0.094 \text{ mol } CO_3^{2-}}{1 \text{ L soln}} = 0.14 \text{ mol } CO_3^{2-}$$

Calculating the mass (g) of Na_2CO_3 needed:

$$\text{Mass (g) of } Na_2CO_3 = 0.14 \text{ mol } Na_2CO_3 \times \frac{105.99 \text{ g } Na_2CO_3}{1 \text{ mol } Na_2CO_3} = \boxed{15 \text{ g } Na_2CO_3}$$

We dissolve 15 g of Na_2CO_3 into 1.3 L of 0.20 M $NaHCO_3$ and add 0.20 M $NaHCO_3$ to make 1.5 L. Using a pH meter, we adjust the pH to 10.00 with strong acid or base.

Check For a useful buffer range, the acidic component, $[HCO_3^-]$, must be within a factor of 10 of the basic component, $[CO_3^{2-}]$. We have (1.5 L)(0.20 M HCO_3^-), or 0.30 mol of HCO_3^-, and 0.14 mol of CO_3^{2-}; 0.30:0.14 = 2.1, which seems fine. Make sure the relative amounts of components seem reasonable: since we want a pH lower than the pK_a of HCO_3^- (10.33), it makes sense that we have more of the acidic than the basic species.

FOLLOW-UP PROBLEM 19.2 How would you prepare a benzoic acid/benzoate buffer with pH = 4.25, starting with 5.0 L of 0.050 M sodium benzoate (C_6H_5COONa) solution and adding the acidic component? K_a of benzoic acid (C_6H_5COOH) is 6.3×10^{-5}.

Another way to prepare a buffer is to form one of the components during the final mixing step by *partial neutralization* of the other component. For example, you can prepare an $HCOOH/HCOO^-$ buffer by mixing appropriate amounts of $HCOOH$ solution and $NaOH$ solution. As the OH^- ions react with the $HCOOH$ molecules, neutralization of part of the total $HCOOH$ present produces the $HCOO^-$ needed:

$HCOOH$ (HA total) + OH^- (amt added) $\longrightarrow$
$\qquad$ $HCOOH$ (HA total − OH^- amt added) + $HCOO^-$ (OH^- amt added) + H_2O

This method is based on the same chemical process that occurs when a weak acid is titrated with a strong base, as you'll see in the next section.

SECTION SUMMARY

A buffered solution exhibits a much smaller change in pH when H_3O^+ or OH^- is added than does an unbuffered solution. A buffer consists of relatively high concentrations of the components of a conjugate weak acid-base pair. The buffer-component concentration ratio determines the pH, and the ratio and pH are related by the Henderson-Hasselbalch equation. As H_3O^+ or OH^- is added, one buffer component reacts with it and is converted into the other component; therefore, the buffer-component concentration ratio, and consequently the free $[H_3O^+]$ (and pH), changes only slightly. A concentrated buffer undergoes smaller changes in pH than a dilute buffer. When the buffer pH equals the pK_a of the acid component, the buffer has its highest capacity. A buffer has an effective range of $pK_a \pm 1$ pH unit. When preparing a buffer, you choose the conjugate acid-base pair, calculate the ratio of buffer components, determine the buffer concentration, and adjust the final solution to the desired pH.

19.2 ACID-BASE TITRATION CURVES

In Chapter 4, we discussed the acid-base titration as an analytical method. Let's re-examine it, this time tracking the change in pH with an **acid-base titration curve,** a plot of pH vs. volume of titrant added. The behavior of an acid-base indicator and its role in the titration are described first. To better understand the titration process, we apply the principles of the acid-base behavior of salt solutions (Section 18.7) and, later in the section, the principles of buffer action.

Monitoring pH with Acid-Base Indicators

The two common devices for measuring pH in the laboratory are pH meters and acid-base indicators. (We discuss the operation of pH meters in Chapter 21.) An *acid-base indicator* is a weak organic acid (denoted as HIn) that has a different color than its conjugate base (In⁻), with the color change occurring over a specific and relatively narrow pH range. Typically, one or both of the forms are intensely colored, so only a tiny amount of indicator is needed, far too little to affect the pH of the solution being studied (see also p. 756).

Indicators are used for estimating the pH of a solution and for monitoring the pH in acid-base titrations and in reactions. Figure 19.4 shows the colors and the pH ranges for the color changes of some common acid-base indicators. Selecting an indicator requires that you know the approximate pH of the titration end point, which in turn requires that you know the ionic species present. You'll see how this is known in our discussion of acid-base titration curves.

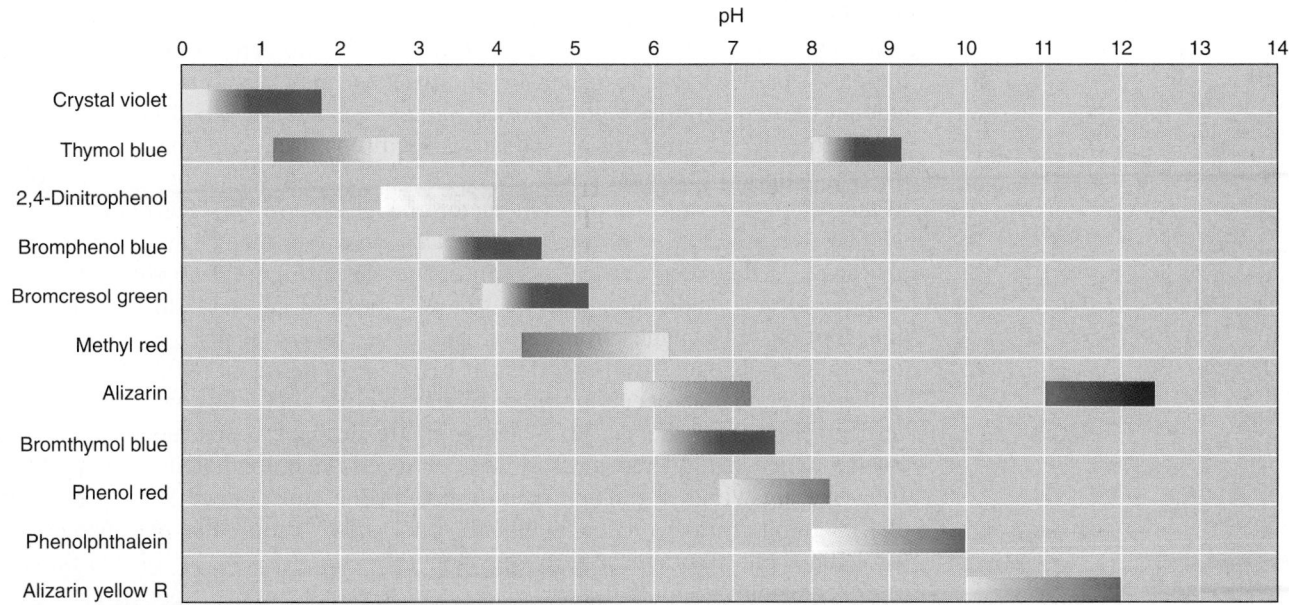

Figure 19.4 **Colors and approximate pH range of some common acid-base indicators.** Most indicators have a range of about 2 pH units, in keeping with a useful buffer range of about 2 pH units ($pK_a \pm 1$). (The specific pH range depends to some extent on the solvent used to prepare the indicator.)

Figure 19.5 The color change of the indicator bromthymol blue. Indicators are weak organic acids whose acidic and basic forms have different colors. The color change takes place over a pH range of about 2 units. The acidic form of bromthymol blue is yellow (*left*) and the basic form is blue (*right*). Over the pH range when the indicator is changing, both forms are present, so the mixture appears greenish (*center*).

Because the indicator molecule is a weak acid, the ratio of the two forms is governed by the $[H_3O^+]$ of the test solution:

$$HIn(aq) + H_2O(l) \rightleftharpoons H_3O^+(aq) + In^-(aq) \qquad K_a \text{ of } HIn = \frac{[H_3O^+][In^-]}{[HIn]}$$

Therefore,
$$\frac{[HIn]}{[In^-]} = \frac{[H_3O^+]}{K_a}$$

How we perceive colors has a major influence on the use of indicators. Typically, the experimenter will see the HIn color if the $[HIn]/[In^-]$ ratio is 10:1 or greater and the In^- color if the $[HIn]/[In^-]$ ratio is 1:10 or less. Between these extremes, the colors of the two forms are merged into an intermediate hue. Therefore, an indicator has a *color range* that reflects a 100-fold range in the $[HIn]/[In]$ ratio, which means that an *indicator changes color over a range of about 2 pH units*. For example, as you can see in Figure 19.4, bromthymol blue has a pH range of about 6.0 to 7.6, and as Figure 19.5 shows, it is yellow below that range, blue above it, and greenish in between.

Strong Acid–Strong Base Titration Curves

A typical curve for the titration of a strong acid with a strong base appears in Figure 19.6, along with the data used to construct it. There are three distinct regions of the curve, which correspond to three major changes in slope:

1. The pH starts out low, reflecting the high $[H_3O^+]$ of the strong acid, and increases gradually as acid is neutralized by the added base.
2. Suddenly, the pH rises steeply. This rise begins when the moles of OH^- that have been added nearly equal the moles of H_3O^+ originally present in the acid. An additional drop or two of base neutralizes the final tiny excess of acid and introduces a tiny excess of base, so the pH jumps 6 to 8 units.
3. Beyond this steep portion, the pH increases slowly again as more base is added.

The **equivalence point,** which occurs within the nearly vertical portion of the curve, is the point at which the *number of moles of added OH⁻ equals the number of moles of H₃O⁺ originally present*. At the equivalence point of a strong acid–strong base titration, *the solution consists of the anion of the strong acid and the cation of the strong base.* Recall from Chapter 18 that *these ions do not react with water, so the solution is neutral, pH = 7.00*. The volume and concentration of base needed to reach the equivalence point allow us to calculate the amount of acid originally present (see Sample Problem 4.5, p. 144).

Before the titration begins, we add a few drops of an appropriate indicator to the acid solution to signal when we reach the equivalence point. The **end point** of the titration occurs when the indicator changes color. *We choose an indicator with an end point close to the equivalence point,* one that changes color in the pH range on the steep vertical portion of the curve. Figure 19.6 shows the color changes for two indicators that are suitable for a strong acid–strong base titration. Methyl red changes from red at pH 4.2 to yellow at pH 6.3, whereas phenolphthalein changes from colorless at pH 8.3 to pink at pH 10.0. Even though neither color change occurs *at* the equivalence point (pH 7.00), both occur on the vertical portion of the curve, where a single drop of base causes a large pH change: when methyl red turns yellow, or when phenolphthalein turns pink, we know we are within a fraction of a drop from the equivalence point. For example, in going from 39.90 to 39.99 mL, one to two drops, the pH changes one whole unit. For all practical purposes, then, the *visible* change in color of the indicator (end point) signals the *invisible* point at which moles of added base equal the original moles of acid (equivalence point).

By knowing the chemical species present during the titration, we can calculate pH at various points along the way. In Figure 19.6, 40.00 mL of 0.1000 *M*

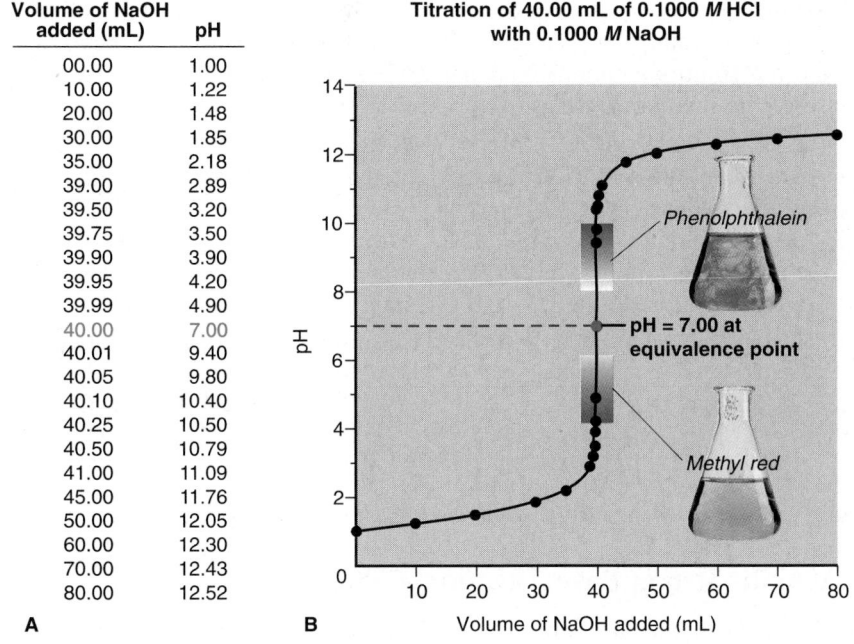

Volume of NaOH added (mL)	pH
00.00	1.00
10.00	1.22
20.00	1.48
30.00	1.85
35.00	2.18
39.00	2.89
39.50	3.20
39.75	3.50
39.90	3.90
39.95	4.20
39.99	4.90
40.00	7.00
40.01	9.40
40.05	9.80
40.10	10.40
40.25	10.50
40.50	10.79
41.00	11.09
45.00	11.76
50.00	12.05
60.00	12.30
70.00	12.43
80.00	12.52

A

Titration of 40.00 mL of 0.1000 *M* HCl with 0.1000 *M* NaOH

Phenolphthalein

pH = 7.00 at equivalence point

Methyl red

B

Volume of NaOH added (mL)

Figure 19.6 Curve for a strong acid–strong base titration. **A,** Data obtained from the titration of 40.00 mL of 0.1000 *M* HCl with 0.1000 *M* NaOH are plotted in **B** to give an acid-base titration curve. The pH increases gradually at first. When the amount (mol) of OH^- added is slightly less than the amount (mol) of H_3O^+ originally present, a large pH change accompanies a small addition of OH^-. The equivalence point occurs when amount (mol) of OH^- added = amount (mol) of H_3O^+ originally present. Note that, for a strong acid–strong base titration, pH = 7.00 at the equivalence point. Added before the titration begins, either methyl red or phenolphthalein is a suitable indicator in this case because each changes color on the steep portion of the curve, as shown by the color strips. Photos showing the color changes from 1–2 drops of indicator appear nearby. Beyond this point, added OH^- causes a gradual pH increase again.

HCl is titrated with 0.1000 *M* NaOH. A strong acid is completely dissociated, so [HCl] = $[H_3O^+]$ = 0.1000 *M*. Therefore, the initial pH is*

$$pH = -\log [H_3O^+] = -\log (0.1000) = 1.00$$

As soon as we start adding titrant, two changes occur that we must incorporate into the pH calculations: (1) some acid is neutralized, and (2) the volume of solution increases. To find the pH at various points during the titration, we find the *amount (mol)* of H_3O^+ (or OH^-) present and then use the change in volume to calculate the *concentration*, $[H_3O^+]$ (or $[OH^-]$), and convert to pH. As an example, let's calculate the pH at three points along the titration curve:

1. *Before the equivalence point.* After adding 20.00 mL of 0.1000 *M* NaOH:

- *Find the moles of H_3O^+ remaining.* Subtracting the number of moles of H_3O^+ reacted from the number originally present gives the number remaining. Moles of H_3O^+ reacted equals moles of OH^- added, so

  Initial moles of H_3O^+ = 0.04000 L × 0.1000 *M* = 0.004000 mol H_3O^+
  −Moles of OH^- added = 0.02000 L × 0.1000 *M* = 0.002000 mol OH^-

  Moles of H_3O^+ remaining =   0.002000 mol H_3O^+

- *Calculate $[H_3O^+]$, taking the total volume into account.* We add one solution to another in a titration, so *the volume increases*. To find the ion concentrations, we use the *total volume* because the water of one solution dilutes the ions of the other:

$$[H_3O^+] = \frac{\text{amount (mol) of } H_3O^+ \text{ remaining}}{\text{original volume of acid + volume of added base}}$$

$$= \frac{0.002000 \text{ mol } H_3O^+}{0.04000 \text{ L} + 0.02000 \text{ L}} = 0.03333 \ M \qquad pH = 1.48$$

Given the moles of OH^- added, we are halfway to the equivalence point; but we are still on the initial slow rise of the curve, so the pH is still very low. Similar calculations give values up to the equivalence point.

---

*In acid-base titrations, volumes and concentrations are usually known to four significant figures, but pH is generally reported to no more than two digits to the right of the decimal point.

2. *At the equivalence point.* After adding 40.00 mL of 0.1000 M NaOH: When the equivalence point is reached, all the H_3O^+ from the acid has been neutralized, and the solution contains Na^+ and Cl^-, neither of which reacts with water. Because of the autoionization of water, however,

$$[H_3O^+] = 1.0 \times 10^{-7} \, M \qquad pH = 7.00$$

In this example 0.004000 mol of OH^- reacted with 0.004000 mol of H_3O^+ to reach the equivalence point.

3. *After the equivalence point.* After adding 50.00 mL of 0.1000 M NaOH: From the equivalence point on, the pH calculation is based on the moles of *excess* OH^- present. After adding 50.00 mL of NaOH, we have

$$\begin{array}{l} \text{Total moles of } OH^- \text{ added} = 0.05000 \text{ L} \times 0.1000 \, M = 0.005000 \text{ mol } OH^- \\ \underline{-\text{Moles of } H_3O^+ \text{ consumed} = 0.04000 \text{ L} \times 0.1000 \, M = 0.004000 \text{ mol } H_3O^+} \\ \qquad \text{Moles of excess } OH^- = \qquad\qquad\qquad\qquad\qquad\qquad 0.001000 \text{ mol } OH^- \end{array}$$

$$[OH^-] = \frac{0.001000 \text{ mol } OH^-}{0.04000 \text{ L} + 0.05000 \text{ L}} = 0.01111 \, M \qquad pOH = 1.95$$

$$pH = pK_w - pOH = 14.00 - 1.95 = 12.05$$

ACID-BASE TITRATION

Weak Acid–Strong Base Titration Curves

Now let's turn to the titration of a weak acid with a strong base. Figure 19.7 shows the curve obtained when we use 0.1000 M NaOH to titrate 40.00 mL of 0.1000 M propanoic acid, a weak organic acid (CH_3CH_2COOH; $K_a = 1.3 \times 10^{-5}$). (We abbreviate the acid as HPr and the conjugate base, $CH_3CH_2COO^-$, as Pr^-.) When we compare this weak acid–strong base titration curve with the strong acid–strong base titration curve (dotted curve portion in Figure 19.7 corresponds to the bottom half of the curve in Figure 19.6), three major differences appear:

- *The initial pH is higher.* Because the weak acid (HPr) dissociates only slightly, less H_3O^+ is present than with a strong acid of the same concentration.
- *A gradually rising portion of the curve, called the buffer region, appears before the steep rise to the equivalence point.* As HPr reacts with the strong base, a significant amount of the conjugate base (Pr^-) forms, which creates an HPr/Pr^- buffer. Note the midpoint of the buffer region. At this point, *half the original HPr has reacted, so [HPr] = [Pr^-], or [Pr^-]/[HPr] = 1.* Therefore, *at the midpoint of the buffer region, the pH equals the pK_a:*

$$pH = pK_a + \log\left(\frac{[Pr^-]}{[HPr]}\right) = pK_a + \log 1 = pK_a + 0 = pK_a$$

In fact, observing the pH at the midpoint of the buffer region is a common experimental method for estimating the pK_a of an unknown acid.
- *The pH at the equivalence point is greater than 7.00.* The solution contains the strong-base cation Na^+, which does not react with water, and the weak-acid anion Pr^-, which acts as a weak base to accept a proton from H_2O and yield OH^-.

Our choice of indicator is more limited here than in a strong acid–strong base titration because the steep rise occurs over a smaller pH range. Phenolphthalein is suitable because its color change lies within this range (Figure 19.7). However, the figure shows that methyl red, our other choice for the strong acid–strong base titration, changes color earlier and slowly over a large volume ($\sim$10 mL) of titrant, thereby giving a vague and false indication of the equivalence point.

The calculation procedure for the weak acid–strong base titration is different from the previous case because we have to consider the partial dissociation of the weak acid and the reaction of the conjugate base with water. There are four key regions of the titration curve, each of which requires a different type of calculation to find [H_3O^+]:

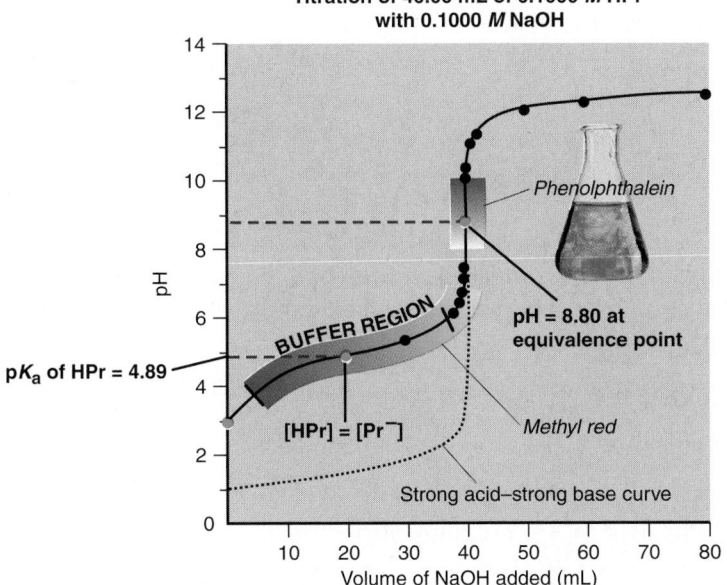

Titration of 40.00 mL of 0.1000 *M* HPr
with 0.1000 *M* NaOH

pK_a of HPr = 4.89

BUFFER REGION

[HPr] = [Pr$^-$]

Phenolphthalein

pH = 8.80 at equivalence point

Methyl red

Strong acid–strong base curve

Figure 19.7 Curve for a weak acid–strong base titration. The curve for the titration of 40.00 mL of 0.1000 *M* CH$_3$CH$_2$COOH (HPr) with 0.1000 *M* NaOH is shown and compared with that for the strong acid HCl (*dotted curve portion*). The initial pH is higher than for a strong acid because HPr dissociates only slightly. At the midpoint of the buffer region, [HPr] = [Pr$^-$], so pH = pK_a. The pH at the equivalence point is above 7.00 because the solution contains the weak base Pr$^-$. Phenolphthalein (*photo*) is a suitable indicator for this titration, but methyl red is not because its color changes over a large volume range.

1. *Solution of HA.* Before base is added, the [H$_3$O$^+$] is that of a weak-acid solution, so we find [H$_3$O$^+$] as in Section 18.4: we set up a reaction table with x = [HPr]$_{dissoc}$, assume [H$_3$O$^+$] = [HPr]$_{dissoc}$ << [HPr]$_{init}$, and solve for x:

$$K_a = \frac{[\text{H}_3\text{O}^+][\text{Pr}^-]}{[\text{HPr}]} \approx \frac{x^2}{[\text{HPr}]_{init}}$$

Therefore, $x = [\text{H}_3\text{O}^+] \approx \sqrt{K_a \times [\text{HPr}]_{init}}$

2. *Solution of HA and added base.* As soon as we add NaOH, it reacts with HPr to form Pr$^-$. This means that up to the equivalence point, we have a mixture of acid and conjugate base, and a buffer solution exists over much of that interval. Therefore, we find [H$_3$O$^+$] from the relationship

$$[\text{H}_3\text{O}^+] = K_a \times \frac{[\text{HPr}]}{[\text{Pr}^-]}$$

(Of course, we can find pH directly with the Henderson-Hasselbalch equation, which is just an alternative form of this relationship.) Note that in this calculation we do *not* have to consider the new total volume because *the volumes cancel in the ratio of concentrations.* That is, [HPr]/[Pr$^-$] = moles of HPr/moles of Pr$^-$, so we need not calculate concentrations.

3. *Equivalent amounts of HA and added base.* At the equivalence point, the original amount of HPr has reacted, so the flask contains a solution of Pr$^-$, a weak base that reacts with water to form OH$^-$:

$$\text{Pr}^-(aq) + \text{H}_2\text{O}(l) \rightleftharpoons \text{HPr}(aq) + \text{OH}^-(aq)$$

Therefore, as mentioned previously, in a weak acid–strong base titration, the solution at the equivalence point is slightly basic, pH > 7.00. We calculate [H$_3$O$^+$] as in Section 18.5: we first find K_b of Pr$^-$ from K_a of HPr, set up a reaction table (assume [Pr$^-$] >> [Pr$^-$]$_{reacting}$), and solve for [OH$^-$]. We need a single concentration, $[Pr^-]$, to solve for $[OH^-]$, so we *do* need the total volume. Then, we convert to $[H_3O^+]$. These two steps are

(1) $[OH^-] \approx \sqrt{K_b \times [Pr^-]}$, where $K_b = \dfrac{K_w}{K_a}$ and $[Pr^-] = \dfrac{\text{moles of HPr}_{init}}{\text{total volume}}$

(2) $[H_3O^+] = \dfrac{K_w}{[OH^-]}$

Combining them into one step gives

$$[H_3O^+] \approx \dfrac{K_w}{\sqrt{K_b \times [Pr^-]}}$$

4. *Solution of excess added base.* Beyond the equivalence point, we are just adding excess OH^- ion, so the calculation is the same as for the strong acid–strong base titration:

$$[H_3O^+] = \dfrac{K_w}{[OH^-]}, \qquad \text{where } [OH^-] = \dfrac{\text{moles of excess } OH^-}{\text{total volume}}$$

Sample Problem 19.3 shows the overall approach.

SAMPLE PROBLEM 19.3 Calculating the pH During a Weak Acid–Strong Base Titration

Problem Calculate the pH during the titration of 40.00 mL of 0.1000 M propanoic acid (HPr; $K_a = 1.3 \times 10^{-5}$) after adding the following volumes of 0.1000 M NaOH:
(a) 0.00 mL   (b) 30.00 mL   (c) 40.00 mL   (d) 50.00 mL
Plan (a) 0.00 mL: No base has been added yet, so this is a weak-acid solution. Thus, we calculate the pH as we did in Section 18.4. **(b)** 30.00 mL: A mixture of Pr^- and HPr is present. We find the amount (mol) of each, substitute into the K_a expression to solve for $[H_3O^+]$, and convert to pH. **(c)** 40.00 mL: The amount (mol) of NaOH added equals the initial amount (mol) of HPr, so a solution of Na^+ and the weak base Pr^- exists. We calculate the pH as we did in Section 18.5, except that we need *total* volume to find $[Pr^-]$. **(d)** 50.00 mL: Excess NaOH is added, so we calculate the amount (mol) of excess OH^- in the total volume and convert to $[H_3O^+]$ and then pH.
Solution (a) 0.00 mL of 0.1000 M NaOH added. Following the approach used in Sample Problem 18.7 and just described in the text, we obtain

$$[H_3O^+] \approx \sqrt{K_a \times [\text{HPr}]_{init}} = \sqrt{(1.3 \times 10^{-5})(0.1000)} = 1.1 \times 10^{-3} \ M$$

$$\boxed{pH = 2.96}$$

(b) 30.00 mL of 0.1000 M NaOH added. Calculating the ratio of moles of HPr to Pr^-:

Original moles of HPr $= 0.04000 \ L \times 0.1000 \ M = 0.004000$ mol HPr
Moles of NaOH added $= 0.03000 \ L \times 0.1000 \ M = 0.003000$ mol OH^-

For 1 mol of NaOH that reacts, 1 mol of Pr^- forms, so we construct the following reaction table for the stoichiometry:

Amount (mol)	HPr(*aq*)	+	OH^-(*aq*)	$\longrightarrow$	Pr^-(*aq*)	+	H_2O(*l*)
Before addition	0.004000		—		0		—
Addition	—		0.003000		—		—
After addition	0.001000		0		0.003000		—

The last line of this table shows the new initial amounts of HPr and Pr^- that will react to attain a new equilibrium. However, with x very small, we assume that the $[HPr]/[Pr^-]$ ratio at equilibrium is essentially equal to the ratio of these new initial amounts (see Comment in Sample Problem 19.1). Thus,

$$\dfrac{[\text{HPr}]}{[\text{Pr}^-]} = \dfrac{0.001000 \text{ mol}}{0.003000 \text{ mol}} = 0.3333$$

Solving for $[H_3O^+]$:

$$[H_3O^+] = K_a \times \dfrac{[\text{HPr}]}{[\text{Pr}^-]} = (1.3 \times 10^{-5})(0.3333) = 4.3 \times 10^{-6} \ M$$

$$\boxed{pH = 5.37}$$

(c) 40.00 mL of 0.1000 M NaOH added. Calculating $[Pr^-]$ after all HPr has reacted:

$$[Pr^-] = \frac{0.004000 \text{ mol}}{0.04000 \text{ L} + 0.04000 \text{ L}} = 0.05000 \text{ } M$$

Calculating K_b:

$$K_b = \frac{K_w}{K_a} = \frac{1.0 \times 10^{-14}}{1.3 \times 10^{-5}} = 7.7 \times 10^{-10}$$

Solving for $[H_3O^+]$ as described in the text:

$$[H_3O^+] \approx \frac{K_w}{\sqrt{K_b \times [Pr^-]}} = \frac{1.0 \times 10^{-14}}{\sqrt{(7.7 \times 10^{-10})(0.05000)}} = 1.6 \times 10^{-9} \text{ } M$$

$$\boxed{pH = 8.80}$$

(d) 50.00 mL of 0.1000 M NaOH added.

Moles of excess $OH^- = (0.1000 \text{ } M)(0.05000 \text{ L} - 0.04000 \text{ L}) = 0.001000 \text{ mol}$

$$[OH^-] = \frac{\text{moles of excess } OH^-}{\text{total volume}} = \frac{0.001000 \text{ mol}}{0.09000 \text{ L}} = 0.01111 \text{ } M$$

$$[H_3O^+] = \frac{K_w}{[OH^-]} = \frac{1.0 \times 10^{-14}}{0.01111} = 9.0 \times 10^{-13} \text{ } M$$

$$\boxed{pH = 12.05}$$

Check As expected from the continuous addition of base, the pH increases through the four stages. Be sure to round off and check the arithmetic along the way.

FOLLOW-UP PROBLEM 19.3 A chemist titrates 20.00 mL of 0.2000 M HBrO ($K_a = 2.3 \times 10^{-9}$) with 0.1000 M NaOH. What is the pH **(a)** before any base is added; **(b)** when $[HBrO] = [BrO^-]$; **(c)** at the equivalence point; **(d)** when the moles of OH^- added are twice the moles of HBrO originally present? **(e)** Sketch the titration curve.

Weak Base–Strong Acid Titration Curves

In the previous case, we titrated a weak acid with a strong base. The opposite process is the titration of a weak base (NH_3) with a strong acid (HCl), shown in Figure 19.8. Note that *the curve has the same shape as the weak acid–strong base*

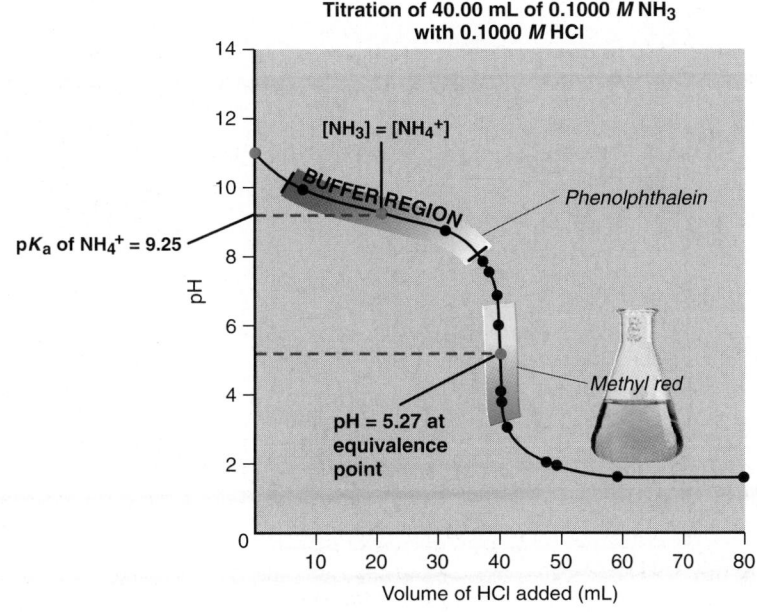

Titration of 40.00 mL of 0.1000 M NH_3 with 0.1000 M HCl

$[NH_3] = [NH_4^+]$

BUFFER REGION

Phenolphthalein

pK_a of $NH_4^+ = 9.25$

Methyl red

pH = 5.27 at equivalence point

Volume of HCl added (mL)

Figure 19.8 **Curve for a weak base–strong acid titration.** Titrating 40.00 mL of 0.1000 M NH_3 with a solution of 0.1000 M HCl leads to a curve whose shape is the same as that of the weak acid–strong base curve in Figure 19.7 but inverted. The midpoint of the buffer region occurs when $[NH_3] = [NH_4^+]$; the pH at this point equals the pK_a of NH_4^+. The pH of the equivalence point is below 7.00 because the solution contains the weak acid NH_4^+. Methyl red (*photo*) is a suitable indicator here, but phenolphthalein is not because its color changes over a large volume range.

curve (Figure 19.7), *but it is inverted.* Thus, the regions of the curve have the same features, but *the pH decreases* throughout the process:

1. The initial solution is that of a weak base, so *the pH starts out above 7.00.*
2. The pH decreases gradually in the buffer region, where significant amounts of base (NH_3) and conjugate acid (NH_4^+) exist. At the midpoint of the buffer region, *the pH equals the pK_a of the ammonium ion.*
3. After the buffer region, the curve drops vertically to the equivalence point, at which all the NH_3 has reacted and the solution contains only NH_4^+ and Cl^-. Note that *the pH at the equivalence point is below 7.00* because Cl^- does not react with water and NH_4^+ is acidic:

$$NH_4^+(aq) + H_2O(l) \rightleftharpoons NH_3(aq) + H_3O^+(aq)$$

4. Beyond the equivalence point, the pH decreases slowly as *excess H_3O^+* is added.

For this titration also, we must be more careful in choosing the indicator than for a strong acid–strong base titration. Phenolphthalein changes color too soon and too slowly to indicate the equivalence point; but methyl red lies on the steep portion of the curve and straddles the equivalence point, so it is a perfect choice.

Titration Curves for Polyprotic Acids

As we discussed in Section 18.4, polyprotic acids have more than one ionizable proton. Except for sulfuric acid, the common polyprotic acids are all weak. The successive K_a values for a polyprotic acid differ by several orders of magnitude, which means that the first H^+ is lost much more easily than subsequent ones, as these values for sulfurous acid show:

$$H_2SO_3(aq) + H_2O(l) \rightleftharpoons HSO_3^-(aq) + H_3O^+(aq)$$
$$K_{a1} = 1.4 \times 10^{-2} \quad \text{and} \quad pK_{a1} = 1.85$$

$$HSO_3^-(aq) + H_2O(l) \rightleftharpoons SO_3^{2-}(aq) + H_3O^+(aq)$$
$$K_{a2} = 6.5 \times 10^{-8} \quad \text{and} \quad pK_{a2} = 7.19$$

In a titration of a diprotic acid such as H_2SO_3, two OH^- ions are required to completely remove both H^+ ions from each molecule of acid. Figure 19.9 shows the titration curve for sulfurous acid with strong base. Because of the large dif-

Figure 19.9 Curve for the titration of a weak polyprotic acid. Titrating 40.00 mL of 0.1000 M H_2SO_3 with 0.1000 M NaOH leads to a curve with two buffer regions and two equivalence points. Because the K_a values are separated by several orders of magnitude, the titration curve looks like two weak acid–strong base curves joined end to end. The pH of the first equivalence point is below 7.00 because the solution contains HSO_3^-, which is a stronger acid than it is a base (K_a of $HSO_3^- = 6.5 \times 10^{-8}$; K_b of $HSO_3^- = 7.1 \times 10^{-13}$).

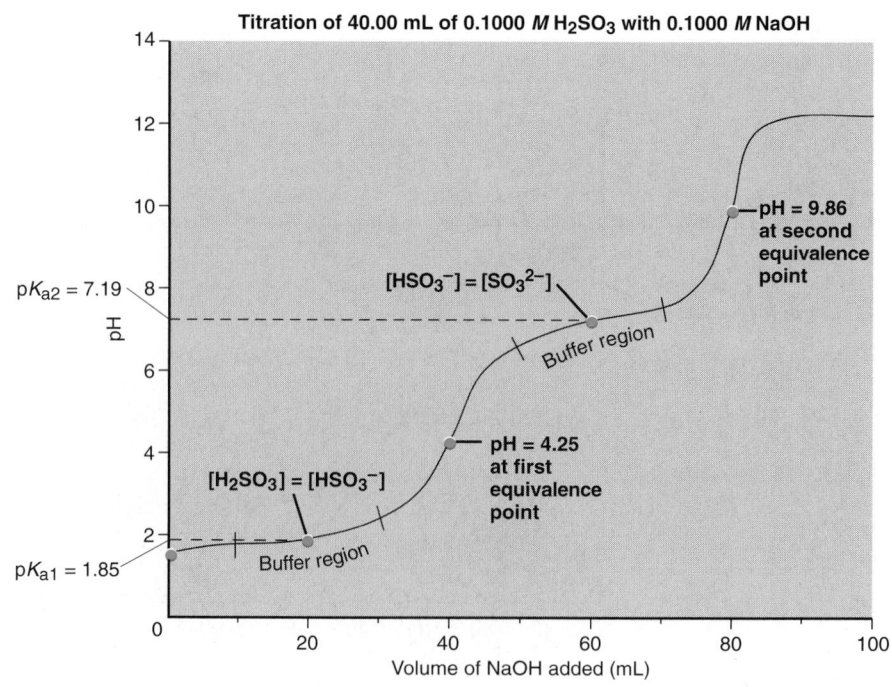

Titration of 40.00 mL of 0.1000 M H_2SO_3 with 0.1000 M NaOH

$pK_{a2} = 7.19$

$[HSO_3^-] = [SO_3^{2-}]$

Buffer region

pH = 9.86 at second equivalence point

pH = 4.25 at first equivalence point

$[H_2SO_3] = [HSO_3^-]$

$pK_{a1} = 1.85$

Buffer region

Volume of NaOH added (mL)

ference in K_a values, we assume that each mole of H^+ is titrated separately; that is, all H_2SO_3 molecules lose one H^+ before any HSO_3^- ions lose one:

$$H_2SO_3 \xrightarrow{\text{1 mol OH}^-} HSO_3^- \xrightarrow{\text{1 mol OH}^-} SO_3^{2-}$$

As you can see from the curve, the loss of each mole of H^+ shows up as a separate equivalence point and buffer region. Like the curve for a weak monoprotic acid, the pH at the midpoint of the buffer region is equal to the pK_a of that acidic species. Note also that the same volume of added base (in this case, 40.00 mL of 0.1000 M OH^-) is required to remove each mole of H^+.

Amino Acids as Biological Polyprotic Acids

Amino acids, the compounds that link together to form proteins (Section 15.6), have the general formula $NH_2-CH(R)-COOH$, where R can be one of about 20 different groups. At low pH, both the amino group ($-NH_2$) and the acid group ($-COOH$) are protonated: $^+NH_3-CH(R)-COOH$. Thus, in this form the amino acid behaves like a polyprotic acid. In the case of glycine (R = H), the simplest amino acid, the dissociation reactions and pK_a values are

$$^+NH_3CH_2COOH(aq) + H_2O(l) \rightleftharpoons {}^+NH_3CH_2COO^-(aq) + H_3O^+(aq) \quad pK_{a1} = 2.35$$
$$^+NH_3CH_2COO^-(aq) + H_2O(l) \rightleftharpoons NH_2CH_2COO^-(aq) + H_3O^+(aq) \quad pK_{a2} = 9.78$$

The pK_a values show that the $-COOH$ group is much more acidic than the $-NH_3^+$ group. As we saw with H_2SO_3, the protons are titrated separately, so virtually all the $-COOH$ protons are removed before any $-NH_3^+$ protons are:

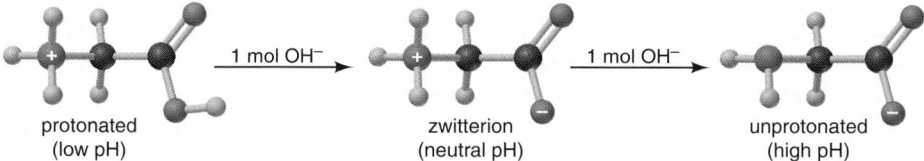

| protonated | zwitterion | unprotonated |
| (low pH) | (neutral pH) | (high pH) |

Thus, at physiological pH (~7), *glycine exists predominantly as a zwitterion* (German *zwitter,* "double"), a species with opposite charges on the same molecule: $^+NH_3CH_2COO^-$. Among the 20 different R groups of amino acids in proteins, several have *additional* $-COO^-$ or $-NH_3^+$ groups at pH 7 (see Figure 15.30).

    When amino acids link into a protein, charged R groups give the protein its overall charge and often play a role in its function. A widely studied example occurs in sickle cell anemia. Normal red blood cells are packed tightly with molecules of hemoglobin, the oxygen-carrying protein. Two amino acids in hemoglobin are critical to the mobility of the molecules, and this mobility is critical to the shape of the cells (Figure 19.10). These two amino acids have negatively charged ($-COO^-$) R groups. The abnormal hemoglobin molecules in sickle cell anemia have uncharged ($-CH_3$) groups replacing the two charged ones. This change in just 2 of hemoglobin's 574 amino acids lowers the charge repulsions between hemoglobin molecules, and they clump together in fiber-like structures, which leads to the sickle shape of the red blood cells. The misshapen cells block capillaries, and the painful course of sickle cell anemia usually ends in early death.

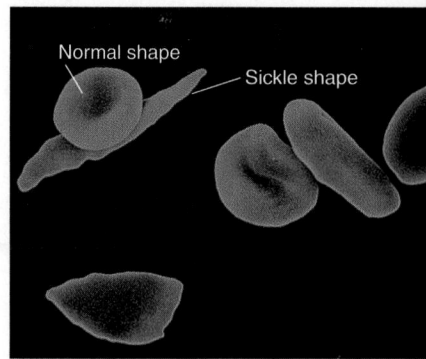

Figure 19.10 Sickle shape of red blood cells in sickle cell anemia.

SECTION SUMMARY

An acid-base (pH) indicator is a weak acid that has differently colored acidic and basic forms and changes color over about 2 pH units. In a strong acid–strong base titration, the pH starts out low, rises slowly, then shoots up near the equivalence point (pH = 7). In a weak acid–strong base titration, the pH starts out higher than in the strong acid titration, rises slowly in the buffer region (pH = pK_a at the midpoint), then rises more quickly near the equivalence point (pH > 7). A weak base–strong acid titration is the inverse of this, with its pH decreasing to the equivalence point (pH < 7). Polyprotic acids have two or more acidic protons, and each is titrated separately. Amino acids exist in different charged forms that depend on the pH of the solution.

19.3 EQUILIBRIA OF SLIGHTLY SOLUBLE IONIC COMPOUNDS

In this section, we explore the aqueous equilibria of slightly soluble ionic compounds. In Chapter 13, we found that most solutes, even those called "soluble," have a limited solubility in a particular solvent. Add more than this amount, and some solute remains undissolved. In a saturated solution at a particular temperature, equilibrium exists between the undissolved and dissolved solute. Slightly soluble (often called "insoluble") ionic compounds have a relatively low solubility, so they reach equilibrium with relatively little solute dissolved. At this point, it would be a good idea for you to review the solubility rules that are listed in Table 4.1 (p. 139).

When a soluble ionic compound dissolves in water, it dissociates completely into ions. In this discussion, we will assume that the small amount of a slightly soluble ionic compound that dissolves in water also dissociates completely into ions. In reality, however, this is not the case. Many slightly soluble salts, particularly those of transition metals and heavy main-group metals, have significant covalent character in their metal-nonmetal bonding, and their solutions often contain other species that are partially dissociated or undissociated. For example, when lead(II) chloride is thoroughly stirred in water (see photo), the solution contains not only the expected $Pb^{2+}(aq)$ and $Cl^-(aq)$ ions, but also undissociated $PbCl_2(aq)$ molecules and $PbCl^+(aq)$ ions. In solutions of some other salts, such as $CaSO_4$, there are no molecules, but pairs of ions exist, such as $Ca^{2+}SO_4{}^{2-}(aq)$. These species increase the solubility above what we calculate assuming complete dissociation. More advanced courses address these complexities, but we will only hint at some of them in the Comments of several sample problems. For these reasons, it is best to keep in mind that our simple calculations should be treated as first approximations.

PbCl$_2$, a slightly soluble ionic compound.

The Ion-Product Expression (Q_{sp}) and the Solubility-Product Constant (K_{sp})

If we make the assumption that there is complete dissociation of a slightly soluble ionic compound into its component ions, then *equilibrium exists between solid solute and aqueous ions.* Thus, for example, for a saturated solution of lead(II) sulfate in water, we have

$$PbSO_4(s) \rightleftharpoons Pb^{2+}(aq) + SO_4{}^{2-}(aq)$$

As with all the other equilibrium systems we've looked at, this one can be expressed by a reaction quotient:

$$Q_c = \frac{[Pb^{2+}][SO_4{}^{2-}]}{[PbSO_4]}$$

As in previous cases, we combine the constant concentration of the solid, $[PbSO_4]$, with the value of Q_c and eliminate it. This gives the *ion-product expression, Q_{sp}:*

$$Q_{sp} = Q_c[PbSO_4] = [Pb^{2+}][SO_4{}^{2-}]$$

And, when solid $PbSO_4$ attains equilibrium with Pb^{2+} and $SO_4{}^{2-}$ ions, that is, when the solution reaches saturation, the numerical value of Q_{sp} attains a constant value. This new equilibrium constant is called the **solubility-product constant, K_{sp}.** The K_{sp} for $PbSO_4$ at 25°C, for example, is 1.6×10^{-8}.

As we've seen with other equilibrium constants, a given K_{sp} value depends only on the temperature, not on the individual ion concentrations. Suppose, for example, you add some lead(II) nitrate, a soluble lead salt, to increase the solution's $[Pb^{2+}]$. The equilibrium position shifts to the left, and $[SO_4{}^{2-}]$ goes down as more $PbSO_4$ precipitates; so the K_{sp} value is maintained.

The form of Q_{sp} is identical to that of the other reaction quotients we have written: each ion concentration is raised to an exponent equal to the coefficient in the balanced equation, which in this case also *equals the subscript of each ion in the compound's formula*. Thus, in general, for a saturated solution of a slightly soluble ionic compound, M_pX_q, composed of the ions M^{n+} and X^{z-}, the equilibrium condition is

$$Q_{sp} = [M^{n+}]^p[X^{z-}]^q = K_{sp} \qquad (19.2)$$

Of course, at saturation, the concentration terms represent equilibrium concentrations, so from here on, we write the ion-product expression directly with the symbol K_{sp}. For example, the equation and ion-product expression that describe a saturated solution of $Cu(OH)_2$ are

$$Cu(OH)_2(s) \rightleftharpoons Cu^{2+}(aq) + 2OH^-(aq) \qquad K_{sp} = [Cu^{2+}][OH^-]^2$$

Insoluble metal sulfides present a slightly different case. The sulfide ion, S^{2-}, is so basic that it is not stable in water and reacts completely to form the hydrogen sulfide ion (HS^-) and the hydroxide ion (OH^-):

$$S^{2-}(aq) + H_2O(l) \longrightarrow HS^-(aq) + OH^-(aq)$$

For instance, when manganese(II) sulfide is shaken with water, the solution contains Mn^{2+}, HS^-, and OH^- ions. Although the sulfide ion does not exist as such in water, you can imagine the dissolution process as the sum of two steps, with S^{2-} occurring as an intermediate that is consumed immediately:

$$MnS(s) \rightleftharpoons Mn^{2+}(aq) + S^{2-}(aq)$$
$$\underline{S^{2-}(aq) + H_2O(l) \longrightarrow HS^-(aq) + OH^-(aq)}$$
$$MnS(s) + H_2O(l) \rightleftharpoons Mn^{2+}(aq) + HS^-(aq) + OH^-(aq)$$

Therefore, the ion-product expression is

$$K_{sp} = [Mn^{2+}][HS^-][OH^-]$$

SAMPLE PROBLEM 19.4 Writing Ion-Product Expressions for Slightly Soluble Ionic Compounds

Problem Write the ion-product expression for each of the following:
(a) Magnesium carbonate (b) Iron(II) hydroxide
(c) Calcium phosphate (d) Silver sulfide
Plan We write an equation that describes a saturated solution and then write the ion-product expression, K_{sp}, according to Equation 19.2, noting the sulfide in part (d).
Solution (a) Magnesium carbonate:

$$MgCO_3(s) \rightleftharpoons Mg^{2+}(aq) + CO_3^{2-}(aq) \qquad K_{sp} = [Mg^{2+}][CO_3^{2-}]$$

(b) Iron(II) hydroxide:

$$Fe(OH)_2(s) \rightleftharpoons Fe^{2+}(aq) + 2OH^-(aq) \qquad K_{sp} = [Fe^{2+}][OH^-]^2$$

(c) Calcium phosphate:

$$Ca_3(PO_4)_2(s) \rightleftharpoons 3Ca^{2+}(aq) + 2PO_4^{3-}(aq) \qquad K_{sp} = [Ca^{2+}]^3[PO_4^{3-}]^2$$

(d) Silver sulfide:

$$Ag_2S(s) \rightleftharpoons 2Ag^+(aq) + S^{2-}(aq)$$
$$\underline{S^{2-}(aq) + H_2O(l) \longrightarrow HS^-(aq) + OH^-(aq)}$$
$$Ag_2S(s) + H_2O(l) \rightleftharpoons 2Ag^+(aq) + HS^-(aq) + OH^-(aq) \qquad K_{sp} = [Ag^+]^2[HS^-][OH^-]$$

Check Except for part (d), you can check by reversing the process to see if you obtain the formula of the compound from K_{sp}.
Comment In part (d), we include H_2O as reactant to obtain a balanced equation.

FOLLOW-UP PROBLEM 19.4 Write the ion-product expressions for each of the following:
(a) Calcium sulfate (b) Chromium(III) carbonate
(c) Magnesium hydroxide (d) Arsenic(III) sulfide

Table 19.2 Solubility-Product Constants (K_{sp}) of Selected Ionic Compounds at 25°C

Name, Formula	K_{sp}
Aluminum hydroxide, Al(OH)$_3$	3×10^{-34}
Cobalt(II) carbonate, CoCO$_3$	1.0×10^{-10}
Iron(II) hydroxide, Fe(OH)$_2$	4.1×10^{-15}
Lead(II) fluoride, PbF$_2$	3.6×10^{-8}
Lead(II) sulfate, PbSO$_4$	1.6×10^{-8}
Mercury(I) iodide, Hg$_2$I$_2$	4.7×10^{-29}
Silver sulfide, Ag$_2$S	8×10^{-48}
Zinc iodate, Zn(IO$_3$)$_2$	3.9×10^{-6}

The magnitude of K_{sp} is a measure of how far to the right the dissolution proceeds at equilibrium (saturation). We'll use it later to compare solubilities. Table 19.2 presents some representative K_{sp} values of slightly soluble ionic compounds. (Appendix C is a much more extensive list.) Note that, even though the values are all quite low, they range over many orders of magnitude.

Calculations Involving the Solubility-Product Constant

In Chapters 17 and 18, we described two types of equilibrium problems. In one type, we use concentrations to find K, and in the other, we use K to find concentrations. Here we encounter the same two types.

Determining K_{sp} from Solubility The solubilities of ionic compounds are determined experimentally, and several chemical handbooks tabulate them. Most solubility values are given in units of grams of solute dissolved in 100 grams of H_2O. Because the mass of compound in solution is small, a negligible error is introduced if we assume that "100 g of water" is equal to "100 mL of solution." We then convert the solubility from grams of solute per 100 mL of solution to **molar solubility,** the amount (mol) of solute per liter of solution (that is, the molarity of the solute). Next, we use the balanced equation to find the molarity of each ion and substitute into the ion-product expression to find the value of K_{sp}.

SAMPLE PROBLEM 19.5 Determining K_{sp} from Solubility

Problem **(a)** Lead(II) sulfate (PbSO$_4$) is a key component in lead-acid car batteries. Its solubility in water at 25°C is 4.25×10^{-3} g/100 mL solution. What is the K_{sp} of PbSO$_4$?
(b) When lead(II) fluoride (PbF$_2$) is shaken with pure water at 25°C, the solubility is found to be 0.64 g/L. Calculate the K_{sp} of PbF$_2$.
Plan We are given the solubilities in various units and must find K_{sp}. For each compound, we write an equation for its dissolution to see the number of moles of each ion, and then write the ion-product expression. We convert the solubility to molar solubility, find the molarity of each ion, and substitute into the ion-product expression to calculate K_{sp}.
Solution **(a)** For PbSO$_4$. Writing the equation and ion-product (K_{sp}) expression:

$$PbSO_4(s) \rightleftharpoons Pb^{2+}(aq) + SO_4^{2-}(aq) \qquad K_{sp} = [Pb^{2+}][SO_4^{2-}]$$

Converting solubility to molar solubility:

$$\text{Molar solubility of PbSO}_4 = \frac{0.00425 \text{ g PbSO}_4}{100 \text{ mL soln}} \times \frac{1000 \text{ mL}}{1 \text{ L}} \times \frac{1 \text{ mol PbSO}_4}{303.3 \text{ g PbSO}_4}$$
$$= 1.40 \times 10^{-4} \ M \text{ PbSO}_4$$

Determining molarities of the ions: Because 1 mol of Pb^{2+} and 1 mol of SO_4^{2-} form when 1 mol of PbSO$_4$ dissolves, $[Pb^{2+}] = [SO_4^{2-}] = 1.40 \times 10^{-4} \ M$.
Calculating K_{sp}:

$$K_{sp} = [Pb^{2+}][SO_4^{2-}] = (1.40 \times 10^{-4})^2 = \boxed{1.96 \times 10^{-8}}$$

(b) For PbF$_2$. Writing the equation and K_{sp} expression:

$$PbF_2(s) \rightleftharpoons Pb^{2+}(aq) + 2F^-(aq) \qquad K_{sp} = [Pb^{2+}][F^-]^2$$

Converting solubility to molar solubility:

$$\text{Molar solubility of PbF}_2 = \frac{0.64 \text{ g PbF}_2}{1 \text{ L soln}} \times \frac{1 \text{ mol PbF}_2}{245.2 \text{ g PbF}_2} = 2.6 \times 10^{-3} \ M \text{ PbF}_2$$

Determining molarities of the ions: Since 1 mol of Pb^{2+} and 2 mol of F^- form when 1 mol of PbF$_2$ dissolves,

$$[Pb^{2+}] = 2.6 \times 10^{-3} \ M \qquad \text{and} \qquad [F^-] = 2(2.6 \times 10^{-3} \ M) = 5.2 \times 10^{-3} \ M$$

Calculating K_{sp}:

$$K_{sp} = [Pb^{2+}][F^-]^2 = (2.6 \times 10^{-3})(5.2 \times 10^{-3})^2 = \boxed{7.0 \times 10^{-8}}$$

Check The low solubilities are consistent with K_{sp} values being small. In part (a), the molar solubility seems about right: $\sim \dfrac{4 \times 10^{-2} \text{ g/L}}{3 \times 10^2 \text{ g/mol}} \approx 1.3 \times 10^{-4} \ M$. Squaring this number gives 1.7×10^{-8}, close to the calculated K_{sp}. In part (b), we check the final step: $\sim (3 \times 10^{-3})(5 \times 10^{-3})^2 = 7.5 \times 10^{-8}$, close to the calculated K_{sp}.

Comment 1. In part (b), the formula PbF_2 means that $[F^-]$ is twice $[Pb^{2+}]$. Then we square this value of $[F^-]$. Always follow the ion-product expression explicitly.

2. The tabulated K_{sp} values for these compounds (Table 19.2) are lower than our calculated values. For PbF_2, for instance, the tabulated value is 3.6×10^{-8}, but we calculated 7.0×10^{-8} from solubility data. The discrepancy arises because we assumed that the PbF_2 in solution dissociates completely to Pb^{2+} and F^-. Here is an example of the complexity pointed out at the beginning of this section. Actually, about a third of the PbF_2 dissolves as $PbF^+(aq)$ and a small amount as undissociated $PbF_2(aq)$. The solubility (0.64 g/L) is determined experimentally and includes these other species, which we did not include in our simple calculation. This is why we treat such calculated K_{sp} values as approximations.

FOLLOW-UP PROBLEM 19.5 When powdered fluorite (CaF_2; see photo) is shaken with pure water at 18°C, 1.5×10^{-4} g dissolves for every 10.0 mL of solution. Calculate the K_{sp} of CaF_2 at 18°C.

Fluorite.

Determining Solubility from K_{sp} The reverse of the previous type of problem involves finding the solubility of a compound based on its formula and K_{sp} value. An approach similar to the one we used for weak acids in Sample Problem 18.7 is to define the unknown molar solubility as S, define the ion concentrations in terms of this unknown in a reaction table, and solve for S.

SAMPLE PROBLEM 19.6 Determining Solubility from K_{sp}

Problem Calcium hydroxide (slaked lime) is a major component of mortar, plaster, and cement, and solutions of $Ca(OH)_2$ are used in industry as a cheap, strong base. Calculate the solubility of $Ca(OH)_2$ in water if the K_{sp} is 6.5×10^{-6}.

Plan We write the dissolution equation and the ion-product expression. We know K_{sp} (6.5×10^{-6}); to find molar solubility (S), we set up a reaction table that expresses $[Ca^{2+}]$ and $[OH^-]$ in terms of S, substitute into the ion-product expression, and solve for S.

Solution Writing the equation and ion-product expression:

$$Ca(OH)_2(s) \rightleftharpoons Ca^{2+}(aq) + 2OH^-(aq) \quad K_{sp} = [Ca^{2+}][OH^-]^2 = 6.5 \times 10^{-6}$$

Setting up a reaction table, with S = molar solubility:

Concentration (M)	$Ca(OH)_2(s)$	$\rightleftharpoons$	$Ca^{2+}(aq)$	+	$2OH^-(aq)$
Initial	—		0		0
Change	—		$+S$		$+2S$
Equilibrium	—		S		$2S$

Substituting into the ion-product expression and solving for S:

$$K_{sp} = [Ca^{2+}][OH^-]^2 = (S)(2S)^2 = (S)(4S^2) = 4S^3 = 6.5 \times 10^{-6}$$

$$S = \sqrt[3]{\frac{6.5 \times 10^{-6}}{4}} = \boxed{1.2 \times 10^{-2} \ M}$$

Check We expect a low solubility from a slightly soluble salt. If we reverse the calculation, we should obtain the given K_{sp}: $4(1.2 \times 10^{-2})^3 = 6.9 \times 10^{-6}$, close to 6.5×10^{-6}.

Comment 1. Note that we did not double and *then* square $[OH^-]$. $2S$ *is* the $[OH^-]$, so we just squared it, as the ion-product expression required.

2. Once again, we assumed that the solid dissociates completely. Actually, the solubility is increased to about 2.0×10^{-2} M by the presence of $CaOH^+(aq)$ formed in the reaction $Ca(OH)_2(s) \rightleftharpoons CaOH^+(aq) + OH^-(aq)$. Our calculated answer is only approximate because we did not take this other species into account.

FOLLOW-UP PROBLEM 19.6 A suspension of $Mg(OH)_2$ in water is sold as "milk of magnesia" to alleviate minor stomach disorders by neutralizing stomach acid. The $[OH^-]$ is too low to harm the mouth and throat, but the suspension dissolves in the acidic stomach juices. What is the molar solubility of $Mg(OH)_2$ ($K_{sp} = 6.3 \times 10^{-10}$) in pure water?

Using K_{sp} Values to Compare Solubilities K_{sp} values are a guide to *relative* solubility, as long as we compare compounds whose formulas contain the *same total number of ions*. In such cases, *the higher the K_{sp}, the greater the solubility.* Table 19.3 shows this point for several compounds. Note that for compounds that form three ions, the relationship holds whether the cation:anion ratio is 1:2 or 2:1, because the mathematical expression containing S is the same ($4S^3$) in the calculation (see Sample Problem 19.6).

Table 19.3 Relationship Between K_{sp} and Solubility at 25°C

No. of Ions	Formula	Cation:Anion	K_{sp}	Solubility (M)
2	$MgCO_3$	1:1	3.5×10^{-8}	1.9×10^{-4}
2	$PbSO_4$	1:1	1.6×10^{-8}	1.3×10^{-4}
2	$BaCrO_4$	1:1	2.1×10^{-10}	1.4×10^{-5}
3	$Ca(OH)_2$	1:2	6.5×10^{-6}	1.2×10^{-2}
3	BaF_2	1:2	1.5×10^{-6}	7.2×10^{-3}
3	CaF_2	1:2	3.2×10^{-11}	2.0×10^{-4}
3	Ag_2CrO_4	2:1	2.6×10^{-12}	8.7×10^{-5}

The Effect of a Common Ion on Solubility

The presence of a common ion decreases the solubility of a slightly soluble ionic compound. As we saw in the case of acid-base systems, Le Châtelier's principle helps explain this effect. Let's examine the equilibrium condition for a saturated solution of lead(II) chromate:

$$PbCrO_4(s) \rightleftharpoons Pb^{2+}(aq) + CrO_4^{2-}(aq) \qquad K_{sp} = [Pb^{2+}][CrO_4^{2-}] = 2.3\times10^{-13}$$

At a given temperature, K_{sp} depends only on the product of the ion concentrations. If the concentration of either ion goes up, the other must go down to maintain the constant K_{sp}. Suppose we add Na_2CrO_4, a very soluble salt, to the saturated $PbCrO_4$ solution. The concentration of the common ion, CrO_4^{2-}, increases, and some of it combines with Pb^{2+} ion to form more solid $PbCrO_4$ (Figure 19.11). The overall effect is a shift in the position of equilibrium to the left:

$$PbCrO_4(s) \overset{\longleftarrow}{\rightleftharpoons} Pb^{2+}(aq) + CrO_4^{2-}(aq; \text{added})$$

Figure 19.11 The effect of a common ion on solubility. When a common ion is added to a saturated solution of an ionic compound, the solubility is lowered and more of the compound precipitates. **A,** Lead(II) chromate, a slightly soluble salt, forms a saturated aqueous solution. **B,** When Na_2CrO_4 solution is added, the amount of $PbCrO_4(s)$ increases, indicating a lower solubility in the presence of the common ion, CrO_4^{2-}.

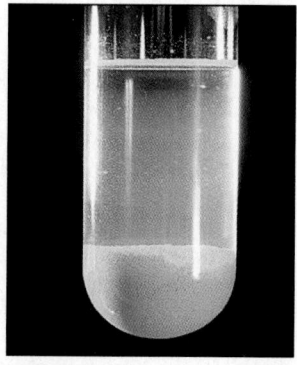

$PbCrO_4(s) \rightleftharpoons Pb^{2+}(aq) + CrO_4^{2-}(aq)$

A

$PbCrO_4(s) \overset{\longleftarrow}{\rightleftharpoons} Pb^{2+}(aq) + CrO_4^{2-}(aq; \text{added})$

B

After the addition, $[CrO_4^{2-}]$ is higher, but $[Pb^{2+}]$ is lower. In this case, $[Pb^{2+}]$ represents the amount of $PbCrO_4$ dissolved; thus, in effect, the solubility of $PbCrO_4$ has decreased. The same result is obtained if we dissolve $PbCrO_4$ in a Na_2CrO_4 solution. We also obtain this result by adding a soluble lead(II) salt, such as $Pb(NO_3)_2$. The added Pb^{2+} ion combines with some $CrO_4^{2-}(aq)$, thereby lowering the amount of dissolved $PbCrO_4$.

SAMPLE PROBLEM 19.7 Calculating the Effect of a Common Ion on Solubility

Problem In Sample Problem 19.6, we calculated the solubility of $Ca(OH)_2$ in water. What is its solubility in 0.10 M $Ca(NO_3)_2$? K_{sp} of $Ca(OH)_2$ is 6.5×10^{-6}.
Plan From the equation and the ion-product expression for $Ca(OH)_2$, we predict that the addition of Ca^{2+}, the common ion, will lower the solubility. We set up a reaction table with $[Ca^{2+}]_{init}$ coming from $Ca(NO_3)_2$ and S equal to $[Ca^{2+}]_{from\ Ca(OH)_2}$. To simplify the math, we assume that, because K_{sp} is low, S is so small relative to $[Ca^{2+}]_{init}$ that it can be neglected. Then we solve for S and check the assumption.
Solution Writing the equation and ion-product expression:

$$Ca(OH)_2(s) \rightleftharpoons Ca^{2+}(aq) + 2OH^-(aq) \qquad K_{sp} = [Ca^{2+}][OH^-]^2 = 6.5 \times 10^{-6}$$

Setting up the reaction table, with $S = [Ca^{2+}]_{from\ Ca(OH)_2}$:

Concentration (M)	$Ca(OH)_2(s)$	$\rightleftharpoons$	$Ca^{2+}(aq)$	$+$	$2OH^-(aq)$
Initial	—		0.10		0
Change	—		$+S$		$+2S$
Equilibrium	—		$0.10 + S$		$2S$

Making the assumption: K_{sp} is small, so $S \ll 0.10\ M$; thus, $0.10\ M + S \approx 0.10\ M$.
Substituting into the ion-product expression and solving for S:

$$K_{sp} = [Ca^{2+}][OH^-]^2 = 6.5 \times 10^{-6} \approx (0.10)(2S)^2$$

Therefore, $4S^2 \approx \dfrac{6.5 \times 10^{-6}}{0.10}$ so $S \approx \sqrt{\dfrac{6.5 \times 10^{-5}}{4}} = \boxed{4.0 \times 10^{-3}\ M}$

Checking the assumption: $\dfrac{4.0 \times 10^{-3}\ M}{0.10\ M} \times 100 = 4.0\% < 5\%$

Check In Sample Problem 19.6, the solubility of $Ca(OH)_2$ was 0.012 M, but here, it is 0.0040 M, so the solubility *decreased* in the presence of added Ca^{2+}, the common ion, as we predicted.

FOLLOW-UP PROBLEM 19.7 To improve the quality of x-ray photos used in the diagnosis of intestinal disorders, the patient drinks an aqueous suspension of $BaSO_4$ before the x-ray procedure (see photo). The Ba^{2+} in the suspension is opaque to x-rays, but it is also toxic, so the Ba^{2+} concentration is lowered by the addition of dilute Na_2SO_4. What is the solubility of $BaSO_4$ ($K_{sp} = 1.1 \times 10^{-10}$) in each of the following:
(a) Pure water
(b) 0.10 M Na_2SO_4

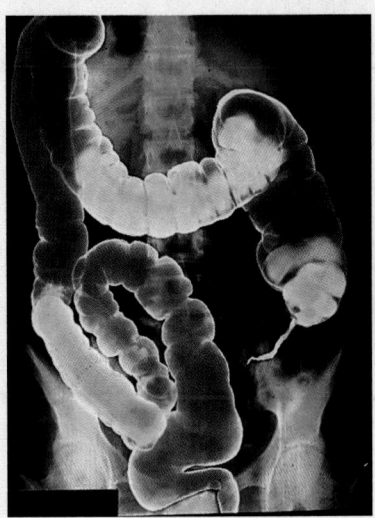

BaSO$_4$ imaging of a human large intestine.

The Effect of pH on Solubility

The hydronium ion concentration can have a profound effect on the solubility of an ionic compound. *If the compound contains the anion of a weak acid, addition of H_3O^+ (from a strong acid) increases its solubility.* Once again, Le Châtelier's principle explains why. In a saturated solution of calcium carbonate, for example, we have

$$CaCO_3(s) \rightleftharpoons Ca^{2+}(aq) + CO_3^{2-}(aq)$$

Figure 19.12 Test for the presence of a carbonate. When a mineral that contains carbonate ion is treated with strong acid, the added H_3O^+ shifts the equilibrium position of the carbonate solubility. More carbonate dissolves, and the carbonic acid that is formed breaks down to water and gaseous CO_2.

Adding some strong acid introduces a large amount of H_3O^+, which immediately reacts with CO_3^{2-} to form the weak acid HCO_3^-:

$$CO_3^{2-}(aq) + H_3O^+(aq) \longrightarrow HCO_3^-(aq) + H_2O(l)$$

If enough H_3O^+ is added, further reaction occurs to form carbonic acid, which decomposes immediately to H_2O and CO_2, which escapes the container:

$$HCO_3^-(aq) + H_3O^+(aq) \longrightarrow H_2CO_3(aq) + H_2O(l) \longrightarrow CO_2(g) + 2H_2O(l)$$

Thus, the net effect of added H_3O^+ is a shift in the equilibrium position to the right, and more $CaCO_3$ dissolves:

$$CaCO_3(s) \overrightarrow{\rightleftharpoons} Ca^{2+} + CO_3^{2-} \xrightarrow{H_3O^+} HCO_3^- \xrightarrow{H_3O^+} H_2CO_3 \longrightarrow$$
$$CO_2(g) + H_2O + Ca^{2+}$$

This particular case illustrates a qualitative field test for carbonate minerals because the CO_2 bubbles vigorously (Figure 19.12).

In contrast, adding H_3O^+ to a saturated solution of a compound with a strong-acid anion, such as silver chloride, has no effect on the equilibrium position:

$$AgCl(s) \rightleftharpoons Ag^+(aq) + Cl^-(aq)$$

Because Cl^- ion is the conjugate base of a strong acid (HCl), it can coexist in solution with high $[H_3O^+]$. The Cl^- does not leave the system, so the equilibrium position is not affected.

SAMPLE PROBLEM 19.8 Predicting the Effect on Solubility of Adding Strong Acid

Problem Write balanced equations to explain whether addition of H_3O^+ from a strong acid affects the solubility of these ionic compounds:
(a) Lead(II) bromide   **(b)** Copper(II) hydroxide   **(c)** Iron(II) sulfide
Plan We write the balanced dissolution equation and note the anion: weak-acid anions react with H_3O^+ and shift the equilibrium position toward more dissolution. Strong-acid anions do not react, so added H_3O^+ has no effect.
Solution (a) $PbBr_2(s) \rightleftharpoons Pb^{2+}(aq) + 2Br^-(aq)$
No effect. Br^- is the anion of HBr, a strong acid, so it does not react with H_3O^+.
(b) $Cu(OH)_2(s) \rightleftharpoons Cu^{2+}(aq) + 2OH^-(aq)$
Increases solubility. OH^- is the anion of H_2O, a very weak acid, so it reacts with the added H_3O^+:

$$OH^-(aq) + H_3O^+(aq) \longrightarrow 2H_2O(l)$$

(c) $FeS(s) + H_2O(l) \rightleftharpoons Fe^{2+}(aq) + HS^-(aq) + OH^-(aq)$
Increases solubility. We noted earlier that the S^{2-} ion reacts immediately with water to form HS^-. The added H_3O^+ reacts with both weak-acid anions, HS^- and OH^-:

$$HS^-(aq) + H_3O^+(aq) \longrightarrow H_2S(aq) + H_2O(l)$$
$$OH^-(aq) + H_3O^+(aq) \longrightarrow 2H_2O(l)$$

FOLLOW-UP PROBLEM 19.8 Write balanced equations to show how addition of $HNO_3(aq)$ affects the solubility of **(a)** calcium fluoride; **(b)** zinc sulfide; **(c)** silver iodide.

Many principles of ionic equilibria are manifested in natural formations, as the upcoming Chemical Connections essay illustrates.

Predicting the Formation of a Precipitate: Q_{sp} vs. K_{sp}

In Chapter 17, we compared the values of Q and K to see if a reaction had reached equilibrium and, if not, in which net direction it would move until it did. Now we use the same approach to see if a precipitate will form and, if not, what concentrations of ions will cause it to do so.

Chemical Connections   Chemistry in Geology
Creation of a Limestone Cave

Limestone caves and the detailed structures within them provide striking evidence of the workings of aqueous ionic equilibria (Figure B19.1). The spires and vaults of these natural cathedrals are the products of reactions between carbonate rocks and the water that has run through them for millennia. Limestone is predominantly calcium carbonate ($CaCO_3$), a slightly soluble ionic compound with a K_{sp} of 3.3×10^{-9}. This rocky material began accumulating in the Earth over 400 million years ago, and a relatively young cave, such as Howe Caverns in eastern New York State, began forming about 800,000 years ago.

Two key facts help us understand how limestone caves form:

1. Gaseous CO_2 is in equilibrium with aqueous CO_2 in natural waters:

$$CO_2(g) \xrightleftharpoons{H_2O(l)} CO_2(aq) \qquad \text{[equation 1]}$$

The concentration of CO_2 in the water is proportional to the partial pressure of $CO_2(g)$ in contact with the water (Henry's law; Section 13.3):

$$[CO_2(aq)] \propto P_{CO_2}$$

Because of the continual release of CO_2 from within the Earth (outgassing), the P_{CO_2} in soil-trapped air is *higher* than the P_{CO_2} in the atmosphere.

2. As we just discussed in the text, the presence of $H_3O^+(aq)$ increases the solubility of ionic compounds that contain the anion of a weak acid. The reaction of CO_2 with water produces H_3O^+:

$$CO_2(aq) + 2H_2O(l) \rightleftharpoons H_3O^+(aq) + HCO_3^-(aq)$$

Thus, the presence of $CO_2(aq)$ leads to the formation of H_3O^+, which increases the solubility of $CaCO_3$:

$$CaCO_3(s) + CO_2(aq) + H_2O(l) \rightleftharpoons$$
$$Ca^{2+}(aq) + 2HCO_3^-(aq)$$
$$\text{[equation 2]}$$

Here is an overview of the cave-forming process. As surface water trickles through cracks in the ground, it meets soil-trapped air with its high P_{CO_2}. As a result, $[CO_2(aq)]$ increases (equation 1 shifts to the right), and the solution becomes more acidic. When this CO_2-rich water contacts limestone, more $CaCO_3$ dissolves (equation 2 shifts to the right). As a result, more rock is carved out, more water flows in, more rock is carved out, and so on. Centuries pass as a cave slowly begins to form.

Eating its way through underground tunnels, some of the aqueous solution, largely dissolved $Ca(HCO_3)_2$, passes through the ceiling of the growing cave. As it drips, it meets air, which has a lower P_{CO_2} than the soil, so some $CO_2(aq)$ comes out of solution (equation 1 shifts to the left). This causes some $CaCO_3$ to precipitate on the ceiling and on the floor below, where the drops land (equation 2 shifts to the left). Decades pass, and the ceiling bears an "icicle" of $CaCO_3$, called a *stalactite,* while a spike of $CaCO_3$, called a *stalagmite,* grows upward from the cave floor. Given enough time, they meet to form a column of precipitated limestone.

The same chemical process can lead to many different shapes. Standing pools of $Ca(HCO_3)_2$ solution form limestone "lily pads" or "corals." Cascades of solution form delicate limestone "draperies" on a cave wall, with fabulous colors arising from trace metal ions, such as iron (reddish brown) or copper (bluish green).

Figure B19.1 A view inside Carlsbad Caverns, New Mexico. The marvelous formations within this limestone cave (shown under colored lights) result from subtle shifts in carbonate ionic equilibria acting over millions of years.

As you know, $Q_{sp} = K_{sp}$ when the solution is saturated. If Q_{sp} is greater than K_{sp}, the solution is momentarily supersaturated, and some solid precipitates until the remaining solution becomes saturated ($Q_{sp} = K_{sp}$). If Q_{sp} is less than K_{sp}, the solution is unsaturated, and no precipitate forms at that temperature (more solid can dissolve). To summarize,

- $Q_{sp} = K_{sp}$: solution is saturated and no change occurs.
- $Q_{sp} > K_{sp}$: precipitate forms until solution is saturated.
- $Q_{sp} < K_{sp}$: solution is unsaturated and no precipitate forms.

SAMPLE PROBLEM 19.9 Predicting Whether a Precipitate Will Form

Problem A common laboratory method for preparing a precipitate is to mix solutions of the component ions. Does a precipitate form when 0.100 L of 0.30 M $Ca(NO_3)_2$ is mixed with 0.200 L of 0.060 M NaF?

Plan First, we decide which slightly soluble salt could form and look up its K_{sp} value in Appendix C. To see whether mixing these solutions will form the precipitate, we find the initial ion concentrations by calculating the amount (mol) of each ion from its concentration and volume, and then dividing by the *total* volume, since one solution dilutes the other. Then we write the ion-product expression, calculate Q_{sp}, and compare it with K_{sp}.

Solution The ions present are Ca^{2+}, Na^+, F^-, and NO_3^-. All sodium and all nitrate salts are soluble (Table 4.1), so the only possibility is CaF_2 ($K_{sp} = 3.2 \times 10^{-11}$).

Calculating the ion concentrations:

$$\text{Moles of } Ca^{2+} = 0.30 \; M \; Ca^{2+} \times 0.100 \; L = 0.030 \; \text{mol } Ca^{2+}$$

$$[Ca^{2+}]_{init} = \frac{0.030 \; \text{mol } Ca^{2+}}{0.100 \; L + 0.200 \; L} = 0.10 \; M \; Ca^{2+}$$

$$\text{Moles of } F^- = 0.060 \; M \; F^- \times 0.200 \; L = 0.012 \; \text{mol } F^-$$

$$[F^-]_{init} = \frac{0.012 \; \text{mol } F^-}{0.100 \; L + 0.200 \; L} = 0.040 \; M \; F^-$$

Substituting into the ion-product expression and comparing Q_{sp} with K_{sp}:

$$Q_{sp} = [Ca^{2+}]_{init}[F^-]^2_{init} = (0.10)(0.040)^2 = 1.6 \times 10^{-4}$$

Because $Q_{sp} > K_{sp}$, CaF$_2$ will precipitate until $Q_{sp} = 3.2 \times 10^{-11}$.

Check Don't forget to round off and quickly check the math. For example, $Q_{sp} = (1 \times 10^{-1})(4 \times 10^{-2})^2 = 1.6 \times 10^{-4}$. With K_{sp} so low, CaF_2 must have a low solubility, and given the sizable concentrations being mixed, we would expect CaF_2 to precipitate.

FOLLOW-UP PROBLEM 19.9 Phosphate in natural waters often precipitates as insoluble salts, such as $Ca_3(PO_4)_2$. If $[Ca^{2+}]_{init} = [PO_4^{3-}]_{init} = 1.0 \times 10^{-9}$ M in a given river, will $Ca_3(PO_4)_2$ precipitate? K_{sp} of $Ca_3(PO_4)_2$ is 1.2×10^{-29}.

Precipitation of CaF$_2$.

As the upcoming Chemical Connections essay demonstrates, the principles of ionic equilibria often help us understand the chemical basis of complex environmental problems and may provide ways to solve them.

SECTION SUMMARY

As a first approximation, the dissolved portion of a slightly soluble salt dissociates completely into ions. In a saturated solution, the ions are in equilibrium with the solid, and the product of the ion concentrations, each raised to the power of its subscript in the formula, has a constant value ($Q_{sp} = K_{sp}$). The value of K_{sp} can be obtained from the solubility, and vice versa. Adding a common ion lowers an ionic compound's solubility. Adding H_3O^+ (lowering the pH) increases the compound's solubility if the anion of the compound is that of a weak acid. If $Q_{sp} > K_{sp}$ for a compound, a precipitate forms when two solutions are mixed. Limestone caves result from shifts in the $CaCO_3$-CO_2 equilibrium system. Lakes with limestone bedrock form a buffer system that prevents harmful acidification.

The Acid-Rain Problem

The conflict between industrial society and the environment is very clear in the problem of *acid rain*, acids resulting from human activity that occur as wet deposition in rain, snow, or fog, and as dry deposition on solid particles. Acidic precipitation has been recorded in all parts of the United States, in Canada, Mexico, the Amazon basin, throughout Europe, Russia, and many parts of Asia, and even at the North and South Poles. We've addressed several aspects of this problem in earlier chapters; now, let's examine some effects of acidic precipitation on aqueous biological and mineral systems and see how to prevent them. There are several troublesome chemical culprits:

1. *Sulfurous acid.* Sulfur dioxide (SO_2) formed primarily by the burning of high-sulfur coal, forms sulfurous acid (H_2SO_3) when it comes into contact with water. Oxidants, such as hydrogen peroxide and ozone, that are present as pollutants in the atmosphere also dissolve in water and convert the sulfurous acid to sulfuric acid:

$$H_2O_2(aq) + H_2SO_3(aq) \longrightarrow H_2SO_4(aq) + H_2O(l)$$

2. *Sulfuric acid.* Sulfur trioxide forms through the atmospheric oxidation of SO_2. Sulfur trioxide forms sulfuric acid (H_2SO_4) on contact with water, including that in the atmosphere.

3. *Nitric acid.* Nitrogen oxides (collectively known as NO_x) form in the reaction of N_2 and O_2. NO is formed during combustion, primarily in car engines, and it forms NO_2 and HNO_3 in air in the process that creates smog (p. 578). At night NO_x is converted to N_2O_5, which hydrolyzes to HNO_3 in the presence of water. The strong acids H_2SO_4 and HNO_3 cause the greatest concern (Figure B19.2).

How does the pH of acidic precipitation compare with that of natural waters? Normal rainwater is weakly acidic because it contains dissolved CO_2 from the air:

$$CO_2(g) + 2\,H_2O(l) \rightleftharpoons H_3O^+(aq) + HCO_3^-(aq)$$

Based on the volume percent of CO_2 in air, the solubility of CO_2 in water, and the K_{a1} of H_2CO_3, the pH of normal rainwater is about 5.6 (see Problem 19.148 at the end of the chapter). In stark contrast, the average pH of rainfall in many parts of the United States was 4.2 as recently as 1984, representing 25 times as much H_3O^+. Worldwide, rain in Sweden and Pennsylvania shared second prize with a pH of 2.7, about the same as vinegar. Rain in Wheeling, West Virginia, won first prize with a pH of 1.8, between that of lemon juice and stomach acid! Acidic fog in California sometimes has a pH of 1.6 because evaporation of water from the particles concentrates the acid.

These 10- to 10,000-fold excesses of [H_3O^+] are very destructive to living things. Some fish and shellfish die at pH values between 4.5 and 5.0. The young of most species are generally most sensitive. At pH 5, most fish eggs cannot hatch. With tens of thousands of rivers and lakes around the world becoming acidified, the loss of fish became a major concern long ago. In addition, acres of forest have been harmed by the acid, which removes nutrients and releases toxic substances from the soil.

continued

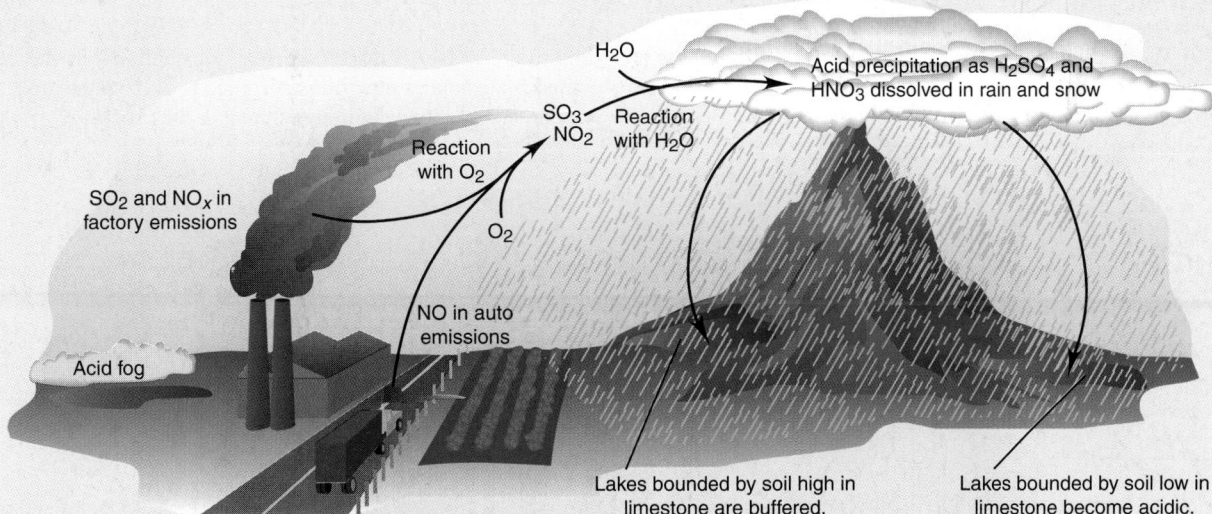

Figure B19.2 Formation of acidic precipitation. A complex interplay of human activities, atmospheric chemistry, and environmental distribution leads to acidic precipitation and its harmful effects. Car exhaust and factory waste gases produce lower oxides of nitrogen and sulfur. These are oxidized in the atmosphere by O_2 (or O_3, not shown) to higher oxides (NO_2, SO_3), which react with moisture to form acidic rain, snow, and fog. In contact with acidic precipitation, many lakes become acidified, whereas limestone-bounded lakes form a carbonate buffer that prevents acidification.

Even if the soil is buffered, acidic fog and clouds can remove essential nutrients from leaf surfaces. (Figure B19.3).

Many principles of aqueous equilibria bear directly on the effects of acid rain. The aluminosilicates that make up most soils are extremely insoluble in water. In these materials, the Al^{3+} ion is bonded to OH^- and O^{2-} ions in complex structures (see the Gallery in Chapter 14, p. 570). Continual contact with the H_3O^+ in acid rain causes these ions to react, and some of the bound Al^{3+}, which is toxic to fish, dissolves. Along with dissolved Al^{3+} ions, the acid rain carries away ions that serve as nutrients for plants and animals.

Acid rain also dissolves the calcium carbonate in the marble and limestone of buildings and monuments (Figure B19.4). Ironically, the same chemical process that destroys these structures is responsible for saving those lakes that lie on or are bounded by limestone-rich soil. As acid rain falls, the H_3O^+ reacts with dissolved carbonate ion in the lake to form bicarbonate:

$$CO_3^{2-} + H_3O^+ \rightleftharpoons HCO_3^- + H_2O$$

In essence, limestone-bounded lakes function as enormous HCO_3^-/CO_3^{2-} buffer solutions, absorbing the additional H_3O^+ and maintaining a relatively stable pH. In fact, lakes, rivers, and groundwater in limestone-rich soils actually remain mildly basic.

For lakes and rivers in contact with limestone-poor soils, expensive remediation methods are needed. A direct attack on the symptoms is the so-called liming (treating with limestone) of lakes and rivers. Sweden spent tens of millions of dollars during the 1990s to neutralize slightly more than 3000 lakes by adding limestone. This approach is, at best, only a stopgap because the lakes are acidic again within several years.

As we pointed out earlier (see Chemical Connections, Chapter 6, p. 243), the principal means of controlling sulfur dioxide is by "scrubbing" the emissions from power plants with limestone. Both dry and wet scrubbers are used. Another method reduces

1944   1994

Figure B19.4 The effect of acid rain on marble statuary. Calcium carbonate, which is the major component of marble, is slowly decomposed by acid rain. These photos of the same statue of George Washington in New York City were taken 50 years apart.

some of the SO_2 with methane or coal to H_2S, and the mixture is catalytically converted to sulfur, which is sold:

$$16H_2S(g) + 8SO_2(g) \longrightarrow 3S_8(s) + 16H_2O(l)$$

Burning low-sulfur coal to reduce SO_2 formation is sometimes an option, but such coal deposits are rare and expensive to mine. Coal can also be converted into gaseous and liquid low-sulfur fuels (see Chemical Connections, p. 243). The sulfur is removed (as H_2S) in an acid-gas scrubber after gasification.

Through the use of a catalytic converter in an auto exhaust system, NO_x species are reduced to N_2 and NH_3. In power plants, the amount of NO_x is decreased by adjusting combustion conditions, but it also can be removed from the hot stack gases by treatment with ammonia:

$$4NO(g) + 4NH_3(g) + O_2(g) \longrightarrow 4N_2(g) + 6H_2O(g)$$

Emissions of NO_x must be curbed substantially in the eastern United States under new rules designed to help states meet ozone standards, and HNO_3 will be reduced in the process.

Figure B19.3 A forest damaged by acid rain.

19.4 EQUILIBRIA INVOLVING COMPLEX IONS

The final type of aqueous ionic equilibrium we consider involves a different type of ion than we've examined up to now. Simple ions, such as Na^+ or SO_4^{2-}, consist of one or a few bound atoms, with an excess or deficit of electrons. A **complex ion** consists of a central metal ion covalently bonded to two or more anions or molecules, called **ligands.** Hydroxide, chloride, and cyanide ions are some ionic ligands; water, carbon monoxide, and ammonia are some molecular ligands. In the complex ion $Cr(NH_3)_6{}^{3+}$, for example, Cr^{3+} is the central metal ion and six NH_3 molecules are the ligands, giving an overall 3+ charge (Figure 19.13).

As we discussed in Section 18.9, *all complex ions are Lewis adducts.* The metal ion acts as a Lewis acid (accepts an electron pair) and the ligand acts as a Lewis base (donates an electron pair). The acidic hydrated metal ions that we discussed in Section 18.6 are complex ions with water molecules as ligands. In Chapter 23, we discuss the transition metals and the structures and properties of the numerous complex ions they form. Our focus here is on equilibria of hydrated ions with ligands other than water.

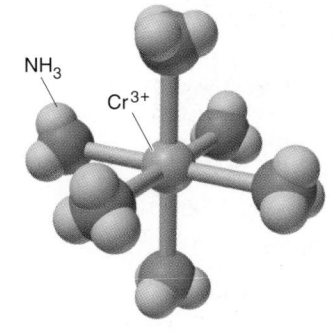

Figure 19.13 $Cr(NH_3)_6{}^{3+}$, **a typical complex ion.** A complex ion consists of a central metal ion, such as Cr^{3+}, covalently bonded to a specific number of ligands, such as NH_3.

Formation of Complex Ions

Whenever a metal ion enters water, a complex ion forms, with water as the ligand. In many cases, when we treat this hydrated cation with a solution of another ligand, the bound water molecules exchange for the other ligand. For example, a hydrated M^{2+} ion, $M(H_2O)_4{}^{2+}$, forms the complex ion $M(NH_3)_4{}^{2+}$ in aqueous NH_3:

$$M(H_2O)_4{}^{2+}(aq) + 4NH_3(aq) \rightleftharpoons M(NH_3)_4{}^{2+}(aq) + 4H_2O(l)$$

At equilibrium, this system is expressed by a ratio of concentration terms whose form follows that of any other equilibrium expression:

$$K_c = \frac{[M(NH_3)_4{}^{2+}][H_2O]^4}{[M(H_2O)_4{}^{2+}][NH_3]^4}$$

Once again, because the concentration of water is essentially constant in aqueous reactions, we incorporate it into K_c and obtain the expression for a new equilibrium constant, the **formation constant, K_f:**

$$K_f = \frac{K_c}{[H_2O]^4} = \frac{[M(NH_3)_4{}^{2+}]}{[M(H_2O)_4{}^{2+}][NH_3]^4}$$

At the molecular level, the actual process is stepwise, with ammonia molecules replacing water molecules one at a time to give a series of intermediate species, each with its own formation constant:

$$M(H_2O)_4{}^{2+}(aq) + NH_3(aq) \rightleftharpoons M(H_2O)_3(NH_3)^{2+}(aq) + H_2O(l)$$

$$K_{f1} = \frac{[M(H_2O)_3(NH_3)^{2+}]}{[M(H_2O)_4{}^{2+}][NH_3]}$$

$$M(H_2O)_3(NH_3)^{2+}(aq) + NH_3(aq) \rightleftharpoons M(H_2O)_2(NH_3)_2{}^{2+}(aq) + H_2O(l)$$

$$K_{f2} = \frac{[M(H_2O)_2(NH_3)_2{}^{2+}]}{[M(H_2O)_3(NH_3)^{2+}][NH_3]}$$

$$M(H_2O)_2(NH_3)_2{}^{2+}(aq) + NH_3(aq) \rightleftharpoons M(H_2O)(NH_3)_3{}^{2+}(aq) + H_2O(l)$$

$$K_{f3} = \frac{[M(H_2O)(NH_3)_3{}^{2+}]}{[M(H_2O)_2(NH_3)_2{}^{2+}][NH_3]}$$

$$M(H_2O)(NH_3)_3{}^{2+}(aq) + NH_3(aq) \rightleftharpoons M(NH_3)_4{}^{2+}(aq) + H_2O(l)$$

$$K_{f4} = \frac{[M(NH_3)_4{}^{2+}]}{[M(H_2O)(NH_3)_3{}^{2+}][NH_3]}$$

The *sum* of the equations gives the overall equation, so the *product* of the individual formation constants gives the overall formation constant:

$$K_f = K_{f1} \times K_{f2} \times K_{f3} \times K_{f4}$$

Table 19.4 Formation Constants (K_f) of Some Complex Ions at 25°C

Complex Ion	K_f
$Ag(CN)_2^-$	3.0×10^{20}
$Ag(NH_3)_2^+$	1.7×10^7
$Ag(S_2O_3)_2^{3-}$	4.7×10^{13}
AlF_6^{3-}	4×10^{19}
$Al(OH)_4^-$	3×10^{33}
$Be(OH)_4^{2-}$	4×10^{18}
CdI_4^{2-}	1×10^6
$Co(OH)_4^{2-}$	5×10^9
$Cr(OH)_4^-$	8.0×10^{29}
$Cu(NH_3)_4^{2+}$	5.6×10^{11}
$Fe(CN)_6^{4-}$	3×10^{35}
$Fe(CN)_6^{3-}$	4.0×10^{43}
$Hg(CN)_4^{2-}$	9.3×10^{38}
$Ni(NH_3)_6^{2+}$	2.0×10^8
$Pb(OH)_3^-$	8×10^{13}
$Sn(OH)_3^-$	3×10^{25}
$Zn(CN)_4^{2-}$	4.2×10^{19}
$Zn(NH_3)_4^{2+}$	7.8×10^8
$Zn(OH)_4^{2-}$	3×10^{15}

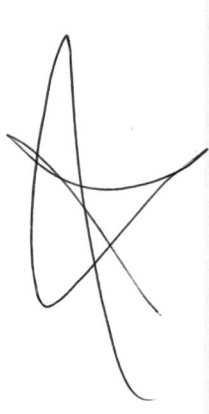

Figure 19.14 summarizes the process on the molecular level. In this case, the K_f for each step is much larger than 1 because ammonia is a stronger Lewis base than water. Therefore, if we add excess ammonia to the $M(H_2O)_4^{2+}$ solution, the H_2O ligands are replaced and essentially all the M^{2+} ion exists as $M(NH_3)_4^{2+}$.

Table 19.4 shows the formation constants of some complex ions. Notice that the K_f values are all 10^6 or greater, which means that these ions form readily. Because of this behavior, complex-ion formation is used to retrieve a metal from its ore, eliminate a toxic or unwanted metal ion from a solution, or convert the metal ion to a different form, as Sample Problem 19.10 shows for the zinc ion.

SAMPLE PROBLEM 19.10 Calculating the Concentration of a Complex Ion

Problem An industrial chemist converts $Zn(H_2O)_4^{2+}$ to the more stable $Zn(NH_3)_4^{2+}$ by mixing 50.0 L of 0.0020 M $Zn(H_2O)_4^{2+}$ and 25.0 L of 0.15 M NH_3. What is the final $[Zn(H_2O)_4^{2+}]$? K_f of $Zn(NH_3)_4^{2+}$ is 7.8×10^8.

Plan We write the equation and the K_f expression and use a reaction table to calculate the equilibrium concentrations. To set up the table, we must first find $[Zn(H_2O)_4^{2+}]_{init}$ and $[NH_3]_{init}$. We are given the individual volumes and molar concentrations, so we find the moles and divide by the *total* volume, because the solutions are mixed. With the large excess of NH_3 and high K_f, we assume that almost all the $Zn(H_2O)_4^{2+}$ is converted to $Zn(NH_3)_4^{2+}$. Because $[Zn(H_2O)_4^{2+}]$ at equilibrium is very small, we use x to represent it.

Solution Writing the equation and the K_f expression:

$$Zn(H_2O)_4^{2+}(aq) + 4NH_3(aq) \rightleftharpoons Zn(NH_3)_4^{2+}(aq) + 4H_2O(l)$$

$$K_f = \frac{[Zn(NH_3)_4^{2+}]}{[Zn(H_2O)_4^{2+}][NH_3]^4}$$

Finding the initial reactant concentrations:

$$[Zn(H_2O)_4^{2+}]_{init} = \frac{50.0\,L \times 0.0020\,M}{50.0\,L + 25.0\,L} = 1.3\times10^{-3}\,M$$

$$[NH_3]_{init} = \frac{25.0\,L \times 0.15\,M}{50.0\,L + 25.0\,L} = 5.0\times10^{-2}\,M$$

Setting up a reaction table: We assume that nearly all the $Zn(H_2O)_4^{2+}$ is converted to $Zn(NH_3)_4^{2+}$, so we set up the table with $x = [Zn(H_2O)_4^{2+}]$ at equilibrium. Because 4 mol of NH_3 are needed per mole of $Zn(H_2O)_4^{2+}$, the change in $[NH_3]$ is

$$[NH_3]_{reacted} \approx 4(1.3\times10^{-3}\,M) = 5.2\times10^{-3}\,M$$

and

$$[Zn(NH_3)_4^{2+}] \approx 1.3\times10^{-3}\,M$$

Concentration (M)	$Zn(H_2O)_4^{2+}(aq)$	$+\ 4NH_3(aq)$	$\rightleftharpoons$ $Zn(NH_3)_4^{2+}(aq)$	$+\ 4H_2O(l)$
Initial	1.3×10^{-3}	5.0×10^{-2}	0	—
Change	$\sim(-1.3\times10^{-3})$	$\sim(-5.2\times10^{-3})$	$\sim(+1.3\times10^{-3})$	—
Equilibrium	x	4.5×10^{-2}	1.3×10^{-3}	—

Solving for x, the $[Zn(H_2O)_4^{2+}]$ remaining at equilibrium:

$$K_f = \frac{[Zn(NH_3)_4^{2+}]}{[Zn(H_2O)_4^{2+}][NH_3]^4} = 7.8\times10^8 \approx \frac{1.3\times10^{-3}}{x(4.5\times10^{-2})^4}$$

$$x = [Zn(H_2O)_4^{2+}] \approx 4.1\times10^{-7}\,M$$

Check The K_f is large, so we expect the $[Zn(H_2O)_4^{2+}]$ remaining to be very low.

FOLLOW-UP PROBLEM 19.10 Cyanide ion is toxic because it forms stable complex ions with the Fe^{3+} ion in certain iron-containing proteins engaged in energy production. To study this effect, a biochemist mixes 25.5 mL of 3.1×10^{-2} M $Fe(H_2O)_6^{3+}$ with 35.0 mL of 1.5 M NaCN. What is the final $[Fe(H_2O)_6^{3+}]$? K_f of $Fe(CN)_6^{3-}$ is 4.0×10^{43}.

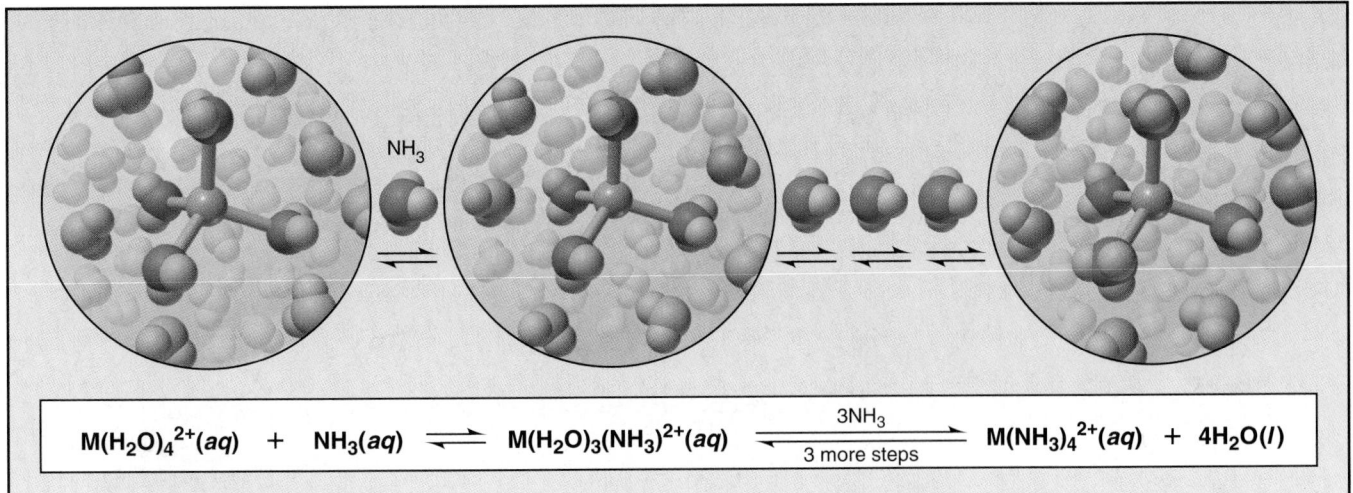

Figure 19.14 The stepwise exchange of NH$_3$ for H$_2$O in M(H$_2$O)$_4$$^{2+}$. The ligands of a complex ion can exchange for other ligands. When ammonia is added to a solution of the hydrated M^{2+} ion, M(H$_2$O)$_4$$^{2+}$, NH$_3$ molecules replace the bound H$_2$O molecules one at a time to form the M(NH$_3$)$_4$$^{2+}$ ion. The molecular-scale views show the first exchange and the fully ammoniated ion.

Complex Ions and the Solubility of Precipitates

In Section 19.3, you saw that H$_3$O$^+$ increases the solubility of a slightly soluble ionic compound if its anion is that of a weak acid. Similarly, *a ligand increases the solubility of a slightly soluble ionic compound if it forms a complex ion with the cation.* For example, zinc sulfide is very slightly soluble:

$$ZnS(s) + H_2O(l) \rightleftharpoons Zn^{2+}(aq) + HS^-(aq) + OH^-(aq) \qquad K_{sp} = 2.0\times10^{-22}$$

When we add some 1.0 *M* NaCN, the CN$^-$ ions act as ligands and react with the small amount of Zn^{2+}(*aq*) to form the following complex ion:

$$Zn^{2+}(aq) + 4CN^-(aq) \rightleftharpoons Zn(CN)_4{}^{2-}(aq) \qquad K_f = 4.2\times10^{19}$$

To see the effect of complex ion formation on the solubility of ZnS, we add the equations and, therefore, multiply their equilibrium constants:

$$ZnS(s) + 4CN^-(aq) + H_2O(l) \rightleftharpoons Zn(CN)_4{}^{2-}(aq) + HS^-(aq) + OH^-(aq)$$
$$K_{overall} = K_{sp} \times K_f = (2.0\times10^{-22})(4.2\times10^{19}) = 8.4\times10^{-3}$$

The overall equilibrium constant increased by more than a factor of 10^{19} in the presence of the ligand; this reflects the increased amount of ZnS in solution.

SAMPLE PROBLEM 19.11 Calculating the Effect of Complex-Ion Formation on Solubility

Problem In black-and-white film developing (see photo), excess AgBr is removed from the film negative by "hypo," an aqueous solution of sodium thiosulfate (Na$_2$S$_2$O$_3$), through formation of the complex ion Ag(S$_2$O$_3$)$_2$$^{3-}$. Calculate the solubility of AgBr in (a) H$_2$O; (b) 1.0 *M* hypo. K_f of Ag(S$_2$O$_3$)$_2$$^{3-}$ is 4.7×10^{13} and K_{sp} of AgBr is 5.0×10^{-13}.
Plan (a) After writing the equation and the ion-product expression, we use the given K_{sp} to solve for S, the molar solubility of AgBr. (b) In hypo, Ag$^+$ forms a complex ion with S$_2$O$_3$$^{2-}$, which shifts the equilibrium and dissolves more AgBr. We write the complex-ion equation and add it to the equation for dissolving AgBr to obtain the overall equation for dissolving AgBr in hypo. We multiply K_{sp} by K_f to find $K_{overall}$. To find the solubility of AgBr in hypo, we set up a reaction table, with $S = [Ag(S_2O_3)_2{}^{3-}]$, substitute into the expression for $K_{overall}$, and solve for S.

Developing the image in "hypo."

Solution (a) Solubility in water. Writing the equation for the saturated solution and the ion-product expression:

$$\text{AgBr}(s) \rightleftharpoons \text{Ag}^+(aq) + \text{Br}^-(aq) \qquad K_{sp} = [\text{Ag}^+][\text{Br}^-]$$

Solving for solubility (S) directly from the equation: We know that

$$S = [\text{AgBr}]_{dissolved} = [\text{Ag}^+] = [\text{Br}^-]$$

Thus, $\quad K_{sp} = [\text{Ag}^+][\text{Br}^-] = S^2 = 5.0 \times 10^{-13}, \quad$ so $S = \boxed{7.1 \times 10^{-7}\ M}$

(b) Solubility in 1.0 M hypo. Writing the overall equation:

$$\text{AgBr}(s) \rightleftharpoons \text{Ag}^+(aq) + \text{Br}^-(aq)$$
$$\underline{\text{Ag}^+(aq) + 2\text{S}_2\text{O}_3{}^{2-}(aq) \rightleftharpoons \text{Ag}(\text{S}_2\text{O}_3)_2{}^{3-}(aq)}$$
$$\text{AgBr}(s) + 2\text{S}_2\text{O}_3{}^{2-}(aq) \rightleftharpoons \text{Ag}(\text{S}_2\text{O}_3)_2{}^{3-}(aq) + \text{Br}^-(aq)$$

Calculating $K_{overall}$:

$$K_{overall} = \frac{[\text{Ag}(\text{S}_2\text{O}_3)_2{}^{3-}][\text{Br}^-]}{[\text{S}_2\text{O}_3{}^{2-}]^2} = K_{sp} \times K_f = (5.0 \times 10^{-13})(4.7 \times 10^{13}) = 24$$

Setting up a reaction table, with $S = [\text{AgBr}]_{dissolved} = [\text{Ag}(\text{S}_2\text{O}_3)_2{}^{3-}]$:

Concentration (M)	AgBr(s) +	2S$_2$O$_3{}^{2-}$(aq) $\rightleftharpoons$	Ag(S$_2$O$_3$)$_2{}^{3-}$(aq) +	Br$^-$(aq)
Initial	—	1.0	0	0
Change	—	$-2S$	$+S$	$+S$
Equilibrium	—	$1.0 - 2S$	S	S

Substituting the values into $K_{overall}$ and solving for S:

$$K_{overall} = \frac{[\text{Ag}(\text{S}_2\text{O}_3)_2{}^{3-}][\text{Br}^-]}{[\text{S}_2\text{O}_3{}^{2-}]^2} = \frac{S^2}{(1.0\ M - 2S)^2} = 24$$

Taking the square root of both sides gives

$$\frac{S}{1.0\ M - 2S} = \sqrt{24} = 4.9 \qquad [\text{Ag}(\text{S}_2\text{O}_3)_2{}^{3-}] = \boxed{S = 0.45\ M}$$

Check (a) From the number of ions in the formula of AgBr, we know that $S = \sqrt{K_{sp}}$, so the order of magnitude seems right: $\sim \sqrt{10^{-14}} = 10^{-7}$. **(b)** The $K_{overall}$ seems correct: the exponents cancel, and $5 \times 5 = 25$. Most importantly, the answer makes sense because the photographic process requires the remaining AgBr to be washed off the film and the large $K_{overall}$ confirms that. We can check S by rounding and working backward to find $K_{overall}$: from the reaction table, we find that

$$[(\text{S}_2\text{O}_3)^{2-}] = 1.0\ M - 2S = 1.0\ M - 2(0.45\ M) = 1.0\ M - 0.90\ M = 0.1\ M$$

so $K_{overall} \approx (0.45)^2/(0.1)^2 = 20$, within rounding of the calculated value.

FOLLOW-UP PROBLEM 19.11 How does the solubility of AgBr in 1.0 M NH$_3$ compare with its solubility in hypo? K_f of Ag(NH$_3$)$_2{}^+$ is 1.7×10^7.

Complex Ions of Amphoteric Hydroxides

The same metals that form amphoteric oxides (Section 8.5 and Interchapter Topic 4) also form slightly soluble *amphoteric hydroxides,* compounds that dissolve very little in water but to a much greater extent in both acidic and basic solutions. Aluminum hydroxide is an example:

$$\text{Al(OH)}_3(s) \rightleftharpoons \text{Al}^{3+}(aq) + 3\text{OH}^-(aq)$$

It is insoluble in water ($K_{sp} = 3 \times 10^{-34}$), but

- It dissolves in acid because H$_3$O$^+$ reacts with the OH$^-$ anion (Section 19.3),
$$3\text{H}_3\text{O}^+(aq) + 3\text{OH}^-(aq) \longrightarrow 6\text{H}_2\text{O}(l)$$
  giving the overall equation
$$\text{Al(OH)}_3(s) + 3\text{H}_3\text{O}^+(aq) \longrightarrow \text{Al}^{3+}(aq) + 6\text{H}_2\text{O}(l)$$
- It dissolves in base through the formation of a complex ion:
$$\text{Al(OH)}_3(s) + \text{OH}^-(aq) \longrightarrow \text{Al(OH)}_4{}^-(aq)$$

Figure 19.15 shows this amphoteric behavior but includes some unexpected formulas for the species. Let's see how these species arise. When we dissolve a soluble aluminum salt, such as $Al(NO_3)_3$, in water and then slowly add a strong base, a white precipitate first forms and then dissolves as more base is added. What reactions are occurring? The formula for the hydrated Al^{3+} ion is $Al(H_2O)_6^{3+}(aq)$. It acts as a weak polyprotic acid and reacts with added OH^- ions in a stepwise manner. In each step, one bound H_2O loses a proton and becomes a bound OH^- ion, so the number of bound H_2O molecules is reduced by 1:

$$Al(H_2O)_6^{3+}(aq) + OH^-(aq) \rightleftharpoons Al(H_2O)_5OH^{2+}(aq) + H_2O(l)$$
$$Al(H_2O)_5OH^{2+}(aq) + OH^-(aq) \rightleftharpoons Al(H_2O)_4(OH)_2^+(aq) + H_2O(l)$$
$$Al(H_2O)_4(OH)_2^+(aq) + OH^-(aq) \rightleftharpoons Al(H_2O)_3(OH)_3(s) + H_2O(l)$$

At this point, the white precipitate has formed. It is the insoluble hydroxide $Al(H_2O)_3(OH)_3(s)$, often written more simply as $Al(OH)_3(s)$. Now you can see that the precipitate is actually the hydrated Al^{3+} ion with an H^+ removed from each of three bound H_2O molecules. Addition of H_3O^+ protonates the OH^- ions and re-forms the hydrated Al^{3+} ion.

Further addition of OH^- removes a fourth H^+ and forms the soluble ion $Al(H_2O)_2(OH)_4^-(aq)$, which we usually write as $Al(OH)_4^-(aq)$:

$$Al(H_2O)_3(OH)_3(s) + OH^-(aq) \rightleftharpoons Al(H_2O)_2(OH)_4^-(aq) + H_2O(l)$$

In other words, this complex ion is not created by ligands substituting for bound water molecules but through an acid-base reaction in which added OH^- ions titrate bound water molecules.

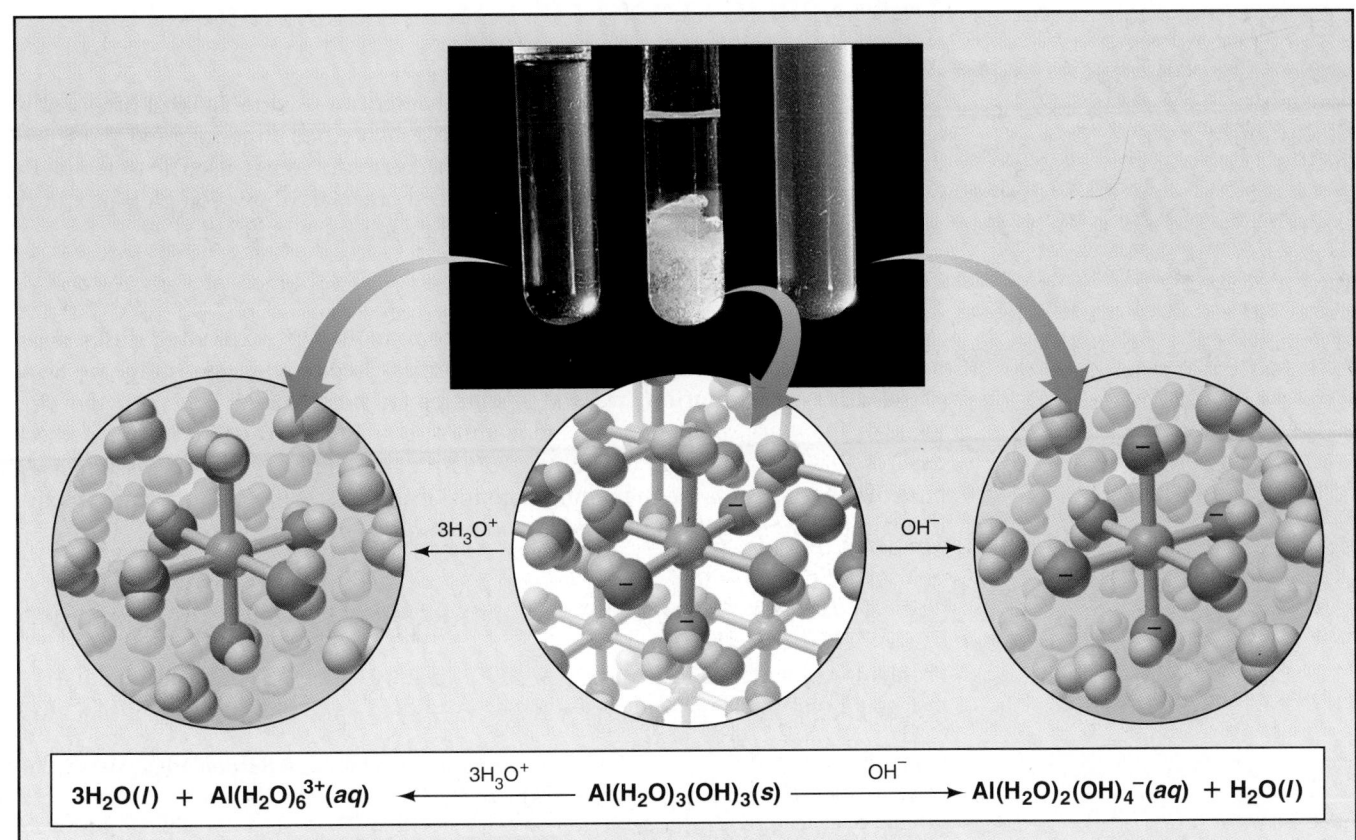

Figure 19.15 The amphoteric behavior of aluminum hydroxide. When solid $Al(OH)_3$ is treated with H_3O^+ (*left*) or with OH^- (*right*), it dissolves as a result of the formation of soluble complex ions. The molecular views show that $Al(OH)_3$ is actually the neutral species $Al(H_2O)_3(OH)_3$. (The extensive OH bridging that occurs between Al^{3+} ions throughout the solid is not shown.) Addition of OH^- (*right*) forms the soluble $Al(H_2O)_2(OH)_4^-$ ion; addition of H_3O^+ (*left*) forms the soluble $Al(H_2O)_6^{3+}$ ion.

Several other slightly soluble hydroxides, such as those of Cr^{3+}, Zn^{2+}, Pb^{2+}, and Sn^{2+}, are amphoteric and exhibit similar reactions:

$$Zn(H_2O)_2(OH)_2(s) + OH^-(aq) \rightleftharpoons Zn(H_2O)(OH)_3^-(aq) + H_2O(l)$$

In contrast, the slightly soluble hydroxides of Fe^{2+}, Fe^{3+}, and Ca^{2+} dissolve in acid, but *not* in base, because the remaining bound water molecules are not acidic enough:

$$Fe(H_2O)_3(OH)_3(s) + 3H_3O^+(aq) \longrightarrow Fe(H_2O)_6^{3+}(aq) + 3H_2O(l)$$

$$Fe(H_2O)_3(OH)_3(s) + OH^-(aq) \longrightarrow \text{no reaction}$$

The difference in solubility in base between $Al(OH)_3$ and $Fe(OH)_3$ is the key to an important separation step in the production of aluminum metal, so we'll consider it again in Section 22.4. We also employ it in the next section in analyzing a mixture of ions.

SECTION SUMMARY

A complex ion consists of a central metal ion covalently bonded to two or more negatively charged or neutral ligands. Its formation is described by a formation constant, K_f. A hydrated metal ion is a complex ion with water molecules as ligands. Other ligands can displace the water in a stepwise process. In most cases, the K_f value of each step is large, so the fully substituted complex ion forms almost completely in the presence of excess ligand. A ligand solution increases the solubility of an ionic precipitate if the cation forms a complex ion with the ligand. Amphoteric metal hydroxides dissolve in acid and base due to acid-base reactions that form soluble complex ions.

19.5 APPLICATION OF IONIC EQUILIBRIA TO CHEMICAL ANALYSIS

Many of the ideas we've discussed in this chapter are used to analyze the ions in a mixture. In this brief introduction to an extensive and time-honored field, we discuss how control of the precipitating-ion concentration is used to selectively precipitate one metal ion in the presence of another and how complex mixtures of ions are separated into smaller groups and each ion identified.

Selective Precipitation

We can often select one ion in a solution from another by exploiting differences in the solubility of their compounds with a given precipitating ion. In the process of **selective precipitation,** we add a solution of precipitating ion until the Q_{sp} value of the *more soluble* compound is almost equal to its K_{sp} value. This method ensures that the K_{sp} value of the *less soluble* compound is exceeded as much as possible. As a result, the maximum amount of the less soluble compound precipitates, but none of the more soluble compound does.

SAMPLE PROBLEM 19.12   Separating Ions by Selective Precipitation

Problem A solution consists of 0.20 M $MgCl_2$ and 0.10 M $CuCl_2$. Calculate the $[OH^-]$ that would separate the metal ions as their hydroxides. K_{sp} of $Mg(OH)_2$ is 6.3×10^{-10}; K_{sp} of $Cu(OH)_2$ is 2.2×10^{-20}.

Plan The two hydroxides have the same formula type (1:2) (see Section 19.3), so we can compare their K_{sp} values to see which is more soluble: $Mg(OH)_2$ is about 10^{10} times more soluble than $Cu(OH)_2$. Thus, $Cu(OH)_2$ precipitates first. We want to precipitate as much $Cu(OH)_2$ as possible without precipitating any $Mg(OH)_2$. We solve for the $[OH^-]$ that will just give a saturated solution of $Mg(OH)_2$ because this $[OH^-]$ will precipitate the greatest amount of Cu^{2+} ion. As confirmation, we calculate the $[Cu^{2+}]$ remaining to see if the separation was accomplished.

Solution Writing the equations and ion-product expressions:

$$Mg(OH)_2(s) \rightleftharpoons Mg^{2+}(aq) + 2OH^-(aq) \qquad K_{sp} = [Mg^{2+}][OH^-]^2$$
$$Cu(OH)_2(s) \rightleftharpoons Cu^{2+}(aq) + 2OH^-(aq) \qquad K_{sp} = [Cu^{2+}][OH^-]^2$$

Calculating the $[OH^-]$ that gives a saturated $Mg(OH)_2$ solution:

$$[OH^-] = \sqrt{\frac{K_{sp}}{[Mg^{2+}]}} = \sqrt{\frac{6.3 \times 10^{-10}}{0.20}} = \boxed{5.6 \times 10^{-5} \, M}$$

This is the maximum $[OH^-]$ that will *not* precipitate Mg^{2+} ion.
Calculating the $[Cu^{2+}]$ remaining in the solution with this $[OH^-]$:

$$[Cu^{2+}] = \frac{K_{sp}}{[OH^-]^2} = \frac{2.2 \times 10^{-20}}{(5.6 \times 10^{-5})^2} = 7.0 \times 10^{-12} \, M$$

Since the initial $[Cu^{2+}]$ was 0.10 M, virtually all the Cu^{2+} ion is precipitated.

Check Rounding, we find the $[OH^-]$ seems right: $\sim\sqrt{(6 \times 10^{-10})/0.2} = 5 \times 10^{-5}$. The $[Cu^{2+}]$ remaining also seems correct: $(200 \times 10^{-22})/(5 \times 10^{-5})^2 = 8 \times 10^{-12}$.

Comment Often, more than one approach will accomplish the same analytical step. Another possibility in this case is to add excess ammonia to make the solution basic enough to precipitate both hydroxides:

$$Mg^{2+}(aq) + 2NH_3(aq) + 2H_2O(l) \rightleftharpoons Mg(OH)_2(s) + 2NH_4^+(aq)$$
$$Cu^{2+}(aq) + 2NH_3(aq) + 2H_2O(l) \rightleftharpoons Cu(OH)_2(s) + 2NH_4^+(aq)$$

Then, the $Cu(OH)_2$ dissolves in excess NH_3 by forming a soluble complex ion:

$$Cu(OH)_2(s) + 4NH_3(aq) \rightleftharpoons Cu(NH_3)_4^{2+}(aq) + 2OH^-(aq)$$

FOLLOW-UP PROBLEM 19.12 A solution containing two alkaline earth metal ions is made from 0.050 M $BaCl_2$ and 0.025 M $CaCl_2$. What concentration of SO_4^{2-} must be present to leave 99.99% of only one of the cations in solution? K_{sp} of $BaSO_4$ is 1.1×10^{-10}, and K_{sp} of $CaSO_4$ is 2.4×10^{-5}.

Sometimes two or more types of ionic equilibria are controlled simultaneously to precipitate ions selectively. This approach is used commonly to separate ions as their sulfides, so the HS^- ion is the precipitating ion. In the manner used in Sample Problem 19.12, we control the $[HS^-]$ to exceed the K_{sp} value of one metal sulfide but not another. We exert this control on the $[HS^-]$ by controlling H_2S dissociation because H_2S is the source of HS^-; we control H_2S dissociation by adjusting the $[H_3O^+]$:

$$H_2S(aq) + H_2O(l) \rightleftharpoons H_3O^+(aq) + HS^-(aq)$$

These interactions are controlled in the following way: If we add strong acid to the solution, $[H_3O^+]$ is high, so H_2S dissociation shifts to the left, which decreases $[HS^-]$. With a low $[HS^-]$, the *less* soluble sulfide precipitates. Conversely, if we add strong base, $[H_3O^+]$ is low, so H_2S dissociation shifts to the right, which increases $[HS^-]$, and then the *more* soluble sulfide precipitates. In essence, what we have done is shift one equilibrium system (H_2S dissociation) by adjusting a second (H_2O ionization) to control a third (metal sulfide solubility).

Qualitative Analysis: Identifying Ions in Complex Mixtures

The practice of inorganic **qualitative analysis,** the separation and identification of the ions in a mixture, was once an essential part of a chemist's skills. Today, many of these wet chemical techniques have been replaced by instrumental methods. Nevertheless, they provide an excellent means for studying ionic equilibria, including some of the very ones utilized by modern analytical instruments. In this discussion, we apply the principles of solubility and complex-ion equilibria to separate and characterize a mixture of cations; similar procedures exist for anions. Rather than being a fixed series of steps for identifying a set of ions, the design of the procedure often depends on the analyst's ingenuity.

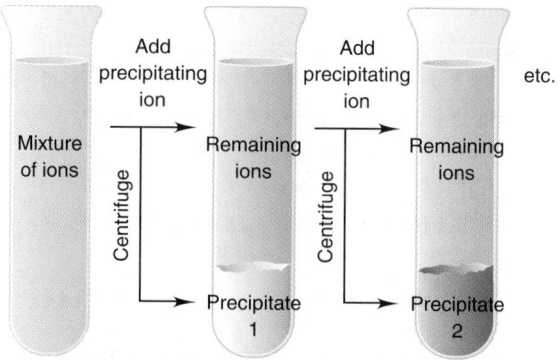

Figure 19.16 The general procedure for separating ions in qualitative analysis. A precipitating ion is added to a mixture of ions, the precipitate is separated by centrifugation, and the remaining dissolved ions are treated with another precipitating ion. The process is repeated until the ions are separated into ion groups.

Separation into Ion Groups The general approach begins by separating the unknown solution into *ion groups*. (Ion groups have *nothing* to do with periodic table groups.) Figure 19.16 depicts the laboratory steps. The mixture of metal ions is treated with a solution that precipitates a certain group of them and leaves the others in solution. Filtration or centrifugation (the rapid spinning of the tube to collect the solid in a compact pellet at the bottom) separates the compounds containing the precipitated metal ions. The solution containing the remaining ions is decanted (poured off) and treated with a solution that precipitates a different group of ions. These steps are repeated until the original mixture has been separated into specific ion groups. In an actual analysis, a *known solution* (one that contains all the ions under study) and a *blank* of distilled water are treated in exactly the same way as the unknown solution, thereby making it much easier to judge a positive or negative result.

Figure 19.17 shows one scheme for separating cations into ion groups.

Ion group 1: Insoluble chlorides. The entire mixture of soluble ions is treated with 6 *M* HCl. Most metal chlorides are soluble, so only those few ions that form insoluble chlorides precipitate in this step. If a precipitate appears, it consists of chlorides of any of the following:

$$Ag^+, Hg_2^{2+}, Pb^{2+}$$

If none appears, no ions from ion group 1 are present in the mixture. If a precipitate forms, the tube is centrifuged, and the solution is carefully decanted.

Ion group 2: Acid-insoluble sulfides. The decanted solution is already acidic from the previous treatment with HCl. The pH is adjusted to 0.5 and the solution treated with aqueous H_2S. The H_3O^+ present keeps the $[HS^-]$ very low, so any precipitate contains one or more of the least soluble sulfides, which are those of

$$Cu^{2+}, Cd^{2+}, Hg^{2+}, As^{3+}, Sb^{3+}, Bi^{3+}, Sn^{2+}, Sn^{4+}, Pb^{2+}$$

($PbCl_2$ is slightly soluble in water, so a small amount of Pb^{2+} remains in solution after addition of HCl and appears in ion groups 1 *and* 2.) If no precipitate forms under these conditions, no members of ion group 2 are present. The tube is centrifuged and the solution decanted.

Ion group 3: Base-insoluble sulfides and hydroxides. The decanted solution is made slightly basic with a buffer of NH_3/NH_4^+. The OH^- present increases the $[HS^-]$, which causes precipitation of the more soluble sulfides and some hydroxides. The cations included are

$$Zn^{2+}, Mn^{2+}, Ni^{2+}, Fe^{2+}, Co^{2+} \text{ as sulfides, and } Al^{3+}, Cr^{3+} \text{ as hydroxides}$$

(Fe^{3+} is reduced to Fe^{2+} in this step.) If no precipitate appears, none of these ions is present. Centrifuging and decanting gives the next solution.

Ion group 4: Insoluble phosphates. To the slightly basic solution, $(NH_4)_2HPO_4$ is added, which precipitates any alkaline earth ions as phosphates. Alternatively, Na_2CO_3 is added, and the precipitate contains alkaline earth carbonates. In either case, any of the following ions are precipitated:

$$Mg^{2+}, Ca^{2+}, Ba^{2+}$$

If no precipitate appears, none of these ions is present.

Ion group 5: Alkali metal and ammonium ions. After centrifuging and decanting, the final solution contains any of the following ions:

$$Na^+, K^+, NH_4^+$$

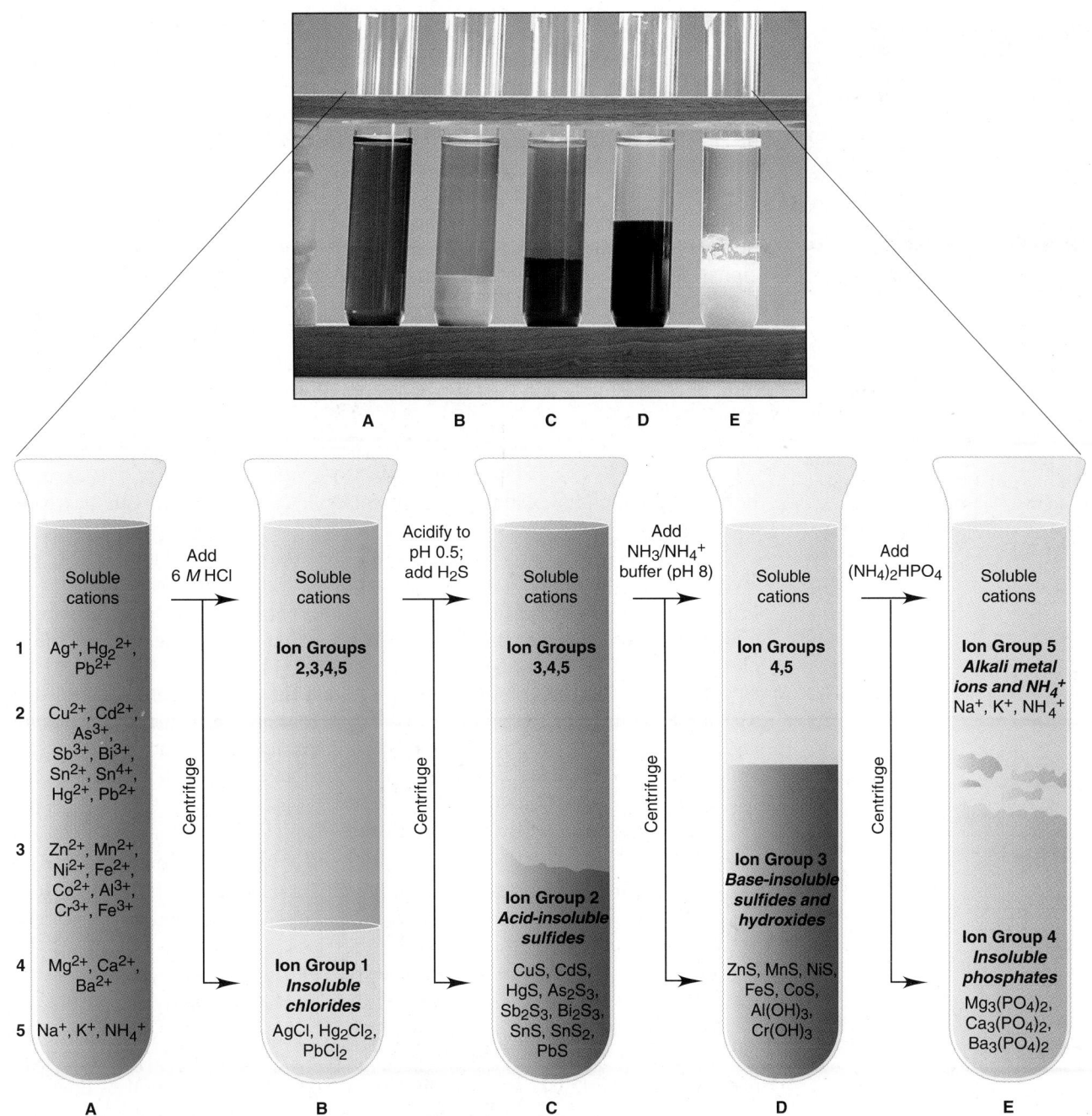

Figure 19.17 A qualitative analysis scheme for separating cations into five ion groups. The first tube contains a solution of ions (listed according to ion group). It is treated with the first precipitating solution (6 *M* HCl) and the procedure continues, as shown in Figure 19.16.

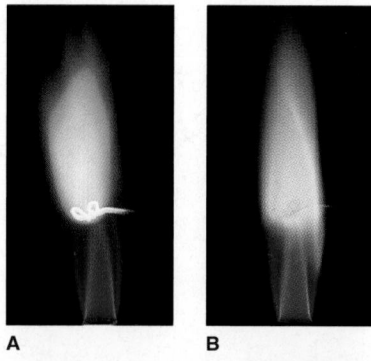

A   **B**

C

Figure 19.18 Tests to determine the presence of cations in ion group 5. Ion group 5 is soluble through all the steps in Figure 19.17. It consists of some alkali metal ions and NH_4^+ ion. Flame tests give characteristic colors for Na^+ ion **(A)** and K^+ ion **(B)**. Adding OH^- to NH_4^+ forms gaseous NH_3, which turns moistened red litmus paper blue **(C)**.

With the ions separated into ion groups, the analyst then devises schemes to identify each ion in a group. For example, identification of the ions in ion group 5 is usually done through flame and color tests (Figure 19.18). (Because NH_4^+ ion is added during earlier steps and Na^+ is a common contaminant in ammonium salts, the analyst performs tests for these ions on the original ion mixture.) In flame tests, sodium produces a characteristic yellow-orange color, and potassium gives a violet color. Acid-base behavior is used to identify NH_4^+. The solution is made basic with NaOH and moist red litmus paper is held over it. If the paper turns blue, NH_4^+ is present because the OH^- reacts to form NH_3, which reacts with the H_2O on the paper:

$$NH_4^+(aq) + OH^-(aq; \text{ added}) \longrightarrow NH_3(g) + H_2O(l)$$
$$NH_3(g) + H_2O \rightleftharpoons NH_4^+(aq) + OH^- \text{ (turns moist red litmus blue)}$$

A Simple "Qual" Scheme Qualitative analyses often include the formation of complex ions in addition to precipitates. Figure 19.19 diagrams a "qual" scheme to identify four ions from several ion groups. Suppose the solution contains some or all of the following: Ag^+, Al^{3+}, Cu^{2+}, and Fe^{3+}. We can separate and identify them by the following steps:

Step 1. Al^{3+} and Fe^{3+} form insoluble hydroxides, and Ag^+ and Cu^{2+} form soluble complex ions with NH_3, so we add aqueous NH_3, which provides both the NH_3 molecules that act as ligands and a small amount of OH^- ions. A precipitate forms that may be $Fe(OH)_3$ (brown), $Al(OH)_3$ (white), or both. The reactions written with the actual formulas of the species are

$$Al(H_2O)_6^{3+}(aq) + 3NH_3(aq) \longrightarrow Al(H_2O)_3(OH)_3(s) \text{ [white]} + 3NH_4^+(aq)$$
$$Fe(H_2O)_6^{3+}(aq) + 3NH_3(aq) \longrightarrow Fe(H_2O)_3(OH)_3(s) \text{ [brown]} + 3NH_4^+(aq)$$

Simplified versions, with the OH^- ions coming from the reaction of NH_3 in water, are

$$Al^{3+}(aq) + 3OH^-(aq) \rightleftharpoons Al(OH)_3(s) \text{ [white]}$$
$$Fe^{3+}(aq) + 3OH^-(aq) \rightleftharpoons Fe(OH)_3(s) \text{ [brown]}$$

In the solution, complex ions of Ag^+(colorless), Cu^{2+}(blue), or both may be present:

$$Ag^+(aq) + 2NH_3(aq) \rightleftharpoons Ag(NH_3)_2^+(aq) \text{ [colorless]}$$
$$Cu^{2+}(aq) + 4NH_3(aq) \rightleftharpoons Cu(NH_3)_4^{2+}(aq) \text{ [blue]}$$

(The Cu^{2+} also forms an insoluble hydroxide, but the high $[NH_3]$ dissolves the precipitate and forms the complex ion.) We centrifuge to separate the soluble complex ions from the precipitate.

Step 2. To separate the Ag^+ and Cu^{2+} that may be present as complex ions, we take advantage of the ability of Cl^- to act as both a ligand and a precipitating ion. We add HCl to the solution of complex ions and centrifuge. The H_3O^+ ions react with the NH_3 in equilibrium with the complex ions to form NH_4^+, which cannot act as a ligand:

$$Ag(NH_3)_2^+(aq) + 2H_3O^+(aq; \text{ added}) \longrightarrow Ag^+(aq) + 2NH_4^+(aq) + 2H_2O(l)$$
$$Cu(NH_3)_4^{2+}(aq) + 4H_3O^+(aq; \text{ added}) \longrightarrow Cu^{2+}(aq) + 4NH_4^+(aq) + 4H_2O(l)$$

Now that the metal ions are hydrated, the Cl^- ion present in the added HCl can react with them to form a white precipitate with $Ag^+(aq)$ and a green complex ion with $Cu^{2+}(aq)$:

$$Ag^+(aq) + Cl^-(aq; \text{ added}) \longrightarrow AgCl(s) \text{ [white]}$$
$$Cu^{2+}(aq) + 4Cl^-(aq; \text{ added}) \longrightarrow CuCl_4^{2-}(aq) \text{ [green]}$$

Step 3. To separate the aluminum and iron compounds that may be in the precipitate of step 1, recall that $Al(OH)_3$ is amphoteric, but $Fe(OH)_3$ is not. We add NaOH to the solid. The amphoteric $Al(OH)_3$ dissolves, but $Fe(OH)_3$ does not:

$$Al(OH)_3(s) + OH^-(aq; \text{ added}) \longrightarrow Al(OH)_4^-(aq)$$
$$Fe(OH)_3(s) + OH^-(aq; \text{ added}) \longrightarrow \text{ no reaction}$$

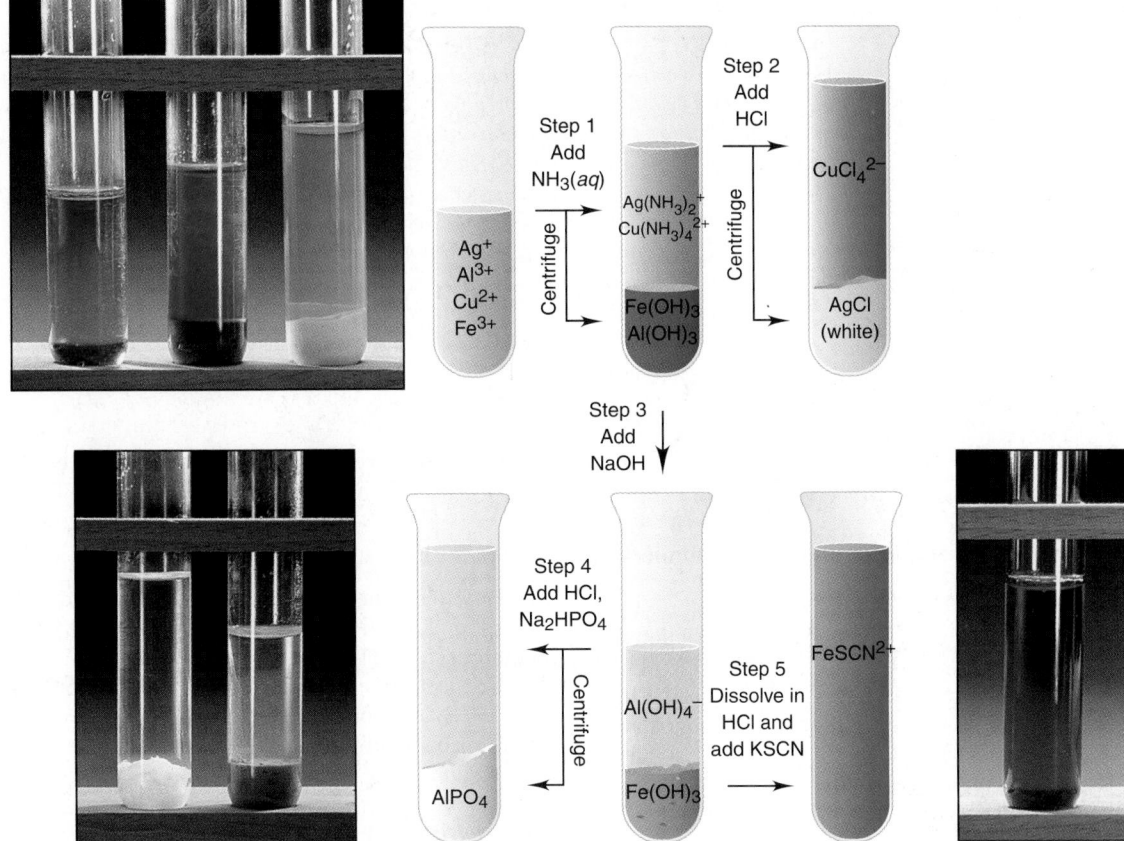

Figure 19.19 **A qualitative analysis scheme for Ag^+, Al^{3+}, Cu^{2+}, and Fe^{3+}.**

Step 4. To confirm the presence of Al^{3+} in the solution of step 3, we decant the solution and make it somewhat acidic with HCl, which first precipitates the Al^{3+} ion as the hydroxide and then converts it to hydrated Al^{3+}. Then we add Na_2HPO_4, which forms a white precipitate of $AlPO_4$:

$$Al(OH)_4^-(aq) \xrightarrow{H_3O^+} Al(OH)_3(s) \xrightarrow{H_3O^+} Al^{3+}(aq)$$
$$Al^{3+}(aq) + HPO_4^{2-}(aq; \text{ added}) + H_2O(l) \longrightarrow H_3O^+(aq) + AlPO_4(s) \text{ [white]}$$

Step 5. To confirm that the remaining solid in step 3 is, in fact, $Fe(OH)_3$, we dissolve it in dilute HCl and add potassium thiocyanate solution (KSCN), which forms a red complex ion with Fe^{3+}:

$$Fe(OH)_3(s) + 3H_3O^+(aq) \longrightarrow Fe^{3+}(aq) + 6H_2O(l)$$
$$Fe^{3+}(aq) + SCN^-(aq; \text{ added}) \longrightarrow FeSCN^{2+}(aq) \text{ [red]}$$

Thus, each ion in the original mixture is separated, identified, and confirmed.

SECTION SUMMARY

Ions are precipitated selectively by adding a precipitating ion until the K_{sp} of one compound is exceeded as much as possible without exceeding the K_{sp} of the other. An extension of this approach is to control the equilibrium of the slightly soluble compound by simultaneously controlling an equilibrium system that contains the precipitating ion. Qualitative analysis of ion mixtures involves adding precipitating ions to separate the unknown ions into ion groups. The groups are then analyzed further through precipitation and complex ion formation.

Chapter Perspective

This chapter is the last of three that explore the nature and variety of equilibrium systems. In Chapter 17, we discussed the central ideas of equilibrium in the context of gaseous systems. In Chapter 18, we extended our understanding to acid-base equilibria. In this chapter, we highlighted three types of aqueous ionic systems and examined their role in the laboratory and the environment. The equilibrium constant, in all its forms, is a number that provides a limit to changes in a system, whether a chemical reaction, a physical change, or the dissolution of a substance. You now have the skills to predict whether a change will take place and to calculate its result, but you still do not know why the change occurs in the first place, or why it stops when it does. In Chapter 20, you'll find out.

For Review and Reference (Numbers in parentheses refer to pages, unless noted otherwise.)

Learning Objectives

Relevant section and/or sample problem (SP) numbers appear in parentheses.

Understand These Concepts

1. How the presence of a common ion suppresses a reaction that forms it (Section 19.1)
2. Why the concentrations of buffer components must be high to minimize the change in pH from addition of small amounts of H_3O^+ or OH^- (Section 19.1)
3. How buffer capacity depends on buffer concentration and on the pK_a of the acid component; why buffer range is within ± 1 pH unit of the pK_a (Section 19.1)
4. The nature of an acid-base indicator as a conjugate acid-base pair with differently colored acidic and basic forms (Section 19.2)
5. The distinction between equivalence point and end point in an acid-base titration (Section 19.2)
6. Why the shapes of strong acid–strong base, weak acid–strong base, and strong acid–weak base titration curves differ (Section 19.2)
7. How the pH at the equivalence point is determined by the species present; why the pH at the midpoint of the buffer region equals the pK_a of the acid (Section 19.2)
8. How the titration curve of a polyprotic acid has a buffer region and equivalence point for each ionizable proton (Section 19.2)
9. How a slightly soluble ionic compound reaches equilibrium in water, expressed by an equilibrium (solubility-product) constant K_{sp} (Section 19.3)
10. Why incomplete dissociation of an ionic compound leads to approximate calculated values for K_{sp} and solubility (Section 19.3)
11. Why a common ion in a solution decreases the solubility of its compounds (Section 19.3)
12. How pH affects the solubility of a compound that contains a weak-acid anion (Section 19.3)
13. How precipitate formation depends on the relative values of Q_{sp} and K_{sp} (Section 19.3)
14. How complex-ion formation is stepwise and expressed by an overall equilibrium (formation) constant K_f (Section 19.4)
15. Why addition of a ligand increases the solubility of a compound whose metal ion forms a complex ion (Section 19.4)

16. How the aqueous chemistry of amphoteric hydroxides involves precipitation, complex-ion formation, and acid-base equilibria (Section 19.4)
17. How selective precipitation and simultaneous equilibria are used to separate ions (Section 19.5)
18. How qualitative analysis is used to separate and identify the ions in a mixture (Section 19.5)

Master These Skills

1. Using stoichiometry and equilibrium problem-solving techniques to calculate the effect of added H_3O^+ and OH^- on buffer pH (SP 19.1)
2. Using the Henderson-Hasselbalch equation to calculate buffer pH (Section 19.1)
3. Choosing the components of a buffer of a given pH and calculating their quantities (SP 19.2)
4. Calculating the pH at any point in an acid-base titration (Section 19.2 and SP 19.3)
5. Choosing an appropriate indicator based on the pH at various points in a titration (Section 19.2)
6. Writing K_{sp} expressions for slightly soluble ionic compounds (SP 19.4)
7. Calculating a K_{sp} value from solubility data (SP 19.5)
8. Calculating solubility from a K_{sp} value (SP 19.6)
9. Using K_{sp} values to compare solubilities for compounds with the same total number of ions (Section 19.3)
10. Calculating the decrease in solubility caused by the presence of a common ion (SP 19.7)
11. Predicting the effect of added H_3O^+ on solubility (SP 19.8)
12. Using ion concentrations to calculate Q_{sp} and compare it with K_{sp} to predict whether a precipitate forms (SP 19.9)
13. Calculating the concentration of metal ion remaining after addition of excess ligand forms a complex ion (SP 19.10)
14. Using an overall equilibrium constant ($K_{sp} \times K_f$) to calculate the effect of complex-ion formation on solubility (SP 19.11)
15. Comparing K_{sp} values in order to separate ions by selective precipitation (SP 19.12)
16. Identifying the ions in a mixture through sequences of precipitation and complex-ion formation (Section 19.5)

Key Terms

Section 19.1
acid-base buffer (806)
common-ion effect (806)
Henderson-Hasselbalch
equation (811)

buffer capacity (812)
buffer range (812)

Section 19.2
acid-base titration curve (815)
equivalence point (816)
end point (816)

Section 19.3
solubility-product constant
(K_{sp}) (824)
molar solubility (826)

Section 19.4
complex ion (835)

ligand (835)
formation constant (K_f) (835)

Section 19.5
selective precipitation (840)
qualitative analysis (841)

Key Equations and Relationships

19.1 Finding the pH from known concentrations of a conjugate acid-base pair (Henderson-Hasselbalch equation) (811):

$$pH = pK_a + \log\left(\frac{[base]}{[acid]}\right)$$

19.2 Defining the equilibrium condition for a saturated solution of a slightly soluble compound, M_pX_q, composed of M^{n+} and X^{z-} ions (825):

$$Q_{sp} = [M^{n+}]^p[X^{z-}]^q = K_{sp}$$

Highlighted Figures and Tables

These figures (F) and tables (T) provide a quick review of key ideas. Entries in color contain frequently used data.

F19.2 How a buffer works (808)
F19.3 Buffer capacity and pH change (812)
F19.4 Colors and pH ranges of acid-base indicators (815)
F19.6 A strong acid–strong base titration (817)

F19.7 A weak acid–strong base titration (819)
F19.8 A weak base–strong acid titration (821)
T19.4 K_f of some complex ions at 25°C (836)

Brief Solutions to Follow-up Problems

19.1 (a) Before addition:
Assuming x is small enough to be neglected,

[HF] = 0.50 M and   [F$^-$] = 0.45 M

$$[H_3O^+] = K_a \times \frac{[HF]}{[F^-]} \approx (6.8\times10^{-4})\left(\frac{0.50}{0.45}\right) = 7.6\times10^{-4}\ M$$

pH = 3.12

(b) After addition of 0.40 g of NaOH (0.010 mol of NaOH) to 1.0 L of buffer,

[HF] = 0.49 M and   [F$^-$] = 0.46 M

$$[H_3O^+] \approx (6.8\times10^{-4})\left(\frac{0.49}{0.46}\right) = 7.2\times10^{-4}\ M;\ pH = 3.14$$

19.2 $[H_3O^+] = 10^{-pH} = 10^{-4.25} = 5.6\times10^{-5}$

$$[C_6H_5COOH] = \frac{[H_3O^+][C_6H_5COO^-]}{K_a}$$

$$= \frac{(5.6\times10^{-5})(0.050)}{6.3\times10^{-5}} = 0.044\ M$$

Mass (g) of C_6H_5COOH

$$= 5.0\ L\ soln \times \frac{0.044\ mol\ C_6H_5COOH}{1\ L\ soln}$$

$$\times \frac{122.12\ g\ C_6H_5COOH}{1\ mol\ C_6H_5COOH}$$

$$= 27\ g\ C_6H_5COOH$$

Dissolve 27 g of C_6H_5COOH in 4.9 L of 0.050 M C_6H_5COONa and add solution to make 5.0 L. Adjust pH to 4.25 with strong acid or base.

19.3 (a) $[H_3O^+] \approx \sqrt{(2.3\times10^{-9})(0.2000)}$
$$= 2.1\times10^{-5}\ M;\ pH = 4.68$$

(b) $[H_3O^+] = K_a \times \frac{[HBrO]}{[BrO^-]} = (2.3\times10^{-9})(1) = 2.3\times10^{-9}\ M$

pH = 8.64

(c) $[BrO^-] = \dfrac{moles\ of\ BrO^-}{total\ volume} = \dfrac{0.004000\ mol}{0.06000\ L} = 0.06667\ M$

$$K_b\ of\ BrO^- = \frac{K_w}{K_a\ of\ HBrO} = 4.3\times10^{-6}$$

$$[H_3O^+] = \frac{K_w}{\sqrt{K_b \times [BrO^-]}}$$

$$\approx \frac{1.0\times10^{-14}}{\sqrt{(4.3\times10^{-6})(0.06667)}} = 1.9\times10^{-11}\ M$$

pH = 10.72

(d) Moles of OH$^-$ added = 0.008000 mol
Volume (L) of OH$^-$ soln = 0.08000 L

$$[OH^-] = \frac{moles\ of\ OH^-\ unreacted}{total\ volume}$$

$$= \frac{0.008000\ mol - 0.004000\ mol}{(0.02000 + 0.08000)\ L} = 0.04000\ M$$

$$[H_3O^+] = \frac{K_w}{[OH^-]} = 2.5\times10^{-13}$$

pH = 12.60

Brief Solutions to Follow-up Problems (continued)

(e)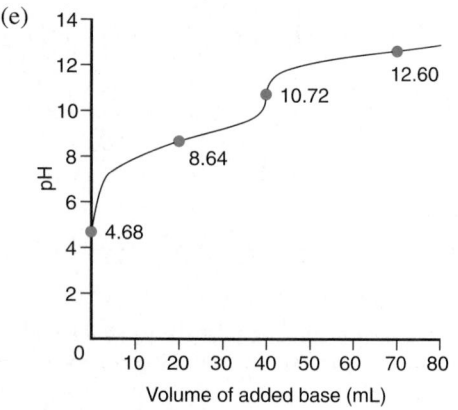

19.4 (a) $K_{sp} = [Ca^{2+}][SO_4^{2-}]$
(b) $K_{sp} = [Cr^{3+}]^2[CO_3^{2-}]^3$
(c) $K_{sp} = [Mg^{2+}][OH^-]^2$
(d) $K_{sp} = [As^{3+}]^2[HS^-]^3[OH^-]^3$

19.5 $[CaF_2] = \dfrac{1.5\times10^{-4} \text{ g } CaF_2}{10.0 \text{ mL soln}}\times\dfrac{1000 \text{ mL}}{1 \text{ L}}\times\dfrac{1 \text{ mol } CaF_2}{78.08 \text{ g } CaF_2}$

$= 1.9\times10^{-4} M$

$$CaF_2(s) \rightleftharpoons Ca^{2+}(aq) + 2F^-(aq)$$

$[Ca^{2+}] = 1.9\times10^{-4} M$ and $[F^-] = 3.8\times10^{-4} M$

$K_{sp} = [Ca^{2+}][F^-]^2 = (1.9\times10^{-4})(3.8\times10^{-4})^2 = 2.7\times10^{-11}$

19.6 From the reaction table, $[Mg^{2+}] = S$ and $[OH^-] = 2S$

$K_{sp} = [Mg^{2+}][OH^-]^2 = 4S^3 = 6.3\times10^{-10}$

$S = 5.4\times10^{-4} M$

19.7 (a) In water: $K_{sp} = [Ba^{2+}][SO_4^{2-}] = S^2 = 1.1\times10^{-10}$;
$S = 1.0\times10^{-5}$

(b) In 0.10 M Na_2SO_4: $[SO_4^{2-}] = 0.10 M$

$K_{sp} = 1.1\times10^{-10} \approx S \times 0.10; \qquad S = 1.1\times10^{-9} M$

S decreases in presence of the common ion, SO_4^{2-}.

19.8 (a) Increases solubility.
$$CaF_2(s) \rightleftharpoons Ca^{2+}(aq) + 2F^-(aq)$$
$$F^-(aq) + H_3O^+(aq) \longrightarrow HF(aq) + H_2O(l)$$
(b) Increases solubility.
$$ZnS(s) + H_2O(l) \rightleftharpoons Zn^{2+}(aq) + HS^-(aq) + OH^-(aq)$$
$$HS^-(aq) + H_3O^+(aq) \longrightarrow H_2S(aq) + H_2O(l)$$
$$OH^-(aq) + H_3O^+(aq) \longrightarrow 2H_2O(l)$$
(c) No effect. $I^-(aq)$ is conjugate base of strong acid, HI.

19.9 $Ca_3(PO_4)_2(s) \rightleftharpoons 3Ca^{2+}(aq) + 2PO_4^{3-}(aq)$

$Q_{sp} = [Ca^{2+}]^3[PO_4^{3-}]^2 = (1.0\times10^{-9})^5 = 1.0\times10^{-45}$

$Q_{sp} < K_{sp}$, so $Ca_3(PO_4)_2$ will not precipitate.

19.10 $[Fe(H_2O)_6^{3+}]_{init} = \dfrac{(0.0255 \text{ L})(3.1\times10^{-2} M)}{0.0255 \text{ L} + 0.0350 \text{ L}}$

$= 1.3\times10^{-2} M$

Similarly, $[CN^-]_{init} = 0.87 M$. From the reaction table,

$K_f = \dfrac{[Fe(CN)_6^{3-}]}{[Fe(H_2O)_6^{3+}][CN^-]^6} = 4.0\times10^{43} \approx \dfrac{1.3\times10^{-2}}{x(0.79)^6}$

$x = [Fe(H_2O)_6^{3+}] \approx 1.3\times10^{-45}$

19.11 $AgBr(s) + 2NH_3(aq) \rightleftharpoons Ag(NH_3)_2^+(aq) + Br^-(aq)$

$K_{overall} = K_{sp} \text{ of AgBr} \times K_f \text{ of } Ag(NH_3)_2^+$

$= 8.5\times10^{-6}$

From the reaction table,

$\dfrac{S}{1.0 - 2S} = \sqrt{8.5\times10^{-6}} = 2.9\times10^{-3}$

$S = [Ag(NH_3)_2^+] = 2.9\times10^{-3} M$

Solubility in 1 M hypo is greater than in 1 M NH_3.

19.12 Both 1:1 salts, so K_{sp} values show that $CaSO_4$ is more soluble:

$[SO_4^{2-}] = \dfrac{K_{sp}}{[Ca^{2+}]} = \dfrac{2.4\times10^{-5}}{(0.025)(0.9999)}$

$= 9.6\times10^{-4} M$

Problems

Problems with **colored** numbers are answered at the back of the text. Sections match the text and provide the number(s) of relevant sample problems. Most offer Concept Review Questions, Skill-Building Exercises (in similar pairs), and Problems in Context. Then Comprehensive Problems, based on material from any section or previous chapter, follow.

Note: Unless specifically stated otherwise, all of the problems for this chapter refer to aqueous solutions at 298 K (25°C).

Equilibria of Acid-Base Buffer Systems
(Sample Problems 19.1 and 19.2)

● **Concept Review Questions**

19.1 What is the purpose of an acid-base buffer?
19.2 How do the acid and base components of a buffer function? Why are the components typically a conjugate acid-base pair of a weak acid?

19.3 What is the common-ion effect? How is it related to Le Châtelier's principle? Explain with equations that include HF and NaF.

19.4 When a small amount of H_3O^+ is added to a buffer, does the pH remain constant? Explain.

19.5 What is the difference between buffers with high and low capacities? Will adding 0.01 mol of HCl produce a greater pH change in a buffer with a high or a low capacity? Explain.

19.6 Choose the factors that determine the capacity of a buffer from among the following and explain your choices:
(a) Conjugate acid-base pair
(b) pH of the buffer
(c) Concentration of buffer components
(d) Buffer range
(e) Buffer-component concentration ratio
(f) pK_a of the acid component
19.7 What is the relationship between buffer range and buffer-component concentration ratio?

19.8 A chemist needs a pH 3.5 buffer. Should she use NaOH with formic acid ($K_a = 1.8 \times 10^{-4}$) or with acetic acid ($K_a = 1.8 \times 10^{-5}$)? Why? What is the disadvantage of choosing the other acid? What is the role of the NaOH?

19.9 Explain the increase or decrease in the pH of a buffer and in the buffer-component concentration ratio, [NaA]/[HA], for each of the following cases:
(a) Add 0.1 M NaOH to the buffer
(b) Add 0.1 M HCl to the buffer
(c) Dissolve pure NaA in the buffer
(d) Dissolve pure HA in the buffer

19.10 Does the pH increase or decrease, and does it do so to a large or small extent, with each of the following additions:
(a) 5 drops of 0.1 M NaOH to 100 mL of 0.5 M acetate buffer
(b) 5 drops of 0.1 M HCl to 100 mL of 0.5 M acetate buffer
(c) 5 drops of 0.1 M NaOH to 100 mL of 0.5 M HCl
(d) 5 drops of 0.1 M NaOH to distilled water

● **Skill-Building Exercises** *(paired)*

19.11 What are the [H_3O^+] and the pH of a propanoate buffer that consists of 0.25 M CH_3CH_2COONa and 0.15 M CH_3CH_2COOH (K_a of propanoic acid $= 1.3 \times 10^{-5}$)?

19.12 What are the [H_3O^+] and the pH of a benzoate buffer that consists of 0.33 M C_6H_5COOH and 0.28 M C_6H_5COONa (K_a of benzoic acid $= 6.3 \times 10^{-5}$)?

19.13 What are the [H_3O^+] and the pH of a buffer that consists of 0.50 M HNO_2 and 0.65 M KNO_2 (K_a of $HNO_2 = 7.1 \times 10^{-4}$)?

19.14 What are the [H_3O^+] and the pH of a buffer that consists of 0.20 M HF and 0.25 M KF (K_a of HF $= 6.8 \times 10^{-4}$)?

19.15 What is the pH of a buffer that consists of 0.55 M HCOOH and 0.63 M HCOONa (pK_a of HCOOH $= 3.74$)?

19.16 What is the pH of a buffer that consists of 0.95 M HBrO and 0.68 M KBrO (pK_a of HBrO $= 8.64$)?

19.17 What is the pH of a buffer that consists of 1.0 M sodium phenolate (C_6H_5ONa) and 1.2 M phenol (C_6H_5OH) (pK_a of phenol $= 10.00$)?

19.18 What is the pH of a buffer that consists of 0.12 M boric acid (H_3BO_3) and 0.82 M sodium borate (NaH_2BO_3) (pK_a of boric acid $= 9.24$)?

19.19 What is the pH of a buffer consisting of 0.20 M NH_3 and 0.10 M NH_4Cl (pK_b of $NH_3 = 4.75$)?

19.20 What is the pH of a buffer consisting of 0.50 M methylamine (CH_3NH_2) and 0.60 M CH_3NH_3Cl (pK_b of $CH_3NH_2 = 3.35$)?

19.21 A buffer consists of 0.25 M $KHCO_3$ and 0.32 M K_2CO_3. Carbonic acid is a diprotic acid with $K_{a1} = 4.5 \times 10^{-7}$ and $K_{a2} = 4.7 \times 10^{-11}$. (a) Which K_a value is more important to this buffer? (b) What is the buffer pH?

19.22 A buffer consists of 0.50 M NaH_2PO_4 and 0.40 M Na_2HPO_4. Phosphoric acid is a triprotic acid ($K_{a1} = 7.2 \times 10^{-3}$, $K_{a2} = 6.3 \times 10^{-8}$, and $K_{a3} = 4.2 \times 10^{-13}$). (a) Which K_a value is most important to this buffer? (b) What is the buffer pH?

19.23 What is the buffer-component concentration ratio, [Pr^-]/[HPr], of a buffer that has a pH of 5.11? See Problem 19.11 for K_a of HPr.

19.24 What is the buffer-component concentration ratio, [NO_2^-]/[HNO_2], of a buffer that has a pH of 2.95? See Problem 19.13 for K_a of HNO_2.

19.25 What is the buffer-component concentration ratio, [BrO^-]/[HBrO], of a buffer that has a pH of 7.88 (K_a of HBrO $= 2.3 \times 10^{-9}$)?

19.26 What is the buffer-component concentration ratio, [CH_3COO^-]/[CH_3COOH], of a buffer that has a pH of 4.39? See Problem 19.8 for K_a of CH_3COOH.

19.27 A buffer containing 0.2000 M of acid, HA, and 0.1500 M of its conjugate base, A^-, has a pH of 3.35. What is the pH after 0.0015 mol of NaOH is added to 0.5000 L of this solution?

19.28 A buffer that contains 0.40 M base, B, and 0.25 M of its conjugate acid, BH^+, has a pH of 8.88. What is the pH after 0.0020 mol of HCl is added to 0.25 L of this solution?

19.29 A buffer that contains 0.110 M HY and 0.220 M Y^- has a pH of 8.77. What is the pH after 0.0010 mol of $Ba(OH)_2$ is added to 0.750 L of this solution?

19.30 A buffer that contains 1.05 M B and 0.750 M BH^+ has a pH of 9.50. What is the pH after 0.0050 mol of HCl is added to 0.500 L of this solution?

19.31 A buffer is prepared by mixing 184 mL of 0.442 M HCl and 0.500 L of 0.400 M sodium acetate. (See Table 18.2.) (a) What is the pH? (b) How many grams of KOH must be added to 0.500 L of the buffer to change the pH by 0.15 units?

19.32 A buffer is prepared by mixing 50.0 mL of 0.050 M sodium bicarbonate and 10.7 mL of 0.10 M NaOH. (See Table 18.5.) (a) What is the pH? (b) How many grams of HCl must be added to 25.0 mL of the buffer to change the pH by 0.07 units?

19.33 Choose specific acid-base conjugate pairs from Tables 18.2, 18.5, and 18.6 that are suitable for preparing the following buffers: (a) pH ≈ 4.0; (b) pH ≈ 7.0.

19.34 Choose specific acid-base conjugate pairs from Tables 18.2, 18.5, and 18.6 that are suitable for preparing the following buffers: (a) [H_3O^+] $\approx 1 \times 10^{-9}$ M; (b) [OH^-] $\approx 3 \times 10^{-5}$ M.

19.35 Choose specific acid-base conjugate pairs from Tables 18.2, 18.5, and 18.6 that are suitable for preparing the following buffers: (a) pH ≈ 2.5; (b) pH ≈ 5.5.

19.36 Choose specific acid-base conjugate pairs from Tables 18.2, 18.5, and 18.6 that are suitable for preparing the following buffers: (a) [OH^-] $\approx 1 \times 10^{-6}$ M; (b) [H_3O^+] $\approx 4 \times 10^{-4}$ M.

● **Problems in Context**

19.37 An industrial chemist studying the effect of pH on bleaching and sterilizing processes prepares several hypochlorite buffers. Calculate the pH of the following buffers:
(a) 0.100 M HClO and 0.100 M NaClO
(b) 0.100 M HClO and 0.150 M NaClO
(c) 0.150 M HClO and 0.100 M NaClO
(d) One liter of the solution in part (a) after 0.0050 mol of NaOH has been added.

19.38 Ions of P are key buffer components in blood. For a Na_2HPO_4/KH_2PO_4 solution with pH 7.40 (average pH of normal arterial blood), what is the buffer-component concentration ratio?

Acid-Base Titration Curves
(Sample Problem 19.3)

● **Concept Review Questions**

19.39 Acid-base indicators are weak acids. What information must you have to estimate the range of an indicator's color change? Why do some indicators have two separate pH ranges?

19.40 Why does the color change of an indicator take place over a range of about 2 pH units?

19.41 Why doesn't the addition of an acid-base indicator affect the pH of the test solution?

19.42 What is the difference between the end point of a titration and the equivalence point? Is the equivalence point always reached first? Explain.

19.43 Some automatic titrators measure the slope of the titration curve to determine when the equivalence point is reached. What happens to the slope of the titration curve that enables the instrument to recognize this point?

19.44 Explain how *strong acid*–strong base, *weak acid*–strong base, and *weak base*–strong acid titrations using the same concentrations differ in terms of (a) the initial pH and (b) the pH at the equivalence point. (The component in italics is in the flask.)

19.45 What species are present in the buffer region of a weak acid–strong base titration? How are they different from the species present at the equivalence point? Are they different from the species present in the buffer region of a weak base–strong acid titration? Explain.

19.46 Why is the center of the buffer region of a weak acid–strong base titration significant?

19.47 How does the titration curve of a monoprotic acid differ from that of a diprotic acid?

● **Skill-Building Exercises (paired)**

19.48 The indicator cresol red has $K_a = 5.0 \times 10^{-9}$. Over what approximate pH range does it change color?

19.49 The indicator thymolphthalein has $K_a = 7.9 \times 10^{-11}$. Over what approximate pH range does it change color?

19.50 Use Figure 19.4 to find an indicator for these titrations:
(a) 0.10 M HCl with 0.10 M NaOH
(b) 0.10 M HCOOH (Table 18.2) with 0.10 M NaOH

19.51 Use Figure 19.4 to find an indicator for these titrations:
(a) 0.10 M CH_3NH_2 (Table 18.6) with 0.10 M HCl
(b) 0.50 M HI with 0.10 M KOH

19.52 Use Figure 19.4 to find an indicator for these titrations:
(a) 0.5 M $(CH_3)_2NH$ (Table 18.6) with 0.5 M HBr
(b) 0.2 M KOH with 0.2 M HNO_3

19.53 Use Figure 19.4 to find an indicator for these titrations:
(a) 0.25 M C_6H_5COOH (Table 18.2) with 0.25 M KOH
(b) 0.50 M NH_4Cl (Table 18.6) with 0.50 M NaOH .

19.54 Calculate the pH during the titration of 50.00 mL of 0.1000 M HCl with 0.1000 M NaOH solution after the following additions of base: (a) 0 mL; (b) 25.00 mL; (c) 49.00 mL; (d) 49.90 mL; (e) 50.00 mL; (f) 50.10 mL; (g) 60.00 mL.

19.55 Calculate the pH during the titration of 30.00 mL of 0.1000 M KOH with 0.1000 M HBr solution after the following additions of acid: (a) 0 mL; (b) 15.00 mL; (c) 29.00 mL; (d) 29.90 mL; (e) 30.00 mL; (f) 30.10 mL; (g) 40.00 mL.

19.56 What is the pH during the titration of 20.00 mL of 0.1000 M butanoic acid, $CH_3CH_2CH_2COOH$ ($K_a = 1.54 \times 10^{-5}$), with 0.1000 M NaOH solution after the following additions of titrant: (a) 0 mL; (b) 10.00 mL; (c) 15.00 mL; (d) 19.00 mL; (e) 19.95 mL; (f) 20.00 mL; (g) 20.05 mL; (h) 25.00 mL.

19.57 What is the pH during the titration of 20.00 mL of 0.1000 M triethylamine, $(CH_3CH_2)_3N$ ($K_b = 5.2 \times 10^{-4}$), with 0.1000 M HCl solution after the following additions of titrant:

(a) 0 mL; (b) 10.00 mL; (c) 15.00 mL; (d) 19.00 mL; (e) 19.95 mL; (f) 20.00 mL; (g) 20.05 mL; (h) 25.00 mL.

19.58 Using Tables 18.2 and 18.5, find the pH and volume (mL) of 0.0372 M NaOH needed to reach the equivalence point(s) in titrations of
(a) 42.2 mL of 0.0520 M CH_3COOH
(b) 18.9 mL of 0.0890 M H_2SO_3 (two equivalence points)

19.59 Using Tables 18.2 and 18.5, find the pH and volume (mL) of 0.0588 M KOH needed to reach the equivalence point(s) in titrations of
(a) 23.4 mL of 0.0390 M HNO_2
(b) 17.3 mL of 0.130 M H_2CO_3 (two equivalence points)

19.60 Find the pH and the volume (mL) of 0.135 M HCl needed to reach the equivalence point(s) in titrations of
(a) 55.5 mL of 0.234 M NH_3
(b) 17.8 mL of 1.11 M CH_3NH_2

19.61 Find the pH and volume (mL) of 0.447 M HNO_3 needed to reach the equivalence point(s) in titrations of
(a) 2.65 L of 0.0750 M pyridine (C_5H_5N)
(b) 0.188 L of 0.250 M NH_3

Equilibria of Slightly Soluble Ionic Compounds
(Sample Problems 19.4 to 19.9)

● **Concept Review Questions**

19.62 The molar solubility of the slightly soluble ionic compound M_2X is 5×10^{-5} M. What is the molarity of each of the two ions? How do you set up the calculation to find K_{sp}? What assumption must you make about the dissociation of M_2X into ions? Why is the calculated K_{sp} higher than the actual value?

19.63 Why does pH affect the solubility of CaF_2 significantly, but not that of $CaCl_2$?

19.64 A list of K_{sp} values like that in Appendix C can be used to compare the solubility of silver chloride directly with that of silver bromide but not with that of silver chromate. Explain.

19.65 In gaseous equilibria, the reverse reaction predominates when $Q_c > K_c$. What occurs in an aqueous ionic solution when $Q_{sp} > K_{sp}$?

● **Skill-Building Exercises (paired)**

19.66 Write the ion-product expressions for (a) silver carbonate; (b) barium fluoride; (c) copper(II) sulfide.

19.67 Write the ion-product expressions for (a) iron(III) hydroxide; (b) barium phosphate; (c) tin(II) sulfide.

19.68 Write the ion-product expressions for (a) calcium chromate; (b) silver cyanide; (c) nickel(II) sulfide.

19.69 Write the ion-product expressions for (a) lead(II) iodide; (b) strontium sulfate; (c) cadmium sulfide.

19.70 The solubility of silver carbonate is 0.032 M at 20°C. Calculate the K_{sp} of silver carbonate.

19.71 The solubility of zinc oxalate is 7.9×10^{-3} M at 18°C. Calculate the K_{sp} of zinc oxalate.

19.72 The solubility of silver dichromate at 15°C is 8.3×10^{-3} g/100 mL solution. Calculate the K_{sp} of silver dichromate.

19.73 The solubility of calcium sulfate at 30°C is 0.209 g/100 mL solution. Calculate the K_{sp} of calcium sulfate.

19.74 Find the molar solubility of $SrCO_3$ ($K_{sp} = 5.4 \times 10^{-10}$) in (a) pure water and (b) 0.13 M $Sr(NO_3)_2$.

19.75 Find the molar solubility of $BaCrO_4$ ($K_{sp} = 2.1\times10^{-10}$) in (a) pure water and (b) 1.5×10^{-3} M Na_2CrO_4.

19.76 Calculate the molar solubility of $Ca(IO_3)_2$ in (a) 0.060 M $Ca(NO_3)_2$ and (b) 0.060 M $NaIO_3$. (See Appendix C.)

19.77 Calculate the molar solubility of Ag_2SO_4 in (a) 0.22 M $AgNO_3$ and (b) 0.22 M Na_2SO_4. (See Appendix C.)

19.78 Which of the following is more soluble in water?
(a) Magnesium hydroxide or nickel(II) hydroxide
(b) Lead(II) sulfide or copper(II) sulfide
(c) Silver sulfate or magnesium fluoride

19.79 Which of the following is more soluble in water?
(a) Strontium sulfate or barium chromate
(b) Calcium carbonate or copper(II) carbonate
(c) Barium iodate or silver chromate

19.80 Which of the following is more soluble in water?
(a) Barium sulfate or calcium sulfate
(b) Calcium phosphate or magnesium phosphate
(c) Silver chloride or lead(II) sulfate

19.81 Which of the following is more soluble in water?
(a) Manganese(II) hydroxide or calcium iodate
(b) Strontium carbonate or cadmium sulfide
(c) Silver cyanide or copper(I) iodide

19.82 Write balanced equations to show whether either of the following compounds has a different solubility with a change in pH: (a) AgCl; (b) $SrCO_3$.

19.83 Write balanced equations to show whether either of the following compounds has a different solubility with a change in pH: (a) CuBr; (b) $Ca_3(PO_4)_2$.

19.84 Write balanced equations to show whether either of the following compounds has a different solubility with a change in pH: (a) $Fe(OH)_2$; (b) CuS.

19.85 Write balanced equations to show whether either of the following compounds has a different solubility with a change in pH: (a) PbI_2; (b) $Hg_2(CN)_2$.

19.86 Does any solid $Cu(OH)_2$ form when 0.075 g of KOH is dissolved in 1.0 L of 1.0×10^{-3} M $Cu(NO_3)_2$?

19.87 Does any solid $PbCl_2$ form when 3.5 mg of NaCl is dissolved in 0.250 L of 0.12 M $Pb(NO_3)_2$?

19.88 Does any solid $Ba(IO_3)_2$ form when 6.5 mg of $BaCl_2$ is dissolved in 500. mL of 0.033 M $NaIO_3$?

19.89 Does any solid Ag_2CrO_4 form when 2.7×10^{-5} g of $AgNO_3$ is dissolved in 15.0 mL of 4.0×10^{-4} M K_2CrO_4?

● **Problems in Context**

19.90 Calcium ion triggers blood clotting, so when blood is donated, the receiving bag contains sodium oxalate solution to precipitate the Ca^{2+} and prevent clotting. A 104-mL sample of blood contains 9.7×10^{-5} g Ca^{2+}/mL. A technologist treats the sample with 100.0 mL of 0.1550 M $Na_2C_2O_4$. Calculate $[Ca^{2+}]$ after the treatment. (See Appendix C for K_{sp} of $CaC_2O_4\cdot H_2O$.)

Equilibria Involving Complex Ions
(Sample Problems 19.10 and 19.11)

● **Concept Review Questions**

19.91 How is it possible for a positively charged metal ion to be at the center of a negatively charged complex ion?

19.92 Write equations to show the stepwise reaction of $Cd(H_2O)_4{}^{2+}$ in aqueous KI to form $CdI_4{}^{2-}$. Show that $K_{f(overall)} = K_{f1}\times K_{f2}\times K_{f3}\times K_{f4}$.

19.93 Consider the dissolution of PbS in water:

$$PbS(s) + H_2O(l) \rightleftharpoons Pb^{2+}(aq) + HS^-(aq) + OH^-(aq)$$

Adding aqueous NaOH causes more PbS to dissolve. Does this violate Le Châtelier's principle? Explain.

● **Skill-Building Exercises (paired)**

19.94 Write a balanced equation for the reaction of $Hg(H_2O)_4{}^{2+}$ in aqueous KCN.

19.95 Write a balanced equation for the reaction of $Zn(H_2O)_4{}^{2+}$ in aqueous NaCN.

19.96 Write a balanced equation for the reaction of $Ag(H_2O)_2{}^+$ in aqueous $Na_2S_2O_3$.

19.97 Write a balanced equation for the reaction of $Al(H_2O)_6{}^{3+}$ in aqueous KF.

19.98 Potassium thiocyanate, KSCN, is often used to determine the presence of Fe^{3+} ions in solution by the formation of the red $Fe(H_2O)_5SCN^{2+}$ (or, more simply, $FeSCN^{2+}$). What is the concentration of Fe^{3+} when 0.50 L each of 0.0015 M $Fe(NO_3)_3$ and 0.20 M KSCN are mixed? K_f of $FeSCN^{2+} = 8.9\times10^2$.

19.99 What is the concentration of Ag^+ when 25.0 mL each of 0.044 M $AgNO_3$ and 0.57 M $Na_2S_2O_3$ are mixed (K_f of $Ag(S_2O_3)_2{}^{3+} = 4.7\times10^{13}$)?

19.100 When 0.82 g of $ZnCl_2$ is dissolved in 255 mL of 0.150 M NaCN, what are $[Zn^{2+}]$, $[Zn(CN)_4{}^{2-}]$, and $[CN^-]$ (K_f of $Zn(CN)_4{}^{2-} = 4.2\times10^{19}$)?

19.101 When 2.4 g of $Co(NO_3)_2$ is dissolved in 0.350 L of 0.22 M KOH, what are $[Co^{2+}]$, $[Co(OH)_4{}^{2-}]$, and $[OH^-]$ (K_f of $Co(OH)_4{}^{2-} = 5\times10^9$)?

19.102 Find the solubility of AgI in 2.5 M NH_3 (K_{sp} of AgI = 8.3×10^{-17}; K_f of $Ag(NH_3)_2{}^+ = 1.7\times10^7$).

19.103 Find the solubility of $Cr(OH)_3$ in a buffer of pH 13.0 (K_{sp} of $Cr(OH)_3 = 6.3\times10^{-31}$; K_f of $Cr(OH)_4{}^- = 8.0\times10^{29}$).

Application of Ionic Equilibria to Chemical Analysis
(Sample Problem 19.12)

● **Problems in Context**

19.104 A 50.0-mL volume of 0.50 M $Fe(NO_3)_3$ is mixed with 125 mL of 0.25 M $Cd(NO_3)_2$.
(a) If aqueous NaOH is added to the mixture, which ion precipitates first? (See Appendix C for the appropriate K_{sp} values.)
(b) Describe how the metal ions can be separated using NaOH.
(c) Calculate the $[OH^-]$ that will accomplish the separation.

19.105 In a qualitative analysis procedure, a chemist adds 0.3 M HCl to a group of ions and then saturates the solution with H_2S.
(a) What is the $[HS^-]$ of the solution?
(b) If 0.01 M of each of the following ions is in the solution, which will form a precipitate: Pb^{2+}, Mn^{2+}, Hg^{2+}, Cu^{2+}, Ni^{2+}, Fe^{2+}, Ag^+, K^+?

19.106 Name a reagent and describe the results that would distinguish between the following pairs of substances:
(a) NH_4Cl and $NiCl_2$
(b) AgCl and $PbCl_2$
(c) $Al(NO_3)_3$ and $Fe(NO_3)_3$

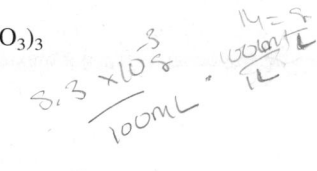

Comprehensive Problems

Problems with an asterisk (*) are more challenging.

19.107 A microbiologist is preparing a medium on which to culture *E. coli* bacteria. She buffers the medium at pH 7.00 to minimize the effect of acid-producing fermentation, which can occur in such cultures. What volumes of equimolar aqueous solutions of K_2HPO_4 and KH_2PO_4 must she combine to make 100. mL of the pH 7.00 buffer?

19.108 As an FDA physiologist, you need 0.600 L of formate buffer with a pH of 3.74.
(a) What is the required buffer-component concentration ratio?
(b) How do you prepare this solution from stock solutions of 1.0 M HCOOH and 1.0 M NaOH? (*Hint:* Mix x mL of HCOOH with $600 - x$ mL of NaOH.)
(c) What is the final concentration of HCOOH in this solution?

19.109 Tris(hydroxymethyl)aminomethane $[(HOCH_2)_3CNH_2$, known as TRIS or THAM] is a weak base widely used in biochemical experiments to make buffer solutions in the pH range of 7 to 9. A certain TRIS buffer has a pH of 8.10 at 25°C and a pH of 7.80 at 37°C. Why does the pH change with temperature?

19.110 Gout is an extremely painful disease caused by an error in nucleic acid metabolism that leads to a buildup of uric acid in body fluids. It is deposited as slightly soluble sodium urate $(C_5H_3N_4O_3Na)$ in the soft tissues of joints. If the extracellular $[Na^+]$ is 0.15 M and the solubility in water of sodium urate is 0.085 g/100. mL, what is the minimum urate ion concentration (abbreviated $[UR^-]$) that will cause a deposit of sodium urate?

19.111 Cadmium ion in solution is analyzed by precipitation as the sulfide, a yellow compound used as a pigment in everything from artists' oil paints to glass and rubber. Calculate the molar solubility of cadmium sulfide at 25°C.

19.112 To achieve maximum accuracy in an acid-base titration, there should be a large change in pH around the equivalence point for a small addition of titrant.
(a) Which will give more accurate results in the titration of a strong acid with a strong base: 0.100 mol/L solutions or 0.000100 mol/L solutions? Explain.
(b) Assuming equal concentrations of 0.100 mol/L, which of the following three will give the most accurate and least accurate results: weak acid titrated with a weak base; weak acid with a strong base; strong acid with a strong base? Explain.

19.113 Phosphate systems form essential buffers in organisms. Use Table 18.5 to calculate the pH of a buffer made by dissolving 0.80 mol of NaOH in 0.50 L of 1.0 M H_3PO_4.

*19.114 A student was given a solution containing one or more of the following cations: K^+, Ag^+, Al^{3+}, Cu^{2+}, and Fe^{3+}. From the following laboratory results, determine the cations present:
1. Added $NH_3(aq)$ to ion mixture $\Longrightarrow$ brown precipitate.
2. Centrifuged $\Longrightarrow$ colorless solution.
3. Treated precipitate with $NaOH(aq) \Longrightarrow$ some solid remained, but some may have dissolved.
4. Neutralized solution from step 3 with $NH_4Cl(aq) \Longrightarrow$ no precipitate.
5. Added $HCl(aq)$ to solution from step 2 $\Longrightarrow$ white precipitate.
6. Flame test on original mixture $\Longrightarrow$ ordinary flame.

19.115 It is possible to detect NH_3 gas over 10^{-2} M NH_3. To what pH must 0.15 M NH_4Cl be raised to form detectable NH_3?

19.116 Manganese(II) sulfide is one of the compounds found in the nodules on the ocean floor that may eventually be a primary source of many transition metals. The solubility of MnS is 4.7×10^{-4} g/100 mL solution. Estimate the K_{sp} of MnS.

19.117 The normal pH of blood is 7.40 ± 0.05 and is controlled in part by the H_2CO_3/HCO_3^- buffer system.
(a) Assuming that the K_a value for carbonic acid at 25°C applies to blood, what is the $[H_2CO_3]/[HCO_3^-]$ ratio in normal blood?
(b) In acidosis, a condition that can have many causes, the blood is too acidic. What is the $[H_2CO_3]/[HCO_3^-]$ ratio in a patient whose blood pH is 7.20 (severe acidosis)?

19.118 Consider the following reaction:

$$Fe(OH)_3(s) + OH^-(aq) \rightleftharpoons Fe(OH)_4^-(aq); K_c = 4 \times 10^{-5}$$

(a) What is the solubility of $Fe(OH)_3$ in 0.010 M NaOH?
(b) How is this result used in qualitative analysis?

19.119 A bioengineer preparing cells for cloning bathes a small piece of rat epithelial tissue in a TRIS buffer (see Problem 19.109). The buffer is made by dissolving 43.0 g of TRIS $(pK_b = 5.91)$ in enough 0.095 M HCl to make 1.00 L of solution. What is the molarity of TRIS and the pH of the buffer?

19.120 Sketch a qualitative curve for the titration of ethylenediamine, $H_2NCH_2CH_2NH_2$, with 0.1 M HCl.

*19.121 A solution contains 0.10 M $ZnCl_2$ and 0.020 M $MnCl_2$. Given the following information, how would you adjust the pH to separate the ions as their sulfides ($[H_2S]$ of a saturated aqueous solution at 25°C = 0.10 M; $K_w = 1.0 \times 10^{-14}$ at 25°C)?

$MnS + H_2O \rightleftharpoons Mn^{2+} + HS^- + OH^-$		$K_{sp} = 3 \times 10^{-11}$
$ZnS + H_2O \rightleftharpoons Zn^{2+} + HS^- + OH^-$		$K_{sp} = 2 \times 10^{-22}$
$H_2S + H_2O \rightleftharpoons H_3O^+ + HS^-$		$K_{a1} = 9 \times 10^{-8}$

*19.122 Amino acids [general formula $NH_2CH(R)COOH$] can be considered polyprotic acids. In many cases, the R group contains additional amine and carboxyl groups.
(a) Can an amino acid dissolved in pure water have a protonated COOH group and an unprotonated NH_2 group (K_a of COOH group = 4.47×10^{-3}; K_b of NH_2 group = 6.03×10^{-5})? Use glycine, NH_2CH_2COOH, to explain why.
(b) Calculate $[^+NH_3CH_2COO^-]/[^+NH_3CH_2COOH]$ at pH 5.5.
(c) The R group of lysine is $—CH_2CH_2CH_2CH_2NH_2$ ($pK_b = 3.47$). Draw the structure of lysine at pH 1, physiological pH (~ 7), and pH 13.
(d) The R group of glutamic acid is $—CH_2CH_2COOH$ ($pK_a = 4.07$). Four ionic forms of glutamic acid (A to D) are shown below. Select the form that predominates at pH 1, at physiological pH (~ 7), and at pH 13.

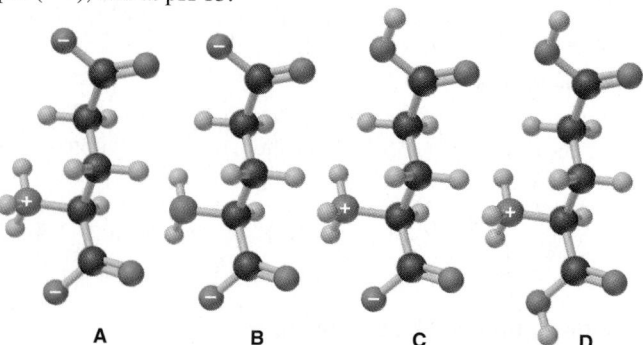

A   **B**   **C**   **D**

19.123 Tooth enamel is composed of hydroxyapatite, $Ca_5(PO_4)_3OH$ ($K_{sp} = 6.8 \times 10^{-37}$). Many water treatment plants add F^- ion to drinking water, which reacts with $Ca_5(PO_4)_3OH$ to form the more tooth decay–resistant fluorapatite, $Ca_5(PO_4)_3F$ ($K_{sp} = 1.0 \times 10^{-60}$). Fluoridated water has resulted in a dramatic decrease in the number of cavities among children. Calculate the solubility of $Ca_5(PO_4)_3OH$ and of $Ca_5(PO_4)_3F$ in water.

19.124 The color of the acid-base indicator ethyl orange turns from red to yellow over the pH range 3.4 to 4.8. Estimate K_a for ethyl orange.

19.125 An acid-base titration of a weak acid with a strong base has an end point at pH = 9.0. What indicator is suitable for the titration?

19.126 Use the values obtained in Problem 19.54 to sketch a curve of $[H_3O^+]$ vs. mL of added titrant. Are there advantages or disadvantages to viewing the results in this form? Explain.

*19.127 Instrumental acid-base titrations use a pH meter rather than a chemical indicator to monitor the changes in pH and volume. The equivalence point is estimated by finding the volume at which the curve has the steepest slope.
(a) Use the data in Figure 19.6 to calculate the slope $\Delta pH/\Delta V$ for all pairs of adjacent points and the average volume (V_{avg}) for each interval.
(b) Plot $\Delta pH/\Delta V$ vs. V_{avg} to find the steepest slope, and thus the volume at the equivalence point. (For example, the first pair of points gives $\Delta pH = 0.22$, $\Delta V = 10.00$ mL; hence, $\Delta pH/\Delta V = 0.022$ mL^{-1}, and $V_{avg} = 5.00$ mL.)

19.128 A student wants to make a basic solution of $Ca(OH)_2$, but its solubility in water is very low ($K_{sp} = 6.5 \times 10^{-6}$). What is the pH of a solution of 6.5×10^{-9} mol of $Ca(OH)_2$ in 10.0 L of water?

19.129 Lactic acid [$CH_3CH(OH)COOH$] is an important bio-molecule. Muscle physiologists study its accumulation during exercise. Food chemists study its occurrence in sour milk products, fruit, beer, wine, and tomatoes. Industrial microbiologists study its formation by various bacterial species from carbohydrates. A biochemist prepares a lactate buffer by mixing 225 mL of 0.85 M lactic acid ($K_a = 1.38 \times 10^{-4}$) with 435 mL of 0.68 M sodium lactate. What is the pH of the buffer?

19.130 A student wants to dissolve the maximum amount of CaF_2 ($K_{sp} = 3.2 \times 10^{-11}$) to make 1 L of aqueous solution.
(a) Into which of the following should she dissolve the salt?
(I) Pure water   (II) 0.01 M HF   (III) 0.01 M NaOH
(IV) 0.01 M HCl   (V) 0.01 M $Ca(OH)_2$
(b) Which would dissolve the least amount of salt?

19.131 A 500.-mL solution consists of 0.050 mol of solid NaOH and 0.13 mol of hypochlorous acid (HClO; $K_a = 3.0 \times 10^{-8}$) dissolved in water. (a) Aside from water, what is the concentration of each species present? (b) What is the pH of the solution? (c) What is the pH after adding 0.0050 mol of HCl to the flask?

19.132 The Henderson-Hasselbalch equation gives a relationship for obtaining the pH of a buffer solution consisting of HA and A^-. Derive an analogous relationship for obtaining the pOH of a buffer solution consisting of B and BH^+.

19.133 Calculate the molar solubility of $Hg_2C_2O_4$ ($K_{sp} = 1.75 \times 10^{-13}$) in 0.13 M $Hg_2(NO_3)_2$.

19.134 The well water in an area is "hard" because it is in equilibrium with $CaCO_3$ in the surrounding rocks. What is the concentration of Ca^{2+} in the well water (assuming the water's pH is such that the CO_3^{2-} ion is not hydrolyzed)? (See Appendix C for K_{sp} of $CaCO_3$.)

19.135 Four samples of an unknown solution are tested individually by adding HCl, H_2S in acidic solution, H_2S in basic solution, or Na_2CO_3, respectively. No precipitate forms in any of the tests. When NaOH is added to the original solution, moistened

litmus paper held above the solution turns blue. A drop of the original solution turns the color of a flame violet. Which ions are present in the unknown solution?

19.136 Human blood contains one buffer system based on phosphate species and one on carbonate species. Assuming that blood has a normal pH of 7.4, what are the principal phosphate and carbonate species present? What is the ratio of the two phosphate species? (In the presence of the dissolved ions and other species in blood, K_{a1} of $H_3PO_4 = 1.3 \times 10^{-2}$, $K_{a2} = 2.3 \times 10^{-7}$, and $K_{a3} = 6 \times 10^{-12}$; K_{a1} of $H_2CO_3 = 8 \times 10^{-7}$ and $K_{a2} = 1.6 \times 10^{-10}$).

*19.137 Most qualitative analysis schemes start with the separation of ion group 1, Ag^+, Hg_2^{2+}, and Pb^{2+}, by precipitating AgCl, Hg_2Cl_2, and $PbCl_2$. Develop a procedure for confirming the presence or absence of these ions from the following:
- $PbCl_2$ is soluble in hot water, but AgCl and Hg_2Cl_2 are not.
- $PbCrO_4$ is a yellow solid that is insoluble in hot water.
- AgCl forms colorless $Ag(NH_3)_2^+(aq)$ in aqueous NH_3.
- When a solution containing $Ag(NH_3)_2^+$ is acidified with HCl, AgCl precipitates.
- Hg_2Cl_2 forms an insoluble mixture of Hg(l; black) and $HgNH_2Cl$(s; white) in aqueous NH_3.

19.138 An environmental technician collects a sample of rainwater. A light on her portable pH meter indicates low battery power, which might result in inaccurate readings, so she uses indicator solutions to estimate the pH. A piece of litmus paper turns red, indicating acidity, so she divides the sample into thirds and obtains the following results: thymol blue turns yellow; bromphenol blue turns green; and methyl red turns red. Estimate the pH of the rainwater.

*19.139 A 0.050 M H_2S solution contains 0.15 M $NiCl_2$ and 0.35 M $Hg(NO_3)_2$. What pH is required to precipitate the maximum amount of HgS but none of the NiS? (See Appendix C.)

*19.140 Quantitative analysis of Cl^- ion is often performed by a titration with silver nitrate, using sodium chromate as an indicator. As standardized $AgNO_3$ is added, both white AgCl and red Ag_2CrO_4 precipitate where the drop of $AgNO_3$ lands, but so long as some Cl^- remains, the Ag_2CrO_4 redissolves as the mixture is stirred. When the red color is permanent, the equivalence point has been reached.
(a) Calculate the equilibrium constant for the reaction
$$2AgCl(s) + CrO_4^{2-}(aq) \rightleftharpoons Ag_2CrO_4(s) + 2Cl^-(aq)$$
(b) Explain why the silver chromate redissolves.
(c) If 25.00 cm^3 of 0.1000 M NaCl is mixed with 25.00 cm^3 of 0.1000 M $AgNO_3$, what is the concentration of Ag^+ remaining in solution? Is this sufficient to precipitate any silver chromate?

*19.141 An ecobotanist separates the components of a tropical bark extract by liquid chromatography. She discovers a large proportion of quinidine, a dextrorotatory stereoisomer of quinine that is a cardiac muscle depressant used for control of arrhythmic heartbeat. Quinidine has two basic nitrogens ($K_{b1} = 4.0 \times 10^{-6}$ and $K_{b2} = 1.0 \times 10^{-10}$). To measure the concentration, she carries out a titration. Because of the low solubility of quinidine, she first protonates both nitrogens with excess HCl and titrates the acidified solution with standardized base. A 33.85-mg sample of quinidine ($M = 324.41$ g/mol) was acidified with 6.55 mL of 0.150 M HCl.
(a) How many milliliters of 0.0133 M NaOH are needed to titrate the excess HCl?
(b) How many additional milliliters of titrant are needed to reach

the first equivalence point of quinidine dihydrochloride?

(c) What is the pH at the first equivalence point?

19.142 Why was rainwater acidic in the preindustrial age? Why did its acidity not erode marble statuary? What solutes make acid rain much more acidic now? How does automobile exhaust contribute to this acidity?

*19.143 A biochemist must prepare a solution for use as a medium in an experiment involving acid-producing bacteria. The pH of the medium must not change by more than 0.05 pH units for every 0.0010 mol of H_3O^+ generated by the organisms per liter of medium. A buffer consisting of 0.10 M HA and 0.10 M A^- is included in the medium to control its pH. What volume of this buffer must be included in 1.0 L of medium?

19.144 A 35.00-mL solution of 0.2500 M HF is titrated with a standardized 0.1532 M solution of NaOH at 25°C.

(a) What is the pH of the HF solution before titrant is added?

(b) How many milliliters of titrant are required to reach the equivalence point?

(c) What is the pH at 0.50 mL before the equivalence point?

(d) What is the pH at the equivalence point?

(e) What is the pH at 0.50 mL after the equivalence point?

*19.145 A student is given a solution containing one or more of the following cations: K^+, Ag^+, Al^{3+}, Cu^{2+}, Fe^{3+}. From the following laboratory results, determine the cations present:

1. Added $NH_3(aq) \Longrightarrow$ white precipitate in colored solution.
2. Centrifuged $\Longrightarrow$ deep-blue solution.
3. Treated precipitate with $NaOH(aq) \Longrightarrow$ dissolved completely. Acidified and obtained white precipitate with $Na_2HPO_4(aq)$.
4. Added $HCl(aq)$ to step 2 solution $\Longrightarrow$ turned green, and white precipitate formed.
5. Centrifuged precipitate from step 4 and washed it to remove excess HCl. Treated it with $NH_3(aq) \Longrightarrow$ dissolved to form colorless solution.
6. Flame test on original solution $\Longrightarrow$ green flame with no red visible.

*19.146 A lake with a surface area of 10.0 acres (1 acre = 4.840×10^3 yd^2) receives 1.00 in. of rain of pH 4.20. (Assume that the acidity of the rain is due to a strong, monoprotic acid.)

(a) How many moles of H_3O^+ are in the rain falling on the lake?

(b) If the lake is unbuffered (pH = 7.00) and its average depth is 10.0 ft before the rain, calculate the pH after the rain has been throughly mixed with lake water. (Ignore runoff from the surrounding land.)

(c) If the lake contains hydrogen carbonate ions (HCO_3^-) and the reaction is

$$HCO_3^-(aq) + H_3O^+(aq) \longrightarrow 2H_2O(l) + CO_2(g)$$

what mass of HCO_3^- would neutralize the acid in the rain?

19.147 A 35.0-mL solution of 0.075 M $CaCl_2$ is mixed with 25.0 mL of 0.090 M $BaCl_2$.

(a) If aqueous KF is added, which fluoride precipitates first?

(b) Describe how the metal ions can be separated using KF to form the fluorides.

(c) Calculate the fluoride ion concentration that will accomplish the separation.

*19.148 Even before the industrial age, rainwater was slightly acidic. Use the following data to calculate pH of unpolluted rainwater at 25°C: vol % in air of CO_2 = 0.033 vol %; solubility of CO_2 in pure water at 25°C and 1 atm = 88 mL CO_2/100 mL H_2O; K_{a1} of H_2CO_3 = 4.5×10^{-7}.

*19.149 Seawater at the surface has a pH of about 8.5.

(a) Which of the following species has the highest concentration at this pH: H_2CO_3; HCO_3^-; CO_3^{2-}? Explain.

(b) What are the concentration ratios $[CO_3^{2-}]/[HCO_3^-]$ and $[HCO_3^-]/[H_2CO_3]$ at this pH?

(c) In the deep sea, light levels are low, and the pH of the water is around 7.5. Suggest a reason for the lower pH at the greater ocean depth. (*Hint:* Consider the presence or absence of plant and animal life, and their effects on carbon dioxide concentrations.)

*19.150 Ethylenediaminetetraacetic acid is a tetraprotic acid, $(HOOCCH_2)_2NCH_2CH_2N(CH_2COOH)_2$ (that is abbreviated as H_4EDTA). Its salts are used to treat toxic metal poisoning through the formation of soluble complex ions that are then excreted. Because $EDTA^{4-}$ also binds essential calcium ions, it is often administered as the calcium disodium salt. For example, when $Na_2Ca(EDTA)$ is given to a patient, the $[Ca(EDTA)]^{2-}$ ion reacts with circulating Pb^{2+} ions and exchanges metal ions:

$$[Ca(EDTA)]^{2-}(aq) + Pb^{2+}(aq) \Longrightarrow$$
$$[Pb(EDTA)]^{2-}(aq) + Ca^{2+}(aq) \qquad K_c = 2.5 \times 10^7$$

A child is found to have a dangerously high blood lead level of 120 µg/100 mL. If the child is administered 100. mL of 0.10 M $Na_2Ca(EDTA)$, assuming that the exchange reaction and excretion process are 100% efficient, what is the final concentration of Pb^{2+} in µg/100 mL blood? (Total blood volume is 1.5 L.)

*19.151 Sodium chloride is purified for use as table salt by adding HCl to a saturated solution of NaCl (317 g/L). When 25.5 mL of 7.85 M HCl is added to 0.100 L of saturated solution, how many grams of purified NaCl precipitate?

*19.152 The solubility of Ag(I) in aqueous solutions containing different concentrations of Cl^- is based on the following equilibria:

$$Ag^+(aq) + Cl^-(aq) \Longrightarrow AgCl(s) \qquad K_{sp} = 1.8 \times 10^{-10}$$
$$Ag^+(aq) + 2Cl^-(aq) \Longrightarrow AgCl_2^-(aq) \qquad K_f = 1.8 \times 10^5$$

When solid AgCl is shaken with a solution containing Cl^-, Ag(I) is present as both Ag^+ and $AgCl_2^-$. The solubility of AgCl is the sum of the concentrations of Ag^+ and $AgCl_2^-$.

(a) Show that $[Ag^+]$ in solution is given by

$$[Ag^+] = 1.8 \times 10^{-10}/[Cl^-]$$

and that $[AgCl_2^-]$ in solution in given by

$$[AgCl_2^-] = (3.2 \times 10^{-5}) \, ([Cl^-])$$

(b) Find the $[Cl^-]$ at which $[Ag^+] = [AgCl_2^-]$.

(c) Explain the shape of the curve in the plot of AgCl solubility vs. $[Cl^-]$.

(d) Find the solubility of AgCl at the $[Cl^-]$ in part (b), which is the minimum solubility of AgCl in the presence of Cl^-.

CHAPTER 20

THERMODYNAMICS: ENTROPY, FREE ENERGY, AND THE DIRECTION OF CHEMICAL REACTIONS

CHAPTER OUTLINE

20.1 The Second Law of Thermodynamics: Predicting Spontaneous Change
Limitations of the First Law
The Sign of ΔH and Spontaneous Change
Disorder and Entropy
Entropy and the Second Law
Standard Molar Entropies and the Third Law

20.2 Calculating the Change in Entropy of a Reaction
The Standard Entropy of Reaction
Entropy Changes in the Surroundings
Entropy Change and the Equilibrium State
Spontaneous Exothermic and Endothermic Reactions

20.3 Entropy, Free Energy, and Work
Free Energy Change and Reaction Spontaneity
Standard Free Energy Changes
ΔG and Work
Temperature and Reaction Spontaneity
Coupling of Reactions

20.4 Free Energy, Equilibrium, and Reaction Direction

Figure: Thermodynamics and the steam engine. The principles of thermodynamics with regard to the relationship between heat and work were developed in the 19th century to improve on the function of the newly invented steam engine. In this chapter, we examine this relationship and focus on the real-world limitations to the work that can be obtained from a chemical or physical system. In the process, we discover *why* reactions occur.

CONCEPTS & SKILLS

to review before you study this chapter

- internal energy, heat, and work (Section 6.1)
- state functions (Section 6.1) and standard states (Section 6.6)
- enthalpy, ΔH, and Hess's law (Sections 6.2 and 6.5)
- entropy and disorder (Section 13.2)
- comparing Q and K to find reaction direction (Section 17.4)

Iron rusts spontaneously.

In the last few chapters, we've posed and answered some extremely important questions about chemical and physical change: How fast does the change occur, and how is this rate affected by concentration and temperature? How much product will be present when the net change ceases, and how is this yield affected by concentration and temperature? We've explored these questions for systems ranging from the stratosphere to a limestone cave and from the cells of your body to the lakes of Sweden.

Now it's time to stand back and ask the most profound question of all: Why does a change occur in the first place? You know that some everyday chemical and physical changes seem to have a natural direction and to happen by themselves, whereas others do not. Methane burns in oxygen with a vigorous burst of heat to yield carbon dioxide and water vapor, but these products will not remake methane and oxygen no matter how long they interact. A steel chain left outside slowly rusts, but a rusty one will not become shiny (see photo). A cube of sugar dissolves in a cup of coffee after a few seconds of stirring, but stir for another century and the cube will not reappear.

All our investigations of nature allow us to define a *spontaneous* process as one that occurs by itself. The opposite of this, a nonspontaneous process, does not occur unless we make it happen. There appears to be a driving force built into a spontaneous process that propels it in a given direction. If we want a process to occur against that natural drive, we must expend energy to accomplish it. "Spontaneous" does *not* mean instantaneous and has nothing to do with how long a process takes to occur; it means that, given enough time, the process will happen by itself. Many processes are spontaneous but slow—ripening, rusting, and (happily!) aging.

A chemical reaction proceeding toward equilibrium is an example of a spontaneous change. As you learned in Chapter 17, we can predict the net direction of the reaction—its spontaneous direction—by comparing the reaction quotient (Q) with the equilibrium constant (K). But why is there a drive to attain equilibrium? What determines the value of the equilibrium constant? And why do some spontaneous processes absorb energy whereas others release it? The principles of thermodynamics were developed in the early 19th century to help utilize the newly invented steam engine more efficiently. Despite that rather narrow focus, the principles apply, as far as we know, to every system in the universe! In this chapter, we apply these principles to chemical and physical change.

The chapter opens with a review of the first law of thermodynamics. As you'll see, the first law accounts for the size of an energy change but *not* its spontaneous direction. The enthalpy change also is not an adequate criterion for predicting direction. Rather, we find this criterion in the change in *entropy,* that is, in the natural tendency of systems to become disordered. This idea leads us to the second law of thermodynamics and its quantitative application to both exothermic and endothermic changes. Then we examine the concept of *free energy,* which simplifies the criterion for spontaneous change, and see how it relates to the useful work a system can perform. Finally, we investigate the key relationship between the free energy change of a reaction and its equilibrium constant.

20.1 THE SECOND LAW OF THERMODYNAMICS: PREDICTING SPONTANEOUS CHANGE

A **spontaneous change** of a system, whether a chemical or physical change or just a change in location, is one that occurs by itself under specified conditions, without an ongoing input of energy from outside the system. The freezing of water, for example, is spontaneous at 1 atm and −5°C. A spontaneous process such as burning or falling may need a little "push" to get started—a spark to ignite gasoline, a shove to knock a book off your desk—but once the process begins, it continues without external aid because the system releases enough energy to keep the process going.

In contrast, for a nonspontaneous change to occur, the system must be supplied with a continuous input of energy. A book falls spontaneously, but it rises only if something else, such as a human hand (or a hurricane-force wind), supplies energy in the form of work. Under a given set of conditions, *if a change is spontaneous in one direction, it is **not** spontaneous in the other.*

How can we tell the direction of a spontaneous change in cases that are not as familiar as burning gasoline or falling books? Can we predict the net direction of any reaction? Does the criterion for spontaneous change involve the change in energy or enthalpy of the reacting system? Because energy changes seem to be involved, let's begin addressing these questions by reviewing the idea of conservation of energy to see if it helps explain spontaneity.

Limitations of the First Law of Thermodynamics

In Chapter 6, we discussed the first law of thermodynamics (the law of conservation of energy) and its application to chemical and physical systems. The first law states that the internal energy (E) of a system, the sum of the kinetic and potential energy of all its particles, changes through the addition or removal of heat (q) and/or work (w):

$$\Delta E = q + w$$

Whatever is not part of the system (sys) is part of the surroundings (surr), so the system and surroundings together constitute the universe (univ):

$$E_{univ} = E_{sys} + E_{surr}$$

Heat and/or work gained by the system is lost by the surroundings, and vice versa:

$$(q + w)_{sys} = -(q + w)_{surr}$$

It follows from these ideas that *the total energy of the universe is constant:*[*]

$$\Delta E_{sys} = -\Delta E_{surr} \quad \text{therefore} \quad \Delta E_{sys} + \Delta E_{surr} = 0 = \Delta E_{univ}$$

Is the first law sufficient to explain why a natural process takes place as it does? It certainly accounts for the energy involved. When a book that was resting on your desk falls to the floor, the first law guides us through the conversion from the potential energy of the resting book to the kinetic energy of the falling book to the heat dispersed in the floor near the point of contact. When gasoline burns in your car's engine, the first law explains that the potential energy difference between the chemical bonds in the fuel mixture and those in the exhaust gases is converted to the kinetic energy of the moving car and its parts plus the heat released to the environment. When an ice cube melts in your hand, the first law tells that energy from your hand was transferred to the ice to change it to a liquid. If you could measure the work and heat involved in each case, you would find that the energy is conserved as it is converted from one form to another.

However, the first law does not help us make sense of the *direction* of the change. Why doesn't the heat in the floor near the fallen book change to kinetic energy in the book and move it back onto your desk? Why doesn't the heat released in the car engine convert exhaust fumes back into gasoline and oxygen? Why doesn't the pool of water in your cupped hand transfer the heat back to your hand and refreeze? None of these events would violate the first law—energy would still be conserved—but they never happen. The first law by itself tells nothing about the direction of a spontaneous change, so we must search elsewhere for a way to predict that direction.

---

[*]Any modern statement of conservation of energy must take into account mass-energy equivalence and the processes in stars, which convert enormous amounts of matter into energy. Therefore, we qualify the statement to be that the total *mass-energy* of the universe is constant.

The Sign of ΔH Cannot Predict Spontaneous Change

In the mid-19th century, some thermodynamicists thought the sign of the enthalpy change (ΔH), the heat added or removed at constant pressure (q_P), was the criterion for spontaneity. (Recall from Section 6.2 that, for most reactions, ΔH is either equal to or very close to ΔE.) They thought that exothermic processes ($\Delta H < 0$) were spontaneous and endothermic ones ($\Delta H > 0$) were nonspontaneous. This hypothesis had experimental support; after all, many spontaneous processes *are* exothermic. All combustion reactions are spontaneous and exothermic; for example,

$$CH_4(g) + 2O_2(g) \longrightarrow CO_2(g) + 2H_2O(g) \qquad \Delta H^0_{rxn} = -802 \text{ kJ}$$

Iron rusts spontaneously and exothermically:

$$2Fe(s) + \tfrac{3}{2}O_2(g) \longrightarrow Fe_2O_3(s) \qquad \Delta H^0_{rxn} = -826 \text{ kJ}$$

Ionic compounds form spontaneously from their elements with a large release of heat; for example,

$$Na(s) + \tfrac{1}{2}Cl_2(g) \longrightarrow NaCl(s) \qquad \Delta H^0_{rxn} = -411 \text{ kJ}$$

However, in many other cases, the sign of ΔH is no help. An exothermic process may occur spontaneously under certain conditions, whereas the opposite, endothermic, process may occur spontaneously under other conditions. For example, at ordinary pressure, water freezes below 0°C but melts above 0°C. Both changes are spontaneous, but the first is exothermic and the second endothermic:

$$H_2O(l) \longrightarrow H_2O(s) \qquad \Delta H^0_{rxn} = -6.02 \text{ kJ } (exo\text{thermic; spontaneous at } T < 0°C)$$
$$H_2O(s) \longrightarrow H_2O(l) \qquad \Delta H^0_{rxn} = +6.02 \text{ kJ } (endo\text{thermic; spontaneous at } T > 0°C)$$

At ordinary pressure and room temperature, liquid water vaporizes spontaneously in dry air, another endothermic change:

$$H_2O(l) \longrightarrow H_2O(g) \qquad \Delta H^0_{rxn} = +44.0 \text{ kJ}$$

In fact, all melting and vaporizing are endothermic changes that are spontaneous under the prevailing conditions.

There are many other spontaneous endothermic physical changes. Recall from Chapter 13 that water-soluble salts dissolve spontaneously, even though most have a positive ΔH^0_{soln}:

$$NaCl(s) \xrightarrow{H_2O} Na^+(aq) + Cl^-(aq) \qquad \Delta H^0_{soln} = +3.9 \text{ kJ}$$
$$RbClO_3(s) \xrightarrow{H_2O} Rb^+(aq) + ClO_3^-(aq) \qquad \Delta H^0_{soln} = +47.7 \text{ kJ}$$
$$NH_4NO_3(s) \xrightarrow{H_2O} NH_4^+(aq) + NO_3^-(aq) \qquad \Delta H^0_{soln} = +25.7 \text{ kJ}$$

Some endothermic chemical changes are also spontaneous:

$$N_2O_5(s) \longrightarrow 2NO_2(g) + \tfrac{1}{2}O_2(g) \qquad \Delta H^0_{rxn} = +109.5 \text{ kJ}$$
$$Ba(OH)_2 \cdot 8H_2O(s) + 2NH_4NO_3(s) \longrightarrow$$
$$Ba^{2+}(aq) + 2NO_3^-(aq) + 2NH_3(aq) + 10H_2O(l) \qquad \Delta H^0_{rxn} = +62.3 \text{ kJ}$$

In the second reaction, the released waters of hydration solvate the ions, but the reaction mixture cannot absorb heat from the surroundings quickly enough, and the container becomes so cold that a wet block of wood freezes to it (Figure 20.1).

Figure 20.1 **A spontaneous endothermic chemical reaction. A,** When the crystalline solid reactants barium hydroxide octahydrate and ammonium nitrate are mixed, a slurry soon forms as waters of hydration are released. **B,** The reaction mixture absorbs heat so quickly from the surroundings that the beaker becomes covered with frost, and a moistened block of wood freezes to it.

A   B

Is there any feature common to these endothermic processes that can give us a clue as to why they occur spontaneously? A close look shows that in all these examples, matter changes from a more ordered to a less ordered state, *from a state with relatively few ways of arranging the particles to one with more ways of doing so*. The phase changes lead from particles fixed in relatively few arrangements to their random motion and separation into many more arrangements:

$$\text{more order} \xrightarrow{\hspace{3cm}} \text{less order}$$
$$\text{solid} \longrightarrow \text{liquid} \longrightarrow \text{gas}$$

The dissolving of salts leads from crystalline solid and pure liquid to dispersed ions in a solution—once again from fewer ways to more ways of arranging the particles:

$$\text{more order} \xrightarrow{\hspace{3cm}} \text{less order}$$
$$\text{crystal} + \text{liquid} \longrightarrow \text{ions in solution}$$

In the chemical reactions, fewer moles of crystalline solids produce more moles of gases and/or dispersed ions; here, too, more particles have more ways of being arranged:

$$\text{more order} \xrightarrow{\hspace{3cm}} \text{less order}$$
$$\text{crystal} + \text{crystal} \longrightarrow \text{gases} + \text{ions in solution}$$

In thermodynamic terms, a change in order is a change in the number of ways of arranging the particles, and it is a key factor in determining the direction of a spontaneous process.

Disorder and Entropy

As you know from many familiar examples, there is a natural tendency for a system to become disordered, that is, for the parts of the system to have more ways of being arranged. Drop a dinner plate, and it shatters into many pieces. Imagine what would happen to the clothes hanging in your closet if you left everything wherever it fell, or to the books in the college library if the staff didn't continuously reshelve them. On the other hand, creating order—gluing the pieces of plate together, hanging up your clothes, reshelving library books—requires work. ● In Chapter 13, we discussed the importance of disorder as a key factor in the solution process; here, we give it a quantitative dimension and examine its central role in other chemical and physical changes.

Let's analyze this idea further in terms of a deck of playing cards. In a new deck, the cards look very ordered, arranged by suit and increasing face value (Figure 20.2A). But someone may perceive more order if the cards were arranged

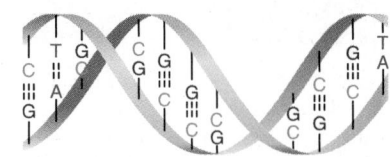

● Vital Orderly Information Searching for an address in a city whose houses are not in numerical order is the stuff of fitful dreams. A close relationship exists between the order of a system and the information the system conveys. One system in which the order and corresponding information are vitally related is the double helix of nucleotide bases in DNA (Section 15.6). The sequence of four bases—G, C, T, and A—in each strand is reproduced during cell division so the genetic makeup of each cell is conserved. A break in a strand, or an insertion or deletion of a base, introduces disorder in the genetic information and creates mutations—most insignificant, some harmful, a few helpful. Of particular relevance here is that DNA becomes disordered naturally. There are a host of chemical repair mechanisms that, like molecular librarians, do work to reorganize the bases into their original sequence.

A   **B**

Figure 20.2 The number of ways to arrange a deck of playing cards. A, A new deck of cards is arranged in one of 52! (52 factorial) possible sequences. **B,** For a deck scattered on the floor, there are many more ways of arranging the cards; thus, the disorder of the system has increased.

Poker and Probability In a game of poker, a rare combination has more value than a common one because it has a lower probability of being dealt. In other words, the poker hand that has fewer possible arrangements wins. There are only four possible royal-flush hands (A, K, Q, J, 10 of the same suit), but 1,098,240 hands with one pair.

Figure 20.3 Spontaneous expansion of a gas. The container consists of two flasks connected by a stopcock. **A,** With the stopcock closed, neon gas at a pressure of 1 atm occupies one flask and the other is evacuated. **B,** With the stopcock open, the gas expands spontaneously and fills both flasks uniformly to a pressure of 0.5 atm.

with all the 2's, followed by all the 3's, followed by all the 4's, and so on. And someone else may perceive another arrangement to be more ordered. In fact, no matter how any arrangement may *look*, in thermodynamic terms, the deck is equally ordered, because there are still only 52 cards and 52 positions for those cards, and therefore the number of ways of arranging them stays the same. And while that number may be high—52! (52 factorial) or about 10^{68} arrangements—there are many more ways of arranging the cards if you scatter them on the floor (Figure 20.2B). Thus, scattering the cards initiates a spontaneous process that produces disorder because it *increases the number of possible arrangements*. In physical and chemical systems, too, disorder increases when a process results in more ways for the atoms, ions, or molecules in the system to be arranged.

Now let's see how the tendency toward disorder appears in a physical system. Figure 20.3 shows an evacuated container consisting of two flasks connected by a stopcock. You fill the left flask with neon to a pressure of 1 atm and connect the other flask. When you open the stopcock, the gas eventually becomes distributed uniformly, and the pressure is 0.5 atm in each flask. This result is what you expect on the basis of experience, but *why* does the gas behave this way? Why don't all the neon atoms rush to the other flask, or just stay where they are?

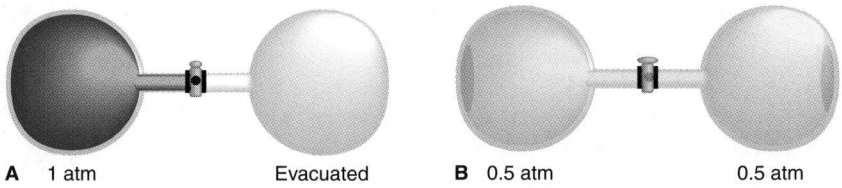

A 1 atm   Evacuated   **B** 0.5 atm   0.5 atm

The randomly moving gas particles spread throughout both flasks because twice the original volume affords each particle twice as many places it can be, which vastly increases the number of ways of arranging the particles. Therefore, it is vastly more probable that the particles will fill the entire volume than a limited part of it. Imagine how unlikely it would be for all the particles in the right flask to move back to the left. Even for 10 Ne atoms, the odds against this happening are about 1000 to 1; for 1 mol of gas, the odds are about $10^{2 \times 10^{23}}$ to 1! Like scattering a deck of cards, opening the stopcock starts a spontaneous change from more to less order—from fewer to more arrangements of the particles.

The scattering of cards on the floor or the redistribution of the gas particles in the container occurred with *no change in the total internal energy of the system,* so the various arrangements of the components are *equivalent.* The number of ways (W) of arranging the components of a system without changing its energy is directly related to the **entropy (S)** of the system. In 1877, the Austrian mathematician Ludwig Boltzmann described this relationship quantitatively:

$$S = k \ln W \tag{20.1}$$

where k, the *Boltzmann constant,* is the universal gas constant divided by Avogadro's number, R/N_A, and equals 1.38×10^{-23} J/K. Because W is unitless, S has units of joules/kelvin (J/K). Recall that Chapter 13 introduced *entropy as a measure of the disorder of the system:*

- A system with relatively few equivalent ways to arrange its components (smaller W), such as a crystalline solid or a stacked deck of cards, has relatively *less disorder and low entropy.*
- A system with many equivalent ways to arrange its components (larger W), such as a gas or a scattered deck of cards, has relatively *more disorder and high entropy.*

If the entropy of a system goes up, the components of the system have more ways of being arranged, and the same total energy of the system becomes distributed over a wider range: $S_{disorder} > S_{order}$.

Like internal energy and enthalpy, *entropy is a state function,* one that depends only on the present state of the system, not on how it arrived at that state. Therefore, the change in entropy of the system, ΔS_{sys}, depends only on its initial and final values:

$$\Delta S_{sys} = S_{final} - S_{initial}$$

Because the gas particles in the flasks in Figure 20.3 ended up with more ways to be arranged, the entropy of the system increased; thus, ΔS_{sys} was positive ($S_{final} > S_{initial}$). If the gas particles in both flasks moved to one flask, they would have fewer ways to be arranged, and so the entropy of the system would decrease: ΔS_{sys} would be negative ($S_{final} < S_{initial}$). How the change in order comes about—what paths the particles take as they become rearranged—does not affect the sign or magnitude of ΔS_{sys}.

Entropy and the Second Law of Thermodynamics

Now back to our original question: what determines the direction of a spontaneous change? As you've seen, there is a natural tendency toward higher entropy. This is the criterion for which we've been searching, but to apply it correctly, we have to consider more than just the system. After all, some systems become *less* ordered (higher entropy) spontaneously, such as ice melting or a crystal dissolving, whereas others become *more* ordered (lower entropy) spontaneously, such as water freezing or a crystal forming. If we consider changes in *both* the system *and* its surroundings, however, we find that *all processes occur spontaneously in the direction that increases the entropy of the universe (system plus surroundings).* This is one way to state the **second law of thermodynamics.**

Notice that the second law places no limitations on the entropy change of the system *or* the surroundings: either may be negative; that is, either system *or* surroundings may become more ordered (lower entropy). The law does state, however, that for a spontaneous process, the *sum* of the entropy changes must be positive. When the entropy of the system decreases, the entropy of the surroundings increases even more to offset this system decrease, and so the entropy of the universe (system *plus* surroundings) increases. A quantitative statement of the second law is that, for any spontaneous process,

$$\Delta S_{univ} = \Delta S_{sys} + \Delta S_{surr} > 0 \qquad (20.2)$$

Standard Molar Entropies and the Third Law

Both entropy and enthalpy are state functions, but the nature of their values differs in a fundamental way. Recall that we cannot determine absolute enthalpies because we have no starting point, that is, no baseline value for the enthalpy of a substance. Therefore, only enthalpy *changes* can be measured. In contrast, we *can* determine the absolute entropy of a substance. To do so requires application of the **third law of thermodynamics,** which states that *a perfect crystal has zero entropy at a temperature of absolute zero:* $S_{sys} = 0$ at 0 K. "Perfect" means that all the particles are aligned flawlessly in the crystal structure, with no defects of any kind. At absolute zero, all particles in the crystal have their minimum energy, and there is only one way they can be arranged: thus, in Equation 20.1, $W = 1$, so $S = k \ln 1 = 0$. If we warm the crystal, the total energy of the system increases, which gives the particles a larger range of energies, meaning that they can be arranged in more ways (Figure 20.4). Thus, $W > 1$, $\ln W > 0$, and $S > 0$. The entropy of a substance at a given temperature is therefore an *absolute* value, equal to the entropy increase it would undergo if heated from 0 K to that temperature.

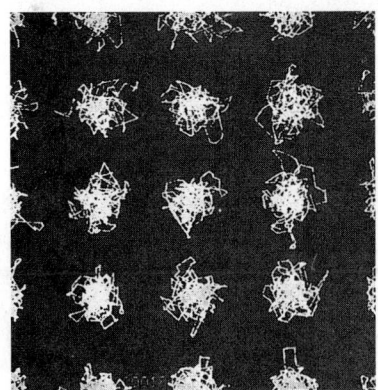

Figure 20.4 Random motion in a crystal. This computer simulation shows the paths of the particle centers in a crystalline solid. At any temperature greater than 0 K, each particle moves about its lattice position. The greater the temperature, the more vigorous the movement. Adding thermal energy increases the total energy, so the particle energies are distributed over a wider range. This increases the number of possible arrangements, so the entropy increases.

As with other thermodynamic variables, we usually compare entropy values for substances in their *standard states* at the temperature of interest. Recall that the standard states we use throughout the text are *1 atm for gases, 1 M for solutions, and the pure substance in its most stable form for solids or liquids.* Entropy is an *extensive* property, one that depends on the amount of substance, so we are interested in the **standard molar entropy (S^0)** in units of J/mol·K (that is, $J \cdot mol^{-1} \cdot K^{-1}$). A list of S^0 values at 298 K (25°C) for many elements, compounds, and ions appears, with other thermodynamic variables, in Appendix B.

Predicting Relative S^0 Values of a System   Based on events at the molecular level, we can predict how the entropy of a substance is affected by temperature, physical state, dissolution, and atomic or molecular complexity. (All S^0 values in the following discussion have units of J/mol·K and, unless stated otherwise, refer to the system at 298 K.)

1. *Temperature changes.* For a given substance, S^0 *increases as the temperature rises.* Consider these typical values for copper metal:

T (K):	273	295	298
S^0:	31.0	32.9	33.2

The temperature increase represents an increase in average kinetic energy of the particles. Recall from Figure 5.14 (p. 198) that the kinetic energies of gas particles in a sample are distributed over a range, which becomes wider as the temperature rises. The same general behavior occurs for liquids and solids. There are more ways to distribute the energy of the substance at the higher temperature, so the entropy of the substance goes up.

2. *Physical states and phase changes.* When a more ordered phase changes to a less ordered one, the entropy change is positive. For a given substance, S^0 *increases as the substance changes from a solid to a liquid to a gas:*

	Na	**H_2O**	**C(graphite)**
S^0(s or l):	51.4(s)	69.9(l)	5.7(s)
S^0(g):	153.6	188.7	158.0

Figure 20.5 shows the entropy of a typical substance as it is heated and undergoes a phase change. Note the pattern of *gradual increase within a phase* as the temperature rises and a *large, sudden increase at the phase change.* The solid is relatively ordered and has the lowest entropy, its particles vibrating about their positions but, on average, remaining fixed. As the temperature rises, the entropy gradually increases with the increase in kinetic energy. When the solid melts, the particles move freely between and around each other in the liquid, so there is an abrupt increase in entropy. Further heating increases the speed of the particles in the liquid, and the entropy increases gradually. Finally, freed from intermolecular forces, the particles undergo another abrupt entropy increase and move chaotically as a gas. Note that *the increase in entropy from liquid to gas is much larger than from solid to liquid.*

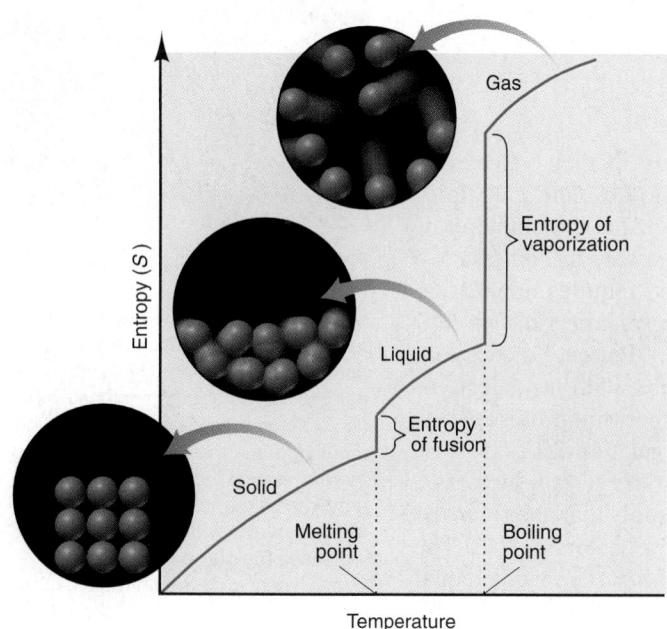

Figure 20.5 **The increase in entropy from solid to liquid to gas.** A plot of entropy vs. temperature shows the gradual increase in entropy within a phase and the abrupt increase with a phase change. The molecular-scale views depict the increase in randomness of the particles as the solid melts and, even more so, as the liquid vaporizes.

3. *Dissolution of a solid or liquid.* The entropy of a dissolved solid or liquid solute is usually *greater* than the entropy of the pure solute, but the type of solute *and* solvent and the nature of the dissolution process affect the overall entropy change (Figure 20.6):

	NaCl	AlCl₃	CH₃OH
S^0(s or l):	72.1(s)	167(s)	127(l)
S^0(aq):	115.1	−148	132

When an ionic solid dissolves in water, the highly ordered crystal mixes with the pure liquid and becomes separate, hydrated ions dispersed randomly in a solution. We expect the entropy of the ions themselves to be greater in the solution than in the crystal. However, some of the water molecules become organized around the ions (see Figure 13.2, p. 487), which makes a negative contribution to the overall entropy change. In fact, for small, multiply charged ions, the solvent becomes so highly ordered that this negative contribution can dominate and lead to negative S^0 values for the ion in solution. For example, the Al^{3+}(aq) ion has an S^0 value of −313 J/mol·K.* Therefore, when $AlCl_3$ dissolves in water, Cl^-(aq) has a positive entropy, but Al^{3+}(aq) has such a negative entropy that the overall entropy of dissolved $AlCl_3$ is lower than that of the solid.

For molecular solutes, the increase in entropy upon dissolving is typically much smaller than for ionic solutes. For a solid, such as glucose, there is no separation into ions, and for a liquid, such as ethanol, the breakdown of a crystal structure is absent as well. Furthermore, in pure ethanol and in pure water, the molecules form many H bonds, so there is relatively little change in their freedom of movement when they are mixed (Figure 20.7). The small increase in the disorder of the dissolved ethanol arises from the random mixing of the two types of molecules.

4. *Dissolution of a gas.* A gas is so disordered to begin with that it becomes *more* ordered when it dissolves in a liquid or solid. Therefore, the entropy of a solution of a gas in a liquid or a solid is always *less* than the entropy of the pure gas. When O_2 gas [S^0(g) = 205.0 J/mol·K] dissolves in water, its entropy

---

*An S^0 value for a hydrated ion can be negative because it is relative to the S^0 value of the hydrated proton, H^+(aq), which is assigned a value of 0. In other words, Al^{3+}(aq) has a lower entropy than H^+(aq).

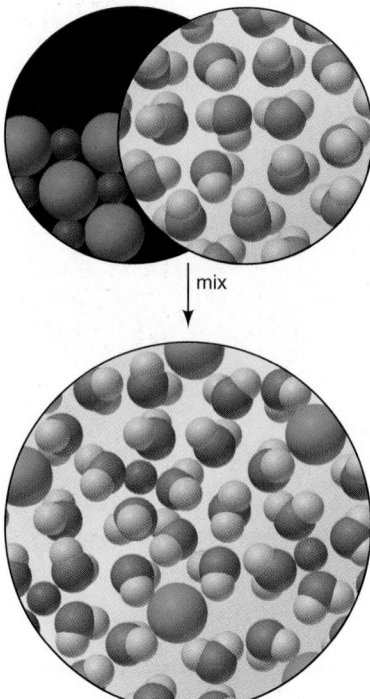

Figure 20.6 The entropy change accompanying the dissolution of a salt. When a crystalline salt and pure liquid water mix to form a solution, the entropy change has two contributions— a positive contribution as the crystal separates into ions and the pure liquid disperses them, and a negative contribution as water molecules become organized around each ion. The relative magnitudes of these positive and negative contributions determine the overall entropy change. In the majority of cases, the entropy of the solution is *greater* than that of the ionic solid and solvent.

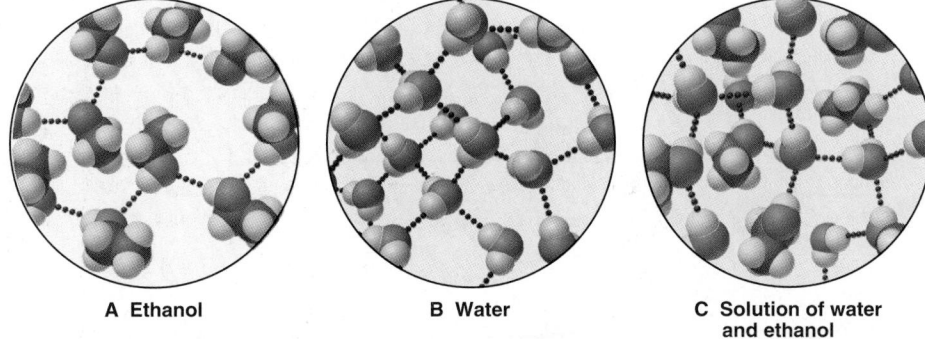

A Ethanol	**B** Water	**C** Solution of water and ethanol

Figure 20.7 The small increase in entropy when ethanol dissolves in water. Pure ethanol **(A)** and pure water **(B)** have many intermolecular H bonds. **C,** When they form a solution, the molecules form H bonds to one another, so their freedom of motion does not change significantly. Thus, the entropy increase is relatively small and is due solely to random mixing.

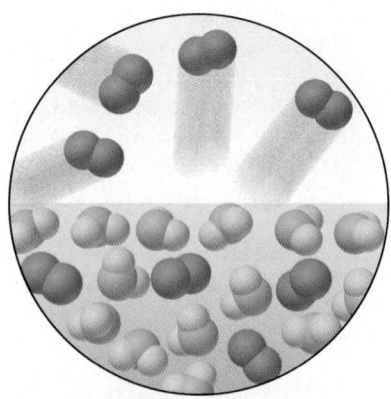

Figure 20.8 The large *decrease* in entropy of a gas when it dissolves in a liquid. The chaotic movement and high entropy of O_2 gas are reduced greatly when it dissolves in water.

decreases sharply [$S^0(aq) = 110.9$ J/mol·K] (Figure 20.8). When a gas dissolves in another gas, however, the entropy increases because of the mixing of the two types of molecules.

5. *Atomic size or molecular complexity.* In general, differences in entropy values for substances in the same phase are based on atomic size and molecular complexity. For elements within a periodic group, atomic size reflects molar mass, and entropy increases down the group:

	Li	Na	K	Rb	Cs
Atomic radius (pm):	152	186	227	248	265
Molar mass (g/mol):	6.941	22.99	39.10	85.47	132.9
$S^0(s)$:	29.1	51.4	64.7	69.5	85.2

The same trend of increasing entropy down a group holds for similar compounds:

	HF	HCl	HBr	HI
Molar mass (g/mol):	20.01	36.46	80.91	127.9
$S^0(g)$:	173.7	186.8	198.6	206.3

For an element that occurs in different forms (allotropes), the entropy is *higher* in the form with bonds that allow the atoms more movement. For example, the S^0 of graphite is 5.69 J/mol·K, whereas the S^0 of diamond is 2.44 J/mol·K. In diamond, covalent bonds extend in three dimensions, allowing the atoms little movement; in graphite, covalent bonds extend only within a sheet, and motion of the sheets is relatively easy, so the entropy is higher.

For compounds, entropy increases with chemical complexity, that is, with the number of atoms in the compound. This trend holds for both ionic and covalent compounds, as long as the substances are in the same physical state:

	NaCl	AlCl₃	P₄O₁₀	NO	NO₂	N₂O₄
$S^0(s)$:	72.1	167	229			
$S^0(g)$:				211	240	304

The trend is based on the types of movement available to the atoms (or ions) in each compound, which relates to the number of equivalent ways in which their energy can be distributed. For example, as Figure 20.9 shows, among the nitrogen oxides listed above, the two atoms of NO can vibrate in only one way, toward and away from each other. The three atoms of NO_2 have several vibrational motions, and the six atoms of N_2O_4 have many more.

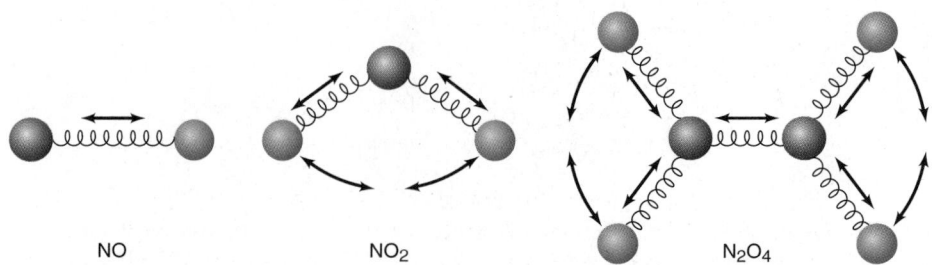

NO   NO₂   N₂O₄

Figure 20.9 Entropy and vibrational motion. A diatomic molecule, such as NO, can vibrate in only one way. NO_2 can vibrate in more ways, and N_2O_4 in even more. Thus, as the number of atoms increases, a molecule can distribute its vibrational energy in more ways, and so has higher entropy.

For larger molecules, we also consider how parts of the molecule move relative to other parts. A long hydrocarbon chain can rotate and vibrate in more ways than a short one can, so entropy increases with chain length. A ring compound, such as cyclopentane (C_5H_{10}), has lower entropy than the chain compound of the same molar mass, in this case pentene (C_5H_{10}), because many of the random molecular motions are restricted by the ring:

$CH_4(g)$	$C_2H_6(g)$	$C_3H_8(g)$	$C_4H_{10}(g)$	$C_5H_{10}(g)$	$C_5H_{10}(cyclo, g)$	$C_2H_5OH(l)$
S^0: 186	230	270	310	348	293	161

Remember that these trends hold only for *substances in the same physical state*. Note, for instance, that gaseous methane (CH_4) has a greater entropy than liquid ethanol (C_2H_5OH), even though the ethanol molecules are much more complex. When gases are compared with liquids, *the effect of physical state usually dominates that of molecular complexity.*

SAMPLE PROBLEM 20.1 Predicting Relative Entropy Values

Problem Choose the member with the higher entropy in each of the following pairs, and justify your choice [assume constant temperature, except in part (e)]:
(a) 1 mol of $SO_2(g)$ or 1 mol of $SO_3(g)$
(b) 1 mol of $CO_2(s)$ or 1 mol of $CO_2(g)$
(c) 3 mol of oxygen gas (O_2) or 2 mol of ozone gas (O_3)
(d) 1 mol of KBr(s) or 1 mol of KBr(aq)
(e) Seawater in midwinter at 2°C or in midsummer at 23°C
(f) 1 mol of $CF_4(g)$ or 1 mol of $CCl_4(g)$
Plan In general, we know that less ordered systems have higher entropy than more ordered systems and that higher temperature increases entropy. We apply the general categories described in the text to choose the member with the higher entropy.
Solution (a) 1 mol of $SO_3(g)$. For equal numbers of moles of substances with the same types of atoms in the same physical state, the more atoms in the molecule, the more types of motion available, and thus the higher the entropy.
(b) 1 mol of $CO_2(g)$. For a given substance, entropy increases in the sequence $s < l < g$.
(c) 3 mol of $O_2(g)$. The two samples contain the same number of oxygen atoms but different numbers of molecules. Despite the greater complexity of O_3, the greater number of molecules dominates in this case because there are many more ways to arrange three moles of particles than two moles.
(d) 1 mol of KBr(aq). The two samples have the same number of ions, but they are highly ordered in the solid and randomly dispersed in the solution.
(e) Seawater in summer. Entropy increases with rising temperature.
(f) 1 mol of $CCl_4(g)$. For similar compounds, entropy increases with molar mass.

FOLLOW-UP PROBLEM 20.1 For 1 mol of substance at a given temperature, select the member in each pair with the higher entropy and give the reason for your choice:
(a) $PCl_3(g)$ or $PCl_5(g)$; (b) $CaF_2(s)$ or $BaCl_2(s)$; (c) $Br_2(g)$ or $Br_2(l)$.

SECTION SUMMARY

A change is spontaneous if it occurs in a given direction under specified conditions without a continuous input of energy. Neither the first law of thermodynamics nor the sign of ΔH predicts the direction. All spontaneous processes involve a change in the universe from more to less order, that is, from fewer ways of arranging the particles to more ways. Entropy is a measure of disorder and is directly related to the number of ways of arranging the components of a system. The second law of thermodynamics states that, in a spontaneous process, the entropy of the universe (system plus surroundings) increases. Entropy values are absolute because perfect crystals have zero entropy at 0 K (third law). Standard molar entropy S^0 (J/mol·K) is affected by temperature, phase changes, dissolution, and atomic size or molecular complexity.

20.2 CALCULATING THE CHANGE IN ENTROPY OF A REACTION

Beyond understanding trends in S^0 values for different substances or for the same substance in different states, chemists are especially interested in learning to predict *and* calculate the change in entropy as a reaction occurs.

Entropy Changes in the System: The Standard Entropy of Reaction (ΔS^0_{rxn})

Based on the ideas we discussed in the previous section, we predict that when a given amount (mol) of more ordered reactants yields the same amount of less ordered products, the reaction has a positive entropy change. For example, when cyclopropane is heated to 500°C, the ring opens and propene is formed. The open chain has more freedom of motion than the ring, so the entropy of the product is greater than that of the reactant; that is, the entropy increases during the reaction:

$$H_2C\!\!\overset{\displaystyle CH_2}{\underset{\diagdown\diagup}{\rule{0pt}{0pt}}}\!\!CH_2(g) \longrightarrow CH_3\!-\!CH\!=\!CH_2(g) \qquad \Delta S^0 = S^0_{products} - S^0_{reactants} > 0$$

Such rearrangements are relatively uncommon reactions, however. Far more common are reactions in which the amount (mol) of substances, especially when they are gases, does change during the reaction. For example, when ammonia forms from its elements, 4 mol of gas produces 2 mol of gas. Since gases have such high molar entropies, we predict that the entropy of the products is less than that of the reactants and the entropy decreases during the reaction:

$$N_2(g) + 3H_2(g) \rightleftharpoons 2NH_3(g) \qquad \Delta S^0 = S^0_{products} - S^0_{reactants} < 0$$

Recall that by applying Hess's law, we can combine ΔH^0_f values to find the standard heat of reaction, ΔH^0_{rxn}. Similarly, we can combine standard molar entropies to find the **standard entropy of reaction, ΔS^0_{rxn}**:

$$\Delta S^0_{rxn} = \Sigma m S^0_{products} - \Sigma n S^0_{reactants} \qquad \text{(20.3)}$$

where m and n are the amounts of the individual species, represented by their coefficients in the balanced equation. For the synthesis of ammonia, we have

$$\Delta S^0_{rxn} = [(2\ mol\ NH_3)(S^0\ of\ NH_3)] - [(1\ mol\ N_2)(S^0\ of\ N_2) + (3\ mol\ H_2)(S^0\ of\ H_2)]$$

From Appendix B, we find the appropriate S^0 values:

$$\Delta S^0_{rxn} = (2\ mol)(193\ J/mol\!\cdot\!K)$$
$$- [(1\ mol)(191.5\ J/mol\!\cdot\!K) + (3\ mol)(130.6\ J/mol\!\cdot\!K)]$$
$$= -197\ J/K$$

As we predicted, $\Delta S^0 < 0$.

SAMPLE PROBLEM 20.2 Calculating the Standard Entropy of Reaction, ΔS^0_{rxn}

Problem Calculate ΔS^0_{rxn} for the combustion of 1 mol of propane at 25°C:

$$C_3H_8(g) + 5O_2(g) \longrightarrow 3CO_2(g) + 4H_2O(l)$$

Plan To determine ΔS^0_{rxn}, we apply Equation 20.3. We predict the sign of ΔS^0_{rxn} from the change in the number of moles of gas: 6 mol of gas yields 3 mol of gas, so the entropy will decrease ($\Delta S^0_{rxn} < 0$).

Solution Calculating ΔS^0_{rxn}. Using Appendix B values,

$$\Delta S^0_{rxn} = [(3\ mol\ CO_2)(S^0\ of\ CO_2) + (4\ mol\ H_2O)(S^0\ of\ H_2O)]$$
$$- [(1\ mol\ C_3H_8)(S^0\ of\ C_3H_8) + (5\ mol\ O_2)(S^0\ of\ O_2)]$$
$$= [(3\ mol)(213.7\ J/mol\!\cdot\!K) + (4\ mol)(69.9\ J/mol\!\cdot\!K)]$$
$$- [(1\ mol)(269.9\ J/mol\!\cdot\!K) + (5\ mol)(205.0\ J/mol\!\cdot\!K)]$$
$$= \boxed{-374\ J/K}$$

Check $\Delta S^0 < 0$, so our prediction is correct. Rounding gives $[3(200) + 4(70)] - [270 + 5(200)] = 880 - 1270 = -390$, close to the calculated value.

Comment We based our prediction on the fact that S^0 values of gases are much greater than those of solids or liquids. (This is usually true even when the condensed phases consist of more complex molecules.) When there is no change in the amount (mol) of gas, however, you *cannot* confidently predict the sign of ΔS_{rxn}^0.

FOLLOW-UP PROBLEM 20.2 Balance the following equations, predict the sign of ΔS_{rxn}^0 when possible, and calculate its value at 25°C:
(a) $NaOH(s) + CO_2(g) \longrightarrow Na_2CO_3(s) + H_2O(l)$
(b) $Fe(s) + H_2O(g) \longrightarrow Fe_2O_3(s) + H_2(g)$

Entropy Changes in the Surroundings: The Other Part of the Total

In many spontaneous reactions, such as those we just examined for the synthesis of ammonia and the combustion of propane, the system becomes *more* ordered at the given conditions ($\Delta S_{rxn}^0 < 0$). The second law dictates that *decreases in the entropy of the system can occur only if **increases** in the entropy of the surroundings outweigh them*. Let's examine the influence of the surroundings on the *total* entropy change.

The essential role of the surroundings is to *either add heat to the system or remove heat from it*. In essence, the surroundings function as an enormous heat source or heat sink, one so large that its temperature remains constant, even though its entropy changes through the loss or gain of heat. The surroundings participate in the two possible types of enthalpy changes as follows:

1. *Exothermic change*. Heat lost by the system is gained by the surroundings. The gain of heat increases the random motion of particles in the surroundings, so the entropy of the surroundings increases:
   For an exothermic change,
   $$q_{sys} < 0, \qquad q_{surr} > 0 \qquad \text{and} \qquad \Delta S_{surr} > 0$$
2. *Endothermic change*. Heat gained by the system is lost by the surroundings. The loss of heat reduces the disorder of the surroundings, so the entropy of the surroundings decreases:
   For an endothermic change,
   $$q_{sys} > 0, \qquad q_{surr} < 0 \qquad \text{and} \qquad \Delta S_{surr} < 0$$

The change in entropy of the surroundings is proportional to the quantity of heat transferred to or from the surroundings and, most importantly, *to an opposite change in heat of the system*.

The *temperature* of the surroundings before heat is transferred to or from them also affects ΔS_{surr}. Consider the effect of an exothermic reaction at a low and at a high temperature. At a low temperature, such as 20 K, there is very little random motion in the surroundings, so they are relatively ordered. Therefore, transferring heat to the surroundings has a large effect on their orderliness. At a higher temperature, such as 298 K, the surroundings are already relatively disordered, so transferring the same amount of heat has a smaller effect on their orderliness. In other words, the change in entropy of the surroundings is greater when heat is added at a lower temperature. Thus, putting these ideas together, *the change in entropy of the surroundings is directly related to an opposite change in the heat of the system and inversely related to the temperature of the surroundings before the heat is transferred:*

$$\Delta S_{surr} \propto -q_{sys} \qquad \text{and} \qquad \Delta S_{surr} \propto \frac{1}{T}$$

A Checkbook Analogy for Heating the Surroundings A monetary analogy may clarify the relative changes in disorder that arise from heating the surroundings at different initial temperatures. If you have $10 in your checking account, a $10 deposit represents a 100% increase in your net worth; that is, a given change to a low initial state has a large impact. If, however, you have a $1000 balance, a $10 deposit represents only a 1% increase. That is, the same change to a high initial state has a smaller impact.

Combining these gives

$$\Delta S_{surr} = -\frac{q_{sys}}{T} \tag{20.4}$$

Recall that for a process at *constant pressure*, the heat (q_P) is ΔH, so

$$\Delta S_{surr} = -\frac{\Delta H_{sys}}{T}$$

This means that we can calculate ΔS_{surr} by measuring ΔH_{sys} and the temperature T at which the change takes place.

To restate the central point, if a spontaneous reaction has a negative ΔS_{sys} (the system becomes more ordered), ΔS_{surr} must be positive enough (the surroundings must become sufficiently disordered) that ΔS_{univ} is positive (the universe becomes more disordered). Sample Problem 20.3 illustrates this situation for one of the reactions we considered earlier.

SAMPLE PROBLEM 20.3 Determining Reaction Spontaneity

Problem At 298 K, the formation of ammonia has a negative ΔS_{sys}^0:

$$N_2(g) + 3H_2(g) \longrightarrow 2NH_3(g) \qquad \Delta S_{sys}^0 = -197 \text{ J/K}$$

Calculate ΔS_{univ}^0, and state whether the reaction occurs spontaneously at this temperature.
Plan For the reaction to occur spontaneously, $\Delta S_{univ}^0 > 0$, so ΔS_{surr}^0 must be greater than $+197$ J/K. To find ΔS_{surr}^0, we need ΔH_{sys}^0, which is the same as ΔH_{rxn}^0. We use ΔH_f^0 values from Appendix B to find ΔH_{rxn}^0. Then, we use this calculated value of ΔH_{rxn}^0 and the given T (298 K) to find ΔS_{surr}^0. To find ΔS_{univ}^0, we take the sum of the calculated ΔS_{surr}^0 and the given ΔS_{sys}^0 (-197 J/K).
Solution Calculating ΔH_{sys}^0:

$$\begin{aligned} \Delta H_{sys}^0 &= \Delta H_{rxn}^0 \\ &= (2 \text{ mol NH}_3)(-45.9 \text{ kJ/mol}) - [(3 \text{ mol H}_2)(0 \text{ kJ/mol}) + (1 \text{ mol N}_2)(0 \text{ kJ/mol})] \\ &= -91.8 \text{ kJ} \end{aligned}$$

Calculating ΔS_{surr}^0:

$$\Delta S_{surr}^0 = -\frac{\Delta H_{sys}^0}{T} = -\frac{-91.8 \text{ kJ} \times \dfrac{1000 \text{ J}}{1 \text{ kJ}}}{298 \text{ K}} = 308 \text{ J/K}$$

Determining ΔS_{univ}^0:

$$\Delta S_{univ}^0 = \Delta S_{sys}^0 + \Delta S_{surr}^0 = -197 \text{ J/K} + 308 \text{ J/K} = \boxed{111 \text{ J/K}}$$

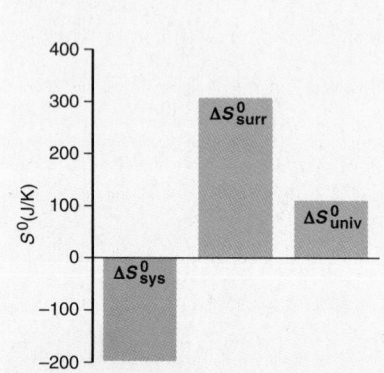

$\Delta S_{univ}^0 > 0$, so the reaction occurs spontaneously at 298 K (see figure in margin).
Check Rounding to check the math, we have

$$\begin{aligned} \Delta H_{rxn}^0 &\approx 2(-45 \text{ kJ}) = -90 \text{ kJ} \\ \Delta S_{surr}^0 &\approx -(-90,000 \text{ J})/300 \text{ K} = 300 \text{ J/K} \\ \Delta S_{univ}^0 &\approx -200 \text{ J/K} + 300 \text{ J/K} = 100 \text{ J/K} \end{aligned}$$

Given the negative ΔH_{rxn}^0, Le Châtelier's principle predicts that low temperature should favor NH$_3$ formation, and so the answer is reasonable (see Chemical Connections essay on the Haber process, Section 17.6).
Comment 1. Note that ΔH^0 has units of kJ, whereas ΔS^0 has units of J/K, so don't forget to convert kJ to J, or you'll introduce a large error.
2. This example highlights the distinction between thermodynamic and kinetic considerations. Even though NH$_3$ forms spontaneously, it does so slowly; the chemical industry expends great effort to catalyze its formation so that it can be produced at a practical rate.

FOLLOW-UP PROBLEM 20.3 Does the oxidation of FeO(s) to Fe$_2$O$_3$(s) occur spontaneously at 298 K?

As Sample Problem 20.3 shows, taking the surroundings into account is crucial to determining reaction spontaneity. Moreover, it clarifies the relevance of thermodynamics to biology, as the Chemical Connections essay shows.

Chemical Connections **Chemistry in Biology**
Do Living Things Obey the Laws of Thermodynamics?

Organisms can be thought of as chemical machines that evolved by extracting energy as efficiently as possible from the environment. Such a mechanical view holds that all processes, whether they involve living or nonliving systems, are consistent with thermodynamic principles. When applied to animals and plants, however, these ideas seem far from obvious. Let's examine the first and second laws of thermodynamics to see if they apply to living systems.

Organisms certainly comply with the first law. The chemical bond energy in food is converted into the mechanical energy of sprouting, crawling, swimming, and countless other movements; the electrical energy of nerve conduction; the thermal energy of warming the body; and so forth. Many experiments have demonstrated that in all these energy conversions, the total energy is conserved. Some of the earliest studies of this question were performed by Lavoisier, who included animal respiration in his new theory of combustion (Figure B20.1). He was the first to show that "animal heat" was produced by a slow combustion process occurring continually in the body. In experiments with guinea pigs, he measured the intake of food and O_2 and the output of CO_2 and heat, for which he invented a calorimeter based on the melting of ice, and he established the principles and methods for measuring the basal rate of metabolism. Modern experiments using a calorimeter large enough to contain an exercising human continue to confirm the conservation of energy (Figure B20.2).

It may not seem as clear, however, that an organism, or for that matter, the whole parade of life, complies with the second law. Mature humans are far more ordered and complex than the simple egg and sperm cells from which they develop. Modern organisms are far more orderly, complex systems than the one-celled ancestral specks from which they evolved. Are the growth of an organism and the evolution of life exceptions to the tendency of natural processes to become more disordered? Does biology violate the second law? Not at all, if we examine the system *and* its surroundings. For an organism to grow or for a species to evolve, uncountable moles of food molecules—carbohydrates, proteins, and fats—and oxygen molecules undergo combustion to form many more moles of gaseous CO_2 and H_2O. The formation and discharge of these waste gases represent a tremendous net increase in the entropy of the surroundings, as does the heat released from the exothermic redox reactions involved. Thus, the increase in *order* apparent in the spontaneous growth of organisms and their evolution occurs at the expense of a far greater increase in *disorder* in the surroundings—Earth and the Sun. When system and surroundings are considered together, the entropy of the universe, as always, increases.

Figure B20.2 A whole-body calorimeter. In this room-sized apparatus, a subject exercises while respiratory gases, energy input and output, and other physiological variables are monitored.

Figure B20.1 Lavoisier studying human respiration as a form of combustion. Lavoisier (standing at right) measured substances consumed and excreted in addition to changes in heat to understand the chemical nature of respiration. The artist (Mme. Lavoisier) shows herself taking notes (seated at right).

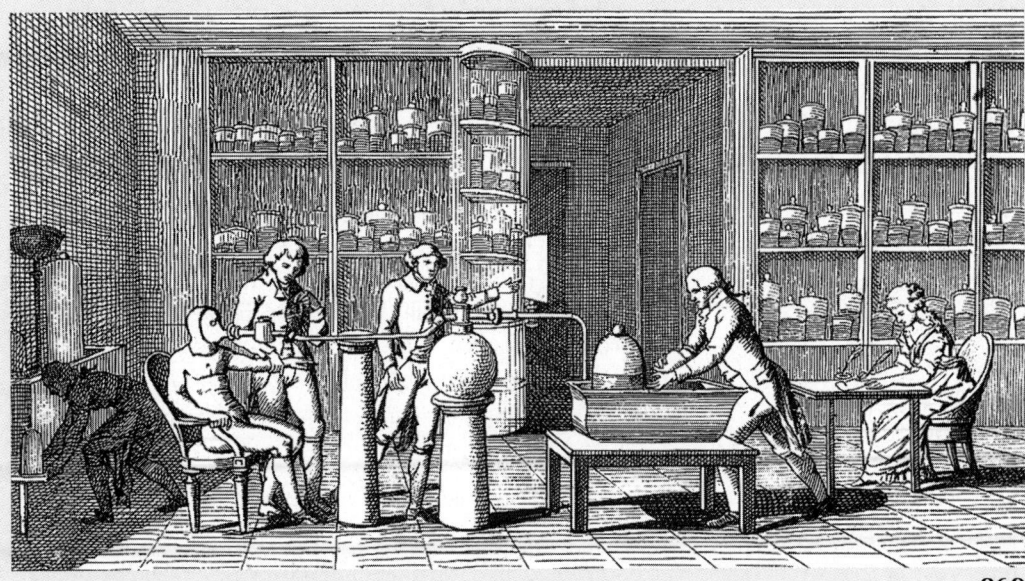

The Entropy Change and the Equilibrium State

A process proceeds toward equilibrium spontaneously, so $\Delta S_{univ} > 0$. When the process reaches equilibrium, there is no longer any driving force to proceed further and, thus, no net change in either direction; that is, $\Delta S_{univ} = 0$. At that point, any entropy change in the system is exactly balanced by an opposite entropy change in the surroundings:

$$\text{At equilibrium: } \Delta S_{univ} = \Delta S_{sys} + \Delta S_{surr} = 0 \quad \text{or} \quad \Delta S_{sys} = -\Delta S_{surr}$$

For example, let's calculate ΔS^0_{univ} for a phase change. For the vaporization-condensation of 1 mol of water at 100°C (373 K),

$$H_2O(l; 373 \text{ K}) \rightleftharpoons H_2O(g; 373 \text{ K})$$

First, we find ΔS^0_{univ} for the forward change (vaporization) by calculating ΔS^0_{sys}:

$$\Delta S^0_{sys} = \Sigma m S^0_{products} - \Sigma n S^0_{reactants} = S^0 \text{ of } H_2O(g; 373 \text{ K}) - S^0 \text{ of } H_2O(l; 373 \text{ K})$$
$$= 195.9 \text{ J/K} - 86.8 \text{ J/K} = 109.1 \text{ J/K}$$

As we expect, the system becomes more disordered ($\Delta S^0_{sys} > 0$) as the liquid changes to a gas.

For ΔS^0_{surr}, we have

$$\Delta S^0_{surr} = -\frac{\Delta H^0_{sys}}{T}$$

where $\Delta H^0_{sys} = \Delta H^0_{vap}$ at 373 K = 40.7 kJ/mol = 40.7×10^3 J/mol. For 1 mol of water, we have

$$\Delta S^0_{surr} = -\frac{\Delta H^0_{vap}}{T} = -\frac{40.7 \times 10^3 \text{ J}}{373 \text{ K}} = -109 \text{ J/K}$$

The negative sign indicates that the surroundings become more ordered because they lose heat. The two entropy changes have the same magnitude but opposite sign:

$$\Delta S^0_{univ} = 109 \text{ J/K} + (-109 \text{ J/K}) = 0$$

For the reverse change (condensation), ΔS^0_{univ} also equals zero, but ΔS^0_{sys} and ΔS^0_{surr} have signs opposite those for vaporization. A similar treatment of a chemical change shows the same result: the entropy change of the forward reaction is *equal in magnitude but opposite in sign* to the entropy change of the reverse reaction. Thus, *when a system reaches equilibrium, neither the forward nor the reverse reaction is spontaneous,* and so neither proceeds any further.

Spontaneous Exothermic and Endothermic Reactions: A Summary

We can now understand why both exothermic and endothermic spontaneous reactions occur. No matter what its *enthalpy* change, a reaction occurs because the total *entropy* of the reacting system and its surroundings increases. The two possibilities are

1. *For an exothermic reaction* ($\Delta H_{sys} < 0$), heat is released, which increases the entropy of the surroundings ($\Delta S_{surr} > 0$).

- If the reacting system yields products that are more disordered than the reactants ($\Delta S_{sys} > 0$), the total entropy change ($\Delta S_{sys} + \Delta S_{surr}$) will be positive, as depicted in Figure 20.10A. For example, in the oxidation of glucose, which is an essential reaction for all higher organisms,

$$C_6H_{12}O_6(s) + 6O_2(g) \longrightarrow 6CO_2(g) + 6H_2O(g) + \textit{heat}$$

6 mol of gas yields 12 mol of gas; thus, $\Delta S_{sys} > 0$, $\Delta S_{surr} > 0$, and $\Delta S_{univ} > 0$.

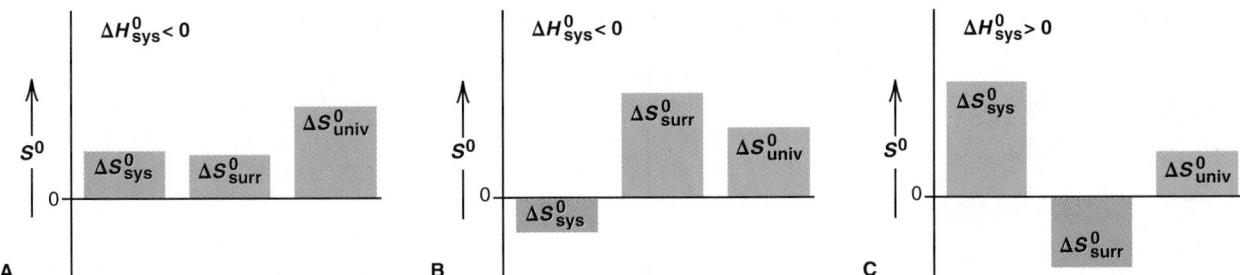

Figure 20.10 **Components of ΔS^0_{univ} for spontaneous reactions.** For a reaction to occur spontaneously, ΔS^0_{univ} must be positive. **A,** An exothermic reaction in which the system becomes more disordered; the size of ΔS^0_{surr} is not important. **B,** An exothermic reaction in which the system becomes more ordered; ΔS^0_{surr} must be larger than ΔS^0_{sys}. **C,** An endothermic reaction in which the system becomes more disordered; ΔS^0_{surr} must be smaller than ΔS^0_{sys}.

- If, on the other hand, the system becomes more ordered as the reaction occurs ($\Delta S_{sys} < 0$), the surroundings must become even more disordered ($\Delta S_{surr} >> 0$) to make the total ΔS positive, as illustrated in Figure 20.10B. For example, when calcium oxide and carbon dioxide gas form calcium carbonate,

$$CaO(s) + CO_2(g) \longrightarrow CaCO_3(s) + \textit{heat}$$

the system becomes more ordered because the amount (mol) of gas decreases. However, the heat released disorders the surroundings even more; thus, $\Delta S_{sys} < 0$, but $\Delta S_{surr} >> 0$, so $\Delta S_{univ} > 0$.

2. *For an endothermic reaction* ($\Delta H_{sys} > 0$), the heat lost by the surroundings lowers the entropy of the surroundings ($\Delta S_{surr} < 0$). Therefore, the only way an endothermic reaction can occur spontaneously is if the system becomes so disordered ($\Delta S_{sys} >> 0$) that its entropy increase outweighs the negative ΔS_{surr}, as shown in Figure 20.10C.

- In the solution processes for many ionic compounds, heat is absorbed to form the solution, so the surroundings become more ordered ($\Delta S_{surr} < 0$). However, the crystalline solid is so highly ordered and the dispersed ions so disordered that the entropy increase ($\Delta S_{sys} >> 0$) outweighs the negative ΔS_{surr}. Thus, ΔS_{univ} is positive.
- Spontaneous endothermic chemical reactions have similar features. For example, in the reaction between barium hydroxide octahydrate and ammonium nitrate (see Figure 20.1),

$$\textit{heat} + Ba(OH)_2 \cdot 8H_2O(s) + 2NH_4NO_3(s) \longrightarrow$$
$$Ba^{2+}(aq) + 2NO_3^-(aq) + 2NH_3(aq) + 10H_2O(l)$$

3 mol of crystalline solids absorbs heat from the surroundings ($\Delta S_{surr} < 0$) and yields 15 mol of dissolved ions and molecules, thereby greatly increasing the disorder of the system ($\Delta S_{sys} >> 0$).

SECTION SUMMARY

The standard entropy of reaction, ΔS^0_{rxn}, is calculated from S^0 values. When the number of moles of gas (Δn_{gas}) increases in a reaction, $\Delta S^0_{rxn} > 0$. The value of ΔS^0_{surr} is related directly to ΔH^0_{sys} and inversely to the temperature at which the enthalpy change occurs. In a spontaneous change, a system can become more ordered only if the surroundings become even more disordered. For a system at equilibrium, $\Delta S_{univ} = 0$.

20.3 ENTROPY, FREE ENERGY, AND WORK

By making *two* separate measurements, ΔS_{sys} and ΔS_{surr}, we can predict whether a reaction will be spontaneous at a particular temperature. It would be useful, however, to have *one* criterion for spontaneity that applies only to the system. The Gibbs free energy, or simply **free energy (G),** is a function that combines the system's enthalpy and entropy:

$$G = H - TS$$

This function, named for Josiah Willard Gibbs, who proposed it and laid the foundation for chemical thermodynamics, is the criterion we've been seeking.

Free Energy Change and Reaction Spontaneity

The free energy change (ΔG) is a measure of the spontaneity of a process and of the useful energy available from it. Let's see how the free energy change is derived from the second law. By definition, the entropy change of the universe is the sum of the entropy changes of the system and the surroundings:

$$\Delta S_{univ} = \Delta S_{sys} + \Delta S_{surr}$$

At constant pressure,

$$\Delta S_{surr} = -\frac{\Delta H_{sys}}{T}$$

Substituting for ΔS_{surr} gives a relationship that lets us focus solely on the system:

$$\Delta S_{univ} = \Delta S_{sys} - \frac{\Delta H_{sys}}{T}$$

Multiplying both sides by $-T$ gives

$$-T\Delta S_{univ} = \Delta H_{sys} - T\Delta S_{sys}$$

Now we can introduce the new free energy quantity to replace the enthalpy and entropy terms. From $G = H - TS$, the *Gibbs equation* shows us the change in free energy of the system (ΔG_{sys}) at constant temperature and pressure:

$$\Delta G_{sys} = \Delta H_{sys} - T\Delta S_{sys} \qquad \textbf{(20.5)}$$

Combining this equation with the previous one shows that

$$-T\Delta S_{univ} = \Delta H_{sys} - T\Delta S_{sys} = \Delta G_{sys}$$

*The **sign** of ΔG tells if a reaction is spontaneous.* The second law dictates

- $\Delta S_{univ} > 0$ for a spontaneous process
- $\Delta S_{univ} < 0$ for a nonspontaneous process
- $\Delta S_{univ} = 0$ for a process at equilibrium

Of course, absolute temperature is always positive, so

$$T\Delta S_{univ} > 0 \qquad \text{or} \qquad -T\Delta S_{univ} < 0 \text{ for a spontaneous process}$$

Because $\Delta G = -T\Delta S_{univ}$, we know that

- $\Delta G < 0$ for a spontaneous process
- $\Delta G > 0$ for a nonspontaneous process
- $\Delta G = 0$ for a process at equilibrium

An important point to keep in mind is that if a process is *nonspontaneous* in one direction ($\Delta G > 0$), it is *spontaneous* in the opposite direction ($\Delta G < 0$). By using ΔG, we have not incorporated any new ideas, but we have developed a way of predicting reaction spontaneity from one variable (ΔG_{sys}) rather than two (ΔS_{sys} and ΔS_{surr}).

As we noted at the beginning of the chapter, the degree of spontaneity of a reaction—that is, the sign *and* magnitude of ΔG—tells us nothing about its rate. Remember that some spontaneous reactions are extremely slow. For example, the reaction between $H_2(g)$ and $O_2(g)$ at room temperature is highly spontaneous—ΔG has a large negative value—but in the absence of a catalyst or a flame, the reaction doesn't occur to a measurable extent because its rate is so low.

Greatness and Obscurity of J. Willard Gibbs Even today, one of the most remarkable minds in science is barely known outside chemistry and physics. In 1878, Josiah Willard Gibbs (1839–1903), a professor of mathematical physics at Yale, completed a 323-page paper that virtually established the science of chemical thermodynamics and included major principles governing chemical equilibrium, phase-change equilibrium, and the energy changes in electrochemical cells. The European scientists James Clerk Maxwell, Wilhelm Ostwald, and Henri Le Châtelier appreciated the significance of Gibbs's achievements long before those in his own country did. In fact, he was elected to the Hall of Fame of Distinguished Americans only in 1950, because he had not received enough votes until then!

Calculating Standard Free Energy Changes

Because free energy (G) combines three state functions, H, S, and T, it is also a state function. As with enthalpy, only free energy *changes* (ΔG) concern us.

The Standard Free Energy Change As with the other thermodynamic variables, to compare the free energy changes of different reactions, we calculate the ***standard* free energy change (ΔG^0),** which occurs when all components of the system are in their standard states. Adapting the Gibbs equation (20.5), we have

$$\Delta G^0_{sys} = \Delta H^0_{sys} - T\Delta S^0_{sys} \qquad (20.6)$$

This important relationship is used frequently to find any one of these three central thermodynamic variables, given the other two. Let's apply this relationship in a sample problem.

SAMPLE PROBLEM 20.4 Calculating ΔG^0 from Enthalpy and Entropy Values

Problem Potassium chlorate, one of the common oxidizing agents in explosives, fireworks (see photo), and matchheads, undergoes a solid-state redox reaction when heated. In this reaction, note that the oxidation number of Cl in the reactant is higher in one of the products and lower in the other (disproportionation):

$$4\overset{+5}{\text{KClO}_3}(s) \xrightarrow{\Delta} 3\overset{+7}{\text{KClO}_4}(s) + \overset{-1}{\text{KCl}}(s)$$

Use ΔH^0_f and S^0 values to calculate ΔG^0_{sys} (ΔG^0_{rxn}) at 25°C for this reaction.

Plan To solve for ΔG^0, we need values from Appendix B. We use ΔH^0_f values to calculate ΔH^0_{rxn} (ΔH^0_{sys}), use S^0 values to calculate ΔS^0_{rxn} (ΔS^0_{sys}), and then apply Equation 20.6.

Solution Calculating ΔH^0_{sys} from ΔH^0_f values (with Equation 6.8):

$$\begin{aligned}
\Delta H^0_{sys} = \Delta H^0_{rxn} &= \Sigma m \Delta H^0_{f(products)} - \Sigma n \Delta H^0_{f(reactants)} \\
&= [(3 \text{ mol KClO}_4)(\Delta H^0_f \text{ of KClO}_4) + (1 \text{ mol KCl})(\Delta H^0_f \text{ of KCl})] \\
&\quad - [(4 \text{ mol KClO}_3)(\Delta H^0_f \text{ of KClO}_3)] \\
&= [(3 \text{ mol})(-432.8 \text{ kJ/mol}) + (1 \text{ mol})(-436.7 \text{ kJ/mol})] \\
&\quad - [(4 \text{ mol})(-397.7 \text{ kJ/mol})] \\
&= -144 \text{ kJ}
\end{aligned}$$

Calculating ΔS^0_{sys} from S^0 values (with Equation 20.3):

$$\begin{aligned}
\Delta S^0_{sys} = \Delta S^0_{rxn} &= [(3 \text{ mol KClO}_4)(S^0 \text{ of KClO}_4) + (1 \text{ mol KCl})(S^0 \text{ of KCl})] \\
&\quad - [(4 \text{ mol KClO}_3)(S^0 \text{ of KClO}_3)] \\
&= [(3 \text{ mol})(151.0 \text{ J/mol·K}) + (1 \text{ mol})(82.6 \text{ J/mol·K})] \\
&\quad - [(4 \text{ mol})(143.1 \text{ J/mol·K})] \\
&= -36.8 \text{ J/K}
\end{aligned}$$

Calculating ΔG^0_{sys} at 298 K:

$$\Delta G^0_{sys} = \Delta H^0_{sys} - T\Delta S^0_{sys} = -144 \text{ kJ} - \left[(298 \text{ K})(-36.8 \text{ J/K})\left(\frac{1 \text{ kJ}}{1000 \text{ J}}\right)\right] = \boxed{-133 \text{ kJ}}$$

Check The reaction *is* spontaneous, which is consistent with $\Delta G^0 < 0$. Rounding to check the math:

$\Delta H^0 \approx [3(-433 \text{ kJ}) + (-440 \text{ kJ})] - [4(-400 \text{ kJ})] = -1740 \text{ kJ} + 1600 \text{ kJ} = -140 \text{kJ}$

$\Delta S^0 \approx [3(150 \text{ J/K}) + 85 \text{ J/K}] - [4(145 \text{ J/K})] = 535 \text{ J/K} - 580 \text{ J/K} = -45 \text{ J/K}$

$\Delta G^0 \approx -140 \text{ kJ} - 300 \text{ K}(-0.04 \text{ kJ/K}) = -140 \text{ kJ} + 12 \text{ kJ} = -128 \text{ kJ}$

All values are close to the calculated ones. Another way to calculate ΔG^0 for this reaction appears in Sample Problem 20.5.

Comment Although the negative ΔG^0 means the reaction is spontaneous, the rate is extremely low because the required transfer of an O atom does not occur easily in the solid. When KClO$_3$ is heated slightly above its melting point, the ions are free to move and the reaction occurs readily.

Potassium chlorate is the oxidizing agent in fireworks.

FOLLOW-UP PROBLEM 20.4 Determine the standard free energy change at 298 K for the reaction $2\text{NO}(g) + \text{O}_2(g) \longrightarrow 2\text{NO}_2(g)$.

The Standard Free Energy of Formation Another way to calculate ΔG^0_{rxn} is with values for the **standard free energy of formation (ΔG^0_f)** of the components; ΔG^0_f is the free energy change that occurs when 1 mol of compound is made *from its elements,* with all components in their standard states. Because free energy is a state function, we can combine ΔG^0_f values of reactants and products to calculate ΔG^0_{rxn} without regard for the reaction path:

$$\Delta G^0_{rxn} = \Sigma m \Delta G^0_{f(products)} - \Sigma n \Delta G^0_{f(reactants)} \qquad \textbf{(20.7)}$$

ΔG^0_f values have properties similar to ΔH^0_f values:

- ΔG^0_f of an element in its standard state is zero.
- An equation coefficient (*m* or *n* above) multiplies ΔG^0_f by that number.
- Reversing a reaction changes the sign of ΔG^0_f.

Many ΔG^0_f values appear with those for ΔH^0_f and S^0 in Appendix B.

SAMPLE PROBLEM 20.5 Calculating ΔG^0_{rxn} from ΔG^0_f Values

Problem Use ΔG^0_f values to calculate ΔG^0_{rxn} for the reaction in Sample Problem 20.4:
$$4KClO_3(s) \longrightarrow 3KClO_4(s) + KCl(s)$$
Plan We apply Equation 20.7 to calculate ΔG^0_{rxn}.
Solution

$$\begin{aligned}
\Delta G^0_{rxn} &= \Sigma m \Delta G^0_{f(products)} - \Sigma n \Delta G^0_{f(reactants)} \\
&= [(3 \text{ mol } KClO_4)(\Delta G^0_f \text{ of } KClO_4) + (1 \text{ mol } KCl)(\Delta G^0_f \text{ of } KCl)] \\
&\quad - [(4 \text{ mol } KClO_3)(\Delta G^0_f \text{ of } KClO_3)] \\
&= [(3 \text{ mol})(-303.2 \text{ kJ/mol}) + (1 \text{ mol})(-409.2 \text{ kJ/mol})] \\
&\quad - [(4 \text{ mol})(-296.3 \text{ kJ/mol})] \\
&= \boxed{-134 \text{ kJ}}
\end{aligned}$$

Check Rounding to check the math:
$$\begin{aligned}
\Delta G^0_{rxn} &\approx [3(-300 \text{ kJ}) + 1(-400 \text{ kJ})] - 4(-300 \text{ kJ}) \\
&= -1300 \text{ kJ} + 1200 \text{ kJ} = -100 \text{ kJ}
\end{aligned}$$
Comment The slight discrepancy between this value and that in Sample Problem 20.4 is within experimental error. As you can see, when ΔG^0_f values are available for a reaction taking place at 25°C, this method is simpler than the calculation in Sample Problem 20.4.

FOLLOW-UP PROBLEM 20.5 Use ΔG^0_f values to calculate the free energy change at 25°C for each of the following reactions:
(a) $2NO(g) + O_2(g) \longrightarrow 2NO_2(g)$ (from Follow-up Problem 20.4)
(b) $2C(graphite) + O_2(g) \longrightarrow 2CO(g)$

ΔG and the Work a System Can Do

Recall that the science of thermodynamics was born soon after the invention of the steam engine, and one of the most practical relationships in the field is that between the free energy change and the work a system can do:

- For a spontaneous process, ΔG is the *maximum work obtainable* **from** the system as the process takes place:
$$\Delta G = w_{max}$$

- For a nonspontaneous process, ΔG is the *minimum work that must be done* **to** the system to make the process take place.

What do we mean by the "maximum work obtainable," and what determines this limit? The free energy change is the maximum work the system can *possibly* do. But the work the system *actually* does depends on how the free energy is released. To understand this, let's first consider a nonchemical process. Suppose a gas is confined within a cylinder, at $V_{initial}$, by a piston attached to a 1-kg weight (see margin). As the gas expands and lifts the weight, its pressure becomes just balanced by the weight at some final volume, V_{final}. The gas lifted the weight in one step, thus doing a certain quantity of work. The expanding gas can do more work by lifting a 2-kg weight to one-half V_{final}; the 2-kg weight is then replaced by the 1-kg weight, which the gas lifts the rest of the way to V_{final}—that is, V_{final} is reached in two steps. Similarly, it can do even more work by lifting a 3-kg weight to one-third V_{final}, the 2-kg weight to one-half V_{final}, and then the 1-kg weight the rest of the way to V_{final}—that is, in three steps. As the number of steps increases, the quantity of work done by the gas increases. Therefore, in the limit, the gas does the maximum possible work because it lifts the weights in an infinite number of steps. Note that, in such a hypothetical process, an infinitesimal increase in the weight at any step would reverse the expansion. In general, *a reversible process is one that can be changed in either direction by an infinitesimal change in a variable. The maximum work from a spontaneous process is obtained only if the work is carried out reversibly.*

Of course, in any *real* process, work is performed *irreversibly*, that is, in a finite number of steps, so *we can never obtain the maximum work.* The free energy not used for work appears as heat lost to the surroundings. This "unharnessed" energy is a consequence of any real spontaneous process.

Let's consider two examples of systems that use a chemical reaction to do work—a car engine and a battery. When gasoline (represented by octane) is burned in a car engine,

$$C_8H_{18}(l) + \tfrac{25}{2}O_2(g) \longrightarrow 8CO_2(g) + 9H_2O(g),$$

a large amount of energy is given off as heat ($\Delta H_{sys} < 0$), and because the number of moles of gas increases, the entropy of the system increases ($\Delta S_{sys} > 0$). This reaction is spontaneous ($\Delta G_{sys} < 0$) at all temperatures. The free energy available does work turning wheels, moving belts, regenerating the battery, and so on. However, only if it is released reversibly, that is, in a series of infinitesimal steps, do we obtain the maximum work available from this reaction. In reality, the process is carried out irreversibly and much of the total free energy just warms the engine and the outside air, which makes the motions of the particles in the universe more chaotic, in accord with the second law.

A battery is essentially a packaged spontaneous redox reaction that releases free energy to the surroundings (flashlight, radio, motor, etc.). If we connect the battery terminals to each other through a short piece of wire, the free energy change is released all at once but does no work—it just heats the wire and battery. If we connect the terminals to a motor, the free energy is released more slowly, and a sizeable portion of it runs the motor, but some is still converted to heat in the battery and the motor. If we discharge the battery still more slowly, more of the free energy change does work and less is converted to heat, but only when the battery discharges infinitely slowly can we obtain the maximum work.

This is the compromise that all engineers and machine designers must face— *in the real world, some free energy is always changed to heat and is, thereby, unharnessed:* no real process uses all the available free energy to do work.

Let's summarize the relationship between the free energy change of a reaction and the work it can actually do:

- A spontaneous reaction ($\Delta G_{sys} < 0$) will occur *and* can do work on the surroundings. In any real machine, however, the work obtained from the reaction is *always less than the maximum* because some of the ΔG released is lost as heat.
- A nonspontaneous reaction ($\Delta G_{sys} > 0$) will not occur unless the surroundings do work on it. In any real machine, however, the work needed to make the reaction occur is *always more than the minimum* because some of the ΔG added is lost as heat.
- A reaction at equilibrium ($\Delta G_{sys} = 0$) can no longer do any work.

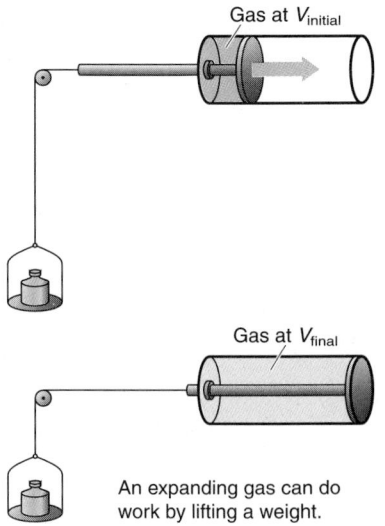

An expanding gas can do work by lifting a weight.

The Wide Range of Energy Efficiency One definition for the efficiency of a device is the percentage of the energy input that results in work output. The range of efficiencies is enormous for the common "energy-conversion systems" of society. For instance, a common incandescent lightbulb converts only 5% of incoming electrical energy to light, the rest is given off as heat. At the other extreme, a large electrical generator converts 99% of the incoming mechanical energy to electricity. Although improvements are continually being made, here are some approximate efficiency values for other devices: home oil furnace, 65%; hand tool motor, 63%; liquid fuel rocket, 50%; car engine, <30%; fluorescent lamp, 20%; solar cell, ~15%.

The Effect of Temperature on Reaction Spontaneity

In most cases, the enthalpy contribution (ΔH^0) to the free energy change (ΔG^0) is much *larger* than the entropy contribution ($T\Delta S^0$). For this reason, most exothermic reactions are spontaneous: the negative ΔH^0 helps make ΔG^0 negative. However, the *temperature at which a reaction occurs influences the magnitude of the $T\Delta S^0$ term,* so, in many cases, the overall spontaneity depends on the temperature.

By scrutinizing the signs of ΔH^0 and ΔS^0, we can predict the effect of temperature on the sign of ΔG^0 and thus on the spontaneity of a process at any temperature. (In this discussion, we assume that ΔH^0 and ΔS^0 change very little with temperature, which is usually correct as long as no phase changes occur.) Let's examine the four combinations of positive and negative ΔH^0 and ΔS^0; two combinations do not depend on temperature and two do:

* *Temperature-independent cases (opposite signs).* When ΔH^0 and ΔS^0 have *opposite* signs, the reaction occurs spontaneously either at all temperatures or at none.

   1. *Reaction is spontaneous at all temperatures: $\Delta H^0 < 0$, $\Delta S^0 > 0$.* Both contributions favor the process occurring spontaneously: ΔH^0 is negative and ΔS^0 is positive, so $-T\Delta S^0$ is negative; thus, ΔG^0 is always negative. Many combustion reactions are in this category, including those of glucose and octane that we considered earlier. The decomposition of hydrogen peroxide, a common disinfectant, is also spontaneous at all temperatures:

   $$2H_2O_2(l) \longrightarrow 2H_2O(l) + O_2(g) \qquad \Delta H^0 = -196 \text{ kJ and } \Delta S^0 = 125 \text{ J/K}$$

   2. *Reaction is nonspontaneous at all temperatures: $\Delta H^0 > 0$, $\Delta S^0 < 0$.* Both contributions oppose the reaction being spontaneous. ΔH^0 is positive and ΔS^0 is negative, so $-T\Delta S^0$ is positive; thus, ΔG^0 is always positive. The formation of ozone from oxygen is not spontaneous at any temperature:

   $$3O_2(g) \longrightarrow 2O_3(g) \qquad \Delta H^0 = 286 \text{ kJ and } \Delta S^0 = -137 \text{ J/K}$$

   This reaction occurs only if enough energy is supplied from the surroundings, as when ozone is synthesized by passing an electrical discharge through pure O_2 (see photo).

* *Temperature-dependent cases (same signs).* When ΔH^0 and ΔS^0 have the *same* sign, the relative magnitudes of the $-T\Delta S^0$ and ΔH^0 terms determine the sign of ΔG^0, so the magnitude of T is crucial.

   3. *Reaction is spontaneous at higher temperatures: $\Delta H^0 > 0$ and $\Delta S^0 > 0$.* In these cases, ΔS^0 favors spontaneity ($-T\Delta S^0 < 0$), but ΔH^0 does not. For example,

   $$2N_2O(g) + O_2(g) \longrightarrow 4NO(g) \qquad \Delta H^0 = 197.1 \text{ kJ and } \Delta S^0 = 198.2 \text{ J/K}$$

   With a positive ΔH^0, the reaction will occur spontaneously only when $-T\Delta S^0$ is large enough to make ΔG^0 negative, which will happen at higher temperatures. The oxidation of N_2O occurs spontaneously at $T > 994$ K.

   4. *Reaction is spontaneous at lower temperatures: $\Delta H^0 < 0$ and $\Delta S^0 < 0$.* In these cases, ΔH^0 favors spontaneity, but ΔS^0 does not ($-T\Delta S > 0$). For example,

   $$2Na(s) + Cl_2(g) \longrightarrow 2NaCl(s) \qquad \Delta H^0 = -822.2 \text{ kJ and } \Delta S^0 = -181.7 \text{ J/K}$$

   With a negative ΔH^0, the reaction will occur spontaneously only if the $-T\Delta S^0$ term is smaller than the ΔH^0 term, and this happens at lower temperatures. Common examples are the formation of ammonium halides from ammonia and a hydrogen halide or the formation of metal oxides, fluorides, and chlorides from their elements. The production of sodium chloride occurs spontaneously at $T < 4525$ K.

Table 20.1 summarizes these four possible combinations of ΔH^0 and ΔS^0.

An industrial ozone generator.

Table 20.1 Reaction Spontaneity and the Signs of ΔH^0, ΔS^0, and ΔG^0

ΔH^0	ΔS^0	$-T\Delta S^0$	ΔG^0	Description
−	+	−	−	Spontaneous at all T
+	−	+	+	Nonspontaneous at all T
+	+	−	+ or −	Spontaneous at higher T; nonspontaneous at lower T
−	−	+	+ or −	Spontaneous at lower T; nonspontaneous at higher T

As you saw in Sample Problem 20.4, one way to calculate ΔG^0 is from enthalpy and entropy changes. Because ΔH^0 and ΔS^0 usually change little with temperature, we can use their values at 298 K to examine the effect of temperature on ΔG^0 and thus on reaction spontaneity.

SAMPLE PROBLEM 20.6 Determining the Effect of Temperature on ΔG^0

Problem An important reaction in the production of sulfuric acid is the oxidation of $SO_2(g)$ to $SO_3(g)$:

$$2SO_2(g) + O_2(g) \longrightarrow 2SO_3(g)$$

At 298 K, $\Delta G^0 = -141.6$ kJ; $\Delta H^0 = -198.4$ kJ; and $\Delta S^0 = -187.9$ J/K.
(a) Use the data to decide if this reaction is spontaneous at 25°C, and predict how ΔG^0 will change with increasing T.
(b) Assuming ΔH^0 and ΔS^0 are constant with increasing T, is the reaction spontaneous at 900.°C?
Plan (a) We note the sign of ΔG^0 to see if the reaction is spontaneous and the signs of ΔH^0 and ΔS^0 to see the effect of T. (b) We use Equation 20.6 to calculate ΔG^0 from the given ΔH^0 and ΔS^0 at the higher T (in K).
Solution (a) $\Delta G^0 < 0$, so the reaction is spontaneous at 298 K: a mixture of $SO_2(g)$, $O_2(g)$, and $SO_3(g)$ in their standard states (1 atm) will spontaneously yield more $SO_3(g)$. With $\Delta S^0 < 0$, the term $-T\Delta S^0 > 0$ and becomes more positive at higher T. Therefore, ΔG^0 will be less negative, and the reaction less spontaneous, with increasing T.
(b) Calculating ΔG^0 at 900.°C ($T = 273 + 900. = 1173$ K):

$\Delta G^0 = \Delta H^0 - T\Delta S^0 = -198.4$ kJ $- [(1173$ K$)(-187.9$ J/K$)(1$ kJ/1000 J$)] = 22.0$ kJ

$\Delta G^0 > 0$, so the reaction is nonspontaneous at the higher T.
Check The answer in part (b) seems reasonable based on our prediction in part (a). The arithmetic seems correct, given the following considerable rounding:

$$\Delta G^0 \approx -200 \text{ kJ} - [(1200 \text{ K})(-200 \text{ J/K})/1000)] = +40 \text{ kJ}$$

FOLLOW-UP PROBLEM 20.6 A reaction is nonspontaneous at room temperature but *is* spontaneous at −40°C. What can you say about the signs and relative magnitudes of ΔH^0, ΔS^0, and $-T\Delta S^0$?

The Temperature at Which a Reaction Becomes Spontaneous As you have just seen, when the signs of ΔH^0 and ΔS^0 are the same, some reactions that are non-spontaneous at one temperature become spontaneous at another, and vice versa. It would certainly be useful to know the temperature at which a reaction becomes spontaneous. This is the temperature at which a positive ΔG^0 switches to a negative ΔG^0 because of the changing magnitude of the $-T\Delta S^0$ term. We find this crossover temperature by setting ΔG^0 equal to zero and solving for T:

$$\Delta G^0 = \Delta H^0 - T\Delta S^0 = 0$$

Therefore, $\Delta H^0 = T\Delta S^0$ and $T = \dfrac{\Delta H^0}{\Delta S^0}$ (20.8)

Consider the reaction of copper(I) oxide with carbon, which does *not* occur at lower temperature but is used at higher temperature in a step during the extraction of copper metal from chalcocite (Sample Problems 3.8 and 3.9):

$$Cu_2O(s) + C(s) \longrightarrow 2Cu(s) + CO(g)$$

We predict that this reaction has a positive entropy change because of the increase in number of moles of gas; in fact, $\Delta S^0 = 165$ J/K. Furthermore, with the reaction *non*spontaneous at lower temperatures, it must have a positive ΔH^0 (58.1 kJ). As the $-T\Delta S^0$ term becomes more negative at higher temperatures, it will eventually outweigh the positive ΔH^0 term, and the reaction will occur spontaneously.

Let's calculate ΔG^0 for this reaction at 25°C and then find the temperature above which the reaction is spontaneous. At 25°C (298 K),

$$\Delta G^0 = \Delta H^0 - T\Delta S^0 = 58.1 \text{ kJ} - \left(298 \text{ K} \times 165 \text{ J/K} \times \frac{1 \text{ kJ}}{1000 \text{ J}}\right) = 8.9 \text{ kJ}$$

Because ΔG^0 is positive, the reaction will not proceed on its own at 25°C. At the crossover temperature, $\Delta G^0 = 0$, so

$$T = \frac{\Delta H^0}{\Delta S^0} = \frac{58.1 \text{ kJ} \times \dfrac{1000 \text{ J}}{1 \text{ kJ}}}{165 \text{ J/K}} = 352 \text{ K}$$

At any temperature above 352 K (79°C), a moderate temperature for recovering a metal from its ore, the reaction occurs spontaneously. Figure 20.11 depicts this result. The line for $T\Delta S^0$ increases steadily (and thus the $-T\Delta S^0$ term becomes more negative) with rising T. It crosses the relatively constant ΔH^0 line at 352 K. At any higher T, the $-T\Delta S^0$ term is greater than the ΔH^0 term, so ΔG^0 is negative.

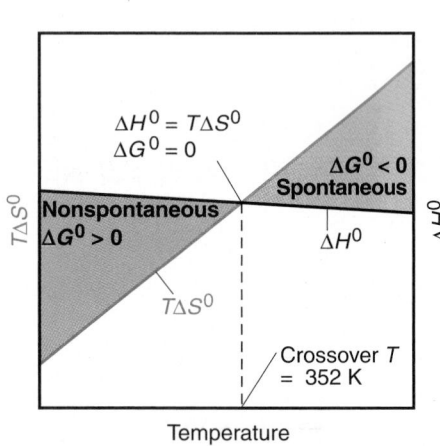

Figure 20.11 **The effect of temperature on reaction spontaneity.** The two terms that make up ΔG are plotted against T. The figure shows a relatively constant ΔH^0 and a steadily increasing $T\Delta S^0$ (and thus more negative $-T\Delta S^0$) for the reaction between Cu_2O and C. At low T, the reaction is nonspontaneous ($\Delta G^0 > 0$) because the positive ΔH^0 term has a greater magnitude than the negative $T\Delta S^0$ term. At 352 K, $\Delta H^0 = T\Delta S^0$, so $\Delta G^0 = 0$. At any higher T, the reaction becomes spontaneous ($\Delta G^0 < 0$) because the $-T\Delta S^0$ term dominates.

Coupling of Reactions to Drive a Nonspontaneous Change

When we break down a spontaneous multistep reaction to its component steps, we often find that a nonspontaneous step is driven by a spontaneous step in a **coupling of reactions.** One step supplies enough free energy for the other to occur, just as the combustion of gasoline supplies enough free energy to move a car.

Consider the reaction we just discussed for the reduction of copper(I) oxide by carbon. We found that the *overall* reaction becomes spontaneous at any temperature above 352 K. When we divide the reaction into two steps, however, we find that even at a higher temperature, such as 375 K, copper(I) oxide does not spontaneously decompose to its elements:

$$Cu_2O(s) \longrightarrow 2Cu(s) + \tfrac{1}{2}O_2(g) \qquad \Delta G^0_{375} = 140.0 \text{ kJ}$$

However, the oxidation of carbon to CO at 375 K is quite spontaneous:

$$C(s) + \tfrac{1}{2}O_2(g) \longrightarrow CO(g) \qquad \Delta G^0_{375} = -143.8 \text{ kJ}$$

Coupling these reactions means having the carbon in contact with the Cu_2O, which allows the reaction with the larger negative ΔG^0 to "drive" the one with the smaller positive ΔG^0. Adding the reactions together and canceling terms gives an overall negative ΔG^0:

$$Cu_2O(s) + C(s) \longrightarrow 2Cu(s) + CO(g) \qquad \Delta G^0_{375} = -3.8 \text{ kJ}$$

Many biochemical reactions are also nonspontaneous. Key steps in the synthesis of proteins and nucleic acids, the formation of fatty acids, the maintenance of ion balance, and the breakdown of nutrients are among essential processes with positive ΔG^0 values. Driving these energetically unfavorable steps by coupling them to a spontaneous one is a life-sustaining strategy common to all organisms—animals, plants, and microbes—as we discuss in the Chemical Connections essay.

The Universal Role of ATP

One of the most remarkable features of living organisms, as well as a strong indication of a common biological ancestry, is the utilization of the same few biomolecules for all the reactions of life. Despite their bewildering diversity of appearance and behavior, virtually all organisms use the same 20 amino acids to make their proteins, the same four or five nucleotides to make their nucleic acids, and the same carbohydrate (glucose) to provide energy.

In addition, *all organisms use the same spontaneous reaction to provide the free energy needed to drive a wide variety of nonspontaneous ones.* This reaction is the hydrolysis of a high-energy molecule called **adenosine triphosphate (ATP)** to adenosine diphosphate (ADP):

$$ATP^{4-} + H_2O \rightleftharpoons ADP^{3-} + HPO_4^{2-} + H^+$$
$$\Delta G^0 = -30.5 \text{ kJ}$$

In the metabolic breakdown of glucose, for example, the initial step is the addition of a phosphate group to a glucose molecule in a dehydration-condensation reaction:

$$\text{Glucose} + HPO_4^{2-} + H^+ \rightleftharpoons [\text{glucose phosphate}]^- + H_2O$$
$$\Delta G^0 = 13.8 \text{ kJ}$$

By coupling this nonspontaneous reaction to ATP hydrolysis, the overall reaction becomes spontaneous. If we add these two reactions, HPO_4^{2-}, H^+, and H_2O cancel, and we obtain

$$\text{Glucose} + ATP^{4-} \rightleftharpoons [\text{glucose phosphate}]^- + ADP^{3-}$$
$$\Delta G^0 = -16.7 \text{ kJ}$$

Like the reactions to extract copper that we just discussed, these two reactions cannot affect each other if they are physically separated. Coupling of the reactions is accomplished through an enzyme, a biological catalyst (Section 16.8) that simultaneously binds glucose and ATP such that the phosphate group of ATP to be transferred lies next to the particular —OH group of glucose

that will accept it (Figure B20.3). Enzymes play similar catalytic roles in all the reactions driven by ATP hydrolysis.

The ADP formed in these energy-releasing reactions is combined with phosphate to regenerate ATP in energy-absorbing reactions catalyzed by other enzymes. In fact, the underlying biochemical reason an organism eats and breathes is to make ATP, so that it has the energy to move, grow, reproduce—and study chemistry, of course. Thus, there is a continuous cycling of ATP to ADP and back to ATP again to supply energy to the cells (Figure B20.4).

(continued)

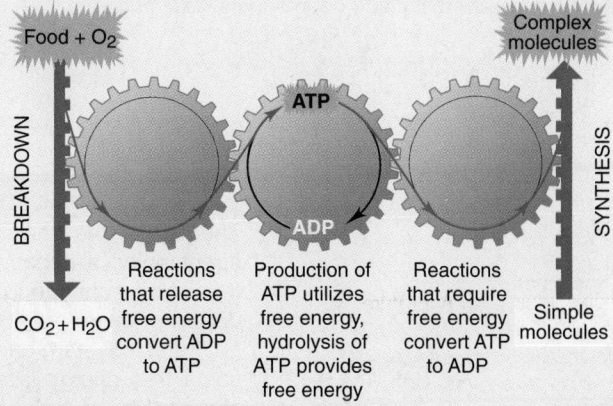

Figure B20.4 The cycling of metabolic free energy through ATP. Processes that release free energy are coupled to the formation of ATP from ADP, whereas those that require free energy are coupled to the hydrolysis of ATP to ADP.

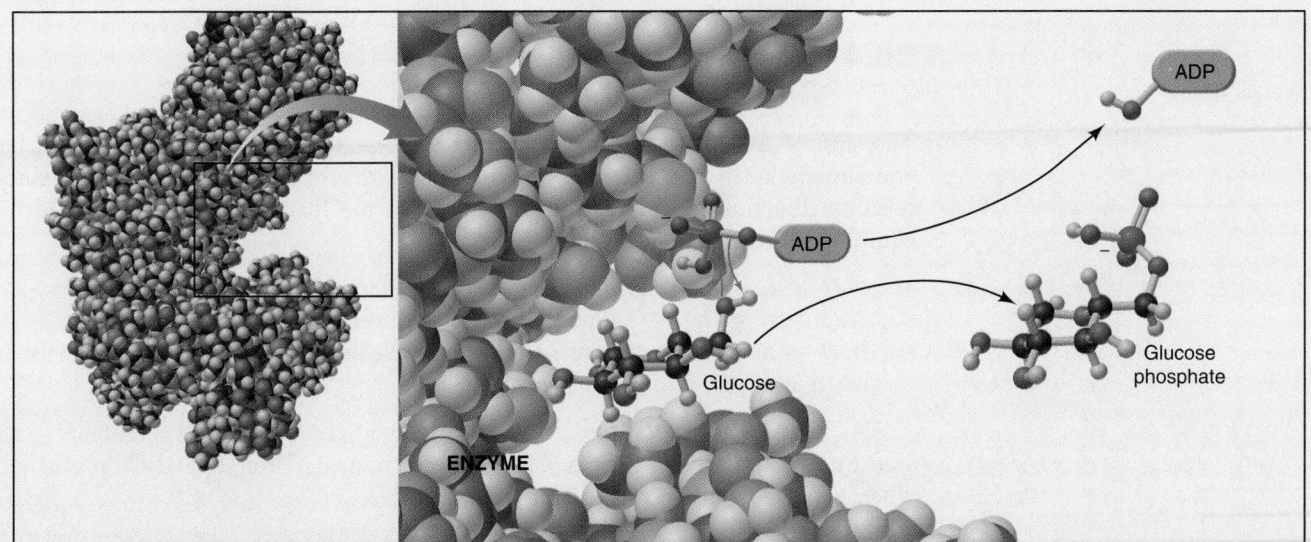

Figure B20.3 The coupling of a nonspontaneous reaction to the hydrolysis of ATP. The glucose molecule must lie next to the ATP molecule (shown as ADP—O—PO₃H) in the enzyme's active site for the correct atoms to form bonds. ADP (shown as ADP—OH) and glucose phosphate are released.

What makes ATP hydrolysis such a good supplier of free energy? By examining the phosphate portions of ATP, ADP, and HPO_4^{2-}, we can see two reasons (Figure B20.5). The first is that, at physiological pH ($\sim$7), the triphosphate portion of the ATP molecule has an average of four negative charges grouped closely together. As a result, the molecule experiences a built-in *high charge repulsion*. In ADP, however, some of these repulsions are relieved.

The second reason relates to the greater delocalization of π electrons in the hydrolysis products, which we can see by using resonance structures. For example, once ATP is hydrolyzed, the π electrons in HPO_4^{2-} can be more readily delocalized, which stabilizes the ion. Thus, greater charge repulsion and less electron delocalization make ATP higher in energy than the sum of ADP and HPO_4^{2-}. When ATP is hydrolyzed, some of this additional energy is released to be harnessed by the organism to drive the metabolic reactions that could not otherwise take place.

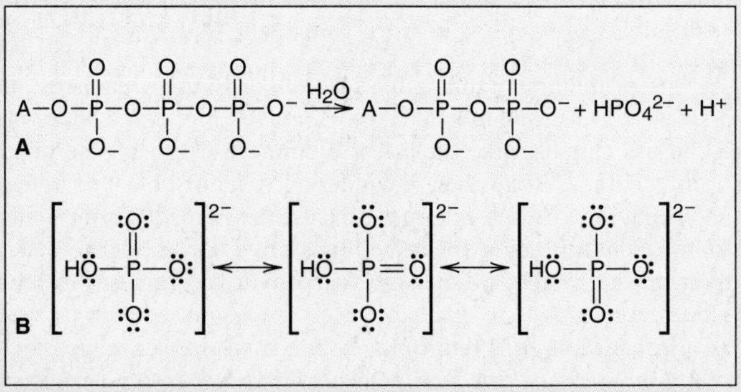

Figure B20.5 Why is ATP a high-energy molecule? The ATP molecule releases a large amount of free energy when it is hydrolyzed because **(A)** the high charge repulsion in the triphosphate portion of ATP is partially relieved, and **(B)** the free HPO_4^{2-} ion is stabilized by delocalization of its π electrons, as shown in several resonance forms.

SECTION SUMMARY

The sign of the free energy change, $\Delta G = \Delta H - T\Delta S$, is directly related to reaction spontaneity: a negative ΔG corresponds to a positive ΔS_{univ}. We use the standard free energy change (ΔG^0) to evaluate a reaction's spontaneity, and we use the standard free energy of formation (ΔG_f^0) to calculate ΔG_{rxn}^0 at 25°C. The maximum work (ΔG) that a system can do is never obtained from a real (irreversible) process because some free energy is always converted to heat. The magnitude of T influences the spontaneity of temperature-dependent reactions (same signs of ΔH and ΔS) by affecting the size of $T\Delta S$. For such reactions, the T at which the reaction becomes spontaneous can be found by setting $\Delta G = 0$. A nonspontaneous reaction ($\Delta G > 0$) can be made to occur by being coupled to a more spontaneous one ($\Delta G \ll 0$). For example, in organisms, the hydrolysis of ATP drives many reactions with a positive ΔG.

20.4 FREE ENERGY, EQUILIBRIUM, AND REACTION DIRECTION

The sign of ΔG allows us to predict reaction spontaneity and thus direction, but you already know that it is not the only way to do so. In Chapter 17 we predicted reaction direction by comparing the values of the reaction quotient (Q) and the equilibrium constant (K). Recall that

- If $Q < K$ ($Q/K < 1$), the reaction as written proceeds to the right.
- If $Q > K$ ($Q/K > 1$), the reaction as written proceeds to the left.
- If $Q = K$ ($Q/K = 1$), the reaction has reached equilibrium, and there is no net reaction in either direction.

As you might expect, these two ways of predicting reaction spontaneity—the sign of ΔG and the magnitude of Q/K—are related. Their relationship emerges when we compare the signs of $\ln Q/K$ with ΔG:

- If $Q/K < 1$, then $\ln Q/K < 0$: reaction proceeds to the right ($\Delta G < 0$).
- If $Q/K > 1$, then $\ln Q/K > 0$: reaction proceeds to the left ($\Delta G > 0$).
- If $Q/K = 1$, then $\ln Q/K = 0$: reaction is at equilibrium ($\Delta G = 0$).

Note that the signs of ΔG and $\ln Q/K$ are identical for a given reaction direction. In fact, ΔG and $\ln Q/K$ are proportional to each other and made equal through the constant RT:

$$\Delta G = RT \ln \frac{Q}{K} = RT \ln Q - RT \ln K \qquad (20.9)$$

What does this central relationship mean? As you know, Q represents the concentrations (or pressures) of a system's components at any time during the reaction, whereas K represents them when the reaction has reached equilibrium. Therefore, Equation 20.9 states that the free energy change of the system is the difference between the free energy of the system in some initial state, Q, and the free energy of the system in its final state, K. For a system at equilibrium, Q has become equal to K, so $\Delta G = 0$. In other words, at equilibrium, no further free energy change occurs; in effect, the system has released all its free energy in the process of attaining equilibrium.

The size of ΔG depends on the difference in the sizes of Q and K. By choosing standard-state values for Q, we obtain the standard free energy change (ΔG^0). When all concentrations are 1 M (or all pressures 1 atm), ΔG equals ΔG^0 and Q equals 1:

$$\Delta G^0 = RT \ln 1 - RT \ln K$$

We know that $\ln 1 = 0$, so the $RT \ln Q$ term drops out, and we have

$$\Delta G^0 = -RT \ln K \qquad (20.10)$$

This very important relationship allows us to calculate the standard free energy change of a reaction (ΔG^0) from its equilibrium constant, or vice versa. Because ΔG^0 is related logarithmically to K, even a small change in the value of ΔG^0 has a large effect on the value of K. Table 20.2 shows the K values that correspond to a range of ΔG^0 values. Note that as ΔG^0 becomes more positive, the equilibrium constant becomes smaller, which means the reaction reaches equilibrium with less product and more reactant. Similarly, as ΔG^0 becomes more negative, the reaction reaches equilibrium with more product and less reactant. For example, if $\Delta G^0 = +10$ kJ, $K \approx 0.02$, which means that the product terms are about 1/50th as large as the reactant terms; whereas, if $\Delta G^0 = -10$ kJ, they are 50 times as large.

Of course, most reactions do not begin with all components in their standard states. By substituting the relationship between ΔG^0 and K (Equation 20.10) into

Table 20.2 **The Relationship Between ΔG^0 and K at 298 K**

ΔG^0 (kJ)	K	Significance		
200	9×10^{-36}	Essentially no forward reaction; reverse reaction goes to completion	FORWARD REACTION	REVERSE REACTION
100	3×10^{-18}			
50	2×10^{-9}			
10	2×10^{-2}			
1	7×10^{-1}	Forward and reverse reactions proceed to same extent		
0	1			
-1	1.5			
-10	5×10^{1}			
-50	6×10^{8}			
-100	3×10^{17}			
-200	1×10^{35}	Forward reaction goes to completion; essentially no reverse reaction		

the expression for ΔG (Equation 20.9), we obtain a relationship that applies to any starting concentrations:

$$\Delta G = \Delta G^0 + RT \ln Q \qquad \text{(20.11)}$$

Sample Problem 20.7 illustrates how Equations 20.10 and 20.11 are applied.

SAMPLE PROBLEM 20.7 Calculating ΔG at Nonstandard Conditions

Problem The oxidation of SO_2, which we considered in Sample Problem 20.6,

$$2SO_2(g) + O_2(g) \longrightarrow 2SO_3(g)$$

is too slow at 298 K to be useful in the manufacture of sulfuric acid. To overcome this low rate, the process is conducted at an elevated temperature.
(a) Calculate K at 298 K and at 973 K. ($\Delta G^0_{298} = -141.6$ kJ/mol of reaction as written; using ΔH^0 and ΔS^0 values at 973 K, $\Delta G^0_{973} = -12.12$ kJ/mol of reaction as written.)
(b) In experiments to determine the effect of temperature on reaction spontaneity, two sealed containers are filled with 0.500 atm of SO_2, 0.0100 atm of O_2, and 0.100 atm of SO_3 and kept at 25°C and at 700.°C. In which direction, if any, will the reaction proceed to reach equilibrium at each temperature?
(c) Calculate ΔG for the system in part (b) at each temperature.
Plan (a) We know ΔG^0, T, and R, so we can calculate the K's from Equation 20.10.
(b) To determine if a net reaction will occur at the given pressures, we calculate Q with the given partial pressures and compare it with each K from part (a). **(c)** Because these are not standard-state pressures, we calculate ΔG at each T from Equation 20.11 with the values of ΔG^0 (given) and Q [found in part (b)].
Solution (a) Calculating K at the two temperatures:

$$\Delta G^0 = -RT \ln K \qquad \text{so} \qquad K = e^{-(\Delta G^0/RT)}$$

At 298 K, the exponent is

$$-(\Delta G^0/RT) = -\left(\frac{-141.6 \text{ kJ/mol} \times \dfrac{1000 \text{ J}}{1 \text{ kJ}}}{8.314 \text{ J/mol·K} \times 298 \text{ K}} \right) = 57.2$$

So

$$K = e^{-(\Delta G^0/RT)} = e^{57.2} = \boxed{7 \times 10^{24}}$$

At 973 K, the exponent is

$$-(\Delta G^0/RT) = -\left(\frac{-12.12 \text{ kJ/mol} \times \dfrac{1000 \text{ J}}{1 \text{ kJ}}}{8.314 \text{ J/mol·K} \times 973 \text{ K}} \right) = 1.50$$

So

$$K = e^{-(\Delta G^0/RT)} = e^{1.50} = \boxed{4.5}$$

(b) Calculating the value of Q:

$$Q = \frac{P_{SO_3}^2}{P_{SO_2}^2 \times P_{O_2}} = \frac{0.100^2}{0.500^2 \times 0.0100} = 4.00$$

Because $Q < K$ at both temperatures, the denominator will decrease and the numerator increase—more SO_3 will form—until Q equals K. However, the reaction will go far to the right at 298 K before reaching equilibrium, whereas it will move only slightly to the right at 973 K.

(c) Calculating ΔG, the nonstandard free energy change, at 298 K:

$$\Delta G_{298} = \Delta G^0 + RT \ln Q$$

$$= -141.6 \text{ kJ/mol} + \left(8.314 \text{ J/mol·K} \times \frac{1 \text{ kJ}}{1000 \text{ J}} \times 298 \text{ K} \times \ln 4.00 \right)$$

$$= \boxed{-138.2 \text{ kJ/mol}}$$

Calculating ΔG at 973 K:

$$\Delta G_{973} = \Delta G^0 + RT \ln Q$$

$$= -12.12 \text{ kJ/mol} + \left(8.314 \text{ J/mol·K} \times \frac{1 \text{ kJ}}{1000 \text{ J}} \times 973 \text{ K} \times \ln 4.00 \right)$$

$$= \boxed{-0.9 \text{ kJ/mol}}$$

Check Note that in parts (a) and (c) we made energy units in free energy changes (kJ) consistent with those in R (J). Based on the rules for significant figures in addition and subtraction, we retain one digit to the right of the decimal place in part (c).

Comment For these starting gas pressures at 973 K, the process is barely spontaneous ($\Delta G = -0.9$ kJ/mol), so why use a higher temperature? As in the synthesis of NH_3 (Section 17.6), this process is carried out at a higher temperature *with a catalyst* to attain a higher *rate*, even though the *yield* is greater at a lower temperature. We discuss these details of the industrial production of sulfuric acid in Chapter 22.

FOLLOW-UP PROBLEM 20.7 At 298 K, hypobromous acid (HBrO) dissociates in water with a K_a of 2.3×10^{-9}.
(a) Calculate ΔG^0 for the dissociation of HBrO.
(b) Calculate ΔG if $[H_3O^+] = 6.0 \times 10^{-4}$ M, $[BrO^-] = 0.10$ M, and $[HBrO] = 0.20$ M.

Another Look at the Meaning of Spontaneity At this point, let's reconsider what we mean by the terms *spontaneous* and *nonspontaneous*. Consider the general reaction A $\rightleftharpoons$ B, for which $K = [B]/[A] > 1$; therefore, the reaction proceeds largely from left to right (Figure 20.12A). From pure A to the equilibrium point, $Q < K$ and the reaction is spontaneous ($\Delta G < 0$). From there on, the reaction is nonspontaneous ($\Delta G > 0$). From pure B to the equilibrium point, $Q > K$ and the reaction is also spontaneous ($\Delta G < 0$), but not thereafter. In either case, *the free energy decreases as the reaction proceeds, until it reaches a minimum at the equilibrium mixture*. If we start with pure A, the reaction stops with mostly B present, so we say the forward reaction A $\longrightarrow$ B is "spontaneous." Likewise, if we start with pure B, the reaction also stops with mostly B present. We say the reverse reaction B $\longrightarrow$ A is "nonspontaneous" because it proceeds very little in that direction. For the overall reaction A $\rightleftharpoons$ B (starting with all components in their standard states), G_B^0 is smaller than G_A^0, so ΔG^0 is negative, which corresponds to $K > 1$ and, therefore, to a spontaneous reaction as written.

Now consider the opposite situation, a general reaction C $\rightleftharpoons$ D for which $K = [D]/[C] < 1$: the reaction proceeds only slightly from left to right (Figure 20.12B). Here, too, whether we start with pure C or pure D, the reaction is spontaneous ($\Delta G < 0$) until the equilibrium point. But in this case, the equilibrium mixture contains mostly C (the reactant), so we say the forward reaction, C $\longrightarrow$ D, is nonspontaneous and the reverse reaction, D $\longrightarrow$ C, is spontaneous.

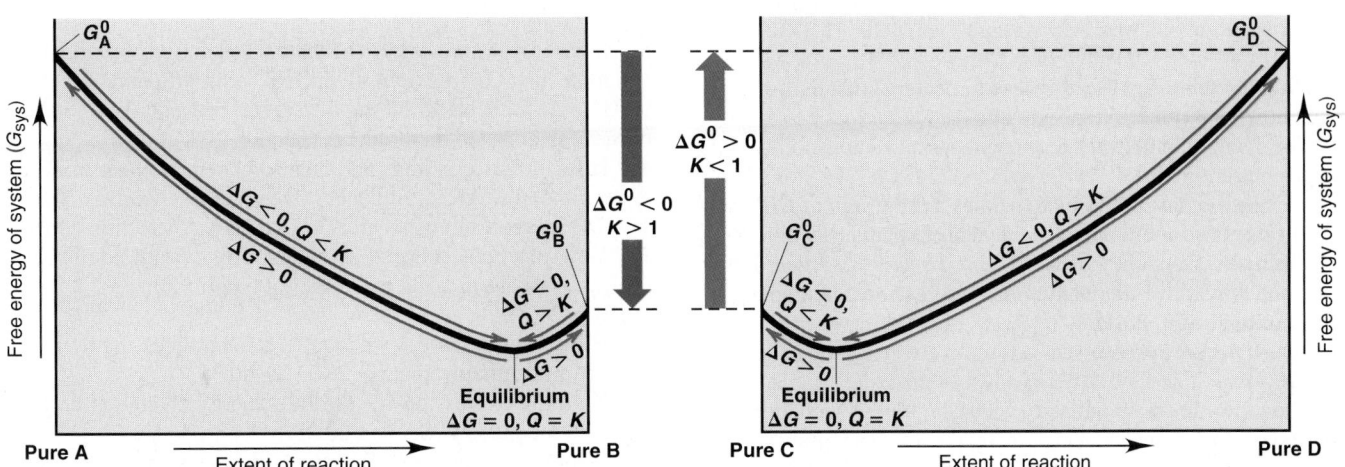

Figure 20.12 **The relation between free energy and the extent of reaction.** The free energy of the system is plotted against the extent of reaction. Each reaction proceeds spontaneously ($Q \neq K$ and $\Delta G < 0$; *curved green arrows*) from either pure reactants (A or C) or pure products (B or D) *to* the equilibrium mixture, at which point $\Delta G = 0$. The reaction *from* the equilibrium mixture to either pure reactants or products is nonspontaneous ($\Delta G > 0$; *curved red arrows*). **A,** For the reaction A $\rightleftharpoons$ B, $G_A^0 > G_B^0$, so $\Delta G^0 < 0$ and $K > 1$. **B,** For the reaction C $\rightleftharpoons$ D, $G_D^0 > G_C^0$, so $\Delta G^0 > 0$ and $K < 1$.

Moreover, G_D^0 is *larger* than G_C^0, so ΔG^0 is *positive*, which corresponds to $K < 1$ and, therefore, to a nonspontaneous reaction as written. The point is that the term *spontaneous reaction* refers to one that goes *predominantly, not necessarily completely, to product* (see Table 20.2, p. 881).

SECTION SUMMARY

Two ways of predicting reaction direction are from the value of ΔG and from the relation of Q to K. These variables represent different aspects of the same phenomenon and are related to each other by $\Delta G = RT \ln Q/K$. When $Q = K$, the system can release no more free energy. Beginning with Q at the standard state, the free energy change is ΔG^0, and it is related to the equilibrium constant by $\Delta G^0 = -RT \ln K$. For nonstandard conditions, ΔG has two components, ΔG^0 and $RT \ln Q$. Any nonequilibrium mixture of reactants and products moves spontaneously ($\Delta G < 0$) toward the equilibrium mixture, but only when $K > 1$ is the reaction spontaneous for the components in their standard states ($\Delta G^0 < 0$).

Chapter Perspective

As processes move toward equilibrium, some of the released energy becomes degraded to unusable, dispersed heat, and the entropy of the universe increases. This unalterable necessity built into all natural processes has led to speculation about the "End of Everything," when all possible processes have stopped occurring, no free energy remains in any system, and nothing but waste heat is dispersed evenly throughout the universe—the final equilibrium! Even if this grim future is in store, however, it is many billions of years away, so you have plenty of time to appreciate more hopeful applications of thermodynamics. In Chapter 21, you'll see how spontaneous reactions can generate electricity and how electricity supplies free energy to drive nonspontaneous ones.

For Review and Reference (Numbers in parentheses refer to pages, unless noted otherwise.)

Learning Objectives

Relevant section and/or sample problem (SP) numbers appear in parentheses.

Understand These Concepts

1. How the tendency of a process to occur by itself is distinct from how long it takes to occur (Intro)
2. The distinction between a spontaneous and a nonspontaneous change (Section 20.1)
3. Why the first law of thermodynamics and the sign of ΔH^0 cannot predict the direction of a spontaneous process (Section 20.1)
4. The natural tendency of a system to become disordered over time and why a disordered system is more probable than an ordered one (Section 20.1)
5. How disorder is expressed quantitatively by entropy (S) (Section 20.1)
6. The criterion for spontaneity according to the second law of thermodynamics: that a change increases S_{univ} (Section 20.1)
7. How absolute values of standard molar entropies (S^0) can be obtained because the third law of thermodynamics provides a "zero point" (Section 20.1)

8. How temperature, physical state, dissolution, atomic size, and molecular complexity influence S^0 values (Section 20.1)
9. How ΔS_{rxn}^0 is based on the difference between the S^0 values of the reactants and the products (Section 20.2)
10. How the surroundings add heat to or remove heat from the system and how ΔS_{surr}^0 influences overall ΔS_{rxn}^0 (Section 20.2)
11. The relationship between ΔS_{surr} and ΔH_{sys} (Section 20.2)
12. How reactions proceed spontaneously toward equilibrium ($\Delta S_{univ} > 0$) but proceed no further at equilibrium ($\Delta S_{univ} = 0$) (Section 20.2)
13. How the free energy change (ΔG) combines the system's entropy and enthalpy changes (Section 20.3)
14. How the expression for the free energy change is derived from the second law (Section 20.3)
15. The relationship between ΔG and the maximum work a system can perform and why this quantity of work is never performed in a real process (Section 20.3)

16. How temperature determines spontaneity for reactions in which ΔS^0 and ΔH^0 have the same sign (Section 20.3)
17. Why the temperature at which a reaction becomes spontaneous occurs when $\Delta G^0 = 0$ (Section 20.3)
18. How a spontaneous change can be coupled to a nonspontaneous change to make it occur (Section 20.3)
19. Why ΔG is the free energy change from the start of a reaction (expressed by Q) to its end (expressed by K) (Section 20.4)
20. The meaning of ΔG^0 and its relation to K (Section 20.4)
21. The relation of ΔG to ΔG^0 and Q (Section 20.4)
22. Why ΔG decreases, no matter what the starting concentrations, as the reacting system moves spontaneously toward equilibrium (Section 20.4)

Master These Skills

1. Predicting relative S^0 values of systems (Section SP 20.1)
2. Calculating ΔS^0_{rxn} for a chemical change (SP 20.2)
3. Finding reaction spontaneity from ΔS^0_{surr} and ΔH^0_{sys} (SP 20)
4. Calculating ΔG^0_{rxn} from ΔH^0_f and S^0 values (SP 20.4)
5. Calculating ΔG^0_{rxn} from ΔG^0_f values (SP 20.5)
6. Calculating the effect of temperature on ΔG^0 (SP 20.6)
7. Calculating the temperature at which a reaction becomes spontaneous (Section 20.3)
8. Calculating K from ΔG^0 (Section 20.4 and SP 20.7)
9. Using ΔG^0 and Q to calculate ΔG at any temperature (SP 20.7)

Key Terms

Section 20.1
spontaneous change (856)
entropy (S) (860)
second law of thermodynamics (861)
third law of thermodynamics (861)

standard molar entropy (S^0) (862)

Section 20.2
standard entropy of reaction (ΔS^0_{rxn}) (866)

Section 20.3
free energy (G) (872)
standard free energy change (ΔG^0) (873)
standard free energy of formation (ΔG^0_f) (874)

coupling of reactions (878)
adenosine triphosphate (ATP) (879)

Key Equations and Relationships

20.1 Quantifying entropy in terms of the number of ways (W) a system can be arranged (860):
$$S = k \ln W$$

20.2 Stating the second law of thermodynamics, for a spontaneous process (861):
$$\Delta S_{univ} = \Delta S_{sys} + \Delta S_{surr} > 0$$

20.3 Calculating the standard entropy of reaction from standard molar entropies of reactants and products (866):
$$\Delta S^0_{rxn} = \Sigma m S^0_{products} - \Sigma n S^0_{reactants}$$

20.4 Relating the entropy change in the surroundings to the heat of the system and the temperature (868):
$$\Delta S_{surr} = -\frac{q_{sys}}{T}$$

20.5 Expressing the free energy change of the system in terms of its component enthalpy and entropy changes (Gibbs equation) (872):
$$\Delta G_{sys} = \Delta H_{sys} - T\Delta S_{sys}$$

20.6 Calculating the standard free energy change from standard enthalpy and entropy changes (873):
$$\Delta G^0_{sys} = \Delta H^0_{sys} - T\Delta S^0_{sys}$$

20.7 Calculating the standard free energy change from standard free energies of formation (874):
$$\Delta G^0_{rxn} = \Sigma m \Delta G^0_{f(products)} - \Sigma n \Delta G^0_{f(reactants)}$$

20.8 Finding the temperature at which a reaction becomes spontaneous (877):
$$T = \frac{\Delta H^0}{\Delta S^0}$$

20.9 Expressing the free energy change as the difference between an initial state Q and a final state K (881):
$$\Delta G = RT \ln \frac{Q}{K} = RT \ln Q - RT \ln K$$

20.10 Expressing the free energy change when Q is evaluated at the standard state (881):
$$\Delta G^0 = -RT \ln K$$

20.11 Expressing the free energy change for a nonstandard initial state (882):
$$\Delta G = \Delta G^0 + RT \ln Q$$

Highlighted Figures and Tables

These figures (F) and tables (T) provide a quick review of key ideas.

F20.5 Entropy and phase changes (862)
F20.10 Components of ΔS_{univ} for spontaneous reactions (871)

T20.1 The signs of ΔH^0, ΔS^0, and ΔG^0 (877)
T20.2 The relationship of ΔG^0 and K (881)
F20.12 Free energy and extent of reaction (883)

Follow-up Problems

...lar mass and more complex mol-
...olar mass; (c) $Br_2(g)$: gases have

$$\longrightarrow Na_2CO_3(s) + H_2O(l);$$

..., $\text{so } \Delta S_{rxn}^0 < 0$

$\Delta S_{rxn}^0 = [(1 \text{ mol } H_2O)(69.9 \text{ J/mol·K})$
$\quad + (1 \text{ mol } Na_2CO_3)(139 \text{ J/mol·K})]$
$\quad - [(1 \text{ mol } CO_2)(213.7 \text{ J/mol·K})$
$\quad + (2 \text{ mol } NaOH)(64.5 \text{ J/mol·K})]$
$\quad = -134 \text{ J/K}$

(b) $2Fe(s) + 3H_2O(g) \longrightarrow Fe_2O_3(s) + 3H_2(g)$
$\Delta n_{gas} = 0$, so cannot predict sign of ΔS_{rxn}^0
$\Delta S_{rxn}^0 = [(1 \text{ mol } Fe_2O_3)(87.4 \text{ J/mol·K})$
$\quad + (3 \text{ mol } H_2)(130.6 \text{ J/mol·K})]$
$\quad - [(2 \text{ mol } Fe)(27.3 \text{ J/mol·K})$
$\quad + (3 \text{ mol } H_2O)(188.7 \text{ J/mol·K})]$
$\quad = -141.5 \text{ J/K}$

20.3 $2FeO(s) + \frac{1}{2}O_2(g) \longrightarrow Fe_2O_3(s)$
$\Delta S_{sys}^0 = (1 \text{ mol } Fe_2O_3)(87.4 \text{ J/mol·K})$
$\quad - [(2 \text{ mol } FeO)(60.75 \text{ J/mol·K})$
$\quad + (\frac{1}{2} \text{ mol } O_2)(205.0 \text{ J/mol·K})]$
$\quad = -136.6 \text{ J/K}$
$\Delta H_{sys}^0 = (1 \text{ mol } Fe_2O_3)(-825.5 \text{ kJ/mol})$
$\quad - [(2 \text{ mol } FeO)(-272.0 \text{ kJ/mol})$
$\quad + (\frac{1}{2} \text{ mol } O_2)(0 \text{ kJ/mol})]$
$\quad = -281.5 \text{ kJ}$
$\Delta S_{surr}^0 = -\dfrac{\Delta H_{sys}^0}{T} = -\dfrac{(-281.5 \text{ kJ} \times 1000 \text{ J/kJ})}{298 \text{ K}} = +945 \text{ J/K}$
$\Delta S_{univ}^0 = \Delta S_{sys}^0 + \Delta S_{surr}^0 = -136.6 \text{ J/K} + 945 \text{ J/K}$
$\quad = 808 \text{ J/K}$; reaction is spontaneous at 298 K.

20.4 Using ΔH_f^0 and S^0 values from Appendix B,
$\Delta H_{rxn}^0 = -114.2 \text{ kJ and } \Delta S_{rxn}^0 = -146.5 \text{ J/K}$
$\Delta G_{rxn}^0 = \Delta H_{rxn}^0 - T\Delta S_{rxn}^0 = -114.2 \text{ kJ}$
$\quad - [(298 \text{ K})(-146.5 \text{ J/K})(1 \text{ kJ}/1000 \text{ J})]$
$\quad = -70.5 \text{ kJ}$

20.5 (a) $\Delta G_{rxn}^0 = (2 \text{ mol } NO_2)(51 \text{ kJ/mol})$
$\quad - [(2 \text{ mol } NO)(86.60 \text{ kJ/mol})$
$\quad + (1 \text{ mol } O_2)(0 \text{ kJ/mol})]$
$\quad = -71 \text{ kJ}$

(b) $\Delta G_{rxn}^0 = (2 \text{ mol } CO)(-137.2 \text{ kJ/mol}) - [(2 \text{ mol } C)(0 \text{ kJ/mol})$
$\quad + (1 \text{ mol } O_2)(0 \text{ kJ/mol})]$
$\quad = -274.4 \text{ kJ}$

20.6 ΔG^0 becomes negative at lower T, so $\Delta H^0 < 0$, $\Delta S^0 < 0$, and $-T\Delta S^0 > 0$. At lower T, the negative ΔH^0 value becomes larger than the positive $-T\Delta S^0$ value.

20.7 (a) $\Delta G^0 = -RT \ln K$

$\quad = -8.314 \text{ J/mol·K} \times \dfrac{1 \text{ kJ}}{1000 \text{ J}} \times 298 \text{ K}$
$\quad \times \ln(2.3 \times 10^{-9})$
$\quad = 49 \text{ kJ/mol}$

(b) $Q = \dfrac{[H_3O^+][BrO^-]}{[HBrO]} = \dfrac{(6.0 \times 10^{-4})(0.10)}{0.20} = 3.0 \times 10^{-4}$

$\Delta G = \Delta G^0 + RT \ln Q$
$\quad = 49 \text{ kJ/mol}$
$\quad + \left[8.314 \text{ J/mol·K} \times \dfrac{1 \text{ kJ}}{1000 \text{ J}} \times 298 \text{ K} \times \ln(3.0 \times 10^{-4})\right]$
$\quad = 29 \text{ kJ/mol}$

Problems

Problems with colored numbers are answered at the back of the text. Sections match the text and provide the number(s) of relevant sample problems. Most offer Concept Review Questions, Skill-Building Exercises (in similar pairs), and Problems in Context. Then Comprehensive Problems, based on material from any section or previous chapter, follow. *Note:* Unless stated otherwise, all problems refer to systems at 298 K (25°C).

The Second Law of Thermodynamics: Predicting Spontaneous Change
(Sample Problem 20.1)

● **Concept Review Questions**

20.1 Distinguish between the terms *spontaneous* and *instantaneous*. Give an example of a process that is spontaneous but very slow, and one that is very fast but not spontaneous.
20.2 Distinguish between the terms *spontaneous* and *nonspontaneous*. Can a nonspontaneous process occur? Explain.
20.3 State the first law of thermodynamics in terms of (a) the energy of the universe; (b) the creation or destruction of energy; (c) the energy change of system and surroundings. Does the first law reveal the direction of spontaneous change? Explain.
20.4 State qualitatively the relationship between entropy and probability. Use this idea to explain why you have probably never (a) seen a large number of coins (>6) flipped into the air land with all heads up; (b) been suffocated because all the air near you moved to the other side of the room; (c) seen half the water in your cup of tea freeze while the other half boiled.
20.5 Why is ΔS_{vap} of a substance always larger than ΔS_{fus}?
20.6 How does the entropy of the surroundings change during the course of an exothermic reaction? An endothermic reaction? Give an example, other than those cited in text, of an endothermic process that occurs spontaneously.
20.7 (a) What is the entropy of a perfect crystal at 0 K?
(b) Does entropy increase or decrease as the temperature rises?
(c) Why is $\Delta H_f^0 = 0$ but $S^0 > 0$ for an element?
(d) Why does Appendix B list ΔH_f^0 values but not ΔS_f^0 values?

● **Skill-Building Exercises (paired)**

20.8 Which of the following processes are spontaneous?
(a) Water evaporating from a puddle in summer
(b) A lion chasing an antelope
(c) An unstable isotope undergoing radioactive disintegration
20.9 Which of the following processes are spontaneous?
(a) The Earth moving around the Sun
(b) A boulder rolling up a hill
(c) Sodium metal and chlorine gas reacting to form solid sodium chloride

20.10 Which of the following processes are spontaneous?
(a) Methane burning in air
(b) A teaspoonful of sugar dissolving in a cup of hot coffee
(c) A soft-boiled egg becoming raw
20.11 Which of the following processes are spontaneous?
(a) A satellite falling to Earth
(b) Water decomposing to H_2 and O_2 at 298 K and 1 atm
(c) Average car prices increasing

20.12 Predict the sign of ΔS_{sys} for each process:
(a) A piece of wax melting
(b) Silver chloride precipitating from solution
(c) Dew forming
20.13 Predict the sign of ΔS_{sys} for each process:
(a) Gasoline vapors mixing with air in a car engine
(b) Hot air expanding
(c) Breath condensing in cold air

20.14 Predict the sign of ΔS_{sys} for each process:
(a) Alcohol evaporating
(b) A solid explosive converting to a gas
(c) Perfume vapors diffusing through a room
20.15 Predict the sign of ΔS_{sys} for each process:
(a) A pond freezing in winter
(b) Atmospheric CO_2 dissolving in the ocean
(c) An apple tree bearing fruit

20.16 Without referring to Appendix B, predict the sign of ΔS^0 for each process:
(a) $2K(s) + F_2(g) \longrightarrow 2KF(s)$
(b) $NH_3(g) + HBr(g) \longrightarrow NH_4Br(s)$
(c) $NaClO_3(s) \longrightarrow Na^+(aq) + ClO_3^-(aq)$
20.17 Without referring to Appendix B, predict the sign of ΔS^0 for each process:
(a) $H_2S(g) + \frac{1}{2}O_2(g) \longrightarrow \frac{1}{8}S_8(s) + H_2O(g)$
(b) $HCl(aq) + NaOH(aq) \longrightarrow NaCl(aq) + H_2O(l)$
(c) $2NO_2(g) \longrightarrow N_2O_4(g)$

20.18 Without referring to Appendix B, predict the sign of ΔS^0 for each process:
(a) $CaCO_3(s) + 2HCl(aq) \longrightarrow$
$\qquad CaCl_2(aq) + H_2O(l) + CO_2(g)$
(b) $2NO(g) + O_2(g) \longrightarrow 2NO_2(g)$
(c) $2KClO_3(s) \longrightarrow 2KCl(s) + 3O_2(g)$
20.19 Without referring to Appendix B, predict the sign of ΔS^0 for each process:
(a) $Ag^+(aq) + Cl^-(aq) \longrightarrow AgCl(s)$
(b) $FeCl_3(s) \longrightarrow FeCl_3(aq)$
(c)
$$CH_3CH=CH_2(g) \longrightarrow H_2\overset{CH_2}{\underset{}{C-CH_2}}(g)$$

20.20 Predict the sign of ΔS for each process:
(a) $C_2H_5OH(g)$ (350 K and 500 torr) $\longrightarrow$
$\qquad C_2H_5OH(g)$ (350 K and 250 torr)
(b) $N_2(g)$ (298 K and 1 atm) $\longrightarrow N_2(aq)$ (298 K and 1 atm)
(c) $O_2(aq)$ (303 K and 1 atm) $\longrightarrow O_2(g)$ (303 K and 1 atm)
20.21 Predict the sign of ΔS for each process:
(a) $O_2(g)$ ($V = 1.0$ L, $P = 1$ atm) $\longrightarrow O_2(g)$
$\qquad (V = 0.10$ L, $P = 10$ atm)
(b) $Cu(s)$ (350°C and 2.5 atm) $\longrightarrow$
$\qquad Cu(s)$ (450°C and 2.5 atm)
(c) $Cl_2(g)$ (100°C and 1 atm) $\longrightarrow Cl_2(g)$ (10°C and 1 atm)

20.22 Predict which has the greater molar entropy.
(a) Butane $CH_3CH_2CH_2CH_3(g)$
or 2-butene $CH_3CH=CHCH_3(g)$
(b) $Ne(g)$ or $Xe(g)$
(c) $CH_4(g)$ or $CCl_4(l)$
20.23 Predict which has greater molar entropy. Explain.
(a) $NO_2(g)$ or $N_2O_4(g)$ (b) $CH_3OCH_3(l)$ or $CH_3CH_2OH(l)$
(c) $HCl(g)$ or $HBr(g)$

20.24 Predict which has greater molar entropy. Explain.
(a) $CH_3OH(l)$ or $C_2H_5OH(l)$ (b) $KClO_3(s)$ or $KClO_3(aq)$
(c) $Na(s)$ or $K(s)$
20.25 Predict which has greater molar entropy. Explain.
(a) $P_4(g)$ or $P_2(g)$ (b) $HNO_3(aq)$ or $HNO_3(l)$
(c) $CuSO_4(s)$ or $CuSO_4 \cdot 5H_2O(s)$

20.26 Without consulting Appendix B, arrange each group in order of *increasing* standard molar entropy (S^0). Explain.
(a) Graphite, diamond, charcoal
(b) Ice, water vapor, liquid water (c) O_2, O_3, O atoms
20.27 Without consulting Appendix B, arrange each group in order of *increasing* standard molar entropy (S^0). Explain.
(a) Glucose ($C_6H_{12}O_6$), sucrose ($C_{12}H_{22}O_{11}$), ribose ($C_5H_{10}O_5$)
(b) $CaCO_3$, $Ca + C + \frac{3}{2}O_2$, $CaO + CO_2$
(c) $SF_6(g)$, $SF_4(g)$, $S_2F_{10}(g)$

20.28 Without consulting Appendix B, arrange each group in order of *decreasing* standard molar entropy (S^0). Explain.
(a) $ClO_4^-(aq)$, $ClO_2^-(aq)$, $ClO_3^-(aq)$
(b) $NO_2(g)$, $NO(g)$, $N_2(g)$ (c) $Fe_2O_3(s)$, $Al_2O_3(s)$, $Fe_3O_4(s)$
20.29 Without consulting Appendix B, arrange each of the following groups in order of *decreasing* standard molar entropy (S^0), and explain your choice:
(a) Mg metal, Ca metal, Ba metal
(b) Hexane (C_6H_{14}), benzene (C_6H_6), cyclohexane (C_6H_{12})
(c) $PF_2Cl_3(g)$, $PF_5(g)$, $PF_3(g)$

Calculating the Change in Entropy of a Reaction
(Sample Problems 20.2 and 20.3)

● Concept Review Questions

20.30 What property of entropy allows Hess's law to be used in the calculation of entropy changes?
20.31 Describe the equilibrium condition in terms of the entropy changes of a system and its surroundings. What does this description mean about the entropy change of the universe?
20.32 Consider the reaction
$$H_2O(g) + Cl_2O(g) \longrightarrow 2HClO(g)$$
Given ΔS^0_{rxn} and S^0 of $HClO(g)$ and of $H_2O(g)$, write an expression to determine S^0 of $Cl_2O(g)$.

● Skill-Building Exercises (paired)

20.33 For each reaction, predict the sign of ΔS^0 and use Appendix B to calculate its value:
(a) $3NO(g) \longrightarrow N_2O(g) + NO_2(g)$
(b) $3H_2(g) + Fe_2O_3(s) \longrightarrow 2Fe(s) + 3H_2O(g)$
(c) $P_4(s) + 5O_2(g) \longrightarrow P_4O_{10}(s)$
20.34 For each reaction, predict the sign of ΔS^0 and use Appendix B to calculate its value:
(a) $3NO_2(g) + H_2O(l) \longrightarrow 2HNO_3(l) + NO(g)$
(b) $N_2(g) + 3F_2(g) \longrightarrow 2NF_3(g)$
(c) $C_6H_{12}O_6(s) + 6O_2(g) \longrightarrow 6CO_2(g) + 6H_2O(g)$

e combustion of ethane (C_2H_6) to
eous water. Is the answer what you

combustion of methane to carbon
e answer what you would expect

... ΔS^0 for the reduction of nitric oxide with hydrogen to form ammonia and water vapor. Is the answer what you would expect qualitatively?

20.38 Calculate ΔS^0 for the combustion of ammonia to form nitrogen dioxide and water vapor. Is the answer what you would expect qualitatively?

20.39 Find ΔS^0 for the formation of $Cu_2O(s)$ from its elements.

20.40 Find ΔS^0 for the formation of $HI(g)$ from its elements.

20.41 Find ΔS^0 for the formation of $CH_3OH(l)$ from its elements.

20.42 Find ΔS^0 for the formation of $PCl_5(g)$ from its elements.

● **Problems in Context**

20.43 Sulfur dioxide is released in the combustion of coal. Scrubbers use lime slurries containing calcium hydroxide to remove much of the SO_2 from flue gases. Write a balanced equation for this reaction and calculate ΔS^0 at 298 K [S^0 of $CaSO_3(s)$ = 101.4 J/mol·K].

20.44 Oxyacetylene welding is a common method for repairing metal structures, including bridges, buildings, and even the Statue of Liberty. Calculate ΔS^0 for the combustion of 1 mol of acetylene (C_2H_2).

Entropy, Free Energy, and Work
(Sample Problems 20.4 to 20.6)

● **Concept Review Questions**

20.45 What is the advantage of calculating free energy changes rather than entropy changes to determine reaction spontaneity?

20.46 Given that $\Delta G_{sys} = -T\Delta S_{univ}$, explain how the sign of ΔG_{sys} correlates with reaction spontaneity.

20.47 Is an endothermic reaction more likely to be spontaneous at higher temperatures or lower temperatures? Explain.

20.48 With its components in their standard states, a certain reaction is spontaneous only at high temperatures. What can you conclude about the signs of ΔH^0 and ΔS^0 for the reaction? Describe a process for which this is true.

20.49 How can ΔS^0 be relatively independent of T if S^0 of each reactant and product increases with T?

● **Skill-Building Exercises (paired)**

20.50 Calculate ΔG^0 for each reaction using ΔG_f^0 values from Appendix B:
(a) $2Mg(s) + O_2(g) \longrightarrow 2MgO(s)$
(b) $2CH_3OH(g) + 3O_2(g) \longrightarrow 2CO_2(g) + 4H_2O(g)$
(c) $BaO(s) + CO_2(g) \longrightarrow BaCO_3(s)$

20.51 Calculate ΔG^0 for each reaction using ΔG_f^0 values from Appendix B:
(a) $H_2(g) + I_2(s) \longrightarrow 2HI(g)$
(b) $MnO_2(s) + 2CO(g) \longrightarrow Mn(s) + 2CO_2(g)$
(c) $NH_4Cl(s) \longrightarrow NH_3(g) + HCl(g)$

20.52 Calculate ΔG^0 for the reactions in Problem 20.50 using ΔH_f^0 and S^0 values from Appendix B.

20.53 Calculate ΔG^0 for the reactions in Problem 20.51 using ΔH_f^0 and S^0 values from Appendix B.

20.54 Consider the oxidation of carbon monoxide to carbon dioxide:
$$CO(g) + \tfrac{1}{2}O_2(g) \longrightarrow CO_2(g)$$
(a) Predict the signs of ΔS^0 and ΔH^0. Explain.
(b) Calculate ΔG^0 by two different methods.

20.55 Consider the combustion of butane gas:
$$C_4H_{10}(g) + \tfrac{13}{2}O_2(g) \longrightarrow 4CO_2(g) + 5H_2O(g)$$
(a) Predict the signs of ΔS^0 and ΔH^0. Explain.
(b) Calculate ΔG^0 by two different methods.

20.56 For the gaseous reaction of xenon and fluorine to form xenon hexafluoride:
(a) Calculate ΔS^0 at 298 K ($\Delta H^0 = -402$ kJ/mol and $\Delta G^0 = -280.$ kJ/mol).
(b) Assuming that ΔS^0 and ΔH^0 change little with temperature, calculate ΔG^0 at 500. K.

20.57 For the gaseous reaction of carbon monoxide and chlorine to form phosgene $(COCl_2)$:
(a) Calculate ΔS^0 at 298 K ($\Delta H^0 = -220.$ kJ/mol and $\Delta G^0 = -206$ kJ/mol).
(b) Assuming that ΔS^0 and ΔH^0 change little with temperature, calculate ΔG^0 at 450. K.

20.58 One reaction used to produce small quantities of pure H_2 is
$$CH_3OH(g) \rightleftharpoons CO(g) + 2H_2(g)$$
(a) Determine ΔH^0 and ΔS^0 for the reaction at 298 K.
(b) Assuming that these values are relatively independent of temperature, calculate ΔG^0 at 38°C, 138°C, and 238°C.
(c) What is the significance of the different values of ΔG^0?

20.59 An important chemical reaction that occurs in the internal combustion engine is
$$N_2(g) + O_2(g) \rightleftharpoons 2NO(g)$$
(a) Determine ΔH^0 and ΔS^0 for the reaction at 298 K.
(b) Assuming that these values are relatively independent of temperature, calculate ΔG^0 at 100.°C, 2560.°C, and 3540.°C.
(c) What is the significance of the different values of ΔG^0?

20.60 The temperature at which the following process reaches equilibrium at 1 atm is the normal boiling point of bromine:
$$Br_2(l) \rightleftharpoons Br_2(g)$$
Use ΔH^0 and ΔS^0 values to determine this temperature.

20.61 The temperature at which the following process reaches equilibrium at 1 atm is the transition temperature for these two allotropes of crystalline sulfur:
$$S(rhombic) \rightleftharpoons S(monoclinic)$$
Use ΔH^0 and ΔS^0 values to determine this temperature.

● **Problems in Context**

20.62 Many researchers have investigated the use of H_2 as an energy source for vehicles because it produces only nonpolluting $H_2O(g)$ when it burns. Moreover, when H_2 combines with O_2 in a fuel cell (Chapter 21), the reaction provides electrical energy.
(a) Calculate ΔH^0, ΔS^0, and ΔG^0 per mol of H_2 at 298 K.
(b) Is the spontaneity of this reaction dependent on T? Explain.
(c) At what temperature does the reaction become spontaneous?

20.63 The U.S. government now requires automobile fuels to contain a renewable component. The fermentation of glucose from corn to produce ethanol for addition to gasoline is a key process in fulfilling this requirement:

$$C_6H_{12}O_6(s) \longrightarrow 2C_2H_5OH(l) + 2CO_2(g)$$

Calculate ΔH^0, ΔS^0, and ΔG^0 for the reaction at 25°C. Is the spontaneity of this reaction dependent on T? Explain.

Free Energy, Equilibrium, and Reaction Direction
(Sample Problem 20.7)

● **Concept Review Questions**

20.64 (a) If $K \ll 1$ for a reaction, what do you know about the sign and magnitude of ΔG^0?
(b) If $\Delta G^0 \ll 0$ for a reaction, what do you know about the magnitude of K? Of Q?

20.65 How is the free energy change of a process related to the work that can be obtained from the process? Is this quantity of work obtainable in practice? Explain.

20.66 In what form is the portion of reaction enthalpy that cannot do work given off? What happens to a portion of the useful energy as it does work?

20.67 What is the difference between the terms ΔG^0 and ΔG? Under what circumstances does $\Delta G = \Delta G^0$?

● **Skill-Building Exercises (paired)**

20.68 Use Appendix B to calculate K at 298 K for
(a) $NO(g) + \frac{1}{2}O_2(g) \rightleftharpoons NO_2(g)$
(b) $2HCl(g) \rightleftharpoons H_2(g) + Cl_2(g)$
(c) $2C(graphite) + O_2(g) \rightleftharpoons 2CO(g)$

20.69 Use Appendix B to calculate K at 298 K for
(a) $MgCO_3(s) \rightleftharpoons Mg^{2+}(aq) + CO_3^{2-}(aq)$
(b) $2HCl(g) + Br_2(l) \rightleftharpoons 2HBr(g) + Cl_2(g)$
(c) $H_2(g) + O_2(g) \rightleftharpoons H_2O_2(l)$

20.70 Use Appendix B to calculate K at 298 K for
(a) $2H_2S(g) + 3O_2(g) \rightleftharpoons 2H_2O(g) + 2SO_2(g)$
(b) $H_2SO_4(l) \rightleftharpoons H_2O(l) + SO_3(g)$
(c) $HCN(aq) + NaOH(aq) \rightleftharpoons NaCN(aq) + H_2O(l)$

20.71 Use Appendix B to calculate K at 298 K for
(a) $SrSO_4(s) \rightleftharpoons Sr^{2+}(aq) + SO_4^{2-}(aq)$
(b) $2NO(g) + Cl_2(g) \rightleftharpoons 2NOCl(g)$
(c) $Cu_2S(s) + O_2(g) \rightleftharpoons 2Cu(s) + SO_2(g)$

20.72 Use Appendix B to determine the K_{sp} of Ag_2S.
20.73 Use Appendix B to determine the K_{sp} of CaF_2.

20.74 For the reaction $I_2(g) + Cl_2(g) \rightleftharpoons 2ICl(g)$, calculate K_p at 25°C [ΔG_f^0 of $ICl(g) = -6.075$ kJ/mol].
20.75 For the reaction $CaCO_3(s) \rightleftharpoons CaO(s) + CO_2(g)$, calculate P_{CO_2} at 25°C.

20.76 The K_{sp} of $PbCl_2$ is 1.7×10^{-5} at 25°C. What is the value of ΔG^0 at this temperature? Is it possible to prepare a solution that has $Pb^{2+}(aq)$ and $Cl^-(aq)$, each at the standard-state concentration? (Neglect any reactions of the ions with water.)
20.77 The K_{sp} of ZnF_2 is 3.0×10^{-2} at 25°C. What is the value of ΔG^0 at this temperature? Is it possible to prepare a solution that has $Zn^{2+}(aq)$ and $F^-(aq)$ each at the standard-state concentration? (Neglect any reactions of the ions with water.)

20.78 The equilibrium constant for the reaction
$$2Fe^{3+}(aq) + Hg_2^{2+}(aq) \rightleftharpoons 2Fe^{2+}(aq) + 2Hg^{2+}(aq)$$
is $K_c = 9.1\times10^{-6}$ at 298 K.
(a) What is ΔG^0 at this temperature?

(b) If standard-state concentrations of the reactants and products are mixed, in which direction does the reaction proceed?
(c) Calculate ΔG when $[Fe^{3+}] = 0.20$ M, $[Hg_2^{2+}] = 0.010$ M, $[Fe^{2+}] = 0.010$ M, and $[Hg^{2+}] = 0.025$ M. In which direction will the reaction proceed to achieve equilibrium?

20.79 The formation constant for the reaction
$$Ni^{2+}(aq) + 6NH_3(aq) \rightleftharpoons Ni(NH_3)_6^{2+}(aq)$$
is $K_f = 5.6\times10^8$ at 25°C.
(a) What is ΔG^0 at this temperature?
(b) If standard-state concentrations of the reactants and products are mixed, in which direction does the reaction proceed?
(c) Determine ΔG when $[Ni(NH_3)_6^{2+}] = 0.010$ M, $[Ni^{2+}] = 0.0010$ M, and $[NH_3] = 0.0050$ M. In which direction will the reaction proceed to achieve equilibrium?

● **Problems in Context**

20.80 High levels of ozone (O_3) gas in the lower atmosphere makes rubber deteriorate, green plants turn brown, and persons with respiratory disease have difficulty breathing.
(a) Is the formation of O_3 from O_2 favored at all T, no T, high T, or low T?
(b) Calculate ΔG^0 for this reaction at 298 K.
(c) Calculate ΔG at 298 K for this reaction in urban smog where $[O_2] = 0.21$ M and $[O_3] = 5\times10^{-7}$ M.

20.81 A $BaSO_4$ slurry is ingested just before an x-ray of the gastrointestinal tract is taken because the precipitate is opaque to x-rays and thus defines the contours of the tract. The Ba^{2+} ion is toxic, but the compound is nearly insoluble. If ΔG^0 at 37°C (body temperature) is 59.1 kJ/mol for the process
$$BaSO_4(s) \rightleftharpoons Ba^{2+}(aq) + SO_4^{2-}(aq)$$
what is $[Ba^{2+}]$ in the intestinal tract? (Assume the only source of SO_4^{2-} is the ingested slurry.)

Comprehensive Problems
Problems with an asterisk (*) are more challenging.

20.82 According to the advertisement, "a diamond is forever."
(a) Calculate ΔH^0, ΔS^0, and ΔG^0 at 298 K for the phase change
$$\text{Diamond} \longrightarrow \text{graphite}$$
(b) Given the conditions under which diamond jewelry is normally kept, argue for and against the statement in the ad.
(c) Given the answers in part (a), what would need to be done to make synthetic diamonds from graphite?
(d) Assuming that ΔH^0 and ΔS^0 do not change with temperature, can graphite be converted to diamond spontaneously at 1 atm?

20.83 Supply the missing information (question mark) in each row of the following table:

	ΔS_{rxn}	ΔH_{rxn}	ΔG_{rxn}	Comment
(a)	+	−	−	?
(b)	?	0	−	Spontaneous
(c)	−	+	?	Not spontaneous
(d)	0	?	−	Spontaneous
(e)	?	0	+	?
(f)	+	+	?	$T\Delta S > \Delta H$

20.84 A complex ion (Lewis adduct) is formed when a ligand (Lewis base) is bonded to a metal ion (Lewis acid) by a pair of electrons. Of the many complex ions of cobalt are the following:
$$Co(NH_3)_6^{3+}(aq) + 3en(aq) \rightleftharpoons Co(en)_3^{3+}(aq) + 6NH_3(aq)$$

where *en* stands for ethylenediamine, $H_2NCH_2CH_2NH_2$. Six Co—N bonds are broken and six Co—N bonds are formed in this reaction, so $\Delta H^0_{rxn} \approx 0$; yet $K > 1$. What are the signs of ΔS^0 and ΔG^0? What drives the reaction?

20.85 What is the change in entropy when 0.200 mol of potassium freezes at 63.7°C ($\Delta H_{fus} = 2.39$ kJ/mol)?

20.86 Is each statement true or false? If false, correct it.
(a) All spontaneous reactions occur quickly.
(b) If a reaction is spontaneous, the reverse reaction is nonspontaneous.
(c) All spontaneous processes release heat.
(d) The boiling of water at 100°C and 1 atm is a spontaneous process.
(e) If a process increases the randomness of the particles of a system, the entropy of the system decreases.
(f) The energy of the universe is constant; the entropy of the universe decreases toward a minimum.
(g) All systems become disordered spontaneously.
(h) Both ΔS_{sys} and ΔS_{surr} equal zero at equilibrium.

*20.87 The hemoglobin molecule in red blood cells carries O_2 from the lungs to tissue cells, where the O_2 is released for metabolic processes. The molecule can be represented as Hb in its unoxygenated form and as Hb·O_2 in its oxygenated form. One reason CO is toxic is that it competes with O_2 for binding to Hb:

$$Hb \cdot O_2(aq) + CO(g) \rightleftharpoons Hb \cdot CO(aq) + O_2(g)$$

(a) If $\Delta G^0 \approx -14$ kJ at 37°C (body temperature), what is the ratio of [Hb·CO] to [Hb·O_2] at 37°C with [O_2] = [CO]?
(b) Use Le Châtelier's principle to suggest how to treat a victim of CO poisoning.

20.88 The most important ore of lithium is spodumene, $LiAlSi_2O_6$. The first step in the extraction of the metal is a conversion of the α form of spodumene into the less dense β form in preparation for subsequent leaching and washing steps. Use the following data to calculate the lowest temperature at which the α $\longrightarrow$ β conversion is feasible:

	ΔH^0_f (kJ/mol)	S^0 (J/mol·K)
α-spodumene	−3055	129.3
β-spodumene	−3027	154.4

20.89 Adenosine triphosphate (ATP) is essential for muscle contraction, protein building, nerve conduction, and numerous other energy-requiring metabolic processes. These nonspontaneous processes are "coupled" to the spontaneous hydrolysis of ATP to ADP (see Chemical Connections essay, Section 20.3). In a similar manner, ATP is regenerated by coupling its synthesis to other energy-yielding reactions, one of which is

Creatine phosphate $\longrightarrow$ creatine + phosphate
$$\Delta G^0 = -43.1 \text{ kJ/mol}$$
ADP + phosphate $\longrightarrow$ ATP $\quad \Delta G^0 = +30.5 \text{ kJ/mol}$

Calculate ΔG^0 for the overall reaction that regenerates ATP.

20.90 Nuclear fuel is prepared by converting U_3O_8 to $UO_2(NO_3)_2$ through treatment with concentrated HNO_3. The uranyl nitrate is then converted into UO_3 and finally UO_2 ("yellow cake"). At this point, the fuel is enriched (that is, the proportion of ^{235}U is increased) by converting the UO_2 into UF_6, a highly volatile solid, in preparation for a gaseous-diffusion separation of isotopes. The chemical conversion occurs in two steps:

$$UO_2(s) + 4HF(g) \longrightarrow UF_4(s) + 2H_2O(g)$$
$$UF_4(s) + F_2(g) \longrightarrow UF_6(s)$$

Calculate ΔG^0 for the overall process at 85°C:

	ΔH^0_f (kJ/mol)	S^0 (J/mol·K)	ΔG^0_f (kJ/mol)
$UO_2(s)$	−1085	77.0	−1032
$UF_4(s)$	−1921	152	−1830.
$UF_6(s)$	−2197	225	−2068

20.91 Methanol, one of the most important industrial feedstocks, is made by several catalyzed reactions, one of which is $CO(g) + 2H_2(g) \longrightarrow CH_3OH(l)$.
(a) Demonstrate that this reaction is thermodynamically feasible.
(b) Is it favored at a low or at a high temperature?
(c) One concern raised about using CH_3OH as a fuel in cars is that it is partially oxidized in air to yield formaldehyde, $CH_2O(g)$, which poses a health hazard. Calculate ΔG^0 at 100.°C for this oxidation.

20.92 Consider the following equilibrium system:

$$H_2(g) + I_2(g) \rightleftharpoons 2HI(g) \qquad \Delta G^0 = -16.77 \text{ kJ}$$

At 298 K, HI is injected into an evacuated container. At equilibrium, the partial pressure of H_2 is 0.103 atm. What is the partial pressure of HI?

*20.93 Calculate the equilibrium constants for decomposition of the hydrogen halides at 298 K:

$$2HX(g) \rightleftharpoons H_2(g) + X_2(g)$$

What do these values indicate about the extent of decomposition of HX at 298 K? Suggest a plausible reason for this trend.

20.94 Hydrogenation is the addition of H_2 to double (or triple) carbon-carbon bonds. Margarine, peanut butter, and most commercial baked goods include hydrogenated vegetable oils. What are the changes in enthalpy, entropy, and free energy for the hydrogenation of ethene (C_2H_4) to ethane (C_2H_6) at 25°C?

*20.95 Banks use coin stackers to group large numbers of coins.
(a) What is the probability that 3 dimes will stack heads-up if you throw them in a coin stacker?
(b) Based on the answer to part (a), develop a general equation for the number of ways, W, that n identical coins can be stacked.
(c) Use this equation to find the probability that 8 dimes will stack heads-up.

20.96 The key process in a blast furnace during the production of iron is the reaction of Fe_2O_3 and carbon to Fe and CO_2.
(a) Calculate ΔH^0 and ΔS^0. [Assume C(graphite).]
(b) Is the reaction spontaneous at low or at high T? Explain.
(c) Is the reaction spontaneous at 298 K?
(d) At what temperature does the reaction become spontaneous?

*20.97 Solid dinitrogen pentaoxide (N_2O_5) undergoes reaction with water to form liquid nitric acid.
(a) Is the reaction spontaneous at 25°C?
(b) The solid decomposes to NO_2 and O_2 at 25°C. Is the decomposition spontaneous at this temperature? At what temperature does the decomposition become spontaneous?
(c) At what temperature does the decomposition of *gaseous* N_2O_5 become spontaneous? Rationalize the difference between this temperature and that in part (b).

20.98 From the following reaction and data, find (a) S^0 of $SOCl_2$ and (b) T at which the reaction becomes nonspontaneous:

$$SO_3(g) + SCl_2(l) \longrightarrow SOCl_2(l) + SO_2(g) \, \Delta G^0_{rxn} = -75.2 \text{ kJ}$$

	$SO_3(g)$	$SCl_2(l)$	$SOCl_2(l)$	$SO_2(g)$
ΔH^0_f (kJ/mol)	−396	−50.0	−245.6	−296.8
S^0 (J/mol·K)	256.7	184	—	248.1

20.99 The catalysts in an automobile exhaust system remove pollutants such as carbon monoxide and nitrogen oxides, as in the following example:

$$2CO(g) + 2NO(g) \longrightarrow 2CO_2(g) + N_2(g)$$

Show that this reaction is spontaneous at standard conditions. Given this fact, what is the role of the catalyst?

20.100 Chemists often use the phrase "thermodynamically unstable, but kinetically stable" to describe a compound such as acetylene. What do you think this phrase means?

20.101 The oxidation of a metal in air is called *corrosion*. Write equations for the corrosion of iron and aluminum. Use ΔG_f^0 values from Appendix B to determine whether either process is spontaneous at 25°C.

20.102 The components of a chemical reaction can reach equilibrium at virtually any temperature. For example, the formation of HI from its elements has a K_c of 50 at 700 K:

$$H_2(g) + I_2(g) \rightleftharpoons 2HI(g) \qquad K_c = 50 \text{ at } 700 \text{ K}$$

In contrast, at a given pressure, the solid and liquid phases of a substance are in equilibrium only at the melting point. For example, at 1 atm, solid and liquid water are in equilibrium only at 273 K:

$$H_2O(s) \rightleftharpoons H_2O(l) \qquad (273 \text{ K and 1 atm})$$

(a) Which of the graphs depicts how G_{sys} changes for the chemical reaction? Explain.
(b) Which of the graphs depicts how G_{sys} changes as ice melts at 1°C and 1 atm? Explain.

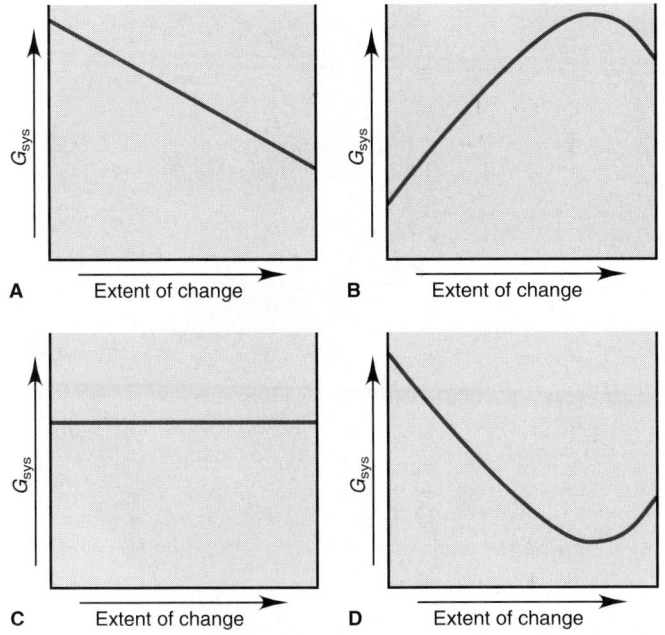

A Extent of change

B Extent of change

C Extent of change

D Extent of change

*20.103 A key step in the metabolism of glucose for energy is the isomerization of glucose-6-phosphate (G6P) to fructose-6-phosphate (F6P): G6P $\rightleftharpoons$ F6P. At 298 K, the equilibrium constant for the isomerization is 0.510.
(a) Calculate ΔG^0 at 298 K.
(b) Calculate ΔG when Q, the [F6P]/[G6P] ratio, equals 10.0.
(c) Calculate ΔG when $Q = 0.100$.
(d) Calculate Q in the cell if $\Delta G = -2.50$ kJ/mol.

*20.104 When heated, the DNA double helix separates into two random-coil single strands. When cooled, the random coils reform the double helix: double helix $\rightleftharpoons$ 2 random coils.
(a) What is the sign of ΔS for the forward process? Why?
(b) Energy must be added to the system to overcome the attractive forces due to H bonds between the base pairs and dispersion forces between the stacked bases. What is the sign of ΔG for the forward process when $T\Delta S$ is smaller than ΔH?
(c) Write an expression that shows T in terms of ΔH and ΔS when the reaction is at equilibrium. (This temperature is called the *melting temperature* of the nucleic acid.)

20.105 Consider the formation of ammonia:

$$N_2(g) + 3H_2(g) \rightleftharpoons 2NH_3(g)$$

(a) Assuming that ΔH^0 and ΔS^0 are constant with temperature, find the temperature at which $K_p = 1.00$.
(b) Find K_p at 400.°C, a typical industrial temperature for ammonia production.
(c) Given the lower K_p at the higher temperature, why are these conditions used industrially?

*20.106 Kyanite, sillimanite, and andalusite are minerals with the formula Al_2SiO_5, and each is stable under different conditions of pressure and temperature (see phase diagram).

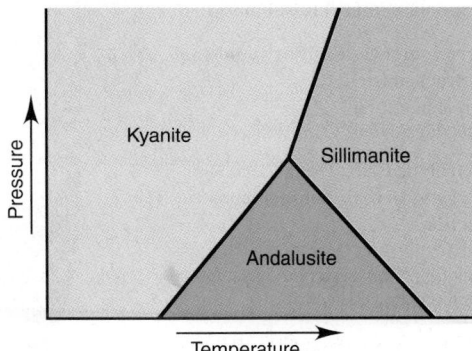

All three phases are in equilibrium where the lines intersect. At this point:
(a) Which mineral, if any, has the lowest free energy?
(b) Which mineral, if any, has the lowest enthalpy?
(c) Which mineral, if any, has the highest entropy?
(d) Which mineral, if any, has the lowest density?

CHAPTER 21

ELECTROCHEMISTRY: CHEMICAL CHANGE AND ELECTRICAL WORK

CHAPTER OUTLINE

21.1 Half-Reactions and Electrochemical Cells
Review of Oxidation-Reduction Concepts
Half-Reaction Method for Balancing
   Redox Reactions
Electrochemical Cells

21.2 Voltaic Cells: Using Spontaneous Reactions to Generate Electrical Energy
Construction and Operation
Cell Notation
Why Does the Cell Work?

21.3 Cell Potential: Output of a Voltaic Cell
Standard Cell Potentials
Strengths of Oxidizing and Reducing
   Agents

21.4 Free Energy and Electrical Work
Standard Cell Potential and K
Effect of Concentration on E_{cell}
E_{cell} and the Relation Between Q and K
Concentration Cells

21.5 Electrochemical Processes in Batteries

21.6 Corrosion: A Case of Environmental Electrochemistry
Corrosion of Iron
Protecting Against Corrosion

21.7 Electrolytic Cells: Using Electrical Energy to Drive a Nonspontaneous Reaction
Construction and Operation
Predicting Electrolysis Products
Stoichiometry of Electrolysis

Figure: Everyday electrochemistry. Today, the electrochemical cells in common batteries power a TV remote, a wristwatch, or a toy; tomorrow, they'll power our cars and household appliances. In this chapter, you'll confront the two faces of electrochemistry—reactions in cells, such as those in batteries, that do electrical work and reactions in cells that require electrical work for them to occur. Both are indispensable to our way of life.

If you think thermodynamics relates only to steam engines, games of chance, and the like, but has few practical, everyday applications, just look around you. Some of these applications are probably within your reach right now in the form of battery-operated devices—laptop computer, palm organizer, DVD remote, and, of course, the old standbys, the wristwatch and calculator—or in the form of a piece of metal-plated jewelry or silverware. The operation and creation of these objects, and the many similar ones you use daily, involve the principles of electrochemistry, certainly one of the most important areas of applied thermodynamics.

Electrochemistry is the study of the relationship between chemical change and electrical work. It is typically investigated through the use of **electrochemical cells,** systems that incorporate a redox reaction to produce or utilize electrical energy. The common objects just mentioned display the essential difference between the two types of electrochemical cells:

- One type of cell *does work by releasing free energy from a spontaneous reaction to produce electricity*; a battery houses such a cell. This application is nothing new; the automotive battery, for instance, was invented over 135 years ago.

- The other type of cell *does work by absorbing free energy from a source of electricity to drive a nonspontaneous reaction.* Such cells are used to plate a thin layer of metal on objects, and, in major industrial processes, they are used to produce some important compounds and nonmetals and to recover many metals from their ores.

In this chapter, we focus on both types of cell and the thermodynamic principles that explain why they work. We start with a review of redox concepts and revisit a method for balancing redox equations, mentioned briefly in Chapter 4, that is particularly useful for electrochemical cells. Then we provide an overview of the two cell types. We focus first on the type that releases free energy to do electrical work. We examine the free energy change and equilibrium nature of the cell's redox reaction and how these relate to its electrical output. We describe concentration cells and batteries, two important examples of this type of cell. Then, we discuss *corrosion,* a destructive, spontaneous electrochemical process similar in principle to the useful ones in these cells. Next, we switch our focus to cells that absorb free energy to do electrical work. We see how they are used to isolate elements from their compounds, how to predict the element formed from aqueous solutions and from pure compounds, and how the quantity of current flowing through the cell determines the amount of product formed. Finally, we examine the redox system that generates energy in living cells.

CONCEPTS & SKILLS

to review before you study this chapter
- redox terminology (Section 4.5 and Interchapter Topic 5)
- balancing redox reactions (Section 4.5)
- activity series of the metals (Section 4.6)
- free energy, work, and equilibrium (Sections 20.3 and 20.4)
- Q vs. K (Section 17.4) and ΔG vs. ΔG^0 (Section 20.4)

The Electrochemical Future Is Here As the combustion of coal and gasoline continues to threaten our atmosphere, a new generation of electrochemical devices is being developed. Battery-gasoline hybrid cars are already common, doubling and often tripling the mileage of traditional car engines. Soon, electric cars, powered by banks of advanced fuel cells, will reduce the need for gasoline to a minimum. Virtually every car company has a fuel-cell prototype in operation. And these devices, once used principally on space missions, are now being produced for everyday residential and industrial applications as well.

21.1 HALF-REACTIONS AND ELECTROCHEMICAL CELLS

Whether an electrochemical process releases or absorbs free energy, it always involves the *movement of electrons from one chemical species to another* in an oxidation-reduction (redox) reaction. In this section, we review the redox process and describe the half-reaction method of balancing redox reactions that was mentioned in Section 4.5. Then we take a preliminary look at how such reactions are used in the two types of electrochemical cells.

A Quick Review of Oxidation-Reduction Concepts

In electrochemical reactions, as in any redox process, *oxidation* is the loss of electrons, and *reduction* is the gain of electrons. An *oxidizing agent* is the species that performs the oxidation, taking electrons from the substance being oxidized. A *reducing agent* is the species that performs the reduction, giving electrons to the substance being reduced. After the reaction, the oxidized substance has a higher

Figure 21.1 **A summary of redox terminology.** In the reaction between zinc and hydrogen ion, Zn is oxidized and H^+ is reduced.

PROCESS	$Zn(s) + 2H^+(aq) \longrightarrow Zn^{2+}(aq) + H_2(g)$	
OXIDATION • One reactant loses electrons. • Reducing agent is oxidized. • Oxidation number increases.	Zinc **loses** electrons. Zinc is the reducing agent and becomes **oxidized**. The oxidation number of Zn **increases** from 0 to +2.	
REDUCTION • Other reactant gains electrons. • Oxidizing agent is reduced. • Oxidation number decreases.	Hydrogen ion **gains** electrons. Hydrogen ion is the oxidizing agent and becomes **reduced**. The oxidation number of H **decreases** from +1 to 0.	

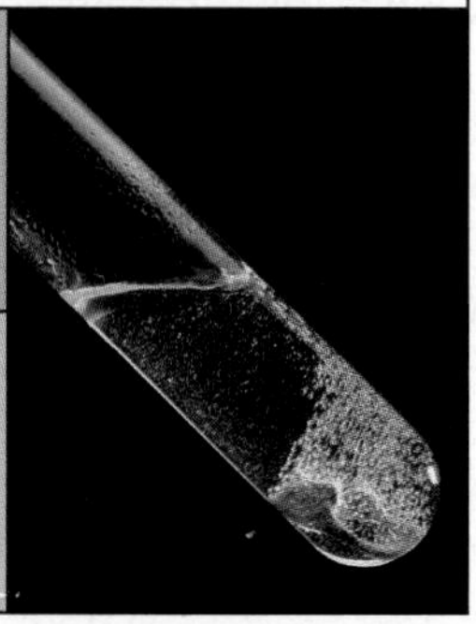

(more positive or less negative) oxidation number, and the reduced substance has a lower (less positive or more negative) one. Keep in mind three key points:

- Oxidation (electron loss) always accompanies reduction (electron gain).
- The oxidizing agent is reduced, and the reducing agent is oxidized.
- The number of electrons gained by the oxidizing agent always equals the number lost by the reducing agent.

Figure 21.1 presents these ideas in the context of the aqueous reaction between zinc metal and a strong acid. Your ability to identify the oxidation and reduction parts of a redox process is essential to understanding the topics in this chapter. If you still find this aspect of redox reactions unclear, review the full discussion in Chapter 4 and the summary in Topic 5 of the Interchapter.

Half-Reaction Method for Balancing Redox Reactions

In Chapter 4, two methods for balancing redox reactions were mentioned—the oxidation number method and the half-reaction method—but only the first method was discussed in detail. Recall that the oxidation number method assigns an oxidation number (O.N.) to each atom in the reactants and products to find those that change O.N. Then it joins the oxidized and reduced atom (or ion) and uses multipliers as coefficients to make the electrons lost equal the electrons gained (see Sample Problem 4.8, p. 151).

The essential difference between the two balancing methods is that the **half-reaction method** *divides the overall redox reaction into oxidation and reduction half-reactions.* Each half-reaction is balanced for mass (atoms) and charge. Then, one or both are multiplied by some integer to make electrons gained equal electrons lost, and the half-reactions are recombined to give the balanced redox equation. The half-reaction method offers several advantages for studying electrochemistry:

- It separates the oxidation and reduction steps, which reflects their actual physical separation in electrochemical cells.
- It makes it easier to balance redox reactions that take place in acidic or basic solution, which is common in these cells.
- It (usually) does *not* require assigning O.N.s. (In cases where the half-reactions are not obvious, we assign O.N.s to determine which atoms undergo a change and write half-reactions with the species that contain those atoms.)

In general, we begin with a "skeleton" ionic reaction, which shows only the species that are oxidized and reduced. *If the oxidized form of a species is on the left side of the skeleton reaction, the reduced form is on the right, and vice versa.* (Unless H_2O, H^+, and OH^- are among those species being oxidized or reduced, they do not appear in the skeleton reaction.) The following steps are used in balancing a redox reaction by the half-reaction method:

Step 1. Divide the skeleton reaction into two half-reactions, each of which contains the oxidized and reduced forms of one of the species. (Which half-reaction is the oxidation and which the reduction becomes clear after we balance the charges in the next step.)

Step 2. Balance the atoms and charges in each half-reaction.

- Atoms are balanced in the following order: atoms other than O and H, then O, and then H.
- Charge is balanced by *adding electrons* (e^-). They are added *to the left in the reduction half-reaction* because the reactant gains them; they are added *to the right in the oxidation half-reaction* because the reactant loses them.

Step 3. If necessary, multiply one or both half-reactions by an integer to make the number of e^- gained in the reduction equal the number of e^- lost in the oxidation.

Step 4. Add the balanced half-reactions, and include states of matter.

Step 5. Check that the atoms and charges are balanced.

We balance a redox reaction that occurs in acidic solution first and then go through Sample Problem 21.1 for balancing one in basic solution.

Balancing Redox Reactions in Acidic Solution When a redox reaction occurs in acidic solution, H_2O molecules and H^+ ions are available for use in the balancing process. Even though we've used H_3O^+ so far to indicate the proton in water, we use H^+ in this chapter because it makes the balanced equations less complex. After using H^+ in this example, we'll balance the equation with H_3O^+ just to show you that the only difference between using H^+ or H_3O^+ is in the number of water molecules needed.

Let's balance the redox reaction between dichromate ion and iodide ion to form chromium(III) ion and solid iodine, which occurs in acidic solution (Figure 21.2). The skeleton ionic reaction shows only the oxidized and reduced species:

$$Cr_2O_7^{2-}(aq) + I^-(aq) \longrightarrow Cr^{3+}(aq) + I_2(s) \qquad \text{[acidic solution]}$$

Step 1. Divide the reaction into half-reactions, each of which contains the oxidized and reduced forms of one species. The two chromium species make up one half-reaction, and the two iodine species make up the other:

$$Cr_2O_7^{2-} \longrightarrow Cr^{3+}$$
$$I^- \longrightarrow I_2$$

Step 2. Balance atoms and charges in each half-reaction. We use H_2O to balance O atoms, H^+ to balance H atoms, and e^- to balance positive charges.

- For the $Cr_2O_7^{2-}/Cr^{3+}$ half-reaction:

  a. *Balance atoms other than O and H.* We balance the two Cr on the left with a coefficient 2 on the right:

  $$Cr_2O_7^{2-} \longrightarrow 2Cr^{3+}$$

  b. *Balance O atoms by adding H_2O molecules.* Each H_2O has one O atom, so we add seven H_2O on the right to balance the seven O in $Cr_2O_7^{2-}$:

  $$Cr_2O_7^{2-} \longrightarrow 2Cr^{3+} + 7H_2O$$

  c. *Balance H atoms by adding H^+ ions.* Each H_2O contains two H, and we added seven H_2O, so we add 14 H^+ ions on the left:

  $$14H^+ + Cr_2O_7^{2-} \longrightarrow 2Cr^{3+} + 7H_2O$$

  d. *Balance charge by adding electrons.* Each H^+ ion has a 1+ charge, and 14 H^+ plus $Cr_2O_7^{2-}$ gives 12+ on the left. Two Cr^{3+} give 6+ on the right. There is an excess of 6+ on the left, so we add six e^- on the left:

  $$6e^- + 14H^+ + Cr_2O_7^{2-} \longrightarrow 2Cr^{3+} + 7H_2O$$

This half-reaction is balanced, and we see it is the *reduction* because electrons are added on the *left*: the reactant $Cr_2O_7^{2-}$ gained electrons (was reduced), so $Cr_2O_7^{2-}$ is the *oxidizing agent*. (Note that the O.N. of Cr decreased from +6 on the left to +3 on the right.)

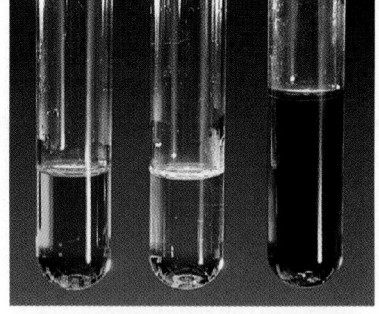

Figure 21.2 The redox reaction between dichromate ion and iodide ion. When $Cr_2O_7^{2-}$ (*left*) and I^- (*center*) are mixed in acid solution, they react to form Cr^{3+} and I_2 (*right*).

• For the I^-/I_2 half-reaction:

a. *Balance atoms other than O and H.* Two I atoms on the right requires a coefficient 2 on the left:

$$2I^- \longrightarrow I_2$$

b. *Balance O atoms with H_2O.* Not needed; there are no O atoms.

c. *Balance H atoms with H^+.* Not needed; there are no H atoms.

d. *Balance charge with e^-.* To balance the $2-$ on the left, we add two e^- on the right:

$$2I^- \longrightarrow I_2 + 2e^-$$

This balanced half-reaction is the *oxidation* because electrons are added on the *right:* the reactant I^- lost electrons (was oxidized), so I^- is the *reducing agent.* (Note that the O.N. of I increased from -1 to 0.)

Step 3. Multiply each half-reaction, if necessary, by an integer so that the number of e^- lost in the oxidation equals the number of e^- gained in the reduction. Two e^- are lost in the oxidation and six e^- are gained in the reduction, so we multiply the oxidation by 3:

$$3(2I^- \longrightarrow I_2 + 2e^-)$$
$$6I^- \longrightarrow 3I_2 + 6e^-$$

Step 4. Add the half-reactions together, canceling substances that appear on both sides, and include states of matter. In this example, only the electrons cancel:

$$6e^- + 14H^+ + Cr_2O_7^{2-} \longrightarrow 2Cr^{3+} + 7H_2O$$
$$6I^- \longrightarrow 3I_2 + 6e^-$$
$$6I^-(aq) + 14H^+(aq) + Cr_2O_7^{2-}(aq) \longrightarrow 3I_2(s) + 7H_2O(l) + 2Cr^{3+}(aq)$$

Step 5. Check that atoms and charges balance:

Reactants (6I, 14H, 2Cr, 7O; 6+) $\longrightarrow$ products (6I, 14H, 2Cr, 7O; 6+)

Balancing a Redox Reaction Using H_3O^+ Now let's balance the reduction half-reaction with H_3O^+ as the supplier of H atoms. In step 2, balancing Cr atoms and balancing O atoms with H_2O are the same as before, so we'll start with

$$Cr_2O_7^{2-} \longrightarrow 2Cr^{3+} + 7H_2O$$

Now we use H_3O^+ instead of H^+ to balance H atoms. Because each H_3O^+ is an H^+ bonded to an H_2O, we balance the 14 H on the right with 14 H_3O^+ on the left and immediately balance the H_2O that are part of those 14 H_3O^+ by adding 14 more H_2O on the right:

$$14H_3O^+ + Cr_2O_7^{2-} \longrightarrow 2Cr^{3+} + 7H_2O + 14H_2O$$

Simplifying the right side by taking the sum of the H_2O molecules gives

$$14H_3O^+ + Cr_2O_7^{2-} \longrightarrow 2Cr^{3+} + 21H_2O$$

None of this affects the redox change, so balancing the charge still requires six e^- on the left, and we obtain the balanced reduction half-reaction:

$$6e^- + 14H_3O^+ + Cr_2O_7^{2-} \longrightarrow 2Cr^{3+} + 21H_2O$$

Adding the balanced oxidation half-reaction gives the balanced redox equation:

$$6I^-(aq) + 14H_3O^+(aq) + Cr_2O_7^{2-}(aq) \longrightarrow 3I_2(s) + 21H_2O(l) + 2Cr^{3+}(aq)$$

Note, as we said, the only difference is the number of H_2O molecules: there are 14 more H_2O (21 instead of 7) from the 14 H_3O^+.

To depict the reaction even more accurately, we could show the metal ion in its hydrated form, which would also affect only the total number of H_2O:

$$6I^-(aq) + 14H_3O^+(aq) + Cr_2O_7^{2-}(aq) \longrightarrow 3I_2(s) + 9H_2O(l) + 2Cr(H_2O)_6^{3+}(aq)$$

Although these added steps present the species in solution more accurately, they only change the number of water molecules, and make balancing more difficult; thus, they detract somewhat from the key chemical event—the redox change. Therefore, we'll employ the method that uses H^+ to balance H atoms.

Balancing Redox Reactions in Basic Solution As you just saw, in acidic solution, H_2O molecules and H^+ (or H_3O^+) ions are available for balancing. As Sample Problem 21.1 shows, when a reaction occurs in basic solution, H_2O molecules and OH^- ions are available. Only one additional step is needed to balance a redox equation that takes place in basic solution. It appears after both half-reactions have first been balanced *as if they took place in acidic solution* (steps 1 and 2), the e^- lost have been made equal to the e^- gained (step 3), and the half-reactions have been combined (step 4). At this point, *we add one OH^- ion* **to both sides of the equation** *for every H^+ ion present.* (We label this step 4 Basic.) The H^+ ions on one side are combined with the added OH^- ions to form H_2O, and OH^- ions appear on the other side of the equation. Excess H_2O is canceled and states of matter are identified. Finally, we check that atoms and charges balance (step 5).

SAMPLE PROBLEM 21.1 Balancing Redox Reactions by the Half-Reaction Method

Problem Permanganate ion is a strong oxidizing agent, and its deep purple color makes it useful as an indicator in redox titrations (see Figure 4.13, p. 152). It reacts in basic solution with the oxalate ion to form carbonate ion and solid manganese dioxide. Balance the skeleton ionic reaction that occurs between $NaMnO_4$ and $Na_2C_2O_4$ in basic solution:

$$MnO_4^-(aq) + C_2O_4^{2-}(aq) \longrightarrow MnO_2(s) + CO_3^{2-}(aq) \quad \text{[basic solution]}$$

Plan We proceed through step 4 as if this took place in acidic solution. Then, we add the appropriate number of OH^- ions and cancel excess H_2O molecules (step 4 Basic).

Solution

1. Divide into half-reactions.

$$MnO_4^- \longrightarrow MnO_2 \qquad\qquad C_2O_4^{2-} \longrightarrow CO_3^{2-}$$

2. Balance.

  a. Atoms other than O and H,   a. Atoms other than O and H,

     Not needed $C_2O_4^{2-} \longrightarrow 2CO_3^{2-}$

  b. O atoms with H_2O,   b. O atoms with H_2O,

    $MnO_4^- \longrightarrow MnO_2 + 2H_2O$ $2H_2O + C_2O_4^{2-} \longrightarrow 2CO_3^{2-}$

  c. H atoms with H^+,   c. H atoms with H^+,

    $4H^+ + MnO_4^- \longrightarrow MnO_2 + 2H_2O$ $2H_2O + C_2O_4^{2-} \longrightarrow 2CO_3^{2-} + 4H^+$

  d. Charge with e^-,   d. Charge with e^-,

    $3e^- + 4H^+ + MnO_4^- \longrightarrow MnO_2 + 2H_2O$ $2H_2O + C_2O_4^{2-} \longrightarrow 2CO_3^{2-} + 4H^+ + 2e^-$

                  [reduction]   [oxidation]

3. Multiply each half-reaction, if necessary, by some integer to make e^- lost equal e^- gained.

$$2(3e^- + 4H^+ + MnO_4^- \longrightarrow MnO_2 + 2H_2O) \qquad 3(2H_2O + C_2O_4^{2-} \longrightarrow 2CO_3^{2-} + 4H^+ + 2e^-)$$
$$6e^- + 8H^+ + 2MnO_4^- \longrightarrow 2MnO_2 + 4H_2O \qquad 6H_2O + 3C_2O_4^{2-} \longrightarrow 6CO_3^{2-} + 12H^+ + 6e^-$$

4. Add half-reactions, and cancel substances appearing on both sides.

    The e^- cancel, eight H^+ cancel to leave four H^+ on the right, and four H_2O cancel to leave two H_2O on the left:

$$\require{cancel}\cancel{6e^-} + \cancel{8}H^{\cancel{+}} + 2MnO_4^- \longrightarrow 2MnO_2 + \cancel{4}H_2O$$
$$2\,6H_2O + 3C_2O_4^{2-} \longrightarrow 6CO_3^{2-} + 4\,\cancel{12}H^+ + \cancel{6e^-}$$
$$\overline{2MnO_4^- + 2H_2O + 3C_2O_4^{2-} \longrightarrow 2MnO_2 + 6CO_3^{2-} + 4H^+}$$

4 Basic. Add OH^- to both sides to neutralize H^+, and cancel H_2O.

    Adding four OH^- to both sides forms four H_2O on the right, two of which cancel the two H_2O on the left, leaving two H_2O on the right:

$$2MnO_4^- + 2H_2O + 3C_2O_4^{2-} + 4OH^- \longrightarrow 2MnO_2 + 6CO_3^{2-} + [4H^+ + 4OH^-]$$
$$2MnO_4^- + \cancel{2H_2O} + 3C_2O_4^{2-} + 4OH^- \longrightarrow 2MnO_2 + 6CO_3^{2-} + 2\,\cancel{4}H_2O$$

    Including states of matter gives the final balanced equation:

$$2MnO_4^-(aq) + 3C_2O_4^{2-}(aq) + 4OH^-(aq) \longrightarrow 2MnO_2(s) + 6CO_3^{2-}(aq) + 2H_2O(l)$$

5. Check that atoms and charges balance.

$$(2Mn, 24O, 6C, 4H; 12-) \longrightarrow (2Mn, 24O, 6C, 4H; 12-)$$

Comment As a final step, we can obtain the balanced *molecular* equation for this reaction by noting the number of moles of each anion in the balanced ionic equation and adding the correct number of moles of spectator ions (in this case, Na^+) to obtain neutral compounds. Thus, for instance, balancing the charge of 2 mol of MnO_4^- requires 2 mol of Na^+, so we have $2NaMnO_4$. The balanced molecular equation is

$$2NaMnO_4(aq) + 3Na_2C_2O_4(aq) + 4NaOH(aq) \longrightarrow$$
$$2MnO_2(s) + 6Na_2CO_3(aq) + 2H_2O(l)$$

FOLLOW-UP PROBLEM 21.1 Write a balanced molecular equation for the reaction between $KMnO_4$ and KI in basic solution. The skeleton reaction is

$$MnO_4^-(aq) + I^-(aq) \longrightarrow MnO_4^{2-}(aq) + IO_3^-(aq) \qquad \text{[basic solution]}$$

The half-reaction method reveals a great deal about redox processes and is essential to understanding electrochemical cells. The major points to remember are

- Any redox reaction can be treated as the sum of a reduction and an oxidation half-reaction.
- Mass (atoms) and charge are conserved in each half-reaction.
- Electrons lost in one half-reaction are gained in the other.
- Although the half-reactions are treated separately, electron loss and electron gain occur simultaneously.

An Overview of Electrochemical Cells

We distinguish two types of electrochemical cells based on the general thermodynamic nature of the reaction:

1. A **voltaic cell** (or **galvanic cell**) uses a spontaneous reaction ($\Delta G < 0$) to generate electrical energy. In the cell reaction, the difference in chemical potential energy between higher energy reactants and lower energy products is converted into electrical energy. This energy is then used to operate the load, such as a flashlight bulb, CD player, car starter motor, or other electrical device. In other words, *the system does work on the surroundings.* All batteries contain voltaic cells.

2. An **electrolytic cell** uses electrical energy to drive a nonspontaneous reaction ($\Delta G > 0$). In the cell reaction, electrical energy from an external power supply converts lower energy reactants into higher energy products. Thus, *the surroundings do work on the system.* Electroplating and recovering metals from ores involve electrolytic cells.

The two types of cell have certain design features in common (Figure 21.3). Two **electrodes,** the objects that conduct the electricity between the cell and the surroundings, are dipped into an **electrolyte,** the mixture of ions (usually in aqueous solution) that are involved in the reaction or that carry the charge. An electrode is identified as the **anode** or the **cathode,** according to the half-reaction that takes place there:

- *The oxidation half-reaction takes place at the anode.* Electrons are given up by the substance being oxidized (reducing agent) and *leave the cell* at the anode.
- *The reduction half-reaction takes place at the cathode.* Electrons are taken up by the substance being reduced (oxidizing agent) and *enter the cell* at the cathode.

Which Half-Reaction Occurs at Which Electrode? If you sometimes forget which half-reaction occurs at which electrode, you're not alone. Here are some memory aids to help:
1. The words *anode* and *oxidation* start with vowels; the words *cathode* and *reduction* start with consonants.
2. Alphabetically, the *A* in anode comes before the *C* in cathode, and the *O* in oxidation comes before the *R* in reduction.
3. Look at the first syllables and use your imagination:
ANode, OXidation;
REDuction, CAThode ⇒
AN OX and a RED CAT

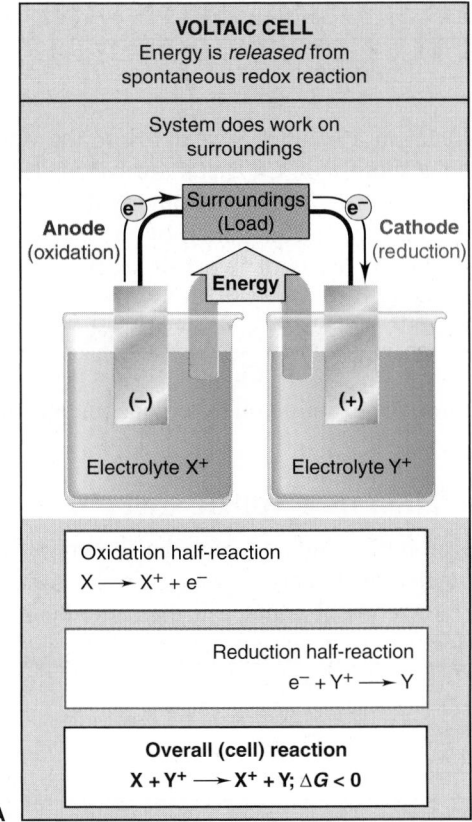

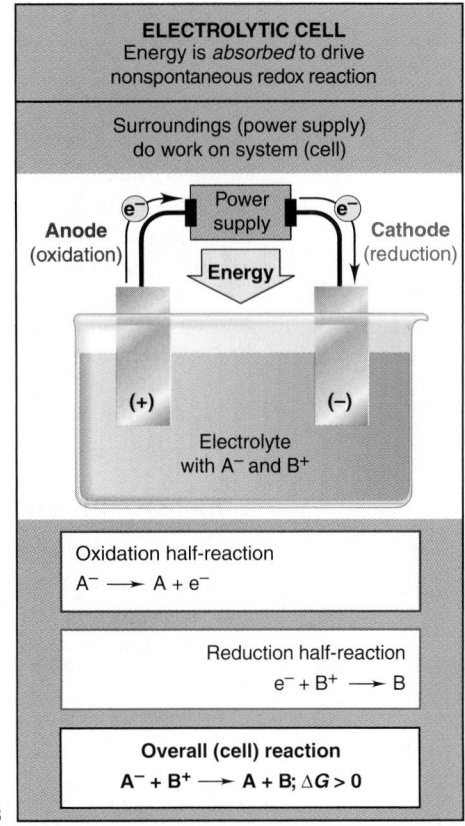

Figure 21.3 **General characteristics of voltaic and electrolytic cells.** A voltaic cell **(A)** generates energy from a spontaneous reaction ($\Delta G < 0$), whereas an electrolytic cell **(B)** requires energy to drive a nonspontaneous reaction ($\Delta G > 0$). In both types of cell, two electrodes dip into electrolyte solutions, and an external circuit provides the means for electrons to flow between them. Most important, notice that oxidation takes place at the anode and reduction takes place at the cathode, but the relative electrode charges are opposite in the two cells.

As shown in Figure 21.3, the relative charges of the electrodes are *opposite* in the two types of cell. As you'll see in the following sections, the opposite charges of the electrodes result from the different phenomena that cause the electrons to flow.

SECTION SUMMARY

Oxidation-reduction processes involve the transfer of electrons from a reducing agent to an oxidizing agent. The half-reaction method of balancing divides the overall reaction into half-reactions that are balanced separately and then are recombined. The two types of electrochemical cells are both based on redox reactions: in one cell it generates electricity, and in the other cell it utilizes electricity. In a voltaic cell, a spontaneous reaction does work on the surroundings (flashlight, starter motor, etc.); in an electrolytic cell, the surroundings (power supply) do work to drive a nonspontaneous reaction. Both types of cell contain two electrodes that dip into electrolyte solutions. In both types, oxidation occurs at the anode, and reduction occurs at the cathode.

21.2 VOLTAIC CELLS: USING SPONTANEOUS REACTIONS TO GENERATE ELECTRICAL ENERGY

If you put a strip of zinc metal in a solution of Cu^{2+} ion, the blue color of the solution fades as a brown-black crust of Cu metal forms on the Zn strip (Figure 21.4). Judging from what we see, the reaction involves the reduction of Cu^{2+} ion to Cu metal, which must be accompanied by the oxidation of Zn metal to Zn^{2+} ion. The overall reaction consists of two half-reactions:

$$Cu^{2+}(aq) + 2e^- \longrightarrow Cu(s) \qquad \text{[reduction]}$$
$$\underline{Zn(s) \longrightarrow Zn^{2+}(aq) + 2e^- \qquad \text{[oxidation]}}$$
$$Zn(s) + Cu^{2+}(aq) \longrightarrow Zn^{2+}(aq) + Cu(s) \qquad \text{[overall reaction]}$$

In the remainder of this section, we examine this spontaneous reaction as the basis of a voltaic (galvanic) cell.

Construction and Operation of a Voltaic Cell

Electrons are being transferred in the Zn/Cu^{2+} reaction shown in Figure 21.4, but the system does not generate electrical energy because the oxidizing agent (Cu^{2+}) and reducing agent (Zn) are in physical contact in the same beaker. If, however, the half-reactions are physically separated and connected by an external circuit, the electrons are transferred by traveling through the circuit, thus producing an electric current.

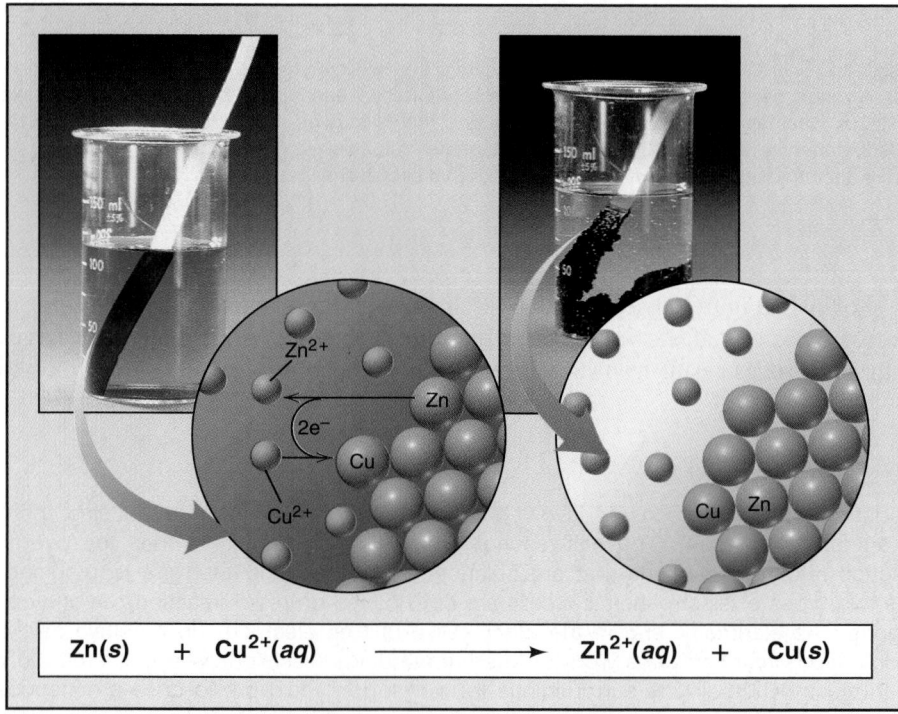

Figure 21.4 The spontaneous reaction between zinc and copper(II) ion. When a strip of zinc metal is placed in a solution of Cu^{2+} ion, a redox reaction begins (*left*) in which the zinc is oxidized to Zn^{2+} and the Cu^{2+} is reduced to copper metal. As the reaction proceeds (*right*), the deep blue color of the solution of hydrated Cu^{2+} ion lightens, and the Cu "plates out" on the Zn and falls off in chunks. (The Cu appears black because it is very finely divided.) At the atomic scale, each Zn atom loses two electrons, which are gained by a Cu^{2+} ion. The process is summarized with symbols in the balanced equation.

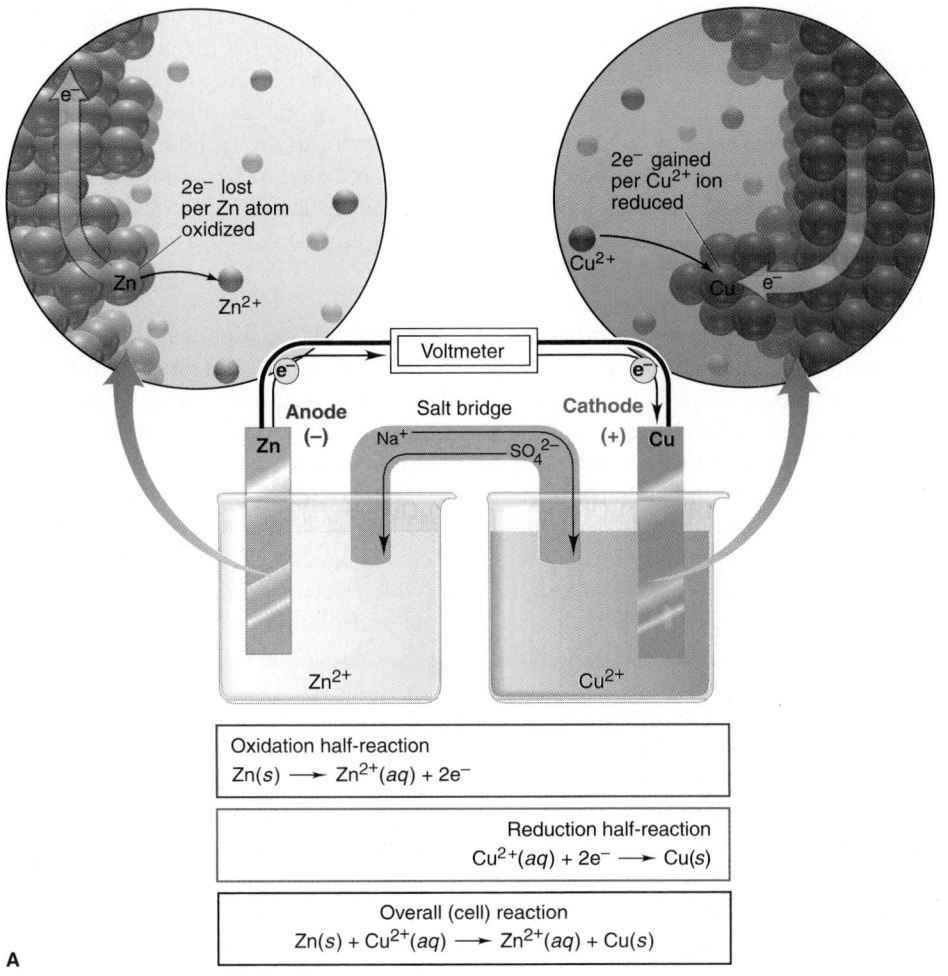

Oxidation half-reaction
$$Zn(s) \longrightarrow Zn^{2+}(aq) + 2e^-$$

Reduction half-reaction
$$Cu^{2+}(aq) + 2e^- \longrightarrow Cu(s)$$

Overall (cell) reaction
$$Zn(s) + Cu^{2+}(aq) \longrightarrow Zn^{2+}(aq) + Cu(s)$$

A

B

Figure 21.5 **A voltaic cell based on the zinc-copper reaction. A,** The anode half-cell (oxidation) consists of a Zn electrode dipping into a Zn^{2+} solution. The two electrons generated in the oxidation of each Zn atom move through the Zn bar and the wire, and into the Cu electrode, which dips into a Cu^{2+} solution in the cathode half-cell (reduction). There, the electrons reduce Cu^{2+} ions. Thus, electrons flow left to right through electrodes and wire. A salt bridge contains unreactive Na^+ and SO_4^{2-} ions that maintain neutral charge in the electrolyte solution: anions in the salt bridge flow to the left, and cations flow to the right. The voltmeter registers the electrical output of the cell. **B,** After the cell runs for several hours, the Zn anode weighs less because Zn atoms have been oxidized to aqueous Zn^{2+} ions, and the Cu cathode weighs more because aqueous Cu^{2+} ions have been reduced to Cu metal.

This separation of half-reactions is the essential idea behind a voltaic cell (Figure 21.5A). The components of each half-reaction are placed in a separate container, or **half-cell,** which consists of an electrode dipping into an electrolyte solution. The two half-cells are joined by the circuit, which consists of a wire and a salt bridge (the inverted U tube in the figure; we will discuss its function shortly). In order to measure the voltage generated by the cell, a voltmeter is inserted in the path of the wire connecting the electrodes. A switch (not shown) is included to close (complete) or open (break) the circuit. By convention, *the oxidation half-cell (anode compartment) is shown on the left and the reduction half-cell (cathode compartment) on the right*. Here are the key points about this voltaic cell:

1. *The oxidation half-cell.* In this case, the anode compartment consists of a zinc metal bar (the anode) immersed in a Zn^{2+} electrolyte (such as a solution of zinc sulfate, $ZnSO_4$). In addition to actively participating as the reactant in the oxidation half-reaction, the zinc bar conducts the released electrons *out* of its half-cell.

2. *The reduction half-cell.* In this case, the cathode compartment consists of a copper metal bar (the cathode) immersed in a Cu^{2+} electrolyte [such as a solution of copper(II) sulfate, $CuSO_4$]. In addition to actively participating as the product in the reduction half-reaction, the copper bar conducts electrons *into* its half-cell.

3. *Relative charges on the electrodes.* The electrode charges are determined by the *source of electrons* and the *direction of electron flow* through the circuit. In this cell, zinc metal is oxidized at the anode to Zn^{2+} ions and electrons. The Zn^{2+} ions enter and flow throughout the solution, while the electrons enter the wire. *The electrons flow left to right* through the wire to the cathode. There, Cu^{2+} ions in the solution accept the electrons and are reduced to Cu atoms. As the cell operates, electrons are continuously generated at the anode and consumed at the cathode. Therefore, the anode has an excess of electrons and a negative charge *relative* to the cathode. *In any* **voltaic** *cell, the anode is negative and the cathode is positive.*

4. *The purpose of the salt bridge.* The cell can't operate until the circuit is complete. The oxidation half-cell originally contains a neutral solution of Zn^{2+} and SO_4^{2-} ions, but if Zn atoms in the bar lost electrons, the solution would develop a net positive charge from the addition of newly formed Zn^{2+} ions. Similarly, in the reduction half-cell, the neutral solution of Cu^{2+} and SO_4^{2-} ions would develop a net negative charge if Cu^{2+} ions left the solution to form Cu atoms. A charge imbalance would arise and stop cell operation if the neutral charge of the half-cells were not maintained. To enable the cell to operate, the two half-cells are joined by a **salt bridge,** which acts as a "liquid wire," allowing ions to flow through both compartments and complete the circuit. The salt bridge shown in Figure 21.5A is an inverted U tube containing a solution of non-reacting ions—in this case, Na^+ and SO_4^{2-}—in a gel. Thus, the solution cannot pour out, but ions can diffuse through it into and out of the half-cells.

To maintain neutrality in the reduction half-cell (right; cathode compartment) as Cu^{2+} ions change to Cu atoms, Na^+ ions move from the salt bridge into the solution (and some SO_4^{2-} ions move from the solution into the salt bridge). Similarly, to maintain neutrality in the oxidation half-cell (left; anode compartment) as Zn atoms change to Zn^{2+} ions, SO_4^{2-} ions move from the salt bridge into that solution (and some Zn^{2+} ions move from the solution into the salt bridge). Thus, as Figure 21.5A shows, the circuit is completed *as electrons move left to right through the wire, while anions move right to left and cations move left to right through the salt bridge.*

5. *Active vs. inactive electrodes.* The electrodes in the Zn/Cu^{2+} cell are *active* because the metals themselves are components of the half-reactions. As the cell operates, the mass of the zinc electrode gradually decreases, and the [Zn^{2+}] in the anode half-cell increases. At the same time, the mass of the copper electrode increases, and the [Cu^{2+}] in the cathode half-cell decreases; we say that the Cu^{2+} "plates out" on the electrode. Figure 21.5B shows the electrodes removed from their half-cells after several hours of operation.

For many redox reactions, however, there are no reactants or products capable of serving as electrodes. In these cases, *inactive* electrodes are used. Most commonly, inactive electrodes are rods of *graphite* or *platinum*; they conduct electrons into or out of the cell but cannot take part in the half-reactions. In a voltaic cell based on the following half-reactions, for instance, the reacting species cannot act as electrodes:

$$2I^-(aq) \longrightarrow I_2(s) + 2e^- \qquad \text{[anode; oxidation]}$$
$$MnO_4^-(aq) + 8H^+(aq) + 5e^- \longrightarrow Mn^{2+}(aq) + 4H_2O(l) \qquad \text{[cathode; reduction]}$$

Therefore, each half-cell consists of inactive electrodes immersed in an electrolyte solution that contains *all the species involved in that half-reaction* (Figure 21.6). In the anode half-cell, I^- ions are oxidized to solid I_2. The electrons released flow into the graphite anode, through the wire, and into the graphite cathode. From there, the electrons are consumed by MnO_4^- ions, which are reduced to Mn^{2+} ions. (A KNO_3 salt bridge is used.)

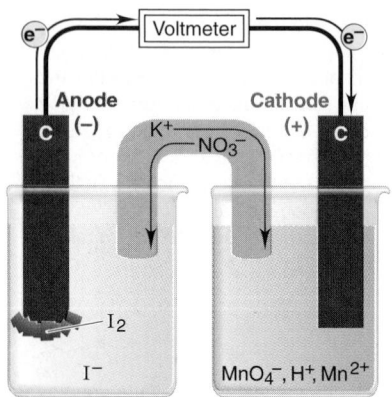

Oxidation half-reaction

$2I^-(aq) \longrightarrow I_2(s) + 2e^-$

Reduction half-reaction

$MnO_4^-(aq) + 8H^+(aq) + 5e^- \longrightarrow$
$\qquad Mn^{2+}(aq) + 4H_2O(l)$

Overall (cell) reaction

$2MnO_4^-(aq) + 16H^+(aq) + 10I^-(aq) \longrightarrow$
$\qquad 2Mn^{2+}(aq) + 5I_2(s) + 8H_2O(l)$

Figure 21.6 A voltaic cell using inactive electrodes. A cell based on the reaction between I^- and MnO_4^- in acidic solution does not have species that can be used as electrodes, so inactive graphite (C) electrodes are used.

As Figures 21.5A and 21.6 illustrate, there are certain consistent features in the *diagram* of any voltaic cell. The physical arrangement includes the half-cell containers, electrodes, wire, and salt bridge, and the following details are shown:

- Components of the half-cells: electrode materials, electrolyte ions, and other substances involved in the reaction
- Electrode name (anode or cathode) and charge. By convention, the anode compartment always appears *on the left*.
- Each half-reaction with its half-cell and the overall cell reaction
- Direction of electron flow in the external circuit
- Nature of ions and direction of ion flow in the salt bridge

You'll see how to specify these details and diagram a cell shortly.

GALVANIC CELL

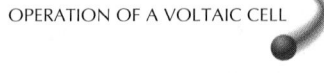
OPERATION OF A VOLTAIC CELL

Notation for a Voltaic Cell

There is a useful shorthand notation for describing the components of a voltaic cell. For example, the notation for the Zn/Cu^{2+} cell is

$$Zn(s) \mid Zn^{2+}(aq) \parallel Cu^{2+}(aq) \mid Cu(s)$$

Key parts of the notation are

- The components of the anode compartment (oxidation half-cell) are written *to the left* of the components of the cathode compartment (reduction half-cell).
- A vertical line represents a phase boundary. For example, $Zn(s) \mid Zn^{2+}(aq)$ indicates that the *solid* Zn is a *different* phase from the *aqueous* Zn^{2+}. A comma separates the half-cell components that are in the *same* phase. For example, the notation for the voltaic cell housing the reaction between I^- and MnO_4^- that is shown in Figure 21.6 is

  $$\text{graphite} \mid I^-(aq) \mid I_2(s) \parallel H^+(aq), MnO_4^-(aq), Mn^{2+}(aq) \mid \text{graphite}$$

  That is, in the cathode compartment, H^+, MnO_4^-, and Mn^{2+} ions are all in aqueous solution with solid graphite immersed in it. Often, we specify the concentrations of dissolved components; for example, if the concentrations of Zn^{2+} and Cu^{2+} are 1 *M*, we write

  $$Zn(s) \mid Zn^{2+}(1\ M) \parallel Cu^{2+}(1\ M) \mid Cu(s)$$

- Half-cell components usually appear in the same order as in the half-reaction, and electrodes appear at the far left and right of the notation.
- A double vertical line separates the half-cells and represents the phase boundary on either side of the salt bridge (the ions in the salt bridge are omitted because they are not part of the reaction).

SAMPLE PROBLEM 21.2 Diagramming Voltaic Cells

Problem Diagram, show balanced equations, and write the notation for a voltaic cell that consists of one half-cell with a Cr bar in a $Cr(NO_3)_3$ solution, another half-cell with an Ag bar in an $AgNO_3$ solution, and a KNO_3 salt bridge. Measurement indicates that the Cr electrode is negative relative to the Ag electrode.

Plan From the given contents of the half-cells, we can write the half-reactions. We must determine which is the anode compartment (oxidation) and which is the cathode (reduction). To do so, we must find the direction of the spontaneous redox reaction, which is given by the relative electrode charges. Since electrons are released into the anode during oxidation, it has a negative charge. We are told that Cr is negative, so it must be the anode and, therefore, Ag is the cathode.

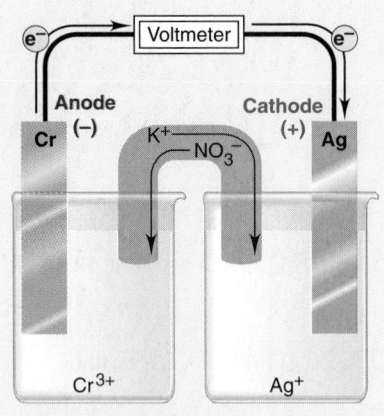

Oxidation half-reaction
$Cr(s) \longrightarrow Cr^{3+}(aq) + 3e^-$

Reduction half-reaction
$Ag^+(aq) + e^- \longrightarrow Ag(s)$

Overall (cell) reaction
$Cr(s) + 3Ag^+(aq) \longrightarrow Cr^{3+}(aq) + 3Ag(s)$

Solution Writing the balanced half-reactions. Since the Ag electrode is positive, the half-reaction consumes e^-:

$$Ag^+(aq) + e^- \longrightarrow Ag(s) \quad \text{[reduction; cathode]}$$

Since the Cr electrode is negative, the half-reaction releases e^-:

$$Cr(s) \longrightarrow Cr^{3+}(aq) + 3e^- \quad \text{[oxidation; anode]}$$

Writing the balanced overall cell reaction. We triple the reduction half-reaction in order to balance e^-, and then we combine the half-reactions to obtain the overall spontaneous redox reaction:

$$Cr(s) + 3Ag^+(aq) \longrightarrow Cr^{3+}(aq) + 3Ag(s)$$

Determining direction of electron and ion flow. The released e^- in the Cr electrode (negative) flow through the external circuit to the Ag electrode (positive). As Cr^{3+} ions enter the anode electrolyte, NO_3^- ions enter from the salt bridge to maintain neutrality. As Ag^+ ions leave the cathode electrolyte and plate out on the Ag electrode, K^+ ions enter from the salt bridge to maintain neutrality. The diagram of this cell is shown in the margin. Writing the cell notation:

$$Cr(s) \mid Cr^{3+}(aq) \parallel Ag^+(aq) \mid Ag(s)$$

Check Always be sure that the half-reactions and cell reaction are balanced, the half-cells contain *all* components of the half-reaction, and the electron and ion flow are shown. You should be able to write the half-reactions from the cell notation as a check.

Comment The key to diagramming a voltaic cell is to use the spontaneous reaction to identify the oxidation (anode) and reduction (cathode) half-reactions.

FOLLOW-UP PROBLEM 21.2 In one compartment of a voltaic cell, a graphite rod dips into an acidic solution of $K_2Cr_2O_7$ and $Cr(NO_3)_3$; in the other, a tin bar dips into a $Sn(NO_3)_2$ solution. A KNO_3 salt bridge joins the half-cells. The tin electrode is negative relative to the graphite. Diagram the cell, write balanced equations, and show the cell notation.

Why Does a Voltaic Cell Work?

By placing a lightbulb in the circuit or looking at the voltmeter, we can see that the Zn/Cu^{2+} cell generates electrical energy. But what principle explains *how* the reaction takes place, and *why* do electrons flow in the direction shown?

Let's examine what is happening when the switch is open and no reaction is occurring. In each of the half-cells, we can consider the metal electrode to be in equilibrium with the metal ions in the electrolyte, with the electrons residing in the metal:

$$Zn(s) \rightleftharpoons Zn^{2+}(aq) + 2e^-(\text{in Zn metal})$$
$$Cu(s) \rightleftharpoons Cu^{2+}(aq) + 2e^-(\text{in Cu metal})$$

From the direction of the overall spontaneous reaction (shown on p. 900), we know that Zn gives up its electrons more easily than Cu does, so it is a stronger reducing agent. Therefore, the equilibrium position of the Zn half-reaction lies farther to the right: Zn produces more electrons than Cu does. You might think of the electrons in the Zn electrode as having a greater electron "pressure" than those in the Cu electrode, a greater potential energy (referred to as *electrical potential*) ready to "push" the electrons through the circuit. Close the switch and electrons flow from the Zn to the Cu electrode to equalize this difference in electrical potential. The flow disturbs the equilibrium at each electrode. The Zn half-reaction shifts to the right to restore the electrons flowing out, and the Cu half-reaction shifts to the left to remove the electrons flowing in. Thus, *the spontaneous reaction occurs as a result of the different abilities of these metals to give up their electrons and the ability of the electrons to flow through the circuit.*

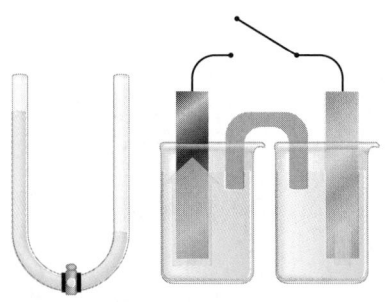

Electron Flow and Water Flow Consider this analogy between electron "pressure" and water pressure. A U-tube (cell) is separated into two arms (two half-cells) by a stopcock (switch), and the two arms contain water at different heights (the half-cells contain half-reactions with different electrical potentials). Open the stopcock (close the switch) and the different heights (potential difference) become equal as water flows (electrons flow and current is generated).

A voltaic cell consists of oxidation (anode) and reduction (cathode) half-cells, connected by a wire to conduct electrons and a salt bridge to provide ions that maintain charge neutrality as the cell operates. Electrons move from anode (left) to cathode (right), while cations move from the salt bridge into the cathode half-cell and anions from the salt bridge into the anode half-cell. The cell notation uses a comma to separate components in the same phase within a half-cell, a single vertical line to separate components in different phases, and a double vertical line to separate the half-cells. A voltaic cell operates because the two half-cells differ in their tendency to lose electrons.

21.3 CELL POTENTIAL: OUTPUT OF A VOLTAIC CELL

The purpose of a voltaic cell is to convert the free energy change of a spontaneous reaction into the kinetic energy of electrons moving through an external circuit (electrical energy). This electrical energy can do work and is proportional to the *difference in electrical potential between the two electrodes.* This difference in the electrical potential of the electrodes is the **cell potential (E_{cell}),** also called the **voltage** of the cell or the **electromotive force (emf).**

Electrons are negatively charged, so they flow spontaneously from the negative to the positive electrode, that is, toward the electrode with the more positive electrical potential. Thus, when the cell operates *spontaneously,* the difference in the electrical potential of the electrodes is positive; that is, there is a *positive* cell potential:

$$E_{cell} > 0 \text{ for a spontaneous process} \qquad (21.1)$$

The more positive E_{cell} is, the more work the cell can do, and the farther the reaction proceeds to the right as written. A *negative* cell potential, on the other hand, is associated with a *nonspontaneous* cell reaction. If $E_{cell} = 0$, the reaction has reached equilibrium and the cell can do no more work. (There is a clear relationship between E_{cell}, K, and ΔG that we discuss in Section 21.4.)

How are the units of cell potential related to those of energy available to do work? As you've seen, work is done when charge moves between electrode compartments that differ in electrical potential. The SI unit of electrical potential is the **volt (V),** and the SI unit of electrical charge is the **coulomb (C).** By definition, for two electrodes that differ by 1 volt of electrical potential, 1 joule of energy is released (that is, 1 joule of work can be done) for each coulomb of charge that moves between the electrodes. Thus,

$$1 \text{ V} = 1 \text{ J/C} \qquad (21.2)$$

Table 21.1 lists the voltages of some commercial and natural voltaic cells. Let's see how to measure cell potential.

Table 21.1   Voltages of Some Voltaic Cells

Voltaic Cell	Voltage (V)
Common alkaline battery	1.5
Lead-acid car battery (6 cells = 12 V)	2.0
Calculator battery (mercury)	1.3
Electric eel (~5000 cells in 6-ft eel = 750 V)	0.15
Nerve of giant squid (across cell membrane)	0.070

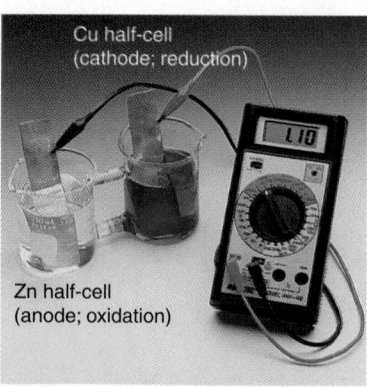

Cu half-cell
(cathode; reduction)

Zn half-cell
(anode; oxidation)

The zinc-copper cell operating at 298 K under standard-state conditions.

Standard Cell Potentials

The measured potential of a voltaic cell is affected by changes in concentration as the reaction proceeds and by energy losses due to heating of the cell and the external circuit. Therefore, in order to compare the output of different cells, we obtain a **standard cell potential (E^0_{cell}),** the potential measured at a specified temperature (usually 298 K) with no current flowing* and *all components in their standard states:* 1 atm for gases, 1 M for solutions, the pure solid for electrodes. When the zinc-copper cell that we diagrammed in Figure 21.5 operates under standard state conditions, that is, when $[Zn^{2+}] = [Cu^{2+}] = 1$ M, the cell produces 1.10 V at 298 K (see photo):

$$Zn(s) + Cu^{2+}(aq, 1\ M) \longrightarrow Zn^{2+}(aq, 1\ M) + Cu(s) \qquad E^0_{cell} = 1.10\ V$$

Standard Electrode (Half-Cell) Potentials Just as each half-reaction makes up part of the overall reaction, the potential of each half-cell makes up a part of the overall cell potential. The **standard electrode potential ($E^0_{half-cell}$)** is the potential associated with a given half-reaction (electrode compartment) when all the components are in their standard states.

By convention, *a standard electrode potential always refers to the half-reaction written as a **reduction.*** For the zinc-copper reaction, for example, the standard electrode potentials for the zinc half-reaction (E^0_{zinc}, anode compartment) and for the copper half-reaction (E^0_{copper}, cathode compartment) refer to the processes written as reductions:

$$Zn^{2+}(aq) + 2e^- \longrightarrow Zn(s) \qquad E^0_{zinc}\ (E^0_{anode}) \qquad \text{[reduction]}$$
$$Cu^{2+}(aq) + 2e^- \longrightarrow Cu(s) \qquad E^0_{copper}\ (E^0_{cathode}) \qquad \text{[reduction]}$$

The overall cell reaction involves the *oxidation* of zinc at the anode, not the *reduction* of Zn^{2+}, so we reverse the zinc half-reaction:

$$Zn(s) \longrightarrow Zn^{2+}(aq) + 2e^- \qquad \text{[oxidation]}$$
$$Cu^{2+}(aq) + 2e^- \longrightarrow Cu(s) \qquad \text{[reduction]}$$

The overall redox reaction is the sum of these half-reactions:

$$Zn(s) + Cu^{2+}(aq) \longrightarrow Zn^{2+}(aq) + Cu(s)$$

Because electrons flow spontaneously toward the copper electrode (cathode), it must have a more positive $E^0_{half-cell}$ than the zinc electrode (anode). Therefore, to obtain a positive E^0_{cell}, we subtract E^0_{zinc} from E^0_{copper}:

$$E^0_{cell} = E^0_{copper} - E^0_{zinc}$$

We can generalize this result for any voltaic cell: *the standard cell potential is the difference between the standard electrode potential of the cathode (reduction) half-cell and the standard electrode potential of the anode (oxidation) half-cell,*

$$E^0_{cell} = E^0_{cathode} - E^0_{anode} \qquad \textbf{(21.3)}$$

Determining $E^0_{half-cell}$: The Standard Hydrogen Electrode What portion of E^0_{cell} for the zinc-copper reaction is contributed by the anode half-cell (oxidation of Zn) and what portion by the cathode half-cell (reduction of Cu^{2+})? That is, how can we know half-cell potentials if we can only measure the potential of the complete cell? Half-cell potentials, such as E^0_{zinc} and E^0_{copper}, are not absolute quantities, but rather are values *relative* to that of a standard. Chemists have chosen *a standard reference half-cell and defined its standard electrode potential as zero* ($E^0_{reference} \equiv 0.00$ V). The **standard reference half-cell** is a **standard hydrogen electrode,** which consists of a specially prepared platinum electrode immersed in

---

*The current required to operate modern digital voltmeters makes a negligible difference in the value of E^0_{cell}.

a 1 M aqueous solution of a strong acid, $H^+(aq)$ [or $H_3O^+(aq)$], through which H_2 gas at 1 atm is bubbled. Thus, the reference half-reaction is

$$2H^+(aq, 1\ M) + 2e^- \rightleftharpoons H_2(g, 1\ atm) \qquad E^0_{reference} = 0.00\ V$$

Now we can construct a voltaic cell consisting of the reference half-cell and another half-cell whose potential we want to determine. With $E^0_{reference}$ defined as zero, the overall E^0_{cell} allows us to find the unknown standard electrode potential, $E^0_{unknown}$. When H_2 is oxidized, the reference half-cell is the anode, so reduction occurs at the unknown half-cell:

$$E^0_{cell} = E^0_{cathode} - E^0_{anode} = E^0_{unknown} - E^0_{reference} = E^0_{unknown} - 0.00\ V = E^0_{unknown}$$

When H^+ is reduced, the reference half-cell is the cathode, so oxidation occurs at the unknown half-cell:

$$E^0_{cell} = E^0_{cathode} - E^0_{anode} = E^0_{reference} - E^0_{unknown} = 0.00\ V - E^0_{unknown} = -E^0_{unknown}$$

Figure 21.7 shows a voltaic cell that has the Zn/Zn^{2+} half-reaction in one compartment and the H^+/H_2 (or H_3O^+/H_2) half-reaction in the other. The zinc electrode is negative relative to the hydrogen electrode, so we know that the zinc is being oxidized and is the anode. The measured E^0_{cell} is $+0.76$ V, and we use this value to find the unknown standard electrode potential, E^0_{zinc}:

$$2H^+(aq) + 2e^- \longrightarrow H_2(g) \qquad E^0_{reference} = 0.00\ V \qquad \text{[cathode; reduction]}$$
$$Zn(s) \longrightarrow Zn^{2+}(aq) + 2e^- \qquad E^0_{zinc} = ?\ V \qquad \text{[anode; oxidation]}$$

$$Zn(s) + 2H^+(aq) \longrightarrow Zn^{2+}(aq) + H_2(g) \qquad E^0_{cell} = 0.76\ V$$

$$E^0_{cell} = E^0_{cathode} - E^0_{anode} = E^0_{reference} - E^0_{zinc}$$
$$E^0_{zinc} = E^0_{reference} - E^0_{cell} = 0.00\ V - 0.76\ V = -0.76\ V$$

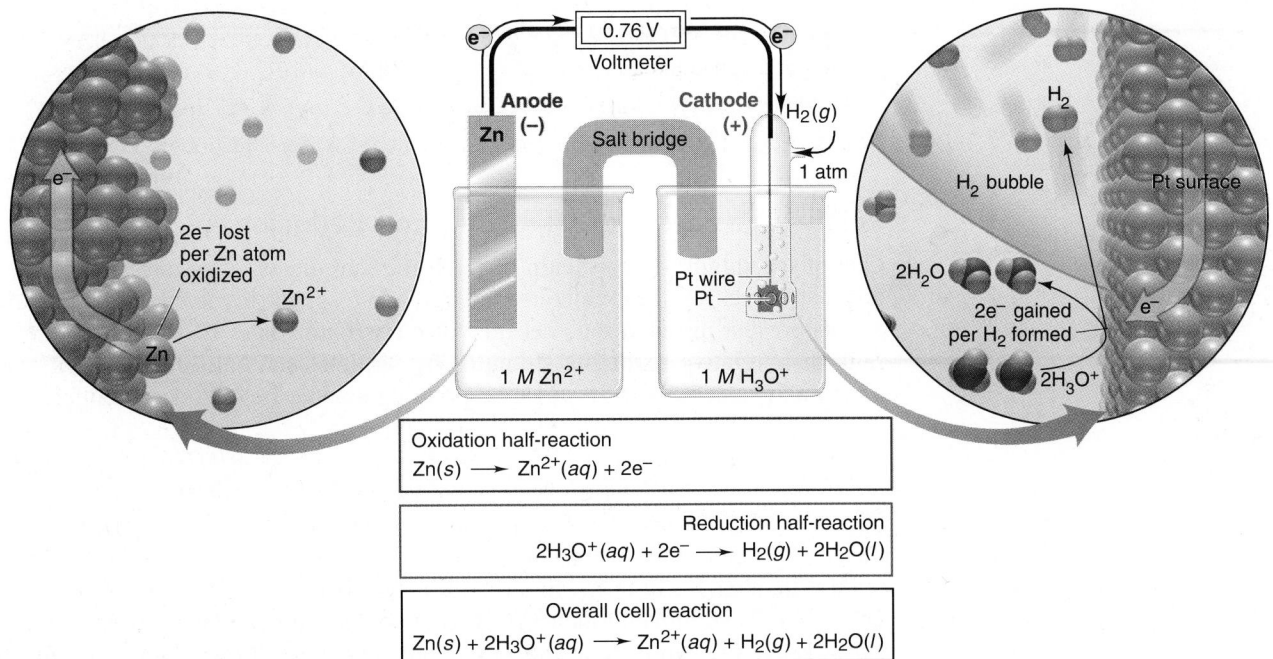

Oxidation half-reaction
$$Zn(s) \longrightarrow Zn^{2+}(aq) + 2e^-$$

Reduction half-reaction
$$2H_3O^+(aq) + 2e^- \longrightarrow H_2(g) + 2H_2O(l)$$

Overall (cell) reaction
$$Zn(s) + 2H_3O^+(aq) \longrightarrow Zn^{2+}(aq) + H_2(g) + 2H_2O(l)$$

Figure 21.7 Determining an unknown $E^0_{half\text{-}cell}$ with the standard reference (hydrogen) electrode. A voltaic cell has the Zn half-reaction in one half-cell and the hydrogen reference half-reaction in the other. The magnified view of the hydrogen half-reaction shows two H_3O^+ ions being reduced to two H_2O molecules and an H_2 molecule, which enters the H_2 bubble. The Zn/Zn^{2+} half-cell potential is negative (anode), and the cell potential is 0.76 V. The potential of the standard reference electrode is defined as 0.00 V, so the cell potential equals the negative of the anode potential; that is, 0.76 V = 0.00 V − E^0_{zinc}, so $E^0_{zinc} = -0.76$ V.

Now let's return to the zinc-copper cell and use the measured value of E^0_{cell} (1.10 V) and the value we just found for E^0_{zinc} to calculate E^0_{copper}:

$$E^0_{cell} = E^0_{cathode} - E^0_{anode} = E^0_{copper} - E^0_{zinc}$$
$$E^0_{copper} = E^0_{cell} + E^0_{zinc} = 1.10 \text{ V} + (-0.76 \text{ V}) = 0.34 \text{ V}$$

By continuing this process of constructing cells with one known and one unknown electrode potential, we can find many other standard electrode potentials. Let's go over these ideas once more with a sample problem.

SAMPLE PROBLEM 21.3 Calculating an Unknown $E^0_{half\text{-}cell}$ from E^0_{cell}

Problem A voltaic cell houses the reaction between aqueous bromine and zinc metal:

$$Br_2(aq) + Zn(s) \longrightarrow Zn^{2+}(aq) + 2Br^-(aq) \qquad E^0_{cell} = 1.83 \text{ V}$$

Calculate $E^0_{bromine}$, given $E^0_{zinc} = -0.76$ V.

Plan E^0_{cell} is positive, so the reaction is spontaneous as written. By dividing the reaction into half-reactions, we see that Br_2 is reduced and Zn is oxidized; thus, the zinc half-cell contains the anode. We use Equation 21.3 to find $E^0_{unknown}$ ($E^0_{bromine}$).

Solution Dividing the reaction into half-reactions:

$$Br_2(aq) + 2e^- \longrightarrow 2Br^-(aq) \qquad E^0_{unknown} = E^0_{bromine} = ? \text{ V}$$
$$Zn(s) \longrightarrow Zn^{2+}(aq) + 2e^- \qquad E^0_{zinc} = -0.76 \text{ V}$$

Calculating $E^0_{bromine}$:

$$E^0_{cell} = E^0_{cathode} - E^0_{anode} = E^0_{bromine} - E^0_{zinc}$$
$$E^0_{bromine} = E^0_{cell} + E^0_{zinc} = 1.83 \text{ V} + (-0.76 \text{ V}) = \boxed{1.07 \text{ V}}$$

Check A good check is to make sure that calculating $E^0_{bromine} - E^0_{zinc}$ gives E^0_{cell}: 1.07 V − (−0.76 V) = 1.83 V.

Comment Keep in mind that, whichever is the unknown half-cell, reduction is the cathode half-reaction and oxidation is the anode half-reaction. Always subtract E^0_{anode} from $E^0_{cathode}$ to get E^0_{cell}.

FOLLOW-UP PROBLEM 21.3 A voltaic cell based on the reaction between aqueous Br_2 and vanadium(III) ions has $E^0_{cell} = 1.39$ V:

$$Br_2(aq) + 2V^{3+}(aq) + 2H_2O(l) \longrightarrow 2VO^{2+}(aq) + 4H^+(aq) + 2Br^-(aq)$$

What is $E^0_{vanadium}$, the standard electrode potential for the reduction of VO^{2+} to V^{3+}?

Relative Strengths of Oxidizing and Reducing Agents

One of the things we can learn through measurements of voltaic cells is the relative strengths of the oxidizing and reducing agents involved. Three oxidizing agents present in the voltaic cell just discussed are Cu^{2+}, H^+, and Zn^{2+}. We can rank their relative oxidizing strengths by writing each half-reaction as a gain of electrons (reduction), with its corresponding standard electrode potential:

$$Cu^{2+}(aq) + 2e^- \longrightarrow Cu(s) \qquad E^0 = 0.34 \text{ V}$$
$$2H^+(aq) + 2e^- \longrightarrow H_2(g) \qquad E^0 = 0.00 \text{ V}$$
$$Zn^{2+}(aq) + 2e^- \longrightarrow Zn(s) \qquad E^0 = -0.76 \text{ V}$$

The more positive the E^0 value, the more readily the reaction (as written) occurs; thus, Cu^{2+} gains two electrons more readily than H^+, which gains them more readily than Zn^{2+}. In terms of strength as an oxidizing agent, therefore, $Cu^{2+} > H^+ > Zn^{2+}$. Moreover, this listing also ranks the strengths of the reducing agents: $Zn > H_2 > Cu$. Notice that this list of half-reactions in order of *decreasing* half-cell potential shows, *from top to bottom*, the oxidizing agents (reactants) *decreasing* in strength and the reducing agents (products) *increasing* in strength; that is, Cu^{2+} (top left) is the strongest oxidizing agent, and Zn (bottom right) is the strongest reducing agent.

By combining many pairs of half-cells into voltaic cells, we can create a list of reduction half-reactions and arrange them in *decreasing* order of standard electrode potential (from most positive to most negative). Such a list, called an *emf*

Table 21.2 Selected Standard Electrode Potentials (298 K)

Half-Reaction	E^0 (V)
$F_2(g) + 2e^- \rightleftharpoons 2F^-(aq)$	+2.87
$Cl_2(g) + 2e^- \rightleftharpoons 2Cl^-(aq)$	+1.36
$MnO_2(s) + 4H^+(aq) + 2e^- \rightleftharpoons Mn^{2+}(aq) + 2H_2O(l)$	+1.23
$NO_3^-(aq) + 4H^+(aq) + 3e^- \rightleftharpoons NO(g) + 2H_2O(l)$	+0.96
$Ag^+(aq) + e^- \rightleftharpoons Ag(s)$	+0.80
$Fe^{3+}(aq) + e^- \rightleftharpoons Fe^{2+}(aq)$	+0.77
$O_2(g) + 2H_2O(l) + 4e^- \rightleftharpoons 4OH^-(aq)$	+0.40
$Cu^{2+}(aq) + 2e^- \rightleftharpoons Cu(s)$	+0.34
$2H^+(aq) + 2e^- \rightleftharpoons H_2(g)$	0.00
$N_2(g) + 5H^+(aq) + 4e^- \rightleftharpoons N_2H_5^+(aq)$	−0.23
$Fe^{2+}(aq) + 2e^- \rightleftharpoons Fe(s)$	−0.44
$2H_2O(l) + 2e^- \rightleftharpoons H_2(g) + 2OH^-(aq)$	−0.83
$Na^+(aq) + e^- \rightleftharpoons Na(s)$	−2.71
$Li^+(aq) + e^- \rightleftharpoons Li(s)$	−3.05

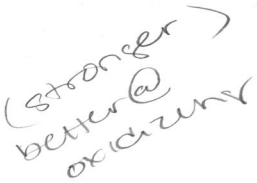

series or *table of standard electrode potentials,* appears in Appendix D, with a few examples in Table 21.2. Keep several key points in mind as you use this list:

- All values are relative to the standard hydrogen (reference) electrode:

$$2H^+(aq, 1\ M) + 2e^- \rightleftharpoons H_2(g, 1\ atm) \qquad E^0_{reference} = 0.00\ V$$

- By convention, the half-reactions are written as *reductions,* which means that *only reactants are oxidizing agents and only products are reducing agents.*
- The more positive the $E^0_{half\text{-}cell}$, the more readily the half-reaction occurs.
- Half-reactions are shown with an equilibrium arrow because each can occur as a reduction or as an oxidation (that is, take place at the cathode or anode, respectively), depending on the conditions and the standard electrode potential of the other half-reaction.
- As Appendix D (and Table 21.2) is arranged, the strength of the oxidizing agent (reactant) *increases up (bottom to top),* and the strength of the reducing agent (product) *increases down (top to bottom).*

Thus, $F_2(g)$ is the strongest oxidizing agent (has the largest positive E^0), so $F^-(aq)$ is the weakest reducing agent. Similarly, $Li^+(aq)$ is the weakest oxidizing agent (has the most negative E^0), so $Li(s)$ is the strongest reducing agent. You may have noticed an analogy to conjugate acid-base pairs: a strong acid forms a weak conjugate base, and vice versa, just as a *strong oxidizing agent forms a weak reducing agent,* and vice versa. If you forget the ranking in the table, just rely on your chemical knowledge of the elements. You know that F_2 is very electronegative and typically occurs as F^-. It is easily reduced (gains electrons), so it must be a strong oxidizing agent (high, positive E^0). Similarly, Li metal has a low ionization energy and typically occurs as Li^+. Therefore, it is easily oxidized (loses electrons), so it must be a strong reducing agent (low, negative E^0).

Writing Spontaneous Redox Reactions Appendix D can be used as a guide to writing spontaneous redox reactions, which can be useful in laboratory analyses or construction of electrochemical cells.

Every redox reaction is the sum of two half-reactions, so there is a reducing agent and an oxidizing agent on each side. In the zinc-copper reaction, for instance, Zn and Cu are the reducing agents, and Cu^{2+} and Zn^{2+} are the oxidizing agents. The stronger oxidizing and reducing agents react spontaneously to form the weaker oxidizing and reducing agents:

$$Zn(s) \quad + \quad Cu^{2+}(aq) \quad \longrightarrow \quad Zn^{2+}(aq) \quad + \quad Cu(s)$$

stronger reducing agent	stronger oxidizing agent	weaker oxidizing agent	weaker reducing agent

Here, too, note the similarity to acid-base chemistry. The stronger acid and base spontaneously form the weaker base and acid, respectively. The members of a conjugate acid-base pair differ by a proton: the acid has the proton and the base does not. The members of a redox pair, or *redox couple,* such as Zn and Zn^{2+}, differ by one or more electrons: the reduced form (Zn) has the electrons and the oxidized form (Zn^{2+}) does not. In acid-base reactions, we compare acid and base strength with K_a and K_b values. In redox reactions, we compare oxidizing and reducing strength with E^0 values.

Based on the order of the E^0 values in Appendix D, *the stronger oxidizing agent (species on the left) has a half-reaction with a larger (more positive or less negative) E^0 value, and the stronger reducing agent (species on the right) has a half-reaction with a smaller (less positive or more negative) E^0 value.* Therefore, a spontaneous reaction ($E^0_{cell} > 0$) will occur between an oxidizing agent and a reducing agent that lies *below* it in the list. For instance, Cu^{2+} (left) and Zn (right) react spontaneously, and Zn lies below Cu^{2+}. In other words, a spontaneous reaction involves one half-reaction that proceeds as a reduction (that is, as written in the table) and a second half-reaction, listed lower in the table, that proceeds as an oxidation (that is, written in reverse). This pairing ensures that the stronger oxidizing agent (higher on the left) and stronger reducing agent (lower on the right) will be the reactants.

However, if we know the electrode potentials, we can write a spontaneous redox reaction even if Appendix D is not available. Let's choose a pair of half-reactions from the appendix and, without referring to their relative positions in the list, arrange them into a spontaneous redox reaction:

$$Ag^+(aq) + e^- \longrightarrow Ag(s) \qquad E^0_{silver} = 0.80 \text{ V}$$
$$Sn^{2+}(aq) + 2e^- \longrightarrow Sn(s) \qquad E^0_{tin} = -0.14 \text{ V}$$

There are two steps involved:

1. Reverse one of the half-reactions into an oxidation step such that the difference of the electrode potentials (cathode *minus* anode) gives a *positive* E^0_{cell}.
2. Add the rearranged half-reactions to obtain a balanced overall equation. Be sure to multiply by coefficients so that e^- lost equals e^- gained and to cancel species common to both sides.

(You may be tempted in this case to add the two half-reactions as written, because you obtain a positive E^0_{cell}, but you would then have two oxidizing agents forming two reducing agents, which cannot occur.)

We want to pair the stronger oxidizing and reducing agents as reactants. The larger (more positive) E^0 value for the silver half-reaction means that Ag^+ is a stronger oxidizing agent (gains electrons more readily) than Sn^{2+}, and the smaller (more negative) E^0 value for the tin half-reaction means that Sn is a stronger reducing agent (loses electrons more readily) than Ag. Therefore, we reverse the tin half-reaction:

$$Sn(s) \longrightarrow Sn^{2+}(aq) + 2e^- \qquad E^0_{tin} = -0.14 \text{ V}$$

Subtracting $E^0_{half\text{-}cell}$ of the tin half-reaction (anode, oxidation) from $E^0_{half\text{-}cell}$ of the silver half-reaction (cathode, reduction) gives a positive E^0_{cell}; that is, 0.80 V − (−0.14 V) = 0.94 V).

With the half-reactions written in the correct direction, we must then be sure that the *number of electrons lost in the oxidation equals the number gained in the reduction.* In this case, we double the silver (reduction) half-reaction and add the half-reactions to obtain the balanced equation:

$2Ag^+(aq) + 2e^- \longrightarrow 2Ag(s)$	$E^0 = 0.80$ V	[reduction]
$Sn(s) \longrightarrow Sn^{2+}(aq) + 2e^-$	$E^0 = -0.14$ V	[oxidation]
$Sn(s) + 2Ag^+(aq) \longrightarrow Sn^{2+}(aq) + 2Ag(s)$	$E^0_{cell} = E^0_{silver} - E^0_{tin} = 0.94$ V	

With the reaction spontaneous as written, the stronger oxidizing and reducing agents are reactants, which confirms that Sn is a stronger reducing agent than Ag, and Ag^+ is a stronger oxidizing agent than Sn^{2+}.

A very important point to note is that, when we doubled the coefficients of the silver half-reaction to balance the number of electrons, we did *not* double its E^0 value—it remained 0.80 V. That is, *changing the balancing coefficients of a half-reaction does **not** change the E^0 value*. The reason is that a standard electrode potential is an *intensive* property, one that does *not* depend on the amount of substance present. The potential is the *ratio* of energy to charge. When we change the coefficients, thus increasing the amount of substance, the energy *and* the charge increase proportionately, so their ratio stays the same. (In a similar sense, density, which is also an intensive property, does not change with the amount of substance because the mass *and* the volume increase proportionately.)

SAMPLE PROBLEM 21.4 Writing Spontaneous Redox Reactions and Ranking Oxidizing and Reducing Agents by Strength

Problem **(a)** Combine the following three half-reactions into three spontaneous, balanced equations (A, B, and C), and calculate E_{cell}^0 for each. **(b)** Rank the relative strengths of the oxidizing and reducing agents:

(1) $NO_3^-(aq) + 4H^+(aq) + 3e^- \longrightarrow NO(g) + 2H_2O(l)$ $E^0 = 0.96$ V

(2) $N_2(g) + 5H^+(aq) + 4e^- \longrightarrow N_2H_5^+(aq)$ $E^0 = -0.23$ V

(3) $MnO_2(s) + 4H^+(aq) + 2e^- \longrightarrow Mn^{2+}(aq) + 2H_2O(l)$ $E^0 = 1.23$ V

Plan **(a)** To write the redox equations, we combine the possible pairs of half-reactions: (1) and (2), (1) and (3), and (2) and (3). They are all written as reductions, so the oxidizing agents appear as reactants and the reducing agents appear as products. In each pair, we reverse the reduction half-reaction that has the smaller (less positive or more negative) E^0 value to an oxidation to obtain a positive E_{cell}^0. We make e^- lost equal e^- gained, without changing the magnitude of the E^0 value, add the half-reactions together, and then apply Equation 21.3 to find E_{cell}^0. **(b)** Because the equation is spontaneous as written, the stronger oxidizing and reducing agents are the reactants. To obtain the overall ranking, we first rank the relative strengths within each equation and then compare them.

Solution **(a)** Combining half-reactions (1) and (2) gives equation (A). The E^0 value of (1) is larger (more positive) than (2), so we reverse (2) to obtain a positive E_{cell}^0:

(1) $NO_3^-(aq) + 4H^+(aq) + 3e^- \longrightarrow NO(g) + 2H_2O(l)$ $E^0 = 0.96$ V

(rev 2) $N_2H_5^+(aq) \longrightarrow N_2(g) + 5H^+(aq) + 4e^-$ $E^0 = -0.23$ V

To make e^- lost equal e^- gained, we multiply (1) by four and the reversed (2) by three. Then we add the half-reactions and cancel the appropriate numbers of common species (H^+ and e^-):

$4NO_3^-(aq) + 16H^+(aq) + 12e^- \longrightarrow 4NO(g) + 8H_2O(l)$ $E^0 = 0.96$ V

$3N_2H_5^+(aq) \longrightarrow 3N_2(g) + 15H^+(aq) + 12e^-$ $E^0 = -0.23$ V

(A) $3N_2H_5^+(aq) + 4NO_3^-(aq) + H^+(aq) \longrightarrow 3N_2(g) + 4NO(g) + 8H_2O(l)$

$E_{cell}^0 = 0.96$ V $- (-0.23$ V$) = 1.19$ V

Combining half-reactions (1) and (3) gives equation (B). Half-reaction (1) must be reversed:

(rev 1) $NO(g) + 2H_2O(l) \longrightarrow NO_3^-(aq) + 4H^+(aq) + 3e^-$ $E^0 = 0.96$ V

(3) $MnO_2(s) + 4H^+(aq) + 2e^- \longrightarrow Mn^{2+}(aq) + 2H_2O(l)$ $E^0 = 1.23$ V

We multiply reversed (1) by two and (3) by three, then add and cancel:

$2NO(g) + 4H_2O(l) \longrightarrow 2NO_3^-(aq) + 8H^+(aq) + 6e^-$ $E^0 = 0.96$ V

$3MnO_2(s) + 12H^+(aq) + 6e^- \longrightarrow 3Mn^{2+}(aq) + 6H_2O(l)$ $E^0 = 1.23$ V

(B) $3MnO_2(s) + 4H^+(aq) + 2NO(g) \longrightarrow 3Mn^{2+}(aq) + 2H_2O(l) + 2NO_3^-(aq)$

$E_{cell}^0 = 1.23$ V $- 0.96$ V $= 0.27$ V

Combining half-reactions (2) and (3) gives equation (C). Half-reaction (2) must be reversed:

(rev 2) $\qquad$ $N_2H_5^+(aq) \longrightarrow N_2(g) + 5H^+(aq) + 4e^-$ $\qquad$ $E^0 = -0.23$ V

(3) $MnO_2(s) + 4H^+(aq) + 2e^- \longrightarrow Mn^{2+}(aq) + 2H_2O(l)$ $\qquad$ $E^0 = 1.23$ V

We multiply reaction (3) by two, add the half-reactions, and cancel:

$$N_2H_5^+(aq) \longrightarrow N_2(g) + 5H^+(aq) + 4e^- \qquad E^0 = -0.23 \text{ V}$$
$$2MnO_2(s) + 8H^+(aq) + 4e^- \longrightarrow 2Mn^{2+}(aq) + 4H_2O(l) \qquad E^0 = 1.23 \text{ V}$$

(C) $N_2H_5^+(aq) + 2MnO_2(s) + 3H^+(aq) \longrightarrow N_2(g) + 2Mn^{2+}(aq) + 4H_2O(l)$

$E^0_{cell} = 1.23$ V $- (-0.23$ V$) = 1.46$ V

(b) Ranking the oxidizing and reducing agents within each equation:

Equation (A): Oxidizing agents: $NO_3^- > N_2$ $\qquad$ Reducing agents: $N_2H_5^+ > NO$
Equation (B): Oxidizing agents: $MnO_2 > NO_3^-$ $\qquad$ Reducing agents: $NO > Mn^{2+}$
Equation (C): Oxidizing agents: $MnO_2 > N_2$ $\qquad$ Reducing agents: $N_2H_5^+ > Mn^{2+}$

Determining the overall ranking of oxidizing and reducing agents. Comparing the relative strengths from the three balanced equations gives

Oxidizing agents: $MnO_2 > NO_3^- > N_2$

Reducing agents: $N_2H_5^+ > NO > Mn^{2+}$

Check As always, check that atoms and charge balance on each side of the equation. A good way to check the ranking and equations is to list the given half-reactions in order of decreasing E^0 value:

$MnO_2(s) + 4H^+(aq) + 2e^- \longrightarrow Mn^{2+}(aq) + 2H_2O(l)$ $\qquad$ $E^0 = 1.23$ V
$NO_3^-(aq) + 4H^+(aq) + 3e^- \longrightarrow NO(g) + 2H_2O(l)$ $\qquad$ $E^0 = 0.96$ V
$N_2(g) + 5H^+(aq) + 4e^- \longrightarrow N_2H_5^+(aq)$ $\qquad$ $E^0 = -0.23$ V

Then the oxidizing agents (reactants) decrease in strength down the list, so the reducing agents (products) decrease in strength up. Moreover, each of the three spontaneous equations you obtained in the Solution should combine a reactant with a product from lower down on this list.

FOLLOW-UP PROBLEM 21.4 Is the following equation spontaneous?

$$3Fe^{2+}(aq) \longrightarrow Fe(s) + 2Fe^{3+}(aq)$$

If not, write the spontaneous equation, calculate E^0_{cell}, and rank the three species of iron in order of decreasing reducing strength.

Relative Reactivities of Metals In Chapter 4, you saw that a more active metal can reduce the ion of a less active metal to its elemental form, thus "displacing" that ion from aqueous solution. The activity series of the metals (see Figure 4.19, p. 160) ranks metals on the basis of this relative reactivity. Now you'll see *why* this displacement occurs, why many, but not all, metals react with acid to form H_2, and why a few metals form H_2 even in water. All these variations in metal reactivity become clear by examining standard electrode potentials.

1. *Metals that can displace H_2 from acid.* The standard hydrogen half-reaction represents the reduction of H^+ ions from an acid to H_2:

$$2H^+(aq) + 2e^- \longrightarrow H_2(g) \qquad E^0 = 0.00 \text{ V}$$

To see which metals reduce H^+ (referred to as "displacing H_2") from acids, choose a metal, write its half-reaction as an oxidation, combine this half-reaction with the hydrogen half-reaction, and see if E^0_{cell} is positive. What you find is that the metals Li through Pb, those that lie *below* the standard hydrogen (reference) half-reaction in Appendix D, give a positive E^0_{cell} when reducing H^+. Iron, for example, reduces H^+ from an acid to H_2:

$Fe(s) \longrightarrow Fe^{2+}(aq) + 2e^-$ $\qquad$ $E^0 = -0.44$ V $\quad$ [anode; oxidation]
$2H^+(aq) + 2e^- \longrightarrow H_2(g)$ $\qquad$ $E^0 = 0.00$ V $\quad$ [cathode; reduction]

$Fe(s) + 2H^+(aq) \longrightarrow H_2(g) + Fe^{2+}(aq)$ $\qquad$ $E^0_{cell} = 0.00$ V $- (-0.44$ V$) = 0.44$ V

$E_{cell} = $ non-spontaneous

The lower the metal in the table, the stronger it is as a reducing agent; therefore, the more positive its half-cell potential when the half-reaction is reversed, and the higher the E^0_{cell} for its reduction of H^+ to H_2. *If E^0_{cell} of metal A for the reduction of H^+ is more positive than E^0_{cell} of metal B, metal A is a stronger reducing agent than metal B and a more "active" metal.*

2. *Metals that cannot displace H_2 from acid.* Metals that are *above* the standard hydrogen (reference) half-reaction *cannot* reduce H^+ from acids. When we reverse the metal half-reaction, the E^0_{cell} is negative, and therefore the reaction is not spontaneous. For example, the coinage metals [Group 1B(11)]— copper, silver, and gold—are not strong enough reducing agents to reduce H^+ from acids:

$$Ag(s) \longrightarrow Ag^+(aq) + e^- \qquad E^0 = 0.80 \text{ V} \quad \text{[anode; oxidation]}$$
$$2H^+(aq) + 2e^- \longrightarrow H_2(g) \qquad E^0 = 0.00 \text{ V} \quad \text{[cathode; reduction]}$$

$$2Ag(s) + 2H^+(aq) \longrightarrow 2Ag^+(aq) + H_2(g) \qquad E^0_{cell} = 0.00 \text{ V} - 0.80 \text{ V} = -0.80 \text{ V}$$

The higher the metal in the table, the more negative is its E^0_{cell} for the reduction of H^+ to H_2, the lower is its reducing strength, and the less active it is. Thus, gold is less active than silver, which is less active than copper.

3. *Metals that can displace H_2 from water.* Metals active enough to displace H_2 from water lie *below the half-reaction for the reduction of water:*

$$2H_2O(l) + 2e^- \longrightarrow H_2(g) + 2OH^-(aq) \qquad E = -0.42 \text{ V}$$

(The value that is shown here is the *nonstandard* electrode potential because, in pure water, $[OH^-]$ is 1.0×10^{-7} M, not the standard-state value of 1 M.) Consider the reaction of sodium in water (with the Na half-reaction reversed and doubled):

$$2Na(s) \longrightarrow 2Na^+(aq) + 2e^- \qquad E^0 = -2.71 \text{ V} \quad \text{[anode; oxidation]}$$
$$2H_2O(l) + 2e^- \longrightarrow H_2(g) + 2OH^-(aq) \qquad E = -0.42 \text{ V} \quad \text{[cathode; reduction]}$$

$$2Na(s) + 2H_2O(l) \longrightarrow 2Na^+(aq) + H_2(g) + 2OH^-(aq)$$
$$E_{cell} = -0.42 \text{ V} - (-2.71 \text{ V}) \; 2.29 \text{ V}$$

The alkali metals [Group 1A(1)] and the larger alkaline earth metals [Group 2A(2)] can displace H_2 from H_2O (Figure 21.8).

4. *Metals that can displace other metals from solution.* We can also predict whether one metal can reduce the aqueous ion of another metal. Any metal that is lower in the list in Appendix D can reduce the ion of a metal that is higher up, and thus displace that metal from solution. For example, zinc can displace iron from solution:

$$Zn(s) \longrightarrow Zn^{2+}(aq) + 2e^- \qquad E^0 = -0.76 \text{ V} \quad \text{[anode; oxidation]}$$
$$Fe^{2+}(aq) + 2e^- \longrightarrow Fe(s) \qquad E^0 = -0.44 \text{ V} \quad \text{[cathode; reduction]}$$

$$Zn(s) + Fe^{2+}(aq) \longrightarrow Zn^{2+}(aq) + Fe(s) \qquad E^0_{cell} = -0.44 \text{ V} - (-0.76 \text{ V}) = 0.32 \text{ V}$$

This particular reaction has tremendous economic importance in protecting iron from rusting, as you'll see shortly. The reducing power of metals has other, more personal, consequences, as the margin note points out. ●

SECTION SUMMARY

The output of a cell is called the cell potential (E_{cell}) and is measured in volts (1 V = 1 J/C). When all substances are in their standard states, the output is the standard cell potential (E^0_{cell}). $E^0_{cell} > 0$ for a spontaneous reaction at standard-state conditions. By convention, standard electrode potentials ($E^0_{half-cell}$) refer to the *reduction* half-reaction. E^0_{cell} equals $E^0_{half-cell}$ of the cathode *minus* $E^0_{half-cell}$ of the anode. Using a standard hydrogen (reference) electrode, other $E^0_{half-cell}$ values can be measured and used to rank oxidizing (or reducing) agents (see Appendix D). Spontaneous redox reactions combine stronger oxidizing and reducing agents to form weaker ones. A metal can reduce another species (H^+, H_2O, or an ion of another metal) if the E^0_{cell} for the reaction is positive.

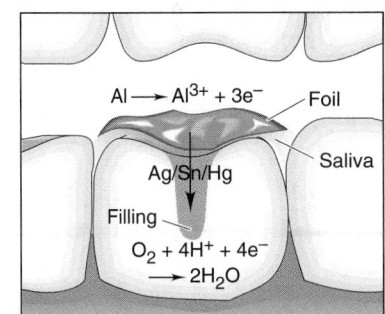

Oxidation half-reaction
$Ca(s) \longrightarrow Ca^{2+}(aq) + 2e^-$

Reduction half-reaction
$2H_2O(l) + 2e^- \longrightarrow H_2(g) + 2OH^-(aq)$

Overall (cell) reaction
$Ca(s) + 2H_2O(l) \longrightarrow Ca(OH)_2(aq) + H_2(g)$

Figure 21.8 **The reaction of calcium in water.** Calcium is one of the metals active enough to displace H_2 from H_2O.

The Pain of a Dental Voltaic Cell Have you ever felt a jolt of pain when biting down with a filled tooth on a scrap of foil left on a piece of food? Here's why. The aluminum foil acts as an active anode (E^0 of Al = -1.66 V), saliva as the electrolyte, and the filling (usually a silver/tin/mercury alloy) as an inactive cathode. O_2 is reduced to water, and the short circuit between the foil in contact with the filling creates a current that is sensed by the nerve of the tooth.

21.4 FREE ENERGY AND ELECTRICAL WORK

In Chapter 20, we discussed the relationship of useful work, free energy, and the equilibrium constant. In this section, we examine this central relationship in the context of electrochemical cells and see the effect of concentration on cell potential.

Standard Cell Potential and the Equilibrium Constant

As you know from Section 20.3, a spontaneous reaction has a *negative* free energy change ($\Delta G < 0$), and now you've just seen that a spontaneous electrochemical reaction has a *positive* cell potential ($E_{cell} > 0$). Note that *the signs of ΔG and E_{cell} are opposite for a spontaneous reaction*. These two indications of spontaneity are proportional to each other:

$$\Delta G \propto -E_{cell}$$

Let's determine the proportionality constant. The cell potential (E_{cell}, in volts) is the work (w, in joules) done *by* the system per unit of charge (in coulombs) that flows through the circuit. If no current flows, no energy is lost to heating the cell components, so the potential represents the *maximum* work the cell can do. The work is done *on* the surroundings, so its sign is negative:

$$E_{cell} = \frac{-w_{max}}{\text{charge}} \qquad \text{or} \qquad w_{max} = -\text{charge} \times E_{cell}$$

The charge that flows through the cell is equal to the number of moles of electrons (n) transferred in the balanced redox equation times the charge of 1 mol of electrons (symbol F):

$$\text{Charge} = \text{moles of e}^- \times \frac{\text{charge}}{\text{mol e}^-} \qquad \text{or} \qquad \text{charge} = nF$$

Substituting for charge in the equation for maximum work, we have

$$w_{max} = -\text{charge} \times E_{cell} = -nFE_{cell}$$

The charge of 1 mol of electrons is the **Faraday constant (F),** in honor of Michael Faraday, the 19th-century British scientist who pioneered the study of electrochemistry:

$$F = \frac{96{,}485 \text{ C}}{\text{mol e}^-}$$

Because 1 V = 1 J/C, we have 1 C = 1 J/V, and

$$F = 9.65 \times 10^4 \; \frac{\text{J}}{\text{V} \cdot \text{mol e}^-} \qquad \text{(3 sf)} \tag{21.4}$$

Recall from Chapter 20 that the free energy change is the maximum work that can be obtained from a spontaneous process:

$$\Delta G = w_{max}$$

Combining these expressions for maximum work, we determine the proportionality constant for the earlier relationship and obtain the key equation in the thermodynamics of electrochemical cells:

$$\Delta G = -nFE_{cell} \tag{21.5}$$

When all of the components are in their standard states, we have the following relationship:

$$\Delta G^0 = -nFE^0_{cell} \tag{21.6}$$

Using this relationship, we can relate the standard cell potential to the equilibrium constant of the redox reaction. Recall that

$$\Delta G^0 = -RT \ln K$$

Substituting for ΔG^0 in Equation 21.6, we obtain

$$-nFE^0_{cell} = -RT \ln K$$

Solving for E^0_{cell} gives

$$E^0_{cell} = \frac{RT}{nF} \ln K \qquad (21.7)$$

Figure 21.9 presents a summary of the interconnections among the standard free energy change, the equilibrium constant, and the standard cell potential. The procedures that we examined previously for determining K required that we know ΔG^0, either from ΔH^0 and ΔS^0 values or from ΔG^0_f values. For redox reactions, we now have a direct experimental method for determining K *and* ΔG^0: measure E^0_{cell}.

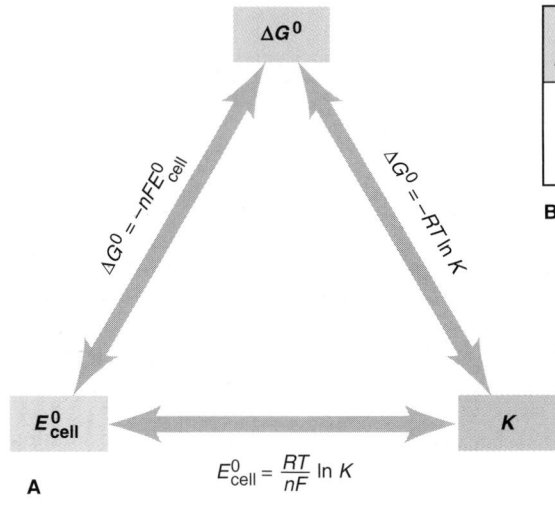

ΔG^0	K	E^0_{cell}	Reaction at standard-state conditions
<0	>1	>0	Spontaneous
0	1	0	At equilibrium
>0	<1	<0	Nonspontaneous

B

Figure 21.9 The interrelationship of ΔG^0, E^0, and K. **A,** Any one of these three central thermodynamic parameters can be used to find the other two. **B,** The signs of ΔG^0 and E^0_{cell} determine the reaction direction at standard-state conditions.

A

If we substitute values for the constants R and F and run the cell at 25°C (298 K), we obtain a simplified form of Equation 21.7 that relates E^0 to K when n moles of e^- are transferred per mole of reaction:

$$E^0_{cell} = \frac{RT}{nF} \ln K = \frac{8.314 \frac{J}{mol\ rxn \cdot K} \times 298\ K}{\frac{n\ mol\ e^-}{mol\ rxn} \left(9.65 \times 10^4 \frac{J}{V \cdot mol\ e^-}\right)} \ln K = \frac{0.0257\ V}{n} \ln K$$

Multiplying by 2.303 in order to obtain the common (base-10) logarithm then gives

$$E^0_{cell} = \frac{0.0592\ V}{n} \log K \quad \text{or} \quad \log K = \frac{nE^0_{cell}}{0.0592\ V} \quad \text{(at 25°C)} \quad (21.8)$$

SAMPLE PROBLEM 21.5 Calculating K and ΔG^0 from E^0_{cell}

Problem Lead can displace silver from solution:

$$Pb(s) + 2Ag^+(aq) \longrightarrow Pb^{2+}(aq) + 2Ag(s)$$

As a consequence, silver is a valuable by-product in the industrial extraction of lead from its ore. Calculate K and ΔG^0 at 25°C for this reaction.

Plan We divide the spontaneous redox equation into the half-reactions and use values from Appendix D to calculate E^0_{cell}. Then, we substitute this result into Equation 21.8 to find K and into Equation 21.6 to find ΔG^0.

Solution Writing the half-reactions and their E^0 values:

(1) $Ag^+(aq) + e^- \longrightarrow Ag(s)$ $E^0 = 0.80$ V

(2) $Pb^{2+}(aq) + 2e^- \longrightarrow Pb(s)$ $E^0 = -0.13$ V

Calculating E^0_{cell}: We double (1), reverse (2), add the half-reactions, and subtract E^0_{lead} from E^0_{silver}:

$$2Ag^+(aq) + 2e^- \longrightarrow 2Ag(s) \qquad\qquad E^0 = 0.80 \text{ V}$$
$$\underline{Pb(s) \longrightarrow Pb^{2+}(aq) + 2e^- \qquad\qquad E^0 = -0.13 \text{ V}}$$
$$Pb(s) + 2Ag^+(aq) \longrightarrow Pb^{2+}(aq) + 2Ag(s) \quad E^0_{cell} = 0.80 \text{ V} - (-0.13 \text{ V}) = 0.93 \text{ V}$$

Calculating K (Equation 21.8): The adjusted half-reactions show that 2 mol of e^- are transferred per mole of reaction as written, so $n = 2$:

$$E^0_{cell} = \frac{0.0592 \text{ V}}{n} \log K = \frac{0.0592 \text{ V}}{2} \log K = 0.93 \text{ V}$$

So, $\log K = \dfrac{0.93 \text{ V} \times 2}{0.0592 \text{ V}} = 31.42$ and $\boxed{K = 2.6 \times 10^{31}}$

Calculating ΔG^0 (Equation 21.6):

$$\Delta G^0 = -nFE^0_{cell} = -\frac{2 \text{ mol } e^-}{\text{mol rxn}} \times \frac{96.5 \text{ kJ}}{\text{V} \cdot \text{mol } e^-} \times 0.93 \text{ V} = \boxed{-1.8 \times 10^2 \text{ kJ/mol rxn}}$$

Check The three variables are consistent with the reaction being spontaneous: $E^0_{cell} > 0$, $\Delta G^0 < 0$, and $K > 1$. Be sure to round and check the order of magnitude: in the ΔG^0 calculation, for instance, $\Delta G^0 \approx -2 \times 100 \times 1 = -200$, so the overall math seems right. Another check would be to obtain ΔG^0 directly from its relation with K:

$$\Delta G^0 = -RT \ln K = -8.314 \text{ J/mol rxn} \cdot \text{K} \times 298 \text{ K} \times \ln (2.6 \times 10^{31})$$
$$= -1.8 \times 10^5 \text{ J/mol} = -1.8 \times 10^2 \text{ kJ/mol rxn}$$

FOLLOW-UP PROBLEM 21.5 When cadmium metal reduces Cu^{2+} in solution, Cd^{2+} forms in addition to copper metal. If $\Delta G^0 = -143$ kJ, calculate K at 25°C. What is E^0_{cell} in a voltaic cell that uses this reaction?

The Effect of Concentration on Cell Potential

So far, we have considered cells that operate with all components in their standard states and have determined standard cell potential (E^0_{cell}) from standard half-cell potentials ($E^0_{half\text{-}cell}$). However, most cells do not start with all components in their standard states, and even if they did, the concentrations change after a few moments of cell operation. Moreover, in all practical voltaic cells, such as batteries, reactant concentrations are far from standard-state values. Clearly, we must be able to determine E_{cell}, the potential under nonstandard conditions.

We can derive an expression for the relation between cell potential and concentration that is based on the relation between free energy and concentration. Recall from Chapter 20 (Equation 20.11) that ΔG equals ΔG^0 (the free energy change when the system moves from standard-state concentrations to equilibrium) *plus* $RT \ln Q$ (the free energy change when the system moves from nonstandard-state to standard-state concentrations):

$$\Delta G = \Delta G^0 + RT \ln Q$$

ΔG is related to E_{cell} and ΔG^0 to E^0_{cell} (Equations 21.5 and 21.6), so we substitute for them and get

$$-nFE_{cell} = -nFE^0_{cell} + RT \ln Q$$

Dividing both sides by $-nF$, we obtain the **Nernst equation,** developed by the German chemist Walther Hermann Nernst in 1889:

$$E_{cell} = E^0_{cell} - \frac{RT}{nF} \ln Q \qquad (21.9)$$

The Nernst equation says that the cell potential under any conditions depends on its potential at standard-state concentrations *and* a term for the potential at non-standard-state concentrations. How do changes in Q affect cell potential? From Equation 21.9, we see that

- When $Q < 1$ and thus [reactant] > [product], $\ln Q < 0$, so $E_{cell} > E^0_{cell}$.
- When $Q = 1$ and thus [reactant] = [product], $\ln Q = 0$, so $E_{cell} = E^0_{cell}$.
- When $Q > 1$ and thus [reactant] < [product], $\ln Q > 0$, so $E_{cell} < E^0_{cell}$.

As before, when we substitute known values for R and F, operate the cell at 25°C (298 K), and convert to common (base-10) logarithms, we obtain

$$E_{cell} = E^0_{cell} - \frac{0.0592 \text{ V}}{n} \log Q \qquad \text{(at 25°C)} \qquad (21.10)$$

Remember that the expression for Q contains only those species with concentrations (or pressures) that can vary; thus, solids do not appear, even though they may act as electrodes. For example, in the reaction between cadmium and silver ion, the Cd and Ag electrodes do not appear in the expression for Q:

$$Cd(s) + 2Ag^+(aq) \longrightarrow Cd^{2+}(aq) + 2Ag(s) \qquad Q = \frac{[Cd^{2+}]}{[Ag^+]^2}$$

Walther Hermann Nernst (1864–1941) The great German physical chemist was only 25 years old when he developed the equation for the relationship between cell voltage and concentration. His career, which culminated in the Nobel Prize in 1920, included formulating the third law of thermodynamics and the principle of the solubility product; he also contributed key ideas to photochemistry and to the development of the Haber process. He is shown here in later years making a point to a few of his celebrated colleagues: from left to right, Nernst, Albert Einstein, Max Planck, Robert Millikan, and Max von Laue.

SAMPLE PROBLEM 21.6 Using the Nernst Equation to Calculate E_{cell}

Problem In a test of a new reference electrode, a chemist constructs a voltaic cell consisting of a Zn/Zn^{2+} half-cell and an H$_2$/H$^+$ half-cell under the following conditions:

$$[Zn^{2+}] = 0.010 \text{ M} \qquad [H^+] = 2.5 \text{ M} \qquad P_{H_2} = 0.30 \text{ atm}$$

Calculate E_{cell} at 25°C.

Plan To apply the Nernst equation and determine E_{cell}, we must know E^0_{cell} and Q. We write the spontaneous reaction, calculate E^0_{cell} from standard electrode potentials (Appendix D), and use the given pressure and concentrations to find Q. (Recall that the ideal gas law allows us to use P at constant T as another way of writing concentration, n/V.) Then we substitute into Equation 21.10.

Solution Determining the cell reaction and E^0_{cell}:

$$\begin{array}{ll} 2H^+(aq) + 2e^- \longrightarrow H_2(g) & E^0 = 0.00 \text{ V} \\ Zn(s) \longrightarrow Zn^{2+}(aq) + 2e^- & E^0 = -0.76 \text{ V} \\ \hline 2H^+(aq) + Zn(s) \longrightarrow H_2(g) + Zn^{2+}(aq) & E^0_{cell} = 0.00 \text{ V} - (-0.76 \text{ V}) = 0.76 \text{ V} \end{array}$$

Calculating Q:

$$Q = \frac{P_{H_2} \times [Zn^{2+}]}{[H^+]^2} = \frac{0.30 \times 0.010}{2.5^2} = 4.8 \times 10^{-4}$$

Solving for E_{cell} at 25°C (298 K), with $n = 2$:

$$E_{cell} = E^0_{cell} - \frac{0.0592 \text{ V}}{n} \log Q = 0.76 \text{ V} - \left[\frac{0.0592 \text{ V}}{2} \log (4.8 \times 10^{-4}) \right]$$

$$= 0.76 \text{ V} - (-0.0982 \text{ V}) = \boxed{0.86 \text{ V}}$$

Check After you check the arithmetic, reason through the answer: $E_{cell} > E^0_{cell}$ (0.86 > 0.76) because the log Q term was negative, which is consistent with $Q < 1$; that is, the amounts of products, P_{H_2} and [Zn^{2+}], are smaller than the amount of reactant, [H$^+$].

FOLLOW-UP PROBLEM 21.6 Consider a cell based on the following reaction: Fe(s) + Cu^{2+}(aq) $\longrightarrow$ Fe^{2+}(aq) + Cu(s). If [Cu^{2+}] = 0.30 M, what must [Fe^{2+}] be to increase E_{cell} by 0.25 V above E^0_{cell} at 25°C?

Cell Potential and the Relation Between Q and K

Let's reconsider the zinc-copper cell to see how the potential changes as concentrations change during cell operation:

$$Zn(s) + Cu^{2+}(aq) \longrightarrow Zn^{2+}(aq) + Cu(s) \qquad Q = \frac{[Zn^{2+}]}{[Cu^{2+}]}$$

The positive E^0_{cell} (1.10 V) means that this reaction proceeds *spontaneously* from the point at which [product] = [reactant] = 1 M ($Q = 1$) to some point at which [product] > [reactant] ($Q > 1$). Now, suppose we start operating the cell when $Q < 1$, for example, $[Zn^{2+}] = 1.0 \times 10^{-4}\ M$ and $[Cu^{2+}] = 2.0\ M$. In this case, the cell potential is *higher* than the standard cell potential:

$$E_{cell} = E^0_{cell} - \frac{0.0592\ V}{2} \log \frac{[Zn^{2+}]}{[Cu^{2+}]} = 1.10\ V - \left(\frac{0.0592\ V}{2} \log \frac{1.0 \times 10^{-4}}{2.0} \right)$$

$$= 1.10\ V - \left[\frac{0.0592\ V}{2} (-4.30) \right] = 1.10\ V + 0.127\ V = 1.23\ V$$

As the cell operates, $[Zn^{2+}]$ increases (as the Zn electrode deteriorates) and $[Cu^{2+}]$ decreases (as Cu plates out on the Cu electrode). Although the process changes smoothly, if we keep Equation 21.10 in mind, we can identify four general stages in the operation of the cell. Figure 21.10 shows the first three. The main point to note is *as the cell operates, its potential decreases:*

- $E_{cell} > E^0_{cell}$ when $Q < 1$: When the cell begins operation, $[Cu^{2+}] > [Zn^{2+}]$, so the $[(0.0592\ V/n) \log Q]$ term < 0 and $E_{cell} > E^0_{cell}$. As operation continues, $[Zn^{2+}]$ increases while $[Cu^{2+}]$ decreases, so Q becomes larger, the $[(0.0592\ V/n) \log Q]$ term becomes less negative (more positive), and E_{cell} decreases.

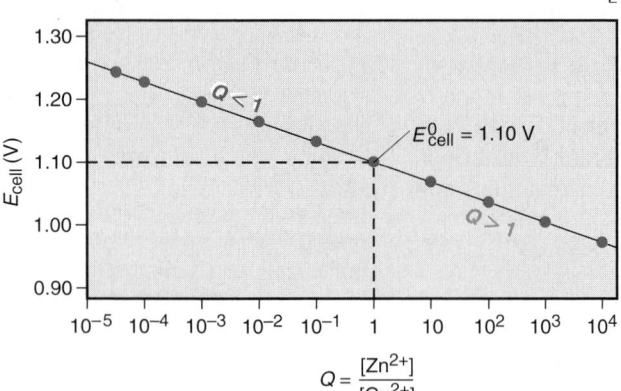

Figure 21.10 The relation between E_{cell} and log Q for the zinc-copper cell. A plot of E_{cell} vs. Q (on a logarithmic scale) for the zinc-copper reaction shows a linear decrease. When $Q < 1$ (*left*), [reactant] is relatively high, and the cell can do relatively more work. When $Q = 1$, $E_{cell} = E^0_{cell}$. When $Q > 1$ (*right*), [reactant] is relatively low, and the cell can do relatively less work.

- $E_{cell} = E^0_{cell}$ when $Q = 1$: At the point when $[Cu^{2+}] = [Zn^{2+}]$, $Q = 1$, so the $[(0.0592\ V/n) \log Q]$ term $= 0$ and $E_{cell} = E^0_{cell}$.
- $E_{cell} < E^0_{cell}$ when $Q > 1$: As the $[Zn^{2+}]/[Cu^{2+}]$ ratio continues to increase, the $[(0.0592\ V/n) \log Q]$ term > 0, so $E_{cell} < E^0_{cell}$.
- $E_{cell} = 0$ when $Q = K$: Eventually, the $[(0.0592\ V/n) \log Q]$ term becomes so large that it equals E^0_{cell}, which means that E_{cell} is zero. This occurs *when the system reaches **equilibrium**: no more free energy is released, so the cell can do no more work.* At this point in the life of a battery, we say it is "dead."

At equilibrium, Equation 21.10 becomes

$$0 = E^0_{cell} - \left(\frac{0.0592\ V}{n} \right) \log K$$

Solving for E^0_{cell}, we have

$$E^0_{cell} = \frac{0.0592\ V}{n} \log K$$

Note that this equation is identical to Equation 21.8, which we obtained from ΔG^0. Solving for K for the zinc-copper cell ($E^0_{cell} = 1.10\ V$),

$$\log K = \frac{2 \times E^0_{cell}}{0.0592\ V}, \qquad \text{so} \qquad K = 10^{(2 \times 1.10\ V)/0.0592\ V} = 10^{37.16} = 1.4 \times 10^{37}$$

Thus, the zinc-copper cell can do work until the $[Zn^{2+}]/[Cu^{2+}]$ ratio is very high.

To conclude, let's examine cell potential in terms of the *starting Q/K* ratio:

- If $Q/K < 1$, E_{cell} is positive for the reaction as written. The smaller the Q/K ratio, the greater the value of E_{cell}, and the more electrical work the cell can do.
- If $Q/K = 1$, $E_{cell} = 0$. The cell is at equilibrium and can no longer do work.
- If $Q/K > 1$, E_{cell} is negative for the reaction as written. The cell will operate in reverse—the reverse reaction will take place—and do work until Q/K equals 1 at equilibrium.

Concentration Cells

You know that if you place a concentrated solution of a salt in contact with a dilute solution of the salt, they spontaneously mix and their final concentrations become equal at an intermediate value. A **concentration cell** employs this spontaneous tendency as a way to generate electrical energy. The two solutions are in separate half-cells, so they do not physically mix; rather, their concentrations become equal through the operation of the cell. Chemists, biologists, and environmental scientists use concentration cells in a host of applications.

How a Concentration Cell Works Suppose both compartments of a voltaic cell house the Cu/Cu^{2+} half-reaction. The cell reaction is the sum of identical half-reactions, written in opposite directions, so the standard half-cell potentials cancel ($E^0_{copper} - E^0_{copper}$) and E^0_{cell} is zero. This occurs because *standard* electrode potentials are based on concentrations of 1 M for each dissolved component. *In a concentration cell, however, the half-reactions are the same but the concentrations are different.* As a result, even though E^0_{cell} equals zero, the nonstandard cell potential, E_{cell}, does *not* equal zero because it depends on the ratio of ion concentrations.

Consider Figure 21.11A, in which a concentration cell has 0.10 M Cu^{2+} in the anode half-cell and 1.0 M Cu^{2+}, a 10-fold higher concentration, in the cathode half-cell:

$$Cu(s) \longrightarrow Cu^{2+}(aq; 0.10\ M) + 2e^- \qquad \text{[anode; oxidation]}$$
$$Cu^{2+}(aq, 1.0\ M) + 2e^- \longrightarrow Cu(s) \qquad \text{[cathode; reduction]}$$

The overall cell reaction is the sum of the half-reactions:

$$Cu^{2+}(aq, 1.0\ M) \longrightarrow Cu^{2+}(aq; 0.10\ M) \qquad E_{cell} = ?$$

The cell potential at the initial concentrations of 0.10 M (dilute) and 1.0 M (concentrated), with $n = 2$, is obtained from the Nernst equation:

$$E_{cell} = E^0_{cell} - \frac{0.0592\ V}{2} \log \frac{[Cu^{2+}]_{dil}}{[Cu^{2+}]_{conc}} = 0\ V - \left(\frac{0.0592\ V}{2} \log \frac{0.10\ M}{1.0\ M} \right)$$

$$= 0\ V - \left[\frac{0.0592\ V}{2} (-1.00) \right] = 0.0296\ V$$

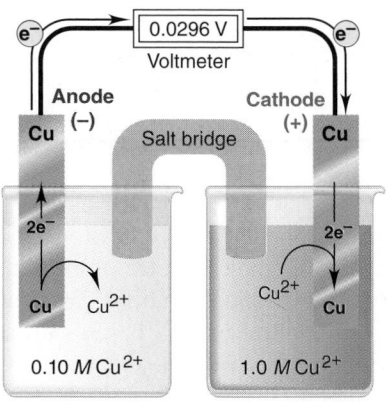

Oxidation half-reaction
$Cu(s) \longrightarrow Cu^{2+}(aq, 0.10\ M) + 2e^-$

Reduction half-reaction
$Cu^{2+}(aq, 1.0\ M) + 2e^- \longrightarrow Cu(s)$

Overall (cell) reaction
$Cu^{2+}(aq, 1.0\ M) \longrightarrow Cu^{2+}(aq, 0.10\ M)$

A

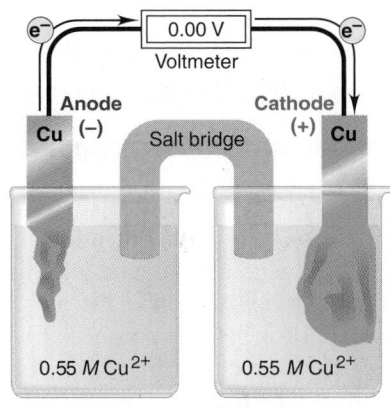

B

Figure 21.11 A concentration cell based on the Cu/Cu^{2+} half-reaction. A, The half-reactions involve the same components, which makes $E^0_{cell} = 0$. Nevertheless, the cell operates because the half-cell concentrations are different, which makes $E_{cell} > 0$ in this case. **B,** The cell operates spontaneously until the half-cell concentrations are equal. Note the change in electrodes (exaggerated here for clarity) and the identical color of the solutions.

As you can see from this result, E_{cell} for a concentration cell depends entirely on the $[(0.0592 \text{ V}/n) \log Q]$ term for nonstandard conditions because E^0_{cell} equals zero.

What is actually going on as this cell operates? In the half-cell with dilute electrolyte (anode), the Cu atoms in the electrode give up electrons and become Cu^{2+} ions, which enter the solution and make it *more* concentrated. The electrons released at the anode flow to the cathode compartment. There, Cu^{2+} ions in the concentrated solution pick up the electrons and become Cu atoms, which plate out on the electrode, thereby making that solution *less* concentrated. As in any voltaic cell, E_{cell} decreases until equilibrium is attained, which happens when the Cu^{2+} concentrations in the half-cells are equal (Figure 21.11B). The same final concentration would result if we mixed the two solutions, but no electrical work would be done.

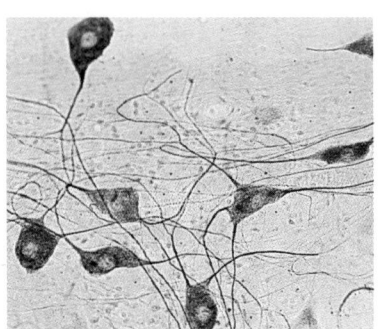

Concentration Cells in Your Nerve Cells The nerve cells that mediate every thought, movement, and other bodily process function by the principle of a concentration cell. The nerve membrane is imbedded with a host of specialized enzyme "gates" that use energy from ATP hydrolysis to separate an *interior* solution of low $[Na^+]$ and high $[K^+]$ from an *exterior* solution of high $[Na^+]$ and low $[K^+]$. As a result of the differences in $[Na^+]$ and $[K^+]$, the membrane is more positive outside than inside. One third of the body's ATP is used to create and maintain these concentration differences. (The 1997 Nobel Prize in chemistry was shared by Jens C. Skou for elucidation of this enzyme mechanism.) When the nerve membrane is stimulated, Na^+ ions spontaneously rush in, and in 0.001 s, the inside of the membrane becomes more positive than the outside. This stage is followed by K^+ ions spontaneously rushing out; after another 0.001 s, the membrane is again more positive outside. These large changes in charge in one region of the membrane stimulate the neighboring region and the electrical impulse moves down the length of the cell.

SAMPLE PROBLEM 21.7 Calculating the Potential of a Concentration Cell

Problem A concentration cell consists of two Ag/Ag^+ half-cells. In half-cell A, electrode A dips into 0.010 M $AgNO_3$; in half-cell B, electrode B dips into 4.0×10^{-4} M $AgNO_3$. What is the cell potential at 298 K? Which electrode has a positive charge?

Plan The standard half-cell reactions are identical, so E^0_{cell} is zero, and we calculate E_{cell} from the Nernst equation. Because half-cell A has a higher $[Ag^+]$, Ag^+ ions will be reduced and plate out on electrode A. In half-cell B, Ag will be oxidized and Ag^+ ions will enter the solution. As in all voltaic cells, reduction occurs at the cathode, which is positive.

Solution Writing the spontaneous reaction: The $[Ag^+]$ decreases in half-cell A and increases in half-cell B, so the spontaneous reaction is

$$Ag^+(aq, 0.010 \text{ } M) \text{ [half-cell A]} \longrightarrow Ag^+(aq; 4.0\times10^{-4} \text{ } M) \text{ [half-cell B]}$$

Calculating E_{cell}, with $n = 1$:

$$E_{cell} = E^0_{cell} - \frac{0.0592 \text{ V}}{1} \log \frac{[Ag^+]_{dil}}{[Ag^+]_{conc}}$$

$$= 0 \text{ V} - \left(0.0592 \text{ V} \log \frac{4.0\times10^{-4}}{0.010}\right) = 0.0828 \text{ V}$$

Reduction occurs at the cathode, electrode A: $Ag^+(aq; 0.010 \text{ } M) + e^- \longrightarrow Ag(s)$. Thus, electrode A has a positive charge due to a relative electron deficiency.

FOLLOW-UP PROBLEM 21.7 A concentration cell is built using two Au/Au^{3+} half-cells. In half-cell A, $[Au^{3+}] = 7.0\times10^{-4}$ M, and in half-cell B, $[Au^{3+}] = 2.5\times10^{-2}$ M. What is E_{cell}, and which electrode is negative?

Applications of Concentration Cells The principle of a concentration cell has major commercial and biological significance. The most important commercial application is the measurement of unknown concentrations, particularly $[H^+]$ (or pH). Suppose that we construct a concentration cell based on the H_2/H^+ half-reaction, in which the cathode compartment houses the standard hydrogen electrode and the anode compartment has the same apparatus dipping into an unknown $[H^+]$ in solution. The half-reactions and overall reaction are

$$H_2(g, 1 \text{ atm}) \longrightarrow 2H^+(aq; \text{ unknown}) + 2e^- \quad \text{[anode; oxidation]}$$
$$\underline{2H^+(aq, 1 \text{ } M) + 2e^- \longrightarrow H_2(g; 1 \text{ atm})} \quad \text{[cathode; reduction]}$$
$$2H^+(aq, 1 \text{ } M) \longrightarrow 2H^+(aq; \text{ unknown}) \quad E_{cell} = ?$$

As we saw for the Cu/Cu^{2+} concentration cell, E^0_{cell} is zero; however, the two half-cells differ in $[H^+]$, so E_{cell} is *not* zero. From the Nernst equation, with $n = 2$, we obtain

$$E_{cell} = E^0_{cell} - \frac{0.0592 \text{ V}}{2} \log \frac{[H^+]^2_{unknown}}{[H^+]^2_{standard}}$$

Substituting 1 *M* for $[H^+]_{standard}$ and zero for E^0_{cell} gives

$$E_{cell} = 0 \text{ V} - \frac{0.0592 \text{ V}}{2} \log \frac{[H^+]^2_{unknown}}{1^2} = -\frac{0.0592 \text{ V}}{2} \log [H^+]^2_{unknown}$$

Because $\log x^2 = 2 \log x$ (see Appendix A), we obtain

$$E_{cell} = -\left[\frac{0.0592 \text{ V}}{2} (2 \log [H^+]_{unknown}) \right] = -0.0592 \text{ V} \times \log [H^+]_{unknown}$$

Substituting $-\log [H^+] = pH$, we have

$$E_{cell} = 0.0592 \text{ V} \times pH$$

Thus, by measuring E_{cell}, we can find the pH.

In the routine measurement of pH, a concentration cell incorporating two hydrogen electrodes is too bulky and difficult to maintain. Instead, as was pointed out in Chapter 18, a pH meter is used. As shown in Figure 21.12, the two electrodes of a pH meter dip into the solution being tested. One of them is a *glass electrode*. This electrode consists of an Ag/AgCl half-reaction immersed in an HCl solution of fixed concentration (usually 1.000 *M*) and enclosed by a thin (~0.05 mm) membrane made of a special glass that is highly sensitive to the presence of H^+ ions. The other electrode is typically a *saturated calomel electrode*. This reference electrode consists of a platinum wire that is immersed in a paste of Hg_2Cl_2 (calomel), liquid Hg, and saturated KCl solution. The glass electrode monitors the solution's $[H^+]$ relative to its own fixed internal $[H^+]$, and the instrument converts the potential difference between the two electrodes into a measure of pH.

The pH electrode is one example of an *ion-selective (or ion-specific) electrode*. Electrodes have been designed with highly specialized membranes in order

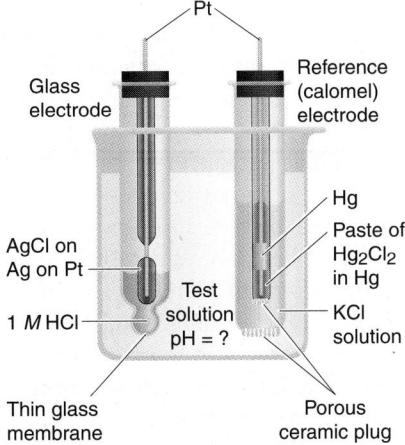

Figure 21.12 The laboratory measurement of pH. The glass electrode (*left*) is a self-contained Ag/AgCl half-cell immersed in an HCl solution of known concentration and enclosed by a thin glass membrane. It monitors the external $[H^+]$ in the solution relative to its own fixed internal $[H^+]$. The saturated calomel electrode (*right*) acts as a reference.

Table 21.3 Some Ions Measured with Ion-Specific Electrodes

Species Detected	Typical Sample
NH_3/NH_4^+	Industrial wastewater, seawater
CO_2/HCO_3^-	Blood, groundwater
F^-	Drinking water, urine, soil, industrial stack gases
Br^-	Grain, plant tissue
I^-	Milk, pharmaceuticals
NO_3^-	Soil, fertilizer, drinking water
K^+	Blood serum, soil, wine
H^+	Laboratory solutions, soil, natural waters

to measure selectively the concentrations of many different ions in industrial, environmental, and biological samples. Recent advances allow measurement in the picomolar (10^{-12} M) range. Table 21.3 shows some of the many ions for which ion-selective electrodes are available.

SECTION SUMMARY

A spontaneous process is indicated by a negative ΔG or a positive E_{cell}, which are related: $\Delta G = -nFE_{cell}$. The ΔG of the cell reaction represents the maximum amount of electrical work the cell can do. Because the standard free energy change, ΔG^0, is related to E_{cell}^0 and to K, we can use E_{cell}^0 to determine K. At nonstandard conditions, the Nernst equation shows that E_{cell} depends on E_{cell}^0 and a correction term based on Q. E_{cell} is high when Q is small (high [reactant]), and it decreases as the cell operates. At equilibrium, ΔG and E_{cell} are zero, which means that $Q = K$. Concentration cells use identical half-reactions, with solutions of differing concentration, that generate electrical energy as the concentrations become equal. Ion-specific electrodes, such as the pH electrode, measure the concentration of one species.

Minimicroanalysis   Electronic miniaturization has greatly broadened the applications of electrochemistry. Environmental chemists use pocket-sized pH meters in the field to study natural waters and soil. Physiologists and biochemists employ electrodes so small that they can be surgically implanted in a single biological cell to study ion channels and receptors. In the photo, a microelectrode (filled with a fluorescent dye for better visibility) is shown penetrating an egg cell from the African toad *Xenopus laevis*.

21.5 ELECTROCHEMICAL PROCESSES IN BATTERIES

Because of their compactness and mobility, batteries have a major influence on our way of life. In industrialized countries, an average of 10 batteries are used by each person every year. A **battery,** strictly speaking, is a self-contained group of voltaic cells arranged in series (plus-to-minus-to-plus, and so on), so that their individual voltages are added together. In everyday speech, however, the term may also be applied to a single voltaic cell.

Batteries are ingeniously engineered devices that house rather unusual half-reactions and half-cells, but they operate through the same electrochemical principles we've been discussing. There are several classes of batteries. A *primary battery* cannot be recharged, so it is thrown away when the battery is "dead," that is, when the components have reached their equilibrium concentrations. In contrast, when a *secondary,* or *rechargeable, battery* runs down, it is recharged by supplying electrical energy to reverse the cell reaction and re-form reactant. In other words, in this type of battery, the voltaic cells are periodically converted to electrolytic cells to restore *nonequilibrium* concentrations. A **fuel cell,** or *flow battery,* is one that is not self-contained. The reactants (usually a combustible fuel and oxygen) enter the cell and the products leave, generating electricity through the controlled oxidation of the fuel. The upcoming multipage Gallery presents some important examples of primary, secondary, and flow batteries, including several newer designs.

Batteries and Their Applications

Primary (Nonrechargeable) Batteries

Dry Cell

Invented in the 1860s, the common dry cell, or *Leclanché cell,* has become a familiar household item. A zinc anode in the form of a can houses a mixture of MnO_2 and an electrolyte paste, consisting of NH_4Cl, $ZnCl_2$, H_2O, and starch. Powdered graphite improves conductivity. The cathode is an inactive graphite rod.

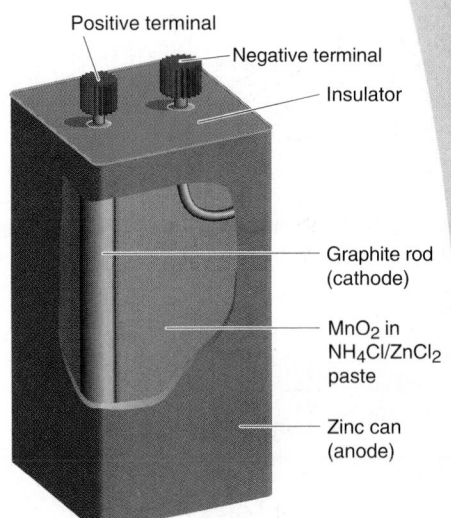

Anode (oxidation):

$$Zn(s) \longrightarrow Zn^{2+}(aq) + 2e^-$$

Cathode (reduction): The cathode half-reaction is complex and not completely understood. $MnO_2(s)$ is reduced to $Mn_2O_3(s)$ through a series of steps that may involve the presence of Mn^{2+} and an acid-base reaction between NH_4^+ and OH^-:

$$2MnO_2(s) + 2NH_4^+(aq) + 2e^- \longrightarrow Mn_2O_3(s) + 2NH_3(aq) + H_2O(l)$$

The ammonia, some of which may be gaseous, forms a complex ion with Zn^{2+}, which crystallizes in contact with Cl^- ion:

$$Zn^{2+}(aq) + 2NH_3(aq) + 2Cl^-(aq) \longrightarrow Zn(NH_3)_2Cl_2(s)$$

Overall (cell) reaction:

$$2MnO_2(s) + 2NH_4Cl(aq) + Zn(s) \longrightarrow Zn(NH_3)_2Cl_2(s) + H_2O(l) + Mn_2O_3(s)$$

$$E_{cell} = 1.5 \text{ V}$$

Uses: Portable radios, toys, flashlights.
Advantages: Inexpensive, safe, available in many sizes.
Disadvantages: At high current drain, $NH_3(g)$ builds up, causing voltage drop. Dry cells have a short shelf life because the zinc anode reacts with the acidic NH_4^+ ions.

Positive terminal
Negative terminal
Insulator
Graphite rod (cathode)
MnO_2 in $NH_4Cl/ZnCl_2$ paste
Zinc can (anode)

Alkaline Battery

The alkaline battery is an improved dry cell. The half-reactions are essentially the same, but the electrolyte is a KOH paste. The basic electrolyte eliminates the buildup of gases and maintains the Zn electrode.

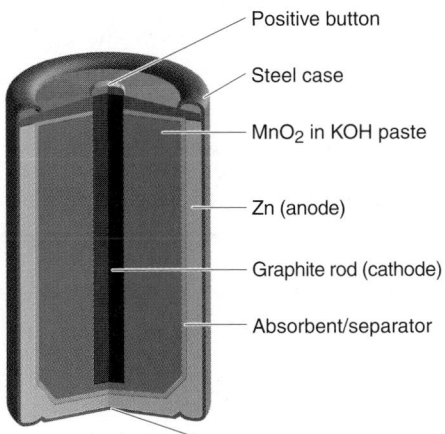

Positive button
Steel case
MnO_2 in KOH paste
Zn (anode)
Graphite rod (cathode)
Absorbent/separator
Negative end cap

Anode (oxidation):

$$Zn(s) + 2OH^-(aq) \longrightarrow ZnO(s) + H_2O(l) + 2e^-$$

Cathode (reduction):

$$MnO_2(s) + 2H_2O(l) + 2e^- \longrightarrow Mn(OH)_2(s) + 2OH^-(aq)$$

Overall (cell) reaction:

$$Zn(s) + MnO_2(s) + H_2O(l) \longrightarrow ZnO(s) + Mn(OH)_2(s) \qquad E_{cell} = 1.5 \text{ V}$$

Uses: Same as dry cell.

Advantages: No voltage drop and longer shelf life than dry cell. Safe, many sizes.
Disadvantages: More expensive than common dry cell.

Mercury and Silver (Button) Batteries

The mercury battery and the silver battery are quite similar. Both use a zinc anode (reducing agent) in a basic medium. One employs HgO as the oxidizing agent, the other Ag_2O, and both use a steel cathode. The solid reactants are compacted separately with KOH, with moist paper for a salt bridge.

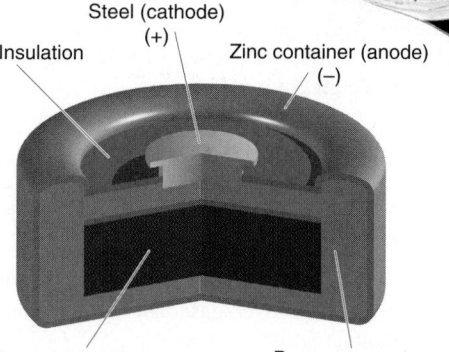

Anode (oxidation):

$$Zn(s) + 2OH^-(aq) \longrightarrow ZnO(s) + H_2O(l) + 2e^-$$

Cathode (reduction) (mercury):

$$HgO(s) + H_2O(l) + 2e^- \longrightarrow Hg(l) + 2OH^-(aq)$$

Cathode (reduction) (silver):

$$Ag_2O(s) + H_2O(l) + 2e^- \longrightarrow 2Ag(s) + 2OH^-(aq)$$

Overall (cell) reaction (mercury):

$$Zn(s) + HgO(s) \longrightarrow ZnO(s) + Hg(l) \quad E_{cell} = 1.3 \text{ V}$$

Overall (cell) reaction (silver):

$$Zn(s) + Ag_2O(s) \longrightarrow ZnO(s) + 2Ag(s) \quad E_{cell} = 1.6 \text{ V}$$

Steel (cathode) (+)

Insulation

Zinc container (anode) (−)

Paste of Ag_2O on electrolyte KOH and $Zn(OH)_2$

Porous separator

Uses: The mercury cell is used in watches and calculators; the silver cell in cameras, heart pacemakers, and hearing aids.

Advantages: Both are very small, with relatively large voltage. The silver cell has a very steady output and is nontoxic.

Disadvantages: Discarded mercury cells release the toxic metal. Silver cells are expensive.

Secondary (Rechargeable) Batteries

Lead-Acid Battery

A typical 12-V lead-acid car battery has six cells connected in series, each of which delivers about 2.1 V. Each cell contains two lead grids packed with the electrode materials: the anode is spongy, powdered Pb, and the cathode is powdered PbO_2.

The grids are immersed in an electrolyte solution of $\sim$4.5 M H_2SO_4. Fiberglass sheets between the grids prevent shorting due to chance physical contact. When the cell discharges, it generates electrical energy as a voltaic cell:

Anode (oxidation):

$$Pb(s) + HSO_4^-(aq) \longrightarrow PbSO_4(s) + H^+ + 2e^-$$

Cathode (reduction):

$$PbO_2(s) + 3H^+(aq) + HSO_4^-(aq) + 2e^- \longrightarrow PbSO_4(s) + 2H_2O(l)$$

Both half-reactions produce Pb^{2+} ion, one through oxidation of Pb, the other through reduction of PbO_2. The Pb^{2+} forms $PbSO_4(s)$ at both electrodes by reaction with HSO_4^-.

Overall (cell) reaction (discharge):

$$PbO_2(s) + Pb(s) + 2H_2SO_4(aq) \longrightarrow 2PbSO_4(s) + 2H_2O(l) \quad E_{cell} = 2.1 \text{ V}$$

When the cell recharges, it uses electrical energy as an electrolytic cell, and the half-cell and overall reactions are reversed.

Overall (cell) reaction (recharge):

$$2PbSO_4(s) + 2H_2O(l) \longrightarrow PbO_2(s) + Pb(s) + 2H_2SO_4(aq)$$

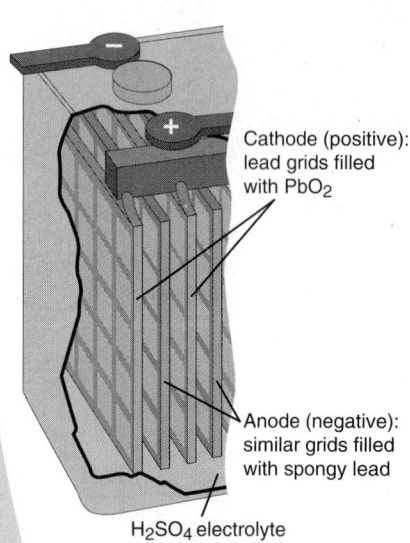

Cathode (positive): lead grids filled with PbO_2

Anode (negative): similar grids filled with spongy lead

H_2SO_4 electrolyte

Uses: In automobiles and trucks.

Advantages: Provides a large burst of current to the engine starter motor; is reliable and has a long life; is effective at low temperature.

Disadvantages: (1) Loss of capacity: $PbSO_4$, which is required in the recharging stage, coats the battery grids after the battery discharges. Mechanical strain and normal bumping can dislodge $PbSO_4$ and reduce battery capacity. If enough $PbSO_4$ is lost, the cell cannot be recharged.
(2) Safety hazard: Older batteries had a cap on each cell to monitor electrolyte density and replace water lost during discharging. During recharging, some water could electrolyze to H_2 and O_2, which could explode if sparked, and splatter H_2SO_4. Modern batteries use a lead alloy that inhibits electrolysis and reduces water loss, so the cells are sealed.

GALLERYGALLERYGALLERYGALLERY

Nickel–Metal Hydride (Ni-MH) Battery

The nickel–metal hydride battery has started to replace the nickel-cadmium battery. The anode half-reaction oxidizes the hydrogen adsorbed in metal alloys (designated M; e.g., $LaNi_5$) in a basic (KOH) electrolyte, while nickel(III) as NiO(OH) is reduced at the cathode.

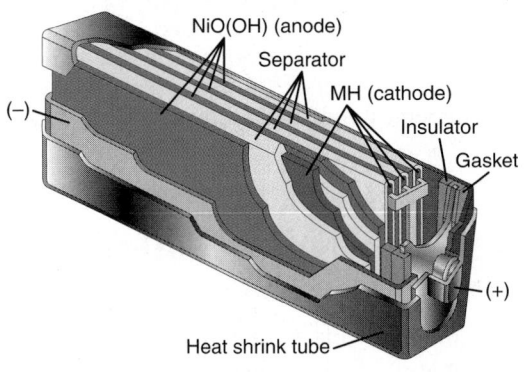

Anode (oxidation):
$$MH(s) + OH^-(aq) \longrightarrow M(s) + H_2O(l) + e^-$$
Cathode (reduction):
$$NiO(OH)(s) + H_2O(l) + e^- \longrightarrow Ni(OH)_2(s) + OH^-(aq)$$
Overall (cell) reaction:
$$MH(s) + NiO(OH)(s) \longrightarrow M(s) + Ni(OH)_2(s); \qquad E_{cell} = 1.4\ V$$

The cell reaction is reversed during recharging.
Uses: Cordless razors, photo flash units, and power tools.
Advantages: Lightweight; high power; avoids the toxicity of cadmium in Nicad battery.
Disadvantages: Discharge upon storage.

Lithium-Ion Battery

The modern lithium-ion battery has an anode of Li atoms intercalated with (lying between) planes of graphite [designated $Li_x(gr)$]. The cathode is a lithium–metal oxide, such as $LiMn_2O_4$, and a typical electrolyte is 1 *M* $LiClO_4$ in ethylene carbonate, an organic solvent. Electrons flow through the circuit while Li^+ ions flow from anode to cathode.

Anode (oxidation): $Li_x(gr) \longrightarrow xLi^+ + x\,e^-$
Cathode (reduction): $Li_{1-x}Mn_2O_4(s) + xLi^+ + x\,e^- \longrightarrow LiMn_2O_4(s)$
Overall (cell) reaction: $Li_x(gr) + Li_{1-x}Mn_2O_4(s) \longrightarrow LiMn_2O_4(s) \qquad E_{cell} = 3.7\ V$

The cell reaction is reversed when recharging.
Uses: Laptop computers, cell phones, and camcorders.
Advantages: Extremely high energy/mass ratio: 1 mol of e^- (1 *F*) produced from < 7 g of metal (*M* of Li = 6.941 g/mol).
Disadvantages: Relatively expensive; organic solvent may be flammable.

Flow Batteries (Fuel Cells)

Fuel cells use combustion reactions to produce electricity. The fuel does not burn because, as in other batteries, the reactants undergo separated half-reactions, and the electrons are transferred through an external circuit. A fuel cell being designed for cars oxidizes H_2 to H_2O. It has graphite electrodes surrounded by a layer of platinum-impregnated catalyst. At the catalyst layer in the anode, H_2 is split, and the e^- enter the circuit. The electrolyte in the new proton-exchange-membrane (PEM) cell has a perfluoroethylene polymer backbone $-\!\!\!-[F_2C\!\!-\!\!CF_2]_n\!\!\!-\!\!\!-$ with attached sulfonic acid groups (RSO_3^-) that ferry H^+ from anode to cathode. At the catalyst layer in the cathode, O_2 is split, reduced, and combined with H^+.

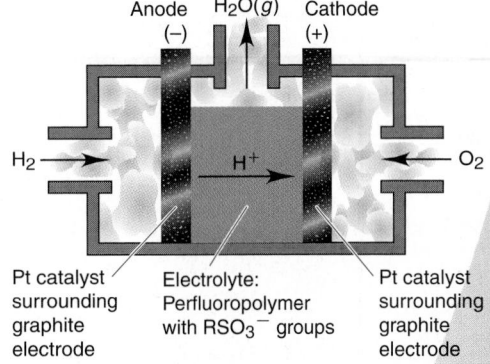

Anode (oxidation): $H_2(g) \longrightarrow 2H^+(aq) + 2e^-$
Cathode (reduction): $\frac{1}{2}O_2(g) + 2H^+(aq) + 2e^- \longrightarrow H_2O(g)$
Overall (cell) reaction: $H_2(g) + \frac{1}{2}O_2(g) \longrightarrow H_2O(g) \qquad E_{cell} = 1.2\ V$

Uses: In the very near future, for transportation, residential, and business electric power; currently provide electricity and pure water (after condensing the water vapor) during space flights.
Advantages: (1) Clean and portable; produce no pollutants. (2) Convert about 75% of bond energy in fuel into electricity, in contrast to 40% for an electric power plant and 25% for a car engine.
Disadvantages: (1) Fuel cells operate with a continuous flow of reactants, so they can't store electrical energy. (2) Electrode catalysts are expensive.

Batteries contain several voltaic cells in series and are classified as primary (such as dry cell, alkaline, mercury, or silver), rechargeable (such as lead acid, nickel–metal hydride, and lithium ion), or flow (fuel cells). Supplying electricity to a rechargeable battery reverses the redox reaction, forming more reactant for further use. Fuel cells release electrons and generate a current through the controlled oxidation of a combustible fuel, such as H_2.

21.6 CORROSION: A CASE OF ENVIRONMENTAL ELECTROCHEMISTRY

By now, you may be thinking that spontaneous electrochemical processes are always beneficial, but consider the problem of **corrosion,** the natural redox process that oxidizes metals to their oxides and sulfides. In chemical terms, corrosion is the reverse of isolating a metal from its oxide or sulfide ore; in electrochemical terms, the process shares many similarities with voltaic cells. Damage from corrosion to cars, ships, buildings, and bridges runs into tens of billions of dollars annually, so it is a major problem in much of the world. Although we focus here on the corrosion of iron, many other metals, such as copper and silver, also corrode.

The Corrosion of Iron

The most common and economically destructive form of corrosion is the rusting of iron. About 25% of the steel produced in the United States is made just to replace steel already in use that has corroded. Contrary to the simplified view given in some earlier discussions, rust is *not* a direct product of the reaction between iron and oxygen but arises through a complex electrochemical process. Let's look at the facts of iron corrosion and then use the features of a voltaic cell to explain them:

1. Iron does not rust in dry air: moisture must be present.
2. Iron does not rust in air-free water: oxygen must be present.
3. The loss of iron and the depositing of rust often occur at *different* places on the *same* object.
4. Iron rusts more quickly at low pH (high $[H^+]$).
5. Iron rusts more quickly in contact with ionic solutions.
6. Iron rusts more quickly in contact with a less active metal (such as Cu) and more slowly in contact with a more active metal (such as Zn).

Picture the magnified surface of a piece of iron or steel (Figure 21.13). Strains, ridges, and dents in contact with water are typically the sites of iron loss

Figure 21.13 **The corrosion of iron. A,** Close-up view of an iron surface. Corrosion usually occurs at a surface irregularity. **B,** A schematic depiction of the surface enlarged, showing the steps in the corrosion process.

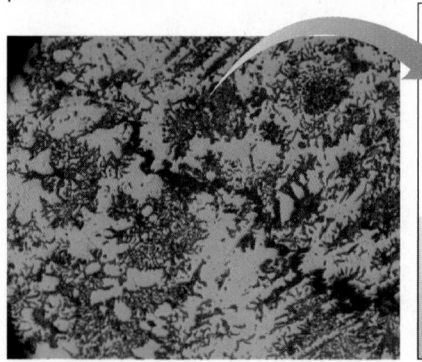

A

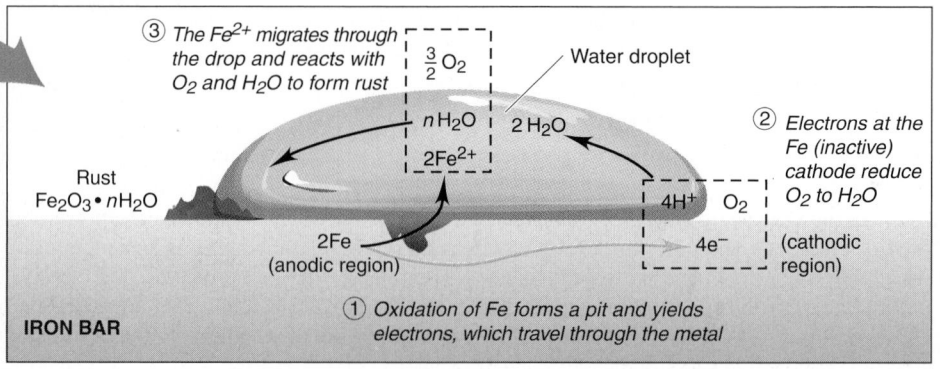

③ The Fe^{2+} migrates through the drop and reacts with O_2 and H_2O to form rust

$\frac{3}{2} O_2$ Water droplet

$n H_2O$ $2 H_2O$

② Electrons at the Fe (inactive) cathode reduce O_2 to H_2O

$2Fe^{2+}$

Rust $Fe_2O_3 \cdot nH_2O$

$4H^+$ O_2

$2Fe$ (anodic region) $4e^-$ (cathodic region)

IRON BAR

① Oxidation of Fe forms a pit and yields electrons, which travel through the metal

B

(fact 1). These sites are called *anodic regions* because the following half-reaction occurs there:

$$Fe(s) \longrightarrow Fe^{2+}(aq) + 2e^- \qquad \text{[anodic region; oxidation]}$$

Once the iron atoms lose electrons, the damage to the object has been done, and a pit forms where the iron is lost.

The freed electrons move through the external circuit—the piece of iron itself—until they reach a region of relatively high O_2 concentration (fact 2), near the surface of a surrounding water droplet, for instance. At this *cathodic region,* the electrons released from the iron atoms reduce O_2 molecules:

$$O_2(g) + 4H^+(aq) + 4e^- \longrightarrow 2H_2O(l) \qquad \text{[cathodic region; reduction]}$$

Notice that this overall redox process is complete, and thus the iron loss has occurred, without any rust forming:

$$2Fe(s) + O_2(g) + 4H^+(aq) \longrightarrow 2Fe^{2+}(aq) + 2H_2O(l)$$

Rust forms through another redox reaction in which the reactants contact each other directly. The Fe^{2+} ions formed originally at the anodic region disperse through the surrounding water and react with O_2, often at some distance from the pit (fact 3). The overall reaction for this step is

$$2Fe^{2+}(aq) + \tfrac{1}{2}O_2(g) + (2+n)H_2O(l) \longrightarrow Fe_2O_3 \cdot nH_2O(s) + 4H^+(aq)$$

[The inexact coefficient n for H_2O in the above equation appears because rust, $Fe_2O_3 \cdot nH_2O$, is a form of iron(III) oxide with variable numbers of waters of hydration.] The rust deposit is really incidental to the damage caused by loss of iron—a chemical insult added to the original injury.

Adding the previous two equations together shows the overall equation for the rusting of iron:

$$2Fe(s) + \tfrac{3}{2}O_2(g) + nH_2O(l) + 4H^+(aq) \longrightarrow Fe_2O_3 \cdot nH_2O(s) + 4H^+(aq)$$

We've shown the canceled H^+ ions to emphasize that they act as a catalyst; that is, they are used up in one step of the overall reaction and created in another. As a result of this action, rusting is faster at low pH (high $[H^+]$) (fact 4). Ionic solutions speed rusting by improving the conductivity of the aqueous medium near the anodic and cathodic regions (fact 5). The effect of ions is especially evident on ocean-going vessels (Figure 21.14) and on the underbodies and around the wheel wells of cars driven in cold climates, where salts are used to melt ice on slippery roads.

In many ways, the components of the corrosion process resemble those of a voltaic cell:

- Anodic and cathodic regions are separated in space.
- The regions are connected via an external circuit through which the electrons travel.
- In the anodic region, iron behaves like an active electrode, whereas in the cathodic region, it is inactive.
- The moisture surrounding the pit functions somewhat like a salt bridge, a means for ions to ferry back and forth and keep the solution neutral.

Figure 21.14 Enhanced corrosion at sea. The high ion concentration of seawater leads to its high conductivity, which enhances the corrosion of iron in the hulls and anchors of ocean-going vessels.

Protecting Against the Corrosion of Iron

A common approach to preventing or limiting corrosion is to eliminate contact with the corrosive factors. The simple act of washing off road salt removes the ionic solution from auto chassis parts. Iron objects are frequently painted to keep out O_2 and moisture, but if the paint layer chips, rusting proceeds. More permanent coatings include chromium on plumbing fixtures. "Blueing" of gun barrels, wood stoves, and other steel objects bonds an adherent coating of Fe_3O_4 (magnetite) to the surface.

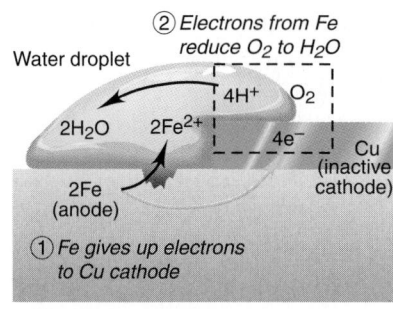
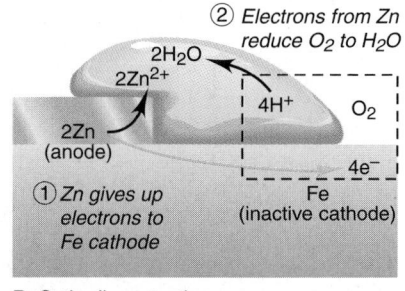

A Enhanced corrosion   **B** Cathodic protection

Figure 21.15 **The effect of metal-metal contact on the corrosion of iron. A,** When iron is in contact with a less active metal, such as copper, the iron loses electrons more readily (is more anodic), so it corrodes faster. **B,** When iron is in contact with a more active metal, such as zinc, the zinc acts as the anode and loses electrons. Therefore, the iron is cathodic, so it does not corrode. The process is known as *cathodic protection.*

The only fact regarding corrosion that we have not yet addressed concerns the relative activity of other metals in contact with iron (fact 6), which leads to the most effective way to prevent corrosion. The essence of the idea is that *iron functions as both anode and cathode in the rusting process, but it is lost only at the anode.* Therefore, it makes sense that anything that makes iron behave more like the anode increases corrosion. As you can see in Figure 21.15A, when iron is in contact with a *less* active metal (weaker reducing agent), such as copper, its anodic function is enhanced. As a result, when iron plumbing is connected directly to copper plumbing with no electrical insulation between them, the iron pipe corrodes rapidly.

On the other hand, anything that makes iron behave more like the cathode prevents corrosion. Application of this principle is called *cathodic protection.* For example, if the iron makes contact with a *more* active metal (stronger reducing agent), such as zinc, the iron becomes cathodic and remains intact, while the zinc acts as the anode and loses electrons (Figure 21.15B). Coating steel with a "sacrificial" layer of zinc is the basis of the *galvanizing* process. In addition to blocking physical contact with H_2O and O_2, the zinc is "sacrificed" (oxidized) instead of the iron.

Sacrificial anodes are employed to protect iron and steel structures (pipes, tanks, oil rigs, and so on) in marine and moist underground environments. The most frequently used metals for this purpose are magnesium and aluminum, elements that are much more active than iron, so they act as the anode while iron acts as the cathode. An example is illustrated in Figure 21.16. Another advantage of these metals is that they form adherent oxide coatings, which slows their own corrosion.

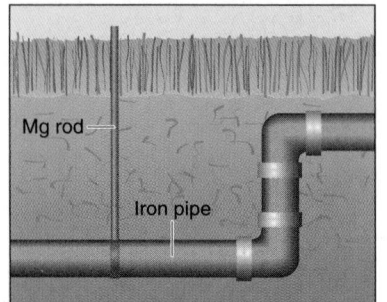

Figure 21.16 **The use of sacrificial anodes to prevent iron corrosion.** Active metals, such as magnesium and aluminum, are connected to underground iron pipes to prevent their corrosion through cathodic protection. The active metal is sacrificed instead of the iron.

SECTION SUMMARY

Corrosion damages metal structures through a natural electrochemical change. Iron corrosion occurs in the presence of oxygen and moisture and is increased by high $[H^+]$, high [ion], or contact with a less active metal, such as Cu. Fe is oxidized and O_2 is reduced in one redox reaction, while rust is formed in another reaction that often takes place at a different location. Since Fe functions as both anode and cathode in the process, the object can be protected by physically covering its surface or joining it to a more active metal (such as Zn, Mg, or Al), which acts as the anode in place of the Fe.

21.7 ELECTROLYTIC CELLS: USING ELECTRICAL ENERGY TO DRIVE A NONSPONTANEOUS REACTION

Up to now, we've been considering voltaic cells, those that generate electrical energy from a spontaneous redox reaction. The principle of an electrolytic cell is exactly the opposite: *electrical energy from an external source drives a nonspontaneous reaction.*

Construction and Operation of an Electrolytic Cell

Let's examine the operation of an electrolytic cell by constructing one from a voltaic cell. Consider the tin-copper voltaic cell in Figure 21.17A. The Sn anode will gradually become oxidized to Sn^{2+} ions, and the Cu^{2+} ions will gradually be reduced and plate out on the Cu cathode because the cell reaction is spontaneous in that direction:

For the voltaic cell

$$Sn(s) \longrightarrow Sn^{2+}(aq) + 2e^- \qquad \text{[anode; oxidation]}$$
$$\underline{Cu^{2+}(aq) + 2e^- \longrightarrow Cu(s) \qquad \text{[cathode; reduction]}}$$
$$Sn(s) + Cu^{2+}(aq) \longrightarrow Sn^{2+}(aq) + Cu(s) \qquad E^0_{cell} = 0.48 \text{ V and } \Delta G^0 = -93 \text{ kJ}$$

Therefore, the *reverse* cell reaction is *non*spontaneous and never happens of its own accord, as the negative E^0_{cell} and positive ΔG^0 indicate:

$$Cu(s) + Sn^{2+}(aq) \longrightarrow Cu^{2+}(aq) + Sn(s) \qquad E^0_{cell} = -0.48 \text{ V and } \Delta G^0 = 93 \text{ kJ}$$

However, we can make this process happen by supplying from an external source an electric potential *greater than* E^0_{cell}. In effect, we have converted the voltaic cell into an electrolytic cell and changed the nature of the electrodes—anode is now cathode, and cathode is now anode (Figure 21.17B):

For the electrolytic cell

$$Cu(s) \longrightarrow Cu^{2+}(aq) + 2e^- \qquad \text{[anode; oxidation]}$$
$$\underline{Sn^{2+}(aq) + 2e^- \longrightarrow Sn(s) \qquad \text{[cathode; reduction]}}$$
$$Cu(s) + Sn^{2+}(aq) \longrightarrow Cu^{2+}(aq) + Sn(s) \qquad \text{[overall (cell) reaction]}$$

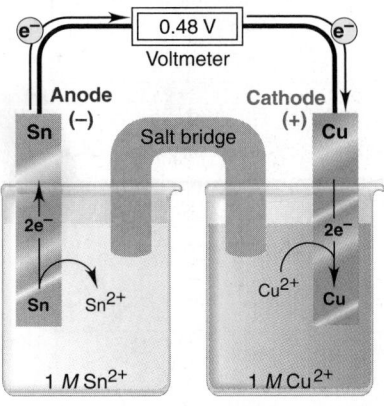

A Voltaic cell

0.48 V — Voltmeter

Anode (−) Sn — Salt bridge — **Cathode (+)** Cu

1 M Sn²⁺   1 M Cu²⁺

Oxidation half-reaction
$Sn(s) \longrightarrow Sn^{2+}(aq) + 2e^-$

Reduction half-reaction
$Cu^{2+}(aq) + 2e^- \longrightarrow Cu(s)$

Overall (cell) reaction
$Sn(s) + Cu^{2+}(aq) \longrightarrow Sn^{2+}(aq) + Cu(s)$

B Electrolytic cell

External source *greater than* 0.48 V

Cathode (−) Sn — Salt bridge — **Anode (+)** Cu

1 M Sn²⁺   1 M Cu²⁺

Oxidation half-reaction
$Cu(s) \longrightarrow Cu^{2+}(aq) + 2e^-$

Reduction half-reaction
$Sn^{2+}(aq) + 2e^- \longrightarrow Sn(s)$

Overall (cell) reaction
$Cu(s) + Sn^{2+}(aq) \longrightarrow Cu^{2+}(aq) + Sn(s)$

Figure 21.17 The tin-copper reaction as the basis of a voltaic and an electrolytic cell. **A,** The spontaneous reaction between Sn and Cu^{2+} generates 0.48 V in a voltaic cell. **B,** If more than 0.48 V is supplied, the same apparatus is changed to an electrolytic cell, and the nonspontaneous reaction between Cu and Sn^{2+} occurs. Note the changes in electrode charges and direction of electron flow.

Note that in an electrolytic cell, as in a voltaic cell, *oxidation takes place at the anode and reduction takes place at the cathode, but the direction of electron flow and the signs of the electrodes are reversed.* To understand these changes, keep in mind the *cause* of the electron flow:

• In a voltaic cell, electrons are generated at the anode, so it is negative, and electrons are consumed at the cathode, so it is positive.

• In an electrolytic cell, the electrons come from the external power source, which *supplies* them *to* the cathode, so it is negative, and *removes* them *from* the anode, so it is positive.

A rechargeable battery functions as a voltaic cell when it is discharging and as an electrolytic cell when it is recharging, so it provides a good way to compare these two cell types and the changes in the functions of the electrodes. Figure 21.18 shows these two functions in the lead-acid battery. In the discharge mode (voltaic cell), oxidation occurs at electrode I, thus making the *negative* electrode the anode. In the recharge mode (electrolytic cell), oxidation occurs at electrode II, thus making the *positive* electrode the anode. Similarly, the cathode is *positive* during discharge (electrode II), and it is *negative* during recharge (electrode I). To reiterate, regardless of the mode, oxidation occurs at the anode and reduction occurs at the cathode.

Table 21.4 summarizes the processes and signs in the two types of electrochemical cells.

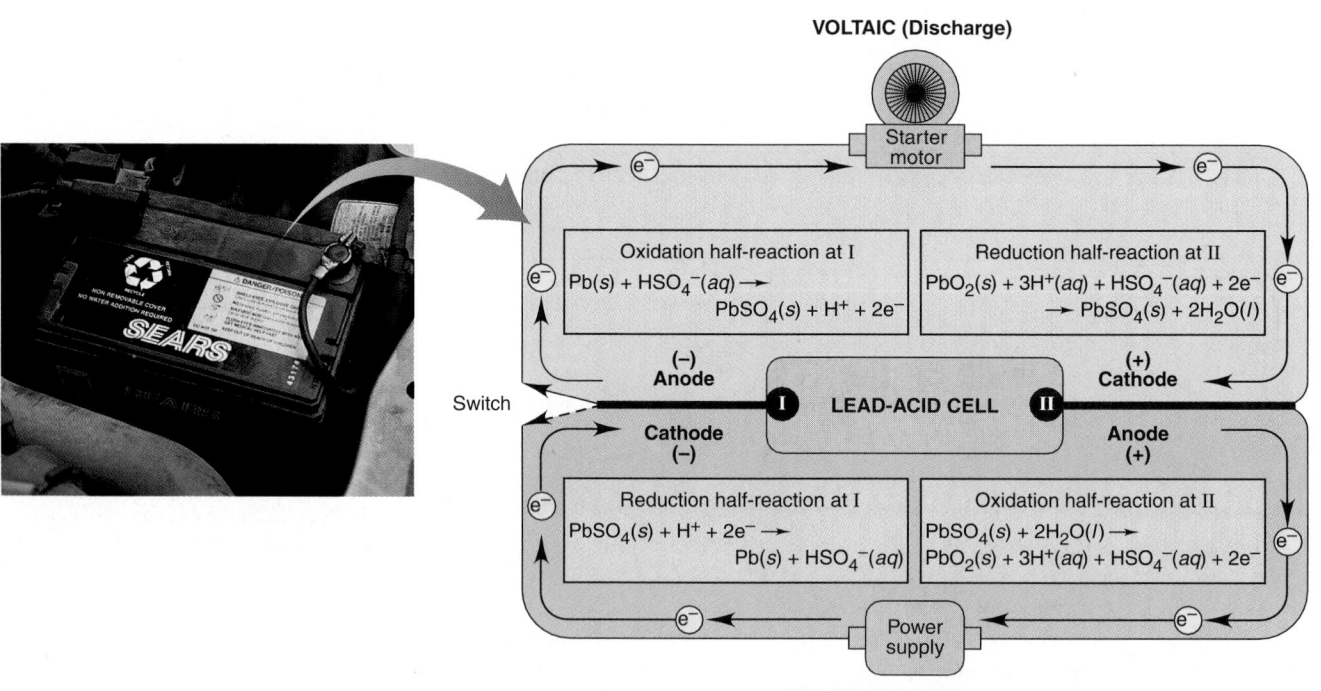

Figure 21.18 The processes occurring during the discharge and recharge of a lead-acid battery. When the lead-acid battery is discharging (*top*), it behaves like a voltaic cell: the anode is negative (electrode I), and the cathode is positive (electrode II). When it is recharging (*bottom*), it behaves like an electrolytic cell: the anode is positive (electrode II), and the cathode is negative (electrode I).

Table 21.4 Comparison of Voltaic and Electrolytic Cells

| Cell Type | ΔG | E_{cell} | Electrode | | |
			Name	Process	Sign
Voltaic	<0	>0	Anode	Oxidation	−
Voltaic	<0	>0	Cathode	Reduction	+
Electrolytic	>0	<0	Anode	Oxidation	+
Electrolytic	>0	<0	Cathode	Reduction	−

Predicting the Products of Electrolysis

Electrolysis, the splitting (lysing) of a substance by the input of electrical energy, is often used to decompose a compound into its elements. Electrolytic cells are involved in key industrial production steps for some of the most commercially important elements, including chlorine, aluminum, and copper, as you'll see in the next chapter. The first laboratory electrolysis of H_2O to H_2 and O_2 was performed in 1800, and the process is still used to produce these gases in ultrahigh purity. The electrolyte in an electrolytic cell can be the pure compound (such as H_2O or a molten salt), a mixture of molten salts, or an aqueous solution of a salt. The products obtained depend on atomic properties and several other factors, so let's examine some actual cases.

Electrolysis of Pure Molten Salts Many electrolytic applications involve isolating a metal or nonmetal from a molten salt. Predicting the product at each electrode is simple if the salt is pure because *the cation will be reduced and the anion oxidized.* The electrolyte is the molten salt itself, and the ions move through the cell attracted by the oppositely charged electrodes.

Consider the electrolysis of molten (fused) calcium chloride. The two species present are Ca^{2+} and Cl^-, so the Ca^{2+} ion is reduced and the Cl^- ion is oxidized:

$$2Cl^-(l) \longrightarrow Cl_2(g) + 2e^- \qquad \text{[anode; oxidation]}$$
$$\underline{Ca^{2+}(l) + 2e^- \longrightarrow Ca(s)} \qquad \text{[cathode; reduction]}$$
$$Ca^{2+}(l) + 2Cl^-(l) \longrightarrow Ca(s) + Cl_2(g) \qquad \text{[overall]}$$

Metallic calcium is prepared industrially this way, as are several other active metals, such as Na and Mg, and the halogens Cl_2 and Br_2. We examine the details of these and several other electrolytic processes in Chapter 22.

Electrolysis of Mixed Molten Salts More typically, the electrolyte is a mixture of molten salts, which is then electrolyzed to obtain a particular metal. When we have a choice of product, how can we tell which species will react at which electrode? The general rule for all electrolytic cells is that *the more easily oxidized species (stronger reducing agent) reacts at the anode, and the more easily reduced species (stronger oxidizing agent) reacts at the cathode.*

It's important to realize that for electrolysis of mixtures of molten salts we *cannot* use tabulated E^0 values to tell the relative strength of the oxidizing and reducing agents. Those values refer to the *change from aqueous ion to free element,* $M^{n+}(aq) + ne^- \longrightarrow M(s)$, under standard-state conditions, but there are no aqueous ions in a molten salt. Instead, we rely on our knowledge of periodic atomic trends to predict which of the ions present gains or loses electrons more easily (Sections 8.4 and 9.4).

SAMPLE PROBLEM 21.8 Predicting the Electrolysis Products of a Molten Salt Mixture

Problem A chemical engineer melts a naturally occurring mixture of NaBr and $MgCl_2$ and decomposes it in an electrolytic cell. Predict the substance formed at each electrode, and write balanced half-reactions and the overall cell reaction.

Plan We have to determine which metal and nonmetal will form more easily at the electrodes. We first list the ions as oxidizing or reducing agents. If a metal holds its electrons more tightly than another, it has a higher ionization energy (IE). Therefore, as a cation, it gains electrons more easily; it is the stronger oxidizing agent and is reduced at the cathode. Similarly, if a nonmetal holds its electrons less tightly than another, it has a lower electronegativity (EN). Therefore, as an anion, it loses electrons more easily; it is the stronger reducing agent and is oxidized at the anode.

Solution Listing the ions as oxidizing or reducing agents:

The possible oxidizing agents are Na^+ and Mg^{2+}.
The possible reducing agents are Br^- and Cl^-.

Determining the cathode product (more easily reduced cation): Mg is to the right of Na in Period 3. IE increases from left to right, so Mg has a higher IE. It takes more energy to remove an e^- from Mg than from Na, so it follows that Mg^{2+} has a greater attraction for e^- and thus is more easily reduced (stronger oxidizing agent):

$$Mg^{2+}(l) + 2e^- \longrightarrow Mg(l) \qquad \text{[cathode; reduction]}$$

Determining the anode product (more easily oxidized anion): Br is below Cl in Group 7A(17). EN decreases down the group, so Br has a lower EN than Cl. Therefore, it follows that Br^- holds its e^- less tightly than Cl^-, so Br^- is more easily oxidized (stronger reducing agent):

$$2Br^-(l) \longrightarrow Br_2(g) + 2e^- \qquad \text{[anode; oxidation]}$$

Writing the overall cell reaction:

$$Mg^{2+}(l) + 2Br^-(l) \longrightarrow Mg(l) + Br_2(g) \qquad \text{[overall]}$$

Comment The cell temperature must be high enough to keep the salt mixture molten. In this case, the temperature is greater than the melting point of Mg, so it appears as a liquid in the equation, and greater than the boiling point of Br_2, so it appears as a gas.

FOLLOW-UP PROBLEM 21.8 A sample of $AlBr_3$ contaminated with KF is melted and electrolyzed. Determine the electrode products and the overall cell reaction.

Electrolysis of Water and Nonstandard Half-Cell Potentials Before we can analyze the electrolysis products of aqueous salt solutions, we must examine the electrolysis of water itself. Extremely pure water is difficult to electrolyze because very few ions are present to conduct a current. If a small amount of a nonreacting salt (such as Na_2SO_4) is added, however, electrolysis proceeds rapidly. A glass electrolytic cell with separated gas compartments is used to keep the H_2 and O_2 gases from mixing (Figure 21.19). At the anode, water is oxidized as the O.N. of O changes from -2 to 0:

$$2H_2O(l) \longrightarrow O_2(g) + 4H^+(aq) + 4e^- \qquad E = 0.82 \text{ V} \qquad \text{[anode; oxidation]}$$

At the cathode, water is reduced as the O.N. of H changes from $+1$ to 0:

$$2H_2O(l) + 2e^- \longrightarrow H_2(g) + 2OH^-(aq) \qquad E = -0.42 \text{ V} \qquad \text{[cathode; reduction]}$$

After doubling the cathode half-reaction to equate e^- loss and gain, adding the half-reactions (which involves combining the H^+ and OH^- into H_2O and canceling e^- and excess H_2O), and calculating E_{cell}, the overall reaction is

$$2H_2O(l) \longrightarrow 2H_2(g) + O_2(g) \qquad E_{cell} = -0.42 \text{ V} - 0.82 \text{ V} = -1.24 \text{ V} \qquad \text{[overall]}$$

Oxidation half-reaction
$2H_2O(l) \longrightarrow O_2(g) + 4H^+(aq) + 4e^-$

Reduction half-reaction
$2H_2O(l) + 2e^- \longrightarrow H_2(g) + 2OH^-(aq)$

Overall (cell) reaction
$2H_2O(l) \longrightarrow 2H_2(g) + O_2(g)$

Figure 21.19 The electrolysis of water. Oxygen forms through oxidation of H_2O at the anode (*right*), and twice its volume of hydrogen forms through reduction of H_2O at the cathode (*left*).

Notice that these electrode potentials are not written with a superscript zero because they are *not* standard electrode potentials. The $[H^+]$ and $[OH^-]$ are 1.0×10^{-7} *M* rather than the standard-state value of 1 *M*. These *E* values are obtained by applying the Nernst equation. For example, the calculation for the anode potential (with $n = 4$) is

$$E_{cell} = E^0_{cell} - \frac{0.0592\ V}{4} \log (P_{O_2} \times [H^+]^4)$$

The standard potential for the *oxidation* of water is -1.23 V (from Appendix D) and $P_{O_2} \approx 1$ atm in the half-cell, so we have

$$E_{cell} = -1.23\ V - \left\{ \frac{0.0592\ V}{4} \times [\log 1 + 4 \log (1.0\times10^{-7})] \right\} = -0.82\ V$$

In aqueous ionic solutions, $[H^+]$ and $[OH^-]$ are approximately 10^{-7} *M* also, so we use these nonstandard E_{cell} values to predict electrode products.

Electrolysis of Aqueous Ionic Solutions and the Phenomenon of Overvoltage

Aqueous salt solutions are mixtures of ions *and* water, so we have to compare the various electrode potentials to predict the electrode products. When two half-reactions are possible at an electrode,

• *The reduction with the less negative (more positive) electrode potential occurs.*
• *The oxidation with the less positive (more negative) electrode potential occurs.*

What happens, for instance, when a solution of potassium iodide is electrolyzed? The possible oxidizing agents are K^+ and H_2O, and their reduction half-reactions are

$$K^+(aq) + e^- \longrightarrow K(s) \qquad\qquad E^0 = -2.93\ V$$
$$2H_2O(l) + 2e^- \longrightarrow H_2(g) + 2OH^-(aq) \qquad E = -0.42\ V \qquad \text{[reduction]}$$

The less *negative* electrode potential for water means that it is much easier to reduce than K^+, so H_2 forms at the cathode. The possible reducing agents are I^- and H_2O, and their oxidation half-reactions are

$$2I^-(aq) \longrightarrow I_2(s) + 2e^- \qquad\qquad E^0 = 0.53\ V \qquad \text{[oxidation]}$$
$$2H_2O(l) \longrightarrow O_2(g) + 4H^+(aq) + 4e^- \qquad E = 0.82\ V$$

The less *positive* electrode potential for I^- means that a lower potential is needed to oxidize it than to oxidize H_2O, so I_2 forms at the anode.

However, the products predicted from this type of comparison of electrode potentials are not always the actual products. For gases such as $H_2(g)$ and $O_2(g)$ to be produced at metal electrodes, an additional voltage is required. This increment over the expected voltage is called the **overvoltage,** and it is 0.4 to 0.6 V for these gases. The overvoltage results from kinetic factors, such as the large activation energy (Section 16.6) required for gases to form at the electrode.

Overvoltage has major practical significance. A multibillion-dollar example is the industrial production of chlorine from concentrated NaCl solution. Water is easier to reduce than Na^+, so H_2 forms at the cathode even with an overvoltage of 0.6 V:

$$Na^+(aq) + e^- \longrightarrow Na(s) \qquad\qquad E^0 = -2.71\ V$$
$$2H_2O(l) + 2e^- \longrightarrow H_2(g) + 2OH^-(aq) \qquad E = -0.42\ V\ (\approx -1\ V\ \text{with overvoltage})$$
$$\text{[reduction]}$$

But Cl_2 forms at the anode, even though the electrode potentials themselves would lead us to predict that O_2 should form:

$$2H_2O(l) \longrightarrow O_2(g) + 4H^+(aq) + 4e^- \qquad E = 0.82\ V\ (\sim 1.4\ V\ \text{with overvoltage})$$
$$2Cl^-(aq) \longrightarrow Cl_2(g) + 2e^- \qquad\qquad E^0 = 1.36\ V \qquad \text{[oxidation]}$$

An overvoltage of ~0.6 V makes the potential needed to form O_2 slightly above that for Cl_2. Keeping the $[Cl^-]$ high also favors Cl_2 formation. Thus, as you'll see in Chapter 22, chlorine, which is one of the 10 most heavily produced industrial chemicals, can be formed from plentiful natural sources of aqueous sodium chloride.

From these and other examples, we can determine which elements can be prepared electrolytically from aqueous solutions of their salts:

1. Cations of less active metals *are* reduced to the metal, including gold, silver, copper, chromium, platinum, and cadmium.
2. Cations of more active metals *are not* reduced, including those in Groups 1A(1) and 2A(2), and Al from 3A(13). Water is reduced to H_2 and OH^- instead.
3. Anions that *are* oxidized, because of overvoltage from O_2 formation, include the halides ($[Cl^-]$ must be high), except for F^-.
4. Anions that *are not* oxidized include F^- and common oxoanions, such as SO_4^{2-}, CO_3^{2-}, NO_3^-, and PO_4^{3-}, because the central nonmetal in these oxoanions is already in its highest oxidation state. Water is oxidized to O_2 and H^+ instead.

SAMPLE PROBLEM 21.9 Predicting the Electrolysis Products of Aqueous Ionic Solutions

Problem What products form during electrolysis of aqueous solutions of the following salts: **(a)** KBr; **(b)** $AgNO_3$; **(c)** $MgSO_4$?

Plan We identify the reacting ions and compare their electrode potentials with those of water, taking the 0.4 to 0.6 V overvoltage into consideration. The reduction half-reaction with the less negative electrode potential, and the oxidation half-reaction with the less positive electrode potential occurs at that electrode.

Solution

(a)
$$K^+(aq) + e^- \longrightarrow K(s) \qquad\qquad E^0 = -2.93 \text{ V}$$
$$2H_2O(l) + 2e^- \longrightarrow H_2(g) + 2OH^-(aq) \qquad E = -0.42 \text{ V}$$

Despite the overvoltage, which makes E for the reduction of water between -0.8 and -1.0 V, H_2O is still easier to reduce than K^+, so $H_2(g)$ forms at the cathode.

$$2Br^-(aq) \longrightarrow Br_2(l) + 2e^- \qquad\qquad E^0 = 1.07 \text{ V}$$
$$2H_2O(l) \longrightarrow O_2(g) + 4H^+(aq) + 4e^- \qquad E = 0.82 \text{ V}$$

Because of the overvoltage, which makes E for the oxidation of water between 1.2 and 1.4 V, Br^- is easier to oxidize than water, so $Br_2(l)$ forms at the anode (see photo).

(b)
$$Ag^+(aq) + e^- \longrightarrow Ag(s) \qquad\qquad E^0 = 0.80 \text{ V}$$
$$2H_2O(l) + 2e^- \longrightarrow H_2(g) + 2OH^-(aq) \qquad E = -0.42 \text{ V}$$

As the cation of an inactive metal, Ag^+ is a better oxidizing agent than H_2O, so Ag forms at the cathode. NO_3^- cannot be oxidized, because N is already in its highest (+5) oxidation state. Thus, O_2 forms at the anode:

$$2H_2O(l) \longrightarrow O_2(g) + 4H^+(aq) + 4e^-$$

(c)
$$Mg^{2+}(aq) + 2e^- \longrightarrow Mg(s) \qquad E^0 = -2.37 \text{ V}$$

Like K^+ in part (a), Mg^{2+} cannot be reduced in the presence of water, so H_2 forms at the cathode. The SO_4^{2-} ion cannot be oxidized because S is in its highest (+6) oxidation state. Thus, H_2O is oxidized, and O_2 forms at the anode:

$$2H_2O(l) \longrightarrow O_2(g) + 4H^+(aq) + 4e^-$$

FOLLOW-UP PROBLEM 21.9 Write half-reactions for the products you predict will form in the electrolysis of aqueous $AuBr_3$.

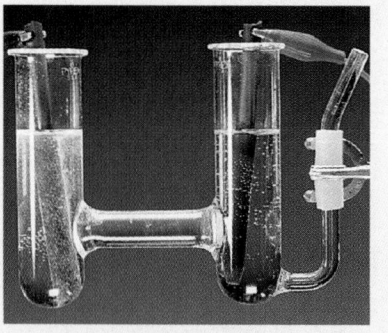

Electrolysis of aqueous KBr.

The Stoichiometry of Electrolysis: The Relation Between Amounts of Charge and Product

As you've seen, the charge flowing through an electrolytic cell yields products at the electrodes. In the electrolysis of molten NaCl, for example, the power source supplies electrons to the cathode, where Na^+ ions migrate to pick them up and become Na metal. At the same time, the power source pulls from the anode the electrons that Cl^- ions release as they become Cl_2 gas. It follows that the more electrons picked up by Na^+ ions and released by Cl^- ions, the greater the amounts of Na and Cl_2 that form. This relationship was first determined experimentally by Michael Faraday and is referred to as *Faraday's law of electrolysis: the amount of substance produced at each electrode is directly proportional to the quantity of charge flowing through the cell.*

Each balanced half-reaction shows the amounts (mol) of reactant, electrons, and product involved in the change, so it contains the information we need to answer such questions as "How much material will form as a result of a given quantity of charge?" or, conversely, "How much charge is needed to produce a given amount of material?" To apply Faraday's law,

1. Balance the half-reaction to find the number of moles of electrons needed per mole of product.
2. Use the Faraday constant ($F = 9.65 \times 10^4$ C/mol e^-) to find the corresponding charge.
3. Use the molar mass to find the charge needed for a given mass of product.

In practice, to supply the correct amount of electricity, we need some means of finding the charge flowing through the cell. We cannot measure charge directly, but we *can* measure current, the charge flowing per unit time. The SI unit of current is the **ampere (A),** which is defined as 1 coulomb flowing through a conductor in 1 second:

$$1 \text{ ampere} = 1 \text{ coulomb/second} \quad \text{or} \quad 1 \text{ A} = 1 \text{ C/s} \qquad \textbf{(21.11)}$$

Thus, the current multiplied by the time gives the charge:

$$\text{Current} \times \text{time} = \text{charge} \quad \text{or} \quad \text{A} \times \text{s} = \frac{\text{C}}{\text{s}} \times \text{s} = \text{C}$$

Therefore, we find the charge by measuring the current *and* the time during which the current flows. This, in turn, relates to the amount of product formed. Figure 21.20 summarizes these relationships.

Problems based on Faraday's law often ask you to calculate current, mass of material, or time. The electrode half-reaction provides the key because it is related to the mass for a certain quantity of charge. Here is a typical problem in practical electrolysis: How long does it take to produce 3.0 g of $Cl_2(g)$ from the electrolysis of aqueous NaCl using a power supply with a current of 12 A? The problem asks for the time needed to produce a certain mass, so let's first relate mass to number of moles of electrons to find the charge needed and then relate charge to current to find the time.

The Father of Electrochemistry and Much More The investigations by Michael Faraday (1791–1867) into the mass of an element that is equivalent to a given amount of charge established electrochemistry as a quantitative science, but his breakthroughs in physics are even more celebrated. On Christmas Day of 1821, Faraday developed the precursor of the electric motor, and his later studies of currents induced by electric and magnetic fields eventually led to the development of the electric generator and the transformer. After his death, this self-educated blacksmith's son was rated by Albert Einstein as the peer of Newton, Galileo, and Maxwell. He is shown here delivering one of his famous lectures on the "Chemical History of a Candle."

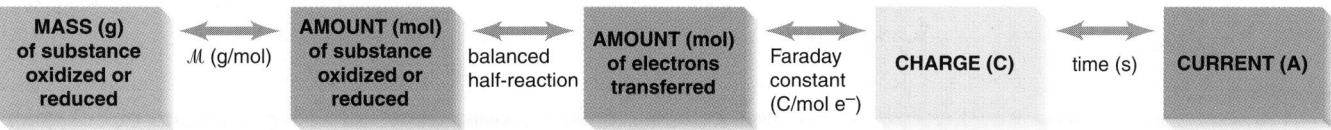

Figure 21.20 A summary diagram for the stoichiometry of electrolysis.

We know the mass of Cl_2 produced, so we can find the amount (mol) of Cl_2. The half-reaction tells us that the loss of 2 mol of electrons produces 1 mol of chlorine gas:

$$2Cl^-(aq) \longrightarrow Cl_2(g) + 2e^-$$

We use this relationship as a conversion factor, and multiplying by the Faraday constant gives us the total charge:

$$\text{Charge (C)} = 3.0 \text{ g } Cl_2 \times \frac{1 \text{ mol } Cl_2}{70.90 \text{ g } Cl_2} \times \frac{2 \text{ mol } e^-}{1 \text{ mol } Cl_2} \times \frac{9.65\times10^4 \text{ C}}{1 \text{ mol } e^-} = 8.2\times10^3 \text{ C}$$

Now we use the relationship between charge and current to find the time needed:

$$\text{Time (s)} = \frac{\text{charge (C)}}{\text{current (A, or C/s)}} = 8.2\times10^3 \text{ C} \times \frac{1 \text{ s}}{12 \text{ C}} = 6.8\times10^2 \text{ s } (\sim11 \text{ min})$$

Note that the entire calculation follows Figure 21.20 until the last step:

grams of Cl_2 $\Rightarrow$ moles of Cl_2 $\Rightarrow$ moles of e^- $\Rightarrow$ coulombs $\Rightarrow$ seconds

Sample Problem 21.10 demonstrates the steps as they appear in Figure 21.20.

SAMPLE PROBLEM 21.10 Applying the Relationship Among Current, Time, and Amount of Substance

Problem A technician is plating a faucet with 0.86 g of Cr from an electrolytic bath containing aqueous $Cr_2(SO_4)_3$. If 12.5 min is allowed for the plating, what current is needed?
Plan To find the current, we divide the charge by the time; so we need to find the charge. First we write the half-reaction for Cr^{3+} reduction. From it, we know the number of moles of e^- required per mole of Cr. As the roadmap shows, to find the charge, we convert the mass of Cr needed (0.86 g) to amount (mol) of Cr. The balanced half-reaction gives the amount (mol) of e^- transferred. Then, we use the Faraday constant (9.65×10^4 C/mol e^-) to find the charge and divide by the time (12.5 min, converted to s) to obtain the current. 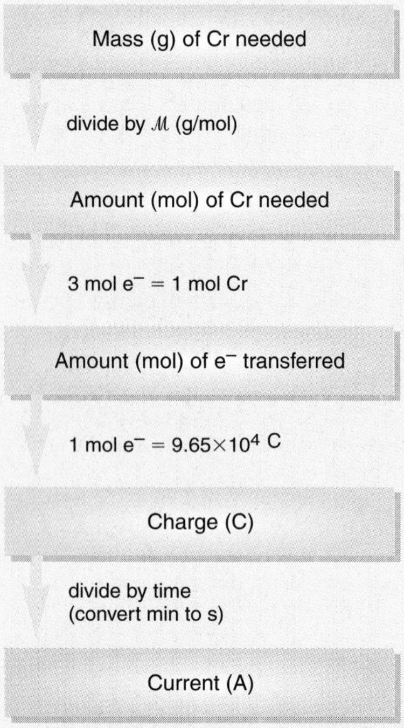
Solution Writing the balanced half-reaction:

$$Cr^{3+}(aq) + 3e^- \longrightarrow Cr(s)$$

Combining steps to find amount (mol) of e^- transferred for mass of Cr needed:

$$\text{Moles of } e^- \text{ transferred} = 0.86 \text{ g Cr} \times \frac{1 \text{ mol Cr}}{52.00 \text{ g Cr}} \times \frac{3 \text{ mol } e^-}{1 \text{ mol Cr}} = 0.050 \text{ mol } e^-$$

Calculating the charge:

$$\text{Charge (C)} = 0.050 \text{ mol } e^- \times \frac{9.65\times10^4 \text{ C}}{1 \text{ mol } e^-} = 4.8\times10^3 \text{ C}$$

Calculating the current:

$$\text{Current (A)} = \frac{\text{charge (C)}}{\text{time (s)}} = \frac{4.8\times10^3 \text{ C}}{12.5 \text{ min}} \times \frac{1 \text{ min}}{60 \text{ s}} = 6.4 \text{ C/s} = \boxed{6.4 \text{ A}}$$

Check Rounding gives

$$(\sim0.9 \text{ g})(1 \text{ mol Cr}/50 \text{ g})(3 \text{ mol } e^-/1 \text{ mol Cr}) = 5\times10^{-2} \text{ mol } e^-$$

then

$$(5\times10^{-2} \text{ mol } e^-)(\sim1\times10^5 \text{ C/mol } e^-) = 5\times10^3 \text{ C}$$

and

$$(5\times10^3 \text{ C}/12 \text{ min})(1 \text{ min}/60 \text{ s}) = 7 \text{ A}$$

Comment For the sake of introducing Faraday's law, the details of the electroplating process have been simplified here. Actually, electroplating chromium is only 30% to 40% efficient and must be run at a particular temperature range for the plate to appear bright. Nearly 10,000 metric tons (2×10^8 mol) of chromium are used annually for electroplating.

FOLLOW-UP PROBLEM 21.10 Using a current of 4.75 A, how many minutes does it take to plate 1.50 g of Cu onto a sculpture from a $CuSO_4$ solution?

Roadmap (left margin):

Mass (g) of Cr needed

divide by $\mathcal{M}$ (g/mol)

Amount (mol) of Cr needed

3 mol e^- = 1 mol Cr

Amount (mol) of e^- transferred

1 mol e^- = 9.65×10^4 C

Charge (C)

divide by time (convert min to s)

Current (A)

The following Chemical Connections essay links several themes of this chapter in the setting of a living cell.

Chemical Connections Chemistry in Biological Energetics
Cellular Electrochemistry and the Production of ATP

Biological cells apply the principles of electrochemical cells to generate energy. The complex multistep process can be divided into two parts:

1. Bond energy in food is used to generate an electrochemical potential.
2. The potential is used to create the bond energy of the high-energy molecule adenosine triphosphate (ATP; see Chemical Connections, Section 20.3).

The redox species that accomplish these steps are part of the *electron-transport chain* (ETC), which lies on the inner membranes of *mitochondria,* the subcellular particles that produce the cell's energy (Figure B21.1).

The ETC is a series of large molecules (mostly proteins), each of which contains a *redox couple* (the oxidized and reduced forms of a species), such as Fe^{3+}/Fe^{2+}, that passes electrons down the chain. At three points along the chain, large potential differences supply enough free energy to convert adenosine diphosphate (ADP) into ATP.

Bond Energy to Electrochemical Potential

Cells utilize the energy in food by releasing it in controlled steps rather than all at once. The reaction that ultimately powers the ETC is the oxidation of hydrogen to form water:

$$H_2 + \tfrac{1}{2}O_2 \longrightarrow H_2O$$

However, instead of H_2 gas, which does not occur in organisms, the hydrogen takes the form of two H^+ ions and two e^-. A bio-logical oxidizing agent called NAD^+ (*n*icotinamide *a*denine *d*inucleotide) acquires these protons and electrons in the process of oxidizing the molecules in food. To show this process, we use the following half-reaction (without canceling the H^+ on both sides):

$$NAD^+(aq) + 2H^+(aq) + 2e^- \longrightarrow NADH(aq) + H^+(aq)$$

At the mitochondrial inner membrane, the NADH and H^+ transfer the two e^- to the first redox couple of the ETC and release the two H^+. The electrons are transported down the chain of redox couples, where they finally reduce O_2 to H_2O. The overall process, with standard electrode potentials,* is

$$NADH(aq) + H^+(aq) \longrightarrow NAD^+(aq) + 2H^+(aq) + 2e^-$$
$$E^{0'} = -0.315 \text{ V}$$
$$\tfrac{1}{2}O_2(aq) + 2e^- + 2H^+(aq) \longrightarrow H_2O(l) \qquad E^{0'} = 0.815 \text{ V}$$

$$NADH(aq) + H^+(aq) + \tfrac{1}{2}O_2(aq) \longrightarrow NAD^+(aq) + H_2O(l)$$
$$E^{0'}_{\text{overall}} = 0.815 \text{ V} - (-0.315 \text{ V}) = 1.130 \text{ V}$$

Thus, for each mole of NADH that enters the ETC, the free-energy equivalent of 1.14 V is available:

$$\Delta G^{0'} = -nFE^{0'}$$
$$= -(2 \text{ mol e}^-/\text{mol NADH})(96.5 \text{ kJ/V·mol e}^-)(1.130 \text{ V})$$
$$= -218 \text{ kJ/mol NADH}$$

(continued)

---

*In biological systems, standard potentials are designated $E^{0'}$ and the standard states include a pH of 7.0 ($[H^+] = 1 \times 10^{-7}$ *M*).

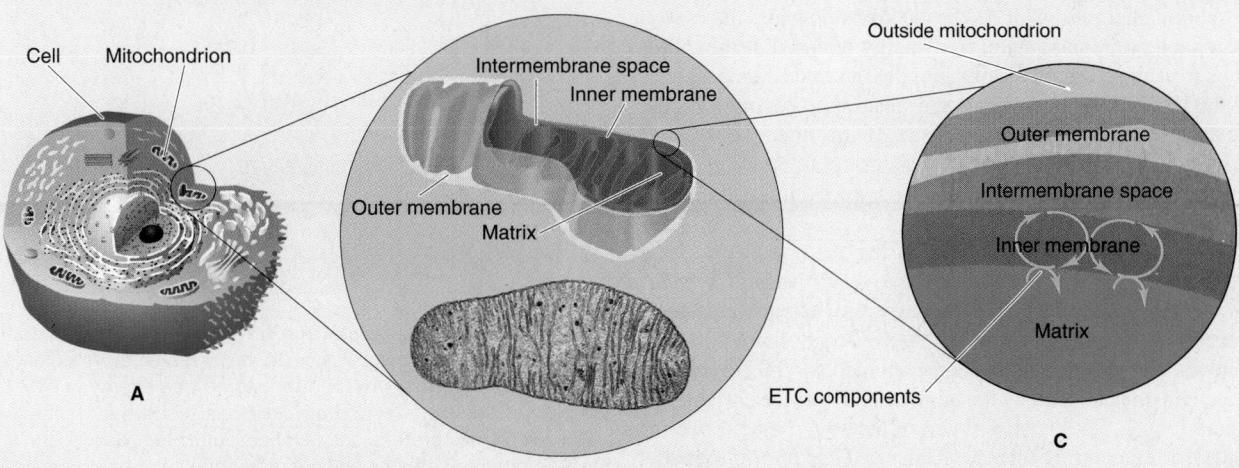

Figure B21.1 The mitochondrion. A, Mitochondria are subcellular particles outside the cell nucleus. **B,** They have a smooth outer membrane and a highly folded inner membrane, shown schematically and in an electron micrograph. **C,** The components of the electron-transport chain are attached to the inner membrane.

Note that this aspect of the process functions like a voltaic cell: a spontaneous reaction, the reduction of O_2 to H_2O, is used to generate a potential. In contrast to a laboratory voltaic cell, in which the overall process occurs in one step, this process occurs in many small steps. Figure B21.2 is a greatly simplified diagram of the three key steps in the ETC that generate the high potential to produce ATP. In the cell, each of these steps is part of a complex consisting of several components, most of which are proteins. Electrons are passed from one redox couple to the next down the chain, such that the reduced form of the first couple reduces the oxidized form of the second, and so forth. Most of the ETC components are iron-containing proteins, and the redox change involves the oxidation of Fe^{2+} to Fe^{3+} in one component caused by the reduction of Fe^{3+} to Fe^{2+} in another:

$$Fe^{2+} \text{ (in A)} + Fe^{3+} \text{ (in B)} \longrightarrow Fe^{3+} \text{ (in A)} + Fe^{2+} \text{ (in B)}$$

In other words, *metal ions within the ETC proteins are the actual species undergoing the redox reactions.*

Electrochemical Potential to Bond Energy

At the three points shown in Figure B21.2, the large potential difference is used to form ATP:

$$ADP^{3-}(aq) + HPO_4^{2-}(aq) + H^+(aq) \longrightarrow ATP^{4-}(aq) + H_2O(l)$$
$$\Delta G^{0'} = 30.5 \text{ kJ/mol}$$

Note that the free energy that is *released* at each of the three ATP-producing points exceeds 30.5 kJ, the free energy that must be *absorbed* to form ATP. Thus, just as in an electrolytic cell, an electrochemical potential is supplied to drive a nonspontaneous reaction.

So far, we've followed the flow of *electrons* through the members of the ETC, but where have the released *protons* gone? The answer is the key to *how* electrochemical potential is converted to bond energy in ATP. As the electrons flow and the redox couples change oxidation state, free energy released at the three key steps is used to force H^+ ions into the intermembrane space, so that the $[H^+]$ of the intermembrane space soon becomes higher than that of the matrix (Figure B21.3). In other words, the electrolytic-cell portion functions by using the free energy supplied by the three steps to *create an H^+ concentration cell across the membrane.*

When the $[H^+]$ difference across the membrane reaches about 2.5-fold, it triggers the membrane to let H^+ ions flow back through spontaneously (in effect, closing the switch and allowing the concentration cell to operate). The free energy released in this spontaneous process drives the nonspontaneous ATP formation via a mechanism catalyzed by the enzyme ATP synthase. (Paul D. Boyer and John E. Walker shared part of the 1997 Nobel Prize in chemistry for elucidating this mechanism.) Thus, the mitochondrion uses the "electron-motive force" of redox couples on the membrane to generate a "proton-motive force" across the membrane, which converts a potential difference to bond energy.

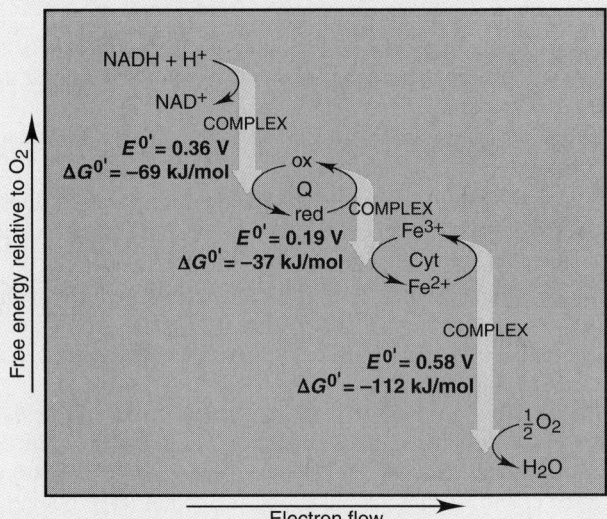

Figure B21.2 The main energy-yielding steps in the electron-transport chain (ETC). The electron carriers in the ETC undergo oxidation and reduction as they pass electrons to one another along the chain. At the three points shown, the difference in potential $E^{0'}$ (or free energy, $\Delta G^{0'}$) is large enough to be used for ATP production. (A complex consists of many components, mostly proteins; Q is a large organic molecule; and Cyt is the abbreviation for a cytochrome, a protein that contains a metal-ion redox couple, such as Fe^{3+}/Fe^{2+}, as the electron carrier.)

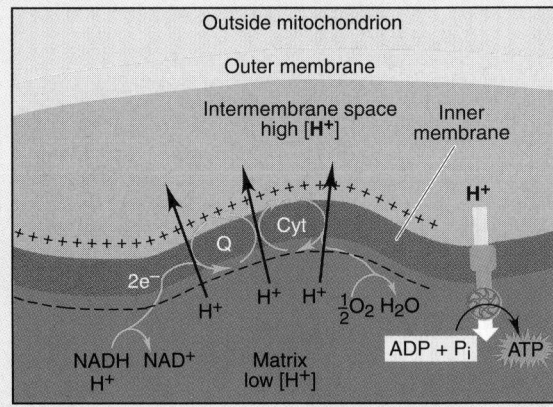

Figure B21.3 Coupling electron transport to proton transport to ATP synthesis. The purpose of the ETC is to convert the free energy released from food molecules into the stored free energy of ATP. It accomplishes this by transporting electrons along the chain *(curved yellow line)*, while protons are pumped out of the mitochondrial matrix. This pumping creates an $[H^+]$ difference and generates a potential across the inner membrane (in effect, a concentration cell). When this potential reaches a "trigger" value, H^+ flows back into the inner space, and the free energy released drives the formation of ATP.

SECTION SUMMARY

Electrolytic cells use electrical energy to drive a nonspontaneous reaction. Oxidation occurs at the anode and reduction at the cathode, but the direction of electron flow and the charges of the electrodes are opposite those in voltaic cells. When two products can form at each electrode, the more easily oxidized substance reacts at the anode and the more easily reduced at the cathode; that is, the half-reaction with the less negative (more positive) E^0 occurs. The reduction or oxidation of water takes place at nonstandard conditions. Overvoltage causes the actual voltage to be unexpectedly high and can affect the electrode product that forms. The amount of product that forms depends on the quantity of charge flowing through the cell, which is related to the magnitude of the current and the time it flows. Cellular redox systems combine aspects of voltaic, concentration, and electrolytic cells to convert bond energy in food into electrochemical potential and then into the bond energy of ATP.

Chapter Perspective

The field of electrochemistry is one of the many areas in which the principles of thermodynamics lead to practical benefits. As you've seen, electrochemical cells can use a reaction to generate energy or use energy to drive a reaction. Such processes are central not only to our mobile way of life, but also to our biological existence. In Chapter 22, we examine the electrochemical (and other) methods used by industry to convert raw natural resources into some of the materials modern society finds indispensable.

For Review and Reference (Numbers in parentheses refer to pages, unless noted otherwise.)

Learning Objectives

Relevant section and/or sample problem (SP) numbers appear in parentheses.

Understand These Concepts

1. The meanings of oxidation and reduction; why an oxidizing agent is reduced and a reducing agent is oxidized (Section 21.1; also Section 4.5)
2. How the half-reaction method is used to balance redox reactions in acidic or basic solution (Section 21.1)
3. The distinction between voltaic and electrolytic cells in terms of the sign of ΔG (Section 21.1)
4. How voltaic cells use a spontaneous reaction to release electrical energy (Section 21.2)
5. The physical makeup of a voltaic cell: arrangement and composition of half-cells, relative charges of electrodes, and purpose of a salt bridge (Section 21.2)
6. How the difference in reducing strength of the electrodes determines the direction of electron flow (Section 21.2)
7. How a positive E_{cell} corresponds to a spontaneous cell reaction (Section 21.3)
8. The usefulness and significance of standard electrode potentials ($E^0_{half-cell}$) (Section 21.3)
9. How $E^0_{half-cell}$ values are combined to give E^0_{cell} (Section 21.3)
10. How the standard reference electrode is used to find an unknown $E^0_{half-cell}$ (Section 21.3)
11. How an emf series (e.g., Table 21.2 and Appendix D) is used to write spontaneous redox reactions (Section 21.3)
12. How the relative reactivity of a metal is determined by its reducing power and is related to the negative of its $E^0_{half-cell}$ (Section 21.3)

13. How E_{cell} (the nonstandard cell potential) is related to ΔG (maximum work) and the charge (moles of electrons times the faraday) flowing through the cell (Section 21.4)
14. The interrelationship of ΔG^0, E^0_{cell}, and K (Section 21.4)
15. How E_{cell} changes as the cell operates (Q changes) (Section 21.4)
16. Why a voltaic cell can do work until $Q = K$ (Section 21.4)
17. How a concentration cell does work until the half-cell concentrations are equal (Section 21.4)
18. The distinction between primary (nonrechargeable) and secondary (rechargeable) batteries (Section 21.5)
19. How corrosion occurs and is prevented; the similarities between a corroding metal and a voltaic cell (Section 21.6)
20. How electrolytic cells use a nonspontaneous redox reaction that is driven by an external source of electricity (Section 21.7)
21. How atomic properties (ionization energy and electronegativity) determine the products of the electrolysis of molten salt mixtures (Section 21.7)
22. How the electrolysis of water influences the products of aqueous electrolysis; the importance of overvoltage (Section 21.7)
23. The relationship between the quantity of charge flowing through the cell and the amount of product formed (Section 21.7)

Learning Objectives *(continued)*

Master These Skills

1. Balancing redox reactions by the half-reaction method (Section 21.1 and SP 21.1)
2. Diagramming and notating a voltaic cell (Section 21.2 and SP 21.2)
3. Combining $E^0_{half-cell}$ values to obtain E^0_{cell} (Section 21.3)
4. Using E^0_{cell} and known $E^0_{half-cell}$ to find an unknown $E^0_{half-cell}$ (SP 21.3)
5. Manipulating half-reactions to write a spontaneous redox reaction and calculate its E^0_{cell} (SP 21.4)
6. Ranking the relative strengths of oxidizing and reducing agents in a redox reaction (SP 21.4)

7. Predicting whether a metal can displace hydrogen or another metal from solution (Section 21.3)
8. Using the interrelationship of ΔG^0, E^0_{cell}, and K to calculate one of the three given the other two (Section 21.4 and SP 21.5)
9. Using the Nernst equation to calculate the nonstandard cell potential (E_{cell}) (SP 21.6)
10. Calculating E_{cell} of a concentration cell (SP 21.7)
11. Predicting the products of the electrolysis of molten salts (SP 21.8)
12. Predicting the products of the electrolysis of aqueous salts (SP 21.9)
13. Calculating the current (or time) needed to produce a given amount of product, and vice versa (SP 21.10)

Key Terms

electrochemistry (893)
electrochemical cell (893)

Section 21.1

half-reaction method (894)
voltaic (galvanic) cell (898)
electrolytic cell (898)
electrode (898)
electrolyte (898)
anode (898)
cathode (898)

Section 21.2

half-cell (901)
salt bridge (902)

Section 21.3

cell potential (E_{cell}) (905)
voltage (905)
electromotive force (emf) (905)
volt (V) (905)
coulomb (C) (905)

standard cell potential (E^0_{cell}) (906)
standard electrode (half-cell) potential ($E^0_{half-cell}$) (906)
standard reference half-cell (standard hydrogen electrode) (906)

Section 21.4

Faraday constant (F) (914)
Nernst equation (917)
concentration cell (919)

Section 21.5

battery (922)
fuel cell (922)

Section 21.6

corrosion (926)

Section 21.7

electrolysis (931)
overvoltage (933)
ampere (A) (935)

Key Equations and Relationships

21.1 Relating a spontaneous process to the sign of the cell potential (905):

$$E_{cell} > 0 \text{ for a spontaneous process}$$

21.2 Relating electric potential to energy and charge in SI units (905):

$$\text{Potential} = \text{energy/charge} \quad \text{or} \quad 1\text{ V} = 1\text{ J/C}$$

21.3 Relating standard cell potential to standard electrode potentials in a voltaic cell (906):

$$E^0_{cell} = E^0_{cathode} - E^0_{anode}$$

21.4 Defining the Faraday constant (914):

$$F = 9.65\times10^4 \frac{\text{J}}{\text{V·mol e}^-} \quad (3\text{ sf})$$

21.5 Relating the free energy change to electrical work and cell potential (914):

$$\Delta G = w_{max} = -nFE_{cell}$$

21.6 Finding the standard free energy change from the standard cell potential (914):

$$\Delta G^0 = -nFE^0_{cell}$$

21.7 Finding the equilibrium constant from the standard cell potential (915):

$$E^0_{cell} = \frac{RT}{nF} \ln K$$

21.8 Substituting known values of R, F, and T into Equation 21.7 and converting to common logarithms (915):

$$E^0_{cell} = \frac{0.0592\text{ V}}{n} \log K \quad \text{or} \quad \log K = \frac{nE^0_{cell}}{0.0592\text{ V}} \quad (\text{at } 25°\text{C})$$

21.9 Calculating the nonstandard cell potential (Nernst equation) (917):

$$E_{cell} = E^0_{cell} - \frac{RT}{nF} \ln Q$$

21.10 Substituting known values of R, F, and T into the Nernst equation and converting to common logarithms (917):

$$E_{cell} = E^0_{cell} - \frac{0.0592\text{ V}}{n} \log Q \quad (\text{at } 25°\text{C})$$

21.11 Relating current to charge and time (935):

$$\text{Current} = \text{charge/time} \quad \text{or} \quad 1\text{ A} = 1\text{ C/s}$$

Highlighted Figures and Tables

These figures (F) and tables (T) provide a quick review of key ideas. Entries in color contain frequently used data.

F21.1 A summary of redox terminology (894)
F21.3 Voltaic and electrolytic cells (899)
F21.5 A voltaic cell based on the zinc-copper reaction (901)
F21.9 The interrelationship of ΔG^0, E^0, and K (915)
F21.10 E_{cell} and log Q for the zinc-copper cell (918)

F21.13 The corrosion of iron (926)
F21.17 Tin-copper reaction in voltaic and electrolytic cells (929)
T21.4 Comparison of voltaic and electrolytic cells (931)
F21.20 Summary of the stoichiometry of electrolysis (935)

Brief Solutions to Follow-up Problems

21.1 $6KMnO_4(aq) + 6KOH(aq) + KI(aq) \longrightarrow$
$$6K_2MnO_4(aq) + KIO_3(aq) + 3H_2O(l)$$

21.2 $Sn(s) \longrightarrow Sn^{2+}(aq) + 2e^-$ [anode; oxidation]
$6e^- + 14H^+(aq) + Cr_2O_7^{2-}(aq) \longrightarrow 2Cr^{3+}(aq) + 7H_2O(l)$
                                 [cathode; reduction]

$3Sn(s) + Cr_2O_7^{2-}(aq) + 14H^+(aq) \longrightarrow$
$$3Sn^{2+}(aq) + 2Cr^{3+}(aq) + 7H_2O(l) \quad \text{[overall]}$$
Cell notation:
$Sn(s) \mid Sn^{2+}(aq) \parallel H^+(aq), Cr_2O_7^{2-}(aq), Cr^{3+}(aq) \mid$ graphite

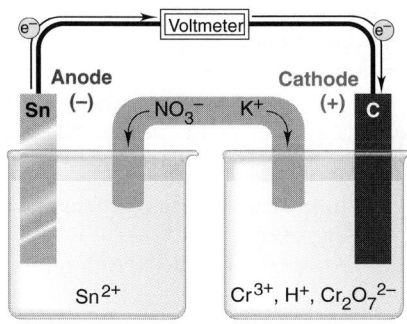

21.3 $Br_2(aq) + 2e^- \longrightarrow 2Br^-(aq)$ $E^0_{bromine} = 1.07$ V
                                       [cathode]
$2V^{3+}(aq) + 2H_2O(l) \longrightarrow 2VO^{2+}(aq) + 4H^+(aq) + 2e^-$
                         $E^0_{vanadium} = ?$ [anode]
$E^0_{vanadium} = E^0_{bromine} - E^0_{cell} = 1.07$ V $- 1.39$ V $= -0.32$ V

21.4 $Fe^{2+}(aq) + 2e^- \longrightarrow Fe(s)$ $E^0 = -0.44$ V
$2[Fe^{2+}(aq) \longrightarrow Fe^{3+}(aq) + e^-]$ $E^0 = 0.77$ V

$3Fe^{2+}(aq) \longrightarrow 2Fe^{3+}(aq) + Fe(s)$

    $E^0_{cell} = -0.44$ V $- 0.77$ V $= -1.21$ V
The reaction is nonspontaneous. The spontaneous reaction is
$2Fe^{3+}(aq) + Fe(s) \longrightarrow 3Fe^{2+}(aq)$ $E^0_{cell} = 1.21$ V
$Fe > Fe^{2+} > Fe^{3+}$

21.5 $Cd(s) + Cu^{2+}(aq) \longrightarrow Cd^{2+}(aq) + Cu(s)$
    $\Delta G^0 = -RT \ln K = -8.314$ J/mol·K $\times 298$ K $\times \ln K$
       $= -143$ kJ; $K = 1.2 \times 10^{25}$

$E^0_{cell} = \dfrac{0.0592 \text{ V}}{2} \log(1.2 \times 10^{25}) = 0.742$ V

21.6 $Fe(s) \longrightarrow Fe^{2+}(aq) + 2e^-$ $E^0 = -0.44$ V
$Cu^{2+}(aq) + 2e^- \longrightarrow Cu(s)$ $E^0 = 0.34$ V

$Fe(s) + Cu^{2+}(aq) \longrightarrow Fe^{2+}(aq) + Cu(s)$ $E^0_{cell} = 0.78$ V
So $E_{cell} = 0.78$ V $+ 0.25$ V $= 1.03$ V

1.03 V $= 0.78$ V $- \dfrac{0.0592 \text{ V}}{2} \log \dfrac{[Fe^{2+}]}{[Cu^{2+}]}$

$\dfrac{[Fe^{2+}]}{[Cu^{2+}]} = 3.6 \times 10^{-9}$

$[Fe^{2+}] = 3.6 \times 10^{-9} \times 0.30 \; M = 1.1 \times 10^{-9} \; M$

21.7 $Au^{3+}(aq, 2.5 \times 10^{-2} \; M)$ [B] $\longrightarrow$
                           $Au^{3+}(aq, 7.0 \times 10^{-4} \; M)$ [A]

$E_{cell} = 0$ V $- \left(\dfrac{0.0592 \text{ V}}{3} \times \log \dfrac{7.0 \times 10^{-4}}{2.5 \times 10^{-2}} \right) = 0.0306$ V

A is negative, so it is the anode.

21.8 Oxidizing agents: K^+ and Al^{3+}. Reducing agents: F^- and Br^-
Al is above and to the right of K, so it has a higher IE:
$Al^{3+}(l) + 3e^- \longrightarrow Al(s)$ [cathode; reduction]
Br is below F, so it has a lower EN:
$2Br^-(l) \longrightarrow Br_2(g) + 2e^-$ [anode; oxidation]
$2Al^{3+}(l) + 6Br^-(l) \longrightarrow 2Al(s) + 3Br_2(g)$ [overall]

21.9 The reduction with the more positive electrode potential is
$Au^{3+}(aq) + 3e^- \longrightarrow Au(s); E^0 = 1.50$ V
                                   [cathode; reduction]
Because of overvoltage, O_2 will not form at the anode, so Br_2 will form:
$2Br^-(aq) \longrightarrow Br_2(l) + 2e^-; E^0 = 1.07$ V
                                   [cathode; oxidation]

21.10 $Cu^{2+}(aq) + 2e^- \longrightarrow Cu(s)$; therefore,
2 mol $e^-/1$ mol $Cu = 2$ mol $e^-/63.55$ g Cu

Time (min) $= 1.50$ g $Cu \times \dfrac{2 \text{ mol } e^-}{63.55 \text{ g } Cu}$

$\times \dfrac{9.65 \times 10^4 \text{ C}}{1 \text{ mol } e^-} \times \dfrac{1 \text{ s}}{4.75 \text{ C}} \times \dfrac{1 \text{ min}}{60 \text{ s}} = 16.0$ min

Problems

Problems with colored numbers are answered at the back of the text. Sections match the text and provide the number(s) of relevant sample problems. Most offer Concept Review Questions, Skill-Building Exercises (in similar pairs), and Problems in Context. Then Comprehensive Problems, based on material from any section or previous chapter, follow.

Note: Unless stated otherwise, all problems refer to systems at 298 K (25°C).

Half-Reactions and Electrochemical Cells
(Sample Problem 21.1)

● **Concept Review Questions**

21.1 Define *oxidation* and *reduction* in terms of electron transfer and change in oxidation number.

21.2 Why must an electrochemical process involve a redox reaction?

21.3 Can one half-reaction in a redox process take place independently of the other? Explain.

21.4 Water is used to balance O atoms in the half-reaction method. Why can't O^{2-} ions be used instead?

21.5 During the balancing process, what step is taken to ensure that e^- loss equals e^- gain?

21.6 How are protons removed when balancing a redox reaction in basic solution?

21.7 Are spectator ions used to balance the half-reactions of a redox reaction? At what stage might spectator ions enter the balancing process?

21.8 Which type of electrochemical cell has a $\Delta G_{sys} < 0$? Which type has an increase in the free energy of the cell?

21.9 Which statements are true? Correct any that are false.
(a) In a voltaic cell, the anode is negative relative to the cathode.
(b) Oxidation occurs at the anode of either type of cell.
(c) Electrons flow into the cathode of an electrolytic cell.
(d) In a voltaic cell, the surroundings do work on the system.
(e) If a metal is plated out of an electrolytic cell, it appears on the cathode.
(f) The cell electrolyte provides a solution of mobile electrons.

● **Skill-Building Exercises (*paired*)**

21.10 Consider the following balanced redox reaction:
$$16H^+(aq) + 2MnO_4^-(aq) + 10Cl^-(aq) \longrightarrow$$
$$2Mn^{2+}(aq) + 5Cl_2(g) + 8H_2O(l)$$
(a) Which species is being oxidized?
(b) Which species is being reduced?
(c) Which species is the oxidizing agent?
(d) Which species is the reducing agent?
(e) From which species to which does electron transfer occur?
(f) Write the balanced molecular equation, with K^+ and SO_4^{2-} as the spectator ions.

21.11 Consider the following balanced redox reaction:
$$2CrO_2^-(aq) + 2H_2O(l) + 6ClO^-(aq) \longrightarrow$$
$$2CrO_4^{2-}(aq) + 3Cl_2(g) + 4OH^-(aq)$$
(a) Which species is being oxidized?
(b) Which species is being reduced?
(c) Which species is the oxidizing agent?
(d) Which species is the reducing agent?

(e) From which species to which does electron transfer occur?
(f) Write the balanced molecular equation, with Na^+ as the spectator ion.

21.12 Balance the following skeleton reactions and identify the oxidizing and reducing agents:
(a) $ClO_3^-(aq) + I^-(aq) \longrightarrow I_2(s) + Cl^-(aq)$ [acidic]
(b) $MnO_4^-(aq) + SO_3^{2-}(aq) \longrightarrow$
$$MnO_2(s) + SO_4^{2-}(aq)$$ [basic]
(c) $MnO_4^-(aq) + H_2O_2(aq) \longrightarrow Mn^{2+}(aq) + O_2(g)$ [acidic]

21.13 Balance the following skeleton reactions and identify the oxidizing and reducing agents:
(a) $O_2(g) + NO(g) \longrightarrow NO_3^-(aq)$ [acidic]
(b) $CrO_4^{2-}(aq) + Cu(s) \longrightarrow$
$$Cr(OH)_3(s) + Cu(OH)_2(s)$$ [basic]
(c) $AsO_4^{3-}(aq) + NO_2^-(aq) \longrightarrow$
$$AsO_2^-(aq) + NO_3^-(aq)$$ [basic]

21.14 Balance the following skeleton reactions and identify the oxidizing and reducing agents:
(a) $Cr_2O_7^{2-}(aq) + Zn(s) \longrightarrow Zn^{2+}(aq) + Cr^{3+}(aq)$ [acidic]
(b) $Fe(OH)_2(s) + MnO_4^-(aq) \longrightarrow$
$$MnO_2(s) + Fe(OH)_3(s)$$ [basic]
(c) $Zn(s) + NO_3^-(aq) \longrightarrow Zn^{2+}(aq) + N_2(g)$ [acidic]

21.15 Balance the following skeleton reactions and identify the oxidizing and reducing agents:
(a) $BH_4^-(aq) + ClO_3^-(aq) \longrightarrow$
$$H_2BO_3^-(aq) + Cl^-(aq)$$ [basic]
(b) $CrO_4^{2-}(aq) + N_2O(g) \longrightarrow Cr^{3+}(aq) + NO(g)$ [acidic]
(c) $Br_2(l) \longrightarrow BrO_3^-(aq) + Br^-(aq)$ [basic]

21.16 Balance the following skeleton reactions and identify the oxidizing and reducing agents:
(a) $Sb(s) + NO_3^-(aq) \longrightarrow Sb_4O_6(s) + NO(g)$ [acidic]
(b) $Mn^{2+}(aq) + BiO_3^-(aq) \longrightarrow$
$$MnO_4^-(aq) + Bi^{3+}(aq)$$ [acidic]
(c) $Fe(OH)_2(s) + Pb(OH)_3^-(aq) \longrightarrow$
$$Fe(OH)_3(s) + Pb(s)$$ [basic]

21.17 Balance the following skeleton reactions and identify the oxidizing and reducing agents:
(a) $NO_2(g) \longrightarrow NO_3^-(aq) + NO_2^-(aq)$ [basic]
(b) $Zn(s) + NO_3^-(aq) \longrightarrow Zn(OH)_4^{2-}(aq) + NH_3(g)$ [basic]
(c) $H_2S(g) + NO_3^-(aq) \longrightarrow S_8(s) + NO(g)$ [acidic]

21.18 Balance the following skeleton reactions and identify the oxidizing and reducing agents:
(a) $As_4O_6(s) + MnO_4^-(aq) \longrightarrow$
$$AsO_4^{3-}(aq) + Mn^{2+}(aq)$$ [acidic]
(b) $P_4(s) \longrightarrow HPO_3^{2-}(aq) + PH_3(g)$ [acidic]
(c) $MnO_4^-(aq) + CN^-(aq) \longrightarrow$
$$MnO_2(s) + CNO^-(aq)$$ [basic]

21.19 Balance the following skeleton reactions and identify the oxidizing and reducing agents:
(a) $SO_3^{2-}(aq) + Cl_2(g) \longrightarrow SO_4^{2-}(aq) + Cl^-(aq)$ [basic]
(b) $Fe(CN)_6^{3-}(aq) + Re(s) \longrightarrow$
$$Fe(CN)_6^{4-}(aq) + ReO_4^-(aq)$$ [basic]
(c) $MnO_4^-(aq) + HCOOH(aq) \longrightarrow$
$$Mn^{2+}(aq) + CO_2(g)$$ [acidic]

● **Problems in Context**

21.20 In many residential water systems, the aqueous Fe^{3+} concentration is high enough to stain sinks and turn drinking water light brown. The analysis of iron content occurs by first reducing the Fe^{3+} to Fe^{2+} and then titrating with MnO_4^- in acidic solution. Balance the skeleton reaction of the titration step:

$$Fe^{2+}(aq) + MnO_4^-(aq) \longrightarrow Mn^{2+}(aq) + Fe^{3+}(aq)$$

21.21 *Aqua regia,* a mixture of concentrated HNO_3 and HCl, was developed by alchemists as a means to "dissolve" gold. The process is actually a redox reaction with the following simplified skeleton reaction:

$$Au(s) + NO_3^-(aq) + Cl^-(aq) \longrightarrow AuCl_4^-(aq) + NO_2(g)$$

(a) Balance the reaction by the half-reaction method.
(b) What are the oxidizing and reducing agents?
(c) What is the function of HCl in aqua regia?

Voltaic Cells: Using Spontaneous Reactions to Generate Electrical Energy
(Sample Problem 21.2)

● **Concept Review Questions**

21.22 Consider the following general voltaic cell:

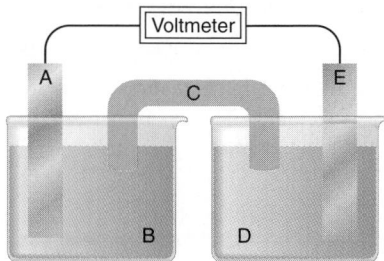

Identify the (a) anode; (b) cathode; (c) salt bridge; (d) electrode at which e^- leave the cell; (e) electrode with a positive charge; (f) electrode that gains mass as the cell operates (assuming that a metal plates out).

21.23 Why does a voltaic cell not operate to any significant extent unless the two compartments are connected through an external circuit?

21.24 What purpose does the salt bridge serve in a voltaic cell and how does it accomplish this purpose?

21.25 What is the difference between an active and an inactive electrode? Why are inactive electrodes used? Name two substances commonly used for inactive electrodes.

21.26 When a piece of metal A is placed in a solution of metal B ions, metal B plates out on the piece of A.
(a) Which metal is being oxidized?
(b) Which metal is being "displaced"?
(c) Which metal would you use as the anode in a voltaic cell incorporating these two metals?
(d) If bubbles of H_2 form when B is placed in acid, will they form if A is placed in acid? Explain.

● **Skill-Building Exercises** *(paired)*

21.27 A voltaic cell is constructed with an Sn/Sn^{2+} half-cell and a Zn/Zn^{2+} half-cell. Measurement shows the zinc electrode is negative.
(a) Write balanced half-reactions and the overall spontaneous reaction.

(b) Diagram the cell, labeling electrodes with their charges and showing the directions of electron flow in the circuit and of cation and anion flow in the salt bridge.

21.28 A voltaic cell is constructed with an Ag/Ag^+ half-cell and a Pb/Pb^{2+} half-cell. Measurement shows that the silver electrode is positive.
(a) Write balanced half-reactions and the overall spontaneous reaction.
(b) Diagram the cell, labeling electrodes with their charges and showing the directions of electron flow in the circuit and of cation and anion flow in the salt bridge.

21.29 Consider the following voltaic cell:

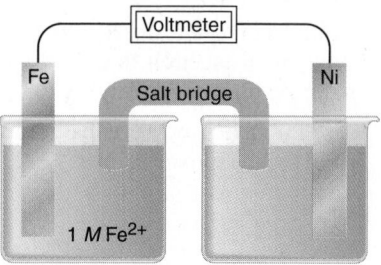

(a) In which direction do electrons flow through the external circuit?
(b) In which half-cell does oxidation occur?
(c) In which half-cell do electrons enter the cell?
(d) At which electrode are electrons consumed?
(e) Which electrode is negatively charged?
(f) Which electrode decreases in mass during operation of the cell?
(g) Suggest a solution for the cathode electrolyte.
(h) Suggest a pair of ions for the salt bridge.
(i) For which electrode could you use an inactive material?
(j) In which direction do anions within the salt bridge move to maintain charge neutrality?
(k) Write balanced half-reactions and an overall cell reaction.

21.30 Consider the following voltaic cell:

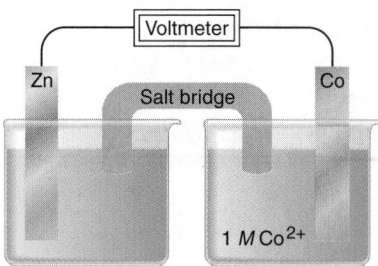

(a) Which direction do electrons flow in the external circuit?
(b) In which half-cell does reduction occur?
(c) In which half-cell do electrons leave the cell?
(d) At which electrode are electrons generated?
(e) Which electrode is positively charged?
(f) Which electrode increases in mass during operation of the cell?
(g) Suggest a solution for the anode electrolyte.
(h) Suggest a pair of ions for the salt bridge.
(i) For which electrode could you use an inactive material?
(j) In which direction do cations within the salt bridge move to maintain charge neutrality?
(k) Write balanced half-reactions and an overall cell reaction.

21.31 A voltaic cell is constructed with an Fe/Fe^{2+} half-cell and an Mn/Mn^{2+} half-cell. Measurement shows that the iron electrode is positive.
(a) Write balanced half-reactions and the overall spontaneous reaction.
(b) Diagram the cell, labeling electrodes with their charges and showing the directions of electron flow in the circuit and of cation and anion flow in the salt bridge.

21.32 A voltaic cell is constructed with a Cu/Cu^{2+} half-cell and an Ni/Ni^{2+} half-cell. Measurement shows that the nickel electrode is negative.
(a) Write balanced half-reactions and the overall spontaneous reaction.
(b) Diagram the cell, labeling electrodes with their charges and showing the directions of electron flow in the circuit and of cation and anion flow in the salt bridge.

21.33 Write the cell notation for the voltaic cell that incorporates each of the following redox reactions:
(a) $Al(s) + Cr^{3+}(aq) \longrightarrow Al^{3+}(aq) + Cr(s)$
(b) $Cu^{2+}(aq) + SO_2(g) + 2H_2O(l) \longrightarrow$
$$Cu(s) + SO_4^{2-}(aq) + 4H^+(aq)$$

21.34 Write a balanced equation from each cell notation:
(a) $Mn(s) \mid Mn^{2+}(aq) \parallel Cd^{2+}(aq) \mid Cd(s)$
(b) $Fe(s) \mid Fe^{2+}(aq) \parallel NO_3^-(aq) \mid NO(g) \mid Pt(s)$

Cell Potential: Output of a Voltaic Cell
(Sample Problems 21.3 and 21.4)

● **Concept Review Questions**

21.35 How is a standard reference electrode used to determine unknown $E^0_{half-cell}$ values?

21.36 What does a negative E^0_{cell} indicate about a redox reaction? What does it indicate about the reverse reaction?

21.37 The standard cell potential is a thermodynamic state function. How are E^0 values treated similarly to ΔH^0, ΔG^0, and S^0 values? How are they treated differently?

● **Skill-Building Exercises (paired)**

21.38 In basic solution, selenide and sulfite ions react spontaneously:
$$2Se^{2-}(aq) + 2SO_3^{2-}(aq) + 3H_2O(l) \longrightarrow$$
$$2Se(s) + 6OH^-(aq) + S_2O_3^{2-}(aq) \quad E^0_{cell} = 0.35 \text{ V}$$
(a) Write balanced half-reactions for the process.
(b) If $E^0_{sulfite}$ is -0.57 V, calculate $E^0_{selenium}$.

21.39 In acidic solution, ozone and manganese(II) ion react spontaneously:
$$O_3(g) + Mn^{2+}(aq) + H_2O(l) \longrightarrow$$
$$O_2(g) + MnO_2(s) + 2H^+(aq) \quad E^0_{cell} = 0.84 \text{ V}$$
(a) Write the balanced half-reactions.
(b) Using Appendix D to find E^0_{ozone}, calculate $E^0_{manganese}$.

21.40 Use the emf series (Appendix D) to arrange the species
(a) In order of *decreasing* strength as *oxidizing* agents: Fe^{3+}, Br_2, Cu^{2+}
(b) In order of *increasing* strength as *oxidizing* agents: Ca^{2+}, $Cr_2O_7^{2-}$, Ag^+

21.41 Use the emf series (Appendix D) to arrange the species
(a) In order of *decreasing* strength as *reducing* agents: SO_2, $PbSO_4$, MnO_2

(b) In order of *increasing* strength as *reducing* agents: Hg, Fe, Sn

21.42 Balance each skeleton reaction, calculate E^0_{cell}, and state whether the reaction is spontaneous:
(a) $Co(s) + H^+(aq) \longrightarrow Co^{2+}(aq) + H_2(g)$
(b) $Mn^{2+}(aq) + Br_2(l) \longrightarrow MnO_4^-(aq) + Br^-(aq)$ [acidic]
(c) $Hg_2^{2+}(aq) \longrightarrow Hg^{2+}(aq) + Hg(l)$

21.43 Balance each skeleton reaction, calculate E^0_{cell}, and state whether the reaction is spontaneous:
(a) $Cl_2(g) + Fe^{2+}(aq) \longrightarrow Cl^-(aq) + Fe^{3+}(aq)$
(b) $Mn^{2+}(aq) + Co^{3+}(aq) \longrightarrow MnO_2(s) + Co^{2+}(aq)$ [acidic]
(c) $AgCl(s) + NO(g) \longrightarrow$
$$Ag(s) + Cl^-(aq) + NO_3^-(aq) \text{ [acidic]}$$

21.44 Balance each skeleton reaction, calculate E^0_{cell}, and state whether the reaction is spontaneous:
(a) $Ag(s) + Cu^{2+}(aq) \longrightarrow Ag^+(aq) + Cu(s)$
(b) $Cd(s) + Cr_2O_7^{2-}(aq) \longrightarrow Cd^{2+}(aq) + Cr^{3+}(aq)$
(c) $Ni^{2+}(aq) + Pb(s) \longrightarrow Ni(s) + Pb^{2+}(aq)$

21.45 Balance each skeleton reaction, calculate E^0_{cell}, and state whether the reaction is spontaneous:
(a) $Cu^+(aq) + PbO_2(s) + SO_4^{2-}(aq) \longrightarrow$
$$PbSO_4(s) + Cu^{2+}(aq) \text{ [acidic]}$$
(b) $H_2O_2(aq) + Ni^{2+}(aq) \longrightarrow O_2(g) + Ni(s)$ [acidic]
(c) $MnO_2(s) + Ag^+(aq) \longrightarrow MnO_4^-(aq) + Ag(s)$ [basic]

21.46 Use the following half-reactions to write three spontaneous reactions, calculate E^0_{cell} for each reaction, and rank the oxidizing and reducing agents:
(1) $Al^{3+}(aq) + 3e^- \longrightarrow Al(s) \quad E^0 = -1.66 \text{ V}$
(2) $N_2O_4(g) + 2e^- \longrightarrow 2NO_2^-(aq) \quad E^0 = 0.867 \text{ V}$
(3) $SO_4^{2-}(aq) + H_2O(l) + 2e^- \longrightarrow SO_3^{2-}(aq) + 2OH^-(aq)$
$$E^0 = 0.93 \text{ V}$$

21.47 Use the following half-reactions to write three spontaneous reactions, calculate E^0_{cell} for each reaction, and rank the oxidizing and reducing agents:
(1) $Au^+(aq) + e^- \longrightarrow Au(s) \quad E^0 = 1.69 \text{ V}$
(2) $N_2O(g) + 2H^+(aq) + 2e^- \longrightarrow N_2(g) + H_2O(l)$
$$E^0 = 1.77 \text{ V}$$
(3) $Cr^{3+}(aq) + 3e^- \longrightarrow Cr(s) \quad E^0 = -0.74 \text{ V}$

21.48 Use the following half-reactions to write three spontaneous reactions, calculate E^0_{cell} for each reaction, and rank the oxidizing and reducing agents:
(1) $2HClO(aq) + 2H^+(aq) + 2e^- \longrightarrow Cl_2(g) + 2H_2O(l)$
$$E^0 = 1.63 \text{ V}$$
(2) $Pt^{2+}(aq) + 2e^- \longrightarrow Pt(s) \quad E^0 = 1.20 \text{ V}$
(3) $PbSO_4(s) + 2e^- \longrightarrow Pb(s) + SO_4^{2-}(aq) \quad E^0 = -0.31 \text{ V}$

21.49 Use the following half-reactions to write three spontaneous reactions, calculate E^0_{cell} for each reaction, and rank the oxidizing and reducing agents:
(1) $I_2(s) + 2e^- \longrightarrow 2I^-(aq) \quad E^0 = 0.53 \text{ V}$
(2) $S_2O_8^{2-}(aq) + 2e^- \longrightarrow 2SO_4^{2-}(aq) \quad E^0 = 2.01 \text{ V}$
(3) $Cr_2O_7^{2-}(aq) + 14H^+(aq) + 6e^- \longrightarrow$
$$2Cr^{3+}(aq) + 7H_2O(l) \quad E^0 = 1.33 \text{ V}$$

● **Problems in Context**

21.50 When metal A is placed in a solution of metal B salt, the surface of metal A changes color. When metal B is placed in acid solution, gas bubbles form on the surface of the metal. When metal A is placed in a solution of metal C salt, no change is observed in the solution or on the metal A surface. Will metal C cause formation of H_2 when placed in acid solution? Rank metals A, B, and C in order of *decreasing* reducing strength.

21.51 When a clean iron nail is placed in an aqueous solution of copper(II) sulfate, the nail immediately begins to turn a brown-black color. In a few minutes, the nail is completely coated with a material of this color.
(a) What is the material coating the iron?
(b) What are the oxidizing and reducing agents?
(c) Can this reaction be made into a voltaic cell?
(d) Write the balanced equation for the reaction.
(e) Calculate E^0_{cell} for the process.

Free Energy and Electrical Work
(Sample Problems 21.5 to 21.7)

● **Concept Review Questions**

21.52 (a) How do the relative magnitudes of Q and K relate to the signs of ΔG and E_{cell}? Explain.
(b) Can a cell do work when $Q/K > 1$ or when $Q/K < 1$? Explain.
21.53 A voltaic cell consists of a metal A/A^+ electrode and metal B/B^+ electrode, with the A/A^+ electrode negative. The initial $[A^+]/[B^+]$ is such that $E_{cell} > E^0_{cell}$.
(a) How do $[A^+]$ and $[B^+]$ change as the cell operates?
(b) How does E_{cell} change as the cell operates?
(c) What is $[A^+]/[B^+]$ when $E_{cell} = E^0_{cell}$? Explain.
(d) Is it possible for E_{cell} to be less than E^0_{cell}? Explain.
21.54 Explain whether E_{cell} of a voltaic cell increases or decreases from each of the following changes:
(a) Decrease in cell temperature
(b) Increase in concentration of an active ion in the anode compartment
(c) Increase in concentration of an active ion in the cathode compartment
(d) Increase in pressure of a gaseous reactant in the cathode component
21.55 In a concentration cell, is the more concentrated electrolyte in the cathode or the anode compartment? Explain.

● **Skill-Building Exercises (paired)**

21.56 What is the value of the equilibrium constant at 25°C for the reaction between each pair?
(a) $Ni(s)$ and $Ag^+(aq)$ (b) $Fe(s)$ and $Cr^{3+}(aq)$
21.57 What is the value of the equilibrium constant at 25°C for the reaction between each pair?
(a) $Al(s)$ and $Cd^{2+}(aq)$ (b) $I_2(s)$ and $Br^-(aq)$

21.58 What is the value of the equilibrium constant at 25°C for the reaction between each pair?
(a) $Ag(s)$ and $Mn^{2+}(aq)$ (b) $Cl_2(g)$ and $Br^-(aq)$
21.59 What is the value of the equilibrium constant at 25°C for the reaction between each pair?
(a) $Cr(s)$ and $Cu^{2+}(aq)$ (b) $Sn(s)$ and $Pb^{2+}(aq)$

21.60 Calculate ΔG^0 for each of the reactions in Problem 21.56.
21.61 Calculate ΔG^0 for each of the reactions in Problem 21.57.

21.62 Calculate ΔG^0 for each of the reactions in Problem 21.58.
21.63 Calculate ΔG^0 for each of the reactions in Problem 21.59.

21.64 What are E^0_{cell} and ΔG^0 at 25°C of a redox reaction for which $n = 1$ and $K = 5.0 \times 10^3$?
21.65 What are E^0_{cell} and ΔG^0 at 25°C of a redox reaction for which $n = 1$ and $K = 5.0 \times 10^{-6}$?

21.66 What are E^0_{cell} and ΔG^0 at 25°C of a redox reaction for which $n = 2$ and $K = 75$?
21.67 What are E^0_{cell} and ΔG^0 at 25°C of a redox reaction for which $n = 2$ and $K = 0.075$?

21.68 A voltaic cell consists of a standard hydrogen electrode in one half-cell and a Cu/Cu^{2+} half-cell. Calculate $[Cu^{2+}]$ when E_{cell} is 0.25 V.
21.69 A voltaic cell consists of an Mn/Mn^{2+} half-cell and a Pb/Pb^{2+} half-cell. Calculate $[Pb^{2+}]$ when $[Mn^{2+}]$ is 1.3 M and E_{cell} is 0.42 V.

21.70 A voltaic cell consists of Ni/Ni^{2+} and Co/Co^{2+} half-cells with the following initial concentrations: $[Ni^{2+}] = 0.80\ M$; $[Co^{2+}] = 0.20\ M$.
(a) What is the initial E_{cell}?
(b) What is E_{cell} when $[Co^{2+}]$ reaches 0.45 M?
(c) What is $[Ni^{2+}]$ when E_{cell} reaches 0.025 V?
(d) What are the equilibrium concentrations of the ions?
21.71 A voltaic cell consists of Mn/Mn^{2+} and Cd/Cd^{2+} half-cells with the following initial concentrations: $[Mn^{2+}] = 0.090\ M$; $[Cd^{2+}] = 0.060\ M$.
(a) What is the initial E_{cell}?
(b) What is E_{cell} when $[Cd^{2+}]$ reaches 0.050 M?
(c) What is $[Mn^{2+}]$ when E_{cell} reaches 0.055 V?
(d) What are the equilibrium concentrations of the ions?

21.72 A concentration cell consists of two H_2/H^+ half-cells. Half-cell A has H_2 at 0.90 atm bubbling into 0.10 M HCl. Half-cell B has H_2 at 0.50 atm bubbling into 2.0 M HCl. Which half-cell houses the anode? What is the voltage of the cell?
21.73 A concentration cell consists of two Sn/Sn^{2+} half-cells. The electrolyte in compartment A is 0.13 M $Sn(NO_3)_2$. The electrolyte in B is 0.87 M $Sn(NO_3)_2$. Which half-cell houses the cathode? What is the voltage of the cell?

Electrochemical Processes in Batteries

● **Concept Review Questions**

21.74 What is the direction of electron flow with respect to the anode and the cathode in a battery? Explain.
21.75 In the everyday batteries used for flashlights, toys, etc., no salt bridge is evident. What is used in these cells to separate the anode and cathode compartments?
21.76 Both a D-sized and an AAA-sized alkaline battery have an output of 1.5 V. What property of the cell potential allows this to occur? What is different about these two batteries?

● **Problems in Context**

21.77 Many common devices require more than one battery for operation.
(a) How many alkaline batteries must be placed in series to light a flashlight with a 6.0-V bulb?
(b) What is the voltage requirement of a camera that uses six silver batteries?
(c) How many volts can a car battery deliver if two of its anode/cathode cells are shorted?

Corrosion: A Case of Environmental Electrochemistry

● **Concept Review Questions**

21.78 During the reconstruction of the Statue of Liberty, Teflon spacers were placed between the iron skeleton and the copper plates that cover the statue. What purpose do these spacers serve?

21.79 Why do steel bridge-supports rust at the waterline but not above or below it?

21.80 Since the 1930s, chromium has replaced nickel for corrosion resistance and appearance on car bumpers and trim. How does the chromium protect steel from corrosion?

21.81 Which of the following metals are suitable for use as sacrificial anodes to protect against the corrosion of underground iron pipes? If any are not suitable, explain why: (a) aluminum; (b) magnesium; (c) sodium; (d) lead; (e) nickel; (f) zinc; (g) chromium.

Electrolytic Cells: Using Electrical Energy to Drive a Nonspontaneous Reaction
(Sample Problems 21.8 to 21.10)

● **Concept Review Questions**

Note: Unless stated otherwise, assume that the electrolytic cells in the following problems operate at 100% efficiency.

21.82 Consider the following general electrolytic cell:

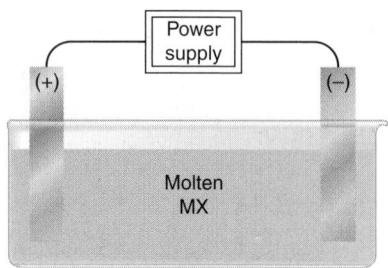

(a) At which electrode does oxidation occur?
(b) At which electrode does elemental M form?
(c) At which electrode are electrons being released by ions?
(d) At which electrode are electrons entering the cell?

21.83 A voltaic cell consists of Cr/Cr^{3+} and Cd/Cd^{2+} half-cells with all components in their standard states. After 10 minutes of operation, a thin coating of cadmium metal has plated out on the cathode. Describe what happens if you attach the negative terminal of a dry cell (1.5 V) to the cell cathode and the positive terminal to the cell anode.

21.84 Why are $E_{half-cell}$ values for the oxidation and reduction of water different from $E^0_{half-cell}$ values for the same processes?

21.85 In an aqueous electrolytic cell, nitrate ions never react at the anode, but nitrite ions do. Explain.

21.86 How does overvoltage influence the products in the electrolysis of aqueous salts?

● **Skill-Building Exercises** *(paired)*

21.87 In the electrolysis of molten NaBr,
(a) What product forms at the anode?
(b) What product forms at the cathode?

21.88 In the electrolysis of molten BaI_2,
(a) What product forms at the negative electrode?
(b) What product forms at the positive electrode?

21.89 In the electrolysis of a molten mixture of KI and MgF_2, identify the product that forms at the anode and at the cathode.

21.90 In the electrolysis of a molten mixture of CsBr and $SrCl_2$, identify the product that forms at the negative electrode and at the positive electrode.

21.91 In the electrolysis of a molten mixture of NaCl and $CaBr_2$, identify the product that forms at the anode and at the cathode.

21.92 In the electrolysis of a molten mixture of RbF and $CaCl_2$, identify the product that forms at the negative electrode and at the positive electrode.

21.93 Identify those elements that can be prepared by electrolysis of their aqueous salts: copper, barium, aluminum, bromine.

21.94 Identify those elements that can be prepared by electrolysis of their aqueous salts: strontium, gold, tin, chlorine.

21.95 Identify those elements that can be prepared by electrolysis of their aqueous salts: lithium, iodine, zinc, silver.

21.96 Identify those elements that can be prepared by electrolysis of their aqueous salts: fluorine, manganese, iron, cadmium.

21.97 What product forms at each electrode in the aqueous electrolysis of the following salts: (a) LiF; (b) $SnSO_4$?

21.98 What product forms at each electrode in the aqueous electrolysis of the following salts: (a) $ZnBr_2$; (b) $Cu(HCO_3)_2$?

21.99 What product forms at each electrode in the aqueous electrolysis of the following salts: (a) $Cr(NO_3)_3$; (b) $MnCl_2$?

21.100 What product forms at each electrode in the aqueous electrolysis of the following salts: (a) FeI_2; (b) K_3PO_4?

21.101 Electrolysis of molten $MgCl_2$ is the final production step in the isolation of magnesium from seawater by the Dow process (Section 22.4). Assuming that 35.6 g of Mg metal forms,
(a) How many moles of electrons are required?
(b) How many coulombs are required?
(c) How many amps are required to produce this amount in 2.50 h?

21.102 Electrolysis of molten NaCl in a Downs cell is the major isolation step in the production of sodium metal (Section 22.4). Assuming that 215 g of Na metal forms,
(a) How many moles of electrons are required?
(b) How many coulombs are required?
(c) How many amps are required to produce this amount in 9.50 h?

21.103 How many grams of radium can be formed by the passage of 215 C through an electrolytic cell containing a molten radium salt?

21.104 How many grams of aluminum can be formed by the passage of 305 C through an electrolytic cell containing a molten aluminum salt?

21.105 How many seconds does it take to deposit 85.5 g of Zn on a steel gate when 23.0 A is passed through a $ZnSO_4$ solution?

21.106 How many seconds does it take to deposit 1.63 g of Ni on a decorative drawer handle when 13.7 A is passed through a $Ni(NO_3)_2$ solution?

● **Problems in Context**

21.107 A professor adds Na_2SO_4 to water to facilitate its electrolysis in a lecture demonstration. (a) What is the purpose of the Na_2SO_4? (b) Why is the water electrolyzed instead of the salt?

21.108 Subterranean brines in parts of the United States are rich in iodides and bromides and serve as an industrial source of these elements. In one recovery method, the brines are evaporated to dryness and then melted and electrolyzed. Which halogen is more likely to form from this treatment? Why?

21.109 Zinc plating (galvanizing) is an important means of corrosion protection. Although the process is done customarily by dipping the object into molten zinc, the metal can also be electroplated from aqueous solutions. How many grams of zinc can be deposited on a steel tank from a $ZnSO_4$ solution when a 0.755-A current flows for 2.00 days?

Comprehensive Problems

Problems with an asterisk (*) are more challenging.

21.110 In a dry cell, MnO_2 occurs in a semisolid electrolyte paste and is reduced at the cathode. The MnO_2 used in dry cells can itself be produced by an electrochemical process of which one half-reaction is

$$Mn^{2+}(aq) + 2H_2O(l) \longrightarrow MnO_2(s) + 4H^+(aq) + 2e^-$$

If a current of 25.0 A is used, how many hours are needed to produce 1.00 kg of MnO_2? At which electrode is the MnO_2 formed?

21.111 Automobile manufacturers are currently developing cars that will use hydrogen as fuel. In Iceland, Sweden, and other parts of Scandinavia, where hydroelectric plants produce relatively inexpensive electric power, the hydrogen may be made industrially by the electrolysis of water.
(a) How many coulombs are needed to produce 2.5×10^6 L of H_2 gas at 10.0 atm pressure and 25°C? (Assume that the ideal gas law applies.)
(b) If the coulombs are supplied at 1.24 V, how many joules are produced?
(c) If the combustion of oil yields 4.0×10^4 kJ/kg, what mass of oil must be burned to yield the number of joules in part (b)?

21.112 The overall cell reaction occurring in an alkaline battery is

$$Zn(s) + MnO_2(s) + H_2O(l) \longrightarrow ZnO(s) + Mn(OH)_2(s)$$

(a) How many moles of electrons flow per mole of reaction?
(b) If 2.50 g of zinc is oxidized, how many grams of manganese dioxide and of water are consumed?
(c) What is the total mass of reactants consumed in part (b)?
(d) How many coulombs are produced in part (b)?
(e) In practice, voltaic cells of a given capacity (coulombs) are heavier than the calculation in part (c) indicates. Explain.

21.113 An inexpensive and accurate method of measuring the quantity of electricity passing through a circuit is to pass it through a solution of a metal ion and weigh the metal deposited. A silver electrode immersed in an Ag^+ solution weighs 1.7854 g before the current passes and weighs 1.8016 g after the current has passed. How many coulombs have passed?

***21.114** Brass, an alloy of copper and zinc, can be produced by simultaneously electroplating the two metals from a solution containing their 2+ ions. If exactly 70.0% of the total current is used to plate copper, while 30.0% goes to plating zinc, what is the mass percent of copper in the brass?

21.115 Compare and contrast a voltaic cell and an electrolytic cell with respect to each of the following:
(a) Sign of the free energy change
(b) Nature of the half-reaction at the anode
(c) Nature of the half-reaction at the cathode

(d) Charge on the electrode labeled "anode"
(e) Electrode from which electrons leave the cell

21.116 A thin disk earring 5.00 cm in diameter is plated with a coating of gold 0.20 mm thick from an Au^{3+} bath.
(a) How many days does it take to deposit the gold if the current used is 0.010 A (d of gold = 19.3 g/cm³)?
(b) If the price of gold is $320 per troy ounce (31.10 g), what is the cost of the gold plating?

21.117 (a) How many minutes does it take to form 10.0 L of O_2 measured at 99.8 kPa and 28°C from water if a current of 1.3 A passes through the electrolytic cell?
(b) What mass of H_2 forms at the same time?

***21.118** Trains powered by electricity, including subways, use direct current. One conductor is the overhead wire (or "third rail" for subways), and the other is the rails upon which the wheels run. The rails are on supports in contact with the ground. To minimize corrosion, should the overhead wire or the rails be connected to the positive terminal? Explain.

21.119 A simple laboratory test to identify the relative charge of the two leads from a DC power supply uses a piece of filter paper moistened with NaCl solution containing phenolphthalein indicator (colorless in acid, pink in base). The two wires from the power supply are touched to the paper about an inch apart. What do you expect to observe? What conclusion can you reach about the relative charges of the leads?

21.120 Like any piece of apparatus, an electrolytic cell operates at less than 100% efficiency. A cell depositing Cu from a Cu^{2+} bath operates for 10 h with an average current of 5.8 A. If 53.4 g of copper is deposited, at what efficiency is the cell operating?

21.121 Commercial electrolysis is performed on both molten NaCl and aqueous NaCl solutions. Identify the anode product, cathode product, species reduced, and species oxidized for the (a) molten electrolysis and (b) aqueous electrolysis.

***21.122** To examine the effect of ion removal on cell voltage, a chemist constructs two voltaic cells, each with a standard hydrogen electrode in one compartment. One cell also contains a Pb/Pb^{2+} half-cell; the other contains a Cu/Cu^{2+} half-cell.
(a) What is E^0 of each cell at 298 K?
(b) Which electrode in each cell is negative?
(c) When Na_2S solution is added to the Pb^{2+} electrolyte, solid PbS forms. What happens to the cell voltage?
(d) When sufficient Na_2S is added to the Cu^{2+} electrolyte, CuS forms and $[Cu^{2+}]$ is lowered to 1×10^{-16} M. What is the cell voltage?

21.123 Electrodes used in electrocardiography are disposable, and many incorporate silver. The metal is deposited in a thin layer on a small plastic "button," and then some is converted to AgCl:

$$Ag(s) + Cl^-(aq) \rightleftharpoons AgCl(s) + e^-$$

(a) If the surface area of the button is 2.0 cm² and the thickness of the silver layer is 7.0×10^{-6} m, calculate the volume of Ag used in one electrode, in cm³.
(b) The density of silver metal is 10.5 g/cm³. How many grams of silver are used per electrode?
(c) If the silver is electroplated on the button from a solution of Ag^+ with a current of 10.0 mA, how many minutes does the electroplating take?
(d) If bulk silver costs $5.50 per troy ounce (31.10 g), what is the cost (in cents) of the silver in one disposable electrode?

21.124 The commercial production of aluminum is done by the electrolysis of a bath containing Al_2O_3 dissolved in molten Na_3AlF_6. Why isn't it done by electrolysis of an aqueous $AlCl_3$ solution?

*__21.125__ Comparing the standard electrode potentials (E^0) of the Group 1A(1) metals Li, Na, and K with the negative of their first ionization energies reveals a discrepancy:

Ionization process reversed: $M^+(g) + e^- \rightleftharpoons M(g)$ $(-IE)$

Electrode reaction: $M^+(aq) + e^- \rightleftharpoons M(s)$ (E^0)

Metal	$-IE$ (kJ/mol)	E^0 (V)
Li	−520	−3.05
Na	−496	−2.71
K	−419	−2.93

Note that the changes in electrode potentials do not follow the decrease in ionization energies down the group. You might expect that if it is more difficult to remove an electron from an atom to form a gaseous ion (larger IE), then it would be less difficult to add an electron to an aqueous ion to form an atom (smaller E^0), yet $Li^+(aq)$ is *more* difficult to reduce than $Na^+(aq)$. Applying Hess's law, use an approach similar to that for a Born-Haber cycle to break down the process occurring at the electrode into three steps and label the energy involved in each step. How can you account for the discrepancy?

21.126 To improve conductivity in the electroplating of automobile bumpers, a thin coating of copper separates the steel from a heavy coating of chromium.
(a) What mass of Cu is deposited on an automobile trim piece if plating continues for 1.25 h at a current of 5.0 A?
(b) If the area of the trim piece is 50.0 cm^2, what is the thickness of the Cu coating (d of Cu = 8.95 g/cm^3)?

21.127 In Appendix D, standard electrode potentials range from about +3 to −3 V. Thus, by using a half-cell from each end of this range, it might seem possible to construct a cell with a voltage of approximately 6 V. However, most commercial aqueous voltaic cells have E^0 values of 1.5 to 2 V. Why are there no aqueous cells with significantly higher potentials?

21.128 Tin is used to coat "tin" cans used for food storage. If the tin is scratched and the iron of the can exposed, will the iron corrode more or less rapidly than if the tin were not present? On the inside of the can, the tin coating is itself coated with a transparent varnish. Explain.

*__21.129__ Commercial electrolytic cells for producing aluminum operate at 5.0 V and 100,000 A.
(a) How long does it take to produce exactly 1 metric ton (1000 kg) of aluminum?
(b) How much electrical power (in kilowatt-hours, kW·h) is used [1 W = 1 J/s; 1 kW·h = 3.6×10^3 kJ]?
(c) Assuming that electricity costs 0.90 cents per kW·h and cell efficiency is 90.%, what is the cost of producing exactly 1 lb of aluminum?

21.130 Magnesium bars are connected electrically to underground iron pipes to serve as sacrificial anodes.
(a) Do electrons flow from the Mg to the pipe or the reverse?
(b) A 12-kg Mg bar is attached to an iron pipe, and it takes 8.5 yr

for the Mg to be consumed. What is the average current flowing between the Mg and the Fe during this period?

21.131 Bubbles of H_2 form when metal D is placed in hot H_2O. No reaction occurs when D is placed in a solution of a salt of metal E, but D is discolored and coated immediately when placed in a solution of a salt of metal F. What happens if E is placed in a solution of a salt of metal F? Rank metals D, E, and F in order of *increasing* reducing strength.

21.132 Calcium is obtained industrially by electrolysis of molten $CaCl_2$ and used in aluminum alloys. How many coulombs are needed to produce 10.0 g of Ca metal? If the cell runs at 15 A, how many minutes will it take to produce the 10.0 g of Ca(s)?

21.133 In addition to reacting with gold (see Problem 21.21), aqua regia is used to bring other precious metals into solution. Balance the skeleton reaction for the reaction with Pt:

$$Pt(s) + NO_3^-(aq) + Cl^-(aq) \longrightarrow PtCl_6^{2-}(aq) + NO(g)$$

*__21.134__ The following three reactions are all used in voltaic cells or batteries:

I $2H_2(g) + O_2(g) \longrightarrow 2H_2O(l)$ $E_{cell} = 1.23$ V
II $Pb(s) + PbO_2(s) + 2H_2SO_4(aq) \longrightarrow$
                $2PbSO_4(s) + 2H_2O(l)$ $E_{cell} = 2.04$ V
III $2Na(l) + FeCl_2(s) \longrightarrow 2NaCl(s) + Fe(s)$
                                $E_{cell} = 2.35$ V

Reaction I is used in fuel cells, II is used in the automobile lead-acid battery, and III is used in an experimental high-temperature battery for powering electric vehicles. In many applications, the aim is to obtain as much work as possible from a cell, while keeping its weight to a minimum.
(a) Deduce the number of moles of electrons being transferred and calculate ΔG in each cell.
(b) Calculate the ratio, in kJ/g, of w_{max} to mass of reactants for each of the cells. Which has the highest ratio, which the lowest, and why? (*Note:* For simplicity, ignore the masses of cell components that do not appear in the cell as reactants, including electrode materials, electrolytes, separators, cell casing, wiring, etc.)

21.135 A current is applied to two electrolytic cells connected in series. In the first cell, silver is deposited. In the second cell, a zinc electrode is consumed. How much Ag is plated out if 1.2 g of Zn dissolves?

21.136 You are working in a laboratory and investigating a particular chemical reaction. State all the types of data available in standard tables that enable you to calculate the equilibrium constant for the reaction at 298 K.

*__21.137__ The maintenance personnel at an electric power plant must pay careful attention to corrosion of the turbine blades. Why does iron corrode faster in steam and hot water than in cold water?

21.138 A voltaic cell using Cu/Cu^{2+} and Sn/Sn^{2+} half-cells is set up at standard conditions, and each compartment has a volume of 245 mL. The cell delivers 0.15 A for 54.0 h.
(a) How many grams of Cu(s) are deposited?
(b) What is the [Cu^{2+}] remaining?

21.139 The $E_{half-cell}$ for the reduction of water is very different from the $E^0_{half-cell}$ value. Calculate the $E_{half-cell}$ value when H_2 is in its standard state.

*21.140 From the skeleton equations below, create a list of balanced half-reactions that has the strongest oxidizing agent on top and the weakest on the bottom:

$$U^{3+}(aq) + Cr^{3+}(aq) \longrightarrow Cr^{2+}(aq) + U^{4+}(aq)$$
$$Fe(s) + Sn^{2+}(aq) \longrightarrow Sn(s) + Fe^{2+}(aq)$$
$$Fe(s) + U^{4+}(aq) \longrightarrow \text{no reaction}$$
$$Cr^{3+}(aq) + Fe(s) \longrightarrow Cr^{2+}(aq) + Fe^{2+}(aq)$$
$$Cr^{2+}(aq) + Sn^{2+}(aq) \longrightarrow Sn(s) + Cr^{3+}(aq)$$

21.141 In the process of electrorefining (Chapter 22), impure copper is purified by using it as the anode and already purified copper as the cathode of an electrolytic cell.
(a) Write the equation for the reaction at the anode.
(b) Write the equation for the reaction at the cathode.
(c) Assuming 100% yield, how much electricity (in C) is needed to produce 1 metric ton of pure copper?
(d) Judging from its position in the periodic table, what other metals might be present in the impure copper?
(e) Why wouldn't these other metals be electrolyzed and contaminate the purified copper?

21.142 Use the half-reaction method to balance the reaction for the conversion of ethanol to acetic acid:

$$CH_3CH_2OH + Cr_2O_7^{2-} \longrightarrow$$
$$CH_3COOH + Cr^{3+} \text{ [acid solution]}$$

21.143 When zinc is refined by electrolysis, the desired half-reaction at the cathode is

$$Zn^{2+}(aq) + 2e^- \longrightarrow Zn(s)$$

A competing reaction, which lowers the yield, is the formation of hydrogen gas:

$$2H^+(aq) + 2e^- \longrightarrow H_2(g)$$

If 91.50% of the current flowing results in zinc being deposited, while 8.50% produces hydrogen gas, how many liters of H_2, measured at STP, form per kilogram of zinc?

*21.144 In an experiment to determine the effect of ions on the solubility of a precipitate, a chemist designs an ion-specific probe for measuring $[Ag^+]$ in an NaCl solution saturated with AgCl. One half-cell is an Ag-wire electrode immersed in the unknown AgCl-saturated NaCl solution. It is connected through a salt bridge to a calomel reference electrode, a platinum wire immersed in a paste of mercury and calomel (Hg_2Cl_2) in a saturated KCl solution. The measured E_{cell} is 0.060 V.
(a) Given the following standard half-reactions, calculate $[Ag^+]$.
Calomel: $Hg_2Cl_2(s) + 2e^- \longrightarrow$
$$2Hg(l) + 2Cl^-(aq) \qquad E^0 = 0.24 \text{ V}$$
Silver: $Ag^+(aq) + e^- \longrightarrow Ag(s) \qquad E^0 = 0.80 \text{ V}$
(*Hint:* Assume that $[Cl^-]$ is so high it is essentially constant.)
(b) A mining engineer sends a silver ore sample to the chemist for analysis with the Ag^+-selective probe. After pretreating the

ore sample, the chemist measures the cell voltage as 0.57 V. What is $[Ag^+]$?
*21.145 Use Appendix D to calculate the K_{sp} of AgCl.
*21.146 Black-and-white photographic film is coated with silver halides. Because the silver is expensive and may harm the environment, the film manufacturer monitors the Ag^+ content of the waste stream, $[Ag^+]_{waste}$, from the plant with an Ag^+-selective electrode kept at 25°C. A stream of known Ag^+ concentration, $[Ag^+]_{standard}$, is regularly passed over the electrode in turn with the waste stream, and the data are recorded by a computer.
(a) Write the equations relating the nonstandard cell potential to the standard cell potential and $[Ag^+]$ for each solution.
(b) Combine these into a single equation to find $[Ag^+]_{waste}$.
(c) Rewrite the equation from part (b) to find $[Ag^+]_{waste}$ in ng/L.
(d) If E_{waste} is 0.003 V higher than $E_{standard}$, and the standard solution contains 1000. ng/L, what is $[Ag^+]_{waste}$?
(e) Rewrite the equation from part (b) to find $[Ag^+]_{waste}$ for a system in which T changes and in which T_{waste} and $T_{standard}$ may be different.
*21.147 Calculate the K_f of $Ag(NH_3)_2^+$ from

$$Ag^+(aq) + e^- \rightleftharpoons Ag(s) \qquad\qquad E^0 = 0.80 \text{ V}$$
$$Ag(NH_3)_2^+(aq) + e^- \rightleftharpoons Ag(s) + 2NH_3(aq) \quad E^0 = 0.37 \text{ V}$$

*21.148 Recently, the toxicity of cadmium became a concern, but nickel-cadmium (nicad) batteries are still used commonly in many devices. The overall cell reaction is

$$Cd(s) + 2NiO(OH)(s) + 2H_2O(l) \longrightarrow$$
$$2Ni(OH)(s) + Cd(OH)_2(s)$$

A certain nicad battery weighs 13.3 g and has a capacity of 300. mA·h (that is, the cell can store charge equivalent to a current of 300. mA flowing for 1 h).
(a) What is the capacity of this cell in coulombs?
(b) What mass of reactants is needed to deliver 300. mA·h?
(c) What percentage of the cell mass consists of reactants?
21.149 The zinc-air battery is a less expensive alternative for silver batteries in hearing aids. The cell reaction is

$$2Zn(s) + O_2(g) \longrightarrow 2ZnO(s)$$

A new battery weighs 0.275 g. The zinc accounts for exactly one-tenth of the mass, and the oxygen does not contribute to the mass because it is supplied by the air.
(a) How much electricity (in C) can the battery deliver?
(b) How much free energy (in J) is released if the cell voltage is 1.3 V?
21.150 Use Appendix D to create an activity series of Mn, Fe, Ag, Sn, Cr, Cu, Ba, Al, Na, Hg, Ni, Li, Au, Zn, and Pb. Rank these metals in order of decreasing reducing strength, and divide them into three groups: those that can displace H_2 from water, those that can displace H_2 from acid, and those that cannot displace H_2.

CHAPTER 22

THE ELEMENTS IN NATURE AND INDUSTRY

CHAPTER OUTLINE

22.1 How the Elements Occur in Nature
Earth's Structure and the Abundance of the
  Elements
Sources of the Elements

**22.2 The Cycling of Elements Through the
Environment**
The Carbon Cycle
The Nitrogen Cycle
The Phosphorus Cycle

**22.3 Metallurgy: Extracting a Metal
from Its Ore**
Pretreating the Ore
Converting Mineral to Element
Refining and Alloying

**22.4 Tapping the Crust: Isolation and Uses
of the Elements**
Sodium and Potassium
Iron, Copper, and Aluminum
Magnesium and Bromine
Hydrogen
Sources, Isolation, and Uses of the Elements

**22.5 Chemical Manufacturing:
Two Case Studies**
Sulfuric Acid
The Chlor-Alkali Process

Figure: Metallurgy and Music. The copper and zinc that make up the brass of this magnificent horn began in rocky deposits that were mined, the elements isolated from their ores, refined, and alloyed before being processed further. In this chapter, you'll apply the principles from the last six chapters to the practical side of chemistry—how we obtain and use the elements and their compounds.

In this chapter we return to the periodic table, but from a new perspective. In Chapter 14, we considered the atomic, physical, and chemical behavior of the main-group elements. Here, we examine the elements from a practical viewpoint to learn how much of an element is present in nature, in what form it occurs, how organisms affect its distribution, and how we obtain it for our own use.

Remember that chemistry is, above all, a practical science, and its concepts were developed to address real-life problems: How can we isolate and recycle aluminum? How can we make iron stronger? How can we lower energy costs and increase the yield of sulfuric acid? How does fossil-fuel combustion affect the environmental cycling of carbon? Here, we apply concepts that you learned from the chapters on kinetics, equilibrium, thermodynamics, and electrochemistry to explain chemical processes in nature and industry that influence our lives.

We first discuss the abundances and sources of elements in the Earth, then consider how three essential elements—carbon, nitrogen, and phosphorus—cycle through the environment. After seeing where and in what form elements are found, we focus on their isolation and utilization. We discuss general procedures for extracting an element from its ore and examine in detail the isolation methods and uses of certain key elements. Then we quickly survey major isolation methods and uses for many of the other elements. The chapter ends with a close look at two of the most important processes in chemical manufacturing.

22.1 HOW THE ELEMENTS OCCUR IN NATURE

To begin our examination of how we use the elements, let's take inventory of our elemental stock—the distribution and relative amounts of the elements in the Earth, especially that thin outer portion of the planet that we can reach.

Earth's Structure and the Abundance of the Elements

Any attempt to isolate an element must begin with a knowledge of its **abundance,** the amount of the element in a particular region of the natural world. The abundances of the elements on the Earth and in its various regions are the result of the specific history of our planet's evolution.

Formation and Layering of the Earth About 4.5 billion years ago, vast clouds of cold gases and interstellar debris from exploded older stars gradually coalesced into the Sun and planets. At first, the Earth was a cold, solid sphere of uniformly distributed elements and simple compounds. In the next billion years or so, heat from radioactive decay and virtually continuous meteor impacts raised the planet's temperature to around 10^4 K, sufficient to form an enormous molten mass. Any remaining gaseous elements, such as the cosmically abundant hydrogen and helium, were ejected into space.

As the Earth cooled, chemical and physical processes resulted in its **differentiation,** the formation of regions of different composition and density. Differentiation gave the Earth its layered internal structure, which consists of a dense (10–15 g/cm^3) **core,** composed of a molten outer core and a solid Moon-sized inner core. (Recent evidence has revealed remarkable properties of the inner core: it is nearly as hot as the surface of the Sun and spins within the molten outer core slightly faster than does the Earth itself!) Around the core lies a thick homogeneous **mantle,** consisting of lower and upper portions, with an overall density of 4–6 g/cm^3. All the comings and goings of life take place on the thin heterogeneous **crust** (average density = 2.8 g/cm^3). Figure 22.1 on the next page shows these layers and compares the abundances of some key elements in the universe, the whole Earth (actually core plus mantle only, which account for more than 99% of its mass), and the Earth's three regions. Because the deepest terrestrial sampling can penetrate only a few kilometers into the crust, some of these data

CONCEPTS & SKILLS

to review before you study this chapter
- catalysts and reaction rate (Section 16.8)
- Le Châtelier's principle (Section 17.6)
- acid-base equilibria (Sections 18.3 and 18.9)
- solubility and complex-ion equilibria (Sections 19.3 and 19.4)
- temperature and reaction spontaneity (Section 20.3)
- free energy and equilibrium (Section 20.4)
- standard electrode potentials (Section 21.3)
- electrolysis of molten salts and aqueous solutions (Section 21.7)

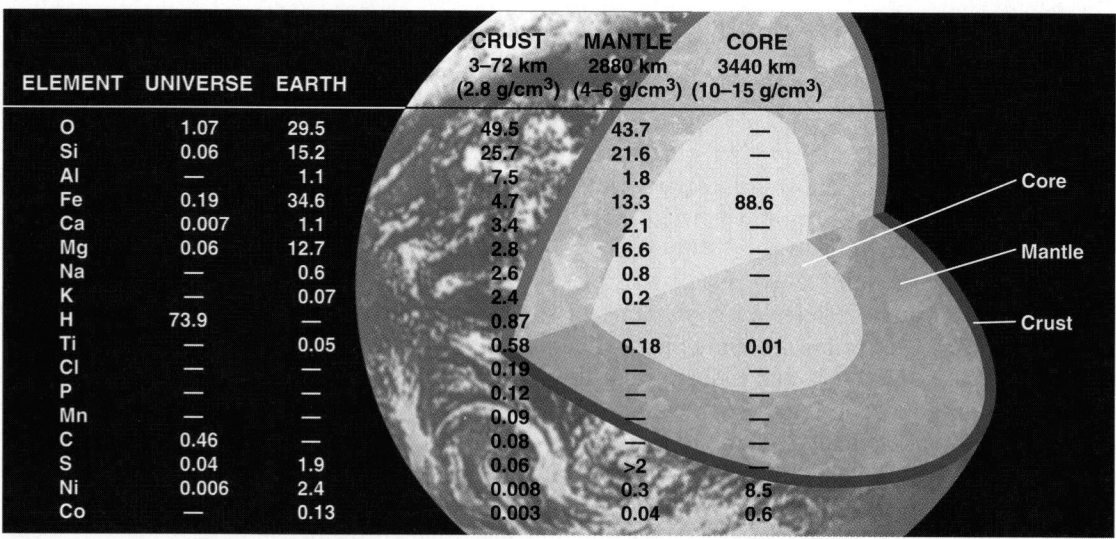

ELEMENT	UNIVERSE	EARTH	CRUST 3–72 km (2.8 g/cm³)	MANTLE 2880 km (4–6 g/cm³)	CORE 3440 km (10–15 g/cm³)
O	1.07	29.5	49.5	43.7	—
Si	0.06	15.2	25.7	21.6	—
Al	—	1.1	7.5	1.8	—
Fe	0.19	34.6	4.7	13.3	88.6
Ca	0.007	1.1	3.4	2.1	—
Mg	0.06	12.7	2.8	16.6	—
Na	—	0.6	2.6	0.8	—
K	—	0.07	2.4	0.2	—
H	73.9	—	0.87	—	—
Ti	—	0.05	0.58	0.18	0.01
Cl	—	—	0.19	—	—
P	—	—	0.12	—	—
Mn	—	—	0.09	—	—
C	0.46	—	0.08	—	—
S	0.04	1.9	0.06	>2	—
Ni	0.006	2.4	0.008	0.3	8.5
Co	—	0.13	0.003	0.04	0.6

Core

Mantle

Crust

Figure 22.1 **Cosmic and terrestrial abundances of selected elements (mass %).** The mass percentages of the elements are superimposed on a cutaway of the Earth. The internal structure consists of a large core, thick mantle, and thin crust. The densities of these regions are indicated in the column heads. Blank abundance values mean either that reliable data are not available or that the value is less than 0.001 mass %. (Helium is not listed because it is not abundant on Earth, but it accounts for 24.0 mass % of the universe.)

represent extrapolations from meteor samples and from seismic studies of earthquakes. Several points stand out:

1. Cosmic and whole-Earth abundances are very different, particularly for H.
2. The elements O, Si, Fe, and Mg are abundant both cosmically and on Earth. Together, they account for more than 90% of the Earth's mass.
3. The core is particularly rich in the dense Group 8B metals: Co, Ni, and especially Fe, the most abundant element in the whole Earth.
4. Crustal abundances are very different from whole-Earth abundances. The crust makes up only 0.4% of the Earth's mass, but has the largest share of nonmetals, metalloids, and light, active metals: Al, Ca, Na, and K. The mantle contains smaller proportions of these, and the core has none. Oxygen is the most abundant element in the crust and mantle but is absent from the core.

These compositional differences in the Earth's major layers, or *phases,* arose from the effects of thermal energy. When the Earth was molten, gravity and convection caused more dense materials to sink and less dense materials to rise. Most of the Fe sank to form the core, or *iron phase.* In the light outer phase, oxygen combined with Si, Al, Mg, and some Fe to form silicates, the material of rocks. This *silicate phase* later separated into the mantle and crust. The *sulfide phase,* intermediate in density and insoluble in the other two, consisted mostly of iron sulfide and mixed with parts of the silicate phase above and the iron phase below. Outgassing, the expulsion of trapped gases, produced a thin, primitive atmosphere, probably a mixture of water vapor (which gave rise to the oceans), carbon monoxide, and nitrogen (or ammonia).

The distribution of the remaining elements was controlled by their chemical affinity for one of the three phases. In general terms, as Figure 22.2 shows (and we discuss using the new group numbers), elements with low or high electronegativity—active metals (Groups 1 through 5, Cr, and Mn) and nonmetals (O, lighter members of Groups 13 to 15, and all of Group 17)—tended to congregate in the silicate phase as ionic compounds. Metals with intermediate electronegativities (many from Groups 6 to 10) dissolved in the iron core. Lower-melting transition metals and many metals and metalloids in Groups 11 to 16 became concentrated in the sulfide phase.

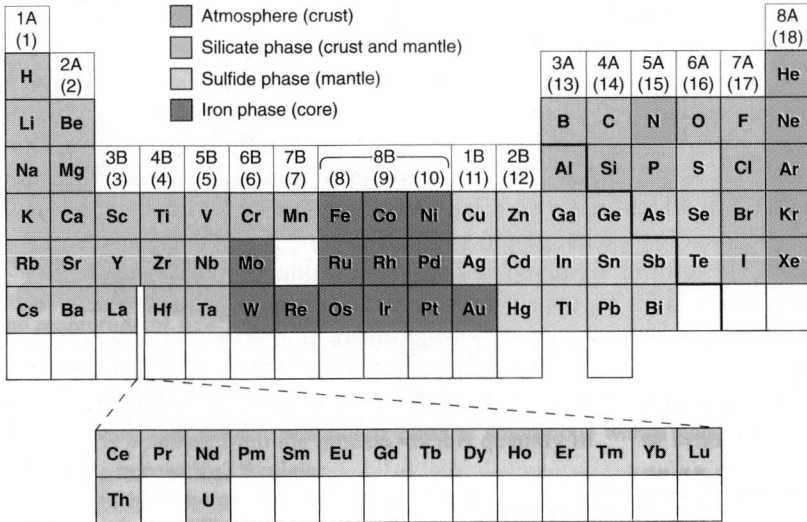

Figure 22.2 **Geochemical differentiation of the elements.** During Earth's formation, elements became concentrated primarily in the atmosphere or one of three phases: silicate, sulfide, or iron.

The Impact of Life on Crustal Abundances At present, the crust is the only accessible portion of our planet, so only crustal abundances have practical significance. The crust is divided into solid, liquid, and gaseous portions called the **lithosphere, hydrosphere,** and **atmosphere.** Over billions of years, weathering and volcanic upsurges dramatically altered the composition of the crust. The **biosphere,** which consists of the living systems that have inhabited the planet, has been another *major influence on crustal element composition and distribution.*

When the earliest rocks were forming and the ocean basins filling with water, binary inorganic molecules in the atmosphere were reacting to form first simple, and then more complex, organic molecules. The energy for these (mostly) endothermic changes was supplied by lightning, solar radiation, geologic heating, and meteoric impact. In an amazingly short period of time, probably no more than 500 million years, the first organisms appeared. It took perhaps another billion years for these to evolve into simple algae that could derive metabolic energy from photosynthesis, converting CO_2 and H_2O into organic molecules and releasing O_2 as a by-product. The importance of this process to crustal chemistry cannot be overstated. Over the next 300 million years, the atmosphere gradually became richer in O_2, and *oxidation became the major source of free energy in the crust and biosphere.* Geologists mark this period by the appearance of Fe(III)-containing minerals in place of the Fe(II)-containing minerals that had predominated (Figure 22.3A). Paleontologists see in it an explosion of O_2-utilizing lifeforms, a few of which evolved into the organisms of today (Figure 22.3B).

Figure 22.3 **Ancient effects of an O_2-rich atmosphere. A,** With the appearance of photosynthetic organisms and the resulting oxidizing environment, much of the iron(II) in minerals was oxidized to iron(III). These ancient banded-iron formations (red beds) from Michigan contain hematite, Fe_2O_3. **B,** A fossil of *Hallucigenia sparsa,* one of the multicellular organisms found in the Burgess Shale, British Columbia, Canada, whose appearance coincides with the increase in O_2. Many of these organisms have no known modern descendants, having disappeared in a great mass extinction at the end of the Cambrian Period about 505 million years ago.

A

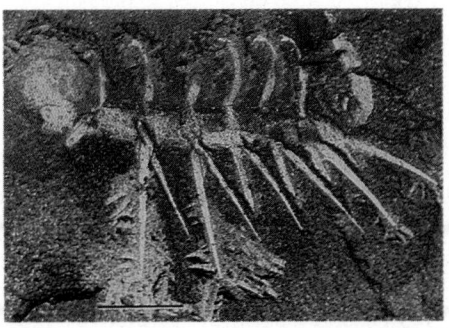

B

The white cliffs of Dover, England, are composed of chalk.

In addition to creating an oxidizing environment, organisms have had profound effects on the *distribution* of specific elements. Why, for instance, is the K^+ concentration of the oceans so much lower than the Na^+ concentration? Forming over eons from outgassed water vapor condensing into rain and streaming over the land, the oceans became complex ionic solutions of dissolved minerals with 30 times as much Na^+ as K^+. There are two principal reasons for this difference. Clays, which make up a large portion of many soils, bind K^+ preferentially through an ion-exchange process. Even more importantly, plants require K^+ for growth, absorbing dissolved K^+ that would otherwise wash down to the sea.

As another example, consider the enormous subterranean deposits of organic carbon. Buried deeply and decomposing under high pressure and temperature in the absence of free O_2, ancient plants gradually turned into coal, and animals buried under similar conditions turned into petroleum. Crustal deposits of this organic carbon provide the fuels that move our cars, heat our homes, and electrify our cities. Moreover, carbon *and* calcium, the fifth most abundant element in the crust, occur together in vast sedimentary deposits all over the world as limestone, dolomite, marble, and chalk, the fossilized skeletal remains of early marine organisms (see photo).

Table 22.1 compares selected elemental abundances in the whole crust, the three crustal regions, and the human body, a representative portion of the biosphere. Note the quantities of the four major elements of life—O, C, H, and N. Oxygen is either the first or second most abundant element in all cases. The biosphere contains large amounts of carbon in its biomolecules and large amounts of hydrogen as water. Nitrogen is abundant in the atmosphere as free N_2 and in organisms combined in proteins. Phosphorus and sulfur occur in exceptionally high amounts in organisms, too.

One striking difference among the lithosphere, hydrosphere, and biosphere (human) is the abundances of the transition metals vanadium through zinc. Their

Table 22.1 **Abundance of Selected Elements in the Crust, Its Regions, and the Human Body as Representative of the Biosphere (Mass %)**

| Element | Crust | Crustal Regions | | | Human |
		Lithosphere	Hydrosphere	Atmosphere	
O	49.5	45.5	85.8	23.0	65.0
C	0.08	0.018	—	0.01	18.0
H	0.87	0.15	10.7	0.02	10.0
N	0.03	0.002	—	75.5	3.0
P	0.12	0.11	—	—	1.0
Mg	1.9	2.76	0.13	—	0.50
K	2.4	1.84	0.04	—	0.34
Ca	3.4	4.66	0.05	—	2.4
S	0.06	0.034	—	—	0.26
Na	2.6	2.27	1.1	—	0.14
Cl	0.19	0.013	2.1	—	0.15
Fe	4.7	6.2	—	—	0.005
Zn	0.013	0.008	—	—	0.003
Cr	0.02	0.012	—	—	3×10^{-6}
Co	0.003	0.003	—	—	3×10^{-6}
Cu	0.007	0.007	—	—	4×10^{-4}
Mn	0.09	0.11	—	—	1×10^{-4}
Ni	0.008	0.010	—	—	3×10^{-6}
V	0.015	0.014	—	—	3×10^{-6}

relatively insoluble oxides and sulfides make them much scarcer in water than on land. Yet organisms, which evolved in the seas, developed the ability to concentrate these elements from the trace amounts in their aqueous environment. In every case, the concentration increases at least 100-fold, with Mn increasing about 1000-fold, and Cu, Zn, and Fe even more. Each of these elements performs an essential role in living systems (see Chemical Connections, Section 23.5).

Sources of the Elements

Given an element's abundance in a region of the crust, we must next determine its **occurrence,** or **source,** the form(s) in which the element exists. Practical considerations often determine the commercial source. Oxygen, for example, is abundant in all three crustal regions, but the atmosphere is its primary industrial source because it occurs there as the free element. Nitrogen and the noble gases (except helium) are also obtained from the atmosphere. Several other elements occur uncombined, formed in large deposits by prehistoric biological action; examples are sulfur in caprock salt domes and nearly pure carbon in coal. The relatively unreactive elements gold and platinum also occur in an uncombined (*native*) state.

The overwhelming majority of elements, however, occur in **ores,** natural compounds or mixtures of compounds from which an element can be extracted by economically feasible means. The phrase "economically feasible" refers to an important point: *the financial costs of mining, isolating, and purifying an element must be considered when choosing a process to obtain it.*

Figure 22.4 shows the most useful sources of the elements. Alkali metal halides are ores for both of their component groups of elements. One example of how cost influences recovery concerns the feasibility of using silicates as ores. Even though many elements occur as silicates, most of these sources are very stable thermodynamically. Thus, the cost of the energy required to process them prohibits their use—aside from silicon, only lithium and beryllium are obtained from their silicates. The Group 2A(2) metals occur as carbonates in the marble and limestone of mountain ranges, although magnesium's great abundance in seawater makes that the preferred source, as we discuss later.

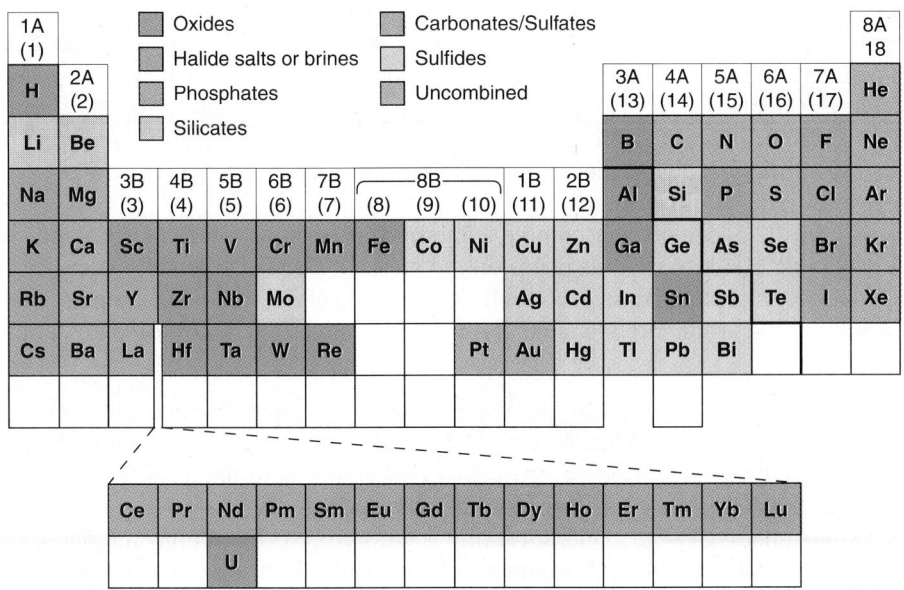

Figure 22.4 **Sources of the elements.** The major sources of most of the elements are shown. Although a few elements occur uncombined, most are obtained from their oxide or sulfide ores.

The ores of most industrially important metals are either *oxides,* which dominate for the left half of the transition series, or *sulfides,* which dominate for the right half of the transition series and a few of the main groups beyond. The reasons for the prominence of oxides and sulfides are complex and include processes of weathering, selective precipitation, and relative solubilities. Nevertheless, some atomic properties are relevant as well. Elements in the left half have lower ionization energies and electronegativities, so they tend to give up electrons or hold them loosely in bonds. The O^{2-} ion is small enough to approach a metal cation closely, which results in a high lattice energy for the oxide. In contrast, the elements toward the right side have higher ionization energies and electronegativities. Thus, they tend to form bonds that are more covalent, which suits the larger, more polarizable S^{2-} ion.

SECTION SUMMARY
As the young Earth cooled, the elements became differentiated into a dense, metallic core, silicate-rich mantle, and low-density crust. High abundances of light metals, metalloids, and nonmetals are concentrated in the crust, which consists of the lithosphere (solid), hydrosphere (liquid), and atmosphere (gaseous). The biosphere (living systems) has affected crustal chemistry primarily by producing free O_2, and thus an oxidizing environment. Some elements occur in their native state, but most are combined in ores. The most important ores of metallic elements are oxides and sulfides.

22.2 THE CYCLING OF ELEMENTS THROUGH THE ENVIRONMENT

The distributions of many elements are in flux, changing at widely differing rates. The physical, chemical, and biological paths that the atoms of an element take on their journey through regions of the crust constitute the element's **environmental cycle.** In this section, we consider the cycles for three important elements—carbon, nitrogen, and phosphorus—and highlight the effects of humans on them.

The Carbon Cycle

Carbon is one of a handful of elements that appear in all three portions of the Earth's crust. In the lithosphere, it occurs as elemental graphite and diamond; combined in fully oxidized form as carbonate minerals and in fully reduced form as petroleum hydrocarbons; and in complex mixtures, such as coal and living matter. In the hydrosphere, it occurs as living matter, as carbonate minerals formed by the action of coral-reef organisms, and as dissolved CO_2. In the atmosphere, it occurs principally as gaseous CO_2, a minor but essential component that exists in equilibrium with the aqueous fraction.

Figure 22.5 depicts the complex interplay of these carbon sources and the considerable effect of the biosphere on the element's environmental cycle. It provides an estimate of the size of each source and, where reliable numbers exist, the amount of carbon moving annually between sources. Here are some key points to note:

- The portions of the cycle are linked by the atmosphere—one link between oceans and air, the other between land and air. The 2.6×10^{12} metric tons of CO_2 in air cycles through the oceans and atmosphere about once every 300 years but spends a much longer time in carbonate minerals.
- Atmospheric CO_2 accounts for a tiny fraction (0.003%) of crustal carbon, but the atmosphere is the prime mover of carbon through the other regions. Estimates indicate that a CO_2 molecule spends, on the average, 3.5 years in the atmosphere.

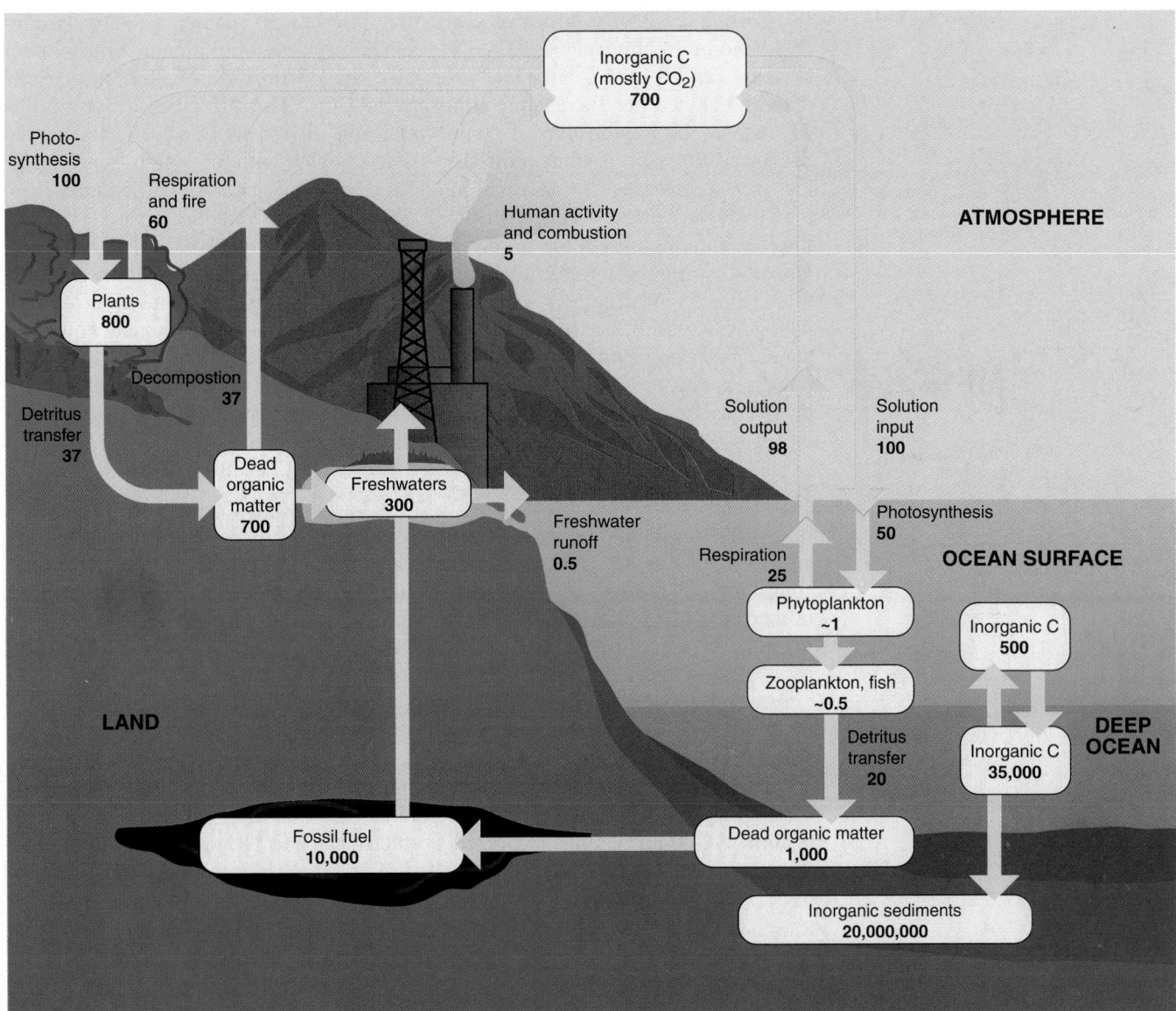

Figure 22.5 **The carbon cycle.** The major sources and changes in carbon distribution are shown. Note that the atmosphere is the major conduit of carbon between land and sea and that CO_2 becomes fixed through the action of organisms. Numbers in boxes refer to the size of the source; numbers along arrows refer to the annual movement of the element from source to source. Values are in 10^9 metric tons of C (1 metric ton = 1000 kg = 1.1 tons).

- The land and oceans are also in contact with the largest sources of carbon, which are effectively immobilized as carbonates, coal, and oil in the rocky sediment beneath the soil.
- The biological processes of *photosynthesis, respiration,* and *decay* are major factors in the cycle. **Fixation** is the process of converting a gaseous substance into a condensed, more usable form. Photosynthesis by marine plankton and terrestrial plants uses sunlight to fix atmospheric CO_2 into carbohydrates. Plants release CO_2 by respiration at night and, much more slowly, when they decay. Animals eat the plants and release CO_2 by respiration and in their decay.
- Natural fires and volcanoes also release CO_2 into the air.

For hundreds of millions of years, the cycle has maintained a relatively constant amount of atmospheric CO_2. The tremendous growth of human industry over the past century and a half, and especially since World War II, has shifted this natural balance by increasing atmospheric CO_2. The principal cause of this increase is the combustion of coal, wood, and oil for fuel and for the thermal decomposition of limestone to make cement, coupled with the simultaneous clearing of vast stretches of forest and jungle for lumber, paper, and agriculture. With the 1990s the hottest decade ever recorded, evidence is mounting that these activities are creating one of the most dire environmental changes in human history—global warming through the greenhouse effect (see Chemical Connections, Section 6.6). Higher temperatures, alterations in dry and rainy seasons, melting of polar ice, and increasing ocean acidity with associated changes in carbonate equilibria are some of the results currently being assessed. Most environmental and science policy experts are pressing for the immediate adoption of a program that combines conservation of carbon-based fuels, an end to deforestation, extensive planting of trees, and development of alternative energy sources.

The Nitrogen Cycle

In contrast to the carbon cycle, the nitrogen cycle includes a direct interaction of land and sea (Figure 22.6). All nitrites and nitrates are soluble, so rain and runoff contribute huge amounts of nitrogen to lakes, rivers, and oceans. Human activity, in the form of fertilizer production and use, plays a major role in this cycle.

Like carbon dioxide, atmospheric nitrogen must be fixed to be used by organisms. However, whereas CO_2 can be incorporated by plants in either its gaseous or aqueous form, the great stability of N_2 prevents plants from using it directly. Fixation of N_2 requires a great deal of energy and occurs through atmospheric, industrial, and biological processes:

1. *Atmospheric fixation.* Lightning and fires (and, to a much smaller extent, car engines) bring about the high-temperature endothermic reaction of N_2 and O_2 to form NO, which is then oxidized exothermically to NO_2:

$$N_2(g) + O_2(g) \longrightarrow 2NO(g) \qquad \Delta H^0 = 180.6 \text{ kJ}$$
$$2NO(g) + O_2(g) \longrightarrow 2NO_2(g) \qquad \Delta H^0 = -114.2 \text{ kJ}$$

Rain converts the NO_2 to nitric acid, which enters both sea and land as $NO_3^-(aq)$ to be utilized by plants:

$$3NO_2(g) + H_2O(l) \longrightarrow 2H^+(aq) + 2NO_3^-(aq) + NO(g)$$

2. *Industrial fixation.* Most human-caused fixation occurs industrially during ammonia synthesis via the Haber process (see Chemical Connections, Section 17.6). The process takes place on an enormous scale: on a mole basis, NH_3 ranks first among compounds in amount produced. Some of this NH_3 is converted to HNO_3 in the Ostwald process (Section 14.7), but most is used as fertilizer, either directly or in the form of urea and ammonium salts (sulfate, phosphate, and nitrate), which enter the biosphere as they are taken up by plants (item 3 below). In recent decades, the internal combustion engine has become an important contributor to total fixed nitrogen, because of the enormous rise in automobile and truck traffic. High engine operating temperatures mimic lightning and fires to form NO from the air taken in to burn the hydrocarbon fuel. The NO in exhaust gases reacts to form nitric acid in the atmosphere, which adds to the nitrate load that reaches the ground. The overuse of fertilizers and automobiles presents an increasingly serious water pollution problem in many areas. Leaching of the land by rain causes fertilizers and car-borne nitrate to enter natural waters (lakes, rivers, and coastal estuaries) and cause *eutrophication,* the depletion of O_2 and the death of aquatic animal life from excessive algal and plant growth and decay. Excess nitrate also spoils nearby drinkable water.

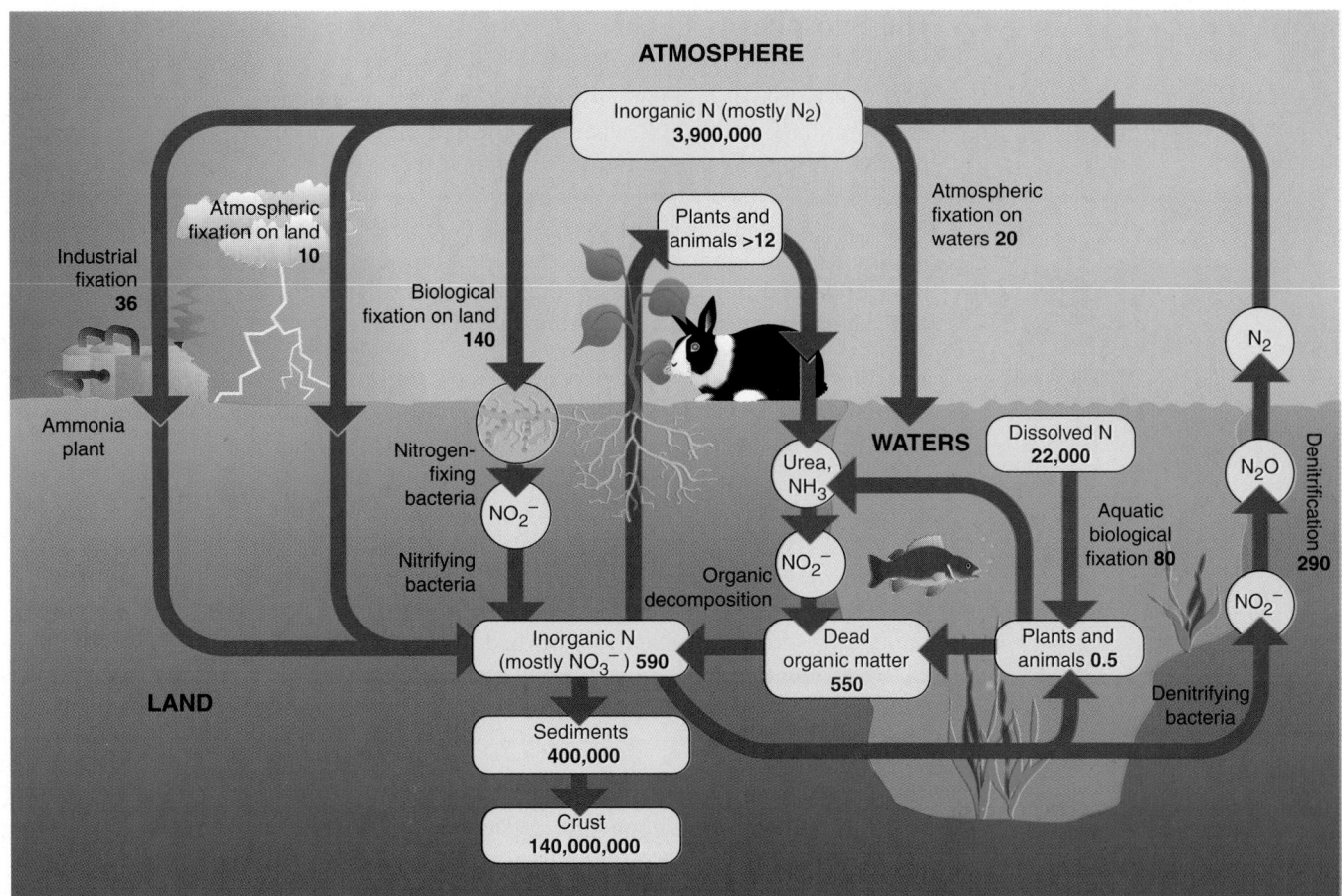

Figure 22.6 **The nitrogen cycle.** The major source of nitrogen, N_2, is fixed atmospherically, industrially, and biologically. Various types of bacteria are indispensable throughout the cycle. Numbers in boxes are in 10^9 metric tons of N and refer to the size of the source; numbers along arrows are in 10^6 metric tons of N and refer to the annual movement of the element between sources.

3. *Biological fixation.* The biological fixation of atmospheric N_2 occurs in marine blue-green algae and in nitrogen-fixing bacteria that live on the roots of leguminous plants (e.g., peas, alfalfa, and clover). On a planetary scale, these microbial processes dwarf the other two, fixing more than seven times as much nitrogen as the atmosphere and six times as much as industry. Root bacteria fix N_2 by reducing it to NH_3/NH_4^+ through the action of enzymes that contain the transition metal molybdenum at their active site. Enzymes in other soil bacteria catalyze the multistep oxidation of the NH_4^+ to NO_2^- and finally NO_3^-, which the plants reduce again to make their proteins. When the plants die, still other soil bacteria oxidize the proteins to NO_3^-.

Animals eat the plants, use the plant proteins to make their own proteins, and excrete nitrogenous wastes, such as urea [$(H_2N)_2C{=}O$]. The nitrogen in the proteins is released when the animals die and decay, and is converted by soil bacteria to NO_2^- and NO_3^- again. This central pool of inorganic nitrate has three main fates: some enters marine and terrestrial plants, some enters the enormous sediment store of mineral nitrates, and some is reduced by denitrifying bacteria to NO_2^- and then to N_2O and N_2, which re-enter the atmosphere to complete the cycle.

The Phosphorus Cycle

Virtually all the mineral sources of phosphorus contain the phosphate group, PO_4^{3-}. The most commercially important ores are **apatites,** compounds of general formula $Ca_5(PO_4)_3X$, where X is usually F, Cl, or OH. The cycling of phosphorus through the environment involves three interlocking subcycles (Figure 22.7). Two rapid biological cycles—a land-based cycle completed in a matter of years and a water-based cycle completed in weeks to years—are superimposed on an inorganic cycle that takes millions of years to complete. Unlike the carbon and nitrogen cycles, the phosphorus cycle has no gaseous component and thus does *not* involve the atmosphere.

The Inorganic Cycle Most phosphate rock formed when the Earth's lithosphere solidified. ● The most important components of this material are nearly insoluble phosphate salts:

$$K_{sp} \text{ of } Ca_3(PO_4)_2 \approx 10^{-29}$$
$$K_{sp} \text{ of } Ca_5(PO_4)_3OH \approx 10^{-51}$$
$$K_{sp} \text{ of } Ca_5(PO_4)_3F \approx 10^{-60}$$

Weathering slowly leaches phosphates from the soil and carries the ions through rivers to the sea. Plants in the land-based biological cycle speed this process. In the ocean, some phosphate is absorbed by organisms in the water-based biological cycle, but the majority is precipitated again by Ca^{2+} ion and deposited on the continental shelf. Geologic activity lifts the continental shelves, returning the phosphate to the land.

The Land-Based Biological Cycle The biological cycles involve the incorporation of phosphate into organisms (biomolecules, bones, teeth, and so forth) and the release of phosphate through excretion and decay. In the land-based cycle, plants continually remove phosphate from the inorganic cycle. Recall that there are three phosphate oxoanions, which exist in equilibrium in the aqueous environment:

$$H_2PO_4^- \rightleftharpoons HPO_4^{2-} \rightleftharpoons PO_4^{3-}$$
$$+ H^+ \qquad + H^+$$

In topsoil, phosphates occur as insoluble compounds of Ca^{2+}, Fe^{3+}, and Al^{3+}. Because plants can absorb only the soluble dihydrogen phosphates, they have evolved the ability to secrete acids near their roots to convert the insoluble salts gradually into the soluble ion:

$Ca_3(PO_4)_2(s; \text{ soil}) + 4H^+(aq; \text{ secreted by plants}) \longrightarrow$
$$3Ca^{2+}(aq) + 2H_2PO_4^-(aq; \text{ absorbed by plants})$$

Animals that eat the plants excrete soluble phosphate, which is used in turn by newly growing plants. As the plants and animals excrete, die, and decay, some phosphate is washed into rivers and from there to the ocean. Most of the 2 million metric tons of phosphate that washes from land to sea each year comes from biological decay. Thus, the biosphere greatly increases the movement of phosphate from lithosphere to hydrosphere.

The Water-Based Biological Cycle Various phosphorus oxoanions continually enter the aquatic environment. Studies that incorporate trace amounts of radioactive phosphorus into $H_2PO_4^-$ and monitor its uptake show that, within 1 minute, 50% of the ion is taken up by photosynthetic algae, which use it to synthesize

●**Phosphorus from Outer Space** Although most phosphate occurs in the rocks formed by geological activity on Earth, a sizeable amount arrived here from outer space and continues to do so, as this time-lapse photo of the 1966 Leonid meteor shower shows. About 100 metric tons of meteorites enter our atmosphere each day. With an average P content of 0.1 mass %, they contribute about 36 metric tons of P per year. Over the 4.6 billion-year lifetime of the Earth, about 10^{11} metric tons of P has arrived from extraterrestrial sources!

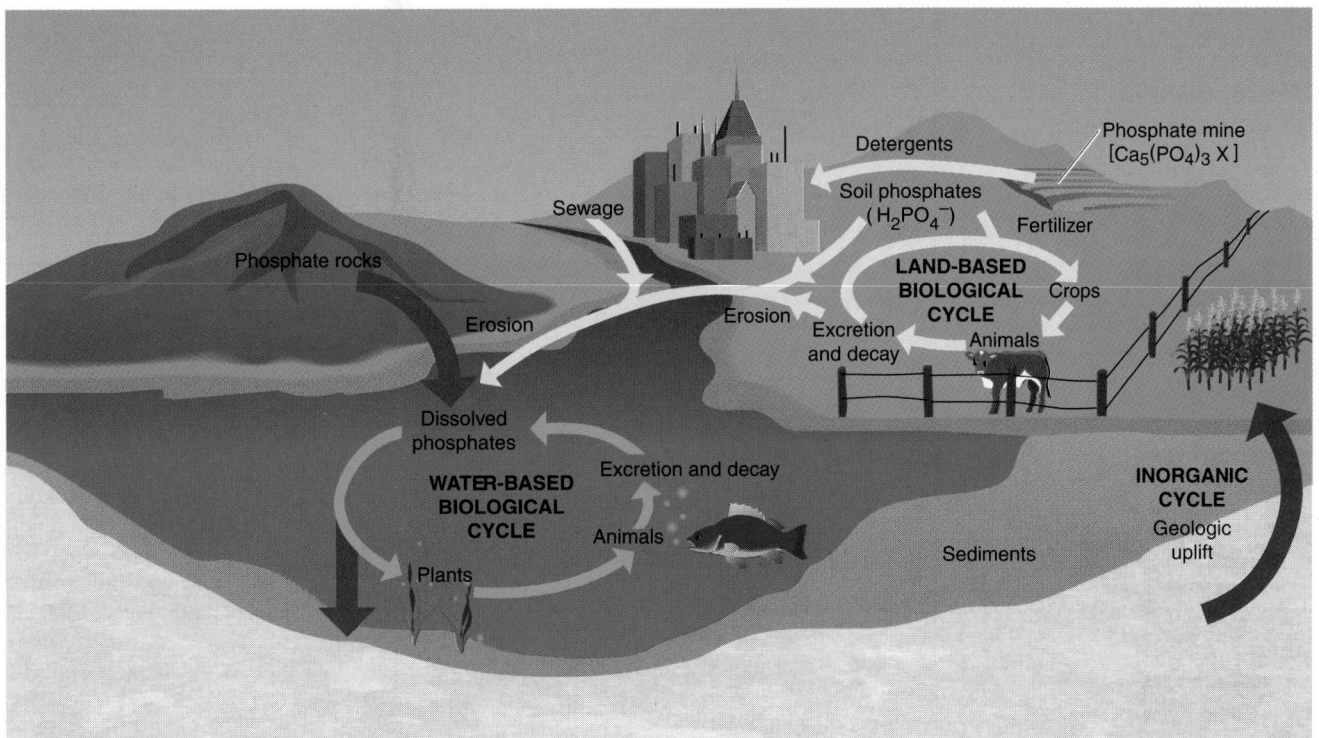

Figure 22.7 **The phosphorus cycle.** The inorganic *(purple arrows)*, water-based biological *(blue arrows)*, and land-based biological *(yellow arrows)* subcycles interact through erosion and through the growth and decay of organisms. The phosphorus cycle has no gaseous component.

their biomolecules (this synthesis is denoted by the top curved arrow in the upcoming equation). The overall process might be represented by the following equation:

$$106CO_2(g) + 16NO_3^-(aq) + H_2PO_4^-(aq) + 122H_2O(l) + 17H^+(aq) \underset{\text{decay}}{\overset{\text{synthesis}}{\rightleftharpoons}}$$

$$C_{106}H_{263}O_{110}N_{16}P(aq; \text{ in algal cell fluid}) + 138O_2(g)$$

(The complex formula on the right represents the total composition of algal biomolecules, not some particular compound.) As on land, animals eat the plants and are eaten by other animals, and they all excrete, die, and decay (denoted by the bottom curved arrow). Some of the released phosphate is used by other aquatic organisms, and some returns to the land when fish-eating animals (mostly humans and birds) excrete phosphate, die, and decay. In addition, some aquatic phosphate precipitates with Ca^{2+}, sinks to the seabed, and returns to the long-term mineral deposits of the inorganic cycle.

Human Effects on Phosphorus Movement From prehistoric through preindustrial times, these interlocking phosphorus cycles were balanced. Modern human activity, however, alters the movement of phosphorus considerably. Our influence on the inorganic cycle is minimal, despite our annual removal of more than

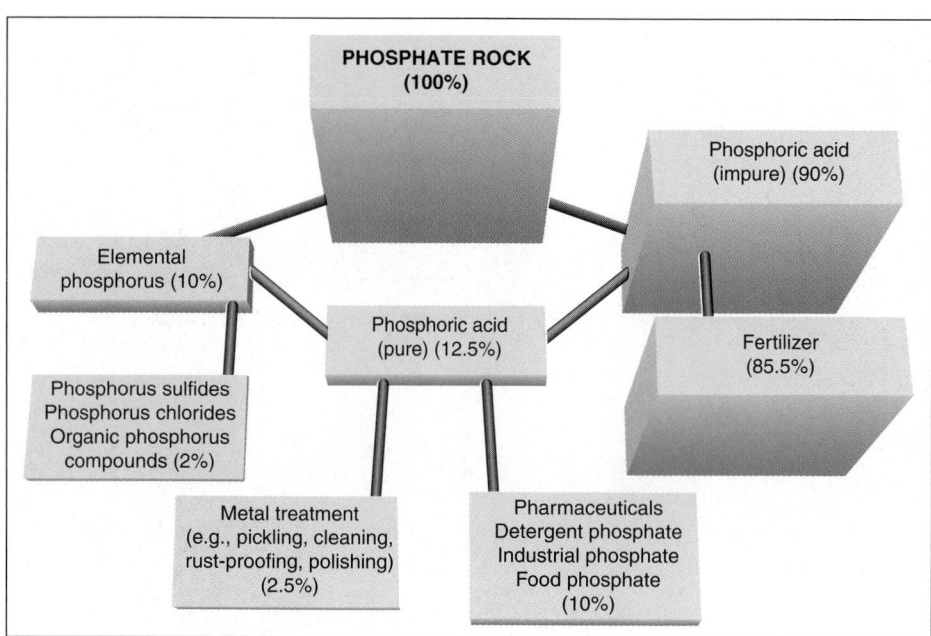

Figure 22.8 **Industrial uses of phosphorus.** The major end products from phosphate rock are shown, with box heights proportional to the amounts of phosphate rock used annually (shown as mass %) to make the end product. Over 100 million metric tons of phosphate rock is mined each year (assigned a value of 100%).

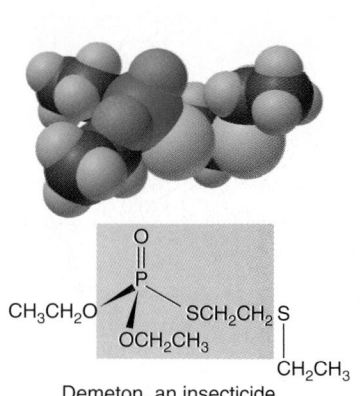

Demeton, an insecticide

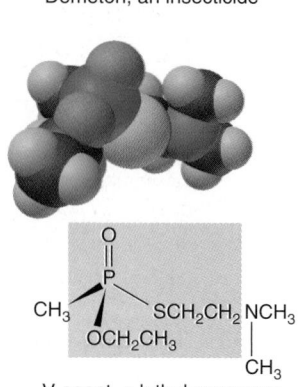

V agent, a lethal nerve gas

Phosphorus Nerve Poisons Because phosphorus is essential for all organisms, its compounds have many biological actions. In addition to their widespread use as fertilizer, phosphorus compounds are also used as pesticides *and* as military nerve gases. In fact, these two types of substances often have similar structures (see models and formulas) and modes of action. Both types affect the target organism by inactivating an enzyme involved in muscle contraction. They bind an amino acid critical to the enzyme's function, allowing the muscle to contract but not relax; the effect on the organism—insect or human—is often fatal. Chemists have synthesized a nerve-gas antidote whose shape mimics the binding site of the enzyme. The drug binds the poison, removing it from the enzyme, and is excreted.

100 million metric tons of phosphate rock for a variety of uses (Figure 22.8). The end products of this removal, however, have unbalanced the two biological cycles.

The imbalance arises from our overuse of *soluble phosphate fertilizers,* such as $Ca(H_2PO_4)_2$ and $NH_4H_2PO_4$, the end products of about 85% of phosphate rock mined. Some fertilizer finds its way into rivers, lakes, and oceans to enter the water-based cycle. Much larger pollution sources, however, are the crops grown with the fertilizer and the detergents made from phosphate rock. The great majority of this phosphate arrives eventually in cities as crops, as animals that were fed crops, and as consumer products, such as the tripolyphosphates in detergents. Human garbage, excrement, detergent wash water, and industrial wastewater containing this phosphate return through sewers to the aquatic system. This human contribution equals the natural contribution—another 2 million metric tons of phosphate per year. The increased concentration in rivers and lakes causes eutrophication, which robs the water of O_2 so that it cannot support life. Such "dead" rivers and lakes are no longer usable for fishing, drinking, or recreation.

As early as 1912, a chemical solution to this problem was tried in Switzerland, where the enormous Lake Zurich had been choked with algae and become devoid of fish because of the release of human sewage. Treatment with $FeCl_3$ precipitated enough phosphate to return the lake gradually to its natural state. Soluble aluminum compounds have the same effect:

$$FeCl_3(aq) + PO_4^{3-}(aq) \longrightarrow FePO_4(s) + 3Cl^-(aq)$$
$$KAl(SO_4)_2 \cdot 12H_2O(aq) + PO_4^{3-}(aq) \longrightarrow$$
$$AlPO_4(s) + K^+(aq) + 2SO_4^{2-}(aq) + 12H_2O(l)$$

After several lakes in the United States became polluted in this way, phosphates in detergents were drastically reduced, and stringent requirements for sewage treatment, including tertiary treatments to remove phosphate (see Chemical Connections, pp. 520–521), were put into effect. Lake Erie, which was severely polluted by the late 1960s and early 1970s, underwent a program to reduce phosphates and is today relatively clean and restocked with fish.

The environmental distribution of many elements changes cyclically with time and is affected in major ways by organisms. Carbon occurs in all three regions of the Earth's crust, with atmospheric CO_2 linking the other two regions. Photosynthesis and decay of organisms alter the amount of carbon in the land and oceans. Human activity has increased atmospheric CO_2. Nitrogen is fixed by lightning, by industry, and primarily by microorganisms. When plants decay, bacteria decompose organic nitrogen and eventually return N_2 to the atmosphere. Through extensive use of fertilizers, humans have added excess nitrogen to freshwaters. The phosphorus cycle has no gaseous component. Inorganic phosphates leach slowly into land and water, where biological cycles interact. When plants and animals decay, they release phosphate to natural waters, where it is then absorbed by other plants and animals. Through overuse of phosphate fertilizers and detergents, humans double the amount of phosphorus entering aqueous systems.

22.3 METALLURGY: EXTRACTING A METAL FROM ITS ORE

Metallurgy is the branch of materials science concerned with the extraction and utilization of metals. In this section, we discuss the general procedures for isolating metals and, occasionally, the nonmetals that are found in their ores. The extraction process applies one or more of these three types of metallurgy: *pyrometallurgy* uses heat to obtain the metal, *electrometallurgy* employs an electrochemical step, and *hydrometallurgy* relies on the metal's aqueous solution chemistry.

The extraction of an element begins with *mining the ore*. Most ores consist of mineral and gangue. The **mineral** contains the element; it is defined as a naturally occurring, homogeneous, crystalline inorganic solid, with a well-defined composition. The **gangue** is the portion of the ore with no commercial value, such as sand, rock, and clay attached to the mineral. Table 22.2 lists the common mineral sources of some metals.

Humans have been removing metals from the Earth for thousands of years, so most of the known concentrated sources are long gone. In many cases, ores containing very low mass percentages of a metal are all that remain, and these require elaborate—and ingenious—extraction methods. Nevertheless, the general procedure for extracting most metals (and many nonmetals) involves a few basic steps, described in the upcoming subsections and previewed in Figure 22.9.

Table 22.2 Common Mineral Sources of Some Elements

Element	Mineral, Formula
Al	Gibbsite (in bauxite), $Al(OH)_3$
Ba	Barite, $BaSO_4$
Be	Beryl, $Be_3Al_2Si_6O_{18}$
Ca	Limestone, $CaCO_3$
Fe	Hematite, Fe_2O_3
Hg	Cinnabar, HgS
Na	Halite, $NaCl$
Pb	Galena, PbS
Sn	Cassiterite, SnO_2
Zn	Sphalerite, ZnS

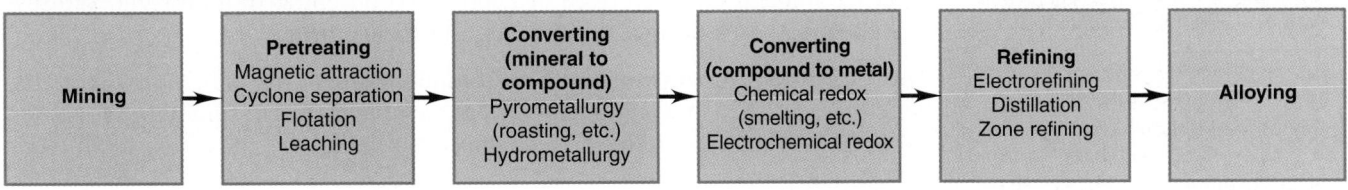

Figure 22.9 Steps in metallurgy.

Pretreating the Ore

Following a crushing, grinding, or pulverizing step, which can be very expensive, pretreatment usually takes advantage of some physical or chemical difference to separate mineral from gangue. For magnetic minerals, such as magnetite (Fe_3O_4), a magnet can remove the mineral and leave the gangue behind. Where large density differences exist, a *cyclone separator* is used; it blows high-pressure air through the pulverized mixture to separate the particles. The lighter silicate-rich

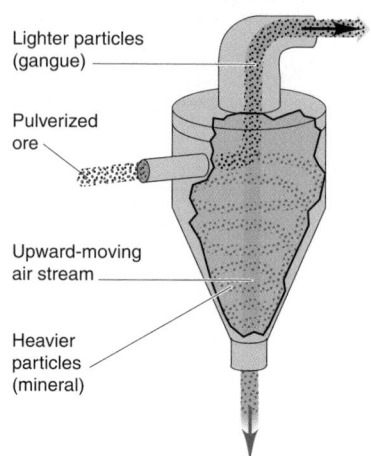

Lighter particles (gangue)

Pulverized ore

Upward-moving air stream

Heavier particles (mineral)

Figure 22.10 The cyclone separator. The pulverized ore enters at an angle, with the more dense particles (mineral) hitting the walls and spiraling downward, while the less dense particles (silicate-rich gangue) are carried upward in a stream of air.

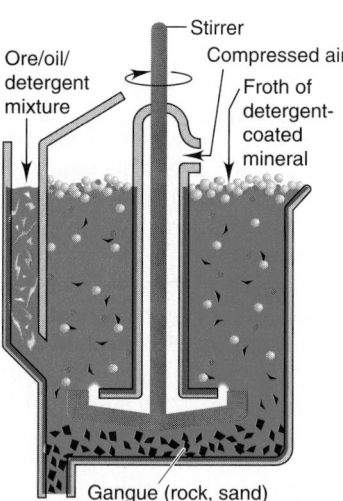

Stirrer

Ore/oil/ detergent mixture

Compressed air

Froth of detergent-coated mineral

Gangue (rock, sand)

Figure 22.11 The flotation process. In flotation, the pulverized ore is treated with an oil-detergent mixture, then stirred and aerated to create a froth that separates the detergent-coated mineral from the gangue. The mineral-rich froth is collected for further processing.

gangue is blown away, while the denser mineral-rich particles hit the walls of the separator and fall through the open bottom (Figure 22.10).

In the process of **flotation**, an oil-detergent mixture is stirred with a pulverized sulfide ore in water to form a slurry (Figure 22.11). Separation by flotation takes advantage of the different abilities of the surfaces of mineral and of gangue to become wet in contact with water and detergent. Rapid mixing moves air bubbles through the mixture and produces an oily, mineral-rich froth that floats, while the silicate particles sink. Skimming, followed by solvent removal of the oil-detergent mixture, isolates the concentrated mineral fraction. Flotation is a key step in copper recovery, as you'll see later.

Leaching is a hydrometallurgic process that selectively extracts the metal, usually by forming a complex ion. The modern extraction of gold is a good example of this technique, because nuggets of the precious metal are found only in museums nowadays.  The crushed ore, which often contains as little as 25 ppm of gold, is contained in a plastic-lined pool, treated with a cyanide ion solution, and aerated. In the presence of CN^-, the O_2 in air oxidizes gold metal to gold(I) ion, Au^+, which forms the soluble complex ion, $Au(CN)_2^-$:

$$4Au(s) + O_2(g) + 8CN^-(aq) + 2H_2O(l) \longrightarrow 4Au(CN)_2^-(aq) + 4OH^-(aq)$$

(The method has become controversial in some areas because cyanide lost from the contained area enters streams and lakes, where it poisons fish and birds.)

Converting Mineral to Element

After the mineral has been freed of debris and concentrated, it may undergo several chemical steps during its conversion to the element.

Converting the Mineral to Another Compound First, the mineral is often converted to another compound, one that has more appropriate solubility properties, is easier to reduce, or is free of a troublesome impurity. Conversion to an oxide is common because oxides can be reduced easily. Carbonates are heated to convert them to the oxide:

$$CaCO_3(s) \xrightarrow{\Delta} CaO(s) + CO_2(g)$$

Metal sulfides, such as ZnS, are converted to oxides by **roasting** in air:

$$2ZnS(s) + 3O_2(g) \xrightarrow{\Delta} 2ZnO(s) + 2SO_2(g)$$

In most countries, hydrometallurgic methods are used now to avoid the atmospheric release of SO_2 during roasting. One way of processing copper, for example, is by bubbling air through an acidic slurry of insoluble Cu_2S, which completes both the pretreatment and conversion steps:

$$2Cu_2S(s) + 5O_2(g) + 4H^+(aq) \longrightarrow 4Cu^{2+}(aq) + 2SO_4^{2-}(aq) + 2H_2O(l)$$

Panning and Fleecing for Gold
The picture of the prospector, alone but for his faithful mule, hunched over the bank of a mountain stream patiently panning for gold, is mostly a relic of old movies. However, the process is valid. It is based on the very large density difference between gold ($d = 19.3$ g/cm^3) and sand ($d \approx 2.5$ g/cm^3). In ancient times, river sands were washed over a sheep's fleece to trap the gold grains, a practice historians think represents the origin of the Golden Fleece of Greek mythology.

Converting the Compound to the Element Through Chemical Redox The next step converts the new mineral form (usually an oxide) to the free element by either chemical or electrochemical methods. In chemical redox, a reducing agent reacts directly with the compound. The most common reducing agents are carbon, hydrogen, and an active metal.

1. *Reduction with carbon.* Carbon, in the form of coke (a porous residue from incomplete combustion of coal) or charcoal, is a very common reducing agent. Heating an oxide with a reducing agent such as coke to obtain the metal is called **smelting.** Many metal oxides, such as zinc oxide and tin(IV) oxide, are smelted with carbon to free the metal, which may need to be condensed and solidified:

$$ZnO(s) + C(s) \longrightarrow Zn(g) + CO(g)$$
$$SnO_2(s) + 2C(s) \longrightarrow Sn(l) + 2CO(g)$$

Several nonmetals that occur in positive oxidation states in a mineral can be reduced with carbon as well. Phosphorus, for example, is produced from calcium phosphate:

$$2Ca_3(PO_4)_2(s) + 10C(s) + 6SiO_2(s) \longrightarrow 6CaSiO_3(s) + 10CO(g) + P_4(s)$$

(Metallic calcium is a much stronger reducing agent than carbon, so it is not formed.)

Low cost and ready availability explain why carbon is such a common reducing agent, and thermodynamic principles explain why it is such an effective one in many cases. Consider the standard free energy change (ΔG^0) for the reduction of tin(IV) oxide:

$$SnO_2(s) + 2C(s) \longrightarrow Sn(s) + 2CO(g) \quad \Delta G^0 = 245 \text{ kJ at } 25°C \text{ (298 K)}$$

The magnitude and sign of ΔG^0 indicate a highly *non*spontaneous reaction at 25°C. However, because solid C becomes gaseous CO, the standard molar entropy change is very positive ($\Delta S^0 \approx 380$ J/K). Therefore, *the $-T\Delta S^0$ term of ΔG^0 becomes more negative with higher temperature* (Section 20.3). As the temperature increases, ΔG^0 decreases, and at a high enough temperature, the reaction becomes spontaneous ($\Delta G^0 < 0$). At 1000°C (1273 K), for example, $\Delta G^0 = -62.8$ kJ. The actual process is carried out at a temperature of around 1250°C (1523 K), so ΔG^0 is even more negative.

As you know, overall reactions often mask several intermediate steps. For instance, the reduction of tin(IV) oxide may occur by first forming tin(II) oxide:

$$SnO_2(s) + C(s) \longrightarrow \overset{\text{---}}{SnO(s)} + CO(g)$$
$$\overset{\text{---}}{SnO(s)} + C(s) \longrightarrow Sn(l) + CO(g)$$

[Molten tin (mp = 232 K) is obtained at this temperature.] Also, the second step may even be composed of others, in which CO, not C, is the actual reducing agent:

$$SnO(s) + \overset{\text{---}}{CO(g)} \longrightarrow Sn(l) + \overset{\text{---}}{CO_2(g)}$$
$$\overset{\text{---}}{CO_2(g)} + C(s) \rightleftharpoons \overset{\text{---}}{2CO(g)}$$

Other pyrometallurgical processes are similarly complex. Shortly, you'll see that the overall reaction for the smelting of iron is

$$2Fe_2O_3(s) + 3C(s) \longrightarrow 4Fe(l) + 3CO_2(g)$$

whereas the actual process is a multistep one with CO as the reducing agent. The advantage of CO over C is the far greater contact the gaseous reducing agent can make with the other reactant, which speeds the process.

2. *Reduction with hydrogen.* For oxides of some metals, especially some members of Groups 6B(6) and 7B(7), reduction with carbon forms metal carbides. These carbides are difficult to convert further, so other reducing agents are used. Hydrogen gas is used for less active metals, like tungsten:

$$WO_3(s) + 3H_2(g) \longrightarrow W(s) + 3H_2O(g)$$

One step in the purification of the metalloid germanium uses hydrogen:

$$GeO_2(s) + 2H_2(g) \longrightarrow Ge(s) + 2H_2O(g)$$

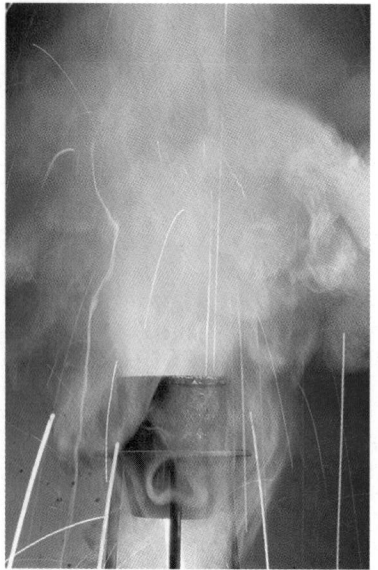

Figure 22.12 The thermite reaction.
This highly exothermic redox reaction
uses powdered aluminum to reduce
metal oxides.

THERMITE REACTION

3. *Reduction with an active metal.* When a metal might form an undesirable hydride, its oxide is reduced by a more active metal. In the *thermite reaction,* aluminum powder reduces the metal oxide in a spectacular exothermic reaction to give the molten metal (Figure 22.12). The reaction for chromium is

$$Cr_2O_3(s) + 2Al(s) \longrightarrow 2Cr(l) + Al_2O_3(s) \quad \Delta H^0 << 0$$

Reduction by a more active metal is also used in situations that do not involve an oxide. In the extraction of gold, after the pretreatment by leaching, the gold(I) complex ion is reduced with zinc to form the metal and a zinc complex ion:

$$2Au(CN)_2^-(aq) + Zn(s) \longrightarrow 2Au(s) + Zn(CN)_4^{2-}(aq)$$

In some cases, an active metal, such as calcium, is used to recover an even more active metal, such as rubidium, from its molten salt:

$$Ca(l) + 2RbCl(l) \longrightarrow CaCl_2(l) + 2Rb(g)$$

A similar process is used to recover sodium, as detailed in the next section.

4. *Oxidation with an active nonmetal.* Just as chemical *reduction* of a mineral is used to obtain the metal, chemical *oxidation* of a mineral is sometimes used to obtain a nonmetal. A stronger oxidizing agent is used to remove electrons from the nonmetal anion to give the free nonmetal, as in the industrial production of iodine from concentrated brines by oxidation with chlorine gas:

$$2I^-(aq) + Cl_2(g) \longrightarrow 2Cl^-(aq) + I_2(s)$$

Converting the Compound to the Element Through Electrochemical Redox In these processes, the mineral components are converted to the elements in an electrolytic cell (Section 21.7). Sometimes, the *pure* mineral, in the form of the molten halide or oxide, is used to prevent unwanted side reactions. The cation is reduced to the metal at the cathode, and the anion is oxidized to the nonmetal at the anode:

$$BeCl_2(l) \longrightarrow Be(s) + Cl_2(g)$$

High-purity hydrogen gas is prepared by electrochemical reduction:

$$2H_2O(l) \longrightarrow 2H_2(g) + O_2(g)$$

Specially designed cells separate the products to prevent their recombination. Cost is a major factor in the use of electrolysis, and an inexpensive source of electricity is essential for large-scale methods. As always, the current and voltage requirements depend on the electrochemical potential and any overvoltage effects.

Figure 22.13 shows, in periodic table format, the most common step for obtaining each free element from its mineral.

☐ Uncombined in nature

☐ Reduction of molten halide (or oxide)
   electrolytically

☐ Reduction of halide with active metal
   (e.g., Na, Mg, Ca)

☐ Reduction of oxide/halide with Al or H_2

☐ Reduction of oxide with C (coke or charcoal)

☐ Oxidation of anion (or oxoanion) chemically
   and/or electrolytically

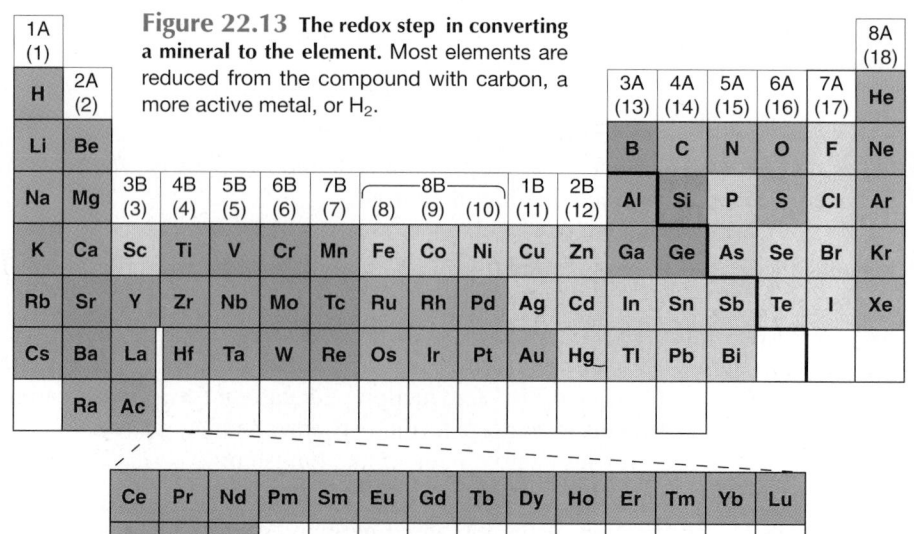

Figure 22.13 The redox step in converting a mineral to the element. Most elements are reduced from the compound with carbon, a more active metal, or H_2.

Refining and Alloying the Element

At this point in the isolation process, the element typically contains impurities from previous steps, so it must be refined. Then, once purified, metallic elements are often alloyed to improve their properties.

Refining (Purifying) the Element Refining is a purification procedure, and three common methods are mentioned here. An electrolytic cell is employed in **electrorefining,** in which the impure metal acts as the anode and a sample of the pure metal acts as the cathode. As the reaction proceeds, the anode slowly disintegrates, and the metal ions are reduced to deposit on the cathode. In some cases, the impurities, which fall beneath the disintegrating anode, are the source of several valuable and less abundant elements. (We examine the electrorefining of copper shortly.)

Metals with relatively low boiling points, such as zinc and mercury, are refined by *distillation.*

In the process of **zone refining** (see the Gallery, p. 512), impurities are removed from a bar of the element by concentrating them in a thin molten zone, while the purified element recrystallizes. Metalloids used in electronic semiconductors, such as silicon and germanium, must be zone refined to greater than 99.999999% purity.

Alloying the Purified Element An **alloy** is a metal-like mixture consisting of solid phases of two or more pure elements, a solid solution, or, in some cases, distinct intermediate phases (these alloys are sometimes referred to as *intermetallic compounds*). The separate phases are often so finely divided that they can only be distinguished microscopically. Alloying a metal with other metals (and, in some cases, nonmetals) is done to alter the metal's melting point and to enhance properties such as luster, conductivity, malleability, ductility, and strength.

Iron, probably the most important metal, is used only when it is alloyed. In pure form, it is soft and corrodes easily. However, when alloyed with carbon and other metals, such as Mo for hardness and Cr and Ni for corrosion resistance, it forms the various steels. Copper stiffens when zinc is added to make brass. Mercury solidifies when alloyed with sodium, and vanadium becomes extremely tough with some carbon added. Table 22.3 shows the composition and uses of some common alloys.

Table 22.3 Some Familiar Alloys and Their Composition

Name	Composition (Mass %)	Uses
Stainless steel	73–79 Fe, 14–18 Cr, 7–9 Ni	Cutlery, instruments
Nickel steel	96–98 Fe, 2–4 Ni	Cables, gears
High-speed steels	80–94 Fe, 14–20 W (or 6–12 Mo)	Cutting tools
Permalloy	78 Ni, 22 Fe	Ocean cables
Bronzes	70–95 Cu, 1–25 Zn, 1–18 Sn	Statues, castings
Brasses	50–80 Cu, 20–50 Zn	Plating, ornamental objects
Sterling silver	92.5 Ag, 7.5 Cu	Jewelry, tableware
14-Carat gold	58 Au, 4–28 Ag, 14–28 Cu	Jewelry
18-Carat white gold	75 Au, 12.5 Ag, 12.5 Cu	Jewelry
Typical tin solder	67 Pb, 33 Sn	Electrical connections
Dental amalgam	69 Ag, 18 Sn, 12 Cu, 1 Zn (dissolved in Hg)	Dental fillings

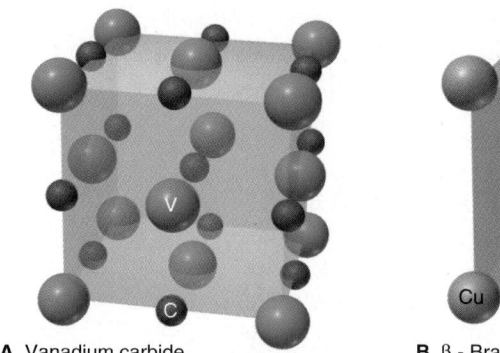

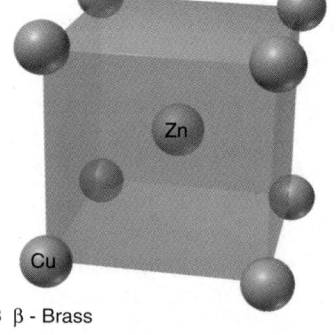

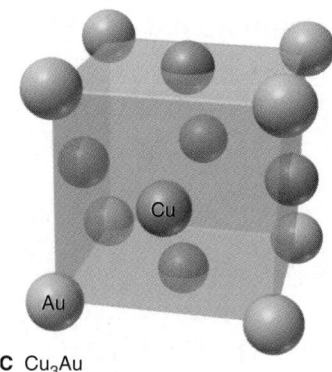

A Vanadium carbide   **B** β - Brass   **C** Cu₃Au

Figure 22.14 Three binary alloys.

The simplest alloys, called *binary alloys,* contain only two elements. In some cases, *the added element enters interstices,* spaces between the parent metal atoms in the crystal structure (Section 12.6). In vanadium carbide, for instance, carbon atoms occupy holes of a face-centered cubic vanadium lattice (Figure 22.14A). In other cases, the *added element substitutes for parent atoms* in the unit cell. In many types of brass, for example, relatively few zinc atoms substitute randomly for copper atoms in the face-centered cubic copper lattice (see Figure 13.4A). On the other hand, β-brass is an intermediate phase that crystallizes as the Zn:Cu ratio approaches 1:1 in a structure with a Zn atom surrounded by eight Cu atoms (Figure 22.14B). In one alloy of copper and gold, Cu atoms occupy the faces of a face-centered cube, and Au atoms lie at the corners (Figure 22.14C).

Atomic properties, including size, electron configuration, and number of valence electrons, determine which metals form stable alloys. Quantitative predictions indicate that transition metals having few *d* electrons often form stable alloys with transition metals having many *d* electrons. Thus, metals from the left half of the *d* block (Groups 3 to 5) often form alloys with metals from the right half (Groups 9 to 11). For example, electron-poor Zr, Nb, and Ta form strong alloys with electron-rich Ir, Pt, or Au; an example is the very stable ZrPt₃.

SECTION SUMMARY

Metallurgy involves mining an ore, separating it from debris, pretreating it to concentrate the mineral source, converting the mineral to another compound that is easier to process further, reducing this compound to the metal, purifying the metal, and in many cases, alloying it to obtain a more useful material.

22.4 TAPPING THE CRUST: ISOLATION AND USES OF THE ELEMENTS

Once an element's abundance and sources are known and the metallurgical processes chosen, the isolation process begins. Obviously, the process depends on the physical and chemical properties of the source: we isolate an element that occurs uncombined in the air differently from one dissolved in the sea or one found in a rocky ore. In this section, we detail methods for recovering some important elements; pay special attention to the application of ideas from earlier chapters to these key industrial processes.

Producing the Alkali Metals: Sodium and Potassium

The alkali metals are among the most reactive elements and thus are always found as ions in nature, either in solid minerals or in aqueous solution. In the laboratory, the free elements must be protected from contact with air (for example, by being stored under mineral oil) to prevent their immediate oxidation and to minimize the possibility of fires. The two most important alkali metals are sodium and potassium. Their abundant, water-soluble compounds are used throughout industry and research, and the Na^+ and K^+ ions are essential to organisms.

Industrial Production of Sodium and Potassium The sodium ore is *halite* (largely NaCl), which is obtained either by evaporation of concentrated salt solutions (brines) or by mining vast salt deposits formed from the evaporation of prehistoric seas. The Cheshire salt field in Britain, for example, is 60 km by 24 km by 400 m thick and contains more than 90% NaCl. Other large deposits occur in New Mexico, Michigan, New York, and Kansas.

The brine is evaporated to dryness, and the solid is crushed and fused (melted) for use in an electrolytic apparatus called the **Downs cell** (Figure 22.15). To reduce heating costs, the NaCl (mp = 801°C) is mixed with $1\frac{1}{2}$ parts $CaCl_2$ to form a mixture that melts at only 580°C. Reduction of the metal ions to Na and Ca takes place at a cylindrical steel cathode, with the molten metals floating on the denser molten salt mixture. As they rise through a short collecting pipe, the liquid Na is siphoned off, while a higher melting Na/Ca alloy solidifies and falls back into the molten electrolyte. Chloride ions are oxidized to Cl_2 gas at a large anode within an inverted cone-shaped chamber. The cell design separates the metals from the Cl_2 to prevent their explosive recombination. The Cl_2 gas is collected, purified, and sold as a valuable by-product.

Sylvite (mostly KCl) is the major ore of potassium. The metal is too soluble in molten KCl to be prepared by a method similar to that for sodium. Instead, chemical reduction of K^+ ions by liquid Na is used. The method is based on atomic properties and the nature of equilibrium systems. Recall that an Na atom is smaller than a K atom, so it holds its outer electron more tightly, as the first ionization energies indicate: IE_1 of Na = 496 kJ/mol; IE_1 of K = 419 kJ/mol.

A Limitless Oceanic Supply of NaCl Will we ever run out of sodium chloride? Not likely. The rock-salt (halite) equivalent of NaCl in the world's oceans has been estimated at 1.9×10^7 km³, about 150% of the volume of North America above sea level. Or, looking at this amount another way: a row of 1-km³ cubes of NaCl would stretch from the Earth to the Moon 47 times! The photo shows evaporative salt beds near San Francisco Bay.

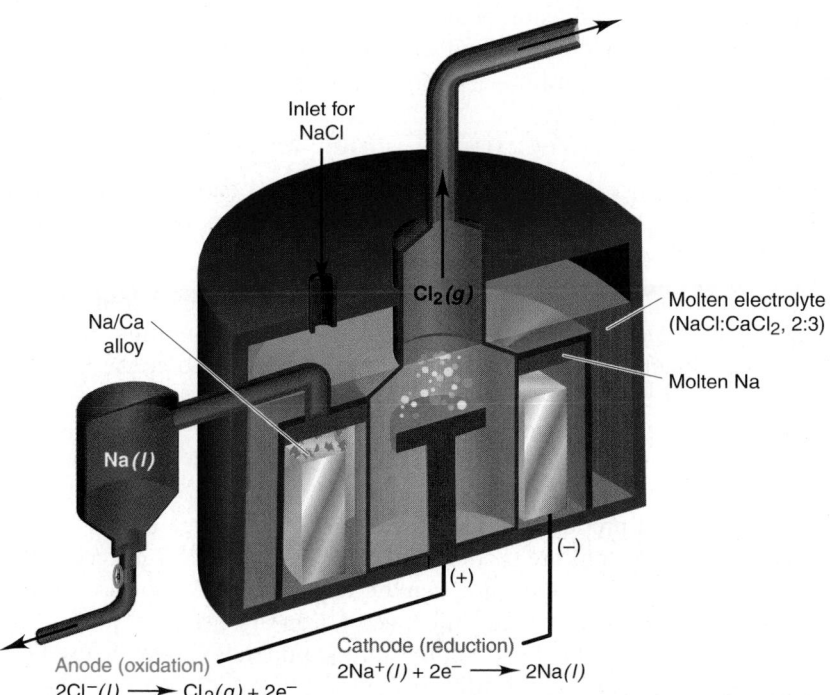

Inlet for
NaCl

$Cl_2(g)$

Na/Ca
alloy

Molten electrolyte
(NaCl:CaCl₂, 2:3)

Molten Na

Na (l)

(−)

(+)

Anode (oxidation)
$2Cl^-(l) \longrightarrow Cl_2(g) + 2e^-$

Cathode (reduction)
$2Na^+(l) + 2e^- \longrightarrow 2Na(l)$

Figure 22.15 The Downs cell for production of sodium. The mixture of solid NaCl and CaCl₂ forms the molten electrolyte. Sodium and calcium are formed at the cathode and float, but due to its higher melting point, an Na/Ca alloy solidifies and falls back into the bath. Chlorine gas forms at the anode.

Thus, if the isolation method for K were based solely on this atomic property, Na would not be very effective at reducing K^+, and the reaction would not move very far toward products. An ingenious approach, which applies Le Châtelier's principle, overcomes this drawback. The reduction is carried out at 850°C, which is above the boiling point of K, so the equilibrium mixture contains *gaseous* K:

$$Na(l) + K^+(l) \rightleftharpoons Na^+(l) + K(g)$$

As the K gas is removed, the system shifts to produce more K. The gas is then condensed and purified by fractional distillation. The same general method (with Ca as the reducing agent) is used to produce rubidium and cesium.

Uses of Sodium and Potassium The compounds of Na and K (and indeed, of all the alkali metals) have many more uses than the elements themselves [see Group 1A(1) Family Portrait, p. 551]. Nevertheless, there are some interesting uses of the metals that take advantage of their strong reducing power. Large amounts of Na were used as an alloy with lead to make gasoline antiknock additives, such as tetraethyllead:

$$4C_2H_5Cl(g) + 4Na(s) + Pb(s) \longrightarrow (C_2H_5)_4Pb(l) + 4NaCl(s)$$

The toxic effects of environmental lead have made this use virtually nonexistent in the United States. Moreover, although leaded gasoline is still used in parts of Europe, recent legislation promises to eliminate this product early in this decade.

If certain types of nuclear reactors, called *breeder reactors* (Chapter 24), become a practical way to generate energy in the United States, Na production would increase enormously. Its low melting point, viscosity, and absorption of neutrons, combined with its high thermal conductivity and heat capacity, make it perfect for cooling the reactor and exchanging heat to the steam generator.

The major use of potassium at present is in an alloy with sodium for use as a heat exchanger in chemical and nuclear reactors. Another application is in the production of its superoxide, which it forms by direct contact with O_2:

$$K(s) + O_2(g) \longrightarrow KO_2(s)$$

This material is employed as an emergency source of O_2 in breathing masks for miners, divers, submarine crews, and firefighters (see photo):

$$4KO_2(s) + 4CO_2(g) + 2H_2O(g) \longrightarrow 4KHCO_3(s) + 3O_2(g)$$

The Indispensable Three: Iron, Copper, and Aluminum

Their familiar presence, innumerable applications, and enormous production levels make three metals—iron (in the form of steel), copper, and aluminum—stand out as indispensable materials of industrial society.

Metallurgy of Iron and Steel Although people have practiced iron smelting for more than 3000 years, it was only a little over 225 years ago that iron assumed its current dominant role. In 1773, an inexpensive process to convert coal to carbon in the form of coke was discovered, and the material was used in a blast furnace. The coke process made iron smelting cheap and efficient and led to large-scale iron production, which ushered in the Industrial Revolution.

Modern society rests, quite literally, on the various alloys of iron known as **steel.** Although steel production has grown enormously since the 18th century— more than 700 million tons are produced annually—the process of recovering iron from its ores still employs the same general approach: *reduction by carbon in a blast furnace.* The most important minerals of iron are listed in Table 22.4. The first four are used for steelmaking, but the sulfide minerals cannot be used because traces of sulfur make the steel brittle.

The process of converting iron ore to iron metal involves a series of overall redox and acid-base reactions that appear quite simple, although the detailed chemistry is very complex and not entirely understood even today. A modern **blast furnace** (Figure 22.16), such as those used in South Korea and Japan, is a tower

Firefighter with KO_2 breathing mask

Table 22.4 Important Minerals of Iron

Mineral Type	Mineral, Formula
Oxide	Hematite, Fe_2O_3
	Magnetite, Fe_3O_4
	Ilmenite, $FeTiO_3$
Carbonate	Siderite, $FeCO_3$
Sulfide	Pyrite, FeS_2
	Pyrrhotite, FeS

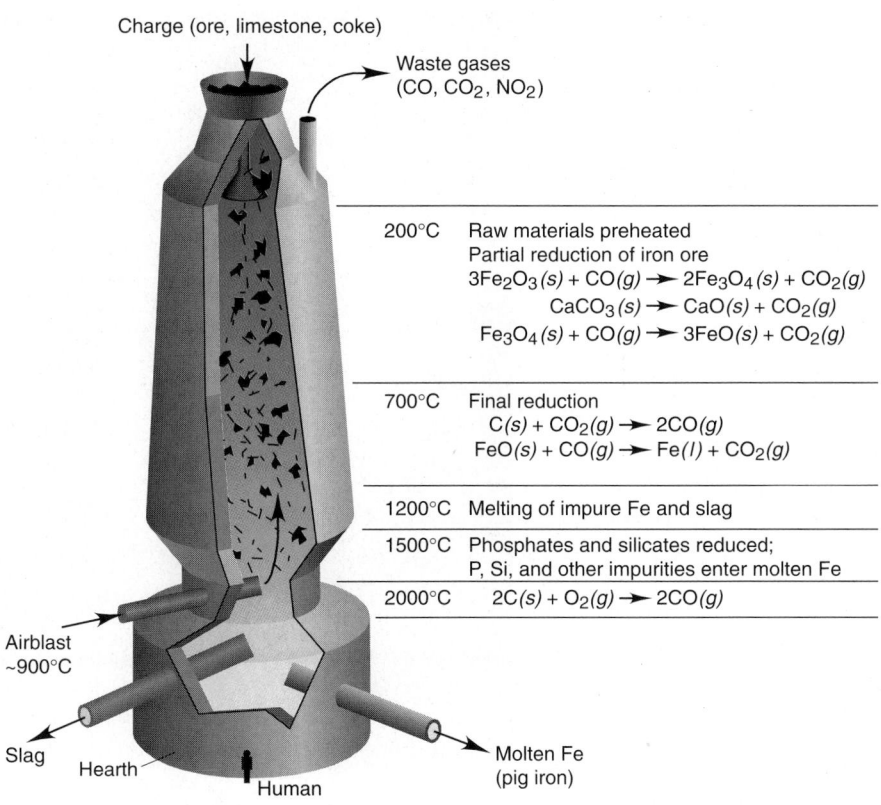

Charge (ore, limestone, coke)

Waste gases
(CO, CO_2, NO_2)

200°C	Raw materials preheated
	Partial reduction of iron ore
	$3Fe_2O_3(s) + CO(g) \rightarrow 2Fe_3O_4(s) + CO_2(g)$
	$CaCO_3(s) \rightarrow CaO(s) + CO_2(g)$
	$Fe_3O_4(s) + CO(g) \rightarrow 3FeO(s) + CO_2(g)$
700°C	Final reduction
	$C(s) + CO_2(g) \rightarrow 2CO(g)$
	$FeO(s) + CO(g) \rightarrow Fe(l) + CO_2(g)$
1200°C	Melting of impure Fe and slag
1500°C	Phosphates and silicates reduced;
	P, Si, and other impurities enter molten Fe
2000°C	$2C(s) + O_2(g) \rightarrow 2CO(g)$

Airblast
~900°C

Slag

Hearth

Human

Molten Fe
(pig iron)

Figure 22.16 The major reactions in a blast furnace. Iron is produced from iron ore, coke, and limestone in a complex process that depends on the temperature at different heights in the furnace. (A human is shown for scale.)

IRON SMELTING

about 14 m wide by 40 m high, made of a brick material that can withstand intense heat. The *charge,* which consists of an iron ore (usually hematite) containing the mineral, coke, and limestone, is fed through the top. More coke is burned in air at the bottom. The charge falls and meets a *blast* of rapidly rising hot air created by the burning coke:

$$2C(s) + O_2(g) \longrightarrow 2CO(g) + \textit{heat}$$

At the bottom of the furnace, the temperature exceeds 2000°C, while at the top, it reaches only 200°C. As a result, different stages of the overall reaction process occur at different heights as the charge descends:

1. In the upper part of the furnace (200°C to 700°C), the charge is preheated, and a *partial reduction* step occurs. The hematite is reduced to magnetite and then to iron(II) oxide (FeO) by CO, the actual reducing agent for the iron oxides. Carbon dioxide is formed as well. Because the blast of hot air passes through the entire furnace in only 10 s, the various gas-solid reactions do *not* reach equilibrium, and many intermediate products form. Limestone also decomposes to form CO_2 and the basic oxide CaO, which is important later in the process.

2. Lower down, at temperatures of 700°C to 1200°C, a *final reduction* step occurs, as some of the coke reduces CO_2 to form more CO, which reduces the FeO to Fe.

3. At still higher temperatures, between 1200°C and 1500°C, the iron melts and drips to the bottom of the furnace. Acidic silica particles from the gangue react with basic calcium oxide in a Lewis acid-base reaction to form a molten waste product called **slag:**

$$CaO(s) + SiO_2(s) \longrightarrow CaSiO_3(l)$$

The siliceous slag drips down and floats on the denser iron (much as the Earth's silicate-rich mantle and crust float on its molten iron core). ●

Was It Slag That Made the Great Ship Go Down? The direct observation of the sunken luxury liner Titanic shows that six small slits caused by collision with an iceberg, rather than an enormous gash, sank the ship. The slits appear between sections of steel plates in the lower hull and suggest weakness in the rivets that connected the plates. Metallurgical analysis of the rivets shows three times as much slag content as modern rivets. Too much of this silicate-rich material makes the wrought iron used in the rivets brittle and easier to break under stress. Moreover, rather than being evenly distributed, the slag particles within the iron, viewed microscopically, appear to be clumped, which causes weak spots. Could low-grade rivets with too much slag have caused this tragedy? Follow-up analysis of more rivets and a historical search of rivet standards in the early 1900s may finally solve the mystery of what brought down the unsinkable Titanic and caused the deaths of 1500 passengers.

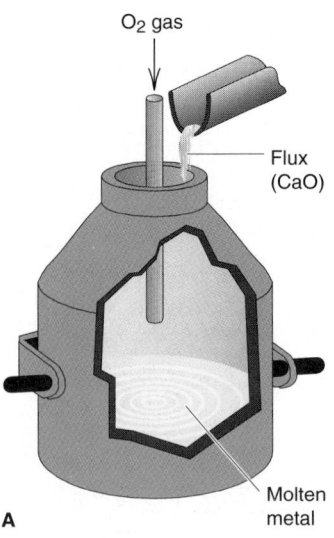

O₂ gas

Flux
(CaO)

Molten
metal

A

Figure 22.17 The basic-oxygen process for making steel. A, Jets of pure O_2 in combination with a basic flux (CaO) are used to remove impurities and lower the carbon content of pig iron in the manufacture of steel. **B,** Impure molten iron is added to a basic-oxygen furnace.

B

4. Some unwanted reactions occur at this hottest stage, between 1500°C and 2000°C. Any remaining phosphates and silicates are reduced to P and Si, and some Mn and traces of S dissolve into the molten iron along with carbon. The resulting impure product is called *pig iron* and contains about 3% to 4% C. A small amount of pig iron is used to make *cast iron,* but most is purified and alloyed to make various kinds of steel.

Pig iron is converted to steel in a separate furnace by means of the **basic-oxygen process,** as shown in Figure 22.17. High-pressure O_2 is blown over and through the molten iron, so that impurities (C, Si, P, Mn, S) are oxidized rapidly. The highly negative heats of formation of their oxides (such as ΔH_f^0 of CO_2 = -394 kJ/mol and ΔH_f^0 of SiO_2 = -911 kJ/mol) raise the temperature, which speeds the reaction. A lime (CaO) flux is added, which converts the oxides to a molten slag [primarily $CaSiO_3$ and $Ca_3(PO_4)_2$] that is decanted from the molten steel. The product is **carbon steel,** which contains 1% to 1.5% C and other impurities. It is alloyed with metals that prevent corrosion and increase its strength or flexibility.

Isolation and Electrorefining of Copper After many centuries of being mined to make bronze and brass articles, copper ores have become less plentiful and less rich in copper, so the metal is more expensive to extract. ● Despite this, more than 2.5 billion pounds of copper is produced in the United States annually. The most common copper ore is chalcopyrite, $CuFeS_2$, a mixed sulfide of FeS and CuS. Most remaining deposits contain less than 0.5% Cu by mass. To "win" this small amount of copper from the ore requires several metallurgical steps, including a final refining to achieve the 99.99% purity needed for electrical wiring, copper's most important application.

● **The Dawns of Three New Ages**
Copper has been known since prehistoric times because it occurs naturally in uncombined form. Its ores were probably first reduced by charcoal fires as early as 3500 BC, which marked the beginning of the Copper Age. With another 500 years of experience, people of India and Greece added molten tin to obtain a harder material, which ushered in the Bronze Age. Some 1800 years later, brass, a copper-zinc alloy, was made in Palestine and used extensively by the Romans for well over 2000 years.

The low copper content in chalcopyrite must be enriched by removing the iron. The first step in copper extraction is pretreatment by flotation (see Figure 22.11, p. 964), which concentrates the ore to around 15% Cu by mass. The next step in many processing plants is a controlled roasting step, which oxidizes the FeS but not the CuS:

$$2FeCuS_2(s) + 3O_2(g) \longrightarrow 2CuS(s) + 2FeO(s) + 2SO_2(g)$$

To remove the FeO and convert the CuS to a more convenient form, the mixture is heated to 1100°C with sand and more of the concentrated ore. Several reactions occur in this step. The FeO reacts with sand to form a molten slag:

$$FeO(s) + SiO_2(s) \longrightarrow FeSiO_3(l)$$

The CuS is thermodynamically unstable at the elevated temperature and decomposes to yield Cu_2S, which is drawn off as a liquid.

In the final smelting step, the Cu_2S is roasted in air, which converts some of it to Cu_2O:

$$2Cu_2S(s) + 3O_2(g) \longrightarrow 2Cu_2O(s) + 2SO_2(g)$$

The two copper(I) compounds then react, with sulfide ion acting as the reducing agent:

$$Cu_2S(s) + 2Cu_2O(s) \longrightarrow 6Cu(l) + SO_2(g)$$

The copper obtained at this stage is usable for plumbing, but it must be purified further for electrical applications by removing unwanted impurities (Fe and Ni) as well as valuable ones (Ag, Au, and Pt). Purification is accomplished by *electrorefining*, which involves the oxidation of Cu and the formation of Cu^{2+} ions in solution, followed by their reduction and the plating out of Cu metal (Figure 22.18). The impure copper obtained from smelting is cast into plates to be used as anodes, and cathodes are made from already purified copper. The electrodes are immersed in acidified $CuSO_4$ solution, and a controlled voltage is applied that accomplishes two tasks simultaneously:

1. Copper and the more active impurities (Fe, Ni) are oxidized to their cations, while the less active ones (Ag, Au, Pt) are not. As the anode slabs react, these unoxidized metals fall off as a valuable "anode mud" and are purified separately. Sale of the precious metals in the anode mud nearly offsets the cost of electricity to operate the cell, which is the main reason Cu wire is so inexpensive.

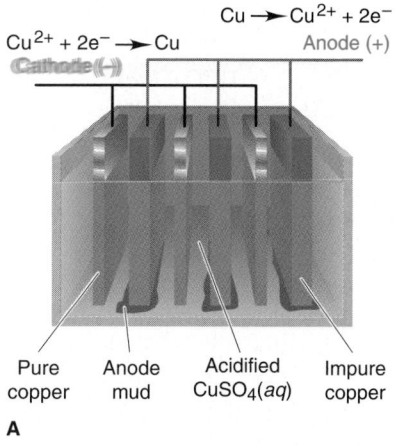

$Cu^{2+} + 2e^- \longrightarrow Cu$
Cathode (−)

$Cu \longrightarrow Cu^{2+} + 2e^-$
Anode (+)

Pure copper | Anode mud | Acidified $CuSO_4(aq)$ | Impure copper

A

B

Figure 22.18 The electrorefining of copper. A, Copper is refined electrolytically, using impure slabs of copper as anodes and sheets of pure copper as cathodes. The Cu^{2+} ions released from the anode are reduced to Cu metal and plate out at the cathode. The "anode mud" contains valuable metal by-products. **B,** A small section of an industrial facility for electrorefining copper.

2. Because Cu is much less active than the Fe and Ni impurities, Cu^{2+} ions are reduced at the cathode, but Fe^{2+} and Ni^{2+} ions remain in solution:

$$Cu^{2+}(aq) + 2e^- \longrightarrow Cu(s) \quad E^0 = 0.34 \text{ V}$$
$$Ni^{2+}(aq) + 2e^- \longrightarrow Ni(s) \quad E^0 = -0.25 \text{ V}$$
$$Fe^{2+}(aq) + 2e^- \longrightarrow Fe(s) \quad E^0 = -0.44 \text{ V}$$

Isolation, Uses, and Recycling of Aluminum Aluminum is the most abundant metal in the Earth's crust by mass and the third most abundant element (after O and Si). It is found in numerous aluminosilicate minerals (Section 14.6), such as feldspars, micas, and clays, and in the rare gems garnet, beryl, spinel, and turquoise. Corundum, pure aluminum oxide (Al_2O_3), is extremely hard; mixed with traces of transition metals, it exists as ruby and sapphire. Impure Al_2O_3 is used in sandpaper and other abrasives.

Through eons of weathering, certain clays became *bauxite,* the major ore of aluminum. This mixed oxide-hydroxide occurs in enormous surface deposits in Mediterranean and tropical regions (see photo), but with world aluminum production approaching 100 million tons annually, it may someday be scarce. In addition to hydrated Al_2O_3 (about 75%), industrial-grade bauxite also contains Fe_2O_3, SiO_2, and TiO_2, which are removed during the extraction.

The overall two-step process combines hydro- and electrometallurgical techniques. In the first step, Al_2O_3 is separated from bauxite; in the second, it is converted to the metal.

1. *Isolating Al_2O_3 from bauxite.* After mining, bauxite is pretreated by extended boiling in 30% NaOH in the *Bayer process,* which involves acid-base, solubility, and complex-ion equilibria. The acidic SiO_2 and the amphoteric Al_2O_3 dissolve in the base, but the basic Fe_2O_3 and TiO_2 do not:

$$SiO_2(s) + 2NaOH(aq) + 2H_2O(l) \longrightarrow Na_2Si(OH)_6(aq)$$
$$Al_2O_3(s) + 2NaOH(aq) + 3H_2O(l) \longrightarrow 2NaAl(OH)_4(aq)$$
$$Fe_2O_3(s) + NaOH(aq) \longrightarrow \text{no reaction}$$
$$TiO_2(s) + NaOH(aq) \longrightarrow \text{no reaction}$$

Further heating slowly precipitates the $Na_2Si(OH)_6$ as an aluminosilicate, which is filtered out with the insoluble Fe_2O_3 and TiO_2 ("red mud").

Acidifying the filtrate precipitates Al^{3+} as $Al(OH)_3$. Recall from our discussion of complex-ion equilibria that the aluminate ion, $Al(OH)_4^-(aq)$, is actually the complex ion $Al(H_2O)_2(OH)_4^-$, in which four of the six water molecules surrounding Al^{3+} have each lost a proton (see Figure 19.15, p. 839). Weakly acidic CO_2 is added to produce a small amount of H^+ ion, which reacts with this complex ion. Cooling supersaturates the solution, and the solid forms and is filtered:

$$CO_2(g) + H_2O \rightleftharpoons H^+(aq) + HCO_3^-(aq)$$
$$Al(H_2O)_2(OH)_4^-(aq) + H^+(aq) \longrightarrow Al(H_2O)_3(OH)_3(s)$$

[Recall that we usually write $Al(H_2O)_3(OH)_3$ more simply as $Al(OH)_3$.]

Drying at high temperature converts the hydroxide to the oxide:

$$2Al(H_2O)_3(OH)_3(s) \xrightarrow{\Delta} Al_2O_3(s) + 9H_2O(g)$$

2. *Converting Al_2O_3 to the free metal.* Aluminum is an active metal, much too strong a reducing agent to be formed at the cathode from aqueous solution (Section 21.7), so the oxide itself must be electrolyzed. However, the melting point of Al_2O_3 is very high (2030°C), so it is dissolved in molten *cryolite* (Na_3AlF_6) to give a mixture that is electrolyzed at ~1000°C. Obviously, the use of cryolite provides a major energy (and cost) savings. The only sizeable cryolite mines are in Greenland, however, and they cannot supply enough natural mineral to meet the demand. Therefore, production of synthetic cryolite has become a major subsidiary industry in aluminum manufacture.

Mining bauxite.

The electrolytic step, called the *Hall-Heroult process*, takes place in a graphite-lined furnace, with the lining itself acting as the cathode. Anodes of graphite dip into the molten Al_2O_3/Na_3AlF_6 mixture (Figure 22.19). The cell typically operates at a moderate voltage of 4.5 V, but with an enormous current flow of 1.0×10^5 to 2.5×10^5 A.

The process is complex and its details are still not entirely known. Therefore, the specific reactions shown below are chosen from among several other possibilities. Molten cryolite contains several ions (including AlF_6^{3-}, AlF_4^{-}, and F^{-}), which react with Al_2O_3 to form fluoro-oxy ions (including $AlOF_3^{2-}$, $Al_2OF_6^{2-}$, and $Al_2O_2F_4^{2-}$) that dissolve in the mixture. For example,

$$2Al_2O_3(s) + 2AlF_6^{3-}(l) \longrightarrow 3Al_2O_2F_4^{2-}(l)$$

Al forms at the cathode (reduction), shown here with AlF_6^{3-} as reactant:

$$AlF_6^{3-}(l) + 3e^{-} \longrightarrow Al(l) + 6F^{-}(l) \quad \text{[cathode (reduction)]}$$

The graphite anode itself is oxidized and forms carbon dioxide gas. Using one of the fluoro-oxy species as an example, the anode reaction is

$$Al_2O_2F_4^{2-}(l) + 8F^{-}(l) + C(graphite) \longrightarrow 2AlF_6^{3-}(l) + CO_2(g) + 4e^{-}$$

[anode (oxidation)]

Thus, the anodes are consumed in this half-reaction and must be replaced frequently.

Combining the three previous equations and making sure that e^{-} gained at the cathode equal e^{-} lost at the anode gives the overall reaction:

$$2Al_2O_3(\text{in } Na_3AlF_6) + 3C(graphite) \longrightarrow 4Al(l) + 3CO_2(g) \quad \text{[overall (cell) reaction]}$$

The Hall-Heroult process uses an enormous quantity of energy: aluminum production accounts for more than 5% of total U.S. electrical usage.

Energy Received and Returned

Aluminum manufacture in the United States uses more electricity in 1 day than a city of 100,000 uses in 1 year! The reason for the high energy needs of Al manufacture is the electron configuration of Al, ([Ne] $3s^2 3p^1$). Each Al^{3+} ion needs $3e^{-}$ to form an Al atom, and the atomic mass of Al is so low (~27 g/mol) that 1 mol of e^{-} produces only 9 g of Al. Compare this with the masses of Mg or Ca (two other lightweight structural metals) that 1 mol of e^{-} produces:

$$\tfrac{1}{3}Al^{3+} + e^{-} = \tfrac{1}{3}Al(s), \sim 9 \text{ g}$$

$$\tfrac{1}{2}Mg^{2+} + e^{-} = \tfrac{1}{2}Mg(s), \sim 12 \text{ g}$$

$$\tfrac{1}{2}Ca^{2+} + e^{-} = \tfrac{1}{2}Ca(s), \sim 20 \text{ g}$$

To turn this disadvantage around, we need an aluminum battery. Once produced, the Al in the battery represents a concentrated form of electrical energy that can deliver 1 mol of e^{-} (96,500 C) for every 9 g of Al consumed. Thus, its electrical output per gram of metal is high. In fact, aluminum-air batteries are now being produced.

ALUMINUM PRODUCTION

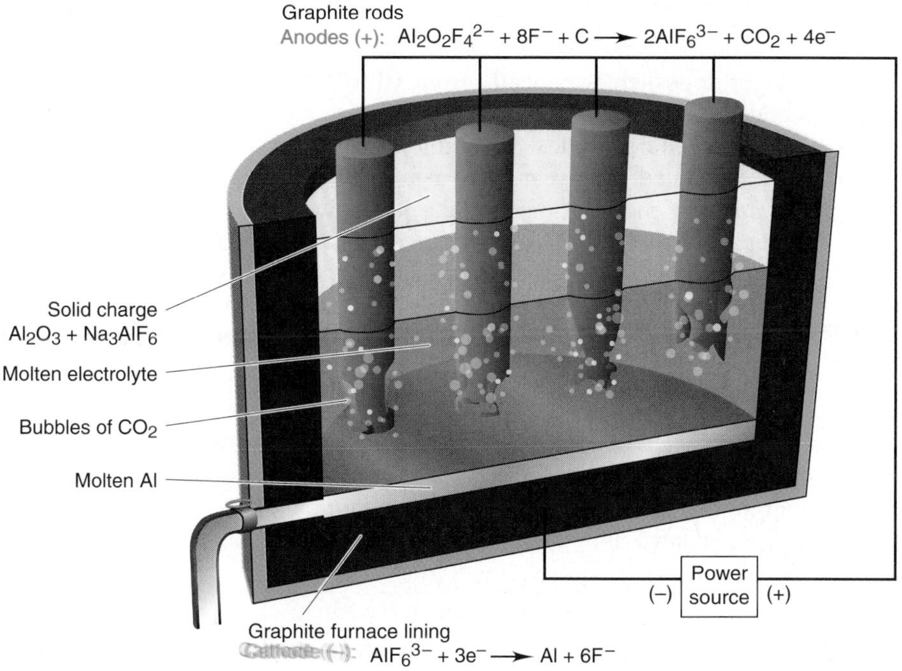

Graphite rods
Anodes (+): $Al_2O_2F_4^{2-} + 8F^{-} + C \longrightarrow 2AlF_6^{3-} + CO_2 + 4e^{-}$

Solid charge
$Al_2O_3 + Na_3AlF_6$

Molten electrolyte

Bubbles of CO_2

Molten Al

Power source   (−)   (+)

Graphite furnace lining
Cathode (−): $AlF_6^{3-} + 3e^{-} \longrightarrow Al + 6F^{-}$

Figure 22.19 The electrolytic cell in the manufacture of aluminum. Purified Al_2O_3 is mixed with cryolite (Na_3AlF_6) and melted. Reduction at the graphite furnace lining (cathode) gives molten Al. Oxidation at the graphite rods (anodes) slowly converts them to CO_2, so they must be replaced periodically.

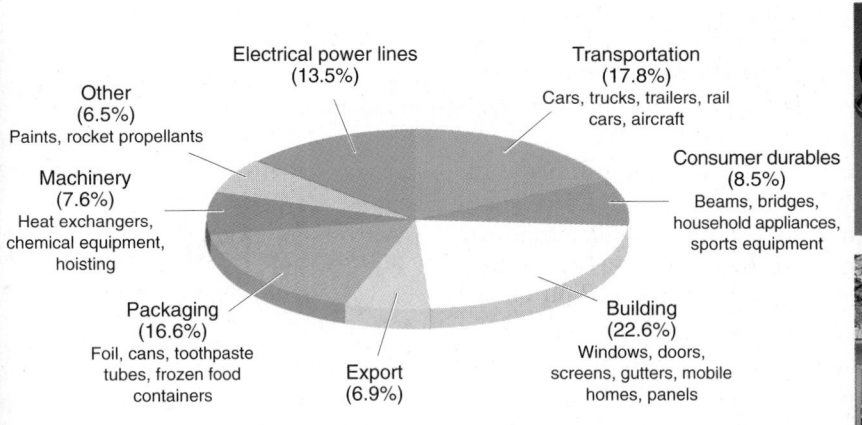

Other
(6.5%)
Paints, rocket propellants

Electrical power lines
(13.5%)

Transportation
(17.8%)
Cars, trucks, trailers, rail
cars, aircraft

Machinery
(7.6%)
Heat exchangers,
chemical equipment,
hoisting

Consumer durables
(8.5%)
Beams, bridges,
household appliances,
sports equipment

Packaging
(16.6%)
Foil, cans, toothpaste
tubes, frozen food
containers

Export
(6.9%)

Building
(22.6%)
Windows, doors,
screens, gutters, mobile
homes, panels

Figure 22.20 **The many familiar and essential uses of aluminum.**

Aluminum is a superb decorative, functional, and structural metal. It is lightweight, attractive, easy to work, and forms strong alloys, many of which are all around us (Figure 22.20). Although Al is very active, it does not corrode readily because of an adherent oxide layer that forms rapidly in air and prevents more O_2 from penetrating. Nevertheless, when it is in contact with less active metals such as Fe, Cu, and Pb, aluminum becomes the anode and deteriorates rapidly (Section 21.6). To prevent this, aluminum objects are often *anodized,* that is, made to act as the anode in an electrolysis that coats them with an oxide layer. The object is immersed in a 20% H_2SO_4 bath and connected to a graphite cathode:

$$6H^+(aq) + 6e^- \longrightarrow 3H_2(g) \qquad \text{[cathode; reduction]}$$
$$\underline{2Al(s) + 3H_2O(l) \longrightarrow Al_2O_3(s) + 6H^+(aq) + 6e^-} \qquad \text{[anode; oxidation]}$$
$$2Al(s) + 3H_2O(l) \longrightarrow Al_2O_3(s) + 3H_2(g) \qquad \text{[overall (cell) reaction]}$$

The Al_2O_3 layer deposited is typically from 10 to 100 μ thick, depending on the object's intended use.

More than 3.5 billion pounds (1.5 million metric tons) of aluminum cans and packaging are discarded each year—a waste of one of the most useful materials in the world *and* the energy used to make it. A quick calculation of the energy needed to prepare 1 mol of Al from *purified* Al_2O_3, compared with the energy needed for recycling, conveys a clear message. The overall cell reaction in the Hall-Heroult process has a ΔH^0 of 2272 kJ and a ΔS^0 of 635.4 J/K. Considering *only* the free energy change of the reaction at 1000.°C, for 1 of mol Al, we obtain

$$\Delta G^0 = \Delta H^0 - T\Delta S^0 = \frac{2272\ kJ}{4\ mol\ Al} - \left(1273\ K \times \frac{0.6354\ kJ/K}{4\ mol\ Al}\right) = 365.8\ kJ/mol\ Al$$

The molar mass of Al is nearly twice the mass of a soft-drink or beer can, so the electrolysis step requires nearly 200 kJ of energy for each can!

When aluminum is recycled, the major energy input, for melting the cans and foil, has been calculated as ~26 kJ/mol Al. The ratio of these energy inputs is

$$\frac{\text{Energy to recycle 1 mol Al}}{\text{Energy for electrolysis of 1 mol Al}} = \frac{26\ kJ}{365.8\ kJ} = 0.071$$

Based on just these portions of the process, recycling uses about 7% as much energy as electrolysis. Recent energy estimates for the entire manufacturing process (including mining, pretreating, maintaining operating conditions, electrolyzing, and so forth) are about 6000 kJ/mol Al, which means recycling requires less than 1% as much energy as manufacturing! The economic advantages, not to mention the environmental ones, are obvious, and recycling of aluminum has become common in the United States (see photo).

Crushed aluminum cans ready for recycling.

Mining the Sea: Magnesium and Bromine

In the not too distant future, as terrestrial sources of certain elements become scarce or too costly to mine, the oceans will become an important source. Despite the abundant distribution of magnesium on land, it is already being obtained from the sea, and its ocean-based production is a good example of the approach we might one day use for other elements. Bromine, too, is recovered from the sea as well as from inland brines.

Isolation and Uses of Magnesium The *Dow process* for the isolation of magnesium from the sea involves steps that are similar to the procedures used for rocky ores (Figure 22.21):

1. *Mining.* Intake of seawater and straining the debris are the "mining" steps. No pretreatment is needed.
2. *Converting to mineral.* The dissolved Mg^{2+} ion is converted to the mineral $Mg(OH)_2$ with $Ca(OH)_2$, which is generated on-site (at the plant). Seashells $(CaCO_3)$ are crushed, decomposed with heat to CaO, and mixed with water to make slaked lime $[Ca(OH)_2]$. This is pumped into the seawater intake tank to precipitate the dissolved Mg^{2+} as the hydroxide ($K_{sp} \approx 10^{-9}$):
$$Ca(OH)_2(aq) + Mg^{2+}(aq) \longrightarrow Mg(OH)_2(s) + Ca^{2+}(aq)$$
3. *Converting to compound.* The solid $Mg(OH)_2$ is filtered and mixed with excess HCl, which is also made on-site, to form aqueous $MgCl_2$:
$$Mg(OH)_2(s) + 2HCl(aq) \longrightarrow MgCl_2(aq) + 2H_2O(l)$$
The water is evaporated in stages to give solid, hydrated $MgCl_2 \cdot nH_2O$.
4. *Electrochemical redox.* Heating above 700°C drives off the water of hydration and melts the $MgCl_2$. Electrolysis gives chlorine gas and the molten metal, which floats on the denser molten salt:
$$MgCl_2(l) \longrightarrow Mg(l) + Cl_2(g)$$
The Cl_2 that forms is recycled to make the HCl used in step 3.

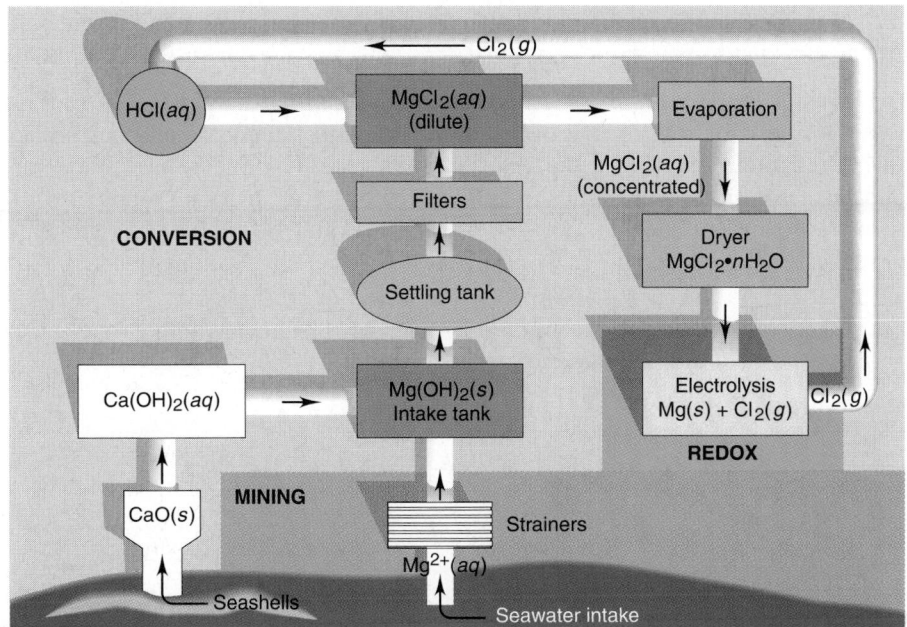

Figure 22.21 The production of elemental Mg from seawater. The multistage Dow process involves obtaining Mg^{2+} ion from seawater, converting the Mg^{2+} to $Mg(OH)_2$ and then $MgCl_2$, which is evaporated to dryness and then electrolytically reduced to Mg metal and Cl_2 gas.

Magnesium is the lightest structural metal available (about one-half the density of Al and one-fifth that of steel). Although Mg is quite reactive, it forms an extremely adherent, high-melting oxide layer (MgO); thus, it finds many uses in metal alloys and can be machined into any form. Magnesium alloys occur in everything from aircraft bodies to camera bodies, and from luggage to auto engine blocks. The pure metal is a strong reducing agent, which makes it useful for sacrificial anodes (Section 21.6) and in the metallurgical extraction of other metals, such as Be, Ti, Zr, Hf, and U. For example, titanium is made from its major ore, ilmenite, in two steps:

$$2FeTiO_3(s) + 7Cl_2(g) + 6C(s) \longrightarrow 2TiCl_4(l) + 2FeCl_3(s) + 6CO(g)$$
$$TiCl_4(l) + 2Mg(l) \longrightarrow Ti(s) + 2MgCl_2(l)$$

Isolation and Uses of Bromine The largest source of bromine is the oceans, where it occurs as Br^- at a concentration of 0.065 g/L (65 ppm). However, the added cost of concentrating such a dilute solution often shifts the choice of source, when available, to much more concentrated salt lakes and natural brines. Arkansas brines, the most important source in the United States, contain about 4.5 g Br^-/L (4500 ppm). The Br^- is readily oxidized to Br_2 with aqueous Cl_2:

$$2Br^-(aq) + Cl_2(aq) \longrightarrow Br_2(l) + 2Cl^-(aq) \qquad \Delta G^0 = -61.5 \text{ kJ}$$

The Br_2 is removed by passing steam through the mixture and then cooling and drying the liquid bromine.

World production of Br_2 is about 1% that of Cl_2. Its major use is in the preparation of organic chemicals, such as ethylene dibromide, which was added to leaded gasoline to prevent lead oxides from forming and depositing on engine parts. Newer products requiring bromine include the flame retardants in rugs and textiles. Bromine is also needed in the synthesis of inorganic chemicals, especially silver bromide for photographic emulsions.

The Many Sources and Uses of Hydrogen

Although hydrogen accounts for 90% of the atoms in the universe, it makes up only 15% of the atoms in the Earth's crust. Moreover, whereas it occurs in the universe mostly as H_2 molecules and free H atoms, virtually all hydrogen in the crust is in combination with other elements, either oxygen in natural waters or carbon in biomass, petroleum, and coal.

Properties of the Hydrogen Isotopes Hydrogen has three naturally occurring isotopes. Ordinary hydrogen (^1H), or protium, is the most abundant and has no neutrons. Deuterium (^2H or D), the next in abundance, has one neutron, and rare, radioactive tritium (^3H or T) has two. All three occur as diatomic molecules, H_2, D_2 (or $^2H_2$), and T_2 (or $^3H_2$), as well as HD, DT, and HT. Table 22.5 compares

Table 22.5 Some Molecular and Physical Properties of Diatomic Protium, Deuterium, and Tritium

Property	H_2	D_2	T_2
Molar mass (g/mol)	2.016	4.028	6.032
Bond length (pm)	74.14	74.14	74.14
Melting point (K)	13.96	18.73	20.62
Boiling point (K)	20.39	23.67	25.04
ΔH^0_{fus} (kJ/mol)	0.117	0.197	0.250
ΔH^0_{vap} (kJ/mol)	0.904	1.226	1.393
Bond energy (kJ/mol at 298 K)	432	443	447

some molecular and physical properties of H_2, D_2, and T_2. As you can see, the heavier the isotope is, the higher the molar mass of the molecule, and the higher its melting point, boiling point, and heats of phase change; it also effuses at a lower rate (Section 5.6).

Because hydrogen is so light, the relative difference in mass of its isotopes is enormous compared to isotopes of other common elements. (For example, the mass of ^{13}C is only 8% greater than that of ^{12}C.) Note, especially, that the mass difference leads to different bond energies, which affects reactivity. Because the mass of D is twice the mass of H, H atoms bonded to a given atom vibrate at a higher frequency than do D atoms, so their bonds are higher in energy. As a result, the *bond to H is weaker and thus quicker to break*. Therefore, any reaction that includes breaking a bond to hydrogen in the rate-determining step occurs *faster with H than with D*. This behavior is called a *kinetic isotope effect,* and we'll see an example of it shortly. No other element displays a kinetic isotope effect nearly as large.

Industrial Production of Hydrogen Hydrogen gas (H_2) is produced on an industrial scale worldwide: 250,000 metric tons (3×10^{12} L at STP; equivalent to $\sim1\times10^{11}$ mol) are produced annually in the United States alone. All production methods are energy intensive, so the choice is determined by energy costs. In Scandinavia, where hydroelectric power is plentiful, electrolysis is the chosen method; on the other hand, where natural gas from oil refineries is plentiful, as in the United States and Great Britain, thermal methods are used to produce hydrogen.

The most common thermal methods use water and a simple hydrocarbon in two steps. Modern American plants use methane, which has the highest H:C ratio of any hydrocarbon. In the first step, the reactants are heated to around 1000°C over a nickel-based catalyst in the endothermic *steam-reforming process:*

$$CH_4(g) + H_2O(g) \longrightarrow CO(g) + 3H_2(g) \qquad \Delta H = 206 \text{ kJ}$$

Heat is supplied by burning methane at the refinery. To generate more H_2, the product mixture (called *water gas*) is heated with steam at 400°C over an iron or cobalt oxide catalyst in the exothermic *water-gas shift reaction,* and the CO reacts:

$$H_2O(g) + CO(g) \rightleftharpoons CO_2(g) + H_2(g) \qquad \Delta H = -41 \text{ kJ}$$

The reaction mixture is recycled several times, which decreases the CO to around 0.2% by volume. Passing the mixture through liquid water removes the more soluble CO_2 (solubility = 0.034 mol/L) from the H_2 (solubility < 0.001 mol/L). Calcium oxide can also be used to remove CO_2 by formation of $CaCO_3$. By removing CO_2, these steps shift the equilibrium position to the right and produce H_2 that is about 98% pure. To attain greater purity ($\sim$99.9%), the gas mixture is passed through a *synthetic zeolite* (see the Gallery, p. 571) selected to filter out nearly all molecules larger than H_2.

In regions where inexpensive electricity is available, very pure H_2 is prepared through electrolysis of water with Pt (or Ni) electrodes:

$$2H_2O(l) + 2e^- \longrightarrow H_2(g) + 2OH^-(aq) \qquad E = -0.42 \text{ V} \quad \text{[cathode; reduction]}$$
$$\underline{H_2O(l) \longrightarrow \tfrac{1}{2}O_2(g) + 2H^+(aq) + 2e^- \qquad E = 0.82 \text{ V} \quad \text{[anode; oxidation]}}$$
$$H_2O(l) \longrightarrow H_2(g) + \tfrac{1}{2}O_2(g) \qquad E_{cell} = -0.42 \text{ V} - 0.82 \text{ V} = -1.24 \text{ V}$$

Overvoltage makes the cell potential about -2 V (Section 21.7). Therefore, under typical operating conditions, it takes about 400 kJ of energy to produce 1 mol of H_2:

$$\Delta G = -nFE = (-2 \text{ mol e}^-)\left(\frac{96.5 \text{ kJ}}{\text{V·mol e}^-}\right)(-2 \text{ V}) = 4\times10^2 \text{ kJ/mol } H_2$$

High-purity O_2 is a valuable by-product that offsets some of the costs.

Calcium reacts with water to form H_2.

Laboratory Production of Hydrogen You may already have produced small amounts of H_2 in the lab by one of several methods. A characteristic reaction of the very active metals in Groups 1A(1) and 2A(2) is the reduction of water to H_2 and OH^- (see photo):

$$Ca(s) + 2H_2O(l) \longrightarrow Ca^{2+}(aq) + 2OH^-(aq) + H_2(g)$$

Alternatively, the strongly reducing hydride ion can be used:

$$NaH(s) + H_2O(l) \longrightarrow Na^+(aq) + OH^-(aq) + H_2(g)$$

Less active metals reduce H^+ in acids:

$$Zn(s) + 2H^+(aq) \longrightarrow Zn^{2+}(aq) + H_2(g)$$

In view of hydrogen's great potential as a fuel, chemists are seeking ways to decompose water by lowering the overall activation energy of the process. More than 10,000 *water-splitting* schemes have been devised; one of the more promising of these schemes was described in Chapter 6 (see the Chemical Connections essay, p. 243).

However, at 298 K, ΔG^0 for the decomposition of 1 mol of $H_2O(g)$ is 229 kJ, so all water-splitting schemes require energy. Given the increase in the number of moles of gas, ΔS^0 is positive; therefore, this is an endothermic process ($\Delta H^0 > 0$) and requires heat to occur spontaneously. One promising source of this heat is focused sunlight.

Industrial Uses of Hydrogen Typically, a plant produces H_2 to make some other end product. In fact, more than 95% of H_2 produced industrially is consumed on-site in ammonia or petrochemical facilities. In a plant that synthesizes NH_3 from N_2 and H_2, the reactant gases are formed through a series of reactions that involve methane, including the steam-reforming and water-gas shift reactions we discussed previously. For this reason, the cost of NH_3 is closely correlated with the cost of CH_4. Here is a typical example of how the reaction series in an ammonia plant works.

The steam-reforming reaction is performed with excess CH_4, which depletes the reaction mixture of H_2O:

$$CH_4(g;\ excess) + H_2O(g) \longrightarrow CO(g) + 3H_2(g)$$

In the next step, an excess of the product mixture (CH_4, CO, and H_2) is burned in an amount of air ($N_2 + O_2$) that is insufficient to effect complete combustion, but is just enough to consume the O_2, heat the mixture to 1100°C, and form additional H_2O:

$$4CH_4(g;\ excess) + 7O_2(g) \longrightarrow 2CO_2(g) + 2CO(g) + 8H_2O(g)$$
$$2H_2(g;\ excess) + O_2(g) \longrightarrow 2H_2O(g)$$
$$2CO(g;\ excess) + O_2(g) \longrightarrow 2CO_2(g)$$

Any remaining CH_4 reacts by the steam-reforming reaction, the remaining CO reacts by the water-gas shift reaction ($H_2O + CO \rightleftharpoons CO_2 + H_2$) to form more H_2, and then the CO_2 is removed with CaO. The amounts are carefully adjusted to produce a final mixture that contains a 1:3 ratio of N_2 (from the added air) to H_2 (with traces of CH_4, Ar, and CO). This mixture is used directly in the synthesis of ammonia, described in Chapter 17 (see the Chemical Connections essay, pp. 774–775).

A second major use of H_2 is the *hydrogenation* of the C=C bonds in liquid oils to form the C—C bonds in solid fats and margarine. The process uses H_2 in contact with transition metal catalysts, such as powdered nickel (see Figure 16.24,

p. 699). Solid fats are used not only as spreads, but also in commercial baked goods. Look at the list of ingredients on most packages of bread, cake, and cookies, and you'll see the "partially hydrogenated vegetable oils" made by this process.

Hydrogen is also essential in the manufacture of numerous "bulk" chemicals, those that are produced in large amounts because they have many further uses. One application that has been gaining great attention is the production of methanol. In this process, carbon monoxide reacts with hydrogen over a copper–zinc oxide catalyst:

$$CO(g) + 2H_2(g) \xrightarrow{\text{Cu-ZnO catalyst}} CH_3OH(l)$$

Many automotive engineers expect methanol to be used increasingly as a gasoline additive and, if an inexpensive source of hydrogen becomes available, used directly as one of several fuel alternatives.

Production and Uses of Deuterium Deuterium and its compounds are produced from D_2O (heavy water), which is present as a minor component (0.016 mol % D_2O) in normal water and is isolated on the multiton scale by *electrolytic enrichment*. This process is based on the kinetic isotope effect for hydrogen noted earlier, specifically on the *higher rate of bond breaking for O—H bonds compared to O—D bonds*, and thus on the higher rate of electrolysis of H_2O compared with D_2O.

For example, using Pt electrodes, H_2O is electrolyzed about 14 times faster than D_2O. As some of the liquid decomposes to the elemental gases, the remainder becomes enriched in D_2O. Thus, by the time the volume of water has been reduced to 1/20,000 of its original volume, the remaining water is around 99% D_2O. By combining samples and repeating the electrolysis, more than 99.9% D_2O is obtained.

Deuterium gas is produced by electrolysis of D_2O or by any of the chemical reactions that produce hydrogen gas from water, such as

$$2Na(s) + 2D_2O(l) \longrightarrow 2Na^+(aq) + 2OD^-(aq) + D_2(g)$$

Similarly, compounds containing deuterium (or tritium) are produced from reactions that give rise to the corresponding hydrogen-containing compound; for example,

$$SiCl_4(l) + 2D_2O(l) \longrightarrow SiO_2(s) + 4DCl(g)$$

Compounds with acidic protons undergo hydrogen/deuterium exchange:

$$CH_3COOH(l) + D_2O(l; \text{ excess}) \longrightarrow CH_3COOD(l) + DHO(l; \text{ small amount})$$

Notice that only the acidic H atom, the one in the COOH group, is exchanged, not any of those attached to carbon. We discuss the natural and synthetic formation of tritium in Chapter 24.

A Group at a Glance: Sources, Isolation, and Uses of the Elements

From a large-scale industrial perspective, certain elements may be more essential than others, but each has its own uses—in many cases, a critical one in a specialty industry. You can survey some of the practical aspects of the elements by consulting the multipart Table 22.6 on the following five pages, which presents sources, isolation methods, and uses of the main-group elements and Period 4 transition elements.

Table 22.6A Sources, Isolation, and Uses of Group 1A(1): The Alkali Metals

Despite their great chemical similarity, the alkali metals rarely occur together in terrestrial minerals because of large differences in ionic size and thus crystal structure. Lithium salts occur with those of magnesium, an example of their diagonal relationship (Section 14.4). Sodium and potassium are the fifth and sixth most abundant metals in the crust. All alkali metals are isolated from their molten salts, through either electrolytic or chemical reduction. Most uses rely on their low densities or great reducing power.

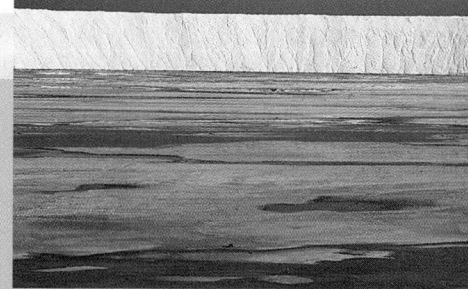

Sea salt harvest, France

Element	Source	Isolation	Uses
Lithium	Spodumene [$LiAl(Si_2O_6)$]	Preparation and electrolysis of molten LiCl	In strong, low-density Mg and Al alloys for armor and aerospace parts; in Li batteries for computers, electric cars
Sodium	NaCl in rock salt (halite); $NaNO_3$ (saltpeter)	Electrolysis of molten NaCl (Downs cell; see text)	Reducing agent for isolation of Ti, Zr, and others; heat exchanger in nuclear reactors
Potassium	KCl (sylvite) in seawater	Na reduction of molten KCl (see text)	Reducing agent; production of KO_2 (see text)
Rubidium	Minor component of Li ores	Ca reduction of molten RbCl; by-product of Li isolation	Reducing agent
Cesium	Minor component of Li ores; pollucite ($Cs_4Al_4Si_9O_{26}\cdot H_2O$)	Ca reduction of molten CsCl; by-product of Li isolation	Reducing agent
Francium	Minute traces from ^{235}U decay		

Table 22.6B Sources, Isolation, and Uses of Group 2A(2): The Alkaline Earth Metals

The alkaline earth metals are extracted mostly from rocky carbonates and sulfates (except for magnesium, which is also isolated from seawater). Like the alkali metals, they are strong reducing agents, so the final isolation step is either electrolytic reduction of the molten chloride or chemical reduction of the oxide with an active metal. The lighter members are used in alloys, the heavier ones as scavengers of nonmetal impurities through formation of ionic compounds.

Dolomite Mountains, Italy

Element	Source	Isolation	Uses
Beryllium	Beryl ($Be_3Al_2Si_6O_{18}$)	Electrolysis of molten $BeCl_2$; reduction of BeF_2 with Mg	In high-strength alloys of Cu and Ni for aerospace engines and electronics; neutron moderator and reflector in nuclear reactors (Section 24.7); window in x-ray tubes
Magnesium	Magnesite ($MgCO_3$), dolomite ($MgCO_3\cdot CaCO_3$), seawater	Electrolysis of molten $MgCl_2$ (see text); silicothermal method, $2(MgO\cdot CaO) + FeSi \longrightarrow 2Mg + Ca_2SiO_4 + Fe$	Lightweight alloys
Calcium	Limestone and aragonite ($CaCO_3$)	Electrolysis of molten $CaCl_2$ formed by HCl on $CaCO_3$	Strengthener in Al alloys; reducing agent to produce Cr, Zr, and U; scavenger of traces of O_2, P, and S in steel
Strontium	Strontianite ($SrCO_3$); celestite ($SrSO_4$)	Thermal decomposition, then Al reduction of SrO	Scavenger of O_2 and N_2 in electronic devices
Barium	Barite ($BaSO_4$)	Al reduction of BaO	Scavenger of O_2 and N_2 in electronic devices
Radium	Minor (0.1 ppb) component in pitchblende (uranium ore)	Electrolysis of molten $RaCl_2$ after extensive extraction (Section 24.1)	Formerly used in cancer therapy

Table 22.6C Sources, Isolation, and Uses of the Period 4 Transition Metals [Groups 3B(3) to 2B(12)]

Several Period 4 transition elements—iron, titanium, and manganese—are among the most abundant metals in the crust and occur together in many ores. Extraction methods vary depending on the source, but all include a reduction step with either carbon or more active metals (magnesium or aluminum). Most of the uses of these metals involve their alloys.

Nodules of Cu, Ni, and Co

Element	Source	Isolation	Uses
Scandium	Thortveitite (40% Sc_2O_3); by-product of U extraction	Reduction of Sc_2O_3 with C	None
Titanium	Rutile (TiO_2); ilmenite ($FeTiO_3$)	Conversion to $TiCl_4$, then reduction with Mg	Very abundant; stronger than steel but half as dense; high-temperature, lightweight alloys for rocket and jet engines and for train and car parts
Vanadium	Carnotite [$K(UO_2)(VO_4)\cdot1.5H_2O$]	Conversion to $NaVO_3$, then reduction with Al (thermite) or FeSi	Combines with C in steel to make very strong alloy for truck springs and axles
Chromium	Chromite ($FeCr_2O_4$)	Conversion to Cr_2O_3, then reduction with Al	Nonferrous alloys; chrome plating; stainless steels
Manganese	Pyrolusite (MnO_2); many other ores; in future, Mn "nodules" on ocean floor	Conversion to Mn_3O_4, then reduction with Al; conversion to $MnSO_4$, then electrolysis of aqueous Mn(II)	Scavenger of O and S in steel; high-strength steel alloys of excavators, rail crossings
Iron	Hematite (Fe_2O_3); magnetite (Fe_3O_4)	Reduction using C (see text)	Steel (see text)
Cobalt	Smaltite ($CoAs_2$); many sulfides with Ni, Cu, and Pb	Roasting in O_2, leaching with H_2SO_4, precipitating $Co(OH)_3$ with ClO^-, heating to form CoO, and reducing with C	Cobalt blue glass and pottery; pigments for paints and inks; catalysts for organic reactions; specialty alloys with Cr and W for drill bits, lathe tools, and surgical instruments; magnetic alloys (Alnico)
Nickel	Pentlandite [$(Ni,Fe)_9S_8$]	Roasting in O_2 to NiO, reducing with C; Mond process, $Ni(CO)_4(g) \rightleftharpoons Ni(s) + 4CO(g)$	Nickel steels for armor; stainless steel and Alnico; nonferrous alloys (tableware) with Ag; Monel with Cu for handling F_2; nichrome; undercoat for chrome plating; hydrogenation catalyst
Copper	Chalcopyrite ($CuFeS_2$)	See text	Wiring, plumbing, coins (see text)
Zinc	Zinc blende (ZnS); sphalerite	Roasting in O_2 to form ZnO, then reducing with C	Brasses (50% to 80% Cu); galvanizing steel to prevent corrosion; batteries (Section 21.5)

Jet engine assembly

Mining chalcopyrite

Table 22.6D Sources, Isolation, and Uses of Group 3A(13): The Boron Family

Most sources are rocky oxides and sulfides. Boron is rare but concentrated in large deposits; the largest (6.4 km × 1.6 km × 30 m thick) is in California's Mojave Desert. Aluminum is the most abundant crustal metal, but bauxite is its only ore. Alloys and semiconductors are the principal end uses.

Borate deposit (tufa), California

Element	Source	Isolation	Uses
Boron	Borax ($Na_2[B_4O_5(OH)_4]\cdot 8H_2O$); kernite ($Na_2[B_4O_5(OH)_4]\cdot 2H_2O$)	Mg reduction of B_2O_3; electrolysis of KBF_4	M_xB_y turbine blades, rocket nozzles, heat shields
Aluminum	Bauxite [contains gibbsite, $Al(OH)_3$]	Electrolysis of Al_2O_3 in Na_3AlF_6 (see text)	Many familiar alloys; electric transmission lines (see text)
Gallium	Trace element in bauxite	Obtained as trace impurity in Al purification	High-speed semiconductors; in photovoltaic solar panels
Indium	Trace element in Zn/Pb sulfide ores	Recovered from sulfide roasting flue dusts	Semiconductors with P or Sb; low-melting alloys in sprinklers
Thallium	Trace element in Zn/Pb sulfide ores	Recovered from sulfide roasting flue dusts	Very toxic; few uses

Table 22.6E Sources, Isolation, and Uses of Group 4A(14): The Carbon Family

Carbon is more abundant cosmically than on Earth; the reverse is true for silicon, the second most abundant element in the crust. Lead, the final product of most radioactive decay processes, is the crust's most abundant heavy metal. The isolation procedures vary, but those for Sn and Pb require milder conditions than those for Si and Ge. Except for Ge, each member has essential large-scale applications.

Purified silicon

Element	Source	Isolation	Uses
Carbon	Diamond; graphite; petroleum; coal; carbonates; CO_2	Used as found, or isolated from petroleum or coal by heating in the absence of air	Graphite: composites; electrodes; control rods in nuclear reactors Diamond: jewelry; abrasives; films Coke, carbon black: reductant in metallurgy; rubber tire strengthener; pigment; decolorizer for sugar
Silicon	Silica (SiO_2); silicate minerals	Reduction of K_2SiF_6 with Al; reduction of SiO_2 with Mg, then zone refining	Semiconductors; glass; ceramics
Germanium	Germanite (mixture of Cu, Fe, and Ge sulfides)	Roasting in O_2, then reducing GeO_2 with H_2, and zone refining	Semiconductors; infrared spectrometer windows and lenses
Tin	Cassiterite (SnO_2)	Thermal reduction of SnO_2 with C	Prevents corrosion in steel cans; alloys (e.g., solder, bronze, pewter)
Lead	Galena (PbS)	Roasting in O_2 to PbO, then reducing with C, and electrorefining	Automotive batteries; solder; ammunition

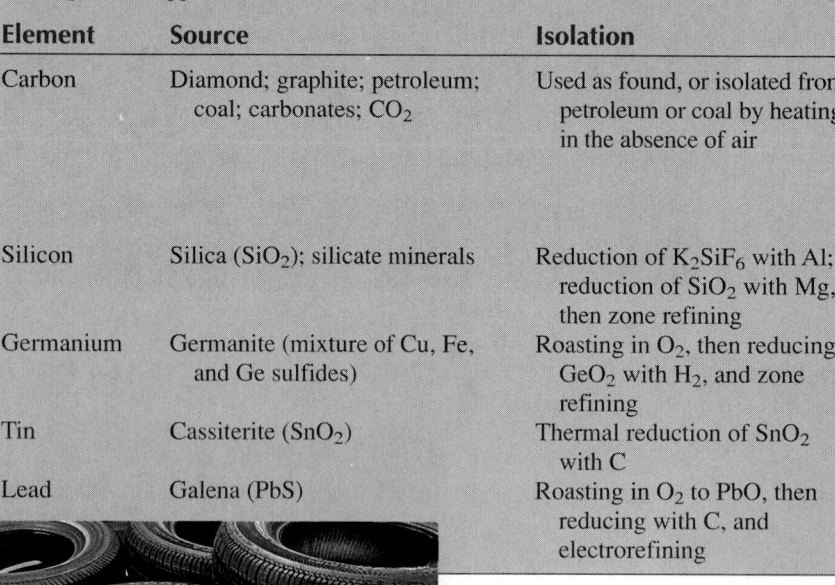

Rubber tires

Pewter mugs

Table 22.6F Sources, Isolation, and Uses of Group 5A(15): The Nitrogen Family

The high abundances of nitrogen and phosphorus contrast with the low abundances of the other group members, but all occur in concentrated sources. N_2 is obtained from liquefied air and the others by thermal reduction of sulfide ores, usually with coke or iron. The major applications of the three larger elements are in lead alloys, but arsenic and antimony are also used as dopants in semiconductors.

N_2 freeze-dried rose

Element	Source	Isolation	Uses
Nitrogen	Air	Fractional distillation of liquefied air	$N_2(g)$: inert atmosphere in metallurgical and petrochemical processing; reactant in NH_3 production $N_2(l)$: freeze-drying food; grinding meat for hamburger; biological preservation: future use in superconductors
Phosphorus	Phosphate ores, e.g., fluoroapatite [$Ca_5(PO_4)_3F$] (see text)	Reducing phosphate rock with C	Starting material for synthesis of H_3PO_4 (90%), PCl_3, P_4S_3 (matches), and P_4S_{10} (pesticide) (see text)
Arsenic	Arsenopyrite (FeAsS); flue dust in Cu and Pb extraction	Heating in absence of air	Films, light-emitting (photo) diodes; lead alloys
Antimony	Stibnite (Sb_2S_3); flue dust in Cu and Pb extraction	Roasting in air to Sb_2O_3, then reducing with C	In lead-acid batteries (5% Sb)
Bismuth	Bismuthinite (Bi_2S_3)	Roasting to Bi_2O_3, then reducing with C or Fe	Alloys; medicines

Table 22.6G Sources, Isolation, and Uses of Group 6A(16): The Oxygen Family

Nearly one of every two atoms in the Earth's crust (and the Moon's) is oxygen, but O_2 is obtained from liquefied air. Its critical application in steelmaking makes its production the third highest of any substance. Sulfur is extracted from large native deposits, and 90% of it is destined for H_2SO_4 production (see text). The other elements are obtained as by-products and, except for selenium's application in photocopying, have minor uses.

Manufacturing sulfuric acid

Element	Source	Isolation	Uses
Oxygen	Air	Fractional distillation of liquefied air	Oxidizing agent in steelmaking (see text), sewage treatment, paper-pulp bleaching, rocket fuel; medical applications
Sulfur	Underground S deposits; sour natural gas or petroleum	Frasch process (see text); catalytic oxidation of H_2S	Production of H_2SO_4 (see text); vulcanization of rubber; chemicals for pharmaceuticals, textiles, and pesticides
Selenium	Impurity in sulfide ores; anode muds of Cu refining	Reducing H_2SeO_3 with SO_2	Electronics; xerography; cadmium pigments
Tellurium	Mixed tellurides and sulfides of Group 8 to 11 metals; anode muds of Cu refining	Oxidizing to Na_2TeO_3, then electrolysis	Steelmaking
Polonium	Pitchblende; trace element formed in radium decay	Isolated in trace amounts	Future use as heat source in space satellites and lunar stations

Table 22.6H   Sources, Isolation, and Uses of Group 7A(17): The Halogens

Because of their high reactivity, the halogens never occur free in nature. In rocks, their abundances decrease down the group, but chlorine dominates by its enormous abundance in the ocean (1.9 mass %). All halogens are produced by oxidation of their halides, chlorine being the oxidizing agent for bromine and iodine. Chlorine's uses, especially in the formation of monomers for the plastics industry, make its production among the 10 highest in the United States.

Fluorite

Element	Source	Isolation	Uses
Fluorine	Fluorite; fluorspar (CaF_2)	Electrolysis of KF in molten anhydrous HF	Synthesis of UF_6 (for nuclear fuel) and SF_6 (electrical insulator); fluorinating agents; Teflon monomer
Chlorine	Halite (NaCl); seawater	Electrolysis of molten NaCl (see text); electrolysis of concentrated seawater (Section 22.5)	Oxidizing agent in bleach and disinfectant; production of poly(vinyl chloride) monomer; major biological anion
Bromine	Brine wells; seawater	Oxidation of Br^- salts by Cl_2 (see text)	Preparation of organic bromides; AgBr in photography
Iodine	Brine wells; Chilean saltpeter ($NaIO_3$)	Oxidation of I^- salts by Cl_2; reduction of IO_3^- with HSO_3^-	In table salt as essential trace element for thyroid hormones; disinfectant
Astatine	Extremely rare radioisotope	Obtained only in trace quantities	None

Teflon cookware

PVC hose

Neon sign

Table 22.6I   Sources, Isolation, and Uses of Group 8A(18): The Noble Gases

Helium and other noble gases are rare on Earth because of outgassing during the planet's formation. Helium is a product of radioactive decay of uranium ores and is obtained from natural gas. The high abundance of argon in air (~0.9 mol %) is due to radioactive decay of crustal ^{40}K.

Element	Source	Isolation	Uses
Helium	In natural gas (>0.4 mass %)	Distillation of condensed natural gas or differential diffusion of natural gas	Coolant for superconducting magnets; substitute for N_2 in deep-sea breathing mixture; mobile phase in gas chromatography
Neon	Air	Fractional distillation of liquefied air	Luminous gas in signs
Argon	Air	Fractional distillation of liquefied air	Arc welding
Krypton	Air	Fractional distillation of liquefied air	None
Xenon	Air	Fractional distillation of liquefied air	None
Radon	Air	Fractional distillation of liquefied air	None; radioactive air pollutant

Production highlights of key elements are as follows:
- Na is isolated by electrolysis of molten NaCl in the Downs process; Cl_2 is a by-product.
- K is produced by reduction with Na in a thermal process.
- Fe is produced through a multistep high-temperature process in a blast furnace. The crude pig iron is converted to carbon steel in the basic-oxygen process and then alloyed with other metals to make different steels.
- Cu is produced by concentration of the ore through flotation, reduction to the metal by smelting, and purification by electrorefining. The metal has extensive electrical and plumbing uses.
- Al is extracted from bauxite by pretreating with concentrated base, followed by electrolysis of the Al_2O_3 in molten cryolite. Al alloys are used throughout home and industry. The total energy required to extract Al from its ore is over 100 times that needed for recycling it.
- The Mg^{2+} in seawater is converted to $Mg(OH)_2$ and then to $MgCl_2$, which is electrolyzed to obtain the metal; Mg forms strong, lightweight alloys.
- Br_2 is obtained from brines by oxidizing Br^- with Cl_2.
- H_2 is produced by electrolysis of water or in the formation of gaseous fuels from hydrocarbons. It is used in NH_3 production and in hydrogenation of vegetable oils. The isotopes of hydrogen differ significantly in atomic mass and thus in the rate at which their bonds to other atoms break. This difference is used to obtain D_2O from water.

22.5 CHEMICAL MANUFACTURING: TWO CASE STUDIES

From laboratory-scale endeavors of the first half of the 19th century, today's chemical industries have grown to multinational corporations producing materials that define modern life: polymers, electronics, pharmaceuticals, consumer goods, and fuels. In this final section, we examine the interplay of theory and practice in two of the most important processes in the inorganic chemical industry: (1) the contact process for the production of sulfuric acid, and (2) the chlor-alkali process for the production of chlorine.

Sulfuric Acid, the Most Important Chemical

The manufacture of sulfuric acid began more than 400 years ago, when the acid was known as *oil of vitriol* and was distilled from "green vitriol" ($FeSO_4 \cdot 7H_2O$). Considering its countless uses, it is not surprising that today sulfuric acid is produced throughout the world on a gigantic scale—more than 150 million tons a year. The green vitriol method and the many production methods used since have all been superseded by the modern **contact process,** which is based on the *catalyzed oxidation of SO_2.* Here are the key steps.

1. *Obtaining sulfur.* In most countries today, the production of sulfuric acid starts with the production of elemental sulfur, often by chemically separating the H_2S in "sour" natural gas and then oxidizing it:

$$2H_2S(g) + 2O_2(g) \xrightarrow{\text{low temperature}} \tfrac{1}{8}S_8(g) + SO_2(g) + 2H_2O(g)$$

$$2H_2S(g) + SO_2(g) \xrightarrow{Fe_2O_3 \text{ catalyst}} \tfrac{3}{8}S_8(g) + 2H_2O(g)$$

Where natural gas is not abundant but natural underground deposits of the element are, sulfur is obtained by the *Frasch process,* a nonchemical method that taps these deposits. A hole is drilled to the deposit, and superheated water (about

Figure 22.22 The Frasch process for mining elemental sulfur. This non-chemical process uses superheated water and compressed air to melt underground sulfur and bring it to the surface. **A,** Superheated water is forced down a drilled hole into the subterranean sulfur deposit, and the sulfur melts. **B,** Compressed air is sent down the drilling apparatus to force up the molten sulfur. **C,** Sulfur obtained from the Frasch process.

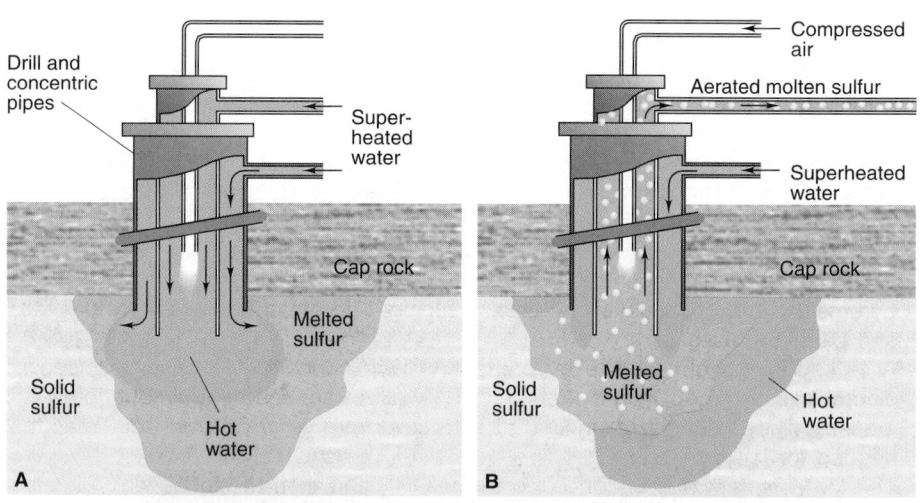

160°C) is pumped down two outer concentric pipes to melt the sulfur (Figure 22.22A). Then, a combination of the hydrostatic pressure in the outermost pipe and the pressure of compressed air sent through a narrow inner pipe forces the sulfur to the surface (Figure 22.22B and C). The costs of drilling, pumping, and supplying water (5×10^6 gallons per day) are balanced somewhat by the fact that the product is very pure (~99.7% S).

2. *From sulfur to sulfur dioxide.* Once obtained, the sulfur is burned in air to form SO_2:

$$\tfrac{1}{8}S_8(s) + O_2(g) \longrightarrow SO_2(g) \qquad \Delta H^0 = -297 \text{ kJ}$$

Some SO_2 is also obtained from the roasting of metal sulfide ores. About 90% of processed sulfur is used in making sulfur dioxide for production of the all-important sulfuric acid. Indeed, this end product from sulfur is central to so many chemical industries that a nation's level of sulfur production is a reliable indicator of its overall industrial capacity: the United States, Russia, Japan, and Germany are the top four sulfur producers.

3. *From sulfur dioxide to trioxide.* The contact process oxidizes SO_2 with O_2 to SO_3:

$$SO_2(g) + \tfrac{1}{2}O_2(g) \rightleftharpoons SO_3(g) \qquad \Delta H^0 = -99 \text{ kJ}$$

The reaction is *exothermic* and very *slow* at room temperature. From Le Châtelier's principle (Section 17.6), we know that the yield of SO_3 can be increased by (1) changing the temperature, (2) increasing the pressure (more moles of gas are on the left than on the right), and (3) adjusting the concentrations (adding excess O_2 and removing SO_3).

First, let's examine the temperature effect. Adding heat (raising the temperature) increases the frequency of SO_2-O_2 collisions and thus increases the *rate* of SO_3 formation. However, because the formation of SO_3 is exothermic, removing heat (lowering the temperature) shifts the equilibrium position to the right and thus increases the *yield* of SO_3. This is a classic situation that calls for use of a catalyst. By lowering the activation energy, *a catalyst allows equilibrium to be reached more quickly and at a lower temperature;* thus, rate *and* yield are optimized (Section 16.8). The catalyst in the contact process is V_2O_5 on inert silica, which is active between 400°C and 600°C.

The pressure effect is small and economically not worth exploiting. The concentration effects are controlled by providing an excess of O_2 in the form of a 5:1 mixture of air:SO_2, or about 1:1 O_2:SO_2, about twice as much O_2 as is called for by the reaction stoichiometry. The mixture is passed over catalyst beds in four stages, and the SO_3 is removed at several points to favor more SO_3 formation. The overall yield of SO_3 is 99.5%.

4. *From sulfur trioxide to acid.* Sulfur trioxide is the anhydride of sulfuric acid, so a hydration step is next. However, SO_3 cannot be added to water because, at the operating temperature, it would first meet water vapor, which catalyzes its polymerization to $(SO_3)_x$, and results in a smoke of solid particles that makes poor contact with water. To prevent this, previously formed H_2SO_4 absorbs the SO_3 and forms pyrosulfuric acid (or disulfuric acid, $H_2S_2O_7$; see margin), which is then hydrolyzed with sufficient water:

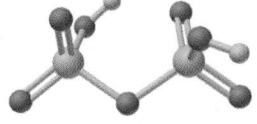

disulfuric acid

$$SO_3(g) + H_2SO_4(l) \longrightarrow H_2S_2O_7(l)$$
$$H_2S_2O_7(l) + H_2O(l) \longrightarrow 2H_2SO_4(l)$$
$$SO_3(g) + H_2O(l) \longrightarrow H_2SO_4(l)$$

The uses of sulfuric acid are legion, as Figure 22.23 indicates.

Sulfuric acid is remarkably inexpensive (about $150/ton), largely because each step in the process is exothermic—burning S ($\Delta H^0 = -297$ kJ/mol), oxidizing SO_2 ($\Delta H^0 = -99$ kJ/mol), hydrating SO_3 ($\Delta H^0 = -132$ kJ/mol)—and the heat is a valuable by-product. Three-quarters of the heat is sold as steam, and the rest of it is used to pump gases through the plant. A typical plant making 825 tons of H_2SO_4 per day produces enough steam to generate 7×10^6 watts of electric power.

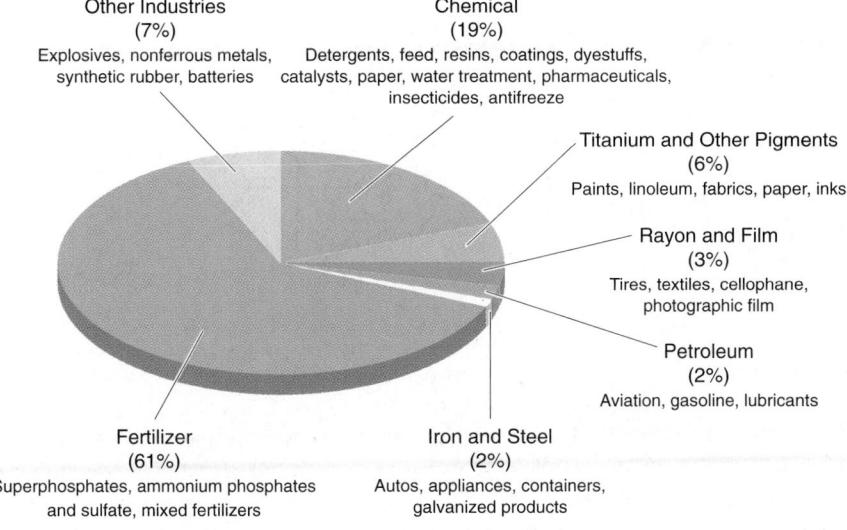

**Other Industries
(7%)**
Explosives, nonferrous metals, synthetic rubber, batteries

**Chemical
(19%)**
Detergents, feed, resins, coatings, dyestuffs, catalysts, paper, water treatment, pharmaceuticals, insecticides, antifreeze

**Titanium and Other Pigments
(6%)**
Paints, linoleum, fabrics, paper, inks

**Rayon and Film
(3%)**
Tires, textiles, cellophane, photographic film

**Petroleum
(2%)**
Aviation, gasoline, lubricants

**Fertilizer
(61%)**
Superphosphates, ammonium phosphates and sulfate, mixed fertilizers

**Iron and Steel
(2%)**
Autos, appliances, containers, galvanized products

Figure 22.23 **The many indispensable applications of sulfuric acid.**

The Chlor-Alkali Process

Chlorine is produced and used in amounts many times greater than all the other halogens combined, ranking among the top 10 chemicals produced in the United States. All of its production methods depend on *the oxidation of Cl⁻ ion from NaCl.*

Earlier, we discussed the Downs process for the isolation of sodium, which yields Cl_2 gas as the other product (Section 22.4). The **chlor-alkali process,** which forms the basis of one of the largest inorganic chemical industries, electrolyzes concentrated aqueous NaCl to produce Cl_2 and several other important chemicals. As you learned in Section 21.7, the electrolysis of aqueous NaCl does not yield both of the component elements. Chloride ions are oxidized at the anode rather than water due to the effects of overvoltage. However, Na^+ ions are not reduced at the cathode because the half-cell potential (-2.71 V) is much more negative than that for reduction of H_2O (-0.42 V), even with the normal overvoltage (around -0.6 V). Therefore, the half-reactions for electrolysis of aqueous NaCl are

$$2Cl^-(aq) \longrightarrow Cl_2(g) + 2e^- \qquad E^0 = 1.36 \text{ V} \quad \text{[anode; oxidation]}$$
$$2H_2O(l) + 2e^- \longrightarrow 2OH^-(aq) + H_2(g) \qquad E \approx -1.0 \text{ V} \quad \text{[cathode; reduction]}$$
$$\overline{2Cl^-(aq) + 2H_2O(l) \longrightarrow 2OH^-(aq) + H_2(g) + Cl_2(g) \qquad E_{cell} = -1.0 \text{ V} - 1.36 \text{ V} = -2.4 \text{ V}}$$

To obtain commercially meaningful amounts of Cl_2, however, a voltage almost twice this value and a current in excess of 3×10^4 A are used.

When we include the Na^+ spectator ion, the total ionic equation shows another important product made by the process:

$$2Na^+(aq) + 2Cl^-(aq) + 2H_2O(l) \longrightarrow 2Na^+(aq) + 2OH^-(aq) + H_2(g) + Cl_2(g)$$

As Figure 22.24 shows, the sodium salts in the cathode compartment exist as an aqueous mixture of NaCl and NaOH; the NaCl is removed by fractional crystallization. Thus, in this version of the chlor-alkali process, which uses an *asbestos diaphragm* to separate the anode and cathode compartments, electrolysis of NaCl brines yields Cl_2, H_2, and industrial-grade NaOH, an important base. Like other reactive products, H_2 and Cl_2 are kept apart to prevent explosive recombination. Note the higher liquid level in the anode compartment. This slight hydrostatic pressure difference minimizes backflow of NaOH, which avoids disproportionation (self–oxidation-reduction) reactions of Cl_2 in the presence of OH^- (Section 14.9), such as

$$Cl_2(g) + 2OH^-(aq) \longrightarrow Cl^-(aq) + ClO^-(aq) + H_2O(l)$$

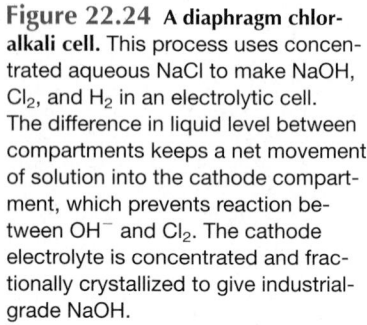

Figure 22.24 A diaphragm chlor-alkali cell. This process uses concentrated aqueous NaCl to make NaOH, Cl_2, and H_2 in an electrolytic cell. The difference in liquid level between compartments keeps a net movement of solution into the cathode compartment, which prevents reaction between OH^- and Cl_2. The cathode electrolyte is concentrated and fractionally crystallized to give industrial-grade NaOH.

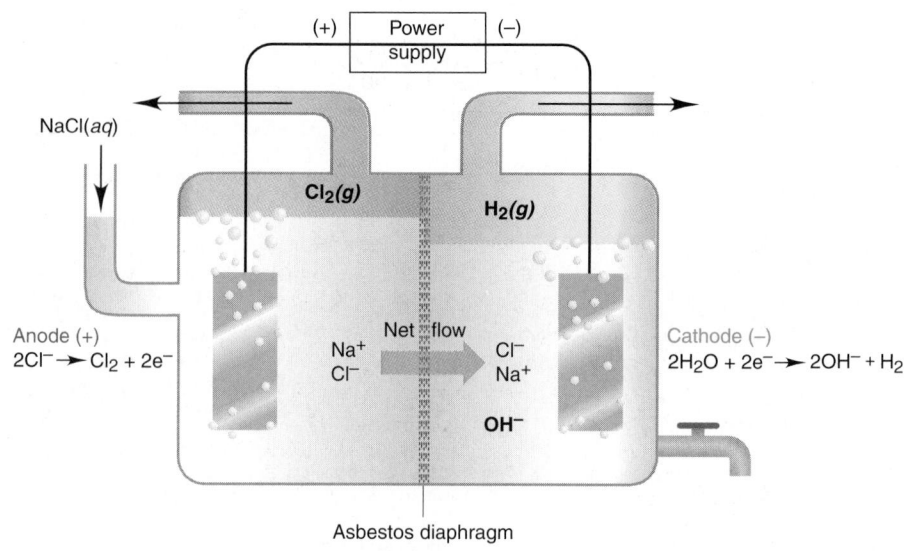

If high-purity NaOH is desired, a slightly different version, called the chlor-alkali *mercury-cell* process, is employed. Mercury is used as the cathode, which creates such a large overvoltage for reduction of H_2O to H_2 that the process *does* favor reduction of Na^+. The sodium dissolves in the mercury to form sodium amalgam, Na(Hg). In the mercury-cell version, the half-reactions are

$$2Cl^-(aq) \longrightarrow Cl_2(g) + 2e^- \qquad \text{[anode; oxidation]}$$
$$2Na^+(aq) + 2e^- \xrightarrow{Hg} 2Na(Hg) \qquad \text{[cathode; reduction]}$$

To obtain sodium hydroxide, the sodium amalgam is pumped out of the system and treated with H_2O, which is reduced by the Na:

$$2Na(Hg) + 2H_2O(l) \xrightarrow{-Hg} 2Na^+(aq) + 2OH^-(aq) + H_2(g)$$

The mercury released in this step is recycled back to the electrolysis bath. Therefore, *the products are the same in both versions,* but the purity of NaOH in the mercury-cell version is much higher.

Despite the formation of purer NaOH, the mercury-cell method is being steadily phased out in the United States and has been eliminated in Japan for more than a decade. Cost is not the major reason for the phase-out, although the method does consume about 15% more electricity than the diaphragm version. Rather, the problem is that as the mercury is recycled, some is lost in the industrial wastewater. On average, 200 g of Hg are lost per ton of Cl_2 produced. In the 1980s, U.S. production was 2.75 million tons of Cl_2 annually via the mercury-cell method; thus, 550,000 kg of this toxic heavy metal was flowing into U.S. waterways each year!

The more recent chlor-alkali *membrane-cell* process replaces the diaphragm with a polymeric membrane to separate the cell compartments. The membrane allows only cations to move through it and only from anode to cathode compartments. Thus, as Cl^- ions are removed at the anode through oxidation to Cl_2, Na^+ ions in the anode compartment move through the membrane to the cathode compartment and form an NaOH solution. In addition to forming purer NaOH than the older diaphragm method, the membrane-cell process uses less electricity and eliminates the problem of Hg pollution. As a result, its application is being established throughout the industrialized world.

SECTION SUMMARY

Sulfuric acid production starts with the extraction of sulfur, either by the oxidation of H_2S or the mining of sulfur deposits. The sulfur is roasted to SO_2, which is oxidized to SO_3 by the catalyzed contact process, which optimizes the yield at lower temperatures. Absorption of the SO_3 into H_2SO_4, followed by hydration, forms sulfuric acid. In the diaphragm-cell chlor-alkali process, aqueous NaCl is electrolyzed to form Cl_2, H_2, and low-purity NaOH. The mercury-cell version produces high-purity NaOH but is being phased out because of mercury pollution. The polymeric membrane–cell process requires less electricity and does not use Hg.

Chapter Perspective

In this chapter, we took a realistic approach to the chemistry of the elements that allowed a glimpse at the intertwining natural cycles in which various elements take part, as well as at the ingenious methods that have been developed to extract them. The impact of living systems in general, and of humans in particular, on the origin and distribution of elements reveals a complex, evolving relationship. Understanding of and respect for that relationship are required for a healthy, productive balance between society's needs and the environment.

In the next chapter, we conclude our exploration of the elements with a close look at the transition metals, a fascinating and extremely useful collection of metals. Then, in the final chapter, we dive down to the atom's nuclear core and investigate its properties and its great potential for society.

For Review and Reference (Numbers in parentheses refer to pages, unless noted otherwise.)

Learning Objectives

Relevant section numbers appear in parentheses.

Understand These Concepts

1. How gravity, thermal convection, and elemental properties led to the silicate, sulfide, and iron phases and the predominance of certain elements in crust, mantle, and core of the Earth (Section 22.1)

2. How organisms affect crustal abundances of elements, especially oxygen, carbon, calcium, and transition metals; the onset of oxidation as an energy source (Section 22.1)

3. How atomic properties influence which elements have oxide ores and which have sulfide ores (Section 22.1)

4. The central role of CO_2 and the importance of photosynthesis, respiration, and decay in the carbon cycle (Section 22.2)

5. The central role of N_2 and the importance of atmospheric, industrial, and biological fixation in the nitrogen cycle (Section 22.2)

6. The absence of a gaseous component, the interactions of the inorganic and biological cycles, and the impact of humans on the phosphorus cycle (Section 22.2)

7. How pyro-, electro-, and hydrometallurgical processes are employed to extract a metal from its ore; the importance of the reduction step from compound to metal; refining and alloying processes (Section 22.3)

8. Functioning of the Downs cell for Na production and the application of Le Châtelier's principle for K production (Section 22.4)

9. How iron ore is reduced in a blast furnace and how pig iron is purified by the basic-oxygen process (Section 22.4)

10. How Fe is removed from copper ore and impure Cu is electrorefined (Section 22.4)

11. The importance of amphoterism in the Bayer process for isolating Al_2O_3 from bauxite; significance of cryolite in the electrolytic step; the energy advantage of Al recycling (Section 22.4)

12. The steps in the Dow process for the extraction of Mg from seawater (Section 22.4)

13. How H_2 production, whether by chemical or electrolytic means, is tied to NH_3 production (Section 22.4)

14. How the kinetic isotope effect is applied to produce deuterium (Section 22.4)

15. How the Frasch process is used to obtain sulfur from natural deposits (Section 22.5)

16. The importance of equilibrium and kinetic factors in H_2SO_4 production (Section 22.5)

17. How overvoltage allows electrolysis of aqueous NaCl to Cl_2 gas in the chlor-alkali process; coproduction of NaOH; comparison of diaphragm-cell, mercury-cell, and membrane-cell methods (Section 22.5)

Key Terms

Section 22.1
abundance (951)
differentiation (951)
core (951)
mantle (951)
crust (951)
lithosphere (953)
hydrosphere (953)
atmosphere (953)
biosphere (953)

occurrence (source) (955)
ore (955)

Section 22.2
environmental cycle (956)
fixation (957)
apatites (960)

Section 22.3
metallurgy (963)
mineral (963)
gangue (963)

flotation (964)
leaching (964)
roasting (964)
smelting (965)
electrorefining (967)
zone refining (967)
alloy (967)

Section 22.4
Downs cell (969)

steel (970)
blast furnace (970)
slag (971)
basic-oxygen process (972)
carbon steel (972)

Section 22.5
contact process (987)
chlor-alkali process (990)

Highlighted Figures and Tables

These figures (F) and tables (T) provide a quick review of key ideas.

F22.1 Cosmic and terrestrial abundances of selected elements (952)
F22.2 Geochemical differentiation of the elements (953)
T22.1 Abundances of elements in crust and biosphere (954)
F22.4 Sources of the elements (955)

F22.5 The carbon cycle (957)
F22.6 The nitrogen cycle (959)
F22.7 The phosphorus cycle (961)
T22.6 Sources, isolation, and uses of the elements (982 to 986)

Problems

Problems with **colored** numbers are answered at the back of the text. Sections match the text and provide the number(s) of relevant sample problems. Most offer Concept Review Questions, and Problems in Context. Then Comprehensive Problems, based on material from any section or previous chapter, follow.

How the Elements Occur in Nature

● Concept Review Questions

22.1 Hydrogen is by far the most abundant element cosmically. In interstellar space, it exists mainly as H_2. In contrast, on Earth, it exists very rarely as H_2 and is ninth in abundance in the crust. Why is hydrogen so abundant in the universe? Why is hydrogen so rare as a diatomic gas in the Earth's atmosphere?

22.2 Metallic elements are recovered primarily from ores that are oxides, carbonates, halides, or sulfides. Give an example for each of these types.

22.3 The location of elements in the regions of the Earth has enormous practical importance.
(a) Define the term *differentiation* and explain which physical property of a substance is primarily responsible for this process.
(b) What are the four most abundant elements in the crust?
(c) Which element is abundant in the crust and mantle but not the core?

22.4 How does the position of a metal in the periodic table relate to whether it exists in nature primarily as an oxide or as a sulfide?

● Problems in Context

22.5 What material is the source for commercial production of each of the following elements: (a) aluminum; (b) nitrogen; (c) chlorine; (d) calcium; (e) sodium?

22.6 Aluminum is widely distributed throughout the world in the form of aluminosilicates. What property of these minerals prevents them from being a source of aluminum?

22.7 Describe two ways in which the biosphere has influenced the composition of the Earth's crust.

The Cycling of Elements Through the Environment

● Concept Review Questions

22.8 Use atomic and molecular properties to explain why life is based on the chemistry of carbon, rather than some other element such as silicon.

22.9 Define the term *fixation*. Name two elements that undergo environmental fixation. What are the natural forms that are fixed?

22.10 Carbon dioxide enters the atmosphere by natural processes and as a result of human activity. Why is the latter source a cause of great concern?

22.11 Diagrams of environmental cycles are simplified to show overall changes and omit relatively minor contributors. For example, the production of lime from limestone is not explicitly shown in the cycle for carbon (Figure 22.5, p. 957). Which labeled category in the figure includes this process? Name two other processes that contribute to this category.

22.12 Describe three pathways for utilization of atmospheric nitrogen. Is human activity a significant factor? Explain.

22.13 Why do the nitrogen-containing species shown in Figure 22.6 (p. 959) not include ring compounds or long-chain compounds with N—N bonds?

22.14 (a) Which region of the Earth's crust is not involved in the phosphorus cycle?
(b) Briefly describe two roles that organisms play in the phosphorus cycle.

● Problems in Context

22.15 Nitrogen fixation requires a great deal of energy because the N_2 bond is strong (activation energy for its breakage is high).
(a) How do the processes of atmospheric and industrial fixation reflect this energy requirement?
(b) How do the thermodynamics of the two processes differ? (*Hint:* Examine the respective heats of formation.)
(c) In view of the mild conditions for biological fixation, what must be the source of the "great deal of energy" in this case?
(d) What would be the most obvious environmental result of a low activation energy for N_2 fixation?

22.16 The following steps are *unbalanced* half-reactions involved in the nitrogen cycle. Balance each half-reaction to show the number of electrons lost or gained, and state whether it is an oxidation or a reduction (all occur in acidic conditions):
(a) $N_2(g) \longrightarrow NO(g)$ (b) $N_2O(g) \longrightarrow NO_2(g)$
(c) $NH_3(aq) \longrightarrow NO_2^-(aq)$ (d) $NO_3^-(aq) \longrightarrow NO_2^-(aq)$
(e) $N_2(g) \longrightarrow NO_3^-(aq)$

22.17 The use of silica to form slag in the production of phosphorus from phosphate rock was introduced by Robert Boyle more than 300 years ago. When fluorapatite [$Ca_5(PO_4)_3F$] is used in phosphorus production, most of the fluorine atoms appear in the slag, but some end up in toxic and corrosive $SiF_4(g)$.
(a) If 15% by mass of the fluorine in 100. kg of $Ca_5(PO_4)_3F$ forms SiF_4, what volume of this gas is collected at 1.00 atm and the industrial furnace temperature of 1450.°C?
(b) In some facilities, the SiF_4 is used to produce sodium hexafluorosilicate (Na_2SiF_6) which is sold for domestic water fluoridation:

$$2SiF_4(g) + Na_2CO_3(s) + H_2O(l) \longrightarrow$$
$$Na_2SiF_6(aq) + SiO_2(s) + CO_2(g) + 2HF(aq)$$

How many cubic meters of drinking water can be fluoridated to a level of 1.0 ppm of F^- from the reaction of the SiF_4 produced in part (a)?

22.18 An impurity sometimes found in $Ca_3(PO_4)_2$ is Fe_2O_3, which is removed during the production of phosphorus as a material called *ferrophosphorus* (Fe_2P).
(a) Why is this impurity troubling from an economic standpoint?
(b) If 50. metric tons of crude $Ca_3(PO_4)_2$ contains 2.0% Fe_2O_3 by mass and the overall yield of phosphorus is 90.%, how many metric tons of P_4 can be isolated?

Metallurgy: Extracting a Metal from Its Ore

● Concept Review Questions

22.19 Define each of the following materials: (a) ore; (b) mineral; (c) gangue; (d) brine.

22.20 Define each of the following processes: (a) roasting; (b) smelting; (c) flotation; (d) refining.

22.21 What factors determine which reducing agent is selected for the production of a specific metal?

22.22 Use atomic properties to explain the reduction of a less active metal by a more active one (a) in aqueous solution; (b) in the molten state. Give a specific example of each process.

22.23 What class of element is typically obtained by oxidation of a mineral? What class of element is typically obtained by reduction of a mineral?

● **Problems in Context**

22.24 Select the group of elemental components that gives each of the following alloys: (a) brass; (b) stainless steel; (c) bronze; (d) sterling silver.

1. Cu, Ag   2. Cu, Sn   3. Ag, Au
4. Fe, Cr, Ni   5. Fe, V   6. Cu, Zn

Tapping the Crust: Isolation and Uses of the Elements

● **Concept Review Questions**

22.25 How are each of the following involved in iron metallurgy: (a) slag; (b) pig iron; (c) steel; (d) basic-oxygen process?

22.26 What are the distinguishing features of each extraction process: pyrometallurgy, electrometallurgy, and hydrometallurgy? Explain briefly how the types of metallurgy are used in the production of (a) Fe; (b) Na; (c) Au; (d) Al.

22.27 What property allows copper to be purified in the presence of iron and nickel impurities? Explain.

22.28 What is the practical reason for using cryolite in the electrolysis of aluminum oxide?

22.29 (a) What is a kinetic isotope effect?
(b) Do compounds of hydrogen exhibit a relatively large or small kinetic isotope effect? Explain.
(c) Carbon compounds also exhibit a kinetic isotope effect. How do you expect it to compare in magnitude with that for hydrogen compounds? Why?

22.30 How is Hess's law involved in water-splitting schemes for the production of H_2?

● **Problems in Context**

22.31 Elemental Li and Na are prepared by electrolysis of a molten salt, whereas K, Rb, and Cs are prepared by chemical reduction.
(a) In general terms, explain why the alkali metals cannot be prepared by electrolysis of their aqueous salt solutions.
(b) Use ionization energies (see Family Portraits, pages 550 and 554) to explain why calcium should *not* be able to isolate Rb from molten RbX (X = halide).
(c) Use physical properties to explain why calcium *is* used to isolate Rb from molten RbX.
(d) Can Ca be used to isolate Cs from molten CsX? Explain.

22.32 A Downs cell operates at 75.0 A and produces 30.0 kg of Na metal.
(a) What volume of $Cl_2(g)$ is produced at 1.0 atm and 580.°C?
(b) How many coulombs were passed through the cell?
(c) How long did the cell operate?

22.33 (a) In the industrial production of iron, what is the reducing substance that is loaded into the blast furnace?
(b) In addition to furnishing the reducing power, what other function does this substance serve?
(c) What is the formula of the active reducing agent in the process?

(d) Write equations for the stepwise reduction of Fe_2O_3 to iron in the furnace.

22.34 One of the substances loaded into the blast furnace is limestone, which produces lime in the furnace.
(a) Give the chemical equation for the reaction forming lime.
(b) Explain the purpose of lime in the furnace. The term *flux* is often used as a label for a substance acting as the lime does. What is the derivation of this word and how does it relate to the function of the lime?
(c) Write a chemical equation describing the action of the lime flux.

22.35 The last step in the Dow process for the production of magnesium metal involves electrolysis of molten $MgCl_2$.
(a) Why isn't the electrolysis carried out in aqueous $MgCl_2$? What are the products of aqueous electrolysis?
(b) Do the high temperatures required to melt $MgCl_2$ favor products or reactants? (*Hint:* Consider the heat of formation of $MgCl_2$.)

22.36 (a) What is the only halogen that occurs in a positive oxidation state in nature (Table 22.6H)?
(b) Is this mode of occurrence consistent with its location in the periodic table? Explain.
(c) Write a balanced equation for the production of the free halogen from this source.

22.37 Selenium is prepared by the reaction of H_2SeO_3 with gaseous SO_2 (Table 22.6G).
(a) What redox process does the sulfur dioxide undergo? What is the oxidation state of sulfur in the product?
(b) Given that the reaction occurs in acidic aqueous solution, what is the formula of the sulfur-containing species?
(c) Write the balanced redox equation for the process.

22.38 The halogens F_2 and Cl_2 are produced by electrolytic oxidation, whereas Br_2 and I_2 are produced by chemical oxidation of the halide in a concentrated aqueous solution (brine) by a more electronegative halogen. State two reasons why Cl_2 is not prepared by this method.

22.39 Silicon is prepared by the reduction of K_2SiF_6 with Al (Table 22.6E). Write the equation for this reaction. (*Hint:* Can F^- be oxidized in this reaction? Can K^+ be reduced?)

22.40 What is the mass percent of iron in each of the following iron ores: Fe_2O_3, Fe_3O_4, FeS_2?

22.41 Phosphorus is one of the impurities present in pig iron that is removed in the basic-oxygen process. Assuming that the phosphorus is present as P atoms, write equations for its oxidation and subsequent reaction in the basic slag.

22.42 The final step in the smelting of $FeCuS_2$ is

$$Cu_2S(s) + 2Cu_2O(s) \longrightarrow 6Cu(l) + SO_2(g)$$

(a) What are the oxidation states of copper in Cu_2S, Cu_2O, and Cu?
(b) What are the oxidizing and reducing agents in this reaction?

22.43 Use balanced equations to explain how acid-base properties are used to separate iron and titanium oxides from aluminum oxide in the Bayer process.

22.44 A piece of Al with a surface area of 2.3 m^2 is anodized to produce a film of Al_2O_3 that is 20. μ ($20.\times10^{-6}$ m) thick.
(a) How many coulombs flow through the cell in this process (assume that the density of the Al_2O_3 layer is 3.97 g/cm^3)?
(b) If this film grows over the course of 15 min, what current must flow through the cell?

22.45 The production of H_2 gas by the electrolysis of water typically requires about 400 kJ of energy per mole.
(a) Use the relationship between work and cell potential (Section 21.4) to calculate the minimum work needed to form 1.0 mol of H_2 gas at a cell potential of 1.24 V.
(b) What is the energy efficiency of the cell operation?
(c) Calculate the cost of producing 500. mol of H_2 if the cost of electrical energy is exactly \$0.05 per kilowatt·hour (1 watt·second = 1 joule).

22.46 (a) What are the components of the reaction mixture following the water-gas shift reaction?
(b) Explain how zeolites (Gallery, p. 571) are used to purify the H_2 formed.

22.47 Typically, metal sulfides are first converted to oxides by roasting in air and then reduced with carbon to produce the metal. Why are the metal sulfides not reduced directly by carbon to yield CS_2? Give a thermodynamic analysis of both processes for a typical case such as ZnS.

Chemical Manufacturing: Two Case Studies

● **Concept Review Questions**

22.48 Explain in detail why a catalyst is used to produce SO_3.
22.49 Among the exothermic steps in the manufacture of sulfuric acid is the process of hydrating SO_3.
(a) Write two chemical reactions that show this process.
(b) Why is the direct reaction of SO_3 with water not feasible?
22.50 What is the principal reason that commercial H_2SO_4 is so inexpensive?
22.51 (a) What are the three commercial products formed in the chlor-alkali process? (b) State an advantage and a disadvantage of the mercury-cell version of this process.

● **Problems in Context**

22.52 (a) Calculate the standard free energy change at 25°C for the oxidation of SO_2 to SO_3 in the contact process. Is the reaction spontaneous at 25°C?
(b) Why is the reaction not performed at 25°C?
(c) Is the reaction spontaneous at 500.°C? (Assume that ΔH^0 and ΔS^0 are constant with temperature.)
(d) How does the value of the equilibrium constant at 500.°C compare with that at 25°C?
(e) What is the highest temperature at which the reaction is spontaneous?
22.53 If a chlor-alkali cell used a current of 3×10^4 A, how many pounds of Cl_2 would be produced in a typical 8-h operating day?
22.54 In the chlor-alkali process, the product chlorine and sodium hydroxide are kept separate from one another.
(a) Why is this precaution necessary when Cl_2 is the desired product?
(b) Hypochlorite or chlorate may be formed by disproportionation of Cl_2 in basic solution. What condition determines which product is formed?
(c) What mole ratio of Cl_2 to OH^- is needed to produce ClO^-? ClO_3^-?

Comprehensive Problems

Problems with an asterisk (*) are more challenging.

22.55 The key step in the manufacture of sulfuric acid is the oxidation of sulfur dioxide in the presence of a catalyst, such as

V_2O_5. At 727°C, 0.010 mol of SO_2 is injected into an empty 2.00-L container ($K_p = 3.18$).
(a) What is the pressure of O_2 at equilibrium that is needed to maintain a 1:1 mole ratio of SO_3:SO_2?
(b) What is the pressure of O_2 at equilibrium that is needed to maintain a 95:5 mole ratio of SO_3:SO_2?

22.56 Sodium carbonate (Na_2CO_3) is an industrial base used in the paper, soap, and detergent industries, and it is indispensable in glassmaking. In the United States, it is mined as *trona* ($Na_2CO_3 \cdot NaHCO_3 \cdot 2H_2O$) from enormous deposits in Wyoming. In many other countries, however, it is still made by the Solvay process, which has the following *overall* reaction:

$$CaCO_3(s) + 2NaCl(aq) \longrightarrow Na_2CO_3(aq) + CaCl_2(aq)$$

(a) Why would you expect the overall reaction not to proceed as written?
(b) In the industrial process, the $CaCO_3$ is heated, and the gas formed reacts with added NH_3 and water. The resulting aqueous bicarbonate salt then undergoes a metathesis reaction with NaCl, and the Na salt formed is heated to produce the desired product. Write balanced equations for these four reactions.
(c) Sodium bicarbonate, a key intermediate in the process, has several uses as a result of its low-temperature (50°C to 100°C) decomposition. If a fire extinguisher contains 150. g of $NaHCO_3$, what volume of CO_2 can be formed at 100.8 kPa and 72°F?

22.57 Tetraphosphorus decaoxide (P_4O_{10}), a compound made from phosphate rock, is commonly used as a drying agent in the laboratory.
(a) Write a balanced equation for its reaction with water.
(b) What is the pH of a solution formed from the addition of 5.0 g of P_4O_{10} in sufficient water to form 0.500 L? (See Table 18.5 for additional information.)

***22.58** Heavy water (D_2O) is used to prepare numerous deuterated chemicals.
(a) What major species, aside from the starting compounds, do you expect to find in a solution formed by mixing CH_3OH and D_2O?
(b) Write equations to explain how these various species arise. (*Hint:* Consider the autoionization of both components.)

22.59 A blast furnace uses Fe_2O_3 to produce 8400. t of Fe per day.
(a) What mass of CO_2 is produced each day?
(b) Compare this amount of CO_2 with that produced by 1.0 million automobiles, each burning 5.0 gal of gasoline a day. Assume that gasoline has the formula C_8H_{18} and a density of 0.74 g/mL, and that it burns completely. (Note that U.S. gasoline consumption is over 4×10^8 gal/day.)

22.60 Several decades ago, various sets of empirical rules were devised to predict the formation of alloys. Two of the rules propose that stable alloys form if the ratio of the number of valence electrons (s plus p) to the number of atoms is 3:2 or 21:13. For example, β-brass (CuZn) is a very stable alloy: Cu has one $4s$ and Zn has two $4s$ electrons, for 3 valence electrons; one Cu atom plus one Zn atom gives 2 atoms; thus, a 3:2 ratio of electrons to atoms. What electron-to-atom ratio does each of the folowing alloys display: (a) Cu_5Sn; (b) Cu_5Zn_8; (c) $Cu_{31}Sn_8$; (d) $AgZn_3$; (e) Ag_3Al?

22.61 In the production of magnesium metal, $Mg(OH)_2$ is precipitated by using $Ca(OH)_2$, which itself is "insoluble."
(a) Use K_{sp} values to show that $Mg(OH)_2$ can be precipitated from seawater in which $[Mg^{2+}]$ is initially 0.051 M.

(b) If the seawater is saturated with $Ca(OH)_2$, what fraction of the Mg^{2+} can be precipitated?

*22.62 The Ostwald process for the production of HNO_3 proceeds as follows:

(1) $4NH_3(g) + 5O_2(g) \xrightarrow{\text{Pt/Rh catalyst}} 4NO(g) + 6H_2O(g)$

(2) $2NO(g) + O_2(g) \longrightarrow 2NO_2(g)$

(3) $3NO_2(g) + H_2O(l) \longrightarrow 2HNO_3(aq) + NO(g)$

(a) Describe the nature of the redox change that occurs in step 3.
(b) Write an overall equation that includes NH_3 and HNO_3 as the only nitrogen-containing species.
(c) Calculate ΔH^0_{rxn} (in kJ/mol N atoms) for this reaction at 25°C.

22.63 Step 1 of the Ostwald process for nitric acid production is

$4NH_3(g) + 5O_2(g) \xrightarrow{\text{Pt/Rh catalyst}} 4NO(g) + 6H_2O(g)$

An unwanted side reaction for this step is

$4NH_3(g) + 3O_2(g) \longrightarrow 2N_2(g) + 6H_2O(g)$

(a) Calculate K_p for these two NH_3 oxidations at 25°C.
(b) Calculate K_p for these two NH_3 oxidations at 900.°C.
(c) The Pt/Rh catalyst is one of the most efficient in the chemical industry, achieving 96% yield in 1 millisecond of contact with the reactants. However, at normal operating conditions (5 atm and 850°C), about 175 mg of Pt is lost per metric ton (t) of HNO_3 produced. If current annual U.S. production of HNO_3 is 7.7×10^6 t and the current market price of Pt is \$385/troy oz, what is the annual cost of the lost Pt (1 kg = 32.15 troy oz)?

22.64 The compounds $(NH_4)_2HPO_4$ and $NH_4(H_2PO_4)$ are water-soluble phosphate fertilizers.
(a) Which has the higher mass % of P?
(b) Why might the one with the lower mass % of P be preferred?

22.65 Several transition metals are prepared by reduction of the metal halide with magnesium metal. Titanium is prepared by the Kroll method, in which ore (ilmenite) is converted to the gaseous chloride, which is then reduced to the metal by molten Mg:

(1) $FeTiO_3(s) + Cl_2(g) + C(s) \longrightarrow$
$\qquad\qquad TiCl_4(g) + FeCl_3(l) + CO(g)$ [unbalanced]

(2) $TiCl_4(g) + Mg(l) \longrightarrow Ti(s) + MgCl_2(l)$ [unbalanced]

Assuming yields of 84% for step 1 and 93% for step 2, and an excess of the other reactants, what mass of Ti metal can be prepared from 17.5 metric tons of ilmenite?

22.66 The production of S_8 from the $H_2S(g)$ found in natural gas deposits occurs through the *Claus process* (Section 22.5):
(a) Use these two unbalanced steps to write an overall balanced equation for this process:

(1) $H_2S(g) + O_2(g) \longrightarrow S_8(g) + SO_2(g) + H_2O(g)$

(2) $H_2S(g) + SO_2(g) \longrightarrow S_8(g) + H_2O(g)$

(b) Write the overall reaction with Cl_2 as the oxidizing agent instead of O_2. Use thermodynamic values to show whether $Cl_2(g)$ can be used to oxidize $H_2S(g)$.
(c) Why is oxidation by O_2 preferred to oxidation by Cl_2?

22.67 Hydrazine, N_2H_4, is produced commercially by the oxidation of NH_3 with sodium hypochlorite, NaClO.
(a) Write a balanced equation for this reaction in base. [*Note:* N_2H_4 and the intermediate NH_2Cl are toxic. Attempts to produce "extra strength" cleaning solutions by mixing bleach (aqueous NaClO) and window cleaner (aqueous NH_3) are dangerous.]
(b) Why is hydrazine not produced by the reduction of N_2? (*Hint:* Consider thermodynamic parameters for the formation of N_2H_4 and of NH_3.)

*22.68 Based on a concentration of 0.065 g Br^-/L, at recent prices the ocean is estimated to have \$244,000,000 of Br_2/mi^3.
(a) What is the price per gram of Br_2 in the ocean?
(b) Find the mass (in g) of $AgNO_3$ needed to precipitate 90.0% of the Br^- in 1.00 L of ocean water.
(c) Use Appendix C to find how low the $[Cl^-]$ must be so that AgCl does not also precipitate?
(d) Given a $[Cl^-]$ of about 0.3 M, is precipitation with silver ion a practical way to obtain bromine from the ocean? Explain.

22.69 Incomplete combustion of C-containing compounds, whether in forest fires, coal-fired power plants, or automobile engines, produces CO as well as CO_2. The binding of CO to transition metal atoms and ions is the reason for its toxicity as well as the basis of several industrial methods to purify metals.
(a) Draw a Lewis structure for CO that has minimal formal charges.
(b) Does CO bond to the metal through the C or the O atom? Explain.

22.70 Why is nitric acid not produced by oxidizing atmospheric nitrogen directly in the following way?

(1) $\quad N_2(g) + 2O_2(g) \longrightarrow 2NO_2(g)$

(2) $3NO_2(g) + H_2O(l) \longrightarrow 2HNO_3(aq) + NO(g)$

(3) $\quad 2NO(g) + O_2(g) \longrightarrow 2NO_2(g)$

Overall: $3N_2(g) + 6O_2(g) + 2H_2O(l) \longrightarrow$
$\qquad\qquad\qquad\qquad\qquad 4HNO_3(aq) + 2NO(g)$

(*Hint:* Evaluate the thermodynamics of each step.)

*22.71 Before the development of the Downs cell, the Castner cell was used for the industrial production of Na metal. The Castner cell was based on the electrolysis of molten NaOH.
(a) Write balanced cathode and anode half-reactions for this cell.
(b) A major problem with this cell was that the water produced at one electrode diffused to the other and reacted with the Na. If all the water produced reacted with Na, what would be the maximum efficiency of the Castner cell expressed as moles of Na produced per mole of electrons flowing through the cell?

*22.72 In the leaching of gold ores with CN^- solutions, gold forms the complex ion $Au(CN)_2^-$.
(a) Calculate E_{cell} for the oxidation in air ($P_{O_2} = 0.21$) of Au to Au^+ in basic (pH 13.55) solution with $[Au^+] = 0.50$ M. Is this reaction spontaneous [$Au^+(aq) + e^- \longrightarrow Au(s)$; $E^0 = 1.68$ V]?
(b) How does the formation of the complex ion change E^0 so that the oxidation can be accomplished?

22.73 Nitric oxide is an important species in the tropospheric nitrogen cycle, but it destroys ozone in the stratosphere.
(a) Write a balanced equation for its reversible reaction with ozone.
(b) Given that the forward and reverse steps are first order in each component, write general rate laws for them.
(c) Calculate ΔG^0 for this reaction at 280. K, the average temperature in the stratosphere. (Assume that the ΔH^0 and S^0 values in Appendix B do not change with temperature.)
(d) Calculate the ratio of rate constants for the reaction at this temperature.

22.74 A key part of the carbon cycle is the fixation of CO_2 by photosynthesis to produce carbohydrates and oxygen gas.
(a) Using the formula $(CH_2O)_n$ to represent a carbohydrate, develop a general balanced equation for the photosynthetic reaction.

(b) If a moderate-sized tree fixes 45 g of CO_2 a day, what volume of O_2 gas measured at 1.0 atm and 80.°F does the tree produce per day?

(c) What volume of air (0.033 mol % CO_2) at the same conditions contains this amount of CO_2?

22.75 In the early part of the 20th century, most sulfuric acid was manufactured by a process using a gaseous NO/NO_2 mixture as the catalyst system to convert SO_2 to SO_3.

(a) What reaction does NO undergo in contact with O_2?

(b) Write reactions to show how the catalyst functions and is re-formed in this process.

(c) Explain how a similar process relates to forming acid rain.

22.76 Small-scale production of chlorine gas is accomplished by the reaction of manganese dioxide and hydrochloric acid. What volume of Cl_2 gas at 755 torr and 29°C can be prepared from 253 g of MnO_2 and 325 mL of 4.2 M HCl? ($MnCl_2$ and H_2O form also.)

22.77 The key reaction (unbalanced) in the manufacture of synthetic cryolite for aluminum electrolysis is

$$HF(g) + Al(OH)_3(s) + NaOH(aq) \longrightarrow$$
$$Na_3AlF_6(aq) + H_2O(l)$$

Assuming a 96.6% yield of dried, crystallized product, what mass (in kg) of cryolite can be obtained from the reaction of 353 kg of $Al(OH)_3$, 1.10 m^3 of 50.0% by mass aqueous NaOH ($d = 1.53$ g/mL), and 225 m^3 of gaseous HF at 315 kPa and 89.5°C? (Assume that the ideal gas law holds.)

*22.78 Because of their different molar masses, H_2 and D_2 effuse at different rates (Section 5.6).

(a) If it takes 14.5 min for 0.10 mol of H_2 to effuse, how long does it take for 0.10 mol of D_2 in the same apparatus at the same T and P?

(b) How many effusion steps does it take to separate an equimolar mixture of D_2 and H_2 to 99 mol % purity?

22.79 The disproportionation of carbon monoxide to graphite and carbon dioxide is thermodynamically favored but slow.

(a) What does this statement mean in terms of the magnitudes of the equilibrium constant K, the rate constant k, and the activation energy E_a?

(b) Write a balanced equation for the disproportionation of carbon monoxide.

(c) Calculate K_c at 298 K.

(d) Calculate K_p at 298 K.

22.80 The overall cell reaction for aluminum production is

$$2Al_2O_3(\text{in } Na_3AlF_6) + 3C(\text{graphite}) \longrightarrow 4Al(l) + 3CO_2(g)$$

(a) Assuming 100% efficiency, how many metric tons (t) of Al_2O_3 are consumed per t of Al produced?

(b) Assuming 100% efficiency, how many metric tons of graphite anode material are consumed per t of Al produced?

(c) Actual conditions in an aluminum plant require 1.89 t of Al_2O_3 and 0.45 t of graphite per t of Al. What is the percent yield with respect to Al_2O_3?

(d) What is the percent yield with respect to graphite?

(e) What volume of CO_2 (in m^3) is produced per t of Al at operating conditions of 960.°C and exactly 1 atm?

22.81 Quinones are biological molecules used extensively in industrial electron-transfer processes, in which the molecules cycle between oxidized (quinone) and reduced (hydroquinone) forms. For example,

1,4-benzoquinone   hydroquinone

Sodium hydrosulfite ($Na_2S_2O_4$) is frequently used to reduce quinones to hydroquinones. Write a balanced equation for this reduction with 1,4-benzoquinone in basic solution. Gaseous SO_2 is produced also.

22.82 The following nitrogen-containing species play roles in the nitrogen cycle: NH_3, N_2O, NO, NO_2, NO_2^-, NO_3^-. Draw a Lewis structure for each species, showing minimal formal charges, and indicate the shape, including ideal bond angles.

*22.83 Even though most metal sulfides are very sparingly soluble in water, their solubilities differ by several orders of magnitude. This difference can sometimes be used to separate the metals in an isolation step by controlling the pH. Use the following data to find the pH at which you can separate 0.10 M Cu^{2+} and 0.10 M Ni^{2+}:

Saturated H_2S = 0.10 M

K_{a1} of H_2S = 9×10^{-8} K_{a2} of H_2S = 1×10^{-17}
K_{sp} of NiS = 1.1×10^{-18} K_{sp} of CuS = 8×10^{-34}

22.84 Ores containing as little as 0.25% by mass of copper are used as sources of the metal.

(a) How many kilograms of such an ore would be needed for another Statue of Liberty, which contains 2.0×10^5 lb of copper?

(b) If the mineral in the ore is chalcopyrite ($FeCuS_2$), what is the mass percent of chalcopyrite in the ore?

*22.85 How does acid rain affect the leaching of phosphate into groundwater from terrestrial phosphate rock? Calculate the solubility of $Ca_3(PO_4)_2$ in each of the following:

(a) Pure water, pH 7.0 (Assume that PO_4^{3-} does not react with water.)

(b) Moderately acidic rainwater, pH 4.5 (*Hint:* Assume that all the phosphate exists in the form that predominates at this pH.)

22.86 The lead(IV) oxide used in car batteries is prepared in place by coating the electrode plate with PbO and then oxidizing this to PbO_2. Despite its formula, PbO_2 has a non-stoichiometric ratio of lead to oxygen of about 1:1.888. In fact, the presence of holes in the PbO_2 crystal structure due to missing O atoms is related to the oxide's conductivity.

(a) What is the mole % of O missing from the PbO_2 structure?

(b) What is the molar mass of the nonstoichiometric compound?

22.87 Chemosynthetic bacteria reduce CO_2 by "splitting" $H_2S(g)$ rather than $H_2O(g)$, which is used by photosynthetic organisms. Compare the free energy change for splitting H_2S with that for splitting H_2O. Is there an advantage to using H_2S instead of H_2O?

CHAPTER 23

THE TRANSITION ELEMENTS AND THEIR COORDINATION COMPOUNDS

CHAPTER OUTLINE

23.1 Properties of the Transition Elements
Electron Configurations
Atomic and Physical Properties
Chemical Properties

23.2 The Inner Transition Elements
The Lanthanides
The Actinides

23.3 Highlights of Selected Transition Metals
Chromium
Manganese
Silver
Mercury

23.4 Coordination Compounds
Structures of Complex Ions
Formulas and Names
Alfred Werner and Coordination Theory
Isomerism

23.5 Theoretical Basis for the Bonding and Properties of Complexes
Valence Bond Theory
Crystal Field Theory

Figure: A transition-metal engine. The usefulness of transition metals is obvious from the composition of a modern jet engine. In addition to aluminum, graphite fibers, and various polymer composites, this engine incorporates numerous alloys that consist of iron, chromium, titanium, and nickel. In this chapter, we examine not only the metals themselves, but also the molecular structures and the chemical and physical properties of their compounds.

Our exploration of the elements so far has, paradoxically, skirted the majority of them. Whereas most important end uses of the main-group elements involve their compounds, the transition elements are remarkably useful in their uncombined form. In Chapter 22, you saw how two of the most important transition elements—copper and iron—are extracted from their ores. In addition to those two, with their countless essential uses, many other transition elements are indispensable as well: chromium in automobile parts, gold and silver in jewelry, tungsten in lightbulb filaments, platinum in automobile catalytic converters, titanium in bicycle frames, and zinc in batteries, to mention just a few of the better known elements. You may be less aware of zirconium in nuclear-reactor liners, vanadium in axles and crankshafts, molybdenum in boiler plates, nickel in coins, tantalum in organ-replacement parts, palladium in telephone-relay contacts—the list goes on and on. As ions, many of these elements also play vital roles in living organisms.

The periodic table in Figure 23.1 shows that the **transition elements** *(transition metals)* occupy the *d* block (B groups) and *f* block *(inner transition elements)*. In the first half of this chapter, we discuss some atomic, physical, and chemical properties of these elements in general and then focus on the chemistry of four familiar ones: chromium, manganese, silver, and mercury. The second half of the chapter is devoted to the most distinctive feature of transition element chemistry, the formation of *coordination compounds,* substances that contain complex ions. We consider two models that explain the striking colors of these compounds, as well as their magnetic properties and structures, and then end with some essential biochemical functions of transition metal ions.

CONCEPTS & SKILLS

to review before you study this chapter

- properties of light (Section 7.1)
- electron shielding of nuclear charge (Section 8.2)
- electron configuration, ionic size, and magnetic behavior (Sections 8.3 to 8.5)
- valence bond theory (Section 11.1)
- constitutional, geometric, and optical isomerism (Section 15.2)
- Lewis acid-base concepts (Section 18.9)
- complex-ion formation (Section 19.4)
- redox behavior and standard electrode potentials (Section 21.3)

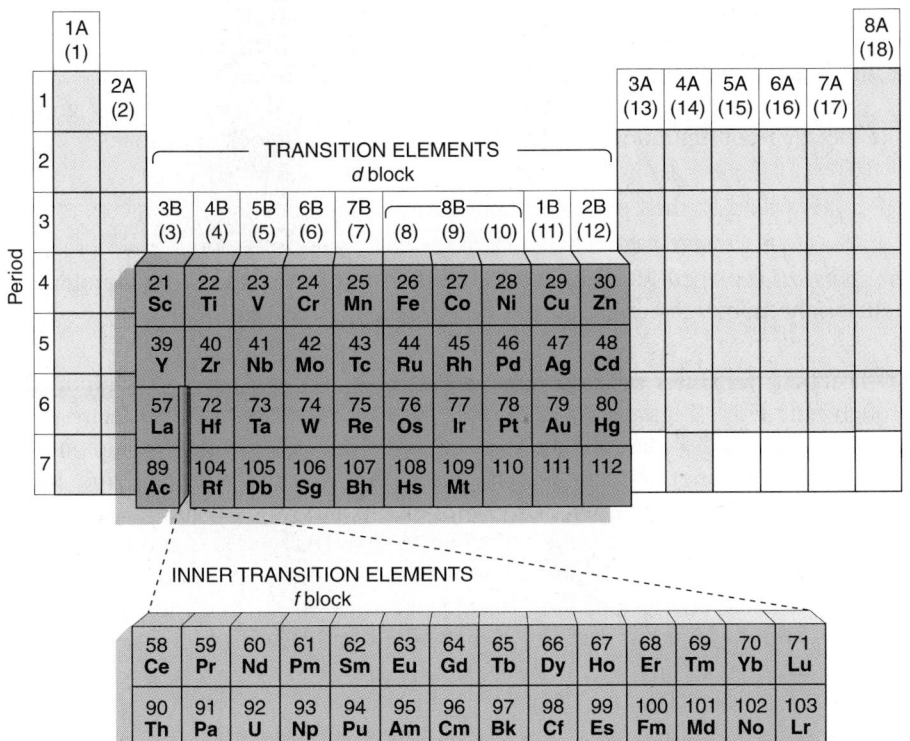

Figure 23.1 The transition elements (*d* block) and inner transition elements (*f* block) in the periodic table.

Scandium, Sc; 3B(3)

Titanium, Ti; 4B(4)

Vanadium, V; 5B(5)

Chromium, Cr; 6B(6)

Manganese, Mn; 7B(7)

Figure 23.2 The Period 4 transition metals. Samples of all ten elements appear as pure metals, in chunk or powder form, in periodic-table order on this and the facing page.

23.1 PROPERTIES OF THE TRANSITION ELEMENTS

The transition elements differ considerably in physical and chemical behavior from the main-group elements. In some ways, they are more uniform: main-group elements in each period change from metal to nonmetal, but *all transition elements are metals.* In other ways, the transition elements are more diverse: most main-group ionic compounds are colorless and diamagnetic, but *many transition metal compounds are highly colored and paramagnetic.* We first discuss electron configurations of the atoms and ions, and then examine certain key properties of transition elements, with an occasional comparison to the main-group elements.

Electron Configurations of the Transition Metals and Their Ions

As with any element, the properties of the transition elements and their compounds arise largely from the electron configurations of their atoms (Section 8.3) and ions (Section 8.5). The *d*-block (B-group) elements occur in four series that lie within Periods 4 through 7 between the last *ns*-block element [Group 2A(2)] and the first *np*-block element [Group 3A(13)]. Each series represents the filling of five *d* orbitals and, thus, contains ten elements. In 1996 and 1997, elements 110 through 112 were synthesized in particle accelerators, so the Period 7 series is complete; thus, all 40 *d*-block transition elements are known. Lying between the first and second members of the *d*-block transition series in Periods 6 and 7 are the inner transition elements, whose *f* orbitals are being filled.

Even though there are several exceptions, in general, the *condensed* ground-state electron configuration for the elements in each *d*-block series is

[noble gas] $ns^2(n-1)d^x$, with $n = 4$ to 7 and $x = 1$ to 10

In Periods 6 and 7, the condensed configuration includes the *f* sublevel:

[noble gas] $ns^2(n-2)f^{14}(n-1)d^x$, with $n = 6$ or 7

The *partial* (valence-level) electron configuration for the *d*-block elements excludes the noble gas core and the filled inner *f* sublevel:

$$ns^2(n-1)d^x$$

The first transition series occurs in Period 4 and consists of scandium (Sc) through zinc (Zn) (Figure 23.2 and Table 23.1). Scandium has the electron configuration [Ar] $4s^2 3d^1$, and the addition of one electron at a time (along with one proton in the nucleus) first half-fills, then fills, the 3*d* orbitals through zinc. Recall that chromium and copper are two exceptions to this general filling pattern: the 4*s* and 3*d* orbitals in Cr are both half-filled to give [Ar] $4s^1 3d^5$, and the 4*s* in Cu is half-filled to give [Ar] $4s^1 3d^{10}$. The reasons for these exceptions involve the change in relative energies of the 4*s* and 3*d* orbitals as electrons are added across the series and the unusual stability of half-filled and filled sublevels.

Transition metal ions form through *the loss of the ns electrons before the (n − 1)d electrons.* Thus, the electron configuration of Ti^{2+} is [Ar] $3d^2$, *not* [Ar] $4s^2$, and Ti^{2+} is referred to as a d^2 ion. Ions of different metals with the same configuration often have similar properties. For example, both Mn^{2+} and Fe^{3+} are d^5 ions; both have pale colors in aqueous solution and, as we'll discuss later, form complex ions with similar magnetic properties.

Table 23.1 shows a general pattern in number of unpaired electrons (or half-filled orbitals) across the Period 4 transition series. Note that the number increases

Iron, Fe; 8B(8)

Cobalt, Co; 8B(9)

Nickel, Ni; 8B(10)

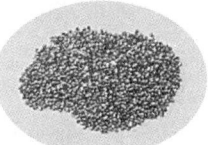
Copper, Cu; 1B(11)

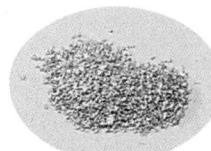
Zinc, Zn; 2B(12)

Table 23.1 **Orbital Occupancy of the Period 4 Transition Metals**

Element	Partial Orbital Diagram			Unpaired Electrons
	$4s$	$3d$	$4p$	
Sc	↑↓	↑		1
Ti	↑↓	↑ ↑		2
V	↑↓	↑ ↑ ↑		3
Cr	↑	↑ ↑ ↑ ↑ ↑		6
Mn	↑↓	↑ ↑ ↑ ↑ ↑		5
Fe	↑↓	↑↓ ↑ ↑ ↑ ↑		4
Co	↑↓	↑↓ ↑↓ ↑ ↑ ↑		3
Ni	↑↓	↑↓ ↑↓ ↑↓ ↑ ↑		2
Cu	↑	↑↓ ↑↓ ↑↓ ↑↓ ↑↓		1
Zn	↑↓	↑↓ ↑↓ ↑↓ ↑↓ ↑↓		0

in the first half of the series and, when pairing begins, decreases through the second half. As you'll see, it is the electron configuration of the transition metal *atom* that correlates with physical properties of the *element,* such as density and magnetic behavior, whereas it is the electron configuration of the *ion* that determines the properties of the *compounds.*

SAMPLE PROBLEM 23.1 Writing Electron Configurations of Transition Metal Atoms and Ions

Problem Write *condensed* electron configurations for the following: **(a)** Zr; **(b)** V^{3+}; **(c)** Mo^{3+}. (Assume that elements in higher periods behave like those in Period 4.)

Plan We locate the element in the periodic table and count its position in the respective transition series. These elements are in Periods 4 and 5, so the general configuration is [noble gas] $ns^2(n-1)d^x$. For the ions, we recall that ns electrons are lost first.

Solution **(a)** Zr is the second element in the $4d$ series: [Kr] $5s^24d^2$.

(b) V is the third element in the $3d$ series: [Ar] $4s^23d^3$. In forming V^{3+}, three electrons are lost (two $4s$ and one $3d$), so V^{3+} is a d^2 ion: [Ar] $3d^2$.

(c) Mo lies below Cr in Group 6B(6), so we expect the same exception as for Cr. Thus, Mo is [Kr] $5s^14d^5$. To form the ion, Mo loses the one $5s$ and two of the $4d$ electrons, so Mo^{3+} is a d^3 ion: [Kr] $4d^3$.

Check Figure 8.12 (p. 301) shows we're correct for the atoms. Be sure charge plus number of d electrons in the ion equals the sum of outer s and d electrons in the atom.

FOLLOW-UP PROBLEM 23.1 Write *partial* electron configurations for the following: **(a)** Ag^+; **(b)** Cd^{2+}; **(c)** Ir^{3+}.

Atomic and Physical Properties of the Transition Elements

The atomic properties of the transition elements contrast in several ways with those of a comparable set of main-group elements.

Trends Across a Period Consider the variations in atomic size, electronegativity, and ionization energy across Period 4 (Figure 23.3):

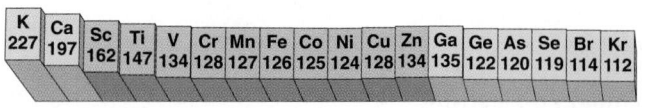

A Atomic radius (pm)

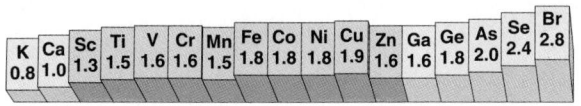

B Electronegativity

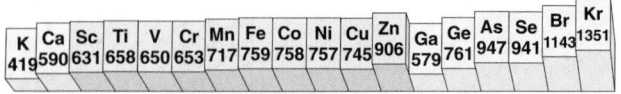

C First ionization energy (kJ/mol)

Figure 23.3 **Horizontal trends in key atomic properties of the Period 4 elements.** The atomic radius, electronegativity, and first ionization energy of the elements in Period 4 are shown as posts of different heights, with darker shades for the transition series. The transition elements exhibit smaller, less regular changes for these properties than do the main-group elements.

- *Atomic size.* Atomic size decreases overall across the period (Figure 23.3A). However, there is a smooth, steady decrease across the main groups because the electrons are added to *outer* orbitals, which shield the increasing nuclear charge poorly. This steady decrease is suspended throughout the transition series, where *atomic size decreases at first but then remains fairly constant.* Recall that the *d* electrons fill *inner* orbitals, so they shield outer electrons from the increasing nuclear charge much more efficiently. As a result, the outer 4*s* electrons are not pulled closer.

- *Electronegativity.* Electronegativity generally increases across the period but, once again, the transition elements exhibit a relatively *small change in electronegativity* (Figure 23.3B), consistent with the relatively small change in size. In contrast, the main groups show a steady, much steeper increase between the metal potassium (0.8) and the nonmetal bromine (2.8). The transition elements all have intermediate electronegativity values, much like the large, metallic members of Groups 3A(13) to 5A(15).

- *Ionization energy.* The ionization energies of the Period 4 main-group elements rise steeply from left to right, more than tripling from potassium (419 kJ/mol) to krypton (1351 kJ/mol), as electrons become more difficult to remove from the poorly shielded, increasing nuclear charge. In the transition metals, however, the *first ionization energies increase relatively little* because the inner 3*d* electrons shield effectively (Figure 23.3C); thus, the outer 4*s* electron experiences only a slightly higher effective nuclear charge. [Recall from Section 8.4 that the drop at Group 3A(13) occurs because it is relatively easy to remove the first electron from the outer *np* orbital.]

Trends Within a Group Vertical trends for transition elements are also different from those in the main groups.

- *Atomic size.* As expected, atomic size increases from Period 4 to 5, as it does for the main-group elements, but there is virtually *no size increase from Period 5 to 6* (Figure 23.4A). Remember that the lanthanides, with their buried 4*f* sublevel, appear between the 4*d* (Period 5) and 5*d* (Period 6) series. Therefore, an element in Period 6 is separated from the one above it in Period 5 by 32 elements (ten 4*d*, six 5*p*, two 6*s*, and fourteen 4*f*) instead of just 18. The extra shrinkage that results from the increase in nuclear charge due to these additional 14 elements is called the **lanthanide contraction.** By coincidence, this *decrease* is about equal to the normal *increase* between periods, so the Periods 5 and 6 transition elements have about the same atomic sizes.

- *Electronegativity.* The vertical trend in electronegativity seen in most transition groups is opposite the trend in main groups. Here, we see an *increase* in electronegativity from Period 4 to Period 5, but then no further increase in Period 6 (Figure 23.4B). The heavier elements, especially gold (EN = 2.4), become quite electronegative, with values exceeding those of most metalloids and even some nonmetals (e.g., EN of Te and of P = 2.1). (In fact, gold forms the saltlike CsAu and the Au⁻ ion, which exists in liquid ammonia.) Although

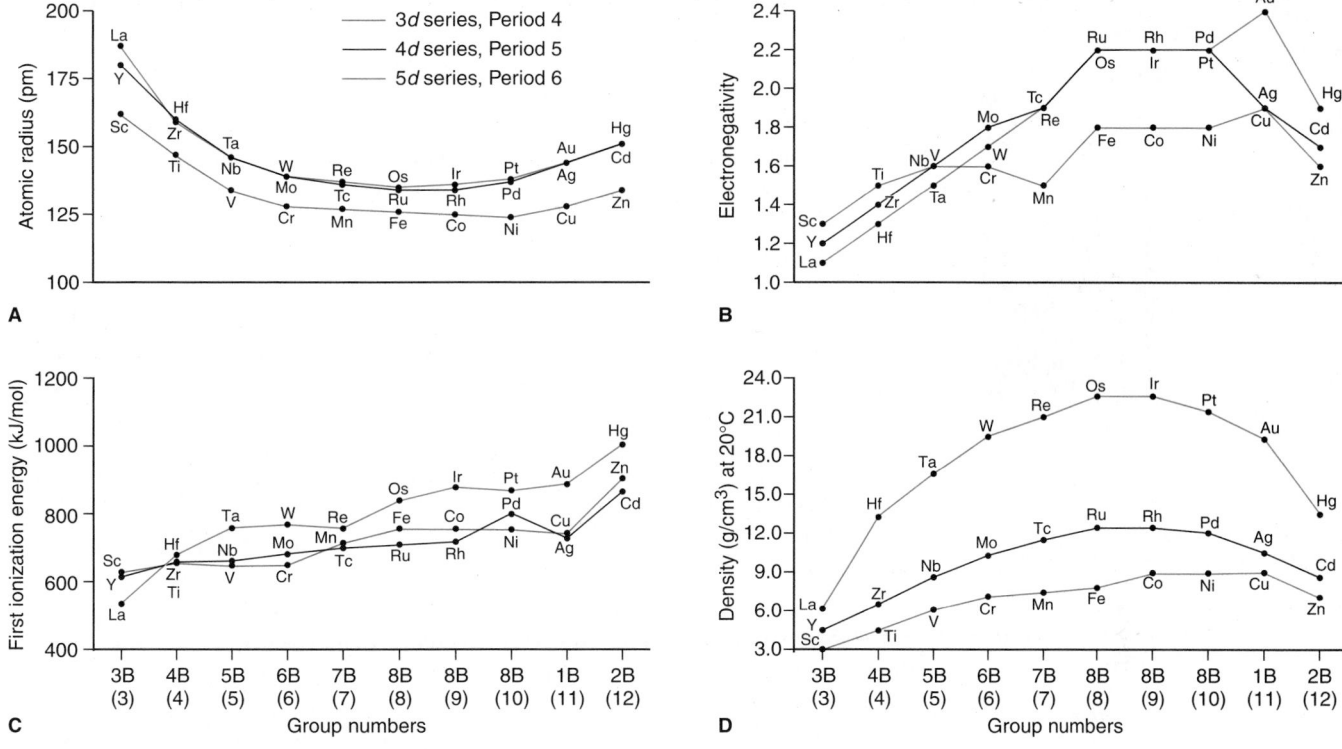

Figure 23.4 **Vertical trends in key properties within the transition elements.** The trends are unlike those for the main-group elements in several ways: **A,** The second and third members of a transition metal group are nearly the same size. **B,** Electronegativity increases down a transition group. **C,** First ionization energies are highest at the bottom of a transition group. **D,** Densities increase down a transition group because mass increases faster than volume.

the atomic size increases slightly from the top to the bottom of a group, the nuclear charge increases much more. Therefore, the heavier members exhibit more covalent character in their bonds and attract electrons more strongly than do main-group metals.

• *Ionization energy.* The relatively small increase in size combined with the relatively large increase in nuclear charge also explains why *the first ionization energy generally increases* down a group (Figure 23.4C). This trend also runs counter to the pattern in the main groups, in which heavier members are so much larger that their outer electron is easier to remove.

• *Density.* Atomic size, and therefore volume, is inversely related to density. Across a period, densities increase, then level off, and finally dip a bit at the end of a series (Figure 23.4D). Down a group, densities increase dramatically because atomic volumes change little from Period 5 to 6, but atomic masses increase significantly. As a result, the Period 6 series contains some of the densest elements known: tungsten, rhenium, osmium, iridium, platinum, and gold have densities about 20 times that of water and twice that of lead.

Chemical Properties of the Transition Metals

Like their atomic and physical properties, the chemical properties of the transition elements are very different from those of the main-group elements. Let's examine the key properties in the Period 4 transition series and then see how behavior changes within a group.

A

B

Figure 23.5 Aqueous oxoanions of transition elements. A, Often, a given transition element has multiple oxidation states. Here, Mn is shown in the +2 (Mn^{2+}, *left*), the +6 (MnO_4^{2-}, *middle*), and the +7 states (MnO_4^-, *right*). **B,** The highest possible oxidation state equals the group number in these oxoanions: VO_4^{3-} (*left*), $Cr_2O_7^{2-}$ (*middle*), and MnO_4^- (*right*).

VANADIUM REDUCTION

Oxidation States One of the most characteristic chemical properties of the transition metals is the occurrence of *multiple oxidation states*. For example, in their compounds, vanadium exhibits two common positive oxidation states, chromium three, and manganese three (Figure 23.5A), and many other oxidation states are seen less often. Since the ns and $(n - 1)d$ electrons are so close in energy, transition elements can involve all or most of these electrons in bonding. This behavior is markedly different from that of the main-group metals, which display one or at most two oxidation states in their compounds.

The highest oxidation state of elements in Groups 3B(3) through 7B(7) is equal to the group number (Table 23.2). These states are seen when the elements combine with highly electronegative oxygen or fluorine. For instance, in the oxoanion solutions shown in Figure 23.5B, vanadium occurs as the vanadate ion (VO_4^{3-}; O.N. of V = +5), chromium occurs as the dichromate ion ($Cr_2O_7^{2-}$; O.N. of Cr = +6), and manganese occurs as the permanganate ion (MnO_4^-; O.N. of Mn = +7). Elements in Groups 8B(8), 8B(9), and 8B(10) exhibit fewer oxidation states, and the highest state is less common and never equal to the group number. For instance, we never encounter Fe in the +8 state and only rarely in the +6 state. The +2 and +3 states are the most common ones for iron* and cobalt, and the +2 state is most common for nickel, copper, and zinc. *The +2 oxidation state is common because ns^2 electrons are readily lost.*

Copper, silver, and gold (the coinage metals) in Group 1B(11) are unusual. Although copper has a fairly common oxidation state of +1, which results from loss of its single $4s$ electron, its most common state is +2. Silver behaves more predictably, exhibiting primarily the +1 oxidation state. Gold exhibits the +3 state and, less often, the +1 state. Zinc, cadmium, and mercury in Group 2B(12) exhibit the +2 state, but mercury also exhibits the +1 state in the Hg_2^{2+} ion, with its Hg—Hg bond.[†]

---

*Iron may seem to have unusual oxidation states in the common ores magnetite (Fe_3O_4) and pyrite (FeS_2), but this is not really the case. In magnetite, one-third of the metal ions are Fe^{2+} and two-thirds are Fe^{3+}, which is equivalent to a 1:1 ratio of $FeO:Fe_2O_3$ and gives an overall formula of Fe_3O_4. Pyrite contains Fe^{2+} combined with the disulfide ion, S_2^{2-}.

[†]Some evidence suggests that the Hg_2^{2+} ion may be considered an Hg atom bonded to an Hg^{2+} ion, with the atom donating its two $6s$ electrons for the covalent bond.

Table 23.2 Oxidation States and *d*-Orbital Occupancy of the Period 4 Transition Metals*

Oxidation State	3B(3) Sc	4B(4) Ti	5B(5) V	6B(6) Cr	7B(7) Mn	8B(8) Fe	8B(9) Co	8B(10) Ni	1B(11) Cu	2B(12) Zn
0	d^1	d^2	d^3	d^5	d^5	d^6	d^7	d^8	d^{10}	d^{10}
+1			d^3	d^5	d^5	d^6	d^7	d^8	d^{10}	
+2		d^2	d^3	d^4	d^5	d^6	d^7	d^8	d^9	d^{10}
+3	d^0	d^1	d^2	d^3	d^4	d^5	d^6	d^7	d^8	
+4		d^0	d^1	d^2	d^3	d^4	d^5	d^6		
+5			d^0	d^1	d^2		d^4			
+6				d^0	d^1	d^2				
+7					d^0					

*Most important in color.

Metallic Behavior and Reducing Strength Atomic size and oxidation state have a major effect on the nature of bonding in transition metal compounds. Like the metals in Groups 3A(13), 4A(14), and 5A(15), the transition elements in their *lower* oxidation states behave chemically more like metals. That is, *ionic bonding is more prevalent for the lower oxidation states, and covalent bonding is more prevalent for the higher states.* For example, at room temperature, $TiCl_2$ is an ionic solid, whereas $TiCl_4$ is a molecular liquid. In the higher oxidation states, the atoms have higher charge densities, so they polarize the electron clouds of the nonmetal ions more strongly and the bonding becomes more covalent. For the same reason, the oxides become less basic as the oxidation state increases: TiO is weakly basic in water, whereas TiO_2 is amphoteric (reacts with both acid and base).

Table 23.3 shows the standard electrode potentials of the Period 4 transition metals in their +2 oxidation state in acid solution. Note that, in general, reducing strength decreases across the series. All the Period 4 transition metals, except copper, are active enough to reduce H^+ from aqueous acid to form hydrogen gas. In contrast to the rapid reaction at room temperature of the Group 1A(1) and 2A(2) metals with water, however, the transition metals have an oxide coating that allows rapid reaction only with hot water or steam.

Color and Magnetism of Compounds *Most main-group ionic compounds are colorless* because the metal ion has a filled outer level (noble gas electron configuration). With only much higher energy orbitals available to receive an excited electron, the ion does not absorb visible light. In contrast, electrons in a partially filled *d* sublevel can absorb visible wavelengths and move to slightly higher energy *d* orbitals. As a result, *many transition metal compounds have striking colors.* Exceptions are the compounds of scandium, titanium(IV), and zinc, which are colorless because their metal ions have either an empty *d* sublevel (Sc^{3+} or Ti^{4+}: [Ar] $3d^0$) or a filled one (Zn^{2+}: [Ar] $3d^{10}$) (Figure 23.6).

Magnetic properties are also related to sublevel occupancy (Section 8.5). Recall that a *paramagnetic* substance has atoms or ions with unpaired electrons, which cause it to be attracted to an external magnetic field. A *diamagnetic* substance has only paired electrons, so it is unaffected (or slightly repelled) by a magnetic field. *Most main-group metal ions are diamagnetic* for the same reason they are colorless: all their electrons are paired. In contrast, *many transition metal compounds are paramagnetic because of their unpaired d electrons.* For example, $MnSO_4$ is paramagnetic, but $CaSO_4$ is diamagnetic. The Ca^{2+} ion has the electron configuraton of argon, whereas Mn^{2+} has a d^5 configuration. Transition metal ions with a d^0 or d^{10} configuration are also colorless and diamagnetic.

Table 23.3 Standard Electrode Potentials of Period 4 M^{2+} Ions

Half-Reaction	E^0 (V)
$Ti^{2+}(aq) + 2e^- \rightleftharpoons Ti(s)$	-1.63
$V^{2+}(aq) + 2e^- \rightleftharpoons V(s)$	-1.19
$Cr^{2+}(aq) + 2e^- \rightleftharpoons Cr(s)$	-0.91
$Mn^{2+}(aq) + 2e^- \rightleftharpoons Mn(s)$	-1.18
$Fe^{2+}(aq) + 2e^- \rightleftharpoons Fe(s)$	-0.44
$Co^{2+}(aq) + 2e^- \rightleftharpoons Co(s)$	-0.28
$Ni^{2+}(aq) + 2e^- \rightleftharpoons Ni(s)$	-0.25
$Cu^{2+}(aq) + 2e^- \rightleftharpoons Cu(s)$	0.34
$Zn^{2+}(aq) + 2e^- \rightleftharpoons Zn(s)$	-0.76

Figure 23.6 Colors of representative compounds of the Period 4 transition metals. Staggered from left to right, the compounds are scandium oxide *(white)*, titanium(IV) oxide *(white)*, vanadyl sulfate dihydrate *(light blue)*, sodium chromate *(yellow)*, manganese(II) chloride tetrahydrate *(light pink)*, potassium ferricyanide *(red-orange)*, cobalt(II) chloride hexahydrate *(violet)*, nickel(II) nitrate hexahydrate *(green)*, copper(II) sulfate pentahydrate *(blue)*, and zinc sulfate heptahydrate *(white)*.

Chemical Behavior Within a Group The *increase* in reactivity seen down a group of main-group metals, as shown by the *decrease* in first ionization energy (IE_1), does *not* appear down a group of transition metals. Consider the chromium (Cr) group [6B(6)], which shows a typical pattern (Table 23.4). IE_1 *increases* down the group, which makes the two heavier metals *less* reactive than the lightest one. Chromium is also a much stronger reducing agent than molybdenum (Mo) or tungsten (W), as shown by the standard electrode potentials.

Table 23.4 Some Properties of Group 6B(6) Elements

Element	Atomic Radius (pm)	IE_1 (kJ/mol)	E^0 (V) for $M^{3+}(aq)|M(s)$
Cr	128	653	−0.74
Mo	139	685	−0.20
W	139	770	−0.11

The similarity in atomic size of Period 5 and 6 members also leads to similar chemical behavior, a fact that has some important practical consequences. Because Mo and W compounds behave similarly, for example, their ores often occur together in nature, which makes the elements very difficult to separate from each other. The same situation occurs between zirconium and hafnium in Group 4B(4), and between niobium and tantalum in Group 5B(5).

SECTION SUMMARY

All transition elements are metals. Atoms of the *d*-block elements have $(n − 1)d$ orbitals being filled, and their ions have an empty *ns* orbital. Unlike the trends in the main-group elements, atomic size, electronegativity, and first ionization energy change relatively little across a transition series. Because of the lanthanide contraction, atomic size changes little from Period 5 to 6 in a transition metal group; thus, electronegativity, first ionization energy, and density *increase* down a group. Transition metals typically have several oxidation states, with the +2 state most common. The elements exhibit more metallic behavior in their lower states. Most Period 4 members are active enough to reduce hydrogen ion from acid solution. Many transition metal compounds are colored and paramagnetic because the metal ion has unpaired *d* electrons.

A Remarkable Laboratory Feat The discovery of the lanthanides is a testimony to the remarkable laboratory skills of 19th-century chemists. The two mineral sources, ceria and yttria, are mixtures of compounds of all 14 lanthanides; yttria also includes oxides of Sc, Y, and La. Purification of these elements was so difficult that many claims of new elements turned out to be mixtures of similar elements. To confuse matters, the periodic table of the time had space for only one element between Ba and Hf, which turned out to be La. Not until 1913, when the atomic number basis of the table was established, did chemists realize that 14 new elements could fit, and in 1918, Niels Bohr proposed that the $n = 4$ level be expanded to include the *f* sublevel.

23.2 THE INNER TRANSITION ELEMENTS

The 14 **lanthanides** (lanthanoids)—cerium (Ce; $Z = 58$) through lutetium (Lu; $Z = 71$)—lie between lanthanum ($Z = 57$) and hafnium ($Z = 72$) in the third *d*-block transition series. Below them are the 14 radioactive **actinides** (actinoids), thorium (Th; $Z = 90$) through lawrencium (Lr; $Z = 103$), which lie between actinium (Ac; $Z = 89$) and rutherfordium (Rf; $Z = 104$). The lanthanides and actinides are called **inner transition elements** because, in most cases, their seven inner $4f$ or $5f$ orbitals are being filled.

The Lanthanides

The lanthanides are sometimes called the *rare earth elements,* a term referring to their presence in unfamiliar oxides, but they are actually not rare at all. Cerium (Ce), for instance, ranks 26th in natural abundance (by mass %) and is five times more abundant than lead. All the lanthanides are silvery, high-melting (800°C to 1600°C) metals. Their chemical properties represent an extreme case of the small variations typical of transition elements in a period or a group, which makes the lanthanides very difficult to separate.

The natural co-occurrence of the lanthanides arises because they exist as M^{3+} ions of very similar radii in their common ores. Most lanthanides have the ground-state electron configuration [Xe] $6s^2 4f^x 5d^0$, where x varies across the series. The three exceptions (Ce, Gd, and Lu) have a single electron in a $5d$ orbital: Ce ([Xe] $6s^2 4f^1 5d^1$) forms a stable 4+ ion with an empty (f^0) sublevel, and the Gd^{3+} and Lu^{3+} ions have a stable half-filled (f^7) or filled (f^{14}) sublevel.

Lanthanide compounds and their mixtures have many uses. Several oxides are used for tinting sunglasses and welder's goggles and for adding color to the fluorescent powder coatings in TV screens. High-quality camera lenses incorporate La_2O_3 because of its extremely high index of refraction. Samarium forms an alloy with cobalt, $SmCo_5$, that is used in the strongest known permanent magnet (see Sample Problem 23.2). Two industrial uses account for more than 60% of rare earth applications. In gasoline refining, the zeolite catalysts (Gallery, Section 14.6) used to "crack" hydrocarbon components into smaller molecules contain 5% by mass rare earth oxides. In steelmaking, a mixture of lanthanides, called *misch metal,* is used to remove carbon impurities from molten iron and steel.

SAMPLE PROBLEM 23.2 Finding the Number of Unpaired Electrons

Problem The alloy $SmCo_5$ forms a permanent magnet because both samarium and cobalt have unpaired electrons. How many unpaired electrons are in the Sm atom ($Z = 62$)?

Plan We write the condensed electron configuration of Sm and then, using Hund's rule and the aufbau principle, place the electrons in a partial orbital diagram and count the unpaired electrons.

Solution Samarium is the eighth element after Xe. Two electrons go into the $6s$ sublevel. In general, the $4f$ sublevel fills before the $5d$ (among the lanthanides, only Ce, Gd, and Lu have $5d$ electrons), so the remaining electrons go into the $4f$. Thus, Sm is [Xe] $6s^2 4f^6$. There are seven f orbitals, so each of the six f electrons enters a separate orbital:

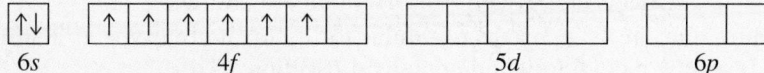

| $6s$ | $4f$ | $5d$ | $6p$ |

Thus, Sm has six unpaired electrons.

Check Six $4f$ e^- plus two $6s$ e^- plus the 54 e^- in Xe gives 62, the atomic number of Sm.

FOLLOW-UP PROBLEM 23.2 How many unpaired electrons are in the Er^{3+} ion?

The Actinides

All actinides are radioactive. Like the lanthanides, they share very similar physical and chemical properties. Thorium and uranium occur in nature, but the transuranium elements, those with Z greater than 92, have been synthesized in high-energy particle accelerators, some in only tiny amounts (Section 24.3). In fact, macroscopic samples of mendelevium (Md), nobelium (No), and lawrencium (Lr) have never been seen. The actinides that have been isolated are silvery and chemically reactive and, like the lanthanides, form highly colored compounds. The actinides and lanthanides have similar outer-electron configurations. Although the +3 oxidation state is characteristic of the actinides, as it is in the lanthanides, other states also occur. For example, uranium exhibits +3 through +6 states, with the +6 state the most prevalent; thus, the most common oxide of uranium is UO_3.

SECTION SUMMARY

There are two series of inner transition elements. The lanthanides ($4f$ series) have a common +3 oxidation state and exhibit very similar properties. The actinides ($5f$ series) are radioactive. All actinides have a +3 oxidation state; several, including uranium, have higher states as well.

23.3 HIGHLIGHTS OF SELECTED TRANSITION METALS

Let's investigate several transition elements, focusing on the general patterns in the aqueous chemistry of each. We examine chromium and manganese from Period 4, silver from Period 5, and mercury from Period 6.

Chromium

Ruby.

Chromium is a very shiny, silvery metal, whose name (from the Greek *chroma*, "color") refers to its many colorful compounds. A solution of Cr^{3+} is deep violet, for example, and a trace of Cr^{3+} in the Al_2O_3 crystal structure gives a ruby its beautiful red hue (see photo). Chromium readily forms a thin, adherent, transparent coating of Cr_2O_3 in air, making the metal extremely useful as an attractive protective coating on easily corroded metals, such as iron. "Stainless" steels often contain as much as 18% chromium by mass and are highly resistant to corrosion.

With six valence electrons ($[Ar]\ 4s^1 3d^5$), chromium occurs in all possible positive oxidation states, but the three most important are +2, +3, and +6 (see Table 23.2). In its three common oxides, chromium exhibits the pattern seen in many elements: *nonmetallic character and oxide acidity increase with metal oxidation state*. Chromium(II) oxide (CrO) is basic and largely ionic. It forms an insoluble hydroxide in neutral or basic solution but dissolves in acidic solution to yield the Cr^{2+} ion:

$$CrO(s) + 2H^+(aq) \longrightarrow Cr^{2+}(aq) + H_2O(l)$$

Chromium(III) oxide (Cr_2O_3) is amphoteric, dissolving in acid to yield the violet Cr^{3+} ion,

$$Cr_2O_3(s) + 6H^+(aq) \longrightarrow 2Cr^{3+}(aq) + 3H_2O(l)$$

and in base to form the green $Cr(OH)_4^-$ ion:

$$Cr_2O_3(s) + 3H_2O(l) + 2OH^-(aq) \longrightarrow 2Cr(OH)_4^-(aq)$$

Thus, chromium in its +3 state is similar to the main-group metal aluminum in several respects, including its amphoterism (Section 19.4).

Deep-red chromium(VI) oxide (CrO_3) is covalent and acidic, forming chromic acid (H_2CrO_4) in water,

$$CrO_3(s) + H_2O(l) \longrightarrow H_2CrO_4(aq)$$

which yields the yellow chromate ion (CrO_4^{2-}) in base:

$$H_2CrO_4(aq) + 2OH^-(aq) \longrightarrow CrO_4^{2-}(aq) + 2H_2O(l)$$

In acidic solution, the chromate ion immediately forms the orange dichromate ion ($Cr_2O_7^{2-}$):

$$2CrO_4^{2-}(aq) + 2H^+(aq) \rightleftharpoons Cr_2O_7^{2-}(aq) + H_2O(l)$$

Because both ions contain chromium(VI), this is not a redox reaction; rather, it is a dehydration-condensation, as you can see from the structures in Figure 23.7. Hydrogen-ion concentration controls the equilibrium position: yellow CrO_4^{2-} predominates at high pH and orange $Cr_2O_7^{2-}$ at low pH. The bright colors of chromium(VI) compounds lead to their wide use in pigments for artist's paints and ceramic glazes. Lead chromate (chrome yellow) is used as an oil-paint color, and also in the yellow stripes that delineate traffic lanes.

Chromium metal and the Cr^{2+} ion are potent reducing agents. The metal displaces hydrogen from dilute acids to form blue $Cr^{2+}(aq)$, which reduces O_2 in air within minutes to form the violet Cr^{3+} ion:

$$4Cr^{2+}(aq) + O_2(g) + 4H^+(aq) \longrightarrow 4Cr^{3+}(aq) + 2H_2O(l) \qquad E^0_{overall} = 1.64\ V$$

Chromium(VI) compounds in acid solution are strong oxidizing agents (concentrated solutions are *extremely* corrosive!), the chromium(VI) being readily reduced to chromium(III):

$$Cr_2O_7^{2-}(aq) + 14H^+(aq) + 6e^- \longrightarrow 2Cr^{3+}(aq) + 7H_2O(l) \qquad E^0 = 1.33\ V$$

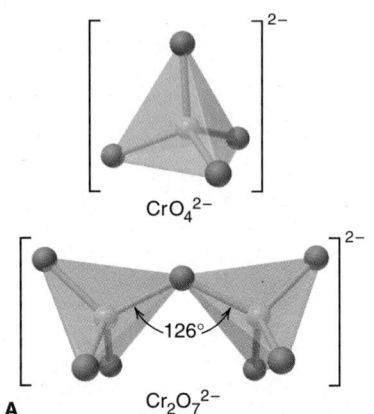

CrO_4^{2-}

$126°$

$Cr_2O_7^{2-}$

A

B

Figure 23.7 The bright colors of chromium(VI) compounds. A, Structures of the chromate (CrO_4^{2-}) and dichromate ($Cr_2O_7^{2-}$) ions. **B,** Samples of K_2CrO_4 (*yellow*) and $K_2Cr_2O_7$ (*orange*).

This reaction is often used to determine the iron content of a water or soil sample by oxidizing Fe^{2+} to Fe^{3+} ion. In basic solution, the CrO_4^{2-} ion, which is a much weaker oxidizing agent, predominates:

$$CrO_4^{2-}(aq) + 4H_2O(l) + 3e^- \longrightarrow Cr(OH)_3(s) + 5OH^-(aq) \qquad E^0 = -0.13 \text{ V}$$

The Concept of Valence-State Electronegativity Why does oxide acidity increase with oxidation state? And how can a metal like chromium form an oxoanion? To answer such questions, we must apply the concept of electronegativity to the various oxidation states of an element. A metal in a higher oxidation state is more positively charged, which increases its attraction for electrons; in effect, its *electronegativity increases*. This effective electronegativity, called *valence-state electronegativity,* also has numerical values. The electronegativity of chromium metal is 1.6, close to that of aluminum (1.5), another active metal. For chromium(III), the value increases to 1.7, still characteristic of a metal. However, the electronegativity of chromium(VI) is 2.3, close to the values of some nonmetals, such as phosphorus (2.1), selenium (2.4), and carbon (2.5). Thus, like P in PO_4^{3-}, Cr in chromium(VI) compounds often occurs covalently bound at the center of the oxoanion of a relatively strong acid.

Manganese

Elemental manganese is hard and shiny and, like vanadium and chromium, is used mostly to make steel alloys. A small amount of Mn (<1%) makes steel easier to roll, forge, and weld. Steel made with 12% Mn is tough enough to be used for naval armor, bulldozer buckets (see photo), and other extremely hard steel objects. Small amounts of manganese are added to aluminum beverage cans and bronze alloys to make them stiffer and tougher as well.

The chemistry of manganese resembles that of chromium in some respects. The free metal is quite reactive and readily reduces H^+ from acids, forming the pale-pink Mn^{2+} ion:

$$Mn(s) + 2H^+(aq) \longrightarrow Mn^{2+}(aq) + H_2(g) \qquad E^0 = 1.18 \text{ V}$$

Like chromium, manganese can use all its valence electrons in its compounds, exhibiting every possible positive oxidation state, with the +2, +4, and +7 states most common (Table 23.5). As the oxidation state of manganese rises, its valence-state electronegativity increases and its oxides change from basic to acidic. Manganese(II) oxide (MnO) is basic, and manganese(III) oxide (Mn_2O_3) is amphoteric. Manganese(IV) oxide (MnO_2) is insoluble and shows no acid-base properties. [It is used in dry-cell and alkaline batteries as the oxidizing agent in a redox reaction with zinc (Gallery, Section 21.5).] Manganese(VII) oxide

Bulldozer parts are made of steel containing manganese.

Table 23.5 Some Oxidation States of Manganese

Oxidation state*	Mn(II)	Mn(III)	Mn(IV)	Mn(VI)	Mn(VII)
Example	Mn^{2+}	Mn_2O_3	MnO_2	MnO_4^{2-}	MnO_4^-
Ion configuration	d^5	d^4	d^3	d^1	d^0
Oxide acidity	BASIC				ACIDIC

*Most common states in **boldface.**

Sharing the Ocean's Wealth At their current rate of usage, known reserves of many key transition metals will be depleted in less than 50 years. Other sources must be found. One promising source is nodules strewn over large portions of the ocean floor. Varying from a few millimeters to a few meters in diameter, these chunks consist mainly of manganese and iron oxides, with oxides of other elements present in smaller amounts. Billions of tons exist, but mining them presents major technical and political challenges. Global agreements have designated the ocean floor as international property, so cooperation will be required to mine the nodules and share the mineral rewards.

(Mn_2O_7), which forms by reaction of Mn with pure O_2, reacts with water to form permanganic acid ($HMnO_4$), which is as strong as perchloric acid ($HClO_4$).

All manganese species with oxidation states greater than $+2$ act as oxidizing agents, but the purple permanganate (MnO_4^-) ion is particularly powerful. Like compounds with chromium in its highest oxidation state, MnO_4^- is a much stronger oxidizing agent in acidic than in basic solution:

$$MnO_4^-(aq) + 4H^+(aq) + 3e^- \longrightarrow MnO_2(s) + 2H_2O(l) \qquad E^0 = 1.68 \text{ V}$$
$$MnO_4^-(aq) + 2H_2O(l) + 3e^- \longrightarrow MnO_2(s) + 4OH^-(aq) \qquad E^0 = 0.59 \text{ V}$$

Unlike Cr^{2+} and Fe^{2+}, the Mn^{2+} ion resists oxidation in air. The Cr^{2+} ion is a d^4 species and readily loses a $3d$ electron to form the d^3 ion Cr^{3+}, which is more stable. The Fe^{2+} ion is a d^6 species, and removing a $3d$ electron yields the stable, half-filled d^5 configuration of Fe^{3+}. Removing an electron from Mn^{2+} disrupts its stable d^5 configuration.

Silver

Silver, the second member of the coinage metals [Group 1B(11)], has been admired for thousands of years and is still treasured for use in jewelry and fine flatware. Because the pure metal is too soft for these purposes, however, it is alloyed with copper to form the harder sterling silver. In former times, silver was used in coins, but it has been replaced almost universally by copper-nickel alloys. Silver has the *highest electrical conductivity of any element* but is not used in wiring because copper is cheaper and more plentiful. In the past, silver was found in nuggets and veins of rock, often mixed with gold, because both elements are chemically inert enough to exist uncombined. Nearly all of those deposits have been mined, so most silver is now obtained from the anode mud formed during the electrorefining of copper (Section 22.4).

The only important oxidation state of silver is +1. Its most important *soluble* compound is silver nitrate, used for electroplating and in the manufacture of the halides used for photographic film. Although silver forms no oxide in air, it tarnishes to black Ag_2S by reaction with traces of sulfur-containing compounds. Some polishes remove the Ag_2S, along with some silver, by physically abrading the surface. An alternative "home remedy" that removes the tarnish and restores the metal involves heating the object in a solution of table salt or baking soda ($NaHCO_3$) in an aluminum pan. Aluminum, a strong reducing agent, reduces the Ag^+ ions back to the metal:

$$2Al(s) + 3Ag_2S(s) + 6H_2O(l) \longrightarrow 2Al(OH)_3(s) + 6Ag(s) + 3H_2S(g) \qquad E^0 = 0.86 \text{ V}$$

The Chemistry of Black-and-White Photography The most widespread use of silver compounds—particularly the three halides AgCl, AgBr, and AgI—is in black-and-white photography, an art that applies transition metal chemistry and solution kinetics. The photographic film itself is simply a flexible plastic support for the light-sensitive emulsion, which consists of AgBr microcrystals dispersed in gelatin. The five steps in obtaining a final photograph are exposing the film, developing the image, fixing the image, washing the negative, and printing the image. Figure 23.8 summarizes the the first four of these steps. The process depends on several key chemical properties of silver and its compounds:

- Silver halides undergo a redox reaction when exposed to visible light.
- Silver chloride, bromide, and iodide are *not* water soluble.
- Ag^+ is easily reduced: $Ag^+(aq) + e^- \longrightarrow Ag(s)$; $E^0 = 0.80$ V.
- Ag^+ forms several stable, water-soluble complex ions.

1. *Exposing the film.* Light reflected from the objects in a scene—more light from bright objects than from dark ones—enters the camera lens and strikes the film. Exposed AgBr crystals absorb photons ($h\nu$) in a very localized redox reac-

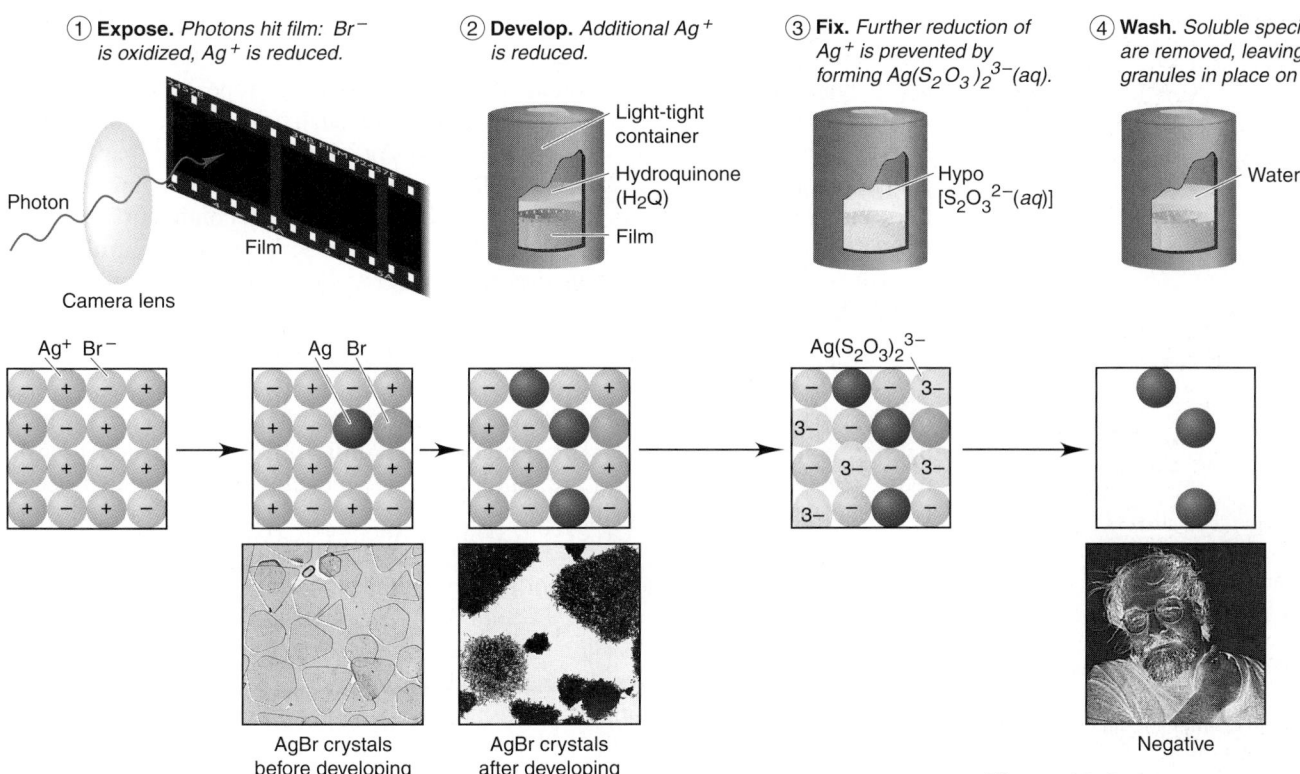

① **Expose.** *Photons hit film: Br⁻ is oxidized, Ag⁺ is reduced.*

② **Develop.** *Additional Ag⁺ is reduced.*

③ **Fix.** *Further reduction of Ag⁺ is prevented by forming Ag(S₂O₃)₂³⁻(aq).*

④ **Wash.** *Soluble species are removed, leaving Ag granules in place on film.*

AgBr crystals before developing   AgBr crystals after developing

Negative

Figure 23.8 **Steps in producing a black-and-white negative.**

tion. A Br^- ion is excited by a photon and oxidized, and the released electron almost immediately reduces a nearby Ag^+ ion:

$$Br^- \xrightarrow{h\nu} Br + e^-$$
$$\underline{Ag^+ + e^- \longrightarrow Ag}$$
$$Ag^+ + Br^- \xrightarrow{h\nu} Ag + Br$$

Wherever more light strikes a microcrystal, more Ag atoms form. The exposed crystals are called a *latent image* because the few scattered atoms of photoreduced Ag are not yet visible. Nevertheless, their presence as crystal defects within the AgBr crystal makes it highly susceptible to further reduction.

2. *Developing the image.* The latent image is developed into the actual image by reducing more of the silver ions in the crystal in a controlled manner. Developing is a rate-dependent step: crystals with many photoreduced Ag atoms react more quickly than those with only a few. The developer is a weak reducing agent, such as the organic substance hydroquinone ($H_2C_6H_4O_2$; H_2Q) (see margin):

$$2Ag^+(s) + H_2Q(aq; \text{ reduced form}) \longrightarrow 2Ag(s) + Q(aq; \text{ oxidized form}) + 2H^+(aq)$$

The reaction rate depends on H_2Q concentration, solution temperature, and the length of time the emulsion is bathed in the solution. After developing, approximately 10^6 as many Ag atoms are present on the film as there were in the latent image, and they form very small, black clusters of silver.

3. *Fixing the image.* After the image is developed, it must be "fixed"; that is, the reduction of Ag^+ must be stopped, or the entire film will blacken on exposure to more light. Fixing involves removing the remaining Ag^+ chemically by converting it to a soluble complex ion with sodium thiosulfate solution ("hypo"):

$$AgBr(s) + 2S_2O_3{}^{2-}(aq) \longrightarrow Ag(S_2O_3)_2{}^{3-}(aq) + Br^-(aq)$$

4. *Washing the negative.* The water-soluble ions are washed away in water. Washing is the final step in producing a photographic *negative,* in which dark objects in the scene appear bright in the image, and vice versa.

hydroquinone   quinone

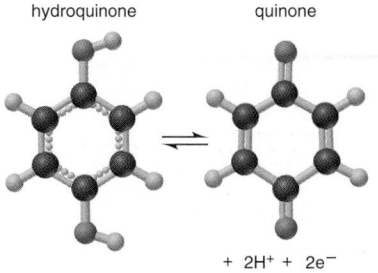

$+ 2H^+ + 2e^-$

5. *Printing the image.* Through the use of an enlarger, the image on the negative is projected onto print paper coated with emulsion (silver halide in gelatin) and exposed to light, and the previous chemical steps are repeated to produce a "positive" of the image. Bright areas on a negative, such as car tires, allow a great deal of light to pass through and reduce many Ag^+ ions on the print paper. Dark areas on a negative, such as clouds, allow much less light to pass through and, thus, much less Ag^+ reduction. Print-paper emulsion usually contains silver chloride, which reacts more slowly than silver bromide, giving finer control of the print image. High-speed film incorporates silver iodide, the most light-sensitive of the three halides.

Mercury

Mercury has been known since ancient times because cinnabar (HgS), its principal ore, is a naturally occurring red pigment (vermilion) that readily undergoes a redox reaction in the heat of a fire. Sulfide ion, the reducing agent for the process, is already present as part of the ore:

$$HgS(s) + O_2(g) \longrightarrow Hg(g) + SO_2(g)$$

The gaseous Hg condenses on cool nearby surfaces.

The Latin name *hydrargyrum* ("liquid silver") is a good description of mercury, the only metal that is liquid at room temperature. Two factors account for this unusual property. First, because of a distorted crystal structure, each mercury atom is surrounded by 6 rather than 12 nearest neighbors. Second, a filled, tightly held d subshell leaves only the two $6s$ electrons for metallic bonding. Thus, interactions among mercury atoms are relatively few and relatively weak and, as a result, the solid form breaks down at $-38.9°C$.

Many of mercury's uses arise from its unusual physical properties. Its liquid range ($-39°C$ to $357°C$) encompasses most everyday temperatures, so Hg is commonly used in thermometers. As you might expect from its position in Period 6 following the lanthanide contraction, mercury is quite dense (13.5 g/mL), which makes it convenient for use in barometers and manometers. Mercury's fluidity and conductivity make it useful for "silent" switches in thermostats. At high pressures, mercury vapor can be excited electrically to emit the bright white light seen in sports stadium and highway lights.

Mercury is a good solvent for metals, and many *amalgams* (alloys of mercury) exist. In mercury batteries, a zinc amalgam acts as the anode and mercury(II) oxide acts as the cathode (see the Gallery in Chapter 21, p. 924). In the mercury-cell version of the chlor-alkali process (Section 22.5), mercury acts as the cathode and as the solvent for the sodium metal that forms. Recall that, when treated with water, the resulting sodium amalgam, Na(Hg), forms two important by-products:

$$2Na(Hg) + 2H_2O(l) \xrightarrow{-Hg} 2NaOH(aq) + H_2(g)$$

The mercury is released in this step and reused in the electrolytic cell. In view of mercury's toxicity, which we discuss shortly, newer methods involving polymeric membranes are replacing the mercury-cell process. Nevertheless, past contamination of wastewater from the chlor-alkali process and the disposal of old mercury batteries remain serious environmental concerns.

The chemical properties of mercury are unique within Group 2B(12). The other members, zinc and cadmium, occur in the $+2$ oxidation state as d^{10} ions, but mercury occurs in the $+1$ state as well, with the condensed electron configuration [Xe] $6s^1 4f^{14} 5d^{10}$. We can picture the unpaired $6s$ electron allowing two Hg(I) species to form the *diatomic ion* [Hg—Hg]$^{2+}$ (written Hg_2^{2+}), one of the first species known with a covalent metal-metal bond (but also see the footnote on p. 1004). Mercury's more common oxidation state is +2. Whereas HgF_2 is largely ionic, many other compounds, such as $HgCl_2$, contain bonds that are predominantly covalent. Most mercury(II) compounds are insoluble in water.

The common ions Zn^{2+}, Cd^{2+}, and Hg^{2+} are biopoisons. Zinc oxide is used as an external antiseptic ointment. The Cd^{2+} and Hg^{2+} ions are two of the so-called toxic heavy-metal ions. The cadmium in solder may be more responsible than the lead for solder's high toxicity. Mercury compounds have been used in agriculture as fungicides and pesticides and in medicine as internal drugs, but these uses have been largely phased out.

Because most mercury(II) compounds are insoluble in water, they were once thought to be harmless in the environment; but we now know otherwise. Microorganisms in sludge and river sediment convert mercury atoms and ions to the methyl mercury ion, $CH_3—Hg^+$, and then to organomercury compounds, such as dimethylmercury, $CH_3—Hg—CH_3$. These toxic compounds are nonpolar and, like chlorinated hydrocarbons, become increasingly concentrated in fatty tissues, as they move up the food chain from microorganisms to worms to fish and, finally, to birds and mammals. Indeed, fish living in a mercury-polluted lake or coastal estuary can have a mercury concentration thousands of times higher than that of the water itself.

The mechanism for the toxicity of mercury and other heavy-metal ions is not fully understood. It is thought that the ions migrate from fatty tissue and bind strongly to thiol (—SH) groups of amino acids in proteins, thereby disrupting the proteins' structure and function. Because the brain has a high fat content, small amounts of lead, cadmium, and mercury ions circulating in the blood become deposited in the brain's fatty tissue, interact with its proteins, and often cause devastating neurological and psychological effects.

Mad as a Hatter The felt top hat and zany manner of the Mad Hatter in *Alice in Wonderland* are literary references to the toxicity of mercury compounds. Mercury(II) nitrate and chloride were used with HNO_3 to make felt for hats from animal hair. Inhalation of the dust generated by the process led to "hatter's shakes," abnormal behavior, and other neurological and psychological symptoms.

SECTION SUMMARY

Chromium and manganese add corrosion resistance and hardness to steels. They are typical of transition metals in having several oxidation states. Valence-state electronegativity refers to the ability of an element in its various oxidation states to attract bonding electrons. It increases with oxidation number, which is the reason elements act more metallic (more ionic compounds, more basic oxides) in lower states and more nonmetallic (more acidic oxides, oxoanions of acids) in higher states. Cr and Mn produce H_2 in acid. Cr(VI) undergoes a pH-sensitive dehydration-condensation reaction. Both Cr(VI) and Mn(VII) are stronger oxidizing agents in acid than in base. The only important oxidation state for silver is +1. The silver halides are light sensitive and are used in photography. Mercury, the only metal that is liquid at room temperature, dissolves many other metals in important applications. The mercury(I) ion is diatomic and has a metal-metal covalent bond. The element and its compounds are toxic and become concentrated as they move up the food chain.

23.4 COORDINATION COMPOUNDS

The most distinctive aspect of transition metal chemistry is the formation of **coordination compounds** (also called *complexes*). These are substances that contain at least one **complex ion,** a species consisting of *a central metal cation (either a transition metal or a main-group metal) that is bonded to molecules and/or anions called **ligands.*** In order to maintain charge neutrality in the coordination compound, the complex ion is typically associated with other ions, called **counter ions.**

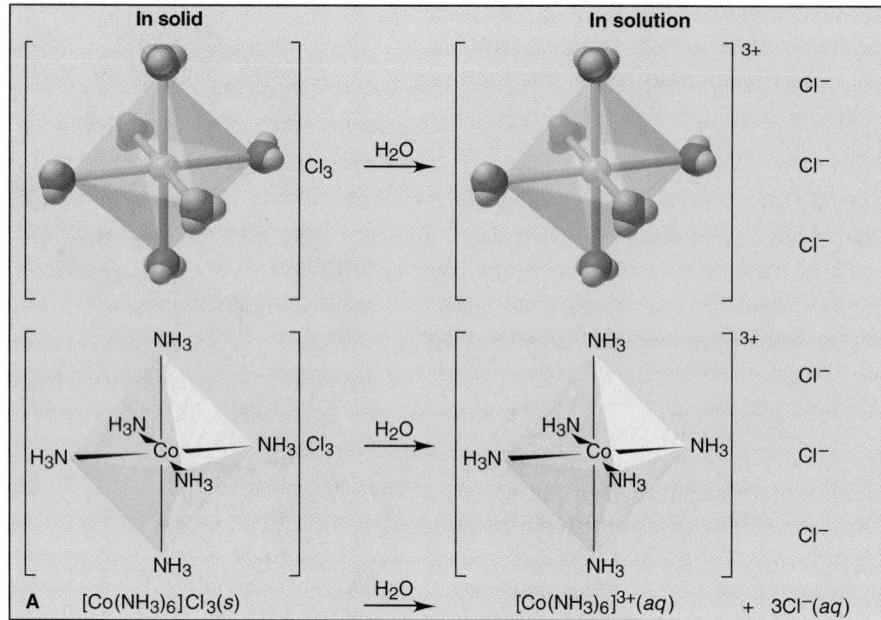

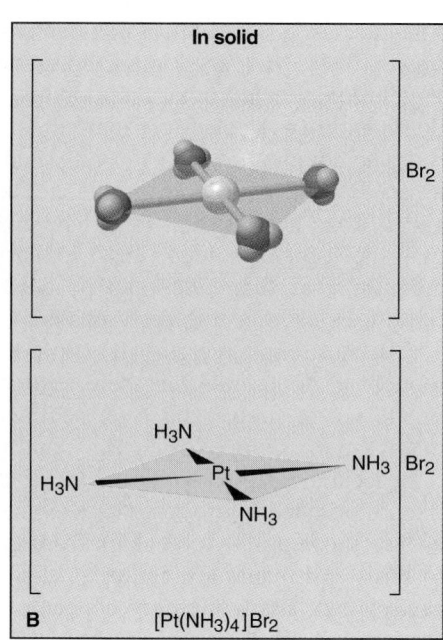

Figure 23.9 Components of a coordination compound. Coordination compounds, shown here as models *(top),* wedge diagrams *(middle),* and chemical formulas *(bottom),* typically consist of a complex ion and counter ions to neutralize the charge. The complex ion has a central metal ion surrounded by ligands. **A,** When solid [Co(NH₃)₆]Cl₃ dissolves, the complex ion and counter ions separate, but the ligands remain bound to the metal ion. Six ligands around the metal ion give the complex ion an octahedral geometry. **B,** Complex ions with a central d^8 metal ion have four ligands and a square planar geometry.

A typical coordination compound appears in Figure 23.9A: the coordination compound is $[Co(NH_3)_6]Cl_3$, the complex ion (always enclosed in square brackets) is $[Co(NH_3)_6]^{3+}$, the six NH_3 molecules bonded to the central Co^{3+} are ligands, and the three Cl^- ions are counter ions.

A coordination compound behaves like an electrolyte in water: the complex ion and counter ions separate from each other. But the complex ion behaves like a polyatomic ion: *the ligands and central metal ion remain attached.* Thus, as Figure 23.9A shows, 1 mol of $[Co(NH_3)_6]Cl_3$ yields 1 mol of $[Co(NH_3)_6]^{3+}$ ions and 3 mol of Cl^- ions.

We discussed the Lewis acid-base properties of hydrated metal ions, which are a type of complex ion, in Section 18.9, and we examined complex-ion equilibria in Section 19.4. In this section, we consider the bonding, structure, and properties of complex ions.

Structures of Complex Ions: Coordination Numbers, Geometries, and Ligands

A complex ion is described by the metal ion and the number and types of ligands attached to it. Its structure is related to three characteristics—coordination number, geometry, and number of donor atoms per ligand:

- *Coordination number.* The **coordination number** is the *number of ligand atoms* that are bonded directly to the central metal ion and is *specific* for a given metal ion in a particular oxidation state and compound. The coordination number of the Co^{3+} ion in $[Co(NH_3)_6]^{3+}$ is 6 because six ligand atoms (N from NH_3) are bonded to it. The coordination number of the Pt^{2+} ion in many of its complexes is 4, whereas that of the Pt^{4+} ion in its complexes is 6. Copper(II) may have a coordination number of 2, 4, or 6 in different complex ions. In general, *the most common coordination number in complex ions is 6,* but 2 and 4 are often seen, and some higher coordination numbers are also known.

- *Geometry. The geometry (shape) of a complex ion depends on the coordination number and nature of the metal ion.* Table 23.6 shows the geometries associated with the coordination numbers 2, 4, and 6, with some examples of each. A complex ion whose metal ion has a coordination number of 2, such as $[Ag(NH_3)_2]^+$, is *linear.* The coordination number 4 gives rise to either of two geometries—square planar or tetrahedral. Most d^8 metal ions form *square planar* complex ions, depicted in Figure 23.9B. The d^{10} ions are among those that form *tetrahedral* complex ions. A coordination number of 6 results in an *octahedral* geometry, as shown by $[Co(NH_3)_6]^{3+}$ in Figure 23.9A. Note the similarity with some of the molecular shapes in VSEPR theory (Section 10.3).

Table 23.6 Coordination Numbers and Shapes of Some Complex Ions

Coordination Number	Shape		Examples
2	Linear		$[CuCl_2]^-$, $[Ag(NH_3)_2]^+$, $[AuCl_2]^-$
4	Square planar		$[Ni(CN)_4]^{2-}$, $[PdCl_4]^{2-}$, $[Pt(NH_3)_4]^{2+}$, $[Cu(NH_3)_4]^{2+}$
4	Tetrahedral		$[Cu(CN)_4]^{3-}$, $[Zn(NH_3)_4]^{2+}$, $[CdCl_4]^{2-}$, $[MnCl_4]^{2-}$
6	Octahedral		$[Ti(H_2O)_6]^{3+}$, $[V(CN)_6]^{4-}$, $[Cr(NH_3)_4Cl_2]^+$, $[Mn(H_2O)_6]^{2+}$, $[FeCl_6]^{3-}$, $[Co(en)_3]^{3+}$

- *Donor atoms per ligand.* The ligands of complex ions are *molecules and/or anions* with one or more **donor atoms** that each *donate a lone pair of electrons* to the metal ion to form a covalent bond. Because they have at least one lone pair, donor atoms often come from Group 5A(15), 6A(16), or 7A(17).

Ligands are classified in terms of the number of donor atoms, or "teeth," that each uses to bond to the central metal ion. *Monodentate* (Latin, "one-toothed") ligands, such as Cl^- and NH_3, use a single donor atom. *Bidentate* ligands have two donor atoms, each of which bonds to the metal ion. *Polydentate* ligands have

Table 23.7 **Some Common Ligands in Coordination Compounds**

Ligand Type	Examples			
Monodentate	$H_2\ddot{O}$: water	$:\!\ddot{F}\!:^-$ fluoride ion	$[:C\!\equiv\!N\!:]^-$ cyanide ion	$[:\ddot{O}\!-\!H]^-$ hydroxide ion
	$:NH_3$ ammonia	$:\!\ddot{C}\ddot{l}\!:^-$ chloride ion	$[:\ddot{S}\!=\!C\!=\!\ddot{N}\!:]^-$ thiocyanate ion $\llcorner$ or $\lrcorner$	$[:\ddot{O}\!-\!N\!=\!\ddot{O}\!:]^-$ nitrite ion $\llcorner$ or $\lrcorner$

Bidentate

$$\begin{array}{c} H_2C-CH_2 \\ H_2\ddot{N} \qquad \ddot{N}H_2 \end{array}$$
ethylenediamine (en)

$$\left[\begin{array}{c} :\ddot{O}: \qquad :\ddot{O}: \\ C-C \\ :\ddot{O}: \qquad :\ddot{O}: \end{array} \right]^{2-}$$
oxalate ion

Polydentate

$$\begin{array}{c} H_2C-CH_2 \quad CH_2-CH_2 \\ H_2\ddot{N} \qquad \ddot{N}H \qquad \ddot{N}H_2 \end{array}$$
diethylenetriamine

$$\left[\begin{array}{ccc} :O: & :O: & :O: \\ \| & \| & \| \\ :\ddot{O}-P-\ddot{O}-P-\ddot{O}-P-\ddot{O}: \\ | & | & | \\ :\ddot{O}: & :\ddot{O}: & :\ddot{O}: \end{array} \right]^{5-}$$
triphosphate ion

$$\left[\begin{array}{c} :O: \qquad\qquad :O: \\ \| \qquad\qquad \| \\ :\ddot{O}-C-CH_2 \qquad CH_2-C-\ddot{O}: \\ :N-CH_2-CH_2-N: \\ :\ddot{O}-C-CH_2 \qquad CH_2-C-\ddot{O}: \\ \| \qquad\qquad \| \\ :O: \qquad\qquad :O: \end{array} \right]^{4-}$$
ethylenediaminetetraacetate ion (EDTA^{4-})

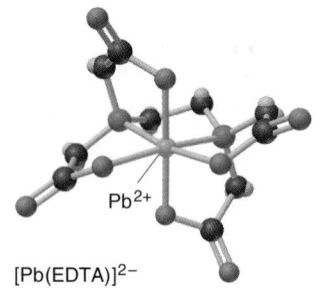

$[Pb(EDTA)]^{2-}$

Grabbing Ions Because it has six donor atoms, the ethylenediamine-tetraacetate (EDTA^{4-}) ion forms very stable complexes with many metal ions. This property makes EDTA useful in treating heavy-metal poisoning. Once ingested by the patient, the ion acts as a scavenger to remove lead and other heavy-metal ions from the blood and other body fluids.

more than two donor atoms. Table 23.7 shows some common ligands in coordination compounds; note that each ligand has one or more donor atoms (colored type), each with a lone pair of electrons to donate. Bidentate and polydentate ligands give rise to *rings* in the complex ion. For instance, ethylenediamine (abbreviated *en* in formulas) has a chain of four atoms (:N—C—C—N:), so it forms a five-membered ring, with the two electron-donating N atoms bonding to the metal atom. Such ligands seem to grab the metal ion like claws, so a complex ion that contains them is also called a **chelate** (pronounced *KEY-late*; Greek *chela*, "crab's claw").

Formulas and Names of Coordination Compounds

There are three important rules for writing the formulas of coordination compounds, the first two being the same for writing formulas of any ionic compound:

1. *The cation is written before the anion.*
2. *The charge of the cation(s) is balanced by the charge of the anion(s).*
3. *In the complex ion, neutral ligands are written before anionic ligands, and the formula for the whole ion is placed in brackets.*

Let's apply these rules as we examine the combinations of ions in coordination compounds. *The whole complex ion may be a cation or an anion.* A complex cation has anionic counter ions, and a complex anion has cationic counter ions. It's easy to find the charge of the central metal ion. For example, in $K_2[Co(NH_3)_2Cl_4]$, two K^+ counter ions balance the charge of the complex anion $[Co(NH_3)_2Cl_4]^{2-}$, which contains two NH_3 molecules and four Cl^- ions as ligands. The two NH_3 are neutral, the four Cl^- have a total charge of $4-$, and the entire complex ion has a charge of $2-$, so the central metal ion must be Co^{2+}:

$$\text{Charge of complex ion} = \text{Charge of metal ion} + \text{ total charge of ligands}$$
$$2- = \text{Charge of metal ion} + [(2 \times 0) + (4 \times 1-)]$$

So,   Charge of metal ion $= (2-) - (4-) = 2+$

In the compound $[Co(NH_3)_4Cl_2]Cl$, the complex ion is $[Co(NH_3)_4Cl_2]^+$ and one Cl^- is the counter ion. The four NH_3 ligands are neutral, the two Cl^- ligands have a total charge of $2-$, and the complex cation has a charge of $1+$, so the central metal ion must be Co^{3+} [that is, $1+ = (3+) + (2-)$]. Some coordination compounds have a complex cation *and* a complex anion, as in $[Co(NH_3)_5Br]_2[Fe(CN)_6]$. In this compound, the complex cation is $[Co(NH_3)_5Br]^{2+}$, with Co^{3+}, and the complex anion is $[Fe(CN)_6]^{4-}$, with Fe^{2+}.

Coordination compounds were originally named after the person who first prepared them or from their color, and some of these common names are still used, but most coordination compounds are named systematically through a set of rules:

1. *The cation is named before the anion.* In naming $[Co(NH_3)_4Cl_2]Cl$, for example, we name the $[Co(NH_3)_4Cl_2]^+$ ion before the Cl^- ion. Thus, the name is

   tetraamminedichlorocobalt(III) chloride

   The only space in the name appears between the cation and the anion.

2. *Within the complex ion, the ligands are named, in alphabetical order, **before** the metal ion.* Note that in the $[Co(NH_3)_4Cl_2]^+$ ion of the compound named in rule 1, the four NH_3 and two Cl^- are named before the Co^{3+}.

3. *Neutral ligands generally have the molecule name,* but there are a few exceptions (Table 23.8). *Anionic ligands drop the -ide and add -o after the root name;* thus, the name *fluoride* for the F^- ion becomes the ligand name *fluoro*. The two ligands in $[Co(NH_3)_4Cl_2]^+$ are *ammine* (NH_3) and *chloro* (Cl^-) with ammine coming before chloro alphabetically.

Table 23.8 Names of Some Neutral and Anionic Ligands

Neutral		Anionic	
Name	**Formula**	**Name**	**Formula**
Aqua	H_2O	Fluoro	F^-
Ammine	NH_3	Chloro	Cl^-
Carbonyl	CO	Bromo	Br^-
Nitrosyl	NO	Iodo	I^-
		Hydroxo	OH^-
		Cyano	CN^-

4. *A numerical prefix indicates the number of ligands of a particular type.* For example, *tetra*ammine denotes *four* NH_3, and *di*chloro denotes *two* Cl^-. Other prefixes are *tri-, penta-,* and *hexa-*. These prefixes do *not* affect the alphabetical order; thus, *tetra*ammine comes before *di*chloro. Some ligand names already contain a numerical prefix (such as ethylene*di*amine), so we use *bis* (2), *tris* (3), or *tetrakis* (4) to indicate the number of such ligands, followed by the ligand name in parentheses. Thus, a complex ion with two ethylenediamine ligands has *bis(ethylenediamine)* in its name.

5. *The oxidation state of the central metal ion is given by a Roman numeral (in parentheses) only* if the metal ion can have more than one state, as in the compound named in rule 1.

6. *If the complex ion is an anion, we drop the ending of the metal name and add -ate.* Thus, the name for $K[Pt(NH_3)Cl_5]$ is

   potassium amminepentachloroplatinate(IV)

   (Note that there is one K^+ counter ion, so the complex anion has a charge of $1-$. The five Cl^- ligands have a total charge of $5-$, so Pt must be in the $+4$ oxidation state.) For some metals, we use the Latin root with the *-ate* ending (Table 23.9). For example, the name for $Na_4[FeBr_6]$ is

   sodium hexabromoferrate(II)

Table 23.9 Names of Some Metal Ions in Complex Anions

Metal	Name in Anion
Iron	Ferrate
Copper	Cuprate
Lead	Plumbate
Silver	Argentate
Gold	Aurate
Tin	Stannate

SAMPLE PROBLEM 23.3 Writing Names and Formulas of Coordination Compounds

Problem (a) What is the systematic name of $Na_3[AlF_6]$?
(b) What is the systematic name of $[Co(en)_2Cl_2]NO_3$?
(c) What is the formula of tetraamminebromochloroplatinum(IV) chloride?
(d) What is the formula of hexaamminecobalt(III) tetrachloroferrate(III)?

Plan We use the rules that were discussed on the preceding page and refer to Tables 23.8 and 23.9.

Solution (a) The complex ion is $[AlF_6]^{3-}$. There are six *(hexa-)* F^- ions *(fluoro)* as ligands, so we have *hexafluoro*. The complex ion is an anion, so the ending of the metal ion (aluminum) must be changed to *-ate:* hexafluoroaluminate. Aluminum has only the $+3$ oxidation state, so we do *not* use a Roman numeral. The positive counter ion is named first and separated from the anion by a space: sodium hexafluoroaluminate.

(b) Listed alphabetically, there are two Cl^- *(dichloro)* and two en [*bis(ethylenediamine)*] as ligands. The complex ion is a cation, so the metal name is unchanged, but we specify its oxidation state because cobalt can have several. One NO_3^- balances the $1+$ cation charge: with $2-$ for two Cl^- and 0 for two en, the metal must be *cobalt(III)*. The word *nitrate* follows a space: dichlorobis(ethylenediamine)cobalt(III) nitrate.

(c) The central metal ion is written first, followed by the neutral ligands and then (in alphabetical order) by the negative ligands. *Tetraammine* is four NH_3, *bromo* is one Br^-, *chloro* is one Cl^-, and *platinate(IV)* is Pt^{4+}, so the complex ion is $[Pt(NH_3)_4BrCl]^{2+}$. Its $2+$ charge is the sum of $4+$ for Pt^{4+}, 0 for four NH_3, $1-$ for one Br^-, and $1-$ for one Cl^-. To balance the $2+$ charge, we need two Cl^- counter ions: $[Pt(NH_3)_4BrCl]Cl_2$.

(d) This compound consists of two different complex ions. In the cation, *hexaammine* is six NH_3 and *cobalt(III)* is Co^{3+}, so the cation is $[Co(NH_3)_6]^{3+}$. The $3+$ charge is the sum of $3+$ for Co^{3+} and 0 for six NH_3. In the anion, *tetrachloro* is four Cl^-, and *ferrate(III)* is Fe^{3+}, so the anion is $[FeCl_4]^-$. The $1-$ charge is the sum of $3+$ for Fe^{3+} and $4-$ for four Cl^-. In the neutral compound, one $3+$ cation is balanced by three $1-$ anions: $[Co(NH_3)_6][FeCl_4]_3$.

Check Reverse the process to be sure you obtain the name or formula asked for in the problem.

FOLLOW-UP PROBLEM 23.3 (a) What is the name of $[Cr(H_2O)_5Br]Cl_2$?
(b) What is the formula of barium hexacyanocobaltate(III)?

A Historical Perspective: Alfred Werner and Coordination Theory

The substances we now call coordination compounds had been known for almost 200 years when the young Swiss chemist Alfred Werner began studying them in the 1890s. He investigated a series of compounds such as the cobalt series shown in Table 23.10, each of which contains one cobalt(III) ion, three chloride ions, and a given number of ammonia molecules. At the time, which was 30 years before the idea of atomic orbitals was proposed, no structural theory could explain how compounds with similar, even identical, formulas could have widely different properties.

Werner measured the conductivity of each compound in aqueous solution to determine the total number of ions that became dissociated. He treated the solutions with excess $AgNO_3$ to precipitate released Cl^- ions as AgCl and thus determine the number of free Cl^- ions per formula unit. Previous studies had

Table 23.10 Some Coordination Compounds of Cobalt Studied by Werner

Traditional Formula	Werner's Data*		Modern Formula	Charge of Complex Ion
	Total Ions	Free Cl^-		
$CoCl_3 \cdot 6NH_3$	4	3	$[Co(NH_3)_6]Cl_3$	3+
$CoCl_3 \cdot 5NH_3$	3	2	$[Co(NH_3)_5Cl]Cl_2$	2+
$CoCl_3 \cdot 4NH_3$	2	1	$[Co(NH_3)_4Cl_2]Cl$	1+
$CoCl_3 \cdot 3NH_3$	0	0	$[Co(NH_3)_3Cl_3]$	—

*Moles per mole of compound.

established that the NH_3 molecules were not free in solution. Werner's data, summarized in Table 23.10, could not be explained by the accepted, traditional formulas of the compounds. Other chemists had proposed "chain" structures, like those of organic compounds, to explain such data. For example, a proposed structure for $[Co(NH_3)_6]Cl_3$ was

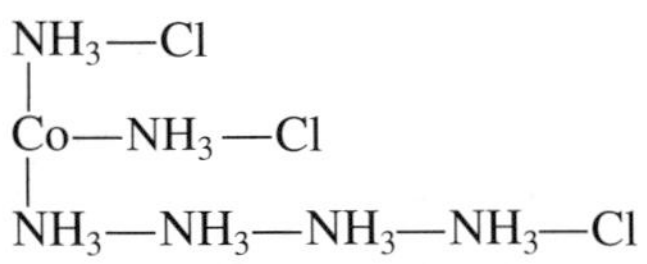

$$NH_3-Cl$$
$$|$$
$$Co-NH_3-Cl$$
$$|$$
$$NH_3-NH_3-NH_3-NH_3-Cl$$

However, these models proved inadequate.

Werner's novel idea was the coordination complex, a central metal ion surrounded by a *constant total number* of covalently bonded molecules and/or anions. The coordination complex could be neutral or charged; if charged, it combined with oppositely charged counter ions, in this case Cl^-, to form the neutral compound.

Werner proposed two types of valence, or *combining ability,* for metal ions. *Primary valence,* now called *oxidation state,* is the positive charge on the metal ion that must be satisfied by an equivalent negative charge. In Werner's cobalt series, the primary valence is +3, and it is always balanced by three Cl^- ions. These anions can be bonded covalently to Co as part of the complex ion and/or associated with it as counter ions. *Secondary valence,* now called *coordination number,* is the constant total number of connections (anionic or neutral ligands) *within* the complex ion. The secondary valence in this series of cobalt compounds is 6.

As you can see, Werner's data are satisfied if the total number of ligands remains the same for each compound, even though the numbers of Cl^- ions and NH_3 molecules in the different complex ions vary. For example, the first compound, $[Co(NH_3)_6]Cl_3$, has a total of four ions: one $[Co(NH_3)_6]^{3+}$ and three Cl^-. All three Cl^- ions are free to form AgCl. The last compound, $[Co(NH_3)_3Cl_3]$, contains no separate ions.

Surprisingly, Werner was an organic chemist, and his work on coordination compounds, which virtually revolutionized his contemporaries' understanding of chemical bonding, was his attempt to demonstrate the unity of chemistry. For these pioneering studies, especially his prediction of optical isomerism (discussed next), Werner received the Nobel Prize in 1913.

Isomerism in Coordination Compounds

Isomers are compounds with the same chemical formula but different properties. We have already discussed many aspects of isomerism in the context of organic compounds in Section 15.2; it may be helpful for you to review that section now. Figure 23.10 presents an overview of the most common types of isomerism in coordination compounds.

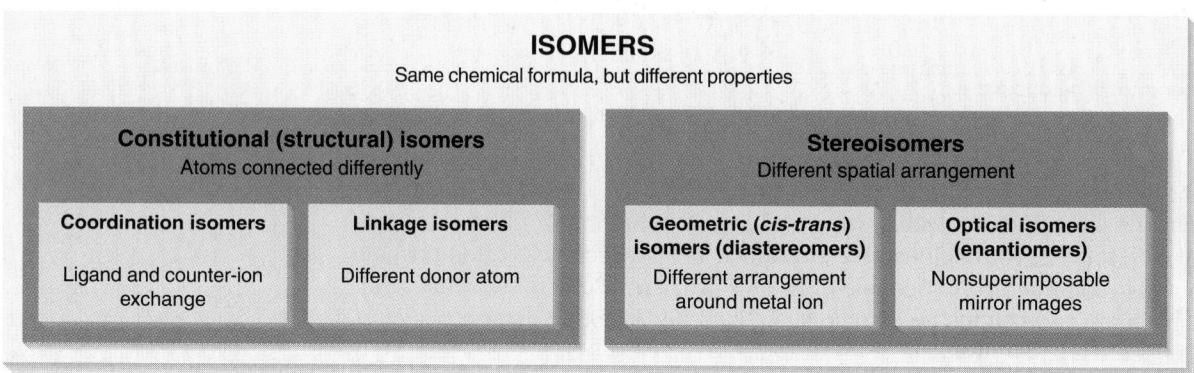

Figure 23.10 Important types of isomerism in coordination compounds.

Constitutional Isomers: Same Atoms Connected Differently Two compounds with the same formula, but with the atoms connected differently, are called **constitutional (structural) isomers.** Coordination compounds exhibit the following two types of constitutional isomers: one involves changes in the composition of the complex ion, the other in the donor atom of the ligand.

1. **Coordination isomers** occur when the composition of the complex ion changes but not that of the compound. One way this type of isomerism occurs is when ligand and counter ion exchange positions, as in $[Pt(NH_3)_4Cl_2](NO_2)_2$ and $[Pt(NH_3)_4(NO_2)_2]Cl_2$. In the first compound, the Cl^- ions are the ligands, and the NO_2^- ions are counter ions; in the second, the roles are reversed. Another way this type of isomerism occurs is in compounds of two complex ions in which the two sets of ligands in one compound are reversed in the other, as in $[Cr(NH_3)_6][Co(CN)_6]$ and $[Co(NH_3)_6][Cr(CN)_6]$; note that NH_3 is a ligand of Cr^{3+} in one compound and of Co^{3+} in the other.

2. **Linkage isomers** occur when the composition of the complex ion remains the same but the attachment of the ligand donor atom changes. Some ligands can bind to the metal ion through *either of two donor atoms*. For example, the nitrite ion can bind through a lone pair on either the N atom (*nitro*, $O_2N\colon$) or one of the O atoms (*nitrito*, $ONO\colon$) to give linkage isomers, as in the orange compound pentaammine*nitro*cobalt(III) chloride $[Co(NH_3)_5(NO_2)]Cl_2$ *(below left)* and its red linkage isomer pentaammine*nitrito*cobalt(III) chloride $[Co(NH_3)_5(ONO)]Cl_2$ *(below right)*:

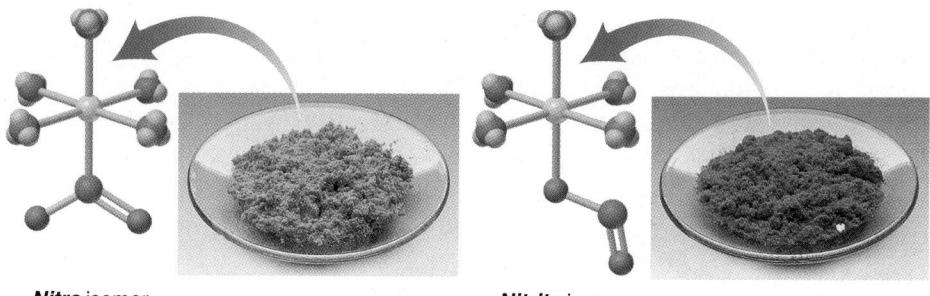

Nitro isomer   *Nitrito* isomer

Another example is the cyanate ion, which can attach via a lone pair on the O atom (*cyanato*, NCO:) or the N atom (*isocyanato*, OCN:); the thiocyanate ion behaves similarly, attaching via the S atom or the N atom:

$$\left[\begin{array}{c} \ddot{O} \\ N: \\ \ddot{O} \end{array} \right]^{-} \qquad [\ddot{O}=C=\ddot{N}:]^{-} \qquad [\ddot{S}=C=\ddot{N}:]^{-}$$

nitrite   cyanate   thiocyanate

Stereoisomers: Different Spatial Arrangements of Atoms Stereoisomers are compounds that have the same atomic connections but different spatial arrangements of the atoms. The two types that we discussed for organic compounds, called *geometric* and *optical* isomers, are seen with coordination compounds as well:

1. **Geometric isomers** (also called ***cis-trans*** **isomers** and, sometimes, *diastereomers*) occur when atoms or groups of atoms are arranged differently in space relative to the central metal ion. For example, the square planar $[Pt(NH_3)_2Cl_2]$ has two arrangements, which give rise to two different compounds (Figure 23.11A). The isomer with identical ligands *next* to each other is *cis*-diamminedichloroplatinum(II), and the one with identical ligands *across* from each other is *trans*-diamminedichloroplatinum(II); their biological behaviors are remarkably different. Octahedral complexes also exhibit *cis-trans* isomerism (Figure 23.11B). The *cis* isomer of the $[Co(NH_3)_4Cl_2]^+$ ion has the two Cl^- ligands next to each other and is violet, whereas the *trans*-isomer has these two ligands across from each other and is green.

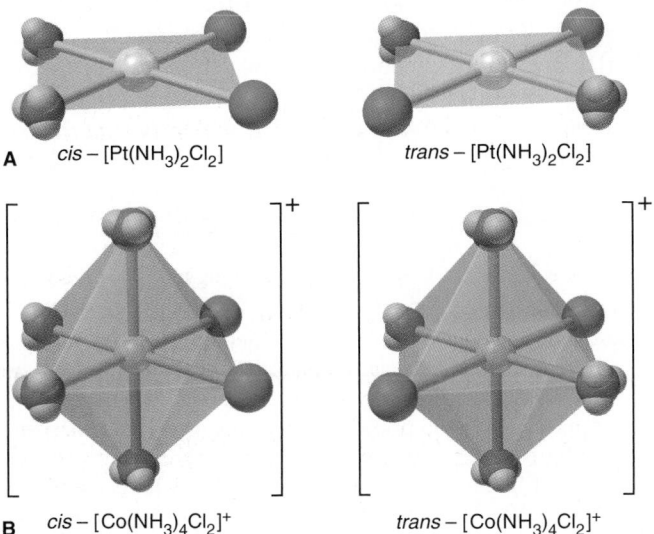

A   *cis* − $[Pt(NH_3)_2Cl_2]$ *trans* − $[Pt(NH_3)_2Cl_2]$

B   *cis* − $[Co(NH_3)_4Cl_2]^+$ *trans* − $[Co(NH_3)_4Cl_2]^+$

Figure 23.11 Geometric (*cis-trans*) isomerism. A, The *cis* and *trans* isomers of the square planar coordination compound $[Pt(NH_3)_2Cl_2]$. **B,** The *cis* and *trans* isomers of the octahedral complex ion $[Co(NH_3)_4Cl_2]^+$. The colored shapes represent the actual colors of the species.

2. **Optical isomers** (also called *enantiomers*) occur when a molecule and its mirror image cannot be superimposed (see Figures 15.8 to 15.10, pp. 617–618). Unlike the other types of isomers, which have distinct physical properties, optical isomers are physically identical in all ways but one: *the direction in which they rotate the plane of polarized light*. Octahedral complex ions show many examples of optical isomerism, which we can observe by rotating one isomer and seeing if it is superimposable on the other isomer (its mirror image). Let's consider the two structures (I and II) of $[Co(en)_2Cl_2]^+$, the *cis*-dichlorobis-(ethylenediamine)cobalt(III) ion; they are mirror images of each other, as shown

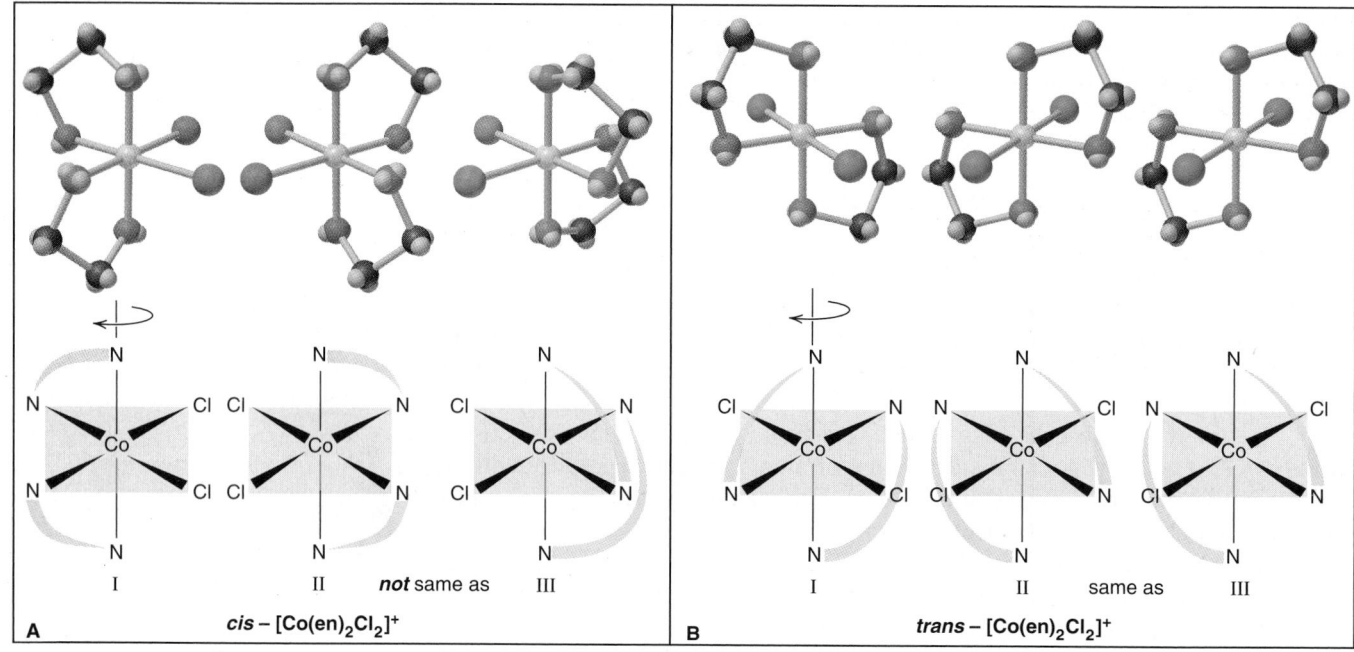

A *cis* – [Co(en)$_2$Cl$_2$]$^+$

B *trans* – [Co(en)$_2$Cl$_2$]$^+$

Figure 23.12 Optical isomerism in an octahedral complex ion. A, Structure I and its mirror image, structure II, are optical isomers of *cis*-[Co(en)$_2$Cl$_2$]$^+$. Rotating I gives III, which is *not* the same as II. (The curved wedges represent the didentate ligand ethylenediamine, H$_2$N—CH$_2$—CH$_2$—NH$_2$.) **B,** The *trans* isomer does *not* have optical isomers. Rotating I gives III, which is *identical* to II, the mirror image of I.

in Figure 23.12A. Rotate structure I 180° around a vertical axis, and you obtain III. The Cl$^-$ ligands of III match those of II, but the en ligands do not: II and III (rotated I) are not superimposable; therefore, they are optical isomers. One isomer is designated *d*-[Co(en)$_2$Cl$_2$]$^+$ and the other is *l*-[Co(en)$_2$Cl$_2$]$^+$, depending on whether it rotates polarized light to the right (*d* for "dextro") or to the left (*l* for "levo-"). (The *d*- or *l*- designation can only be determined experimentally, *not* by examination of the structure.) In contrast, the two structures of the *trans*-dichlorobis(ethylenediamine)cobalt(III) ion (Figure 23.12B) are *not* optical isomers: rotate I 90° around a vertical axis and you obtain III, which *is* superimposable on II.

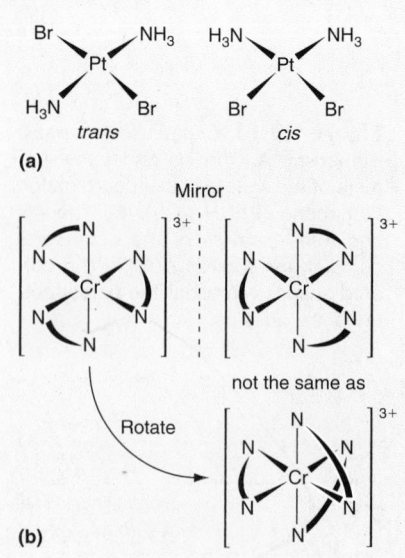

(a)

(b)

SAMPLE PROBLEM 23.4 Determining the Type of Stereoisomerism

Problem Draw all stereoisomers for each of the following and state the type of isomerism:
(a) [Pt(NH$_3$)$_2$Br$_2$] (square planar)   **(b)** [Cr(en)$_3$]$^{3+}$ (en = H$_2$N̈CH$_2$CH$_2$N̈H$_2$)
Plan We first determine the geometry around each metal ion and the nature of the ligands. If there are two different ligands that can be placed in different positions relative to each other, geometric *(cis-trans)* isomerism occurs. Then, we see whether the mirror image of an isomer is superimposable on the original. If it is *not*, optical isomerism occurs.
Solution (a) The Pt(II) complex is square planar, and there are two different monodentate ligands. Each pair of ligands can lie next to or across from each other (see structures in margin). Thus, geometric isomerism occurs. Each isomer *is* superimposable on its mirror image, so there is no optical isomerism.
(b) Ethylenediamine (en) is a bidentate ligand. The Cr^{3+} has a coordination number of 6 and an octahedral geometry, like Co^{3+}. The three bidentate ligands are identical, so there is no geometric isomerism. However, the complex ion has a nonsuperimposable mirror image (see structures in margin). Thus, optical isomerism occurs.

FOLLOW-UP PROBLEM 23.4 What stereoisomers, if any, are possible for the [Co(NH$_3$)$_2$(en)Cl$_2$]$^+$ ion?

SECTION SUMMARY

Coordination compounds consist of a complex ion and charge-balancing counter ions. The complex ion has a central metal ion bonded to neutral and/or anionic ligands, have one or more donor atoms that act as Lewis bases. The most common geometry is octahedral (six ligand atoms bonding). Formulas and names of coordination compounds follow systematic rules. Alfred Werner established the structural basis of coordination compounds. These compounds can exhibit constitutional isomerism (coordination and linkage) and stereoisomerism (geometric and optical).

23.5 THEORETICAL BASIS FOR THE BONDING AND PROPERTIES OF COMPLEXES

In this section, we consider models that address, in different ways, several key features of complexes: how metal-ligand bonds form, why certain geometries are preferred, and why these complexes are brightly colored and often paramagnetic. As you saw with covalent bonding in other compounds (Chapter 11), more than one model is often needed to tell the whole story.

Application of Valence Bond Theory to Complex Ions

Valence bond (VB) theory, which helped explain bonding and structure in main-group compounds (Section 11.1), is also used to describe bonding in complex ions. In the formation of a complex ion, the filled ligand orbital overlaps the empty metal-ion orbital. *The ligand (Lewis base) donates the electron pair, and the metal ion (Lewis acid) accepts it to form one of the covalent bonds of the complex ion (Lewis adduct)* (Section 18.9). Such a bond, in which one atom in the bond contributes both electrons, is called a **coordinate covalent bond,** although, once formed, it is identical to any covalent single bond. Recall that the VB concept of hybridization proposes the mixing of particular combinations of *s*, *p*, and *d* orbitals to give sets of hybrid orbitals, which have specific geometries. Similarly, for coordination compounds, the model proposes that *the number and type of metal-ion hybrid orbitals occupied by ligand lone pairs determine the geometry of the complex ion.* Let's discuss the orbital combinations that lead to octahedral, square planar, and tetrahedral geometries.

Octahedral Complexes The hexaamminechromium(III) ion, $[Cr(NH_3)_6]^{3+}$, illustrates the application of VB theory to an *octahedral complex* (Figure 23.13). The six lowest energy empty orbitals of the Cr^{3+} ion—two $3d$, one $4s$, and three $4p$—mix and become six equivalent d^2sp^3 hybrid orbitals that point toward the corners of an octahedron.* Six NH_3 molecules donate their nitrogen lone electron pairs to form six metal-ligand bonds. The three unpaired $3d$ electrons of the central Cr^{3+} ion ([Ar] $3d^3$), which make the complex ion paramagnetic, remain in unhybridized orbitals.

---

*Note the distinction between the hybrid-orbital designation here and that for octahedral molecules like SF_6. The designation gives the orbitals in energy order within a given n value. In the $[Cr(NH_3)_6]^{3+}$ complex ion, the d orbitals have a *lower* n value than the s and p orbitals, so the hybrid is d^2sp^3. For the orbitals in SF_6, the d orbitals have the *same* n value as the s and p, so the hybrid is sp^3d^2.

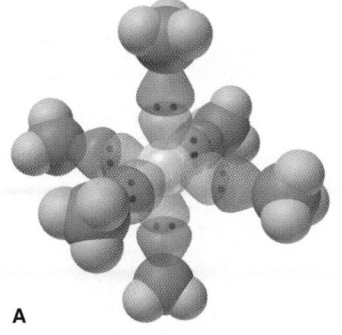

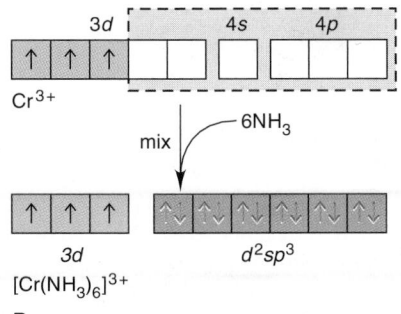

Figure 23.13 Hybrid orbitals and bonding in the octahedral $[Cr(NH_3)_6]^{3+}$ ion. A, VB depiction of the $Cr(NH_3)_6^{3+}$ ion. **B,** The partial orbital diagrams depict the mixing of two $3d$, one $4s$, and three $4p$ orbitals in Cr^{3+} to form six d^2sp^3 hybrid orbitals, and bonding with six NH_3 lone pairs (*red*).

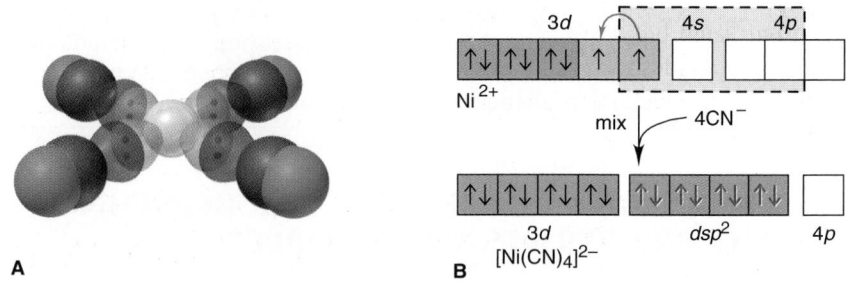

Figure 23.14 **Hybrid orbitals and bonding in the square planar [Ni(CN)$_4$]$^{2-}$ ion. A,** VB depiction of [Ni(CN)$_4$]$^{2-}$. **B,** Two lone $3d$ electrons pair up and free one $3d$ orbital for hybridization with the $4s$ and two $4p$ orbitals to form four dsp^2 orbitals, which become occupied with lone pairs (*red*) from four CN$^-$ ligands.

Square Planar Complexes Metal ions with a d^8 configuration usually form *square planar complexes* (Figure 23.14). In the [Ni(CN)$_4$]$^{2-}$ ion, for example, the model proposes that one $3d$, one $4s$, and two $4p$ orbitals of Ni^{2+} mix and form four dsp^2 hybrid orbitals, which point to the corners of a square and accept one electron pair from each of four CN$^-$ ligands.

A look at the ground-state electron configuration of the Ni^{2+} ion, however, raises a key question: how can the Ni^{2+} ion ([Ar] $3d^8$) offer an empty $3d$ orbital for accepting a bonding pair, if its eight $3d$ electrons lie in three filled and two half-filled orbitals? Apparently, in the d^8 configuration of Ni^{2+}, electrons in the half-filled orbitals pair up and leave one $3d$ orbital empty. This explanation is consistent with the fact that the complex is diamagnetic (no unpaired electrons). Moreover, it requires that the energy *gained* by using a $3d$ orbital for bonding in the hybrid orbital is greater than the energy *required* to overcome repulsions from pairing the $3d$ electrons.

Tetrahedral Complexes Metal ions that have a filled d sublevel, such as Zn^{2+} ([Ar] d^{10}), often form *tetrahedral complexes* (Figure 23.15). For the complex ion [Zn(OH)$_4$]$^{2-}$, for example, VB theory proposes that the lowest available Zn^{2+} orbitals—one $4s$ and three $4p$—mix to become four sp^3 hybrid orbitals that point to the corners of a tetrahedron and are occupied by a lone pair from each of four OH$^-$ ligands.

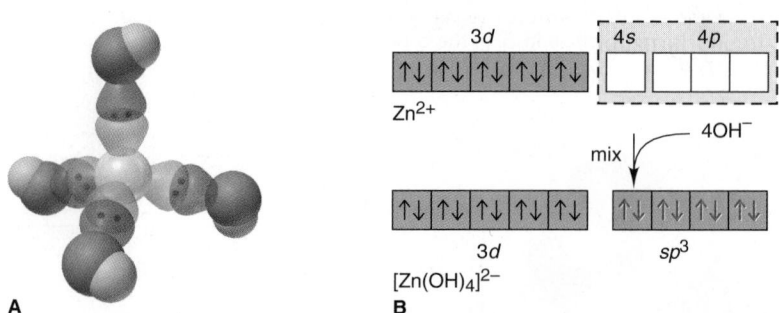

Figure 23.15 **Hybrid orbitals and bonding in the tetrahedral [Zn(OH)$_4$]$^{2-}$ ion. A,** VB depiction of [Zn(OH)$_4$]$^{2-}$. **B,** Mixing one $4s$ and three $4p$ orbitals gives four sp^3 orbitals available for accepting lone pairs (*red*) from OH$^-$ ligands.

Crystal Field Theory

The VB model is easy to picture and rationalizes bonding and shape, but it treats the orbitals as little more than empty "slots" for accepting electron pairs. Consequently, it gives no insight into the colors of coordination compounds and sometimes predicts their magnetic properties incorrectly. In contrast to the VB approach, **crystal field theory** provides little insight about metal-ligand bonding but explains color and magnetism clearly. To do so, it highlights the *effects on the d-orbital energies of the metal ion as the ligands approach.* Before we discuss this theory, let's consider what causes a substance to be colored.

What Is Color? White light is electromagnetic radiation consisting of all wavelengths (λ) in the visible range (Section 7.1). It can be dispersed into a spectrum of colors, each of which has a narrower range of wavelengths. Objects appear colored in white light because they absorb certain wavelengths and reflect or transmit others: an opaque object *reflects* light, whereas a clear one *transmits* it. The reflected or transmitted light enters the eye and the brain perceives a color. If an object absorbs all visible wavelengths, it appears black; if it reflects all, it appears white.

Each color has a *complementary* color. For example, green and red are complementary colors. A mixture of complementary colors absorbs all visible wavelengths and appears black. Figure 23.16 shows these relationships on an artist's color wheel, a circle in which complementary colors appear as wedges opposite each other.

An object has a particular color for one of two reasons:

* It reflects (or transmits) light of *that* color. Thus, if an object absorbs all wavelengths *except* green, the reflected (or transmitted) light enters our eyes and is interpreted as green.
* It absorbs light of the *complementary* color. Thus, if the object absorbs only red, the *complement* of green, the remaining mixture of reflected (or transmitted) wavelengths enters our eyes and is interpreted as green also.

Table 23.11 lists the color absorbed and the resulting color perceived. A familiar example of this phenomenon results in the seasonal colors of deciduous trees. In the spring and summer, leaves contain high concentrations of the photosynthetic pigment *chlorophyll* and lower concentrations of other pigments called *xanthophylls.* Chlorophyll absorbs strongly in the blue and red regions, reflecting mostly green wavelengths into your eyes. In the fall, photosynthesis slows, so the leaf no longer makes chlorophyll. Gradually, the green color fades as the chlorophyll decomposes, revealing the xanthophylls that were present all along but masked by the chlorophyll. Xanthophylls absorb green and blue strongly, reflecting the bright yellows and reds of autumn (see photo).

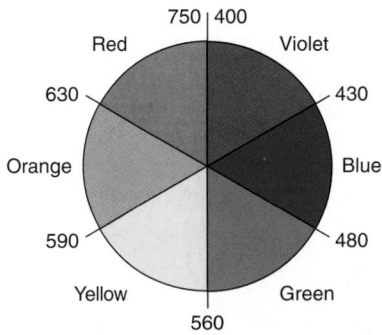

Figure 23.16 An artist's wheel. Colors, with approximate wavelength ranges, are shown as wedges. Complementary colors, such as red and green, lie opposite each other.

Foliage changing color in autumn.

Table 23.11 Relation Between Absorbed and Observed Colors

Absorbed Color	λ (nm)	Observed Color	λ (nm)
Violet	400	Green-yellow	560
Blue	450	Yellow	600
Blue-green	490	Red	620
Yellow-green	570	Violet	410
Yellow	580	Dark blue	430
Orange	600	Blue	450
Red	650	Green	520

Splitting of *d* Orbitals in an Octahedral Field of Ligands The crystal field model explains that the properties of complexes result from the splitting of *d*-orbital energies, which arises from electrostatic interactions between metal ion and ligands. The model assumes that a complex ion forms as a result of *electrostatic attractions between the metal cation and the negative charge of the ligands.* This negative charge is either partial, as in a polar neutral ligand like NH_3, or full, as in an anionic ligand like Cl^-. The ligands approach the metal ion along *x*, *y*, and *z* axes, which minimizes the overall energy of the system.

Picture what happens as the ligands approach. Figure 23.17A shows six ligands moving toward a metal ion to form an octahedral complex. Let's see how the various *d* orbitals of the metal ion are affected as the complex forms. As the ligands approach, their electron pairs repel electrons in the five *d* orbitals. In the isolated metal ion, the *d* orbitals have equal energies despite their different orientations. In the electrostatic field of ligands, however, the *d* electrons are *repelled unequally because they have different orientations.* Because the ligands move along the *x*, *y*, and *z* axes, they approach *directly toward* the lobes of the $d_{x^2-y^2}$ and d_{z^2} orbitals (Figure 23.17B and C) but *between* the lobes of the d_{xy}, d_{xz}, and d_{yz} orbitals (Figure 23.17D to F). As a result, electrons in the $d_{x^2-y^2}$ and d_{z^2} orbitals experience *stronger* repulsions than those in the d_{xy}, d_{xz}, and d_{yz} orbitals.

Figure 23.17 The five *d* orbitals in an octahedral field of ligands. The direction of ligand approach influences the strength of repulsions of electrons in the five metal *d* orbitals. **A,** We assume that ligands approach a metal ion along the three linear axes in an octahedral orientation. **B** and **C,** Lobes of the $d_{x^2-y^2}$ and d_{z^2} orbitals lie *directly in line* with the approaching ligands, so repulsions are stronger. **D** to **F,** Lobes of the d_{xy}, d_{xz}, and d_{yz} orbitals lie *between* the approaching ligands, so repulsions are weaker.

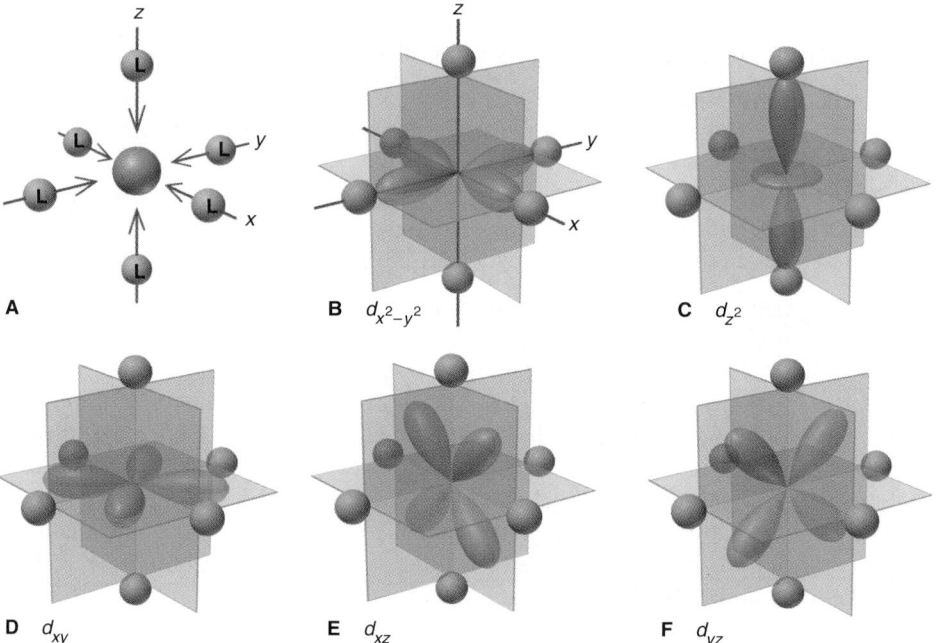

An energy diagram of the orbitals shows that all five *d* orbitals are higher in energy in the forming complex than in the free metal ion because of repulsions from the approaching ligands, but *the orbital energies split, with two d orbitals higher in energy than the other three* (Figure 23.18). The two higher energy orbitals are called ***e_g* orbitals,** and the three lower energy ones are ***t_{2g}* orbitals.** (These designations refer to features of the orbitals that need not concern us here.)

The splitting of orbital energies is called the *crystal field effect,* and the difference in energy between the e_g and t_{2g} sets of orbitals is the **crystal field splitting energy (Δ).** Different ligands create crystal fields of different strength and, thus, cause the *d*-orbital energies to split to different extents. **Strong-field ligands** lead to a *larger* splitting energy (larger Δ); **weak-field ligands** lead to a *smaller*

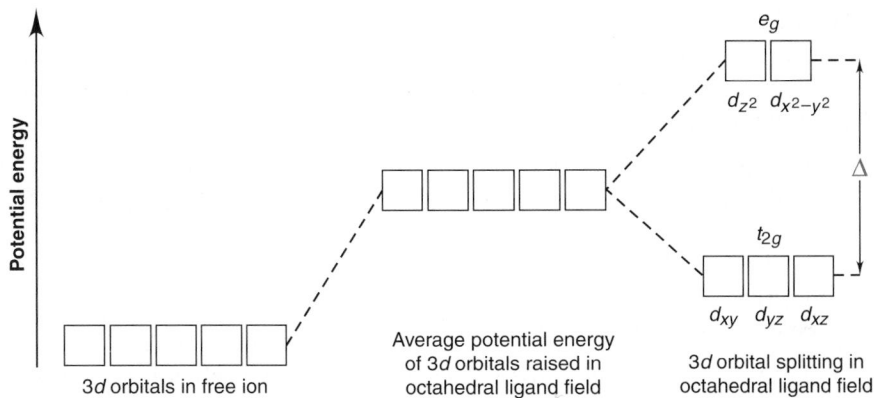

Figure 23.18 **Splitting of *d*-orbital energies by an octahedral field of ligands.** Electrons in the *d* orbitals of the free metal ion experience an *average* net repulsion in the negative ligand field that increases all *d*-orbital energies. Electrons in the d_{xy}, d_{yz}, and d_{xz} orbitals, which form the t_{2g} set, are repelled less than those in the $d_{x^2-y^2}$ and d_{z^2} orbitals, which form the e_g set. The energy difference between these two sets is the crystal field splitting energy, Δ.

splitting energy (smaller Δ). For instance, H_2O is a weak-field ligand, and CN^- is a strong-field ligand (Figure 23.19). The magnitude of Δ relates directly to the color and magnetic properties of complexes.

Explaining the Colors of Transition Metals The remarkably diverse colors of coordination compounds are determined by the energy difference (Δ) between the t_{2g} and e_g orbital sets in their complex ions. When the ion absorbs light in the visible range, electrons are excited ("jump") from the lower energy t_{2g} level to the higher e_g level. Recall that the *difference* between two electronic energy levels in the ion is equal to the energy (and inversely related to the wavelength) of the absorbed photon:

$$\Delta E_{electron} = E_{photon} = h\nu = hc/\lambda$$

The substance has a color because only certain wavelengths of the incoming white light are absorbed.

Consider the $[Ti(H_2O)_6]^{3+}$ ion, which appears purple in aqueous solution (Figure 23.20). Hydrated Ti^{3+} is a d^1 ion, with the d electron in one of the three lower energy t_{2g} orbitals. The energy difference (Δ) between the t_{2g} and e_g orbitals in this ion corresponds to the energy of photons spanning the green and yellow range. When white light shines on the solution, these colors of light are absorbed, and the electron jumps to one of the e_g orbitals. Red, blue, and violet light are transmitted, so the solution appears purple.

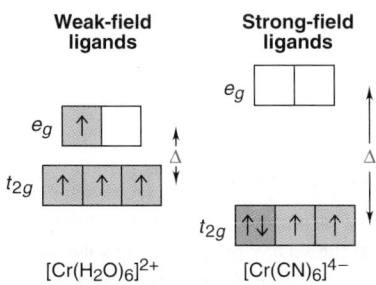

Figure 23.19 **The effect of the ligand on splitting energy.** Ligands interacting strongly with metal-ion *d* orbitals, such as CN^-, produce a larger Δ than those interacting weakly, such as H_2O.

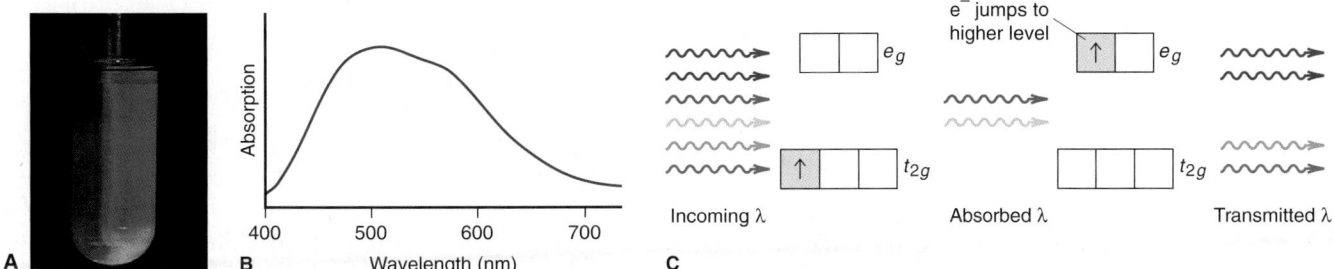

Figure 23.20 **The color of $[Ti(H_2O)_6]^{3+}$.** **A,** The hydrated Ti^{3+} ion is purple in aqueous solution. **B,** An absorption spectrum shows that incoming wavelengths corresponding to green and yellow light are absorbed, whereas other wavelengths are transmitted. **C,** An orbital diagram depicts the colors absorbed in the excitation of the *d* electron to the higher level.

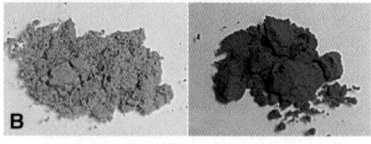

Figure 23.21 **Effects of the metal oxidation state and of ligand identity on color. A,** Solutions of $[V(H_2O)_6]^{2+}$ (*left*) and $[V(H_2O)_6]^{3+}$ (*right*) ions have different colors. **B,** A change in even a single ligand can influence the color. The $[Cr(NH_3)_6]^{3+}$ ion is yellow-orange (*left*); the $[Cr(NH_3)_5Cl]^{2+}$ ion is purple (*right*).

Absorption spectra show the wavelengths absorbed by a given metal ion with different ligands and by different metal ions with the same ligand. From such data, we relate the energy of the absorbed light to the Δ values, and two important observations emerge:

1. *For a given ligand, the color depends on the oxidation state of the metal ion.* A solution of $[V(H_2O)_6]^{2+}$ ion is violet, and a solution of $[V(H_2O)_6]^{3+}$ ion is yellow (Figure 23.21A).
2. *For a given metal ion, the color depends on the ligand.* Even a single ligand substitution can have a major effect on the wavelengths absorbed and, thus, the color, as you can see for two Cr^{3+} complex ions in Figure 23.21B.

Observation 2 allows us to rank ligands into a **spectrochemical series** with regard to their ability to split *d*-orbital energies. An abbreviated series, moving from weak-field ligands (small splitting, small Δ) to strong-field ligands (large splitting, large Δ), is shown in Figure 23.22. Using this series, we can predict the *relative* size of Δ for a series of octahedral complexes of the same metal ion. Although it is difficult to predict the actual color of a given complex, we can determine whether a complex will absorb longer or shorter wavelengths than other complexes in the series.

Figure 23.22 **The spectrochemical series.** As the crystal field strength of the ligand increases, the splitting energy (Δ) increases, so shorter wavelengths (λ) of light must be absorbed to excite electrons. Water is usually a weak-field ligand.

$$I^- < Cl^- < F^- < OH^- < H_2O < SCN^- < NH_3 < en < NO_2^- < CN^- < CO$$

WEAKER FIELD	STRONGER FIELD
SMALLER Δ	LARGER Δ
LONGER λ	SHORTER λ

SAMPLE PROBLEM 23.5 Ranking Crystal Field Splitting Energies for Complex Ions of a Given Metal

Problem Rank the ions $[Ti(H_2O)_6]^{3+}$, $[Ti(NH_3)_6]^{3+}$, and $[Ti(CN)_6]^{3-}$ in terms of the relative value of Δ and of the energy of visible light absorbed.
Plan The formulas show that titanium's oxidation state is +3 in the three ions. From Figure 23.22, we rank the ligands in terms of crystal field strength: the stronger the ligand, the greater the splitting, and the higher the energy of light absorbed.
Solution The ligand field strength is in the order $CN^- > NH_3 > H_2O$, so the relative size of Δ and energy of light absorbed is

$$Ti(CN)_6^{3-} > Ti(NH_3)_6^{3+} > Ti(H_2O)_6^{3+}$$

FOLLOW-UP PROBLEM 23.5 Which complex ion absorbs visible light of higher energy, $[V(H_2O)_6]^{3+}$ or $[V(NH_3)_6]^{3+}$?

Explaining the Magnetic Properties of Transition Metal Complexes The splitting of energy levels influences magnetic properties by affecting the number of *unpaired* electrons in the metal ion's *d* orbitals. Based on Hund's rule, electrons

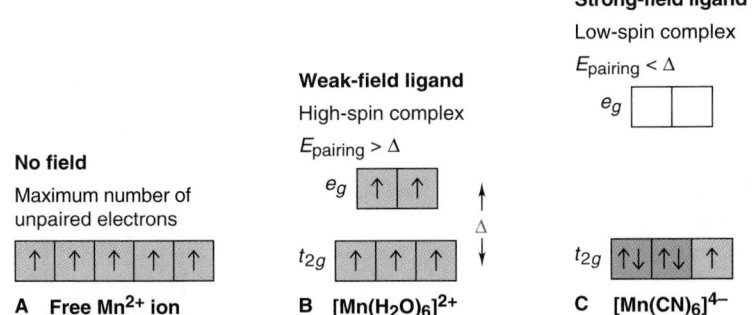

Figure 23.23 High-spin and low-spin complex ions of Mn²⁺. A, The free Mn^{2+} ion has five unpaired electrons. **B,** Bonded to weak-field ligands (smaller Δ), such as H_2O, Mn^{2+} still has five unpaired electrons (high spin). **C,** Bonded to strong-field ligands (larger Δ), such as CN^-, Mn^{2+} has only one unpaired electron (low spin).

occupy orbitals one at a time as long as orbitals of equal energy are available. When all lower energy orbitals are half-filled, the next electron can

- enter a half-filled orbital and pair up by overcoming a repulsive *pairing energy* $(E_{pairing})$, or
- enter an empty, higher energy orbital by overcoming the crystal field splitting energy (Δ).

Thus, *the relative sizes of $E_{pairing}$ and Δ determine the occupancy of the d orbitals.* This, in turn, determines the number of unpaired electrons and, thus, the paramagnetic behavior of the ion.

As an example, the isolated Mn^{2+} ion ([Ar] $3d^5$) has five unpaired electrons in $3d$ orbitals of equal energy (Figure 23.23A). In an octahedral field of ligands, the orbital energies split. The orbital occupancy is affected by the ligand in one of two ways:

- *Weak-field ligands and high-spin complexes.* Weak-field ligands, such as H_2O in $[Mn(H_2O)_6]^{2+}$, cause a *small* splitting energy, so it takes *less* energy for d electrons to jump to the e_g set than to pair up in the t_{2g} set. Therefore, the d electrons remain unpaired (Figure 23.23B). Thus, with weak-field ligands, the pairing energy is *greater* than the splitting energy $(E_{pairing} > \Delta)$; therefore, *the number of unpaired electrons in the complex ion is the **same** as in the free ion.* Weak-field ligands create **high-spin complexes,** those with the *maximum* number of unpaired electrons.
- *Strong-field ligands and low-spin complexes.* In contrast, strong-field ligands, such as CN^- in $[Mn(CN)_6]^{4-}$, cause a *large* splitting of the d-orbital energies, so it takes *more* energy for electrons to jump to the e_g set than to pair up in the t_{2g} set (Figure 23.23C). With strong-field ligands, the pairing energy is *smaller* than the splitting energy $(E_{pairing} < \Delta)$; therefore, *the number of unpaired electrons in the complex ion is **less** than in the free ion.* Strong-field ligands create **low-spin complexes,** those with *fewer* unpaired electrons.

Orbital diagrams for the d^1 through d^9 ions in octahedral complexes show that both high-spin and low-spin options are possible only for d^4, d^5, d^6, and d^7 ions (Figure 23.24). With three lower energy t_{2g} orbitals available, the d^1, d^2, and d^3 ions always form high-spin complexes because there is no need to pair up. Similarly, d^8 and d^9 ions always form high-spin complexes because the t_{2g} set is filled with six electrons, so the two e_g orbitals *must* have either two (d^8) or one (d^9) unpaired electron.

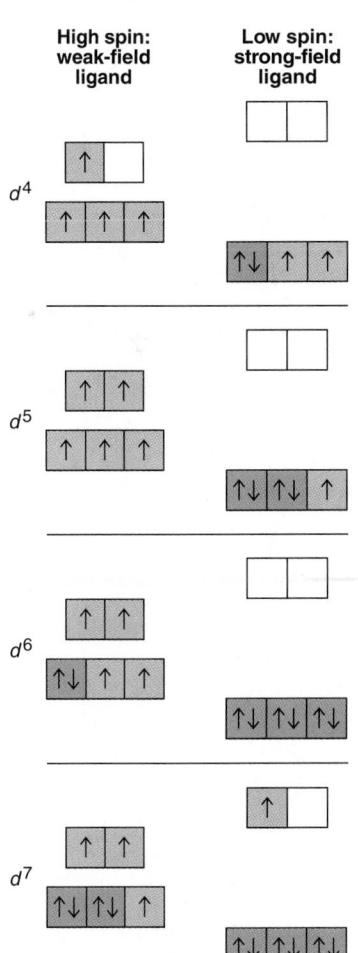

Figure 23.24 Orbital occupancy for high- and low-spin complexes of d^4 through d^7 metal ions.

SAMPLE PROBLEM 23.6 Identifying Complex Ions as High Spin or Low Spin

Problem Iron(II) forms an essential complex in hemoglobin. For each of the two octahedral complex ions $[Fe(H_2O)_6]^{2+}$ and $[Fe(CN)_6]^{4-}$, draw an orbital splitting diagram, predict the number of unpaired electrons, and identify the ion as low or high spin.

Plan The Fe^{2+} electron configuration gives us the number of d electrons, and the spectrochemical series in Figure 23.22 shows the relative strengths of the two ligands. We draw the diagrams, separating the t_{2g} and e_g sets by a greater distance for the strong-field ligand. Then we add electrons, noting that a weak-field ligand gives the *maximum* number of unpaired electrons and a high-spin complex, whereas a strong-field ligand leads to pairing and a low-spin complex.

Solution Fe^{2+} has the [Ar] $3d^6$ configuration. According to Figure 23.22, H_2O produces smaller splitting than CN^-. The diagrams are shown in the margin. The $[Fe(H_2O)_6]^{2+}$ ion has four unpaired electrons (high spin), and the $[Fe(CN)_6]^{4-}$ ion has no unpaired electrons (low spin).

Comment 1. H_2O is a weak-field ligand, so it almost always forms high-spin complexes. **2.** These results are correct, but we cannot confidently predict the spin of a complex without having actual values for Δ and $E_{pairing}$. **3.** Cyanide ions and carbon monoxide are highly toxic because they interact with the iron cations in proteins.

FOLLOW-UP PROBLEM 23.6 How many unpaired electrons do you expect for $[Mn(CN)_6]^{3-}$? Is this a high-spin or low-spin complex ion?

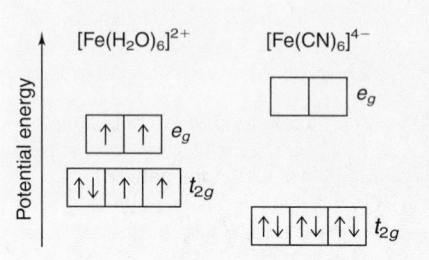

Crystal Field Splitting in Tetrahedral and Square Planar Complexes Four ligands around a metal ion also cause d-orbital splitting, but the magnitude and pattern of the splitting depend on whether the ligands are in a tetrahedral or a square planar arrangement.

- *Tetrahedral complexes.* With the ligands approaching from the corners of a tetrahedron, none of the five d orbitals is directly in their paths (Figure 23.25A). Thus, splitting of d-orbital energies is *less* in a tetrahedral complex than in an octahedral complex having the same ligands:

$$\Delta_{tetrahedral} < \Delta_{octahedral}$$

Minimal repulsions arise if the ligands approach the d_{xy}, d_{yz}, and d_{xz} orbitals closer than they approach the d_{z^2} and $d_{x^2-y^2}$ orbitals. This situation is the *opposite of the octahedral case,* and the relative d-orbital energies are reversed: the d_{xy}, d_{yz}, and d_{xz} orbitals become *higher* in energy than the d_{z^2} and $d_{x^2-y^2}$ orbitals. *Only high-spin tetrahedral complexes are known* because the magnitude of Δ is so small.

- *Square planar complexes.* The effects of the ligand field in the square planar case are easier to picture if we imagine starting with an octahedral geometry and then remove the two z-axis ligands (Figure 23.25B). With no interactions along the z axis, the d_{z^2} orbital energy decreases greatly, and the d_{xz} and d_{yz} orbital energies also decrease. As a result, the two d orbitals in the xy plane interact most strongly with the ligands, and because the $d_{x^2-y^2}$ orbital has its lobes *on* the axes, its energy is highest. As a result of this splitting pattern, square planar complexes with d^8 ions, such as $[PdCl_4]^{2-}$, are diamagnetic because four pairs of d electrons fill the four lowest orbitals. Indeed, *square planar complexes are low spin.*

A final word about bonding theories may be helpful. As you have seen with several other topics, no one model is satisfactory in every respect. The VB approach offers a simple picture of bond formation but does not even attempt to

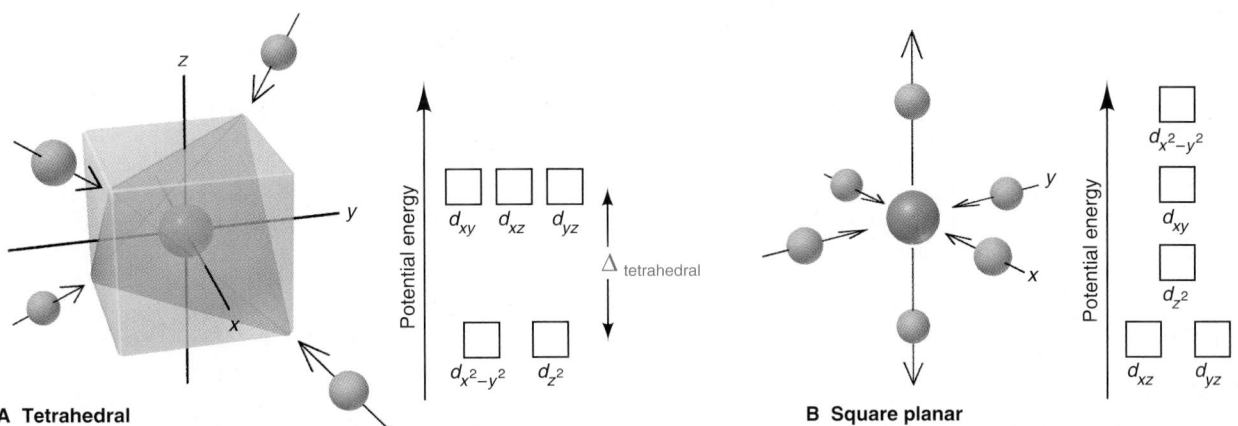

Figure 23.25 Splitting of *d*-orbital energies by a tetrahedral field and a square planar field of ligands. **A,** Electrons in d_{xy}, d_{yz}, and d_{xz} orbitals experience greater repulsions than those in $d_{x^2-y^2}$ and d_{z^2}, so the tetrahedral splitting pattern is the opposite of the octahedral pattern. **B,** In a square planar field, the energies of d_{xz}, d_{yz}, and especially d_{z^2} orbitals decrease relative to the octahedral pattern.

explain color. The crystal field model predicts color and magnetic behavior beautifully but treats the metal ion and ligands as points of opposite charge and thus offers no insight about the covalent nature of metal-ligand bonding. Despite its complexity, chemists now rely on a more refined model, called *ligand field–molecular orbital theory,* which combines aspects of the previous two models with MO theory (Section 11.3). We won't explore it here, but it is a powerful predictive tool, yielding information on bond properties that result from the overlap of metal ion and ligand orbitals as well as information on the spectral and magnetic properties that result from the splitting of metal *d* orbitals.

In addition to their important chemical applications, complexes of the transition elements play vital roles in living systems, as the following Chemical Connections essay describes.

SECTION SUMMARY

Valence bond theory pictures bonding in complex ions as arising from coordinate covalent bonding between Lewis bases (ligands) and Lewis acids (metal ions). Ligand lone pairs occupy hybridized metal-ion orbitals to form complex ions with characteristic shapes. Crystal field theory explains color and magnetism of complexes. As the result of a surrounding field of ligands, the *d*-orbital energies of the metal ion split. The magnitude of this crystal field splitting energy (Δ) depends on the charge of the metal ion and the crystal field strength of the ligand. In turn, Δ influences the energy of the photon absorbed (color) and the number of unpaired *d* electrons (paramagnetism). Strong-field ligands create a large Δ and produce low-spin complexes that absorb light of higher energy (shorter λ); the reverse is true of weak-field ligands. Several transition metals, such as iron and zinc, are essential dietary trace elements that function in complexes within proteins.

Chapter Perspective

Our study of the transition elements, a large group of metals with many essential industrial and biological roles, points up once again that macroscopic properties, such as color and magnetism, have their roots at the atomic and molecular level. In the next and final chapter, we explore the core of the atom and learn how we can apply its enormous power.

Transition Metals as Essential Dietary Trace Elements

Living things consist primarily of water and complex organic compounds of four key elements: carbon, oxygen, hydrogen, and nitrogen. All known organisms also contain seven other elements, known as *macro*nutrients because they occur in fairly high concentrations. In order of increasing atomic number, they are sodium, magnesium, phosphorus, sulfur, chlorine, potassium, and calcium. In addition, organisms contain a surprisingly large number of other elements in much lower concentrations, and most of these *micro*nutrients, or *trace elements,* are transition metals.

With the exception of scandium and titanium, all of the Period 4 transition elements are essential for organisms, and plants require molybdenum (from Period 5) as well. Within the organism, the transition metal ion usually occurs at a bend of a protein chain covalently bonded to surrounding amino acid groups whose N and O atoms act as ligands. Despite the structural complexity of biomolecules, the principles of bonding and *d*-orbital splitting are the same as in simple inorganic systems. Table B23.1 is a list of the Period 4 transition metals that are known, or thought, to be essential in human nutrition. In this discussion we focus on iron and zinc.

Iron plays a crucial role in oxygen transport in all vertebrates. The oxygen-transporting protein hemoglobin (Figure B23.1A) consists of four folded protein chains called *globins,* each cradling the iron-containing complex *heme*. Heme is a porphyrin, a complex derived from a metal ion and the tetradentate ring ligand known as *porphin*. Iron(II) is centered in the plane of the porphin ring through coordinate covalent bonds with four N lone pairs, resulting in a square planar complex. When heme is bound in hemoglobin (Figure B23.1B), the complex is *octahedral,* with the fifth ligand of iron(II) being an N atom from a nearby amino acid (histidine), and the sixth an O atom from either an O_2 (shown) or an H_2O molecule.

Hemoglobin exists in two forms, depending on the nature of the sixth ligand. In the blood vessels of the lungs, where O_2 concentration is high, heme binds O_2 to form *oxyhemoglobin,* which is transported in the arteries to O_2-depleted tissues. At these tissues, the O_2 is released and replaced by an H_2O molecule to form *deoxyhemoglobin,* which is transported in the veins back to the lungs. The H_2O is a weak-field ligand, so the d^6 ion Fe^{2+} in deoxyhemoglobin is part of a high-spin complex. Because of the relatively small *d*-orbital splitting, deoxyhemoglobin absorbs light at the red (low-energy) end of the spectrum and looks purplish blue, which accounts for the dark color of venous blood. On the other hand, O_2 is a strong-field ligand, so it increases the splitting energy, which gives rise to a low-spin complex. For this reason, oxyhemoglobin absorbs at the blue (high-energy) end of the spectrum, which accounts for the bright red color of arterial blood.

The position of the Fe^{2+} ion relative to the plane of the porphin ring also depends on this sixth ligand. Bound to O_2, Fe^{2+} is *in* the porphin plane; bound to H_2O, it moves *out of* the plane slightly. This tiny (~ 60 pm $= 6 \times 10^{-11}$ m) change in the position of Fe^{2+} on release or attachment of O_2 influences the shape of its globin chain, which in turn alters the shape of a neighboring globin chain, triggering the release or attachment of *its* O_2, and so on to the other two globin chains. The very survival of vertebrate life is the result of this cooperative "teamwork" by the four globin chains because it allows hemoglobin to pick up O_2 rapidly from the lungs and unload it rapidly in the tissues.

Carbon monoxide is highly toxic because it binds to the Fe^{2+} ion in heme about 200 times more tightly than O_2, thereby eliminating that heme group from functioning in the circulation. Like O_2, CO is a strong-field ligand and produces a bright red, "healthy" look in the individual. Because heme binding is an equilibrium process, CO poisoning can be reversed by breathing

Table B23.1 Some Transition Metal Trace Elements in Humans

Element	Biomolecule Containing Element	Function of Biomolecule
Vanadium	Protein (?)	Redox couple in fat metabolism (?)
Chromium	Glucose tolerance factor	Glucose utilization
Manganese	Isocitrate dehydrogenase	Cell respiration
Iron	Hemoglobin and myoglobin Cytochrome *c* Catalase	Oxygen transport Cell respiration; ATP formation Decomposition of H_2O_2
Cobalt	Cobalamin (vitamin B_{12})	Development of red blood cells
Copper	Ceruloplasmin Cytochrome oxidase	Hemoglobin synthesis Cell respiration; ATP formation
Zinc	Carbonic anhydrase Carboxypeptidase A Alcohol dehydrogenase	Elimination of CO_2 Protein digestion Metabolism of ethanol

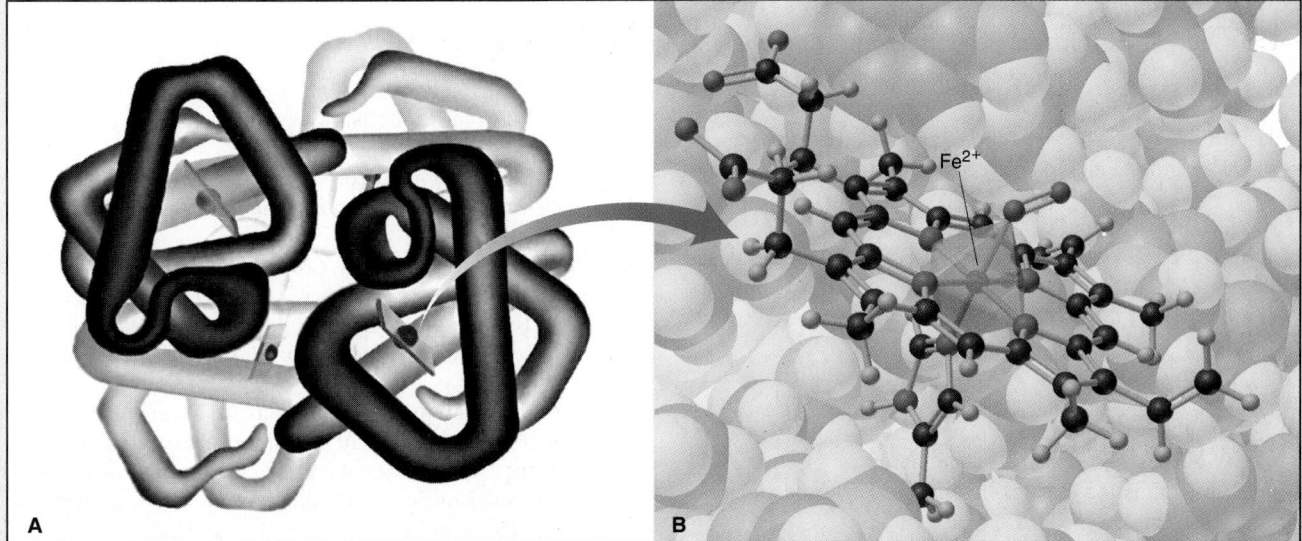

Figure B23.1 Hemoglobin and the octahedral complex in heme. A, Hemoglobin consists of four protein chains, each with a bound heme. (Illustration by Irving Geis. Rights owned by Howard Hughes Medical Institute. Not to be used without permission.) **B,** In oxyhemo-globin, the octahedral complex in heme has iron(II) at the center surrounded by the four N atoms of the porphin ring, a fifth N from histidine *(below)*, and an O_2 molecule *(above)*.

extremely high concentrations of O_2, which effectively displaces CO from the heme:

$$\text{heme}-\text{CO} + O_2 \rightleftharpoons \text{heme}-O_2 + \text{CO}$$

Porphin rings are among the most common biological ligands. Chlorophyll, the photosynthetic pigment of green plants, is a porphyrin with Mg^{2+} at the center of the porphin ring, and vitamin B_{12} has Co^{3+} at the center of a very similar ring system. Heme itself is found not only in hemoglobin, but also in proteins called *cytochromes* that are involved in energy metabolism (see Chemical Connections, pp. 937–938).

The zinc ion occurs in many enzymes, the protein catalysts of cells (see Chemical Connections, pp. 700–701). With its d^{10} configuration, Zn^{2+} is typically surrounded tetrahedrally by the N atoms of three amino acid groups, and the fourth position is free to interact with the molecule whose reaction is being catalyzed (Figure B23.2). In every case studied, the Zn^{2+} ion acts as a Lewis acid, accepting a lone pair from the reactant as a key step in the catalytic process. Consider the enzyme carbonic anhydrase, which catalyzes the essential reaction between H_2O and CO_2 during respiration:

$$CO_2(g) + H_2O(l) \rightleftharpoons H^+(aq) + HCO_3^-(aq)$$

The Zn^{2+} ion at the enzyme's active site binds three histidine N atoms and the H_2O reactant as the fourth ligand. By withdrawing electron density from the O—H bonds, the Zn^{2+} makes the H_2O acidic enough to lose a proton. In the rate-determining step, the resulting bound OH^- ion attacks the partially positive C atom of

CO_2 much more vigorously than could the lone pair of a free water molecule, so the reaction rate is higher. One reason the Cd^{2+} ion is toxic is that it competes with Zn^{2+} for fitting into the carbonic anhydrase active site.

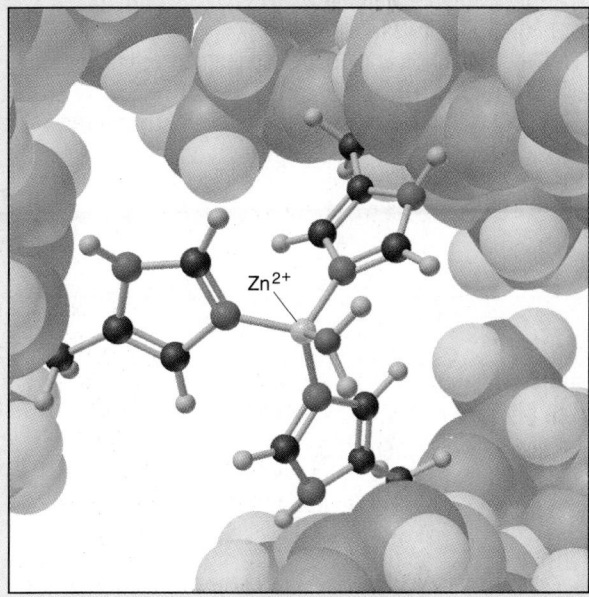

Figure B23.2 The tetrahedral Zn^{2+} complex in carbonic anhydrase.

For Review and Reference (Numbers in parentheses refer to pages, unless noted otherwise.)

Learning Objectives

Relevant section and/or sample problem (SP) numbers appear in parentheses.

Understand These Concepts

1. The position of the *d*- and *f*-block elements and the general forms of their atomic and ionic electron configurations (Section 23.1)

2. How atomic size, ionization energy, and electronegativity vary across a period and down a group of transition elements and how these trends differ from those of the main-group elements; why the densities of Period 6 transition elements are so high (Section 23.1)

3. Why the transition elements often have multiple oxidation states and why the +2 state is common (Section 23.1)

4. Why metallic behavior (prevalence of ionic bonding and basic oxides) decreases as oxidation state increases (Section 23.1)

5. Why many transition metal compounds are colored and paramagnetic (Section 23.1)

6. The common +3 oxidation state of lanthanides and the similarity in their M^{3+} radii; the radioactivity of actinides (Section 23.2)

7. How valence-state electronegativity explains why oxides become more covalent and acidic as the O.N. of the metal increases (Section 23.3)

8. Why Cr and Mn oxoanions are stronger oxidizing agents in acidic than in basic solutions (Section 23.3)

9. The role silver halides play in black-and-white photography (Section 23.3)

10. How the high density and low melting point of mercury account for its common uses; the toxicity of organomercury compounds (Section 23.3)

11. The coordination numbers, geometries, and ligand structures of complex ions (Section 23.4)

12. How coordination compounds are named and their formulas written (Section 23.4)

13. How Werner correlated the properties and structures of coordination compounds (Section 23.4)

14. The types of constitutional isomerism (coordination and linkage) and stereoisomerism (geometric and optical) of coordination compounds (Section 23.4)

15. How valence bond theory uses hybridization to account for the shapes of octahedral, square planar, and tetrahedral complexes (Section 23.5)

16. How crystal field theory explains that approaching ligands cause *d*-orbital energies to split (Section 23.5)

17. How the relative crystal-field strength of ligands (spectrochemical series) affects the *d*-orbital splitting energy (Δ) (Section 23.5)

18. How the magnitude of Δ accounts for the energy of light absorbed and, thus, the color of complexes (Section 23.5)

19. How the relative sizes of pairing energy and Δ determine the occupancy of *d* orbitals and, thus, the magnetic properties of complexes (Section 23.5)

20. How *d*-orbital splitting in tetrahedral and square planar complexes differs from that in octahedral complexes (Section 23.5)

Master These Skills

1. Writing electron configurations of transition metal atoms and ions (SP 23.1)

2. Using a partial orbital diagram to determine the number of unpaired electrons in a transition-metal atom or ion (SP 23.2)

3. Recognizing the structural components of complex ions (Section 23.4)

4. Naming and writing formulas of coordination compounds (SP 23.3)

5. Determining the type of stereoisomerism in complexes (SP 23.4)

6. Correlating complex-ion shape with the number and type of hybrid orbitals of the central metal ion (Section 23.5)

7. Using the spectrochemical series to rank complex ions in terms of Δ and the energy of light absorbed (SP 23.5)

8. Using the spectrochemical series to determine if a complex is high or low spin (SP 23.6)

Key Terms

transition elements (999)

Section 23.1
lanthanide contraction (1002)

Section 23.2
lanthanides (1006)
actinides (1006)
inner transition elements (1006)

Section 23.4
coordination compound (1013)

complex ion (1013)
ligand (1013)
counter ion (1013)
coordination number (1014)
donor atom (1015)
chelate (1016)
isomer (1020)
constitutional (structural) isomers (1020)
coordination isomers (1020)

linkage isomers (1020)
stereoisomers (1021)
geometric *(cis-trans)* isomers (1021)
optical isomers (1021)

Section 23.5
coordinate covalent bond (1023)
crystal field theory (1025)
e_g orbital (1026)
t_{2g} orbital (1026)

crystal field splitting energy (Δ) (1026)
strong-field ligand (1026)
weak-field ligand (1026)
spectrochemical series (1028)
high-spin complex (1029)
low-spin complex (1029)

Highlighted Figures and Tables

These figures (F) and tables (T) provide a quick review of key ideas.

T23.1 Orbital occupancy of Period 4 transition metals (1001)
F23.3 Trends in atomic properties of Period 4 elements (1002)
F23.4 Trends in key properties of the transition elements (1003)
T23.2 Oxidation states and *d*-orbital occupancy of Period 4 transition metals (1004)
T23.6 Coordination numbers and shapes of complex ions (1015)
T23.7 Common ligands in coordination compounds (1016)
T23.8 Names of neutral and anionic ligands (1017)
T23.9 Names of metal ions in complex anions (1017)

F23.10 Isomerism in coordination compounds (1020)
F23.17 *d* Orbitals in an octahedral field of ligands (1026)
F23.18 Splitting of *d*-orbital energies by an octahedral field of ligands (1027)
F23.22 The spectrochemical series (1028)
F23.24 Orbital occupancy for high-spin and low-spin complexes (1029)
F23.25 Splitting of *d*-orbital energies by tetrahedral and square planar fields of ligands (1031)

Brief Solutions to Follow-up Problems

23.1 (a) Ag^+: $4d^{10}$; (b) Cd^{2+}: $4d^{10}$; (c) Ir^{3+}: $5d^6$
23.2 Three; Er^{3+} is $[Xe]\,4f^{11}$:

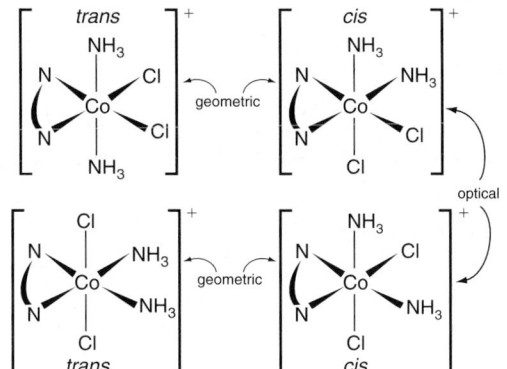

23.3 (a) Pentaaquabromochromium(III) chloride;
(b) $Ba_3[Co(CN)_6]_2$
23.4 Two sets of *cis-trans* isomers, and the two *cis* isomers are optical isomers.

23.5 Both metal ions are V^{3+}; in terms of ligand field energy, $NH_3 > H_2O$, so $[V(NH_3)_6]^{3+}$ absorbs light of higher energy.
23.6 The metal ion is Mn^{3+}: $[Ar]\,3d^4$.

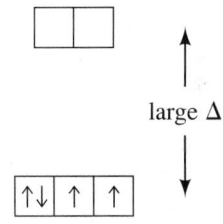

large Δ

Two unpaired *d* electrons; low spin

Problems

Problems with **colored** numbers are answered at the back of the text. Sections match the text and provide the number(s) of relevant sample problems. Most offer Concept Review Questions, Skill-Building Exercises (in similar pairs), and Problems in Context. Then Comprehensive Problems, based on material from any section or previous chapters, follow.
Note: In these problems, the term *electron configuration* refers to the condensed, ground-state electron configuration.

Properties of the Transition Elements
(Sample Problem 23.1)

● **Concept Review Questions**

23.1 How is the *n* value of the *d* sublevel of a transition element related to the period number of the element?
23.2 (a) Write the general electron configuration of a transition element in Period 5.

(b) Write the general electron configuration of a transition element in Period 6.
23.3 What is the general rule concerning the order in which electrons are removed from a transition metal atom to form an ion? Give an example from Group 5B(5). Name two types of measurements used to study electron configurations of ions.
23.4 What is the maximum number of unpaired *d* electrons that an atom or ion can possess? Give an example of an atom and an ion that have this number.
23.5 How does the variation in atomic size across a transition series contrast with the change across the main-group elements of the same period? Why?
23.6 (a) What is the lanthanide contraction?
(b) How does it affect atomic size down a group of transition elements?
(c) How does it influence the densities of the Period 6 transition elements?

23.7 (a) What is the range in electronegativity values across the first (3*d*) transition series?
(b) What is the range across Period 4 of main-group elements?
(c) Explain the difference between the two ranges.

23.8 (a) Explain the major difference between the number of oxidation states of transition elements and that of most main-group elements.
(b) Why is the +2 oxidation state so common among transition elements?

23.9 (a) What difference in behavior distinguishes a paramagnetic substance from a diamagnetic one?
(b) Why are paramagnetic ions common among the transition elements but not the main-group elements?
(c) Why are colored solutions of metal ions common among the transition elements but not the main-group elements?

● **Skill-Building Exercises (paired)**

23.10 Using the periodic table to locate each element, write the electron configuration of (a) V; (b) Y; (c) Hg.

23.11 Using the periodic table to locate each element, write the electron configuration of (a) Ru; (b) Cu; (c) Ni.

23.12 Using the periodic table to locate each element, write the electron configuration of (a) Os; (b) Co; (c) Ag.

23.13 Using the periodic table to locate each element, write the electron configuration of (a) Zn; (b) Mn; (c) Re.

23.14 Give the electron configuration and the number of unpaired electrons in each of the following ions: (a) Sc^{3+}; (b) Cu^{2+}; (c) Fe^{3+}; (d) Nb^{3+}.

23.15 Give the electron configuration and the number of unpaired electrons in each of the following ions: (a) Cr^{3+}; (b) Ti^{4+}; (c) Co^{3+}; (d) Ta^{2+}.

23.16 What is the highest possible oxidation state for each of the following: (a) Ta; (b) Zr; (c) Mn?

23.17 What is the highest possible oxidation state for each of the following: (a) Nb; (b) Y; (c) Tc?

23.18 Which transition metals have a maximum oxidation state of +6?

23.19 Which transition metals have a maximum oxidation state of +4?

23.20 In which compound does Cr exhibit greater metallic behavior, CrF_2 or CrF_6? Explain.

23.21 VF_5 is a liquid that boils at 48°C, whereas VF_3 is a solid that melts above 800°C. Explain this difference in properties.

23.22 Is it more difficult to oxidize Cr or Mo? Explain.

23.23 Is MnO_4^- or ReO_4^- a stronger oxidizing agent? Explain.

23.24 Which oxide, CrO_3 or CrO, forms a more acidic aqueous solution? Explain.

23.25 Which oxide, Mn_2O_3 or Mn_2O_7, displays more basic behavior? Explain.

● **Problems in Context**

23.26 The green patina of copper-alloy roofs of old buildings is the result of the corrosion (oxidation) of copper in the presence of O_2, H_2O, CO_2, and sulfur compounds. Silver and gold—the other members of Group 1B(11)—do not form this patina. Corrosion of copper and silver in the presence of sulfur and its compounds leads to the familiar black tarnish, but gold does not react with sulfur. This pattern is markedly different from that in

Group 1A(1), where ease of oxidation *increases* down the group. What causes the different patterns in the two groups?

The Inner Transition Elements
(Sample Problem 23.2)

● **Concept Review Questions**

23.27 What atomic property of the lanthanides leads to their remarkably similar chemical properties?

23.28 (a) What is the maximum number of unpaired electrons exhibited by an ion of a lanthanide?
(b) How does this number relate to occupancy of the 4*f* subshell?

23.29 Which of the actinides are radioactive?

● **Skill-Building Exercises (paired)**

23.30 Write the electron configurations of the following atoms and ions: (a) La; (b) Ce^{3+}; (c) Es; (d) U^{4+}.

23.31 Write the electron configurations of the following atoms and ions: (a) Pm; (b) Lu^{3+}; (c) Th; (d) Fm^{3+}.

23.32 Only a few of the lanthanides show any oxidation state other than +3. Two of these, europium (Eu) and terbium (Tb), are found near the middle of the series and can be associated with a half-filled *f* subshell.
(a) Write the electron configurations of Eu^{2+}, Eu^{3+}, and Eu^{4+}. Why is Eu^{2+} a common ion, whereas Eu^{4+} is unknown?
(b) Write the electron configurations of Tb^{2+}, Tb^{3+}, and Tb^{4+}. Do you expect Tb to show a +2 or a +4 oxidation state? Explain.

23.33 Cerium (Ce) and ytterbium (Yb) exhibit other oxidation states in addition to +3.
(a) Write the electron configurations of Ce^{2+}, Ce^{3+}, and Ce^{4+}.
(b) Write the electron configurations of Yb^{2+}, Yb^{3+}, and Yb^{4+}.
(c) In addition to the 3+ ions, the ions Ce^{4+} and Yb^{2+} are stable. Suggest a reason for this stability.

● **Problems in Context**

23.34 One of the lanthanides displays the maximum possible number of unpaired electrons both for an atom and for a 3+ ion. Name the element and give the number of unpaired electrons in the atom and the ion.

Highlights of Selected Transition Metals

● **Concept Review Questions**

23.35 What is the chemical reason chromium is so useful for decorative plating on metals?

23.36 What is valence-state electronegativity? Use the concept to explain the change in acidity of the oxides of Mn with changing O.N.

23.37 What property does manganese confer to steel?

23.38 What chemical property of silver leads to its use in jewelry and other decorative objects?

23.39 How is a photographic latent image different from the image you see on a piece of developed film?

23.40 Mercury has an unusual physical property and an unusual 1+ ion. Explain.

● **Problems in Context**

23.41 When a basic solution of $Cr(OH)_4^-$ ion is slowly acidified, solid $Cr(OH)_3$ first precipitates out and then redissolves as excess acid is added. If $Cr(OH)_4^-$ is actually $Cr(H_2O)_2(OH)_4^-$, write equations that represent these two reactions.

23.42 Use the following data to determine if $Cr^{2+}(aq)$ can be prepared by the reaction of $Cr(s)$ with $Cr^{3+}(aq)$:

$$Cr^{3+}(aq) + e^- \longrightarrow Cr^{2+}(aq) \qquad E^0 = -0.41 \text{ V}$$
$$Cr^{3+}(aq) + 3e^- \longrightarrow Cr(s) \qquad E^0 = -0.74 \text{ V}$$
$$Cr^{2+}(aq) + 2e^- \longrightarrow Cr(s) \qquad E^0 = -0.91 \text{ V}$$

23.43 When solid CrO_3 is dissolved in water, the solution is orange rather than the yellow of H_2CrO_4. How does this observation indicate that CrO_3 is an acidic oxide?

23.44 Solutions of $KMnO_4$ are used commonly in redox titrations. The dark purple MnO_4^- ion serves as its own indicator, changing to the almost colorless Mn^{2+} as it is reduced. The end point occurs when a pale purple color remains as the $KMnO_4$ solution is added. If a sample that has reached this end point is allowed to stand for a long period of time, the color fades and a suspension of a small amount of brown, muddy MnO_2 appears. Use standard electrode potentials to explain this result.

Coordination Compounds
(Sample Problems 23.3 and 23.4)

● **Concept Review Questions**

23.45 Describe the makeup of a complex ion, including the nature of the ligands and their interaction with the central metal ion. Explain how a complex ion can be positive or negative and how it occurs as part of a neutral coordination compound.

23.46 What electronic feature must a donor atom of any ligand possess?

23.47 What is the coordination number of a metal ion in a complex ion? How does it differ from oxidation number?

23.48 What structural feature is characteristic of a complex described as a chelate?

23.49 What geometries are associated with the coordination numbers 2, 4, and 6?

23.50 What are the coordination numbers of cobalt(III), platinum(II), and platinum(IV) in complexes?

23.51 In what sense is a complex ion the adduct of a Lewis acid-base reaction?

23.52 What does the ending -*ate* signify in a complex ion name?

23.53 In what order are the metal ion and ligands given in the name of a complex ion?

23.54 Is a linkage isomer a type of constitutional isomer or stereoisomer? Explain.

● **Skill-Building Exercises** *(paired)*

23.55 Give systematic names for the following formulas:
(a) $[Ni(H_2O)_6]Cl_2$ (b) $[Cr(en)_3](ClO_4)_3$ (c) $K_4[Mn(CN)_6]$

23.56 Give systematic names for the following formulas:
(a) $[Co(NH_3)_4(NO_2)_2]Cl$ (b) $[Cr(NH_3)_6][Cr(CN)_6]$
(c) $K_2[CuCl_4]$

23.57 What are the charge and coordination number of the central metal ion(s) in each compound of Problem 23.55?

23.58 What are the charge and coordination number of the central metal ion(s) in each compound of Problem 23.56?

23.59 Give systematic names for the following formulas:
(a) $K[Ag(CN)_2]$ (b) $Na_2[CdCl_4]$ (c) $[Co(NH_3)_4(H_2O)Br]Br_2$

23.60 Give systematic names for the following formulas:
(a) $K[Pt(NH_3)Cl_5]$ (b) $[Cu(en)(NH_3)_2][Co(en)Cl_4]$
(c) $[Pt(en)_2Br_2](ClO_4)_2$

23.61 What are the charge and coordination number of the central metal ion(s) in each compound of Problem 23.59?

23.62 What are the charge and coordination number of the central metal ion(s) in each compound of Problem 23.60?

23.63 Give formulas corresponding to the following names:
(a) Tetraamminezinc sulfate
(b) Pentaamminechlorochromium(III) chloride
(c) Sodium bis(thiosulfato)argentate(I)

23.64 Give formulas corresponding to the following names:
(a) Dibromobis(ethylenediamine)cobalt(III) sulfate
(b) Hexaamminechromium(III) tetrachlorocuprate(II)
(c) Potassium hexacyanoferrate(II)

23.65 What is the coordination number of the metal ion and the number of individual ions per formula unit in each of the compounds in Problem 23.63?

23.66 What is the coordination number of the metal ion and the number of individual ions per formula unit in each of the compounds in Problem 23.64?

23.67 Give formulas corresponding to the following names:
(a) Hexaaquachromium(III) sulfate
(b) Barium tetrabromoferrate(III)
(c) Bis(ethylenediamine)platinum(II) carbonate

23.68 Give formulas corresponding to the following names:
(a) Potassium tris(oxalato)chromate(III)
(b) Tris(ethylenediamine)cobalt(III) pentacyanoiodomanganate(II)
(c) Bromochlorodiaquadiamminealuminum nitrate

23.69 What is the coordination number of the metal ion and the number of individual ions per formula unit in each of the compounds in Problem 23.67?

23.70 What is the coordination number of the metal ion and the number of individual ions per formula unit in each of the compounds in Problem 23.68?

23.71 Which of these ligands can participate in linkage isomerism: (a) NO_2^-; (b) SO_2; (c) NO_3^-? Explain with Lewis structures.

23.72 Which of these ligands can participate in linkage isomerism: (a) SCN^-; (b) $S_2O_3^{2-}$ (thiosulfate); (c) HS^-? Explain with Lewis structures.

23.73 For any of the following that can exist as isomers, state the type of isomerism and draw the structures: (a) $[Pt(CH_3NH_2)_2Br_2]$; (b) $[Pt(NH_3)_2FCl]$; (c) $[Pt(H_2O)(NH_3)FCl]$.

23.74 For any of the following that can exist as isomers, state the type of isomerism and draw the structures: (a) $[Zn(en)F_2]$; (b) $[Zn(H_2O)(NH_3)FCl]$; (c) $[Pd(CN)_2(OH)_2]^{2-}$.

23.75 For any of the following that can exist as isomers, state the type of isomerism and draw the structures: (a) $[PtCl_2Br_2]^{2-}$; (b) $[Cr(NH_3)_5(NO_2)]^{2+}$; (c) $[Pt(NH_3)_4I_2]^{2+}$.

23.76 For any of the following that can exist as isomers, state the type of isomerism and draw the structures: (a) $[Co(NH_3)_5Cl]Br_2$; (b) $[Pt(CH_3NH_2)_3Cl]Br$; (c) $[Fe(H_2O)_4(NH_3)_2]^{2+}$.

● **Problems in Context**

23.77 Chromium(III), like cobalt(III), has a coordination number of 6 in many of its complex ions. Compounds are known that have the traditional formula $CrCl_3 \cdot nNH_3$, where $n = 3$ to 6. Which of the compounds has an electrical conductivity in aqueous solution similar to that of an equimolar $NaCl$ solution?

23.78 When $MCl_4(NH_3)_2$ is dissolved in water and treated with $AgNO_3$, 2 mol of $AgCl$ precipitate immediately for each mole of $MCl_4(NH_3)_2$. Give the coordination number of M in the complex.

23.79 Palladium, like its group neighbor platinum, forms four-coordinate Pd(II) and six-coordinate Pd(IV) complexes. Write modern formulas for the complexes with these compositions: (a) $PdK(NH_3)Cl_3$ (b) $PdCl_2(NH_3)_2$ (c) PdK_2Cl_6 (d) $Pd(NH_3)_4Cl_4$

Theoretical Basis for the Bonding and Properties of Complexes
(Sample Problems 23.5 and 23.6)

● **Concept Review Questions**

23.80 (a) What is a coordinate covalent bond?
(b) Is it involved when $FeCl_3$ dissolves in water? Explain.
(c) Is it involved when HCl gas dissolves in water? Explain.

23.81 According to valence bond theory, what set of orbitals is used by a Period 4 metal ion in forming (a) a square planar complex; (b) a tetrahedral complex?

23.82 A metal ion is described as using a d^2sp^3 set of orbitals when forming a complex. What is the coordination number of the metal ion and the shape of the complex?

23.83 A complex in solution absorbs green light. What is the color of the solution?

23.84 What *two* possibilities of color absorption by a solution could give rise to an observed blue color?

23.85 (a) What is the crystal field splitting energy (Δ)?
(b) How does it arise for an octahedral field of ligands?
(c) How is it different for a tetrahedral field of ligands?

23.86 What is the distinction between a weak-field ligand and a strong-field ligand? Give an example of each.

23.87 Is a complex with the same number of unpaired electrons as the free gaseous metal ion termed high spin or low spin?

23.88 How do the relative magnitudes of $E_{pairing}$ and Δ affect the paramagnetism of a complex?

23.89 Why are there both high-spin and low-spin octahedral complexes but only high-spin tetrahedral complexes?

● **Skill-Building Exercises** *(paired)*

23.90 Give the number of d electrons (n of d^n) for the central metal ion in each of these species: (a) $[TiCl_6]^{2-}$; (b) $K[AuCl_4]$; (c) $[RhCl_6]^{3-}$.

23.91 Give the number of d electrons (n of d^n) for the central metal ion in each of these species: (a) $[Cr(H_2O)_6](ClO_3)_2$; (b) $[Mn(CN)_6]^{2-}$; (c) $[Ru(NO)(en)_2Cl]Br$.

23.92 Give the number of d electrons (n of d^n) for the central metal ion in each of these species: (a) $Ca[IrF_6]$; (b) $[HgI_4]^{2-}$; (c) $[Co(EDTA)]^{2-}$.

23.93 Give the number of d electrons (n of d^n) for the central metal ion in each of these species: (a) $[Ru(NH_3)_5Cl]SO_4$; (b) $Na_2[Os(CN)_6]$; (c) $[Co(NH_3)_4CO_3I]$.

23.94 Sketch the orientation of the orbitals relative to the ligands in an octahedral complex to explain the splitting and the relative energies of the d_{xy} and the $d_{x^2-y^2}$ orbitals.

23.95 The two e_g orbitals are identical in energy in an octahedral complex but have different energies in a square planar complex, with the d_{z^2} orbital being much lower in energy than the $d_{x^2-y^2}$. Explain with orbital sketches.

23.96 Which of these ions *cannot* form both high- and low-spin octahedral complexes: (a) Ti^{3+}; (b) Co^{2+}; (c) Fe^{2+}; (d) Cu^{2+}?

23.97 Which of these ions *cannot* form both high- and low-spin octahedral complexes: (a) Mn^{3+}; (b) Nb^{3+}; (c) Ru^{3+}; (d) Ni^{2+}?

23.98 Draw orbital-energy splitting diagrams and use the spectrochemical series to show the orbital occupancy for each of the following (assuming that H_2O is a weak-field ligand):
(a) $[Cr(H_2O)_6]^{3+}$ (b) $[Cu(H_2O)_4]^{2+}$ (c) $[FeF_6]^{3-}$

23.99 Draw orbital-energy splitting diagrams and use the spectrochemical series to show the orbital occupancy for each of the following (assuming that H_2O is a weak-field ligand):
(a) $[Cr(CN)_6]^{3-}$ (b) $[Rh(CO)_6]^{3+}$ (c) $[Co(OH)_6]^{4-}$

23.100 Draw orbital-energy splitting diagrams and use the spectrochemical series to show the orbital occupancy for each of the following (assuming that H_2O is a weak-field ligand):
(a) $[MoCl_6]^{3-}$ (b) $[Ni(H_2O)_6]^{2+}$ (c) $[Ni(CN)_4]^{2-}$

23.101 Draw orbital-energy splitting diagrams and use the spectrochemical series to show the orbital occupancy for each of the following (assuming that H_2O is a weak-field ligand):
(a) $[Fe(C_2O_4)_3]^{3-}$ ($C_2O_4^{2-}$ creates a weaker field than H_2O does)
(b) $[Co(CN)_6]^{4-}$ (c) $[MnCl_6]^{4-}$

23.102 Rank the following complex ions in order of *increasing* Δ and energy of visible light absorbed: $[Cr(NH_3)_6]^{3+}$, $[Cr(H_2O)_6]^{3+}$, $[Cr(NO_2)_6]^{3-}$.

23.103 Rank the following complex ions in order of *decreasing* Δ and energy of visible light absorbed: $[Cr(en)_3]^{3+}$, $[Cr(CN)_6]^{3-}$, $[CrCl_6]^{3-}$.

23.104 A complex, ML_6^{2+}, is violet. The same metal forms a complex with another ligand, Q, that creates a weaker field. What color might MQ_6^{2+} be expected to show? Explain.

23.105 $[Cr(H_2O)_6]^{2+}$ is violet. Another CrL_6 complex is green. Can ligand L be CN^-? Can it be Cl^-? Explain.

● **Problems in Context**

23.106 Octahedral $[Ni(NH_3)_6]^{2+}$ is paramagnetic, whereas planar $[Pt(NH_3)_4]^{2+}$ is diamagnetic, even though both metal ions are d^8 species. Explain.

23.107 The hexaaqua complex $[Ni(H_2O)_6]^{2+}$ is green, whereas the hexaammonia complex $[Ni(NH_3)_6]^{2+}$ is violet. Explain.

23.108 Three of the complex ions formed by Co^{3+} are $[Co(H_2O)_6]^{3+}$, $[Co(NH_3)_6]^{3+}$, and $[CoF_6]^{3-}$. These ions have the observed colors (listed in arbitrary order) yellow-orange, green, and blue. Match each complex with its color. Explain.

Comprehensive Problems
Problems with an asterisk (*) are more challenging.

23.109 When neptunium (Np) and plutonium (Pu) were discovered, the periodic table did not include the actinides, so these elements were placed in Group 7B(7) and 8B(8). When americium (Am) and curium (Cm) were synthesized, they were placed in Group 8B(9) and 8B(10). However, during chemical isolation procedures Glenn Seaborg and his group, who had synthesized these elements, could not find their compounds among other compounds of these transition series, which led Seaborg to suggest they were part of a new inner transition series. (a) How do the electron configurations of these elements support Seaborg's suggestion? (b) The highest fluorides of Np and Pu are hexafluorides, as is the highest fluoride of uranium. How does this chemical evidence support the placement of Np and Pu as *inner* transition elements rather than transition elements?

23.110 (a) What are the central metal ions in chlorophyll, heme, and vitamin B_{12}?
(b) What similarity in structure do these compounds have?

23.111 At one time, it was common to write the formula for copper(I) chloride as Cu_2Cl_2, instead of CuCl, analogously to Hg_2Cl_2 for mercury(I) chloride. Use electron configurations to explain why Hg_2Cl_2 is correct but so is CuCl.

23.112 Correct each name that has an error.
(a) $Na[FeBr_4]$, sodium tetrabromoferrate(II)
(b) $[Ni(NH_3)_6]^{2+}$, nickel hexaammine ion
(c) $[Co(NH_3)_3I_3]$, triamminetriiodocobalt(III)
(d) $[V(CN)_6]^{3-}$, hexacyanovanadium(III) ion
(e) $K[FeCl_4]$, potassium tetrachloroiron(III)

23.113 For the compound $[Co(en)_2Cl_2]Cl$, give
(a) The coordination number of the metal ion
(b) The oxidation number of the central metal ion
(c) The number of individual ions per formula unit
(d) The moles of AgCl that precipitate immediately when 1 mol of compound is dissolved in water and treated with $AgNO_3$

23.114 Hexafluorocobaltate(III) ion is a high-spin complex. Draw the orbital-energy splitting diagram for its d orbitals.

23.115 Werner prepared two compounds by heating a solution of $PtCl_2$ with triethyl phosphine, $P(C_2H_5)_3$, which is an excellent ligand for Pt. The two compounds gave the same analysis: Pt, 38.8%; Cl, 14.1%; C, 28.7%; P, 12.4%; and H, 6.02%. Write formulas, structures, and systematic names for the two isomers.

23.116 Criticize and correct the following statement: strong-field ligands always give rise to low-spin complexes.

*23.117 Two major bidentate ligands used in analytical chemistry are bipyridyl (bipy) and *ortho*-phenanthroline (*o*-phen):

bipyridyl   *o*-phenanthroline

Draw structures and discuss the possibility of isomers for
(a) $[Pt(bipy)Cl_2]$ (b) $[Fe(o\text{-phen})_3]^{3+}$
(c) $[Co(bipy)_2F_2]^+$ (d) $[Co(o\text{-phen})(NH_3)_3Cl]^{2+}$

*23.118 A shortcut to finding optical isomers is to see if the complex has a "plane of symmetry"—a plane passing through the metal atom such that every atom on one side of the plane is matched by an identical one at the same distance from the plane on the other side. Any planar complex has a plane of symmetry, since all atoms lie in one plane. Use this approach to determine if these exist as optical isomers: (a) $[Zn(NH_3)_2Cl_2]$ (tetrahedral); (b) $[Pt(en)_2]^{2+}$; (c) *trans*-$[PtBr_4Cl_2]^{2-}$; (d) *trans*-$[Co(en)_2F_2]^+$; (e) *cis*-$[Co(en)_2F_2]^+$.

23.119 The metal ion in platinum(IV) complexes, like that in cobalt(III) complexes, has a coordination number of 6. Many of these complexes occur with Cl^- ions and NH_3 molecules as ligands. Consider the following traditional (before the work of Werner) formulas for two coordination compounds:
(a) $PtCl_4 \cdot 6NH_3$ (b) $PtCl_4 \cdot 4NH_3$
For each of these compounds,
(1) Give the modern formula and charge of the complex ion.
(2) Predict moles of ions formed per mole of compound dissolved and moles of AgCl formed immediately with excess $AgNO_3$.

23.120 In 1940, when the elements Np and Pu were prepared, a controversy arose about whether the elements from Ac on were analogs of the transition elements (and related to Y through Mo) or analogs of the lanthanides (and related to La through Nd). The arguments hinged primarily on comparing observed oxidation states to those of the earlier elements. (a) If the actinides were analogs of the transition elements, what would you predict about the maximum oxidation state for U? For Np? (b) If the actinides were analogs of the lanthanides, what would you predict about the maximum oxidation state for U? For Pu?

23.121 The effect of entropy on reactions is evident in the stabilities of certain complexes.
(a) Using the criterion of number of product particles, predict which of the following will be favored in terms of ΔS^0_{rxn}:
$$[Cu(NH_3)_4]^{2+}(aq) + 4H_2O(l) \longrightarrow$$
$$[Cu(H_2O)_4]^{2+}(aq) + 4NH_3(aq)$$
$$[Cu(H_2NCH_2CH_2NH_2)_2]^{2+}(aq) + 4H_2O(l) \longrightarrow$$
$$[Cu(H_2O)_4]^{2+}(aq) + 2en(aq)$$
(b) Given that the Cu—N bond strength is approximately the same in both complexes, which complex will be more stable with respect to ligand exchange in water? Explain.

23.122 For the permanganate ion, draw a Lewis structure that has the lowest formal charges.

23.123 The coordination compound $[Pt(NH_3)_2(SCN)_2]$ displays two types of isomerism. Name the types and give names and structures for the six possible isomers.

23.124 In the sepia "toning" of a black-and-white photograph, the image is converted to a rich brownish violet by placing the finished photograph in a solution of gold(III) ions, in which metallic gold replaces the metallic silver. Use Appendix D to explain the chemistry of this process.

23.125 An octahedral complex with three different ligands (A, B, and C) can have formulas with three different ratios of the ligands:
$[MA_4BC]^{n+}$,   such as $[Co(NH_3)_4(H_2O)Cl]^{2+}$
$[MA_3B_2C]^{n+}$,   such as $[Cr(H_2O)_3Br_2Cl]$
$[MA_2B_2C_2]^{n+}$,   such as $[Cr(NH_3)_2(H_2O)_2Br_2]^+$
For each example, give the name, state the type(s) of isomerism present, and draw all isomers.

23.126 In black-and-white photography, what are the major chemical changes involved in exposing, developing, and fixing?

23.127 In $[Cr(NH_3)_6]Cl_3$, the $[Cr(NH_3)_6]^{3+}$ ion absorbs visible light in the blue-violet range, and the compound is yellow-orange. In $[Cr(H_2O)_6]Br_3$, the $[Cr(H_2O)_6]^{3+}$ ion absorbs visible light in the red range, and the compound is blue-gray. Explain these differences in light absorbed and colors of the compounds.

23.128 Dark green manganate salts contain the MnO_4^{2-} ion. The ion is stable in basic solution but disproportionates in acid to $MnO_2(s)$ and MnO_4^-. (a) What is the oxidation state of Mn in MnO_4^{2-}, MnO_4^-, and MnO_2? (b) Write a balanced equation for the reaction of MnO_4^{2-} in acidic solution.

23.129 Aqueous electrolysis of potassium manganate (K_2MnO_4) in basic solution is used for the production of tens of thousands of tons of potassium permanganate ($KMnO_4$) annually. (a) Write a balanced equation for the electrolysis reaction. (Water is reduced also.) (b) How many moles of MnO_4^- can be formed if 12 A flow through a tank of aqueous MnO_4^{2-} for 96 h?

23.130 The actinides Pa, U, and Np form a series of complex ions, as in the compound Na_3UF_8, in which the central metal ion has an unusual geometry and oxidation state. In the crystal structure, the complex ion can be pictured as resulting from interpenetration of simple cubic arrays of uranium and fluoride ions. (a) What is the coordination number of the metal ion in the complex ion? (b) What is the oxidation state of uranium in the compound? (c) Sketch the complex ion.

CHAPTER 24

NUCLEAR REACTIONS AND THEIR APPLICATIONS

CHAPTER OUTLINE

24.1 Radioactive Decay and Nuclear Stability
Components of the Nucleus
Types of Radioactive Emissions
Types of Radioactive Decay;
    Nuclear Equations
The Mode of Decay

24.2 The Kinetics of Radioactive Decay
Rate of Radioactive Decay
Radioisotopic Dating

**24.3 Nuclear Transmutation: Induced
Changes in Nuclei**
Early Transmutation Experiments
Particle Accelerators

**24.4 The Effects of Nuclear Radiation
on Matter**
Excitation and Ionization
Ionizing Radiation and Living Matter

24.5 Applications of Radioisotopes
Radioactive Tracers
Applications of Ionizing Radiation

24.6 The Interconversion of Mass and Energy
The Mass Defect
Nuclear Binding Energy

24.7 Applications of Fission and Fusion
Nuclear Fission
Nuclear Fusion

Figure: Subatomic fireworks. These colorized tracks show the characteristic spirals and curlicues of charged subatomic particles traveling through the super-heated liquid hydrogen in a bubble chamber. Each particle, under the influence of a strong external magnetic field, leaves a curved trail of tiny bubbles. In this chapter, you'll enter the strange world that lies at the core of every atom to observe its behavior and see how we can put it to use.

Far below the outer fringes of the cloud of electrons lies the atom's tiny, dense core, held together by the strongest force in the universe. For nearly the entire text so far, we have focused on an atom's electrons, treating the nucleus as little more than their electrostatic anchor, while examining the effect of its positive charge on atomic properties and, ultimately, chemical behavior. But, for the scientists probing the structure and behavior of the atomic nucleus itself, there is the scene of real action, one that holds enormous potential benefit and great mystery and wonder.

Society is ambivalent about the applications of nuclear research, however. The promise of abundant energy and treatments for disease comes hand-in-hand with the threat of nuclear waste contamination, reactor accidents, and unimaginable destruction from nuclear war or terrorism. Can the power of the nucleus be harnessed for our benefit, or are the risks too great? In this chapter, we discuss the principles that can help you answer this vital question.

The changes that occur in atomic nuclei are strikingly different from chemical changes. In the reactions you've studied so far, electrons are shared or transferred to form *compounds,* while nuclei sit by passively, never changing their identities. In nuclear reactions, the roles are reversed: electrons in their orbitals are usually bystanders as the nuclei undergo changes that, in nearly every case, form different *elements.* Nuclear reactions may be accompanied by energy changes a million times greater than those in chemical reactions, energy changes so great that changes in mass *are* detectable. Moreover, nuclear reaction yields and rates are typically *not* subject to the effects of pressure, temperature, and catalysis that so clearly influence chemical reactions. Table 24.1 summarizes the general differences between chemical and nuclear reactions.

CONCEPTS & SKILLS

to review before you study this chapter
- discovery of the atomic nucleus (Section 2.4)
- protons, neutrons, mass number, and the $^A_Z X$ notation (Section 2.5)
- half-life and first-order reaction rate (Section 16.4)

Table 24.1 Comparison of Chemical and Nuclear Reactions

Chemical Reactions	Nuclear Reactions
1. One substance is converted into another, but atoms never change identity.	1. Atoms of one element typically are converted into atoms of another element.
2. Orbital electrons are involved as bonds break and form; nuclear particles do not take part.	2. Protons, neutrons, and other particles are involved; orbital electrons rarely take part.
3. Reactions are accompanied by relatively small changes in energy and no measurable changes in mass.	3. Reactions are accompanied by relatively large changes in energy and measurable changes in mass.
4. Reaction rates are influenced by temperature, concentration, catalysts, and the compound in which an element occurs.	4. Reaction rates are affected by number of nuclei, but not by temperature, catalysts, or, normally, the compound in which an element occurs.

We begin this brief treatment of nuclear chemistry with an investigation of nuclear stability—why some nuclei are stable, whereas others are unstable, or *radioactive*. Then, we examine the detection and rate of radioactive decay. Next, we discuss how nuclei formed in particle accelerators continue to extend the periodic table beyond uranium, the last naturally occurring element. We consider the effects of radioactive emissions on matter, especially living matter, focusing on some major applications in science, technology, and medicine. Then, we calculate the energy released in nuclear reactions and discuss current and future engineering attempts to harness this energy. Finally, we end the chapter with a look at the nuclear processes that create the chemical elements in the stars.

24.1 RADIOACTIVE DECAY AND NUCLEAR STABILITY

A stable nucleus remains intact indefinitely, but *the great majority of nuclei are unstable.* An unstable nucleus exhibits **radioactivity:** it spontaneously disintegrates, or *decays,* by emitting radiation. Each type of unstable nucleus has a characteristic *rate* of radioactive decay. For different types of nuclei, this rate ranges from a fraction of a second to many billion years. In this section, we become familiar with important terms and notation for nuclei, discuss some of the key events in the discovery of radioactivity, and consider the various types of radioactive decay.

The Components of the Nucleus: Terms and Notation

The Remarkably Tiny, Massive Nucleus If you could strip all the electrons from the atoms in an object and compress all the nuclei together, the object would lose only a fraction of a percent of its mass, but it would shrink to 0.0000000001% (10^{-10}%) of its volume. An atom the size of the Houston Astrodome would have a nucleus the size of a grapefruit, which would contain virtually all the atom's mass.

Recall from Chapter 2 that the nucleus contains essentially all the atom's mass but is only about 10^{-4} times its diameter (or 10^{-12} times its volume). Obviously, the nucleus is incredibly dense: about 10^{14} g/mL. *Protons* and *neutrons,* the elementary particles that make up the nucleus, are collectively called **nucleons.** The term **nuclide** refers to a variety of nuclei with a particular composition, that is, with specific numbers of the two types of nucleons. Most elements occur in nature as a mixture of **isotopes,** atoms with the characteristic number of protons of the element but different numbers of neutrons. Therefore, each isotope of an element is a single nuclide, and an element generally has a number of different isotopes. The nuclide that is the most abundant isotope of oxygen, for example, contains eight protons and eight neutrons, whereas the nuclide that is the least abundant contains eight protons and ten neutrons.

The relative mass and charge of a particle—nucleon, other elementary particle, or nuclide—is described by the notation $^A_Z X$, where X is the *symbol* for the particle, A is the *mass number,* or the total number of nucleons, and Z is the *charge* of the particle; for nuclides, Z is the same as the *number of protons* (atomic number). Using this notation, we write the three subatomic elementary particles as follows:

$$^{0}_{-1}e \text{ (electron)}, \quad ^{1}_{1}p \text{ (proton), and } ^{1}_{0}n \text{ (neutron)}$$

(In nuclear notation, the element symbol refers to the nucleus only, so a proton is also sometimes represented as $^1_1 H$.) The number of neutrons (N) in a nucleus is the mass number (A) minus the atomic number (Z): $N = A - Z$. The two naturally occurring isotopes of chlorine, for example, have 17 protons ($Z = 17$), but one has 18 neutrons ($^{35}_{17}Cl$, also written ^{35}Cl) and the other has 20 ($^{37}_{17}Cl$, or ^{37}Cl). Nuclides are also designated with the element name followed by the mass number, for example, chlorine-35 and chlorine-37. Despite some small variations, in *naturally occurring* samples of an element or its compounds, the isotopes of the element are present in particular, fixed proportions. Thus, in a sample of sodium chloride (or any Cl-containing substance), 75.77% of the Cl atoms are chlorine-35 and the remaining 24.23% are chlorine-37.

To understand this chapter, it is very important that you are comfortable with nuclear notations, so please take a moment to review Sample Problem 2.2 on p. 53 and Problems 2.37 to 2.44 at the end of Chapter 2.

The Discovery of Radioactivity and the Types of Emissions

In 1896, the French physicist Antoine-Henri Becquerel discovered, quite by accident, that uranium minerals, even when wrapped in paper and stored in the dark, emit a penetrating radiation that can produce bright images on a photographic plate. Becquerel also found that the radiation creates an electric discharge in air, thus providing a means for measuring its intensity. Two years later, a young doctoral student named Marie Sklodowska Curie began a search for other minerals that emitted radiation. She found that thorium minerals also emit radiation and showed that *the intensity of the radiation is directly proportional to the concentration of the element in the mineral, not to the nature of the compound* in which the element occurs. Curie named the emissions *radioactivity* and showed that they are *unaffected by temperature, pressure, or other physical and chemical conditions.*

To her surprise, Curie found that certain uranium minerals were even more radioactive than pure uranium, which implied that they contained traces of one or more as yet unknown, highly radioactive elements. She and her husband, the physicist Pierre Curie, set out to isolate all the radioactive components in pitchblende, the principal ore of uranium. After months of painstaking chemical work, they isolated two extremely small, highly radioactive fractions, one that precipitated with bismuth compounds and another that precipitated with alkaline earth compounds. Through chemical and spectroscopic analysis, Marie Curie was able to show that these fractions contained two new elements, which she named polonium (after her native Poland) and radium. Polonium (Po; $Z = 84$), the most metallic member of Group 6A(16), lies to the right of bismuth in Period 6. Radium (Ra; $Z = 88$), which is the heaviest alkaline earth metal, lies under barium in Group 2A(2).

Purifying radium proved to be another arduous task. Starting with several tons of pitchblende residues from which the uranium had been extracted, Curie prepared compounds of the larger Group 2A(2) elements, continually separating minuscule amounts of radium compounds from enormously larger amounts of chemically similar barium compounds. It took her four years to isolate 0.1 g of radium chloride, which she melted and electrolyzed to obtain pure metallic radium.

During the next few years, Henri Becquerel, the Curies, and P. Villard in France and Ernest Rutherford and his coworkers in England studied the nature of radioactive emissions. Rutherford and his colleague Frederick Soddy observed that elements other than radium were formed when radium decayed. In 1902, they proposed that radioactive emission results in the change of one element into another. To their contemporaries, this idea sounded like a resurrection of alchemy and was met with disbelief and ridicule. We now know it to be true: under most circumstances, *when a nuclide of one element decays, it changes into a nuclide of a different element.*

These studies led to an understanding of the three most common types of radioactive emission:

- **Alpha particles** (symbolized α or ^4_2He) are dense, positively charged particles identical to helium nuclei.
- **Beta particles** (symbolized β, β^-, or more usually $^{\ 0}_{-1}\beta$) are negatively charged particles identified as high-speed electrons. (The emission of electrons from the nucleus may seem strange, but as you'll see shortly, β particles arise as a result of a nuclear reaction.)
- **Gamma rays** (symbolized as γ, or sometimes $^0_0\gamma$) are very high-energy photons, about 10^5 times as energetic as visible light.

The behavior of these three emissions in an electric field is shown in Figure 24.1. Note that α particles bend toward the negative plate, β particles bend toward the positive plate, and γ rays are not affected by the electric field. We'll discuss the effects of these emissions on matter later.

Her Brilliant Career Marie Curie (1867–1934) is the only person to be awarded Nobel Prizes in two *different* sciences, one in physics in 1903 for her research into radioactivity and the other in chemistry in 1911 for the discovery of polonium and the discovery, isolation, and study of radium and its compounds.

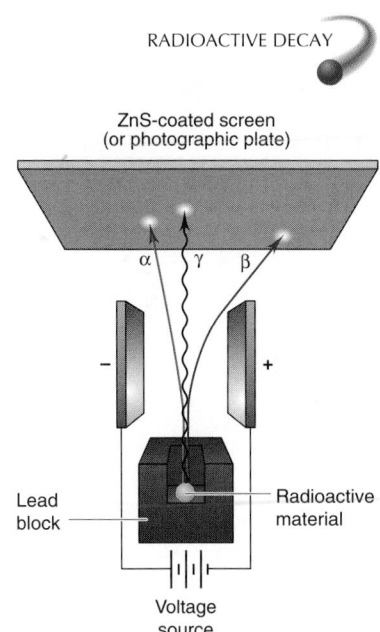

Figure 24.1 The behavior of three types of radioactive emissions in an electric field. Alpha particles (α, ^4_2He) are positively charged, so they bend toward the negative plate. In contrast, beta particles (β, $^{\ 0}_{-1}\beta$) are negatively charged, so they bend toward the positive plate. Note that the curvature is greater for the β particles because they have much lower mass than the α particles. Gamma rays (γ, $^0_0\gamma$) consist of uncharged high-energy photons.

Types of Radioactive Decay; Balancing Nuclear Equations

When a nuclide decays, it forms a nuclide of lower energy, and the excess energy is carried off by the emitted radiation. The decaying, or reactant, nuclide is called the *parent;* the product nuclide is called the *daughter.* Nuclides can decay in several ways. As we discuss the major types of decay, which are summarized in Table 24.2, note the principle used to balance nuclear reactions: *the total Z (charge, number of protons) and total A (sum of protons and neutrons) of the reactants equal those of the products:*

$$\text{Total } A \atop \text{Total } Z} \text{Reactants} = {\text{Total } A \atop \text{Total } Z} \text{Products} \qquad (24.1)$$

1. **Alpha decay** involves the loss of an α particle (^4_2He) from a nucleus. For each α particle emitted by the parent nucleus, *A decreases by 4 and Z decreases by 2.* Every element that is heavier than lead (Pb; Z = 82), as well as a few lighter ones, exhibits α decay. In Rutherford's classic experiment that established the existence of the atomic nucleus (Section 2.4, p. 50), radium was the source of the α particles that were used as projectiles. Radium undergoes α decay to yield radon (Rn; Z = 86):

$$^{226}_{88}\text{Ra} \longrightarrow {}^{222}_{86}\text{Rn} + {}^4_2\text{He}$$

Note that the *A* value for Ra equals the sum of the *A* values for Rn and He (226 = 222 + 4), and that the *Z* value for Ra equals the sum of the *Z* values for Rn and He (88 = 86 + 2).

Table 24.2 Modes of Radioactive Decay*

Mode	Emission	Decay Process	Change in A	Change in Z	Change in N
α Decay	α (^4_2He)	 Reactant (parent) Product (daughter) α expelled	−4	−2	−2
β Decay	$^0_{-1}\beta$	^1_0n in nucleus $\longrightarrow$ ^1_1p in nucleus + $^0_{-1}\beta$ β expelled	0	+1	−1
Positron emission	$^0_1\beta$	$h\nu$ + nucleus with $x\,\text{p}^+$ and $y\,\text{n}^0$ $\longrightarrow$ nucleus with $(x-1)\,\text{p}^+$ and $(y+1)\,\text{n}^0$ + $^0_1\beta$ positron expelled high-energy photon	0	−1	+1
Electron capture	x-ray photon	$^0_{-1}\text{e}$ absorbed from low-energy orbital + ^1_1p in nucleus $\longrightarrow$ ^1_0n in nucleus	0	−1	+1
γ Emission	$^0_0\gamma$	 excited nucleus stable nucleus γ photon radiated	0	0	0

*Neutrinos (ν) are involved in several of these processes but are not shown.

2. **Beta decay** involves the ejection of a β particle ($_{-1}^{0}\beta$) from the nucleus. This change does not involve the expulsion of a β particle that was actually in the nucleus, but rather the *conversion of a neutron into a proton, which remains in the nucleus, and a β particle, which is expelled immediately:*

$$_{0}^{1}n \longrightarrow \,_{1}^{1}p + \,_{-1}^{0}\beta$$

As always, the totals of the *A* and the *Z* values for reactant and products are equal. Radioactive nickel-63 becomes stable copper-63 through β decay:

$$_{28}^{63}Ni \longrightarrow \,_{29}^{63}Cu + \,_{-1}^{0}\beta$$

Another example is the β decay of carbon-14, which we return to in the context of radiocarbon dating:

$$_{6}^{14}C \longrightarrow \,_{7}^{14}N + \,_{-1}^{0}\beta$$

Note that *β decay results in a product nuclide with the same A but with Z one higher (one more proton) than in the reactant nuclide.* In other words, an atom of the element with the next *higher* atomic number is formed.

3. **Positron decay** involves the emission of a positron from the nucleus. A key idea of modern physics is that every fundamental particle has a corresponding *antiparticle,* another particle with the same mass but opposite charge. The **positron** (symbolized $_{1}^{0}\beta$; note the positive *Z*) is the antiparticle of the electron. Positron decay occurs through a process in which *a proton in the nucleus is converted into a neutron, and a positron is expelled.** *Positron decay has the opposite effect of β decay, resulting in a daughter nuclide with the same A but with Z one lower (one fewer proton) than the parent;* thus, an atom of the element with the next *lower* atomic number forms. Carbon-11, a synthetic radioisotope, decays to a stable boron isotope through emission of a positron:

$$_{6}^{11}C \longrightarrow \,_{5}^{11}B + \,_{1}^{0}\beta$$

4. **Electron capture** occurs when the nucleus of an atom draws in an electron from an orbital of the lowest energy level. The net effect is that *a nuclear proton is transformed into a neutron:*

$$_{1}^{1}p + \,_{-1}^{0}e \longrightarrow \,_{0}^{1}n$$

(We use the symbol $_{-1}^{0}e$ to distinguish an orbital electron from a beta particle, symbol $_{-1}^{0}\beta$.) The orbital vacancy is quickly filled by an electron that moves down from a higher energy level, and that energy difference appears as an *x-ray photon.* Radioactive iron forms stable manganese through electron capture:

$$_{26}^{55}Fe + \,_{-1}^{0}e \longrightarrow \,_{25}^{55}Mn + h\nu \text{ (x-ray)}$$

Electron capture has the same net effect as positron decay (Z lower by 1, A unchanged), even though the processes are entirely different.

5. **Gamma emission** involves the radiation of high-energy γ photons from an excited nucleus. Recall that an atom in an excited *electronic* state reduces its energy by emitting photons, usually in the UV and visible ranges. Similarly, a nucleus in an excited state lowers its energy by emitting γ photons, which are of much higher energy (much shorter wavelength) than UV photons. Many nuclear processes leave the nucleus in an excited state, so *γ emission accompanies most other types of decay.* Several γ photons (γ rays) of different frequencies can be emitted from an excited nucleus as it relaxes, usually to the ground state. Many of Marie Curie's experiments involved the release of γ rays, such as

$$_{92}^{238}U \longrightarrow \,_{90}^{234}Th + \,_{2}^{4}He + 2_{0}^{0}\gamma$$

Because γ rays have no mass or charge, *γ emission does not change A or Z.* Gamma rays also result when a particle and an antiparticle annihilate each other, as when an emitted positron meets an orbital electron:

$$_{1}^{0}\beta \text{ (from nucleus)} + \,_{-1}^{0}e \text{ (outside nucleus)} \longrightarrow 2_{0}^{0}\gamma$$

The Little Neutral One A neutral particle called a *neutrino* (ν) is also emitted in many nuclear reactions, including the change of a neutron to a proton:

$$_{0}^{1}n \longrightarrow \,_{1}^{1}p + \,_{-1}^{0}\beta + \nu$$

Theory suggests that neutrinos have a mass much less than 10^{-4} times that of an electron, and that at least 10^{9} neutrinos exist in the universe for every proton. Neutrinos interact with matter so slightly that it would take a piece of lead 1 light-year thick to absorb them. We will not discuss them further, except to mention that experiments in Japan in the 1990s detected neutrinos and obtained evidence that they have mass. Using a cathedral-sized pool containing 50,000 tons of ultrapure water buried 1 mile underground in a zinc mine, an international team of scientists obtained results that suggest that neutrinos may account for a significant portion of the "missing" matter in the universe and may provide enough mass (and, thus, gravitational attraction) to prevent the universe from expanding forever.

---

*The process, called *pair production,* involves energy actually turning into matter. A high-energy ($>1.63\times10^{-13}$ J) photon becomes an electron and a positron simultaneously. The electron and a proton in the nucleus form a neutron, while the positron is expelled.

SAMPLE PROBLEM 24.1 Writing Equations for Nuclear Reactions

Problem Write balanced equations for the following nuclear reactions:
(a) Naturally occurring thorium-232 undergoes α decay.
(b) Chlorine-36 undergoes electron capture.
Plan We first write a skeleton equation that includes the mass numbers, atomic numbers, and symbols of all the particles, showing the unknown particles as $^A_Z X$. Then, because the total of mass numbers and the total of charges on the left side and the right side must be equal, we solve for A and Z, and use Z to determine X from the periodic table.
Solution (a) Writing the skeleton equation:

$$^{232}_{90}\text{Th} \longrightarrow\ ^A_Z X\ +\ ^4_2\text{He}$$

Solving for A and Z and balancing the equation: For A, $232 = A + 4$, so $A = 228$. For Z, $90 = Z + 2$, so $Z = 88$. From the periodic table, we see that the element with $Z = 88$ is radium (Ra). Thus, the balanced equation is

$$^{232}_{90}\text{Th} \longrightarrow\ ^{228}_{88}\text{Ra}\ +\ ^4_2\text{He}$$

(b) Writing the skeleton equation:

$$^{36}_{17}\text{Cl}\ +\ ^0_{-1}\text{e} \longrightarrow\ ^A_Z X$$

Solving for A and Z and balancing the equation: For A, $36 + 0 = A$, so $A = 36$. For Z, $17 + (-1) = Z$, so $Z = 16$. The element with $Z = 16$ is sulfur (S), so we have

$$^{36}_{17}\text{Cl}\ +\ ^0_{-1}\text{e} \longrightarrow\ ^{36}_{16}\text{S}$$

Check Always read across superscripts and then across subscripts, with the yield arrow as an equal sign, to check your arithmetic. In part (a), for example, $232 = 228 + 4$, and $90 = 88 + 2$.

FOLLOW-UP PROBLEM 24.1 Write a balanced equation for the reaction in which a nuclide undergoes β decay and produces cesium-133.

Nuclear Stability and the Mode of Decay

There are several ways that an unstable nuclide *might* decay, but can we predict how it *will* decay? Indeed, can we predict *if* it will decay at all? Our knowledge of the nucleus is much less complete than our knowledge of the atom as a whole, but some patterns emerge when we examine the naturally occurring nuclides.

The Band of Stability and the Neutron/Proton (N/Z) Ratio A key factor that determines the stability of a nuclide is the ratio of the number of neutrons to the number of protons, the **N/Z ratio,** which we calculate from $(A - Z)/Z$. For lighter nuclides, one neutron for each proton ($N/Z \approx 1$) is enough to provide stability. However, for heavier nuclides to be stable, the number of neutrons must exceed the number of protons, and often by quite a significant proportion. If the N/Z ratio is either too high or not high enough, the nuclide is unstable and decays.

Figure 24.2A is a plot of number of neutrons vs. number of protons for the *stable* nuclides. The nuclides form a narrow **band of stability** that gradually increases from an N/Z ratio of 1, near $Z = 10$, to an N/Z ratio slightly greater than 1.5, near $Z = 83$ for ^{209}Bi. Several key points are

- Very few stable nuclides exist with $N/Z < 1$; the only two are $^1_1$H and $^3_2$He. For lighter stable nuclides, $N/Z \approx 1$: $^4_2$He, $^{12}_6$C, $^{16}_8$O, and $^{20}_{10}$Ne are particularly stable.
- The N/Z ratio of stable nuclides gradually increases as Z increases. No stable nuclide exists with $N/Z = 1$ for $Z > 20$. Thus, for $^{56}_{26}$Fe, $N/Z = 1.15$; for $^{107}_{47}$Ag, $N/Z = 1.28$; and for $^{184}_{74}$W, $N/Z = 1.49$.
- All nuclides with $Z > 83$ are unstable. Bismuth-209 is the heaviest stable nuclide. Therefore, the largest members of Groups 1A(1), 2A(2), 6A(16), 7A(17), and 8A(18) are radioactive, as are all the actinides and the elements of the fourth transition series (Period 7).

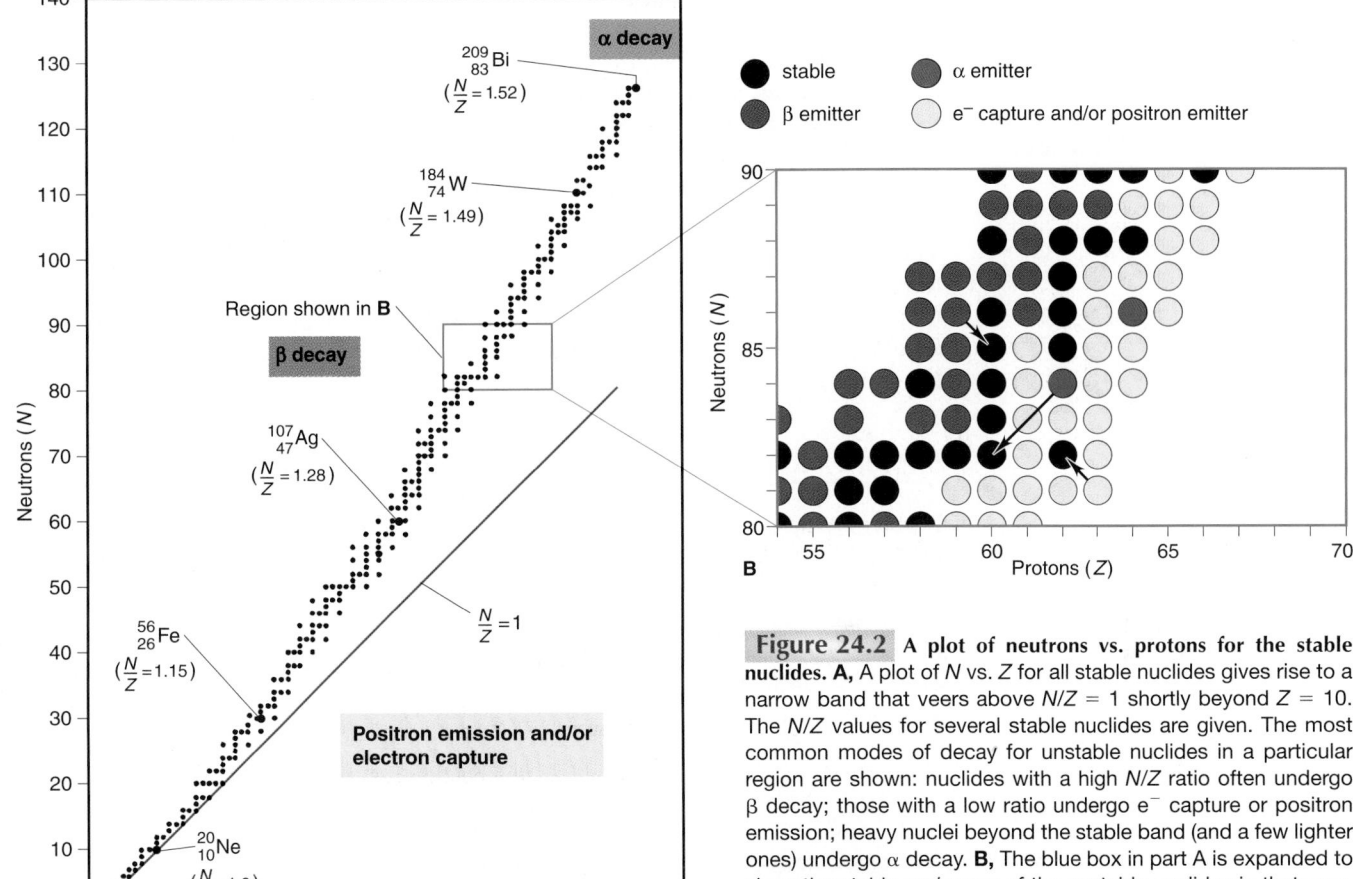

A

B

Figure 24.2 **A plot of neutrons vs. protons for the stable nuclides.** **A,** A plot of N vs. Z for all stable nuclides gives rise to a narrow band that veers above $N/Z = 1$ shortly beyond $Z = 10$. The N/Z values for several stable nuclides are given. The most common modes of decay for unstable nuclides in a particular region are shown: nuclides with a high N/Z ratio often undergo β decay; those with a low ratio undergo e^- capture or positron emission; heavy nuclei beyond the stable band (and a few lighter ones) undergo α decay. **B,** The blue box in part A is expanded to show the stable *and* many of the unstable nuclides in that area. Note the modes of decay: α decay decreases both N and Z by 2; β decay decreases N and increases Z by 1; positron emission and e^- capture increase N and decrease Z by 1.

Stability and Nuclear Structure Given that protons are positively charged and neutrons uncharged, what holds the nucleus together? Nuclear scientists answer this question and explain the importance of the N/Z ratio in terms of two opposing forces. Electrostatic repulsive forces between protons would break the nucleus apart if not for the presence of an attractive force that exists between all nucleons (protons and neutrons) called the **strong force.** This force is about 100 times stronger than the repulsive force but *operates only over the short distances within the nucleus.* Competition between the *attractive* strong force and the *repulsive* electrostatic force determines nuclear stability.

Curiously, the oddness or evenness of N and Z values is related to some important patterns of nuclear stability. Two interesting points appear when we classify the known stable nuclides:

- Elements with an even Z (number of protons) usually have a larger number of stable nuclides than elements with an odd Z. Table 24.3 demonstrates this point for cadmium ($Z = 48$) through xenon ($Z = 54$).
- Well over half the stable nuclides have *both* even N and even Z (Table 24.4, next page). (Only seven nuclides with odd N and odd Z are either stable—$_1^2H$, $_3^6Li$, $_5^{10}B$, $_7^{14}N$—or decay so slowly that their amounts have changed little since the Earth formed—$_{23}^{50}V$, $_{57}^{138}La$, and $_{71}^{176}Lu$.)

Table 24.3 Number of Stable Nuclides for Elements 48 Through 54*

Element	Atomic No. (Z)	No. of Nuclides
Cd	**48**	**8**
In	49	2
Sn	**50**	**10**
Sb	51	2
Te	**52**	**8**
I	53	1
Xe	**54**	**9**

*Even Z shown in boldface.

Table 24.4 An Even-Odd Classification of the Stable Nuclides

Z	N	Number of Nuclides
Even	Even	157
Even	Odd	53
Odd	Even	50
Odd	Odd	7
	TOTAL	267

One model of nuclear structure that attempts to explain these findings postulates that protons and neutrons lie in *nucleon shells,* or energy levels, and that stability results from the *pairing* of like nucleons. This arrangement leads to the stability of even values of N and Z. (The analogy to electron energy levels and the stability that arises from electron pairing is striking.)

Just as the noble gases—the elements with 2, 10, 18, 36, 54, and 86 electrons—are exceptionally stable because of their filled *electron* shells, nuclides with N or Z values of 2, 8, 20, 28, 50, 82 (and $N = 126$) are exceptionally stable as well. These so-called *magic numbers* are thought to correspond to the numbers of protons or neutrons in filled *nucleon* shells. A few examples are $^{50}_{22}\text{Ti}$ ($N = 28$), $^{88}_{38}\text{Sr}$ ($N = 50$), and the nine stable nuclides of tin ($Z = 50$). Some extremely stable nuclides have double magic numbers: $^{4}_{2}\text{He}$, $^{16}_{8}\text{O}$, $^{40}_{20}\text{Ca}$, and $^{208}_{82}\text{Pb}$ ($N = 126$).

SAMPLE PROBLEM 24.2 Predicting Nuclear Stability

Problem Which of the following nuclides would you predict to be stable and which radioactive? Explain.

(a) $^{18}_{10}\text{Ne}$ (b) $^{32}_{16}\text{S}$

(c) $^{236}_{90}\text{Th}$ (d) $^{123}_{56}\text{Ba}$

Plan To evaluate the stability of each nuclide, we find the N/Z ratio from $(A - Z)/Z$, the value of Z, stable N/Z ratios (from Figure 24.2), and whether Z and N are even or odd.

Solution (a) Radioactive. The ratio $N/Z = \dfrac{18 - 10}{10} = 0.8$. The minimum ratio for stability is 1.0; so, despite even N and Z, this nuclide has too few neutrons to be stable.

(b) Stable. This nuclide has $N/Z = 1.0$ and $Z < 20$, with even N and Z. Thus, it is most likely stable.

(c) Radioactive. Every nuclide with $Z > 83$ is radioactive.

(d) Radioactive. The ratio $N/Z = 1.20$. For Z from 55 to 60, Figure 24.2A shows $N/Z \geq 1.3$, so this nuclide probably has too few neutrons to be stable.

Check By consulting a table of isotopes, such as the one in the *CRC Handbook of Chemistry and Physics,* we find that our predictions are correct.

FOLLOW-UP PROBLEM 24.2 Why is $^{31}_{15}\text{P}$ stable but $^{30}_{15}\text{P}$ unstable?

Predicting the Mode of Decay An unstable nuclide generally decays in a mode that shifts its N/Z ratio toward the band of stability. This fact is illustrated in Figure 24.2B on the preceding page, which expands a small region of part A to show all of the stable *and* many of the radioactive nuclides in that region, as well as their modes of decay. Note the following points, and then we'll apply them in a sample problem:

1. *Neutron-rich nuclides.* Nuclides with too many neutrons for stability (a high N/Z) lie above the band of stability. They undergo β *decay,* which converts a neutron into a proton, thus reducing the value of N/Z.

2. *Neutron-poor nuclides.* Nuclides with too few neutrons for stability (a low N/Z) lie below the band. They undergo *positron decay* or *electron capture,* both of which convert a proton into a neutron, thus increasing the value of N/Z.

3. *Heavy nuclides.* Nuclides with $Z > 83$ are too heavy to lie within the band and undergo α *decay,* which reduces their Z and N values by two units per emission. (Several lighter nuclides also exhibit α decay.)

SAMPLE PROBLEM 24.3 Predicting the Mode of Nuclear Decay

Problem Predict the nature of the nuclear change(s) each of the following radioactive nuclides is likely to undergo:
(a) $^{12}_{5}B$ (b) $^{234}_{92}U$
(c) $^{74}_{33}As$ (d) $^{127}_{57}La$

Plan We use the N/Z ratio to decide where the nuclide lies relative to the band of stability and how its ratio compares with others in that region of the band. Then, we predict which of the modes just discussed will yield a product nuclide that is closer to the band.

Solution (a) This nuclide has an N/Z ratio of 1.4, which is too high for this region of the band. It will probably undergo β decay, which will increase Z to 6 and lower the N/Z ratio to 1.

(b) This nuclide is heavier than those in the band of stability. It will probably undergo α decay and decrease its total mass.

(c) This nuclide, with an N/Z ratio of 1.24, lies in the band of stability, so it will probably undergo either β decay or positron emission.

(d) This nuclide has an N/Z ratio of 1.23, which is too low for this region of the band, so it will decrease Z by either positron emission or electron capture.

Comment Both the possible modes of decay are observed for the nuclides in parts (c) and (d).

FOLLOW-UP PROBLEM 24.3 What mode of decay would you expect for (a) $^{61}_{26}Fe$; (b) $^{241}_{95}Am$?

Decay Series A parent nuclide may undergo a series of decay steps before a stable daughter nuclide forms. The succession of steps is called a **decay series,** or **disintegration series,** and is typically depicted on a gridlike display. Figure 24.3 shows the decay series from uranium-238 to lead-206. Numbers of neutrons (N) are plotted against numbers of protons (Z) to form the grid, which displays a series of α and β decays. The zigzag pattern is typical and occurs because α decay decreases both N and Z, whereas β decay decreases N but increases Z. Note that it is quite common for a given nuclide to undergo both types of decay. (Gamma decay accompanies many of these steps, but it does not affect the mass or type of the nuclide.) This decay series is one of three that occur in nature. All end with isotopes of lead whose nuclides all have one ($Z = 82$) or two (^{208}Pb) magic numbers. A second series begins with uranium-235 and ends with lead-207, and a third begins with thorium-232 and ends with lead-208. (Neptunium-237 began a fourth series, but its half-life is so much shorter than the age of the Earth that only traces of it remain today.)

SECTION SUMMARY

Nuclear reactions are not affected by reaction conditions or chemical composition and release much more energy than chemical reactions. A radioactive nuclide is unstable and may emit α particles ($^{4}_{2}He$ nuclei), β particles ($^{0}_{-1}\beta$; high-speed electrons), positrons ($^{0}_{1}\beta$), or γ rays ($^{0}_{0}\gamma$; high-energy photons) or may capture an orbital electron. A narrow band of neutron-to-proton ratios (N/Z) includes those of all the stable nuclides. Radioactive decay allows an unstable nuclide to achieve a more stable N/Z ratio. Certain "magic numbers" of neutrons and protons are associated with very stable nuclides. By comparing a nuclide's N/Z ratio with those in the band of stability, we can predict that, in general, heavy nuclides undergo α decay, neutron-rich nuclides undergo β decay, and proton-rich nuclides undergo positron emission or electron capture. Three naturally occurring decay series all end in isotopes of lead.

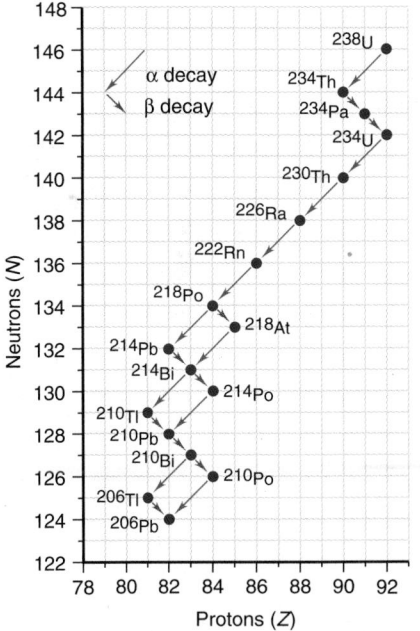

Figure 24.3 The ^{238}U decay series.
Uranium-238 (*top right*) decays through a series of emissions of α or β particles to lead-206 (*bottom left*) in 14 steps.

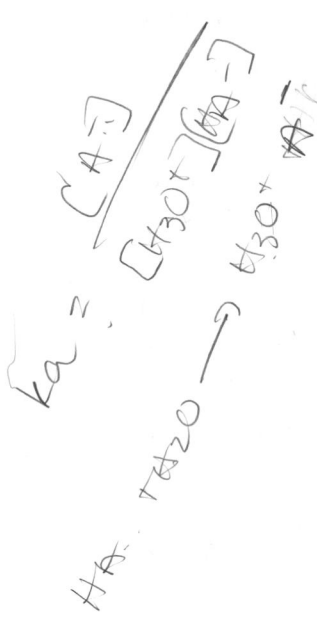

24.2 THE KINETICS OF RADIOACTIVE DECAY

Chemical and nuclear systems both tend toward maximum stability. Just as the concentrations in a chemical system change in a predictable direction to give a stable equilibrium ratio, the type and number of nucleons in an unstable nucleus change in a predictable direction to give a stable N/Z ratio. As you know, however, the tendency of a chemical system to become more stable tells nothing about how long that process will take, and the same holds true for nuclear systems. In this section, we examine the kinetics of nuclear change; later, we'll examine the energetics of nuclear change. To begin, a Tools of the Laboratory essay describes how radioactivity is detected and measured.

The Rate of Radioactive Decay

Radioactive nuclei decay at a characteristic rate, regardless of the chemical substance in which they occur. The *decay rate,* or **activity** ($\mathscr{A}$), of a radioactive sample is the change in number of nuclei ($\mathscr{N}$) divided by the change in time (t). As we saw with chemical reaction rates, because the number of nuclei is *decreasing,* a minus sign precedes the expression:

$$\text{Decay rate } (\mathscr{A}) = -\frac{\Delta \mathscr{N}}{\Delta t}$$

The SI unit of radioactivity is the **becquerel (Bq);** it is defined as one disintegration per second (d/s): 1 Bq = 1 d/s. A much larger and more common unit of radioactivity is the **curie (Ci):** 1 curie equals the number of nuclei disintegrating each second in 1 g of radium-226:

$$1 \text{ Ci} = 3.70 \times 10^{10} \text{ d/s} \tag{24.2}$$

Because the curie is so large, the millicurie (mCi) and microcurie (μCi) are commonly used. We often express the radioactivity of a sample in terms of *specific activity,* the decay rate per gram.

An activity is meaningful only when we consider the large number of nuclei in a macroscopic sample. Suppose there are 1×10^{15} radioactive nuclei of a particular type in a sample and they decay at a rate of 10% per hour. Although any particular nucleus in the sample might decay in a microsecond or in a million hours, the *average* of all decays results in 10% of the entire collection of nuclei disintegrating each hour. During the first hour, 10% of the *original* number, or 1×10^{14} nuclei, will decay. During the next hour, 10% of the remaining 9×10^{14} nuclei, or 9×10^{13} nuclei, will decay. During the next hour, 10% of those remaining will decay, and so forth. Thus, for a large collection of radioactive nuclei, *the number decaying per unit time is proportional to the number present:*

$$\text{Decay rate } (\mathscr{A}) \propto \mathscr{N} \qquad \text{or} \qquad \mathscr{A} = k\mathscr{N}$$

where k is called the **decay constant** and is characteristic of the nuclide. The larger the value of k, the higher is the decay rate.

Combining the two rate expressions just given, we obtain

$$\mathscr{A} = -\frac{\Delta \mathscr{N}}{\Delta t} = k\mathscr{N} \tag{24.3}$$

Note that the activity depends only on $\mathscr{N}$ raised to the first power (and on the constant value of k). Therefore, *radioactive decay is a first-order process* (see Section 16.4). The only difference in the case of nuclear decay is that we consider the *number* of nuclei rather than their concentration.

Half-Life of Radioactive Decay Decay rates are also commonly expressed in terms of the fraction of nuclei that decays over a given time interval. The **half-life ($t_{1/2}$)** of a nuclide is the time it takes for half the nuclei present to decay. *The number of nuclei remaining is halved after each half-life.* Thus, half-life has the same meaning for a nuclear change as for a chemical change (Section 16.4).

Counters for the Detection of Radioactive Emissions

Radioactive emissions interact with atoms in surrounding materials. To determine the rate of nuclear decay, we measure the radioactivity of a sample by observing the effects of these interactions over time. Because these effects can be electrically amplified billions of times, it is even possible to detect the decay of a single nucleus. Ionization counters and scintillation counters are two devices used to measure radioactive emissions.

An *ionization counter* detects radioactive emissions as they ionize a gas. Ionization produces free electrons and gaseous cations, which are attracted to electrodes that conduct a current to a recording device. The most common type of ionization counter is a **Geiger-Müller counter** (Figure B24.1). It consists of a tube filled with argon gas; the tube housing acts as the cathode, and a thin wire in the center of the tube acts as the anode. Emissions from the sample enter the tube through a thin window and strike argon atoms, producing free electrons that are accelerated toward the anode. These electrons collide with other argon atoms and free more electrons in an *avalanche effect*. The current created is amplified and appears as a meter reading and/or an audible click. The initial release of one electron can release 10^{10} electrons in a microsecond, giving the Geiger-Müller counter great sensitivity.

In a **scintillation counter,** radioactive emissions too weak to ionize surrounding atoms are detected by their ability to excite atoms and cause them to emit light. The light-emitting substance in the counter, called a *phosphor,* is coated onto part of a *photomultiplier tube,* a device that increases the original electrical signal. Incoming radioactive particles strike the phosphor, which emits photons. Each photon, in turn, strikes a cathode, releasing an electron through the photoelectric effect (Section 7.1). This electron hits other portions of the tube that release increasing numbers of electrons, and the resulting current is recorded. Liquid scintillation counters employ an organic mixture that contains a phosphor and a solvent (Figure B24.2). This "cocktail" dissolves the sample *and* emits light when excited by the emission. These counters are often used to measure emissions from dissolved radioactive biological samples.

Figure B24.2 **Vials of a scintillation "cocktail" emitting light.** A radioactive substance dissolved in an organic mixture (cocktail) emits particles that excite the phosphor component to emit light. Light intensity is proportional to the concentration of the substance.

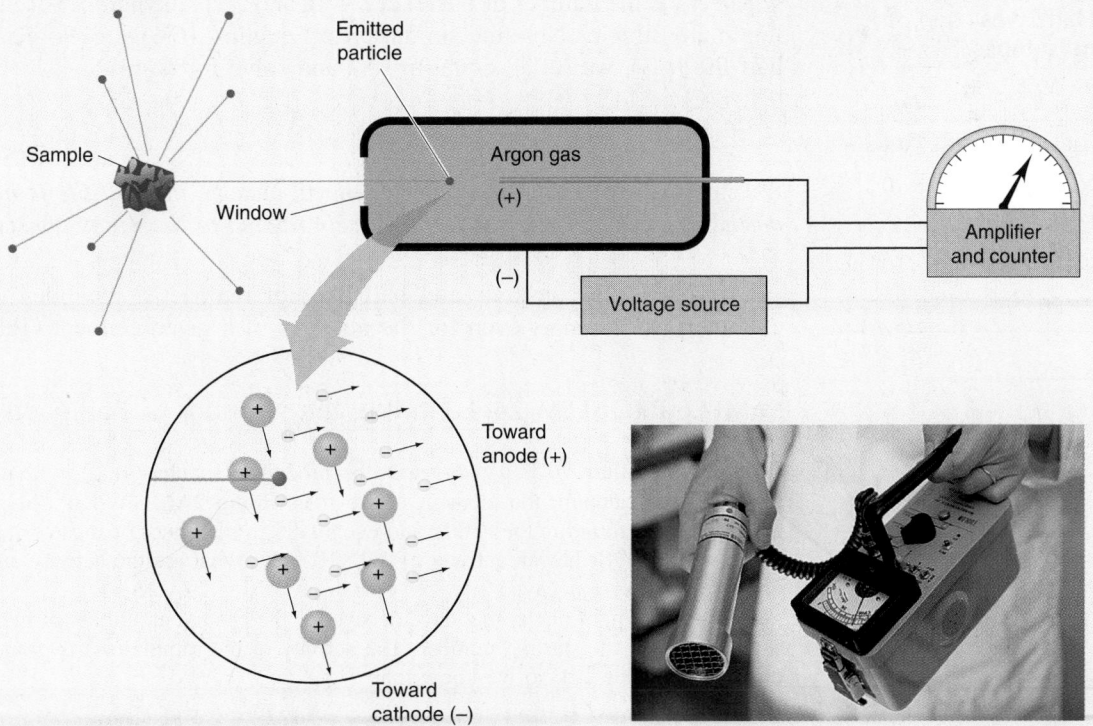

Figure B24.1 Detection of radioactivity by an ionization counter. When an Ar atom absorbs the energy of a radioactive particle *(red)*, it is ionized to an Ar^+ ion *(purple)* and an electron *(yellow)*. The free electron collides with and ionizes another Ar atom. As the process continues, the Ar^+ ions migrate to the negative electrode, and the electrons migrate to the positive electrode, resulting in a current.

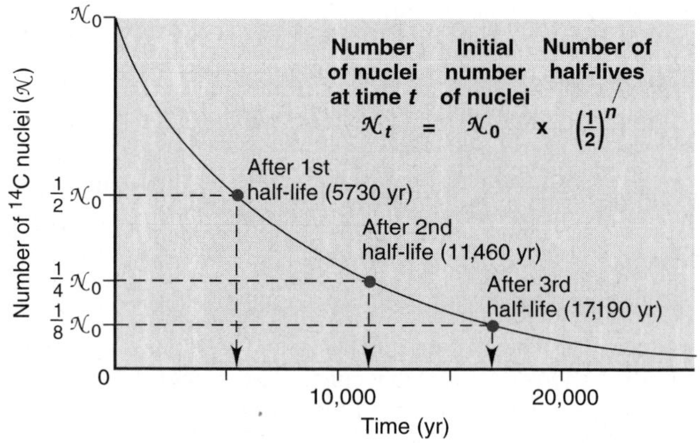

Figure 24.4 Decrease in number of ^{14}C nuclei over time. A plot of number of ^{14}C nuclei vs. time gives a decreasing curve. In each half-life (5730 years), half the ^{14}C nuclei present undergo decay. A plot of mass of ^{14}C vs. time is identical.

Figure 24.4 Decrease in number of ^{14}C nuclei over time. A plot of number of ^{14}C nuclei vs. time gives a decreasing curve. In each half-life (5730 years), half the ^{14}C nuclei present undergo decay. A plot of mass of ^{14}C vs. time is identical.

Figure 24.4 shows the decay of carbon-14, which has a half-life of 5730 years, in terms of number of ^{14}C nuclei remaining:

$$^{14}_{6}C \longrightarrow \, ^{14}_{7}N + \, ^{0}_{-1}\beta$$

We can also consider the half-life in terms of mass of substance. As ^{14}C decays to the product ^{14}N, its mass decreases. If we start with 1.0 g of carbon-14, half that mass of ^{14}C (0.50 g) will be left after 5730 years, half of that mass (0.25 g) after another 5730 years, and so on. The activity depends on the number of nuclei present, so the activity is halved after each succeeding half-life as well.

We determine the half-life of a nuclear reaction from its rate constant. Rearranging Equation 24.3 and integrating over time gives

$$\ln \frac{\mathcal{N}_t}{\mathcal{N}_0} = -kt \qquad \text{or} \qquad \ln \frac{\mathcal{N}_0}{\mathcal{N}_t} = kt \qquad \textbf{(24.4)}$$

where $\mathcal{N}_0$ is the number of nuclei at $t = 0$, and $\mathcal{N}_t$ is the number of nuclei remaining at any time t. (Note the similarity to Equation 16.4, p. 678.) To calculate the half-life ($t_{1/2}$), we set $\mathcal{N}_t$ equal to $\frac{1}{2}\mathcal{N}_0$ and solve for $t_{1/2}$:

$$\ln \frac{\mathcal{N}_0}{\frac{1}{2}\mathcal{N}_0} = kt_{1/2} \qquad \text{so} \qquad t_{1/2} = \frac{\ln 2}{k} \qquad \textbf{(24.5)}$$

Exactly analogous to a first-order chemical change, *the half-life is **not** dependent on the number of nuclei and is inversely related to the decay constant:*

large $k \Rightarrow$ short $t_{1/2}$ and   small $k \Rightarrow$ long $t_{1/2}$

The decay constants and half-lives of radioactive nuclides vary over an extremely wide range, even for the nuclides of a given element (Table 24.5).

Table 24.5 Decay Constants (k) and Half-Lives ($t_{1/2}$) of Beryllium Isotopes

Nuclide	k	$t_{1/2}$
$^{7}_{4}Be$	1.30×10^{-2}/day	53.3 days
$^{8}_{4}Be$	1.0×10^{16}/s	6.7×10^{-17} s
$^{9}_{4}Be$	Stable	
$^{10}_{4}Be$	4.3×10^{-7}/yr	1.6×10^{6} yr
$^{11}_{4}Be$	5.02×10^{-2}/s	13.8 s

SAMPLE PROBLEM 24.4 Finding the Number of Radioactive Nuclei

Problem Strontium-90 is a radioactive by-product of nuclear reactors that behaves biologically like calcium, the element above it in Group 2A(2). When ^{90}Sr is ingested by mammals, it is found in their milk and eventually in the bones of those drinking the milk. If a sample of ^{90}Sr has an activity of 1.2×10^{12} d/s, what are the activity and the fraction of nuclei that have decayed after 59 yr ($t_{1/2}$ of $^{90}Sr = 29$ yr)?

Plan The fraction of nuclei that have decayed is the change in number of nuclei, expressed as a fraction of the starting number. The activity of the sample ($\mathcal{A}$) is proportional to the number of nuclei ($\mathcal{N}$), so we know that

$$\text{Fraction decayed} = \frac{\mathcal{N}_0 - \mathcal{N}_t}{\mathcal{N}_0} = \frac{\mathcal{A}_0 - \mathcal{A}_t}{\mathcal{A}_0}$$

We are given $\mathcal{A}_0$ (1.2×10^{12} d/s), so we find $\mathcal{A}_t$ from the integrated form of the first-order rate equation (Equation 24.4), in which t is 59 yr. To solve that equation, we first need k, which we can calculate from the given $t_{1/2}$ (29 yr).

Solution Calculating the decay constant k:

$$t_{1/2} = \frac{\ln 2}{k} \quad \text{so} \quad k = \frac{\ln 2}{t_{1/2}} = \frac{0.693}{29 \text{ yr}} = 0.024 \text{ yr}^{-1}$$

Applying Equation 24.4 to calculate $\mathcal{A}_t$, the activity remaining at time t:

$$\ln \frac{\mathcal{N}_0}{\mathcal{N}_t} = \ln \frac{\mathcal{A}_0}{\mathcal{A}_t} = kt \quad \text{or} \quad \ln \mathcal{A}_0 - \ln \mathcal{A}_t = kt$$

So,
$$\ln \mathcal{A}_t = -kt + \ln \mathcal{A}_0 = -(0.024 \text{ yr}^{-1} \times 59 \text{ yr}) + \ln (1.2 \times 10^{12} \text{ d/s})$$
$$\ln \mathcal{A}_t = -1.4 + 27.81 = 26.4$$
$$\boxed{\mathcal{A}_t = 2.9 \times 10^{11} \text{ d/s}}$$

(All the data contain two significant figures, so we retained two in the answer.) Calculating the fraction decayed:

$$\text{Fraction decayed} = \frac{\mathcal{A}_0 - \mathcal{A}_t}{\mathcal{A}_0} = \frac{1.2 \times 10^{12} \text{ d/s} - 2.9 \times 10^{11} \text{ d/s}}{1.2 \times 10^{12} \text{ d/s}} = \boxed{0.76}$$

Check The answer is reasonable: t is about 2 half-lives, so $\mathcal{A}_t$ should be about $\frac{1}{4}\mathcal{A}_0$, or about 0.3×10^{12}; therefore, the activity should have decreased by about $\frac{3}{4}$.

Comment An *alternative approach* is to use the number of half-lives ($t/t_{1/2}$) to find the fraction of activity (or nuclei) remaining. By combining Equations 24.4 and 24.5 and substituting ($\ln 2)/t_{1/2}$ for k, we obtain

$$\ln \frac{\mathcal{N}_0}{\mathcal{N}_t} = \left(\frac{\ln 2}{t_{1/2}}\right) t = \frac{t}{t_{1/2}} \ln 2 = \ln 2^{t/t_{1/2}}$$

Thus,
$$\ln \frac{\mathcal{N}_t}{\mathcal{N}_0} = \ln \left(\frac{1}{2}\right)^{t/t_{1/2}}$$

Taking the antilog gives

$$\text{Fraction remaining} = \frac{\mathcal{N}_t}{\mathcal{N}_0} = \left(\frac{1}{2}\right)^{t/t_{1/2}} = \left(\frac{1}{2}\right)^{59/29} = 0.24$$

So,
$$\text{Fraction decayed} = 1.00 - 0.24 = 0.76$$

FOLLOW-UP PROBLEM 24.4 Sodium-24 has a half-life of 15 h and is used to study blood circulation. If a patient is injected with a ^{24}NaCl solution whose activity is 2.5×10^9 d/s, how much of the activity is present in the patient's body and excreted fluids after 4.0 days?

Radioisotopic Dating

The historical record fades rapidly with time and virtually disappears for events of more than a few thousand years ago. Much of our understanding of prehistory comes from a technique called **radioisotopic dating,** which uses **radioisotopes** to determine the age of an object. The method supplies data about the ages of objects in fields as diverse as art history, archeology, geology, and paleontology.

The technique of *radiocarbon dating,* for which the American chemist Willard F. Libby won the Nobel Prize in 1960, is based on measuring the amounts of ^{14}C and ^{12}C in materials of biological origin. The accuracy of the method falls off after about six half-lives of ^{14}C ($t_{1/2} = 5730$ yr), so it is used to date objects up to about 36,000 years old.

Here is how the method works. High-energy neutrons resulting from cosmic ray collisions reach Earth continually from outer space. They enter the atmosphere and cause the slow formation of ^{14}C by bombarding ordinary ^{14}N atoms:

$$^{14}_{7}\text{N} + ^{1}_{0}\text{n} \longrightarrow ^{14}_{6}\text{C} + ^{1}_{1}\text{p}$$

Through the processes of formation and radioactive decay, the amount of ^{14}C has remained nearly constant.*

HALF-LIFE

---

*Cosmic ray intensity does vary slightly with time, which affects the amount of atmospheric ^{14}C. From ^{14}C activity in ancient trees, we know the amount fell slightly about 3000 years ago to current levels. Recently, nuclear testing and fossil fuel combustion have also altered the fraction of ^{14}C slightly. Taking these factors into account improves the accuracy of the dating method.

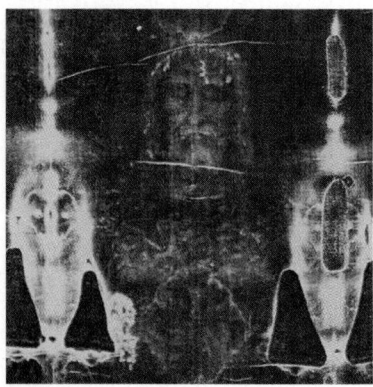

The Case of the Shroud of Turin
One of the holiest relics of the Christian world is the famed Shroud of Turin. It is a piece of linen that bears a faint image of a man's body and was thought to be the burial cloth used to wrap the body of Jesus Christ. In 1988, after much controversy, the Vatican allowed scientific testing of the age of the cloth by radiocarbon dating. Three laboratories in Europe and the United States independently measured the $^{12}C:^{14}C$ ratio of a 50-mg piece of the linen and determined that the flax from which the cloth was made was grown between 1260 AD and 1390 AD. Despite this evidence, the shroud lost none of its fascination: 10 years later, in 1998, when the shroud was again put on display, about 2 million people lined up to view it.

The ^{14}C atoms combine with O_2, diffuse throughout the lower atmosphere, and enter the total carbon pool as gaseous $^{14}CO_2$ and aqueous $H^{14}CO_3^-$. They mix with ordinary $^{12}CO_2$ and $H^{12}CO_3^-$, reaching a constant $^{12}C:^{14}C$ ratio of about $10^{12}:1$. The CO_2 is taken up by plants during photosynthesis, and then taken up and excreted by animals that eat the plants. Thus, the $^{12}C:^{14}C$ ratio of a living organism has the same constant value as the environment. When an organism dies, however, it no longer takes in ^{14}C, so the $^{12}C:^{14}C$ ratio steadily increases because the amount of ^{14}C *decreases as it decays*:

$$^{14}_{6}C \longrightarrow {}^{14}_{7}N + {}^{0}_{-1}\beta$$

The difference between the $^{12}C:^{14}C$ ratio in a dead organism and the ratio in living organisms reflects the time elapsed since the organism died.

As you saw in Sample Problem 24.4, the first-order rate equation can be expressed in terms of a ratio of activities:

$$\ln \frac{\mathcal{N}_0}{\mathcal{N}_t} = \ln \frac{\mathcal{A}_0}{\mathcal{A}_t} = kt$$

We use this expression in radiocarbon dating, where $\mathcal{A}_0$ is the activity in a living organism and $\mathcal{A}_t$ is the activity in the object whose age is unknown. Solving for t gives the age of the object:

$$t = \frac{1}{k} \ln \frac{\mathcal{A}_0}{\mathcal{A}_t} \qquad \textbf{(24.6)}$$

A useful graphical method in radioisotopic dating shows a plot of the natural logarithm of the specific activity vs. time, which gives a straight line with a slope of $-k$, the negative of the decay constant. Using such a plot and measuring the ^{14}C specific activity of an object, we can determine its age; several examples appear in Figure 24.5. To determine the ages of more ancient objects or of objects that do not contain carbon, different radioisotopes are measured. (See the margin note on the opposite page.)

Figure 24.5 Radiocarbon dating for determining the age of artifacts. The natural logarithms of the specific activity of ^{14}C (activity/g ^{14}C) of various artifacts are projected onto a line whose slope equals $-k$, the negative of the ^{14}C decay constant. The age (in years) of an artifact is determined from the horizontal axis.

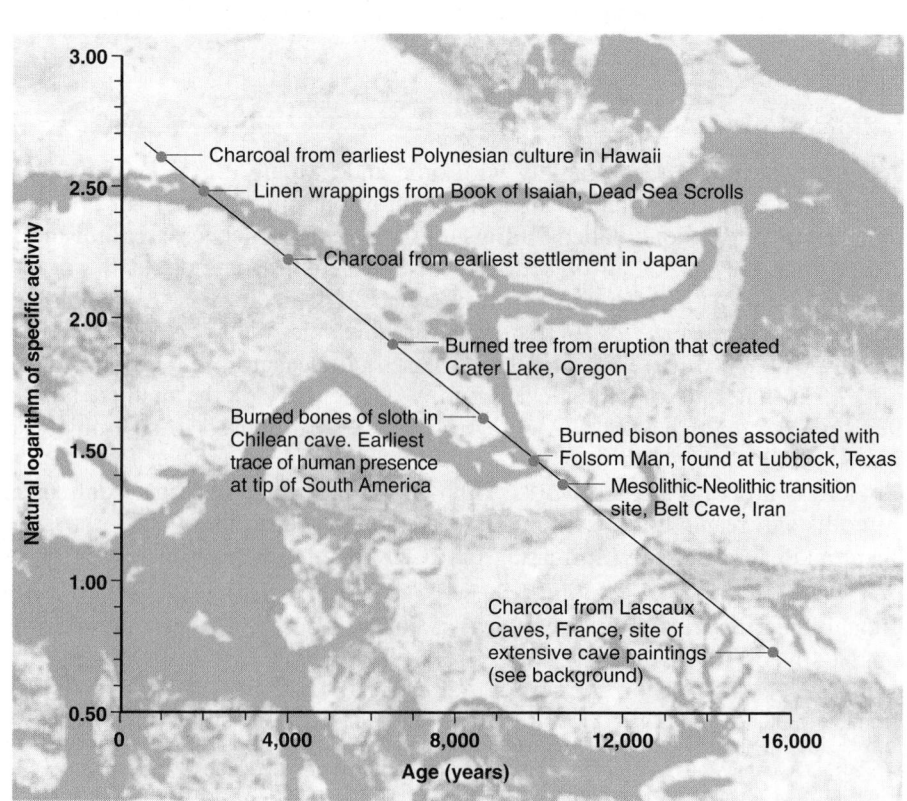

SAMPLE PROBLEM 24.5 Applying Radiocarbon Dating

Problem The charred bones of a sloth in a cave in Chile represent the earliest evidence of human presence in the southern tip of South America. A sample of the bone has a specific activity of 5.22 disintegrations per minute per gram of carbon (d/min·g). If the ratio of $^{12}C:^{14}C$ in living organisms results in a specific activity of 15.3 d/min·g, how old are the bones ($t_{1/2}$ of ^{14}C = 5730 yr)?

Plan We first calculate k from the given $t_{1/2}$ (5730 yr). Then we apply Equation 24.6 to find the age (t) of the bones, using the given activities of the bones ($\mathscr{A}_t$ = 5.22 d/min·g) and of a living organism ($\mathscr{A}_0$ = 15.3 d/min·g).

Solution Calculating k for ^{14}C decay:

$$k = \frac{\ln 2}{t_{1/2}} = \frac{0.693}{5730 \text{ yr}} = 1.21 \times 10^{-4} \text{ yr}^{-1}$$

Calculating the age (t) of the bones:

$$t = \frac{1}{k} \ln \frac{\mathscr{A}_0}{\mathscr{A}_t} = \frac{1}{1.21 \times 10^{-4} \text{ yr}^{-1}} \ln \left(\frac{15.3 \text{ d/min·g}}{5.22 \text{ d/min·g}} \right) = 8.89 \times 10^3 \text{ yr}$$

The bones are about $\boxed{8900 \text{ yr old.}}$

Check The activity of the bones is between $\frac{1}{2}$ and $\frac{1}{4}$ the modern activity, so the age should be between one and two half-lives (5730 to 11,460 yr).

FOLLOW-UP PROBLEM 24.5 A sample of wood from an Egyptian mummy case has a specific activity of 9.41 d/min·g. How old is the case?

SECTION SUMMARY

Ionization and scintillation counters measure the number of emissions from a radioactive sample. The decay rate (activity) of a sample is proportional to the number of radioactive nuclei. Nuclear decay is a first-order process, so the half-life does not depend on the number of nuclei. Radioisotopic methods, such as ^{14}C dating, determine the ages of objects by measuring the ratio of specific isotopes in the sample.

24.3 NUCLEAR TRANSMUTATION: INDUCED CHANGES IN NUCLEI

The alchemists' dream of changing base metals into gold was never realized, but in the early 20th century, atomic physicists found that they *could* change one element into another. Research into **nuclear transmutation,** the *induced* conversion of one nucleus into another, was closely linked with research into atomic structure and led to the discovery of the neutron and to the production of artificial radioisotopes. Later, high-energy bombardment of nuclei in particle accelerators began a scientific endeavor, which continues to this day, of creating many new nuclides and a growing number of new elements.

Early Transmutation Experiments; Discovery of the Neutron

The first recognized transmutation occurred in 1919, when Ernest Rutherford showed that α particles emitted from radium bombarded atmospheric nitrogen to form a proton and oxygen-17:

$$^{14}_{7}N + ^{4}_{2}He \longrightarrow ^{1}_{1}H + ^{17}_{8}O$$

By 1926, experimenters had found that α bombardment transmuted most elements with low atomic numbers to the next higher element, with ejection of a proton.

A shorthand notation commonly used for nuclear bombardment reactions shows the reactant (target) nucleus and product nucleus separated by parentheses, within which a comma separates the projectile particle from the ejected particle(s):

Reactant nucleus (particle in, particle(s) out) product nucleus

Using this notation, the previous reaction is $^{14}N (\alpha,p) ^{17}O$.

How Old Is the Solar System? By comparing the ratio of ^{238}U to its final decay product, ^{206}Pb, geochemists have found that the oldest known surface rocks on Earth—granite in western Greenland—are about 3.7 billion years old. The ratio of $^{238}U:^{206}Pb$ in meteorites gives 4.65 billion years for the age of the Solar System, and thus the Earth. From this and other isotope ratios, such as $^{40}K:^{40}Ar$ ($t_{1/2}$ of ^{40}K = 1.3×10^9 yr) as well as $^{87}Rb:^{87}Sr$ ($t_{1/2}$ of ^{87}Rb = 4.9×10^{10} yr), Moon rocks collected by Apollo astronauts have been shown to be 4.2 billion years old, and they provide evidence for volcanic activity on the Moon's surface about 3.3 billion years ago. That was about the time that, according to these methods, the first organisms were evolving on Earth.

The Joliot-Curies in their laboratory.

An unexpected finding in a transmutation experiment led to the discovery of the neutron. When lithium, beryllium, and boron were bombarded with α particles, they emitted highly penetrating radiation that could not be deflected by a magnetic or electric field. Unlike γ radiation, these emissions were massive enough to eject protons from the substances they penetrated. In 1932, James Chadwick, a student of Rutherford, proposed that these emissions consisted of neutral particles with a mass similar to that of a proton, and he named them *neutrons*. Chadwick received the Nobel Prize in 1935 for his discovery.

In 1933, Irene and Frederic Joliot-Curie (see photo), daughter and son-in-law of Marie and Pierre Curie, created the first artificial radioisotope, phosphorus-30. When they bombarded aluminum foil with α particles, phosphorus-30 and neutrons were formed:

$$^{27}_{13}\text{Al} + {}^{4}_{2}\text{He} \longrightarrow {}^{1}_{0}\text{n} + {}^{30}_{15}\text{P} \qquad \text{or} \qquad {}^{27}\text{Al}\ (\alpha,\text{n})\ {}^{30}\text{P}$$

Since then, other techniques for producing artificial radioisotopes have been developed. In fact, the majority of the nearly 1000 known radionuclides have been produced artificially.

Particle Accelerators and the Transuranium Elements

During the 1930s and 1940s, researchers probing the nucleus bombarded elements with neutrons, α particles, protons, and **deuterons** (nuclei of the stable hydrogen isotope deuterium, ^{2}H). Neutrons are especially useful as projectiles because they have no charge, so they are not repelled as they approach a target nucleus. The other particles are all positive, so early researchers found it difficult to give them enough energy to overcome their repulsion by the target nuclei. Beginning in the 1930s, however, **particle accelerators** were invented to impart high kinetic energies to particles by placing them in an electric field, usually in combination with a magnetic field. In the simplest and earliest design, protons are introduced at one end of a tube and attracted to the other end by a potential difference.

A major advance occurred with the invention of the *linear accelerator,* a series of separated tubes of increasing length that change their charge from positive to negative in synchrony with the movement of the particle through them (Figure 24.6A). A proton, for example, exits the first tube just when that tube becomes positive and the next tube negative. Repelled by the first tube and attracted by the second, the proton accelerates across the gap between them. A 40-ft linear accelerator with 46 tubes, built in California after World War II, accelerated protons to speeds several million times faster than those early accelerators. Later designs, such as the Stanford Linear Accelerator (Figure 24.6B), accelerate

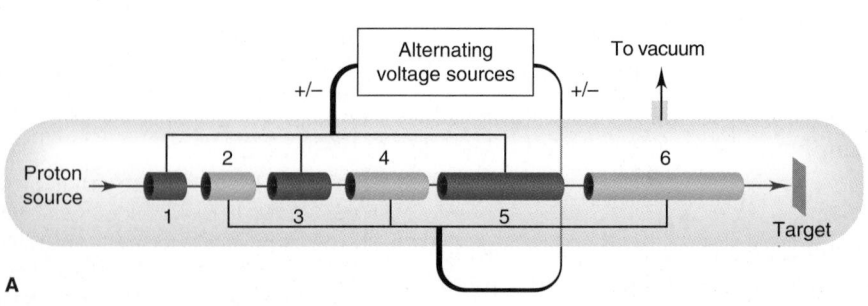

Figure 24.6 A linear accelerator. A, The voltage of each tubular section is alternated, such that the positively charged particle (a proton here) is repelled from the section it is leaving and attracted to the section it is entering. As a result, the particle speed is continually increased. **B,** The linear accelerator operated by Stanford University in California.

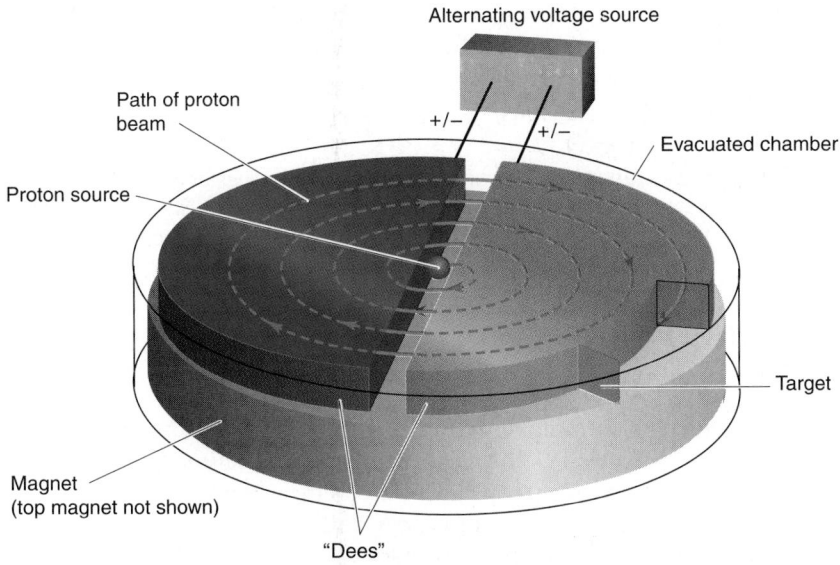

Figure 24.7 The cyclotron accelerator. When the positively charged particle reaches the gap between the two D-shaped electrodes ("dees"), it is repelled by one dee and attracted by the other. The particles move in a spiral path, so the cyclotron can be much smaller than a linear accelerator.

heavier particles, such as B, C, O, and Ne nuclei, several hundred million times faster, with correspondingly greater kinetic energies.

The *cyclotron* (Figure 24.7), invented by E. O. Lawrence in 1930, applies the principle of the linear accelerator but uses electromagnets to give the particle a spiral path, thus saving space. The magnets lie within an evacuated chamber above and below two "dees," open, D-shaped electrodes that function like the tubes in the linear design. The particle is accelerated as it passes from one dee, which is momentarily positive, to the other, which is momentarily negative. Its speed and radius increase until it is deflected toward the target nucleus. The *synchrotron* uses a synchronously increasing magnetic field to make the particle's path circular rather than spiral.

Accelerators have many applications, from producing radioisotopes used in medical applications to studying the fundamental nature of matter. Perhaps their most specific application for chemists is the synthesis of **transuranium elements,** those with atomic numbers higher than uranium, which is the heaviest naturally occurring element. Some reactions that were used to form several of these elements appear in Table 24.6. The transuranium elements include the remaining actinides ($Z = 93$ to 103), in which the $5f$ sublevel is being filled, and the elements in the fourth transition series ($Z = 104$ to 112), in which the $6d$ sublevel

The **Powerful Bevatron** The *bevatron* includes a linear section and a synchrotron section. The instrument at the Lawrence Berkeley Laboratory in California increases the kinetic energy of the particles by a factor of more than 6 billion. A beam of 10^{10} protons makes more than 4 million revolutions, a distance of 300,000 miles, in 1.8 s, attaining a final speed about 90% the speed of light! Even more powerful bevatrons are in use at the Brookhaven National Laboratory in New York and at CERN, outside Geneva, Switzerland. At one of the CERN facilities (see photo), separate bending and focusing magnets accelerate protons along a 4.3-mile circular tunnel.

Table 24.6 Formation of Some Transuranium Nuclides

Reaction						Half-life of Product
$^{239}_{94}\text{Pu}$	$+$	$^{4}_{2}\text{He}$	$\longrightarrow$	$^{240}_{95}\text{Am}$	$+$	$^{1}_{1}\text{H} + 2^{1}_{0}\text{n}$ 50.9 h
$^{239}_{94}\text{Pu}$	$+$	$^{4}_{2}\text{He}$	$\longrightarrow$	$^{242}_{96}\text{Cm}$	$+$	$^{1}_{0}\text{n}$ 163 days
$^{244}_{96}\text{Cm}$	$+$	$^{4}_{2}\text{He}$	$\longrightarrow$	$^{245}_{97}\text{Bk}$	$+$	$^{1}_{1}\text{H} + 2^{1}_{0}\text{n}$ 4.94 days
$^{238}_{92}\text{U}$	$+$	$^{12}_{6}\text{C}$	$\longrightarrow$	$^{246}_{98}\text{Cf}$	$+$	4^{1}_{0}n 36 h
$^{253}_{99}\text{Es}$	$+$	$^{4}_{2}\text{He}$	$\longrightarrow$	$^{256}_{101}\text{Md}$	$+$	$^{1}_{0}\text{n}$ 76 min
$^{252}_{98}\text{Cf}$	$+$	$^{10}_{5}\text{B}$	$\longrightarrow$	$^{256}_{103}\text{Lr}$	$+$	6^{1}_{0}n 28 s

Naming Transuranium Elements

The last naturally occurring element was named after Uranus, thought at the time to be the outermost planet, so the first two artificial elements were named after the more recently discovered Neptune and Pluto. The next few elements were named after famous scientists, as in curium, and places, as in americium. However, conflicting claims of discovery by scientists in different countries led to controversies about names for elements 104 and higher. To provide interim names until the disputes could be settled, the International Union of Pure and Applied Chemistry (IUPAC) adopted a system that uses the atomic number as the basis for a Latin name. Thus, for example, element 104 was named unnilquadium (un = 1, nil = 0, quad = 4, ium = element suffix), with the symbol Unq. After much compromise, the IUPAC has finalized these names: 104, rutherfordium (Rf); 105, dubnium (Db); 106, seaborgium (Sg); 107, bohrium (Bh); 108, hassium (Hs); 109, meitnerium (Mt). Elements with atomic numbers 110 and higher have not yet been named.

is being filled. (In 1999, one research group reported the synthesis of elements 114, 116, and 118, the second, fourth, and sixth members of the $7p$ sublevel. In 2001, the data regarding elements 116 and 118 were retracted.*)

SECTION SUMMARY
One nucleus can be transmuted to another through bombardment with high-energy particles. Accelerators increase the kinetic energy of particles in nuclear bombardment experiments and are used to produce transuranium elements.

24.4 THE EFFECTS OF NUCLEAR RADIATION ON MATTER

In 1986, an accident at the Chernobyl nuclear facility in the former Soviet Union released radioactivity that is estimated to have already caused thousands of cancer deaths. In the same year, isotopes used in medical treatment emitted radioactivity that has prevented thousands of cancer deaths. In this section and the next, we examine the harm and benefit of radioactivity.

The key to both of these outcomes is that *nuclear changes cause chemical changes in surrounding matter.* In other words, even though the nucleus of an atom undergoes a reaction with little or no involvement of the atom's electrons, the emissions *do* affect the electrons of nearby atoms.

The Effects of Radioactive Emissions: Excitation and Ionization
Radioactive emissions interact with matter in two ways, depending on their energies:

- *Excitation.* In the process of **excitation,** a particle of relatively low energy collides with an atom of a substance, which absorbs some of the energy and then re-emits it. Because electrons are not lost from the atom, the radiation that causes excitation is called **nonionizing radiation.** If the absorbed energy causes the atoms to move, vibrate, or rotate more rapidly, the material becomes hotter. Concentrated aqueous solutions of plutonium salts boil because the emissions excite the surrounding water molecules. Polonium has been suggested as a lightweight heat source, with no moving parts, for use on space stations. Particles of somewhat higher energy excite electrons in other atoms to higher energy levels. As the atoms return to their ground state, they emit photons, often in the blue or ultraviolet region (see scintillation counters in the Tools of the Laboratory essay, p. 1051).
- *Ionization.* In the process of **ionization,** radiation collides with an atom energetically enough to dislodge an electron:

$$\text{Atom} \xrightarrow{\text{ionizing radiation}} \text{ion}^+ + \text{e}^-$$

A cation and free electron result, and the number of such *cation-electron pairs* that are produced is directly related to the energy of the incoming radiation. The high-energy radiation that gives rise to this effect is called **ionizing radiation.** The free electron of the pair often collides with another atom and ejects a second electron (see Geiger-Müller counters in the Tools essay, p. 1051).

Effects of Ionizing Radiation on Living Matter

Whereas nonionizing radiation is relatively harmless, ionizing radiation has a destructive effect on living tissue. When the atom that was ionized is part of a biological macromolecule or membrane component, the results can be devastating.

---

*Very recently (unfortunately, too late for inclusion in periodic tables of this text), another group, using different reactant nuclides, synthesized and confirmed the presence of element 116.

Units of Radiation Dose and Its Effects To measure the effects of ionizing radiation, we need a unit for radiation dose. Units of radioactive decay, such as the becquerel and curie, state the number of decay events in a given time but not their energy or absorption by matter. The number of cation-electron pairs produced in a given amount of living tissue is a measure of the energy absorbed by the tissue. The SI unit for such energy absorption is the **gray (Gy)**; it is equal to 1 joule of energy absorbed per kilogram of body tissue: 1 Gy = 1 J/kg. A more widely used unit is the **rad (*r*adiation-*a*bsorbed *d*ose),** which is equal to 0.01 Gy:

$$1 \text{ rad} = 0.01 \text{ J/kg} = 0.01 \text{ Gy}$$

To measure actual tissue damage, we must account for differences in the strength of the radiation, the exposure time, and the type of tissue. To do this, we multiply the number of rads by a *relative biological effectiveness* (RBE) factor, which depends on the effect of a given type of radiation on a given tissue or body part. The product is the **rem (*r*oentgen *e*quivalent for *m*an),** the unit of radiation dosage equivalent to a given amount of tissue damage in a human:

$$\text{no. of rems} = \text{no. of rads} \times \text{RBE}$$

Doses are often expressed in millirems (10^{-3} rem). The SI unit for dosage equivalent is the **sievert (Sv).** It is defined in the same way as the rem but with absorbed dose in grays; thus, 1 rem = 0.01 Sv.

Penetrating Power of Emissions The effect on living tissue of a radiation dose depends on the penetrating power *and* ionizing ability of the radiation. Figure 24.8 depicts the differences in penetrating power of the three common emissions. Note, in general, that *penetrating power is inversely related to the mass and charge of the emission*. In other words, if a particle interacts strongly with matter, it penetrates only slightly, and vice versa:

- *α Particles*. Alpha particles are massive and highly charged, which means that they interact with matter most strongly of the three common types of emissions. As a result, they penetrate so little that a piece of paper, light clothing, or the outer layer of skin can stop α radiation from an external source. However, if ingested, an α emitter such as plutonium-239 causes grave localized damage through extensive ionization.
- *β Particles and positrons*. Beta particles and positrons have less charge and much less mass than α particles, so they interact less strongly with matter. Even though a given particle has less chance of causing ionization, a β (or positron) emitter is a more destructive external source because the particles penetrate deeper. Specialized heavy clothing or a thick (0.5 cm) piece of metal is required to stop these particles.
- *γ Rays*. Neutral, massless γ rays interact least with matter and, thus, penetrate most. A block of lead several inches thick is needed to stop them. Therefore, an external γ ray source is the most dangerous because the energy can ionize many layers of living tissue.

Molecular Interactions How does the damage take place on the molecular level? When ionizing radiation interacts with a molecule, it causes the *loss of an electron from a bond or lone pair*. The resulting charged species go on to form **free radicals,** molecular or atomic species with one or more unpaired (lone) electrons. As we've seen several times already, species with lone electrons are very reactive and tend to form electron pairs by bonding to other species. To do this, they attack bonds in other molecules, sometimes forming more free radicals.

When γ radiation strikes biological tissue, for instance, the most likely molecule to absorb it is water, which forms an electron and a water ion-radical:

$$H_2O + \gamma \longrightarrow H_2O\cdot^+ + e^-$$

α (~0.03 mm)

β (~2 mm)

γ (~10 cm)

Figure 24.8 Penetrating power of radioactive emissions. Penetrating power is often measured in terms of the depth of water that stops 50% of the incoming radiation. Water is used because it is the main component of living tissue. Alpha particles, with the highest mass and charge, have the lowest penetrating power and γ rays the highest. (Average values of actual penetrating distances are shown.)

A Tragic Way to Tell Time in the Dark In the beginning of the 20th century, wristwatch and clock dials were painted by hand with paint containing radium so they would "glow" in the dark. To write numbers clearly, the young women hired to apply the paint "tipped" fine brushes repeatedly between their lips. Small amounts of ingested $^{226}Ra^{2+}$ were incorporated into the bones of these women, along with normal Ca^{2+}, which led to numerous cases of bone fracture and jaw cancer.

The $H_2O\cdot^+$ and e^- collide with other water molecules to form free radicals:

$$H_2O\cdot^+ + H_2O \longrightarrow H_3O^+ + \cdot OH \qquad \text{and} \qquad e^- + H_2O \longrightarrow H\cdot + OH^-$$

These free radicals go on to attack more water molecules and surrounding bio-molecules, whose bonding and structure, as you know (Section 15.6), are intimately connected with their function.

The double bonds in membrane lipids are particularly susceptible to free-radical attack:

$$H\cdot + RCH{=}CHR' \longrightarrow RCH_2{-}\dot{C}HR'$$

In this reaction, one electron of the π bond forms a C—H bond between one of the doubly bonded carbons and the H·, and the other electron resides on the other carbon to form a free radical. Changes to lipid structure cause changes in membrane fluidity and other damage that, in turn, cause leakage of the cell and destruction of the protective fatty tissue around organs. Changes to critical bonds in enzymes lead to their malfunction as catalysts of metabolic reactions. Changes in the nucleic acids and proteins that govern the rate of cell division cause cancer. Genetic damage and mutations may occur when bonds in the DNA of sperm and egg cells are altered by free radicals.

Sources of Ionizing Radiation It is essential to keep the molecular effects of ionizing radiation in perspective. After all, we are continuously exposed to ionizing radiation from natural and artificial sources (Table 24.7). Indeed, life evolved in

Table 24.7 Examples of Typical Radiation Doses from Natural and Artificial Sources

Source of Radiation	Average Adult Exposure
Natural	
Cosmic radiation	30–50 mrem/yr
Radiation from the ground	
From clay soil and rocks	~25–170 mrem/yr
In wooden houses	10–20 mrem/yr
In brick houses	60–70 mrem/yr
In concrete (cinder block) houses	60–160 mrem/yr
Radiation from the air (mainly radon)	
Outdoors, average value	20 mrem/yr
In wooden houses	70 mrem/yr
In brick houses	130 mrem/yr
In concrete (cinder block) houses	260 mrem/yr
Internal radiation from minerals in tap water and daily intake of food (^{40}K, ^{14}C, Ra)	~40 mrem/yr
Artificial	
Diagnostic x-ray methods	
Lung (local)	0.04–0.2 rad/film
Kidney (local)	1.5–3 rad/film
Dental (dose to the skin)	≤ 1 rad/film
Therapeutic radiation treatment	Locally $\leq 10,000$ rad
Other sources	
Jet flight (4 h)	~1 mrem
Nuclear testing	<4 mrem/yr
Nuclear power industry	<1 mrem/yr
TOTAL AVERAGE VALUE	100–200 mrem/yr

the presence of natural ionizing radiation, called **background radiation.** The same radiation that causes harmful mutations also causes beneficial mutations that, over time, allow organisms to adapt and species to change.

Background radiation has several sources. One source is *cosmic radiation,* which increases with altitude because of decreased absorption by the atmosphere. Thus, people in Denver absorb twice as much cosmic radiation as people in Los Angeles; even a jet flight involves measurable absorption. The sources of most background radiation are thorium and uranium minerals present in rocks and soil. Radon, the heaviest noble gas [Group 8A(18)], is a radioactive product of uranium and thorium decay, and its concentration in the air we breathe varies with type of local soil and rocks. About 150 g of K^+ ions is dissolved in the water in the tissues of an average adult, and 0.0118% of that amount is radioactive ^{40}K. The presence of these substances and of atmospheric $^{14}CO_2$ means that all food, water, clothing, and building materials are slightly radioactive.

The largest artificial source of radiation, and the easiest to control, is associated with medical diagnostic techniques, especially x-rays. The radiation dosage from nuclear testing and radioactive waste disposal is miniscule for most people, but exposures for those living near test sites, nuclear energy facilities, or disposal areas may be many times higher.

Assessing the Risk from Ionizing Radiation How much radiation is too much? To approach this question, we must ask several others: How strong is the exposure? How long is the exposure? Which tissue is exposed? Are offspring affected? One reason we lack clear data to answer these questions is that scientific ethical standards forbid the intentional exposure of humans in an experimental setting. However, accidentally exposed radiation workers and Japanese atomic bomb survivors have been studied extensively. Table 24.8 summarizes the immediate tissue effects on humans of an acute single dose of ionizing radiation to the whole body. The severity of the effects increases with dose; a dose of 500 rem will kill about 50% of the exposed population within a month.

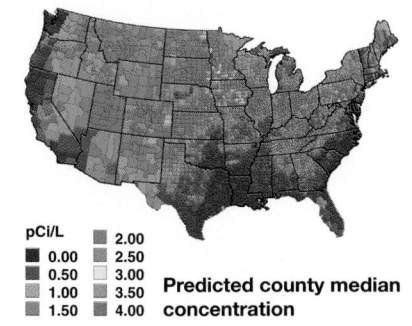

pCi/L
0.00   2.00
0.50   2.50
1.00   3.00
1.50   3.50
       4.00

Predicted county median concentration

The Risk of Radon Radon (Rn; Z = 86), the largest noble gas, is a natural decay product of uranium. Therefore, the uranium content of the local soil and rocks is a critical factor in the extent of the threat, but radon occurs everywhere in varying concentrations. Radon itself decays to radioactive nuclides of Po, Pb, and Bi, through α, β, and γ emission. These processes occur inside the body when radon is inhaled and pose a serious potential hazard. The emissions damage lung tissue, and the heavy-metal atoms formed aggravate the problem. The latest EPA estimates indicate that radon contributes to 15% of annual lung cancer deaths.

Table 24.8 Acute Effects of a Single Dose of Whole-Body Irradiation

Dose (rem)	Effect	Lethal Dose Population (%)	No. of Days
5–20	Possible late effect; possible chromosomal aberrations	—	—
20–100	Temporary reduction in white blood cells	—	—
50+	Temporary sterility in men (100+ rem = 1 yr duration)	—	—
100–200	"Mild radiation sickness": vomiting, diarrhea, tiredness in a few hours Reduction in infection resistance Possible bone growth retardation in children	—	—
300+	Permanent sterility in women	—	—
500	"Serious radiation sickness": marrow/intestine destruction	50–70	30
400–1000	Acute illness, early deaths	60–95	30
3000+	Acute illness, death in hours to days	100	2

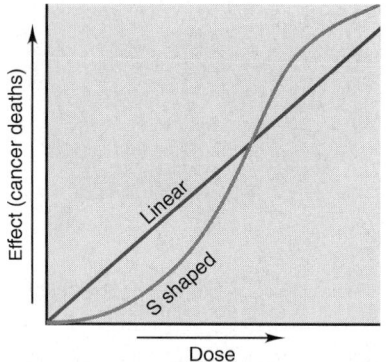

Modeling Radiation Risk There are two current models of effect vs. dose. The *linear response* model proposes that radiation effects, such as cancer risks, accumulate over time regardless of dose and that populations should not be exposed to any radiation above background levels. The *S-shaped response* model assumes an extremely low risk at low doses and advocates concern only at higher doses. If the linear model is more accurate, we should limit all excess exposure, but this would severely restrict medical diagnosis and research, military testing, and nuclear energy production.

Most data come from laboratory animals, whose biological systems may differ greatly from ours. Nevertheless, studies with mice and dogs show that lesions and cancers appear after massive whole-body exposure, with rapidly dividing cells affected first. In an adult animal, these are the cells of the bone marrow, organ linings, and reproductive organs, but many other tissues are affected in an immature animal or fetus. Both animal and human studies show an increase in the incidence of cancer from either a high, single exposure or a low, chronic exposure.

Reliable data on genetic effects are few. Pioneering studies on fruit flies show a linear increase in genetic defects with both dose and exposure time. However, in the mouse, whose genetic system is obviously much more similar to ours than is the fruit fly's, a total dose given over a long period created one-third as many genetic defects as the same dose given over a short period. Therefore, rate of exposure is a key factor. The children of atomic bomb survivors show higher-than-normal childhood cancer rates, implying that their parents' reproductive systems were affected.

SECTION SUMMARY

Relatively low-energy emissions cause excitation of atoms in surrounding matter, whereas high-energy emissions cause ionization. The effect of ionizing radiation on living matter depends on the quantity of energy absorbed and the extent of ionization in a given type of tissue. Radiation dose for the human body is measured in rem. Ionization forms free radicals, some of which proliferate and destroy biomolecular function. All life is exposed to varying quantities of natural ionizing radiation. Studies show that a large acute dose and a chronic small dose are both harmful.

24.5 APPLICATIONS OF RADIOISOTOPES

Our ability to detect minute amounts of radioisotopes makes them powerful tools for studying processes in biochemistry, medicine, materials science, environmental studies, and many other scientific and industrial fields. Such uses depend on the fact that *isotopes of an element exhibit **very** similar chemical and physical behavior.* In other words, except for having a less stable nucleus, a radioisotope has nearly the same chemical properties as a nonradioactive isotope of that element.* For example, the fact that $^{14}CO_2$ is utilized by a plant in the same way as $^{12}CO_2$ forms the basis of radiocarbon dating.

Radioactive Tracers: Applications of Nonionizing Radiation

Just think how useful it could be to follow a substance through the stages of a complex process or from one region of a system to another. A tiny amount of a radioisotope mixed with a large amount of the stable isotope can act as a **tracer,** a chemical "beacon" emitting nonionizing radiation that signals the presence of the substance.

Reaction Pathways Tracers help us choose from among possible reaction pathways. Consider the reaction between periodate and iodide ions:

$$IO_4^-(aq) + 2I^-(aq) + H_2O(l) \longrightarrow I_2(s) + IO_3^-(aq) + 2OH^-(aq)$$

---

*Although this statement is generally correct, differences in isotopic mass *can* influence bond strengths and therefore reaction rates. Such behavior is called a *kinetic isotope effect* and is particularly important for isotopes of hydrogen—^1H, ^2H, and ^3H—because their masses differ by such large proportions. Section 22.4 discusses how the kinetic isotope effect is employed in the industrial production of heavy water, D_2O.

Is IO_3^- the result of IO_4^- reduction or I^- oxidation? When we add "cold" (non-radioactive) IO_4^- to a solution of I^- that contains some "hot" (radioactive) $^{131}I^-$, we find that the I_2 is radioactive, not the IO_3^-:

$$IO_4^-(aq) + 2\,^{131}I^-(aq) + H_2O(l) \longrightarrow\,^{131}I_2(s) + IO_3^-(aq) + 2OH^-(aq)$$

These results show that IO_3^- forms through the reduction of IO_4^-, and that I_2 forms through the oxidation of I^-. To confirm this pathway, we add IO_4^- containing some hot $^{131}IO_4^-$ to a solution of cold I^-. As we expected, the IO_3^- is radioactive, not the I_2:

$$^{131}IO_4^-(aq) + 2I^-(aq) + H_2O(l) \longrightarrow I_2(s) +\,^{131}IO_3^-(aq) + 2OH^-(aq)$$

The tracer acts like a "handle" we can "hold" to follow the changing reactants. Far more complex pathways can be followed with tracers as well. The photosynthetic pathway, the most essential and widespread metabolic process on Earth, in which energy from sunlight is used to form the chemical bonds of glucose, has an overall reaction that looks quite simple:

$$6CO_2(g) + 6H_2O(l) \xrightarrow[\text{chlorophyll}]{\text{light}} C_6H_{12}O_6(s) + 6O_2(g)$$

However, the actual process is extremely complex, requiring 13 enzyme-catalyzed steps for each molecule of CO_2 incorporated, which occurs six times through the pathway for each molecule of $C_6H_{12}O_6$ that forms. Melvin Calvin and his co-workers took seven years to determine the pathway, using ^{14}C in CO_2 as the tracer and paper chromatography as the means of separating the products formed after different times of light exposure. Calvin won the Nobel Prize in 1961 for this remarkable achievement.

Tracers are used in many studies of biological function. Most recently, life in space has required answers to new questions. In an animal study of red blood cell loss during extended space flight, blood plasma volume was measured with ^{125}I-labeled albumin, and ^{51}Cr-labeled red blood cells were used to assess survival of blood cells. In another study, blood flow in skin under long periods of microgravity was monitored using injected ^{133}Xe.

Material Flow Tracers are used in studies of solid surfaces and the flow of materials. Metal atoms hundreds of layers deep within a solid have been shown to exchange with metal ions from the surrounding solution within a matter of minutes. Chemists and engineers use tracers to study material movement in semiconductor chips, paint, and metal plating, in detergent action, and in the process of corrosion, to mention just a few of many applications.

Hydrologic engineers use tracers to study the volume and flow of large bodies of water. By following radionuclides formed during atmospheric nuclear bomb tests (^3H in H_2O, $^{90}Sr^{2+}$, and $^{137}Cs^+$), scientists have mapped the flow of water from land to lakes and streams to oceans. Surface and deep ocean currents that circulate around the globe are also studied, as are the mechanisms of hurricane formation and the mixing of the troposphere and stratosphere. Industries employ tracers to study material flow during the manufacturing process, such as the flow of ore pellets in smelting kilns, the paths of wood chips and bleach in paper mills, the diffusion of fungicide into lumber, and in a particularly important application, the porosity and leakage of oil and gas wells in various geological formations.

Activation Analysis A somewhat different use of tracers occurs in *neutron activation analysis* (NAA). In this method, neutrons bombard a nonradioactive sample, converting a small fraction of its atoms to radioisotopes, which exhibit characteristic decay patterns, such as γ-ray spectra, that reveal the elements present. Unlike chemical analysis, NAA leaves the sample virtually intact, so the

Sojourner on the surface of Mars, heading toward Yogi *(right)*.

Table 24.9 Some Radioisotopes Used as Medical Tracers

Isotope	Body Part or Process
^{11}C, ^{18}F, ^{13}N, ^{15}O	PET studies of brain, heart
^{60}Co, ^{192}Ir	Cancer therapy
^{64}Cu	Metabolism of copper
^{59}Fe	Blood flow, spleen
^{67}Ga	Tumor imaging
^{123}I, ^{131}I	Thyroid
^{111}In	Brain, colon
^{42}K	Blood flow
^{81m}Kr	Lung
^{99}Tc	Heart, thyroid, liver, lung, bone
^{201}Tl	Heart muscle
^{90}Y	Cancer, arthritis

method can be used to determine the composition of a valuable object or a very small sample. For example, a painting thought to be a 16th-century Dutch masterpiece was shown through NAA to be a 20th-century forgery, because a microgram-sized sample of its pigment contained much less silver and antimony than the pigments used by the Dutch masters. Forensic chemists use NAA to detect traces and type of ammunition on a suspect's hand or traces of arsenic in the hair of a victim of poisoning. In early 1998, space scientists incorporated NAA instrumentation in the *Sojourner* robot vehicle to analyze the composition of Martian soils and rocks (see photo).

Automotive engineers employ NAA and γ-ray detectors to measure friction and wear of moving parts without having to take the engine apart. For example, when a steel surface that has been neutron-activated to form some radioactive ^{59}Fe moves against a second steel surface, the amount of radioactivity on the second surface indicates the amount of material rubbing off. The radioactivity appearing in a lubricant placed between the surfaces can demonstrate the lubricant's ability to reduce wear.

Medical Diagnosis The largest use of radioisotopes is in medical science. In fact, over 25% of U.S. hospital admissions are for diagnoses based on data from radioisotopes. Tracers with half-lives of a few minutes to a few days are employed to observe specific organs and body parts. For example, a healthy thyroid gland incorporates dietary I^- into iodine-containing hormones at a known rate. To assess thyroid function, the patient drinks a solution containing a trace amount of $Na^{131}I$, and a scanning monitor follows the uptake of $^{131}I^-$ into the thyroid (Figure 24.9A). Technetium-99 ($Z = 43$) is also used for imaging the thyroid (Figure 24.9B), as well as the heart, lungs, and liver. Technetium does not occur naturally, so the radioisotope (actually a "metastable" form, ^{99m}Tc) is prepared just before use from radioactive molybdenum:

$$^{99}_{42}Mo \longrightarrow \, ^{99m}_{43}Tc + \, ^{0}_{-1}\beta$$

Tracers are also used to measure physiological processes, such as blood flow. The rate at which the heart pumps blood, for example, can be observed by injecting ^{59}Fe, which concentrates in the hemoglobin of blood cells. Several radioisotopes used in medical diagnosis are listed in Table 24.9.

Positron-emission tomography (PET) is a powerful imaging method for observing brain structure and function. A biological substance is synthesized with one of its atoms replaced by an isotope that emits positrons. The substance is injected into a patient's bloodstream, from which it is taken up into the brain. The

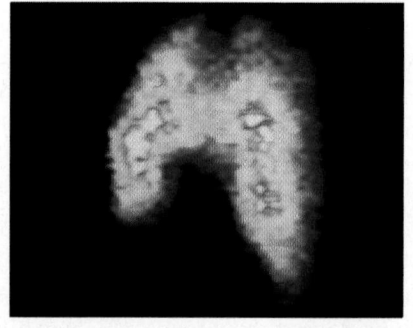

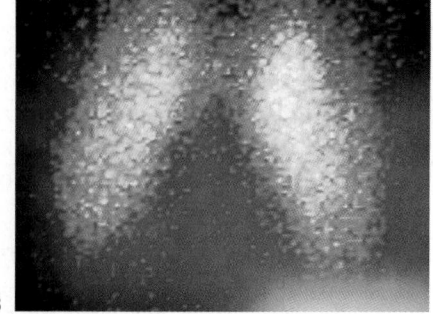

A **B**

Figure 24.9 The use of radioisotopes to image the thyroid gland. Thyroid scanning is used to assess nutritional deficiencies, inflammation, tumor growth, and other thyroid-related ailments. **A,** In ^{131}I scanning, the thyroid gland absorbs $^{131}I^-$ ions whose β emissions expose a photographic film. The asymmetric image indicates disease. **B,** A ^{99}Tc scan of a healthy thyroid.

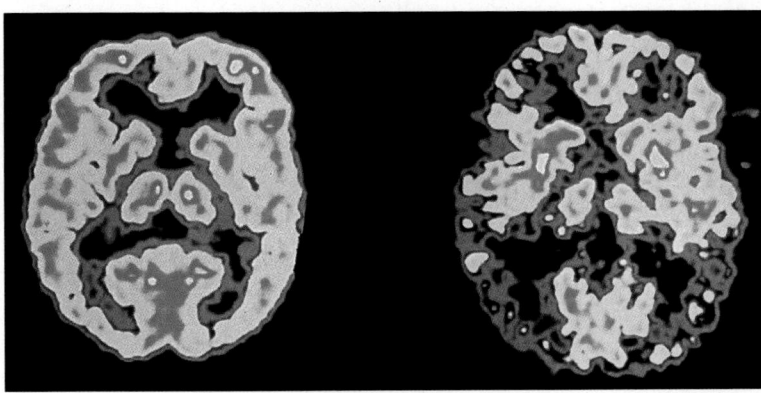

Figure 24.10 PET and brain activity. These PET scans show brain activity in a normal person (*left*) and in a patient with Alzheimer's disease. Red and yellow indicate relatively high activity within a region.

isotope emits positrons, each of which annihilates a nearby electron. In the annihilation process, two γ photons are emitted simultaneously 180° from each other:

$$^{0}_{1}\beta + ^{0}_{-1}e \longrightarrow 2^{0}_{0}\gamma$$

An array of detectors around the patient's head pinpoints the sites of γ emission, and the image is analyzed by computer. Two of the isotopes used are ^{15}O, injected as $H_2{}^{15}$O to measure blood flow, and ^{18}F bonded to a glucose analog to measure glucose uptake, which is a marker for energy metabolism. Among many fascinating PET findings are those that show how changes in blood flow and glucose uptake accompany normal or abnormal brain activity (Figure 24.10). In a recent nonmedical development, substances incorporating ^{11}C and ^{15}O are being investigated by PET to learn how molecules interact with and move along the surface of a catalyst.

Applications of Ionizing Radiation

To be used as a tracer, a radioisotope need emit only low-energy detectable radiation. Many other uses of radioisotopes, however, depend on the effects of high-energy, ionizing radiation.

The interaction between radiation and matter that causes cancer can also be used to eliminate it. Cancer cells divide more rapidly than normal cells, so radioisotopes that interfere with the cell-division process kill more cancer cells than normal ones. Implants of ^{198}Au or of a mixture of ^{90}Sr and ^{90}Y have been used to destroy pituitary and breast tumor cells, and γ rays from ^{60}Co have been used to destroy brain tumors.

Irradiation of food increases shelf life by killing microorganisms that cause food to rot (Figure 24.11), but the practice is quite controversial. Advocates point to the benefits of preserving fresh foods, grains, and seeds for long periods, whereas opponents suggest that irradiation might lower the food's nutritional content or produce harmful by-products. The increased use of antibiotics in animal feed has brought about an increased incidence of illness from newer, more resistant bacterial strains, providing a stronger argument for the use of irradiation. The United Nations has approved irradiation for potatoes, wheat, chicken, and strawberries, and the United States allows irradiation of chicken.

Ionizing radiation has been used to control harmful insects. Captured males are sterilized by radiation and released to mate, thereby reducing the number of offspring. This method has been used to control the Mediterranean fruit fly in California and disease-causing insects, such as the tsetse fly and malarial mosquito, in other parts of the world.

Nonirradiated   Irradiated

Figure 24.11 The increased shelf life of irradiated food.

SECTION SUMMARY
Radioisotopic tracers emit nonionizing radiation and have been used to study reaction mechanisms, material flow, elemental composition, and medical conditions. Ionizing radiation has been used to destroy cancerous tissue, kill organisms that spoil food, and control insect populations.

24.6 THE INTERCONVERSION OF MASS AND ENERGY

Most of the nuclear processes we've considered so far have involved radioactive decay, in which a nucleus emits one or a few small particles or photons to become a slightly lighter nucleus. Two other nuclear processes cause much greater changes. In nuclear **fission,** a heavy nucleus splits into two much lighter nuclei, emitting several small particles at the same time. In nuclear **fusion,** the opposite process occurs as two lighter nuclei combine to form a heavier one. Both fission and fusion release enormous quantities of energy. Let's take a look at the origins of this energy by first examining the change in mass that accompanies the breakup of a nucleus into its nucleons and then considering the energy that is equivalent to this mass change.

The Mass Defect

We have known for most of the 20th century that mass and energy are interconvertible. The traditional mass and energy conservation laws have been combined to state that *the total quantity of mass-energy in the universe is constant*. Therefore, when *any* reacting system releases or absorbs energy, there must be an accompanying loss or gain in mass.

This relation between mass and energy did not concern us earlier because the energy changes involved in breaking or forming chemical bonds are so small that the mass changes are negligible. When 1 mol of water breaks up into its atoms, for example, heat is absorbed:

$$H_2O(g) \longrightarrow 2H(g) + O(g) \qquad \Delta H^0_{rxn} = 2 \times \text{BE of O—H} = 934 \text{ kJ}$$

We find the mass that is equivalent to this energy from Einstein's equation:

$$E = mc^2 \qquad \text{or} \qquad \Delta E = \Delta mc^2 \qquad \text{so} \qquad \Delta m = \frac{\Delta E}{c^2} \qquad \textbf{(24.7)}$$

where Δm is the change in mass between the reactants and the products. Substituting the heat of reaction (in J/mol) for ΔE and the numerical value for c (2.9979$\times$10^8 m/s), we obtain

$$\Delta m = \frac{9.34 \times 10^5 \text{ J/mol}}{(2.9979 \times 10^8 \text{ m/s})^2} = 1.04 \times 10^{-11} \text{ kg/mol} = 1.04 \times 10^{-8} \text{ g/mol}$$

(Units of kg/mol are obtained because the joule includes the kilogram: 1 J = 1 kg·m^2/s^2.) The mass of 1 mol of H$_2$O (reactant) is about 10 ng *less* than the combined masses of 2 mol of H and 1 mol of O (products), a change too small to measure with even the most sophisticated balance. Such minute mass changes when bonds break or form allow us to assume that mass is conserved in *chemical* reactions.

The much larger mass change that accompanies a *nuclear* process is related to the enormous energy required to bind the nucleus together or break it apart. Consider, for example, the change in mass that occurs when one ^{12}C nucleus breaks up into its nucleons: six protons and six neutrons. We calculate this change in mass by combining the mass of six H *atoms* and six neutrons and then subtracting the mass of one ^{12}C *atom*. This procedure cancels the masses of the electrons [six e$^-$ (in six ^{1}H atoms) cancel six e$^-$ (in one ^{12}C atom)]. The mass of one ^{1}H atom is 1.007825 amu, and the mass of one neutron is 1.008665 amu, so we have

$$
\begin{array}{rl}
\text{Mass of six } ^1\text{H atoms} = & 06.046950 \text{ amu} \\
\underline{\text{Mass of six neutrons} =} & \underline{06.051990 \text{ amu}} \\
\text{Total mass} = & 12.098940 \text{ amu}
\end{array}
$$

The mass of one ^{12}C atom is 12 amu (exactly). The difference in mass (Δm) is the total mass of the nucleons minus the mass of the nucleus:

$$
\begin{aligned}
\Delta m &= 12.098940 \text{ amu} - 12.000000 \text{ amu} \\
&= 0.098940 \text{ amu}/^{12}\text{C} = 0.098940 \text{ g/mol } ^{12}\text{C}
\end{aligned}
$$

Note that *the mass of the nucleus is **less** than the combined masses of its nucleons.* The mass decrease that occurs when nucleons are united into a nucleus is called the **mass defect.** The size of this mass change (9.89×10^{-2} g/mol) is nearly 10 million times that of the previous bond breakage (1.04×10^{-8} g/mol) and is easily observed on any laboratory balance.

Nuclear Binding Energy

Einstein's equation for the relation between mass and energy also allows us to find the energy equivalent of a mass defect. For ^{12}C, after converting grams to kilograms, we have

$$
\begin{aligned}
\Delta E = \Delta mc^2 &= (9.8940 \times 10^{-5} \text{ kg/mol})(2.9979 \times 10^8 \text{ m/s})^2 \\
&= 8.8921 \times 10^{12} \text{ J/mol} = 8.8921 \times 10^9 \text{ kJ/mol}
\end{aligned}
$$

This quantity of energy is called the **nuclear binding energy** for carbon-12. In general, the nuclear binding energy is the quantity of energy required to *break up 1 mol of nuclei into their individual nucleons:*

$$\text{Nucleus} + \text{nuclear binding energy} \longrightarrow \text{nucleons}$$

Thus, in a qualitative sense, the nuclear binding energy is analogous to the sum of the bond energies of a compound. However, quantitatively, nuclear binding energies are typically several million times greater.

We use joules to express the binding energy per mole of nuclei, but the joule is an impractically large unit to express the binding energy of a single nucleus. Instead, nuclear scientists use the **electron volt (eV),** the energy an electron acquires when it moves through a potential difference of 1 volt:

$$1 \text{ eV} = 1.602 \times 10^{-19} \text{ J}$$

Binding energies are commonly expressed in millions of electron volts, that is, in *mega–electron volts* (MeV):

$$1 \text{ MeV} = 10^6 \text{ eV} = 1.602 \times 10^{-13} \text{ J}$$

A particularly useful factor converts a given mass defect in atomic mass units to its energy equivalent in electron volts:

$$1 \text{ amu} = 931.5 \times 10^6 \text{ eV} = 931.5 \text{ MeV} \qquad (24.8)$$

Earlier we found the mass defect of the ^{12}C nucleus to be 0.098940 amu. Therefore, the binding energy per ^{12}C nucleus, expressed in MeV, is

$$\frac{\text{Binding energy}}{^{12}\text{C nucleus}} = 0.098940 \text{ amu} \times \frac{931.5 \text{ MeV}}{1 \text{ amu}} = 92.16 \text{ MeV}$$

We can compare the stability of nuclides of different elements by determining the *binding energy per nucleon.* For ^{12}C, we have

$$\frac{\text{Binding energy}}{\text{nucleon}} = \frac{92.16 \text{ MeV}}{12 \text{ nucleons}} = 7.680 \text{ MeV/nucleon}$$

The Force That Binds Us According to current theory, the nuclear binding energy is related to the strong force, which holds nucleons together in a nucleus. There are three other fundamental forces: (1) the weak nuclear force, which is important in β decay, (2) the electrostatic force that we observe between charged particles, and (3) the gravitational force. Toward the end of his life, Albert Einstein tried unsuccessfully to develop a theory to explain how the four forces were really different aspects of one unified force that governs all nature. Through experiments with the next generation of giant accelerators, modern particle physicists will attempt to realize Einstein's dream.

SAMPLE PROBLEM 24.6 Calculating the Binding Energy per Nucleon

Problem Iron-56 is an extremely stable nuclide. Compute the binding energy per nucleon for ^{56}Fe and compare it with that for ^{12}C (mass of ^{56}Fe atom = 55.934939 amu; mass of ^{1}H atom = 1.007825 amu; mass of neutron = 1.008665 amu).

Plan Iron-56 has 26 protons and 30 neutrons in its nucleus. We calculate the mass defect by finding the sum of the masses of 26 ^{1}H atoms and 30 neutrons and subtracting the given mass of 1 ^{56}Fe atom. Then we multiply Δm by the equivalent in MeV (931.5 MeV/amu) and divide by 56 (no. of nucleons) to obtain the binding energy per nucleon.

Solution Calculating the mass defect:

Mass Defect = [(26 × mass ^{1}H atom) + (30 × mass neutron)] − mass ^{56}Fe atom
= [(26)(1.007825 amu) + (30)(1.008665 amu)] − 55.934939 amu
= 0.52846 amu

Calculating the binding energy per nucleon:

$$\text{Binding energy per nucleon} = \frac{0.52846 \text{ amu} \times 931.5 \text{ MeV/amu}}{56 \text{ nucleons}} = \boxed{8.790 \text{ MeV/nucleon}}$$

An ^{56}Fe nucleus would require more energy to break up into its nucleons than would ^{12}C (7.680 MeV/nucleon), so ^{56}Fe is more stable than ^{12}C.

Check The answer is consistent with the great stability of ^{56}Fe. Given the number of decimal places in the values, rounding to check the math is useful only to find a *major* error. The number of nucleons (56) is an exact number, so we retain four significant figures.

FOLLOW-UP PROBLEM 24.6 Uranium-235 is the essential component of the fuel in nuclear power plants. Calculate the binding energy per nucleon for ^{235}U. Is this nuclide more or less stable than ^{12}C (mass of ^{235}U atom = 235.043924 amu)?

Fission or Fusion: Means of Increasing the Binding Energy Per Nucleon

Calculations similar to Sample Problem 24.6 for other nuclides show that the binding energy per nucleon varies considerably. The essential point is that *the greater the binding energy per nucleon, the more stable the nuclide.*

Figure 24.12 shows a plot of the binding energy per nucleon vs. mass number. It provides information about nuclide stability and the two possible processes nuclides can undergo to form more stable nuclides. Nuclides with fewer than 10 nucleons have a relatively small binding energy per nucleon. The ^{4}He nucleus has an exceptionally large value, however, which is why it is emitted intact as an α particle. Above $A = 12$, the binding energy per nucleon varies from about 7.6 to 8.8 MeV.

The most important observation is that *the binding energy per nucleon peaks for elements with $A \approx 60$.* In other words, nuclides become more stable with increasing mass number up to around 60 nucleons and then become less stable with higher numbers of nucleons. The existence of a peak of stability suggests that there are two ways nuclides can increase their binding energy per nucleon:

- *Fission.* A heavier nucleus can *split into lighter ones (closer to $A \approx 60$) by* undergoing fission. The product nuclei have greater binding energy per nucleon (are more stable) than the reactant, and the difference in *energy is released.* Nuclear power plants generate energy through fission, as do atomic bombs (Section 24.7).

- *Fusion.* Lighter nuclei, on the other hand, can *combine to form a heavier one (closer to $A \approx 60$) by* undergoing fusion. Once again, the product is more stable than the reactants, and *energy is released.* The Sun and other stars generate energy through fusion, as do hydrogen bombs. In these examples and all current research efforts for developing fusion as a useful energy source, hydrogen nuclei fuse to form the very stable helium-4 nucleus.

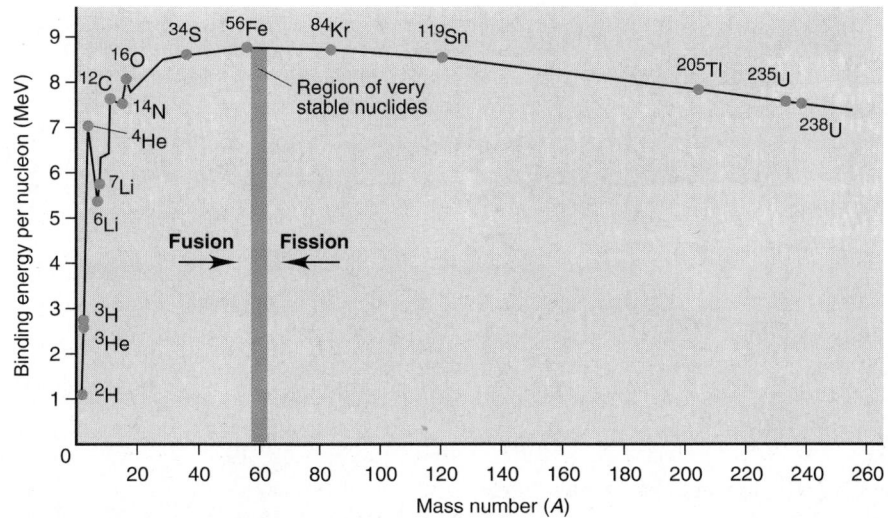

Figure 24.12 **The variation in binding energy per nucleon.** A plot of the binding energy per nucleon vs. mass number shows that nuclear stability is greatest in the region near ^{56}Fe. Lighter nuclei may undergo fusion to become more stable; heavier ones may undergo fission. Note the exceptional stability of ^{4}He among extremely light nuclei.

In the next section, we examine fission and fusion and the industrial energy facilities designed to utilize them.

SECTION SUMMARY

The mass of a nucleus is less than the sum of the masses of its nucleons by an amount called the mass defect. The energy equivalent to the mass defect is the nuclear binding energy, usually expressed in units of MeV. The binding energy per nucleon is a measure of nuclide stability and varies with number of nucleons. Nuclides with $A \approx 60$ are most stable. Lighter nuclides can join (fusion), and heavier nuclides can split (fission) to become more stable.

24.7 APPLICATIONS OF FISSION AND FUSION

Of the many beneficial applications of nuclear reactions, the greatest is the potential for almost limitless amounts of energy, which is based on the multimillion-fold increase in energy yield of nuclear reactions over chemical reactions. Our experience with nuclear energy from power plants in the late 20th century, however, has forced a realization that we must strive to improve ways to tap this energy source safely and economically. In this section we discuss how fission and fusion occur and how we are applying them.

The Process of Nuclear Fission

During the mid-1930s, Enrico Fermi and coworkers bombarded uranium ($Z = 92$) with neutrons in an attempt to synthesize transuranium elements. Many of the unstable nuclides produced were tentatively identified as having $Z > 92$, but other scientists were skeptical. Four years later, the German chemist Otto Hahn and his associate F. Strassmann showed that one of these unstable nuclides was an isotope of barium ($Z = 56$). The Austrian physicist Lise Meitner, a coworker of Hahn, and her nephew Otto Frisch proposed that barium resulted from the *splitting* of the uranium nucleus into *smaller* nuclei, a process they named *fission* because of its similarity to the reproductive fission of a biological cell.

Lise Meitner

 Lise Meitner (1878–1968) Until very recently, this extraordinary physicist received little of the acclaim she deserved. Meitner worked in the laboratory of the chemist Otto Hahn, and she was responsible for the discovery of protactinium (Pa; $Z = 91$) and numerous radioisotopes. After leaving Germany in advance of the Nazi domination, Meitner proposed the correct explanation of nuclear fission. When Hahn received the Nobel Prize in 1944, however, he did not even acknowledge Meitner in his acceptance speech. Today, most physicists believe that Lise Meitner should have received the prize. Despite controversy over names for elements 104 to 109, it was widely agreed that element 109 should be named meitnerium.

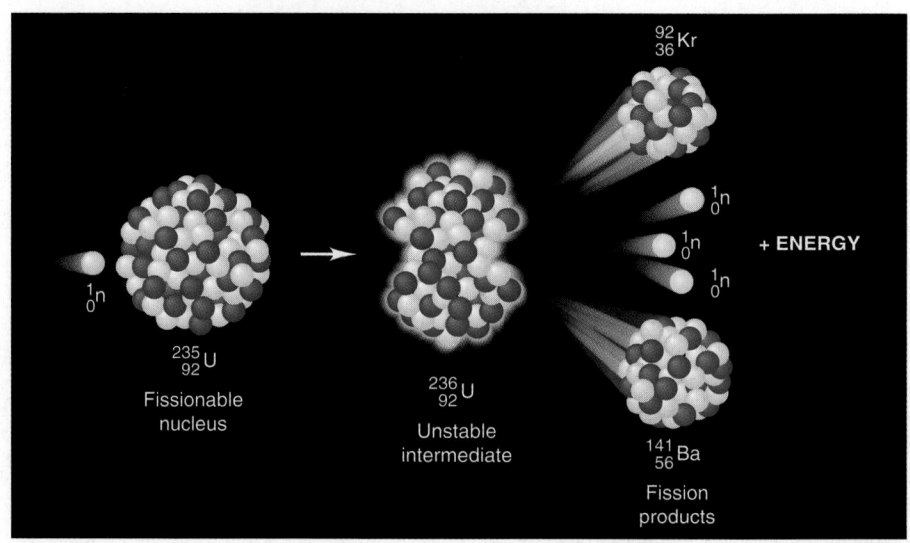

Figure 24.13 Induced fission of ^{235}U. A neutron bombarding a ^{235}U nucleus results in an extremely unstable ^{236}U nucleus, which becomes distorted in the act of splitting. In this case, which shows one of many possible splitting patterns, the products are ^{92}Kr and ^{141}Ba. Three neutrons and a great deal of energy are released also.

The ^{235}U nucleus can split in many different ways, giving rise to various daughter nuclei, but all routes have the same general features. Figure 24.13 depicts one of these fission patterns. Neutron bombardment results in a highly excited ^{236}U nucleus, which splits apart in 10^{-14} s. The products are two nuclei of unequal mass, two or three neutrons (average of 2.4), and a large quantity of energy. A single ^{235}U nucleus releases 3.5×10^{-11} J when it splits; 1 mol of ^{235}U (about $\frac{1}{2}$ lb) releases 2.1×10^{13} J—a billion times as much energy as burning $\frac{1}{2}$ lb of coal (about 2×10^4 J)!

We harness the energy of nuclear fission, much of which appears as heat, by means of a **chain reaction,** illustrated in Figure 24.14: the two to three neutrons that are released by the fission of one nucleus collide with other fissionable nuclei and cause them to split, releasing more neutrons, which then collide with other nuclei, and so on, in a self-sustaining process. In this manner, the energy released increases rapidly because each fission event in a chain reaction releases two to three times as much energy as the preceding one.

Whether a chain reaction occurs depends on the mass (and thus the volume) of the fissionable sample. If the piece of uranium is large enough, the product neutrons strike another fissionable nucleus *before* flying out of the sample, and a chain reaction takes place. The mass required to achieve a chain reaction is called the **critical mass.** If the sample has less than the critical mass (called a *subcritical mass*), most of the product neutrons leave the sample before they have the opportunity to collide with and cause the fission of another ^{235}U nucleus, and thus a chain reaction does not occur.

Uncontrolled Fission: The Atomic Bomb An uncontrolled chain reaction can be adapted to make an extremely powerful explosive, as several of the world's leading atomic physicists suspected just prior to the beginning of World War II. In August 1939, Albert Einstein wrote the president of the United States, Franklin Delano Roosevelt, to this effect, warning of the danger of allowing the Nazi government to develop this power first. It was this concern that led to the Manhattan Project, an enormous scientific effort to develop a bomb based on nuclear

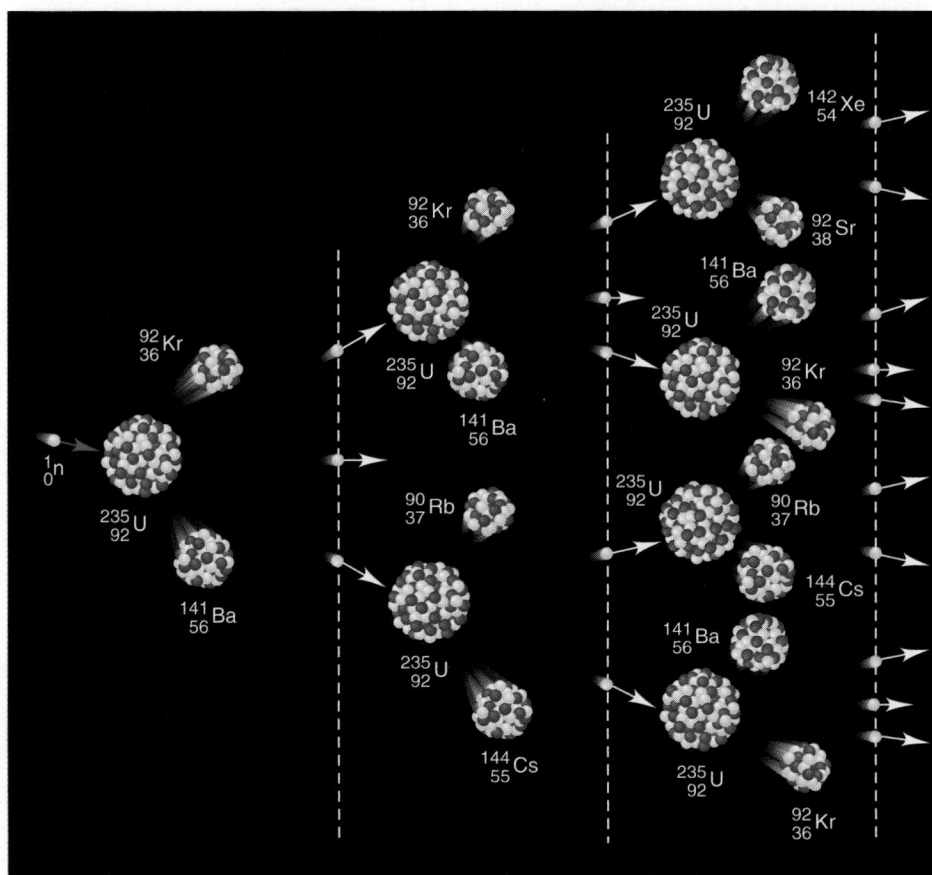

Figure 24.14 **A chain reaction of** ^{235}U. If a sample exceeds the critical mass, neutrons produced by the first fission event collide with other nuclei, causing their fission and the production of more neutrons to continue the process. Note that various product nuclei form. The vertical dashed lines identify succeeding "generations" of neutrons.

fission, which was initiated in 1941.* In August 1945, the United States detonated two atomic bombs over Japan, and the horrible destructive power of these bombs was a major factor in the surrender of the Japanese a few days later.

In an atomic bomb, small explosions of trinitrotoluene (TNT) bring subcritical masses of fissionable material together to exceed the critical mass, and the ensuing chain reaction brings about the explosion (Figure 24.15). The proliferation of nuclear power plants, which use fissionable materials to generate energy for electricity, has increased concern that more countries (and unscrupulous individuals) may have access to such material for making bombs. In light of the devastating terrorist attacks of September 11, 2001 in the United States, this concern has been heightened. After all, only 1 kg of fissionable uranium was used in the bomb dropped on Hiroshima, Japan.

Controlled Fission: Nuclear Energy Reactors Controlled fission can produce electric power more cleanly than burning coal. Like a coal-fired power plant, *a nuclear power plant generates heat to produce steam, which turns a turbine attached to an electric generator.* In a coal plant, the heat is produced by burning coal; in a nuclear plant, it is produced by splitting uranium.

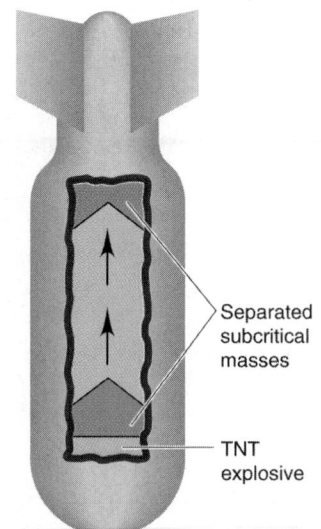

Figure 24.15 **Diagram of an atomic bomb.** Small TNT explosions bring subcritical masses together, and the chain reaction occurs.

---

* For an excellent scientific and historical account of the development of the A-bomb, see R. Rhodes, *The Making of the Atomic Bomb*, New York, Simon and Schuster, 1986.

Heat generation takes place in the **reactor core** of a nuclear plant (Figure 24.16). The core contains the *fuel rods,* which consist of fuel enclosed in tubes of a corrosion-resistant zirconium alloy. The fuel is uranium(IV) oxide (UO_2) that has been *enriched* from 0.7% ^{235}U, the natural abundance of this fissionable isotope, to the 3% to 4% ^{235}U required to sustain a chain reaction. (Enrichment of nuclear fuel is the most important application of Graham's law; see the margin note, p. 202.) Sandwiched between the fuel rods are movable *control rods* made of cadmium or boron (or, in nuclear submarines, hafnium), substances that absorb neutrons very efficiently. When the control rods are moved between the fuel rods, the chain reaction slows because fewer neutrons are available to bombard uranium atoms; when they are removed, the chain reaction speeds up. Neutrons that leave the fuel-rod assembly collide with a *reflector,* usually made of a beryllium alloy, which absorbs very few neutrons. Reflecting the neutrons back to the fuel rods speeds the chain reaction.

Figure 24.16 **A light-water nuclear reactor. A,** Photo of a facility showing the concrete containment shell and nearby water source. **B,** Schematic of a light-water reactor.

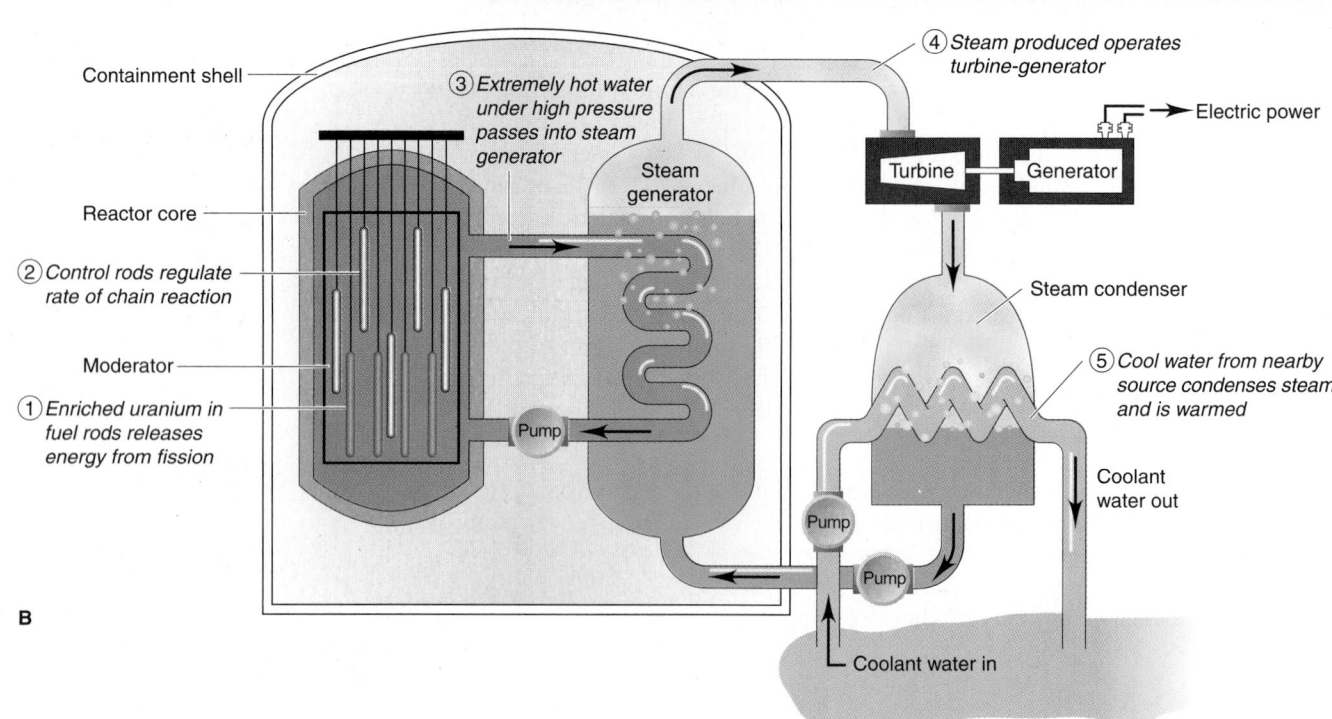

Flowing around the fuel and control rods in the reactor core is the *moderator,* a substance that slows the neutrons, making them much better at causing fission than the fast ones emerging directly from the fission event. In most modern reactors, the moderator also acts as the *coolant,* the fluid that transfers the released heat to the steam-producing region. Because ^{1}H absorbs neutrons, *light-water reactors* use H_2O as the moderator; in heavy-water reactors, D_2O is used. The advantage of D_2O is that it absorbs very few neutrons, leaving more available for fission, so heavy-water reactors can use *unenriched* uranium. As the coolant flows around the encased fuel, pumps circulate it through coils that transfer its heat to the water reservoir. Steam formed in the reservoir turns the turbine that runs the generator. The steam is then condensed in large cooling towers (see Figure 13.11, p. 499) with water from a lake or river and returned to the water reservoir.

Some major accidents at nuclear plants have caused decidedly negative public reactions. In 1979, malfunctions of coolant pumps and valves at the Three-Mile Island facility in Pennsylvania led to melting of some of the fuel, serious damage to the reactor core, and the release of radioactive gases into the atmosphere. In 1986, a million times as much radioactivity was released when a cooling system failure at the Chernobyl plant in Ukraine of the former Soviet Union caused a much greater melting of fuel and an uncontrolled reaction. High-pressure steam and ignited graphite moderator rods caused the reactor building to explode and expel radioactive debris. Carried by prevailing winds, the radioactive particles contaminated vegetables and milk in much of Europe. Health officials have evidence that thousands of people living near the accident have already or may soon develop cancer from radiation exposure. The design of the Chernobyl plant was particularly unsafe because, unlike reactors in the United States and western Europe, the reactor was not enclosed in a massive, concrete containment building.

Despite potential safety problems, nuclear power remains an important source of electricity. In the late 1990s, nearly every European country employed nuclear power, and it is the major power source in some countries—Sweden creates 50% of its electricity this way and France almost 80%. Currently, the United States obtains about 20% of its electricity from nuclear power, and Canada slightly less. As our need for energy grows, new, safer reactors will be designed.

However, even a smoothly operating plant has certain inherent problems. The problem of *thermal pollution* is common to all power plants. Water used to condense the steam is several degrees warmer when returned to its source, which can harm aquatic organisms (Section 13.3). A more serious problem is *nuclear waste disposal.* Many of the fission products formed in nuclear reactors have long half-lives, and no satisfactory plan for their permanent disposal has yet been devised. Proposals to place the waste in containers and bury those in deep bedrock cannot possibly be field-tested for the thousands of years the material will remain harmful. Leakage of radioactive material into groundwater is always a danger, and earthquakes can occur even in geologically stable regions. Despite studies indicating that the proposed disposal site at Yucca Mountain, Nevada may be too geologically active, the U.S. government recently approved the site. It remains to be seen whether we can operate fission reactors *and* dispose of the waste safely and inexpensively.

The Promise of Nuclear Fusion

Nuclear fusion is the ultimate source of nearly all the energy on Earth because nearly all other sources depend, directly or indirectly, on the energy produced by nuclear fusion in the Sun. But the Sun and other stars generate more than energy; in fact, *all the elements larger than hydrogen were formed in fusion and decay processes within stars,* as the upcoming Chemical Connections essay describes.

"Breeding" Nuclear Fuel Uranium-235 is not an abundant isotope. One solution to a potential fuel shortage is a *breeder reactor,* designed to consume one type of nuclear fuel as it produces another. Fuel rods are surrounded by natural U_3O_8, which contains 99.3% *nonfissionable* ^{238}U atoms. As fast neutrons, formed during ^{235}U fission, escape the fuel rod, they collide with ^{238}U, transmuting it into ^{239}Pu, another fissionable nucleus:

$$^{238}_{92}U + ^{1}_{0}n \longrightarrow ^{239}_{92}U$$
($t_{1/2}$ of $^{239}_{92}U$ = 23.5 min)

$$^{239}_{92}U \longrightarrow ^{239}_{93}Np + ^{0}_{-1}\beta$$
($t_{1/2}$ of $^{239}_{93}Np$ = 2.35 days)

$$^{239}_{93}Np \longrightarrow ^{239}_{94}Pu + ^{0}_{-1}\beta$$
($t_{1/2}$ of $^{239}_{94}Pu$ = 2.4×10^4 yr)

Although breeder reactors can make fuel as they operate, they are difficult and expensive to build, and the ^{239}Pu is extremely toxic and long lived. No breeder reactors are in use in the United States, although there are several in Europe and Japan.

Origin of the Elements in the Stars

How did the universe begin? Where did matter come from? How were the elements formed? Every culture has creation myths that address such questions, but only recently have astronomers, physicists, and chemists begun to offer a scientific explanation. The most accepted current model proposes that a sphere of unimaginable properties—diameter of 10^{-28} cm, density of 10^{96} g/mL (density of a nucleus $\approx 10^{14}$ g/mL), and temperature of 10^{32} K—exploded in a "Big Bang," for reasons not yet even guessed, and distributed its contents through the void of space. Cosmologists consider this moment the beginning of time.

One second later, the universe was an expanding mixture of neutrons, protons, and electrons, denser than rock and hotter than an exploding H-bomb (about 10^{10} K). During the next few minutes, it became a gigantic fusion reactor creating the first atomic nuclei: ^{2}H, ^{3}He, and ^{4}He. After 10 minutes, more than 25% of the mass of the universe existed as ^{4}He, and only about 0.025% as ^{2}H. About 100 million years later, or about 15 billion years ago, gravitational forces pulled this cosmic mixture into primitive, contracting stars.

This account of the origin of the universe is based on the observation of spectra from the Sun, other stars, nearby galaxies, and cosmic (interstellar) dust. Spectral analysis of planets and chemical analysis of the Earth, Moon, meteorites, and cosmic-ray particles furnish data about isotope abundance. From these a model has been developed for **stellar nucleogenesis,** the origin of the elements in the stars. The overall process occurs in several stages during a star's evolution, and the entire sequence of steps occurs only in very massive stars, 10 to 100 times the mass of the Sun. Each step involves a contraction of the star that produces higher temperature and heavier nuclei. Such events are forming elements in stars today. Here are the key stages in the process (Figure B24.3):

1. *Hydrogen burning produces He.* The initial contraction of a star heats its core to about 10^7 K, at which point a fusion process called *hydrogen burning* produces helium from the abundant protons:

$$4^{1}_{1}\text{H} \longrightarrow {}^{4}_{2}\text{He} + 2^{0}_{1}\beta + 2\gamma + \text{energy}$$

2. *Helium burning produces C, O, Ne, and Mg.* After several billion years of hydrogen burning, about 10% of the ^{1}H is consumed, and the star contracts further. The ^{4}He forms a dense core, hot enough (2×10^8 K) to fuse ^{4}He. The energy released during *helium burning* expands the remaining ^{1}H into a vast envelope: the star becomes a *red giant,* more than 100 times its original diameter. Within its core, pairs of ^{4}He nuclei (α particles) fuse into unstable ^{8}Be nuclei ($t_{1/2} = 2\times10^{-16}$ s), which collide with another ^{4}He to form stable ^{12}C. Further fusion with ^{4}He creates nuclei up to ^{24}Mg:

$$^{12}\text{C} \xrightarrow{\alpha} {}^{16}\text{O} \xrightarrow{\alpha} {}^{20}\text{Ne} \xrightarrow{\alpha} {}^{24}\text{Mg}$$

3. *Elements through Fe and Ni form.* For another 10 million years, ^{4}He is consumed, and the heavier nuclei form a core. This core contracts and heats, expanding the star to a *supergiant.* Within the hot core (7×10^8 K), *carbon and oxygen burning* occur:

$$^{12}\text{C} + {}^{12}\text{C} \longrightarrow {}^{23}\text{Na} + {}^{1}\text{H}$$
$$^{12}\text{C} + {}^{16}\text{O} \longrightarrow {}^{28}\text{Si} + \gamma$$

Absorption of α particles forms nuclei up to ^{40}Ca:

$$^{12}\text{C} \xrightarrow{\alpha} {}^{16}\text{O} \xrightarrow{\alpha} {}^{20}\text{Ne} \xrightarrow{\alpha} {}^{24}\text{Mg} \xrightarrow{\alpha}$$

$$^{28}\text{Si} \xrightarrow{\alpha} {}^{32}\text{S} \xrightarrow{\alpha} {}^{36}\text{Ar} \xrightarrow{\alpha} {}^{40}\text{Ca}$$

Further contraction and heating to 3×10^9 K allow reactions in which nuclei release neutrons, protons, and α particles and then recapture them. As a result, nuclei with lower binding energies supply nucleons to create those with higher binding energies. This process, which takes only a few minutes, stops at iron ($A = 56$) and nickel ($A = 58$), the nuclei with the highest binding energies.

4. *Heavier elements form.* In very massive stars, the next stage is the most spectacular. With all the fuel consumed, the core collapses within a second. Many Fe and Ni nuclei break down into neutrons and protons. Protons capture electrons to form neutrons, and the entire core forms an incredibly dense *neutron star.* (An Earth-sized star that became a neutron star would fit in the Houston astrodome!) As the core implodes, the outer layers explode in a *supernova,* which expels material throughout space. A supernova occurs an average of every few hundred years in each galaxy; one was observed from the southern hemisphere in 1987, about 160,000 years after the event occurred (Figure B24.4). The heavier elements form during supernovas and are found in *second-generation stars,* those that coalesce from interstellar ^{1}H and ^{4}He and the debris of exploded first-generation stars.

Heavier elements form through *neutron-capture* processes. In the *s-process,* a nucleus captures a neutron and emits a γ ray. Days, months, or even thousands of years later, the nucleus emits a β particle to form the next element, as in this conversion of ^{68}Zn to ^{70}Ge:

$$^{68}\text{Zn} \xrightarrow{n} {}^{69}\text{Zn} \xrightarrow{\beta} {}^{69}\text{Ga} \xrightarrow{n} {}^{70}\text{Ga} \xrightarrow{\beta} {}^{70}\text{Ge}$$

The stable isotopes of most heavy elements form by the s-process.

Less stable isotopes and those with A greater than 230 cannot form by the s-process because their half-lives are too short. These form by the *r-process* during the fury of the supernova. Multiple neutron captures, followed by multiple β decays, occur in a second, as when ^{56}Fe is converted to ^{79}Br:

$$^{56}_{26}\text{Fe} + 23^{1}_{0}\text{n} \longrightarrow {}^{79}_{26}\text{Fe} \longrightarrow {}^{79}_{35}\text{Br} + 9^{0}_{-1}\beta$$

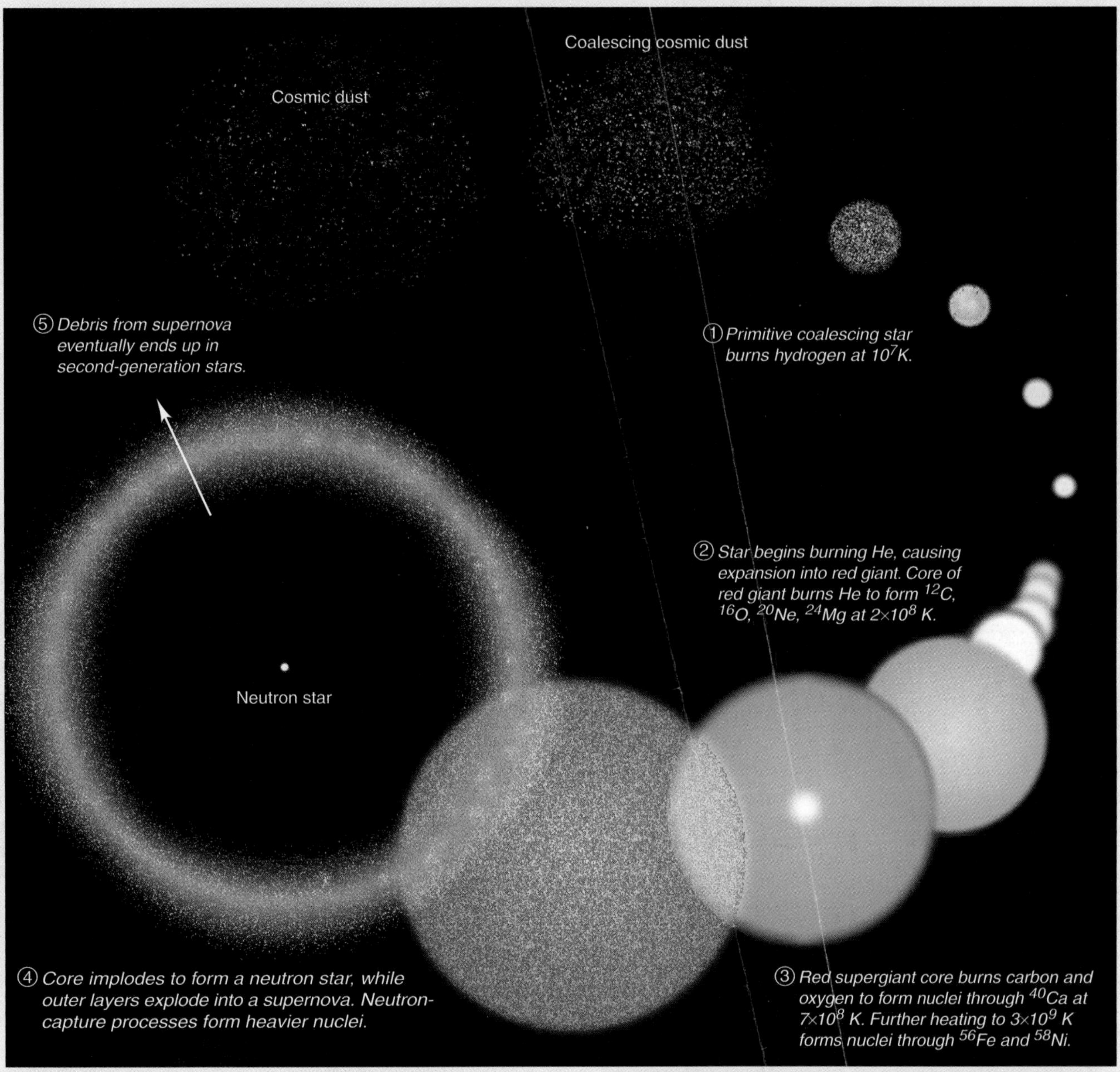

⑤ *Debris from supernova eventually ends up in second-generation stars.*

① *Primitive coalescing star burns hydrogen at 10^7 K.*

Cosmic dust

Coalescing cosmic dust

② *Star begins burning He, causing expansion into red giant. Core of red giant burns He to form ^{12}C, ^{16}O, ^{20}Ne, ^{24}Mg at 2×10^8 K.*

Neutron star

④ *Core implodes to form a neutron star, while outer layers explode into a supernova. Neutron-capture processes form heavier nuclei.*

③ *Red supergiant core burns carbon and oxygen to form nuclei through ^{40}Ca at 7×10^8 K. Further heating to 3×10^9 K forms nuclei through ^{56}Fe and ^{58}Ni.*

Figure B24.3 Element synthesis in the life cycle of a star.

We know from the heavy elements present in the Sun that it is at least a second-generation star presently undergoing hydrogen burning. Together with its planets, it was formed from the dust of exploded stars about 4.6×10^9 years ago. This means that many of the atoms on Earth, including some within you, came from exploded stars and are older than the Solar System itself!

Any theory of element formation must be consistent with the element abundances we observe (Section 22.1). Although local compositions, such as those of the Earth and Sun, differ, large regions of the universe have, on average, similar compositions. Thus, scientists believe that element forming reaches a dynamic equilibrium, which leads to *relatively constant amounts of the isotopes.*

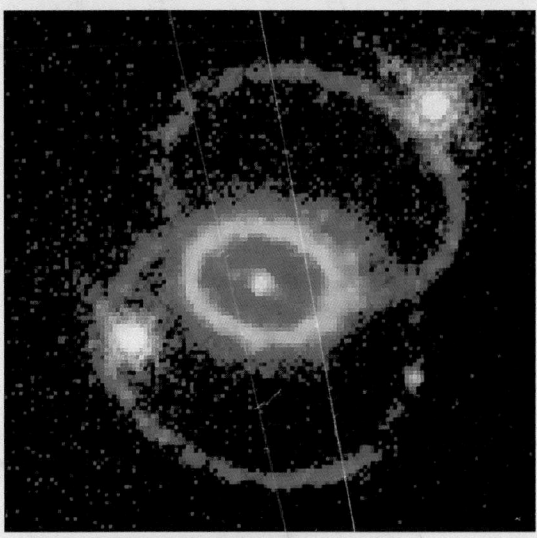

Figure B24.4 A view of Supernova 1987A.

1075

Much research is being devoted to making nuclear fusion a practical, direct source of energy on Earth. To understand the advantages of fusion, let's consider one of the most discussed fusion reactions, in which deuterium and tritium react:

$$_1^2H + {}_1^3H \longrightarrow {}_2^4He + {}_0^1n$$

This reaction produces 1.7×10^9 kJ/mol, an enormous quantity of energy with no radioactive by-products. Moreover, the reactant nuclei are relatively easy to come by. We obtain deuterium from the electrolysis of water (Section 22.4). In nature, tritium forms through the cosmic (neutron) irradiation of ^{14}N:

$$_7^{14}N + {}_0^1n \longrightarrow {}_1^3H + {}_6^{12}C$$

However, this process results in a natural abundance of only $10^{-7}\%$ ^3H. More practically, tritium can be produced in nuclear accelerators by bombarding lithium-6 or by surrounding the fusion reactor itself with material containing lithium-6:

$$_3^6Li + {}_0^1n \longrightarrow {}_1^3H + {}_2^4He$$

Thus, fusion seems very promising, at least in principle. However, some extremely difficult problems exist. Fusion requires enormous energy in the form of heat to give the positively charged nuclei enough kinetic energy to force themselves together. The fusion of deuterium and tritium, for example, occurs at practical rates at about 10^8 K, hotter than the Sun's core! How can such temperatures be achieved? The reaction that forms the basis of a *hydrogen, or thermonuclear, bomb* fuses lithium-6 and deuterium, with an atomic bomb inside the device providing the heat. Obviously, a power plant cannot begin operation by detonating atomic bombs.

Two research approaches are being used to achieve the necessary heat. In one, atoms are stripped of their electrons at high temperatures, which results in a gaseous *plasma*, a neutral mixture of positive nuclei and electrons. Because of the extreme temperatures needed for fusion, no *material* can contain the plasma. The most successful approach to date has been to enclose the plasma within a magnetic field. The *tokamak* design has a donut-shaped container in which a helical magnetic field confines the plasma and prevents it from contacting the walls (Figure 24.17). Scientists at the Princeton University Plasma Physics facility have achieved some success in generating energy from fusion this way. In another approach, the high temperature is reached by using many focused lasers to compress and heat the fusion reactants. In any event, as a practical, everyday source of energy, fusion still seems to be a long way off.

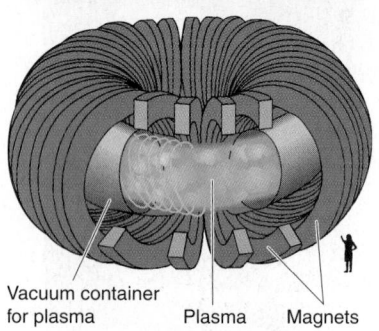

Vacuum container
for plasma   Plasma   Magnets

Figure 24.17 The tokamak design for magnetic containment of a fusion plasma. The donut-shaped chamber of the tokamak (photo, *top;* schematic, *bottom*) contains the plasma within a helical magnetic field.

SECTION SUMMARY

In nuclear fission, neutron bombardment causes a nucleus to split, releasing neutrons that split other nuclei to produce a chain reaction. A nuclear power plant controls the rate of the chain reaction to produce heat that creates steam, which is used to generate electricity. Potential hazards, such as radiation leaks, thermal pollution, and disposal of nuclear waste, remain current concerns. Nuclear fusion holds great promise as a source of clean abundant energy, but it requires extremely high temperatures and is not yet practical. The elements were formed through a complex series of nuclear reactions in evolving stars.

Chapter Perspective

With this chapter, our earlier picture of the nucleus as a static point of positive mass at the atom's core has changed radically. Now we picture a dynamic body, capable of a host of actions that involve incredible quantities of energy. Our attempts to apply the behavior of this minute system to benefit society have created some of the most fascinating and challenging fields in science today.

We began our investigation of chemistry 24 chapters ago, by seeing how the chemical elements and the products we make from them influence nearly every aspect of our material existence. Now we have come full circle to learn that these elements, whose patterns of behavior we marvel at, are continually being born in the countless infernos that twinkle in the night sky.

For you, the end of this course is a beginning—a chance to apply your new abilities to visualize molecular events and solve problems in whatever field you choose. For the science of chemistry, future challenges are great: What new forms of energy can satisfy society's needs without damaging our environment? What new products can feed, clothe, and house the world's people and maintain precious resources? What new medicines can defend against cancer, AIDS, and other dreaded diseases? What new materials and technologies can make life more productive and meaningful? The questions are many, but the science of chemistry will always be one of our most powerful means of answering them.

For Review and Reference (Numbers in parentheses refer to pages, unless noted otherwise.)

Learning Objectives

Relevant section and/or sample problem (SP) numbers appear in parentheses.

Understand These Concepts

1. How nuclear changes differ, in general, from chemical changes (Intro)
2. The meanings of radioactivity, nucleon, nuclide, and isotope (Section 24.1)
3. Characteristics of three types of radioactive emissions: α, β, and γ (Section 24.1)
4. The various forms of radioactive decay and how each changes the values of A and Z (Section 24.1)
5. How the N/Z ratio and the even-odd nature of N and Z correlate with nuclear stability (Section 24.1)
6. How the N/Z ratio correlates with the mode of decay of an unstable nuclide (Section 24.1)
7. How a decay series combines numerous decay steps to reach a stable nuclide (Section 24.1)
8. Why radioactive decay is a first-order process; the meaning of decay rate and specific activity (Section 24.2)
9. The meaning of first-order half-life in the context of radioactive decay (Section 24.2)
10. How the specific activity of an isotope in an object is used to determine its age (Section 24.2)
11. How particle accelerators are used to synthesize new nuclides (Section 24.3)
12. The distinction between excitation and ionization and the extent of their effects on matter (Section 24.4)
13. The units of radiation dose; the effects on living tissue of various dosage levels; the inverse relationship between the mass and charge of an emission and its penetrating power (Section 24.4)
14. How ionizing radiation creates free radicals that damage tissue; sources and risks of ionizing radiation (Section 24.4)
15. How radioisotopes are used in research, analysis, and diagnosis (Section 24.5)

16. Why the mass of a nuclide is less than the sum of its nucleons' masses (mass defect) and how this mass difference is related to the nuclear binding energy (Section 24.6)
17. How nuclear stability is related to binding energy per nucleon (Section 24.6)
18. How unstable nuclides undergo either fission or fusion to increase their binding energy per nucleon (Section 24.6)
19. The current application of fission and potential application of fusion to produce energy (Section 24.7)

Master These Skills

1. Expressing the mass and charge of a particle with the $_Z^A X$ notation (Section 24.1; see also Section 2.5)
2. Using changes in the values of A and Z to write and balance nuclear equations (SP 24.1)
3. Using the N/Z ratio and the even-odd nature of N and Z to predict nuclear stability (SP 24.2)
4. Using the N/Z ratio to predict the mode of radioactive decay (SP 24.3)
5. Converting units of radioactivity (Section 24.2)
6. Calculating specific activity, decay constant, half-life, and number of nuclei (Section 24.2 and SP 24.4)
7. Estimating the age of an object from the specific activity and half-life of carbon-14 (SP 24.5)
8. Writing and balancing equations for nuclear transmutation (Section 24.3)
9. Calculating radiation dose and converting units (Section 24.4)
10. Calculating the mass defect and its energy equivalent in J and eV (Section 24.6)
11. Calculating the binding energy per nucleon and using it to compare stabilities of nuclides (SP 24.6)

Key Terms

Section 24.1
radioactivity (1042)
nucleon (1042)
nuclide (1042)
isotope (1042)
alpha (α) particle (1043)
beta (β) particle (1043)
gamma (γ) ray (1043)
alpha decay (1044)
beta decay (1045)
positron decay (1045)
positron (1045)
electron capture (1045)
gamma emission (1045)
N/Z ratio (1046)
band of stability (1046)

strong force (1047)
decay (disintegration) series (1049)

Section 24.2
activity ($\mathcal{A}$) (1050)
becquerel (Bq) (1050)
curie (Ci) (1050)
decay constant (1050)
half-life ($t_{1/2}$) (1050)
Geiger-Müller counter (1051)
scintillation counter (1051)
radioisotopic dating (1053)
radioisotope (1053)

Section 24.3
nuclear transmutation (1055)

deuteron (1056)
particle accelerator (1056)
transuranium element (1057)

Section 24.4
excitation (1058)
nonionizing radiation (1058)
ionization (1058)
ionizing radiation (1058)
gray (Gy) (1059)
rad (radiation-absorbed dose) (1059)
rem (roentgen equivalent for man) (1059)
sievert (Sv) (1059)
free radical (1059)
background radiation (1061)

Section 24.5
tracer (1062)

Section 24.6
fission (1066)
fusion (1066)
mass defect (1067)
nuclear binding energy (1067)
electron volt (eV) (1067)

Section 24.7
chain reaction (1070)
critical mass (1070)
reactor core (1072)
stellar nucleogenesis (1074)

Key Equations and Relationships

24.1 Balancing a nuclear equation (1044):
$$\text{Total } A \atop \text{Total } Z} \text{ Reactants} = {\text{Total } A \atop \text{Total } Z} \text{ Products}$$
24.2 Defining the unit of radioactivity (curie, Ci) (1050):
$$1 \text{ Ci} = 3.70 \times 10^{10} \text{ disintegrations per second (d/s)}$$
24.3 Expressing the decay rate (activity) for radioactive nuclei (1050):
$$\text{Decay rate } (\mathcal{A}) = -\frac{\Delta \mathcal{N}}{\Delta t} = k\mathcal{N}$$
24.4 Finding the number of nuclei remaining after a given time, $\mathcal{N}_t$ (1052):
$$\ln \frac{\mathcal{N}_0}{\mathcal{N}_t} = kt$$

24.5 Finding the half-life of a radioactive nuclide (1052):
$$t_{1/2} = \frac{\ln 2}{k}$$
24.6 Calculating the time to reach a given specific activity (age of an object in radioisotopic dating) (1054):
$$t = \frac{1}{k} \ln \frac{\mathcal{A}_0}{\mathcal{A}_t}$$
24.7 Using Einstein's equation and the mass defect to calculate the nuclear binding energy (1066 and 1067):
$$\Delta E = \Delta mc^2$$
24.8 Relating the atomic mass unit to its energy equivalent in MeV (1067):
$$1 \text{ amu} = 931.5 \times 10^6 \text{ eV} = 931.5 \text{ MeV}$$

Highlighted Figures and Tables

These figures (F) and tables (T) provide a quick review of key ideas.

T24.1 Chemical vs. nuclear reactions (1041)
T24.2 Modes of radioactive decay (1044)
F24.2 N vs. Z for the stable nuclides (1047)

F24.4 Decrease in number of ^{14}C nuclei over time (1052)
F24.12 The variation in binding energy per nucleon (1069)

Brief Solutions to Follow-up Problems

24.1 $^{133}_{54}\text{Xe} \longrightarrow {}^{133}_{55}\text{Cs} + {}^{0}_{-1}\beta$
24.2 Phosphorus-31 has a slightly higher N/Z ratio and an even N (16).
24.3 (a) N/Z = 1.35; too high for this region of band: β decay
(b) Mass too high for stability: α decay
24.4 $\ln \mathcal{A}_t = -kt + \ln \mathcal{A}_0$
$$= -\left(\frac{\ln 2}{15 \text{ h}} \times 4.0 \text{ days} \times \frac{24 \text{ h}}{1 \text{ day}}\right) + \ln (2.5 \times 10^9)$$
$$= 17.20$$
$$\mathcal{A}_t = 3.0 \times 10^7 \text{ d/s}$$

24.5 $t = \dfrac{1}{k} \ln \dfrac{\mathcal{A}_0}{\mathcal{A}_t} = \dfrac{5730 \text{ yr}}{\ln 2} \ln \left(\dfrac{15.3 \text{ d/min·g}}{9.41 \text{ d/min·g}}\right) = 4.02 \times 10^3 \text{ yr}$
The mummy case is about 4000 years old.
24.6 ^{235}U has 92 ^1_1p and 143 ^1_0n.
$\Delta m = [(92 \times 1.007825 \text{ amu}) + (143 \times 1.008665 \text{ amu})]$
$$- 235.043924 \text{ amu} = 1.9151 \text{ amu}$$
$$\frac{\text{Binding energy}}{\text{nucleon}} = \frac{1.9151 \text{ amu} \times \dfrac{931.5 \text{ MeV}}{1 \text{ amu}}}{235 \text{ nucleons}}$$
$$= 7.591 \text{ MeV/nucleon}$$
Therefore, ^{235}U is less stable than ^{12}C.

Problems

Problems with **colored** numbers are answered at the back of the text. Sections match the text and provide the number(s) of relevant sample problems. Most offer Concept Review Questions, Skill-Building Exercises (in similar pairs), and Problems in Context. Then Comprehensive Problems, based on material from any section or previous chapter, follow.

Radioactive Decay and Nuclear Stability
(Sample Problems 24.1 to 24.3)

● **Concept Review Questions**

24.1 How do chemical and nuclear reactions differ in
(a) Magnitude of the energy change?
(b) Effect on rate of increasing temperature?
(c) Effect on rate of higher reactant concentration?
(d) Effect on yield of higher reactant concentration?
24.2 Sulfur has four naturally occurring isotopes. The one with the lowest mass number is sulfur-32, which is also the most abundant (95.02%).
(a) What percentage of the S atoms in a matchhead are ^{32}S?
(b) The isotopic mass of ^{32}S is 31.972070 amu. Is the atomic mass larger, smaller, or equal to this mass? Explain.
24.3 What evidence led Marie Curie to draw the following conclusions?
(a) Radioactivity is a property of the element and not the compound in which it is found.
(b) A highly radioactive element, aside from uranium, occurs in pitchblende.
24.4 Which of the following types of radioactive decay produce an atom of a *different* element: (a) alpha; (b) beta; (c) gamma; (d) positron; (e) electron capture? Show how Z and N change, if at all, with each type.
24.5 Why is $_2^3$He stable whereas $_2^2$He is so unstable that it has never been detected?
24.6 How do the modes of decay differ for a neutron-rich nuclide and a proton-rich nuclide?
24.7 Why can't you use the position of a nuclide's N/Z ratio relative to the band of stability to predict whether it is more likely to decay by positron emission or by electron capture?

● **Skill-Building Exercises (paired)**

24.8 Write balanced nuclear equations for the following:
(a) Alpha decay of $_{92}^{234}$U
(b) Electron capture by neptunium-232
(c) Positron emission by $_7^{12}$N
24.9 Write balanced nuclear equations for the following:
(a) Beta decay of sodium-26
(b) Beta decay of francium-223
(c) Alpha decay of $_{83}^{212}$Bi

24.10 Write balanced nuclear equations for the following:
(a) Beta emission by magnesium-27
(b) Neutron emission by $_3^9$Li
(c) Electron capture by $_{46}^{103}$Pd
24.11 Write balanced nuclear equations for the following:
(a) Simultaneous β and neutron emission by helium-8
(b) Alpha decay of polonium-218
(c) Electron capture by $_{49}^{110}$In

24.12 Write balanced nuclear equations for the following:
(a) Formation of $_{22}^{48}$Ti through positron emission
(b) Formation of silver-107 through electron capture
(c) Formation of polonium-206 through α decay
24.13 Write balanced nuclear equations for the following:
(a) Production of $_{95}^{241}$Am through β decay
(b) Formation of $_{89}^{228}$Ac through β decay
(c) Formation of $_{83}^{203}$Bi through α decay

24.14 Write balanced nuclear equations for the following:
(a) Formation of ^{186}Ir through electron capture
(b) Formation of francium-221 through α decay
(c) Formation of iodine-129 through β decay
24.15 Write balanced nuclear equations for the following:
(a) Formation of ^{52}Mn through positron emission
(b) Formation of polonium-215 through α decay
(c) Formation of ^{81}Kr through electron capture

24.16 Which nuclide(s) would you predict to be stable? Why?
(a) $_8^{20}$O (b) $_{27}^{59}$Co (c) $_3^9$Li
24.17 Which nuclide(s) would you predict to be stable? Why?
(a) $_{60}^{146}$Nd (b) $_{48}^{114}$Cd (c) $_{42}^{88}$Mo

24.18 Which nuclide(s) would you predict to be stable? Why?
(a) ^{127}I (b) tin-106 (c) ^{68}As
24.19 Which nuclide(s) would you predict to be stable? Why?
(a) ^{48}K (b) ^{79}Br (c) argon-32

24.20 What is the most likely mode of decay for each?
(a) $_{92}^{238}$U (b) $_{24}^{48}$Cr (c) $_{25}^{50}$Mn
24.21 What is the most likely mode of decay for each?
(a) $_{26}^{61}$Fe (b) $_{17}^{41}$Cl (c) $_{44}^{110}$Ru

24.22 What is the most likely mode of decay for each?
(a) ^{15}C (b) ^{120}Xe (c) ^{224}Th
24.23 What is the most likely mode of decay for each?
(a) ^{234}Th (b) ^{141}Eu (c) ^{241}Am

24.24 Why is $_{24}^{52}$Cr the most stable isotope of chromium?
24.25 Why is $_{20}^{40}$Ca the most stable isotope of calcium?

● **Problems in Context**

24.26 Only traces of neptunium-237 remain in nature, but it is the first nuclide in a decay series that starts with emission of an α particle, followed by a β particle, and then two more α particles. Write a balanced nuclear equation for each decay step.
24.27 Why is helium found in deposits of uranium and thorium ores? What kind of radioactive emission produces it?
24.28 In the natural radioactive series that starts with uranium-235, a sequence of α and β emissions ends with lead-207. How many α and β particles are emitted per atom of uranium-235 to result in an atom of lead-207?

The Kinetics of Radioactive Decay
(Sample Problems 24.4 and 24.5)

● **Concept Review Questions**

24.29 What electronic process is the basis for detecting radioactive emissions in (a) a scintillation counter; (b) a Geiger-Müller counter?
24.30 What is the kinetic reaction order of radioactive decay? Explain.

24.31 After 1 minute, half the radioactive nuclei remain from an original sample of six nuclei. Is it valid to conclude that $t_{1/2}$ equals 1 minute? Would this conclusion be valid if the original sample contained 6×10^{12} nuclei? Explain.

24.32 Radioisotopic dating depends on the constant rate of decay and formation of various nuclides in a sample. How is the proportion of ^{14}C kept relatively constant in living organisms?

● **Skill-Building Exercises** *(paired)*

24.33 What is the specific activity (in Ci/g) if 1.55 mg of an isotope emits 1.66×10^6 α particles per second?

24.34 What is the specific activity (in Ci/g) if 2.6 g of an isotope emits 4.13×10^8 β particles per hour?

24.35 What is the specific activity (in Bq/g) if 8.58 μg of an isotope emits 7.4×10^4 α particles per minute?

24.36 What is the specific activity (in Bq/g) if 1.07 kg of an isotope emits 3.77×10^7 β particles per minute?

24.37 One-trillionth of the atoms in a sample of radioactive isotope disintegrate each hour. What is the decay constant of the process?

24.38 Measurement shows that $2.8 \times 10^{-10}\%$ of the atoms in a sample of radioactive isotope disintegrate in 1.0 yr. What is the decay constant of the process?

24.39 If 1.00×10^{-12} mol of ^{135}Cs emits 1.39×10^5 β particles in 1.00 yr, what is the decay constant?

24.40 If 6.40×10^{-9} mol of ^{176}W emits 1.07×10^{15} positrons in 1.00 h, what is the decay constant?

24.41 The isotope $^{212}_{83}Bi$ has a half-life of 1.01 yr. What mass (in mg) of a 2.00-mg sample will not decay after 3.75×10^3 h?

24.42 The half-life of radium-226 is 1.60×10^3 yr. How many hours will it take for a 2.50-g sample to decay until 0.185 g of the isotope remains?

24.43 A rock contains 270 μmol of ^{238}U ($t_{1/2} = 4.5 \times 10^9$ yr) and 110 μmol of ^{206}Pb. Assuming that all the ^{206}Pb comes from decay of the ^{238}U, estimate the rock's age.

24.44 A fabric remnant from a burial site has a $^{14}C{:}^{12}C$ ratio of 0.735 of the original value. How old is the fabric?

● **Problems in Context**

24.45 Due to decay of ^{40}K, cow's milk has a specific activity of about 6×10^{-11} mCi per milliliter. How many disintegrations of ^{40}K nuclei are there per minute in 1.0 qt of milk?

24.46 Plutonium-239 ($t_{1/2} = 2.41 \times 10^4$ yr) represents a serious nuclear waste disposal problem. If seven half-lives are required to reach a tolerable level of radioactivity, how long must the ^{239}Pu be stored?

24.47 A rock that contains 2.1×10^{-15} mol of ^{232}Th ($t_{1/2} = 1.4 \times 10^{10}$ yr) has 9.5×10^4 fission tracks, each representing the fission of one atom of ^{232}Th. How old does the rock appear to be?

24.48 Volcanism was much more common in the distant past than today, and in some cases, specific events can be dated. A volcanic eruption melts a large area of rock, and all gases are expelled. After cooling, $^{40}_{18}Ar$ accumulates from the ongoing decay of $^{40}_{19}K$ in the rock ($t_{1/2} = 1.25 \times 10^9$ yr). When a piece of rock is analyzed, it is found to contain 1.38 mmol of ^{40}K and 1.14 mmol of ^{40}Ar. How long ago did the rock cool?

Nuclear Transmutation: Induced Changes in Nuclei

● **Concept Review Questions**

24.49 Irene and Frederic Curie-Joliot converted $^{27}_{13}Al$ to $^{30}_{15}P$ in 1933. Why was this transmutation significant?

24.50 Early workers mistakenly thought neutron beams were γ radiation. Why were they misled? What evidence led to the correct conclusion?

24.51 Why must the electrical polarity of the tubes in a linear accelerator be reversed at very short time intervals?

24.52 Why does bombardment with protons usually require higher energies than bombardment with neutrons?

● **Skill-Building Exercises** *(paired)*

24.53 Determine the missing species in these transmutations and write a full nuclear equation from the shorthand notation:
(a) ^{10}B (α,n) __
(b) ^{28}Si (d,__) ^{29}P (the deuteron, d, is ^2H)
(c) __ (α,2n) ^{244}Cf.

24.54 Determine the missing species in these transmutations and express the process in shorthand notation:
(a) Bombardment of a nuclide with a γ photon yields a proton, a neutron, and ^{29}Si.
(b) Bombardment of ^{252}Cf with ^{10}B yields five neutrons and a nuclide.
(c) Bombardment of ^{238}U with a particle yields three neutrons and ^{239}Pu.

● **Problems in Context**

24.55 Names for elements 104, 105, and 106 have been approved as rutherfordium (Rf), dubnium (Db), and seaborgium (Sg), respectively. These elements have been synthesized from californium-249 by bombardment with carbon-12, nitrogen-15, and oxygen-18 nuclei, respectively. Four neutrons are formed in each reaction as well. (a) Write balanced nuclear equations for the formation of these elements. (b) Write the equations in shorthand notation.

The Effects of Nuclear Radiation on Matter

● **Concept Review Questions**

24.56 Gamma radiation and ultraviolet radiation cause different processes to occur in matter. What are these processes, and how do they differ?

24.57 What is a cation-electron pair, and how does it form?

24.58 Why is ionizing radiation more dangerous to children than to adults?

24.59 Why is ·OH more dangerous in an organism than OH^-?

● **Skill-Building Exercises** *(paired)*

24.60 A 135-lb person absorbs 3.3×10^{-7} J of energy from radioactive emissions. (a) How many rads does she receive? (b) How many grays (Gy) does she receive?

24.61 A 3.6-kg laboratory animal receives a single dose of 8.92×10^{-4} Gy. (a) How many rads did the animal receive? (b) How many joules did the animal absorb?

24.62 A 70.-kg person exposed to ^{90}Sr absorbs 6.0×10^5 β particles, each with an energy of 8.74×10^{-14} J. (a) How many grays does the person receive? (b) If the RBE is 1.0, how many millirems is this? (c) What is the equivalent dose in sieverts (Sv)?

24.63 A laboratory rat weighs 265 g and absorbs 1.77×10^{10} β particles, each with an energy of 2.20×10^{-13} J. (a) How many rads does the animal receive? (b) What is this dose in Gy? (c) If the RBE is 0.75, what is the equivalent dose in Sv?

● **Problems in Context**

24.64 If 2.50 pCi (1 pCi, picocurie = 1×10^{-12} Ci) of ^{239}Pu stays in a 95-kg human for 65 h, and each disintegration has an energy of 8.25×10^{-13} J, how many grays does the person receive?

24.65 A small portion of a cancer patient's brain is exposed for 27.0 min to 475 Bq of ^{60}Co for treatment of a tumor. If the brain mass exposed is 1.588 g and each β particle emitted has an energy of 5.05×10^{-14} J, what is the dose in rads?

Applications of Radioisotopes

● **Concept Review Questions**

24.66 Describe two ways that radioactive tracers are used in organisms.

24.67 Why is neutron activation analysis (NAA) useful to art historians and criminologists?

24.68 Positrons cannot penetrate matter more than a few atomic diameters, but positron emission of radiotracers can be monitored in medical diagnosis. Explain.

24.69 A steel part is treated to form some iron-59. Oil used to lubricate the part emits 298 β particles (with the energy characteristic of ^{59}Fe) per minute per milliliter of oil. What other information would you need to calculate the rate of removal of the steel from the part during use?

● **Problems in Context**

24.70 The oxidation of methanol to formaldehyde can be accomplished by reaction with chromic acid:

$$6H^+(aq) + 3CH_3OH(aq) + 2H_2CrO_4(aq) \longrightarrow$$
$$3CH_2O(aq) + 2Cr^{3+}(aq) + 8H_2O(l)$$

The reaction can be studied with the stable isotope tracer ^{18}O and mass spectrometry. When a small amount of $CH_3^{18}OH$ is present in the alcohol reactant, $H_2C^{18}O$ forms. When a small amount of $H_2Cr^{18}O_4$ is present, $H_2^{18}O$ forms. Does chromic acid or methanol supply the O atom to the aldehyde? Explain.

The Interconversion of Mass and Energy

(Sample Problem 24.6)

Note: Use the following data to solve the problems in this section: mass of ^{1}H atom = 1.007825 amu; mass of neutron = 1.008665 amu.

● **Concept Review Questions**

24.71 Many scientists at first reacted skeptically to Einstein's equation, $E = mc^2$. Why?

24.72 What is a mass defect, and how does it arise?

24.73 When a nuclide forms from nucleons, is energy absorbed or released? Why?

24.74 What is the binding energy per nucleon? Why is the binding energy per nucleon, rather than per nuclide, used to compare nuclide stability?

● **Skill-Building Exercises (paired)**

24.75 A ^{3}H nucleus decays with an energy of 0.01861 MeV. Convert this energy into (a) electron volts; (b) joules.

24.76 Arsenic-84 decays with an energy of 1.57×10^{-15} kJ per nucleus. Convert this energy into (a) eV; (b) MeV.

24.77 How many joules are released when 1.0 mol of ^{239}Pu decays, if each nucleus releases 5.243 MeV?

24.78 How many MeV are released per nucleus when 3.2×10^{-3} mol of chromium-49 releases 8.11×10^5 kJ?

24.79 Oxygen-16 is one of the most stable nuclides. The mass of an ^{16}O atom is 15.994915 amu. Calculate the binding energy (a) per nucleon in MeV; (b) per atom in MeV; (c) per mole in kJ.

24.80 Lead-206 is the end product of ^{238}U decay. One ^{206}Pb atom has a mass of 205.974440 amu. Calculate the binding energy (a) per nucleon in MeV; (b) per atom in MeV; (c) per mole in kJ.

24.81 Cobalt-59 is the only stable isotope of this transition metal. One ^{59}Co atom has a mass of 58.933198 amu. Calculate the binding energy (a) per nucleon in MeV; (b) per atom in MeV; (c) per mole in kJ.

24.82 Iodine-131 is one of the most important isotopes used in medical diagnosis for thyroid cancer. One atom has a mass of 130.906114 amu. Calculate the binding energy (a) per nucleon in MeV; (b) per atom in MeV; (c) per mole in kJ.

● **Problems in Context**

24.83 The ^{80}Br nuclide decays either by β decay or by electron capture. (a) What is the product of each process? (b) Which process releases more energy? (Masses of atoms: ^{80}Br = 79.918528 amu; ^{80}Kr = 79.916380 amu; ^{80}Se = 79.916520 amu; neglect the mass of the electron involved.)

Applications of Fission and Fusion

● **Concept Review Questions**

24.84 What is the minimum number of neutrons from each fission event that must be absorbed by other nuclei for a chain reaction to occur?

24.85 In what main way is fission different from radioactive decay? Are all fission events in a chain reaction identical? Explain.

24.86 What is the purpose of enrichment in the preparation of fuel rods? How is it accomplished?

24.87 Describe the nature and purpose of the following components of a nuclear reactor:
(a) control rods   (b) moderator   (c) reflector

24.88 State an advantage and a disadvantage of heavy-water reactors compared to light-water reactors.

24.89 What are the expected advantages of fusion reactors over fission reactors?

24.90 Account for the fact that there is a greater mass of iron in the Earth than any other element.

24.91 Why do so many nuclides have isotopic masses that are close to multiples of 4 amu?

24.92 What is the cosmic importance of unstable ^{8}Be?

● **Problems in Context**

24.93 The reaction that will probably power the first commercial fusion reactor is

$$^3_1H + {}^2_1H \longrightarrow {}^4_2He + {}^1_0n$$

How much energy would be produced per mole of reaction? (Masses of atoms: $^3_1$H = 3.01605 amu; $^2_1$H = 2.0140 amu; $^4_2$He = 4.00260 amu; mass of $^1_0$n = 1.008665 amu.)

Comprehensive Problems

Problems with an asterisk (*) are more challenging.

24.94 Some $^{243}_{95}$Am was present when the Earth formed, but it all decayed in the next billion years. The first three steps in this decay are emission of an α particle, emission of a β particle, and emission of another α particle. What other isotopes were present on the young Earth in a rock that contained some $^{243}_{95}$Am?

24.95 Curium-243 undergoes α-decay to plutonium-239:

$$^{243}Cm \longrightarrow {}^{239}Pu + {}^4He$$

(a) Calculate the change in mass, Δm (in kg). (Masses: $^{243}Cm = 243.0614$ amu; $^{239}Pu = 239.0522$ amu; $^4He = 4.0026$ amu; 1 amu $= 1.661 \times 10^{-24}$ g.)
(b) Calculate the energy released in joules.
(c) Calculate the energy released in kJ/mol of reaction, and comment on the difference between this value and a typical heat of reaction for a chemical change of a few hundred kJ/mol.

24.96 Plutonium "triggers" for nuclear weapons were manufactured at the Rocky Flats plant in Colorado. An 85-kg worker inhales a dust particle containing 1.00 μg of $^{239}_{94}$Pu, which resides in his body for 16 h ($t_{1/2}$ of ^{239}Pu $= 2.41 \times 10^4$ yr; each disintegration releases 5.15 MeV).
(a) How many rads does he receive?
(b) How many grays?

*24.97** Archeologists removed pieces of charcoal from a Native American campfire, burned them in O_2, and bubbled the CO_2 formed into $Ca(OH)_2$ solution (limewater). The $CaCO_3$ that precipitated was filtered and oven dried. A 4.38-g sample of the $CaCO_3$ had a radioactivity of 3.2 d/min. How long ago was the campfire?

*24.98** A 5.4-μg sample of $^{226}RaCl_2$ has a radioactivity of 1.5×10^5 Bq. Calculate $t_{1/2}$ of ^{226}Ra.

24.99 How many rads does a 65-kg human receive each year from the approximately 10^{-8} g of $^{14}_6$C naturally present in her body ($t_{1/2} = 5730$ yr; each disintegration releases 0.156 MeV)?

24.100 The major reaction taking place during hydrogen burning in a young star is

$$4{}^1_1H \longrightarrow {}^4_2He + 2{}^0_1\beta + 2{}^0_0\gamma + \text{energy}$$

How much energy (in MeV) is released per He nucleus formed? Per mole of He? (Masses: ^1_1H atom $= 1.007825$ amu; ^4_2He atom $= 4.00260$ amu; positron $= 5.48580 \times 10^{-4}$ amu.)

24.101 A sample of AgCl emits 175 nCi/g. A saturated solution prepared from the solid emits 1.25×10^{-2} Bq/mL due to radioactive Ag^+ ions. What is the molar solubility of AgCl?

24.102 From the burning of fossil fuels, the proportion of total CO_2 in our atmosphere continues to increase. Moreover, as a result of nuclear explosions and similar events, the CO_2 also contains more ^{14}C. How will these factors affect the efforts of future archeologists to determine the apparent ages of our artifacts by radiocarbon dating?

24.103 What fraction of the ^{235}U ($t_{1/2} = 7.0 \times 10^8$ yr) created when the Earth was formed would remain after 2.8×10^9 yr?

*24.104** In the event of a nuclear accident, government and industrial radiation officers must obtain many pieces of data to decide on appropriate action.
(a) If a person ingests radioactive material, which of the following is the *most* important quantity in deciding whether a serious medical emergency has occurred?

1. The number of rems he receives
2. The number of curies he absorbs
3. The length of time he is exposed to the radiation
4. The number of moles of radioisotopes he ingests
5. The energy emitted per disintegration by the radioisotopes
(b) If the drinking water in a town becomes contaminated with radioactive material, what is the *most* important factor in deciding whether drastic and expensive action is warranted?
1. The radioactivity per volume, Ci/m^3
2. How long the water supply has been contaminated
3. $(Ci/m^3) \times$ energy per disintegration
4. The type of radiation emitted
5. The radioisotopes involved

24.105 Suggest a reason the critical mass of a fissionable substance depends on its shape.

*24.106** Technetium-99m is a metastable nuclide used in numerous cancer diagnostic and treatment programs. It is prepared just before use because it decays rapidly through γ emission:

$$^{99m}Tc \longrightarrow {}^{99}Tc + \gamma$$

Use the data below to determine:
(a) The half-life of ^{99m}Tc
(b) The percentage of the isotope that is lost if it takes 2.0 h to prepare and administer the dose

Time (h)	γ Emission (photons/s)
0	5000.
4	3150.
8	2000.
12	1250.
16	788
20	495

*24.107** How many curies are produced by 1.0 mol of ^{40}K ($t_{1/2} = 1.25 \times 10^9$ yr)? How many becquerels?

24.108 The fraction of radioactive isotope remaining at time t is

$$\left(\tfrac{1}{2}\right)^{t/t_{1/2}}$$

where $t_{1/2}$ is the half-life. If the half-life of carbon-14 is 5730 yr, what fraction of carbon-14 in a piece of charcoal remains after (a) 10.0 yr, (b) 10.0×10^3 yr; (c) 10.0×10^4 yr? (d) Why is radiocarbon dating more reliable for the fraction remaining in part (b) than that in part (a) or in part (c)?

*24.109** The isotopic mass of $^{210}_{86}$Rn is 209.989669 amu. When this nuclide decays by electron capture, it emits 2.368 MeV. What is the isotopic mass of the resulting nuclide?

24.110 Exactly 1/10th of the radioactive nuclei in a sample decay each hour. Thus, after n hours, the fraction of nuclei remaining is $(0.900)^n$. Find the value of n equal to one half-life.

24.111 In neutron activation analysis (NAA), stable isotopes are bombarded with neutrons. Depending on the isotope and the energy of the neutron, a wide variety of emissions are then observed. What are the products when the following neutron-activated species decay? Write an overall equation in nuclear shorthand notation for the reaction starting with the stable isotope before neutron activation.

(a) $^{52}_{23}V^* \longrightarrow$ [β emission]

(b) $^{64}_{29}Cu^* \longrightarrow$ [positron emission]

(c) $^{28}_{13}Al^* \longrightarrow$ [β emission]

*24.112 A metastable (excited) form of ^{50}Sc rearranges to its stable form by emitting γ radiation with a wavelength of 8.73 pm. What is the change in mass of 1 mol of the isotope when it undergoes this rearrangement?

24.113 Isotopic abundances are relatively constant throughout the Earth's crust. Could the science of chemistry have developed if, for example, one sample of tin(II) oxide contained mostly ^{112}Sn and another mostly ^{124}Sn? Explain.

*24.114 What volume of radon will be produced per hour at STP from 1.000 g of ^{226}Ra ($t_{1/2}$ = 1599 yr; 1 yr = 8766 h; mass of one ^{226}Ra atom = 226.025402 amu)?

24.115 A sample of ^{90}Kr ($t_{1/2}$ = 32 s) is to be used in a study of a patient's respiration. How soon after being made must it be administered to the patient if the activity must be at least 90% of the original activity?

24.116 Which isotope in each pair would you predict to be more stable? Why?

(a) $^{140}_{55}$Cs or $^{133}_{55}$Cs   (b) $^{79}_{35}$Br or $^{78}_{35}$Br

(c) $^{28}_{12}$Mg or $^{24}_{12}$Mg   (d) $^{14}_{7}$N or $^{18}_{7}$N

24.117 A sample of bone contains enough strontium-90 ($t_{1/2}$ = 29 yr) to emit 8.0×10^4 β particles per month. How long will it take for the emissions to decrease to 1.0×10^4 per month?

*24.118 The 23rd-century starship *Enterprise* uses a substance called "di-lithium crystals" as its fuel.

(a) Assuming this material is the result of fusion, what is the product of the fusion of two ^{6}Li nuclei?

(b) How much energy is released per kilogram of di-lithium formed? (Mass of one ^{6}Li atom is 6.015121 amu.)

(c) When four ^{1}H atoms fuse to form ^{4}He, how many positrons are released?

(d) To determine the energy potential of the fusion processes in parts (b) and (c), compare the changes in mass per kilogram of dilithium and of ^{4}He.

(e) Compare the change in mass in part (b) to that in the formation per kilogram of ^{4}He by the method used in current fusion reactors (Section 24.7). (For masses, see Problem 24.93.)

(f) Using early 21st-century fusion technology, how much tritium can be produced per kilogram of ^{6}Li in the following reaction: $^6_3Li + ^1_0n \longrightarrow ^4_2He + ^3_1H$? When this amount of tritium is fused with deuterium, what is the change in mass? How does this quantity compare with the use of di-lithium in part (b)?

24.119 Uranium and radium are found in differing amounts in many rocky soils throughout the world. Both undergo radioactive decay, and one of the products is radon-222, the heaviest noble gas ($t_{1/2}$ = 3.82 days). Inhalation of basement and room air containing this gas contributes to a large proportion of the lung cancers reported each year. According to current Environmental Protection Agency safety recommendations, the level of radioactivity from radon in homes should not exceed 4.0 pCi/L of air.

(a) What is the safe level of radon in Bq/L of air?

(b) A home has a radon measurement of 43.5 pCi/L. The owner vents the basement air in such a way that no more radon enters the living area. What is the activity of the radon remaining in the room air (in Bq/L) after 8.5 days?

(c) How many more days does it take to reach the EPA recommended level?

24.120 Nuclear disarmament could be accomplished if weapons were not "replenished." The tritium in nuclear warheads decays to helium with a half-life of 12.26 yr, and it must be periodically replaced or the weapon is useless. What fraction of the tritium is lost in 5.50 yr?

24.121 A decay series starts with the synthetic isotope $^{239}_{92}$U. The first four steps are emissions of a β particle, another β, an α particle, and another α. Write a balanced nuclear equation for each step. Which natural radioactive series could be started by this sequence?

24.122 How long can a 48-lb child be exposed to 1.0 mCi of radiation from ^{222}Rn before accumulating 1.0 mrad if the energy of each disintegration is 5.59 MeV?

24.123 The approximate date of an earthquake in the San Francisco area is to be determined by measuring the ^{14}C activity ($t_{1/2}$ = 5730 yr) of parts of a tree that was uprooted during the event. The tree parts have an activity of 12.9 d/min·g C, and a living tree has an activity of 15.3 d/min·g C. How long ago did the earthquake occur?

24.124 Were organisms a billion years ago exposed to more or less ionizing radiation than similar organisms today? Explain.

24.125 Tritium (^{3}H; $t_{1/2}$ = 12.26 yr) is continually formed in the upper troposphere by interaction of solar particles with nitrogen. As a result, natural waters contain a small amount of tritium. Two samples of wine are analyzed, one known to be made in 1941, and another that may have been made as early as 1902, but is suspected of being younger. The water in the 1941 wine has 2.32 times as much tritium as the water in the other. When was the other wine produced?

*24.126 Plutonium-239 ($t_{1/2}$ = 2.41×10^4 yr) is a serious radiation hazard present in spent uranium fuel from nuclear power plants. How many years does it take for 99% of the plutonium-239 in spent fuel to decay?

24.127 Carbon from the most recent remains of an extinct Australian marsupial, called *Diprotodon*, has a specific activity of 0.61 pCi/g. Modern carbon has a specific activity of 6.89 pCi/g. How long ago did the *Diprotodon* apparently become extinct?

24.128 The reaction that allows for radiocarbon dating is the continual formation of carbon-14 in the upper atmosphere:

$$^{14}_{7}N + ^1_0n \longrightarrow ^{14}_{6}C + ^1_1H$$

What is the energy change associated with this process in eV/reaction and in kJ/mol reaction? (Masses of atoms: $^{14}_{7}$N = 14.003074 amu; $^{14}_{6}$C = 14.003241 amu; $^1_1$H = 1.007825 amu; mass of $^1_0$n = 1.008665 amu.)

*24.129 The starship *Voyager*, like many other vessels of the newly designed 24th-century fleet, uses antimatter as fuel.

(a) How much energy is released when 1.00 kg each of antimatter and of matter annihilate each other?

(b) When the antimatter is atomic antihydrogen, a small amount of it is mixed with excess atomic hydrogen (gathered from interstellar space during flight). The annihilation releases so much heat that the remaining hydrogen nuclei fuse to form ^{4}He. If each hydrogen-antihydrogen collision releases enough heat to fuse 1.00×10^5 hydrogen atoms, how much energy (in kJ) is released per kilogram of antihydrogen?

(c) Which produces more energy per kilogram of antihydrogen, the procedure in part (a) or in part (b)?

*24.130 Using early 21st-century technology, hydrogen fusion requires temperatures around 10^8 K, but lower temperatures succeed if the hydrogen is compressed. In the late 24th century, the starship *Leinad* uses such methods to fuse hydrogen at 10^6 K.
(a) What is the kinetic energy of a hydrogen atom at 1.00×10^6 K?
(b) How many hydrogen atoms can be heated to 1.00×10^6 K from the energy released when one hydrogen atom and one antihydrogen atom collide and annihilate each other?
(c) If these hydrogen atoms fuse to form ^{4}He atoms (with the loss of two positrons for each ^{4}He formed), how much energy (in J) is generated?
(d) How much more energy is generated by the fusion in (c) than by the hydrogen-antihydrogen collision in (b)?
(e) Should the chief engineer of the *Leinad* advise the captain to change the technology and produce ^{3}He (mass = 3.01603 amu) instead of ^{4}He?

*24.131 Because radioactive decay is a random process, a random-number generator can be used to simulate the probability of a given atom decaying over a given time. For example, the formula "=RAND()" in the Excel spreadsheet returns a random number between 0 and 1, so for one radioactive atom and a time of one half-life, a random number less than 0.5 means the atom decays and a number greater than 0.5 means it doesn't.
(a) Place the "=RAND()" formula in cells A1 through A10 of an Excel spreadsheet. In cell B1, place "=IF(A1<0.5, 0, 1)". This formula returns 0 if A1 is < 0.5 (the atom decays) and 1 if A1 is > 0.5 (the atom does not decay). Place analogous formulas in cells B2 through B10 (using the "Fill Down" procedure in Excel). To determine the number of atoms remaining after one half-life, sum cells B1 through B10 by placing "=SUM(B1:B10)" in cell B12. To create a new set of random numbers, click on an empty cell (e.g., B13) and hit "Delete." Perform 10 simulations, each time recording the total number of atoms remaining. Do half of the atoms remain after each half-life? If not, why not?
(b) Increase the number of atoms to 100 by placing suitable formulas in cells A1 through A100, B1 through B100, and in cell B102. Perform 10 simulations and record the number of atoms remaining each time. Is this simulation more realistic for radioactive decay? Explain.

*24.132 In the following Excel-based simulation, the fate of 256 atoms is followed over five half-lives. Set up formulas in

columns A and B, as in Problem 24.131, and simulate the fate of the sample of 256 atoms over one half-life. Cells B1 through B256 should contain 1's and 0's. In cell C1, enter "=IF(B1=0, 0, RAND())". This returns 0 if the original atom decayed in the previous half-life or a random number between 0 and 1 if it did not. Fill down the formula in C1 to cell C256. Column D should have formulas similar to those in B, but with modified references, as should columns F, H, and J. Columns E, G, and I should have formulas similar to those in C, but with modified references. In cell B258, enter "=SUM(B1:B256)". This records the number of atoms remaining after the first half-life. Put formulas in cells D258, F258, H258, and J258 to record atoms remaining after subsequent half-lives.
(a) Ideally, how many atoms should remain after each half-life?
(b) Make a table of the atoms remaining after each half-life in four separate simulations. Compare these outcomes to the ideal outcome. How would you make this simulation more realistic?

24.133 What is the nuclear binding energy of a lithium-7 nucleus in units of kJ/mol and eV/nucleus? (Mass of a lithium-7 atom = 7.016003 amu.)

24.134 Uranium-238 undergoes a slow decay step ($t_{1/2} = 4.5 \times 10^9$ yr) followed by a series of relatively fast steps to form the stable isotope ^{206}Pb. Thus, on a timescale of billions of years, ^{238}U effectively decays "directly" to ^{206}Pb, and the relative amounts of these two isotopes are used to determine the age of some rocks (see Margin Note, p. 1055). Students 1 and 2 derive equations relating the number of half-lives (n) since the rock formed to the amounts of the two isotopes:
Student 1:

$$\left(\tfrac{1}{2}\right)^n = \frac{^{238}_{92}U}{^{206}_{82}Pb}$$

Student 2:

$$\left(\tfrac{1}{2}\right)^n = \frac{^{238}_{92}U}{^{238}_{92}U + ^{206}_{82}Pb}$$

(a) Which equation is correct, and why?
(b) If a rock contains exactly twice as much ^{238}U as ^{206}Pb, what is its age in years?

*24.135 Use Einstein's equation, the mass in grams of exactly 1 amu, and the relation between electron volts and joules to find the energy equivalent (in MeV) of a mass defect of exactly 1 amu.

Appendix A

COMMON MATHEMATICAL OPERATIONS IN CHEMISTRY

In addition to basic arithmetic and algebra, four mathematical operations are used frequently in general chemistry: manipulating logarithms, using exponential notation, solving quadratic equations, and graphing data. Each is discussed briefly below.

MANIPULATING LOGARITHMS

Meaning and Properties of Logarithms

A logarithm is an exponent. Specifically, if $x^n = A$, we can say that the logarithm to the base x of the number A is n, and we can denote it as

$$\log_x A = n$$

Because logarithms are exponents, they have the following properties:

$$\log_x 1 = 0$$

$$\log_x (A \times B) = \log_x A + \log_x B$$

$$\log_x \frac{A}{B} = \log_x A - \log_x B$$

$$\log_x A^y = y \log_x A$$

Types of Logarithms

Common and natural logarithms are used frequently in chemistry and the other sciences. For common logarithms, the base (x in the examples above) is 10, but they are written without specifying the base; that is, $\log_{10} A$ is written $\log A$. For example, the common logarithm of 1000 is 3; in other words, you must raise 10 to the 3rd power to obtain 1000:

$$\log 1000 = 3 \quad \text{or} \quad 10^3 = 1000$$

Similarly, we have

$$\log 10 = 1 \quad \text{or} \quad 10^1 = 10$$

$$\log 1{,}000{,}000 = 6 \quad \text{or} \quad 10^6 = 1{,}000{,}000$$

$$\log 0.001 = -3 \quad \text{or} \quad 10^{-3} = 0.001$$

$$\log 853 = 2.931 \quad \text{or} \quad 10^{2.931} = 853$$

The last example illustrates an important point about significant figures with all logarithms: the number of significant figures in the number equals the number of digits to the right of the decimal point in the logarithm. That is, the number 853 has three significant figures, and the logarithm 2.931 has three digits to the right of the decimal point.

To find a common logarithm with an electronic calculator, you simply enter the number and press the LOG button.

For natural logarithms, the base is the number e, which is $2.71828\ldots$, and $\log_e A$ is written $\ln A$. The relationship between the common and natural logarithms is easily obtained: because

$$\log 10 = 1 \quad \text{and} \quad \ln 10 = 2.303$$

we have

$$\ln A = 2.303 \log A$$

To find a natural logarithm with an electronic calculator, you simply enter the number and press the LN button. If your calculator does not have an LN button, enter the number, press the LOG button, and multiply by 2.303.

Antilogarithms

The antilogarithm is the number you obtain when you raise the base to the logarithm:

$$\text{antilogarithm (antilog) of } n \text{ is } 10^n$$

Using two of the earlier examples, the antilog of 3 is 1000, and the antilog of 2.931 is 853. To obtain the antilog with a calculator, you enter the number and press the 10^x button. Similarly, to obtain the natural antilogarithm, you enter the number and press the e^x button. [On some calculators, you enter the number and first press INV and then the LOG (or LN) button.]

USING EXPONENTIAL (SCIENTIFIC) NOTATION

Many quantities in chemistry are very large or very small. For example, in the conventional way of writing numbers, the number of gold atoms in 1 gram of gold is

$$59{,}060{,}000{,}000{,}000{,}000{,}000{,}000 \text{ atoms (to four significant figures)}$$

As another example, the mass in grams of one gold atom is

$$0.00000000000000000000003272 \text{ g (to four significant figures)}$$

Exponential (scientific) notation provides a much more practical way of writing such numbers. In exponential notation, we express numbers in the form

$$A \times 10^n$$

where A (the coefficient) is greater than or equal to 1 and less than 10 (that is, $1 \leq A < 10$), and n (the exponent) is an integer.

If the number we want to express in exponential notation is larger than 1, the exponent is positive ($n > 0$); if the number is smaller than 1, the exponent is negative ($n < 0$). The size of n tells the number of places the decimal point (in conventional notation) must be moved to obtain a coefficient A greater than or equal to 1 and less than 10 (in exponential notation). In exponential notation, 1 gram of gold contains 5.906×10^{22} atoms, and each gold atom has a mass of 3.272×10^{-22} g.

Changing Between Conventional and Exponential Notation

In order to use exponential notation, you must be able to convert to it from conventional notation, and vice versa.

1. To change a number from conventional to exponential notation, move the decimal point to the left for numbers equal to or greater than 10 and to the right for numbers between 0 and 1:

   $$75{,}000{,}000 \text{ changes to } 7.5 \times 10^7 \text{ (decimal point 7 places to the left)}$$

   $$0.006042 \text{ changes to } 6.042 \times 10^{-3} \text{ (decimal point 3 places to the right)}$$

2. To change a number from exponential to conventional notation, move the decimal point the number of places indicated by the exponent to the right for numbers with positive exponents and to the left for numbers with negative exponents:

   $$1.38 \times 10^5 \text{ changes to } 138{,}000 \text{ (decimal point 5 places to the right)}$$

   $$8.41 \times 10^{-6} \text{ changes to } 0.00000841 \text{ (decimal point 6 places to the left)}$$

3. An exponential number with a coefficient greater than 10 or less than 1 can be changed to the standard exponential form by converting the coefficient to the standard form and adding the exponents:

   $$582.3 \times 10^6 \text{ changes to } 5.823 \times 10^2 \times 10^6 = 5.823 \times 10^{(2+6)} = 5.823 \times 10^8$$

   $$0.0043 \times 10^{-4} \text{ changes to } 4.3 \times 10^{-3} \times 10^{-4} = 4.3 \times 10^{[(-3)+(-4)]} = 4.3 \times 10^{-7}$$

Using Exponential Notation in Calculations

In calculations, you can treat the coefficient and exponents separately and apply the properties of exponents (see earlier section on logarithms).

1. To multiply exponential numbers, multiply the coefficients, add the exponents, and reconstruct the number in standard exponential notation:

$$(5.5\times10^3)(3.1\times10^5) = (5.5 \times 3.1)\times10^{(3+5)} = 17\times10^8 = 1.7\times10^9$$

$$(9.7\times10^{14})(4.3\times10^{-20}) = (9.7 \times 4.3)\times10^{[14+(-20)]} = 42\times10^{-6} = 4.2\times10^{-5}$$

2. To divide exponential numbers, divide the coefficients, subtract the exponents, and reconstruct the number in standard exponential notation:

$$\frac{2.6\times10^6}{5.8\times10^2} = \frac{2.6}{5.8} \times 10^{(6-2)} = 0.45\times10^4 = 4.5\times10^3$$

$$\frac{1.7\times10^{-5}}{8.2\times10^{-8}} = \frac{1.7}{8.2} \times 10^{[(-5)-(-8)]} = 0.21\times10^3 = 2.1\times10^2$$

3. To add or subtract exponential numbers, change all numbers so that they have the same exponent, then add or subtract the coefficients:

$$(1.45\times10^4) + (3.2\times10^3) = (1.45\times10^4) + (0.32\times10^4) = 1.77\times10^4$$

$$(3.22\times10^5) - (9.02\times10^4) = (3.22\times10^5) - (0.902\times10^5) = 2.32\times10^5$$

SOLVING QUADRATIC EQUATIONS

A quadratic equation is one in which the highest power of x is 2. The general form of a quadratic equation is

$$ax^2 + bx + c = 0$$

where a, b, and c are numbers. For given values of a, b, and c, the values of x that satisfy the equation are called solutions of the equation. We calculate x with the quadratic formula:

$$x = \frac{-b \pm \sqrt{b^2 - 4ac}}{2a}$$

We commonly require the quadratic formula when solving for some concentration in an equilibrium problem. For example, if $x = [H_3O^+]$, we might have an expression that is rearranged into the quadratic equation

$$4.3x^2 + 0.65x - 8.7 = 0$$

Applying the quadratic formula, with $a = 4.3$, $b = 0.65$, and $c = -8.7$, gives

$$x = \frac{-0.65 \pm \sqrt{(0.65)^2 - 4(4.3)(-8.7)}}{2(4.3)}$$

The "plus or minus" sign ($\pm$) indicates that there are always two possible values for x. In this case, they are

$$x = 1.3 \quad \text{and} \quad x = -1.5$$

In any real physical system, however, only one of the values will have any meaning. In this case, for example, if x were $[H_3O^+]$, the negative value would mean a negative concentration, which has no meaning.

GRAPHING DATA IN THE FORM OF A STRAIGHT LINE

Visualizing changes in variables by means of a graph is a very useful technique in science. In many cases, it is most useful if the data can be graphed in the form of a straight line. Any equation will appear as a straight line if it has, or can be rearranged to have, the following general form:

$$y = mx + b$$

where y is the dependent variable (typically plotted along the vertical axis), x is the independent variable (typically plotted along the horizontal axis), m is the slope of the line, and b is the intercept of the line on the y axis. The intercept is the value of y when $x = 0$:

$$y = m(0) + b = b$$

The slope of the line is the change in y for a given change in x:

$$\text{Slope } (m) = \frac{y_2 - y_1}{x_2 - x_1} = \frac{\Delta y}{\Delta x}$$

The *sign* of the slope tells the *direction* of the line. If y increases as x increases, m is positive, and the line slopes upward with higher values of x; if y decreases as x increases, m is negative, and the line slopes downward with higher values of x. The *magnitude* of the slope indicates the *steepness* of the line. A line with $m = 3$ is three times as steep (y changes three times as much for a given change in x) as a line with $m = 1$.

Consider the linear equation $y = 2x + 1$. A graph of this equation is shown in Figure A.1. In practice, you can find the slope by drawing a right triangle to the line, using the line as the hypotenuse. Then, one leg gives Δy, and the other gives Δx. In the figure, $\Delta y = 8$ and $\Delta x = 4$.

At several places in the text, an equation is rearranged into the form of a straight line in order to determine information from the slope and/or the intercept. For example, in Chapter 16, we obtained the following expression:

$$\ln \frac{[A]_0}{[A]_t} = kt$$

Based on the properties of logarithms, we have

$$\ln [A]_0 - \ln [A]_t = kt$$

Rearranging into the form of an equation for a straight line gives

$$\ln [A]_t = -kt + \ln [A]_0$$
$$y \quad = mx + \quad b$$

Thus, a plot of $\ln [A]_t$ versus t is a straight line, from which you can see that the slope is $-k$ (the negative of the rate constant) and the intercept is $\ln [A]_0$ (the natural logarithm of the initial concentration of A).

At many other places in the text, linear relationships occur that were not shown in graphical terms. For example, the conversion of temperatures scales in Chapter 1 can also be expressed in the form of a straight line:

$$°F = \tfrac{9}{5}°C + 32$$
$$y = mx + b$$

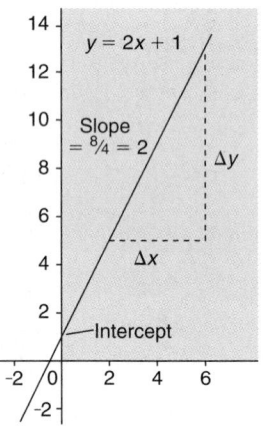

Figure A.1

Appendix B

STANDARD THERMODYNAMIC VALUES FOR SELECTED SUBSTANCES AT 298 K

(NOTE: See the rear endpaper for a list of other important data tables.)

Substance or ion	ΔH_f^0 (kJ/mol)	ΔG_f^0 (kJ/mol)	S^0 (J/mol·K)	Substance or ion	ΔH_f^0 (kJ/mol)	ΔG_f^0 (kJ/mol)	S^0 (J/mol·K)
$e^-(g)$	0	0	20.87	$CaF_2(s)$	−1215	−1162	68.87
Aluminum				$CaCl_2(s)$	−795.0	−750.2	114
$Al(s)$	0	0	28.3	$CaCO_3(s)$	−1206.9	−1128.8	92.9
$Al^{3+}(aq)$	−524.7	−481.2	−313	$CaO(s)$	−635.1	−603.5	38.2
$AlCl_3(s)$	−704.2	−628.9	110.7	$Ca(OH)_2(s)$	−986.09	−898.56	83.39
$Al_2O_3(s)$	−1676	−1582	50.94	$Ca_3(PO_4)_2(s)$	−4138	−3899	263
Barium				$CaSO_4(s)$	−1432.7	−1320.3	107
$Ba(s)$	0	0	62.5	Carbon			
$Ba(g)$	175.6	144.8	170.28	C(graphite)	0	0	5.686
$Ba^{2+}(g)$	1649.9	—	—	C(diamond)	1.896	2.866	2.439
$Ba^{2+}(aq)$	−538.36	−560.7	13	$C(g)$	715.0	669.6	158.0
$BaCl_2(s)$	−806.06	−810.9	126	$CO(g)$	−110.5	−137.2	197.5
$BaCO_3(s)$	−1219	−1139	112	$CO_2(g)$	−393.5	−394.4	213.7
$BaO(s)$	−548.1	−520.4	72.07	$CO_2(aq)$	−412.9	−386.2	121
$BaSO_4(s)$	−1465	−1353	132	$CO_3^{2-}(aq)$	−676.26	−528.10	−53.1
Boron				$HCO_3^-(aq)$	−691.11	587.06	95.0
B(β-rhombo-hedral)	0	0	5.87	$H_2CO_3(aq)$	−698.7	−623.42	191
$BF_3(g)$	−1137.0	−1120.3	254.0	$CH_4(g)$	−74.87	−50.81	186.1
$BCl_3(g)$	−403.8	−388.7	290.0	$C_2H_2(g)$	227	209	200.85
$B_2H_6(g)$	35	86.6	232.0	$C_2H_4(g)$	52.47	68.36	219.22
$B_2O_3(s)$	−1272	−1193	53.8	$C_2H_6(g)$	−84.667	−32.89	229.5
$H_3BO_3(s)$	−1094.3	−969.01	88.83	$C_3H_8(g)$	−105	−24.5	269.9
Bromine				$C_4H_{10}(g)$	−126	−16.7	310
$Br_2(l)$	0	0	152.23	$C_6H_6(l)$	49.0	124.5	172.8
$Br_2(g)$	30.91	3.13	245.38	$CH_3OH(g)$	−201.2	−161.9	238
$Br(g)$	111.9	82.40	174.90	$CH_3OH(l)$	−238.6	−166.2	127
$Br^-(g)$	−218.9	—	—	$HCHO(g)$	−116	−110	219
$Br^-(aq)$	−120.9	−102.82	80.71	$HCOO^-(aq)$	−410	−335	91.6
$HBr(g)$	−36.3	−53.5	198.59	$HCOOH(l)$	−409	−346	129.0
Cadmium				$HCOOH(aq)$	−410	−356	164
$Cd(s)$	0	0	51.5	$C_2H_5OH(l)$	−277.63	−174.8	161
$Cd(g)$	112.8	78.20	167.64	$C_2H_5OH(g)$	−235.1	−168.6	282.6
$Cd^{2+}(aq)$	−72.38	−77.74	−61.1	$CH_3CHO(g)$	−166	−133.7	266
$CdS(s)$	−144	−141	71	$CH_3COOH(l)$	−487.0	−392	160
Calcium				$C_6H_{12}O_6(s)$	−1273.3	−910.56	212.1
$Ca(s)$	0	0	41.6	$CN^-(aq)$	151	166	118
$Ca(g)$	192.6	158.9	154.78	$HCN(g)$	135	125	201.7
$Ca^{2+}(g)$	1934.1	—	—	$HCN(l)$	105	121	112.8
$Ca^{2+}(aq)$	−542.96	−553.04	−55.2	$HCN(aq)$	105	112	129
				$CS_2(g)$	117	66.9	237.79

Substance or ion	ΔH_f^0 (kJ/mol)	ΔG_f^0 (kJ/mol)	S^0 (J/mol·K)	Substance or ion	ΔH_f^0 (kJ/mol)	ΔG_f^0 (kJ/mol)	S^0 (J/mol·K)
$CS_2(l)$	87.9	63.6	151.0	Iron			
$CH_3Cl(g)$	−83.7	−60.2	234	$Fe(s)$	0	0	27.3
$CH_2Cl_2(l)$	−117	−63.2	179	$Fe^{3+}(aq)$	−47.7	−10.5	−293
$CHCl_3(l)$	−132	−71.5	203	$Fe^{2+}(aq)$	−87.9	−84.94	113
$CCl_4(g)$	−96.0	−53.7	309.7	$FeCl_2(s)$	−341.8	−302.3	117.9
$CCl_4(l)$	−139	−68.6	214.4	$FeCl_3(s)$	−399.5	−334.1	142
$COCl_2(g)$	−220	−206	283.74	$FeO(s)$	−272.0	−251.4	60.75
Cesium				$Fe_2O_3(s)$	−825.5	−743.6	87.400
$Cs(s)$	0	0	85.15	$Fe_3O_4(s)$	−1121	−1018	145.3
$Cs(g)$	76.7	49.7	175.5	Lead			
$Cs^+(g)$	458.5	427.1	169.72	$Pb(s)$	0	0	64.785
$Cs^+(aq)$	−248	−282.0	133	$Pb^{2+}(aq)$	1.6	−24.3	21
$CsF(s)$	−554.7	−525.4	88	$PbCl_2(s)$	−359	−314	136
$CsCl(s)$	−442.8	−414	101.18	$PbO(s)$	−218	−198	68.70
$CsBr(s)$	−395	−383	121	$PbO_2(s)$	−276.6	−219.0	76.6
$CsI(s)$	−337	−333	130	$PbS(s)$	−98.3	−96.7	91.3
Chlorine				$PbSO_4(s)$	−918.39	−811.24	147
$Cl_2(g)$	0	0	223.0	Lithium			
$Cl(g)$	121.0	105.0	165.1	$Li(s)$	0	0	29.10
$Cl^-(g)$	−234	−240	153.25	$Li(g)$	161	128	138.67
$Cl^-(aq)$	−167.46	−131.17	55.10	$Li^+(g)$	687.163	649.989	132.91
$HCl(g)$	−92.31	−95.30	186.79	$Li^+(aq)$	−278.46	−293.8	14
$HCl(aq)$	−167.46	−131.17	55.06	$LiF(s)$	−616.9	−588.7	35.66
$ClO_2(g)$	102	120	256.7	$LiCl(s)$	−408	−384	59.30
$Cl_2O(g)$	80.3	97.9	266.1	$LiBr(s)$	−351	−342	74.1
Chromium				$LiI(s)$	−270	−270	85.8
$Cr(s)$	0	0	23.8	Magnesium			
$Cr^{3+}(aq)$	−1971	—	—	$Mg(s)$	0	0	32.69
$CrO_4^{2-}(aq)$	−863.2	−706.3	38	$Mg(g)$	150	115	148.55
$Cr_2O_7^{2-}(aq)$	−1461	−1257	214	$Mg^{2+}(g)$	2351	—	—
Copper				$Mg^{2+}(aq)$	−461.96	−456.01	118
$Cu(s)$	0	0	33.1	$MgCl_2(s)$	−641.6	−592.1	89.630
$Cu(g)$	341.1	301.4	166.29	$MgCO_3(s)$	−1112	−1028	65.86
$Cu^+(aq)$	51.9	50.2	−26	$MgO(s)$	−601.2	−569.0	26.9
$Cu^{2+}(aq)$	64.39	64.98	−98.7	$Mg_3N_2(s)$	−461	−401	88
$Cu_2O(s)$	−168.6	−146.0	93.1	Manganese			
$CuO(s)$	−157.3	−130	42.63	$Mn(s, \alpha)$	0	0	31.8
$Cu_2S(s)$	−79.5	−86.2	120.9	$Mn^{2+}(aq)$	−219	−223	−84
$CuS(s)$	−53.1	−53.6	66.5	$MnO_2(s)$	−520.9	−466.1	53.1
Fluorine				$MnO_4^-(aq)$	−518.4	−425.1	190
$F_2(g)$	0	0	202.7	Mercury			
$F(g)$	78.9	61.8	158.64	$Hg(l)$	0	0	76.027
$F^-(g)$	−255.6	−262.5	145.47	$Hg(g)$	61.30	31.8	174.87
$F^-(aq)$	−329.1	−276.5	−9.6	$Hg^{2+}(aq)$	171	164.4	−32
$HF(g)$	−273	−275	173.67	$Hg_2^{2+}(aq)$	172	153.6	84.5
Hydrogen				$HgCl_2(s)$	−230	−184	144
$H_2(g)$	0	0	130.6	$Hg_2Cl_2(s)$	−264.9	−210.66	196
$H(g)$	218.0	203.30	114.60	$HgO(s)$	−90.79	−58.50	70.27
$H^+(aq)$	0	0	0	Nitrogen			
$H^+(g)$	1536.3	1517.1	108.83	$N_2(g)$	0	0	191.5
Iodine				$N(g)$	473	456	153.2
$I_2(s)$	0	0	116.14	$N_2O(g)$	82.05	104.2	219.7
$I_2(g)$	62.442	19.38	260.58	$NO(g)$	90.29	86.60	210.65
$I(g)$	106.8	70.21	180.67	$NO_2(g)$	33.2	51	239.9
$I^-(g)$	−194.7	—	—	$N_2O_4(g)$	9.16	97.7	304.3
$I^-(aq)$	−55.94	−51.67	109.4	$N_2O_5(g)$	11	118	346
$HI(g)$	25.9	1.3	206.33	$N_2O_5(s)$	−43.1	114	178

Substance or ion	ΔH_f^0 (kJ/mol)	ΔG_f^0 (kJ/mol)	S^0 (J/mol·K)	Substance or ion	ΔH_f^0 (kJ/mol)	ΔG_f^0 (kJ/mol)	S^0 (J/mol·K)
$NH_3(g)$	−45.9	−16	193	Silver			
$NH_3(aq)$	−80.83	26.7	110	$Ag(s)$	0	0	42.702
$N_2H_4(l)$	50.63	149.2	121.2	$Ag(g)$	289.2	250.4	172.892
$NO_3^-(aq)$	−206.57	−110.5	146	$Ag^+(aq)$	105.9	77.111	73.93
$HNO_3(l)$	−173.23	−79.914	155.6	$AgF(s)$	−203	−185	84
$HNO_3(aq)$	−206.57	−110.5	146	$AgCl(s)$	−127.03	−109.72	96.11
$NF_3(g)$	−125	−83.3	260.6	$AgBr(s)$	−99.51	−95.939	107.1
$NOCl(g)$	51.71	66.07	261.6	$AgI(s)$	−62.38	−66.32	114
$NH_4Cl(s)$	−314.4	−203.0	94.6	$AgNO_3(s)$	−45.06	19.1	128.2
Oxygen				$Ag_2S(s)$	−31.8	−40.3	146
$O_2(g)$	0	0	205.0	Sodium			
$O(g)$	249.2	231.7	160.95	$Na(s)$	0	0	51.446
$O_3(g)$	143	163	238.82	$Na(g)$	107.76	77.299	153.61
$OH^-(aq)$	−229.94	−157.30	−10.54	$Na^+(g)$	609.839	574.877	147.85
$H_2O(g)$	−241.826	−228.60	188.72	$Na^+(aq)$	−239.66	−261.87	60.2
$H_2O(l)$	−285.840	−237.192	69.940	$NaF(s)$	−575.4	−545.1	51.21
$H_2O_2(l)$	−187.8	−120.4	110	$NaCl(s)$	−411.1	−384.0	72.12
$H_2O_2(aq)$	−191.2	−134.1	144	$NaBr(s)$	−361	−349	86.82
Phosphorus				$NaOH(s)$	−425.609	−379.53	64.454
$P_4(s, white)$	0	0	41.1	$Na_2CO_3(s)$	−1130.8	−1048.1	139
$P(g)$	314.6	278.3	163.1	$NaHCO_3(s)$	−947.7	−851.9	102
$P(s, red)$	−17.6	−12.1	22.8	$NaI(s)$	−288	−285	98.5
$P_2(g)$	144	104	218	Strontium			
$P_4(g)$	58.9	24.5	280	$Sr(s)$	0	0	54.4
$PCl_3(g)$	−287	−268	312	$Sr(g)$	164	110	164.54
$PCl_3(l)$	−320	−272	217	$Sr^{2+}(g)$	1784	—	—
$PCl_5(g)$	−402	−323	353	$Sr^{2+}(aq)$	−545.51	−557.3	−39
$PCl_5(s)$	−443.5	—	—	$SrCl_2(s)$	−828.4	−781.2	117
$P_4O_{10}(s)$	−2984	−2698	229	$SrCO_3(s)$	−1218	−1138	97.1
$PO_4^{3-}(aq)$	−1266	−1013	−218	$SrO(s)$	−592.0	−562.4	55.5
$HPO_4^{2-}(aq)$	−1281	−1082	−36	$SrSO_4(s)$	−1445	−1334	122
$H_2PO_4^-(aq)$	−1285	−1135	89.1	Sulfur			
$H_3PO_4(aq)$	−1277	−1019	228	$S_8(rhombic)$	0	0	31.9
Potassium				$S_8(monoclinic)$	0.3	0.096	32.6
$K(s)$	0	0	64.672	$S(g)$	279	239	168
$K(g)$	89.2	60.7	160.23	$S_2(g)$	129	80.1	228.1
$K^+(g)$	514.197	481.202	154.47	$S_8(g)$	101	49.1	430.211
$K^+(aq)$	−251.2	−282.28	103	$S^{2-}(aq)$	41.8	83.7	22
$KF(s)$	−568.6	−538.9	66.55	$HS^-(aq)$	−17.7	12.6	61.1
$KCl(s)$	−436.7	−409.2	82.59	$H_2S(g)$	−20.2	−33	205.6
$KBr(s)$	−394	−380	95.94	$H_2S(aq)$	−39	−27.4	122
$KI(s)$	−328	−323	106.39	$SO_2(g)$	−296.8	−300.2	248.1
$KOH(s)$	−424.8	−379.1	78.87	$SO_3(g)$	−396	−371	256.66
$KClO_3(s)$	−397.7	−296.3	143.1	$SO_4^{2-}(aq)$	−907.51	−741.99	17
$KClO_4(s)$	−432.75	−303.2	151.0	$HSO_4^-(aq)$	−885.75	−752.87	126.9
Rubidium				$H_2SO_4(l)$	−813.989	−690.059	156.90
$Rb(s)$	0	0	69.5	$H_2SO_4(aq)$	−907.51	−741.99	17
$Rb(g)$	85.81	55.86	169.99	Tin			
$Rb^+(g)$	495.04	—	—	$Sn(white)$	0	0	51.5
$Rb^+(aq)$	−246	−282.2	124	$Sn(gray)$	3	4.6	44.8
$RbF(s)$	−549.28	—	—	$SnCl_4(l)$	−545.2	−474.0	259
$RbCl(s)$	−435.35	−407.8	95.90	$SnO_2(s)$	−580.7	−519.7	52.3
$RbBr(s)$	−389.2	−378.1	108.3	Zinc			
$RbI(s)$	−328	−326	118.0	$Zn(s)$	0	0	41.6
Silicon				$Zn(g)$	130.5	94.93	160.9
$Si(s)$	0	0	18.0	$Zn^{2+}(aq)$	−152.4	−147.21	−106.5
$SiF_4(g)$	−1614.9	−1572.7	282.4	$ZnO(s)$	−348.0	−318.2	43.9
$SiO_2(s)$	−910.9	−856.5	41.5	$ZnS(s, zinc blende)$	−203	−198	57.7

Appendix C

SOLUBILITY-PRODUCT CONSTANTS (K_{sp}) OF SLIGHTLY SOLUBLE IONIC COMPOUNDS AT 298 K

Name, Formula	K_{sp}	Name, Formula	K_{sp}
Carbonates		Cobalt(II) hydroxide, $Co(OH)_2$	1.3×10^{-15}
Barium carbonate, $BaCO_3$	2.0×10^{-9}	Copper(II) hydroxide, $Cu(OH)_2$	2.2×10^{-20}
Cadmium carbonate, $CdCO_3$	1.8×10^{-14}	Iron(II) hydroxide, $Fe(OH)_2$	4.1×10^{-15}
Calcium carbonate, $CaCO_3$	3.3×10^{-9}	Iron(III) hydroxide, $Fe(OH)_3$	1.6×10^{-39}
Cobalt(II) carbonate, $CoCO_3$	1.0×10^{-10}	Magnesium hydroxide, $Mg(OH)_2$	6.3×10^{-10}
Copper(II) carbonate, $CuCO_3$	3×10^{-12}	Manganese(II) hydroxide, $Mn(OH)_2$	1.6×10^{-13}
Lead(II) carbonate, $PbCO_3$	7.4×10^{-14}	Nickel(II) hydroxide, $Ni(OH)_2$	6×10^{-16}
Magnesium carbonate, $MgCO_3$	3.5×10^{-8}	Zinc hydroxide, $Zn(OH)_2$	3×10^{-16}
Mercury(I) carbonate, Hg_2CO_3	8.9×10^{-17}	Iodates	
Nickel(II) carbonate, $NiCO_3$	1.3×10^{-7}	Barium iodate, $Ba(IO_3)_2$	1.5×10^{-9}
Strontium carbonate, $SrCO_3$	5.4×10^{-10}	Calcium iodate, $Ca(IO_3)_2$	7.1×10^{-7}
Zinc carbonate, $ZnCO_3$	1.0×10^{-10}	Lead(II) iodate, $Pb(IO_3)_2$	2.5×10^{-13}
Chromates		Silver iodate, $AgIO_3$	3.1×10^{-8}
Barium chromate, $BaCrO_4$	2.1×10^{-10}	Strontium iodate, $Sr(IO_3)_2$	3.3×10^{-7}
Calcium chromate, $CaCrO_4$	1×10^{-8}	Zinc iodate, $Zn(IO_3)_2$	3.9×10^{-6}
Lead(II) chromate, $PbCrO_4$	2.3×10^{-13}	Oxalates	
Silver chromate, Ag_2CrO_4	2.6×10^{-12}	Barium oxalate dihydrate, $BaC_2O_4 \cdot 2H_2O$	1.1×10^{-7}
Cyanides		Calcium oxalate monohydrate, $CaC_2O_4 \cdot H_2O$	2.3×10^{-9}
Mercury(I) cyanide, $Hg_2(CN)_2$	5×10^{-40}	Strontium oxalate monohydrate,	
Silver cyanide, $AgCN$	2.2×10^{-16}	$SrC_2O_4 \cdot H_2O$	5.6×10^{-8}
Halides		Phosphates	
Fluorides		Calcium phosphate, $Ca_3(PO_4)_2$	1.2×10^{-29}
Barium fluoride, BaF_2	1.5×10^{-6}	Magnesium phosphate, $Mg_3(PO_4)_2$	5.2×10^{-24}
Calcium fluoride, CaF_2	3.2×10^{-11}	Silver phosphate, Ag_3PO_4	2.6×10^{-18}
Lead(II) fluoride, PbF_2	3.6×10^{-8}	Sulfates	
Magnesium fluoride, MgF_2	7.4×10^{-9}	Barium sulfate, $BaSO_4$	1.1×10^{-10}
Strontium fluoride, SrF_2	2.6×10^{-9}	Calcium sulfate, $CaSO_4$	2.4×10^{-5}
Chlorides		Lead(II) sulfate, $PbSO_4$	1.6×10^{-8}
Copper(I) chloride, $CuCl$	1.9×10^{-7}	Radium sulfate, $RaSO_4$	2×10^{-11}
Lead(II) chloride, $PbCl_2$	1.7×10^{-5}	Silver sulfate, Ag_2SO_4	1.5×10^{-5}
Silver chloride, $AgCl$	1.8×10^{-10}	Strontium sulfate, $SrSO_4$	3.2×10^{-7}
Bromides		Sulfides	
Copper(I) bromide, $CuBr$	5×10^{-9}	Cadmium sulfide, CdS	1.0×10^{-24}
Silver bromide, $AgBr$	5.0×10^{-13}	Copper(II) sulfide, CuS	8×10^{-34}
Iodides		Iron(II) sulfide, FeS	8×10^{-16}
Copper(I) iodide, CuI	1×10^{-12}	Lead(II) sulfide, PbS	3×10^{-25}
Lead(II) iodide, PbI_2	7.9×10^{-9}	Manganese(II) sulfide, MnS	3×10^{-11}
Mercury(I) iodide, Hg_2I_2	4.7×10^{-29}	Mercury(II) sulfide, HgS	2×10^{-50}
Silver iodide, AgI	8.3×10^{-17}	Nickel(II) sulfide, NiS	1.1×10^{-18}
Hydroxides		Silver sulfide, Ag_2S	8×10^{-48}
Aluminum hydroxide, $Al(OH)_3$	3×10^{-34}	Tin(II) sulfide, SnS	1.3×10^{-23}
Cadmium hydroxide, $Cd(OH)_2$	7.2×10^{-15}	Zinc sulfide, ZnS	2.0×10^{-22}
Calcium hydroxide, $Ca(OH)_2$	6.5×10^{-6}		

Appendix D

STANDARD ELECTRODE (HALF-CELL) POTENTIALS AT 298 K*

Half-Reaction	E^0 (V)
$F_2(g) + 2e^- \rightleftharpoons 2F^-(aq)$	+2.87
$O_3(g) + 2H^+(aq) + 2e^- \rightleftharpoons O_2(g) + H_2O(l)$	+2.07
$Co^{3+}(aq) + e^- \rightleftharpoons Co^{2+}(aq)$	+1.82
$H_2O_2(aq) + 2H^+(aq) + 2e^- \rightleftharpoons 2H_2O(l)$	+1.77
$PbO_2(s) + 3H^+(aq) + HSO_4^-(aq) + 2e^- \rightleftharpoons PbSO_4(s) + 2H_2O(l)$	+1.70
$Ce^{4+}(aq) + e^- \rightleftharpoons Ce^{3+}(aq)$	+1.61
$MnO_4^-(aq) + 8H^+(aq) + 5e^- \rightleftharpoons Mn^{2+}(aq) + 4H_2O(l)$	+1.51
$Au^{3+}(aq) + 3e^- \rightleftharpoons Au(s)$	+1.50
$Cl_2(g) + 2e^- \rightleftharpoons 2Cl^-(aq)$	+1.36
$Cr_2O_7^{2-}(aq) + 14H^+(aq) + 6e^- \rightleftharpoons 2Cr^{3+}(aq) + 7H_2O(l)$	+1.33
$MnO_2(s) + 4H^+(aq) + 2e^- \rightleftharpoons Mn^{2+}(aq) + 2H_2O(l)$	+1.23
$O_2(g) + 4H^+(aq) + 4e^- \rightleftharpoons 2H_2O(l)$	+1.23
$Br_2(l) + 2e^- \rightleftharpoons 2Br^-(aq)$	+1.07
$NO_3^-(aq) + 4H^+(aq) + 3e^- \rightleftharpoons NO(g) + 2H_2O(l)$	+0.96
$2Hg^{2+}(aq) + 2e^- \rightleftharpoons Hg_2^{2+}(aq)$	+0.92
$Hg_2^{2+}(aq) + 2e^- \rightleftharpoons 2Hg(l)$	+0.85
$Ag^+(aq) + e^- \rightleftharpoons Ag(s)$	+0.80
$Fe^{3+}(aq) + e^- \rightleftharpoons Fe^{2+}(aq)$	+0.77
$O_2(g) + 2H^+(aq) + 2e^- \rightleftharpoons H_2O_2(aq)$	+0.68
$MnO_4^-(aq) + 2H_2O(l) + 3e^- \rightleftharpoons MnO_2(s) + 4OH^-(aq)$	+0.59
$I_2(s) + 2e^- \rightleftharpoons 2I^-(aq)$	+0.53
$O_2(g) + 2H_2O(l) + 4e^- \rightleftharpoons 4OH^-(aq)$	+0.40
$Cu^{2+}(aq) + 2e^- \rightleftharpoons Cu(s)$	+0.34
$AgCl(s) + e^- \rightleftharpoons Ag(s) + Cl^-(aq)$	+0.22
$SO_4^{2-}(aq) + 4H^+(aq) + 2e^- \rightleftharpoons SO_2(g) + 2H_2O(l)$	+0.20
$Cu^{2+}(aq) + e^- \rightleftharpoons Cu^+(aq)$	+0.15
$Sn^{4+}(aq) + 2e^- \rightleftharpoons Sn^{2+}(aq)$	+0.13
$2H^+(aq) + 2e^- \rightleftharpoons H_2(g)$	0.00
$Pb^{2+}(aq) + 2e^- \rightleftharpoons Pb(s)$	−0.13
$Sn^{2+}(aq) + 2e^- \rightleftharpoons Sn(s)$	−0.14
$N_2(g) + 5H^+(aq) + 4e^- \rightleftharpoons N_2H_5^+(aq)$	−0.23
$Ni^{2+}(aq) + 2e^- \rightleftharpoons Ni(s)$	−0.25
$Co^{2+}(aq) + 2e^- \rightleftharpoons Co(s)$	−0.28
$PbSO_4(s) + H^+(aq) + 2e^- \rightleftharpoons Pb(s) + HSO_4^-(aq)$	−0.31
$Cd^{2+}(aq) + 2e^- \rightleftharpoons Cd(s)$	−0.40
$Fe^{2+}(aq) + 2e^- \rightleftharpoons Fe(s)$	−0.44
$Cr^{3+}(aq) + 3e^- \rightleftharpoons Cr(s)$	−0.74
$Zn^{2+}(aq) + 2e^- \rightleftharpoons Zn(s)$	−0.76
$2H_2O(l) + 2e^- \rightleftharpoons H_2(g) + 2OH^-(aq)$	−0.83
$Mn^{2+}(aq) + 2e^- \rightleftharpoons Mn(s)$	−1.18
$Al^{3+}(aq) + 3e^- \rightleftharpoons Al(s)$	−1.66
$Mg^{2+}(aq) + 2e^- \rightleftharpoons Mg(s)$	−2.37
$Na^+(aq) + e^- \rightleftharpoons Na(s)$	−2.71
$Ca^{2+}(aq) + 2e^- \rightleftharpoons Ca(s)$	−2.87
$Sr^{2+}(aq) + 2e^- \rightleftharpoons Sr(s)$	−2.89
$Ba^{2+}(aq) + 2e^- \rightleftharpoons Ba(s)$	−2.90
$K^+(aq) + e^- \rightleftharpoons K(s)$	−2.93
$Li^+(aq) + e^- \rightleftharpoons Li(s)$	−3.05

*Written as reductions; E^0 value refers to all components in their standard states: 1 M for dissolved species; 1 atm pressure for the gas behaving ideally; the pure substance for solids and liquids.

Appendix E

ANSWERS TO SELECTED PROBLEMS

Chapter 1

1.2 Solids and liquids have fixed volumes; that is, they will not necessarily totally fill a container. Gases take the volume of their container.   (a) Gas   (b) Liquid   (c) Liquid   **1.4** A physical property is one that can be measured without interacting the sample with another substance. A chemical property is one that characterizes how one material reacts with or transforms into another substance.   (a) Color (yellow-green, silvery, white) and physical state (gas, crystals) are physical properties; reaction of sodium with chlorine is chemical.   (b) Magnetism and color (black, white) are both physical properties.   **1.6** (a) Physical; change in temperature only.   (b) Chemical; "browning" process involves change in composition.   (c) Physical; only changes in the size of the pieces.   (d) Chemical; wood and ashes have different composition.   **1.8** (a) Fuel   (b) Wood   **1.13** Lavoisier showed that during combustion the total mass remained constant. He also showed that the volume of air decreased as a component of the air was "absorbed."   **1.16** A well-designed experiment should test the relationship between at least two variables, all but one controlled, and the remaining one free to vary as the others change. It should test the relationship to see if there is any obvious cause and effect. For the results to be meaningful, they need to be reproducible from one trial to another.   **1.19** (a) $(2.54 \text{ cm}/1 \text{ in})^2$ (b) $(10^3 \text{ m}/1 \text{ km})^2$ (c) $(1 \text{ km}/0.6214 \text{ mi}) \times (10^3 \text{ m}/1 \text{ km}) \times (10^2 \text{ cm}/1 \text{ m}) \times (1 \text{ h}/3600 \text{ s})$ (d) $(1000 \text{ g}/2.205 \text{ lb})(30.48 \text{ cm}/1 \text{ ft})^3$ **1.21** An intensive property depends only on the type of material in the sample; an extensive property depends on the quantity of material present.   (a) and (c) are extensive; (b) and (d) are intensive.   **1.23** (a) increases   (b) remains the same   (c) decreases   (d) increases   (e) remains the same   **1.26** 0.128 nm   **1.28** 3.94×10^3 in   **1.30** (a) 1.77×10^{-9} km^2 (b) $8.92   **1.32** 77.1 kg (assuming 170. lb)   **1.34** (a) 5.52×10^3 kg/m^3 (b) 345 lb/ft^3 **1.36** (a) 1.72×10^{-9} mm^3 (b) 2×10^{-10} L   **1.38** (a) 9.626 cm^3 (b) 64.92 g   **1.40** 2.70 g/cm^3 **1.42** (a) 22°C; 295 K   (b) 109 K; -263°F   (c) -273°C; -459°F   **1.45** (a) 5.45×10^{-7} m   (b) 6830 Å   **1.46** 1.53×10^{-6} cm   **1.50** Initial zeros are never significant; internal zeros are always significant; terminal zeros to the right of a decimal point are significant; terminal zeros to the left of a decimal point are significant *only* if they were measured.   **1.52** (a) 0.39   (b) 0.039   (c) 0.0390   (d) 3.0900×10^4 **1.54** (a) 0.00036   (b) 35.83   (c) 22.5   **1.56** 600   **1.58** (a) 1.34 m   (b) 3.350×10^3 cm^3 (c) 4.43×10^2 cm   **1.60** (a) 1.310000×10^5 (b) 4.7×10^{-4} (c) 2.10006×10^5 (d) 2.1605×10^3 **1.62** (a) 5550   (b) 10070.   (c) 0.000000885   (d) 0.003004   **1.64** (a) 8.025×10^4 (b) 1.0098×10^{-3} (c) 7.7×10^{-11} **1.66** (a) 4.06×10^{-19} J   (b) 1.56×10^{24} molecules   (c) 1.82×10^5 J/mol   **1.68** (a) height is

measured, not exact   (b) number of planets is counted, exact   (c) number of grams in a pound is not a unit definition, not exact   (d) a unit definition, exact   **1.70.** 7.50 ± 0.05 cm   **1.73** (a) $I_{avg} = 8.72$ g, $II_{avg} = 8.72$, $III_{avg} = 8.50$, $IV_{avg} = 8.56$ I and II most accurate.   (b) III is the most precise, but the least accurate.   (c) I has the best combination of high accuracy and precision.   (d) IV has both low accuracy and precision.   **1.76** (a)

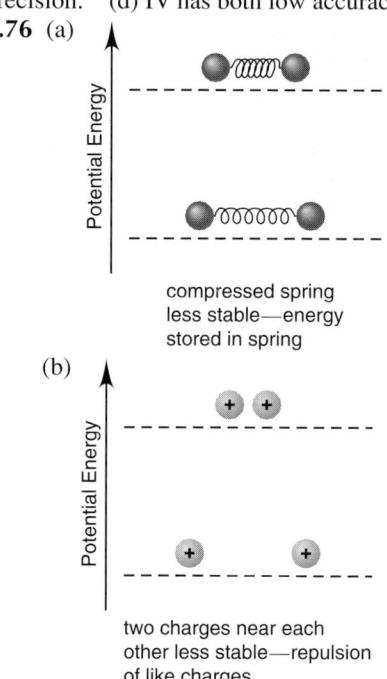

compressed spring
less stable—energy
stored in spring

(b)

two charges near each
other less stable—repulsion
of like charges

1.79 26 minutes   **1.82** (a) 0.21 g/L, will float   (b) CO_2 is more dense than air, will sink   (c) 0.30 g/L, will float   (d) O_2 is more dense than air, will sink   (e) 1.4 g/L, will sink   (f) 0.50 g   **1.84** d = 12 g/cm^3, the crown is not pure gold.   **1.87** (a) -195.79°C   (b) -320.42°F   (c) 5.05 L   **1.89** (a) 2.6 m/s   (b) 16 km   (c) 12:48 PM   **1.96** freezing pt = -3.7°X, boiling point = 63.3°X.

Chapter 2

2.1 An element consists of only one type of atom; a compound consists of more than one type of atom.   **2.4** (a) Since it has constant composition, it is a compound.   (b) Since all the atoms are identical, it is an element.   (c) Since it has variable composition, it is a mixture.   (d) Since it has constant composition, it is a compound.   **2.9** The original sample was a mixture because it was composed of more than one component.   **2.11** (a) all substances   (b) compounds   (c) compounds   **2.13** (a) Law of definite composition: the composition is independent of the source.   (b) Law of conservation of mass: the total quantity of matter does not change.   (c) Law of multiple proportions: both materials are pure, but one pair of elements can combine in two different proportions.   **2.14** (a) No; the composition is independent of amount. Twice as much of Element A will combine with twice as much of element B and the composition of the compound formed will be the same.   (b) Yes; more compound will contain proportionally more of each component element.   **2.16** These two experiments demonstrate the law of definite composition. The unknown blue compound decomposes the same way in both experiments, giving 64% white compound and 36% colorless gas. They also demonstrate the law of conservation of mass, since in both cases the total mass before reaction equals the total mass after reaction.   **2.18** (a) 1.34 g   (b) 0.514 Ca; 0.486 F   (c) 51.4% Ca; 48.6% F   **2.20** (a) 0.603   (b) 262 g Mg   **2.22** 3.498×10^6 g Cu, 1.766×10^6 g S   **2.24** 0.905 g S/g Cl in compound 1 and 0.451 g S/g Cl in compound 2, 0.905/0.451 = 2:1   **2.27** Coal A   **2.29** Dalton postulated that atoms of an element are identical. Dalton also postulated that compounds result from the chemical combination of specific ratios of different elements.   **2.30** If you know the ratio of **any** two quantities, and the value of one of them, the other can always be calculated. In this case, charge/(charge/mass) = mass.   **2.34** The atomic number of an atom is the number of protons in the nucleus; the mass number is the sum of the number of protons and number of neutrons. The mass number can vary without changing the chemical identity of the atom.   **2.37** The mass numbers are 36, 38, and 40, respectively, containing 18, 20, and 22 neutrons. *All* the isotopes have 18 protons and 18 electrons.   **2.39** (a) These have the same number of protons and electrons, but different numbers of neutrons. Same Z.   (b) These have the same number of neutrons, but different numbers of protons and electrons. Same N.   (c) These differ in all three (number of protons, neutrons, and electrons), but have the same A.   **2.41** (a) $^{38}_{18}Ar$ (b) $^{55}_{25}Mn$ (c) $^{109}_{47}Ag$

2.43 (a)   (b)   (c)

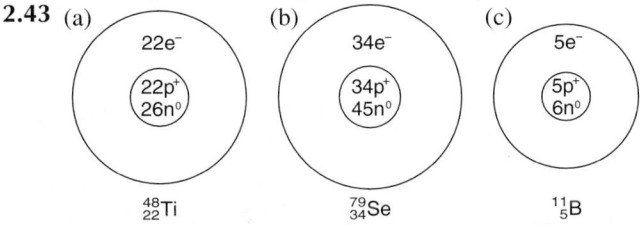

$^{48}_{22}Ti$ $^{79}_{34}Se$ $^{11}_{5}B$

2.45 69.72 amu   **2.47** 75.774% ^{35}Cl and 24.226% ^{37}Cl **2.50** (a) In the modern periodic table, the elements are arranged in order of increasing atomic **number.**   (b) Elements in a **group** (or family) have similar chemical properties.   (c) Elements can be classified as **metals,** metalloids, or nonmetals.   **2.53** The elements in group 1A(1) are metals; they generally lose electrons in chemical reactions. The elements in group 7A(17) are nonmetals; they generally gain electrons in chemical reactions.   **2.54** (a) germanium; Ge; 4A(14); metalloid   (b) sulfur; S; 6A(16); nonmetal   (c) helium; He; 8A(18); nonmetal   (d) lithium; Li; 1A(1); metal   (e) molybdenum; Mo; 6B(6); metal   **2.56** (a) Ra; 88   (b) As; 33   (c) Cu; 63.55   (d) Br; 79.90   **2.58** These atoms will form **ionic** bonds, in which one or more electrons is transferred from the metal atom to the non-metal atom to form a cation and an anion, respectively.   **2.61** Coulomb's law states that the strength of interaction of

two charged particles is directly proportional to the product of the charges and inversely proportional to the distance between them. In this case, the doubly charged ions of MgO (Mg^{2+} and O^{2-}) would attract more strongly than the similarly sized, singly charged ions of LiF (Li^+ and F^-).   **2.64** The Group 1A(1) elements are metals and form cations; the Group 7A(17) elements are nonmetals and form anions.   **2.66** Each of the sulfur atoms would gain two electrons to form an S^{2-} ion. Twice this number of potassium atoms would each lose one electron to form K^+ ions. The K^+ and S^{2-} ions form a regular array of ions in the solid K_2S.   **2.68** K^+, I^- **2.70** (a) oxygen; 17; 6A(16); 2   (b) fluorine; 19; 7A(17); 2   (c) calcium; 40; 2A(2); 4   **2.72** Li forms Li^+ and O forms O^{2-}, so the compound contains half as many O^{2-} (2.6×10^{20}) as Li^+ (5.3×10^{20}).   **2.74** Sodium chloride   **2.76** The empirical formula shows the **simplest** ratio of the atoms involved, while a molecular formula shows the **actual** numbers of the atoms which make up the molecule. The molecular formula is always a whole number multiple of the empirical formula; if the whole number is 1, the empirical and molecular formulas are identical.   **2.78** The two samples are similar in that they both contain 20 billion H atoms and 20 billion O atoms. They are different in that, in the $H_2 + O_2$ mixture, H atoms are bonded only to other H atoms and O atoms are bonded only to other O atoms, while in the H_2O_2 sample, H atoms are bonded to O atoms and O atoms are bonded both to H and to other O atoms. Also, the $H_2 + O_2$ mixture contains a total of 20 billion particles, while the H_2O_2 sample contains 10 billion particles.   **2.82** (a) NH_2 (b) CH_2O **2.84** (a) Li_3N; lithium nitride   (b) SrO; strontium oxide   (c) $AlCl_3$; aluminum chloride   **2.86** (a) magnesium fluoride; MgF_2 (b) sodium sulfide; Na_2S (c) strontium chloride, $SrCl_2$ **2.88** (a) $SnCl_4$ (b) iron(III) bromide   (c) CuBr   (d) manganese(III) oxide   **2.90** (a) cobalt(II) oxide   (b) Hg_2Cl_2 (c) lead(II) acetate trihydrate   (d) Cr_2O_3 **2.92** (a) BaO   (b) $Fe(NO_3)_2$ (c) MgS   **2.94** (a) sulfuric acid, H_2SO_4 (b) iodic acid, HIO_3 (c) hydrocyanic acid, HCN   (d) hydrosulfuric acid, H_2S **2.96** (a) dinitrogen pentaoxide   (b) ClF_3 (c) silicon disulfide   **2.98** S_2F_4; disulfur tetrafluoride   **2.100** (a) carbon monoxide   (b) sulfur dioxide   (c) dichlorine monoxide   **2.102** (a) 12; 342.2 amu   (b) 9; 132.06 amu   (c) 8; 344.6 amu   **2.104** (a) $(NH_4)_2SO_4$; 132.15 amu   (b) NaH_2PO_4; 119.98 amu   (c) $KHCO_3$; 100.12 amu   **2.106** (a) 108.02 amu   (b) 331.2 amu   (c) 72.08 amu   **2.108** (a) SO_3; sulfur trioxide; 80.07   (b) C_3H_8; propane; 44.09   **2.110** Disulfur dichloride; SCl; 135.04   **2.112** Separating the components of a mixture is done by physical methods only; that is, no chemical changes (i.e., changes in composition) take place and the components maintain their chemical identities and properties throughout. Separating the components of a compound requires a chemical change (i.e., change in composition).   **2.115** (a) compound   (b) homogeneous mixture   (c) heterogeneous mixture   (d) homogeneous mixture   (e) homogeneous mixture   **2.117** (a) Add water. Salt will dissolve, pepper won't. Filter off the pepper and evaporate the water to recover the salt.   (b) Add water. Sugar will dissolve, sand won't. Filter off the sand and evaporate the water to recover the sugar.   (c) Heat the mixture to boiling; the water will evaporate first and can be collected and condensed.   (d) Use a separatory funnel to drain off the vinegar (lower level).

Or heat the mixture to boiling; the vinegar (water + acetic acid) will evaporate first and can be collected and condensed. **2.119** (a) filtration   (b) adsorption chromatography   (c) extraction   **2.121** (a) 5.2×10^{-13} (b) 0.999726   **2.124** Magnetite (a) 267.48 amu   (b) 449.5 amu   (c) 363.46 amu   (d) 835.2 amu   **2.126** %Fe = 2.7%, %Mg = 2.4%, %Si = 43.2%, %O = 51.6%   **2.130** In each case, the total mass of reactants equals the total mass of products, so the law of conservation of mass is obeyed. In each case, the product (NaCl) contains 60.66% Cl by mass, so the law of definite composition is obeyed.   **2.131** mg/kg$\times10^{-4}$ = mg/100 g = mass %, so (a) Cl$^-$: 1.8980%, Na$^+$: 1.0560%, SO$_4{}^{2-}$: 0.2650%, Mg^{2+}: 0.1270%, Ca^{2+}: 0.0400%, K$^+$: 0.0380%, HCO$_3{}^-$: 0.0140% (b) 30.716%   (c) alkaline-earth cations = 0.1670% (~15% of alkali metal); alkali-metal cations = 1.0940%   (d) anions: 2.1770% > cations: 1.2610%   **2.132** C$_4$H$_6$O$_4$; C$_2$H$_3$O$_2$; 118.09   **2.134** 52.00 amu   **2.136** 58.091 amu   **2.138** (a) The number of chlorine atoms in 1 liter of chlorine gas is equal to the number of chlorine atoms in 2 liters of hydrogen chloride gas. Therefore, there must be two chlorine atoms per chlorine molecule.   (b) Because equal volumes of hydrogen and chlorine react, there must be one atom of hydrogen per atom of chlorine in a hydrogen chloride molecule.   (c) 35.5 amu   **2.141** %V in carnotite = 11.29%, %V in tyuyamunite = 11.10%, %V in patronite = 28.42%, %V in vanadinite = 10.79%, %V in roscoelite = 22.830%. Patronite has the most vanadium per gram and vanadinite has the least.   **2.143** 37.01% C; 2.219% H; 18.50% N; 42.26% O.

Chapter 3

3.2 (a) 1 mol of C$_{12}$H$_{22}$O$_{11}$ contains 12 mol of C   (b) 7.226×10^{24} C atoms.   **3.6** (a) left   (b) left   (c) left   (d) neither   **3.8** (a) 121.64 g/mol   (b) 44.02 g/mol   (c) 106.44 g/mol   (d) 152.00 g/mol   **3.10** (a) 150.7 g/mol   (b) 175.3 g/mol   (c) 342.17 g/mol   (d) 125.84 g/mol   **3.12** (a) 9.0×10^1 g KMnO$_4$ (b) 0.331 mol O atoms   (c) 1.8×10^{20} O atoms   **3.14** (a) 97 g MnSO$_4$ (b) 4.46×10^{-2} mol Fe(ClO$_4$)$_3$ (c) 1.74×10^{24} N atoms   **3.16** (a) 1.57×10^3 g Cu$_2$CO$_3$ (b) 0.366 g N$_2$O$_5$ (c) 2.85×10^{23} formula units NaClO$_4$ (d) 2.85×10^{23} Na$^+$ ions, 2.85×10^{23} ClO$_4{}^-$ ions, 2.85×10^{23} Cl atoms, 1.14×10^{24} O atoms   **3.18** (a) 6.375% H (b) 71.52% O   **3.20** (a) 0.1252   (b) 0.3428   **3.23** (a) 0.9507 mol Pt(NH$_3$)$_2$Cl$_2$ (b) 3.5×10^{24} H atoms   **3.25** (a) 281 mol rust   (b) 281 mol Fe$_2$O$_3$ (c) 3.14×10^4 g Fe   **3.27** urea > ammonium nitrate > ammonium sulfate > potassium nitrate   **3.29** 4   **3.31** (b) From the mass percentages, determine the empirical formula. Add up the total number of atoms in that, and divide that into the total number of atoms in the molecule. The result tells you the multiplier to make the empirical formula into the molecular formula.

(c) (mass %)$\left(\dfrac{1 \text{ mol}}{\text{molar mass}}\right)$ = moles of each element

(e) Count the numbers of the various types of atoms in the structural formula and put these into a molecular formula.
3.33 (a) CH$_2$; 14.03   (b) CH$_3$O; 31.03   (c) N$_2$O$_5$; 108.02 (d) Ba$_3$(PO$_4$)$_2$; 601.8   (e) TeI$_4$; 635.2   **3.35** (a) C$_3$H$_6$ (b) N$_2$H$_4$ (c) N$_2$O$_4$ (d) C$_5$H$_5$N$_5$ **3.37** (a) Cl$_2$O$_7$ (b) SiCl$_4$

(c) CO$_2$ **3.39** (a) NO$_2$ (b) N$_2$O$_4$ **3.41** (a) 1.20 mol F (b) 24.0 g M   (c) Ca   **3.44** C$_{21}$H$_{30}$O$_5$ **3.46** C$_{10}$H$_{20}$O   **3.47** The balanced equation provides information on the amount and type of reactants and products, in terms of molecules, moles, and mass in grams.   **3.50** (b)
3.51 (a) $16Cu(s) + S_8(s) \longrightarrow 8Cu_2S(s)$
   (b) $P_4O_{10}(s) + 6H_2O(l) \longrightarrow 4H_3PO_4(l)$
   (c) $B_2O_3(s) + 6NaOH(aq) \longrightarrow 2Na_3BO_3(aq) + 3H_2O(l)$
   (d) $4CH_3NH_2(g) + 9O_2(g) \longrightarrow$
$$4CO_2(g) + 10H_2O(g) + 2N_2(g)$$
3.53 (a) $2SO_2(g) + O_2(g) \longrightarrow 2SO_3(g)$
   (b) $Sc_2O_3(s) + 3H_2O(l) \longrightarrow 2Sc(OH)_3(s)$
   (c) $H_3PO_4(aq) + 2NaOH(aq) \longrightarrow$
$$Na_2HPO_4(aq) + 2H_2O(l)$$
   (d) $C_6H_{10}O_5(s) + 6O_2(g) \longrightarrow 6CO_2(g) + 5H_2O(g)$
3.55 (a) $4Ga(s) + 3O_2(g) \xrightarrow{\Delta} 2Ga_2O_3(s)$
   (b) $2C_6H_{14}(l) + 19O_2(g) \longrightarrow 12CO_2(g) + 14H_2O(g)$
   (c) $3CaCl_2(aq) + 2Na_3PO_4(aq) \longrightarrow$
$$Ca_3(PO_4)_2(s) + 6NaCl(aq)$$
3.59 Balance the equation for the reaction. From the relationship moles = g/$\mathcal{M}$, calculate the moles of D and E present. By comparing the ratio of moles present to that needed in the balanced equation, determine the limiting reactant. Based on the number of moles of limiting reactant, proceed as follows:

$$m(g)\ F = mol\ L.R. \times \frac{mol\ F}{mol\ L.R.} \times \frac{g\ F}{mol\ F} = g\ F$$

3.61 (a) 0.455 mol Cl$_2$ (b) 32.3 g Cl$_2$ **3.63** (a) 2.22×10^3 mol KNO$_3$ (b) 2.24×10^5 g KNO$_3$ **3.65** 150.2 g H$_3$BO$_3$ and 14.69 g H$_2$ **3.67** 2.03×10^3 g Cl$_2$
3.69 (a) $I_2(s) + Cl_2(g) \longrightarrow 2ICl(s)$
   $ICl(s) + Cl_2(g) \longrightarrow ICl_3(s)$
   (b) $I_2(s) + 3Cl_2(g) \longrightarrow 2ICl_3(s)$
   (c) 1.71×10^4 g I$_2$
3.71 (a) 0.105 mol CaO   (b) 0.175 mol CaO   (c) Ca   (d) 5.89 g CaO   **3.73** 1.47 mol HIO$_3$, 258 g HIO$_3$, 38.0 g H$_2$O   **3.75** 4.40 g CO$_2$; 4.80 g O$_2$ in excess   **3.77** 6.4 g Al(NO$_2$)$_3$, 0 g NH$_4$Cl, 45.4 g AlCl$_3$, 28.6 g N$_2$, 36.8 g H$_2$O   **3.79** 53% **3.81** 98.2%   **3.83** 24.5 g   **3.85** 52.9 g   **3.88** 68.4 g NaBH$_4$ **3.89** (a) C   (b) B   (c) C   (d) B   **3.91** No. It should read: "Take 100.0 mL of the 10.0 M solution and add sufficient water to make 1.00 L of solution."   **3.92** (a) 5.76 g Ca(C$_2$H$_3$O$_2$)$_2$ (b) 0.254 M (c) 124 mol NaCN   **3.94** (a) 4.65 g   (b) 5.90×10^{-2} M (c) 1.11×10^{20} ions   **3.96** (a) 0.0617 M (b) 0.00363 M (c) 0.0150 M **3.98** (a) 987 g   (b) 15.7 M **3.100** 845 mL HCl(aq)   **3.102** 0.87 g   **3.105** (a) Slowly add 5.7×10^3 mL of concentrated HCl to approximately 3.0 gal of water with constant stirring. Monitor increase in temperature. When all of the concentrated HCl has been added, slowly add enough water to make a final volume of 5.0 gallon. (b) 22.4 mL   **3.109** ethane > propane > cetyl palmitate > ethanol > benzene   **3.110** narceine·3H$_2$O **3.115** (a) Fe$_2$O$_3$ + 3CO $\longrightarrow$ 2Fe + 3CO$_2$ (b) 3.01×10^7 g **3.116** 89.8%   **3.118** (a) 2AB$_2$ + B$_2$ $\longrightarrow$ 2AB$_3$ (b) AB$_2$ (c) 5.0 mol   (d) 0.5 mol   **3.120** 8.9×10^{-7} M **3.122** 0.6855 g   **3.124** (a) C   (b) B   (c) D   **3.128** 0.071 M KBr **3.129** 38.5 mass% B   **3.132** 586 g CO$_2$ **3.134** 10:0.66:1.0 **3.137** 32.7% C   **3.139** (a) 192.12 g/mol   (b) 0.549 mol citric acid

3.140 (a) $N_2(g) + O_2(g) \longrightarrow 2NO(g)$
      $2NO(g) + O_2(g) \longrightarrow 2NO_2(g)$
      $3NO_2(g) + H_2O(l) \longrightarrow 2HNO_3(aq) + NO(g)$
    (b) $2N_2(g) + 5O_2(g) + 2H_2O(l) \longrightarrow 4HNO_3\ (aq)$
    (c) 5.62×10^3 t HNO_3
3.143 14% XeF_4, 86.2% XeF_6 **3.144** (a) 0.027 g heme   (b)
4.4×10^{-5} mol heme   (c) 2.4×10^{-3} g Fe   (d) 0.028 g hemin
3.146 (a) 46.65% N in urea, 31.98% N in arginine, 21.04% N
in ornithine   (b) 30.36 g N   **3.148** (a) 84.3%   (b) 2.39 g
C_2H_4 **3.150** (a) 125 g salt   (b) 65.6 L

Chapter 4

4.2 Ionic compounds or polar covalent compounds.   **4.3** Ions,
which could come from ionic compounds or from other electrolytes such as acids or bases.   **4.6** 2   **4.10** (a) Molecules of
benzene are symmetrical and not much like water, so it would
most likely be insoluble in water.   (b) Sodium hydroxide, an
ionic compound, would be expected to be soluble in water, and
the solubility rules confirm this.   (c) Ethanol (CH_3CH_2OH)
molecules are similar in structure to water molecules (both contain O—H bonds), so it is soluble.   (d) Potassium acetate, an
ionic compound, would be expected to be soluble in water, and
this is confirmed by the solubility rules.   **4.12** (a) Yes; CsI is a
salt.   (b) Yes; HBr is a strong acid.   **4.14** (a) 0.50 mol   (b)
0.251 mol   (c) 5.91×10^{-4} mol   **4.16** (a) 0.60 mol   (b)
0.527 mol   (c) 0.281 mol   **4.18** (a) 0.234 mol Al^{3+},
1.41×10^{23} Al^{3+} ions, 0.702 mol Cl^-, 4.23×10^{23} Cl^- ions   (b)
0.162 mol Na^+, 9.73×10^{22} Na^+ ions, 0.0808 mol SO_4^{2-},
4.86×10^{22} SO_4^{2-} ions   (c) 3.58×10^{-3} mol Mg^{2+},
2.16×10^{21} Mg^{2+} ions, 7.16×10^{-3} mol Br^-, 4.31×10^{21} Br^-
ions   **4.20** (a) 0.35 mol H^+ (b) 3.5×10^{-3} mol H^+ (c)
0.14 mol H^+ **4.24** Spectator ions (i.e., ions which do not
chemically change in the reaction) do not appear, since they are
not involved in the reaction and are only present to balance
charge.   **4.28** Assuming that the left beaker is $AgNO_3$ and the
right is NaCl, the Ag^+ is gray, the NO_3^- is blue, the Na^+ is
brown, and the Cl^- is green.
molecular: $AgNO_3(aq) + NaCl(aq) \longrightarrow$
                   $NaNO_3(aq) + AgCl(s)$
total ionic: $Ag^+(aq) + NO_3^-(aq) + Na^+(aq) + Cl^-(aq) \longrightarrow$
                   $Na^+(aq) + NO_3^-(aq) + AgCl(s)$
net ionic: $Ag^+(aq) + Cl^-(aq) \longrightarrow AgCl(s)$
4.29 (a) no reaction   (b) silver iodide (AgI)   **4.31** (a) no reaction   (b) barium sulfate ($BaSO_4$)
4.33 (a) molecular: $Hg_2(NO_3)_2(aq) + 2KI(aq) \longrightarrow$
                   $Hg_2I_2(s) + 2KNO_3(aq)$
total ionic: $Hg_2^{2+}(aq) + 2NO_3^-(aq) + 2K^+(aq) +$
   $2I^-(aq) \longrightarrow Hg_2I_2(s) + 2K^+(aq) + 2NO_3^-(aq)$
net ionic: $Hg_2^{2+}(aq) + 2I^-(aq) \longrightarrow Hg_2I_2(s)$
Spectator ions are K^+ and NO_3^-.
   (b) molecular: $FeSO_4(aq) + Ba(OH)_2(aq) \longrightarrow$
                   $Fe(OH)_2(s) + BaSO_4(s)$
total ionic: $Fe^{2+}(aq) + SO_4^{2-}(aq) + Ba^{2+}(aq) +$
   $2OH^-(aq) \longrightarrow Fe(OH)_2(s) + BaSO_4(s)$
      net ionic: same as total ionic; no spectator ions
4.35 0.0389 M **4.37** 1.80%   **4.43** (a) formation of a gas and
formation of a nonelectrolyte   (b) formation of a precipitate
and formation of a nonelectrolyte

4.45 (a) molecular: $KOH(aq) + HI(aq) \longrightarrow$
                   $KI(aq) + H_2O(l)$
total ionic: $K^+(aq) + OH^-(aq) + H^+(aq) + I^-(aq)$
          $\longrightarrow K^+(aq) + I^-(aq) + H_2O(l)$
net ionic: $OH^-(aq) + H^+(aq) \longrightarrow H_2O(l)$
Spectator ions are K^+ and I^-.
   (b) molecular: $NH_3(aq) + HCl(aq) \longrightarrow NH_4Cl(aq)$
total ionic: $NH_3(aq) + H^+(aq) + Cl^-(aq) \longrightarrow$
                   $NH_4^+(aq) + Cl^-(aq)$
net ionic: $NH_3(aq) + H^+(aq) \longrightarrow NH_4^+(aq)$
Spectator ion is Cl^-.
4.47 The hydrochloric acid reacts with the CO_3^{2-} bound in the
$CaCO_3$, releasing $CO_2(g)$:
total ionic: $CaCO_3(s) + 2H^+(aq) + 2Cl^-(aq) \longrightarrow$
         $Ca^{2+}(aq) + 2Cl^-(aq) + H_2O(l) + CO_2(g)$
net ionic: $CaCO_3(s) + 2H^+(aq) \longrightarrow$
           $Ca^{2+}(aq) + H_2O(l) + CO_2(g)$
4.49 0.03319 M **4.58** (a) The S in SO_4^{2-} (i.e., H_2SO_4) has
O.N. = +6, and in SO_2, O.N.(S) = +4, so the S has been reduced (and the I^- oxidized), so the H_2SO_4 is an oxidizing
agent.   (b) The oxidation numbers remain constant throughout; H_2SO_4 transfers a proton to F^- to produce HF, so it acts as
an acid.   **4.60** (a) +4   (b) +3   (c) +4   (d) −3   **4.62** (a)
−1   (b) −2   (c) −3   (d) +3   **4.64** (a) −3   (b) +5   (c)
+3   **4.66** (a) +6   (b) +3   (c) +7   **4.68** (a) Oxidizing
agent (O.A.) = MnO_4^-; Reducing agent (R.A.) = $H_2C_2O_4$
(b) O.A. = NO_3^-; R.A. = Cu   **4.70** (a) O.A. = NO_3^-; R.A.
= Sn   (b) O.A. = MnO_4^-; R.A. = Cl^- **4.72** S is in Group
6A (16), so its highest possible O.N. is +6 and its lowest possible O.N. is 6 − 8 = −2.   (a) In S^{2-}, O.N.(S) = −2, so it can
only be oxidized (i.e., act as a reducing agent).   (b) In SO_4^{2-},
O.N.(S) = +6, so it can only be reduced (i.e., act as an oxidizing agent).   (c) In SO_2, O.N.(S) = +4, so it can either be oxidized or reduced (i.e., act as a reducing or oxidizing agent).
4.74 (a) $8HNO_3(aq) + K_2CrO_4(aq) + 3Fe(NO_3)_2(aq) \longrightarrow$
   $2KNO_3(aq) + Cr(NO_3)_3(aq) + 3Fe(NO_3)_3(aq) + 4H_2O(l)$
     K_2CrO_4 is the oxidizing agent (O.N.(Cr) goes from
     +6 to +3) and $Fe(NO_3)_2$ is the reducing agent
     [O.N.(Fe) goes from +2 to +3].
   (b) $8HNO_3(aq) + 3C_2H_6O(l) + K_2Cr_2O_7(aq) \longrightarrow$
   $2KNO_3(aq) + 3C_2H_4O(l) + 7H_2O(l) + 2Cr(NO_3)_3(aq)$
     $K_2Cr_2O_7$ is the oxidizing agent [O.N.(Cr) goes from
     +6 to +3] and C_2H_6O is the reducing agent [O.N.(C)
     goes from −2 to −1].
   (c) $6HCl(aq) + 2NH_4Cl(aq) + K_2Cr_2O_7(aq) \longrightarrow$
      $2KCl(aq) + 2CrCl_3(aq) + N_2(g) + 7H_2O(l)$
     $K_2Cr_2O_7$ is the oxidizing agent [O.N.(Cr) goes from
     +6 to +3], and NH_4Cl is the reducing agent [O.N.(N)
     goes from −3 to 0].
   (d) $KClO_3(aq) + 6HBr(aq) \longrightarrow$
                   $3Br_2(l) + 3H_2O(l) + KCl(aq)$
     $KClO_3$ is the oxidizing agent [O.N.(Cl) goes from +5
     to −1], and HBr is the reducing agent [O.N.(Br) goes
     from −1 to 0].
4.76 (a) 4.54×10^{-3} mol   (b) 0.0113 mol   (c) 0.386 g   (d)
2.80%   (e) H_2O_2 **4.81** $2Mg(s) + O_2(g) \longrightarrow 2MgO(s)$ is a
combination reaction and also a redox. $CaO(s) + H_2O(l) \longrightarrow$
$Ca(OH)_2(s)$ is a combination but not a redox.

4.83 (a) $Ca(s) + 2H_2O(l) \longrightarrow Ca(OH)_2(aq) + H_2(g)$; displacement

(b) $2NaNO_3(s) \longrightarrow 2NaNO_2(s) + O_2(g)$; decomposition

(c) $C_2H_2(g) + 2H_2(g) \longrightarrow C_2H_6(g)$; combination

4.85 (a) $2Sb(s) + 3Cl_2(g) \longrightarrow 2SbCl_3(s)$; combination

(b) $2AsH_3(g) \longrightarrow 2As(s) + 3H_2(g)$; decomposition

(c) $3Mn(s) + 2Fe(NO_3)_3(aq) \longrightarrow$
$$3Mn(NO_3)_2(aq) + 2Fe(s);$$
displacement

4.87 (a) $Ca(s) + Br_2(l) \longrightarrow CaBr_2(s)$

(b) $2Ag_2O(s) \xrightarrow{\Delta} 4Ag(s) + O_2(g)$

(c) $Mn(s) + Cu(NO_3)_2(aq) \longrightarrow Mn(NO_3)_2(aq) + Cu(s)$

4.89 (a) $N_2(g) + 3H_2(g) \longrightarrow 2NH_3(g)$

(b) $2NaClO_3(s) \xrightarrow{\Delta} 2NaCl(s) + 3O_2(g)$

(c) $Ba(s) + 2H_2O(l) \longrightarrow Ba(OH)_2(aq) + H_2(g)$

4.91 (a) $2Cs(s) + I_2(s) \longrightarrow 2CsI(s)$

(b) $2Al(s) + 3MnSO_4(aq) \longrightarrow Al_2(SO_4)_3(aq) + 3Mn(s)$

(c) $2SO_2(g) + O_2(g) \xrightarrow{\Delta} 2SO_3(g)$

(d) $C_3H_8(g) + 5O_2(g) \longrightarrow 3CO_2(g) + 4H_2O(g)$

(e) $2Al(s) + 3Mn^{2+}(aq) \longrightarrow 2Al^{3+}(aq) + 3Mn(s)$

4.93 315 g O_2; 3.95 kg Hg (mercury)   **4.95** (a) O_2 (b) 0.117 mol Li_2O (c) 0 g Li; 4.13 g O_2; 3.50 g Li_2O **4.97** 56.7%

4.99 223 g   **4.100** 99.9 g   **4.104** The reaction $2NO + Br_2 \rightleftharpoons 2NOBr$ can proceed in either direction. If NO and Br_2 are placed in a container, they will react to form NOBr, and NOBr will decompose to form NO and Br_2. Eventually, the concentrations of NO, Br_2, and NOBr adjust so that the rates of the forward and reverse reactions become equal, and equilibrium is reached.

4.106 (a) $Fe(s) + 2H^+(aq) \longrightarrow Fe^{2+}(aq) + H_2(g)$
O.N.:   0   +1   +2   0

(b) 3.1×10^{21} ions

4.110 (a) $C_2O_4^{2-}(aq) + Ca^{2+}(aq) \longrightarrow CaC_2O_4(s)$

(b) $5H_2C_2O_4(aq) + 2MnO_4^-(aq) + 6H^+(aq) \longrightarrow$
$$10CO_2(g) + 2Mn^{2+}(aq) + 8H_2O(l)$$

(c) $KMnO_4$ (d) $H_2C_2O_4$ (e) 55.06% $CaCl_2$

4.111 Ag_2S, $CaSO_4$; 0.02 M Ag^+, 0 M Ca^{2+}, 0.03 M Na^+, 0.03 M NO_3^-, 0.01 M SO_4^{2-}, 0 M S^{2-} **4.114** 3.027%

4.116 (a) 1.79×10^5 g HNO_3 (b) 238 kg O_2 (c) 89.4 kg H_2O (d)

step 1 NH_3 - reducing agent, N is oxidized
        O_2 - oxidizing agent, O is reduced

step 2   NO - reducing agent, N is oxidized
        O_2 - oxidizing agent, O is reduced

step 3 NO_2 is oxidizing and reducing agent, N is both oxidized and reduced

4.120 (a) $2CrO_4^{2-}(aq) + 3HSnO_2^-(aq) + H_2O(l) \longrightarrow$
$$2CrO_2^-(aq) + 3HSnO_3^-(aq) + 2OH^-(aq)$$
CrO_4^{2-} is the oxidizing agent [O.N.(Cr) goes from +6 to +3] and $HSnO_2^-$ is the reducing agent [O.N.(Sn) goes from +2 to +4].

(b) $2KMnO_4(aq) + 3NaNO_2(aq) + H_2O(l) \longrightarrow$
$$2MnO_2(s) + 3NaNO_3(aq) + 2KOH(aq)$$
$KMnO_4$ is the oxidizing agent [O.N.(Mn) goes from +7 to +4] and $NaNO_2$ is the reducing agent [O.N.(N) goes from +3 to +5].

(c) $4I^-(aq) + O_2(g) + 2H_2O(l) \longrightarrow$
$$2I_2(s) + 4OH^-(aq)$$
O_2 is the oxidizing agent [O.N.(O) goes from 0 to −2] and I^- is the reducing agent [O.N.(I) goes from −1 to 0].

4.123 0.75 kg SiO_2, 0.26 kg Na_2CO_3, 0.18 kg $CaCO_3$

4.126 (a) 4 mol   (b) 12 mol   (c) 12.87%   **4.128** (a) 0.0030 mol CO_2 (b) 0.11 L CO_2

4.131 (a) $C_2H_5OH(l) + 3O_2(g) \longrightarrow 2CO_2(g) + 3H_2O(l)$
$$2C_8H_{18}(l) + 25O_2(g) \longrightarrow 16CO_2(g) + 18H_2O(l)$$
(b) 2.50×10^3 g   (c) 1.75×10^3 L   (d) 8.37×10^3 L

4.134 (a) 61.13 g CO   (b) 82.50 g CO   (c) 37.91 g CO

4.138 (a) The second reaction is a redox reaction   (b) 2.00×10^5 g Fe_2O_3, 4.06×10^5 g $FeCl_3$ (c) 2.09×10^5 g Fe, 4.75×10^5 g $FeCl_2$ (d) 0.313 g $FeCl_2$/g $FeCl_3$

Chapter 5

5.1 (a) The volume of liquid remains constant but the volume of gas increases to the volume of the larger container.   (b) The volume of the container containing the gas sample would increase when heated but the volume of the container containing the liquid sample would remain essentially constant when heated.   (c) The volume of the liquid remains essentially constant but the volume of the gas would be reduced.   **5.6** 979 cm H_2O **5.8** 0.9408 atm   **5.10** 0.966 atm   **5.12** (a) 566 mm Hg (b) 1.32 bar   (c) 3.60 atm   (d) 107 kPa   **5.18** $P_g \propto n$ if T, V are constant   **5.20** (a) V reduced to $\frac{1}{3}$ of its original value.   (b) V increased by a factor of 2.5.   (c) V increased by a factor of 3.   **5.22** (a) V reduced by a factor of 2.   (b) V increased by a factor of 1.56.   (c) V reduced by a factor of 4.   **5.24** −42°C **5.26** 35.3 L   **5.28** 0.061 mol   **5.30** 0.675 g   **5.33** no   **5.35** inverted for H_2; upright for CO_2; because $\mathcal{M}(CO_2) > \mathcal{M}(air) > \mathcal{M}(H_2)$.   **5.38** 5.86 g/L   **5.40** 0.00179 mol; 3.48 g/L   **5.42** 51.1 g/mol   **5.44** 1.33 atm   **5.48** C_5H_{12} **5.50** (a) 0.90 mol (b) 6.76 torr   **5.51** 39.3 g   **5.53** 41.2 g   **5.55** 2.49×10^{-2} g **5.57** 286 mL   **5.59** 9.97×10^{-2} atm   **5.63** At STP (or any identical temperature and pressure) the volume occupied by a mole of any gas will be identical. This is due to the fact that, at the same temperature, all gases have the same average kinetic energy, resulting in the same pressure.   **5.66** (a) pressure: A > B > C   (b) $\bar{E}_k$: A = B = C   (c) diffusion rate: A > B > C (d) total E_k: A > B > C   (e) density: A = B = C   (f) collision frequency: A > B > C   **5.67** 13.21   **5.69** (a) curve 1 (b) curve 1   (c) Curve 1; fluorine gas has about the same molar mass as argon.   **5.71** 14.0 min   **5.73** 4   **5.75** Negative deviations. $N_2 < Kr < CO_2$ **5.77** At 1 atm. At high P, intermolecular forces and molecular volume become more important. **5.80** 6.80×10^4 g/mol   **5.83** (a) 22.5 atm   (b) 21.2 atm **5.86** (a) $P_{N_2} = 597$ torr, $P_{O_2} = 159$ torr, $P_{CO_2} = 0.3$ torr, $P_{H_2O} = 3.5$ torr   (b) mol % $N_2 = 74.9\%$, mol % $O_2 = 13.7\%$, mol % $CO_2 = 5.3\%$, mol % $H_2O = 6.2\%$ (c) 1.6×10^{21} molecules O_2 **5.88** (a) 399 mL   (b) 1.2×10^{-2} mol N_2 **5.90** 35.7 L **5.95** Al_2Cl_6 **5.97** 1.63×10^{-2} mol   **5.101** (a) 1.95×10^3 g (b) 3.5×10^4 g Ni   (c) 63.2 m³   **5.103** (a) 9 volumes   (b) CH_5N **5.107** A factor of 4.87. He could ascend 52.4 ft to a depth of 73 ft.   **5.109** 6.07 g   **5.112** 4.93×10^{-3} g   **5.114** (a) Xe   (b) H_2O (c) Hg   (d) H_2O. In all cases, due to stronger intermolecular attractions. **5.117** 17.2 g CO_2; 17.8 g Kr **5.120** 6.76×10^2 m/s for Ne; 4.81×10^2 m/s for Ar; 1.52×10^3 m/s for He **5.122** (a) 0.055 g (b) 1.1 mL

5.123 (a) $\frac{1}{2}m\overline{u^2} = \frac{3}{2}\left(\frac{R}{N_A}\right)T$

$$m\overline{u^2} = 3RT\left(\frac{R}{N_A}\right)T$$

$$\overline{u^2} = \frac{3RT}{mN_A} \text{ and } \mathcal{M} = mN_A$$

$$\overline{u} = \sqrt{\frac{3RT}{\mathcal{M}}}$$

(b) $\overline{E}_k = \frac{1}{2}m_1\overline{u_1^2} = \frac{1}{2}m_2\overline{u_2^2}$

$$m_1\overline{u_1^2} = m_2\overline{u_2^2}$$

$$\frac{m_1}{m_2} = \frac{\overline{u_2^2}}{\overline{u_1^2}}$$

$$\sqrt{\frac{m_1}{m_2}} = \frac{\overline{u_2}}{\overline{u_1}}$$

Since $\frac{m_1}{m_2} = \frac{\mathcal{M}_1}{\mathcal{M}_2}$ and $\overline{u} \propto$ rate, $\sqrt{\frac{\mathcal{M}_1}{\mathcal{M}_2}} = \frac{\text{rate}_2}{\text{rate}_1}$

5.127 (a) 14.8 L (b) P_{H_2O} = 42.2 torr; $P_{O_2} = P_{CO_2}$ = 370 torr **5.131** 334 **5.133** 1.6 **5.135** P_{total} = 0.323 atm, P_{I_2} = 0.0330 atm

Chapter 6

6.4 Increase: eating food, lying in the sun, taking a hot bath Decrease: exercising, taking a cold bath, going outside on a cold day **6.6** The ΔE for the heater and the air conditioner are the same magnitude. $\Delta E_{\text{system}} = -\Delta E_{\text{surroundings}}$ **6.8** zero J **6.10** 1.52×10^3 J **6.12** (a) 3.3×10^7 kJ (b) 7.9×10^6 kcal (c) 3.1×10^7 Btu **6.15** 9.3 hr **6.17** Measuring the heat transfer at constant pressure is more convenient than measuring the heat transfer at constant volume. **6.19** (a), (c), (d), and (g) are exothermic; (b), (e), and (f) are endothermic.

6.22

Reactants

H $\Delta H = (-)$

Products

6.24 (a) $CH_4(g) + 2O_2(g) \longrightarrow CO_2(g) + 2H_2O(g)$ + heat

$CH_4 + 2O_2$

H $\Delta H = (-)$

$CO_2 + 2H_2O$

(b) $H_2O(l) \longrightarrow H_2O(s)$ + heat

$H_2O_{(l)}$

H $\Delta H = (-)$

$H_2O_{(s)}$

6.26 (a) $C_2H_6O(l) + 3O_2(g) \longrightarrow 2CO_2(g) + 3H_2O(g)$ + heat

$C_2H_6O + 3O_2$

H $\Delta H = (-)$

$2CO_2 + 3H_2O$

(b) $\frac{1}{2}N_2(g) + O_2(g)$ + heat $\longrightarrow NO_2(g)$

NO_2

H $\Delta H = (+)$

$\frac{1}{2}N_2 + O_2$

6.28 Both are one-carbon molecules. Since methane contains the fewer C—O bonds, it will have the greater heat of combustion per mole. **6.31** Energy, mass, change in temperature. **6.33** The heat capacity of an object is the amount of energy required to raise its temperature by 1 K (or 1°C). For pure substances, the heat capacity is usually defined for a specified amount of the substance. The specific heat capacity is the amount of energy required to raise the temperature of one gram of the substance by 1 K (or 1°C). (a) Since the chrome-plated brass fixture is not a pure substance, one would use heat capacity. (b) and (c) Since the copper wire and water are pure substances; one would use specific heat capacity. **6.35** 4.0 kJ **6.37** 323°C **6.39** 78°C **6.41** 42°C **6.43** 57.0°C **6.48** $\Delta H > 0$ since energy is required to break the O—O bond. **6.49** $\Delta H < 0$; Opposite in sign and $\frac{1}{2}$ the value for the vaporization of 2 moles of H_2O **6.50** (a) exothermic (b) +20.2 kJ (c) -5.2×10^2 kJ (d) -12.6 kJ **6.52** (a) $\frac{1}{2}N_2(g) + \frac{1}{2}O_2(g) \longrightarrow NO(g)$ $\Delta H = 90.29$ kJ (b) -4.51 kJ **6.54** 2.11×10^6 kJ is released. **6.58** (a) $C_2H_4(g) + 3O_2(g) \longrightarrow 2CO_2(g) + 2H_2O(g)$ $\Delta H = -1411$ kJ (b) 1.39 g **6.62** -110.5 kJ **6.63** -813.4 kJ **6.65** 44.0 kJ **6.67** $N_2(g) + 2O_2(g) \longrightarrow 2 NO_2(g)$; $\Delta H = 66.4$ kJ; A = 1, B = 2, C = 3. **6.70** The standard heat of reaction, ΔH^0_{rxn}, is the enthalpy change for any reaction where all substances are in their standard states. The standard heat of formation, ΔH^0_f, is the enthalpy change that accompanies the formation of *one mole* of a compound in its standard state from elements in their standard states. **6.72** (a) $\frac{1}{2}Cl_2(g) + Na(s) \longrightarrow NaCl(s)$ (b) $H_2(g) + \frac{1}{2}O_2(g) \longrightarrow H_2O(l)$ (c) no change **6.73** (a) $Ca(s) + Cl_2(g) \longrightarrow CaCl_2(s)$ (b) $Na(s) + C(\text{graphite}) + \frac{1}{2}H_2(g) + \frac{3}{2}O_2(g) \longrightarrow$ $NaHCO_3(s)$ (c) $C(\text{graphite}) + 2Cl_2(g) \longrightarrow CCl_4(l)$ (d) $\frac{1}{2}N_2(g) + \frac{1}{2}H_2(g) + \frac{3}{2}O_2(g) \longrightarrow HNO_3(l)$ **6.75** (a) -1036.8 kJ (b) -433 kJ **6.77** -157.3 kJ/mol **6.80** (a) 503.9 kJ (b) $-\Delta H_1 + 2\Delta H_2 = 504$ kJ **6.81** (a) $C_{18}H_{36}O_2(s) + 26O_2(g) \longrightarrow 18CO_2(g) + 18H_2O(g)$

(b) −10,488 kJ   (c) −36.9 kJ/g = −8.81 kcal/g   (d) 11.0 g
fat × 8.81 kcal/g = 96.9 kcal, so the information is consistent.
6.83 (a) 23.6 L initial; 24.9 L final   (b) 187 J   (c)
−1.3×10² J   (d) 320 J   (e) 320 J   (f) $\Delta H = \Delta E + P\Delta V =$
$\Delta E - w = q_p$ **6.84** (a) ΔH^0 for both is −55.90 kJ   (b) All
three are the same since all have the same net ionic equation.
6.86 (a) −65.2 kJ/mol   (b) 1.558×10⁴ kJ/kg SiC   **6.90**
−94 kJ/mol   **6.93** (a) 1054 J/Btu   (b) 1.05×10⁸ J/therm   (c)
1.31×10² mol   (d) $0.0035   (e) $0.53   **6.95** 721 kJ   **6.99**
(a) 4.3×10² kJ/mol O_2 (b) 6.1×10² kJ/mol CO_2 (c) 39 kJ/g
$C_{57}H_{110}O_6$ (d) 139 g $C_{57}H_{110}O_6$ **6.102** 94°C   **6.104** (a)
III > II > I   (b) I > III > II   **6.106** (a) $\Delta H^0_{rxn1} = -657.0$
kJ, $\Delta H^0_{rxn2} = 32.9$ kJ   (b) $\Delta H^0_{rxn2+3} = -106.6$ kJ   **6.108** (a)
6.81×10³ J   (b) 243°C   **6.110** −22.2 kJ   **6.112** (a) 34 kJ
(b) −757 kJ   **6.114** (a) 1.25×10³ kJ   (b) 2.24×10³°C

Chapter 7

7.2 (a) 5 < 2 < 6 < 4 < 1 < 3   (b) 3 < 1 < 4 < 6 < 2 < 5
(c) 3 < 1 < 4 < 6 < 2 < 5   **7.5** The energy of an atom is not
continuous, but quantized. It exists only in certain fixed
amounts called a quantum.   **7.7** 313 m, 3.13×10¹¹ nm,
3.13×10¹² Å   **7.9** 2.4×10⁻²³ J   **7.11** red < yellow < blue
7.13 1.3483×10⁷ nm, 1.3483×10⁸ Å   **7.16** (a) 8.21×10⁻¹⁹ J
(b) $\nu = 1.4×10^{15}$ s⁻¹, $E = 9.0×10^{-19}$ J   **7.18** Bohr's key assumption was that the electron in an atom does *not* radiate energy while in a stationary state, but that the electron can move
to a different orbit only by absorbing or emitting a photon
whose energy is equal to the difference in energy between two
states. These differences in energy correspond to the wavelengths in known line spectra for the hydrogen atom.   **7.20** (a)
absorption   (b) emission   (c) emission   (d) absorption
7.22 Yes, the predicted line spectra are accurate. The energies
for a one-electron system would have the values

$$E_n = \frac{-(Z^2)2.18×10^{-18}\text{ J}}{n^2}$$ where Z is the atomic number for the

atom or ion. The energy levels for Be^{3+} would be greater by a
factor of 16 than those for the hydrogen atom. This means that
the pattern of lines would be similar, but at different wavelengths.   **7.23** 434.17 nm   **7.25** 1875 nm   **7.27**
2.76×10⁵ J/mol   **7.29** (d) < (a) < (c) < (b)   **7.31** $n = 4$
7.34 3.37×10⁻¹⁹ J/photon; 2.03×10² kJ/einstein   **7.37**
Wavelengths associated with objects of ordinary size and mass
are so extremely small that they are impossible to observe.
7.39 (a) 7.56×10⁻³⁷ m   (b) 1×10⁻³⁵ m   **7.41**
2.2×10⁻²⁶ m/s   **7.43** 3.75×10⁻³⁶ kg/photon   **7.47** The total
probability of finding the 1s electron in any direction at a distance r from the nucleus is greatest when the value of r is
0.529 Å. This distance is for the greatest probability of finding
a 1s electron. The distance for a 2s electron would be farther
from the nucleus.   **7.48** (a) principal determinant of the electron's energy or distance from the nucleus   (b) determines the
shape of the orbital   (c) determines the orientation of the orbital   **7.49** (a) 1   (b) 5   (c) 3   (d) 9   **7.51** (a) m_l: −2, −1,
0, +1, +2   (b) m_l: 0   (c) m_l: −3, −2, −1, 0, +1, +2, +3

7.53 (a)

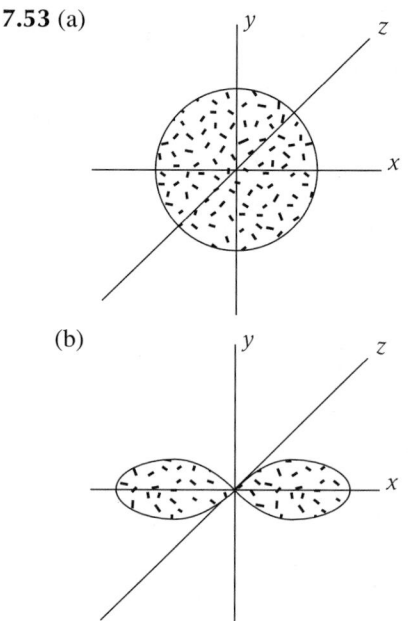

(b)

7.55

	Sublevel	Allowable m_l	No. of Orbitals
(a)	d	−2, −1, 0, +1, +2	5
(b)	p	−1, 0, +1	3
(c)	f	−3, −2, −1, 0, +1, +2, +3	7

7.57 (a) $n = 5$; $l = 0$; 1 orbital   (b) $n = 3$; $l = 1$; 3 orbitals
(c) $n = 4$; $l = 3$; 7 orbitals   **7.59** (a) No; $n = 2$, $l = 0$; $m_l = 0$;
$n = 2$, $l = 1$; $m_l = -1$ (b) OK   (c) OK   (d) No; $n = 5$,
$l = 2$; $m_l = +2$; $n = 5$, $l = 3$; $m_l = +3$ **7.62** (a) $E =$
$-[h^2/(8\pi^2m_ea_0^2)](1/n^2)$; constant = 2.180×10¹⁸ J, as in Bohr's
theory   (b) 3.027×10⁻¹⁹ J   (c) 656.2 nm, yes   **7.63** (a) The
attractive force of the nucleus for the electron must be overcome.   (b) The e⁻ in silver is more tightly held by the nucleus.
(c) Silver   (d) Once the electron is freed from the atom, its energy increases in proportion to the frequency.   **7.66** Li^{2+}
7.69 (a) 2 ⟶ 1   (b) 5 ⟶ 2   (c) 4 ⟶ 2   (d) 3 ⟶ 2
(e) 6 ⟶ 3   **7.71** (a) 462 nm   (b) 278 nm to 292 nm
7.73 8.70×10¹⁴ S⁻¹; 345 nm; ultraviolet   **7.75** (a) $n = 1$,
$l = 1$, $m_l = -1$ (b) $n = 3$, $l = 2$, $m_l = +1$ (c) $n = 7$, $l = 3$,
$m_l = +3$ (d) $n = 4$, $l = 2$, $m_l = -2$

7.77 (a) $\Delta E_{final} = -2.18×10^{-18}$ J/atom $\times \left(\dfrac{-Z^2}{n_i^2}\right)$

(b) 3.28×10⁴ kJ/mol   (c) 205 nm   (d) 22.8 nm
7.81 6.4×10²⁷   **7.83** (a) no overlap   (b) overlap   (c) 2   (d)
At longer wavelengths, the spectrum begins to become a continuous band.   **7.88** (a) 1.87×10⁻¹⁹ J   (b) 3.68×10⁻¹⁹ J
7.90 (a) red, green   (b) 1.18 kJ (Sr), 1.17 kJ (Ba)   **7.92** (a)
This is the wavelength of maximum absorbance, so gives the
highest sensitivity.   (b) UV   (c) 1.93×10⁻² g vitamin A/g
fish-liver oil   **7.96** 1.0×10¹⁸ photons/s   **7.98** 3s ⟶ 2p,
3d ⟶ 2p, 4s ⟶ 2p, 3p ⟶ 2s

Chapter 8

8.1 Elements are listed in the periodic table in an ordered, systematic way that correlates with a periodicity of their chemical
and physical properties. The theoretical basis for the table in
terms of atomic number and electron configuration does not al-

low for an "unknown element" between Sn and Sb. **8.3** (a) predicted 54.23 amu (b) predicted 6.3°C **8.6** m_s relates just to the electron; the others describe the orbital. **8.9** Shielding occurs when inner-shell electrons protect or shield outer-shell electrons from the full attractive force of the nucleus. The effective nuclear charge is the nuclear charge an electron actually experiences. As the number of shielding electrons increases, the effective nuclear charge decreases. **8.11** (a) 6 (b) 10 (c) 2 **8.13** (a) 6 (b) 2 (c) 14

8.16 In degenerate orbitals (those of identical energy), the filling will occur in such a manner that the spins will have the same value. N: $1s^2\,2s^2\,2p^3$

8.18 Main-group elements from the same group have similar outer electron configurations where the (old) group number equals the number of outer electrons. Outer electron configurations vary in a periodic manner within a period, with each succeeding element having one additional electron. **8.20** The maximum number of electrons in any orbital n is $2n^2$. For $n = 4$, the maximum electron capacity is $2(4)^2$ or 32 electrons.

8.21 (a) $n = 5$; $l = 0$; $m_l = 0$; $m_s = +\frac{1}{2}$ (b) $n = 3$; $l = 1$; $m_l = 1$; $m_s = -\frac{1}{2}$ (c) $n = 5$; $l = 0$; $m_l = 0$; $m_s = +\frac{1}{2}$ (d) $n = 2$; $l = 1$; $m_l = +1$; $m_s = -\frac{1}{2}$ **8.23** (a) Rb: $1s^2 2s^2 2p^6 3s^2 3p^6 4s^2 3d^{10} 4p^6 5s^1$ (b) Ge: $1s^2 2s^2 2p^6 3s^2 3p^6 4s^2 3d^{10} 4p^2$ (c) Ar: $1s^2 2s^2 2p^6 3s^2 3p^6$ **8.25** (a) Cl: $1s^2 2s^2 2p^6 3s^2 3p^5$ (b) Si: $1s^2 2s^2 2p^6 3s^2 3p^2$ (c) Sr: $1s^2 2s^2 2p^6 3s^2 3p^6 4s^2 3d^{10} 4p^6 5s^2$

8.27 (a) Ti: $[Ar]4s^2 3d^2$

(b) Cl: $[Ne]3s^2 3p^5$

(c) V: $[Ar]4s^2 3d^3$

8.29 (a) Mn: $[Ar]4s^2 3d^5$

(b) P: $[Ne]3s^2 3p^3$

(c) Fe: $[Ar]4s^2 3d^6$

8.31 (a) O; Group 6A(16); Period 2 (b) P; Group 5A(15); Period 3

8.33 (a)

Cl; 7A(17); 3

(b)

As; 5A(15); 4

8.35 (a) $[Ar]\,4s^2 3d^{10} 4p^1$; Ga; 3A(13); 4 (b) $[He]\,2s^2 2p^6$; Ne, 8A(18); 2

8.37

	Inner Electrons	Outer Electrons	Valence Electrons
(a) O	2	6	6
(b) Sn	46	4	4
(c) Ca	18	2	2
(d) Fe	18	2	8
(e) Se	28	6	6

8.39 (a) B; Al, Ga, In, Tl (b) S; O, Se, Te, Po (c) La; Sc, Y, Ac **8.41** (a) C; Si, Ge, Sn, Pb (b) V; Nb, Ta, Ha (c) P; N, As, Sb, Bi

8.43 Na: $1s^2 2s^2 2p^6 3s^1$
Na (excited): $1s^2 2s^2 2p^6 3p^1$

8.46 Atomic radius increases from top to bottom in a group, whereas ionization energy decreases from top to bottom in a group. The farther away an electron is from the nucleus, the more easily it is removed. **8.48** Successive ionization energies of a particular element increase because each additional electron is being pulled away from a more positive ion. An extremely large jump in ionization energies occurs when all valence electrons have been removed and the first core electron is removed. Thus, by looking at a series of successive IE's, we can determine the number of valence electrons. **8.50** Group 7A(17) has high IE, and negative first electron affinities. These elements tend to form -1 ions. **8.53** (a) K < Rb < Cs (b) O < C < Be (c) Cl < S < K (d) Mg < Ca < K **8.55** (a) Ba < Sr < Ca (b) B < N < Ne (c) Rb < Se < Br (d) Sn < Sb < As **8.57** B: $1s^2 2s^2 2p^1$ **8.59** (a) Na (b) Na (c) Be **8.61** Metals have a characteristic luster, generally have high melting points, and are good electrical conductors. Nonmetals have dull surfaces, typically low melting points, and are poor electrical conductors. **8.62** Metallic character increases from right to left across a period and from top to bottom in a group. This trend is identical to atomic size and opposite to that of ionization energy. **8.64** +2 or +4. The +2 ions form by loss of the outermost two p electrons, while the +4 ions form by loss of these and the outermost two s electrons. **8.68** (a) Rb (b) Ra (c) I **8.70** (a) As (b) P (c) Be **8.72** acidic solution; $SO_3(g) + H_2O(l) \longrightarrow H_2SO_4(aq)$ **8.74** (a) Cl$^-$: $1s^2 2s^2 2p^6 3s^2 3p^6$ (b) Na$^+$: $1s^2 2s^2 2p^6$ (c) Ca^{2+}: $1s^2 2s^2 2p^6 3s^2 3p^6$ **8.76** (a) Al^{3+}: $1s^2 2s^2 2p^6$ (b) S^{2-}: $1s^2 2s^2 2p^6 3s^2 3p^6$ (c) Sr^{2+}: $1s^2 2s^2 2p^6 3s^2 3p^6 4s^2 3d^{10} 4p^6$
8.78 (a) 0 (b) 3 (c) 0 (d) 1 **8.80** (a), (b), and (d) **8.82** (a) V^{3+}: $[Ar]3d^2$, paramagnetic (b) Cd^{2+}: $[Kr]4d^{10}$, diamagnetic (c) Co^{3+}: $[Ar]3d^6$, paramagnetic (d) Ag$^+$: $[Kr]4d^{10}$, diamagnetic

8.84 (a) [Kr] 2 unpaired e$^-$

(b) [Kr] 0 unpaired e$^-$

(c) [Kr] 2 unpaired e$^-$

The electron configuration of (b) is consistent with a diamagnetic palladium atom.

8.86 (a) $Li^+ < Na^+ < K^+$; size increases down a group.   (b) $Rb^+ < Br^- < Se^{2-}$; size increases as atomic number decreases in an isoelectronic series.   (c) $F^- < O^{2-} < N^{3-}$; same reason as (b).   **8.90** (a) K is further left (greater radius) while Sr is further down (also greater radius).   (b) Z_{eff} remains fairly constant after the first 2 or 3 elements across a transition series. (c) Na is farther left (more metallic) but Ca is farther down (also more metallic).   (d) Se is farther right (more acidic oxide) but P is farther up (also more acidic oxide). **8.92** (a) Aluminum: $1s^2 2s^2 2p^6 3s^2 3p^1$
(b) Lutetium: $1s^2 2s^2 2p^6 3s^2 3p^6 4s^2 3d^{10} 4p^6 5s^2 4d^{10} 5p^6 6s^2 4f^{14} 5d^1$
(c) Yttrium: $1s^2 2s^2 2p^6 3s^2 3p^6 4s^2 3d^{10} 4p^6 5s^2 4d^1$
(d) Sulfur: $1s^2 2s^2 2p^6 3s^2 3p^4$
(e) Strontium: $1s^2 2s^2 2p^6 3s^2 3p^6 4s^2 3d^{10} 4p^6 5s^2$
(f) Arsenic: $1s^2 2s^2 2p^6 3s^2 3p^6 4s^2 3d^{10} 4p^3$ **8.95** (a) $SrBr_2$; strontium bromide   (b) CaS; calcium sulfide   (c) ZnF_2; zinc fluoride   (d) LiF; lithium fluoride   **8.97** (a) $\Delta H < 0$; Li has a larger negative electron affinity than Na due to its smaller size (b) $\Delta H < 0$; K has a smaller ionization energy than Na due to its larger size.   **8.101** Wear blue/purple clothing to stay cooler. Blue/purple clothing absorbs orange/red light, which is lower energy light than blue/purple. Blue/purple clothing reflects higher energy blue/purple light.   **8.103** $n = 8, l = 0$; $8s$; 1 orbital.   **8.105** (a) $E_{436nm} = 4.56 \times 10^{-19}$ J, $E_{254nm} = 7.83 \times 10^{-19}$ J   (b) 1.24×10^{-18} J   (c) 160. nm   **8.107** Ce: $[Xe] 6s^2 4f^1 5d^1$; Ce^{4+}: [Xe]; Eu: $[Xe] 6s^2 4f^7$; Eu^{2+}: $[Xe] 4f^7$; Ce^{4+} has a noble-gas configuration and Eu^{2+} has a half-filled f subshell.

Chapter 9

9.1 (a) Larger ionization energy decreases metallic character. (b) Larger atomic radius increases metallic character.   (c) Larger number of outer electrons decreases metallic character. (d) Larger effective nuclear charge decreases metallic character.   **9.4** (a) Cs   (b) Rb   (c) As   **9.6** (a) ionic   (b) covalent   (c) metallic   **9.8** (a) covalent   (b) ionic   (c) covalent

9.10 (a) Rb·   (b) ·S̈i·   (c) :Ï·

9.12 (a) ·Sr·   (b) :P̈·   (c) :S̈:

9.14 (a) 6A(16); $ns^2 np^4$ (b) 3A(13); $ns^2 np^1$ **9.17** Because the lattice energy is the result of electrostatic attractions among the oppositely charged ions, its magnitude depends on several factors, including ionic size and ionic charge. For a particular arrangement of ions, the lattice energy increases as the charges on the ions increase and as their radii decrease.   **9.20** (a) Ba^{2+} = [Xe]; Cl^- = $[Ne] 3s^2 3p^6$; $BaCl_2$ (b) Sr^{2+} = [Kr]; O^{2-} = $[He] 2s^2 2p^6$; SrO   (c) Al^{3+} = [Ne]; F^- = $[He] 2s^2 2p^6$; AlF_3 (d) Rb^+ = [Kr]; O^{2-} = $[He] 2s^2 2p^6$; Rb_2O **9.22** (a) 2A(2) (b) 6A(16)   (c) 1A(1)   **9.24** (a) 3A(13)   (b) 2A(2)   (c) 6A(16)   **9.26** (a) BaS; Ba and S have larger ion charges than Cs and Cl.   (b) LiCl; Li has a smaller radius than Cs.   **9.28** (a) BaS; Ba has a larger radius than Ca.   (b) NaF; Na and F have smaller charges than Mg and O.   **9.30** Lattice energy of NaCl = -788 kJ. This is less than for LiF due to the larger ionic radii of Na^+ and Cl^-.   **9.33** -336 kJ   **9.34** Chlorine gas consists of diatomic molecules. As two chlorine atoms approach each other, each atomic nucleus starts to attract the

other's electron, which lowers the potential energy of the system. The atoms continue to draw each other closer, and the potential energy continues to be lowered. As attractions increase, so do repulsions between the nuclei and between electrons. At some internuclear distance, the maximum attraction is achieved in spite of the repulsion, and the system is at its lowest energy. The mutual attraction of nuclei and electron pair constitutes the covalent bond.
9.35 Bond energy is the change in enthalpy accompanying the dissociation of one mole of the bond in the gaseous state.
$$H—Cl(g) + Energy \longrightarrow H(g) + Cl(g)$$
ΔH is positive, requiring energy or endothermic.
ΔH of bond formation is the same magnitude, but opposite in sign.
9.39 (a) $I—I < Br—Br < Cl—Cl$ (b) $S—Br < S—Cl < S—H$ (c) $C—N < C=N < C\equiv N$ **9.41** $C=O > C—O >$ intermolecular force   **9.43** Electronegativity increases from left to right and decreases from top to bottom. Fluorine and oxygen are the two most electronegative atoms, while cesium and francium are the two least electronegative.   **9.45** The H—O bond in water is polar covalent. A nonpolar covalent bond occurs when the bonding electrons are between two atoms of the same electronegativity. The polar covalent bond is when the bonding electrons are between two different atoms with differing electronegativity, and therefore one atom can attract the bonding electrons more strongly. Ionic bonding would imply that an electron transfer between the atoms has occurred.
9.48 (a) $Si < S < O$ (b) $Mg < As < P$
9.50 (a) $N > P > Si$ (b) $As > Ga > Ca$
9.52 (a) N—B   (b) N—O   (c) C—S (no poles)
(d) S—O   (e) N—H   (f) Cl—O
9.54 (a), (d), (e)   **9.56** (a) nonpolar covalent   (b) ionic   (c) polar covalent   (d) polar covalent   (e) nonpolar covalent   (f) polar covalent; $SCl_2 < SF_2 < PF_3$
9.58 (a) $H—I < H—Br < H—Cl$
(b) $H—C < H—O < H—F$
(c) $S—Cl < P—Cl < Si—Cl$
9.61 (a) shiny, malleable, good electrical conductivity, relatively high melting and boiling points   (b) form basic oxides, form cations by losing electrons   **9.65** (a) $-1220.$ kJ   (b) 2.343×10^4 kJ   (c) 1.690×10^3 g   (d) 65.2 L   **9.67** (a) -125 kJ   (b) Yes, since ΔH_f^0 is negative.   (c) -392 kJ   (d) No, since ΔH for the conversion of MgCl to $MgCl_2$ is negative. **9.70** (a) 406 nm   (b) 2.92×10^{-19} J/atom   (c) 1.87×10^4 m/s   **9.74** (a) 4.04×10^{-19} J/molecule, 6.09×10^{14} s^{-1} (b) 353 nm **9.77** (a) The bond energies are for the formation of gaseous atoms and thus are for homolytic bond breakage.   (b) Since one atom gets both of the bonding electrons in heterolytic bond breakage, this results in the formation of ions.   (c) Since chlorine is more electronegative than carbon, the carbon atoms in $H_3C—CCl_3$ which are bonded to the chlorine effectively have a greater electronegativity than those in $H_3C—CH_3$. This would make that atom hold onto its electrons more tightly and increase the bond energy.   (d) Yes. In addition to breaking the C—C bond, you would have to overcome the energy of attraction of the ions formed.   **9.79** (a) solid silver   (b) The va-

lence electrons in solid metals are delocalized. The delocalized electrons are easier to remove than electrons located on a single gaseous atom. **9.81** for ethane: 413 kJ/mol, for ethene: 410. kJ/mol, for ethyne: 404 kJ/mol

Chapter 10

10.1 (b) He and (d) H cannot serve as a central atom in a Lewis structure. Both have an outermost shell with maximum of two electrons. F cannot serve as a central atom because its valence electrons are in the $n = 2$ level. F cannot have an expanded octet because the $n = 2$ level only contains s and p orbitals, so F can form only one bond.

10.3 a, b, d, e, f and h obey the octet rule; c and g do not.

10.5 (a) $:\ddot{F}:$ (b) $:\ddot{Cl}-\ddot{Se}-\ddot{Cl}:$ (c) $:\ddot{F}-C-\ddot{F}:$

10.7 (a) $:\ddot{F}-\ddot{P}-\ddot{F}:$ (b) $H-\ddot{O}-C-\ddot{O}-H$ (c) $:\ddot{S}=C=\ddot{S}:$

10.9 (a) $:\ddot{O}-\ddot{N}=\ddot{O}: \longleftrightarrow :\ddot{O}=\ddot{N}-\ddot{O}:$

(b) $:\ddot{O}-N=\ddot{O}: \longleftrightarrow :\ddot{O}=N-\ddot{O}:$ with $:\ddot{F}:$

10.11 (a) $[:\ddot{N}=N=\ddot{N}: \longleftrightarrow :\ddot{N}-N\equiv N: \longleftrightarrow :N\equiv N-\ddot{N}:]^-$

(b) $[:\ddot{O}=N-\ddot{O}: \longleftrightarrow :\ddot{O}-N=\ddot{O}:]^-$

10.13 (a) f.c. = 0 for F and I

(b) f.c. = 0 for H; = −1 for Al

10.15 (a) $[:C\equiv N:]^-$ f.c. = −1 for C; = 0 for N
(b) $[:\ddot{O}-\ddot{Cl}:]^-$ f.c. = −1 for O; = 0 for Cl

10.17 (a) $[:\ddot{O}-Br=\ddot{O}]$ f.c. = −1 for O; = 0 for (2)O; = 0 for Br ON = −2 for O; = +5 for Br

(b) $[:\ddot{O}-S=\ddot{O}]^{2-}$ f.c. = −1 for (2)O; = 0 for O; = 0 for S ON = −2 for O; = +4 for S

10.19 (a) $H-B-H$ only 6 e⁻ in outer shell

(b) $[:\ddot{F}-\ddot{As}-\ddot{F}:]^-$ expanded outer shell to 10 e⁻

(c) $:\ddot{Cl}-\ddot{Se}-\ddot{Cl}:$ expanded outer shell to 10 e⁻

10.21 (a) $:\ddot{F}-\ddot{Br}-\ddot{F}:$ expanded outer shell to 10 e⁻

(b) $[:\ddot{Cl}-\ddot{I}-\ddot{Cl}:]^-$ expanded outer shell to 10 e⁻

(c) $:\ddot{F}-Be-\ddot{F}:$ only 4 e⁻ in outer shell

10.23

$:\ddot{Cl}-Be-\ddot{Cl}: + 2:\ddot{Cl}:^- \longrightarrow [:\ddot{Cl}-Be-\ddot{Cl}:]^{2-}$ (with $:\ddot{Cl}:$ above and below)

10.26 Structure A has all atoms with formal charge = 0, so it is the best structure. **10.28** Less energy is required to break weaker bonds. **10.30** −168 kJ

10.32

$H_2C=CH_2 + H-\ddot{O}-H \longrightarrow H-CH_2-CH_2-\ddot{O}-H$

$\Delta H = -37$ kJ

10.34 −22 kJ **10.37** When all electron pairs are involved in bonding (i.e., there are no lone pairs), the name of the molecular shape is the same as the name of the electron-group arrangement.

10.39 tetrahedral AX₄; trigonal pyramidal AX₃E; bent or V shaped AX₂E₂

10.41 (a) AX₂ (b) AX₃

(c) AX₅ (d) AX₃E₂

(e) AX₃E (f) AX₅E

10.43 (a) trigonal planar, bent, 120° (b) tetrahedral, trigonal pyramidal, 109.5° (c) tetrahedral, trigonal pyramidal, 109.5° **10.45** (a) trigonal planar, trigonal planar, 120° (b) trigonal planar, bent, 120° (c) tetrahedral, tetrahedral, 109.5° **10.47** (a) trigonal planar, AX₃, 120° (b) trigonal pyramidal, AX₃E, 109.5° (c) trigonal bipyramidal, AX₅, 90° and 120° **10.49** (a) bent, 109.5°, smaller (b) trigonal bipyramidal, 90° and 120°, none (c) see-saw, 90° and 120°, smaller (d) linear, 180°, none **10.51** (a) C: tetrahedral, 109.5°; O: bent, < 109.5° (due to lone pairs) (b) N: trigonal planar, > 120° (O—N=O) **10.53** (a) C in CH₃: tetrahedral, 109.5°; C (with C=O): trigonal planar, < 120° for C—C—O(H); O (with O—H): bent, < 109.5° (due to lone pairs) (b) O: bent, < 109.5° (due to lone pairs) **10.55** OF₂ < NF₃ < CF₄ < BF₃ < BeF₂

	Angle	Ideal Value	Deviation
10.57 (a)	H—C—H	120°	smaller
	H—C—N	120°	greater
	C—N—O	120°	smaller
	N—O—H	109.5°	smaller
(b)	H—C—H	109.5°	none
	H—C—O	109.5°	none
	C—O—C	109.5°	smaller
(c)	H—O—B	109.5°	smaller
	O—B—O	120°	none

10.60 $2PCl_5(g) \longrightarrow PCl_4^+(s) + PCl_6^-(s)$

  $AX_5 \longrightarrow AX_4$ AX_6

  trigonal   tetrahedral   octahedral
  bipyramidal
  120°, 90°   109.5°   90°

10.61 Polar bonds have magnitude and direction; if equal and opposite, the molecule is nonpolar. If the forces are not balanced the molecule is polar.   **10.64** (a) CF_4 (b) SCl_2 and BrCl   **10.66** (a) SO_2; bent vs. trigonal planar structure   (b) IF; greater ΔEN (c) SF_4; see-saw vs. tetrahedral structure   (d) H_2O; greater ΔEN.

10.68 X: Y: Z:

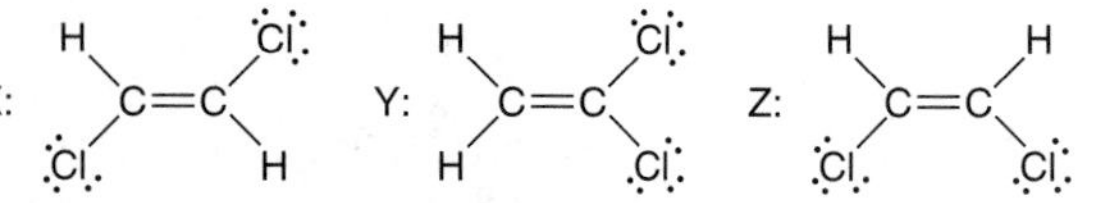

Compound Y does have a dipole moment.

10.70 (a) Bond strength is greatest in N_2. Bond length is greatest in N_2H_4. Bond order = 1 for N_2H_4, 2 for N_2H_2, and 3 for N_2.

(b)

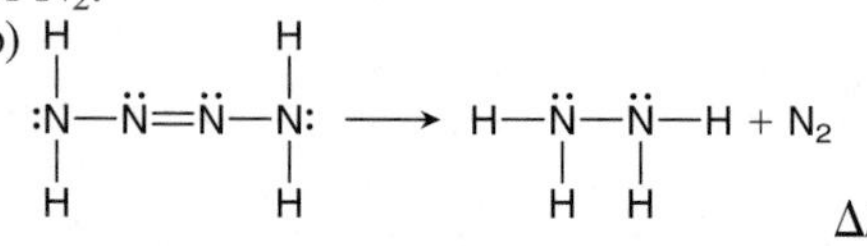

:N—N̈=N̈—N: ⟶ H—N̈—N̈—H + N_2

$\Delta H = -367$ kJ/mol

10.73 (a) f.c. Al = −1, f.c. Cl, ends = 0, f.c. Cl, interior = +1 f.c. I = −1, f.c. Cl, ends = 0, f.c. Cl, interior = +1   (b) I_2Cl_6 is planar, the iodine atoms have four bonding pairs and two non-bonding pairs, which corresponds to a square *planar* shape.

10.76

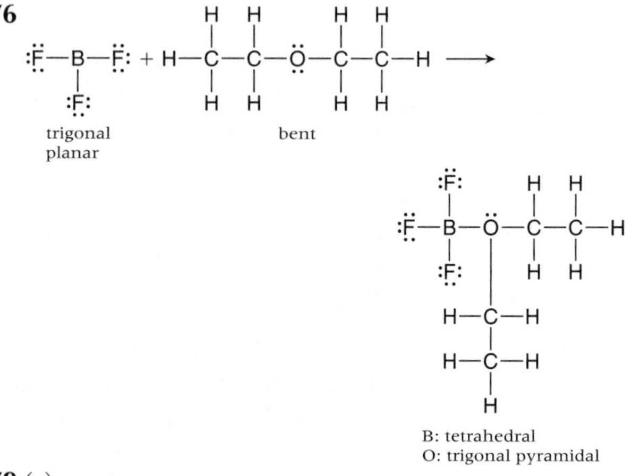

:F̈—B—F̈: + H—C—C—Ö—C—C—H ⟶

  :F̈:

  trigonal   bent
  planar

B: tetrahedral
O: trigonal pyramidal

10.79 (a)

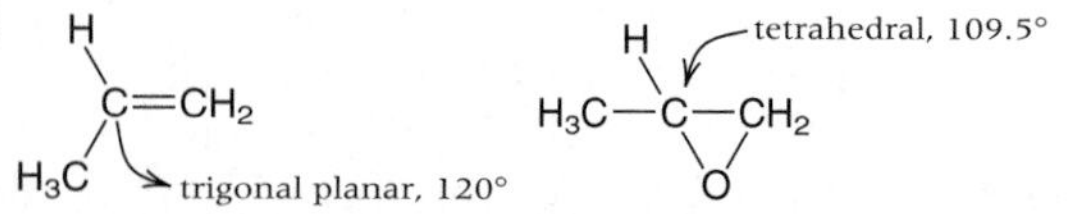

trigonal planar, 120°

tetrahedral, 109.5°

(b) The C—C—H bond angle in propene is compressed (<120°) because the double bond takes up more space than the other bonds. The central C atom in propylene is part of a three-member ring, so the C—C—O bond angle is 60° and the remaining angles are >109.5°.

10.81

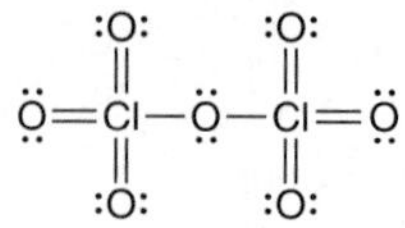

Ö=Cl—Ö—Cl=Ö

The Cl—O—Cl bond is less than 109.5° due to the presence of two lone pairs on oxygen.

10.85 (a) −1267 kJ/mol   (b) −1226 kJ/mol   (c)

−1234.8 kJ/mol, the values differ by less than 1%, so there is good agreement between calculation methods.

10.88 H—Ö—C—C—Ö—H

  :O: :O:

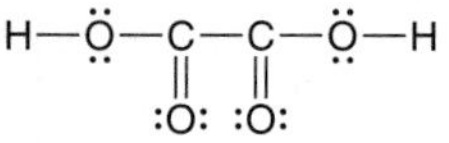

10.92 (a) It has an odd number of electrons.   (b) 426 kJ/mol   (c) 508 kJ/mol

10.96 $H_2C_2O_4$:

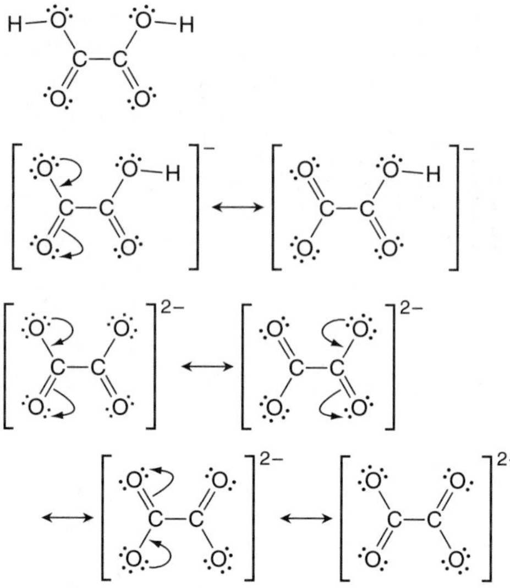

$HC_2O_4^-$:

$C_2O_4^{2-}$:

In $H_2C_2O_4$, the C=O bonds will be shorter and stronger than the C—O bonds. In $HC_2O_4^-$, the O which bonds to the C that has lost the acidic hydrogen will be equivalent and will be longer and weaker than the C=O bond on the other carbon, but shorter and stronger than the C—O bond on the other carbon. In $C_2O_4^{2-}$ all the carbon to oxygen bonds will be the same length (intermediate between C—O and C=O) and strength.

10.100 possible: trigonal pyramidal or T shaped; impossible: trigonal planar

Chapter 11

11.1 (a) sp^2 (b) sp^3d^2 (c) sp (d) sp^3 (e) sp^3d

11.3 C has only 2s and 2p orbitals, allowing for a maximum of four hybrids. Si has 3d orbitals available in addition to the 3s and 3p, so it can form up to six hybrids.   **11.5** (a) six sp^3d^2 orbitals   (b) four sp^3 orbitals   **11.7** (a) sp^2 (b) sp^2 (c) sp^2

11.9 (a) sp^3 (b) sp^3 (c) sp^3

11.11 (a) $sp^3 \longleftarrow s + 3p$ (b) $sp \longleftarrow s + p$

11.13 (a) $sp^3d \longleftarrow s + 3p + d$ (b) $sp^3 \longleftarrow s + 3p$

11.15 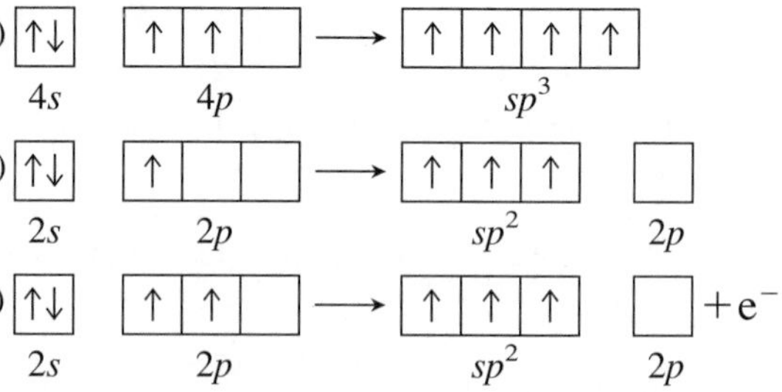 (a)

11.17 (a)

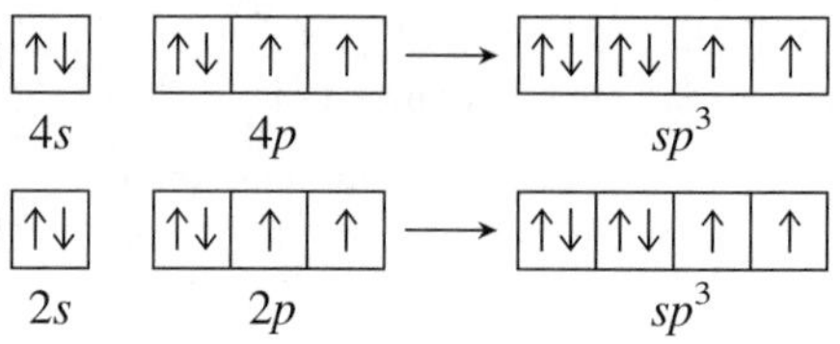

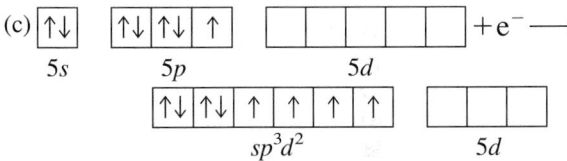

(c) $5s$ $5p$ $5d$ $+e^- \longrightarrow$ sp^3d^2 $5d$

11.20 (a) False. A double bond is one sigma and one pi bond. (b) False. A triple bond consists of one sigma (σ) and two pi (π) bonds. (c) True. (d) True. (e) False. A π bond consists of a second pair of electrons after a σ bond has been previously formed. (f) False. End-to-end overlap results in a bond with electron density along the bond axis.

11.21 (a) $3(\sigma) + 1(\pi)$ about N, sp^2 hybridization (b) $2(\sigma) + 2(\pi)$ about C, sp hybridization (c) $3(\sigma) + 1(\pi)$ about C, sp^2 hybridization

11.23 (a) $:\ddot{F}\!-\!\ddot{N}\!=\!\ddot{O}:$ $2(\sigma) + 1(\pi)$ about N, sp^2 hybridization

(b) 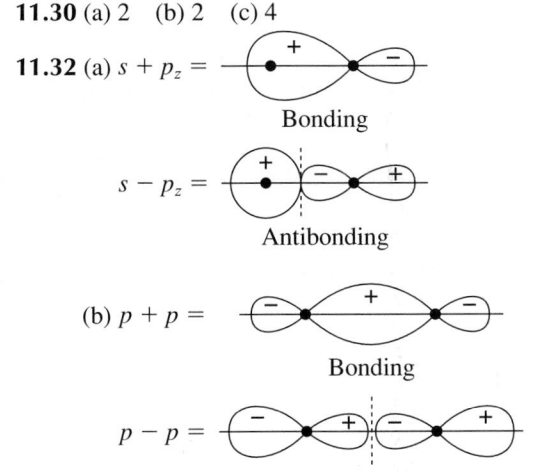 $3(\sigma) + 1(\pi)$ about each C, sp^2 hybridization

(c) $:N\!\equiv\!C\!-\!C\!\equiv\!N:$ $2\sigma + 2(\pi)$ about each C, sp hybridization

11.25

CH₃ H / CH₃ CH₃
C=C (trans)   C=C (cis)
H CH₃ / H H

The single C—H and C—C bonds are σ bonds; the C=C bonds are $1\sigma + 1\pi$.

11.26 Two p AO + two p AO $\longrightarrow$ 4 molecular orbitals. The number of MO is equal to the number of AO from which they are formed.

11.28 Bonding orbitals (BO) *vs.* antibonding orbitals (ABO): (a) energy—BO lower in energy—more stable. (b) nodal plane—BO have none perpendicular to the bond axis. (c) internuclear electron density—the BO has a higher electron density than the ABO.

11.30 (a) 2 (b) 2 (c) 4

11.32 (a) $s + p_z =$ **Bonding**

$s - p_z =$ **Antibonding**

(b) $p + p =$ **Bonding**

$p - p =$ **Antibonding**

11.34 (a) stable (b) paramagnetic (c) $(\sigma_{2s})^2(\sigma^*_{2s})^1$

11.36 (a) Increasing bond energy $C_2^+ < C_2 < C_2^-$ (b) Increasing bond length $C_2^- < C_2 < C_2^+$ **11.39** (a) $9\sigma, 2\pi$ (b) No, since each double bond has two identical substituent groups on one C.

11.41 (a)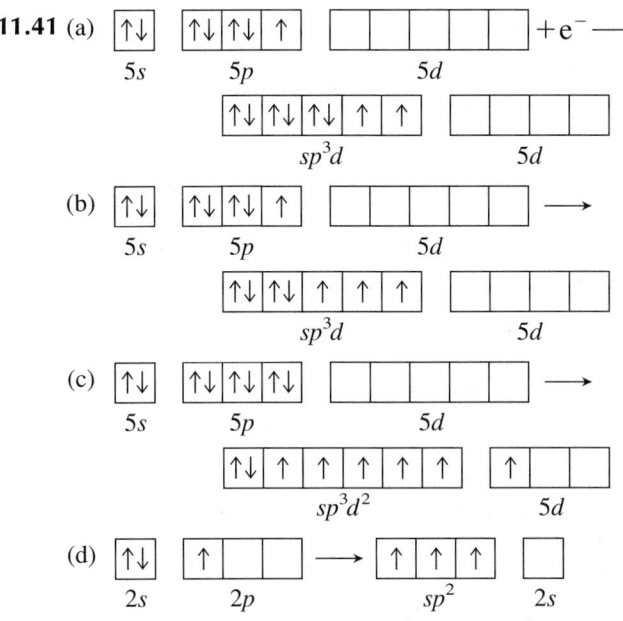
$5s$ $5p$ $5d$ $+e^- \longrightarrow$ sp^3d $5d$

(b) $5s$ $5p$ $5d$ $\longrightarrow$ sp^3d $5d$

(c) $5s$ $5p$ $5d$ $\longrightarrow$ sp^3d^2 $5d$

(d) $2s$ $2p$ $\longrightarrow$ sp^2 $2s$

11.42 (a) 17 sigma bonds (b) all C: sp^2, ring N: sp^2, other N: sp^3 **11.44** (a) B changes from sp^2 to sp^3 hybridization. (b) P changes from sp^3 to sp^3d. (c) C changes from sp to sp^2. (d) Si changes from sp^3 to sp^3d^2. (e) S is sp^2 and remains sp^2.

11.47 (a) the one with two S=O bonds (b) tetrahedral; sp^3 (c) S: d orbitals; O: p orbitals (d)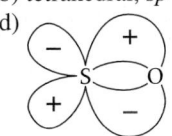

11.54 The LiF molecules are polar due to the large difference in electronegativity between Li and F. The relatively small ionization energy of Li and the relatively large electron affinity of F reinforce this.

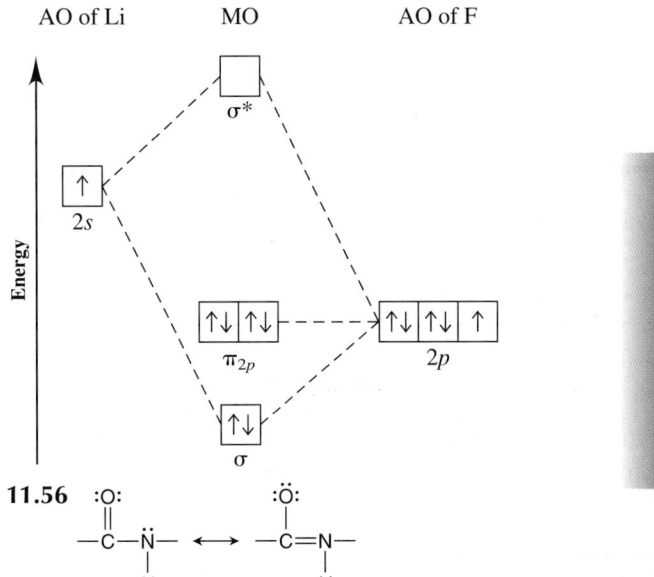

AO of Li   MO   AO of F

11.56

:O: :Ö:
‖ |
—C—Ṅ— ⟷ —C=N—
      | |
      H   H

Resonance interaction gives the C—N bond double bond character which hinders rotation about the C—N bond. The C—N single bond is a σ bond; the resonance interaction exchanges a

C—O π bond for a C—N π bond. **11.58** (a) C in CH_3: sp^3; all other C: sp^2; central O: sp^3 (b) 2 (c) 8; 1

Chapter 12

12.1 In a solid, the energy of attraction is greater than the energy of motion, while in a gas it is less. Gases have high compressibility and the ability to flow, while solids have neither. **12.4** (a) Because the intermolecular forces are only partially overcome in fusion and need to be totally overcome in vaporization. (b) Because greater intermolecular forces are found in solids as opposed to liquids. (c) $\Delta H_{cond} = -\Delta H_{vap}$. **12.5** (a), (b) intermolecular; (c), (d) intramolecular **12.7** (a) condensation (b) fusion (c) evaporation **12.9** The molecules of a gas are moving more rapidly than those in a liquid. When the liquid condenses, the molecules slow down and energy is released to the surroundings. **12.13** At first, the evaporation of liquid molecules from the surface predominates, which increases the number of gas molecules and hence the vapor pressure. Once more molecules enter the gas phase, the probability of a gas molecule hitting the surface of the liquid and "sticking" increases, so the condensation rate increases. When the evaporation and condensation rates become equal, the vapor pressure becomes constant. **12.14** As intermolecular forces increase, the (a) critical temperature increases. (b) boiling point increases. (c) vapor pressure decreases. (d) heat of vaporization increases. **12.18** Because the condensation of steam releases a much greater quantity of heat (40.7 kJ/mol) than does the cooling of water. **12.19** 4.16×10^3 J **12.21** 526 torr **12.23** 18 kJ/mol

12.25

Solid ethylene is more dense than liquid ethylene.
12.28 41 atm **12.32** O is a much smaller atom than Se. The electron density on O is much greater than on Se as a result, and it attracts the H atom much more strongly. **12.34** *All* molecules exhibit dispersion forces. Since these are so weak, however, they are in general overwhelmed if any other intermolecular forces (dipole-dipole, H bonding) exist. **12.37** (a) H bonding (b) dispersion (c) dispersion **12.39** (a) dipole-dipole (b) dispersion (c) H bonding

12.41 (a)

(b) H—$\ddot{F}$:········H—$\ddot{F}$:········H—$\ddot{F}$:

12.43 (a) dispersion (b) H bonding (c) dispersion **12.45** (a) I^- (b) C_2H_4 (c) H_2Se In cases (a) and (c), the larger particle (with more electrons) has the higher polarizability. In case (b), the less tightly held π electrons are more easily distorted. **12.47** (a) C_2H_6; smaller molecules have smaller intermolecular forces. (b) CH_3CH_2F; weaker intermolecular force (dipole-dipole *vs.* H bonding in CH_3CH_2OH). (c) PH_3; weaker intermolecular force (dipole-dipole *vs.* H bonding in NH_3). **12.49** (a) LiCl; ionic bonds (LiCl) *vs.* dipole-dipole (HCl). (b) NH_3; H bonding (NH_3) *vs.* dipole-dipole (PH_3). (c) I_2; larger molecule has larger dispersion forces.
12.51 (a) CH_2—CH_2: more compact molecule.
        CH_2—CH_2

(b) PBr_3; dipole-dipole (PBr_3) *vs.* ionic bonds (NaBr).
(c) HBr; dipole-dipole (HBr) *vs.* H bonding (H_2O).
12.53 The polarity of the X—H bond increases N—H < O—H < F—H. **12.57** The cohesive forces of H_2O (H bonds) and of Hg (metallic bonds) are stronger than the adhesive forces between those drops and the nonpolar wax (held together by van der Waals forces), meaning that these two would remain as spherical drops. The weak cohesive forces of oil (van der Waals forces) are about as strong as the adhesive forces between the oil and the wax, so that the oil drops would spread out more. **12.59** Because it is defined as the amount of energy required to increase the surface area by a specific amount. **12.61** $CH_3CH_2CH_2OH < HOCH_2CH_2OH <$ $HOCH_2CH(OH)CH_2OH$ The number of H bonds, and hence the strength of intermolecular forces, would increase as shown, so also would the surface tension. **12.63** Viscosity and surface tension both increase with the strength of intermolecular forces, so the order would be the same. **12.65** Apple juice (primarily water) is held together by H bonds; cooking oil is held together by dispersion forces. The adhesion of the paper and water would be stronger (H bonds in paper to H bonds in water) than that between paper and cooking oil. **12.68** Water is a good solvent for materials which form ions and/or which can hydrogen bond. The more polar (i.e., like water) the material, the better. Water is a poor solvent for nonpolar materials. **12.69** The two hydrogen atoms covalently bonded to the oxygen can form hydrogen bonds with two other molecules. In addition, two hydrogen atoms from adjoining H_2O molecules can form hydrogen bonds with the two lone pairs of electrons on the oxygen. **12.72** Water has a high capillary action, which allows it to easily be absorbed by the plant's roots and transported to the leaves. **12.78** Simple cubic **12.81** The energy gap is the energy difference between the highest filled energy level (the valence band) and the lowest unfilled energy level (the conduction band). In conductors and superconductors, the gap is zero (the bands overlap); in semiconductors it is small, and in insulators it is large. **12.83** Atomic mass and atomic radius **12.84** (a) face-centered (b) body-centered (c) face-centered **12.86** (a) metallic, since Sn is a metal (b) network covalent, since Si is similar to C (diamond) (c) atomic, since Xe is a monatomic element **12.88** (a) metallic, since Ni is a metal (b) molecular, since this is a molecule (c) molecular,

since this is a molecule   **12.90** 4   **12.92** (a) 4 of each   (b) 577.40 amu or 9.59×10^{-22} g   (c) 1.77×10^{-22} cm^3 (d) 5.61Å   **12.94** (a) insulator   (b) conductor   (c) semiconductor   **12.96** (a) increase   (b) increase   (c) decrease   **12.98** 1.68×10^{-8} cm (1.68Å)   **12.105** Isotropic properties are the same in all directions, while anisotropic properties depend on direction. Liquid crystals show some degree of order, so they are anisotropic.   **12.111** (a) n-type   (b) p-type   **12.113** 3.4×10^3 **12.115** 7.9 nm   **12.118** (a) 19.8 torr   (b) 0.0388 g   **12.120** (a)

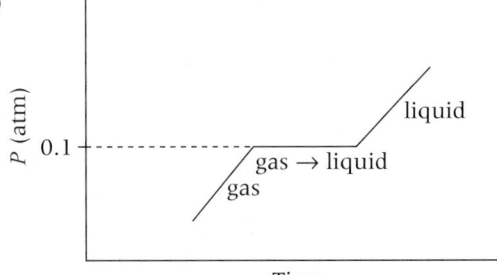

(b)

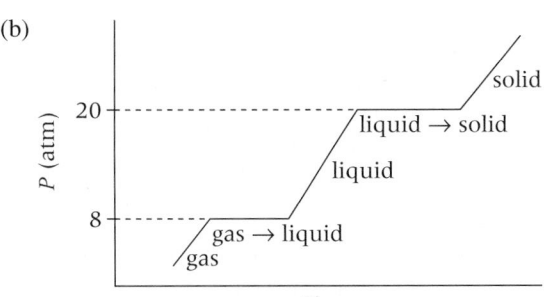

12.122 More energy is needed to sublime ice since the heat of sublimation is equal to the sum of the heats of fusion (to melt the ice) and vaporization (to boil the water). The same would be true of CO_2, but, since liquid CO_2 is not stable at 1 atm pressure, ordinarily sublimation takes place in one step rather than two for this substance.   **12.123** (a) 1.50×10^3 (b) 3.33 g   (c) $SiO_2(s) + 4HF(g) \longrightarrow SiF_4(g) + 2H_2O(g)$ (d) 3.56×10^{-3} mol   **12.126** 259K ($-14°C$)   **12.135** (a)

(b)

12.138 (a) 0.088 g   (b) yes   (c) 0.15 g   **12.140** 2.9 g/m^3 **12.143** (a) 1.1 min   (b) 10. min

(c)

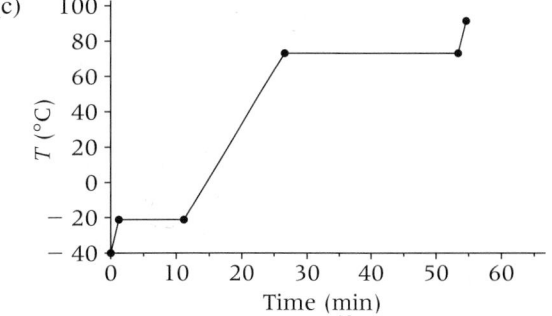

12.148 45.98 amu or 7.635×10^{-23} g   **12.152** (a) 49 metric tons   (b) 1.1×10^8 kJ

Chapter 13

13.2 When a salt such as NaCl dissolves, ion-dipole forces cause the ions to become separated, and more water molecules cluster around them in hydration shells.   **13.4** Sodium stearate is more effective as a soap. The long chain stearate ion has both a hydrophilic carboxylate and a hydrophobic hydrocarbon chain, which allows it to reduce surface tension between oil and water and to act as a soap. The acetate ion is too small to exhibit any hydrophobic qualities.   **13.7** (a) In the polar solvent water, KNO_3 ionizes completely and forms more concentrated solutions. KNO_3, an ionic salt, does not dissolve in or interact well with the nonpolar carbon tetrachloride.   **13.9** (a) ion-dipole   (b) hydrogen bonding   (c) dipole-induced dipole   **13.11** (a) hydrogen bonding   (b) dipole-induced dipole   (c) dispersion forces   **13.13** (a) HCl(g); molecular interactions (dipole-dipole forces) are greater.   (b) CH_3CHO; molecular interactions (dipole-dipole) are greater.   (c) CH_3CH_2MgBr; molecular interactions (dispersion forces) are greater.   **13.16** Gluconic acid is soluble in water due to its extensive hydrogen bonding through its polar hydroxyl groups which caproic acid does not have. The nonpolar end of caproic acid allows it to be soluble in nonpolar hexane.   **13.18** For a general solvent, the energy changes needed to separate solvent into particles ($\Delta H_{solvent}$), and that needed to mix the solvent and solute particles (ΔH_{mix}) would be combined to obtain $\Delta H_{solvation}$.   **13.23** Very soluble. A decrease in energy and an increase in entropy both favor the formation of a solution.   **13.24**

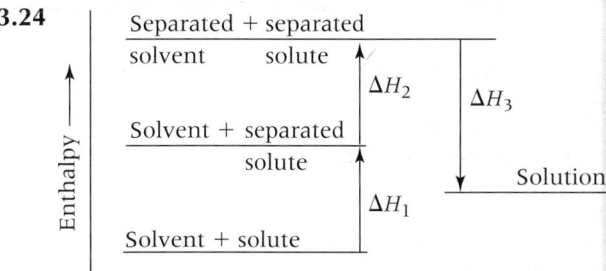

13.26 (a) Na$^+$ has a greater charge density than Cs$^+$. Na$^+$ has a smaller ion volume.   (b) Sr^{2+} has a larger ion charge than Rb$^+$, so will have the larger charge density.   (c) Na$^+$ has a smaller ion volume than Cl$^-$, so will have the larger charge density.   (d) O^{2-} has a larger ion charge than F$^-$, so will have the larger charge density.   (e) OH$^-$ has a smaller ion volume than SH$^-$, so will have the larger charge density.

13.28 (a) Na^+ (b) Sr^{2+} (c) Na^+ (d) O^{2-} (e) OH^-
13.30 (a) -704 kJ/mol (b) probably K^+ due to its smaller size (larger charge density) **13.32** (a) increases (b) decreases (c) increases (d) decreases **13.35** Add a pinch of the solid solute to each solution. The supersaturated solution is unstable and addition of a "seed" crystal of solute causes the excess solute to crystallize immediately, leaving behind a saturated solution. The solution in which the added solid solute dissolves is the unsaturated solution of X. The solution in which the added solid solute remains undissolved is the saturated solution of X. **13.38** (a) increase (b) decrease **13.40** (a) 8.19×10^{-2} g O_2 (b) 1.71×10^{-2} g O_2 **13.43** 0.20 mol/L **13.46** (a) % w/v; molarity (b) % w/w; mole fraction (c) molality **13.48** With just this information, you can interconvert between molarity and molality, but you need to know the identity (or at least the molar mass) of the solvent to convert between either of these and the mole fraction. **13.50** (a) 1.24 M (b) 0.158 M **13.52** (a) 0.0750 M (b) 0.31 M **13.54** (a) Add 4.22 g KH_2PO_4 to enough water to make 355 mL of aqueous solution. (b) Add 107 mL of 1.25 M NaOH to enough water to make 425 mL of aqueous solution. **13.56** (a) Weigh out 45.9 g KBr, then add to enough water to make 1.50 L of solution. (b) Add 139 mL of 0.244 M $LiNO_3$ to enough water to make 355 mL of solution. **13.58** (a) 0.942 m (b) 1.29 m **13.60** 3.09 m **13.62** (a) Add 2.13 g $C_2H_6O_2$ to 298 g H_2O. (b) Add 32.3 g of 62.0 mass % HNO_3 to 968 g H_2O. **13.64** (a) 0.27 (b) 56% (c) 21 m **13.66** 33.7 g; mol fraction $= 7.15 \times 10^{-3}$; 6.31% by mass **13.68** 5.11 m; 4.53 M; 0.0842 mol fraction **13.70** 2.2 ppm Ca^{2+}; 0.66 ppm Mg^{2+} **13.74** A strong electrolyte is a solute that dissociates completely into ions in solution. **13.76** The temperature at which the solution boils is higher and freezing point is lower for the solution compared with the pure solvent. **13.79** A dilute solution of an electrolyte behaves more ideally than a concentrated one. With increasing concentration the "activity" concentration deviates from the molar concentration. Thus, 0.050 m NaF has a boiling point closer to its predicted value. **13.82** (a) strong electrolyte (b) strong electrolyte (c) nonelectrolyte (d) weak electrolyte **13.84** (a) 0.4 mol (b) 0.14 mol (c) 3×10^{-4} mol (d) 0.07 mol **13.86** (a) 10.0 g CH_3OH in 100. g H_2O (b) 10.0 g H_2O in 1.00 kg CH_3OH **13.88** (a) I and II same; III higher (b) I and II increased; III increased by $1\frac{1}{2}$ times (c) I and II decreased; III decreased by $1\frac{1}{2}$ times (d) I and II decreased; III decreased by $1\frac{1}{2}$ times **13.90** 23.36 torr **13.92** $-0.206°C$ **13.94** 79.0°C **13.96** 11.3 kg **13.98** (a) 0.173 m; $i = 1.84$ (b) 0.0837 m; $i = 1.02$ **13.100** 0.240 atm **13.102** 211 torr (CH_2Cl_2); 47.2 torr (CCl_4) **13.103** The fluid inside a cell is both a colloid and a solution. Within a typical cell, colloid-sized protein and nucleic acid molecules are dispersed in an aqueous solution of ions and numerous small molecules. **13.107** Soap molecules from spherical micelles, with the charged "heads" forming the micelle exterior and the nonpolar "tails" the interior. Because the surfaces of the colloidal particles have like charges, repulsions prevent them from adhering to each other when they collide. Soap would be less effective in seawater with its larger concentrations of divalent cations. **13.108** In order to make the cream foam, the gas must have fairly low sol-

ubility in the liquid at 1 atm pressure. If CO_2 does not work as well as N_2O, this would imply that CO_2 is more soluble in cream than is N_2O, so its Henry's law constant would be larger than that for N_2O. **13.113** 3.5×10^9 L **13.115** This would only be true if the density of the solvent is near 1.0 g/mL. **13.120** (a) 90.0 g/mol (b) C_2H_5O; $C_4H_{10}O_2$ (c) $HOCH_2CH_2CH_2CH_2OH$ forms H bonds; $CH_3OCH_2CH_2OCH_3$ does not **13.123** (a) Mg^{2+} (b) Mg^{2+} (c) CO_3^{2-} (d) SO_4^{2-} (e) Fe^{3+} (f) Ca^{2+} **13.125** (a) 68 g/mol (b) 2.0×10^2 g/mol (c) The actual formula mass (164.1 g/mol) is between that calculated assuming that it is a nonelectrolyte and that calculated assuming that it is a strong electrolyte. A strong electrolyte with a formula mass of 164.1 g/mol and which produces three ions per mole should have given an apparent formula mass of $(164.1)/3 = 54.7$ g/mol. This material is highly, but not totally, dissociated into ions in solution. (d) 2.4 **13.129** (a) 1.82×10^4 g/mol (b) $3.41 \times 10^{-5}°C$ **13.130** 9.4×10^5 ng/L **13.134** Add 3.11 g $NaHCO_3$ to 247 g H_2O.

13.138 $M = \dfrac{m \times \text{kg solvent}}{V \text{ of soln}} = \dfrac{m \times \text{kg solvent}}{V \text{ of soln} \times \text{density soln}}$

As density of solution approaches 1 g/mL, $M = m$. If the solvent is water, assuming a density of 1.000 g/mL, one kg of water is equal to a volume of 1.0 liter. In dilute solutions, the volume of solvent and volume of solution are essentially equal. **13.140** (a) CH_4N_2O (b) 60. g/mol; CH_4N_2O **13.142** (a) 0.088 cm^3/g H_2O (b) 1.5 cm^3/g $H_2O \cdot MPa$ (c) 0.74 cm^3/g H_2O, 11% error **13.146** (a) Yes, but the triple point would be at a lower pressure due to the vapor-pressure lowering of the solvent by the dissolved air. (b) A lower temperature, again due to the vapor-pressure lowering. (c) Yes, since you would be below the new triple point. (d) No, the vapor pressures would be lower due to the dissolved air.

Chapter 14

14.2 (a) Z_{eff} increases across a period and decreases down a group. (b) As you move to the right across a period, the atomic size decreases, the IE_1 increases and the EN increases, all as a result of the increased Z_{eff}. **14.3** (a) ICl is polar; Br_2 is nonpolar. The electronegativity of Cl is greater than that of I, so the I end of ICl will be partially positively charged due to the greater "pull" of the Cl on the shared electrons. (b) ICl will have a higher boiling point than Br_2 since dipole-dipole forces in ICl are greater than dispersion forces in Br_2. **14.7** The bonding would change from metallic to (perhaps) very polar covalent to ionic as the "other" element went farther to the right. The electronegativity, ionization energy, and electron affinity all change as you move to the right. The atomic properties of the leftmost element and the "other" element become increasingly different, causing the type of bonding to change in response. **14.9** The elements at the left form cations, which are smaller than the corresponding atoms; the elements at the right form anions, which are larger than the corresponding atoms. **14.11** (a) The higher the electronegativity of the element, the more covalent is the bonding in its oxide. (b) The more electronegative the element, the more acidic its oxide. **14.14** In general, network solids have very high melting and boiling points, and are very hard, while molecular solids have

low melting and boiling points and are soft. To melt, boil, or distort a network solid requires you to break chemical bonds, which requires a great amount of energy. In molecular solids, energy is needed to separate molecules which are held together with weaker intermolecular forces.   **14.15** (a) Mg < Sr < Ba (b) Na < Al < P   (c) Se < Br < Cl   (d) Ga < Sn < Bi
14.17 (b) and (d)
14.19 $:\ddot{O}=C=\ddot{O}:$ **14.21** $CF_4 < SiCl_4 < GeBr_4$ **14.23** (a) $Na^+ < F^- < O^{2-}$ (b) $Cl^- < S^{2-} < P^{3-}$
14.25 $O < O^- < O^{2-}$
14.27 (a) $CaO(s) + H_2O(l) \longrightarrow Ca(OH)_2(aq)$ (more basic)
$SO_3(g) + H_2O(l) \longrightarrow H_2SO_4(aq)$
(b) BaO   (c) CO_2 (d) K_2O
14.29 (a) LiCl   (b) PCl_3 (c) NCl_3 **14.31** (a) Ar < Na < Si
(b) Cs < Rb < Li   **14.33** The outermost electron in Li is much farther away from the nucleus than the electron in H. It is also screened from a large portion of the nuclear charge by the inner electrons.
14.35 (a) $:\ddot{F}-\ddot{N}-\ddot{F}:$ and $H-\ddot{N}-H$ NH_3 will H bond.
   $:\ddot{F}:$ H

(b)
$$H-\overset{\underset{|}{H}}{\underset{|}{C}}-\ddot{O}-\overset{\underset{|}{H}}{\underset{|}{C}}-H \quad \text{and} \quad H-\overset{\underset{|}{H}}{\underset{|}{C}}-\overset{\underset{|}{H}}{\underset{|}{C}}-\ddot{O}-H$$

CH_3CH_2OH will H bond.
14.37 (a) $2Al(s) + 6HCl(aq) \longrightarrow 2AlCl_3(aq) + 3H_2(g)$
(b) $LiH(s) + H_2O(l) \longrightarrow LiOH(aq) + H_2(g)$
14.39 (a) Na = +1, B = +3, H = −1 in $NaBH_4$
Al = +3, B = +3, H = −1 in $Al(BH_4)_3$
Li = +1, Al = +3, H = −1 in $LiAlH_4$

(b)
$$\left[H-\overset{\underset{|}{H}}{\underset{\underset{H}{|}}{B}}-H \right]^- \quad \text{tetrahedral}$$

14.42 For Groups 1A(1)–4A(4), the number of covalent bonds equals the (old) group number. For Groups 5A(15)–7A(17), it equals eight minus the (old) group number. In higher periods, this correlation holds but there are exceptions where the elements in Groups 3A(13) through 7A(17) can form more bonds than this.   **14.45** (a) Groups 3B(3) or 3A(13).   (b) If in Group 3B(3), the oxide and fluoride would be more ionic.
14.48 (a) reducing agent
(b) Their outermost electrons are easily removed due to low ionization energy.
(c) $2Na(s) + Cl_2(g) \longrightarrow 2NaCl(s)$
$2Na(s) + 2H_2O(l) \longrightarrow 2NaOH(aq) + H_2(g)$
14.50 (a) and (b) increase down the group. (c), (d), and (e) decrease down the group.
14.52 $2Na(s) + O_2(g) \longrightarrow Na_2O_2(s)$
14.54 $K_2CO_3(s) + 2HI(aq) \longrightarrow 2KI(aq) + H_2O(l) + CO_2(g)$
14.58 The Group 2A(2) metals have an additional electron available for bonding, so the bonding is stronger and the melting points increase. They are also harder, denser, and have higher boiling points.
14.59 (a) $CaO(s) + H_2O(l) \longrightarrow Ca(OH)_2(s)$
(b) $2Ca(s) + O_2(g) \longrightarrow 2CaO(s)$
14.62 (a) $BeO(s) + H_2O(l) \longrightarrow$ no reaction

(b) $BeCl_2(l) + 2Cl^-$ (solvated) $\longrightarrow$
$BeCl_4^{2-}$ (solvated)
In reaction (b).
14.65 Be: $1s^2 2s^2$ B: $1s^2 2s^2 2p^1$
In B, the outermost electron (i.e., the one removed when the atom is ionized) is in a $2p$ orbital as opposed to the $2s$ in Be. Since the energy of the $2p$ orbital is higher than the $2s$, the electron is more easily removed (i.e., IE_1 is lower) despite the atom's smaller size.
14.66 (a) "Normal" atoms have completed outer shells (i.e., an octet). Atoms with fewer electrons than eight are considered "electron deficient."
(b) $BF_3(g) + NH_3(g) \longrightarrow F_3B-NH_3(g)$
$B(OH)_3(aq) + OH^-(aq) \longrightarrow B(OH)_4^-(aq)$
14.68 $In_2O_3 < Ga_2O_3 < Al_2O_3$
14.70 ON = +3 (apparent); = +1 (actual)
$[:\ddot{I}-\ddot{I}-\ddot{I}:]^-$ class = AX_2E_3; bond angles = 180°
TlI_3 doesn't exist as $(Tl^{3+})(I^-)_3$ because of the low strength of the Tl—I bond.
14.73 (a) $In(s) + As(s) \longrightarrow InAs(s)$
(b) $GaSb(s) + 3H_2O(g) \longrightarrow Ga(OH)_3(s) + SbH_3(g)$

14.75
$$:\ddot{Br}: \quad :\ddot{Br}: \quad :\ddot{Br}:$$
$$\diagdown \diagup \diagdown \diagup$$
$$Ga \quad Ga \qquad 109.5° \text{ bond angles}$$
$$\diagup \diagdown \diagup \diagdown$$
$$:\ddot{Br}: \quad :\ddot{Br}: \quad :\ddot{Br}:$$

14.76 Sn, being the more metallic element, would have the more basic oxide.   **14.78** (a) Ionization energy decreases as you go down a group.   (b) The deviations (increases) from the trend are due to the appearance of the first transition series between Si and Ge and the appearance of the lanthanide elements between Sn and Pb.   (c) Group 3A(13).   **14.81** The sizes of the atoms increase and their ionization energies decrease as you go down the group. If the electrons are more easily ionized, the atoms will behave more metallic.
14.83 (a)

(b)

14.85

14.88 (a) diamond   (b) $CaCO_3$ (c) CO_2 (d) $[(CH_3)_2SiO]_n$
(e) SiC   (f) CO   (g) Pb   **14.92** (a) from −3 to +5.   (b) The highest oxidation state equals the (old) group number; the lowest equals the (old) group number minus eight.   **14.95**
$H_3AsO_4 < H_3PO_4 < HNO_3$

14.97 (a) $4As(s) + 5O_2(g) \longrightarrow 2As_2O_5(s)$
(b) $2Bi(s) + 5F_2(g) \longrightarrow 2BiF_5(s)$
(c) $Ca_3As_2(s) + 6H_2O(l) \longrightarrow$
$\qquad\qquad\qquad 3Ca(OH)_2(s) + 2AsH_3(g)$
14.99 (a) $N_2(g) + 2Al(s) \overset{\Delta}{\longrightarrow} 2AlN(s)$
(b) $PF_5(g) + 4H_2O(l) \longrightarrow H_3PO_4(aq) + 5HF(g)$
14.101 A trigonal bipyramid with F axial and Cl equatorial.
14.105 Yes—it could form H bonds similar to NH_3.

14.107 (a) $2KNO_3(s) \overset{\Delta}{\longrightarrow} 2KNO_2(s) + O_2(g)$
(b) $4KNO_3(s) \overset{\Delta}{\longrightarrow} 2K_2O(s) + 2N_2(g) + 5O_2(g)$
14.109 (a) Their boiling points and conductivities vary in similar ways down the group.   (b) The amount of metallic character and their types of bonding vary in similar ways down the group.   (c) They both react with oxygen to form acidic oxides; both are oxidizing agents.   (d) They are both gaseous nonmetals which form diatomic molecules.   (e) O_2 is more reactive and contains a double bond; N_2 is less reactive and contains a triple bond.
14.111 (a) $NaHSO_4(aq) + NaOH(aq) \longrightarrow$
$\qquad\qquad\qquad Na_2SO_4(aq) + H_2O(l)$
(b) $S_8(s) + 24F_2(g) \longrightarrow 8SF_6(g)$
(c) $FeS(s) + 2HCl(aq) \longrightarrow FeCl_2(aq) + H_2S(g)$
(d) $Te(s) + 2I_2(s) \longrightarrow TeI_4(s)$
14.113 (a) acidic   (b) acidic   (c) basic   (d) amphoteric   (e) basic   **14.115** $H_2O < H_2S < H_2Te$ **14.118** (a) sulfur hexafluoride, SF_6 (b) ozone, O_3 (c) sulfur trioxide, SO_3 (d) sulfur dioxide, SO_2 (e) sulfur trioxide SO_3 (f) sodium thiosulfate pentahydrate, $Na_2S_2O_3 \cdot 5H_2O$ (g) hydrogen sulfide, H_2S
14.120 $S_2F_{10}(g) \longrightarrow SF_4(g) + SF_6(g)$
O.N.: $+5$ $+4$ $+6$
14.122 (a) $-1, +1, +3, +5, +7$ (b) Cl has a $1s^22s^22p^63s^23p^5$ configuration. By adding one electron, the "octet" is achieved. By forming covalent bonds, it can complete (or expand) its octet but it does so in a way maintaining its electrons paired in bonds or lone pairs.   (c) F is an exception because of its small size and high electronegativity.
14.124 (a) Cl—Cl   (b) Br—Br   (c) Cl—Cl
The F—F bond strength is weaker than expected due to the small size of the F atom. This means that the lone electron pairs on the adjacent atoms repel, weakening the bond.
14.126 (a) $3Br_2(l) + 6OH^-(aq) \longrightarrow$
$\qquad\qquad\qquad 5Br^-(aq) + BrO_3^-(aq) + 3H_2O(l)$
(b) $ClF_5(l) + 6OH^-(aq) \longrightarrow$
$\qquad\qquad\qquad 5F^-(aq) + ClO_3^-(aq) + 3H_2O(l)$
14.127 (a) $2Rb(s) + Br_2(l) \longrightarrow 2RbBr(s)$
(b) $I_2(s) + H_2O(l) \longrightarrow HI(aq) + HIO(aq)$
(c) $Br_2(l) + 2I^-(aq) \longrightarrow I_2(s) + 2Br^-(aq)$
(d) $CaF_2(s) + H_2SO_4(l) \longrightarrow CaSO_4(s) + 2HF(g)$
14.129 $HIO < HBrO < HClO < HClO_2$ **14.132** (a) oxidation-reduction   (b) Because I^- is more easily oxidized than Cl^- due to its larger size.   (c) It would have to be a non-oxidizing acid.   **14.134** helium; argon   **14.136** Because they are held together only by dispersion forces.

14.139 (a) $2NaH(s) + 2SO_2(l) \longrightarrow Na_2S_2O_4(s) + H_2(g)$
O.N.: $+1\ -1$ $+4\ -2$ $+1\ +3\ -2$ 0
(b) 1.46×10^5 L
14.141 3.371×10^{-19} J
14.145 (a) $MgCl_2 \cdot 6H_2O(s) + 6SOCl_2(l) \longrightarrow$
$\qquad\qquad\qquad MgCl_2(s) + 6SO_2(g) + 12HCl(g)$
(b)

14.146 (a) Second ionization energies for alkali metals are high because the second electron must be removed from the next lower energy level. The second electron is held tighter.   (b) $2CsF_2(s) \longrightarrow 2CsF(s) + F_2(g)$, -405 kJ/mol
14.148 (a) hyponitrous acid is $H_2N_2O_2$; nitroxyl is HNO.
(b)   and
(c) Both would be bent at the N atoms.
(d)

14.150 (a) $4NH_3(g) + 5O_2(g) \longrightarrow 4NO(g) + 6H_2O(g)$
$2NO(g) + O_2(g) \longrightarrow 2NO_2(g)$
$3NO_2(g) + H_2O(l) \longrightarrow 2HNO_3(l) + NO(g)$
(b) 1.5   (c) 2.7×10^2 mL
14.154 13 metric tons   **14.156** A disproportionation reaction is one in which a substance containing an element in an intermediate oxidation state undergoes simultaneous oxidation and reduction. Reactions (b)–(f) are disproportionations.   **14.158** (a) 5A(15)   (b) 7A(17)   (c) 6A(16)   (d) 1A(1) or 2A(2)   (e) 3A(13)   (f) 8A(18)
14.160 (a)   (b) 3 (all on N)   (c) 1.5

14.163 (a) The increase in size due to the occupation of the next quantum level is offset by the decrease in size due to the 2+ charge.   (b) The lattice energy of CaF_2 is greater than that of NaF.   (c) Be compounds have more covalent character than Ca compounds, which are essentially ionic.
14.166 as oxidizing agent: $2Na(s) + H_2(g) \longrightarrow 2NaH(s)$
as reducing agent: $Cl_2(g) + H_2(g) \longrightarrow 2HCl(g)$
14.168 Both contain 2 mol of N per mole of compound, but N_2H_4 has 0.8742 g N/g N_2H_4 and NH_4NO_3 has 0.3500 g N/g NH_4NO_3.
14.170 (a)   has sp^3 hybridization at each Al

(b) tetrahedral   (c) sp. Since the ion is linear, the central atom must be sp hybridized.   (d) There are no lone pairs on the central atom. Instead the Cl forms double bonds by donating the lone pairs into empty d orbitals of Al.

14.172 The Lewis structures are

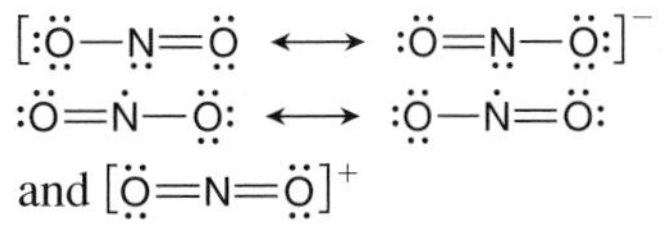

$$[:\ddot{\text{O}}-\text{N}=\ddot{\text{O}} \longleftrightarrow :\ddot{\text{O}}=\text{N}-\ddot{\text{O}}:]^-$$

$$:\ddot{\text{O}}=\dot{\text{N}}-\ddot{\text{O}}: \longleftrightarrow :\ddot{\text{O}}-\dot{\text{N}}=\ddot{\text{O}}:$$

and $[\ddot{\text{O}}=\text{N}=\ddot{\text{O}}]^+$

There are 3, "2.5", and 2 groups of electrons on the N in NO_2^-, NO_2 and NO_2^+, respectively. The bond angles are consistent with the presence of fewer groups at the N giving rise to larger bond angles.   **14.174** 2.29×10^4 g   **14.176** -769 kJ/mol

Chapter 15

15.3 (a) Carbon's position in the periodic table gives it an electronegativity midway between the most metallic and nonmetallic elements of period 2. Therefore, carbon shares electrons to attain a filled shell, bonding covalently in molecules, networks, and polyatomic ions.   (b) Carbon has four electrons in its outer shell, so it needs to form four covalent bonds to fulfill its octet requirement.   (c) To achieve a noble gas configuration, a carbon atom must either lose 4 e⁻ to become C^{4+} or gain 4 e⁻ to become C^{4-}. Both require prohibitively large amounts of energy.   (d) Carbon forms relatively short, strong bonds as a result of its small size, which allows close approach to another atom and thus greater orbital overlap.   (e) The C—C bond is short enough to allow the sideways overlap of unhybridized p orbitals on neighboring carbon atoms. The sideways overlap of p orbitals results in double and triple bonds.   **15.4** (a) Carbon bonds most frequently to C, H, O, N, P, S, and halogens.   (b) Heteroatoms are all atoms other than C or H.   (c) N, O, F, Cl, and Br are more electronegative. H and P are less electronegative. S and I have the same electronegativity as carbon.   (d) Since it can bond to a wide variety of atoms, this means it can form many different compounds.   **15.7** The C—C, C—H, and C—I bonds are relatively unreactive due to fairly equal sharing of electrons. The C=O bond is reactive because of the difference in electronegativity between carbon and oxygen and the electron-rich π bond. The C—Li bond is also reactive because of the electronegativity difference between carbon and lithium.   **15.8** (a) An alkane and a cycloalkane contain only single C—C bonds, while an alkene contains a C—C double bond and an alkyne contains a C—C triple bond. A cycloalkane contains a ring of C atoms.   (b) C_nH_{2n+2}; C_nH_{2n}; C_nH_{2n}; C_nH_{2n-2}, respectively.   (c) Both alkanes and cycloalkanes are saturated.   **15.11** (a), (c), and (f)

15.14 (a)

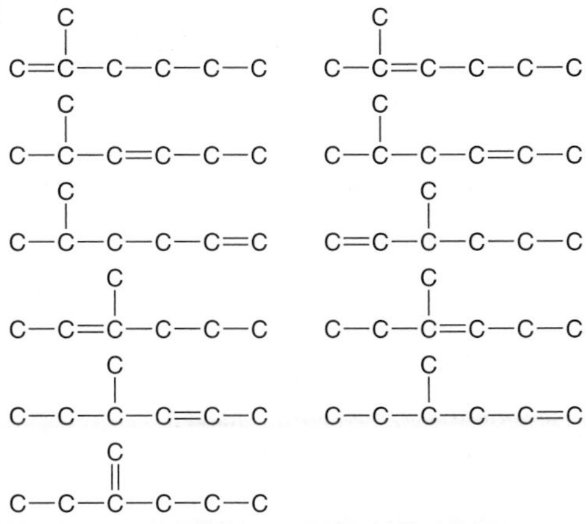

(b)

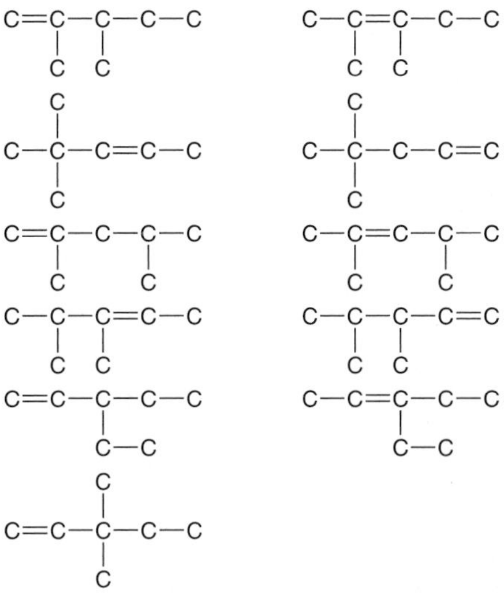

(c)

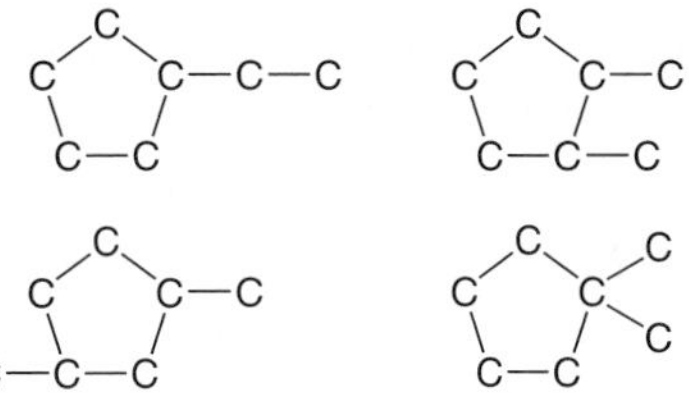

15.16 (a)

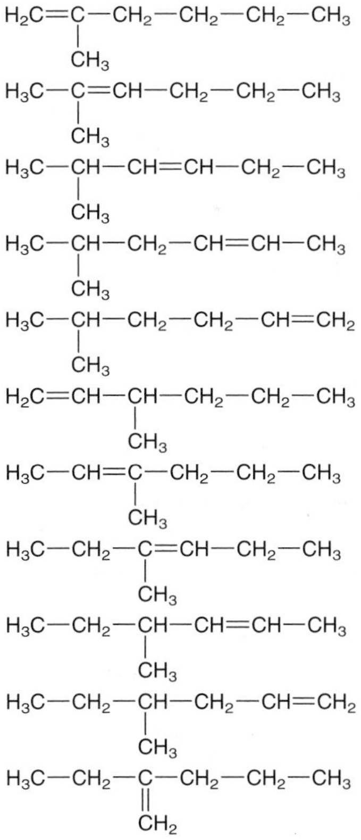

(b) H₂C=C—CH—CH₂—CH₃ — $H_2C=C-CH-CH_2-CH_3$ with CH₃, CH₃ below

H₃C—C=C—CH₂—CH₃ with CH₃, CH₃ below

CH₃ / H₃C—C—CH=CH—CH₃ / CH₃

CH₃ / H₃C—C—CH₂—CH=CH₂ / CH₃

H₂C=C—CH₂—CH—CH₃ with CH₃ and CH₃ below

H₃C—C=CH—CH—CH₃ with CH₃ and CH₃ below

H₃C—CH—C=CH—CH₃ with CH₃ and CH₃ below

H₃C—CH—CH—CH=CH₂ with CH₃ and CH₃ below

H₂C=CH—CH—CH₂—CH₃ with CH₂—CH₃ below

H₃C—CH=C—CH₂—CH₃ with CH₂—CH₃ below

CH₃ / H₂C=CH—C—CH₂—CH₃ / CH₃

(c) cyclopentane structures:

CH₂ / CH₂ CH—CH₂—CH₃ / H₂C—CH₂

CH₂ / CH₂ CH—CH₃ / H₂C—CH—CH₃

CH₂ / H₂C CH—CH₃ / H₃C—HC—CH₂

CH₂ / H₂C C with CH₃, CH₃ / H₂C—CH₂

15.18 (a)
CH₃ / CH₃—C—CH₂CH₃ / CH₃

(b) CH₂=CH—CH₂CH₃

(c) H—C≡C—CH—CH₃ with CH₂ / CH₃ below   (d) OK

15.20 (a)
CH₃ CH₃ / H₃C—CH—CH—CH₂CH₂CH₂CH₂CH₃

(b)
CH₂CH₃ / CH / CH₂ CH₂ / CH₂ CH / H₂C   CH₃

(c) 3,4-dimethylheptane   (d) 2,2-dimethylbutane

15.22 (a) H₃C—CH₂—CH—CH₂—CH₂—CH₃ with CH₃ below
3-methylhexane

(b) H₃C—CH₂—CH—CH₂—CH₂—CH₃ with CH₃ below
3-methylhexane

(c)
CH₂—CH₂ / CH₂   CH—CH₃ / CH₂—CH₂
methylcyclohexane

(d)
CH₃ / H₃C—CH₂—C—CH—CH₂—CH₂—CH₂—CH₃ / CH₃ CH₂CH₃
4-ethyl-3,3-dimethyloctane

15.24 Both can exhibit optical activity.
(a)
CH₂ / H©C ©H—CH₃ / Cl

(b)
CH₃ H / H₃C—CH₂—©—©—CH₃ / H CH₃

15.26 (a) H₃C—CH₂—©H—CH₂—CH₂—CH₃ with Br below   optically active

(b)
CH₃ / H₃C—CH₂—C—CH₂—CH₃ / Cl   not optically active

(c)
CH₃ / BrCH₂—©—CH₂—CH₃ / Br   optically active

15.28 (a)
CH₃CH₂   CH₃ / C=C / H   H
cis-2-pentene

CH₃CH₂   H / C=C / H   CH₃
trans-2-pentene

(b)
H   H / C=C / (cyclohexyl)   CH₃
cis-1-cyclohexylpropene

H   CH₃ / C=C / (cyclohexyl)   H
trans-1-cyclohexylpropene

(c) no geometric isomers

15.30 (a)
H   H / C=C / CH₃   H   no geometric isomers

(b)
H   H / C=C / CH₂   CH₂ / CH₃   CH₃
cis-3-hexene

H   CH₂CH₃ / C=C / CH₂   H / CH₃
trans-3-hexene

(c)
Cl   H / C=C / Cl   H   no geometric isomers

(d)
H   H / C=C / Cl   Cl
cis-1,2-dichloroethene

H   Cl / C=C / Cl   H
trans-1,2-dichloroethene

15.32

benzene ring with Cl at 1,2 positions
1,2-dichlorobenzene
(o-dichlorobenzene)

benzene ring with Cl at 1,3 positions
1,3-dichlorobenzene
(m-dichlorobenzene)

benzene ring with Cl at 1,4 positions
1,4-dichlorobenzene
(p-dichlorobenzene)

15.34

15.35

(*cis*)   (*trans*)

2-methyl-3-hexene

   no geometric isomers

15.39 π bond   **15.41** (a) elimination   (b) addition

15.43 (a)

$$CH_3CH_2CH{=}CHCH_2CH_3 + H_2O \xrightarrow{H^+}$$

$$CH_3CH_2CH_2\underset{\underset{OH}{|}}{C}HCH_2CH_3$$

(b)

$$CH_3{-}\underset{\underset{Br}{|}}{C}H{-}CH_3 + CH_3CH_2OK \xrightarrow{\Delta}$$

$$CH_3{-}CH{=}CH_2 + CH_3CH_2OH + KBr$$

(c) $CH_3CH_3 + 2Cl_2 \xrightarrow{h\nu} Cl{-}\underset{\underset{Cl}{|}}{C}H{-}CH_3 + 2HCl$

15.45 (a) oxidized   (b) reduced   (c) reduced   **15.47** (a) oxidized   (b) oxidized   **15.50** (a) Methylethylamine is more water soluble due to hydrogen bonding between the nitrogen and water.   (b) Assuming that melting point varies like boiling point, 1-butanol has a higher melting point due to intermolecular H bonding.   (c) Propylamine has a higher boiling point than trimethylamine due to intermolecular H bonding of the primary amine not possible in the tertiary amine.

15.53 Both react by addition across the electron-rich trigonal planar π bond. Due to bond polarity, an electron-rich group bonds to the carbonyl carbon and an electron-poor group to the carbonyl oxygen, whereas in the alkene the addition can produce two isomeric products.

15.56 Common to the formation of esters and acid anhydrides is a dehydration-condensation reaction, the other product being water.   **15.58** (a) alkyl halide   (b) nitrile   (c) carboxylic acid (d) aldehyde

15.60 (a)

alkene   alcohol

(b)

haloalkane   carboxylic acid

(c)

alkene   amide

(d)

nitrile   ketone

(e)

ester

15.62

$$H_3C{-}CH_2{-}CH_2{-}CH_2{-}CH_2{-}OH$$

$$H_3C{-}CH_2{-}CH_2{-}\underset{\underset{OH}{|}}{C}H{-}CH_3$$

$$H_3C{-}\underset{\underset{CH_3}{|}}{C}H{-}\underset{\underset{OH}{|}}{C}H{-}CH_3 \qquad H_3C{-}\underset{\underset{CH_3}{|}}{\overset{\overset{OH}{|}}{C}}{-}CH_2{-}CH_3$$

$$H_3C{-}CH_2{-}\underset{\underset{OH}{|}}{C}H{-}CH_2{-}CH_3 \qquad H_3C{-}\underset{\underset{CH_3}{|}}{C}H{-}CH_2{-}CH_2{-}OH$$

$$H_3C{-}CH_2{-}\underset{\underset{CH_3}{|}}{C}H{-}CH_2{-}OH \qquad H_3C{-}\underset{\underset{CH_3}{|}}{\overset{\overset{CH_3}{|}}{C}}{-}CH_2{-}OH$$

15.64

$$H_3C{-}CH_2{-}CH_2{-}CH_2{-}NH_2 \qquad H_3C{-}CH_2{-}\underset{\underset{NH_2}{|}}{C}H{-}CH_3$$

$$H_3C{-}\underset{\underset{CH_3}{|}}{C}H{-}CH_2{-}NH_2 \qquad H_3C{-}\underset{\underset{CH_3}{|}}{\overset{\overset{CH_3}{|}}{C}}{-}NH_2$$

$$H_3C{-}CH_2{-}\underset{\underset{H}{|}}{N}{-}CH_2{-}CH_3 \qquad H_3C{-}CH_2{-}CH_2{-}\underset{\underset{H}{|}}{N}{-}CH_3$$

$$H_3C{-}CH_2{-}\underset{\underset{CH_3}{|}}{N}{-}CH_3 \qquad H_3C{-}\underset{\underset{CH_3}{|}}{C}H{-}\underset{\underset{H}{|}}{N}{-}CH_3$$

15.66 (a) $CH_3{-}\underset{\underset{O}{\|}}{C}{-}CH_2CH_3$ (b)

(c)

15.68 (a)

(b)

(c)

15.70 (a) $CH_3(CH_2)_4COOH + CH_3CH_2OH$
(b) $C_6H_5COOH + CH_3CH_2CH_2OH$
(c) $CH_3CH_2OH + C_6H_5CH_2CH_2COOH$

15.72 (a)

$$CH_3-CH_2-OH; \quad CH_3CH_2-\overset{\overset{\displaystyle O}{\|}}{C}-O-CH_2CH_3$$

(b)

$$CH_3CH_2-\overset{\overset{\displaystyle CN}{|}}{CH}-CH_3; \quad CH_3-CH_2-\overset{\overset{\displaystyle COOH}{|}}{CH}-CH_3$$

15.74 (a)   (b) $CH_3CH_2NH_2$

$$CH_3CH_2\overset{\overset{\displaystyle O}{\|}}{C}-OH; \quad H^+$$

15.76 (a) insoluble   (b) soluble—hydrogen bonding   (c) soluble—hydrogen bonding   (d) soluble—hydrogen bonding and some dissociation (weak acid)   (e) insoluble—alkyl group too hydrophobic   (f) insoluble   **15.78** Addition; condensation
15.81 The dispersion forces between the long unbranched chains HDPE are strong due to the size of the polyethylene chains. LDPE has increased branching which prevents packing and weakens intermolecular dispersion forces.   **15.83** Amine + carboxylic acid; alcohol + carboxylic acid. Specifically in nylon a diamine and dicarboxylic acid; and in polyesters, a diol and a dicarboxylic acid.

15.84 (a)   (b)

15.86

$$nHO-\overset{\overset{\displaystyle O}{\|}}{C}-C_6H_4-\overset{\overset{\displaystyle O}{\|}}{C}-OH + nHOCH_2CH_2OH \longrightarrow$$

$$HO\!\!-\!\!\left[\!CH_2CH_2O\overset{\overset{\displaystyle O}{\|}}{C}-C_6H_4-\overset{\overset{\displaystyle O}{\|}}{C}-O\!\right]_{\!n}\!\!H + nH_2O$$

15.88 (a) condensation   (b) addition   (c) condensation   (d) condensation   **15.90** The amino acid sequence in a protein determines its shape and structure, which determine its function.
15.93 The base sequence in DNA determines the base sequence in RNA, which determines the amino acid sequence in the protein.
15.94 (a) CH_3- (b)

(c) $CH_3-S-CH_2CH_2-$
15.96 (a)

(b)

15.98 (a) AATCGG   (b) TCTGTA   **15.100** ACAATGCCT; three   **15.102** (a) cysteine-cysteine $\longrightarrow$ cystine; disulfide bond   (b) Lysine and aspartic acid by salt link.   (c) Asparagine with serine by hydrogen bonding.   (d) Valine with phenylalanine by hydrophobic forces.   **15.104** (a) Acid-catalyzed dehydration of the alcohol followed by addition of bromine.   (b) Oxidize one mole of the ethanol to acetic acid. React this with a second mole of ethanol to form the ester.
15.106 $CH_3-CH-CHO$

$\quad\quad\quad\quad\overset{|}{CH_3}$
$\quad\quad\quad C_4H_8O$

2-methylpropanal
15.108 $CH_3-CH=CH-CH_3$. Decolorization of bromine.
15.110 Cadaverine (1,5-diaminopentane):

$$H_2N-CH_2CH_2CH_2CH_2CH_2-NH_2$$

Putrescine (1,4-diaminobutane):

$$H_2N-CH_2CH_2CH_2CH_2-NH_2$$

$$BrCH_2CH_2Br + 2\ NaCN \longrightarrow$$

$$N\equiv CCH_2CH_2C\equiv N + 2NaBr$$

$$N\equiv CCH_2CH_2C\equiv N + \text{``4H''} \longrightarrow$$

$$H_2NCH_2CH_2CH_2CH_2NH_2$$

15.111 (a) From left to right, the functional groups are alkene, ester, cyclic unsaturated ketone, and another ester, respectively.
(b) 1-sp^2, 2-sp^3, 3-sp^3, 4-sp^2, 5-sp^3, 6-sp^2, 7-sp^2 (c) carbons 2, 3, and 5
15.113

The C$=$O bond is the shortest because it is a double bond.
15.116

The second structure shows double bond character that inhibits free rotation.
15.120 (a) Hydrolysis involves addition of water to break the peptide bonds.   (b) glycine = 4; alanine = 1; valine = 3; proline = 6; serine = 7; arginine = 49   (c) 10,700 g/mol
15.122 The reaction is not stereospecific and produces a racemic mixture.

Chapter 16

16.2 An increase in pressure increases the reactant concentrations and would result in an increased reaction rate.   **16.3** Additional solvent would decrease the reactant concentrations and would result in a decreased reaction rate.   **16.5** An increase in temperature increases both the number and energy of the collisions and thus would increase the reaction rate.   **16.8** (a) The slope of the line joining any two points on the graph of concentrations *vs.* time provides the average rate over that time period. The shorter this time period, the closer the average rate will equal the instantaneous rate, the rate at a particular instant during the reaction.   (b) The initial rate is the instantaneous rate at the moment the reactants are mixed (at $t = 0$).

16.10

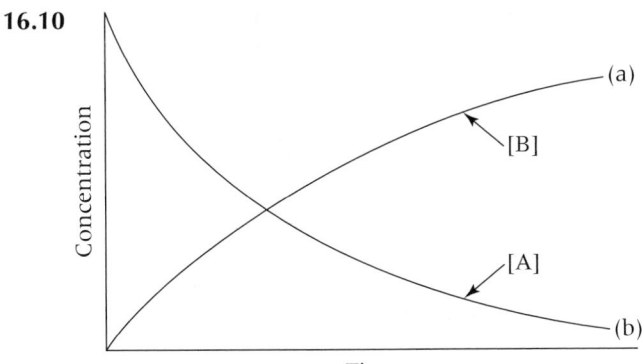

16.12 (a) Rate $= -\dfrac{1\Delta[AX_2]}{2\Delta t}$

$= -\dfrac{(0.0088\ \text{mol/L}) - (0.0500\ \text{mol/L})}{2(20.0 - 0)\ \text{s}}$

$= 1.0\times 10^{-3}\ \text{mol/L·s}$

(b)

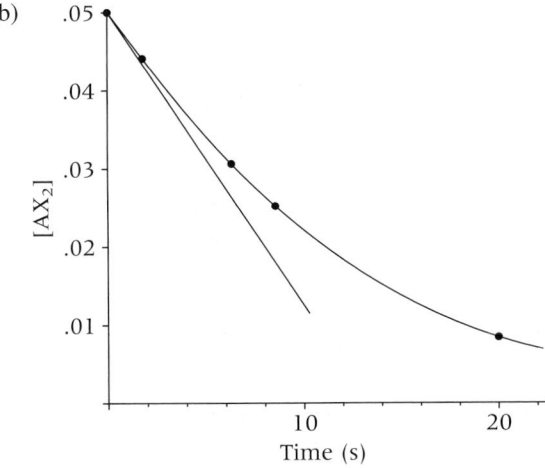

The initial rate is determined by the slope of the tangent to the curve at the point $t = 0$. The initial rate is higher than the rate in part (a). Initial rate $\approx 4\times 10^{-3}\ \text{mol/L·s}$

16.14 Rate $= -\dfrac{1}{2}\dfrac{\Delta[A]}{\Delta t} = \dfrac{\Delta[B]}{\Delta t} = \dfrac{\Delta[C]}{\Delta t}$; 4 mol/L·s

16.16 Rate $= -\dfrac{\Delta[A]}{\Delta t} = -\dfrac{1}{2}\dfrac{\Delta[B]}{\Delta t} = \dfrac{\Delta[C]}{\Delta t}$; 0.25 mol/L·s

16.18 $2N_2O_5 \longrightarrow 4NO_2 + O_2$

16.21 Rate $= -\dfrac{\Delta[N_2]}{\Delta t} = -\dfrac{1}{3}\dfrac{\Delta[H_2]}{\Delta t} = \dfrac{1}{2}\dfrac{\Delta[NH_3]}{\Delta t}$

16.22 (a) Rate $= -\dfrac{1}{3}\dfrac{\Delta[O_2]}{\Delta t} = \dfrac{1}{2}\dfrac{\Delta[O_3]}{\Delta t}$

(b) 1.45×10^{-5} mol/L·s

16.23 (a) k is the rate proportionality constant (usually called the rate constant) and is specific for a particular reaction at a particular temperature. (b) The exponents m and n are called the reaction orders and define how the concentration of each reactant affects the rate. The coefficients a and b in the balanced equation are not necessarily related to the reaction orders m and n. (c) $L^2/\text{mol}^2\cdot\text{min}$ **16.25** (a) doubled (b) decreased by a factor of 4 (c) increased by a factor of 9 **16.26** first order with respect to BrO_3^-; first order with respect to Br^-; second

order with respect to H^+; fourth order overall. **16.28** (a) doubled (b) halved (c) increased by a factor of 16 **16.30** second order with respect to NO_2; first order with respect to Cl_2; third order overall. **16.32** (a) increased by a factor of 9 (b) increased by a factor of 8 (c) halved **16.34** (a) second order in A, first order in B (b) Rate $= k[A]^2[B]$ (c) 5.00×10^3 $L^2/\text{mol}^2\cdot\text{min}$

16.36 (a) time^{-1} (b) $\dfrac{L}{\text{mol·time}}$

(c) $\dfrac{L^2}{\text{mol}^2\cdot\text{time}}$ (d) $\dfrac{L^{3/2}}{\text{mol}^{3/2}\cdot\text{time}}$

16.39 (a) first order (b) second order (c) zero order **16.41** 7 s **16.43** (a) 6.60×10^{-2} min^{-1} (b) 21.0 min

16.45 (a)

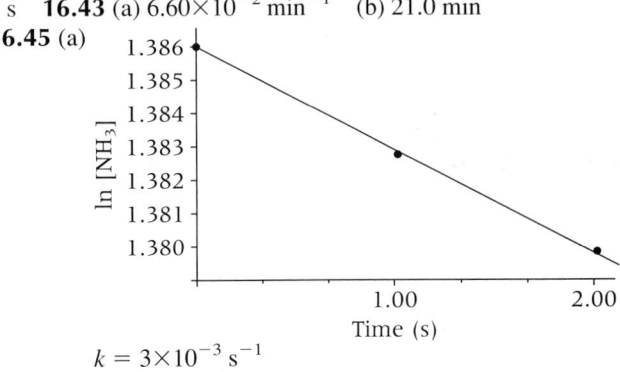

$k = 3\times 10^{-3}\ \text{s}^{-1}$

(b) $t_{1/2} = 2\times 10^2$ s

16.47 Take the logarithm of both sides of the Arrhenius equation. Rearrange the resulting equation, then plot the logarithm of the rate constant *vs.* $1/T$. The slope of this line is $-E_a/R$ (or $-E_a/2.303R$ if common logarithms are used). **16.49** $3.3\times 10^{-2}\ \text{s}^{-1}$ **16.53** No. For most reactions, only a small fraction of collisions actually lead to reaction. If this were not true, every gaseous reaction would be over instantaneously. Only those collisions with energy equal to or greater than the activation energy and with the proper orientation can lead to reaction. **16.56** No. All reactions are reversible and will eventually reach a state where the forward rate equals the reverse rate. At this point without disturbing the reaction, a nonzero constant concentration will exist for all reactants and products. Thus, there will be less than 4×10^{-5} mol EF formed. **16.57** At the same temperature, the NH_3 and $N(CH_3)_3$ will have the same average kinetic energy. However, due to their larger mass, $N(CH_3)_3$ molecules will be traveling more slowly. This will decrease the number of collisions per unit time and reduce the rate of the second reaction. In addition, the higher molecular complexity of the $N(CH_3)_3$ would probably reduce its orientation probability factor, decreasing the rate of the second reaction still more. **16.58** 12 **16.60** 2.96×10^{-18}

16.62 (a)

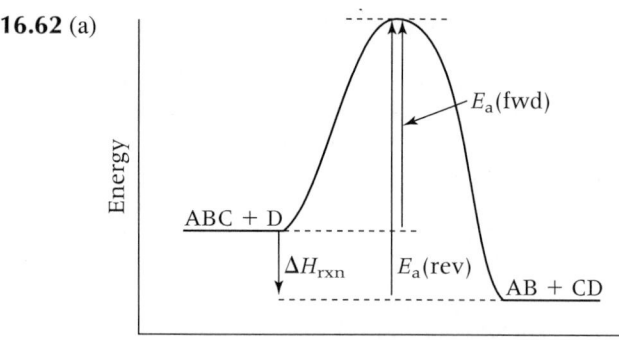

(b) 270. kJ/mol

(c) A
|
B······C······D

16.64 (a)

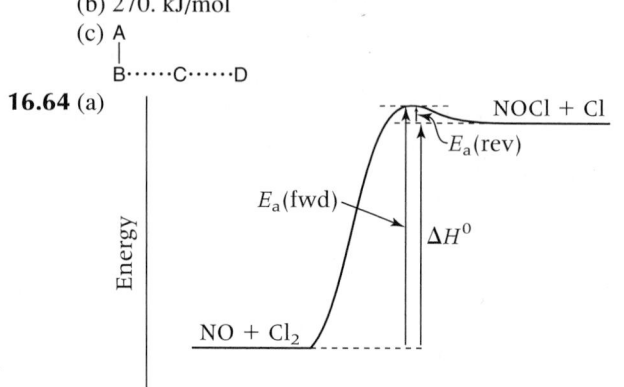

(b) 3 kJ

(c) :Ö=N̈······C̈l······C̈l:

16.65 In many cases, one of the steps is much slower than the others. If this is true, the rate equation for this slowest step represents the rate equation for the overall reaction and the rate would be slower than the average rate of the individual steps.
16.69 The probability of three particles coming together at the same time is much less than that of two coming together.
16.70 No. The overall rate law includes only the species up to and including those in the rate-determining (slow) step.
16.72 (a) $A(g) + B(g) + C(g) \longrightarrow D(g)$
(b) $X(g)$ and $Y(g)$
(c) (1) bimolecular; $Rate_1 = k_1[A][B]$
(2) bimolecular; $Rate_2 = k_2[X][C]$
(3) unimolecular; $Rate_3 = k_3[Y]$
(d) yes
(e) yes
16.74 The proposed mechanism is possible since the elementary reactions which are all physically reasonable add up to the overall equation and the mechanism rate law is consistent with the experimental rate law. **16.77** No. An increase in temperature increases the reaction rate by increasing the fraction of collisions whose energy is greater than the activation energy. A catalyst increases the reaction rate by providing a different reaction mechanism with a lower activation energy. **16.78** (a) No. It supplies an initial amount of heat energy to start the exothermic reaction by allowing some of the molecules to achieve the threshold energy. (b) Yes. The metal catalyzes by chemically adsorbing the H_2 and O_2 molecules onto its surface which provides a low-activation-energy step.

16.82 (a) Water is a solvent and thus its concentration is effectively constant due to the large excess.
(b) (1) $Rate_1 = k_1[(CH_3)_3CBr]$
(2) $Rate_2 = k_2[(CH_3)_3C^+]$
(3) $Rate_3 = k_3[(CH_3)_3COH_2^+]$
(c) $(CH_3)_3C^+$ and $(CH_3)_3COH_2^+$
(d) Rate $= k_1[(CH_3)_3CBr]$, the rate law for the slowest step.
16.83 $O_3(g) + NO(g) \longrightarrow NO_2(g) + O_2(g)$
$$O_3(g) \longrightarrow O_2(g) + O(g)$$
$$\underline{NO_2(g) + O(g) \longrightarrow NO(g) + O_2(g)}$$
$$2O_3(g) \longrightarrow 3O_2(g)$$
NO acts as a catalyst for the decomposition of O_3 to O_2 by providing a different mechanism and is regenerated. **16.85** 46 kJ/mol
16.87 (a)

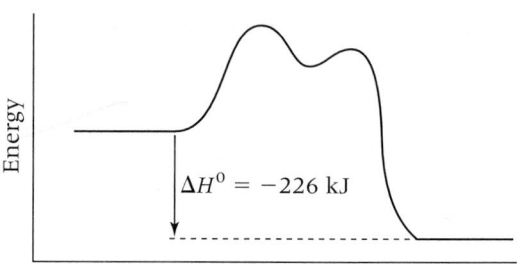

(b) Yes, it is consistent, but the in-text mechanism is more reasonable, since it involves only bimolecular steps.
16.89 (a) decrease (b) increase (c) no effect (assuming this reactant does not also occur earlier) (d) increase (e) increase
16.92 second order **16.94** 7 times faster **16.96** 57 yr
16.99 (a) $k = 0.21$ h^{-1}; $t_{1/2} = 3.3$ h (b) 6.6 h (c) If the concentration of one reactant (in this case, water) is much greater than the other (in this case, sucrose), the change in concentration of the reactant in great excess is negligible, yielding an apparent zero-order behavior with respect to that reactant.
16.101 (a) 0.68 mol/L (b) 0.57 **16.104** 71 kPa
16.108 Rate $= -\dfrac{\Delta[C_2H_6]}{\Delta t} = -\dfrac{1}{4}\dfrac{\Delta[Cl_2]}{\Delta t} = \dfrac{\Delta[C_2H_2Cl_4]}{\Delta t} =$
$\dfrac{1}{4}\dfrac{\Delta[HCl]}{\Delta t}$
16.110 (a) 2.4×10^{-15} M (b) 2.4×10^{-11} mol/L·s
16.114 (a) $Rate_1 = 1.7\times10^{-5}$ M s^{-1};
$Rate_2 = 3.4\times10^{-5}$ M s^{-1};
$Rate_3 = 3.4\times10^{-5}$ M s^{-1}
(b) zero order in $Na_2S_2O_8$; first order in I^-
(c) 4.3×10^{-4} s^{-1} (d) Rate $= (4.3\times10^{-4}$ $s^{-1})[I^-]$

Chapter 17

17.1 $Rate_f = k_f[R]$ $Rate_r = k_r[P]$ $K = \dfrac{k_f}{k_r} = \dfrac{[P]}{[R]}$

If the change is one of concentrations, the product concentrations will increase, while the reactant concentrations will decrease but the ratio (equilibrium constant) will be constant. If the change is one of temperature, the product concentrations and equilibrium constant will increase while the reactant concentrations will decrease. **17.5** This reaction probably would have a large K, since it is very exothermic. **17.7** The equilibrium constant expression is $K = [O_2]$. If the temperature remains constant and the initial amount of Li_2O_2 present was sufficient to reach equilibrium, the amount of O_2 obtained will be constant, regardless of how much $Li_2O_2(s)$ is present.

17.8 (a) $H_2(g) + I_2(g) \rightleftharpoons 2HI(g)$ $Q = \dfrac{[HI]^2}{[H_2][I_2]}$

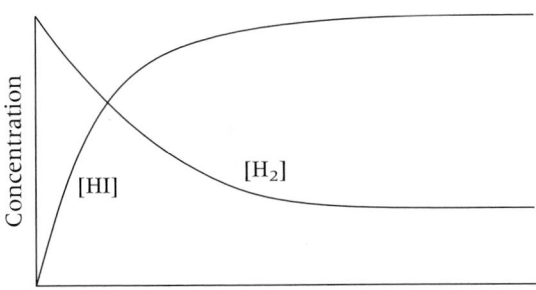

The value of Q increases as a function of time until it reaches the value of K.   (b) No

17.11 Yes; $Q_2 = Q_1^2$, where Q_1 is for the formation of 1 mol of NH_3 and Q_2 is for the loss of 1 mol of N_2.

17.12 (a) $4NO(g) + O_2(g) \rightleftharpoons 2N_2O_3(g)$

$$Q_c = \dfrac{[N_2O_3]^2}{[NO]^4[O_2]}$$

(b) $SF_6(g) + 2SO_3(g) \rightleftharpoons 3SO_2F_2(g)$

$$Q_c = \dfrac{[SO_2F_2]^3}{[SF_6][SO_3]^2}$$

(c) $2SClF_5(g) + H_2(g) \rightleftharpoons S_2F_{10}(g) + 2HCl(g)$

$$Q_c = \dfrac{[S_2F_{10}][HCl]^2}{[SClF_5]^2[H_2]}$$

17.14 (a) $2NO_2Cl(g) \rightleftharpoons 2NO_2(g) + Cl_2(g)$

$$Q_c = \dfrac{[NO_2]^2[Cl_2]}{[NO_2Cl]^2}$$

(b) $2POCl_3(g) \rightleftharpoons 2PCl_3(g) + O_2(g)$

$$Q_c = \dfrac{[PCl_3]^2[O_2]}{[POCl_3]^2}$$

(c) $4NH_3(g) + 3O_2(g) \rightleftharpoons 2N_2(g) + 6H_2O(g)$

$$Q_c = \dfrac{[N_2]^2[H_2O]^6}{[NH_3]^4[O_2]^3}$$

17.16 (a) 7.9   (b) 3.2×10^{-5}

17.18 (a) $2Na_2O_2(s) + 2CO_2(g) \rightleftharpoons 2Na_2CO_3(s) + O_2(g)$

$$Q_c = \dfrac{[O_2]}{[CO_2]^2}$$

(b) $H_2O(l) \rightleftharpoons H_2O(g); Q_c = [H_2O(g)]$
(c) $NH_4Cl(s) \rightleftharpoons NH_3(g) + HCl(g); Q_c = [NH_3][HCl]$

17.20 (a) $2NaHCO_3(s) \rightleftharpoons Na_2CO_3(s) + CO_2(g) + H_2O(g)$

$$Q_c = [CO_2][H_2O]$$

(b) $SnO_2(s) + 2H_2(g) \rightleftharpoons Sn(s) + 2H_2O(g)$

$$Q_c = \dfrac{[H_2O]^2}{[H_2]^2}$$

(c) $H_2SO_4(l) + SO_3(g) \rightleftharpoons H_2S_2O_7(l)$

$$Q_c = \dfrac{1}{[SO_3]}$$

17.23 (a) (1) $Cl_2(g) + F_2(g) \rightleftharpoons 2ClF(g)$
(2) $2ClF(g) + 2F_2(g) \rightleftharpoons 2ClF_3(g)$
overall: $Cl_2(g) + 3F_2(g) \rightleftharpoons 2ClF_3(g)$

(b) $Q_c = \dfrac{[ClF_3]^2}{[Cl_2][F_2]^3}$

$$Q_{c_1} \times Q_{c_2} = \dfrac{[ClF]^2}{[Cl_2][F_2]} \times \dfrac{[ClF_3]^2}{[ClF]^2[F_2]^2} = \dfrac{[ClF_3]^2}{[Cl_2][F_2]^3}$$

17.25 K_c and K_p are **only** equal when $\Delta n_{gas} = 0$.   **17.26** (a) smaller   (b) Assuming that $RT > 1$ (i.e., that $T > 12.2$ K), $K_p > K_c$ if there are more moles of gas in the products than in the reactants, and $K_p < K_c$ if there are fewer moles of gas in the products than in the reactants.   **17.27** (a) 3   (b) -1 (c) 3
17.29 (a) 3.2   (b) 28.5   **17.31** (a) 0.15   (b) 3.6×10^{-7}
17.33 The value Q equals the ratio [products]/[reactants]. When Q is less than K, the concentration of products is less than its equilibrium value, so the reaction proceeds to the right (towards the products) to reach equilibrium.   **17.35** No; to the left.   **17.38** Since the product coefficients are 1:1, equal concentrations of $CFCl_3$ and HCl will form, regardless of the initial reactant ratio.   **17.40** (a) In general, if a reaction has a relatively small equilibrium constant and a relatively large initial reactant concentration, the concentration change (x) can often be ignored.   (b) If the above conditions do not exist, the approximation cannot be made. More precisely, the assumption is not justified if it results in an error of greater than 5% in the calculated equilibrium concentration(s).   **17.41** 50.8

17.43

Concentration (M)	$PCl_5(g)$	$\rightleftharpoons$	$PCl_3(g)$	$+$	$Cl_2(g)$
Initial	0.075		0		0
Change	$-x$		$+x$		$+x$
Equilibrium	$0.075 - x$		x		x

17.45 28 atm   **17.47** 0.33 atm   **17.49** 3.6×10^{-3} M   **17.51** [HI] $= 0.0152$ M; [I_2] $= 0.00328$ M   **17.53** [I_2] $=$ [Cl_2] $= 0.0199$ M; [ICl] $= 0.060$ M   **17.55** 6.01×10^{-6} **17.58** The equilibrium position refers to the specific concentrations (or pressures) that exist when Q equals K, the equilibrium constant. Only the equilibrium position changes in response to a change in concentration.   **17.59** (a) B. Amount of product increases with temperature.   (b) A. Lowest temperature will yield the smallest amount of product.   **17.62** The rate of the forward reaction is increased more than the rate of the reverse reaction, making K_2 larger than K_1.   **17.63** (a) shifts toward products   (b) shifts toward products   (c) no change   (d) shifts toward reactants   **17.65** (a) less F_2; more F   (b) less CH_4; more C_2H_2 and H_2 **17.67** (a) no change   (b) more $H_2O(l)$; less H_2 and O_2 **17.69** (a) amount of product is independent of the volume.   (b) increase volume   **17.71** (a) decrease   (b) increase   (c) increase   (d) decrease   **17.73** 2.0   **17.76** (a) high pressure; low temperature   (b) Q would be decreased, K would be unchanged.   (c) The low temperature needed for increased yields would also lower the reaction rate. A catalyst is employed to increase the rate. Without the catalyst, a much higher temperature would be needed to achieve the same rate and result in a much lower yield.   **17.79** (a) 4.82×10^{-5} atm   (b) 19.6 mg/L   **17.81** 0.204 atm   **17.84** (a) 3×10^{-3} atm   (b) high pressure, low temperature   (c) 2×10^5 (d) No, since H_2O would condense at a higher temperature than C_2H_5OH.   **17.89** (a) 1.6×10^{-2} atm   (b) $K_c = 5.6 \times 10^2$; $P_{SO_2} = 0.16$ atm   **17.91** 12.5 g $CaCO_3$ **17.95** Both concentrations increased by a factor of 2.2.   **17.97** (a) 3.0×10^{-14} atm   (b) 1.3×10^{-2} pg CO/L   **17.100** (a) 97.9%   (b) 99.2%   (c) 259.6 kJ

17.102 (a) $2CH_4(g) + O_2(g) + 2H_2O(g) \rightleftharpoons$
$$2CO_2(g) + 6H_2(g)$$
   (b) 1.76×10^{29} (c) 3.19×10^{23} (d) 48 atm
17.105 (a) 4.0×10^{-21} atm   (b) 5.5×10^{-8} atm
   (c) 29 N atoms; 4.0×10^{14} H atoms
   (d) $N_2(g) + H(g) \longrightarrow NH(g) + N(g)$
17.108 (a) $P_{eq}(N_2) = 0.780$ atm, $P_{eq}(O_2) = 0.210$ atm,
$P_{eq}(NO) = 2.67 \times 10^{-16}$ atm   (b) 0.990 atm   (c) $K_c = K_p$, because there are equal numbers of moles of gaseous products and reactants   **17.111** (a) 5   (b) 8   (c) Once enough F is produced neither branch of the reaction would take place.   (d) Once enough F is produced the second branch of the reaction would not take place.

Chapter 18

18.2 All Arrhenius acids produce H^+ ions and all Arrhenius bases produce OH^- ions in aqueous solution. Neutralization involves the combination of H^+ ions and OH^- ions to produce water molecules. It was this single reaction for all strong acid-base neutralizations which according to Arrhenius produced the identical heat of neutralization reaction value of -56 kJ per mol of water formed.   **18.4** Weak acids exist mainly as undissociated molecules in water. The words "strong" and "weak" refer to the extent of dissociation in water. Strong acids are essentially 100% dissociated in water.   **18.5** (a), (c), and (d)   **18.7** (b) and (d)

18.9 (a) $K_a = \dfrac{[H_3O^+][CN^-]}{[HCN]}$ (b) $K_a = \dfrac{[H_3O^+][CO_3^{2-}]}{[HCO_3^-]}$

   (c) $K_a = \dfrac{[H_3O^+][HCOO^-]}{[HCOOH]}$

18.11 (a) $K_a = \dfrac{[H_3O^+][NO_2^-]}{[HNO_2]}$

   (b) $K_a = \dfrac{[H_3O^+][CH_3COO^-]}{[CH_3COOH]}$

   (c) $K_a = \dfrac{[H_3O^+][BrO_2^-]}{[HBrO_2]}$

18.13 $CH_3COOH < HF < HIO_3 < HI$ **18.15** (a) weak acid (b) strong base   (c) weak acid   (d) strong acid   **18.17** (a) strong base   (b) strong acid   (c) weak acid   (d) weak acid **18.22** (a) $K_a = 4 \times 10^{-5}$. The smaller value of K_a indicates the weaker acid (higher pH).   (b) $pK_a = 3.5$. The larger pK_a indicates the weaker acid.   (c) $0.01\ M$. The lower the concentration, the lower $[H_3O^+]$.   (d) $0.1\ M$ weak acid. Less dissociation will lead to a higher pH value.   (e) $0.1\ M$ base. pH values of bases are larger than pH values for acids.   (f) $pOH = 6.0$. $pH = 14.0 - pOH$ **18.23** (a) 12.045; basic   (b) 11.090; acidic   **18.25** (a) 2.298; acidic   (b) -0.708; basic **18.27** (a) $[H_3O^+] = 1.7 \times 10^{-10}\ M$; $[OH^-] = 6.0 \times 10^{-5}\ M$; $pOH = 4.22$ (b) $[H_3O^+] = 2.7 \times 10^{-4}\ M$; $[OH^-] = 3.7 \times 10^{-11}\ M$; $pH = 3.57$ **18.29** (a) $[H_3O^+] = 1.7 \times 10^{-3}\ M$; $[OH^-] = 5.9 \times 10^{-12}\ M$; $pOH = 11.23$ (b) $[H_3O^+] = 1.5 \times 10^{-9}\ M$; $[OH^-] = 6.6 \times 10^{-6}\ M$; $pH = 8.82$ **18.31** 3.4×10^{-4} mol OH   **18.33** 5.9×10^{-5} mol OH   **18.36** (a) An increase in temperature shifts the autoionization of water to the right which will increase the value of K_w.   (b) $K_w = 2.5 \times 10^{-14}$; $pOH = 6.80$; $[OH^-] = 1.6 \times 10^{-7}\ M$ **18.37** A Brønsted acid is a proton donor; hence, any Brønsted acid must

contain at least one hydrogen atom. An Arrhenius acid donates a proton to water and thus is also an acid under the Brønsted definition. While an Arrhenius base produces OH^- ions in water, a Brønsted base is a substance that accepts a proton, H^+. Ammonia (NH_3) and fluoride ion (F^-) are Brønsted-Lowry bases but not Arrhenius bases. All Brønsted-Lowry acids can be considered potential Arrhenius acids.
18.40 An amphoteric substance can act either as an acid or as a base.
$NH_3 + H_3O^+ \longrightarrow NH_4^+ + H_2O$ (NH_3 acts as a base)
$NH_3 + CH_3^- \longrightarrow NH_2^- + CH_4$ (NH_3 acts as an acid)
18.41 (a) $H_3PO_4(aq) + H_2O(l) \rightleftharpoons$
$$H_3O^+(aq) + H_2PO_4^-(aq)$$

   $K_a = \dfrac{[H_3O^+][H_2PO_4^-]}{[H_3PO_4]}$

   (b) $C_6H_5COOH(aq) + H_2O(l) \rightleftharpoons$
$$H_3O^+(aq) + C_6H_5COO^-(aq)$$

   $K_a = \dfrac{[H_3O^+][C_6H_5COO^-]}{[C_6H_5COOH]}$

   (c) $HSO_4^-(aq) + H_2O(l) \rightleftharpoons H_3O^+(aq) + SO_4^{2-}(aq)$

   $K_a = \dfrac{[H_3O^+][SO_4^{2-}]}{[HSO_4^-]}$

18.43 (a) Cl^- (b) HCO_3^- (c) OH^- **18.45** (a) NH_4^+ (b) NH_3 (c) $C_{10}H_{14}N_2H^+$
18.47 (a) $\underset{acid}{HCl} + \underset{base}{H_2O} \rightleftharpoons \underset{\substack{conjugate \\ base}}{Cl^-} + \underset{\substack{conjugate \\ acid}}{H_3O^+}$
   Conjugate pairs: HCl/Cl^-; H_3O^+/H_2O
   (b) $\underset{acid}{HClO_4} + \underset{base}{H_2SO_4} \rightleftharpoons \underset{\substack{conjugate \\ base}}{ClO_4^-} + \underset{\substack{conjugate \\ acid}}{H_3SO_4^+}$
   Conjugate pairs: $HClO_4/ClO_4^-$; $H_3SO_4^+/H_2SO_4$
   (c) $\underset{base}{HPO_4^{2-}} + \underset{acid}{H_2SO_4} \rightleftharpoons \underset{\substack{conjugate \\ acid}}{H_2PO_4^-} + \underset{\substack{conjugate \\ base}}{HSO_4^-}$
   Conjugate pairs: H_2SO_4/HSO_4^-; $H_2PO_4^-/HPO_4^{2-}$
18.49 (a) $\underset{base}{NH_3} + \underset{acid}{H_3PO_4} \rightleftharpoons \underset{\substack{conjugate \\ acid}}{NH_4^+} + \underset{\substack{conjugate \\ base}}{H_2PO_4^-}$
   Conjugate pairs: $H_3PO_4/H_2PO_4^-$; NH_4^+/NH_3
   (b) $\underset{base}{CH_3O^-} + \underset{acid}{NH_3} \rightleftharpoons \underset{\substack{conjugate \\ acid}}{CH_3OH} + \underset{\substack{conjugate \\ base}}{NH_2^-}$
   Conjugate pairs: NH_3/NH_2^-; CH_3OH/CH_3O^-
   (c) $\underset{base}{HPO_4^{2-}} + \underset{acid}{HSO_4^-} \rightleftharpoons \underset{\substack{conjugate \\ acid}}{H_2PO_4^-} + \underset{\substack{conjugate \\ base}}{SO_4^{2-}}$
   Conjugate pairs: HSO_4^-/SO_4^{2-}; $H_2PO_4^-/HPO_4^{2-}$
18.51 (a) $\underset{base}{OH^-(aq)} + \underset{acid}{H_2PO_4^-(aq)} \rightleftharpoons$
$$\underset{\substack{conjugate \\ acid}}{H_2O(l)} + \underset{\substack{conjugate \\ base}}{HPO_4^{2-}(aq)}$$
   Conjugate pairs: $H_2PO_4^-/HPO_4^{2-}$; H_2O/OH^-
   (b) $\underset{acid}{HSO_4^-(aq)} + \underset{base}{CO_3^{2-}(aq)} \rightleftharpoons$
$$\underset{\substack{conjugate \\ base}}{SO_4^{2-}(aq)} + \underset{\substack{conjugate \\ acid}}{HCO_3^-(aq)}$$
   Conjugate pairs: HSO_4^-/SO_4^{2-}; HCO_3^-/CO_3^{2-}

18.53 $K_c > 1$: $HCl + HS^- \rightleftharpoons Cl^- + H_2S$
$K_c < 1$: $Cl^- + H_2S \rightleftharpoons HCl + HS^-$
18.55 both (a) and (b)   **18.57** both (a) and (b)   **18.59** (a) A strong acid is nearly 100% dissociated, so the acid concentration is very different after dissociation.   (b) A weak acid is only very slightly dissociated, so the acid concentration would be nearly the same after dissociation.   (c) Same as (b), but the extent of dissociation would be greater than in (b).   (d) Same as (a).   **18.60** No. The % dissociation of hydrochloric acid must be much larger than that of acetic acid. Even though hydrochloric acid is much stronger than acetic acid, the $[H_3O^+]$ of these samples would be approximately the same.   **18.63** 1.5×10^{-5} **18.65** $[H_3O^+] = [NO_2^-] = 1.9 \times 10^{-2}\ M$; $[OH^-] = 5.3 \times 10^{-13}\ M$ **18.67** $[H_3O^+] = [ClCH_2COO^-] = 3.8 \times 10^{-2}\ M$; pH = 1.42; $[ClCH_2COOH] = 1.01\ M$ **18.69** (a) $[H_3O^+] = 7.5 \times 10^{-3}\ M$; pH = 2.12; $[OH^-] = 1.3 \times 10^{-12}\ M$; pOH = 11.89   (b) 2.3×10^{-4} **18.71** 3.5×10^{-7} **18.73** (a) 2.47   (b) 11.42   **18.75** (a) 2.378   (b) 12.57   **18.77** 1.6%   **18.79** $[H_3O^+] = [HS^-] = 9 \times 10^{-5}\ M$, pH = 4.0, $[H_2S] = 0.10\ M$, $[OH^-] = 1 \times 10^{-10}\ M$, pOH = 10.0 M, $[S^{2-}] = 1 \times 10^{-17}\ M$ **18.82** 1.9%   **18.83** All Brønsted-Lowry bases contain at least one lone pair of electrons able to bind the H^+ ion and thus allow them to be proton acceptors.
18.86 (a) $C_5H_5N(aq) + H_2O(l) \rightleftharpoons$
$$OH^-(aq) + C_5H_5NH^+(aq)$$
$$K_b = \frac{[OH^-][C_5H_5NH^+]}{[C_5H_5N]}$$
(b) $CO_3^{2-}(aq) + H_2O(l) \rightleftharpoons OH^-(aq) + HCO_3^-(aq)$
$$K_b = \frac{[OH^-][HCO_3^-]}{[CO_3^{2-}]}$$
18.88 (a) $HONH_2(aq) + H_2O(l) \rightleftharpoons$
$$OH^-(aq) + HONH_3^+(aq)$$
$$K_b = \frac{[OH^-][HONH_3^+]}{[HONH_2]}$$
(b) $HPO_4^{2-}(aq) + H_2O(l) \rightleftharpoons$
$$H_2PO_4^-(aq) + OH^-(aq)$$
$$K_b = \frac{[OH^-][H_2PO_4^-]}{[HPO_4^{2-}]}$$
18.90 11.71   **18.92** 11.34   **18.94** (a) 5.6×10^{-10} (b) 2.5×10^{-5} **18.96** (a) 12.05   (b) 10.77   **18.98** (a) 10.95   (b) 5.62   **18.100** (a) 8.73   (b) 4.58   **18.102** $[OH^-] = 4.8 \times 10^{-4}\ M$; pH = 10.68   **18.104** As the nonmetal becomes more electronegative, its bond to hydrogen becomes more polar and the H^+ is more easily lost.   **18.107** The electronegativity of Cl is greater than that of I. The charge on Cl (due to the larger number of O atoms) is greater than the charge on I.   **18.108** (a) H_2SeO_4 (b) H_3PO_4 (c) H_2Te **18.110** (a) H_2Se (b) $B(OH)_3$ (c) $HBrO_2$ **18.112** (a) $0.05\ M\ Al_2(SO_4)_3$ (b) $0.1\ M\ PbCl_2$ **18.114** (a) $0.1\ M\ Ni(NO_3)_2$ (b) $0.1\ M\ Al(NO_3)_3$ **18.117** NaF contains the anion (F^-) of the weak acid HF, while NaCl contains the anion (Cl^-) of the strong acid HCl.
18.119 (a) $KBr(s) + H_2O(l) \longrightarrow K^+(aq) + Br^-(aq)$ neutral
(b) $NH_4I(s) + H_2O(l) \longrightarrow NH_4^+(aq) + I^-(aq)$
$NH_4^+(aq) + H_2O(l) \rightleftharpoons H_3O^+(aq) + NH_3(aq)$ acidic
(c) $KCN(s) + H_2O(l) \longrightarrow K^+(aq) + CN^-(aq)$
$CN^-(aq) + H_2O(l) \rightleftharpoons OH^-(aq) + HCN(aq)$ basic

18.121 (a) $Na_2CO_3(s) + H_2O(l) \longrightarrow 2Na^+(aq) + CO_3^{2-}(aq)$
$CO_3^{2-}(aq) + H_2O(l) \rightleftharpoons$
$$OH^-(aq) + HCO_3^-(aq) \quad \text{basic}$$
(b) $CaCl_2(s) + H_2O(l) \longrightarrow Ca^{2+}(aq) + 2Cl^-(aq)$ neutral
(c) $Cu(NO_3)_2(s) + nH_2O(l) \longrightarrow$
$$Cu(H_2O)_n^{2+}(aq) + 2NO_3^-(aq)$$
$Cu(H_2O)_n^{2+}(aq) + H_2O(l) \rightleftharpoons$
$$Cu(H_2O)_{n-1}OH^+(aq) + H_3O^+(aq) \quad \text{acidic}$$
18.123 (a) $SrBr_2(s) + H_2O(l) \longrightarrow$
$$Sr^{2+}(aq) + 2Br^-(aq) \quad \text{neutral}$$
(b) $Ba(CH_3COO)_2(s) + H_2O(l) \longrightarrow$
$$Ba^{2+}(aq) + 2CH_3COO^-(aq)$$
$CH_3COO^-(aq) + H_2O(l) \rightleftharpoons$
$$CH_3COOH(aq) + OH^-(aq) \quad \text{basic}$$
(c) $(CH_3)_2NH_2Br(s) + H_2O(l) \longrightarrow$
$$(CH_3)_2NH_2^+(aq) + Br^-(aq)$$
$(CH_3)_2NH_2^+(aq) + H_2O(l) \rightleftharpoons$
$$(CH_3)_2NH(aq) + H_3O^+(aq) \quad \text{acidic}$$
18.125 (a) $(NH_4)_3PO_4(s) + H_2O(l) \longrightarrow$
$$3NH_4^+(aq) + PO_4^{3-}(aq)$$
$NH_4^+(aq) + H_2O(l) \rightleftharpoons NH_3(aq) + H_3O^+(aq)$
$PO_4^{3-}(aq) + H_2O(l) \rightleftharpoons HPO_4^{2-}(aq) + OH^-(aq)$
Since $K_b(PO_4^{3-}) > K_a(NH_4^+)$ basic
(b) $Na_2SO_4(s) + H_2O(l) \longrightarrow 2Na^+(aq) + SO_4^{2-}$
$SO_4^{2-}(aq) + H_2O(l) \rightleftharpoons$
$$HSO_4^-(aq) + OH^-(aq) \quad \text{basic}$$
(c) $LiClO(s) + H_2O(l) \longrightarrow Li^+(aq) + ClO^-(aq)$
$ClO^-(aq) + H_2O(l) \rightleftharpoons$
$$HOCl(aq) + OH^-(aq) \quad \text{basic}$$
18.127 (a) $Fe(NO_3)_2 < KNO_3 < K_2SO_3 < K_2S$
(b) $NaHSO_4 < NH_4NO_3 < NaHCO_3 < Na_2CO_3$
18.129 Water acting as an acid makes all strong bases appear **equally** strong by forming OH^-:
$CH_3O^-(aq) + H_2O(l) \longrightarrow OH^-(aq) + CH_3OH(aq)$
$NH_2^-(aq) + H_2O(l) \longrightarrow OH^-(aq) + NH_3(aq)$
18.131 NH_3 as a solvent is more basic than H_2O and as such, weak acids like HF act like strong acids and are 100% dissociated.   **18.133** A Lewis acid is an electron pair acceptor, while a Brønsted-Lowry acid is a proton donor. They are similar in the fact that the proton is an electron pair acceptor and thus is a Lewis acid. A Lewis base is an electron pair donor to any species, while a Brønsted-Lowry base is an electron pair donor specifically for the proton; that is, it is a proton acceptor. Thus, Lewis bases and Brønsted-Lowry bases are essentially the same.
18.134 (a) No.
$$Zn^{2+}(aq) + 6H_2O(l) \rightleftharpoons Zn(H_2O)_6^{2+}(aq)$$
H_2O is a very poor Brønsted base, but a reasonable Lewis base.
(b) CN^- and H_2O (c) CN^-
18.137 (a) acid   (b) base   (c) acid   (d) base   **18.139** (a) acid   (b) base   (c) base   (d) acid
18.141 (a) $Na^+ + 6H_2O \rightleftharpoons Na(H_2O)_6^+$
    acid   base
(b) $CO_2 + H_2O \rightleftharpoons H_2CO_3$
    acid   base
(c) $F^- + BF_3 \rightleftharpoons BF_4^-$
    base   acid

18.143 (a) Lewis　(b) Brønsted-Lowry and Lewis　(c) none (d) Lewis　**18.146** Stronger in seawater.　**18.148** (a) Weak acids will vary in % dissociation (acid strength) in solvents of different acid-base strength.　(b) Methanol is a weaker base than water since carbolic acid dissociates less in methanol than in water.　(c) $C_6H_5OH(solvated) + CH_3OH(l) \rightleftharpoons$ $CH_3OH_2^+(solvated) + C_6H_5O^-(solvated)$　(d) $2CH_3OH$ $\rightleftharpoons CH_3OH_2^+(solvated) + CH_3O^-(solvated)$; $K =$ $[CH_3OH_2^+][CH_3O^-]$　**18.151** (a) Lewis acid: $SnCl_4$; Lewis base: $(CH_3)_3N$　(b) $5d$　**18.152** pH = 5.00, 6.00, 6.79, 6.98, 7.00　**18.155** H_3PO_4　**18.156** (a) Higher. HCl dissociates into ions to a much greater extent than CH_3COOH.　(b) About the same. At these extremely low acid concentrations, the concentration of ions due to the autoionization of water becomes significant and the total ion concentration in each solution is nearly equal. Also, the % dissociation of a weak electrolyte such as CH_3COOH increases with decreasing concentration.
18.159 $5.6\times10^{-8}\ M$　**18.163** 10.43　**18.165** 2.41　**18.168** amylase, $2\times10^{-7}\ M$; pepsin, $1\times10^{-2}\ M$; trypsin, $3\times10^{-10}\ M$
18.173 (a) $Ca(CH_3CH_2COO)_2(s) + H_2O(l) \longrightarrow$
$$Ca^{2+}(aq) + 2CH_3CH_2COO^-(aq)$$
$$CH_3CH_2COO^-(aq) + H_2O(l) \rightleftharpoons$$
$$OH^-(aq) + CH_3CH_2COOH(aq)$$
　　Therefore, it is a basic salt.
　　(b) 9.03
18.175 The small radius of the Be^{2+} ion means its charge to radius ratio is much larger than that of the other Group 2A(2) ions. This results in the hydrated beryllium ion's ability to transfer a proton to water and form an acidic solution.
18.178 4.5×10^{-5}　**18.180** (a) 10.0　(b) $[OH^-]$ from tertiary amine = $1\times10^{-4}\ M$; from aromatic ring N = $6\times10^{-7}\ M$　(c) 4.6　(d) 5.1

Chapter 19

19.2 The weak acid in the system reacts with added base and the weak base in the system reacts with added acid. The acid and base need to be of comparable acid/base strength so that they don't neutralize one another.　**19.4** When H_3O^+ is added to **any** system, the pH will drop. However, the change in pH will be minimized (and perhaps not measurable) with a buffer solution.　**19.7** When the buffer-component ratio is 1:1, the capacity of the buffer for "absorbing" added base and acid are equal, and the buffer range will be symmetrical around the pK_a. As the ratio deviates more from this, the buffer range will be unsymmetrical (e.g., from $pK_a - 0.5$ to $pK_a + 1$).　**19.9** (a) pH would increase; ratio would increase　(b) pH would decrease; ratio would decrease　(c) pH would increase, ratio would increase　(d) pH would decrease; ratio would decrease
19.11 $7.8\times10^{-6}\ M$; 5.11　**19.13** $5.5\times10^{-4}\ M$; 3.26　**19.15** 3.80　**19.17** 9.92　**19.19** 9.55　**19.21** (a) K_{a2}　(b) 10.44
19.23 1.7　**19.25** 0.17　**19.27** 3.37　**19.29** 8.79　**19.31** (a) 4.91　(b) 0.66 g　**19.33** (a) $HCOOH/HCOO^-$, $C_6H_5NH_2/C_6H_5NH_3^+$　(b) $H_2PO_4^-/HPO_4^{2-}$, $H_2AsO_4^-/HAsO_4^{2-}$　**19.35** (a) $H_3AsO_4/H_2AsO_4^-$, $H_3PO_4/H_2PO_4^-$　(b) $C_5H_5N/C_5H_5NH^+$　**19.38** 1.6　**19.40** To see a distinct color in a mixture of two, you need one color to be about 10 times the intensity of the other. For this to take place, the concentration ratio $[HIn]/[In^-]$ needs to be greater than 10:1 or less than 1:10. This will occur when pH = $pK_a -$

1 or pH = $pK_a + 1$, respectively, giving a transition range of about two units.　**19.42** The end point is where you stop titrating, i.e., when the indicator changes color. The equivalence point is when the stoichiometric ratio of reactants is reached. The end point may precede or follow the equivalence point, depending on the relative magnitudes of K_a for the acid being titrated and K_a for the indicator.　**19.44** (a) initial pH: strong acid-strong base < weak acid-strong base < strong acid-weak base　(b) equiv. pt. pH: strong acid-weak base < strong acid-strong base < weak acid-strong base.　**19.46** In the exact center of the buffer region (i.e., where $[HA] = [A^-]$), the pH = the pK_a for the acid.　**19.48** 7.3 − 9.3　**19.50** (a) bromthymol blue　(b) thymol blue, phenol red, or phenolphthalein
19.52 (a) methyl red　(b) bromthymol blue　**19.54** (a) 1.00　(b) 1.48　(c) 3.00　(d) 4.00　(e) 7.00　(f) 10.00　(g) 11.96
19.56 (a) 2.91　(b) 4.81　(c) 5.29　(d) 6.09　(e) 7.41　(f) 8.76　(g) 10.10　(h) 12.05　**19.58** (a) 59.0 mL; 8.54　(b) 45.2 mL; 4.52 and 90.4 mL; 9.69　**19.60** (a) 96.2 mL; 5.16　(b) 146 mL; 5.78　**19.63** Because F^- is the anion (i.e., conjugate base) of a weak acid and Cl^- is the anion of a strong acid. The equilibrium $F^-(aq) + H_2O(l) \rightleftharpoons HF(aq) + OH^-(aq)$ would be shifted by changing the pH (i.e., changing $[H_3O^+]$ and $[OH^-]$), but the corresponding equilibrium is not set up for Cl^-.　**19.65** The solid precipitates.　**19.66** (a) $Q_{sp} = [Ag^+]^2[CO_3^{2-}]$　(b) $Q_{sp} = [Ba^{2+}][F^-]^2$　(c) $Q_{sp} = [Cu^{2+}][HS^-][OH^-]$　**19.68** (a) $Q_{sp} = [Ca^{2+}][CrO_4^{2-}]$　(b) $Q_{sp} = [Ag^+][CN^-]$　(c) $Q_{sp} = [Ni^{2+}][HS^-][OH^-]$　**19.70** 1.3×10^{-4}　**19.72** 2.8×10^{-11}　**19.74** (a) $2.3\times10^{-5}\ M$　(b) $4.2\times10^{-9}\ M$　**19.76** (a) $1.7\times10^{-3}\ M$　(b) $2.0\times10^{-4}\ M$
19.78 (a) $Mg(OH)_2$　(b) PbS　(c) Ag_2SO_4　**19.80** (a) $CaSO_4$　(b) $Mg_3(PO_4)_2$　(c) $PbSO_4$
19.82 (a) $AgCl(s) \rightleftharpoons Ag^+(aq) + Cl^-(aq)$; Cl^- is the anion of a strong acid, so it does not react with H_3O^+. No change in solubility with pH.　(b) $SrCO_3(s) \rightleftharpoons Sr^{2+}(aq) + CO_3^{2-}(aq)$; CO_3^{2-} is the anion of a weak acid, so the equilibria $CO_3^{2-}(aq) + H_2O(l) \rightleftharpoons HCO_3^-(aq) + OH^-(aq)$ and $HCO_3^-(aq) + H_2O(l) \rightleftharpoons H_2CO_3(aq) + OH^-(aq)$ would be set up. The solubility of $SrCO_3$ will increase with decreasing pH.　**19.84** (a) $Fe(OH)_2(s) \rightleftharpoons Fe^{2+}(aq) + 2OH^-(aq)$; Solubility increases with decreasing pH.　(b) $CuS(s) + H_2O(l) \rightleftharpoons Cu^{2+}(aq) + HS^-(aq) + OH^-(aq)$; Solubility increases with decreasing pH.　**19.86** Yes.　**19.88** Yes.　**19.93** This is most likely due to the formation of a complex ion by a reaction like $Pb^{2+}(aq) + nOH^-(aq) \rightleftharpoons Pb(OH)_n^{(2-n)}(aq)$.
19.94 $Hg(H_2O)_4^{2+}(aq) + 4CN^-(aq) \rightleftharpoons$
$$Hg(CN)_4^{2-}(aq) + 4H_2O(l)$$
19.96 $Ag(H_2O)_2^+(aq) + 2S_2O_3^{2-}(aq) \rightleftharpoons$
$$Ag(S_2O_3)_2^{3-}(aq) + 2H_2O(l)$$
19.98 $1\times10^{-5}\ M$　**19.100** $[Zn^{2+}] = 5.9\times10^{-17}\ M$, $[Zn(CN)_4^{2-}] = 0.024\ M$, $[CN^-] = 0.056\ M$　**19.102** $9.4\times10^{-5}\ M$　**19.104** (a) $Fe(OH)_3$　(b) Add enough NaOH to the solution to maintain the pH just below that which will precipitate $Cd(OH)_2$, then separate the precipitate $[Fe(OH)_3]$ and solution (Cd^{2+}).　(c) $2.1\times10^{-7}\ M$　**19.106** (a) 1) Add OH^-: NH_4^+ will give off NH_3, which can be detected with moist red litmus paper, while Ni^{2+} will give a pale green precipitate; or 2) add Na_2CO_3, Na_3PO_4, HS^-, etc: Ni^{2+} will precipitate, NH_4^+ will give no reaction.　(b) Hot water will dissolve $PbCl_2$

but not AgCl.   (c) NaOH will cause a red-brown precipitate with Fe^{3+}, but will form a white precipitate with Al^{3+} which will redissolve in excess OH^-.   **19.108** (a) 0.99   (b) Assuming volumes are additive, mix 400 mL of 1.0 M HCOOH and 200 mL of 1.0 M NaOH, so that the HCOOH will be half-neutralized.   (c) 0.33 M **19.110** 1.3×10^{-4} M **19.113** 7.38   **19.114** Fe^{3+} and Ag^+ **19.117** (a) 0.088   (b) 0.14   **19.119** [TRIS] = 0.260 M, pH = 8.53   **19.121** pH < 6.6   **19.124** 8×10^{-5}

19.127 (a)

V (mL)	pH	$\dfrac{\Delta pH}{\Delta V}$	$V_{average}$ (mL)
0.00	1.00		
10.00	1.22	0.022	5.00
20.00	1.48	0.026	15.00
30.00	1.85	0.037	25.00
35.00	2.18	0.066	32.50
39.00	2.89	0.178	37.00
39.50	3.20	0.620	39.25
39.75	3.50	1.20	39.63
39.90	3.90	2.67	39.83
39.95	4.20	6.00	39.93
39.99	4.90	17.50	39.97
40.00	7.00	210.	40.00
40.01	9.40	240.	40.01
40.05	9.80	10.00	40.03
40.10	10.40	12.00	40.08
40.25	10.50	0.67	40.18
40.50	10.79	1.16	40.38
41.00	11.09	0.600	40.75
45.00	11.76	0.168	43.00
50.00	12.05	0.058	47.50
60.00	12.30	0.025	55.00
70.00	12.43	0.013	65.00
80.00	12.52	0.009	75.00

(b)

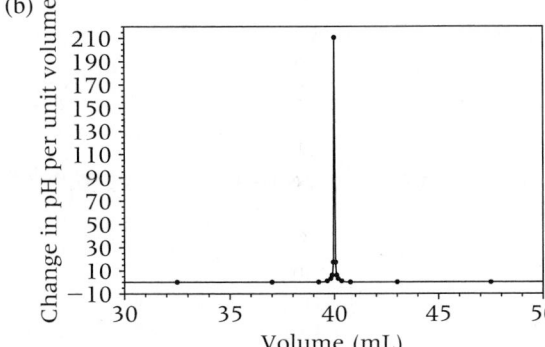

Maximum slope (equivalence point) is at V = 40.00 mL

19.132 $K_b = \dfrac{[BH^+][OH^-]}{[B]}$

$-\log K_b = -\log [OH^-] - \log [BH^+]/[B]$

$pOH = pK_b + \log [BH^+]/[B]$

or $pOH = pK_b - \log [B]/[BH^+]$

19.136 $H_2PO_4^-/HPO_4^{2-}$ and H_2CO_3/HCO_3^-

$[HPO_4^{2-}]/[H_2PO_4^-]$ = 5.8

19.138 3.5 – 4   **19.141** (a) 58.2 mL   (b) 7.84 mL   (c) 6.30   **19.143** 170 mL   **19.146** (a) 65 mol   (b) 6.28   (c) 4.0×10^3 g   **19.148** 5.68   **19.150** 3.9×10^{-9} µg/100 mL

Chapter 20

20.2 A spontaneous process is capable of proceeding as written or described without need of an outside source of energy. Under a specific set of conditions, if a process is spontaneous in one direction, it is not spontaneous in the other. By changing the set of conditions a nonspontaneous process can be made spontaneous.   **20.5** The larger value for ΔS_{vap} results largely from the increased volume in which the molecules may be found. An increase in volume means an increase in randomness. Gases have **much** higher entropy than liquids or solids, due to the random motion of their molecules.   **20.6** $\Delta S_{surr} > 0$ for an exothermic reaction. $\Delta S_{surr} < 0$ for an endothermic reaction. Melting of ice cream at room temperature.   **20.8** (a), (b), and (c)   **20.10** (a) and (b)   **20.12** (a) > 0   (b) < 0   (c) < 0   **20.14** (a) > 0   (b) > 0   (c) > 0   **20.16** (a) < 0   (b) < 0   (c) > 0   **20.18** (a) > 0   (b) < 0   (c) > 0   **20.20** (a) > 0   (b) < 0   (c) > 0   **20.22** (a) Butane (g); 2-butene has restricted rotation about the double bond and would have less flexibility.   (b) Xe(g); greater mass.   (c) $CH_4(g)$; $S(g)$ is greater than $S(l)$.   **20.24** (a) $C_2H_5OH(l)$; greater molecular complexity.   (b) $KClO_3(aq)$; $S(aq)$ is greater than $S(s)$.   (c) K(s); greater mass.   **20.26** (a) diamond < graphite < charcoal; entropy increases from the more ordered crystalline state to the amorphous state.   (b) ice < liquid water < water vapor; entropy increases as substance changes from a solid to a liquid to a gas.   (c) O atoms < O_2 < O_3; entropy increases with molecular complexity.   **20.28** (a) $ClO_4^-(aq) > ClO_3^-(aq) > ClO_2^-(aq)$; entropy decreases with lower molecular complexity.   (b) $NO_2(g) > NO(g) > N_2(g)$; entropy decreases with lower molecular complexity.   (c) $Fe_3O_4(s) > Fe_2O_3(s) > Al_2O_3(s)$; entropy decreases with both lower mass and lower molecular complexity.   **20.31** For a process spontaneously moving to equilibrium: $\Delta S_{sys} + \Delta S_{surr} = \Delta S_{universe} > 0$. For a process at equilibrium: $\Delta S_{sys} + \Delta S_{surr} = \Delta S_{universe} = 0$. For any spontaneous process, the entropy of the universe increases.   **20.32** $S^0_{Cl_2O(g)} = 2S^0_{HClO(g)} - S^0_{H_2O(g)} - \Delta S^0_{rxn}$ **20.33** (a) < 0; $\Delta S^0 = -172.4$ J/K   (b) > 0; $\Delta S^0 = 141.6$ J/K   (c) < 0; $\Delta S^0 = -837$ J/K   **20.35** $\Delta S^0 = 93.1$ J/K. Yes—an increase in the number of mol of gas should give $\Delta S > 0$.   **20.37** $\Delta S^0 = -311$ J/K. Yes—a decrease in the number of mol of gas should give $\Delta S < 0$.   **20.39** -75.6 J/K   **20.41** -242 J/K   **20.44** -97.2 J/K   **20.46** For a spontaneous process, $\Delta S_{universe} > 0$. Since absolute temperature is always positive, it follows that $\Delta G_{sys} < 0$ for a spontaneous process.   **20.48** If $\Delta G^0 = \Delta H^0 - T\Delta S^0 < 0$ at high temperatures $\Delta H^0 > 0$ and $\Delta S^0 > 0$. This would be true for melting processes.   **20.49** Entropy values increase only gradually within a particular phase. Thus, as long as no phase changes occur, the value of ΔS^0 is relatively independent of temperature.   **20.50** (a) -1138 kJ   (b) -1379.4 kJ   (c) -224 kJ   **20.52** (a) $\Delta H^0 = -1202$ kJ; $\Delta S^0 = -216.6$ J/K; $\Delta G^0 = -1138$ kJ   (b) $\Delta H^0 = -1351.9$ kJ; $\Delta S^0 = 91$ J/K; $\Delta G^0 = -1379$ kJ   (c) $\Delta H^0 = -277$ kJ; $\Delta S^0 = -174$ J/K; $\Delta G^0 = -226$ kJ   **20.54** (a) A decrease in the number of moles of gas should result in a negative ΔS^0 value. The combustion of CO(g) (a fuel) will result in a release of energy or a negative ΔH^0 value.   (b) $\Delta G^0 = -257.2$ kJ   **20.56** (a) -0.409 kJ/mol·K   (b) -198 kJ/mol   **20.58** (a) $\Delta H^0 = 90.7$ kJ; $\Delta S^0 = 221$ J/K   (b) $\Delta G = 22.0$ kJ at 38°C, -0.1 kJ at 138°C, -22.2 kJ at 238°C.   (c) The reaction is unfavorable at 38°C, favorable at 238°C, and essentially at equilibrium at 138°C.   **20.60** $\Delta H^0 = 30.91$ kJ; $\Delta S^0 = 93.15$ J/K; $T_b = 331.8$ K   **20.62** (a) $\Delta H^0 = -241.826$ kJ, $\Delta S^0 = -44.4$ J/K, $\Delta G^0 = -228.60$ kJ   (b) Yes, it will become nonspontaneous at high temperatures.   (c) Below 5.45×10^3 K, the reaction becomes spontaneous.   **20.64** (a) ΔG^0 would be a relatively large positive value.   (b) K would have an extremely large value; the value of Q depends on the initial conditions.   **20.67** The standard free energy change, ΔG^0, occurs when all components of the system are in their standard states. Under standard state conditions, $\Delta G = \Delta G^0$.   **20.68** (a) 1.7×10^6 (b) 3.89×10^{-34} (c) 1.26×10^{48} **20.70** (a) 6.57×10^{173} (b) 5×10^{-15} (c) 3.5×10^4 **20.72** 5.0×10^{-51} **20.74** 3.4×10^5 **20.76** $\Delta G^0 = 27$ kJ/mol; no   **20.78** (a) 28.8 kJ/mol   (b) left   (c) $\Delta G = 7.1$ kJ/mol; left   **20.80** (a) no temperature   (b) 326 kJ   (c) 266 kJ   **20.83** (a) spontaneous   (b) $+$ (c) $+$ (d) $-$ (e) $-$; Not spontaneous   (f) $-$ **20.87** (a) 2.3×10^2 (b) Give the victim oxygen-enriched air to breathe.   **20.90** -370 kJ   **20.94** $\Delta H^0 = -137.14$ kJ, $\Delta S^0 = -120.3$ J/K, $\Delta G^0 = -101.25$ kJ   **20.96** (a) $\Delta H^0 = 470.5$ kJ; $\Delta S^0 = 558.4$ J/K   (b) It will be spontaneous at higher temperature, where the $-T\Delta S$ term will be larger in magnitude than ΔH.   (c) No   (d) 842.6 K

20.101 $4Fe(s) + 3O_2(g) \longrightarrow 2Fe_2O_3(s)$
$\Delta G^0 = -1487.2$ kJ; spontaneous
$4Al(s) + 3O_2(g) \longrightarrow 2Al_2O_3(s)$
$\Delta G^0 = -3164$ kJ; spontaneous

Chapter 21

21.1 Oxidation is a loss of electron(s) and an increase in oxidation number. Reduction is a gain of electron(s) and a decrease in oxidation number.   **21.3** An electrochemical process involves an electron flow. At least one substance must lose electron(s) and one substance must gain electron(s) to produce the flow. This electron transfer is a redox process.   **21.6** Add an equal number of OH^- ions to both sides to neutralize H^+ ions and produce H_2O.   **21.8** A voltaic or galvanic cell has a $\Delta G_{sys} < 0$. An electrolytic cell has a $\Delta G_{sys} > 0$, meaning an increase in free energy.

21.10 (a) Cl^- (b) MnO_4^- (c) MnO_4^- (d) Cl^-
(e) From Cl^- to MnO_4^-
(f) $8H_2SO_4(aq) + 2KMnO_4(aq) + 10KCl(aq) \longrightarrow$
$\quad 2MnSO_4(aq) + 6K_2SO_4(aq) + 5Cl_2(g) + 8H_2O(l)$
21.12 (a) $ClO_3^-(aq) + 6I^-(aq) + 6H^+(aq) \longrightarrow$
$\quad Cl^-(aq) + 3I_2(s) + 3H_2O(l)$
oxidizing agent: ClO_3^-; reducing agent: I^-
(b) $2MnO_4^-(aq) + 3SO_3^{2-}(aq) + H_2O(l) \longrightarrow$
$\quad 2MnO_2(s) + 3SO_4^{2-}(aq) + 2OH^-(aq)$
oxidizing agent: MnO_4^-; reducing agent: SO_3^{2-}
(c) $2MnO_4^-(aq) + 5H_2O_2(aq) + 6H^+(aq) \longrightarrow$
$\quad 2Mn^{2+}(aq) + 5O_2(g) + 8H_2O(l)$
oxidizing agent: MnO_4^-; reducing agent: H_2O_2
21.14 (a) $Cr_2O_7^{2-}(aq) + 3Zn(s) + 14H^+(aq) \longrightarrow$
$\quad 2Cr^{3+}(aq) + 3Zn^{2+}(aq) + 7H_2O(l)$
oxidizing agent: $Cr_2O_7^{2-}$; reducing agent: Zn
(b) $3Fe(OH)_2(s) + MnO_4^-(aq) + 2H_2O(l) \longrightarrow$
$\quad MnO_2(s) + 3Fe(OH)_3(s) + OH^-(aq)$
oxidizing agent: MnO_4^-; reducing agent: $Fe(OH)_2$

(c) $5Zn(s) + 2NO_3^-(aq) + 12H^+(aq) \longrightarrow$
$\quad 5Zn^{2+}(aq) + N_2(g) + 6H_2O(l)$
oxidizing agent: NO_3^-; reducing agent: Zn
21.16 (a) $4Sb(s) + 4NO_3^-(aq) + 4H^+(aq) \longrightarrow$
$\quad Sb_4O_6(s) + 4NO(g) + 2H_2O(l)$
oxidizing agent: NO_3^-; reducing agent: Sb
(b) $2Mn^{2+}(aq) + 5BiO_3^-(aq) + 14H^+(aq) \longrightarrow$
$\quad 2MnO_4^-(aq) + 5Bi^{3+}(aq) + 7H_2O(l)$
oxidizing agent: BiO_3^-; reducing agent: Mn^{2+}
(c) $2Fe(OH)_2(s) + Pb(OH)_3^-(aq) \longrightarrow$
$\quad 2Fe(OH)_3(s) + Pb(s) + OH^-(aq)$
oxidizing agent: $Pb(OH)_3^-$
reducing agent: $Fe(OH)_2$
21.18 (a) $5As_4O_6(s) + 8MnO_4^-(aq) + 18H_2O(l) \longrightarrow$
$\quad 20AsO_4^{3-}(aq) + 8Mn^{2+}(aq) + 36H^+(aq)$
oxidizing agent: MnO_4^-; reducing agent: As_4O_6
(b) $P_4(s) + 6H_2O(l) \longrightarrow$
$\quad 2HPO_3^{2-}(aq) + 4H^+(aq) + 2PH_3(g)$
oxidizing agent: P_4; reducing agent: P_4
(c) $2MnO_4^-(aq) + H_2O(l) + 3CN^-(aq) \longrightarrow$
$\quad 2MnO_2(s) + 3CNO^-(aq) + 2OH^-(aq)$
oxidizing agent: MnO_4^-; reducing agent: CN^-
21.21 (a) $Au(s) + 3NO_3^-(aq) + 6H^+(aq) + 4Cl^-(aq) \longrightarrow$
$\quad AuCl_4^-(aq) + 3NO_2(g) + 3H_2O(l)$
(b) oxidizing agent: NO_3^-; reducing agent: Au
(c) The function of HCl is as a source of Cl^- to act as a complexing agent to form highly stable $AuCl_4^-$.
21.22 (a) A   (b) E   (c) C   (d) A   (e) E   (f) E   **21.25** Active electrodes are components of the half-cell reactions. Inactive electrodes are used to conduct electrons but do **not** participate in the half-cell reactions. Two common inactive electrodes are graphite and platinum.   **21.26** (a) metal A   (b) metal B   (c) metal A   (d) The formation of H_2 bubbles means that H^+ was reduced by metal B. Since metal A is a better reducing agent than metal B, H_2 bubbles would form when metal A is placed in acid.
21.27 (a) (red half-rxn) $Sn^{2+}(aq) + 2e^- \longrightarrow Sn(s)$
(ox half-rxn) $\underline{Zn(s) \longrightarrow Zn^{2+}(aq) + 2e^-}$
(overall rxn) $Zn(s) + Sn^{2+}(aq) \longrightarrow Zn^{2+}(aq) + Sn(s)$

(b)

21.29 (a) left to right   (b) left   (c) right   (d) Ni   (e) Fe
(f) Fe   (g) 1 M NiSO$_4$ (h) K$^+$, NO$_3^-$ (i) neither   (j) right to left
(k) reduction: $Ni^{2+}(aq) + 2e^- \longrightarrow Ni(s)$
oxidation: $Fe(s) \longrightarrow Fe^{2+}(aq) + 2e^-$
overall: $Fe(s) + Ni^{2+}(aq) \longrightarrow Fe^{2+}(aq) + Ni(s)$
21.31 (a) reduction: $Fe^{2+}(aq) + 2e^- \longrightarrow Fe(s)$
oxidation: $Mn(s) \longrightarrow Mn^{2+}(aq) + 2e^-$
overall: $Fe^{2+}(aq) + Mn(s) \longrightarrow Mn^{2+}(aq) + Fe(s)$

(b)

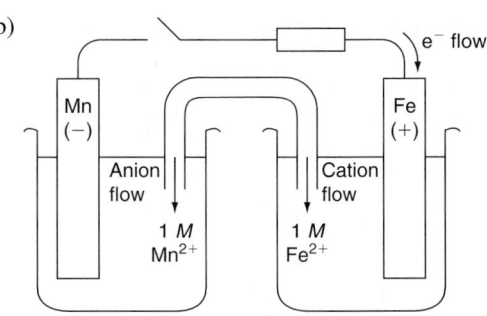

21.33 (a) $Al(s) \mid Al^{3+}(aq) \parallel Cr^{3+}(aq) \mid Cr(s)$
(b) $Pt \mid SO_2(g) \mid SO_4^{2-}(aq), H^+(aq) \parallel Cu^{2+}(aq) \mid Cu(s)$
Note: graphite could be used in place of Pt.
21.36 A negative E_{cell}^0 indicates that the reaction as written is nonspontaneous. The reverse reaction is actually spontaneous.
21.37 Cell potentials (E^0), like other thermodynamic quantities, change sign when the reaction is reversed. Unlike other thermodynamic quantities, cell potentials are intensive properties. Thus, changing the stoichiometric coefficients of a half-cell reaction does **not** affect the value of the cell potential.
21.38 (a) (red half-rxn) $4e^- + 2SO_3^{2-}(aq) + 3H_2O(l) \longrightarrow$
$$6OH^-(aq) + S_2O_3^{2-}(aq)$$
(ox half-rxn) $Se^{2-}(aq) \longrightarrow Se(s) + 2e^-$
(b) $E_{cell}^0 = E_{sulfite}^0 - E_{selenide}^0$
$E_{selenide}^0 = E_{sulfite}^0 - E_{cell}^0$
$= -0.57 \text{ V} - 0.35 \text{ V}$
$= -0.92 \text{ V}$
21.40 (a) $Br_2 > Fe^{3+} > Cu^{2+}$ (b) $Ca^{2+} < Ag^+ < Cr_2O_7^{2-}$
21.42 (a) $Co(s) + 2H^+(aq) \longrightarrow Co^{2+}(aq) + H_2(g)$;
0.28 V; spontaneous
(b) $2Mn^{2+}(aq) + 5Br_2(l) + 8H_2O(l) \longrightarrow$
$$2MnO_4^-(aq) + 10Br^-(aq) + 16H^+(aq);$$
-0.44 V; nonspontaneous
(c) $Hg_2^{2+}(aq) \longrightarrow Hg^{2+}(aq) + Hg(l)$;
-0.07 V; nonspontaneous
21.44 (a) $2Ag(s) + Cu^{2+}(aq) \longrightarrow 2Ag^+(aq) + Cu(s)$;
-0.46 V; nonspontaneous
(b) $3Cd(s) + Cr_2O_7^{2-}(aq) + 14H^+(aq) \longrightarrow$
$$3Cd^{2+}(aq) + 2Cr^{3+}(aq) + 7H_2O(l);$$
1.73 V; spontaneous
(c) $Ni^{2+}(aq) + Pb(s) \longrightarrow Ni(s) + Pb^{2+}(aq)$;
-0.12 V; nonspontaneous
21.46 (I) $3SO_4^{2-}(aq) + 3H_2O(l) + 2Al(s) \longrightarrow$
$$3SO_3^{2-}(aq) + 6OH^-(aq) + 2Al^{3+}(aq)$$
$E_{cell}^0 = 0.93 \text{ V} - (-1.66 \text{ V}) = 2.59 \text{ V}$
oxidizing agents: $SO_4^{2-} > Al^{3+}$
reducing agents: $Al > SO_3^{2-}$
(II) $3N_2O_4(g) + 2Al(s) \longrightarrow 6NO_2^-(aq) + 2Al^{3+}(aq)$
$E_{cell}^0 = 0.867 \text{ V} - (-1.66 \text{ V}) = 2.53 \text{ V}$
oxidizing agents: $N_2O_4 > Al^{3+}$
reducing agents: $Al > NO_2^-$
(III) $SO_4^{2-}(aq) + H_2O(l) + 2NO_2^-(aq) \longrightarrow$
$$SO_3^{2-}(aq) + 2OH^-(aq) + N_2O_4(g)$$
$E_{cell}^0 = 0.93 \text{ V} - (0.867 \text{ V}) = 0.06 \text{ V}$
oxidizing agents: $SO_4^{2-} > N_2O_4$
reducing agents: $NO_2^- > SO_3^{2-}$
oxidizing agents: $SO_4^{2-} > N_2O_4 > Al^{3+}$
reducing agents: $Al > NO_2^- > SO_3^{2-}$

21.48 (I) $2HClO(aq) + 2H^+(aq) + Pb(s) + SO_4^{2-}(aq) \longrightarrow$
$$Cl_2(g) + 2H_2O(l) + PbSO_4(s)$$
$E_{cell}^0 = 1.94 \text{ V}$
oxidizing agents: $HClO > PbSO_4$
reducing agents: $Pb + SO_4^{2-} > Cl_2$
(II) $Pt^{2+}(aq) + Pb(s) + SO_4^{2-}(aq) \longrightarrow$
$$Pt(s) + PbSO_4(s)$$
$E_{cell}^0 = 1.51 \text{ V}$
oxidizing agents: $Pt^{2+} > PbSO_4$
reducing agents: $Pb + SO_4^{2-} > Pt$
(III) $2HClO(aq) + 2H^+(aq) + Pt(s) \longrightarrow$
$$Pt^{2+}(aq) + Cl_2(g) + 2H_2O(l)$$
$E_{cell}^0 = 0.43 \text{ V}$
oxidizing agents: $HClO > Pt^{2+}$
reducing agents: $Pt > Cl_2$
oxidizing agents: $HClO > Pt^{2+} > PbSO_4$
reducing agents: $Pb + SO_4^{2-} > Pt > Cl_2$
21.50 $C > A > B$; $C(s) + 2H^+(aq) \longrightarrow C^{c+}(aq) + H_2(g)$
(Metal C **would** cause formation of H_2 gas.)
21.53 $A(s) + B^+(aq) \longrightarrow A^+(aq) + B(s)$
(a) $[A^+]$ increases and $[B^+]$ decreases.
(b) E_{cell} decreases
(c) $E_{cell} = E_{cell}^0 - \dfrac{0.0592 \text{ V}}{n} \log \dfrac{[A^+]}{[B^+]}$
E_{cell} will equal E_{cell}^0 when $\log \dfrac{[A^+]}{[B^+]} = 0$, which will occur when $\dfrac{[A^+]}{[B^+]} = 1$.
(d) Yes. E_{cell} will be less than E_{cell}^0 when $Q \left(= \dfrac{[A^+]}{[B^+]} \right.$ for this problem$\left. \right)$ is greater than 1.

21.55 In a concentration cell, the redox reaction proceeds in the direction to achieve concentration equality in the half-cells.
cathode rxn: $M^+(aq) + e^- \longrightarrow M(s)$
anode rxn: $M(s) \longrightarrow M^+(aq) + e^-$
The more concentrated electrolyte is in the cathode compartment.
21.56 (a) 3×10^{35} (b) 4×10^{-31} **21.58** (a) 1×10^{-67} (b) 6×10^9 **21.60** (a) -203 kJ (b) 1.73×10^2 kJ **21.62** (a) 382 kJ (b) -55.9 kJ **21.64** $E^0 = 0.22$ V; $\Delta G = -21$ kJ **21.66** $E^0 = 0.055$ V; $\Delta G = -11$ kJ **21.68** 9×10^{-4} M **21.70** (a) 0.05 V (b) 0.03 V (c) 0.40 M (d) $[Ni^{2+}] = 0.09$ M; $[Co^{2+}] = 0.91$ M **21.72** Electrode A is the anode; $E_{cell} = 0.085$ V **21.74** Electrons in a battery flow from the anode to the cathode, since oxidation still takes place at the anode and reduction still takes place at the cathode. **21.76** Since cell potentials are intensive properties, the voltage depends on the chemical reaction taking place, not on the size of the battery. What *is* different is the total charge the battery can produce, with the larger D-size battery producing more charge. **21.78** If the copper plates touch the iron skeleton, the more active iron would corrode very rapidly.
$Fe(s) + Cu^{2+}(aq) \rightleftharpoons Fe^{2+}(aq) + Cu(s)$ $E_{cell}^0 = 0.78$ V
In reality, the copper simply acts as a conductor for electrons and the oxidizing agent is not Cu^{2+} but more probably O_2 or oxides of N or S. In any case, the teflon spacers are designed to keep the metals from touching, to prevent the corrosion of the

iron.   **21.81** A more reactive metal (that is, one which is preferentially oxidized) is used as the sacrificial anode. Only (a), (b), (c), (f), and (g) are suitable for iron (although Na wouldn't last very long!).   **21.83** The dry cell potential of 1.5 V is greater than the voltaic cell voltage of 0.34 V. This is more than sufficient to reverse the reaction so that Cd redissolves and Cr will plate out.   **21.85** The nitrate ion cannot be oxidized (so it can't react at the anode) because it is already in its highest oxidation state.   **21.87** (a) $Br_2(g)$ (b) $Na(l)$ **21.89** anode: $I_2(g)$; cathode: $Mg(l)$ **21.91** anode: $Br_2(g)$; cathode: $Ca(l)$ **21.93** copper and bromine   **21.95** iodine, zinc, and silver   **21.97** (a) cathode: $H_2(g)$ and $OH^-(aq)$; anode: $O_2(g)$ and $H_3O^+(aq)$ (b) cathode: $Sn(s)$; anode: $O_2(g)$ and $H_3O^+(aq)$ **21.99** (a) cathode: $NO(g)$; anode: $O_2(g)$ and $H_3O^+(aq)$ (b) cathode: $H_2(g)$ and $OH^-(aq)$; anode: $Cl_2(g)$ **21.101** (a) 2.93 mol of e^- (b) 2.83×10^5 C   (c) 31.4 A   **21.103** 0.252 g   **21.105** 1.10×10^4 s   **21.107** (a) Pure water has very few ions present to conduct the current needed for electrolysis. The addition of the Na^+ and SO_4^{2-} ions increases the rate of electrolysis.   (b) The reduction of H_2O has a more positive half-potential than the reduction of Na^+; the oxidation of H_2O is the only reaction possible because SO_4^{2-} cannot be oxidized. In other words, it is easier to reduce H_2O than Na^+ and easier to oxidize H_2O than SO_4^{2-}.   **21.109** 44.2 g   **21.111** (a) 2.0×10^{11} C   (b) 2.4×10^{11} J   (c) 6.1×10^6 g   **21.114** 69.4% Cu   **21.116** (a) 26 days   (b) $\$1.6\times10^2$

21.119 anode (positive electrode): $2H_2O(l) \longrightarrow$
$$O_2(g) + 4H^+(aq) + 4e^-$$
cathode (negative electrode): $2H_2O(l) + 2e^- \longrightarrow$
$$H_2(g) + 2OH^-(aq)$$
At the negative lead, the filter paper will turn pink.   **21.122** (a) $E^0_{cell} = 0.13$ V for Pb (cell 1); $E^0_{cell} = 0.34$ V for Cu (cell 2)   (b) In cell 1, the negative electrode is $Pb(s)$. In cell 2, the negative electrode is the inert $Pt(s)$.   (c) An addition of S^{2-} lowers the Pb^{2+} concentration, which will result in an increase in cell voltage.   (d) -0.13 V

21.125 The electrode reaction of $M^+(aq)$ can be divided into three steps:

1. $M^+(aq) \longrightarrow M^+(g)$ $\Delta H = -\Delta H_{hydration}$
2. $M^+(g) + e^- \longrightarrow M(g)$ $\Delta H = -\Delta H_{ionization}$
3. $M(g) \longrightarrow M(s)$ $\Delta H = -\Delta H_{atomization}$

The second step of the electrode reaction is the same as the reversed ionization energy. The small Li^+ ion has the greatest (most negative) hydration energy of the Group 1A(1) elements; thus, step 1 above is much more unfavorable for Li^+ than for either Na^+ or K^+. The atomization energies (the negative of step 3) of the three metals are fairly similar. Overall, in the electrode reaction, the hydration energy is dominant in terms of ordering the three E^0 values.

21.127 The key factor is that the table deals with electrode potentials in aqueous solution. The very high and very low standard potentials involve extremely reactive substances, such as F_2 and Li. These substances react directly with water, rather than according to the desired half-cell reactions. Alternatively (but equivalently), any aqueous cell with a voltage greater than 1.23 volts has the ability to decompose water into H_2 and O_2. When two electrodes with 6 volts across them are placed in water, electrolysis of water will occur.   **21.129** (a) 1.07×10^5 s

(b) 1.5×10^4 kW·h   (c) $\$0.068$/lb Al   **21.130** (a) from the Mg to the pipe   (b) 0.36 A

21.131 F < D < E; $E(s) + F^{f+}(aq) \longrightarrow F(s) + E^{e+}(aq)$ (If E metal were placed in a solution of an F salt, F metal is produced on the surface of E metal.)

21.134 (a) Cell I: 4 mol e^-; $\Delta G = -475$ kJ
      Cell II: 2 mol e^-; $\Delta G = -394$ kJ
      Cell III: 2 mol e^-; $\Delta G = -453$ kJ
  (b) Cell I: -13.2 kJ/g
      Cell II: -0.613 kJ/g
      Cell III: -2.62 kJ/g

The fuel cell (I) gives the highest ratio and the lead-acid cell (II) gives the lowest. This is mostly because the fuel cell uses materials with small molar masses.

21.135 4.0 g Ag   **21.138** (a) 9.6 g Cu   (b) 0.38 M

21.140 $Sn^{2+}(aq) + 2e^- \longrightarrow Sn(s)$
    $Cr^{3+}(aq) + e^- \longrightarrow Cr^{2+}(aq)$
    $Fe^{2+}(aq) + 2e^- \longrightarrow Fe(s)$
    $U^{4+}(aq) + e^- \longrightarrow U^{3+}(aq)$

21.141 (a) $Cu(s) \longrightarrow Cu^{2+}(aq) + 2e^-$
  (b) $Cu^{2+}(aq) + 2e^- \longrightarrow Cu(s)$
  (c) 2.9×10^2 kW·h
  (d) Ag and Au; possibly Ni, Fe, Pd, and Pt
  (e) Copper and the more active impurities (e.g., Ni, Fe) are oxidized to cations but due to differences (less positive E^0_{red}) in their standard electrode potentials, only the Cu^{2+} ions are reduced at the cathode. The less active metals (e.g., Au, Pd, Pt) remain unoxidized in the "anode mud."

21.144 (a) 3.6×10^{-9} M (b) 1.5 M

21.146 (a) $E_{stand} = E^0_{cell} - (0.0592\text{V})\log [Ag^+]_{stand}$
    $E_{waste} = E^0_{cell} - (0.0592\text{V})\log [Ag^+]_{waste}$
(b) antilog $\left(\dfrac{E_{stand} - E_{waste}}{0.0592}\right)[Ag^+]_{stand} = [Ag^+]_{waste}$
(c) antilog $\left(\dfrac{E_{stand} - E_{waste}}{0.0592}\right)C_{Ag\ stand} = C_{Ag\ waste}$
  (where C = concentration in ng/L)
(d) 9×10^2 ng/L
(e) antilog $\left(\dfrac{\dfrac{nF}{2.303R}(E_{stand} - E_{waste}) + T_{stand}\log [Ag^+]_{stand}}{T_{waste}}\right)$
  $= [Ag^+]_{waste}$

Chapter 22

22.2 Fe in Fe_2O_3 and Fe_3O_4; Ca in $CaCO_3$; Na in NaCl; Zn in ZnS, and Pb in PbS.   **22.3** (a) Differentiation is the separating of the materials of the earth into different regions (layers) based primarily on their densities.   (b) O, Si, Al, and Fe.   (c) O. Si, Al, Ca, Na, K, Mg also present in crust and mantle but not core.   **22.7** Plant life has produced O_2, which reacts with most metals. Plants need large amounts of K^+ for growth, so remove that ion from the crust.   **22.9** Fixation is the process of converting an element in a form not directly usable by animals or plants (usually in gaseous form) to a form which is usable (usually in condensed form). Nitrogen is fixed from atmospheric N_2 and carbon is fixed from atmospheric CO_2.   **22.12** Nitrogen is removed from the atmosphere by atmospheric fixation on land

and water, by biological fixation on land, and by industrial fixation. The first involves reaction of N_2 and O_2 induced by lightning, the second involves nitrogen-fixing bacteria, and the third involves reaction of N_2 in chemical plants to produce NH_3 and related compounds. Human activity is a significant factor contributing about 17% of the nitrogen removed. **22.14** (a) the atmosphere (b) Plants absorb phosphate in the form of $H_2PO_4^-$. Animals eat the plants and excrete phosphates, and both animals and plants produce phosphates by decay after death. **22.17** (a) 1.1×10^3 L (b) 4.2×10^2 m^3 **22.18** (a) Because it removes some of the (desired) P and is of limited value. (b) 8.8 tonne **22.20** (a) Heating an ore in the presence of oxygen. (b) Heating an ore in the presence of a reducing agent. (c) Stirring an ore with a detergent-oil mixture to separate the mineral from the gangue. (d) Purifying a material after its initial formation. **22.25** (a) This is a by-product of steelmaking containing the impurity SiO_2. (b) This is impure iron containing 3–4% C plus other impurities. (c) This is iron containing 1–1.5% C and small amounts of other materials. (d) The process used to produce steel from pig iron. **22.27** Fe and Ni are more easily oxidized and less easily reduced than Cu. In the electrorefining process, all three metals are oxidized at the anode, but only Cu^{2+} is reduced at the cathode. **22.30** Hess's law is used to analyze the thermodynamic feasibility of the various steps in a proposed scheme.
22.31 (a) $E_{red}^0 = -3.05$ V, -2.93 V, and -2.71 V for Li^+, K^+, and Na^+, respectively. (Rb^+ and Cs^+ would be expected to be similar.) In all of these cases, it is energetically more favorable to reduce H_2O to H_2 than $M^+(aq)$ to M.
(b) $2RbX + Ca \longrightarrow CaX_2 + 2Rb$
$\Delta H = IE_1(Ca) + IE_2(Ca) - 2IE_1(Rb)$
$= +929$ kJ/mol
Since $\Delta H > 0$, the reaction should be unfavorable.
(c) Rb has a lower boiling point than Ca, so the Rb produced boils out of the reaction mixture, shifting the reaction to the right.
(d) $2CsX + Ca \longrightarrow CaX_2 + 2Cs$
$\Delta H = IE_1(Ca) + IE_2(Ca) - 2IE_1(Cs)$
$= +983$ kJ/mol
This reaction is more unfavorable than for Rb, but Cs has a lower boiling point, making the process reasonable.
22.32 (a) 4.6×10^4 L (b) 1.26×10^8 C (c) 466 h **22.35** (a) Mg^{2+} is more difficult to reduce than H_2O, so H_2 would be produced (not Mg metal), as well as Cl_2. (b) ΔH_f^0 for $MgCl_2$ is negative, so high temperature would favor its decomposition into Mg and Cl_2.
22.37 (a) The SO_2 acts as a reducing agent, and is oxidized to the +6 state.
(b) $HSO_4^- (aq)$
(c) $H_2SeO_3(aq) + 2SO_2(g) + H_2O(l) \longrightarrow$
$Se(s) + 2HSO_4^-(aq) + 2H^+(aq)$
22.42 (a) Cu_2S and Cu_2O: O.N. $= +1$; Cu: O.N. $= 0$ (b) Oxidizing agent: Cu_2O; Reducing agent: Cu_2S **22.44** (a) 1.1×10^6 C (b) 1.2×10^3 A **22.47** ΔG^0 for $2ZnS + C \longrightarrow 2Zn + CS_2$ is $+463$ kJ; for $2ZnO + C \longrightarrow 2Zn + CO_2$, ΔG^0 $+242.0$ kJ. The process for ZnO is less unfavorable. **22.48** The reaction of SO_2 with O_2 to produce SO_3 is thermodynami-

cally favorable, but very slow at ordinary temperatures. A catalyst is used to speed up the reaction. **22.51** (a) H_2, Cl_2, and NaOH (b) The product NaOH is of higher purity, but the Hg is an environmental pollutant. **22.52** (a) $\Delta G^0 = -142$ kJ; spontaneous (b) It is very slow at that temperature. (c) $\Delta G_{500} = -53$ kJ; spontaneous (d) $K_{25} = 7.8\times10^{24} > K_{500} = 3.8\times10^3$ (e) 1.05×10^3 K **22.53** 7×10^2 lb
22.57 (a) $P_4O_{10}(s) + 6H_2O(l) \longrightarrow 4H_3PO_4(l)$ (b) 1.55
22.59 (a) 9.929×10^3 t (b) 4.3×10^4 t from automobiles, so $CO_2(autos) > CO_2(steelmaking)$ **22.61** (a) $[OH^-] >$ 1.1×10^{-4} M (i.e., pH > 10.04) will precipitate $Mg(OH)_2$. (b) 99.998%
22.66 (a) $8H_2S(g) + 4O_2(g) \longrightarrow S_8(g) + 8H_2O(g)$
(b) $8H_2S(g) + 8Cl_2(g) \longrightarrow S_8(g) + 16HCl(g)$
$\Delta G^0 = -1212$ kJ, so the process is thermodynamically spontaneous.
(c) O_2 is considerably less expensive than Cl_2, and H_2O is a very benign by-product (although HCl could be sold profitably).
22.69 (a) $:C \equiv O:$ (b) C has formal charge of -1, so would be the more likely spot for e^- donation.
22.71 (a) cathode: $Na^+ + e^- \longrightarrow Na$
anode: $4OH^- \longrightarrow O_2 + 2H_2O + 4e^-$
(b) 0.5 mol Na/mol e^-
22.73 (a) $NO(g) + O_3(g) \rightleftharpoons NO_2(g) + O_2(g)$
(b) $Rate_f = k_f[NO][O_3]$
$Rate_r = k_r[NO_2][O_2]$
(c) $\Delta G_{280}^0 = -199$ kJ
(d) $k_f/k_r = 1.3\times10^{37}$
22.74 (a) $nCO_2(g) + nH_2O(l) \longrightarrow (CH_2O)_n(s) + nO_2(g)$
(b) 25 L (c) 7.6×10^4 L
22.77 795 kg **22.78** (a) 20.4 min (b) 14 steps **22.80** (a) 1.890 t (b) 0.3339 t (c) 100% (d) 74% (e) 2.813×10^3 m^3 **22.85** Acid rain increases the leaching of PO_4^{3-} into the ground water, due to the protonation of the PO_4^{3-} to form HPO_4^{2-} and $H_2PO_4^-$. (a) 6.4×10^{-7} M (b) 1.1×10^{-2} M

Chapter 23

23.2 (a) $1s^22s^22p^63s^23p^64s^23d^{10}4p^65s^24d^x$
(b) $1s^22s^22p^63s^23p^64s^23d^{10}4p^65s^24d^{10}5p^66s^24f^{14}5d^x$
23.4 Five; $Mn([Ar]4s^23d^5)$ and $Mn^{2+}([Ar]3d^5)$. **23.6** (a) The lanthanide contraction is the "shrinkage" of atoms following the filling of the $4f$ subshell (i.e., the atoms are smaller than expected). (b) The size increases from Period 4 to Period 5, but stays fairly constant (or increases only slightly) from Period 5 to Period 6. (c) The smaller-than-expected sizes lead to very large densities. **23.9** (a) Paramagnetic materials are attracted into a magnetic field; diamagnetic ones are repelled. Frequently, paramagnetic materials are also colored. (b) Most main-group element ions are isoelectronic with a noble gas, meaning that all their electrons are paired. (c) Because main-group metal ions are diamagnetic, most have a filled outer shell (or at least the subshell) and too much energy would be required to promote an electron to a higher energy orbital.
23.10 (a) $[Ar] 4s^23d^3$ (b) $[Kr] 5s^24d^1$ (c) $[Xe] 6s^24f^{14}5d^{10}$
23.12 (a) $[Xe] 6s^24f^{14}5d^6$ (b) $[Ar] 4s^23d^7$ (c) $[Kr] 5s^14d^{10}$
23.14 (a) $[Ar]$; 0 (b) $[Ar] 3d^9$; 1 (c) $[Ar] 3d^5$; 5 (d) $[Kr]$

$4d^2$; 2   **23.16** (a) +5   (b) +4   (c) +7   **23.18** Cr, Mo, and W   **23.20** In CrF_2, since metallic behavior is greater in lower oxidation states.   **23.22** Atomic size decreases down a group, but the nuclear charge increases much more, so the ionization energy increases, making the heavier element (Mo) more difficult to oxidize.   **23.24** CrO_3, because higher oxidation states form more acidic oxides.   **23.28** (a) 7   (b) This corresponds to a half-filled f subshell.   **23.30** (a) [Xe] $6s^2 5d^1$ (b) [Xe] $4f^1$ (c) [Rn] $7s^2 5f^{11}$ (d) [Rn] $5f^2$
23.32 (a) Eu^{2+}: [Xe] $4f^7$
    Eu^{3+}: [Xe] $4f^6$
    Eu^{4+}: [Xe] $4f^5$
    The stability of the half-filled f subshell makes Eu^{2+} most stable. Too much energy is required to remove two more electrons and make Eu^{4+}.
    (b) Tb^{2+}: [Xe] $4f^9$
    Tb^{3+}: [Xe] $4f^8$
    Tb^{4+}: [Xe] $4f^7$
    Tb would show a +4 oxidation state, since that has the stable half-filled f^7 configuration.
23.34 Gd has 8 unpaired electrons ([Xe] $6s^2 5d^1 4f^7$) and Gd^{3+} has 7 unpaired electrons ([Xe] $4f^7$).   **23.36** This is the effective electronegativity of the element in a given oxidation (or valence) state. As oxidation number increases, the atom becomes more electronegative, strengthening its M—O bonds and making the oxide more acidic. MnO is basic, Mn_2O_7 is acidic.   **23.40** It is the only metal which is a liquid at room temperature. Its "1+" ion is really Hg_2^{2+}.   **23.42** E^0 for $Cr(s) + 2Cr^{3+}(aq) \longrightarrow 3Cr^{2+}(aq)$ is +0.50 V, so the process is favorable.   **23.44** $E^0 = 0.28$ V for $3Mn^{2+}(aq) + 2MnO_4^-(aq) + 2H_2O(l) \longrightarrow 5MnO_2(s) + 4H^+(aq)$, so this process occurs to remove MnO_4^- and produce $MnO_2(s)$.
23.47 The coordination number is the number of ligand donor atoms attached to the metal. It is unrelated to the oxidation number, which is determined by the number of electrons the metal has lost to form the ion.   **23.49** 2: linear; 4: tetrahedral, square planar; 6: octahedral   **23.52** The complex ion has a negative charge.   **23.55** (a) hexaaquanickel(II) chloride   (b) tris(ethylenediamine)chromium(III) perchlorate   (c) potassium hexacyanomanganate(II)   **23.57** (a) +2, 6   (b) +3, 6   (c) +2, 6   **23.59** (a) potassium dicyanoargentate(I)   (b) sodium tetrachlorocadmate(II)   (c) triamminehaquabromocobalt(III) bromide   **23.61** (a) +1, 2   (b) +2, 4   (c) +3, 6   **23.63** (a) $[Zn(NH_3)_4]SO_4$ (b) $[Cr(NH_3)_5Cl]Cl_2$ (c) $Na_3[Ag(S_2O_3)_2]$
23.65 (a) 4, 2   (b) 6, 3   (c) 2, 4   **23.67** (a) $[Cr(H_2O)_6]_2(SO_4)_3$ (b) $Ba[FeBr_4]_2$ (c) $[Pt(C_2H_8N_2)_2]CO_3$ **23.69** (a) 6, 5   (b) 4, 3   (c) 4, 2
23.71 (a) $[\ddot{\text{O}}=\ddot{\text{N}}-\ddot{\text{O}}: \longleftrightarrow :\ddot{\text{O}}-\ddot{\text{N}}=\ddot{\text{O}}:]^-$ can form linkage isomers, since two different atoms (O, N) have lone pairs.
(b) $:\ddot{\text{O}}=\ddot{\text{S}}-\ddot{\text{O}}: \longleftrightarrow :\ddot{\text{O}}-\ddot{\text{S}}=\ddot{\text{O}}:$ can also form linkage isomers (S, O).
(c) $\left[:\ddot{\text{O}}-\text{N}-\ddot{\text{O}}: \longleftrightarrow :\ddot{\text{O}}=\text{N}-\ddot{\text{O}}: \longleftrightarrow :\ddot{\text{O}}-\text{N}=\ddot{\text{O}}:\right]^-$
cannot form linkage isomers since only the three identical O atoms have lone pairs.

23.73 (a) Geometric isomers
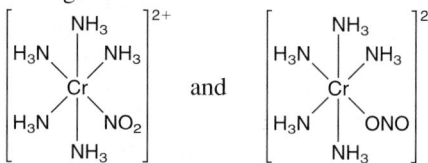
(b) Geometric isomers
(c) Geometric isomers

23.75 (a) Geometric isomers
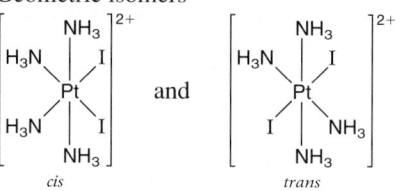
(b) Linkage isomers
(c) Geometric isomers

23.77 $CrCl_3(NH_3)_4$, which is actually $[Cr(NH_3)_4Cl_2]Cl$.   **23.79** (a) $K[Pd(NH_3)Cl_3]$ (b) $[Pd(NH_3)_2Cl_2]$ (c) $K_2[PdCl_6]$ (d) $[Pd(NH_3)_4Cl_2]Cl_2$ **23.81** (a) dsp^2 (b) sp^3 **23.84** yellow and/or orange   **23.85** (a) The energy difference between the upper and lower energy levels after splitting by the crystal field.   (b) The energy of the d_{z^2} and $d_{x^2-y^2}$ orbitals is raised more than that of the d_{xy}, d_{xz} and d_{yz} orbitals.   (c) The splitting is reversed for a tetrahedral field—i.e., the d_{xy}, d_{xz}, and d_{yz} orbitals are higher in energy than the d_{z^2} and $d_{x^2-y^2}$.
23.88 If $\Delta > E_{\text{pairing}}$, the complex will be low-spin and will have a lower paramagnetism than if $\Delta < E_{\text{pairing}}$.   **23.90** (a) 0   (b) 8   (c) 6   **23.92** (a) 5   (b) 10   (c) 7
23.94

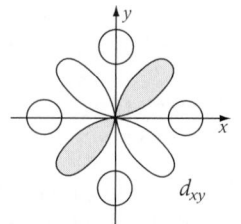

The ligands "point" directly at the electrons in the $d_{x^2-y^2}$ orbital, raising their energy. The ligands are further away from the electrons in the d_{xy} orbital, so their energy is affected less.
23.96 (a) and (d)

23.98 (a)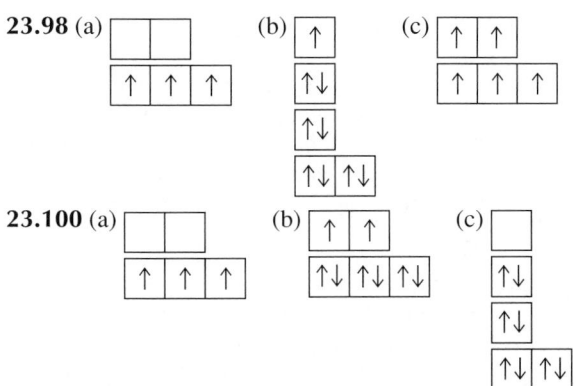

23.100 (a)

23.102 $[Cr(H_2O)_6]^{3+} < [Cr(NH_3)_6]^{3+} < [Cr(NO_2)_6]^{3-}$

23.104 A violet complex will absorb yellow-green light. A complex of a weaker ligand will absorb lower-energy yellow, orange, or red light, making the observed color blue or green.

23.107 NH_3 is a stronger ligand than H_2O, so $[Ni(NH_3)_6]^{2+}$ will absorb higher-energy light than $[Ni(H_2O)_6]^{2+}$. Being green, $[Ni(H_2O)_6]^{2+}$ is probably absorbing primarily red light, while violet $[Ni(NH_3)_6]^{2+}$ is probably absorbing higher-energy yellow-green light. **23.111** Hg^+ is [Xe] $6s^1 4f^{14} 5d^{10}$ but Cu^+ is [Ar] $3d^{10}$. The unpaired electron in Hg^+ would be likely to pair up, forming Hg_2^{2+}. The full subshell configuration of Cu^+ is stable. **23.113** (a) 6 (b) +3 (c) 2 (d) 1

23.115 $PtC_{12}H_{30}P_2Cl_2$ or $Pt[P(C_2H_5)_3]_2Cl_2$

The isomers are

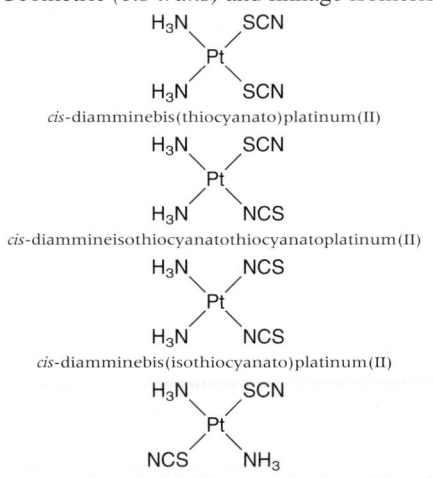

cis-dichlorobis(triethylphosphine)platinum(II)

and

trans-dichlorobis(triethylphosphine)platinum(II)

23.118 (a) no (b) no (c) no (d) no (e) yes **23.121** (a) The first reaction would have $\Delta S \approx 0$; the second, $\Delta S < 0$. Thus, the first is more favorable (or less unfavorable). (b) The second ($Cu(en)_2^{2+}$) would be more stable because of the unfavorable entropy change.

23.123 Geometric (cis-trans) and linkage isomerism.

H₃N, SCN / Pt / H₃N, SCN
cis-diamminebis(thiocyanato)platinum(II)

H₃N, SCN / Pt / H₃N, NCS
cis-diammineisothiocyanatothiocyanatoplatinum(II)

H₃N, NCS / Pt / H₃N, NCS
cis-diamminebis(isothiocyanato)platinum(II)

H₃N, SCN / Pt / NCS, NH₃
trans-diamminebis(thiocyanato)platinum(II)

H₃N, SCN / Pt / SCN, NH₃
trans-diammineisothiocyanatothiocyanatoplatinum(II)

H₃N, NCS / Pt / SCN, NH₃
trans-diamminebis(isothiocyanato)platinum(II)

23.125 $[Co(NH_3)_4(H_2O)Cl]^{2+}$
tetraammineaquachlorocobalt(III)

2 geometric isomers

cis   trans

$[Cr(H_2O)_3Br_2Cl]$ triaquadibromochlorochromium(III)

3 geometric isomers

bromines are *trans*   bromines are *cis* waters are *cis* and *trans*   bromines and waters are *cis*

$[Cr(NH_3)_2(H_2O)_2Br_2]^+$
diamminediaquadibromochromium(III)

6 isomers (5 geometric)

all ligands are *trans*   only NH₃ is *trans*   only Br is *trans*

only H₂O is *trans*   optical isomers all three ligands are *cis*

23.128 (a) MnO_4^{2-}: +6; MnO_4^-: +7; MnO_2: +4

(b) $3MnO_4^{2-}(aq) + 4H^+(aq) \longrightarrow$
$MnO_2(s) + 2MnO_4^-(aq) + 2H_2O(l)$

Chapter 24

24.1 (a) Chemical reactions are accompanied by relatively small changes in energy while nuclear reactions are accompanied by relatively large changes in energy. (b) Rates of chemical reactions are increased by increasing temperature while nuclear reactions are not affected by temperature. (c) Both are increased by higher reactant concentrations. (d) Higher reaction concentrations would increase the yield of both nuclear reactions and chemical reactions. **24.2** (a) 95.02% (b) Larger than this, since ^{32}S is the lightest isotope.

24.4 (a) Z down by 2, N down by 2

(b) Z up by 1, N down by 1

(c) Z and N unchanged

(d) and (e) Z down by 1, N up by 1

A different element is produced in all cases except (c).

24.6 Neutron-rich nuclides, with a high N/Z, undergo β decay. Neutron-poor nuclides, with a low N/Z, undergo positron decay or electron capture.

24.8 (a) $^{234}_{92}\text{U} \longrightarrow ^{4}_{2}\text{He} + ^{230}_{90}\text{Th}$

(b) $^{232}_{93}\text{Np} + ^{0}_{-1}\text{e} \longrightarrow ^{232}_{92}\text{U}$

(c) $^{12}_{7}\text{N} \longrightarrow ^{0}_{1}\beta + ^{12}_{6}\text{C}$

24.10 (a) $^{27}_{12}\text{Mg} \longrightarrow ^{0}_{-1}\beta + ^{27}_{13}\text{Al}$

(b) $^{9}_{3}\text{Li} \longrightarrow ^{1}_{0}\text{n} + ^{8}_{3}\text{Li}$

(c) $^{103}_{46}\text{Pd} + ^{0}_{-1}\text{e} \longrightarrow ^{103}_{45}\text{Rh}$

24.12 (a) $^{48}_{23}\text{V} \longrightarrow ^{48}_{22}\text{Ti} + ^{0}_{1}\beta$

(b) $^{107}_{48}\text{Cd} + ^{0}_{-1}\text{e} \longrightarrow ^{107}_{47}\text{Ag}$

(c) $^{210}_{86}\text{Rn} \longrightarrow ^{206}_{84}\text{Po} + ^{4}_{2}\text{He}$

24.14 (a) $^{186}_{78}\text{Pt} + ^{0}_{-1}\text{e} \longrightarrow ^{186}_{77}\text{Ir}$

(b) $^{225}_{89}\text{Ac} \longrightarrow ^{221}_{87}\text{Fr} + ^{4}_{2}\text{He}$

(c) $^{129}_{52}\text{Te} \longrightarrow ^{129}_{53}\text{I} + ^{0}_{-1}\beta$

24.16 (a) $^{20}_{8}\text{O}$ $N/Z = 12/8 = 1.5$;

unstable, too high for this region of the band

(b) $^{59}_{27}\text{Co}$ $N/Z = 32/27 = 1.2$; stable, N/Z OK

(c) $^{9}_{3}\text{Li}$ $N/Z = 6/3 = 2$; unstable, too high N/Z

24.18 (a) $^{127}_{53}\text{I}$ $N/Z = 74/53 = 1.4$; stable, N/Z OK

(b) $^{106}_{50}\text{Sn}$ $N/Z = 56/50 = 1.1$;

unstable, N/Z too small for this region of the band

(c) $^{68}_{33}\text{As}$ $N/Z = 35/33 = 1.1$; unstable, odd N and Z

24.20 (a) α decay (b) β^+ decay or e^- capture (c) β^+ decay or e^- capture **24.22** (a) β decay (b) β^+ decay or e^- capture (c) α decay **24.24** It lies in the band of stability, with N and Z both even. **24.28** 7α, 4β **24.31** No. A decay rate is an average rate and is meaningful only when we consider a large number of nuclei in a macroscopic sample. The conclusion would be valid if the original sample were large. **24.33** 2.89×10^{-2} Ci/g **24.35** 1.4×10^{8} Bq/g **24.37** 1×10^{-12} h^{-1} **24.39** 2.31×10^{-7} yr^{-1} **24.41** 1.49 mg **24.43** 2.2×10^{9} yr **24.45** 1×10^{2} d/min **24.47** 1.6×10^{6} yr **24.50** Both gamma radiation and neutrons have no charge but the neutron has mass approximately equal to that of a proton. **24.52** Protons experience repulsion from the target nuclei due to their positive charge.

24.53 (a) $^{10}_{5}\text{B} + ^{4}_{2}\text{He} \longrightarrow ^{1}_{0}\text{n} + ^{13}_{7}\text{N}$

(b) $^{28}_{14}\text{Si} + ^{2}_{1}\text{H} \longrightarrow ^{1}_{0}\text{n} + ^{29}_{15}\text{P}$

(c) $^{242}_{96}\text{Cm} + ^{4}_{2}\text{He} \longrightarrow 2^{1}_{0}\text{n} + ^{244}_{98}\text{Cf}$

24.58 Ionizing radiation is more dangerous to children because of their growing, developing body parts and function. **24.60** (a) 5.4×10^{-7} rad (b) 5.4×10^{-9} Gy **24.62** (a) 7.5×10^{-10} Gy (b) 7.5×10^{-5} mrem (c) 7.5×10^{-10} Sv **24.65** 2.45×10^{-3} rad **24.67** In neutron activation analysis (NAA), neutrons bombard a non-radioactive sample, converting a small fraction of its atoms to radioisotopes, which exhibit characteristic decay patterns, such as γ-ray spectra, that reveal the elements present. Unlike chemical analysis, NAA analysis leaves the sample virtually intact. **24.70** The methanol supplies the oxygen to the aldehyde. The oxidation involves the loss of hydrogen atoms from the methanol to form the formaldehyde. The H_2CrO_4 supplies the oxygen to the water. **24.73** Energy is released when nuclides form from their nucleons. The nuclear binding energy is the amount of energy holding the nucleus together. **24.75** (a) 1.861×10^{4} eV (b) 2.981×10^{-15} J **24.77** 5.1×10^{11} J **24.79** (a) 7.976 MeV/nucleon (b) 127.6 MeV/atom (c) 1.2313×10^{10} kJ/mol **24.81** (a) 8.768 MeV/nucleon (b) 517.3 MeV/atom (c) 4.991×10^{10} kJ/mol **24.85** Radioactive decay is a spontaneous process in which nuclides emit small particles and energy. In fission, larger atomic nuclides are bombarded and break into small nuclides in addition to small particle(s) and energy emission.

24.88 Light-water reactors simply use H_2O as the moderator; in heavy-water reactors, D_2O is used. The advantage of D_2O is that it absorbs fewer neutrons, leaving more available for fission, and thus the ^{235}U fuel does not need to be enriched. The disadvantage is in the initial additional cost of D_2O. **24.91** After the initial hydrogen burning process, the star's energy source changes to helium burning where continual heavier nuclei are formed by the fusion of helium nuclei which have a mass of 4 amu. **24.96** (a) 1.28×10^{-4} rad (b) 1.28×10^{-6} Gy **24.98** 5.1×10^{10} s **24.101** 1.35×10^{-5} M **24.103** 6.3×10^{-2} **24.106** (a) 6.05 h (b) 21% **24.108** (a) 0.999 (b) 0.298 (c) 5.58×10^{-6} (d) Radiocarbon dating is more reliable for (b) because a change in the amount of C-14 would be noticeable. In the case of either (a) or (c) a change would be difficult to measure and the error would be large.

24.111 (a) β emission by vanadium-52 produces chromium-52.
$^{52}_{24}\text{Cr}$; $^{51}_{23}\text{V} (n, \beta) ^{52}_{24}\text{Cr}$

(b) Positron emission by copper-64 produces nickel-64.
$^{64}_{28}\text{Ni}$; $^{63}_{29}\text{Cu} (n, \beta^+) ^{64}_{28}\text{Ni}$

(c) β emission by aluminum-28 produces silicon-28.
$^{28}_{14}\text{Si}$; $^{27}_{13}\text{Al} (n, \beta) ^{28}_{14}\text{Si}$

24.114 4.904×10^{-9} L/h **24.117** 87 yr **24.119** (a) 0.15 Bq/L (b) 0.34 Bq/L (c) 4.7 days more; 13.2 days total **24.122** 6.6 s **24.125** 1926 **24.129** (a) 1.80×10^{14} kJ (b) 6.7×10^{16} kJ (c) procedure (b) **24.135** 9.316×10^{2} MeV

GLOSSARY

Numbers in parentheses refer to the page(s) on which a term is introduced and/or discussed.

A

absolute scale (also *Kelvin scale*)   The preferred temperature scale in scientific work, the absolute scale takes absolute zero (0 K, or $-273.15°C$) as the lowest temperature. (24) [See also *kelvin (K)*.]

absorption spectrum   The spectrum produced when atoms absorb specific wavelengths of incoming light as they become excited from lower to higher energy levels. (267)

abundance   The amount of an element in a particular region of the universe. (951)

accuracy   The closeness of a measurement to the actual value. (31)

acid   In common laboratory terms, any species that produces H^+ ions when dissolved in water. (141) [See also *classical (Arrhenius), Brønsted-Lowry,* and *Lewis acid-base definitions.*]

acid anhydride   A compound, sometimes formed by a dehydration-condensation reaction of an oxoacid, that yields two molecules of the acid when it reacts with water. (637)

acid-base buffer (also *buffer*)   A solution that resists changes in pH when a small amount of either strong acid or strong base is added. (806)

acid-base indicator   A species whose color is different in acid and in base, which is used to monitor the equivalence point of a titration or the pH of a solution. (143, 768, 815)

acid-base reaction   Any reaction between an acid and a base. (140) (See also *neutralization reaction.*)

acid-base titration curve   A plot of the pH of a solution of acid (or base) versus the volume of base (or acid) added to the solution. (815)

acid-dissociation (acid-ionization) constant (K_a)   An equilibrium constant for the dissociation of an acid (HA) in H_2O to yield the conjugate base (A^-) and H_3O^+: (761)

$$K_a = \frac{[H_3O^+][A^-]}{[HA]}$$

actinides (also *actinoids*)   The Period 7 elements that form the second inner transition series ($5f$ block), including thorium (Th; $Z = 90$) through lawrencium (Lr; $Z = 103$). (303, 1006)

activated complex   (See *transition state.*)

activation energy (E_a)   The minimum energy with which molecules must collide to react. (683)

active site   The region of an enzyme formed by specific amino acid side chains at which catalysis occurs. (700)

activity ($\mathscr{A}$) (also *decay rate*)   The change in number of nuclei ($\mathscr{N}$) of a radioactive sample divided by the change in time (t). (1050)

activity series of the metals   A listing of metals arranged in order of decreasing strength of the metal as a reducing agent in aqueous reactions. (160)

actual yield   The amount of product actually obtained in a chemical reaction. (112)

addition polymer (also *chain-reaction,* or *chain-growth, polymer*)   A polymer formed when monomers (usually containing $C=C$) combine through an addition reaction. (642)

addition reaction   A type of organic reaction in which atoms linked by a multiple bond become bonded to more atoms. (625)

adduct   The product of a Lewis acid-base reaction characterized by the formation of a new covalent bond. (792)

adenosine triphosphate (ATP)   A high-energy molecule that serves most commonly as a store and source of energy in organisms. (879)

alchemy   An occult study of nature that flourished for 1500 years in northern Africa and Europe and resulted in the development of several key laboratory methods. (8)

alcohol   An organic compound (ending, *-ol*) that contains a

$$-\overset{|}{\underset{|}{C}}-\overset{..}{\underset{..}{O}}-H \text{ functional group. (628)}$$

aldehyde   An organic compound (ending, *-al*) that contains the carbonyl functional group ($C=\overset{..}{\underset{..}{O}}$) in which the carbonyl C is also bonded to H. (633)

alkane   A hydrocarbon that contains only single bonds (general formula, C_nH_{2n+2}). (613)

alkene   A hydrocarbon that contains at least one $C=C$ bond (general formula, C_nH_{2n}). (618)

alkyl group   A saturated hydrocarbon chain with one bond available. (624)

alkyl halide   (See *haloalkane.*)

alkyne   A hydrocarbon that contains at least one $C\equiv C$ bond (general formula, C_nH_{2n-2}). (621)

allotrope   One of two or more crystalline or molecular forms of an element. In general, one allotrope is more stable than another at a particular pressure and temperature. (563)

alloy   A mixture with metallic properties that consists of solid phases of the pure elements, a solid/solid solution, or distinct intermediate phases. (350, 492, 967)

alpha (α) decay   A radioactive process in which an alpha particle is emitted from a nucleus. (1044)

alpha particle (α or ^4_2He)   A positively charged particle, identical to a helium nucleus, that is one of the common types of radioactive emissions. (1043)

amide   An organic compound that contains the $-\overset{\overset{\displaystyle :O:}{\|}}{C}-\overset{|}{\underset{..}{N}}-$ functional group. (636)

amine   An organic compound (general formula, $R-\overset{..}{\underset{|}{N}}-$) derived structurally by replacing one or more H atoms of ammonia with alkyl groups; a weak organic base. (631, 780)

amino acid   An organic compound [general formula, $H_2N-CH(R)-COOH$] with at least one carboxyl and one amine group on the same molecule; the monomer unit of a protein. (646)

amorphous solid   A solid that occurs in different shapes because it lacks a well-defined crystal structure. (445)

ampere (A) The SI unit of electric current; 1 ampere of current results when 1 coulomb flows through a conductor in 1 second. (935)

amphoteric Able to act as either an acid or a base. (313)

amplitude The height of the crest (or depth of the trough) of a wave; related to the intensity of the energy. (256)

angular momentum quantum number (l) (also *orbital-shape quantum number*) An integer from 0 to $n-1$ that is related to the shape of an atomic orbital. (275)

anion A negatively charged ion. (59)

anode The electrode at which oxidation occurs in an electrochemical cell. Electrons are given up by the reducing agent and leave the cell at the anode. (898)

antibonding MO A molecular orbital formed when wave functions are subtracted from each other, which decreases electron density between the nuclei and leaves a node. Electrons occupying such an orbital destabilize the molecule. (404)

apatite A compound of general formula $Ca_5(PO_4)_3X$, where X is generally F, Cl, or OH. (960)

aqueous solution A solution in which water is the solvent. (74)

aromatic hydrocarbon A compound of C and H with one or more rings of C atoms (often drawn with alternating C—C and C=C bonds), in which there is extensive delocalization of π electrons. (622)

Arrhenius acid-base definition [See *classical (Arrhenius) acid-base definition*.]

Arrhenius equation An equation that expresses the exponential relationship between temperature and the rate constant: $k = Ae^{-E_a/RT}$. (682)

atmosphere The mixture of gases that extends from a planet's surface and eventually merges with outer space; the gaseous portion of the Earth's crust. (204, 953) (For unit, see *standard atmosphere*.)

atom The smallest particle of an element that retains the chemical nature of the element. A neutral, spherical entity composed of a positively charged central nucleus surrounded by one or more negatively charged electrons. (46)

atomic mass (also *atomic weight*) The average of the masses of the naturally occurring isotopes of an element weighted according to their abundances. (53)

atomic mass unit (amu) (also *dalton, D*) A mass exactly equal to $\frac{1}{12}$ the mass of a carbon-12 atom. (53)

atomic number (Z) The unique number of protons in the nucleus of each atom of an element (equal to the number of electrons in the neutral atom). An integer that expresses the positive charge of a nucleus in multiples of the electronic charge. (52)

atomic orbital (also *wave function*) A mathematical expression that describes the motion of the electron's matter-wave in terms of time and position in the region of the nucleus. The term is used qualitatively to mean the region of space in which there is a high probability of finding the electron. (273)

atomic size A term referring to the atomic radius, one-half the distance between nuclei of identical bonded elements. (304, 533) (See also *covalent radius* and *metallic radius*.)

atomic solid A solid consisting of individual atoms held together by dispersion forces; the frozen noble gases are the only examples. (452)

atomic symbol (also *element symbol*) A one- or two-letter abbreviation for the English, Latin, or Greek name of an element. (52)

aufbau principle (also *building-up principle*) The conceptual basis of a process of building up atoms by adding one proton at a time to the nucleus and one electron around it to obtain the ground-state electron configurations of the elements. (295)

autoionization (also *self-ionization*) A reaction in which two molecules of a substance react to give ions. The most important example is for water: (764)

$$2H_2O(l) \rightleftharpoons H_3O^+(aq) + OH^-(aq)$$

average rate The change in concentration of reactants (or products) divided by a finite time period. (668)

Avogadro's law The gas law stating that, at fixed temperature and pressure, equal volumes of any ideal gas contain equal numbers of particles and, therefore, the volume of a gas is directly proportional to its amount (mol): $V \propto n$. (184)

Avogadro's number A number (6.022×10^{23} to four significant figures) equal to the number of atoms in exactly 12 g of carbon-12; the number of atoms, molecules, or formula units in one mole of an element or compound. (87)

axial group An atom (or group) that lies above or below the trigonal plane of a trigonal bipyramidal molecule, or a similar structural feature in a molecule. (375)

B

background radiation Natural ionizing radiation, the most important form of which is cosmic radiation. (1061)

balancing coefficient (also *stoichiometric coefficient*) A numerical multiplier of all the atoms in the formula immediately following it in a chemical equation. (102)

band of stability The narrow band of stable nuclides that appears on a plot of number of neutrons versus number of protons for all nuclides. (1046)

band theory An extension of molecular orbital (MO) theory that explains many properties of metals, in particular, the differences in electrical conductivity of conductors, semiconductors, and insulators. (456)

barometer A device used to measure atmospheric pressure. Most commonly, a tube open at one end, which is filled with mercury and inverted into a dish of mercury. (176)

base In common laboratory terms, any species that produces OH^- ions when dissolved in water. (141) [See also *classical (Arrhenius), Brønsted-Lowry,* and *Lewis acid-base definitions.*]

base-dissociation (base-ionization) constant (K_b) An equilibrium constant for the reaction of a base (B) with H_2O to yield the conjugate acid (BH^+) and OH^-: (779)

$$K_b = \frac{[BH^+][OH^-]}{[B]}$$

base pair Two complementary bases in mononucleotides that are H bonded to each other; guanine (G) always pairs with cytosine (C), and adenine (A) always pairs with thymine (T) (or uracil, U). (651)

base unit (also *fundamental unit*) A unit that defines the standard for one of the seven physical quantities in the International System of Units (SI). (17)

basic-oxygen process The method used to convert pig iron to steel, in which O_2 is blown over and through molten iron to oxidize impurities and decrease the content of carbon. (972)

battery   A self-contained group of voltaic cells arranged in series. (922)

becquerel (Bq)   The SI unit of radioactivity; 1 Bq = 1 d/s (disintegration per second). (1050)

bent shape (also *V shape*)   A molecular shape that arises when a central atom is bonded to two other atoms and has one or two lone pairs; occurs as the AX_2E shape class (bond angle < 120°) in the trigonal planar arrangement and as the AX_2E_2 shape class (bond angle < 109.5°) in the tetrahedral arrangement. (373)

beta (β) decay   A radioactive process in which a beta particle is emitted from a nucleus. (1045)

beta particle (β or $_{-1}^{0}\beta$)   A negatively charged particle identified as a fast-moving electron that is one of the common types of radioactive emissions. (1043)

bimolecular reaction   An elementary reaction involving the collision of two reactant species. (692)

binary covalent compound   A compound that consists of atoms of two elements in which bonding occurs primarily through electron sharing. (70)

binary ionic compound   A compound that consists of the oppositely charged ions of two elements. (59)

biomass conversion   The process of applying chemical and biological methods to convert plant and/or animal matter into fuels. (243)

biosphere   The living systems that have inhabited the Earth. (953)

blast furnace   A tower-shaped furnace made of brick material in which intense heat and blasts of air are used to convert iron ore and coke to iron metal and carbon dioxide. (970)

body-centered cubic unit cell   A unit cell in which a particle lies at each corner and in the center of a cube. (446)

boiling point (bp or T_b)   The temperature at which the vapor pressure of a gas equals the external (atmospheric) pressure. (428)

boiling point elevation (ΔT_b)   The increase in the boiling point of a solvent caused by the presence of dissolved solute. (508)

bond angle   The angle formed by the nuclei of two surrounding atoms with the nucleus of the central atom at the vertex. (371)

bond energy (BE) (also *bond strength*)   The enthalpy change accompanying the breakage of a given bond in a mole of gaseous molecules. (338, 535)

bond length   The distance between the nuclei of two bonded atoms. (340, 535)

bond order   The number of electron pairs shared by two bonded atoms. (338, 535)

bonding MO   A molecular orbital formed when wave functions are added to each other, which increases electron density between the nuclei. Electrons occupying such an orbital stabilize the molecule. (404)

bonding pair (also *shared pair*)   An electron pair shared by two nuclei; the mutual attraction between the nuclei and the electron pair forms a covalent bond. (338)

Born-Haber cycle   A series of hypothetical steps and their enthalpy changes needed to convert elements to ionic compound and devised to calculate the lattice energy. (332)

Boyle's law   The gas law stating that, at constant temperature and amount of gas, the volume occupied by a gas is inversely proportional to the applied (external) pressure: $V \propto 1/P$. (181)

branch   A side chain appended to a polymer backbone. (472)

bridge bond (also *three-center, two-electron bond*)   A covalent bond in which three atoms are held together by two electrons. (561)

Brønsted-Lowry acid-base definition   A model of acid-base behavior based on proton transfer, in which an acid and a base are defined, respectively, as species that donate and accept a proton. (768)

buffer   (See *acid-base buffer.*)

buffer capacity   A measure of the ability of a buffer to resist a change in pH; related to the concentrations and relative proportions of buffer components. (812)

buffer range   The pH range over which a buffer acts effectively. (812)

C

calibration   The process of correcting for systematic error of a measuring device by comparing it to a known standard. (31)

calorie (cal)   A unit of energy defined as exactly 4.184 joules; originally defined as the heat needed to raise the temperature of 1 g of water 1°C (from 14.5°C to 15.5°C). (225)

calorimeter   A device used to measure the heat released or absorbed by a physical or chemical process taking place within it. (234)

capillarity (or *capillary action*)   A property that results in a liquid rising through a narrow space. (439)

carbon steel   The steel that is produced by the basic-oxygen process, contains about 1% to 1.5% C and other impurities, and is alloyed with metals that prevent corrosion and increase strength. (972)

carbonyl group   The C=O grouping of atoms. (633)

carboxylic acid   An organic compound (ending, *-oic acid*) that contains the $-\overset{\displaystyle :O:}{\underset{\displaystyle ||}{C}}-\ddot{O}H$ group. (635)

catalyst   A substance that increases the rate of a reaction without being used up in the process. (697)

cathode   The electrode at which reduction occurs in an electrochemical cell. Electrons enter the cell and are acquired by the oxidizing agent at the cathode. (898)

cathode ray   The ray of light emitted by the cathode (negative electrode) in a gas discharge tube; travels in straight lines, unless deflected by magnetic or electric fields. (48)

cation   A positively charged ion. (59)

cell potential (E_{cell}) (also *electromotive force*, or *emf; cell voltage*)   The potential difference between the electrodes of an electrochemical cell when no current flows. (905)

Celsius scale (formerly *centigrade scale*)   A temperature scale in which the freezing and boiling points of water are defined as 0°C and 100°C, respectively. (24)

ceramic   A nonmetallic material that is hardened by heating it to high temperatures and, in most cases, consists of silicate microcrystals suspended in a glassy cementing medium. (465)

chain reaction   In nuclear fission, a self-sustaining process in which neutrons released by splitting of one nucleus cause other nuclei to split, which releases more neutrons, and so on. (1070)

change in enthalpy (ΔH)   The change in internal energy plus the product of the constant pressure and the change in volume: $\Delta H = \Delta E + P\Delta V$; the heat lost or gained at constant pressure: $\Delta H = q_P$. (228)

charge density The ratio of the charge of an ion to its volume. (495)

Charles's law The gas law stating that at constant pressure and amount occupied by gas, the volume of a gas is directly proportional to its absolute temperature: $V \propto T$. (182)

chelate A complex ion in which the metal ion is bonded to a bidentate or polydentate ligand. (1016)

chemical bond The force that holds two atoms together in a molecule (or formula unit). (59)

chemical change (also *chemical reaction*) A change in which a substance is converted into a substance with different composition and properties. (3)

chemical equation A statement that uses chemical formulas to express the identities and quantities of the substances involved in a chemical or physical change. (101)

chemical formula A notation of atomic symbols and numerical subscripts that shows the type and number of each atom in a molecule or formula unit of a substance. (64)

chemical kinetics The study of the rates and mechanisms of reactions. (664)

chemical property A characteristic of a substance that appears as it interacts with, or transforms into, other substances. (3)

chemical reaction (See *chemical change*.)

chemistry The scientific study of matter and the changes it undergoes. (3)

chiral molecule One that is not superimposable on its mirror image; an optically active molecule. In organic compounds, a chiral molecule typically contains a C atom bonded to four different groups (asymmetric C). (617)

chlor-alkali process An industrial method that electrolyzes concentrated aqueous NaCl and produces Cl_2, H_2, and NaOH. (990)

chromatography A separation technique in which a mixture is dissolved in a fluid (gas or liquid), and the components are separated through differences in adsorption to (or solubility in) a solid surface (or viscous liquid). (76)

***cis-trans* isomers** (See *geometric isomers*.)

classical (Arrhenius) acid-base definition A model of acid-base behavior in which an acid is a substance that has H in its formula and produces H^+ in water, and a base is a substance that has OH in its formula and produces OH^- in water. (759)

Clausius-Clapeyron equation An equation that expresses the relationship between vapor pressure P of a liquid and temperature T: (428)

$$\ln P = \frac{-\Delta H_{vap}}{R}\left(\frac{1}{T}\right) + C, \text{ where } C \text{ is a constant}$$

coal gasification An industrial process for altering the large molecules in coal to sulfur-free gaseous fuels. (243)

colligative property A property of a solution that depends on the number, not the identity, of solute particles. (506) (See also *boiling point elevation, freezing point depression, osmotic pressure,* and *vapor pressure lowering*.)

collision frequency The average number of collisions per second that a particle undergoes. (203)

collision theory A model that explains reaction rate as the result of particles colliding with a certain minimum energy. (685)

colloid A suspension in which a solute-like phase is dispersed throughout a solvent-like phase. (517)

combustion The process of burning in air, often with release of heat and light. (10)

combustion analysis A method for determining the formula of a compound from the amounts of its combustion products. (98)

common-ion effect The shift in the position of an ionic equilibrium away from an ion involved in the process that is caused by the addition or presence of that ion. (806)

complex (See *coordination compound*.)

complex ion An ion consisting of a central metal ion bonded covalently to molecules and/or anions called ligands. (835, 1013)

composition The types and amounts of simpler substances that make up a sample of matter. (3)

compound A substance composed of two or more elements that are chemically combined in fixed proportions. (42)

concentration A measure of the quantity of solute dissolved in a given quantity of solution. (114)

concentration cell A voltaic cell in which both compartments contain the same components but at different concentrations. (919)

condensation The process of a gas changing into a liquid. (421)

condensation polymer A polymer formed by monomers with two functional groups that are linked together in a dehydration-condensation reaction. (643)

conduction band In band theory, the empty, higher energy portion of the band of molecular orbitals into which electrons move when conducting heat and electricity. (458)

conductor A substance (usually a metal) that conducts an electric current well regardless of the temperature. (458)

conjugate acid-base pair Two species related to each other through the gain or loss of a proton; the acid has one more proton than its conjugate base. (769)

constitutional isomers (also *structural isomers*) Compounds with the same molecular formula but different arrangements of atoms. (615, 1020)

contact process An industrial process for the manufacture of sulfuric acid based on the catalyzed oxidation of SO_2. (987)

controlled experiment An experiment that measures the effect of one variable at a time by keeping other variables constant. (12)

conversion factor A ratio of equivalent quantities that is equal to 1 and used to convert the units of a quantity. (13)

coordinate covalent bond A covalent bond formed when one atom donates both electrons to form the shared pair and, once formed, is identical to any single bond. (1023)

coordination compound (also *complex*) A substance containing at least one complex ion. (1013)

coordination isomers Two or more coordination compounds with the same composition in which the complex ions have different ligand arrangements. (1020)

coordination number In a crystal, the number of nearest neighbors surrounding a particle. (446) In a complex, the number of ligand atoms bonded to the central metal ion. (1014)

copolymer A polymer that consists of two or more types of monomer. (473)

core The dense, innermost region of the Earth. (951)

core electrons (See *inner electrons*.)

corrosion   The natural redox process that results in unwanted oxidation of a metal. (926)

coulomb (C)   The SI unit of electric charge. One coulomb is the charge of 6.242×10^{18} electrons; one electron possesses a charge of 1.602×10^{-19} C. (905)

Coulomb's law   A law stating that the electrostatic force associated with two charges A and B is directly proportional to the product of their magnitudes and inversely proportional to the square of the distance between them: (333)

$$\text{force} \propto \frac{\text{charge A} \times \text{charge B}}{(\text{distance})^2}$$

counter ion   A simple ion associated with a complex ion in a coordination compound. (1013)

coupling of reactions   The pairing of reactions of which one releases enough free energy for the other to occur. (878)

covalent bond   A type of bond in which atoms are bonded through the sharing of two electrons; the mutual attraction of the nuclei and an electron pair that holds atoms together in a molecule. (62, 338)

covalent bonding   The idealized bonding type that is based on localized electron-pair sharing between two atoms with little difference in their tendencies to lose or gain electrons (most commonly nonmetals). (328, 534)

covalent compound   A compound that consists of atoms bonded together by shared electron pairs. (59)

covalent radius   One-half the distance between nuclei of identical covalently bonded atoms. (304)

critical mass   The minimum mass needed to achieve a chain reaction. (1070)

critical point   The point on a phase diagram above which the vapor cannot be condensed to a liquid; the end of the liquid-gas curve. (430)

crosslink   A branch that covalently joins one polymer chain to another. (472)

crust   The thin, light, heterogeneous outer layer of the Earth. (951)

crystal defect   Any of a variety of disruptions in the regularity of a crystal structure. (460)

crystal field splitting energy (Δ)   The difference in energy between two sets of metal-ion d orbitals that results from electrostatic interactions with the surrounding ligands. (1026)

crystal field theory   A model that explains the color and magnetism of coordination compounds based on the effects of ligands on metal-ion d-orbital energies. (1025)

crystalline solid   Solid with a well-defined shape because of the orderly arrangement of the atoms, molecules, or ions. (445)

crystallization   A technique used to separate and purify the components of a mixture through differences in solubility in which a component comes out of solution as crystals. (75)

cubic closest packing   A crystal structure based on the face-centered cubic unit cell in which the layers have an ABCABC . . . pattern. (448)

cubic meter (m^3)   The SI derived unit of volume. (19)

curie (Ci)   The most common unit of radioactivity, defined as the number of nuclei disintegrating each second in 1 g of radium-226; 1 Ci = 3.70×10^{10} d/s (disintegrations per second). (1050)

cyclic hydrocarbon   A hydrocarbon with one or more rings in its structure. (615)

D

d orbital   An atomic orbital with $l = 2$. (280)

dalton (D)   A unit of mass identical to *atomic mass unit*. (53)

Dalton's law of partial pressures   A gas law stating that, in a mixture of unreacting gases, the total pressure is the sum of the partial pressures of the individual gases: $P_{\text{total}} = P_1 + P_2 + \cdots$. (192)

data   Pieces of quantitative information obtained by observation. (12)

de Broglie wavelength   The wavelength of a moving particle obtained from the de Broglie equation: $\lambda = h/mu$. (270)

decay constant   The rate constant k for radioactive decay. (1050)

decay series (also *disintegration series*)   The succession of steps a parent nucleus undergoes as it decays into a stable daughter nucleus. (1049)

degree of polymerization (n)   The number of monomer repeat units in a polymer chain. (468)

dehydration-condensation reaction   A reaction in which H and OH groups on two molecules react to form water as one of the products. (580)

delocalization   (See *electron-pair delocalization.*)

density (d)   An intensive physical property of a substance at a given temperature and pressure, defined as the ratio of the mass to the volume: $d = m/V$. (22)

deposition   The process of changing directly from gas to solid. (422)

derived unit   Any of various combinations of the seven SI base units. (17)

desalination   A process used to remove large amounts of ions from seawater, usually by reverse osmosis. (521)

deuterons   Nuclei of the stable hydrogen isotope deuterium, ^{2}H. (1056)

diagonal relationship   Physical and chemical similarities between a Period 2 element and one located diagonally down and to the right in Period 3. (553)

diamagnetism   The tendency of a species not to be attracted (or to be slightly repelled) by a magnetic field as a result of its electrons being paired. (316, 1005)

differentiation   The geochemical process of forming regions in the Earth based on differences in composition and density. (951)

diffraction   The phenomenon in which a wave striking the edge of an object bends around it. A wave passing through a slit as wide as its wavelength forms a circular wave. (259)

diffusion   The movement of one fluid through another. (202)

dimensional analysis (also *factor-label method*)   A calculation method in which arithmetic steps are accompanied by the appropriate canceling of units. (15)

dipole-dipole force   The intermolecular attraction between oppositely charged poles of nearby polar molecules. (432)

dipole–induced dipole force   The intermolecular attraction between a polar molecule and the oppositely charged pole it induces in a nearby molecule. (487)

dipole moment (μ)   A measure of molecular polarity; the magnitude of the partial charges on the ends of a molecule (in coulombs) times the distance between them (in meters). (381)

disaccharide   An organic compound formed by a dehydration-condensation reaction between two simple sugars (monosaccharides). (645)

disintegration series (See *decay series.*)

dispersion force (also *London force*) The intermolecular attraction between all molecules as a result of instantaneous polarizations of their electron clouds; the force primarily responsible for the condensed states of nonpolar substances. (436)

disproportionation reaction A reaction in which a given substance is both oxidized and reduced. (578)

distillation A separation technique in which a more volatile component of a mixture vaporizes and condenses separately from the less volatile components. (75)

donor atom An atom that donates a lone pair of electrons to form a covalent bond, usually from ligand to metal ion in a complex. (1015)

doping Adding small amounts of other elements into the crystal structure of a semiconductor to enhance a specific property, usually conductivity. (460)

double bond A covalent bond that consists of two bonding pairs; two atoms sharing four electrons in the form of one σ and one π bond. (338, 401)

double helix The two intertwined polynucleotide strands held together by H bonds that form the structure of DNA (deoxyribonucleic acid). (651)

Downs cell An industrial apparatus that electrolyzes molten NaCl to produce sodium and chlorine. (969)

dynamic equilibrium The condition at which the forward and reverse reactions are taking place at the same rate, so there is no net change in the amounts of reactants or products. (162)

E

e_g orbitals The set of orbitals (composed of $d_{x^2-y^2}$ and d_{z^2}) that results when the energies of the metal-ion d orbitals are split by a ligand field. This set is higher in energy than the other (t_{2g}) set in an octahedral field of ligands and lower in energy in a tetrahedral field. (1026)

effective collision A collision in which the particles meet with sufficient energy and an orientation that allows them to react. (687)

effective nuclear charge (Z_{eff}) The nuclear charge an electron actually experiences as a result of shielding effects by other electrons. (293)

effusion The process by which a gas escapes from its container through a tiny hole into an evacuated space. (201)

elastomer A polymeric material that can be stretched and springs back to its original shape when released. (472)

electrochemical cell A system that incorporates a redox reaction to produce or use electrical energy. (893)

electrochemistry The study of the relationship between chemical change and electrical work. (893)

electrode The part of an electrochemical cell that conducts the electricity between the cell and the surroundings. (898)

electrolysis The nonspontaneous lysing (splitting) of a substance, often to its component elements, by supplying electrical energy. (931)

electrolyte A substance that conducts a current when it dissolves in water. (132, 506) A mixture of ions, in which the electrodes of an electrochemical cell are immersed, that conducts a current. (898)

electrolytic cell An electrochemical system that uses electrical energy to drive a nonspontaneous chemical reaction ($\Delta G > 0$). (898)

electromagnetic (EM) radiation (also *electromagnetic energy* or *radiant energy*) Oscillating, perpendicular electric and magnetic fields moving simultaneously through space as waves and manifested as visible light, x-rays, microwaves, radio waves, and so on. (255)

electromagnetic spectrum The continuum of wavelengths of radiant energy. (257)

electromotive force (emf) (See *cell potential.*)

electron (e^-) A subatomic particle that possesses a unit negative charge (1.602×10^{-19} C) and occupies the space around the atomic nucleus. (52)

electron affinity (EA) The energy change accompanying one mole of electrons being added to one mole of gaseous atoms or ions. (310)

electron capture A type of radioactive decay in which a nucleus draws in an orbital electron, usually one from the lowest energy level, and releases energy. (1045)

electron cloud An imaginary representation of an electron's rapidly changing position over time around the nucleus. (274)

electron configuration The distribution of electrons within the orbitals of the atoms of an element; also the notation for such a distribution. (289, 532)

electron deficient Referring to a bonded atom, such as Be or B, that has fewer than eight valence electrons. (365)

electron density diagram The pictorial representation for a given energy sublevel of the quantity ψ^2 (the probability that the electron lies within a particular tiny volume) as a function of r (distance from the nucleus). (274)

electron-pair delocalization (also *delocalization*) The process by which electron density is spread over several atoms rather than remaining between two. (362)

electron-sea model A qualitative description of metallic bonding proposing that metal atoms pool their valence electrons into a delocalized "sea" of electrons in which the metal cores (metal ions) are submerged in an orderly array. (349)

electron volt (eV) The energy (in joules, J) that an electron acquires when it moves through a potential difference of 1 volt; 1 eV $= 1.602\times10^{-19}$ J. (1067)

electronegativity (EN) The relative ability of a bonded atom to attract shared electrons. (344, 533)

electronegativity difference (ΔEN) The difference in electronegativities between the atoms in a bond. (347)

electrorefining An industrial electrolytic process in which a sample of impure metal acts as the anode and a sample of the pure metal acts as the cathode. (967)

element The simplest type of substance with unique physical and chemical properties. An element consists of only one kind of atom, so it cannot be broken down into any simpler substances. (41)

elementary reaction (also *elementary step*) A simple reaction that describes a single molecular event in a proposed reaction mechanism. (692)

elimination reaction A type of organic reaction in which C atoms in the product are bonded to fewer atoms than in the reactant, which leads to multiple bonding. (626)

emission spectrum The line spectrum produced when excited atoms emit photons characteristic of the element. (267)

empirical formula A chemical formula that shows the lowest relative number of atoms of each element in a compound. (64)

enantiomers (See *optical isomers.*)

end point The point in a titration at which the indicator changes color. (143, 816)

end-to-end overlap Overlap of *s, p,* or hybrid atomic orbitals that yields a sigma (σ) bond. (535)

endothermic Occurring with an absorption of heat from the surroundings and therefore an increase in the enthalpy of the system ($\Delta H > 0$). (229)

energy The capacity to do work, that is, to move matter. (6) [See also *kinetic energy* (E_k) and *potential energy* (E_p).]

enthalpy (*H*) A thermodynamic quantity that is the sum of the internal energy plus the product of the pressure and volume. (228)

enthalpy diagram A graphic depiction of the enthalpy change of a system. (229)

enthalpy of hydration (ΔH_{hydr}) (See *heat of hydration.*)

enthalpy of solution (ΔH_{soln}) (See *heat of solution.*)

entropy (*S*) A thermodynamic quantity that is a measure of the disorder of a system; related to the number of ways the components of a system can be arranged without changing the system's energy. (496, 860)

environmental cycle The physical, chemical, and biological paths through which the atoms of an element move within the Earth's crust. (956)

enzyme A biological macromolecule (usually a protein) that acts as a catalyst. (700)

enzyme-substrate complex (ES) The intermediate in an enzyme-catalyzed reaction, which consists of enzyme and substrate(s) and whose concentration determines the rate of product formation. (701)

equatorial group An atom (or group) that lies in a trigonal plane of a trigonal bipyramidal molecule, or a similar structural feature in a molecule. (375)

equilibrium constant (*K*) The value obtained when equilibrium concentrations are substituted into the reaction quotient. (716)

equivalence point The point in a titration when the number of moles of the added species is stoichiometrically equivalent to the original number of moles of the other species. (143, 816)

ester An organic compound that contains the $-\overset{\displaystyle :O:}{\underset{\displaystyle |}{\overset{\displaystyle \|}{C}}}-\overset{..}{\underset{..}{O}}-\overset{|}{\underset{|}{C}}-$ group. (636)

exact number A quantity, usually obtained by counting or based on a unit definition, that has no uncertainty associated with it and, therefore, contains as many significant figures as a calculation requires. (30)

excitation The process by which a substance absorbs energy from low-energy radioactive particles causing its electrons to move to higher energy levels. (1058)

excited state Any electron configuration of an atom or molecule other than the lowest energy (ground) state. (264)

exclusion principle A principle developed by Wolfgang Pauli stating that no two electrons in an atom can have the same set of four quantum numbers. The principle arises from the fact that an orbital has a maximum occupancy of two electrons and their spins are paired. (291)

exothermic Occurring with a release of heat to the surroundings and therefore a decrease in the enthalpy of the system ($\Delta H < 0$). (229)

expanded valence shell A valence level that can accommodate more than 8 electrons by using available *d* orbitals; occurs only for elements in Period 3 or higher. (366)

experiment A clear set of procedural steps that tests a hypothesis. (12)

extensive property A property, such as mass, that depends on the quantity of substance present. (23)

extraction A method of separating the components of a mixture based on differences in their solubility in a nonmiscible solvent. (75)

F

face-centered cubic unit cell A unit cell in which a particle occurs at each corner and in the center of each face of a cube. (446)

Faraday constant (*F*) The physical constant representing the charge of 1 mol of electrons: $F = 96{,}485$ C/mol e$^-$. (914)

fatty acid A carboxylic acid with a long hydrocarbon chain that is derived from a natural source. (636)

filtration A method of separating the components of a mixture on the basis of differences in particle size. (75)

first law of thermodynamics (See *law of conservation of energy.*)

fission The process by which a heavier nucleus splits into lighter nuclei with the release of energy. (1066)

fixation A chemical/biochemical process that converts a gaseous substance in the environment into a form that can be used by organisms. (957)

flame test A procedure for identifying the presence of metal ions in which a granule of a compound or a drop of its solution is placed in a flame to observe a characteristic color. (267)

flotation A metallurgical process in which oil and detergent are mixed with pulverized ore in water to create a slurry that separates the mineral from the gangue. (964)

formal charge The hypothetical charge on an atom in a molecule or ion. The number of valence electrons minus the sum of all the unshared and half the shared valence electrons. (364)

formation constant (K_f) An equilibrium constant for the formation of a complex ion from the hydrated metal ion and ligands. (835)

formation equation An equation in which 1 mol of a compound forms from its elements. (240)

formula unit The chemical unit of a compound that contains the number and type of atoms (or ions) expressed in the chemical formula. (65)

fossil fuel Any fuel, including coal, petroleum, and natural gas, derived from the products of the decay of dead organisms. (243)

fraction by mass (also *mass fraction*) The portion of a compound's mass contributed by an element; the mass of an element in a compound divided by the mass of the compound. (44)

fractional distillation A process involving numerous vaporization-condensation steps used to separate two or more volatile components. (515)

free energy (*G*) A thermodynamic quantity that is the difference between the enthalpy and the product of the absolute temperature and the entropy: $G = H - TS$. (872)

free radical A molecular or atomic species with one or more unpaired electrons, which typically make it very reactive. (365, 1059)

freezing   The process of cooling a liquid until it solidifies. (421)

freezing point depression (ΔT_f)   A lowering of the freezing point of a solvent caused by the presence of dissolved solute particles. (510)

frequency (ν)   The number of cycles a wave undergoes per second. (256)

frequency factor (A)   The product of the collision frequency Z and an orientation probability factor p specific for a reaction. (687)

fuel cell (also *flow battery*)   A battery that is not self-contained and in which electricity is generated by the controlled oxidation of a fuel. (922)

functional group   A specific combination of atoms, typically containing a carbon-carbon multiple bond and/or carbon-heteroatom bond, that reacts in a characteristic way no matter in what molecule it occurs. (609)

fundamental unit   (See *base unit*.)

fusion   (See *melting*.)

fusion (nuclear)   The process by which light nuclei combine to form a heavier nucleus with the release of energy. (1066)

G

galvanic cell   (See *voltaic cell*.)

gamma emission   The type of radioactive decay in which gamma rays are emitted from an excited nucleus. (1045)

gamma (γ) ray   A very high-energy photon. (1043)

gangue   In an ore, the debris, such as sand, rock, and clay, attached to the mineral. (963)

gas   One of the three states of matter. A gas fills its container regardless of the shape. (4)

Geiger-Müller counter   An ionization counter that detects radioactive emissions through their ionization of gas atoms within the instrument. (1051)

genetic code   The set of three-base sequences that is translated into specific amino acids during the process of protein synthesis. (652)

geometric isomers (also *cis-trans isomers* or *diastereomers*)   Stereoisomers in which the molecules have the same connections between atoms but differ in the spatial arrangements of the atoms. The *cis-* isomer has similar groups on the same side of a structural feature; the *trans-* isomer has them on opposite sides. (619, 1021)

Graham's law of effusion   A gas law stating that the rate of effusion of a gas is inversely proportional to the square root of its density (or molar mass): (201)

$$\text{rate} \propto \frac{1}{\sqrt{\mathcal{M}}}$$

gray (Gy)   The SI unit of absorbed radiation dose; 1 Gy = 1 J/kg tissue. (1059)

ground state   The electron configuration of an atom or ion that is lowest in energy. (264)

group   A vertical column in the periodic table. (57)

H

Haber process   An industrial process used to form ammonia from its elements. (744)

half-cell   A portion of an electrochemical cell in which a half-reaction takes place. (901)

half-life ($t_{1/2}$)   In chemical processes, the time required for half the initial reactant concentration to be consumed. (680). In nuclear processes, the time required for half the initial number of nuclei to decay. (1050)

half-reaction method   A method of balancing redox reactions by treating the oxidation and reduction half-reactions separately. (894)

haloalkane (also *alkyl halide*)   A hydrocarbon with one or more halogen atoms(X) in place of H; contains a $-\overset{|}{\underset{|}{C}}-\ddot{X}:$ group. (630)

hard water   Water that contains large amounts of divalent cations, especially Ca^{2+} and Mg^{2+}. (520)

heat (q)   The energy transferred between objects because of differences in their temperatures only; thermal energy. (24, 223)

heat capacity   The quantity of heat required to change the temperature of an object by 1 K. (233)

heat of combustion (ΔH_{comb})   The enthalpy change occurring when 1 mol of a substance combines with oxygen in a combustion reaction. (230)

heat of formation (ΔH_f)   The enthalpy change occurring when 1 mol of a compound is produced from its elements. (230)

heat of fusion (ΔH_{fus})   The enthalpy change occurring when 1 mol of a solid substance melts. (230, 422)

heat of hydration (ΔH_{hydr}) (also *enthalpy of hydration*)   The enthalpy change occurring when 1 mol of a gaseous species is hydrated. The sum of the enthalpies of separating water molecules and mixing the gaseous solute with them. (494)

heat of reaction (ΔH_{rxn})   The enthalpy change of a reaction. (229)

heat of solution (ΔH_{soln}) (also *enthalpy of solution*)   The enthalpy change occurring when a solution forms from solute and solvent. The sum of the enthalpies from separating solute and solvent molecules and mixing them. (494)

heat of sublimation (ΔH_{subl})   The enthalpy change occurring when 1 mol of a solid substance changes directly to a gas. The sum of the heats of fusion and vaporization. (422)

heat of vaporization (ΔH_{vap})   The enthalpy change occurring when 1 mol of a liquid substance vaporizes. (230, 422)

heating-cooling curve   A plot of temperature vs. time for a substance when heat is absorbed or released by the system at a constant rate. (423)

Henderson-Hasselbalch equation   An equation for calculating the pH of a buffer system: (811)

$$pH = pK_a + \log\left(\frac{[\text{base}]}{[\text{acid}]}\right)$$

Henry's law   A law stating that the solubility of a gas in a liquid is directly proportional to the partial pressure of the gas above the liquid: $S_{gas} = k_H \times P_{gas}$. (500)

Hess's law of heat summation   A law stating that the enthalpy change of an overall process is the sum of the enthalpy changes of the individual steps of the process. (238)

heteroatom   Any atom in an organic compound other than C or H. (609)

heterogeneous catalyst   A catalyst that occurs in a different phase from the reactants, usually a solid interacting with gaseous or liquid reactants. (699)

heterogeneous mixture　A mixture that has one or more visible boundaries among its components. (74)

hexagonal closest packing　A crystal structure based on the hexagonal unit cell in which the layers have an ABAB . . . pattern. (448)

high-spin complex　Complex ion that has the same number of unpaired electrons as in the isolated metal ion; contains weak-field ligands. (1029)

homogeneous catalyst　A catalyst (gas, liquid, or soluble solid) that exists in the same phase as the reactants. (698)

homogeneous mixture (also *solution*)　A mixture that has no visible boundaries among its components. (74)

homologous series　A series of organic compounds in which each member differs from the next by a —CH_2— (methylene) group. (613)

homonuclear diatomic molecule　A molecule composed of two identical atoms. (407)

Hund's rule　A principle stating that when orbitals of equal energy are available, the electron configuration of lowest energy has the maximum number of unpaired electrons with parallel spins. (296)

hybrid orbital　An atomic orbital postulated to form during bonding by the mathematical mixing of specific combinations of nonequivalent orbitals in a given atom. (394)

hybridization　A postulated process of orbital mixing to form hybrid orbitals. (394)

hydrate　A compound in which a specific number of water molecules are associated with each formula unit. (68)

hydration　Solvation in water. (494)

hydration shell　The oriented cluster of water molecules that surrounds an ion in aqueous solution. (486)

hydrocarbon　An organic compound that contains only H and C atoms. (610)

hydrogen bond (H bond)　A type of dipole-dipole force that arises between molecules that have an H atom bonded to a small, highly electronegative atom with lone pairs, usually N, O, or F. (434, 541)

hydrogenation　The addition of hydrogen to a carbon-carbon multiple bond to form a carbon-carbon single bond. (699)

hydrolysis　Cleaving a molecule by reaction with water, in which one part of the molecule bonds to the water —OH and the other to the water H. (638)

hydronium ion (H_3O^+)　A proton covalently bonded to a water molecule. (758)

hydrosphere　The liquid portion of the Earth's crust. (953)

hypothesis　A testable proposal made to explain an observation. If inconsistent with experimental results, a hypothesis is revised or discarded. (12)

I

ideal gas　A hypothetical gas that exhibits linear relationships among volume, pressure, temperature, and amount (mol) at all conditions; approximated by simple gases at ordinary conditions. (180)

ideal gas law (also *ideal gas equation*)　An equation that expresses the relationships among volume, pressure, temperature, and amount (mol) of an ideal gas: $PV = nRT$. (185)

ideal solution　A solution whose vapor pressure equals the mole fraction of the solvent times the vapor pressure of the pure solvent; approximated only by very dilute solutions. (507) (See also *Raoult's law*.)

indicator　(See *acid-base indicator*.)

induced-fit model　A model of enzyme action that pictures the binding of the substrate as inducing the active site to change its shape and become catalytically active. (701)

infrared (IR)　The region of the electromagnetic spectrum between the microwave and visible regions. (257)

infrared (IR) spectroscopy　An instrumental technique for determining the types of bonds in a covalent molecule by measuring the absorption of IR radiation. (343)

initial rate　The instantaneous rate occurring as soon as the reactants are mixed, that is, at $t = 0$. (669)

inner electrons (also *core electrons*)　Electrons that fill all the energy levels of an atom except the valence level; electrons also present in atoms of the previous noble gas and any completed transition series. (302)

inner transition elements　The elements of the periodic table in which *f* orbitals are being filled; the lanthanides and actinides. (303, 1006)

instantaneous rate　The reaction rate at a particular time, given by the slope of a tangent to a plot of reactant concentration versus time. (668)

insulator　A substance (usually a nonmetal) that does not conduct an electric current. (459)

integrated rate law　A mathematical expression for reactant concentration as a function of time. (677)

intensive property　A property, such as density, that does not depend on the quantity of substance present. (23)

interhalogen compound　A compound consisting entirely of halogens. (592)

intermolecular forces (also *interparticle forces*)　The attractive and repulsive forces among the particles—molecules, atoms, or ions—in a sample of matter. (420)

internal energy (E)　The sum of the kinetic and potential energies of all the particles in a system. (222)

interstitial hydride　The substance formed when metals absorb H_2 into the spaces between their atoms. (543)

ion　A charged particle that forms from an atom (or covalently bound group of atoms) when it gains or loses one or more electrons. (59)

ion-dipole force　The intermolecular attractive force between an ion and a polar molecule (dipole). (432)

ion exchange　A process of softening water by exchanging one type of ion (usually Ca^{2+}) for another (usually Na^+) by binding on a specially designed resin. (520)

ion–induced dipole force　The intermolecular attractive force between an ion and the dipole it induces in the electron cloud of a nearby particle. (486)

ion pair　A pair of ions that form a gaseous ionic molecule; sometimes formed when a salt boils. (336)

ion-product constant for water (K_w)　The equilibrium constant for the autoionization of water: (764)

$$K_w = [H_3O^+][OH^-]$$

ionic atmosphere　A cluster of ions of net opposite charge surrounding a given ion in solution. (516)

ionic bonding The idealized bonding type based on the attraction of oppositely charged ions that arises through electron transfer between atoms with large differences in their tendencies to lose or gain electrons (typically between metals and nonmetals). (328, 534)

ionic compound A compound that consists of oppositely charged ions. (59, 541)

ionic radius The size of an ion as measured by the distance between the centers of adjacent ions in a crystalline ionic compound. (318)

ionic solid A solid whose unit cell contains cations and anions. (453)

ionization The process by which a substance absorbs energy from high-energy radioactive particles and loses an electron to become ionized. (1058)

ionization energy (IE) The energy required to remove completely one mole of electrons from one mole of gaseous atoms or ions. (307, 533)

ionizing radiation The high-energy radiation that forms ions in a substance by causing electron loss. (1058)

isoelectronic Having the same number and configuration of electrons as another species. (314)

isomer One of two or more compounds with the same molecular formula but different properties, generally as a result of different arrangements of atoms. (615, 1020)

isotopes Atoms of a given atomic number (that is, of a specific element) that have different numbers of neutrons and therefore different mass numbers. (52, 1042)

isotopic mass The mass (in amu) of an isotope relative to the mass of carbon-12. (53)

J

joule (J) The SI unit of energy; $1\text{ J} = 1\text{ kg·m}^2/\text{s}^2$. (225)

K

kelvin (K) The SI base unit of temperature. The kelvin is the same size as the Celsius degree. (24)

Kelvin scale (See *absolute scale.*)

ketone An organic compound (ending, *-one*) that contains a carbonyl group bonded to two other C atoms, $-\overset{|}{\underset{|}{C}}-\overset{\overset{\displaystyle :O:}{||}}{C}-\overset{|}{\underset{|}{C}}-$. (634)

kilogram (kg) The SI base unit of mass. (21)

kinetic energy (E_k) The energy an object has because of its motion. (6)

kinetic-molecular theory The model that explains gas behavior in terms of particles in random motion whose volumes and interactions are negligible. (197)

L

lanthanide contraction The additional decrease in atomic and ionic size, beyond the expected trend, caused by the poor shielding of the increasing nuclear charge by f electrons in the elements following the lanthanides. (1002)

lanthanides (also *rare earth*) The Period 6 ($4f$) series of inner transition elements that includes cerium (Ce; $Z = 58$) through lutetium (Lu; $Z = 71$). (303, 1006)

lattice The three-dimensional arrangement of points created by choosing each point to be at the same location within each particle of a crystal; thus, the lattice consists of all points with identical surroundings. (446)

lattice energy ($\Delta H^0_{\text{lattice}}$) The enthalpy change that occurs when gaseous ions coalesce into a solid ionic compound. (331)

law (See *natural law.*)

law of chemical equilibrium (also *law of mass action*) The law stating that when a system reaches equilibrium at a given temperature, the ratio of quantities that make up the reaction quotient has a constant numerical value. (717)

law of conservation of energy (also *first law of thermodynamics*) A basic observation that the total energy of the universe is constant: $\Delta E_{\text{universe}} = \Delta E_{\text{system}} + \Delta E_{\text{surroundings}} = 0$. (225)

law of definite (or constant) composition A mass law stating that, no matter what its source, a particular chemical compound is composed of the same elements in the same parts (fractions) by mass. (44)

law of mass action (See *law of chemical equilibrium.*)

law of mass conservation A mass law stating that the total mass of substances does not change during a chemical reaction. (43)

law of multiple proportions A mass law stating that if elements A and B react to form two compounds, the different masses of B that combine with a fixed mass of A can be expressed as a ratio of small whole numbers. (45)

Le Châtelier's principle A principle stating that if a system in a state of equilibrium is disturbed, it will undergo a change that shifts its equilibrium position in a direction that reduces the effect of the disturbance. (736)

leaching A hydrometallurgical process that extracts a metal selectively, usually through formation of a complex ion. (964)

level (also *shell*) A specific energy state of an atom given by the principal quantum number n. (276)

leveling effect The inability of a solvent to distinguish the strength of an acid (or base) that is stronger than the conjugate acid (or conjugate base) of the solvent. (790)

Lewis acid-base definition A model of acid-base behavior in which acids and bases are defined, respectively, as species that accept and donate an electron pair. (791)

Lewis electron-dot symbol A notation in which the element symbol represents the nucleus and inner electrons, and surrounding dots represent the valence electrons. (329)

Lewis structure (also *Lewis formula*) A structural formula consisting of electron-dot symbols, with lines as bonding pairs and dot pairs as lone pairs. (358)

ligand A molecule or anion bonded to a central metal ion in a complex ion. (835, 1013)

like-dissolves-like rule An empirical observation stating that substances having similar kinds of intermolecular forces dissolve in each other. (487)

limiting reactant (also *limiting reagent*) The reactant that is consumed when a reaction occurs and therefore the one that determines the maximum amount of product that can form. (110)

line spectrum A series of separated lines representing photons whose wavelengths are characteristic of an element. (263) (See also *emission spectrum.*)

linear arrangement The geometric arrangement obtained when two electron groups maximize their separation around a central atom. (372)

linear shape   A molecular shape formed by three atoms lying in a straight line: bond angle of 180° (shape class AX_2 or AX_2E_3). (372, 376)

linkage isomers   Coordination compounds with the same composition but with different ligand donor atoms linked to the central metal ion. (1020)

lipid   Any of a class of biomolecules, including fats and oils, that are soluble in nonpolar solvents. (637)

liquid   One of the three states of matter. A liquid fills a container to the extent of its own volume and thus forms a surface. (4)

liquid crystal   A substance that flows like a liquid but packs like a crystalline solid at the molecular level. (462)

liter (L)   A non-SI unit of volume equivalent to 1 cubic decimeter (0.001 m^3). (19)

lithosphere   The solid portion of the Earth's crust. (953)

lock-and-key model   A model of enzyme function that pictures rigid shapes of the enzyme active site and the substrate fitting together as a lock and key, respectively. (701)

London force   (See *dispersion force*.)

lone pair (also *unshared pair*)   An electron pair that is part of an atom's valence shell but not involved in covalent bonding. (338)

low-spin complex   Complex ion that has fewer unpaired electrons than in the free metal ion because of the presence of strong-field ligands. (1029)

M

macromolecule   (See *polymer*.)

magnetic quantum number (m_l) (also *orbital-orientation quantum number*)   An integer from $-l$ through 0 to $+l$ that specifies the orientation of an atomic orbital in the three-dimensional space about the nucleus. (275)

manometer   A device used to measure the pressure of a gas in a laboratory experiment. (178)

mantle   A thick homogeneous layer of Earth's internal structure that lies between the core and the crust. (951)

mass   The quantity of matter an object contains. Balances are designed to measure mass. (21)

mass-action expression   (See *reaction quotient*.)

mass defect   The mass decrease that occurs when nucleons combine to form a nucleus. (1067)

mass fraction   (See *fraction by mass*.)

mass number (A)   The total number of protons and neutrons in an atom. (52)

mass percent (Also *mass % percent by mass*.)   The fraction by mass expressed as a percentage (44). A concentration term [% (w/w)] expressed as the mass in grams of solute dissolved per 100. g of solution. (502)

mass spectrometry   An instrumental method for measuring the relative masses of particles in a sample by creating charged particles and separating them according to their mass-charge ratio. (53)

matter   Anything that possesses mass and occupies volume. (3)

mean free path   The average distance a molecule travels between collisions at a given temperature and pressure. (203)

melting (also *fusion*)   The change of a substance from a solid to a liquid. (421)

melting point (mp or T_f)   The temperature at which the solid and liquid forms of a substance are at equilibrium. (429)

metabolic pathway   A biochemical reaction sequence that flows in one direction and in which each reaction is enzyme catalyzed. (745)

metal   A substance or mixture that is relatively shiny and malleable and is a good conductor of heat and electricity. In reactions, metals tend to transfer electrons to nonmetals and form ionic compounds. (58)

metallic bonding   An idealized type of bonding based on the attraction between metal ions and their delocalized valence electrons. (328, 534) (See also *electron-sea model*.)

metallic radius   One-half the distance between the nuclei of adjacent individual atoms in a crystal of an element. (304)

metallic solid   A solid whose individual atoms are held together by metallic bonding. (455)

metalloid (also *semimetal*)   An element with properties between those of metals and nonmetals. (58)

metallurgy   The branch of materials science concerned with the extraction and utilization of metals. (963)

metathesis reaction (also *double-displacement reaction*)   A reaction in which atoms or ions of two compounds exchange bonding partners. (139)

meter (m)   The SI base unit of length. The distance light travels in a vacuum in 1/299,792,458 second. (19)

milliliter (mL)   A volume (0.001 L) equivalent to 1 cm^3. (19)

millimeter of mercury (mmHg)   A unit of pressure based on the difference in the heights of mercury in a barometer or manometer. Redefined as 1 torr. (178)

mineral   The portion of an ore that contains the element of interest, a mineral is a naturally occurring, homogeneous, crystalline inorganic solid, with a well-defined composition. (963)

miscible   Soluble in any proportion. (486)

mixture   A group of two or more elements and/or compounds that are physically intermingled. (42, 481)

MO bond order   One-half the difference between the number of electrons in bonding and antibonding MOs. (406)

model (also *theory*)   A simplified conceptual picture based on experiment that explains how an aspect of nature occurs. (12)

molality (m)   A concentration term expressed as the moles of solute dissolved in 1000 g (1 kg) of solvent. (502)

molar heat capacity (C)   The quantity of heat required to change the temperature of 1 mol of a substance by 1 K. (233)

molar mass ($\mathcal{M}$) (also *gram-molecular weight*)   The mass of 1 mol of entities (atoms, molecules, or formula units) of a substance, in units of g/mol. (89)

molar solubility   The solubility expressed in terms of amount (mol) of dissolved solute per liter of solution. (826)

molarity (M)   A concentration term expressed as the moles of solute dissolved in 1 L of solution. (114)

mole (mol)   The SI base unit for amount of a substance. The amount that contains a number of objects equal to the number of atoms in exactly 12 g of carbon-12. (87)

mole fraction (X)   A concentration term expressed as the ratio of moles of one component of a mixture to the total moles present. (192, 503)

molecular compounds   Compounds that consist of individual molecules and whose physical state depends on intermolecular forces. (541)

molecular equation A chemical equation showing a reaction in solution in which reactants and products appear as intact, undissociated compounds. (137)

molecular formula A formula that shows the actual number of atoms of each element in a molecule. (64, 98)

molecular mass (also *molecular weight*) The sum (in amu) of the atomic masses of a formula unit of a compound. (71)

molecular orbital (MO) An orbital of given energy and shape that extends over a molecule and can be occupied by no more than two electrons. (404)

molecular orbital (MO) diagram A depiction of the relative energy and number of electrons in each MO, as well as the atomic orbitals from which the MOs form. (405)

molecular orbital (MO) theory A model that describes a molecule as a collection of nuclei and electrons in which the electrons occupy orbitals that extend over the entire molecule. (404)

molecular polarity The overall distribution of electronic charge in a molecule, determined by its shape and bond polarities. (381)

molecular shape The three-dimensional structure defined by the relative positions of the atomic nuclei in a molecule. (371)

molecular solid A solid held together by intermolecular forces between individual molecules. (452)

molecularity The number of reactant particles involved in an elementary step. (692)

molecule A structure consisting of two or more atoms that are chemically bound together and behave as an independent unit. (42)

monatomic ion An ion derived from a single atom. (59)

monomer A small molecule that is linked covalently to form a polymer and on which the repeat unit of the polymer is based. (468)

mononucleotide A monomer unit of a nucleic acid, consisting of an N-containing base, a sugar, and a phosphate group. (650)

monosaccharide A simple sugar; a polyhydroxy ketone or aldehyde with three to nine C atoms. (645)

N

nanotechnology The science and engineering of nanoscale (1–50 nm) systems. (473)

natural law (also *law*) A summary, often in mathematical form, of a universal observation. (12)

Nernst equation An equation stating that the cell voltage under any conditions depends on the standard cell voltage and the concentrations of the cell components: (917)

$$E_{cell} = E_{cell}^0 - \frac{RT}{nF} \ln Q$$

net ionic equation A chemical equation of a reaction in solution in which spectator ions have been eliminated to show the actual chemical change. (138)

network covalent solid A solid in which all the atoms are bonded covalently. (455, 541)

neutralization reaction An acid-base reaction that gives water and a solution of a salt; when a strong acid reacts with a stoichiometrically equivalent amount of a strong base, the solution is neutral. (140, 759)

neutron (n^0) An uncharged subatomic particle found in the nucleus, with a mass slightly greater than that of a proton. (52)

nitrile An organic compound containing the —C≡N: group. (640)

node A region of an orbital where the probability of finding the electron is zero. (279)

nonbonding MO A molecular orbital that is not involved in bonding. (412)

nonelectrolyte A substance whose aqueous solution does not conduct an electric current. (135, 506)

nonionizing radiation Radiation that does not cause loss of electrons. (1058)

nonmetal An element that lacks metallic properties. In reactions, nonmetals tend to bond with each other to form covalent compounds or accept electrons from metals to form ionic compounds. (58)

nonpolar covalent bond A covalent bond between identical atoms such that the bonding pair is shared equally. (346)

nuclear binding energy The energy required to break 1 mole of nuclei of an element into individual nucleons. (1067)

nuclear magnetic resonance (NMR) spectroscopy An instrumental technique used to determine the molecular environment of a given type of nucleus, most often ^{1}H, from its absorption of radio waves in a magnetic field. (624)

nuclear transmutation The induced conversion of one nucleus into another by bombardment with a particle. (1055)

nucleic acid An unbranched polymer consisting of mononucleotides that occurs as two types, DNA and RNA (deoxyribonucleic and ribonucleic acids), which differ chemically in the nature of the sugar portion of the mononucleotides. (650)

nucleon A subatomic particle that makes up a nucleus; a proton or neutron. (1042)

nucleus The tiny central region of the atom that contains all the positive charge and essentially all the mass. (51)

nuclide A nuclear species with specified numbers of protons and neutrons. (1042)

N/Z ratio The ratio of the number of neutrons to the number of protons, a key factor that determines the stability of a nuclide. (1046)

O

observation A fact obtained with the senses, often with the aid of instruments. Quantitative observations provide data that can be compared objectively. (12)

occurrence (also *source*) The form(s) in which an element exists in nature. (955)

octahedral arrangement The geometric arrangement obtained when six electron groups maximize their space around a central atom; when all six groups are bonding groups, the molecular shape is octahedral (AX_6; ideal bond angle = 90°). (376)

octet rule The observation that when atoms bond, they often lose, gain, or share electrons to attain a filled outer shell of eight electrons. (330)

optical isomers (also *enantiomers*) A pair of stereoisomers that arises when a molecule and its mirror image cannot be superimposed on each other. (617, 1021)

optically active A molecule that rotates the plane of polarized light. (618)

orbital diagram A depiction of electron number and spin in an atom's orbitals by means of arrows in a series of small boxes, lines, or circles. (295)

ore A naturally occurring mixture of gangue and mineral from which an element can be profitably extracted. (955)

organic compound A compound in which carbon is nearly always bonded to itself, to hydrogen, and often to other elements. (607)

organometallic compound An organic compound in which carbon is bonded covalently to a metal atom. (634)

osmosis The process by which solvent flows through a semipermeable membrane from a dilute to a concentrated solution. (511)

osmotic pressure (Π) The pressure that results from the inability of solute particles to cross a semipermeable membrane. The pressure required to prevent the net movement of solvent across the membrane. (511)

outer electrons Electrons that occupy the highest energy level (highest n value) and are, on average, farthest from the nucleus. (302)

overall (net) equation A chemical equation that is the sum of two or more sequential equations. (108)

overvoltage The additional voltage, usually associated with gaseous products, that is required above the standard cell voltage to accomplish electrolysis. (933)

oxidation The loss of electrons by a species, accompanied by an increase in oxidation number. (148)

oxidation number (O.N.) (also *oxidation state*) A number equal to the number of charges an atom would have if its shared electrons were held completely by the atom that attracts them more strongly. (148)

oxidation number method A method for balancing redox reactions in which the change in oxidation numbers is used to determine balancing coefficients. (151)

oxidation-reduction reaction (also *redox reaction*) A process in which there is a net movement of electrons from one reactant (reducing agent) to another (oxidizing agent). (147)

oxidation state (See *oxidation number.*)

oxidizing agent The substance that accepts electrons in a reaction and undergoes a decrease in oxidation number. (148)

oxoanion An anion in which an element is bonded to one or more oxygen atoms. (67)

P

p block The main-group elements that lie to the right of the transition elements in the periodic table whose p orbitals are being filled. (532)

p orbital An atomic orbital with $l = 1$. (279)

packing efficiency The percentage of the available volume occupied by atoms, ions, or molecules in a unit cell. (448)

paramagnetism The tendency of a species with unpaired electrons to be attracted by an external magnetic field. (316, 1005)

partial ionic character An estimate of the actual charge separation in a bond (caused by the electronegativity difference of the bonded atoms) relative to complete separation. (347)

partial pressure The portion of the total pressure contributed by a gas in a mixture of gases. (192)

particle accelerator A device used to impart high kinetic energies to nuclear particles. (1056)

pascal (Pa) The SI unit of pressure; $1 \text{ Pa} = 1 \text{ N/m}^2$. (178)

penetration The process by which an outer electron moves through the region occupied by the core electrons to spend part of its time closer to the nucleus; penetration increases the average effective nuclear charge for that electron. (294)

percent by mass (mass %) (See *mass percent*)

percent yield (% yield) The actual yield of a reaction expressed as a percent of the theoretical yield. (112)

period A horizontal row of the periodic table. (57)

periodic law A law stating that when the elements are arranged by atomic number, they exhibit a periodic recurrence of properties. (289)

periodic table of the elements A table in which the elements are arranged by atomic number into columns (groups) and rows (periods). (56)

pH The negative common logarithm of $[H_3O^+]$. (765)

phase A physically distinct portion of a system. (420)

phase change A physical change from one phase to another, usually referring to a change in physical state. (420)

phase diagram A diagram used to describe the stable phases and phase changes of a substance as a function of temperature and pressure. (430)

phlogiston theory An incorrect theory of combustion proposing that a burning substance releases the undetectable material phlogiston. (10)

photoelectric effect The observation that when light of sufficient energy shines on a metal, an electric current is produced. (261)

photon A quantum of electromagnetic radiation. (261)

photovoltaic cell A device capable of converting light directly into electricity. (245)

physical change A change in which the physical form (or state) of a substance, but not its composition, is altered. (3)

physical property A characteristic shown by a substance itself, without interacting with or changing into other substances. (3)

pi (π) bond A covalent bond formed by sideways overlap of two atomic orbitals that has two regions of electron density, one above and one below the internuclear axis. (401)

pi (π) MO A molecular orbital formed by combination of two atomic (usually p) orbitals whose orientations are perpendicular to the internuclear axis. (408)

Planck's constant (h) A proportionality constant relating the energy and the frequency of a photon, equal to 6.626×10^{-34} J·s. (260)

plastic A material that, when deformed, retains its new shape. (472)

polar covalent bond A covalent bond in which the electron pair is shared unequally, so the bond has partially negative and partially positive poles. (346)

polar molecule A molecule with an unequal distribution of charge as a result of its polar bonds and shape. (134)

polarimeter A device used to measure the rotation of plane-polarized light by an optically active compound. (618)

polarizability The ease with which a particle's electron cloud can be deformed. (436)

polyatomic ion An ion in which two or more atoms are bonded covalently. (63)

polymer (also *macromolecule*) An extremely large molecule that results from the covalent linking of many simpler molecular units (monomers). (468)

polyprotic acid   An acid with more than one ionizable proton. (776)

polysaccharide   A macromolecule composed of many simple sugars linked covalently. (645)

positron ($^0_1\beta$)   The antiparticle of an electron. (1045)

positron decay   A type of radioactive decay in which a positron is emitted from a nucleus. (1045)

potential energy (E_p)   The energy an object has as a result of its position relative to other objects or because of its composition. (6)

precipitate   The insoluble product of a precipitation reaction. (138)

precipitation reaction   A reaction in which two soluble ionic compounds form an insoluble product, a precipitate. (138)

precision (also *reproducibility*)   The closeness of a measurement to other measurements of the same phenomenon in a series of experiments. (31)

pressure (P)   The force exerted per unit of surface area. (176)

pressure-volume work (PV work)   A type of work in which a volume change occurs against an external pressure. (224)

principal quantum number (n)   A positive integer that specifies the energy and relative size of an atomic orbital. (275)

probability contour   A shape that defines the volume around an atomic nucleus within which an electron spends a given percentage of its time. (275)

product   A substance formed in a chemical reaction. (102)

property   A characteristic that gives a substance its unique identity. (3)

protein   A natural, linear polymer composed of any of about 20 types of amino acid monomers linked together by peptide bonds. (646)

proton (p^+)   A subatomic particle found in the nucleus that has a unit positive charge (1.602×10^{-19} C). (52)

proton acceptor   A substance that accepts an H^+ ion; a Brønsted-Lowry base. (768)

proton donor   A substance that donates an H^+ ion; a Brønsted-Lowry acid. (768)

pseudo–noble gas configuration   The $(n-1)d^{10}$ configuration of a *p*-block metal atom that empties its outer energy level. (314)

pure substance   (See *substance*.)

Q

qualitative analysis   The separation and identification of the ions in an aqueous mixture. (841)

quantum   A packet of energy equal to $h\nu$. The smallest quantity of energy that can be emitted or absorbed. (260)

quantum mechanics   The branch of physics that examines the wave motion of objects on the atomic scale. (273)

quantum number   A number that specifies a property of an orbital or an electron. (260)

R

rad (*radiation-absorbed dose*)   The quantity of radiation that results in 0.01 J of energy being absorbed per kilogram of tissue; 1 rad = 0.01 J/kg tissue = 10^{-2} Gy. (1059)

radial probability distribution plot   The graphic depiction of the total probability distribution (sum of ψ^2) of an electron in the region near the nucleus. (275)

radioactivity   The emissions resulting from the spontaneous disintegration of an unstable nucleus. (1042)

radioisotope   An isotope with an unstable nucleus that decays through radioactive emissions. (1053)

radioisotopic dating   A method for determining the age of an object based on the rate of decay of a particular radioactive nuclide. (1053)

radius of gyration (R_g)   A measure of the size of a coiled polymer chain as the average distance from the center of mass of the chain to its outside edge. (470)

random coil   The shape adopted by most polymer chains and caused by the random rotation of each repeat unit. (469)

random error   Human error that occurs in all measurements and results in values *both* higher and lower than the actual value. (31)

Raoult's law   A law stating that the vapor pressure of a solution is directly proportional to the mole fraction of solvent: $P_{solvent} = X_{solvent} \times P^0_{solvent}$. (507)

rare earth   (See *lanthanides*.)

rate constant (k)   The proportionality constant that relates the reaction rate to reactant (and product) concentrations. (671)

rate-determining step (also *rate-limiting step*)   The slowest step in a reaction mechanism and therefore the step that limits the overall rate. (693)

rate law (also *rate equation*)   An equation that expresses the rate as a function of reactant (and product) concentrations. (671)

reactant   A starting substance in a chemical reaction. (102)

reaction energy diagram   A graph that shows the potential energy of a reacting system as it progresses from reactants to products. (689)

reaction intermediate   A substance that is formed and used up during the overall reaction and therefore does not appear in the overall equation. (691)

reaction mechanism   A series of elementary steps that sum to the overall reaction and is consistent with the rate law. (691)

reaction order   The exponent of a reactant concentration in a rate law that shows how the rate is affected by changes in that concentration. (671)

reaction quotient (Q) (also *mass-action expression*)   A ratio of terms for a given reaction consisting of product concentrations multiplied together divided by reactant concentrations multiplied together, each raised to the power of their balancing coefficient. The value of Q changes until the system reaches equilibrium, at which point it equals K. (717)

reaction rate   The change in the concentrations of reactants (or products) with time. (667)

reactor core   The part of a nuclear reactor that contains the fuel rods and generates heat from fission. (1072)

redox reaction   (See *oxidation-reduction reaction*.)

reducing agent   The substance that donates electrons in a redox reaction and undergoes an increase in oxidation number. (148)

reduction   The gain of electrons by a species, accompanied by a decrease in oxidation number. (148)

refraction   A phenomenon in which a wave changes its speed and therefore its angle as it passes through a phase boundary. (258)

rem (*roentgen equivalent for man*)   The unit of radiation dosage for a human based on the product of the number of rads and a factor related to the biological tissue; 1 rem $= 10^{-2}$ Sv. (1059)

resonance hybrid The weighted average of the resonance structures of a molecule. (362)

resonance structure (also *resonance form*) One of two or more Lewis structures for a molecule that cannot be adequately described by a single structure. Resonance structures differ only in the position of electron pairs. (362)

reverse osmosis A process for preparing drinkable water that uses an applied pressure greater than the osmotic pressure to remove ions from an aqueous solution, typically seawater. (521)

rms (root-mean-square) speed (u_{rms}) The speed of a molecule having the average kinetic energy; very close to the most probable speed. (200)

roasting A pyrometallurgical process in which metal sulfides are converted to oxides. (964)

round off The process of removing digits based on a series of rules to obtain an answer with the proper number of significant figures (or decimal places). (28)

S

s **block** The main-group elements that lie to the left of the transition elements in the periodic table whose *s* orbitals are being filled. (532)

s **orbital** An atomic orbital with $l = 0$. (278)

salt An ionic compound that results from a classical acid-base reaction. (142)

salt bridge An inverted U tube that connects the compartments of a voltaic cell and contains a solution of nonreacting electrolyte that maintains neutrality by allowing ions to flow between compartments. (902)

saturated hydrocarbon A hydrocarbon in which each C is bonded to four other atoms. (613)

saturated solution A solution that contains the maximum amount of dissolved solute at a given temperature in the presence of undissolved solute. (498)

scanning tunneling microscopy An instrumental technique that uses electrons moving across a minute gap to observe the topography of a surface on the atomic scale. (451)

Schrödinger equation An equation that describes how the electron matter-wave changes in space around the nucleus. Solutions of the equation provide allowable energy levels of the H atom. (273)

scientific method A process of creative thinking and testing aimed at objective, verifiable discoveries of the causes of natural events. (12)

scintillation counter A device used to measure radioactivity through its excitation of atoms and their subsequent emission of light. (1051)

second (s) The SI base unit of time. (26)

second law of thermodynamics A law stating that a process occurs spontaneously in the direction that increases the entropy of the universe. (861)

seesaw shape A molecular shape caused by the presence of one equatorial lone pair in a trigonal bipyramidal arrangement (AX_4E). (375)

selective precipitation The process of separating ions through differences in the solubility of their compounds with a given precipitating ion. (840)

self-ionization (See *autoionization*.)

semiconductor A substance whose electrical conductivity is poor at room temperature but increases significantly with temperature. (459)

semimetal (See *metalloid*.)

semipermeable membrane A membrane that allows solvent, but not solute, to pass through. (511)

shared pair (See *bonding pair*.)

shell (See *level*.)

shielding The ability of other electrons, especially inner ones, to lessen the nuclear attraction on an outer electron. (293)

SI unit A unit composed of one or more of the base units of the Système International d'Unités, a revised metric system. (17)

side reaction A different chemical reaction that consumes some of the reactant and reduces the overall yield of the desired product. (112)

side-to-side overlap Overlap of *p* with *p* (or sometimes with *d*) atomic orbitals that leads to a pi (π) bond. (535)

sievert (Sv) The SI unit of human radiation dosage; 1 Sv = 100 rem. (1059)

sigma (σ) bond A type of covalent bond that arises through end-to-end orbital overlap and has most electron density along the bond axis. (401)

sigma (σ) MO A molecular orbital that is cylindrically symmetrical about an imaginary line that runs through the nuclei of the component atoms. (405)

significant figures The digits obtained in a measurement. The greater the number of significant figures, the greater the certainty of the measurement. (27)

silicate A type of compound found throughout rocks and soil and consisting of —Si—O groupings and, in most cases, metal cations. (568)

silicone A type of synthetic polymer containing —Si—O chains, with organic groups and crosslinks. (568)

simple cubic unit cell A unit cell in which a particle occurs at each corner of a cube. (446)

single bond A bond that consists of one electron pair. (338)

slag A molten waste product formed in a blast furnace by the reaction of acidic silica with a basic metal oxide. (971)

smelting Heating a mineral with a reducing agent, such as coke, to obtain a metal. (965)

soap The salt of a fatty acid and usually a Group 1A(1) or 2A(2) hydroxide. (490)

solid One of the three states of matter. A solid has a fixed shape that does not conform to the container shape. (4)

solubility (*S*) The maximum amount of solute that dissolves in a fixed quantity of a particular solvent at a specified temperature when excess solute is present. (486)

solubility-product constant (K_{sp}) An equilibrium constant for the dissolving of a slightly soluble ionic compound in water. (824)

solute The substance that dissolves in the solvent. (114, 486)

solution (See *homogeneous mixture*.)

solvated Surrounded by solvent molecules. (132)

solvation The process of surrounding a solute particle with solvent particles. (132, 494)

solvent The substance in which the solute(s) dissolve. (114, 486)

source (See *occurrence*.)

sp **hybrid orbital** An orbital formed by the mixing of one *s* and one *p* orbital in a central atom. (394)

sp^2 hybrid orbital   An orbital formed by the mixing of one s and two p orbitals in a central atom. (396)

sp^3 hybrid orbital   An orbital formed by the mixing of one s and three p orbitals in a central atom. (396)

sp^3d hybrid orbital   An orbital formed by the mixing of one s, three p, and one d orbital in a central atom. (397)

sp^3d^2 hybrid orbital   An orbital formed by the mixing of one s, three p, and two d orbitals in a central atom. (398)

specific heat capacity (c)   The quantity of heat required to change the temperature of 1 gram of a substance by 1 k. (233)

spectator ion   An ion that is present as part of a reactant but is not involved in the chemical change. (138)

spectrochemical series   A ranking of ligands in terms of their ability to split d-orbital energies. (1028)

spectrophotometry   A group of instrumental techniques that create an electromagnetic spectrum to measure the atomic and molecular energy levels of a substance. (267)

speed of light (c)   A fundamental constant giving the speed at which electromagnetic radiation travels in a vacuum: $c = 2.9979 \times 10^8$ m/s. (256)

spin quantum number (m_s)   A number, either $+\frac{1}{2}$ or $-\frac{1}{2}$, that indicates the direction of electron spin. (291)

spontaneous change   A change that occurs by itself, that is, without an ongoing input of energy. (856)

square planar shape   A molecular shape (AX_4E_2) caused by the presence of two axial lone pairs in an octahedral arrangement. (376)

square pyramidal shape   A molecular shape (AX_5E) caused by the presence of one lone pair in an octahedral arrangement. (376)

standard atmosphere (atm)   The average atmospheric pressure measured at sea level, defined as 1.01325×10^5 Pa. (178)

standard cell potential (E^0_{cell})   The potential of a cell measured with all components in their standard states and no current flowing. (906)

standard electrode potential ($E^0_{half\text{-}cell}$) (also *standard half-cell potential*)   The standard potential of a half-cell, with the half-reaction written as a reduction. (906)

standard entropy of reaction (ΔS^0_{rxn})   The entropy change that occurs when all components are in their standard states. (866)

standard free energy change (ΔG^0)   The free energy change that occurs when all components are in their standard states. (873)

standard free energy of formation (ΔG^0_f)   The standard free energy change that occurs when 1 mol of a compound is made from its elements. (874)

standard half-cell potential   (See *standard electrode potential*.)

standard heat of formation (ΔH^0_f)   The enthalpy change that occurs when 1 mol of a compound forms from its elements, with all substances in their standard states. (240)

standard heat of reaction (ΔH^0_{rxn})   The enthalpy change that occurs during a reaction, with all substances in their standard states. (240)

standard hydrogen electrode   (See *standard reference half-cell*.)

standard molar entropy (S^0)   The entropy of 1 mol of a substance in its standard state. (862)

standard molar volume   The volume of 1 mol of an ideal gas at standard temperature and pressure: 22.414 L. (184)

standard reference half-cell (also *standard hydrogen electrode*)   A specially prepared platinum electrode immersed in 1 M $H^+(aq)$ through which H_2 gas at 1 atm is bubbled. $E^0_{half\text{-}cell}$ is defined as 0 V. (906)

standard states   A set of specifications used to compare thermodynamic data: 1 atm for gases behaving ideally, 1 M for dissolved species, or the pure substance for liquids and solids. (240)

standard temperature and pressure (STP)   The reference conditions for a gas: (184)

$$0°C \ (273.15 \ K) \ and \ 1 \ atm \ (760 \ torr)$$

state function   A property of the system determined by its current state, regardless of how it arrived at that state. (226)

state of matter   One of the three physical forms of matter: solid, liquid, or gas. (4)

stationary state   In the Bohr model, one of the allowable energy levels of the atom in which it does not release or absorb energy. (263)

steel   An alloy of iron with small amounts of carbon and usually other metals. (970)

stellar nucleogenesis   The process by which elements are formed in the stars through nuclear fusion. (1074)

stereoisomers   Molecules with the same connections of atoms but different orientations of groups in space. (617, 1021) (See also *geometric isomers* and *optical isomers*.)

stoichiometric coefficient   (See *balancing coefficient*.)

stoichiometry   The study of the mass-mole-number relationships of chemical formulas and reactions. (87)

strong-field ligand   A ligand that causes larger crystal field splitting energy and therefore is part of a low-spin complex. (1026)

strong force   An attractive force that exists between all nucleons and is about 100 times stronger than the electrostatic repulsive force. (1047)

structural formula   A formula that shows the actual number of atoms, their relative placement, and the bonds between them. (64)

structural isomers   (See *constitutional isomers*.)

sublevel (also *subshell*)   An energy substate of an atom within a level. Given by the n and l values, the sublevel designates the size and shape of the atomic orbitals. (276)

sublimation   The process of a solid changing directly into a gas. (422)

substance (also *pure substance*)   A type of matter, either an element or a compound, that has a fixed composition. (41)

substitution reaction   An organic reaction that occurs when an atom (or group) from one reactant substitutes for one in another reactant. (626)

substrate   A reactant that binds to the active site in an enzyme-catalyzed reaction. (700)

superconductivity   The ability to conduct a current with no loss of energy to resistive heating. (459)

supersaturated solution   An unstable solution in which more solute is dissolved than in a saturated solution. (498)

surface tension   The energy required to increase the surface area of a liquid by a given amount. (439)

surroundings　All parts of the universe other than the system. (221)

suspension　A heterogeneous mixture containing particles that are distinct from the surrounding medium. (517)

syngas　Synthesis gas; a combustible mixture of CO and H_2. (243)

synthetic natural gas (SNG)　A gaseous fuel mixture, mostly methane, formed from coal. (243)

system　The defined part of the universe under study. (221)

systematic error　A type of error producing values that are either higher or lower than the actual value and often caused by faulty equipment or a consistent fault in technique. (31)

T

t_{2g} orbitals　The set of orbitals (composed of d_{xy}, d_{yz}, and d_{xz}) that results when the energies of the metal-ion d orbitals are split by a ligand field. This set is lower in energy than the other (e_g) set in an octahedral field and higher in energy in a tetrahedral field. (1026)

T shape　A molecular shape caused by the presence of two equatorial lone pairs in a trigonal bipyramidal arrangement (AX_3E_2). (375)

temperature (*T*)　A measure of how hot or cold a substance is relative to another substance. (24)

tetrahedral arrangement　The geometric arrangement formed when four electron groups maximize their separation around a central atom; when all four groups are bonding groups, the molecular shape is tetrahedral (AX_4; ideal bond angle 109.5°). (374)

theoretical yield　The amount of product predicted by the stoichiometrically equivalent molar ratio in the balanced equation. (112)

theory　(See *model*.)

thermochemical equation　A chemical equation that shows the heat of reaction for the amounts of substances specified. (236)

thermochemistry　The branch of thermodynamics that focuses on the heat involved in chemical reactions. (221)

thermodynamics　The study of heat (thermal energy) and its interconversions. (221)

thermometer　A device for measuring temperature that contains a fluid that expands or contracts within a graduated tube. (24)

third law of thermodynamics　A law stating that the entropy of a perfect crystal is zero at 0 K. (861)

titration　A method of determining the concentration of a solution by monitoring its reaction with a solution of known concentration. (143)

torr　A unit of pressure identical to 1 mmHg. (178)

total ionic equation　A chemical equation of an aqueous reaction that shows all the soluble ionizable substances dissociated into ions. (138)

tracer　A radioisotope that signals the presence of the species of interest by emitting nonionizing radiation. (1062)

transition element (also *transition metal*)　An element that occupies the *d* block of the periodic table; one whose *d* orbitals are being filled. (300, 999)

transition state (also *activated complex*)　An unstable species formed in an effective collision of reactants that exists momen-

tarily when the system is highest in energy and that can either form products or re-form reactants. (688)

transition state theory　A model that explains how the energy of reactant collision is used to form a high-energy transitional species that can change to reactant or product. (685)

transuranium element　An element with atomic number higher than that of uranium ($Z = 92$). (1057)

trigonal bipyramidal arrangement　The geometric arrangement formed when five electron groups maximize their separation around a central atom. When all five groups are bonding groups, the molecular shape is trigonal bipyramidal (AX_5); ideal bond angles, axial-center-equatorial 90° and equatorial-center-equatorial 120°. (375)

trigonal planar arrangement　The geometric arrangement formed when three electron groups maximize their separation around a central atom. (372)

trigonal planar shape　A molecular shape (AX_3) formed when three atoms around a central atom lie at the corners of an equilateral triangle; ideal bond angle 120°. (372)

trigonal pyramidal shape　A molecular shape (AX_3E) caused by the presence of one lone pair in a tetrahedral arrangement. (374)

triple bond　A covalent bond that consists of three bonding pairs, two atoms sharing six electrons; one σ and two π bonds. (338)

triple point　The pressure and temperature at which three phases of a substance are in equilibrium. In a phase diagram, the point at which three phase-transition curves meet. (431)

Tyndall effect　The scattering of light by a colloidal suspension. (518)

U

ultraviolet (UV) region　The region of the electromagnetic spectrum between the visible and the x-ray regions. (257)

uncertainty　A characteristic of every measurement that results from the inexactness of the measuring device and the necessity of estimating when taking a reading. (27)

uncertainty principle　The principle stated by Werner Heisenberg that it is impossible to know simultaneously the exact position and velocity of a particle; the principle becomes important only for particles of very small mass. (272)

unimolecular reaction　An elementary reaction that involves the decomposition or rearrangement of a single particle. (692)

unit cell　The smallest portion of a crystal that, if repeated in all directions, gives the crystal. (446)

universal gas constant (*R*)　A proportionality constant that relates the energy, amount of substance, and temperature of a system; $R = 0.0820578$ atm·L/mol·K $= 8.31447$ J/mol·K. (185)

unsaturated hydrocarbon　A hydrocarbon with a carbon-carbon multiple bond; one in which C is bonded to fewer than four atoms. (618)

unsaturated solution　A solution in which more solute can be dissolved at a given temperature. (498)

unshared pair　(See *lone pair*.)

V

V shape　(See *bent shape*.)

valence band   In band theory, the lower energy portion of the band of molecular orbitals, which is filled with valence electrons. (458)

valence bond (VB) theory   A model that attempts to reconcile the shapes of molecules with those of atomic orbitals through the concepts of orbital overlap and hybridization. (393)

valence electrons   The electrons involved in compound formation; in main-group elements, the electrons in the valence (outer) level. (302)

valence-shell electron-pair repulsion (VSEPR) theory   A model explaining that the shapes of molecules and ions result from minimizing electron-pair repulsions around a central atom. (370)

van der Waals constants   Experimentally determined positive numbers used in the van der Waals equation to account for the molecular interactions and molecular volume of real gases. (209)

van der Waals equation   An equation that accounts for the behavior of real gases. (209)

van der Waals radius   One-half of the closest distance between the nuclei of identical nonbonded atoms. (432)

vapor pressure (also *equilibrium vapor pressure*)   The pressure exerted by a vapor at equilibrium with its liquid in a closed system. (426)

vapor pressure lowering (ΔP)   The lowering of the vapor pressure of a solvent caused by the presence of dissolved solute particles. (507)

vaporization   The process of changing from a liquid to a gas. (421)

variable   A quantity that can have more than a single value. (12) (See also *controlled experiment*.)

viscosity   A measure of the resistance of a liquid to flow. (440)

volatility   The tendency of a substance to become a gas. (75)

volt (V)   The SI unit of electric potential: 1 V = 1 J/C. (905)

voltage   (See *cell potential*.)

voltaic cell (also *galvanic cell*)   An electrochemical cell that uses a spontaneous reaction to generate electric energy. (898)

volume (V)   The space occupied by a sample of matter. (19)

volume percent [% (v/v)]   A concentration term defined as the volume of solute in 100. volumes of solution. (503)

W

wastewater (also *sewage*)   Used water, usually containing industrial and/or residential waste, that is treated before being returned to the environment. (521)

water softening   The process of removing hard-water ions, such as Ca^{2+} and Mg^{2+}, from water. (520)

wave function   (See *atomic orbital*.)

wave-particle duality   The principle stating that both matter and energy have wavelike and particle-like properties. (272)

wavelength (λ)   The distance between any point on a wave and the corresponding point on the next wave, that is, the distance a wave travels during one cycle. (256)

weak-field ligand   A ligand that causes smaller crystal field splitting energy and therefore is part of a high-spin complex. (1026)

weight   The force exerted by a gravitational field on an object. (21)

work (w)   The energy transferred when an object is moved by a force. (223)

X

x-ray diffraction analysis   An instrumental technique used to determine spatial dimensions of a crystal structure by measuring the diffraction patterns caused by x-rays impinging on the crystal. (451)

Z

zone refining   A process used to purify metals and metalloids in which impurities are removed from a bar of the element by concentrating them in a thin molten zone. (508, 967)

CREDITS

Chapter 1

Figure 1.1A: © Paul Morrell/Stone; *1.1B, p. 4 (top left):* © McGraw-Hill Higher Education/Stephen Frisch Photographer; *p. 4 (bottom left):* © Ruth Melnick; *p. 4 (3 photos on right):* © McGraw-Hill Higher Education/Stephen Frisch Photographer; *p. 5:* © NASA Media Resource Center; *1.4:* © Giraudon/Art Resource, NY; *p. 10:* © Culver Pictures; *1.5A:* © Science Library/Photo Researchers, Inc.; *1.5B:* © PhotoDisc; *1.5C:* © Kyocera Industrial Ceramics Corp.; *1.5D:* © Reuters NewMedia Inc./Corbis; *p. 13:* © Loren Santow/Stone; *1.7:* © Kevin Schafer/Stone; *1.9A:* © McGraw-Hill Higher Education/Stephen Frisch Photographer; *1.9B:* © BrandTech Scientific, Inc; *p. 21:* © National Institute of Standards & Technology; *p. 24:* © McGraw-Hill Higher Education/Stephen Frisch Photographer; *1.13:* © Steve Jefferts (Time and Frequency)/National Institute of Standards & Technology; *p. 27:* © North Wind Picture Archives; *1.14B:* © Hart Scientific; *1.15 (both):* © McGraw-Hill Higher Education/Stephen Frisch Photographer; *B1.1A:* ©Bruce Forster/Stone; *B1.1B:* © Bob Daemmrich/Daemmrich Photography; *B1.1C:* © Cindy Yamanaka/National Geographic Image Collection; *B1.1D:* © Tom Pantages

Chapter 2

Opener: © Mark A. Schneider/Visuals Unlimited; *p. 43 (all):* © McGraw-Hill Higher Education/Stephen Frisch Photographer; *p. 44 (top):* © Sleeping Children, after 1859–1874; William Henry Rinehart, American (1825–1874), Marble; overall 15×18×37 in. (38.1×45.7×94 cm), Gift of Daniel and Jessie Lie Farver, Roy Taylor, and Mary E. Moore Gift, 1984.268/Museum of Fine Arts Boston; *p. 44 (bottom):* © Wards Earth Science Catalog; *p. 46:* © University of Cincinnati-Oesper Collection; *p. 48:* © Brad Jirka/American School of Neon; *2.6:* © PhysicsDept/Cavendish Laboratory at University of Cambridge; *B2.2D:* © James Holmes/Oxford Centre for Molecular Sciences/Science Photo Library/Photo Researchers, Inc.; *B2.2E:* © Curtis Hendrickson/National High Magnetic Field Laboratory; *p. 56:* © Hulton Getty/Stone; *2.12 (all), 2.13 (all):* © McGraw-Hill Higher Education/Stephen Frisch Photographer; *2.18:* © Dane S. Johnson/Visuals Unlimited; *p. 68(both), 2.21A, B, B2.3, B2.4:* © McGraw-Hill Higher Education/Stephen Frisch Photographer; *B2.9:* © Tom Pantages; *2.22:* © Dr. Carla Montgomery

Chapter 3

Opener: © McGraw-Hill Higher Education/Fundamental Photography; *3.1A, B3.2:* © McGraw-Hill Higher Education/Stephen Frisch Photographer; *p. 95:* © McGraw-Hill Higher Education; *p. 97:* © Wedgworth/Custom Medical Stock Photo; *p. 98:* © Ruth Melnick; *3.7, 3.8, 3.12 (all), 3.13 (both):* © McGraw-Hill Higher Education/Stephen Frisch Photographer

Chapter 4

Figure 4.1, 4.5, 4.6, p. 141, 4.7 (all), 4.9, p. 151, 4.13 (both), 4.14 (all), 4.15 (all), 4.16 (all), 4.17, 4.18 (both), p. 172: © McGraw-Hill Higher Education/Stephen Frisch Photographer

Chapter 5

Opener: © Richard Megna/Fundamental Photographs; *5.1 (all), 5.2 (both), 5.9:* © McGraw-Hill Higher Education/Stephen Frisch Photographer; *p. 189:* © Deborah Davis/PhotoEdit; *p. 190:* © North Wind Picture Archives; *p. 196:* © McGraw-Hill Higher Education/Stephen Frisch Photographer; *p. 206:* © Julian Baum/Science Photo Library/Photo Researchers, Inc.; *p. 217:* © Romilly Lockyer/Image Bank

Chapter 6

Opener: © Yoav Levy/Phototake; *p. 221:* © David Malin/David Malin Images; *6.1:* © McGraw-Hill Higher Education/Stephen Frisch Photographer; *p. 225:* © AIP Emilio Serge Visual Archives, E. Scott Barr Collection/American Institute of Physics

Chapter 7

Page 257: © Ralph Wetmore/Stone; *p. 259:* © Michael Marten/Science Photo Library/Photo Researchers, Inc.; *7.6A,B:* © Dr. E.R. Degginger/Color-Pic Inc.; *7.6C:* © Paul Dance/Stone; *B7.1A:* © McGraw-Hill Higher Education/Stephen Frisch Photographer; *B7.1B:* © Washington DC Convention & Visitors Bureau; *p. 269:* © Michael Marten/Science Photo Library/Photo Researchers, Inc.; *p. 270:* © Jean Claude Revy/Phototake; *7.14A,B:* © PSSC Physics © 1965, Education Development Center, Inc.; D.C. Heath & Company/Educational Development Center; *p. 273:* © Photo by Mark Oliphant, AIP Emilio Segre Visual Archives, Margrethe Bohr Collection/American Institute of Physics

Chapter 8

Opener: © Renee Lynn/Stone; *p. 289:* Corbis; *p. 290:* © University of Cincinnati-Oesper Collection; *8.11A, B, p. 312, 8.23 (all):* © McGraw-Hill Higher Education//Stephen Frisch Photographer

Chapter 9

Opener: © Japack Company/Corbis; © Bancroft Library, University of California; *9.5A, B9.8A, 9.9, 9.14, 9.21:* © McGraw-Hill Higher Education/Stephen Frisch Photographer; *9.22:* © Richard Megna/Fundamental Photographs; *9.24:* © McGraw-Hill Higher Education/Stephen Frisch Photographer; *p. 351:* © Diane Hirsch/Fundamental Photographs

Chapter 10

Opener: © McGraw-Hill Higher Education/Richard Megna/Fundamental Photographs; *p. 365:* © Richard Megna/Fundamental Photographs; *10.4, 10.9:* © McGraw-Hill Higher Education/Stephen Frisch Photographer

Chapter 11

Figure 11.21: © McGraw-Hill Higher Education/Richard Megna/Fundamental Photographs

Chapter 12

Page 421: © Phil Jude/Science Photo Library/Photo Researchers, Inc.; *p. 422:* © Jill Birschbach Photo Services; *12.8:* © McGraw-Hill Higher Education/Richard Megna/Fundamental Photographs; *p. 439:* © Yoav Levy/Phototake; *12.20A,B:* © McGraw-Hill Higher Education/Stephen Frisch Photographer; *p. 441 (leaf):* © Rob Planck/Photo Researchers, Inc.; *p. 441 (oil):* © Chris Sorensen Photography; *p. 441 (water strider):* © Nuridsany & Perennou/Photo Researchers, Inc.; *p. 441 (bubbles):* © Phil Jude/Science Photo Library/Photo Researchers, Inc.; *p. 441 (child):* © Ruth Melnick; *12.22B:* © Richard C. Walters/Visuals Unlimited; *12.23:* © Tom Pantages; *12.25 (all):* © Paul Silverman/Fundamental Photographs; *p. 448:* © Michal Heron/Woodfin Camp & Associates; *B12.3:* © Physics Dept., Imperial College/Science Photo Library/Photo Researchers, Inc.; *12.38:* © AT&T Bell Labs/Science Photo Library/Photo Researchers, Inc.; *p. 460:* © NASA/Science Photo Library/Photo Researchers, Inc.; *12.44A:* © Jan Robert Factor/Science Source/Photo Researchers, Inc.; *12.44B:* © James Dennis/Phototake; *p. 472:* © Sue Klemens/Stock Boston; *12.50A:* © Felice Frankel; *12.50B:* © NASA Ames Research Center

Chapter 13

Opener: © Richard Megna/Fundamental Photographs; *p. 485:* © A. B. Dowsett/Science Photo Library/Photo Researchers, Inc.; *13.4A:* © Ruth Melnick; *13.4B:* © E. R. Degginger/Color-Pic Inc.; *p. 493:* © Root Resources; *13.9A–C:* © McGraw-Hill Higher Education/Stephen Frisch Photographer; *13.11:* © Shiela Terry/Science Photo Library/Photo Researchers, Inc.; *p. 500:* © John Forsythe/Visuals Unlimited; *13.13:* © Darwin Dale/Photo Researchers, Inc.; *13.14A–C:* © McGraw-Hill Higher Education/Stephen Frisch Photographer; *p. 512 (top):* © Maxwell Mackenzie; *p. 512 (center left):* © OSF/Doug Allen/Animals Animals; *p. 512 (center right):* © Chris Sorensen Photography; *p. 512 (bottom):* © Westinghouse/Visuals Unlimited; *p. 513 (top three):* © David M. Phillips/Photo Researchers, Inc.; *p. 513 (center left):* © John Lei/Stock Boston; *p. 513 (center right):* © Root Resources; *p. 513 (bottom):* © Osentoski & Zoda/Envision; *13.21A:* © McGraw-Hill Higher Education/Color-Pic Inc.; *13.21B:* © E. R. Degginger/Color-Pic Inc.; *p. 519:* © Earth Satellite Corporation/Science Photo Library/Photo Researchers, Inc.; *B13.4A:* © McGraw-Hill Higher Education/Stephen Frisch Photographer; *B13.5A:* © Robert Essel/Corbis/Stock Market

Interchapter

Opener: © Richard Megna/Fundamental Photographs; *p. 538:* © McGraw-Hill Higher Education/Stephen Frisch Photographer

Chapter 14

Opener: © Aaron Haupt/Photo Researchers, Inc.; *p. 546, p. 547, p. 550, p. 551 (top):* © McGraw-Hill Higher Education/Stephen Frisch Photographer; *p. 551 (bottom):* © Visuals Unlimited; *p. 554, p. 555 (top):* © McGraw-Hill Higher Education/Stephen Frisch Photographer; *p. 555 (center):* © Scovil Photography; *p. 555 (bottom right):* © Scott Camazine/Photo Researchers, Inc.; *p. 558:* © McGraw-Hill Higher Education/Stephen Frisch Photographer; *p. 559 (top):* © Scott Camazine/Photo Researchers, Inc.; *p. 559 (center):* © E. R. Degginger/Color-Pic Inc.; *p. 559 (bottom left):* © Phil Degginger/Color-Pic Inc.; *p. 559 (bottom right):* © AT&T Bell Labs/Science Photo Library/Photo Researchers, Inc.; *p. 560:* © McGraw-Hill Higher Education/Stephen Frisch Photographer; *14.13A:* © Courtesy MER Corporation; *14.13B:* © S. C. Tsang/

Science Photo Library/Photo Researchers, Inc.; *p. 564:* © McGraw-Hill Higher Education/Stephen Frisch Photographer; *p. 565 (top):* © C. Winters/Photo Researchers, Inc.; *p. 565 (center):* © Courtesy IBM Microelectronics; *p. 565 (bottom left):* © Phil Jude/Science Photo Library/Photo Researchers, Inc.; *p. 565 (bottom right):* © McGraw-Hill Higher Education/Pat Watson, photographer; *p. 566:* © McGraw-Hill Higher Education/Stephen Frisch Photographer; *p. 570, p. 571 (top and center):* © Scott Camazine/Photo Researchers, Inc.; *p. 571 (bottom):* © Tom Pantages; *p. 572 (top):* © Photri-Microstock; *p. 572 (center):* © Mark Richards/PhotoEdit; *p. 572 (bottom):* © Rare Books Department, Courtesy of the Trustees, Boston Public Library; *p. 574:* © McGraw-Hill Higher Education/Stephen Frisch Photographer; *p. 575 (top left):* © Stone; *p. 575 (top right, bottom):* © McGraw-Hill Higher Education/Pat Watson photographer; *p. 582:* © McGraw-Hill Higher Education/Stephen Frisch Photographer; *p. 583 (top):* © Chris Sorensen Photography; *p. 583 (bottom):* © Gregory Dimijian/Photo Researchers, Inc.; *p. 586:* © Bruce Forster/Viewfinders; *14.26:* © McGraw-Hill Higher Education/Stephen Frisch Photographer; *14.27:* © Wards Natural Science Establishment; *14.30B, p. 590:* © McGraw-Hill Higher Education/Stephen Frisch Photographer; *p. 591 (left):* © Scovil Photography; *p. 591 (right):* © NASA/Science Photo Library/Photo Researchers, Inc.; *14.33:* © E. R. Degginger/Color-Pic Inc.; *p. 596:* © McGraw-Hill Higher Education/Stephen Frisch Photographer;

Chapter 15

Opener: © Biophoto Associates/Photo Researchers, Inc.; *p. 607:* © Science Photo Library/Photo Researchers, Inc.; *B15.5:* © Scott Camazine/Photo Researchers, Inc.; *15.13A, B:* © McGraw-Hill Higher Education/Richard Megna/Fundamental Photographs; *p. 638:* © McGraw-Hill Higher Education/Pat Watson, photographer; *15.28:* © E. R. Degginger/Color-Pic Inc.; *p. 645:* © McGraw-Hill Higher Education/Pat Watson, photographer; *15.33A:* © J. Gross/Science Photo Library/Photo Researchers, Inc.; *15.33B:* © E. R. Degginger/Color-Pic Inc.

Chapter 16

Opener: © Joe McDonald/McDonald Wildlife Photography; *16.2A:* © Stone; *16.2B:* © Ruth Melnick; *16.2C:* © Paul Silverman/Fundamental Photographs; *16.2D:* © Bruce Ayers/Stone; *16.3A,B:* © McGraw-Hill Higher Education/Stephen Frisch Photographer; *p. 667:* © Gary Holscher/Stone; *B16.1:* © Varian Associates; *B16.2, B16.3:* © McGraw-Hill Higher Education/Stephen Frisch Photographer; *p. 693:* © Andie Norwood Browne/Stone; *B16.6:* © NASA/RDF/Visuals Unlimited

Chapter 17

Opener: © David Noton/FPG International; *17.1:* © McGraw-Hill Higher Education/Stephen Frisch Photographer; *p. 720:* © Deborah Davis/PhotoEdit; *p. 721:* © Noranda Inc. Fonderie Horne; *p. 737:* © Kennan Ward Photography; *B17.1:* © Grant Heilman Photography

Chapter 18

18.1A: © Jill Birschbach Photo Services; *18.1B:* © Michael W. Davidson/Photo Researchers, Inc.; *p. 758, 18.4, 18.8A,B:* © McGraw-Hill Higher Education/Stephen Frisch Photographer; *p. 774:* © Tony Freeman/PhotoEdit; *p. 779:* © Louis Mazzatenta/National Geographic Image Collection; *p. 788:* © McGraw-Hill Higher Education/Stephen Frisch Photographer.

Chapter 19

Opener: © Jason Childs/FPG International; *19.1, 19.5:* © McGraw-Hill Higher Education/Stephen Frisch Photographer; *19.10:* © Sr. Gopal Murti/Science Photo Library/Photo Researchers, Inc.; *p. 824:* © McGraw-Hill Higher Education/Stephen Frisch Photographer; *p. 827:* © Scovil Photography; *19.11:* © McGraw-Hill Higher Education/Stephen Frisch Photographer; *p. 829:* © CNRI/SPL/Science Source/Photo Researchers, Inc.; *19.12:* © McGraw-Hill Higher Education/Stephen Frisch Photographer; *B19.1:* © Ned Haines/Photo Researchers, Inc.; *p. 832:* © McGraw-Hill Higher Education/Stephen Frisch Photographer; *B19.3:* © Clyde H. Smith/Peter Arnold, Inc.; *B19.4 (left):* © NYC Parks Archive/Fundamental Photographs; *B19.4 (right):* © Kristen Brochmann/Fundamental Photographs; *p. 837:* © Phil Degginger/Color-Pic Inc.; *19.15, 19.17, 19.18, 19.19:* © McGraw-Hill Higher Education/Stephen Frisch Photographer

Chapter 20

Opener: © Milepost 92 1/2/Corbis; *p. 856:* © Richard Megna/Fundamental Photographs; *20.1:* © McGraw-Hill Higher Education/Stephen Frisch Photographer; *20.2A:* © Chris Sorensen Photography; *20.2B:* © Kristen Brochmann/Fundamental Photographs; *p. 860:* © Paul Silverman/Fundamental Photographs; *20.4:* © Alder and Wainwright, "Molecular Motion", October 1959, Scientific American, Inc.; *B20.1:* © Bettmann/Corbis; *B20.2:* © Vanderbilt University-Clinical Nutrition Research; *p. 872:* © Science Photo Library/Photo Researchers, Inc.; *p. 873:* © Karl Weatherly/PhotoDisc; *p. 876:* © Tom Pantages

Chapter 21

Opener: © David Wrobel/Visuals Unlimited; *p. 893:* © AP/Wide World Photos; *21.1, 21.2, 21.5B:* © McGraw-Hill Higher Education/Stephen Frisch Photographer; *p. 906:* © Richard Megna/Fundamental Photographs; *21.8:* © McGraw-Hill Higher Education/Stephen Frisch Photographer; *p. 917:* © AIP Visual Archives/American Institute of Physics; *p. 920:* © Biophoto Association/Photo Researchers, Inc.; *p. 922:* © Courtesy of Dr. Gerhard Dahl/University of Miami School of Medicine; *p. 923 (both):* © Chris Sorensen Photography; *p. 924:* © McGraw-Hill Higher Education/Pat Watson, photographer; *p. 925 (drill):* © McGraw-Hill Higher Education/Stephen Frisch Photographer; *p. 925 (laptop):* © AP/Wide World Photos; *21.13:* © F. J. Dias/Photo Researchers, Inc.; *21.14:* © David Weintraib/Photo Researchers, Inc.; *21.18:* © Chris Sorensen Photography; *21.19, p. 934:* © McGraw-Hill Higher Education/Stephen Frisch Photographer; *p. 935:* © Faraday Museum/Royal Institute of Great Britain; *B21.1B:* © K. R. Porter/Photo Researchers, Inc.

Chapter 22

Opener: © Aaron Haupt/Photo Researchers, Inc.; *22.3A:* © Doug Sherman; *22.3B:* S. Conway Morris/Cambridge University-Dept. of Earth Science; *p. 954:* © Ric Ergenbright Photography; *p. 960:* © Dennis Milon/Science Photo Library/Photo Researchers, Inc.; *p. 964:* © Patrick W. Grace/Photo Researchers, Inc.; *22.12:* ©

Richard Megna/Fundamental Photographs; *p. 969:* © Christine L. Case/Visuals Unlimited; *p. 970:* © Remi Benali/Getty/Liaison Agency; *22.17B:* © Bethlehem Steel Corporation; *22.18B:* © Tom Hollyman/Photo Researchers, Inc.; *p. 974:* © Laine Remi/Getty/Liaison Agency; *p. 976 (top left):* © Rob Crandall/Rainbow; *p. 976 (bottom left):* © Hank Morgan/Rainbow; *p. 976 (top right):* © E. R. Degginger/Color-Pic Inc.; *p. 976 (bottom right):* © Alcoa Technical Center; *p. 980:* © McGraw-Hill Higher Education/Stephen Frisch Photographer; *p. 982 (top):* © Helga Lade/Peter Arnold, Inc.; *p. 982 (bottom):* © Ruth Melnick; *p. 983 (top):* © Tom McHugh/Photo Researchers, Inc.; *p. 983 (bottom left):* © Ken Whitmore/Stone; *p. 983 (bottom right):* © Lester Lefkowitz/Tech Photo, Inc.; *p. 984 (top):* © Robert Holmes/Corbis; *p. 984 (center):* © Westinghouse/Visuals Unlimited; *p. 984 (bottom two):* © Chris Sorensen Photography; *p. 985 (top):* © McGraw-Hill Higher Education/Stephen Frisch Photographer; *p. 985 (bottom):* © Noranda Inc. Fonderie Horne; *p. 986 (top):* © Scott Camazine/Photo Researchers, Inc.; *p. 986 (center two):* © Chris Sorensen Photography; *p. 986(bottom):* © Brad Jirka and St. Elmo's; *22.22C:* © Nathan Benn/Woodfin Camp & Associates

Chapter 23

Opener: © Michael Melford Inc./Image Bank; *23.2, 23.5, 23.6:* © McGraw-Hill Higher Education/Stephen Frisch Photographer; *p. 1008:* © E. R. Degginger/Color-Pic Inc.; *23.7B:* © Richard Megna/Fundamental Photographs; *p. 1009 (top):* © E. R. Degginger/Color-Pic Inc.; *p. 1009 (bottom):* © McGraw-Hill Higher Education/Stephen Frisch Photographer; *p. 1010:* © Institute of Oceanographic Sciences/NERC/Science Photo Library/Photo Researchers, Inc.; *23.8 (right):* © Kathy Shorr; *p. 1013:* © The Granger Collection; *p. 1020:* © Richard Megna/Fundamental Photographs; *p. 1025:* © Ruth Melnick; *23.20A:* © McGraw-Hill Higher Education/Pat Watson, photographer; *23.21:* © McGraw-Hill Higher Education/Stephen Frisch Photographer

Chapter 24

Opener: © Cern, P. Loiez/Science Photo Library/Photo Researchers, Inc.; *p. 1043:* © W. F. Meggers Collection/American Institute of Physics; *B24.1:* © Hank Morgan/Rainbow; *B24.2:* © Hewlett Packard; *p. 1054:* © Getty/Liaison Agency; *p. 1055:* © Monique Salaber/Getty/Liaison Agency; *p. 1056:* © Hulton Archive/Getty Images; *24.6B, p. 1057:* © Stanford Linear Accelerator Center; *p. 1061:* © E. O. Lawrence Berkeley National Laboratory; *p. 1064:* © NASA/Photo Researchers, Inc.; *24.9A:* © Scott Camazine/Photo Researchers, Inc.; *24.9B:* © T. Youssef/Custom Medical Stock Photo; *24.10:* © Dr. Robert Friedland/Science Photo Library/Photo Researchers, Inc.; *24.11:* © Dr. Dennis Olson/Meat Lab- Iowa State University, Ames, IA; *p. 1069:* © Emilio Segre Visual Archives/American Institute of Physics; *24.16A:* © Albert Copley/Visuals Unlimited; *B24.4:* © Space Telescope Science Institute/NASA/Science Photo Library/Photo Researchers, Inc.; *24.17 (top):* © Dietmar Krause/Princeton Plasma Physics Lab

INDEX

Page numbers followed by *f* indicate figures; *mn,* marginal notes; *n,* footnotes; and *t,* tables. Page numbers preceded by "G–" indicate pages in the *Glossary.*

A

Absolute temperature scale, 24, G–1
Absolute zero, 24
Absorption spectrum, 267–268, 267*f,* 268*f,* G–1
Abundance, of elements, 951–955, 952*f,* G–1
Accuracy, in measurement, 31, 31*f,* G–1
Acetaldehyde, molecular structure, 629*t,* 634*f*
Acetamide, molecular structure, 629*t*
Acetaminophen, molecular structure, 638*f*
Acetate ion, formula, 67*t*
Acetic acid
  boiling point elevation, 509*t*
  formula, 100*t,* 758*t*
  freezing point depression, 509*t*
  molecular model, 73*f*
  molecular structure, 629*t*
  reaction with baking soda, 146, 146*f*
  uses, 100*t,* 758*t*
Acetone
  molecular structure, 629*t,* 634*f*
  NMR spectrum, 624, 625*f*
Acetonitrile, molecular structure, 629*t*
Acetylene, molecular structure of, 629*t*
Acid(s)
  Arrhenius acid, 141, 759, 763, 794
  Brønsted-Lowry acid, 145, 768–770, 790–791, 794
  buffers, 806–815
  concentration, finding from titration, 143
  conjugate acid-base pair, 769–770, 782–784, 806, 811
  defined, 141, G–1
  dissociation, 759–761, 760*f,* 776
  equilibria *see* Acid-base equilibria
  household uses, 758*t*
  indicators, 142, 768, 815–816, 823
  leveling effect, 790–791
  Lewis acid, 791–795, 795
  monoprotic, 762, 762*t*
  naming, 69–70, 72
  pH, 765–768, 920
  polyprotic, 776–777, 777*t,* 822–823
  reactions, 140–146, 141*f,* 141*t,* 145*f*
  strength *see* Acid strength
  strong acids, 140–141, 141*t,* 143, 145*f,* 146, 760, 762, 763, 791
  titrations, 143–144, 143*f,* 816–823
  in water, 758–763
  weak acids, 140–141, 141*t,* 145*f,* 146, 760, 763
    weak-acid equilibria, 772–779
    weak bases and, 779–784
Acid anhydride, 637, G–1
Acid-base behavior, element oxides, 312–313, 537
Acid-base buffer systems, 806–815
  buffer capacity, 812, 812*f*

buffer range, 812, 812*f*
common-ion effect, 806–811, 807*t*
conjugate acid-base pair, 813
defined, 806, G–1
essential features, 808–809, 808*f*
Henderson-Hasselbalch equation, 811
preparing, 813–814
Acid-base equilibria, 756–795
  acid-base indicators, 768
  acid-dissociation constant (K_a), 759–762, 760*f,* 761*f,* 762*t,* 766*t*
    weak-acid equilibria problem solving, 773–776, 777*t*
    weak bases and weak acids, 782–784
  autoionization of water, 764–768, 765*f*
  base-dissociation constant (base-ionization constant) (K_b), 779, 782–784
  Brønsted-Lowry acid-base definition *see* Brønsted-Lowry definition, acids and bases
  conjugate acid-base pairs, 782–784
  hydrated metal ions, 786, 786*f,* 786*t*
  hydronium ion, 758–759
  ion-product constant for water (K_w), 764–768, 765*f,* 782–784
  Lewis acid-base definition *see* Lewis acid-base definition
  pH, 765–768, 766*t,* 766*t,* 767*f,* 768*f*
  polyprotic acids, 776–779, 777*t*
  salt solutions, 787, 790
    acidic solution yields, 787–788
    basic solution yields, 788–789, 788*t*
    neutral solution yields, 787
    weakly acidic cations, 789–790
    weakly basic anions, 789–790
  variation in acid strength, 759–762, 760*f,* 761*f,* 762*t*
  weak-acid equilibria problem solving, 772–773, 779
    acid-dissociation constant, 773–776, 777*t*
    concentrations, 773–776
    notation system, 773
    polyprotic acids, 776–779, 777*t*
Acid-base indicators
  acid-base reactions, 143
  acid-base titration curves, 815–816, 815*f,* 816*f*
  defined, 768, G–1
Acid-base properties, of salt solutions, 787–790
Acid-base (neutralization) reactions, 140–141, 146
  acid-base indicator, 143
  acid-base titrations, 143–144, 143*f*
  acids, 141, 141*t*
  bases, 141, 141*t*
  defined, 140, G–1, G–12
  driving force and net change, 141–142
  end point, 143
  equivalence point, 143
  forming a gaseous product, 146, 146*f*
  H_2O formation, 141–142
  ionic equations, 142, 146*f,* 146

molecular equations, 146*f*
net ionic equations, 146*f*
as proton-transfer reaction, 144–146
salts, 142
strong acid with strong base, 145–146, 145*f*
titrations, 143–144, 143*f*
total ionic equations, 146*f*
of weak acids, 146, 146*f*
of weak bases, 141, 141*t*
Acid-base titration, 143–144, 143*f,* 816–823
Acid-base titration curves, 815–823
  acid-base indicators, 815–816, 815*f,* 816*f*
  defined, 815, G–1
  polyprotic acids, 822–823
  strong acid-strong base titration curves, 816–818, 823
  weak acid-strong base titration curves, 818–821, 823
  weak base-strong acid titration curves, 821–822, 823
Acid-dissociation constant (K_a)
  defined, 760, G–1
  meaning of, 759–762, 760*f,* 761*f,* 762*t,* 766*t*
  relation with K_b in conjugate pair, 782–784
  weak-acid equilibria problem solving, 773–776, 777*t*
Acid-ionization constant *see* Acid-dissociation constant (K_a)
Acid rain, 32, 33*f,* 833–834, 833*f,* 834*f*
Acid strength, 763, 786
  acid-dissociation constant and, 759–761, 762*t*
  direction of reaction and, 770–772
  hydrated metal ions, 786, 786*f,* 786*t*
  molecular properties and, 784–786
  of nonmetal hydrides, 784–785, 786
  of oxoacids, 785, 786
  strong acids, 140–141, 141*t,* 145–146, 145*f,* 146, 760, 762, 763, 791
  weak acids, 140–141, 141*t,* 146, 146*f,* 760, 763
Acrylonitrile
  infrared spectrum, 343
  molecular structure, 566
Actinides, 1007
  defined, 1006, G–1
  electron configuration, 303
  periodic table, 57, 57*f*
Activated complex, 688, G–1
Activation energy (E_a), 683, G–1
Active metal, reduction of, 966, 966*f*
Active nonmetal, oxidation of, 966
Active site, enzyme, 700*f,* G–1
Activity, 1050–1053, G–1
Activity series of metals, 160, 160*f,* 912–913, G–1
Actual yield, 112, G–1
Addition, significant figures, 29
Addition polymers, 642–643, 642*f,* 643*t,* G–1
Addition reaction
  with benzene, 633
  defined, 625, G–1
  organic, 625

Adduct *see* Lewis acid-base definition
Adenine, molecular structure, 631*f*
Adenosine triphosphate *see* ATP
Airplanes, de-icing, 512
Air pollution
  acid rain, 32, 33*f,* 833–834, 833*f,* 834*f*
  CFCs, 702
  power plant emissions, 33*f*
  sulfur dioxide, 586*mn*
-al (suffix), 629*t*
Alanine, structure, 646*f*
Alchemy, 8–9, 9*f,* G–1
Alcohol, grain, density, 23*t*
Alcohols, 628, 629*t,* 630, G–1
Aldehydes, 629*t,* 633–635, 634*f,* G–1
Alkali metals
  electron configuration, 299
  industrial production, 969–970
  natural sources, 982*t*
  periodic table, 67
  properties, 548–551, 552
  reactions, 551
  reaction with halogen, 155, 155*f*
  reactivity, 548–549
  uses, 982*t*
Alkaline battery, 923, 923*f*
Alkaline earth metals
  isolation, 982*t*
  natural occurrence, 982*t*
  periodic table, 67
  properties, 552–554
  reactions, 555
  reactivity, 552
  uses, 982*t*
Alkanes, 613
  compounds, 71, 71*t*
  condensed formula, 614, 614*f*
  covalent compounds, 71, 71*t*
  cyclic hydrocarbon, 615, 615*f*
  cycloalkanes, 615, 615*f*
  defined, 613, G–1
  depicting with formulas and models, 614–615, 614*f*
  expanded formula, 614, 614*f*
  homologous series, 613
  naming, 613–614, 613*t,* 614*t*
  saturated hydrocarbons, 613
Alkenes, 618–619, 629*t,* 633, G–1
Alkoxide, 628
Alkyl group, 624, G–1
Alkyl halides, 630, 632, G–1
Alkylmagnesium halides, 555
Alkynes, 621, 623, 629*t,* G–1
Allotropes
  carbon, 563
  defined, 563, G–1
  oxygen, 584
  phosphorus, 573, 573*f*
  selenium, 584
  sulfur, 584
  tin, 566
Alloys, 492, 492*f,* 967
  common alloys, 967*t*
  defined, 350, 492, 967, G–1
  dental alloys, 421*mn*
  metallic bonding, 350
  metallurgy, 967–968, 967*t,* 968*f*
  solutions, 492
Alpha (α) decay, 1044, 1044*t*

Alpha (α) particles
  de Broglie wavelength, 270t
  defined, 1043, G–1
  nuclear reactions, 1043, 1043f,
      1044t, 1059
  Rutherford's scattering experiment,
      50–51, 50f, 56mn
Aluminum
  bond type, 562t
  electrochemical energy, 892f
  electron configuration, 298t, 299f
  ion formation, 62
  isolation, 974–976, 984t
  melting point, 562t
  metallurgy, 974–976, 987
  mineral source, 963t, 984t
  properties, 312, 313f, 558, 562
  recycling, 975
  specific heat capacity, 233t
  uses, 976f, 984t
Aluminum chloride, 348, 557f
Aluminum hydroxide, complex ions
      with, 838–839, 839f
Aluminum ion, 65t
Aluminum oxide, 313, 559
Aluminum salts, 758t
Americium, name, origin of, 52mn
-amide (suffix), 629t
Amide group, 636
Amides, 629t, 636, 638–639, 638f, G–1
-amine (suffix), 629t
Amines, 629t, 631–632, 779–781,
      780f, G–1
Amino acids, 646–647, 646f, G–1
Ammonia, 575
  Haber process, 744–745
  history, 779mn
  Lewis structure, 780t
  molecular shape, 374
  specific heat capacity, 233t
  as weak base, 141
  as weak base in water, 141–142
Ammonium carbonate, moles and
      number of formula units,
      calculating, 92
Ammonium ion, 67t
Ammonium perchlorate, 591
Amontons's law, 183
Amorphous solids, 445, 456–459, G–1
Amount, unit of, 17t, 87
Ampere (A) (unit), 17t, 935, G–2
Amphoteric, 313, G–2
Amphoteric hydroxides, complex ions,
      838–840
Amplitude, of wave, 256, 256f
Analytical
  electrochemistry, 893–939
  history, 8–11, 41, 757mn
  inorganic chemistry, 58
  organic chemistry, 58, 566–567,
      607–608
  qualitative analysis, 841–845
  real-world problems, 32–33, 33f
  scientific method, 11–13, 12f, 13mn
  thermochemistry, 221
  units, 13
Analytical balance, 21
-ane (suffix), 71
Angstrom (Å), conversion to meters,
      19, 258
Angular momentum quantum number
      (l), 275, 291t, G–2
Aniline, 780t
Anions, 782
  compounds, 59, 65t
  defined, 59, G–2
  weakly basic anions, 789–790

Anode, 898, 898mn, G–2
Antarctic ozone hole, 702, 702f
Antibiotics, mode of action, 487
Antibonding MO, 404–405, 405f, G–2
Anticancer drugs, 1021mn
Antifreeze, 512
Antimony
  bond type, 562t
  isolation, 985t
  melting point, 562t
  natural sources, 985t
  properties, 312, 313f, 574
  uses, 985t
Apatites, 960, G–2
Aqueous equilibria
  acid-base buffer system, 806–815
  acid-base titration curves, 815–823
  chemical analysis and, 840–845
  complex ions, 835–840
  slightly soluble salts, 824–832
Aqueous ionic reactions, 137–138
  acid-base reactions see Acid-base
      (neutralization) reactions
  molecular equations, 137, 137f
  net ionic equation, 137f, 138
  precipitation reactions, 137f
  spectator ions, 138
  total ionic equation, 137f, 138
Aqueous solutions, 74
  defined, 74, G–2
  electrolysis, 933–934
  of salts, 787–790
  standard state, 240
Arginine, structure, 646f
Argon
  cubic closest packing, 452f
  electron configuration, 298t, 299f
  isolation, 986t
  natural occurrence, 59, 986t
  properties, 595, 596
  uses, 986t
Aristotle, nature of matter, 41
Aromatic carcinogens, 623
Aromatic compounds, reactivity, 633
Aromatic hydrocarbons, 622–623, G–2
Arrhenius, Svante, 141
Arrhenius acid-base definition see
      Classical (Arrhenius) acid-base
      definition
Arrhenius equation, 682, 685, G–2
Arsenic
  bond type, 562t
  electron configuration, 300t
  isolation, 985t
  melting point, 562t
  natural sources, 985t
  orbital diagrams, 300t
  properties, 312, 313f, 574
  uses, 985t
Arsenic acid, acid-dissociation
      constants, 777t
Asbestos, 570
Ascorbic acid, equilibrium
      concentrations, 778–779
Asparagine, structure, 646f
Aspartic acid, structure, 646f
Aspirin, chemical formula, 73f
Astatine
  isolation, 986t
  natural sources, 986t
  properties, 590
  uses, 986t
-ate (suffix), 68
Atmosphere
  defined, 204, 953, G–2
  of earth, 174, 204
    composition, 205, 205t

gases, diffusion and flow, 421mn
ozone, 205–206
ozone depletion, 206, 584,
      702, 702f
pressure variations, 204, 204f
primitive atmosphere, 206
regions of, 205–206
sulfur dioxide, 586mn
temperature variations,
      204–205, 204f
planetary, 206, 206t
Atmosphere (atm) (unit), 178, 179, 179t
Atmospheric fixation, 958
Atom, 46, 77, G–2
  history of concept, 41, 46
  notation, 52, 56
  nucleus see Nucleus
  structure see Atomic structure
Atomic bomb, 1070–1071
Atomic clock, 26f
Atomic mass, 52–55, 56, 89n
  calculating, 55
  defined, 53, 89t, G–2
  of element, 88–89
Atomic mass unit (amu), 53, 55,
      88, G–2
Atomic nucleus see Nucleus
Atomic number (Z)
  defined, 52, G–2
  periodic table, 290
Atomic orbital, 273, 281
  aufbau principle, 295, 304
  defined, 273, G–2
  d orbital, 280
  filling order, 302
  f orbital, 280, 280f
  hybridization, 394
  Pauli exclusion principle,
      291–292, 304
  p orbital, 279, 279f
  quantum numbers, 275–277, 276t
  s orbital, 278–279, 278f
Atomic properties
  alkali metals, 550
  alkaline earth metals, 554
  atomic size see Atomic size
  boron family, 558
  carbon family, 564
  electron configuration see Electron
      configuration
  electronegativity see
      Electronegativity (EN)
  halogens, 590
  ionization energy see Ionization
      energy (IE)
  nitrogen family, 574
  noble gases, 596
  oxygen family, 582
  transition elements/metals,
      1002–1003, 1002f, 1003f
Atomic radius, 304–306, 304f–306f,
      306mn, 318f
  crystal structure, determination
      from, 450
Atomic size, 304, 304f, 306, 311,
      311f, 533
  alkali metals, 550
  alkaline earth metals, 554
  boron family, 558
  carbon family, 564
  covalent radius, 304
  defined, 533, G–2
  electronegativity, 345, 345f
  halogens, 590
  main-group elements, 305, 305f
  metallic radius, 304
  monatomic ions, 318–319

nitrogen family, 574
noble gases, 596
oxygen family, 582
periodic trends in, 304–307, 533
thermodynamics, 864
transition elements/metals,
      305–306, 305f, 306f, 1002,
      1002f, 1003f
Atomic solids, 452, 452f, 453t,
      459, G–2
Atomic spectra, 255, 262–263, 266
  Bohr, Niels, 263
  Bohr model of the hydrogen atom,
      263–264, 265f
  energy states of the hydrogen atom,
      264–266, 265f
  excited state, 264
  ground state, 264
  Janssen, Pierre, 266
  line spectrum, 262f, 263
  quantum staircase, 264f
  Ramsay, William, 266
  Rydberg equation, 263
  stationary state, 263–264
Atomic structure, 51–52, 51f, 56 see
      also Electron (e⁻); Neutron
      (n⁰); Proton (p⁺)
  Bohr model, 263–264
  chemical reactivity and, 311–319
  many-electron atoms, 290–294
  orbitals, 278–280, 278f–280f
Atomic symbol, 52, 56, G–2
Atomic theory
  atomic nucleus, discovery of,
      50–51, 50f
  Dalton's theory, 43, 46–47
  electron, discovery of, 48–49, 49f
  history, 41, 43–47, 255
  law of definite composition, 43,
      44–45, 44f, 47
  law of mass conservation, 43–44,
      43f, 46
  many-electron atoms, 290–294
  Millikan's oil drop experiment,
      49, 49f
  modern theory, 51–56
  nuclear atom model, 48–51
  relative masses of atoms, 47
  Rutherford's scattering experiment,
      50–51, 50f, 56mn
Atomic weight see Atomic mass
ATP (adenosine triphosphate), 575, 581,
      879–880, 879f, 880f, G–1
Aufbau principle, 295, 304, G–2
Autoionization, water, 764–765,
      768, G–2
Automobiles
  antifreeze, 512
  car wax, 493mn
  catalytic converter, 699mn
Average reaction rate, 668, G–2
Avogadro, Amedeo, 87
Avogadro's law, 180, 183–184, 183f,
      184mn, 185, 185f, 189, 197,
      200, 200f, G–2
Avogadro's number, 87, 201
  as conversion factor, 90, 91, 92f,
      95, 120f
  defined, 87, G–2
Axial groups, VSEPR theory, 375, G–2

B

Background radiation, 1061, G–2
Baking soda, reaction with vinegar,
      146, 146f

Balanced equation, 113
  information in, 105, 105*t*
  process of balancing, 102–104
  redox reaction, 150–152, 154,
     894–898
  states of matter, 103
  stoichiometrically equivalent molar
     ratios, 106–108, 113
Balancing coefficient, 102, G–2
Ball-and-stick models, 73*f*
Balloon, hot-air, 190*mn*
Ballpoint pen, action of, 441
Band of stability, 1046, G–2
Band theory, 456–459, G–2
Bar (unit), 178, 179*t*
Barium
  isolation, 982*t*
  mineral source, 963*t*, 982*t*
  properties, 554
  uses, 982*t*
Barometer, 176–177, 177*f*, G–2
Base(s) *see also* Acid-base
     (neutralization) reactions;
     Acid-base titration; Acid-base
     titration curves
  anions of weak acids as, 782
  Arrhenius base, 141, 759, 763, 794
  Brønsted-Lowry base, 145, 768–770,
     790–791, 794
  buffers, 806–815
  conjugate acid-base pair, 769–770,
     782–784, 806, 811
  defined, 141, G–2
  household uses, 758*t*
  indicators, 143, 768, 815–816, 823
  leveling effect, 790–791
  Lewis base, 791, 835, 1023
  strong bases, 140–141, 141*t*,
     145–146, 145*f*, 146, 763, 791
  titrations, 143–144, 143*f*, 816–823
  in water, 758–763
  weak bases, 140–141, 141*t*, 763
    weak acids and, 779–784
Base-ionization constant (K_b), 779, G–2
Base pairing, 651, G–2
Base units, in SI, 17, 26, G–2
Basic-oxygen process, 972, 972*f*, G–2
Batteries, 922, 926
  alkaline, 923
  aluminum, 975
  button batteries, 924
  defined, 922, G–3
  dry cell, 923
  flow batteries, 922, 925
  fuel cells, 922, 925
  lead-acid batteries, 924
  lithium-ion batteries, 925
  mercury and silver batteries, 924
  nickel-metal hydride (Ni-MH)
     batteries, 925
  nonrechargeable batteries, 923–924
  primary batteries, 923–924
  rechargeable batteries, 924–925
  secondary batteries, 924–925
Battery-gasoline hybrid cars, 893
Bauxite, 974
Becquerel, Antoine-Henri, 1042, 1043
Becquerel (bq) (unit), 1050, G–3
Beeswax, 493*mn*
"Bends," 500*mn*
Bent shape, VSEPR theory, 373, G–3
Benzaldehyde, molecular
     structure, 634*f*
Benzene, 622, 622*f*
  bonding, 413, 413*f*
  Lewis structure, 362–363
Benzo[*a*]pyrene, 623*mn*

Benzoic acid, molecular structure, 636*f*
Benzylcetyldimethylammonium
     chloride, molecular
     structure, 632*f*
Beryl, 555
Beryllium
  bonding, 329
  diberyllium, 407, 407*f*
  electron configuration, 296, 297*f*
  ionization energy, 309*t*
  isolation, 982*t*
  natural sources, 963*t*, 982*t*
  properties, 312, 546*t*, 553, 554, 562
  uses, 546*t*, 982*t*
Beryllium chloride
  hybrid orbitals, 394, 395*f*
  Lewis structure, 365
  molecular shape, 372
Beryllium compounds, 365n
Beta (β) decay, 1044*t*, 1045, G–3
Beta (β) particles, 1043, 1043*f*, 1044*t*,
     1059, G–3
Bevatron, 1057*mn*
Bhopal, 189
Bicarbonate ion, 67*t*
Bidentate ligands, 1016*t*
Big Bang theory, 1074–1075, 1075*f*
Bile salts, 518*mn*
Bimolecular reaction, 692, G–3
Binary acid, naming, 69
Binary alloys, 968, 968*f*
Binary covalent compounds
  defined, 70, G–3
  names and formulas, 70–71
  numerical prefixes for, 68*f*
Binary ionic compounds, 59–61,
     65–66, G–3
Binding energy *see* Nuclear
     binding energy
Binnig, Gerd, 452
Biological antifreeze, 512
Biological catalysts, 700–701,
     700f, 701*f*
Biological energetics, 879–880,
     879*f*, 880*f*
  electrochemistry, 937–938, 937*f*, 938*f*
  thermodynamics, 879–880,
     879*f*, 880*f*
Biological fixation, 959
Biological macromolecules, 644, 653
  amino acids, 646–649, 646*f*, 647*f*
  base pairs, 651
  basic amino acids, 646*f*
  cellulose, 645
  chromosomes, 651
  disaccharides, 645, 645*f*
  disulfide bridge, 648
  DNA, 650–653, 650*f*, 651*f*, 653*f*
  DNA replication, 653, 653*f*
  double helix, 651, 651*f*
  fibrous proteins, 648–649, 649*f*
  genetic code, 652
  globular proteins, 649
  glycogen, 646
  mononucleotides, 650, 650*f*
  monosaccharides, 645
  mRNA, 652
  natural polymers, 644
  nonpolar amino acids, 646*f*
  nucleic acids, 650–653
  nucleotides, 650–653
  peptide bond, 647, 647*f*
  polar amino acids, 646*f*
  polysaccharides, 644–646
  proteins, 646–649, 647*f*, 648*f*
  protein synthesis, 652, 652*f*
  ribosome, 652

RNA, 650, 650*f*
  starch, 645
  sugars, 644–646
  tRNA, 652
Biomass conversion, 243, G–3
Biopolymers *see* Biological
     macromolecules
Biosphere, 953, G–3
Bismuth
  bond type, 562*t*
  isolation, 985*t*
  melting point, 562*t*
  natural sources, 985*t*
  properties, 312, 313*f*, 574
  uses, 985*t*
Bismuth subsalicylate, 575
Bisulfate ion, 67*t*
Blackbody radiation, 260, 260*f*, 262
Blast furnace, 970, 971*f*, G–3
Block copolymer, 473
Body-centered cubic cell, 448, G–3
Body-centered cubic unit cell, 446, 446*f*
Body temperature, 26
Bohr, Niels, 263, 266, 273*mn*
Bohr model, 263–264, 265*f*, 266
Boiling point
  defined, 428, G–3
  dipole moment and, 434, 435*f*
  hydrogen bonding and, 434, 435*f*
  ionic compounds, 336*t*
  metals, 350*t*
  molar mass and, 437, 437*f*
  molecular shape and, 437, 437*f*
  phase changes, 428–429
  solubility and, 491*t*
  vapor pressure and, 428–429
  water, 25*f*
Boiling point elevation, 508–509, 509*t*,
     510, G–3
Bomb calorimeter, 235, 235*f*
Bond angle, 371, 371*f*, G–3
Bond energy (BE), 338, 339*t*, 341*t*, 535
  calculating heat of reaction from,
     368–370, 368*f*
  defined, 338, 535, G–3
  electrochemical potential to, 938
  in foods, 232
  in fuels, 232
Bond enthalpy *see* Bond energy
Bonding, 327–329, 328*f*, 330, 534–535
  alkali metals, 550
  alkaline earth metals, 554
  atomic properties and, 327–330
  bond strength, 230
  boron family, 558
  breaking bonds, 231
  chemical bond defined, 59, G–4
  covalent *see* Covalent bonds
  delocalized electron-pair bonding,
     362–363
  electronegativity, 344–346
  electron-sea model, 328,
     349–350, 456
  forming bonds, 231
  ionic *see* Ionic bonding
  Lewis electron-dot symbols,
     329, 329*f*
  metallic, 328–329, 328*f*, 349–351,
     433*t*, 534
  octet rule, 330
  oxygen family, 582
  partial ionic character, 347–348, 347*f*
  polar covalent bonds, 346–348
  types, 327–329, 534
Bonding MO, 404–405, 405*f*, G–3
Bonding pairs, 338, G–3

Bond length, 340*f*, 340*t*, 341*t*, 535, G–3
Bond-line formula, 73*f*
Bond order, 338, 535, G–3
Bond polarity, 346
Bond strength, 230, 338
Boranes, 561
Borates, in labware, 560*mn*
Borax, 559
Boric acid, 559, 758*t*
Born-Haber cycle, 332–333, 336,
     494, G–3
Boron
  bond type, 562*t*
  diagonal relationship, 569
  diatomic molecule, orbitals, 410*f*
  electron configuration, 296, 297*f*
  ionization energy, 309*t*
  isolation, 984*t*
  melting point, 562*t*
  natural sources, 984*t*
  properties, 546*t*, 558
  uses, 546*t*, 984*t*
Boron family
  alkaline earth metals compared, 553
  isolation, 984*t*
  properties, 556–562
  sources, 984*t*
  uses, 984*t*
Boron oxide, 559
Boron trifluoride
  hybrid orbitals, 396, 396*f*
  Lewis structure, 365
  molecular shape, 372
Boyle, Robert, 41, 180, 757*mn*
Boyle's law, 180–181, 180*f*, 184*mn*,
     185, 185*f*, 197, 198, 198*f*, G–3
Bragg, W.H., 451
Bragg, W.L., 451
Bragg equation, 451
Branch, of polymer chain, 472, G–3
Brass, 492, 967*t*
Breathing, gas laws and, 184*mn*
Bridge bond, 561, G–3
British thermal unit (Btu) (unit), 226
Bromine
  electron configuration, 300*t*
  isolation, 978, 986*t*, 987
  natural sources, 986*t*
  orbital diagrams, 300*t*
  properties, 590
  uses, 978, 986*t*
Brønsted, J. N., 145
Brønsted-Lowry acid-base
     reactions, 592
Brønsted-Lowry definition, acids and
     bases, 145, 768–770, 790–791,
     794, G–3
Bronze, 967*t*
Brownian motion, 518
Buckminsterfullerene ("bucky ball"),
     380, 380*f*, 563, 563*f*
Buffer *see* Acid-base buffer systems
Buffer capacity, 812, G–3
Buffer range, 812, G–3
Butane
  boiling point, 617*f*
  chemical formula, 73*f*
  combustion of, 161
  formula, 71*t*
  properties, 616*t*
  structural isomers, 616*t*
  uses, 73*f*
Butanoic acid, molecular
     structure, 636*f*
Butanol, solubility, 488, 488*t*
2-Butanone, molecular structure, 634*f*

Butene, geometric isomers, 619, 619*t*
Button battery, 924
Buytric acid, molecular structure, 636*f*

C

Cadmium, toxicity, 1012
Calcium
electron configuration, 300, 300*t*
ion formation, 62
isolation, 982*t*
natural occurrence, 954*t*, 963*t*, 982*t*
orbital diagrams, 300*t*
properties, 554
standard heat of formation, 240*t*
uses, 982*t*
Calcium oxide, 552*mn*
Calculators, significant figures and, 30
Calibration, in measurement, 31, 31*f*, G–3
Calorie (cal) (unit), 225, G–3
Calorimeter, 234, G–3
Calorimetry, 234, 236
bomb calorimeter, 235, 235*f*
coffee-cup calorimeter, 234, 234*f*
constant-pressure calorimetry, 234
constant-volume calorimetry, 235–236
heat capacity of calorimeter, 233
whole-body calorimeter, 869*f*
Cameron, 189
Cancer drugs, 1021*mn*
Candela (cd) (unit), 17*t*
Capillarity, 439–440, 440*f*, 441, 442, 443, G–3
Carbohydrates, heat of combustion of, 232*t*
Carbon see also Diamond; Graphite; Organic chemistry; Organic compounds; Organic molecules
allotropes, 563
bond type, 562*t*
chemistry, 566–567, 607–608
diatomic molecule, orbitals, 410*f*
electron configuration, 296, 297*f*
ionization energy, 309*t*
isolation, 984*t*
isotopes, 52–53
melting point, 562*t*
molecular stability, 608
name, origin of, 52*mn*
natural occurrence, 59, 954*t*, 984*t*
properties, 311, 546*t*, 564, 608
standard heat of formation, 240*t*
uses, 546*t*, 984*t*
Carbonate, reaction with acid, 145
Carbonate ion, 67*t*
Carbonates
limestone cave, 831, 831*f*
metal, 567
thermal decomposition, 156
Carbon atoms
atomic number, 52, 53*f*
mass number, 52, 53*f*
Carbon compounds, naming, 613*t*
Carbon cycle, 956–958, 957*f*
Carbon dioxide, 565
dry ice, 429
formation in acid-base reaction, 145
global warming and, 244, 244*f*
law of multiple proportions, 47, 47*f*
phase diagram, 430*f*
states of, 422
Carbon disulfide, 509*t*
Carbon family
allotropes, 563
isolation, 984*t*

natural sources, 984*t*
properties, 562–572
uses, 984*t*
Carbonic acid, acid-dissociation constants, 777*t*
Carbon monoxide, 565
boiling point, 491*t*
law of multiple proportions, 47, 47*f*
molecular model, 73*f*
solubility, 491*t*
toxicity, 1033
Carbon skeleton see Organic compounds, naming
Carbon steel, 492, 972, G–3
Carbon tetrachloride
boiling point elevation, 509*t*
freezing point depression, 509*t*
specific heat capacity, 233*t*
Carbonyl group, 633, 634*f*, G–3
Carboxylic acids, 635–636
defined, 635, G–3
functional group, 629*t*, 635
Carcinogens, aromatic compounds, 623*mn*
Car engines, combustion, 226
Carnauba wax, 493*mn*
Cars see Automobiles
Catalysis, 697–699, 703
enzymes, 700–701, 700*f*, 701*f*
heterogeneous, 699
homogeneous, 698, 702
ozone depletion and, 702, 702*f*
Catalysts
biological, 700–701, 700*f*, 701*f*
defined, 697, 703, G–3
equilibrium system, 743
heterogeneous, 699
homogeneous, 698, 702
stereoselective, 642
Catalytic converter, 699*mn*
Catalytic hydrogenation, 699, 699*f*
Catenated inorganic hydrides, 623
Catenation, 566, 608
Cathode, 898, 898*mn*, G–3
Cathode rays, 48, 48*f*, 48*mn*, 51, G–3
Cathode ray tube, 48*mn*
Cation, 59
common cations, 65*t*
defined, 59, G–3
metal cations as Lewis acids, 793–794, 793*f*
polyatomic, 67*t*
weakly acidic cations, 789–790
Caves, carbonates, 831, 831*f*
Cell potential (E_{cell})
concentration and, 916–917
defined, 905, G–3
voltaic cells, 906–913
Cell shape, and osmosis, 513
Cellular electrochemistry, 937–938, 937*f*, 938*f*
Cellular metabolism, metabolic pathway, 745–746
Cellular respiration, as combustion process, 161
Cellulose, 645
Celsius, Anders, 24
Celsius/Fahrenheit conversion, 24–26
Celsius scale, 24, 27, G–3
Cement, specific heat capacity, 233*t*
Centi- (prefix), 18*t*
Centigrade scale see Celsius scale
Centimeter (cm) (unit)
conversion to inches, 16
English equivalent, 18*t*
Ceramic materials, 465–466
defined, 466, G–3
preparation, 466

structures, 466–467, 467*f*
uses, 466–467, 466*t*
Cesium
isolation, 982*t*
natural sources, 982*t*
properties, 311
uses, 982*t*
Cesium atomic clock, 26*f*
Cetyl palmitate, molecular structure, 637*f*
CFCs (chlorofluorocarbons), 567*mn*, 702
Chadwick, James, 51
Chain-growth polymers, 642
Chain reactions, nuclear reactions, 1070, G–3
Chain silicates, 570, 570*f*
Chain silicones, 571, 571*f*
Chalcocite, 107
Chalcopyrite, 972
Change in enthalpy see Enthalpy, change in (ΔH)
Charge density, 495, G–4
Charge-induced dipole force, solubility and, 486–487
Charles, J.A.C., 181, 190*mn*
Charles's law, 180, 181–183, 182*f*, 184*mn*, 185, 185*f*, 189, 197, 199, 199*f*, G–4
Chelate, 1016, G–4
Chemical analysis
infrared spectroscopy, 343, 343*f*
ionic equilibria, 840–845
mass spectrometry, 54, 54*f*, 267–268, 267*f*, 268*f*
minimicroanalysis, 922*mn*
nuclear magnetic resonance (NMR) spectroscopy, 624–625, 624*f*
qualitative, 841–845
radioactive emissions, detection, 1051, 1051*f*
reaction rates, measuring, 672–673
scanning tunneling microscopy, 452, 452*f*
selective precipitation, 840–841, 845
X-ray diffraction analysis, 451–452, 451*f*
Chemical bonds see Bonding
Chemical change, 3, 3*f*, 5, 8, 41–42, 56, 77*f*, 86*f*, G–4 see also Chemical reactions
Chemical elements see Element
Chemical equations
actual yield, 112
balancing, 102
defined, 101–102, G–4
information contained in balanced equations, 105, 105*t*
ionic equation, 137–138, 137*f*
limiting reactants, 110–112
molar ratios from balanced equations, 106, 107*f*
molecular equation, 137, 146
net ionic equation, 138
overall, 98
percent yield (% yield), 112–113
products, 102, 107–109, 111–112
reactants, 102, 107–109, 111–112
reaction sequences, 108–109
side reactions, 112
theoretical yield, 112
thermochemical equations, 236–237
total ionic equation, 138, 146
Chemical equilibrium, 162–163 see also Equilibrium
dynamic equilibrium, 162–163
law of, 717

Chemical formulas, 72, 73*f*
by combustion analysis, 98–99, 101
covalent compounds, 70–71
defined, 64, G–4
determining, 95–99
empirical formula, 64, 66–67, 72, 95–96, 99, 100*t*, 101
ionic compounds
binary, 65–66
hydrates, 68, 72
monatomic, 64*f*, 65, 65*t*
oxoanions, 67–68, 68*f*
polyatomic, 67, 67*t*, 68–69
mass percent from, 93–95
molecular formula, 64, 72, 100, 100*t*, 101
molecular masses from, 71–72
molecular structure and, 99–100, 100*t*
structural formula, 64, 72, 100
Chemical kinetics, 664–671
defined, 664, G–4
reaction rate, 664–671, 672, 682–684
Chemical names see Nomenclature
Chemical potential energy, 7, 8
Chemical properties, 3, 4*t*, 56 see under individual chemical elements
defined, 3, G–4
periodic table, 59, 60–61, 61*f*
Chemical reactions see also Chemical equations
acid-base reactions, 140–146, 141*f*, 141*t*, 143*f*, 145*f*, 146, 146*f*
aqueous, 137–138
catalysis, 697–699, 703
combination reactions, 154–156, 155*f*, 162
coupling, 879
decomposition, 156–157, 157*f*, 162
electrolytic decomposition, 157
defined, 3, 3*f*, 154, G–4
dehydration-condensation, 580–581
displacement, 158–160, 162
double-displacement, 158
metathesis reactions, 158
disproportionation, 578
dynamic equilibrium, 162–163
endothermic, 229
equilibrium, 162–163, 162*f*
exothermic, 229
heat of reaction, 229, 230–231
Hess's law of heat summation, 238–239, 425
ionic reactions see Aqueous ionic reactions
law of mass conservation, 43–44, 43*f*, 46
law of multiple proportions, 43, 47
limiting reactant, 110–112, 110*f*, 111*mn*
mass-mole-number relationships, 107*f*, 120*f*
moles, 87–93, 101, 120*f*
multistep reactions, 108–109
neutralization, 141, 146, 757*f*, 759, 763
nuclear see Nuclear reactions
organic see Organic reactions
overall (net) equation, 108
precipitation, 137*f*, 138–140, 139*t*
reaction rate see Reaction rate
redox reactions see Redox reactions
reversibility, 162–163
spontaneity, 876–878
standard heat of, 240
stoichiometry see Stoichiometry
titration see Titration

Chemical reactions *(continued)*
  water
    polar nature of, 134–136, 134*f*–136*f*
    as a solvent, 132–136
    yields, 112–113
Chemical reactivity, atomic structure and, 311–319
Chemical shift, 624
Chemistry
  analytical *see* Chemical analysis
  defined, 3, G–4
Chemistry problems *see* Problem solving
Chirality, 617, 617*f*, 618, 618*mn see also* Optical isomers
Chiral molecule, 617, G–4
Chitin, 645*mn*
Chlor-alkali process, 990–991, 990*f*, G–4
Chlorate ion, 67*t*, 68
Chlorine
  diatomic, 348
  electron configuration, 298*t*, 299*f*
  industrial production, 990–991, 990*f*
  isolation, 986*t*
  name, origin of, 52*mn*
  natural occurrence, 954*t*, 986*t*
  properties, 43*t*, 312, 313*f*, 590
  uses, 986*t*
Chlorine dioxide, 593, 593*f*
Chlorine oxides, 593
Chlorite ion, 67*t*, 68
Chlorofluorocarbons (CFCs), 702
Chloroform, 509*t*
Chloromethane, molecular structure, 629*t*
Chlorophyll, 272*f*, 1033
Cholesteric phase, 464, 464*f*
Cholesterol, molecular structure, 628*f*
Chromate ion, 67*t*
Chromatography, 76, 76*f*, G–4
Chromium
  chromium(II) and (III), 66*t*
  electron configuration, 300, 300*t*
  in humans, 1032*t*
  isolation, 983*t*
  natural occurrence, 954*t*, 983*t*
  orbital diagrams, 300*t*
  orbitals, 1001, 1001*t*
  properties, 1008–1009, 1013
  uses, 983*t*
Chromosomes, 651
Cisplatin, 1021*mn*
cis-trans isomers, 619, 623, 1021
Citric acid
  acid-dissociation constants, 777*t*
  formula, 758*t*
  uses, 758*t*
Classical (Arrhenius) acid-base definition, 759, 794, G–4
Clausius-Clapeyron equation, 428, 431, G–4
Closed-end manometer, 178, 178*f*
Coal, 243
Coal gasification, 243, G–4
Cobalt
  cobalt(II) and (III), 66*t*
  coordination compounds, 1018–1019, 1019*t*
  electron configuration, 300*t*
  in humans, 1032*t*
  isolation, 983*t*
  natural occurrence, 954*t*, 983*t*
  orbital diagrams, 300*t*
  orbitals, 1001, 1001*t*
  uses, 983*t*

Cocaine, molecular structure, 631*f*
Coffee-cup calorimeter, 234, 234*f*
Cold packs, 496*mn*
Colligative properties, 506–511, 514–517
  antifreeze, 512
  boiling point elevation, 508–509, 509*t*, 510
  defined, 506, G–4
  electrolytes, 506
  electrolyte solutions, 516, 516*f*
  freezing point depression, 509–510, 512, 512*f*
  nonelectrolytes, 506, 506*f*
    nonvolatile, 506–511, 514–515
    volatile, 515
  osmotic pressure *see* Osmotic pressure (P)
  plane de-icing, 512
  road salt, 512
  solute molar mass from, 514–515
  uses
    in biology, 513, 513*f*
    in industry, 512, 512*f*
  vapor pressure lowering, 507–508, 507*f*
  zone refining, 512
Collision frequency, 203, 203*mn*, G–4
Collisions, gases, behavior, 197
Collision theory, 685–688, G–4
Colloids, 485, 517–519, 517*t*, 519*f*
  Brownian motion, 518
  defined, 517, G–4
  suspension, 517
  Tyndall effect, 518, 518*f*
  water purification, 520–521, 520*f*, 521*f*
Color, 1005, 1025, 1025*f*
Column chromatography, 76*f*
Combination reactions, 154–156, 155*f*, 162
Combined gas law, 183
Combustion, 160–161 *see also* Fuels
  car engines, 226
  defined, 10, G–4
  fuel cells, 160*mn*
  gasoline, 226
  heat of combustion, 232*t*
  of octane, 104–105
  phlogiston theory, 10–11, 13
  of propane, 105, 105*t*, 106
Combustion analysis, to determine chemical formula, 98–99, 101, G–4
Common-ion effect, 806, 828–829, G–4
Complexes *see* Coordination compounds
Complex ions, 835, 835*f*, 840
  amphoteric hydroxides, 838–840, 839*f*
  coordination compounds, 1013, 1014–1016, 1014*f*, 1015*t*
  defined, 835, 1013, G–4
  formation, 835–836, 836*t*, 837*f*
  formation constant, 835, 836*t*
  ligands, 835
  precipitates, 837–838
  valence bond theory, 1023, 1031
    coordinate covalent bond, 1023
    octahedral complexes, 1023, 1023*f*
    square planar complexes, 1024, 1024*f*
    tetrahedral complexes, 1024, 1024*f*
Composition, 3, G–4
Compounds, 77
  chemical formulas, 63–72
  covalent *see* Covalent compounds

defined, 42, 43, G–4
  formation, 56, 59
    covalent compounds, 62–63
    ionic compounds, 59–61, 60*f*
  formed from polyatomic ions, 67, 67*t*, 68–69, 146
  ionic *see* Ionic compounds
  mixture different from, 74, 74*f*
  models, 73*f*
  molar mass, 90
  molecular mass, 71–72
  moles, converting, 92
  naming, 63–72
  organic *see* Organic compounds
  physical properties, 541
  salts, 142, 145*f*
  specific heat capacity, 233*t*
  standard state, 240
Compton, Arthur, 271
Computer chips, 461
Concentration, 501–505
  from acid-base titration, 144
  calculating equilibrium constant from, 729
  cell potential and, 916–917
  collision theory *see* Collision theory
  of complex ions, 836
  defined, 114, 501, G–4
  Le Châtelier's principle, 737–739, 738*f*, 738*t*, 743*t*
  mass percent [% (w/w)], 502
  molality, 501–502, 501*t*
  molarity, 501–502, 501*t*
  mole fraction, 501*t*, 503, 504
  parts by mass, 501*t*, 502–503, 504
  parts by volume, 501*t*, 503, 504
  reaction rate and, 665
  from redox titration, 153–154
  units of, 501
    converting, 504–505
  volume percent [% (v/v)], 503
Concentration cells, 919–922, G–4
Condensation, 421, 423*f*, G–4
Condensation polymers, 643, G–4
Condensed electron configuration, 298
Condensed organic formula, 614, 614*t*
Condensed phases, 420
Conduction band, 458, G–4
Conductometric methods, reaction rates, measuring, 672
Conductor, 458, 458*f*, G–4
Conjugate acid-base pair, 782–784
  acid-base buffer systems, 813
  Brønsted-Lowry acid-base definition, 769–770, 770*t*, 771*f*
  defined, 769, G–4
Conservation of energy, law of, 225
Constant-pressure calorimetry, 234, 234*f*
Constant-volume calorimetry, 235
Constitutional isomers, 615, 1020–1021, G–4
Contact-lens rinses, 513
Contact process, 987, G–4
Controlled experiments, 12
  defined, 12, G–4
  variable defined, G–18
Convective mixing, 205
Conversion factors, 13–15, 17
  concentrations, 504–505
  defined, 13, G–4
  derived units, 17
  English equivalents to SI units, 18*t*
  Fahrenheit/Celsius temperatures, 24–26
  moles/molar mass, 91–93, 92*f*, 120*f*
  pressure, units of, 179

SI units to English equivalents, 18*t*
  wavelength and frequency, 257–258
Coordinate covalent bond, 1023, G–4
Coordination compounds, 1013–1014, 1014*f*, 1022–1023 *see also* Transition metal complexes (coordination compounds)
  bidentate ligands, 1015, 1016*t*
  chelate, 1016
  complex ions, 1013, 1014–1016, 1014*f*, 1015*t*
  constitutional isomers, 1020–1021
  coordination isomers, 1020
  coordination numbers, 1014–1015, 1015*t*
  coordination theory, 1018–1019, 1019*t*
  counter ions, 1013, 1014*f*
  defined, 1013, G–4
  donor atoms, 1015
  enantiomers, 1021–1022, 1022*f*
  formulas and names, 1016–1018, 1017*t*
  geometric isomers, 1021, 1021*f*
  isomers, 1020–1022, 1020*f*, 1021*f*, 1022*f*
  ligands, 1013, 1014*f*, 1015–1016, 1016*t*, 1017, 1017*t*
  linkage isomers, 1020–1021
  monodentate ligands, 1015, 1016*t*
  polydentate ligands, 1015–1016, 1016*t*
  stereoisomers, 1021–1022, 1021*f*, 1022*f*
  Werner, Alfred, 1018–1019, 1019*t*
Coordination isomers, 1020–1021, G–4
Coordination number, 446, 1014, 1015*t*, G–4
Coordination theory, 1018–1019
Copolymer, 473, G–4
Copper
  copper (I) and (II), 66*t*
  displacement reactions, 159, 159*f*
  electron configuration, 300–301, 300*t*
  extraction from ore, 109
  in humans, 1032*t*
  isolation and electrorefining, 972–974, 973*f*, 983*t*, 987
  natural occurrence, 59, 954*t*, 983*t*
  orbital diagrams, 300*t*
  orbitals, 1001, 1001*t*
  specific heat capacity, 233*t*
  from sulfide ores, 107
  uses, 972*mn*, 983*t*
Copper(I) sulfide, "roasting," 107–108
Coral, 442
Core, Earth, 951, G–4
Core electrons, 302, G–9
Corrosion, 926–928, G–5
CO-shift reaction, 243
Cosmic ray intensity, 1053n
Cosmology, element synthesis, 1074–1075
Coulomb (C) (unit), 905, G–5
Coulomb's law, 60, 333, 420, G–5
Counter ions, 1013, G–5
Counters for detection of radioactive emissions, 1051, 1051*f*
Coupling of reactions, 878, 879, G–5
Covalent bonds, 328, 328*f*, 337–342, 339*t*–340*t*, 341–342, 393–414, 433*t*, 534
  bond energy, 338, 339*t*, 341*t*
  bonding pairs, 338
  bond length, 340, 340*f*, 340*t*, 341*t*
  bond order, 338

coordinate covalent bond, 1023
defined, 62, 338, G–5
in diborane, 561
double bond, 338
electronegativity difference, 347, 347f
formation, 62–63, 337–338, 337f
intermolecular forces compared, 431–432, 433t
lone pairs, 338
molecular orbital theory *see* Molecular orbital theory
nonpolar covalent bond, 346
polar covalent bond, 346–348
polyatomic ions, 63, 63f
single bond, 338
triple bond, 338, 402
valence bond (VB) theory *see* Valence bond (VB) theory
Covalent compounds, 77f
defined, 59, G–5
formation, 62–63, 62f
names and formulas, 70–71
properties, bonding and, 341–342
in water, 135–136
Covalent hydrides, 544
Covalent radius, 304, G–5
Critical mass, 1070, G–5
Critical point, 430–431, G–5
Critical pressure, 430
Critical temperature, 430
Crosslinks, in polymers, 472, G–5
Crust, Earth, 951, G–5
Crutzen, Paul, 702
Crystal defects, in electronic materials, 460, G–5
Crystal field splitting energy, 1026, G–5
Crystal field theory, 1025, 1031
chlorophyll, 1025
color and, 1025, 1025f, 1025t, 1027–1028, 1027f
complementary color, 1025
crystal field splitting energy, 1026
crystal field splitting in tetrahedral and square planer complexes, 1030–1031, 1031f
defined, 1025, G–5
e_g orbitals, 1026
high-spin complexes, 1029–1030, 1029f
ligand field-molecular orbital theory, 1031
low-spin complexes, 1029–1030, 1029f
magnetic properties, 1028–1030, 1029f
octahedral complexes, 1026–1027, 1026f, 1027f
spectrochemical series, 1028
splitting *d* orbitals in an octahedral field of ligands, 1026–1027, 1026f, 1027f
square planar complexes, 1030, 1031f
strong-field ligands, 1026, 1027f
tetrahedral complexes, 1030, 1031f
t_{2g} orbitals, 1026
weak-field ligands, 1026–1027, 1027f
Crystal lattice, 446, 446f
Crystalline solids
atomic, 452, 452f, 453t, 459
crystal lattice, 446, 446f
defined, 445, G–5
ionic, 453–455, 453t, 454f, 459
metallic, 453t, 455, 455f
molecular, 452–453, 452f, 453t
network covalent, 453t, 455–456, 459

types, 452, 453t
unit cell, 446, 447f, 448, 450
Crystallinity, of polymers, 470–471, 470f
Crystallization, 75, 75f, G–5
Crystals, 445–446, 445f
crystal lattice, 446, 446f
unit cell, 446, 447f, 448, 450
Crystal structure
determination from atomic radius, 450
fluorite, 454–455, 455f
metals, 455, 455f
silicon dioxide, 456, 457f
sodium chloride, 453–454, 454f
zinc blende, 454, 454f
Crystal systems, 446
Cubanes, 380, 380f
Cubic centimeter (cm³) (unit), English equivalent, 18t
Cubic closest packing, 448, 449t, 452f, G–5
Cubic decimeter (dm³) (unit), 19
English equivalent, 18t
Cubic foot (ft³) (unit), conversion to SI units, 18t
Cubic meter (m³) (unit), 19, 26
defined, 19, G–5
English equivalent, 18t
Cubic system, crystals, 446, 447f
Curie, Marie, 1043, 1043mn
Curie, Pierre, 1043
Curie (Ci) (unit), 1050, G–5
Curium, name, origin of, 52mn
Cyclic hydrocarbons, 615, 615f, 623, G–5
Cycloalkanes, 615, 615f
Cyclobutane, molecular structure, 615f
Cyclohexane, molecular structure, 615f
Cyclone separator, 964f
Cyclopentane, molecular structure, 615f
Cyclopropane, molecular structure, 615f
Cyclo-S_8, 584
Cyclotron, 1057, 1057f
Cysteine, structure, 646f
Cytochromes, 1033

D

d orbital, 280, 280f, 1026–1027, 1027f, G–5
Dalton, John, 46
atomic theory, 45, 46–47
career highlights, 46mn
Dalton (D) (unit), 53, G–5
Daltonism, 46mn
Dalton's law of partial pressures, 192, 197, 199, 199f, G–5
Data, 12, G–5
Davisson, C., 270
de Broglie, Louis, 269
de Broglie wavelength, 269, 270t, 270, 273, G–5
Debye, Peter, 381
Debye (unit), 381
Decane
boiling point, 617f
formula, 71t
Decay constant, 1050, G–5
Decay series, 1049, G–5
Deci- (prefix), 18t
Decimal places, significant figures, 28
Decimal prefixes, 17, 18t
Decomposition reaction, 156–157, 157f, 162
Decompression sickness, 500mn

Definite composition, 43, 44–45, 44f, 47
Degree of polymerization, 468, G–5
Dehydration-condensation reaction, 580–581, G–5
Deka- (prefix), 18t
Delocalized electron-pair bonding, 362–363
Demeton, molecular structure, 962mn
Democritus, nature of matter, 41, 46
Dendrimers, 380
Density *(d)*
alkali metals, 550
alkaline earth metals, 554
boron family, 558
calculation from mass and length, 23–24
carbon family, 564
of common substances, 23t
defined, 22, G–5
of gases, 175
halogens, 590
nitrogen family, 574
noble gases, 596
oxygen family, 582
transition elements, 1003
unit of, 23, 27
water, 23t, 443–445, 443f, 444f
Dental alloys, 421mn, 967t
Dental anesthetic, 578
Deoxyhemoglobin, 1032
Deoxyribonucleic acid *see* DNA (Deoxyribonucleic acid)
Deposition, 422, 423f, G–5
Derived units, 17, 17t, 26, G–5
Desalination, 521, G–5
Desulfurization devices, 243
Detergents, 490, 632
Deuterium, production and uses, 981
Deuterons, 1056, G–5
Dextrorotatory isomer, 618
Diagonal relationship, in periodic table, 553, 555, G–5
Diamagnetism, 316–317, 1005, 1008, G–5
Diamond, 563
physics of, 259mn
properties, 455–456, 456t
Diastereomers, 1021
Diatomic molecules, 42, 62
heteronuclear, 412–413
homonuclear, 407–412
Diberyllium, 407, 407f
Diborane, 559, 561f
Dichlorodifluoromethane, molecular shape, 373
Dichromate ion, 67t, 1004
Dietary trace elements, 1032–1033, 1032t, 1033f
Diethylamine, 780t
Diethyl ether
boiling point elevation, 509t
freezing point depression, 509t
solubility, 489
Differentiation, geochemical, 951, 953f, G–5
Diffraction, 259, 262
defined, 259, G–5
X-ray diffraction analysis, 451–452, 451f
Diffraction pattern, 259, 259f, 273
electrons, 270, 271f
Diffusion, 202–203, 203f
in biosphere, 421mn
defined, 202, G–5
Dihydrogen phosphate ion, 67t
Dilithium, 407, 407f

Dimensional analysis, 15, G–5
Dimethoxymethane, NMR, 624–625, 625f
Dimethylamine, 780t
Dimethylbenzenes, structure, 622, 622f
Dimethyl ether, properties, 100t
Dimethylformamide, molecular structure, 638f
2,2-Dimethylpropane, 616t
Dinitrogen monoxide, 577t, 578
Dinitrogen pentaoxide, 577t
Dinitrogen tetraoxide, 111, 577t
Dinitrogen trioxide, 71, 577t
Dioxin *see* TCDD
Dioxygen, 89
Diphosphate ion, 581, 581f
Diphosphoric acid, 639
Dipole-dipole force, 432, 433f, 433t, 434, 438, G–5
Dipole–induced dipole force, 433t, 436, 441, 487, G–5
Dipole moment
boiling point and, 434, 434f
defined, 381, G–5
Direction of reaction, 725–727, 770–772, 880–884
Disaccharides, 645, G–5
Disintegration series, 1049
Disorder
entropy and, 859–861
probability, 859–861, 859f
solution process and, 496–497
thermodynamics, 859–861
vapor-pressure lowering, 507
Dispersion (London) forces, 433t, 436–437, 437f, 487, G–6
Displacement reaction, 158–160
Disproportionation, 578, G–6
Dissociation, 135–136, 136
acids, 757–761, 760f, 776
Dissolution
entropy change, 862–863, 862f
of a gas, 863–864
Distillation, 75, 75f *see also* Fractional distillation
defined, 75, G–6
laboratory tools, 75, 75f, 77f
Disulfide bridge, 648
Disulfuric acid, 639
Divers, decompression sickness, 500mn
Division, significant figures, 29
DNA (deoxyribonucleic acid)
double helix of, 651, 651f
molecular model, 73f
protein synthesis, 652
replication, 653, 653f
Döbereiner, Johann, 289
Donor atoms, 1015, G–6
Doped semiconductor, 460
Doping, 460, G–6
Double bonds, 338
color test for, 626f
defined, 338, 405, G–6
functional groups with, 633–635
functional groups with single and double bonds, 635–639
hydrocarbons with, 618–619
inorganic compounds with, 635
Double-displacement reactions, 158
Double helix, DNA, 651, 651f, G–6
Downs cell, 969, 969f, G–6
Dow process, 977
Drugs, chirality of molecules, 618mn
Dry cell batteries, 923
Dry ice, 429
Dumas, J.B.A., 191
Dynamic equilibrium, 162, 162f, G–6

E

e_g orbitals, 1026, G–6
Earth
   abundance of elements, 951–955
   atmosphere *see* Atmosphere
   biosphere, 953, 954*t*
   core, 951
   crust, 951, 953–955, 953*f*, 954*t*
   differentiation, 951
   formation and layering of, 951
   hydrosphere, 953
   lithosphere, 953
   mantle, 951
EDTA, 1016*mn*
Effective collisions, 687, G–6
Effective nuclear charge (Z_{eff}), 293,
   305, 532, G–6
Effusion, 201–202, 202*mn*, G–6
Einstein, Albert, 44*mn*, 261, 269,
   269*mn*, 273*mn*, 917*mn*
Einsteinium, name, origin of, 52*mn*
Eka silicon, 289, 290*t*
Elastomers, 472, G–6
Electrical conductance, ion mobility
   and, 335*f*
Electrical conductivity, 458
   ionic solutions, 132–134, 133*f*
   semiconductors, 458*f*, 459
   superconductivity, 459
Electric cars, 893
Electric current, unit of, 17*t*
Electrochemical cells, 893,
   898–899, G–6
Electrochemistry
   batteries, 922–926, 930
   biological energetics, 937–938,
      937*f*, 938*f*
   cellular, 937–938
   corrosion, 926–928
   defined, 893, G–6
   electrochemical cells, 893, 898–899
   electrolysis, 157, 931–936
   electrolytic cells *see* Electrolytic cells
   free energy and electrical work,
      914–922
   galvanic cells *see* Voltaic cells
   half reactions, 893–895
   voltaic cells *see* Voltaic cells
Electrodes, 898, G–6
Electrolysis *see* Electrolytic cells
Electrolytes, 506
   defined, 132, G–6
   strong, weak, and non- electrolytes,
      141, 141*f*
Electrolytic cells, 898, 899*f*, 929,
   931, 939
   ampere, 935
   aqueous ionic solutions, 933–934
   construction and operation,
      929–930, 929*f*
   defined, 898, G–6
   Faraday, Michael, 935
   lead-acid battery, 930, 930*f*
   mixed molten salts, 931–932
   nonstandard half-cell potentials,
      932–933
   overvoltage, 933–934
   pure molten salts, 931
   stoichiometry of, 935–936, 935*f*
   voltaic cells compared, 929, 929*f*,
      930, 931*t*
   water, 932–933, 932*f*
Electrolytic decomposition, 157
Electromagnetic radiation, 255, 257*mn*,
   262, G–6
Electromagnetic spectrum, 257, 257*f*,
   262, G–6

Electromagnetic waves, 262
Electrometallurgy, 963
Electromotive force (emf), 905, G–6
Electron (e⁻), 51, 51*mn*, 56
   compounds, formation of, 59–63,
      60*f*–63*f*
   de Broglie wavelength, 270*t*
   defined, 52, G–6
   diffraction pattern, 270, 271*f*
   inner (core), 302
   Millikan's oil drop experiment,
      49, 49*f*
   movement of in redox reaction, 147,
      147*f*, 149*f*, 154
   outer, 302
   valence, 302
   wave nature of, 269–272
Electron affinity (EA), 344*n*
   defined, 310, G–6
   trends in, 310–311, 311*f*
Electron capture, 1044*t*, 1045, G–6
Electron cloud, 274, G–6
Electron configuration, 289, 295, 532
   alkali metals, 550
   alkaline earth metals, 554
   atomic properties and, 304
      atomic size, 304–306, 304*f*–306*f*,
         306*mn*, 311
      electron affinity, 310–311, 311*f*
      ionization energy, 307–310, 307*f*,
         308*f*, 309*t*, 311
   aufbau principle, 295, 304
   boron family, 558
   carbon family, 564
   chemical reactivity and, 311–319
   condensed electron
      configuration, 298
   defined, 289, 532, G–6
   halogens, 590
   magnetic properties, 316–317
   main-group ions, 314–315
   nitrogen family, 574
   noble gases, 596
   oxygen family, 582
   Period 3, 298, 298*t*, 299*f*
   Period 4, 299–301, 300*t*, 301*f*
   Periods 1 and 2, 295–297, 297*f*
   principles, 301–302
   transition elements, 302–304,
      1000–1001
   transition metal ions, 315–316
   within groups, 299, 299*f*, 301, 304,
      307, 311
Electron deficient, 365, G–6
Electron-deficient molecules, 365
Electron density diagram, 274,
   274*f*, G–6
Electron-density models, 73*f*
Electron-dot formula, 73*f*
Electronegativity (EN), 344–346, 533
   alkali metals, 550
   alkaline earth metals, 554
   atomic size and, 345, 345*f*
   bonding, 344–346, 344*f*
   boron family, 558
   carbon family, 564
   defined, 344, 533, G–6
   halogens, 590, 594
   nitrogen family, 574
   noble gases, 596
   oxidation number and, 346
   oxygen family, 582
   Pauling scale, 344–345, 344*f*
   periodic trends in, 345, 533
   transition elements, 1002–1003
   valence-state electronegativity,
      1009, 1013

Electronegativity difference, 347, G–6
Electronic balance, 21
Electronic calculators, significant
   figures and, 30
Electronic materials, 460
   crystal defects, 460
   doped semiconductors, 460–461, 461*f*
   doping, 460
   manufacturing p-n junctions,
      462, 462*f*
   solar cells, 460
Electron microscope, 270*mn*
Electron-pair delocalization,
   362–363, G–6
Electron pair donation, acids and bases,
   791–795
Electron pooling, metallic bonding
   and, 328
Electron-sea model, 349–351, G–6
Electron sharing, covalent bonding
   and, 328
Electron-spin quantum number (m_s),
   290–291, 291*f*
Electron transfer, ionic bonding
   and, 328
Electron-transport chain, 937–938, 938*f*
Electron volt (eV) (unit), 1067, G–6
Electrorefining, 967, 972–974, G–6
Electrostatic effect, splitting of energy
   levels, 292–294
Element, 52–53, 77 *see also* Individual
   elements
   abundance, 951–955
   atmosphere, 953
   atomic mass, 52–55, 56, 88–89
   atomic number, 52
   atomic size, 304–306, 304*f*–306*f*,
      306*mn*, 311
   atomic theory, 56
   biosphere, 953, 954*t*
   compounds, formation, 59–63,
      60*f*–63*f*
   Dalton's atomic theory, 46–47
   defined, 41–42, 43, G–6
   electron affinity, 310–311, 311*f*
   electron configuration *see* Electron
      configuration
   environmental cycles, 956, 963
      carbon cycle, 956–958
      nitrogen cycle, 958–959
      phosphorus cycle, 960–962
   forming more than one ion, 66–67, 67*t*
   geochemical differentiation, 953*f*
   history, 56
   in industry, 950*f*, 969–991
   ionization energy, 307–310, 307*f*,
      308*f*, 309*t*
   isotopes, 53–55, 56
   line spectra, 262*f*
   mass number, 52, 53*f*, 56
   mass percent, 93–95
   metallurgy, 963–968
   molar mass, 89
   moles, converting, 90–91, 92*f*
   names of, 52*mn*
   in nature, 951–956
   ores, 955
   origin in the stars, 1074–1075, 1075*f*
   periodic table *see* Periodic table
   phase changes, 541
   physical states of, 540
   sources of, 955–956, 955*f*
   specific heat capacity, 233*t*
   standard state, 240
   symbol, 52, 56
   transuranium elements,
      1056–1058, 1058*mn*

Elemental substances, 41
Elementary reactions (steps),
   692–693, G–6
Element oxides, acid-base behavior,
   312–313, 313*f*, 537
Element symbol, 52, 56
Elimination reactions, organic, 626, G–6
Emission spectrum, 266*mn*, 267, G–6
Empirical formula, 64, 66–67, 72, 101
   defined, 64, 99, 100*t*, G–6
   determining, 95–96
Enantiomers, 617, 1021
Endothermic reactions, 229, 229*f*, 616
   defined, 229, G–7
   dissolution, 494, 494*f*
   spontaneous, 871
End point, of titration, 143, 816, G–7
End-to-end overlap, 401, 535, G–7
-ene (suffix), 629*t*
Energy
   activation energy, 683–684, 683*f*,
      685–687, 686*f*, 686*t*, 687*f*
   blackbody radiation and the
      quantization of energy,
      260, 260*f*
   bond energy *see* Bond energy
   conservation of energy, 8, 225, 857
   crystal field splitting energy, 1026
   defined, 6, G–7
   electromagnetic energy, 255
   first ionization energy variations,
      307–309, 307*f*, 308*f*
   flow to and from a system, 222, 222*f*
   forms and interconversion, 221, 227
   free energy *see* Free energy *(G)*
   free energy change *see* Free energy
      change (Δ*G*)
   Gibbs free energy, 872, 879–880
   internal energy *E*, 222*f*, 230–231
   ionic bonding
      importance of lattice energy,
         331–333
      periodic trends in lattice energy,
         333–335, 334*f*
   ionization energy *see* Ionization
      energy (IE)
   kinetic energy, 6–8, 7*f*, 231, 231*f*
   lattice energy, 331–335
   law of conservation of energy, 8,
      225, 857
   mass-energy, 44
   nuclear binding energy,
      1067–1069, 1069*f*
   nuclear energy *see* Nuclear fission;
      Nuclear fusion
   potential energy, 6–8, 7*f*, 231, 231*f*
   radiant energy, 255
   solar energy, 245
   system energy, 221–222, 222*f*
   thermal energy, 223, 223*f*, 224*t*
   units of, 225–226
   wave-particle duality of matter and
      energy *see* Wave-particle
      duality
Energy efficiency, 875*mn*
Energy-level splitting, 292–294
Energy transfer, heat and work,
   223–224, 223*f*, 224*t*
Enthalpy, change in (Δ*H*), 228–229,
   230, 232
   defined, 228, G–3
   Hess's law of heat summation,
      238–239, 425
   phase changes, 421–423
   standard heat of formation, 240
   standard heat of reaction, 240–242
Enthalpy diagram, 229, G–7

Enthalpy (H), 228–232
  bond strength changes, 230, 232
  breaking and forming bonds, 231
  defined, 228, 228f, G–7
  endothermic processes, 229–230, 229f
  enthalpy diagram, 229, 229f
  exothermic processes, 229–230, 229f
  Hess's law of heat summation, 238–239, 425
  types of change, 230
Enthalpy of hydration, G–7
Enthalpy of solution, G–7
Entropy (S), 496, 497
  defined, 496, 860, G–7
  disorder and, 859–861
  second law of thermodynamics, 861
  solutes, 862f
  solutions, 496
  standard entropy of reaction (ΔS^0_{rxn}), 866–867
  standard molar entropy, 861–865, 862f, 863f, 864f, 876–878, 877t, 878f
  standard molar entropy (S^0), 859–865, 861f, 862f, 876–878, 877t, 878f
Environmental cycles, 956, G–7
  carbon cycle, 956–958
  nitrogen cycle, 958–959
  phosphorus cycle, 960–962
Environmental issues
  air pollution
    acid rain, 32, 33f, 833–834, 833f, 834f
    CFCs, 702
    power plant emissions, 33f
    sulfur dioxide, 586mn
  chemistry and, 32–33, 33f
  defined, G–7
  fuels, 243–245
  global warming, 244, 244f
  green chemistry see Green chemistry
  "greenhouse" effect, 244–245, 244f
  noise pollution, 765mn
  ozone, 73, 205, 413–414, 413f, 702, 702f
  PCBs, 566, 591
  pollutants
    in the food chain, 630mn
    ultralow concentrations, 503mn
Enzymes see also Catalysis
  binding site, 619f
  as catalysts, 700–701, 700f, 701f
  defined, 700, G–7
  zinc ion in, 1033
Enzyme-substrate complex (ES), 701, G–7
Epinephrine, molecular structure, 631f
Equations see Chemical equations
Equatorial groups, VSEPR theory, 375, G–7
Equilibrium, 714
  acid-base, 756–795
  acid-base titration curves, 815–823
  complex ions, 835–840
  dynamic nature of, 714–717
  entropy change and, 870
  ionic, 805–845
  law of chemical equilibrium, 717
  Le Châtelier's principle, 736–743, 746
  metabolic pathway, 745–746
  phase changes
    liquid-gas, 426–427, 426f
    solid-gas, 429, 429f
    solid-liquid, 429
  phase changes and, 425–431
  pressure units, 724–725

reaction direction, 725–727
reaction quotient, 717–724
solubility and, 501–505
solving problems, 727–736
Equilibrium constant (K), 716f, 716n, 717, 724
  based on concentration (K_c), 718
  based on pressure (K_p), 724–725
  defined, 716, G–7
  reaction quotient and, 717–723
  standard cell potential and, 914–916
  standard free energy change and, 881, 915
Equivalence point, 143, 816, G–7
Erbium, name, origin of, 52mn
Error
  random, 31
  systematic, 31
Erythrose, 100t
Ester, functional group, 629t, 636
Esterification, 637–638
Esters, 636, 637–638, 639–640, G–7
Ethanal, molecular structure, 629t, 634f
Ethanamide, molecular structure, 629t
Ethane
  boiling point, 617f
  formula, 71t
  molecular shape, 400–401, 401f
1,2-Ethanediol, 628f
Ethanenitrile, molecular structure, 629t
Ethanoic acid, molecular structure, 629t
Ethanol
  boiling point elevation, 509t
  freezing point depression, 509t
  properties, 100t
  solubility, 488, 488t
  specific heat capacity, 233t
Ethanolamine, 780t
Ethene see Ethylene
Ethers, 628
Ethylamine, molecular structure, 629t
Ethylene
  bonds in, 400–401, 401f
  free-radical polymerization, 642, 642f
  Lewis structure for, 361
  molecular structure, 629t
Ethylene glycol
  molecular structure, 628f
  solubility, 489
  specific heat capacity, 233t
Ethyllithium, 634
Eutrophication, 958, 962
Exact numbers, 30, G–7
Excitation, radioactive emissions, 1058, G–7
Excited state, 264, G–7
Exclusion principle, 291–292, 295, G–7
Exosphere, 204
Exothermic reactions, 229, 229f
  defined, 229, G–7
  dissolution, 494, 494f
  spontaneous, 870–871
Expanded formula, 614, 614t
Expanded valence shell, 366, G–7
Experiment, in scientific method, 12, 12f, 13, G–7
Explosives, 594mn
Extensive properties, 23, 27, G–7
Extraction, 75, 75f, G–7

F

f orbital, 280, 280f
Face-centered cubic unit cell, 446, 447f, 448, G–7

Factor-label method, 15
Fahrenheit/Celsius conversion, 24–26
Fahrenheit scale, 24
Faraday, Michael, 935mn
Faraday constant (F), 914, G–7
Fats
  digestion, 518mn
  heat of combustion of, 232t
Fatty acids, 636, G–7
Femto- (prefix), 18t
Ferrous chlorate, naming, 68
Fertilizers, 962
Fibrous proteins, 648–649, 649f
Filtration, 75, 75f, G–7
Fire extinguishers, 189
Fireworks, 267f, 594mn
First ionization energy, 307–308, 309f, 311
First law of thermodynamics, 225, 857, 865, G–7
First-order reactions, 682t
  half-life, 680–682
  integrated rate laws for, 677–678
  rate law, 672, 674
  zero- and second-order reactions compared, 682t
Fission see Nuclear fission
Fixation
  carbon, 957
  defined, 957, G–7
  nitrogen, 958–959
Flame tests, 267, 267f, G–7
Flashbulb, chemical action in, 102, 103f, 147
Flotation process, 964, 964f, G–7
Flow, in biosphere, 421mn
Flow battery, 922, 925, 925f
Fluid ounce (unit), conversion to SI units, 18t
Fluorine
  diatomic molecule, orbitals, 410f
  electron configuration, 296
  ionization energy, 309t
  isolation, 986t
  natural sources, 986t
  oxidation number, 148t
  properties, 547t, 590
  uses, 547t, 986t
Fluorite
  crystal structure, 454–455, 455f, 591f
  uses, 591
Fluorspar, uses, 591
Food
  energy in bonds of, 232
  hypertonic preservation, 513
  irradiation of, 1065
Food chain, pollutants in, 630mn
"Fool's gold," 587, 587f
Formal charge, 364–365, 368, G–7
Formaldehyde, 100t, 634f
Formation constant (K_f), 835, 836t, G–7
Formation equation, 240, G–7
Formic acid, molecular structure, 636f
Formula mass, 71–72, 89t
Formulas see Chemical formulas
Formula unit, 65, G–7
Fossil fuels, 243, G–7
Fractional distillation, 515, 515f, G–7
Fraction by mass, 44, G–7
Fragrance industry, 638mn
Framework silicates, 571, 571f
Framework silicones, 572, 572f
Francium, 982t
  properties of, 550
Frasch process, 987–988, 988f
Free energy (G), 872–878, 880
  electrochemistry and, 914–916, 915f

reaction direction and, 880–884
reversible reactions, 163
Free energy change (ΔG), 872–878
  defined, G–7
  equilibrium, 880–884, 881t
  extent of reaction, 883–884, 883f
  reaction direction, 880–884, 881t
  standard free energy change (ΔG^0), 873
  standard free energy of formation (ΔG^0_f), 874
  temperature and, 876–878, 877t, 878f
  work and, 874–875
Free radical, 642
  biological systems, 365mn
  defined, 365mn, 1059, G–7
  Lewis structure, 365–366
Free-radical polymerization, ethene, 642, 642f
Freezing, 421, 423f, G–8
Freezing point, 25f
Freezing point depression, 509–510, 512, 512f, G–8
Frequency, of wave (ν), 256, 256f, G–8
Frequency factor, 687, G–8
Frisch, Otto, 1069
Fructose, 645, 645f
Fuel cell, 160mn, 922, 925, 925f, G–8
Fuels see also Combustion
  bonds in, 232
  cleaner fuels, 243–245
  coal, 243
  combustion, 160mn, 243
  fossil fuels, 243
  "greenhouse" effect, 244–245, 244f
  hydrogen fuels, 243–244, 545mn
  nuclear fuel, 202, 1073mn
  rocket fuel, 111–112
  wood, 243
Fuller, R. Buckminster, 380, 563
Fullerenes, 380, 380f
Functional groups, organic, 609, 627–628, 629t
  defined, 609, G–8
  with double bonds, 633–635
  interconnections among, 641f
  with single and double bonds, 635–639
  with single bonds, 628–633
  with triple bonds, 640–641
Fusion, 421
  heat of, 230, 422
  nuclear see Nuclear fusion
  phase changes, 421

G

g orbital, 280
Galileo, 177
Galileo spacecraft, 5
Gallium
  bond type, 562t
  electron configuration, 300t
  in Ga-As semiconductors, 556mn
  isolation, 984t
  melting point, 350f, 562t
  natural sources, 984t
  orbital diagrams, 300t
  properties, 311, 558
  uses, 984t
Gallon (gal) (unit), conversion to SI units, 18t
Galvanic cell see Voltaic cells
Gamma (γ) emission, 1044t, 1045, G–8
Gamma (γ) ray, 1043, 1043f, 1044t, 1059, G–8
Gangue, 963, G–8

Gas, 4f
  chemical behavior, 174
  defined, 4, G–8
  density, 175
    human disasters and, 189mn
    ideal gas law and, 189–190,
      189mn, 190mn, 194
  deviations from ideal behavior,
    207–210
  extreme conditions and, 207–209,
    207f–209f
  gas laws, 180, 198–201
    Amotion's law, 183
    Avogadro's law, 180, 183–184,
      183f, 184mn, 185, 185f, 189,
      200, 201f
    Boyle's law, 180–181, 180f,
      184mn, 185, 185f, 197,
      198, 198f
    breathing and, 184mn
    Charles's law, 180, 181–183, 182f,
      184mn, 185, 185f, 189, 197,
      199, 199f
    combined gas law, 183
    Dalton's law of partial pressures,
      192, 197, 199, 199f
    ideal gas law, 180, 185, 185f, 189,
      194, 195–196
    solving problems, 186–188
  Graham's law, 201–202, 202mn, 203
  Henry's law, 500
  kinetic-molecular theory,
    197–203, 421
  molar mass, 191, 191f
  molar volume of common gases, 207t
  partial pressure, 192–194
  physical behavior, 174
  pressure, 176–179
  properties, 174–176, 421t
  solubility, 500
  standard molar volume, 184, 184f
  standard temperature and pressure
    (STP), 184
  universal gas constant, 185, 683mn
  van der Waals constants for common
    gases, 209t
  van der Waals equation, 209
  viscosity, 175
Gas-gas solutions, 492
Gasification, of coal, 243
Gas-liquid chromatography (GLC),
  76, 76f
Gas-liquid solutions, 491
Gasohol, 243
Gasoline, combustion, 226
Gas-solid solutions, 492
Gas solutions, 492
Gay-Lussac, Joseph, 181, 190mn
Geiger-Müller counter, 1051, G–8
Genes, 651
Genetic code, 652, G–8
Geology
  elements in Earth's crust, 951–955
  formation of Earth, 951
  limestone cave, 831, 831f
Geometric isomers, 619, 623, 1021
  chemistry of vision, 620, 620f
  defined, 619, 1021, G–8
Geraniol, 95mn
Germanium
  bond type, 562t
  electron configuration, 300t
  isolation, 984t
  melting point, 562t
  name, origin of, 52mn
  natural sources, 984t

  properties, 564
  uses, 984t
Germer, L., 270
Gibbs, J. Willard, 872mn
Gibbs free energy, 872
Giga- (prefix), 18t
Glass, specific heat capacity, 233t
Glass transition temperature, 472
Glauber, Johann, 757mn
Glauber's salt, 757mn
Global warming, 244, 244f
Globins, 1032
Globular proteins, 649
Glucosamine, 645mn
Glucose, 100t, 645, 645f
  chemical formula, 90t
  mass percents, calculating, 93–94
  metabolic pathway, 109
Glutamic acid, structure, 646f
Glutamine, structure, 646f
Glycine, structure, 646f
Glycogen, 646
Gold
  density, 23t
  malleability, 351mn
  natural occurrence, 59
  panning, 964mn
  specific heat capacity, 233t
Graham's law of effusion, 201–202,
  202mn, 203, G–8
Grain alcohol, density, 23t
Gram (g) unit
  atomic mass units and, 88
  English equivalent, 18t
Gram-molecular weight see Molar mass
Granite, specific heat capacity, 233t
Graphite, 563
  properties, 455–456, 456t, 563
  specific heat capacity, 233t
Gray (Gy) (unit), 1059, G–8
Gray tin, 566
Greases, silicones, 571
Green chemistry, 109, 243 see also
  Environmental issues
"Greenhouse" effect, 244–245, 244f
  carbon cycle and, 958
Grignard reagents, 555
Ground state, 264, G–8
Group(s)
  defined, 57, G–8
  periodic table, 57, 57f, 59, 149, 149f,
    299, 299f, 301, 304, 307, 311
Group 1A(1) see Alkali metals
Group 2A(2) see Alkaline earth metals
Group 3A(13) see Boron family
Group 4A(14) see Carbon family
Group 5A(15) see Nitrogen family
Group 6A(16) see Oxygen family
Group 7A(17) see Halogens
Group 8A(18) see Noble gases
Group number, 302
Groups 3B(3) to 2B(12) see Transition
  elements
Guldberg, Cato, 717

H

Haber process, 744–745, G–8
Hahn, Otto, 1069, 1069mn
Half-cell, 901, G–8
Half-cell potential, 906–908, 907f
Half-life
  defined, 680, 1050, G–8
  first-order reaction of, 682t
  radioactive decay, 680–681,
    1050–1053

  reaction half-life, 680–682
  second-order reaction of, 682t
  zero-order reaction of, 682t
Half-reaction method, for balancing
  redox equations, 151,
    894–898, G–8
Halic acid, 594t
Hall-Heroult process, 975
Halo- (prefix), 629t
Haloalkanes, 629t, 630, G–8
Halogen oxoacids, 594, 594t
Halogens
  chemistry, 592–594
  displacement of halide ion from
    solution by, 160
  electron affinity, 311
  electron configuration, 299
  ionization energy, 311
  isolation, 986t
  name, origin of, 60f
  natural sources, 986t
  periodic table, 59
  properties, 588–594
  reactivity, 588–589
  uses, 986t
Halomethanes, 567
Halous acid, 594t
Hard water, 520, G–8
Hard water treatment, 520, 521f
"Hatter's shakes," 1013mn
H bond see Hydrogen bond (H bond)
Heat, 27
  defined, 24, 223, G–8
  endothermic reactions, 229
  exothermic reactions, 229
  phase changes and, 423–425
Heat capacity, 233, G–8
Heat capacity, liquid water, 236mn
Heating-cooling curve, 423–424, 431,
  G–8
Heat of atomization, 548
Heat of combustion
  calculating, 236
  of carbon compounds, 232t
  defined, 230, G–8
  of fats and carbohydrates, 232t
Heat of formation, 230
  defined, 230, G–8
  standard heat of formation
    (ΔH_f^0), 240
Heat of fusion, 230, 422, G–8
Heat of hydration, 494, 495, G–8
Heat of reaction, 229, 230–231
  calculation from Lewis structures and
    bond energies, 368–370, 368f
  calorimetry, 233, 234–236, 235f
  defined, 229, G–8
  to find amounts, 237
  standard heat of reaction (ΔH_{rxn}^0),
    240–242
Heat of solution, 493–494, 497, G–8
Heat of sublimation, 422, G–8
Heat of vaporization, 230, 422, 422f,
  443, G–8
Heat transfer, in refrigerator, 223mn
Heavy-metal ion toxicity,
  1013, 1013mn
Hecto- (prefix), 18t
Heisenberg, Werner, 272
Heisenberg uncertainty principle,
  272–273
Helium
  boiling point, 491t
  dihelium, MO diagram of, 406, 406f
  effusion rate, 202
  isolation, 986t
  natural occurrence, 59, 986t

  properties, 595, 596
  solubility, 491t
  uses, 986t
Heme, molecular model, 73f
Hemoglobin, 451f, 1032–1033, 1033f
Henderson-Hasselbalch equation,
  811, G–8
Henry's law, 500, G–8
Heptane
  boiling point, 617f
  formula, 71t
Hess's law of heat summation,
  238–239, 425, G–8
Heteroatoms, 609, G–8
Heterogeneous catalysis, 699, G–8
Heterogeneous mixture, 74, 77f, G–9
Heteronuclear diatomic molecules,
  412–413
Heterosphere, 205
Hexagonal closest packing, 448,
  449t, G–9
Hexagonal unit cell, 448
Hexane
  boiling point, 617f
  formula, 71t
  solubility, 487
  as solvent, 488–489, 488t
Hexanol, solubility, 488, 488t
High-performance liquid
  chromatography (HPLC),
  76, 76f
High-spin complexes, 1029, G–9
Histidine, structure, 646f
Homogeneous catalysis, 698, 702, G–9
Homogeneous mixture, 74, 77f, G–9
Homologous series, organic
  compounds, 613, G–9
Homonuclear diatomic molecules,
  407–412, G–9
Homopolymer, 473
Homosphere, 205
Hooke, Robert, 10
Hot-air balloon, 190mn
Hot packs, 496mn
Hund's rule, 296, 304, G–9
Hybridization, 394, G–9
Hybrid orbitals
  defined, 394, G–9
  sp hybridization, 394, 395f, 396
  sp^2 hybridization, 396, 396f
  sp^3 hybridization, 396, 397f
  sp^3d hybridization, 397, 397f
  sp^3d^2 hybridization, 398
Hydrate, 68, G–9
Hydrated metal ions, 786
Hydrates, 68, 68f, 68t, 72
Hydration, 494, 497, G–9
Hydration shells, 486, 487f, G–9
Hydrazine, 111, 575, 576mn
Hydride bridge bond, 561
Hydrides, types of, 544–545, 545mn
Hydrobromic acid, 70, 114
Hydrocarbons, 610, 623
  alkanes, 71, 71t, 613–617
  alkenes, 618–619, 629t
  alkynes, 621, 623, 629t
  aromatic, 622–623
  cyclic, 615, 615f, 623
  defined, 610, G–9
  double bonds, 618–619
  isomers, 615
    constitutional isomers,
      615–617, 616t
    geometric isomers, 619, 619t,
      620, 620f
    optical isomers, 617–618, 617f,
      618f, 619f

naming, 71, 613–614
saturated, 613
structure, 610–613, 611*f*
triple bonds, 621
unsaturated, 618
Hydrochlorofluorocarbons, 702
Hydrocyanic acid, 70
Hydrofluoric acid, 592, 592*mn*, 757*f*
Hydrogen
chemistry, 544–545
density, 23*t*
deuterium, 981
electron configuration, 295, 297*f*
as fuel, 243–244
green chemistry, 243–244
hydrides *see* Hydrides, types of
industrial production, 979, 987
isotopes, 978–979
laboratory production, 980
molecular orbitals, 405, 405*f*
name, origin of, 52*mn*
natural occurrence, 954*t*
oxidation number, 148*t*
periodic table, position in, 57*f*,
543–544, 543*f*
spectral lines, 263*f*
standard heat of formation, 240*t*
uses, 980–981
Hydrogenation, catalytic, 699,
699*f*, G–9
Hydrogen atom
Bohr model of, 263–264, 265*f*
energy levels of, 281
energy states of, 264–266, 265*f*
Hydrogen bond (H bond)
defined, 434, 541, G–9
intermolecular forces, 433*t*, 434–436,
435*f*, 436*f*, 541
solutions, 486
Hydrogen carbonate ion, 67*t*
Hydrogen chloride, 62, 591, 592
Hydrogen chloride gas, dissolution in
water, 136
Hydrogen fluoride, 591
bonding, 412, 412*f*
formation, 101, 101*f*
Hydrogen fuels, 545*mn*
Hydrogen gas
bonding, 62
fuel cells, 160*mn*
Hydrogen ion, 65*t*
Hydrogen molecule
covalent bond formation in, 337*f*
electron distribution in,
134–136, 134*f*
Hydrogen peroxide, 583, 585*mn*
Hydrogen phosphate ion, 67*t*
Hydrogen sulfate ion, 67*t*
Hydrogen sulfide, 583
Hydrohalic acid, 592
Hydrolysis, 638, 782*n*, G–9
Hydrolysis constant, 782*n*
Hydronium ion, 67*t*, 136, 136*f*,
758–759
defined, 758, G–9
pH scale, 765–768
in redox balancing, 896
water and, 136, 136*f*, 758–759
Hydrosphere, 953, G–9
Hydrosulfuric acid, acid-dissociation
constants, 777*t*
Hydroxide ion, 67*t*
Hydroxides, amphoteric *see*
Amphoteric hydroxides,
complex ions
Hypertonic food preservation, 513
Hypertonic solution, 513

Hypo- (prefix), 69
Hypochlorite ion, 67*t*, 68
Hypohalites, 594
Hypohalous acid, 594*t*
Hypothesis, in scientific method, 12,
12*f*, 13, G–9

I

-ic (suffix), 67, 69
Ice, 443, 443*f*
-ide (suffix), 65
Ideal gas, 180, G–9
Ideal gas law, 180, 185, 185*f*, 189
defined, 185, G–9
gas density, 189–190, 189*mn*, 190*mn*
individual gas laws and, 185*f*
limiting-reactant problem and, 196
molar mass, 191
reaction stoichiometry and, 195–196
van der Waals equation and, 209–210
Ideal solution, 507, G–9
Indicators *see* Acid-base indicators
Indium
bond type, 562*t*
electron configuration, 315
isolation, 984*t*
melting point, 562*t*
natural sources, 984*t*
properties, 558
uses, 984*t*
Induced-fit model, enzyme action,
701, G–9
Industrial fixation, 958
Industrial gases, 174*t*
Inert-pair effect, 314, 557
Infrared (IR), 257, G–9
Infrared region, 257
Infrared spectroscopy (IR), 343,
343*f*, G–9
Infrared spectrum, 343
Inhibitor, 642
Initial rate, 669, G–9
Initial reaction rate, 669, 672
Initiator, in polymerization, 642
Inner (core) electrons, 302, G–9
Inner transition elements, 57, 57*f*,
302–303, 999, 999*f*
actinides, 1006, 1007
compounds, 63–72
defined, 303, 1006, G–9
lanthanides, 1006–1007
Inorganic chemistry, 58
Inorganic oxoacids, *versus* organic,
639–640
Instantaneous reaction rate, 668,
669*f*, G–9
Instrument calibration, in measurement,
31, 31*f*
Insulators, 458*f*, 459, G–9
Integrated rate law, 677, 682
defined, 677, G–9
first-order reactions, 677–679,
680–682, 682*t*
half-life, 680–682, 681*f*, 682*t*
reaction order, 679–680, 679*f*, 680*f*
second-order reactions,
677–679, 682*t*
zero-order reactions, 677–679, 682*t*
Intensive properties, 23, 27, G–9
Interference, 259
Interhalogen compounds, 592, G–9
Intermetallic compounds, 967
Intermolecular attractions, gas pressure
and, 208
Intermolecular forces, 420, 431–432
advanced materials, 460, 475

defined, 420, G–9
dipole-dipole forces, 432, 433*f*, 433*t*,
434, 438
dipole-induced dipole forces, 433*t*,
436, 441
dispersion (London) forces, 433*t*,
436–437, 437*f*
hydrogen bonds, 433*t*, 434–436, 435*t*
ion-dipole forces, 432, 433*t*, 438
ion-induced dipole forces, 433*t*, 486
polarizability, 436
solutions and, 486–487, 486*f*
van der Waals radius, 432, 432*f*
vapor pressure and, 427–428, 427*f*
Internal energy, 222
change in, 226
components of, 231*f*
defined, 222, G–9
International System of Units *see*
SI units
Interstitial alloys, 492
Interstitial hydrides, 543, G–9
Intramolecular forces, 420
Io, 5
Iodine
electron configuration, 315
ion formation, 62
isolation, 986*t*
natural sources, 986*t*
properties, 311, 590
sublimation, 429*f*
uses, 986*t*
Iodite, naming, 69
Iodous acid, naming, 69
Ion(s)
in aqueous solution, 133–134
complex ions *see* Complex ions
defined, 59, G–9
formation, 60–61, 330
Lewis structures, 358–368
naming
monatomic, 64*f*, 65, 65*t*
oxoanions, 67–68, 68*f*, 72
polyatomic, 67, 67*t*, 68–69
ranking by size, 318–319
spectator ions, 138
transition metals, 1000
Ion-dipole forces, 432, 433*t*, 438
defined, 432, G–9
solubility of ionic compounds, 486
Ion-exchange water softening, 520,
521*f*, G–9
Ion groups, qualitative analysis,
842–844, 842*f*, 843*f*, 844*f*
Ionic acid, naming, 70
Ionic atmosphere, 516, G–9
Ionic bonding, 59–61, 60*f*, 61*f*, 328,
328*f*, 330–336, 433*t*, 534
Born-Haber cycle, 332–333, 332*f*
Coulomb's law, 333
defined, 328, 534, G–10
ion pairs, 336, 336*f*
lattice energy
importance, 331–333
periodic trends, 333–335, 334*f*
octet rule, 330
Ionic compounds, 60
boiling point, 336*t*
components of, 62–63
defined, 59, 77*f*, 541, G–10
electrostatic forces, 335*f*
electrostatic forces in, 335*f*
formation, 59–61, 61*f*, 63, 155, 155*f*
formulas, 64*f*, 65, 65*t*
hydrates, 68*f*, 72
melting point, 336*t*
molar mass, 90

naming, 72
binary, 65–66
physical properties, 541
properties, ionic bonding model and,
335–336, 335*f*, 336*f*
slightly soluble *see* Slightly soluble
ionic compounds
solubility
ion-dipole forces in, 486
in water, 139, 139*t*
vaporization, 336*f*
in water, 135, 135*f*
Ionic equations, 137–138, 137*f*
acid-base reactions, 142
aqueous ionic reactions *see* Aqueous
ionic reactions
Ionic equilibria, 805–845
acid-base buffer systems *see* Acid-
base buffer systems
acid-base titration curves *see* Acid-
base titration curves
chemical analysis and, 840–845
complex ions *see* Complex ions
qualitative analysis, 841, 845
selective precipitation, 840–841
slightly soluble ionic compounds *see*
Slightly soluble ionic
compounds
Ionic hydrides, 544
Ionic oxides, 141–142
Ionic radius, 318, 318*f*, G–10
Ionic reaction equations, aqueous *see*
Aqueous ionic reactions
Ionic solids, 453–455, 453*t*, 454*f*,
459, G–10
Ionic solutions
electrical conductivity, 132–134, 133*f*
electrolysis, 933–934
Ion-induced dipole forces, 433*t*,
486, G–9
Ionization, radioactive emissions,
1058, G–10
Ionization counter, 1051, 1055
Ionization energy (IE), 533
alkali metals, 550
alkaline earth metals, 554
boron family, 558
carbon family, 564
defined, 307, 533, G–10
electron configuration and, 307–310,
307*f*, 308*f*, 309*t*, 311
halogens, 590
nitrogen family, 574
noble gases, 596
oxygen family, 582
periodic trends, 307–309, 533
successive, 309–310, 309*f*, 309*t*
transition elements, 1002, 1003
Ionizing radiation, 1065
applications, 1065
background radiation, 1061
defined, 1058, G–10
risk from, 1061–1062, 1061*t*
sources, 1060–1061, 1060*t*
units of radiation dose, 1059
Ion mobility, electrical conductance
and, 336*f*
Ion pairs, 336, 336*f*, G–9
Ion-product constant for water (K_w),
764–765, G–9
Ion-product expression (Q_{sp}), 824–828
Iron
corrosion, 926–928
dietary need for, 1032
electron configuration, 300*t*
ferrous/ferric, 66*t*, 67
in humans, 1032*t*

Iron (continued)
  isolation, 983t
  metallurgy of, 970–972, 987
  natural occurrence, 954t, 963t,
    970t, 983t
  orbital diagrams, 300t
  orbitals, 1001, 1001t
  specific heat capacity, 233t
  uses, 983t
Iron(III) oxide, naming, 67
Irradiated food, 1065, 1065f
Isoelectronic, 314, 567, G–10
Isoleucine, structure, 646f
Isomers, 623
  chirality, 617, 617f, 618, 618mn
  cis-trans isomerism, 619, 623, 1021
  constitutional isomers, 615, 616t,
    1020–1021
  coordination isomers, 1020–1021
  defined, 615, 1020, G–10
  diastereomers, 1021
  enantiomers, 617, 1021
  geometric isomers, 619, 623, 1021
  linkage isomers, 1020
  optical isomers, 617, 617f, 618, 623,
    1021–1022
  stereoisomers, 617, 1021–1022
  structural isomers, 100
Isotonic saline, 116, 513
Isotopes, 52–53, 56, 1042
  of carbon, 53
  defined, 52, 1042, G–10
  of hydrogen, 978–979
  mass spectrometry, 53, 54, 54f
  nuclear reactions, 1042
  of silicon, 53–55
Isotopic mass, 53, 89t, G–10
-ite (suffix), 68

J

Janssen, Pierre, 266mn
Joliot-Curie, Frederic, 1056
Joliot-Curie, Irene, 1056
Joule, James, 225mn
Joule (J) (unit), 225, G–10
Jupiter (planet), atmosphere, 206t

K

K see Equilibrium constant (K)
K_a see Acid-dissociation constant (K_a)
K_b see Base-ionization constant (K_b)
K_f see Formation constant (K_f)
K_{sp} see Solubility-product
  constant (K_{sp})
K_w see Ion-product constant for
  water (K_w)
Kelvin, Lord see Thomson, William
  (Lord Kelvin)
Kelvin (K) (unit), 17t, 24, 27, G–10
Kelvin scale, 24, G–1, G–10
Ketones, 629t, 633–635, 634f, G–10
Kilo- (prefix), 18t
Kilocalorie (kcal) (unit), 225
Kilogram (kg) (unit), 17t, 21, 27
  conversion from English equivalent,
    21–22
  defined, 21, G–10
  English equivalent, 18t
  master kilogram, 21mn
Kinetic energy, 6–8, 7f, 231, G–10
Kinetic-molecular theory, 203
  collision frequency, 203, 203mn
  defined, 197, G–10
  diffusion, 202–203, 203f
  effusion, 201–202, 202mn

gas laws and, 197–203, 198f–201f
mean free path, 203
molecular speeds, distribution of, 203
molecular view of the gas laws, 198,
  200–201
origin of pressure, 197, 198
phase changes, 423–425
postulates of, 197–198
rms speed, 200
states of matter, 420–421
temperature, 201
Kinetics see Chemical kinetics
Kolbe, Adolf, 47mn
Krypton
  electron configuration, 300t
  isolation, 986t
  natural occurrence, 59, 986t
  orbital diagrams, 300t
  properties, 595, 596
  uses, 986t
Kyoto conference on climate
  change, 245

L

Laboratory balance, 21
Laboratory glassware, 20, 20f, 560mn
Lactic acid, 100t
Lactose, 645, 645f
Lanthanide contraction, 556,
  1002, G–10
Lanthanides, 1006–1007, 1006mn, 1007
  defined, 303, 1006, G–10
  electron configuration, 303–304
  periodic trends in, 57, 57f
Lattice, 446, 446f, G–10
Lattice energy, 331–335, 331f, 332f,
  334f, 336
  defined, 331, G–10
  periodic trends in, 333–335, 334f
Lavoisier, Antoine, 10–11, 10mn, 13, 17
  and calorimetry, 869, 869f
  and chemical elements, 56
  periodic table of elements, 56
  scientific method, 13
Law of chemical equilibrium,
  717, G–10
Law of conservation of energy,
  225, G–10
Law of definite (constant) composition,
  43, 44–45, 44f, 47, G–10
Law of mass action, 717
Law of mass conservation, 43–44, 43f,
  46, 47, G–10
Law of multiple proportions, 43, 45,
  47, G–10
Lawrence, E.O., 1057
Lawrencium, 1007
Laws of thermodynamics
  first, 225, 857, 865
  living things and, 869
  second, 856, 861
  third, 861–862, 917mn
Layer silicates, 570, 570f
Leaching, 964, G–10
Lead
  bond type, 562t
  density, 23t
  electron configuration, 303–304
  isolation, 984t
  lead(II) and (IV), 66t
  melting point, 562t
  mineral source, 963t, 984t
  properties, 564
  uses, 984t
Lead-acid battery, 924, 924f, 930f
Le Châtelier, Henri, 872mn

Le Châtelier's principle, 736–743, 746
  concentration changes, 737–739,
    738f, 738t, 743t
  defined, 736, G–10
  Haber process for synthesis of
    ammonia, 744–745, 744f,
    744t, 745f
  pressure changes, 739–741, 740f, 743t
  temperature changes, 741–743, 743t
Lecithin, molecular structure, 637f
Leclanché cell, 923
Length
  common values of, 22f
  conversion to English
    equivalents, 18t
  unit of, 17t, 18t, 19, 27
Leucine, structure, 646f
Level, 276, G–10
Leveling effect, of solvent,
  790–791, G–10
Levorotatory isomer, 618
Lewis, G.N., 329, 330mn
Lewis acid-base definition, 791–792,
  794–795
  adduct, 792, G–1
  defined, 791, G–10
  Lewis acids with electron-deficient
    atoms, 792–793
  Lewis acids with polar multiple
    bonds, 793
  metal cations as Lewis acids,
    793–794, 793f
  molecules as Lewis acids, 792–793
Lewis electron-dot symbols, 329,
  329f, G–10
Lewis structures (formulas), 358, 368
  defined, 358, G–10
  electron-deficient molecules, 365
  electron-pair delocalization, 362–363
  expanded valence shells, 366–367
  formal charge, 364–365
  free radicals, 365–366
  heats of reaction from, 368–370
  molecules with multiple bonds, 361
  molecules with single bonds,
    358–360, 358f
  octet rule, 358–361, 358f
    exceptions to, 365–367
  odd-electron molecules, 365–366
  resonance hybrid, 362
  resonance structures, 362–363
  valence-shell electron-pair repulsion
    theory, 374, 374f
Libby, Willard F., 1053
Life, origin of, 442
Ligand field-molecular orbital
  theory, 1031
Ligands
  common types of, 1014–1018, 1016t
  complex ions, 835
  coordination compounds, 1013,
    1014f, 1015–1016, 1016t,
    1017, 1017t
  defined, 835, 1013, G–10
  splitting d orbitals in an octahedral
    field of ligands, 1026–1027,
    1026f, 1027f
  strong-field ligands, 1026, 1027f
  weak-field ligands, 1026–1027, 1027f
Light
  as electromagnetic radiation, 255
  particle nature of, 260–262, 260f
  photon theory, 260–261
  from stars, 269mn
  wave nature of, 256, 256f, 262
Light scattering see Tyndall effect
Like-dissolves-like rule, 487,
  489f, G–10

Lime, 552mn
Limestone cave, 831, 831f
Limiting reactant, 110–112, 110f,
  111mn, 113, G–10
Limiting-reactant problem, 111, 113
  ideal gas law and, 196
  reactions in solution, 118–119
Linear accelerator, 1056, 1056f
Linear arrangement
  defined, 372, G–10
  hybrid orbitals, 398t
  VSEPR, 371f, 372
Linear shape, 372, 376, G–11
Line spectrum, 263, 266
  defined, 263, G–10
  elements, 262f
Linkage isomers, 1020, G–11
Lipids, 637, G–11
Liquid, 4, 4f, 175f, G–11
Liquid crystals, 462–464, 463f
  application, 464–465
  cholesteric phase, 464, 464f
  defined, 462, G–11
  liquid crystal displays, 465, 465f
  lyotropic phase, 463
  nematic phase, 464, 464f
  polymers, 465
  smectic phase, 464, 464f
  thermotropic phase, 463
Liquid-gas equilibria, 426–427, 426f
Liquid-liquid solutions, 487–489
  capillarity, 439–440, 440f, 441,
    442, 443
  freezing, 421mn, 423f
  kinetic-molecular view, 421
  properties, 421t, 439–442
  surface tension, 439, 439f, 441,
    442, 443
  viscosity, 441
Liquid solutions and molecular polarity,
  487–491
Liter (L) (unit), 19, 27, G–11
Lithium
  density, 23–24, 23f
  diagonal relationships, 553
  dilithium, 407, 407f
  electron configuration, 295, 297f
  ionization energy, 309t
  isolation, 982t
  molecular orbitals, 457, 457f
  natural sources, 982t
  properties, 546t, 549
  uses, 546t, 982t
Lithium fluoride
  Born-Haber cycle, 332–333, 332f
  ionic bonding, 331–333, 332f
Lithium-ion batteries, 135mn
Lithosphere, 953, G–11
Lock-and-key model, enzyme action,
  701, G–11
London, Fritz, 436
London forces, 433t, 436–437,
  437f, 438
Lone pairs, 338, G–11
Lowry, Thomas, 145
Low-spin complexes, 1029, G–11
LSD, molecular structure, 638f
Luminous intensity, unit of, 17t
Lysine, molecular structure, 566,
  631f, 646f

M

Mach, Ernst, 47mn
Macromolecules, 468 see also Polymers
  biological macromolecules see
    Biological macromolecules

Macroscopic properties, 6
"Mad hatter," 1013*mn*
Magnesium
  diagonal relationships, 553
  electron configuration, 298*t*, 299*f*
  isolation, 977, 982*t*
  metallurgy, 977, 987
  natural sources, 954*t*, 982*t*
  properties, 312, 313*f*, 554
  uses, 552*mn*, 977, 982*t*
Magnetic properties
  crystal field theory and, 1028–1030
  transition metal complexes,
    1028–1029
  transition metal compounds, 1005
  transition metal ions, 316–317
Magnetic quantum number (m_l), 275,
    291*t*, G–11
Magnetic resonance imaging (MRI),
    625, 625*f*
Main-group elements, 57, 57*f see also*
    specific elements
  atomic size, 305, 305*f*
  ionization energy, 308, 308*f*
  ions of, 64*f*, 314–315
  oxidation states, 149, 149*f*, 539
  oxides of, 537
Manganese
  electron configuration, 300*t*
  in humans, 1032*t*
  isolation, 983*t*
  natural occurrence, 954*t*, 983*t*
  orbital diagrams, 300*t*
  orbitals, 1001, 1001*t*
  properties, 1009–1010, 1013
  uses, 983*t*
Manometers, 178, 178*f*, G–11
Manometric methods, reaction rate,
    measuring, 673
Mantle, Earth, 951, G–11
Many-electron atoms, 290–294
Mars (planet), atmosphere, 206*t*
Mass
  common values of, 22*f*
  conversion to English equivalents,
    18*t*, 21–22
  defined, 21, G–11
  mass-mole-number conversions,
    91–93, 92*f*, 120*f*
  measuring, 21
  unit of, 17*t*, 21–22, 21*mn*, 27
Mass action, law of, 717
Mass-action expression *see* Reaction
    quotient *(Q)*
Mass and energy, interconversion, 1066
Mass conservation, 43–44, 43*f*, 46, 47
Mass defect, 1066–1067, G–11
Mass fraction, 44–45, 95
Mass laws, 43–45, 47
Mass number *(A)*, 52, 53*f*, 56, G–11
Mass percent, 44–46, 93–95, 502, G–11
Mass spectrometry, 53–54, 54*f*, 56,
    267–268, 267*f*, 268*f*, G–11
Materials science, 460
Matter
  ancient Greek view of, 41
  atomic theory *see* Atomic theory
  compounds *see* Compounds
  defined, 3, G–11
  different from energy, 258–259
  electric properties, 48–49, 48*mn*
  elemental substances *see* Element
  energy changes, 6–8
  mixtures *see* Mixtures
  nature of, 41
  nuclear radiation, effect on,
    1058–1062

phase changes, 420, 421–431
properties, 3–4
states of *see* States of matter
wave-particle duality, 266, 269–273
Maxwell, James Clerk, 872*mn*
Mean free path, 203, G–11
Measurement
  accuracy in, 31, 31*f*
  calibration, 31, 31*f*
  conversion factors, 13–15, 17, 18*t*
  decimal prefixes, 17, 18*t*
  history, 17, 17*mn*
  instrument calibration, 31, 31*f*
  laboratory glassware, 20, 20*f*, 560*mn*
  of mass, 21
  metric system, 17
  precision, 31, 31*f*
  random error, 31, 32–33
  significant figures, 27–33
  SI units, 17–27
  systematic error, 31, 32–33
  uncertainty in, 27–33
Medical science, radioisotopes,
    1064–1065, 1064*f*, 1064*t*
Medical tradition of chemistry, 9
Mega- (prefix), 18*t*
Meitner, Lise, 1069, 1069*mn*
Meitnerium, 1069*mn*
Melting, 421, 423*f*, G–11
Melting point
  alkaline earth metals, 554
  defined, 429, G–11
  gallium, 350*f*
  Groups 1A and 2A elements, 350*f*
  ionic compounds, 336*t*
  metals, 350*t*
  noble gases, 596
Mendeleev, Dmitri, 56, 289
Mendelevium, 1007
Mercury, 440, 440*f*
  barometer, 177
  mercury(I) and (II), 66*t*
  mineral source, 963*t*
  properties, 1012–1013
  toxicity, 1013, 1013*mn*
  uses, 119, 1013
Mercury battery, 924
Mercury compounds, removal from
    wastewater, 119
Mercury (planet), atmosphere, 206*t*
Mesosphere, 204
Messenger RNA, 652
Metabolic pathway, 109,
    745–746, G–11
Metal(s)
  activity series of, 160, 160*f*
  atomic properties, 327, 328*f*
  boiling point, 350*t*
  crystal structures, 455, 455*f*
  defined, 58, G–11
  displacement reactions, 158–160,
    158*f*, 159*f*
  electron-sea model, 328,
    349–350, 456
  forming more than one ion, 66–67,
    66*t*, 72
  melting point, 350*t*
  organometallic compounds, 634
  periodic table, 57*f*, 58, 58*f*, 59
  properties, 311, 319, 350–351, 536
Metal carbonates, 567
Metallic behavior, 536
  acid-base behavior of element oxides,
    312–313, 313*f*, 537
  molecular orbitals and, 457
  transition metals, 1005

Metallic bonding, 328–329, 328*f*,
    349–351, 433*t*, 534, G–11
  alkali metals, 550
  alkaline earth metals, 554
  band theory, 456–459, G–2
  electron-sea model, 349–351
Metallic hydrides, 545
Metallic luster, molecular orbitals
    and, 457
Metallic radius, 304, G–11
Metallic solids, 453*t*, 455, 455*f*, G–11
Metalloids, 58, 458, 458*f*, G–11
  oxidation number, 148, 149*f*
  periodic table, 57*f*, 58, 58*f*, 59
  properties, 311, 536
Metallurgy, 963–968, G–11
  aluminum, 974–976
  copper, 972–974, 973*f*
  iron, 970–972
  magnesium, 977
  steel, 970–972
Metal oxides, reaction with water,
    158–159
Metathesis reactions, 139, 158, G–11
Meter (m) (unit), 17*t*, 19, 19*mn*, 27
  defined, 19, G–11
  English equivalent, 18*t*
Methanal, molecular structure, 634*f*
Methane
  boiling point, 617*f*
  bonding, 394
  cubic closest packing, 452*f*
  hybrid orbitals, 396, 396*f*
  molecular shape, 373
  uses, 565
Methanoic acid, molecular
    structure, 636*f*
Methanol
  hybrid orbitals, 399
  Lewis structure, 360–361
  molecular structure, 628*t*, 629*t*
  solubility, 488, 488*t*
Methionine, structure, 646*f*
Methyl ethyl ketone, molecular
    structure, 634*f*
Methyl red, 815*f*
  in acid-base titrations, 816, 818, 822
  structure, 757–758
Metric system, 17
Meyer, Julius Lothar, 289
Mica, 570
Micro- (prefix), 18*t*
Microelectrode, 922*mn*
Milli- (prefix), 18*t*
Millikan, Robert, 49, 51, 345*n*, 917*mn*
Millikan's oil-drop experiment, 49, 49*f*
Milliliter (mL) (unit), 19, G–11
Millimeter (mm) (unit), English
    equivalent, 18*t*
Millimeter of mercury (mmHg) (unit),
    178, 179, 179*t*, G–11
Mineral, as element source, 963, G–11
Minimicroanalysis, 922*mn*
Miscible solutions, 486, G–11
Mitochondrion, 937, 937*f*
Mixtures, 42, 43, 74, 74*f*, 77, 481, G–11
    see also Colloids; Solution
Mobile phase, chromatography, 76
MO bond order, 406, G–11
Model, in scientific method, 12, 12*f*,
    13, G–11
Molality *(m)*, 501–502, 501*t*, 505, G–11
Molar heat capacity *(C)*, 233, G–11
Molarity *(M)*
  defined, 114, 120, 501*t*, 505, G–11
  mole-mass-volume conversions and,
    115, 115*f*, 120*f*

Molar mass, 89–90, 89*n*, 89*t*, 95, 438
  boiling point and, 437, 437*f*
  converting to grams, 90
  converting to moles, 90
  defined, 89, G–11
  elements, 89
  gases, 191, 191*f*, 202
  Graham's law, 202, 203
  from osmotic pressure, 514–515
  of solutes, from colligative
    properties, 514–515
Molar ratios from balanced equations,
    106, 107*f*
Molar solubility, 826, G–11
Molar solutions, laboratory preparation,
    116–117, 116*f*, 117*f*
Molar volume, of common gases, 207*t*
Mole (mol) (unit), 17*t*, 87–89, 89*f*,
    89*mn*, 90, 95
  chemical reactions and, 101
  defined, 87, G–11
  mass-mole-number conversions,
    91–93, 92*f*, 120*f*
Molecular architecture, of polymers,
    472–473
Molecular compounds, 541, G–11
Molecular equations
  acid-base reactions, 146*f*
  aqueous ionic reaction equations,
    137, 137*f*
  defined, 137, G–12
Molecular formulas, 64, 72, 100,
    100*t*, 101
  converting to hybrid orbitals, 398
  converting to Lewis structure,
    358–361, 358*f*
  defined, 64, 98, G–12
  determining, 96–97
Molecular hydrides, 544
Molecularity, reaction mechanism,
    692–693, G–12
Molecular mass, 71–72, 89*t*, G–12
Molecular orbital
  defined, 404, G–12
  formation, 404
  of hydrogen, 405, 405*f*
  of lithium, 457, 457*f*
Molecular orbital band theory, 456–459
Molecular orbital diagram, 405,
    405*f*, G–12
Molecular orbital theory, 404, 414
  antibonding MO, 404–405, 405*f*
  atomic 2*p*-orbital combinations,
    408–409, 408*f*, 409*f*
  benzene, 413–414, 413*f*
  bonding in the *s*-block diatomic
    molecules, 407, 407*f*
  bonding MO, 404–405, 405*f*
  defined, 404, G–12
  formation of molecular orbitals
    (MOs), 404–405, 404*f*
  heteronuclear diatomic
    molecules, 412
    bonding in HF, 412, 412*f*
    bonding in NO, 413, 413*f*
  homonuclear diatomic molecules,
    407, 407*f*, 409, 409*f*
  MO bond order, 406, 406*f*
  nonbonding MOs, 412
  ozone, 413–414, 413*f*
  pi (π) MOs, 408, 408*f*
  sigma (σ) MOs, 405, 408*f*
Molecular polarity
  bond angle, 381–382
  bond polarity, 381–382
  defined, 381, G–12

Molecular polarity (continued)
  liquid solutions and, 487–491
  molecular shape and, 381–383, 381f
Molecular rotation, 403, 404
Molecular shape
  boiling point and, 437, 437f
  defined, 371, G–12
  molecular polarity and, 381–383, 381f
  number of bonds and, 535
  valence-shell electron-pair repulsion (VSEPR), 370–379, 370f–379f, 379
Molecular solids, 452–453, 452f, 453t, G–12
Molecular speed, 203, 203f
Molecular structure, reaction rate and, 687–688
Molecular volume, gases and, 208–209
Molecular weight see Molecular mass
Molecules
  chemical formulas, 63–72
  defined, 42, 43, G–12
  Lewis structures, 358–368
  models, 73f
  naming, 63–72
  orbitals see Molecular orbital
  organic see Organic compounds
  polarity see Molecular polarity
  rotation, 403, 404
  shape see Molecular shape
  speed, 203, 203f
  structure, reaction rate and, 687–688
Mole fraction, 192, 503, G–11
Mole-mass-number conversions, 115, 115f
Molina, Mario J., 702
Molybdenum, electron configuration, 303–304
Monatomic elements, molar mass, 89
Monatomic ions, 59, 63
  common ions, 64f, 65t
  compounds formed from, 65
  defined, 59, G–12
  electron configurations of, 314–316, 314f, 315f
  magnetic properties of transition metal ions, 316–317, 316f
  properties, 314–319
  radii, 318, 318f
Monodentate ligands, 1015, 1016t
Monomers, 468, G–12
Mononucleotides, 650, 650f, G–12
Monoprotic acids, 762, 762t
Monosaccharides, 645, G–12
Moseley, Henry G.J., 290, 290mn
Motor oil, viscosity, 441
Multiple bonds, VB theory, 400–403
Multiplication, significant figures, 29

N

NAA see Neutron activation analysis (NAA)
Nano- (prefix), 18t
Nanometer (nm) (unit), 27
Nanotechnology, 473–474
  consolidated materials, 475
  defined, 473, G–12
  dispersions and coatings, 474
  functional devices, 474–475, 475f
  high-surface-area materials, 474
Nanotubes, 380, 380f, 563, 563f
Naphthalene, 623mn
Naproxen, 618mn
Natta, Giulio, 642
Natural law, 12, G–12

Nematic phase, 464, 464f
Neon
  boiling point, 491t
  diatomic molecule, orbitals, 410f
  electron configuration, 296
  ionization energy, 309t
  isolation, 986t
  molar mass, 89
  natural occurrence, 59, 986t
  positively charged neon particle, 54
  properties, 547t, 596
  solubility, 491t
  uses, 547t, 986t
Neon signs, 48mn
Neopentane, 616t
Neptune (planet), atmosphere, 206t
Nernst, Walther Hermann, 917mn
Nernst equation, 917, G–12
Nerve cells, as concentration cells, 920mn
Nerve poisons, 962mn
Net ionic equation, 138, G–12
Network covalent compounds, physical properties, 541
Network covalent solids, 453t, 455–456, 459, 541, G–12
Neutralization reaction see Acid-base (neutralization) reactions
Neutrino, 1045mn
Neutron (n⁰), 51mn, 52, 56
  defined, 52, G–12
  discovery, 1055–1056
  mass number, 52
Neutron activation analysis (NAA), 1063–1064
Neutron/proton ratio (N/Z ratio), 1046, G–12
Nickel
  electron configuration, 300, 300t
  isolation, 983t
  natural occurrence, 954t, 983t
  orbital diagrams, 300t
  uses, 983t
Nickel-cadmium (nicad) battery, 925
Nickel-metal hydride batteries, 925
Nitrate ion, 67t, 68, 363
Nitric acid, 578–579, 833
  molarity, calculating, 136
  structure, 579f
  uses, 136
Nitric oxide, 575, 578
  boiling point, 491t
  bonding and MO diagram, 413, 413f
  properties, 577t
  solubility, 491t
  structure, 577t
Nitrile, 640, G–12
Nitrile, functional group, 629t
-nitrile (suffix), 629t
Nitrite ion, 67t, 68
Nitrogen
  boiling point, 491t
  bond type, 562t
  chemistry, 577–579
  decompression sickness, 500mn
  diatomic molecule, orbitals, 410f
  electron configuration, 296
  ionization energy, 309t
  isolation, 985t
  Lewis structure for, 361
  melting point, 562t
  natural occurrence, 59, 954t, 985t
  properties, 312, 313f, 547t, 574
  solubility, 491t
  standard heat of formation, 240t
  uses, 547t, 985t

Nitrogen cycle, 958–959, 959f
Nitrogen dioxide, 578
  Lewis structure, 366
  molecular model, 73f
  properties, 577t
  structure, 577t
Nitrogen family
  allotropes, 573, 573f
  isolation, 985t
  natural sources, 985t
  properties, 573–581
  uses, 985t
Nitrogen monoxide see Nitric oxide
Nitrogen oxides, 577–578, 577t
Nitrogen oxoacids, 578
Nitrous acid, 578, 581
  naming, 70
  structure, 579f
Nitrous oxide, 577t, 578
NMR see Nuclear magnetic resonance (NMR) spectroscopy
Nobelium, 1007
Noble gases
  alkali metals compared, 595, 595f
  as atomic solids, 452
  electron affinity, 311
  halogens compared, 595, 595f
  ionization energy, 311
  ions formed and nearest noble gas, 59, 61f, 62, 63
  isolation, 986t
  natural occurrence, 59
  natural sources, 986t
  periodic table, 59, 61, 61f
  properties, 595–596
  uses, 986t
Node, in electron probability diagram, 279, G–12
Noise pollution, and p scales, 765mn
Nomenclature
  acids, 69–70, 72
  alkanes, 71, 613–614
  aromatic compounds, 622
  carbon chains and branches, 613t
  coordination compounds, 1016–1017
  covalent compounds, 70–71, 72
  ionic compounds, 65–66, 72
  ions, 64–69
  organic compounds, 613t, 614t
  transuranium elements, 1058mn
Nonbonding MO, 412, G–12
Nonelectrolytes, 135
  colligative properties, 506–511, 514–515
  defined, 135, 506, G–12
Nonionizing radiation, 1058
  defined, 1058, G–12
  radioactive tracers, 1062–1065
Nonmetal, 458, 458f
  atomic properties, 327, 328f
  covalent compounds, 62
  defined, 58, G–12
  oxidation number, 149
  periodic table, 57f, 58, 58f, 59
  properties, 311, 536
  reaction with oxygen, 156
Nonmetal halides, reaction with additional halide, 156
Nonmetal hydrides, acid strength, 784–785, 786
Nonmetal oxides, reaction with additional oxygen, 156
Nonpolar covalent bond, 346, G–12
Nonstandard half-cell potentials, 932–933
Nonvolatile nonelectrolytes, colligative properties, 506–511, 514–515

Nuclear atom model, 48–51
Nuclear binding energy, 1067–1069, G–12
Nuclear energy reactors, 1071–1073, 1072f
Nuclear equations, balancing, 1044–1045
Nuclear fission, 1066, 1069–1073, 1076, G–7
Nuclear fuel, 202, 1073mn
Nuclear fusion, 1066, 1069, 1073, 1076, G–8
Nuclear magnetic resonance (NMR) spectroscopy, 624–625, 624f, 625f, G–12
Nuclear radiation, matter, effect on, 1058–1062, 1062mn
Nuclear reactions, 1040–1076
  atomic bomb, 1070–1071
  balancing, 1044–1045
  chain reaction, 1070
  chemical reactions compared, 1041, 1041t
  mass defect, 1066–1067
  nuclear binding energy, 1067–1069
  nuclear radiation, effect on matter, 1058–1062
  radioactive decay, 1042–1049
    decay series, 1049
    kinetics, 1050–1055
    nuclear equations, 1044–1045
    radioisotopic dating, 1053–1055
  radioisotopes see Radioisotopes
  transmutation see Nuclear transmutation
  transuranium elements, 1057
Nuclear structure, 1047–1048, 1047t, 1048t
Nuclear transmutation, 1055
  bevatron accelerator, 1057
  cyclotron accelerators, 1057, 1057f
  defined, 1055, G–12
  linear accelerators, 1056, 1056f
  particle accelerators, 1056–1058, 1056f, 1057f
  synchrotron accelerator, 1057
  transuranium elements, 1056–1058, 1057t
Nucleic acids, 650–653, G–12
Nucleons, 1042, G–12
Nucleotides, 650
Nucleus, 52 see also Nuclear reactions
  components, 51–52, 1042, 1042mn
  defined, 51, G–12
  Rutherford's scattering experiment, 50–51, 50f, 56mn
Nuclides, 1042, 1048, 1057t, G–12
Numbers
  exact, 30
  rounding off, 28–29, 30
Nylon-66, molecular models, 73f
N/Z ratio see Neutron/proton ratio (N/Z ratio)

O

-oate (suffix), 629t
Observations, in scientific method, 12, 12f, 13, G–12
Occurrence, 955, G–12
Oceans, 805f, 806, 1010mn
  diffusion and flow in, 421mn
  vaporization, 443
Octahedral arrangement
  defined, 376, G–12
  hybrid orbitals, 398t
  VSEPR, 371f, 376, 376f

Octahedral complexes, 1023, 1023*f*
  d-orbital splitting and, 1026–1030
  VB theory, 1023
Octane
  boiling point, 617*f*
  combustion reaction, 104–105
  formula, 71*t*
Octanitrocubane, 380
Octet rule, 330
  defined, 330, G–12
  Lewis structures, 358–361, 365
Odd-electron molecules, 365–366
Odor, theory of, 384–385
Odors, organic compounds, 638*mn*
-oic acid (suffix), 629*t*
Oils, silicones, 571
-ol (suffix), 629*t*
Olfactory receptors, 384*f*
-one (suffix), 629*t*
Open-end manometer, 178, 178*f*
Optical isomers, 617, 1021–1022, G–12
Optically active, 618, G–12
Orbit, 273
Orbital, 273 *see also* Atomic orbital;
    Molecular orbital
Orbital diagrams, 295, 296, G–13
Orbital energies, electrostatic effects
    and, 292–294
Orbital-orientation quantum number
    (l), 275
Orbital overlap, 400–404, 535
Ores, 955, 956, 963–968, G–13
Organic chemistry, 58
  chemistry of carbon, 566–567,
      607–608
  history, 607, 607*mn*
Organic compounds, 607, 610,
      627–628, 629*t*, 640–641, 641*f*
      see also individual compounds
  alcohols, 628, 628*f*, 630, 632–633
  aldehydes, 633–635, 634*f*
  alkenes, 633
  alkynes, 640
  amides, 636, 638–640, 638*f*, 640*f*
  amines, 631–633, 631*f*, 632*f*
  aromatic compounds, 633
  biological macromolecules *see*
      Biological macromolecules
  carbonyl group, 633, 634*f*
  carboxylic acid, 635–637, 636*f*,
      639, 639*f*
  combustion analysis, 98–99, 98*f*
  defined, 607, G–13
  esters, 636, 637–638, 637*f*,
      639–640, 640*f*
  haloalkanes, 630, 632–633
  hydrocarbons *see* Hydrocarbons
  hydrolysis, 638
  ketones, 633–635, 634*f*
  naming, 613*t*, 614*t*
  nitriles, 640
  organic molecules *see* Organic
      molecules
  organic synthesis, 635
  organometallic compounds, 634
  reactions *see* Organic reactions
  vitalism, 607
  Wöhler, Friedrich, 607
Organic molecules
  bond properties, 608, 609
  chemical diversity, factors in,
      608–609, 609*f*
  functional groups, 609
  heteroatoms, 609
  structural complexity, factors in, 608
Organic reactions, 624, 627
  addition reactions, 625
  alkyl group, 624

elimination reactions, 626
  oxidation-reduction reactions, 627
  substitution reactions, 626
  types, 624–626
Organic synthesis, 635
Organometallic compounds, 634, G–13
Organotin compounds, 565
Orthosilicate group, 568
Osmosis, 511, 521, 521*f*, G–13
Osmotic pressure (Π), 511, 511*f*, 513,
    G–13 *see also* Colligative
    properties
Ostwald, Wilhelm, 47*mn*, 872*mn*
Ostwald process, 578
-ous (suffix), 67, 69
Outer electrons, 302, G–13
Overall (net) equation, 108, G–13
Overvoltage, 933, G–13
Oxalic acid, acid-dissociation
    constants, 777*t*
Oxidation, 148, G–13
Oxidation number (O.N.), 148–150,
      148*t*, 149*f*, 154, 538
  defined, 148, G–13
  electronegativity and, 346
  halogens, 592–593
  main-group elements, 539
  rules for assigning, 148*t*
  transition metals, 1003–1004, 1004*t*
Oxidation number method, for
    balancing redox equations,
    151–152, G–13
Oxidation-reduction reactions *see*
    Redox reactions
Oxidation states *see* Oxidation number
Oxides *see* Element oxides, acid-base
    behavior
Oxidizing agent, 149, 150, 154,
    538, 893
  defined, 148, G–13
  redox titrations, 152–153, 152*f*
  relative strength, 908–909
Oxoacids, 632
  acid strength, 785, 786
  halogen oxoacids, 594*t*
  inorganic, 639–640
  naming, 69
  nitrogen oxoacids, 578
  of nonmetals, 639–640
Oxoanions, 67–68
  defined, 67, G–13
  naming, 68*f*, 72
Oxygen
  allotropes, 584
  chemistry, 586
  combustion, 10–11
  density, 23*t*
  diatomic molecule, orbitals, 410*f*
  electron configuration, 296
  isolation, 985*t*
  natural occurrence, 59, 954*t*, 985*t*
  oxidation number, 148*t*
  properties, 547*t*, 582
  solubility, 491*f*
  uses, 547*t*, 985*t*
Oxygen family
  isolation, 985*t*
  natural sources, 985*t*
  periodic table, 59
  properties, 581–588
  uses, 985*t*
Oxyhemoglobin, 1032
Ozone
  bonding, 413, 413*f*
  depletion from stratosphere, and
      ozone hole, 205–206, 584,
      702, 702*f*

Lewis structure, 362
  molecular model, 73*f*
  stratospheric, 205

P

p block, 532, G–13
p orbital, 279, 279*f*, G–13
Packing efficiency, unit cells and, 448,
    449*f*, 450, G–13
PANs (peroxyacylnitrates), 578
Paracelsus, 9
Paramagnetism, 316–317, 319,
    1005, G–13
Partial ionic character, 347–348,
    347*f*, G–13
Partial pressure
  collecting a gas over water, 193–194
  Dalton's law of partial pressures,
      192, 197, 199, 199*f*
  defined, 192, G–13
  of gas in mixture of gases, 192–194
Particle accelerators, 1056–1058, G–13
Particle nature of light, 260, 262
Particles, compared to waves, 258*f*
Parts by mass, 501*t*, 502–503
Parts by volume, 501*t*, 503
Pascal (Pa) (unit), 178, 179, 179*t*, G–13
Path independence of energy change,
    226–227
Pauli, Wolfgang, 291 *see also*
    Exclusion principle
Pauling, Linus, 394
Pauling electronegativity scale,
    344*f*, 345
PCBs, 566, 591
Penetration, 294, 294*f*, G–13
Pentane
  boiling point, 617*f*
  formula, 71*t*
  properties, 616*t*
  structural isomers, 616*t*
Peptide bond, 647, 647*f*
Per- (prefix), 69
Percent by mass, 44, 93–94, 502, G–13
Percent yield, 112–113, G–13
Perhalic acid, 594*t*
Period, 57, G–13
Periodic law, 289, G–13
Periodic table, 56–59, 57*f*, 543, 952,
    953*f*, 955–956, 955*f see also*
    specific elements and specific
    groups
  atomic number, 290, 290*mn*
  atomic properties and, 327, 328*f*
  defined, 56, G–13
  diagonal relationships, 553, 569
  electron configuration
    atomic properties and, 304–311
    chemical reactivity and, 311–319
    within groups, 299, 299*f*, 301, 304,
        307, 311
    period 3, 298, 298*t*, 299*f*
    period 4, 299–301, 300*t*, 301*f*
    periods 1 and 2, 295–297, 297*f*
    principles, 301–302
    transition elements, 303–304
  group number, 302
  groups, 57, 57*f*, 59, 299, 299*f*, 301,
      304, 307, 311
  halogens, 588–594
  history, 56–59, 289–290
  inner transition elements, 57, 57*f*,
      999, 999*f*
  oxidation number, 148*t*, 149, 149*f*
  Pauling electronegativity scale,
      344*f*, 345

periods, 57, 57*f*, 59, 302
  quantum-mechanical model, 295–304
  transition elements, 57, 57*f*, 983*t*,
      999, 999*f*
Permalloy, 967*t*
Permanganate ion, 67*t*, 152, 152*f*
Peroxide ion, 67*t*
Peroxyacylnitrates *see* PANs
    (Peroxyacylnitrates)
pH, 765–768, 766*f*, 766*t*, 767*f*, 768*f*
  defined, 765, G–13
  pH meter, 768, 921
  pH paper, 768
  slightly soluble ionic compounds,
      829–830, 830*f*
Phase, 420, G–13
Phase changes, 420, 429, 541
  boiling point, 428–429
  condensation, 421, 423*f*
  critical point, 430
  defined, 420, G–13
  deposition, 422, 423*f*
  elements, 541
  equilibrium nature of, 425–431
  freezing, 421, 423*f*
  fusion, 421
  intermolecular forces, 420
  kinetic-molecular approach, 423–425
  liquid-gas equilibria, 426–427, 426*f*
  melting, 421, 423*t*
  quantitative aspects, 423–431
  solid-gas equilibria, 429, 429*f*
  solid-liquid equilibria, 429, 431
  sublimation, 422, 423*t*
  triple point, 431
  types, 421–423
  vaporization, 421, 423*f*
  vapor pressure, 427–428, 428*f*
Phase diagrams, 430–431, 430*f*, G–13
Phenolphthalein, 815*f*
  in acid-base titrations, 816, 818, 822
Phenylalanine, structure, 646*f*
Pheromones, 503*f*
Phlogiston theory, 10–11, 13, G–13
Phosgene, 189*mn*
Phosphate, uses, 580*mn*, 962
Phosphate ester, 639
Phosphate fertilizers, 962
Phosphate ion, 67*t*
Phosphoric acid, 580
  acid-dissociation constants, 777*t*
  formula, 758*t*
  uses, 758*t*
Phosphorous acid, 580, 777*t*
Phosphorus
  allotropes, 573, 573*f*
  bond type, 562*t*
  chemistry, 580–581
  electron configuration, 298*t*, 299*f*
  isolation, 985*t*
  melting point, 562*t*
  natural occurrence, 954*t*, 985*t*
  nerve poisons, 962*mn*
  occurrence in outer space, 960*mn*
  properties, 312, 313*f*, 574
  standard state, 240
  uses, 985*t*
Phosphorus compounds, uses, 581*mn*
Phosphorus cycle, 960–962, 961*f*
Photochemical smog, 189*mn*, 578, 668
Photodissociation, 205
Photoelectric effect, 260–261,
    261*f*, G–13
Photography
  flashbulb, chemical action in, 102,
      103*f*, 147
  silver compounds, 1010–1012, 1011*f*

Photoionization, 205
Photons, 261, 262, 269, G–13
Photon theory of light, 260–261, G–13
Photosynthesis, 957
Photovoltaic cells, 245, G–13
Physical change, 3, 3f, 4–5, 5, 77f, G–13
Physical properties, 3, 4t, G–13 see also individual groups
Physical states, 540, 666, 666f
    gas see Gas
        kinetic-molecular view, 420–421, 421t
    liquids see Liquid
    solids see Solid
    thermodynamics, 862
Pico- (prefix), 18t
Picometer (pm) (unit), 27
Pi (π) bond, 401, G–13
Pi (π) MO, 408, G–13
Pitchblende, uranium in, 45
pK scale, 766, 766f
Planck, Max, 260, 917mn
Planck's constant (h), 260, G–13
Planets, atmosphere, 206t, 207
Plasma, 1076
Plastic, 472, G–13
Platinum, natural occurrence, 59
Pluto (planet), atmosphere, 206t
p-n junction, 460–462, 461f
Polar covalent bonds, 346–348, G–13
Polarimeter, 618, 618f, G–13
Polarity, molecular see Molecular polarity
Polarizability, 436, G–13
Polar molecule, 134, G–13
Pollutants
    in the food chain, 630mn
    ultralow concentrations, 503mn
Polonium, 581, 582, 985t
Polyatomic ions, 63, 63f, 146
    compounds formed from, 67, 68–69
    defined, 63, G–13
    naming, 67t
    oxoanions, 67–68, 68f
Polychlorinated biphenyls (PCBs), 566, 591
Polydentate ligands, 1016t
Polyethylene, 643t
Poly(ethylene oxide), as solid solvent, 135mn
Polymer, 465, 468 see also Macromolecules
    addition polymers, 642–643, 642f, 643t
    biopolymers, 468
    branch, 472
    chain dimensions, 468–471
    condensation polymers, 643
    copolymer, 473
    crosslinks, 472
    crystallinity, 470–471, 470f
    defined, 468, G–13
    degree of polymerization (n), 468
    dendrimers, 472
    elastomers, 472, 473t
    flow behavior, 471–472, 471f
    glass transition temperature, 472
    homopolymer, 472
    mass, 468–469, 468t
    molecular architecture, 472–473
    monomers, 468
    nucleic acids, 650–653
    nucleotides, 650
    plastic, 472
    polysaccharides, 645–646, 653
    proteins, 646–649, 653

radius of gyration, 469f, 470, 470n
    random coil, 469
    silicone, 568–569, 571–572
    thermoplastic elastomers, 473
    thermoplastic polymer, 472
    thermoset polymer, 472
polymeric materials, 468–473
Polymeric viscosity index improvers, 441
Polypeptide chain, 646–647, 647f
Polyphosphates, 580, 581f
Polyprotic acids, 776–777, 777t
    acid-base titration curves, 822–823, 823
    defined, 776, G–14
Polysaccharides, 645–646, 645mn, 653, G–14
Polysulfanes, 623
Porphin ring, 1033
Positron, 1045, G–14
Positron decay, 1045, G–14
Positron emission, 1044t, 1045, 1045n
Positron-emission tomography (PET), 1064–1065, 1065t
Potassium
    electron configuration, 300, 300t, 303–304, 315
    industrial production, 969–970, 969f, 987
    isolation, 982t
    natural occurrence, 954t, 982t
    orbital diagram, 300t
    uses, 970, 982t
Potential energy, 6–8, 7f, 231, 231f, G–14
Precipitate, 138, G–14
Precipitation reactions, 138, 140
    aqueous ionic reaction equations, 137f
    defined, 138, G–14
    driving force for, 138–139, 139f
    ionic equations, 140
    metathesis reactions, 139
    predicting, 139–140, 139t
Precision, in measurement, 31, 31f, G–14
Prefixes, decimal, 17, 18t
Pressure
    Amontons's law, 183
    atmosphere, 204, 204f
    critical pressure, 430
    Dalton's law of partial pressures, 192, 197, 199, 199f
    defined, 176, G–14
    equilibrium constant, 724–725
    gases, 207, 207f
    Henry's law, 500
    Le Châtelier's principle, 739–740
    measuring, 176–178, 177f, 179f
    partial, 192–194
    solubility and, 500
    units of, 178–179, 179t
Pressure-volume work (PV work), 224, 228f, G–14
Priestley, Joseph, 11mn
Primary amines, 631, 631f
Primary sewage treatment, 521
Principal quantum number (n), 275, 291t, 305, G–14
Probability contour, 275, G–14
Probability factor (p), 687
Problem solving
    chemistry in the real world, 32–33, 33f
    conversion factors, 13–15, 17
    systematic approach to, 15–16
    units of measurement, 13
Products, defined, 102, G–14

Property, 3, G–14
Protactinium, 1069mn
Proteins, 646–649, 653
    amino acids, 646–647, 646f
    composition, 648–649
    defined, 646, G–14
    fibrous proteins, 648–649, 649f
    globular proteins, 649
    size, 647
    structure, 647–648
    synthesis, 652, 652f
Proton (p+), 51mn, 52, 56
    defined, 52, G–14
    hydronium ion, 67t, 136, 136f, 758–759, 765–768
    mass number and, 52
Proton acceptor, 768, G–14
Proton donor, 768, G–14
Proton transfer, acids and bases, 768–770
Pseudo-noble gas configuration, 314, G–14
PV/RT curve, 207, 207f
Pyrometallurgy, 963
Pyrotechnics, 594mn

Q

Qualitative analysis, 841–845, G–14
Quantum, 260, 262, G–14
Quantum mechanics, 255, 273, 281
    atomic orbital (wave function), 273–275, 277
        quantum numbers, 275–277, 276t
        shapes, 278–280, 278f, 279f
    defined, 273, G–14
    electron cloud, 274
    electron density diagram, 274, 274f
    energy levels in the hydrogen atom, 281, 281f
    level (shell), 276
    nodes, 279
    periodic table see Periodic table
    probability contour, 274f, 275
    radial probability distribution plot, 274f, 275
    Schrödinger, Erwin, 273
    Schrödinger equation, 273
    sublevel (subshell), 276, 277
Quantum numbers, 260, 275–277, 276t, 290–291, G–14
Quantum theory, 254f, 255 see also Quantum mechanics
Quartz, crystal structure, 456
Quaternary ammonium salt, 632, 632f

R

R see Universal gas constant
Rad see Radiation-absorbed dose
Radial probability distribution plot, 275, 275mn, G–14
Radiant energy, 255
Radiation see Nuclear radiation, matter, effect on
Radiation-absorbed dose (rad), 1059, G–14
Radiation dose, units of, 1059
Radioactive decay, 680–681, 1042–1049
    decay series, 1049
    half-life, 680–681, 1050–1053
    kinetics, 1050–1055
    nuclear equations, 1044–1045
    nuclear stability and, 1046–1049
    radioisotopic dating, 1053–1055

rate of, 1050
    types of, 1044–1046
Radioactive emissions
    alpha particles, 270t, 1043, 1043f, 1044t, 1059
    beta particles, 1043, 1043f, 1044t, 1059
    detection, 1051, 1051f
    excitation, 1058
    gamma rays, 1043, 1043f, 1044t, 1059
    ionization, 1058
    positron emission, 1044t, 1045, 1059
Radioactive tracers, 1062–1065
Radioactivity see also Nuclear reactions
    defined, 1042, G–14
    discovery, 50, 1042–1043, 1043mn
Radiocarbon dating, 1053–1055, 1054f, 1054mn
Radioisotopes, 1053, 1062–1066, G–14
    applications, 1062–1066
    dating, 1053–1055
    ionizing radiation, 1065
    tracers, 1062–1065
Radioisotopic dating, 1053, G–14
Radium, 1043
    history, 1043
    isolation, 982t
    natural occurrence, 982t
    toxicity, 1059mn
    uses, 982t
Radius of gyration, 469f, 470, 470n, G–14
Radon
    isolation, 986t
    natural occurrence, 59, 986t
    properties, 554
    toxicity, 1061mn
    uses, 986t
Rainbows, dispersion, 259mn
Ramsey, William, 266mn
Random coil of polymer chain, 469, G–14
Random error, 31, G–14
Raoult's law, 507, G–14
Rare earth elements, electron configuration, 303–304
Rate constant (k), 671, 677, G–14
Rate-determining step, 693–694, G–14
Rate law (rate equation), 671, 677
    collision theory see Collision theory
    defined, 671, G–14
    elementary steps, 692t
    first order, 672, 674
    initial rate, 672
    integrated, 677–682
    rate constant, 671, 677, 677t
    reaction mechanism and, 694–696
    reaction orders, 671, 675–676
    reaction order terminology, 672–674
    second order, 672
    zero order, 673
Rate-limiting step see Rate determining step
Rate of reaction see Reaction rate
Reactants
    defined, 102, G–14
    limiting, 110–112, 110f, 111mn
Reaction direction
    Brønsted-Lowry acid-base definition, 770–772, 771f
    equilibrium, 725–727, 726f, 734–736
    free energy change, 880–884, 881t
    reaction quotient, 725–727, 726f, 734–736
Reaction energy diagrams, 689–690, G–14

Reaction half-life, 680–682
Reaction intermediate, 691–692, G–14
Reaction mechanism, 696–697
    defined, 691, G–14
    elementary reaction, 692–693
    molecularity, 692–693
    rate-determining step, 693–694
    rate law and, 694–696
Reaction order
    defined, 671, G–14
    integrated rate law, 679–680,
        679f, 680f
    rate law, 671, 675–676
    terminology, 672–674
Reaction quotient (Q), 717–718, 718f,
        723f, 723t, 724, 918
    defined, 717, G–14
    law of mass action, 717
    reaction direction, 725–727, 726f,
        734–736
    relation to equilibrium constant,
        717–723
    variations in form, 719–723
    writing, 718–719
Reaction rate, 664–665, 664f, 665f,
        667, G–14
    average, 668
    concentration and, 665
    expressing, 667–671
    initial, 669
    instantaneous, 668
    measurement of, 672–673
    molecular structure and, 687–688
    physical state and, 666
    surface area and, 666, 666f
    temperature and, 666, 682–684,
        685–687, 686f
Reaction(s) see Chemical reactions;
        Nuclear reactions; Organic
        reactions
Reactor core, 1072, G–14
Real gases, 207, 210
    extreme conditions, 207–209
    intermolecular attractions, 208, 208f
    molecular volume, 208–209, 209f
    van der Waals constants, 209, 209t
    van der Waals equation, 209–210
Receptor sites, 384–385, 384f
Rechargeable battery, 924–925, 930
Recycling, aluminum, 974–976,
        975f, 976f
Redox couple, 937
Redox reactions, 146, 147, 147f, 154,
        538, 893–898
    activity series of the halogens, 160
    activity series of the metals,
        158–160, 158f, 159f, 160f
    balancing, 150–152, 894–898
        half-reaction method, 894–899
        half-reaction method, hydronium
            ion in, 896
        oxidation number method, 151–152
    in biosphere, 174mn
    combination reactions as, 154
    combustion reactions, 160–161
    decomposition reactions as, 154
    defined, 147, G–13
    displacement reactions as, 154
    disproportionation, 578
    driving force for, 147
    electrochemistry and, 893–894,
        894f, 899
        acidic solutions, 895–896, 895f
        basic solutions, 897
    electrolytic cells, 893–899
    electron movement, 147, 147f, 148,
        149f, 154

equations, 150–152
    in metallurgy, 965–966
    organic reactions, 627
    oxidation, 148
    oxidation numbers see Oxidation
        number
    oxidizing agents, 148, 150, 538
    redox titrations, 152–154, 152f
    reducing agents, 148, 150, 538
    reduction, 148
    terminology, 148, 149f
    voltaic cells, 909–912
        relative reactivities of metals,
            912–913, 913f
        spontaneous redox reactions,
            909–912
Red phosphorus, 573f
Reducing agent, 148, G–14
Reduction, 148, G–14
Refining, metallurgy, 967–968
Refraction, 258, 262, G–14
Relative atomic mass, 89n
Rem see Roentgen equivalent for man
Representative elements, 57
Reproducibility, 31, 31f, 32
Resins, ion-exchange, 572
Resonance, 624
Resonance hybrid, 362, 362mn,
        363, G–15
Resonance structures
    defined, 362, G–15
    formal charge and selecting,
        364–365, 368
    Lewis structures, 362–365
Retinal, in vision, 620, 620f
Reverse osmosis, 521, 521f, G–15
Reversible reactions, 162–163
    dynamic equilibrium, 162, 162f
    free energy, 163
Rhodopsin, 620, 620f
Ribonucleic acid (RNA), 650, 650f
Ribosome, 652
Roasting, of sulfides, 964, G–15
Roentgen equivalent for man (rem)
        (unit), 1059, G–14
Rohrer, Heinrich, 452
Root-mean-square (rms) speed,
        200, G–15
Rosenberg, Barnett, 1021mn
Rounding off, 28–29, 30, G–15
Rowland, F. Sherwood, 702
Rubidium, 982t
Rusting of iron, 926–928
Rutherford, Ernest, 50, 51, 273mn,
        1043, 1055
Rutherford's scattering experiment,
        50–51, 50f, 56mn
Rydberg constant, 266
Rydberg equation, 266

S

s block, 532, G–15
s orbital, 278–279, 278f, G–15
Safety matches, 594
Salt, 142, G–15
Salt bridge, 902, G–15
Saltlike hydrides, 544
Salt link, 648
Salt solutions, acid-base properties,
        787–790
Salt (table salt)
    density, 23t
    formation, 60, 60f
Salt (type of compound), 142, 145f
    yielding acidic solutions, 787–788

    yielding basic solutions, 788
    yielding neutral solutions, 787
Saponification, 638
Saturated hydrocarbon, 613, G–15
Saturated solutions, 497–498,
        498mn, G–15
Saturn (planet), atmosphere, 206t
Scandium
    electron configuration, 300t
    isolation, 983t
    natural sources, 983t
    orbital diagrams, 300t
    orbitals, 1001, 1001t
    uses, 983t
Scanning tunneling microscopy, 451,
        452, 452f, G–15
SCF see Supercritical fluid (SCF)
Schrödinger equation, 273, 273n,
        290, G–15
Scientific method, 11–13, 13mn, G–15
Scintillation counter, 1051, 1055, G–15
Scrubbers, 243
Second (s) (unit), 17t, 26, G–15
Secondary amines, 631, 631f
Secondary batteries, 924–925
Secondary sewage treatment, 521
Second law of thermodynamics, 856,
        861, G–15
Second order reactions
    first- and zero-order compared, 682t
    integrated rate law, 677–679, 682t
    rate law, 672
Seesaw shape, VSEPR theory, 375, G–15
Seismology, logarithmic scale, 765mn
Selective precipitation, 840–841,
        845, G–15
Selenium
    allotropes, 584
    electron configuration, 300t
    isolation, 985t
    natural sources, 985t
    orbital diagrams, 300t
    properties, 582
    uses, 584mn, 985t
    xerography, 584mn
Self-assembly, 473
Semiconductor, 458f, 459, 460–461,
        475, G–15
Semicrystalline regions, in
        polymers, 470
Semimetals see Metalloids
Semipermeable membrane, 511, G–15
Sense of smell, 384–385
Separation techniques, 75–76, 75f
Serine, 628f, 646f
Sewage treatment, 521
Shared electron pair, 338
Sheet silicates, 570, 570f
Sheet silicones, 572
Shell see Level
Shielding, 293, G–15
Shroud of Turin, radiocarbon
        dating, 1054mn
Side reactions, 112, 112f, 113, G–15
Side-to-side overlap, 535, G–15
Sievert (Sv) (unit), 1059, G–15
Sigma (σ) bond, 401, G–15
Sigma (σ) bonding, 401
Sigma (σ) molecular orbital, 405, G–15
Significant figures, 27–33
    in calculations, 28–30
    defined, 27, G–15
    determining, 28
    electronic calculators and, 30
    exact figures, 30
    measuring devices and, 27, 27f, 30, 30f
    rounding off, 28–29, 30

Silanes, 623
Silicate, 568, G–15
Silicate minerals, 570
Silicon
    bond type, 562t
    chemistry, 568–569
    diagonal relationship, 569
    electron configuration, 298t, 299f
    isolation, 984t
    isotopes, 53–55
    melting point, 562t
    properties, 312, 313f, 564
    uses, 984t
Silicon carbide, 565
Silicon dioxide, 313, 565
    crystal structure, 456, 457f
Silicone, 568, G–15
Silicone polymers, 568–569,
        571–572
Silicon hexafluoride, hybrid orbitals,
        298, 398f
Silicon tetrachloride, properties, 348
Silver
    atomic mass, 55
    mass-mole-number conversion, 91
    natural occurrence, 59
    photography, 1010–1012, 1011f
    properties, 1010–1012
    standard heat of formation, 240t
Silver alloys, 967t
Silver battery, 924
Silver ion, 65t
Silver nitrate, reaction with sodium
        chromate, 137f
Simple cubic unit cell, 446, 447f,
        448, G–15
Single bonds, 338
    defined, 338, G–15
    functional groups with, 628–633
    functional groups with single and
        double bonds, 635–639
    Lewis structure for molecules with,
        358–361
    valence bond (VB) theory,
        400–402
Single-displacement reactions,
        158–160, 158f, 159f, 162
SI units, 17–27
    base units, 17, 26
    conversion factors to, 18t
    conversion to English
        equivalents, 18t
    decimal prefixes, 17, 18t
    defined, 17, G–15
    derived units, 17, 17t, 26
    features, 17
Slag, 971, G–15
Slightly soluble ionic compounds,
        824, 832
    common ions, 828–829, 828f
    ion-product expression, 824–826,
        830, 832
    molar solubility, 826
    pH, 829–830, 830f
    precipitate, 830, 832
    solubility-product constant, 824–828,
        826t, 828t, 830, 832
Smectic phase, 464, 464f
Smell, 384–385
Smelting, in metallurgy, 965, G–15
Smog, 156, 189mn, 578, 668
SNG see Synthetic natural gas (SNG)
Snowshoes, 176mn
Soaps
    defined, 490, G–15
    dual polarity of, 490, 490f
    saponification, 638

Sodium
  electron configuration, 298t, 299f
  industrial production, 969–970, 969f, 987
  ionization energy, 309t
  isolation, 982t
  natural occurrence, 954t, 963t, 982t
  properties, 43t, 312, 313f
  reaction with water, 531f
  standard heat of formation, 240t
  uses, 970, 982t
Sodium atoms, emission and absorption spectra, 267f
Sodium carbonate, 551, 758t
Sodium chloride, 551
  crystal structure, 453–454, 454f
  formation, 60, 60f
  isotonic saline, 116
  properties, 43t, 348
  solubility, 489
Sodium chromate, reaction with silver nitrate, 137f
Sodium hydrogen carbonate, 551, 758t
Sodium hydroxide, 551, 758t
Sodium hypochlorite, 591
Sodium iodide, reaction of lead nitrate with, 139, 139f
Sodium ion, 65t
Sodium stearate, 490
Sodium sulfate, reaction with strontium nitrate, 140
Sodium tripolyphosphate, 575
Solar cells, 460mn
Solar emission, 266mn
Solar energy, 245
Solar system
  age, 1055mn
  planetary atmosphere, 206t, 207
Solder, 967t, 1013
Solid, 4, 4f, 175f
  advanced materials, 460
    ceramics, 465–467, 466t, 467f, 475
    liquid crystals, 462–465, 463f–465f, 475
    nanotechnology, 473–475, 475
    semiconductors, 458f, 459, 460–461, 475
  amorphous, 445, 456–459
  bonding in, molecular orbital band theory, 456–459
  crystalline, 445–446, 445f–449f, 448, 450, 452–456
  defined, 4, G–15
  kinetic-molecular view, 421
  properties, 421t
Solid-gas equilibria, 429, 429f
Solid-liquid equilibria, 429, 431
Solid-liquid solutions, 487–489
Solid-solid solutions, 492–493, 492f
Solubility, 486
  boiling point and, 491t
  chromatography, 76, 76f
  common-ion effect, 828–829
  complex ion formation and, 837–838
  concentration, 501–505
  crystallization, 75, 75f
  defined, 75, 486, G–15
  equilibrium and, 501–505
  extraction, 75, 75f
  gases, 175
  gas-gas solutions, 492
  gas-liquid solutions, 491
  gas-solid solutions, 492
  of ionic compounds, 132–134, 133f, 139, 139t
  liquid-liquid solutions, 487–489
  pH and, 829–830

pressure and, 500
solid-liquid solutions, 487–489
solid-solid solutions, 492–493, 492f
temperature and, 498–499
in water
  covalent compounds, 135–136, 136
  ionic compounds, 135, 135f, 136, 139, 139t
Solubility-product constant (K_{sp}), 824–828, 826t, 828t, 830, 832, G–15 see also Appendix C
Soluble compound, 135
Solutes
  defined, 114, 486, G–15
  entropy, 862–863, 862f
  molar mass from colligative properties, 514–515
  vapor pressure and, 507f
Solution, 486 see also Solubility; Solution process; Solvents
  chemical reactions, stoichiometry, 118–119
  colligative properties, 506–511, 514–517
    in biology, 513, 513f
    boiling point elevation, 508–509, 509t, 510
    electrolyte solutions, 516, 516f
    freezing point depression, 509–510, 512, 512f
    in industry, 512, 512f
    osmotic pressure, 512
    solute molar mass from, 514–515
    vapor pressure lowering, 507–508, 507f
  concentration, 501–505
  defined, 74, 114, 492
  electrolysis, 933–934
  gas-gas, 492
  gas-liquid, 491
  gas-solid, 492
  ideal, 507
  intermolecular forces in, 486–487, 486f
  liquid-liquid, 487–489
  salt, 787–790
  saturated, 498, 498mn
  slightly soluble salts, 824–832
  solid-liquid, 487–489
  solid-solid, 492, 492f
  stoichiometry see Solution stoichiometry
  supersaturated, 498, 498f
  unsaturated, 498
Solution process see also Solubility; Solution
  disorder and, 496–497
  energy changes in, 501–505
  heat of hydration, 494, 495
  heat of solution, 493–494, 497
Solution stoichiometry, 114
  chemical reactions in solution, 118–119
  concentration, 114
  molarity, 114
  molar solutions, laboratory preparation, 116–117, 116f, 117f
Solvated, 132, G–15
Solvation, 132, 136, 494, G–15
Solvents, 486, 487
  defined, 114, 486, G–15
  solid solvents for ions, 135mn
  water as, 442
Sound, logarithmic scale, 765mn
sp hybridization, 394, 395f, 396, G–15

sp^2 hybridization, 396, 396f, G–16
sp^3 hybridization, 396, 397f, G–16
sp^3d hybridization, 397, 397f, G–16
sp^3d^2 hybridization, 398, G–16
Space-filling models, 73f
Space Shuttle, fuel cells, 160mn
Specific activity, 1050
Specific heat capacity (c), 233–234, 233t, G–16
Spectator ions, 137f, 138, G–16
Spectra
  absorption
    absorption spectra, 267–268, 267f, 268f
    atomic spectra, 262–263
    emission spectra, 267, 267f
    line spectrum, 263
Spectrochemical series, 1028, G–16
Spectrometry, reaction rates, measuring, 672
Spectrophotometer, 268, 268f
Spectrophotometry, 267–268, G–16
Speed of light (c), 256, G–16
Spin, 290
Spin quantum number (m_s), 291, 291t, G–16
Spontaneous change, 856, 865
  defined, 856, G–16
  second law of thermodynamics, 856–865
  temperature and, 876–878
Square planar complexes, 1024, 1024f, 1030–1031
Square planar shape, VSEPR theory, 376, G–16
Square pyramidal shape, VSEPR, 376, G–16
Stainless steel, 967t
Standard atmosphere (unit), 178, 179, 179t, G–16
Standard cell potential (E^0_{cell}), 906, 914–916, G–16
Standard electrode potential ($E^0_{half-cell}$), 906, 909t, G–16 see also Appendix D
Standard entropy of reaction (ΔS^0_{rxn}), 866–867, G–16
Standard free energy change, 873, G–16
Standard free energy of formation (ΔG^0_f), 874, G–16 see also Appendix B
Standard half-cell potentials see Standard electrode potential
Standard heat of formation (ΔH^0_f), 240, 240t, G–16 see also Appendix B
Standard heat of reaction (ΔH^0_{rxn}), 240–242, G–16
Standard hydrogen electrode, 906, G–16
Standard molar entropy (S^0), 861–862, G–16 see also Appendix B
Standard molar volume, 184, 184f, G–16
Standard reference half-cell see Standard hydrogen electrode
Standard states, 240, 240n, G–16
Standard temperature and pressure (STP), 184, G–16
Stannous fluoride, naming, 67
Stars, life cycle, 1074–1075, 1075f
State functions, 226–227, G–16
States of matter, 4–5, 8, 420
  balanced equation and, 103
  changes of state, 540–541
  defined, 4, G–16
  gases, 4, 4f, 174–210
  kinetic-molecular view, 420–421

liquids, 4, 4f, 175f
solids, 4, 4f, 175f
Stationary phase, chromatography, 76
Stationary states, 263, G–16
Steam-carbon reaction, 243
Stearic acid, molecular structure, 636f
Steel
  defined, 970, G–16
  metallurgy of, 970–972
  specific heat capacity, 233t
Steel alloys, 967t, 970
Stellar nucleogenesis, 1074–1075, 1075f, G–16
Stereoisomers, 617, 1021–1022, G–16
Stereoselective catalysts, 642
Sterling silver, 967t
Stoichiometrically equivalent molar ratios, 106–108, 113
Stoichiometric coefficient, 102
Stoichiometry, 87–120
  acid-base titration, 143–144
  chemical equations, 101–105
  chemical formulas, 95–101
  defined, 87, G–16
  electrolysis, 935–936
  ideal gas law and, 195–196
  mass percent, 93–95
  moles and molar mass, 87–93, 101, 120f
  reactants and products, 105–113
  solution stoichiometry, 114–120
  thermochemical equations, 236–237
STP see Standard temperature and pressure
Strassmann, F., 1069
Stratosphere, 204
Strong acids, 140–141, 141t, 145–146, 145f, 146, 760, 762, 763, 791
Strong acid-strong base titration curves, 816–818, 823
Strong bases, 140–141, 141t, 145–146, 145f, 146, 763, 791
Strong electrolyte, 141, 141f
Strong-field ligands, 1026, G–16
Strong force, 1047, G–16
Strontium
  isolation, 982t
  natural occurrence, 982t
  properties, 554
  uses, 982t
Strontium ion, 65t
Structural formula, 64, 72, 100, G–16
Structural isomers, 100 see also Constitutional isomers
Subatomic particles see also Electron (e^-); Neutron (n^0); Proton (p^+)
  isotopes, 53–55, 56
  properties, 52t
Sublevel, 276, G–16
Sublimation, 422, 423t, G–16
Submicroscopic properties, 6
Substance
  defined, 3, 41, 42, 43, 77f, G–16
  standard state, 240
Substitutional alloys, 492
Substitution reaction, organic, 626, G–16
Substrate, enzyme, 700, G–16
Subtraction, significant figures, 29
Suction pump, 177mn
Sugars, 645
Sulfate ion, 67t, 68
Sulfide ion, 65t
Sulfite ion, 67t, 68
Sulfonamides, 640
Sulfur
  allotropes, 584
  chemistry, 586–587

electron configuration, 298t, 299f
isolation, 985t
molar mass, 89
natural occurrence, 59, 954t
natural sources, 985t
properties, 312, 313f, 582
standard heat of formation, 240t
uses, 985t
Sulfur atoms, atomic mass, 88
Sulfur dioxide, 73f, 583, 586
Sulfur hexafluoride, 585, 586f
Lewis structure, 366
molecular shape, 376
Sulfur hydrides, 623
Sulfuric acid, 586–587, 587f
chemical manufacturing, 987–990
Lewis structure, 366–367
naming, 70
reaction with barium hydroxide, 142
uses, 989f
Sulfurous acid, acid-dissociation
constants, 777t
Sulfur tetrafluoride, 585, 586f
hybrid orbitals, 399
naming, 71
Sulfur trioxide, 583, 586
Superconductivity, 459, G–16
Superconductors, 559
Supercritical fluid (SCF), 431mn
Supernova, 1074
Supersaturated solution, 498,
498f, G–16
Surface tension, 439, 439f, 441, 442,
443, G–16
Surfactants, 490
Surroundings, 221, G–17
Suspensions, 517, G–17
Synchrotron, 1057
Syngas, 243, G–17
Synthesis gas, 243
Synthetic natural gas (SNG), 243, G–17
System, 221, G–17
System, in thermodynamics, 221
Systematic error, 31, G–17

T
t_{2g} orbitals, 1026, G–17
T shape, VSEPR theory, 375, G–17
Tachenius, Otto, 757mn
TCDD, 503mn
Tellurium
isolation, 985t
natural sources, 985t
properties, 582
uses, 985t
Temperature, 201
common values, 24f
critical, 430
defined, 24, G–17
Fahrenheit/Celsius conversion, 24–26
measuring, 24
normal body, 26
reaction rate and, 666, 682–684,
685–687, 686f
solubility and, 498–499
spontaneous change and, 876–878
unit of, 17t, 26
vapor pressure and, 427–428, 428f
Temperature scales, 24, 27, 181–182
Tera- (prefix), 18t
Terbium, name, origin of, 52mn
Termolecular reaction, 692
Tertiary amines, 631f
Tertiary sewage treatment, 521
Tetraethyl lead, 565

Tetrahedral arrangement
defined, 374, G–17
hybrid orbitals, 398t
VSEPR, 371f, 373–375, 374f
Tetrahedral complexes, 1024, 1024f,
1030–1031
Tetranitrocubane, 380, 380f
Tetraphosphorus decaoxide, 575, 580
Tetraphosphorus hexaoxide, 580
Thalidomide, 618mn
Thallium
bond type, 562t
isolation, 984t
melting point, 562t
natural sources, 984t
properties, 558
uses, 984t
Theoretical yield, 112, G–17
Theory, in scientific method, 12, 12f, 13
Thermal decomposition, 156, 157f
Thermal pollution, 499mn
Thermochemical equations,
stoichiometry, 236–237, G–17
Thermochemistry, 221, G–17
Thermodynamics, 221, 225
calorimetry, 234–236, 234f, 235f
defined, 221, G–17
energy, 221–227
law of energy conservation, 225
state functions, 226–227
units of, 225–226
enthalpy, 228–232
Hess's law of heat summation,
238–239, 425
laws of
first, 225, 857, 865
living things and, 869
second, 856, 861
third, 861–862, 917mn
specific heat capacity, 233–234
standard heat of formation, 240
Thermometer, 24, G–17
Thermosphere, 204
Thiosulfuric acid, 587
Third law of thermodynamics,
861–862, 917mn, G–17
Thomson, J.J., 49, 49f, 51
Thomson, William (Lord Kelvin),
27mn, 182
Thorium, 1007
Threonine, structure, 646f
Threshold frequency, 261
Thymol blue, 815f
Time, 26
unit of, 17t, 26
Tin
allotropes, 566
bond type, 562t
displacement of hydrogen from acid
by, 159, 159f
isolation, 984f
melting point, 562t
mineral source, 963t, 984t
organotin compounds, 565
properties, 564
tin(II) and (IV), 66t
uses, 984t
Tin(II) chloride, molecular shape, 373
Tin(II) ion, 314
Titanium
electron configuration, 300t
isolation, 983t
natural sources, 983t
orbital diagrams, 300t
orbitals, 1001, 1001t
paramagnetism, 316–317
uses, 983t

Titration
acid-base titrations, 143–144, 143f
defined, 143, G–17
redox titrations, 152–154, 152f
Titration curves, acid-base, 815–823
TNT, 622
Tokamak, 1040f, 1076f
Toluene, 622
Torr (unit), 178, 179, 179t, G–17
Torricelli, Evangelista, 176, 178
Total ionic equation, 138, G–17
Tracers, radioactive, 1062–1065, G–17
trans- isomer, 619
Transistors, 461
Transition elements
actinides, 57, 57f, 303–304, 1007
complexes see Transition metal
complexes (coordination
compounds)
coordination theory, 1018–1019
defined, 300, 999, G–17
as dietary trace elements, 1032–1033
electron configuration, 302–303,
303–304, 304, 1000–1001
inner see Inner transition elements
ions of, 64f
lanthanide contraction, 1002
lanthanides, 57, 57f, 303–304,
1006–1007
periodic table, 57, 57f, 999, 999f
properties, 1000–1006
sources, isolation, uses, 983t
Transition metal complexes
(coordination compounds),
1013–1023
crystal field theory, 1025–1031, 1031
formulas and names, 1016–1018
isomerism, 1020–1023
structure, 1014–1016
valence bond theory,
1023–1024, 1031
Transition metal ions
electron configuration, 315–316
magnetic properties, 316–317
Transition state, 688, G–17
Transition state theory, 685,
688–690, G–17
Transmutation, 1055–1058
rejected by Dalton, 46
Rutherford's scattering experiment,
50–51, 50f, 56mn
Transuranium elements,
1056–1058, G–17
Triethylamine, 780t
Triglycerides, 637
Trigonal bipyramidal arrangement
defined, 375, G–17
hybrid orbitals, 398t
VSEPR, 371f, 375–376, 375f
Trigonal bipyramidal shape, 375, G–17
Trigonal planar arrangement
defined, 372, G–17
hybrid orbitals, 398t
VSEPR, 371f, 372–373, 372f
Trigonal planar shape, 372, G–17
Trigonal pyramidal shape, VSEPR
theory, 374, G–17
Trinitrotoluene (TNT), 622
Triple bonds, 338, 402
defined, 332, G–17
functional groups with, 640–641
hydrocarbons with, 621
inorganic compounds with, 640
Triple point, 431, G–17
Trisilylamine, bonding, 568, 568f
Trisodium phosphate, 758t
Tristearin, molecular structure, 637f

Troposphere, 204
Tryptophan, structure, 646f
Tyndall effect, 518, 518f, G–17
Tyrosine, structure, 646f

U
Ultraviolet (UV) radiation, 257, G–17
Uncertainty, in measurement,
27–33, G–17
Uncertainty principle, 272–273, G–17
Unidentate ligands, 1016t
Unimolecular reaction, 692, G–17
Unit cell, 446, 447f, 448
defined, 446, G–17
packing efficiency and, 448,
449f, 450
Units of measurement, 13 see also
Measurement; individual units
of measurement
conversion factors, 13–15, 17
decimal prefixes, 17, 18t
history, 17, 17mn
metric system, 17
SI units, 17–27
Universal gas constant (R), 185,
683mn, G–17
Unsaturated hydrocarbons, 618, G–17
Unsaturated solutions, 498, G–17
Unshared electron pair, 338
Uranium, 1007
atomic bomb, 1070–1071
mass fraction, calculating, 45
nuclear fission, 1069–1073
as nuclear fuel, 202mn
Uranus (planet), atmosphere, 206t

V
V agent, molecular structure, 962mn
V shape, VSEPR theory, 373
Vacuum, 176f, 177
Valence band, 458, G–18
Valence bond (VB) theory, 393–400
complexes, 1023–1024, 1030–1031
hybrid orbitals in, 1026–1027
defined, 393, G–18
hybrid orbitals, 394, 395f, 396–400
molecular rotation, 403, 404
orbital overlap, 403, 404
single and multiple bonds, 400–402
Valence electrons, 302, G–18
Valence-shell electron-pair repulsion
(VSEPR) theory, 370–379,
370f–379f, 379 see also
individual shapes
defined, 370, G–18
linear arrangement, 371f, 372
octahedral arrangement, 371f,
376, 376f
square planar shape, 376
square pyramidal shape, 376
tetrahedral arrangement,
373–375, 374f
trigonal bipyramidal arrangement,
371f, 375–376, 375f
trigonal planar arrangement,
372–373, 372f
Valence-state electronegativity,
1009, 1013
Valine, structure, 646f
Vanadium
electron configuration, 300, 300t
in humans, 1032t
isolation, 983t
natural occurrence, 954t, 983t
orbital diagrams, 300t

Vanadium *(continued)*
orbitals, 1001, 1001*t*
uses, 983*t*
Vanadium carbide, 968*f*
van der Waals, Johannes, 209
van der Waals constant, 209,
209*t*, G–18
van der Waals equation, 209, G–18
van der Waals forces, 432
van der Waals radius, 432, 432*f*,
438, G–18
van't Hoff equation, 742
Vaporization, 421, 423*f*
defined, 421, G–18
heat of, 230, 422, 422*f*
Vapor pressure, 193, 431
boiling point and, 428–429
defined, 426, G–18
temperature and intermolecular
forces, effect on,
427–428, 427*f*
of water, 193*t*
Vapor pressure lowering, 507–508,
507*f*, G–18
Variable *see* Controlled experiments
VB theory *see* Valence bond
(VB) theory
Venus (planet), atmosphere, 206*t*
Villard, P., 1043
Vinegar, reaction with baking soda,
146, 146*f*
Viscosity, 440, G–18
of gases, 175
of liquids, 441
of motor oil, 441
of water, 440*t*
Vision, chemistry of, 620, 620*f*
Vitamin A, 620
Vitamin B$_{12}$, 1032*t*
Vitamin C, 98–99, 778
Vitamin E, 365*mn*
Volatile nonelectrolytes, colligative
properties, 515
Volatility, 75, 75*f*, G–18
Volt (V) (unit), 905, G–18
Voltage *see* Cell potential (E_{cell})
Voltaic cells, 898, 899*f*, 904–905, 913
action of, 904–905
cell potential, 905, 905*t*
construction and operation, 900–903,
900*f*, 901*f*, 902*f*
defined, 898, G–18
diagramming, 903–904
electrodes, 902, 902*f*
electrolytic cells compared, 929,
929*f*, 930, 931*t*
half-cell, 901
notation, 903

oxidizing and reducing agents,
908–909, 909*t*
relative reactivities of metals,
912–913, 913*f*
spontaneous redox reactions,
909–912
spontaneous redox reactions,
909–912
standard cell potential, 906–908
standard electrode potentials,
906–908, 907*f*
standard hydrogen electrode,
906–908, 907*f*
Volume
amount of gas and, 180, 183–184,
183*f*, 184*mn*, 185, 185*f*, 186
common values of, 22*f*
conversion to English
equivalents, 18*t*
defined, 19, G–18
determining by displacement of
water, 20
gases
Avogadro's law, 180, 183–184,
183*f*, 184*mn*, 185, 185*f*, 186,
189, 197, 200, 200*f*
Boyle's law, 180–181, 180*f*,
184*mn*, 185, 185*f*, 186, 197,
198, 198*f*
Charles's law, 180, 181–183, 182*f*,
184*mn*, 185, 185*f*, 186, 189,
197, 199, 199*f*
pressure and, 180–181, 180*f*,
184*mn*, 185, 185*f*, 186
temperature and, 180, 181–183,
182*f*, 184*mn*, 185, 185*f*, 186
Le Châtelier's principle, 739–740
unit of, 19–20, 19*f*, 20*f*, 27
Volume percent, 503, G–18
Volumetric glassware, 20, 20*f*
von Laue, Max, 451, 917*mn*
von Mayer, J.R., 225*mn*
VSEPR theory *see* Valence-shell
electron-pair repulsion
(VSEPR) theory

W

Waage, Peter, 717
Wastewater, 521, G–18
Wastewater treatment, 521
Water, 583
autoionization, 764–765
boiling point, 25*f*
boiling point elevation, 509*t*
Celsius scale in, 24
covalent compounds dissolving in,
135–136

decomposition, 157
density, 23*t*, 443–445, 444*f*
electrolysis, 932–933
electron distribution in,
134–136, 134*f*
freezing point, 25*f*
freezing point depression, 509*t*
heat capacity, 236*mn*
heating-cooling curve, 424*f*
hybrid orbitals, 397, 397*f*
hydrogen bonding, 442
hydronium ion, 67*t*, 136, 136*f*,
758–759, 765–768
ice, 443, 443*f*
ionic compounds dissolving in,
135, 135*f*
ion-product constant, 764–765
life processes, role in, 442
melting, 423
phase changes, 423–425
phase diagram, 430*f*
polar nature of, 134–136, 134*f*–136*f*
properties, 3, 3*f*
reaction of metal oxides with,
158–159
as a solvent, 132–136
solvent properties, 442
specific heat capacity, 233*t*
states, 4–5, 423
surface properties, 443
thermal properties, 442–443
vaporization, 423, 423*f*
vapor pressure, 193*t*
viscosity, 440*t*
Water-gas reaction, 243
Water purification, 520–521, 520*f*, 521*f*
Waters of hydration, 68
Water softening, 520, 521*f*, G–18
Water softening agent, 575
Water treatment plants, 520, 520*f*
Wave function, 273, 281
Wavelength (λ), 256, 256*f*, G–18
Wave-particle duality, 266,
269–273, G–18
Waves, behaviors of, 258*f*
Waxes, 493, 493*mn*
Weak acid-strong base titration curves,
818–821, 823
Weak acids, 140–141, 141*t*, 146, 146*f*,
760, 763
weak-acid equilibrium, 772–779
weak bases and, 779–784
Weak base-strong acid titration curves,
821–822, 823
Weak bases, 140–141, 141*t*, 763
weak acids and, 779–784
Weak-field ligands, 1026, G–18
Weight, defined, 21, 27, G–18
Werner, Alfred, 1018

"Whiskers," in ceramic composites, 466
White phosphorus, 573*f*
White tin, 566
Whole-body irradiation, 1061*t*
Wöhler, Friedrich, 607
Wood
as fuel, 243
specific heat capacity, 233*t*
Work
defined, 223, G–18
heat and, 223–224, 223*f*, 224*f*
thermodynamics, 874–875
Wristwatch, radium-containing, 1059*mn*

X

Xanthophylls, 1025
Xenon
isolation, 986*t*
natural occurrence, 59, 986*t*
properties, 595, 596
uses, 986*t*
Xerography, 584*mn*
X-ray diffraction analysis, 451–452,
451*f*, G–18
X-rays
diffraction pattern, 270, 271*f*
and electrons in diffraction, 270, 271*f*
Xylene, structure, 622, 622*f*

Y

Yields, chemical reactions, 112–113
-yne (suffix), 629*t*
Ytterbium, name, origin of, 52*mn*
Yttrium, name, origin of, 52*mn*

Z

Zeolites, 571
Zero order reactions
first- and second-order
compared, 682*t*
integrated rate law, 677–679, 682*t*
rate law, 673
Ziegler, Karl, 642
Ziegler-Natta catalysts, 642
Zinc
electron configuration, 300*t*
in humans, 1032*t*
isolation, 983*t*
natural occurrence, 954*t*, 963*t*, 983*t*
orbital diagrams, 300*t*
orbitals, 1001, 1001*t*
uses, 983*t*
Zinc blende, crystal structure, 454, 454*f*
Zone refining, 512, 512*f*, 967, G–18

Fundamental Physical Constants (six significant figures)

Avogadro's number	N_A	$= 6.02214 \times 10^{23}$/mol
atomic mass unit	amu	$= 1.66054 \times 10^{-27}$ kg
charge of the electron (or proton)	e	$= 1.60218 \times 10^{-19}$ C
Faraday constant	F	$= 9.64853 \times 10^4$ C/mol
mass of the electron	m_e	$= 9.10939 \times 10^{-31}$ kg
mass of the neutron	m_n	$= 1.67493 \times 10^{-27}$ kg
mass of the proton	m_p	$= 1.67262 \times 10^{-27}$ kg
Planck's constant	h	$= 6.62608 \times 10^{-34}$ J·s
speed of light in a vacuum	c	$= 2.99792 \times 10^8$ m/s
standard acceleration of gravity	g	$= 9.80665$ m/s^2
universal gas constant	R	$= 8.31447$ J/(mol·K)
		$= 8.20578 \times 10^{-2}$ (atm·L)/(mol·K)

SI Unit Prefixes

p	n	μ	m	c	d	k	M	G
pico-	nano-	micro-	milli-	centi-	deci-	kilo-	mega-	giga-
10^{-12}	10^{-9}	10^{-6}	10^{-3}	10^{-2}	10^{-1}	10^3	10^6	10^9

Conversions and Relationships

Length
SI unit: meter, m

1 km	$= 1000$ m
	$= 0.62$ mile (mi)
1 inch (in)	$= 2.54$ cm
1 m	$= 1.094$ yards (yd)
1 pm	$= 10^{-12}$ m $= 0.01$ Å

Volume
SI unit: cubic meter, m^3

1 dm^3	$= 10^{-3}$ m^3
	$= 1$ liter (L)
	$= 1.057$ quarts (qt)
1 cm^3	$= 1$ mL
1 m^3	$= 35.3$ ft^3

Pressure
SI unit: pascal, Pa

1 Pa	$= 1$ N/m^2
	$= 1$ kg/m·s^2
1 atm	$= 1.01325 \times 10^5$ Pa
	$= 760$ torr
1 bar	$= 1 \times 10^5$ Pa

Mass
SI unit: kilogram, kg

1 kg	$= 10^3$ g
	$= 2.205$ lb
1 metric ton (t)	$= 10^3$ kg

Energy
SI unit: joule, J

1 J	$= 1$ kg·m^2/s^2
	$= 1$ coulomb·volt (1 C·V)
1 cal	$= 4.184$ J
1 eV	$= 1.602 \times 10^{-19}$ J

Math relationships

$$\pi = 3.1416$$

volume of sphere $= \frac{4}{3}\pi r^3$

volume of cylinder $= \pi r^2 h$

Temperature
SI unit: kelvin, K

0 K	$= -273.15°$C
mp of H_2O	$= 0°$C (273.15 K)
bp of H_2O	$= 100°$C (373.15 K)
T (K)	$= T$ (°C) $+ 273.15$
T (°C)	$= [T$ (°F) $- 32]\frac{5}{9}$
T (°F)	$= \frac{9}{5}T$ (°C) $+ 32$